ENCYCLOPEDIA OF
ANIMAL BEHAVIOR

ENCYCLOPEDIA OF
ANIMAL BEHAVIOR

EDITORS-IN-CHIEF

PROFESSOR MICHAEL D. BREED
University of Colorado, Boulder, CO, USA

PROFESSOR JANICE MOORE
Colorado State University
Fort Collins, CO, USA

AMSTERDAM • BOSTON • HEIDELBERG • LONDON • NEW YORK • OXFORD
PARIS • SAN DIEGO • SAN FRANCISCO • SINGAPORE • SYDNEY • TOKYO
Academic Press is an imprint of Elsevier

ACADEMIC
PRESS

Academic Press is an imprint of Elsevier
32 Jamestown Road, London NWI 7BY, UK
30 Corporate Drive, Suite 400, Burlington, MA 01803, USA
525 B Street, Suite 1900, San Diego, CA 92101-4495, USA

British Library Cataloguing in Publication Data
A catalogue record for this book is available from the British Library

Library of Congress Catalog Number: 2010922487

ISBN: 978-0-08-045333-0

For information on all Elsevier publications
visit our website at books.elsevier.com

10 11 12 13 14 10 9 8 7 6 5 4 3 2 1

Cover Photo: An iguana, Costa Rica, photograph by Michael D. Breed

Printed and Bound in Spain

In Memoriam

Christopher J. Barnard

Ross H. Crozier

PREFACE

Ancient drawings on the walls of caves speak for the ageless intrigue that animal behavior holds for human beings. In those days, the fascination was certainly motivated in part by survival; our ancestors were both predators and prey. There is some evidence that early humans also found animal behavior to be intrinsically interesting; the myths and stories that come down to us from prehistory contain elements of what animals do in the world and what they mean to people. These are the oldest statements of human relationship with the natural world and the living things that inhabit it.

Our ancestors would not recognize the far-flung universe of the modern science of animal behavior. Only 14 decades (approximately) have passed since Darwin first published *The Expression of Emotions in Man and Animals* – generally acknowledged as the starting point for the scientific study of animal behavior – and behavioral biologists now ask questions about topics ranging from the relationship of immunological phenomena and behavioral disorders in dogs, rats, and people to the integration of animal behavior and conservation. Experts in animal behavior provide commentary on the mating displays of rare primates, on television, where entire channels are devoted to the sensory worlds of insects and the ability of octopus to disappear in plain sight. In short, human fascination with animal behavior has produced a field that is rich beyond imagination, and frustratingly beyond the full embrace of any one person.

The almost hopelessly dispersed primary literature of animal behavior reflects the reticulated evolution of the field, which comes to us from field studies; from laboratory experiments; from our understanding of nerves, muscles, and hormones; and from our grasp of social interactions and ecology. It is difficult to think of a major area of biological inquiry that has not been touched by a behavioral tendril or two. A temptation exists to surrender to this fragmentation – allowing our intellectual landscape to reflect increasingly small and disjunct patches of thought and discovery.

Such surrender is, of course, distasteful to any scholar, but there is a more penetrating reason that makes it unacceptable: Anthropogenic change is occurring at a higher rate than ever before, and if we are to preserve our own habitat – the world that the ancients felt compelled to explain in their stories about animals – we must not fail in our attempts to understand its inhabitants. Those residents sustain our own habitat, and their requirements are varied, going far beyond calories and oxygen. They migrate and forage, choose mates, and defend territories, and all this behavior is influenced by hormones, external physical stimuli, trophic and social interactions, and eons of fitness outcomes. A fully integrated knowledge of animal behavior will be indispensable as scientists analyze changing populations, communities, ecosystems, and landscapes. Indeed, it will be indispensable for anyone who seeks to be an honest custodian of nature.

This encyclopedia offers over 300 authoritative and accessible synopses of topics ranging from dolphin signature whistles to game theory. As library reference material, the encyclopedia serves a public that is increasingly challenged to be aware of scientific advances. It is designed as a first stop for the curious advanced undergraduate or graduate student, as well as for the researcher desiring to learn about developments in fields related to his or her own study or to enter a new phase of inquiry.

In compiling this work, we contacted internationally known scientists in the broad array of fields that inform animal behavior. These accomplished men and women are the section editors, and they, in turn, invited some of the best scholars and rising stars in their subject areas to write for the encyclopedia. Thus, every contribution has been reviewed by experts. In short, the articles approach the best that our field has to offer, written by people whose passion for animal behavior is equaled only by their expertise.

Creating the list of sections was as daunting as it was enjoyable. Of course, we included traditional, major areas like foraging, predator–prey interactions, mate choice, and social behavior, along with endocrinology, methods, and neural processes, to name a few. You will see those and more as you survey these volumes. We also included areas that have recently captured the attention of an increasing number of behavioral biologists; these include infectious disease, cognition, conservation, and animal welfare. Looking to the future, we invited contributors from robotics and applied areas. We realized that we could not do the study of animal behavior justice without some exploration of the model systems – the landmark studies – that have molded and continue to guide the development of the field.

In general, you will not find human behavioral studies in this collection, although some articles are tangentially related to such work. That exclusion was a difficult decision, but was motivated not so much by some parochial commitment to a human/non-human divide as by the fact that the non-human literature itself is rich beyond description. Limiting the collection to non-human studies in no way removed the danger of intellectual gluttony.

We remember two eminent behavioral biologists in the dedication of this work – Professors Christopher J. Barnard and Ross H. Crozier. Both of these men played important roles in the creation of this work, and neither lived to see it come to fruition. Professor Barnard (1952–2007) was the first editor-in-chief of the encyclopedia and developed the initial overview of topics, but had to step back from the process because of the illness that eventually took his life. Professor Crozier (1943–2009) was the section editor of Genetics until his untimely death in late 2009. Their immense and varied contributions enriched our knowledge of animal behavior and are cataloged in numerous locations. Those contributions are remarkable in their scope and influence, but they are nonetheless dwarfed by the legions of students, friends, and family members who feel fortunate to have known these scientists and who will carry their legacy forward.

We are grateful to Dr. Andrew Richford, formerly the Senior Acquisitions Editor, Life Sciences Books, Academic Press, who guided us through the formative part of this project. His expertise in and enthusiasm for animal behavior provided significant momentum, not to mention some good fun. Simon Wood, Major Reference Works Development Editor, was indispensable to the project. He answered an amazing variety of questions from contributors and editors, kept the project organized and moving forward, and did all this without losing his fine sense of humor. Nicky Carter, Project Manager, guided us through the completion of the project, providing a pleasantly seamless interface between the scientific scribblers and other publishing professionals. We thank Kristi Gomez and Will Smaldon, also of Academic Press, for their roles in bringing the project to completion. Finally, working with the section editors (see pp. ix) was a real treat; their expertise and devotion to animal behavior is reflected in every page of this work. We particularly thank James Ha, Joan Herbers, James Serpell, and David Stephens for attending an organizational meeting to set the stage for the development of the project.

We are pleased to see this culmination of effort on the part of hundreds of authors and co-authors. Each article is the distillation of expert understanding, acquired over many years. We are excited to be part of such a remarkable collaboration, one that opens so many doors to the future of animal behavior for undergraduates and professionals alike.

Michael Breed, Boulder, CO
Janice Moore, Fort Collins, CO
August 2010

SECTION EDITORS

Bonnie Beaver

Bonnie is internationally recognized for her work in the normal and abnormal behaviors of animals. She has given over 250 scientific presentations to veterinary and veterinary student audiences on subjects of animal behavior, animal welfare, and the human–animal bond, as well as discussed many areas of veterinary medicine for the public media. In addition, she has authored over 150 scientific articles and has nine published books, including *The Veterinarian's Encyclopedia of Animal Behavior* (Blackwell Press), *Feline Behavior: A Guide for Veterinarians* (Saunders), and the newly released second edition of *Canine Behavior: Insights and Answers* (Saunders).

Bonnie is a member of numerous local, state, and national professional organizations and has served as president or chair of several organizations, including the American Veterinary Society for Animal Behavior, the American College of Veterinary Behaviorists, Phi Zeta, and the Texas Veterinary Medical Association. She is board certified by the American College of Veterinary Behaviorists and currently serves as its Executive Director. In addition, Bonnie is the President of the Organizing Committee for the American College of Animal Welfare.

Bonnie is a past president of the American Veterinary Medical Association and has served as Chair of the AVMA Executive Board. She has also served on several AVMA committees, including the Animal Welfare Committee, Council on Education, Committee on the Human–Animal Bond, and American Board on Veterinary Specialties.

In addition, she chaired the AVMA's Canine Aggression and Human–Canine Interactions Task Force, and the Panel on Euthanasia.

Professionally, Bonnie has been honored by being elected as a Distinguished Practitioner of the National Academies of Practice, named as the recipient of the 1996 AVMA Animal Welfare Award, awarded the 2001 Friskies PetCare Award in Animal Behavior, and received the 2001 Leo K. Bustad Companion Animal Veterinarian of the Year Award. She has been recognized for outstanding professional achievement in more than 150 editions of over 50 publications, including *Who's Who in America*, *The World Who's Who of Women*, *Who's Who in the World*, and *American Men and Women of Science*.

Michael Breed

After receiving his PhD from the University of Kansas in 1977, Michael came to Colorado to work as a faculty member at the University of Colorado, Boulder, where he has been ever since. He is currently a Professor in the Department of Ecology and Evolutionary Biology, and teaches courses in general biology, animal behavior, insect biology, and tropical biology. Michael's research program focuses on the behavior and ecology of social insects, and he has worked on ants, bees, and wasps. He has studied the nestmate recognition, the genetics of colony defense, the behavior of defensive bees, and communication, during colony defense. He was the Executive Editor of *Animal Behaviour* from 2006 to 2009.

Jae Chun Choe

After receiving his PhD from Harvard University in 1990, Jae became a Junior Fellow at the Michigan Society of Fellows. He then returned to his home country, Korea, to work in the School of Biological Sciences at Seoul National University. In 2006, he moved to Ewha Womans University to take the post of university chair professor and the director of its natural history museum. He served as the president of the Ecological Society of Korea and is currently serving as the co-president of the Climate Change Center. Since his return to Korea, he has been conducting a long-term ecological research of magpies while continuing to study insects. Quite recently, he began a field study of Javan Gibbons in the Gunuung Halimun-Salak National Park of Indonesia.

Nicola Clayton

Nicola is Professor of Comparative Cognition in the Department of Experimental Psychology at the University of Cambridge, and a Fellow of Clare College. She received her undergraduate degree in Zoology at the University of Oxford and her doctorate in animal behavior at St. Andrews University. In 1995, she moved to the University of California Davis where she gained her first Chair in Animal Behaviour in 2000. She moved to Cambridge and was appointed a personal Chair in 2005. She has 185 publications to her credit.

Nicola studies the development and evolution of intelligence. For example, she addresses the question of whether animals can plan for the future and what they remember about the past, as well as when these abilities develop in children. She is also interested in social and physical intelligence, such as whether animals can differentiate between what they know and what other individuals know. Nicola's work deals mainly with the members of the crow family (e.g., rooks and jays), and comparisons between crows, nonhuman apes, and young children.

Jeff Galef

After receiving his Ph.D. from the University of Pennsylvania in 1968, Jeff moved as an Assistant Professor to McMaster University in Hamilton, Ontario where, for 38 years, his research focused on understanding social influences on the feeding behavior of Norway rats and the mate choices of Japanese quail. Empirical work in his laboratory on social learning in animals has resulted in the publication of more than 100 scientific articles, (www.sociallearning.info) and his scholarly pursuits have produced three co-edited volumes (*Social Learning: Psychological and Biological Perspectives* (with TR Zentall), *Social Learning and Imitation: the Roots of Culture* (with CM Heyes), and *The Question of Animal Culture* (with KN Laland)) as well as a special issue of the journal *Learning & Behavior* (2004, 32(1) (with CM Heyes)). He was honored with the Lifetime Contribution Award of the Social Learning Group, St. Andrews University, Scotland, in 2005, and in 2009, was elected a Fellow of the Royal Society of Canada.

Sidney Gauthreaux

Sidney received his PhD in 1968 and did a post-doctorate at the Institute of Ecology at the University of Georgia in the following 2 years. He joined the zoology faculty at Clemson University in 1970 and retired as Centennial Professor of Biological Sciences in 2006. In 1959, he began working with weather surveillance radar at National Weather Service installations in an effort to detect, quantify, and monitor migrating birds in the atmosphere.

His research has focused on radar studies of bird migration across the Gulf of Mexico and over much of the United States in spring and fall. Since 1992, modern Doppler weather radar has 'revolutionized' the study of bird migration, and he has used it to monitor the flight behavior of birds in the surveillance areas of approximately 150 weather radar stations throughout the United States and explore the interrelationships of bird movements at different spatial scales in relation to geography, topography, habitat, weather, and climatic factors. Recent work with high-resolution surveillance radar (modified marine radar) and thermal imaging and vertically pointing radar (TI-VPR) has greatly enhanced his capability to work at small spatial scales and explore the behavior of migrating birds within 12 km of the radar.

Sidney was President of the Animal Behavior Society from 1987 to 1988 and was elected a Fellow in the American Association for the Advancement of Science in 1988. In October 2006, he received the William Brewster Memorial Award of the American Ornithologists' Union, and in April 2009, the Margaret Morse Nice Medal of the Wilson Ornithological Society.

Deborah M. Gordon

After receiving her PhD from Duke University in 1984, Deborah joined the Harvard Society of Fellows. She did her postdoctoral research at Oxford and at the Centre for Population Biology at Silwood Park, University of London. She came to Stanford in 1991 and is currently a Professor in the Department of Biology. She teaches courses in ecology and behavioral ecology. Deborah's research program focuses on the organization and ecology of ant colonies, and how colonies, without central control, use interaction networks to regulate colony behavior. Her projects include a long-term study of a population of harvester ant colonies in Arizona, studies of the invasive Argentine ant in northern California, and ant–plant mutualisms in Central America.

Patricia Adair Gowaty

Patricia is a Distinguished Professor of Ecology and Evolutionary Biology – UCLA and a Distinguished Research Professor *Emerita* of Ecology at the University of Georgia. After receiving her PhD in 1980, she supported herself with funding from NSF and NIH, until her first tenure track job as an Associate Professor of Zoology in 1993 at the University of Georgia. She studied social behavior, demography, and ecology of eastern bluebirds in the field for 30 years. She pioneered studies of extra-pair paternity in socially monogamous species. She studied fitness outcomes of reproduction under experimentally imposed social constraints in flies, mice, ducks, and cockroaches. Her theoretical work includes papers on the evolution of social systems, forced copulation, compensation, and sex role evolution. Currently, she is completing studies in the genetic mating system of eastern bluebirds, experiments on the fitness variation of males and females in the three species of *Drosophila*, and a book on reproductive decisions under ecological and social constraints. She was President of the Animal Behavior Society in 2001. She is a Fellow of the American Association for the Advancement of Science, the American Ornithologists' Union, the International Ornithologists' Union, and the Animal Behavior Society.

James Ha

James has a 1989 Ph.D. in Zoology/Animal Behavior from Colorado State University and has been on the faculty of the University of Washington since 1992. He is actively involved in research on the social behavior of Old World monkeys and their management in captivity, Pacific Northwest killer whales, local and Pacific island crows, and domestic dogs. He is also certified as an Applied Animal Behaviorist by the Animal Behavior Society and has his own private practice in dealing with companion animal behavior problems in the Puget Sound area.

Joan M. Herbers

Joan is a Professor of Evolution, Ecology, and Organismal Biology at The Ohio State University in Columbus Ohio. She has studied social evolution in ants for many years, with contributions to queen-worker conflict, sex ratio theory, and coevolution. She is currently serving as the Secretary-General of the International Union for the Study of Social Insects and also as the President of the Association for Women in Science.

Jeffrey Lucas

Jeffrey received a Ph.D. from the University of Florida in 1983, studying under Dr. H. Jane Brockmann. He then took a postdoc position in Dr. John Kreb's lab at Oxford University. After teaching at the College of William & Mary and Redlands University, he came to Purdue University in 1987, where he is currently a professor of Biological Sciences. Jeffrey teaches courses in ecology, animal behavior, sensory ecology, and animal communication. His research program focuses on the chick-a-dee call of chickadees and a comparison of auditory physiology in a variety of birds. He has worked on seed dispersal, antlions, and fish, and has published dynamic programming models of a number of systems. He is a past Executive Editor of *Animal Behaviour* and is a fellow of the Animal Behavior Society.

Constantino Macías Garcia

Constantino has been interested in animal behavior ever since he joined Hugh Drummond's laboratory to study the feeding habits of snakes for his BSc and MSc. His main research has been on sexual selection and the evolution of ornaments, which he has studied mainly in Goodeid fish. He was careless not to follow the early forays of his PhD supervisor, Bill Sutherland, into the hybrid field of behavior and conservation. But time, as well as the increasingly grim reality of Mexican fauna, has led him to investigate the links between behavior and conservation in fish, frogs, and birds.

Justin Marshall

Justin's interest in biology and the sea came from his parents, both marine biologists and keen communicators of the ocean realm. He was then fortunate to begin learning about sensory biology in aquatic life during his undergraduate degree in Zoology at The University of St Andrews. The Gatty Marine Laboratory and its then director, Mike Laverack, introduced him to the diversity of marine life and the challenges of different sensory environments under water. Enjoying the cold clear waters of Scotland, he also began to take interest in tropical biodiversity and traveled to Australia and The Great Barrier Reef toward the end of his undergraduate degree. Currently, he is the President of The Australian Coral Reef Society and lives in Australia working at The University of Queensland. He holds a position of Professor at The Queensland Brain Institute and is an Australian Research Council Professorial Research Fellow. Before moving into the superb sensory environment of Jack Pettigrew's Vision Touch and Hearing Research Centre, he did his D.Phil and spent his initial postdoctoral years at The University of Sussex in the U.K., Mike Land and The University of Maryland's

Tom Cronin were his mentors during these years and Justin developed an enthusiasm for the amazing world of invertebrate vision only because of them. His work now focuses on the visual ecology of a variety of animals, mostly aquatic, and has branched out to include fish, reptiles, and birds. Animal behavior and questions, such as 'why are animals colorful?', form a large section of his current research.

Janice Moore

As an undergraduate, Janice was inspired by parasitologist Clark P. Read to think about the ecology and evolution of parasites in new ways. She was especially excited to learn that parasites affected animal behavior, another favorite subject area. Most biologists outside the world of parasitology were not interested in parasites; they were relegated to a nether world between the biology of free-living organisms and medicine. After peregrination through more than one graduate program, she completed her PhD studying parasites and behavior at the University of New Mexico. Janice did postdoctoral work on parasite community ecology with Dan Simberloff at Florida State University, and then accepted a faculty position at Colorado State University, where she has remained since 1983. She is currently a Professor in the Department of Biology where she teaches courses in invertebrate zoology, animal behavior, and the history of medicine. She studies a variety of aspects of parasite ecology and host behavior ranging from behavioral fever and transmission behavior to the ecology of introduced parasite species.

Daniel Papaj

After receiving his PhD from the Duke University in 1984, Daniel engaged in postdoctoral research at the University of Massachusetts at Amherst and at Wageningen University in The Netherlands. He joined the faculty of the Department of Ecology and Evolutionary Biology at the University of Arizona in 1991, where he has been ever since. His research focuses on the reproductive dynamics of insects, with special attention to the role of learning by the insect in its interactions with plants. Daniel's focal organisms have included butterflies, tephritid fruit flies, parasitic wasps, and more recently, bumble bees. Recent projects in the lab include the costs of learning in butterflies, the dynamics of social information use in bumble bees, the thermal ecology of host preference in butterflies, ovarian dynamics in fruit flies, multimodal floral signaling, and bumblebee learning. He teaches courses in animal behavior, behavioral ecology, and introductory biology.

Ted Stankowich

Ted grew up in suburban Southern California where opportunities to observe macrofauna in nature were few, but still found ways to observe and enjoy the animals that he could find in his own backyard. While his initial interests in biology were in biochemistry and genetics, after taking introductory courses at Cornell University, he quickly realized that these disciplines were not his calling. He developed interests in ecology and evolution after taking introductory courses and working in George Lauder's functional morphology lab for a summer at the University of California, Irvine, but he took an abiding interest in animal behavior after taking a course as a junior at Cornell and joined Paul Sherman's naked mole-rat lab, where he completed an honors thesis on parental pup-shoving behavior. Ted entered the Animal Behavior graduate program at the University of California, Davis to work with Richard Coss. He spent three field seasons working on predator recognition, flight decisions, and antipredator behavior, in Columbian black-tailed deer, and completed his dissertation in 2006. Ted served as the Darwin Postdoctoral Fellow at the University of Massachusetts, Amherst from 2006 to 2008, investigating escape behavior in jumping spiders. Since completing his tenure, he has continued to work as a postdoc and teach at UMass.

David W. Stephens

David received his PhD from Oxford University in 1982. Currently, David is a Professor at the University of Minnesota in the Twin Cities. His research takes a theoretical and experimental approach to behavior ecology. His research focuses on the connections between evolution and animal cognition, especially the evolutionary forces that have shaped animal learning and decision-making. His work makes connections with many disciplines within the behavioral sciences, and he has presented his work to groups of psychologists, economists, anthropologists, mathematicians, and neuroscientists. He is the author, with John Krebs, of the well-cited book *Foraging Theory*, and the editor (with Joel Brown and Ronald Ydenberg) of *Foraging: behavior and ecology*. He served as an editor of *Animal Behaviour* from 2006 to 2009.

John C. Wingfield

John's undergraduate degree was in Zoology (special honors program) from the University of Sheffield and he did his Ph.D. in Comparative Endocrinology and Zoology from the University College of North Wales, UK. Although John is trained as a comparative endocrinologist, he has always interacted with behavioral ecologists and has strived to integrate ecology and physiology down to cellular and molecular levels. The overarching question is how animals cope with a changing environment – basic biology of how environmental signals are perceived, transduced into endocrine secretions that then regulate morphological, physiological, and behavioral responses. The diversity of mechanisms is becoming more and more apparent and how these evolved is another intriguing question. He was an Assistant Professor at the Rockefeller University in New York and then spent over 20 years as a Professor at the University of Washington. Currently, he is a Professor and Chair in Physiology at the University of California at Davis.

Harold Zakon

Harold received a B.S. degree from Marlboro College in Vermont. He worked as a research technician at Harvard Medical School for 2 years and realized his love for doing research. He earned a Ph.D. from the Neurobiology & Behavior program at Cornell University, working with Robert Capranica, studying the regeneration of the frog auditory nerve. He did postdoctoral work at the University of - California, San Diego with Theodore Bullock and Walter Heiligenberg. There, he began working on weakly electric fish. He established his laboratory at the University of Texas in Austin, Texas where he has been studying communication in electric fish, and the regulation and evolution of ion channels in electric fish and other organisms. He was the first chairman of the then newly established Section of Neurobiology at UT. He has been Chairman for Gordon Research Conference on Neuroethology and organizer for the International Congress in Neuroethology. His hobbies include playing guitar and piano. He, his wife Lynne (mandolin), and son Alex (banjo), have a band called Red State Bluegrass. Their goal is to perform on Austin City Limits one day.

CONTRIBUTORS

J. S. Adelman
Princeton University, Princeton, NJ, USA

E. Adkins-Regan
Cornell University, Ithaca, NY, USA

J. F. Aggio
Neuroscience Institute and Department of Biology,
Atlanta, GA, USA

M. Ah-King
University of California, Los Angeles, CA, USA

I. Ahnesjö
Uppsala University, Uppsala, Sweden

J. Alcock
Arizona State University, Tempe, AZ, USA

L. Angeloni
Colorado State University, Fort Collins, CO, USA

B. R. Anholt
University of Victoria, Victoria, BC, Canada; Bamfield
Marine Sciences Centre, Bamfield, BC, Canada

C. J. L. Atkinson
University of Queensland, St Lucia, QLD, Australia

F. Aureli
Liverpool John Moores University, Liverpool, UK

A. Avarguès-Weber
CNRS, Université de Toulouse, Toulouse, France; Centre
de Recherches sur la Cognition Animale, Toulouse,
France

K. L. Ayres
University of Washington, Seattle, WA, USA

J. Bakker
University of Liège, Liège, Belgium

G. F. Ball
Johns Hopkins University, Baltimore, MD, USA

J. Balthazart
University of Liège, Liège, Belgium

L. Barrett
University of Lethbridge, Lethbridge, AB, Canada

A. H. Bass
Cornell University, Ithaca, NY, USA

D. K. Bassett
University of Auckland, Auckland, New Zealand

M. Bateson
Newcastle University, Newcastle upon Tyne, UK

G. Beauchamp
University of Montréal, St. Hyacinthe, QC, Canada

B. V. Beaver
Texas A&M University, College Station, TX, USA

P. A. Bednekoff
Eastern Michigan University, Ypsilanti, MI, USA

M. Beekman
University of Sydney, Sydney, NSW, Australia

J. A. Bender
Case Western Reserve University, Cleveland,
OHIO, USA

G. E. Bentley
University of California, Berkeley, CA, USA

A. Berchtold
University of Lausanne, Lausanne, Switzerland

I. S. Bernstein
University of Georgia, Athens, GA, USA

S. Bevins
Colorado State University, Fort Collins, USA

D. T. Blumstein
University of California, Los Angeles, CA, USA

C. R. B. Boake
University of Tennessee, Knoxville, TN, USA

R. A. Boakes
University of Sydney, Sydney, NSW, Australia

W. J. Boeing
New Mexico State University, Las Cruces, NM, USA

N. J. Boogert
McGill University, Montréal, QC, Canada

T. Boswell
Newcastle University, Newcastle upon Tyne, UK

A. Bouskila
Ben-Gurion University of the Negev, Beer Sheva, Israel

R. M. Bowden
Illinois State University, Normal, IL, USA

E. M. Brannon
Duke University, Durham, NC, USA

M. D. Breed
University of Colorado, Boulder, CO, USA

M. R. Bregman
University of California, San Diego, CA, USA

J. Brodeur
Université de Montréal, Montréal, QC, Canada

E. D. Brodie, III
University of Virginia, Charlottesville, VA, USA

A. Brodin
Lund University, Lund, Sweden

D. M. Broom
University of Cambridge, Cambridge, UK

J. L. Brown
University at Albany, Albany, NY, USA

J. S. Brown
University of Illinois at Chicago, Chicago, IL, USA

M. J. F. Brown
Royal Holloway University of London, Egham, UK

H. Brumm
Max Planck Institute for Ornithology, Seewiesen, Germany

R. Buffenstein
University of Texas Health Science Center at San Antonio, San Antonio, TX, USA

J. D. Buntin
University of Wisconsin-Milwaukee, Milwaukee, WI, USA

J. Burger
Rutgers University, Piscataway, NJ, USA

G. M. Burghardt
University of Tennessee, Knoxville, TN, USA

R. W. Burkhardt, Jr.
University of Illinois at Urbana-Champaign, Urbana, IL, USA

N. T. Burley
University of California, Irvine, CA, USA

S. S. Burmeister
University of North Carolina, Chapel Hill, NC, USA

D. S. Busch
Northwest Fisheries Science Center, National Marine Fisheries Service, Seattle, WA, USA

R. W. Byrne
University of St. Andrews, St. Andrews, Fife, Scotland, UK

R. M. Calisi
University of California, Berkeley, CA, USA

J. Call
Max Planck Institute for Evolutionary Anthropology, Leipzig, Germany

U. Candolin
University of Helsinki, Helsinki, Finland

J. F. Cantlon
Rochester University, Rochester, NC, USA

C. E. Carr
University of Maryland, College Park, MD, USA

C. S. Carter
University of Illinois at Chicago, Chicago, IL, USA

F. Cézilly
Université de Bourgogne, Dijon, France

E. S. Chang
University of California-Davis, Bodega Bay, CA, USA

J. W. Chapman
Rothamsted Research, Harpenden, Hertfordshire, UK

J. C. Choe
Ewha Womans University, Seoul, Korea

J. A. Clarke
University of Northern Colorado, Greeley, CO, USA

N. S. Clayton
University of Cambridge, Cambridge, UK

B. Clucas
University of Washington, Seattle, WA, USA; Humboldt University, Berlin, Germany

R. B. Cocroft
University of Missouri, Columbia, MO, USA

J. H. Cohen
Eckerd College, St. Petersburg, FL, USA

S. P. Collin
University of Western Australia, Crawley, WA, Australia

L. Conradt
University of Sussex, Brighton, UK

W. E. Cooper, Jr.
Indiana University Purdue University Fort Wayne, Fort Wayne, IN, USA

R. G. Coss
University of California, Davis, CA, USA

J. T. Costa
Western Carolina University, Cullowhee, NC, USA; Highlands Biological Station, Highland NC, USA

I. D. Couzin
Princeton University, Princeton, NJ, USA

N. J. Cowan
Johns Hopkins University, Baltimore, MD, USA

R. M. Cox
Dartmouth College, Hanover, NH, USA

J. Crast
University of Georgia, Athens, GA, USA

S. Creel
Montana State University, Bozeman, MT, USA

W. Cresswell
University of St. Andrews, St. Andrews, Scotland, UK

D. Crews
University of Texas, Austin, TX, USA

K. R. Crooks
Colorado State University, Fort Collins, CO, USA

J. D. Crystal
University of Georgia, Athens, GA, USA

S. R. X. Dall
University of Exeter, Cornwall, UK

D. Daniels
University at Buffalo, State University of New York, Buffalo, NY, USA

J. M. Davis
Vassar College, Poughkeepsie, NY, USA

K. Dean
University of Maryland, College Park, MD, USA

J. Deen
University of Minnesota, St. Paul, MN, USA

R. J. Denver
University of Michigan, Ann Arbor, MI, USA

C. D. Derby
Neuroscience Institute and Department of Biology, Atlanta, GA, USA

M. E. Deutschlander
Hobart and William Smith Colleges, Geneva, NY, USA

F. B. M. de Waal
Emory University, Atlanta, GA, USA

D. A. Dewsbury
University of Florida, Gainesville, FL, USA

A. Dickinson
University of Cambridge, Cambridge, UK

J. L. Dickinson
Cornell University, Ithaca, NY, USA

A. G. Dolezal
Arizona State University, Tempe, AZ, USA

B. Doligez
Université de Lyon, Villeurbanne, France

R. H. Douglas
City University, London, UK

K. B. Døving
University of Oslo, Oslo, Norway

V. A. Drake
University of New South Wales at the Australian Defence Force Academy, Canberra, ACT, Australia

L. C. Drickamer
Northern Arizona University, Flagstaff, AZ, USA

H. Drummond
Universidad Nacional Autónoma de México, México

J. P. Drury
University of California, Los Angeles, CA, USA

J. E. Duffy
Virginia Institute of Marine Science, Gloucester Point, VA, USA

R. Dukas
McMaster University, Hamilton, ON, Canada

F. C. Dyer
Michigan State University, East Lansing, MI, USA

W. G. Eberhard
Smithsonian Tropical Research Institute; Universidad de Costa Rica, Ciudad Universitaria, Costa Rica

N. J. Emery
Queen Mary University of London, London, UK; University of Cambridge, Cambridge, UK

C. S. Evans
Macquarie University, Sydney, NSW, Australia

S. E. Fahrbach
Wake Forest University, Winston-Salem, NC, USA

F. Fernández-Juricic
Purdue University, West Lafayette, IN, USA

J. R. Fetcho
Cornell University, Ithaca, NY, USA

J. H. Fewell
Arizona State University, Tempe, AZ, USA

G. Fleissner
Goethe-University Frankfurt, Frankfurt, Germany

G. Fleissner
Goethe-University Frankfurt, Frankfurt, Germany

T. H. Fleming
University of Miami, Coral Gables, FL, USA

A. Florsheim
Veterinary Behavior Solutions, Dallas, TX, USA

E. S. Fortune
Johns Hopkins University, Baltimore, MD, USA

R. B. Forward, Jr.
Duke University Marine Laboratory, Beaufort, NC, USA

S. A. Foster
Clark University, Worcester, MA, USA

D. M. Fragaszy
University of Georgia, Athens, GA, USA

O. N. Fraser
University of Vienna, Vienna, Austria

P. J. Fraser
University of Aberdeen, Aberdeen, Scotland, UK

T. M. Freeberg
University of Tennessee, Knoxville, TN, USA

K. A. French
University of California, San Diego, La Jolla, CA, USA

A. Frid
Vancouver Aquarium, Vancouver, BC, Canada

C. B. Frith
Private Independent Ornithologist, Malanda, QLD, Australia

D. J. Funk
Vanderbilt University, Nashville, TN, USA

L. Fusani
University of Ferrara, Ferrara, Italy

C. R. Gabor
Texas State University-San Marcos, San Marcos, TX, USA

R. Gadagkar
Indian Institute Science, Bangalore, India

B. G. Galef
McMaster University, Hamilton, ON, Canada

C. M. Garcia
Instituto de Ecología, UNAM, México

S. A. Gauthreaux, Jr.
Clemson University, Clemson, SC, USA

F. Geiser
University of New England, Armidale, NSW, Australia

T. Q. Gentner
University of California, San Diego, CA, USA

H. C. Gerhardt
University of Missouri, Columbia, MO, USA

M. D. Ginzel
Purdue University, West Lafayette, IN, USA

L.-A. Giraldeau
Université du Québec à Montréal, Montréal, QC, Canada

M. Giurfa
CNRS, Université de Toulouse, Toulouse, France; Centre de Recherches sur la Cognition Animale, Toulouse, France

J.-G. J. Godin
Carleton University, Ottawa, ON, Canada

J. Godwin
North Carolina State University, Raleigh, NC, USA

E. Goodale
Field Ornithology Group of Sri Lanka, University of Colombo, Colombo, Sri Lanka

M. A. D. Goodisman
Georgia Institute of Technology, Atlanta, GA, USA

C. J. Goodnight
University of Vermont, Burlington, Vermont, USA

P. A. Gowaty
University of California, Los Angeles, CA, USA; Smithsonian Tropical Research Institute, USA

W. Goymann
Max Planck Institute for Ornithology, Seewiesen, Germany

P. Graham
University of Sussex, Brighton, UK

T. Grandin
Colorado State University, Fort Collins, CO, USA

M. D. Greenfield
Université François Rabelais de Tours, Tours, France

G. F. Grether
University of California, Los Angeles, CA, USA

A. S. Griffin
University of Newcastle, Callaghan, NSW, Australia

M. Griggio
Konrad Lorenz Institute for Ethology, Vienna, Austria

T. G. G. Groothuis
University of Groningen, Groningen, Netherlands

R. Grosberg
University of California, Davis, CA, USA

C. M. Grozinger
Pennsylvania State University, University Park, PA, USA

R. D. Grubbs
Florida State University Coastal and Marine Laboratory, St. Teresa, FL, USA; George Mason University, Fairfax, VA, USA

R. R. Ha
University of Washington, Seattle, WA, USA

J. P. Hailman
University of Wisconsin, Jupiter, FL, USA

I. M. Hamilton
Ohio State University, Columbus, OH, USA

R. R. Hampton
Emory University, Atlanta, GA, USA

I. C. W. Hardy
University of Nottingham, Loughborough, Leicestershire, UK

B. L. Hart
University of California, Davis, CA, USA

L. I. Haug
Texas Veterinary Behavior Services, Sugar Land, TX, USA

M. Hauser
Harvard University, Cambridge, MA, USA

L. S. Hayward
University of Washington, Seattle, WA, USA

S. D. Healy
University of St. Andrews, St. Andrews, Fife, Scotland, UK

E. A. Hebets
University of Nebraska, Lincoln, NE, USA

M. R. Heithaus
Florida International University, Miami, FL, USA

H. Helanterä
University of Sussex, Brighton, UK; University of Helsinki, Helsinki, Finland

J. M. Hemmi
Australian National University, Canberra, ACT, Australia

L. M. Henry
University of Oxford, Oxford, UK

J. M. Herbers
Ohio State University, Columbus, OH, USA

M. R. Heupel
James Cook University, Townsville, QLD, Australia

H. Hoi
Konrad Lorenz Institute for Ethology, Vienna, Austria

K. E. Holekamp
Michigan State University, East Lansing, MI, USA

R. A. Holland
Max Planck Institute for Ornithology, Radolfzell, Germany

A. G. Horn
Dalhousie University, Halifax, NS, Canada

L. Huber
University of Vienna, Vienna, Austria

M. A. Huffman
Kyoto University, Inuyama, Aichi Prefecture, Japan

H. Hurd
Keele University, Staffordshire, UK

P. L. Hurd
University of Alberta, Edmonton, AB, Canada

A. Jacobs
University of California, Riverside, CA, USA

V. M. Janik
University of St. Andrews, St. Andrews, Fife, Scotland, UK

K. Jensen
Queen Mary University of London, London, UK

C. Jozet-Alves
University of Caen Basse-Normandie, Caen, France

J. Kaminski
Max Planck Institute for Evolutionary Anthropology, Leipzig, Germany

L. Kapás
Washington State University, Spokane, WA, USA

A. S. Kauffman
University of California, San Diego, La Jolla, CA, USA

J. L. Kelley
University of Western Australia, Crawley, WA, Australia

A. J. King
Zoological Society of London, London, UK; University of Cambridge, Cambridge, UK

S. L. Klein
Johns Hopkins Bloomberg School of Public Health, Baltimore, MD, USA

M. J. Klowden
University of Idaho, Moscow, ID, USA

J. Komdeur
University of Groningen, Groningen, Netherlands

M. Konishi
California Institute of Technology, Pasadena, CA, USA

J. Korb
University of Osnabrueck, Osnabrück, Germany

I. Krams
University of Daugavpils, Daugavpils, Latvia

R. T. Kraus
Florida State University Coastal and Marine Laboratory, St. Teresa, FL, USA; George Mason University, Fairfax, VA, USA

W. B. Kristan, Jr.
University of California, San Diego, La Jolla, CA, USA

J. M. Krueger
Washington State University, Spokane, WA, USA

C. W. Kuhar
Cleveland Metroparks Zoo, Cleveland, OH, USA

C. P. Kyriacou
University of Leicester, Leicester, UK

F. Ladich
University of Vienna, Vienna, Austria

K. N. Laland
University of St Andrews, St Andrews, Fife, Scotland, UK

P. H. L. Lamberton
Imperial College Faculty of Medicine, London, UK

A. V. Latchininsky
University of Wyoming, Laramie, WY, USA

L. Lefebvre
McGill University, Montréal, QC, Canada

J. E. Leonard
Hiwassee College, Madisonville, TN, USA

M. L. Leonard
Dalhousie University, Halifax, NS, Canada

G. R. Lewin
Max-Delbrück Center for Molecular Medicine, Berlin, Germany

F. Libersat
Institut de Neurobiologie de la Méditerranée, Parc Scientifique de Luminy, Marseille, France

A. E. Liebert
Framingham State College, Framingham, MA, USA

C. H. Lin
University of British Columbia, Vancouver, BC, Canada

J. A. Linares
Texas A&M University, Gonzales, TX, USA

J. Lind
Stockholm University, Stockholm, Sweden

T. A. Linksvayer
University of Copenhagen, Copenhagen, Denmark

C. List
London School of Economics, London, UK

N. Lo
Australian Museum, Sydney, NSW, Australia; University of Sydney, Sydney, NSW, Australia

C. M. F. Lohmann
University of North Carolina, Chapel Hill, NC, USA

K. J. Lohmann
University of North Carolina, Chapel Hill, NC, USA

Y. Lubin
Ben-Gurion University of the Negev, Beer Sheva, Israel

J. Lucas
Purdue University, West Lafayette, IN, USA

S. K. Lynn
Boston College, Chestnut Hill, MA, USA

K. E. Mabry
New Mexico State University, Las Cruces, NM, USA

D. Maestripieri
University of Chicago, Chicago, IL, USA

D. L. Maney
Emory University, Atlanta, GA, USA

T. G. Manno
Auburn University, Auburn, AL, USA

S. W. Margulis
Canisius College, Buffalo, NY, USA

L. Marino
Emory University, Atlanta, GA, USA

T. A. Markow
University of California at San Diego, La Jolla, CA, USA

C. A. Marler
University of Wisconsin, Madison, WI, USA

P. P. Marra
Smithsonian Migratory Bird Center, National Zoological Park, Washington, DC, USA

L. B. Martin
University of South Florida, Tampa, FL, USA

M. Martin
North Carolina State University, Raleigh, NC, USA

J. A. Mather
University of Lethbridge, Lethbridge, AB, Canada

K. Matsuura
Okayama University, Okayama, Japan

T. Matsuzawa
Kyoto University, Kyoto, Japan

K. McAuliffe
Harvard University, Cambridge, MA, USA

E. A. McGraw
University of Queensland, Brisbane, QLD, Australia

N. L. McGuire
University of California, Berkeley, CA, USA

N. J. Mehdiabadi
Smithsonian Institution, Washington, DC, USA

R. Menzel
Freie Universität Berlin, Berlin, Germany

J. C. Mitani
University of Michigan, Ann Arbor, MI, USA

J. C. Montgomery
University of Auckland, Auckland, New Zealand

J. Moore
Colorado State University, Fort Collins, CO, USA

J. Morand-Ferron
Université du Québec à Montréal, Montréal, QC, Canada

J. Moreno
Museo Nacional de Ciencias Naturales, Madrid, Spain

K. Morgan
University of St. Andrews, St. Andrews, Fife, Scotland, UK

R. Muheim
Lund University, Lund, Sweden

C. A. Nalepa
North Carolina State University, Raleigh, NC, USA

D. Naug
Colorado State University, Fort Collins, CO, USA

D. A. Nelson
Ohio State University, Columbus, OH, USA

R. J. Nelson
Ohio State University, Columbus, OH, USA

I. Newton
Centre for Ecology & Hydrology, Wallingford, UK

K. Nishimura
Hokkaido University, Hakodate, Japan

J. E. Niven
University of Cambridge, Cambridge, UK; Smithsonian Tropical Research Institute, Panamá, República de Panamá

P. Nonacs
University of California, Los Angeles, CA, USA

A. J. Norton
Imperial College Faculty of Medicine, London, UK

B. P. Oldroyd
University of Sydney, Sydney, NSW, Australia

T. J. Ord
University of New South Wales, Sydney, NSW, Australia

M. A. Ottinger
University of Maryland, College Park, MD, USA

D. H. Owings
University of California, Davis, CA, USA

J. M. Packard
Texas A&M University, College Station, TX, USA

A. Pai
Spelman College, Atlanta, GA, USA

T. J. Park
University of Illinois at Chicago, Chicago, IL, USA

L. A. Parr
Yerkes National Primate Research Center, Atlanta, GA, USA

Y. M. Parsons
La Trobe University, Bundoora, VIC, Australia

G. L. Patricelli
University of California, Davis, CA, USA

M. M. Patten
Museum of Comparative Zoology, Cambridge, MA, USA

A. Payne
Tufts University, Medford, MA, USA

I. M. Pepperberg
Harvard University, Cambridge, MA, USA

M.-J. Perrot-Minnot
Université de Bourgogne, Dijon, France

S. Perry
University of California-Los Angeles, Los Angeles, CA, USA

K. M. Pickett
University of Vermont, Burlington, VT, USA

N. Pinter-Wollman
Stanford University, Stanford, CA, USA

D. Plachetzki
University of California, Davis, CA, USA

G. S. Pollack
McGill University, Montréal, QC, Canada

G. D. Pollak
University of Texas at Austin, Austin, TX, USA

R. Poulin
University of Otago, Dunedin, New Zealand

S. C. Pratt
Arizona State University, Tempe, AZ, USA

V. V. Pravosudov
University of Nevada, Reno, NV, USA

G. H. Pyke
Australian Museum, Sydney, NSW, Australia; Macquarie University, North Ryde, NSW, Australia

D. C. Queller
Rice University, Houston, TX, USA

M. Ramenofsky
University of California, Davis, CA, USA

C. H. Rankin
University of British Columbia, Vancouver, BC, Canada

F. L. W. Ratnieks
University of Sussex, Brighton, UK

D. Raubenheimer
Massey University, Auckland, New Zealand

S. M. Reader
Utrecht University, Utrecht, Netherlands

H. K. Reeve
Cornell University, New York, NY, USA

J. Reinhard
University of Queensland, Brisbane, QLD, Australia

L. Rendell
University of St Andrews, St Andrews, Fife, Scotland, UK

A. N. Rice
Cornell University, Ithaca, NY, USA

J. M. L. Richardson
University of Victoria, Victoria, BC, Canada

H. Richner
University of Bern, Bern, Switzerland

T. Rigaud
Université de Bourgogne, Dijon, France

R. E. Ritzmann
Case Western Reserve University, Cleveland, OHIO, USA

A. J. Riveros
University of Arizona, Tucson, AZ, USA

D. Robert
University of Bristol, Bristol, UK

G. E. Robinson
University of Illinois at Urbana-Champaign, Urbana, IL, USA

I. Rodriguez-Prieto
Museo Nacional de Ciencias Naturales, Madrid, Spain

B. D. Roitberg
Simon Fraser University, Burnaby, BC, Canada

L. M. Romero
Tufts University, Medford, MA, USA

T. J. Roper
University of Sussex, Brighton, UK

G. G. Rosenthal
Texas A&M University, College Station, TX, USA

C. Rowe
Newcastle University, Newcastle upon Tyne, UK

L. Ruggiero
Barnard College and Columbia University, New York, NY, USA

G. D. Ruxton
University of Glasgow, Glasgow, Scotland, UK

M. J. Ryan
University of Texas, Austin, TX, USA

R. Safran
University of Colorado, Boulder, CO, USA

W. Saltzman
University of California, Riverside, CA, USA

R. M. Sapolsky
Stanford University, Stanford, CA, USA

L. S. Sayigh
Woods Hole Oceanographic Institution, Woods Hole, MA, USA

A. Schmitz
University of Bonn, Bonn, Germany

H. Schmitz
University of Bonn, Bonn, Germany

J. Schulkin
Georgetown University, Washington, DC, USA; National Institute of Mental Health, Bethesda, MD, USA

H. Schwabl
Washington State University, Pullman, WA, USA

A. M. Seed
Max Planck Institute for Evolutionary Anthropology, Leipzig, Germany

M. R. Servedio
University of North Carolina, Chapel Hill, NC, USA

J. C. Shaw
University of California, Santa Barbara, CA, USA

S.-F. Shen
Cornell University, New York, NY, USA

B. L. Sherman
North Carolina State University, Raleigh, NC, USA

T. N. Sherratt
Carleton University, Ottawa, ON, Canada

D. M. Shuker
University of St. Andrews, St. Andrews, Fife, Scotland, UK

R. Silver
Barnard College and Columbia University, New York, NY, USA

B. Silverin
University of Göteborg, Göteborg, Sweden

A. M. Simmons
Brown University, Providence, RI, USA

S. J. Simpson
University of Sydney, Sydney, NSW, Australia

U. Sinsch
University Koblenz-Landau, Koblenz, Germany

H. Slabbekoorn
Leiden University, Leiden, Netherlands

P. J. B. Slater
University of St. Andrews, St. Andrews, Fife, Scotland, UK

C. N. Slobodchikoff
Northern Arizona University, Flagstaff, AZ, USA

A. R. Smith
Smithsonian Tropical Research Institute, Balboa, Ancon, Panamá

G. T. Smith
Indiana University, Bloomington, IN, USA

J. E. Smith
Michigan State University, East Lansing, MI, USA

B. Smuts
University of Michigan, Ann Arbor, MI, USA

E. C. Snell-Rood
Indiana University, Bloomington, IN, USA

C. T. Snowdon
University of Wisconsin, Madison, WI, USA

R. B. Srygley
USDA-Agricultural Research Service, Sidney, MT, USA

T. Stankowich
University of Massachusetts, Amherst, MA, USA

P. T. Starks
Tufts University, Medford, MA, USA

C. A. Stern
Cornell University, Ithaca, NY, USA

J. R. Stevens
Max Planck Institute for Human Development, Berlin, Germany

P. K. Stoddard
Florida International University, Miami, FL, USA

J. E. Strassmann
Rice University, Houston, TX, USA

C. E. Studds
Smithsonian Migratory Bird Center, National Zoological Park, Washington, DC, USA

L. Sullivan-Beckers
University of Nebraska, Lincoln, NE, USA

R. A. Suthers
Indiana University, Bloomington, IN, USA

J. P. Swaddle
College of William and Mary, Williamsburg, VA, USA

R. Swaisgood
San Diego Zoo's Institute for Conservation Research, Escondido, CA, USA

É. Szentirmai
Washington State University, Spokane, WA, USA

M. Taborsky
University of Bern, Hinterkappelen, Switzerland

Z. Tang-Martínez
University of Missouri-St. Louis, St. Louis, MO, USA

E. Tauber
University of Leicester, Leicester, UK

D. W. Thieltges
University of Otago, Dunedin, New Zealand

F. Thomas
Génétique et Evolution des Maladies Infectieuses, Montpellier, France; Université de Montréal, Montréal, QC, Canada

C. V. Tillberg
Linfield College, McMinnville, OR, USA

M. Tomasello
Max Planck Institute for Evolutionary Anthropology, Leipzig, Germany

A. L. Toth
Pennsylvania State University, University Park, PA, USA

B. C. Trainor
University of California, Davis, CA, USA

J. Traniello
Boston University, Boston, MA, USA

K. Tsuji
University of the Ryukyus, Okinawa, Japan

G. W. Uetz
University of Cincinnati, Cincinnati, OH, USA

M. Valentine
University of Vermont, Burlington, VT, USA

A. Valero
Instituto de Ecología, UNAM, México

J. L. Van Houten
University of Vermont, Burlington, VT, USA

M. A. van Noordwijk
University of Zurich, Zurich, Switzerland

C. P. van Schaik
University of Zurich, Zurich, Switzerland

S. H. Vessey
Bowling Green State University, Bowling Green, OH, USA

G. von der Emde
University of Bonn, Bonn, Germany

H. G. Wallraff
Max Planck Institute for Ornithology, Seewiesen, Germany

R. R. Warner
University of California, Santa Barbara, CA, USA

E. Warrant
University of Lund, Lund, Sweden

R. Watt
University of Edinburgh, Edinburgh, Scotland, UK

J. P. Webster
Imperial College Faculty of Medicine, London, UK

M. Webster
Cornell Lab of Ornithology, Ithaca, NY, USA

N. Wedell
University of Exeter, Penryn, UK

E. V. Wehncke
Biodiversity Research Center of the Californias, San Diego, CA, USA

M. J. West-Eberhard
Smithsonian Tropical Research Institute, Costa Rica

G. Westhoff
Tierpark Hagenbeck gGmbH, Hamburg, Germany

C. J. Whelan
Illinois Natural History Survey, University of Illinois at Chicago, Chicago, IL, USA

A. Whiten
University of St. Andrews, St. Andrews, Fife, Scotland, UK

A. Wilkinson
University of Virginia, Charlottesville, VA, USA

D. M. Wilkinson
Liverpool John Moores University, Liverpool, UK

S. P. Windsor
University of Auckland, Auckland, New Zealand

J. C. Wingfield
University of California, Davis, CA, USA

K. E. Wynne-Edwards
University of Calgary, Calgary, AB, Canada

D. D. Yager
University of Maryland, College Park,
MD, USA

R. Yamada
University of Queensland, Brisbane, QLD,
Australia

J. Yano
University of Vermont, Burlington, VT, USA

K. Yasukawa
Beloit College, Beloit, WI, USA

J. Zeil
Australian National University, Canberra, ACT,
Australia

T. R. Zentall
University of Kentucky, Lexington, KY, USA

E. Zou
Nicholls State University, Thibodaux, LA, USA

M. Zuk
University of California, Riverside, CA, USA

GUIDE TO USE OF THE ENCYCLOPEDIA

Structure of the Encyclopedia

The material in the Encyclopedia is arranged as a series of articles in alphabetical order.

There are four features to help you easily find the topic you're interested in: an alphabetical contents list, a subject classification index, cross-references and a full subject index.

1. Alphabetical Contents List

The alphabetical contents list, which appears at the front of each volume, lists the entries in the order that they appear in the Encyclopedia. It includes both the volume number and the page number of each entry.

2. Subject Classification Index

This index appears at the start of each volume and groups entries under subject headings that reflect the broad themes of Animal Behavior. This index is useful for making quick connections between entries and locating the relevant entry for a topic that is covered in more than one article.

3. Cross-references

All of the entries in the Encyclopedia have been extensively cross-referenced. The cross-references which appear at the end of an entry, serve three different functions:

i. To indicate if a topic is discussed in greater detail elsewhere
ii. To draw the readers attention to parallel discussions in other entries
iii. To indicate material that broadens the discussion

Example

The following list of cross-references appears at the end of the entry Landmark Studies: Honeybees

See also: Communication: Social Recognition; Invertebrate Social Behavior: Ant, Bee and Wasp Social Evolution; Invertebrate Social Behavior: Caste Determination in Arthropods; Invertebrate Social Behavior: Collective Intelligence; Invertebrate Social Behavior: Dance Language; Invertebrate Social Behavior: Developmental Plasticity; Invertebrate Social Behavior: Division of Labor; Invertebrate Social Behavior: Queen-Queen Conflict in Eusocial Insect Colonies; Invertebrate Social Behavior: Queen-Worker Conflicts Over Colony Sex Ratio.

4. Index

The index includes page numbers for quick reference to the information you're looking for. The index entries differentiate between references to a whole entry, a part of an entry, and a table or figure.

5. Contributors

At the start of each volume there is list of the authors who contributed to the Encyclopedia.

SUBJECT CLASSIFICATION

Anti-Predator Behavior

Section Editor: *Ted Stankowich*

Antipredator Benefits from Heterospecifics
Co-Evolution of Predators and Prey
Conservation and Anti-Predator Behavior
Defensive Avoidance
Defensive Chemicals
Defensive Coloration
Defensive Morphology
Ecology of Fear
Economic Escape
Empirical Studies of Predator and Prey Behavior
Games Played by Predators and Prey
Group Living
Life Histories and Predation Risk
Predator Avoidance: Mechanisms
Parasitoids
Predator's Perspective on Predator–Prey Interactions
Risk Allocation in Anti-Predator Behavior
Risk-Taking in Self-Defense
Trade-Offs in Anti-Predator Behavior
Vigilance and Models of Behavior

Applications

Section Editor: *Michael D. Breed* and *Janice Moore*

Conservation and Animal Behavior
Robot Behavior
Training of Animals

Arthropod Social Behavior

Section Editor: *Jae Chun Choe*

Ant, Bee and Wasp Social Evolution
Caste Determination in Arthropods
Collective Intelligence
Colony Founding in Social Insects
Crustacean Social Evolution
Dance Language

Developmental Plasticity
Division of Labor
Kin Selection and Relatedness
Parasites and Insects: Aspects of Social Behavior
Queen–Queen Conflict in Eusocial Insect Colonies
Queen–Worker Conflicts Over Colony Sex Ratio
Recognition Systems in the Social Insects
Reproductive Skew
Sex and Social Evolution
Social Evolution in 'Other' Insects and Arachnids
Spiders: Social Evolution
Subsociality and the Evolution of Eusociality
Termites: Social Evolution
Worker–Worker Conflict and Worker Policing

Behavioral Endocrinology

Section Editor: *John C. Wingfield*

Aggression and Territoriality
Aquatic Invertebrate Endocrine Disruption
Behavioral Endocrinology of Migration
Circadian and Circannual Rhythms and Hormones
Communication and Hormones
Conservation Behavior and Endocrinology
Experimental Approaches to Hormones and Behavior: Invertebrates
Female Sexual Behavior and Hormones in Non-Mammalian Vertebrates
Field Techniques in Hormones and Behavior
Fight or Flight Responses
Food Intake: Behavioral Endocrinology
Hibernation, Daily Torpor and Estivation in Mammals and Birds: Behavioral Aspects
Hormones and Behavior: Basic Concepts
Immune Systems and Sickness Behavior
Invertebrate Hormones and Behavior
Male Sexual Behavior and Hormones in Non-Mammalian Vertebrates
Mammalian Female Sexual Behavior and Hormones
Maternal Effects on Behavior
Memory, Learning, Hormones and Behavior

Cognition

Section Editor: *Nicola Clayton*

Communication

Section Editor: *Jeffery Lucas*

Conservation

Section Editor: *Constantíno Macías Garcia*

Decision Making by Individuals

Section Editor: *David W. Stephens*

Evolution

Section Editor: *Joan M. Herbers*

Evolution: Fundamentals
Isolating Mechanisms and Speciation
Levels of Selection
Microevolution and Macroevolution in Behavior
Nervous System: Evolution in Relation to Behavior
Phylogenetic Inference and the Evolution of Behavior
Reproductive Success
Specialization

Foraging

Section Editor: *David W. Stephens*

Caching
Digestion and Foraging
Foraging Modes
Habitat Selection
Hormones and Breeding Strategies, Sex Reversal, Brood
    Parasites, Parthenogenesis
Hunger and Satiety
Internal Energy Storage
Kleptoparasitism and Cannibalism
Optimal Foraging and Plant-Pollinator Co-Evolution
Optimal Foraging Theory: Introduction
Patch Exploitation
Wintering Strategies: Moult and Behavior

Genetics

Section Editor: *Ross H. Crozier*

Caste in Social Insects: Genetic Influences Over Caste
    Determination
Dictyostelium, the Social Amoeba
Drosophila Behavior Genetics
Genes and Genomic Searches
Kin Recognition and Genetics
Marine Invertebrates: Genetics of Colony Recognition
Nasonia Wasp Behavior Genetics
Orthopteran Behavioral Genetics
Parmecium Behavioral Genetics
Social Insects: Behavioral Genetics
Unicolonial Ants: Loss of Colony Identity

History

Section Editor: *Michael D. Breed*

Animal Behavior: Antiquity to the Sixteenth Century
Animal Behavior: The Seventeenth to the Twentieth
    Centuries
Behavioral Ecology and Sociobiology
Comparative Animal Behavior – 1920–1973
Ethology in Europe

Future of Animal Behavior: Predicting Trends
Integration of Proximate and Ultimate Causes
Neurobiology, Endocrinology and Behavior
Psychology of Animals

Infectious Disease and Behavior

Section Editor: *Janice Moore*

Avoidance of Parasites
Beyond Fever: Comparative Perspectives on Sickness
    Behavior
Conservation, Behavior, Parasites and Invasive Species
Ectoparasite Behavior
Evolution of Parasite-Induced Behavioral Alterations
Intermediate Host Behavior
Parasite-Induced Behavioral Change: Mechanisms
Parasite-Modified Vector Behavior
Parasites and Sexual Selection
Propagule Behavior and Parasite Transmission
Reproductive Behavior and Parasites: Vertebrates
Reproductive Behavior and Parasites: Invertebrates
Self-Medication: Passive Prevention and Active Treatment
Social Behavior and Parasites

Landmark Studies

Section Editor: *Michael D. Breed*

Alex: A Study in Avian Cognition
Aplysia
Barn Swallows: Sexual and Social Behavior
Betta Splendens
Boobies
Bowerbirds
Chimpanzees
Cockroaches
Domestic Dogs
Hamilton, William Donald
Herring Gulls
Honeybees
Locusts
Lorenz, Konrad
Norway Rats
Octopus
Pheidole: Sociobiology of a Highly Diverse Genus
Pigeons
Rhesus Macaques
Sharks
Spotted Hyenas
Swordtails and Platyfishes
Threespine Stickleback
Tinbergen, Niko
Tribolium

Túngara Frog: A Model for Sexual Selection and
   Communication
Turtles: Freshwater
White-Crowned Sparrow
Wolves
Zebra Finches
Zebrafish

Learning and Development

Section Editor: *Daniel Papaj*

Costs of Learning
Decision-Making and Learning: The Peak Shift
   Behavioral Response
Habitat Imprinting
Mate Choice and Learning
Play
Spatial Memory

Methodology

Section Editor: *James Ha*

Cost–Benefit Analysis
Dominance Relationships, Dominance Hierarchies and
   Rankings
Endocrinology and Behavior: Methods
Ethograms, Activity Profiles and Energy Budgets
Experiment, Observation, and Modeling in the Lab and
   Field
Experimental Design: Basic Concepts
Game Theory
Measurement Error and Reliability
Neuroethology: Methods
Playbacks in Behavioral Experiments
Remote-Sensing of Behavior
Robotics in the Study of Animal Behavior
Sequence Analysis and Transition Models
Spatial Orientation and Time: Methods

Migration, Orientation, and Navigation

Section Editor: *Sidney Gauthreaux*

Amphibia: Orientation and Migration
Bat Migration
Bats: Orientation, Navigation and Homing
Bird Migration
Fish Migration
Insect Migration
Insect Navigation
Irruptive Migration

Magnetic Compasses in Insects
Magnetic Orientation in Migratory Songbirds
Maps and Compasses
Migratory Connectivity
Pigeon Homing as a Model Case of Goal-Oriented
   Navigation
Sea Turtles: Navigation and Orientation
Vertical Migration of Aquatic Animals

Networks – Social

Section Editor: *Deborah M. Gordon*

Consensus Decisions
Disease Transmission and Networks
Group Movement
Life Histories and Network Function
Nest Site Choice in Social Insects

Neuroethology

Section Editor: *Harold Zakon*

Acoustic Communication in Insects: Neuroethology
Bat Neuroethology
Crabs and Their Visual World
Insect Flight and Walking: Neuroethological Basis
Leech Behavioral Choice: Neuroethology
Naked Mole Rats: Their Extraordinary Sensory World
Nematode Learning and Memory: Neuroethology
Neuroethology: What is it?
Parasitoid Wasps: Neuroethology
Predator Evasion
Sociogenomics
Sound Localization: Neuroethology
Vocal–Acoustic Communication in Fishes:
   Neuroethology

Reproductive Behavior

Section Editor: *Patricia Adair Gowaty*

Bateman's Principles: Original Experiment and Modern
   Data For and Against
Compensation in Reproduction
Cryptic Female Choice
Differential Allocation
Flexible Mate Choice
Forced or Aggressively Coerced Copulation
Helpers and Reproductive Behavior in Birds and
   Mammals
Infanticide

CONTENTS

B

C

D

E

F

VOLUME 2

G

H

I

K

L

M

VOLUME 3

Q

R

W

Z

Game Theory

K. Yasukawa, Beloit College, Beloit, WI, USA

Introduction

Modeling is a way to test behavioral hypotheses, but it does not involve observing or experimentally manipulating the behavior of animals. Instead, we study the behavior of a model. Like a simple, hand-drawn map, a model is a representation of reality, but it is not meant to be real. Despite their simplified view of behavior, mathematical models can be used to study animal behavior by presenting our understanding of an aspect of behavior in a formal and testable set of equations. The equations can be solved mathematically to examine how behavior might operate under very clearly described circumstances, which are called the model assumptions. Despite the mathematical nature of these methods, they are used to test predictions of hypotheses, so they are relevant to us. In animal behavior, a commonly used modeling method is game theory.

Game theory, a branch of applied mathematics, was developed to study decision making in conflict situations. It describes behavior strategically in situations where one individual's success depends on the choices of others. It was initially developed to analyze zero-sum contests in which one individual does better at another's expense, but it has been expanded to cover many other situations. In many applications, the goal is to identify equilibria or best solutions. A well-known example is the Nash equilibrium, which was devised by John Nash (made famous in the movie, *A Beautiful Mind*) for economic systems. Game theory formally began in 1944 with the publication of John von Neumann and Oskar Morgenstern's *Theory of Games and Economic Behavior.* The application of game theory to biology did not occur immediately, despite the prevailing view that evolution is a contest between alternate entities (genes, individuals, groups, species). John Maynard Smith is credited for applying game theory to animal behavior, and he was awarded the 1999 Crafoord Prize for this work.

The Hawk–Dove Game

Maynard Smith and Price (1973) first used game theory to analyze contests between rivals who are competing for an important resource such as food, territory, or mates. Maynard Smith and Price were trying to answer a question that had been puzzling animal behaviorists for many years: Why do animals use conventional methods such as display (like disputing neighbors shaking their fists at each other) rather than more violent means to settle disputes? At one time, the answer was, because fighting would produce lots of injuries, which would be bad for the species. Explanations that rely on advantages to the species or other groups of individuals are called group selection hypotheses, but evolutionary analyses in the 1960s and 1970s showed that these hypotheses are often inadequate – they cannot explain how advantages to a group could overcome disadvantages to individuals. If a displayer comes up against a fighter, the fighter would win every time, even if fighting were disadvantageous for the species as a whole. Maynard Smith and Price's solution was the now-classic hawk–dove game.

As with all modeling studies, the hawk–dove game starts with assumptions.

- Animals engage in contests over a resource, and in each contest there is a winner that takes possession of the contested resource.
- Winning the resource item increases the fitness (survival/mating success/reproductive success) of the winner.
- An injury sustained in a contest reduces fitness.
- If a contest continues for a long time, both contestants experience a reduction in fitness as a result of the time wasted (i.e., time that could have been used to acquire other resource items).
- Finally, each animal always employs (plays) a particular strategy (a method of competing) in all contests.

Before we look at the game, the meaning of the term strategy in this context must be clear. A behavioral strategy is simply a fixed and predictable way of behaving in a contest. It does not imply that contesting animals make conscious decisions. Although contests involve two contestants, the purpose of a game-theory model is to compare alternate strategies with each other to see if one is better. In this case, we compare the contest strategies of hawk and dove.

A dove (i.e., an animal that always plays the dove strategy) uses threat display in a contest but never fights. If the opponent also displays, then the dove continues to display as well, but if the opponent attacks, the dove retreats immediately, losing the contest but avoiding injury. Thus, a contest between two doves is protracted and wastes a lot of time for both contestants, although neither contestant is injured. In contrast, a hawk (an animal that always plays the hawk strategy) attacks immediately. If a hawk plays against a dove, the hawk always wins and the dove always loses because the dove retreats immediately. On the other hand, if a hawk plays another hawk, a fight ensues and both contestants risk injury as a result.

These written descriptions of what happens in particular contests can be stated formally in a payoff matrix, which is usually presented in the form of a table of fitness payoffs to one strategy when confronted by either the same or the other strategy (**Table 1**).

What do all these letters mean? The fitness payoff of winning a contest is V (for victory). V measures the amount by which the winner's fitness increases as a result of gaining the contested resource. The fitness loss as a result of injury when two hawks fight is W (for wound). T represents the fitness loss caused by wasting time in a protracted contest between doves. When a hawk plays a dove, the hawk wins the resource, so gets a payoff of V. When a dove plays a hawk, the dove loses the resource but avoids injury, so the payoff to the dove is 0 (dove neither gains nor loses fitness as a result of the contest). If two doves play each other, the payoff to the dove is only $V/2$ because each wins only half the time, but because the contests are protracted, there is a cost of T for the wasted time. Finally, if two hawks play each other, the fitness gain of victory (V) is devalued by the cost of injury (W). The difference $V-W$ is halved because each wins only half the time. With the payoff matrix formally specified, we can try to find the best strategy.

The concept that makes game theory models useful in the study of animal behavior is the evolutionarily stable strategy (ESS). An ESS cannot be invaded by (is stable against) any other strategy. Let us assume that a population of animals is comprised of individuals that all play a particular strategy. What would happen if a new individual with a different strategy joins this population? If the new strategy wins against the old one, it will begin to spread in the population (i.e., this new individual will be successful in reproducing, so its offspring will become more and more prevalent over time). Eventually, the new strategy will become so common that most contests are between two contestants playing that strategy. At this point, if the new strategy still has higher fitness than the old one, which is now rare, the new strategy will continue to spread. Eventually, all animals in the population will play the new strategy. In this case, the old strategy is clearly not an ESS.

What about the hawk–dove game – is there an ESS? To answer this question, we must compare the average fitness of each of the two competing strategies. To do that calculation, we assume that we have a mixed population (i.e., some animals play hawk and some play dove). Let us denote the proportion of the mixed population that plays hawk as H and the proportion that plays dove as D. As there are only two strategies in our population, their proportions must add (sum) to 1 ($H+D=1$, and thus $D=1-H$). If we assume that encounters occur at random (i.e., doves do not seek out doves, for example), then we can calculate the mean fitness of doves and hawks by weighting the payoffs of each strategy by its frequency, as follows.

$$\text{Mean fitness of doves} = (0 \times H) + ([V/2 - T] \times D) \quad [1]$$

$$\text{Mean fitness of hawks} = \{([V-W]/2) \times H\} + VD \quad [2]$$

The term $(0 \times H)$ in eqn [1] is simply the payoff to dove when playing against hawk (0 because dove loses but avoids injury) times the proportion of hawks in the population (how often that interaction will occur). Likewise, the term $([V/2 - T] \times D)$ in the same equation is the payoff of to a dove that plays another dove times the likelihood that the dove's opponent is a dove. (Equation [2] has a similar construction.)

So, how do we decide whether dove or hawk is an ESS? We start with a population that is entirely one strategy and then we see what happens when an individual that plays the other strategy arrives. Suppose that we have a population of doves. If hawk cannot invade this population, then dove is an ESS. Incidentally, this situation is the one that group selection (for the good of the species) explanations would predict, so our game theory model allows us to test the group selection hypothesis. It should be clear to you, however, that dove cannot resist invasion by hawk. Try using eqns [1] and [2] with values of $D=0.99$ and $H=0.01$. As long as both V and T are greater

Table 1   Payoff matrix for the hawk–dove game. Fitness payoffs accrue to the strategies on the left when each plays the strategies at the top

	Hawk	*Dove*
Hawk	(V − W)/2	V
Dove	0	(V/2) − T

than 0, which is how we have defined them, hawk has higher fitness than dove and therefore, hawk will spread. If we start with an all-hawk population, we get a similar result. Dove can invade because it plays against hawk almost all of the time initially (dove is rare and hawk is common) and under these conditions, dove does better than hawk because dove does not pay the cost of injury. Our game-theory model predicts that neither strategy is an ESS, at least as long as $V < W$ (i.e., the cost of injury is high).

If neither strategy is an ESS, what will happen to our all-dove and all-hawk populations? Your intuition might suggest that a mixture of the two strategies might result, but a mathematical analysis of our equations will show us that a particular mixture is a stable equilibrium. What is a stable equilibrium? In this case, it is the mixture (proportions) of hawks and doves at which the fitnesses of the two strategies are equal, or the proportions at which neither strategy does better than the other on average. To calculate this stable mixture, we simply set eqn [1] equal to eqn [2] and solve for the equilibrium values of H and D, which we call H_{eq} and D_{eq}. A bit of algebra yields the following equilibrium solutions.

Equilibrium proportion of doves $(D_{eq}) = (W - V)/(2T + W)$

[3]

Equilibrium proportion of hawks $(H_{eq}) = (2T + V)/(2T + W)$

[4]

In game-theory terms, this equilibrium is called a mixed ESS. Recall that our last assumption when we began this modeling exercise was that each animal always plays the same strategy. If we keep that assumption, then, regardless of the starting mixture, at equilibrium our population will have some individuals (H_{eq}) that play hawk and others (D_{eq}) that play dove. One important aspect of modeling, however, is to examine what happens when we relax assumptions. If we drop the fixed strategy assumption, we end up with a population of individuals that play hawk H_{eq} of the time and dove the rest. As in the fixed strategy version, eqns [3] and [4] describe these equilibrium proportions.

The Hawk–Dove–Retaliator Game

Hawk and dove are certainly not the only ways that an animal might behave in a contest. Another possibility might be called retaliator, which displays against a dove but fights (retaliates against) hawk. To illustrate how this new strategy changes the analysis, I will present a simplified payoff matrix (**Table 2**).

The new payoff matrix uses hypothetical fitness values rather than algebraic expressions, but is analyzed in the same way as the more general version of **Table 1**. These values would result from parameter values of $V = 2$, $W = 4$, and $T = 0$. A few other differences must be

Table 2 Simplified payoff matrix for the hawk–dove–retaliator game. Fitness payoffs accrue to the strategies on the left when each plays the strategies at the top

	Hawk	*Dove*	*Retaliator*
Hawk	−1	2	−1
Dove	0	1	0.9
Retaliator	−1	1.1	1

explained as well. When hawk plays retaliator, both fight immediately, so the payoff to hawk (−1) is the same as if hawk played another hawk. If retaliator always displayed against dove, then these two strategies would look identical, so we make a minor modification to the retaliator strategy. When playing against a dove, retaliator occasionally 'probes' with an attack. Such probes result in the immediate retreat of the dove, so retaliator does slightly better than dove when playing against that strategy (1.1), and dove does slightly worse when playing against retaliator (0.9). For this reason, a better name for this new strategy is prober–retaliator.

Is there an ESS for this game? Using the ESS analysis methods of the previous model, it should be clear that there are two. As in the original hawk–dove game, there is a mixed ESS, in this case a 50–50 mix of hawk and doves (or individuals that play hawk half the time and dove half the time), but retaliator is also an ESS because it can resist invasion by either dove or hawk.

These classic game theory models of contests have been expanded considerably over the years. For example, other models have considered games between individuals that differ in ability (asymmetric games), games involving repeated interactions between the same contestants (iterated games) or between relatives, games in which a strategy is compared to the population as a whole (playing the field), and games that are dynamic rather than static.

A study of bowerbirds provides an example of mathematical modeling using game theory. In many species of bowerbirds, the males build and decorate amazing structures (bowers) that females use to choose their mates. Females are able to assess the quality of a male's bower, so a male needs a good one to reproduce. Females are known to visit many bowers before they choose the best one, so of course there is tremendous competition among the males. Just a bit of thought about this system suggests that a male bowerbird might do one of three things to be successful. He could spend lots of time building and decorating a great bower and then defend it against raiding by other males (defender). Or, he could split his time between building and defending his own bower and visiting other bowers to steal the decorations placed there by neighboring males (stealer). Or, he could split his time between building and defending his own bower and visiting other bowers to destroy them (destroyer). By measuring the costs and benefits of these strategies in terms of access to females, the game-theory model shows that both destroyer and stealer are stable against defender under most circumstances.

The Prisoner's Dilemma

Game theory models can be used to analyze many competitive aspects of animal behavior, including habitat selection, foraging, predator–prey interactions, communication, parent–offspring interactions, and sibling interactions. They have also been used to study cooperative behavior, perhaps most famously in the prisoner's dilemma.

Cooperative behavior is another topic that seemed difficult to explain without resorting to group advantages. The prisoner's dilemma game has provided a means of investigating the fitness consequences of cooperation on the basis of reciprocity. The model assumes that interactions between pairs of individuals occur on a probabilistic basis, and the results of a computer tournament show how cooperation can spread in an asocial world, can thrive while interacting with a wide range of other strategies, and can resist invasion once fully established.

The prisoner's dilemma is a symmetric, two-player game with two alternate strategies, 'cooperate' and 'defect.' The payoff matrix is shown in **Table 3**.

Imagine that the police arrest two suspected thieves, who are immediately placed in separate rooms. The police do not have enough evidence to convict either of them, but each is offered the following deal. If one testifies (defects) against the other and the other remains silent (cooperates), the defector goes free and the cooperator receives 10-year sentence. If both remain silent (cooperate), each is sentenced to only 6 months a minor charge. If each betrays the other, each receives a 5-year sentence. Thus, there is a strong temptation to defect (T) and if the other suspect cooperates, he gets a sucker's payoff (S). If both cooperate, each gets the reward for mutual cooperation (R), and if both defect, they pay the punishment for mutual defection (P). Each player does better by defecting than cooperating ($T > R$ and $P > S$), but the combined payoff to cooperation is greater than the combined payoff for cheating ($R > P$), which produces the dilemma. Each suspect must then choose to defect or to cooperate. What should the suspects do?

Alexrod and Hamilton (1981) answered this question for an iterated (repeated interactions) prisoner's dilemma

in two ways: by soliciting strategies and then playing the 14 that were submitted in a round-robin computer tournament, and by determining mathematically whether one strategy is an ESS. One strategy, 'tit for tat' (TFT), which was submitted by Anatol Rapoport, won the tournament and was shown to be an ESS when the probability of interacting with the same player on the next move of the game was high enough. An animal (or suspect) playing TFT cooperates when first meeting an opponent, and subsequently does whatever the opponent does. The three characteristics that make TFT a winning strategy are as follows: it is nice initially, it retaliates, and it forgives immediately.

Since the original use of the prisoner's dilemma as a model for the evolution of reciprocity (one route to cooperation), many modifications have been developed, including changing the number of players, number of strategies, relatedness of players, and degree of stochasticity.

One possible example of reciprocation by TFT is predator inspection by guppies. When guppies and other fish first encounter a potential predator, individuals often approach it, perhaps to gather information about the identity and motivation of the predator. It is very likely that the payoff for inspecting in a group is greater than the payoff if no fish inspects, so $R > P$. In addition, although having no inspectors is dangerous, it is more dangerous to be the lone inspector, so $P > S$. Thus, guppies engaging in predator inspection seem to experience a prisoner's dilemma. A reciprocal strategy such as TFT would ensure that the advantages of inspection exceed those of keeping a safe distance. Guppies are capable of recognizing and remembering the inspection behavior of partners and may employ a conditional approach strategy in which a fish swims toward a predator (inspect) on the first move of a game and subsequently only moves forward if the other fish swims beside it. Inspectors thus appear to be nice (starts to inspect), retaliatory (ceases inspecting if partner stops inspecting), and forgiving (resumes inspecting if partner resumes inspecting).

Honest Communication

Communication occurs when the behavior or some other cue of one animal affects the behavior of another. Signals and displays are traits that function specifically for communication and have evolved by natural selection for that function. They are thus products of signaler/ receiver coevolution, but communication is not necessarily cooperative because the interests of signaler and receiver are not identical. In many cases signalers and receivers have conflicting interests, so that communication has the potential to be manipulative. In such situations, game theory can provide powerful insights into signal evolution and mechanisms that maintain signal honesty or reliability.

Table 3 Payoff matrix for the prisoner's dilemma game. Fitness payoffs accrue to the strategies on the left when each plays the strategies at the top

	Cooperate	Defect
Cooperate	$R = 3$	$S = 0$
Defect	$T = 5$	$P = 1$

Many game theory models have been developed to study interactions in which communication occurs, such as between parents and offspring, contestants for important resources, prey and predators, and males and females. One question they have in common is, How is honest communication possible when signalers can benefit from deceiving receivers?

Maynard Smith (1974) considered this question when he wondered why accurate information is ever transferred during contests. He viewed the coevolution of sender and receiver as analogous to an arms race, with senders attempting to manipulate receivers and receivers attempting the resist the manipulation of senders. But because empirical studies showed clearly that animals communicate information honestly, he concluded that lying may be impossible or may be punished in some cases.

Subsequent studies showed that displays that are physically constrained (e.g., large frogs produce deep croaks, but small ones cannot) are honest. After considerable controversy, game theory models showed that signals that confer a handicap on the signaler are also honest (the handicap principle). Signals can be costly to signalers because they are energetically expensive to produce, or they are disadvantageous in some other way such as interfering with locomotion or the immune system. They can also be costly because signals attract unwanted eavesdroppers such as predators or receivers that attack (punish) dishonest signalers.

Honest signaling models indicate that reliable information transfer is common even if deceit occurs occasionally and that the form of the communication is influenced by selection for signal efficacy and for signal reliability because unreliable signals are not evolutionarily stable.

Game Theory and Animal Behavior

Many researchers consider the application of game theory to animal behavior to be one of the two most important theoretical developments since the modern synthesis of evolution and genetics. In many respects, game theory has changed the thinking of those who study animal behavior. The fundamental principle of game theory that the behavior of one animal affects the fitness of others and that these effects must be understood when explaining the evolution of behavior, and the concept of an ESS, have become a fundamental principle of animal behavior in particular and of biology more generally. The concept of ESS has also invaded psychology, political science, and even mathematics itself.

Despite the ability of game theory models to address many aspects of the social behavior of animals, there have been relatively few empirical tests of game theory models, especially in comparison with other classes of models such as optimal foraging. In addition to the usual objections to theoretical approaches to biology, game theory models seem to face some particular difficulties, including that they are unnecessary or that they are irrelevant because they ignore the underlying genetic structure and constraints.

On the other hand, game theory models have yielded a rigorous evolutionary understanding of social behavior that is otherwise difficult to explain, such as settling contests conventionally, communicating honestly, cooperation in the face of the temptation to cheat, and the maintenance of behavioral polymorphisms. It is likely that empirical testing will become more common as game theory models make more realistic assumptions and more explicit predictions, which will make these models more accessible to empiricists.

If you are interested in learning how to develop game theory models, try Gamebug, a teaching and learning resource.

See also: Agonistic Signals; Alarm Calls in Birds and Mammals; Communication: An Overview; Cooperation and Sociality; Decision-Making: Foraging; Experiment, Observation, and Modeling in the Lab and Field; Games Played by Predators and Prey; Honest Signaling; Optimal Foraging and Plant–Pollinator Co-Evolution; Parent–Offspring Signaling; Vigilance and Models of Behavior; Wintering Strategies: Moult and Behavior.

Further Reading

Axelrod, R and Hamilton, WD (1981). The evolution of cooperation. *Science* 211: 1390–1396.

Dugatkin, LA and Reeve, HK (1998). *Game Theory and Animal Behavior.* Oxford, UK: Oxford University Press.

Gintis, H (2009). *Game Theory Evolving: A Problem-Centered Introduction to Modeling Strategic Interaction,* 2nd edn. Princeton, NJ: Princeton University Press.

Maynard Smith, J (1974). The theory of games and the evolution of animal conflicts. *Journal of Theoretical Biology* 47: 209–221.

Maynard Smith, J (1982). *Evolution and the Theory of Games.* Cambridge, UK: Cambridge University Press.

Maynard Smith, J and Price, GR (1973). The logic of animal conflict. *Nature* 246: 15–18.

Searcy, WA and Nowicki, S (2005). *The Evolution of Animal Communication: Reliability and Deception in Signaling Systems.* Princeton, NJ: Princeton University Press.

von Neumann, J and Morgenstern, O (1944). *Theory of Games and Economic Behavior.* Princeton, NJ: Princeton University Press.

von Neumann, J and Morgenstern, O (2007). *Theory of Games and Economic Behavior,* Commemorative edition. Princeton, NJ: Princeton University Press.

Relevant Websites

hoylab.cornell.edu/gamebug/ – GameBug Software.
www.gametheory.net/ – Game Theory.net.
www.gametheorysociety.org/ – Game Theory Society.

Games Played by Predators and Prey

A. Bouskila, Ben-Gurion University of the Negev, Beer Sheva, Israel

Introduction

Game theory has been very useful in the understanding of the behavior of animals. Game theory models and the concept of evolutionarily stable strategy (ESS) provided sound explanations to a variety of phenomena that could not otherwise be fully understood. A game theoretic approach should be used to understand the behavior of animals whenever there are reasons to believe that the strategy or the behavior of one organism is affected by the behavior of the other and vice versa. The mathematical tools used to solving game theory problems generate predictions regarding the best response of each player to the strategy of the opponent. Initially, most game theory models dealt with different individuals within a species. With time, asymmetric games were analyzed, and later on, this was expanded to include games between individuals of different species. Predator–prey games are a special type of asymmetric games, in which the players are engaged in a predator–prey relationship and often belong to different species. The players do not necessarily have to belong to different species (e.g., as in cannibalistic relationships), but the examples in this study only refer to predator–prey game models between different species.

As in other types of games between animals, one can investigate the predator–prey game on two different time scales: the game may describe situations and life-history strategies that were selected for at the evolutionary time scale, or it could describe behaviors and strategies operating within the life of the individual, often termed ecological time scale.

The Types of Predator–Prey Games Considered

Games of Temporal Distribution

Predators and prey may be involved in games of temporal nature, the most general of which are games that address the question of when should prey and predators choose to be active outside their shelter or roost. A different type of game, often termed the 'waiting game' or 'shell game,' also belongs here: a prey animal escapes into a shelter from which it cannot know whether the predator is still waiting outside the shelter. These models calculate simultaneously both the optimal emergence time of the prey from the shelter and the appropriate length of time for the predator to wait for the prey outside the shelter, before it moves on in search of alternative prey. Hugie showed that the distributions for the waiting times of the predator and the prey have different shapes, and only rarely was the waiting time of the predator longer than that of the prey.

Games of Spatial Distribution

Habitat selection games involve the physical location in which the players spend their activity time. In one of the first predator–prey games described, Iwasa analyzed the vertical migration of zooplankton species and their predators in lakes or in the sea (great depth during daytime and near the surface at night). Previous explanations included effects of the physical environment or biotic relationship between zooplankton and phytoplankton. However, a habitat selection game based on predator avoidance at the time of high predator efficiency better explained many observed characteristics of the vertical migration.

Intuitively, in habitat-selection games, the prey is expected to concentrate its activity where its food is abundant, while the predator seeks the habitat or microhabitat where it can capture the most prey. In fact, these considerations are much more complicated due to the involvement of trade-offs in the strategy of each player, and one of the most obvious trade-offs is between food and safety: often, the habitat that has the highest abundance of food exposes the animal to higher predation risk. Habitat selection games are the most common predator–prey games modeled and discussed so far. As we shall see later, due to the nature of the stable solutions, at times, we find the results of these games quite unintuitive.

Games of Vigilance and Search Intensities

Even when the time of activity and its locations have been determined, the predator and the prey may be involved in games that determine how much effort each one should invest in detection of the other. The prey may invest time in vigilance, thereby sacrificing foraging time or forging efficiency. The predator may determine how much (in terms of energy and time) it should invest in search activities. Search activities often involve much more energy expenditure than resting, expose the predator to risks of injury or risks from its own predators, and reduce the time available for social interactions.

Games of Pursuit and Escape Behaviors

After the prey has been detected, both predator and prey still need to determine how much effort to invest in pursuit and in escape, respectively. Before predator–prey games were investigated specifically, Stewart used a genetic model to find the appropriate search strategy of a predator and the corresponding escape behavior of the prey, after it had encountered cues of the predator. Later on, Vega-Redondo and Hasson investigated pursuit deterrence: which behaviors can a prey animal use to manipulate and reduce the pursuit motivation of the predator. In this case, the investigators were seeking an honest signal by the prey that would clarify to the predator that the pursuit will not lead to a successful capture.

Games of Life-History Parameters: Growth, Birth, and Death Rates

While the previous types of games represent the evolution of behavioral decisions of predators and prey in situations formed while they are engaged in a game, there are games between predators and prey in which their life-history parameters are assumed to be determined at a larger evolutionary scale. Such games may involve growth rate decisions or other growth decisions. For example, Bouskila and colleagues modeled the timing of the switch of a prey animal from one growth phase to another, considering the ability of predators to evolve and modify their search strategy to adjust for this change. In addition, games have been proposed to address the co-evolution of characters that affect birth and death rates of predators and prey (such as body size) toward a stable solution that maintains coexisting populations of the two types of organisms.

Games of Traits: Thermal Physiology

Another set of predator–prey games at a large evolutionary scale concern the co-evolution of physiological and morphological traits. The evolution of physiological adaptation under a situation of a game was recently entered into a framework of a game between competing conspecifics, through an Ideal Free Distribution game. This concept was expanded to a predator–prey game by Mitchell and Angiletta, leading to conclusions regarding the effects of the predators on the evolution of physiological specialty. For example, under severe predation pressure, prey animals are predicted to evolve toward being generalists in thermal preference, rather than specialists. Prey that specialize on a narrow range of temperatures spend time in specific patches that maintain this range of temperatures and facilitate predation. Thermal specialization among prey animals can be a stable solution only when predation pressure is mild.

The Type of Models Used

Various approaches have been used to model predator–prey games. Analytical solutions are often sought to solve the simplest games. Models are formulated as a set of equations, including the fitness functions of predators and prey, sometimes in matrix form. Equilibrium points are found analytically through the simultaneous calculation of the derivatives of the functions. These models often necessitate simplifying assumptions, such as assuming that all players within a category (predators or prey) are identical. While this assumption has been useful in simplifying the models, there are cases in which more complicated elements need to be included in order to capture the essence of the system described by the model. In such cases, computing-intensive simulations are employed for their solution by using a state-variable dynamic game or an evolutionary algorithm. The former calculates the best response of a mutant in a group and then allows the rest of the group to copy the state-dependent solution found by the mutant. This process is cycled as many times as needed to reach a stable solution, in which the best reply to the group's strategy is the same strategy. In order to adjust this game to a predator–prey game, the state-dependent solutions are found simultaneously both for the predator and the prey. The evolutionary algorithms have a game concept embedded in their structure, because new genomes are formed either by genetic mutations or recombinations, and they compete against all other genomes. A stable solution is reached when a genome proliferates and cannot be invaded by new genomes, and here too, the process is run simultaneously for genomes of prey and predators. Mitchell combined an evolutionary algorithm with an Individual-Based Model, to determine the fitness consequences of the actions taken by the different genomes. All together, these techniques enable the incorporation of such concepts as the states of individual animals and their spatial distribution, which have been shown in other disciplines as very useful tools for reaching solutions of complex evolutionary problems. As it is sometimes done, Alonzo analyzed each of the games among prey individuals and among predators separately, and then the full game between predators and prey was analyzed and compared to the partial games.

Empirical Work Used

Difficulties

The theory of predator–prey games was mainly developed in recent years. Empirical studies have been conducted only rarely in the past in a way that demonstrates that animals indeed use a game-theory solution. Apart from the fact that the theory that might have led to such empirical studies was not very well developed, there are objective difficulties that stem from the fact that predation events are involved here. Predation events in general, not necessarily in game situations, are quite rare to observe in nature. It is thus quite complicated to design a study in a natural system that includes predation rates or predation events. One solution to this problem has been to transfer such studies to seminatural conditions or even to laboratory conditions. An additional difficulty stems from the nature of the game itself. When animals are engaged in a game situation, simple observation cannot easily verify if a game is going on. Observing the end result of the game may hide the behavioral options that were not chosen or led to unstable solutions. In order to test the existence of the game, quite often, the game needs to be perturbed, that is, animals need to get cues for a different situation, and only then may their behaviors be evaluated under the framework of game. Here, too, the ability to change the conditions of the game are limited when done in nature and are more likely to be done under more controlled conditions. Thus, occasionally, the empirical work mentioned in games of predators and prey is compatible with the concept of such games but cannot always demonstrate the existence of the game as the best or only explanation for the observed pattern. A few examples of field studies, as well as studies in seminatural arenas and in the lab, are listed as follows. Such studies are often mentioned in theoretical work, and in some cases, have contributed to the development of the theory of predator–prey games.

Empirical Work Used

Field Observations

Quinn and Cresswell found that predators preferentially attack prey according to their vulnerability, rather than according to their numbers. They performed experiments with model birds at different vulnerability positions and recorded the attacks on these birds. This result is compatible with game theory models that predict the common interest of nonvulnerable prey and predators, both against vulnerable prey individuals. Bouskila manipulated presence and absence of snakes in the Mojave Desert and used seasonal changes in activity of the snakes to study the habitat selection of rodents and their predators. Rodents avoided habitats in which snakes were placed, and also habitats in which snakes are likely to choose for ambush, even when they were not placed there. Results were compatible with a model that described the simultaneous habitat selection of a predator and its prey. Additionally, the same model was also compatible with observations of snake movements in a rich oasis embedded in a dry desert matrix.

Experiments in Seminatural Field Conditions

Altwegg manipulated simultaneously the state of prey and predators in order to analyze the effectiveness of antipredator behavior of tadpoles against their invertebrate predators while they are at different states. The study was done outdoors, in standard tubs, and it demonstrated that predatory rates depended both on the behavior of the prey and the predators. A somewhat similar study was performed by Berger-Tal and Kotler, but with vertebrates both as prey and as predators, in a large aviary (**Figure 1**): the energetic state of owls and their prey (desert gerbils) were simultaneously manipulated in order to shed some light on the game between these players. Unlike the previous study, the predators in this study were not very sensitive to the state of the prey, while the prey definitely modified their forging behavior depending on the hunger level of the predators.

An important consideration that predators need to consider while they pursue prey is their own exposure to risk of injury or risk from their own predators. One of the empirical studies that addressed the game between foxes and their prey involved experiments in the same aviary mentioned earlier. Berger-Tal and colleagues manipulated the risk of injury to foxes when they foraged in food trays. Foxes demonstrated that they consider the risk of injury in their decisions of time allocation, and this has important implications for predator–prey games in which the prey spends time in microhabitats that may impose high risk of injury to the predator.

Laboratory Experiments

Hammond and colleagues studied habitat selection of tadpoles and dragonflies in laboratory experiments.

Figure 1 An example of a two-compartment large enclosure suitable for observing and manipulating elements of predator–prey games. The enclosure includes infrared sensors to monitor transitions of foxes between the compartments as well as electronic seed trays, to monitor individual rodent visits at various stations. Photo: A. Bouskila.

They recorded choice of habitat (rich or poor in prey resource) under three treatments: prey alone, predator alone, and both species together. These experiments tested and confirmed some of the predictions generated by predator–prey game models, such as the preference of predators for patches with abundance of prey food, which the predators do not consume. In addition, the prediction that prey animals are not sensitive to the density of their own resources was confirmed too. In this study, a model selection approach was used to choose among factors that could potentially explain the patterns of space use that were observed.

In some cases, experiments are performed in situations where the predators and the prey are likely to be in a game situation, but the experiment itself was not meant to demonstrate or verify if the results fit any theory based on games. For example, Dangles and colleagues describe the optimal velocities of a spider for approaching and capturing a cricket and found that there are two speeds in which the vulnerability of the cricket was maximized, and thus, this approach speed was utilized by the spiders. Although this study was performed without an underlying model to test, it deals with simultaneous decisions of predators and prey and may provide a basis for such a model.

Common Themes in Models and Experiments

Cases in Which Predators Respond to Prey Resource

In many cases, it has been found that predators should distribute themselves according to the distribution of prey resource, rather than according to the parameters that are supposed to affect the predators directly. This nonintuitive result is one of the most consistent results that emerged from several predator–prey game models. This effect of one player's parameters on the second player is especially pronounced when there is no competition or other intraspecific interactions within the population of each player. In some of the models, when intraspecific interactions are included, the predators are still affected by prey resource distribution, but the prey too is affected by its food distribution. In such cases, prey distribution does not match the distribution of resource, as we would expect according to the Ideal Free Distribution model, rather it undermatches the resources, that is, the proportion of animals in the rich patches is smaller than the resource proportion. Other considerations emerged when Hugie and Dill included metabolic and foraging costs, and found that these caused undermatching to the resource too.

Alonzo included the state of the players in the model and found that another consideration emerged and caused undermatching of prey resources by prey animals: individuals at lower states were forced to forage in the risky and

rich food patch. Thus, the predicted resource matching was not achieved in this game too.

Habitat preferences due to food distribution are often traded off with safety considerations, if safety differs among habitats. When the number of shelters or the level of safety in a habitat is manipulated in models, another general result emerges, namely, the prey is strongly associated with the habitat that provided safety, while the predators often concentrate in the habitat where the predator has a higher success rate.

Efficient Predators Make Predator Avoidance Ineffective

Predator prey game models have also shown that when a predator is able to efficiently react to the change in the prey distribution, the predator may render the prey antipredator behavior inefficient. In such cases, the prey cannot escape predation and as a result, the prey should abandon all predator-avoidance strategies. Avoiding predation usually comes at a cost of foraging efficiency, due to the food and safety trade-off mentioned earlier. In those situations where the prey will gain nothing by antipredator behaviors, the prey should attempt to collect food as much as possible and at least grow as fast as possible.

Equally, during a thermal game where patches with different temperatures were provided as the resource, prey animals became indifferent to temperatures (i.e., evolved into generalists) when predation pressure was high.

Another case in which antipredator behavior is predicted to be ineffective was described by Wolf and Mangel. They analyzed a situation in which the prey selects the antipredator behavior, while the predator chooses the attack rate. At high attack rates, the prey loses so much time following attacks that they are forced to forage in the rich and dangerous habitat to avoid starvation, basically abandoning their antipredator behavior. An interesting situation is thus predicted in such systems: the predators should make many false attacks (undistinguishable by the prey from true attacks) in order to induce the prey to abandon its antipredator behavior.

Implications and Importance of Predator–Prey Games

Individual Behavior Implications

The recent development of models of predator–prey games has demonstrated that the incorporation of the game approach into the analysis of the predator–prey relationship has important implications. For many years, studies of predators and prey assumed, often for simplicity, that the prey is faced with a constant level of predation, at least at a given time and habitat. Predators were considered unresponsive to prey distribution or to prey behavior. However, as in other disciplines within animal behavior in which the game approach has been incorporated, the development of predator–prey games has allowed us to capture a much more precise image of the situation by allowing predators and prey to choose their behavior for the best strategy to both players. Removing the assumption of rigid behavior of predators has opened up the development of new hypotheses and a different way to view the behavior of prey and, obviously, predators too. The use of this approach is beginning to change the view of predator–prey behavior in all situations in which a game is likely to take place.

Population and Community-Level Implications

Predator–prey models are usually calculated at the individual level. Nevertheless, the implications from the predicted behaviors of the individuals have been extended at times to populations and communities. For example, Brown and colleagues demonstrated that a predator–prey game that predicts the activity time of each player given that the resource for the prey is provided as a pulse, increases the stability of population dynamics. At a larger evolutionary scale, Brown and Vincent address co-evolution of predators and prey in a community and allow the number of predator and prey species to be evaluated at the ESS point, as an emergent property of the stable strategy. Depending upon conditions, the predators may either be keystone species (whose removal may lead to prey species extinction) or may have insignificant impact on current populations of prey. Not only may the number of species in the community be determined by a predator–prey game, but in some cases, such a game may lead to selection of certain characters, and in extreme cases, even to speciation, such as toward two different species with different thermal preference.

Future Challenges

Multitrophic-Levels Games

Most of the current predator prey games deal with two trophic levels. In one of the few exceptions, Rosenheim added the consideration of a top predator and demonstrated that under such circumstances a predator–prey game reaches very different predictions. Models have yet to be developed in order to describe the simultaneous decisions of several species within one of the trophic levels. The more alternative prey species that are involved in the game (or the more predator species), the more complicated the results may become. However, in such models, the decisions of one player are likely to be less coupled to those of any one player from the second trophic level.

Incorporating Realistic Assumptions

As in other types of models, the assumptions of the predator–prey game models are introduced for simplification and to keep the models tractable. There is a well-known trade-off between generality and specificity in models: in general models, there might be simplifying assumptions, but they might not be realistic for many specific systems, or even not for any of the systems. Now that some of the general patterns are beginning to emerge from the current models, it seems useful to modify some of the assumptions and make them more realistic (at the cost of generality, in some cases).

Experimental Design

In spite of the wave of game theory models of predator and prey, there still is a great lack of empirical studies that deal with such situations and even fewer studies that can demonstrate that a game situation indeed exists between predators and prey. As mentioned before, there are inherent difficulties in the design of experiments to test the existence of a game between predators and prey. The game often needs to be perturbed by either supplying misleading cues or limiting the freedom of choice from one of the players in order to check whether the reactions of the other player are compatible with the existence of game considerations in the interactions. For example, Sih suggested five treatments to fully analyze a predator prey habitat selection game: (1) prey alone, (2) predator alone, (3) prey with restricted predator, (4) predator with restricted prey, and (5) predators and prey free to move between both habitats. This requirement is not impossible to achieve, but it requires specific studies in controlled environments, such as in the lab and in seminatural arenas and enclosures. Such designs are beginning to be more prevalent, and the best demonstrations will be achieved when specific experiments will be coupled with specific predator–prey game models, designed or adjusted to the system in question.

See also: Empirical Studies of Predator and Prey Behavior; Game Theory.

Further Reading

Alonzo SH (2002) State-dependent habitat selection games between predators and prey: The importance of behavioural interactions and expected lifetime reproductive success. *Evolutionary Ecology Research* 4: 759–778.

Altwegg R (2003) Hungry predators render predator-avoidance behavior in tadpoles ineffective. *Oikos* 100: 311–316.

Berger-Tal O, Mukherjee S, Kotler BP, and Brown JS (2009) Look before you leap: Is risk of injury a foraging cost? *Behavioral Ecology and Sociobiology* 63: 1821–1827.

Bouskila A (2001) A habitat selection game of interactions between rodents and their predators. *Annales Zoologici Fennici* 38: 55–70.

Bouskila A, Robinson ME, Roitberg BD, and Tenhumberg B (1998) Life-history decisions under predation risk: Importance of a game perspective. *Evolutionary Ecology* 12: 701–715.

Brown JS, Kotler BP, and Bouskila A (2001) Ecology of fear: Foraging games between predators and prey with pulsed resources. *Annales Zoologici Fennici* 38: 71–87.

Cresswell W and Quinn J (2004) Faced with a choice, sparrowhawks more often attack the more vulnerable prey group. *Oikos* 104: 71.

Hammond JI, Luttbeg B, and Sih A (2007) Predator and prey space use: Dragonflies and tadpoles in an interactive game. *Ecology* 88: 1525–1535.

Hugie DM and Dill LM (1994) Fish and game: A game theoretic approach to habitat selection by predators and prey. *Journal of Fish Biology* 45: 151–169.

Iwasa Y (1982) Vertical migration of zooplankton: A game between predator and prey. *American Naturalist* 120: 171–180.

Mitchell WA (2009) Multi-behavioral strategies in a predator–prey game: An evolutionary algorithm analysis. *Oikos* 118: 1073–1083.

Mitchell WA and Angilletta MJ Jr (2009) Thermal games: Frequency-dependent models of thermal adaptation. *Functional Ecology* 23: 510–520.

Rosenheim JA (2004) Top predators constrain the habitat selection games played by intermediate predators and their prey. *Israel Journal of Zoology* 50: 129–138.

Sih A (2005) Predator–prey space use as an emergent outcome of a behavioral response race. In: Barbosa P and Castellanos I (eds.) *Ecology of Predator–Prey Interactions*, pp. 240–255. New York, NY: Oxford University Press.

Stewart FM (1971) Evolution of dimorphism in a predator–prey model. *Theoretical Population Biology* 2: 493–506.

Vega-Redondo F and Hasson O (1993) A game-theoretic model of predator–prey signaling. *Journal of Theoretical Biology* 162: 309–319.

Wolf N and Mangel M (2007) Strategy, compromise, and cheating in predator–prey games. *Evolutionary Ecology Research* 9: 1293–1304.

Genes and Genomic Searches

C. P. Kyriacou and E. Tauber, University of Leicester, Leicester, UK

Introduction

Behavioral genetics can be argued to be the oldest branch of genetics and can loosely trace its ancestry back to Francis Galton's book 'Hereditary Genius,' which was first published in 1869. In this epic study, Galton (1822–1911) examined the male relatives of highly distinguished Victorian men and observed that the larger the genetic distance between family members, the lower the frequency of outstanding mental abilities. Galton's work became central to the 'eugenics' movement of the first half of the twentieth century, which was later so tragically perverted by the Nazis. Nor did the American psychiatric establishment distinguish itself in this regard, with thousands of people sterilized and institutionalized, often on the flimsiest 'evidence' of genetic mental 'inferiority.' From this extreme genetic determinism of the 1920s sprang behaviorism, the extreme environmentalism of the psychologist John Broadus Watson (1878–1958) that carried almost everything before it, but with the odd exception, notably the studies on the genetic basis of learning in rodents from the laboratories of Edward Chase Tolman (1886–1959) and Robert Chaote Tryon (1901–1967). Perhaps it is from this period that behavioral genetics, as an experimental discipline, was finally born. However, it was not until the dust of the Second World War settled that a handful of zoologists and psychologists began serious work on the genetic basis of animal behavior. A subgroup of these, the ethologists Konrad Lorenz (1903–1989), Niko Tinbergen (1907–1988), and Karl von Frisch (1886–1982) studied instinctive species-specific behavior in vertebrates and insects, with the implication that these motor programs had an underlying genetic basis. They were to share a Nobel Prize for Medicine and Physiology in 1973. Yet it was many years before the first 'behavioral' gene was identified at the molecular level (see Hay's (1985) textbook for more on the history of this subject).

The behavioral geneticists of the 1950s, using inbred lines or selection experiments, studied the genetic architecture underlying behavioral phenotypes such as mating or open field activity, by making a series of genetic crosses, usually in mice, rats, or flies. They could even map differences in behavior between strains of flies to specific chromosomes. Studies, such as those of Fulker, used the methods of quantitative genetics to provide some information about the evolutionary history of the behavioral trait in question. However, it was not until this kind of formal genetic analysis was blended with molecular biology in the 1990s that progress was made in identifying individual genes that contributed, at least partially, to complex behavioral phenotypes. Thus, in the 1960s, the best one could do if one wished to study single gene effects on behavior was to take a morphological mutant in the fly such as *ebony* or *yellow, or* a neurological mutant mouse such as *waltzer* or *twirler,* and study various behavioral phenotypes in the hope that something interesting might emerge. Sometimes it did and sometimes not.

The Birth of Neurogenetics – Genetic Screens for Behavioral Phenotypes

In the mid- 1960s, Seymour Benzer (1921–2007) suggested a novel 'bottom up' approach whereby a single mutation was made randomly within the genome of a model organism, and then behavior was screened for interesting phenotypes. His organism of choice was the fruitfly. Not only did it have a life cycle of only 10 days, making genetic analysis relatively rapid (compared to several months in mice), but the genetic map of the fly was already well understood, and the behavior of a fly seemed genuinely interesting. Benzer's idea was to feed the flies a powerful mutagen, and then screen for behavioral mutants using various fly-specific genetic tricks. The underlying mutation would alter one nucleotide base pair, and if that altered a codon, the amino acid change might generate a phenotypic difference. Using simple yet ingenious behavioral screens involving flight, movement, vision, courtship, etc., Benzer's students soon identified many mutants that would do strange things, not fly, not mate, not see, shake violently, and they were given colorful names like *drop-dead, coitus interruptus, ether-a-go-go,* etc. These behavioral mutations could be mapped to the genome, thereby identifying the corresponding gene as a position on a chromosome. In addition, 'fate mapping,' a technique that Benzer extended to behavior, allowed an approximate identification of the likely neuronal (or otherwise) tissues in which the mutated behavioral gene was having its primary effect. The molecular analysis of these genes came much later, with the advent of cloning and germline transformation techniques in the 1980s.

Benzer's students soon progressed to identifying mutations in more complex and interesting behaviors, for example in learning and memory and circadian rhythms. Flies learn to associate specific odors with electric shocks,

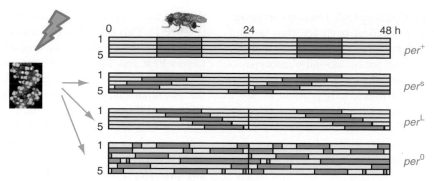

Figure 1 Identification of the *period* gene in *Drosophila melanogaster*. Chemical mutagenesis of DNA resulted in three single nucleotide changes within the *per* gene giving rise to short, long period and arrhythmic animals. The locomotor activity of a fly is double-plotted on the horizontal axis, for 5 days (vertical axis). Wild-type *per⁺* flies, are active (blue) or sleep (yellow) in 24 h rhythms, so they start activity and end it at the same time each day. The short, *perˢ* mutant, has fast 19 h rhythms so activity begins and ends 5 h earlier on every successive day (the 'actogram' moves to the left). The *perᴸ* mutant has long 29 h cycles, so the activity trace moves to the right, while the *per⁰* mutant is arrhythmic. Adapted from Konopka RJ and Benzer S (1971) Clock mutants of *Drosophila melanogaster*. *Proceedings of the National Academy of Sciences of the United States of America* 68: 2112–2116, with permission from · · ·.

and avoid these in future. Mutations in genes *dunce, amnesiac, rutabaga, turnip, zucchini* fail these memory tests. One of Benzer's students, Ron Konopka, developed a method to measure circadian (24 h) locomotor activity rhythms in flies, and subsequent mutagenesis identified three alternative mutant alleles of a single gene termed *period (per)*, which produced short- or long-period, or arrhythmic behavioral cycles (**Figure 1**). Sometime later, it was discovered that *per* could be deleted entirely from the genome, yet the fly appeared happy, healthy, but arrhythmic – in other words, *per* was not a vital gene – it was as true a behavioral gene as any could be. Thus, the identification of *per* comes at the birth of the field known as neurogenetics, which has flourished ever since. Indeed, the fly story mentioned here has been significantly enhanced by similar studies of phenotype-driven 'forward genetics' in mice (for further reading on flies, see Nitabach and Taghert (2008)). In 1994, Joe Takahashi and his group used chemical mutagenesis to identify a variant that disrupted circadian locomotor behavior. They called this mutant mouse *Clock* (*Clk*) and it provided the entree into the molecular basis of the vertebrate circadian mechanism, which incidentally turns out to be highly conserved between flies and mice.

Transposon Mutagenesis

Chemical mutagenesis usually changes one base pair at a time, but mutagenesis can also be accomplished by hopping a mobile piece of DNA (a transposable element, TE) into another gene and disrupting it to cause a behavioral phenotype. Many behavioral mutants in flies are caused by such TEs, including some clock mutants. These types of approaches have one neat advantage over chemical methods in that they can be used as molecular tags to

clone the surrounding areas (the behavioral gene into which they have hopped). These flanking DNA sequences surrounding the transposon can be identified and entered into the fly genome database to find the disrupted fly gene (http://flybase.org/). One disadvantage however is that transposons, unlike chemical mutagens, do not interrogate the genome randomly, but tend to prefer certain sequence compositions for their insertion. On the other hand, chemical mutagenesis has the disadvantage that time-consuming genetic mapping followed by positional cloning is usually the only way to identify the molecular lesion, as in the case of mouse *Clk*.

RNAi – RNA Interference, a Revolution in Genome Screening

A few years ago, small double-stranded RNA molecules (dsRNA) were discovered by Fire and Mello, which was to earn them a Nobel Prize in 2006. These molecules have the ability to interfere with the translation of any mRNA that has a similar sequence and provides a means for 'knocking down' gene expression. To downregulate a particular mRNA, a double-stranded RNA molecule corresponding to the gene must be made and introduced into the organism. For example, a short sequence from *per* could be used to make an inverted repeat of that sequence that will allow the two complementary sequences to base pair and form a dsRNA molecule. This can then be transformed into the fly in a way that will target it to cells that express *per*. These short molecules will then pair with the endogenous *per* mRNA and block translation. A number of centers around the world have generated dsRNA molecules providing RNAi for every gene in the fly genome, all 14 000. One can order a fly strain that carries a dsRNA of interest, and then by crossing this line to another strain that carries an activator of this dsRNA, fused to a sequence that targets the activator to a tissue of choice, your favorite gene can be knocked down tissue-specifically. Systematic screening of all RNAi lines for behavioral phenotypes is usually too laborious, unless the behavioral screen can be made 'high throughput.' Instead, it is possible to use a cellular model as a behavioral readout, as was done by Amita Sehgal, who was able to screen a molecular RNAi library within cell lines in order to identify new genes that were important for entraining the cellular circadian clock to light–dark cycles.

RNAi experiments provide an example of gene product driven 'reverse genetic' approaches. Another reverse method is to knock out or eliminate a gene. In circadian biology, most of the murine clock genes that are homologous to the fly genes were identified by sequence similarity and then targeted by gene knockouts (KOs) to examine any phenotypes. Thus, a KO of the mouse homolog of fly *cyc* (called *Bmal1*) made by Bradfield and colleagues gives complete arrhythmicity, revealing the striking functional conservation of the two species genes.

What Do Gene Sequences Tell Us?

From the mid-1980s, it became possible to molecularly clone fly genes, identify their DNA sequences, and translate them into their putative proteins *in silico* using the genetic code. When the *per* gene was first sequenced in the mid-1980s, it looked like nothing else in the databases – it encoded a 'pioneer protein.' Over the years, a number of other proteins were identified in various organisms that shared a particular sequence domain with PER called PAS. This domain was important for protein–protein interactions and was found in many proteins that were environmental sensors, and particularly responsive to light, oxygen, and voltage. This makes a certain sense as PER must have evolved in response to environmental light–dark cycles. This PAS domain of PER was used in a reverse genetic approach as a trap to identify a protein partner of PER called TIMELESS (TIM). At about the same time, a forward genetics mutagenesis produced a *tim* mutant which was arrhythmic. It turns out that PER and TIM are partner molecules in the fly clock mechanism. They are transcribed into mRNA early at night in clock cells and then translated into proteins in the cytoplasm during the night (**Figure 2**). Late at night they dimerise via the PAS domain of PER and move into the nucleus. There, they (PER–TIM) interact with the transcription factor CLK (see above – it is found in the fly as well as the mouse) and negatively regulate their own genes by sequestering CLK and its partner CYCLE (CYC, also initially defined by mutagenesis, both CYC and CLK have PAS domains). Later on, around dawn, PER and TIM degrade, releasing their block on CLK and CYC, which are now free

to move back onto *per* and *tim* and reactivate transcription (**Figure 2**). This relentless molecular cycle of *per* and *tim* mRNA and their proteins thus requires the two negative factors PER and TIM, and the two positive factors CLK and CYC, within the negative feedback loop that underlies the circadian mechanism, both in flies, and with some minor modifications, in mammals.

Gene sequences can be translated *in silico* into a protein sequence, which can then be compared with thousands of other sequences of known function in a protein database. For example, if the protein has a kinase domain, it will phosphorylate another protein, possibly leading to changes in its stability. If it is a transcription factor, it will be turning on or off other downstream genes. If it is a signaling protein, it will be involved in a transduction cascade, and so on. This information is crucial for understanding the underlying functional biology of the behavioral phenotype and informs and guides future experimentation.

Cellular Biology of Behavior

We linger on biological rhythms as they provide the best example we have of forward genetics being used to identify clock components. However, once a gene is identified, so is the protein, and using reagents such as antibodies or hybridization probes for the endogenous mRNA, a precise determination of exactly which tissues express the gene and protein, and when, can be made. This opens up the cellular as well as the molecular biology of the behavioral phenotype, and needless to say, in circadian rhythms, or learning, and courtship in flies, these approaches have been refined to an art form. Almost any gene can be expressed or misexpressed in almost any tissue of the fly, and this permits a panoramic exploration of the biology of behavior. So, for example, the critical clock neurons in the fly have been identified, and misexpressing clock genes or apoptotic genes (that cause cell death), within subsets of these neurons has revealed separate oscillators that control 'morning' and 'evening' behavior. In courtship, misexpression of a male-specific splice form of the gene *fruitless* (*fru*), in different neurons within the antennal regions, can convert a phenotypic female into a 'she-male,' who will inappropriately court other females. Careful examination of these regions of the central and peripheral nervous system by Billeter and colleagues reveal sex-specific anatomical differences in the shapes and the numbers of some of these *fru*- expressing neurons.

Neurogenetic Disease Models

It is also possible to subvert the fly and use it to study behavior indirectly. For example, Huntington's disease

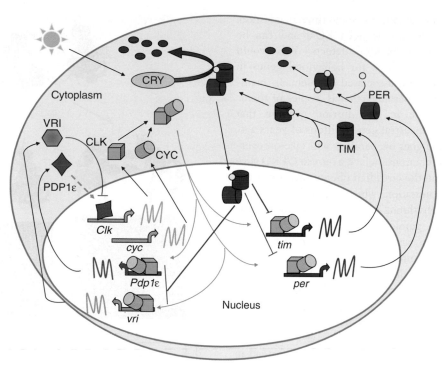

Figure 2 Forward genetics defines the molecular basis for the fly's intracellular circadian clock. The genes (italics) and corresponding proteins (Roman) are color coded. The *period* and *timeless* genes are activated by the CLK and CYC transcription factors (green arrows). The mRNAs (shown as single stranded squiggles) are exported to the cytoplasm where PER and TIM are translated. PER is phosphorylated (small yellow circles) by DBT kinase (encoded by the *doubletime* gene, not shown), which earmarks it for degradation (small blue circles). TIM is also phosphorylated (small yellow circles) by the kinase encoded by the *shaggy* (*sgg*) gene (not shown). Late at night TIM prevents DBT from phosphorylating PER so PER levels build up, and TIM–PER enter the nucleus and negatively regulate the CLK–CYC dimer (red lines), thereby repressing *per/tim* transcription as well. Thus *per* and *tim* mRNAs and proteins cycle in abundance during the circadian cycle. The *Clk* gene is itself positively (green arrows) and negatively (red lines) regulated by VRI and possible PDP1ε (dotted green arrow) leading to cycles in *Clk* and CLK abundance. Thus the *Clk* and *per/tim* feedback loops are interconnected, leading to additional stability. Both *vri* and *Pdp1ε* genes are also positively regulated by CLK–CYC (green arrows) and negatively by PER–TIM (red arrows). The blue-light photoreceptor Cryptochrome (CRY) is activated by light at dawn, and physically interacts with TIM, causing its degradation (small red circles). PER is thus exposed to DBT and degraded, thereby releasing the repression on the *per* and *tim* genes (this also occurs in constant darkness via another molecular route not involving CRY). The CLK–CYC dimer can now restart the molecular cycle by activating *per* and *tim* transcription. The roles of all these genes in the circadian clock were initially identified by forward genetics (i.e., mutagenesis) except for *Pdp1ε*, which was identified initially as a cycling transcript in fly heads. The *vri* and *sgg* genes were identified via a clever transposon mutagenesis whereby a specially constructed TE landed close to each gene, and was activated to overexpress the adjacent *vri* or *sgg* mRNA in clock neurons, revealing disruptive effects on circadian behavior.

(HD) is caused by an expansion of a polyglutamine tract (polyQ) within the huntingtin protein that is toxic to the human nervous system and causes devastating neurobehavioral impairments. When this expanded mammalian polyQ region is expressed in the eye of a fly, the eye degenerates, providing a cellular model for HD. Benzer screened 7000 TE lines and found several that could suppress the mutant Huntington's eye phenotype. Two of these lines had TEs inserted into genes that encoded chaperone domains, which are found in proteins that can prevent the misfolding of proteins that are under stress, be it mutational or environmental. Thus, the fly eye can be used as a substitute for more laborious behavioral screening and implicate gene products that might be used in future therapeutic interventions. Indeed, the fly has provided a surprisingly good model for dissecting

neurodegenerative disorders, and not just those related to expanded polyQ repeats (there are nine polyQ diseases known in humans). Alzheimer's, Parkinson's, Fragile-X, and Angelman's syndrome are just some of the other neurogenetic diseases that are being studied with the fly.

Mammalian Screens

Obviously, systematic genomic searches for behavioral genes are time consuming and expensive, and hence the prominence of *Drosophila* as the major model system. Nevertheless, large-scale mouse screens for many behavioral phenotypes such as learning and memory, circadian rhythms, psychostimulant responses, vision, and stress responses have been underway for some time. In addition, many mouse genes have now been KO'd and can be directly screened for behavioral defects. This would seem the perfect way to look for behavioral genes in mammals, but there is an associated problem. In mammals, many genes have paralogs, that is, copies of themselves somewhere else in the murine genome that duplicated from the ancestral gene millions of years ago. As evolutionary time goes by, paralogs will take on overlapping but related functions. In fact, a mouse *Clk* KO gave very subtle effects on the circadian phenotype, compared to the original *Clk* mutation, which, when homozygous, was dramatically arrhythmic. The *Clk* KO phenotype was compensated by a paralog, whereas the chemically induced *Clk* mutation was a dominant gain-of-function allele that basically gummed up the clockworks. It is interesting to speculate that if the *Clk* KO had been the only means to screen interesting circadian genes in the mouse, the *Clk* gene would probably have remained undiscovered in this context. This kind of result in which the KO mutant is not as dramatic as a chemically induced mutant may turn out to be quite widespread in mammals. In flies, most genes are single copy, so this problem of compensating paralogs in fly gene KOs does not usually rear its ugly head.

Transcriptomics

There are other ways of screening genomes for behavioral genes, and all are based on reverse genetics approaches. Transcriptomics is a popular method for detecting change in mRNA levels that correlate with altered behavior. For example, Dierick and Greenspan selected for highly aggressive male flies over a number of generations from a base population. They then isolated the head mRNA from the aggressive and the control males, and after copying it into cRNA, hybridized it to a commercial gene chip or microarray. On this microarray were placed the DNA corresponding to the entire fly transcriptome (~13 500 sequences, **Figure 3**). If any one of these sequences (arrayed as DNA spots) gave a higher or lower intensity hybridization signal in the aggressive compared to the control flies, it would suggest that the mRNA for that particular gene was up or downregulated. About 80 genes were differentially expressed in the aggressive flies, one of them, *Cyp6a20a*, encoded a cytochrome P450. To validate the microarray results, a mutant strain for this gene was obtained and was found to be significantly more aggressive, consistent with the microarray observation that the selected aggressive flies were downregulated in this particular mRNA species (**Figure 3**). Thus, a transcriptomic screen had identified a gene for aggression, which was subsequently found to be expressed in nonneuronal cells that are associated with pheromone receptors, indicating

Figure 3 Transcriptomic screen for aggression genes in *Drosophila*. A base population was selected for highly aggressive flies (black arrow) or simply maintained as neutral flies (blue arrow). Microarrays were independently interrogated with mRNA from the heads of aggressive and neutral flies, and a number of genes were differentially regulated (seen as dark or light spots, each spot corresponding to a particular gene sequence). From these candidate aggression genes, one, Cyp6a20, is downregulated in aggressive flies and a mutation in this gene which reduces mRNA levels, gives increased aggression (loosely based on Dierick and Greenspan (2006) *Nature Genetics* 38: 1023-1031; cartoon of fighting flies reprinted with permission from Dierick H (2008) *Curr Biol* 18: R161–163.).

that olfaction plays a prominent role in these agonistic encounters.

Similar transcriptomic analyses have been used to identify ~150 genes whose mRNAs cycle in abundance with a circadian period in the fly's head, or several hundred similarly cycling genes from the suprachiasmatic nuclei of the mouse, the organ that determines murine behavioral rhythms. Unlike the example of aggression, it is not differences in behavior that are being assayed here, but a molecular phenotype that has behavioral implications.

These types of studies require considerable statistical aplomb in order to separate false hits from real ones, and validation of candidate mRNAs is required, either by independent molecular methods or with the use of mutants, as in the fly aggression example cited earlier. However, as an entrée into the molecular basis of a behavioral phenotype, transcriptomics have the added flexibility that even nonmodel organisms can be studied. What is required in these cases is the generation of the microarray (gene chip) carrying thousands of cDNA spots, each one corresponding to a different gene made from the RNA of the relevant organism (see **Figure 3**). This is followed by the interrogation of the chip with the RNA from the individuals that show differences in the phenotype, be it behavioral or molecular. Any positive hits on the slide can then be sequenced to identify the corresponding differentially expressed gene.

Applying Molecular Genetics to Identify Natural Genetic Variation

Generating mutants by forward genetics approaches is rather like hitting the animal on the head. The screen usually involves a drastic change in the phenotype for the new mutant to be noticed. However, the gene sequences that are identified by mutagenesis can then become the focus for studies of natural genetic variation. Thus, a natural polymorphism in the *tim* gene of *D. melanogaster* was shown to be spreading from southern Europe into northern Europe, under directional selection. The new mutation had originally occurred a few thousand years ago in a single fly in southern Italy and the new mutation had spread slowly northwards. This new *tim* allele provided the fly with a more adaptive behavioral response to the seasonal environments experienced in Europe, compared to the ancestral *tim* allele, which had evolved in sub-Saharan Africa, in which there is much less of a seasonal challenge. Without the *tim* sequence (identified by forward genetics and mutagenesis), there could have been no reverse genetics whereby the natural polymorphism in *tim* could be placed within a functional and evolutionary context.

Natural genetic variation can also be used to dispense with mutagenesis completely, and provide a gentler approach for searching for behavioral genes. The exponential increase in the available DNA sequence data and the identification of specific sequence regions ('markers') make this a tractable proposition. Quantitative Trait Loci (QTL) mapping is a natural continuation of the types of studies that biometrical behavioral geneticists were doing in the 1980s, before the molecular revolution really took off. This method can commence with two inbred parental lines, ideally (but not necessarily) showing a different phenotype (**Figure 4**). The two lines are crossed, and recombination in the F1 is captured in the F2 generation, which is then itself inbred for a number of generations. Each recombinant inbred line is therefore a unique mosaic of the two parental strains, and various algorithms are available to correlate the behavior of each recombinant inbred line with the genetic marker information (**Figure 4**). This method can also be directly applied to

individuals from an F2 or a backgross generation without further inbreeding.

The development of molecular markers does not necessarily require a sequenced genome, so QTL mapping can be extended to nonmodel organisms. Most of the studies, however, are in model-organisms such as *Drosophila* and mouse. Not only are there stable recombinant inbred lines available for this type of work, but also the available genome sequences permit detailed mapping of QTLs and potentially may identify single loci mediating the behavior. For example, an extensive QTL study of circadian behavior in the mouse revealed 14 loci that were involved in regulating various rhythmic parameters. However, most of these QTLs did not include known circadian clock genes.

QTL mapping is limited to loci that are variable (polymorphic) between the parental lines. Genes encoding critical components for Darwinian fitness will probably be under strong directional selection, which reduces genetic variation, and so these loci may not be uncovered by QTL mapping. However, a powerful aspect of QTL mapping, which is unrivalled by the other methods we have described earlier, is the opportunity to scan simultaneously for the interaction (epistasis) between different loci: indeed, using our circadian example, a substantial amount of epistasis across the mouse genome among circadian loci was revealed.

Identifying the causative genes within a QTL is still a major challenge, as these genomic regions are large (tens to hundreds of kilobases) and typically include many genes. Finer mapping of these large regions can be extremely painful, but this process can be accelerated if some candidate genes are lurking therein. As yet, few behavioral QTL studies have revealed the underlying gene(s). In a study of emotionality in mice, a modified QTL screen using outbred stocks indicated that a regulator of G-protein signaling, *Rgs2* contributed a small proportion to the behavioral variation (~5%). Although this might seem less than overwhelming, *Rgs2* null mutants studied by Willis-Owen and Flint did show altered anxiety responses, thereby validating the QTL. Thus, QTL's may provide the candidate genes through subtle natural variation, but for validation, mutants (KOs, RNAi knockdowns or chemically or transposon induced, see above) will be required.

One potentially informative approach is to apply microarrays to the kinds of genetic crosses that we have discussed, and correlate behavior with differential gene expression in the segregating generations. The net result can be described as a gene network in which large numbers of genes interact to produce the phenotype. This kind of analysis has become popular recently, and some believe that the future of behavioral genetics may lie in understanding these networks. One possible problem may be that these networks may not be very robust, in that a

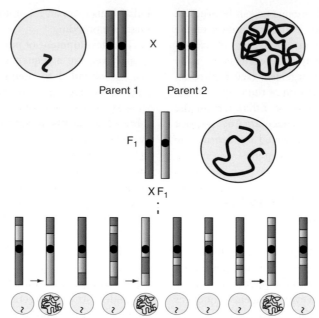

Figure 4   Mapping quantitative trait loci (QTL) involved in foraging behavior in *Drosophila* larvae (hypothetical example). Two parental strains that behave like 'sitters' (left) and do not move around very much on an agar plate, or 'rovers' (right), which do move considerably more, are crossed. The F1 progeny show intermediate behavior. The F1 are crossed for a few generations and then recombinant inbred lines (RILs) are generated by full-sib crosses. Each RIL is a mosaic of the parental genome, which can be identified by molecular markers. The behavior of each RIL is scored. The arrow indicates RILs that inherit a fragment from parent 2, and show the parental rover phenotype. This fragment is likely to carry a QTL affecting foraging behavior. Modified from Mauricio R (2001) *Nature Reviews Genetics* 2: 370–381, with apologies to Marla Sokolowski.

slight change in the conditions in which the behavior is measured ('noise') could significantly affect the overall topography of the network, recruiting new genes or losing others.

Human Molecular Neurogenetics

We cannot end without some comment on the development of neurogenetics with that most difficult of model organisms, *Homo sapiens*. The major tool that is used here is the linkage study. Briefly, if we take a family pedigree within which is segregating a behavioral phenotype of interest and consider the underlying causative behavioral mutation, the flanking genomic regions will likely contain another variant (perhaps a SNP, single nucleotide polymorphism, in a nearby gene) that always cosegregates with the behavioral mutation because the two loci are so close together that they remain undisturbed by genetic recombination. The two loci are thus in linkage and the two variants are in linkage disequilibrium, and thus the SNP in this case becomes a marker for the behavioral mutation. This is the basic principle behind linkage studies, and they have had their successes in human behavioral genetics.

A classic case involves that of a large Dutch family in which some of the boys showed unusually high levels of violence and antisocial behavior, including arson, attempted rape, and other impulsive displays such as exhibitionism. The mutation was tracked down by linkage analysis to an X-linked gene encoding monoamine oxidase A (MAOA), an enzyme that is used to break down neurotransmitters. Later studies in other families were to show that boys carrying milder mutations that produced less active versions of MAOA would not show any of these problems unless they had been subjected to abuse during childhood. These studies show beautifully how the social environment can modulate the expression of a mutant phenotype. Indeed, in the field of maternal behavior in rodents, there exists some stunning work that documents how environment can alter the heritable expression of a gene. Rat pups that receive minimal maternal care from their mothers do the same to their offspring because their gene encoding the receptor for the steroid stress response hormone, glucocorticoid, has been epigenetically modified through methylation of the DNA sequence (Fish et al., 2004). This environmentally triggered modification of the gene is passed on to the next generation, providing, superficially, a quasi-Lamarckian type of inheritance.

Another remarkable linkage study showed that a family that was segregating a dominant, autosomal disorder in which the affected individuals would wake up early and also fall asleep extremely early (Advanced Sleep Phase Syndrome or 'larks') contained a mutation in one of the four copies of the human *Per* gene (*hPer2*). In fact, this human clock mutation was very similar to the original *per^s* mutation found by Konopka in his short-period fly mutant. In both the fly and the human variants, a key Serine amino acid that is phosphorylated had been replaced, and mutants of both species showed fast-running clocks. In a 24 h world, both the fast-running mutants adapted by advancing their sleep–wake cycles by several hours and becoming 'larks.'

These two spectacular and successful examples are rarities within the behavioral field, because an enormous and largely unsuccessful effort has been mounted over the past two decades in identifying some of the genes that contribute to common complex phenotypes, particularly those involving psychpathology, schizophrenia, uni- and bi-polar depression, alcoholism, etc. The net result of hundreds of such studies, many large scale and expensive, has been disappointing. A number of studies have found associations between genes such as *neuregulin*, *dysbindin*, and the gene encoding COMT (catechol-o-methyltransferase), and schizophrenia, for example, yet for every study that identifies such a candidate gene, there appear to be several others than cannot confirm this association. This has led some to question whether this kind of approach will ever be successful in isolating these loci, and some imaginative alternative hypotheses about the genetic and evolutionary basis of schizophrenia have been proposed, particularly by Tim Crow. He has suggested that epigenetic, not genetic, modifications are responsible, thereby explaining why no genetic factors have been consistently identified. This epigenetic modification is invoked to involve the *protocadherin* genes (encoding cell surface adhesion molecules) located on the X and Y chromosomes within a chromosomal rearrangement that distinguishes humans from the great apes and other primates. This rearrangement may have played a role in both the evolution of language and in its distortion in schizophrenia (hallucinations and delusions, i.e. hearing voices). Crow's ingenious epigenetic theory fits in well with the known environmental modulation of this pathology, yet a stringent experimental molecular analysis is difficult with human subjects.

Future Prospects

Neurogenetics is now a mature discipline that straddles behavior, evolution, neurobiology, and genetics. Technical developments such as RNAi have extended the field beyond the model organisms of fly, mouse, zebrafish, and nematode worm. A marine biologist, for example, might be interested in using RNAi to knock down a *per* homolog in a crab to study whether this manipulation disrupted the crustacean's 12 h tidal rhythms. Perhaps a gene originally identified within the fly that affects memory, if knocked down in a honeybee, might affects the workers ability to associate the sun compass with a food source? The technology now exists to potentially manipulate genes in organisms that have no formal genetics, so these rather more interesting eco-behavioral phenotypes will become open for neurogenetic analysis. Natural genetic variation will continue to be studied through QTL-type approaches, although many challenges still remain in dissecting out loci that contribute small yet significant components of behavioral variation, via reverse-genetics, where a gene sequence originally identified through mutagenesis then becomes the substrate for examination of natural polymorphisms. Complex behavior in humans as well as animals will have complex underlying genetic architectures, so whether the QTL or linkage and association approaches will make major contributions to dissecting out natural genetic variation remains to be seen. One prediction is that the epigenetic modification of behavioral genes that we have touched on briefly will become a major field of study in the ensuing years. Needless to say, those will be very exciting times.

See also: Honeybees; *Nasonia* Wasp Behavior Genetics.

Further Reading

Bilen J and Bonini NM (2005) *Drosophila* as a model for human neurodegenerative disease. *Annual Review of Genetics* 39: 153–171.

Billeter J-C, Rideout EJ, Dornan AJ and Goodwin SF (2006) Control of male sexual bhehaviour in *Drosophila* by the sex determination pathway. *Current Biology* 16: R766–R776.

Brunner HG, Nelen M, Breakefield XO, Ropers HH and van Oost BA (1993) Abnormal behavior associated with a point mutation in the structural gene for monoamine oxidase A. *Science* 262: 578–580.

Caspi A, McClay J, Moffitt TE et al. (2002) Role of genotype in the cycle of violence in maltreated children. *Science* 297: 851–854.

Crow TJ (2007) How and why genetic linkage has not solved the problem of psychosis: Review and hypothesis. *American Journal of Psychiatry* 164: 13–21.

Dierick HA and Greenspan RJ (2006) Molecular analysis of flies selected for aggressive behavior. *Nature Genetics* 38: 1023–1031.

Fish EW, Shahrokh D, Bagot R et al. (2004) Epigenetic programming of stress responses through variations in maternal care. *Annals of the New York Academy of Sciences* 1036: 167–180.

Fulker DW (1966) Mating speed in male *Drosophila melanogaster*: A psychogenetic analysis. *Science* 153: 203–205.

Hay DA (1985) *Essentials of Behaviour Genetics*. Victoria: Blackwell.

Kyriacou CP, Peixoto AA, Sandrelli F, Costa R and Tauber E (2008) Clines in clock genes: Fine-tuning circadian rhythms to the environment. *Trends in Genetics* 24: 124–132.

Nitabach MN and Taghert PH (2008) Organisation of the *Drosophila* circadian control circuit. *Current Biology* 18: R84–R93.

Panda S, Hogenesch JB and Kay SA (2002) Circadian rhythms from flies to human. *Nature* 417: 329–335.

Sathyanarayanan S, Zeng X, Kumar S et al. (2008) Identification of novel genes involved in light-dependent CRY degradation through a genome-wide RNAi screen. *Genes & Development* 22: 1522–1533.

Shimomura K, Low-Zeddies SS, King DP et al. (2001) Genome-wide epistatic interaction analysis reveals complex genetic determinants of circadian behavior in mice. *Genome Research* 11: 959–980.

Takahashi JS, Hong H-K, Ko CH, and McDearmon EL (2008) The genetics of mammalian circadian order and disorder: Implications for physiology and disease. *Nature Reviews Genetics* 9: 764–775.

Willis-Owen SAJ and Flint J (2006) The genetic basis of emotional behaviour in mice. *European Journal of Human Genetics* 14: 721–728.

Group Living

G. Beauchamp, University of Montréal, St. Hyacinthe, QC, Canada

Introduction

Living in groups is widespread in animals, ranging from invertebrate swarms to mammalian herds and including other well-known aggregations such as fish schools and bird flocks. Group living has been thought to increase foraging efficiency and reduce predation risk, and these benefits often outweigh the negative consequences of living in groups such as increased food competition and disease transmission.

In this article, I will explore the various ways in which group living enhances survival through a reduction in predation risk. I will focus mostly on foraging groups, but many of the concepts reviewed here apply to other types of groups such as colonies and communal breeding groups. In illustrating the various ways group living can reduce predation risk, I will cover a large range of taxa, although it should be clear that not all mechanisms necessarily apply to any one species.

Group living can enhance survival at several stages during the predatory attack sequence. Here, I focus on adaptations that reduce the attack rate by predators and those that reduce the capture rate once an attack is launched. I then move on to contentious issues regarding how to disentangle the mechanisms that are purported to reduce predation risk.

Adaptations to Reduce Attack Rate

Several mechanisms are known to reduce the attack rate by predators, and as such, these mechanisms can decrease predation risk and increase survival. I will discuss how aggregation by prey and the use of mobbing and group defenses work to reduce the attack rate by predators.

Spatial Aggregation

When predators must search through space for their prey, spatial aggregation by prey will increase the amount of time between prey encounters and may lead to increased survival for aggregated prey when aggregations are no more detectable than solitary individuals and when the predator does not capture all individuals within the group upon attack. In some cases, larger groups can be less detectable than solitary prey if, for instance, a predator will abandon an area with aggregated prey because of low prey encounter rate. But generally, larger groups are probably more detectable visually due to the larger area that they occupy in space and because such groups may also be noisier. Therefore, reduced encounter rate with a group is often not considered sufficient on its own to account for the formation of groups. While predators are often attracted to large aggregations, large groups of prey that use conspicuous morphological traits, such as warning colors, to signal unprofitability are often avoided. The use of such aposematic displays is covered in detail in other articles.

Group Defenses

Groups of animals can also reduce the attack rate by resisting attacks through group defense, or deterring attacks through active pursuit of predators. In the face of predatory threat, many species bunch together in defensive formation. The pinwheel formation in musk oxens upon attacks by wolves represents a well-known example. A defensive formation multiplies individual defenses, such as horns or hooves, and may thus act as an attack deterrent. Perhaps more impressive are active pursuits of predators by potential prey as seen in many species of birds and monkeys. Such mobbing seems counterintuitive at first given that it decreases the distance between a predator and a prey. Indeed, mobbing animals have been known to be captured by predators. However, mobbing in groups reduces individual risk and can, in fact, be more effective in driving the predator away. In the end, a predator that is mobbed cannot rely on surprise to launch an attack, may be forced to move on, or can even be killed, thus decreasing, at least temporarily, predation risk for individuals in the group.

Adaptations to Reduce Capture Rate

Once a predator has launched an attack, animals in groups can reduce the capture rate using several mechanisms which are described below.

Improved Detection

As far back as the nineteenth century, naturalists wondered about the adaptive value of living in groups and suggested that by being in groups, rather than alone, individuals would be less likely to be approached undetected by a predator due to the presence of many eyes and ears tuned to predation threats. By increasing detection of threats, living in groups may allow animals to deploy escape maneuvers more quickly and thus reduce the risk of capture.

Intuitively, the odds of detection should increase quite rapidly with group size given that many eyes and ears are available to detect predators. Therefore, in a large group, any individual could reduce its own investment in vigilance against predators at no increased risk to itself, given that the odds of detection by the group are still far superior to those of a single individual. For the cheaters, the time thus freed from vigilance activities could be used for other fitness-enhancing activities such as feeding and resting, which are largely incompatible with vigilance. In response to this threat of unilateral cheating, the best response by other group members is to reduce their own vigilance level. The end result of this game amongst group members is a level of individual vigilance that is much lower than the vigilance that would provide the most benefit for the group. The game-theory model of vigilance leads to the prediction that vigilance, when aimed at predation threats, ought to decline with group size. Nevertheless, the odds of detection at the group level, even in large groups with low individual vigilance levels, are expected to be higher than what a solitary individual can achieve, meaning that group living can still provide extra protection. In addition, it may be costly for an individual to maintain vigilance levels that are too low since predators can aim their attacks at the least vigilant group members. Also, least vigilant group members may be amongst the last to escape from an attack and thus be more vulnerable to capture.

The advantages from improved detection rely on the ability of group members that have not directly detected the predation threat to use information gleaned from others to initiate their own response to an attack. This use of information from the group, which is known as collective detection, assumes that once an individual has detected a predator, all other individuals in the group that have not detected the threat directly are alerted about the threat very rapidly. Collective detection can be based on alarm calls or obvious escape behavior by the detectors, but in practice, signals of threat detection can be ambiguous and collective detection can thus lead to frequent false alarms in a group.

That groups, as opposed to solitary individuals, benefit from improved predator detection has been corroborated in many species. For instance, larger groups of prey are quicker to detect an approaching threat and can detect these threats at greater distances. The prediction that vigilance ought to decline with group size has also been documented in many species, but the group-size effect on vigilance is not as strong as predicted in many species, including many primate species. One possibility to explain these unexpected findings is that vigilance can be aimed not only at predation threats but also at other group members, for instance, to monitor aggressive companions or detect foraging opportunities. Given that such intragroup monitoring often increases with group size, the end result is that the overall vigilance may not decrease with group size as rapidly as expected. Other reasons why vigilance may not decline with group size include the fact that large groups may be targeted more often by predators or that vigilance in large groups may not be as effective as predicted due to visual interference from other group members.

While the above model assumes that adjustments in vigilance are a response to variation in group size, it is clear that the effect of group size can be confounded by several ecological factors, most notably, food density. Given that large groups often aggregate in areas of higher food density and that vigilance may be reduced when animals feed more, the group-size effect of vigilance may be confounded by the direct effect of food density on vigilance. Other factors that can obscure the effect of group size on vigilance include scramble competition in larger groups, which on its own would also predict a decrease in vigilance with group size, and individual phenotypic attributes that vary with group size, such as sex or satiation level, which could on their own explain part of the decline in vigilance with group size. Disentangling the effect of these various factors often requires experimental manipulation of group size or statistical control of confounding factors.

Dilution of Predation Risk

In many predator–prey systems, predators attacking a group can at most capture one individual. Therefore, the presence of many group members dilutes predation risk for everyone. Specifically, if groups of different sizes are attacked at the same rate, each individual in a group of n foragers only experiences a $1/n$ risk of being attacked by the predator. An added assumption is that the risk of being targeted by the predator is similar for all group members. Put simply, the dilution effect is a decrease in predation risk due to the presence of alternative targets in a group.

The dilution effect can, on its own, explain why vigilance decreases with group size. Most models of vigilance in animals incorporate collective detection and dilution effects, which are then seen as being additive, although dilution is expected to play a greater role in larger groups. The assumption of equal risk is probably questionable in larger groups where some individuals are more likely to be targeted by predators due to a more exposed position on the edges of the group.

Confusion Effect

Prey aggregation can also reduce predator success through the confusion effect. With the confusion effect, a predator attacking a group becomes disoriented by the flight reaction in the group and thus experiences difficulties in singling out one individual. Flight reaction can confuse and disorient a visually guided predator providing extra time for the prey to flee. The confusion effect can be enhanced by conspicuous body markings and erratic flight behavior. The confusion effect predicts a decrease in the capture rate with group size and this has been most convincingly shown in laboratory experiments with fish and more recently in experiments with human subjects trying to locate objects on computer monitors. Exactly how a predator can be confused remains largely unexplored, but recent studies imply increased spatial targeting errors when attacking larger groups.

Selection against oddity would seem to follow from the confusion effect, since any individual that would deviate from the norm when fleeing may represent an easier target for the predator. This may account for the observation that upon attack, odd fish in mixed-species fish shoals abandon the group.

Selfish-Herd Effect

Selfish behavior by individuals in avoiding a predator has been thought to lead to spatial aggregation. This scenario is based on the assumption that a predator will strike the nearest prey individual. Therefore, a solitary forager is more likely to be targeted because it has a larger domain of danger, defined as the space closer to that individual than to any other group members. Predation risk for an individual in such a group should be proportional to the ratio of its domain of danger to the area occupied by the group. Consequently, to reduce predation risk, an individual could move closer to neighbors to reduce its own domain of danger at the expense of others. As the group becomes more compact, all domains of dangers necessarily become smaller, but the key point is to have a relatively smaller domain of danger than the neighbors.

The concept of dilution is often confused with the selfish herd effect since both rely on predation risk dilution in groups. However, with dilution, an individual can only reduce its predation risk by foraging in a larger group. With the selfish-herd effect, greater safety can be achieved within the confines of the same group by altering position. Key predictions from the selfish-herd effect include a tendency to bunch when attacked and a lower predation risk for individuals with a smaller domain of danger.

The selfish-herd effect means that individuals at the edges of groups, which have greater domains of danger, will usually be more at risk from predators attacking from the outside of the group. The observation that vigilance against predators is usually higher at the edges of groups is compatible with position-sensitive risk. However, foraging opportunities are often greater at the edges of groups, and in the end, the spatial position may reflect a trade-off between food gains and predation risk.

Empirically, studies have shown increased spatial cohesion upon attack and a preference for central locations, which are more buffered from attacks originating from the outside of the group. One issue that confounds many empirical studies is the probable correlation between phenotypic attributes related to predation risk, such as hunger, which could force an individual to forage away from the group, and the size of the domain of danger or position within the group. Selfish-herd effects assume, however, that all individuals are intrinsically equally at risk controlling for the size of the domain of danger.

Recent theoretical emphasis has been on determining which movement rules could account for the observed bunching given that simply moving to the nearest neighbor does not necessarily produce spatial cohesion. Rules whereby individuals move to the more crowded part of the group appear to produce bunching as predicted by the original model of the selfish-herd effect. Further work is needed to assess these issues.

Disentangling Collective Detection and Predation Risk Dilution

The contribution of collective detection and risk dilution to the group-size effect on vigilance has been difficult to disentangle because the two mechanisms make very similar predictions. For instance, in addition to the predicted decrease in vigilance with group size, the two mechanisms also predict an increase in vigilance with higher interindividual spacing. With collective detection, the behavior of distant neighbors can be more difficult to monitor, and with risk dilution, a wider spacing increases the domain of danger of each individual and thus their vulnerability to predation.

Disentangling the two mechanisms requires varying the contribution of one mechanism while maintaining the other constant. One example of this occurs when predators are more likely to attack from one side of the group. Collective detection can be assumed constant if all individuals in the group can detect the predator equally well. In the case of risk dilution acting alone or with constant collective detection, individuals occurring on the riskier side of the group are unlikely to be protected to the same extent as group members buffered further inside the group. The prediction that vigilance should be higher on the riskier side of the group was corroborated recently indicating that collective detection alone could not account for changes in vigilance within the group. Other scenarios holding risk dilution constant and varying collective detection in order to disentangle the two mechanisms are also promising.

Disentangling Scramble Competition Effects and Predation Risk Management

The cause for the decline in vigilance with group size has been usually attributed to predation risk management involving mechanisms such as collective detection and risk dilution. However, it has become clear that mechanisms unrelated to predation risk may also cause a decrease in vigilance with group size. For instance, increased competition in large groups may induce a decrease in individual vigilance levels as foragers scramble for a greater proportion of limited resources. Although the potential influence of scramble competition on vigilance has long been recognized, theoretical and empirical interest in the matter has only risen recently.

Scramble models assume that resources such as food are limited and that foragers jostle to obtain a greater share of resources. Scrambling for food should lead foragers to adopt more risky behavior such as a decrease in vigilance. To obtain a greater share of resources, individuals must forage more quickly than their competitors, since the best response to an increase in exploitation speed by companions is a further increase in speed. If vigilance is traded-off against foraging gains, foragers could decrease individual investment in vigilance to increase exploitation speed. As competition increases with group size, scrambling for resources can induce a decline in vigilance with group size.

The empirical evidence in support of the scramble hypothesis comes from recent work indicating that vigilance decreases when individuals feed on rapidly disappearing food items. In another experiment, foragers decreased their vigilance levels when other group members were foraging to a greater extent. In both cases, group size was maintained constant to ensure that predation risk dilution also remained constant.

Conclusion

Several mechanisms allow groups to reduce predation risk either by avoiding predators or by reducing the capture rate once an attack is launched. While the theoretical underpinnings of these mechanisms is rather well known, empirical studies have not always been successful in disentangling their effects, and support for some mechanisms, such as the selfish-herd effect and the confusion effect, is still scant.

See also: Antipredator Benefits from Heterospecifics; Defensive Coloration; Trade-Offs in Anti-Predator Behavior; Vigilance and Models of Behavior.

Further Reading

Beauchamp G and Ruxton GD (2008) Disentangling risk dilution and collective detection in the antipredator vigilance of semipalmated sandpipers in flocks. *Animal Behaviour* 75: 1837–1842.

Bednekoff PA and Lima SL (1998) Randomness, chaos and confusion in the study of antipredatory vigilance. *Trends in Ecology and Evolution* 13: 284–287.

Caro TM (2005) *Antipredator defenses in birds and mammals.* Chicago: The University of Chicago Press.

Elgar MA (1989) Predator vigilance and group size in birds and mammals. *Biological Reviews* 64: 13–33.

Foster WA and Treherne JE (1981) Evidence for the dilution effect in the selfish herd from fish predation on a marine insect. *Nature* 293: 466–467.

Hamilton WD (1971) Geometry of the selfish herd. *Journal of Theoretical Biology* 31: 295–311.

Ioannou CC, Tosh CR, Neville L, and Krause J (2008) The confusion effect from neural networks to reduced predation risk. *Behavioral Ecology* 19: 126–130.

Krause J and Ruxton GD (2002) *Living in Groups.* Oxford: Oxford University Press.

Lima SL (1995) Back to the basics of anti-predatory vigilance: The group size effect. *Animal Behaviour* 49: 11–20.

Pulliam HR, Pyke GH, and Caraco T (1982) The scanning behavior of juncos: A game-theoretical approach. *Journal of Theoretical Biology* 95: 89–103.

Quinn JL and Cresswell W (2006) Testing domains of danger in the selfish herd: Sparrowhawks target widely spaced Redshanks in flocks. *Proceedings of the Royal Society London B: Biological Sciences* 273: 2521–2526.

Roberts G (1996) Why vigilance declines as group size increases. *Animal Behaviour* 51: 1077–1086.

Group Movement

I. D. Couzin, Princeton University, Princeton, NJ, USA
A. J. King, Zoological Society of London, London, UK; University of Cambridge, Cambridge, UK

Introduction

Why Group?

Individuals from almost any animal species will be found in association with others at certain points in their lives. At one end of the spectrum, solitary (sexually reproducing) individuals, if successful enough to find a mate, will have temporarily belonged to a pair. At the other extreme, individuals can spend their entire lives, from the moment they are born until the moment they die, in close proximity with many other individuals. Most animals, however, fall somewhere in between, forming and breaking groups with remarkable frequency. Current ideas on the evolution and ecology of group living are therefore the result of researchers scrutinizing the interactions that occur when individuals come into proximity with one another, and trying to understand the short- and long-term consequences of such interactions.

Where these costs and benefits of grouping dictate that individuals are better off acting together, individuals should be expected to coordinate their movements. Imagine a pair of hungry individuals that have to stick together for protection and have to choose between two available food patches – patch A or patch B – and these patches are of exactly the same size and quality. Whichever individual makes the move to patch A or B first will leave the other individual no option but to follow. Since they get the same food reward at each food patch, a failure to coordinate their behavior, and to move together, will mean that they forfeit the reduced risk of predation they gain from sticking together.

Moving as a Group

Since the action of movement is preceded by the group decision to move, we will look at group movements associated with two types of decisions – when to move and where to move. Often, these decisions will have to be made together; for example, a honeybee colony choosing a new nest site is required to agree both when they move, and the location of the new site, before departure. Nevertheless, we make this distinction to allow us to better illustrate the different mechanisms animals within groups adopt to achieve collective group movements, and the functional costs and benefits to realizing these movements. We begin with a discussion of theoretical models that have been developed over recent years.

Theoretical Models of Group Movement

Group movements are dependent on the social interactions of its members. It is easy to make verbal arguments for groups of two (like our above foraging example), but difficult, or impossible, to think through the consequences of a large number of social interactions by verbal argument alone. Therefore, mathematical models have proved to be very useful not only for structuring our thinking about group movement but also for generating testable predictions.

Where to Go?

The precise rules by which animals interact within mobile groups are still poorly understood, largely due to a dearth in appropriate experimental data. However, individual-based models that simulate explicitly local interactions such as avoidance of collisions, attraction and alignment to others, have helped to explore the principles by which coordinated motion can merge, giving us a better understanding of how group cohesion and structure results from these interactions (see **Figure 1**).

These types of models have also been an important starting point for investigations into how individuals within mobile groups decide where to go when there are differences in informational status among them. For example, individuals can differ in their directional preference which may be used on an expectation of the location and quality of a remembered resource, or on sensory information available to individuals as they move through their environment. Individuals with a preferred direction must therefore reconcile this with their need to maintain group cohesion (if they are to accrue the benefits from group membership). If there are few costs to leaving group-mates, and/or there is a large benefit to moving in the preferred direction, individuals are expected to strongly bias their movement in their preferred direction, and leave the group. However, if maintaining group membership is important, individuals with a preferred direction of travel may be able to exert their influence on the group by exhibiting a less strong (but existent) bias. Thus, groups can be guided spontaneously without requiring individual recognition or signaling. Furthermore, the models reveal that as group size gets larger, the proportion of informed individuals required to accurately guide the group actually gets much smaller, and only a very small

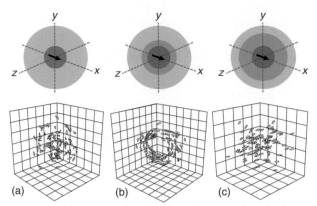

Figure 1 Principles of individual-based models. Shown are sets of rules followed by 'agents' in the individual-based models we describe, and the emergent properties at the level of the group structure. (a)–(c) represent three different model scenarios, where an agent is represented as an arrow, and the shaded volumes represent its interaction zones with its neighbors. The red zone at the center (consistent across all scenarios) represents an agents' 'zone of repulsion'; individuals try to maintain personal space and to avoid collisions by being repelled from conspecifics that enter this zone. The largest area, shown in light blue (also consistent) is an agent's 'zone of attraction.' This represents the volume in which individuals are attracted to one another, which is responsible for maintaining group cohesion. Finally, the dark blue intermediate zone, which is not present in (a), is small in (b), and larger in (c), represents the 'zone of orientation' – the zone in which agents tend to orientate themselves in the average direction of their neighbors within this zone. With very low, or no, zone of orientation (a) agents in the models form a 'swarm.' In this swarm state, even though individuals rotate around the group center, they do so in different directions. As the size of the zone or orientation is increased, (b) groups spontaneously form a 'torus' in which individuals perpetually rotate around an empty core, the direction of rotation being random. Finally, if the zone is increased further, the group adopts a 'dynamic parallel' structure in which the agent's movements are aligned in a single direction, but individuals move throughout the group and the group itself can spontaneously change the direction of travel. Adapted from Couzin ID, Krause J, James R, Ruxton GD, and Franks NR (2002) Collective memory and spatial sorting in animal groups. *Journal of Theoretical Biology* 218(1): 1–11.

proportion of individuals can effectively determine the group direction.

This example of a few 'leaders' guiding many 'followers' could represent the case where only one, or a few, individuals are socially dominant; if subordinate individuals do not express directional preferences (or are punished from doing so), then spontaneously it will be the dominant individual(s) that guide groups. Similarly, if differences in the temperaments of individuals exist, bolder individuals can more often take the initiative and bias group movements. In other cases, it may be that leadership is transferable among group members. Perhaps only one, or a few, individuals are knowledgeable about a food source, or a migration route, and others are naive. Under such circumstances, the same model can explain how leadership will emerge according to informational status.

In the case of conflict within the group, such as when individuals that are exerting a directional bias are not in agreement about where to go, it has been shown that these types of social interactions can facilitate consensus decision-making, even if the individuals in disagreement only constitute a small proportion of the overall group membership. The models predict that if there is no majority (that there are an equal number of individuals wanting to go in each preferred direction), and that if the angle between the preferred directions of travel is small, the group will tend to split the difference and go in the average direction. Above a critical difference in opinion, however, it is predicted that the group will collectively select one or other direction (see **Figure 2**). Consensus can also be made in favor of the direction preferred by the numerical majority and sometimes by those individuals that mostly depend on moving in their preferred direction, even if they are in a (slight) minority, as they are most willing to pay the costs of leaving the group and thus risk isolation by exerting a strong directional preference (this has been termed 'leading according to need').

When to Go?

Models exploring the functional costs and benefits of leading and following also abound, but tend to concentrate on 'when' problems, which require a group to agree on the timing of a particular activity or event (e.g., when to start foraging). These game theoretical models predict that leadership (where a single individual's preference is followed), as well as scenarios where an average or median 'compromise' between all individual's preferences is employed, can each be evolutionarily stable in groups of all sizes, but that the averaging of preferences – as in the individual-based models described above – should evolve under a wider range of social and ecological conditions.

Signals, Rules, and the Brains that Implement Them

The social environment and a need to coordinate clearly impose strong selective pressures on animals, but does complexity in group movements necessitate cognitive complexity? We have already stated that the precise rules by which animals interact within mobile groups are poorly understood, but we can speculate: animals can use simple, local (and error-prone) rules during the decision-making process and are unaware of whether other group members are in agreement or disagreement with them, or even if there are any other individuals within the group that have a directional preference. Furthermore, if members of groups integrate their directional preferences with one another, individuals cannot only make collective decisions, but by determining whether they can assert their own influence on the group motion infer something about

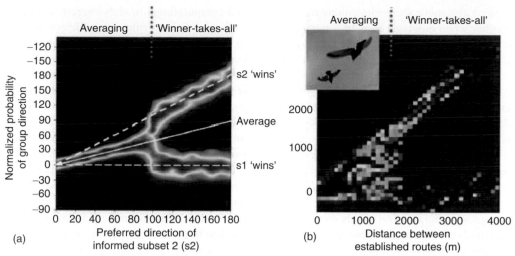

Figure 2 Outcome of decision-making in homing pigeons. (a) Predictions of a theoretical model of group movement under conflict of interest. In this model, there are two conflicting subsets of informed individual (s1 and s2; here, five individuals in each subset), and the whole group adopts the average preferred direction below a critical difference in opinion. Above this, the group enters a consensus phase in which, given a symmetrical conflict, the whole group goes in one preferred direction or the other with equal probability (here, the group size was 100). In cases of an asymmetry, such that s1 does not equal s2, the group will select, collectively, the majority direction with high probability. (b) When faced with a similar conflict, pairs of homing pigeons also tend to compromise when the difference in the preferred route is small, but do tend to select one direction or the other when this becomes large. The figure shows the point-by-point distances between each bird's established route and its route taken during experimental trials. Adapted from Couzin ID (2009) Collective cognition in animal gourps. *Trends in Cognitive Sciences* 13(1): 36–43.

their informational status with respect to other group members. This latter behavior, although more complicated, still requires only simple rules. This does not mean that individuals are incapable of more cognitively demanding decision-making, but rather that apparently sophisticated inference may often be based on relatively simple integration of local information.

It is often difficult to distinguish between simple and more cognitively complex mechanisms of collective decision-making associated with group movements. For instance, ungulates and primates are described as participating in 'voting' behavior whereby individuals use body orientation or initiation movements to come to a consensus on group movement directions. This cognitively demanding hypothesis relies on an estimate of the relative number of votes. Note, however, that the motion characteristics described may also look, to an observer, like voting behavior since individuals with a directional preference tend to orient in their preferred direction. It may be that individuals respond to near-neighbors with only relatively simple interaction rules. Further work now needs to focus on this issue, in order that we can discriminate between these hypotheses, and describe these interactions more precisely.

Testing the Models

The theoretical models described here have been developed with the intention of their predictions being tested across different animal taxa. Yet the models have advanced far more rapidly than empirical studies, often without validation of basic assumptions and without tests of their predictions. In fact, the need for empirical studies to test these model predictions has become increasingly obvious – and this challenge has been taken up with fervor in recent times by researchers studying in animal behavior, with researchers considering both the 'when' and 'where' problems (and those that combine both problems).

Empirical studies have not only begun to describe interaction patterns but have also emphasized how the relative differences between individual group-mates' physiology or temperament, as well as their position within networks of social alliances and dominance, can all contribute to the process by which animal groups are able to coordinate their actions and move as a collective.

We will now provide a series of empirical examples which describe group movement patterns in a variety of species. We will follow a taxonomic structure, beginning with the insects and fish, and then the birds and mammals.

Swarming Insects

Some of the most sophisticated and best-studied group movement decisions are made by the social insects (species of bees, ants, and termites). Here, the close genetic relatedness among individuals often results in individuals working together to achieve a common goal in daily activities such as foraging, nest building, and brood care.

Social insects must also move the entire colony from one location to another when moving home. In some species, this important movement occurs only when a colony outgrows its existing home, or when it becomes irreversibly damaged. In other species, such as the nomadic army and driver ants, colonies must regularly move to new foraging grounds when they have decimated the local invertebrate (prey) population. In the case of nomadic species, the timing of movement is determined by the reproductive status of the colony, while their destination appears to have been selected during the daily raid of prey that precedes the move. However, the practical difficulties encountered when working with such wide-ranging species limit our current understanding of this process.

Much better understood are the habits of honeybees, *Apis mellifera*, and a small cavity-dwelling rock ant, *Temnothorax albipennis*. In both species, independent scouts assess the quality of potential nest sites, using a range of criteria, including an estimate of its size as well as the light-level, humidity, and structural integrity. If considered acceptable, the scouts will begin to actively recruit others to visit the nest. The ants do so by attempting to lead a single follower to the site (**Figure 3(a)**), whereas honeybees express their preference by increasing the duration with which they perform special 'waggle dances' for favored sites (**Figure 3(b)**). Recruits independently assess the site and may then begin recruiting others themselves. Once a threshold (quorum) of individuals is detected at the nest site, the insects increase their rate of recruitment, which allows a compromise between the speed and accuracy of decision-making.

In honeybees, a further process is required; once honeybee scouts have come to a consensus among themselves, the colony (or part of it, if the move is related to reproductive fissioning) must relocate and be guided to the new site. The problem here is that the collective movement of the bee colony can comprise tens of thousands of individuals, of which typically less than 5% (i.e., the scouts) have information on the newly chosen site location. This appears to pose a considerable problem, and yet whilst the precise mechanism is currently unknown, experiments and computational models suggest that once again, relatively simple local rules are sufficient to guide the colony to their new home, not unlike the process of leading a mobile group as we have already described.

The above examples are from eusocial insects, but one of the most dramatic examples of group movement among insects are the massive migratory swarms formed by insects such as locusts and crickets. The Desert locust, *Schistocerca gregaria*, is the most notorious and best-studied example. Swarms of this species can cover several hundred square kilometers, and contain billions of insects. They can be devastating, with the locusts' range expanding in plague years to cover in excess of 20% of the Earth's land surface, and they are estimated to damage the livelihood of one in ten people on the planet.

Locust aggregates typically form when insects are juveniles, and before they grow wings. These mobile groups are called 'marching bands' and inevitably precede the flying swarms. Although highly coordinated, the drive for such collective motion is far from cooperative. Locusts, and other band-forming insects such as the Mormon cricket, *Anabrus simplex*, are in fact engaged in a 'forced march' driven by cannibalism resulting from urgent nutritional needs. When critical resources become scarce, specifically protein, salt, or water, the insects are forced to prey upon each other, since conspecifics represent the only source of these essential nutrients. Perhaps surprisingly, rather than creating chaos, these cannibalistic interactions establish order in the movements of the bands: insects move away from those attempting to cannibalize them, and towards those whose vulnerable abdomen may become the important next meal. Individual-based models that represent these types of interactions in mathematical terms show that if the local population is above a critical density, this can lead to an auto-catalytic process where cannibalism drives the formation of marching bands, and these predictions have been verified experimentally. Since it is cannibalistic interactions that drive these movements, it is likely that individuals with the greatest nutritional needs most strongly exert an influence on the onset and maintenance of marching within bands.

Figure 3 (a) Tandem running in ants; the ant on the left is following the other to a known food source, and is led via tactile communication. Image courtesy of Tom Richardson and Nigel Franks. (b) Honeybee 'waggle' dance (indicated by the white lines) signaling the location and quality of potential nest sites to colony members. Image courtesy of Jürgen Tautz and Marco Kleinhenz.

The self-driven motion of these locust marching bands has also been shown to be inherently unpredictable, making control efforts difficult. Cannibalistic interactions may not only provide locusts with essential nutrients, but may drive the band to cover substantially greater distances than individual insects, thus effectively searching larger areas for new sources of food. Additionally, traveling with conspecifics as a potential food source may decrease predation risk and enable locusts in bands to persist longer while traversing unfavorable parts of a patchy nutritional environment. Thus, a large number of different selection pressures may underlie the motion of migrating insect swarms.

Schooling Fish

Experiments on schooling fish have revealed that individuals with a directional bias can spontaneously lead uninformed individuals to resources, or through a maze. The degree to which an individual appears to be influenced by the directed motion of others depends on its informational status. If individuals lack personally acquired information about their environment, or when it is dangerous to acquire personal information, they tend to follow the movement decisions of others.

Social interactions have also been shown to be important to the timing of such movement events. Stickleback fish, *Gasterosteus aculeatus*, have been shown to exhibit variation in their propensity to leave a 'safe zone' (deep water with vegetation), and to move to look for food in a 'risky zone' (shallow with no shelter) in their tank. Randomly pairing fish that lie on this bold–shy continuum has revealed that the temperament of both fish plays a role in the exploratory behavior of the pair. Both fish in a pair respond to each other's movements – each more likely to leave the safe zone if the other was already out and to return if the other returns. Bolder fish display a greater initiative to move to risky zones, and are less responsive to partners, whereas shyer fish displayed less initiative but follow their partners more faithfully. Notably, the shy fish, when following, also elicited greater leadership tendencies in their bold partners – showing a sort of social feedback that ensured both fish fed, and importantly that they journeyed into the risky zones together.

Complimentary to these studies, it has also been revealed that the number of individuals appearing to choose a common direction of travel also plays an important role in fish school movement decisions. It may be ill-advised to indiscriminately copy the decisions of others when decisions are based on uncertain information (as in most biological scenarios). Stickleback fish have evolved a simple, but elegant, solution. They largely disregard the movement decisions of a single neighbor, but strongly increase their probability of copying the movement decisions as more neighbors (a 'quorum') commit to a give direction of travel. This response was found not only to dramatically improve the accuracy of collective decision-making by allowing individuals to integrate their own estimation with that of others, but does so with minimal costs in terms of the time taken to make the decision. Using this functional response, it has also been shown that individual-level accuracy in decision-making increases as group size increases.

Under other circumstances, such as during collective migration, fish may form very large aggregates. This may not only be a consequence of synchronized timing of the event (e.g., individuals collectively respond to an environmental cue indicating a change in season), but allow schools to locate and climb weak and noisy long-range gradients (such as resource or thermal gradients) that individual fish may have only a poor capability of detecting. Specifically, by grouping, fish form a large 'sensor-array' which both covers a large area, and allows individuals to average their estimates with a large number of others (see **Figure 4**). Thus, groups may be able to respond to structure in the environment, which is not possible for individuals. Such principles may also underlie the migration abilities of other organisms such as wildebeest and flocking birds.

Flocking Birds

Experiments on homing pigeons, *Columba livia*, have tested the theoretical prediction, outlined in our earlier model section, that the degree to which individuals differ in directional preference may play an important role. As predicted by the model, pairs of homing pigeons tend to compromise when the difference in preferred directions among the group members is small, but to select one direction or the other when this difference becomes large (**Figure 2**). Navigational accuracy in homing pigeons has also been shown to depend on group size, with larger groups being more accurate, as predicted by models in which the individual interactions spontaneously act to average individual error (**Figure 4**).

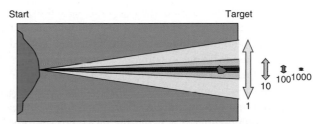

Figure 4   The advantageous effect of large group size on navigational accuracy. The figure shows the navigational accuracy for a theoretical migrating bird flock. The triangles depict 95% confidence intervals of trajectories for a single bird and flocks of 10, 100, and 1000 individuals of equal navigational ability. Figure adapted from Simons AM (2004) Many wrongs: The advantage of group navigation. *Trends in Ecology & Evolution* 19: 453–455.

Grouping Mammals

Many ungulate species, such as plains zebra, *Equus burcellii*, live in fission–fusion societies (Couzin and Laidre, 2009). In populations where group membership is fluid, movement initiation tends to depend on the individual state at the time of the decision. Specifically, in the case of zebra, female reproductive state (lactation) alters the water and energy needs for females, creating a greater motivation to seek out resources, and, consequently, becomes a crucial determinant of leadership and group movements, similar to the 'leading to according to need' hypothesis.

There have also been a number of group movement studies in primates in recent times. Getting a large sample to study (i.e., multiple groups) is a challenge, since primates typically range over large distances and are shy of observation. This means that researchers, for the moment, need to be content with small sample sizes or to resort to studies on semi-free ranging and captive populations in which the social and ecological conditions are not always ideal for studies of movement patterns. Nevertheless, a number of studies have been successful.

One such study took place on wild chacma baboons, *Papio ursinus*, in Namibia. Researchers presented baboon groups with experimental food patches within their home ranges, in addition to natural patches that they regularly forage in. There was an important difference between these two patch types. At natural patches, dominant baboons could rarely monopolize the food, and so, food intake was relatively evenly spread across group members. At experimental food patches, the food was at a higher density, allowing dominant baboons to monopolize almost the entire patch. If you consider the individual costs and benefits for each of these movement decisions (to visit natural or experimental patches), movements to natural patches are expected to represent movements based on the average of all group members' preferences. In contrast, movements to the experimental patches should represent movements based on the preferences of a few dominant individuals.

The baboon groups studied consistently visited experimental patches in preference to natural patches (**Figure 5**), and the dominant male appeared to lead groups to these locations in both cases. Despite not coercing subordinates, they followed nevertheless – despite getting less food than had they foraged independently to natural resources. Why did the subordinates follow those that were dominant, and why did the group not split up? As we have already discussed, individuals will be expected to leave a group only when the benefits of leaving outweigh the benefits of staying. Since the subordinates do not benefit in food acquisition, they should be expected to do so in other ways. It may be that dominants provide subordinates increased protection from predators, as well as protection for their infants from infanticidal males from other groups – all risks they would run by leaving. These hypotheses are backed up by the fact that the individuals that followed the dominant most closely in these movements were the dominant's closest 'associates' who most frequently groomed one another.

The baboon experiment has a very simple design, but was difficult to implement as it required observers to monitor the movements of groups over many kilometers, rather than within experimental tanks like in the stickleback fish experiments described above. For that reason, semi-free ranging conditions – where the movements of groups are somewhat restricted – provide scope for more detailed investigations into the mechanisms for group movement.

In a recent study, researchers examined group movements in a group of 11 captive white-faced capuchins, *Cebus capucinus*, moving between a resting and foraging zone in their semi-captive enclosure. They specifically investigated what happened after an individual began moving in its preferred direction (getting up after resting and moving a short distance towards the foraging zone). Using a modeling approach, they demonstrated that the capuchin monkeys' group movements were determined by two important and complementary phenomena. First, the frequency with which capuchins followed an individual that had proposed moving (the initiator), and second, the willingness of the propensity of the initiator to give up (i.e., cancellation rate). Interestingly, this cancellation rate appeared to be completely reliant on the number of followers an initiator attracted; if an initiator elicited more than three followers, the chances were that the whole group would move toward the feeding zone. But if the initiator was unsuccessful in attracting three followers, or less, the chances are (s)he would give up on leading the group, and return to resting. Such a quorum response (see **Figure 6**) is very similar to that we previously described for stickleback fish. Any individual could act as an initiator, and their success was not found to correlate to age, sex, or dominance. Under this captive setting though, the capuchins only need to choose between a resting or foraging zone. This makes the problem a true 'when' problem, seeing as they have no alternate choices of where to move. It will be interesting to see if these results hold where these two problems are interacting – when capuchins have to decide both when and where to move simultaneously.

Conclusions

Organisms that live in interactive groups have to cope with the ever-changing social demands – preserving bonds, forging alliances, tracking cooperation, detecting cheaters,

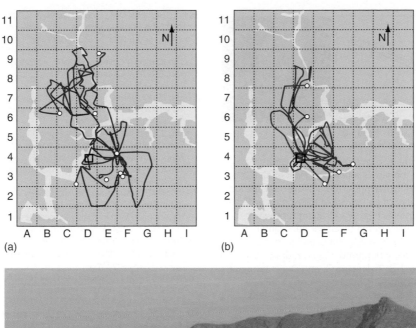

Figure 5 Baboon group movements and leadership. Group movements for a baboon group at the edge of the Namib Desert in Namibia studied by King et al. (2008). (a) Shows the GPS-tracked movements (collected by observers on foot following the baboons) of the group under regular conditions, over 13 days. (b) Shows the GPS-tracked movements of the group when an experimental food patch was presented at gird location D4, marked with a black square, over $n = 16$ days, when the alpha male led subordinate individuals to the experimental food patch, resulting in drastic changes in collective group movements. Sleeping site locations, where the group started and ended each day, are shown by white-filled circles. Light shaded areas represent a (dry) river and its tributaries that run through the baboon home-range. Grid cells represent 1 km by 1 km. (c) Shows the baboons at King et al.'s field location, moving through their home-range in a single file. Photo credit: ZSL Tsaobis Baboon Project/Hannah Peck. Figures from King AJ, Douglas CMS, Huchard E, Isaac NJB and Cowlishaw G (2008) Dominance and affiliation mediate despotism in a social primate. *Current Biology* 18: 1833–1838.

communicating information, and manipulating competitors (discussed in these volumes). All of these play a crucial role in shaping local movement patterns, and consequently, group cohesion and collective group movements.

We have shown in this article that such a complicated social environment does not necessarily require complicated cognition. Instead, simple rules-of-thumb may suffice, and excel in generating extremely complex patterns of group movements. We have described fundamental

shared mechanisms that operate across insects, fish, and mammals, and are even known to work in the studies of human crowd behavior. Indeed, a surprising degree of common themes and organizational principles are beginning to emerge from research in animal group movements. Tests between alternate (but not necessarily mutually exclusive) hypotheses concerning how individual interactions scale to group movements are now revealing specific decision rules (like the stickleback fish and capuchin

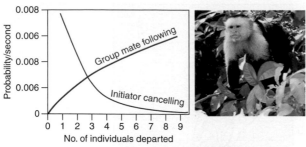

Figure 6 Capuchin monkey group movements. Group movements in capuchin monkey's studies by Petit et al. (2009). The figure shows the results of a model fitted to the data. It shows the probability (per second) of an individual following a proposer (capuchin monkey who moves first) to a foraging zone within their enclosure and the probability of the proposer canceling the movement and staying in their resting zone, as function of the number of individuals already departed. Figure redrawn from Petit O, Gautrais J, Leca J-B, Theraulaz G, and Deneubourg JL (2009) Collective decision-making in white-faced capuchins. *Proceedings of the Royal Society, Series B* 276: 3495–3503. Photograph courtesy of Odile Petit.

monkey examples we discussed). There are now clear predictions to test, and the problems of measurement can be overcome by a careful selection of variables and methodology.

Our understanding of the mechanisms and ultimate function of animal group movements is important not only because of the relevance to grouping behavior in social animals, but because the same principles might have wider application, such as to coordinated control of autonomous grouping robots, and may give us deep insight into the evolution of the highly sophisticated group movements and decision-making that established the evolutionary success and expansion of our own species.

See also: Collective Intelligence; Consensus Decisions; Dance Language; Locusts; Queen–Queen Conflict in Eusocial Insect Colonies; Social Learning: Theory.

Further Reading

Bazazi S, Buhl J, Hale JJ, et al. (2008) Collective motion and cannibalism in locust marching bands. *Current Biology* 18(10): 735–739.

Biro D, Sumpter DJT, Meade J, and Guilford T (2006) From compromise to leadership in pigeon homing. *Current Biology* 16: 2123–2128.

Buhl J, Sumpter DJT, Couzin ID, et al. (2006) From disorder to order in marching locusts. *Science* 312: 1402–1406.

Conradt L and Roper TJ (2005) Consensus decision making in animals. *Trends in Ecology & Evolution* 20: 449–456.

Couzin ID (2007) Collective minds. *Nature* 445: 715.

Couzin ID (2009) Collective cognition in animal groups. *Trends in Cognitive Sciences* 13(1): 36–43.

Couzin ID, Krause J, Franks NR, and Levin SA (2005) Effective leadership and decision-making in animal groups on the move. *Nature* 433: 513–516.

Couzin ID, Krause J, James R, Ruxton GD, and Franks NR (2002) Collective memory and spatial sorting in animal groups. *Journal of Theoretical Biology* 218(1): 1–11.

Couzin ID and Laidre ME (2009) Fission-fusion populations. *Current Biology* 19(15): R633–R635.

Fischhoff IR, Sundaresan SR, Cordingley J, Larkin HM, Sellier M-J, and Rubenstein DI (2007) Social relationships and reproductive state influence leadership roles in movements of plains zebra, *Equus burchellii. Animal Behaviour* 73: 825–831.

Gigerenzer G and Tood PM and ABC Research Group (1999) *Simple Heuristics That Make Us Smart. Evolution and Cognition Series.* Oxford University Press.

Harcourt JL, Ang TZ, Sweetman G, Johnstone RA, and Manica A (2009) Social feedback and the emergence of leaders and followers. *Current Biology* 19: 248–252.

King AJ and Cowlishaw G (2007) When to use social information: The advantage of large group size in individual decision-making. *Biology Letters* 3: 137–139.

King AJ and Cowlishaw G (2009) Leaders, followers and group decision-making. *Communicative and Integrative Biology* 2: 147–150.

King AJ, Douglas CMS, Huchard E, Isaac NJB, and Cowlishaw G (2008) Dominance and affiliation mediate despotism in a social primate. *Current Biology* 18: 1833–1838.

King AJ, Johnson DDP, and van Vugt M (2009) The origins and evolution of leadership. *Current Biology* 19: R911–R916.

Petit O, Gautrais J, Leca J-B, Theraulaz G, and Deneubourg JL (2009) Collective decision-making in white-faced capuchins. *Proceedings of the Royal Society, Series B* 276: 3495–3503.

Seeley TD (1995) *The Wisdom of the Hive: Social Physiology of Honey Bee Colonies.* Cambridge, MA: Harvard University Press.

Sumpter DJT, Krause J, James R, Couzin ID, and Ward A (2008) Consensus decision-making by fish. *Current Biology* 18(22): 1773–1777.

Ward AJ, Sumpter DJT, Couzin ID, Hart PJB, and Krause J (2008) Quorum decision-making facilitates information transfer in fish shoals. *Proceedings of the National Academy of Sciences USA* 105 (19): 6948–6953.

Relevant Websites

http://www.zsl.org; http://www.zoo.cam.ac.uk – Andrew King's homepages.

http://www.princeton.edu/nicouzin – Couzin Lab of Collective Animal Behaviour.

http://www.sussex.ac.uk – Larissa Conradt's homepage.

http://angel.elte.hu – Starlings in Flight, Starflag website.

http://sols.asu.edu – Stephen Pratt's homepage.

http://www.nbb.cornell.edu – Tom Seeley's homepage.

Habitat Imprinting

J. M. Davis, Vassar College, Poughkeepsie, NY, USA

'Habitat imprinting' is one of several terms used to describe a tendency of animals to use or settle in habitats containing stimuli experienced early in life. As a process that influences the environment, mates, and other selective pressures that animals experience after natal dispersal, it is of particular interest to ecologists and evolutionary biologists interested in meta-population dynamics, host race formation, and speciation. Furthermore, it is of interest to researchers involved in animal reintroduction because these efforts rely on animals accepting habitats that may be very different from the habitats in which they were reared.

Habitat 'Imprinting' and Natal Habitat Preference Induction

The term 'imprinting' was originally used to describe the acquisition of social preferences early in life. Most agree that imprinting is not caused by any unique learning mechanism, but rather by some combination of associative conditioning and perceptual learning that results in two distinct characteristics: (1) the existence of a 'sensitive period' during which experience has a particularly strong effect on future preference, and (2) the persistence of that preference even when the individual is provided alternative experiences later in life.

The majority of studies purporting to demonstrate habitat imprinting do not provide experimental evidence of a sensitive period or of persistence. For this reason, Davis and Stamps used the term 'Natal Habitat Preference Induction' (NHPI) to describe any case where 'experience with stimuli in an individual's natal habitat increases the probability that the individual will, following dispersal, select a habitat that contains comparable stimuli.' NHPI is, thus, an umbrella concept which includes habitat imprinting, but is not necessarily characterized by a sensitive period.

Three mechanisms can cause animals to select habitats similar to their natal habitat. First, individuals from low-quality habitats may suffer physiological deficits that cause them to become less choosy during dispersal, and more likely to accept a low-quality habitat, than individuals that developed in high-quality habitats. Second, individuals can develop behavioral, morphological, and physiological traits prior to natal dispersal that make them particularly adept at utilizing resources and avoiding predators found in their natal habitat type. Therefore, when sampling the environment during dispersal, animals are more likely to have positive (i.e., rewarding) experiences in a particular habitat when that habitat is similar to their natal habitat. Habitat preferences learned during dispersal will therefore be biased toward the natal habitat. Third, prior to natal dispersal, animals may learn to prefer stimuli present in their natal habitat. NHPI and habitat imprinting fall under this final category. A number of learning mechanisms can be responsible. These include habituation to typically aversive stimuli present in the natal habitat, a learned association between some unconditioned stimuli (e.g., food and hosts) and stimuli unique to that habitat type, or perceptual learning, in which the animal learns to quickly discriminate its natal habitat type from other environments.

More research is needed, but it is likely that in many species, NHPI exhibits imprinting-like properties. That is, experience early in life probably has a stronger influence on habitat preferences than later experiences. NHPI will bias the habitats sampled during dispersal toward those that resemble the natal habitat, thus preempting the possibility of learning about alternatives. Animals often move long distances during natal dispersal relative to dispersal events later in life, when they are less likely to encounter alternative habitat types. As a result, preferences shaped in the natal habitat will have disproportionate influence on the habitat types that animals will live in. In many species, individuals spend more time in their natal habitat than they will in any habitat they encounter prior to settlement, allowing them more time to recognize and respond to the stimuli present in that habitat.

Because the natal habitat is experienced first, and for a relatively long time, NHPI can share qualities with imprinting even if there is no selective pressure favoring a particular sensitive period in neural development. That being said, in some cases, there are selective advantages to developing preferences in the natal habitat that cannot be altered by later experience. In these cases, preferences formed in the natal habitat may persist through life even if alternative experiences are experimentally forced upon the animal.

Evidence of Habitat Imprinting

The experiments demonstrating NHPI vastly outnumber those that clearly demonstrate habitat imprinting. Below, the evidence for each is reviewed for some major groups of animals.

Vertebrates

Mammals

While several studies on mammalian dispersal suggest that habitat preferences are influenced by NHPI or habitat imprinting, it has been difficult to eliminate the confounding influence of other sources of variation in habitat preference. In most studies on mammals to date, juveniles are marked while still in their natal habitat and tracked or recaptured in the field to determine their habitat selection decisions. Studies on mice and squirrels conducted in this way demonstrate that while individuals tend to settle in habitats similar to the habitats where they were captured, but they do not control for the possibility that individuals captured in particular habitats are there because their parents had a heritable preference for that habitat – a preference that the juvenile experimental subjects will also express. NHPI can only be convincingly shown if juveniles (or mothers) are randomly divided into natal experience treatments. Wecker found that prairie deer mice (*Peromyscus maniculatus bairdi*) from lab strains, but not those from recently wild-caught strains, showed a preference for their natal habitat type when tested in outdoor enclosures.

There is no clear-cut evidence of a sensitive period in the acquisition of mammalian habitat preferences. Population genetic and behavioral studies of rodents, canines, and ungulates suggest that habitat preferences tend to be retained over long periods of time.

Birds

NHPI has been shown to influence perching, nesting, or foraging habitat preferences in nine families of birds. Of particular interest are studies on the host–nest preferences of brood parasitic species, as habitat or host imprinting has long been thought to be a potential mechanism by which races or gentes of brood parasitic birds, such as cuckoos, maintain host-specific adaptations (e.g., egg and

nestling appearances that closely mimic those of the host). Cuckoos (*Cuculus canorus*) demonstrate individual-level variation in habitat preferences, but the evidence that those preferences are shaped by juvenile experience is mixed. Female indigobirds (*Vidua chalybeata*) show a strong preference for the nests and songs of the host species they were reared with. Male indigobirds mimic their natal hosts' song. As a consequence of host and sexual imprinting, reproductively isolated lineages of indigobirds specializing on different hosts have evolved.

A few studies on birds address sensitive periods in the formation of habitat preference. Grünberger and Leisler demonstrated that a coal tit's (*Periparus ater*) decision to perch in deciduous or coniferous tree branches is influenced strongly by natal experience, and that forcing the birds to perch on the nonnatal tree for 7 weeks reduced, but did not eliminate, the preference for the natal branch-type. A study on Barn owls (*Tyto alba*) indicated that owls preferred the type of habitat they were reared in from 4 to 10 weeks of age over the habitat they were kept in from 10 weeks to 10 months of age. On the other hand, a study on cuckoos indicated that the small effect of NHPI seen during the first breeding season was lost by the second breeding season.

Fish, reptiles, and amphibians

The extreme home stream fidelity seen in some anadromous fishes clearly demonstrates that early experience can influence a fish's response to habitat cues. To date, however, the effect of natal experience on fish preferences for *generalized* habitat-types has been shown only for the anemone host preferences of the anemonefish (*Amphiprion* sp.). The few studies conducted on reptiles and amphibians indicate that early experience with odor, visual, and tactile cues can affect later response to those cues in some species. Little work has been done on the existence of sensitive periods for the development of habitat preferences in these groups. Snapping turtles (*Chelydra serpentina*) exhibit preferences for food items experienced early in life that persist despite later experience with alternative food types.

Invertebrates

Perhaps because of the ease of manipulating early experience, most studies on NHPI have been conducted on insects. On average, the influence of natal experience on insect habitat preferences is not as large as is seen in birds; however, in some species, the effect is very strong. For example, Jaisson demonstrated that when worker ants (from the species *Formica polyctena*) reared in artificial nests with thyme sprigs were forced to colonize new nests, all chose nests with thyme over nests without thyme. All but one group of ants reared without any plant cues avoided nests containing thyme, and the group that selected the nest with thyme made quick work of removing the thyme from the nest!

The host preferences of some caterpillars are strongly affected by their previous diet. In some of these cases, after feeding on one host plant during the first and/or second stage of larval life, the caterpillars refuse to eat alternative hosts during later instars and, indeed, will starve before doing so. This pattern was originally described as 'imprinting' because the strong host preferences are formed during a sensitive period. However, the preferences observed as a consequence of larval food imprinting have only rarely been shown to carry over into the host preferences of adults. Because larval food preference induction does not necessarily influence the adult behavior, it is not considered a form of NHPI.

The question of whether larval experience can influence adult preferences (a concept referred to as 'Hopkins' Host Selection Principle (HHSP) by entomologists) has been a major focus of insect NHPI studies. While the majority of published experiments have failed to support HHSP, a handful of recent studies on moths and parasitoid wasps have provided some evidence. One mechanism for HHSP is that chemical stimuli may be trapped on the pupae of metamorphic individuals, allowing emerging adults to learn those stimuli and use them during host search. This mechanism is referred to as the 'chemical legacy hypothesis.' HHSP could be caused by the retention of memories formed through preimaginal conditioning (learning prior to adulthood). The neural circuitry of holometabolous insects is greatly reorganized during metamorphosis, so this mechanism seems unlikely (or at least costly). Preimaginal aversive conditioning (training larval insects to avoid stimuli associated with punishment treatments such as electrical shock) has been shown to generate adult aversion to chemical stimuli in moths. It is at least possible for the memory of chemical stimuli to be maintained across metamorphosis. Perhaps preimaginal conditioning has a potential role in the development of host preferences, but because of neural costs, is only rarely supported by selection.

Very little work has been conducted on noninsect invertebrates. Preference induction for food items has been shown in gastropods, crustaceans, and spiders.

Adaptive Value of Habitat Imprinting

There are several hypotheses explaining how an increased preference for the natal habitat type might be favored by selection. First, as discussed above, animals develop morphological and physiological phenotypes that allow them to deal with environmental challenges particular to the natal habitat type. When this occurs, there is a clear selective advantage for individuals to more readily accept

their natal habitat type relative to individuals that developed in other habitat types. This has been called the habitat-training hypothesis. If development in the natal habitat results in relatively permanent changes in the phenotype, selection may favor natal habitat preferences that are not altered by postdispersal experience, that is, habitat imprinting.

Another adaptive explanation for NHPI is that survival to dispersal age indicates that the natal habitat is at least minimally suitable for development. On average, individuals that increase the acceptance of the natal habitat will perform better than those that do not alter their preferences in response to natal experience. This is the habitat cuing hypothesis. In this case, experience in the natal habitat provides information to dispersers that, prior to settling in the postdispersal habitat and attempting to breed, cannot be gathered elsewhere, specifically information about the ability of the natal habitat to support juvenile development. The value of early experience should favor a sensitive period such that an induced preference for a habitat (in response to mere survival) will only occur prior to natal dispersal.

Habitat cuing and habitat training hypotheses suggest that NHPI provides fitness benefits after settlement in a new habitat. In addition to the potential postsettlement advantages of settling in the natal habitat, NHPI may increase efficiency during search. Elizabeth Bernay's 'neural hypothesis' of specialization argues that animals (in particular, insects) that specialize on particular resources are able to make decisions faster and with lower neural cost than generalists. If this is the case, then imprinting may allow generalist animals to enjoy the search efficiencies of specialists by attending only to stimuli from one habitat type. This fitness advantage would only accrue if, as a consequence of early experience, the difference in attractiveness between the natal habitat and alternative habitats increased. For example, if an animal reared in a typically avoided habitat becomes just as likely to accept that habitat as the most preferred habitat, it has exhibited NHPI, but has become more, not less, generalized, and thus, is expected to pay costs in terms of search efficiency.

Ecological and Evolutionary Implications of Habitat Imprinting

Habitat imprinting is a source of individual variation in habitat preferences, and therefore, has important ecological and evolutionary consequences. If mothers are occasionally forced to rear their offspring in habitats not previously used by that particular species, habitat imprinting may allow that species to rapidly invade and thrive in empty niches. The ongoing recovery of peregrine falcon (*Falco peregrinus*) populations is partially a consequence of the falcons' ability to learn to recognize urban environments as potential habitats. Because the falcons return to their natal habitat type, the success of falcons in urban environments has not resulted in the recolonization of some of their traditional breeding grounds (i.e., Appalachian cliffs).

While habitat imprinting may be the first step in niche expansion, it is not clear how readily entirely new habitat preferences can emerge. Many tested species demonstrate biases in what habitats they will develop strong preferences for. In anemonefish (*Amphiprion melanopus*), a strong preference for the natural anemone host requires early experience with that host. However, providing experience with an alternative host does not result in a strong preference for that alternative.

Once a population begins using more than one habitat type, habitat imprinting may create a correlation between the performance of individuals in a particular habitat and their preference for that habitat. This is because those individuals that have genotypes particularly well suited to their natal habitat are more likely to survive to dispersal age. Individuals dispersing from a particular habitat are better suited to that habitat type than the population as a whole. If, as a consequence of NHPI, those individuals prefer to settle in their natal habitat type, a correlation between habitat preference and performance will form. Such correlations have been shown to be important in allowing environmental heterogeneity to maintain genetic variation in populations. The correlation between preference and performance will reduce gene flow between populations breeding in different types of habitats. Reduction of gene flow, in turn, facilitates local adaptation. If the strength of preferences is strong enough, gene flow between populations breeding in different habitats could become so low that populations evolve into distinct species.

Applications in Conservation and Pest Management

The importance of habitat imprinting to conservation efforts is most evident in translocation and captive-release programs. Such efforts frequently fail when animals move long distances after release, indicating that they may perceive postrelease habitat as unsuitable. For species in which habitat preferences are shaped by imprinting, successful reintroduction will depend on the extent to which salient habitat cues present in the rearing environment and the reintroduction environment are similar. Release efforts with ferrets and lynx indicate that animals tend to disperse shorter distances upon release into the wild when they are reared in naturalistic environments, as compared to when they are reared in traditional cages. Ensuring that individuals are reared in the type of habitat in which they will be released into has the additional advantage that

animals are more likely to develop phenotypic characters important for success in that habitat type (e.g., the ability to recognize predators and the ability to recognize and digest food). If natal experience influences the habitat preferences of a candidate for reintroduction, it is useful to know whether the mechanism resembles imprinting. If there is no sensitive period during which persistent habitat preferences are formed, conservation biologists can acclimatize soon-to-be-released animals by maintaining animals in proximity with their new habitat for some amount of time prior to release. Such efforts have been shown to reduce postrelease dispersal distances.

In addition to conservation efforts, knowledge of NHPI can be used to improve pest and weed management strategies. Parasitoid wasps lay their eggs in the eggs and larvae of other insect species. They are often used as a biological control agent for insect pests. Numerous studies have shown that these wasps use chemical stimuli experienced in their natal habitat to track down potential hosts. Thus, rearing these wasps in environments similar to the ones they will be released into, can improve the effectiveness of this control strategy. Browsing animals such as goats can be trained to forage on particular weed species and then released into areas where that species needs to be controlled.

See also: Avian Social Learning; Foraging Modes; Habitat Selection.

Further Reading

Arvedlund M, McCormick MI, Fautin DG, and Bildsoe M (1999) Host recognition and possible imprinting in the anemonefish *Amphiprion melanopus* (Pisces: Pomacentridae). *Marine Ecology Progress Series* 188: 207–218.

Barron AB (2001) The life and death of Hopkins' host-selection principle. *Journal of Insect Behavior* 14: 725–737.

Beltman JB and Haccou P (2005) Speciation through the learning of habitat features. *Theoretical Population Biology* 67: 189–202.

Beltman JB, Haccou P, and Ten Cate C (2004) Learning and colonization of new niches: A first step toward speciation. *Evolution* 58: 35–46.

Bernays EA and Weiss MR (1996) Induced food preferences in caterpillars: The need to identify mechanisms. *Entomologia Experimentalis Et Applicata* 78: 1–8.

Burghardt GM and Hess EH (1966) Food imprinting in the snapping turtle *Chelydra serpentina*. *Science* 151: 108–109.

Corbet SA (1985) Insect chemosensory responses: A chemical legacy hypothesis. *Ecological Entomology* 10: 143–153.

Davis JM (2008) Patterns of variation in the influence of natal experience on habitat choice. *Quarterly Review of Biology* 83: 363–380.

Grünberger S and Leisler B (1990) Innate and learned components in the habitat selection of coal tits (*Parus ater*). *Journal fuer Ornithologie* 131: 460–464.

Immelmann K (1975) Ecological significance of imprinting and early learning. *Annual Review of Ecology and Systematics* 6: 15–37.

Jaenike J (1988) Effects of early adult experience on host selection in insects: Some experimental and theoretical results. *Journal of Insect Behavior* 1: 3–15.

Jaisson P (1980) Environmental preference induced experimentally in ants (Hymenoptera: Formicidae). *Nature* 286: 388–389.

Janz N, Soderlind L, and Nylin S (2009) No effect of larval experience on adult host preferences in *polgonia c-album* (Lepidoptera: Nymphalidae): On the persistence of Hopkins' host selection principle. *Ecological Entomology* 34: 50–57.

Mabry KE and Stamps JA (2008) Dispersing brush mice prefer habitat like home. *Proceedings of the Royal Society B: Biological Sciences* 275: 543–548.

Stamps JA and Davis JM (2006) Adaptive effects of natal experience on habitat selection by dispersers. *Animal Behaviour* 72: 1279–1289.

Stamps JA and Swaisgood RR (2007) Someplace like home: Experience, habitat selection and conservation biology. *Applied Animal Behaviour Science* 102: 392–409.

Wecker SC (1963) The role early experience in habitat selection by the Prairie Deer Mouse, *Peromyscus maniculatus bairdi. Ecological Monographs* 33: 307–325.

Habitat Selection

I. M. Hamilton, Ohio State University, Columbus, OH, USA

Habitat selection refers to the rules used by organisms to choose among patches or habitats that differ in one or more variables, such as food availability or predation risk, that influence its fitness. These rules determine the spatial distribution of organisms, and may thereby influence population and community-level processes. For example, differences in habitat selection rules between species or competitive classes may allow these groups to coexist. Habitat selection rules influence the growth of populations, particularly if some habitats act as population sinks – habitats in which fewer individuals are produced than die, but which are maintained by migration. Changes in habitat use in response to changes in resources or perceived risk of predation can result in a variety of direct and indirect effects to predator and prey species, with implications for community structure and dynamics and management.

The theoretical framework of habitat selection at small scales (the microhabitat or foraging scale) is closely related to that of foraging theory, particularly models of patch selection, patch residence time, and social foraging. Throughout this article, I define patches as relatively homogeneous areas that differ in some way from other parts of the landscape; I refer to patches and habitats interchangeably. Much of this chapter focuses on the ideal free distribution (IFD) and its derivatives, which form a general model of microhabitat selection. This social foraging model is used to predict the distribution of foragers when the benefits of foraging in a patch or habitat are density-dependent. Many more recent empirical and theoretical models of microhabitat selection have included other fitness influencing factors, such as social interactions, risk of predation or physiological tolerances to the basic IFD framework, as well as incorporating costs of movement within and among different patches.

History

The first influential models of habitat selection were introduced in the late 1960s and early 1970s. The IFD and ideal despotic distribution model habitat use decisions by ideal animals; that is, those with perfect information about the relative qualities of different habitats at all times. The IFD further assumes that animals can freely move among habitats, while the ideal despotic distribution assumes that some patch residents are able to despotically control access to those habitats.

In the 1970s, Rosenzweig introduced a graphical tool, the isoleg, to understand how competition between species influences habitat use. Isolegs are the sets of points of equal fitness for an organism that is selective in terms of habitat and one that does not actively select among habitats. Isolegs demonstrated the link between individual habitat use decisions and patterns of community structure. For example, isolegs allowed exploration of how past competitive interactions could explain habitat segregation and competitive coexistence (the ghost of competition past).

Starting in the mid-1980s, a wide variety of new models were developed which relaxed or changed some of the assumptions of the basic IFD. These included models of interference competition, models in which competitive abilities varied among consumers, models incorporating perceptual constraints and costs of movement, and many others. Also in the 1980s, another graphical tool, the isodar, was introduced. Isodars are the sets of points of equal fitness for animals choosing between habitats on a plot of density in each habitat, and are used to understand patterns of density-dependent habitat use. Isodars provide information about how individuals value different habitats on the landscape. Although isodars and isolegs measure different things, the two are conceptually related, and isolegs can be derived from isodars. Similarly, expected isodars can be derived from ideal distribution models.

While the original ideal distribution models and many of their descendants assume that animals maximize the long-term rate of net energy intake, several more recent models investigate how animals trade off energetic intake against other factors, such as predation risk or temperature. Deviation from the expectations of IFD and other changes in foraging behavior have been used to measure the risk of predation and other habitat-specific fitness costs. The isodar framework has also been used extensively to understand how animals value their landscape in terms of both energy and mortality.

Theory

The Ideal Free Distribution

The IFD describes the expected distribution of foragers when resource patches differ in quality and foragers compete for resources. This model assumes that foragers are 'ideal': they have perfect information about relative patch quality and the densities of foragers in each patch, and 'free': foragers are able to move between patches without cost or time delay and are not excluded from entering patches by the current inhabitants. As an example, consider two people at opposite ends of a small pond feeding pieces of bread to a large population of ducks. Suppose that one person provides twice as many pieces of bread per minute as the other. Where should ducks feed if they seek to maximize the amount of bread received? A piece of bread eaten by one duck is not available to the others, and so if all the ducks congregated at the more productive location, each individual duck would receive few pieces of bread. At the other end of the pond, the other feeder provides fewer pieces of bread overall, but there is also no competition for that bread. Therefore, some ducks could increase the amount of bread received by moving to that end. By doing so, however, they would increase competition for food at the new location. How many ducks, then, should occupy the more productive location and how many should occupy the less productive location? The answer depends on how food arrives in a patch and how the density of competitors influences the rate at which ducks ingest their food.

Continuous input

The continuous-input IFD model assumes that food items arrive randomly at different rates in different patches and foragers consume them immediately. If these assumptions are met, the long-term rate of net energy intake in patch i is the per capita encounter rate with food items, R_i. This rate is the ratio of the rate of resource (food) input (Q_i) to the density of competitors in that patch (d_i) (i.e., $R_i = Q_i/d_i$). The model also assumes that foragers should maximize their long-term rate of net energy intake. By assumption, ideal free consumers can identify and move to any habitat that will provide them with a greater intake rate. When all competitors are equal, there is a unique equilibrium distribution by which no foragers can improve their intake rate by switching habitats unilaterally (in game theory, this is referred to as a Nash equilibrium; see Glossary). If there are two patches that differ in quality, then this point occurs when the ratio of competitor densities in patches 1 and 2, d_1/d_2, equals the ratio of resource inputs Q_1/Q_2 (input matching).

The continuous-input IFD model makes two testable predictions. The first is that all patches provide the same intake rate at equilibrium. The second is that, at equilibrium, consumers should be distributed so that the ratio of consumer densities across patches equals the ratio of patch resource input rates, that is, there should be input matching. However, empirical tests of the IFD rarely find support for the quantitative predictions of input matching and equal intake rates across patches. In many experiments, the proportion of foragers in high-quality patches is less than the theoretical expectation,

a phenomenon known as 'undermatching.' When this is so, foragers in the high-quality patch often have higher intake rates than do occupants of lower-quality patches.

Interference

In the previous model, food items arrive in the patch and are consumed immediately. However, many resources do not deplete so rapidly. Nevertheless, the fitness payoffs to foraging in non- or slowly depleting habitats may still be density dependent because consumers interfere with one another. Interference can result from aggressive defense of resources, kleptoparasitism (the theft of resources from others), changes in the behavior or defenses of prey, or any other interaction that leads to a reversible decrease in intake rate with increasing consumer density. One way to incorporate interference into an analogous model to the continuous-input IFD is through the addition of an 'interference constant' (m) on searching rate, so that the intake rate when handling time is zero is:

$$R_i = \frac{Q_i}{d_i{}^m}$$

Here, Q_i is the foraging rate of a solitary forager. When m is high, changes in density have a strong effect on intake rate; when m is low, changes in density have little effect on intake rate. The equilibrium point in the two-patch interference model is reached when the ratio of the densities of competitors in patches 1 and 2, d_1/d_2, equals $(Q_1/Q_2)^{1/m}$. When $m = 1$, this model makes the same predictions as the continuous-input model. If $m > 1$, and the effects of interference are strong, this model predicts undermatching at equilibrium. This is because the effects of competition are particularly strong in the more productive, and more densely occupied, patch. If $m < 1$, competition is weak and the interference IFD model predicts overmatching. Again, intake rates are equal across patches at equilibrium.

'Mechanistic' interference models of the IFD assign foragers to states, such as searching for food, handling or processing food, and fighting over food, and define the density-dependent transition rates between states. Such models are solved analytically or through simulation to find the equilibrium distribution of individuals among patches, if one exists. The predictions of mechanistic models depend strongly on their assumptions. For example, if handling individuals do not experience interference (e.g., food cannot be stolen while being handled) then mechanistic models of equal competitors predict overuse of high-quality patches relative to resource abundance. If, on the other hand, handlers remain susceptible to interference (e.g., food can be stolen during handling), then such models predict input matching. These predictions change when competitors differ in ability to search for food or engage in interference (see section 'Differences in Competitive Ability').

Differences in competitive ability

One of the assumptions of the continuous-input and interference models introduced earlier is that all competitors are equal. That is, the probability of encountering, capturing, and ingesting food items and the effects of interference on intake are the same for all individuals. This is unlikely to be the case in many, if not most, foraging situations where competitive ability may vary with age, experience, size, and many other factors. Continuous-input ideal free models that include unequal competitors predict input matching, where each individual is weighted by its relative competitive ability. In a large population, a very large number of distributions exist that can satisfy input matching of competitive weights.

Unequal-competitors interference models often predict a *truncated phenotype distribution* in which the best competitors exploit the most productive habitats, and poorer competitors are restricted to poorer habitats. Average intake rates are not equal across habitats at equilibrium. Rather, there is a positive correlation between habitat quality and intake rate because a small number of good competitors have exclusive use of the most productive patches. This distribution is also predicted when competitive weights themselves are a function of habitat quality – competitors are predicted to use the habitat in which their relative competitive ability is the greatest. As in other ideal free models, perceptual constraints and the need to sample multiple patches in order to choose among them may prevent achievement of the equilibrium distribution and result in overuse of less profitable patches (see Empirical tests, later). Some mechanistic models of interference predict a similar distribution, known as a *semi-truncated phenotype* distribution. In this distribution, better competitors (those with competitive ability exceeding the boundary phenotype) are found only in the high-quality patches, but poorer competitors occur in all patches.

Other Ideal Distributions

When competitors differ in some way that allows some individuals to preempt or displace others, the IFD does not apply, because preempted or displaced individuals are no longer free to choose among patches. These effects can occur because dominant individuals defend territories that maximize their fitness, interfering with habitat selection by nonterritorial or more subordinate individuals. The expected distribution in this case is referred to as the 'ideal despotic distribution.' Preemption may also occur if early arrivals stake out the best foraging locations. This is referred to as an 'ideal preemptive distribution.' The expected distributions of foragers in both these models are similar; at equilibrium, individuals in higher-quality habitats will have higher fitness than individuals in lower-quality habitats.

Isolegs

The IFD and related models predict the distribution of organisms under specific sets of assumptions about how organisms interact with one another and their environment. In contrast, researchers can use the observed distribution or behavior of animals to interpret how animals value habitats and the effects of competition within and between species on habitat use. For example, in the Negev Desert, two species of gerbils, *Gerbillus andersoni allenbyi* and *Gerbillus pyramidum*, compete for seeds. Both species prefer to forage in the same semistabilized habitat. Early in the night, the larger *G. pyramidum* is able to exclude its competitor from these patches. However, later in the night, *G. pyramidum* stops foraging and *G. andersoni allenbyi* switches to partially using the preferred habitat. These effects – shared preferred habitats and exclusion of the smaller *G. andersoni allenbyi* from preferred habitats when *G. pyramidum* are common – have been replicated in populations established in enclosures. The set of densities of *G. pyramidum* at which *G. andersoni allenbyi* switches from using both habitats to using a single habitat, across a variety of competitor densities is referred to as its *isoleg* (**Figure 1**). Isolegs are particularly useful in identifying density-dependent competition effects that influence community structure. For example, observed patterns of different habitat use may result from past competitive

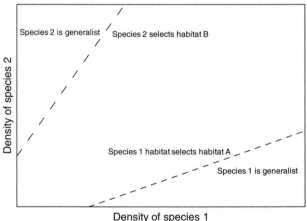

Figure 1 The ghost of competition past, using isolegs to illustrate coexistence of competing species when two habitats (habitat A and habitat B) are available. Species 1 is a generalist below and to the right of its isoleg (dotted line), and selects habitat A to above and to the left of this line. The isoleg represents the set of points for which the fitness benefits of using both habitats and using only one habitat are equal. Species two is a generalist above and to the left of its isoleg (the dashed line) and selects habitat B below and to the right of this line. In the region between the two isolegs, coexistence of the two species is possible because they use different habitats. Differential habitat use is a result of competitive interactions, however. Competition results in the observed patterns of habitat use even though these species no longer compete in this region of species density space.

interactions (the ghost of competition past), even if competition is no longer apparent because of specialization in different habitats.

Isodars

In the Rocky Mountains of Alberta, Canada, small rodents, such as deer mice, pine chipmunks, and red-backed voles, can choose between dry open habitats and moister forest. For each species, plotting the density of that species in one of these habitats against its density in the other habitat reveals different patterns of habitat preference. For example, deer mice are found at much higher densities in dry habitats relative to forest habitats; deer mice continue to prefer dry habitats even when no deer mice are present in the alternative habitat. Similarly, red-backed voles prefer forest habitats, and their abundance in these habitats has little correlation with their abundance in the alternative habitat. In contrast, chipmunks use both habitats and there is a strong correlation between densities in the two habitats. The set of points describing the relationship between density in each habitat for each species is known as an *isodar* (**Figure 2**). Assuming that animals can freely move among habitats, the isodar is also the set of points on a plot of densities in each habitat for which the fitness effects of choosing each habitats are equal; therefore, isodars can be derived from foraging models. The slopes and intercepts of isodars provide information on the value of different habitats to foragers. The slopes of isodars indicate the relative quality of different habitats. In the continuous-input IFD model, the slope of the isodar is the ratio of resource input rates. The intercepts of isodars also yield information about

preferences. Nonzero intercepts indicate that habitats differ in fitness returns at low densities; as a consequence, habitat selection rules vary with population size. Zero intercepts indicate that habitat selection rules do not vary with population size (if the slope is linear), as in the continuous-input IFD.

Costs of Foraging and Movement

Moving among and within patches may be costly. How such costs influence habitat selection depend on the scale of movement, because costs of dispersal (rare, long-distance movements between patches) are different than costs of moving within patches or short distances between patches. If short-distance movement is costly, individuals may avoid some of those costs by being nonselective (in other words, selecting the first habitat they encounter). If encounters with habitats are random, this is likely to result in overuse of low-quality patches relative to the IFD. However, at the dispersal scale, the decision to leave one's current home range should only be undertaken if the expectation of finding a sufficiently high-quality patch to compensate for costs of movement is high. Thus, costs of dispersal result in a decrease in the value of low-quality patches and an increase in the value of high-quality patches. Dispersal costs alone should result in increased use of high-quality patches, while short-distance costs alone should result in decreased use of high-quality patches over no cost. In reality, both costs are likely present for a habitat-selecting animal, but the importance of each varies with spatial scale.

Moving Beyond Energetics

In nature, foragers often must trade off the long-term rate of net energy intake against other influences on fitness, such as risk of mortality or injury from predators, exposure to pathogens and parasites, risk of starvation, opportunities to engage in or avoid social interactions, mating opportunities, nutrient acquisition, and osmoregulatory, thermoregulatory, or other physiological considerations, among others. Several models investigate how animals trade off net energy intake rate and predation risk when selecting habitats. The predictions of models of habitat selection under predation risk differ greatly depending on assumptions regarding relative riskiness of patches and how predation risk is influenced by the density of prey. For example, when patches differ in riskiness, fewer individuals are expected in the risky patch than expected based only on resource input rates. If predation risk decreases with increasing density of prey, this can result in complete abandonment of the risky patch. If both patches are equally risky but differ in productivity, prey are expected to exclusively use the most productive patch

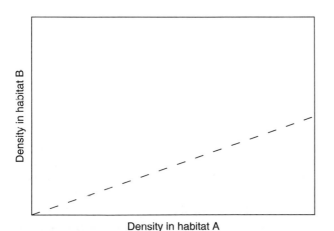

Figure 2 An illustration of the isodar framework. The isodar is the set of points at which the fitness payoffs for a consumer selecting one habitat equal those for a consumer selecting the other habitat. The example shown here is consistent with the continuous-input IFD. Because the slope of the isodar is linear and the intercept is zero, habitat selection does not change with density (i.e., the ratio of consumers in habitat A/habitat B is the same for all densities).

if risk decreases with increasing prey density. Incorporating differences in competitive ability or vulnerability to predators leads to further complexity in predator–prey models.

Further complexity arises if predation risk depends on the density of predators, and predators are also able to move among patches. When this is so, habitat use becomes a game between predators and prey. Models of the IFD that include mobile predators and prey predict that the distribution of prey should be a function of the relative riskiness of patches, while that of predators should be a function of the abundance of the prey's resource, rather than the distribution of the prey itself.

One common way of measuring how predation risk or other costs influence habitat use is to examine *giving up densities*. In a patch that depletes without renewal, the rate at which a randomly searching forager encounters new prey items will decline with increased time spent in the patch as more and more items are eaten. At some point, the encounter rate with prey in the patch will fall below that expected in the environment as a whole. An optimal forager should quit the patch when this is so. If the cost of remaining in a patch is relatively high, for example, because the risk of predation is high, then the forager should give up on this patch earlier, leaving a higher remaining density of prey. A researcher can use giving up densities to estimate differences in the cost of foraging in different habitats by creating artificial patches that have equal initial densities of food and then returning after animals have foraged to measure the density of prey remaining in the patch.

Empirical Tests of Habitat Selection Theory

Tests of the Ideal Free Distribution

Experiments designed to test the two main predictions of the continuous-input IFD, input matching and equal intake rates across habitats, have focused on a small number of systems that closely match the assumptions of the continuous-input model, particularly the assumption that food continually enters the habitat and is immediately consumed or removed. One such system is ducks feeding on small pieces of bread thrown into the pond, as introduced previously. Another classical system is stream fish feeding on small invertebrates drifting in the current. These studies have found that the qualitative prediction of increased densities in high-quality patches (those with a high resource input rate) is met, but that there are typically lower densities than expected in high-quality patches and higher densities than expected in low-quality patches. As this pattern of undermatching is common, it suggests that the assumptions of this simple continuous-input model are often violated. Interference models can

lead to such patterns when interference is very high, but also commonly predict the opposite effect, overmatching of resource inputs in high-quality patches. Violation of the assumption of perfect information can lead to undermatching and differences in intake rate between patches. Animals may be limited by cognitive or perceptual abilities from distinguishing between alternatives that are very similar in payoffs. If this is so, there may be some minimal necessary difference in quality, below which animals choose habitats randomly. Such perceptual constraints will always lead to overuse of low-quality patches and underuse of high-quality patches. Deviation from input matching may also result from stochastic variation in input rates. If resource input rates are variable and there is some risk of starvation, then the variance of resource inputs may be important influences on habitat use. In one risk-sensitive model of the IFD (risk sensitivity here referring to the classical behavioral ecological definition of risk as variance), underuse of high-quality patches is predicted when the risk of starvation is great and overuse of high-quality patches is expected when the risk of starvation is slight.

Predation Risk and Habitat Use

A pair of laboratory experiments used guppies (*Poecilia reticulata*) and coho salmon (*Oncorhynchus kisutch*) to test both the predictions of IFD models and the effects of predation risk on habitat use. In these experiments, fish were first allowed to choose between habitats that differed in the food input rates. In both cases, observed habitat use matched the input matching predictions of the IFD (or the unequal competitors IFD in the case of coho salmon). Overhead cover was then added to one patch. For many fish, predation risk from birds is significant; the addition of overhead cover should therefore increase the safety of that patch. This addition resulted in a shift in the distribution of fish to greater use of the covered patch than before. The investigators then added food to the uncovered (riskier) patch until the distribution matched that in the absence of cover. Thus, the investigators were able to estimate the amount of energy required to make up for the difference in safety between patches, allowing risk and energy intake rate to be expressed using the same currency (although this interpretation has been criticized for only holding if there is no effect of group size on predation risk).

Experimental manipulation of predation risk may not be possible in many field situations; however, variation in the presence and density of predators may allow evaluation of the effects of predators on habitat use. The shallow waters of Shark Bay, Western Australia, support extensive sea grass beds which, in turn, provide food and shelter for a diverse array of marine vertebrates. The presence and density of a top predator, tiger sharks (*Galeocerdo cuvier*), varies seasonally within the bay, and from year to year within seasons. For several large marine vertebrates, such as bottlenose dolphins (*Tursiops aduncus*), dugongs (*Dugong dugon*), and cormorants (*Phalacrocorax varius*), shallow seagrass beds represent risky but productive foraging habitats, while deeper channels between the seagrass beds are less productive but safer from sharks. These species appear to alter their patterns of habitat use depending on the presence of tiger sharks. When sharks are absent, these herbivores and piscivores are found at higher densities in productive shallow seagrass beds. When sharks are present, these species shift to greater use of deeper, less productive patches. These habitat use decisions have cascading effects in Shark Bay communities, with reduced grazing pressure in risky areas and increased grazing pressure in safer areas.

See also: Ecology of Fear; Endocrinology and Behavior: Methods; Foraging Modes; Game Theory; Hormones and Breeding Strategies, Sex Reversal, Brood Parasites, Parthenogenesis; Hunger and Satiety; Optimal Foraging Theory: Introduction; Parasitoids; Trade-Offs in Anti-Predator Behavior; Wintering Strategies: Moult and Behavior.

Further Reading

Abrahams MV and Dill LM (1989) A determination of the energetic equivalence of risk of predation. *Ecology* 70: 999–1007.

Abramsky Z, Rosenzweig ML, Pinshow B, Brown JS, Kotler B, and Mitchell WA (1990) Habitat selection: An experimental field test with two gerbil species. *Ecology* 71: 2358–2369.

Fretwell SD (1972) Populations in a seasonal environment. *Monographs in Population Biology* 5: 1–217.

Fretwell SD and Lucas HL (1969) On territorial behavior and other factors influencing habitat distribution in birds. *Acta Biotheoretica* 19: 16–36.

Giraldeau L-A and Caraco T (2000) *Social Foraging Theory*. NJ, USA: Princeton University Press.

Grand TC and Dill LM (1997) The energetic equivalence of cover to juvenile coho salmon (*Oncorhynchus kisutch*): Ideal free distribution theory applied. *Behavioral Ecology* 8: 437–447.

Grand TC and Dill LM (1999) Predation risk, unequal competitors and the ideal free distribution. *Evolutionary Ecology Research* 1: 389–409.

Heithaus MR, Frid A, Wirsing AJ, and Worm B (2008) Predicting ecological consequences of marine top predator declines. *Trends in Ecology & Evolution* 23: 202–210.

Holmgren N (1995) The ideal free distribution of unequal competitors: Predictions from a behaviour-based functional response. *Journal of Animal Ecology* 64: 197–212.

Hugie DM and Dill LM (1994) Fish and game: A game theoretic approach to habitat selection by predators and prey. *Journal of Fish Biology* 45: (A), 151–169.

Kennedy M and Gray RD (1993) Can ecological theory predict the distribution of foraging animals? A critical analysis of experiments on the ideal free distribution. *Oikos* 68: 158–166.

Morris DW (1992) Scales and costs of habitat selection in heterogeneous landscapes. *Evolutionary Ecology* 6: 412–432.

Morris DW (1996) Coexistence of specialist and generalist rodents via habitat selection. *Ecology* 77: 2352–2364.

Morris DW (2003) Toward an ecological synthesis: A case for habitat selection. *Oecologia* 136: 1–13.

Rosenzweig ML (1981) A theory of habitat selection. *Ecology* 62: 327–335.

Sutherland WJ (1996) *From Individual Behaviour to Population Ecology*. Oxford: Oxford University Press.

William Donald Hamilton

M. J. West-Eberhard, Smithsonian Tropical Research Institute, Costa Rica

William D. 'Bill' Hamilton (1936–2000) was a friend and a hero to many twentieth-century students of animal behavior. The pages of this encyclopedia refer often to his work, especially his 1964 papers on the genetics of the evolution of social behavior, commonly called 'kin selection theory.' This essay is based on one written for a memorial symposium at a meeting of the Human Behavior and Evolution Society (Amherst College, June 11, 2000). The editors of this encyclopedia encouraged me to submit the essay essentially as it was, so I have not tried to make it a balanced treatment of Hamilton's contributions to the field of animal behavior, or to completely remove the personal tone.

A coarse-grained view of history – especially one written by a human evolutionist – might suggest that Hamilton's famous 1964 papers were buried in oblivion until the rise of sociobiology in the middle and late 1970s. And geneticists sometimes suggest that Hamilton's theory was nothing new, having been outlined years before by Haldane, Fisher, and even Darwin. So it is important for readers to appreciate the importance of Hamilton and his theory for animal behavior as a field and to see how the early interest of Hamilton (and readers of his papers) in the behavior of social insects was decisive for the establishment of his ideas within biology.

Ethologists, especially entomologists interested in the evolution of insect sociality, were the first to pay serious attention to Hamilton's ideas. This was due in part to the effectiveness of Hamilton's writing, which kept theory tied to examples from living nature. In this essay, I refer especially to correspondence with Hamilton in the early days of his career, because they show how kin selection theory got established and reveal the charismatic personality of a man who had a profound influence on the science of animal behavior. His writings on behavior included not only social collaboration and its limits, but, later, sexual behavior, aggregations as selfish herds, and the polymorphisms and tactics of competing males, exemplified in fig wasps he studied in Brazil. Here, I focus on a period earlier than that covered by most Hamilton biographers. As Jeremy Leighton John, head of the Hamilton Archives has written (2005), eminent scientists, Hamilton included, often do their most important work when they are young, yet "many people (especially those who did not know the person in his or her younger days) find it difficult to imagine the distinguished professor as a younger scientist." He then features a photograph of Hamilton taken in 1967, the year my correspondence with Hamilton began.

William D. 'Bill' Hamilton was born in 1936 in Cairo, Egypt, the second oldest of six children. His father was Archibald Milne Hamilton, a noted New Zealand–born engineer known for building the 'Hamilton Road' through Kurdistan, and for designing the Callender-Hamilton bridge, a precursor of the Bailey metal bridge. His mother, Bettina Matraves Collier, was a physician who encouraged her son's interest in natural history. He was educated at Tonbridge School and St. John's College, Cambridge, before becoming a lecturer in genetics at Imperial College, London, in 1964. He moved to the University of Michigan in 1978, and since 1984, he was Royal Society Research Professor in the zoology department at Oxford.

Bill described the history of his thoughts on Darwinian evolution and social behavior in a long letter [22/ii/79] that began with his childhood and student days. He felt lonely and unappreciated as a student, both as an undergraduate at Cambridge and as a graduate student at the Galton Laboratories in London. As a teenager, he had read Darwin and a number of semi-popular books on evolution. He was encouraged in this by his remarkable mother, who roughly explained natural selection to him at about age 11 or 12, so that at age 14, he asked for the Origin of Species as the book for a school prize. Then, he was disillusioned when textbooks gave a different picture of evolution, one full of "social idealism, soft-heartedness" and benefit to the species, a view that contrasted with the "strongly individualistic" version he had read on his own. This led to a very early mistrust of what he termed "the intellectual dishonesty of mainstream biology" for "not having its fundamental tenets sorted out." Bill was determined even as an undergraduate to set Darwinism back on track. He wanted to be Darwin's new Bulldog, a twentieth-century Huxley who would defend and clarify Darwin's ideas in the face of criticisms and misinterpretations. As he said in another letter [29/xi/70]: "I have immense admiration for Darwin and almost equal admiration for T.H. Huxley ... if you ask me what sort of a scientist I am I have to reply that I am trying to be a latter-day Huxley and show that Darwin was even more right than biologists think!"

Hamilton was characterized by a modest manner, evident in his response [5/x/67] to my appreciation of his 1964 papers, where he refers to earlier theoretical research on kin selection: "It is indeed nice to know that someone thinks so highly of my work, although I am sure that you rate it too highly. The idea of the involvement of genetical relationship in social evolution did not originate with me, as you know. ... I certainly feel myself subject to Fisher's guidance in all my evolutionary thought, so much have I been impressed with his standards of thought and his biological insight ever since I first read – or tried to read – 'the Genetical Theory of Natural Selection'." The title of his 1964 papers – 'The genetical theory of social behavior' – was clearly inspired by that of Fisher's book, a landmark of the genetic theory of evolution that importantly influenced twentieth-century Darwinian thought.

In retrospect, Bill did not have much success as Darwin's new bulldog until he began to use Darwin's tactic of presenting his theories alongside convincing facts, especially facts on social insects. Bill said that the emphasis on social insects in his 1964 papers was an 'afterthought,' but it was an extremely important one. The first people who paid attention to those papers were entomologists interested in social insects, and there is no better audience for a theoretician. The social insects have always appealed to philosophically inclined, intellectually adventurous, and broadly synthetic biologists – William Morton Wheeler, Alfred Emerson, Caryl Haskins, and Edward O. Wilson to mention a few – scientists who sought to understand complexity in nature, and who liked to generalize about the evolution of sociality in animals, including humans. Among the first to cite Hamilton's papers were Ed Wilson and Caryl Haskins (both students of ants) and George Williams, an ichthyologist with some of the positive traits of an entomologist, having written, along with his wife Doris, a paper on social insects that anticipated some of Bill's ideas. Bill told me that the first people who corresponded with him about kin selection were Richard Alexander, then an entomologist primarily interested in speciation and animal behavior who later became a pioneer in applying modern evolutionary theory to humans; Ed Wilson, who was to expand and name the field of sociobiology with his famous book on that subject; and the ecologist Gordon Orians. But the most important entomologist for Bill in his early career was O.W. Richards, an eminent student of social wasps and a fine evolutionary biologist. Richards appointed Bill to his first job, and encouraged his excursions into both theory and the tropics. Richards was soon to retire as head of the Department at Imperial College, but his influence and support were crucial early in Hamilton's career.

Many factors contributed to the success of Bill's classic 1964 papers, including his provision of a detailed genetic model that specified how self-costing the costs and benefits of social aid were. But he did not depend on theoretical genetics to propagate his ideas about how self-costing aid, or 'altruism,' could evolve, as Haldane and Fisher had in their brief and relatively superficial discussions. Hamilton was raised from obscurity and frustration on the wings of wasps and bees (and to a lesser degree, the feeble and disposable wings of ants and termites), insects with extremely self-sacrificing individuals in the form of sterile workers devoted to helping reproductive relatives rear their young. The second of the 1964 papers is entirely devoted to applying the theory to organisms, and 24 of its 35 pages are on social insects. Hamilton cited work by

Wilson, Michener, Richards, and Haskins, all influential students of social insects who would become his early champions. Hamilton also featured the three-fourth relatedness idea, an insight that effectively dramatized the possible importance of relatedness for social evolution. He pointed out that in the Hymenoptera (wasps, ants, and bees), a female helper can potentially get a larger genetic payoff than she would by rearing her own offspring: due to the haplodiploid sex determination of these insects, hymenopteran females are related to their sisters by three-fourth, whereas they are related to their own daughters by only one half. Even though this particular aspect of kin selection theory has been questioned due to its failure to hold for various wasp species at the threshold of sociality (the critical species for testing this idea), the three-fourth relatedness idea dramatized the main point of the general theory, and placed the social insects at the center of discussions about the genetics of the evolution of social life.

As soon as the manuscript for his 1964 papers was submitted for publication in 1963, Bill made the first of many trips to Brazil to study the social Hymenoptera. The final revisions of those papers were done in Brazil, and Bill inserted his own observations on the genus *Polistes*. By lucky coincidence, I went to Cali, Colombia, in 1964, to study tropical *Polistes* wasps as part of my doctoral research, having already observed a temperate-zone species in detail. As soon as Bill's papers appeared, Richard Alexander (who was my advisor) sent copies to me in Colombia, along with a letter saying that he thought the papers might be extremely important for my research. This wonderfully understated and timely advice enabled me to be the first person, after Bill himself, to apply kin-selection theory in the field.

Bill did not use the term 'kin selection' in the 1964 papers. He said [3/i/73], in some comments on a manuscript of mine, that "When writing my 1964 paper I actually considered inventing a name for the kind of selection I was considering with 'kin selection' reviewed as one possibility but eventually decided that invention of the term 'inclusive fitness' would be better as not having a subtle implication of a process *competing* with individual selection. However, I think I may later have forgotten about the trap that I then clearly saw" (he ended up freely using the term kin selection). This is an interesting comment, for it reveals that Hamilton regarded his idea as an extension of the 'strongly individualistic' ideas of Darwin, which he had admired as a boy, and not, as some have interpreted the idea, as a kind of selection on groups of kin.

My correspondence with Bill began in 1967, when I was a postdoctoral fellow at Harvard. He invited my husband Bill and me to join him and his wife Christine on a field trip to Brazil, organized by the Royal Geographical Society, but conceded [9/iv/68] that Belem is "probably not the best place" for the baby we were expecting at that time. The Hamiltons traveled to Brazil in nineteenth-century style, on a cargo ship.

We did not actually meet Bill Hamilton until he visited the US in 1969. Bob Trivers was a graduate student at Harvard then, working on his now famous ideas about reciprocal altruism, and he was very anxious to meet Bill, so we invited them both to dinner in our apartment on Cambridge Street, in a picturesque tenement district, one floor above Aram's pizza. The humble setting probably had some of the same appeal for Bill as his trips to other poverty-stricken areas of the world.

Some of Bill's best writing was in the tiny script of his letters. He would cram the whole account of a trip to the Amazon onto one or two postcards – the aquatic plants and density of mosquitoes on a white-water river; his fight with a would-be robber who slashed him on the elbow with a broken bottle; his efforts to keep a sinking boat afloat by diving beneath it to afix a bedsheet over the leaks; and his dreamy (and also endearing) thought that the headwaters of the river where the boat was about to sink originated on the slopes of a snow-capped volcano in Colombia, where he had stood with us 15 years before. All this was typical of Bill – his love of tropical nature, of drama, of risk, and of odd connections between events. The unpublished novel he wrote in the late 1980s was dominated by these same themes.

'Love of risk' may be better termed a love of physical contest combined with a kind of bumbling intensity. There is a frustrated warrier-athlete just beneath the skin of many a scientist, sublimated into the deft raquet-twirl of an insect net or the virtuoso changing of a flat tire. I saw the Olympic side of Bill Hamilton one time on a trip to the London airport. We started late, and then got stuck in a traffic jam, so by the time we reached Heathrow, my plane was about to leave. Bill grabbed all of my bags, including one very heavy suitcase full of papers and books, and started a gangling sprint to the gate with me running as fast as I could behind. Something about the way Bill would slide into turns gave me an attack of the giggles that compromised my ability to run. The whole performance must have amazed the many onlookers, though I was too worried about keeping up with Bill to observe their faces. During the intense cold of a Michigan winter, he would chop wood outdoors for hours, to see what it was like to experience physical exertion in very cold conditions.

Bill's longing for the tropics was not satisfied by two early trips to Brazil. When he and his wife Christine bought their first house, Bill added a small glassed-in conservatory in order to recreate a piece of tropical forest at home. Once when I was in England, the Hamiltons decided to invite the chairman of Bill's department to dinner. Bill was very nervous about this because at the time he was worried about losing his job. While Christine frantically worked on the dinner, I tried to help with other last-minute preparations. And where was Bill in this hour of need? He was in the glass conservatory, warming up some butterflies, which he had reared on their tropical foodplants, then carefully chilled so as to release them just as the guests were about to arrive.

The butterflies at the dinner party were symptoms of a romantic idealism that spilled over into some more ambitious projects. One was to take two promising rural Brazilian teenagers, whose father was a gifted naturalist, with him to England, in order to give them the opportunity for an education that they would never be able to afford in their own land. Bill [1/1/68] described the man's daughter as "a walking handbook on the fauna and flora of her area" and hoped that she and her brother would be welcome company for his parents, for "my youngest brother died in a rock-climbing accident a little over a year ago and my parents had been expecting to have him around . . . for a few years more." Like the Fuegians who traveled to England on the Voyage of the Beagle, they had trouble adjusting to their new surroundings, first staying with Bill's parents and then in his own home. They eventually returned to Brazil without changing their prospects as Bill, perhaps somewhat naively, had hoped.

You might wonder why an eminent scientist would ever have worried about keeping his job. That tense period was in the early seventies. Bill was a junior faculty member at Imperial College, and the main problem seemed to be that he was unsuccessful as an undergraduate lecturer. Even after Bill died, an obituary in the London Times mentioned that he was "not a charismatic lecturer," and illustrated this by describing how Bill "once stopped in the middle of a talk, staring into space for some 2 min while he tried to think out the answer to a question he had just raised." I witnessed that same incident but interpreted it in quite a different way. I thought it showed perfectly why Bill was such a charismatic character – why he was such an admirable and likeable man. He was worried that he had made a mistake in a lecture, and was willing to admit it publically, try to set it right, and go on. He could make his mouth stop while his brain was still working. His brand of charisma was not based on smooth talking one-upmanship, or slashing verbal duels. It grew out of humility, out of taking a moment to think, out of a soft voice, out of making others feel at home. It was worth more than any public praise to have Bill listen quietly to an intense expression of your heartfelt ideas, and then raise his bushy-browed rather simian face and say, with an earnestness that seemed to match your own: "Yes, quite."

I was often grateful for Bill's kindness in the face of naive enthusiasm, sentimentality, or outright mistakes. Just to provide a concrete example: in the first letter that I wrote to Bill [6/ix/67], I said that his 1964 papers were "Next to *Huckleberry Finn* . . . the most important things

I have ever read" – a risky literary comparison to write to an Englishman you don't even know. I also carried on effusively about wasps in a way that some high-flown intellectuals would consider daft. I began to worry about this when the letter was already in the mail. But Bill's kind reply [5/x/67] put me at ease. He said that he was "surprised but also ... very pleased that those papers should ever have been considered in any context along with 'Huckleberry Finn'," a book that he often thought of while studying *Polistes* in Brazil. "Whether this was because of the human associations connected with the ramshackle buildings where I used to watch the wasps or whether it was because of an indescribable quality of the wasp life itself – wayward, mysterious, almost human – I can't clearly remember. I know I often had the feeling that the wasps themselves must be in a great dilemma how to act, just as mankind seems to be, both as to what was expedient, as in the sense of the theory I was trying to test, and also perhaps as to what was 'right'!"

Bill showed no signs of ever being embittered over the sociobiology wars that followed the 1975 publication of Wilson's *Sociobiology* – attacks on evolutionary-genetic applications of theories like kin selection to humans. This may have been due in part to having his own bulldog in the form of Richard Dawkins. Or it may have been due to his mellow view of enemies, expressed one time when we were discussing how to deal with an irrational attack on kin selection theory. He said [6/viii/78]:

> I suppose we must need enemies: I always feel sad or irritated when supposed enemies seem to want to be friends or turn out to be much weaker than one had imagined...
> I like to see and take part in fights that are fairly and evenly matched. I always liked [my] mother's story of the Moors who when they had a British garrison penned up in a fort sent up to ask if they needed food or water so that they could have a good fight the following day.

Bill Hamilton did original research on wasps and many other insects that was never published. He was an accomplished naturalist, especially knowledgeable about plants and insects in the field, and always anxious to meet biologists expert in particular groups of organisms, especially those of rare or poorly studied taxa, and he knew about their biology in surprising detail. Throughout his life, he continued to encourage and participate in the field studies of others. He maintained an interest in social insects, especially through his 'Italian connection,' with Stefano Turillazzi and his students, including Laura Beani, Rita Cervo, and their collaborators (e.g., Joan Strassmann and David Queller) who have continued a distinguished tradition of work on *Polistes* that began with Leo Pardi in the 1940s; this connection was strengthened through his relationship with Luisa Bozzi, the Italian science writer who was his devoted companion at the end of his life and whom he met while attending meetings in

Italy on social behavior (in 1988) and *Polistes* (in 1993). He wrote: "sometimes [I think] that having somehow become a theorist I should remain one – that the cobbler should stick to his last and avoid trying to be cattle farmer as well. ... At the same time, I would much like to show that technique is not beyond me, that I do love and study the living world."

Hamilton died in 2000, following a bout with the complications of malaria, having been rushed to a London hospital from Africa. That final episode echoed other events in his life. In a series of letters written in the early 1970s, Bill told us of a family tragedy, when the brilliant young son of his physician sister fell from a tree and remained paralyzed and unconscious for a prolonged period. Bill spent hours talking and reading to him, sure that he could see signs of awareness and recovery. He felt triumphant when his nephew, who eventually recovered, began to laugh and show signs of being able to move. In his own last illness, Bill, the risk-taking uncle, ended his days in a coma, in the care of his sister, Mary Bliss – that boy's mother – herself known as the inventor of devices to alleviate the suffering of bed-ridden patients. The relentless intellectual curiosity, humane creativity, and thirst for adventure that seemed to characterize the Hamilton family, had taken him into one of the most disease- and violence-ridden corners of Africa, in search of the origin of the AIDS virus among polio-stricken non-human primates.

Of Hamilton's many new insights, perhaps his contribution to the understanding of costly social aid was the most important, for it is one of the towering landmarks in the progressive disillusionment of humankind that bring us closer to a realistic idea of our place in nature. We have had to learn that we live on a planet that is not the center of the universe, that we are organisms not much different from other animals, and that our conscious will can be subverted by our subconscious mind. Hamilton's insights about altruism, summarized in the simple expression now called 'Hamilton's Rule,' forced us to realize, in addition, that our beneficent feelings may have originated in an underlying self-interest. This insight has illuminated studies of animal behavior, including that of humans, showing how evolutionary biology can be a tool of human self-knowledge. In his life, he showed some other things that are just as important: that a deep knowledge of natural history can be crucially useful to a theoretician; and that a Darwinian understanding of human nature need not condemn us to complete selfishness and lack of concern for others, something I like to think of as *Hamilton's Second Rule*.

See also: Ant, Bee and Wasp Social Evolution; Kin Selection and Relatedness; Parasites and Sexual Selection; Social Selection, Sexual Selection, and Sexual Conflict.

Further Reading

Grafen A (2005) William Donald Hamilton. In: Hamilton WD and Ridley M (eds.) *Narrow Roads of Gene Land,* vol. 3, pp. 423–458. Oxford: Oxford University Press.

Hamilton WD (1964) The genetical theory of social behavior. I, II. *Journal of Theoretical Biology* 7: 1–16, 17–52.

Hamilton WD (1996) *Narrow Roads of Gene Land,* vol. 1. New York, NY: W.H. Freeman at Macmillan Press, Limited.

Hamilton WD (2000) My intended burial and why. *Ethology Ecology & Evolution* 12: 111–122.

Hamilton WD (2000) *Narrow Roads of Gene Land,* vol. 3. Oxford: Oxford University Press.

Hamilton WD (2001) *Narrow Roads of Gene Land,* vol. 2. Oxford: Oxford University Press.

Leighton John J (2005) Because topics often fade: Letters, essays, notes, digital manuscripts and other unpublished works. In: Hamilton WD and Ridley M (eds.) *Narrow Roads of Gene Land,* vol. 3, pp. 399–422. Oxford: Oxford University Press.

Moran N, Pierce N, and Seger J (2000) W.D. Hamilton, 1936–2000. *Nature Medicine* 6: 367.

Queller DC (2001) W.D. Hamilton and the evolution of sociality. *Behavioral Ecology* 12(3): 261–264.

Segerstråle U (2009) *Nature's Oracle: A Life of W.D. Hamilton*. Oxford: Oxford University Press.

Summers A and Leighton John J (2001) The W.D. Hamilton archive at the British library. *Ethology Ecology & Evolution* 13: 373–384.

Trivers RL (2000) William Donald Hamilton (1936–2000). *Nature* 404: 828.

Williams GC (2000) Some thoughts on William D. Hamilton (1936–2000). *Trends in Ecology & Evolution* 15(7): 302.

Hearing: Insects

D. Robert, University of Bristol, Bristol, UK

Introduction

Focusing on insects, this article presents their remarkable capacity to adapt their sensory morphology to unconventional and sometimes severe evolutionary constraints. As morphological diversity is also a hallmark of arthropods in general, the notion that auditory mechanisms may be present in many more arthropod species than reported so far is also discussed. As our understanding of hearing acquires depth, it is likely that this sensory modality will be discovered in more arthropod species, unveiling more diversity in this spectacularly successful and creative phylum.

The sense of hearing in insects has been the subject of several reviews covering different aspects, such as evolution and development, structure and diversity, and function. The work edited by Hoy, Popper, and Fay gathers a series of authoritative studies of insect hearing, and several reviews covering each of these aspects are also included in further reading.

Hearing is defined as the detection of acoustic energy borne in the fluid medium surrounding the animal, air, or water. Therefore, an essential aspect of the auditory process relies on the efficient coupling between sensory structures and sound energy in the physical environment. This coupling is the first step in the chain of events that converts acoustic energy into mechanical energy, and its transduction into neural activity and auditory information. Hearing, as a matter of definition, is concerned with the detection of low levels of air-borne or water-borne vibrations. An acoustic stimulus is considered to be adequate if it elicits a specific response in the auditory organ, as opposed to nonspecific high-level vibrations that may impart mechanical energy to tissue and organs other than the auditory structures, or for that matter the entire animal.

Behaviorally relevant sound waves or acoustic signals are usually inherently low in energy. Hearing, as a mechanical sense, is therefore a delicate act of mechanoreception. For insects, as for many other auditory animals, specialized mechanosensory cells can detect mechanical vibrations that induce motions in the range of just a few nanometers (10^{-9} m), or sometimes a fraction thereof. Such motions are commensurate with, for instance, the thickness of the cell membrane, the diameter of an ion channel, or the length of a glucose molecule. The exact molecular and cellular mechanisms supporting the detection of such small mechanical inputs are still not completely known. Yet, significant progress has been achieved by studying fruit flies and their particular amenability to genetic analysis. Additionally, the range of stimulation magnitude, the dynamic range, of hearing organs in insects is remarkable as they are capable of responding to eight orders of magnitudes of acoustic or vibration inputs. Such range of detection is comparable to that reported for vertebrate ears.

Auditory functions are diverse and nonexclusive. Insects use sounds for the detection and recognition of conspecifics, and in some species, they engage in two-way acoustic communication. Female crickets detect and localize the songs of the conspecific males in acoustically complex environments. Although the sound emissions of male crickets are loud in comparison with other insect sounds, such as the wing noise of passing flies, one challenge has been to understand the mechanisms subtending the directional detection of male songs. The mechanism involved relies on the acoustic coupling between the ears, whereby the tympanal membrane is driven by external and internal sound pressure. Acting as pressure difference receivers, the ears of crickets gain their directionality by virtue of constructive and destructive interference between the internal and external pressure inputs.

Insect ears have also evolved in response to selective pressures exerted by important acoustic predators such as bats. Well-documented examples come from the capacity of moths to detect the echolocation calls of bats and engage in aerial maneuvers attempting to avoid predation. A third function found in insect ears is prey localization, a task that has been documented in parasitoid flies that acoustically seek hosts for their larval progeny.

Outstanding issues remain to fully understand the exact functions of physiological and behavioral auditory processes in insects, in relation to, for instance, a physical environment generating deleterious signal distortions. How does a female cricket auditory system process temporally and spectrally distorted signals to assess the presence and quality of a singing male? How robust is signal processing in insect auditory systems?

Insects are small animals. The physics of sound propagation in the atmosphere is such that most insects are smaller than the wavelength of sound they emit or hear. Consequently, the acoustic cues usually used by larger animals such as most, but not all, vertebrates can become very small. Illustrating such size constraints, a small parasitoid fly *Ormia* relies on finding singing male crickets as a food source for her larvae. Yet, because the fly is only a couple of millimeters in size, the difference in the time of arrival of a sound wave at the two ears, a major cue for directional hearing, is no larger than a couple of microseconds. Such time difference is admittedly too short for neural encoding, which usually takes place in the millisecond range. Yet, a peripheral mechanism has been reported that allows the fly ears to be directionally sensitive to the songs of their cricket host.

Another example pertains to the remarkable acoustic behavior of mosquitoes. Male mosquitoes have been long known, since Johnston's work in 1856, to use their antennae to hear passing females; the complexity of their auditory mechanisms and behavior has only recently been revealed. In effect, it turns out that both male and female mosquitoes can hear flight sounds and can alter the sound emissions generated by their flapping wings. In addition, recent work has demonstrated that hearing in mosquitoes is an active process similar to that found in vertebrate ears. This process is deemed to boost the mosquito's sensitivity to faint sounds and enhance frequency selectivity, thus improving overall fidelity in signal detection.

The Sophisticated Small Ears of Insects

Insects have a diffusion-limited respiratory system that operates efficiently only on a small scale (cm range). This physiological constraint is considered to limit the body size of insects. Insect ears are therefore necessarily physically close to each other, sampling the sound field at adjacent points in space. From all insects known to have tympanal membranes on each side of their body, on their thorax, abdomen, or legs, the distance between the ears is typically less than 1 cm, and sometimes in the range of a few millimeters. This makes target location, based on the time differences between the sound reaching the two ears, hard. A spatial separation of 1 cm between ears, such as that of a field cricket, generates a maximal time difference, the interaural time difference, of about 30 ms. Such time intervals are very short in terms of neural processing, as neurons usually operate at time scales of milliseconds. However, some smaller insects, namely parasitoid flies, with an interaural distance as small as 0.5 mm, are capable of detecting the direction of incident sound waves, using vanishingly small time cues and produce appropriately oriented phonotactic behaviors.

Hearing in insects is made possible by two fundamentally different types of auditory organs. Anatomically, insect ears can present an eardrum – as in tympanal ears – or take the form of an antennal or a hair shaft – as in flagellar ears. Interestingly, tympanal ears have been found and characterized in insects only to date, while flagellar sound receivers have been identified in several classes of invertebrates, including insects, arachnids, and crustaceans. Morphologically, tympanal ears are made of a thin cuticular membrane stretching over a cavity,

a modified tracheal air sac. In addition, multicellular mechanosensory structures, the scolopidia serve to convert mechanical energy into neural signals. Scolopidia can directly or indirectly attach to the tympanal membrane and are composed of attachments cells, support cells, and ciliated mechanosensory neurons. Noteworthy is the fact that both hearing (flagellar and tympanal) and somatosensory organs are of the chordotonal type, employing similar histoarchitectures of multicellular scolopidia containing ciliated neurones.

Functionally, tympanal ears are sensitive to variations in acoustic pressure in the order of pascals to micropascals (atmospheric pressure typically being 100 kPa). Pressure waves propagate well across the atmosphere and therefore carry information many wavelengths away from the sound source. Another physical component of the energy constitutive of a sound wave is the particle velocity in the medium, a quantity expressed in m/s, which retains usable magnitude only close to the sound source. In effect, the part of the acoustic energy contained in the particle velocity component of the propagating sound wave becomes very weak, and therefore hardly detectable many wavelengths away from the source. Typical sensitivity is in the range of mm/s. An important example of such auditory apparatus is the fruitfly *Drosophila* for which sound is part of signaling during mating behavior. Remarkably, hearing in *Drosophila* is an active process, whereby mechanosensory neurons endowed with axonemal cilia are actively feeding back mechanical energy in the oscillations of the auditory system. The molecular mechanisms enabling active and passive mechanoreception are currently being investigated in great detail using the powerful genetic analytical tools available in the fruitfly. This is set to yield exciting future results.

Many organisms have evolved such sensitive detectors that capture the velocity component of the sound field. Flagellar ears are typically antennae (**Figure 1**) or single filiform hairs borne on the surface of the insect body. Notably, antennal structures are found in many, if not all, insect species. In fact, 99% of insect species (100%

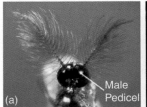

Figure 1 The sexually dimorphic antennal hearing organs of *Aedes aegypti* mosquitoes. (a) Male antenna, with numerous fine setae borne on the antennal shaft and a large second antennal segment, the pedicel at the base of the antenna. (b) Female auditory organs presenting fewer antennal setae, a smaller pedicel. Both antennae respond very well to incident acoustic waves, an indication that both sexes are endowed with a sense of hearing.

of winged species) are endowed with an organ rich in mechanosensory neurons (Johnston's organ) at the base of their antenna. This does not mean that all antennal structures are organs that are mechanically sufficiently sensitive (μm/s–mm/s range) to constitute functional auditory organs. On the other hand, hair-like sensory systems are widespread among arthropods. Spiders can have thousands of such protruding hair sensors, the trichobothria, endowed with several mechanosensory neurons. The function of these sensors has been related to both prey and predator detection. In field crickets, hundreds of innervated mechanosensory hairs can be found on the caudal appendages, the cerci. Their behavioral function is primarily linked to predator avoidance, but an involvement of short-range detection of air currents and sounds emitted by singing males has not been excluded. Most arthropods have mechanosensitive hairs on the surface of their cuticle; it is therefore quite possible that both harmonic and bulk oscillations of the medium (air or water) elicit hair vibrations and deflections, and therefore neural activity. The informational value of such activity is not trivial to investigate, yet evidence is plentiful that arthropods make use of this sensory modality to detect the presence and direction of a source of sound, and therefore a mate, a prey, or a predator in their nearby environment. Caterpillars (*Barathra* sp.) have been shown to react to the flight tones of approaching wasps, while jumping spiders can locate and pounce on passing flies in complete darkness. The specialized sensory system spiders use to detect motion of the air medium is superbly adapted to the task, as revealed by Friedrich Barth's research on hunting spiders. From sensory and behavioral ecological, and evolutionary points of view, it would be very interesting to investigate the possibility of such hearing and sensing capacity in other arthropods such as crustaceans. Crabs are endowed with numerous arborized hairs on their cuticle. Many other examples dot the literature and indicate that this form of capture of air-borne acoustic and water-borne energy is widespread across the phylum. Evidence of the sensory mechanisms and adaptive behaviors involved still seems, however, to be sparse.

It is in mosquitoes that the most elaborate flagellar auditory organs have been found and described to date. Johnston's organ in male and female mosquitoes contains 16 000 and 8000 mechanosensory cells (**Figure 2**), respectively by comparison, the human cochlea contains many hair cells, an indication that the metabolic investment made by mosquitoes in their Johnston's organ must be under sustained selection pressure.

Mate Finding Using Auditory Cues

Acoustically driven behavior in insects is very diverse and reflects the variety of sensory ecologies in insects and the

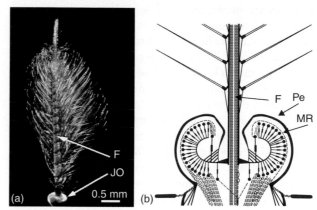

Figure 2 The ear of *Toxorhynchites brevipalpis*. (a) The verticillate antennal flagellum and the pedicel harboring the mechanoreceptive Johnston's organ (JO). (b) Schematic cross-section of Johnston's organ, showing the organization of the mechanoreceptive neurons (MR) inside the pedicel (Pe). Modified from Clements AN (1999) The Biology of Mosquitoes, vol. 2. Oxford, UK: CAB International. The ~60-fold radial symmetry of JO suggests a capacity for directional hearing in this type of auditory organ that is inherent to the vectorial nature of the particle velocity acoustic stimulus.

superb capacity of adaptation. A complete list of key references to foundational work done on praying mantises, field and bush crickets, moths, flies, and cicadas can be found in Hoy et al. (1998).

Acoustic communication in insects is particularly well illustrated by the acoustic signaling of male crickets, which advertise their presence and quality to their conspecifics. Acoustic signals fulfill several functions, helping orientation and localization for mate finding, courtship and mate selection, and competition between males. In some other orthopteran species, such as the bow-winged grasshopper (*Chorthippus biguttulus*), females respond to male signals by emitting very brief signals that are deemed to constitute clues for males, yet remain cryptic to potential predators. A rich documentation of the mechanisms, neurobiology, and behavior of acoustic communication, and its comparative analysis in insects and frogs can be found in a seminal book by Gerhardt and Huber. Insect communication systems have allowed the investigation of the role of acoustic conditions in scattering and nonscattering environments, complex phonotactic behaviors, and the neurobiology of signal detection and recognition, quite remarkably in both laboratory and field settings.

Recent research has revisited the role played by acoustic cues in the mating behavior of mosquitoes, revealing a complex dynamic acoustic interplay between male and female in midair. Elegant experiments have put flying males and females tethered to a thin wire in each other's presence. Surprisingly, the flight tone (e.g., the wing beat frequency) of both sexes varied as to reduce, but not nullify, the difference between them. The behavior is remarkable in that male–female interactions lead to convergence in frequency, while male–male interactions lead to a divergence of tones in these conditions. More importantly, this behavior constitutes the first observation of an acoustically mediated response in female mosquitoes. Further research is likely to establish whether or not mosquito swarms rely on acoustic interactions of that sort for their formation and cohesion. Acoustics may be one aspect so far neglected in the search for agents of mosquito control.

Hearing in mosquitoes is an active process that resembles that found in vertebrates and in *Drosophila*. The mechanical response of the antennal flagellum was shown to vary nonlinearly with the amplitude of the stimulus acoustic stimulus (**Figure 3**). This nonlinearity results in a variable mechanical sensitivity, whereby faint stimuli are amplified (excess of mechanical energy) and loud ones attenuated. The mechanical activity of the ciliary mechanosensory neurons is deemed to be involved. The active mechanisms in Johnston's organ contribute to enhancing frequency selectivity and high-fidelity signal detection, likely contributing to the observed acoustic interplay behavior.

A Particular Problem for Insects: Directional Detection

As we have seen, size is the major constraint insects face in their task to reliably detect sound and be directionally sensitive to incident sound waves. Remarkably though,

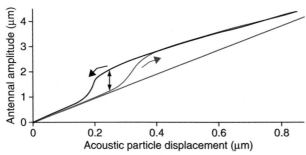

Figure 3 The nonlinear mechanical response of the antenna of the mosquito *Toxorhynchites brevipalpis* to varying particle velocity magnitude. The response is measured using microscanning laser Doppler vibrometry, assessing the velocity of the antenna in a calibrated sound field. The thin blue light depicts a linear proportional response of the antenna to increasing sound particle displacement stimulus (x-axis). The red line shows the antennal response as the amplitude of the stimulus increases to its maximum (red arrow). Past ~0.3 μm, the antennal response exceeds that of the stimulus. The black curve indicates the response to a decreasing stimulus amplitude (black arrow). Notably, during the decreasing amplitude regime the antennal response remains high to displacements that otherwise would generate a smaller antennal response (hysteresis). The vertical double arrow highlights the difference in the level of mechanical energy that the antennal system dissipates. Modified from Jackson JC, Wndmill JFC, Pook VG, and Robert D (2009) Synchrony through twice-frequency forcing for sensitive and selective auditory processing. *PNAS* 106: 10177–10182.

this constraint seems to apply to tympanal receivers only, as particle velocity detectors are inherently directional. Inherent directionality stems from the fact the particle velocity is a vectorial quantity that will apply physical forcing of the receiver along the vector of sound propagation. Although acoustic vector fields can change substantially in the close vicinity of objects, it is generally accepted that flagellar or hair-like receivers can determine the direction of incidence directly from the orientation of their own oscillation in the sound field.

Small tympanal ears, however, face the dual problems that sound pressure variations over short distances are negligible, and that time differences between the ears are very short. An extreme example of such size constraint is the parasitoid flies that use the acoustic emissions (the calling songs) of their hosts to locate them. Such flies, some tachinids hunting field crickets and sarcophagids homing in on cicadas, have evolved ears that can use minute directional cues from the sound field. As true evolutionary innovations, the mechanisms used have been found thus far only in these flies and rely on the mechanical coupling between the two bilateral but adjacent ears. In tachinids, the mechanisms of mechanical coupling allow of minuscule 1.7-μs time difference at the ears to be converted into some 300 μs at the level of the primary sensory neurons. The phonotactic behavior of one of the flies *Ormia ochracea* has been investigated in free flight, using two infrared cameras mounted on computer-controlled pantilt servomotors. Video image analysis was used to provide command signals to the motors to keep the flying fly within the video frame, like one's eyes would follow a fly across the flight arena. The 3-D trajectory of the fly was recorded as she approached a loudspeaker broadcasting an attractive cricket song in infrared darkness (**Figure 4**). The phonotactic parasitoid flies are remarkably accurate in their localization of the sound source. Surprisingly, they are also accurate when the attractive song is present for only part of the phonotactic flight bout (**Figure 4**). In effect, the transient presence of acoustic information is sufficient for the fly to successfully find the source of the cricket song. This simple experiment suggests the presence of 3-D acoustic detection in this fly, and the possible memorization of sufficient auditory spatial information for a cue-free, yet oriented flight toward the target. The mechanisms allowing this behavior are still unknown.

How Ubiquitous Could Hearing Be in Arthropods?

In particular, with respect to the very widespread capacity of insects to harbor numerous long, thin mechanoreceptive sensillae, it is perhaps worth calling for a renewed attention to the equally impressive morphologies of arachnids and crustaceans to build such elaborate sensory structures.

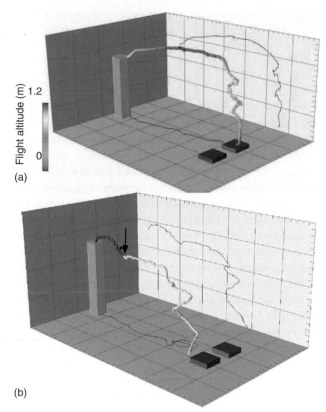

Flight altitude (m)

1.2

0

(a)

(b)

Figure 4 The phonotactic behavior of the acoustic parasitoid fly *Ormia ochracea*. (a) A naive fly never flown in that experimental arena (length 6.8 m, width 4.9 m, height 4.0 m) is deposited on a starting platform (green column). A loudspeaker (blue square) located on the floor some 3 m from the standing fly is switched on. The fly takes off, cruises at relatively constant altitude before engaging in a descending spiral to land on the loudspeaker. Measured for $n = 80$ trajectories from $N = 10$ flies, landing accuracy was 8.2 ± 0.6 cm. (b) When tested to the other (red square) loudspeaker, the same fly initiates a similar flight trajectory (yellow trace). As the loudspeaker is switched off (black arrow), the fly is left with no extrinsic information about the location of the source. Yet, the fly continues her flight (red trace) and initiates her descent to the target with remarkable accuracy. Modified from Müller P and Robert D (2001) A shot in the dark: The silent quest of a free-flying phonotactic fly. *Journal of Experimental Biology* 204: 1039–1052.

Research on hunting spiders clearly indicates the use of air velocity to detect passing prey, a strong evidence that the sense of hearing, as defined by the detection of harmonic or bulk air velocity, in arthropods in general may be vastly more widespread than previously thought.

See also: Hearing: Vertebrates.

Further Reading

Barth FG (2000) How to catch the wind: Spider hairs specialized for sensing the movement of air. *Naturwissenschaften* 87: 51–58.

Cator LJ, Arthur BJ, Harrington L, and Hoy RR (2009) Harmonic convergence in the love songs of the dengue vector mosquito. *Science* 323: 1077–1079.

Field LH and Matheson T (1998) Chordotonal organs in insects. *Advances in Insect Physiology* 27: 1–28.

Gerhardt HC and Huber F (2002) *Acoustic Communication in Insects and Anurans. Common Problems and Diverse Solutions*, p. 542. Chicago: University of Chicago Press.

Göpfert MC and Robert D (2007) Active auditory mechanics in insects. In: Manley G, Popper AN, and Fay RR (eds.) *Active Auditory Mechanics*, pp. 191–209. New York: Springer-Verlag Springer.

Hoy RR, Popper AN, and Fay RR (eds.) (1998) *Comparative Hearing: Insects*. New York: Springer-Verlag.

Jackson JC, Windmill JFC, Pook VG, and Robert D (2009) Synchrony through twice-frequency forcing for sensitive and selective auditory processing. *PNAS* 106: 10177–10182.

Mason AC, Oshinsky MI, and Hoy RR (2001) Hyperacute directional hearing in a microscale auditory system. *Nature* 410: 686–690.

Müller P and Robert D (2001) A shot in the dark: The silent quest of a free-flying phonotactic fly. *Journal of Experimental Biology* 204: 1039–1052.

Robert D (2005) Directional hearing in insects. In: Hoy RR, Popper AN, and Fay RR (eds.) *Sound Source Localization*, pp. 6–35. New York: Springer-Verlag.

Robert D and Göpfert MC (2002) Novel schemes for hearing and acoustic orientation in insects. *Current Opinion in Neurobiology* 12: 715–720.

Robert D and Hoy RR (2007) Auditory systems in insects. In: Greenspan R and North G (eds.) *Invertebrate Neurobiology*, pp. 155–183. Cold Spring Harbour, NY: Cold Spring Harbour Laboratory Press.

Warren B, Gibson G, and Russell IJ (2009) Sex recognition through midflight mating duets in *Culex* mosquitoes is mediated by acoustic distortion. *Current Biology* 19: 485–491.

Yack J (2004) The structure and function of chordotonal organs in insects. *Microscopy Research and Technique* 63: 315–337.

Yager D (1999) Structure, development, and evolution of insect auditory systems. *Microscopy Research and Technique* 47: 380–400.

Hearing: Vertebrates

F. Ladich, University of Vienna, Vienna, Austria

Introduction

Animals are able to detect sound if they can couple the sound energy in some ways to receptor organs. Receptors can respond either to the motion of molecules in a sound field (acoustic particle motion) or to the sound pressure, the force acting on a given area. All tetrapods (amphibian, reptiles, birds, and mammals) are sound pressure sensitive, whereas fish, the largest groups of vertebrates comprising about half of all extant vertebrates (~28 000 out of 55 000 species in total), detect the motion component in a sound wave (a large number of fishes can detect both). Because of different physical constraints, sound detectors and their periphery have to be constructed differently to pick up sound energy and excite sensory cells.

Particle motion detectors oscillate in phase with the particles of the surrounding medium. In aquatic animals wherein the density of the medium and the body are quite similar, it is difficult to get relative movement between a sensory cell structure and the main body of the animal. In addition, at a given sound energy, particle displacement is much smaller in water than in air, which makes it even more difficult to evolve a sensor that responds to these tiny particle movements. In order to solve this problem, the sensory cells have to be coupled to an object that is denser than the surrounding tissue. This object then oscillates either with a lag or a lower amplitude relative to the sensory hairs, creating a relative motion, which excites sensory cells. If the denser object is made of calcium carbonate, which is the case in fishes (e.g., the otolith), then the particle motion detectors are sensitive only to lower frequencies.

Particle motion detectors have one advantage compared to sound pressure detectors but several properties that make them impractical for solving many hearing and communication tasks. The main advantage is that motion detectors are inherently directional and thus can be used to determine the location of sound sources. Disadvantages of motion detectors include the fact that they respond only to low frequencies of a few hundred hertz (<1 kHz). Moreover, they react only to sound close to the sound source, which limits sound detection to rather short distances and to higher amplitudes of sounds.

Sound pressure detection evolved successfully in all classes of vertebrates and in arthropods as well. In order to detect sound pressure, animals need a thin membrane (often called 'tympanum') over an air-filled cavity, which moves in and out when outside pressure is higher than in the cavity and vice versa. This membrane movement is then transmitted to sensory cells. Sound pressure has no inherent direction. Thus, in order to allow animals to localize sound sources such as vocalizing mates or territory intruders, two ears are necessary, and differences between the ears in arrival time or intensity of sound can be used to extract information about the sound source's location. Depending on the wavelength and size of the animal's head, there exist different ways for how sound pressure can be used for sound source localization. If the ears are connected to each other and sound pressure acts on both sides of the tympanum, the pressure difference will bend the tympanum toward the side with the lower sound pressure. Such pressure-gradient receivers are inherently directional similar to particle movement detectors.

Utilizing pressure detectors, vertebrates are able to detect sound frequencies of up to several or even tens of kilohertz with upper frequency limit beyond 100 kHz at very low sound intensities. Sound pressure detecting membranes can be found either on the outside of the head, as in all tetrapods, or within the body as in numerous fishes, called 'hearing specialists.' Sound pressure detection allows animals to communicate over distances of up to several hundred meters or even kilometers such as in baleen whales. This allows the establishment of large territories by acoustically advertising ownership, attracting mates, or performing mate choice over large distances.

Ears and Hearing in Fishes

Fishes do not possess external or middle ears similar to tetrapods. They possess inner ears (or labyrinth) which consist of three semicircular canals with ampullary enlargements, two oriented vertically and one horizontally (these are similar to all other vertebrates with the exception of jawless agnathans) and three otolithic end organs, the utricle, saccule, and lagena. In the following, we focus on the structure and function of bony fishes because, contrary to cartilaginous fishes, many species use sound for intraspecific acoustic communication in addition to orientation. Very often, the saccule is the largest otolithic end organ and the lagena the smallest. Each otolithic end organ consists of a sensory epithelium, or macula, and an overlying calcium carbonate otolith (ear stone). Each macula is built up of a large number of sensory hair cells, which are usually oriented in different direction. The shapes of the maculae of the three otolithic endorgans differ in that they could be round, moon shaped, or L-shaped. Sensory epithelia are smaller than the overlying otolith especially within the saccules where the ovoid otoliths (called 'sagittae') only contact the sensory maculae along an L-shaped shallow indentation.

According to our current knowledge, fishes do not possess a morphological structure solely devoted to hearing. The majority of fishes seem to utilize the saccule for hearing. To what degree the saccule serves vestibular task such as gravity detection in addition to hearing needs to be explored. In a few taxa such as herrings and marine catfishes, the utricle might be the main auditory end organ. Thus, fishes are clearly distinguished from frogs, reptiles, birds, and mammals in not having a sensory structure exclusively for hearing and in not being able to detect sound pressure but particle motion, as mentioned in the introduction. Subsequently, their hearing is limited to low frequencies of a few hundred hertz.

Interestingly, while all fishes are sensitive to motion of molecules, at least a third of all species developed accessory morphological structures, termed 'hearing specializations,' which enable them to detect sound pressure. Sound pressure detectors need a tympanum and an air-filled chamber to work properly. External tympana do not work when the animals and the medium have the same density and vibrate in phase, so some fishes have found another way to circumvent this problem. Their solution is to use internal tympana, by connecting gas-filled chambers, which undergo volume changes in an acoustic field, to the inner ear. Oscillations of the epithelia of such chambers are transmitted in various ways in nonrelated taxa to the inner ear and thus evolved independently several times in fishes. Squirrelfishes (soldierfishes, holocentrids), drums (sciaenids), and herrings (clupeids) possess tubelike anterior extension of the gas-filled swim bladder which are more or less in contact with the inner ear and enhance the hearing sensitivity of these families. The African freshwater family Mormyridae, a group of weakly electric fishes, have an air bubble very close to the inner ear (otic bulla) which fulfils the same task. The Southeast Asian and African group of labyrinth fishes (Anabantoidei) use an air-breathing chamber, the suprabranchial organ, which is located close to the inner ear for hearing sound pressure waves. The most sophisticated sound pressure receiving mechanism evolved in otophysines, a group of fishes comprising carps and minnows (cypriniforms), catfishes (siluriforms), characins (characiforms), and knifefishes (gymnotiforms). This group includes ~8000 species and make up about two-thirds of all freshwater fish species. In otophysines, a series of tiny ossicles transmit oscillations of the anterior part of the swimbladder to the inner ear. These ossicles, called 'Weberian ossicles,' are functionally similar to the middle ear ossicles in mammals, but

they differ phylogenetically because they did not derive from parts of jaw elements but from the anterior vertebrae (**Figure 1**).

The aforementioned groups are often called 'hearing specialists' because they possess special structures for hearing. The hearing sensitivity in these specialists extends

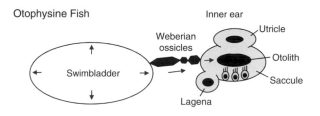

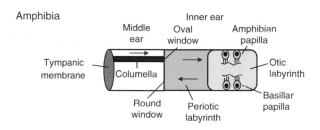

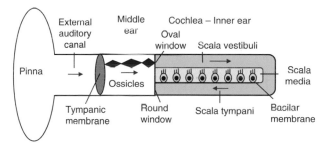

Figure 1 Schematic drawings of the ears of otophysine fish, amphibia, and a higher vertebrate. Otophysine fish belong to hearing specialists and possess a connection between the inner ear (utricle, saccule, lagena) and an air-filled cavity (swim bladder). Oscillations of the swim bladder wall in the sound field are transmitted to the inner ear via a chain of tiny ossicles (Weberian ossicles). In the amphibian ear, movements of the tympanic membrane are transferred by the columella to the oval window of the inner ear. This causes fluid movements within the periotic labyrinth and furthermore in the otic labyrinth, which results in motion of the tectorial membrane relative to the hair cells. The bird, reptile, and mammal ear is built up of an outer ear (with or without a pinna), one (birds and reptiles), or three (mammals) middle ear ossicles and the cochlear duct of the inner ear. The cochlea is stretched in birds and reptiles, coiled in mammals, and consists of three fluid-filled canals. Movements of the fluids cause bending of the basilar membrane and hair cells relative to the tectorial membrane and stimulation of the hair cells. Amphibia and higher vertebrate, modified from Bradbury JW and Vehrencamp SL (1998) *Principles of Animal Communication*. Sunderland, MA: Sinauer Associates, Inc.

up to several kilohertz and to lower sound levels; thus, they possess much better hearing abilities than groups lacking accessory hearing structures (called 'hearing non-specialists' or 'generalists') (**Figure 2**). Some herrings (clupeids) such as shads are able to detect ultrasound up to 200 kHz, higher than bats or dolphins. This hearing may have evolved to detect dolphin predators.

What were the selective forces leading to the evolution of sound pressure detection and thus hearing enhancement in certain fish taxa? Fishes are a vocal group of vertebrates and it is intriguing to assume that many hearing refinements evolved to facilitate acoustic communication. However, the ability to produce sounds and communicate acoustically is not limited to hearing specialists. Many hearing nonspecialists such as toadfishes (batrachoidids), cichlids, sunfishes (centrarchids), or gobies (gobiids) emit sounds in various behavioral contexts such as territorial defense, courtship, and spawning. Obviously, they can compensate for their relatively poor hearing ability by increasing the intensity of their sounds or by shortening the communication distance under given ambient noise levels. Communication often takes place over a distance of a few centimeters, where even faint sounds are detectable by conspecifics. On the other hand, sound production is not a common feature of all hearing specialists. Among otophysines, carps and minnows (cypriniforms) are the most primitive groups. Interestingly, only a few representatives of this large order are known to vocalize and in no case has there been a vocal organ described. Hearing sensitivity

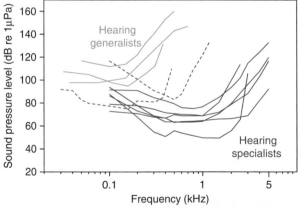

Figure 2 Hearing thresholds for various fishes (hearing generalists and hearing specialists). Hearing thresholds or audiograms show the softest sound an animal can hear at different frequencies. Dashed lines illustrate hearing curves of a damselfish and a cod which have a less extended frequency range and the swim bladder is not connected to the inner ear, though it is believed to play a role in hearing. Modified from Ladich F and Popper AN (2004) Parallel evolution in fish hearing organs. In: Manley G, Popper AN, and Fay RR (eds.) *Evolution of the Vertebrate Auditory System. Springer Handbook of Auditory Research*, pp. 95–127. New York, NY: Springer-Verlag. Note different *x*-axis ranges in **Figure 2** versus **Figures 3** and **4**.

and fine structure of hearing organs do not seem to differ between vocal and nonvocal genera among otophysines, labyrinth fishes, or mormyrids. Both groups of fishes seem to use the temporal patterning of sounds to code information about species identity, sex, and mate assessment. Dominant frequencies of sounds can extend to higher frequencies (above 1 kHz) in hearing specialists than in generalists. In summary, data suggest that acoustic communication was not the driving force for the selection of hearing enhancement in fishes.

Which other constraints cause the ancestors of otophysines and other specialists to enhance their hearing sensitivity? Interestingly, specializations are often found in taxa, which initially arose in shallow freshwater habitats such as lakes, rivers, and flood plains, and they are seldom found in taxa from upper regions of rives or in fishes that evolved in marine environments except probably for the deep sea. The majority of freshwater fishes (more than two-thirds) are hearing specialists such as otophysines, mormyrids, and labyrinth fishes, while only a small portion of marine species possess hearing specializations such as a few squirrelfishes (holocentrids), drums (sciaenids), and probably deep-sea fishes. One explanation may be related to the limited range of propagation of low-frequency sounds in shallow water. As a result, to detect sound that comes from any distance, there would have been selective pressure for detection of the higher frequencies. Another explanation might have been that the detection of low-level sound in particular in quiet stagnant habitats was advantageous. What types of sound sources did freshwater specialists detect if not communication sounds? Important sound sources for any animals are sounds that emanate from predators. Many fishes show a characteristic startle behavior so that the fish turns away from the direction of the attack of predators (C-start). Predator avoidance through the development or improvement of specific hearing abilities has been the major selective pressure in the evolution of ultrasonic hearing in flying insects. A similar response has been described in many species of herrings (clupeids) in response to dolphin-like ultrasonic pulses. Another important aspect of improving survival is prey or food detection. Feeding or swimming noises from conspecifics as well as all kinds of noise generated by prey species might be very important for the localization of prey and food items.

Ears and Hearing in Amphibia

The ear of adult amphibians as well as other tetrapods possesses an external tympanum (and thus acts as sound pressure receiver), a middle ear, and auditory endorgans in the inner ear, which are solely devoted to hearing. This is in contrast to fishes, which generally lack such structures (**Figure 1**).

The tympanum or ear drum is easy to see from outside because it is a large round membrane located immediately behind the eyes in frogs and toads. The middle ear consists of an air-filled cavity and a column-like ossicle, the columella (which includes a small extracolumella), which transmits the vibrations of the tympanum to the oval window, a thin epithelium of the inner ear. This columellar system primarily conducts airborne (and sometimes waterborne) sound in adult anurans and urodeles. Besides this columellar pathway, amphibians possess a second sound transmission pathway, the opercular subsystem for the detection of seismic signals borne through the ground. The opercular system consists of a muscle connecting the shoulder girdle to the operculum, a moveable skeletal element which is located in the oval window. Thus, vibrations of the ground are picked up by the forelimbs and transmitted via the shoulder girdle to the inner ear. Tadpoles do not possess ear drums (tympana) – these develop during metamorphosis – but use extratympanic pathways such as lungs for sound detection similar to the way certain fishes use their swim bladders for hearing. Thus, amphibia have the ability to hear efficiently underwater, underground and in air, an ability few if any other vertebrates have (**Figure 1**). Frogs and toads listen to the ground and to the air at the same time, whereas urodeles (newts and salamanders) mainly listen to the ground.

Besides these various sound-conducting pathways, the amphibian ear is unusual in the number of end organs the inner ear contains. All anurans and some urodeles contain eight endorgans, three of them (saccule, amphibian papilla, and basilar papilla) seem to serve entirely acoustic functions (auditory and/or seismic/ground vibrations). Acoustic signals arriving at the oval window by the tympanic or extratympanic pathways are channeled via the periotic labyrinth to the acoustic sensors in the otic labyrinth (saccule, amphibian, and basilar papillae). Movement of the oval window causes fluid flow in the periotic labyrinth and via contact membranes movement in the endolymph of the otic labyrinth. Contact membranes are thin membranes where epithelia of the periotic and the otic labyrinth fuse and where acoustic energy is transmitted to the endolymph. The saccule, the amphibian, and basilar papillae differ in general structure. The saccule is an otolithic end organ where calcium carbonate crystals called 'otoconia' cover the apical part of the sensory hair cells, while the amphibian and basilar papillae lack otoconia but possess tectorial membranes instead.

The acoustic frequency range seems to be divided among the saccule, amphibian papilla, and basilar papilla. The anuran saccule exhibits the best excitatory frequency typically below 100 Hz and is exquisitely sensitive to substrate vibrations (seismic signals). Many frogs are thus able to detect enough low-frequency energy to be very sensitive to approaching predators. The amphibian papilla responds to frequencies ranging from 100 Hz to ~1200 Hz. The basilar papilla responds to frequencies higher than those detected by the amphibian papilla and is generally tuned to some component of the animal's calls at several kilohertz. The saccule and amphibian papilla seem to be more general-purpose acoustic sensors providing tuning over a broad frequency range. The American green tree frog *Hyla cinerea* shows maximal sensitivity to a frequency of 900 Hz representing a spectral peak in its advertisement call. Moreover, the audiogram shows a second dip in threshold around 3000 Hz, near the dominant high-frequency peaks in its call.

In general, amphibian hearing curves show maximal sensitivity between 600 and 1000 Hz similar to hearing specialists in fishes. Unlike fish, hearing in amphibia is limited to frequencies below 6000 Hz, with sensitivity somewhat less than that measured in other vertebrates (**Figure 4**). Under certain circumstances such as high ambient noise level near waterfalls, some Asian frogs switch to ultrasonic acoustic communication.

Ears and Hearing in Birds and Reptiles

Among vertebrates, birds are one of the most vocal groups. Many birds rely strongly on their sense of hearing for communication in social contexts and for alarm signals. Among reptiles, crocodiles and geckos are also vocal, a rare trait in reptiles.

The ears of birds are built up of an outer ear consisting of an external auditory meatus, a middle ear, and an inner ear. On the tympanum inserts a columella (or stapes similar to amphibians), which crosses the middle ear cavity and ends with its footplate into the oval window of the cochlear duct next to the round window. The cochlea is not coiled as in mammals but bent and somewhat twisted. It consists of three fluid-filled canals and one single sensory epithelium entirely devoted to acoustical functions. Adjacent to the oval window follows the scala vestibuli and the scala tympani which ends in the round window. Both scalae are connected with each other at the apical end of the cochlear duct, and both are filled with perilymph (**Figure 1**). Between both fluid-filled canals lies the endolymphatic scala media which is separated from the scala tympani by the basilar membrane.

In the avian auditory sensory epithelium (basilar papilla), hair cells are not arranged in rows as in mammals but form a mosaic. The sensory epithelium is shorter (4–5 mm, maximum: 11 mm) and wider than that of a typical mammal. A typical cross section shows up to 50 hair cells, as against four to six in mammals. There are no structural grounds to distinguish discrete types of hair cells in the avian papilla. Hair cell height is roughly correlated with response frequency, being generally shorter in birds hearing higher frequencies, and shows a reduction in height toward the base of the papilla. The basilar membrane is tonotopically organized with low frequencies encoded at the apical end and high frequencies at the base of the papilla. The tectorial membrane covers the entire papilla in all species and all hair cells are firmly attached to it.

Birds are sensitive to frequencies from 100 Hz up to 10 kHz and hear best at frequencies between 1 and 5 kHz. The barn owl *Tyto alba* has exceptional high-frequency sensitivity that exceeds 10 kHz, while the dove *Columba livia* is sensitive to infrasound (10 Hz). Based on phylogeny and hearing, three groups of birds can be defined. Median audibility curves illustrate general trends (**Figure 3**). Song birds (Passeriformes such as sparrows, tits, starlings, and crows) hear better at higher frequencies than nonsongbirds such as chicken (Galliformes), ducks and geese (Anseriformes), doves (Columbiformes), and falcons (Falconiformes). Therefore, high-frequency 8-kHz alarm calls of great tits, and related songbirds are mostly inaudible for their predators such as sparrow hawks. On the other hand, nocturnal predators such as owls hear better than both former groups over the entire frequency range.

The main energy of bird vocalizations falls within the frequency region of their best hearing sensitivity. Thus, intraspecific communication was most likely the main selective force in the evolution of their hearing abilities. Owls constitute an exception because their absolute sensitivity is driven by their predatory lifestyle and less by the correlation between the distribution of the energy in their vocal repertoire and the hearing sensitivity. Interestingly, oilbirds *Steatornis* that produce clicks and detect their echoes did not evolve high-frequency ultrasonic hearing such as bats and dolphins, but have auditory curves similar to other birds because the main energy of echolocation pulses falls in the 1.5–2.5 kHz range. Oilbirds use echolocation only for orientation in the darkness of caves, where they roost and nest and not primarily for hunting small prey species such as echolocating mammals (bats and dolphins).

Hearing in reptiles depends on the presence or absence of external ear openings and tympana. In reptiles with external ear structures, the tympanic membrane is visible, either nearly contiguous to the surface of the skin (as with iguanids such as the green iguana), or recessed deeper into the head. Crocodilians are the only reptiles with an outer ear that moves. A mobile flap of skin allows the crocodilians to close their external ears to a thin slit when they are under water. Snakes have no tympanum and the columella is attached to the quadrate bone which is linked to the lower jaw. Snakes are more sensitive to vibrations in the ground than to airborne sounds.

Our knowledge of reptile hearing comes almost exclusively from physiological recordings (either from the ear or from nerve recordings) contrary to that of birds whose hearing was studied primarily by behavioral training methods. Reptiles are sensitive to frequencies from 100 Hz up to 4–8 kHz. As a rule, reptiles without an external ear opening show absolute sensitivities roughly 20–30 dB higher than other groups. Snakes reveal best hearing at 100–500 Hz. Lizards and crocodilians, two relatively derived groups, exhibit better hearing at higher frequencies. In crocodilians, best frequencies are between 200 and 1000 Hz with a level of sensitivity approaching that of most birds (**Figure 3**). In crocodiles, some lizards such as geckos, and some turtles, hearing serves intraspecific acoustic communication during territorial contests, mating and parent–offspring communication (crocodiles). In nonvocal reptiles, hearing probably functions in predator avoidance and prey detection partly through the detection of ground vibration.

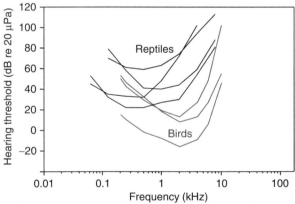

Figure 3 Median hearing thresholds curves from reptiles (black) and birds (birds). In reptiles, the uppermost curve is from snakes that have no external ear opening, the lowest curve showing highest sensitivity from crocodiles. In between, curves from lizards and turtles are located. Lizards and crocodiles possess the best hearing sensitivity at higher frequencies. In birds, the lowest hearing curve and thus best hearing sensitivity is found in nocturnal predators such as owls. The curve in between is from songbirds. Worst high-frequency hearing sensitivity is found in the evolutionary older nonsongbirds such as geese, falcons, and doves. Modified from Dooling RJ, Lohr B, and Dent ML (2000) Hearing in birds and reptiles. In: Dooling RJ, Fay RR, and Popper AN (eds.) *Comparative Hearing: Birds and Reptiles. Springer Handbook of Auditory Research*, pp. 308–359. New York, NY: Springer-Verlag.

Ear and Hearing in Mammals

Mammals differ from fishes, amphibian, reptiles, and birds in the morphology of the ear as well as in the audible frequency range, which is significantly broader than in other vertebrate classes due to the ability to detect higher frequency sound.

The mammalian ear differs in the external, middle, and inner ear morphology from other vertebrates. The external ear consists of large pinna, which is often moveable and which enhanced sound intensity at the tympanum and provides frequency-dependent directional cues. A moveable tragus (bats) may function as a valve to control sound intensities. The pinna surrounds the tube-like ear canal which ends at the tympanum. The middle ear consists of three articulated tiny ossicles (malleus/hammer, incus/anvil, stapes/stirrup) (**Figure 1**). The leverage and the small size of the bones make the whole system very sensitive to higher frequencies. The inner ear contains the cochlear duct, which consists of three fluid-filled canals (scalae vestibuli, media, and tympani). The cochlea is two to three times longer and narrower than in birds and coiled. The sensory structure known as 'the organ of Corti' possesses two classes of hair cells: one row of inner hair cells and 3–5 rows of outer hair cells separated by tunnels all of which are covered by a tectorial membrane. Outer hair cells are innervated from the brain, undergo length changes in response to stimulation, and in some way, control and amplify the sensitivity of the inner hair cells. Sound pressure changes arriving at the tympanum are transmitted by the middle ear ossicles to the oval window and propagate in the scala vestibuli. They generate a traveling wave in the basilar membrane and exhibit a peak of maximum displacement at a specific location that is determined by the frequency of the acoustic stimulus. High-frequency stimulation causes maximum displacement near the base of the cochlea, and low frequencies near the apex (tonotopic arrangement).

Special adaptations to aquatic life are found in aquatic mammals The pinnae may be reduced (seals) or completely missing (whales). The opening of the ear canal is missing in whales or wax-filled and the tympanic membrane unrecognizable but middle ear ossicle are existent. It is assumed that sound is transmitted to the middle ear in various ways. In dolphins, echolocation sound is picked up by the lower jaw and transmitted in a fat-filled canal to the middle ear.

The audible frequency range of mammals is generally broader than that in other vertebrates, extends up to 180 kHz, and reveals a large diversity. Some species are able to detect low frequencies and even infrasound such as large terrestrial and aquatic mammals, for example, elephants and baleen (plankton feeding) whales. Infrasound detection enables animals to communicate over long distances. Elephants can detect sounds up to kilometers and baleen whales up to dozens of kilometers. A majority of mammals, including cats, dogs, horses, and rodents, are able to detect ultrasound either for intraspecific communication purposes or for the detection of predators and prey. Hearing of ultrasound up to 100 kHz and higher is particularly well developed in echolocating mammals such as bats and dolphins (**Figure 4**). High-frequency hearing is linked to their predatory lifestyle, because they need to create sound waves with wavelengths shorter than the body size of their prey species (e.g., moth in bats or fish in

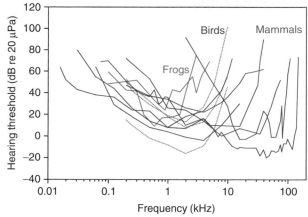

Figure 4 Hearing threshold curves from some frogs and several mammals. For comparative purposes, hearing thresholds of birds (dotted lines) have been added to this comparison. Hearing extends up to higher frequencies in birds as compared to frogs but cannot compete with that of mammals, a vertebrate class where the audible frequency range regularly extends into the ultrasonic region. The lowest detectable frequencies are found in the elephant (left most curve), the highest in bats and dolphin (both right most curves). Data from Fay RR (1988) *Hearing in Vertebrates: A Psychophysics Data Book*. Winnetka, IL: Fay-Hill Associates; Dooling RJ, Lohr B, and Dent ML (2000) Hearing in birds and reptiles. In: Dooling RJ, Fay RR, and Popper AN (eds.) *Comparative Hearing: Birds and Reptiles. Springer Handbook of Auditory Research*, pp. 308–359. New York, NY: Springer-Verlag.

dolphins) in order to discriminate objects reasonably well. Besides ultrasonic clicks, toothed whales produce social sounds at much lower frequencies, which are outside their best hearing range.

The upper frequency limit of hearing in a wide range of mammals is correlated with the effective head size. As distances between ears decrease, the sound frequencies required to generate detectable interaural intensity differences increase. Therefore, high-frequency hearing and subsequently acoustic communication seem to be an adaptation for the localization of sound sources in small-sized species. Mammals lack connections between both ears. Such connections enable small animals, such as frogs, birds, and reptiles, to create a pressure gradient at a single sound detector and improve directional hearing.

Acknowledgments

I would like to thank A.N. Popper for carefully reading the first part of the manuscript.

See also: Acoustic Signals; Alarm Calls in Birds and Mammals; Hearing: Insects; Sound Localization: Neuroethology; Sound Production: Vertebrates; Vocal–Acoustic Communication in Fishes: Neuroethology.

Further Reading

Bradbury JW and Vehrencamp SL (1998) *Principles of Animal Communication.* Sunderland, MA: Sinauer Associates, Inc.

Dooling RJ, Lohr B, and Dent ML (2000) Hearing in birds and reptiles. In: Dooling RJ, Fay RR, and Popper AN (eds.) *Comparative Hearing: Birds and Reptiles. Springer Handbook of Auditory Research*, pp. 308–359. New York, NY: Springer-Verlag.

Echteler SM, Fay RR, and Popper AN (1994) The structure of the mammalian cochlea. In: Fay RR and Popper AN (eds.) *Comparative Hearing: Mammals. Springer Handbook of Auditory Research*, pp. 134–171. New York, NY: Springer-Verlag.

Fay RR (1988) *Hearing in Vertebrates: A Psychophysics Data Book.* Winnetka, IL: Fay-Hill Associates.

Fay RR and Megela-Simmons A (1998) The sense of hearing of fishes and amphibians. In: Fay RR and Popper AN (eds.) *Comparative Hearing: Fish and Amphibians. Springer Handbook of Auditory Research*, pp. 269–318. New York, NY: Springer-Verlag.

Feng AS, Narins PM, Xu CH, et al. (2006) Ultrasonic communication in frogs. *Nature* 440: 333–336.

Gleich O and Manley GA (2000) The hearing organ of birds and crocodilia. In: Dooling RJ, Fay RR, and Popper AN (eds.) *Comparative Hearing: Birds and Reptiles. Springer Handbook of Auditory Research*, pp. 70–138. New York, NY: Springer-Verlag.

Ladich F and Myrberg AA (2006) Agonistic behaviour and acoustic communication. In: Ladich F, Collin SP, Moller P, and Kapoor BG (eds.) *Communication in Fishes,* vol. 1, pp. 121–148. Enfield, NH: Science Publishers, Enfield.

Ladich F and Popper AN (2004) Parallel evolution in fish hearing organs. In: Manley G, Popper AN, and Fay RR (eds.) *Evolution of the Vertebrate Auditory System. Springer Handbook of Auditory Research*, pp. 95–127. New York, NY: Springer-Verlag.

Lewis ER and Narins P (1999) The acoustic periphery of amphibians: Anatomy and physiology. In: Fay RR and Popper AN (eds.) *Comparative Hearing: Fish and Amphibians. Springer Handbook of Auditory Research*, pp. 101–154. New York, NY: Springer-Verlag.

Smotherman M and Narins P (2004) Evolution of the amphibian ear. In: Manley G, Popper AN, and Fay RR (eds.) *Evolution of the Vertebrate Auditory System. Springer Handbook of Auditory Research*, pp. 164–199. New York, NY: Springer-Verlag.

Helpers and Reproductive Behavior in Birds and Mammals

J. Komdeur, University of Groningen, Groningen, Netherlands

Introduction

Animals rarely act in isolation. The vast majority of animals live within a social environment, their lives being affected by the presence and activities of the other individuals around them. For example, foraging, mating, rearing young, predator defense, and practically every aspect of an individual's behavior will be influenced in some way by others around them. Many of us are genuinely fascinated by the wide spectrum of social interactions we see across the diversity of animal life. This may be so because everybody has a social life, and much of the interest probably emerges from comparisons with our own social situation. On the other hand, our fascination may be in understanding the widespread existence of social behaviors between individuals. These social interactions, which can be in the form of cooperation or conflict, can occur over a wide range of strategies and at various levels. Individuals cooperate in hunting, food sharing, fending off enemies, and migrating from one site to another.

Many animals live and breed in colonies, and males and females interact when mating and/or caring for offspring. Behavior that provides a benefit to another individual and importantly, has evolved at least partly because of this benefit can be defined as cooperative. But why should an individual carry out a cooperative behavior that appears costly to perform, but benefits other individuals? This question seems to be hard to answer from an evolutionary perspective. Evolutionary theory states that selection favors individuals who efficiently translate resources into survival and reproductive success, thus maximizing their genetic contribution to future generations. Evolutionary theory predicts that individuals are adapted to maximize their own fitness even if it leads to a decrease in fitness of their partner or of other group or family members. This would appear to lead to a world dominated by selfish behavior. As such, cooperative behaviors pose a problem to evolutionary theory because they appear to reduce the relative fitness of the performer and hence should be selected against. As cooperation is apparent throughout the natural world, there must be a solution to this paradox. As a consequence, the evolution of investment should be driven by the relative costs and benefits of investment. For example, the amount of care provided by the male to offspring may be adjusted in line with his confidence of genetic parentage.

In the last few decades, detailed long-term studies, which use modern molecular techniques to accurately determine genetic relationships, have revealed the true diversity and complexity of breeding systems and have changed our perception of monogamy radically. For example, in passerines (less accurately known as 'songbirds' with around 5000 species), true genetic monogamy (i.e., all offspring are fertilized within the social pair-bond) occurs only in few studied species, while extrapair paternity (i.e., offspring resulting from fertilizations outside the social pair-bond) occurs regularly in most studied species.

While many instances of apparent cooperation are, after detailed investigation, explicable as the selfish motives of individuals, other forms of cooperation have proved more difficult to explain (for the hierarchical structure leading to cooperative breeding and for the different forms of cooperative breeding, see **Figure 1**). For example, in some species, individuals live and breed in bisexual groups of three or more adults and share in providing parental care at a single breeding attempt. Some of these adults are mature individuals that do not breed independently but instead act as helpers that care for young that are not their own genetic offspring, as in Florida scrub-jays (*Aphelocoma coerulescens*; **Figure 2**).

Typically, such cooperative breeding systems comprise family groups – for example, a breeding pair and their offspring – that live together on permanent, stable, all-purpose territories. However, cooperative breeding systems may also comprise individuals that help unrelated dominants to raise offspring. Systems may include a breeding pair assisted by nonbreeding helpers, as in gray wolves (*Canis lupus*) and naked mole-rats (*Heterocephalus glaber*; **Figure 2**). Although in most species cooperative breeding groups contain only a small number of helpers, in several species they may contain dozens, or even, as in the naked mole-rat, hundreds of helpers. Alternatively, a dominant pair may be accompanied by subordinates, of one or both sexes, that share reproduction with the dominant of the opposite sex, as in Lake Tanganyika cichlids (*Neolamprologus pulcher*; **Figure 2**), dwarf mongoose (*Helogale parvula*), meerkats (*Suricata suricatta*), superb fairy-wrens (*Malurus cyaneus*), and Seychelles warblers (*Acrocephalus sechellensis*; **Figure 2**). Among vertebrates, cooperative breeding is found in at least 9% of birds, 3% of mammals, and in some fish species, with a particularly high frequency of 19% in oscine passerine species, that is, the true songbird species, and in Australian birds and in primate species. Whatever the structure of the system (for the various forms of helping, see **Figure 1**), the fact that cooperative breeding systems consist of individuals that spend part,

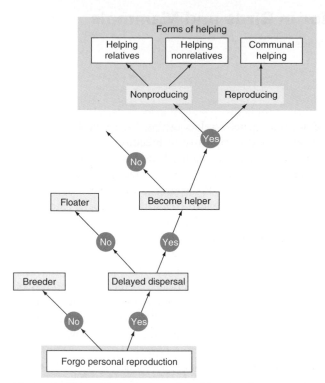

Figure 1 Hierarchical structure of decisions leading to cooperative breeding.

or all (i.e., naked mole rats), of their lives helping others to reproduce provides an intriguing evolutionary paradox. For example, when one of these subordinates has found food, why should it hand it over to one of the offspring of the dominant pair? How can we reconcile this behavior with selfish interests and the struggle for survival and reproduction?

The aim of this review is to discuss the evolution of cooperative breeding in vertebrates and to investigate how ecological and environmental factors may lead to such behavior. We discuss cooperative breeding behavior in relation to habitat use and quality, indirect and direct fitness benefits; resource competition and social interactions; social behavior and population dynamics; and interspecific interactions. Throughout, we use the term 'subordinates' for mature individuals that cohabit with the dominant breeding pair either as nonhelpers, nonreproducing helpers, or breeding helpers.

Evolution of Cooperative Breeding Systems: Kin-Selected Benefits

Although the existence of helpers within cooperative breeding systems has been known for many years, it was not until Hamilton and Maynard Smith developed the theory now referred to as 'kin selection,' that there was

a firm foundation for the empirical study of cooperative breeding. These authors argued that the total number of genes, identical by descent to its own, that are present in subsequent generations, determines the fitness of each individual. Consequently, nonbreeding subordinates can increase their fitness indirectly through enhancing the reproductive success of related individuals (which carry the same genes), the increased survival of recipient young, and through increased parental survival or future breeding success. There are multiple lines of evidence that the presence of helpers increases fledging success, food delivery rates, nestling growth rates, and parental survival. By helping a close relative reproduce, an individual is still passing copies of its genes on to the next generation indirectly. Cooperative behaviors that are costly to the actor and beneficial to the recipient are termed altruistic and indirect fitness effects are the best explanation for their evolution. Hamilton realized that their evolution would be affected by relatedness. In other words, alloparental care will be favored by selection only if $rb - c > 0$, where r is the genetic relatedness between the helper and the offspring helped, b is the fitness benefit to the offspring helped, and c is the fitness cost of helping. Hamilton's rule therefore predicts greater levels of cooperation when the benefits (b) or relatedness (r) is higher, and lower levels when the costs of behaving in such a manner are higher. These indirect, or kin-selected, benefits had been widely regarded as being of fundamental importance in the evolution of cooperative breeding systems, but such benefits can be accrued only if helpers assist their relatives. Helpers can accrue the largest indirect fitness gains by providing aid to the closest genetic relatives ('kin discrimination'). As a consequence, Komdeur and Hatchwell suggested that the ability to discriminate between individuals or groups of individuals and to assess the relatedness of social partners plays a major role in the evolution of social behavior. Only by responding differently to kin and nonkin can an individual maximize the fitness benefits delineated by inclusive fitness. Individuals should, therefore, possess some mechanism or rules for identifying their kin and assessing their relatedness to social partners.

It is clear that a degree of relatedness between the helpers and the helped is essential if indirect benefits are to be accrued. High relatedness between interacting individuals within groups can be a result of active kin discrimination – where an individual distinguishes relatives from nonrelatives and preferentially helps relatives – or merely a consequence of limited dispersal (see Section 'Access to Reproduction: Role of Ecological Constraints'). Active kin discrimination has been demonstrated in several cooperatively breeding vertebrates, such as long-tailed tits (*Aegithalos caudatus*; **Figure 2**). In this species, all mature individuals try to breed independently each year, but if their breeding attempt fails, these failed breeders may help raise offspring at another nest. Observations showed that helpers usually assist at the nest of a relative, and their help has a significant effect on nestling recruitment so that helpers accrue a substantial kin-selected indirect fitness benefit from their cooperation. However, an apparent kin preference could emerge simply by failed breeders becoming helpers at the closest available nest in a kin-structured population. In fact, this is not the case; an active choice of kin was demonstrated in an experiment where the success of breeding attempts was manipulated to offer potential helpers (i.e., failed breeders) the choice between equidistant broods belonging to kin and those belonging to nonkin: helpers preferentially helped kin.

Playback experiments by Sharp and colleagues have shown that long-tailed tits can discriminate between kin and nonkin on the basis of their vocalizations. The calls that provide cues for discrimination develop during the last few days of the nestling period and as adults, calls are individually distinctive. Moreover, the calls of siblings are more similar than those of nonsiblings. Partial cross-fostering of nestlings among unrelated broods showed that cross-fostered birds that survived to adulthood had calls that were more similar to their foster siblings and foster parents than to their true siblings reared apart or their true parents. Thus, the characteristics of calls that can be used as kin recognition cues are learned during development rather than being genetically determined. Furthermore, the recognition template must develop in a similar way because foster siblings did not discriminate between related and unrelated brood mates when deciding which to help as adults. This leads to a situation in which

Figure 2 Continued

Figure 2 Six species used in studies investigating the evolution of delayed dispersal and cooperative breeding (a–f) and the evolution of delayed dispersal and the absence of cooperative breeding (f) in vertebrates: (a) naked mole-rat (*Heterocephalus glaber*, photo: R.A. Mendez); (b) Lake Tanganyika cichlid (*Neolamprologus pulcher*). A breeding group defending their territory against the predatory fish *Lepidiolamprologus elongatus* (photo: M. Taborsky); (c) Seychelles warbler (*Acrocephalus sechellensis*, photo: D. Ellinger); (d) long-tailed tit (*Aegithalos caudatis*, photo: A. MacColl); (e) white-winged choughs (*Corcorax melanorhamphos*, photo: P. Thistle); (f) Florida scrub-jay (*Aphelocoma coerulescens*, photo: Mwanner); (g) Siberian jay (*Perisoreus infaustus*). Offspring do not provide alloparental care while in the family groups (photo: J. Ekman).

individuals tend to help relatives which they have been associated with during the nestling phase.

However, in most species helpers usually accrue lower mean indirect fitness returns by helping than they do by breeding independently. Indeed, in some systems where subordinates are unrelated to the dominant breeders, such indirect benefits are not possible at all. Thus, cooperative behavior can be seen as a best-of-a-bad-job strategy, adopted when opportunities for independent breeding are limited. Therefore, understanding why mature individuals do not, or cannot, breed independently is the key to understanding the evolution of cooperative breeding. As such, the evolution of vertebrate cooperative breeding systems can be viewed as a two-step process; first, the decision by mature individuals to join a group and forgo independent breeding, and second, the decision by subordinates in a group, to become helpers (**Figure 1**).

The first step should be the key to the formation of family units and is usually attributed to the existence of ecological constraints, such as a shortage of breeding territories or mates, that prevent offspring from independent breeding ('ecological constraints' hypothesis). The second step envisages that individuals that have already delayed dispersal can gain a net benefit through helping ('benefits of philopatry' hypothesis).

Access to Reproduction: Role of Ecological Constraints

The 'ecological constraints' hypothesis has been widely accepted as explanation for the evolution of delayed dispersal and group living. The logic behind the role of environmental conditions comes from the fact that remaining and not breeding paves the way for the formation of family groups while offspring may chose to wait for a breeding opportunity at their birth site. Offspring are expected to delay dispersal if the benefits they receive due to increased survival or increased probability of current or future reproduction exceed the benefits they would receive if they were to float or attempt to disperse to another site. A critical prediction of the constraints hypothesis is that helpers are making the best of a bad job and would become breeders if given the chance. Numerous observational studies have found a positive association between the severity of constraints – such as shortage of adequate breeding territories, high dispersal costs, a shortage of breeding partners – and the postponement of dispersal. However, these observational studies do not demonstrate causation. Cogent support for a causal role of 'ecological constraints' in delay of the reproductive debut is found in experimental studies in which constraints are artificially relaxed.

Experimental removal of breeders results in helpers of the same sex as the removed birds abandoning helping and moving to occupy the vacant breeding opportunities. For example, Walters and colleagues found that in red-cockaded woodpeckers (*Picoides borealis*) the provision of new nest sites induced helpers to leave home and establish breeding territories. In acorn woodpeckers (*Melanerpes formicivor*), experimental removal of the sole breeder of one sex created a 'power struggle' over the resulting reproductive vacancy involving a large number of philopatric offspring from other territories. The contests often lasted for several days. Similarly, in Seychelles warblers, the number of territories with subordinates increased as the habitat became saturated. Seychelles warblers were previously endangered because their range is restricted to a few small Seychelles islands. By 1940, anthropogenic disturbance had pushed this species to the verge of extinction and less than 29 individuals remained on the island of Cousin. In 1968, when the population consisted of just 26 individuals, habitat restoration programs were implemented. Over the following 30 years, the population grew impressively. By 1982, the population had grown to nearly 320 birds. No cooperative breeding was reported among Seychelles warblers until 1973, roughly the time at which all suitable breeding habitats become occupied. In essence, young Seychelles warblers delayed dispersal and stayed in their natal territories as the habitat became saturated with territories. To enhance the numbers of this endangered species, birds were introduced onto three nearby, previously unoccupied islands. To obtain birds for these transfers, breeding adults were removed from occupied territories on the original island. By this it was possible to create breeding opportunities. In the Seychelles warbler, removal of a breeder from a territory resulted in subordinates of the same sex rapidly moving in from other groups to fill the breeding opportunity.

Dispersal not only requires vacant breeding territories, but more generally it also requires other resources like breeding mates and access to food. For example, families of superb fairy wrens (*M. cyaneus*) consist predominantly of parents and grown sons. The creation of male breeding vacancies through removal of breeding males caused the dissolution of family groups, with mature-nonbreeding sons leaving home to fill these vacancies. However, when vacant territories without a breeding partner were created by removal of the breeding pair, male helpers did not disperse. Therefore, habitat and mate availability are both important constraints in these species. Dispersal as well as natal philopatry can also be induced through manipulations of food levels. In a unique experiment, the natural food resources were depleted in family territories of western bluebird (*Sialis mexicana*). In winter, western bluebirds are dependent mainly on berries of the oak mistletoe (*Phoradendron villosum*) for food. After removal of half of the mistletoe by volume, sons (the philopatric sex) left the depleted territory. Likewise offspring in a Spanish population of carrion crows (*Corvus corone corone*) that were given additional food (dog food) in their birth territory were more likely to delay dispersal. These experiments show that offspring are more likely to stay when territories are of higher quality in terms of food levels.

Cooperative Breeding: Adaptive Responses Under Individual Control

Yet, despite the aforementioned evidence for a causal role (referred to as proximate role) of 'ecological constraints' on limiting immediate dispersal, there should also be adaptive motives of staying (referred to as ultimate role). As such, Emlen further developed the 'ecological constraints' hypothesis. He suggested that dispersal and independent reproduction should be delayed only when it carries compensating fitness gains over the lifetime. In other words, the seeming contradiction between the facts that a postponement of dispersal is constrained by environmental conditions while it should simultaneously be an adaptive choice can be reconciled by the difference in perspective.

The 'ecological constraint' hypothesis was later refined by Stacey and Ligon with a new perspective on the importance of variation in habitat quality for the evolution of cooperative breeding. The 'benefits of philopatry' hypothesis they put forward predicts that young will stay on high-quality territories because the direct benefits of increased survivorship and access to high-quality territories, combined with indirect benefits due to helping, exceed the fitness expectations for individuals dispersing to breed independently on available low-quality territories. The idea was important because it focused on individual assessment and demonstrated that the decision to delay dispersal and help should be based not on the average fitness of helpers versus breeders, nor on the absolute availability of breeding habitat, but on the relative fitness consequences of the helping and breeding options available to an individual at any given point in time. In the 'benefits of philopatry' hypothesis, the benefits of helping and staying are no longer viewed separately. An individual that gains inclusive fitness benefits by helping on a high-quality territory should not move to a low-quality territory where its inclusive fitness benefits, through direct reproduction, will be comparatively low. Therefore, the delay in dispersal should be seen as a trade-off decision under individual control.

Direct comparisons of lifetime reproductive success of individuals delaying dispersal with that of individuals that do not delay dispersal provide the strongest evidence for an adaptive delayed dispersal. An understanding of the role of 'ecological constraints' for natal philopatry has been severely hampered by the fact that the consequences of behavior have rarely been assessed in a lifetime perspective. This is so because in most studies calculations of the total lifetime reproductive success of each individual are hampered because survival and reproductive estimates are confounded by dispersal outside the study population and the complex patterns of shared parentage of broods and extrapair parentage which may go unnoticed. However, those studies capable of addressing fitness values through the relative difference in the lifetime reproductive success rather than the proximate control of immediate dispersal show that the prospects from delaying dispersal may indeed not be bleak at all.

The early 1990s saw the first experimental evidence for the simultaneous importance of both habitat saturation and variation in habitat quality with work by Komdeur on the Seychelles warbler. In this species, the benefits, in terms of future breeding success and survival, of remaining on high-quality territories (high insect food abundance) as nonbreeders outweigh the benefits of independent breeding on lower quality territories (lower insect food abundance). Consequently, nonbreeding Seychelles warblers did disperse to take up a breeding vacancy only in territories classified as of high quality, and conversely nonbreeding individuals from high-quality territories rarely dispersed to fill vacancies in territories classified as of low quality. The translocations of individuals to previously unoccupied islands allowed this result to be experimentally verified. First, all of the offspring initially produced by translocated birds on 'unsaturated' Aride and Cousine dispersed from their natal territories as yearlings, and none became helpers. Thus, in the absence of habitat saturation, delayed dispersal and cooperative breeding simply did not occur. Only later, when all of the high-quality areas on Aride and Cousine became occupied, did young birds from the best

territories begin to remain on their natal territories and act as helpers. This occurred, even though there was abundant space in lower quality areas to establish territories and to breed independently. Second, in the original population on Cousin breeding vacancies created by breeder removals were filled immediately (some within hours) by formerly nonbreeding helpers, which dispersed from territories of the same or lower quality (**Figure 3**). In other words, nonbreeders from high-quality territories dispersed to fill vacancies on other high-quality territories, but did not fill medium- or low-quality vacancies because in the latter case helping on high-quality territories remained a better option than breeding on superior-quality territories. Similarly, nonbreeders from medium-quality territories never filled vacancies on low-quality territories (**Figure 3**). These results clearly demonstrate that dispersal decisions by warblers are influenced by the relative quality of both the natal and vacant territory. That the offspring can actually do better over life by delaying dispersal was also demonstrated for the Siberian jay (*Perisoreus infaustus*; **Figure 2**), the mountain gorilla (*Gorilla beringei beringei*), and the green woodhoopoe (*Phoeniculus purpureus*) (see Komdeur and Ekman, 2009).

These studies confirm that the 'benefits of philopatry' hypothesis can be accommodated within the 'ecological constraints' hypothesis by recognizing that these hypotheses differ only in the emphasis they place on either the costs of leaving or the benefits of staying. This realization has resulted in a more inclusive approach to investigating the evolution of delayed dispersal and helping, with

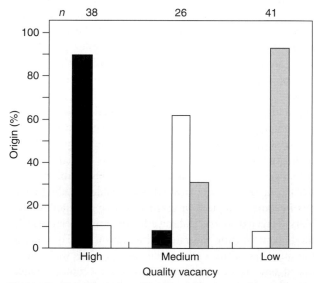

Figure 3 The effect of quality of breeding vacancies on Cousin Island on the origin of individual Seychelles warblers that filled the vacancies (*n*, number of vacancies; from Komdeur, 1992). Mean quality territory: gray bars, low; white bars, medium; black bars, high. Komdeur J (1992) Importance of habitat saturation and territory quality for evolution of cooperative breeding in the Seychelles warbler. *Nature* 358: 493–495.

Cooperative Breeding in the Absence of Kin-Selected Benefits

Until the late 1990s, the indirect, or kin-selected, benefits have been widely regarded as being of fundamental importance for the evolution of cooperative breeding systems. However, the case for kin selection in the evolution of cooperative breeding systems is not as strong as once considered, especially for vertebrates. First, although most social vertebrate groups consist of relatives, it is not clear that relatedness is consistently higher in cooperative breeders than in species that live in stable, but noncooperative, groups. Second, in many systems the amount of help given does not vary with the relatedness of the subordinates and in some systems totally unrelated helpers often occur within groups. Third, the magnitude of indirect fitness benefits relative to direct fitness benefits may have been overestimated because of factors such as a failure to recognize the costs of competing with kin, the confounding effect of individual or territory quality, or an underestimation of the extent of cobreeding by subordinates. Finally, extragroup paternity (with young sired by males from outside the group) occurs in many cooperative systems. This complicates our understanding of cooperative breeding as extragroup parentage will reduce relatedness between subordinates and the offspring they help to raise and consequently, the potential indirect benefits of helping.

Given that the case for kin selection for helping behavior is not as strong as once considered, there has been a trend of shifting the focus of family living away from relatedness, r, in Hamilton's rule toward the effects of costs, c, and benefits, b. As a consequence, an adaptive explanation is still required for those cooperative breeding societies where related subordinates do not help, or where helpers are unrelated to the young but still invest as heavily as close relatives. In some cases, helpers actively compete for access to unrelated offspring, as shown in the meerkat, the dwarf mongoose, the Lake Tanganyika cichlid, the African wild dog (*Lycaon pictus*), the superb fairy-wren, and the stripe-backed wren (*Campylorynchus nuchalis*). This competition to feed unrelated young may result in adoption, involving recruitment and care of dependent young from another group.

Adoption has been seen in the Florida scrub jay (*A. coerulescens*) and the Arabian babbler (*Turdoides squamiceps*). An extreme form of adoption, 'kidnapping,' occurs when adults herd young from another territory into their own territory. This has been seen in white-winged choughs (*Corcorax melanorhamphos*; **Figure 2**).

Adult 'kidnappers' fed the fledglings they adopted, and these young later become unrelated helpers in their new groups. Another oft-cited explanation for care being given by helpers to nonkin is that helpers may foster the formation of 'social bonds' with recipient young, bonds that later benefit the helper either by increasing the probability that the young will return the favor, or by promoting the development of coalitions beneficial in competing for breeding positions. For example, white-winged choughs require helpers to reproduce successfully; selection may have favored kidnapping because the resulting 'special bonds' cause kidnapped young to help their kidnappers (Heinsohn, 1991). Thus, by feeding unrelated offspring, helpers may parasitize a kin-recognition mechanism based on associative learning; the deceived offspring recognize those that care for them as kin and later help rear their provisioner's offspring.

An alternative hypothesis to explain why helpers might provision unrelated offspring is that helpers increase their own survival and future reproduction by cooperating with others (direct fitness gains). For example, helping may lead to an increase in overall group size which, because larger groups are better at competing with other groups or deterring predators, increases the survival of all group members, including the subordinates. Such group augmentation does not require kinship within cooperative groups. Therefore, it pays to recruit new members by increasing group productivity or even by 'kidnapping' members of other groups. Another direct benefit that may be gained by helping to raise offspring is the accumulation of breeding experience, which allows individuals to be more productive when they gain a breeding position themselves (as observed in the Seychelles warbler). Alternatively, individuals may help only to avoid being evicted by the dominants ('pay to stay' hypothesis), whereby they would lose the benefits associated with remaining in a group while waiting for future breeding opportunities. Indeed prolonged residency within a group can also enhance the probability that an individual ascends to the dominant breeding position itself, or can allow subordinates to gain resources through 'budding off' of a portion of the territory.

The important point is that in situations where helpers gain some direct fitness through their cooperative behavior, the ability to recognize and discriminate kin from nonkin is not necessarily a prerequisite for the evolution of helping behavior. For example, in several societies, subordinates not only maintain social relationships and helping activities with members of their own group but also (temporarily) leave groups to join other unrelated groups nearby and become helpers there. In such cases, the goal of helping may be the establishment of familiarity and social relationships with individuals from other territories. Subordinates use these neighboring groups' territories as safe havens when the risk of staying in the home territory increases, and may successfully migrate into other groups. This suggests that subordinates may be prepared to risk expulsion because other groups are available to disperse to, and they may strategically choose which groups to join and which breeders to help. If dominants gain fitness by accepting additional helpers, helpers might trade their helping contribution for being accepted in a territory that provides beneficial conditions.

Remaining in the Group: The Role of Dominants

Although the assumption is that higher numbers of subordinates have a positive effect on the reproduction or survival of the group, there may also be disadvantages of living in larger groups. The decision to remain or disperse is not, however, a unilateral one. Subordinate group members may compete with dominant group members for mates or food. Large groups may attract more predators, and may have a higher risk of parasite or disease infection and increased competition for food, which may lead to reduced survival and reproduction of group members. On the other hand, dominant group members may force subordinates to disperse when the costs of having subordinates in the group exceed the benefits. Whether or not eviction occurs, and who is targeted is, at least in part, dependent on relatedness between group members.

A number of studies of parent behavior in family groups have documented that parents are more tolerant toward offspring than toward nonkin group members. Dominant individuals may be less tolerant toward independent young to which they are less related. In other words, dominants that are unrelated to the group young are more likely to evict young. This is most obvious when territories get taken over by new dominants, as for example in African lions (*Panthera leo*) and white-faced capuchins (*Cebus capucinus*), where takeovers result in the immediate killing or eviction of unrelated young. This is not to say that parent–offspring relations are without conflict or that related dominant individuals are always tolerant. For example, in the superb fairy-wren, any female offspring still in the natal territory at the start of the next breeding season are forced by the mother to disperse. Overall it is clear that in cooperative breeding systems those individuals that remain on natal territories may only be able to do so because the dominant group members allow them to. Yet, data on such parental tolerance and its implications for family living are still limited in comparison with the effort devoted to communal breeding. One reason for this paucity of data could be that such tolerance is taken for granted as a part of parental care, but it is also nonbehavior being characterized by the absence of aggression and as such easily overlooked.

That members of cooperatively breeding groups are often related may also affect dispersal in another way. When a dominant individual dies or disappears, group young of the same sex may attempt to occupy the vacant breeding position. If the parent of the opposite sex is still alive and dominant, this pattern of inheritance would result in incestuous mating, which may lead to decreased fitness. To avoid this, young should be selective when it comes to territory inheritance. In Florida scrub-jays, territory inheritance occurs more often when the surviving breeder is a stepparent of the potential heir rather than its natural parent. Similarly, males of *Antechinus agilis*, a small-sized dimorphic carnivorous marsupial, are more likely to remain philopatric if the mother is removed at the time of weaning. In acorn woodpeckers, red-cockaded woodpecker and superb fairy-wrens, dominants have even been observed to give up their breeder positions when all the opposite sex members of the group are closely related to them. These results show that incest avoidance can lead to individuals giving up breeding opportunities and therefore, to increased dispersal in an attempt to find other breeding vacancies with unrelated individuals. At a less facultative level, the evolution of general sex-biased patterns of dispersal may also have evolved to avoid inbreeding. Indeed although sex-biased dispersal is almost ubiquitous in vertebrates, it appears to be especially pronounced in cooperative breeders.

An alternative way that females can avoid incestuous matings with related group members is through extragroup paternity, which, as stated earlier, can be relatively common. However, this means that in cooperative groups there can be a conflict of interest between dominant females that wish to gain extragroup paternity, and the dominant males and helping subordinates of either sex that may gain kin-selected helping benefits by protecting within-group paternity. How these conflicting interests are resolved within groups remains undetermined.

Conclusions and Future Directions

This review has described and discussed empirical studies to identify the proximate and ultimate causes underlying the occurrence of cooperative breeding and their evolutionary consequences. Association in family groups arises as the offspring delay dispersal and in doing so, they are often assumed to incur an evolutionary cost in lost personal reproduction. Selection for associating in families has been studied extensively for philopatric offspring participating in brood rearing, and the current evidence indicates that alloparental care is a selective trait offering inclusive fitness gains. However, there are many species in which young delay dispersal and live in groups but do not provide alloparental care. Brown suggested that the greatest insights into cooperative breeding would come from comparisons of species in which delaying dispersal and delaying breeding on the one hand and helping on the other are uncoupled. So far, species exhibiting delayed dispersal of offspring that do not help have attracted little attention. There are only a few in depth studies that have revealed that the offspring forego personal reproduction to associate with breeding parents and yet do not provide alloparental care.

Parents sometimes actively prevent offspring from approaching the nest. For example, in the Siberian jay, breeding success is poor because of high risk of predation by nest predators, mainly corvids. Unlike cooperative breeders, Siberian jays, which may stay with their family for a couple of years, do not gain indirect benefits or take part in the care of younger siblings hatched from consecutive broods. This is so because given the high risk of nest predation, activity around the nest is a main threat to reproductive success and retained offspring are actively prevented from approaching the nest. These retained offspring do receive other benefits. While in company with their parents, retained offspring enjoy survival benefits because parents provide protection against predators and tolerance in sharing of food. Furthermore, once retained offspring become breeders, they obtain better territories and enjoy higher lifetime reproductive success than their siblings that did not stay in the family unit. In general, species with delayed dispersal of offspring that do not provide alloparental care have received little attention, because such young do not exhibit the apparent altruism that has been the focus of research into cooperative breeding in the past decades. Any inclusive fitness benefits of helping would certainly augment benefits of delayed dispersal but they appear to be neither necessary nor sufficient to explain why dispersal should be delayed.

The two questions 'why stay?' and 'why help?' represent two independent behavioral decisions, though the costs and enefits of one decision may affect the potential fitness consequences of other decisions. Studies of species with delayed dispersal that do not help allow investigators to analyze the specific fitness consequences of delayed dispersal without the confounding fitness consequences of helping behavior. This topic should receive more attention.

See also: Cooperation and Sociality; Group Living; Kin Selection and Relatedness; Subsociality and the Evolution of Eusociality.

Further Reading

Bergmüller R, Heg D, Peer K, and Taborsky M (2005) Extended safe havens and between-group dispersal of helpers in a cooperatively breeding cichlid. *Behaviour* 142: 1643–1667.

Brown JL (1987) *Helping and Communal Breeding in Birds.* Princeton, NJ: Princeton University Press.

Clutton-Brock TH (2002) Breeding together: Kin selection and mutualism in cooperative vertebrates. *Science* 296: 69–72.

Cockburn A (1998) Evolution of helping behavior in cooperatively breeding birds. *Annual Review of Ecology and Systematics* 29: 141–177.

Dickinson JL and McGowan A (2005) Winter resource wealth drive's delayed dispersal and family-group living in western bluebirds. *Proceedings of the Royal Society B: Biological Sciences* 272: 2423–2428.

Ekman J, Bylin A, and Tegelstrom H (1999) Increased lifetime reproductive success for Siberian jay (*Perisoreus infaustus*) males with delayed dispersal. *Proceedings of the Royal Society of London Series B: Biological Sciences* 266: 911–915.

Ekman J, Sklepkovych B, and Tegelstrom H (1994) Offspring retention in the Siberian jay (*Perisoreus infaustus*) – the prolonged brood care hypothesis. *Behavioural Ecology* 5: 245–253.

Emlen ST (1982) The evolution of helping. 1. An ecological constraints model. 2. The role of behavioral conflict. *American Naturalist* 119: 29–39.

Emlen ST (1997) Predicting family dynamics in social vertebrates. In: Krebs JR and Davies NB (eds.) *Behavioural Ecology: An Evolutionary Approach*, pp. 228–253. Oxford: Blackwell Scientific Publications.

Griffith SC, Owens IPF, and Thuman KA (2002) Extra-pair paternity in birds: A review of interspecific variation and adaptive function. *Molecular Ecology* 11: 2195–2212.

Hamilton WD (1964) The genetical evolution of social behaviour. *Journal of Theoretical Biology* 7: 1–52.

Hatchwell BJ and Komdeur J (2000) Ecological constraints, life history traits and the evolution of cooperative breeding. *Animal Behaviour* 59: 1079–1086.

Hawn AT, Radford AN, and Du Plessis MA (2007) Delayed breeding affects lifetime reproductive success differently in male and female green woodhoopoes. *Current Biology* 17: 844–849.

Heinsohn RG (1991) Kidnapping and reciprocity in cooperatively breeding white-winged choughs. *Animal Behaviour* 41: 1097–1110.

Koenig WD and Dickinson JL (eds.) (2004) *Ecology and Evolution of Cooperative Breeding in Birds.* Cambridge: Cambridge University Press.

Komdeur J (1992) Importance of habitat saturation and territory quality for evolution of cooperative breeding in the Seychelles warbler. *Nature* 358: 493–495.

Komdeur J and Ekman J (in press) Adaptations and constraints in the evolution of delayed dispersal. In: Székely T, Moore AJ, and Komdeur J (eds.) *Social Behaviour: Genes, Ecology and Evolution.* Cambridge: Cambridge University Press.

Maynard Smith J (1964) Group selection and kin selection. *Nature* 201: 1145–1147.

Maynard Smith J (1977) Parental investment: A prospective analysis. *Animal Behaviour* 25: 1–9.

Robbins AM and Robbins MM (2005) Fitness consequences of dispersal decisions for male mountain gorillas (*Gorilla beringei beringei*). *Behavioural Ecology and Sociobiology* 58: 295–309.

Russell AF and Hatchwell BJ (2001) Experimental evidence for kin-biased helping in a cooperatively breeding vertebrate. *Proceedings of the Royal Society B-Biological Sciences* 268: 2169–2174.

Sharp SP, McGowan A, Wood MJ, and Hatchwell BJ (2005) Learned kin recognition cues in a social bird. *Nature* 434: 1127–1130.

Solomon NG and French JA (eds.) (1997) *Cooperative Breeding in Mammals.* Cambridge: Cambridge University Press.

Stacey PB and Koenig WD (eds.) (1990) *Cooperative Breeding in Birds: Long-Term Studies of Ecology and Behaviour.* Cambridge: Cambridge University Press.

Stacey PB and Ligon JD (1987) Territory quality and dispersal options in the acorn woodpecker, and a challenge to the habitat saturation model of cooperative breeding. *American Naturalist* 130: 654–676.

Walters JR, Copeyon CK, and Carter JH (1992) Test of the ecological basis of cooperative breeding in red-cockaded woodpeckers. *Auk* 109: 90–97.

Woolfenden GE and Fitzpatrick J (eds.) (1984) *The Florida Scrub Jay: Demography of a Cooperative-breeding Bird.* Princeton, NJ: Princeton University Press.

Herring Gulls

J. Burger, Rutgers University, Piscataway, NJ, USA

Introduction

In this article, Herring Gulls serve as a model for the study of behavior, ecology, and the effects of human disturbance and contaminants on the species. Increasingly, public policy-makers, managers, biologists, environmentalists, and the public are interested in assessing the health of our environment, including species, populations, and communities. Yet most habitats are composed of hundreds or thousands of species, making it impossible to determine the health status and populations trends of even a small proportion of the species within any habitat. While it is possible to describe the behavior, ecology, and population biology of an array of different species, it is often useful to examine a range of attributes in one or two species within a habitat, using them as bioindicators of the overall health of the ecosystem. Bioindicators should serve as indicators of the health of the individuals and populations of that species, as well as of its prey, competitors, and predators.

Bioindicators that provide information that is useful for assessing both ecological health and human health, and that can be models for anthropogenic effects are particularly useful. Ideally, bioindicators are sufficiently common so that their use does not jeopardize their populations, and are sufficiently widespread that they can be used to provide information over a wide geographical range.

Herring Gulls as a Model

Herring Gulls are ideal as bioindicators of ecosystem and environmental health, and as model organisms to study ecology, behavior, and the effects of environmental variables and contaminants because they are diurnal, common, abundant, large, and long-lived, as well as they nest in colonies over a wide geographical distribution. Because they are diurnal, common, large, and nest in colonies, they can be studied easily without unduly disrupting their populations; because they are long-lived they make ideal subjects for studies of age-related differences in behavior, reproductive success, and biomagnification of contaminants; and because they have wide geographical distribution they can be used to assess differential exposures and effects of environmental and anthropogenic factors (**Figure 1**).

Long-term and comparative studies of Herring Gulls are being conducted in many parts of the world, including in the Great Lakes, New Jersey, and in Europe. Studies by my research group in New Jersey will be used to illustrate the kinds of studies that can be conducted with Herring Gulls to examine behavior, ecology, and the effects of human disturbance and contaminants on behavior and ecology. These studies provide paradigms for the study of Herring Gulls elsewhere, other birds and other vertebrates elsewhere, and in some cases, serve as models and sentinels for understanding the exposure and effects of contaminants on human behavior.

Relevant Life History

Herring Gulls are large, white-headed gulls that nest in colonies ranging in size from only a few individuals to many hundreds of pairs, although occasionally pairs nest solitarily. Pierotti and Good's account in the Cornell Laboratory of Ornithology's *Birds of North America* provides basic information about the biology of this species. They primarily live along shorelines of oceans, seas, large rivers, and lakes, such as the Great Lakes of North America and migrate south in the winter. Herring Gulls generally nest near water in places that are sheltered from inclement weather and storms, and are safe from predators, such as islands, offshore rocks, abandoned piers, cliff ledges, and even on the roofs of buildings in some cities. They are highly territorial, and some birds even prey on the eggs or chicks of neighbors, of their own and other species. This species was nearly extirpated in some parts of North America by plumage hunters and eggers during the nineteenth century, but they have recovered their populations due mainly to protection and the food provided by garbage dumps in the twentieth century.

Its circumboreal breeding range includes much of North America, Central Asia, and Europe. They are not long-distance migrants, some breeding birds remain relatively near their breeding colonies throughout the year, while others migrate a few hundred miles south. In Europe, they are considered nonmigratory. There is, however, postbreeding dispersal away from the breeding colonies, presumably to decrease foraging competition. Although there is size dimorphism within pairs (male on average weigh 1050–1250 g; female on average weigh 800–980 g), the overlap often makes sexual identification

Figure 1   Picture of Herring Gull with young. Photo by Joanna Burger.

Table 1   Effect of vegetation type on nesting behavior of Herring Gulls in New Jersey (6 years of study)

Vegetation type	Mean date of laying (depending upon year)	Hatching success (fledging success/nest)
Dense bushes	20 April to 2 June	2.4 ± 0.8 eggs (1.6 ± 0.9 chicks)
Sparse bushes	23 April to 5 June	2.3 ± 0.8 eggs (1.7 ± 0.5 chicks
Dense grass	25 April to 6 June	2.0 ± 1.0 eggs (1.0 ± 1.3 chicks)
Sparse or low grass	1 May to 12 June	1.8 ± 1.3 eggs (1.2 ± 0.9 chicks)

Source: Burger, J and Shisler, J (1980). The process of colony formation among Herring Gulls *Larus argentatus* nesting in New Jersey. *Ibis* 122: 15–26.
Burger, J (1984). Pattern, mechanism, and adaptive significance of territoriality in Herring Gulls (*Larus argentatus*). *Ornithological Monographs* 34: 1–92.

difficult. Color-marking or banding are essential to identify individuals, and morphometric measurements (weight, bill length, and width) are usually sufficient to identify sex.

Herring Gulls are opportunistic foragers that feed primarily on fishes and invertebrates. Foraging habitat typically is spatially separate from nesting habitat; they nest on land and forage in nearby bays, estuaries, lakes, or the ocean. They forage at sea, in intertidal, on sandy beaches and mudflats, in refuse dumps and ploughed fields, and around picnic areas or fish-processing plants. In general, there are age-related differences in foraging success, with young of the year being less efficient and foraging in less difficult situations.

Habitat Selection

Herring Gulls normally nested in the northern regions of North America, but in the early 1900s they moved into Massachusetts and then into Long Island. The first Herring Gull nested in New Jersey in 1948. The expanding population, which has continued to the present, allowed for an examination of habitat selection and interactions with native species. Herring Gulls nest in a range of habitats, but in coastal New Jersey they have moved into salt marshes, nesting in the higher areas of *Spartina* or on wrack (dead *Spartina* and eelgrass strews in windrows on the high marsh). The use of salt marshes was a new adaptation as a result of lack of other suitable habitats, such as rocky islands or cliffs, although they nest on sandy islands where they are available. On salt marsh islands there is a premium for nesting on the highest spots, because these are the last to flood during high or storm tides (**Table 1**).

Behavioral observations over a 4-year period indicate that there is competition between Herring Gulls and other species that nest with them. In general, larger species (e.g., Great Black-backed Gull, *Larus marinus*) win over Herring Gulls, and they in turn can win over smaller species (such as Laughing Gulls, *Larus atricilla*). The success of Herring Gulls is partly due to their arrival earlier on the breeding grounds than Laughing Gulls, but is also a function of the larger species winning in aggressive encounters. Selecting the best sites for nests often involved choosing the highest places, which reduces the chance of reproductive losses due to high tides and storm tides. Thus, Herring Gulls succeed in using the highest locations, which are located in *Spartina patens* or near bushes, which has forced Laughing Gulls into lower sites, which increased their losses due to high tides. Over the last 30 years, salt marsh islands in Barnegat Bay (New Jersey) once inhabited only by Laughing Gulls have been taken over completely by Herring Gulls, and Laughing Gull populations have declined.

Natural changes in habitat due to storms or high tides often resulted in decreasing available habitat, especially on vulnerable salt marsh islands, or in small sandy islands. If the changes are not severe, however, the gulls simply nest more densely. Unfortunately, sandy islands, or sandy beaches on salt marsh islands, are preferred by recreationists, and such disturbance is detrimental to nesting gulls by keeping them from incubating their eggs or protecting young. With increased human use, the gulls eventually abandon these sites, or are crowded into smaller places.

Anthropogenic changes to salt marshes, such as ditching for mosquito control, resulted in Herring Gulls avoiding these islands in favor of natural marshes. The continued ditching and open marsh management (a procedure where ditches and pools are dug to mimic the natural marsh) allowed for a natural experiment of the effects of habitat manipulation on Herring Gull nesting behavior. Pairs that had previously used islands that were later ditched continued to do so, but new prospecting pairs avoided the ditched islands, perhaps because the spoil piles were unvegetated, and there were fewer bushes (the preferred habitat because such areas are the highest ones on the marshes). Colony formation usually involves the formation of epicenters where a few pairs nest densely, and subsequent pairs choose to nest near these clusters. Initially, the gulls avoided the unvegetated spoil piles (dirt thrown up on the marsh by ditching) in favor of other nearby islands that had bushes and no bare soil. However, in successive years, *Iva* bushes moved onto the spoil piles, and these then became preferred places, because they were both higher than other places on the marsh, and had bushes that provided cover from predators and from inclement weather.

Territory

Herring Gulls are territorial, and their territory size depends on season, reproductive stage, habitat, number of birds in the colony (and the available space), and the presence of other species (and whether they are larger or smaller). They maintain three types of territories: (1) a unique territory, the smallest one, that is defended against all intruders at all times; (2) a primary territory that is defended against all conspecifics (and some other species); and (3) a secondary territory that is defended against neighbors. Territory size is elastic, in that it is large in the preincubation phase when they are initially defending space, is smallest when they are incubating eggs, and expands during the chick phase to be larger than it was during the preincubation phase. Thus to some extent, territory size reflects what they are protecting; it is possible to enlarge their territory size after incubation because some pairs fail, leaving open space.

There is a complex relationship between aggressive interactions, territory size, and reproductive success in Herring Gulls. The most successful pairs (those raising some chicks), had intermediate-sized territories, and engaged in less aggression that those raising no chicks (**Figure 2**). That is, pairs that spent much of their time defending the territory either had smaller ones, because there was such intense competition that they spent all their time defending it (mainly from close neighbors), or their territory was so large that many different pairs tried to usurp some of the space. When pairs were very aggressive toward other gulls, they did not adequately defend their eggs or chicks from predators (other gulls, crows) or protect them from inclement weather. In general, pairs that raised three chicks engaged in an average of less than one aggressive interaction per hour, those that raised three chicks engaged in less than 1.3 aggressive interactions per hour, and those that raised one chick engaged in up to 2.6 aggressive interactions per hour.

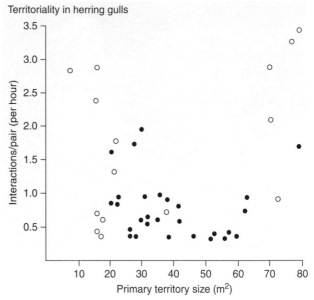

Territoriality in herring gulls

Figure 2 Relationship of reproductive success (open circle = nonfledged, solid circle = some fledged) to primary territory size and rate of aggression for 43 Herring Gull pairs observed in intermediate bush habitat on Calm Island. Reproduced from Burger, J (1984) Pattern, mechanism, and adaptive significance of territoriality in Herring Gulls (*Larus argentatus*). *Ornithological Monographs* 34, 1–92, with permission from Elsevier.

Effects of Human Disturbance on Behavior and Reproductive Success

Because Herring Gulls are so common, and nest in colonies (with and without other species), it is possible to examine the effects of people on behavior and reproductive success. As with other aspects of Herring Gulls behavior and ecology, such studies can be used to test hypotheses that help manage them, as well as other species, including threatened and endangered species. Upon approaches by people, Herring Gulls usually become alert, then stand, and eventually fly from the nest or chicks when people approach too closely. Herring Gulls then engage in active defense of their eggs and chicks by mobbing and dive-bombing intruders, both humans and predators. Leaving the nest exposes vulnerable eggs and chicks to predation and weather stresses (either cold or hot), and thus can lead directly to nest failure. When, for example, two or three Crows are in a colony, while the gulls are mobbing one, another may successfully eat eggs or chicks. Similarly, when gulls are mobbing people or a predator, other Herring Gulls may eat eggs or chicks.

In Herring Gulls, continued exposure leads to habituation, where gulls respond less quickly to people, fewer leave the nest, and with time, gulls mob humans more intensely, even hitting them on the head. Habituation to predators, however, such as dogs, does not occur, mainly because the direct threat has not decreased. That is, people usually cause no direct harm to the gulls, their eggs or chicks, whereas a dog may eat or trample eggs and chicks (and might kill an adult that remained). It should be remembered that this is not always the case. Egging was common in North America and Europe in the late 1800s and early 1990s, and is still common today in some places, especially by Native populations or by immigrants to the United States who collected eggs in their native countries.

Remarkably, gulls are able to discriminate a direct versus a tangential approach. Gulls respond when a person is greater distance away if the person is facing them and on a trajectory that would take them directly to a nest, in comparison to a person that is on a path to take them by the nest, even if they will only be a meter away. This fine discrimination allows them to remain on the nest, and not waste energy flying when there is no direct threat from people.

Bioindicators of Environmental Contamination

Because Herring Gulls are so common and widespread, they can be used as bioindicators of environmental contaminants, and have been used extensively to track seasonal, yearly, and geographical differences in a wide range of contaminants, including organochlorines and heavy metals. Contaminants have been examined in eggs, feathers, and a wide range of internal tissues. Feathers and eggs have proven particularly useful as bioindicators of contaminants, because they can be collected with little effect on population dynamics. Since gulls normally lay three eggs, but fledge only one or two chicks, the removal of one egg for contaminant analysis does not affect reproductive success. Similarly, feathers can be collected without injuring the bird. Because there are age-related differences in plumage patterns, differential accumulation can be examined in young, juveniles, and adults.

Selection of the tissue for analysis of contaminants provides information on local contamination versus distant contamination and the season of exposure, as well as information about individuals and pairs. For example, contaminants in eggs reflect exposure of the female parent and usually reflects rather local exposure if the female was on the breeding grounds for a few weeks prior to egg-laying. Metals in the feathers of young birds in the nest reflect some exposure from the egg as well as local exposure since the parents had to collect the food they fed them locally, feathers from fledglings reflect local exposure because as recently (or almost-fledged) chicks they grew their feathers while their parents were feeding them, and these feathers (as opposed to down) reflect little exposure from the egg (the chick has grown so much since hatchling that any residual has been swamped by growth). Metals in the feathers of adults reflect where they were when they last molted (often on the wintering ground), and metal levels in primaries can be used to identify temporal patterns of exposure, because they mold them in sequence. Internal tissues can be used to evaluate a wide range of contaminants, and each tissue is known to integrate exposure over different time frames. That is, blood usually represents recent exposure, while internal tissues reflect exposure over a longer period of time.

Effects of Contaminants on Behavior

Herring Gulls are ideal models to study the effect of contaminants on behavior and reproductive success, because young birds can be maintained in the laboratory, and experiments can be conducted both in the laboratory and in the field. One of the most important tools in understanding the effect of contaminants on wildlife is to be able to determine contaminant levels in the field, determine effects that specific doses cause in the laboratory, and correlate the specific doses that cause effects with levels in specific tissues. Understanding these three aspects allows managers, biologists, and the public to effectively determine the significance of levels found in wild populations.

Understanding the effects of contaminants on behavior and reproductive success thus requires controlled laboratory experiments, correlating dose and associated effects with levels in tissues, and conducting controlled field experimentation to validate the laboratory studies. Herring Gulls proved an ideal model for the studies described in the following section, conducted mainly with lead, a contaminant of some concern in many estuarine and coastal waters. Further, exposure to lead remains a significant concern for people residing in cities, partly as a function of the past use of lead paint and lead in gasoline.

In the laboratory experiments, the level of lead used was selected on the basis of examining lead levels found in feathers of young in nature in New York and New Jersey, and dosing young in the laboratory until the dose produced the levels found in nature. Therefore, unlike most laboratory studies with contaminants where higher doses are given to obtain an effect so that dose–response curves can be determined, these studies used lead levels that were directly relevant to the exposure of birds in the wild. Thus lethal effects were not expected. On the contrary, sublethal effects might occur in nature, and these might directly affect survival or reproductive success. These experiments were then followed by field experiments where one chick in each nest was injected with a dose of lead that was used for the laboratory experiments, one was injected with a saline solution (an injection control), and the third was not injected. Herring Gulls normally lay three eggs, and treatment was random with respect to laying order. Behavior was then observed by an observer blind to treatment, and the gulls were generally not disturbed.

The sublethal behavioral deficits and morphological effects of lead exposure in Herring Gulls in laboratory studies included growth abnormalities, begging response, feeding behavior, feeding response time, righting response, avoidance of heat stress, sibling recognition, exercise ability, endurance, perception, learning, and expression of synaptic neural cell adhesion molecules in the brain; these findings are summarized in a review by Burger and Gochfeld. All of these characteristics affect the survival of young in nature. This is one of the primary advantages of working with behavioral and morphological effects of contaminants in birds – the behaviors that can be examined both in a laboratory and a field situation relate directly to survival. While these experiments are time consuming and labor intensive, they nonetheless provide information on important behavior, ecological, and evolutionary traits leading to survival, and eventually to reproductive success differences.

In the laboratory, young chicks exposed to lead are less able than control birds to accomplish the following: (1) right themselves quickly when turned on their backs, which directly affects their ability to respond quickly to falling down sand dunes or logs in the wild; (2) locate and remain in shade when exposed to full sun, which would expose them to potentially damaging heat stress; (3) avoid falling off a cliff, exposing them to injury when faced with an incline or cliff face (Herring Gulls sometimes nest on rock cliffs or buildings); (4) identify food as early, thus rendering them less competitive for limited food resources; (5) beg as vigorously to stimulate food provisioning by parents, or to obtain the food parents bring back, which could lead to starvation; (6) walk as steadily or walk on a narrow balance beam, which would affect their ability to remain on narrow cliff ledges during development; (7) learn in a series of trials where the food was located when it was covered with a cup, which affects their ability to quickly find food in a competitive situation; (8) learn to walk on a treadmill or to maintain endurance, which would affect their ability to run from predators, avoid humans, or follow parents that were changing territory locations; and (9) identify siblings, which in nature, would allow them to return correctly to their own nest rather than that of a neighbor. These experiments suggested that, in nature, lead-exposed chicks would be less able to survive to fledging (see Burger and Gochfeld, 2000, 2005, and references therein).

One of the key questions in ecotoxicology is whether effects identified in the laboratory exist in the wild, whether they are as severe, and whether they directly affect survival. In the field, the same doses of lead resulted in similar effects, and the effects were still evident 3 weeks

after injection, although some effects were less severe with time (**Figure 3**). For example, the differences in begging ability between experimental and control chicks remained relatively constant, while they lead-injected chicks improved in their walking ability and their failure to beg from their parents.

In comparing the results from laboratory experiments with those conducted in nature revealed that field effects were more severe for some things (number of hits on their parents bill while begging for food, walking and number of falls), they were less severe for the percent of misses

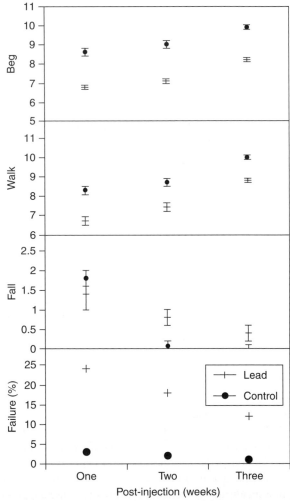

Figure 3 Behavioral responses of herring gulls in the wild. Shown are mean scores (± standard error) of lead-injected and control chick (all significantly different, dose = 100 mg kg^{-1}. Reproduced from Burger, J and Gochfeld, M (1994). Behavioral impairments of lead-injected young Herring Gulls in nature. *Fundamental and Applied Toxicology* 23: 553–561; Burger, J and Gochfeld, M (1997). Lead and neurobehavioral development in gulls: A model for understanding effects in the laboratory and the field. *NeuroToxicology* 18: 279–287. Walking and begging ability were scored on a scale of 10 (the highest score). Fall refers to the number of times a chick fell when walking a 1-m distance; failure refers to the percent of times a chick missed its parent's bill when pecking for food.

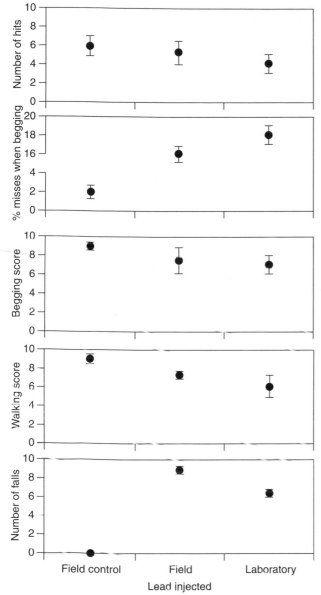

Figure 4 Comparison of behavior herring gulls raised in the field (lead-injected and controls) with lead-injected herring gulls raised in the laboratory (all injections were at 2 days of age, dose = 100 mg kg^{-1}). In all cases the lead-injected chicks differed from the field controls. Reproduced from Burger, J and Gochfeld, M (1994). Behavioral impairments of lead-injected young Herring Gulls in nature. *Fundamental and Applied Toxicology* 23: 553–561; Burger, J and Gochfeld, M (1997). Lead and neurobehavioral development in gulls: A model for understanding effects in the laboratory and the field. NeuroToxicology 18: 279–287. Hits are successful pecks at parent's bill.

while begging (**Figure 4**). On the whole, however, the experiments indicated that the effects found in the laboratory did indeed occur in the wild, at about the same intensity. These deficits in the field clearly indicated that the young were impaired, which in the laboratory resulted in the lead-injected chicks being severely underweight at

fledging, despite being individually fed. However, in the field, the adults compensated for the inability of lead-injected chicks to beg and feed as well as their siblings by splitting up the brood for feeding. That is, one parent would feed the lead-impaired chick, which the other fed the other two chicks, which were begging furiously and able to obtain the food rapidly from their parent's bills. This behavioral compensation by the parents was surprising, and resulted in there being very little difference in weight of the lead-injected and control chicks at fledging in nature. However, these experiments were conducted in a year when food was abundant, in a bad food year, parents may be unable to conduct split feeding of their brood.

The inability to recognize siblings on the part of lead-injected chicks, however, led directly to survival differences. In Herring Gulls, parental recognition by chicks occurs at the time they begin to walk freely around their territory, and to wander into those of their neighbors. Such wandering occurs with regularity when the colony is disturbed by people or predators, because the parents give a warning call when they fly from the nest. The chicks respond by running to cover (which may not be their nest). The chicks that wandered into the territories of neighbors, and were consequently killed by them, were lead-injected chicks. Thus, deficits in sibling recognition (which might also apply to parental recognition) resulted in direct differences in survival.

These experiments demonstrate the utility of Herring Gulls as a model for the study of morphological, behavioral, and physiological effects of contaminants. Further, abnormalities and deficits identified in the laboratory were also found in the wild, indicating that laboratory studies can be extrapolated to the field. The ability to perform similar experiments in the field is a real advantage because it allows for the examination of the biological significance of deficits or abnormalities caused by lead or other contaminants.

Acknowledgments

Over the years, several people have contributed to this work, and I thank them, including M. Gochfeld, F. Lesser, T. Shukla, S. Shukla, and C. Jeitner. My research has been supported by NIMH, NIEHS (P3OES005022), EPA, U.S. Fish & Wildlife Service, National Wildlife Foundation, N.J. Endangered and NonGame Species Program, Consortium for Risk Evaluation with Stakeholder Participation (CRESP) through the Department of Energy cooperative agreement (AI # DE-FC01-95EW55084, DE-FC01-06EW07053), and the Environmental and Occupational Health Sciences Institute. The views expressed are my own, and do not necessarily represent those of any funding agency.

See also: Conservation and Behavior: Introduction; Pigeons.

Further Reading

Becker PH, Conrad B, and Sperveslage H (1980) Comparison of quantities of chlorinated hydrocarbons and PCBs in Herring Gull eggs. *Vogelwarte* 30: 294–296.

Burger J (1979) Competition and predation: Herring Gulls versus Laughing Gulls. *Condor* 81: 269–277.

Burger J (1983) Competition between two species of nesting gulls: On the importance of timing. *Behavioral Neuroscience* 97: 492–501.

Burger J (1984) Pattern, mechanism, and adaptive significance of territoriality in Herring Gulls (*Larus argentatus*). *Ornithological Monographs* 34, 1–92.

Burger J (1988) Foraging behavior in gulls: Differences in method, prey, and habitat. *Colonial Waterbirds* 11: 9–23.

Burger J and Gochfeld M (1981a) Unequal sex ratios and their consequences in Herring Gulls. *Behavioral Ecology and Sociobiology* 8: 125–128.

Burger J and Gochfeld M (1981b) Discrimination of the threat of direct versus tangential approach to the nest by incubating Herring and Great Balck-backed Gulls. *Journal of Comparative Physiology and Psychology* 95: 676–684.

Burger J and Gochfeld M (1994) Behavioral impairments of lead-injected young Herring Gulls in nature. *Fundamental and Applied Toxicology* 23: 553–561.

Burger J and Gochfeld M (2000) Effects of lead on birds (Laridae): A review of laboratory and field studies. *J. Toxicol. Environ. Health* 3: 59–78.

Burger J and Gochfeld M (2005) Effects of lead on learning in Herring Gulls: An avian wildlife model for neurobehavioral deficits. *NeuroToxicology* 26: 615–624.

Burger J and Shisler J (1980) The process of colony formation among Herring Gulls. *Larus argentatus* nesting in New Jersey. *Ibis* 122: 15–26.

Fox GA (1990) Epidemiological and pathobiological evidence of contaminant induced alterations in sexual development in free-living wildlife. In: Colborn T and Clement C (eds.) *Chemically-Induced Alterations in Sexual and Functional Development: The Wildlife/ Human Connection. Advances in Modern Environmental Toxicology*, vol. 21, p. 147.

Mineau P, Fox GA, Norstrom RJ, Weseloh DV, Hallet DJ, and Ellenton JA (1984) Using the herring gull to monitor levels and effects of organochlorine contamination in the Canadian Great Lakes. *Advances in Environmental Science and Technology* 14: 425–453.

Pierotti R (1982) Habitat selection and its effect on reproductive output in the Herring Gull in Newfoundland. *Ecology* 63: 854–868.

Pierotti R (1987) Behavioral consequences of habitat selection in the Herring Gull. *Studies on Avian Biology* 10: 119–128.

Pierotti R and Annett CA (1991) Diet choice in the Herring Gull: Constraints imposed by reproductive and ecological factors. *Ecology* 72(1): 319–328.

Pierotti R and Good TP (1994) Herring Gull *(Larus argentatus)*. In: Poole A (ed.) *The Birds of North America Online.* Ithaca: Cornell Lab of Ornithology. Retrieved from Birds of North America. http://bna.birds,cornell.edu.bnaproxy.birds.cornell.edu/bna/species/124.

Hibernation, Daily Torpor and Estivation in Mammals and Birds: Behavioral Aspects

F. Geiser, University of New England, Armidale, NSW, Australia

Introduction

Mammals and birds are endothermic (within heating). They differ from ectothermic organisms, which rely on external heat for thermoregulation and comprise most animals and plants, primarily in their ability to regulate body temperature (T_b) by a generally high but adjustable internal production of heat generated by the combustion of fuels. Because the surface area in relation to the volume of heat-producing tissues of animals increases with decreasing size, many small endotherms must produce an enormous amount of heat to compensate for heat loss over their relatively large body surface. While heat loss is especially pronounced during cold exposure, even exposure to mild ambient temperatures (T_a) of 25–30 °C, considered to be warm by humans, causes mild cold stress in many small species.

Obviously, prolonged periods of high metabolic rates (MR) for heat production can only be sustained by regular food intake. During adverse environmental conditions and/or food shortages, energetic costs for thermoregulation may exceed those that can be obtained via food uptake. High energy expenditure and food uptake also require substantial foraging times and consequently exposure to predators even when food is abundant. Therefore, not all mammals and birds are permanently homeothermic (i.e., maintain a constant high normothermic T_b), but many, especially small species, enter a state of torpor during certain times of the day or the year. Torpor in these 'heterothermic endotherms' is characterized by a controlled reduction of T_b, energy expenditure, and other physiological processes and functions.

Torpor is by far the most effective means for energy conservation available to mammals and birds. Torpor conserves energy because no thermoregulatory heat for maintenance of a high normothermic T_b of around 37–40 °C is required. Moreover, because many torpid animals are thermoconforming over a wide range of T_a, T_b falls with T_a, and the substantial fall of T_b reduces MR via temperature effects. Further, in some species, inhibition of metabolism (in addition to temperature effects) can substantially lower energy expenditure to only a small fraction of the basal metabolic rate (BMR) or maintenance MR of normothermic, resting individuals under thermoneutral conditions.

Although MR and T_b during torpor in heterothermic endotherms are very low and often similar to those in ectotherms, torpid endotherms can rewarm from low T_b during torpor by using internally generated heat, whereas ectotherms, such as lizards, must rely on uptake of heat from external sources for raising T_b. Moreover, unlike in ectotherms, T_b in torpid endotherms is regulated at or above a species-specific minimum by a proportional increase in heat production that compensates for heat loss to prevent T_b from falling to critically low levels, likely to prevent tissue or organ damage, or to maintain the ability for endothermic arousal.

Torpor is often confused with 'hypothermia,' which also is characterized by reduced T_b and MR. However, torpor is a precisely controlled physiological state, whereas hypothermia is pathological and nothing but a failure of thermoregulation often due to depletion of energy reserves, excessive cold exposure, or from the influence of certain drugs.

Hibernation and Daily Torpor

The two most common patterns of torpor are hibernation (prolonged multiday torpor) and daily torpor. Hibernation often is seasonal and usually lasts from autumn to spring; however, 'hibernators' do not remain torpid continuously throughout the hibernation season (**Figure 1**). Bouts of torpor, during which T_b are low and bodily functions are reduced to a minimum, last for several days or weeks, but are interrupted by periodic rewarming and brief (usually <1 day) resting periods with high normothermic T_b and high energy turnover, apparently to recuperate from the prolonged time at low T_b. Hibernating mammals (with the exception of bears and some other large carnivores that reduce T_b only by about 5–8 °C) are generally small (<10 kg), and most weigh between 10 and 1000 g with a median mass of 85 g. Many hibernators fatten extensively before the hibernation season and rely to a large extent on stored fat for an energy source in winter, whereas fewer species store food for the hibernation season.

Hibernating species usually reduce their T_b to below 10 °C, with a minimum of −3 °C in arctic ground squirrels and most have minimum T_b of around 5 °C. The MR in torpid hibernators is on average reduced to about 5% of the BMR, but can be as low as 1–2% of BMR in small hibernators and often is <1% of that in

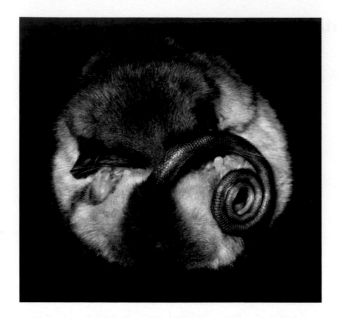

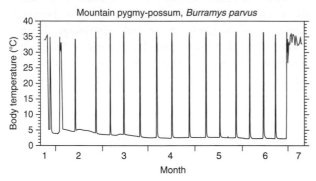

Figure 1   Body temperature (T_b) fluctuations during the hibernation season of a free-ranging mountain pygmy possum, *Burramys parvus* (body mass: 50 g). The hibernation season lasts for ~6 months, but torpor bouts when T_b and other functions are reduced to a minimum last only for up to about 2 weeks and are interrupted by periodic endothermic arousals.

active individuals. Energy expenditure during the mammalian hibernation season is reduced by ~85–95% in comparison to that of an animal that would have remained normothermic throughout winter, even if the high cost of periodic arousals is considered, which consume most of the energy required during the hibernation season. This enormous reduction in energy expenditure is perhaps best illustrated by the fact that many hibernating mammals can survive for 6–8 months or even longer entirely on body fat that has been stored prior to the hibernation season.

Daily torpor is the other widely used pattern of torpor in mammals and also in birds. This form of torpor in the 'daily heterotherms' is usually not as deep as hibernation, lasts only for hours rather than days or weeks, and is usually interrupted by daily foraging and feeding. Daily heterotherms are unable to express multiday torpor bouts. Many daily heterotherms are less seasonal than hibernators, may employ torpor throughout the year, although

torpor use often increases in winter. While daily torpor in many species occurs predominantly as a response to acute energy shortages, in other species, it appears to be used regularly to balance energy budgets even when environmental conditions appear favorable. For example, in hummingbirds, daily torpor is not only used to lower energy expenditure during adverse conditions, but may be employed to conserve energy during migration when birds are relatively fat. The marsupial Mulgara (*Dasycercus cristicauda*) appears to use daily torpor during pregnancy to store fat for the energetically demanding period of lactation. On average, daily heterotherms are even smaller than hibernators and most weigh between 5 and 100 g with a median of 19 g.

In contrast to hibernators, many daily heterotherms do not show extensive fattening before the season torpor is most commonly employed, and often only enter torpor when their body mass is low. The main energy supply of daily heterotherms even in their main torpor season remains food, often gathered during daily foraging, rather than stored body fat. In daily heterotherms, T_b usually fall to 10–20 °C with an average minimum T_b of 18 °C. However, in some hummingbirds, T_b below 10 °C have been reported, whereas in other, mainly large species, such as Tawny frogmouths (*Podargus strigoides*), T_b just below 30 °C are maintained. The MR during daily torpor are on average reduced to about 30% of the BMR although this percentage is strongly affected by body mass and other factors. When the energy expenditure at low T_a is used as point of reference, reductions of MR during daily torpor to about 10–20% of that in normothermic individuals at the same T_a are common. Overall, daily energy expenditure is usually reduced by 10–50% and in extreme cases by up to 90% on days when daily torpor is employed in comparison to days when no torpor is used, primarily depending on the species, the duration of the torpor bout, torpor depth, whether or not basking is employed during rewarming and rest, and how long animals are active.

As stated earlier, torpor bouts in the daily heterotherms are shorter than 1 day, independent of food supply or prevailing ambient conditions. Hibernators also can show brief torpor bouts lasting <1 day early and late in the hibernation season or at high T_a. However, it appears that physiologically these are nothing but brief bouts of hibernation with MR well below those of the daily heterotherms even at the same T_b. Thus, the term 'daily torpor' seems inappropriate for describing short torpor bouts of hibernators.

Often contrasted with hibernation and daily torpor, 'estivation' describes a period of torpor in summer or under warm conditions, which appears to be induced to a large extent by a reduced availability of water and consequently lack of food due to high T_a. In some ground squirrels, the hibernation season begins in the hottest part of the year and therefore qualifies as estivation. Many hibernating bats enter short bouts of torpor in summer and therefore estivate. Small arid zone marsupials regularly express daily torpor in summer, which not only reduces energy expenditure, but also water loss. However, it appears that there is no physiological difference between estivation and hibernation/daily torpor, apart from the higher T_b and thus MR during estivation because of the relative high T_a experienced in summer. It also appears that estivation is often not directly induced by heat because animals still employ torpor during the coolest part of the day, and it is likely that heat affects torpor use indirectly by reducing supply of food and water. However, reduced foraging opportunities because of heat may also result in torpor use.

Torpor in Mammals and Birds

Over recent years, the number of known heterothermic mammals and birds has increased substantially. Contrary to what was widely believed in the past, these are found in a wide diversity of taxa and in a variety of climatic regions ranging from arctic and alpine areas to the tropics.

In mammals, hibernation occurs in many species from all three mammalian subclasses. Hibernators include the egg-laying short-beaked echidna (*Tachyglossus aculeatus*) of Australia (Monotremata) and several marsupials including the Chilean opossum (*Dromiciops gliroides*, Microbiotheriidae), several pygmy-possums (Burramyidae), and feathertail glider (*Acrobates pygmaeus*, Acrobatidae). In the placental mammals, hibernation occurs in rodents (dormice, marmots, chipmunks, ground squirrels, hamsters), armadillos (*Zaedyus pichiy*), perhaps in some elephant shrews (Macroscelidea), some small primates (*Cheirogaleus medius*, fat-tailed lemur; *Microcebus murinus*, mouse lemur), many bats (Microchiroptera), and the insectivores (e.g., *Erinaceus europaeus*, hedgehogs; *Echinops telfairi*, tenrecs).

The 'winter sleep' of the large Carnivores (bears; European badger, *Meles meles*) appears to differ somewhat from deep hibernation in small mammals since T_b falls only by ~5 °C rather than by >30 °C, as in most of the small species. This type of dormancy, especially in bears, is often referred to as 'winter anorexia.'

Daily torpor (**Figure 2**) is known from a very large number of small marsupial and placental mammals. It occurs in several marsupial families from Australia (e.g., insectivorous/carnivorous marsupials, e.g., *Sminthopsis* spp. or *Pseudantechinus macdonnellensis*, Dasyuridae; small possums, *Petaurus breviceps*, Petauridae; honey possum, *Tarsipes rostratus*, Tarsipedidae) and South America (e.g., mouse opossums, *Thylamys elegans*, Didelphidae). In placentals, daily torpor occurs in rodents (deermice, *Peromyscus* spp., gerbils, *Gerbillus* spp., small siberian hamsters, *Phodopus sungorus*), some elephant shrews (*Elephantulus* spp., Macroscelidea), some primates (*Microcebus* mouse lemurs), bats (some Microchiroptera and small Megachiroptera), the

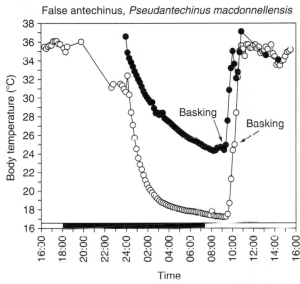

False antechinus, *Pseudantechinus macdonnellensis*

Figure 2 Daily torpor in two free-ranging false antechinus, *Pseudantechinus macdonnellensis* (31 g). The species is generally active with high T_b of about 36 °C for the first half of the night, enters torpor around midnight (characterized by a rapid reduction of T_b), and in the morning employs basking in the sun (arrows indicate visual observations) for rewarming from torpor (rapid rise of T_b) to minimize energy expenditure.

insectivores (shrews, e.g., *Crocidura* spp.), and some small carnivores (e.g., skunk, *Mephitis mephitis*).

In birds, daily torpor also is common. Many birds have normothermic T_b around 40 °C, whereas during daily torpor, T_b are usually in the range of 10–30 °C, depending on the species. In diurnal birds, daily torpor occurs at night. In nocturnal birds, daily torpor often commences in the second part of the night or early in the morning. Daily torpor is known from several avian orders including todies (*Todus mexicanus*) and kookaburras (*Dacelo novaeguineae*, Coraciiformes), mouse birds (*Colius* spp., Coliiformes), swifts (e.g., *Apus apus*, Apodiformes), hummingbirds (e.g., *Calypte* spp., Trochiliformes), nightjars (e.g., *Caprimulgus* spp., Caprimulgiformes), pigeons (e.g., *Drepanoptila holosericea*, Columbiformes), and martins (*Delicon urbica*), woodswallows (*Artamus cyanopterus*), chickadees (*Parus* spp.), and sunbirds (*Nectarinia* spp.) (Passeriformes). The largest bird presently known to enter daily torpor is the Australian tawny frogmouth (500 g), a nightjar relative. In contrast to mammals, multiday hibernation is presently known only for one bird, the common poorwill (*Phalaenoptilus nuttallii*) from North America.

Torpor is characterized by reduced activity, and traditionally, lack of movement or poor coordination has even been used for defining torpor. However, in recent years, new evidence has emerged showing that even torpid individuals can move at low T_b, for example, to basking sites, expressing behaviors that minimize energy expenditure for rewarming from torpor. Other species employ social torpor in groups and must behaviorally interact to maximize energetic outcomes. Some species are reproductively active even during the torpor season, which obviously will entail some behavior. Moreover, before they commence to employ torpor, many species prepare for the torpor season by selecting sites for torpor use or hibernacula, hoarding food, or accumulating fat, and this usually is accompanied by a change in behavior. The aim of this summary is to synthesize these behaviors.

Preparation for Hibernation

Shortening of photoperiod in late summer or autumn initiates physiological and behavioral changes of many species in preparation for hibernation. In other species, as for example ground squirrels, a strong innate circannual rhythm controls the seasons of activity and torpor use largely irrespective of photoperiod. Other species, for example, those from unpredictable habitats, may show opportunistic hibernation and seem to enter prolonged torpor irrespective of season or photoperiod, but at any time of the year when environmental conditions deteriorate, or perhaps to avoid predation.

As many daily heterotherms enter torpor throughout the year, they have to be able to do so without major preparation. However, in those species that express seasonal changes in torpor use, photoperiod, food availability, and T_a appear the major factors that affect the seasonal adjustments in physiology.

Selection of Hibernacula and Torpor Sites

Selection of an appropriate hibernaculum or torpor site is of vital importance. Hibernators often use underground burrows, boulder fields, piles of wood or leaves, tree hollows, caves, or mines. Hibernacula do not only provide shelter from potential predators, but also from temperature extremes and potential desiccation. Most hibernacula show temperatures a few degrees above the freezing point

of water even when outside T_a are well below freezing. Snow often acts as additional thermal blanket.

The selection of thermally appropriate hibernacula or sites where torpor is expressed is important, because at T_a close to the minimum T_b that is defended during torpor, MR are lowest and arousals are least frequent and therefore energy expenditure is minimal. Selection of a hibernaculum with a T_a below the minimum T_b for much of the hibernation season can be detrimental for small and solitary species because of the increased thermoregulatory energy expenditure and more frequent arousals. Therefore, it is likely that the minimum T_b is subject to strong selective pressure for adjustments that result in approximating the minimum T_b to the minimum T_a experienced. Arctic ground squirrels (*Spermophilus parryii*) do hibernate solitary at a T_a that is well below their minimum T_b (-2 to $-3\,^\circ$C); however, this species supercools and is rather large ($\sim$1 kg), is likely to use insulated nests, and obviously there is a limit how far T_b can be reduced, without causing freezing of tissues.

There is also some evidence that the selection of hibernacula may change during winter during periodic arousals apparently when the thermal conditions change due to, for example, rainfall or seasonal T_a change in caves. Mountain pygmy possums select different torpor sites after rain, which decreases T_a in sub-nivean spaces, whereas bats are known to select appropriate hibernacula sites along thermal gradients in caves that generally change with season.

Some species such as bats or fat-tailed lemurs enter torpor under bark, in trees hollows, or even under leaves with little physical protection. In blossom bats, roosts selection changes with season with bats selecting forest centers in summer to avoid heat exposure and forest edges in winter likely to allow bats to rewarm passively from torpor with the increasing T_a in the late morning. Long-eared bats (*Nyctophilus* spp.) enter torpor under bark that will be exposed to sun on the following morning to also take advantage of passive rewarming.

Fat Stores and Dietary Lipids

Preparation for hibernation involves primarily fattening and/or hoarding of food. Prehibernation fattening is common in many hibernators. Fat stores are important quantitatively because in many species they are the main source of energy throughout the prolonged hibernation season. Some species approximately double their body mass largely due to fat storage in autumn, but increases in body mass the order of 10–30% are more common. Fattening often is achieved by a combination of hyperphagy and a reduction in activity.

While the quantity of fat is important as it is the main energy source in fat-storing hibernators during winter, patterns of hibernation are also affected by the composition of dietary fats and body lipids. Function at low T_b during mammalian torpor obviously requires some physiological adjustments. In ectothermic organisms, increases of unsaturated polyunsaturated fatty acids in tissues and cell membranes form an important role in facilitating function at low T_b because unsaturated fatty acids lower the melting point of depot fats and increase the fluidity of cell membranes at low temperatures. In torpid endotherms, polyunsaturated fatty acids increase only slightly in depot fat and some membrane fractions. However, dietary polyunsaturated fatty acids have been shown to enhance torpor in hibernators as well as in daily heterotherms. Ground squirrels and chipmunks fed on a diet rich in polyunsaturated fatty acids, in comparison to conspecifics on a diet rich in saturated fatty acids, have lower T_b and MR and longer torpor bouts substantially reducing energy requirements during winter. In the wild, ground squirrels and marmots select food rich in polyunsaturated fatty acids during prehibernation fattening apparently to enhance winter survival, and perhaps to retain fat for the mating season immediately after hibernation. Further, recent evidence shows that selection of dietary fats is affected by photoperiod exposure with hamsters exposed to short photoperiod increasing their preference for diets rich in unsaturated fatty acids. These observations support the view that uptake of appropriate dietary fats form part of the winter preparation of many heterothermic mammals.

Behavior During Torpor or the Torpor Season

Social Torpor

Interestingly, alpine marmots (*Marmota marmota*) do successfully hibernate at T_a below their minimum T_b for some of the winter despite increased thermoregulatory energy expenditure. However this species, like other marmots, is very large and uses social hibernation to enhance the chance of winter survival. To achieve this, individuals huddle closely in their hibernacula to decrease the exposed surface area and to limit heat loss. Moreover, torpid marmots synchronize entry and arousals from torpor to minimize heat loss to the environment. Especially during endothermic rewarming, large adults typically commence to rewarm first and their endogenously produced heat can be shared by other, particularly small juveniles, minimizing their rewarming costs. Other species that are known to enter torpor socially are sugar gliders and feathertail gliders; however, although it is known that arousals in the former can be highly synchronized and that huddling in normothermic groups reduces energy expenditure, there are no data on whether or not huddling reduces rewarming costs.

Reproduction and Torpor

Social torpor and also solitary torpor do not prevent some heterothermic species to undertake some important reproductive behaviors. Whereas torpor and reproduction often are seen to be mutually exclusive, recent evidence has shown that both pregnant and lactating females may use torpor for energy conservation. For example, hoary bats (*Lasiurus cinereus*) employ torpor during pregnancy during cold spells in spring, which in addition to energy conservation may be used to reduce growth rates of young to delay parturition until conditions are more favorable for lactation and neonatal survival. Moreover, mating may occur during the hibernation season in bats and echidnas. Short-beaked echidnas mate in late winter when promiscuous males appear to seek out torpid females and mate with these either while torpid or during brief normothermic periods of females. A large proportion of torpid females were pregnant, suggesting that mating during hibernation is a common practice in this species, and perhaps may be employed by females for selection of males. Another group of mammals that is known to mate during the hibernation season are bats. In little brown bats (*Myotis lucifugus*), mating occurs frequently early in the hibernation season when bats are active. However, later in the hibernation season when the number of torpid females increases, adult males often force copulation with torpid individuals (females and males).

Basking During Rewarming from Torpor

Endothermic rewarming from torpor is energetically expensive and reduces the savings accrued from daily torpor and often results in death of light individuals during hibernation if they arouse too frequently. Small insectivorous/carnivorous dasyurid marsupials living in Australian deserts use daily torpor in winter frequently in the field and frequently employ basking during rewarming apparently to lower energy expenditure during arousal. Elephant shrews are the only other group of mammals for which data strongly suggest they may bask when rewarming from torpor. Basking during rewarming from torpor in dasyurids can reduce rewarming costs by up to 85% and consequently is highly significant to small mammals with high thermoregulatory energy expenditure, especially for those living in resource-poor environments such as deserts.

Arid zone dasyurids known to employ basking during rewarming are two dunnarts and a planigale living on sandy or clay substrate, and the rock-dwelling false antechinus (**Table 1**). In these species, torpor occurs frequently ($\sim$60–100% of days) in the wild in autumn and winter. Basking was observed in individuals that moved from their rest site, where they apparently had entered torpor, to a basking site in the sun at T_b ranging between 13.8 and 19.7 °C, well below the T_b often used for defining torpor

Table 1   Basking in torpid mammals

Species	Basking T_b minimum (°C)	Source
Marsupials		
Planigale *Planigale gilesi* (8 g)	13.8	Geiser et al. (2008)
Dunnart *Sminthopsis crassicaudata* (10 g)	14.6	Warnecke et al. (2008)
Dunnart *Sminthopsis macroura* (15 g)	19.3	Geiser et al. (2008)
False Antechinus *Pseudantechinus macdonnellensis* (31 g)	19.3	Geiser et al. (2002)
Placentals		
Elephant shrew *Elephantulus myurus* (55 g)	–	Mzilikazi et al. (2002)

in mammals (i.e., $T_b < 30\,°C$). Basking often commenced about 3 h after sunrise when the sun reached openings of rock crevices or soil cracks employed by torpid individuals to expose themselves to sun. Basking often lasted throughout the entire rewarming process. When an observer approached too closely, animals rapidly retreated into their shelter, demonstration that these individuals are fully alert and well enough coordinated even at low T_b.

In Elephant shrews, rewarming from torpor was tightly linked with changes in T_a. The T_b rose from low values at the same time T_a in the sun increased, strongly suggesting that the animals were basking, as basking was also observed independently in normothermic individuals.

Detailed behavioral observation on basking are available for free-ranging false antechinus, which typically basked with the back oriented toward the sun and less commonly with the flanks facing the sun. Both exposure of the entire body and exposure of only parts of the body were observed. Basking animals did not remain stationary for long periods and regularly changed body position or posture. These changes included altering orientation toward the sun by as much as 180°, and small movements of the body to increase or decrease the amount of sun exposure. Usually, a change of body posture or position occurred every ~5 min, suggesting that the animals were seeking to maximize heat uptake depending on their internal thermal condition. In the laboratory, torpid dunnarts (*Sminthopsis crassicaudata*) actively moved from a shaded area where they entered torpor to a heat lamp on 100% of observations to passively rewarm from low T_b.

Summary

While in the past lack of activity and movement were widely used to define periods of torpor, it is now clear that torpid animals at low T_b often are well aware of their

surroundings and even express a number of complex behaviors either while torpid or during normothermic periods between torpor bouts. These behaviors likely contribute to maximize survival of the torpor season or the survival of a species in general as heterothermic species are more resistant to extinction than homeotherms. It thus appears that torpor, although widely viewed as the 'physiological option,' in contrast the 'behavioral option' as for example migration, for survival/avoidance of adverse conditions, also involves several important behavioral components that form an crucial role in enhancing a species' fitness.

See also: Behavioral Endocrinology of Migration; Caching; Circadian and Circannual Rhythms and Hormones; Conservation and Anti-Predator Behavior; Female Sexual Behavior and Hormones in Non-Mammalian Vertebrates; Group Living; Habitat Selection; Internal Energy Storage; Seasonality: Hormones and Behavior.

Further Reading

Arnold W (1993) Energetics of social hibernation. In: Carey C, Florant GL, Wunder BA, and Horwitz B (eds.) *Life in the Cold: Ecological, Physiological, and Molecular Mechanisms*, pp. 65–80. Boulder, CO: Westview Press.

Barnes BM and Carey HV (eds.) (2004) Life in the cold: Evolution, mechanisms, adaptation, and application. *Twelfth International Hibernation Symposium. Biological Papers of the University of Alaska #27.* Institute of Arctic Biology, University of Alaska, Fairbanks.

Bieber C and Ruf T (2009) Summer dormancy in edible dormice (*Glis glis*) without energetic constraints. *Naturwissenschaften* 96: 165–171.

Boyer BB and Barnes BM (1999) Molecular and metabolic aspects of mammalian hibernation. *Bioscience* 49: 713–724.

Brigham RM (1992) Daily torpor in a free-ranging goatsucker, the common poorwill (*Phalaenoptilus nuttallii*). *Physiological Zoology* 65: 457–472.

Carey C, Florant GL, Wunder BA, and Horwitz B (eds.) (1993) *Life in the Cold: Ecological, Physiological, and Molecular Mechanisms.* Boulder, CO: Westview Press.

Carpenter FL and Hixon MA (1988) A new function for torpor: Fat conservation in a wild migrant hummingbird. *Condor* 90: 373–378.

Dausmann KH (2008) Hypometabolism in primates: Torpor and hibernation. In: Lovegrove BG and McKechnie AE (eds.) *Hypometabolism in animals: Torpor, hibernation and cryobiology. 13th International Hibernation Symposium*, pp. 327–336. University of KwaZulu-Natal, Pietermaritzburg.

Florant GL (1999) Lipid metabolism in hibernators: The importance of essential fatty acids. *American Zoologist* 38: 331–340.

Frank CL (1994) Polyunsaturate content and diet selection by ground squirrels (*Spermophilus lateralis*). *Ecology* 75: 458–463.

French AR (1985) Allometries of the duration of torpid and euthermic intervals during mammalian hibernation: A test of the theory of metabolic control of the timing of changes in body temperature. *Journal of Comparative Physiology B* 156: 13–19.

Geiser F (2004) Metabolic rate and body temperature reduction during hibernation and daily torpor. *Annual Review of Physiology* 66: 239–274.

Geiser F, Christian N, Cooper CE, et al. (2008) Torpor in marsupials: Recent advances. In: Lovegrove BG and McKechnie AE (eds.) *Hypometabolism in animals: Torpor, hibernation and cryobiology. 13th International Hibernation Symposium*, pp. 297–306. University of KwaZulu-Natal, Pietermaritzburg.

Geiser F, Goodship N, and Pavey CR (2002) Was basking important in the evolution of mammalian endothermy? *Naturwissenschaften* 89: 412–414.

Geiser F, Hulbert AJ, and Nicol SC (eds.) (1996) Adaptations to the Cold. *Tenth International Hibernation Symposium.* Armidale, Australia: University of New England Press.

Geiser F and Kenagy GJ (1987) Polyunsaturated lipid diet lengthens torpor and reduces body temperature in a hibernator. *American Journal of Physiology* 252: R897–R901.

Geiser F and Ruf T (1995) Hibernation versus daily torpor in mammals and birds: Physiological variables and classification of torpor patterns. *Physiological Zoology* 68: 935–966.

Geiser F and Turbill C (2009) Hibernation and daily torpor minimize mammalian extinctions. *Naturwissenschaften* 96: 1235–1240.

Guppy M and Withers PC (1999) Metabolic depression in animals: Physiological perspectives and biochemical generalizations. *Biological Reviews* 74: 1–40.

Heldmaier G and Klingenspor M (eds.) (2000) Life in the Cold. *Eleventh International Hibernation Symposium.* Heidelberg: Springer Verlag.

Hiebert SM (1993) Seasonality of daily torpor in a migratory hummingbird. In: Carey C, Florant GL, Wunder BA, and Horwitz B (eds.) *Life in the Cold: Ecological, Physiological and Molecular Mechanisms*, pp. 25–32. Boulder, CO: Westview.

Körtner G and Geiser F (2000) The temporal organisation of daily torpor and hibernation: Circadian and circannual rhythms. *Chronobiology International* 17: 103–128.

Lovegrove BG and McKechnie AE (eds.) (2008) Hypometabolism in animals: Torpor, hibernation and cryobiology. *13th International Hibernation Symposium.* University of KwaZulu-Natal, Pietermaritzburg.

Lovegrove BG, Raman J, and Perrin MR (2001) Heterothermy in elephant shrews, *Elephantulus* spp. (Macroscelidea): Daily torpor or hibernation? *Journal of Comparative Physiology B* 171: 1–10.

Lyman CP, Willis JS, Malan A, and Wang LCH (1982) *Hibernation and Torpor in Mammals and Birds.* New York, NY: Academic Press.

McKechnie AE and Lovegrove BG (2002) Avian facultative hypothermic responses: A review. *Condor* 104: 705–724.

Morrow G and Nicol SC (2009) Hibernation and reproduction overlap in the echidna. *PLoS ONE* 4(6): e6070.

Mzilikazi N, Lovegrove BG, and Ribble DO (2002) Exogenous passive heating during torpor arousal in free-ranging rock elephant shrews, *Elephantulus myurus.* *Oecologia* 133: 307–314.

Stawski C and Geiser F (2010) Fat and fed: Frequent use of summer torpor in a subtropical bat. *Naturwissenschaften* 97: 29–35.

Storey KB and Storey JM (1990) Metabolic rate depression and biochemical adaptation in anaerobiosis, hibernation and estivation. *The Quarterly Review of Biology* 65: 145–174.

Wang LCH (ed.) (1989) *Animal Adaptation to Cold.* Heidelberg: Springer Verlag.

Warnecke L, Turner JM, and Geiser F (2008) Torpor and basking in a small arid zone marsupial. *Naturwissenschaften* 95: 73–78.

Willis CKR, Brigham RM, and Geiser F (2006) Deep, prolonged torpor by pregnant, free-ranging bats. *Naturwissenschaften* 93: 80–83.

Honest Signaling

P. L. Hurd, University of Alberta, Edmonton, AB, Canada

Introduction

Classical ethologists largely thought of communication in terms of senders broadcasting information indicating their internal motivational state, and receivers upon whom these signals then acted as releaser stimuli. The receiver's response to these signals was assumed to be a behavioral response appropriate to the signaler's state. The outcome of this interaction would then be of mutual benefit to both signaler and receiver. Evolution was thought to act to make signals less ambiguous about the signaler state, and more efficient and effective at longer and longer ranges. Unfortunately, the logical basis of this classical approach has a potential flaw.

The outline of a challenging problem to this classical view emerged during the 1970s with the rise of explicitly 'selfish' gene-centric views of evolution. The application of game theoretical thinking to social behavior highlighted the importance of conflicting interests in signaling interactions. Most conspicuous communication occurs between individuals in some sort of conflict. Fighting animals often communicate their strength to one another. A prospective suitor will show-off to a prospective mate using courtship displays. Dependent young beg their parents for food. If honest, then all these situations seem to involve individuals giving away information to their disadvantage: some individuals informing a receiver that they are weak, unfit mates, or less hungry than their siblings. Such submaximally escalated signals ought to encourage the opponent to continue fighting, the wooed to spurn, parents to provision less. On the other hand, if all fighting animals signal that they are incredibly strong, suitors always signal that they are the most conceivably deserving mates, and offspring signal that they are starving then, presumably the receivers of these signals would evolve to ignore them entirely.

Do Displays Transfer Information?

An early hypothesis advanced to solve the dilemma of communication with conflicting interests was to question whether signals actually conveyed any information at all. The prediction that there ought to be no information transmitted follows from the game theoretical model of an auction as applied to the use of threat displays. If an agonistic interaction is seen as a bidding process in which the individual who is willing to escalate the most wins, then a threat display that announces how much the signaler is going to bid guarantees losing. A signal that correctly identifies the signaler's state, that is, the intended bid, allows the opponent to out-bid by a small amount every time. In such a situation, it is clearly best to keep one's bid secret. In a 1979 analysis of avian threat display use, Peter Caryl concluded that there was very little evidence to support the traditional view that there was bid-like information in the signal. Few, if any, threat displays were followed by an attack with more than even probability; different threat displays in the same species did not precede attacks with remarkably different probabilities; and a single threat display may 'predict' attack with the same likelihood of predicting abandonment of the contest. In the early 2000s, several authors critically reviewed these conclusions, pointing out that threat displays did consistently predict, albeit with low predictability, subsequent aggressive escalation on the part of the signaler. Caryl's conclusion that the quality of the information transmitted is poor, remains true. These are, at best, quite ambiguous signals.

Is Communication an Arms Race?

A wider case against honest communication was advanced in a number of highly influential articles by John Krebs and Richard Dawkins in the late 1970s and early 1980s. Dawkins and Krebs argued that the exaggerated signals seen in communication between individuals with conflicting interests were attempts to manipulate receivers into acting against their own self-interests. Such hypnotic signals overwhelmed the receiver's senses, hypnotizing them into acting as the signaler's agent. This manipulation of the receiver's behavior was suggested to evolve in concert with ever increasing sales resistance on the receiver's part in an evolutionary arms race.

In contrast to this view of evolutionarily unstable spiralling co-evolution of manipulation and scepticism is the view that signals actually convey useful information, that is, they are basically honest and their use is an evolutionarily stable strategy. Reconciling the latter view with the criticisms leveled against the classical ethological perspective has been a very active research topic, with game theoretical models playing a prominent role. We can address this problem by considering how different types of signals can be used to convey information. For each of these signal types, there is a different reason that receivers may believe the information communicated is reliable.

Types of Signals

Several distinct types of signal have been proposed to be evolutionarily stable against the corrupting pressure of deception or manipulation.

Handicaps

By far the most influential honest signaling hypothesis is the handicap principle, advanced by Amotz Zahavi in a pair of papers in the mid-1970s. Zahavi's verbal model proposed that signal reliability was maintained by the inherently wasteful costliness of a signal. Signalers advertising a desirous or fearsome ability or state could only do so credibly if they use up some of that ability or state. A signal that was wastefully costly to produce could be afforded only by the most able signalers. This verbal model, likening biological communication to a signal of wealth by means of conspicuous consumption, was met with a great deal of scepticism by theoreticians such as John Maynard Smith.

Zahavi's hypothesis gained widespread support only after a formal game theoretical model by Alan Grafen demonstrated that the idea could work, in principle. Zahavi has gone on to suggest that the mechanism underlying handicapped signaling applies to a far wider class of phenomena, claims that have not been widely embraced by researchers in the fields involved. Grafen's models do continue to have widespread support, but some notable criticisms have been made. For example, Tom Getty has questioned whether the critical assumption in Grafen's model, and others like it, that the costs and benefits of different signals are linearly separable, is justified in the biological cases the model is typically applied to. Linearly separable means that, as far as the signaler is concerned, the sources of signal costs are independent of the benefits that a signal will bring. For example, if a courting male uses a handicapping courtship display, then a more attractive display must be more costly when the male is of lower quality. The traditional assumption is that the benefit of a successful courtship is equal for males of all quality. But it may be that males of higher quality are able to turn a successful courtship into greater reproductive success than lower-quality males. If the costs vary with signaler quality, and the benefits do as well, then the two are not linearly separable. Getty concludes that this critical assumption is not justified, and, moreover, that the whole idea of a handicap principle is distractingly unhelpful, and the metaphor ought to be boycotted in favor of less loaded language. The costs of producing and bearing intense signals may prevent some signalers from using them, thereby maintaining honesty, in a process quite like Grafen's models, but without necessarily conforming to Zahavi's larger view of handicaps. Plausible models of signals without linearly separable costs and benefits may produce evolutionarily stable, honest signals in which the cost of producing more attractive signals is prohibitive to signalers of lower quality (the cost of increasing signal intensity prevents signalers of all qualities from exaggerating) yet the absolute cost paid by signalers is zero. Whether this situation can be described as 'handicapping' is debatable. The cost that prevents dishonest exaggeration looks just like a handicap, but that cost is not actually paid. At some point, debating whether it is a handicap or not is less useful than asking how close this model matches what is seen in real biological signals.

Indices

The handicap continues to be a very influential idea, and while it may be the most talked-about form of signal, it is not the only one. Another form of signal that has a much longer, less controversial history is the index. Indices are signals that are honest by way of physical constraint. A large toad makes a deeper croak than a smaller toad and is constrained to do so because of the physics of sound production. If a larger male toad is more attractive to females, and more intimidating to other males, then a deep croak serves as an index of that desirable, fearsome dimension. Whether a specific signal is best described as a handicap, or an index, may be debatable. Some authors, such as Maynard Smith and Harper, see indices as a more widespread, and important, class of signal than do some others. For example, it may be that the depth of the toads croak is exaggerated to a maximum across the population, and the cost of further overcoming the physical constraint on call frequency prevents any further exaggeration. Empirical models of index use such as the sequential assessment game, a model of escalating threat display use based on indices of size, are better supported by empirical data than are handicaps.

Conventional Signals

A more controversial alternative to indices and handicaps are conventional signals, signals without the wasteful cost of a handicap or the physically constrained honesty of an index. A signal is said to be conventional when the meaning of the signal can, theoretically at least, be exchanged with that of another signal. For example, a human in a bar could extend their middle finger upwards and show the back of the hand to another human. This is called 'giving the finger' and involves no more inherent, wasteful, cost than would a similar display using an upraised thumb. One can easily imagine a culture in which a 'thumbs-up' and 'the finger' have their meanings reversed. If the costs and benefits of these signals can be reversed, then their meaning is established by convention; they are conventional signals. Several authors, Zahavi included, have dismissed conventional signals as impossibilities. Game theoretical models of conventional signal use show that they work, at least in principle. The cost imposed by receivers acting on their conventional interpretation of a bluffed signal of strength or desperation in a threat display game can make cheating far less successful than honesty. Some signals, such as the threat displays used by birds, seem more like conventional signals than handicaps or indices.

Some signals do incorporate aspects of both handicapping signals and conventional signals. Vulnerability, or interaction, handicaps are signals that have no inherent cost of production but may bear costs that depend entirely on the receiver's response to the signal, as in conventional signals. Interaction handicaps share with the more classic handicaps described earlier, the property that the form of the signal influences the cost. No cost is paid to produce the signal per se; instead, a cost is paid only if the receiver makes a specific response. The best example of such a signal is a threat display the form of which makes the individual performing the display vulnerable to counterattack from the receiver. The classic example of a vulnerability handicap is the use of a threat display by a newly molted arthropod. Waving their large claw functions as a threat display but entails a great risk of cost, since the claw is effectively useless and vulnerable to damage if (and only if) the receiver reacts to the threat with a counterattack. Although these signals do blur the distinction between conventional signal and handicaps, it is important to realize that conventional signals and handicaps do not grade into one another as general classes of signals.

The costs which maintain honesty in handicaps, if imposed on a game theoretical model in which conventional signals are evolutionarily stable and honest, produce counterintuitive results. Simply adding an arbitrary cost to a conventional signal will not make it extra-resistant to dishonest use. In a model of conventional threat display, if one of the conventional signals is made to be costly in a handicapping sense, then it will be used by the stronger, not weaker, individual. Both the original and posthandicap versions are evolutionarily stable and produce honest signaling, but signalers gain higher fitness in the former, in which signaler quality could be associated with either signal. In the latter, the lower-quality signalers use the handicap and they gain lower payoffs, while everyone else's payoffs remain unchanged. This handicap-enhanced outcome seems unstable in the long run, in that the handicapped signal will not be used at all if a new, costless, signal is made available for use to replace it. The handicap will be used only if the number of signals that can be used is so small that signalers have no option but to include the costly one in their repertoire.

All the three of these signal types: handicaps, indices, and conventional signals, can maintain honesty between individuals with conflicting interests, at least in theory. All have some degree of support from empirical studies. Most likely, all the three do function in stabilizing communication between animals, to some degree, in some cases. However, the debate over the relative importance of these three signal types is far from settled.

Signaling in Biological Contexts: The Degree of Conflict

While animals signal to each other in a wide variety of situations, there are several specific communication scenarios that have interested researchers. Begging signals, courtship displays, threat displays and signals about, or directed to, predators are some of the best-studied examples of biological signaling. These examples may all be placed on a continuum of conflicting interests, as has been done by William Searcy and Stephen Nowicki. Searcy and Nowicki classify the interests of signalers and receivers as identical, overlapping, divergent or opposing, corresponding roughly to interactions within an individual, between kin, between potential mates, and between individuals in aggressive interactions, respectively. The idea that the form of signals used in a social interaction is influenced by the degree of conflict between the participants traces back at least as far as Krebs and Dawkins. The assumption has been that more escalated, or costly, signals will be used in those interactions in which there is a greater degree of conflict. While the idea makes a great deal of intuitive sense, a number of counterexamples can be made. Aposematic signals, in which a toxic signaler and a potential predator have little conflict, show considerable signal exaggeration nonetheless. Conventional signals, where the cost is least, seem to be most likely used in aggressive interactions where the conflicting interests seem rather more stark. In fact, it may be that animals involved in an aggressive interaction have interests that are far more in common than it seems on first inspection. While both individuals prefer that they prevail and the opponent concede, they are united in preferring not to have a potentially injurious escalated physical battle. This sort of pattern of common interest across some possible outcomes, and conflicting interests across others, not only drives conventional signaling models, but may also make it hard to predict large-scale patterns of signal property from the overall degree of conflict.

Honesty and Deception, or Ambiguity?

Various game theoretical models show that signaling can be honest in the face of conflicting interests between signaler and receiver. Empirical studies show that animals do communicate to receivers information that they then put to use. But this does not mean that signals are necessarily honest. Communication between animals seems remarkably ambiguous and imprecise. With the possible exception of alarm calls, where one short call may be used to provide warning to the audience, biological signals seem nothing like the maximally informative, 'say it once, and all is revealed,' traits predicted by most game theory models, or the acme of signals the classical ethologists suggested that evolution selects for. The imprecision of signals may simply reflect some external constraint acting on signals that prevents evolution from removing a large 'noise' component from the signal. According to this view, signalers would benefit if the signals could be made more precise and informative.

An alternative view is that these noisy signals are not failed attempts at maximum honesty, but best thought of as honest-on-average. Alan Grafen and Rufus Johnstone modeled an evolutionarily stable blend of honest and deceptive signalers in a handicapped begging game. The idea that signalers may sporadically exploit receivers while the proportion of honest signals remains high enough to maintain the response in the receivers makes intuitive sense and is widely accepted. The Grafen and Johnstone model, like that of stomatopod threat displays by Eldridge Adams and Michael Mesterton-Gibbons, produces deceptive signaling within a larger population of honest signalers by combining the honest and dishonest signalers together through the use of the same signal. The receiver has to treat the class of all signalers using the same signal as the average of both honest and dishonest types using that signal, in effect never being able to adjust to the types differentially. It is as if the receiver were in a city where 5% of all the $20 bills were counterfeit, choosing to be paid in $20 bills from various sources, but always receiving $19 whenever spending any $20 bill. The receiver chooses to accept $20 bills, without fear of getting stuck with a counterfeit, but never expecting the bill to be worth anything more than $19. Thus, honest and dishonest $20 bills coexist, but the receiver does not have to be described as being 'deceived.'

An alternative to the view that signals are either 'honest' or 'deceptive' is taken by Tom Getty and Peter Hurd, among others. This view proposes that signals might be better characterized as simply 'ambiguous' but still informative. Consider the case of a sentry giving an alarm call. Empirical studies of alarm calling in birds have found examples in which about half the alarm calls are given in the absence of predators, but provide the sentry with the opportunity to take food uncovered by the flock while they take cover from the predator that isn't there. This signal clearly meets the definition of a deceptive signal (below) when given in the absence of a predator. Whenever the alarm is raised, the receivers may either be deceived (by inferring that they were not aware of the possibility that the 'alarm call' signal might be a false alarm), or they may have the correct expectation that a predator has a 50% probability of being present. In the latter case, the receivers may not be deceived, so much as

gambling with known odds. This same logic may be extended more productively to situations with more variation. Imagine that a territorial male is either more or less likely to attack an intruder because of some variation in subjective resource value, such as whether a female nesting on his territory is fertile or not. The male has a repertoire of several different threat displays. In each of these different levels of resource value, the male has different probabilities of attacking after each of the different threat displays. If evolution has led to a stable pair of signaler and receiver strategies, then the receiver must be working with an accurate expectation of these probabilities. Outcomes may be better or worse, but the signaler plays the game knowing the odds and cannot properly be described as cheating. If one accepts that evolution has led to an evolutionarily stable pairing of signaler and receiver states, then any signaling system will be honest in this sense.

Definitions

A word about definitions. Perhaps more than any other area within animal behavior, the study of communication has a long history of making liberal use of important terms, with poor, or multiple conflicting, definitions. This may indicate that the central concepts are so intuitively clear that reasonable progress can be made without universally agreed upon formal definitions or that the concepts are so hoary and vague that ideas appearing to be simple require redefining at each use, resulting in a pile of incommensurate crosstalk obstructing the resolution of any of the central issues. The following definitions have been paraphrased from Searcy and Nowicki, and may be subject to all the criticisms mentioned earlier, but still function adequately.

Signal: A signal is a character or behavior that has evolved so as to provide information to other organisms. Signals are usually defined so as to exclude traits that convey information to the detriment of the signaler, such as the rustling noise of a mouse in the grass which is heard by a nearby owl. Some of the more extreme applications of the handicap idea may include such apparently detrimental signals. For example, it could be argued that such life-threatening grass rustling serves as a signal of his quality because of the handicap it imposes on him.

Honesty: For the purpose of this article, honesty will be usually mean 'reliable,' that is to say: there is some variable state that the signaler is in, or aware of, that the receiver cannot know directly. The receiver would benefit from knowing this state. The signal chosen by the signaler allows the receiver to choose a response that is appropriate for the actual state.

Deception: A signal X is said to be deceptive if it elicits a response Y from the receiver which benefits the signaler, and the response Y would be appropriate, beneficial to the receiver, if the state that the signaler is in, or aware of, were a different one from the actual state.

See also: Agonistic Signals; Alarm Calls in Birds and Mammals; Game Theory; Mating Signals; Parent–Offspring Signaling.

Further Reading

Adams ES and Mesterton-Gibbons M (1995) The cost of threat displays and the stability of deceptive communication. *Journal of Mathematical Biology* 175: 405–421.

Bradbury JW and Vehrencamp SL (1998) *Principles of Animal Communication.* Sunderland, MA: Sinauer.

Caryl PG (1979) Communication by agonistic displays: What can games theory contribute? *Behaviour* 68: 136–169.

Dawkins R and Krebs JR (1978) Animal signals: Information or manipulation? In: Krebs JR and Davies NB (eds.) *Behavioural Ecology: An Evolutionary Approach,* 1st edn., pp. 282–309. Oxford: Blackwell.

Getty T (1997) Deception: The correct path to enlightenment? *Trends in Ecology and Evolution* 12: 159–160.

Getty T (2006) Sexually selected signals are not similar to sports handicaps. *Trends in Ecology and Evolution* 21: 83–88.

Hurd PL and Enquist M (2005) A strategic taxonomy of biological communication. *Animal Behaviour* 70: 1155–1170.

Johnstone RA and Grafen A (1993) Dishonesty and the handicap principle. *Animal Behaviour* 46: 759–764.

Krebs JR and Dawkins R (1984) Animal signals: Mind-reading and manipulation. In: Krebs JR and Davies NB (eds.) *Behavioural Ecology: An Evolutionary Approach,* 2nd edn., pp. 380–402. New York, NY: Sinauer.

Maynard Smith J and Harper DGC (2003) *Animal Signals.* Oxford: Oxford University Press.

Searcy WA and Nowicki S (2005) *The Evolution of Animal Communication: Reliability and Deception in Signaling Systems.* Princeton, NJ: Princeton University Press.

Zahavi A (1975) Mate selection: A selection for a handicap. *Journal of Theoretical Biology* 53: 205–214.

Zahavi A (1977) The cost of honesty (further remarks on the handicap principle). *Journal of Theoretical Biology* 67: 603–605.

Honeybees

M. D. Breed, University of Colorado, Boulder, CO, USA

Background

The honeybees, all members of the genus *Apis*, are one of the most familiar flying animals of terrestrial habitats. The center of diversity of the genus is Southeast Asia, where several species are found. Most of these species are limited in range to tropical and montane zones in Southeast and South Asia, but two species have far broader ranges. *Apis cerana*, sometimes called the Eastern hive bee, occurs as far north as Japan and into the Middle East. *Apis mellifera*, or the Western hive bee, is native to Africa. It expanded its range into Europe and Asia as the ice-age glaciers retreated, and has been spread by humans to the Americas, Australia, and Hawaii (**Figure 1**). *A. mellifera* has also been introduced through much of the range occupied by *A. cerana*, including Japan and mainland China. In this article, I review the basic biology of the genus *Apis*, with a focus behavior.

Human associations with honeybees are deeply rooted in prehistory. Honey hunting was likely an important source of food for early humans living in Asia, Africa, and Europe. Cave paintings in Spain record prehistoric honey hunting by humans, and honey remains an important human food source. An image of the honeybee forms a key element in the ancient Egyptian hieroglyph symbolizing Lower Egypt. Roman ruins sometimes feature niches in walls that were designed to hold honeybee colonies, and numerous such walls constructed in the seventeenth and eighteenth centuries are found in the French and British countryside. Currently, honeybees are highly important pollinators, providing pollination services in manipulated agroecosystems in which native bee populations are reduced and for nonnative crop plants.

Most of our scientific knowledge of honeybees comes from studies of *A. mellifera*. Scientists' fascination with the honeybee dates back at least to Aristotle, and the maintenance of bee colonies by medieval religious communities fostered knowledge of bee management and behavior. Early publications, such as Charles Bulter's *Feminine Monarchie* (1609) held a great amount of detail about honeybee biology, some correct and some incorrect, and form the foundation for modern knowledge of honeybees. The discovery by Jan Dzierzon in 1845 that male honeybees arise from unfertilized (haploid) eggs and females from fertilized (diploid) eggs is an excellent example of how the tradition of beekeeping by educated clergymen led to intellectual explorations of bee biology and behavior. Haplodiploidy, the system of sex determination discovered by Dzierzon and known as Dzierzon's

rule, is now understood to be a characteristic of nearly all Hymenoptera.

Twentieth century studies of honeybees are rooted in the work of Karl von Frisch (1886–1982), who shared the 1973 Nobel Prize in Physiology or Medicine with Niko Tinbergen and Konrad Lorenz. Von Frisch's discoveries of color vision in bees, the dance language of the honeybee, and the ability of bees to use the polarization of light in their orientation were key to the development of sensory physiology and animal behavior as scientific fields. Generations of scientists have followed in von Frisch's footsteps, refining and further developing concepts first proposed by von Frisch and his students.

The *Apis* Family Tree

Honeybees lie in the subfamily Apinae of the family Apidae (**Figure 2**) although it should be noted that there has been considerable controversy over the phylogeny of these bees. Their branch of the apine family tree includes other eusocial taxa, the bumblebees and the stingless bees. In addition to eusociality, honeybees share the use of wax in nest construction with bumblebees and stingless bees, although honeybees are the only taxa to use wax as the sole construction element of their combs.

Two major sets of differences align with the evolution of species in *Apis* – size and nesting habit. The genus divides among dwarf bees, giant bees, and bees that we would regard as more 'normal' in size, like *A. mellifera*. The overall appearance of all *Apis* species is quite similar, but size varies manyfold between the dwarf species, like *A. florea*, and giant species like *A. dorsata*. The dwarf honeybees, subgenus *Micrapis*, which includes *A. florea* (**Figure 3**) and *A. andreniformis*, are generally considered the basal, or most primitive, group within *Apis*. The giant honeybees are first derived from the dwarf honeybees and then the cavity-nesting honeybees are derived. The giant honeybees, subgenus *Macrapis*, include *A. dorsata* (**Figure 4**) and *A. laboriosa*. *Micrapis* and *Macrapis* species build nests of a single exposed comb suspended from a branch, cave roof or cliff. Bees in the subgenus *Apis*–*Apis mellifera*, *A. cerana*, *A. nigrocincta*, *A. nuluensis*, and *A. koschevnikovi* – all nest in cavities, such as hollow trees, and build nests of multiple combs.

What features are held in common among all species in the genus? While the level of documentation varies among species, from the very well-known *A. mellifera* to the almost entirely unknown *A. nuluensis*, some generalizations about

Figure 1 Western honeybee, *Apis mellifera*, colony in a small cave. The use of multiple combs is typical of cavity nesting honeybees, subgenus Apis. Photo by Thomas Ranker.

the genus are widely accepted. All *Apis* species are highly eusocial, meaning that queens and their daughter workers live together in colonies, there is a strong morphological differentiation between queens and workers, and the workers cooperate to rear additional workers, males (drones), and new queens from among their sibs. All members of the genus build nests of wax combs, have workers with barbed, autotomous stings, and have queens that, after multiple matings, store sperm for use throughout their life. Workers in all honeybee species communicate about food resources using a dance language and release alarm pheromones to alert nestmates to threats. Honeybees reproduce by swarming, with the queen in a colony leaving the nest with about half the workers and establishing a new nest; she is replaced in the old nest by one of her daughter queens.

As with other eusocial Hymenoptera (ants, some wasps, as well as some other types of bee), all honeybee workers are female. *Apis* workers are highly differentiated from queens, with larger eyes, hairier bodies, small ovaries, no capacity to mate, and autotomous stings. Drones have yet stouter bodies and even larger eyes, probably adaptations for both flight and spotting potential mates.

Honeybees, Agriculture, and Ecosystems Services

A. mellifera is the champion honey producer in the genus. While all *Apis* species store some concentrated nectar (honey) in their combs, most honeybee species abscond

from nests if conditions are poor and seek out a more favorable nesting site, or are seasonally migratory, moving to accommodate variation in rain and temperature. *A. mellifera* is the only species which is successful across a vast expanse of the northern temperate zone (in Europe and Asia, and introduced into temperate North America). Escaping temperate winters is beyond the migratory ability of honeybees and it is likely that storage of large amounts (50 kg or more) of honey evolved as a way of having food stores for overwintering. Temperatures inside *A. mellifera* nests (and to the extent *A. cerana* lives in temperate climates, in *cerana* nests) are maintained near 30 °C through the unfavorable season.

Storing food creates an attractive resource for animals like skunks, raccoons, bears, and, obviously, humans. Additionally, the larvae and pupae, which are reared in cells in the comb, have high nutritional value. It is no surprise that honeybees have intense defensive methods available to them, and these defenses are highly effective against vertebrate predators. In many human cultures, bee larvae or pupae are consumed in addition to honey. Beeswax is regarded as a high-grade base for cosmetics and candles. Propolis (plant resins collected by bees to seal cracks and supplement wax as a structural material) was used as a base for varnishes prior to the development of petroleum-based wood finishes. Propolis was reputedly a key ingredient in finishes used on fine violins, such as those made by Stradivarius. Recently, hive products such as pollen, royal jelly, and propolis have been touted for their medicinal properties.

Globally, most agriculturally managed honeybees are *A. mellifera*, but *A. cerana* colonies are maintained in parts of Asia. The techniques used to keep these two species are similar. Until the nineteenth century, bees were typically maintained in sections of hollow logs (**Figure 5**) or in woven straw hives, called 'skeps.' The skep is a state symbol for Utah, appearing on highway signs and at the center of the state seal. These arrangements had the disadvantage of not allowing inspection and management of the combs, and required at least partial destruction of the colony when honey was harvested. Early honeyhunters and beekeepers learned that smoke seemed to pacify bees, allowing for collection of honey with less risk of stinging.

Much of beekeeping is based in the management of bee behavior. Beekeeping underwent a major advancement with the invention of moveable frames; these wooden (or now, plastic) supports hold the comb and allow for removal of individual combs from a hive. Also important in artificial hive construction is the concept of 'bee space'; if the right-sized gaps – about 1.0 cm – are left between frames, and between the frames and the hive box, the bees use the spaces as passages, rather than filling them with wax and propolis. Keeping bee space means that frames are not cemented into place, facilitating their easy removal. The American beekeeper, Lorenzo Langstroth,

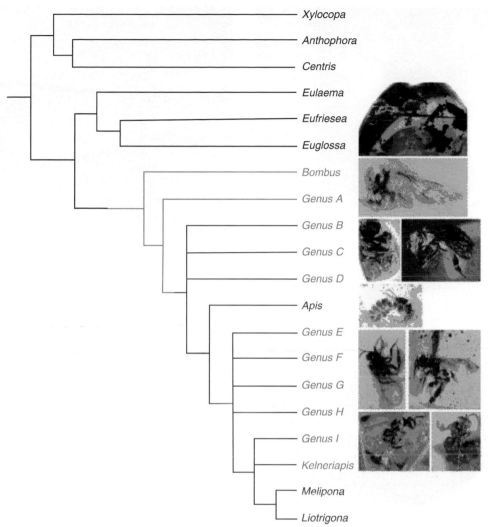

Figure 2 A cladogram of the bee subfamily Apinae places the genus *Apis* on a branch with other eusocial bees, Bombus (the bumblebees) and the stingless bees (Melipona/Liotrigona). Genera labeled in green are extinct. *Bombus*, labeled in yellow, is primitively eusocial, with relatively little morphological differentiation between queens and workers. The genera labeled in red, including *Apis* and the stingless bees, are highly eusocial, with strong morphological differentiation between queens and workers. Reproduced with permission from Engel M S (2001) Monophyly and extensive extinction of advanced eusocial bees: Insights from an unexpected Eocene diversity. *Proceedings of the National Academy of Sciences USA* 1661–1664. Copyright 2001 National Academy of Sciences, USA.

is generally credited with developing the concept of bee space and linking it with moveable frames in the 1850s, resulting in the basis for modern beekeeping. The perfection of techniques for artificial insemination of honeybee queens, particularly by Harry Laidlaw, has allowed beekeepers and scientists to make controlled crosses among honeybees to enhance honeybee agricultural performance and for experimental purposes.

Diseases and Parasites of Honeybees

Honeybees have attracted their share of diseases and parasites. Because of the commercial value of honeybees,

more is probably known about honeybee diseases than diseases afflicting any other insect. Viruses, such as bee paralysis viruses (BPV), bacteria, including the spore-forming American foulbrood, *Paenibacillus larvae*, protists such as *Nosema apis*, and the chalkbrood fungus, *Pericystis apis*, can all impair or kill workers (larvae in the cases of foulbrood and chalkbrood, adults for BPVs and Nosema). Mites, such as the tracheal mite, *Acarapis woodi*, and the Varroa mite, *Varroa destructor*, are also significant threats to honeybees. For most of these diseases and parasites, it is not clear in which species of *Apis* they originated, although all are now problems for *A. mellifera* populations. The *Varroa* mite evolved as a pest of *A. cerana* in Southeast Asia and its recent switch to *A. mellifera*, facilitated by

Figure 3 An *Apis florea* colony. This dwarf honeybee nests lower in the vegetation and on slimmer branches than giant honeybees (see **Figure 4**). The comb, which is obscured by the bees, extends over the branch and a flattened portion of comb above the branch serves as a platform for dances in this species. Photo by Xiaobao Deng.

Figure 5 An *Apis cerana* colony maintained in a log hive in southern China (Yunnan province). The ends are removable for access to the combs. Photo by Michael Breed.

Figure 4 An *Apis dorsata* colony. Note that the comb does not extend up and over the tree branch. *Apis dorsata* often nests in aggregations, with many colonies occupying a large tree. Photo by Michael Breed.

global commerce in honeybees, has been disastrous for *A. mellifera* populations, whose natural defenses are only weakly developed.

Recent public attention has focused on colony collapse disorder (CCD), a disease or complex of diseases that has greatly impacted global *A. mellifera* populations. Colony losses have exceeded 90% in some locations and loss of pollination services have had major impacts on some growers of fruits and vegetables. A variety of diseases have been suggested as causes of CCD, including paralysis viruses and *Nosema ceranae*, which has switched hosts from *A. cerana* to *A. mellifera*, but other factors, such as use of the neonicotinoid insecticide imadocloprid and global climate change have also been implicated. Resolution of

the cause or causes of CCD is a matter of extreme urgency, given the importance of ecosystem services provided by honeybees.

Hygienic behavior is very much a part of honeybee responses to disease and parasites. Bees may respond to diseased or dying larvae and pupae by removing them from their colony. Carl Rothenbuhler's famous studies in the 1960s of hygienic behavior in response to brood infections stood for many years as classic examples in behavioral genetics of how simple Mendelian models could explain behavioral variation. While we now understand that the genetics underlying hygienic behavior are more complex than what Rothenbuhler proposed, his basic finding of genetic variation for hygienic behavior should be recognized as a key stimulus for the development of the field of behavioral genetics. Genetic lines of *A. mellifera* selected for hygienic behavior have good levels of resistance to *Varroa* mites because infested larvae are removed from the colony before the mites reproduce. Bees may also groom themselves or their nestmates (allogrooming) to remove ectoparasites; this behavior is not genetically related to classic hygienic behavior but may enhance resistance to *Varroa* mites, as well.

The Honeybee Life Cycle

Honeybees have no solitary phase in their life cycle. This differentiates them many, perhaps most, other eusocial insects. Queens depend completely on workers for their survival, and workers are nearly completely dependent on queens for reproduction. (Workers may occasionally lay a male egg.) New colonies are started by swarms, which consist of the reigning queen from a colony and a large number of her workers. Swarms settle usually on a tree limb or fence, but sometimes at inconvenient spots like a parked car, and scouts from the swarm search for appropriate nest sites. Returning scouts dance to indicate possible nest locations and within a few hours or days, the colony arrives at a consensus and moves to the new site, where they initiated comb construction and the queen, once comb is available, starts to lay. In temperate habitats, spring is the most promising time to swarm, as the colony will then have ample time to provision its new nest for the upcoming winter. Hence the Mother Goose rhyme:

A swarm of bees in May
Is worth a load of hay;
A swarm of bees in June
Is worth a silver spoon;
A swarm of bees in July
Is not worth a fly.

In the days prior to swarming, colonies rear several new queens in larger than typical cells that extend from the bottom edges of the comb. Queen larvae receive a diet of royal jelly that apparently stimulates development of the queen morphology but all female eggs have the potential to develop into either queens or workers. While typically only one new queen is needed, the extra queens provide a degree of insurance. The first queen emerging from her cocoon may kill the other queens; sometimes, large colonies produce secondary swarms headed by new queens.

Colonies also produce drones, beginning in the spring and extending well into the summer. Drones fly in aggregations several meters in the air, and queens on mating flights seek those aggregations, with drones then engaging in a chase to actually mate with the queen. Drone genitalia are ripped out and left in the queen's reproductive tract when a drone mates, providing an ineffective block against subsequent matings by the queen.

The Comb and Behavior

One of the most unique and intriguing aspects of honeybee biology is the use of wax for the construction of the nest (**Figure 6**). The mechanical and architectural properties of the comb are the result of an enchanting interplay between an outstanding construction material and perfect design. Honeybees produce wax from glands located in the intersegmental membranes on the sternal surface of the abdomen. Wax scales from the glands are chewed, secretions from glands in the head are added, and then wax is formed into comb. Wax is a metabolically costly product; 6–7 kg of honey are required to make 1 kg of wax. Wax consists of hundreds of compounds. Alkanes, like those found in petroleum-based waxes, are but a small component of beeswax. Wax esters, and, notably, fatty acids, which add strength and resilience, are

Figure 6 *Apis mellifera* comb, showing the beautiful symmetry and efficient use of materials by honeybees in comb construction. Photo by Michael Breed.

important parts of the blend. Even though beeswax has superficial similarities to paraffin wax, its unique properties have made it a preferred material for cosmetics, applications in art such as batik, and high-quality candles. Behaviorally, the fatty acids and alkenes (which are present in smaller amounts) serve as important cues for bees in the colony in social recognition.

Under natural conditions the bees shape the comb. In managed colonies apiarists often give the bees a foundation, made of wax or plastic, that guides comb construction within moveable frames. The hexagonal cells (**Figure 6**) are a marvelous extraction of maximum strength and storage space from a minimum of material. The bees appear to be guided in their construction by innate knowledge and the ability to build new comb following the pattern of existing comb. The comb is a result of the collective behavior of many individual bees, each adding a few wax scales secreted from their wax glands. It may appear that complex guiding forces drive comb-building. Hexagonal construction is, however, not unique to honeybees; it appears in the paper nests of many eusocial wasps and in some parts of stingless bee nests, and may be a simple result of close packing of developing larvae. As the comb is used for rearing larvae, the silk larval cocoons are incorporated, and the old comb has a much darker color than the new comb.

Details of construction, though, are remarkably consistent across species. The size of most of the cells in the comb matches the size of worker larvae; dwarf species produce the comb with small cells and giant bees have very large cells. Cells intended for rearing drones are slightly larger than worker cells, and tend to be located around the margins of the comb. As a unifying feature of the genus *Apis*, the use of the wax comb for nests is, indeed, a marvel of nature.

Division of Labor

All honeybee workers are alike, morphologically, and it seems that all are equally capable of contributing to tasks within the colony. Age is the most important variable in determining the behavior of a bee at any given moment, with the youngest adult workers performing tasks deep inside the colony, such as cleaning and nursing brood. Slightly older bees engage in comb construction. Between 10 and 20 days of adult age workers shift to the periphery of the colony, guarding (**Figure 7**), fanning to circulate air, and removing dead bees. Yet older bees forage or stand ready to fly in colony defense, serving as soldiers. In the summer, workers live roughly 4 weeks. This pattern of age-related shifts in activity is termed as temporal polyethism.

Physiologically, the picture is more complex, as workers vary in the likelihood that they will become nectar or pollen foragers. Genetic differences among workers, at least in part, underly fates of bees in becoming nectar or pollen foragers. The interactions between age, demand for task performance, and genetics in determining worker activities within colonies remain a fascinating topic for future study.

The role of queens is simpler and much more easily defined. After emergence as adults, queens may enter a short phase of competition with other potential queens, which can include fights to the death. Queens then leave the colony for one or more mating flights, after which they return. Having mated 10–20 times (at least in *A. mellifera*), the queen soon begins laying eggs, her only task until she dies; queens may live 5–7 years. Egg-laying by the queen is interrupted only by unfavorable seasons – winter in the temperate zone, dry seasons in the tropics – and times when the colony is absconding (all bees leave the nest and

search for a new location) and swarming (the queen leaves with about half the workers).

Male honeybees, drones, do not work within the colony. Their role is to fly and attempt to mate with queens. Late in summer, drones may be forcibly removed from colonies by the workers.

Queen Pheromones

Queens release pheromones throughout their lives. 9-Oxodecenoic acid (9-ODA) serves as a sex pheromone in *A. mellifera*, attracting males to queens on their mating flight, and a related compound, 10-hydroxydecenoic acid (10-HDA) serves a similar function in *A. florea*. A mixture of compounds from the mandibular glands of queens (queen mandibular pheromone, or QMP) appears to inhibit production of new queens and to maintain some aspects of worker-like behavior. In *A. mellifera*, QMP includes 9-ODA, 9-HDA, methyl-*p*-hydroxybenzoate (HOB), and 4-hydroxy-3-methoxy phenylethanol (HVA). Queens continuously produce QMP; if the queen is removed or QMP production is blocked, workers begin rearing emergency replacement queens. QMP attracts a retinue of workers to surround the queen; QMP is transmitted via retinue workers throughout the colony, but it is volatile and disappears quickly from the colony if the queen is not present.

Other Pheromones

Pheromones probably play a role in almost every aspect of honeybee behavior. Brood pheromone, for example, has received attention as a possible stimulant of foraging behavior. Nasonov pheromone, produced from the Nasonov glands on the dorsum of the abdomen, is composed of highly volatile oils, such as geraniol and citral, and serves as an assembly pheromone during swarming. Footprint pheromones may mark locations that foragers have visited. Alarm pheromones are discussed below. The rich chemical life of honeybees is only partly understood.

Colony Defense

Honeybee colony defense can be loosely divided into three categories of responses – defense against other bee colonies, defense against invertebrate predators and honey thieves, and defense against vertebrates. The primary weapon of the honeybee is the sting, which is supplemented by biting with the mandibles. The sting is autotomous, meaning the tissue attaching the sting to the bee's body is weak and the sting pulls out of the bee easily. Barbs on the shaft of the sting catch in the skin of

Figure 7 A guard honeybee, *Apis mellifera*. Photo by Michael Breed.

the victim, the sting pulls loose, and muscles remain with the sting pump venom into the skin. The queen has a sting, but it is not barbed. Alarm pheromones, highly volatile chemicals released from a gland associated with the sting, serve to alert other bees in the colony to threats.

Honeybee nests have tempting concentrations of honey, and when nectar is low in availability, workers may turn to robbing other colonies. Robbing and defenses against robbing are best studied in *A. mellifera*, which, as pointed out above, tends to store large amounts of honey. The mixture of fatty acids in the comb of any given colony is unique to that colony, and gives the workers the means to discriminate colony mates from noncolony mates. Guard bees, which patrol near hive entrances, examine approaching bees with their antennae and bite and sting bees from other colonies. Guarding is more intense when nectar supplies are low. Less is known about this type of colony defense in other species of honeybee.

Invertebrate predators on honeybee colonies have generated a set of interesting adaptations. In Asia and the Middle East, hornets feed on adult bees and can decimate colonies. Bees forming the blanket on the surface of *A. florea* and *A. dorsata* nests (**Figures 3** and **4**) 'shimmer' – move their wings to produce a wave-like effect radiating through the bees – in response to the presence of a hornet. *A. cerana* workers shimmer at the entrance of their nest. If the bees capture a hornet, large numbers may surround it; the bees then produce heat by shivering and can actually bake a hornet to death.

For most people, the pain caused by the defensive abilities of honeybees against vertebrates determines their most immediate impression of bees. The sting of a single worker can cause local pain, itching, and, in hypersensitive individuals a catastrophic anaphylactic reaction. Honeybee venom consists of two nonenzyme proteins, apamin and melittin, an enzyme, hyalonuridase, and the biogenic amines, dopamine and histamine. Apamin is a neurotoxin which may associate with the pain that accompanies a bee sting. Melittin lyses cell walls, probably enhancing local inflammation. Hyalonuridase breaks down connective tissues, perhaps enhancing the spread of the venom, and dopamine and histamine probably increase circulation in the area of the sting, also improving venom spread. Apamin and melittin are hyperallergens, so vertebrates with repeated exposure may be at risk for developing strong immune reactions to bee venom. Major disturbances of a honeybee nest, by a human or other vertebrate, result in hundreds or thousands of bees flying near the nest. These bees orient to dark colors and movement, and tend to concentrate around the eyes and ankles of the victim. Alarm pheromone released from stings stimulates even more bees to fly.

Defensive behavior against vertebrates is, again, best studied in *A. mellifera*, but the giant honeybees also have a reputation of being aggressive defenders of their nest and fierce stingers. Some ecotypes of *A. mellifera*, sometimes called 'African' or 'Africanized' bees, have particularly intense colony defense, with their defended area extending many meters from the nest and large numbers of bees responding to threats. Following the release of these more strongly defensive bees into Brazil, thousands of people have died in stinging events. 'African' bees are more successful in tropical climates than 'European' bees and have spread to their southern climatic limits in Argentina and near their northern limits in California, Arizona, and Texas. In addition to their defensiveness, 'African' bees are noted for their high likelihood of absconding under poor living conditions, production of large numbers of swarms, and long-distance movement of swarms. Beekeepers in North America attempt to genetically manage their bees to reduce the defensive characteristics of their colonies, while beekeepers in Central and South America have, over time, adapted to working with these more difficult bees.

Foraging Behavior

Honeybees are usually thought of as floral generalists, exploiting a wide range of flowers for their nectar and pollen. Honeybees sometimes collect food from wind pollinated plants, like pollen cattails, or from extrafloral nectaries. Honeybees collecting sugary fluid from soda cans or from unscreened kitchens may be viewed as pests. Within this overall pattern of generalist foraging, though, individual worker bees can remain quite faithful to a given plant species.

Here I highlight a few interesting points about the large topic of communication about food resources. A small subset of foragers work as scouts for a colony. The mechanism for determining which bees scout and how many bees scout is poorly understood. These bees, when they find nectar or pollen, dance upon return to the nest, and recruit other bees to the same food source. As noted above, some foragers specialize in pollen collection, others in nectar, and some collect both types of food. Another class of foragers collects water, which on hot days is brought into the colony and evaporated for cooling; without evaporative cooling extreme temperatures would quickly melt the comb wax. In times of severe forage shortage in the surrounding landscape, honeybee colonies send out few scouts; low activity may conserve food reserves.

Honeybee foragers can range up to several kilometers from their nest, but most of the foraging activity from a given colony is concentrated within a few hundred meters of the nest. Honeybees are not territorial on flowers, so bees from many colonies can forage in the same area without direct aggression occurring at flowers.

This contrasts with some stingless bees, which are highly territorial at flowers.

Honeybees are efficient pollinators; the fidelity of a given bee to a plant species results in effective cross-pollination. Flowers with open corollas and accessible nectaries, such as apples, are easily worked by honeybees, while honeybees have difficulty with deep, narrow corollas, such as honeysuckle, or complicated flowers, like alfalfa. The manipulability of honeybee behavior and their broad range of acceptable flowers makes them the prime agricultural pollinator in many ecosystems.

Genomics and the Future of Honeybee Research

The honeybee genome has been sequenced, with the first release of sequence data in 2003. The availability of genomic data opens important future pathways for learning how gene expression relates to eusocial behavior and caste differentiation. The basic groundplan for reproductive biology in insects, and in Hymenoptera in particular, has been modified in bees to support the existence of distinct reproductive and nonreproductive castes; genomic research may ultimately help to explain how animals with the same genes – queens and workers – can have such different morphology and behavior, and what genetic modifications are required to evolutionarily move from the solitary ancestors of *Apis* to the sophisticated social behavior of this genus.

See also: Ant, Bee and Wasp Social Evolution; Caste Determination in Arthropods; Collective Intelligence; Dance Language; Developmental Plasticity; Division of Labor; Queen–Queen Conflict in Eusocial Insect Colonies; Queen–Worker Conflicts Over Colony Sex Ratio; Social Recognition.

Further Reading

Buchwald R, Greenberg AR, and Breed MD (2005) A biomechanical perspective on beeswax. *American Entomologist* 51: 39–41.
Engel MS (2001) Monophyly and extensive extinction of advanced eusocial bees: Insights from an unexpected Eocene diversity. *Proceedings of the National Academy of Sciences USA* 98: 1661–1664.
Hepburn HR (1986) *Honeybees and Wax*, p. 205. Berlin: Springer-Verlag.
Hepburn HR and Radloff SE (1998) *Honeybees of Africa*, p. 377. Berlin: Springer.
Oldroyd BP and Wongsiri S (2006) *Asian Honey Bees: Biology, Conservation, and Human Interactions*, p. 360. Cambridge: Harvard University Press.
Seeley TD (1996) *The Wisdom of the Hive: The Social Physiology of Honey Bee Colonies*, p. 318. Cambridge: Harvard University Press.
Winston ML (1991) *The Biology of the Honey Bee*, p. 294. Cambridge: Harvard University Press.

Hormones and Behavior: Basic Concepts

R. J. Nelson, Ohio State University, Columbus, OH, USA

Introduction

Behavioral endocrinology is the scientific study of the interaction between hormones and behavior. This interaction is bidirectional: hormones can affect behavior, and behavior can feedback to influence hormone concentrations. Hormones are chemical messengers released from endocrine glands that influence the nervous system to regulate the physiology and behavior of individuals. Over evolutionary time, hormones regulating physiological processes have been co-opted to influence behaviors linked to these processes. For example, hormones associated with gamete maturation such as estrogens are now broadly associated with the regulation of female sexual behaviors. Such dual hormonal actions ensure that mating behavior occurs when animals have mature gametes available for fertilization. Generally speaking, hormones change gene expression or cellular function, and affect behavior by increasing the likelihood that specific behaviors occur in the presence of precise stimuli. Hormones achieve this by affecting individuals' sensory systems, central integrators, and/or peripheral effectors. To gain a full understanding of hormone–behavior interactions, it is important to monitor hormone values, as well as receptor interactions in the brain. Because certain chemicals in the environment can mimic natural hormones, these chemicals can have profound effects on the behavior of humans and other animals.

Behavioral Endocrinology Techniques

A number of methods are used to gather the evidence needed to establish hormone–behavior relationships. Much of the recent progress in behavioral endocrinology has resulted from technical advances in the tools that allow us to detect, measure, and probe the functions of hormones and their receptors. These techniques, with a brief description, are listed in **Table 1**. Several of these techniques are the result of advanced research in behavioral endocrinology, including the time-honored ablation-replacement techniques, bioassays, as well as modern assays that utilize the concept of competitive binding of antibodies that include radioimmunoassay (RIA), enzyme-linked immunosorbent assay (ELISA; enzyme-linked immunoassay (EIA)), autoradiography, and immunocytochemistry. Other techniques commonly used in behavioral endocrinology include neural stimulation and single-unit recording, techniques that activate or block hormone receptors with drugs, and gene arrays and genetic manipulations including interfering with RNA and use of viral gene vectors to deliver novel genes directly into the brain. Because hormones must interact with specific receptors to evoke a response, many of these techniques are used to influence or measure hormone secretion, hormone binding, or the physiological and behavioral effects that ensue after hormones bind to their respective receptors.

Hormones

Hormones are organic chemical messengers produced and released by specialized glands called 'endocrine glands.' *Endocrine* is derived from the Greek root words *endon,* meaning 'within,' and *krinein,* meaning 'to release,' whereas the term *hormone* is based on the Greek word *hormon,* meaning 'to excite.' Hormones are released from these glands into the bloodstream (or the tissue fluid system in invertebrates), where they act on target organs (or tissues) generally at some distance from their origin. Hormones coordinate the physiology and behavior of an animal by regulating, integrating, and controlling its bodily function. Hormones are similar in function to other chemical mediators including neurotransmitters and cytokines. Indeed, the division of chemical mediators into categories mainly reflects the need by researchers to organize biological systems into endocrine, nervous, and immune systems, rather than real functional differences among these chemical signals. Hormones often function locally as neurotransmitters and also interact with neurotransmitters and cytokines to influence behavior.

Hormones can be grouped into four classes: (1) peptides or proteins, (2) steroids, (3) monoamines, and (4) lipid-based hormones. Generally, only one class of hormone is produced by a single endocrine gland, but there are some notable exceptions. It is important and useful to discriminate among the four types of hormones because they differ in several important characteristics, including their mode of release, how they move through the blood, the location of their target tissue receptors, and the manner by which the interaction of the hormone with its receptor results in a biological response. The major vertebrate hormones and their primary biological actions are listed in **Table 2**.

Although exceptions always exist, the endocrine system has several general features: (1) endocrine glands are ductless, (2) endocrine glands have a rich blood supply, (3) hormones, the products of endocrine glands, are

Table 1   Common techniques in behavioral endocrinology

Ablation (removal or extirpation) of the suspected source of a hormone to determine its function is a classic technique in endocrinology. Suspected brain regions that may regulate the behavior in question can also be ablated. Typically, four steps are required: (1) a gland that is suspected to be the source of a hormone affecting a behavior is surgically removed; (2) the effects of removal are observed; (3) the hormone is replaced, by reimplanting the removed gland, by injecting a homogenate or extract from the gland, or by injecting a purified hormone; and (4) a determination is made whether the observed consequences of ablation have been reversed by the replacement therapy.

Radioimmunoassay (RIA) is based on the principle of competitive binding of an antibody to its antigen. An antibody produced in response to any antigen, in this case a hormone, has a binding site that is specific for that antigen. Antigen molecules can be 'labeled' with radioactivity, and an antibody cannot discriminate between an antigen that has been radiolabeled (or 'hot') and a normal, nonradioactive ('cold') antigen. A standard curve is produced with several tubes, each containing the same measured amount of antibody, the same measured amount of radiolabeled hormone, and different amounts of cold purified hormone of known concentrations. The radiolabeled hormone and unlabeled hormone compete for binding sites on the antibody, so the more cold hormone that is present in the tube, the less hot hormone will bind to the antibody. The quantity of bound hormone can be determined by precipitating the antibody and measuring the associated radioactivity resulting from the radiolabeled hormone that remains bound. The unknown concentration of hormone in a sample can then be determined by subjecting it to the same procedure and comparing the results with the standard curve.

Enzymoimmunoassay (EIA), as RIA, works on the principle of competitive binding of an antibody to its antigen. EIAs do not require radioactive tags; instead, the antibody is tagged with a compound that changes optical density (color) in response to binding with antigen. Other than home pregnancy tests, most EIAs are developed to provide quantitative information. A standard curve is generated so that different known amounts of the hormone in question provide a gradient of color that can be read on a spectrometer. The unknown sample is then added, and the amount of hormone is interpolated by the standard curves. A similar technique is called 'enzyme-linked immunosorbent assay' (ELISA).

Immunocytochemistry (ICC) techniques use antibodies to determine the location of a hormone in cells. Antibodies linked to marker molecules, such as those of a fluorescent dye, are usually introduced into dissected tissue from an animal, where they bind with the hormone or neurotransmitter of interest. For example, if a thin slice of brain tissue is immersed in a solution of antibodies to a protein hormone linked to a fluorescent dye, and the tissue is then examined under a fluorescent microscope, concentrated spots of fluorescence will appear, indicating where the hormone is located.

Autoradiography is typically used to determine hormonal uptake and indicate receptor locations. Radiolabeled hormone is injected into an individual or into dissected tissue. Suspected target tissues are sliced into several very thin sections; adjacent sections are then subjected to different treatments. One section of the suspected target tissue is stained in the usual way to highlight various cellular structures. The next section is placed in contact with photographic film or emulsion for some period of time, and the emission of radiation from the radiolabeled hormone develops an image on the film. The areas of high radioactivity on the film can then be compared with the stained section to determine how the areas of highest hormone concentration correlate with cellular structures. This technique has been very useful in determining the sites of hormone action in nervous tissue, and consequently has increased our understanding of hormone–behavior interactions.

Blot tests use electrophoresis to determine in which cells specific DNA, RNA, or proteins are located. Homogenized tissue of interest is placed on a nitrocellulose filter, which is subjected to electrophoresis that involves application of an electric current through a matrix or gel that results in a gradient of molecules separating out along the current on the basis of size (smaller molecules move farther than larger molecules during a set time period). The filter is then incubated with a labeled substance that can act as a tracer for the protein or nucleic acid of interest: radiolabeled complementary deoxyribonucleic acid (cDNA) for a nucleic acid assay, or an antibody that has been radiolabeled or linked to an enzyme for a protein assay. If radiolabeling is used, the filter is then put over film to locate and measure radioactivity. In enzyme-linked protein assays, the filter is incubated with chromogenic chemicals, and standard curves reflecting different spectral densities are generated. Southern blots assay DNA; Northern blots assay RNA, whereas Western blots test for proteins.

In situ hybridization is used to identify cells or tissues in which mRNA molecules for a specific protein (e.g., a peptide hormone) are produced. The tissue is fixed, mounted on slides, and either dipped into emulsion or placed over film and developed with photographic chemicals. Typically, the tissue is also counterstained to identify specific cellular structures. A radiolabeled cDNA probe is introduced into the tissue. If the mRNA of interest is present in the tissue, the cDNA will form a tight association (i.e., hybridize) with it. The tightly bound cDNA, and hence the messenger RNA (mRNA), will appear as dark spots. This technique can be used to determine whether a particular substance is produced in a specific tissue.

Pharmacological techniques are used to identify hormones and neurotransmitters involved in specific behaviors. Some specific chemical agents can act to stimulate or inhibit endocrine function by affecting hormonal release; these agents are called 'general agonists' and 'antagonists,' respectively. Other drugs act directly on receptors, either enhancing or negating the effects of the hormone under study; these drugs are referred to as 'receptor agonists' and 'antagonists,' respectively.

Brain imaging techniques reveal brain activation during behaviors. Paired with endocrine manipulations or monitoring, imaging can provide important information about hormone–behavior interactions. Positron emission tomography (PET) scanning permits detailed measurements of real-time functioning of specific brain regions of people who are conscious and alert. PET gives a dynamic representation of the brain at work. Computer-assisted tomography (CT) scanner shoots fine beams of X-rays into the brain from several directions. The emitted information is fed into a computer that constructs a composite picture of the anatomical details within a 'slice' through the brain of the person. Magnetic resonance imaging (MRI) does much the same thing, but uses nonionizing radiation formed by the excitation of protons by radiofrequency energy in the presence of large magnetic fields. Functional MRI (fMRI) uses a very high spatial (∼1 mm) and temporal resolution to detect changes in brain activity during specific tasks or conditions. When neurons become more active, they use more energy, and require additional blood flow to deliver glucose and oxygen. The fMRI scanner detects this change in cerebral blood flow by detecting changes in the ratio of oxyhemoglobin and deoxyhemoglobin.

Continued

Table 1   Continued

Gene manipulations. In behavioral endocrinology research, common genetic manipulations include the insertion (transgenic or knockin) or removal (knockout) of the genetic instructions encoding a hormone or the receptor for a hormone. In knockout mice, behavioral performance can then be compared among wild-type (+/+), heterozygous (+/−), and homozygous (−/−) mice, in which the gene product is produced normally, produced at reduced levels, or completely missing, respectively. Inducible knockouts when specific genes are inactivated in adulthood promise to become important tools in behavioral endocrinology. An alternative approach involved gene silencing via RNA interference (RNAi), which is used to deplete protein products made in cells.

Gene arrays can be used to determine relative gene expression during the onset of a behavior, or a change in developmental state, or among individuals that vary in the frequency of a given behavior or hormonal state. Essentially, a miniscule spot of nucleic acid of known sequence is attached to a glass slide (or occasionally nylon matrix) in a precise location often by high speed robotics. This identified, attached nucleic acid is called 'the probe,' whereas the sample nucleic acid is the target. The identification of the target is revealed by hybridization, the process by which the nucleotides link to their base pair.

secreted into the bloodstream, (4) hormones can travel in the blood to virtually every cell in the body and can thus potentially interact with any cell that has appropriate receptors, and (5) hormone receptors are rather specific binding sites, embedded in the cell membrane or located elsewhere in the cell that interact with a particular hormone or class of hormones. As mentioned, the products of endocrine glands are secreted directly into the blood, whereas other glands, called 'exocrine glands,' have ducts into which their products are secreted (e.g., salivary, sweat, and mammary glands). Some glands have both endocrine and exocrine structures (e.g., the pancreas). Recently, the definition of an endocrine gland also had to be reconsidered. For example, adipose tissue produces the hormone, leptin, and the stomach produces a hormone called 'ghrelin.' Probably the most active endocrine organ, and the one that produces the most diverse types of hormones is the brain.

As single cells evolved into multicellular organisms, chemical communication within and between cells, as well as between individuals and populations, developed. The endocrine system evolved to become a key component of this complicated intra- and intercellular communication system, although other systems of chemical mediation exist. For example, chemical mediation of intracellular events is called 'intracrine mediation.' Some intracrine mediators may have changed their function over the course of evolution and now serve as hormones or pheromones. Autocrine cells secrete products that may feed back to affect processes in the cells that originally produced them. For example, many steroid–hormone-producing cells possess receptors for their own secreted products. Chemical mediators released by one cell that induce a biological response in nearby cells are called 'paracrine agents'; nerve cells are well-known paracrine cells. In several cases, a single hormone (especially peptides) can have autocrine, paracrine, or endocrine functions. For example, leptin stimulates expression of itself and its receptor. Generally, leptin is produced in adipose tissue and it functions as a hormone when released into the blood by regulating energy balance at the level of the hypothalamus. However, leptin is also produced in the

anterior pituitary gland where it diffuses locally to influence thyroid-stimulating hormone (TSH) secretion (paracrine). Many chemical mediators display similar diversity in function.

Some hormones are water-soluble proteins or small peptides that are stored in the endocrine cell in secretory granules, or vesicles. In response to a specific stimulus for secretion, the secretory vesicle fuses its membrane with the cellular membrane, an opening develops, and the hormones diffuse into the extracellular space via a process called 'exocytosis.' The expelled hormones then enter the blood system from the extracellular space. Other hormones, such as steroid hormones, are lipid soluble (i.e., fat soluble), and because they can move easily through the cell's membrane, they are not stored in the endocrine cells. Instead, a signal to an endocrine gland to produce steroid hormones essentially serves as a signal to release them into the blood as soon as they are produced by the cellular machinery.

Hormone receptors, that are either embedded in the cell membrane or located elsewhere within the cell, interact with a particular hormone or class of hormones. Receptor proteins bind to hormones with high affinity and generally high specificity. As a result of the high affinity of hormone receptors, hormones can be very potent in their effects, despite their very dilute concentrations in the blood (e.g., $1\ \mathrm{ng\ ml}^{-1}$ of blood). However, when blood concentrations of a hormone are high, binding with receptors that are specific for other related hormones can occur in sufficient numbers to cause a biological response (i.e., crossreaction). Also, many hormones are structurally similar so that antibodies designed to attach to one hormone may cross react with other similarly shaped molecules (e.g., growth hormone and prolactin, the sex steroids, and the glycoproteins, viz., luteinizing hormone, follicle-stimulating hormone, and thyroid-stimulating hormone).

But generally, hormones can directly influence only cells that have specific receptors for that particular hormone and served as target cells. The interaction of a hormone with its receptor begins a series of cellular events that either eventually lead to activation of enzymatic pathways

Table 2   Vertebrate hormones

Glands/hormone	Abbreviation	Source	Major biological action
Adrenal glands			
Mineralocorticoids			
Aldosterone		Zona glomerulosa of adrenal cortex	Sodium retention in kidney
11-Deoxycorticosterone	DOC	Zona glomerulosa of adrenal cortex	Sodium retention in kidney
Glucocorticoids			
Cortisol (hydrocortisone)	F	Zona fasciculata and z. reticularis of adrenal cortex	Increases carbohydrate metabolism; antistress hormone
Corticosterone	B	Zona fasiculata and z. reticularis of adrenal cortex	Increased carbohydrate metabolism; antistress hormone
Dehydroepiandro-sterone	DHEA	Zona reticularis of adrenal cortex	Weak androgen; primary secretory product of fetal adrenal cortex
Ovaries			
Estradiol		Follicles	Uterine and other female tissue development
Estriol		Follicles	Uterine and mammary tissue development
Estrone		Follicles	Uterine and mammary tissue development
Progesterone	P	Corpora lutea, placenta	Uterine development; mammary gland development; maintenance of pregnancy
Testes			
Androstenedione		Leydig cells	Male sex characters
Dihydrotestosterone	DHT	Seminiferous tubules and prostate	Male secondary sex characters
Testosterone	T	Leydig cells	Spermatogenesis; male secondary sex characters

Peptide and protein hormones

Hormone	Abbreviation	Source	Major biological action
Adipose tissue			
Leptin (Ob protein)		Adipocytes	Regulation of energy balance
Adiponectin		Adipocytes	Modulates endothelial adhesion molecules
Plasminogen activator inhibitor-1	PAI-1	Adipocytes	Regulation of vascular hemostasis
Adrenal glands			
Met-enkephalin		Adrenal medulla	Analgesic actions in CNS
Leu-enkephalin		Adrenal medulla	Analgesic actions in CNS
Gut			
Bombesin		Neurons and endo-crine cells of gut	Hypothermic hormone; increases gastrin secretion
Cholecystokinin (pancreozymin)	CCK	Duodenum and CNS	Stimulates gallbladder contraction and bile flow; affects memory, eating behavior
Gastric inhibitory polypeptide	GIP	Duodenum	Inhibits gastric acid secretion
Gastrin		G-cells of midpyloric glands in stomach antrum	Increases secretion of gastric acid and pepsin
Gastrin-releasing peptide	GRP	GI tract	Stimulates gastrin secretion
Ghrelin		Stomach mucosa/GI tract	Regulation of energy balance
Glucogon-like peptide-1	GLP-1	L cells of intestine	Regulates insulin secretion
Motilin		Duodenum, pineal gland	Alters motility of GI tract
Secretin		Duodenum	Stimulates pancreatic acinar cells to release bicarbonate and water
Vasoactive intestinal polypeptide	VIP	GI tract, hypothalamus	Increases secretion of water and electrolytes from pancreas and gut; acts as neurotransmitter in autonomic nervous system
Peptide YY	PPY	GI tract	Regulation of energy balance/food intake
Heart			
Atrial naturetic factor	ANF	Atrial myocytes	Regulation of urinary sodium excretion
Hypothalamus			
Agouti-related protein	AGRP	Arcuate nucleus	Regulation of energy balance

Continued

Table 2   Continued

Glands/hormone	Abbreviation	Source	Major biological action
Arg-vasotocin	AVT	Hypothalamus and pineal gland	Regulates reproductive organs
Corticotropin-releasing hormone	CRH	Paraventricular nuclei, anterior periventricular nuclei	Stimulates release of ACTH and β-endorphin from anterior pituitary
Gonadotropin-releasing hormone	GnRH	Preoptic area; anterior hypothalamus; suprachiasmatic	Stimulates release of FSH and LH from anterior pituitary
Gonadotropin-inhibiting hormone	GnIH	Species-dependent loci	Inhibits release of LH (in birds)
Kisspeptin	KISS	Arcuate and anteroventral periventricular nuclei	Critical for normal puberty
Luteinizing hormone-releasing hormone	LHRH	Nuclei; medial basal hypothalamus (rodents and primates); arcuate nuclei (primates)	
Somatostatin (growth hormone-inhibiting hormone)		Anterior periventricular nuclei	Inhibits release of GH and TSH from anterior pituitary inhibits release of insulin and glucagon from pancreas
Somatocrinin (growth hormone-releasing hormone)	GHRH	Medial basal hypothalamus; arcuate nuclei	Stimulates release of GH from anterior pituitary
Melanotropin-release inhibitory factor (Dopamine)	MIF (DA)	Arcuate nuclei	Inhibits the release of MSH (no evidence of this peptide in humans)
Melanotropin-releasing factor	MRF	Paraventricular nuclei	Stimulates the release of MSH from anterior pituitary (no evidence of this peptide in humans)
Neuropeptide Y	NPY	Arcuate nuclei	Regulation of energy balance
Neurotensin		Hypothalamus; intestinal mucosa	May act as a neurohormone
Orexin A and B		Lateral hypothalamic area	Regulation of energy balance/food intake
Prolactin-inhibitory factor (Dopamine)	PIF (DA)	Arcuate nuclei	Inhibits PRL secretion
Prolactin-releasing hormone		Paraventricular nuclei	Stimulates release of PRL from anterior pituitary
Substance P	SP	Hypothalamus, CNS, intestine	Transmits pain; increases smooth muscle contractions of GI tract
Thyrotropin-releasing hormone	TRH	Paraventricular nuclei	Stimulates release of TSH and PRL from anterior pituitary
Urocortin		Lateral hypothalamus	CRH-related peptide
Liver			
Somatomedins		Liver, kidney	Cartilage sulfation, somatic cell growth
Angiotensinogen		Liver, blood	Precursor of angiotensins, which affect blood pressure
Ovaries			
Relaxin		Corpora lutea	Permits relaxation of various ligaments during parturition
Inhibin (folliculostatin)		Follicles	Inhibits FSH secretion
Gonadotropin surge-attenuating factor	GnSAF	Follicles	Control of LH secretion during menstruation
Activin		Sertoli cells	Stimulates FSH secretion
Pancreas			
Glucagon		α-cells	Glycogenolysis in liver
Insulin		β-cells	Glucose uptake from blood; glycogen storage in liver
Somatostatin		δ-cells	Inhibits insulin and glucagon secretion
Pancreatic polypeptide	PP	Peripheral cells of pancreatic islets	Effects on gut in pharmacological doses
Pituitary			
Adrenocorticotropic hormone	ACTH	Anterior pituitary	Stimulates synthesis and release of glucocorticoids
Vasopressin (antidiuretic hormone)	ADH or AVP	Posterior pituitary	Increases water reabsorption in kidney
β-endorphin		Intermediate lobe of pituitary	Analgesic actions

Continued

Table 2   Continued

Glands/hormone	Abbreviation	Source	Major biological action
Follicle-stimulating hormone	FSH	Anterior pituitary	Stimulates development of ovarian follicles and secretion of estrogens; stimulates spermatogenesis
Growth hormone	GH	Anterior pituitary	Mediates somatic cell growth
Lipotropin	LPH	Anterior pituitary	Fat mobilization; precursor of opioids
Luteinizing hormone	LH	Anterior pituitary	Stimulates Leydig cell development and testosterone production in males; stimulates corpora lutea development and production of progesterone in females
Melanocyte-stimulating hormone	MSH	Anterior pituitary	Affects memory; affects skin color in amphibians
Oxytocin		Posterior pituitary	Stimulates milk letdown and uterine contractions during birth
Prolactin	PRL	Anterior pituitary	Many actions relating to reproduction, water balance, etc.
Thyroid-stimulating hormone (thyrotropin)	TSH	Anterior pituitary	Stimulates thyroid hormone secretion
Placenta			
Chorionic gonadotropin	CG	Placenta	LH-like functions; maintains progesterone production during pregnancy
Chorionic somatomammotropin (placental lactogen)	CS (PL)	Placenta	Acts like PRL and GH
Testes			
Müllerian inhibitory hormone	MIH	Fetal Sertoli cells of testes	Mediates regression of Müllerian duct system
Inhibin (folliculostatin)		Seminiferous tubules (and ovaries)	Inhibits FSH secretion
Activin		Sertoli cells	Stimulates FSH secretion
Thyroid/parathyroid			
Calcitonin	CT	C-cells of thyroid	Lowers serum Ca^{2+} levels
Parathyroid hormone	PTH	Parathyroid gland	Stimulates bone resorption; increases serum Ca^{2+} levels
Thyroxine (tetraiodothyronine)	T_4	Follicular cells	Increases oxidation rates in tissue
Triiodothyronine	T_3	Follicular cells	Increases oxidation rates in tissue
Parathyroid-related peptide	PTHrP	Parathyroid gland (and other tissues)	Regulation of bone/skin development
Thymus			
Thymosin		Thymocytes	Proliferation/differentiation of lymphocytes
Thymostatin		Thymocytes	Proliferation/differentiation of lymphocytes

Monoamine hormones
Adrenal glands

Hormone	**Abbreviation**	**Source**	**Major biological action**
Epinephrine (adrenaline)	EP	Adrenal medulla (and CNS)	Glycogenolysis in liver; increases blood pressure
Norepinephrine (noradrenaline)	NE	Adrenal medulla (and CNS)	Increases blood pressure
Central nervous system			
Dopamine	DA	Arcuate nuclei of hypothalamus	Inhibits prolactin release (and other actions)
Serotonin	5-HT	CNS (also pineal)	Stimulates release of GH, TSH, ACTH; inhibits release of LH
Pineal gland			
Melatonin		Pineal gland	Affects reproductive functions

Lipid-based hormones (eicosanoids)

Hormone	**Abbreviation**	**Source**	**Major biological action**
Leukotrienes	LT	Lung	Long-acting bronchoconstrictors
Prostaglandins E_1 and E_2	PGE_1 and PGE_2	Variety of cells	Stimulates cAMP

Continued

Table 2   Continued

Glands/hormone	Abbreviation	Source	Major biological action
Prostaglandins $F_{1\alpha}$ and $F_{2\alpha}$	$PGF_{1\alpha}$ and $PGF_{2\alpha}$	Variety of cells	Active in dissolution of corpus luteum and in ovulation
Prostaglandin A_2	PGA_2	Kidney	Hypotensive effects
Prostacyclin I	PGI_2	Variety of cells	Increased second messenger formation
Thromboxane A_2	TX_2	Variety of cells	Increased second messenger formation

Reproduced from Nelson RJ (2005) *An Introduction to Behavioral Endocrinology*. Sunderland, MA: Sinauer Associates.

or to a genomic response wherein the hormone acts directly or indirectly to activate genes that regulate protein synthesis. The newly synthesized proteins may activate or deactivate other genes, causing yet another cascade of cellular events see section 'Steroid Hormone' below.

When sufficient receptors are unavailable because of a clinical condition, or because previous high concentrations of a hormone have occupied all the available receptors and new ones have yet to be made, a biological response may not be sustained (see later). Such a reduction in the numbers of receptors may lead to a so-called endocrine deficiency despite normal or even supernormal levels of circulating hormones. For example, a deficiency of androgen receptors can prevent the development of male traits despite normal circulating testosterone concentrations. Conversely, elevated receptor numbers may produce clinical manifestations of endocrine excess despite a normal blood concentration of the hormone. Thus, in order to understand hormone–behavior interactions, it is sometimes necessary to characterize target tissue sensitivity (i.e., the number and type of receptors possessed by the tissue in question) in addition to measuring hormone concentrations.

Protein Hormones

Most vertebrate hormones are proteins. Protein hormones that comprise only a few amino acids in length are called 'peptide hormones,' whereas larger ones are called 'protein' or 'polypeptide hormones.' Protein and peptide hormones include insulin, the glucagons, the neurohormones of the hypothalamus, the tropic hormones of the anterior pituitary, inhibin, calcitonin, parathyroid hormone, the gastrointestinal (gut) hormones, ghrelin, leptin, adiponectin, and the posterior pituitary hormones. Protein and peptide hormones can be stored in endocrine cells and are released into the circulatory system by means of exocytosis. Protein and peptide hormones are soluble in blood, and therefore, do not require a carrier protein to travel to their target cells, as do steroid hormones. However, protein and peptide hormones may bind with other plasma proteins, which slow their metabolism by peptidases in the blood. Hormones are removed from the blood

via degradation or excretion. The metabolism of a hormone is reported in terms of its biological half-life, which is the amount of time required to remove half of the hormone (radioactively tagged) from the blood. Generally, larger protein hormones have longer half-lives than smaller peptide hormones (e.g., growth hormone has 200 amino acids and a biological half-life of 20–30 min; thyroid-releasing hormone has three amino acids and a biological half-life of <5 min in humans). Again, a gut hormone such as cholecystokine (CCK) may function in a paracrine manner when released locally in the brain to affect behavior.

Steroid Hormones

The adrenal glands, the gonads, and the brain are the most common sources of steroid hormones in vertebrates. Vertebrate steroid hormones have a characteristic chemical structure that includes three six-carbon rings plus one conjugated five-carbon ring. In the nomenclature of steroid biochemistry, substances are identified by the number of carbon atoms in their chemical structure. The precursor to all vertebrate steroid hormones is cholesterol. The cholesterol molecule contains 27 carbon atoms (a C_{27} substance), although cholesterol itself is not a true steroid and can be stored within lipid droplets inside cells.

Because steroid hormones are fat soluble and move easily through cell membranes, they are never stored, but leave the cells in which they were produced almost immediately. A signal to produce steroid hormones is also a signal to release them. The range of responses can be a rather slow one: the delay between stimulus and response in biologically significant steroid production may be hours, although ACTH stimulates corticoid secretion within a few minutes and LH acts quickly to affect progesterone production during the periovulatory surge. In most cases, however, the signal to produce steroids is relatively slow; steroid hormones are not very water soluble. In the circulatory system, steroid hormones must generally bind to water-soluble carrier proteins that increase the solubility of the steroids and transport them through the blood to their target tissues. These carrier proteins also protect the steroid hormones from being degraded prematurely. The target tissues have receptors for steroid hormones and accumulate steroids against a concentration gradient.

Upon arrival at the target tissues, steroid hormones dissociate from their carrier proteins and either interact with receptors embedded in the membrane or diffuse through the cell membrane into the cytoplasm or nucleus of the target cell, where they bind to cytoplasmic receptors. The amino acid sequence of steroid hormone receptors is highly conserved among vertebrates. Each steroid hormone receptor comprises three major domains: the steroid hormone binds to the C-terminal domain, the central domain is involved in DNA binding, and the N-terminal domain interacts with other DNA-binding proteins to affect transcriptional activation. Steroid receptors are kept inactive by the presence of corepressors (consisting mainly of heat-shock proteins (HSP)), which bind to the internal receptors and keep them inactive. It is the release of these HSP after formation of the hormone–receptor complex that activates the steroid receptor, and if not there already, the activated steroid–receptor complex is transported into the cell nucleus, where it binds to DNA sequences called 'hormone response elements' and stimulates or inhibits the transcription of specific mRNA. The effects of environmental, social, or other extrinsic or intrinsic factors on the regulation of specific coactivators have been understudied and represent yet another process by which individual variation in hormone–behavior interactions may be mediated. Environmental factors such as day length can determine whether photoperiod regulates whether steroids affect physiological and behavioral processes via slow (hours to days) genomic or fast (seconds to minutes) nongenomic pathways. For instance, in beach mice, estrogen rapidly (<15 min) increases aggression in short- but not long-day mice. This suggests that estrogen increases aggression via nongenomic actions on short days, but not on long days. Moreover, gene chip analyses indicated that estrogen-dependent expression of genes containing estrogen response elements in their promoters was decreased in the brains of short-day mice compared with that of long-day mice suggesting that the environment regulates the effects of steroid hormones on aggression in by determining the molecular pathways that are activated by steroid receptors. Transcribed mRNA migrates to the cytoplasmic rough endoplasmic reticulum, where it is translated into specific structural proteins or enzymes that produce the physiological response.

Thus, the actions of steroids on target tissues are based on three factors: (1) the steroid hormone concentrations in the blood, (2) the number of available receptors in the target tissue, and (3) the availability of appropriate coactivators. Blood concentrations of steroid hormones are also dependent on three factors: (1) the rate of steroid biosynthesis; (2) the rate of steroid inactivation by catabolism, which occurs mainly in the liver; and (3) the 'tenacity' (affinity) with which the steroid hormone is bound to its plasma carrier protein. Recently, it has been determined that different 'types' of steroid receptors exist. For example, three versions of the estrogen receptor (α, β, and γ) are currently recognized. Multiple versions of steroid receptors represent another mechanism by which responsiveness to steroid hormones can be regulated. Also, it appears that the brain can produce steroid hormones de novo and that the local effects of these steroids can have dramatic behavioral effects without altering blood concentrations of these hormones. These paracrine effects of steroids in the brain present special challenges to assessing hormone–behavior interactions.

How Might Hormones Affect Behavior?

All behavioral systems, including animals, comprise three interacting components: (1) input systems (sensory systems), (2) integrators (the central nervous system), and (3) output systems, or effectors (e.g., muscles). Again, hormones do not cause behavioral changes. Rather, hormones influence these three systems so that specific stimuli are more likely to elicit certain responses in the appropriate behavioral or social context. In other words, hormones change the probability that a particular behavior will be emitted in the appropriate situation. This is a critical distinction that affects conceptualization of hormone–behavior relationships. For example, female rodents must adopt a rigid mating posture (called 'lordosis') for successful copulation to occur. Females only show this posture when blood estrogen concentrations are high coincident with the maturing ova. Females adopt the lordosis posture in repsonse to tactile stimuli provided by a mounting male. Estrogens affect sensory input by increasing the receptive field size in sensory cells in the flanks. Estrogen affects protein synthesis, the electrophysiological responses of neurons, and the appearance of growth-like processes on neurons in the central nervous system, thus altering the speed of processing and connectivity of neurons. Finally, estrogen affects the muscular output that results in lordosis, as well as chemosensory stimuli important in attracting a mating partner.

How Might Behavior Affect Hormones?

The female rodent mating posture example demonstrates how hormones can affect behavior, but, as noted previously, the reciprocal relation also occurs, that is, behavior can affect hormone concentrations. For example, chemosensory cues from males may elevate blood estradiol concentrations in females, and thereby stimulate proceptive or male-seeking behaviors. Similarly, male mammals that lose an aggressive encounter decrease circulating testosterone concentrations for several days or even weeks afterward. Similar results have also been reported in humans. Human testosterone concentrations are affected not only in those involved in physical combat, but also in those involved in simulated battles. For example, testosterone concentrations are elevated in winners and reduced in losers of regional chess tournaments.

Types of Evidence for Establishing Hormone–Behavior Interactions

What sort of evidence would be sufficient to establish that a particular hormone affected a specific behavior or that a specific behavior changed hormone concentrations? Experiments to test hypotheses about the effects of hormones on behavior must be carefully designed, and, generally, three conditions must be satisfied by the experimental results to establish a causal link between hormones and behavior: (1) a hormonally dependent behavior should disappear when the source of the hormone is removed or the actions of the hormone are blocked, (2) after the behavior stops, restoration of the missing hormonal source or its hormone should reinstate the absent behavior, and (3) finally, hormone concentrations and the behavior in question should be covariant, that is, the behavior should be observed only when hormone concentrations are relatively high and never or rarely observed when hormone concentrations are low.

The third class of evidence has proved difficult to obtain because hormones may have a long latency of action, and because many hormones are released in a pulsatile manner. Also, some pharmaceutical grades of steroids (e.g., esterfied steroids) have been altered to remain in circulation longer than endogenous steroids. Pulsatile secretion of hormones presents difficulties with making hormone–behavior inferences. For example, if a pulse of hormone is released into the blood, and then is not released for an hour or so, a single blood sample will not provide an accurate picture of the endocrine status of the animal under study. Completely different conclusions about the effect of a hormone on behavior could be obtained if hormone concentrations were assessed at their peak rather than at their nadir. This problem can be overcome by obtaining measures in several animals or by taking several sequential blood samples from the same animal and averaging across peaks and valleys. Another problem is that biologically effective amounts of hormones are vanishingly small and difficult to measure accurately. Effective concentrations of hormones are usually measured in micrograms (μg, 10^{-6} g), nanograms (ng, 10^{-9} g), or picograms (pg, 10^{-12} g). The development

of techniques, such as the radioimmunoassay, has increased the precision with which hormone concentrations can be measured, but because of the multiple difficulties associated with obtaining reliable covariant hormone–behavior measures, obtaining the first two classes of evidence usually has been considered sufficient to establish a causal link in hormone–behavior relations.

The unique conditions of the laboratory environment may themselves cause changes in an animal's hormone concentrations and behavior that may confound the results of experiments; thus, it has become apparent that hormone–behavior relationships established in the laboratory should be verified in natural environments. The verification of hormone–behavior relationships in natural environments is challenging, but useful for differentiating laboratory artifacts from true biological phenomena. Establishing hormone–behavior interactions in the field presents other challenges including difficulties in reliability, treatment with exogenous hormones, and recaptures for hormone determinations. These difficulties can be overcome by noninvasive hormone determinations (e.g., fecal steroid assays), but again, coordination between lab and field studies is needed for a full appreciation of hormone–behavior interactions.

See also: Field Techniques in Hormones and Behavior.

Further Reading

Balthazart J and Ball GF (2006) Is brain estradiol a hormone or a neurotransmitter? *Trends in Neurosciences* 29: 241–249.

Brosens JJ, Tullet J, Varshochi R, and Lam EW (2004) Steroid receptor action. *Best Practices in Research and Clinical Obstetrics and Gynaecology* 18: 265–283.

Chen C, Chang YC, Liu CL, Chang KJ, and Guo IC (2006b) Leptin-induced growth of human ZR-75-1 breast cancer cells is associated with up-regulation of cyclin D1 and c-Myc and down-regulation of tumor suppressor p53 and p21WAF1/CIP1. *Breast Cancer Research and Treatment* 98: 121–132.

Hadley M and Levine JE (2007) *Endocrinology.* New York, NY: Benjamin Cummings.

Kiran SK, Scotti M-AL, Newman AEM, Charlier TD, and Demas GE (2008) Novel mechanisms for neuroendocrine regulation of aggression. *Frontiers in Neuroendocrinology* 29: 476–489.

Nelson RJ (2005) *An Introduction to Behavioral Endocrinology.* Sunderland, MA: Sinauer Associates.

Nelson RJ and Trainor BC (2007) Neural mechanisms of aggression. *Nature Reviews Neuroscience* 8: 536–546.

Norris DO (2007) *Vertebrate Endocrinology.* San Diego, CA: Elsevier Academic Press.

Trainor BC, Lin S, Finy MS, Rowland MR, and Nelson RJ (2007) Photoperiod reverses the effects of estrogen on male aggression via genomic and non-genomic pathways. *Proceedings of the National Academy of Sciences of the United States of America* 104: 9840–9845.

Hormones and Breeding Strategies, Sex Reversal, Brood Parasites, Parthenogenesis

C. Rowe, Newcastle University, Newcastle upon Tyne, UK

Introduction

Antipredator defenses provide us with some of the most fascinating and puzzling adaptations in the natural world. For a moment, just think of the spines of a porcupine, the vivid warning colors of lethally toxic poison dart frogs, the hot acidic spray of bombardier beetles, or the color changes in the skin of chameleons. They have all evolved as strategies to avoid being eaten by predators.

Traditionally they have been classified into 'primary' and 'secondary' defenses. Primary defenses are those which deter predators from actively pursuing prey, while secondary defenses are those which come into play once a prey has been attacked and caught, and which enhance the prey's probability of escape and survival. While this classification can be useful, it is confusing when a defense can function both as a primary and a secondary defense. For example, the conspicuous spines of some caterpillars are highly visible to deter attack (a primary defense) but also vicious enough to act as a repellent toward any predator (a secondary defense).

Because of this, defense strategies are now often classified according to the stage of predation at which they enhance survival. There are four stages of predation that can be clearly defined. First, prey are detected and identified as a profitable source of nutrients. Once prey have been located, predators initiate their attack on the prey and actively pursue their quarry. The third stage is capture, when the predator secures the prey and prevents its escape. And finally, predators consume their prey.

At any of these four stages of detection, pursuit, capture, and consumption, prey can have defenses which enable them to escape predation. In the first part of this article, I will describe some of these incredible adaptations. In the second part of the article, I will use recent research on crypsis and aposematism to demonstrate the importance of sensory and cognitive processes of predators in the evolution of prey defenses. The main selection pressure for prey defenses comes from the predators themselves, and we can begin to understand the nature and the benefits of prey defenses by studying their predators. Finally, I will briefly review how predators can evolve counterstrategies in an attempt to overcome prey defenses in the struggle for existence between predators and prey.

Avoiding and Surviving an Attack

There is a bewildering array of defensive strategies that prey have evolved to increase their survival chances. While it is impossible to provide an exhaustive review here, I will describe some of the amazing adaptations that have evolved as antipredator defenses at each of the four stages of attack.

How Do Prey Avoid Detection?

Animals can make themselves hard to find in many ways, and there has been a long history of studying animal camouflage. Many animals make themselves highly cryptic by matching their body coloration to that of their backgrounds. Where animals rest predominantly against a monomorphic background, they can simply match the dominant color of their environment. For example, green aphids can be hard to see on plant stems, and sand-colored fish resting on the seabed can be near-impossible to detect. However, other species are more mobile, or live in changing environments, and need to be able to match different environmental backgrounds. Mobile animals need a different strategy to avoid detection, since a color that is cryptic in one part of the environment may stand out in another. For example, some species of tropical butterflies, like the glasswing butterfly (*Greta oto*), have transparent wings which allow them to match the color of any background when resting.

Another solution to the problem of being cryptic in multiple environments is to change coloration. Some animals will change their coloration according to the season, like the ptarmigan (*Lagopus muta*) and arctic hare (*Lepus arcticus*), which are both white to match the snow in winter, and brown in the summer when the ground is bare. An extreme example of this can be found in cuttlefish and octopuses, which can rapidly adopt the color pattern of their background within seconds. Experiments with *Sepia officinalis* have shown that they can even match the pattern of a checkerboard. Chromatophores located in the skin make these dynamic color changes possible. These chromatophores are small sacs of different color pigments that the animal can contract or expand to produce different color patterns. Octopuses can even change the texture of their skin according to the substrate that they currently rest on, for example seaweed or rock, making them even harder to find.

While some color patterns reduce detection via background matching, others reduce detection through other means, for example, through the use of conspicuous and contrasting markings, a phenomenon known as 'disruptive coloration.' Disruptive coloration is an important type of camouflage that makes use of strongly contrasting elements which create false edges and break up the body outline. This can be achieved by either creating false boundaries within the body of the animal or using punctuated bold patterns along the body edge. Although there are many potential examples of this in nature, disruptive coloration has only been shown to be important feature in the design of animal concealment patterns in a single species of marine isopod (*Idotea baltica*). However, this may be because of the technical skills required to quantify color patterns rather than rarity in nature.

Another good way to hide the body is to minimize shadowing, which may be a useful cue for hunting predators. Some prey flatten their bodies against a surface to reduce their shadows, while others use a strategy known as 'countershading.' Countershading occurs when the dorsal surface is darker than the ventral surface; when lit from above, such an animal's shadow is hidden. We commonly find this type of concealment in both terrestrial and aquatic species, and good examples would include many species of fish, such as mackerel or sardines, as well as many species of birds, including penguins and wading species.

Finally, and perhaps most impressively, some animals have the most exquisite resemblance to objects in their environment that do not interest hunting predators. For example, stick insects mimic the stems of their host plants, sea dragons disguise themselves as pieces of seaweed, and some swallowtail larvae look remarkably similar to bird droppings. This is known as 'masquerade,' and is aimed at deceiving the predators' abilities to recognize them as edible prey. Therefore, this strategy differs slightly from crypsis. While crypsis reduces the likelihood of detection, predators detect masquerading pray, but the predator does not see them as food. Predators may be completely unaware that prey are present, even when they can see them.

These strategies have all considered how prey avoid detection by visually hunting predators. However, prey can reduce the probability of being detected in other sensory modalities too. Perhaps the simplest case of reducing detectability in another sensory modality is silence, particularly when the predation risk is high. It may also be possible to mask odors or perhaps smell like another aspect of the environment. Research in this area is rather limited, probably because vision is such an important sensory system in humans. However, it is possible that future research in this area may find equivalents to visual defensive strategies in other sensory domains.

If Detected, How Do Prey Avoid Being Attacked?

Once detected, if prey become threatened by the proximity or investigative behavior of a predator, there are a number of behavioral responses that can help prevent a prolonged or persistent attack. For example, prey can make a dash for cover, or brandish weaponry as a threat. Prey can also attempt to startle a predator by a sudden display of conspicuous coloration. For example, many species of arctiid moths have highly conspicuous hindwings, which they normally cover with their camouflaged forewings. If detected, they flash these conspicuous markings, which are often highly contrasting eye spots; this may give them the time to escape. Prey can also alter their appearance to look like a more dangerous species that would be a potential threat to a predator. Perhaps one of the most impressive examples of this is found in the fifth instar of the spicebush swallowtail (*Papilio troilus*), which swells upon investigation by a predator to take on the appearance of a small snake.

Another way in which animals can reduce the chances of being pursued by a predator is to signal that an attack is unlikely to be profitable. For example, the prey may just signal to the predator that it has been seen, meaning that an attack has lost any element of surprise. Stotting in Thomson gazelle (*Eudorcas thomsoni*) has been one of the most cited examples of signaling between predators and prey. When the gazelle detect a predator, such as a lion or hyena, they leap high in the air, keeping their legs pointing straight down in a highly stereotyped display. This not only signals to the predator that it has been seen, but it may provide information about the health of the individual, and that it is fast and fit enough to escape any subsequent attack.

Some prey have chemical or physical defenses, and signal this to predators in order to prevent an attack. Of course, some defenses are easy to detect, such as spines or hairs, which can cause physical damage to the predator, or just take longer to handle, reducing the time a predator has to find more edible prey. However, toxins are often stored inside the body and therefore need to be signaled to the predator in advance. 'Aposematism' is the term used to describe the phenomenon where toxic prey use conspicuous warning color patterns to advertise them to predators. This defensive strategy has been well studied in a range of species, including monarch butterflies (*Danaus plexippus*), coral snakes (*Micrurus* spp.), and poison dart frogs (*Epipedobates* spp.). Prey can also employ olfactory and auditory warning signals, which are produced either separately or in addition to warning coloration. For example, at night, when coloration is ineffective, moths produce clicks to signal to foraging bats. These signals are effective deterrents because the predator learns to associate the signal with unprofitability and reduces its attack probability on signaling prey.

If Attacked, How Can Prey Avoid Being Caught?

Aside from running fast or finding a place to hide, prey have evolved a number of adaptations to avoid capture. Living in a group can be an important way for an individual to avoid being caught. First, if a predator attacks a group, the chances of a single individual being selected is much lower than if the animal were alone. This is known as the 'dilution effect' because the presence of conspecifics reduces the risk of a single individual being caught. This kind of defense is particularly important in environments which do not provide many opportunities for escape, and may explain the large schools of fish found in the open sea, and the herding behavior of herbivores on the African savannah. Animals in a group can also reduce the probability of being caught by combining individual defenses. One of the best examples of this comes from musk oxen (*Ovibus moschatus*), which form a circle, facing outward to defend all angles from attack from predators, such as wolves. In this way, the combined efforts of the group provide a more effective defense than a single individual could achieve.

Weaponry is also a common way to reduce vulnerability to capture. For example, ungulates have horns and antlers that help them fend off predators, while spines protect a range of animals, including Daphnia, sticklebacks, and porcupines. These defenses make the animals difficult to grasp and consume, allowing prey to escape. Other adaptations also make it difficult for predators to hold on to the prey. For example, some prey can readily shed parts of their body if grasped by a predator, a phenomenon known as 'autotomy.' Many lizards can drop their tails, which continue to move after being shed, and may distract a predator away from the fleeing prey. While this may seem an extreme adaptation, the lizard can regrow its tail in just a few weeks. This is a relatively small price to pay to prevent capture.

If Caught, How Can Prey Avoid Being Eaten?

Prey employ a number of 'last resort' defenses to prevent predators from eating them. Some prey, such as caterpillars, have hairs that make them difficult to handle and ingest. Chemical defenses can also be employed upon capture and some can be extremely unpleasant for the predator. For example, bombardier beetles can spray hot acid in the direction of an attacker, enabling them to make good their escape. However, other secretions and regurgitations that prey produce may allow predators to taste the presence of toxins and avoid prey with high toxin content. Chemical defenses are effective deterrents, and although injury may occur during the attack, chemically defended prey tend to be quite robust and can have high postattack survival rates.

Another interesting strategy to prevent consumption is that of immobility or feigning death. This is an effective strategy for prey because eating prey that is already dead can be dangerous for predators, and consequently, the predators often lose interest in nonmoving prey. This phenomenon has been well documented across the animal world and can be particularly effective in the presence of other prey that can distract the predator's attention.

The Evolution of Prey Defensive Strategies

So far, I have cataloged some of the amazing antipredator strategies found in the natural world, but how did they evolve? Of course, the defenses are adaptations to avoiding predation and, therefore, the capabilities of their predators have been crucial in the evolutionary process. In this section, I first outline recent experiments that show how the perceptual and cognitive mechanisms underlying predatory behavior have been a powerful selection force on prey. I will then consider how predators have evolved counterstrategies to overcome prey defenses and also how the costs of defenses also influence their evolution.

Perceptual Processes

Given that crypsis reduces the probability of detection, we would predict that it will have been under strong selection pressures from the perceptual processes of predators. What pattern features make an animal cryptic and how do cryptic color patterns fool a predator's visual system? Recent experiments using artificial moths have begun to reveal how cryptic patterns can interact with the avian visual system.

In one laboratory-based approach, artificial digital 'moths' have been constructed as pixilated triangles and presented on touchscreens to blue jays (*Cyanocitta cristata*). The experimenter controls the background and the visual properties of the virtual moth, and the birds receive a food reward for correctly detecting a moth on the background. In this way, it is possible to see which moth patterns birds detect easily and which are more difficult. Experiments simulate the evolution of color patterns by starting with a parental population of digital moths (see **Figure 1(a)**) and then allowing surviving patterns (i.e., those not detected by the jays) to 'reproduce' and generate the next generation of digital moths. At the end of these experiments, by comparing the patterns that have been selected by jays to those that have been selected randomly in a control condition, it is easy to see that the jays select for more cryptic prey (see **Figure 1(b)** and **1(c)**). These experiments also show that the birds select for moths with polymorphic or asymmetric patterns, and that they tend to use commonly identifiable or symmetrical features to search for prey. It is therefore likely that moths in the wild will be selected for the same kinds of features when faced

Parental population (P$_0$)

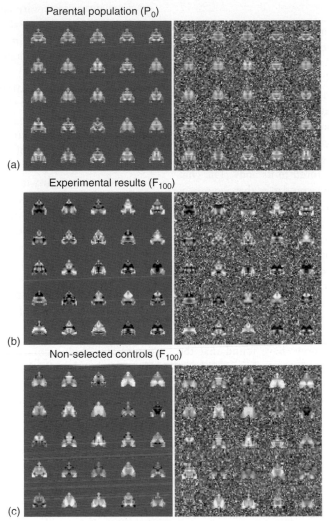

Experimental results (F$_{100}$)

(a)

(b)

Non-selected controls (F$_{100}$)

(c)

Figure 1 Samples of digital moths on a uniform gray (left) and textured (right) backgrounds from (a) a parental population, (b) a population that was selected by jays, and (c) a population that was not under selection. In this experiment, the moths selected by the jays are more polymorphic than the parental population or the nonselected controls. Reproduced from Bond AB and Kamil AC (2001) Visual predators select for crypticity and polymorphism in virtual prey. *Nature* 415: 609–614.

with avian predators, which is indeed what we see in a many moth species. These kinds of experiments illustrate how perceptual abilities and search strategies of predators can explain the color patterns that we see in nature.

A second, successful field-based approach has involved pinning mealworms with patterned triangular paper wings to tree trunks in the forest to make 'artificial moths.' The color patterns of these paper wings have been calibrated to ensure that they match the background in 'avian color space.' This means that the colors of the wings and the bark appear the same to birds, which have different spectral sensitivities to our own and can even see colors that we can't see. By manipulating the pattern on the wings and measuring the rates at which mealworms are found and removed, it is possible to compare the effectiveness of different patterns in hiding the prey from foraging avian predators. This technique has been used to show the importance of disruptive coloration in enhancing prey survival, for example, by manipulating the degree to which pattern elements fall against the body edge (see **Figure 2**). Similar experiments have also confirmed the role of contrasting pattern elements and pattern asymmetry as important components in evading detection by predators.

Taken together, these experiments have led the way in demonstrating the importance of the predator's visual detection system in the evolution of cryptic color patterns. They not only provide valuable insights into how predators see the world, but also measure the impact of their search strategies on the defenses of their prey. Indeed, there is now a renewed interest in cryptic coloration and how different types of pattern are effective at reducing detection.

Cognitive Processes

The importance of cognition in the evolution of prey defenses has been best studied in the context of aposematism, a defense strategy that is taxonomically widespread, and is found across terrestrial and aquatic habitats. Aposematism is the signals between conspicuous coloration and the presence of defenses, such as toxins or spines. Perhaps one of the better known examples of aposematism is the monarch butterfly (*D. plexippus*), which has a distinctive orange and black pattern to warn predators of the toxins (cardiac glycosides) that are stored in its body. The conspicuous coloration is a signal to predators that the prey is dangerous, which the predator can use to avoid attacking it. However, one of the main challenges in understanding the evolution of aposematism is understanding the advantage of being conspicuous: why do prey advertise their presence to predators by being so conspicuous, when surely it is better to maintain crypsis?

The answer to this question has come from understanding aversion learning in predators. Over the last 30 years, experiments using avian predators foraging on prey (both live and artificial prey) have found that the conspicuous coloration can be beneficial when considered from a predator's point of view. In particular, birds can learn to avoid toxic prey faster when they are conspicuously colored compared to when they are cryptically colored. Therefore, conspicuous coloration is advantageous to toxic prey because it ensures that predators learn quickly about their signal, making it a more effective deterrent. This means that fewer individuals are killed during the learning process, and warning signals enjoy a selective advantage. However, experiments have also found additional benefits to being conspicuous. For example, warning signals may allow prey to be detected earlier, giving predators more time to recognize the prey as aposematic and not make mistakes in identification. Conspicuously colored prey may also be approached more cautiously, prompting predators to 'go slow' and carefully inspect prey before committing to an attack. These benefits to being conspicuous are thought to outweigh the costs of being more detectable.

Predator cognition has also been studied in order to explain the evolution of other common features of warning signals, for example, that they are often 'multimodal.' Multimodal signals are those that occur in multiple sensory modalities, and many warningly colored prey produce sounds or odors upon attack. Recent studies show that these additional signal components can also enhance the speed with which predators learn to avoid aposematically colored prey. Therefore, studying the cognitive processes underlying how animals combine sensory information can also help explain the complex nature of warning signals.

Finally, learning and memory have also been important for understanding the evolution of mimicry, where prey species share the same warning signals. In Müllerian mimicry, two sympatric and defended species share the same warning signal, but what are the benefits to that? The proposed benefits again come through predator education and how predators learn to avoid toxic prey. The original hypothesis, proposed more than 100 years ago, suggested that if predators need to attack and ingest a certain number of toxic prey before the predator learns to avoid them, then it is better for species to share the same color pattern, thus reducing the number of prey killed in each species. Although recent experiments have shown that mimicry does reduce mortality in each species during predator learning, many questions remain about how factors such as toxin content and prey density affect predator learning and the evolutionary dynamics of mimicry. Only by understanding these cognitive processes will be able to fully understand the evolution of aposematism and mimicry.

Co-evolution of Predators and Prey: Overcoming Prey Defenses

Of course, as prey evolve defenses against predators, predators also evolve counterstrategies to overcome them. We must therefore think of the evolution of prey defenses as a coevolutionary arms race, where strategies and counterstrategies are constantly evolving as both predator and prey struggle to survive.

For example, predators have evolved search strategies that help them detect cryptic prey. As the work with the blue jays shows, predators seem to be more effective at detecting prey when searching for monomorphic prey compared to prey that have a more varied appearance. It has been suggested that this phenomenon could explain the evolution of prey polymorphism, where prey populations are not only cryptic, but also variable in their appearance. This variation could be a counteradaptation on the part of the prey to prevent predators from finding them so easily.

Some predators have evolved specialized adaptations and behaviors to overcome different types of prey defenses. For example, birds will often taste aposematic prey before eating them in order to assess the amount of toxin they contain and reject them if the concentration is too high. This behavior allows them to discriminate between prey that have differing amounts of toxin and reduce their intake. Predators might also invest in detoxification enzymes to reduce the potency of any toxin ingested, or perhaps eat clay to actively bind to toxins and prevent them from passing from the gut into the body. The investment by predators in these countermeasures may require prey to invest more heavily in toxins in order to maintain an effective deterrent. The relationship between predators and prey is dynamic, and will also be influenced by other predators and prey in the environment. We have to remember that what we see today is just a snapshot in a coevolutionary process.

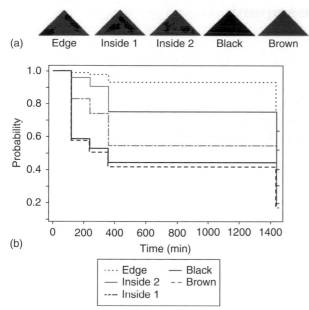

Figure 2 An example of how disruptive color patterns can enhance the survival of artificial moths in the wild. (a) Patterns were manipulated to have contrasting edge elements (Edge), the same elements but presented within the body outline (Inside 1), random elements within the body outline (Inside 2), or be monochrome (Black and Brown). (b) The probability of survival over time was much greater for prey with Edge patterns, confirming that the edge patterning makes prey harder to detect. Reproduced from Cuthill IC, Stevens M, Sheppard J, Maddocks T, Parraga CA, and Troscianko TS (2005) Disruptive coloration and background pattern matching. *Nature* 434: 72–74.

Costs of Defense

Up to this point, I have only considered the benefits of defenses to prey, but like any other trait, defenses can be costly. Possessing defenses, particularly chemical and physical defenses, requires an investment of resources. Although toxins do not always carry significant costs to prey, they have been found to limit growth rates and size in some insect species. Prey may also need to invest in anatomical specializations in order to keep toxins in their bodies, or detoxification pathways to ensure that they are not poisoned by their own defenses. Physical defenses, such as spines and hairs, also require energy and may also lead to slower growth. Defenses may also inflict opportunity costs on prey. For example, crypsis may actually constrain animals to particular environments or niches. If prey cannot evolve color patterns that give them protection in different types of habitat, then they may be limited in the environmental resources that they can exploit. Because of the costs involved in having a defense, some prey only invest in defenses when the perceived risk of predation is high. For example, some tadpoles only start

producing distasteful chemicals when conspecifics have been attacked by a predator. This plasticity is further evidence for prey needing to allocate resources between defenses and other traits that enhance their fitness.

Conclusions

The interaction between predators and prey over evolutionary time has led to some of the most striking adaptations seen across the animals. This article has highlighted some of the evolutionary selection pressures acting on prey defenses, and the coevolutionary relationship between predators and prey. Future work will continue to unearth spectacular antipredator adaptations and fill fundamental gaps in our knowledge. Perhaps the most important of these is to understand why some prey avoid predation by reducing detection, while others have defenses to reduce consumption: what are the relative costs and benefits of each type of defense, and what is it about their predators that might make one strategy more effective than another?

See also: Co-Evolution of Predators and Prey; Defensive Avoidance; Defensive Chemicals; Defensive Coloration; Defensive Morphology; Group Living; Predator's Perspective on Predator–Prey Interactions.

Further Reading

Bond AB and Kamil AC (2002) Visual predators select for crypticity and polymorphism in virtual prey. *Nature* 415: 609–614.

Bowers MD (1992) The evolution of unpalatability and the cost of chemical defense in insects. In: Roitberg BD and Isman MB (eds.) *Insect Chemical Ecology: An Evolutionary Approach*, pp. 216–244. New York, NY: Chapman & Hall.

Caro T (2005) *Antipredator Defenses in Birds and Mammals*. Chicago, IL: The University of Chicago Press.

Cott HB (1940) *Adaptive Coloration in Animals*. London: Methuen.

Cuthill IC, Stevens M, Sheppard J, Maddocks T, Parraga CA, and Troscianko TS (2005) Disruptive coloration and background pattern matching. *Nature* 434: 72–74.

Gittleman JL and Harvey PH (1980) Why are distasteful prey not cryptic? *Nature* 286: 149–150.

Guilford T (1988) The evolution of conspicuous coloration. *American Naturalist* 131: S7–S21.

Rowe C (ed.) (2001) *Special Issue: Warning Signals and Mimicry Evolutionary Ecology* 13: 601–827.

Ruxton GD, Sherratt TN, and Speed MP (2005) *Avoiding Attack: The Evolutionary Ecology of Crypsis, Warning Signals and Mimicry.* Oxford: Oxford University Press.

Sherratt TN (2008) The evolution of Müllerian mimicry. *Naturwissenschaften* 95: 681–695.

Stevens M and Merilaita S (eds.) (2009) *Special Issue: Animal Camouflage: Current Issues and New Perspectives Philosophical Transactions of the Royal Society* 364: 423–557.

Horses: Behavior and Welfare Assessment

B. V. Beaver, Texas A&M University, College Station, TX, USA

Horse welfare is often evaluated in terms of the behavior the animal shows, just as it occurs for other species. Getting information about such a large animal is challenging, so the tendency is to observe behavior and use that information as a guide relative to the animal's welfare state without looking for ways to validate the connection. There are times when behavior is a good monitor of a horse's welfare, and there are times when it is not. In addition, there are a number of other things that have a direct impact on welfare that need to be mentioned to understand the breadth of this subject. Because animal welfare has a philosophical component, it is important to understand as much as we can about the scientific validation of a behavioral assessment of welfare. There is still much to be learned.

Behavior as a Good Indicator of Welfare

In looking at the myriads of welfare issues that affect horses, it becomes clear that perceptions of what a horse is vary considerably around the world and that these perceptions do change over time. Horses have always been considered a prime mode of transportation, whether it is by physically carrying a person from one place to the next, or by pulling carts and wagons. The native people of North America and Mongolia used horses to move their possessions to new campsites and to carry warriors. Mare's milk is a nutritious beverage in some cultures, while meat is used as food in others. Racing can be added to the list of uses in most cultures. Over time, however, things change, especially as a country becomes more affluent, and the population shifts from rural to urban areas. People have the luxury of worrying about things besides where their next meal might come from, and they lose the latently learned understanding of what is normal horse behavior. The focus of attention shifts from self and family, and so animal welfare becomes more significant. With regard to horses, this has meant that in peoples' minds their role has changed more to that of a typical pet.

Stereotypies

The behaviors most commonly associated with poor welfare are stereotypies and the discussions about them are extensive. It has been argued that the presence of a stereotypic behavior indicates that the horse has had poor welfare and has developed the stereotypy as a coping mechanism. Others argue that a stereotypy is not a sign of poor welfare, but rather a successful mechanism used to cope with a less than perfect environment. A lot of recent research on horses suggests that the picture is even more complicated than just environmental imperfection.

Cribbing (aerophagia, wind sucking)

Cribbing is probably the best studied of all the stereotypies, at least in horses, and various studies have suggested that the incidence of cribbing is ~4% of the horse population, at least in the United States. This is a behavior where a horse will bite on an object without actually chewing it. During the time it contacts the object, the neck muscles tighten. (It should be mentioned that a few horses will show the tightening of the neck muscles and other associated oral behaviors without actually having contact with an object.) Eventually most horses will begin to make a 'gulping' sound and actually swallow air.

For a long time, the behavior has been considered to be a learned vice, and so horse owners were concerned that other horses in a stable would start showing the behavior if one of the horses was already showing it. Other people argued that the behavior 'spread' between horses in a barn because all the horses tended to be of the same breed and personality type and/or were subjected to the same environmental conditions. What has recently been shown is that the tendency to show this behavior is about 60% heritable in thoroughbred horses, indicating that there is a genetic propensity for the behavior to occur, at least in certain breeds.

Several other findings connect cribbing to conditions of the gastrointestinal tract. The peak amount of cribbing behavior is found to occur about the same time that ingesta reach the cecum. Horses that crib have also been shown to have a fecal pH lower than that of unaffected horses, although the reason is not known. A high number of affected horses have gastric ulcers, and cribbers tend to improve – showing fewer and shorter bouts of the cribbing behavior – when fed antacid medications. There might be a connection between ulcers and ingesta reaching the cecum. The stomach is more likely to be empty and susceptible to damage from acid in those horses stalled all day and fed only a couple of times a day, and the associated pain would be expressed the most about the time the ingesta reached the large intestine and the next meal was due. There has also been an interesting finding that for some reason, over 50% of horses that have epiploic foramen entrapment of intestines are cribbers. The relationship of these gastrointestinal findings is not known, but it is interesting that they exist at all.

Other recent research findings suggest certain individuals may be predisposed to cribbing, and perhaps other stereotypies, because of physiological differences. Cribbers have a higher baseline cortisol level, suggesting they are more stress-susceptible. This stress-prone personality can be shown as having behavioral components too. Horses that are more aggressive to neighboring horses are more likely to develop cribbing behavior. Cribbers also have lower levels of β-endorphins, so the endorphin released while doing the cribbing behavior may have a greater-than-normal reward value.

Self-mutilation syndrome

A unique problem in horses is the tendency for some to traumatize themselves (**Figure 1**). This behavior is most common in stallions (about 70% of the cases), but does occur in geldings and mares. The stereotypic form of the behavior will involve other repetitive behaviors such as spinning and head tossing, bucking, and being hypersensitive to touch, in about 40% of horses each. Vocalization occurs in 32%. The typical horse looks at its side, nips at the region (anywhere from its chest to its flank), turns about 180°, and kicks out with one hind foot. Some will then squeal. We do not know if there are internal predisposing factors, such as those that exist for cribbing, but we do know that many of these horses are racing-bred animals. These animals get injured while in racing training and are stalled for several weeks to months but remain on full feed during that time.

There are a couple of factors thought to be associated with this condition in animals that have been injured. Stallions are often isolated even more than usual from contact or sight of other horses, which is an atypical situation for a social species. They are consuming a lot of energy but not expending as much while their injuries are healing, so over time this imbalance can lead to the behavior. Whether other physiological or genetic factors are involved is not known. Management while the horse is healing is the current general approach to treatment of self-mutilation. Reducing the energy level of the food is done by putting them on a high-quality hay-only diet. Social factors are addressed by putting them with another horse, pony, or donkey if possible. A goat or dog may also work as a social contact when another equid cannot be used. Drug therapy, as described later, has been tried but is expensive.

Other stereotypies

Horses express a variety of other stereotypies (**Table 1**), but these have not been as well studied as cribbing. Wood chewing occurs in ~12% of horses, weaving in 3%, and stall walking in 2%. We do not have data about the frequency of other behaviors such as head bobbing, pawing, and noise making.

Even though wood chewing is listed as a stereotypy, that is not always the case. A direct connection has been shown between this behavior and the low intake of dietary roughage and with the lack of dietary salt, too. And while 'boredom' is difficult to prove in animals, horses confined to stalls without a great deal of environmental enrichments or social interactions will often show the behavior in a nonstereotypical form.

Drug therapy has been used on stereotypic behavior, particularly drugs that are classified as selected serotonin

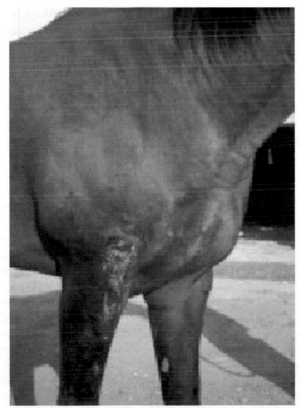

Figure 1 Lesions on a racing-bred Quarter Horse stallion that began the self-directed aggression when confined after a training injury.

Table 1 Common stereotypies in horses

Head-oriented stereotypies	Locomotor stereotypies
Cribbing/wind sucking	Head tossing, circling, shaking, nodding
Lip/tongue flapping	Kicking
Lip licking	Noise making (door banging, kicking)
Licking	Pacing (with head flipping)
Playing in water	Pawing
Rubbing of teeth	Rearing
Self-mutilation	Running away
Wood chewing	Self-rubbing
	Stall walking
	Tail swishing
	Weaving

reuptake inhibitors and tricyclic antidepressants. From work done with dogs, it is expected that these drugs would be helpful in about 50% of cases, and then not necessarily at high levels of success. Narcotic antagonists also have some success in helping relieve stereotypies, probably by blocking the β-endorphin effects. Unfortunately in both cases, the drugs are expensive, hard to administer, and not always available in long-acting formulations, thereby limiting their use in horses.

Obsessive–Compulsive Disorders

There has also been discussion about the development of obsessive–compulsive disorders in horses. The obvious problem is that without being able to read a horse's mind, we will never be sure if there is really an obsessive component. Separating the obsessive–compulsive behaviors with a repetitive component from true stereotypies may not be possible. Our understanding of obsessive–compulsive disorders is rudimentary at this time.

Pain and Behavior

Behavior changes are commonly seen with painful conditions because horses are not considered to be particularly stoic animals. The two most common behavioral expressions of pain are lameness and colic. There are, however, many other times when behavior changes indicate pain. In a study done at Texas A&M University, horses were videotaped before a dental surgery and then afterward, so they served as their own control. Postsurgical behaviors of anorexia, restlessness or reduced movements, and a lowered head posture were indicators that pain medications were needed. Return to normal, presurgical behavior indicated the successful application of these medications and eventually that medication was no longer needed. One of the problems that was noted had to do with the extreme variation in severity of behavioral changes that occurred. Some of the horses showed changes that would be obvious to the most casual observer, regardless of whether that person knew what presurgical behaviors were. Other horses showed relatively few obvious signs of pain, but these signs were more noticeable if the pre- and postsurgical behaviors were compared.

Horse owners often remark that the animal 'just isn't acting right' even before an obvious problem exists. We know now that the stress of traveling and competition can be associated with gastric ulcers, and these could certainly be part of the subtle changes an owner recognizes. There are other painful conditions, such as sore backs caused by poor fitting or dirty equipment, that can also result in subtle changes.

Horse Show Competition

Competition in the horse industry has led to a number of unique problems of welfare. Competitors in most breed shows have a number of practices unique to that type of horse that is intended to enhance the desirability of the horse while it is showing. This practice is often associated with poor welfare for the individual horse. As an example, the soring of Tennessee Walking Horses is an ongoing problem that persists even with very strict measures intended to prevent it. The special walking gait of this breed is very animated compared to the typical walk of most horses. The exaggeration of this gait is highly desirable in the show ring, so trainers/owners try to promote as much animation as possible through genetics and artificial techniques. One such technique is to place chains around the fetlock, loose enough to hit the sensitive area at the top of the hoof. In addition, people may actually damage the skin in that area to create a painful sore, which the chains then keep active. This practice is an irritation at best and probably very painful. Because it is bilateral, lameness is not obvious. In the Quarter Horse breed, a few horses that show in western pleasure classes will have had the nerves to their tail deadened so that they cannot swish the tail. Horses that are trained in a rush, so that they can perform well at a very young age, may develop this behavior from being booted in the side too heavily and often. Arabian horses that show in halter classes may have the pupils of the eyes dilated so that the bright lights of the arena frighten them into a more high-strung behavior. The list goes on and on. Horse associations, and in the case of soring, the federal government, try to discourage the practices that negatively impact welfare, but these attempts have not been totally successful. These are cases where the human desire to win can result in the application of techniques to change a behavior but negatively impact welfare.

Unique Problems in Assessing Welfare by Behavior

The biggest problem in assessing the welfare of the horse is anthropomorphism by the general public. Horses and other equids continue to be work animals in much of the world, especially outside North America and Europe. As such, there are welfare concerns about their proper use and injuries that can result. In First World countries, horses tend to be thought of as large dogs by the majority of the population because most people in those areas only see horses in the riding-for-pleasure environment. They are uneducated about how horses behave and what their actual welfare needs might be. Any injury that occurs or stereotypy that develops must, by their reasoning, be the human's fault and detrimental to the animal's welfare. They fail to understand that race horses run for the love of running and are not beaten to force them to run. They fail to appreciate that occasionally bones will break in a running horse because of the amount of pressure carried on a relatively small surface area called a foot. Then, they do not understand the difficulty in trying to treat a broken bone in a horse or how significant the individual's behavior is in order for the healing process to go well.

Another example of the misunderstanding of horse behavior and welfare is commonly associated with rodeo stock. 'Bucking broncos' buck, not because the flank straps are hurting them, but because they are genetically bred to do so or are individuals who preferentially show the bucking behavior. For thousands of years, the wild horse had to react with bucking to fend off mountain lions and other predators to survive. Thus, there is a strong tendency to want to keep things off its back. Training can reduce that inborn fear from most horses, but some retain the behavior. Anthropomorphic viewers want the horse free to roam on the prairies or in lush green pastures and be treated with lumps of sugar or carrots. They forget about the dangers that the mustangs and other feral horses still face.

Lameness, as previously mentioned, is associated with pain. Broken bones, tendon inflammation, navicular disease, bone cysts, and many other conditions do cause pain and lameness. However, a unique challenge to the horse owner and the veterinarian is to differentiate these conditions from lameness that is not associated with pain. Medications are not particularly helpful in the latter cases. Mechanical lameness results when a limb cannot work correctly, such as when a joint is fused, a muscle becomes calcified, the patella gets stuck abnormally in an upward position, or a ligament was ruptured in the past. This lameness may be viewed by others as being inhumane, and yet the horse is not experiencing pain or distress.

Housing situations can also be called into question regarding welfare. Wild horses are capable of finding shelter from the elements. Domesticated horses are protected by their owners with shelter that ranges from rudimentary things such as wind barriers, to elaborate barns with extensively bedded stalls. Even the fanciest barn and paddock setup may not meet all the behavioral needs of the horse, even though the picture of the welfare looks perfect. On a breeding farm, for example, the stallions may be kept in one barn and mares in another. The paddock area where the stallions exercise is extensive, yet careful observation shows that the stallions spend most of their exercise time in an area near the mare barn, often going back and forth along the fence (**Figure 2**).

Areas Where the Interrelation of Behavior and Welfare Need More Research

Effects of Roundups

The mustangs that are free-roaming in the western United States and the ponies living on the coastal islands experience some problems where welfare and behavior are comingled. Each of these groups has to have the herd

Figure 2 A large stallion exercise arena at the Lippizan Piber breeding farm in Austria shows that the stallions limit their activities to areas near the mare barn located to the left of the picture. The solid black arrows show the extent of the most used path. The hollow white arrows point to the top rail where it is chewed (left) and relatively untouched (right).

size artificially managed so that they do not overgraze areas where they live. To manage herd size, there are periodic roundups of the herd and then selection of which animals will be returned to the wild and which ones will be sold to the public or cared for in managed environments. When this type of artificial selection occurs, it results in stress behaviors and injuries by forcing harem bands to come together into large groups of horses that would not ordinarily be in such close proximity. In addition to all the other implications of stress, aggression is a common occurrence. A second consequence is that the existing harem groups are disrupted so there is a need to reshape the social structures for those horses that are returned to the wild. For those kept in holding facilities, they are usually forced to live in a crowded situation that does not respect personal space needs, so the amount of agonistic behavior is greater than should be seen. A great deal of research needs to be done to understand the implications of breaking up of harem groups and of social crowding on behavior and on the health of individual horses. Ulcers are commonly associated with the stress of showing; could they be associated with these artificial management situations too?

The 'Unwanted Horse'

One of the biggest challenges to the welfare of the horses has come to light in the United States with what is being called 'the unwanted horse.' In recent years, activists have effectively shut down the few remaining horse slaughter plants in this country, claiming that the transport methods used are inhumane and that the method of stunning is not humane. Previously, the horse slaughter facilities in the United States would process between 60 000 and 120 000 horses a year, with much of the meat going for human consumption in other countries. The majority of the horses that went to these plants were those that owners could not find other homes for – such as those that were chronically lame or dangerous to handle. As a result of plant closures, there has been a buildup of ~100 000 horses each year that are no longer wanted. Added to this picture has been the downturn of the global economy, which has resulted in more people who can no longer afford to take care of their horses nor take on care of additional ones. Horses are being turned loose to fend for themselves, and owners are neglecting the horses that they do have. More and more horses are starving. The traditional city or county animal shelter or humane society has difficulty raising enough money to care for owner-surrendered and captured free-roaming dogs and cats. The expense of caring for just one horse is far greater than that for several of the smaller animals. The few horse sanctuaries that do exist are overflowing already and do not have funding, space, or facilities to handle more animals. In addition, there is no consistent regulatory oversight of these sanctuaries to ensure proper care of the horses kept there. In fact, some have actually been charged with the inhumane care of the horses they were supposed to be helping. Alternative humane euthanasia and carcass disposal options are also expensive and limited for an animal as large as a horse.

The quandary is what to do with these unwanted animals. One proposed solution is to bring them together on large sections of land in the western United States; however, funding for the overall management and care has not been forthcoming from the private sources that were promised, and no consideration has been given for what will happen when the grasslands on which they would roam are overgrazed or for ensuring that the care for them is as promised. Some people are proposing that breeders should simply decrease the number of foals born each year by ~100 000 to solve the problem. That tactic has not worked for dogs and cats, so it is not expected to work for horses either unless there is significant governmental control on breeding – something not likely to happen in the United States. In addition, the problem is not as simple as too many horses being born each year, but rather that there are horses that are not behaviorally or physically appropriate to keep.

Horse welfare and its relationship to behavior is an important area for additional study. To do so will require a scientific evaluation of carefully designed projects. Care must be taken to ensure that the emotional component of the horse-as-a-pet not be allowed to override the scientific conclusions. At the same time, it is important to ensure good welfare and the ability of horses to express natural behaviors.

See also: Group Living; Neural Control of Sexual Behavior; Pair-Bonding, Mating Systems and Hormones; Parent–Offspring Signaling; Reproductive Success; Social Selection, Sexual Selection, and Sexual Conflict; Stress, Health and Social Behavior.

Further Reading

Fraser AF and Broom DM (1990) *Farm Animal Behaviour and Welfare,* 3rd edn. London: Baillière Tindall.
Houpt KA and Lieb S (1993) Horse handling and transport. In: Grandin T (ed.) *Livestock Handling and Transport*, pp. 233–252. Oxon: CAB International.
Kiley-Worthington M (1997) *Equine Welfare.* London: JA Allen & Co.
Waring GH (2003) *Horse Behavior,* 2nd edn. Norwich, NY: Noyes Publications.

Hunger and Satiety

D. Raubenheimer, Massey University, Auckland, New Zealand
S. J. Simpson, University of Sydney, Sydney, NSW, Australia

Introduction

Hunger and satiety have played a conspicuous role in shaping human prehistory and history, and persist as a dominant influence at scales from international politics to the rhythm of our daily lives. Given their salience in our personal experience, it is no wonder that these concepts also permeate our interpretations of the behaviors of non-human animals, both informally and in the context of scientific research.

But what are 'hunger' and 'satiety'? This is a deceptively complex question. In the context of human behavior, these terms refer to the subjective sensations associated with eating: hunger is what we experience when we need food, and satiety is the persisting sensation of repletion that results from eating A related concept, 'satiation,' is the feeling of fullness at the end of a meal. These concepts fit comfortably with our everyday experience of food and eating, and also work in the scientific study of nutrition in a species that can articulate subjective sensations. But these terms are deeply associated with the study of nutrition in non-human animals for which we have no way of measuring subjective sensations, or even knowing for sure whether these sensations are experienced. Indeed, 'hunger' and 'satiety' are among the most widely used concepts in animal behavior in relation to feeding and nutrition.

How can unmeasurable and unknowable things form the bedrock of the study of feeding and nutrition in animal behavior? Our aim in this chapter is to address this question by providing an overview of the scientific development of these concepts in animal behavior, ranging from conceptual difficulties with their use to the constructive role they have played in behavioral science. We will first examine hunger and satiety from the behavioral viewpoint, and then provide an overview of what is known about the physiological mechanisms. We end with a model that enables the hunger–satiety spectrum to be linked to animal performance and evolution.

The Behavioral View

The difficulties of measuring subjective sensations like 'hunger' and 'satiety' in non-human animals do not necessarily mean that these cannot play a useful role in understanding nutritional behavior. It does, however, mean that we need to be extra vigilant in how we incorporate these concepts into scientific thinking. We begin this section with a brief overview of some of the philosophical dangers of using subjective notions like hunger and satiety, and thereafter explain the constructive role they can play in scientific thinking about animal behavior.

Conceptual Pitfalls

At one level, it seems obvious that when an animal sets out in search of prey, and invests energy and risks injury in the pursuit and attack, it is being driven by the deep-seated sensation that would cause us to do the same. This interpretation is reinforced by the observation that, as is true in our personal experience, the levels of effort and risk that animals are prepared to take to acquire food increase with hunger.

Unfortunately, compelling as it may seem, there are hidden dangers with this view. It is based on a logical error called 'anthropomorphism,' and can infect scientific reasoning with other logical errors called affirming the consequent and circular reasoning. It also raises the specter of the scientifically debatable doctrine of mentalism.

Anthropomorphism

Anthropomorphism is the attribution of human characteristics to non-human things – for example, the beliefs that earthquakes strike when the earth is angry, love bonds frogs to their mating partners, and the feeling of hunger causes horses to eat.

Anthropomorphism can be useful in forming hypotheses about the similarities between humans and other animals, but presents the risk that the compelling reality of our own subjective experience may distort the objectivity with which we observe and interpret the behavior of other animals. At worst, anthropomorphism can fool us into accepting as scientific truth something for which there exists no objective evidence.

Indeed, even if our anthropomorphic inference is true – if, for example, horses do feel hunger – problems nonetheless remain with attributing human feelings to non-human animals. The fact that these states cannot be observed in other animals, but must be inferred, makes their use in animal behavior vulnerable to the logical fallacies called affirming the consequent and circular reasoning.

Affirming the consequent

Affirming the consequent is a logical error of the form:

If H then F

Observe F

Conclude H.

Consider, for example, the case where H is 'hunger,' and F 'feeding.' Since we cannot observe hunger directly, we need to rely on feeding (which we can observe) to infer hunger. 'If H then F' now becomes 'if the animal is hungry then it will feed.' The logical error arises if we then observe an animal feeding (Observe F), and from this conclude that it is hungry (Conclude H).

The error here, of course, is that 'If H then F' states that the animal will feed when it is hungry, but it does not state that the animal will feed *only* when it is hungry. The feeding animal might not be hungry at all, but thirsty, or feeding to provision its hungry offspring, or to self-medicate against parasites and pathogens. Therefore, our ability to infer the state (hunger) from the behavior (feeding) is limited, and vulnerable to the logical fallacy of affirming the consequent.

Circular reasoning

Circular reasoning is the logical fallacy we commit when we assume the proposition to be proved in our premise. For example, an experimenter might observe that a rat in a cage ate more than its identical twin. If the experiment was properly controlled to rule out other causes of eating (e.g., thirst, provisioning of offspring, and self-medication), then the experimenter could reasonably infer that the animal that ate more was the hungrier of the two. In this case, because hunger could not be observed, the scientist took the acceptable step of using the amount eaten as an operational definition of hunger.

However, the experimenter could not then conclude that the animal ate more *because* it was the hungrier. This would be circular reasoning: hunger has been defined in terms of food intake, and so to say that the animal ate more because it was hungrier (=ate more) is to say that it ate more because it ate more. This is a pseudoexplanation, which does not move beyond the original observation that food intake was greater for one animal than for the other.

Mentalism

To break the circularity in the aforementioned example and legitimately explain the observed difference in food intake, the experimenter would need to link food intake to another observable variable, such as gut fullness. In this case, a noncircular explanation would be that the one animal ate more than the other because it was hungrier, and it was hungrier because it had an emptier gut. Here, however, all the explanatory work is done by the association between the two observables – gut fullness and amount eaten – and to insert a causal role for the inferred variable, 'hunger,' adds nothing to the explanation.

Unless, that is, the experimenter subscribes to the doctrine of mentalism. Mentalism is the view that mental states, such as hunger and satiety, have causal power in their own right, and thus provide a level of explanation that is different from the mechanisms regulating the behavior and from the behavior itself. Many scientists studying animal behavior stridently contest this view, most notably those that have leanings toward the behaviorist school of animal psychology.

The relative merit or otherwise of mentalism as a valid scientific paradigm is a huge field that has been debated at library-length by philosophers and psychologists better qualified for the task than we are. For now, suffice to say that mentalism clearly is a useful model for the study of human behavior (on it rests much of clinical psychology), and some headway has been made in extrapolating to apes and possibly also to nonprimate mammals such as rats. But at which point does it cease to be useful? What about fish, frogs, or pubic lice – is their feeding behavior also caused by mental states? At best, this remains an empirical question (to the extent that it can be addressed empirically at all). At worst, even in humans, mental states such as hunger may be no more than transitory illusions we experience as bodily mechanisms affect behavior. For the general case, therefore, to answer the unknown 'what caused the animal to feed?' by invoking a mental state such as hunger is to substitute one unknown for another.

Hunger and Satiety as Intervening Variables

We would excuse the reader for having concluded at this point that the concepts hunger and satiety are disreputable obstacles to the study of animal behavior. This is not our view: these concepts have played an important historical role, and will continue to do so provided we remain vigilant of the pitfalls discussed above. We now explain the sense in which hunger and satiety can make a useful contribution to the study of nutrition and foraging.

In the nonanthropomorphic, nonmentalist sense, 'hunger' and 'satiety' are considered in behavior theory to be *'intervening variables.'* Intervening variables are hypothetical internal states that are used as a framework to explain the relationships between two real and observable events, but are themselves not assumed to be either observable or real. They cannot, therefore, in any reasonable sense be said to 'cause' behavior.

For example, if a properly replicated and controlled experiment showed that rats move toward an air source (one observable event) when it emits a food odor (the second observable event) but not when it emits air alone, then it can reasonably be concluded that food odor stimulates (causes) the rats to move toward the source.

However, the behavior of animals is more complex than is implied by this simple stimulus–response relationship. Thus, a rat that responded to the stimulus on one occasion might on an identical subsequent trial need a stronger odor to elicit the same behavior, might not respond at all, or might even respond aversively to the odor. In this case, something clearly has intervened in the relationship between the stimulus and response that accounts for the difference between the behaviors on the successive trials. Since the environment has not changed between the trials, this something must be intrinsic to the animal – it might have learned that the odor is not associated with a food reward, it might have become ill, or it might have fed shortly before the second trial and so had no incentive to pursue food-related cues. *Intervening variable* is the general term used to refer to such changes in the state of the animal that are associated with this kind of behavioral variability. Examples of intervening variables are personality, learning, motivation, intelligence, libido, and . . . yes . . . hunger and satiety.

The useful thing about the concept 'intervening variable' is that it enables researchers to take account of factors that are clearly important in the animal's interactions with its environment without specifically knowing the details. For example, in the scenario we gave in the previous section of the twin rats, the observation that the rat with the emptier stomach ate the most is important, because it links two observable events (gut fullness and amounts eaten) and thus furthers the researcher's ability to predict and understand rat behavior. This is true whether or not the researcher knows anything about *how* gut fullness influences rat behavior, or indeed has any interest in the messy details of how the stomachs, nervous systems, and muscles of rats are linked. Instead, the researcher can construct a framework according to which gut fullness influences behavior via a general – and in this case poorly understood – state called 'hunger,' and proceed with the study of how rats relate to their nutritional environment.

If that is where it ended, then adopting the term 'hunger' as shorthand for 'gut fullness influences feeding but how I do not know' would be useful, but not particularly exciting. On the other hand, the concept of hunger could be a more substantial component of the model that is developed to explain the animal's nutritional behavior. For instance, in the twin rat example discussed earlier, we operationally defined hunger in relation to gut fullness. Research shows, however, that in many cases a complex mix of several factors influences the amount eaten – including gut fullness, levels of nutrients in the blood, body water stores, etc. In such cases, we can move beyond the operational definition of 'hunger = gut fullness,' and adopt a model according to which gut fullness is one component among many in the hunger of the animal. This model can then be used as a framework to investigate the factors that affect feeding. Researchers have used various forms of systems analysis from engineering to construct highly sophisticated models of hunger, which are then refined using controlled experiments.

The key point in the new model is that hunger is no longer synonymous with gut fullness, and therefore 'hunger' plays a role in the model that is different from gut fullness. But does 'hunger' now contribute anything that is not contributed by the second observable, feeding behavior? The answer is 'yes,' it can contribute something extra. Imagine that we performed an experiment in which the two rats were treated identically, and so had equal gut fullness, levels of nutrients in the blood, state of hydration, etc. According to our model, the rats would be equally hungry and should eat similar amounts of the experimental food. But if the two rats were offered different foods (e.g., one got the original food and the other the same food but combined with a flavor that it associated with toxins), then it is likely that the rats would eat dissimilar amounts – even though they were identical in terms of the hypothetical variable hunger. This demonstrates that feeding behavior is *not* synonymous with hunger, because the same level of hunger can be associated with different behaviors if there are differences in the environment (e.g., food flavors differ).

One Hunger or Many?

Hunger and satiety become especially important devices in the study of behavior when they are linked not only to the behavior they are postulated to control, as in the earlier examples, but also to the *consequences* for the animal of performing that behavior. In this section we consider one category of consequences, namely the consequences for the homeostatic regulation of the animal's physiological condition. In the final section we take this a step further, and consider also the longer-term, evolutionary consequences.

From the viewpoint of homeostasis, feeding should be linked to the nutritional status of the animal – the animal should feed when reserves of a nutrients drop below a certain threshold and cease to feed when they are replenished. It is clearly the case that animals do regulate their feeding in this way, and even the most basic models of hunger should take this into account. However, the details of the way that nutritional status is linked to feeding are potentially very complex. This is because animals need, and foods contain, not just a single nutrient, but a particular balance of many nutrients (carbohydrates, fats, amino acids, calcium, sodium, etc.), and an effective homeostatic system should take this multiplicity of nutrients into account. But how could it do this?

The simplest mechanism would be to have 'unitary hunger' – a generalized hunger that paid attention only to a single food component. Many models of nutrition make this assumption, most commonly assuming that feeding is linked only to the energetic status of the animal. Such a system would serve the animal's energetic needs, but would not enable it to balance its intake of different energetic nutrients (carbohydrates, proteins, and fats) or nonenergetic nutrients such as minerals and vitamins. A slightly more complex model would link hunger to a range of nutrients, where the animal feeds when any one of these drops below some threshold. Such a system would help the animal to avoid shortages not just of energy but of all nutrients to which it was linked. In this regard, it would be an improvement on the previous model, but it would present problems of its own. Specifically, it would not enable the animal to feed selectively on foods that contain high levels of the particular nutrient that is deficient at a given time. Such nonselective feeding would result in the animal not only redressing the current nutrient shortage, but also ingesting excesses of other nutrients that it already has in sufficient quantities. This is an important problem, because experiments have shown that ingested excesses of nutrients can be as costly as nutrient deficiencies, or even more costly.

To get around this problem, the animal would need to select foods with a balance of nutrients that suits its current nutritional state. For example, if deficient in carbohydrate but not protein, it would need to feed specifically on foods that are high in carbohydrate and low in protein (e.g., honey); conversely, when protein deficient but carbohydrate replete it would feed not on honey but on foods with a high ratio of protein:carbohydrates (e.g., meat). This could be achieved if the animal had separate hunger systems for different nutrients – that is, it would seek honey when 'carbohydrate hungry' but 'protein satiated,' and meat when hungry for protein but not carbohydrate.

Considerable evidence shows that many animals can indeed mix their intake in this way to obtain a nutritionally balanced diet – a phenomenon known as *dietary self-selection*. This has been observed not only in laboratory experiments, but also in wild animals in their natural habitats (**Figure 1**). Some studies have linked dietary self-selection to physiological mechanisms in a way that clearly demonstrates a role for nutrient-specific hungers. For example, over half a century ago, the experimentalist Curt Richter showed that rats could compensate for surgically induced losses of a nutrient from the body by increasing their intake of that nutrient. When Richter disrupted salt homeostasis in rats by removing their adrenal glands, the animals compensated for the resulting salt loss by drinking large volumes of saline; likewise, rats in which calcium homeostasis had been disrupted by removing their parathyroid glands specifically increased their calcium intake. In both cases, the altered feeding preferences of the experimental animals rescued them from what otherwise would have been certain death. Moreover, in neither case did the intake of other nutrients increase,

Figure 1 Mountain gorillas in Bwindi Impenetrable National Park, Uganda, eat dead and decaying wood as a source of sodium. Photo: J. Rothman.

demonstrating that feeding is linked to physiological homeostasis in a nutrient-specific way. More recent studies have shown that in addition to calcium and sodium, a wide range of animals have nutrient-specific appetites for macronutrients (protein, carbohydrates, and fat), as discussed further below.

The advantages of nutrient-specific appetites are thus clear: they provide a framework for understanding how animals choose foods and combinations of foods that satisfy their complex and ever-changing nutrient needs. In the following section we discuss the physiological mechanisms involved in nutrient-specific feeding, and in the final section we consider the links between mechanisms, behavior, and evolution.

Mechanisms for Regulating Nutrient Intake

Maintaining a balanced intake of energy and nutrients requires three things: (1) an ability to assess the nutritional quality of foods, (2) the capacity to measure current nutritional state, and (3) a means of integrating these two to produce appropriate feeding responses. We will consider each of these in turn.

Detecting Nutrients in Foods

The most fundamental and direct way of assessing the nutritional value of food is to detect nutrients by tasting them. Receptors for the detection of amino acids, sugars, and essential minerals such as sodium and calcium evolved well before the evolution of multicelled organisms and direct the chemotaxic behavior of single-celled organisms. Both invertebrate and vertebrate animals possess such taste receptors on external appendages (e.g., the feet and external mouthparts of insects and the barbels of some fish), within the oral cavity, and along the gastrointestinal tract, providing the opportunity to assess the nutritional value of food before, during, and after ingestion.

The balance of positive and negative chemosensory inputs to the central nervous system influences whether feeding is initiated, and if so how much is eaten within a meal. Other chemoreceptors, such as the bitter receptor of vertebrates and the deterrent receptors of herbivorous insects, are stimulated by potentially toxic compounds in food and elicit food avoidance responses. An evolutionary twist on this pattern is seen in those herbivorous insects which have evolved to specialize on particular plant taxa and are able to detoxify the associated secondary metabolites in their host plants. In these herbivores, secondary plant compounds that would be deterrent to other insects stimulate feeding and serve as a means of recognizing host plants.

Whereas evolutionarily ancient taste receptor mechanisms provide the basic framework for detecting nutrients in food, animals cannot taste all nutrients so the chemical consequences of eating a food may not become apparent until the meal is processed and absorbed. Such postingestive feedbacks provide the basis for learned associations which affect subsequent responses to food-related sensory cues. Two basic types of association between food cues and postingestive consequences are found: learned aversions and positive learned associations. Food aversion learning was first reported in 1955 by John Garcia in rats, and has since been found to be widespread among vertebrates and invertebrates. It is unusual among associatively learned responses, in that an aversive association is made between the flavor of a food (the conditioned stimulus, CS) and deleterious consequences of its ingestion (the unconditioned stimulus, US) that may manifest themselves hours later – rather than the CS–US interval being a few seconds as in other types of associative learning. Deleterious consequences can include toxicity and nutritional imbalances such as absence of essential amino acids.

Learned positive associations, such as flavor preference learning, occur when previously ineffective food cues paired with nutritionally rewarding foods come to stimulate feeding, and have been demonstrated in a wide range of vertebrates and invertebrates. A special but rarely reported case is known as a 'learned specific appetite.' In such cases, an animal learns to associate a cue with the specific nutrient content of a food and is subsequently stimulated by that cue when in a state of deprivation for that nutrient. Such nutrient-specific learned associations have been reported for protein and carbohydrates in rodents and insects. For example, locusts which have

been specifically deprived of protein for only 4 h become attracted to odor or color cues previously paired with high-protein foods.

In addition to associatively learned feeding behaviors, nonassociative responses such as neophilia and neophobia, in which novel foods become more or less acceptable, respectively, according to nutritional state provide a means of maintaining within acceptable bounds the intake of nutrients that are not directly sensed, for example, micronutrients such as vitamins.

Sources of Information about Nutritional State

Assessment of current nutritional state requires systemic nutrient and energy sensors. Many, if not all, cells in the body have the capacity to sense their nutrient status, but at the level of the whole animal the major coordinating role is played by the brain, in neural and chemical dialog with other organs, notably the liver and pancreas in vertebrates and the fat body in insects. There are three major sources of nutritional information. The first are the concentrations of nutrients such as glucose, amino acids, free fatty acids, and mineral ions circulating in the blood system; whether the closed blood system of vertebrates or the open hemocoel of invertebrates. Circulating nutrients potentially provide an instantaneous measure of nutritional state in respect of specific nutrient groups. Second is the level of energy in cells, as indicated for example by the ratio of adenosine monophosphate (AMP) to adenosine triphosphate (ATP). The third source of information about nutritional state comes from levels of reserves in storage organs such as adipose tissue and liver, as signaled by secretion of hormones such as insulin from the pancreas and the vertebrate hormones leptin and adiponectin from adipose tissue.

The two best known (but not the only) intracellular nutrient sensing pathways are the evolutionarily conserved protein kinase systems, AMPK (AMP-activated kinase) and TOR (target of rapamycin), which are expressed in a range of tissues in vertebrates and invertebrates. The AMPK pathway responds to a decline in circulating levels of glucose, amino acids, and fatty acids, and to energy depletion as indicated by an increased AMP/ATP ratio, and triggers a cascade of metabolic changes that are broadly catabolic, involving the breakdown of complex molecules such as fats and proteins, thus liberating stored nutrients and inhibiting growth and reproduction. In contrast, high levels of nutrients stimulate TOR, notably branch chain amino acids such as leucine, and glucose (in an amino -acid-dependent manner). TOR signaling coordinates a series of anabolic metabolic responses, involving synthesis of fats and proteins, and enhanced cell division, growth, and reproduction.

Integrating Food Nutrients and Nutritional State

Taste and other inputs associated with food nutritional composition need to be integrated with systemic signals indicating nutritional state to produce regulatory feeding behavior. The brain coordinates feeding behavior, and it is here that nutritional integration must ultimately be mediated. We will first consider the example of insects, where a primary site of nutritional integration occurs at the sensory neurones that detect nutrients in food, leaving the brain with the simple task of responding to already integrated, nutrient specific inputs.

Insects possess chemoreceptors within porous cuticular pegs and hairs (called 'sensilla') located on various body parts, including antennae and external mouthparts, within the preoral cavity, on the tarsi (feet) and ovipositor. Chemicals on and within foods enter the terminal pore on gustatory sensilla and stimulate the dendrites of neurones within the lumen of the sensilla. Experiments on locusts have shown that concentrations of nutrients and the signaling molecule nitric oxide in the blood directly modulate the responsiveness of these gustatory neurones in a nutrient-specific manner. Hence, when the insect is protein deprived but sugar replete, levels of free amino acids in the blood fall and sugar levels rise, the responsiveness of gustatory receptors to amino acids in the food is elevated, and responsiveness to sugar falls. As a result, the insect becomes highly responsive to foods containing amino acids and ignores foods containing sugar. Sugar, amino acid, and salt responses are modulated independently by blood composition, resulting in the animal tasting what it needs and eating what it tastes. Hence, sophisticated nutritional choices are made without the brain having to do more than summate incoming gustatory impulses, with the peripheral taste organs serving as the integrator of information about the quality of food and the animal's instantaneous nutritional state. Learned responses (see earlier) help to fine-tune this feedback system and direct efficient foraging behavior. Additionally, volumetric feedbacks from the gut and body wall play a role in modulating feeding behavior, determining when meals end and influencing when they next begin.

In mammals, the major point of nutritional integration is within the brain itself. Gustatory inputs from the mouth project to the hindbrain, as do stretch receptor inputs from the upper gastrointestinal tract, nutrient receptor inputs from the gastrointestinal tract and liver (carried via the vagus nerve), and neural signals from the forebrain (discussed further below). The other source of signals associated with food in the gut is a suite of hormones that are released from the gastrointestinal tract and act variously on gut motility, the vagus nerve, and directly on the brain. Among these hormones are peptide YY (PYY), glucagon-like peptide-1 (GLP-1), and cholecystokinin (CKK), which are stimulated by food intake and depress feeding behavior, and ghrelin, secretion of which is inhibited by food intake, rises with food deprivation and stimulates food intake.

Within the forebrain, the hypothalamus plays a central role in integrating nutritional signals, in particular the arcuate nucleus. The arcuate nucleus contains two major groups of neurones that control feeding behavior. One of these subpopulations is termed 'POMC/CART' because they coexpress pro-opiomelacortin (POMC) and cocaine- and amphetamine-related transcript (CART). These neurones release α-melanocyte-stimulating hormone, which binds to melanocortin-4 receptors and inhibits food intake. The second population of neurones in ARC is termed 'NPY/AgRP neurones' since they release neuropeptide Y (NPY) and agouti-related protein (AgRP), which increase food intake. Systemic nutritional signals modulate the levels of activity in these two neural subpopulations. Hence, elevated levels of amino acids, glucose, fatty acids, insulin, and leptin inhibit NPY/AgRP neurones and stimulate POMC/CART neurones, as does low AMP/ATP ratio and low ghrelin concentrations. By contrast, the opposite nutritional and hormonal trends have the reverse effect, exciting NPY/AgRP neurones and inhibiting POMC/CART neurones. The sensing of these nutritional variables by the ARC neurones is mediated via AMPK and TOR signaling pathways (see earlier), providing an elegant link between cellular nutrient sensing mechanisms and the regulation of nutrient intake and utilization at the level of the whole animal.

There is crosstalk between the arcuate nucleus and hindbrain integrative centers, as well as input from higher brain areas dealing with other sensory modalities and food reward, allowing learned responses to be integrated into the control of feeding behavior. Quite how specific appetites for different macronutrients are controlled within these hormonal and neural pathways is not understood as yet, but it is worth noting that other forebrain areas such as the anterior piriform cortex have been implicated in regulation of amino acid intake.

Linking Mechanisms, Behavior, and Evolution

Above, we have shown how the concepts 'hunger' and 'satiety,' if used with due caution, can provide a useful framework that guides research into the links between behavior, physiological homeostasis, and regulatory physiology. The Holy Grail for nutrition research is to extend this framework to deal also with evolutionary aspects. In this section, we present an approach that has been developed for this purpose, known as the 'geometric framework for nutrition.'

We have already mentioned that the hunger–satiety spectrum has been represented using sophisticated systems models derived from engineering. The geometric framework is an example, based on a method known as 'state-space geometry.' It differs from other system models of feeding in combining three key features. (1) It represents several nutrients simultaneously, and can therefore distinguish between unitary and multiple hungers. (2) While in most models of hunger and satiety the primary observables are behavior and regulatory physiology, the geometric framework is centered on the *outcomes* of behavior and physiology, in terms of nutrient gains, losses, and expenditures. (3) Focusing on nutrients in this way provides a common currency that enables direct links to be made between the environment (foods), behavior (food selection and feeding), physiology (nutrient extraction, utilization, and excretion), animal performance (health, survival, growth rates, reproduction, etc.), and, ultimately, evolution (how natural selection has shaped the behavior and physiology of the animal to achieve favorable nutritional outcomes).

The first step in building a model using the geometric framework is to identify the nutrients of interest in a given situation, then represent each nutrient on an axis to define a geometrical space called a 'nutrient space' (see **Figure 2**). This space can have as many dimensions as there are relevant nutrients, but for illustration we will assume a two-dimensional nutrient space comprising axes for protein (P) and carbohydrate (C). A nutrient space so constructed provides the setting within which foods, feeding, nutritional physiology, and animal performance are interrelated.

To see the value of nutritional space, consider a food type that contains 30% protein and 70% carbohydrate. An animal that eats Y grams of this food obtains $0.3Y$ grams of protein and $0.7Y$ grams of carbohydrate. If the animal eats more, it obtains more protein and more carbohydrate, but the amount of carbohydrate gained increases more quickly than the amount of protein. We can represent this relationship in nutritional space using a straight line, which will be steeper for foods with more carbohydrate and shallower for foods with more protein. A hypothetical food type with equal amounts of protein and carbohydrate would have a slope of 1. All such lines go through the origin ($P = C = 0$), because eating amount zero necessarily yields zero carbohydrate and zero protein, regardless of the mixture in a given food. Following the geometric framework, we call lines like these 'nutritional rails.'

An animal can therefore be viewed as 'moving' along a nutritional rail as it feeds, over a distance that is proportional to the amount of the food that it eats. In **Figure 2(a)** for example, the unfilled circle on the nutritional rail represents the current nutritional state of the animal (the amounts of P and C gained) having fed on the food for some hypothetical period at the beginning of which its state was taken to be $0P\ 0C$ (no gain of either nutrient).

Also depicted within the nutrient space is a point or region representing the optimal position that the animal could move to (i.e., the amounts of the two nutrients that would benefit it most), known as the 'intake target' (shown in **Figure 2(a)** as the pink region labeled IT). The intake target in reality moves with time, because the animal's

nutrient requirements change with development, health, etc., but it can also be considered to be static if the model pertains to a snapshot in time. For simplicity, in the examples that follow, we will assume the static case, although they could equally be developed in dynamic models. In either case, the 'goal' of the animal is to move through nutrient space, reaching or approaching as close as possible to the intake target.

In the scenario shown in **Figure 2(a)**, this is a straightforward task because the nutritional rail passes through the intake target (i.e., the food contains the same balance of P and C as is needed by the animal) and the animal needs only to ensure that it eats enough of the food to reach the target. In Foods 2 and 3 (**Figure 2(b)**), by contrast, the $P{:}C$ ratio is too low and too high, respectively, so an animal assigned either of those foods could not reach the intake target. However, if it had access to both these foods, then it *could* reach the target by mixing its intake from the two. It could, for example, first feed on Food 2 until it reached the point S1 and then switch to Food 3; alternatively it could first feed on Food 3 and at point S2 switch to Food 2; or it could switch between the foods several times in one of a very large number of possible patterns (e. g., the route to point S3). The scenario depicted in this figure is the phenomenon of nutrient self-selection that we discussed earlier in the section titled 'Hunger and satiety as intervening variables.'

An important point that this hypothetical example illustrates is how geometric models provide a framework for representing internal nutritional states in a quantitative way and relating these to behavior and regulatory mechanisms. Thus, the distance and direction from the intake target of the animal's nutritional state can be used as a measurable, geometric quantity, to make a prediction about the animal's feeding behavior. Consider, for example, the three states represented in **Figure 2(b)** as S1, S2, and S3. That all three of these states fall on the negatively sloped dashed line shows that the animals in these states have all eaten the same total amount of macronutrient ($P + C$), and might well have done so by eating the same total amount of food (if the macronutrient density of Food 2 and 3 are equal). And yet their different positions in nutrient space lead to the prediction that the three animals will show very different feeding behavior from that point on. The animal in state S1 would be more likely to feed on Food 3 (as this will take it to the intake target), while the animal in state S2 would be more likely to feed on Food 2. The animal in state S3 would, like the animal at S1, probably feed on Food 3, because this is the food that would take it closer to the target balance of nutrients. It would, nonetheless, differ from the animal at S1, because its incentive to select Food 3 would be lower. This is because if the animal in S1 continued to feed on Food 2 (e.g., to S4), then it would have veered too far off course and could no longer subsequently reach the target by

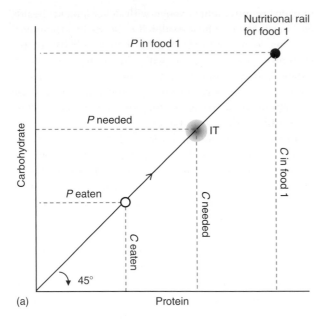

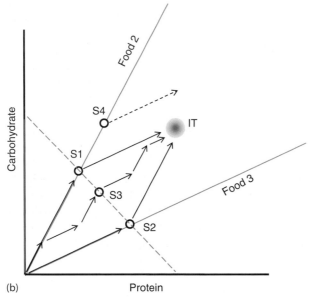

Figure 2 The geometric framework for nutrition. (a) Focal food components, in this case protein (P) and carbohydrate (C), are plotted as axes which define a 'nutrient space.' The solid dot shows the amount of the two nutrients in a food item, and the line radiating from the origin through this point gives the balance of the two nutrients in the food. In this case, the P/C balance is 1:1, and the line thus has a slope of 45°. Such lines representing the nutrient balance of foods are called 'nutritional rails.' The amount of the nutrients needed by the animal (the 'intake target') is also shown (labeled IT). As the animal eats the food, it gains nutrients in the same proportion they are present in the food (i.e., P eaten/ C eaten = P in food/C in food), and can thus be viewed as 'moving' along the nutritional rail (hollow circle and arrow). In panel A, the rail passes directly through the intake target (because the balance P in food/C in food = P needed/C needed), and the animal can thus reach its target by eating this food – that is, this food is balanced with respect to the animal's requirements for the two nutrients. (b) Here the animal does not have access to nutritionally balanced Food 1, but to two nutritionally imbalanced

eating F3 – at best, it could occupy some position along the broken arrow. The animal at S3, by contrast, would have more leeway to correct its course before reaching the point where the target is no longer accessible. Experiments have shown that animals do indeed 'navigate' through nutrient space in this way (e.g., **Figure 3**) and, as shown in the previous section, the mechanisms are partly understood.

Geometric models also enable nutritional state to be linked to functional outcomes and ultimately evolutionary fitness. To do this, you construct a performance surface which maps onto nutrient space the consequences for the animal of achieving different nutrient intakes. A performance surface is equivalent to a topographical map, but with isolines that represent not altitude but the levels achieved of whatever performance variables are of interest (e.g., survival, growth rates, fecundity, etc.). Performance surfaces enable the fitness consequences to be compared between different feeding strategies – for example, the two food selection options open to the animal at S1 in **Figure 2(b)**, as discussed earlier. In **Figure 4**, we give a real example, showing that female fruit flies select the balance of nutrients that maximizes lifetime egg production. Such performance surfaces provide a powerful approach for understanding the costs and benefits for animals of achieving different nutritional states, thereby clarifying the factors that underlie the evolution of nutritional regulatory systems.

Conclusions

Nutrition-related behavior gives the strong impression of goal-directedness: animals become increasingly focused on nutrition as food deprivation progresses, and cease to do so once sated. Hunger and satiety are useful concepts in animal behavior because they provide a framework that captures this property of goal-directedness. Not only does this represent an important characteristic of the system being modeled, but the property of goal-directedness also enhances the predictive power of a framework. However,

foods with rails that do not pass through the intake target: Food 2 has excess C relative to P, and Food 3 excess P relative to C. However, since these rails lie on opposite sides of the intake target, the animal can nonetheless reach the intake target by mixing its intake from the two foods. Several scenarios of such food mixing are given: first eat Food 2, and when in states S1 switch to Food 3; first eat Food 3 then switch to Food 2 at state S2; or pass through state S3 by switching repeatedly between the two foods. If the animal stays on Food 2 until it reaches S4 it cannot subsequently reach the target, because by then switching to Food 3 it will be confined to the trajectory given by the dashed arrow. The diagonal dashed red line is a nutritional isoline, any point on which represents the same total gain of the two nutrients (i.e., $P + C$ = constant).

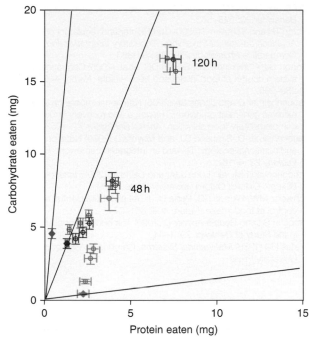

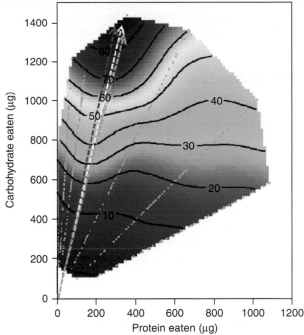

Figure 3 Results of an experiment in which male cockroaches (*Blatella germanica*) were given one of three foods for 48 h, and then for the following 120 h allowed access to all three foods. The solid black lines are nutritional rails representing the protein/carbohydrate balance of the artificial foods, which were identical except for the *P/C* ratio. Each point represents mean + bivariate standard error intake of protein and carbohydrate by > 15 insects, with the colors distinguishing treatment groups (green = high *P/C*, blue − intermediate *P/C*, and red = low *P/C*). The first mean in the series for each treatment (i.e., that which falls on the respective nutritional rail) represents the nutrient intake following 48 h in which the animals were confined to the respective food, while subsequent means show the point of cumulative intake over subsequent 4 h periods in which the insects were able to self-select a diet from all three foods. All treatment groups headed in different directions, so as to compensate for their period of constrained imbalance, and by 48 h had converged on the same intake point. They maintained this grouping, and continued to select the same macronutrient balance, until the experiment was terminated at 120 h. Data from Raubenheimer D and Jones SA (2006) Nutritional imbalance in an extreme generalist omnivore: Tolerance and recovery through complementary food selection. *Animal Behaviour* 71: 1253–1262.

Figure 4 Experiment investigating the link between macronutrient selection and performance in fruit flies (*Drosophila melanogaster*). Female flies were allocated to one of 28 dietary treatments, which resulted in a wide spread of macronutrient intakes (the small gray dots represent intakes of 1000 individual flies). The lifetime egg production of these flies was then plotted as a performance landscape, where red represents high values and blue low values. In a second experiment, three groups of 25 flies were each given a different combination of nutritionally complementary foods from which they could self-select a diet. Dashed arrows show how flies in the three self-selecting treatments adjusted their intake of separate yeast and sugar solutions to converge on a common nutrient intake trajectory that maximized lifetime egg production. In each case, flies had access to a 180 g l^{-1} solution of sugar, and a yeast hydrolysate solution at either 180 g l^{-1} (green arrow), 90 g l^{-1} (white arrow), or 45 g l^{-1} (pink arrow). This experiment demonstrates dietary self-selection, as seen for cockroaches in **Figure 3**, and also that self-selected macronutrient balance corresponds with that which maximizes fitness. Data are from Lee KP, Simpson SJ, Clissold FJ, et al. (2008) Lifespan and reproduction in Drosophila: New insights from nutritional geometry. *Proceedings of the National Academy of Sciences of the United States of America* 105: 2498–2503.

the use of goal-directed frameworks in animal behavior calls for particular vigilance against the risk of teleology − the anthropomorphic, mentalistic notion of conscious intent. Teleology can be avoided, however, by adopting the working hypothesis that 'hunger' and 'satiety' are not things that actually exist within animals, but 'intervening variables.' These are hypothetical internal states that allow researchers better to understand those aspects of animals that we can observe − their physiology, behavior, and the functional consequences of these. Such an approach has been labeled 'teleonomic,' to distinguish it

from teleology. Teleonomic frameworks that use objective, quantitative terms of reference, such as Cartesian geometry, are better insulated against teleology than are those couched in the language of human psychology. We believe that the field is fast progressing to the point where teleonomic frameworks can confidently be used in the study of nutritional behavior, while avoiding the risks of teleology. Central to this progress is rapid growth in our understanding of the mechanisms that control feeding, and the development of models for exploring how these mechanisms interact in the regulation of behavior.

See also: Adaptive Landscapes and Optimality; Norway Rats; Self-Medication: Passive Prevention and Active Treatment; Taste: Invertebrates; Taste: Vertebrates.

Further Reading

Chaudri OB, Salem V, Murphy KG, et al. (2008) Gastrointestinal satiety signals. *Annual Review of Physiology* 70: 239–255.

Cota D, Proulx K, and Seeley RJ (2007) The role of CNS fuel sensing in energy and glucose regulation. *Gastroenterology* 132: 2158–2168.

Dethier VG (1976) *The Hungry Fly*. Princeton, NJ: Princeton University Press.

Finger TE (1997) Evolution of taste and solitary chemoreceptor cell systems. *Brain, Behavior and Evolution* 50: 234–243.

Kennedy JS (1992) *The New Anthropomorphism*. Cambridge: Cambridge University Press.

Lee KP, Simpson SJ, Clissold FJ, et al. (2008) Lifespan and reproduction in Drosophila: New insights from nutritional geometry. *Proceedings of the National Academy of Sciences of the United States of America* 105: 2498–2503.

McFarland DJ and Sibly RM (1972) 'Unitary drives' revisited. *Animal Behaviour* 20: 548–563.

Moran TH and Schulkin J (2000) Curt Richter and regulatory physiology. *American Journal of Physiology-Regulatory Integrative and Comparative Physiology* 279: R357–R363.

Morton GJ, Cummings DE, Baskin DG, et al. (2006) Central nervous system control of food intake and body weight. *Nature* 443: 289–295.

Raubenheimer D and Jones SA (2006) Nutritional imbalance in an extreme generalist omnivore: Tolerance and recovery through complementary food selection. *Animal Behaviour* 71: 1253–1262.

Raubenheimer D, Simpson SJ, and Mayntz D (2009) Nutrition, ecology and nutritional ecology: Toward an integrated framework. *Functional Ecology* 23: 4–16.

Schoonhoven LM, van Loon JJA, and Dicke M (2005) *Insect–Plant Biology*. Oxford: Oxford University Press.

Schwartz MW, Woods SC, Porte D, Jr., et al. (2000) Central nervous control of food intake. *Nature* 404: 661–671.

Simpson SJ and Raubenheimer D (2000) The hungry locust. *Advances in the Study of Behavior* 29: 1–44.

Toates FM (1986) *Motivational Systems*. Cambridge: Cambridge University Press.

Imitation: Cognitive Implications

T. R. Zentall, University of Kentucky, Lexington, KY, USA

Introduction

Definitions of concepts relevant to social learning and influence presented in this article have borrowed freely from earlier attempts by Galef, Whiten and Hamm, and Zentall to distinguish among the various mechanisms thought to affect the behavior of animals when exposed to others of their species.

Some researchers who have studied social learning have been interested in its adaptive value. Others have been more interested in the mechanisms by which animals are able to integrate the information that is available when they observe another animal perform a response, and then perform the observed response themselves. In the following sections, I will try to describe the various mechanisms that have been thought to play a role in social learning, and examine the evidence for imitation in animals, a kind of social learning that is thought to imply higher cognitive functioning.

Social Influence

Species Typical Influences: Contagion

Predisposed tendencies to match specific behaviors of a conspecific are called contagious behaviors, mimesis, response facilitation, or response priming. Contagion can be used to describe certain courtship displays, antipredator behavior (such as mobbing), and social eating. For example, a chicken that has eaten its fill and then stopped eating will begin eating again if placed with another chicken that is eating.

Motivational Influences

Factors that affect the arousal or motivation of an animal may affect its general activity, which may, in turn, lead to faster learning.

Social facilitation

The mere presence of another animal, irrespective of the other animal's behavior, may increase (or decrease) arousal, a phenomenon called social facilitation. Increased arousal can lead to increased activity leading to increased contact with environmental contingencies. For example, if the presence of another animal increases an observer rat's general activity, the rat may discover (on its own) the lever that when pressed leads to reward.

Socially induced incentive motivation

Being in the presence of a conspecific that is eating may further increase the arousal of an observing animal. Thus, if a rat sees another rat eating, it may result in a greater increase in the activity of an observer than just seeing another animal, leading the observer to discover the rewarding lever.

Observation of aversive conditioning

Being in the presence of a demonstrator that is acquiring or performing a response motivated by avoidance of painful stimulation (e.g., electric shock) can provide the observer with an additional source of motivation. For example, emotional cues provided by another animal either escaping from or avoiding shock may evoke fear in an observer, and increased fear may lead to faster learning.

Perceptual Factors

Observation of a performing demonstrator also may draw attention to the place where the demonstrator is performing (local enhancement) or to the object that the demonstrator is manipulating (stimulus enhancement). The added attention may increase the likelihood that the observer will explore the place or the object and learn individually about the object. For example, the movement of a lever pressed by a demonstrator may draw attention to the lever and cause the observer to approach the moving lever (or a lever that looks like the moving lever) more readily.

Simple Social Learning

Social influence results in behavior that makes it more likely that an observer will be motivated to learn or will notice cues that are necessary for learning. However, the learning is individual.

Discriminated Following or Matched Dependent Behavior

Rats can be trained to follow a trained conspecific to food in a **T** maze in the absence of any other discriminative stimulus. Although the leader rat in such experiments is clearly a social stimulus, the data are parsimoniously interpreted as demonstrating simple discrimination learning with the leader serving as a salient, discriminative stimulus. For example, a rat could easily be trained to follow a block of wood pulled through a maze by a string. It may be easier for the rat to learn to follow a social stimulus, because it is more salient than an inanimate object. However, a leader rat is not conceptually different from a block of wood.

Observational Conditioning

The observation of a performing demonstrator may do more than draw attention to the object being manipulated (e.g., a lever or joystick). Because the observer's orientation to the object is often followed immediately by presentation of food to the demonstrator, a (higher order) Pavlovian association called observational conditioning may be established between the object, towards which the observer orients, and food. If sight of the moving object subsequently serves as a conditioned stimulus, the animal may be drawn to it, contact it, and thus attain a reward.

Socially transmitted food preferences of the sort studied by Galef and his colleagues represent a special case of observational conditioning. In these studies, rats that have interacted with another rat eating a novel food item will be more likely to eat that same food than another equally novel food item. Although food preference may appear to fall into the category of unlearned behavior, consuming food that has a *novel* taste is an acquired behavior that can be learned through its association with another animal.

Vocal Mimicry (Bird Song)

A special case of matching behavior is the acquisition of regional variations in bird song that appear to depend on the bird's early experience with conspecifics. Bird song consists of auditory stimuli, and the sounds produced by a singing demonstrator and those produced by a listener learning to sing a regional dialect can be similar. Thus, stimulus matching by trial and error could mediate the listeners' acquisition of a song, similar to that of its demonstrator.

Visual Matching

The preceding analysis of the imitation of vocal behavior can also be applied to certain examples of visual imitation. Any behavior that produces a clear change in the environment, such that, from the perspective of the observer, there is a match between the result of the action of a demonstrator and that produced by an action of an observer, may allow for stimulus matching. Consider what happens when you see someone turn up the volume of a radio. When the knob turns to the right, the volume increases. Similarly, when you turn the knob, you see it turn to the right and hear the volume increase. You could learn to turn the knob simply by learning to match the movement of the knob in response to your action with the movement of the knob in response to the demonstrated action. Such cases of visual stimulus matching can be distinguished from the, perhaps, more abstract and interesting case that I will consider in more detail later, opaque imitation, in which no visual stimulus match is possible, for example, imitating the body position of a person who has his hands clasped behind his back.

Emulation (Object Movement Reenactment)

When observation of a demonstrator allows an animal to learn how the environment functions, a sophisticated form of learning is certainly involved. However, because learning how the environment works may not require observation of the behavior of another animal, one may not want to view such learning as social learning. Learning about how the environment works has been referred to as the emulation of affordances or object movement reenactment.

Emulation may be involved in a procedure used with chimpanzees in which they learned to open a box to obtain a reward. Some demonstrator chimpanzees were trained to poke a bolt to open a box, whereas other demonstrators were trained to twist and pull the bolt to achieve the same result. Observers given access to the box tended to remove the bolt the same way that they had seen it removed. However, because the bolt moved differently in the two cases, it is possible that the observers learned how the bolt moved (by emulation) rather than to copy the actions of their demonstrators (by imitation).

Although emulation may not be considered social learning, because it is learning about the movement of objects rather than the actions of demonstrators, emulation is a phenomenon of interest in its own right. Understanding how things work has implications for cognitive learning. Interestingly, although there is some evidence that pigeons can emulate the direction of movement of a screen permitting access to food, chimpanzees have shown little evidence of emulation.

Complex Social Learning

To researchers interested in cognitive mechanisms involved in social learning, the most interesting forms of social learning are those that cannot be explained easily by any of the previously described mechanisms.

Opaque Imitation

The term imitation is used to indicate behavior of an observer that matches the behavior of a demonstrator that cannot be accounted for with any of the motivational, attentional, or simple learning mechanisms described so far. Such imitation may be referred to as opaque imitation because, under these conditions, the observer's own movements cannot be readily seen by the observer. Under appropriate conditions, the bidirectional control and two-action procedures are accepted methods for demonstrating opaque imitation.

The bidirectional control procedure
When a screen that could be pushed to the left or right was placed in front of a feeder, observer pigeons tended to push the screen in the same direction that they saw a demonstrator push it. To distinguish this learning from emulation, a 'ghost control' is needed in which the screen appears to move by itself without a demonstrator pushing it.

The two-action procedure
The two-action procedure controls for emulation by having the demonstrator produce the same environmental outcome in one of two different ways. For example, quail were trained to respond to a treadle (a metal plate near the floor of the chamber) for food either by pecking at the treadle or by stepping on it. When given access to the treadle, observers used the same part of their body to make the response as had their respective demonstrators.

It is important to note, first, that the environmental consequences of stepping and pecking were essentially the same (i.e., everything was the same except the response topographies of the demonstrators), and second, that there was little if any similarity between the visual stimulus the observer saw during observation and the visual stimulus it saw during its own performance of either response. That is, the appearance of the demonstrator's beak on the treadle must have appeared quite different to the observer from the sight of its own beak on the treadle. Similarly, although perhaps not so obviously, when the quail stepped on the treadle (located near the corner of the chamber), it pulled

its head back and thrust its chest forward. Thus, it could not see its foot making contact with the treadle. Once again, to the observer, the demonstrator's response to the treadle must have appeared quite different from the observer's own response to the treadle. Thus, in such an experiment, any account of imitation based on stimulus matching is quite implausible.

Variables That May Influence Opaque Imitation
Demonstrator reinforcement
A cognitive account of imitation suggests that the observer understands what the demonstrator is doing and perhaps why it is doing it. If such an interpretation is correct, whether evidence of imitation is found may depend on the consequences of the demonstrated response for the demonstrator. Consistent with such a hypothesis, Zentall found that quail imitated only when they observed demonstrators receiving a reward after they pecked or stepped on a treadle.

It is possible to explain the effect of demonstrator reward on observer imitation by appealing to observational conditioning, a simpler form of learning described above. In observational conditioning, an observer's attention is drawn to a stimulus (in this case, the demonstrator quail depressing the treadle) because this action precedes reward. Although observational conditioning might account for the difference in the effect on observer imitation of reward, it cannot account for the correspondence between the observer's and demonstrator's response topographies. Thus, demonstrator reward may act as a catalyst to bring out imitative learning in an observer.

The effect of the outcome for the demonstrator may play an even more important role when more complex responses than pressing a treadle are demonstrated. For example, humans performed a task in the presence of chimpanzees in which, first, one response (poking a stick in a hole in the top of a box) did not lead to obtaining food but another response (poking the stick in a hole in the side of the box) did. Next, subjects were given access to the box and the stick. When the box was opaque, so that subjects could not see that the top hole did not provide access to food, the subjects often started by poking the stick in the top hole. However, when the box was transparent, and subjects could see that the top hole did not provide access to food, the subjects generally avoided poking the stick in the top hole and instead poked the stick directly into the side hole that had produced food. Horner and Whiten proposed that when the box was transparent, subjects recognized the causal structure of the task and avoided the response that did not lead to reward, however, when the box was opaque, it was not clear that inserting the stick in the top hole was not a necessary prerequisite to inserting the stick in the side hole. Thus, chimpanzees could acquire the entire sequence of responses through observation, but omitted part of the sequence when it was apparent that one of the demonstrated responses was not necessary to achieve the goal.

Observer motivation

The hypothesis that observer motivation affects imitation was tested by comparing imitative learning by quail that were either hungry or sated at the time of observation. It was found that the hungry quail matched the demonstrator's reinforced behavior, whereas sated quail did not. It is possible, however, that the observers that were hungry paid more attention to what the demonstrators were doing than observers that were not hungry.

Deferred imitation

Bandura proposed that there is an important cognitive difference between immediate imitation (that Bandura called imitation) and deferred imitation (that he called observational learning) where some time passes between observation and observer performance. For Bandura, immediate imitation may be a reflexive response akin to contagious behavior, whereas deferred imitation indicates a more cognitive process where an observer has to *represent* the response at the time of observation for later retrieval when performance is assessed.

To determine if animals are capable of deferred imitation, hungry quail observed either treadle pecking or treadle stepping. When they were tested 30 min later, they imitated as frequently as observers tested immediately following observation. If, as Bandura proposed, deferred imitation is evidence of a more cognitive process, then quail show good evidence of cognitive representation.

More Complex Forms of Imitation

Generalized gestural imitation

A form of imitative learning that is conceptually related to the two-action procedure involves copying the gestures of a model on command (e.g., 'Do this!'). Successful do-as-I-do performance has been reported in chimpanzees (Custance et al.), dolphins (Herman), and parrots (Pepperberg). Remarkably, because the imitated models were humans, there was little similarity between corresponding body parts of observer dolphins and parrots, and the human demonstrator.

Custance et al. found that chimpanzees learned to respond to the command 'Do this!' by imitating a broad class of behaviors demonstrated by humans including touching the back of their heads and other actions that could not be seen as they were performed. Such imitated opaque actions cannot be explained by some form of visual stimulus matching. Furthermore, because objects are not involved in this kind of imitation, local and stimulus enhancement are irrelevant. Finally, each imitated gesture serves as a control for other imitated

gestures, and the broad range of gestures that have been imitated within a few seconds of demonstration suggests that differential motivation plays no role. Success in such do-as-I-do experiments shows not only that chimpanzees can imitate, but also that they are capable of forming a generalized *concept* of imitation, because they selectively imitate any of a broad class of gestures when cued to do so.

Intentionality

Interest in imitation research can be traced, at least in part, to the possibility that imitation involves some degree of intentionality. Intentionality is surely involved in many higher order forms of imitation, such as the human dancer who repeats the movements of a teacher. Unfortunately, because of its indirect nature, intentionality in animals can only be inferred, and evidence for intentionality appears most often in the form of anecdote rather than experiment.

Mitchell provides a number of examples of imitation at these higher levels. For example, Mitchell discusses observations of a young female rhesus monkey who, after seeing her mother carrying a sibling, walked around carrying a coconut shell at the same location on her own body. If there were any way to carry out experiments involving the manipulation of intentionality, the credibility of these anecdotes would be greatly increased.

Understanding the intentions of others

Recent evidence suggests that 14-month-old children are able to understand the intentions of another and use this understanding to mediate their imitative behavior. When children watched a demonstrator whose hands were occupied turn on a light by touching it with her forehead, they subsequently turned on the light more efficiently by using their hands. However, when the demonstrator's hands were not occupied, so that observing children might assume that it was necessary to use their forehead to turn on the light, children copied the demonstrator in using their forehead.

More surprising, there is evidence that dogs may be able to make similar inferences. When dogs watched a dog demonstrator with a ball in its mouth pull a rod with its paw to obtain a treat, the observer dogs pulled the rod more efficiently with their mouth. However, if the demonstrator's mouth was not occupied and it pulled the rod with its paw, the observers also pulled the rod with their paw, suggesting that dogs, like human children cannot only imitate but can understand intentions of a demonstrator.

Program-level imitation

Byrne has distinguished action level imitation, for example, pressing a lever or poking at a bolt, from program-level imitation that involves learning of a coordinated sequence of actions leading to reward. As already noted,

Horner and Whiten have shown that when a human demonstrated a response sequence (poking a stick into the top hole of an opaque box and then the side hole), chimpanzees copied both responses. Even pigeons can show evidence of the imitation of a sequence of two quite different response alternatives (stepping on a treadle vs. pecking at the treadle and then pushing a screen blocking a feeder either to the right or to the left). Whether program level imitation is conceptually different from imitation of a single response is yet to be shown.

Associative Learning Accounts of Imitation

Heyes and Ray have recently proposed an associative learning account of imitation. According to their associative sequence learning (ASL) model, specific responses, such as stepping or pecking, have been reinforced in the past when they have occurred in the presence of other animals engaged in similar behavior. So, for example, the theory assumes that before participating in any experiment, observers ate at the same time as others and consequently learned to peck when others were pecking, and they fed from a similar feeder and consequently learned to step toward the feeder when others were doing so. As a result, seeing others pecking or stepping would become a discriminative stimulus for engaging in the same behavior.

There are some difficulties with this account. First, laboratory birds are often fed at different times, so typically pecking and stepping at the same time as other birds would not be reinforced. Second, the theory requires that the home-cage context generalizes to the experimental context, despite the fact that the experimental environment is typically quite different from the home-cage environment. Third, in the two-action procedure, the treadle is not located near the feeder. Finally, it is not clear how ASL can account for results of bidirectional control experiments in which a screen encountered for the first time is pushed in the same direction as a demonstrator pushed it. As Whiten suggests, if the mechanisms responsible for imitation involve basic learning processes that are present in many animal species, why is it that only humans, certain great apes, and birds, show clear evidence imitation learning?

Possible Biological Mechanisms

Response Facilitation

Byrne proposed that response facilitation could account for much of the imitation reported in animals. As noted above, response facilitation implies that observation of a response elicits a similar response in an observer. By this account, the observed behavior is already in the repertoire of the observer, and observation of it automatically primes the representation of the behavior in the brain, increasing the probability that the behavior will occur. Although this view provides a noncognitive mechanism for the kinds of imitation most frequently studied in animals, it does not provide a particularly convincing account of imitation of behaviors that an animal encounters for the first time in an experimental setting, for example, pushing a screen to left or right to access food.

The Mirror System

Neurons found in the premotor cortex of monkeys are activated not only when the monkey picks up an object but also when it sees either a human or another monkey pick up an object. These 'mirror' neurons are thought to be responsible for imitation, and their presence in the premotor cortex rather than the visual cortex suggests that they may have a preparatory cognitive function. Although the mirror neurons may be involved in stimulus matching, it is not clear whether they can account for perceptually opaque imitation when there is little similarity between the visual input animals receive from watching the behavior of another and what they can see of their own behavior.

Conclusions

Procedures have recently been developed that separate imitation from other forms of social influence and social learning, and the results of initial studies indicate that species from chimpanzees to quail can imitate. Such findings should not be surprising because social learning, whether by imitation or some other process, often provides greater benefits than genetically based behavior or trial-and-error learning. However, the mechanisms enabling animals to match their behavior to that of a demonstrator are poorly understood. Imitation may involve some form of coordination of visual and tactile sensory modalities, perspective taking, or response facilitation. However, the role of such processes in opaque imitation is not yet well understood. A reasonable strategy to better understand the mechanisms involved in imitation would be to determine the necessary and sufficient conditions for opaque imitation to occur and to explore the range of behaviors that animals can imitate.

See also: Apes: Social Learning; Vocal Learning.

Further Reading

Bandura A (1969) Social learning theory of identificatory processes. In: Goslin DA (ed.) *Handbook of Socialization Theory and Research*, pp. 213–262. Chicago: Rand-McNally.
Byrne R (1994) The evolution of intelligence. In: Slater PJB and Halliday TR (eds.) *Behavior and Evolution*, pp. 223–265. Cambridge: Cambridge University Press.

Custance DM, Whiten A, and Bard KA (1995) Can young chimpanzees imitate arbitrary actions? Hayes and Hayes revisited. *Behaviour* 132: 839–858.

Galef BG, Jr (1988a) Communication of information concerning distant diets in a social, central-place foraging species: *Rattus norvegicus*. In: Zentall TR and Galef BG, Jr (eds.) *Social Learning: Psychological and Biological Perspectives*, pp. 119–139. Hillsdale, NJ: Erlbaum.

Galef BG, Jr (1988b) Imitation in animals: History, definition, and interpretation of data from the psychological laboratory. In: Zentall TR and Galef BG, Jr (eds.) *Social Learning: Psychological and Biological Perspectives*, pp. 3–28. Hillsdale, NJ: Erlbaum.

Herman LM, Matus D, Herman EYK, Ivancio M, and Pack AA (2001) The bottlenosed dolphin's (*Tursiops truncatus*) understanding of gestures as symbolic representations of body parts. *Animal Learning & Behavior* 29: 250–264.

Heyes CM and Dawson GR (1990) A demonstration of observational learning in rats using a bidirectional control. *Quarterly Journal of Experimental Psychology* 42B: 59–71.

Heyes CM and Ray ED (2000) What is the significance of imitation in animals. *Advances in the Study of Behavior* 29: 215–245.

Hopper LM, Lambeth SP, Schapiro SJ, and Whiten A (2007) Observational learning in chimpanzees and children studied through 'ghost' conditions. *Proceedings of the Royal Society B*, doi: 10.1098/rsbv.2007.1542.

Horner V and Whiten A (2004) Causal knowledge and imitation/emulation switching in chimpanzees (*Pan troglodytes*) and children (*Homo sapiens*). *Animal Cognition* 8: 164–181.

Janik VM and Slater PJB (2003) Traditions in mammalian and avian vocal communication. In: Fragaszy DM and Perry S (eds.) *The Biology of Tradition: Models and Evidence*, pp. 213–235. Cambridge, UK: Cambridge University Press.

Mitchell RW (1987) A comparative-developmental approach to understanding imitation. In: Bateson PPG and Klopfer PH (eds.) *Perspectives in Ethology,* Vol. 7, pp. 183–215. New York: Plenum.

Pepperberg IM (1988) The importance of social interaction and observation in the acquisition of communicative competence: Possible parallels between avian and human learning. In: Zentall TR and Galef BG Jr. (eds.) *Social Learning: Psychological and Biological Perspectives*, pp. 279–299. Hillsdale, NJ: Erlbaum.

Rizzolatti G, Fadiga L, Fogassi L, and Gallese V (2002) From mirror neurons to imitation: Facts and speculations. In: Meltzoff A and Prinz W (eds.) *The Imitative Mind: Development, Evolution, and Brain Bases*, pp. 247–266. Cambridge, UK: Cambridge University Press.

Whiten A (2005) The imitative correspondence problem: Solved or sidestepped? In: Hurley S and Chater N (eds.) *Perspectives on Imitation: From Neuroscience to Social Science,* Vol. 1, pp. 220–222. Cambridge, MA: MIT Press.

Whiten A and Ham R (1992) On the nature and evolution of imitation in the animal kingdom: Reappraisal of a century of research. In: Slater PJB, Rosenblatt JS, Beer C, and Milinski M (eds.) *Advances in the Study of Behavior,* Vol. 21, pp. 239–283. New York: Academic Press.

Zentall TR (2006) Imitation: Definitions, evidence, and mechanisms. *Animal Cognition* 9: 355–367.

Immune Systems and Sickness Behavior

J. S. Adelman, Princeton University, Princeton, NJ, USA
L. B. Martin, University of South Florida, Tampa, FL, USA

Introduction

The vertebrate neuroendocrine system responds to changes in social and physical environments, adjusting an animal's phenotype to match current or impending conditions. A classic example is the prebreeding rise in testosterone among seasonally breeding males, which increases the likelihood of exhibiting courtship, mating, and territorial behaviors. Another vital physiological process affected by hormones is immune function. While immune defenses are crucial to individual health and survival, they entail considerable metabolic and opportunity costs (other behaviors or processes that must be sacrificed to mount the response). Organisms must balance these costs against competing processes, which require information about the value of defense against infection versus other fitness priorities (e.g., breeding). Endocrine hormones provide a reliable, integrated source of such information.

Exemplar endocrine–immune interactions include the behavioral responses to infection, or sickness behaviors. These behavioral changes include lethargy, anorexia, adipsia (decreased water intake), somnolence, anhedonia (reduced performance of pleasurable behaviors), and decreased libido. In 1988, Benjamin Hart suggested that sickness behaviors are likely to increase fitness by conserving energy for use on costly physiological immune responses, such as fever and defensive protein synthesis, both of which increase pathogen clearance. Moreover, anorexia can limit a pathogen's access to nutrients crucial to replication. On the other hand, as sickness behaviors impose considerable opportunity costs, infected animals must reduce some adaptive behaviors (e.g., foraging, territorial defense, parental care, and mating effort) to deal with an infection. Understanding how and why animals balance these various costs to achieve protection against infection remains an ongoing challenge and comes under the purview of the emerging field of ecological immunology.

In this article, we examine some of the ways in which hormones can influence immune responses in general, and sickness behaviors in particular. To begin with, we provide a brief introduction to the vertebrate immune system and discuss how studying immunology in the context of life history can help elucidate why immune responses vary and when (and how) we would expect endocrine modulation to occur. We then introduce cytokines and immune system-signaling molecules as key links among the endocrine, immune, and nervous systems. Finally, we demonstrate how several specific hormones enable the appropriate expression of sickness behaviors (see **Figure 1**), and we briefly discuss several directions for future research.

The Vertebrate Immune System

The vertebrate immune system protects against infection via a network of proteins, cells, and tissues, classically divided into innate and adaptive arms. The innate arm includes rapidly mobilized immune cells and defensive proteins that bind, kill, or engulf foreign invaders. The adaptive arm becomes fully active several days after pathogen infiltration and includes antibody production by B-lymphocytes and/or proliferation of T-lymphocytes that eliminate infected host cells, regulate B-cell function, or coordinate innate immune activity. Adaptive immunity has the capacity for memory of specific antigens, in that a second infection by the same pathogen induces a more efficient response. Innate immunity lacks this capacity, but it is much more broadly effective.

Historically, there has been a bias towards understanding the adaptive immune system. However, it is becoming clearer that the dichotomy between these two arms is oversimplified; the majority of adaptive immune processes coordinate innate immune function. For these reasons, and because innate immunity is the only defense that most organisms possess, the innate immune system is receiving greater attention today.

Sickness behaviors are innate immune defenses and part of the acute phase response, a systemic enhancement of the immune system engaged early in the course of an infection. At present, we do not understand how initial local, low-grade inflammation escalates into a full-scale acute phase response. Nevertheless, once acute phase responses are induced, they place the host in an emergency life history state in which survival from infection dominates physiology and behavior. Acute phase responses entail heterothermia (i.e., fever in most vertebrates), copious release of defense-oriented hepatic proteins (which control foreign invaders through a variety of mechanisms), increased leukocyte activity (e.g., phagocytotic capacity), and sickness behaviors. These defenses come

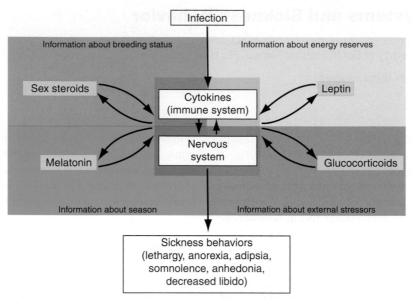

Figure 1 Upon infection, a vertebrate's immune cells and tissues release cytokines, signaling molecules that promote subsequent immune responses and the sickness behaviors listed above. To express these sickness behaviors in a context-dependent manner, animals must physiologically integrate information about breeding status (purple), season (green), energy reserves (yellow), and external stressors (pink). In this article, we highlight how bidirectional interactions among cytokines, endocrine hormones (blue), and the nervous system facilitate this integration and the adaptive expression of sickness behaviors.

at considerable costs: fever increases the metabolic rate by 10–15% for each degree Celsius increase in body temperature, defensive protein production requires both energy and at least 5% of daily amino acid intake, and sickness behaviors carry the substantial opportunity costs discussed above. Moreover, because of their broad, nonspecific effects, these defenses can also induce considerable collateral damage to host tissue.

Acute phase responses are coordinated predominantly by the proinflammatory cytokines, interleukin-1β (IL-1β), interleukin-6 (IL-6), and tumor necrosis factor-α (TNF-α). These signaling molecules are typically induced once leukocyte receptors (pattern recognition receptors, PRRs) that detect molecules shared by numerous pathogens (pathogen-associated molecular patterns, PAMPs) are activated. Thus, it has been possible to study sickness behaviors (and other acute phase response components) simply by exposing organisms to microbial fragments. This approach has been critical for distinguishing the costs associated with true pathogens from the costs of the immune defenses engaged to combat them. One particularly commonly used molecule is lipopolysaccharide (LPS), a part of Gram-negative bacterial cell walls.

Immune Variation Reflects Costs and Benefits in an Ecological Context

In spite of their putative protective values, sickness behaviors are variable. Some sickness behaviors are diminished

greatly and some eliminated altogether in certain species in specific contexts. The emerging field of ecological immunology proposes that the costs and benefits of sickness behaviors drive this variability. In other words, immune variability may be explained by the fitness benefits of defending oneself from infection versus maximizing current reproductive success. A unifying paradigm that directs this field is the following trade-off: if immune responses require metabolic resources or time to resolve infections, other priorities may have to be sacrificed in the meantime or over some period thereafter. Extensive recent work in a host of organisms has demonstrated the costs of immune function sufficient to affect other fitness-related traits. Whether this cost-benefit context is useful in explaining the variation in sickness behaviors is only now being considered.

Unlike other acute phase components such as fever and hepatic protein production, sickness behaviors impose few direct resource costs. Rather, opportunity costs represent the greatest challenges to fitness for sickness behaviors. Indeed, animals appear to dampen sickness behaviors when opportunity costs become too high. For instance, during a study in early spring, when territory defense is critical to annual reproductive success, male song sparrows (*Melospiza melodia morphna*) showed little or no reduction in aggression after simulated infection. Later in the year, when the reproductive benefit of territoriality was presumably lower than the benefit of surviving to the next breeding season, territorial aggression was greatly attenuated when birds were infected. Opportunity costs also influence how sick female mice care for their pups. Females treated with LPS at low ambient temperature suppressed sickness behaviors, retrieved pups quickly, and ensured that nests were well maintained. At room temperature, when pups were not thermally challenged, LPS-treated females took longer to retrieve pups and did not maintain nests of the same quality. A similar explanation was proposed for differences in sickness behaviors between male and female rats. Sexual behaviors of ill male rats were no different than untreated controls, whereas in females, lordosis (a characteristic mating posture) and other reproductive behaviors were dampened in response to simulated infection. For males, the costs of missed reproductive opportunities outweigh the transient reduction in immune function that is mandated to engage in copulations. For females, conception during sickness may increase the risk of complications during pregnancy, so the benefits of reducing sexual behaviors during illness outweigh the costs.

Cost–benefit perspectives also apply to species or population-level differences in sickness behaviors. LPS-treated song sparrows from populations with small clutch sizes and a long breeding season reduced both territoriality and locomotor activity more than populations with larger clutch sizes and shorter breeding seasons. This outcome suggests that when time is limited, as during a short breeding season, reproduction takes priority over sickness behaviors. In a study on *Peromyscus* mice, similar patterns emerged: reproductively, prolific species showed little lethargy when infected, whereas reproductively, conservative species displayed pronounced lethargy. Moreover, reproductively, prolific species did not alter food intake when sick but conservative ones ate less when infected. In a study of captive white-crowned sparrows, the pattern was reversed: during the breeding season, a population with greater reproductive effort per unit time showed exaggerated anorexia compared to the one with lower effort.

Given that two different classes of vertebrates (mammals and birds), some wild and some captive, some breeding and some reproductively quiescent, were used in the above studies, it is difficult to make generalizations about the underlying causes of alterations in anorexia or other sickness behaviors. The extensive variability, however, warrants explanation. Evaluating the importance of the opportunity costs associated with sickness behaviors versus the costs associated with other components of acute phase responses explains some inconsistencies. As sickness behaviors are regulated by the same mechanisms (e.g., proinflammatory cytokines) as other acute phase components, they may not always be dissociable. In such cases, variation in sickness behaviors may be best explained through concurrent consideration of other acute phase response components and their respective costs.

Cytokines Facilitate Neuroendocrine–Immune Interactions and the Modulation of Sickness Behaviors

In order to understand variation in sickness behaviors among individuals, populations, and species, it is critical to identify how animals physiologically integrate diverse environmental information. Because endocrine hormones relay information about external and internal environments, these signals can help modulate immune responses in accordance with current conditions. Endocrine interactions with the immune system are widespread and diverse in nature. Hormones can act directly upon immune cells to influence their proliferation and function. In addition, endocrine hormones can alter the production and efficacy of cytokines. Cytokines have a broad array of functions, including control of local and systemic inflammation, promotion of antibody or T-cell responses, and modulation of sickness behaviors. In this section, we present several of the best-studied examples of endocrine–immune interactions, paying particular attention to sickness behaviors and the role of cytokines in these relationships.

Androgens

Androgens, most notably testosterone, promote male reproductive physiology and behavior. In seasonal breeders, androgen levels rise in response to environmental cues such as changes in day length, which are transduced through photoreceptors in the brain to the hypothalamic–pituitary–gonadal axis. Additionally, social instability and the presence of receptive females elevate testosterone. Thus, androgens serve as integrated mediators of reproductive and social state and could be important in facilitating trade-offs between breeding success and immune defense. Androgens have long been portrayed as immunosuppressants. Even before the isolation and purification of testosterone in the 1930s, Calzolari found that castrated rabbits had larger thymuses. More recently, Folstad and Karter proposed a physiological variant of the handicap hypothesis of sexual selection involving the immunosuppressive effects of androgens. Their immunocompetence handicap hypothesis suggests that because testosterone is crucial in developing some secondary sexual traits and also in suppressing immune function, only the highest quality males could afford (honest) secondary sexual signals.

Although androgens indeed depress some immune responses, the effects are inconsistent across species and parts of the immune system. Additionally, testosterone-mediated immunosuppression cannot explain seasonal differences in immune function commonly found in females, which have low levels throughout the year, or sex-differences in immune function in invertebrates (females stronger) that do not possess androgens. On the other hand, androgen receptors exist on or in several types of leukocytes. Moreover, androgens can decrease synthesis and the release of inflammatory cytokines, including TNF-α and IL-1β, in cultured human immune cells.

Such findings support a link between androgens and the immune system, but is likely to be a context- or tissue-dependent one. Further, the link between androgens and the immune system is not unidirectional. In mammals and birds, proinflammatory cytokines decrease the circulating levels of luteinizing hormone and testosterone. Such changes may help facilitate decreased libido during sickness in many animals. In general, though, the role of androgens in modulating sickness behaviors is relatively unknown. Consistent with the ability to down-regulate certain proinflammatory cytokines, testosterone would be predicted to diminish sickness behaviors. This appeared to be the case in a study of white-crowned sparrows, as exogenous testosterone decreased anorexia. However, few other studies focus on wild vertebrates, necessitating more research in this area.

Estrogens and Progestins

As with testosterone in males, in seasonally breeding females, estrogens and progestins vary across the year. Additionally, pregnancy in mammals involves transiently elevated estrogen, suppression of T-cell functions, and increases in antibody production and innate immune responses. Studies on adult mice have shown that at least some of these changes can be attributed directly to estrogen. Progesterone can compromise cell-mediated immunity too, inhibiting T-lymphocyte activation in cultured cells and decreasing the likelihood of skin graft or exogenous tissue rejection in mammals. Both progesterone and estrogen concentrations correlate with the production of pro- and anti-inflammatory cytokines in humans, which may help explain how the immune system becomes biased towards certain immune responses during pregnancy. As with testosterone, more research is needed about the roles of estrogens and progestins in regulating sickness behaviors. In one study, progesterone enhanced the lethargy induced by the proinflammatory cytokine IL-1β, suggesting that this hormone helps modulate sickness behaviors across the estrous cycle. To our knowledge, however, no work has been done in an ecological context on the effects of either estrogens or progestins on sickness behaviors.

Melatonin

In mammals, melatonin integrates information about seasonal changes in environmental conditions. The pineal gland and retina secrete this hormone during darkness; as day-length changes with season, the changing duration of melatonin secretion informs an animal of the time of year. In general, melatonin augments immune function. In at least some species, however, melatonin diminishes the components of the acute phase response, including the intensity and duration of fever. In terms of sickness behaviors, the links between melatonin in the immune system have been best studied in Siberian hamsters. When housed under long day lengths (shorter nocturnal melatonin secretion duration), this species shows greater lethargy and anorexia than under short day lengths; these differences are largely dependent upon pineal melatonin. Interestingly, differences in sickness behaviors between seasons correlate highly with both production of and sensitivity to inflammatory cytokines. This suggests that cytokines and melatonin work in concert to integrate seasonal information into appropriate immune responses.

Leptin

Leptin is produced in adipose tissue and reflects the status of current energy reserves and functions as a satiety factor. These functions make this hormone an attractive candidate for modulating both immune function and behavior in accordance with an animal's energetic reserves. Indeed, leptin can influence a variety of immune responses, including fever and T-cell proliferation. With regard to sickness behaviors, impairing leptin signaling in LPS-treated mice caused exaggerated lethargy. This result is consistent with severely diminished locomotor activity as a means of conserving limited supplies of energy, as when fat reserves are extremely low. While increased leptin generally decreases feeding in healthy animals, its role in sickness-induced anorexia is somewhat less clear; disrupting leptin signaling in rats and mice can enhance, suppress, or have no effect on feeding behavior. Such variable results may reflect the numerous connections between leptin and cytokines. During the early stages of an acute phase response, peripheral proinflammatory cytokines increase circulating leptin. Leptin may then modulate anorexia by affecting cytokine signaling within the hypothalamus, which can influence glucocorticoid secretion (see section 'Glucocorticoids'). Additionally, leptin can diminish the production of proinflammatory cytokines by peripheral monocytes, which may provide a negative feedback on systemic inflammation as an infection resolves.

Glucocorticoids

Glucocorticoids are steroid hormones that modulate metabolism, salt balance, development, reproductive processes, and immune function. While acute elevations of glucocorticoids enhance some immune functions, such as leukocyte infiltration at the sites of injury, chronic elevations induce leukocyte apoptosis, reduce proinflammatory cytokine release, and generally suppress immune activity. In terms of sickness behaviors, glucocorticoids reduce sickness-induced lethargy and anorexia, and also dampen immune defenses as an infection resolves. These outcomes are regulated at multiple levels: glucocorticoids decrease the efficacy of proinflammatory cytokine signaling in brain regions associated with sickness behaviors, and they down-regulate proinflammatory cytokine synthesis in the periphery. These connections between cytokine and glucocorticoid signaling pathways are also bidirectional, since the induction of an acute phase response increases glucocorticoid production. Such effects support the notion that in the wild, glucocorticoids may help minimize sickness behaviors after an acute stressor, when time must be allotted to other behaviors crucial to immediate survival. As of yet, however, the role of glucocorticoids in modulating sickness behaviors in wild animals remains unstudied.

Conclusions and Future Directions

Hormones physiologically encode endogenous and exogenous resource levels and threats and have manifold effects on immune responses, including sickness behaviors. Through interactions with cytokines, these hormones link the immune and nervous systems and help animals express sickness behaviors appropriate to their environment. Numerous examples of such interactions exist and ecological immunologists are developing frameworks to understand how and when these connections vary among individuals, populations, and species.

Several avenues of research will augment our understanding of the endocrine modulation of sickness behaviors in the coming years. First, few studies have examined how such connections change when resources are experimentally manipulated and how and why effects might vary across species. Such studies will help determine which hormones are crucial in relating information about energetic, nutritional, or temporal resources, and what species-specific characteristics may enhance these relationships. Additionally, examining the effects of exogenous hormones on sickness behaviors in free-living animals is crucial for understanding the importance of endocrine–immune interactions when natural conditions constrain an animal's response. Finally, because lethargy, somnolence, and decreased libido can reduce the contact rates and interactions among individuals, being able to predict which individuals or groups are most likely to express sickness behaviors will have important consequences for epidemiology. Progress will thus require an integrative and multidisciplinary study of diverse animals.

See also: Field Techniques in Hormones and Behavior; Hormones and Behavior: Basic Concepts; Male Sexual Behavior and Hormones in Non-Mammalian Vertebrates; Stress, Health and Social Behavior.

Further Reading

Aubert A, Goodall G, Dantzer R, and Gheusi G (1997) Differential effects of lipopolysaccharide on pup retrieving and nest building in lactating mice. *Brain, Behavior, and Immunity* 11: 107–118.

Avitsur R and Yirmiya R (1999) The immunobiology of sexual behavior: Gender differences in the suppression of sexual activity during illness. *Pharmacology, Biochemistry, and Behavior* 64: 787–796.

Besedovsky H and del Rey A (2001) Cytokines as mediators of central and peripheral immune-neuroendocrine interactions. In: Ader R, Felten DL, and Cohen N (eds.) *Psychoneuroimmunology*, 3rd edn, vol. 1, pp. 1–17. San Diego, CA: Academic Press.

Dantzer R (2004) Cytokine-induced sickness behaviour: A neuroimmune response to activation of innate immunity. *European Journal of Pharmacology* 500: 399–411.

Folstad I and Karter AJ (1992) Parasites, bright males, and the immunocompetence handicap. *American Naturalist* 139: 603–622.

Grossman CJ (1984) Regulation of the immune system by sex steroids. *Endocrine Reviews* 5: 435–455.

Hart BL (1988) Biological basis of the behavior of sick animals. *Neuroscience and Biobehavioral Reviews* 12: 123–137.

Lee KA, Martin LB, and Wikelski MC (2005) Responding to inflammatory challenges is less costly for a successful avian invader, the house sparrow (*Passer domesticus*), than its less-invasive congener. *Oecologia* 145: 244–251.

Lochmiller RL and Deerenberg C (2000) Trade-offs in evolutionary immunology: Just what is the cost of immunity? *Oikos* 88: 87–98.

Martin LB, Weil ZM, and Nelson RJ (2008a) Fever and sickness behaviour vary among congeneric rodents. *Functional Ecology* 22: 68–77.

Martin LB, Weil ZM, and Nelson RJ (2008b) Seasonal changes in vertebrate immune activity: Mediation by physiological trade-offs. *Philosophical Transactions of the Royal Society B-Biological Sciences* 363: 321–339.

Owen-Ashley NT and Wingfield JC (2007) Acute phase responses of passerine birds: Characterization and seasonal variation. *Journal of Ornithology* 148: S583–S591.

Roberts ML, Buchanan KL, and Evans MR (2004) Testing the immunocompetence handicap hypothesis: A review of the evidence. *Animal Behaviour* 68: 227–239.

Steiner AA and Romanovsky AA (2007) Leptin: At the crossroads of energy balance and systemic inflammation. *Progress in Lipid Research* 46: 89–107.

Wen JC and Prendergast BJ (2007) Photoperiodic regulation of behavioral responsiveness to proinflammatory cytokines. *Physiology & Behavior* 90: 717–725.

Infanticide

C. P. van Schaik and M. A. van Noordwijk, University of Zurich, Zurich, Switzerland

Introduction

Infanticide refers to the killing of dependent offspring, or more formally, to any form of lethal curtailment of parental investment in offspring brought about by conspecifics. As is evident from this broad definition, it is a behavior that is as diverse in its forms and functions as it is disturbing to human observers. The victims can be embryos, newborns, or older but still dependent offspring; the perpetrators can be mothers, fathers, unrelated adults, or even other, larger immatures; and the actions involved can range from neglect or abandonment to active killing and eating. What unites this seemingly disparate array of actions and contexts is the fact that the species involved have parental care, dependent offspring are killed, and that it usually improves the perpetrator's access to limiting resources, be it food, nest sites, help, or mates.

Especially when the perpetrators kill but do not eat the immatures, the function of this behavior is not immediately apparent. Little wonder, then, that early observers tended to dismiss their observations of infanticide in mammals as pathological aberrations evoked by unusual conditions, rather than adaptations honed by natural selection. Yet, even though infanticide is often rare in the species in which it is found and therefore often overlooked and incompletely documented, there is increasingly abundant and systematic evidence that perpetrators gain some fitness benefit from their act. Indeed, many biologists are convinced that the threat of infanticide has acted as the selective agent for the evolution of a slew of counteradaptations ranging from biparental care to permanent male–female association to unusual mating behavior or reproductive physiology. The presence of these counteradaptations can explain why infanticide is often rare: it happens only when these defenses fail.

Infanticide overlaps with cannibalism, which refers to the consumption of conspecifics, regardless of whether they are young. We also follow common usage by regarding it as distinct from siblicide, in which the perpetrators themselves are immatures. In the first major review of the topic in 1979, Sarah Hrdy recognized several functionally distinct forms of infanticide (i.e., kinds of contexts where infanticide improves the perpetrator's fitness), and her categories will largely serve as the section headers of this article. Nonparents can kill conspecific young (1) to eat them, (2) to reduce competition for critical resources to them or their own offspring, (3) to avoid misdirecting their parental care, or (4) to be able to sire offspring sooner with the victim's parent, in a bizarre twist to sexual selection. Under some conditions, (5) even parents can benefit from eliminating their own offspring. In all these forms, the killing of the infant is usually deliberate, not accidental, although occasionally infants may die because they got in the way of escalated fights among others.

These functional categories are not necessarily exclusive, such as when an infant killed to improve resources available to one's own young is also eaten. But each functional benefit arises only under a strictly delineated set of biological conditions and thus is expected only in some lineages but not in others. Each comes with a set of testable predictions about the kinds of situations in which infants are at risk and from whom. Each is also likely to have favored the evolution of counterstrategies. **Table 1** provides an overview of these forms of nonparental infanticide.

Finally, each functional category is also likely to have its own set of psychological mechanisms that regulate the behavior, specifying the stimuli that elicit attacks on infants or the rules that animals use to identify potential victims. For obvious reasons, there is less experimental evidence for the regulation of infanticide than for that of other behaviors, so that much of our insight into its proximate control is anecdotal. Moreover, the experimental insights obtained for one functional category cannot necessarily be generalized to others. Yet, we know enough in general of the stimuli and conditions eliciting infanticidal attacks not to expect that each and every instance of infanticide be clearly adaptive to the perpetrator, merely that cases should be so on average. This consideration is important to delineate functional infanticide from the possibility that infanticide is an expression of (6) pathological behavior.

Cannibalism of Young

The most obvious functional benefit of the killing of conspecific immatures is that they serve as food. The younger, and thus smaller the immatures, the easier it should be for adults to capture and eat them. This advantage may account for many cases among species where parents do not attend to young. Cannibalism of young has been recorded among many invertebrates, amphibians, and fishes.

The only precondition is that the perpetrator not eat its own offspring (although even that can occasionally be adaptive: see the following section) and thus that some kin

Table 1 Nonparental infanticide: perpetrators, conditions, counterstrategies

	Male	Female	Condition	Counterstrategy parent
Cannibalism	Yes	Yes	Infant unprotected & recognized as nonkin	(Bi)parental care
Resource: nest site	No	Yes	Nest sites limiting & infant unprotected	Cooperative defense with other relatives; biparental care
Resource: allo-care	No	Yes	Allo-care limiting & infant unprotected	Synchronize breeding if infants share nest; desynchronize breeding if infants are carried
Avoid misdirected care	Yes	Rare	Perpetrator is provisioning sex; females kill if exploited by foreign young	Infant: hide identity of parent
Mating competition	Yes	Sometimes	Infant loss advances mating access to other sex; female kills only in role-reversed species	Reproductive physiology, producing paternity confusion and illusions; associate with protector

recognition mechanism ensures that there is a strict separation between the treatment of own young and other small conspecifics. A potential cost is that the perpetrator may contract (species-specific) communicable diseases; this probably happens quite rarely, but when it happens, the costs may be dramatic.

It is likely that the risk of cannibalism of unguarded young served as one of the major selective benefits for the evolution of parental care and in some cases, even of biparental care. In Australian lizards of the genus *Egernia*, for instance, the presence of parents in experimental encounters strongly reduced attacks by conspecific adults on the young. Similarly, the presence of both parents seems to be most effective in preventing infanticide in burying beetles.

Among species with parental care, cannibalism either complements other benefits, thus lowering the threshold for such behaviors being selectively favored by increasing the benefits, or may, alternatively, be a byproduct of other benefits to killing infants. The distinction depends on whether infants are consistently eaten. Thus, while in carnivores, such as bears or hyenas, and some rodents, such as ground squirrels, infants killed are generally cannibalized as well, in many primates and pinnipeds, infants that are killed are rarely eaten.

Resource Competition

Offspring can be vulnerable because killing them improves access to resources such as food or nest sites abandoned by the victim's parent, for either the perpetrator or its offspring. Since there are some costs to such actions, natural selection reasoning does not predict infanticide due to resource competition whenever the benefits might equally accrue to many others as well (in effect creating a public good), because those who receive them for free have a net gain relative to the perpetrators. Indeed, the best examples come from organisms in which

only a few broods share a single resource. Thus, among burying beetles (*Nicrophorus*), when two unrelated females breed communally on the same carcass, one female may kill the eggs laid by the other one, thus eliminating competitors for her own offspring. Similarly, in ground squirrels, female victims abandon nesting sites that are in short supply.

Perhaps the most common resource over which there is competition is care for the young: killing someone's dependent offspring turns this former parent into a helper to the perpetrator, providing the latter with the double benefit of having less competition and more support. These conditions should thus largely overlap with those favoring cooperative breeding. In social insects and a variety of cooperatively breeding birds and mammals, dominant females kill the newborns of other females in their unit, which subsequently help the dominant female's offspring (and in mammals sometimes even allo-nurse). Similar behavior is sometimes seen among communal breeders, in which females jointly rear their young. It can be speculated that communal breeding (in which cohabiting females all reproduce and care for each other's young) is found only where protective mechanisms exist to prevent frequent infanticide by dominants.

Psychologically, the governing mechanisms are expected to be condition dependent, because dominants kill young only if the timing is such that their care would compete with the care for their own young. If a subordinate female gives birth at other times, the young may get spared. However, because dominants have high reproductive rates, such opportunities may arise only rarely, and in many of the cooperative breeders vulnerable to this kind of infanticide, subordinate females will not show ovarian activity, which can be seen as an adaptation to preventing certain infanticide or eviction. Where the animals have communal nests, as in meerkats (*Suricata suricatta*) and banded mongooses (*Mungos mungo*), subordinate females may be able to prevent infanticide of their young by synchronizing themselves with the dominant female, adjusting their gestation length to give birth on the same day as the dominant female. This works because females cannot pick out their own young unless there are major size differences. Not surprisingly, dominant females tend to evict pregnant subordinates.

Social counterstrategies apart from emigration will not work in cooperative breeders, but in others, one way to reduce the risk of being victimized is maintaining strict nesting territories against conspecifics, even when food is superabundant. In yet others, females find a protective associate, often the other parent. In burying beetles mentioned earlier, for instance, new females rarely try to intrude on a carcass occupied by both a male and a female, thus providing the female with a benefit to pair bonding.

A puzzling instance is the killing of infants, especially those of recent immigrants, by female chimpanzees (*Pan troglodytes*) seen at some sites with long-term observations. Because single females cannot easily kill another female's baby, the precondition seems to be that females, though basically solitary, can occasionally form a coalition stable enough to overwhelm a female with an infant and together kill her infant. Cannibalism cannot be the main benefit because it does not happen systematically enough, but it is possible that the killers benefit from the fact that the affected mothers will in future avoid the area in which they currently range. However, the perpetrators must have somehow overcome the problem befuddling all collective action, namely that free riders benefit more than those that participate, which often prevents the collective action from happening. Because such cases are quite rare, they may be dismissed as flukes, but they may also represent adaptive responses that can only be shown in very specific and therefore rare conditions, as suggested by the targeted and persistent nature of the attacks. Continued detailed long-term study may help to reveal these conditions. Moreover, similar cases may happen in other species, but may not be recorded because studies are not intensive enough.

Adoption Avoidance

Where a parent engages in extensive postnatal care of its offspring, it is important that this care is directed at the correct offspring lest valuable time should be lost that could be spent investing in their own offspring. Various mechanisms ensure that the offspring is recognized as such, and when they suggest that the offspring are foreign, the young may be removed or killed, and in some cases eaten. In many birds, foreign eggs may be ejected from nests, provided they are recognized as such. Female guira cuckoos, which have communal nests, eject eggs laid by their co-habiting females (in this case, this functional benefit may overlap with the elimination of competition discussed earlier). Males in some cichlid fish that guard nests sites with a limited capacity will try to remove broods produced by females they did not spawn with. Male blue-footed boobies that had been experimentally removed from their mate for less than a day during the week before egg laying started evicted the first egg laid by their female, whereas separation before that time had no effect.

Surprisingly, high parental investment does not automatically lead to elimination of foreign young. Thus, in many Neotropical primates, males placed into new groups will immediately care for the young. The function of this restraint is unclear, but female control over male behavior may be a major factor. Human males, too, often care for infants not their own, although they may also kill them (see the following section).

Gaining Access to Mates: The Sexual Selection Hypothesis

Another nonobvious condition in which infanticide is adaptive, much studied in mammals and especially carnivores and primates, is where by killing a mate's offspring the perpetrator (usually a male) gains more rapid mating access to the parent. This idea is called the sexual selection hypothesis, because it contains elements of both competition for mates and mating conflict. Infanticide may be an adaptive male strategy whenever four conditions are met: (1) the male is unlikely to have sired the offspring, (2) the loss of dependent offspring advances the timing of renewed receptivity; (3) the male is more likely than before to father the female's next offspring; and (4) the costs of killing the infant, either as injury or as time lost to other more valuable pursuits, are low enough. A compilation of observed cases in primates indicated that the average infanticide leads to a fitness benefit for the male perpetrator, because he was probably not the sire of the infant he killed, the female returned to receptivity on average about 30% earlier than she would have otherwise, and, being newly dominant, he had very good chances of siring the next infant, while never sustaining any serious injuries.

The key condition, then, is whether female reproductive condition is affected by the loss of dependent offspring. In most mammals, females become pregnant again soon after giving birth, during their so-called postpartum estrus, so loss of the current offspring will not affect her reproductive status. However, as life-history pace slows down, for example, because body size increases, lactation becomes relatively longer. Once its duration begins to exceed that of gestation, postpartum estrus disappears, and so-called postpartum amenorrhea ensues. Only species with postpartum amenorrhea are vulnerable to this form of infanticide. Thus, the lactation/gestation ratio predicts whether a species is vulnerable to sexually selected infanticide. An additional risk factor is the litter size, in part because large litters usually mean longer lactation, but also because the size of the current litter may affect the future performance of the gestated litter, both in terms of embryo survival and growth rate. These rather unusual conditions explain why the vulnerability to infanticide by males shows such a restricted taxonomic distribution among mammals, being limited largely to precocial taxa with slow life history or altricial ones with very large litters and prolonged lactation.

Male mammals cannot usually directly recognize their immature offspring and must rely on a set of indirect indicators, based on the quality, timing, and degree of monopolization of copulations with the female, in addition to being locally dominant and thus likely to mate with the female later. Experiments with rodents showed that mating played a critical role and that timing mattered: male mice experience their maximum inhibition to kill a female's infants at around one mouse gestation length (3 weeks), which declines to zero at around the end of what would be the lactation period. In primates, anecdotal evidence suggests a less strict clock mechanism, but in all cases mating history is critical.

The male assessment mechanisms provided natural selection with the opportunity to evolve changes in female reproductive physiology and behavior that manipulate these recognition mechanisms. Female mammals in species vulnerable to infanticide almost universally use polyandrous mating to cause paternity dilution (by inciting sperm competition) and create paternity illusions (due to mating when conception is highly unlikely or impossible, as during pregnancy). Females usually have a conflict of interest with the dominant male about paternity, however, and in a subset of these vulnerable species, the Old World primates, dominant males tend to mate guard females so assiduously that the latter have limited control over which male they mate with. In these species, additional features serve to reduce the hold of the dominant males over the female's preferences for balanced polyandry: females have lengthened attractive periods per cycle (accompanied by unpredictable ovulation) and thus extended follicular phases of their ovarian cycles, have multiple cycles per conception, and often continue mating during pregnancy. Many even show exaggerated sexual swellings and copulation calls, both of which may serve to attract subordinate males intimidated by the dominant male. Some of these characteristics are even plastic within individuals. In various Old World monkeys, the arrival of unfamiliar males induces females to extend their follicular phase well beyond their regular length or to become receptive again during pregnancy or even during early lactation.

Social counterstrategies are more straightforward: the likely sire is interested in protecting the offspring, but he must be around to do so. Among primates and equids, species vulnerable to infanticide show year-round association between males and females, whereas others do not.

Where adult females can disperse between social units, decisions whether and when they do so often serve to reduce the risk of infanticide.

A third class of counterstrategies is curtailment in the investment in the current offspring. The most dramatic version of this is the Bruce effect, where a pregnant female experimentally exposed to a novel dominant male (implying the elimination of the previous local dominant) resorbs (or more rarely, aborts) her fetuses. This is interpreted as cutting her losses in light of near-certain loss of the offspring upon birth. Although some naturalistic observations on rodents suggest that this may be a laboratory artifact, some primate observations fit nicely. Less dramatic but probably more widespread is that females prematurely wean their offspring and develop sexual receptivity to novel dominant males even though they were still fully lactating just before. Here, the interpretation is that the infants' survival prospects are compromised less when weaned early than when not weaned at all.

Among birds, infanticide by males should be favored when mated males die in the middle of the breeding season (which is not so common), or when displacing a female's mate followed by destruction of her clutch or nestlings would cause her to mate with the perpetrator and start laying again, and there is enough time in the breeding season to do so. The second condition may appear quite general, yet infanticide is not very common in birds, perhaps because the usual combination of monogamy and seasonal breeding means that unmated males are both rare and rarely strong enough to beat breeding males. In fact, a long-term study of barn swallows found increased infanticide as the proportion of unmated males increased and as the condition of male breeders decreased.

Sexually selected infanticide is even found among invertebrates. In eresid spiders, females provide suicidal care for their offspring, allowing the hatchlings to kill and eat her. A male can remove the female's egg sac containing the fertilized eggs, forcing her to mate again with him.

Although it is usually males that reap the mating benefits by killing infants, logic suggests that in species with reversed sex roles (exclusive paternal care) females would be seen to commit infanticide and thus gain a mate. Indeed, in jacanas, polyandrous lake and swamp birds, females are the ones that kill clutches of other females and then mate with the males, which would otherwise have been occupied with brood care duties.

Parental Manipulation

In all the functional categories examined so far, the perpetrators were not related to the immatures they killed. At first sight, the idea that parents that kill their own offspring derive a fitness benefit from doing so seems very far-fetched. Yet, under some circumstances, this may turn out to be adaptive as shown by the following examples. First, fathers in some mouth-brooding fishes sometimes swallow their own young when in poor condition. This allows them to gain enough energy to care for the remaining young until independence, or to care more successfully for a next brood. In effect, they are using some of the female's eggs to sustain their brooding activity. Second, many rodent mothers, when they experience a sudden deterioration in food abundance while trying to rear a large litter, lose pups (especially males are vulnerable to starvation). Although in many cases these young may simply die and then be consumed, there are also reports of mothers actively removing or killing young. Third, there are reports of mothers killing deformed or runty young, which have poor survival prospects. Similar reports abound in humans, where also babies of the nonpreferred sex (usually females), babies born too early, or the smaller of twins may be killed. Because in humans these decisions are often culturally sanctioned and thus geographically variable, this form of infanticide is presumably controlled differently in humans than in other species.

Cooperative breeders constitute a special case. In them, mothers frequently abandon, reject, or even kill, their newborns when help is absent or insufficient. In callitrichid monkeys, for instance, newborns are twice as likely to be rejected when the female has no helpers; and when triplets are born, rather than the usual twins, the weakest tends to be rejected as well. Something similar is seen in humans, where (usually young) women without resources and social support, abandon or kill their newborn babies. Psychologically, some women tend to hide or deny their pregnancy and have a detached attitude toward their baby.

Pathology and Byproduct

The pathology interpretation could explain sexually selected infanticide by males and particularly in primates. However, the pathology prediction is not met, because perpetrators are otherwise behaviorally normal individuals, who target their victims quite specifically under the conditions in which infanticidal behavior is expected and then turn into protective parents of their own offspring, once these are born (see **Figure 1**, for male long-tailed macaques). This high conditionality of the behavior is indicative of adaptive responses rather than pathology. The case of humans may form an exception, where committing infanticide has in some cases been linked to other behavioral abnormalities.

Of course, this does not mean that the generally adaptive responses are perfect in that they will always prove to increase the perpetrator's inclusive fitness. Thus, several cases in primates have been reported of infants getting killed that belong to a neighboring group or that had already been orphaned. Such cases can easily be understood as a reflection of the stimuli and conditions that elicit these attacks, even though the actual outcome was not adaptive. However, pointing out such misfiring of an adaptive system is not the same as arguing that the system as a whole is pathological, that is, systematically producing maladaptive results because a behavior system that evolved for a different function is triggered under unusual circumstances. A pathological interpretation would be plausible when such mistaken infanticides make up a considerable proportion of all cases and add up to a net loss to the perpetrator's fitness.

Some behaviors may look pathological but are merely a byproduct of other motivations that are by themselves adaptive. A case in point is the 'aunting to death' seen in

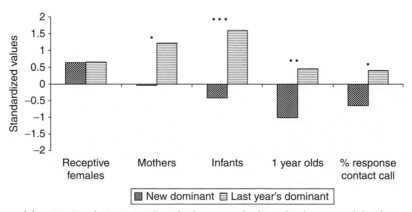

Figure 1 Difference in social parameters between males who have acquired top dominance rank but have not yet sired offspring and males (usually the same individuals in subsequent years) who were already dominant during the previous mating season, and thus were the likely sires of the infants, in a group of long-tailed macaques (*Macaca fascicularis*) in a natural population (Ketambe, Sumatra, Indonesia): % time male spends within 5 m of receptive females, mothers with dependent infants, infants and 1-year old juveniles, as well as the tendency to answer contact calls from other group members (based on data in van Noordwijk MA and van Schaik CP (1988) Male careers in Sumatran longtailed macaques (*Macaca fascicularis*). *Behaviour* 107: 24–43.

female primates and social carnivores such as hyenas. The infant eventually dies as a result of younger or infantless females directing inadequate mothering skills or simply keeping the mother from reaching and nursing her own infant. This is functional behavior if it turns out that the young female actually gains experience that improves her own rearing success, but the evidence for this functional benefit is mixed at best. Most work suggests instead that infant abuse is simply a byproduct of mothering instincts directed at another female's infant because the female is too young to have her own infant or just lost her own. This handling may be costly for the infant involved but not costly for the handler for selection to be able to eliminate it. On the other hand, high-ranking females might even benefit from this abuse, because they can more easily acquire the infants to handle and may eliminate rivals for their own infants or those of kin. Thus, careful study is needed of the behaviors, the triggering stimuli, and their demographic consequences to evaluate where on the gradient from adaptive to byproduct 'aunting to death' is located.

Consequences of Infanticide

Infanticide may have consequences at the level of the population. In a comparative study of blue monkeys (*Cercopithecus mitis*), for instance, unmated males collected in the most productive habitat, leading to such short tenures of males in the harem groups there and thus such a poor infant survival that the local population density was far lower than elsewhere in the same forest. Other things being equal, population growth rates are reduced as infanticide increases in prevalence. It is possible that high rates of infanticide have driven some populations or even species extinct and therefore that species selection has weeded out lineages with social and life-history characteristics that generated exaggerated infanticide rates.

Infanticide can also have immediate consequences for population management and conservation efforts. In both lions and various bear species, elimination of breeding males by trophy hunters has led to increased frequency of takeovers by other males, followed by infanticide or other costs to females because of avoidance behavior. Thus, managers of vulnerable species must be aware of this risk.

Even though elaborate social norms govern much of human behavior, the empirical study of human infanticide

with an evolutionary perspective has proved invaluable in improving our understanding of the circumstances in which a human dependent child faces a heightened risk of being killed, abused, or abandoned. This example shows that the study of animal behavior can have unanticipated benefits for humanity.

See also: Compensation in Reproduction; Cooperation and Sociality; Differential Allocation; Forced or Aggressively Coerced Copulation; Risk Allocation in Anti-Predator Behavior; Social Selection, Sexual Selection, and Sexual Conflict.

Further Reading

Agrell J, Wolff JO, and Ylöonen H (1998) Counter-strategies to infanticide in mammals: Costs and consequences. *Oikos* 83: 507–517.
Daly M and Wilson M (1998) *The Truth About Cinderella: A Darwinian View of Parental Love.* New Haven, CT: Yale University Press.
Ebensperger LA (1998) Strategies and counterstrategies to infanticide in mammals. *Biological Reviews* 73: 231–346.
Hausfater G and Hrdy SB (eds.) (1984) *Infanticide: Comparative and Evolutionary Perspectives.* New York, NY: Aldine Publishing Company.
Hrdy SB (1979) Infanticide among animals: A review, classification, and examination of the implications for the reproductive strategies of females. *Ethology and Sociobiology* 1: 13–40.
O'Connor DE and Shine R (2004) Parental care protects against infanticide in the lizard *Egernia saxatilis* (Scincidae). *Animal Behaviour* 68: 1361–1369.
Parmigiani S and Vom Saal FS (eds.) (1994) *Infanticide and Parental Care.* Chur, Switzerland: Harwood Academic Publishers.
Schneider JM and Lubin Y (1996) Infanticidal male eresid spiders. *Nature* 381: 655–656.
Trumbo ST (2006) Infanticide, sexual selection and task specialization in a biparental burying beetle. *Animal Behaviour* 72: 1159–1167.
van Schaik CP and Janson CH (eds.) (2000) *Infanticide by Males and Its Implications.* Cambridge: Cambridge University Press.

Relevant Websites

http://www.psych.upenn.edu/~seyfarth/Baboonresearch/infanticide01-web.jpg
http://www.bio.davidson.edu/people/vecase/Behavior/Spring2004/shelburne/infanticide.html
www.arkive.org/hippopotamus/hippopotamus-amphibius/image-G20281.html
http://www.arkive.org/hippopotamus/hippopotamus-amphibius/video-12b.html
http://www.youtube.com/watch?v=xZRwOIYdf3g&eurl=http://www.itsnature.org/videos/wild-videos/lions-and-cubs-infanticide/

Information Content and Signals

J. P. Hailman, University of Wisconsin, Jupiter, FL, USA

Introduction

Communication is the transfer of information from a sender to a receiver by means of signals. This succinct definition raises many questions, such as what signals are, what information is, how information is measured, and how it is transferred by the sender and used by the receiver. These issues are considered before I turn to the major focus of the chapter: the principles by which information is embedded in signals.

Physical Structure of Signals

The physical nature of signals is highly diverse. Animal signals may be visible light, ultraviolet radiation, infrared radiation, sound, ultrasound, vibration, air- or waterborne chemical molecules, molecules deposited on a substrate, mechanical contact, electricity, and perhaps even others.

The physical signals used by a given species in intraspecific communication are determined by such factors as its sensory systems, its ability to generate signals, the environment in which the signaling takes place, and constraints imposed by the evolutionary ancestry of the species. In communication between species, the sensory systems of the receiver are also an important factor in decoding information carried by the signal.

In some cases, the kind of information to be communicated may put additional constraints on the kind of physical signals employed. This factor, however, appears to be a relatively minor one because information of all kinds is at heart simply variety. So long as a system of signals can transfer the requisite variety needed for the kind of information communicated, the physical structure of the signals is unimportant. For example, depending upon the species, an individual may signal its sex by its coloration, the sounds that it makes, the chemical signals it releases, or other types of signals. The amount and kind of information (sex identity) are the same but the signals used to communicate this information vary widely among species.

Information, Entropy, and Capacity

The usual quantitative concern about communication is the average amount of information encoded by a given system, as opposed to the information associated with a particular signal. This average amount may fall short of the theoretical capacity of the system, in part because not all of the information encoded is necessarily transferred effectively to the intended receiver. In other words, there are three related quantities: (1) the amount of information that the system could theoretically encode in signals, called the channel capacity; (2) the average amount of information actually encoded in the signals, called the source entropy; and (3) the actual amount of information passed to the receiver, called the transferred information. These terms are often shortened to capacity, entropy, and information, respectively. As entropy is the potential amount of information that could be transferred, it is often called information when confusion between the latter two quantities would be unlikely.

Channel capacity and simple entropy

As information is variety or diversity, it is a measurable quantity. For various reasons, a logarithmic measure is particularly useful. So instead of designating the information in sex signals as 2 (male and female), it is expressed as log 2. Different logarithm bases have been employed for expressing variety of various kinds, but communication studies use base 2. Thus the entropy of a sex-signaling system is $\log_2 2 = 1$, the unit being called the bit, a contraction of the term binary digit. The entropy of a sex-signaling system is 1 bit only if the two alternative signals are equally likely to occur. Most species do in fact have approximately equal numbers of males and females.

The expression $\log_2 x$ is general for all values of x so long as the x alternative signals are equiprobable. For example, there are four suits in a deck of playing cards, with equal numbers of clubs, diamonds, hearts, and spades. So if the suits were used as signals, the entropy would be $\log_2 4 = 2$ bits per suit (or bits/suit). Technically, units should be bits per something, in this case per signal. When rates of information are under scrutiny, the quantities would be in bits per unit time. When bits/signal is intended, the per-signal part is often omitted.

The bit unit is the common currency of communication. Channel capacity, source entropy, and information transferred are all expressed in bits. Furthermore, signaling systems from two different species using physically different signals could be compared quantitatively. Comparison is useful in answering such questions as to whether species A or species B communicates the location of a food source more precisely.

Realistic entropy

Alternative signals are rarely equiprobable in the real world so that $\log_2 x$ merely sets an upper limit to actual entropy. That limit is in fact the aforementioned channel capacity of the system. When the alternative signals are not equiprobable, the proportional contribution of each must be calculated and summed over all the signals. In the simple case of a one-bit system such as signaling sex, the alternative signals for both male and female have the same probability of 0.5. Therefore, the entropy is calculated as $0.5 \log_2 2 + 0.5 \log_2 2$ or 1 bit/signal as previously calculated. This is the maximum entropy a two-signal system can have, and it can be appreciated intuitively that the minimum approaches zero as one of the components increases to a probability of 1 with the other component decreasing to a probability of zero.

The foregoing calculation can be written more symbolically. Begin by expressing the probability of occurrence (0.5 in the example) as p, which is 1 divided by the number of alternative signals (2) or n. Therefore, we can write each equiprobable component as $p \log_2 n$. As $p = 1/n$, it is also true that $n = 1/p$, so the component can be written as $p \log_2 1/p$. Finally, another mathematical trick for simplification is remembering that the log $1/x$ is $-\log x$ (this is true for any base). So bringing the negative sign to the start, we can write an entropy component as $-p \log_2 p$. It turns out that this works not only when the alternative signals are equiprobable but is perfectly generalizable to any value of p. The sum of the individual p's is always unity and the sum of the $-p \log_2 p$ components is the entropy.

Information transferred

The last of the three informational quantities is the average information transferred to the receiver. The theoretical capacity of a signaling system is $\log_2 n$ and the average source entropy (or encoded information) is the sum of the $-p \log_2 p$ components. It is obvious by definition that if all the entropy that is encoded in signals gets through to the receiver, then the information transferred is quantitatively identical with the entropy. That is another reason why the entropy is often called the information of a system.

It need not be the case, however, that all the information is transferred. For simplicity, return to the silly example of using the suits of playing cards as a signal. Suppose a card is seen so briefly that the shape of the suit icon could not be made out but the color was detected as red. Now the observer knows that the signal is a diamond or heart so the new entropy becomes $\log_2 2 = 1$ bit/signal. The initial entropy was $\log_2 4 = 2$ bits, and the difference between the initial and new entropy is the information transferred, in this case 1 bit. Put differently, the receiver is uncertain about something, as measured by the source entropy. The amount by which that uncertainty is reduced by receipt of a signal is the amount of information transferred.

Effect of Information on the Receiver

We can tell that information has been transferred if the receiver does something different from what it would have done in the absence of communication. In practice, detecting such a difference is often quite difficult. It is an easy call only when the receiver makes an immediate, obvious, and unexpected change in behavior upon receipt of the signal.

One reason why a signal's effect may be difficult to detect is that a receiver might have changed its behavior in the absence of communication but upon receipt of a signal does not alter its behavior. For example, many birds have a special alarm call that causes nestlings to freeze in silence. Eventually, they will resume begging, but if hearing another alarm note will instead continue to be motionless and silent. Sometimes a given signal causes the receiver to change behavior and sometimes to maintain ongoing behavior.

Another reason for the difficulty of detecting a signal's effect is that it may be delayed. These cases are often extremely hard to establish. Suppose a parent bird sees a snake and gives an alarm call. The nearby fledglings may not see the snake, but they associate the place with alarm. The fledgling may then avoid that place in future movements or be especially vigilant when near it.

The most general reason why detecting a signal's effect on the receiver is tough is the simple difficulty of establishing what would have happened in the absence of communication. Many repetitions of the same context with and without receipt of a signal are required to demonstrate the effect the signal has on the receiver.

Coding and Information Content

More is known about the source entropy and how signals encode it than about the information actually passed to the recipient. This difference occurs because it is difficult to determine the effect of communication on receivers, and therefore the amount of information transferred. In general, the more complex the encoding, the greater the information content. Types of signaling codes can be grouped into three major categories: binary codes, multivalued codes, and a heterogeneous category of multivariate codes.

Binary Codes

Binary codes have two alternative signals (values of the encoding variable). Binary codes are probably the most prevalent category of animal signals and are used to convey all sorts of information about species identity, sex, group identity, territorial boundaries, predator danger, reproductive state, and so on.

Types of binary codes are nearly as numerous as the kinds of information they encode or the physical structures of the signals used. Indeed, there are more kinds than can be enumerated here. One common type has already been mentioned: two distinctive signals like those often used for sex identification. If there are equal numbers of males and females, the system has an entropy of 1 bit. The sex ratio of adults of several species may be highly skewed, however, in which case the binary signals encode far less than the maximum of 1 bit.

Another common type is the encounter sign. Here, one of the alternative values of the coding variable is some signal and the other is its absence. For example, a territorial species of mammal may encounter a deposited scent indicating the boundary of another individual's territory. Much of the time an animal may encounter no scent at all so the binary system encodes only some fraction of a bit per unit time. It is true, though, that the rare encounter is a very informative signal. It is possible to express the magnitude of this information as $-\log_2 p$, where p is the encounter probability. This quantity has been named surprisal.

Periodic reports are another common class of binary codes. While defending a territory in the spring, for example, a male song bird may utter a song every few minutes. The signal indicates that a state is being constantly maintained, in this case the territory.

Closely related to periodic reports are state indicators when a signal is chronically on while some state is maintained. For example, females of some lizards sport a unique color pattern while in reproductive condition.

Of various other kinds of binary signaling systems perhaps the least obvious is one in which the outside observer could incorrectly conclude that more than two values are involved. These systems have been called cryptically binary and they are common signals of species identity. An experienced bird observer can identify instantly many species of puddle ducks in flight by their species-specific wing markings. The birds themselves appear to cue on the same signals, but they make only a binary distinction: conspecific or not. A duck in flight tries to join others of its kind, but makes no distinctions among other species.

Multivalued Codes

Obviously a simple binary code can convey at most only 1 bit of information. It follows that the informational content can be increased by using three or more values of the coding variable. With equiprobable signals, a three-valued code has a maximum of $\log_2 3 = 1.58$ bits/signal, a four-valued code $\log_2 4 = 2$ bits, and so on. For every doubling of the number of coding values, the maximum entropy increases by a bit. Many more kinds of multivalued codes exist than can be mentioned here.

Most straightforward are simple multivalued codes which have three or more different signals. For example, in several species of damselflies three or more body colorations occur. 'Ordinary' males and females have one type, but there is also a male-like coloration of some females, believed to be a deceptive signal to deter males from harassing such females. In some damselfly species, even more colorations exist.

A number of avian and mammalian species use multivalued event markers when detecting a predator. One value of the variable is the absence of any signal and the two or more others are different signals for different types of predators detected. This kind of event marker is known from a number of birds, vervet monkeys, tree squirrels, and other animals.

The aforementioned codes are discrete, with a countable number of signal types, but graded signals also exist. The brightness of feathers in many male birds varies continuously among individuals, communicating to a potential mate their health and general fitness. Coloration varies quantitatively in other animals such as the head coloration of cichlid fish and the facial skin of mandrills.

A related kind of signaling is by variations in performance rates. Honeybees give more dances per unit time when communicating the nearness of a food source. Gull chicks beg at an increasing rate indicating their level of hunger. Chickadees encode levels of danger in their chick-a-dee call. The smaller the size of a predator, the more 'dee' notes chickadees give in their mobbing calls, as small owls are their most dangerous predators.

Multivariate Codes

The defining characteristic of multivariate codes is that the information-laden signal is constructed from values of two or more underlying variables that are transmitted simultaneously or successively to form a decodable message. If this definition seems vague or overly general, it is so because types of multivariate codes are so different from one another. Indeed, each subcategory could be considered as conceptually equal in level to binary and multivalued codes. Five such subcategories can be distinguished.

Composite signals

The most straightforward type of multivariate code is composed of values of two or more variables taken together. It does not matter whether these variables are binary or multivalued.

Composite signals are well illustrated with a slightly involved but clear example, namely the species-specific wing patterns of puddle ducks. As noted earlier, this is also an example of a cryptically binary signal in that a given duck will make only the distinction between its species and all remaining species collectively. As this binary distinction is true for every species of duck, every species must be different. Therefore, the question is how to encode this variety.

The critical wing pattern used for species recognition in flight is called the speculum (plural specula). The speculum is a rectangular area on the trailing half of the inner wing, and it may be divided into seven parts (**Figure 1**). Black, white, gray, and five chromatic colors are used for these parts in the common puddle duck species of eastern North America.

With seven parts, each of which could be one of eight colors, an enormous number of different specula are theoretically possible. Any given sample of species, such as the nine illustrated in **Figure 1**, need to use only a few of these possible patterns in order to render it unique. This composite signal can be characterized by the color of each patch in the order numbered in the figure. Using B for blue, G green, K black, L light blue, N brown, P purple, W white, x for absent, and Y for gray,

the species' specular patterns can be represented as a seven-letter 'word' of sorts (**Table 1**). In the parlance of coding theory, these seven-letter designations constitute a code word.

Most of the many examples of composite signals in the literature employ many fewer than seven variables. Even if there are merely two, however, they still constitute a code word, which is the smallest unit having a decodable meaning. Nor do composite signals have to be visual. Two closely related American treefrogs, green treefrog (*Hyla cinerea*) and barking treefrog (*H. gratiosa*), for example, use a two-variable code word in their calls. Each has a high-frequency band and a low-frequency band, but the frequencies used are different in the two species. Experiments in female choices showed that both bands were needed to elicit a full species recognition.

Compound signals

For lack of a better term, signals made up of two or more interacting variables can be called compound. The difference between these and the foregoing composite signals is that compound signals cannot be described by values of a finite list of coding variables. Put differently, compound signals do not have code words. Think of the world's national flags: a huge array of distinct patterns but differing in all sorts of ways. The whole is more than the sum of individual parts. The entirety is perceived as an organized unit, or Gestalt (to borrow a term from psychology).

Compound signals are very common in animals and are often signals of species, social group, or individual identity. Compound signals have been described in all sorts of physical forms, such as octopus display colorations, sex pheromones of insects and mammals, electric fish discharge patterns, and calls of Australian frogs. In fact acoustic signals provide abundant examples from

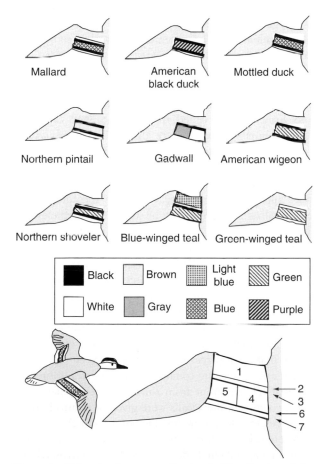

Figure 1 Example of a composite signal with a fixed spatial word length of seven: wing specula of nine species of puddle ducks. The spatial words are diagrammed at lower right and arbitrarily numbered, with a female mallard at lower left showing the position of the specula. Reproduced from Hailman, JP (2008). *Coding and Redundancy: Man-Made and Animal-Evolved Signals*, p. 122. Cambridge, MA: Harvard University Press.

Table 1   Speculum code words

Species	Code-word
Mallard	xWKBBKW
American black duck	xxKPPKW
Mottled duck	xxKBBKW
Northern pintail	xxKxxKW
Gadwall	xxKWYxx
American wigeon	xxKGGKx
Northern shoveler	xWKGGKW
Blue-winged teal	LWKGGxx
Green-winged teal	xWxGGxW

Source
Hailman, JP (2008). *Coding and Redundancy: Man-Made and Animal-Evolved Signals*, p. 123. Cambridge, MA: Harvard University Press.
The spatial code – composed of the seven elements shown in **Figure 1** (lower right) – is tabulated here for all the species shown in **Figure 1** (upper).

birds and mammals: vocalizations of finches, terns, Guinea fowl, reindeer calves, piglets, mother seals, and individually specific calls of monkeys.

Concurrent signals

Some kinds of unitary signals are made of two or more variables that encode different kinds of information. A simple analogy is the wind sock, where its orientation in a circle encodes the direction of the wind and the degree to which the sock is inflated encodes the wind's speed.

It seems likely that many signals of animals that we see as unitary in fact have different aspects that encode different kinds of information. The classical case is the figure-8 dance of the honeybee. The orientation of the straight part between the loops indicates the direction to a food source, whereas the rate of several components all encode the food source's distance. Odor trails of fire ants have similar dual meanings.

Concurrent signals are common in vocalizations. Domestic fowl give different alerting calls to objects on the ground and in the sky, and at least the aerial call encodes the degree to which the caller sees the object as dangerous. Dual aspects have also been described in cricket calls, frog songs, and chimpanzee vocalizations.

To provide a visual example, consider the coloration of birds, especially displaying males. As mentioned, the brightness of colorful aspects signals health, because the colors are produced by carotenoid pigments. Carotenoids must be taken in the diet as they cannot be synthesized from scratch, and the production of these pigments is negatively correlated with immune system activity. In other words, dull carotenoid-based coloration indicates poor nutrition or infection. The dark red, brown, gray, and black colorations, on the other hand, are based on synthesized melanin, a biochemical process under genetic control. Good dark colors thus indicate good genes, so the complex coloration of birds concurrently encodes both good health and good genes.

Alternation signals

A type of multivariate code using on and off states of an apparent single variable in fact involves two variables, because in alternation patterns both the on and off states vary in duration to form patterns. Put differently, the on state and off state are both coding variables in their own rights.

The classic example of an alternation signal is the bioluminescent flashing patterns of small beetles known as fireflies or lightning bugs. The male of each species has a different pattern, which is a series of on and off states. In fact, most of the species in eastern North America were not even distinguished until the patterns of their flashing revealed various types that were repeatedly consistent. The flashing is done in flight so that the display in many species also describes a characteristic shape, somewhat like

shapes children make with chemically luminescent wands. The soft courtship buzzing of *Drosophila* flies is another example of a species' distinctive alternation pattern.

Alternation patterns are not just distinctive for species, but can also be individual recognition signals. The lower frequencies of the 'kuyrriet' calls of different crested terns are not noticeably different, but the upper frequencies (5–10 kHz) show alternation patterns of sound and silence. In a sonographic display of frequency as a function of time, the patterns somewhat resemble the familiar UPC bar codes that have become ubiquitous on all items for sale. It turns out that upper frequency patterns are very similar in the calls of one individual but consistently different among individuals.

Combinatorial signals

Finally, the most complicated multivariate encoding so far described for animals begins to take on language-like characteristics. In combinatorial signals, a meaningful message is a combination of elements. The elements used in any given signal are a sample from a finite group of different elements. These elements, however, can be combined in more ways than there are types of elements. Only a few such systems have been described for animals but many more probably await discovery.

An interesting example is the call system of a mischievous little monkey called the wedge-capped capuchin. The combinable elements are all used by themselves as well as in combinations, and were named chirps, huhs, screams, squaws (sic), trills, and whistles. The trills, though, vary and four types could be distinguished reliably. Sometimes apparent intermediate calls were recorded, and a few vocalizations could not be typed. Nevertheless, as biological systems go, this is a fairly clean one.

From the use of calls in a large variety of social contexts, the investigator identified three quantitative emotional variables underlying the vocalizations: the levels of aggressiveness, submission, and seeking of physical contact. The call types could then be plotted as unique loci in three-dimensional space. The combinations of these calls – nearly 300 doublets, eight triplets, and three quadruplets – were plotted in places intermediate between their component calls. Thus, the combinations encoded information quantitatively different from the lone uncombined calls. When combined, these calls were given in relatively fixed sequences, but the reason for this syntax is unclear.

A somewhat similar system is the 'chick-a-dee' calls of the black-capped chickadee and its close relatives. Different species may have different numbers of combinable note types, the black-capped chickadee having four (unimaginatively termed A, B, C, and D). Like the capuchin calls, occasional intermediate and unclassified notes are uttered. Also like the capuchin calls, notes are almost always given in a fixed sequence, namely A–B–C–D.

Unlike capuchin calls, though, chickadee calls can be combined in a virtually limitless number of ways. This potentially huge call vocabulary results from repeating the component notes a variable number of times. Also unlike capuchin calls, chickadee calls are only rarely of one note type, except long trains of D notes do occur in special contexts such as mobbing predators.

Overview

The information content of signals is not determined by their physical structure so long as the sender's alternative signals can encode sufficient information that, in turn, can be decoded by the receiver. Information is variety or diversity and can be expressed quantitatively when the relative probabilities of occurrence of the alternative signals are known. The simplest systems have two alternative signals (binary codes), more complex systems have three or more alternative signals (multivalued codes), and the most complicated systems employ two or more signal variables (multivariate codes). Of these last, using different combinations from the same set of alternative signals to mean different things (combinatorial codes) begins to approach the complexity of human language.

See also: Adaptive Landscapes and Optimality; Anthropogenic Noise: Impacts on Animals; Communication Networks; Decision-Making: Foraging; Multimodal Signaling; Syntactically Complex Vocal Systems.

Further Reading

Bradbury JW and Vehrencamp SL (1998) *Principles of Animal Communication*. Sunderland, MA: Sinauer Associates.

Darwin C (1872) *The Expression of the Emotions in Man and Animals*. New York: D. Appleton and Company.

Gerhardt HC and Huber F (2002) *Acoustic Communication in Insects and Anurans: Common Problems and Diverse Solutions*. Chicago: University of Chicago Press.

Hailman JP (2008) *Coding and Redundancy: Man-Made and Animal-Evolved Signals*. Cambridge, MA: Harvard University Press.

Hailman JP, Ficken MS, and Ficken RE (1985) The 'chick-a-dee' calls of Parus atricapillus: A recombinant system of animal communication compared with written English. *Semiotica* 56: 191–224.

Haldane JBS and Spurway H (1954) A statistical analysis of communication in 'Apis mellifera' and a comparison with communication in other animals. *Insectes Sociaux* 1: 247–283.

Hauser MD (1996) *The Evolution of Communication*. Cambridge, MA: MIT Press.

Hill GE (2002) *A Red Bird in a Brown Bag*. Oxford: Oxford University Press.

Maynard Smith J and Harper D (2003) *Animal Signals*. Oxford: Oxford University Press.

Owings DH and Morton ES (1998) *Animal Vocal Communication: A New Approach*. Cambridge: Cambridge University Press.

Robinson JG (1983) Syntactic structures in the vocalizations of wedge-capped capuchin monkeys, *Cebus olivaceus. Behaviour* 90: 46–79.

Sebeok TA (ed.) (1968) *Animal Communication: Techniques of Study and Results of Research*. Bloomington: Indiana University Press.

Sebeok TA (ed.) (1977) *How Animal Communicate*. Bloomington: Indiana University Press.

von Frisch K (1967) *The Dance Language and Orientation of Bees*. Cambridge, MA: Harvard University Press.

Innovation in Animals

K. N. Laland, University of St Andrews, St Andrews, Fife, Scotland, UK
S. M. Reader, Utrecht University, Utrecht, Netherlands

Introduction

In Swaythling, a village in southern Britain, in 1921 the first observation was made of a blue tit opening milk bottles left outside homes and drinking the cream. Ornithologists and members of the public noted repeated occurrence of the behavior, and over the next 30 years, milk-bottle opening was reported at numerous sites across the United Kingdom and mainland Europe. Milk-bottle opening has become the best-characterized and perhaps best-known example of the spread of a novel behavior pattern, or innovation, in non-human animals. Experimental investigation determined that the rate of innovation was likely quite high and that the behavior probably spread by social learning in localities where milk bottles were introduced. However, it is just one example of thousands of innovations reported in non-human animals, with examples ranging from novel elements in the song of birds to novel tool use in primates. Many innovations appear to be responses to changed circumstances, such as human impacts. However, innovations are also produced in stable environments, where an animal discovers a new method of exploiting the environment, an example of niche construction. It would seem that a large number of animals invent new behavior patterns, modify existing behavior to a novel context, or respond to social and ecological stresses in an appropriate and novel manner. Such behavior can usefully be termed 'innovation,' and can be distinguished from related processes such as exploration and learning.

When a novel learned behavior spreads through an animal population as individuals learn from one another, typically a single individual will have initiated the process. Such diffusion requires two processes: the initial inception of the behavioral variant, termed innovation, and the spread of the novel trait between individuals, a process called social learning. In recent years, animal innovation has begun to be recognized as an important component of animal social learning research. However, not all innovations spread by social learning. A common observation, in both primates and birds, is that many innovations fail to spread through animal groups, even though they are of apparent utility to the inventor. This phenomenon has also been noted in humans. Possible explanations for this limited diffusion include differences in ability or motivation, noninventors being able to benefit (scrounge) from the innovations of others without learning the innovation, and that innovators may not be attended to (e.g., they may be on the periphery of the group). In birdsong innovation, female preferences can act in some circumstances against the spread of novelties and is thus a conservative force (e.g., in swamp sparrows), while in other circumstances a preference for large vocal repertoires may promote the spread of innovations (e.g., sedge warblers).

Innovation is widespread in non-humans, and evidence is mounting for its functional importance. For example, innovation appears to play an important role in ecology (for instance, facilitating range expansion), in evolution (for instance, driving subspecies diversification), and as the vital first step of social learning and cultural diversification. Innovations appear in both the social domain (e.g., some instances of tactical deception) and the nonsocial domain (e.g., novel tool use or foraging techniques). Innovation is also key to cumulative cultural evolution, where a careful balance must be struck between faithful social transmission (to minimize loss of previous innovations) and innovation (to minimize stagnation and allow adaptive change). Although clear evidence of cumulative cultural evolution is lacking in animals, the use of hammers and anvils by common chimpanzees to crack nuts, and leaf tool manufacture in New Caledonian crows, have been suggested as possible cases. Innovation may have also played an important role in primate and avian brain evolution. A strong case can be made for the assertion that non-human animals innovate, although the consanguinity of animal and human innovation is a matter of debate.

Defining Animal Innovation

In 2003, Reader and Laland put forward a definition of animal innovation in the first book on the topic. They proposed two operational definitions, relating to the use of innovation as a newly invented behavior pattern (the 'product') or as the process of devising inventions (the 'process'). Innovation sensu product is a new or modified learned behavior not previously found in the population. Innovation sensu process is a process that introduces novel behavioral variants into a population's repertoire and results in new or modified learned behavior. Thus, the key features of innovations are that they are learned and that they are novel to a population. Innovations can spread by social learning; however, the introduction of a novel behavior into a population by social learning is not considered innovation (this contrasts with some definitions of human innovation, which refer to acquisition of a novel act by any route as innovation, and the initial inception as 'invention'). Other researchers add another definitional qualification, that an innovation should not be environmentally induced, that is, not predictably produced in a particular environment. However, the utility of this extra qualifier in identifying innovations has been questioned.

Ecological and Evolutionary Significance

Behavioral scientists have long noted that species differ in their tendency to innovate. In 1912, Lloyd Morgan speculated that behavior may be composed of a repetitive component that has occurred before and a smaller proportion of novel behavior, found particularly in so-called 'higher' organisms, that can be regarded as a creative departure from routine. Although there are long traditions of research into related topics, such as neophilia, exploration, and insight learning in animals, and while innovation in humans has been subject to considerable investigation, animal innovation has only relatively recently begun to receive attention. Field studies of primates, particularly in common chimpanzees and Japanese macaques, were critical in drawing attention to the potentially significant role of innovation. For example, in Kummer and Goodall's (1985) landmark review of primate innovation, they suggested that some innovations derive from the ability of the individual to profit from an accidental happening, while others result from the ability to use existing behavior patterns for new purposes. A third kind are completely new behavior patterns. Innovation may be prompted by need (such as a period of drought or a social challenge), or alternatively an excess of resources, which will allow animals to bear the costs of exploration. Recent experimentation and meta-analyses of published cases of innovation suggest that necessity is probably the dominant factor prompting animal innovation, consistent with the saying 'necessity is the mother of invention.'

In the last decade, two major surveys of innovation have collated reports from published literature, using keywords such as 'novel' or 'never seen before' to classify behavior patterns as innovations. The first and most extensive survey, by Louis Lefebvre and colleagues, documented over 2200 examples of foraging innovations in birds. We conducted a similar survey in primates, where over 500 innovations were documented. Although these kinds of analysis are vulnerable to reporting biases – well-studied animals will be reported to exhibit more innovation than poorly studied animals – researchers have devised statistical methods for evaluating and counteracting these biases. Accordingly, there are reasonable grounds to be confident that these innovation data are reliable, allowing the rate of innovations for a given species to be used as one measure of its behavioral

flexibility. Confidence in such measures is further enhanced by the observation that a species' innovation rate correlates with its performance in laboratory tests of learning, reinforcing the idea that innovation is a cognitive measure. In both birds and primates, innovation rate has also been shown to correlate positively with the forebrain size. This parallel finding in two independent taxa raises the possibility that increased innovativeness may have given a selective advantage driving brain enlargement over evolutionary time, although since the data are correlational, causality is difficult to establish. The phylogenetic distribution of innovation suggests that selection may have favored innovativeness as part of a cognitive suite of traits in particular lineages, with innovation part of a survival strategy based on flexibility to cope with unpredictable, changing socioecological environments. Thus, particular taxa appear unusually prone to innovation.

There are numerous reports of specific innovations apparently facilitating survival in changed circumstances, while innovation is also thought to be of critical importance to those endangered or threatened species forced to adjust to impoverished environments. To test the idea that innovations may facilitate survival in novel circumstances, biologist Daniel Sol and colleagues took advantage of a series of natural experiments, where humans have introduced bird species into new habitats. Species innovative in their natal habitat were found to be more likely to survive and establish themselves when introduced to new locations than other species. This supports the hypothesis that the ability to innovate may aid survival in changed environmental conditions. Consistent with this, nonmigratory birds are known to innovate most in the harsher winter months. Conversely, migratory species have been found to be less innovative than nonmigrants, suggesting that migratory birds may be forced to migrate because of an inability to adjust behaviorally to the changed winter months.

Innovation has also been proposed to play a key influence on the tempo and course of evolution. For instance, in 1985, molecular evolutionist Allan Wilson proposed a 'behavioral drive' hypothesis, which argued that innovation combined with cultural transmission led animals to exploit the environment in new ways, exposing them to novel selection pressures and increasing the rate of genetic evolution. Support for this idea comes from the aforementioned studies of bird innovation. Innovative species and innovative taxa contain more subspecies and species, respectively, compared to less innovative taxa, suggesting that the rate of evolutionary diversification has been greater in innovative groups.

Individual Differences in Animal Innovation

Observational and experimental studies show that individuals, like species, differ in their propensity to innovate.

Innovation is influenced by variables such as social rank, age, sex, competitive ability, and motivational state, and is correlated with both behavioral (e.g., learning rate, tool use) and hormonal measures (e.g., testosterone). There is even evidence for innovative 'personalities' in some species (e.g., guppies), although consistent individual differences in the propensity to innovate have not been found in other species (e.g., capuchin monkeys).

Some of these studies rely on natural observations of innovation, and recognition criteria have been developed to identify 'true' innovations. For example, in wild orangutans, behavior patterns were classified as innovations if they were not universally expressed across populations and if their absence did not have a clear cause (e.g., an observational artifact, such as insufficient observation time, or an ecological cause, such as lack of a particular resource). This process identified 43 innovations, such as the manufacture of branch cushions or 'throat-scraping' sounds made by mothers with young infants before moving. Although such recognition criteria cannot provide incontrovertible evidence that a particular act is or is not an innovation, they can identify potential innovations for further investigation. Such investigation may have surprising results. For example, grackles dunk hard food in water, a relatively rare behavior originally described as innovative, even insightful. However, the vast majority of the population will dunk food if placed in ideal circumstances, suggesting dunking is rarely expressed because the costs (e.g., kleptoparasitism) usually exceed the benefits (softened food). This implies that rarity per se cannot be used to identify innovation.

As an alternative or addition to observational studies, innovation can be studied experimentally, in both captivity and the wild, by presenting animals with novel challenges, such as puzzle boxes that they must open to access food, and exploring the factors influencing innovation. One of the most marked examples of innovative tool manufacture was observed when a female New Caledonian crow named Betty bent a wire to manufacture a hooked tool and obtain a food reward. Further experiments investigated the technical understanding of this individual in solving such tasks. Innovative tool use has also been recorded in wild-living New Caledonian crows, such as individuals enhancing their usual leaf tools by bending them.

Experiments have documented a number of behavioral correlates of innovation. For example, studies of several species of birds and of callitrichid monkeys (marmosets and tamarins) have established that those individuals least reluctant to approach novel objects (i.e., showing low levels of object neophobia) are fastest to solve novel foraging tasks. Thus, differences in innovative tendency need not be ascribed to differences in cognitive ability, but can be at least partially explained by the willingness to engage with novel stimuli. The ability to inhibit previously learned responses may be another important

correlate of innovation. In feral pigeons *Columba livia*, cut-throat finches *Amadina fasciata*, and zebra finches *Taeniopygia guttata*, animals that performed well in innovation tasks were also superior in social learning tasks. Thus, in these species at least, innovators also tend to be those individuals most able to make use of social information.

A number of studies have focused on age and innovation. Perhaps influenced by a small number of high-profile cases, the prevailing assumption among many primatologists is that young or juvenile primates are more innovative than adult individuals. This innovative tendency among the young is often thought to be a consequence or side effect of their increased rates of exploration and play. However, a recent meta-analysis of the primate innovation literature challenges this view. A greater incidence of innovation was found in adults than in nonadults, which the researchers interpreted, in part, as a reflection of the greater experience and competence of older individuals.

These findings are supported by a detailed experimental analysis of innovation in callitrichid monkeys. Researchers presented novel extractive foraging tasks to family groups of monkeys in 26 zoo populations in order to examine whether youth or experience most facilitates innovation. Exploration and innovation were found to be positively correlated with age, perhaps reflecting adults' greater experience, manipulative competence, or cognitive ability. Younger monkeys, particularly subadults and young adults, were disproportionately likely to first contact the tasks, but adults were disproportionately first to solve the tasks. Thus, older individuals were significantly more likely than younger individuals to turn task manipulations into solutions. Subsequent statistical analyses provided evidence that at least some of the methods of box opening subsequently spread through the group through social learning. Another study, this time of brown-mantled tamarins, also found that adults acquire information more efficiently and that they can recognize and classify objects more quickly, than nonadults. Such experiments suggest that experience and competence allow older individuals to solve novel problems more effectively than younger individuals. However, other developmental factors, such as improvements in manipulative skills, increased strength, and maturity with age, may also play a role. Further investigations of species differences in innovativeness among monkeys suggest that certain life-history characteristics, particularly a diet reliant on extractive foraging, may favor enhanced innovation.

If dominant individuals monopolize resources, or if low-status individuals are driven by lack of success in other regards to devise new solutions, then social rank orders might predict who innovates. A number of avian studies, as well as observations of macaques and other primates, have demonstrated that subordinates are more likely to innovate, but are often usurped by dominants. In primate groups, low-ranking monkeys may acquire a novel behavior but not express it in order to avoid the attention of dominants. Conspecifics may speed or slow approach to novel objects (e.g., ravens approach novel objects more rapidly when alone than when in groups, but will spend more time investigating the objects when in groups). Thus, there may be social constraints and influences on the invention and expression of novel behavior patterns.

Studies of guppies demonstrate that motivational state can be a critical determinant of innovation. Small groups of fish were presented with novel maze tasks containing food, and the first individual to solve the task was characterized as an innovator. Females were found to be more likely to innovate than males, food-deprived fish were more likely to innovate than nonfood-deprived fish, and smaller fish were more likely to innovate than were larger fish. Innovators were neither the most active fish (males) nor those with the greatest swimming speed (large fish). Here, the most parsimonious explanation for the observed individual differences in problem solving is that innovators do not need to be particularly intelligent or creative, but are driven to find novel solutions to foraging problems by hunger or by the metabolic costs of growth or pregnancy.

To further investigate how motivational state affects innovation, researchers monitored the relation between past foraging success and foraging innovation, again using guppies. Groups of fish were fed food items one at a time, and thus had to compete for food. Poor competitors – fish that had gained the least weight and obtained the least food during the scramble competition – were predicted to be more likely to innovate when presented with the novel foraging tasks. In male but not female guppies, this prediction was upheld. Females appeared more motivated to solve the foraging tasks than males, regardless of how they had fared during the scramble competition. In many vertebrate species, female parental investment exceeds that of males, so male reproductive success is most effectively maximized by prioritizing mating whereas female reproductive success is limited by access to food resources. This is particularly true in guppies, since females can store sperm, are viviparous, and unlike males, have indeterminate growth, with a correlation between energy intake and female fecundity. Consequently, finding high-quality food has greater marginal fitness value for females than for males, which may explain why females should be more investigative than males and are constantly searching for new food sources, whereas males begin searching for food only when they become food-deprived.

A study of the spread of innovations in small captive groups of starlings (*Sturnus vulgaris*) investigated whether the pattern of spread could be predicted by the knowledge of relevant variables. The researchers presented small groups of starlings with a series of novel extractive foraging tasks. Object neophobia and social-rank measures best characterized which animal was the first of the group to contact the novel foraging tasks. However, asocial learning performance, measured in isolation, was the best predictor of the first-solvers of the novel foraging tasks in the group. In other words, one can predict how innovative a starling will be on the basis of its previously measured learning performance in isolation. The solutions to these tasks appeared to spread through social learning, since individuals acquiring the behavior later in the diffusion exhibited shorter learning times. This pattern would be expected if the subjects learn socially, since later solvers have more demonstrators than individuals that acquire the behavior early. However, perhaps surprisingly, association patterns did not predict the spread of solving: birds were no more likely to learn from close associates than birds with which they spent little time. Similar results were found in studies of novel foraging behavior in guppies. This may reflect the relatively small size of the groups and enclosures in both studies, and innovations may be more likely to spread along networks of association in larger groups living in more naturalistic environments.

Summary

Numerous animals acquire novel skills and information from others, and behavioral innovations can diffuse through natural and captive populations by social learning processes. It is instructive to refer to the initial inception of such behavioral variants as innovation and to investigate the factors that underlie and predict variation in innovation within and between species. Innovation can be studied in animals experimentally by presenting novel tasks to captive or natural populations and carefully monitoring the spread of the solution. Experimental studies of animal innovation, in various vertebrates including fish, birds, and primates, suggest that the adage 'necessity is the mother of invention' explains a lot of data. Hungry, small individuals and individuals of low status disproportionately engage in innovative behavior. Adults perform a disproportionate amount of innovation, and the greater experience, strength, and maturity of elder individuals may be necessary to translate exploration into successful exploitation. Sex differences in innovation can be interpreted, and to some extent predicted, using conventional behavioral-ecology theory, such as parental-investment and sexual-selection theory. Comparative data demonstrate a potential role for innovation in brain evolution, introduction success, and evolutionary diversification. Animal innovation has received relatively little empirical attention, but there is growing evidence that it is functionally important.

See also: Social Cognition and Theory of Mind.

Further Reading

Biro D, Inoue-Nakamura N, Tonooka R, Yamakoshi G, Sousa C, and Matsuzawa T (2003) Cultural innovation and transmission of tool use in wild chimpanzees: Evidence from field experiments. *Animal Cognition* 6: 213–223.

Boogert NJ, Reader SM, Hoppitt W, and Laland KN (2008) The origin and spread of innovations in starlings. *Animal Behaviour* 75: 1509–1518.

Fisher J and Hinde RA (1949) The opening of milk bottles by birds. *British Birds* 42: 347–357.

Kendal RL, Coe RL, and Laland KN (2005) Age differences in neophilia, exploration, and innovation in family groups of callitrichid monkeys. *American Journal of Primatology* 66: 167–188.

Kummer H and Goodall J (1985) Conditions of innovative behaviour in primates. *Philosophical Transactions of the Royal Society of London Series B* 308: 203–214.

Lefebvre L, Whittle P, Lascaris E, and Finkelstein A (1997) Feeding innovations and forebrain size in birds. *Animal Behaviour* 53: 549–560.

Lloyd Morgan C (1912) *Instinct and Experience.* London: Methuen.

Ramsey G, Bastian ML, and van Schaik C (2007) Animal innovation defined and operationalized. *Behavioral and Brain Sciences* 30: 393–437.

Reader SM and Laland KN (2002) Social intelligence, innovation and enhanced brain size in primates. *Proceedings of the National Academy of Sciences of the United States of America* 99: 4436–4441.

Reader SM and Laland KN (eds.) (2003) *Animal Innovation.* Oxford: Oxford University Press.

Sol D, Duncan RP, Blackburn TM, Cassey P, and Lefebvre L (2005) Big brains, enhanced cognition and response of birds to novel environments. *Proceedings of the National Academy of Sciences of the United States of America* 102: 5460–5465.

Sol D, Lefebvre L, and Rodríguez-Teijeiro JD (2005) Brain size, innovative propensity and migratory behaviour in temperate palaearctic birds. *Proceedings of the Royal Society of London Series B* 272: 1433–1441.

Wilson AC (1985) The molecular basis of evolution. *Scientific American* 253: 148–157.

Wyles JS, Kunkel JG, and Wilson AC (1983) Birds, behaviour, and anatomical evolution. *Proceedings of the National Academy of Sciences of the United States of America* 80: 4394–4397.

Insect Flight and Walking: Neuroethological Basis

R. E. Ritzmann and J. A. Bender, Case Western Reserve University, Cleveland, OHIO, USA

Introduction

Insects are among the most agile and adaptable animals in the world. In addition to stable movements on relatively flat, horizontal substrates, they can readily walk through tortuous terrain that requires climbing, tunneling, or turning movements to deal with obstacles. Beyond that, many insects can also walk on water or fly through the air. What is it about insects that make them so successful? Certainly, much of their success in locomotion is due to mechanical factors. The hexapod body plan allows insects to maintain passive stability over a variety of movements, and a sprawled leg posture adds to that stability. However, beyond their body plan, insects take advantage of a remarkable array of sensors that provide a wealth of information to a sophisticated central nervous system (CNS) that in turn acts through an efficient neuromuscular system to alter movements appropriately. These neural and muscular systems, working with their lightweight exoskeletons, allow them to adjust to a wide range of terrain.

Sensors fall into two groups. Local sensors located on appendages such as legs and wings monitor immediate properties of the limb such as joint position and load. These proprioceptors contribute to reflex circuits in the thoracic ganglia and also impact pattern generators that control the timing of joint movements. Exteroceptors located primarily on the head are associated with a sophisticated brain. These sensors provide a remarkable amount of information on the insect's surroundings (e.g., chemical, mechanical, auditory, and visual cues). The sensory information gained by these structures is processed in primary sensory regions, and then used by association centers to generate descending commands, which in turn alter local movement parameters in the thoracic ganglia. Many researchers have learned a tremendous amount about the local reflex circuits of the thoracic ganglia and the primary sensory regions of the brain. However, we are only beginning to examine the brain circuitry that influences locomotion.

Interestingly, this hierarchical locomotion control scheme parallels that of vertebrate animals, in spite of the fact that they evolved legged locomotion independently. In vertebrates, local control similar to that found in insect thoracic ganglia is found in the spinal cord. Vertebrate exteroceptors are also processed in sophisticated brain regions, leading to descending commands. The convergence between insect and vertebrate control systems underscores the notion that the hierarchical design used by both these successful animal groups represents a particularly good system that was selected for at least twice for negotiating a range of complex terrains.

An Overview of Walking

Before examining the neural basis of walking, it is important to understand the leg structures and the basic walking movements that are made by most insects. Insect legs are made up of a series of leg segments surrounded by hard cuticle and connected by soft joints. Starting with the first leg segment that attaches to the thorax and moving distally, the leg segments are the coxa, trochanter, femur, and tibia (**Figure 1**). Beyond that, a series of tarsal segments make up the foot, which typically ends in a segment containing claws and/or attachment pads. Most of these joints are simple hinges that flex in one plane. In contrast, the most proximal thoracic-coxal (ThC) joint that connects the leg to the thoracic cuticle is a mechanical wonder, a compound joint that allows rotation around three different axes. Effective leg movements that propel the body forward require coordination of these joints, both within a single leg and between the various different legs of the body.

Being hexapods, adult insects typically walk in a tripod gait, in which the front and rear legs on one side of the body step simultaneously with the middle leg on the opposite side. This tripod alternates with the remaining three legs, maintaining continuous static stability under most conditions. The period of the walking cycle during which legs are in contact with the ground and propel the body forward is called the *stance* phase. This period alternates with the *swing* phase when the legs are lifted off the ground to either return to their starting position or move to a new starting position in anticipation of a turn or other maneuver. At slower speeds, legs move in a metachronal pattern with one leg stepping at a time on each side of the body, starting from the rear and moving forward. This can grade into a tripod gait if, for example, the hind leg begins another step before the front leg finishes moving.

Although the three legs of a tripod set down and lift off together, their actions are not the same. Rather, the rear, middle, and front legs move through very different joint actions and interact differently with the ground, generating unique ground-reaction forces. The rear legs make piston-like movements, with the two distal joints extending through similar angles to produce a powerful

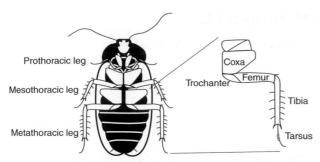

Figure 1 Diagram of a cockroach from the ventral side showing the location of the three pairs of legs. One metathoracic leg is expanded to the right with leg segments labeled.

rearward-directed motion at the tarsus. That movement generates accelerating horizontal ground-reaction forces that propel the body forward. In contrast, the front legs make variable movements that typically extend the tarsus in front of the insect's head. On contact, these legs actually decelerate the forward motion of the body. The middle legs make intermediate sweeping movements that counteract the lateral forces of the other two legs. Their ground-reaction forces initially decelerate and then accelerate the body. The details of these mechanical properties of insect locomotion and their responses to perturbations have been studied extensively by Robert J. Full and his collaborators.

Local Control Circuits in Thoracic Ganglia

Insects require a sophisticated neural control system to coordinate the movements of each of these legs. Recent neurobiological studies in several laboratories are beginning to describe the local control circuits that lead to these movements.

A technical advantage to studying insect motor control stems from the relatively small number of motor neurons that control each muscle. Indeed, at several joints, one can monitor motor activity with simple electromyogram wires anchored to the cuticle and thereby examine activity from a specific, identified motor neuron. For example, in the cockroach, the main muscle that extends the femur-tibia (FTi) joint is controlled by only two motor neurons (the slow and the fast extensors of the tibia or SETi and FETi, respectively), and FETi is only active when the insect is running at relatively high speed.

The timing of extension and flexion in each joint is controlled by a separate pattern generator located within the thoracic ganglia. These small neural circuits are found throughout the animal kingdom and can act independent of sensory activity to provide basic timing commands for oscillatory behaviors. However, in normal behavioral situations, pattern generators typically act in concert with associated sensory structures to produce specific actions that are appropriate for the conditions that the

animal is experiencing at any given moment. In insects, the transitions between extension and flexion of individual joints are influenced by sensory signals from leg proprioceptors. These sensors include chordotonal organs that monitor angle and direction of movement in each joint, strain detectors called 'campaniform sensilla' that provide information on load and muscle forces, and hair plates that detect maximal joint flexion. Thus, for example, a transition between stance and swing movements typically requires both pattern-generator activity and unloading of the leg.

How do leg proprioceptors exert their influence on leg movements? Their activation typically evokes reflex responses in specific motor neurons. In the simplest situation, these are resistance reflexes. For example, stretch of a chordotonal organ that normally occurs when the FTi joint is flexed will excite the motor neuron that, in turn, activates the muscle that counteracts the movement (SETi). Increases in load are detected by the campaniform sensilla and evoke similar reflexes to counteract imposed load and maintain posture.

When the insect is walking, these reflexes become more complex. Many reflexes are both quantitatively and qualitatively altered from the situation found in a standing insect. Moreover, they must now interact with the pattern generators that control timing of joint behavior in order to produce different leg movements that are appropriate for walking on horizontal surfaces, inclines, walls, or ceilings. Changes in joint coordination can also lead to movements associated with turning, climbing, or backward walking.

A particularly interesting finding from Ansgar Büschges' laboratory in Cologne, Germany, revolves around the roles of proprioceptors in coordinating joint movements. After eliminating sensory signals from leg proprioceptors, the outputs of the pattern generators continue to burst, but their individual phase patterns are unrelated, indicating independence among the actual pattern generation circuitry. That is, there does not appear to be a master pattern generator present for each leg that would coordinate the various joints of that leg. However, coordination of joints can occur via interjoint reflexes. It is now apparent that proprioceptors project to neurons that control adjacent joints as well as those that control the joint that they monitor. These weaker interjoint reflexes influence transitions between flexion and extension of adjacent joints.

The interjoint reflexes that control stick insect legs have now been documented in great detail. Moreover, the pattern of these reflexes was used by Ekeberg, Blümel, and Büschges to control a sophisticated, dynamic simulation of a stepping leg. By simply progressing through the various chains of reflexes, the simulated leg moves realistically through stance and swing phases of walking. Interestingly, hind legs tend to move very differently than front and middle legs, yet the simulation can replicate those movements also by simply reversing some of the reflexes. Similar alterations in leg movement occur in cockroach middle legs when they transition between forward walking and turning behaviors. Robotic models of both the stick insect and the cockroach leg, developed in the Ritzmann laboratory based upon the Ekeberg model, have confirmed that a transition from forward walking to turning can occur through changes in reflex sign.

These observations and models suggest a hypothetical mechanism for redirecting leg movement. Under this model, descending commands from the brain alter locally coordinated actions in each leg by modifying various opposing reflex circuits in the thoracic ganglia. Laiyong Mu, working in the Ritzmann laboratory, demonstrated that proprioceptive leg reflexes could, in fact, be influenced by activity that descends from the brain. His experiments examined an interjoint reflex generated by stretching and relaxing the femoral chordotonal organ (FCo) that monitors the position of the FTi joint. In addition to the intra-joint reflex described earlier for SETi, this organ also excites the slow depressor motor neuron that controls extensor muscles of the coxa-trochanter joint (Ds). In intact animals, relaxation of the FCo inhibits Ds. However, in animals with descending activity from the brain eliminated by cutting both cervical connectives, the reflex reverses and FCo stretch generates excitation in Ds. Similarly, Turgay Akay, working in the Büschges laboratory, has demonstrated that reflexes in stick insects reverse sign when the animal changes from forward to backward walking.

In addition to coordination within each leg, effective movements must also be coordinated among all six legs. The manner in which this occurs has been formalized by a set of rules described by investigators in Holk Cruse's laboratory in Bielefeld, Germany. These rules stem from examination of various insect locomotion behaviors. They describe effects that the state of one leg has upon adjacent leg movements. For example, one rule states that a leg should not be lifted off the substrate for return while the next posterior leg is still in swing. Another rule states that as an anterior leg progresses through the stance power stroke, it increasingly influences the next posterior leg to enter into its swing return phase.

The total set of 'Cruse rules' can explain much of the interleg coordination that is seen in insect locomotion. They have also been used to control several robotic devices and to guide neurobiological studies that seek to understand the neural basis of these properties.

Influence of Brain Circuits on Locomotion

One might conclude from the previous section that walking movements are controlled exclusively by local circuits, with higher centers serving merely to initiate movement. However, the situation becomes more complex when considering movements through unpredictable, natural terrain. Upon encountering a block, a cockroach rears up to an appropriate height, so that it can place its front legs on the top surface, and then climbs over the barrier. A shelf-like obstacle can evoke either tunneling or climbing. Detailed behavioral analysis by Cynthia Harley in the Ritzmann laboratory indicates that this decision is largely dictated by the manner in which antennae contact the obstacle. Contact from above evokes climbing, while contact from below yields tunneling. Of course, other options are possible. Insects readily turn and walk around obstacles, follow complex walls, and span gaps in substrates, and several laboratories have described these behaviors in a variety of insect species.

Any of these obstacles requires a directed change in the actions that are typically made by the legs as the insect walks forward. While some barriers can be dealt with through local reflex effects, the redirection of leg movements associated with large barriers requires that the animal become aware of the barrier, evaluate it, and then redirect leg movements accordingly. In order to accomplish these adjustments, the insect takes advantage of the various sensory systems located on its head. These include tens of thousands of mechanoreceptors on each antenna and on the maxillary palps, chemoreceptors also on those structures, visual cues detected by either compound eyes or simple ocelli, and in some insects, auditory or vibration signals. Many of these sensors project to primary sensory regions and then to association regions of the brain. This neural processing of sensory information must ultimately lead to descending commands that interact with the local control circuits in the thoracic ganglia and thereby redirect leg movements appropriately. Indeed, insects that have experienced damage within various brain regions fail to negotiate barriers appropriately.

Two main association regions exist in the insect brain. They are the mushroom bodies (MB) and the central complex (CC). The MBs are perhaps the most studied region of the insect brain. They are involved in learning and memory as well as olfactory processing. The CC was originally described by Les Williams. It is made up of several midline neuropils, including the fan-shaped body, ellipsoid body, protocerebral bridge, and two paired nodules. The morphology of these neuropils has been described in detail for several insects. Several have a distinctly columnar organization, and fibers project between the left and right columns in the protocerebral bridge and columns of the fan-shaped body. This anatomy suggests that turning decisions may be made in the CC. In earlier experiments, Franz Huber found that stimulation within the CC of crickets evoked increases in locomotion and turning movements. More recently, both genetic and mechanical lesions within the CC have been associated with motor deficits.

Physiological studies have only recently focused upon the CC. Perhaps the most remarkable finding is a topographic map of polarized light space that was described by Uwe Homberg's laboratory. Observations in the Ritzmann laboratory describe multisensory CC units that respond to visual and antennal stimulation. Some of these units are biased to one antenna or to a preferred direction of movement of either antenna. Thus, it is possible that sensory information on barriers as well as surrounding ambient conditions acts through the circuits of the CC to ultimately influence the local control circuits of the thoracic ganglia.

Descending activity, influenced by sensory information processed in the CC, could represent the commands that are required to alter local reflexes and redirect leg movements, as described earlier. Neurons leaving the CC project to a region of the brain called 'the lateral accessory lobe,' and several neurons that ultimately descend to the thoracic ganglia also pass through this region. Thus, it is possible that the remarkable agility of insects in the face of complex terrain arises in large part from descending commands that are influenced by head-based sensors through the actions of CC circuits. Appropriate modifications on basic walking patterns would then occur when these descending commands alter local reflexes within the thoracic ganglia. By allowing the local reflex circuits to perform the moment-by-moment control of actual leg movements, insects achieve both remarkable stability and amazing flexibility.

Insect Flight

Of course, insect locomotion is not limited to walking and running. Many insects are among the most maneuverable flyers in the animal kingdom. How is the hierarchical control system modified to suit this form of locomotion? Conceptually, the issues faced by flying insects are similar to those confronted by walking ones. The wings beat up and down with an intrinsic rhythm, and descending commands from the brain based on sensory information must be processed and executed within the context of this ongoing locomotor pattern. In many ways, flight control can be viewed as an extreme example of terrestrial locomotor control, as many of the same sensors and neural circuits are involved. The extreme limits of a behavior, however, are often informative regarding what factors limit an animal's ability to perform, survive, and thrive.

One prominent control problem, differing between walking and flying, is that flight is only dynamically stable, meaning that an insect cannot simply stop in place to gather or process more sensory information before implementing an action. Further complicating this situation is the fact that motor commands must be executed at an appropriate phase of the wing beat cycle, just as a walking animal must contract its muscles at the right times to influence movement. Because the wing beat frequencies are generally higher than stepping rates, this means that descending control mechanisms must include extremely precise methods for gating and synchronization with locomotor rhythms. To make matters even worse from a control standpoint, in some insects the powerful muscles which drive the wings up and down are not under direct control of the nervous system, as the upstroke and the downstroke muscles excite each other mechanically through stretch activation. Therefore, the animal must use sensory information to determine the state of its own muscles and body, combining this appropriately with information about the outside world to produce and execute a turning movement through precisely timed contractions of the wings' small steering muscles.

The true (dipteran or two-winged) flies have specialized sense organs which are extremely well suited to monitor motions of their bodies. In these insects, the hind pair of wings has evolved into very small, club-shaped organs called 'halteres' (**Figure 2**). The halteres beat back and forth at the same time as the wings, but have no aerodynamic function and instead are tightly packed with mechanosensors. Most of the sensors on the halteres are dedicated to measuring the back-and-forth motions, which presumably serve to report the wings' positions to the central nervous system. In combination with similar signals from the wings themselves, information from the halteres could be used to calculate not only the state of each wing power muscle, whether contracting or relaxing, but may also contain clues as to whether the wings are moving as they should be, in phase with the halteres, or if wind, load, or other external forces have derailed the wings from their normal pattern.

Beyond these uses, the halteres also have a more celebrated function as onboard gyroscopes, recently studied in depth by Michael Dickinson's lab. Some of the halteres' mechanosensors do not respond phasically to each stroke,

Figure 2   Crane fly showing prominent halteres behind wings. Photo: Armin Hinterwirth.

but rather are activated when the halteres are deflected from their stroke plane. Since the halteres are so tiny, such a deflection would not result from a swat or even from wind blowing the haltere, but only by Coriolis forces corresponding to a rotation of the fly's body, and then only during a particular part of the haltere's stroke phase. Afferent neurons from the deflection-sensitive fields are electrically coupled to motor neurons of the wing steering muscles, thereby producing very fast and precisely timed contractions designed to compensate for the imposed rotation of the body. This ability to sense body rotations seems to be very important to insect flight, possibly because the inherent instability of flight in small animals makes the maintenance of equilibrium a task of primary importance. Recent studies suggest that moths and butterflies may use their antennae as rotation sensors, in a similar manner to the halteres of flies.

In line with the hypothesis that ongoing patterns of movement are controlled by descending commands working indirectly through reflex arcs, the halteres are equipped with their own set of steering muscles. Thus, a neuron from the brain might cause a contraction in the haltere steering muscles, in addition to or instead of modifying the wing motions directly. These muscles could act to deflect the haltere from its stroke plane in the same way that Coriolis forces do, thus triggering a reflex cascade that leads to the fly turning. Work in the Dickinson laboratory has demonstrated that visual stimuli are capable of eliciting activity in the haltere steering muscles, but the rest of this pathway remains conjectural. One especially appealing part of this hypothesis is that the haltere reflexes, just like those acting through leg mechanosensors, normally have precise and effective influence over locomotor activity, so co-opting those mechanisms to produce directed movements takes advantage of this built-in robustness.

Mechanosensation is not the only way that insects detect the motions of their bodies during flight. Bees, beetles, ants, locusts, crickets, and even cockroaches can fly without any known mechanical ability to sense body rotations independently from their wings, but exceedingly few insect species can fly stably in complete darkness. Dragonflies, in fact, are so dependent on vision for flight control that they cannot fly in normal indoor lighting. Even nocturnal insects, such as moths, make extensive use of vision to navigate and to sustain flight. Many fundamental visual processes, such as the delay-and-correlate model for retinal motion detection, were derived from experiments on flies by Werner Reichardt and subsequent investigators. Karl Götz, in particular, advanced the use of what might now be called 'insect psychophysics' to probe the neural processes underlying visually mediated flight behaviors. These experiments involve tethering a fly to a sensitive torque meter and using the forces produced by the insect to close the

feedback loop, rotating a patterned cylinder in place around the animal. This artificial-closed-loop sensory environment allows experimenters to manipulate parameters of both the stimulus and the feedback to determine how visual information is processed in a behaving animal. For example, researchers such as Axel Borst and Martin Egelhaaf have shown that, rather than parsing the world into collections of objects like trees or rocks, basic insect flight control is based only on optic flow – the amount and direction of movement seen by the retina. This is partially due to the poor spatial resolution of the insect eye (something like us viewing the world through a 40×40-pixel camera), but also reflects the minimum requirements to keep an insect airborne. Martin Heisenberg has suggested some forms of shape learning in flies and Mandyam Srinivasan has shown memory for color and texture in bees, but simply maintaining steady flight is a much lower-level goal and can seemingly be achieved using motion cues alone.

Summary

Insects take advantage of both local reflexes and sensors mounted on their heads to generate adaptable movements. Sensory information is processed both locally within the thoracic ganglia and in sophisticated brain regions. Interactions within this hierarchical neural control system then allow insects to quickly adjust their movements according to immediate demands. As a result, they can successfully navigate a wide range of terrestrial terrain and, for many insects, fly with an agility that is unmatched by manmade devices. These control systems, coupled with efficient body mechanics, certainly contribute to their unparalleled success within the animal kingdom.

See also: Insect Navigation; Nervous System: Evolution in Relation to Behavior; Robot Behavior; Robotics in the Study of Animal Behavior; Vision: Invertebrates.

Further Reading

Borst A and Haag J (2002) Neural networks in the cockpit of the fly. *Journal of Comparative Physiology A: Neuroethology, Sensory, Neural, and Behavioral Physiology* 188(6): 419–437.

Büschges A, Akay T, Gabriel JP, and Schmidt J (2008) Organizing network action for locomotion: Insights from studying insect walking. *Brain Research Reviews* 57: 162–171.

Büschges A and Gruhn M (2008) Mechanosensory feedback in walking: from joint control to locomotory patterns. *Advances in Insect Physiology* 34: 194–234.

Frye MA and Dickinson MH (2004) Closing the loop between neurobiology and flight behavior in Drosophila. *Current Opinion in Neurobiology* 14(6): 729–736.

Ritzmann RE and Büschges A (2007) Insect walking: From reduced preparations to natural terrain. In: North G and Greenspan RJ (eds.) *Invertebrate Neurobiology*, pp. 229–250. Cold Spring Harbor, NY: Cold Spring Harbor Laboratory Press.

Ritzmann RE and Büschges A (2007) Adaptive motor behavior in insects. *Current Opinion in Neurobiology* 17: 629–636.

Zill SN, Schmitz J, and Büschges A (2004) Load sensing and control of posture and locomotion. *Arthropod Structure & Development* 33: 273–286.

Zill SN and Seyfarth E-A (1996) Exoskeletal sensors for walking. *Scientific American* 275: 86–90.

Insect Migration

J. W. Chapman, Rothamsted Research, Harpenden, Hertfordshire, UK
V. A. Drake, University of New South Wales at the Australian Defence Force Academy, Canberra, ACT, Australia

Introduction

Although invasive movements of desert locusts were recorded in biblical times, insect migration received relatively little scientific attention before the 1920s and 1930s when systematic observations of the highly visible low-altitude migrations of butterflies and dragonflies by field naturalists began. At about the same time, the development of aviation revealed an unexpected abundance of insects at altitudes of hundreds, and even thousands, of meters during warm weather. It was soon realized that these movements well above the surface were intense enough, and occurred on a large enough scale, to account for the annual reappearance of some species and the infestation each year of newly planted crops by specialist pests. In the 1940s and 1950s, research on the migrations and migratory behavior of agricultural pests, most notably locusts in Africa and aphids in the United Kingdom, bloomed. By the 1960s, this work had led to the development of a conceptual basis, founded in biology rather than natural history, that remains extant and generally valid today. At the end of that decade, which can be seen now to have been a pivotal period for this field, C.G. Johnson published a large and comprehensive volume presenting thematic syntheses that distinguished the individual and collective aspects of migration and related migratory activity to life history, weather and wind systems, and habitat condition. Just a year earlier, in 1968, the first special-purpose insect-detecting radar had been deployed in Africa and had revealed some remarkable and unforeseen phenomena (**Figure 1**). The new capability that radar provided – of direct observation of insects flying at high altitude and at night – opened a new era for insect migration science. The 1970s and 1980s saw a radar-driven surge in field observations that demonstrated the importance of night-time movements and further emphasized significance of the meteorological factors and phenomena on migration distances, directions, and intensities. In parallel with the field observations, the 1980s especially saw the development of laboratory programs to investigate the flight performance of migrant species and the physiological traits enabling these. This research extended, through selection experiments, to establishing that these traits have a genetic basis. In the 1990s and 2000s, the navigational capabilities of insect migrants have been receiving particular attention.

Migration occurs in all major insect taxonomic groups and has evidently evolved multiple times. Insect orders with numerous and well-studied (and often economically important) migrant species include Orthoptera (locusts and grasshoppers), Hemiptera (particularly aphids and planthoppers), and Lepidoptera (butterflies and moths, the latter being the adult forms of various crop pests known as armyworms, cutworms, bollworms, etc.). Some migrants are vectors of disease, and some are parasites of migrant pest species and therefore, beneficial to agriculture. Insect migration has been most studied in Europe (especially Britain and northern Europe), Africa (the desert and sahel zones across the north, and the east-African savannah), east Asia (China and Japan), Australia, North America (especially the southern and eastern United States and eastern Canada), and Central America (Panama). It occurs commonly in these regions, and probably over all larger land areas, when temperatures are high enough to support insect development and flight.

Entomologists have come to view migration as, first and foremost, a behavioral phenomena. The behaviors are understood as being adaptive and as leading to outcomes (displacements) that have significant consequences for individual survival and population development. This article reflects this perspective. It also emphasizes aspects of the migration phenomenon that are peculiar to, or particularly evident in, insects. Finally, it aims to provide some context, both by relating migratory behavior to lower (e.g., physiological) and higher (e.g., ecological) organizational levels and by recognizing that it has evolved through, and is still subject to, natural selection.

What Distinguishes Migratory Movement?

Although movements that are recognizably migratory occur throughout the animal kingdom, their character differs somewhat from one taxonomic group to the next. These differences likely arise from broad constraints imposed by the animal's general form: its size, its lifetime, its mode of locomotion, its cognitive and sensory capacities, and the medium in which it lives. Insects probably sit toward one end of the spectrum on most of these counts, and definitions or assumptions about what constitutes migration that derive from the studies of higher animals

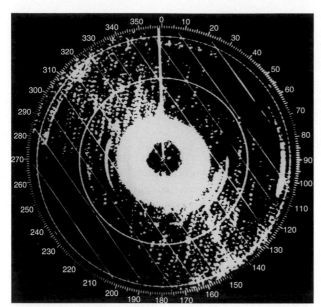

Figure 1 Display of a scanning entomological radar showing a migration of moths at altitudes of 100–200 m over inland New South Wales, Australia. The triple exposure photograph shows steady movement to just east of north, which was the downwind direction. The uneven pattern of echoes around the screen indicates a high degree of common orientation to the northeast. (Targets to the northeast and southwest are head- or tail-on to the radar and produce such a weak echo that they are not detected.) The range rings are at 463-m intervals and the radar beam was pointing upward at an angle of 8°. The ring of white near the center is artefactual. Reproduced from Drake VA and Farrow RA (1985) A radar and aerial-trapping study of an early spring migration of moths (Lepidoptera) in inland New South Wales. *Australian Journal of Ecology* 10: 223–235.

are not always applicable. What perhaps most differentiates insect migration is that there are only a very few cases where a single generation completes an entire round trip. These special characters of insect migration have provoked some insights, including the recognition that behavioral criteria may best distinguish migration from other forms of movement and that the process of migration should be examined primarily at the individual level while its function is to be understood more in terms of population processes and the changing distribution of resources. This latter dichotomy was well summarized by the title of a landmark 1985 paper by J.S. Kennedy: "Migration, Behavioral and Ecological."

Aphids are small insects and have airspeeds so low that their horizontal displacement due to locomotion is essentially negligible. Yet, on warm days they, and many other types of microinsect, are numerous at altitudes of several hundred meters over many terrestrial habitats. They are carried along on the wind and travel tens, even hundreds, of kilometers in a single day. As agriculturalists well know, these movements are very effective at colonizing new habitats (e.g., spring cereal crops), and they likely also provide a means to escape from exhausted or otherwise

deteriorating ones. By the 1960s, the notion that these movements were accidental was no longer being seriously entertained. It was established that the insects initiated their flights by launching themselves into the air and that, while convective up-currents may have helped them gain altitude, they maintained themselves in the wind stream by continuous wing flapping. These sustained flights were therefore, clearly an adaptation, one that enabled exploitation of short-lived habitats and that perhaps also allowed populations to grow to sizes that, in a sedentary species, would have been overtaken by epidemic disease or parasites. Still, movement direction was determined by the wind and, at least over the United Kingdom where these pioneering studies took place, the seasonal back-and-forth (usually approximately north–south) population movements typical of vertebrate migration were not especially evident; some workers, therefore, preferred the term 'dispersal' to describe these microinsect flights.

Some key behavioral experiments undertaken by J.S. Kennedy, and his inspired interpretation of the results, firmly established that not only is migration in insects an active rather than a passive process, but that migratory flight can be recognized by its persistence and by being undistracted. Migrants do not respond to appetitive cues (e.g., food items or sex pheromones) that would normally cause them to stop or change course, and migration trajectories are therefore, steady and straightened out. They are quite distinct from the erratic and intermittent pattern traced out by a foraging individual: that is, one that does respond to localized cues. Migration ends when its characteristic inhibition of responsiveness to appetitive stimuli declines (perhaps as a direct physiological consequence of the act of migrating itself) and the migrant settles. From this behavioral perspective, widely adopted by migration entomologists, the important distinction is between migratory and appetitive (sometimes termed trivial) movement. The term 'dispersal' is now usually reserved for movements that either incorporate some spreading out of a population or that are internal to a population's range and lead to mixing and the occupation of vacant habitat patches.

Migration Syndromes

Migratory behavior depends upon, and must have evolved alongside, a suite of enabling morphological, physiological, and life-history traits. This was first recognized by C.G. Johnson who described the oogenesis-flight syndrome, in which, for the great majority of insect migrants, migratory activity is confined to the short period following metamorphosis into the adult form and before sexual maturation. Flight is by far the most efficient means of long-distance movement available to insects, and only the final adult life stage has wings; but females can be much heavier when gravid than when newly developed and presumably are then much less able to undertake sustained flight. Whether for this or other reasons, for most species, the two primary tasks of a migrant adult female – relocation to a habitat in a suitable condition for the next generation to develop, and fertilization and laying of eggs – are undertaken sequentially. Not uncommonly, the flight apparatus is autolyzed after migration and its materials and energy content converted to egg mass.

What has come to be known as the migration syndrome (a concept developed especially by H. Dingle through his studies of seed-feeding bugs) comprises the entire suite of traits enabling migration – or, when both migrant and nonmigrant individuals occur within the same species, the suite of traits that differ between these forms. In addition to the timing of reproductive maturation, the syndrome includes wing and wing-muscle size, wing polymorphism, deposition and metabolism of flight fuels, flight duration and propensity (and modulation of the latter by environmental cues), and the hormonal processes that largely control all these. Selection experiments have shown that these traits are variable and that variations are genetically correlated in migrant species. Laboratory studies have established that this correlation arises in part through mediation by a common hormonal pathway, that involving juvenile hormone (JH) and its esterase (JHE). Traits not directly involved in migration, such as size, development rate, and fecundity, may also exhibit correlations, many of which presumably arise through trade-offs.

One of the seed-feeding bugs studied by Dingle is also interesting because its populations are generally sedentary, and flight is required only for locating scattered host plants or moving within and between stands of massfruiting hosts. Thus, a migration syndrome is not necessarily associated with movement over long distances or with region-wide deterioration of habitat. Another interesting case peculiar to insects is the autumn movement of the sexual forms (gynoparae) of alternating-host aphids to the woody plants on which they pass the winter. This movement may be over short distances and within the population's range, but it certainly requires flight and some capacity for finding the specific winter host. A focus on the migration syndrome, therefore, leads to a broader perspective on what constitutes migration, or perhaps alternatively leads to the recognition that adaptations required for ecological processes that are commonly regarded as distinct may, in fact, be very similar.

Migration as a Spatial Process

Even if we set aside short-distance flights and confine the term 'migration' to movements that lead to a significant change in the boundary of the region that a population occupies, migration can still take on a variety of spatial forms: to and fro, loop, or erratic – the latter often being

termed 'nomadism.' Because of the short lifetimes of most insects, each generation typically completes only a one-way movement, or even only a segment of one. Thus, the migration of Monarch Butterflies through the United States to their overwintering sites in Mexico is completed by one generation, while the return movement, the following spring, extends over two or even three. One counterexample is provided by the Bogong moth of Australia: in this univoltine species, the adults fly to the mountains and estivate there and then return to the plains (though probably not to the particular region from which they originated) to initiate the next generation on winter-growing host plants.

Nomadic movement patterns have seemed more prevalent in insects than in vertebrate migrants, perhaps because much research on insect migration has been undertaken in semiarid regions where patchy and unpredictable rainfall, rather than regular seasonal changes of temperature, determines host-plant availability. The pioneering work of L.R. Taylor, using a network of traps and map-based analyses, showed that annual patterns of aphid and moth distributions are also sometimes irregular in temperate regions. This led to the perception of a reticulate population structure, in which temporarily distinct subpopulations form and, perhaps only after a number of generations of near isolation, coalesce. However, at higher latitudes, a north–south pattern is usually present, with poleward spring movements taking the migrants out of lower-latitude regions with dry summers and into favorable habitats (most obviously growing crops). The return movement in autumn is often less apparent, but many instances have now been established.

In insects, movements over distances sufficiently great to allow the exploitation of climatic differences almost always require wind assistance, and prevailing seasonal airflows play as important a role in sustaining a migratory population as do seasonal changes of temperature and rainfall. This dependence on wind led to the concept of rectification (the term recognizes an analogy with electrical circuit theory), whereby poleward winds in spring are favorable for migration as they are warm, but the return movement in autumn is suppressed because the equatorward winds required then are cool, with temperatures often below the flight threshold of these poikilothermic organisms. The notion that insects carried to high latitudes and breeding successfully on summer crops (or other seasonal vegetation) there might constitute dead-end populations unable to escape fatal winter cold – termed 'the Pied Piper effect' – has largely been discounted. However, the autumn migration is probably a more chancy process than that in spring and natural selection can be expected to act particularly vigorously on it, fine-tuning the migration syndrome to exploit the few available transport opportunities. A pioneering study by M. Wikelski and colleagues using radiotracking (a technology that at present can be employed only with the largest insect species) has shown that the green darner dragonfly *Anax junius* is stimulated to fly to lower latitudes in autumn by two successive nights of falling temperatures. In subtropical climates, temperature is not limiting and the seasonal back-and-forth pattern, sometimes evident only as a trend in predominantly erratic and reticulate population trajectories, is between winter and summer rainfall regions.

At a more fundamental level, and as recognized particularly by C. Solbreck and T.R.E. Southwood, migration can be understood as an adaptation to resources that are ephemeral and moreover that arise in different places at different times. In a study of 35 species of wing-dimorphic planthoppers, R.F. Denno and colleagues found that long-winged forms were predominant (>50%) in species inhabiting temporary habitats like crops but were infrequent (<1%) in species using freshwater and salt marshes that have existed for thousands of years. As insects often complete several generations in a year, each resource development episode may support breeding; this leads to much higher reproductive rates than will arise from the alternative strategy of a sedentary existence and dormancy until favorable conditions recur. Both strategies, of course, have trade-offs in terms of material and energy expended and risks taken. Migration will often need to be preemptive, with habitats abandoned before they have deteriorated and the Pied Piper trap is closed: such preemption depends on behavioral responses and the inhibition of responsiveness to appetitive stimuli so emphasized by J.S. Kennedy.

The Biometeorology of Insect Migration

Insect airspeeds in the sustained flight required for migration are typically $3-6 \, m \, s^{-1}$ for larger species (grasshoppers, butterflies, and larger moths) and $\sim 1 \, m \, s^{-1}$ for microinsects, and above the vegetation canopy, these are often less than the wind speed. Friction due to the surface leads to an increase in wind speed with height through the lowest few hundred meters of the atmosphere. L.R. Taylor recognized that for each species there will be a flight boundary layer (FBL), within which individuals are able to make their way in any direction (albeit only slowly as the wind speed approaches the insect's airspeed), but above which only movement with a downwind component is possible. Movement with an upwind component is best effected at as low an altitude as practicable, to take maximal advantage of the frictional slowing of the air. Movement in an approximately downwind direction will be most efficient above the FBL, where the migrant may need to expend only the energy needed to keep itself aloft. Wind speeds of $10-20 \, m \, s^{-1}$ occur frequently at altitudes of only a few hundred meters and can carry insects between climate zones – that is, over distances of the order of 1000 km – over a few days or nights of flight. Nevertheless, some strong-flying and longer-lived species,

notably butterflies, do migrate in opposing winds, often in daylight, and their low-altitude flights are highly visible and amenable to study. The more opportunistic downwind movements above the FBL, whether by day or by night, are much less apparent and have been investigated mainly through trapping and with entomological radar. Microinsects flying at altitude are sometimes referred to as 'aerial plankton,' but these flights are not continuous and are actively initiated (and probably also actively terminated), so the analogy with oceanic drift is not close and the term is probably better avoided.

An early biometeorological contribution to insect migration science was the realization, by R.C. Rainey, that flights by locusts in the winds north and south of the intertropical convergence zone would carry the locusts to, and maintain them within, a region receiving rain. Although horizontal inflows of air into a convergence are approximately balanced by an ascending outflow, insects generally will not rise far, either through an active behavioral response or due to passive cessation of locomotory activity when ascent takes them to altitudes where temperatures (which normally decrease quite rapidly with height) are below their threshold for flight. Thus, the migrants will become concentrated, and in locusts, this can lead to the development of swarms. Radar observations have shown that concentrations also form at more localized and shorter-lived convergences, such as those occurring in small-scale wind systems like such as sea breezes and thunderstorm outflows; again, some behavioral response presumably inhibits ascent and causes the insects (but not the air that is conveying them) to accumulate at low altitudes near the front (that is, the interface of the two flows). On sunny days, convective air movements usually develop, to a depth of about 1 km, over land areas. Most daytime migration begins around mid-morning just as these vertical flows are becoming strong (**Figure 2**). Insects are carried into updrafts by the convergent airflows around their bases, and become concentrated in the updraft zone through some response that prevents them from being carried up to the region of divergent flow around the updraft's top. Remaining in an updraft obviously conveys an advantage in reducing the wing-flapping effort, and thus, the energy expenditure, required to remain aloft. In models, behavioral responses to ascent and descent seem to reproduce observed concentration patterns and mean profiles (variation with height) better than a simple cut-off of locomotion at the height of the species' temperature threshold.

In the fair-weather conditions generally favorable for migration, convection ceases before nightfall and the surface becomes cool and in turn cools the air immediately above it. The night-time lower atmosphere is thus, stratified, with temperatures increasing with height (an inversion) to a maximum somewhere between 100 and 400 m; wind speed is also reduced in the inversion zone, to which the slowing effect of surface friction is confined. Except when an ephemeral disturbance (e.g., a thunderstorm outflow) passes through, vertical air movements are essentially absent. In these circumstances, radar observations almost invariably show some degree of aggregation of insect migrants around particular altitudes, sometimes in quite narrow layers (**Figure 2**). Layers often form near the top of the inversion, where temperatures are highest; in warmer conditions, when temperatures are apparently not limiting, they also occur at greater heights where winds are faster or in different directions. These observations indicate behavioral responses to environmental variables and possibly also to visual cues from the surface, but the details remain to be elucidated. The adaptive benefit, however, appears clear: maximal utilization of the transport opportunity and possibly some control over the direction.

Night-time migration is generally favored by the larger species that exploit wind transport – principally orthopterans (grasshoppers and locusts) and noctuid and pyralid moths. Migration commences at dusk with an ascent flight to the cruising altitude, which is usually between 100 and 1500 m (**Figure 2**). Flight durations vary from an hour or two through to the whole night, but – except when the migrants find themselves over the sea – nearly always cease at dawn. Termination of flight is probably often due to poor condition (low fuel reserves) or environmental inclemency (e.g., low temperatures). However, behavioral responses, either to external cues or internal physiological changes arising from the act of migration itself, may also be involved. This stage of migration is particularly difficult to study in the field, and little is known about termination behaviors even though their importance for postmigration survival and reproductive fitness seems obvious.

Orientation and Navigation

The low-altitude daytime migrations of butterflies and dragonflies are typically highly directed and occur in predictable seasonal directions. This indicates that these insects possess one or more compass orientation mechanisms

and a preferred heading that changes with the season in a beneficial manner. The best-known example is the Monarch Butterfly of eastern North America, which migrates up to 3500 km southwestward in autumn to its communal hibernation site in central Mexico. While Monarchs have been observed ascending to considerable altitudes when winds are favorable, much of this migration, and the return movements in spring, occur within the FBL. Experiments by H. Mouritsen and B.J. Frost with tethered Monarchs in a flight simulator have established that during the autumn migration, a time-compensated solar compass is used to select and maintain the preferred migratory heading; whether this species also possesses a magnetic compass that would allow them to keep oriented in completely overcast conditions, is a subject of debate.

A time-compensated solar compass has also been demonstrated for two neotropical butterflies, which move between forests on the Atlantic and Pacific coasts of the Isthmus of Panama around the onset of the rainy season. R.B. Srygley and R. Dudley tracked the flight of these and other insect migrants over a large expanse of water by following them in a boat and measuring both flight and wind parameters. They have shown that some species can fully compensate for wind drift by using visual landmarks on the horizon. Females of one species flew higher when benefiting from a tailwind, and for four species, airspeeds were lower when relative lipid content (i.e., fuel load) was lower, which is consistent with a strategy of regulating energy expenditure to maximize the distance covered per unit of fuel. This research has revealed that butterflies have evolved specific migratory behaviors that are reliant on sophisticated sensory systems.

Radar observations of nocturnal migrations of larger insects flying at altitudes of hundreds of meters very frequently show that the insects are exhibiting some degree of common orientation (**Figure 1**). Convective updrafts and downdrafts may disrupt this behavior during the day, but the stable night-time atmosphere clearly allows an insect to maintain its orientation. The spread

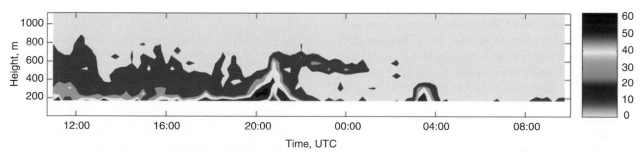

Figure 2   Time/height plot of the number of windborne insect migrants recorded by a vertical-looking entomological radar in southern UK from 11:00 h on 25 June to 10:00 h on 26 June 2003. The color key refers to the number of resolvable insects detected at each of the 15 sampling altitudes between 190 and 1100 m above the ground, during 5-min periods scheduled every 15 min. Numbers of diurnal migrants decline with altitude during the daytime, but in contrast at night, a dense altitudinal layer at about 600 m forms from the intense dusk (around 20.30 h) take-off of nocturnal migrants. There is also a discrete dawn (around 03.30 h) take-off, which did not lead to prolonged migration on this occasion.

of orientations is often quite broad, but there is usually a clear maximum. The mean direction often varies with height, and is sometimes closely aligned to that of the wind, which results in the insects adding their airspeed to the windspeed and achieving the maximal displacement distance possible from flight at that particular height. As an organism carried along in a stream cannot determine the flow speed and direction directly, these migrants must be able either to assess their movement relative to the ground (in darkness, from altitudes of hundreds of meters) or to determine it somehow from anisotropies within the flow itself, most likely small-scale turbulence associated with wind shear. As wind shear is sometimes concentrated in shallow zones, it may also be related to some forms of layering (see earlier); common orientation is indeed often present in layers, though this is unsurprising as both phenomena are of frequent occurrence. The average distance between migrants is usually so great that visual interactions with other migrants can be discounted as an orientation cue.

Using season-long observations from an entomological radar, J.W. Chapman and colleagues have shown that the moth *Autographa gamma*, when migrating over the United Kingdom at night, uses a sophisticated suite of behavioral mechanisms to achieve long-distance movements in seasonally adaptive directions. Firstly, the moths use a compass (probably magnetic) to select favorable tailwinds, and secondly, they migrate predominantly at the height of the fastest airstreams (200–1000 m). Thirdly, they orientate approximately downwind, and lastly, they partially compensate for crosswind drift from their preferred headings (toward the north in spring and the south in autumn). These strategies in combination can result in movements of 600 km per night, which is sufficient to carry them between a summer-breeding area in the British Isles and a winter-breeding area in the Mediterranean Basin.

Concluding Comments

Insect migration researchers today perceive that migration encompasses all levels of biological organization, from the genetic through the developmental/morphological, physiological, and behavioral to the ecological and evolutionary. They view migratory activity, and the capacity to undertake it, as an adaptation that has evolved through, and moreover is maintained by, the action of natural selection on migration outcomes (and nonattempts). This perspective indicates a central role for behavior in the phenomena they study, because it constitutes the expression at the individual level of the broader migratory adaptation which includes the morphological and physiological traits that enable it. It is behavior – migratory activity and its modulation in response to environmental cues – that directly influences migration outcomes and thus, is acted on most immediately

by natural selection. However, despite this appreciation of its importance, and of the impact that J.S. Kennedy's pioneering flight-chamber experiments on aphids have had on our understanding of migration, direct investigation of migratory behavior through stimulus-response experiments has been minimal. The recent development, by H. Mouritsen and B.J. Frost, of flight simulators that can be used to investigate the compass of butterflies migrating in the FBL is therefore, welcome. Most laboratory investigation of migratory flight has been concerned with biomechanics or aerodynamics, or with determining maximum durations. The greater part of our knowledge of insect migratory behavior has been inferred from the observations of migration patterns in natural populations, and while this is valid and often insightful, it needs to be complemented by experimentation in controlled conditions.

See also: Insect Navigation; Magnetic Compasses in Insects.

Further Reading

Chapman JW, Reynolds DR, Mouritsen H, et al. (2008) Wind selection and drift compensation optimize migratory pathways in a high-flying moth. *Current Biology* 18: 514–518.

Dingle H (1996) *Migration: The Biology of Life on the Move.* New York: Oxford University Press.

Dingle H and Drake VA (2007) What is migration? *Bioscience* 57: 113–121.

Drake VA and Farrow RA (1988) The influence of atmospheric structure and motions on insect migration. *Annual Review of Entomology* 33: 183–210.

Drake VA and Gatehouse AG (eds.) (1995) *Insect Migration: Tracking Resources through Space and Time.* Cambridge: Cambridge University Press.

Gatehouse AG (1997) Behavior and ecological genetics of wind-borne migration by insects. *Annual Review of Entomology* 42: 475–502.

Johnson CG (1969) *Migration and Dispersal of Insects by Flight.* London: Methuen.

Kennedy JS (1985) Migration: Behavioral and ecological. In: Rankin MA (ed.) *Migration: Mechanisms and Adaptive Significance. Contributions in Marine Science,* vol. 27 (supplement), pp. 5–26. Port Aransas, TX: Marine Science Institute, The University of Texas at Austin.

Mouritsen H and Frost BJ (2002) Virtual migration in tethered flying monarch butterflies reveals their orientation mechanisms. *Proceedings of the National Academy of Sciences of the United States of America* 99: 10162–10166.

Rainey RC (1989) *Migration and Meteorology. Flight Behaviour and the Atmospheric Environment of Locusts and other Migrant Pests.* Oxford: Oxford University Press.

Roff DA and Fairbairn DJ (2007) The evolution and genetics of migration in insects. *Bioscience* 57: 155–164.

Srygley RB and Dudley R (2008) Optimal strategies for insects migrating in the flight boundary layer: Mechanisms and consequences. *Integrative and Comparative Biology* 48: 119–133.

Wikelski M, Moskowitz D, Adelman JS, et al. (2006) Simple rules guide dragonfly migration. *Biology Letters* 2: 325–329.

Woiwod IP, Reynolds DR, and Thomas CD (eds.) (2001) *Insect Movement: Mechanisms and Consequences.* Wallingford, UK: CABI.

Wood CR, Chapman JW, Reynolds DR, et al. (2006) The influence of the atmospheric boundary layer on nocturnal layers of noctuids and other moths migrating over southern Britain. *International Journal of Biometeorology* 50: 193–204.

Insect Navigation

P. Graham, University of Sussex, Brighton, UK

Introduction

The possibility that insects learn the spatial layout of their environment had been dismissed by some as mere 'anthropomorphic delusion,' before experiments by early twentieth century ethologists, such as Jean-Henri Fabre and George Romanes, showed that insects were indeed capable of learning about and navigating around a familiar environment. One elegant example from Romanes involved taking a hive of bees, relocating them to a house, and then allowing them to forage freely from that location. After a period during which bees foraged, he captured a cohort of foragers and transported them ~250 m to a flowerless cliff top. From this location, where bees were unlikely to have foraged, no individuals found their way home. In contrast, all bees released from a flowered garden, also 250 m from the hive, successfully returned. Romanes had shown simply that rather than any arcane 'spatial sense,' it was the experience of places that was necessary for successful homing. When an animal moves to a goal location in this way, it can be classified as true navigation, because the animal must calculate the current location relative to the goal before it can move accordingly. In contrast, equally remarkable spatial behaviors such as long-distance migration or chemotaxis, need only involve the moment-to-moment alignment of an animal's body using sensory feedback and are classified as orientation.

In this article, we focus on the mechanisms that underpin true navigation in insects. We begin with a discussion of path integration, an innate navigational strategy that enables an animal to return to the starting point of a route. We then discuss how insects learn about the landmarks within their environment and use this knowledge to guide complex routes. Insects are capable of impressive feats of navigation using only this simple toolkit of innate behaviors and learned landmark information. Yet, the smooth operation of these navigational strategies requires sophisticated cognitive mechanisms, and we end the article with a discussion of how insects organize the large set of memories required for navigation.

As is evident from this article, our knowledge of navigation in insects is almost exclusively drawn from the study of central place foragers, predominantly, the hymenopteran social insects. Unfortunately, we know much less about the navigation of other insects, but there are strong suggestions that the general mechanisms discussed here are likely widely applicable.

Path Integration

In order to fully exploit environmental resources, an animal must leave her nest and occasionally venture into new territory, from where she must safely return to the starting point of the journey. This is a basic requirement for an animal navigator, and the general mechanism, which is shared by most animals, is called path integration (known to sailors as dead reckoning). Path integration (PI) involves monitoring the orientation and length of journey segments and integrating this information to maintain a continuous estimate of the distance and direction of the direct line back to the starting point of the route (**Figure 1(a)**). Therefore, at any time, such as when a food item is located or a predator attacks, the animal can take the quickest route home. In addition to guiding a direct route home, information acquired using PI can be used to inform others of the location of a food source. This is seen in the remarkable waggle dance of the honeybee, which is performed by a forager upon returning from a profitable food source. The dances (**Figure 1(b)**) were decoded by Karl von Frisch, who discovered that the orientation of the waggle runs, relative to gravity, indicates the direction of the goal relative to the azimuthal position of the sun, while the distance to the goal is strongly correlated with the waggle duration. Following von Frisch, scientists were able to read the dance of a returning forager, thus giving an insight into an insect's mind. We will see in subsequent sections how reading the dance has been an important tool for investigations of the mechanisms of navigation.

To perform path integration, an insect needs three things: a compass to measure orientation, an odometer to measure distance or speed, and neural machinery to iteratively perform the path integration calculation. As yet, we know little about how insects' brains perform the PI calculation, though we do know about the compass and odometer mechanisms used by walking and flying insects.

Odometric Mechanisms

There are three possible ways for an insect to measure the distance it has traveled: a proprioceptive mechanism that monitors the movement of the insect's legs or wings; a system that monitors energy usage during a route; or, a sensory mechanism monitoring the consequences of movement, for instance, the optic flow experienced

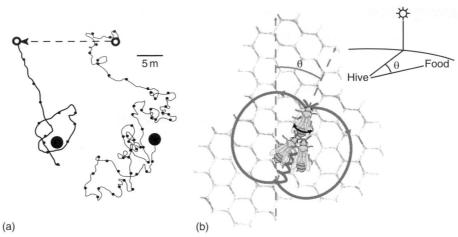

(a)   (b)

Figure 1 (a) Path integration in desert ants. The outward foraging path of a desert ant is shown until she finds food. Before the ant can make a return trip, it is displaced. The homeward path is parallel to the route from food to nest indicating the use of an egocentric estimate of position rather than landmark information. Filled circle denotes nest and fictive nest, open circle food Reproduced from Wehner R, Boyer M, Loertscher F, Sommer S, and Menzi U (2006) Ant navigation: One-way routes rather than maps. *Current Biology* 16(1): 75–79. (b) The waggle dance of the honeybee. A returning forager performs a waggle dance on the vertical honeycomb in order to communicate food location to other bees. The dance is made up of a waggle runs followed by loops of alternating direction. The direction to the food is signaled by the orientation of the waggle run relative to gravity which represents the bearing the recruit should take relative to the current azimuthal position of the sun (inset). Distance to the goal is signaled by the duration of the 'waggling' during the waggle run.

during a route. The most appropriate odometric mechanism for a given species will vary as a consequence of their mode of locomotion.

For a walking insect that is in contact with the ground, a reliable estimate of the distance traveled can be retrieved from monitoring its own leg movements. This 'step-counting' hypothesis was verified for ants by manipulating the lengths of their legs before they were allowed to make a PI guided return to their nest. Foragers were allowed to find a feeder at the end of a long channel and from there they were transported to a test channel. When control ants are released in this channel, they walk the distance that would have ordinarily taken them back to their nest, before starting to search. Two further groups of ants were subjected to a delicate surgery before being placed in the test channel. Ants with shortened legs searched at a reduced distance and ants with their legs lengthened with pig bristle, overshot the fictive nest position. In each of these three conditions, the ants' return journey consisted of approximately the same number of strides. So, we can conclude that ants use a proprioceptive odometer as also used by other terrestrial invertebrates such as spiders and crabs.

For flying insects, the influence of air movement means that the attempts to measure distance in terms of motor output or energy usage may be inaccurate, although, following the ideas of von Frisch for a long time, the predominant theory of odometry for flying insects was one based on energy consumption. It took a simple experiment by Harald Esch and John Burns to overturn the energy hypothesis and suggest an alternative. They

trained bees to forage at a feeder in an open field and then recorded the distance signaled by the dances of returning bees. Over time, the feeder was raised above the ground, so bees had to fly further and use more energy flying against gravity. However, the distance signaled in their dances reduced, leading Esch and Burns to suggest that bees' distance estimates may depend on the degree of visual motion generated by their flight. Visual motion, also known as optic flow, is generated by an animal's movement as the images of environmental objects move across the retina. The degree of visual motion depends on the animal's speed and the distance between the animal and environmental objects. As Esch and Burns forced bees to fly higher and higher to the elevated feeder, the amount of perceived visual motion dropped because bees were flying further from the ground. This led to the 'optic flow' hypothesis: that a bee's estimate of distance depends on the amount of perceived visual motion rather than its energy usage.

The optic flow hypothesis has been rigorously tested by Mandyam Srinivasan and colleagues who trained bees to fly in small tunnels with high contrast stripes on the walls. One of their earliest experiments shows the influence of the optic flow on the bees' perception of distance. Bees were trained to find food at a fixed distance along a tunnel which had a radial (perpendicular to the direction of flight) stripe pattern on the walls and floor. Bees were able to learn the food distance and would search persistently at the correct distance when tested in a feederless fresh tunnel. However, when the radial stripes were replaced with axial stripes (aligned with the direction of flight), bees showed no focused search pattern. The axial pattern provides no optic flow relative to the direction of flight, making distance estimation impossible.

The fact that flying insects use optic flow to measure distance creates a problem for our perception of the waggle dance, as the dance cannot signal an absolute distance but only the amount of optic flow experienced along a route. This will depend on the proximity and density of objects along the flight path, so the relationship between the waggle duration and the absolute distance varies as a function of the environment within which a particular colony forages. However, this is not a problem as long as the bee following the dance takes a similar path through the environment as did the dancer.

Compass Mechanisms

For many insects, the sun plays a major role in providing compass information, as is evident from the waggle dance of honeybees. To use the sun as a compass, insects must solve two fundamental issues: Firstly, the sun moves during the day, and secondly, the sun is often hidden by clouds. Insects can compensate for the daily movement of the sun by learning its position relative to the time of day. Insects have some innate knowledge of the sun's movements, they know that the sun's position changes slowly near dawn and dusk and more quickly around noon. Combining this knowledge with observed sun positions relative to stable environmental landmarks allows an insect to learn an accurate function describing how the sun moves throughout the day. We know about the innate knowledge of bees following experiments where new foragers have their experience of sun position restricted to the morning. In orientation tests in the afternoon, the insects show that they have fitted their morning experience of the sun's position to a step-shaped template which includes information about the rapid sun movement around noon and slower sun movement late in the day.

When the sun is obscured by clouds, but portions of blue sky remain, insects are able to derive compass information from the polarization patterns created by the scattering of sunlight in the upper atmosphere. The orientation of polarized light forms concentric circles around the sun's position, and if an insect knows the time of day, she can retrieve compass information from any patch of blue sky. Insects are, therefore, able to use the sun as a compass even when it is not in view. To detect these polarization patterns, most diurnal insects' compound eyes have special dorsal areas that are sensitive to the direction of polarized light. Of course, there will be overcast days when no celestial compass information is available. On such days, bees use the same prominent landmarks they used as references when learning about the movement of the sun. Other insects are able to take a

more leisurely approach on overcast days, the desert ant *Cataglyphis* remains inside the nest on those rare days when there is no sun.

Path integration allows insects to explore unfamiliar terrain while being connected to the starting point of their journey by the distance and the direction information required for a direct route back to the start. What is more, insects can store the PI co-ordinates of a profitable location and use PI to guide a subsequent return or signal that location to a nestmate. With such an elegant mechanism to guide insects between important locations, why should they use any other navigational strategy? The answer lies in the fact that PI is an egocentric estimate of position and so, small errors will accumulate throughout a route. Therefore, upon the completion of a PI-guided homeward trajectory, the insect may not be at the goal as expected. An inexperienced animal has no choice but to search systematically for the goal. However, if an insect is familiar with a location, it can use terrestrial landmarks to guide its search and correct for any errors accrued during the path integration process. In the next section, we look at the mechanisms and uses of landmark guidance.

Using Landmarks to Pinpoint a Goal

Terrestrial landmarks provide a stable geocentric reference by which animals can define a location. This was demonstrated by Niko Tinbergen in his famous digger wasp experiment (**Figure 2(a)**). Tinbergen identified a digger wasp nest and placed a ring of pine cones around the entrance. When the digger wasp departed, she inspected the nest surrounds before leaving on her foraging trip. While the wasp was away, Tinbergen relocated the pine cone ring, and on her return, the wasp searched at the center of the relocated pine cone ring even though the real nest entrance was only centimeters away. Tinbergen concluded that the information learned about the nest entrance's position relative to the pine cones dominates over any directly perceptible odor or visual cues from the entrance itself. In natural situations, because landmarks are usually stable, learning how a goal location relates to the surrounding landmarks mitigates the risk that cumulative errors from PI will lead to missing the goal.

View-Based Navigation

We now know much about how insects use visual landmarks to define a goal location. Evidence from ants, flies, and solitary and social bees and wasps suggests that places are represented as 2D retinotopic images of the world as seen from that place. We can illustrate the basic phenomenon with recent data from wood ants (**Figure 2(b)**). Ants were trained to find food at a location defined solely by two cylindrical landmarks. In tests, with no feeder present,

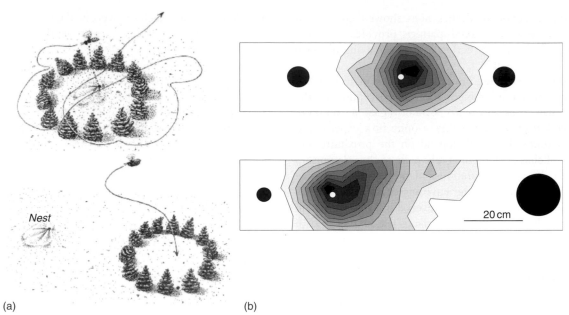

(a)   (b)

Figure 2   (a) Tinbergen's digger wasp. Upper: Tinbergen placed a circle of pine cones around the nest of a solitary digger wasp. This change in surroundings triggers a period of observation before she finally departs the nest locale. Lower: During the wasp's foraging trip, Tinbergen moved the ring of pine cones, resulting in the wasp searching for the entrance in the wrong location. Reproduced from Tinbergen N (1951) *The Study of Instinct*. Oxford: Clarendon Press. Reprinted with the permission of Oxford University Press. (b) Goal localization using snapshots. Upper: Ants are trained to find food midway between two cylinders (white circle) and their search distribution is concentrated on that spot when the feeder is missing. Darker areas denote the regions where ants spent the most time during search. Lower: Training cylinders are replaced by one cylinder that is smaller (in height and width) and one that is larger. Ants' searches are focused at the location where the cylinders look the same as they did from the feeder in training (white circle). Reproduced from Graham et al. (2004).

ants search mostly at the training location. In further tests, with one small and one large cylinder, ants show a search distribution biased towards the small cylinder. The peak of this new search distribution is at the location where the small and large cylinders have the same apparent size as did the regular cylinders from the food location during training. Using view-based matching to find a goal is an economical navigational strategy as it does not require the computation of the absolute distance to objects.

Stored retinotopic views of the world from goal locations are commonly known as snapshots. Snapshots are made up of the retinal positions of a set of visual features such as apparent size, orientation of high contrast edges, color, and vertical center of gravity. Importantly, snapshots are not only used to identify when an insect is at the goal location. Insects also use the difference between their current view of the world and their stored snapshot to derive a movement direction. Identifying the differences between the current view of the world and the stored snapshot is made easier if the insect faces in the same direction as when the snapshot was stored. Bees and wasps can fly in any direction relative to their line of sight and can therefore fix a body orientation, using their compass mechanisms, thus aligning current view with the stored snapshot before moving to reduce the discrepancy between the two. The process of snapshot alignment is

less clear for walking insects because they can only translate in the direction of their body axis and cannot maintain a fixed orientation. There is some evidence that ants align snapshots by fixating conspicuous landmarks or using compass information. However, these mechanisms can align snapshots only temporarily, and it is not yet clear what ants do when the current view and goal snapshot are not aligned.

Learning About a Goal Location

When Tinbergen's Digger wasp left her nest, she initiated a period of observation of the nest locale prompted by the conspicuous change in nest surroundings. This was an example of a learning flight, a type of predictive learning, where insects anticipate what information will be useful in the future and use stereotyped learning behavior to acquire that information. Tinbergen's rough sketch did not capture the detailed structure of this flight, but using film and video technology, further experiments have revealed the fine structure of the similar learning behaviors of other wasp and bee species.

When an individual bee or wasp leaves an important location, be it nest, feeder, or even a parasitized host, she turns to face that location and moves backwards and upwards while flying in arcs of increasing size (**Figure 3(a)**).

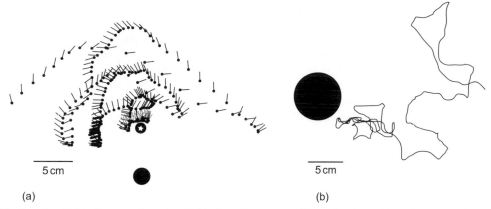

Figure 3   (a) The learning flight of a wasp. Learning flight of a solitary wasp. (*Cerceris rybyensis*) as she leaves the nest Circles with tails denote the wasp's position and orientation every 20 ms. The star shows the nest entrance, and the filled black circle indicates a small cylinder placed nearby. Reproduced from Collett TS and Zeil J (1996) Flights of learning. *Current Directions in Psychological Science* 5(5): 149–155. (b) The learning walk of a wood ant. Departure walk of an ant after visiting a feeder at the base of a black cone. Reproduced from Nicholson et al. (1999).

This provides a prolonged period of observation from the point of view that she will adopt when returning to the goal. Because bees do not have stereoscopic vision, they have to generate 3D depth structure from the motion parallax generated by translatory movements. The arcing structure of the flight is, therefore, suited to determining the relative distance of the landmarks surrounding the goal and identifying those close landmarks which are the most reliable for navigation. The apparent paradox between a snapshot, which does not contain any absolute depth or distance information, and the learning flight, which seems ideally structured to extract depth information, was addressed by Miriam Lehrer and Tom Collett. They trained bees to collect a reward from a feeder that was located in a fixed position relative to a cylinder. Probe tests demonstrated that bees learned both apparent size (a snapshot) and also the absolute distance between the landmark and the feeder. Following further tests, it was shown that the absolute distance provides the primary cue during the initial phase of learning, with the absolute size becoming more important later on. Later, when the close-by landmarks have been reliably identified, the bee can rely on a snapshot which contains only the apparent size of those landmarks. As the insect gains experience, the duration of learning flights drops off and eventually, the insect will fly directly away from the goal. However, flights will recommence at the start of each day, if local landmarks are changed or following a difficult inbound trip.

Early in their foraging careers, bees also gain experience of the large-scale environment around their nest by undertaking a series of flights in which they do not collect nectar or pollen. The so-called survey flights seem to consist of a series of loops with relatively direct trajectories away from and returning to the hive. This structure would seem suited to learning about the routes back to the hive from surrounding areas, and it has been shown that

after a single survey flight, bees are able to fly directly home from release sites within a direct line of sight of the hive. The time and effort invested in learning and survey flights highlights the importance of gaining accurate information about local landmarks. This advantage also applies to ants which have been observed performing a similar behavior (**Figure 3(b)**), referred to as learning walks. These maneuvers are similar to learning flights in terms of inspecting the goal area and surroundings, but as yet, we do not have a functional analysis of learning walk behavior.

Habitual Routes

Insects could theoretically navigate using a toolkit of path integration augmented by snapshot guidance near a goal. However, they do not restrict themselves to this simple procedure, but also build extensive knowledge of their environment by learning sets of instructions that can guide long and complex idiosyncratic routes. As with place learning, route learning depends on the sophisticated interaction of innate behavior and predictive learning. Additionally, the organization of the large sets of memories required for route navigation highlights the cognitive abilities of insects.

The consummation of the route learning process is a set of procedural instructions associated with visual landmark information. **Figure 4(a)** gives an elegant demonstration of the properties of such a learned route. Martin Kohler and Rudiger Wehner allowed individual Australian desert ants to learn a route between their nest and a food source. Individual foragers showed idiosyncratic and stable routes through the scrub and grass tussocks. Experienced ants were then taken from the feeder or from near their nest and relocated to the midpoint of their habitual route.

In both cases, they accurately reproduce the second half of the route. From this demonstration, we can infer several of the properties of visually guided routes. Firstly, even with its low resolution eyes in a world of similar objects, the ant is able to identify its location. Secondly, this knowledge has to be accessible independently of path integration or sequence information. Finally, the ant knows which way she is going and reproduces the second half of her homeward rather than her foodward route. These properties of routes come about because of the way navigational memories are organized in the insects' brain, and this is discussed in the section 'Organization of Spatial Memories.' In the remainder of this section, we look at the types of procedural information used to guide routes as well as the strategies that insects have for the rapid learning of route information.

Mechanisms for Route Following

Experienced foragers demonstrate in their habitual routes a variety of ways of utilizing visual landmark information. In the section 'Using Landmarks to Pinpoint a Goal,' we showed how snapshots can be used to navigate to a single goal location, and snapshot guidance can similarly be used to guide insects to subgoals along a route. However, routes do not generally require the same level of precision along their entire length as they do at the end. Therefore, simple procedural instructions, such as taking the correct direction at a recognized location, will ensure that an insect stays on course.

Two simple procedural mechanisms have been identified from the studies of the North African desert ant *Cataglyphis*. Firstly, if navigating in cluttered terrain, routes can be described as a series of detours around recognized landmarks.

Studies have shown that ants will learn the appearance of landmarks close to a route. They will then make appropriate detours to put the landmark on the same side of their path as they experienced during route learning, although they do not take a precise route to exactly match the retinotopic appearance of the landmark experienced in training. The second procedural mechanism for route guidance comes from associating compass directions with defined locations. In the experimental example from **Figure 4(b)**, ants were trained to take an L-shaped route from a permanent feeder back to their nest. The first part of the route was along a channel, and from the end of the channel, ants would head due south to the nest. Experienced ants have learned to associate the end of the channel with the southerly direction habitually taken there. This association of a direction with a salient location is called a local vector. If the channel is shortened, ants will have conflicting information at the end of the channel. Their local vector will point due south but their path integration system will point in a south westerly direction. We see that the procedural information 'wins out' and ants follow their local vector, though the path integrator continues to calculate the homeward direction, and after a while, the ants switch to following the direction set by PI.

The Scaffolding of Route Learning

In order to accurately and quickly learn route information, insects should establish a consistent route shape as soon as possible. As the shapes of routes are determined by path integration and an insect's innate responses to objects, innate behaviors play an important role in learning by ensuring that a naïve individual takes similar routes on her early foraging trips. We can thus say that the

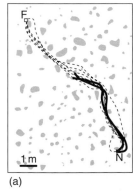

(a)

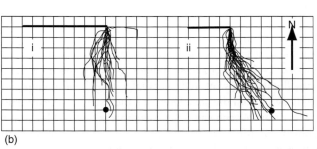

(b)

Figure 4   (a) Idiosyncratic foraging routes. Australian desert ants (*Melophorus bagoti*) complete their habitual route when placed at the midpoint after being taken from the feeder (F) or close to the nest (N) Dotted lines show normal homeward trajectories. Black lines show test paths. Gray areas depict grass tussocks Reproduced from Kohler M and Wehner R (2005) Idiosyncratic route-based memories in desert ants, *Melophorus bagoti*: How do they interact with pathintegration vectors? *Neurobiology of Learning and Memory* 83(1): 1–12. Reprinted with the permission of Elsevier. (b) Local vectors. Ants were trained on an L-shaped homeward path consisting of 8 m in an open topped channel (thick line) and 8 m over open ground to the nest (●). i: The trajectories of ants taken from the feeder and placed at the end of a test channel on a novel test ground. ii: Trajectories from ants taken from the feeder to the end of a 4 m channel. Reproduced from Collett M, Collett TS, Bisch S, and Wehner R (1998) Local and global vectors in desert ant navigation. *Nature* 394: 269–271.

innate behaviors act as a scaffold for learning, providing a consistent route shape allowing the insect to confidently learn landmark information from along the route. The key outcome of this process is that experienced foragers can guide their routes independently of the original innate behavior which determined the route shape.

Path integration provides one obvious mechanism for ensuring a consistent route across multiple foraging runs. Another is the innate attraction that some insects show towards conspicuous visual landmarks (beacons). In an illustrative experiment with wood ants, individual foragers were allowed to learn routes with a large landmark placed laterally to the direct path between the start-point and the feeder. Mature routes were biased towards this beacon, which became an intermediate goal on the route to the feeder. In tests, however, experienced ants performed the same routes even when the beacon was removed or displaced, showing that they had learned other visual cues to guide their route. Although the beacon determined the shape of the route, routes were robust to the removal of the beacon.

Organization of Spatial Memories

An insect's navigational repertoire consists of a set of simple behaviors, including path integration, view-based homing, and local vectors. In combination, these simple behaviors enable insects to learn and robustly perform complex foraging routes of many thousands of body lengths. What is more, individuals can learn multiple routes as necessitated by cyclical or seasonal changes in resource availability. Therefore, an individual forager not only has to organize the large set of procedural instructions that define a single route but must also have multiple sets of route instructions that lead to different locations. The efficient organization and accurate recall of route memories is critical for navigation, and the studies of these processes highlight the impressive cognitive performance of an insect's small brain.

Organization of Memories for a Single Route

A single route through cluttered terrain is built from many instructions associated with specific locations or landmarks (route marks) of which accurate identification is essential. Yet, natural landmarks, such as trees, shrubs, bushes, and grass tussocks, may look similar to insects' low-resolution eyes. To lessen the chance of misidentification leading to the recall of an inappropriate route instruction, insects bind together memories of route marks with contextual information from larger and more distant landmarks. As an insect moves along a route, the appearance of distant landmarks changes more gradually than the appearance of local route marks. These distant landmarks can therefore be used as contextual cues, reducing the set of possibilities about the insect's current position and simplifying the task of recalling the appropriate route instruction for the current route mark. The binding together of information about route marks with information from larger landmarks has been shown for both ants and bees when a spatial task requires an individual to treat identical landmarks differently in two contexts.

One alternative mechanism for recalling the appropriate instruction for the currently perceived landmark would be to store route memories as a rigid sequence of instructions where the performance of one action primes the recall of the next instruction in the sequence. An internal mechanism of this type may not be necessary when considering spatial behavior as the route sequence already resides in the external environment. Additionally, we have seen that the Australian desert ant is able to perform her routes independently of a rigid sequence (**Figure 4(a)**). However, experiencing route landmarks in the correct sequence does seem to have some effect. Lars Chittka and colleagues trained bees to fly along a route marked by a series of conspicuous tents. Compressing the distance between the tents prompted bees to search for the feeder after a shorter distance than usual. This showed that the bee's expectation of finding the feeder is not just triggered by landmarks close to the goal but by perceiving those goal landmarks after experiencing the route landmarks in the correct sequence.

Organization of Multiple Route Memories

Although insects show flexibility in their ability to access route memories out of sequence, the individual must ensure that only those memories associated with the current route are available to be accessed. In the simplest instance, a central place forager with experience of its environment will have two routes through it: an outward route to a food source and an inward route back to the central place. An elegant experiment by Rudiger Wehner shows how the information for guiding outward routes is insulated from that for guiding inward routes. Using barriers, Wehner was able to spatially separate outward and inward routes so that ants took a looped route from the nest to the feeder and back. Experienced ants on their inbound route were captured either from the feeder, along the inbound route, or near the nest and then displaced to a point on their habitual outward route. Despite this being a familiar location, ants behaved as if they were lost and only managed to return home if their systematic search led them by chance to discover their familiar inbound route. Further experiments with bees and ants have confirmed that an insect's internal motivational state can act as a contextual cue for priming appropriate memories for outbound or inward routes even in identical spatial contexts.

Other examples of contextual signals that can prime particular route memories are the time of day and odor. Bees will readily learn routes to two locations if nectar is available in one location in the morning and another in the afternoon. Similarly, if bees are trained to forage at two locations with differently scented feeders, simply introducing one of the scents into the hive is enough to motivate bees to recall the route instructions that will take them to the feeder carrying that scent. In summary, we see that the use of multiple contextual cues allows an insect to organize memories for multiple routes. These mechanisms give an insect the flexibility to choose different routes for different purposes and yet ensure that only the instructions for the current route are used.

Cognitive Maps

One of the most persistent debates within navigation research is whether insects are able to organize their large set of spatial memories into a single representation of the world; a so-called 'cognitive map.' In principle, the path integration system would allow co-ordinates to be allocated to key goals and landmarks so that locations would share a common frame of reference. While much energy has been wasted on debates surrounding the definition of the term 'cognitive map,' there is a general consensus that the behavioral signature of such a map would be the ability to take novel shortcuts between familiar locations.

Here, we look at the evidence from ants and bees to determine whether insects construct cognitive maps. For ants, the simple experiment by Rudiger Wehner (section 'Organization of Multiple Route Memories') shows how sets of route instructions are insulated from each other rather than integrated into a single map. Moreover, despite the indirect interaction between landmark learning and path integration (section 'The Scaffolding of Route Learning'), there is no evidence that familiar locations or prominent landmarks can become associated with the metric information acquired from path integration. In contrast, the debate over whether bees might hold a cognitive map has been more intense.

The idea that bees might be able to develop a unitary representation of the world, while ants do not, may be based on the bees' elevated perspective of the world and their ability to cover large distances during foraging trips. This would, in principle, make it easier to integrate information from different locations and routes. So, when James Gould reported that bees did, in fact, build a cognitive map, the finding was not considered controversial. Gould had trained bees to find one feeder (A) located in a wooded area and then proceeded to catch regular foragers as they left the hive and transported them to a new location (B). The ground sloped up from B to A such that bees could not see A from the new location. Yet, bees

flew directly from B to A suggesting that they had knowledge of the world, enabling them to take direct paths to important locations; this would satisfy many people of the existence of a cognitive map in bees. However, this finding has proved difficult to replicate. For instance, Fred Dyer reports an experiment similar to Gould's original. Two feeders, A and B, were established at equal distances from a hive. The terrain meant that feeder A was higher than B. Bees departing the hive for either A or B were transported to the other feeder and released; bees familiar with B were able to fly directly to B from A; however, bees familiar with A could not reach it from B. Only when bees from A could see the landmarks surrounding B, could they fly directly there, suggesting that bees were using simple landmark guidance rather than a map-like representation. Dyer's experiment had seemingly quietened the cognitive map debate; yet, recent experiments using radar tracking have shown bees appearing to take novel shortcuts to their hive. So far, the data is inconclusive, but the interest and debate surrounding cognitive maps is sure to continue.

Concluding Remarks

True navigation requires a combination of physical, sensory, and cognitive adaptations, and insects tell a fascinating story in all these three dimensions. The scale over which some insects navigate is astounding, and the mechanisms they use to do this, represent a paradoxical combination of simplicity and sophistication. We have seen how insect navigation is built on the interaction of innate strategies with learned information. Throughout its life, an individual gains experience of the world and thus, develops a reportoire of stored snapshots and procedural vector instructions linked to familiar places. These memories are bound together into contextually labeled routes. Representing knowledge of the world in this way, as a series of routes, may be the limit of an insect's navigational ability. There is no evidence that insects are capable of building a map-like representation of the world as vertebrates appear to do, and perhaps, this is a fundamental limitation of the small brains of insects. Either way, with ever-improving technologies, the continued study of social and solitary insects is sure to reveal more examples of elegant behavioral and cognitive solutions to the problem of navigating through the real-world.

See also: Magnetic Compasses in Insects.

Further Reading

Chittka L, Geiger K, and Kunze J (1995) The influences of landmarks on distance estimation of honey bees. *Animal Behaviour* 50(1): 23–31.
Collett M, Collett TS, Bisch S, and Wehner R (1998) Local and global vectors in desert ant navigation. *Nature* 394: 269–271.

Collett TS and Zeil J (1996) Flights of learning. *Current Directions in Psychological Science* 5(5): 149–155.

Esch H and Burns J (1996) Distance estimation by foraging honeybees. *Journal of Experimental Biology* 199(1): 155–162.

Frisch K (1967) *The Dance Language and Orientation of Bees*. Cambridge, MA: Harvard University Press.

Gould J (1986) The locale map of honey bees: Do insects have cognitive maps? *Science* 232(4752): 861–863.

Graham P, Fauria K, and Collett TS (2003) The influence of beacon-aiming on the routes of wood ants. *Journal of Experimental Biology* 206(3): 535–541.

Kohler M and Wehner R (2005) Idiosyncratic route-based memories in desert ants, *Melophorus bagoti*: How do they interact with path-integration vectors? *Neurobiology of Learning and Memory* 83(1): 1–12.

Nicholson DJ (1999) Learning walks and landmark guidance in wood ants (*Formica rufa*). *Journal of Experimental Biology* 202: 1831–1838.

Srinivasan M, Zhang S, Lehrer M, and Collett T (1996) Honeybee navigation en route to the goal: Visual flight control and odometry. *Journal of Experimental Biology* 199(1): 237–244.

Tinbergen N (1951) *The Study of Instinct*. Oxford: Clarendon Press.

Wehner R (2003) Desert ant navigation: How miniature brains solve complex tasks. *Journal of Comparative Physiology A* 189(8): 579–588.

Wehner R, Boyer M, Loertscher F, Sommer S, and Menzi U (2006) Ant navigation: One-way routes rather than maps. *Current Biology* 16(1): 75–79.

Insect Social Learning

R. Dukas, McMaster University, Hamilton, ON, Canada

Introduction

Our understanding of insect learning has changed dramatically in the past few decades and it is now well established that learning affects all major insect activities including feeding, predator avoidance, social interactions, and sexual behavior. Although we understand individual learning in insects rather well, the role of social learning in this diverse group is still not clear. Until recently, there have been no research programs devoted to examining insect social learning. Furthermore, students of insect behavior who worked on topics relevant to social learning typically did not relate their research to the literature on social learning, which has focused on vertebrates. Examples detailed in this article include studies of intergenerational transfer of substrate preference in a variety of insects and communication about food in eusocial insects.

This brief review of insect social learning begins with general considerations of how insect characteristics could affect the prevalence of social learning in this large and diverse group. The analysis of what is known about socially influenced learning is then divided into two parts, one devoted to the majority of insects that are solitary, and the other to the well-studied minority that are highly social.

Insect Life History and Social Learning

Social learning has a fitness advantage over individual learning only under a restricted set of conditions. Such conditions may not be widespread among insects, suggesting that social learning may occur only in a small proportion of insect taxa. In this article, two key life history traits are discussed that may limit the use of social learning in insects.

Lack of Parental Care

Social learning is perhaps most beneficial for young, inexperienced individuals that are cared for by their more experienced parents. The life history of many vertebrates requires a period of parental care, during which the dependent young can acquire reliable information from their parents. For example, in most song birds (oscines), young males must hear their father sing in order to sing properly when sexually mature, and both sexes sexually imprint on their opposite-sex parent. However, vertebrate-like parental care is rare in insects. Even in the exceptional cases where an insect cares for her young, for example, in some sand wasps (tribe Bembicini), parental care is limited to the adult providing food for her larvae inside the underground burrow. This limits the repertoire of information that can be socially transmitted because the parents and offspring do not spend time together in the above-ground settings most relevant to the adult. Furthermore, only little information may survive metamorphosis from larvae to adults.

Another type of interaction between kin that is widespread among social insects involves siblings, but siblings typically do not exhibit the same dichotomy in experience as parents and offspring do. Hence learning from siblings may not be as beneficial as learning from parents. Nevertheless, social learning among siblings is known among social insects and is discussed in a following section.

Nonoverlapping Generations

As previously mentioned, social learning is most beneficial when inexperienced individuals can gain relevant information from experienced ones. In animals with overlapping generations, there are typically distinct age groups that differ in their levels of experience so the younger, inexperienced generations can acquire social information from the older, more experienced ones. Many insects, however, have nonoverlapping generations. For example, many solitary bees emerge in the spring, provision their nest with floral reward, and lay eggs. The bees typically live for only a few weeks, whereas their offspring emerge in the following spring. The lack of age-related experience hierarchies in insects with nonoverlapping generations could limit the occurrence of social learning. Furthermore, unlike individual learning, social learning allows for information transmition across generations but this is less likely to occur in insects with nonoverlapping generations. This means that insects with nonoverlapping generations are less likely to possess socially transmitted traditions.

Social Learning in Solitary Insects

Social learning is more likely to occur if there are frequent interactions among individuals who gain from sharing valuable information. This is clearly the case among social insects, which typically share a nest with numerous closely related conspecifics. Because social insects are unique among animals and more likely to possess social learning than solitary insects, they will be discussed separately. Here, the limited knowledge we have about social learning in solitary insects is addressed, dividing the discussion between insects that interact little with other individuals and insects that typically occur in aggregations.

Truly Solitary Insects

In many insects, interactions among individuals are limited to courtship and mating. Otherwise, there may be little between-individuals contact, which could allow for information transfer. Currently, no social learning in the context of sexual behavior is known in solitary insects. Similarly, although a variety of insects are territorial and many meet incidentally at food sites, no social learning in these contexts has as yet been described in solitary insects.

Perhaps the best opportunity for social learning to occur among solitary insects involves the indirect interactions between mothers and offspring. Many insect species specialize on a single food source and possess an innate repertoire of behaviors and physiology linked to that source. At the other end of the spectrum, generalist insects may exploit a variety of food sources. In insects that are not extreme specialists, offspring could acquire indirect information about appropriate food from their mother. In many insects, the mother either lays her eggs on a food substrate or delivers food to her nest, where she has laid her eggs. Although the larvae typically do not interact with their mother, they do consume the food she has chosen. If these larvae succeed in maturing into adults, the newly eclosed adults who know their mothers' food choice possess socially acquired information that this food is adequate.

How can newly eclosed adults know what they have eaten as larvae? There are three scenarios, all apparently occurring among insects. First, the larvae can simply learn the characteristics of the food they consume. Individual learning has been well studied in several insect larvae, most notably fruit flies (*Drosophila melanogaster*). The larval memory, however, would have to survive the massive cellular reorganization that accompanies metamorphosis. Neurobiological work in fruit flies indicates that parts of the mushroom body, the brain part involved in olfactory learning and memory, remains intact throughout metamorphosis, suggesting that memory transfer from larvae to adults is mechanistically feasible. Empirical studies of this scenario are somewhat inconclusive. The first careful test documenting survival of memory from larvae to adults in fruit flies involved associating one odor with electric shock and another odor with safety. The fly larvae exhibited avoidance learning of the odor associated with shock and the same individuals tested as adult flies

exhibited similar odor-specific avoidance. Although attempts to replicate this finding have failed, studies in other insect taxa including moths and parasitoid wasps have also suggested that specific memories can be transferred from larvae to adults. These studies await replication.

The other two mechanisms allowing for social transmission of information from larvae to adults do not require survival of memory through metamorphosis. In some insects, the pupal case may contain odors of the larval food, which the newly eclosed adults can learn. Finally, various insect species pupate either on the larval food substrate or close to it, so the young adults can learn about their larval food upon emergence. Indeed a few studies in *D. melanogaster* indicate that adults prefer either odors of their larval food remaining on the pupal case or simply the food substrate upon which they have eclosed.

Regardless of the mechanism involved, it is clear that intergenerational transfer of information can take place in a variety of insects. Although this constitutes a simple form of social learning, it can have dramatic ecological and evolutionary implications as it can influence patterns of host-plant use by herbivores and processes leading to speciation.

Insect Aggregations

Many solitary insects live in aggregations, which vary greatly in the frequency of interactions among members. Some aggregations are formed owing to either individuals' tendency to stay where they were born or some desirable substrate features, which independently attract many individuals. Such aggregations seem to have little social interactions. For example, many solitary bees and wasps nest in aggregations in which each female appears to have minimal contact with her neighbors. No social learning among adults is known in these species. Other aggregations are created as a result of active recruitment and attraction of conspecifics, which often rely on species-specific aggregation pheromones. Examples include many fruit flies (*Drosophila* spp.), bark beetles (most species of *Dendroctonus* and *Ips*), and locust (e.g., *Schistocerca gregaria*). To date, no study has documented social learning in actively aggregating species, but such taxa would be prime candidates for relying on socially acquired information, given the frequent social interactions among conspecifics.

Social Learning in Social Insects

Until recently, research on social-insect behavior was disassociated from the literature on vertebrate social learning. Consequently, cases of probable social learning in social insects were typically described as communication and no critical tests for social learning were conducted. Nevertheless, some behaviors of social insects have been so well studied that it is obvious that they constitute social learning.

Learning About Distant Food

Some social insects forage on plentiful but ephemeral food sources. For example, a flower patch can provide nectar and pollen for many bees, but may end blooming within a couple weeks. Locating new flower patches is a difficult task so bees could benefit from informing their hive mates, which are typically closely related, about a rich source they have discovered. Indeed many social bees, wasps and ants (hymenoptera) possess means of communication about distant food sources. The most celebrated case of such social transfer of information is the waggle dance of honeybees (*Apis mellifera*). Foragers (models) returning with nectar from a rich patch of flowers regurgitate their load to workers in the hive. If a forager senses high demand for her nectar, she performs the waggle dance on the vertical comb inside the dark nest cavity. The waggle portion of the dance involves the bee moving in a certain direction while waggling her body from side to side and vibrating her wings to produce a buzzing sound. At the end of each waggle run, the bee circles back to her starting point, alternating between clockwise and counter clockwise turns such that each two successive rounds create a figure eight. The angle of the waggle run relative to the upward direction indicates the angle of the flower patch relative to the sun's position in the sky and the duration of the waggle is positively correlated with the distance to the flowers. Finally, the overall number of waggle runs is positively correlated with relative food quality. Typically, a few observer bees closely attend to the dancer's movements and the floral odors carried on her body.

A variety of experiments as well as recent observations using harmonic radar indicate that observer bees indeed learn the direction and distance information encoded in the waggle dance and rely on that knowledge to arrive in the general vicinity of the flowers. The bees are further assisted by olfactory and visual cues from the flowers and perhaps also by directly following model bees and pheromones. Honeybees also rely on waggle dances to inform hive mates about other resources as well as new potential nest sites when the colony swarms.

Although the waggle dance has not originally described as such, it meets the commonly agreed upon definition of teaching. First, the teacher should incur some cost. The model bee spends time and energy on the dance and also pays the opportunity cost involved in delaying her departure to the rich flower patch. Second, teaching of a given task should be performed selectively only in the presence of individuals not familiar with that task. A returning forager performs the waggle dance based

on her assessment of the patch and colony needs, and the sole function of the dance is to recruit bees unfamiliar with a given flower patch to that patch. Third, the pupil should benefit from the teaching. Critical experiments indeed indicate that inexperienced bees can find a given flower patch much faster after attending to waggle dances coding its location. Finally, an implicit assumption about teaching is that the pupil has learned new information that would guide its future behavior in the absence of the teacher. In honeybees, new recruits first learn the dance information, which enhances their initial arrival in the flower patch. They then learn the landmarks associated with the patch and can locate the patch in the future on their own. Furthermore, the recruits can also code this newly acquired information in their own waggle dances if they choose to perform recruitment dances.

Unlike the waggle dance, tandem running, which is among the simplest means of conveying social information, has been formally described as teaching. Tandem running occurs in some ant species and involves a successful forager leading a recruit from their nest to the food site. Here the teacher adjusts her behavior to ensure that the recruit follows her and this lengthens the teacher's travel time. The recruit, however, arrives in the food faster than she would on her own and she can find the food independently in further trips.

Between the extremes of advanced waggle dances and simple tandem running, social hymenoptera exhibit a variety of means for conveying information to nestmates. Many but not all of these mechanisms of social communication may be classified as social learning. For example, stingless bees (tribe Meliponini), which are among the closest relatives of honeybees (tribe Apini), consist of over 450 species found mostly in the Neotropics. In most of the species that have been examined, successful foragers display behaviors somewhat similar to those of honeybee dances. The dances are followed by recruits leaving the nest in search of food. Some species rely heavily on scent trails leading to the food, whereas others, such as *Melipona panamica*, seem to communicate the distance and height of the food via sound. In *M. panamica*, no feature of the dance is correlated with the food direction, leading to the suggestion that observers directly watch the departing model and follow her in the direction of the food. All social hymenoptera studied exhibit excellent individual learning, so it is likely that when observers respond to social signals, they also learn that information.

Choice and Handling of Flowers

In addition to recruiting nestmates to distant food sources, social bees can also copy the flower choice of experienced foragers. In one study with bumblebees (*Bombus terrestris*), each trial was initiated by allowing a demonstrator bee to forage on an inflorescence consisting of artificial flowers in an arena containing four yellow and four blue inflorescences that were equally rewarding. Then an inexperienced observer bee was introduced to the arena. The observer bees showed significant preference for landing on the occupied inflorescence over the others and exhibited significant preference for that inflorescence color on subsequent foraging trips in the absence of the demonstrator.

Social learning may also influence bees' handling of flowers. Nectar robbing means that bees either punch a hole at the base of a flower or use previously punched holes to extract nectar rather than access the flowers legitimately, in the way that facilitates pollination. Observer bees (*B. terrestris*) that extracted nectar from flowers with holes previously punched by model bees were later more likely to punch holes in intact flowers than control bees with no prior nectar robbing experience.

Conclusion

Research on insect social learning is still in its infancy but it is already clear that some insect species rely on social learning to guide their behavior. The types of information learned from others can be rather minimal as in the case of odor cues remaining from the larval period, which can help newly eclosed adult insects choose their own egg-laying substrate, or sophisticated as in the honeybee waggle dance, which involves symbolic coding of environmental features. It is likely that many cases of insect social learning have not been described, whereas others, such as the forms of social learning about food sources within the numerous species of stingless bees (Meliponini), require further study. Some features of insect life history, including lack of parental care and nonoverlapping generations, could limit the prevalence of social learning. Other insect attributes, most notably their sheer diversity and the social behavior of some species, could allow for intriguing forms of socially acquired information.

See also: Dance Language; Group Movement; Honeybees; Insect Navigation; Robotics in the Study of Animal Behavior; Social Learning: Theory.

Further Reading

Dukas R (2008) Evolutionary biology of insect learning. *Annual Review of Entomology* 53: 145–160.

Dukas R and Ratcliffe J (eds.) (2009) *Cognitive Ecology II.* Chicago, IL: University of Chicago Press.

Dyer FC (2002) The biology of the dance language. *Annual Review of Entomology* 47: 917–949.

Holldobler B and Wilson EO (1990) *The Ants.* Cambridge, MA: Harvard University Press.

Leadbeater E and Chittka L (2007) Social learning in insects – from miniature brains to consensus building. *Current Biology* 17: R703–R713.

O'Neill KM (2001) *Solitary Wasps: Natural History and Behavior.* Ithaca, NY: Cornell University Press.

Prokopy RJ and Roitberg BD (2001) Joining and avoidance behavior in nonsocial insects. *Annual Review of Entomology* 46: 631–665.

Seeley TD (1996) *The Wisdom of the Hive.* Cambridge, MA: Harvard University Press.

von Frisch K (1967) *The Dance Language and Orientation of Bees.* Cambridge, MA: Harvard University Press.

Integration of Proximate and Ultimate Causes

S. H. Vessey, Bowling Green State University, Bowling Green, OH, USA
L. C. Drickamer, Northern Arizona University, Flagstaff, AZ, USA

Introduction

A discussion of proximate and ultimate causation in relation to animal behavior must begin with some definitions. Proximate causation refers to the underlying endocrine system, nervous system, immune system, and developmental processes that result in observed behavior patterns. Ultimate causation refers to the effects of behavior on fitness, through an understanding of the ecology of the organism and its evolution. Consequently, the integration of these two concepts would involve an examination of the ways in which evolutionary selection pressures shape the various internal mechanisms that regulate behavior.

These terms, and their definitions, should not be confused with the nature–nurture debates. The observed behavior, part of the organism's phenotype, is a product of its genetic blueprint unfolding under the influence of all the experiences and environmental effects beginning from fertilization. In this way, we perceive genetics as setting up limits for potential phenotypic traits and the experiences as shaping the actual phenotype. For example, feeding behavior is constrained genetically by several traits, including the ability to forage for and ingest certain foods, the digestion of those foods in terms of enzymes and other features of the digestive tract, and the capacity of the animal to shift its diet seasonally whenever that is necessary. The actual, phenotypic, food preferences are shaped by where the animal lives, the foods that are available, and the experiences that it has had with different foods.

Our purpose in this article is to provide a historical perspective on the ideas behind proximate and ultimate causation to give the reader some context for where we are today. As discussed in another article on the future of animal behavior, scientists are moving forward with studies that integrate the proximate and ultimate causation concepts. The link between these two major ideas is, of course, genetics. For evolution to occur, via differential reproductive success, there must be changes in the genetics of the organism. Each organism 'faces' the world with its phenotype as its set of tools. The success of the organism depends on how its internal mechanisms meet the ecological challenges it faces on a daily basis; that is, how well the animal functions can be measured in terms of the number of progeny, which is a dual product of genetics and environmental influences. Studies of both proximate and ultimate causation occur in the context of the interaction of the genetics and environment of the organism. The article concludes with an up-to-date coverage of the ongoing integration of proximate and ultimate causation.

One term in particular, mechanism, is typically defined in one way by those who study primarily proximate mechanisms, and in another way by those who study ultimate mechanisms. For those who explore proximate causation, the meaning of mechanism incorporates the animal's physiology, immunology, endocrinology, nervous system, and development. These are basically internal processes. To an individual examining ultimate causation of behavior, mechanism generally refers to the ways in which an animal's phenotype functions in its ecological context. These are usually considered as traits best studied in a natural or wild setting. We attempt to be explicit when we use the term 'mechanism' in this essay.

Ultimate Causation: Historical Perspective

The easiest way to distinguish between proximate and ultimate causation is to consider the answers one might get when asking *why* a particular behavior pattern occurs. For instance, when asked why dogs wag their tails, we might give an answer based on proximate causation, in terms of the nerves and muscles involved, the role of the central nervous system, and so on. Alternatively, we might answer that it is based on ultimate causation, in terms of the function of the pattern (communicating aggression perhaps) and the evolutionary history of the pattern.

A second example is to ask why male rhesus monkeys invariably leave their natal social group around the time of puberty (4 years) and join a new social group. The transition process can be difficult, as they are likely to be rejected by members of any different group. While in their natal group, they share their mother's social rank, but in the new group they assume the lowest dominance rank, where rank is based on group tenure. Why would an adolescent male give up the security and status of his mother's group and risk rejection and possible injury by joining a new group? The proximate causes of this behavior are not clearly understood, but are manifested as an incest taboo, as is seen in human cultures. Either the males are somehow repulsed by the proximity of their mothers and other female relatives, or they are rejected by them as possible mates. The male sex hormone testosterone is likely involved because the departure occurs during the mating season and castrated males do not emigrate. The ultimate causes are likely the avoidance of inbreeding and its resulting accumulation of deleterious recessive genes. A male that fails to emigrate potentially could mate with the large number female relatives of his mother and thus could have inbred offspring of lower fitness.

In this section, we focus on the study of ultimate causation, that is, the evolutionary forces responsible for behavior patterns. Ultimate causation of behavior is mediated by the environment, through such negative pressures as climate, predators, and competitors, as well as through positive opportunities such as new food sources or new habitats. These environmental variables create pressures for genetic evolution by means of natural selection. In responding to these pressures, the species is constrained by the amount of genetic variability present in a population and the existence of mutations that might increase fitness. This process of adaptation takes many generations.

One of the earliest experimental studies exploring ultimate causation of behavior was Tinbergen's observations of black-headed gulls removing the eggshells from around the nest after their young hatched. Tinbergen wondered what the function of such behavior was, given the apparent cost of such a behavior pattern. Although the eggs are mottled-brown colored on the outside, inside they are bright white and highly visible, suggesting that egg predators might cue in on broken egg shells. Tinbergen tested this notion by placing broken eggs at varying distance from intact eggs. As predicted, the closer the intact eggs were to the shells, the more likely they were to be taken by predators. From this manipulation, Tinbergen inferred that the tendency to remove egg shells was a heritable trait that had become typical of this and other gull species through natural selection.

Tinbergen's approach to questions of ultimate causation derives directly from Darwin's theory of evolution by means of natural selection, whereby members of a species differ in characteristics, some of which increase survival and reproduction. Although Darwin did not understand the mechanism of inheritance, he realized that these traits tended to be passed on to offspring, so individuals with such traits tended to increase in the population from one generation to the next. In the gull example, if an individual happened to remove eggs shells from the vicinity of the nest, its chicks would be more likely to survive. If the tendency were heritable, their chicks in turn would be more likely to show that behavior pattern, and egg shell removal would spread throughout the population.

Study of the evolution of behavior patterns by means of natural selection had received relatively little attention up to this point, in spite of much progress being made on the role of genetics in animal behavior. A series of publications in the 1960s and 1970s focused attention on

the level, individual versus group, that natural selection acts on. Darwin had argued that selection acts at the level of the individual, since the individual is what natural selection 'sees' in terms of survival and offspring production. He was troubled by instances of seemingly altruistic behavior in social insects, where workers sacrifice their reproductive interests to help the queen raise more offspring. His way around this problem was to consider the colony or hive a 'superorganism,' such that selection would act on the colony as a whole.

In the early twentieth century, many, if not most, biologists accepted the notion that traits could evolve for the good of the group or species and paid little attention to the level upon which selection was supposed to act. Survival of individuals and groups was thought to increase as a function of the degree to which they harmoniously adjusted themselves to their physical and social environment. This line of thinking was especially prevalent among many plant ecologists, who spoke of plant 'sociology.' Studies of overcrowding and stress in animal populations revealed that hormonal changes can lead to reduced reproductive activity and increased morbidity. It was argued that this behavioral-endocrine feedback loop serves as a group-level adaptation for regulating population density; as populations reach the carrying capacity of the environment, aggression increases, followed by increases in stress hormones and the associated loss of fitness. Once the population declines, aggression declines and individual fitness increases.

The idea that individuals sacrifice reproduction for the good of the group was expounded in two books by Wynne-Edwards. He argued that many species have evolved behavior patterns he called epideictic displays that function to tune the birth rate to the available resources. Groups or populations that lack such patterns are more likely to exceed the carrying capacity of the habitat and potentially go extinct. If the unit of selection were the population, then selection might occur for individuals that sacrifice their own reproductive interests for the good of the group. In other words, groups that avoid overexploiting the environment are more likely to survive and may later colonize habitats left vacant by imprudent groups that have become extinct.

This train of thought was vigorously challenged by George C. Williams, who argued (as had Darwin) that group selection, as proposed by Wynne-Edwards and others, was much less likely than individual selection to be a potent force for evolutionary change. He pointed out that the conditions necessary for group selection to override selection at the individual level were stringent, given that groups must maintain their integrity for relatively long periods, that groups must differ genetically for traits that affect the groups' survival, and that group extinction rates must be relatively high. In most cases, traits that lower fitness of individuals will be selected against, even if they favor group survival. Williams argued that selection at higher levels should be invoked only if lower-level selection, that is, the individual and its offspring, cannot explain the evolution of the observed traits. This view has dominated the thinking of most behavioral ecologists as they explore the ultimate causes of behavior, and group selection, although considered theoretically possible, is generally discounted or ignored altogether.

In the meantime, several other ideas of how altruistic behavior could evolve were put forth that were argued not to involve group selection. One is the concept of inclusive fitness, that is, the sum of an individual's direct fitness, as measured by the reproductive success of one's own offspring, and indirect fitness, as measured by the reproductive success of one's nondescendent relatives, for example, siblings or cousins. For instance, in the classic study of prairie dogs by Sherman, he argued that although individuals risked their own lives by giving alarm calls, thus revealing their whereabouts to predators, they increased their inclusive fitness by warning close relatives, a process referred to as kin selection. Other models based on games theory were developed, such as reciprocal altruism, where one individual performs an altruistic act with the expectation of being repaid, with interest, at a later time.

In 1975, E. O. Wilson published his compendium entitled *Sociobiology, The New Synthesis*. Wilson drew heavily on population biology in shaping the emerging fields of sociobiology and behavioral ecology. Evolutionary history and environmental factors that determine the ecological niche affect population growth and dispersal rates, the focus of evolutionary ecology. Behavioral and population parameters, such as birth and death schedules and gene flow between populations feed into the theory of sociobiology. This theory's goal is to enable predictions of behavior from knowledge of population parameters and the behavioral constraints imposed by the gene pool. Although several models of group selection were presented in the book, it was generally assumed by Wilson that individual selection is main force for behavioral evolution.

The year after the publication of Wilson's tome, Richard Dawkins published *The Selfish Gene*. The basic argument was that the unit of selection is the gene, rather than the individual organism or group. Genes are referred to as replicators that typically help their temporary host, the organism, survive, and reproduce, thus improving the gene's own chances of being passed on. In some instances, however, the gene's interests may not coincide with the host's, and hence the term selfish gene.

Most biologists still think of the organism, or phenotype, as the main unit of selection because that is what natural selection 'sees.' Genes can perhaps be better thought of as the unit of evolution, since evolution results from changes in gene frequency. Thinking of genes as replicators makes it somewhat easier to understand how altruistic traits might be maintained in a population via kin selection: Organisms might act against their individual interests to help related organisms reproduce because genes set 'helping' copies of themselves in other bodies to replicate. Thus, the 'selfish' actions of genes might lead to unselfish actions by organisms.

The study of ultimate causation in behavioral ecology and sociobiology mushroomed in the 1970s and continues to this day, testing the fitness consequences of behavior patterns and social groupings. The vast majority of these studies follows Williams's law of parsimony and assumes that selection occurs at the level of the individual and no higher. Most instances of helping behavior or other potentially altruistic acts not explained by classic Darwinian selection are explained by kin selection, reciprocal altruism, or some other model not involving selection above the level of the individual.

Although group selection was dismissed by most behavioral ecologists, as noted earlier, new modeling techniques and empirical data suggest that it may play a more important role than previously thought. D. S. Wilson has argued that adaptations can potentially evolve at any level, from genes to ecosystems. In a joint paper with E. O. Wilson, they dismiss what they call the naïve group selection arguments of early workers, including Wynne-Edwards, who assumed that group selection would easily prevail over selection at the individual level. But Wilson and Wilson go on to argue that the theoretical foundation of sociobiology needs to be reformulated to include multilevel selection, including selection at the level of the group. Whether theoretical and empirical evidence will continue to build in support of higher-level selection remains to be seen.

Proximate Causation: Historical Perspective

The study of proximate mechanisms dates back to antiquity, in a general sense, with initial interest in the 'how' questions of animal behavior with regard to potential food sources and predators. In the first millennium, anatomists learned a great deal about animal structure through their extensive dissections. When, after stagnation during the Middle Ages, scientific inquiry resumed in the sixteenth and seventeenth centuries, new discoveries were made. These included Harvey's findings on circulation and Borelli's contributions on form, function, and muscular physiology. These works and others provided the basis for the emergence of studies of how internal and developmental processes influence behavior.

One important distinction between studies of ultimate and proximate causation of behavior involves the ability to 'see' the subject matter. Generally, as covered in the previous section, behavior in the functional and evolutionary sense can be observed directly, whether in a field or laboratory environment. On the other hand, mechanisms that involve the nervous system, endocrine system, and underlying genetics and development, all take place away from our normal visual world. To be sure, these events can be observed with a variety of techniques, but most people do not see behavior in this way. This distinction could be a partial reason for the divergence between those who study proximate mechanisms and those who explore ultimate causation.

In a modern sense, the investigation of proximate causes of behavior begins with three threads, all of which emerged in the latter half of the nineteenth century and continued into the first half of the twentieth century. First, psychologists, primarily in North America, explored the possible relationships between human and non-humans in terms of their mental processes. This progressively led to interest in species-specific behavior and the functional aspects of observed phenomena. Second, American zoologists began formulating explanations about behavior mechanisms on the basis of both natural history and physiology. Early writings in natural history by colonists and later explorations westward provided information that led to questions about how the behavior patterns were controlled and concerning their functions. During the first half of the twentieth century, this blossomed into studies relating physiological mechanisms to observed behavior. Last, the ethology tradition in Europe was initiated with studies of natural history and attempts to explain, via models, the internal processes underlying behavior. So, for example, Lorenz and others developed terminology including 'innate releasing mechanism' and 'sign stimulus' to explain behavior that was under significant genetic control. European work also included aspects of physiology, such as the studies by von Holst, bridging ethology and emerging neurobiology.

During the last half of the twentieth century, several individuals provided overall schemes for categorizing the way scientists posed questions about animal behavior. Niko Tinbergen's 1963 paper 'On the Aims and Methods of Ethology,' provided such a scheme, one that is used even today. He proposed four types of questions: two concerning proximate mechanisms and two about ultimate mechanisms. His scheme involved causation (control), ontogeny (development), survival value (function), and evolution. Frank Beach provided a similar scheme, which included historical determinants, direct and indirect determinants, and organismal determinants. More recently, Donald Dewsbury proposed a structure involving three categories of questions: the genesis of the behavior, its control, and the consequences of the actions. The common elements in these schemes, from a proximate causation perspective, encompass physiological mechanisms and development.

Physiological studies of the neural bases for behavior, and explorations of endocrine functions and behavior have their roots in early American comparative psychology,

augmented into the mid-twentieth century by work in zoology. Karl Lashley was a key pioneer in the exploration of the neural bases for behavior, sensory systems, and brain function. Others who made significant contributions to this emerging field were Hermann von Helmholtz for work on visual systems; Donald O. Hebb, who worked on connections between the brain and learning; James Olds, who co-discovered the pleasure centers in the brain; and Rita Levi-Montalcini, who discovered the nerve growth factor. Signaling the growing importance of neurobiology in relation to behavior, Nobel Prizes for Physiology or Medicine were awarded to Roger Sperry, David Hubel, Torsten Wiesel, and Eric Kandel for their research and findings on vision and cognitive neuroscience.

Many journals involving various aspects of neurobiology began publication during the decades of the 1970s and 1980s as the field expanded and diversified. Until the past few years, there were almost no papers published in *Animal Behaviour*, the primary journal in this field, with neural aspects of behavior as a major focus. Even now, the vast majority of papers are concerned with ultimate causation issues. This decades-long emphasis in the journal no doubt was a significant stimulus for the many neurobiology journals that emerged.

Frank Beach was an early proponent of examining the endocrine bases for behavior. His work on hormones and reproduction served as the primary basis for launching a number of careers and lines of inquiry. One key principle Beach championed was that the endocrine-behavior link worked both ways. That is, hormones can affect behavior, but also behavior can influence hormone levels.

Both field and laboratory environments are used for investigations of endocrines and behavior. Much of the work on female sexual receptivity in birds and mammals, maternal and paternal behavior, male aggression, and the interplay between behavior and endocrine systems has been laboratory-based research. Lehrman's elegant work on coordinated activities and hormones in the breeding cycle of the ring dove is particularly fascinating in this regard.

Field studies have involved the use of artificial hormone doses to test effects in wild or free-living animals. Collection of blood, urine, or feces provides a way to measure various hormone levels. However, the stress associated with capture and restraint, or even just being in a laboratory setting, can compromise hormonal information gathered in this manner. With the advent of new hormone assay technologies, investigators can now gather samples of urine or feces without the necessity of capturing the animal and providing a picture of hormone levels based not on a point in time, but representative of a longer period, up to a day or more. Monitoring hormone levels in wild animals makes it possible to examine variations in both sex steroids and stress hormones for animals under different conditions and in different social situations.

Research on hormones and behavior led to its inclusion in classes and textbooks in physiological psychology by the 1950s. Further growth in this arena spawned a new journal, *Hormones and Behavior.* There have been at least three textbooks devoted to the subject of endocrines and behavior.

In the last three decades, a new horizon has emerged: the investigation of relationships between the immune system and behavior. This topic was addressed by Hamilton and Zuk in their work on blood parasites in birds. Work on the Major Histocompatibility Complex (MHC) in mice and its role in odor preference and mate choice also relate to the role of the immune system in affecting behavior. More recent work by Wingfield draws connections between stress, sickness, and immune system function in birds and mammals. Though the interplay between an animal's immune system and behavior has received modest attention for several decades, this relationship is currently being more thoroughly investigated.

The interrelationships among neural, hormonal, and immunological systems provide a stimulus for future exploration of the two-way roles between these systems and behavior. The key to understanding the connections between proximate mechanisms of behavior and the functional or ultimate causation is genetics. This is a key theme in the section on integration at the conclusion of this article.

Behavior development received considerable attention in the early days of comparative psychology, probably because of the connections drawn between human and non-human animal learning during growth and maturation. The processes of behavior development are often divided along a chronological timeline beginning with prenatal or prehatching events, followed by early postnatal considerations, and aspects of behavior during juvenile stages. Play behavior is often considered as part of the investigation of behavior development.

Prenatal influences on behavior include such things as exposure to hormones in utero, effects of stress on the mother on later offspring behavior, and ways in which both external and internal stimulation contribute to the maturation and refinement of the brain, sensory systems, and motor development. Early postnatal events include imprinting, emergence of food preferences, and the beginning of some forms of play behavior. Among the extensive investigations of juvenile and adolescent behavior are those on bird song, the form and functions of play behavior, sexual maturation and changes in behavior, and connections between early behavior actions, prior to birth or hatching and the period immediately following those events with later behavior.

Imprinting refers to the formation of either filial ties involving formation of an attachment to a parent or object, or the establishment of strong tendencies to court and mate with individuals of the same kind. The phenomena associated with imprinting encompass many subtopics among which are the notions of critical and sensitive periods, the importance of different sensory modalities, and variations in the timing and strength of the imprinting experience.

A favorite procedure for the early investigators of behavior development was the isolation (deprivation) experiment. Can we isolate an organism from particular sorts of stimulation and discover specific deficits in later actions? These studies provided useful information, but various confounds rendered them less important as the field progressed. The opposite manipulation, providing and enriched environment with added stimulation, sometimes specific and other times general, was used to explore ways in which environmental impacts enhanced the learning of various organisms. Many of these studies were, of course, directed at understanding the processes occurring in human development. Relating differences in environment to learning and other endpoints served as the basis for changes in such areas as early childhood education and orphanages.

Synthesis and Integration

Categorization of causes of behavior patterns as either proximate or ultimate is an arbitrary distinction, though convenient at some levels. The term 'ultimate' is also problematic in this context, since it conveys the notion of an absolute end point; 'proximate,' on the other hand, is a relative term. A more appropriate antonym for 'proximate' is 'distal.' Consider the study of the role of natural selection by observing behavior in the field, as Tinbergen did with the egg shell removal experiment, versus documenting changes in gene frequencies due to selection using new molecular and statistical techniques. Both approaches ask evolutionary questions, but the latter is clearly getting at proximate causes of evolution. The same can, no doubt, be said for examples of proximate causation discussed earlier: some are more proximal than others. A full understanding of any behavior pattern requires study at all levels, from its selective advantage or disadvantage in the field, that is, how and why it affects fitness, through all levels of organization down to the mechanisms of gene action. This line of thinking might predict a trend toward studies taking a multilevel approach to causation, but this has not always been the case.

In his 1975 treatise on sociobiology, E. O. Wilson depicted the relative sizes of the different fields dealing with animal behavior in 1950, 1975, and 2000. On the left, or proximal end, was neurophysiology and its close connections with cellular biology. On the right, or ultimate end, were behavioral ecology and sociobiology, with connections to population biology. In 1950, connecting the two ends of this 'dumbbell' were the two large traditional branches of animal behavior, ethology and comparative psychology. Wilson predicted that, rather than the expansion of these connecting disciplines to unify the ends, neurophysiology would cannibalize ethology and comparative psychology from one end, and behavioral ecology would cannibalize them from the other. To some extent this has come true, given the explosion of research in both cellular/molecular biology at one end and in population biology/behavioral ecology at the other.

Countering the trend toward increasing dichotomy between proximate and ultimate causation is the appearance of a number of academic departments dedicated to integrative biology. Many of these, however, incorporate evolution and ecology to the extent that they deal mainly with ultimate causation, leaving proximate causation to departments of cellular and molecular biology. Funding agencies tend to follow the same pattern.

Perhaps more important than names for channeling academic programs and grant proposals, however, is the increasing number of research projects in animal behavior that span many levels of causation. For example, the study of mating systems in rodents has been approached at a number of levels, from the environmental conditions favoring the evolution of monogamy versus polygamy in field experiments to the differences in the DNA that regulate these mating systems. At the proximal level, not only have researchers identified a gene that controls the number of hormone receptors in the forebrain of the polygynous meadow vole, but they have also been able to transfer extra copies of this gene into the brain and cause meadow voles to behave like monogamous prairie voles. At the same time, others are exploring the long-term fitness consequences of these manipulated animals in seminatural field enclosures.

A number of other research teams are studying behavior at multiple levels of causation, including those of both Ketterson and Wingfield on bird physiology, Houck on salamander mating pheromones, Robinson on honeybee genomics and social behavior, and Strassmann and Queller on genomics and the evolution of slime mold sociality. Although much of this research gets published in specialized journals, the teams work at multiple levels and ask questions about both proximate and ultimate causation. The success of such efforts to demonstrate how evolutionary selection pressures shape the various internal mechanisms that regulate behavior depends in no small part on project leaders that have a broad vision and the ability to coordinate the activities of researchers with different specializations.

See also: Behavioral Ecology and Sociobiology; Comparative Animal Behavior – 1920–1973; Ethology in Europe; Future of Animal Behavior: Predicting Trends; Neurobiology, Endocrinology and Behavior; Psychology of Animals.

Further Reading

Beach FA (1960) Experimental investigations of species-specific behavior. *American Psychologist* 15: 1–18.

Darwin C (1859) *On the Origin of Species.* London: Murray.

Dawkins R (1976) *The Selfish Gene.* New York, NY: Oxford University Press.

Dewsbury DA (1992) On the problems studied in ethology, comparative psychology, and animal behavior. *Ethology* 92: 89–107.

Hamilton WD (1964) The genetical evolution of social behavior I, II. *Journal of Theoretical Biology* 7: 1–52.

Hamilton WD and Zuk M (1982) Heritable true fitness and bright birds: A role for parasites? *Science* 218: 384–387.

Lehrman DS (1955) The physiological basis of parental feeding behavior in ring doves (*Streptopelia risoria*). *Behaviour* 28: 337–369.

Sherman PW (1977) Nepotism and the evolution of alarm calls. *Science* 197: 1246–1253.

Thorpe WH (1979) *The Origins and Rise of Ethology.* London: Praeger.

Tinbergen N (1963a) The shell menace. *Natural History* 72: 28–35.

Tinbergen N (1963b) On aims and methods in ethology. *Zeitschrift für Tierpsychologie* 20: 410–433.

Williams GC (1966) *Adaptation and Natural Selection: A Critique of Some Current Evolutionary Thought.* Princeton, NJ: Princeton University Press.

Wilson DS and Wilson EO (2007) Rethinking the theoretical foundation of sociobiology. *The Quarterly Review of Biology* 82: 327–348.

Wilson EO (1975) *Sociobiology: The New Synthesis.* Cambridge: Harvard University Press.

Wingfield JC (2005) The concept of allostasis: Coping with a capricious environment. *Journal of Mammalogy* 86: 248–254.

Wynne-Edwards VC (1962) *Animal Dispersion in Relation to Social Behavior.* Edinburgh: Oliver and Boyd.

Wynne-Edwards VC (1986) *Evolution Through Group Selection.* Boston, MA: Blackwell Scientific Publications.

Relevant Websites

http://findarticles.com/p/articles/mi_qa3746/is_199802/ai_n8801814/ – A symposium on integration of proximate and ultimate causation.

http://www.ias.ac.in/currsci/oct102005/1180.pdf – An article exploring differences between those studying proximate and ultimate causation.

http://cas.bellarmine.edu/tietjen/Ethology/introduction_and_history_of_anim.htm – Historical perspective on many aspects of animal behavior.

http://www.sciencedirect.com/science?_ob=ArticleURL&_udi=B6VJ1-3Y86861-G&_user=109269&_rdoc=1&_fmt=&_orig=search&_sort=d&_docanchor=&view=c&_searchStrId=1012941289&_rerunOrigin=google&_acct=C000059546&_version=1&_urlVersion=0&_userid=109269&md5=7a129b63b5c0c3ee43d09427ee55dd53 – Examination of shifts toward studies of proximate mechanisms.

Intermediate Host Behavior

J. Moore, Colorado State University, Fort Collins, CO, USA

Introduction

For internal parasites, hosts are both habitat and nutritional resource. Because the host is also ephemeral, transmission between hosts is essential for parasite survival and reproduction. Parasite transmission falls into three general categories: transmission of propagules, transmission by living vectors, and transmission by intermediate hosts.

Some parasites have direct life cycles; these parasites produce propagules that colonize new hosts and produce yet more propagules. Other parasites use vectors – commonly hematophagous insects – to transmit infective stages to new hosts. In the case of yet other parasites, immature stages develop to an infective stage in intermediate hosts. When the intermediate host and the parasite it contains are eaten by an appropriate predator (i.e., the final, or definitive, host), the immature parasite develops into an adult in the intestine or the intestinal diverticulae of that final host. The parasite then produces eggs, which usually exit with feces, becoming available to other potential intermediate hosts.

Of the parasites that are transmitted by intermediate hosts, the best known are cestodes (tapeworms), trematodes (flukes), nematodes (roundworms), and acanthocephalans (thorny-headed worms), along with some protists. There are many variations on this intermediate host–final host life cycle, but no matter what happens in the remainder of the life cycle, the intermediate host must be eaten for transmission to occur. Because of this, the behavior of the intermediate host can influence transmission success in unexpected ways. These predator–prey interactions involving intermediate and final hosts are critical to the survival and reproduction of the parasite; without predation by the final host, the parasite does not develop or produce offspring. Changes in intermediate host behavior as a result of parasitism can therefore be seen as potentially adaptive for the parasite if they aid transmission. Indeed, this adaptive-for-parasite scenario is a major hypothesis that can explain some of these behavioral changes, although it is not the only hypothesis.

In fact, parasites are known to alter host geotaxis, phototaxis, humidity responses, antipredator behavior, habitat selection, activity levels, and a variety of other behaviors. The adaptive significance of these alterations is more difficult to demonstrate than their occurrence, which seems to be ubiquitous.

In this article, the history of this field is summarized and some recent findings are introduced that are at once exciting and that point to the complexity of this little understood field. Keep in mind that a generation ago, topics such as pathogens, disease, and parasites were considered inappropriate and/or uninteresting for ecological and behavioral study. Times have changed.

A Brief History of Parasites and Intermediate Host Behavior

Parasitologists have long suspected that intermediate host behavior is influenced by parasites. In the 1930s, Eloise Cram noticed that grasshoppers parasitized by a nematode of chickens were sluggish, and hypothesized that this might make these grasshoppers easier prey for chickens. Around the same time, Wesenberg-Lund watched snails that contained brightly striped, pulsating trematode (*Leucochloridium* sp.) broodsacs in their tentacles. These broodsacs have been mistaken for caterpillars by humans, and it is reasonable to expect birds (the final host) to make the same error. However, Wesenberg-Lund had to admit that despite hours of observation, he never witnessed avian predation on the trematode and, in a statement that (regrettably) would never be allowed in a modern scientific paper, expressed his sympathy for the hardworking trematode. *Leucochloridium* is difficult to maintain in the laboratory, and that might be part of the reason why such a spectacular parasite has not been the subject of more investigation.

When we think of predatory final hosts, large animals with big teeth probably come to mind. However, as we have seen with *Leucochloridium*, the final host does not have to be a 'typical' predator, that is, a carnivore; herbivores or insectivores are frequently final hosts if they accidentally consume the intermediate host, say, a small arthropod. Such a life cycle characterizes another trematode, *Dicrocoelium dendriticum*, a parasite that is as well known as *Leucochloridium*, and just as recalcitrant when it comes to definitive studies. The adult worm lives in sheep and other ungulates. Eggs pass out with feces and are eaten by pulmonates. 'Slime balls' – mucous balls containing cercariae, a trematode larval stage – are released from the mantle cavity of the infected snail and picked up by foraging ants. When an ant feeds on this material, the ingested parasites begin an anterior migration that ceases when one of them reaches a depression in the anterior part of the subesophageal ganglion near the nerves to the mouthparts. It produces a thin-walled cyst there, while the remaining metacercariae (the trematode stage following cercariae) remain in the ant hemocoel. The infected ant behaves like other ants during the day, but as the cool of the evening approaches, it does not return to the nest. Instead, it ascends a blade of grass and attaches there with its mandibles until temperatures rise the next day. Ungulates that graze in the evening or early morning may ingest this ant, along with its 'brainworm.' Clearly, the parasite's influence on ant behavior is very likely to increase risk of predation from a source (an ungulate) that is rarely a danger to uninfected ants; nonetheless, quantifying this risk, or even showing that it exists, is a nontrivial undertaking.

It is for reasons like these – uncooperative snails, uncountable ants, and in general, the constraints that face ecologists who study predation – that the examples of parasite-induced behavioral alterations far exceed the evidence that such alterations actually cause a change in transmission. However, the evidence does exist, some of which comes from surprising corners of the animal world. In fact, over 50 years ago, W. H. van Dobben used ejecta from cormorants to show that the birds caught an unusually large number of roach (*Leuciscus rutilus*) infected with a tapeworm, *Ligula intestinalis*, compared to what would be expected on the basis of samples from fishermen's catches. The cestode was making the roach more available to cormorants (final hosts) than to fishermen. The immature cestodes in the roach become quite large, and the 'fat' fish are probably more visible – and perhaps closer to the surface – than uninfected roach.

In the 1970s, William Bethel and John Holmes focused on acanthocephalans of ducks in some of the first laboratory demonstrations of parasite-induced behavioral alterations and associated increases in predation risk. These worms use amphipod intermediate hosts (e.g., *Polymorphus paradoxus* in *Gammarus lacustris*); infected amphipods exhibited normal antipredator behavior until the parasite reached an infective stage. Amphipods containing immature parasites behaved like uninfected conspecifics; they avoided light, diving and burrowing into the substrate if disturbed. Once the parasite became infective, these responses changed dramatically. The gammarids became photophilic and positively phototaxic, and they skimmed across the surface or clung to objects on the surface – including final hosts! In laboratory tests, naïve mallards (a final host) ate nearly four times as many gammarids infected with *P. paradoxus* as they did uninfected gammarids, in large part because the ducks seemed to be intrigued by the skimming organisms and the floating vegetation to which they frequently clung. Muskrats, an unsuitable host, also ate some of these gammarids.

Other duck parasites had different effects on amphipods. *Corynosoma constrictum* (another parasite of mallards) in *Hyalella azteca* induced increases in both photophilia and phototaxis. *Polymorphus marilis*, a parasite of diving ducks (e.g., scaup), induced photophilia in its intermediate host (*G. lacustris*), but the parasitized crustacean did not go to the surface, and as a result, did not get consumed by mallards. Being in well-lit portions of the water column, albeit not at the surface, might have enhanced parasite transmission to diving ducks. (Scaup are reluctant foragers in the laboratory, so tests of this hypothesis could not be conducted.) In the end, it seemed to Bethel and Holmes that the parasites altered amphipod behavior in ways that were likely to increase encounters with final hosts – thus increasing transmission – even if some of the amphipods were noticed by unsuitable predators (e.g., muskrats).

As noted earlier, assessing predation in the field, much less differential predation on infected and uninfected animals, is no simple task. In addition, observing behavioral changes themselves in the field is fraught because although we may see that infected and uninfected animals differ behaviorally, we do not know whether they have always shown those differences, that is, if infected animals are infected simply *because* they behaved differently in the first place. In the 1980s, Moore combined field and laboratory approaches to show that behavioral changes observed in the laboratory can have transmission consequences in the field. She studied an acanthocephalan (*Plagiorhynchus cylindraceus*) that uses terrestrial isopods (*Armadillidium vulgare*) as intermediate hosts and European Starlings as final hosts. She infected isopods in the laboratory and learned that infected animals were more likely than uninfected conspecifics to spend time in areas of relatively low humidity and on light-colored substrate, and they were less likely to spend time under overhanging shelter. In addition, the infected isopods had a tendency to be more active than uninfected ones. In the field, Moore measured acanthocephalan prevalence in isopods collected from areas where parent starlings foraged; simultaneously, she calculated rates of isopod delivery to nestling starlings. She discovered that there were significantly more infected nestlings than were predicted on the basis of the proportion of infected isopods in the isopod population; in other words, the parents were not taking isopods randomly, but were feeding an unusually high proportion of infected isopods to their nestlings, probably because they were encountering these isopods more frequently, given the behavioral changes induced by the parasite. This 'preference' was also demonstrated in an aviary setting.

An Abbreviated Review of Behavioral Changes

The list of behaviors that parasites alter is amazingly long and ranges from odor production to changes in predator avoidance. Some of the behaviors for which there are long lists of examples include altered phototaxis/photophilia and elevation seeking, as well as changes in activity, reproductive behavior, and feeding behavior. Increased movement toward light, increased tolerance of lighted areas, or decreased photophobia can make intermediate hosts more conspicuous to predators. In the case of elevation seeking, intermediate hosts may climb up to high places, swim high in the water column, crawl high on beaches, fail to burrow, and do any of a number of other things that would put them in the way of predators. Activity changes are more puzzling. They are probably the most commonly reported of all parasite-induced behavioral changes and perhaps among the least understood of those changes. Although decreased activity seems a more pervasive response to parasitism, there are numerous examples of increased activity as well, and researchers have argued that decreased activity (depending on the nature of the change in activity) makes intermediate hosts less likely to escape predators or that increased activity makes intermediate hosts more noticeable to predators. A large literature has emerged around parasites and host reproductive behavior; common interpretations include the possibility that reduced reproductive activity might enhance host survival or might result in reallocation of resources to parasites. As for modified feeding, this is most typically reported in vectors, but intermediate hosts are also known to alter their selection of food items and feeding times when parasitized. It is both intriguing and intimidating to speculate upon the variety of behavioral changes that may occur given the fact that there is no parasite-free organism in nature.

Keep in mind that these changes are just that – changes in behavior. We do not know whether they might be associated with parasite transmission until appropriate transmission studies have been performed, and in many cases, those are very difficult to do. Indeed, changes in reactions to light, elevation seeking, activity and foraging preferences may also be consistent with host defensive behavior. Moreover, recall that humans are a visual species. We are much more likely to record visually salient behaviors (e.g., activity, photophilia, elevation seeking, etc.) than other phenomena. The sensory-perceptual world of other animals extends far beyond these stimuli and we are barely conscious of what is probably an immense array of parasite-induced auditory or olfactory alternations, for instance.

'Targeted' Transmission?

There are other parasite-induced behavioral modifications that might seem to be more 'targeted' at transmission. In the case of intermediate hosts, reduced antipredator behavior would probably top this list. Indeed, aquatic isopods (*Caecidotea intermedius*) that are

hosts to acanthocephalans (*Acanthocephalus dirus*) are more likely to be near the fish final host predator than uninfected isopods are, and sticklebacks infected with cestodes (*Schistocephalus solidus*) recover from frightening stimuli faster than uninfected conspecifics. Rats infected with a protist (*Toxoplasma gondii*) that is spread in cat feces are unafraid of cat odor. However, even these behaviors can be complicated. Another protist (*Eimeria vermiformis*) induces fearlessness in mice, and in this case, predation would spell the end of both mouse and parasite; however, the same phenomenon can also increase social interaction on the part of the infected mouse, thus directly transmitting this parasite to other mice.

This raises the question of 'mistakes' in manipulated hosts. Even if parasites benefit from transmission as a result of predation upon behaviorally aberrant intermediate hosts, what about the cost of possibly getting into the wrong host? Although this does happen, there is some evidence that in some systems the probability of this outcome is minimized. Habitat shifts that result in increased encounters with some hosts may decrease encounters with other predators (see Bethel and Holmes' work). In the case of a trematode (*Diplostomum spathaceum*) that lives in the eyes of fish and affects vision, avoidance responses to avian final host predators are diminished, but the infected fish continues to avoid piscine predators; its nonvisual senses (e.g., lateral line, olfaction) remain intact. In general, we might expect behavioral modifications that lead to transmission to reflect the general constraints and proclivities of the host–parasite association. Thus, parasites in intermediate hosts that are cryptic species might increase visual conspicuousness, especially if the final host is a visual predator, while parasites in intermediate hosts that escape predation by fleeing might affect stamina or speed. Such a prediction has yet to be tested in a comparative manner and requires a substantial database; however, it is not an unreasonable expectation.

Timing of Behavioral Change

Another expectation of behavioral alterations in intermediate hosts has to do with timing: if the intermediate host engages in risky, predation-prone behavior induced by a parasite and favored by natural selection acting on that parasite, we expect this behavior to occur when the parasite is infective for the next host, and not before that time. Tests of this prediction are relatively rare because they are time-consuming; the same behavioral tests must be performed at intervals throughout the infection, during parasite development, so that the behavioral effects of immature and infective stages can be compared. Nonetheless, the experiments that have been performed have largely upheld the expectation. One of the first such demonstrations was that of Bethel and Holmes, who showed that the acanthocephalan in question did not change gammarid behavior until it was infective to ducks. Similar results have been found with some trematodes and cestodes in their intermediate hosts. In yet other host–parasite associations, changes in behavior seem to intensify as the infection develops; these are more difficult to interpret and may involve some pathological responses.

In addition to the ontogeny of the altered behavior, timing may also be important in the daily expression of the parasite-induced behavior. In the case of the brain-worm of ants, recall that the infected ants were trapped atop blades of grass at dusk and remained until dawn. If not consumed by morning, they were able to release their grasp and resume normal behavior. This meant that while the ants were in position to be eaten by ungulate final hosts during the foraging times of those hosts, they were not exposed to heat and desiccation during the day. Likewise, snails (*Potamopyrgus antipodarum*) containing infective stages of trematodes (*Microphallus sp.*) shifted feeding (i.e., exposed) times to coincide with the foraging schedule of the final host.

Appearance and Behavior

Although changes in appearance (e.g., size, color) do not strictly qualify as behavior, they are common among intermediate hosts, and some are among the few parasite-induced attributes that are amenable to experimental examination because they can be isolated from a broader array of changes that a given parasite might induce. Gigantism is common in snails parasitized by trematodes, and across parasite taxa, a veritable rainbow of host alterations has been reported. For instance, the juvenile of the acanthocephalan *Pomphorhynchus laevis* looks like an orange dot under the exoskeleton of the amphipod *Gammarus pulex*, which also becomes more positively phototactic when infected. To see what effect this orange dot had on the likelihood that the gammarid would be eaten by sticklebacks, Theo Bakker and colleagues painted orange dots on uninfected gammarids, and covered up the orange parasite in the case of infected gammarids. In this experiment, they discovered that both altered behavior and altered appearance increased gammarid predation risk. Again, comparative studies would be most helpful in understanding both the evolutionary history and the current role of color changes in parasitized hosts. Parasite-induced changes in the substrate color preference of hosts can have the same overall effect as the color change of the host itself, that is, increased conspicuousness as a result of color contrast; this is probably important in Moore's study of terrestrial isopods, which do not change their dark color but do spend more time on light substrates when infected.

Some Current Questions About Behavioral Change

Given the rich diversity of behavioral alterations in parasitized hosts, it is all too tempting to envision adaptive significance where none has been demonstrated. Faced with an expanding literature about behavioral alteration, in the 1990s Moore & Gotelli and Poulin suggested more rigorous approaches to these studies, especially when addressing evolutionary explanations. They emphasized the role of phylogeny and the importance of testing alternative hypotheses. For instance, a behavior that seems to predispose a host to predation might actually function in host defense; the fitness consequences of the altered behavior for either host or parasite must be investigated. Since then, workers have attempted to pay attention to such nuances while struggling with host–parasite systems that are not always amenable to all the questions one would like to ask.

Of course, if one parasite can modify behavior, what happens if several species share a host, as is often the case? Theoretically, it should depend on whether or not the interests of the participating parasites coincide or conflict. For instance, when two manipulating parasites have the same life cycle and alter behavior in ways that have additive transmission benefits for both, we might expect them to co-occur more frequently than a random model would predict, that is, natural selection should favor co-occurrence. When one of these parasites is a manipulator and one is not, we expect natural selection to favor the nonmanipulator that can 'hitch-hike' with a manipulator, taking advantage of the enhance transmission offered by the manipulation. The scenarios become even more intriguing in the case of parasites that share intermediate hosts but have different final hosts. In these cases, successful manipulation on the part of one parasite increases the probability that the other reaches a dead end. These parasites are in conflict, and different outcomes can be expected: potentially conflicting parasites may avoid parasitized intermediate hosts, that is, avoid co-occurrence, or one may kill or out-manipulate the other. Indeed, some parasites such as microsporidians are often overlooked by students of parasite-induced behavioral alterations, but recent studies indicate that these ubiquitous organisms can themselves affect the behavioral outcome in hosts with other co-occurring parasites. While the literature on behavioral effects of co-occurring parasites in intermediate hosts is still spare, it is growing, and offers a rich source of evolutionary understanding.

Some of the predictions about the outcomes of co-occurrence are based on the assumption that it is costly to manipulate host behavior. To date, these hypothesized costs have not been measured, but in general are thought to be closely linked to the mechanisms that cause the behavioral shifts. Mechanisms range from neurotransmitter changes found in some acanthocephalan infections to mechanical interference that occurs when trematodes invade eyes or when cestodes grow so large that a streamlined fish becomes bulky and visible. The costs of behavioral alterations will be as variable as the vast range of mechanisms that we are only beginning to identify, but without some measure of these costs, our understanding of the evolution and distribution of behavioral manipulation will be greatly compromised.

In some host–parasite systems, the parasite that alters behavior acts almost like an on–off switch. This is true in the gammarid–amphipod system that Bethel and Holmes studied. Fully developed cystacanths always induced the same response – reversed phototaxis, surface-skimming, and clinging. In many other host–parasite systems, however, all hosts are not altered in the same way, or to the same extent. The source of this variation is poorly understood. Might it be some mitigating effect of host age or resistance to manipulation? Do parasites themselves vary in their ability to induce alteration? Is this an indication of genetic variation in host or parasite, or some environmental factor? For that matter, we understand very little about how parasite-induced alterations vary over the range of the parasite or host. For example, it is reasonable to expect different responses if we compare phototaxis in hosts from humid environments to those from dry environments. Will this in turn affect how their parasites might alter phototaxis? What constraints limit the ability of parasites to alter host behavior and where do we expect to find them? At the root of all this is the assumption that the life cycles that we study are those that actually function in nature. In fact, while many organisms may support some generalist parasites in experimental contexts, it is much more difficult to determine which of these hosts is responsible for the majority of that parasite's survival and reproduction in nature. These and other questions are central to our understanding of how parasite-induced behaviors have come to be.

They are also central to our understanding of how these behaviors affect communities and ecosystems. If parasitized hosts demonstrate altered habitat preferences, react differently to stimuli, forage at different times of day or prefer different foods or unusual quantities of food or interact differently with predators – to name a few common alterations – if they do all this and more, then what is the effect on their participation in ecological interactions? We know from 'simple' two-species competition experiments that a protistan parasite can reverse the outcome of competition between two species of flour beetles; the consequences that are possible in a complex ecosystem challenge the imagination.

See also: Beyond Fever: Comparative Perspectives on Sickness Behavior; Evolution of Parasite-Induced Behavioral Alterations; Parasite-Induced Behavioral Change: Mechanisms; Parasite-Modified Vector Behavior; Reproductive Behavior and Parasites: Vertebrates.

Further Reading

Haine ER, Boucansaud K, and Rigaud T (2005) Conflict between parasites with different transmission strategies infecting an amphipod host. *Proceedings of the Royal Society B* 272: 2505–2510.

Leung TLF and Poulin R (2007) Recruitment rate of gymnophallid metacercariae in the New Zealand cockle *Austrovenus stutchburyi*: An experimental tests of the hitch-hiking hypothesis. *Parasitology Research* 101: 281–287.

Moore J (2002) *Parasites and the Behavior of Animals Oxford Series in Ecology and Evolution*, p. 315. New York: Oxford University Press.

Seppala O, Karvonnen A, and Valtonen FT (2006) Host manipulation by parasites and risk of non-host predation: Is manipulation costly in an eye fluke-fish interaction? *Evolutionary Ecology Research* 8: 871–879.

Thomas F, Adamo S, and Moore J (2005) Parasitic manipulation: Where are we and where should we go? *Behavioural Processes* 68: 185–199.

Internal Energy Storage

A. Brodin, Lund University, Lund, Sweden

Introduction

If you hear the expression 'she eats like a bird' you might imagine a thin woman who eats very little in a delicate way. It would be more appropriate, however, to think about the Monty Python character, Mr. Creosote from the film 'The meaning of life.' He is a disgustingly fat character who eats an incredible amount of food before he finally explodes. Birds do not eat until they explode, but a small bird in a cold environment will have to eat so much food that it can gain 10% of its lean body mass in fat – on a daily basis! Not all food can be transformed to fat – such a gain will require an awful lot of eating.

For most animals, food comes in bouts even though they need energy continuously. This means that energy supply to the cells in the body must be evened out relative to how food is ingested. In animals with developed digestive tracts, food will pass through these during digestion, but it cannot be saved for long here, because digestion will soon break it down. Still, food under digestion may be an important energy buffer. Seed-eating birds such as grouse and finches may fill their digestive tract with food in the evening and use this for energy expenditure over night.

Energy from digested food will be released into the bloodstream, for example, as glucose or free fatty acids. If more energy is available than required for immediate metabolism, the surplus will be stored as carbohydrates or triglycerides. Carbohydrates are stored as the polysaccharide glycogen. A triglyceride is an alcohol glycerol esterified with three fatty acids. A triglyceride that is fluid in room temperature is called an 'oil,' whereas one that is solid is called a 'fat.' Usually, plants store triglycerides as oils, whereas animals store them as fats. The difference between these makes most vegetable oils better for human health than animal fat. The former are less saturated (meaning that have more double carbon bounds), and do not contain cholesterol. In marine animals and in insect cuticles, fatty acids can be stored as wax esters. Proteins are also a form of stored energy, but have other primary purposes.

Cellular respiration metabolizes glucose and releases energy. This process starts with the citric acid cycle. This well-known metabolic pathway transforms the glucose molecule into acetyl coenzyme A, and other pathways transform fat and proteins into this coenzyme. Via anabolic (builds up) and catabolic (breaks down) reactions, fat and carbohydrates (and proteins) can be seen as strategic alternatives for energy storing. For example, surplus glucose may be stored as fat rather than the more natural polymer carbohydrate, glycogen. This explains why humans can get fat from eating carbohydrate-rich food.

Storing Fat or Carbohydrates?

Glycogen occurs in the muscles or in the liver. Since carbohydrates are more oxygen efficient than fat, they are suitable for short-term high power work. Fat, on the other hand, is hydrophobic and therefore less heavy. Since each fat molecule contains three long fatty acid chains, fat is more energy efficient per gram than carbohydrates and thus suitable for long-term energy storing in the body. A person who expects a long period of starvation should build up large fat reserves. Someone who is going to participate in a marathon run, on the other hand, should build up glycogen reserves a day or two before the race.

Flying animals such as birds build up and metabolize fat deposits on a daily basis. The reason that they have developed the ability to use fat efficiently in such a short-term perspective is that they need to stay light, but flying animals will also store energy in the form of carbohydrates. At least in insects, long-distance flyers tend to store energy as fat, whereas short-distance flyers store energy as glycogen.

Amphibians provide an example of this temporal difference in storage strategies. Species that breed explosively mainly rely on glycogen, whereas species with prolonged breeding periods mainly rely on fat as a stored energy source.

Strategic or Environmental Control?

Since flying organisms are sensitive to increases in body mass, natural selection should act strongly to favor an optimal level of fat storage for these animals. Investigators have explored the idea of an optimal and precisely regulated fat storage system in at least three avian study systems: (1) small passerines in cold winter environments, (2) fuelling and energy regulation in migrating birds, and, (3) overnight energy regulation in hummingbirds.

It is important to realize, however, that constraints will often prevent animals from regulating their fat reserves optimally. In a study in South-central Sweden, subordinate willow tits carried larger fat reserves than dominant birds. This may make sense because dominant individuals get to eat first when a foraging flock encounters food. So, dominants experience a more predictable food supply and they can afford to carry less fat than a subordinate that experiences less predictability. The dominant willow tits were minimizing predation risk by staying lighter than the subordinates.

This study has later been replicated independently twice. Both replicates got the opposite result; dominant individuals carried larger fat deposits than subordinates. One might ask who was right then, should dominants carry larger or smaller fat deposits than subordinates? The answer is that both were right! The replicates were done in field sites further north than the first study. In these regions, the climate is harsher. Under such conditions, dominants used their priority access to food to grab a larger proportion of the food. The subordinates would probably have carried even more fat than the dominants if they had been able to, but the limited amount of food in the environment prevented this. In the first study, the birds had strategic control of their fuel reserves, while the subordinates in the replicates carried suboptimal fat deposits due to environmental restrictions.

Fat in Birds 1: The Little Bird in Winter

For a 10-g bird, simply surviving the winter in a cold northern forest is a remarkable feat. Since surface increases with the square of length while body mass increases with the cube, small homeothermic animals must metabolize more energy per unit body mass than large ones just to maintain body temperature. This means that small passerines must spend most of their time foraging during daylight hours in midwinter. At high latitudes, the short winter days make this especially hard. Furthermore, temperatures may stay below −20 °C for long periods. To survive a cold night under such conditions, species such as willow tits, Siberian tits, and boreal chickadees must gain almost 10% of their lean body mass in fat each day. At night, they will metabolize this fat to maintain body temperature. To repeat this mass gain on a daily basis means that these birds must find food items such as nuts and insects almost continuously. A small mammal that inhabits the same forest faces a less difficult problem because it can forage under the insulating snow cover. Also, most small mammals can extend foraging into the night, but a diurnal bird cannot.

Starvation Risk and Predation Risk

Random fluctuations in overnight temperatures mean that a bird cannot know beforehand how much fuel it will need. Furthermore, the next day's foraging conditions will also vary unpredictability, meaning that the bird cannot accurately predict how much fat it will need in the morning. The only safe option is to carry sufficient reserves for the worst possible case. However, starvation is not the only threat. Predators, especially airborne ones, comprise an ever-present risk. If a predator such as an owl or a hawk attacks, acceleration and maneuverability may be critical to the bird's survival. A light bird that carries small fat deposits will have a better chance of escape than a heavy bird. The small bird in winter thus faces a trade-off between two sources of mortality; avoiding predators favors a lean body but avoiding starvation favors fat storage.

To optimize survival, a bird should minimize total mortality from both these sources. As a bird gains fat, the predation risk increases at the same time as the starvation risk decreases. When the slope of the increase balances the slope of the decrease, the result will be a straight, horizontal line. This line is the derivative of the total mortality, which then must be at its minimum since the slope is zero. This will usually occur where death from predation is much more frequent than death from starvation.

Some evidence suggests that birds can reduce mass-dependent predation risk by increasing flight muscle mass in parallel with body fat. In an experiment, starlings were equipped with extra weights and tested in an obstacle course. Initially, the weights clearly reduced their speed and maneuverability. After carrying the weights for a week, however, they could fly as well as before. Somehow, the starlings compensated for this extra mass, probably by gaining flight muscle mass.

Fat Regulation in Food Hoarders and Nonhoarders

Considering that it is dangerous to carry large fat reserves, one might believe that a food-hoarding species should delay the gain of night time fuel deposits compared to a nonhoarding species. Since a hoarder has access to reliable food (caches), it could wait longer before it builds up its night time fuel. In this way, a hoarder could minimize the time it has to carry this fat load. However, both models and empirical data show that it is the other way around, hoarders gain more fat in the morning than nonhoarders, but not later in the day. This occurs because hoarders always carry smaller body fat deposits than nonhoarders. After a night, fuel reserves will be at their daily low and the hoarder will have to build up a buffer of fat. Nonhoarders, in contrast, do not have access to caches and must therefore hedge for the worst possible night by carrying sufficient reserves. Most nights are not as bad as the worst one, meaning that nonhoarders will experience a carryover of fat until the morning.

Fat in Birds 2: Migration and Fat

When we think of migration, we typically think of high-flying bird flocks, surprising orientation abilities, spectacular long-distance flights, etc. Most of the time, however, a migrating bird will be eating. A migrating bird spends up to 90% of migration time on the ground, refueling, and resting. Migrants of many species will gain considerable store of fat before a long-distance flight, sometimes more than 100% of their lean body mass. This means that some birds may double their body mass before they take off. There is a ceiling, however, to how much fat a flying animal can carry. The maximum flight distance follows a curve of diminishing returns as the animal gets fatter, because the cost of flying accelerates with increasing body mass.

Yet, some birds do make spectacularly long flights. According to records from satellite tracking, for example, a female bar-tailed godwit flew directly from Alaska to New Zealand, a distance of more than 11 000 km. Judged from her even speed, she did not land once during this 8-day flight. The Black brant, an American subspecies of the brent goose, regularly fly overseas from the Alaskan peninsula to the coast of Baja Californica. This will be shorter than a shorter crossing from Alaska to British Columbia followed by southward migration along the coast. However, the direct passage to Baja California is potentially dangerous since the geese cannot feed in the open ocean. Still, most geese prefer this route in autumn, even though they follow the coast when they return north in the spring. The reason for this difference is probably that there are reliable southward tailwinds in autumn.

On migration flights, birds usually burn both proteins and fat; we can think of the energy stored for migration as fuel rather than just fat. Several factors affect how much fuel migratory birds need before a long flight. If arriving quickly is important, birds should carry large fuel deposits. This may occur in the spring when migrants need to reach their breeding territories before competitors. In other cases, it may be more important to minimize energy expenditure. In such cases, a migrant should migrate in shorter steps, taking on smaller amounts of fuel at each stop. In cases when food is abundant and the bird is not in hurry, it may be more important to minimize predation risk.

Fuelling before migration is usually referred to as 'premigratory fattening,' a phenomenon that has been studied in some detail. Not surprisingly, birds respond adaptively to variation in the energy requirements of migration, for example, migrants behave as if they 'know' how much fuel they will need to make their next flight. It has been demonstrated that they will carry more fuel before a long nonstop flight than before other flights. Especially long flights are necessary to pass over barriers such as seas or deserts, where foraging and/or resting is hard or impossible.

Even first-time migrants load more fuel in anticipation of long flights, so this cannot be learned from prior experience. Environmental factors such as regional differences in the earth magnetism might provide cues for this. For example, thrush nightingale's winter in tropical Africa and the long passage across Sahara requires larger fuel deposits than average-distance laps on the European continent. First-year thrush nightingales caught and tested in Sweden increased fuel loading when investigators exposed them to a magnetic field that simulated a location in Egypt.

Even if premigratory fattening has primarily been studied in birds, it also occurs in other migrating animals. Many species of large whales migrate between cold areas where food is abundant and warm calving areas. Many baleen whales prefer prey such as krill that they can find in high concentrations only in Arctic and Antarctic waters. Newborn calves have relatively thin blubber layers meaning that they do not possess sufficient insulation for cold water. In the calving areas, however, there is not much food for baleen whales. For many species, the fat they have stored before migration will be the main energy source until they return to their Arctic/Antarctic foraging areas. Especially pregnant females must build up large premigratory fat deposits since they will need to sustain not only themselves but also their calves for many months before they can resume serious feeding.

The migration of the gray whales along the American west coast is well known since the whales in many places can be observed from land. In summer, the gray whales feed in the Chukchi and Bering seas off Alaska. In October, the whales will start their journey south. Compared to birds, the whale migration proceeds slowly, the bulk of the whales will reach the lagoons they prefer in Baja California in mid-February to mid-March. (The brant goose makes roughly the same trip in 3 days!) These lagoons will then be filled with mating and calving whales. Since the whales follow the coast, this will be the longest known regular migration in any mammal.

Hibernation and Fat

Hibernation is a hypothermic condition when animals save energy during cold seasons by lowering body temperature, metabolism, breathing, and heart rate. This makes it possible for an animal to survive the winter on a fat deposit that would not last the winter. Animals usually hibernate in a protected, well-insulated site. There are several forms of hibernation: true hibernation, torpor, and denning. The term 'hypothermia' is general, because it can occur in a range of situations from hibernation to cooling during accidents.

This terminology is not always agreed on, I will use 'hibernation' for the deep dormant state from which animals cannot wake up quickly. Torpor has sometimes been used for true hibernation, but here I will use 'torpor' to refer to more temporary hypothermic conditions, for example an overnight reduction in body temperature. Denning describes the situation, common in bears and badgers, in which animals enter a state similar to deep normal sleep, with only a slight reduction in body temperature. People commonly call this 'hibernation,' but this is strictly not correct.

Truly hibernating animals may lower body temperatures to a few degrees above freezing. At such low temperatures, heart rates and breathing are lowered to a minimum. In many homeothermic animals, hibernation will be interrupted by arousals when the animal restores normal body temperature before entering hibernation again. Examples of true hibernators are bats, hedgehogs, dormice, marsupials, and snakes. Among birds, some nightjars such as the common poorwill may enter true hibernation.

In autumn, before they go into winter hibernation, bats may use temporary torpor as a preparation for hibernation. At such occasions, they will roost in relative cool locations even if warmer ones are available. The lower ambient temperature makes it easier for them to decrease body temperature and metabolism during daytime roosting, thereby saving body fat for the winter hibernation.

While true hibernation and torpor primarily are strategies to save energy, denning can be seen as a strategy to escape food shortage. Before entering a winter den, bears and badgers will build up large body fat stores. These fat deposits will be sufficient for the winter if the animal stays near its BMR, basal metabolic rate, that is, if it sleeps. They will not be sufficient otherwise, because an active animal's metabolism would be several times larger than the BMR. The advantage of this condition compared to true hibernation is obvious. A bear or a badger will have a body core temperature that almost is normal. This makes it possible to care for cubs, react to danger, etc. when they are in the den. That's why you should never poke a bear – even if it is 'hibernating.'

Regular fat is white but there is also brown fat (BAT, brown adipose tissue). This type of fat will not only insulate, but also generate heat. Brown fat contains many more mitochondria and capillaries than white fat. Brown fat occurs in newborns, but it has also been found in hibernating animals, for example bats. Brown fat will generate heat when the bat is leaving its hypothermic state. It is believed to be important for arousal.

The benefit of the hypothermic condition in a hibernating animal is obviously to reduce energy expenditure. Hypothermia can therefore be seen as another form of energy storing, the balance can be improved not just by storing, but also through saving. Chickadees and titmice in the boreal winter forest are not only large-scale food hoarders and finely tuned fat regulators. They will also exhibit a night time torpor that is better developed than in close relatives living in temperate climates. On a cold night, a willow tit or a black-capped chickadee will reduce its body temperature down to 35° C (from 41° C), thereby saving a substantial amount of energy. This strategy is not without a cost, however. At dawn, it will take a hypothermic

bird about 15 min to regain normal temperature. Meanwhile, it will be stiff and immobile, and this may be fatal if a predator passes by.

The ability of 10-g birds to spend the winter in the boreal forests is impressive, but a hummingbird's abilities to regulate energy may be even more impressive. Many species have a lean body mass of 3–4 g. Due to their tiny size, they have an even higher metabolism than chickadee-sized birds. Hummingbirds need to eat several times their own body mass each day. Maintaining their daytime metabolic rate during a cool night would require larger fat deposits than they can carry. To solve this problem, hummingbirds have developed special torpor adaptations. They can decrease night time body temperatures to a state only a few degrees above the environment. They will not, however, lower it to near freezing, as hibernating bats do. If the ambient temperature drops below around 7°, the hummingbird will increase metabolism so that its temperature does not drop below this.

Reproduction and Fat

Reproduction is energetically costly. Reproducing individuals must be in good condition and have sufficient fat reserves to breed successfully. Female lizards, for example, store fat in their tail that they can use during egg production. Female geese breeding in the far north carry larger fat deposits than conspecific breeders in southern regions. For geese migrating to arctic breeding grounds, early arrival is an imperative. Fat may be critical for these early arriving geese, because food in the form of fresh plants may not be available when they arrive. Also, fat females can spend more time in the nest than a lean female who has to spend more time foraging. This is important in the far north, where bad weather is dangerous for eggs and nestlings that are not covered by a female.

Female mammals, for example humans, will gain fat during pregnancy and this will be an important energy buffer for lactation. Female seals provide an extreme example of this, in some species, such as the gray seal, the mother will fast and stay with the pup during the whole lactation period. A gray seal pup will gain almost 3 kg of fat each day during the 16-day lactation. The mother may lose almost 40% of her body mass during lactation.

The king penguin has a breeding period that lasts longer than a year. It breeds on sub-Antarctic islands. During the southern hemisphere summer, the parents will feed the chick intensively. In autumn, both parents leave the island and depart for winter foraging areas. The chick is still downy and it cannot forage or enter the water. Cuddled together with other chicks, it will spend the winter standing on land, living on stored body fat. In spring, the parents will return and resume provisioning. After a couple of months, the chick will lose its down, fledge, and become independent. The amount of body fat in autumn is crucial, because lean chicks experience very high winter mortality.

Food Hoarding Versus Fat Storing

As discussed earlier, we expect food-hoarding birds to carry smaller fat reserves than nonhoarding species. The extent to which animals can substitute stored food for body fat, however, depends on how accessible the stored food is. Food in a larder or a granary can substitute for fat effectively, because the hoarding animal can easily access it. Scatter hoarded food, on the other hand, may not always be a good alternative to body fat, because it take more time to recover the individually hoarded items. In this case, a good spatial memory (that will speed up retrieval) can make it easier to substitute hoarded food for fat. In contrast, a hoarder that does not remember exact caching positions may have to spend time searching for the caches before they can be eaten. In some cases, hoarded food might thus serve as an alternative to body fat, but there are other cases when it instead should be seen as a way of increasing food availability.

Some Current Questions

The question of how migrating birds can adjust fuel levels to local conditions, even when they are inexperienced, continues to attract interest. The topic under study here is the relationship between migrative fattening and geomagnetic cues/latitude. Another field discussed in this article that still attracts much interest is adaptations to cold climates in small animals, especially small birds in winter.

In some current studies, fat is seen as one aspect of a larger question: energy regulation in general. Fat deposits cannot be considered in isolation; optimal energy regulation includes energy-saving behavior and hypothermia.

Body fat deposits shows distinctly at magnetic resonance scanning images (MRI). This will make it possible to measure fat contents on living animals with higher precision than in older studies.

It is not known how fast birds can gain flight muscles in parallel to fat deposits, but it has been suggested that it may occur on a daily basis. This would explain why bird flight in some cases seems less affected by increased fat deposits than predicted by physical laws.

Summary

Energy is stored in the body either as fat or carbohydrates. In mammals, fat is primarily long-term fuel, whereas it may serve also as short-term fuel in birds. There is a strategic aspect of fat; especially in flying animals the amount of body fat carried is under strong selection. Lean animals may starve to death, fat animals may be easy to capture for attacking predators. Several study systems in birds have been thoroughly studied, for example small birds in cold environments and optimal fuel loading during migration.

See also: Amphibia: Orientation and Migration; Bat Migration; Bats: Orientation, Navigation and Homing; Bird Migration; Caching; Fish Migration; Insect Migration; Insect Navigation; Irruptive Migration; Magnetic Compasses in Insects; Magnetic Orientation in Migratory Songbirds; Maps and Compasses; Migratory Connectivity; Pigeon Homing as a Model Case of Goal-Oriented Navigation; Sea Turtles: Navigation and Orientation; Vertical Migration of Aquatic Animals.

Further Reading

Bednekoff PA (2007) Foraging in the face of danger. In: Stephens DW, Ydenberg RC, and Brown JL (eds.) *Foraging*, pp. 305–329. Chicago: Chicago University Press.

Brodin A (2007) Theoretical models of adaptive energy management in small wintering birds. *Philosophical Transactions of the Royal Society of London Series B: Biological Sciences* 362: 1857–1871.

Brodin A and Clark CW (2007) Resource storage and expenditure. In: Stephens DW, Ydenberg RC, and Brown JL (eds.) *Foraging*, Ch. 7, pp. 221–269. Chicago: Chicago University Press.

Witter MS and Cuthill IC (1993) The ecological cost of avian fat storage. *Philosophical Transactions of the Royal Society of London Series B: Biological Sciences* 340: 73–92.

Interspecific Communication

I. Krams, University of Daugavpils, Daugavpils, Latvia

Introduction

Interspecific communication usually deals with signals that provide a receiver with information about the heterospecific sender, which includes the sender's aggressive intentions in the case of interspecific aggression, territorial status, which can be important in the case of interspecific competition (undoubtedly a common interaction among animals), suitability as a mate in the case of interspecific hybridization, and current actions which are important signals for the coordinated activities of group members to maintain group cohesion. Interspecific communication may also deal with signals in which the sender provides the heterospecific receiver with referential information about objects external to itself. Such objects could be other individuals of the same trophic level, accessible resources such as food, or predators.

Most often interspecific signaling occurs between ecologically related species which usually belong to the same guild, to alert both conspecifics and heterospecific members of mixed-species groups of potential danger and recruit nearby individuals for mobbing defense. Examples include mixed troops of monkeys, mixed-species flocks of birds, and fish schools in the tropics. However, interspecific communication can also occur between nonadjacent trophic levels to warn a predator that it has been spotted, or involve prey individuals signaling to the predators of their predators. In some cases, signal use occurs between species of the same or similar trophic levels, but under circumstances that do not involve individuals of the same social group. This is called eavesdropping, also known as information parasitism, which occurs when an individual other than the intended receiver cues in on public sensory information.

Within-Trophic-Level Interspecific Communication

Within-Trophic-Level Social Integration Signals of Heterospecific Animals

There is a variety of social contexts in which individuals belonging to two or more species must integrate their activities to achieve a common goal. Protection against predators is among the most important reasons for individuals to become members of a multispecies group. The integration of members of group-living heterospecifics to maintain group cohesion, to coordinate movement, and to organize communal activities of all members of multispecies groups should be synchronized in such a way that they enhance predator protection. Social integration requires recognition of key target individuals (such as conspecific and heterospecific group members), appreciation of both interspecific and intraspecific dominance hierarchies, and honest signals. For example, mixed-species aggregations of breeding birds and mixed-species foraging groups may result from an active search for heterospecific companions. Recognition of neighboring species is involved in these assemblages. It is well established that birds actively aggregate in mixed-species foraging groups to gain protection against predators and/or enhance feeding efficiency, supporting the view that positive ecological interactions are important in structuring some breeding bird communities.

Signaling an internal state to heterospecifics

Many animals stay as members of mixed-species groups where interspecific competition restricts access both to food and to the safest and most preferred feeding sites. In such groups, the position of the individual in the group's intraspecific and interspecific dominance hierarchy affects its resource access and survival. Larger animals are usually more dominant in interspecific dominance hierarchies. Given the stability of interspecific dominance hierarchies, the individual's ability to recognize aggressive intentions by heterospecific group members allows a potential threat posed by more dominant individuals of other species to be more easily recognized. The ability to understand the aggressive signaling by other species makes the groups consisting of heterospecific individuals more coherent and decreases the probability of overall aggressiveness and being injured during the interference competition between social species.

Interspecific signaling of mating and territorial status

Animals living in complex communities need to be able to identify other species and retain memories of species identities for long periods of time to make optimal choices. The ability to discriminate between conspecifics and heterospecifics may be crucially important in the refinement of mating preferences during speciation and for premating isolation of sympatric species. Various species defend territories not only against conspecifics but also against certain other, usually congeneric, species. In birds, interspecific interactions over territoriality are effected through interspecific reactions to territorial songs. Where *Regulus* species are interspecifically territorial in Spain, both goldcrest (**Figure 1**), *Regulus regulus*, and firecrest, *R. ignicapillus*, can sing species-specific or mixed songs, and most males react aggressively to the song of the other species. This example shows how song convergence between two species of songbirds may play an important role in resource partitioning by signaling to a heterospecific competitor. Recently, it has been shown that male blackcaps (**Figure 2**) *Sylvia atricapilla*, can associate species-specific songs with species-specific plumage and that they can retain the memory of this association for an 8-month period without contact with heterospecific rivals, indicating that signals from two different sensory modalities (visual and auditory) are important for distinguishing conspecifics from heterospecifics.

Multispecies group integration

Group living serves a variety of adaptive functions and social animals need special signals to coordinate their movements and other activities to facilitate social benefits such as warning and defense against predators. The level of group integration depends largely on mechanisms used to recognize group members and signals used to maintain group cohesion. When group membership is unstable, a group label is meaningless and groups are more appropriately called aggregations. Examples include swarms of insects, migratory flocks of birds, and herds of antelopes.

In contrast, group labels can be critical in socially cohesive mixed-species assemblages. Individual recognition based on familiarity with group members is a common mechanism used to distinguish group members from nonmembers. Groups must be relatively small and stable in composition for this mechanism to operate. In multispecies groups, individual recognition facilitates the development of stable dominance hierarchies within groups because each individual can remember the outcome of prior encounters with other members. In mixed groups of the tits (*Paridae*), crested tits (**Figure 3**), *Lophophanes cristatus*, usually dominates over willow tit (**Figure 4**), *Poecile montanus*, individuals. In some cases, willow tits may rise in rank above crested tits by demonstrating aggressive postures and giving aggressive calls, and these changes in dominance hierarchy are permanently accepted by all crested tits.

Figure 1 Goldcrest, *Regulus regulus*.

Figure 3 Crested tit, *Lophophanes cristatus*.

Figure 2 Blackcap, *Sylvia atricapilla*.

Figure 4 Willow tit, *Poecile montanus*.

Coordination of multispecies group movement

For the majority of group-living species, the advantages of grouping can accrue only when group members remain in close proximity. When groups move over large areas as a unit, mechanisms for maintaining group cohesion are required. The signals used to coordinate group movement vary depending on the function of grouping, the precision required, and the mode of locomotion. In a wide variety of social species, three types of vocalizations are found: (1) soft, frequently repeated contact calls with individual signature, (2) louder and longer separation calls by individuals that have become separated from the group, and (3) movement initiation calls, which are the loudest of the three types and are structured to be easily localizable. The existence of such vocalizations is consistent with the ideas of Marler about signal design features and motivation–structure rules proposed by Morton suggesting that selection would favor the use of tonal, high-frequency vocalizations in fearful and friendly contexts because these sounds symbolize smaller size or juvenile age class, thus reducing the likelihood of attack by the receiver of the call.

Within-Trophic-Level Interspecific Referential Communication About External Objects

Mixed-species groups are often observed in birds, primates, and cetaceans. However, direct measurements of the fitness value of gregariousness and the understanding of principles of interspecific communication in these animal species are still in their infancy. Predation represents a strong selective factor in prey populations, resulting in the development of complicated and unique behavioral defenses. The trade-off between maintaining antipredator vigilance and devoting resources to other activities has been ameliorated in many group-living species by the development of calls accessible to members of a local community. The most important signals about external objects used in heterospecific company include resource-recruitment signals, distress signals, and alarm signals. Alarm signals can be further categorized as low-risk warning signals, which may be followed by on-guard signals, inspection or mobbing signals, and high-risk warning signals.

Resource-recruitment signals

A variety of social primates and birds produce a food call while at the location of a newly found food source. Call rates usually increase with food preferences and quantity found. In primates, an individual out of sight of other group members is more likely to produce a discovery call than one with others nearby. This suggests that the calls are used to ensure sufficient group sizes for predator surveillance during feeding. However, actual field data on food calls among individuals belonging to different species are surprisingly rare, despite the fact that social

foraging should reduce the risk of energetic shortfall, and that resource-recruitment signals are well known among members of groups consisting of a single species. Nonetheless, such calls might be expected to evolve because some associations composed of heterospecific animals (such as mixed-species flocks of temperate tits and mixed-species associations of dolphins) are most convincingly explained by the foraging advantage hypothesis.

Alarm calls

Alarm calls are among those signals that are usually shared by ecologically related sympatric species. Mixed-species associations can reduce group members' risk of predation through dilution effects, predator deterrence, and improved detection. However, a positive association between the group size and predator detection efficiency relies on the ability of the members to communicate. For this reason, the evolution of social complexity is often associated with the evolution of a large vocal alarm repertoire. Not surprisingly, such animals as passerine birds, primates, and cetaceans having large alarm call repertoires are predisposed to grouping with heterospecifics and responding to heterospecific alarm calls. Defining sociality as group living in which members interact and form relationships, several researchers have found that recognition of heterospecific alarm vocalizations is an essential component of antipredator behavior, especially in primates. Their alarm calls are very similar structurally, and all members of a mixed troop respond to any species' alarms.

Several studies have shown that the forest wintering tits save vigilance time not only by flocking with conspecifics but also by associating with other tit species, goldcrests and treecreepers, *Certhia familiaris*. Heterospecifics in such mixed-species flocks are generally considered to substitute for conspecifics as predator protection at lower competition cost. However, this poses several important questions related to the causes of alarm calling when in the company of heterospecific individuals. Will members of multispecies groups give alarm calls irrespective of the absence of conspecifics? Will an individual belonging to one species warn individuals belonging to the other species – in other words, how reliable are heterospecific alarm calling? It is possible that even within the same flock, the compensatory benefits of emitting warning signals may vary with the participant, which suggests that the field of alarm calling among heterospecifics and the origin and evolution of warning communication are still puzzling and largely unanswered.

Mobbing behavior

A simple definition of mobbing is the development of an assemblage of individuals around a potentially dangerous predator. In 1960s, the eminent Austrian ethologist Konrad Lorenz used the English term mobbing to describe the behavior that animals use to scare away a stronger, predatory enemy: 'a number of weaker individuals crowd together and display attacking behavior, such as geese scaring away a fox.' This well-defined behavioral pattern occurs in a wide diversity of animals, especially in birds and mammals. If two territorial neighbors cooperate during mobbing, they have an increased opportunity to drive the predator from their breeding area. Under natural conditions, mobbing often occurs in a company of heterospecifics and they benefit from the antipredator behavior of other species. This implies that individuals belonging to different species should recognize mobbing calls of other species and that there is a possibility of reciprocity among heterospecifics. This is so because the emission of mobbing signals puts the mobber in jeopardy, and an altruistic act helping a nonrelative only pays the altruist if it is directed at a particular heterospecific individual that on a later occasion reciprocates. Although human behavior abounds with reciprocal altruism, few examples exist documenting reciprocal altruism in animals, while experimental evidence on interspecific mobbing as a reciprocity-based behavior is so far lacking.

Within-trophic-level distress signals

Prey individuals often give distress calls in the final stages of predator attack or being caught in a trap. These signals are fundamentally different from alarm calls and mobbing calls. Among avian species, distress calls are notably similar in structure, generally consisting of short, screaming repeated bursts of unmusical sound covering a wide range of frequencies, characteristics that increase the effective distance and localization of the call. The calls seem to be convergent among different bird species. Parents of many animal species will respond to distress calls of their own offspring. However, intraspecific responses by the vespertilionid bats, *Pipistrellus pygmaeus*, showed that they respond to the distress calls of unrelated conspecifics, thus rejecting the hypothesis that distress calls request aid from kin. Moreover, some authors have observed interspecific responses to the distress calls of the phyllostomid bat, *Artibeus jamaicensis*.

A number of hypotheses have been put forward to explain the function of distress calls. Some of the explanations require strong reciprocity links between sender and receiver before it pays a receiver to respond. For example, an individual giving distress calls may warn other heterospecific individuals of the presence of a predator, provide other individuals with information about a predator (and thus reduce the chance of those individuals falling prey to a predator in the future), or attract other individuals that will mob a predator (which may in turn facilitate the caller's escape). Distress signals may therefore resemble a request of aid from reciprocal altruists. While there is some evidence for these hypotheses, within-trophic-level distress calling has not, to date, been studied in detail.

Within-trophic-level interspecific eavesdropping

Animal communication networks can be extremely complex, consisting of many signalers and many receivers. The idea about communication network follows from the observation that many signals travel further than the average spacing between animals. This is self evidently true for long-range signals, but at a high density the same is true for short-range signals such as begging calls of nestling birds. Growing evidence clearly shows that alarm calls elicit responses not only from conspecifics but also from eavesdropping heterospecifics. If an individual intercepts information from the alarm calls directed toward another individual, this is termed 'interceptive eavesdropping.' This differs from 'social eavesdropping,' which involves obtaining information from an interaction between two signalers. Numerous studies have shown that mammals, birds, amphibians, and fish recognize alarm signals of other species, and interceptive eavesdropping can even occur between taxa that are not closely related. For example, vervet monkeys, *Cercopithecus aethiops*, respond to the alarm calls of superb starlings, *Spreo superbus*, dwarf mongooses, *Helogale parvula*, recognize hornbill, *Tockus* spp., alarm calls; and red squirrels, *Sciurus vulgaris*, respond to the alarm calls of jays (**Figure 5**), *Garrulus glandarius*. Although these examples show the ability of many animals to recognize the alarm calls produced by other species, the amount of information they glean from these eavesdropped signals is largely unknown.

Instead of using a general type of alarm call, some species have categorically distinct vocalizations that are each associated with a different type of predator encounter (e.g., different calls that refer to aerial vs. terrestrial predators). There are a few studies that show that birds and mammals discriminate among categorically distinct types of heterospecific vocalizations. For example, the closely related Diana monkeys, *C. diana*, and Campbell's monkeys, *C. campbelli*, recognize each other's acoustically distinct leopard and eagle alarm calls and treat them

Figure 5 Northern (or Eurasian) jay, *Garrulus glandarius*.

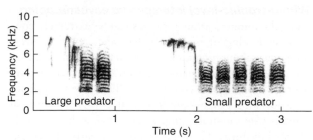

Figure 6 Chickadee (*Poecile atricapillus*) alarm calls encode information about predator size and risk.

similar to conspecific alarm calls. A sympatric bird, the yellow-casqued hornbill, *Ceratogymna elata*, also differentiates between these two types of monkey alarm calls. Similarly, vervet monkeys discriminate among the aerial and terrestrial predator alarm calls of superb starlings.

Instead of or in addition to using acoustically distinct vocalizations, some animals encode considerable information about predators in more subtle variations of a single type of call. It was recently shown that *chick-a-dee* alarm calls (**Figure 6**) of black-capped chickadees, *P. atricapillus*, encode a surprising amount of information about the size and threat of a given predator through variations in the acoustic structure of the calls, demonstrating that chickadees have one of the most sophisticated mobbing call systems yet discovered.

Complex information, including predator size and type, can potentially be gleaned from these calls by heterospecific individuals. This idea is supported by experimental work revealing that red-breasted nuthatches, *Sitta canadensis*, a temporary flock-mate of black-capped chickadees, respond appropriately to subtle variations of these 'chick-a-dee' calls, thereby showing that they have gained important information about potential predators in their environment. These findings clearly demonstrate that our knowledge about complexity and adaptiveness of communication between species is surprisingly superficial.

Intertrophic-Level Interspecific Communication

Signals exchanged by animals within the same trophic level are likely to be cooperative in the sense that the sender is providing useful information to the receiver, whereas signals exchanged between trophic levels occur when conflicts of interest between sender and receiver may be really high. This guarantees signal honesty, which is required before the receivers attend to the signal. Many animals give signals often to individuals belonging to species of different trophic levels (prey individuals vs. predators). Such signals can be divided into those that cause a receiver to approach the sender and those that encourage the receiver to repel.

Interspecific Attraction Signals

Distress signals

A number of animals give loud distress calls when attacked by predators. Some may release pheromones from damaged tissues as soon as they are attacked. Growing evidence suggests that most animal distress signals function to attract predators other than the one attacking the distressed individual. If the additional predators distract or interfere with the first, the prey may have a chance to escape. A more complex function of distress signals is the attraction of predators of the attacking predators. This could be viewed as behavioral manipulation, although there are no definitive examples of this manipulation. However, such interactions have been described in a system consisting of plant, herbivore and herbivore's parasites or predators.

Interspecific Repellent Signals

Notification of predator detection

A signal emitted in the presence of a predator could be not only an alarm intended to warn other prey individuals, but it can also be designed to deter the predator from attack. Stalking predators rely on surprise, and if a prey individual detects a predator and signals this fact to it, the predator usually gives up its current hunt, since it is now likely to be unsuccessful. Loud alarms of a variety of birds and primates, foot stamping by kangaroo rats, and tail wagging by many lizards, all appear to be signals to notify a predator that it has been detected. Ground squirrels have recently been shown to signal to potential predatory rattlesnakes by waving their tails that they have heated through increased blood flow. Clearly, a predator will attend to such signals only if they are largely honest. Prey individuals that perform the signal repeatedly, whether a predator is stalking or not, risk having their signal ignored by the predator. Therefore, many prey animals give just a few calls, which are usually enough to signal to the predator that it has been spotted.

Predator inspection and mobbing

In many animals, calls of notification of detection may be followed by predator inspection. Predator inspection ensues when the assembled prey conspicuously monitor the predator's subsequent activities, which is often followed by mobbing in some species. Mobbing alarm calls are produced in response to perched raptors or terrestrial predators. Mobbing behavior may have several different functions, but one of the most important is that the mobbing is likely to harass the predator enough to drive it from the area so that it does not surprise the birds later. This view is supported by anecdotal observations suggesting that when mobbing perched raptors, passerine birds tend to approach more dangerous stationary predators more closely than less dangerous ones, probably in order to ensure its desertion from the vicinity.

Notification of condition

Prey in good enough condition to escape pursuit may benefit by signaling this fact to a predator when it has already begun an attack. Prey that failed to signal would therefore be at a disadvantage, thus providing a strong incentive for prey in poor condition to cheat and produce the signal regardless of condition. Predators, in turn, should rely on such signals only if they are honest indicators of condition. One mechanism that would ensure signal honesty is the 'handicap principle': honesty is ensured if the signal is sufficiently difficult or costly to produce that only high-quality prey are able to produce it. This appears to be the case for the jumping gait called stotting used by a number of African antelopes. Once wild dogs begin a chase, some Thompson's gazelles, *Gazella thomsoni*, stot and some do not, and those that do, stot at different rates. The stotting itself neither hinders nor helps the gazelle to escape once it is chased by a dog. However, it appears to be difficult for a gazelle in poor condition to stot at high rates. Wild dogs preferentially focus their chases on gazelles with lower stotting rates. In contrast, Thompson's gazelles rarely stot for stalking cheetah, in which the attack is quick and there is little time for targeted prey to respond before the predator is upon them.

Aposematic signals

Animals armed with unpalatable and even toxic spines often exhibit conspicuous colors or behavioral patterns. Such aposematic signals are clearly aimed at discouraging predators from attacking the senders of such signals; these senders are usually highly conspicuous against the relevant background. Many salamanders, butterflies and other insects, frogs, and skunks are well-known examples.

Eavesdropping of prey signaling by predators

Interspecific eavesdropping by predators and parasites is common and occurs in all sensory modalities. Many mating systems are characterized by conspicuous male sexual displays, where males aggregate and advertise acoustically to attract mates. Chorusing as an example of male aggregation and sexual display is especially prevalent in anurans. Males in aggregations might have advantages in attracting females over males advertising alone. There are also disadvantages to males that join choruses. Several studies have shown that acoustically advertising anuran males have a higher probability of attracting predators such as frog-eating bats, *Trachops cirrhosus*, and philander opossums, *Philander opossum*. Recently, it was confirmed that white storks, *Cionia ciconia*, could discriminate the

Figure 7 (a) *Carothella* blood-sucking flies attacking tungura frogs. (b) Closeup of a fly on a frog's nose.

signals of moor frogs, *Rana arvalis*. This suggests that such predators as frog-eating bats, storks, opossums, and even blood-sucking flies (**Figure 7**), *Carothrella*, are acoustically orienting predators during the reproductive season of frogs and that chorusing is costly in terms of predation.

Learning of Heterospecific Signals

Alarm signals contain some of the most pertinent information an animal can learn about its environment: the presence of a predator. Thus, an animal that is able to eavesdrop on the alarm signals of another species may obtain considerable, potentially even life-saving, information. This response could be innate and may be triggered by acoustical properties shared between alarm calls. Playback experiments with apostlebirds, *Struthidea cinera*, suggested that experience with a particular species' call is not essential to elicit mobbing; rather, intrinsic aspects of the calls themselves may explain heterospecific recognition.

The heterospecific response can result also from a learned association between predator presence and alarm calls. This was supported while examining the role of learning in the discrimination of heterospecific vocalizations by wild bonnet macaques, *Macaca radiata*, in southern India. The bonnet macaques' flight and scanning responses to playbacks of their own alarm vocalizations were compared with their responses to playbacks of vocalizations of Nilgiri langurs, *Trachypithecus johnii*; Hanuman langurs, *Semnopithecus entellus*; and sambar deer, *Cervus unicolor*. Age and experience appeared to be important factors in heterospecific call recognition by bonnet macaques.

Some other studies of heterospecific eavesdroppers such as the Galapagos marine iguana, *Amblyrhynchus cristatus* (that does not emit any kind of vocalization or auditory alarm signal); the golden-mantled ground squirrel, *Spermophilus lateralis*; the red squirrel, *S. vulgaris*; and Gunther's dik-dik, *Madoqua guentheri* (which do not live socially) also support the idea that the learning of heterospecific alarm signals is important and suggest that sociality is not always a necessary prerequisite for the evolution of flexible associative learning abilities.

See also: Agonistic Signals; Alarm Calls in Birds and Mammals; Communication Networks; Dolphin Signature Whistles; Dominance Relationships, Dominance Hierarchies and Rankings; Ethology in Europe; Food Signals; Group Movement; Honest Signaling; Konrad Lorenz; Referential Signaling; Rhesus Macaques; Social Learning: Theory; Túngara Frog: A Model for Sexual Selection and Communication; Vocal Learning; Wintering Strategies: Moult and Behavior.

Further Reading

Bernal XE, Rand AS, and Ryan MJ (2006) Acoustic preferences and localization performance of blood-sucking flies (*Corethrella* Coquillett). *Behavioral Ecology* 17: 709–715.

Bradbury JW and Vehrencamp SL (1998) *Principles of Animal Communication*. Sunderland, MA: Sinauer Associates.

Caro T (2005) *Antipredator Defenses in Birds and Mammals (Interspecific Interactions)*. Chicago: University of Chicago Press.

Hamilton WD (1964) The genetical evolution of social behaviour. *Journal Theoretical Biology* 7: 1–52.

Krams I and Krama T (2002) Interspecific reciprocity explains mobbing behaviour of the breeding chaffinches, *Fringilla coelebs*. *Proceedings of the Royal Society of London, Series B* 269: 2345–2350.

Lea AJ, Barrera JP, Lauren TM, and Blumstein DT (2008) Heterospecific eavesdropping in a nonsocial species. *Behavioral Ecology* 19: 1041–1046.

Marler PR and Slabbekoorn H (eds.) (2004) *Nature's Music: The Science of Birdsong*. St Louis, MO: Academic Press.

McGraw WS, Zuberbühler K, and Nöe R (eds.) (2007) *Monkeys of the Taï Forest: An African Primate Community Cambridge Studies in Biological and Evolutionary Anthropology*. Cambridge: Cambridge University Press.

McGregor PK (ed.) (2005) *Animal Communication Networks Cambridge Studies in Biological and Evolutionary Anthropology*. New York: Cambridge University Press.

Mönkkönen M, Forsman J, and Helle P (1996) Mixed-species foraging aggregations and heterospecific attraction in boreal bird communities. *Oikos* 77: 127–136.

Morton ES (1977) On the occurrence and significance of motivation-structural rules in some bird and mammal sounds. *American Naturalist* 111: 855–869.

Ryan MJ (1985) *The Túngara Frog, A Study in Sexual Selection and Communication*. Chicago: University of Chicago Press.

Searcy WA and Nowicki S (2005) *The Evolution of Animal Communication: Reliability and Deception in Signaling Systems*. Princeton, NJ: Princeton University Press.

Templeton CN and Greene E (2007) Nuthatches eavesdrop on variations in heterospecific chickadee mobbing alarm calls. *Proceedings of the National Academy of Sciences of the United States of America* 104: 5479–5482.

Trivers RL (1971) The evolution of reciprocal altruism. *Quarterly Review of Biology* 46: 35–57.

Intertemporal Choice

J. R. Stevens, Max Planck Institute for Human Development, Berlin, Germany

Introduction

A Clark's nutcracker (*Nucifraga columbiana*) flies to the top of a pine tree and selects one of the many cones. She twists and pecks at the stem until the cone breaks free from the branch and, with one foot, holds the cone in the crotch of a branch. She then repeatedly hammers her long bill in between the scales of the cone. After forcing out one of the seeds, she tips her head back a bit, clicks the seed in her bill a few times, and closes her bill. A few minutes later, the shredded cone drops to the snow, the nutcracker having extracted several dozen seeds. Each time she closes her bill, the nutcracker makes a choice: she either swallows the seed or places it in a small pouch of skin under her tongue. The seeds placed in her pouch are destined for a site several kilometers away, where she buries them under a bit of dirt or leaf litter. In the autumn, she may bury 33 000 seeds in this manner, only to return to uncover them a few months later during the harsh mountain winter. This form of food storing or caching typifies an interesting and wide-ranging set of decisions faced by animals: intertemporal choices.

The term 'intertemporal' choice refers to decisions in which the benefits associated with different outcomes occur at different times. For instance, for each seed, the nutcracker must choose between eating it now versus waiting to consume it in the winter. Often, there exists a trade-off between the size of the benefit and the cost (time delay), such that larger benefits accrue after longer delays. Thus, the decisions of interest are between obtaining immediate or short-term rewards and investing in a grander future.

Intertemporal Choice in the Wild

Although not usually framed in this way, many decisions that animals face involve a temporal trade-off. The life history trade-off of growth versus reproduction offers an example of balancing the immediate, competitive benefits of growing larger with the delayed benefits of investing in offspring. Even plants and other organisms face these kinds of temporal trade-offs. Although these provide perfectly reasonable examples of temporal trade-offs, most work in this area has explored more active intertemporal choice decisions that often reflect the particular ecology of the individual species. The food caching example provides a nice illustration of species-specific choices

between immediate and delayed consumption in a foraging context. Caching provides a remarkable example of intertemporal choice because of the long delay until food recovery. Many other foraging-based intertemporal choices involve rather short delays: for instance, continuing feeding in the current food patch versus moving on to another patch. In fact, patch exploitation offers a classic example of intertemporal choice from behavioral ecology that is well studied both theoretically and empirically.

Imagine a bird eating berries from a bush. Every berry consumed depletes the patch and increases the average time required to find the next berry. When should the bird stop searching in that patch (after all, there may be no more berries in the bush) and move to the next bush? Staying too long can waste time better used in searching for food elsewhere. Leaving too early can waste opportunities to obtain a quick meal. Optimal foraging theory predicts a simple patch-leaving rule when patches are similar: leave when the foraging rate in the patch drops below the average foraging rate in the environment. We can calculate this foraging rate:

$$\frac{A}{\tau + t + h} \qquad [1]$$

where A represents the amount of food, τ represents the time required to travel between patches, t represents the delay to finding food within a patch, and h represents the time required to process and consume the food. Maximizing this foraging rate results in an optimal solution to the question of intertemporal choice in the patch-foraging situation.

Foraging decisions often involve the temporal trade-offs characterizing intertemporal choices, from patch foraging to caching to decide between a smaller, easier-to-process food item and a larger, more difficult one. Yet intertemporal choice extends far beyond foraging. Returning to the category of life history examples, parental investment exemplifies an important temporal trade-off. Investment in current offspring reduces potential investment in future offspring. How should a parent distribute investment over time? Also, mating decisions have a critical temporal component – along the lines of choosing between Mr. Right and Mr. Right Now. Should an individual accept the currently available mate or continue looking for a higher quality mate?

Cooperative situations may also involve intertemporal choices. Reciprocal altruism, for instance, requires that animals trade the immediate benefits of defection against the delayed benefits of reciprocated cooperation. Clearly, many behaviors fall under the umbrella of intertemporal choices, although they are not typically analyzed with this framework. Even when viewing these decisions as intertemporal choices, researchers use various terms such as temporal discounting, self-control, impulsivity, patience, and delayed gratification. Because of the broad nature of intertemporal choices, they have attracted the attention of many disciplines, including economics and psychology, as well as biology.

Economics of Intertemporal Choice

One of the first disciplines to investigate intertemporal choice was economics. Economists study the consumption of goods and services; thus, they want to know how agents manage streams of benefits that accrue over time. How do they balance the consumption of a small amount of a commodity now compared to a larger amount later? The economic perspective focuses on a 'rational' account of how agents should respond to choices over time. That is, assuming that agents have all information about the goods, what response provides the optimal return? Economists noticed that, when given a choice between the same benefit immediately or in the future (say, one dollar now vs. in 20 years), people preferred the immediate payoff. This effect suggests that people temporally discount or devalue future payoffs. In other words, a delay reduces the

subjective value of receiving a benefit. Why should we discount the future?

Economic theory proposes a number of reasons to discount the future. First, inflation literally makes money less valuable. One dollar will buy more lollipops today than it will after 20 years. Also, individuals can invest currently available benefits. That is, there are opportunity costs associated with not being able to use or invest benefits that are locked away during the delay. Investing one dollar now will yield much more in 20 years than one dollar. Finally, the future is uncertain. A bird in the hand is worth three in the bush because the three in the bush may never be in hand. Future rewards run the risk of not being realized; instead, some force may interrupt their consumption.

These three reasons for discounting are related, and economists have developed a model to account for intertemporal choice. In this model, each option yields a present value:

$$V = \delta^t A \qquad [2]$$

where δ represents a discount factor and t represents the delay to receiving reward amount A. The discount factor δ accounts for the remaining value after a single unit of delay (thus, from δ we can calculate the rate of discounting – the proportional rate of decrease in value). This 'exponential discounting model' has a special feature: the rate of discounting remains constant across the delay (**Figure 1**). So, a reward delayed 1 day after another reward is available will lose the same value if the first

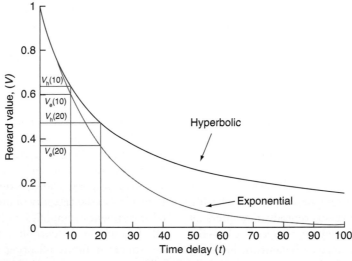

Figure 1   Models of temporal discounting describe how the subjective value of a reward at the present time decreases with the delay to receiving that reward. The exponential model promoted by economists predicts that the rate of discounting is constant over time, whereas, the hyperbolic model promoted by psychologists predicts a decreasing rate of discounting. As an example, we can compare the difference in values for the two models over two time frames: from $t = 0$ to $t = 10$ and from $t = 10$ to $t = 20$. At $t = 0$, both the hyperbolic model ($V_h(0)$) and the exponential model ($V_e(0)$) start with a value of 1. At $t = 10$, $V_h(10) = 0.65$ and $V_e(10) = 0.61$, thus the hyperbolic value decreased by 0.35 in 10 time units, and the exponential value decreased by 0.39. At $t = 20$, however, $V_h(20) = 0.48$ and $V_e(20) = 0.37$. The relative decrease from $t = 10$ is 0.26 for hyperbolic and again 0.39 for exponential. Thus, the decrease remained the same for exponential value but diminished for hyperbolic value.

reward is available today or in a year. In other words, value decreases at the same rate across time.

This constant rate of discounting makes sense when organisms discount because of future uncertainty, that is, when the risk of interruption makes a delayed reward less valuable. If random events interrupt the receipt of delayed payoffs, then decision makers face a constant probability of loss, making discounting to match the environmental loss rate beneficial. Thus, discounting may closely relate to uncertainty and risk.

Psychology of Intertemporal Choice

For decades, psychologists have acted as the fly in the ointment for the elegant economic models of decision making. The psychological approach focuses on describing the actual behavior of decision makers rather than creating models of omniscient, godlike agents. In many cases, the models do not hold up well – behavior deviates substantially from the rational predictions. So, how can we measure the temporal preferences of animals to test the models?

Experimental Methods

Animal experiments often use the 'self-control' paradigm to explore intertemporal choice (**Figure 2(a)**). This typically involves offering a subject the choice between a smaller amount of food after a shorter delay (smaller–sooner option) and a larger amount of food available after a longer delay (larger–later option). Subjects often start with a fixed set of options, then the experimenter adjusts either the long delay or the larger amount to titrate an

indifference point, that is, to find a pair of options between which the subjects choose equally. For example, a subject may first face the choice between two food items available immediately and six food items available immediately. Assuming that the subject prefers the larger amount, a one-second delay is added to the larger option. The experimenter will continue to add one second increments to the large amount until the subject chooses the two immediate food items as often as she chooses the six delayed items. This indifference point then indicates how long a subject will wait for three times as much food. Many psychologists interpret these data as a kind of discounting: the delayed food loses value relative to the immediate food. As discussed later, biologists have another interpretation that does not invoke discounting. Pigeons have been the workhorse for self-control experiments, but rats and primates have been tested using this technique as well.

The 'delayed gratification' technique provides a second method to study intertemporal choice in animals (**Figure 2(b)**). This method mirrors Walter Mischel's pioneering work on delayed gratification in children. In the animal version, a stream of food rewards accumulates over a period of time. For instance, a grape appears in front of the subject every 5 s. The catch is, once a subject interrupts this stream by reaching for or eating the food, the stream stops. So if subjects can delay their gratification, they will receive all of the rewards in the stream; however, they constantly face the temptation to consume the available rewards. Rather than choosing between two options, in the delayed gratification paradigm, subjects choose when to stop waiting for the reward. Researchers have primarily used this method with primates but occasionally with pigeons as well.

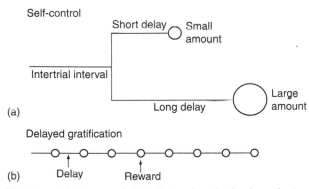

(a)

(b) Delay   Reward

Figure 2 (a) The most frequently used method to investigate intertemporal choice in pigeons and rats is the self-control paradigm. In this technique, subjects experience an intertrial interval in which nothing happens. Following this interval, subjects choose between receiving a small amount of food after a short delay and a large amount of food after a long delay. After consuming the food, another intertrial interval begins. (b) In the delayed gratification paradigm, a certain amount of food accumulates at fixed rate, say, one food reward per 10 s. Interrupting the stream of food results in stopping the accumulation of rewards.

Hyperbolic Discounting

Most work on animal intertemporal choice uses the self-control paradigm and assumes that this tests temporal discounting in their subjects. With a series of indifference points, one can derive a discounting function that quantitatively describes how reward values decrease with delays. Recall that exponential discounting (eqn [2]) implies a constant rate of discounting. Unfortunately, experiments in both humans and other animals show little support for this prediction. Instead, the discount rate decreases with time, showing high discounting at short delays and a lower rate at longer delays (**Figure 1**). This type of discounting is termed 'hyperbolic discounting.' Psychologists favor a particular hyperbolic function that describes how the present value of a reward amount A decreases with the delay t:

$$V = \frac{A}{1 + kt} \qquad [3]$$

where k represents a discounting parameter that accounts for the steepness of the slope. This model has a declining discounting rate and describes data from pigeons, rats, and humans quite well.

The actual rate of discounting also violates the economic model. If viewed as a proxy for the interest rate in humans or the interruption rate in animals, the rate of discounting should be rather small. A rational investor should have a discount rate that matches available interest rates (around, say, 5% per year). In experiments and 'field studies' in humans, the estimated discounting rates often range between 10 and 30% per year for delays beyond 1 year (and are much higher for shorter delays). Humans, therefore, choose much more impulsively than predicted by economic analysis because they have strong preferences for sooner outcomes. Animals also exhibit impulsive preferences, but on an even shorter time scale, typically only waiting for seconds or minutes for delayed payoffs. This would imply implausibly high interruption rates (up to four interruptions per minute!) to discount the future at this level. Thus, from a psychological perspective, high levels of impulsivity remain a puzzle. Rather than offer ultimate explanations of behavior, the psychological perspective emphasizes the cognitive variables underlying behavior.

Cognitive Variables

The psychological study of intertemporal choice often highlights how individuals overcome temptation for short-term gratification. A number of cognitive variables play key roles in trade-offs between short- and long-term rewards.

Commitment

One way to avoid the temptation of immediate gratification is to use external commitment devices that force an individual to choose the delayed option. Examples of commitment devices in humans include automatically transferring salary into a retirement account to save money for the future, throwing away a pack of cigarettes to avoid smoking, placing the alarm clock across the room to avert the draw of the snooze button, and Ulysses lashing himself to the mast of his ship to resist the Sirens' songs. Although little or no evidence suggests that animals actively pursue commitment devices, they can use them when available. For instance, Howard Rachlin and Leonard Green conducted a series of experiments in which pigeons faced an additional choice before experiencing the standard self-control choice between a smaller–sooner and larger–later option. In one version of the task, the subjects could choose between experiencing a delay, then continuing on to the standard self-control choice or experiencing a delay, then automatically receiving the larger–later option. This second choice represents a form of commitment because the pigeons can

commit themselves in advance to the larger–later option. Interestingly, the pigeons did use the commitment device, and most subjects significantly preferred it when a long delay separated to two sets of choices. Therefore, the pigeons used commitment if the temptation was far enough in the future.

Reward magnitude

Both the exponential and hyperbolic models predict that the absolute magnitude of the rewards should not matter; only the relative magnitudes should matter. So, the choice between one food item now and three items tomorrow should be devalued the same way as 20 items now and 60 items tomorrow. Only the threefold increase in reward amount should matter. The absolute magnitude does, however, influence choices in humans: the discounting rate decreases as the magnitude increases. So, human subjects choose more patiently (meaning that they opt for the larger–later reward more often) when assessing rewards in hundreds or thousands of dollars compared to tens of dollars. Interestingly, this 'magnitude effect' does not appear in animals. The ratio of rewards influences choices rather than the absolute magnitude. Testing the magnitude effect, however, proves difficult in animals because the magnitudes cannot scale to the same degree as in humans. Experiments in humans can vary hypothetical monetary payoffs over several orders of magnitude, whereas animal food rewards can only vary over a single order of magnitude. Thus, as we will see in the next section, the currency of the reward is a key aspect of intertemporal choice.

Currency

Food is the most commonly used reward currency in studies of animal intertemporal choice because it is easy to manipulate, highly motivating, and slow to cause satiation. Most experimental studies show that animals will wait for seconds or minutes for food rewards. Water provides another primary reward (a reward needed for survival) used in studies of intertemporal choice. Self-control studies rewarding deprived rats with water show similar patterns as when using food: rats only wait for a few seconds and their discounting function matches the hyperbolic model. Unfortunately, we do not have good experimental data on other currencies such as mating opportunities or social contact, but this provides an important avenue of future research. If intertemporal choices are adaptive in animals, we might expect that different currencies vary in how they lose value over different time scales, and animal intertemporal choices might match this variation. Food may elicit a strong preference for immediacy because it often does not persist long in the environment – competitors will take it if you do not. Also, food is, of course, something animals constantly need, so discounting of food may result from close ties to metabolic rates.

Currency effects appear more prominently in humans. Food, money, and health options all seem to show hyperbolic discounting but over different time scales. In fact, when tested with food in a similar way as other animals, humans also show very impulsive choices. So, food seems to be a universally impulsive currency. However, money and health options allow for much longer-term delays, even if they are shorter than those predicted by economic models.

Attention

Mischel's work on delayed gratification in children highlights the role of attention. In his design, an experimenter placed a single treat (cookie or marshmallow) in front of a child and said that she would leave the room. If the child waited and did not eat the treat until she returned, the child could have two additional treats. The experimenter would then leave the room and measure how long the children would wait (up to 15–20 min). Mischel and colleagues manipulated attention in several ways. First, they simply varied whether the children could see the treat. When the treat was hidden, the children waited significantly longer than when it was visible. Next, the experimenters drew the children's attention to the treat in different ways. They either focused the children's attention on the delicious properties of the treat (e.g., 'the marshmallow sure looks like a yummy, sweet treat, doesn't it?') or had them divert their attention by thinking of the treat as something else (e.g., 'imagine the marshmallow is a soft, fluffy cloud'). Again, diverting attention from the treat as food increased patience. Similar studies in animals have used the delayed gratification paradigm. Pigeons, for instance, can wait longer when their food is not visible, and chimpanzees can wait longer when experimenters provide toys to distract subjects from the accumulating food. The availability of distraction therefore can increase patience – reducing attention to waiting makes delays more tolerable.

Mechanisms of control

One of the most interesting and controversial topics in psychology is the nature of mechanisms of control over behavior. Are behaviors consciously or reflexively controlled? Are they genetically determined, learned, or reasoned out? These questions certainly apply to intertemporal choice, but unfortunately we have not begun to address them systematically. Claiming that plants make intertemporal choices suggests that strong genetic mechanisms with relevant environmental input can generate intertemporal choices. Of course, the same can be said of animals. Parasitoid wasps, for instance, can detect cues associated with a short life expectancy (such as lower barometric pressure indicating an impending storm). When detecting these cues,

they lay more eggs in lower quality hosts than in the absence of the cues. They therefore accept a lower reproductive output when responding to a shortened temporal horizon. Caching also likely falls under the category of intertemporal choices with strong genetic components. Caching species probably do not weigh the current and future benefits of the seeds in front of them – foresight months into the future seems unlikely. Nevertheless, caching species show extreme flexibility in their behavior, and foresight into a much shorter future seems perfectly reasonable. Experiments with scrub jays show that they attend to the decay of food, the time since caching, the presence of possible cache thieves, and future need. Thus, they have an extraordinarily flexible system for dealing with delayed rewards, although we do not fully understand how they represent the future. The abstract representation of time in humans allows us extreme flexibility in anticipating future payoffs – we can mentally travel in time. Although other animals can plan for the short term (hours, maybe days), the full scope of their mental time horizons remains unclear.

Evolution of Intertemporal Choice

The psychological approach offers insight into the mechanisms of intertemporal choice, but it does not offer a satisfying explanation of the circumstances under which animals should choose patiently or impulsively. An evolutionary account, however, can make specific predictions about temporal preferences and the change in discount rate over time. The evolutionary view stresses the fit between the decision mechanisms used to make temporal trade-offs and the environment in which these mechanisms evolved. Thus, natural selection favors a good fit between the decisions and the ecology of organisms – temporal preferences should be 'ecologically rational' rather than economically rational. This perspective leads to predictions that can account for some of the variation in species differences in patience and impulsivity.

The ecological rationality perspective suggests that decision mechanisms should fit the environment in which they operate. Thus, intertemporal choices should match the kinds of problems often faced by animals. This may explain animal impulsivity in the self-control paradigm. Rather than discounting, the rats and pigeons in these experiments may use simple rate-maximizing rules that are adapted to foraging in patches (maximizing intake also results in a hyperbolic discounting function). David Stephens and his colleagues have proposed that actual foraging situations rarely have the property of simultaneous choice used in the self-control paradigms (**Figure 2(a)**). Instead, animals typically choose when to leave a patch. A rule that maximizes the short-term intake rate:

$$\frac{A}{t+h} \qquad [4]$$

where t represents the delay and b represents the time required to process the food, makes similar predictions as the long-term rule (eqn [1]) in the patch situation. In the self-control situation, however, it predicts impulsive choice. Experiments with blue jays suggest that they make appropriate decisions in a patch situation, but choose more impulsively than expected in a self-control situation. This short-term rule is ecologically rational because it works well in a more naturalistic environment in which animals forage in patches.

Ecological rationality can also make predictions about species differences in intertemporal choice because species differ in their ecologies. Although relatively few animals have been tested systematically, interesting patterns emerge in the data across species. Comparing species can pose difficulties, especially with phylogenetically distant species. With more closely related species, however, the comparative method can yield interesting insights. For instance, chimpanzees and bonobos are sister taxa that share many morphological, ecological, and behavioral similarities. Yet, they differ in key aspects of their foraging ecologies. Although their diets overlap substantially, chimpanzees often hunt for food, whereas bonobos spend more time consuming the abundant terrestrial herbaceous vegetation in their habitat. This means that chimpanzees frequently face delays in food consumption: they decide to hunt and then must wait until capturing food before consuming it. Bonobos, in contrast, rarely hunt, instead feeding on the plentiful vegetation that is virtually immediately accessible. Ecological rationality would predict that these differences in foraging ecology should translate into different decision mechanisms and preferences between the two species. In fact, chimpanzees are more patient in the self-control task than bonobos, reflecting the differences in natural foraging. Although chimpanzees and bonobos differ, they wait longer than any other species systematically tested so far. Macaques wait for an intermediate length of time, and capuchin monkeys, tamarins, and marmosets wait as long as pigeons and rats. Yet differences still exist between these species, some of which may result from foraging ecology. The comparative study of intertemporal choice remains in its infancy, and testing more species can help reveal the underlying nature of temporal preferences.

See also: Animal Arithmetic; Caching; Mental Time Travel: Can Animals Recall the Past and Plan for the Future?; Optimal Foraging Theory: Introduction; Patch Exploitation; Rational Choice Behavior: Definitions and Evidence.

Further Reading

Evans TA and Beran MJ (2007) Chimpanzees use self-distraction to cope with impulsivity. *Biology Letters* 3: 599–602.

Frederick S, Loewenstein G, and O'Donoghue T (2002) Time discounting and time preference: A critical review. *Journal of Economic Literature* 40: 351–401.

Green L, Myerson J, Holt DD, Slevin JR, and Estle SJ (2004) Discounting of delayed food rewards in pigeons and rats: Is there a magnitude effect? *Journal of the Experimental Analysis of Behavior* 81: 39–50.

Kacelnik A (2003) The evolution of patience. In: Loewenstein G, Read D, and Baumeister RF (eds.) *Time and Decision: Economic and Psychological Perspectives on Intertemporal Choice*, pp. 115–138. New York: Russell Sage Foundation.

Loewenstein G, Read D, and Baumeister R (2003) *Time and Decision: Economic and Psychological Perspectives on Intertemporal Choice*. New York: Russell Sage Foundation.

Madden GJ and Bickel WK (2009) *Impulsivity: The Behavioral and Neurological Science of Discounting*. Washington, DC: American Psychological Association.

Mazur JE (1987) An adjusting procedure for studying delayed reinforcement. In: Commons ML, Mazur JE, Nevin JA, and Rachlin H (eds.) *Quantitative Analyses of Behavior: The Effect of Delay and of Intervening Events on Reinforcement Value*, vol. 5, pp. 55–73. Hillsdale, NJ: Lawrence Erlbaum Associates.

Mischel W, Shoda Y, and Rodriguez ML (1989) Delay of gratification in children. *Science* 244: 933–938.

Rachlin H and Green L (1972) Commitment, choice and self-control. *Journal of the Experimental Analysis of Behavior* 17: 15–22.

Read D (2004) Intertemporal choice. In: Koehler D and Harvey N (eds.) *Blackwell Handbook of Judgment and Decision Making*, pp. 424–443. Oxford: Blackwell.

Rosati AG, Stevens JR, Hare B, and Hauser MD (2007) The evolutionary origins of human patience: Temporal preferences in chimpanzees, bonobos, and adult humans. *Current Biology* 17: 1663–1668.

Stephens DW and Anderson D (2001) The adaptive value of preference for immediacy: When shortsighted rules have farsighted consequences. *Behavioral Ecology* 12: 330–339.

Stephens DW, Brown JS, and Ydenberg RC (2007) *Foraging: Behavior and Ecology*. Chicago: University of Chicago Press.

Stephens DW and Krebs JR (1986) *Foraging Theory*. Princeton: Princeton University Press.

Stevens JR and Stephens DW (2009) The adaptive nature of impulsivity. In: Madden GJ and Bickel WK (eds.) *Impulsivity: The Behavioral and Neurological Science of Discounting*, pp. 361–387. Washington, DC: American Psychological Association.

Relevant Websites

http://www.eva.mpg.de/3chimps/movies/chimpanzee_discounting-web.mpg – Chimpanzee Intertemporal Choice Video.

Invertebrate Hormones and Behavior

E. S. Chang, University of California-Davis, Bodega Bay, CA, USA

Introduction

It is difficult to provide an overview of just about any topic dealing with all the invertebrate animals found in the biosphere. Indeed, about 98% of all animal species are invertebrates, comprising approximately 30 phyla. This is especially true about hormones and behavior. There is much known about some topics in some species (such as molting in arthropods or egg laying in molluscs), but comparisons are complex due to differences observed among related species (e.g., crustaceans vs. insects or caterpillars vs. fruit flies). Given these constraints, I attempt to provide an overview of the topic. This topic is of basic and applied interest to a wide range of readers due to the importance of invertebrates as follows: (1) cultured and wild-caught species (crustaceans and molluscs) that form some of the world's most valuable food commodities; (2) keystone species in various ecosystem food webs (including agricultural systems), which in some cases are predators and in others prey species; and (3) major disease vectors (especially insects) and parasites that greatly influence human endeavors. In addition to these practical reasons for the promotion of research into invertebrates, they provide very useful model systems for the study of the chemical mediation of behavior. Invertebrates are well suited to endocrine-related experiments on behavior. They are especially tolerant of experiments using classical endocrinological techniques (e.g., ablation, injection, reimplantation, and organ culture).

Arthropoda

Crustacea

Molting

Crustaceans, like all arthropods, must periodically shed their external, confining exoskeletons and take up air or water to expand their new and larger exoskeletons in order to grow in size. The problem of how to increase in body size is even more formidable for crustaceans (compared to insects) due to their mineralized, relatively rigid exoskeletons. **Figure 1** shows the molting of the overlying old exoskeleton in the lobster *Homarus americanus*. This ecdysial process lasts about 30 min. However, the entire molt cycle occurred over many weeks, and ecdysis was simply the culmination of this process.

The arthropod molting hormone was identified as the steroid 20-hydroxyecdysone (20E; see **Table 1**). 20E is the principal active form of the molting hormone found in both insects and crustaceans. It is a member of a class of steroids, ecdysteroids, that show molting hormone activity. In most species examined, ecdysone is the prohormone secreted by the molting gland (Y-organ in crustaceans), and it is hydroxylated by target tissues to 20E. Although 20E is the predominant molting hormone in all decapod species examined to date, other ecdysteroids have been characterized in hemolymph and tissues of various crustacean species.

Like vertebrate steroid hormones, ecdysteroids recognize target tissues by binding with nuclear receptors. The ecdysteroid receptor has been isolated and characterized from the fiddler crab *Uca pugilator*. It has been sequenced and has homologies with insect ecdysteroid receptors. Transcripts for the receptor were isolated from crab limb buds and developing ovaries.

Hemolymph ecdysteroid concentration fluctuates dramatically during the molt cycle (e.g., from $<10 \, \text{ng ml}^{-1}$ in postmolt to $>350 \, \text{ng ml}^{-1}$ in premolt lobsters). These changes mediate the various biochemical and physiological processes that occur during the cycle. The rates of synthesis and/or secretion of ecdysone by the Y-organ vary during the molt cycle and partially account for these hemolymph fluctuations in ecdysteroid titer. Just prior to the substage of premolt in which the highest concentration of ecdysteroids is found in the hemolymph, explanted Y-organs secrete the greatest amount of ecdysone. Low hemolymph concentrations correlate with low Y-organ secretory rates.

Removal of both stalked eyes of *U. pugilator* results in a shortening of the molt interval. This observation led to the postulation of an endocrine factor present in the eyestalks that normally inhibits molting – a molt-inhibiting hormone (MIH). Detailed microscopical examinations resulted in the discovery of a neurohemal organ in the eyestalk of several decapod crustaceans. This neurohemal organ is called 'the sinus gland' and serves as a storage site for neurosecretory products. It consists of the enlarged endings of a group of neurosecretory neurons collectively called 'the X-organ.'

The shortened molt interval observed in eyestalk-ablated decapods is likely due to a rapid elevation in the concentration of circulating ecdysteroids, which is a result of X-organ/sinus gland removal. However, recent evidence indicates that the regulation of the Y-organ is more complex than simple inhibition by MIH. No overt changes in the hemolymph levels of MIH were observed

Figure 1 Molting of a juvenile lobster (*Homarus americanus*). The entire sequence took about 30 min from start to finish. (a) The lobster has resorbed much of its old exoskeleton's mineralization. The exoskeleton splits at the junction of the thorax and abdomen. (b) The anterior of the animal retracts from its old exoskeleton by pulling posteriorly. (c) The posterior of the lobster pulls anteriorly. (d) The lobster is almost free of its old exoskeleton and begins to take up water to expand its new, flexible, larger exoskeleton. (e) The animal is above its shed exoskeleton. It will take several days for the exoskeleton to deposit layers of chitin, protein, and calcium carbonate. Photos by staff of the Bodega Marine Laboratory.

Table 1   Overview of the hormones affecting invertebrate behavior discussed in the article

Taxonomic group	Hormone	Glandular source	General function(s)
Crustacea	20-Hydroxyecdysone	Y-organ	Molting, pheromone, allelochemical (in the chelicerate pycnogonids)
	Molt-inhibiting hormone	X-organ/sinus gland	Molt inhibition
	Androgenic gland hormone	Androgenic gland	Reproduction, aggression
Insecta	20-Hydroxyecdysone	Prothoracic gland	Molting
	Prothoracicotropic hormone	Brain	Molt promotion
	Corazonin	Brain	Ecdysial behavior
	Pre-ecdysis-triggering hormone	Inka cells	Ecdysial behavior
	Ecdysis-triggering hormone	Inka cells	Ecdysial behavior
	Eclosion hormone	Brain	Ecdysial behavior
	Juvenile hormone	Corpus allatum	Development, reproduction, migration, social behavior
Mollusca	Egg-laying hormone	Bag cell neurons	Egg laying
	Caudodorsal cell hormone-I	Caudodorsal cells	Egg laying
	Peptides A, B	Apical gland	Egg-laying promotion
Echinodermata	Radial nerve factor	Radial nerves	Spawning promotion
	1-Methyl adenine	Follicle cells	Spawning

in a crab over the molt cycle, except for a large increase during late premolt immediately before ecdysis. This peak of premolt MIH likely mediates the sudden drop in circulating ecdysteroids just prior to ecdysis. The MIH

from the green crab *Carcinus maenas* was among the initial MIHs to be characterized. It is a member of a novel neuropeptide family, representatives of which have so far been found only in arthropods. This neuropeptide family regulates such diverse functions as molting, reproduction, and metabolism.

Pheromones and allelochemicals

The hypothesis that hormones can serve as pheromones has been tested with positive results in only a few instances in the animal kingdom. Perhaps the best known examples are from fish. Compared to fish, relatively little evidence exists for a pheromonal role of crustacean hormones.

Ecdysteroids are not only mediators of molting, but also act as gonadotropins. Since molting is usually a prerequisite for insemination in many female decapod crustaceans, it is not unreasonable to hypothesize that ecdysteroids may act as mating pheromones. There are some early studies in support of this hypothesis. A premating stance is typically displayed by male crabs in the presence of chemical cues released by a premolt female. This stance involves elevation of the cephalothorax with the anterior margin tilted up. Males walk on the tips of their dactyls of the first three pairs of walking legs with the fourth pair extended backward. The chelipeds are partially extended in a lowered position. In the presence of 20E, male shore crabs (*Pachygrapsus crassipes*) display the premating stance. A dose response curve for the activity of 20E was published showing a range of effects from 10^{-13} to 10^{-5} M. Researchers observed that lower doses of 20E resulted in longer reaction times before the premating stance was displayed. These researchers then hypothesized that 20E was released by premolt female crabs (presumably in their urine) and initiated mating behavior.

The proposal that crustacean release pheromones through urine is supported by several experiments. There is evidence that urine contains ecdysteroids and that their amounts vary during the molt cycle. Urinary concentrations of ecdysteroids are highest in late premolt, corresponding to an observed dramatic decline in hemolymph ecdysteroids immediately prior to ecdysis.

There have been challenges, however, to the proposal that ecdysteroids are in fact pheromones. For example, other research failed to demonstrate any pheromonal activity of various tested ecdysteroids on *C. maenas* and did not find changes in mating behavior when lobsters, *H. americanus*, were bioassayed with 20E or ecdysteroid metabolites. These latter experiments indicated the initiation of alert responses to some of the compounds tested. No mating responses were observed when blue crabs, *Callinectes sapidus*, were tested with 20E. Perhaps the major difficulty in proposing a pheromonal role of ecdysteroids is the lack of species specificity. Since all arthropods examined so far synthesize, secrete, and use ecdysteroids for molting, it is difficult to envision how an ecdysteroid acting as a pheromone can be species specific. One possibility is that only a very limited number of species evolved receptors capable of binding extraorganismal ecdysteroids.

The hypothesis that ecdysteroids have some pheromonal activity has recently been revisited. Male *C. maenas* had decreased feeding responses during the summer reproductive season. During these months, postmolt females are soft shelled. Males are observed to decrease their foraging activities and are less likely to cannibalize these postmolt females. The application of exogenous 20E deterred male crabs from feeding on bivalve prey items. Female crabs were not deterred. Thus, it appears that 20E may act as a sex-specific feeding-deterrent pheromone. Although postmolt female crabs have low circulating ecdysteroid concentrations, presumably they excrete sufficient quantities via the urine or from exoskeletal pores to be perceived by males in very close proximity (as in the mating embrace).

Another recent study indicates a role for 20E as a chemical mediator of aggressive interactions in premolt lobsters. When physiologically relevant concentrations of 20E were released near the antennules of premolt lobsters, their level of aggression increased relative to controls. The controls consisted of the release of the prohormone ecdysone, artificial seawater, or the use of lobsters that could not smell. Presumably, the increased aggression will chase off other lobsters and ensures that the premolt lobster will deter a subsequent physical confrontation between its future postmolt, defenseless self and any nearby cannibalistic conspecifics.

As described earlier, the primary function of ecdysteroids is to mediate the molting process. Recent work demonstrates that molting hormones have been additionally used by pycnogonids (sea spiders) as defensive compounds against crustacean predators. Pycnogonids are not crustaceans; they are related chelicerate marine arthropods. Eight different ecdysteroids have been isolated from the pycnogonid *Pycnogonum litorale*. Much of these ecdysteroids exist as acetate or glycolate conjugates. These conjugates may affect metabolism of the ecdysteroids and/or their biological activities within the host pycnogonid. These combined ecdysteroids are found in very high concentrations in *P. litorale* – as much as 2–3 orders of magnitude higher than in other arthropods.

In feeding choice experiments, crabs were deterred from eating either powdered extracts of pycnogonids or food pellets containing high levels of ecdysteroids. Since all arthropods to date have been shown to be sensitive to the molt-promoting effects of ecdysteroids, *P. litorale* must have a mechanism to isolate their allelochemical compounds from their circulating hemolymph. *P. litorale* sequesters its defensive ecdysteroids in epidermal glands. These glands can be selectively stimulated to release their contents. Ecdysteroids apparently act as antipredation compounds because potential crustacean predators of pycnogonids are susceptible to the molt-inducing effects of exogenously applied ecdysteroids. The effect of exogenously applied ecdysteroids is frequently an early entry into premolt followed by an unsuccessful ecdysis.

Just as some terrestrial plants have high concentrations of ecdysteroids that act as antifeeding allelochemicals against herbivorous insects, it will be interesting to determine if marine algae have high levels of ecdysteroids that act as deterrents against herbivorous crustaceans. The evolution of plant ecdysteroids appears to have occurred multiple times throughout the plant kingdom. There is at least one example of ecdysteroid-like molecules in a red alga.

Aggression

Ecdysteroids also appear to have neuromodulatory activities. With in vitro neuromuscular preparations, increased amplitudes and frequency of the excitatory potentials were observed in the opener muscle of the lobster claw in the presence of 20E. In abdominal muscle, these potentials were significantly smaller in the presence of 20E compared to control abdominal muscle. These observations were consistent with the changes observed in vivo in which premolt lobsters (with high circulating levels of ecdysteroids) had increased aggressiveness relative to other molt stages. In support of these observations on alterations in neuromuscular activity, experiments were conducted on lobsters using 20E injections. Intermolt females injected with 20E displayed increased aggressiveness relative to saline-injected controls.

Reproduction

The androgenic gland (AG) is located on the distal portion of the sperm duct in male crustaceans. In genetic males, the AG develops and begins to secrete AG hormone. In the absence of a developed, active AG (i.e., in females), there is an absence of AG hormone and female structures develop. Following removal of the AG from a male, spermatogenesis waned and in some cases oocytes appeared. More recent experiments have been conducted in the giant freshwater prawn *Macrobrachium rosenbergii* in which complete sex reversal followed the removal of the AGs at an early immature stage. This operation resulted in complete female differentiation, complete with ovaries and oviducts including reproductive behavior, successful mating, and production of offspring. Implantation of AGs into females resulted in the development of male copulatory organs with reported cases of functional sex reversal and progeny obtained when fertile sex-reversed animals were crossed with normal prawns.

M. rosenbergii males progress through a succession of male morphotypes beginning with small males. These small males have relatively short claws and minimal reproductive capacity. Orange-claw males are a larger, intermediate morphotype. This progression of male morphotypes ends with the dominant blue-claw males. These very large males have long claws and maximal reproductive capacity. The AG is necessary for this morphotypic progression in males. When implanted into females, the AG induces male-like reproductive and aggressive behavior.

These observations have interesting implications for the aquaculture industry. Most economically important cultured crustaceans show sexually bimodal growth patterns in which males grow faster than females or vice versa. Monosex culture has been recently suggested as one of the most promising ways to improve production efficiencies of crustacean aquaculture. An example of the application of endocrine research in crustaceans is the potential for the production of male monosex *M. rosenbergii*. This would be the result of AG removal from immature males; this results in sex reversal with complete and functional female differentiation (neofemale). Neofemale prawns display female mating behavior and are capable of mating with normal males to produce all-male offspring. These crosses produce all-male progeny because of the homogametic (ZZ) nature of the males. Selective harvest of the largest males would have to be conducted in order to reduce aggression.

Insecta

Ecdysis

Like the crustaceans, insects use 20E as their predominant molting hormone. There are, however, several distinct differences between the two groups of arthropods. Instead of the inhibitory control of the molting gland by MIH as in crustaceans, insects employ a stimulatory hormonal regulation of their molting glands (prothoracic glands). This stimulatory peptide is prothoracicotropic hormone that is produced in brain neurons and released from neurohemal organs called 'the corpora cardiaca.' The morphological stage produced as a result of an insect molt is regulated by juvenile hormone (JH). This terpenoid hormone is produced by the corpora allata and promotes the developmental status quo. For example, in the presence of high concentrations of JH, most insect taxa retain their juvenile or larval characteristics when stimulated to molt by 20E. In the presence of reduced amounts of JH, ecdysteroid-mediated molting proceeds toward mature stages.

Details of the actual mechanisms involved in the shedding of the old exoskeleton, or ecdysis, have only been elucidated in the insects. Several peptide hormones mediate this ecdysial behavior. A brain hormone (corazonin) stimulates the secretion of pre-ecdysis-triggering hormone (PETH). PETH is produced by the Inka cells, which are located near the spiracles. PETH acts on the abdominal ganglia to promote the initial phase of pre-ecdysial behavior. This may include searching for an appropriate location in which to molt and body movements of rhythmic muscular contractions that serve to loosen the connections between the old overlying exoskeleton and the new one lying beneath it.

Ecdysis-triggering hormone (ETH) is also secreted by the Inka cells. It mediates the next phase of the pre-ecdysis motor pattern. In the tobacco hornworm (*Manduca sexta*), this involves rhythmic cycles of muscular constriction and relaxation along the lateral margin of the abdomen. Additionally, the prolegs display waves of extension and relaxation. The next targets are brain neurons that secrete eclosion hormone (EH). EH mediates the actual ecdysis behavior. In *M. sexta*, one primary activity of EH is the initiation of waves of peristaltic movements that result in the sliding of the old cuticle to the posterior. The details of ecdysis in insects are highly complex and are still being worked out. There are several other peptide hormones involved and reciprocal interactions cloud the situation. For example, there is evidence that EH forms a positive feedback loop upon the secretion of ETH. ETH also induces the secretion of other hormones that have further downstream effects. Study of ecdysis behavior permits an elegant blend of research that uses the tools of neural physiology, molecular biology, and comparative endocrinology.

Reproduction

As with many aspects of the endocrine control of insect behavior, it is difficult to make generalizations across the diverse and numerous insect orders. The primary mediator of female reproductive behavior is JH. However, some insects continue to display apparently normal reproductive behavior in the absence of functioning corpora allata, while there are examples of the loss of sexual receptivity upon removal of these glands in individual species of various groups. Often, reproductive behavior can be reinstated with JH replacement therapy. Functioning corpora allata or JH injections stimulate male mating behavior in some species.

A great deal of research has been conducted on the isolation and characterization of insect pheromones (see Further Reading). There is no evidence for insect hormones acting as pheromones. However, in the mosquito *Anopheles gambiae*, males transfer large amounts of 20E into the female upon copulation. This transferred hormone is pheromone-like in that it is a chemical messenger produced by one member of a species and transmitted to another member of the same species, resulting in a physiological change. In this case, the change is enhanced production of yolk by the female.

Migration

The milkweed bug (*Oncopeltus fasciatus*) presents one of the best models for migratory behavior. Under the proper photoperiod (long days) and high temperatures, the female initiates reproductive processes after a short dispersal flight. This behavior is mediated by a large and rapid rise in circulating JH. Short days and lower temperatures result in intermediate titers of JH. This results in a long migratory flight. If the insects do not find sufficient food and environmental conditions, the JH concentration drops further and they enter into a quiescent resting stage called 'diapause.'

Social behavior

In social insects, different castes perform specific functions. The proportion of the members of each caste is often determined by colonial and environmental conditions. In some termites, transformation of workers to soldiers is mediated by elevated levels of JH. In some ant species, application of JH to juveniles will result in the formation of a higher proportion of queens. And in honeybees, low JH titers in adults may promote nursing and hive-related activities, while high JH titers may promote foraging behavior.

Mollusca

The neuroendocrine control of egg laying is well understood in two species of gastropods – the sea hare *Aplysia californica* and the pond snail *Lymnaea stagnalis*. Egg-laying behavior of *A. californica* is controlled by two clusters of peptidergic neurons called 'bag cells.' These bag cell neurons of the abdominal ganglion synthesize an egg-laying hormone (ELH) precursor protein. This precursor hormone undergoes posttranslational processing that results in multiple peptides in addition to ELH. These additional peptides include α-, β-, γ-, δ-, and ε-bag cell peptides and an acidic peptide, which have various activities upon the animal's physiology and behavior. Some of these peptides can mimic the action of ELH. As the name implies, the primary behavior mediated by ELH is egg laying. Other behaviors associated with egg laying are mediated by ELH. These include inhibition of locomotion and feeding, increased respiratory pumping, and initiation of stereotypical head movements.

The ultimate control of egg laying has not been completely elucidated. Other factors, such as the apical gland, which is associated with the reproductive tract, may be involved. It produces several factors, including peptides A and B that stimulate ELH secretion. Another apical gland peptide ('egg-laying release hormone') can stimulate egg laying in the absence of the abdominal ganglion.

Egg-laying behavior in the pond snail *L. stagnalis* is regulated by the caudodorsal cells located in the cerebral ganglia. Like the bag cells of *A. californica*, the caudodorsal cells produce a family of peptides through a series of posttranscriptional and posttranslational modifications. The major peptide of these neurosecretory cells is caudodorsal cell hormone-I (CDCH-I). The CDCH-I precursor displays some homology to the ELH precursor (about 28%). CDCH-I mediates a suite of behaviors related to egg laying. In addition to ovulation, this hormone initiates a resting behavior in the snail followed by a turning behavior. During the turning phase, the snail scrapes the substrate with its buccal mass. This scraped substrate is the future location of egg deposition. Oviposition then occurs and is followed by the snail's inspection of the egg mass.

Both these molluscan neurohormonal systems have provided valuable insight into gene structure and posttranscriptional and posttranslational processing. These species have relatively few cells in their relatively accessible central nervous systems and well-defined behaviors that are responsive to the relatively small peptide hormones.

Echinodermata

Most of the work on hormones in echinoderms has been conducted on sea stars. This research includes the separation of follicle cells from the oocyte (ovulation), the induction of germinal vesicle breakdown, and the release of maturing oocytes into seawater (spawning). Fertilization then occurs in the aquatic medium. Pioneering studies demonstrated that the radial nerves, components of the sea star central nervous system, contain a peptide that is able to mediate these events. This hormonal factor was initially called 'gonad-stimulating substance' and is now called 'radial nerve factor' (RNF). Subsequent work identified the target cells of RNF as the follicle cells (though more recent work indicates that nonfollicular cells may also be targets). The primary action of RNF is to stimulate the release of maturation-inducing substance that actually induces the biological activities observed with RNF. Maturation-inducing substance has been identified as 1-methyladenine (1-MA). Some of the actions of 1-MA may be due to the subsequent induction of a locally acting factor called 'maturation-promoting substance.'

Of particular relevance to this article's central topic, injection of 1-MA induces release of oocytes from their surrounding follicle cells (**Figure 2**) and generates spawning posture and brooding behavior. During spawning, sea stars usually become stationary and assume a characteristic posture of raising their central disk off of the substrate. This may be accompanied with rhythmic waves of contraction that start at the arm tips and progress medially. This is followed by spawning of gametes (and occurs in both females and males). In some brooding sea stars, 1-MA injection mediates the formation of a brood pouch under the arms and central disk of females. Normally this pouch is used for the brooding of embryos.

Conclusions

This is an exciting time for research on the interface of endocrinology and behavior in invertebrates. I think that some key areas of future research on this topic include the following:

1. I have only discussed data from three invertebrate phyla. There are dozens of other invertebrate phyla that undoubtedly have hormones that regulate behavior.

Figure 2 An ovarian lobe dissected from an arm of the sea star *Pisaster ochraceus* and placed into filtered seawater. The lobe in the bottom panel was incubated with 1.0 μM of 1-methyladenine for 60 min. Mature, fertilizable oocytes have been released from their surrounding follicle cells. The lobe in the top panel did not have anything added to the seawater.

Those hormones and their effects upon behavior are waiting to be characterized.

2. The further identification and characterization of ecdysteroid and neuropeptide receptors in arthropods will be achieved using modern molecular techniques. This area of research will identify putative target tissues and quantify receptor activities during molt and reproductive cycles.

3. The molluscs that display the most complex behaviors are the cephalopods. Although classical hormonal manipulations (ablation and implantation experiments) have been conducted on squid and octopus, little is known about the endocrines in this class of animals. With advances in the successful culture of cephalopods, we can expect new findings on the hormonal regulation of behavior.

4. The sea urchin genome has been sequenced. This database will permit informative comparisons between hormones and behavior in these invertebrate deuterostomes and the chordates.

5. There is growing interest in the field of invertebrate endocrine disrupters. There will likely be much more research devoted to the determination of the effects

of exogenous chemicals upon invertebrate endocrine systems. Behavioral assays will be useful indicators of environmental contamination.

Acknowledgment

I thank Sharon A. Chang for editorial assistance.

See also: Aggression and Territoriality; Aquatic Invertebrate Endocrine Disruption; Communication and Hormones; Experimental Approaches to Hormones and Behavior: Invertebrates; Hormones and Behavior: Basic Concepts; Vertebrate Endocrine Disruption.

Further Reading

Bernheim SM and Mayeri E (1995) Complex behavior induced by egg-laying hormone in *Aplysia*. *Journal of Comparative Physiology A* 176: 131–136.

Blomquist GJ, Jurenka R, Schal C, and Tittiger C (2005) Biochemistry and molecular biology of pheromone production. In: Gilbert LI, Iatrou K, and Gill SS (eds.) *Comprehensive Molecular Insect Science*, vol. 3, pp. 705–751. Oxford: Elsevier B.V.

Chang ES and Kaufman WR (2005) Endocrinology of Crustacea and Chelicerata. In: Gilbert LI, Iatrou K, and Gill SS (eds.) *Comprehensive Molecular Insect Science*, vol. 3, pp. 805–842. Oxford: Elsevier B.V.

Chang ES and Sagi A (2008) Male reproductive hormones. In: Mente E (ed.) *Reproductive Biology of Crustaceans: Case Studies of Decapod Crustaceans*, pp. 299–317. Enfield: Science Publishers.

Cogllanese DL, Cromarty SI, and Kass-Simon G (2008) Perception of the steroid hormone 20-hydroxyecdysone modulates agonistic interactions in *Homarus americanus*. *Animal Behaviour* 75: 2023–2034.

Conn PJ and Kaczmarek LK (1989) The bag cell neurons of *Aplysia*. *Molecular Neurobiology* 3: 237–273.

Cromarty SI and Kass-Simon G (1998) Differential effects of a molting hormone, 20-hydroxyecdysone, on the neuromuscular junctions of the claw opener and abdominal flexor muscles of the American lobster. *Comparative Biochemistry and Physiology* 120: 289–300.

Denlinger DL, Yocum GD, and Rinehart JP (2005) Hormonal control of diapause. In: Gilbert LI, Iatrou K, and Gill SS (eds.) *Comprehensive Molecular Insect Science*, vol. 3, pp. 615–650. Oxford: Elsevier B.V.

Geraerts WPM, Ter Maat A, and Vreugdenhil E (1988) The petidergic neuroendocrine control of egg-laying behavior in *Aplysia* and *Lymnaea*. In: Laufer H and Downer RGH (eds.) *Endocrinology of Selected Invertebrate Types*, pp. 141–231. New York: Alan R. Liss.

Hartfelder K and Emlen DJ (2005) Endocrine control of insect polyphenism. In: Gilbert LI, Iatrou K, and Gill SS (eds.) *Comprehensive Molecular Insect Science*, vol. 3, pp. 651–703. Oxford: Elsevier B.V.

Nijhout HF (1994) *Insect Hormones*. Princeton: Princeton University Press.

Pondeville E, Maria A, Jacques J-C, Bourgouin C, and Dauphin-Villemant C (2008) *Anopheles gambiae* males produce and transfer the vitellogenic steroid hormone 20-hydroxyecdysone to females during mating. *Proceedings of the National Academy of Sciences USA* 105: 19631–19636.

Sagi A and Aflalo ED (2005) The androgenic gland and monosex culture of freshwater prawn *Macrobrachium rosenbergii* (De Man): A biotechnological perspective. *Aquaculture Research* 36: 231–237.

Schuetz AW (2000) Extrafollicular mediation of oocyte maturation by radial nerve factor in starfish *Pisaster ochraceus*. *Zygote* 8: 359–368.

Shirai H and Walker CW (1988) Chemical control of asexual and sexual reproduction in echinoderms. In: Laufer H and Downer RGH (eds.) *Endocrinology of Selected Invertebrate Types*, pp. 453–476. New York: Alan R. Liss.

Strumwasser F (1984) The structure of the command for a neuropeptide-mediated behavior, egg-laying, in an opisthobranch mollusc. In: Hoffmann J and Porchet M (eds.) *Biosynthesis, Metabolism and Mode of Action of Invertebrate Hormones*, pp. 36–43. Berlin: Springer-Verlag.

Tomaschko K-H (1994) Ecdysteroids from *Pycnogonum litorale* (Arthropoda, Pantopoda) act as chemical defense against *Carcinus maenas* (Crustacea, Decapoda). *Journal of Chemical Ecology* 20: 1445–1455.

Truman JW (2002) Invertebrate systems for the study of hormone brain-behavior relationships. In: Becker JB, Breedlove SM, Crews D, and McCarthy MM (eds.) *Behavioral Endocrinology*, 2nd edn., pp. 652–686. Cambridge: The MIT Press.

Truman JW (2005) Hormonal control of insect ecdysis: Endocrine cascades for coordinating behavior with physiology. *Vitamins and Hormones* 73: 1–30.

Žitňan D and Adams ME (2005) Neuroendocrine regulation of insect ecdysis. In: Gilbert LI, Iatrou K, and Gill SS (eds.) *Comprehensive Molecular Insect Science*, vol. 3, pp. 1–60. Oxford: Elsevier B.V.

Žitňan D, Kingan TG, Hermesman J, and Adams ME (1996) Identification of ecdysis-triggering hormone from an epitracheal endocrine system. *Science* 271: 88–91.

Invertebrates: The Inside Story of Post-Insemination, Pre-Fertilization Reproductive Interactions

T. A. Markow, University of California at San Diego, La Jolla, CA, USA

Introduction

In organisms with internal fertilization, females serve as arenas for postinsemination, prefertilization interactions between the sexes. Although these interactions are not visible to the naked eye, from those species in which they have been studied, females are clearly the site of a rich intersexual dialog that considerably impacts the fitness of both sexes.

Among and within various invertebrate taxa are species with either external or internal fertilization (**Table 1**). Even where fertilization is internal, it is not necessarily preceded by copulation or by penetration of the female by an intromittent organ. For example, in some marine and freshwater invertebrates, sperm enter the females via the water column, while in others, females collect spermatophores and place them into their own reproductive tracts. Cases also exist in which insemination is referred to as 'traumatic' as it involves piercing of the female's body in order to deliver the ejaculate. This article will not concern itself with copulatory organs or copulation itself. Instead, the focus, regardless of how females come to be inseminated, will be the nature of intersexual interactions that take place inside the female. The intent is not to provide a complete review of all invertebrate taxa with internal fertilization, but instead to highlight some of the variability that exists in reproductive tract interactions in invertebrates with internal fertilization. This article is organized to first present what is known of the components interacting inside the mated female reproductive tract, including the sperm and any accompanying ejaculate as well as the structural and chemical features of the female. Following this, intersexual interactions, starting with the arrival of the ejaculate until the moment of fertilization, will be treated. The final section will address the evolutionary implications of the reproductive structures, substances, and processes that occur inside females.

Components

Male Ejaculate

What is the nature of the material inseminated females acquire from males? Typically, the ejaculate consists of the sperm and the chemical cocktail accompanying them.

In cases where there is a spermatophore, or packet of sperm, the chemical cocktail includes the covering of the packet. Both sperm and nonsperm constituents of ejaculates can be highly variable within and between species. Sperm can vary morphologically, chemically, and quantitatively. Nonsperm components also exhibit qualitative and quantitative variability. Furthermore, some of the ejaculate variability will be nongenetic: nutrition, age, mating status, and mating type (genotypes of each member of the pair) may influence the qualitative and quantitative features of what males pass to females.

Sperm

Sperm size and shape and surface chemistry are all features capable of interacting with the female. While we typically think of sperm as having a head, a midpiece, and tail, in some taxa, sperm have multiple or no flagella. In others, they are amoeboid. Among those species with tailed sperm, such as certain *Drosophila*, sperm can be many times the length of the male's body. The number of sperm transferred to females can vary within and among individuals or species by over a 100-fold.

The sperm surface contains substances that interact directly with the female or with the oocyte surface. Sperm plasma membrane glycosidases are implicated in gamete recognition in a range of invertebrates, including molluscs, ascidians, and insects, where they are considered as candidate proteins in binding with the egg surface at fertilization. In a detailed comparative study of 11 *Drosophila* species, the Perotti laboratory at the University of Milan has examined four sperm glycosidases, candidates for sperm–egg binding. While expression of all four glycosidases was observed in the 11 species, their distribution on sperm surfaces as well their activities were found to vary among species.

Nonsperm

In many invertebrates, including worms, crabs, beetles, crickets, and butterflies, sperm are transferred in a spermatophore, or a discrete package. A spermatophore's contents may be compartmentalized into portions with and without sperm, depending upon the species. Furthermore, not all spermatophores are directly deposited into the female. In some species, females themselves insert spermatophores into their own reproductive tracts. Because nonsperm components of

Table 1   Internal fertilization in the invertebrates

Taxon	Fertilization	Copulation
Poriferans	Both	No
Colenterates	Both	No
Platyhelminthes	Both	Yes
Roundworms nematode	Yes	Yes
Echinoderms	Rarely internal But some stars and sea cucumbers	Rare if at all
Molluscs	Both	Yes
Crustaceans	Both	Yes
Arachnids	Internal	
Insects	Internal	Yes and no

spermatophores are in discrete packages, they lend themselves more easily to comprehensive chemical analyses than does seminal fluid passed directly, without any sort of membrane, to the female. Spermatophores vary tremendously in their size and composition. In some Orthoptera, spermatophores can be up to one-third of the male's body weight. Bush cricket spermatophores are mostly water and protein, with small amounts of glycogen and lipid, primarily hydrocarbons. Crab spermatophores are largely mucopolysaccharide, lacking glycogen. Specialized compounds such as alkaloids and carotenoids also are found in spermatophores of certain species, and may provide benefits to females or their eggs in the form of protection against parasitism or oxidative stress. In butterflies, spermatophore composition varies with adult diet.

In invertebrates lacking spermatophores, chemical analyses of seminal fluid are less advanced. Ejaculate quantities are small and the fluid itself is difficult to obtain in pure form. Some studies have utilized radioactive labeling of males to track and identify male-derived substances in females. Labeling studies have revealed the transfer of elements such as sodium and phosphorus. Studies of invertebrate seminal fluid, however, have tended to focus on seminal fluid proteins and the genes that code for them. Genomic and proteomic approaches have led to the characterization of male accessory gland and other seminal proteins in the honeybee *Apis melifera*, in the malarial vector *Anopheles gambiae*, and in several species of *Drosophila*. These studies, despite using different approaches, have revealed over a 100 different male proteins in each of the foregoing taxa. Some proteins appear to be conserved across these diverse invertebrate taxa, and even share, in the case of the honeybee, considerable overlap with human seminal fluid. Other proteins, in *Drosophila*, for example, are so rapidly evolving that they cannot even be identified in other members of the same genus. Seminal fluid proteins fall into a number of functional categories, indicating their involvement in processes such as sperm energetics and immune defense.

Female Arena

Morphology

Some species such as poriferans and coelenterates have internal fertilization but typically lack specialized female tissue for receiving or storing sperm. In some sponges, however, oocytes develop in association with a cluster of nurse cells, one of which is specialized to capture sperm. Certain bivalve molluscs store sperm in gill chambers. At the other extreme, many molluscs and arthropods have highly developed female reproductive tracts, often with multiple types of structures. Reproductive tracts typically include a uterine structure or bursa where sperm enter and may or may not remain. Sperm storage organs can vary in shape, size, and number. Frequently, there are tubular receptacles and paired spermathecae, some of which have sphincter valves at the base and muscular coverings. Spermathecae and seminal receptacles also exhibit high levels of interspecific variability in shape and size. A wide variety of accessory glands occur in female tracts, although their specific functions typically remain unidentified. Lampyrid beetle female reproductive tracts, for example, have a certain structure that is assumed to be a spermatophore-digesting gland.

Chemistry

Reproductive tract morphology of female invertebrates has been studied far more extensively than has female chemistry. In those species in which female reproductive tract chemistry has been examined, however, it is proving to be complex and variable. Our information primarily derives from proteomic and genomic studies of economically or medically important insects and of *Drosophila*. In bloodsucking sandflies, spermathecae are filled with a mucopolysaccharide secretory mass. Honeybee spermathecal fluid has been found to have over a 100 proteins. In other cases, chemical interactions can be inferred from structural or ultrastructural studies. For example, in the lampyrid beetle, the spermatophore-digesting gland presumably contains digestive enzymes of some sort. In female houseflies, the posterior reproductive tract has at least six proteins with acid phosphatase activity, each of which exhibits specificity with respect to particular structures of the reproductive tract. Female reproductive proteins in several *Drosophila* species and in honeybees have been the most extensively studied of any invertebrates. Several hundred candidate female reproductive genes turned up in expression studies of whole *Drosophila* female lower reproductive tracts, while ~40 proteins were found in the spermathecae alone. Proteomic studies of the seminal fluid of honeybee queens revealed over a 100 proteins in this storage organ, most of which have energetic or antioxidant function.

Eggs

Outer membranes of freshwater and marine invertebrates, because desiccation is not a problem, are usually different from those of terrestrial species. Oocytes of terrestrial arthropods typically are covered not only with the vitelline membrane, but also a tough chorion, or shell, on the outside. In the vast majority of insect species, because of the impenetrable outer coverings, oocytes have a tubular structure or micropyle, through which sperm must enter. Oocytes of species in the Hemipteran family of Cemicoidea, where fertilization takes place in the ovariole even before the egg's vitelline membrane and chorion are complete, lack a micropyle. In most insects, oocytes have just one micropyle, but in some species there are two and even up to 70 micropyles, often arranged in a circle. In related taxa where the micropyle number differs, it is usually constant for a given species. Micropyles have unique surface carbohydrates that interact with sperm, purportedly the glycosidases mentioned earlier. Consistent with this idea, in *Drosophila*, the Perotti group has shown interspecific differences in the distribution of these carbohydrates in and around the micropyle. Spiders and other invertebrates, for the most part, lack any specialized sperm-guidance structure on the egg surface.

Processes

Once the ejaculate enters the female, one or more prezygotic processes occur, including sperm storage, various female physiological responses, and sperm retrieval and fertilization. These processes may happen quickly or last several years, depending upon the species, but all rely on interactions among the structural and molecular components discussed in the previous section.

Reaching Their Destination

Whether the immediate destination inside the female is the oocyte or a longer-term storage location, some mechanism (s) must guide sperm to their destination. In certain species, for example, the sponges previously mentioned, sperm are not stored. Rather, upon flowing through the body cavity with the seawater, a sperm is captured by a specialized female cell and quickly delivered to the oocyte where fertilization takes place. Nematode sperm are amoeboid and crawl from the uterus to the spermatheca where fertilization takes place. Oocytes of *Caenorhabditis elegans* use polyunsaturated fatty acids to control directional sperm motility within the uterus.

In species where sperm have tails, the storage process is assumed to be at least partially a function of their motility. Contraction of female reproductive tract musculature moves sperm in many insect species. As in *C. elegans*, however, chemical signaling is likely to play a role in where sperm go. Even in taxa with flagellated, highly motile sperm, such as *Drosophila*, the Wolfner laboratory at Cornell University has shown using RNA interference and antibodies that certain seminal fluid proteins control the sperm storage process. In some cases, male proteins localize to specific regions of the female reproductive tract. Such specificity would be difficult without corresponding localized biochemical differences in the female reproductive tract. In another Dipteran, the sand fly, the mucopolysaccharide mass of the spermatheca appears to activate the sperm in the spermatophore, promoting their uptake into storage. In certain insects, females appear to dump or kill sperm, although the underlying mechanisms are unknown.

Traumatic and hypodermic insemination represent special cases of sperm delivery. A number of invertebrate taxa, including molluscs, insects, and spiders, have traumatic insemination. In most cases of traumatic insemination, sperm are delivered directly to the female reproductive tract. Hypodermic insemination, in which the females' body is pierced and sperm are delivered to the female hemolymph, requires some mechanism by which sperm navigate through the female's body to her reproductive tissue. In the Dysderid spider, *Harpactea sadistica*, for example, where sperm storage organs are atrophied, sperm travel directly to the ovary where fertilization occurs. In this and other cases, the mechanisms guiding sperm to their destination remain unknown.

Maintenance in Storage

In those invertebrates where sperm are stored, what does maintenance of sperm in storage require and who provides it? Duration of sperm storage can last less than an hour or several years as in some species of hymenoptera. In honeybees and leaf cutter ants, both the seminal fluid and spermathecal secretions are important in the long-term viability of sperm in the sperm storage organs. In many invertebrates in which females store sperm, they also mate multiply. The presence of multiple ejaculates in the storage organs sets the stage for interejaculate interactions as well as more complex, multiindividual interactions that include the multiply inseminated female and influence the fitness of all parties.

Physiological Changes in the Female

Male materials produce a variety of changes not only in the female reproductive tract but in somatic tissues as well, either by mechanical or chemical means or both. Innervation of the female reproductive tract can detect the presence of sperm and signal responses leading to ovulation and fertilization. In isopods, mating itself triggers remodeling of the sperm storage organs. In some *Drosophila* species, especially those of the mulleri complex of the repleta group, a large mass forms in the uterus following mating that can last up to 10 h in intraspecific and indefinitely in interspecific matings. *Drosophila* species have proved highly informative regarding the action of specific seminal proteins in triggering oogenesis, ovulation, and delaying remating. While the well-known sex peptide is found in many *Drosophila* species, other proteins produce similar responses in mated females. Microarray and other experiments have revealed postmating changes in the expression of many functional categories of genes in females. Immune responses to mating may include upregulation of immune-response genes as in *Drosophila* as well as their downregulation as in *Tribolium*.

Females of several taxa take up seminal fluid substances, including proteins, sodium, and phosphorus, to incorporate them into their somatic tissues and ovaries. In *Drosophila melanogaster*, some seminal proteins appear to be associated with reduced female lifespan, while in the majority of *Drosophila* examined, no lifespan reduction seems to result from mating. In certain ant species, however, queens mating with either fertile or sterile males experience significantly longer lifespans than virgin queens.

Retrieval/Fertilization

Fertilization in invertebrates with stored sperm requires (1) the retrieval of sperm from storage, (2) ovulation, and (3) fusion of the sperm with the oocyte. In the majority of species studied, females release from their ovaries mature oocytes, which then are fertilized by stored sperm. What triggers the processes leading to fertilization? Signals may consist of environmental cues, internal cues, and/or their interaction. Environmental cues may be abiotic, such as light or temperature, or biotic, such as the availability of suitable oviposition sites. Internal factors can include the nutritional state of females, such as a recent blood meal in hematophagous species, and its influence on the production of mature oocytes. Female reproductive tracts frequently are innervated, by stretch or proprioceptors, signaling to a female that she is inseminated. Chemical features of seminal fluid, particularly male accessory gland proteins, also can initiate the steps toward fertilization via the female nervous system. Oviposition decisions, therefore, while they may appear to be female controlled, can be under the morphological or biochemical influences of both sexes long after copulation. In *Rhodnius*, male secretions cause contractions of the female oviduct.

An unidentified product of the paired sex accessory glands of the posterior reproductive tract of female houseflies allows penetration of the eggs by sperm, either by 'activation' of the sperm or alteration of the egg membrane. *Tribolium* females exert muscular control over sperm storage, although there is no evidence to date that females use this to differentiate among mates. The development of sperm storage organs allows females control over sperm storage and subsequent utilization.

Molecular mechanisms are implied to be the male's way of controlling the same processes, but in ways that benefit the male.

Evolutionary Implications of Reproductive Tract Interactions

A major challenge remains to understand the evolutionary significance of the observed intra- and interspecific variability in the processes that take place within inseminated females. Postmating control over reproduction frequently is discussed in different contexts such as cryptic female choice or sexually antagonistic co-evolution. These processes presumably occur inside the mated female and ultimately influence the genotypes of the fertilized eggs, embryos, or even older stages of the progeny that subsequently issue from the female. If we are to understand the evolution of these processes at their most fundamental levels, their underlying genetic architecture first must be identified. For example, if multiply mated females are able to selectively store and or use the sperm from one male over another, some heritable characteristic of the ejaculate must be not only variable among males, but it must also be detectable to females. Females also must have heritable mechanisms by which to detect male ejaculate variants and in order to differentially respond to them. Sperm uptake, storage, and retrieval each represent critical control points if female sperm choice is a reality. A firm grasp of the genetic underpinnings of each step will allow stringent and manipulative tests of the evolutionary processes hypothesized to explain the particular intersexual interaction.

Currently, we know only parts of the story. For example, while there is strong evidence for selection on genes encoding both male and female reproductive proteins, especially in *Drosophila* and there is evidence of assortative fertilization in several taxa, including *Drosophila*, no connection currently exists between these different bodies of evidence. What is the role of precopulatory factors, such as whether a mating was preferred by one or both partners, on postcopulatory processes? Are certain steps in the postcopulatory sequence of interactions more probable than others to influence fertilization success? In other words, is selection stronger on particular types of internal interactions than on others? Are certain components more variable than others either at a genetic or nongenetic, including the environmental level?

What is the relationship between fertilization success within a species and the evolution of postcopulatory-prezygotic reproductive isolation?

Much of our information on proximate mechanisms comes from laboratory studies of model organisms such as *Caenorhabditis*, *Drosophila*, *Tribolium*, or *Apis*. These model organisms exist in nature as well, and are not especially difficult to study in the wild. Furthermore, with next-generation sequencing technologies and other molecular approaches becoming affordable, a larger range of ecologically interesting taxa can be added to the pioneering studies on model organisms. Perusal of **Table 1** shows that in the porifera, coelenterata, platyhelminthes, mollusca, and crustea both internal and external fertilization occurs. How many times within one of these groups has internal fertilization evolved and what are the ecological, behavioral, and internal correlates? Carefully selecting new taxa to study should generate hypothesis-driven investigations of the relationship between mating system, reproductive tract interactions, and fitness.

See also: Cryptic Female Choice; Forced or Aggressively Coerced Copulation; Social Selection, Sexual Selection, and Sexual Conflict; Sperm Competition.

Further Reading

Baer B, Armitage SA, and Boomsma JJ (2006) Sperm storage induces an immunity cost in ants. *Nature* 441: 872–875.

Bauer RT and Martin JW (1991) *Crustacean Sexual Biology.* New York, NY: Columbia University Press.

Gowaty PA (2008) Reproductive compensation. *Journal of Evolutionary Biology* 21: 1189–1200.

Markow TA and O'Grady PM (2005) Evolutionary genetics of reproductive behavior in *Drosophila*: Connecting the dots. *Annual Review of Genetics* 39: 263–291.

Markow TA and O'Grady PM (2008) Reproductive ecology of *Drosophila. Functional Ecology* 22: 747–759.

Ravi Ram K and Wolfner MF (2007) Seminal influences: *Drosophila* Acps and the molecular interplay between male and female during reproduction. *Integrative and Comparative Biology* 47: 427–445.

Rogers DW, Whitten MMA, Thallayil J, Solchot J, Levashina E, and Catteruccia F (2008) Molecular and cellular components of the mating machinery in *Anopheles gambiae* females. *Proceedings of the National Academy of Sciences of the United States of America* 105: 19390–19395.

Schmitt V, Anthes N, and Michiels NK (2007) Mating behaviour in the sea slug *Elysia timida* (Opisthobranchia, Sacoglossa): Hypodermic injection, sperm transfer and balanced reciprocity. *Frontiers in Zoology* 4: 17. doi: 10.1186/1742-9994-4-17.

Singson A, Hang JS, and Parry JM (2008) Genes required for the common miracle of fertilization in *Caenorhabditis elegans*. *International Journal of Developmental Biology* 52: 647–656.

Irruptive Migration

I. Newton, Centre for Ecology & Hydrology, Wallingford, UK

Introduction

Most migrations in birds are regular, taking place at the same seasons each year, with individuals moving between fixed breeding and wintering areas. In irruptive migrations, the proportions of birds leaving the breeding range, and the distances they travel, vary greatly from year to year. While regular migrations are associated with regular and predictable food supplies, irruptions are associated with sporadic food supplies, which vary in abundance and distribution from 1 year to the next. Typical irruptive migrants are all food specialists for at least part of the year, and all can breed or winter in widely separated areas in different years. Strictly, the term 'irruption' (or 'invasion') is applicable only to the region receiving the birds, whereas 'eruption' is often applied to the region losing them, but for general discussion it is convenient to use a single term.

Typical irruptive migrants of northern regions include: (1) boreal finches and others that depend on fluctuating tree-seed crops; and (2) owls and others that depend on cyclically fluctuating prey populations. This chapter is concerned with both groups, but in some parts of the world waterfowl and other birds of arid regions show irruptive movements, in association with sporadic rainfall patterns that influence their habitats and food supplies.

Boreal Seedeaters

The Food Base

Typically, northern tree-seed crops vary greatly in size from year to year, and in some years fail completely (**Figure 1**). Fruiting depends partly on the natural rhythm of the trees themselves and partly on the weather. Trees of most species require more than 1 year to accumulate the nutrient reserves necessary to produce fruit. For a good crop, the weather must ideally be fine and warm in the preceding autumn when the fruit buds form, and again in the spring when the flowers set. Otherwise, the crop is reduced or delayed for another year. In any one area, most of the trees of a species fruit in phase with one another partly because they come under the same weather, and often those of several different species also crop in phase. The result is a great profusion of tree fruits in some years, and practically none in others. Good crops almost never occur in consecutive years, and nor do poor crops, but each good crop is usually followed by a poor one (**Figure 1**).

The trees in widely separated areas may be on different fruiting regimes, partly because of regional variations in weather, so that good crops in some areas may coincide with poor crops in others. Nevertheless, good crops may occur in many more areas in some years than in others, so the total continental seed production also varies greatly from year to year. An analysis by W.D. Köenig and J.M.H. Knops of the fruiting patterns of various boreal conifer species in many localities in North America and Eurasia revealed high synchrony in seed production in localities 500–1000 km apart, depending on tree species. The synchrony declined at greater distances, and by 5000 km no correlation was apparent within particular tree species.

The various seed-eating species of irruptive migrants are listed in **Table 1** (in which scientific names are given). Some seedeaters eat mainly seeds the year-round, but different types of seeds at different seasons, as exemplified by the Common Redpoll and Pine Siskin. Others eat mainly insects in summer and seeds (or fleshy fruits) in winter, as exemplified by the Bohemian Waxwing, Brambling, and Evening Grosbeak. The latter tend to concentrate on breeding in areas with insect outbreaks – the Brambling in areas with high densities of the Autumn Moth *Epirrita autumnata* and the Evening Grosbeak in areas with Spruce Budworm *Choristoneura fumiferana* – but both species seek tree seeds in winter. Other species also exploit the same foods as irruptive species, but do not depend so heavily on them, so are less affected by their fluctuations.

Density Fluctuations

In parallel with their local food supplies, the densities of many irruptive seedeaters have been found to fluctuate greatly from year to year (**Figure 2**). Local breeding densities can vary from nil or almost nil in poor food years to hundreds or thousands of pairs per 100 km^2 in good food years. Local increases in numbers from 1 year to the next are often far greater than could be explained by high survival and reproduction from the previous year, so must also involve immigration. Such species contrast with other seedeaters, whose densities typically vary by less than threefold from year to year.

Movements

One line of evidence suggesting frequent shifts between breeding localities in different years comes from measured rates of turnover in the occupants of particular study areas. In regular migrants, if the birds occupying a particular study area are trapped and banded in the breeding season, large proportions of the same individuals are usually found nesting in the same area next year. Most are found on the same territories, and those that change territories usually move over relatively short distances, staying within the general area. The proportions of individuals that return to the same area are usually within the range 30–60% for passerines and 60–90% for nonpasserines. Allowing for expected mortality, such high figures imply that most surviving individuals return to breed in the same limited area year after year. In some such regular migrants, the same holds for wintering areas.

Among irruptive migrants, in contrast, return rates to the same study area are much lower. For example, among Bramblings trapped in the breeding season in various areas, individuals were seldom or never caught in the same locality in a later year, so that each year's occupants were different from those the year before. In one study, Lindström and colleagues reported that only seven (0.6%) of 1238 adult Bramblings were retrapped in the same area in a later year, and none of 1806 juveniles, despite a regular annual trapping program over many years.

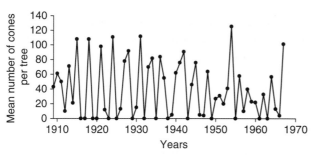

Figure 1 Annual fluctuations in the size of Norway Spruce *Picea abies* cone crops in south Sweden, 1909–1967. Almost complete fruiting failure occurred in 15 different years out of the 59 years covered. Reproduced from Newton I (2008) *The Migration Ecology of Birds*. London: Academic Press.

Table 1 Established year-to-year correlations between bird abundance and food supply in seed-eating and fruit-eating birds

	Preferred winter food[a]	Summer densities	Winter densities	Autumn emigration
Great spotted woodpecker *Dendrocopos major* (P)	Spruce, pine, and other seeds	•	•	•
Bohemian waxwing *Bombycilla garrulus* (H)	Rowan and other berries			•
Fieldfare *Turdus pilaris* (P)	Rowan and other berries			•
Coal tit *Parus ater* (P)	Spruce seeds, insects			•
Black-capped chickadee *Parus atricapillus* (N)	Conifer seeds, insects			•
Great tit *Parus major* (P)	Beech seeds		•	•
Blue tit *Parus caeruleus* (P)	Beech seeds		•	•
Wood nuthatch *Sitta europaea* (P)	Spruce seeds			•
Red-breasted nuthatch *Sitta canadensis* (N)	Pine and other conifer seeds			•
Brambling *Fringilla montifringilla* (P)	Beech seeds	•	•	•
Eurasian siskin *Carduelis spinus* (P)	Birch, alder, and conifer seeds	•	•	•
Pine siskin *Carduelis pinus* (N)	Conifer, birch, and alder seeds			•
Common redpoll *Carduelis flammea* (H)	Birch and alder seeds	•		•
Arctic (Hoary) redpoll *Carduelis hornemanni* (H)		•		•
Eurasian bullfinch *Pyrrhula pyrrhula* (P)	Various tree seeds and berries			•
Pine grosbeak *Pinicola enucleator* (H)			•	•
Evening grosbeak *Hesperiphona vespertina* (N)	Maple and other tree seeds	•		•
Purple finch *Carpodacus purpureus* (N)	Various tree seeds			•
Common (Red) crossbill *Loxia curvirostra* (H)	Spruce and other conifer seeds	•	•	•
Two-barred (white-winged) crossbill *Loxia leucoptera* (H)	Larch and other conifer seeds	•	•	•
Parrot crossbill *Loxia pytyopsittacus* (P)	Scots pine seeds	•	•	•
Eurasian Jay *Garrulus glandarius* (P)	Oak seeds			•
Thick-billed nutcracker *Nucifraga c. macrorhynchos* (P)	Hazel and Swiss stone pine seeds			•
Thin-billed nutcracker *Nucifraga c. caryocatactes* (P)	Siberian stone pine seeds, brush pine seeds			•
Clark's nutcracker *Nucifraga columbiana* (N)	Whitebark pine and other conifer seeds			•

[a]Scientific names of trees: alder *Alnus*, Beech *Fagus sylvatica*, birch *Betula*, Hazel *Corylus avellana*, larch *Larix*, maple *Acer*, oak *Quercus*, Rowan *Sorbus aucuparia*, Scots Pine *Pinus sylvestris*, Siberian Stone Pine *Pinus sibirica*, spruce *Picea*, Swiss Stone (Arolla) Pine *Pinus cembra*, Brush Pine *P. pumila*, Whitebark Pine *Pinus albicaulis*, Chihuahua Pine *Pinus chihuahuana*. Where several species in the same genus are involved, only the generic name is given.
H, Holarctic; N, Nearctic; P, Palearctic.
Source: Newton I (2008) *The Migration Ecology of Birds*. London: Academic Press.

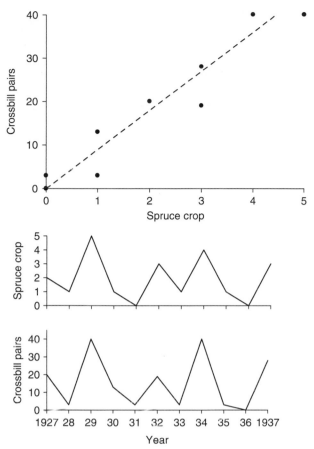

Figure 2 Annual fluctuations in the breeding densities of the common crossbill *Loxia curvirostra* in relation to Norway Spruce cone crops in an area of Finland. Crossbills in number of pairs per 120 km transect; spruce crop ranked in five categories. Redrawn from Reinikainen A (1937) The irregular migrations of the crossbill, *Loxia c. curvirostra*, and their relation to the cone-crop of the conifers. *Ornis Fennica* 14: 55–64.

Similar findings have come from studies on other irruptive species elsewhere.

Return to specific wintering sites was even lower than return to breeding sites. Despite some very large numbers banded, return rates of irruptive seedeaters were mostly nil or less than 1%. Evidently, extremely few individuals of such species returned to the same areas in subsequent years. They thus showed little or no site fidelity, summer or winter, in striking contrast to more regular migrants.

For some species, band recoveries have given some idea of how far individuals can shift their breeding site from one year to another. The main breeding season of the spruce-feeding Common Crossbill in Europe is in January–April, when the cones begin to open. Several adults trapped and banded in this period one year were recovered in the same period in a later year. Apart from one bird which had moved only 28 km, the rest had moved distances of 790–3170 km, with none at lesser distances (**Figure 3**). Similarly, three Crossbills banded as chicks or recently fledged juveniles were recovered in later breeding seasons

at distances of 1100–2950 km, with none at lesser distances (**Figure 3**). In North America, a study by Brewer and colleagues showed that two Red Crossbills were recovered at places 1288 km apart (November 1969–October 1971) and 1409 km apart (May 1991–May 1992); these dates could have fallen within the autumn and spring breeding periods of the species on this continent. Although relevant band recoveries are few, other irruptive finches have also been found in different breeding seasons at localities up to several hundred kilometers apart.

Individuals of irruptive species have also been found in widely separated localities in different winters, sometimes on opposite sides of a continent. Extreme examples include a Pine Siskin banded in Quebec in one winter and recovered in California in a later winter, an Evening Grosbeak banded in Maryland in one winter and recovered in Alberta in a later winter, and a Common Redpoll banded in Belgium in one winter and recovered in China in a later one. Another Common Redpoll was recorded in North America in one winter and Eurasia in a later one, having been banded in Michigan and recovered near Okhotsk in Siberia, some 10 200 km to the northeast. All these birds are likely to have returned to the breeding range in the interim and taken a markedly different migration direction in the second year. Although they are extreme examples, dozens of individuals of various irruptive species have now been recorded at places 500–1500 km apart in different winters.

The extent to which irruptive finches wander for food is well shown by the Evening Grosbeak, which breeds in conifer forests and moves south or southeast in autumn. This species feeds in winter mainly on large, hard tree fruits, but also visits garden feeding trays, a habit which makes it easy to catch. D.H. Speirs analyzed the recoveries of 17 000 individuals banded over 14 winters at a site in Pennsylvania. Of these, only 48 (0.003%) were recovered in the same place in subsequent winters, yet 451 others were scattered among 17 American States and four Canadian Provinces. Another 348 birds that had been banded elsewhere were caught at this same locality, and these had come from 14 different States and four Provinces. These recoveries show how widely individual Grosbeaks range, and how weak is their tendency to return to the same place in later years.

Irruptions

The extent of autumn emigration from the breeding range has been related to food supplies in almost all seed-eating and fruit-eating species discussed here, with the biggest emigration in the years of crop failure (**Table 1**). Moreover, the various species that depend heavily on the same seed or fruit crops tend to irrupt in the same years: examples include north European Blue Tits and Great Tits, both of which feed heavily on beech mast, and Common

Figure 3 Banding and recovery sites of common crossbills *Loxia curvirostra* that were both banded and recovered in different breeding seasons (taken as January–April in areas of Norway Spruce *Picea abies*). Continuous lines – banded as adults (representing breeding dispersal); dashed lines – banded as juveniles (representing natal dispersal). Reproduced from Newton I (2006) Movement patterns of common crossbills *Loxia curvirostra* in Europe. *Ibis* 148: 782–788.

Crossbills and Great Spotted Woodpeckers, both of which feed heavily on spruce seeds. Where different tree species fruit in phase with one another, the number of participating species is increased further. Over much of the boreal region of North America, conifer and other seed crops tend to fluctuate biennially, and in alternate years of poor crops several species that depend on them migrate to lower latitudes. Analyses by Bock & Lepthien and Köenig & Knops show that for much of the twentieth century, at least eight species of boreal seedeaters tended to irrupt together, in response to a widespread, synchronized pattern of seed crop fluctuations (namely Common Redpoll, Pine Siskin, Purple Finch, Evening Grosbeak, Red Crossbill, White-winged Crossbill, Red-breasted Nuthatch, and Black-capped Chickadee). These species vary in the proportions of conifer and broad-leaved tree seeds in their diets. Over periods of years, different species of trees in the same area can drift in and out of synchrony with one another, affecting the movements of birds. During the period 1921–1950, Larson and Bock found that the biennial pattern and synchrony between the various North American seedeaters was less marked than before or after this period.

In 1954, David Lack noted that the food shortage that leads to a long and heavy migration is accentuated if the birds themselves are especially numerous at the time, as a result of good survival and breeding in the previous years, when food was plentiful. It is not, therefore, merely poor food supplies that stimulate large-scale emigration, but poor supplies relative to the number of birds present.

In 1962, Staffan Ulfstrand showed that when poor food supplies are associated with low bird numbers, little emigration occurs, while at other times moderate food supplies are associated with exceptionally high bird numbers and large-scale emigration. It is the ratio of birds to food that seems to count: the greater the imbalance, the greater the proportion of birds that leaves.

Other Aspects

In some irruptive species, the timing of autumn emigration appears much more variable than among regular migrants. Roos found that the dates for peak passage of Eurasian Siskins through Falsterbo Bird Observatory in Sweden during 1949–1988 varied from 15 August (in 1988) to 17 November (in 1958), the last date the station was manned that year, and elsewhere Svärdson noted that heavy southward movements have been seen as late as December–January. The birds tended to pass in largest numbers, and at the earliest dates, in years when their food crops were poor. The tendency for birds to linger longer in the north in years of good tree-seed crops has been noted in many other irruptive seedeaters, as has their tendency to arrive earlier in more southern wintering areas in invasion years.

Compared to regular migrants in autumn, irruptive seedeaters also tend to show more spread in their departure directions, as is evident both from observations at particular watch sites and from subsequent band recoveries. Whereas recoveries of typical migrants from particular sites normally fall on a relatively narrow route toward wintering areas, those of irruptive migrants are often spread over an arc of more than 45°, say, from east to south, or southwest to southeast.

Another difference concerns the frequency of movements within a winter, for irruptive migrants do not necessarily remain in the same localities throughout a winter, but move on repeatedly. Banding has revealed rapid turnover in the individuals present at particular sites, and the occurrence of the same individuals at widely separated sites within the same winter. The implication is that, in the nonbreeding period, individuals frequently make long moves, perhaps in continual search for good feeding areas. As Haila, Tiainen, and Vepsäläinen have shown in their work in Finland, many seedeaters travel progressively further from their breeding areas during the nonbreeding period, stripping food crops as they go.

Owls and Other Rodent Eaters

The Food Base

Similar aspects of behavior are shown by those species of owls and raptors that specialize on cyclically fluctuating prey populations, with mass emigration occurring in years when prey is scarce. Two main cycles are recognized: (a) an approximately 4-year cycle in small rodents in temperate, boreal, or tundra environments and (b) an approximately 10-year cycle of Snowshoe Hares *Lepus americanus* in the boreal forests of North America. Some grouse-like birds are also involved, but whereas in some areas they follow the 4-year rodent cycle, in others they follow the 10-year hare cycle. As with tree seeds, the peaks in rodent numbers are usually synchronized over hundreds or thousands of square kilometers, but out of phase with those in other regions. However, Chitty noted that peak populations may occur simultaneously over many more areas in some years than in others, giving a measure of synchrony, for example, to lemming cycles over large parts of northern Canada, with few regional exceptions. In most places, the increase phase of the rodent cycle usually takes 2–3 years, and the crash phase occurs within 1 year. The hare cycle, in contrast, occurs more or less simultaneously over most of the North American boreal region, with little regional variation, the increase phase occurring over several years and the crash phase over 1–2 years.

Density Fluctuations

Among the rodent eaters, local breeding densities can vary from nil in low rodent years to several tens of pairs per $100 \, km^2$ in intermediate (increasing) or high rodent years. For example, in a 47-km^2 area of western Finland, over an 11-year period, Korpimäki and Norrdahl found

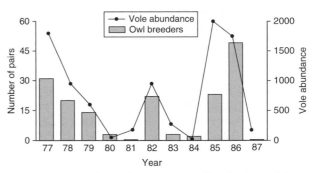

Figure 4 Annual fluctuations in the breeding densities of short-eared owls *Asio flammeus* in relation to field vole *Microtus arvalis* densities in an area of western Finland. Reproduced from Korpimäki E and Norrdahl K (1991) Numerical and functional responses of Kestrels, short-eared owls and long-eared owls to vole densities. *Ecology* 72: 814–825.

that numbers of short-eared Owls varied between 0 and 49 pairs, numbers of long-eared Owls *Asio otus* between 0 and 19 pairs, and Eurasian Kestrels *Falco tinnunculus* between 2 and 46 pairs, all in accordance with spring densities of *Microtus* voles (**Figure 4**). All these predators were summer visitors to the area concerned and settled according to vole densities at the time. Similar fluctuations in these and other rodent specialists were recorded elsewhere. In each case, the year-to-year increases were so great that they could be explained only by massive immigration. Their fluctuations contrast with findings from other owls and raptors that depend on a wider range of prey species and show much more stable breeding densities from year to year.

Movements

The importance of movements has again been confirmed by banding studies. For example, among Eurasian Kestrels, of 146 individual breeders trapped and banded in a 63-km^2 area in Finland over an 11-year period, Korpimäki and Norrdahl reported that only 13% of males and 3% of females were found back in the same area in a later year. In this and other species, return rates were extremely low, compared to what would be expected from their annual survival rates. The implication is again that a large proportion of breeders changed their nesting localities from year to year; and as in the seedeaters, long-distance moves between the breeding sites of different years have again been shown by banding or radiotracking studies.

Mark Fuller and coinvestigators followed some remarkable movements by four adult Snowy Owls *Nyctia scandiaca* that were radiotagged while nesting near Point Barrow in Alaska (long. 150° W), and tracked by satellite over the next 1–2 years. These birds mostly stayed in the arctic but dispersed widely in different directions from Point Barrow, reaching west as far as 147° E and east as far as 116° W, a geographical spread encompassing nearly one-third of the species Holarctic breeding range. Two birds that bred at Point Barrow in 1999 were present during the next breeding season in northern Siberia (147° E and 157° E, respectively), up to 1928 km east of Point Barrow, and then in the following breeding seasons they were on Victoria Island (116° W) and Banks Island (122° W), respectively, in northern Canada. The two birds that bred at Point Barrow in 2000 were present on Victoria and Banks Islands in the breeding season of 2001. The successive summering areas of these four birds were thus separated by distances of 628–1928 km. From the dates they were present, some could have bred successfully, while others were unlikely to have done so, having arrived too late or left too early. None returned to the same breeding or wintering site used in a previous year, but three passed through Point Barrow in 2001.

Extensive data from banding are available for Boreal Owls *Aegolius funereus*, which nest readily in boxes and have been studied at many localities in northern Europe. In this species, the males are mainly resident and the females highly dispersive. Both sexes tend to stay in the same localities if vole densities remain high, moving no more than about 5 km between nest boxes used in successive years; but if vole densities crash, females move much longer distances, with many having moved 100–600 km between nest sites in different years (**Figure 5**). In contrast, fewer long movements were recorded from males, with only two at more than

100 km. The greater residency of males was attributed to their need to guard cavity nest sites, which are scarce in their conifer-forest nesting-habitat, while their smaller size makes them better able than females to catch birds and hence survive (without breeding) through low vole conditions.

Fewer records are available for other nomadic owl species, because the chances of recording marked individuals at places far apart are low. However, movements between breeding sites in different years of hundreds or thousands of kilometers have been recorded from several other species, including Northern Hawk Owl *Surnia ulula*, Great Grey Owl *Strix nebulosa*, short-eared Owl *Asio flammeus*, and long-eared Owl *Asio otus*. Again, however, these various irruptive owls contrast greatly with more sedentary populations, which exploit more stable food supplies. Analyses by Newton and studies by Saurola show that in such species, adults usually remain in their territories year after year, with only small proportions moving to other territories, usually nearby.

Few band recoveries from different winters are yet available for any species of irruptive owl. However, the satellite-tagged Snowy Owls mentioned earlier were present in widely separated localities in different winters and often moved long distances within a winter. Another Snowy Owl was banded near Edmonton in January 1955 and recovered 330 km to the southeast in Saskatchewan in January 1957. There is also an intriguing record of a long-eared Owl banded in California in April and recovered in Ontario in October of the same year. In general, therefore, banding and radiotracking evidence bear out the inference from local counts that many individual owls both breed and winter in widely separated areas in different years.

Irruptions

Among rodent-eating species, irruptions of Snowy Owls from the tundra to the boreal and temperate regions of North America have been documented since about 1880. For the next 120 years, Newton noted that irruptions into northeastern regions occurred every 3–5 years, at a mean interval of 3.9 years. Moreover, in periods when information on lemmings was available from the breeding areas, both Lack and Chitty found that mass southward movements of owls coincided with widespread crashes in lemming numbers. In western North America, irruptions of Snowy Owls were not well synchronized with those in the east, presumably reflecting asynchrony in lemming cycles between breeding regions. An analysis by Kerlinger, Lein, and Sevick documented that irruptions were also less regular and less pronounced in the west than in the east, with some birds appearing on the northern prairies every year, and some of the same marked individuals appearing on the same territories in different (not necessarily consecutive) winters.

Both Lack and Newton have reported that in eastern North America, two other vole eaters, the Rough-legged

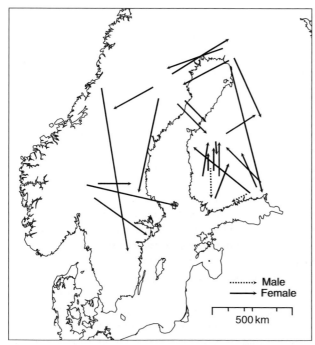

Figure 5 Banding and recovery sites of adult boreal owls *Aegolius funereus* that were identified in different breeding seasons. Continuous lines – females; dashed lines – males. Only movements greater than 100 km are shown. Reproduced from Newton I (2008) *The Migration Ecology of Birds*. London: Academic Press.

Hawk *Buteo lapopus* and Northern Shrike *Lanius excubitor*, have irrupted at similar 3–5 year intervals, mostly (but not always) in the same years as Snowy Owls. Perfect synchrony between the three species would not be expected, because their breeding ranges only partly overlap. Irruptions at similar intervals have been recorded in other vole-eating species in both North America and Eurasia.

Irruptions of Goshawks *Accipiter gentilis* in North America tend to occur roughly every 10 years, corresponding with the cyclic crashes in their main prey, the Snowshoe Hare. Irruptions may occur in only one autumn or in two successive autumns if prey remains scarce. Great Horned Owls *Bubo virginianus*, which also feed on hares, tend to irrupt in the same years as Goshawks, but are less well documented.

Discussion

Given all this information, how do irruptive migrants differ from regular migrants? In the first place, their numbers at particular localities seem to fluctuate much more from year to year than do those of regular migrants, and in most species, these fluctuations have been clearly linked to fluctuations in local food supplies (**Table 2**). The fact that these birds can go from absence or near absence to abundance in less than 1 year strongly suggests the role of movements in influencing local densities, giving greatly varying patterns of distribution across the range from year to year. Movement has been confirmed by the high turnover rates in the occupants of particular areas, and by the long breeding dispersal distances found for a small number of individuals from band recoveries. It is now clear that irruptive seedeaters and rodent eaters

Table 2 Comparison between typical regular and typical irruptive migration

	Regular (obligate) migrants	Irruptive (facultative) migrants
Habitat/food	Predictable	Unpredictable
Breeding areas	Fixed	Variable
Wintering areas	Fixed	Variable
Site fidelity	High	Low
Migration		
Proportion migrating	Constant	Variable
Timing	Consistent	Variable
Distance	Consistent	Variable
Direction	Consistent	Variable
Main presumed ultimate stimulus	Food supply	Food supply
Main presumed proximate stimulus	Daylength	Food supply

can travel hundreds or thousands of kilometers between the breeding areas used in different years. They can also spend the winter in widely separated areas in different years, up to several thousand kilometers apart, and often on an east–west axis. This behavior is strikingly different from that of regular migrants, which usually return to the same breeding localities year after year, and often also to the same wintering localities, migrating more or less directly between the two.

To judge from the variable extent and timing of their migrations, many irruptive species – while genetically equipped to migrate – must presumably respond directly to food conditions at the time. Only in this way could they show the level of flexibility in movement patterns recorded. Food shortage apparently acts not only as the ultimate causal factor to which migration is supposedly adapted as suggested by David Lack in 1954, but also as the main proximate factor delaying or promoting departure from particular localities. The influence of any other factors (such as an endogenous rhythm or daylength response) in triggering movement seems less in irruptive than in more regular migrants.

The ease with which particular individuals could gain food is likely to depend, not only on food abundance, but on the density of other birds, which could reduce feeding rates through depletion and interference competition. In this way, the likelihood of an individual migrating is a function of population density as well as of food supply. In a study by Tyrväinen, it was noted that the autumn emigration of Fieldfares from southern Finland occurred when their main food crop (Rowan berries *Sorbus aucuparia*) had been reduced to an average of about two fruits per inflorescence. The date at which this occurred depended on both the initial crop size and the number of consumers. The role of competitive interactions in reducing feeding rates of individuals is well known from field studies of wild birds, and their role in the development of migratory condition was demonstrated experimentally by Scott Terrill in captive birds. In both situations, the effects of competition fell most strongly on the subordinate (usually younger) individuals. One might speculate, therefore, that competition for food, and its effect on body condition, might be the major proximate stimulus to eruptive migration, with greater proportions of individuals affected in poor food years than in good ones.

In considering the proximate control of migration, a distinction has been drawn between 'obligate migration' (formerly called 'instinct' or 'calendar' migration) and 'facultative migration' (formerly called 'weather' migration). According to Peter Berthold, in obligate migration, all main aspects are viewed as under firmer endogenous (genetic) control, mediated by daylength changes. This internal control gives a high degree of year-to-year consistency in the timing, directions, and distances of movements. For the most part, each individual behaves in the same way year after year, being much less influenced by prevailing conditions. In response to endogenous stimuli and daylength change, obligate migrants often leave their breeding areas each year well before food supplies collapse, and while they still have ample opportunity to accumulate body reserves for the journey. Most long-distance migrants that breed in temperate and boreal regions but winter in the tropics are in this category.

In contrast, facultative migration is viewed as a direct response to prevailing conditions, especially food supplies, and the same individual may migrate in some years, but not in others. Within a population, the proportions of individuals that leave the breeding range, the dates they leave, and the distances they travel can vary greatly from year to year, as can the rate of progress on migration, all depending on conditions at the time. In consequence, facultative migrants have been seen on migration at almost any date in the nonbreeding season, at least into January, and their winter distributions vary greatly from year to year. Although in such facultative migrants, the timing and distance of movements may vary with individual circumstances, other aspects must presumably be under firmer genetic control, notably the directional preferences and the tendency to return at appropriate dates in spring.

In general, it seems that obligate migration occurs in populations that breed in areas where food is predictably absent in winter, whereas facultative migration occurs in populations that breed in areas where food is plentiful in some winters and scarce or lacking in others. The distinction between obligate and facultative migrants is useful because it reflects the degree to which individual behavior is sensitive to prevailing external conditions, and hence varies from year to year. The advantage of strong endogenous control, as in obligate migrants, is that it can permit anticipatory behavior, allowing birds to prepare for an event, such as migration, before it becomes essential for survival, and facilitating fat deposition before food becomes scarce. But such a fixed control system is likely to be beneficial only in predictable circumstances, in which food supplies change in a consistent manner, and at about the same dates, from year to year. It is not suited to populations, which have to cope with a large degree of spatial and temporal unpredictability in their food supplies. It is these aspects of food supply, which probably result in irruptive migrants showing greater variations in autumn timing, directions, and distances, selection having imposed less precision on these aspects than in regular migrants. Both regular and irregular systems are adaptive, but to different types of food supplies. Nevertheless, obligate and facultative migrants are perhaps best regarded, not as distinct categories, but as opposite ends of a continuum, with predominantly endogenous control (=rigidity) at one end and predominantly external control (=flexibility) at the other. Irruptive migrants, with their facultative behavior, belong to the latter category.

See also: Bird Migration; Migratory Connectivity.

Further Reading

Berthold P (2001) *Bird Migration: A General Survey*, 2nd edn. Oxford: Oxford University Press.

Bock CE and Lepthien LW (1976) Synchronous eruptions of boreal seed-eating birds. *The American Naturalist* 110: 559–571.

Brewer D, Diamond A, Woodsworth EJ, Collins BT, and Dunn EH (2000) *Canadian Atlas of Bird Banding*, vol. 1. Ottawa: Canadian Wildlife Service.

Chitty H (1950) Canadian arctic wildlife enquiry, 1943–49, with a summary of results since 1933. *Journal of Animal Ecology* 19: 180–193.

Fuller M, Holt D, and Schueck L (2003) Snowy owl movements: Variation on a migration theme. In: Berthold P, Gwinner E, and Sonnenschein E (eds.) *Avian Migration*, pp. 359–366. Berlin: Springer.

Haila Y, Tiainen J, and Vepsäläinen K (1986) Delayed autumn migration as an adaptive strategy of birds in northern Europe: Evidence from Finland. *Ornis Fennica* 63: 1–9.

Kerlinger P, Lein MR, and Sevick BJ (1985) Distribution and population fluctuations of wintering Snowy owls (*Nyctea scandiaca*) in North America. *Ecology* 63: 1829–1834.

Koenig WD and Knops JMH (1998) Scale of mast seeding and tree-ring growth. *Nature* 396: 225–226.

Koenig WD and Knops JMH (2000) Patterns of annual seed production by northern hemisphere trees: A global perspective. *The American Naturalist* 155: 59–69.

Koenig WD and Knops JMH (2001) Seed crop size and eruptions of North American boreal seed-eating birds. *Journal Animal Ecology* 70: 609–620.

Korpimäki E and Norrdahl K (1991) Numerical and functional responses of Kestrels, short-eared owls and long-eared owls to vole densities. *Ecology* 72: 814–825.

Lack D (1954) *The Natural Regulation of Animal Numbers*. Oxford: University Press.

Larson DL and Bock CE (1986) Eruptions of some North American boreal seed-eating birds, 1901–1980. *Ibis* 128: 137–140.

Lindström Å, Enemar A, Andersson G, von Proschwitz T, and Nyholm NEI (2005) Density-independent reproductive output in relation to a drastically varying food supply: Getting the density measure right. *Oikos* 110: 155–163.

Newton I (1972) *Finches*. London: Collins.

Newton I (1979) *Population Ecology of Raptors*. Berkhamsted, UK: Poyser.

Newton I (1998) *Population Limitation in Birds*. London: Academic Press.

Newton I (2002) Population limitation in Holarctic owls. In: Newton I, Kavanagh R, Olson J, and Taylor IR (eds.) *Ecology and Conservation of Owls*, pp. 3–29. Collingwood, Victoria: CSIRO Publishing.

Newton I (2006) Movement patterns of common crossbills *Loxia curvirostra* in Europe. *Ibis* 148: 782–788.

Newton I (2008) *The Migration Ecology of Birds*. London: Academic Press.

Reinikainen A (1937) The irregular migrations of the crossbill, *Loxia c. curvirostra*, and their relation to the cone-crop of the conifers. *Ornis Fennica* 14: 55–64.

Roos G (1991) Visible bird migration at Falsterbo in autumn 1989, with a summary of the occurrence of six *Carduelis* species in 1973–90. *Anser* 30: 229–253.

Saurola P (1989) Ural owl. In: Newton I (ed.) *Lifetime Reproduction in Birds*, pp. 327–345. London: Academic Press.

Saurola P (2002) Natal dispersal distances of Finnish owls: Results from ringing. In: Newton I, Kavanagh R, Olsen J, and Taylor I (eds.) *Ecology and Conservation of Owls*, pp. 42–55. Collingwood, Victoria: CSIRO Publishing.

Terrill SB (1990) Ecophysiological aspects of movements by migrants in the wintering quarters. In: Gwinner E (ed.) *Bird Migration. Physiology and Ecophysiology*, pp. 130–143. Berlin: Springer.

Tyrväinen H (1975) The winter irruption of the fieldfare *Turdus pilaris* and the supply of rowan-berries. *Ornis Fennica* 52: 23–31.

Ulfstrand S (1962) On the non-breeding ecology and migratory movements of the Great Tit (*Parus major*) and the Blue Tit (*Parus caeruleus*) in southern Sweden. Vår Fågelvärld Suppl 3.

Ulfstrand S, Roos G, Alerstam T, and Osterdahl L (1974) Visible bird migration at Falsterbo, Sweden. Vår Fågelvärld Suppl. 8.

Isolating Mechanisms and Speciation

M. R. Servedio, University of North Carolina, Chapel Hill, NC, USA

Introduction

Whenever two species overlap geographically, there is the opportunity for behavioral isolating mechanisms to play a large role in maintaining species identity. The degree to which these mechanisms are involved in the initial process of speciation is more open to debate. In particular, the opportunity for behavior to play a role in speciation exists primarily when there is geographic overlap between incipient species at some point during the speciation process, a factor that is often unknown. Nevertheless, what we know about behavioral isolating mechanisms that are involved in speciation is often inferred from the mechanisms that we observe keeping species apart in the present day.

Behavioral isolating mechanisms can act at several stages during the life cycle of an animal (the degree to which behavioral mechanisms apply to plants is a semantic question). These include the premating, the postmating-prezygotic, and the postzygotic stages (**Figure 1**). The importance of isolating mechanisms in general is integrally tied to the life stage at which they act. If a strong isolating mechanism is in place that acts before mating, it renders mechanisms that act after mating less relevant, simply because they rarely come into effect. For this reason, the premating isolating mechanisms described in the following section may often be more biologically relevant than the postmating-prezygotic or postzygotic mechanisms, although all may be important in certain circumstances.

Behaviors that serve as isolating mechanisms can either evolve adaptively for that purpose or can act in that capacity as a byproduct of adaptation for other purposes (an exaptation). It is generally believed that selection for behavioral isolating mechanisms occurs most frequently when they act as a form of premating isolation. When behavioral isolating mechanisms evolve in true allopatry (without any migration between populations), they are not adaptations because they cannot be a response to selection imposed by the other population. Rather, we can describe isolating mechanisms as adaptive only when the speciating populations contact one another. There are two syndromes that fit such a scenario: populations become species while sympatric, or they come into secondary contact after initial divergence.

The Biological Species Concept

The discussion of processes involved in speciation can be complicated by the fact that there is little agreement over what constitutes a species. Essentially all species concepts that have been developed by biologists have some conditions in which they do not apply, or have limited utility in the practice of describing a species. In 1942, Ernst Mayr developed what is known as the Biological Species Concept, which states that "Species are groups of actually or potentially interbreeding populations, which are reproductively isolated from other such groups." Most biologists that study the process of speciation claim to subscribe to the Biological Species Concept, and accordingly stress the development of mechanisms that prevent interbreeding (isolating mechanisms). At the same time, these biologists sometimes call different groups species even if they are known to produce fertile hybrids; theory rarely translates into practice where species concepts are concerned. For the purpose of this article, we will consider the process of speciation to be occurring if isolating mechanisms are being strengthened, that is, are under active selection (see section Is Speciation Complete).

Premating Behavioral Isolating Mechanisms

One of the most thoroughly studied ways in which behavior can serve to isolate different species is the prevention of mating across species boundaries, leading to premating isolation (**Figure 1**). There are three basic categories of mechanisms by which this can occur. Two of these, temporal isolation and habitat choice, prevent individuals of different species from encountering one another during the time of breeding, despite sympatry of the incipient species. The other, mate choice, prevents hybridization upon an encounter. All of these mechanisms can be either complete, resulting in full isolation, or partial, allowing some heterospecific matings.

Temporal Isolation

Temporal isolation occurs when the timing of mating differs between two species. The scale of temporal isolation can vary widely. Some closely related species differ in whether they are diurnal or nocturnal, so individuals are looking for mates at different times of the day. In other cases, the timing of the breeding season is shifted, reducing overlap in the time of year during which individuals look for mates. Temporal isolation is one of the clearest mechanisms for behavior to apply to plants; there is clear evidence in several systems of shifts in flowering time that serve as isolating mechanisms between species.

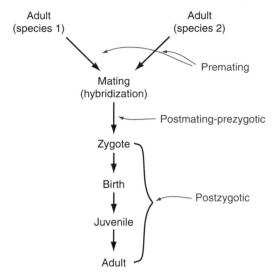

Figure 1 Life stages at which behavioral isolating mechanisms can act.

Habitat Choice

Different species can be prevented from hybridizing in sympatry if they occupy different habitats. As with temporal isolation, the scale over which this type of isolation occurs can vary. In some cases, individuals may overlap only in a dispersal stage and not encounter one another at other times because of habitat choice. In other cases, habitat choice leading to isolation may be more subtle.

Striking cases of habitat choice leading to isolation between species occur in phytophagous insects. Certain groups of these plant-eating insects have a high degree of host fidelity and lay eggs upon their host plant. In these cases, species or races may be effectively isolated despite the fact that individuals of each species remain within cruising range of one another. Indeed, habitat choice by phytophagous insects is one of the most commonly cited potential examples of sympatric speciation. Work by Sarah Via has demonstrated that regardless of whether initial differentiation was sympatric, current maintenance of host races in pea aphids is made possible by a high degree of host fidelity of different populations on alfalfa versus clover. In this case, host races are maintained and perhaps strengthened by both disruptive selection on habitat choice, in which more extreme specialists on each host are selectively favored, and ecological selection against hybrids that are intermediate in phenotype (see section Postzygotic Isolation).

Mate Choice

Even when adults of different species encounter one another during the breeding season, they may be effectively isolated by their choice of mates. Although the physical possibility of interbreeding may exist because of proximity of individuals, mate choice can be just as effective as habitat choice and temporal isolation in keeping species apart in certain cases. The set of behaviors used by individuals to discriminate between species are termed species recognition mechanisms. Such mechanisms consist of preferences for conspecific traits over heterospecific traits in the opposite sex and can be expressed by both males and females. Traits that act as the targets of species recognition can be morphological cues, vocalizations, olfactory cues, and behaviors. It is important here to distinguish species recognition from sexual selection. The extent to which species recognition traits overlap with cues that evolved in the context of sexual selection within species remains unknown. Examples have been found of species recognition both reinforcing and opposing sexual selection on display traits in nature.

The strength or direction of preferences for conspecifics can sometimes be influenced by prior experiences. Cross-fostering experiments have shown that in some cases, preferences for conspecifics are acquired by sexual imprinting. In other instances, individuals learn to avoid heterospecifics after experience as adults with courting or being courted by members of the opposite species. It is also possible for either innate or learned preferences to be based upon traits that are themselves partly learned (e.g., birdsong).

Postmating-Prezygotic Behavioral Isolating Mechanisms

Postmating-prezygotic isolating mechanisms take place after mating but before zygote production (**Figure 1**). Incompatibilities between species can occur, for example, in the travel of the sperm through the female, in sperm storage, or in egg–sperm interactions. Most of these mechanisms are inherently physiological, with behavior playing a minor role. Egg–sperm recognition, for example, is mediated by proteins on each cell; a mismatch can lead to the failure of zygote formation. Some researchers consider behaviors that occur during copulation that result in lower fertility, such as a mismatch in the timing or placement of egg and sperm release in animals with external fertilization, to be forms of postmating-prezygotic isolation.

Another postmating-prezygotic interaction mediated by behavior is cryptic female choice resulting in conspecific sperm precedence. When a female has mated with both conspecific and heterospecific males, the stage is set for the possibility of conspecific sperm precedence, the disproportional fertilization of the female's eggs with conspecific rather than heterospecific sperms. The mechanisms by which conspecific sperm precedence occurs are largely still being determined. It is not yet known to what extent or how often conspecific sperm precedence occurs via a mechanism that can be thought to be controlled by the female (in other words, as a form of cryptic female choice). This mechanism, like the other postmating-prezygotic mechanisms described earlier, may generally be thought of as more physiological than behavioral per se, although the distinction may be semantic.

Postzygotic Behavioral Isolating Mechanisms

Postzygotic isolating mechanisms between species occur in the form of incompatibilities that cause low fitness in hybrids (**Figure 1**). In many cases, these incompatibilities cause low viability or sterility because of genetic, morphological, or physiological problems (intrinsic incompatibilities) or because hybrids have phenotypes that are not well adapted to the environment (extrinsic incompatibilities). In some cases, these viability or fertility problems in hybrids can have a behavioral component. The foraging behavior of hybrids, for example, may not be optimized, especially if the parental species rely on different types of food. Habitat choice, antipredator defenses, or other behaviors could also theoretically be suboptimal. Likewise, hybrids may lack the ability to produce the appropriate courtship behaviors to find mates, especially in taxa in which such behaviors are highly ritualized. Although such mechanisms are straightforward to imagine, they have received little empirical attention, and we do not know how important they are in nature.

The Evolution of Behavioral Isolating Mechanisms in Allopatry

If populations are separated by geographic barriers that prevent contact between individuals, then behavioral isolating mechanisms may evolve rather easily. Behaviors may evolve, adaptively or otherwise, in each population that, when and if the populations come into contact, will result in isolation. However, because the populations are allopatric, these behaviors do not evolve as isolating mechanisms per se, but for other purposes. Their role as behavioral isolating mechanisms, should the populations come into secondary contact, can thus be viewed as a byproduct or an exaptation.

Premating behavioral isolating mechanisms may evolve in allopatry through several pathways. Behaviors shifting the time or season of mating (resulting in temporal isolation) or behaviors shifting the habitat in which mating occurs (which may result in isolation via habitat choice) can result from environmental differences that two species experience in allopatry. Sexual selection occurring in allopatric populations can also result in isolation, in this case via mate choice, if these populations were to come into contact. The traits that are favored by the opposite sex may be preferred to different extents in different populations or

may switch completely, via any of the many proposed mechanisms driving preference and trait evolution during sexual selection (e.g., Fisherian runaway, good genes, sensory drive, sexual conflict). In cases where species recognition cues differ from sexually selected traits, genetic drift may create differences between populations; species recognition cues that serve no other function are neutral in allopatry, so unless the populations are in contact it is difficult to see how these can evolve adaptively. One example of a mating cue that could potentially evolve via drift is the specific syntax of bird song or bird calls. In some cases, however, natural selection may be involved in call or song evolution because of differences in the way songs or calls are transmitted in different physical environments or with different types of ambient noise present.

Postmating-prezygotic behavioral isolating mechanisms can evolve adaptively in allopatry if cryptic female choice between conspecifics and heterospecifics reflects cryptic female choice among conspecifics. Egg and sperm signal molecules can evolve very rapidly because of sexual conflict (as may signals and preferences involved in premating isolation), but we still know little about mechanisms producing cryptic choice between species in nature.

Postzygotic behavioral isolating mechanisms may also arise as a byproduct of adaptive divergence in behavior in allopatry. As with physiological or morphological traits, behavioral traits involved in postzygotic isolation can result from intrinsic or extrinsic isolation, although the differences can be subtle. Local adaptation to the environment may result in new behaviors, involved in foraging, predator avoidance, etc., that differ between populations. This evolution can in turn cause hybrids to express extrinsic isolation by behaving in a way that is maladaptive in the environment in which the two come into secondary contact. Likewise, inappropriate courtship on the part of hybrids can be thought of as a type of extrinsic isolation if courtship behaviors have diverged between incipient species; local female mating preferences act as a biotic environment that makes inappropriate courting behavior maladaptive. In contrast, maladaptive behavior may also result from hybrids being in too poor a shape physically, because of intrinsic incompatibilities that reduce fitness, to perform appropriate behaviors for survival or mating, despite having the genes that code for these behaviors. For example, a hybrid may have the genes to perform an appropriate elaborate courtship ritual, but may not be agile enough to perform it well because of intrinsic incompatibilities that lead to poor condition. In practice, these cases of intrinsic isolation leading to maladaptive behavior in hybrids may be functionally indistinguishable from the extrinsic cases described earlier.

In certain cases, populations can be allopatric but still be geographically close enough to exchange occasional migrants. In this situation, many of the mechanisms discussed in the section on evolution in sympatry can apply.

The Evolution of Behavioral Isolating Mechanisms in Sympatry

When populations of incipient species are in contact with one another, isolating mechanisms can evolve under direct selection. In true sympatry, defined here as the situation in which organisms live within each other's common range of movement at some point in the life cycle (cruising range), selection can act on isolating mechanisms per se. This is true both when sympatry has existed from the beginning of the speciation event (sympatric speciation) and when sympatry follows secondary contact between two populations that have begun to diverge in allopatry.

Mate Choice

The evolution of mating preferences in sympatry has been both highly studied and, in some cases, strongly contested. Many researchers have speculated that sympatric speciation could potentially occur by sexual selection alone: an idea that has received mixed theoretical and empirical support. Sympatric speciation is exceedingly difficult to demonstrate conclusively, so the relative role of various forces that may potentially drive sympatric speciation in nature is, to date, inferred largely through a small number of compelling cases, indirect evidence, or theoretical analyses. Many sympatric species differ widely from one another in characteristics used for mate choice. This does not mean, however, that these differences arose in sympatry as the sole mechanism keeping species apart. In order to make this claim, researchers must rule out past episodes of allopatry and establish that no other form of isolation between the species was already present before divergent mating preferences evolved. Establishing conclusively that there were no prior episodes of allopatry is often impossible, although in some cases the probability of prior allopatry can be shown to be very low.

In other cases, premating isolation can evolve in sympatry after secondary contact between species that have diverged to some extent in allopatry. If incipient species can hybridize but hybrids have low (but nonzero, see below) fitness, selection is strong for the evolution of conspecific mating preferences and/or divergence of species-specific mating cues to avoid hybridization. This scenario, which was first described by Theodosius Dobzhansky in 1937, has come to be called reinforcement. Reinforcement could occur in populations that exhibit no premating isolation or the process could serve to strengthen premating isolation that has already begun to evolve in allopatry. The precise definition of reinforcement has been expanded by some researchers to also include the evolution of premating isolation driven by fitness costs from hybridization writ large (e.g., immediate viability or fertility costs to the hybridizing parents,

forms of postmating-prezyogtic isolation), instead of solely from the formation of low-fitness hybrids per se (postzygotic isolation).

Reinforcement is far from inevitable. The incompatibilities that drive reinforcement are countered evolutionarily by gene flow between the incipient species; gene flow tends to homogenize mating preferences between populations, making conspecific preferences difficult to evolve. Because by definition gene flow must exist in order to create the opportunity for hybridization, selection for reinforcement must always be accompanied by gene flow tending to prevent it.

Scientific consensus on reinforcement has waxed and waned in recent decades. Like sympatric speciation, reinforcement is difficult to prove empirically because of several alternative explanations that must be ruled out. Because of this lack of conclusive examples and theoretical arguments against it, based largely on the issue of gene flow, during the 1980s reinforcement was generally considered unlikely to occur. In the 1990s, however, reinforcement started to garner a significant amount of theoretical and empirical support, including a number of convincing case studies. The primary remaining question about this process is not whether it occurs, but how often it occurs, and thus how important it is in speciation.

In cases where postzygotic or postmating-prezygotic isolation is complete before secondary contact, conspecific mating preferences can evolve easily because of the absence of gene flow. Some scientists would call this process reproductive character displacement, defined as the divergent evolution of mating preferences or traits used as cues for mating. This term is meant to suggest a reproductive parallel for the process of ecological character displacement, whereby species diverge in their use of an ecological resource because of interspecies competition. These researchers draw a distinction between reinforcement and reproductive character displacement based upon whether isolation between the species is complete prior to secondary contact; if isolation is complete, the divergence of mating characters would be called reproductive character displacement, but if there is still gene flow then the divergence of mating characters would be called reinforcement. Reinforcement is thus considered part of the process of speciation, while reproductive character displacement occurs after speciation.

The argument can be made, though, that distinguishing reinforcement from reproductive character displacement on the basis of whether there is still gene flow is problematic; for example, if one fertile hybrid out of a million survives, is the process reinforcement or reproductive character displacement? A second group of researchers instead draw a distinction between the two phenomena based upon pattern versus process. For them, reproductive character displacement describes the pattern of divergence of characteristics such as mating preferences or mating traits in sympatry, where populations encounter one another, versus in allopatry, where they do not. Reinforcement instead describes the process of divergence when gene flow is involved. Reproductive character displacement thus has two competing definitions. Reproductive character displacement defined as a pattern does not necessarily imply anything about the selective forces that may have caused the differences between sympatric and allopatric populations. This pattern can, however, serve as evidence that the process of reinforcement may have occurred. This long-standing contention over definitions is not likely to be resolved by consensus, so definitions should be carefully laid out and considered when the term reproductive character displacement is used.

The pattern of differences in reproductive characters in sympatry but not in allopatry (the pattern definition of reproductive character displacement) has been observed across many taxa in characteristics as diverse as bird songs, frog and cricket calls, lizard push-up displays, flower color and flower size in plants, and female preferences for a variety of characters, to name a few. The fact that these patterns of divergence are observed so frequently in nature may indicate that the process of reinforcement is common, but as discussed earlier, a number of alternative explanations may also account for this data.

Temporal Isolation and Habitat Choice

Like mate choice, both temporal isolation and habitat choice can produce behavioral premating isolation in situations of sympatric speciation and reinforcement. The mechanisms causing divergence include selection against hybrids and reproductive competition when speciation is complete (i.e., reproductive character displacement when this terms is used to represent an evolutionary process). The best evidence for these processes comes from the latter case and includes examples of temporal isolation due to flowering time divergence in plants.

Other Behavioral Mechanisms of Premating Isolation

Another behavioral mechanism leading to enhanced isolation is reduced migration between populations. If premating isolation is incomplete, high levels of migration between populations produce more hybrids. If individuals that hybridize have low fitness because of partial or complete postmating-prezygotic or postzygotic isolation, then individuals that do not move between populations on average have higher fitness than migrants. As a result, selection for more sedentary behavior can reinforce other forms of isolation. An interesting interpretation of this understudied process is that allopatry itself is the focus of selection.

Postmating-Prezygotic Isolation

The evolution of conspecific sperm precedence occurring via cryptic female choice could be driven by selection against hybrids in the same way as would the evolution of premating isolation, in a process analogous to reinforcement. As is the case for premating isolation, divergence leading to postmating-prezyotic isolation by this mechanism may be difficult to tell from the evolution of similar mechanisms in allopatry, before secondary contact. It is not known whether there is a mechanism whereby conspecific gamete precedence could arise as a primary means of speciation in sympatry.

Postzygotic Isolation

Although there have been several recent suggestions in the scientific literature for ways in which low viability of hybrids can evolve as an adaptive response to selection, these have thus far only been developed in verbal and mathematical models and have not yet been applied specifically to behavior. It may be likely that the most common way in which behavioral postzygotic isolation can evolve is that most studied to date, as a byproduct of independent evolution of diverging populations in allopatry.

Is Speciation Complete?

Evolution of behavioral isolating mechanisms serves to strengthen species boundaries. In the case of sympatric speciation, premating isolation could potentially evolve at the very beginning of the speciation process. Assortative mating between different types could theoretically arise without any prior isolating mechanism present. In other instances, either later in the process of sympatric speciation or in the case of reinforcement, some isolation could be present at the time at which a behavioral isolating mechanism starts to evolve. In this case, it is interesting to consider whether speciation could be completed by the evolution of a behavioral mechanism. One of the primary arguments against reinforcement during the 1980s was that as premating isolation strengthened and hybridization become rare, the source of selection for the further evolution of premating divergence, namely low hybrid fitness, would become ineffectually weak. By this logic, the true completion of speciation would not be expected to be driven by selection against hybrids under the classic definition of reinforcement. There are other mechanisms driving the evolution of assortment, considered primarily in theoretical models of sympatric speciation, that do show that complete assortative mating can evolve under certain conditions. Some researchers consider this discussion to be largely semantic, since many pairs of species that are universally considered to be good can produce fertile hybrids. The disconnect between the definitions of species in theory and those used to distinguish species in practice may thus cause conflicts over which questions about speciation are considered important to pursue in the future.

See also: Cryptic Female Choice; Evolution: Fundamentals; Invertebrates: The Inside Story of Post-Insemination, Pre-Fertilization Reproductive Interactions; Mate Choice in Males and Females; Sexual Selection and Speciation; Social Selection, Sexual Selection, and Sexual Conflict.

Further Reading

Berlocher SH and Feder JL (2002) Sympatric speciation in phytophagous insects: Moving beyond controversy? *Annual Review of Entomology* 47: 773–815.

Coyne JA and Orr HA (2004) *Speciation*. Sunderland, MA: Sinauer Associates Inc.

Dobzhansky T (1937) *Genetics and the Origin of Species*. New York: Columbia University Press.

Kirkpatrick M and Ravigné V (2002) Speciation by natural and sexual selection: Models and experiments. *The American Naturalist* 159: S22–S35.

Mayr E (1942) *Systematics and the Origin of Species*. New York: Columbia University Press.

Ritchie MG (2007) Sexual selection and speciation. *Annual Review of Ecology, Evolution and Systematics* 28: 79–102.

Servedio MR and Noor MAF (2003) The role of reinforcement in speciation: Theory and data meet. *Annual Review of Ecology and Systematics* 34: 339–364.

Via S (2001) Sympatric speciation in animals: The ugly duckling grows up. *Trends in Ecology and Evolution* 16: 381–390.

Kin Recognition and Genetics

A. Payne and P. T. Starks, Tufts University, Medford, MA, USA
A. E. Liebert, Framingham State College, Framingham, MA, USA

Introduction: A Gene's Eye View of Altruism

At first glance, altruistic behaviors – those that benefit others at the actor's expense – are not well suited to the natural world. Within a population, individuals compete for access to food, mates, and habitats, and only those that are best able to secure those resources survive and reproduce. Given this struggle for resources – what Darwin called the 'dreadful, but quiet war of organic beings' – it is reasonable to assume that selfish and exploitive behaviors would always prevail.

This does not, however, appear to be the case. While nature has more than its share of selfishness, it proves surprisingly hospitable to cooperation, sharing, and self-sacrifice. Studies of animal behavior provide numerous examples. When faced with famine, some unicellular slime moulds sacrifice themselves to help others escape. Vervet monkeys put themselves at increased risk to warn others of an approaching predator. Perhaps most spectacularly, the eusocial insects have evolved sterile worker castes, the members of which give up their own reproduction for the sake of their nestmates'.

Darwin was particularly impressed with the last example and saw it as a formidable challenge to natural selection: he went so far as to call the evolution of worker ants 'by far the most serious special difficulty which my theory has encountered.' His solution largely ignored for over a century – was to recast natural selection in terms of competition between families; Darwin reasoned that if a sterile ant worker increased its family's overall fitness, then it might also increase the chances that its own traits would recur in the next generation. In the early 1960s, William Hamilton revisited this idea with mathematical rigor in a series of papers that challenged the way behavioral ecologists thought about altruism. In them, he introduced and formalized the concept of inclusive fitness, a framework for viewing natural selection from the perspective of genes rather than of individuals. Such a paradigm shift has profound implications for the study of altruism: from a gene's point of view, no behavior is selfless as long as it results in an increased number of copies of that gene in the next generation. According to Hamilton, even if an individual sacrifices its life, the genes responsible for that sacrifice can still spread through the population so long as $rb - c > 0$ (Hamilton's rule, where b is the benefit to the recipient of a behavior, c is the cost incurred by the actor, and r is the proportion of genes that the two share by common descent). Altruism is therefore advantageous only under certain conditions and only when it is preferentially directed toward close relatives, where r is highest and the number of shared genes is greatest.

Forty years on, it is hard to overestimate the impact that inclusive fitness theory has had on the study of animal behavior. Gene-level explanations help us to understand why some wasps forgo reproduction and instead raise sisters, why scrub jays sometimes devote their youths to defending their parents' nests, and why baboons often risk their lives to protect their kin groups. But these explanations also generate new questions and new directions for research. One such direction – research into the existence and nature of kin recognition systems – has proved particularly fruitful. The central question arises directly from Hamilton's insights: If we expect organisms to behave in ways that preferentially benefit their close relatives, should we not also expect them to identify those relatives? If they do discriminate between individuals on the basis of relatedness, what mechanisms might they use to do so? More importantly for our purposes, to what extent are those mechanisms influenced by genetics?

Since the late 1970s, researchers have produced convincing evidence that many animals do in fact discriminate between relatives and nonrelatives. Organisms as diverse as sweat bees, bank swallows, sea squirts, and macaques – even some plants and microorganisms – all demonstrate some form of kin recognition and discrimination when directing care or choosing mates. Biologists have also come a long way in describing the mechanisms behind this recognition, but much remains unknown about the relative contributions of genetic factors. In this article, we review our current understanding of the genetics of kin recognition systems and describe some of the mechanisms animals use to recognize relatives.

A Recognition Lexicon: Describing the Components of a Recognition System

The terms used to describe recognition and communication systems have fluctuated considerably over the last 25 years: what some authors have called recipients, others have called signalers; and what some call receivers, others call actors. For consistency and clarity, we suggest that the following terms be used to describe recognition behaviors.

All recognition systems, whether they identify kin, evaluate mates, or discover predators, require at least two participants: an agent that gets recognized and another that does the recognizing. We use the terms cue-bearer and evaluator, respectively, to clarify these different roles. Evaluators must also possess some set of criteria against which to compare a cue-bearer and make a decision about its identity. This identification key is called a template, and it can be either fixed at birth or learned later during development. Some of these learned templates are formed early on and quickly become resistant to further modification; others, meanwhile, are subject to constant revision and may change dramatically throughout an organism's lifetime. In either case, the learned template is established with reference to some external model, the so-called referent. Although the referent is most often another individual – usually a nestmate, a parent, or some other relative – it may also be the organism itself, or, as in some social insects, the organism's natal nest.

Researchers have found it useful to divide the recognition process into three essential and discrete parts: the so-called expression, perception, and action components. Expression refers to the production or acquisition of identity cues by the cue-bearer; perception entails the recognition and interpretation of those cues by the evaluator; and action describes all those behaviors elicited on the part of the evaluator by the act of recognition. Organisms demonstrate varying degrees of genetic involvement in each of these components, and in many cases the relative contributions of genes remains unclear.

Finally, it is worth pointing out the fine distinction between kin discrimination and kin recognition. Although the terms are sometimes used interchangeably, the former properly refers to an observed behavioral change in the presence of kin, while the latter refers to processes occurring inside the evaluator. We can easily imagine an organism that, while fully aware of who is who in its social group, does not modify its behavior in response to that information. Such an instance would resist typical empirical analysis and would belie the presence of recognition. Keeping the terms separate not only helps us design more rigorous analyses but also provides a humbling reminder of what we can and cannot know about animal minds.

The Gull and the Egg-Shaped Rock: Is a Genetic Component Necessary?

Genes for altruism thrive only if they direct help toward other individuals carrying those genes. As such, we might expect successful recognition systems to evaluate genotypes directly by looking for unambiguous phenotypes that reveal matches at specific loci. However, instances of such direct allelic recognition are exceedingly rare; much more common are systems that look not for specific alleles, but rather for indicators of overall relatedness. These 'best guess' mechanisms use those indicators to approximate the degree of kinship and thus the likelihood that any given allele is shared. If, for instance, you can safely assume that another individual is your brother, then there is a 50% chance that you and he will share any given allele; if the individual is only your half-brother, then the chance falls to 25%. Recognition systems that determine approximate kinship can therefore help organisms make appropriate behavioral choices even in the absence of detailed genotypic information. While many of these mechanisms depend on matching newly encountered phenotypes to recognition templates, some do not evaluate phenotypes at all. These are the so-called context-dependent mechanisms, and they are the focus of this section.

Imagine a couple driving home from the grocery store with their infant in the backseat. Knowing that they had strapped their child in just minutes before, they would be surprised – and more than a little nonplussed – to turn around and find a different infant smiling back. As long as certain situations are always associated with kinship (e.g., the backseat of the couple's car normally contains only their child), they may serve as reliable indicators of relatedness. In his 1964 paper, Hamilton cites Niko Tinbergen's example of the herring gull: Through controlled experiments, Tinbergen found that these birds were incapable of differentiating between their own eggs and any egg-shaped objects placed inside their nests. In fact, anything that approximated the correct shape and color was treated with the care due to offspring. 'This is what we would expect,' Hamilton concludes, 'in view of the fact that eggs do not stray at all.' In other words, the gull has little reason to suspect that an egg-shaped object sitting in its nest is anything other than its offspring.

Similar context-dependent recognition is found in some swallows, dunnocks, and prairie dogs, among others.

The bank swallow (*Riparia riparia*) is a particularly useful example, as its kin recognition system shifts to match conditions found at different periods of parental care. Shortly after they hatch, the altricial offspring are confined to natal nests, and parents have good reason for supposing that any young found in those nests are their offspring. At this stage, the swallows operate under a reasonable-assumption, context-dependent recognition rule: if a chick is sitting in your nest, treat it as though it were your own. As the chicks age, however, they leave the nest more frequently and begin to intermingle more freely with other individuals in their age group. This creates more opportunities for accidental switches and misdirection of care, but parents compensate by adopting new mechanisms for recognizing their offspring. Around the time that they begin to venture out from the nest, young swallows also develop individually distinct calls, which parents increasingly rely on for recognition. Once individual calls are established as the criteria for recognition, it is no longer enough to be in the right place at the right time; if they expect to be fed, young swallows must also be the right individuals.

Although context-dependent mechanisms can be effective at helping organisms direct altruism, they are also particularly susceptible to invasion by parasites. The cuckoo bird provides a classic example. Cuckoos lay their eggs in the nests of other bird species and then abandon them to the care of their unwitting foster parents. The cuckoo chick hatches from its egg and, in most cases, quickly dispatches its adopted brothers and sisters; when the parents return, they find the young cuckoo alone, assume that it is their last remaining offspring, and raise it to maturity. Such parasitic behavior clearly exploits parental assumptions about kinship identity and therefore provides a strong selective force against the maintenance of context-dependent recognition.

Because inter- and intraspecific parasitism exert such strong pressures on context-dependent systems, we might also expect organisms to supplement those systems with increasingly sophisticated recognition templates. Indeed, it appears that some bird species frequently parasitized by cuckoos have evolved mechanisms to discriminate between species-specific egg coloration patterns. Of course, this level of detection in turn exerts pressures on the eggs of the parasites themselves, leading to an evolutionary arms race that favors ever more subtle deception and detection. At least one species of brood parasite, the brown-headed cowbird, *Molothrus ater*, has evolved a dramatic strategy for dealing with this increased scrutiny. Adult cowbirds revisit parasitized nests on a regular basis, checking up on the eggs they left behind. If a host bird has been keen enough to discover a cowbird egg and eject it from its nest, the adult cowbird retaliates by destroying that nest and any eggs left inside. At this point, the behavior does not directly benefit the cowbird's

offspring, which have already been killed. It does, however, impose sharp costs on any host species capable of evolving more precise kin recognition systems. This not only slows the evolution of antiparasitic behaviors, but may also lead to behavioral changes within a single season: studies have shown that when some host birds rebuild their nests following retaliations, they become less discriminatory the second time around. In this way, the retaliation behavior is not just spiteful toward the host, but also directly beneficial to the individual cowbird's reproduction.

As these examples make clear, context-dependent recognition is far too precarious a strategy for some organisms, and this is especially true for those that live in large social groups without clear spatial boundaries. Most of these species have instead evolved template-based mechanisms for discriminating between kin and nonkin. One such mechanism, phenotype matching, is the subject of our next section.

The Persistence of Memory: Inferring Genetic Relationships Through Phenotype Matching

Many organisms recognize their kin through phenotype matching: a system in which individuals learn to associate specific sets of traits with varying degrees of kinship. The templates used in phenotype matching are established early in life, with reference either to family members encountered in the natal nest (where r is likely to be high) or to the individual's own phenotype (in which case $r = 1$). The latter is a special case known as self-referent phenotype matching or, as Dawkins memorably put it, the armpit effect. We will return to it in more detail shortly.

Nestmate-based phenotype matching has some obvious advantages over the context-dependent mechanisms discussed earlier. For one thing, parasites hoping to escape detection must do more than simply show up in the right place at the right time; they must also evolve to mimic the genetically influenced traits required to pass for kin. Perhaps more importantly, phenotype matching allows individuals to assess the kin status of previously unencountered individuals. In social insects with large colonies and highly mobile workers, for example, this ability to assess never-before-encountered individuals is essential to determining who should be treated with care and who should be attacked. Obviously, neither context-dependent, spatially based recognition nor individual recognition would be sufficient in such large and complex populations.

While initial template formation may occur under similar circumstances in both context-dependent and phenotype matching recognition systems (e.g., 'this individual is in my nest and therefore must be kin'), the durability of the latter allows recognition to take place long after the natal nest is abandoned. In fact, once a template is established, it is likely to remain intact for some time, often guiding that organism's recognition decisions throughout its life. Errors may still occur if templates are formed in the presence of unrelated individuals, but the window of opportunity for such mistakes is considerably smaller and limited to early development.

Organisms can presumably escape even this difficulty if they use the self-referent phenotype matching system mentioned earlier. If an individual can identify particular aspects of its own phenotype, use those aspects as referents for a kin recognition template, and then discriminate between other individuals on the basis of their similarities to that template, then that organism will possess a highly accurate means for identifying kin. Dawkins suggested the name 'armpit effect' because humans appear to generate individual-specific armpit odors which could, presumably, serve as referents for template formation. Although there is no direct evidence that this occurs in humans – dogs appear to be much better at smelling our individual differences than we are – the armpit effect has been successfully demonstrated in a handful of species. In one of the first studies of its kind, researchers tested whether or not pig-tailed macaques (*Macaca nemistrina*) could discriminate between previously unencountered paternal half-brothers and unrelated individuals. Since these paternal half-siblings neither grew up together nor had the same mothers, they could not have identified each other on the basis of phenotypes learned during early development; nevertheless, individual macaques showed a significant preference for half-siblings over unrelated individuals. This strongly implies that these monkeys were able to recognize paternally derived aspects of their own phenotype in their half-brothers and sisters, and the most likely mechanism for that recognition is self-referent phenotype matching.

The armpit effect has also been found in the golden hamster, *Mesocricetus auratus*, and in the honeybee, *Apis mellifera*. The latter case is especially interesting, as honeybee workers have strong incentives to match other individuals to their own phenotypes. Workers are nonreproductive and, through a peculiarity of Hymenopteran sex determination (haplodiploidy, in which females have two parents and males are fatherless), they are more related to their full-sisters ($r = 0.75$) than they are to either their potential offspring or their mothers (both $r = 0.5$). As such, honeybee society is particularly amenable to altruism between full-sister nestmates. This situation is complicated, however, by the polyandrous nature of the honeybee queen and by the multiple lines of paternity that exist in each hive. Again due to haplodiploid sex determination, workers are considerably less related to their half-sisters ($r = 0.25$) than they are to either their full sisters or their mothers. Thus, altruistic behaviors

directed toward these half-sisters would not be favored and honeybees capable of discriminating between full- and half-sisters during nestmate care and queen rearing should be at an inclusive fitness advantage. Since these bees grow up surrounded by both full- and half-sisters, nestmate-based phenotype matching would not give individual workers the tools to make fine discriminations between patrilines; it is not surprising, then, to find self-referent phenotype matching at work in these insects.

In the absence of carefully controlled studies, it is difficult to determine whether a particular phenotype matching system is based on nestmate- or self-referent formation. It is even more difficult, however, to discriminate between instances of the armpit effect and of so-called recognition alleles, genes that simultaneously code for both a phenotypic difference and an ability to recognize that difference in others. We will return to recognition alleles and to the closely related 'green-beard gene' concept shortly; first, we look more closely at the relationship between genes and phenotype matching in two well-studied systems: insects and mammals.

Smells of Home: Nest and Nestmate Odors in Insect Phenotype Matching

The first evidence for genetically influenced kin recognition in insects came in the late 1970s, when Greenberg demonstrated discrimination on the part of a primitively eusocial sweat bee, *Lasioglossum zephyrum*. In this study, bees standing guard at a nest entrance were introduced to previously unencountered individuals that varied by degree of kinship with the guards. Greenberg found that an individual's likelihood of making it past the guard was directly proportional to the coefficient of relatedness r between the two, regardless of a lack of past encounters; this implied that the bees were somehow able to determine relatedness in the absence of individual recognition or previous association. Drawing on previous studies, Greenberg concluded that the kinship signal was chemical in nature and that the bees were able to pick up on genetically determined odor differences between relatives and nonrelatives, presumably through phenotype matching.

In the years since, much of the research into insect kin recognition has focused on the role of olfaction and chemical cues, particularly those associated with cuticular hydrocarbons, or CHCs. The epicuticles of most insects are covered by these antidesiccation molecules, which vary between individuals and species in terms of both structure and relative abundance. This high variability makes CHCs obvious candidates for the expression component of kin recognition.

An impressive body of evidence does, in fact, point to a central role for CHCs in insect communication and recognition, and most researchers now view them as the primary cues necessary for detecting kin. But this raises an interesting question: Are these chemical profiles intrinsic (i.e., specified by an individual's genes), environmentally acquired from nests or relatives, or some complex combination of the two? While studies have shown that some components of CHC profiles are heritable in male crickets, honeybees, and fruit flies, much remains unknown about other insect groups. Most likely, many insects develop their profiles through a complex interplay of both genetic and environmental factors.

In several groups of ants, for instance, each individual contributes its personal, intrinsic scent to a colony-wide, or 'gestalt,' odor that serves as a referent for template formation. Meanwhile, each ant acquires this complex gestalt odor on its cuticle, thus labeling itself as a member of its home colony. In some other species, colony-specific odors come exclusively from the queen, while in still others, the gestalt odor is based on a mixture of exclusively worker-derived odors. Some species use exogenous, environmental odors from foods or nest materials to supplement the chemical cues produced by the ants themselves, while others may rely exclusively on those intrinsic cues. The important point is that the relative contributions of genetics and of the social and physical environments vary widely between species and that, in most of those species, the details remain a mystery.

The use of chemical recognition cues has, however, been well studied in paper wasps of the genus *Polistes* and serves as an excellent example of the complex role of CHCs in recognition. Paper wasps are primitively eusocial insects that live in colonies dominated by a single reproductive female assisted by nonreproductive workers. According to Hamilton's rule, such altruistic division of labor is advantageous for worker genes only when helping behaviors are preferentially directed toward close relatives. It is in a worker's interest, then, to possess an accurate and effective kin recognition system.

Paper wasps appear to discriminate between kin and nonkin with the help of CHC profiles borne on the epicuticular surfaces of each individual. There is evidence to suggest that these profiles are influenced by both genetic and environmental factors and that the natal nest is the primary source of these odors. Wasps are much more readily accepted by individuals from the nest on which they eclosed (emerged from pupation) than by individuals from other nests; this holds true even when wasps are reared in nests where the average relatedness is low and nestmates are less-related to each other than to nonnestmates. The mechanism underlying this ability appears to be the acquisition of chemical cues embedded in the nest material itself. Indeed, studies have shown that paper wasps can discriminate between out-of-context fragments of their own and foreign nests, even in the absence of adults and brood. Recognition cues appear, then, to be homogenized across nestmates and to employ nest of origin as an imperfect proxy for genetic relatedness.

Nest of origin also plays a critical role in the formation of *Polistes* recognition templates. Wasps tend to accept individuals raised on the same nest comb as themselves, while rejecting those raised on foreign nests. Again, this rule applies regardless of actual genetic relationship; in studies in which preeclosion wasps were switched to the nests of nonrelated females, wasps discriminated against their genetic relatives in favor of individuals from the same natal nest. The wasps apparently develop their recognition templates through a phenotype matching system using early, postemergence environmental cues as referents. It is worth pointing out, however, that individual wasps still appear to identify and use genetic cues when nest-origin cues are absent, hinting at a possible role for self-referent phenotype matching.

Polistes wasps respond to complex signals when evaluating kinship, but interestingly, do not appear to make subtle distinctions between kinship categories. During a specific recognition event – say an encounter between two previously unacquainted wasps – the evaluator compares the cue-bearer's odor profile to the colony-specific template. If the cue-bearer sufficiently matches the template, then she is accepted as a nestmate; if not, the evaluator responds to the perceived intruder with aggression. These recognition interactions appear to follow an 'all or none' rule, with cue-bearers neatly divided into kin and nonkin categories. Getz's 'kingram' concept provides a helpful framework for thinking about such dichotomous decision making in the face of highly variable cue information (see **Box 1**).

While CHC-based kin recognition requires a great deal of specificity on the part of cue-bearers, it is still susceptible to sophisticated mimicry by parasites. A recently discovered example involves chemical deception in a cuckoo wasp, *Hedychrum rutilans*, a highly specialized brood parasite of the European beewolf, *Philanthus triangulum*. The beewolf is itself a predator of honeybees; it attacks and paralyzes worker bees in the field, carries them to its underground nest, and lays eggs on their cuticles. These egg-carrying honeybees are temporarily stored just within the nest entrance, then deposited deeper in the underground burrow and sealed up inside.

Females of *H. rutilans* have evolved to take advantage of this food provisioning through a sophisticated suite of behaviors. Before the honeybee is taken inside the nest, the cuckoo wasp quickly deposits her own eggs on the bee's surface. When the beewolf returns to its nest, it carries the bee, its own egg, and the egg of its parasite into the brood chamber. The cuckoo wasp larvae emerge first and immediately set to devouring both the beewolf offspring and the paralyzed honeybee, thus increasing their own fitness at a considerable cost to that of their hosts. Normally, we might expect *P. triangulum* females to evolve mechanisms to detect the presence of

Box 1   Kingrams and the optimal acceptance threshold

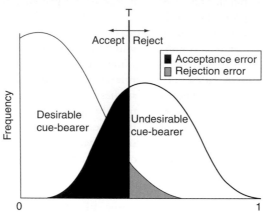

Dissimilarity between an evaluator's recognition template
and a cue-bearer's recognition cues.

Kingrams help us to visualize the recognition process and to better understand the role of thresholds in decision making. Organisms are often forced to make binary decisions in the face of complex and highly variable information. For instance, when deciding whether or not to accept a newly-encountered individual as kin, an organism must confront a great deal of intra-group variation. Not all relatives will share precisely the same set of identifying labels and some will no doubt match the evaluator's template more closely than others. Likewise, there will probably be some overlap between relatives and nonrelatives in terms of template dissimilarity. How do organisms make good decisions in the face of this uncertainty?

Many appear to do so through an optimal acceptance threshold, a point (T on the diagram above) at which increasing cue dissimilarity triggers rejection behavior. All individuals falling to the right of T are rejected as nonrelatives, while all those on the left of T are accepted. We expect organisms to position T such that fitness is maximized with respect to the relative costs of acceptance and rejection errors. For instance, when rejecting true kin is more costly than accepting imposters, the optimum acceptance threshold should move to the right.

H. rutilans-specific CHC profiles and tag 'contaminated' honeybees for removal. But this is apparently made considerably more difficult by chemical mimicry on the part of the cuckoo wasp: analysis of *H. rutilans* and *P. triangulum* CHC profiles found that the cuckoo wasps share more of their chemical signatures in common with beewolves than they do with other *Hedychrum* species!

The use of CHCs as species- or individual-recognition cues, while common, is not universal in the insect world. For example, CHCs, while present in diurnal fireflies, appear to be absent in their nocturnal cousins. These night-flying species not only have a lower risk of desiccation (presumably making dehydration-resistant CHCs unnecessary), but also rely much more heavily on visual cues for mate evaluation and species recognition. Indeed, the flash signals in *Photinus* fireflies are surprisingly complicated and information-rich, and apparently compensate well for a lack of chemical cues.

To Smell a Rat: Phenotype Matching and the Genes of the MHC

As we have seen, phenotype matching is not limited to insect systems; many other organisms use similar mechanisms to establish and maintain recognition templates. In this section, we explore the likely role of phenotype matching in mammalian kin recognition, discuss its implications for inbreeding avoidance and nepotism, and focus on a likely source of referent cues, the family of genes known as the major histocompatibility complex or MHC.

Previous sections have emphasized the importance of kin recognition to the evolution of altruism and nepotism and, indeed, much of the early work on recognition focused on these behaviors. There is, however, at least one instance in which animals might benefit from avoiding close kin and from associating instead with unrelated individuals: that instance, of course, is mate choice.

Inbreeding between closely related individuals often exacts a heavy toll on fitness, as rare and deleterious recessive alleles are expressed to disastrous effect in the offspring. As such, organisms that recognize kin do well to avoid mating with close relations. Several decades of research have now shown that inbreeding avoidance is indeed an important consequence of kin recognition and a powerful force in the maintenance of cue diversity.

Much of the research on inbreeding avoidance in mammals has focused on the highly polymorphic genes of the MHC. This family of genes has long been the subject of intensive study, though more for its central role in the vertebrate immune system than for its use in kin recognition. The genes of the complex code for cell surface glycoproteins that help immune T cells differentiate between self and nonself; if a cell has been invaded by a pathogen, small fragments of the invader's protein are presented to T cells via a pocket on the MHC protein's surface, marking that cell for destruction. Medical researchers have an obvious interest in exploring the functions of these genes and so routinely keep inbred strains of laboratory mice differing only at the MHC. It was through a fortuitous observation of these mice's mating behavior that researchers first thought to investigate the role of such immunity genes in kin recognition.

Yamazaki and colleagues first conducted studies on MHC and mate choice in the late 1970s and found that mice preferentially mate with individuals that differ from them at these loci. Subsequent studies showed that mice could be trained to discriminate between urine samples from MHC-different individuals, indicating olfactory perception of either the gene products themselves or of secondary compounds linked to those gene products. These researchers also discovered a fascinating role for the MHC in the phenomenon known as pregnancy block, the spontaneous abortion of pregnancy that occurs when female mice are exposed to the scent of an unfamiliar male. Consistent with kin selection theory, the risk of pregnancy block is considerably increased when the second male varies from the first at MHC loci.

The importance of these genes for mouse kin recognition goes beyond inbreeding avoidance to a role in cooperation and altruism. Female mice often nest together, sharing nursing duties and other parental responsibilities. One study found that females preferred to nest with MHC-similar individuals, thus increasing the likelihood that helping behaviors will benefit close relatives and increase inclusive fitness. This sets up an interesting contest between opposing selection pressures, with sexual selection promoting MHC diversity (because individuals tend to prefer genetically dissimilar mates) and kin selection favoring greater homogeneity (through benefits accruing to MHC-similar individuals): a situation known as Crozier's Paradox. The extreme polymorphism found

at the MHC – Brown and Eklund suggest that in mice over three billion phenotypes are possible at just three principal loci – seems to imply that the effects of mate choice and sexual selection outweigh the selective forces applied by nepotism.

One area of research that has received particular attention and generated considerable controversy focuses on the role that MHC (also known as HLA) genes might play in human mate choice. Wedekind and colleagues' well-known 1995 study of odor preference in humans required females to smell a series of T-shirts worn by various males and to rank these shirts' scents according to pleasantness. The researchers found that the women were significantly more likely to deem an odor 'pleasant' if the shirt was worn by a man who differed at HLA loci. They were also more likely to report that such shirts reminded them of a current or former romantic partners, suggesting that pleasantness of odor and mate choice are related. Interestingly, preferences for HLA dissimilarity were reversed in women taking oral contraceptives at the time of the study.

Recent research seems to support the conclusion that HLA genotypes influence mate choice in humans. A large-scale study of Hutterite marriage patterns found that individuals were significantly more likely to marry partners with different HLA haplotypes than would be expected by chance. Similarly, a study testing odor preferences among Brazilian college students found that females preferred the smell of sweat from HLA-dissimilar males significantly more often than that from HLA-similar males; this study did not, however, find a similar effect when males ranked female sweat odors or when females ranked odors originating in male urine.

Much of the available evidence points to early childhood-based phenotype matching as the most likely source of these mating preferences. Yamazaki reversed odor preferences by rearing mice with MHC-different individuals: a situation unlikely to occur in nests in the wild. Several other studies and decades of anecdotal evidence indicate that humans learn relatives' phenotypes during childhood and later avoid those phenotypes when choosing mates.

Although a growing body of evidence supports a link between MHC genes and kin recognition, some researchers remain skeptical. They point out that the majority of mouse studies use highly inbred laboratory strains living under artificial conditions and thus may not shed much light on the behaviors of genetically diverse, wild populations. A recent study investigated the effects of MHC and MUP (major urinary protein) genotypes on inbreeding avoidance in mice living under natural conditions. Mice sharing both alleles at the MUP locus were significantly less likely to mate with each other than they were with mice who shared one or neither of their MUP alleles; the sharing of MHC alleles, however, had no significant effect on mating.

Another study of an unmanaged sheep (*Ovis aries*) population found no evidence of dissortative mating on the basis of MHC haplotype, suggesting that such genetic kin recognition may not be ubiquitous among mammals.

Not Easy Being Green: Super-Genes in Theory and in Nature

Workers of the invasive red fire ant, *Solenopsis invicta*, are not known for their civility. Fierce colony defenders, these tiny insects are quick to bite, grapple with, and sting, often to the point of death, any organism unlucky enough to wander into their nest mounds. Even so, it comes as a shock to watch a group of these workers descend on one of their own queens, overwhelm her with well-placed stings and bites, and literally rip her apart. What could possibly lead worker ants – organisms that depend on their queens as the sole sources of colony reproduction and inclusive fitness – to commit regicide?

The answer, it turns out, is a green beard. In his 1964 paper on the genetics of social behavior, William Hamilton described a hypothetical 'super-gene' (i.e., a tightly linked complex of several genes) that would simultaneously code for (1) a conspicuous phenotypic cue, (2) an ability to detect this cue in others, and (3) a propensity to direct care exclusively toward other cue-bearing individuals. Dawkins memorably dubbed such super-genes 'green-beards,' after an imaginary example in which beard color served as the recognition cue. It is just such a green-beard gene, and its recognition by *S. invicta* workers, that leads these ants to dispatch their queens with such surprising violence.

Colonies of *S. invicta* come in two forms, those with single queens (monogynous colonies) and those with multiple queens (polygynous colonies), and this difference in social structure corresponds to a genetic difference at a specific locus known as *Gp-9*. All queens and workers in monogynous colonies are dominant homozygotes (*BB*) at *Gp-9*, while all queens and most of the workers in the polygynous colonies are heterozygotes (*Bb*); homozygous recessive females are not found in the colonies because the *bb* genotype inevitably leads to death early in development.

The monogyne and polygyne populations are distinct and are connected genetically only through males who move between the two groups. In the polygyne colonies, matings between *Bb* females and B males (males are haploid) inevitably result in some new *BB* queens in every generation; without intervention, these *BB* queens would grow up to become reproductives and, year by year, their genotype would increase its representation in the colony. In reality, however, these new *BB* queens do not stand a chance: workers, with their penchant for political assassination, quickly intervene to maintain genetic disequilibrium. *Bb* workers can, through pheromonal cues, detect a

new queen's genotype and modify their behavior accordingly. If she is heterozygous like them, she receives the typical protection due to a colony's queen, but if she carries two copies of the *B* allele, she is marked for execution. The *Bb* workers surround the *BB* queen and swiftly eliminate her from the gene pool.

This is not exactly what biologists expected from a green-beard in nature. Recall that Hamilton's original conditions specified altruism, not lethal violence, as the behavior associated with this degree of direct genetic recognition. But even though heterozygous *S. invicta* workers respond to the absence rather than the presence of a shared genotype, the effect remains the same. To return to Dawkins's metaphor, imagine a set of closely linked alleles specifying both the possession of a green beard and an unusually high degree of aggression directed toward any individuals with purple beards. When green beards encounter other green beards, they treat each other with the normal degree of cooperation expected in the species. When they encounter purple beards, however, the latter suffers a significant reduction in fitness. This situation is analogous to that of the *S. invicta* workers and is functionally equivalent to the classic green-beard scenario described earlier.

The most surprising thing about the green-beard gene *Gp-9* is not that it results in aggressive behavior, but rather that it exists at all. Both Hamilton and Dawkins thought the likelihood of finding a green-beard in nature very small, because either (1) such super-alleles would be so successful that they would quickly fix themselves in a population and become invisible to analysis, or (2) once everyone in a population carried the conspicuous marker of genetic altruism, the system would become highly vulnerable to cheating. Individuals born with the marker (the green beard, say), but without the associated altruism, would benefit from the help of everyone else in the population without making sacrifices of their own. With such a fitness advantage, these individuals would quickly come to dominate the gene pool, and the benefits associated with the green beard would turn into handicaps. *Gp-9* avoids this fate at least in part because the *b* allele responsible for aggressive behavior can never become fully fixed in the population; its lethality in homozygous recessives prevents it from becoming the predominant allele and so the polygynous cycle of *BB* birth and elimination continues generation after generation.

So, contrary to predictions, green-beard genes do exist in nature. But are they common? As of now, the consensus answer is a qualified 'no' – very few examples have been discovered, the theoretical obstacles to their evolution still apply, and it is not immediately obvious how best to differentiate them from instances of the armpit effect mentioned earlier. But in the wake of research into the role of *Gp-9* in fire ants and of a possible green-beard effect at placental interfaces, researchers are more willing than ever to consider a possible role for these super-genes in recognition systems.

One fairly unequivocal example comes from a species of unicellular slime mold, *Dictyostelium discoideum*. These social amoebas spend most of their time living as single-celled predators of bacteria, but when food gets scarce, they coalesce to form a multicellular colony known as a 'motile slug.' This cellular aggregation moves as a single unit until it comes to a suitable location on which to form a fruiting body, a distinctively shaped structure with reproductive spores sitting atop a long, thin stalk. Only those cells that find themselves at the top of this structure get the opportunity to escape into a better environment; the cells that make up the stalk inevitably die, sacrificing themselves for colony mates further up in the spores.

Since *D. discoideum* fruiting bodies are almost always composed of more than one clonal line and therefore contain individuals that vary widely in relatedness, the stage is set for kin selection. It appears that *D. discoideum* individuals bearing wild-type copies of a gene known as *csA* are able to respond to the presence of this allele in others and to preferentially direct cooperative behavior toward those individuals. *csA* codes for a self-binding (homophilic) protein that sits on the surface of *D. discoideum* cells, and this self-binding property causes wild-type individuals to adhere more readily to each other than they do to mutants bearing a different *csA* allele. The increased binding between these cells allows them to preferentially pull each other upward toward the top of the fruiting body, increasing their chances of survival and leaving mutant cells to die in the stalk.

The wild-type *csA* allele thus fulfills all the requirements for a green-beard gene. First, each cell bears a conspicuous cue in the form of a cell-surface adhesion protein; second, the homophilic nature of these proteins allows cells to recognize the genotype of another individual; finally, because these wild-type *csA* cells preferentially bind to and pull at genotypically similar individuals, the allele is able to direct care toward other copies of itself in the population. Indeed, the *csA* gene product may be only the first of many cell-surface adhesion proteins found to play a role in kin recognition. Because of their highly developed ability to discriminate between self and nonself, cell-surface adhesion proteins hold out particular promise as green-beard mechanisms.

Researchers have even speculated that green-beard genes and their surface adhesion products could influence sperm behavior in New World marsupials. In these animals, sperm cells swim in pairs, attaching to each other by their head segments and beating their flagella in concert. Together, they consistently out-swim any un-partnered rivals, but the alliance is an uneasy one because only one member of the pair typically goes on to fertilize an egg. If a gene could cause a sperm cell to partner exclusively with other cells carrying that same gene, then it could

ensure that at least one copy of itself would enter the egg. From this gene's point of view, the eventual competition between the two sperm cells would become no competition at all. While no evidence has yet been found to support the suggestion that green beards play a role in sperm partner choice, recent discoveries in *Solenopsis* and *Dictyostelium* show that such a mechanism is far from improbable.

It is important to point out that not all recognition alleles are green-beards *sensu stricto*. The former definition only requires that a gene create a conspicuous cue and the ability to detect that cue in others; the latter, meanwhile, requires both of those conditions plus behaviors favoring the cue-bearers. The key difference is that a green-beard gene carries all three phenotypic effects at a single locus or tightly linked set of loci, while some recognition alleles may rely on behaviors coded for at other locations in the genome. In the case of *Botryllus schlosseri*, a marine tunicate, individual colonies are able to recognize kin on the basis of genetic similarity at a single locus, the *FuHC* gene. Sessile colonies often grow into each other across a shared substrate. If individuals of the colonies share one or both of their alleles (out of literally hundreds available), then the colonies easily fuse together; when they share neither, the colonies reject each other with an aggressive inflammatory response. This example, along with similar ones from hydrozoans and plants, does not imply the existence of green-beards, as the accompanying behavior is not a result of the *FuHC* gene itself. Instead, this gene serves as a source of recognition cues and as a mechanism for comparing those cues to self, in other words, as a recognition allele.

Finally, it is worth asking whether or not green-beard alleles are really kin recognition mechanisms at all. To return once more to Dawkins's beards, a *DD* green-bearded individual may well grow up in a household with a *dd* red-bearded sibling (if parental cross is *Dd* × *Dd*). Given the choice between helping a green-bearded second cousin or a red-bearded brother, the individual in question would prefer to assist the former every time. In this way, green-beard super-genes help themselves not only at the expense of the individual's fitness, but also at the expense of the other genes with which they share a body.

Conclusions and Future Directions

Forty years ago, Hamilton's insights lay the groundwork for an explosion of kin recognition research; today, the rapid development of modern genetic techniques is building a new foundation, this time for understanding how genes interact with the environment to build kin recognition mechanisms. The examples given in this article illustrate both the interdisciplinary nature of the work and the breadth of strategies and mechanisms that animals use to recognize kin. Future contributions from the fields of animal behavior, neurobiology, and cellular biology will only expand this understanding and may well challenge many of our current assumptions.

See also: Kin Selection and Relatedness; Recognition Systems in the Social Insects; Social Recognition.

Further Reading

Blaustein AR (1983) Kin recognition mechanisms: Phenotypic matching or recognition alleles? *The American Naturalist* 121: 749–754.

Brown JL and Eklund A (1994) Kin recognition and the major histocompatibility complex: An integrative review. *The American Naturalist* 143: 435–461.

Crozier RH and Dix MW (1979) Analysis of two genetic models for the innate components of colony odor in social hymenoptera. *Behavioral Ecology and Sociobiology* 4: 217–224.

Dawkins R (1982) *The Extended Phenotype.* Oxford: Oxford University Press.

Fletcher DJC and Michener CD (eds.) (1987) *Kin Recognition in Animals.* New York, NY: John Wiley & Sons.

Greenberg L (1979) Genetic component of bee odor in kin recognition. *Science* 206: 1095–1097.

Hamilton WD (1964) The genetic evolution of social behaviour I, II. *Journal of Theoretical Biology* 7: 1–52.

Hepper PG (ed.) (1991) *Kin Recognition.* Cambridge: Cambridge University Press.

Keller L and Ross KG (1998) Selfish genes: A green beard in the red fire ant. *Nature* 394: 573–575.

Mateo JM and Johnston RE (2000) Kin recognition and the 'armpit effect': Evidence of self-referent phenotype matching. *Proceedings of the Royal Society of London B* 267: 695–700.

Rousset F and Roze D (2007) Constraints on the origin and maintenance of genetic kin recognition. *Evolution* 61: 2320–2330.

Sherman PW, Reeve HK, and Pfennig DW (1997) Recognition systems. In: Krebs J and Davies N (eds.) *Behavioural Ecology: An Evolutionary Approach.* Oxford: Wiley-Blackwell.

Starks PT (ed.) (2004) Recognition systems (special issue). *Annales Zoologici Fennici* 41: 689–892.

Strohm E, Kroiss J, Herzner G, et al. (2008) A cuckoo in wolves' clothing? Chemical mimicry in a specialized cuckoo wasp of the European beewolf (Hymenoptera, Chrysididae and Crabronidae). *Frontiers in Zoology* 5: 2.

Wedekind C, Seeback T, Bettens F, and Paepke AJ (1995) MHC-dependent mate preferences in humans. *Proceedings of the Royal Society of London B* 260: 245–249.

Kin Selection and Relatedness

D. C. Queller, Rice University, Houston, TX, USA

Introduction

Animals often treat their kin differently from other individuals. Parental care is the most obvious manifestation of this. Parents will often take great risks to give their offspring help, and will not do the same for other young. The same phenomenon sometimes occurs with other kinds of relatives. A ground squirrel will give a warning call if its relatives are nearby, but not when it is surrounded by nonrelatives. A worker honeybee allows its sisters into the hive, but tries to sting nonrelatives.

The workers of honeybees and many other social insects take kin preference to an extreme. They have evolved to forego reproduction, instead devoting their lives to rearing sisters and brothers. Darwin was troubled over how worker traits could evolve by natural selection if their bearers did not reproduce: How would good traits be favored and bad ones be disfavored? He settled on family selection, in which the good traits of a sterile or nonreproducing individual could be passed on by its kin. He noted that it is analogous to selecting for well-marbled beef. Even though we have to kill the cow to assess its quality, we can breed relatives of the good ones and have them transmit the good qualities.

Inclusive Fitness and Kin Selection

In the 1960s, William D. Hamilton formalized this notion in his concept of inclusive fitness, also known as kin selection. He showed that when an individual's behavior affects multiple individuals, selection favors the behavior according to the sum of its effects on their fitness effect, with each fitness multiplied by relatedness. A behavior is favored if this sum, termed the inclusive fitness effect, exceeds zero:

$$w + \sum_i r_i \Delta w_i > 0 \qquad [1]$$

where w is its effect on its own fitness, each w_i is its fitness effect on another individual, and r_i is its relatedness to that individual. This inequality is known as Hamilton's rule. The most familiar example is when an individual, such as a social insect worker, altruistically benefits a relative by amount b, at cost c to its own fitness, in which case formula [1] yields

$$-c + rb > 0 \text{ or } c < rb \qquad [2]$$

The formula shows that, if the benefit to kin is high enough compared to the altruist's cost, altruism can be favored, and it is most easily favored if relatedness is high. From the standpoint of an altruism gene, it can decrease its replication via the altruist if it sufficiently increases its replication through a relative. The chief initial appeal of this theory was the solution to the problem of altruism, but the concept of inclusive fitness is generally applicable to all kinds of interactions.

Hamilton's rule is a concise summary of the outcome of formal models of gene frequency change. These models make some assumptions, such as weak selection and additive fitness effects, but Hamilton's rule appears widely applicable.

Genetic Relatedness

Kin selection works because relatives share genes. But all individuals share some genes (we even share genes with chimpanzees), so what is the appropriate measure? The answer comes from mathematical population genetic models, but the idea is easy to understand. For selection to increase the frequency of an allele, that allele has to do better than the average allele in the population. Thus, a behavior that helps random individuals in the population will not spread, even if some of those individuals possess the helping allele. Benefits will change allele frequency only to the extent that the helping allele is present above random levels in the beneficiaries. The proper relatedness for Hamilton's inclusive fitness rule must therefore capture this deviation from random expectation. A regression coefficient does exactly that. It measures the probability, given that an individual has an allele, and that its partner has the same allele above random expectation. Thus, regression coefficients are widely used to measure relatedness in kin selection studies.

Relatedness can be measured in several ways. One method involves the use of variable neutral genetic markers. Various kinds of genetic markers will do, but microsatellites have proved particularly useful and popular. Microsatellites are short tandem repeats of DNA, such as the doublet CA repeated nine times in succession. Such repeat sequences occasionally misalign during replication, resulting in the mutational gain or loss of a repeat. As a result, microsatellites are often quite variable. This means that random identity is low and they therefore provide a lot of information about similarity above random levels. Computer programs are used to measure regression-like relatedness, and when information is combined from enough loci, these estimates become increasingly precise and accurate.

Relatedness can also be estimated though pedigree methods. In the simplest example, an offspring gets half its genes directly from its mother and is therefore related by $1/2$ to her. That is, each allele in the offspring has a $1/2$ chance of having been inherited directly from the mother. We say that those alleles are identical by descent. Of course, there is usually some additional total identity between offspring and mother; for example, an offspring allele derived from the father might also, by chance, be identical to one in the mother. But these additional identities are random identities (known as identical by state) and therefore do not count or, more accurately, they are exactly cancelled out by random nonidentities. More complicated cases can be calculated by pedigree diagrams and tracing all the connections between classes of relatives. Relatedness to a maternal half sibling is $1/4$, the average of $1/2$ for genes identical by descent through the mother, and 0 for genes identical by descent through the father. Full siblings are related by $1/2$ because now both maternally and paternally inherited genes are identical by descent to that degree. Relatedness to grandparents, grandoffspring, aunts, uncles, nephews, and nieces is $1/4$. Cousins are $1/8$. Inbreeding will alter these coefficients by providing additional paths for nonrandom identity by descent.

These two methods, statistical and pedigree, generally give the same answers, although statistical estimates based on molecular markers include some sampling variance. For example, unrelated individuals will sometimes have negative estimates. The two kinds of estimates can diverge if pedigrees are taken back too far. In the limit, all copies of an allele in the present generation can be traced back to a single allele in the remote past and in this sense they are all identical by descent. Here, the statistical method is the only one that captures the essential feature of similarity above random levels.

Both of these methods measure relatedness across the genome, whereas what really matters in the models of kin selection is relatedness at a locus that affects the behavior. This is not usually a problem because probabilities of identity by descent are typically uniform across different loci. However, there are two potential reasons for this not to be true. First, strong selection at either the locus under selection or at a marker locus can cause relatednesses to be altered somewhat, so kin selection predictions are on strongest ground when selection is not too strong. Second, similarity at the social locus could be due to factors other than average kinship, such as the green-beard genes discussed in the following sections.

Kin Recognition

Animals are not expected to calculate relatedness, let alone conduct the full fitness calculations of Hamilton's

rule. Instead, those that happen to behave in ways that most accord with Hamilton's rule are favored over those that act differently. An important component of this is how information about relatedness is used. Three types of recognition cues are possible: spatial cues, genetic kinship cues, and green-beard cues.

The simplest, and often most reliable, kinship cues are environmental or spatial. For example, the individuals that hatch out of the eggs you laid are usually your offspring. The individual that nurses you when you are young is your mother. The other occupants of your natal nest are your siblings, as are individuals in later nests of your mother. Therefore, behaviors that are always directed toward nestmates are selected according to Hamilton's rule for siblings. Recognition occurs simply by the performance and selection of behaviors toward such classes of environmentally defined recipients.

However, care must be taken to consider the full context of the behavior. Increasing a nestmate's fitness may not be beneficial, even if $c < rb$, if there is strong competition among animals within the nest. If increasing the fitness of one nestmate decreases that of another by an equal amount, then there is no net benefit. There are two ways for us to adjust the calculations to take this into account. One is to add in the fitness of the harmed nestmate to Hamilton's rule. The other is to adjust the relatedness coefficient so that it is measured on the scale of competition. Because relatedness is similarity above and beyond the random levels of similarity in the competing population and because here competition occurs on the scale of the nest, relatedness could be estimated with respect to that population. However, in a great many natural situations, these kinds of adjustments are unnecessary because competition occurs at greater scales; nestmates that are helped go out and compete against the larger population.

A second kind of recognition is phenotype matching. Spatial and environmental cues alone are unlikely to be useful in identifying unfamiliar relatives. For this purpose, you need genetic cues, or more accurately, phenotypic cues based on genetic variation. A problem is that there can be no universal cue of kinship. Instead, you need variable markers and a way of knowing which ones mark your particular kin. This is achieved through learning, starting with known relatives. That means phenotype matching is built upon the kinds of spatial cues discussed earlier. So, an individual could learn odor cues from its mother, or from its siblings, known by spatial context, and then match those cues to later unfamiliar individuals. Those that smell most like siblings are likely to be relatives. Such phenotype matching involves three components: the production of variable cues, the perception of cues, and the taking of appropriate action in response.

The cues might have evolved for the purpose of recognition, but there is one reason to believe that they evolved for other reasons and were subsequently exploited for kin recognition. Common cue alleles will match more often and will therefore be more likely to be on the receiving end of kin-direct altruism and less likely to be on the receiving end of other-directed selfishness. Thus, kin selection tends to favor and fix common cue alleles, and it tends to remove the cue variation required to identify kin. This is puzzling because some animals clearly do have genetic kin recognition systems. One possible resolution is that they use cues where variability is maintained for other reasons, such as variation in the major histocompatibility locus for disease resistance.

The final form of recognition involves green-beard genes. Here, it is not kin that are recognized, but actual bearers of the social allele. The name comes from a thought experiment of Richard Dawkins, who imagined an allele with three effects: it codes for a green beard, recognizes green beards in others, and acts favorably only toward possessors of green beards.

Green-beard recognition has two problems. The first is that it requires quite complex effects resulting from expression of a single locus. However, it must be remembered that the allele need not build all the three traits by itself; it needs to create only the necessary differences in the traits. Green-beard systems have been found. In a social amoeba, *Dictyostelium discoideum*, a single gene affecting cell adhesion causes greater altruistic behavior and ensures that it goes to like individuals by preferential binding to them. And in bacteria, poison-antidote systems known as bacteriocins function to kill cells that lack the relevant alleles.

The other problem with green-beard systems, or at least green-beard altruism, is that it may act against the interests of the rest of the genome. Under Hamilton's rule, a green-beard locus that sacrifices itself to benefit another green-beard bearer is favored if $c < b$ (because r at this locus, is 1). However, the rest of the genome may not be related by 1, so such extreme altruism would not be favored at these loci. There may therefore be selection at other loci to suppress the green-beard effects (and especially to suppress the altruism part, perhaps leaving the green beard itself so that carriers can still be recipients of altruism). This does not mean that green-beards are impossible – we have seen that they exist – but it may result in quite complex and unstable selective dynamics involving multiple loci.

Relatedness Affects Behavior

Hamilton's inclusive fitness theory pointed out the importance of genetic relatedness. It also served as an explanation for the general preference for kin pointed out at the beginning of this article, and it has led to many testable predictions.

Oddly though, the theory's initially most spectacular success is no longer viewed as definitive. Hamilton noted

that most social insects – specifically the ants, bees, and wasps of the insect order Hymenoptera – shared an unusual genetic system that results in some odd relatednesses. The system is called haplodiploidy because males are haploid and females are diploid, the difference arising according to whether they come from an unfertilized or fertilized egg. In normal diploid species (which include the social termites) full siblings are related by 1/2, whereas in haplodiploids, it varies by sex. Specifically, a female is related to her brother by 1/4 (because they share no paternal alleles) and related to her sisters by 3/4 (because the paternal allele is always shared, the father having only one to give). Hamilton suggested that this relatedness pattern explained why altruistic sociality was so common in haplodiploid species and also why, because males do not have this high 3/4 relatedness, altruistic workers are always female in haplodiploid species. One problem with this idea is that workers rear both sisters and brothers, and the average relatedness to a sister and brother is 1/2, just as in diploids. A smaller, temporary relatedness advantage can be obtained if workers invest more in sisters, but other advantages based more on benefits than relatedness (still kin selection explanations) seem more powerful.

However, the special haplodiploid relatednesses pointed out by Hamilton did lead to very successful tests of kin selection theory. When workers are related by 3/4 to sisters and 1/4 to brothers, they should do better by raising sisters (here, there is no personal fitness term w in formula 1, but w_i effects on two kinds of relatives). The workers of ants, bees, and wasps do appear to favor sisters and bias the sex ratio away from 1:1, which is best for the queen. This is clearest in species in which workers in some colonies have 3/4 relatedness to females and workers in other colonies do not. For example, in the ant *Formica truncorum*, colonies with singly mated queens (hence 3/4 sister relatedness) produce mostly females while those with multiply mated queens (hence many half sisters) produce more males. Queens in the full-sister colonies do lay many male eggs, but workers kill most of the males in order to get the benefits of rearing highly related sisters.

Other comparative tests have shown the importance of relatedness. In many species, workers will begin laying eggs that will become males when their queen dies. In species in which relatedness among the workers within colonies is low, a high fraction of them selfishly lay eggs, while in species with higher worker relatedness, more of them opt to altruistically rear the sons of their sisters.

Microbes also have cooperation and even altruism, and their short generation times have allowed researchers to go beyond comparative tests and do evolutionary experiments. The bacterium *Pseudomonas aeruginosa* secretes siderophores, molecules that bind iron and can then be recaptured, or captured by another bacterium. Siderophore production is therefore partly cooperative and vulnerable to exploitation by nonproducers who avoid the cost, but benefit from the production of others. Experiments have shown that high-relatedness conditions favor the spread of producers, while low-relatedness conditions favor selfish nonproducers. Even viruses have cooperation that is favored by high relatedness – when a single clone infects a host – than by low relatedness.

It should also be noted that multicellular bodies can be viewed as societies of cells that have been selected to be highly altruistic because their common descent from a single cell makes relatedness maximal.

Costs and Benefits Affect Behavior

It is important to remember that inclusive fitness is more than just relatedness. Some investigators have erroneously concluded that if the 3/4 relatedness hypothesis does not explain the social insects, then kin selection has failed. Some degree of relatedness is essential for the evolution of altruism (formula [2] cannot be satisfied with $r = 0$) but there is no magic number. The required relatedness drops if costs to donors are low and benefits to recipients are high.

Thus, the reason that eusociality has evolved multiple times in Hymenoptera may have more to do with special benefits than special relatedness. One special feature of Hymenoptera is the unusual prevalence of parental care relative to other insects. Care of one's young is a useful preadaptation for care of another's young. Moreover certain kinds of parental situations promote helping. When the mother rears her young in an enclosed space that provides both protection and food, such as a plant gall, the benefits of staying and helping may be high, relative to the risks of leaving the protected site. Social insects of this type, such as certain aphids and thrips and perhaps termites, are sometimes called fortress defenders. They build up colonies inside their protected site and workers generally specialize in defense. Other social insects, called life insurers, exploit a different set of benefits. For parental insects that must conduct dangerous foraging for their dependent young, short adult life expectancy makes it difficult for a single adult to provide extended care of offspring. Even one or a few workers help to ensure that some one survives long enough to provide sufficient care. Neither of these paths to sociality works without some degree of relatedness among colony members, but unusually high relatedness is not required.

Evidence on the importance of costs and benefits of social behavior comes from variation in individual behavior. In the hover wasp *Liostenogaster flavolineata*, there is a queen and a small number of workers in each colony. If the queen dies, a worker succeeds her according to an age-based rule, with the oldest worker becoming queen. Workers adjust their working effort according to their cue positions. Those near the top of the cue forage comparatively

little, unwilling to risk dying when they have reasonable prospects of reproduction. Those farther back in the queue work harder and take more risks for the colony. Because they have less chance or reproducing, the cost of altruism is lower.

Relative costs and benefits mold differences between species. We saw earlier that relatedness determines the amount of egg-laying by workers in queenless colonies. When the queen is present, it is determined more by the extent to which worker-laid eggs are often eaten by other workers. In other words, if the benefit of laying is reduced in this way, fewer workers try to lay eggs.

Benefits interact with relatedness in cooperatively breeding vertebrates. When the benefit of helping is high, as in pied kingfishers, the helpers discriminate strongly between kin and nonkin. But when it is low, as in superb fairly-wrens, there is less discrimination.

Group Selection and Kin Selection

An alternative way of explaining social behaviors, including cooperative and altruistic ones, is group selection. This argument has been made most strongly by David Sloan Wilson. If groups reproduce to different degrees and if they have varying characteristics that are heritable, then selection can operate at the group level to favor traits, such as altruism, that are good for the group. Of course, such selection can be opposed by selection within groups – selfish individuals do better than altruists in the same group. The outcome depends therefore on the balance. For some years, group selection was generally viewed as unlikely, partly because some early proponents of the idea paid insufficient attention to genetic structure – the exact feature that inclusive fitness highlights as essential. But it turns out that group selection can work under the right population structures, particularly when groups consist of close kin. In fact, inclusive fitness and group selection are not really alternative explanations; they are alternative ways of slicing up the way selection works. Either approach can be successful if used with care.

Selfishness and Conflicts

Inclusive fitness is not just about cooperation and altruism. It can be used to analyze any kind of behavior that affects others within a population. Inclusive fitness theory predicts, for example, that selection favors harming relatives if it results in sufficient fitness gain to oneself. Although worker honeybees give extraordinary help to kin, queen honeybees do the extreme opposite. When the time comes for a colony to reproduce, the old queen departs with much of the worker force and food stores, leaving the home colony for a daughter queen. For some reason, perhaps insurance, they usually leave behind several immature daughter queens. When these young queens emerge, they fight to the death, even though they are sisters. But this makes sense under Hamilton's rule. If a queen kills her half-sister, she gets the whole colony, let us say w fitness units. If instead, she splits the colony output with her sister (leaving aside the possibility that her sister kills her), they both get $w/2$. Neither is equipped to be a helper, nor can they leave and reproduce elsewhere. So by killing her sister, a queen gains $w/2$, while her sister, related by $1/4$, loses $w/2$. Hamilton's rule favors killing if $w/2 - (1/4 \times w/2) > 0$, which is always true. In other words, in this case the benefit equals the cost, and under that condition, each queen would prefer to reproduce rather than let her sister do so. The other queen, of course, favors the opposite result, and therefore they fight to the death.

Potential conflicts exist whenever two nonidentical individuals interact. Take the example of a mother horse caring for her offspring, which one might think is purely cooperative. In fact, as Robert Trivers first showed, such relationships do involve conflict, for example over the timing of weaning. A colt may continue to try to nurse at a point when the mother ignores it, chases it away, and even bites it. This can be analyzed as follows. The mother horse produces milk helping her colt by an amount b, but at cost c to her own future reproduction. For the mother, giving milk is favored if $-c + b/2 > 0$. However, from the colt's point of view, the $r = 1/2$ now goes on the mother's cost, and it favors more milk for itself if $-c/2 + b > 0$ (for simplicity, we assume here that the mother's future offspring would be half siblings to the current one; otherwise we would need to add in effects on the colt's father's reproduction too). The colt's condition is less strict; it is selected to favor more milk for itself than the mother is selected to give.

Genomic Imprinting

Let us return to conflict within individuals. We already saw how green-beard genes can be too altruistic from the point of view of other loci in the genome. There is also a means to have conflict at a single locus. Normally, when we calculate relatedness, we average over a diploid individual's two alleles, one inherited from its mother (I will call it the matrigene) and the other inherited from its father (the patrigene). This can be justified in two ways. First, for many relationships, the matrigene and patrigene are equally related to relatives, for example by $1/2$ to offspring and also by $1/2$ to full siblings (in diploid species). But what if they are differently related? Only the matrigene is related to the mother and only the patrigene to the father. And, as we have already noted,

the $1/4$ relatedness to maternal half siblings is really the average of $1/2$ for matrigenes and 0 for patrigenes. Averaging the two relatednesses can still be justified if there is no way for the offspring's alleles to assess their matrigene/patrigene status; in the absence of information, there is no way to take action. Relatedness to a maternal half sibling is $1/4$, the average of $1/2$ for genes identical by descent through the mother and 0 for genes identical by descent through the father.

However, there is a way in which matrigenes and patrigenes can be distinguished for some genes in some species. Parents can mark the DNA that they put into gametes by adding methyl groups to certain sites (usually the cytosine of specific CG doublets), and DNA in sperm is methylated differently from DNA in eggs. The zygote then comes with some of its genes differentially marked by parental origin and the cytosine methlyation can be inherited through the mitotic cell divisions that produce the cells in the offspring's body. Methylation can affect the DNA winding and unwinding, which in turn affects gene expression. The result is that matrigenes and patrigenes can be differentially expressed, sometimes with one of them being entirely silenced (which one depends on both the gene and the tissue). Such genes are said to be 'imprinted.'

David Haig has argued that for imprinted genes, we should calculate separate relatedness coefficients for matrigenes and patrigenes. For example, consider an offspring locus that affects milk acquisition in the colt discussed earlier. If, as assumed previously, the mother's future offspring will usually have different fathers (half siblings to the colt), then the colt's patrigene is unrelated to these future offspring and should be selected to take as much milk as it can use. Its matrigene is related to these future siblings, so it will be less selected to harm the mother's future output. Indeed, Haig argues that the mother might be selected to methlyate and silence her alleles at this locus to decrease overall expression. Fathers would methylate and silence loci when it would have the opposite effect of increasing milk acquisition.

There is now considerable comparative evidence in agreement with this hypothesis. Imprinting is particularly common in mammals and flowering plants, both of which are taxa in which offspring take resources from mothers, and imprinted genes are commonly expressed in embryos or in the placenta or endosperm tissues that are specialized to acquire nutrients from the mother. In one case, two loci appear to battle against each other in exactly the manner predicted. The IGF2 locus in mice helps mouse embryos acquire nutrients from the mother and is expressed only from the patrigene. Another locus acts to degrade the IGF2 protein, thereby limiting growth, and it is expressed only from the matrigene. How common imprinting is and how often it generates conflict within and between loci are important questions for future work.

See also: Behavioral Ecology and Sociobiology; Caste in Social Insects: Genetic Influences Over Caste Determination; Cooperation and Sociality; *Dictyostelium*, the Social Amoeba; Honeybees; Kin Recognition and Genetics; Levels of Selection; Marine Invertebrates: Genetics of Colony Recognition; Wolves.

Further Reading

Bourke AFG and Franks NR (1995) *Social Evolution in Ants*. Princeton, NJ: Princeton University Press.

Crozier RH and Pamilo P (1996) *Evolution of Social Insect Colonies: Sex Allocation and Kin Selection*. Oxford: Oxford University Press.

Dawkins R (1979) Twelve misunderstandings of kin selection. *Zeitschrift für Tierpsychologie* 51: 184–200.

Field J, Cronin A, and Bridge C (2006) Future fitness and helping in social queues. *Nature* 441: 214–217.

Frank SA (1998) *Foundations of Social Evolution*. Princeton, NJ: Princeton University Press.

Griffin AS and West SA (2003) Kin discrimination and the benefit of helping in cooperatively breeding vertebrates. *Science* 302: 634–636.

Griffin AS, West SA, and Buckling A (2004) Cooperation and competition in pathogenic bacteria. *Nature* 430: 1024–1027.

Haig D (2002) *Genomic Imprinting and Kinship*. New Brunswick, NJ: Rutgers University Press.

Hamilton WD (1964) The genetical evolution of social behaviour. I–II. *Journal of Theoretical Biology* 7: 1–52.

Queller DC and Goodnight KF (1989) Estimating relatedness using genetic markers. *Evolution* 43: 258–275.

Queller DC and Strassmann JE (1998) Kin selection and social insects. *Bioscience* 48: 165–175.

Sherman PW and Holmes WG (1983) Kin recognition in animals. *American Scientist* 71: 46–55.

Trivers RL (1974) Parent-offspring conflict. *American Zoologist* 14: 249–264.

Wenseleers T and Ratnieks FLW (2006) Enforced altruism in insect societies. *Nature* 444: 50.

Kleptoparasitism and Cannibalism

K. Nishimura, Hokkaido University, Hakodate, Japan

Introduction

Animals acquire resources in many ways, and this diversity allows for many types of interactions between individuals. Students of behavior commonly use the idea of resource ownership to classify these. When, for example, a group of feeding animals contests a limited amount of ownerless resources, we call this interaction 'exploitative competition.' In many situations, however, animals obtain exclusive control of resources in a manner analogous to human ownership. These acquired and defended resources present an opportunity for thieves. Kleptoparasitism is an interaction in which one individual takes a resource from its owner by stealth or aggressive conflict. The Greek word *kleptes* literally means 'thief,' so kleptoparasitism means 'parasitism by theft.' In the end, the food resources that animals acquire are assimilated into their bodies, so killing and eating another animal is, in a way, the ultimate theft.

Cannibalism is one form of this ultimate theft. To the more squeamish human readers, cannibalism – killing and eating a member of your own species – makes the 'theft' implicit in kleptoparasitism seem neighborly. The word 'cannibalism' comes from the Spanish word *Canibales*, which is the variant of the English word *Caribes*, the name of a West Indian people reputed to eat humans.

Figure 1 shows kleptoparasitism and cannibalism in a schematic space of the biological interaction of resource exploitation. Cannibalism is exploitation of the resource with full ownership, that is, the body, in which the resources are already assimilated and stored. Thus, the act is lethal to the victim. Kleptoparasitism is in the wide spectrum of varieties of parasitism. The victims of kleptoparasitism may be members of the same or different species, but the act of the kleptoparasite does not kill the victim. The resources 'stolen' in an act of kleptoparasitism may be food or objects such as nest material.

Kleptoparasitism

Kleptoparasitism takes many forms and occurs in many situations. Reports of kleptoparasitism have often used words with anthropomorphic connotations, such as 'usurpation,' 'robbing,' 'stealing,' 'pilfering,' and 'sponging,' and even phrases such as 'using others as truffle pigs.' In their authoritative book, Giraldeau and Caraco recognized three distinct forms of kleptoparasitism: overt aggression, competitive scramble, and stealth. An aggressive kleptoparasite uses force or the threat of force to gain exclusive access to resource. In scramble kleptoparasitism, one or a few individuals actively hunt for resource, but nonhunters can exploit discovered resources in an open scramble. A stealthy kleptoparasite takes the resource, but avoids interaction with the host.

Distribution Among Animal Groups

A recent review by Iyengar found that most reports of kleptoparasitism involve birds. The preponderance of records from birds probably reflected research effort and visibility more than a true pattern in nature. Investigators have reported kleptoparasitism in insects, spiders, mollusks, fishes, birds, and mammals. The following paragraphs give several concrete examples of kleptoparasitism.

Spotted hyenas are masters of piracy in African savanna. Group of hyenas steal kills made by wild dogs, cheetah, and lions. Kleptoparasitism by spotted hyenas profoundly affects the energy acquisition of wild dogs and cheetah. Chipmunks take seeds from their neighbors by entering the burrows of absent conspecifics and pilfering seeds from the larders.

Observers frequently see intra- and interspecific kleptoparasitism in seabird colonies. Jaegers and skuas rely exclusively on kleptoparasitizing other seabirds, such as terns, kittiwakes, and gulls. These birds take food from others in the air, during courtship feeding, and when adults regurgitate food to their chicks. Ornithologists have also reported kleptoparasitism in several species of waterfowl, passerines, egrets, and bird of prey. In feeding flocks of passerine species, we often see some individuals actively searching for food, while others search for opportunities to exploit the food discoveries of the others (see the discussion of producer–scrounger systems in Elsevier, Encyclopedia.

In fishes, there are a few reports of kleptoparasitism. Among territorial reef fishes that gather food algae at a fixed site (termed as 'garden'), theft from gardens occurs, and theft is not only when the territory holder is absent. Blue tang surgeonfish and striped parrotfish forage in large roving groups, feeding from the algal turf defended by damselfish.

Thrips and flies create shelters (galls) on host plants. Invading individuals sometimes actively evict occupants from their galls. Some parasitic wasps steal hosts that another wasp has located previously. In some water striders, males, which are usually smaller than females, ride on their mate's back for long periods; during this time, they often take food items that their mate catches.

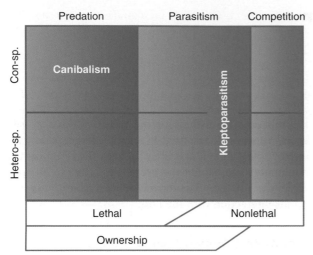

Figure 1 Kleptoparasitism and cannibalism in a schematic space of the biological interaction of resource exploitation.

Kleptoparasitism among mates is common in spiders when males live on or near the female's web for extended periods. These 'cohabiting' males will feed on prey caught in the female's web. Several spiders of genus *Argyrodes* steal prey from other spiders. They can move on the webs of their hosts without being detected, yet they can detect the position of prey trapped in the web. Some spiders engage in a unique form of kleptoparasitism called 'silk stealing.' Silk thieves cut silk out of the hosts' orb webs and eat it.

We have a few records of kleptoparasitism of terrestrial and marine gastropods. Carnivorous plants capture large quantities of high-quality food. A slug species is known to take over the food resource. Some species of conches (a marine snail) steal food from tube-dwelling polychaete worms.

Extension of the Concept of Kleptoparasitism

Outright expropriation of a food resource from its owner is the fundamental phenomenon we call kleptoparasitism. Kleptoparasites, however, also take inanimate objects. Several warbler species engage in nest material stealing, and each of these species can act both as a perpetrator and a victim.

Furthermore, kleptoparasites can also exploit intangible quantities. A kleptoparasite may exploit the searching behavior of another individual and usurp food discoveries before the discoverer can consume them. This type of parasitism has significance for the value of group living, because within a group, some members may produce information for themselves while others scrounge information from the producers. Little brown bats, for example, eavesdrop on echolocation calls of others to find prey and other resources.

Brood parasitism occurs in some birds, fishes, and, rarely, insects, and some authors interpret this to be a

type of kleptoparasitism in which the parasite uses several services, including nest-building labor and parental care. In mating situations, peripheral males may obtain matings by parasitizing the displays and other female-attracting activities of dominant males. One can view this well-known stealthy mating tactic as a form of kleptoparasitism, and investigators sometimes call it 'kleptogamy.' Thus, we realize that kleptoparasites can exploit a wide range of nonfood resources such as nest materials, domicile, parental care, mating partners, and information.

Evolutionary Ecology of Kleptoparasitism

Kleptoparasitism represents an adaptive strategy that may pay off in some situations but not in others. Taking resources from another eliminates the need to search for and handle food items, and it may therefore, save time and energy. The fitness value of kleptoparasitism depends on the relative cost obtaining resources on your own, and the ease with which the kleptoparasite can steal food from others.

In some cases, kleptoparasites work in groups, and we predict that this will only happen if the resource in question can be divided. Of course, in a world composed entirely of kleptoparasites, no one would eat. The value of kleptoparasitism hinges on the presence of individuals who find and capture their own food. In the language of behavioral game theory, kleptoparasites are 'scroungers' who depend on the presence of 'producers.' Therefore, the mix of producers and scroungers in a population determines the relative payoffs of the two strategies.

The producer–scrounger game, a model proposed by Barnard and Sibly in 1981, addresses this question. **Figure 2** demonstrates an envisaged fitness change of producer individual and scrounger individual in the population of a given proportion of scroungers. When the proportion of scroungers is low, scroungers do well because there are many producers to exploit. As scroungers become more common, encounters with producers become less common and the fitness of scrounger ultimately falls below the fitness of producers. These fitness relationships stabilize a mixture composition of producer/scrounger in a population.

The producer–scrounger framework helps us to understand how and why kleptoparasitism can evolve and be maintained in a population. As with most adaptive arguments, however, one needs to consider other factors to achieve a deep understanding of kleptoparasitism. Probably, internal state, growth and developmental history of the individual, and genetic and epigenetic constraints — all play a role in kleptoparasitism. Iyengar reviewed a wide range of distribution patterns of kleptoparasitism among the animal kingdom and argued some explanations of the distribution pattern among taxonomic groups. Morand-Ferron et al. also review the distribution pattern of food-stealing in birds and offer some explanations of this phenomenon.

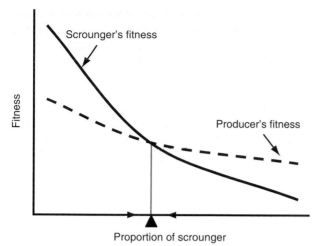

Figure 2 The relative fitness of producer and scrounger (kleptoparasite).

Kleptoparasitism and Ecology

Kleptoparasitism has some implications that go beyond its effects on the perpetrators and their victims. For example, large groups of African wild dogs have a higher rate of food intake, because large groups can defend carcasses against kleptoparasitic hyenas. So, 'defense against kleptoparasites' may play an important role in animal group size. To take another example, in Zeus bugs, male Zeus bugs ride on the backs of their mates and steal food secured by females. In response, females produce a glandular secretion that males feed on, which reduces the extent to which the males kleptoparasitize the female's food.

Kleptoparasites exploit various types of resources including food, inanimate objects, domicile, parental care, mating partner, and information. Perpetrator and victims may be solitary or in groups, and they may be conspecifics or heterospecifics. The amount of damage that kleptoparasites cause also varies. This diversity makes it difficult to draw general conclusions about the implications of the ecological consequences of kleptoparasitism. However, we do often find that interspecific kleptoparasites are often fairly close phylogenetic relatives of their victims, which in turn suggests that kleptoparasitism will often have important implications for the dynamics of guilds within ecological communities.

Cannibalism

In earlier discussions, investigators dismissed cannibalistic behavior as an anomaly. Zoos and animal-rearing facilities commonly observed cannibalism in captive situations, but it was dismissed as an artifact of crowding and stress. In addition, animals sometimes eat their relatives. Parents eat their babies, juveniles eat their siblings, and so on. This odd behavior seemed to preclude an evolutionary explanation, and further suggested some type of mistake or behavior out of context.

Notwithstanding this preconception, comprehensive surveys by several authors show that cannibalism occurs in nature in many groups of including: protozoa, planaria, rotifers, gastropods (snails and slugs), ciliates, copepods, centipedes, mites, insects, fish, amphibians, birds, and mammals. In short, cannibalism seems nearly ubiquitous, so much so that it is not even restricted to carnivorous species; we commonly find cannibalism in herbivores and detritivores.

Variety of Cannibalism

Relative size and vulnerability dependence

As noted earlier, cannibalism can take many forms. The relative sizes, ages, and developmental stages of the consumed and the consumer can vary. Crudely speaking, however, the consumers tend to be larger and more aggressive, while the consumed are small and vulnerable.

In many situations, differences in relative size create opportunities for cannibalism. Predators that swallow their prey whole – like many species of fish – can only open their mouths so much, so for these animals, cannibalism can only occur when large individuals attack victims small enough to fit in their mouths.

An animal can vary in size for many reasons. An individual may be smaller because it is younger or because it is a different gender. Yet, we find size variation even within cohorts of the same age and sex. This can occur because of differences in resource allocation during development or it may be due to random variation during an individual's development. Size variation, regardless of its source, sets the stage for cannibalism, even within cohorts of the same age and sex.

In addition, developing animals often pass through vulnerable life-history stages such as ecdysis and pupation. Cannibalism often occurs during these vulnerable periods. Indeed, younger and smaller individuals may cannibalize older and larger individuals during these vulnerable stages. We find situations where smaller individuals cannibalize larger individuals in some insects, crustaceans, and amphibians.

Parental cannibalism of progeny

Filial cannibalism, where parents eat their own eggs or infants, is widely observed in mammals (e.g., chimpanzees, lions, hyenas, and baboons), birds (e.g., several bird of prey, house finches, and house sparrows), amphibians (e.g., salamanders), reptiles (e.g., skink and boas), insects (e.g., assassin bug and burying beetles), and spiders (e.g., wolf spiders). Studies of teleost fish suggest that filial cannibalism is especially prevalent in this group (e.g., bullheads, damselfish, cichlid, flagfish, goby, and stickleback). Filial cannibalism occurs most frequently while parents are caring for their eggs and young.

Filial cannibalism presents an evolutionary conundrum, because consuming your own offspring surely decreases your current net reproductive success. One explanation is that filial cannibalism represents a 'decision' to redirect the resource away from current reproductive output and toward survival and future reproduction. Cannibalism by fathers occurs more frequently than cannibalism by mothers. Presumably, this pattern arises because males typically have less invested in offspring than females.

Another explanation of filial cannibalism is that cannibalism removes failed offspring. Parents eat diseased or parasitized eggs from their clutches. We can view this as a form of parental care that prevents diseases and parasites from spreading to the entire clutch.

Sibling cannibalism

Cannibalism among siblings occurs when sibling groups aggregate. This may be important for some species in which juveniles commonly pass through an aggregating developmental stage. Investigators have observed sibling cannibalism in birds (e.g., several birds of prey, several sea birds, and house sparrows), teleost fishes (e.g., pike, perch, and cods), selachian fishes (sharks), reptiles (snakes), amphibians (salamanders), mites (predatory mites), insects (e.g., ant lions, lady beetles, and water bugs), gastropods (snails and slugs), Spionidae (segment worms), and echinoderms (viviparous sea stars). In this form of sibling cannibalism, the developmentally advanced individuals in a clutch consume eggs, embryos, or newborn siblings within their clutch. Asymmetric development among sibs makes differential vulnerability among offspring and facilitates sibling cannibalism. Small or stunted offspring experience a greater risk of cannibalism. In many bird species, asynchronous hatching, which creates a size and age difference within the next, sets the stage for sibling cannibalism.

In some amphibians, insects, spiders, and gastropods, newborns eat eggs from their clutch as the first food of their life. In some cases, these eggs are 'nurse' or 'trophic' eggs that are sterile or have stopped developing at any early stage. In some live-bearing sharks, embryonic offspring eat their embryonic siblings while they are still within their mother's body.

Cannibalism of parents (matriphagy) and mate cannibalism

In some species of spiders, scorpions, and insects, offspring eat their mother. We interpret this behavior as a form of parental care, because the mother's body provides resources that promote the growth and development of her offspring. In one species of earwig, the mother-eating

(or matriphagy) delays offspring dispersal and increases their survival. From an evolutionary perspective, offspring cannibalizing parents is less surprising than parents cannibalizing young, because the consumed parents are typically postreproductive and decrepit.

Sexual cannibalism occurs when one sexual partner eats another. Sexual cannibalism frequently occurs as part of courtship and mating, and we see it in mantids, scorpions, and spiders. In most situations, females eat their male partners, and not the other way around. People often think of courtship and mating as a harmonious and cooperative reproductive partnership, so sexual cannibalism reminds us that mating can be fraught with conflict. In cases where the female consumes the male after copulation, we can interpret the male's 'sacrifice' as parental investment. Yet, females sometimes consume males before insemination, and this suggests sexual conflict.

Cannibalism and Ecology

Cannibalism is synonymous with 'intraspecific predation.' It has implications for population and community ecology that go beyond its importance in behavior, physiology, and life history.

Cannibalism directly eliminates conspecific individuals, so it inevitably lowers population density. In some cases, we have evidence that high densities lead to increased cannibalism and hence, to greater reductions in population size via cannibalism. The relationship between cannibalism and conspecific density is direct and immediate. Thus, cannibalism can help regulate population size.

Classical models of predator–prey dynamics suggest that predator and prey population may exhibit couple oscillations. Allowing cannibalism with the predator can reduce or eliminate these oscillations and stabilize predator-prey dynamics in the following way. When the prey population is low, predators cannot obtain enough energy from prey, and they will engage in some cannibalism. This quickly reduces the predator numbers and hence, the effect of predators on prey. The net effect is that when predators increase their rate of cannibalism, this stabilizes predator and prey population densities.

On the contrary, cannibalism can also destabilize population dynamics. Consider a situation, for example, in which older individuals cannibalize younger and more vulnerable age classes. If cannibalism eliminates a high proportion of a given cohort, this age class will be a small group throughout its life history. The resulting group of cannibalistic adults will, because they are small in number, have a smaller effect on cohorts that follow them. This 'less cannibalized' cohort will, in turn, have a larger effect on the cohorts younger than themselves, and so on. This multistep chain inference implies that the intercohort cannibalism may cause violent population fluctuations. Cannibalism can have various implications for ecological communities. For example, cannibalistic species often have complex food habits, where young animals feed on resources that adults do not eat, and adults cannibalize the young. Thus, a single cannibalistic species can connect multiple trophic levels, and it can influence a community's food web in a complex manner.

Evolutionary Arguments

As with other traits, we would like to understand the evolution and adaptive significance of cannibalism. As this review shows, cannibalism takes many forms and we cannot offer a single comprehensive explanation for all cannibalistic phenomena. The following paragraphs outline current thinking about the evolution of cannibalism.

The basic principles of adaptation
Individual fitness
Among free-living animals, cannibalism is usually facultative (meaning that it only occurs in some conditions). Cannibalism should occur when populations are crowded or alternative preys are rare or difficult to obtain. This, of course, broadly agrees with observed facts: crowding and poor access to alternatives do increase cannibalism. We can apply foraging theory's diet model to further understand the conditions that favor cannibalism. The diet model assumes that the forager makes choices that maximize its own energy acquisition or the probability of survival under fear of death from hunger.

Typically, we would expect that conspecific prey would have a lower rank as potential diet items than other food types, because attacking and handling conspecifics (who have similar size and defensive abilities) will be costly in line for potential menu items. The diet model predicts that when the abundance of relatively high-ranking food types decreases, the lower-ranked prey should be included in the diet menu, and thus, cannibalism occurs. The widespread observation that cannibalism increases when the availability of alternative foods declines is consistent with this argument.

We could explain the facultative parental cannibalism of offspring as a conditional decision to give up the current reproductive output in order in increase future reproduction. We would expect to observe parental cannibalism, therefore, in harsh environmental conditions where a parent must choose between eating its offspring and starvation.

Parental manipulation
As explained earlier, differences in size and development can set the stage for sibling cannibalism. In some cases, however, parents may pull the strings behind the scene. For example, parents may produce embryos asynchronously, differentially partition resources among their embryos, or simply feed some offspring more than others.

All the mechanisms can generate asymmetries among offspring and create the potential of sibling cannibalism.

One rather surprising interpretation of sibling cannibalism is that it represents a parental food storage strategy. According to this view, we see the bodies of vulnerable offspring as food stores for larger, older, or more viable offspring. And, of course, the larger offspring must cannibalize the smaller to exploit this 'stored food.' This may seem fanciful to some readers, but it clearly happens in some cases, where parents produce trophic eggs (eggs that others eat) that offspring need for survival. In cases like these, sibling cannibalism may be an integral part of a parent's reproductive strategy.

Evolutionary game theoretical view
Consider the question 'why is cannibalism relatively rare?' or even the reverse 'why is cannibalism relatively common?' How can the energy-based optimization argument answer these questions? Explaining this variation via simple optimization requires pre-existing differences in size or vulnerability and treats victims as merely food items.

Suppose, however, that all individuals are the same, and there is no energetic advantage to including conspecifics in the diet. Can cannibalism occur? Can we imagine conditions where some individuals act as cannibals while others do not? To answer this question, we turn to evolutionary game theory. According to this body of theory, we need to ask how the frequency of individuals 'playing' the cannibal and noncannibal strategies influences the fitness value of the two strategies.

In a mixed population of cannibals and noncannibals, each individual is threatened not only by starvation but also by cannibalism. Cannibal types are more likely to survive attacks from others than noncannibals. Fitness of both cannibals and noncannibals depends on the relative encounter rate to cannibal-type individuals in the population.

Figure 3 shows a hypothetical example of the fitness curves for the two types. When the proportion of cannibals is low, the risk of death via cannibalism is low for both cannibals and noncannibals, and so, noncannibals have higher fitness. As cannibals become more common, encounters with cannibals become more common and the fitness of noncannibal falls below the cannibal's fitness. As the figure shows, a critical proportion of cannibals exists; below this, we expect a population of noncannibals to evolve, and above this, we expect a pure cannibal population of cannibals. This argument shows how evolutionary game theory can help us understand the evolutionary origins and maintenance of cannibalism.

Evolutionary implications
Disease transmission
Evidence from several species (including several mammals, reptiles, amphibian, insects, crustaceans, and fishes)

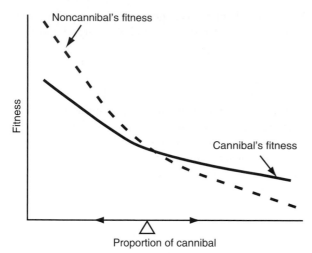

Figure 3 The relative fitness of cannibal and noncannibal.

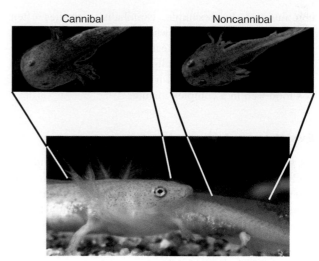

Figure 4 Cannibalism in larval salamander, *Hynobius retardatus*.

shows that cannibalism can transmit diseases such as viral, bacterial, and parasitic infections. So, a possible answer to the question 'why is cannibalism relatively uncommon?' is that cannibalism incurs a potential cost of pathogen transmission from conspecifics. Cannibals experience a greater risk than other predators because they are genetically similar to their prey, and hence, susceptible to the same kinds of pathogens. Studies by Pfennig document the enhanced risk cannibals experience. Using tiger salamanders, these studies showed that eating diseased conspecifics (cannibalism) caused infections more frequently than eating heterospecifics (normal predation).

Cannibalistic polyphenism
Cannibalistic polyphenism is an intriguing phenomenon known in some flagellates, ciliates, rotifers, and amphibians. In cannibalistic polyphenisms, some individuals in a population are cannibalistic, while others are not even though both types have the same genetic background. Cannibals commonly have modifications to their eating machinery (bigger jaws and sharper teeth) that make them more efficient cannibals. Cannibalistic polyphenism is an instance of developmental plasticity, because cannibalistic morphs develop their enlarged jaws in response to environmental conditions – primarily crowding.

Early developmental stage larvae of the salamander, *Hynobius retardatus*, can become cannibals. In crowded conditions, some individuals become cannibal morphs that prey on noncannibals and have distinct morphological structures such as a broad head and a large jaw and a large body (**Figure 4**). We can observe either of dimorphic or monomorphic local population in natural ponds depending on larval density during the sensitive phase of development. The facultative cannibalism involves important evolutionary questions, such as what conditions maintain this developmental phenotypic plasticity and

what allows the coexistence of both morphs within a given population.

See also: Avoidance of Parasites; Co-Evolution of Predators and Prey; Decision-Making: Foraging; Defensive Avoidance; Defensive Morphology; Foraging Modes; Game Theory; Games Played by Predators and Prey; Hormones and Breeding Strategies, Sex Reversal, Brood Parasites, Parthenogenesis; Infanticide; Optimal Foraging and Plant–Pollinator Co-Evolution; Optimal Foraging Theory: Introduction; Propagule Behavior and Parasite Transmission; Rhesus Macaques; Signal Parasites; Spotted Hyenas; Wintering Strategies: Moult and Behavior.

Further Reading

Barnard CJ and Sibly RM (1981) Producers and scroungers – A general-model and its application to captive flocks of house sparrows. *Animal Behaviour* 29: 543–550.

Elgar MA and Crespi BJ (1992) *Cannibalism.* Oxford: Oxford University Press.

Fox LR (1975) Cannibalism in natural populations. *Annual Review of Ecology and Systematics* 6: 87–106.

Giraldeau L-A and Caraco T (2000) *Social Foraging Theory.* Princeton, NJ: Princeton University Press.

Iyengar EV (2008) Kleptoparasitic interactions throughout the animal kingdom and a re-evaluation, based on participant mobility, of the conditions promoting the evolution of kleptoparasitism. *Biological Journal of the Linnean Society* 93: 745–762.

Morand-Ferron J, Sol D, and Lefebvre L (2007) Food stealing in birds: Brain or brawn? *Animal Behaviour* 74: 1725–1734.

Pfennig DW, Ho SG, and Hoffman EA (1998) Pathogen transmission as a selective force against cannibalism. *Animal Behaviour* 55: 1255–1261.

Polis GA (1981) The evolution and dynamics of intraspecific predation. *Annual Review of Ecology and Systematics* 12: 225–251.

Stephens DW and Krebs JR (1986) *Foraging Theory.* Princeton, NJ: Princeton University Press.

L

Learning and Conservation

A. S. Griffin, University of Newcastle, Callaghan, NSW, Australia

Introduction: History and Definitions

A rat scuffles along a track in the undergrowth and comes across a food source it has not encountered previously. Next day, heading out to feed, it goes straight to that location to forage once again. A hummingbird lands on a pink flower and discovers a rich source of nectar. On subsequent foraging trips, it chooses to land preferentially on pink flowers. A young calf narrowly escapes an attack by a crocodile while drinking at a riverbank. Next time it approaches the river, it does so more cautiously and remains more vigilant while it drinks. In all these examples, animals change their behavior as a consequence of experience. This ability, referred to as 'learning,' is taxonomically widespread and is critical to survival and reproduction.

The study of learning has a long and rich history. Early interest in the phenomenon can be traced back to the great biologist, Charles Darwin, who conducted extensive empirical work to explore whether earthworms learn where to build holes. The study of learning emerged as a modern academic discipline in the nineteenth century when Russian scientist, Ivan Pavlov, began his well-known work on associative learning in dogs. Continuing in Pavlov's tracks, other psychologists, such as Edward Thorndike and Burrhus Skinner, undertook to determine the laws that govern how, when, and under which conditions learning occurs – a field of widespread research even today. For psychologists, the primary motivation has been, and still is, to formulate so-called universal laws that are assumed to govern learning across all species and all situations, an approach known as 'the General Process approach' to the study of learning. Consequently, their highly controlled empirical work has employed only a handful of convenient model laboratory species, such as rats and pigeons, a surprising approach given the motivation to establish general laws of learning, and has focused on asking how these animals learn about arbitrary stimuli with little ecological significance (e.g., single tones). In the 1960s, however, findings from the emerging field of ethology brought with them the awareness that learning occurs in a broad range of contexts outside the laboratory and plays an essential role in the survival of most animal species. Subsequently, zoologists and behavioral ecologists embraced the study of learning, successfully combining experimental control with ecological significance. A particular focus of this work has been on the mechanisms and functions of social learning. Involvement of this scientific community in the study of learning brought with it an increase in taxonomic breadth.

An awareness of how important learning is to conservation emerged in the early 1990s when the number of captive-breeding programs increased dramatically as conservation biologists began to attempt to stall the impending global species extinction crisis. Motivated by the initially poor success rates of wildlife reintroductions, reintroduction biologists began to pay more attention to the role of learning in captive-breeding programs in particular. Although the role of learning has been considered mostly within the captive-breeding context, learning needs to be considered in any wildlife management strategy that isolates an individual from the habitat in which it will ultimately have to survive. Translocation of animals to predator-free islands to increase population numbers with the ultimate aim of returning future generations to their original environment is one such example.

Types of Learning

To understand how animal learning affects conservation, it is necessary to briefly introduce the reader to two classes of learning phenomena, as well as their mechanisms and functions. Of most importance to conservation have been two types of associative learning traditionally known as 'classical' (a.k.a. Pavlovian) conditioning and 'instrumental' (a.k.a. operant) conditioning. The principles that describe when and under which conditions classical and instrumental conditioning occur are well established. Some attention to the huge body of empirical data on associative learning is important for conservation work because it provides the basis for understanding how experience shapes behavior and for designing conservation interventions that produce animals well suited to the postrelease environment.

In classical conditioning, an initially neutral stimulus (conditioned stimulus, CS; e.g., a light) is presented repeatedly together with a biologically significant event, which evokes a spontaneous response (unconditioned stimulus, US; e.g., food). As a result, animals learn an association between the CS and US. Learning of the association is reflected in that the CS acquires the ability to evoke a response that is related to the response evoked by the US (e.g., foraging behavior). It is generally accepted that learning occurs when appearance of the CS predicts, or signals, the subsequent occurrence of the US. Hence, the function, or adaptive significance, of classical conditioning is to allow organisms to prepare themselves for biologically significant events. For instance, young white-tailed ptarmigan chicks (*Lagopus leucurus*) learn to forage on foods high in protein (CS), which they have associated with their mother's food calls (US). In trials designed to reduce egg depredation in endangered species, American crows (*Corvus brachyrhynchos*) learn to avoid green-painted eggs after ingesting similar colored eggs (CS) injected with a chemical that made them ill (US). Crows acquire aggressive responses to humans (CS) who have netted them (US).

Also of importance to conservation is instrumental conditioning. Here, rather than an association between two stimuli, animals learn an association between a behavior and its consequences. Behaviors followed by successful outcomes increase in frequency, while those followed by adverse outcomes decrease in frequency. In many predator species, mothers bring live prey to their offspring, thus creating opportunities for them to practice their capturing and killing techniques. The adaptive value of instrumental conditioning is that it provides a mechanism by which animals increase the effectiveness of their behavior.

Ontogenetic Isolation and Its Effects on Behavior

Captive breeding isolates animals from the environment in which they will ultimately have to survive when they are released into the wild. Comparisons between behavior of captive-reared and wild-born individuals have shown that captive rearing can lead to substantial changes in behavior, and that these changes can affect postrelease survival. Captive-reared northern bobwhites (*Colinus virginianus*) have deficient antipredator skills and as a consequence undergo far greater postrelease predation than wild-born individuals. Juvenile black-tailed prairie dogs (*Cynomys leucurus*) bred in captivity are less vigilant and alarm call less to predators than age-matched wild

individuals, and suffer higher rates of mortality. Captive-bred Attwater's prairie chickens (*Tympanuchus cupido attwateri*) tolerate closer approach by humans and dogs than wild prairie chickens, and suffer high rates of postrelease predation. Captive-bred golden lion tamarins (*Leontopithecus rosalia*) are deficient in their locomotor and foraging skills when compared to their wild-born offspring, and these deficiencies persist several years after release.

There are exceptions, however, to the general rule that captivity has detrimental effects on behavior. For instance, survival rates of captive-reared takahe (*Porphyrio mantelli*) do not differ significantly from those of wild-born takahe. But, in many instances, equivalent survival requires implementing postrelease measures to reduce the impact of deficient behavior. For example, captive bred black and white ruffed lemurs (*Varecia variegata*) survive equally well as their wild counterparts, as do scarlet macaws (*Ara macao*) raised with wild mates, as long as they receive postrelease supplemental feeding.

Captivity-associated alterations in behavior have two sources. First, captive-bred animals do not have the opportunity to acquire lifetime experience with their natural environment. Both individual learning through direct experience with the environment, and social learning through interactions with more experienced individuals (e.g., a parent) are compromised. Taking this process one step further, animals may even adjust their behavior to suit life in captivity in ways that are detrimental to survival in the wild. Captive-held river otters (*Lontra canadensis*) are more prone to predation and accidents with traps than wild otters, perhaps reflecting habituation to captivity.

A second effect that may occur when animals are bred in isolation from their natural habitat over several successive generations is evolutionary loss of behavior. Loss may occur either because behavior well suited to survival in the wild is selected against in captivity, or alternatively because once beneficial behavior is lost under the effects of genetic drift. High numbers of captive-bred Saudi Arabian houbara bustards (*Chlamydotis macqueenii*) die from trauma-related deaths, usually involving collisions with cages by frightened birds. Such mortality can result in selection for individuals whose behavior is more suited to life in cages, but whose predator escape responses are inadequate. Indeed, pen-reared Attwater's prairie chickens fly significantly less far in response to an approaching human or dog than wild greater prairie chickens (*T. cupido*).

Evolutionary loss of behavior is complicated by the fact that genetic predispositions focusing attention on stimuli that are particularly relevant to survival often guide learning. Although this phenomenon has not been studied in the conservation context, several examples can be found in the literature on mechanisms of learning. Snake-naïve rhesus monkeys (*Macaca mulatta*) learn to associate snakes, but not flowers, with fear responses of social companions. Male quail associate an object with quail-like features with copulation and feeding opportunities more quickly than an arbitrary object. Guided learning is also evident when young chicks learn the features of their mother, although the emergence of such preferences is itself dependent upon nonspecific visual experience shortly after birth. Learning predispositions ensure that animals learn quickly and effectively about ecologically significant events. Consequently, reintroduction biologists need to be aware that evolutionary isolation may lead not only to the loss of behavior, but also potentially, and for the same reasons, the loss of learning predispositions. By impacting both the experiential and the genetic underpinnings of learning, captive breeding has the potential to reduce substantially the ability of animals to survive once released into the wild.

To address this problem, a huge effort has been made to design captive-breeding environments that provide animals with enriched learning opportunities. In addition, numerous prerelease preparation programs are implemented to train individuals in the survival skills they lack. These conservation interventions target a diverse range of animal behaviors, three of which are discussed in the following sections.

Specific Research Areas in Learning and Conservation

The Role of Learning in Predator Avoidance

The high incidence of postrelease predation on captive-bred individuals has been, and still is, one of the greatest challenges to wildlife reintroductions. Encouragingly, however, a large body of empirical work has demonstrated that a taxonomically wide range of species exhibit the ability to learn about novel predators. Learning can occur both through direct individual aversive experience with the predator stimulus, and through indirect (social) experience, for example by perceiving predator together with alarm responses of predator-experienced individuals. Both direct and indirect learning engage classical conditioning in which a novel predator plays the role of a CS and inherently aversive stimuli, such as being chased or bitten in direct learning, or social alarm signals in indirect learning, serve as the US. Furthermore, predator avoidance learning is guided by predispositions to learn about predator stimuli more readily than arbitrary (e.g., plastic bottle) or nonpredator stimuli (e.g., goat), and does not take long (one to three exposures to aversive associations are sufficient).

Building on this knowledge, reintroduction biologists have developed a range of predator avoidance training techniques for captive-bred animals. Broadly speaking, these methods all engage classical conditioning, in which opportunities are created for animals to associate novel predator stimuli with aversive events. Both direct learning and indirect social learning training methods have been tested, with tentative evidence that social learning produces greater changes in behavior and greater improvements in survival. For example, several reintroduction programs (e.g., Attwater's prairie chickens, houbara bustards, takahe, bobwhite quail (*C. virginianus*), prairie dogs (*Cynomys ludovicianus*)) have used direct attack, or harassment, by a predator (e.g., fox (live or mounted); dog; stoat; human) to enhance antipredator responses. In bobwhite quail, training improves cover seeking and covey coordination and increases postrelease survival rates. In houbara bustards, harassment by a live fox, but not a fox mount, enhances postrelease survival rates. In prairie dogs, pairing predator stimuli with social alarm vocalizations or the opportunity to observe a predator-experienced prairie dog respond to the predator stimuli enhances antipredator vigilance, alarm call rates, and time in or near shelter. Social training increases postrelease survival to the point that trained prairie dogs survive as often as their wild counterparts.

Social learning can also be triggered by allowing predator-naïve individuals to watch the attack of a conspecific by a predator. For example, in an attempt to reduce postrelease predation of captive-bred Puerto Rican parrots (*Amazona vittata*) by red-tailed hawks (*Buteo jamaicensis*), captive-born individuals are given the opportunity to witness a staged attack of a nonendangered Hispaniolian parrot (*Amazona ventralis*) by the aerial predator. Similarly, takahe chicks watch a model stoat apparently attack and kill a takahe chick. In both cases, there is tentative evidence that such training increases postrelease survival.

It is important to note that firm conclusions about the effects of prerelease predator avoidance training on postrelease mortality rates can really only be made if the content of learning is known. This is because training has the potential to cause general increases in stress, rather than to teach predator recognition or predator-specific antipredator responses, changes in behavior that are unlikely to improve postrelease survival. Consequently, measuring responses to the target predator both before and after training to ensure that learning has occurred, as well as to nontrained control stimuli to ensure that changes in behavior are predator-specific, is important.

The Role of Learning in Social Behavior

Just like antipredator behavior, social behavior is shaped by experience and is of prime importance to conservation. The learning phenomena that have received the most attention in this context are filial and sexual imprinting. During filial imprinting, the young individual learns soon after birth to recognize its mother (or its carer) and becomes socially attached to her. Such learning occurs in precocial animals, such as ducks and guinea pigs, and is conceptualized as a form of classical conditioning in which learning of the carer's visual features (CS) is triggered by association with inherently salient stimuli (US), such as movement. Similar to predator avoidance learning, imprinting is guided by predispositions to learn about some stimuli more readily than others. For example, chicks imprint most readily on a hen-like stimulus. During sexual imprinting, individuals learn the visual attributes of potential mates. The critical experiences for sexual imprinting are different from those involved in filial imprinting and occur later in life.

Filial and sexual imprinting have serious implications for animals reared away from their natural social environment. For instance, widespread breeding techniques for endangered species, such as hand rearing and crossfostering to related species, can produce individuals maladapted to reproduction. While artificial rearing environments (e.g., brooder boxes; commercial incubators) can boost population growth at relatively low cost, benefits may be offset by problems associated with deficient social behavior. For example, from 1986 to 2000, 67% of unsuccessful releases of southern sea otter pups (*Enhydra lutris nereis*) reared using methods that rely heavily on human care were caused by failure of pups to integrate with wild populations and avoid interactions with humans.

An increasing awareness of the interaction between early social environment and later reproductive behavior has triggered a number of strategies to expose animals to natural social contexts immediately after birth and during subsequent development. The Mississippi sandhill crane (*Grus Canadensis pulla*) reintroduction program has been at the forefront of such attempts. Here, extensive efforts are made to expose chicks immediately after hatching to adult cranes that can serve as imprinting models. Taxidermy mounts of adult cranes lying in brood posture are placed beneath the heat lamps and sandhill crane brood calls are played back by tape recorder, while mounts of crane heads are used to teach chicks to feed. Later on, chicks are housed in pens adjacent to adult cranes. Furthermore, during occasional interactions, humans are disguised in amorphous gray costumes. Chicks raised using these techniques survive at least as well as parent-reared birds, which are less wary of approach by humans and predators after release, a behavior attributed to their tendency to approach motor vehicles and uncostumed humans in captivity. Other reintroduction programs have followed in the steps of the sandhill crane reintroduction.

Social interactions between members of the same species can shape social behavior in subtle ways that extend beyond filial and sexual imprinting, however. Cultural transmission of mate choice in female Japanese quail (*Coturnix japonica*) is one example; they remember, and prefer to mate with, males whom they have previously seen court and mate with another female. Social behavior of cowbirds (*Molothrus ater ater*) provides another. Here, females enhance the frequency of specific songs within the male song repertoire by using a wing stroke to indicate their song preferences, and these songs later evoke higher levels of female copulatory behavior. In addition, female cowbird behavior enhances male–male competition. Males that are involved in more male–male competition later receive more copulations, and aviaries containing more competitive males produce more eggs. Although such experiential effects have not been studied in a conservation context, these examples illustrate that learning associated with social interactions can have far-reaching consequences on the genetic composition of a population, on individual breeding success, and, as a result, on the long-term outcome of a reintroduction program.

The Role of Learning in Foraging Behavior

Finally, we turn to the effect of experience on foraging behavior. The fact that postrelease supplemental feeding is a widely recommended practice and so often increases postrelease survival rates provides indirect evidence that many animals reared away from their natural environment have deficient foraging behavior. Instrumental conditioning is critical to the development of adequate food handling techniques. In many predator species, such as cats, mothers bring live prey to their young, thus creating opportunities for inexperienced individuals to practice capturing and killing techniques. Similar opportunities can be created in captivity by exposing captive-reared animals to the foods they will later encounter in the wild. Captive-bred Puerto Rican parrots are fed a range of rainforest fruits, allowing them presumably not only to improve their handling skills, but also through classical conditioning, learn the colors and smells of edible foods. Such learning can start early. Rat pups exhibit preferences for foods, the flavor of which they have experienced through their mother's milk.

Animals also acquire foraging behavior from interacting with more experienced individuals. Ptarmigan chicks learn to forage on foods high in protein they have associated with their mothers' food calls. Southern sea otter pups reared by surrogate mother otters forage independently sooner and have higher survival rates than pups reared using methods that rely heavily on human care.

Individual and social learning shapes not only food preferences, but also food avoidances. Red-winged blackbirds (*Agelaius phoeniceus*) avoid a distinctively colored food after they have observed a conspecific eat the food and subsequently develop toxin-induced illness. Similarly, domestic chicks (*Gallus domesticus*) that peck at a colored bead dipped in a bitter-tasting chemical, or watch other chicks peck at the bead and express a disgust response, subsequently avoid pecking at beads of that color.

But learning about food is not restricted to acquiring handling techniques and recognizing edible and nonedible foods. Both temporal and spatial food-related information can be acquired through both individual and social experience, and can have far-reaching consequences on behavior of individuals after release. For example, young black bears (*Ursus americanus*) reared by mothers accustomed to feeding on anthropogenic food sources tend to maintain these preferences as adults, and are consequently more likely to venture close to humans. As a consequence, where, when, and on what an individual learns to forage early in its life can affect how it distributes its behavior in space and time as an adult, perhaps exposing it to greater risk of predation. Acquisition of food-related information can hence have consequences that extend far beyond the immediate problem of foraging skills and food recognition.

The Role of Learning in Adjusting to Urbanization

As urban environments expand and natural habitats retract, selection for species able to adjust to human-modified habitats intensifies. In recent years, there has been an increasing interest in understanding why some species, but not others, adjust to environmental change. One hypothesis is that environmental change is one of the factors that selects for increases in brain size. It is thought that larger brains afford greater innovation and learning capabilities (a.k.a. *behavioral flexibility*), which allow individuals to modify their behavior in adaptive ways and hence increase survival in modified habitats. Indeed, large-scale analyses of species-specific innovation rates, operationally defined as the number of anecdotal reports of feeding innovations in the wild (e.g., foraging on a novel food) and obtained by reviewing the field-based literature, have revealed a positive relationship in both mammals and birds between brain size and innovation rate once the effects of body size and phylogeny have been removed. Coupled with a small within-species experimental literature pointing to a positive relationship between innovation and learning, these analyses suggest that larger brains afford greater behavioral flexibility.

Furthermore, behavioral flexibility seems to increase survival in harsh, novel, or altered environments. Birds innovate more frequently in winter when resources are scarce. Species with larger brains and higher innovation rates are more likely to become established in novel environments than species with smaller brains, while long-term avian population trends in England indicate that large-brained species are fairing the best. In sum, relationships between brain size, behavioral flexibility, and environmental change point to a potential, but yet untested, relationship between behavioral flexibility and population expansion in disturbed environments.

It should be noted that the comparative correlational literature on brain size has its critics, who call for more experimental work to properly understand the function of large brains. It might also be helpful to properly identify the life history traits that support the evolution of big brains (e.g., extended parental care). Together, this information might allow us to predict whether large-brained species will be better able to adjust to large-scale environmental change, including rampant urbanization and climate change.

Future Research

Reintroduction programs are outcome-driven exercises that aim to restore species to their historical habitats. To date, much reintroduction research has been undertaken in an ad hoc manner, and knowledge regarding the parameters that favor reintroduction success has been gained using an opportunistic and/or a posteriori evaluation of management strategies. Development of methods to offset the effects of captivity is likely to benefit most from an experimental hypothesis-driven approach. Experimental protocols should measure behavior both before and after a controlled learning experience in both trained-experimental and nontrained-control animals to detect changes specifically attributable to the learning experience, use control stimuli to assess to what extent learning is specific to the trained stimulus, and include the release of nontrained animals to ascertain to what extent learning provides a survival benefit. Only in this way will we be sure that management practices provide a measurable benefit and are not simply a matter of faith. In this regard, the fundamental literature on animal learning, behavior, and developmental biology has much to offer in terms of procedures and theory, and should be used to inform reintroduction research. Integration of fundamental and applied research can only be achieved by providing reintroduction biologists with a thorough training in these scientific disciplines and/or through their close collaboration with behavioral scientists.

One of the greatest obstacles facing reintroduction biologists is that hypothesis-driven research designed to understand the effects of experience on behavior requires relatively large sample sizes, which are not always available when working with threatened species. One way to overcome this problem is to use surrogate species as models. For instance, the effects of puppet rearing have been evaluated using common ravens (*Corvus corax*) as a model for the endangered Hawaiian crow (*C. hawaiiensis*) and the Mariana crow (*C. kubaryi*). In Australia, tammar wallabies (*Macropus eugenii*) have been used as a model macropodid marsupial to develop predator avoidance training techniques and systematically assess their effect on behavior. Better planning and coordination of research across reintroduction projects dealing with taxonomically related groups may also assist in this regard. For example, reintroduction programs involving precocial birds could work together to explore the effects of various rearing methods on social integration of wild populations.

Another possible approach is to use data from fundamental work in animal behavior to predict which interventions will be most successful. For example, Griffin and colleagues integrated an understanding of how ontogenetic and evolutionary isolation from predators modify antipredator responses with principles of associative learning to predict that predator avoidance training is likely to be more successful with animals that have undergone ontogenetic isolation from some, but not all, predators. This kind of predictive framework can assist decision makers in allocating limited resources to prerelease training.

In conclusion, it is proposed that integration of reintroduction research and fundamental work in animal behavior, coupled with an experimental hypothesis-driven methodology, will be the most fruitful way forward for research into learning and conservation.

See also: Memory, Learning, Hormones and Behavior; Ontogenetic Effects of Captive Breeding.

Further Reading

Ellis DH, Gee GF, Hereford GH, et al. (2000) Post-release survival of hand-reared and parent-reared Mississippi Sandhill Cranes. *Condor* 102: 104–112.

Griffin AS (2003) Training tammar wallabies (*Macropus eugenii*) to respond to predators: A review linking experimental psychology to conservation. *International Journal of Comparative Psychology* 16: 111–129.

Griffin AS, Blumstein DT, and Evans CS (2000) Training captive-bred or translocated animals to avoid predators. *Conservation Biology* 14: 1317–1326.

Lefebvre L and Sol D (2008) Brains, lifestyles and cognition: Are there general trends? *Brain, Behavior and Evolution* 72: 135–144.

Mazur R and Seher V (2008) Socially learned foraging behaviour in wild black bears, *Ursus americanus. Animal Behaviour* 75: 1503–1508.

Seddon PJ, Armstrong DP, and Maloney RF (2007) Developing the science of reintroduction biology. *Conservation Biology* 21: 303–312.

Ten Cate C (2000) How learning mechanisms might affect evolutionary processes. *Trends in Ecology & Evolution* 15: 179–181.

Valutis LL and Marzluff JM (1999) The appropriateness of puppet-rearing birds for reintroduction. *Conservation Biology* 13: 584–591.

Leech Behavioral Choice: Neuroethology

W. B. Kristan Jr. and K. A. French, University of California, San Diego, La Jolla, CA, USA

Introduction: Using Behavioral Choice to Study Decision-Making

Animals make decisions constantly about how to respond to sensory input from both external and internal sources. Decisions are characteristic of species (e.g., a dog and a cat respond differently to human verbalizations) and of individuals (e.g., George Bush and Barack Obama make very different decisions based on the same set of data). Even within a single individual, the response to the same stimulus varies depending upon such conditions as time of day, state of hunger, developmental stage, and their own previous history.

To study decision making, experimentalists have taken two general approaches. One approach derives from behavioral psychology: recording the activity of individual neurons in the brains of behaving mammals as they respond differentially to different versions of a stimulus, such as the direction of movement of a random dot display, the difference between complex shapes, the rate of vibration of tactile stimuli, or different types of smells. The animal (usually a monkey, sometimes a rat) has been highly trained to perform the task for a reward, and the recording takes place from the time immediately preceding stimulus presentation until the animal indicates its choice (by moving its eyes to one of two locations, for instance, or moving to one of two locations to receive its reward). In some cases, instead of making a discrimination, the animal plays an interactive game with the experimenter, with a computer, or with another animal, so that it needs to take the responses of its competitor into account to optimize its chances of receiving a reward.

The second approach is neuroethological: recording from presumed 'decision-making' neurons in a relatively simple nervous system as the animal makes choices among mutually exclusive behaviors, and its choice is assayed by observing its behavior. To control the choices, the animal simultaneously receives two stimuli, each of which produces a different behavior on its own. Typically, the two simultaneous stimuli elicit a single behavior; in other words, the animal 'chooses' one behavior over the other, rather than producing a qualitatively new behavior or a combination of the two behaviors. This approach can be extended by pairing each of the two stimuli with a third one that, by itself, would produce a distinctly different, third behavior. In this way, one can build a *behavioral hierarchy*, in which behavior A is selected over behavior B, and either one is selected over behavior C. (Usually – but not always – this is a transitive relationship: if A is selected over B, and B is selected over C, then A is selected over C.)

The experimental psychological approach requires a simple response (e.g., movement of the eyes) in a complex nervous system (e.g., that of a monkey or a rat), and the response itself (e.g., an eye movement either to the left or to the right) is usually interesting only as a way to indicate the discrimination that has been made or the abstract strategy that has been selected. The neuroethological approach typically uses more complex responses (e.g., feeding or swimming or egg-laying) in a relatively simple nervous system (e.g., that of a sea slug, an insect, or a leech) in which the behavioral responses themselves are important, because they may be part of the decision-making process. The psychological approach has been reviewed beautifully in recent years by Glimcher, Gold and Shadlen, and by Wang. The neuroethological approach was inspired by the work of ethologists, particularly Niko Tinbergen, who quantified many natural behaviors of a variety of animals. This article focuses on a single organism, the medicinal leech, and shows how the neuroethological approach has revealed several different mechanisms that neuronal circuits use to make behavioral choices.

Behavioral Choices in the European Medicinal Leech

Anatomy and Experimental Approaches

For centuries, Europeans have used several species of the leech genus *Hirudo* to remove blood from sick people in a more or less controlled manner. These 'medicinal leeches' are still used clinically to maintain blood flow through body parts that microsurgeons have stitched back onto people. In addition, neurobiologists have used this robust creature to study the basic properties of neurons, the development and regeneration of the nervous system, and the neuronal basis of such behaviors as swimming, crawling, heartbeat, bending, shortening, and feeding. Leeches are segmented worms that come in different shapes and sizes. *Hirudo* adults that have been used for behavioral studies are about the size of a human adult's index finger. Each of the overt behaviors was first studied in intact animals (**Figure 1(a)**) to determine the temporal and spatial pattern of body movements that constitute the behavior; such a description is called *kinematics*. Remarkably, the kinematics in intact parts of the animal are normal even after large pieces of the animal are removed to expose the nervous system (**Figure 1(b)**, for example). Such an animal is said to be semiintact. Recording from motor neurons – either intracellularly using sharp electrodes or from nerves using suction electrodes – in the exposed parts of the nervous system while the intact regions are performing a behavior has revealed that the motor neurons in these denuded ganglia produce impulse patterns that would be appropriate if the motor neurons were still connected to their muscles and were producing the same behavior. In fact, for many of the behaviors studied, the motor neuronal activity patterns when the semiintact animal is pinched or prodded can also be seen in the isolated nerve cord that has been entirely disconnected and removed from the rest of the body (**Figure 1(c)**), in response to electrical stimulation of a single neuron or a nerve.

In addition to this behavioral robustness, the leech nervous system offers the experimenter other advantages: the somata of its neurons are relatively large (20–90 μm in diameter in an adult), they are readily visualized (they form a monolayer over the surface of the ganglion), and they are identifiable (each of the 21 midbody ganglia in a leech has nearly the same complement of 400 neurons, and they are consistently found from leech to leech). About 30% of the 400 have been identified as individuals, and the activity patterns of nearly 75% of them have been

reliably characterized during various behaviors. These anatomical and physiological features have been exploited to perform the neuroethological studies required to characterize behaviors and the interactions between the behaviors.

Behavioral Choices

When touched lightly anywhere in the *middle* of the body, a *Hirudo* bends locally by contracting longitudinal muscles on the side of the touch and relaxing the muscles on the opposite side (**Figure 2**, lower left). When prodded or firmly pinched in the *posterior half*, *Hirudo* either swims or crawls (**Figure 2**, bottom middle and right). Swimming is a fast (1–3 cycles s^{-1}), up-and-down undulation of the body that moves from front-to-back, propelling the leech forward. To make these undulatory movements, the animal flattens itself dorsoventrally and then alternately contracts longitudinal muscles on the dorsal and ventral surfaces. Crawling is also a rhythmic form of locomotion but it is much slower (each step cycle takes 2–10 s), and the leech uses circular muscle contractions to make its body longer, which alternate with longitudinal muscle contractions all around the body that shorten its body. The lengthening and shortening movements progress from the front segments rearward and are

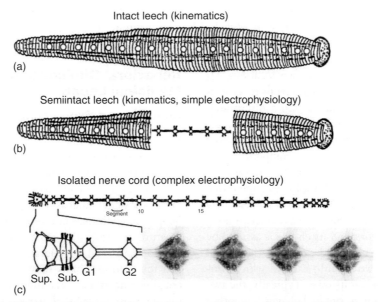

Figure 1 Different types of *Hirudo* preparations used to relate neuronal activity to behavior. (a) Intact leech. Markers are sewn or glued onto the surface of the leech. (In this case, white beads were sewn onto the dorsal surface in the middle of 19 of the 21 midbody segments.) Detailed body movements are analyzed from video-tapes of leeches as they perform a behavior. At rest, a typical adult leech measures 5–10 cm. (b) Semiintact leech. In this example, the body wall and internal organs were removed from five midbody segments, leaving only the ganglia and the interganglionic connectives intact in these segments. (c) Isolated central nervous system (CNS). All the body has been dissected away. The CNS consists of an anterior brain (composed of two parts, the supraesophageal ganglion (sup.) and subesophageal (sub.) ganglion), a nerve cord of 21 nearly identical ganglia (shown in the expanded view as photographs), and a posterior brain. The brains develop from the compression of ganglion-like primordia: the subesophageal ganglion from 4 and the posterior brain from 7. Each ganglion and brain connects to the skin and muscles in the body wall via laterally directed nerves.

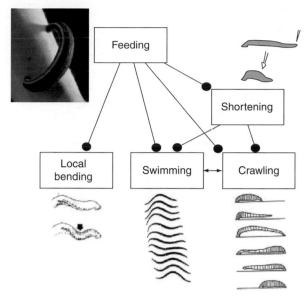

Figure 2 Interactions among five *Hirudo* behaviors: pictures and drawings of leeches feeding, performing local bending, swimming, crawling, and shortening. Swimming and crawling are rhythmic behaviors; a single cycle of each is shown, reading from top to bottom. The lines ending in filled circles indicate inhibitory interactions among behaviors: feeding inhibits all of the others and shortening inhibits swimming and crawling. The two-headed arrow between swimming and crawling indicates complex interactions between these two behaviors: whether a leech swims or crawls in response to a touch applied to its rear end depends upon such things as its state of hunger, its behavioral state (e.g., resting vs. aroused), and the depth of the water.

coordinated with attachment and release of suckers on the front and back ends to pull the animal forward. If touched briskly on the *front end*, a *Hirudo* shortens (**Figure 2**, upper right) by contracting all the longitudinal muscles in its body at once. If presented with appropriate chemical and temperature cues (a warm-blooded mammal is a favored food), *Hirudo* feeds: it latches onto the target with both suckers, everts its rasping teeth through its mouth in the middle of the front sucker, cuts through the skin, and sucks in the blood that oozes from the cut (**Figure 2**, upper left). When stimulated with two stimuli that would elicit any two of these behaviors when presented singly (e.g., stimulating the front and back simultaneously, or presenting a blood meal while prodding the leech in the back), the following hierarchy is seen:

- Feeding turns off all other behaviors.
- Shortening overrides swimming and crawling.
- Swimming and crawling show complex interactions.

The neuronal basis of each interaction is discussed in the following section, but to frame the discussion, we will first present two extreme possibilities that have been proposed to explain how neuronal circuits might accomplish behavioral choice.

Possible Neuronal Mehanisms of Behavioral Choice

One mechanism that has been proposed to explain behavioral choice is competition between reflexive pathways, using inhibition to turn off the alternative behavior(s) (**Figure 3(a)**). In this scheme, a particular set of *sensory receptors* elicits each behavior. In this example, activity in the A sensory receptors elicits shortening and the B sensory receptors activate swimming. The activity of these receptors is processed by one or more levels of *sensory processor* neurons (e.g., in mammalian visual systems, there may be a dozen or more layers of neurons). At the behavioral end of the reflex, *motor neurons* activate muscles to produce shortening or swimming. The timing, intensity, and locations of the muscles activated are determined, not by connections among motor neurons, in general, but by networks of *pattern generator* neurons that turn unpatterned excitatory input into a complex output that is imposed on the motor neurons. (For instance, a scary encounter might produce a temporally unpatterned increase in activity in your brain that would lead to your running away, a highly patterned behavioral output.) Between the sensory processors and pattern generators lie a group of neurons variously called *command neurons*, decision neurons, or more poetically, the sensorimotor watershed. The name 'command neuron' comes from the observation that stimulation of certain individual neurons of this sort in many relatively simple nervous systems produces a program of motor activity that looks like a natural behavior. An extreme hypothesis for the role of command neurons is that they are both necessary and sufficient to activate a particular behavior: they are activated only by appropriate input from sensory processors and, in turn, activate the pattern generators.

In this model of decision-making, the question, 'How does shortening override swimming' comes down to 'How is the pattern generator for shortening activated in preference to the pattern generator for swimming?' The favored – and simplest – hypothesis has been that the command neurons for one behavior (shortening) inhibit the command neurons for the other (swimming). For this hypothesis to be true, the swim-producing command neurons should be inhibited whenever the shortening-producing pathway is activated.

A second possible mechanism for behavioral choice is that the same neurons are responsible for selecting both behaviors, but that the *dynamics* – that is, the temporal pattern of activity in each neuron – of the neuronal circuit are different. A simple example of this possibility is shown in **Figure 3(b)**, which represents the activity of just three neurons (x, y, and z) that function as decision-makers. At rest (the gray cloud), cell x fires at a high rate, cell y fires much less, and cell z fires hardly at all. The arrows rising out of the gray cloud represent two different responses to

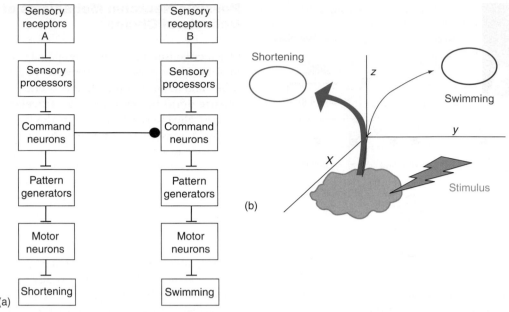

Figure 3 Possible neuronal mechanisms of behavioral choice. (a) Inhibition between command neurons, each dedicated to a different behavior. Two separate reflexive pathways are hypothesized, one that produces shortening and the other that produces swimming. The lines between the boxes indicate the nature of the influence: T-junctions indicate excitation and the filled circle indicates inhibition. Each box represents one or more levels of processing and is labeled by the type of processing performed. (b) Alternative dynamic states of the same set of three command neurons. The activity of each of the three neurons (x, y, and z) is indicated by the three axes of the graph. The gray cloud indicates the activity of these three cells at rest. The lightening bolt represents a stimulus and the two arrows rising from the cloud indicate two activity patterns, which change over time, of the three neurons. The ovals represent locations in the activity space that are attractive: outside these areas, activity tends to change, whereas inside these areas activity persists. When the activity of the three neurons takes the red trajectory and ends up inside the red oval, the leech shortens; when the activity takes the blue trajectory into the blue oval, the leech swims. The relative widths of the two arrows indicates that the pathway to shortening is more likely, so the response to the stimulus is usually shortening.

the same electrical stimulus. The red arrow pointing to the red oval indicates that the activity of cell z has increased, that of cell y has decreased, and cell x activity has stayed relatively constant. The red oval itself indicates that, once this combination of x, y, and z activity is reached, it remains in this region. In the terminology of dynamical systems, the region of activity space enclosed by the red oval is an *attractor*; it is a combination of interconnections and neuronal properties that produce maintained activity that is easy to initiate, but very difficulty to terminate.

The thinner arrow pointing toward the blue oval indicates that the same stimulus occasionally produces increased activity in all three neurons that moves the system toward an attractor for a different behavior: swimming. Again, whenever the activity of the three neurons gets within this region, it stays there, so that swimming tends to be self-sustaining once begun. In this dynamical view of behavioral choice, all three neurons are helping to make both decisions, so the designation of 'command neuron for behavior X' is inappropriate. Instead, all the neurons are *multifunctional*: they are active during more than one behavior. In addition, consider the action of cell z: it shows the same activity in both behaviors. Whether

cell z activity leads to shortening or swimming depends on the context – on what cells x and y are doing. This dependence on context means that the decision-making process is *combinatorial*: which behavior occurs depends upon the *combination* of neurons that are active, not upon which particular one turns on and which are turned off.

In a real animal, there are always more than two behavioral possibilities; animals can produce several qualitatively different responses to similar – or even identical – stimuli. (Consider your response to a tall glass of water when you have just finished a long run, when you have just downed a 32-ounce soda, or when you find it standing next to your favorite microbrew.) Both mechanisms of behavioral choice are readily expanded to handle multiple behaviors. In the 'inhibition between reflex pathways' model, each additional behavior has its own dedicated group of command neurons that inhibit the command neurons for all the other behaviors. In the 'dynamical system' model, adding behaviors would correspond to increasing the number of attractor regions in the activity space. (With many neurons, the activity space gets very large and difficult to picture, but the mathematics accommodates many attractors.) However, no matter how complex they get, the two kinds of strategies can be distinguished by the same experimental procedure: are decision-making neurons that are active during one behavior inhibited during all the dominating behaviors in the hierarchy? If so, the neuronal system is likely to be more like inhibition among reflexive pathways (**Figure 3(a)**); if not, the neuronal system would be better described as multiple attractor states in a dynamical system (**Figure 3(b)**). We have looked at interaction among three sets of behaviors in leeches and found evidence for both kinds of systems.

Feeding Inhibits Other Behaviors

A very hungry medicinal leech can ingest over ten times its body weight in a single meal. During that meal, it will ignore mechanosensory inputs from touches, from pinches, and even from normally painful stimuli. (A feeding leech will continue to feed – it will actually feed longer – after being cut in half!) Remarkably, recording from identified neurons in the back half of the animal as the front half sucks blood showed that none of the command neurons for any of the behaviors appear to be either inhibited or excited during feeding. This suggests that neither mechanism represented in **Figure 3** could explain

how feeding blocks any response to mechanical stimuli. Instead, it turns out that during feeding, presynaptic inhibition chokes off the input from mechanosensory neurons: the presynaptic terminals of these neurons that would otherwise initiate local bending, swimming, crawling, or shortening are kept from releasing their excitatory transmitter onto their target neurons. In effect, this mechanism is very similar to the one shown in **Figure 3(a)**, except that the inhibition from the feeding circuit is directed at the presynaptic terminals of the sensory receptors, rather than at the command neurons for the other behaviors. This is a coarse, but effective, strategy: the leech shuts off all mechanosensory input from the body by acting on a single cell type, the mechanosensory neurons, thus making itself numb.

Shortening Inhibits Swimming

Swimming is initiated and produced by at least five levels of neurons (**Figure 4(a)**): *mechanosensory neurons* (touch, pressure, and pain receptors) activate *trigger neurons* in the subesophageal ganglion, which excite segmental *gating neurons* in the segmental ganglia, which excite – and are

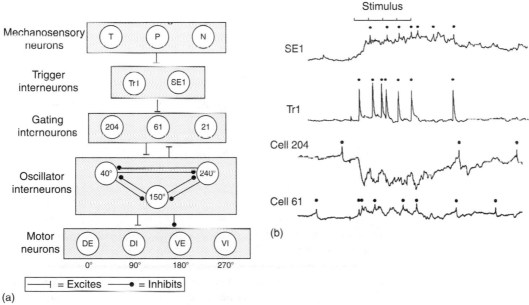

Figure 4 Shortening inhibits swimming, but not by inhibiting all swimming 'command' neurons. (a) The hierarchical circuitry that produces swimming in the leech. The relevant *sensory* neurons are *mechanosensory neurons* in the posterior end of the body that respond to touch (T), pressure (P), and pain (N). The *trigger neurons*, SE1 (swim excitor #1) and Tr1 (trigger neuron #1), are two pairs of neurons found only in the subesophageal ganglion. In many of the segmental ganglia, there are two pairs of *gating neurons* (of which cell 61 is one) and an unpaired one (cell 204). The *pattern-generating* neurons consist of 17 identified neurons whose interconnections produce bursts of impulses when the circuit is activated. The *motor neurons* burst in four different phases (roughly 0°, 90°, 180°, and 270°), in a cycle that repeats every 0.4–1.0 s. The motor neurons are named according to the types of longitudinal muscles they innervate (D = dorsal, V = ventral) and whether they excite (E) or inhibit (I) the muscles. Lines with T-junctions represent excitatory connections and lines ending in filled circles represent inhibitory connections. (b) Intracellular recordings from two trigger neurons (SE1 and Tr1) and two gating neurons (cells 204 and 61) while stimulating the skin at the anterior end of a leech with five pulses at times indicated by the 'stimulus' bar. (Note: The recordings were obtained sequentially; the four traces have been aligned based on the timing of the stimuli delivered during each recording.) Because the action potentials are of different amplitudes and shapes in different neurons, each action potential is identified by a dot above it.

excited by – rhythmic *pattern-generating neurons* that activate appropriate motor neurons in the characteristic swim motor pattern. By analogy with the general circuitry of **Figure 3(a)**, the trigger neurons serve both as sensory processors and as command neurons, whereas the gating neurons have an entirely command-like function. The reason why both trigger and gating neurons are called 'command neurons' is that stimulating any single trigger or gating neuron activates the swimming motor pattern. The functional difference between them is that a short (1 s or less) burst of impulses in a trigger neuron will activate the motor pattern for tens of seconds, whereas depolarization of gating neurons must be maintained to keep the swimming motor pattern active.

To determine whether shortening overrides swimming by inhibiting these command neurons, we delivered a stimulus to the leech's front end that reliably produced shortening while recording activity from each of two trigger and two gating neurons in turn. Surprisingly, three of these four command neurons for swimming were *excited* by the stimulus that elicited shortening (**Figure 4(b)**); the one exception, cell 204, was indeed inhibited (third trace). This pattern means that three of the four neurons that on their

own elicit swimming are also active during shortening, a behavior that strongly suppresses swimming. This result strongly suggests that these decision-making neurons are used in different combinations to produce different behaviors, that is, they use a *combinatorial code* that is more like a dynamical system (**Figure 3(b)**) than it is like the inhibition between reflexive pathways (**Figure 3(a)**).

Swimming and Crawling Have Subtle Interactions

Use of voltage-sensitive dyes to record simultaneously from many of the neurons in a midbody ganglion showed that the neurons that are rhythmically active during swimming are largely a subset of the neurons that are rhythmically active during crawling. When a stimulus was presented that sometimes produced swimming and at other times produced crawling, a small group of neurons were found to be active in one pattern just before swimming and in a different pattern just before crawling ('early individual discriminators' in **Figure 5**). Another small group of neurons had activity patterns that co-varied (i.e., all of them varied in a similar way) in one way before swimming and in another

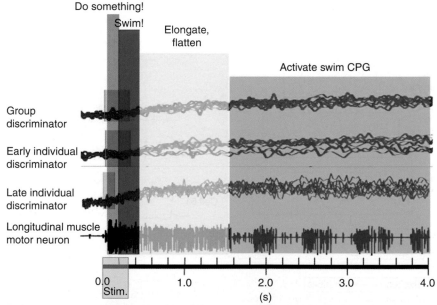

Figure 5   Summary of neuron activity patterns as an isolated leech nerve cord chooses to perform swimming. The lowest of the four traces shows a recording from a peripheral nerve extending from a midbody ganglion. The largest action potentials in this recording are generated by DE-3, an excitatory motor neuron that innervates dorsal longitudinal muscle fibers. In response to electrical stimulation (gray rectangle labeled 'stim') of a more posterior nerve, the activity of DE-3 increases for about 1.5 s, then begins to produce bursts at nearly 2 s⁻¹; these bursts are the swimming motor pattern. Each of the upper three sets of traces are optical recordings from three different neurons. Each set shows nine overlapping recordings: in four of the recordings (blue traces), the nerve cord produced the same kind of swimming motor program shown in the bottom trace; in the other five recordings (blue traces), the nerve cord produced the crawling motor pattern. The top trace ('group discriminator') is from a neuron that produced patterns of activation that were indistinguishable individually, but that were correlated with activity in a small number of other neurons during the time highlighted in red. The other traces are from neurons whose activity patterns were different early (during the time highlighted in yellow) or late (the time highlighted in green). These types of neurons are called 'early individual discriminators' and 'late individual discriminators.' The initial period of increased DE-3 activity, highlighted in violet, may represent preparation for definitive movement. The words above the highlighted areas are explained in the text.

way before crawling, allowing the experimenter to reliably predict, on the basis of the pattern of covariation, whether the response would be swimming or crawling ('group discriminators' in **Figure 5**). When the activity pattern of individual neurons was changed by depolarizing or hyperpolarizing one neuron at a time, the only trials in which the behavioral response could be changed were those in which the test neuron was among the 'group discriminators.' Surprisingly, the previously identified command neurons were either in the 'early individual discriminator' group, or in an entirely separate 'late individual discriminator' population. We conclude that the group discriminators decide between swimming or crawling during the time indicated by red in **Figure 5**, and then the animal prepares to do just one of these behaviors (by elongating and flattening, e.g., when the decision is to swim). When the motor pattern actually begins (green region), the 'late individual discriminators' are activated; these include gating neurons, pattern-generators, and motor neurons.

Notice that there is motor neuronal activity before the decision is detected (violet region in **Figure 5**). This activity could represent a generalized decision, maybe something like 'stimulus detected, prepare to do something!' Alternatively, this activity might indicate that the swim/crawl decision has been made but by neurons, or by neuronal activity patterns, that have not yet been found. In any case, these findings strongly suggest that the swimming versus crawling decision is made by a dynamical mechanism (**Figure 3(a)**) rather than by competition among command neurons each of which is dedicated to a single behavior.

Summary and Conclusions

The hypothesis that inhibition among command neurons produces behavioral choice was first proposed to explain results from experiments on molluscs, and it is currently proposed to be the way that primates decide which direction to move their eyes. The evidence for such a mechanism is strongest in slugs, although the decision-making neurons tend to be multifunctional in these animals, as they are in leeches. In the three different leech behavioral choices described in the previous section, competition between reflexive pathways was found only during feeding, which effectively renders the leech numb to mechanosensory input. This mechanism is a generalized, sledge-hammer strategy, whereby one behavior turns everything else off. Shortening is a bit more subtle: it turns off locomotory responses (leeches stop swimming and crawling when touched on the front end), but does not turn off responses to food or to pain. In this case, some

of the same cells that activate swimming are also activated during shortening. The decision to swim or to crawl is even more subtle: essentially all the neurons active in crawling are active during swimming, and the decision between these behaviors is made by subtle variations in the correlated activity of a group of neurons. In fact, crawling can turn into swimming, and leeches can execute a combined crawling/swimming behavior.

All the behaviors described in this article are for well-fed, alert leeches in a laboratory setting. Many factors can modify the interactions described. For instance, a very hungry leech is very active, even with no apparent external stimulation, and it quickly recovers from a shortening response to investigate a pinch to its anterior end, seemingly to determine whether there might be a food source nearby. A well-sated leech, on the other hand, is quite lethargic and cannot be provoked by any mechanical stimulation to swim (instead, it crawls or shortens to stimuli that previously elicited swimming). In addition, age and reproductive state greatly influence behaviors: a hatchling leech responds to stimuli in a manner very different from that of an adult, and adults respond differently when they are ready to reproduce. Understanding the basic mechanisms of behavioral choice will help to determine how these choices can be modified by factors such as the age and behavioral state of an animal.

See also: Consensus Decisions; Decision-Making: Foraging; Neuroethology: Methods; Rational Choice Behavior: Definitions and Evidence.

Further Reading

Briggman KL, Abarbanel HDI, and Kristan WB Jr (2005) Optical Imaging of neuronal populations during decision-making. *Science* 307: 896–901.

Briggman KL and Kristan WB Jr (2008) Multifunctional pattern generating circuits. *Annual Review of Neuroscience* 31: 271–294.

Esch TE and Kristan WB Jr (2002) Decision-making in the leech nervous system. *Integrative and Comparative Biology* 42: 716–724.

Gaudry Q and Kristan WB Jr (2009) Behavioral choice by presynaptic inhibition of tactile sensory terminals. *Nature Neuroscience* 12: 1450–1457.

Glimcher P (2003) *Decisions, Uncertainty and the Brain: The Science of Neuroeconomics.* Cambridge, MA: MIT Press.

Gold JI and Shadlen MN (2007) The neural basis of decision making. *Annual Review of Neuroscience* 30: 535–574.

Kristan WB Jr, Calabrese RL, and Friesen WO (2005) Neuronal basis of leech behaviors. *Progress in Neurobiology* 76: 279–327.

Kristan WB and Gillette R (2007) Decision-making in small neuronal networks. In: North G and Greenspan R (eds.) *Invertebrate Neurobiology,* pp. 533–554. New York, NY: Cold Spring Harbor Laboratory Press.

Tinbergen N (1951) *The Study of Instinct.* Oxford: Clarendon Press.

Wang X-J (2008) Decision making in recurrent neuronal circuits. *Neuron* 60: 215–234.

Levels of Selection

M. M. Patten, Museum of Comparative Zoology, Cambridge, MA, USA

Causal Explanations and Adaptive Evolution

Rabbits are fast. Let us assume that at some time in the evolutionary past, they were not all that fast. We then have an observation that the rabbit population has evolved increased speed. With that observation in hand, we might then wish to tell a causal story that can account for this change in the average speed from then until now. Reconstructing the causal story of evolutionary change is one of the goals of evolutionary biology.

A first attempt at a causal story for rabbit evolution might run like this: proto-rabbits endowed with greater speed had greater success at escaping predators and thus left more offspring. Each successive offspring generation was therefore enriched for speediness because in the previous generation, the slowest rabbits were the most likely to be captured by predators and the least likely to leave descendants. Due to the reliable inheritance of speed from parents to offspring, the average speed of the offspring generation improves on that of the parental population.

This causal story should look familiar to any student of biology as evolution by natural selection: Traits vary; trait variation is heritable; traits determine their representation in future generations because of a statistical association with fitness. In this example, care is taken to place phenotypic variation at the level of the individual: some individual rabbits are faster than others. But in the depiction of evolution by natural selection offered in this article, there is no requirement that traits or fitness be measured on individuals. The genes within individuals are free to vary in their phenotypic properties and their fitness, as are the groups to which individuals belong.

In the next sections, how selection may act at these different levels is described: alleles within organisms (gene level or genic selection); organisms within groups (individual or organismal selection); and between groups (group selection). The possibility that different levels of selection may contribute to evolution has been granted (though at times debated) for well over a century but a complete theoretical understanding of how each of these levels contributes to evolution has taken almost this whole time to solidify. Our evolving understanding of the levels of selection concept is summarized and our current – and we hope mostly complete – understanding of this topic is presented. The article begins by providing some useful definitions and clarifications. Then it turns to an examination of each level of selection. A special section on kin selection is provided, as this is one of the most important theoretical concepts in behavioral ecology as well as one of the biggest stumbling blocks for students of that topic.

The Selfish Gene, the Gene's Eye View, and Careful Accounting

Dawkins, in *The Selfish Gene*, advises where to keep our focus when thinking about any evolutionary event or process. He suggests that we focus on the gene and take the gene's eye view. Why should we focus our attention on the gene? The reason is simple. Fitness is essentially a measure of the persistence of something through time and the only 'somethings' that persist through time in nature are genes – but not the material versions of the genes, of course. Semiconservative DNA replication ensures that ten generations from now, the material version of any gene – the bases and the sugar-phosphate backbone – will be almost completely diluted away (it would appear in $<0.1\%$ of future copies). What persists is the information encoded by such a gene. Such information is persistent through time, across generations and across rounds of semiconservative DNA replication, and it is the change in the informational content of populations with which any measure of evolution must be concerned. Individuals come and go and populations come and go, but information is that special kind of something that persists. By focusing our attention on genes and the information they contain, we ensure that our accounting of evolution is sensible. Throughout this article, then, the focus is kept on how the informational content of populations is changed by referring to gene/allele frequency change.

Our unit of selection will thus be gene/allele frequencies or informational content, used interchangeably. It is worth taking care to explain what is meant by unit. One of the debates that has plagued multilevel selection theory stems from the many ways in which the word 'unit' can be employed in evolutionary reasoning. Here, unit is taken to mean the standard of measurement (e.g., height is given in units of inches); this is the thing that we are keeping a tally of in each generation. Unit is not meant as an organizational concept (e.g., the army is subdivided into different units: cavalry, infantry, tank platoon, etc.). When the question is asked, "What is the unit of selection?", either notion of unit may be employed as an answer, but we run the risk of talking at cross-purposes and creating needless confusion and debate. As long as we are explicit with our choice, we avoid such problems. For instance, a paleontologist may keep track of the relative proportion of kinds of species through time, whereas a population geneticist may score relative allele frequencies through time. If both state that body size is increasing through time in their studies, it is clear that the former is referring to the frequency of larger species increasing, while the latter is referring to an increase in the frequency of alleles giving rise to large body size. On the other hand, when we are concerned more with the organizational unit that is experiencing selection and is causally responsible for evolutionary change, it is best to use a word different from "unit." 'Target,' 'vehicle,' and 'interactor' are all terms that have been introduced, somewhat interchangeably, to take the place of unit in these discussions. In the body size example, we cannot tell which unit of selection is responsible for increased average size. There are a number of causal explanations available to both the paleontologist and the population geneticist. Fruitful debates stem from the question of which organizational level is the target of natural selection. Indeed, the debate about levels of selection is primarily centered on this one question.

One source of confusion surrounding the gene's eye view of evolution is to misread Dawkins as saying that selection could not act at any level other than the gene. This is certainly not Dawkins' point and the confusion stems from conflating units and targets of selection. Selection occurs at a level if variation in a trait at that level causes fitness variation among the underlying genes, but ultimately, it is the alleles that are changing in frequency. In other words, the information stored in a gene may produce material that causes fitness differences and this material may be proteins, individual bodies, or groups of bodies (it may even be artifacts – though that is beyond the scope of this article). The information in a gene informs the construction of these material objects and it is these material objects that vary in their gene replicating ability. Thus, looking back at a past evolutionary change, we are faced with the question of why the informational content of the population was so modified. We are forced to ask whether the differential fitness of alleles was a consequence of individual differences, gene level differences, or group differences. One can adopt the gene as the unit of selection but may allow selection to act at any level. One's choice of the units that measure selective change does not commit one to ascribe the selective cause of change only to that level of the hierarchy.

With gene/allele frequency or informational content as our unit and these other potential stumbling blocks out of the way, we can now say that a 'Levels of Selection' understanding of evolution simply means the following: the informational content of populations can change as a consequence of gene level, or individual level, or group level trait variation (or some combination of these),

provided the traits at each level have a statistical association with the persistence of the information that informed their construction and a heritable genetic basis. This statement can be used as a template to describe any evolutionary change.

We can reexamine the rabbit example from this new perspective. We would say that there was genetic (informational) variation associated with speed variation among individuals and due to differential survival and reproduction of slower and faster individuals, the relative frequencies of alleles were changed. Alleles build rabbits that are better or worse suited for surviving and making offspring and thereby these alleles find themselves more or less likely to persist.

We will see in later sections, though, that genes have other avenues of causing differential persistence of information besides building individual level traits. They may build groups of individuals that together give rise to differential persistence of information. Likewise, they may build proteins that act within bodies to cause differential persistence of alleles. The variation in the information in genes causes variation in traits, which causes fitness variation, which brings about evolutionary change of the something that we are concerned with accounting for: the informational content of populations.

Gene-Level Selection

As mentioned earlier, it is theoretically permissible for selection to act on the effects of genes at the intraindividual level. Typically, we may think that genes encode proteins that serve some function in the body. But genes that are selected for their intraindividual effects usually produce proteins or other molecules that contribute nothing to bodily function; instead, they make proteins or other molecules that ensure the preferential transmission of the allele that encodes them. For instance, a heterozygous individual is a fifty-fifty mix of alleles, but in certain heterozygotes, instead of transmitting a fifty-fifty mix of those alleles to the next generation, they transmit more of one allele than the other. This is called segregation distortion or meiotic drive. Within that individual, there are fitness differences for the two alleles. One allele produces a protein that allows it to out-compete the other during meiosis. When summed over all heterozygotes, this contributes to allele frequency change in the population.

Gene-level selection likely has little impact on the evolution of animal behaviors, generally speaking. Most selection in the evolution of animal behavior occurs at higher levels. However, gene level selection is worth noting, because many of the selection pressures faced by genes within genomes are analogous to the selection pressures faced by individuals within social groups, as Maynard Smith, Haig, and other authors have described.

A genome is an assemblage of genes that, together, produce a body. We can think of the genes collaborating socially to produce a body for mutual benefit, which is shared among them equally. However, there are rogue alleles that contribute nothing to the functioning of the body and therefore nothing to the success of the genes in the genome. They instead free-ride on the efforts of all the other genes. Thus, they ensure their persistence in the population by out-competing other alleles within the body for transmission.

Individual-Level Selection

Selection at the level of the individual is the easiest to grasp – though it is just as easily misunderstood – because our senses are tuned to detect and emphasize individual differences. However, this perceptual bias may also make it easier to misattribute evolution to individual selection when some other process, like group selection, is actually responsible. This is the first of the common misunderstandings of selection at this level. Some careful language will help avoid this mistake: instead of individual level selection, perhaps it is better to say 'within-group' selection. This linguistic trick helps avoid attributing group level adaptations and group level effects to 'individual' level selection.

To illustrate how such misattribution works, we can use an example that gives two equivalent formulations of the same scenario. Suppose we know that the fitness of an individual is correlated with both its own phenotype and the phenotype of the group to which it belongs. In other words, we know that both individual and group levels of selection are at work. In the first formulation, we can then say that the fitness of a given type of individual is the sum of these two contributions. In the second formulation, we average the fitness of a type of individual across all of the different groups in which it is found and say that this average is the fitness of that type of individual. The first formulation of evolutionary change is causally correct. The second formulation, with its averaging, obscures the true causal pathways. However, because they are mathematically equivalent, both formulations return the same prediction for the amount of change delivered in one generation. But one invokes group selection; the other does not. One gets the causal story right; the other does not. For whatever reason, there is a tendency to overlook the influence of group selection on evolution and to formulate evolution in the second manner, such that individual selection is the sole force for change and group selection evaporates.

The second common misunderstanding surrounding individual level selection is to assume that what is good for individuals is necessarily good for groups. The reasoning might go as follows: since individual selection concentrates the traits that have the highest fitness, then after repeated bouts of individual selection, we should expect our group to be composed of fitter individuals, thus raising the average fitness of the group. Continuing with this reasoning, we find that the average adaptedness of individuals increases and the adaptedness of the group as a whole is therefore increased. It is this last connection that proves flimsy and the mistaken reasoning that gives rise to it can again be avoided by using 'within-group' selection. Within-group selection will ensure that the individuals with the highest relative fitness within a group contribute more of the genetic endowment of subsequent generations. But this says nothing at all about whether groups composed of these individuals will have higher fitness or not. Selection at the level of individuals and selection at the level of groups can be independent of one another. Conflicts between different levels of selection are possible. What makes individuals most adapted to their environment is not necessarily what makes groups most adapted. Overall group fitness may decline as a consequence of selection at the level of the individual. An analog exists at a lower level of the hierarchy. Selfish genetic elements, which are selected for within individuals, may lower the fitness of the bodies that bear them, creating a conflict between different levels over what is selectively favored. Nowhere is the possible conflict between levels of selection more apparent than in the mating arena, a common area of inquiry for animal behavior studies. For instance, traits that are favored at the individual level in males may reduce group fitness. As we see at all adjacent pairs of levels in the hierarchy, selection at one level does not give us any information about what selection at any other level does. Indeed, one favored mathematical approach to multilevel selection, the Price equation, shows how the different levels may be partitioned into separate terms that are then summed in order to account for the total evolutionary change in a population.

Group-Level Selection

Group selection is occurring when the replication of genes differs because of differences between groups that those genes give rise to. This definition permits an understanding of group selection in several ways. The failure to recognize that group selection can be understood and modeled in different ways is likely the source of much of the confusion (and controversy) surrounding group selection. First, the history of this controversy is discussed before delving into a more careful explanation of what group selection is (or, what the different notions of group selection are).

History of an Idea

The proposition that groups may be targets of selection goes back to Darwin. In *The Descent of Man, and Selection in*

Relation to Sex, Darwin suggested that groups of humans may have experienced differential survival and extinction based on their differing morality. In the first half of the twentieth century, the major authors of the Modern Synthesis, perhaps surprisingly, did not contribute anything substantial to a theoretical understanding of group selection. The one lasting contribution by any of the founders of the Modern Synthesis may be Haldane's quip about giving his life to save at least two drowning brothers or eight drowning cousins, foreshadowing the later development of kin selection in the 1960s. Through the middle part of the twentieth century and into the 1960s, the picture of group selection that emerged was one of adaptations for the good of the species, championed by Wynne-Edwards (1962). This picture was challenged convincingly by Williams in his 1966 book *Adaptation and Natural Selection.* But his challenge, coupled with Hamilton's (1963) development of kin selection (a seeming alternative to group selection) in the same decade, may have unintentionally created the idea that group selection in any form was a theoretical impossibility and therefore of perhaps only minor significance in evolutionary history and theory, if at all. In the 1970s, some theoretical developments rescued the group selection baby from the discarded bathwater, so to speak. Some of the zeal for kin selection as an alternative to group selection gave way to an understanding of their theoretical equivalence, beginning with Hamilton's adoption of Price's (1972) approach to modeling evolution. Additionally in the 1970s, David Sloan Wilson's (1975) development of trait group selection offered a different formulation of group selection, one that was not as restrictive as had been dismissed by the critical works of the 1960s. In the 1980s and 1990s, group selection thinking contributed advances to the evolution of individuality, multicellularity, sexual reproduction, macroevolutionary theory, and major transitions in hereditary organization. Most recently, the problems of cooperation and altruism, particularly in humans, have reignited an interest in multilevel selection theory.

Different Senses of Group Selection

An early, and much derided, vision of group selection posited that the design of organisms was that which best suited the group. In other words, adaptations – typically, behavioral ones – occur because they increase the persistence of the species, buttressing them from extinction. This view was championed by Wynne-Edwards (1962), though he was not the only author with this viewpoint. This is the notion of group selection with perhaps the worst reputation. Though it is not theoretically impossible for nature to provide a compelling example of this notion of group selection – and indeed studies of microorganism behaviors may hold promise for such group selected adaptations – this framework of group selection has received very little empirical support.

Another take on group selection allows that different groups have different probabilities of going extinct (an analog of dying) or of duplicating (an analog of reproducing) and that evolutionary change can be driven by these two processes alone. Groups have properties, groups have fitness variation (measured in terms of duplication/extinction), and provided the properties of groups are heritable, then this may cause changes in the frequencies of different types of groups. This take on group selection reflects the currently held picture of species selection that some macroevolutionary theorists prefer. Notice here that the unit of accounting is not the gene but rather the group. Species selectionists and other group selectionists who adopt this notion keep track of the frequency of different kinds of groups rather than the frequency of different kinds of genetic information. Under certain restrictive assumptions, the two accounting approaches give equivalent predictions.

Yet another view of group selection requires neither that adaptations be for the group's persistence nor that groups show differential likelihoods of replicating. This view merely states that there are features of groups that contribute to differential reproduction of genetic information. If this genetic information is responsible for these group features, then the informational content of a population can so evolve. In this view of group selection, the fitness of an allele is partitioned as being due to a within-group effect and a between-group effect. This second effect may capture differences of group productivity or it may simply capture a contextual effect of a group level property on individual reproductive success. Regardless of what this partition is capturing, we can see that it accounts for differential allele replication as a consequence of both individual- and group-level differences.

What Is Kin Selection?

Kin selection is best understood by taking a gene's eye view (which, we should be reminded, is not a commitment to invoking selection at any particular level), though it is most accurately described as a form of group selection. Although mathematically it is possible – and even sometimes heuristically invaluable – to make all fitness variation under kin selection a property of genes or of individuals, this obscures the true causal forces that bring about gene frequency change under kin selection. Kin selection is a way of understanding allele frequency change as a consequence of the actions and interactions amongst individuals who share alleles by recent common descent, that is, kin. As with group selection, it is a consequence of the properties of groups that cause allele frequency change; with kin selection, though, the groups have this special genetic structure.

Kin selection has been used to explain the evolution of cooperation and altruism in animal societies. The evolution of altruistic traits, which is opposed within groups but favored between groups, is facilitated by close kinship within groups. The within-group fitness losses that altruists typically experience are partially offset by the fitness gains of kin who share the same genetic information. In this way, the genes that control behavior can recoup the fitness losses of the donors of altruistic actions. Hamilton specified a useful rule for altruistic acts such as these that determines whether such behaviors are evolutionarily favored: $rb > c$. That is, if the benefits (b) conferred on kin, weighted by the relatedness (r) of the donor to the recipient, is greater than the cost (c) conferred on the donor, then such an action is favored by natural selection (this also puts Haldane's quip on firm theoretical ground).

Hamilton introduced a method of accounting for kin selection called inclusive fitness that assigned all of the fitness effects of an allele to the individual bearer of that allele. This is why kin selection is often used – misleadingly – as an individual-level alternative to group selection. A different way of accounting for all of the effects of genes is the direct fitness approach, which again measures individual fitness variation. This approach accounts for all of the social effects on a gene in a focal individual rather than on the effects of the gene in a focal individual on others. These are two different ways of accounting, both using the individual (or the gene) as the bearer of the fitness variation, both sometimes employed to show that group selection does not actually exist. But both are mathematical instruments that cannot tell us about the causal influences on evolutionary change. Inclusive fitness and direct fitness lead to correct predictions about the direction of evolution under kin selection but obscure the true causal story, which in both cases is, at least in part, group selection.

Conclusion

Rabbits are fast, but proto-rabbits were not all that fast. It may be that alleles for speed outcompeted alleles for sluggishness at meiosis; this would be a gene-level selection explanation. It may be that faster individuals left more descendants than slow individuals; this would be an individual-level selection explanation. Perhaps groups with higher speed were fitter than slower groups; this is one type of group-level selection explanation. Or speed could be an altruistic trait – the benefit of which was directed preferentially towards kin; this is a kin-selection explanation. The levels of selection offer all of these theoretical explanations for the cause of evolutionary change in rabbits and elsewhere, as this article has shown.

See also: Cooperation and Sociality; Evolution: Fundamentals; Microevolution and Macroevolution in Behavior; Morality and Evolution; Social Selection, Sexual Selection, and Sexual Conflict.

Further Reading

Dawkins R (1976) *The Selfish Gene*. Oxford: Oxford University Press.

Haig D (1994) The social gene. In: Krebs JR and Davies NB (eds.) *Behavioural Ecology,* 4th edn., pp. 284–304. Oxford: Blackwell Scientific.

Hamilton WD (1963) The evolution of altruistic behavior. *American Naturalist* 97: 354–356.

Hamilton WD (1998) *Narrow Roads of Gene Land, Vol. 1: Evolution of Social Behaviour*. Oxford: Oxford University Press.

Lewontin RC (1970) The units of selection. *Annual Review of Ecology and Systematics* 1: 1–18.

Lloyd E (2001) Units and levels of selection: An anatomy of the units of selection debate. In: Singh RS, Krimbas CB, Paul DB, and Beatty J (eds.) *Thinking About Evolution: Historical, Philosophical, and Political Perspectives*, vol. 2, pp. 267–291. Cambridge, UK: Cambridge University Press.

Maynard Smith J (1964) Group selection and kin selection. *Nature* 201: 1145–1147.

Maynard Smith J (1976) Group selection. *Quarterly Review of Biology* 51: 277–283.

Okasha S (2006) *Evolution and the Levels of Selection*. Oxford: Oxford University Press.

Price GR (1972) Extension of covariance selection mathematics. *Annals of Human Genetics* 35: 485–490.

Sober E (1984) *The Nature of Selection*. Cambridge, MA: MIT Press.

Williams GC (1966) *Adaptation and Natural Selection*. Princeton, NJ: Princeton University Press.

Wilson DS (1975) A theory of group selection. *Proceedings of the Natural Academy of Sciences of the United States of America* 72: 143–146.

Wilson DS and Wilson EO (2007) Rethinking the theoretical foundations of sociobiology. *Quarterly Review of Biology* 82: 327–348.

Wynne-Edwards VC (1962) *Animal Dispersion in Relation to Social Behaviour*. Edinburgh: Oliver & Boyd.

Life Histories and Network Function

T. G. Manno, Auburn University, Auburn, AL, USA

Introduction

When animals aggregate, they form complex social relationships and structures via social interaction. For instance, discrete social groups or coalitions may result from the association of individuals, providing the basis for the evolution of cooperative behavior via kin selection. Amicable interactions may also allow individuals to become familiar with appropriate breeding partners and to select from among them. A dominance hierarchy is another example of a social relationship that can be formed with agonistic interactions among individuals in an aggregation. By studying the structure of social relationships and interactions, we can understand the causes and consequences of sociality and the role of interaction in shaping the evolution of sociality more thoroughly.

These relationships and interactions can be described by a 'social network' that models the entire system and the way individuals are interconnected. By providing information about individual group members and the entire group, as well as direct and indirect interactions, social network analysis (SNA) offers an alternative definition of animal social groups by visualizing relationships among individuals by collecting and mapping relational data that are organized into a matrix. Historically used to analyze human political connections, food webs, and metabolic pathways, the SNA approach has recently undergone a phoenix-like transformation as a result of animal behavior studies that use software programs to analyze social interaction and colony substructure. For animal studies, SNA examines social behavior by defining groups of animals as collections of individual units that are connected by social ties (usually amicable or hostile social interactions). In the pages that follow, I summarize the history of the field, categorize overarching concepts, identify recent applications, introduce recent findings, and suggest materials for analysis.

Origin and Development of Social Network Theory

The metaphor of a 'social network' has been used for over a century to describe complex sets of relationships between members of social systems. Precursors of social networks were proposed by É. Durkheim and F. Tönnies (c. 1890) when they maintained that social groups could exist as personal and direct social ties and that social phenomena arise when interacting individuals constitute a reality that can no longer be accounted for in terms of the properties of the individuals. G. Simmel began to write about the nature of network size and social interaction, and three schools of thought in network analysis followed – analysis of social interaction in small groups such as classrooms and work groups, interpersonal relations in the workplace, and systematic study of networks (c. 1930).

In the 1950s and 1960s, SNA developed further in Britain following E. Bott's kinship studies and the community network research of M. Gluckman and J. C. Mitchell. S. Milgram then provided evidence of a human social network by stating that any two people in a portion of the US population could be connected via six acquaintances. The subsequent work of S. Borgatti, K. Faust, S. Freeman and L. Freeman, and S. Wasserman, among others, expanded the use of social networks and initiated the availability of SNA to animal behaviorists.

Public interest in SNA studies, including those involving animals, is now widespread with the advent of human social networking tools such as the Internet (e.g., MySpace, Facebook), and unintentional applications to popular culture such as the Kevin Bacon game (where college students in Pennsylvania connected Bacon to other Hollywood colleagues via common film appearances with the same degrees of separation as the Milgram study). The public is also interested in recent findings suggesting that animals form social structures and interact in a similar way to humans.

Concepts and Techniques

Individual Measures

SNA examines behaviors by defining animal social groups as collections of individual units connected by social *ties* (also known as '*edges*'). SNA describes the position of a particular individual in a group, called a *node* (or *vertex*) and the effect that individual has on others. The number of direct ties that a focal individual has to other individuals is its node *degree*, and the ties are shown on a *sociomatrix*, which is translated into a network. For affiliative networks, direct ties are usually the result of amicable social interaction between individuals (e.g., copulation, anal sniffing), and individuals with high degree have many 'friends.' For agonistic networks, direct ties are the result of hostile interactions (e.g., territorial displays,

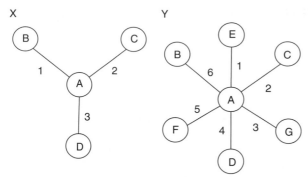

Figure 1   Representation of two theoretical networks, which could be affiliative or agonistic, visualized from sociomatrices *M* and *N*. The networks *X* and *Y* contain four and seven nodes, respectively (represented by circles labeled with letters), as well as 3 and 6 ties, respectively (represented with solid black lines labeled with numbers). In network *X*, node *A* has a degree of 3 because it has ties with nodes *B*, *C*, and *D*. This is in contrast to node *A* in network *Y*, which has a degree of 6 because it has ties with nodes *B*, *C*, *D*, *E*, *F*, and *G*.

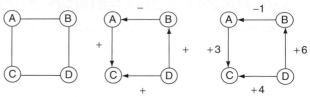

Figure 2   Representation of three networks. The first is unweighted and undirected, and all members have a degree of 2. The second is a directed network, with node *D* having an outdegree of 2, for example, because of the 2 ties emanating from *D* to other nodes. Node *A* has an outdegree of 1 because of 1 tie emanating from *A* to another node, and an indegree of 1 because of 1 tie emanating from another node to node *A*. Since the local component of sociometric net status equals difference between indegree and outdegree, *D* has a net status of 2 and *A* has a net status of 1. The final network is a weighted network, with the strength of the connections (e.g., number of social interactions, number of times sighted together) denoted by a numerical value.

fighting), and individuals with high degree have many 'enemies' (**Figure 1**). Animals with a high degree obtain both the benefits and costs of being socially connected. For example, ground squirrels with a high degree have a better selection of copulatory partners but are probably more likely to transmit infectious diseases. Likewise, in an agonistic network such as a primate aggregation, a high-degree individual may be more likely to form coalitions, but will suffer more injuries as the result of increased aggression with other individuals.

Networks may be *directed* or *nondirected*. *Indegree* reflects the number of relationships in which an individual is the receptor of the relationship; *outdegree* is the number of ties originating from an animal. The difference between indegree and outdegree, which can be a positive or negative number, is the local component of *sociometric net status* (**Figure 2**). For affiliative and agonistic networks, these measures can have different implications. In ground squirrels, a male in an agonistic network with high outdegree likely initiates hostile interactions in a particular area in an attempt to defend it as a territory, and a low-outdegree male may not defend territory. In an affiliative network, a male ground squirrel with high outdegree is probably a dominant individual in an area and looks for opportunities to mate, and a low-outdegree individual may be nonreproductive.

Networks may also be *weighted* or *unweighted*. A weighted network that reflects the strength of the relationships through the number of ties can be constructed, but usually networks are nonweighted and reflect a direct tie if the individuals involved interacted more than would be expected from random interactions. Models for random ties are usually simulated with computer software (see below; **Figure 2**).

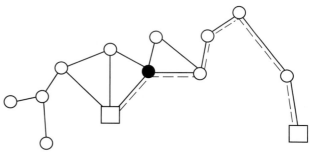

Figure 3   Partial social network of Columbian ground squirrels in Alberta, Canada from a study by Manno (2008). Individual squirrels, or nodes, are represented by circles and squares and vary in degree from 0 to 4. Ties or edges are represented by solid black lines. The central node with degree 4 is shaded black, and also has the highest betweenness. The shortest path between the square nodes, illustrated by the dotted lines, takes four steps.

Centrality defines the structural importance of an individual in a group. *Betweenness centrality* is the number of shortest paths between every other pair of animals in the network on which a focal animal lies (**Figure 3**). Betweenness therefore indicates how important an animal is as a point of social connection. Thus, animal behaviorists using SNA often simulate removal of individuals with high betweenness centrality (called 'targeted removals') to test for *resiliency* of the network if it were to come under attack from some outside force that removed such individuals. The most notable natural examples of targeted removal are selective predation and trophy hunting, since these phenomena typically affect males in the act of breeding, and reproductive males often have high betweenness centrality because of their search for copulatory partners. A network that is resilient does not split during targeted removals of individuals and is therefore more likely to retain the social activity patterns in place

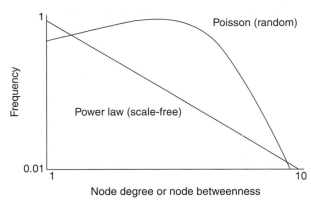

Figure 4 Comparison of the degree distributions produced by models of network formation shown on a log–log plot. Poisson, or random, networks have degree distributions characterized by a modal degree. Scale-free networks have preferential attachment which produces a power-law degree distribution that is linear.

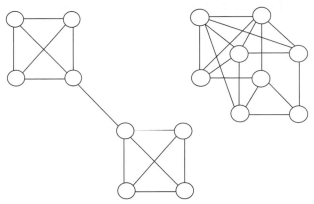

Figure 5 Two 8-node networks. The first has two distinct subgroups or cliques as told by an algorithm or a high clustering coefficient. The second has no distinct subgroups and low cliquishness and clustering.

after such a disaster than a network that is not resilient. The same analysis can be run by removing random individuals, to simulate natural phenomena that eliminate random individuals from a network (e.g., plague, infectious disease, random shooting by humans).

Animal networks have typically been found to possess a few individuals with high betweenness centrality, with most individuals having low betweenness centrality. Thus, for animals with affiliative social interactions such as whales, dolphins, and primates, it is usually only a few individuals that have a large role in the overall cohesion of the network and that dictate the resiliency of the network. Such networks are called '*scale-free*' networks, because when centrality and frequency of individuals with different levels of centrality are plotted on a log–log graph, a straight line with a *power-law* exponent fits the plotted points well. This is as opposed to a random network, where the graph will show a *Poisson* (curved) distribution of plotted points, and most individuals would have a more or less equal betweenness centrality and an equal role in the overall cohesion of the network (**Figure 4**).

Intermediate Measures

For animals such as prairie dogs, primates, and pikas, social groups or substructures within an aggregation may exhibit genetic or behavioral properties that facilitate the evolution of cooperation and group cohesion. The substructures may have different levels of exclusivity among members or different flexibility in the defense of a territory over time. These *subgroups* can be detected because SNA integrates measures that reflect the distribution of direct ties (amicable interactions) between animals and describe relationships beyond a single individual. *Cliquishness* describes to what extent the network is divided into cohesive subgroups (i.e., sets of nodes where each node

is directly tied to each other). The amount of *clustering* in an aggregation of animals is determined by a *coefficient* that quantifies the *density* of relationships among a focal animal's neighbors (i.e., the number of existing ties between neighbors divided by the maximal possible number of such ties). These measures can be compared with a predetermined number of random networks to determine whether a network is cliquish. Algorithms such as those proposed by Girvan and Newman can also be used to detect subgroups of connections within a network (**Figure 5**).

Group Measures

To describe aspects of overall or global network structure, SNA employs several measures. Diameter is the basic measure of *connectivity* in a network and is the longest path length in the network (*n* diameter means that no two individuals are more than n steps away from each other). The average of all path lengths between individuals (previously mentioned) also yields a general idea of a network's overall connectivity. Phenomena such as disease transmission and information transfer would become slower with a high number of paths than with a low number. Diameter was used by Milgram to show 'degrees of separation' between people and illustrate the 'small world phenomenon,' where random humans have a relatively small number of degrees of separation between them.

Cohesion is a more sophisticated way of measuring connectivity. One measure of cohesion is density, which is the number of ties present divided by the number of possible ties in the network (generally only calculated for unweighted networks). A group with higher density has more ties per individual than a group with lower density and therefore is theoretically more cohesive, and less likely to split up during targeted and random attacks. The idea of *transitivity* holds that if X has a relationship with Y, and Y has a relationship with Z, then X has a relationship with Z as well. Along these lines, *reciprocity* reflects how many of the relationships in a network are mutual, thus yielding an idea of how well balanced connections in the network are.

Technology

Computer software is the most popular visualization tool for animal behaviorists using SNA. UCINET, Netminer, and JUNG are convenient programs for calculating SNA measures and simulating random networks for comparisons with visualized networks. GUESS, Ora, Pajek, Netminer, and InFlow are programs typically used for business applications but can also be used by animal behaviorists. For Linux users, SocNetV is an available open source package. Data are usually entered in an Excel database as a sociomatrix, with all of the individuals in a group listed on the outside of rows and columns of the matrix, and the connections (weighted or unweighted) between individuals on the inside. While Excel is sufficient for taking and entering sociomatrix data, Onasurveys.com and Network Genie are also excellent tools for social network survey data collection and are adaptable for animal behavior investigations.

Recent Applications to Animal Societies

Although only recently applied to animal societies, SNA has helped to make breakthroughs in understanding the general principles that govern social life in simpler societies and that have extensions to our own social evolution. The scale-free pattern, for instance, is applicable to human societies and many other human-influenced entities such as the internet, molecular pathways, and terrorist networks. Research on animals, however, has already yielded several examples of scale-free networks, such as individuals in bottle-nosed dolphin social groups, ground squirrel colonies, freetail bat tree-roosting systems, killer whale aggregations, and primate societies.

Thus, further research has shown a tendency for societies to possess individuals with differential influences on the cohesion of the network. This idea has been further supported by resiliency analyses, where scientists use simulated or experimental removals of individuals that resemble genetic knockouts used to identify gene function. Results from a resiliency analysis of macaque network structure indicated that key individuals had disproportionately large effects on network stability and the connectivity of individuals. However, dolphin groups that have shown the 'scale-free' pattern are actually resilient to the removal of those individuals most important to the group's cohesion, as the network stays intact despite the removal of those key individuals.

Milgram's 'small-world phenomenon,' started by his finding that any two random individuals in a portion of the US population can be connected by six handshakes, has also been found in the dolphin networks. The networks have a combination of highly clustered subgroups and a short average path length. This pattern is similar to that of human societies and probably encourages communication and information transfer.

More complex social structures and dyadic interactions have been described in animal societies, using social network theory. The primary example comes from guppies and sticklebacks, two groupings of animals with small-world qualities and social cliquishness where the network structure predicted patterns of cooperation. By visualizing a social network, these organisms were found to be homophilic in that similar-sized individuals engaged in repeated dyadic interactions that satisfy prerequisites for the evolution of reciprocity and cooperation. Individuals of the same sex and age also associated. In human and some animal societies, individuals are also known to associate according to degree (i.e., individuals with many 'friends' associate with others that have many 'friends'), although with animals such as dolphins, this is not always the case.

Most recently, social networks have been used to explore mating systems. When applied to ground squirrels, for instance, SNA revealed a social structure where females were clustered around the male with which they have mated, and showed that males interact with females more directly before they are estrous. The evolution of the relationship between multiple males in lekking mating systems has also been explored using SNA to determine male and female social interactions during breeding, most notably in manakins.

Avenues for Future Research

While network theory has been used effectively to describe social interaction between animals, the fitness consequences of these social relationships are not well known. Animals with high betweenness, for instance, or many connections with other individuals, might have increased fitness because of these qualities. On the other hand, the same animals might suffer from increased exposure to pathogens or be at the forefront of a disease epidemic because of their socially central position. By using SNA to study social interactions, biologists are becoming more aware of fitness tradeoffs and why animals live in social groups.

Further information regarding the different social 'roles' of group members is another avenue that is certain to be explored during future research. In light of debates on how 'sociality,' 'social complexity,' or a 'social group,' can be defined, a network approach may provide a way to describe these terms and to standardize definitions across varied taxa in the future. The measures may provide different ways to define social roles, although issues such as controlling for group size will need to be resolved. Although some algorithms have proved useful in deciphering social divisions, the facilitation of interspecific and intraspecific comparisons of social groups is an area that has been largely unexplored thus far.

By the same token, the future visualization of animal social networks from varied taxa is helping biologists begin to understand the characteristics that are crucial for social group persistence, structure, or the fitness of individuals within the group. For instance, there are probably reasons why some groups are more cohesive or persist longer, and if these groups have individuals with higher fitness, it might suggest a consequence on which natural selection would act to favor social behavior. Perhaps future researchers will use SNA to model social group stability and cohesion, or obtain field data that can be visualized with SNA.

Because SNA has already been used to study human disease networks, understanding the mechanisms of disease transmission continues as another focus of network applications in the future. Besides tracing the original source of a disease (e.g., 'patient zero'), diseases have a variety of modes of transmission and their spread may depend largely on how a network is structured. Identifying socially central animals will suggest which groups members are most influential in disease transfer and describe the degree to which pathogen transmission depends on social relationships.

Limitations of Analysis

SNA is an excellent tool for discovering relationships between animals at a given point in time and with specific reference to the type of ties (e.g., amicable or hostile social interactions) selected by the investigator. Although relationships are generally included in a sociomatrix only if they exceed those that would occur in a simulated null model, schools of thought vary on the issue. Investigators must therefore decide which SNA measures and methods of relationship inclusion are appropriate for their questions, with the understanding that including too few individuals will give a truncated picture of the network and too many may result in a highly fragmented network. Since networks are snapshots of relationships at a particular point in time, it is also a challenge to model dynamic processes such as transmission with SNA, and inferences must therefore be made over long periods of time or permutation methods must be used. Along these lines, network analysis assumes that relationships are relatively stable over time, which must be understood when interpreting network visualizations. While social networks and spatial networks may be similar, SNA does

not consider space constraints on network structure, which limits the ways that animals can construct their social networks.

Professional Associations and Journals

Animal behaviorists can join the International Network for Social Network Analysis, which is the professional association of SNA. Started in 1977 by sociologist B. Wellman (University of Toronto), it has more than 1200 members and is now headed by G. Barnett (University of Buffalo). Netwiki is a scientific wiki devoted to SNA that uses tools from subjects such as graph theory, statistical mechanics, and dynamical systems to study real-world networks in the social sciences, technology, and biology. Several journals are devoted completely to network analysis and sometimes publish articles regarding animal behavior: Social Networks, Connections, the Journal of Social Structure, and the Network Science Report. Mainstream journals such as Animal Behavior, American Naturalist, Proceedings of the Royal Society Series B, and Proceedings of the National Academy of Science also have published many studies involving SNA and animal behavior recently.

See also: Consensus Decisions; Disease Transmission and Networks; Group Movement; Nest Site Choice in Social Insects.

Further Reading

Barabási AL (2002) Linked: The New Science of Networks. Cambridge: Perseus.

Borgatti SP, Carley KM, and Karckhardt D (2006) On the robustness of centrality measures under conditions of imperfect data. Social Networks 28: 124–136.

Connor RC, Mann J, Tyack PL, and Whitehead H (1998) Social evolution in toothed whales. Trends in Ecology & Evolution 13: 228–232.

Croft DP, James R, and Krause J (2007) Exploring Animal Social Networks. Princeton, NJ: Princeton University Press.

Fewell JH (2003) Social insect networks. Science 301: 1867–1870.

Flack JC, Girvan M, de Wall FBM, and Krakauer DC (2006) Policing stabilized construction of social niches in primates. Nature 439: 426–429.

Freeman LC (1979) Centrality in social networks: Conceptual clarification. Social Networks 1: 215–239.

Guare J (1990) Six Degrees of Separation. New York, NY: Random House.

Lusseau D and Newman MEJ (2004) Identifying the role that animals play in their social networks. Proceedings of the Royal Society of London Series B 271: S477–S481.

Manno TG (2008) Social networking in the Columbian ground squirrel. Animal Behaviour 75: 1221–1228.

Milgram S (1967) The small-world problem. Psychology Today 1: 61–67.

Newman MEJ and Girvan M (2004) Finding and evaluating community structure in networks. Physical Review E 69: 026113.

Proulx SR, Promislow DEL, and Phillips PC (2005) Networking thinking in ecology and evolution. Trends in Ecology & Evolution 20: 345–353.

Wasserman S and Faust K (1994) *Social Network Analysis: Methods and Applications.* Cambridge: Cambridge University Press.

Wey TW, Blumstein DT, Shen W, and Jordan F (2008) Social network analysis of animal behaviour: A promising tool for the study of sociality. *Animal Behaviour* 75: 333–344.

Whitehead H (2008) *Analyzing Animal Societies: Quantitative Methods for Vertebrate Social Analysis.* Chicago, IL: University of Chicago Press.

Relevant Websites

http://www.insna.org/ – International Network for Social Network Analysis.

http://www.visualcomplexity.com/vc/ – Visual complexity.

http://netwiki.amath.unc.edu/ – Netwiki.

Life Histories and Predation Risk

P. A. Bednekoff, Eastern Michigan University, Ypsilanti, MI, USA

Introduction

Life histories are schedules of growing, surviving, and reproducing. Organisms differ radically in life histories. For example, blue whales can weigh 160 000 kg as adults; they take a long time to grow to adult size then tend to produce one calf every few years. At the other end of reproductive scheduling among mammals, in the mouse-like marsupial *Antechinus stuartii*, males go into a frenzy of breeding activity as they approach 1 year old. During this frenzy, their immune systems shut down, their fur starts to fall out, and then they die. Females forge on for some months as single mothers, and then most of them die.

In describing life histories, we recognize juvenile and reproductive phases. The juvenile phase includes all growth and development prior to reproduction. The juvenile phase is described by size at birth, growth pattern, and age and size at maturity. The reproductive phase starts at first reproduction and includes the distribution of all periods of potential reproduction thereafter. The reproductive phase is described by age and size-specific reproductive investments, the number and size of offspring, mortality schedules, and length of life. A few animals, including humans, live substantial periods after reproduction has ceased. Overall, we can describe life histories by the age and size at maturity and by survival and fecundity as functions of age.

Life histories are built around tradeoffs, and thus tradeoffs are the key to understanding the diversity in life histories. Tradeoffs occur when two good things are not completely compatible. Tradeoffs exist between all the major components of life histories: growing, surviving, and reproducing. Growing, surviving, and reproducing cannot be maximized simultaneously. In this article, I concentrate on how antipredator behavior interacts with life histories. Tradeoffs with survival often involve predators: behaviors that lead animals to encounter more food or more mates often also lead them to encounter more predators. Tadpoles, for example, need to move to find food, yet moving brings them within the range of dragonfly larvae and other predators.

Tradeoffs between growing, surviving, and reproducing mean that organisms will be selected to 'invest' in different mixes of these three things at different times. In life histories, the basic measure to maximize is the expected reproductive value, $b + SV$, where b is current reproduction, S is survival until the following breeding season, and V is the expected reproductive value for an animal that does survive to the next breeding season. Reproductive value weights the contributions of individuals of different ages to future populations. Reproductive value often increases as individuals grow and survive through the juvenile period, and frequently decreases among adults as they age. As formulated here, the expected reproductive value for a newborn is its expected lifetime reproductive success – the average number of descendants produced by this life history. For stable populations of sexually reproducing organisms, we expect this to be two – enough to replace the parents. Therefore, the interesting part is not the total expected lifetime reproductive success but how survival and reproduction at different ages contribute to this total.

During the nonbreeding season, current reproduction (b) equals zero, so the expected fitness is SV: survival to the next breeding season times the reproductive value expected if our animal survives this long. Activities during the nonbreeding season can serve to increase either S or V. For example, some foraging is needed to avoid starvation. Foraging more than this may contribute to growth or energy reserves that allow greater reproduction in the future. In this life history perspective, the costs of predation are proportional to the potential benefits. Being killed by a predator lowers the expected fitness to zero. Therefore, the cost of being killed is the reproductive success a forager could have had if it had survived. This linkage means that we cannot ask how much risk a forager should take on to produce one additional offspring without knowing how many offspring it would produce without taking the additional risk. For example, a forager that would otherwise expect to produce one offspring might risk a lot to produce a second, while a forager that would otherwise expect to produce three offspring should risk less to produce a fourth, and a forager that would otherwise expect to produce a dozen should risk little to produce a thirteenth. This linkage of costs and benefits sets up an automatic state-dependence: the potential losses from being killed increase with previous success in acquiring resources, so the relative value of further resources is likely to be lower. Even if the fitness gains of resource acquisition are constant, the costs will increase since the expected reproductive value increases, and that entire value would be lost in death.

Predation and Growth Rate (Resource Acquisition Rate and Something Else)

Faster growth can lead to lower survival in several different ways. First, organisms do dangerous things during growth. For example, salmon with genetically engineered growth promoters forage intensely even when predators are around. If these fish escaped into habitats with predators, they would have low survival rates. Thus, faster growth can lead to decreased survival because it requires more intensive foraging and less intensive antipredator behavior. A tradeoff between growth and survival can also happen for reasons beyond antipredator behavior. For example, Atlantic silverside (fish) that grow more quickly also die more rapidly – even with no predators involved. The physiology involved here is not obvious in animals but somehow tissues rapidly thrown together tend to be shoddy constructions that are not built to last. With trees we can actually see and feel the difference between rapidly and slowly grown wood. The tradeoff between growth and survival is important in understanding the great diversity of trees in tropical forests. Seedlings of some tree species grow quickly if exposed to sunlight, while seedlings of other species can survive very well when shaded by a canopy. Data show that no species do well in both environments. While sun-adapted trees grow more quickly in both sun and shade, they are more likely to die in the shade – a tradeoff between growth and mortality. Shade-adapted trees produce dense, tightly grained wood that is often highly coveted as lumber (and harvested at rates faster than these slow-growing trees can sustain).

Reproduction and Growth

Animals face a tradeoff between reproduction and growth. Obviously, the energy used for reproduction cannot be used for growth. Reproduction might also require behaviors that limit energy intake. We are most familiar with animals that finish growth before starting reproduction. Here longer growth leads to later reproduction. This is true for many kinds of plants, lizards, and fish. Other organisms, including turtles, crocodiles, and elephants, have the potential to grow throughout their lives. Such indefinite growth makes it easier to study the tradeoff because more potential combinations of growth and reproduction can be observed. Blue-headed wrasse are coral reef fish that grow indefinitely. By comparing across reefs, we see that wrasse that reproduce more grow less. Different guppy populations from Trinidad that allocate more energy to reproduction grow less quickly, even when kept in the laboratory under standardized conditions. Guppies from low-predation sites mature later and allocate less of their energy to reproduction after they mature.

How do predators affect the tradeoff between reproduction and growth? Within the formulation $b + SV$, growth can increase V, the capacity for future reproduction. Theoretically, reproduction should be delayed only if the reproduction foregone (lower b) is more than made up by future reproduction, discounted by survival in the meanwhile (greater SV). As we shall see, predation risk can alter the age at first reproduction. This effect, however, is more likely due to how predators affect survival than due to how they affect reproductive capacity. Therefore, we will examine the effect later after discussing survival.

Total Reproductive Effort Limits Survival

Reproduction often involves potentially risky activities. Finding a mate often requires searching for a mate, perhaps through unfamiliar territory, or advertising for one to come. Male black-tailed prairie dogs suffer mortality when moving between burrows of different females. The same sorts of activities that advertise to potential mates could attract the attention of predators. Animals may act to establish 'private' channels of communication for mating that predators cannot detect, to the extent that they can. For example, male guppies court by showing off their spotted sides in the light of streams. Guppies look different in areas with different sorts of predators. Those with few predators (including domesticated strains) tend to be colorful. Where fish predators regularly prey on adults, males have little red spots among a pattern that generally blends in with the gravel substrate. Where the major predators are prawns, which do not see red but see black patterns, the red spots are well developed but not edged in black. Thus, male guppies are showy in ways less noticeable to their predators. Furthermore, guppies court at dawn and dusk when their signals are far less likely to be seen by predators, even though somewhat less likely to be seen by potential mates. Thus, guppies act to minimize the tradeoff between showing off and being seen by their predators. At high-predation sites, males also display less and harass females more. Thus, predators increase conflict between the sexes for guppies.

Reproduction itself may involve vulnerable states. Pregnant animals, whether garter snakes or humans, are less mobile than usual and would have greater trouble in evading predators. Predation on a pregnant animal eliminates both current and future reproduction, so we might expect it to be especially selected against. Birds do not get pregnant but starlings escape slowly when carrying eggs. Seahorse males carry young in a pouch and giant waterbugs carry their young on their backs. Such males undertaking greater care might be more vulnerable to predators. If pregnancy leads to steeply increasing predation risk, we might expect animals to give birth to a succession of small broods, rather than a few larger ones. I know of no direct evidence of this, but it matches what guppies do under moderate risk of predation. Nest predation favors multiple attempts at small broods, since the whole brood is often lost when nest predators find it.

Greater current reproduction often decreases survival to breed again. For example, Nazca boobies that raise two chicks are less likely to survive the next year than those that raise one chick. To understand this, we have to consider how size and condition affect survival. I will highlight the potential effects of predators, though disease could act similarly. Reproduction may leave a parent depleted and needing to take risks to restore their condition. The question is what is depleted and what time and activities are needed to restore this. Zebra finches deplete their wing muscles when laying large clutches. They escape more slowly while depleted but can restore the muscles from their diet. Parents may also deplete their body calcium or energetic stores. Restoration of calcium depends on diet and local environment and is more difficult in environments that have suffered from acid rain.

Reproduction may also limit survival by preventing parents from performing other functions. For example, willow tits in Sweden that raise large late broods have to molt more quickly and less effectively. A rushed molt gives them a less insulating set of feathers for the upcoming winter, and their chances of surviving the winter are reduced. Some may starve but likely many more are killed by predators while trying to find the extra food they need to survive.

Now we revisit how predators affect the balance between current and future reproduction. How predators affect this balance depends on who is more vulnerable to the predators. Where large individuals are safest, predators select for individuals that grow to safety. Where big adults are still at risk, predation will tend to select for early reproduction. Across guppy populations, with high adult predation, adults start reproducing at young ages and continue at high rates from then on. Here small adults produce large clutches of small offspring. With moderate predation on juveniles, larger adults produce larger embryos at lower rates. With low predation risk at all ages, big adults produce big embryos on short intervals. Data show that high-predation sites are more dangerous for all sizes and ages of guppies. Surprisingly, guppies have similar survival to maturity at high- and low-predation sites – they just grow faster and mature at smaller sizes at high-predation sites. High growth rates are possible because high-predation sites support lower population densities and more food per guppy. In summary, predators affect reproduction in guppies as predicted from theory, and predators also indirectly affect feeding conditions in a way that was not originally predicted.

Fishing by humans often targets the largest adults. Nets capture everything too big to slip out through the mesh. Angling regulations generally mandate that fish under a size limit must be released. In these ways, fishing has often selected for reproduction at small sizes. This life history response selects for smaller size at maturity and smaller fishing catches. Where large fish would be able to

produce many more offspring, it may have also reduced recruitment of young into the population. Alternative fishing strategies could harness life history responses to produce more large adults. These include allowing fishing only outside a network of reserves and setting a maximum size limit (in addition to a minimum). The second strategy is most feasible for lobsters and other fisheries where each individual caught can be measured.

Parental Care to Reduce Offspring Mortality

Parental care is one way that behavior affects current reproduction and adult survival. Adults are predicted to take on less risk when they can expect greater future reproduction, and take on more when it affects offspring prospects more.

The location where young spend the early part of their lives interacts with the number and kind of offspring that they have. Mammals such as rodents that leave their young protected in nests produce large litters of small newborns. Mammals that give birth in the open on land or at sea tend to produce small numbers of large young. Mammals such as primates that carry their young produce litters of one or a few offspring. Birds that nest in cavities have larger clutches than birds that nest in open nests, and also have offspring that develop at slower rates.

When parents leave offspring alone somewhere, they may return to care at times and in ways that avoid drawing predators to the young. The number of offspring that parents can raise may be limited by the amount of care they can safely provide, and not by the amount of care they can potentially provide. Thus, parents could feed larger broods, but the foraging trips needed to do so would likely attract the attention of nest predators. This topic has been most discussed in considering the number of eggs laid by birds. In particular, many tropical birds lay clutches of one or two eggs and are very likely to lose them to snakes, monkeys, and other nest predators. Compared to temperate birds, parents make very few and secretive visits. Some tropical birds are well known as adults yet their nests have never been observed. By contrast, the nests of some seabirds have long been observed while the lives of adults at sea are just starting to be known in any detail.

Parental care may also be dangerous for adults and the timing of the care may depend on the balance of threats to adults and to young. White-breasted nuthatches have larger clutches and lower survival than (smaller) red-breasted nuthatches. White-breasted nuthatches delay returning to the nest more in response to a nest predator and less in response to an adult predator. In these ways, white-breasted nuthatches value current reproduction more and future reproduction less than do red-breasted nuthatches.

Parents also act to deter predators. Nest defense may force adults to risk themselves (and future reproduction) to preserve current reproduction. Nest defense often increases with brood age. Younger offspring will need more feeding and perhaps more risky defense before becoming independent. When adults have a lower survival rate even if they do not defend, they have less to lose by defending.

Offspring may continue to rely on parental protection and nourishment for some time. In humans, this may last for decades. Thus, human females stop having babies at menopause, but may well provide parental care long afterwards. Parental care over an extended period means that the tradeoff between the number and survival of offspring is likely not to be a one-time allocation decision. Instead, parents may opt for fewer, safer offspring at various points, even if these are not demonstrably better offspring. The effects of parental care are most clearly defined by examining survival to independence and the condition of offspring at independence, but independence is not always simple to identify. A mother red squirrel in Canada can leave her territory, including its nest and food stores, to one offspring while she and other surviving offspring go to live elsewhere for the winter. Independence is far less clear for an offspring inheriting a territory than for one searching for a new one. Where animals live in multigenerational groups, such as hyenas and baboons, parents may invest in their mature offspring for years. With multigenerational groups, it may be more important to examine how adults and grown offspring are interdependent than to try to define a time when offspring become independent. Multigenerational fitness effects are likely needed to explain how humans and some other animals came to have substantial postreproductive periods in their life histories. Menopause may be explained as a foregoing of future reproduction to ensure that current offspring reach maturity, and perhaps to help with the grandchildren as well.

Predation Risk and Rate of Aging (Discounting of Later Reproduction)

Aging or senescence is a late life decline in an individual's fertility or probability of survival. This is measured on a per time period basis, so that a human is more likely to die between ages 60 and 61 than between 30 and 31 (and also less able to reproduce). Although aging is individually unavoidable, a great deal of research, often conducted with fruit flies and *C. elegans* nematodes, shows that we can select for longer-living animals. It is important to note that *C. elegans* and fruit flies are selected to be short lived in their natural environments. Even though we can artificially select for long-lived flies and 'worms,' these strains, when returned to their natural habitats, are outcompeted by wild-type individuals in exploiting patchy and ephemeral resources.

Organisms differ greatly in their rate of aging. Rates of aging best correlate with rates of extrinsic mortality. Extrinsic mortality is mortality that an organism would suffer even if it did not reproduce. Organisms with low extrinsic mortality age slowly. While such organisms may be big, this is secondary to extrinsic mortality. Long-lived organisms have low extrinsic mortality for different reasons – tortoises are safe in their shells while parrots are safe in their watchful flocks – but are united in living safe lives as adults in the wild. Thus, animals age quickly if they have a low probability of surviving even under the best circumstances. Extrinsic mortality, however, is not something we simply observe. Actual mortality is also affected by intrinsic mortality that changes with allocations among reproduction, maintenance, and defensive structures. Aging seems to be due to unrepaired damage and organisms differ both in their rates of damage and their rates of repairing damage.

The force of selection against aging depends on survival to that age multiplied by the residual reproductive value for individuals of that age – that is the number of offspring they can expect to have in the future. Because mortality is a one-way street, the fraction of individuals living cannot increase with age. The residual reproductive value, however, can increase if older females are better at reproducing than are younger ones. If residual reproductive value increases enough to offset loss to mortality, we might expect antiaging. As turtles grow, they become more fecund and more safe. By standard measures of fertility and survival, they show the opposite of aging. Why turtles start to reproduce early is less obvious, since this reproduction limits their growth. Perhaps waiting longer increases the risk that they will not reproduce at all.

We can examine how predators affect aging by comparing high- and low-predation sites. Off the coast of South Carolina, opossums on Sapelo Island have lived for thousands of years without their usual suite of mammalian predators. These opossums often live into their second year and age less quickly from the first year to the second. Even their collagen fibers stay flexible longer than those of opossums on the mainland.

In guppies, females from low-predation sites live longer in the field. In the lab without predators, however, they actually live shorter lives despite starting reproduction later. Although predators clearly eat guppies and shape their life histories, their effects on guppy aging differ for different aspects of performance. In the laboratory, reproductive output slows less quickly with age in fish from high-predation sites. On the other hand, fast start performance, the ability to suddenly swim away, is better in young guppies from high-predation sites, but old guppies from all sites are similarly slow. By definition, guppies from high-predation sites age more rapidly in swimming performance. Thus, this example shows that aging is not necessarily a unified phenomenon, and predators may select on aging differently for different traits.

Value of Experiments and Selection Studies in Studying Life Histories

Life histories contain a cycle of related events and are potentially influenced by many factors. Experiments are enormously helpful in understanding causal relationships in life histories. For example, I have discussed high- and low-predation guppy sites. These sites, however, differ in many ways besides predators. Low-predation sites also have lower parasite diversity and resistance, plus lower overall genetic diversity. Low-predation sites generally occur upstream in drainages where temperatures are lower and productivity is lower. We can be confident that predation causes differences between guppies at these sites because guppies and their predators have been experimentally moved between sites, and compared with experimental studies of selection in the laboratory. In experiments, changing just the predators at the site results in large and rapid changes in life histories that parallel the differences between the different high- and low-predation sites in nature.

Where experiments are not feasible, we can take advantage of situations resembling experiments. We know most about the effects of predators on life histories from islands and northern lakes where predators may naturally be missing. Life histories without predators are generally very different from those with a potent predator or two. At large scales, we can take advantage of changes in predator number that have occurred for other

reasons. Humans have introduced predatory fish to many formerly fishless lakes and removed large carnivores from many areas of the world within historical times. As predators recolonize or are reintroduced to some areas, we can examine changes in life histories of prey. The reintroduction of wolves to Yellowstone National Park has led to major changes not only for their primary prey, elk, but also for coyotes, pronghorn, and riparian woodlands. Here shifts in foraging behavior by elk profoundly affect the ecology of the ecosystem.

See also: Conservation and Anti-Predator Behavior; Reproductive Success; Risk-Taking in Self-Defense; Trade-Offs in Anti-Predator Behavior.

Further Reading

Caro T (2005) *Antipredator Defenses in Birds and Mammals.* Chicago, IL: University of Chicago Press.
Clark CW (1994) Antipredator behavior and the asset-protection principle. *Behavioral Ecology* 5: 159–170.
Gordon SP, Reznick DN, Kinnison MT, et al. (2009) Adaptive changes in life history and survival following a new guppy introduction. *American Naturalist* 174: 34–45.
Magurran AE (2005) *Evolutionary Ecology: The Trinidadian Guppy.* Oxford: Oxford University Press.
Reznick DN, Bryant MJ, Roff D, Ghalambor CK, and Ghalambor DE (2004) Effect of extrinsic mortality on the evolution of senescence in guppies. *Nature* 431: 1095–1099.
Roff D (2002) *Life History Evolution.* Sunderland, MA: Sinauer Associates.
Stearns SC (1992) *The Evolution of Life Histories.* Oxford: Oxford University Press.

Locusts

A. V. Latchininsky, University of Wyoming, Laramie, WY, USA

What Are Locusts and What Are Not?

Locusts (from the Latin '*locus ustus*' = 'burnt place') are short-horned grasshoppers (Orthoptera: Acrididae), distinguished by their density-dependent behavioral, physiological, and phenotypic polymorphism. Under low population densities, locusts exist in the so-called 'solitarious phase.' Solitarious nymphs are characterized by camouflage coloration, infrequent social interactions, and sedentary behavior. When crowded, locusts develop into the 'gregarious phase' with nymphs often strikingly colored in dark brown or black with contrasting yellow, orange, or red. The gregarious nymphs form cohesive groups or 'hopper bands,' capable of long-distance, concerted marching. Adults of the gregarious phase differ from solitarious individuals by having longer wings and shorter hind femora, as well as by some less conspicuous morphological traits.

The most spectacular differences between the phases are in behavior: the solitarious adults avoid each other except for mating, while the gregarious adults pack together in swarms; they migrate, feed, mate, and lay eggs in crowds. The swarms contain from several thousand to 40 billion individuals. Swarms are capable of sustained day-time flights covering distances from several dozen to several thousand kilometers. In 1988, swarms of the Desert locust took off from the coast of West Africa, crossed the Atlantic Ocean and landed in the Caribbean Islands and northern shores of South America, covering over 5000 km in 6–10 days. Solitarious adults also can make long-distance flights although they fly individually and at night. The differences in pigmentation between the adults of the gregarious and solitarious phases are not as striking as in the nymphal stage (**Figures 1** and **2**).

Out of more than 12 000 described grasshopper species in the world, only about a dozen exhibit pronounced behavioral and/or morphological differences between phases of both nymphs and adults, and should be considered locusts. In other words, all locusts are grasshoppers, but not all grasshoppers are locusts. The capacity to produce a swarming phase appeared independently a number of times within the family Acrididae and is considered as a relatively recent trait in grasshopper evolution. The most economically important locust species and their geographic distribution are presented in **Table 1**. The term 'locust' is sometimes erroneously applied to periodic cicadas (e.g., '17-year cicada'), which belong to a different insect order, Homoptera. Another misnomer comes from the plant kingdom (e.g., 'black (or yellow) locust tree' *Robinia pseudoacacia* or 'honey locust tree' *Gleditsia triacanthos*, both from the legume family) (**Figures 3–6**).

Economic Importance

Locusts have been the enemies of humans since the dawn of agriculture. They are mentioned in ancient writings such as the Torah and the Koran. In the Old Testament of the Bible, locusts constitute the infamous Eighth Plague of Egypt. Locust swarms often brought devastation and hunger to entire nations. According to the Roman historian Pliny the Elder, in 125 BC, 800 000 people died in the Roman colonies of Cyrenaica and Numidia (territories of contemporary Libya, Algeria, and Tunisia) from famine caused by a locust plague. In 1958 in Ethiopia, locusts destroyed 167 000 tons of grain, which is enough to feed 1 million people for a year. Locust outbreaks have occurred on all continents except Antarctica and can affect the livelihood of one in ten people on Earth.

Besides the direct economic losses to crops, locust outbreaks may be devastating to ecological processes by destroying vegetative food sources for many animal species. The passages of locust swarms may cause human demographic changes. In contemporary society, subsistence farmers abandon fields that are wrecked by locusts and move into cities, adding to the demand on urban infrastructures in already overpopulated and impoverished settings. Locust-control efforts, which are essentially chemical, can produce negative environmental impacts and continue to be very costly. In 2003–2005, to curtail a Desert locust outbreak that affected 8 million people in Africa, 13 million ha (approximately the area of the state of New York) was treated with broad-spectrum neurotoxic insecticides in 26 countries. The cost of the campaign including food aid to the affected population amounted to half a billion US dollars.

While insect pests destroy annually 14% of crops worldwide, the annual crop losses from locusts are estimated at only 0.2%. The seemingly low figure is misleading because the perception of locust damage is scale-dependent. Locust swarms can be compared to other natural disasters like hurricanes or tornadoes. For an entire national economy, the total crop losses from locusts may seem negligible, but for a given farmer or cooperative, even a brief passage of a locust swarm may result in a complete destruction of the whole season's work. This is particularly relevant to subsistence farmers in developing

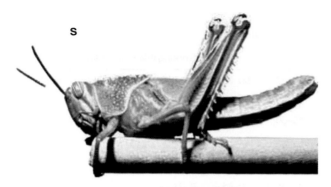

Figure 1 Nymphs of the solitarious (S) and gregarious (G) phases of the Desert locust. Photo courtesy Compton Tucker, NASA GSFC.

Figure 2 Marching hopper band of the Migratory locust nymphs. Photo: A. Latchininsky.

countries. As such, the socio-political consequences of a locust plague are difficult to translate into simple monetary terms.

Life Cycle

The life cycle of any locust species includes a sequence of embryonic (egg), nymphal, and adult stages. Females lay eggs, using short hook-like valves of the ovipositor at the tip of their abdomen to bore into the top layer of the soil.

During oviposition, a female extends her abdomen three to four times its normal length by means of elastic intersegmentary membranes. The duration of oviposition depends on soil properties, particularly compactness, and lasts from 30 min to 2 h. Eggs are laid in packets called egg-pods. The number of eggs in an egg-pod depends on the species and is often correlated with the body size of the females. The smaller, Moroccan and Italian locusts have respectively 18–42 and 20–60 eggs in their egg-pods, while the larger Migratory and Desert locusts have 40–120 and 30–146 eggs in their egg-pods, respectively. Egg-pods of the big Red locusts may contain up to 190 eggs. Each female lays one to four egg-pods with about a 10-day interval between successive ovipositions. The initial egg-pod contains more eggs than the subsequent ones of the same female (**Figures 7–9**).

In the temperate zones, locust embryos develop with an obligatory diapause, meaning that eggs laid in the summer delay hatch the next spring. To enhance survival through long periods of freezing temperatures and to protect from predator attacks, the egg-pods of the temperate locust species, such as Moroccan and Italian locusts, have thick walls made of soil particles cemented by secretions from female's accessory glands and oviduct. The eggs are cemented together as one by the female's secretions and form an egg cluster occupying the lower part of the egg-pod. Its upper part consists of a foamy or spongy mass, which hardens after oviposition, and a lid that allows the hatching nymphs to exit. Eggs of most tropical and subtropical locust species incubate without diapause and hatch 2–3 weeks after oviposition. In such cases (e.g., in the Desert or Red locusts), the egg-pods are just clusters of loosely attached eggs with a foam plug on top of them.

During hatching, the newly born nymphs tunnel through the softened foam plug to reach the surface of the soil. Once out of the egg-pod, they immediately undergo an intermediate molt, shedding their embryonic cuticle and becoming first-instar nymphs. With a mortality rate of up to 90%, the first instar constitutes the critical developmental stage for the survival of the locust population. Nymphal development includes 5 (rarely 4, 6, 7, or even up to 9) successive instars separated by molts. In some species, females have an extra instar compared to males. Later instars are distinguished from the earlier ones by their bigger size and more developed wing pads. The rate of development and duration of the nymphal period depend largely on the prevailing weather, primarily air temperature and humidity. Under optimal conditions, each instar lasts 5–7 days with a total of 25–35 days to reach adulthood. Under unfavorable environmental conditions, the nymphs may require 2–3 months to develop fully. After the final molt, or fledging, the nymphs turn into adults.

Adult locusts are characterized by two pairs of fully grown wings. The front pair, or tegmina, are narrow and

Table 1   Main locust species, their body sizes and geographic distribution

No	Common name	Latin name	Body length (mm)[a] ♂	♀	Geographic distribution
1	Desert locust	*Schistocerca gregaria* (Forskål, 1775)	45–56	50–65	Africa, S. Europe, SW and Central Asia (invasion area)
2	Migratory locust	*Locusta migratoria* (L., 1758)	35–50	45–65	Africa, Eurasia, Australia
3	Moroccan locust	*Dociostaurus maroccanus* (Thunberg, 1815)	15–30	20–40	N. Africa, Europe, Central Asia
4	Italian locust	*Calliptamus italicus* (L., 1758)	15–30	20–42	Europe, Asia
5	Red locust	*Nomadacris septemfasciata* (Audinet-Serville, 1838)	35–50	50–62	Africa
6	Brown locust	*Locustana pardalina* (Walker, 1870)	35–45	40–55	S. Africa
7	Australian Plague locust	*Chortoicetes terminifera* (Walker, 1870)	20–30	30–45	Australia
8	American Bird locust	*Schistocerca americana* (Drury, 1773)	40–50	45–56	S. and Central America
9	Bombay locust	*Nomadacris succinta* (Johansson, 1763)	40–50	55–65	SE Asia
10	Rocky Mountain locust[b]	*Melanoplus spretus* (Walsh, 1866)	15–25	20–30	N. America

[a]Measured from the tip of the head to the end of the abdomen.
[b]Went extinct in the early twentieth century.

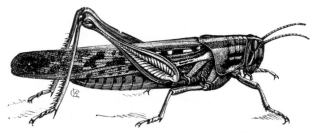

Figure 3   Adult American Bird locust. Hebard (1934) Bull Illinois State, Nat. Hist. 20: 125 279, Fig. 157, after Thomas, 9th Report State Ent Ill. The image is taken from Orthoptera Species file online http://osf2.orthoptera.org/HomePage.aspx: Eades, D.C. & D. Otte. Orthoptera Species File Online. Version 2.0/3.5.

Figure 5   Adult Moroccan locust. Photo: A. Latchininsky.

Figure 4   Adult Migratory locust. Photo: A. Latchininsky.

Figure 6   Adult Italian locust male. Photo: A. Latchininsky.

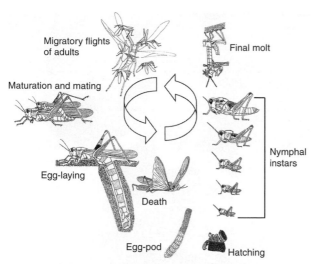

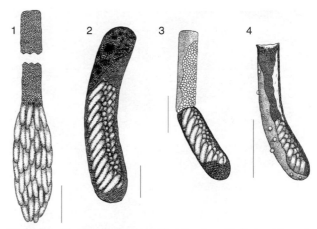

Figure 7   Locust life cycle. Modified from Latchininsky AV, Sergeev MG, Childebaev MK, et al. (2002) Acridids of Kazakhstan; Central Asia and Adjacent Territories Laramie, WY: Association for Applied Acridology International and the University of Wyoming.

Figure 9   Egg-pods of Desert (1), Migratory (2), Italian (3), and Moroccan (4) locusts. Scale bar represents 10mm. Source: Latchininsky AV, Sergeev MG, Childebaev MK, et al. (2002) Acridids of Kazakhstan; Central Asia and Adjacent Territories Laramie, WY: Association for Applied Acridology International and the University of Wyoming.

Figure 8   Egg-laying Moroccan locust female. Photo: A. Foucart (CIRAD).

Figure 10   Immature Desert locusts. Photo: FAO UN.

leathery, concealing the broad membranous hind wings, which are folded along the main veins fanwise while at rest. Newly fledged locusts have a soft cuticle that hardens in several days. Adult locusts make short wandering flights in the first days after fledging but only after their cuticle hardens can they accomplish long-distance migratory flights. Sexual maturation takes from just a few days to several weeks, after which the locusts start mating. In some cases, such as Desert and Red locusts and tropical tree locusts of the genus *Anacridium*, mating can be delayed by unfavorable weather conditions (e.g., low air temperature or insufficient humidity). The adults can remain sexually immature for up to 9 months. They continue to fly and feed but, if unfavorable conditions persist,

they would eventually die without producing offspring. A favorable change of the environmental conditions can trigger sexual maturation and eventual reproduction at any time during this period. In most locust species, maturation is manifest by noticeable changes in pigmentation. Immature Desert locusts are pink and turn bright yellow when sexually mature. Immature Migratory locusts gradually change their coloration from green or brownish to mostly yellowish as they mature. Some other species as the Italian or Australian Plague locust do not exhibit noticeable pigmentation changes associated with sexual maturation. Physiological changes during maturation

Figure 11 Mature Desert locusts. Photo: FAO UN.

Figure 12 Copulating Moroccan locusts. Photo: A. Latchininsky.

include the growth of testes and accessory glands in males and ovarian development and egg growth in females (**Figures 10** and **11**).

Locust males mature from one to several days earlier than females. The presence of the mature males accelerates the maturation of females. Locusts reproduce sexually; cases of parthenogenetic reproduction are rare. Olfactory and acoustic signals are used to ensure the meeting of prospective mates. Locusts (often both sexes) produce stridulating call signals by rubbing the inner surface of the hind femora over the thickened veins on the tegmina. Copulation lasts from 1 to 20 h. During copulation, the male mounts the female, grasps the tip of her abdomen with his and transfers the sperm packet, called spermatophore, into the female's genital opening. The female stores the sperm in a special organ called the spermatheca. Locusts can mate multiple times during their adult lives. However, a single copulation is usually sufficient to fertilize all the eggs produced by the female (**Figure 12**).

In temperate zones, locusts exhibit a univoltine life cycle characterized by only one generation per year and an obligate embryonic diapause. Some subtropical species, such as the Egyptian tree locust *Anacridium aegyptium*, hibernate as sluggish and practically nonfeeding adults during the winter months. Most tropical locusts develop continuously, without embryonic diapause, and under favorable conditions can produce two, three, and rarely even four annual generations.

Phase Transformation

The key event in the biology of locusts is the change from a single-living and mostly sedentary solitarious phase to a gregarious phase in which they live in dense bands or swarms, actively migrate and may devastate

crops and rangeland. This phenomenon, which is nonexistent among nonswarming grasshoppers, is called locust phase transformation. Besides the conspicuous behavioral, morphological, and color changes, the extreme solitarious and gregarious locust phases differ in food selection, nutritional physiology, metabolism, reproductive physiology, neurophysiology, endocrinology, pheromone production, longevity, and molecular biology. Locusts are polymorphic, and the extreme phases are connected by a continuum of intermediate, transitional forms (phase transiens). It takes at least four consecutive generations to complete the phase transformation from a typical solitarious phase to a fully gregarious phase. Phase transformation is a cumulative, but also a reversible, process that requires suitable environmental conditions. The process is density dependent and starts when locust density exceeds a certain threshold. In the Desert locust, behavioral changes first become manifest when the density of the young nymphs exceeds approximately 50 000 individuals per ha or 5 per m^2. For older nymphal instars, the threshold is 5000 per ha, and for adults it is 250–500 per ha. At these density levels, the locusts start switching from their solitarious tendency of avoiding each other to a proclivity for forming sustained cohesive groups and moving in a concerted way. Most other locust species have higher phase transformation thresholds than the Desert locust.

Recent studies have shown that different locust phases are produced by different expressions of genes in response to crowding. Locust species vary in the number of phase traits they exhibit. Some, like the Australian Plague locust, produce only behaviorally different phases. In others, like Desert, Migratory, and Moroccan locusts, in addition to behavior, the solitarious and gregarious phases are distinguished by morphology and pigmentation.

The sight and smell of conspecifics trigger behavioral changes in solitarious locusts after only 4 h of crowding above the phase transformation threshold. When the locusts meet up, their nervous systems release serotonin, an evolutionarily conserved mediator of neuronal plasticity. Serotonin causes the locusts to become mutually attracted, which is a prerequisite for swarming. Once the locusts start to aggregate together, their interactions increase and a positive feedback loop accelerates the changes toward the gregarious phase. At this point, the direct mutual contacts between the individuals become the most powerful gregarizing stimuli. Tactile receptors on external side of the hind femora of nymphs are particularly sensitive to such contacts. Their repeated stimulation by crowded locusts enhances the expression of gregarious phenotypic traits, particularly the contrasting black and orange, yellow, or red pigmentation. However, the locusts may revert to solitarious behavior after 4 h of reisolation.

Phase characteristics not only develop during the lifespan of a locust, but are also being transferred from mothers to offspring. This maternal effect is mediated through certain chemicals secreted by females into the foam substance surrounding the eggs. Such a mechanism allows for maintaining and developing phase status across generations. Behavioral, chromatic, and physiological changes are followed by emerging traits of gregarious morphology, but these changes become noticeable only several (at least four) generations after gregarization starts. Tropical locusts which have multiple generations per year can build up a dense gregarious population and mass migrate in hopper bands and swarms in just 2–3 years. In temperate zones, where locusts have a single annual generation, it usually takes them longer to accomplish the transformation from the solitarious to gregarious phase and to build up a swarming population.

The concept of the locust phases was first put forward in 1921 by the Russian entomologist Boris Petrovich Uvarov (1888–1970), who was the founder and first director of the Anti-Locust Research Centre in London. Uvarov postulated that the two phenotypically very different forms of *Locusta*, which were then considered as separate species *L. danica* and *L. migratoria*, were in fact the two extreme phases of a single species. Similar phase differences were later found in other locust species. Uvarov's 'phase theory of locusts' emphasized the crucial role of phase transformation

in developing locust outbreaks and had important practical applications for locust population management.

Environmental Conditions Leading to Phase Transformation

Most of the time, locusts lead a solitarious life, and this may be considered as the normal state of their populations. At some points in time, however, changes in their environment may initiate the gregarization. In the case of the Desert locust, such changes are triggered by abundant rains that promote lush vegetation growth in an otherwise arid milieu. The locusts start to congregate on the patches of green vegetation, forming loose groups at first, and dense hopper bands later. They feed and march together, and increasingly become phenotypically gregarious. Similarly, rains in the arid zones trigger gregarization of the Australian Plague locust. In the case of the Migratory locust, which inhabits reed stands in wetlands along rivers and lakes, it is the excessive drought that usually initiates the aggregation of locusts and their consequent gregarization. The locusts concentrate on few remaining patches of reeds and start producing hopper bands with gregarious behavior and appearance. Drought is also responsible for initial concentrations of the Moroccan locust in the Mediterranean semi-deserts and Central Asian arid steppes. The hoppers crowd together on few patches of green grasses and forbs which emerge in early spring. Such concentrations may eventually lead to the appearance of the gregarious phase. These examples show that habitat discontinuity or patchiness, which can result from a variety of meteorological events, is the most important condition for initial locust gregarization and phase transformation. If the resources, particularly vegetation, are distributed in a uniform fashion, the chances that the locusts will start to produce gregarious populations are low. For example, locust outbreaks originating from dense forests are unknown. Mosaic habitats which represent a combination of green vegetation clumps and areas of bare ground are most favorable for producing and maintaining the gregarious phase.

Although the ecological conditions leading to the initiation of phase transformation are well understood for most locust species, outbreaks (the spectacular hopper band movements and swarm flights) still often remain 'unexpected.' The main reason for this is that the areas of initial locust aggregations are scattered over a vast territory with difficult access and low human populations. For the Desert locust, the area where incipient gregarious populations may form covers 16 million km^2, which is roughly equal to the areas of the United States and Australia, combined. The total distribution area of the Migratory locust is even larger. Despite efforts to implement efficient locust monitoring using satellite images and automated weather stations, there is always a threat that in some locations locusts may produce an undetected gregarious population, leading to a large-scale outbreak.

Gregarious Behavior

Hopper Bands

Gregarious females oviposit in dense groups, which results in simultaneous hatching of large number of hoppers in close proximity to each other. In the mornings, the hoppers form very dense groups staying on the ground and basking in the sun. Once their body temperature rises, they start marching. The concerted, directional movement of dense hopper bands may represent an antipredator strategy: the grouped nymphs saturate the predators with sheer numbers and are much less likely to become their victims than individual hoppers living on their own. At the same time, the members of the band suffer from intraspecific competition for nutritional resources and

Figure 13   Hopper band of the Moroccan locust basking in the sun. Photo A. Latchininsky.

from cannibalistic pressure. The hopper band migration is a 'forced march' driven by cannibalism. Until a band encounters new nutritional resources, the hoppers need to move to escape attacks from behind by the hungry members of the same band (**Figures 13** and **14**).

The sizes of hopper bands vary from several square yards to many acres. The record figure, $110 \, km^2$, comes from a Moroccan locust band observed in Iraq. Accordingly, the number of hoppers in the band can be astronomically high. The density in the band is highest during early nymphal instars and often reaches thousands per m^2. The record density of the first-instar hoppers of the Migratory locust is known to reach 80 000 per m^2; similar estimates for other species are 37 000 for the desert locust, 28 000 for the Australian Plague locust, and 21 000 for the Moroccan locust (**Figure 15**).

The speed of hopper marching and distances traveled by the bands depend on the age of the hoppers, vegetation, relief, and weather. First- and second-instar hoppers rarely travel more than $200 \, m \, day^{-1}$. Organized marching usually starts in the third instar and continues until adulthood. If the vegetation is sparse, late-instar hopper bands of the Desert and Migratory locusts can travel over a mile per day. The total maximum distances covered by hopper bands during the entire nymphal period are 3 km for the Italian, 10 km for the Red, 17 km for the Moroccan and Brown, and up to 30 km for the Migratory and Desert locusts.

Swarms

Adult gregarious locusts spend nights roosting in very dense aggregations on trees, shrubs, or bare ground. In the first days after fledging, when their cuticle is still soft, individual adults can produce only short erratic or escape flights. Group flights start 10–15 days after fledging. At first,

Figure 14   Cannibalistic Migratory locust nymphs. Photo: A. Latchininsky.

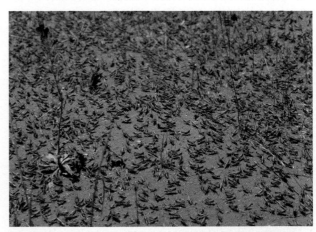

Figure 15   Australian plague locust hopper band. Photo A. Latchininsky.

the swarms fly short distances, from several hundred m to a couple of km. The flight range becomes progressively longer with the age of the adults. Swarms take off when the internal body temperature of the locusts exceeds certain level: 20°C for the Migratory, 25°C for Desert, 28°C for Moroccan locusts, etc. The swarms fly generally downwind and maintain remarkable cohesion. The flights take place in the mornings and afternoons and can last from 9 to 20 h day^{-1}. At night and during the hottest periods of the day, locusts usually roost or feed in the vegetation. The flight speed of the swarms, typically between 8 and 32 km h^{-1}, depends largely on the speed of the wind. Swarms fly at a very wide range of altitudes, from 15 to over 1500 m above the ground. The average distance covered during a day varies from 10 km for the Moroccan locust to 200 km for Desert and Migratory locusts. During the entire adult life, locust swarms can fly hundreds and even thousands of miles. The longest migrations are known for the swarms of the Desert locust which flew across the Atlantic Ocean in 1988 covering 5000 km in 6–10 days. Migratory locust swarms can fly distances of up to 1000 km. Other locust species usually do not fly farther than 300 km during their adult life.

When the swarms are flying, the locust densities in them are relatively low, with a maximum of 10 per m^3. When they land for a night rest or midday roosting, the locust densities can range from 100 to 2000 per m^2. Settled swarm sizes can be enormous, covering areas for up to 800 km^2. Swarms of the Desert locust can contain up to 40 billion locusts, which is the largest terrestrial congregation of animals on Earth. An exceptionally huge swarm of the now extinct Rocky Mountain locust, measured in the 1875, occupied an estimated area of 256 409 km^2, which is larger than the entire states of Wyoming or Oregon (**Figure 16**).

Atmospheric conditions, especially wind force and direction, govern the flight pattern of locust swarms. Swarms of the Desert locust often concentrate in the intertropical convergence zone, where air masses may collide and produce rainfall, generating potentially suitable habitat for locust oviposition and nymphal development. Locust swarming is often considered as an adaptation for searching for nutritional resources and colonizing new habitats. However, the factors that cause swarms to take off and mass migrate are not fully understood. On many occasions, swarms have abruptly taken off from an area covered with lush vegetation and without any external disturbance. Numerous swarms have been recorded flying toward open seas or oceans, and the great majority of them, except some rare cases mentioned earlier, drown.

Food and Feeding Behavior

Locusts are proverbial for their voracity. However, the view that locusts are 'chewing automata,' that is, they devour everything in their way, appears to be far from being accurate. In the solitarious phase, locusts exhibit marked food preferences and feed only on a limited number of preferred plant species. The Desert locust prefers the foliage of trees and shrubs to the herbaceous plants, and among the latter, it clearly prefers forbs to grasses. The Migratory locust, on the contrary, has a preference for grasses and related families of sedges and rushes. The forbivorous Italian locust favors sage shrubs from the genus *Artemisia* and legumes. Food selection includes finding suitable plants, first using visual and then olfactory stimuli. Host-plant choice may be limited by the distribution of deterrent compounds (glucosides, alkaloids, essential oils, organic acids) in nonhost plants. Plants with mechanical defenses (hooks, spines, trichomes) are often avoided. The situation changes when the locusts undergo phase transformation. Crowded locusts are less selective in food searching because they often live under the stress of food and water shortage. Among the factors separating unpalatable from palatable food, the water content of the food plays a significant role. This is one of the reasons why locusts sometimes attack not only habitually rejected plants but also many nonplant substances like textiles, dung, woodwork, wool – even on live sheep! After a long migratory flight, the need to compensate for water losses becomes overwhelming, and upon landing, locust swarms consume literally anything that holds the slightest moisture. The degree of polyphagy in swarming locusts is inversely proportioned to the water requirements of the organism (**Figure 17**).

Although locusts are essentially herbivorous, they can be cannibalistic or necrophagic, especially if crowded. The impending cannibalism is considered to be one of the driving forces of the concerted hopper marching behavior.

The proverb 'each locust can eat its weight in plants each day' holds true only for the nymphal instars, while for adults the ratio of daily consumption to the body weight is lower, about 0.5. The amount of food consumed

Figure 16 Migratory locust swarm. Photo: A. Latchininsky.

Figure 17 Moroccan locusts feeding on fresh dung. Photo: A. Latchininsky.

Figure 18 Group egg-laying by Moroccan locust females. Phtoto: A. Latchininsky.

by locusts varies among the species and developmental stages and is proportionate to the insect's size. Locust hoppers fast before and after each molt, and the duration of this fasting can reach 10–15% of the total nymphal period. At some periods of their life, such as sexual maturation and egg production, locusts become more voracious. Energy expenditures during hopper band marching or swarm flying is often compensated for by increased feeding.

Egg Laying

Females of the gregarious phase are well known for laying their eggs in dense groups. Arriving females are visually attracted to those that have already started egg-laying. Furthermore, the soil in which they oviposit attracts other females because of the pheromones that are contained in the secretions surrounding the eggs in an egg-pod. Group egg laying ensures maintaining and enhancing the gregarious status of the population as the resulting hatchlings form dense groups from the first days of their lives. Average egg-pod density is 5–10 per m^2 and maximum densities range from 500 per m^2 for the Desert locust to 8000 and 10 000 per m^2 for the Moroccan and Italian locusts, respectively. In such extreme cases, the soil in which the egg-pods were laid resembles a honeycomb. Multiplying these numbers by the average number of eggs in an egg-pod, it is possible to estimate the number of hoppers which will hatch per unit of area. However, even in the large and fecund locust species which can lay multiple pods containing more than 100 eggs each, the rate of multiplication from one generation to the next does not usually exceed 20-fold. This relatively low multiplication rate is explained by high mortality of eggs and early hopper instars due to predation and unfavorable environmental conditions (**Figure 18**).

Conclusion: Locusts as Models

For decades, locusts have been used as model organisms to study different aspects of individual and group behavior. Their collective migrations are of particular interest because the patterns of hopper band movement and swarm flight are similar to those observed in other animals. Apparently unifying laws and mechanisms exist that govern group movement in animals. This underlying framework may be so general that the individual locusts can be considered as analogs to interacting, inanimate particles. Self-propelled particle models have recently been used to account for the emerging density-dependent transition from wandering solitarious individuals to concerted hopper band marching. On the other hand, the amazing capability of locusts to avoid crashing into each other when flying together in a swarm fascinated car makers in their attempts to create a crash-proof car. This ability appears to be due to the fact that the visual input is transmitted directly to the wings of the locust, seemingly bypassing the brain.

See also: Collective Intelligence; Group Living; Group Movement; Insect Migration; Insect Navigation; Klepto-parasitism and Cannibalism; Orthopteran Behavioral Genetics.

Further Reading

Anstey ML, Rogers SM, Swidbert RO, Burrows M, and Simpson SJ (2009) Serotonin mediates behavioral gregarization underlying swarm formation in Desert locusts. *Science* 323: 627–630.
Bazazi S, Buhl J, Hale JJ, et al. (2008) Collective motion and cannibalism in locust migratory bands. *Current Biology* 18: 1–5.
Buhl J, Sumpter DJT, Couzin IC, et al. (2006) From disorder to order in marching locusts. *Science* 312: 1402–1406.
COPR (1982) *The Locust and Grasshopper Agricultural Manual*. London: Centre for Overseas Pest Research.

FAO (2001) *Desert Locust Guidelines*. vols. 1–7. Rome: Food and Agriculture Organization of the United Nations.

Kang L, Chen XY, Zhou Y, et al. (2004) The analysis of large-scale gene expression correlated to the phase changes of the migratory locust. *Proceedings of the National Academy of Sciences, USA* 101: 17611–17615.

Latchininsky AV, Sergeev MG, Childebaev MK, et al. (2002) *Acridids of Kazakhstan, Central Asia and Adjacent Territories*. Laramie, WY: Association for Applied Acridology International and the University of Wyoming (in Russian with English Summary).

Lockwood JA (2004) *Locust. The Devastating Rise and Mysterious Disappearance of the Insect That Shaped the American Frontier*. NewYork, NY: Basic Books.

Simpson SJ, Despland E, Haegele BF, and Dodgson T (2001) Gregarious behaviour in desert locusts is evoked by touching their back legs. *Proceedings of the National Academy of Sciences, USA* 98: 3895–3897.

Simpson SJ, McCaffery AR, and Haegele B (1999) A behavioural analysis of phase change in the desert locust. *Biological Reviews* 74: 461–480.

Simpson SJ and Sword GA (2007) Phase polyphenism in locusts: Mechanisms, population consequences, adaptive significance and evolution. In: Whitman D and Ananthakrishnan TN (eds.) *Phenotypic Plasticity of Insects: Mechanisms and Consequences*, pp. 93–135. Plymouth: Science Publishers Inc.

Simpson SJ and Sword GA (2008) Locusts. *Current Biology* 18: R364–R366.

Simpson SJ, Sword GA, and De Loof A (2005) Advances, controversies and consensus in locust phase polyphenism research. *Journal of Orthoptera Research* 14(2): 213–222.

Steedman A (1988) *Locust Handbook*. 2nd edn. London: Overseas Development Natural Resource Institute.

Uvarov BP (1966) *Grasshoppers and Locusts. A Handbook of General Acridology, Vol. I: Anatomy, Physiology, Development, Phase Polymorphism, Introduction to Taxonomy*. Cambridge: Anti-Locust Research Centre, University Press.

Uvarov BP (1977) *Grasshoppers and Locusts. A Handbook of General Acridology, Vol. II.: Behavior, Ecology, Biogeography, Population Dynamics*. London: Centre for Overseas Pest Research, University Press.

Relevant Websites

http://www.fao.org/ag/locusts/en/info/info/index.html – FAO UN locusts.
http://140.247.119.138/OS_Homepage/ – Orthopterists Society.
http://www.daff.gov.au/animal-plant-health/locusts – Australian Plague Locust Communication (APLC)
http://locust.cirad.fr/ – CIRAD (France).
http://www.pestinfo.org/ – International Society for Pest Information (ISPI).
http://www.schistocerca.org/ – Schistocerca Information Site.
http://www.sdvc.uwyo.edu/grasshopper/ghwywfrm.htm – Grasshoppers of Wyoming and the West.
http://www.bio.usyd.edu.au/staff/simpson/simpson.htm – Steve Simpson (University of Sydney) website.

Konrad Lorenz

R. W. Burkhardt, Jr., University of Illinois at Urbana-Champaign, Urbana, IL, USA

Life and Scientific Career

Born on 7 November 1903, Lorenz was the second and last child of Emma Lorenz and Dr. Adolf Lorenz, a distinguished and wealthy orthopedic surgeon. Growing up in comfortable surroundings at the family home in the village of Altenberg, on the outskirts of Vienna, the young Lorenz was allowed to pursue his enthusiasms as an animal lover. His interest in animals and evolution as an adolescent led him to think of becoming a zoologist or paleontologist, but his father wanted him to be a physician instead. After one semester of premedical studies at Columbia University, Lorenz enrolled in 1923 as a medical student in the Second Anatomical Institute of the University of Vienna. There he came under the influence of the distinguished comparative anatomist Ferdinand Hochstetter, who taught him how comparative anatomists use physical structures to reconstruct evolutionary lineages (**Figure 1**).

Lorenz's receipt of his doctorate of medicine in 1928 seems to have satisfied his father's desire that he receive a medical education. With his MD in hand, he enrolled at the University of Vienna's Zoological Institute, receiving his PhD in zoology in 1933 for a study of bird flight and wing form. In the meantime, he had continued to raise birds and observe their behavior, and his observations had brought him to the attention of Germany's leading ornithologists, Erwin Stresemann and Oskar Heinroth. They, along with Hochstetter and the psychologist, Karl Bühler, at the University of Vienna, encouraged Lorenz to pursue a career combining zoology and animal psychology. His talents in this regard were displayed in a series of papers he published on bird behavior, culminating in his path-breaking 'Kumpan,' monograph of 1935, entitled 'Der Kumpan in der Umwelt des Vogels: der Artgenosse als auslösendes Moment sozialer Verhaltungsweisen' ('The Companion in the Bird's World: Fellow Members of the Species as Releasers of Social Behavior').

Lorenz's rise in scientific visibility was not followed immediately, however, by gainful academic employment. As of 1937, his only position was that of *Dozent* (unpaid lecturer) in Bühler's Psychological Institute. By then, he had already been married for a decade (to Margarethe Gebhardt, his child sweetheart), and he had two children (Thomas, b. 1928; Agnes, b. 1930) (a third child, Dagmar, would be born in 1941). He and his family lived with his parents in Altenberg. He came to fear that his chances for professional advancement in Catholic Austria were slim, given his Protestant background and his firmly held belief that human behavior should be understood in the context of biological evolution. This contributed to his enthusiasm in March 1938 for the *Anschluss*, the incorporation of Austria into Germany. He expected that his chances of scientific support would be greater under the Third Reich than they had been under the Austrian clerico-fascists. His greatest hope was that the Kaiser Wilhelm Gesellschaft (KWG), Germany's primary organization for supporting scientific research, would establish an institute for him in Altenberg.

Not hesitant about signaling his enthusiasm for the new regime, Lorenz in May 1938 applied for membership in the Nazi Party. In July 1938, at a joint meeting of the German societies for psychology and animal psychology, and then over the next several years in other papers and addresses, he argued that animal behavior studies could shed light on matters of racial hygiene. Breakdowns in the innate social behavior patterns of domesticated animals, he claimed, were strictly analogous to the 'signs of decay' in civilized man. He expressed support for Nazi race purity laws. In addition, in an article in 1940, he argued that Darwinism, properly understood, led not to communism or socialism but instead to National Socialism.

A Kaiser Wilhelm Institute never materialized for Lorenz. The KWG Senate reviewed favorably the idea of providing him with an institute, but the funds for it were not forthcoming. In 1940, he was named Professor of Psychology at the University of Königsberg. This professorship traced back to Immanuel Kant. The post inspired him to develop his philosophical interests and recast Kant's categorical imperative in an evolutionary context. His time at Königsberg was brief, however. He was drafted into the military in 1941, serving successively as a psychologist, psychiatrist, and then troop physician. In June 1944, the Russians captured him on the eastern front. He spent the next three and a half years in Russian prisoner-of-war camps. He did not return to Austria until February 1948.

Back in Austria, Lorenz found himself once again without an academic position. The Austrian Academy of Sciences provided him with modest support for his research station at the family home in Altenberg, where he and his family continued to live (both of his parents were now deceased). He wrote his popular book *King Solomon's Ring* (published originally in German in 1949)

Figure 1 Konrad Lorenz lecturing student research assistants about the principles of ethology during observations of hand-raised geese at the Max-Planck-Institut für Verhaltensphysiologie in Seewiesen in 1971. Photo by Jane Packard.

as a means of making money. In 1950, he appeared to be the top choice to replace Karl von Frisch for the professorship of zoology at the University of Graz (Frisch was returning to his earlier post at Munich), but political and ideological considerations scuttled his candidacy. This would not be the last time that allegations of earlier Nazi sympathies on his part caused him difficulties. Concluding that he had no chance of ever getting a professorial appointment in Austria, he appealed to colleagues in Britain to find a position for him there. As they went about this task, Lorenz's friend, the German behavioral physiologist, Erich von Holst, persuaded the Max Planck Gesellschaft (MPG) (the KWG's successor) to work to keep Lorenz in Germany. The MPG quickly set up an institute for Lorenz in Buldern, Westphalia, under the auspices of Holst's Max Planck Institute for Marine Biology in Wilhelmshaven. Lorenz gladly took up the new post. In 1956, the MPG established for him and Holst a new Max Planck Institute for Behavioral Physiology in Seewiesen, near Starnberg, in Bavaria. Lorenz remained there until his retirement in 1973. He then returned home to Austria and Altenberg, where he continued his researches. In the course of his long career, he received many honors, including the 1973 Nobel Prize. He died on 27 February 1989.

Lifelong Scientific Practices

Lorenz prided himself on being an animal lover. The scientific value of being an animal lover, he liked to explain, was that without the love of an animal, one would never have the patience to watch it long enough to become familiar with its entire set of behaviors. His own favored method of research was to raise wild birds in a state of semicaptivity and observe them over the course of months and even years, thereby allowing himself to come to know the whole of a bird's normal behavior patterns. This also permitted him to witness rare but instructive behavioral events that a field observer might never see, as for example when an instinct 'misfired' in a situation where the proper stimuli for releasing it were not present. In addition, by raising different species side by side, he was able to make comparative observations that again would not have been possible for a field observer. On the other hand, he never developed the keen ecological sense of a field biologist. Nor did he develop strong skills as an experimenter. He credited himself with an intuitive understanding of animals, on the basis of his years of close observation of how animals behaved.

Given his predilection for raising animals, it is not surprising that Lorenz developed a special admiration for two of his predecessors in particular, the American biologist Charles Otis Whitman (1842–1910) and the German ornithologist Oskar Heinroth (1871–1945), both of whom raised birds and observed their behavior closely. Lorenz portrayed these two scientists, with some exaggeration, as animal lovers who were content to watch their pigeons and ducks in a completely unbiased way, unburdened by any hypotheses. However, he also appreciated their ideas. He credited Whitman with having discovered what he called the 'Archimedean point' on which the new science of ethology revolved. This was the idea that, as Whitman expressed it in 1898, 'Instinct and structure are to be studied from the common standpoint of phyletic descent.' Heinroth became a model for Lorenz for his studies of how instincts function in avian social life.

Lorenz's first scientific publication (in 1927) was an empirical study reporting his observations on the behavior of a tame, pet jackdaw. His experiences with this bird led him to want to understand how its instinctive behaviors functioned in the life of a jackdaw colony. To this end, he established a colony of jackdaws in the attic of the family home, marked the birds for identification, and began studying the social life of jackdaws. His successes in this regard led him to study night herons and then graylag geese (along with a host of other species). He promoted his practices as the key to advancing animal psychology. In his Kumpan paper of 1935, he wrote that the proper method for the animal psychologist in studying any species was to begin with 'an extensive period of general observation' prior to any experimentation, and furthermore to focus on instincts before tackling learning. The investigator unwilling to begin by gaining a thorough familiarity of the full behavioral repertoire of his subject species, Lorenz admonished, 'should leave animal psychology well alone.'

The Conceptual Foundations of Ethology

Lorenz's publications became increasingly theoretical in the 1930s, as he addressed the nature of instincts and the role they play in the social life of birds. In a 1932 paper on instinct, he argued that instincts and learning are wholly distinct from each other, even when they are 'intercalated' in complex, coordinated chains of behavior. In 1935, in his remarkable '*Kumpan*' monograph, he advanced his theorizing further by employing the concept of the 'releaser' (an idea previously enunciated by the theoretical biologist Jakob von Uexküll, with whom Lorenz had been interacting and to whom Lorenz dedicated the monograph). By Lorenz's account, lower animals such as birds are adapted to their environments not very much through acquired knowledge (as are humans) but instead through highly differentiated instinctive motor patterns, created over time by natural selection. To function effectively, they need to be released only by a very few stimuli emanating from the thing to which the animal is responding. These stimuli, however, must characterize the object sufficiently well that the animal does not respond to similar stimuli coming from an inappropriate object. Like a key fitting a lock, the proper combination of stimuli evokes a response from an 'innate schema' (later to be called the innate releasing mechanism or IRM), releasing the performance of its associated instinctive motor pattern.

The interrelations of stimuli and innate schema, Lorenz proceeded to explain, were subject to even greater refinement when the sender of the stimuli and their recipient were members of the same species. Then the releasing stimuli and innate schema could be mutually fine tuned over time by natural selection so as to make the fit between them ever more precise, resulting in combinations of such overall improbability that an animal's instinctive reactions would only rarely be triggered by stimuli from the 'wrong' object. Lorenz used the word 'releasers' (*Auslöser*) for characters that serve to activate the innate schemata of conspecifics. Releasers could be morphological structures or conspicuous behavior patterns or, most often, a combination of both.

Lorenz went on to describe how the highly organized social life of jackdaws depends on a surprisingly small number of instinctive reactions to releasers provided by fellow jackdaws. Borrowing the idea of the 'companion' from Uexküll, who had used the word in the first place to describe what Lorenz had told him about the social life of jackdaws, Lorenz maintained that every jackdaw has a number of social drives with respect to which other jackdaws serve as 'companions.' As 'parental,' 'infant,' 'sexual,' 'social,' or 'sibling' companions they provide stimuli that release the innate behavior patterns appropriate to the jackdaw's drives.

Lorenz's Kumpan paper was also the site in which he called attention to the phenomenon he called 'imprinting' (*Prägung*). Whitman and Heinroth, among others, had

been familiar with the phenomenon, but Lorenz was the first to focus scientific attention upon it. He reported that in most bird species, the newly hatched baby bird does not have an innate ability to recognize its own kind; rather, the object of its instinctive behavior patterns is imprinted upon it in a brief, early period in its life. Thus, if a baby gosling sees a human before it sees a mother goose, the gosling will follow the human, directing toward this foster parent the instinctive behavior patterns that would under normal circumstances have been directed toward members of its own species. Lorenz distinguished imprinting from learning, likening it instead to embryological induction. He maintained that imprinting was irreversible.

Lorenz's *Kumpan* monograph evoked a strong, appreciative response among behaviorally oriented ornithologists, including Julian Huxley and Henry Eliot Howard in Britain and Margaret Morse Nice and Wallace Craig in the United States. Lorenz had not yet, however, arrived at his final explanation of how instincts work physiologically. Up to this time, he had endorsed a chain-reflex theory of instinct. Between 1935 and 1937, he decided that that theory was wrong. His interactions with the American Wallace Craig and especially the German Erich von Holst led him to jettison it in favor of a theory involving the internal build up of instinctive energies. Holst's studies of the endogenous generation and central coordination of nervous impulses led Lorenz to conclude that instincts involve some kind of energy (later called 'action-specific-energy) that builds up in the organism until it is released or it overflows. This new theory made sense of what Craig had called 'appetitive behavior,' where the animal seems internally motivated to seek the stimuli that will elicit its instinctive motor patterns. It also served to explain two phenomena that were apparently related to each other: 'threshold lowering' and 'vacuum activities.' Threshold lowering described the finding that the longer it had been since an instinctive action was last performed, the easier it became for the behavior to be released. A 'vacuum activity' was when an instinctive behavior pattern was performed 'in vacuo,' that is, it 'went off' without any apparent or appropriate releasing stimulus and thus without serving its proper biological function. These findings made no sense if one viewed instincts as chains of reflexes set in motion by external stimuli. They did make sense, Lorenz decided, if instincts were understood to be internally generated.

While Craig and Holst were of special help to Lorenz in his theory building, the arrival of Niko Tinbergen on the scene provided Lorenz with an ally who gave Lorenz's key concepts critical experimental support. The two men first met at a conference on instinct held in Leiden in November 1936. In the presence of older animal psychologists who seemed primarily interested in gaining insights into the animal mind, Lorenz and Tinbergen found themselves sharing a different commitment. They both wanted to put animal behavior studies on a much firmer, objectivistic, physiological foundation. Tinbergen was impressed by Lorenz's insights and ambition as a theorist. Lorenz was ecstatic to learn of the experiments that Tinbergen and Tinbergen's students at Leiden had been doing on the instinctive behavior of the three-spined stickleback. Their analysis of the stimuli eliciting the sticklebacks' instinctive movements struck Lorenz as precisely what he needed. The following spring Tinbergen was given a leave of absence from his department at Leiden to go to Austria to study with Lorenz in Altenberg. There the two men worked together for three and a half months, conducting, among other projects, their classic study of the egg-rolling behavior of the graylag goose. And there too they established a firm friendship. This friendship, which survived the strains of war and lasted for the rest of their lives, was of major importance for the development of ethology as a scientific discipline.

Lorenz's publications during the war varied considerably in nature. They included his writings about domestication and racial degeneration and his paper arguing that evolutionary biology was consistent with National Socialism; an early paper on evolutionary epistemology; an extended comparison of the instinctive behavior patterns of different duck species as a means of assessing their genetic affinities; and a major monograph on 'the inborn forms of possible experience.' He offered his duck study as a confirmation of the idea that the comparative method could be applied successfully to instincts in reconstructing phylogenies. His 'inborn forms' monograph was a sweeping synthesis of his recent thinking in which he addressed such topics as instinctive behavior, domestication phenomena and the threat these posed to racial hygiene, the reinterpretation of Kantian epistemology in evolutionary terms, and what man might make of himself in the future.

Lorenz in the Postwar Period

The rebuilding of ethology immediately after the war fell to Tinbergen rather than Lorenz, since Lorenz did not return from the war until 1948. Lorenz's first major occasion to present his ideas again after the war occurred at a special conference on physiological mechanisms in behavior, held in Cambridge, England, in 1949. There he offered a visual representation of the instinct theory he had developed. His 'hydro-mechanical' or 'psycho-hydraulic' model, as he called it, featured a reservoir containing a fluid, a spring valve connected by way of a pulley to a scale, and a weight on a scale (**Figure 2**). In this model, the fluid building up in the reservoir represents action-specific-energy; the spring, pulley, and scale represent the innate releasing mechanism; the weight on the scale represents the stimuli serving to trigger the innate releasing mechanism;

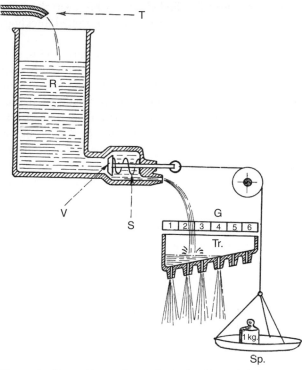

Figure 2 Konrad Lorenz's psycho-hydraulic model of instinctive action. Reproduced from Lorenz KZ (1950) The comparative method in studying innate behaviour patterns. *Symposia of the Society for Experimental Biology* 4: 221–268.

and the instinctive reaction itself is represented by the jet of liquid coming through the valve, producing different results according to its strength. Although Lorenz acknowledged the 'extreme crudeness and simplicity' of this model, he insisted that the model represented 'a surprising wealth of facts really encountered in the reactions of animals.' The model stimulated considerable debate and experimentation over the next decade. Although it came to be generally discredited by the end of the 1950s, Lorenz remained attached to it, and he presented a revised version of it two decades later.

There is no doubt that Lorenz did his most creative work before and during the war, not after it. In 1950, prior to being given his first Max Planck institute, he complained to the British ethologist W. H. Thorpe that he was not gaining any new knowledge but rather simply using up his capital of old knowledge. But this picture did not change all that much even after he had special institutional resources at his disposal. His postwar intellectual activity consisted primarily of recycling, developing, and defending ideas he had formulated earlier. He did this, however, with great gusto, and he continued to be a powerful, charismatic leader of the field. He attracted students to study with him, he energized ethology's international congresses, he challenged psychologists to put behavior in an evolutionary perspective, and he provided the public with an attractive view of the science of ethology, frequently highlighted by his own charming image as the foster mother of some imprinted ducklings or goslings.

As ethology began to flourish in the early 1950s, several of Lorenz's key concepts drew criticism, both from inside the discipline and from other quarters. Among ethology's own new recruits, Robert Hinde in particular called into question behavioral models involving fluids flowing. Meanwhile, from outside the discipline, the American comparative psychologist Daniel Lehrman launched a multipronged attack on Lorenzian ethology. Lehrman insisted, among other things, that Lorenz's sharp distinction between innate and learned behavior stood in the way of a better understanding of how behavior develops in the individual. Much to Lorenz's disappointment, some of his colleagues, including Tinbergen, came to feel that Lehrman had a point. Lorenz himself, however, was not inclined to make concessions. Although his counterattacks on American behaviorists were not all that successful in addressing Lehrman's actual complaints, Lorenz did in the course of these debates introduce an instructive concept, which he playfully designated 'the innate schoolmarm.' This, as he expressed it in 1965, is the idea that an organism's ability to learn particular things is itself a function of mechanisms that natural selection has built into that organism. In brief, innate mechanisms determine what a species can learn.

Although always considering himself a good Darwinian and always insisting on the importance of bringing evolutionary perspectives to bear on animal behavior, Lorenz was better at applying the methods of comparative anatomy to behavior than he was at thinking about the mechanisms by which evolution operates. His remarks in the latter regard simply reflected his confidence that natural selection typically works for 'the good of the species,' a view that came to be regarded as old-fashioned in the 1960s and 1970s as evolutionary biologists, behavioral ecologists, and sociobiologists promoted ideas of individual selection or kin selection instead of group selection. In contrast, Lorenz's efforts to understand human cognitive processes in evolutionary terms have been viewed as much more farseeing in nature, and he is regarded as a pioneer in evolutionary epistemology. His book *Behind the Mirror* (published first in German in 1973) represents his mature thinking on the philosophical ideas he began developing in the 1940s, when he found himself in his professorial chair descending from Kant.

Lorenz from early in his career was eager to explore the broader human implications of his studies of behavior. He enjoyed playing the role of the scientist-prophet bringing the lessons of biology to a society in peril. This motif appeared in his prewar and wartime warnings about genetic deterioration in civilized man. It reappeared in his first popular book, *King Solomon's Ring*. Though that book is best known for Lorenz's charming accounts of

his experiences and observations as an animal-raiser, Lorenz concluded the book with a somber claim. The human species, he maintained, is unique among higher animals in that it lacks innate inhibitions against killing its own kind. He returned to the theme of human nature in his bestseller, *On Aggression*. There he portrayed aggression as an instinct that builds up naturally in humans as in animals and ultimately needs release. The problem of civilized man, Lorenz argued, is that he does not have sufficient outlets for his aggressive drive. In the 1970s, in his slender volume entitled *Civilized Man's Eight Deadly Sins*, Lorenz became ever more pontifical, reciting a whole litany of dangers threatening humankind, including overpopulation, environmental destruction, genetic deterioration, and nuclear warfare.

Lorenz's Legacy

As early as the 1930s, Lorenz planned to write a general textbook on the study of animal behavior. He did not succeed in doing so until 1978, when he published his *Vergleichende Verhaltensforschung: Grundlagen der Ethologie* (the English translation appeared 3 years later as *The Foundations of Ethology*). By then, he was not trying to write an up-to-date textbook on ethology. His emphasis instead was on the founding concepts of ethology, which he felt modern ethologists were forgetting, to their detriment. In the book's preface, and with some bitterness, he likened the recent development of ethology to the way that the tips of a coral reef grow quickly away from its foundations, sometimes breaking off from where they started, and then dying or failing to develop in any clear direction. Most of the reviewers of the book found it disappointing. They saw Lorenz as clinging to concepts that had outlived their usefulness. Lorenz's text included, among other things, a revised version of his old psychohydraulic model of instinctive action.

Although many of Lorenz's specific concepts did not remain central to the field, his historical significance for the field's development should not be understated. When Lorenz began his researches, zoologists showed only marginal interest in behavior, European animal psychologists tended to endorse quasi-vitalistic or subjectivistic approaches to behavior, and American comparative psychologists had little appreciation of interspecific differences in behavior or the value of looking at behavior from an evolutionary perspective. Lorenz was the key figure in transforming this landscape. He demanded that the student of behavior gain, through assiduous and detailed observation, a knowledge of the whole range of behaviors of multiple species, and that biological questions – questions in particular of evolutionary history, survival value, and physiology – be brought to bear on this material. He provided ethology with its early conceptual foundations; he attracted talented researchers to his cause; and he served as a highly visible and popular promoter of the ideas and practices of his field. Although his model of human aggression was disputed, his insistence that human behavior be considered in its broader, evolutionary context remains of fundamental importance.

See also: Behavioral Ecology and Sociobiology; Comparative Animal Behavior – 1920–1973; Ethology in Europe; Future of Animal Behavior: Predicting Trends; Integration of Proximate and Ultimate Causes; Neurobiology, Endocrinology and Behavior.

Further Reading

Burkhardt RW (2005) *Patterns of Behavior: Konrad Lorenz, Niko Tinbergen, and the Founding of Ethology*. Chicago: University of Chicago Press.

Lehrman DS (1953) A critique of Konrad Lorenz's theory of instinctive behavior. *The Quarterly Review of Biology* 298: 337–363.

Lorenz KZ (1935) Der Kumpan in der Umwelt des Vogels: der Artgenosse als auslösendes Moment sozialer Verhaltungsweisen. *Journal für Ornithologie* 83: 37–215; 289–413.

Lorenz KZ (1941) Vergleichende Bewegungsstudien an Anatiden. *Journal für Ornithologie* 89. Ergänzungsband 3: 194–293.

Lorenz KZ (1943) Die angeborenen Formen möglicher Erfahrung. *Zeitschrift für Tierpsychologie* 5: 235–409.

Lorenz KZ (1950) The comparative method in studying innate behaviour patterns. *Symposia of the Society for Experimental Biology* 4: 221–268.

Lorenz KZ (1952) *King Solomon's Ring*. London: Methuen.

Lorenz KZ (1965) *Evolution and Modification of Behavior*. Chicago: University of Chicago Press.

Lorenz KZ (1966) *On Aggression*. New York: Harcourt Brace and World.

Lorenz KZ (1970–1971) In: *Studies in Animal and Human Behaviour* 2 vols). Cambridge, MA: Harvard University Press.

Lorenz KZ (1974) *Civilized Man's Eight Deadly Sins*. New York and London: Harcourt Brace Jovanovich.

Lorenz KZ (1977) *Behind the Mirror: A Search for a Natural History of Human Knowledge*. London: Methuen.

Lorenz KZ (1981) *The Foundations of Ethology*. New York and Vienna: Springer-Verlag.

Lorenz KZ and Tinbergen N (1938) Taxis und Instinkthandlung in der Eirollbewegung der Graugans, I. *Zeitschrift für Tierpsychologie* 2: 1–29.

Taschwer K and Föger B (2003) *Konrad Lorenz: Biographie*. Paul Zolnay Verlag: Vienna.

Whitman X (1898) Animal behavior. *Biological Lectures from the Marine Biological Laboratory Wood's Holl, Mass* 1898: 285–338.

Magnetic Compasses in Insects

A. J. Riveros, University of Arizona, Tucson, AZ, USA
R. B. Srygley, USDA-Agricultural Research Service, Sidney, MT, USA

Introduction

The use of magnetic information as a compass is among the most intriguing mechanisms used by animals to orient and navigate. Part of our fascination with the use of magnetism comes from our inability to perceive it relying only on our sensory machinery. In recent decades, we have seen a burst of interest and research on how animals detect and use the Earth's magnetic field. This article focuses on our current knowledge on the use of magnetic information as a compass for orientation and navigation in insects.

The use of magnetic information in insects was first recognized in the late 1950s, with alignment of the body axis in termites relative to the Earth's magnetic field. During the 1960s, interest in the alignment behavior increased, and several other species belonging to taxonomic insect Orders as diverse as Diptera, Coleoptera, Orthoptera, and Hymenoptera were added to the list of insects with magnetic sensitivity. Although it has been difficult to interpret the biological meaning of such alignments, their discovery initiated further studies on the use of magnetic compasses. The discovery of bacteria, with magnetite crystals causing them to move in alignment with the Earth's magnetic field, stimulated the search for magnetic compasses in a diversity of vertebrates and invertebrates, based on similar principles. Also, the continued analysis since the 1970s of the use of magnetic information by model insect species, such as the honeybee *Apis mellifera* and the fruit fly *Drosophila melanogaster*, not only showed that magnetic fields could be used under diverse contexts but also motivated the exploration of such capacities in other insect species. Thus, during the 1980s and 1990s, research turned toward the search of a magnetic compass for navigation. Of particular interest were the species exhibiting long-distance migrations, which were predicted to rely on prominent compasses available across the unknown terrains of their migratory routes. This interest was further supported by additional findings of magnetic particles in insect tissues, which could become the substrate for the compass.

However, determining the use of a magnetic compass has not been an easy task. Part of the problem is that magnetic compasses do not seem to be the primary tools within the multimodal systems of navigation. Thus, the role of the magnetic compass is often uncovered only after other sources of information, such as the sun or other significant landmarks, are unavailable or unreliable. On the other hand, experimental manipulations are constrained by our lack of understanding of the mechanisms underlying the magnetic compass. These two main problems have become the focus of research in the new millennium. First, the analysis of central place foragers, such as honeybees and ants, has allowed for controlled manipulations uncovering interactions of the magnetic compass with other navigational mechanisms. Second, model species, such as cockroaches, honeybees, and fruit flies are becoming essential for the understanding of proximate causes. Particularly remarkable in this respect is the use of genetic manipulations in the fruit fly *D. melanogaster*.

In general, understanding the use and role of magnetic compasses requires comprehension at different levels, from the nature and source of the magnetic cues to the mechanisms of perception, including the nature of the compass and interactions with other cues during the process of decision-making. Our current knowledge of these levels appears as a puzzle, where behavioral studies have contributed with most of the pieces.

What Is Special About Magnetic Information?

The primary source of magnetic information is the iron-rich molten core of the Earth, which makes it hold an enormous magnetic field with lines of force running from magnetic South toward the North Pole (conventionally named north and south relative to the geographic north and south, respectively). The magnetic lines of force vary in inclination, pointing upward in the southern hemisphere, parallel to the Earth's surface at the magnetic equator,

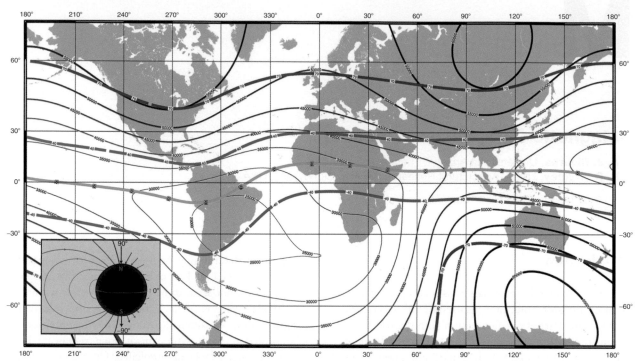

Figure 1   A projection of the Earth derived from the World Magnetic Model (http://www.ngdc.noaa.gov/geomag) showing the intensity of magnetic flux in shades of red and the magnetic inclination from the magnetic equator to the poles in shades of blue. The North and South magnetic poles are not shown because the map ends at the eightieth parallels. Inset: A diagram showing the trajectory of the magnetic field from the southern geomagnetic pole to the northern geomagnetic pole. The equatorial line is the geographic equatorial line and so the inclination is zero on the line in only two locations (i.e., where the light blue line crosses the equatorial line at ca. 315 degrees and at ca. 200 degrees).

and downward in the northern hemisphere (**Figure 1**). Thus, these lines of force have both horizontal and vertical components, which can potentially be used by compasses. The horizontal parameter is easily described by the North–South polarity, whereas the vertical component is called inclination and depends on the angle at which the line of field at a particular location meets with the horizontal plane of the Earth's surface. Inclination roughly correlates with latitude, decreasing from the poles where it is 90° (+90° at the magnetic north pole and −90° at the south pole) to 0° at the magnetic equator which roughly corresponds with the geographic equator. A third component of the Earth's magnetic field is its intensity, with its maxima at the magnetic poles and minimum at the equator. Although two positions are at the same latitude, it is likely that they will have unique inclinations and intensities, such that an animal with the ability to sense both of these features may be able to use them as beacons identifying its position. Importantly, in addition to the global pattern of the Earth's magnetic field, local deposits of iron may interfere with the Earth's magnetic field, creating magnetic anomalies, which serve as beacons for orientation, navigation, and the construction of maps of local magnetic information.

From the previous description, several particularities of the magnetic information can be drawn. First, it is continuously available everywhere on Earth. This marks

a difference with, for instance, solar information, which is intermittently available above the ground and never available below the ground. Second, it is intrinsically directional in one of its components, with such directionality being relatively stable overtime (even after considering the changes in declination from 1 year to the next and the flipping of the magnetic poles in geological time). And third, it is spatially variable in two of its parameters (vertical component and intensity).

For What Purposes Do Insects Use a Magnetic Compass?

Body Alignment and Nest Construction

Body alignment refers to the preference of individuals to orient their body axes relative to the lines of force of the magnetic field. In most cases, insects align themselves parallel or perpendicular to the field, but some intermediate orientations have been reported. Early studies found body alignment in flies (Diptera) and termites (Isoptera) at rest. Body alignment has also been observed in bees and wasps (Hymenoptera), beetles (Coleoptera), crickets, and cockroaches (Orthoptera). Body alignments are evaluated by rotating or canceling the local magnetic field (a magnetic coil design that might be used to reverse the

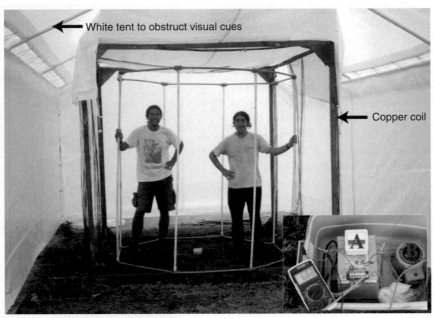

Figure 2 A wood frame constructed with brass screws holds the copper wires wound in a Merritt 4-coil construction. The coil is positioned around the geomagnetic North–South axis. We transformed AC to DC electricity to power the coil with sufficient current to reverse the horizontal component of the Earth's magnetic field. A white tent surrounds the coils to obstruct celestial cues and landmarks on the horizon. The PVC frame can hold a nylon net to prevent insects from escaping, and in the floor of the arena is a camera to observe the insects from a remote location. *Inset:* (a) AC–DC transformer; (b) Variac.

horizontal component of the Earth's magnetic field is shown in **Figure 2**). Under the experimental condition, an animal preferring a particular orientation aligns according to the artificial field; if the field is compensated, the preferences disappear. Following the manipulation, the animal realigns to the natural field.

Although typically associated with resting behavior, body alignments may also be involved in more complex contexts. In some cases, body alignments may be over-ridden by other cues and may represent an adaptive, rather than a passive, response. Three remarkable examples are nest construction in honeybees and termites, alignment of waggle dances by honeybee foragers, and nomadic movements of the nomadic ant *Pachycondyla marginata*.

During nest construction, compass termites (*Amitermes meridionalis*) orient their gigantic mounds of northern Australia on the North–South axis (with some regional variation reported). Similarly, honeybees seem to rely on magnetic information during comb construction as suggested by the irregular combs built when magnetic alterations are experimentally introduced. In both species, the use of magnetic information during nest construction has been associated with the absence of directional cues in the dark and the need to coordinate many individuals involved in the task. Furthermore, in compass termites, the preference for the North–South axis may benefit nest thermoregulation, maximizing exposure to the sun during mornings and afternoons and minimizing it at noon. Indeed, regional variation in orientation is related to the environment in such a way as to suggest that nest alignments with the magnetic field are adaptive.

A preference for a magnetic axis is also evident during the waggle dance of the honeybee. Foragers dance to communicate to their hive mates the direction of a resource by transposing the angle between the sun and the resource to the angle between the gravity and the axis of their dance. The directions communicated exhibit systematic errors that vary across the day (with the position of the sun). Such 'misdirections' disappear if the magnetic field is compensated, or if the direction toward the resource coincides with the cardinal directions. Thus, the waggle dance seems to reflect the preference of the honeybees to align with the magnetic field, as also observed during comb construction or when they are at rest.

A colony of nomadic ants *P. marginata* relocates its nest with a preference to move on a North–South axis. Relocation of nests is associated with the capture of their only prey, the termite *Neocapritermes opacus*. Following relocation, they forage on either side across their migratory axis, enabling colonies to minimize overlap with areas that were previously searched.

Tropotactic-Based Navigation

In tropotactic behavior, insects move toward or away from a stimulus, such as light, humidity, or temperature. Such movements typically lead the animals to more favorable conditions within relatively short distances. In cases where the directional stimulus and the magnetic field are spatially associated, an animal may substitute use of the latter as a directional cue if the primary stimulus is absent or lacks directional information.

For example, fruit fly (*D. melanogaster*: Diptera) larvae exhibit negative phototaxis during their first three developmental instars, but when they begin to search for a pupation site, they switch to positive phototaxis. In the right experimental setup, the light's direction can be associated with the direction of the magnetic field. After experimental shifts of the field, fruit fly larvae reorient as predicted.

As another example, mealworms (*Tenebrio molitor*: Coleoptera) show positive phototaxis in low or high relative humidity, and negative phototaxis in intermediate conditions. Similar to fruitflies, the light's direction can be associated with the direction of the magnetic field. This has been verified by shifting the magnetic field under homogeneous lighting or in the absence of light. Under homogeneous light conditions, adult mealworms moved toward the predicted direction of greater or lesser light that was indicated by the rotated magnetic field, whereas they oriented randomly in darkness.

Central Place Foraging

Central place foraging is a challenging task that involves displacements between home and particular resources, typically food and mating areas. This implies that movements occur within a more restricted area than migrations, which allow animals to rely on their memory to recognize particularities of the terrain (e.g., visual landmarks). Within insects, research on the use of a magnetic compass during central place foraging has focused on the species of Hymenoptera, particularly ants and bees, and Isoptera (termites).

The use of a magnetic compass during foraging is suggested by the ability of certain species to navigate home even when cues, such as sunlight and landmarks, are absent. Often, central place foragers may be evaluated in a natural context in which the location and nature of the goal is specified by the experimenter. Under these conditions, foragers are typically trained to visit a feeder where either the external state is manipulated (e.g., exposure to reversed fields or training to experimentally produced magnetic anomalies) or their internal state is altered with strong disruptive fields (e.g., a brief magnetic pulse, **Figure 3** or a strong bar magnet, **Figure 4**).

Training in a discriminative paradigm has been used in the honeybee *A. mellifera*. Honeybee foragers can be trained to recognize the location of a food source based on differences between two locally produced magnetic fields. The repeatedly successful replication of this paradigm has proved that honeybees can rely on magnetic cues during foraging; yet this is not the only proof of

Figure 3 Pulse magnet treatment of a migrating *Urania* moth captured flying across the Panama Canal. Should the compass be composed of single domain magnetite and arranged in a similar way to that in the magnetotactic bacteria, this treatment will reverse the geographic orientation of the moths upon release.

such capacity or the only context in which honeybees use magnetic information.

On the other hand, manipulations of the horizontal component may include a total reversal (180°) or a partial shift (typically 90°) of the magnetic polarity. Between these two variants, the partial switch is preferred because the total reversal may lead to axial distributions that are more difficult to interpret. Nevertheless, both manipulations have successfully provided evidence for the use of a magnetic compass in insects. Carpenter ants and honeybees turn their orientations in experimental arenas according to the artificial magnetic shifts (magnetic field turned 90°). Similar changes are observed in termites when fields are shifted even less than 45°. On the other hand, under complete reversal of the magnetic polarity, foragers of weaver ants and leaf-cutter ants shift their homing orientation ~180° relative to bearings of control individuals, demonstrating that they can rely on a magnetic compass for homing when other cues are absent.

However, insects do not rely on magnetic compasses only when information from other compasses is not available or reliable. This seems to be true particularly for central place foragers, which, as mentioned before, navigate within a familiar area. In this case, alternative mechanisms, such as landmark navigation, pheromone

Figure 4 Strong magnet treatment of migrating butterflies captured flying across the Panama Canal. *Inset:* Detail of a butterfly during exposure to a strong magnet bar.

trails, or path integration, may be suitable enough for foraging and homing; yet, those mechanisms may interact with, or be supported by, a magnetic compass. Indeed, a path-integrated vector can be updated by the use of a magnetic compass in leaf-cutter ants, as demonstrated by the shift in the vector home after a reversal in the magnetic polarity or by the inability to orient homeward after exposure to a strong magnet.

The use of landmarks may also be supported by a magnetic compass. During route learning, honeybee foragers should recognize the location of the landmarks on the terrain based on a directional framework. Although a sun/sky compass may efficiently provide such a framework, a magnetic compass may act as an alternative if other cues are not available. The magnetic compass also has the advantage of providing an unambiguous system when compared, for instance, with a polarized sky. Magnetic compasses may be of major importance for species living in the forest or in the dark, where landmarks are certainly available but neither the sun nor the sky can be reliably used. Indeed, termite foragers use a magnetic compass in conjunction with pheromones, in order to determine the trail's polarity and indicate the goal's direction.

Long-Distance Migrations

All the features described before have made the magnetic field a recurrent candidate to be a directional cue for long-distant migrants. Within insects, long-distant migrants include species of dragonflies, beetles, butterflies, moths, and locusts. In some of these insects, observations of

directed migrations in the absence of celestial cues, such as sunlight, have been used to suggest the use of a magnetic compass. This is the case for migratory butterflies, such as the monarch butterfly *Danaus plexippus* or the sulphur butterflies *Aphrissa statira* and *Phoebis argante*, all of which can orient with a sun compass, but are also observed migrating directionally under overcast skies. It is also the case for some migratory moths, such as the silver Y *Autographa gamma*, which maintain migratory directions on moonless nights.

Results from experiments manipulating the magnetic environment and from experiments disrupting the compass support the hypothesis that insects may use a magnetic compass for long-distant migrations. In Neotropical butterflies, natural migratory orientation is altered after exposure to a strong magnet (**Figures 4** and **6**). Also, orientation relative to migration can be reversed when the magnetic polarity is experimentally reversed (by a coil such as in **Figure 2**). Although in these manipulations, control groups do not always follow their natural directions of migration, the significant differences with treated groups suggest sensitivity of the compass to the experimentally manipulated magnetic field. In addition, magnetite crystals have been detected in the body of monarch butterflies and at least one of the Neotropical migrating butterflies (*A. statira*, see 'Properties of the Insect Magnetic Compass').

Of course, magnetic information is not necessarily the only or the primary mechanism that migrants may rely on. This fact makes the study of interactions between compasses an exciting field of research but complicates the experimental evaluation of a magnetic compass, especially in the field, where factors, such as weather and alternative navigational cues (e.g., landmarks, sun), are difficult to control. Therefore, field studies are often combined with laboratory manipulations; yet more controlled environments are not completely safe from confounded effects or lack of motivation of the animals.

Properties of the Insect Magnetic Compass

The use of a magnetic compass requires several steps, from the acquisition (perception) of the information to its transduction and subsequent use during the process of decision-making. Our current understanding of these levels includes mainly behavioral evidence for the perception of the magnetic information and, in some cases, how such information interacts with other cues. To date, we lack a complete understanding of the mechanistic processes underlying the perception of magnetism and its integration into multimodal strategies of navigation. For example, it has only rarely been tested whether insects detect polarity from the horizontal component of the Earth's magnetic field or from its inclination. Honeybees and ants obtain polarity information from the horizontal component of the Earth's magnetic field, but recently, flour beetles *Tenebrio* sp. were shown to use the inclination of the magnetic field relative to gravity for short distance movements.

The discovery that magnetotactic bacteria use chains of single-domain magnetite to cause them to move along the lines of magnetic flux stimulated the search for magnetite in animals. The mechanism for a compass based on single-domain magnetite is similar to our anthropogenic compass. The magnetite crystals are rotated to align with the magnetic field, providing the animal with information on the field's polarity. Although the mechanisms for neural transduction have never been verified, it is hypothesized that magnetite chains are attached to ion channels so that magnetically induced realignments would lead to opening of the channels and cell depolarization.

Support for the use of a magnetite-based compass comes from both behavioral assays and the presence of particles of magnetite in different species. Interestingly, the presence of magnetite is not exclusively associated with any organ in particular, and within an individual, it is not limited to a particular area. Its presence has been shown in diverse body areas such as the abdomen (e.g., honeybee *A. mellifera*), the thorax (e.g., monarch butterfly *D. plexippus*), the antenna (e.g., nomadic ant *P. marginata*, stingless bee *Schwarziana quadripunctata*), and the head (e.g., fire ant *Solenopsis invicta*).

Recently, a second magnetite-based configuration for sensing magnetic fields has been proposed for honeybees. In this system, variations in the field intensity are associated with changes in the size of magnetic granules (located inside iron deposition vesicles of trophocytes). Increases in field intensity lead to shrinking of magnetic granules in a direction parallel to the applied field and to their expansion perpendicular to the applied field. Furthermore, such changes in the magnetic granule's size are associated with intracellular release of calcium from the trophocytes. As iron deposition vesicles are attached to the internal cytoskeleton, it is proposed that changes in the magnetic granules' sizes induce relaxation or contraction of the cytoskeleton, which in turn, lead to the release of calcium ions for signal transduction.

The radical pair compass is a mechanism proposed for sensing magnetic fields without magnetite. Photosensitive molecules are excited by the incidence of light, and an electron is elevated to the singlet excited state. Singlet radical pairs form with antiparallel spin, and there is a reversible conversion of singlet radical pairs with antiparallel spin to triplet radical pairs with parallel spin. The equilibrium state of the reversible reaction forming the two radical pairs depends on the alignment of the sensory system to the earth's magnetic field. Presumably, the animal could sense the orientation of the magnetic field by comparing the amount of conversion from singlet to triplet radical pairs in different orientations. The conversion of singlet to triplet radical pairs would be symmetrical around the axial vector of the magnetic field, and thus serve as an inclination compass. Since photopigments, such as opsins, do not form radical-pairs in reaction to light, it has been proposed that other molecules, specifically cryptochromes, may be involved in the magnetoreception. Within insects, cryptochromes are found in the fruit fly *D. melanogaster*, which, accordingly, has a light-dependent magnetic compass.

Indeed, a direct connection between cryptochromes and magnetic sensitivity was recently determined in fruit-flies. In a T-maze paradigm, fruit fly adults can be trained to associate the presence of a magnetic alteration with a food reward. Wild strains thoroughly learned the association, whereas Cry mutants (i.e., cryptochrome-lacking mutants) failed in the task. Wild strains that were trained in light spectra that do not activate the cryptochromes also failed in the learning task.

Future Challenges and Prospects

We have emphasized throughout this article that in order to understand orientation by insects with magnetic compasses, we need to integrate how the information is sensed, how it is perceived and processed, and how the animal uses and responds to the magnetic information. The integration of those levels is probably one of the most interesting challenges for the near future.

First, we are in need of developing and refining the methodologies for the evaluation of the magnetic sensory system. We currently lack replicable behavioral paradigms enabling the simultaneous evaluation of neural activity while magnetic information is experimentally manipulated. Recent attempts aiming to standardize behavioral assays have relied on alignment behavior in cockroaches, which might be used under more restrained conditions (e.g., for electrophysiology recordings). However, a magnetic compass involved in body alignment may involve completely different sensory organs and neural wiring relative to that involved in goal orientation. We have not yet found a specialized organ for sensing magnetic information in insects. Concentration of magnetite in the head and antennae indicate that these sections might play a role in the magnetic perception of certain species. Research that focuses on the antennae and other body parts accessible for electrophysiological recordings will hopefully improve the chance of isolating the relevant sensory tissues.

For example, the protein-stringing magnetite vesicles onto the bacterial cell filament is known to be mamJ. Researchers have suggested that a general survey of the Animal Kingdom be conducted for other species that express mamJ. Similarly, an antibody to mamJ could be used as a marker to structurally link magnetite crystals to neural cells. For example, magnetite has been observed with electron microscopy in the antennae of the nomadic ant *P. marginata.* Is mamJ also expressed in the same general region and can it be associated with neural synapses?

In the previous section, we discussed the lack of research on a compass based on the polarity of the magnetic field versus a compass based on inclination. If only the horizontal component of the magnetic field is experimentally changed, one cannot distinguish the two types of compasses. A complete experimental protocol would include a natural control, changing the horizontal component without changing the inclination, and reversing the inclination without changing the horizontal component (**Figure 5**). The experimental setup can be accomplished with two overlapping coils – one oriented about the polarity vector and another oriented about the vertical flux.

Second, as the magnetic compass is part of a multimodal system of navigation, research on the hierarchical and supportive interactions with other mechanisms warrants further experimental efforts. Many animals rely on more than one compass, with the sun being a typical reference for diurnal insects. African dung beetles orient with polarized moonlight. When more than one compass is invoked, they may conflict with one another. For example, both the sun and moon compasses are based on the rotation of the Earth about its geological poles, whereas

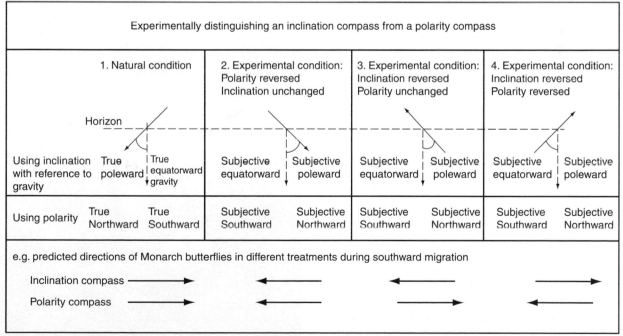

Figure 5 Manipulations of the local magnetic field to distinguish a polarity compass from an inclination compass. Experiments would include a natural control, changing the horizontal component without changing the inclination (which will alter insect orientations whether a polarity compass or an inclination compass is operating), reversing the inclination without changing the horizontal component (which will alter insect orientations if they use an inclination compass but not if they use a polarity compass), and reversing both the inclination and the horizontal component (which will alter insect orientations if they use a polarity compass but not if they use an inclination compass).

the magnetic compass is based on the geomagnetic poles (true North lies in the Arctic Ocean, whereas magnetic North is offset 11° latitude onto the Canadian island of Ellsmere). The difference in orientation between the geological and geomagnetic axes is called declination and varies with location on the Earth. In short range movements, declination will not be an issue, but over long distances, animals that use both a sun and a magnetic compass must calibrate one with the other.

Over the longer term, the geomagnetic poles are not stable points. Declination changes on an ecological timescale and the poles may reverse on a geological timescale (on the order of one-half million years). The most recent reversal was 750 000 years ago. For long-distance migrants, the Earth's magnetic poles are stable within a generation, but an insect, such as a monarch butterfly, must have a means to reach its winter destination site in central Mexico encoded genetically. Thus, whether the insect's preferred magnetic compass headings are plastic or hard-wired will be important to its success at reaching its destination. Finally, sunspot activity can disrupt the Earth's magnetic field creating magnetic anomalies that ebb and wane on a 11-year cycle. For example, orientation of stingless bees *S. quadripunctata* at their nest entrance was altered by a magnetic storm in 2001. How do animals cope with changes in declination and magnetic anomalies?

Third, how might insects use the magnetic field? Monarch butterflies east of the Rocky Mountains must carry the genetic blueprint necessary to fly from natal grounds to an overwintering site in the mountains of Michoacan, Mexico, as far as 5000 km away. Elaborate hypotheses for how they navigate en route involve the use of geomagnetic information. One particularly interesting feature is a magnetic anomaly at their destination that may guide their final approach like a beacon. On a more local scale, honeybees have been trained to detect spatial anomalies associated with nectar rewards, which can be used to measure their sensitivity to differences in magnetic fields or to set up experiments where orientation cues conflict with one another.

We need to make experiments as natural as possible to investigate how insects use the magnetic field. Tethering of insects confounds body alignment with goal orientation, orientations that may involve different sensory tissues and neural processing. Arenas that obscure celestial cues and landmarks cause insects to lose motivation to move toward a goal and attempt to escape instead. Insects that migrate for long distances or at high altitudes are notoriously difficult to study in these artificial settings. We have successfully tracked naturally migrating butterflies, *Urania* moths, and dragonflies, as they flew over bodies of water, by following them in a boat. We have also conducted releases of butterflies and day-flying moths over water in order to conduct experiments in an open environment that is as homogeneous as possible

(**Figure 6**). However, the boat has its limitations of distance, and for nonmigratory species, a body of water could be an unnatural setting. Radio transmitters are becoming lighter in weight to the point that dragonflies and other migratory insects can be followed remotely (**Figure 7**), and the launch of low-orbiting satellites to track insects with satellite transmitters is on the horizon in project ICARUS (International Cooperation for Animal Research Using Space).

The use of animal models, such as the fruitfly *D. melanogaster* or the honeybee *A. mellifera*, will certainly be of major relevance. This is exemplified by the recent genetic manipulations affecting the magnetic compass in *Drosophila* adults and providing us with a more detailed understanding of the mechanistic perception of magnetic information. Having genetic tools to integrate levels from sensory input to information processing and behavior output seem very promising. Nevertheless, it is also very important to consider the limitations of traditional models

Butterfly ⟶

Photo courtesy of Christian Ziegler

Figure 6   A butterfly released over the Panama Canal. Following manipulation of its internal state with a strong magnet, we measure the compass orientation on the horizon at which the insect vanishes (a vanishing bearing).

Photo courtesy of U.S. department of agriculture-agriculture research service

Figure 7   A migrating Mormon cricket *Anabrus simplex* (Orthoptera) with a radiotransmitter glued to its thorax.

in aspects such as experimental manipulations and the extent to which they enable generalizations (i.e., at what degree they are really models). This is particularly true as our knowledge of the diversity of magnetism-associated behaviors increases. An example is the difference between magnetite and light-based magnetic compasses. Thus research on species, such as *Drosophila*, may shed light on the processing of magnetic information, but it may remain limited to those insects with a light-based magnetic compass.

Therefore, the exploration of magnetic compasses in other species is warranted. Increasing the range of species not only enables the evaluation of the taxonomic distribution of magnetic compasses but, importantly, it might provide us with more suitable models for specific questions. In this respect, the use of nocturnal species appears promising, since other relevant mechanisms of navigation, such as a sun compass or landmark navigation, are naturally controlled and, of course, those species might adopt the magnetic compass as the primary mechanism. A complete understanding of orientation by the magnetic compass will also provide a better comprehension of insect cognition such as decision-making, learning, and memory.

See also: Insect Migration; Insect Navigation; Magnetic Orientation in Migratory Songbirds.

Further Reading

Banks AN and Srygley RB (2003) Orientation by magnetic field in leaf-cutter ants, *Atta colombica* (Hymenoptera: Formicidae). *Ethology* 109: 835–846.

Capaldi EA, Robinson GE, and Fahrbach SE (1999) Neuroethology of spatial learning: The birds and the bees. *Annual Review of Psychology* 50: 651–682.

Gegear RJ, Casselman A, Waddell S, and Reppert SM (2008) Cryptochrome mediates light-dependent magnetosensitivity in *Drosophila*. *Nature* 454: 1014–1018.

Gould JL (2008) Animal navigation: The evolution of magnetic orientation. *Current Biology* 18: R482–R484.

Hsu C-Y, Ko F-Y, Li C-W, Fann K, and Lue J-T (2007) Magnetoreception system in honeybees (*Apis mellifera*). *PloS One* 2: e395.

Kirschvink JL, Walker MM, and Diebel CE (2001) Magnetite-based magnetoreception. *Current Opinion in Neurobiology* 11: 462–467.

Riveros AJ and Srygley RB (2008) Do leaf-cutter ants *Atta colombica* orient their path-integrated, home vector with a magnetic compass? *Animal Behaviour* 75: 1273–1281.

Srygley RB, Dudley R, Oliveira EG, and Riveros AJ (2006) Experimental evidence for a magnetic sense in Neotropical migrating butterflies (Lepidoptera: Pieridae). *Animal Behaviour* 71: 183–191.

Walker MM (1997) Magnetic orientation and the magnetic sense in arthropods. In: Lehrer M (ed.) *Orientation and Communication in Arthropods*, pp. 187–213. Basel: Birkhäuser.

Williams JED (1994) *From Sails to Satellites: The Origin and Development of Navigational Science*. Oxford: Oxford University Press.

Wiltschko R and Wiltschko W (1995) *Magnetic Orientation in Animals*. Berlin: Springer.

Magnetic Orientation in Migratory Songbirds

M. E. Deutschlander, Hobart and William Smith Colleges, Geneva, NY, USA
R. Muheim, Lund University, Lund, Sweden

Introduction

The navigational challenges faced by migratory songbirds are both immense and complex. Whether migrating short or long distances, within or between continents, or to reach wintering or breeding sites, migratory songbirds must navigate across diverse landscapes, often facing large ecological barriers and adverse weather conditions, to reach habitats appropriate for their needs; indeed, many individuals are capable of migrating to the same breeding site and/or wintering site year after year. Such navigational feats impress amateur birders, naturalists, and scientists alike, and beg for questions about how small birds, often weighing only a few grams or tens of grams, manage to find their way. Moreover, while their final destinations have captured our attention for centuries, scientists have only begun to realize the importance of stopover sites, habitats where birds can rest and 'refuel' to continue migration. The success of the seasonal trips of these birds relies not only on reaching their destination at appropriate times but also by following 'historical' routes that provide adequate habitat along the way.

To find their way, birds require a complement of navigation mechanisms and strategies, which allow them to cope with changing habitats and information as they move between equatorial and polar latitudes. Orientation cues available to songbirds, most of which migrate during the night, include celestial information, such as the stars and sunset (**Figure 1**), as well as the Earth's magnetic field. While the use of the geomagnetic field is the focus of this chapter, birds must integrate or calibrate the direction information they obtain from different cues. Much like humans and other animals that integrate information from their visual and vestibular systems to provide a sense of position and movement, migratory birds must use information from their visual system (about the position of the stars, setting sun, etc.) along with magnetic information in order to obtain their overall 'sense of direction.'

Our goal is to provide an overview of what is currently known about magnetic orientation in songbirds. We will address both physiological mechanism(s) for sensing the Earth's magnetic field, as well as ecological, or functional, uses of magnetic information. While we have known that songbirds are capable of orienting using the Earth's magnetic field for over half a century, there is much more to learn about magnetic sensing, how the nervous system encodes and processes magnetic information, and how birds use magnetic information in different ecological contexts and in combination with other directional cues. We hope this chapter provides an impetus for students of animal behavior to address these mysteries in the decades to come.

A Brief History

Even though the discovery of sensitivity to the geomagnetic field in animals is rather recent, the notion that animals might make use of geomagnetic information for orientation tasks is quite old. As early as the mid-to-late 1800s, scientists suggested that the geomagnetic field might underlie the extraordinary navigational capabilities of birds and insects; Charles Darwin suggested that it might be worth investigating the effect of attaching small magnets to bees to try and manipulate their orientation behavior. However, it was not until the 1960s that Wolfgang Wiltschko, under the tutelage of Friedrich Merkel, was able to demonstrate that the orientation response of a caged migratory bird, the European robin (*Erithacus rubecula*), was affected by the direction of an Earth-strength magnetic field. Although initial evidence for magnetic orientation was met with much skepticism, the body of evidence for magnetoreception in birds and other animals, including many species of invertebrates, fish, amphibians, reptiles, and some mammals, has grown quite rapidly in the past 45 years. Now even a skeptical reviewer of the literature would have to conclude that magnetoreception is a widespread sense among animals.

Wiltschko and Merkel's experimental design to test for magnetic orientation in birds is essentially still the method of choice for examining migratory orientation, although the technique has been modified slightly over the years. The technique is based on observations made by Gustav Kramer in the late 1940s; Kramer observed that captive (i.e., caged) migratory birds exhibit migratory, or nocturnal, restlessness (in German known as 'Zugunruhe'). More importantly, Kramer noticed that the direction of the birds' activity, which is indicated by increased hopping in their cages, corresponds with their seasonal migratory direction. To assess a bird's orientation, Wiltschko and Merkel used radially positioned, recording 'event' perches in an octagonal-shaped cage, which allowed them to monitor each bird's position (i.e., directional preference) and activity. Currently, the most prevalent cage type used by researchers is the circular 'Emlen' funnel (**Figure** 2), first used by Stephen Emlen to study stellar orientation in indigo buntings (*Passerina cyanea*).

The key to demonstrate magnetic orientation is to show that birds change their absolute, or geographic, direction when the direction of the magnetic field is no longer pointing toward geographic North. 'Magnetic coils' are used to rotate the direction of the magnetic field; these three-dimensional coils are often cubical or octagonal in shape and wrapped with copper wire to which electric current can be applied to create an artificial

magnetic field (**Figure** 3). With proper current and positioning of the coil(s), Earth-strength magnetic fields can be created that differ from the actual Earth's magnetic field only in direction; that is, a magnetic field can be created to point toward geographic East, South, West, or a variety of positions in between. Wiltschko and Merkel used such a magnetic coil to show that European robins, when given access only to magnetic cues, will consistently migrate to the north–north east; when the direction of the magnetic field was rotated, the birds changed their orientation to reflect the position of the magnetic field.

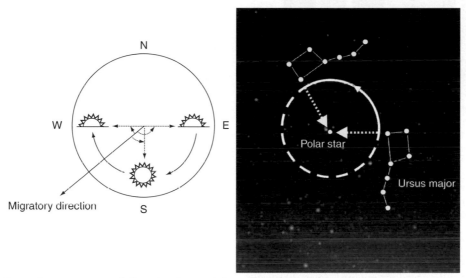

Figure 1   Illustrations of a sun compass (left) and a star, or stellar, compass (right). Celestial compasses each rely on the rotation of the Earth, which causes relative movement of the sun and the stars, to provide information about geographic, or true, North in the northern hemisphere. Left: Although most songbirds migrate at night, some are diurnal (or day) migrants. Diurnal migrants (and homing pigeons) can use the sun's position along with an internal circadian clock to determine a migratory direction. In order to maintain a constant geographic heading, the angle between the migratory direction and the sun's position changes with time of day. Hence, the need to coordinate direction with an internal clock is required to use a sun compass. Nocturnal migrants, which take off and land at sunset and sunrise, can use the position of the sun at sunset or sunrise to determine a direction just before departure or landing, respectively. Right: During the night, the Polar star (in the northern hemisphere) can be used by nocturnal birds to steer a heading. Birds can learn which star is the pole star by the rotation of other constellations around this immobile star in the night sky.

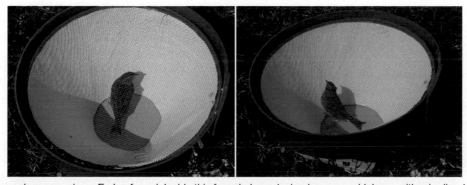

Figure 2   A Savannah sparrow in an Emlen funnel. Inside this funnel-shaped, circular cage, which can either be lined with a recording paper (such as typewriter correction paper or thermal paper) or blank newspaper and an inkpad at the bottom, songbirds will hop. Either method results in marks (scratch marks or ink blotches, respectively) on the paper, which indicate the direction the bird was hopping. During migration, songbirds will tend to hop in the direction in which they would be flying when placed in Emlen funnels during dusk or evening. To test for solely magnetic orientation, visual cues can be blocked by covering the funnel with opaque, white Plexiglas.

Figure 3 A magnetic cube coil for manipulation of the Earth's magnetic field. On the table in the center of this coil are several Emlen funnels for testing bird orientation. Each is covered with opaque, translucent Plexiglas, which will block out visual cues but allow some light to enter the funnel. The coil is doubly wrapped with copper wires, which allow two magnetic vectors to be created in order to change the magnetic field so that it can point toward geographic north, east, south, or west depending on which wire(s) electric current is applied to.

Since these earliest experiments, researchers have not only demonstrated that the ability to use geomagnetic cues for orientation is widespread in songbirds (over 20 species have been examined to date), but similar methods have been used to provide the evidence for most of the ideas we present in this chapter. While technological advances in tracking devices (such as miniaturized telemetry devices and geolocators) are providing new methods to explore orientation choices in free-flying migrants, experiments on caged migrants have provided the bulk of our current knowledge about magnetic orientation. Indeed, Wolfgang Wiltschko, along with his wife Roswitha, have continued to explore magnetic orientation since the first experiments on European robins, providing a wealth of hypotheses, experiments, and data that have always been at the forefront of research on magnetic orientation in birds. No review of magnetic orientation would be complete without acknowledging their lifetime achievements and their clear influence on their own students, their collaborators, and all of us who have worked on magnetic orientation.

The Geomagnetic Field

The Earth's magnetic field is analogous to a field produced by a bar magnet. However, the geomagnetic field is actually produced by a self-generating geodynamo, where fluid motion in the core of the Earth moves electrically

conducting material across an existing field. The magnetic field lines leave the Earth in the southern hemisphere and enter the Earth in the northern hemisphere (**Figure 4**). The intensity of the geomagnetic field ranges from about $68 \, \mu T$ at the magnetic poles, where the field lines stand vertically (known as an inclination, or dip, angle of $90°$), to about $23 \, \mu T$ around the magnetic equator, where the field lines are parallel to the Earth's surface (inclination angle is $0°$). The two magnetic poles are not static; rather they constantly drift or 'wander.' Moreover, the magnetic poles do not coincide with the geographic poles, which are defined by the axis of rotation of the Earth. The difference between a magnetic pole and its corresponding geographic pole (known as 'declination') is measured as an angle from any reference point on the Earth. Currently, the poles are wandering several tenths of a degree annually, which is called 'secular' variation, and the total intensity of the Earth's magnetic field has decreased by about 10% since 1900.

Functions of the Earth's Magnetic Field for Avian Migration

The Avian Magnetic Inclination Compass

The magnetic compass of migratory birds functions differently from the industrial compasses that humans use for orienteering. The needle of most commercially available compasses points toward the magnetic North pole, which is why this type of compass is called a 'polarity compass.' The magnetic compass of birds is insensitive to the polarity of the magnetic field. Rather birds sense only the axis of the magnetic field and they must rely on magnetic inclination, or the dip angle, to determine direction. Therefore, the avian magnetic compass is called an 'inclination compass.' Birds use the inclination of the magnetic field lines to determine which of the two sides of the magnetic axis leads toward the magnetic equator or toward the closer of the two magnetic poles (**Figure 4**). The side of the magnetic axis where the magnetic field lines meet with the horizon always leads polewards, in both the northern and the southern hemispheres, and the side where the field lines and the horizon diverge always leads toward the magnetic equator.

An experimental method to determine the type of compass (inclination compass or polarity compass) that an animal possesses is to artificially invert only the vertical component of the magnetic field with a magnetic coil surrounding a testing apparatus (such as birds placed in an Emlen funnel). Inverting the vertical component flips the magnetic field vector, that is, reverses inclination, but leaves polarity unchanged. Consequently, animals, such as birds, that possess an inclination compass will reverse their direction of orientation when the vertical component of the magnetic field is inverted, even though the polarity (i.e., N–S axis) of the magnetic field is

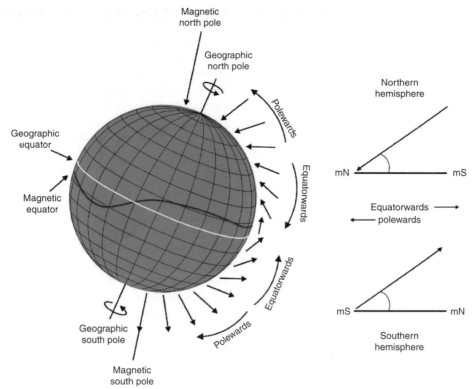

Figure 4 The Earth's magnetic field (left) and function of the avian inclination compass (right). Left: The arrows near the Earth's surface indicate the intensity (lengths of arrows), direction (direction of arrowhead), and angle of inclination (steepness of the arrows in relation to the surface of the Earth) of the magnetic field at a particular site. Right: The birds' inclination compass is not sensitive to the direction of the magnetic field, but rather its alignment and sign of inclination. Birds do not distinguish between 'north' and 'south,' but between 'equatorwards' and 'polewards.' The side of the magnetic axis where the magnetic field lines meet with the horizon always leads toward the pole, in both the northern and the southern hemispheres, and the side where the field lines and the horizon diverge always leads toward the magnetic equator.

unchanged. A polarity-based magnetic compass, such as the one commonly used by humans for orienteering, will not respond to an inversion of the vertical field (i.e., it would continue to point toward magnetic North).

Similar to other sensory systems (such as the visual system), the functional range of the avian magnetic inclination compass also appears to be adaptable to different magnetic field intensities. Experiments with European robins demonstrated that birds are disoriented when tested in artificial magnetic fields weaker (16 and $34\,\mu T$) or stronger ($60-105\,\mu T$) than the Earth's magnetic field. However, preexposure to such unnatural magnetic fields for 1 h resulted in seasonally appropriate orientation, implying that the functional range of the magnetoreceptor is flexible and allows adjustment (although relatively slowly compared to other senses) to new magnetic conditions.

Determining direction: a flexible migratory compass program

Possessing a physiological magnetic compass provides birds only with a directional reference. Determining which direction to migrate (such as 'equatorwards' during

autumn migration) requires birds to 'know' the appropriate direction to fly for their species. Andreas Helbig, Peter Berthold, Eberhard Gwinner, and others have shown that the general direction and distance (or length) of migration is, at least in part, determined by an inherited (i.e., genetic) migratory program in birds. Because this genetic program, available to juvenile songbirds on their first migration, encodes information about length of migration and direction, it is called a 'clock and compass' migration strategy.

The migratory programs of songbirds are similar in both hemispheres; species that breed toward the poles migrate 'equatorwards' after the breeding season in autumn when day length decreases, and 'polewards' in spring when day length increases. However, some species (or even populations within a species) may fly in one direction for part of their migration (such as southwest) and then change to a different direction for the remaining part of their migration (such as more southerly directions). Therefore, determining the correct direction for successful migration can be more complicated than just to fly 'equatorwards' or 'polewards,' and even slight species-specific or population-specific differences in migratory direction appear to be at least partially determined by genetic information. For example, when Helbig crossbred male and female blackcaps (*Sylvia atricapilla*) from two populations with different autumn migratory directions in Europe, the offspring oriented in a direction intermediate to the two population-specific directions.

Crossing the magnetic equator using an inclination compass is a challenging task for extremely long-distance migrants, such as bobolinks (*Dolichonyx oryzivorus*) and garden warblers (*Sylvia borin*). The horizontal alignment of magnetic field lines at the magnetic equator prevents the determination of direction with an axial, inclination compass. While crossing the magnetic equator, birds would thus have to rely on other compass cues, such as stellar patterns. Moreover, transequatorial migrants have to change their migratory program from 'fly equatorwards' to 'fly polewards' during fall migration, while at the same time the inclination compass information is ambiguous. Experimental evidence indeed suggests that exposure to the horizontal magnetic field at the magnetic equator triggers this change in migratory program.

Energetic condition and local geography can influence orientation

The migratory program of songbirds can be quite flexible during a single migratory journey. For example, birds can and will adjust their migratory direction to reflect their own energetic condition and/or local ecological features. Songbirds must be physically prepared for migration, which takes enormous amounts of energy, which they store largely as fat. Rather than carry excessive fat loads during migration, birds optimize migration speed and time, at least in part, by periodically arresting migration to replenish fat stores at suitable stopover sites *en route*. Especially important stopover sites are located just prior to or after crossing expansive ecological barriers, such as large bodies of water or deserts, where feeding and refueling are difficult or impossible. At stopover sites near these ecological barriers, birds can gain significant fat to prepare for, or recover from, crossing the barrier. Orientation preferences of individual birds at these sites are dependent on both season and energetic condition of the birds. For example, upon encountering a large body of water (such as the Gulf of Mexico, one of the Great Lakes, or the Baltic Sea), fat migrants usually cross the barrier by exhibiting 'forward' migration in a seasonally appropriate direction. In contrast, lean birds often orient in opposite directions (i.e., reverse orientation) of fat birds when they encounter these same barriers, or they at least discontinue migration temporarily until their energy reserves are sufficient enough to continue migration. Reverse orientation may lead leaner individuals to more suitable stopover areas for refueling, with better food sources, less competition for food, and/or less predation pressure.

Comparing compasses – cue calibration

Migratory songbirds use both celestial and geomagnetic information for compass orientation. Celestial patterns, such as a stellar compass, provide birds with information about true or geographic North or South (**Figure 1**). Having multiple compass mechanisms during migration can be advantageous. Under overcast weather, for example, birds cannot use their sun and star compasses, but need to rely on the magnetic compass. Likewise, the wandering of the magnetic poles makes magnetic information less reliable than geographically based compass mechanisms. The directional information from these two types of compass systems changes during migration because of the spatially changing relationship between geographic and magnetic North (i.e., magnetic declination). Birds migrating at high arctic areas close to the magnetic North pole are exposed to particularly large changes in magnetic declination, because the differences between magnetic and geographic North are maximal there (**Figure 5**).

Both before the start and during the migratory journey, birds can correct for magnetic declination by calibrating their magnetic compass with a celestial compass, thus prioritizing the information from the celestial compass (i.e., geographic, or true, North) over magnetic compass information. Although a controversial idea, polarization patterns of skylight near the horizon at sunrise and sunset may serve as the primary calibration reference for the magnetic compass in many migrants, such as Savannah sparrows (*Passerculus sandwichensis*) and white-throated sparrows (*Zonotrichia albicollis*). These are the two times of day when the skylight polarization pattern is seen as a band of maximum polarization (BMP) across the zenith at a 90° angle relative to the position of the sun (**Figure 6**). The BMP intersects the horizon vertically; thus, detection is independent of horizon height.

Geographic Positioning and Use of Magnetic Information for Noncompass Behaviors

The occurrence of global geomagnetic gradients has led to several hypotheses for a magnetic 'map,' or geographic-positioning, sense. From the equator, both the intensity and the angle of inclination of the geomagnetic field increase toward the poles (as mentioned previously, see **Figures 4–7**). Map-based (or 'true') navigation requires nonparallel gradients of two or more geophysical features to determine one's position relative to a goal in order to return to a familiar area following displacement. A bicoordinate map (one that would provide the equivalent of latitude and longitude) based on the geomagnetic field would require that an animal perceives at least two components of the magnetic field that vary geographically, such as intensity and inclination, and that the two gradients be nonparallel. Geomagnetic intensity and

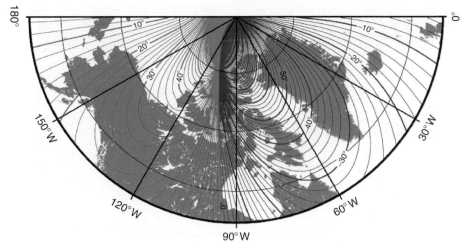

Figure 5 Illustration of magnetic declination at high Northern latitudes according to the World Magnetic Model of the Epoch 2000 (http://geomag.usgs.gov/). Declination isolines in green and red denote positive (deviations to the east of geographic North) and negative values (deviations to the west of geographic North), respectively. Note that in some areas a magnetic compass and a celestial compass could be more than 50° off. Therefore, calibration at high latitudes is essential for birds to make use of their magnetic compass.

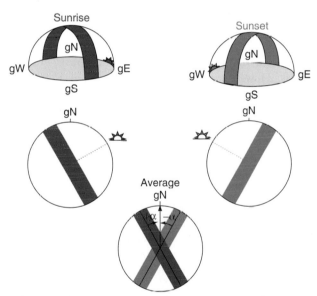

Figure 6 The band of maximum polarization (BMP) of skylight at sunrise and sunset. Top: Three-dimensional view of the BMP that intersects the horizon at a 90° angle at sunrise and sunset. Middle: This pattern is always symmetrical to geographic North, independent of time of year and latitude. Bottom: Averaging of the BMP vectors available at sunrise and sunset would provide birds with a true geographic reference for calibration of the magnetic compass and corrections of magnetic declination.

inclination mostly vary concomitantly along a north-south axis in the Americas and Europe and Africa and may provide only a unicoordinate map limited to latitudinal information for migrants on these continents. However, in several regions (e.g., the Indian and South Atlantic Oceans, and parts of Europe), these two gradients are not parallel to each other, making bicoordinate geomagnetic navigation theoretically feasible (**Figure 7**).

Because of local, regional variation in the alignment and steepness of geomagnetic gradients and temporal (such as daily or annual) geomagnetic variation, most hypotheses about map-based navigation presume that animals would have to learn the pattern of magnetic gradients within their home range or along their migratory route. Juveniles, that have not yet learned these gradients, would have to rely on other orientation strategies (i.e., the inherited, clock and compass strategy for migration as mentioned earlier). Displacement and recovery experiments with free-flying migrants, as well as laboratory experiments, consistently support a purely compass-based orientation strategy in juvenile birds. Age-dependent recoveries of geographically displaced

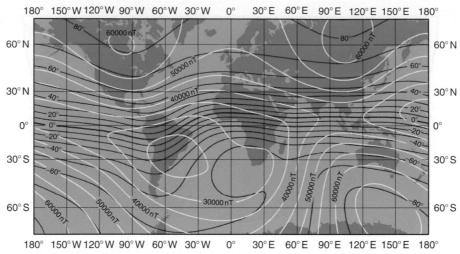

Figure 7 Total intensity (yellow lines) and inclination (red lines) of the Earth's magnetic field according to the World Magnetic Model (WMM) 2000 (http://geomag.usgs.gov/). The total intensity is shown in 5000 nT steps and the inclination is shown in 10° steps.

migratory European starlings (*Sturnus vulgaris*) by Albert Perdeck first suggested that adults use a different orientation strategy than do juveniles. After displacing thousands of banded starlings to the southwest of their autumn migratory route, adults were recovered in their usual population-specific wintering areas. However, juvenile birds, which had never migrated before, were recovered to the southwest of their population-specific wintering grounds. In other words, adults compensated for the geographic displacement, whereas juveniles continued to orient in a fixed compass direction without compensation.

Direct tests of magnetic map hypotheses, where geographic displacements are simulated by altering the intensity and/or inclination of the geomagnetic field, are few and in most cases involve 'homing behavior' in species other than songbirds. Eastern red-spotted newts (*Notophthalmus viridescens*), spiny lobsters (*Panulirus argus*), and green sea turtles (*Chelonia mydas*) orient toward a home or capture site when exposed to magnetic field values that simulate disparate geographical locations. Also consistent with map-based geomagnetic navigation, temporal variation and spatial anomalies in the geomagnetic field also affect homing orientation in pigeons (*Columba livia*) and alligators (*Alligator mississippiensis*). In songbirds, only Australian silvereyes (*Zosterops l. lateralis*) have been directly tested for a magnetic map sense by examining their orientation responses to magnetically simulated geographic displacements. Silvereyes that breed in Tasmania migrate northwards to wintering sites on the Australian mainland and then southwards to Tasmania during spring migration. Fischer and others (including the authors of this article, in unpublished studies) have found that adult silvereyes, but not juveniles, became disoriented or reoriented during autumn migration when exposed to magnetic field values that simulate a northerly displacement (i.e., beyond their normal wintering range). Although other

explanations are possible for the orientation behavior observed in these experiments, silvereyes may learn to use gradients in the geomagnetic field for at least a unicoordinate geomagnetic 'map' sense, which provides latitudinal information (see Freake et al., 2006).

Although it is unclear whether songbirds possess a magnetic 'map' sense, specific values of the geomagnetic field have been shown to serve as innate 'sign posts' (or sign stimuli) for locations along a migratory route. Genetically encoded geomagnetic values may stimulate an 'innate releasing mechanism,' causing migrants to change migratory behavior at appropriate locations, such as at stopover sites or migratory boundaries. When exposed to gradually decreasing values of magnetic intensity and inclination, juvenile pied flycatchers (*Ficedula hypoleuca*) shift their autumn orientation from southwest to southeast in magnetic fields that simulate those of southern Spain, as would freely migrating birds. Southeast reorientation toward Africa prevents the birds from migrating over the Atlantic Ocean. This example does not require birds to determine their position; instead, specific geomagnetic field values elicit an appropriate 'programmed' change in the bird's behavior, orientation, or otherwise. Likewise, juvenile thrush nightingales (*Luscinia luscinia*) increase feeding rates in a magnetic field simulating a known stopover site in northern Egypt.

Identifying the Avian Magnetoreceptor(s)

Early hypothetical biophysical models for magnetoreception have led to the discovery of two candidate magnetoreception systems in birds: (1) a light-dependent mechanism located in the eye and (2) an iron-based mechanism associated with the trigeminal nerve. Magnetic compass orientation of both juvenile and adult birds is light dependent, affected by both wavelength and intensity, and in at least two species, it is lateralized to the right eye. In addition, pulse remagnetization experiments and neurophysiological studies suggest that an iron-based mechanism innervated by the trigeminal nerve provides adult birds with magnetic information other than simply compass direction, possibly geomagnetic-positioning information.

Light-Dependent Magnetoreception and Compass Orientation

Magnetic compass orientation of migratory songbirds and homing pigeons is influenced when they are tested under different wavelengths (i.e., colors) of light. Experiments with songbirds, both adults and juveniles, in Emlen funnels illuminated with monochromatic lights demonstrate that birds tested under nearly monochromatic blue, turquoise, or green light (all relatively short wavelengths) were well oriented toward their seasonally expected migratory direction. Birds tested under longer wavelengths (i.e., yellow and red), however, either became disoriented or showed approximately 90° shifted orientation. Experiments with European robins tested under green and green-yellow lights, which differed by only 8 nm, showed that the transition from oriented behavior to disorientation occurred very abruptly. Light-dependent magnetoreception varies in birds not only with wavelength, but also with light intensity (i.e., brightness) at the same wavelength, leading to shifts in orientation, disorientation, or axial alignment along the migratory direction. The interactions of wavelength and intensity of light on compass orientation are complex and still not well understood.

Peter Semm demonstrated some of the first neurophysiological recordings on magnetically sensitive neurons in bird brains. In the nucleus of the basal optic root (nBOR) and the optic tectum, Semm showed a clear involvement of the visual system in light-dependent magnetoreception. His recordings demonstrated magnetic responsiveness to changes in the direction of a magnetic field and to slow inversions of the vertical component of the magnetic field, and thus strongly implied that light-dependent magnetoreception takes place at locations innervated by the optic nerve, with the eyes as likely candidates for the locations of receptor cells. Experiments testing the magnetic orientation of birds (i.e., European robins and Australian silvereyes) with one eye covered with light-proof caps suggest that light-sensitive magnetoreception is actually lateralized; mainly the right eye is involved in magnetoreception. Birds were well oriented and reacted to an inversion of the magnetic field when tested with the right eye open, but were disoriented when tested with the right eye covered.

Currently, the most accepted magnetoreception model for the light-dependent magnetic compass, originally proposed by Klaus Schulten, suggests that an external magnetic field can modulate photon-induced processes in specialized photoreceptors. In this process, radical pairs of light-sensitive molecules are formed by photon excitation through light absorption similar to the photosynthetic reactions. The interconversion between the two excited state products can be modified by an external magnetic field, resulting in the formation of different yields of singlet and triplet products (the triplet products being the signaling state). Cryptochromes, candidates for such a magnetoreception molecule, have been found in retinas of two migratory bird species, European robins and garden warblers. Photoreceptors containing such magnetosensitive molecules arranged in an ordered array in the eye would respond differently, depending on their relative alignment to the magnetic field. Birds would be able to 'see' the magnetic field lines as a three-dimensional pattern of light irradiance (i.e., brightness) or color variation in their visual field or through a dedicated parallel pathway in the brain (**Figure 8**).

The use of low-intensity oscillating radiofrequency magnetic fields (RF fields) in the lower MHz range (0.1–100 MHz) has been established as an important tool to test whether a radical pair mechanism is involved in the primary magnetoreception process of an orientation response. RF fields of distinct frequencies interfere with the interconversion between the excited states of the molecule(s) and can mask or alter the magnetic field effects produced by the Earth's magnetic field. RF interferences can lead to either disorientation or change in orientation, depending on the amount and type of change, and how the animals integrate the information into a migratory direction. Experiments with European robins exposed to such RF showed that birds become disoriented

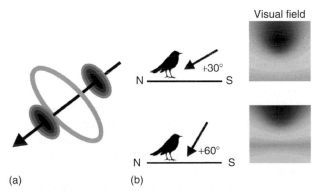

Figure 8 Illustration of light-dependent magnetic compass perception through magnetosensitive photoreceptors.
(a) Magnetic field vector (arrow) and three-dimensional pattern which birds are thought to perceive, consisting of a dark area on each side of the magnetic field axis and a ring in the center.
(b) Visual pattern perceived by a bird, depending on the relative alignment of the magnetoreceptors and the magnetic field; in this example, the bird is facing toward magnetic North with the eyes horizontally aligned at two different latitudes (i.e., magnetic field inclinations of 30° and 60°, respectively).

when tested under either a broadband RF field or distinct single frequencies in the lower MHz range. Iron-based magnetoreceptors, in contrast, would not be affected by RF fields, because the rotation of iron oxide particles, such as magnetite, would be too slow and the ferromagnetic resonance frequency is expected to be in the GHz rather than MHz range.

Iron-Based Magnetoreception

A magnetoreceptor based on a direct interaction with the magnetic field is fairly easy to imagine if one considers coupling a tiny biological 'bar magnet' with a sensory neuron; pull on or rotation of such a biological 'micromagnet' could, in theory, be detected by a mechanoreceptor-like neuron. Magnetite (Fe_3O_4) is a biogenically produced compound that exhibits ferromagnetic properties, which could give rise to such a 'bar magnet' based magnetoreceptor. In fact, particles of magnetite have been shown to be responsible for geomagnetic alignment of some anaerobic bacteria. In the mid-1970s, Richard Blakemore and Richard Frankel found that both living and dead marine bacteria from the North Atlantic passively oriented parallel to the magnetic field lines; the anterior end of each bacterium was pointed northward and downward (as are the lines of inclination in the northern hemisphere; **Figure 4**). In living bacteria, flagellar motion at the posterior end of the organism results in movement of the organism along the field lines toward the anaerobic areas of sediment at the water–substrate boundary. A variety of magnetotactic bacteria and algae have been found in both fresh and marine waters of the northern and southern hemispheres. In each case, long chains of (single domain) magnetite particles or, in some cases, greigite (an iron sulfide) are present within the bacteria and cause passive alignment with the geomagnetic field lines.

To confirm the role of magnetite in the orientation of these microorganisms, Blakemore (and his colleague, Adrianus Kalmijn) remagnetized the chains of magnetite in bacteria with a strong magnetic pulse oriented antiparallel to the orientation of the bacteria. This technique, known as 'pulse remagnetization,' will remagnetize (i.e., reverse the polarity of) any permanently magnetic particles; however, paramagnetic particles such as radical pairs will not be permanently affected. After pulse remagnetization, the magnetotactic bacteria oriented in the opposite direction showing that the magnetic pulse had reversed their 'behavior' by reversing the polarity of their magnetite chains.

The findings in bacteria triggered the search for a magnetite-based magnetoreception mechanism in animals. Magnetite is a fairly ubiquitous biogenic compound in animals and has been reported in insects, birds, fish, and mammals; it occurs in a variety of tissue types including the nervous system. In order to be useful for magnetic orientation, however, a magnetite-based magnetoreceptor would need to be associated with a directionally selective

sensory system. Physiological, anatomical, and behavioral studies have all provided evidence for an iron-based mechanism associated with the ethmoid, or nasal, region in birds and fish. Using traditional neurophysiological techniques, Robert Beason and Peter Semm first demonstrated that the ophthalmic branch of the trigeminal nerve, which innervates the ethmoid region of songbirds, is sensitive to changes in Earth-strength magnetic fields.

Recently, Gerta and Guenther Fleissner and others more fully described iron-containing structures in the dendrites of the ophthalmic branch of the trigeminal nerve in the upper beak of homing pigeons. The complex structures were found to contain both platelets of maghemite (another ferromagnetic iron oxide) and small round intracellular 'bullets' of magnetite, which are influenced by local magnetic fields around the sensors to detect the magnetic field and likely also amplify it so that sensory transduction can take place. Thus, geomagnetic transduction may work similar to other senses such as hearing, where the stimulus is amplified to increase detection. These iron-containing sensory neurons were found in three pairs, bilaterally arranged within the upper beak. Each pair is aligned along a different axis, so that when taken together, they could act as a three-dimensional magnetometer to detect multiple components of the magnetic field, analogous to a human-made three-axis magnetometer. With this structure, birds could sense both the direction and the intensity of the surrounding magnetic field.

Pulse remagnetization experiments similar to those conducted on magnetotactic bacteria have provided evidence that some aspect of magnetic orientation in birds is mediated by an iron-based magnetoreception mechanism. However, the characteristics of this trigeminal magnetic system suggest that it may mediate a magnetic 'map' sense rather than a magnetic compass sense. When birds are exposed to a strong magnetic pulse designed to remagnetize magnetite particles, a change in the direction of migratory orientation is observed in adult birds, but only when the trigeminal nerve is intact. If information from the trigeminal nerve is blocked with anesthesia, bobolinks can still show magnetic orientation, but the effect of the pulse (i.e., a shift in their orientation) is no longer evident. In other words, pulse remagnetization does not influence the adult's compass sense. Likewise, juvenile songbirds captured prior to their first migration, which should not have a map sense, are unaffected by pulse remagnetization. Interestingly, trigeminal neurons exhibit the requisite sensitivity to extremely small magnetic field changes that would be expected for precise geographic positioning.

Conclusion

Since the discovery of magnetic orientation in the European robin and other songbirds, researchers have begun to investigate many proximate questions about the genetics and development of magnetic orientation and the sensory 'rules' and processing of magnetic information. Despite almost 50 years of research, we still do not understand many of the basic rules of operation of this sensory system or its ecological functions; even the elusive magnetoreceptor(s) and magnetoreception mechanism(s) in birds have yet to be conclusively identified. Like many other senses, however, magnetoreception appears predisposed to be 'tuned' to Earth-strength magnetic fields, able to adapt to changes in the magnetic field, and to provide more than one type of information (e.g., birds appear to be able to detect both quantity, or magnetic field strength, and quality, such as magnetic inclination). Moreover, magnetoreception is not used alone for navigation. Rather, songbirds are capable of multimodal processing (i.e., combining magnetic and visual cues) in order to determine a direction for orientation during migration. However, how birds combine information from different compass types in their nervous system is still largely unknown. Furthermore, the functional role of geographic variation in the geomagnetic field for map-based navigation, geographic positioning, or sign post navigation needs to be more fully explored in species with different migratory pathways and constraints. Another 50 years of research on avian magnetoreception, including behavioral studies on caged and free-flying migrants, physiological and anatomical investigations of neurological mechanisms, and collaborations of biologist and physicists will likely lead to some of these answers and to the inclusion of this important sensory system in textbooks on animal behavior, physiology, and sensory systems, where it has largely been ignored.

See also: Amphibia: Orientation and Migration; Bird Migration; Magnetic Compasses in Insects; Maps and Compasses; Sea Turtles: Navigation and Orientation.

Further Reading

Able KP and Able MA (1996) The flexible migratory orientation system of the Savannah sparrow (*Passerculus sandwichensis*). *The Journal of Experimental Biology* 199: 3–8.

Beason RC (2005) Mechanisms of magnetic orientation in birds. *Integrative and Comparative Biology* 45: 565–573.
Berthold P (1996) *Control of Bird Migration*. London: Chapman & Hall.
Berthold P, Gwinner G, and Sonnenschein E (2003) *Avian Migration*. Berlin: Springer-Verlag.
Blakemore RP and Frankel RB (1981) Magnetic navigation in bacteria. *Scientific American* 245: 58–65.
Deutschlander ME and Muheim R (2009) Fuel reserves affect migratory orientation of thrushes and sparrows both before and after crossing an ecological barrier near their breeding grounds. *The Journal of Avian Biology* 40: 85–89.
Emlen ST and Emlen JT (1966) A technique for recording migratory orientation in captive birds. *Auk* 83: 361–367.
Fleissner G, Stahl B, and Falkenberg G (2007) Iron-mineral-based magnetoreception in birds: The stimulus conducting system. *Journal of Ornithology* 148: S643–S648.
Freake MJ, Muheim R, and Phillips JB (2006) Magnetic maps in animals: A theory comes of age. *The Quarterly Review of Biology* 81: 327–347.
Gwinner E (1996) Circadian and cirannual programmes in avian migration. *The Journal of Experimental Biology* 199: 39–48.
Helbig AJ (1996) Genetic basis, mode of inheritance, and evolutionary changes of migratory directions in palearctic warblers (Aves: Sylviidae). *The Journal of Experimental Biology* 199: 49–55.
Mouritsen H and Ritz T (2005) Magnetoreception and its use in bird navigation. *Current Opinion in Neurobiology* 15: 406–414.
Muheim R, Åkesson S, and Alerstam T (2003) Compass orientation and possible migratory routes of passerine birds at high arctic latitudes. *Oikos* 103: 341–349.
Muheim R, Moore FR, and Phillips JB (2006) Calibration of magnetic and celestial compass cues by migratory birds – a review of cue conflict experiments. *The Journal of Experimental Biology* 209: 2–17.
Perdeck AC (1958) Two types of orientation in migratory starlings, *Sturnus vulgaris* L., and chaffinches, *Fringilla coelebs* L., as revealed by displacement experiments. *Ardea* 46: 1–37.
Ritz T, Adem S, and Schulten K (2000) A model for photoreceptor-based magnetoreception in birds. *Biophysical Journal* 78: 707–718.
Ritz T, Thalau P, Phillips JB, Wiltschko R, and Wiltschko W (2004) Resonance effects indicate a radical-pair mechanism for avian magnetic compass. *Nature* 429: 177–180.
Rozhok A (2008) *Orientation and Navigation in Vertebrates*. Berlin: Springer-Verlag.
Wiltschko R and Wiltschko W (2009) Avian navigation. *The Auk* 126: 717–743.
Wiltschko R and Wiltschko W (1995) *Magnetic Orientation in Animals*. Berlin: Springer-Verlag.

Magnetoreception

G. Fleissner and G. Fleissner, Goethe-University Frankfurt, Frankfurt, Germany

A vast amount of studies can be found on the phenomenon of magnetic-field-guided behavior. In nearly all animal phyla, species have been shown to use parameters of the Earth magnetic field for navigation and orientation (**Figure 1**). Also, various physiological processes make use of information of the local magnetic field: for example, migratory birds grow fat before they cross the Mediterranean Sea and the Sahara desert and lose weight on their way back – under the control of the correct geomagnetic coordinates. Therefore, in laboratory studies, these processes can only be observed when the geomagnetic landmarks are simulated. Even in plants, phenomena have been demonstrated to depend on special geomagnetic parameters. So, it is no surprise that magnetic-field-guided behavior is frequently reported in additional organisms.

Another topic, which has gained great interest among researchers and especially practitioners, is man-made electromagnetic radiation and its putative impact on body and brain, especially at the biochemical and tissue level. These studies deal with conditions different from the natural geomagnetic field and are not concerned with the analysis of the basic processes involved in the magnetoreception of animals.

Despite much interest in the area, the mechanisms underlying magnetoreception are still enigmatic and obviously pose a challenge for researchers from different scientific areas. Several hypotheses and myths have evolved, and up to now, seem to dominate – and sometimes distort – the discussion of upcoming solutions to this fascinating question (**Box 1**). In order to find sound concepts for animals' magnetoreception, we will try to recall the essential physiological and physical background knowledge, which has to be respected before you may name a structure a magnetoreceptor. Once a promising structure has been found, studies can be planned to verify its function (within biologically relevant magnetic field conditions) and all steps of the related information processing pathways controlling behavior and/or physiology.

Although a relatively poorly understood sense, here we open the 'black box' and follow the steps inside between the magnetic field – as input to the organism, and magnetic-field-controlled behavior as output phenomenon, recalling on the way the cascade of physiological principles required (**Figure 2**). Examples will be given at each level of magnetoreception and magnetic field perception.

Parameters of the Earth Magnetic Field Relevant for Magnetoreception

The Earth's magnetic field is omnipresent. It is never hidden behind clouds or invisible in a daily cycle like the planets or transient like sound and smell. The Earth has many of the features of a huge dipolar magnet (**Figure 3**), where each geographic position is characterized by the local magnetic vector: the *vector direction* derived from the orientation and inclination of the local magnetic field lines, and the *vector length* as field intensity. The field lines run from the South Pole toward the North Pole (the magnetic South Pole next to the geographic North Pole and the magnetic North Pole near the geographic South Pole). This shows the inclination of the field lines to be maximal near the poles and parallel to the earth surface at the equator. The total intensity of the geomagnetic field is lower near the equator and higher poleward.

Some remarkable deviations of the smooth and stable dipole magnet model Earth may occur depending on: (1) *The local magnetic 'topography,'* which may show anomalies like magnetic hills and valleys, mainly originating from ancient volcanic activities. Detailed magnetic maps are available to be consulted as reference, for example, when selecting a release site for homing studies. (2) *Slow continuous modulations* of up to 500...nT. (3) *Time of day*: depending on the sun spot activity and the rotation of the Earth, the mean magnetic flux is lower at noon than in the evening or morning, especially during the summer, and especially on the sun-facing side of the earth. These parameters (1–3) must be controlled, when the Earth magnetic field is used as a reference in magnetic orientation experiments. (4) *Magnetic storms*, solar flares, or approaching thunderstorms may induce generally short but strong changes of the geomagnetic fields. They provide severe interference with electronic equipment (the last great geomagnetic storm happened 13 March 1989 affecting voltage regulation relay operations in Canada and North America) and also animal orientation (reported for migratory and homing and long-distance marine travelers such as dolphins and whales). Such animal reactions may indirectly hint toward the physiological limits of biological magnetic receptors, that is, which intensities and slopes of magnetic stimuli are used for the magnetic-field-controlled behavior or body functions. (5) *Long-term variations* far beyond the lifespan of any animal: the field intensity

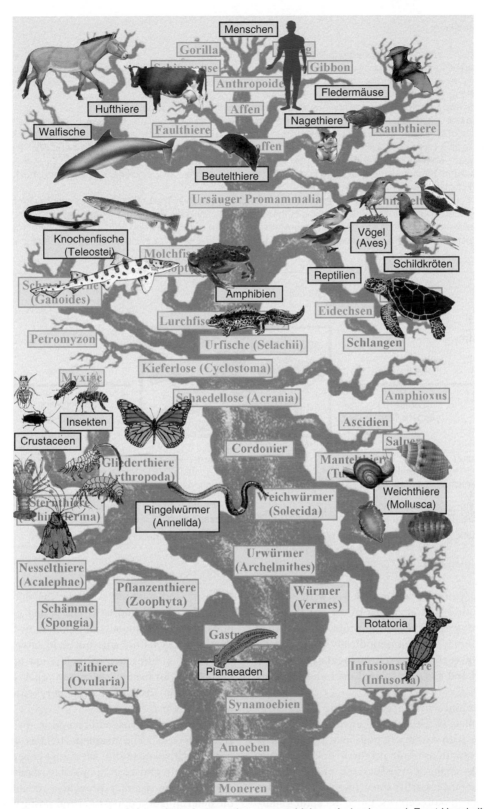

Figure 1 Schematic overview on animals investigated concerning magnetobiology. As background, Ernst Haeckel's tree of life (*The Evolution of Man*, 1874) provides a rough survey on an old systematic affiliation of the studied organisms (German terms according to Haeckel's original presentation). The intensity and the quality of the respective magnetobiological analyses are not respected. Data mainly according to listings in Wiltschko W and Wiltschko R (1995); pictures: private or public domain.

Box 1 Myths and Facts

- *Myth*: Magnetite incrustations hint toward putative magnetoreceptors: WRONG.
  Metal oxides can be found in hard body appendages like the radula of snails or the sting and the mouthparts of arthropods. Here, the metal inclusions serve to harden the delicate structures; sensory innervations are missing, which excludes their involvement in magnetoreception.
- *Myth*: Magnetite accumulations in central nervous tissue hint to a magnetoreceptor site: WRONG.
  Magnetite in the brain often occurs as leftover of the iron metabolism, with an increased amount along with inflammation, degeneration or neoplasmic processes. The central nervous system in vertebrates is no sensory organ and therefore cannot monitor parameters of the geomagnetic field. Likewise, magnetite in different organ systems is rather a by-product of the iron metabolism than an indicator of magnetosensation – unless it is localized within or next to sensory nerve endings.
- *Myth*: All metazoan magnetoreceptors originate from endosymbiontic magnetotactic bacteria, and therefore, must be composed of chains of single domain magnetite crystals: WRONG.
  Magnetotactic bacteria are not universal model systems for magnetoreceptors. Their magnetosomes are no component parts of a magnetomechanical transducer process. By means of their magnetosomal chain, bacteria are passively aligned in the geomagnetic field. The bacterial magnetosomes can contain magnetite or maghemite or Greigite or no iron at all. Even here, magnetite is not an obligatory magnetic material. This myth is persistently defended in order to antagonize 'noniron'-based magnetoreceptor mechanisms in metazoans.

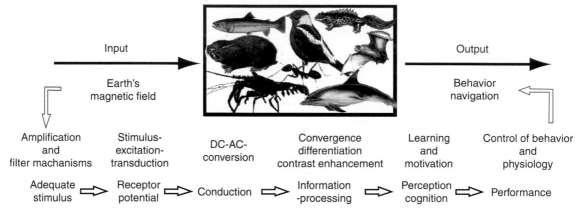

Figure 2 Schematic overview of the different steps from magnetoreception to magnetic-field-guided behavior.

varies by 0.05% per year and has a westward drift of 0.2° per year with a period length of several thousand years before it returns to the past position. (6) *Reversal of the polarity*, which may happen periodically at an average interval of 250 000 years. About 730 000 years ago, the last change occurred and can be detected in studies of magnetic rock.

We will refer only to the geomagnetic parameters, available for animals in their natural environment. Laboratory conditions with extremely high field intensities or high-frequency variations of the magnetic field, magnetic pulses, may evoke changes in tissues and molecules, but the results do not necessarily show these elements are components of magnetoreceptors. Good receptor candidates must match the limits and dynamics of detectable natural field conditions, derived from thorough behavioral and receptor physiological studies. Astonishingly small and slow changes of magnetic flux, field direction, and inclination may serve as a magnetic map and compass.

Intensity and inclination of the magnetic field vary by about $3\,nT\,km^{-1}$ and $0.01°\,km^{-1}$, and experiments with birds, reptiles, mammals, amphibians, or fish indicate that they may react to changes of field intensity of less than $50\,nT$ and a difference of inclination of about $1°$. This means that, for example, they can reliably recognize minute deviations from their migratory route and intended landing place.

For our topic, magnetoreception, two characteristics are central: (1) The magnetic field is omnipresent and perpetual; this means life has always experienced it, and in common with gravity most probably cannot exist without it. Both features together provide a reliable reference for the determination of position and posture of an organism. (2) The geomagnetic field vector cannot be detected by a 'far sense.' Only the local field conditions count, detectable by a 'near sense.' Simply expressed, the magnetic vector behind the next corner or hill cannot be 'seen.' But studies (with birds) have shown a gate open for about

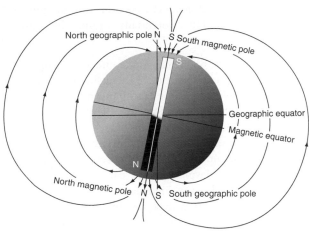

Figure 3 The Earth's magnetic field appears to be formed by a huge bipolar magnet with its magnetic South Pole next to the geographic North Pole, and the magnetic North Pole next to the geographic South Pole. The magnetic field lines run from the magnetic North Pole to the magnetic South Pole. By 'tradition,' the magnetic North Pole is named South Pole and vice versa; thus, the compass needle is said to point to the 'North.'

Figure 4 The Chinese 'South Pointer,' replication of a Chinese compass, probably constructed during the Han dynasty (202 BC to AD 220). The handle of the magnetic spoon always points to the South. The nonmagnetic bronze plate has the eight celestial directions engraved.

half a year for imprinting the landmarks near the home place: experience and learning offer a continuously growing 'mental' magnetic map, which then can serve as reference for the evaluation of the current magnetic parameters.

Physical Principles and Technical Devices for Measuring Magnetic Field Parameters

Geographic positioning during navigation requires devices, which can measure the local magnetic vector, that is, the direction, inclination and intensity of the magnetic field, determined by a magnetic compass and a three-axial magnetometer.

The Magnetic Compass

Long before the Earth was understood as a dipole magnet, in the third century BC, Chinese engineers had discovered the magnetic compass, a tool always pointing in the same direction independent of its spatial position. This followed from finding of magnetic rocks, iron ore, later named 'loadstone.' Metal rubbed with this stone became magnetic. This natural phenomenon was an attraction for the public, for example, when prognosticators used it as pointer on their telling boards. But soon magnetism was applied for navigation, as the geographic orientation of magnetic material was found to be constant, a prototype of a magnetic compass. The first device for practical use was the 'wet compass,' a magnetized needle floating on water. Later, the Chinese used a little spoon of magnetite,

representing the constellation 'Great Bear,' placed on a nonmagnetic bronze plate (**Figure 4**). After a short 'dance' on the plate, the handle of the spoon reliably pointed to the South, 'South pointer.' Engraved in the plate were the heaven as central circle, surrounded by the earth with eight main directions, and fine gradations for 28 lunar houses.

Prior its use for navigation, the Chinese compass was an important tool for divination and geomancy (aligning buildings according to the traditional rules of Feng Shui). This can be deduced from the alignment of ancient temples – also in Europe: they face magnetic north, not geographic north. In medieval times, the magnetic compass found its way to Europe by traders and was developed mainly by Arab scientists, to various shapes matching the requirements of marine and terrestrial navigation. Even today, we use the magnetic needle compass.

The Magnetometer

Accurate positioning depends on the knowledge of the local magnetic vector, the total intensity of the magnetic field, and its direction, that is, declination and inclination of the magnetic field lines. Additional to the magnetic compass, two measuring devices are in use: a dip needle, a freely suspended compass, for determination of the inclination; and a magnetometer to determine the magnitude, which at least in principle includes the direction of the magnetic field. Different from the 'simple tools,' magnetic compass and dip needle, magnetometers are complex technical machines. In principle, magnetic probes oriented in the three spatial directions (X – north, Y – east, Z – downward) measure the respective field intensity, a combination of these Cartesian values results in the magnetic vector. Alternative types of magnetometers following various physical principles are realized depending on their specific application, for example, near the surface of the Earth or in the magnetosphere, the aural zone surrounding the Earth far into space. By means of satellites, the earth magnetic field, the magnetosphere and their modulations are surveyed. The current detailed magnetic maps shown by these programs can be consulted via internet.

The description of basic features of the earth's magnetic field can be attributed to William Gilbert, English physician, who had built a 'terrella,' a little model of the Earth, consisting of magnetite in order to demonstrate the changing declination of magnetic field lines, when a dip needle is moved around the sphere. His distinction between the use of a dip needle and the horizontal compass is one of the topics of his book 'De magnete.' Several scientists involved in the exploration of the electric field, for example, Ampere, Gauss, Maxwell, Oersted, Tesla, Weber, recognized its close relation to magnetism; their names can be found as units of electromagnetic parameters (**Table 1**).

The First Step of Magnetoreception

The Basic Principles of Transducing the Magnetic Field Parameters into a Receptor Potential

A good first question for animal systems is which material or biophysical process is able to react to the low magnetic flux of the natural geomagnetic field and its minute changes, and serve their magnetic sense? This 'material' is essential for the transduction from the external magnetic field parameters into a receptor potential that may conduct information to the brain. Only this primary process is specific for magnetoreception, all following steps of information processing and output control can be principally found in other sensory systems (**Box 2**).

Three models of magnetoreceptor transduction are currently considered potentially valid in various organisms.

- *Electromagnetism*: Electromagnetic effects are studied in detail in physical experiments and found in multiple practical applications, for example, when constructing electro motors. For decades, electromagnetic induction evoked by moving conducting objects within an electric field has attracted researchers, who tried to find this principle in fish. The electroreceptors ('Lorenzini ampullae') of elasmobranchia can sense the voltage drop induced by the environmental magnetic field. But, up to now, despite some attempts and ideas, no sound evidence of the Lorenzini ampullae functioning as magnetoreceptors has been published.

- *Microbiology/iron based*: About 40 years ago, bacteria with small magnetic inclusions were found in fresh water lakes and also in seawater. Magnetite crystals form an intracellular chain, which enables these so-called magnetotactic bacteria to behave like little compass needles and move into water zones of optimal oxygen concentration. Since then, researchers have tried to find magnetoreceptors in metazoan organisms according to this magnetic crystal model. The bacteria themselves are not performing any sensory processes. Magnetite nanocrystals assumed to be the key molecules have been found everywhere in the body and brain, and a large variety of candidate magnetoreceptors have been proposed. Much of the remainder of this study discusses whether or not these structures deserve the notion magnetoreceptor, and we will show a promising receptor system based on magnetic iron minerals in the avian beak, apparently serving as a most sensitive biological GPS system.

- *Biochemistry/molecular based*: Several magnetic field effects on biological molecules have been discussed as putatively hazardous and also as a basic mechanism of magnetoreception. The greatest problem is obviously that several of these reported processes are improbable at the naturally low level of the Earth's magnetic field.

Table 1   Magnetic units of measurements and symbols according to the three systems in use

Parameter	Symbol	Unit of measurement		
		CGS	*SI*	*English*
Field force	mmf	Gilbert (Gb)	Amp-turn	Amp-turn
Field flux	Φ	Maxwell (Mx)	Weber (Wb)	Line
Field intensity	H	Oersted (Oe)	Amp-turns per meter	Amp-turns per inch
Flux density	B	Gauss (G)	Tesla (T)	Lines per square inch
Reluctance	ℝ	Gilberts per Maxwell	Amp-turns per Weber	Amp-turns per line
Permeability	μ	Gauss per Oersted	Tesla-meters per Amp-turn	Lines per inch-Amp-turns

For more information and the conversion refer to, for example, Roche JJ (1998) *The Mathematics of Measurement: A Critical History*. Berlin: Springer.

But at a smaller scale, such reactions, which involve an electron transfer between different orbitals, are now proposed to be a clue to a biochemical magnetoreceptor. In plant photopigments, for example, chlorophylls and flavins, light energy enhances the electron transfer between neighboring molecules, leading to unpaired electrons, 'radical pairs' in both partners. This process has been described in detail in green plants, which show a different growth rate depending on magnetic field conditions. The main question remaining is 'what are the partner molecules'. Cryptochrome, a photolyase, with otherwise 'cryptic' function and oxygen seem to be involved in this radical pair mechanism. This effect in plants is not magnetoreception. A hint toward the involvement of a biochemical process such as that in animal magnetoreception was first derived from experiments with robins. These birds were found to navigate by means of a magnetic compass mediated, oddly enough, through vision. Although initially independent of photosensitive or magnetoreceptive function, the pigment cryptochrome has now been found in distinct areas of the avian retina. Since high-frequency magnetic field stimuli are known to generally interfere with radical pair processes, similar experiments were performed with birds during orientation experiments. As expected, these stimuli disturbed the magnetic orientation and are thought to be convincing evidence for a photo-pigment-based magnetic compass.

The Structural Site of Transduction, the Magnetoreceptive Cell

To date, there is some good evidence that the latter two of these three possibilities magnetoreceptor mechanisms are found in various species localized in different tissues (**Table 2**). Iron-based magnetoreception is suggested present in sensory dendritic terminals of the trigeminal nerve (in fish and birds, possibly also in mammals), and the photopigment-based magnetoreception happens in retinal cells (birds) or in the pineal (amphibians). Both the photopigment- and the iron-based magnetoreceptor systems may occur side by side within the same organism, possibly serving different functions of magnetic orientation: in birds, there is evidence for a magnetic compass in the eye, and an iron-based magnetometer for map information in the upper beak.

Where is the site of the magnetoreceptive transduction in animals?

Biochemical magnetoreception. Here, we can simply state that so far the magnetoreceptive cells are not yet known. We do not know which type of retinal cells are involved, and therefore, it is not known, whether the observed photopigment-based magnetoreception is a primary receptor process or whether the magnetic field modifies the visual transduction or information processing. Behavioral experiments show an effect but not the basis of this reaction. Immunohistology at the electron microscopic level would be a more promising approach, which must be combined with traditional receptor physiological investigation in the periphery.

Iron-based magnetoreception. So far, the magnetoreceptive unit is convincingly shown in birds only. A candidate structure exists, which fulfils all biophysical and receptor physiological prerequisites of an iron-based magnetoreceptor. According to mathematical simulations, its sensitivity also matches the behaviorally defined threshold values of the magnetic sense.

Magnetic material is concentrated in the upper beak skin within distinct sensory dendrites of the median branch of the Ramus ophthalmicus, a part of the trigeminal nerve. Derived from light and electron microscopical analyses of various avian species, these dendrites have the similar shape and size and their subcellular components are well ordered (**Figure 5(a)**). In each dendrite (20–30 μm long, 5-μm diameter), little bullets (1-μm diameter) composed of nanomagnets (6–7 nm) adhere to the cell membrane and may trigger sensitive membrane channels. Via dense fiber scaffolding, they are connected to chains of platelets ($1 \times 12 \times 0.1$ μm). In the midst of the dendrite lies a vesicle (5-μm diameter) surrounded by an iron crust. By means of X-ray analyses, the iron minerals inside the dendrites have been identified as mostly maghemite and some magnetite, both strongly magnetic iron oxides[15]. Based on these data, a first sound hypothesis of the magnetomechanical transduction process in these dendrites

Table 2   Overview of the mostly fragmentary knowledge of magnetoreception

Animal	Key principle/ molecule	Receptor structure	System features	Information processing pathways	CNS representation	Critical test paradigms	Biological function (in behaviour)
Birds	Radical-pair-process Cryptochrome	Retinal cells	Best during migratory restlessness Unilaterally organized	Optic nerve	Nucleus of basal optic root	Dim light of short wavelength (from UV up to bluegreen)	Inclination non-polar compass
	Partner ??	Cones?? Displaced retinal ganglion cells					
Birds	Nano-sized iron crystals in three different subcellular configurations	In the upper beak: distinct nervous terminals of the median ophthalmic branch	Bilaterally three dendritic fields with a 3D-alignment of dendrites	Trigeminal ramus ophthalmicus medialis: recording of action potentials	Cluster N: immunohistology Trigeminal terminal regions: immunohistology, recording of action potentials	Blocked by RF magnetic fields Can be disturbed by high intensity magnetic pulses parallel to aligned dendrites	3-Axial-magnetometer Map factor
Mammals	Magnetite ?? Light-dependent process ??	Corneal cells ?? Retinal cells ?? Pinealocytes ??	??	??	Superior colliculus?? : immunohistology	Can be disturbed by strong magnetic pulses	Polarity compass
Turtles	Magnetite ??	??	??	??	??	Can be disturbed by strong magnetic pulses Needs light	Inclination compass
Amphibia	Cryptochrome ??	Pinealocytes ??	??	??	??	??	Axial compass 3D-magnetometer
Fish	Magnetite Iron crystal chain	Cells in the ethmoid ?? Cells in the olfactory lamellae ??	??	Ramus ophthalmicus superfacialis: recording of action potentials	??	??	Polarity compass Magnetometer
Crustacea	??	??	??	??	??	??	Polarity compass
Insects	Cryptochrome ?? Magnetite	??	??	??	??	??	Compass ??

It is evident that – so far – only in case of birds a distinct molecular basis (cryptochrome, magnetite) and a sound structural candidate (dendrites in the beak) are known. This knowledge is essential to nail down the relevant neurophysiologic steps between magnetic field input and behavioral output functions. Shaded areas indicate missing evidence; for more details see Fleissner et al. (2007b), Johnsen and Lohmann (2005), Mouritsen and Ritz (2005), Wiltschko and Wiltschko (2006).

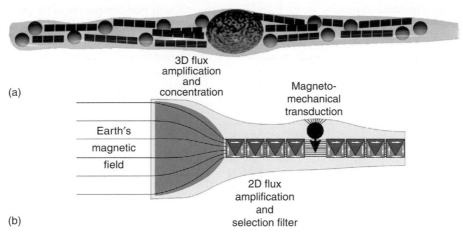

(a)

3D flux
amplification
and
concentration

Magneto-
mechanical
transduction

Earth's

magnetic

field

2D flux
amplification
and
selection filter

(b)

Figure 5 Candidate structure of the iron-mineral-based magnetoreceptor in the avian beak. (a) Semischematic drawing of a single dendrite according to electron microscopic serial sections. (b) Hypothesis of magnetomechanical transduction, meanwhile principally verified by mathematical simulations.

was developed (**Figure 5(b)**). Mathematical model calculations have shown:

- The little bullets may be deformed or dislocated when the Earth's magnetic field is turned or has changed its flux: The crystalline nanomagnets inside the bullets will be polarized when exposed to a strong magnetic field and then commonly attracted according to the strength of this field. They also regain their 'untidy' state immediately when the field strength decreases. This means that the bullets may reversibly exert a graduated pull at the membrane, which is an ideal basis for a transducer monitoring temporal variations of the magnetic flux. Multiple bullets attracted simultaneously may induce a primary receptor potential of a single dendrite via the excitation of mechanoreceptive membrane channels.

The natural Earth's magnetic field, however, is not strong enough to have any impact on either shape or place of the isolated bullets; hence, the magnetic field must be locally amplified. This amplification is probably achieved by the two other iron-containing subcellular components, the vesicle and the platelet chains.

- The iron-crusted vesicle may serve as a tiny Mu-metal chamber. As a result, the surrounding magnetic field lines may be 'compressed,' which means a flux concentration relative to the difference between the field inside and outside the chamber. Thus, the vesicle may serve as a 3D amplifier of the geomagnetic field by about 2 log units.
- When a chain of platelets is parallel to the magnetic field, the net magnetization of each platelet is cooperatively enhanced by the neighboring particles, and the stray field is maximal at both ends of the chain. Since the bullets occur exactly at these sites, they may become attracted by the locally enhanced magnetic field. Mathematical simulations have predicted that

this effect induces a mechanic strain to the cell membrane of several pN, enough to open known mechanoreceptor membrane channels.

- As the net magnetization of the platelet band drops, when it is turned out of the ideal alignment with the surrounding magnetic field, each dendrite can clearly provide unidirectional information on field intensity and direction along this one vector component. The dendrites arranged in another orientation will not sense it.

Conclusion: Each iron-containing dendrite in the beak could be a sensor with a specific orientation in the magnetic field. The adequate stimulus for this dendrite is the momentary intensity of the parallel component of the local magnetic field vector.

System Features of the Proposed Avian Magnetoreceptor Candidates

In analogy to sensory systems in general, a magnetoreceptor usually is not only a single cell, but rather a complex system composed of multiple receptor units. The mutual spatial and functional interrelation of these components and their information coupling with other types of sensory systems is essential to understand the physiological function and context.

The two magnetoreceptor principles seem to follow different strategies.

Iron-Based Magnetometer

The microarchitecture of the iron-containing dendrites in the avian beak provides evidence for the biological function of this magnetoreceptor candidate. The dendrites do not occur randomly distributed over the entire skin; they are concentrated in six fields, three on each side near the lateral margin of the beak. The dendrites are nearly uniformly aligned in a distinct direction in each field perpendicular to each other: in frontocaudal direction in the caudal fields, mediolaterally in the median fields and dorsoventrally in the frontal fields. Since each dendrite can sense only one component of the geomagnetic field, the array of all dendrites provides the local magnetic vector, when recomposed in the central nervous system and thus is a perfect candidate for a three-axis magnetometer.

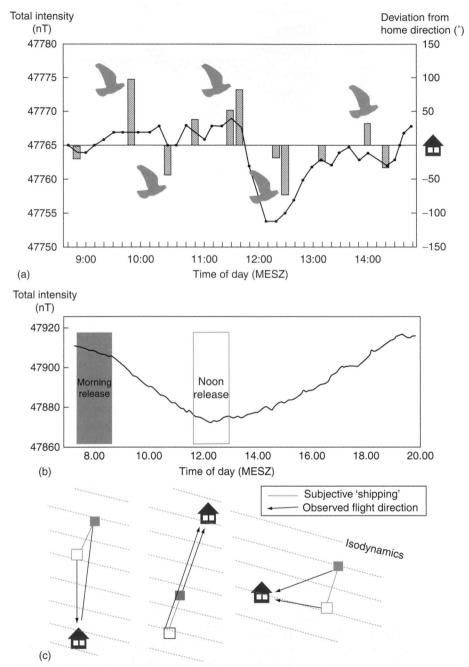

Figure 6 Evidence for a gradual sensing of the magnetic flux. Schematic representation of release experiments with free-flying homing pigeon showing the direct influence of geomagnetic field intensity on homing. (a) Natural low variations of the local magnetic field intensity (left *y*-axis) change the flight direction (right *y*-axis) in each bird according to the actual value at the time of release (*x*-axis time of day). (b) Scheme of experiments to test the influence of the noon dip of magnetic field intensity: the same flock of pigeons is released from a familiar place twice a day: in the morning and again – after they returned to their home loft – around noon (*x*-axis time of day, *y*-axis local field intensity). (c) The results of these experiments depend on the geographic position of the release site relative to the home loft. During the noon release, when magnetic field intensity is low, the birds behave according to a subjective shift to a position further south. This can be monitored as changed vanishing bearings, when the site of the loft is eastward or westward of the release localization. (a) Reproduced from Holtkamp-Rötzler E (1999) Ph.D. Thesis, University Frankfurt; (b, c) Reproduced from Becker M (2000) Ph.D. Thesis University, Frankfurt; both with courtesy of the authors.

Prior electrophysiological recordings from the ophthalmic nerve and the site of its first terminal regions in the CNS have corroborated this hypothesis, matching recent histochemical tracing experiments.

Conclusion: The iron-based magnetoreceptor might provide information on the magnetic vector, that is, field intensity and direction, as a sound basis for magnetic map information (**Figure 6**).

Photopigment-Based Magnetic Compass in the Eye

For a magnetic compass, it is essential to have a clear directionality of sensing the magnetic field. Here, the focusing apparatus of the eye cannot help, as the magnetic field simply penetrates the eye without being focused, for example, to certain areas on the retina. The biochemical magnetic compass has been shown to function in dim light and short wavelengths (**Figure 7(a)**). Since neither the cellular nor the subcellular site of the assumed 'key

molecule' cryptochrome is known, models propose its colocalization with photopigments in the highly ordered membrane stacks of the outer segments of retinal photoreceptor cells. It has been hypothesized that light sensitivity might be modified in certain retinal areas or distinct receptor cells, when they are aligned with the magnetic fields lines, but the cellular and subcellular localization of cryptochrome has not yet been found. Receptor physiological analyses are still missing, and it is not clear how the signal is disambiguated from vision. Few – partially contradicting – details concerning the neuronal wiring have been published, showing a common cortical representation of magnetic and photic stimuli.

Conclusion: The model of a photopigment-based magnetic compass in the eye relies mainly on results from behavioral experiments and physical simulations and still waits for its receptor structure and physiological verification. So far, behavioral data have shown that the biochemical compass may monitor the magnetic field direction, with an accuracy of about ±15°.

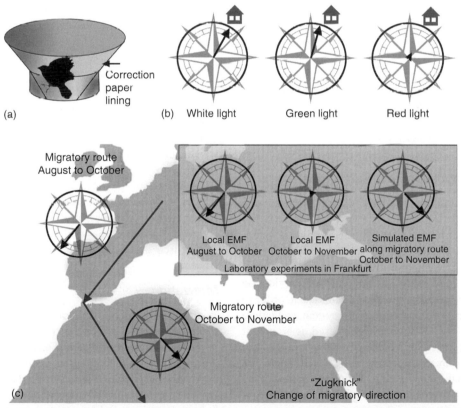

Figure 7 The magnetic compass of birds. (a) Schematic view of an Emlen funnel, which is traditionally used in the lab to test the flight direction of birds during migratory restlessness. (b) Experiments with Australian silvereyes under white and monochromatic light. The birds are clearly oriented under white and short wavelength light (UV to blue green). Under light of longer wavelength, the birds are disoriented. (c) On their autumnal journey to African feeding places, flycatchers are guided by an inborn compass and head to a Southwestern direction until they reach Southern Spain. Then, they change their migratory route and turn to the Southeast. (Inset) This effect can be observed in the lab, too, when the geomagnetic parameters of Southern Spain are presented after a time span matching the natural timing program (EMF, Earth's magnetic field). Reproduced from Wiltschko W and Wiltschko R (2004) Avian magnetoreception: A radical pair and a magnetite-based mechanisms. In: *Proceedings 60th American Meeting of the Institute of Navigation Dayton*, pp. 138–147.

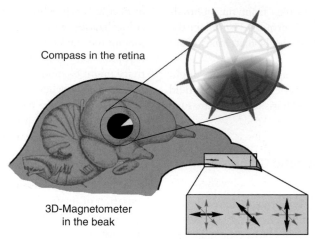

Compass in the retina

3D-Magnetometer
in the beak

Figure 8 Hypothesis on the interaction of the two different magnetoreceptor systems in birds. *Compass*: Due to the assumed cryptochrome influenced magnetic field interaction with retinal cells, the visual sensitivity is reduced in a sector of the retina. The animal can 'see' the compass direction. The acuity is not high, and the exact cellular localization of this effect is still uncertain. *Magnetometer*: Iron-containing dendrites in the inner lining of the upper beak are arranged in six fields, which cover the three spatial directions. For the cellular details of the dendrites see **Figure 6**.

Adaptation, Habituation, Learning, and Motivation

For any receptor system, the stimulus-response function must be derived, and then, it must be tested against processes such as adaptation, habituation, gating, learning, and motivation. Here, we can demonstrate few examples, as only few electrophysiological data are available. Receptor physiological aspects were rarely applied well enough to interpret behavioral experiments.

Threshold, Saturation, and Refractoriness

The intensity-response function clearly outlines the physiological working range of a receptor. It does not only show the minimum, but also the maximum values that should be regarded in order to avoid damage of the receptor structure. Though this functional context is not fully known, the following provides some evidence: (1) Magnets were glued to the head of animals in order to influence the 'magnetite'-based receptor system, and failure of an impact of this manipulation was used as argument against such a system. Here, the experimenters overlooked receptor adaptation as effects were observed, when the test was performed immediately, but diminished after some hours or days. (2) In electrophysiological experiments, recordings of action potentials from the trigeminal axons were inconsistent because the peripheral anatomy was still unknown. Possibly the most efficient stimulus direction for subpopulation of receptor cells was not determined. (3) High-intensity magnetic pulses applied to the head of a bird or a turtle shift the homeward orientation. When the head is bent in the solenoid, the indicated direction is turned. The effect may last several days. This result can

best be interpreted by the three-axis receptor system, where one axis, one dendritic field, has been damaged by a too high magnetic flux.

Learning

During long-distance migration of birds, the compass direction, which they follow on their route, seems to be innate. This is only to a certain extent, as can be shown by a comparison between young and experienced travelers, which can easily compensate for a wind drift or after shift experiments – by means of a magnetic sense (**Figure 7(b)**).

Another aspect is the finding with young birds that they explore the landmarks around their home place and that star or sun compass are calibrated by the magnetic compass. Afterward, these different mechanisms and cues can be used as the circumstances demand.

Motivation

With respect to the photopigment-based system, it is still a matter of debate whether the cryptochrome-based process in the eye is a magnetoreceptor or rather a magnetic-field-induced process that modifies vision. Experiments with birds have provided evidence for the latter interpretation: Opaque but translucent 'spectacles,' which allowed both light and magnetic field to reach the retina, but destroyed imaging, inhibited the magnetic orientation.

Another interpretation is possible. Motivation might have got lost by this manipulation. For birds, vision seems to be the most important sensory input, and in the dark, their motivation might simply be decreased. Motivation has also been discussed as a putative reason for the selection of several types of mechanisms, which are used for orientation by different animals. Especially the distinction between magnetic field, optic landmarks, olfactory and infrasound environment seems to depend on the momentarily best available key feature rather than a systematic or species-specific decision.

To date, only in birds are details of the receptor candidate structures and their wiring known along with behavior. Partial stories do exist in fish and amphibian; in some organism, magnetic sense is apparently coupled with olfaction. In birds, probably two different magnetoreceptor systems share the task of recognition of the magnetic vector: a magnetic compass in the eye and a magnetometer in the upper beak (**Figure 8**). Hypotheses on a general concept of metazoan magnetoreception following the avian model are still premature. Obviously, only interdisciplinary studies following the receptor physiological principles combined with sound biophysical analyses will help describe the magnetoreceptor system features as background for the interpretation of behavioral findings.

See also: Electroreception in Vertebrates and Invertebrates.

Further Reading

Fleissner Ge, Fleissner Gue, Stahl B, and Falkenberg G (2007a) Iron-mineral-based magnetoreception in birds: The stimulus conducting system. *Journal of Ornithology* 148(supplement 2): S643–S648.

Fleissner Ge, Stahl B, Thalau P, Falkenberg G, and Fleissner Gue (2007b) A novel concept of Fe-mineral based magnetoreception: Histological and physicochemical data from the upper beak of homing pigeons. *Naturwissenschaften* 94: 631–642.

Fransson T, Jakobsson S, Johansson P, Kullberg C, Lind J, and Vallin A (2001) Magnetic cues trigger extensive refuelling. *Nature* 94: 631–635.

Gilbert W (1600) *De Magnete, Magneticisque Corporibus, et de Magno Magnete Tellure.* London: Petrus Short.

Heyers D, Manns M, Luksch H, Güntürkün O, and Mouritsen H (2007) A visual pathway links brain structures active during magnetic compass orientation in migratory birds. *PLoS ONE* 2(9): e937. PMID: 17895978.

Holland RA, Thorup K, Vonhof MJ, Cochran WW, and Wikelski M (2006) Navigation: Bat orientation using Earth's magnetic field. *Nature* 444: 702.

Johnsen S and Lohmann KJ (2005) The physics and neurobiology of magnetoreception. *Nature Reviews Neuroscience* 6: 703–712.

Kalmijn AD (2000) Detection and processing of electromagnetic and near-field acoustic signals in elasmobranch fishes. *Philosophical Transactions of the Royal Society of London Series B: Biological Sciences* 355: 1135–1141.

Korhonen JV, Fairhead JD, Hamoudi M, et al. (2007) Magnetic Anomaly Map of the World, Scale 1:50 000 000, Commission for the Geological Map of the World. Supported by UNESCO, Helsinki, GTK. (URL: http://projects.gtk.fi/WDMAM/project/perugia/index.html).

Mann S, Sparks NH, Walker MM, and Kirschvink JL (1988) Ultrastructure, morphology and organization of biogenic magnetite from sockeye salmon, *Oncorhynchus nerka*: Implications for magnetoreception. *The Journal of Experimental Biology* 140: 35–49.

Mouritsen H and Ritz T (2005) Magnetoreception and its use in bird navigation. *Current Opinion in Neurobiology* 15: 406–414.

Phillips JB, Schmidt-Koenig K, and Muheim R (2006) True navigation: Sensory bases of gradient maps. In: Brown MF and Cook RG (eds.) *Animal Spatial Cognition: Comparative, Neural, and Computational Approaches.* www.pigeon.psy.tufts.edu/asc/phillips [on-line].

Schüler D (ed.) (2007) *Magnetoreception and Magnetosomes in Bacteria.* Berlin: Springer.

Schulten K and Windemuth A (1986) Model for a physiological magnetic compass. In: Maret G, Boccara N, and Kiepenheuer J (eds.) *Biophysical Effects of Steady Magnetic Fields*, pp. 99–106. Berlin: Springer.

Solov'yov IA and Greiner W (2007) Theoretical analysis of an iron mineral-based magnetoreceptor model in birds. *Biophysical Journal* 93: 1493–1509.

Wiltschko W and Wiltschko R (2004) Avian magnetoreception: A radical pair and a magnetite-based mechanisms. In: *Proceedings 60th American Meeting of the Institute of Navigation Dayton*, pp. 138–147.

Wiltschko W and Wiltschko R (2005) Magnetic orientation and magnetoreception in birds and other animals. *Journal of Comparative Physiology A* 191: 675–693.

Wiltschko R and Wiltschko W (1995) Magnetic orientation in animals. In: *Zoophysiology*, vol. 33, p. 297. Berlin: Springer. ISBN 3-540-59257 1

Wiltschko R and Wiltschko W (2006) Magnetoreception. *BioEssays* 28(2): 157–168.

Male Ornaments and Habitat Deterioration

U. Candolin, University of Helsinki, Helsinki, Finland

Introduction

Animal communication requires that signals are efficiently transmitted and that the information they convey is reliable. Since habitats vary in characteristics that influence signal transmission and reliability, such as light conditions, background, and acoustic properties, individuals have to adjust their signals to prevailing conditions to be able to efficiently communicate. Species, or populations, that occupy different habitats can therefore vary in the design of their signals.

To attract mates, many animals use ornaments, behavioral displays, or vocalizations that advertise their quality and draw the attention of the other sex. These sexual signals evolve if their benefit in mate attraction is higher than the cost of signaling. For the receiver to pay attention to the signals, the signals have to reflect direct or indirect benefits to the receiver. Direct benefits are, for instance, parental care or resources that increase the number of offspring produced. Indirect benefits are the inheritance of advantageous genes that increase the quality of the offspring. Since the transmission of sexual signals depends on environmental conditions, in the same way as other signals, their expression also needs to be adjusted to the environment to ensure efficient communication. For instance, a species pair of lizards, *Anolis cooki* and *Anolis cristatellus*, that occupy distinct local light environments differ in the reflectance spectra of their throat fan, a colorful dewlap. The dewlap is used by males to repel other males from their territory and to attract females. Leal and Fleishman showed that the dewlaps had reflectance spectra that increased their contrast to the prevailing background. Thus, differences in signal expression between habitats enhanced the transmission of the signal in each habitat.

Presently, the habitats of the earth are changing more quickly than before because of human activities. If the changes occur more rapidly than the speed at which signals can be adjusted, then the signals may become less well adapted to the environment. They may become more difficult to judge, or the link between the signal and the sought benefit may be disrupted, resulting in dishonest signaling. Hampered transmission of information or reduced reliability of signals can lead to maladaptive mate choices. This can reduce the fitness of the individuals born in the altered environment, and hence, reduce the viability of the population. Moreover, signaling may be costlier in the new environment, due to, for instance, increased predation risk. This may further reduce population viability. Thus,

effects of human-induced environmental changes on the costs and benefits of sexual signals can contribute to declines in biodiversity in disturbed areas.

Here, I review the effects that sudden, human-induced changes of the environment can have on sexual signals through effects on their costs and benefits. I discuss the consequences that alterations of sexual signals and their information may have on adaptive mate choice, and hence, on the viability of populations.

Environment Dependence of Signals

The optimum expression of a signal depends on the costs and benefits of signaling under the prevailing conditions. Sexual signaling is beneficial to the signaler if it increases mating success in terms of the production of more offspring or offspring that are better adapted to prevailing environmental conditions. For example, male barn swallows *Hirundo rustica* perform an elaborate courtship display to females, showing off their tail feathers. Males with longer tail feathers acquire mates sooner and have a higher reproductive success than those with shorter tail feathers, as documented by Møller in 1994. However, signaling also incurs costs that reduce the fitness of the signaler, such as the expenditure of energy, increased mortality risk, or the attraction of unwanted receivers. In the case of the barn swallow, longer tails are costly to produce and increases mortality risk. It is therefore expected that only males in the best condition will be able to carry the cost of long tail feathers and that this will ensure that the trait reflects male quality.

Over evolutionary time, signal expression is expected to evolve to reach a balance between costs and benefits and maximize the net benefit of signaling. However, sudden changes in environmental conditions could disrupt the balance, which would influence the fitness of the signaler. Common effects of environmental changes are modifications of the energetic or mortality costs of signaling, the distance over which signals are detected, the ability of receivers to evaluate signals, and the honesty of signals as indicators of direct and indirect benefits. For instance, an increase in the carotenoid content of the food reduces the value of red colors as indicators of an individual's ability to find carotenoid-rich food. This can relax sexual selection for conspicuous red colors and lead to populations where red colors are less commonly used in signaling.

Sexual displays vary in complexity from single traits to multicomponent signals. In general, more complex displays, which contain several information rich components, convey more information. However, complex displays can suffer higher rates of environmental attenuation than simpler signals. They can therefore be more severely affected by environmental degradation. This implies that a preponderance of complex displays in sexual signaling systems can result in large negative effects of environmental change on the costs and benefits of sexual signaling.

Adjustment of Signals to New Conditions

There are two main pathways by which signals can be adjusted to environmental change. The first is phenotypic plasticity, where the signaler adjusts the expression of the signal to the environment according to a genetically determined reaction norm. The other possibility is genetic change, in which case, the effects are apparent in the following generations. Since genetic changes require time, the primary response of individuals to sudden environmental change is often plastic alterations of signaling. How individuals adjust their signaling depends on their genetically determined reaction norm. These reaction norms have evolved in the old environment, which differs from the new one, and the adjustment can be either adaptive or maladaptive.

If plastic adjustment results in signals that are adaptive in the new environment, then genetic changes are no longer needed. However, if the adjustment results in signals that are better adapted to the new conditions but still displaced from the optimum, that is, incomplete adaptive plasticity, then genetic differentiation may be required for the population to survive in the long run. On the other hand, if adjustment of signaling results in signals that deviate much from the optimum, or in signals that are ever further displaced from the optimum, then the viability of the population may decline. This can, under a worst-case scenario, contribute to the extinction of the population.

Deteriorating Visibility

Vision is an important sensory channel for many animals. The transmission and reception of visual signals relies on light conditions, the attenuation and degradation of the signals, the background, and the sensory properties of the receiver. Since human activities, such as forest management, human settlement, and eutrophication alter these factors, humans often have a large impact on the transmission and reliability of visual signals.

Detection and Mate Encounter Rate

The detection of visual signals depends much on the background and the light conditions. For instance, many animals use body movements to attract mates and deter rivals. The detection of these movements is hampered if the background is moving, such as wind-blown plants. Since reduced detection can decrease mate encounter rate, changed background can reduce the opportunity for mate choosiness.

Recently, the effect of human-induced eutrophication of natural waters on sexual signaling in fishes has gained much attention. Due to increased input of nutrients into aquatic systems, primary production has grown and led to reduced visibility. This is now hampering the ability of several organisms to use visual signals in mate attraction. In particular, fishes that use conspicuous colors and elaborate courtship displays to both draw the attention of the other sex and to advertise their quality, are negatively affected. Reduced visibility impairs the transmission and reliability of their visual signals, which results in more random mating.

An area that is heavily affected by eutrophication is the Baltic Sea. A fish species that spawns along the coast of this sea is the threespine stickleback *Gasterosteus aculeatus*. Males establish territories and build nests in shallow water and then attempt to attract females to spawn in their nests. One male can collect eggs from several females. The male cares alone for the eggs and the newly hatched offspring by defending them against predators and by fanning the eggs to provide adequate oxygenation. The competition among males for favorable territories and for females is fierce and males develop conspicuous nuptial coloration to attract females and deter rivaling males. The coloration, which is exposed to the female through a conspicuous courtship dance, consists of a red ventral side and blue eyes. Field observations and manipulation of algae growth show that increased density of filamentous algae impairs the ability of females to detect males. This reduces mate encounter rates, which results in more random mating. Similarly, increased water turbidity due to the growth of phytoplankton reduces visibility and the ability of females to detect males. To counteract these negative effects of visibility on mate encounter rate, males enhance their courtship activity. However, this increases the time and energy spent on courtship and mate attraction. Increased courtship activity also increases the mortality risk of any eggs that the male already has collected in his nest, as more time spent courting reduces the time spent on parental care. Thus, impaired visibility induces more costly courtship in males, which can have negative effects on male viability and egg survival.

To increase detection under reduced visibility, animals could add an alert to their visual signal. Simple signal components suffer lower rates of environmental degradation than more complex components and the alert could enhance signal detection under adverse signaling conditions. The most efficient thing to do is to begin signaling with a simple conspicuous component, an alert, and then produce a more detailed component that contains more information.

Evaluation

After a potential mate has been detected, its quality is usually evaluated. Reduced visibility can hinder the evaluation of visual signals, and thereby, influence mate choice. A classical example of this is the cichlid fishes of the Great lakes of Africa. Increased turbidity of the water due to human activities has caused a decrease in light penetration and a narrowing of the light spectrum. Seehausen and coworkers found this to constrain color vision of the cichlids in the lakes, and hence, to interfere with their mate choice, which is largely based on interspecific differences in male color patterns. Deteriorating visibility consequently broke down reproductive barriers among species. This has most likely contributed to the recent erosion of species diversity in the lakes.

Similarly, increased turbidity of waters due to the growth of phytoplankton has been found to hamper mate evaluation in the sand goby *Pomatoschistus minutus*. When Järvenpää and Lindström allowed goby females to choose among males in clear and turbid water, the distribution of eggs among males was less skewed toward larger males in turbid water. Since male body size is an important mate choice cue under clear water conditions, this implies that eutrophication relaxes selection for larger males.

In threespine sticklebacks, the evaluation of visual sexual signals is similarly hampered under reduced visibility. Female sticklebacks that evaluate males in dense growth of algae spend more time assessing them before they make a choice than females evaluating males in more open habitats. Thus, the cost of mate choice in terms of the expenditure of time and energy is higher when visibility is low. Females also pay less attention to visual signals when making their decision, which results in mate choices that often differ from those under good visibility. Thus, selection on visual sexual signals is relaxed under poor visibility and other mates are chosen.

Honesty

When signal transmission is hampered, the reliability of a signal can decrease. A negative effect of reduced visibility on the reliability of visual signals has been shown for threespine sticklebacks spawning in eutrophied waters. While visual traits, such as the red nuptial coloration and the courtship activity, reflect condition and dominance status under good visibility, they do not do it under poor visibility. This is due to male competition no longer controlling signal expression under poor visibility. Under good visibility, males adjust their signaling to their dominance status in relation to other males. Dominant males become colorful and court at a high rate, while subdominant males signal their subdominant status by fading their colors and becoming passive. However, when visibility is reduced, due to, for instance, phytoplankton

blooms, the social control of signaling relaxes and the visual displays become a less honest index of male dominance and condition. A subdominant male in poor condition can then express bright red colors and court intensively without being punished by dominant males.

Dominance is an important predictor of male parental ability in sticklebacks. Dominant males are better at defending the offspring against predators than subdominant males, particularly conspecific predators. They are usually also in better condition and have a higher probability of surviving the parental phase. The reduced ability of females to tell the dominance status of males under poor visibility can, hence, result in maladaptive mate choices.

Noise

Anthropogenic noise arising from urbanization and traffic influences the transmission of auditory sexual signals and restricts acoustic communication. A good example of this is the song of great tits *Parus major* that is masked by human-induced low frequency noise in urban areas. Slabbekorn and Peet found that birds that nest at noisy locations have to sing with a higher minimum frequency to prevent their song from being masked.

Underwater noise pollution from shipping can similarly influence communication in aquatic environments. For instance, noise from ferry boats lies within the most sensitive hearing range of the Lusitanian toadfish *Halobatrachus didactylus*. Vasconcelos and coworkers found the noise to increase the auditory threshold of the toadfish. This hampered the ability of the fish to detect acoustic signals from conspecifics. Since the auditory signals are essential during agonistic encounters and mate attraction, ferry boat noise could influence mate choice in the species. Similarly, Foote and coworkers found acoustic communication in whales to be restricted by the engine noise of whale watcher boats. To adjust for the anthropogenic noise, whales increase the duration of their primary calls in the presence of boats.

Chemical Pollution

Influx of untreated sewage and agricultural waste is disturbing many water bodies and changing the chemical environment. This can influence the detection of olfactory sexual signals, and thereby influence mate evaluation and mate choice in aquatic environments. An example of this is the swordtail fish *Xiphophorus birchmanni*. Fisher and coworkers found that the exposure of the fish to sewage effluent and agricultural runoff removes the ability of females to recognize male conspecifics. This is most likely due to high concentrations of humic acid, a natural product that is elevated to high levels by anthropogenic processes, causing females to lose their preference for the odor cues of conspecific males. Disturbances of the chemical environment thus hinder chemically mediated species recognition, which ultimately can cause hybridization between swordtail fishes.

Another human-induced problem is the acidification of oceans. This arises from the increased concentration of carbon dioxide in the atmosphere, which decreases the saturation of oceans with calcium carbonate. An acidification of water changes the value and quality of olfactory signals, which can influence mate choice. For instance, several fishes are less able to detect chemical signals when the pH is reduced.

Consequences

What are the consequences of human-induced environmental changes on sexual signaling, or on the reception of the signals, for individuals and for populations? If females cannot properly evaluate males, they may end up doing maladaptive choices that reduce their fitness. For instance, threespine stickleback females are less able to tell the condition and dominance status of potential mates in turbid water. They may therefore mate with males that are poor fathers and have a low hatching success, or males that are of poor genetic quality and will sire offspring with a low fitness. Moreover, the cost of sexual signaling may increase and reduce the fitness of the displayer due to increased mortality risk or due to the allocation of energy and resources away from other fitness-enhancing traits.

A reduction in individual fitness, that is, a reduction in the number or quality of offspring produced over the lifetime of an individual, can result in a declining population size or in a population consisting of maladapted individuals. This will reduce population viability and increase the risk of extinction. Moreover, hampered mate evaluation due to deteriorating environmental conditions can complicate the recognition of conspecifics, and hence, cause hybridization between species. An example of this is the cichlids of the African great lakes that are hybridizing due to increased turbidity of the water. Similarly, a species pair of sympatric threespine sticklebacks in the Canadian lakes, a larger benthic and a smaller limnetic species, are collapsing into a hybrid swarm, probably due to the destruction of aquatic vegetation and increased water turbidity. In a Mexican stream, two species of swordtail

fishes are merging into a hybrid swarm due to impaired chemically mediated species recognition. Consequently, a reduction in signal transmission or reliability, or an increase in the cost of signaling could have negative effects on population viability and persistence, and influence biodiversity.

The effects of altered costs and benefits of sexual signals on population viability depend on the relative importance of sexual selection in comparison to natural selection during adaptation to environmental change. The mentioned studies on sticklebacks, gobies, and cichlids show that mate choice is often more random and sexual selection therefore relaxed under new conditions. If the intensity of natural selection then increases and compensates for the reduction in the intensity of sexual selection, then natural selection could drive the population toward the optimum phenotype in the new environment and rescue the population. More investigations are presently needed on the relative importance of sexual and natural selection during environmental change and the ability of populations to adapt quickly enough to changed selection pressures.

See also: Anthropogenic Noise: Implications for Conservation.

Further Reading

Andersson M (1994) *Sexual Selection*. Princeton, NJ: Princeton University Press.

Candolin U (2003) The use of multiple cues in mate choice. *Biological Reviews* 78: 575–595.

Candolin U (2009) Population responses to anthropogenic disturbance: Lessons from three-spined sticklebacks *Gasterosteus aculeatus* in eutrophied hapitats. *Journal of Fish Biology* 75: 2108–2121.

Candolin U and Heuschele J (2008) Is sexual selection beneficial during adaptation to environmental change? *Trends in Ecology and Evolution* 23: 446–452.

Endler JA (1992) Signals, signal conditions, and the direction of evolution. *American Naturalist* 139: S125–S153.

Espmark Y, Amundsen T, and Rosenqvist G (2000) *Animal Signals: Signalling and Signal Design in Animal Communication*. Trondheim: Tapir Academic Press.

Maynard-Smith J and Harper D (2003) *Animal Signals*. Oxford: Oxford University Press.

Møller AP (1994) *Sexual Selection and the Barn Swallow*. Oxford: Oxford University Press.

Seehausen O (2006) Conservation: Losing biodiversity by reverse speciation. *Current Biology* 16: R334–R337.

West-Eberhard MJ (2003) *Developmental Plasticity and Evolution*. New York, NY: Oxford University Press.

Male Sexual Behavior and Hormones in Non-Mammalian Vertebrates

J. Balthazart, University of Liège, Liège, Belgium
G. F. Ball, Johns Hopkins University, Baltimore, MD, USA

Introduction: Scope of the Article

This article will review the hormonal regulation of sexual behavior in male nonmammalian vertebrates. Sexual behavior encompasses male-typical courtship behaviors as well as copulatory behavior but does not include other reproductive behaviors such as parental behaviors. The term 'hormones' of course includes many different types of chemical messengers, but gonadal sex steroids play the key role in the hormonal regulation of male sexual behaviors. Therefore, the emphasis in this article will be on androgens secreted by the testis such as testosterone as well as the androgenic and estrogenic metabolites of this hormone such as dihydrotestosterone and estradiol. In terms of species, our focus will be on piscine, amphibian, reptilian, and avian species, but some reference will be made to mammalian studies to help illustrate general principles. It is important to note that a consideration of a wide range of species is not only important because these species can serve as a 'model system' for questions relevant to mammals including humans but also to construct general theories about the evolution of neuroendocrine control mechanisms.

Organization of Male Sexual Behavior

Most vertebrate species employ sexual reproduction to maximize individual fitness. Parthenogenetic reproduction has been described, but it is relatively rare. The topography of gamete transfer and other aspects of sexual reproduction varies considerably among species. For example, in many aquatic species (various taxa of fish as well as some amphibians), the gametes are released into the environment and external fertilization occurs. In amniotic vertebrates, internal fertilization is the most common pattern. Of course, there are many variations on this theme given that some species are oviparous or ovoviviparous, while others, such as the eutherian mammals, have a specialized organ in the mother (the uterus) where the zygote migrates to and develops into the fetus.

In order to reproduce successfully, males must not only be able to produce gametes but also must be able to attract a female, ensure that she is sexually receptive, and then copulate with her (or release gametes in a coordinated manner into the external environment) so that zygote

formation can occur. These different aspects of the interaction with the female correspond to different phases of sexual behavior that have been described by terms such as 'attractivity,' 'appetitive sexual behavior' (ASB), and 'consummatory sexual behavior' (CSB). CSB is usually followed by a period of variable duration during which the male's interest in the female is greatly reduced or nonexistent (refractory period). The distinction between the appetitive and the consummatory phases of 'instinctive' (motivated) behaviors was originally made by Charles Sherrington and the European Ethologists of the first generation (Konrad Lorenz and Niko Tinbergen) and was introduced to the field of behavioral endocrinology by Frank Beach in the 1950s (see Ball and Balthazart, 2008 for more information). However, most of the research analyzing the neuroendocrine bases of male sexual behavior during the past 50 years has been devoted to the consummatory aspects of the behavior (intromission and ejaculation in species with internal fertilization, ejaculation in other species). Less attention has been devoted to the endocrine control of other aspects of sexual behavior with a few exceptions that will be considered in the following sections. Broad generalizations that presumably apply to most, if not all, vertebrate species are thus available only for consummatory (copulatory) behavior.

Testosterone as Key Endocrine Signal

From a reductionist point of view, sexual behaviors can be viewed as a suite of muscular contractions triggered by an organized set of nerve impulses. In most cases, hormones modulate the expression of behaviors by changing the electrical activity in specific circuits of brain neurons. One important way through which hormones achieve this goal is by modifying the concentration and activity of neurotransmitters and neuropeptides and by modulating the concentration of their receptors. The sex steroid hormone testosterone, produced mainly in the testes and to a lesser extent the adrenal glands, is in almost all vertebrates the key hormone mediating these changes in neurotransmission that activate the expression of male sexual behavior.

The origin of behavioral endocrinology is generally traced to early studies of Arnold Adolph Berthold who identified, in 1849, the critical role of the testes in the expression of copulatory behavior of domestic fowl

(*Gallus domesticus*). Surgical removal of the testes in young male chicks produced adult males whose appearance (secondary sexual characters such as the comb and wattles) and behavior (crowing and mounting females) did not develop as in normal males. In contrast, castrated males whose testes were reimplanted, or who received a testis from another bird developed both rooster-typical plumage and behavior. Because the testes could be grafted anywhere in the abdominal cavity, these experiments suggested that a blood-borne substance was responsible for induction of behavioral changes observed in adults.

This substance was identified as the steroid hormone testosterone at the beginning of the twentieth century and pure testosterone, first purified from animal testes and soon thereafter chemically synthesized, became broadly available for experimentation. It could thus be experimentally demonstrated that, in a broad range of vertebrate species belonging to all classes from fishes to mammals, castration eliminates or at least markedly reduces the expression of male copulatory behaviors, which are restored by a treatment with exogenous testosterone.

Few exceptions to this rule have, to date, been identified. One of the best-documented example concerns the red-sided garter snakes (*Thamnophis sirtalis*) that copulate in the early spring almost immediately after they emerge from the den where they hibernated (Crews, 2005). It has been suggested that no endocrine signal is required to activate copulation. Indeed, castration in the fall before hibernation does not prevent spring copulation and treatment with exogenous testosterone does not affect the rate of behavior expression. These snakes exhibit a 'dissociated pattern of reproduction' in which spermatogenesis and steroid secretion take place during the summer months and precede by at least three-quarters of a year the period when copulation actually takes place. Once these summer endocrine events have taken place, no endocrine or neuroendocrine signal seems to be required for triggering copulatory behavior. The only identified controlling event is a period of 'vernalization' (exposure to cold) of at least a few weeks that must take place before the behavior is expressed. A similar dissociated pattern of reproduction has also been identified in some bat species.

It is important to note that, in many cases, testosterone is not the chemical signal that will by itself induce the changes in neurotransmission leading to behavior. Often, testosterone must be first metabolized into another steroid before acting at the cellular level. For example, in many species of birds and mammals, testosterone is transformed by specialized neurons into estradiol, a sex steroid hormone erroneously considered as only a 'female' hormone, and it is estradiol produced locally in the brain that produces at the cellular level the neurochemical changes that result in the activation of male sexual behavior.

In many fish species, the main androgenic steroid found in the blood that plays a major role in the activation

of male sexual behavior is not testosterone itself but a related compound derived from testosterone called '11-ketotestosterone' (11KT). Many experiments on a variety of fish species coming from diverse families demonstrate that 11KT circulates in larger concentrations than testosterone, and treatment of castrated subjects with this exogenous steroid shows that 11KT is more powerful than testosterone for restoring sexual (and aggressive) behaviors. In these two examples, testosterone thus acts as a prohormone in the control of sexual behavior. However, in the first case, testosterone is the circulating hormone and the transformation into the active metabolite (estradiol) takes place in the target organ (specific parts of the brain), whereas in the second case, the critical transformation into 11KT already takes place in the testis and the active steroid is also the steroid found in the circulation.

Another key notion to keep in mind is that steroid hormones such as testosterone, estradiol, or 11KT do not by themselves induce behavior. They simply modify the activity of neural circuits so that an individual is more likely to react to relevant sexual stimuli with the expression of sexual behavior. At the phenomenological level, hormones thus change the probability that a given behavior will be produced in response to specific stimuli as well as the intensity of the elicited behavioral response. In the accepted terminology, it is often said that hormones activate behaviors, but it is important to keep in mind that they do not induce or trigger a particular behavior as a chemical pheromone would trigger a behavior in invertebrates.

Additionally, the relationship between hormones and behavior is not unidirectional. Hormones activate behavior, but the behavior of a congener (as well as other environmental stimuli) can also alter the endocrine state of an animal. For example, in seasonal breeders, hormones secreted by the pituitary gland and gonads in many species respond to environmental stimuli such as the change in photoperiod to synchronize the onset of reproductive behavior with the time of year most conducive for high reproductive success. In these species, the secretion of gonadal hormones is also influenced by other environmental stimuli such as the behavior of conspecifics that can enhance or retard reproductive development. For example, the view of and interaction with a female will in many species acutely increase plasma testosterone concentration in the male, and conversely the sexual displays of the male will promote steroid hormone release and the maturation of oocytes in females. The effects of behavior on hormone secretion will ensure an optimal synchronization between partners of a pair, as elegantly demonstrated already in the 1960s by a suite of experiments on two avian species, the ring dove (*Streptopelia risoria*) studied by Daniel Lehrman and his group at Rutgers University and the canary (*Serinus canaria*) that was studied by Robert Hinde and his collaborators at Cambridge University, UK.

Central Versus Peripheral Actions of Testosterone

The brain is the major site of hormone action for the expression of sexual behaviors and a large part of this article reviews the way in which steroids act at this level to regulate behavioral expression. However, hormones have widespread effects in the entire organism and there are at least three other ways that hormones act on peripheral tissues that are relevant to behavior control.

Steroids can first affect sensory inputs to the brain. For example, androgens have marked effects on secondary sexual characters such as the penis in mammals and some species of birds (e.g., ratites, some waterfowl species) or associated structures such as the cloacal gland of quail (*Coturnix japonica*). In male rats, androgens influence ejaculation by enhancing the sensitivity of the penis and a similar effect may take place in birds that possess a penis although this has never been tested. A recent study in quail indicated that anesthesia of the cloacal region has no detectable effect on the copulatory frequency or latency but the related brain activation, as measured by the expression of the immediate early gene *c-fos*, was decreased suggesting that some subtle sensory feedback from the cloacal region had been modified.

Secondly, androgens are known to have anabolic (trophic) effect on muscles that are ultimately the effector organs of behavior. For example, the mass of muscles controlling the syrinx (the primary vocal production organ in birds) is increased by androgens. Castration also decreases cholinergic activity in the songbird syrinx and this effect is reversed by a treatment with androgens. These morphological and biochemical changes in the syrinx represent one way that androgens can modify singing behavior. Similarly, the androgen-dependent development of the thumb pad in male frogs during the reproductive season is clearly needed to facilitate egg fertilization by males. During mating, the male grasps a female with his front legs (amplexus) and then releases his sperm in synchrony with female oviposition. Appropriate amplexus is only possible in these aquatic animals following development of the thumb pad.

Finally, steroids also alter social signals that will indirectly affect behavior expression. Many birds exhibit a marked steroid-dependent sex dimorphism in their plumage and in a number of integumentary derivatives and skin appendages such as beak, comb, wattles, and cloacal gland. These structures play an important role as social signals during sexual interactions. By influencing these visual signals, hormones can therefore have profound effects on the expression of sexual behaviors. Steroid-dependent olfactory signals are also produced in males and females of many vertebrate species from fishes to mammals and directly influence the expression of sexual behaviors.

It is thus clear that a hormone-induced modification in sexual behavior usually represents more than the consequence of the action of the hormone in the brain. Peripheral effects of these hormones are also likely to be involved in many cases.

Seasonal Changes in Testosterone Concentrations and Male Sexual Behavior: Variations on a Theme

Based on castration and testosterone replacement studies, it has been established that testosterone plays a critical role in the expression of sexual behaviors in a wide variety of vertebrate species. It is therefore reasonable to expect that seasonal changes in the intensity and frequency of sexual behavior should follow changes in plasma testosterone concentration. This expectation has been documented in many species, including species that display very different patterns of seasonal change. Demonstration of this temporal correlation became reasonably easy with the advent of radioimmunoassays for steroids in the early 1970s. Studies of domestic ducks (*Anas platyrhynchos*) provided one of the first clear illustrations of a clear coincidence between the vernal peak in plasma testosterone and the maximal frequencies of male sexual behavior (**Figure 1**).

Such a close positive correlation is not always observed. Studies by Wingfield and collaborators of field-caught wild songbirds illustrate several variations in the pattern of testosterone and male reproductive behaviors (**Figure 2**). For example, in song sparrows (*Melospiza melodia*) during the breeding cycle, there are usually two periods when plasma testosterone is high: the time just after the male arrives on the breeding grounds when he is initially establishing a territory and the period later in the season when females are laying eggs and males copulate with them frequently and also deter other males who are trying to copulate with their mates (i.e., mate guard). The study of three subspecies/populations of white-crowned sparrows

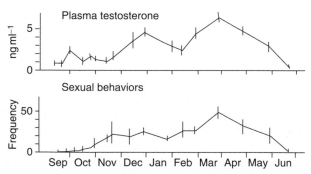

Figure 1   Seasonal changes in plasma testosterone concentrations and in the frequency of male sexual behaviors in a group of captive male ducks (*A. platyrhynchos*). Redrawn from data in Balthazart J and Hendrick JC (1976) Annual variation in reproductive behavior, testosterone, and plasma FSH levels in the Rouen duck, *Anas platyrhynchos*. *General and Comparative Endocrinology* 28: 171–183.

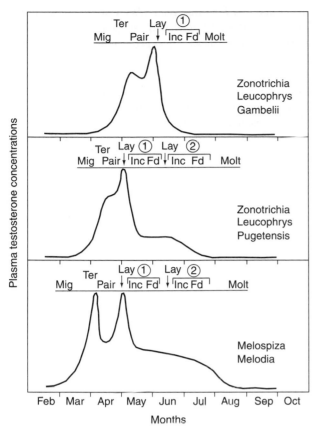

Figure 2 Changes in testosterone plasma concentrations during the breeding season in two subspecies of white-crowned sparrows (*Z. leucophrys gambelii* or *pugetensis*) and in song sparrows (*M. melodia*). The different stages of the reproductive cycle are indicated above the graphs. Mig: spring migration, Ter: establishing a territory; Pair: pair formation; Lay: egg-laying period; Inc: incubation; Fd: feeding nestlings or fledglings; Molt: postnuptial molt. The numbers 1 and 2 indicate successive broods. Redrawn from Wingfield JC and Moore MC (1987) Hormonal, social, and environmental factors in the reproductive biology of free-living male birds. In: Crews D (ed.) *Psychobiology of Reproductive Behavior: An Evolutionary Perspective*, pp. 148–175. Englewood Cliffs, NJ: Prentice-Hall.

(*Zonotrichia leucophrys*) producing one or two broods during a single reproductive season interestingly demonstrated that there are also exceptions to this rule. The establishment of the territory was always associated with a rise in plasma testosterone concentrations, but during female egg laying, testosterone concentrations were high only during the production of the first clutch. In populations with two clutches, the subsequent period of egg laying did not correlate with an increase in plasma testosterone. This dissociation was observed within a single species, *Z. leucophrys*, in which the subspecies *gambelii* lays a single clutch and egg laying is associated with a rise of testosterone, while the subspecies *pugetensis* lays two clutches but only the first one correlates with increased plasma testosterone concentrations.

Testosterone is still detectable in the plasma of male *Z. leucophrys pugetensis* as well as *M. melodia* when females lay their second clutch so that it can still be argued that copulatory behavior of males is activated by the steroid during this period. However, there is no clear correlation and Wingfield has argued that social competition among males is the variable most responsible for the rise in testosterone during egg laying, and this appears to be reduced during second clutches in these species.

This observation about the complexity of correlations between plasma testosterone concentrations and different aspects of male reproductive behavior illustrates the important notion previously mentioned that testosterone cannot be held responsible for 'inducing' a particular behavior. It is rather changing the probability of its occurrence when proper stimuli are encountered. There is no simple stimulus-response chain. Various aspects of male reproductive behavior have indeed been shown to be somewhat independent of testosterone. As an example, courtship singing directed at females in common starlings (*Sturnus vulgaris*) is clearly enhanced by exogenous testosterone but at the same time does not disappear in castrated birds. Testosterone in this case affects the stimulus conditions that will affect singing. Castrated males still sing relatively frequently in isolation or in all-male groups, but the addition of a female will increase the singing rate only in males that have high concentration of testosterone.

The Process of Sexual Differentiation

Another important dissociation between testosterone concentrations and sexual behaviors is also frequently observed when comparing the sexes. A variety of behaviors in animals, including humans, are preferentially or exclusively exhibited in one sex. Due to the actions of sexual selection, sex biases in behavioral expression seem more apt to occur among sexual and parental behaviors than other aspects of an animal's behavioral repertoire. It was initially thought that these sex differences resulted from the presence of a different hormonal milieu in adult males and females, such as high plasma testosterone in the former and high plasma estrogen and progesterone in the latter. More than 50 years of research have, however, demonstrated that in many cases this interpretation is not correct. Estrogens often cannot activate female-typical behaviors in adult males, and conversely testosterone often fails to activate male-typical behaviors in adult females.

This discrepancy between effects of testosterone on male-typical sexual behavior in males and females reflects the sexual differentiation of the brain that takes place during ontogeny. This process has been identified in all classes of tetrapods, but although the general principle remains the same in all species (early action of steroids organize in an irreversible manner the responsiveness of the adult brain to steroid hormones), there are again in this case many variations on this general theme.

In mammals, early exposure to testosterone (or its metabolite estradiol derived from local aromatization in the brain) produces a masculine phenotype. Behavioral characteristics of the male can be enhanced (masculinization), and more or less independent of this process, the capacity of males to display female-typical behavior is diminished or lost (defeminization; **Figure 3**). The female phenotype in contrast develops in the absence of (high) concentrations of sex steroids. Recent evidence suggests that small amounts of estrogens are required during ontogeny to allow the development of a female brain that will be able to support in adulthood the expression of the full complement of female behaviors. The development of the masculine and feminine phenotypes of responsiveness to sex steroids in adulthood can thus be obtained irrespective of the genetic sex of the subjects by neonatal gonadectomy or injection of testosterone.

In avian species that have been studied in detail, such as chicken (*G. domesticus*), quail (*Co. japonica*), and ducks (*A. platyrhynchos*), exposure to steroids early in ontogeny also affects the responsiveness to steroids in adulthood, but this sexual differentiation process exclusively affects male-typical behaviors. A variety of male-typical sexual behaviors have been described that cannot be activated in females even after treatment with high doses of testosterone. Female reproductive behaviors (e.g., sexual receptivity postures such as squatting) are in contrast not sexually differentiated and can be activated in both sexes provided

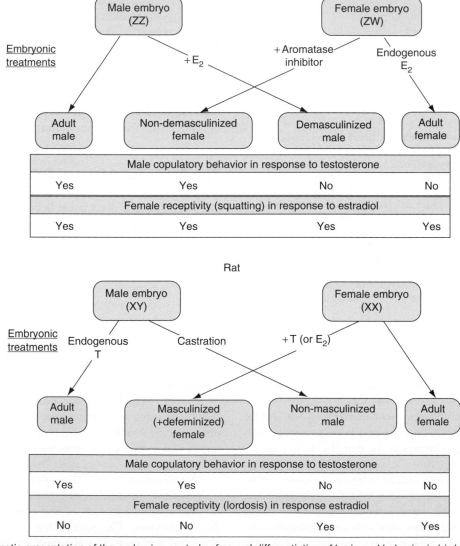

Figure 3   Schematic presentation of the endocrine controls of sexual differentiation of brain and behavior in birds and mammals. These principles are derived from studies in quail and rats, respectively, but seem applicable to a broad range of species in the corresponding classes. See text for additional details.

an adequate treatment with estrogens is administered. In these avian species, the sex difference in responsiveness to adult testosterone is the result of the early exposure of female embryos to ovarian estrogens (as opposed to exposure of males to testicular androgen as is the case in mammals). In the absence of embryonic steroids, the male phenotype of responsiveness to testosterone develops. In the presence of estrogens, the brain and the behavior are demasculinized: birds lose the capacity of express male-typical copulatory behavior in response to testosterone.

Appropriate manipulations of the embryonic endocrine environment can, as in mammals, produce a male or female behavioral phenotype independent of the genetic sex. The processes are simply mirror images of one another in the two classes (**Figure 3**). These conclusions are derived from studies in which the embryonic hormonal milieu was manipulated by the injection of estrogens in males and blockade of estrogen production (i.e., injection of an aromatase inhibitor) or action (i.e., injection of an antiestrogen) in females. These manipulations clearly explain why adult females do not show male-typical behavior when injected with testosterone and also why females with intact gonads do not show these behaviors. Testosterone concentrations, even though they are on average lower in females than in males, overlap between males and females and appear to be high enough in many females to activate male-typical copulatory behavior if their brains were organized in an appropriate manner to respond to such treatments.

These contrasting patterns of differentiation in birds and mammals (i.e., relative absence of endocrine stimulation during early life results in male behavioral phenotype in birds and in female behavioral phenotype in mammals) may be related to the fact that females are the homogametic sex in mammals (female XX and male XY) while males are homogametic in birds (male ZZ and female ZW). This observation could then highlight a more general rule according to which the phenotype of the heterogametic sex (male mammals [XY] and female birds [ZW]) would always develop in response to hormonal stimulation during ontogeny while the behavioral phenotype of the homogametic sex would be the 'default' sex observed in the absence of early endocrine stimulation (also referred to as the 'neutral' sex). This principle must be accepted cautiously at the present time because the proximate mechanisms that might explain this connection between the nature of the sex chromosomes in males and females and the process of behavioral differentiation have still not been identified. In particular, the process of sexual differentiation of the gonads is reasonably well established in mammals (the *sry* gene of the Y chromosome induces formation of the testes) but not in birds where the differentiation of the gonad seems to be the result of the presence on one or two Z chromosomes of gene DMRT1 (gene dosage effect).

In other vertebrate classes including fish, amphibians, and reptiles, many species do not appear to have sex chromosomes, and their brain and behavior differentiate by endocrine mechanisms that appear to be driven by the physical (e.g., temperature) or social environment (presence of congeners of the other sex, position in a dominance-subordinate hierarchy). For example, in reptiles, the first identified event signaling sexual differentiation is the increase in aromatase expression in the female gonad that leads to an increase in estrogen concentration that will by itself transform the undifferentiated gonad into an ovary. Whether gonadal steroids have an early and irreversible effect on the brain and its responsiveness to steroids in adulthood is not broadly established in fish, amphibians, and reptiles. There are, however, indications that such a phenomenon may take place at least in some species. For example, in tree lizards (*Urosaurus ornatus*), concentrations of progesterone and testosterone during development determine, in an apparently irreversible manner, whether the adult males will be territorial during their entire life or will switch from nomadic to satellite male as a function of environmental conditions. In species displaying alternative mating strategies such as the midshipman fish (*Porichthys notatus*), the physiological 'decision' to become a territorial male who builds a nest to attract females or a sneaker (satellite male) is made early in development and is apparently irreversible, but the endocrine bases of this 'decision' are unclear at present.

Given the plasticity in the expression of sex-typical behavior that can be observed by adult fish species, it seems likely that, in many species, the brain and the behavioral phenotype are not determined in an irreversible manner by the early endocrine environment. Variable situations are likely to be observed in amphibians and reptiles and these issues certainly deserve more attention in these taxa.

In mammals, birds, reptiles, and amphibians, multiple sex differences in brain structures implicated in the control of male sexual behavior have been observed. These differences can be morphological in nature (e.g., groups of neurons in the preoptic area (POA), size of the neurons or of cell nuclei larger in males, dendritic arborization more developed in ventromedial hypothalamus of females), and also biochemical. For example, the concentration and turnover of various neurotransmitters or neuropeptides as well as the density of their receptors can vary according to the sex. Some of these anatomical or biochemical differences persist in adult gonadectomized animals placed in similar hormonal environments. It is therefore likely that they are causally related to behavioral differences between sexes and induced by differences in the early endocrine environment. Overall, the organizational effects of steroids on behavior are the consequence of changes in gene expression that result in the multiplication, migration, or death of neurons, as well as the modulation of their functional differentiation (e.g., expression of neurotransmitters, neuropeptides, and their receptors). Many aspects of these effects remain, however, to be uncovered.

Cellular Mechanisms of Testosterone Action in the Brain

Sex steroids including testosterone act on the brain to activate male sexual behavior by molecular mechanisms that are for the most part similar to those described based on studies of peripheral organs (i.e., hormones binding to specific receptors and modifying cellular physiology including protein synthesis). More recent work indicates, however, that the central effects of steroids are also achieved through other brain-specific mechanisms involving, for example, direct effects on neuronal membranes.

Two distinct types of interactions between steroids and their target cells have been described to date. Steroid hormones, and other similar lipophilic compounds such as the thyroid hormones, can more or less freely enter target cells and produce their biological effects by binding to specific intracellular receptors. The complex formed by the hormone and its receptor then acts, in the cell nucleus, as a transcription factor leading to changes in the transcription of new messenger RNA (mRNA) and new proteins that ultimately alter cell function. These effects are usually fairly slow and take hours to days to develop. Many effects of steroids including testosterone and its metabolite estradiol are, however, too rapid to be produced by these mechanisms and appear to result from direct effects at the level of the cell membranes and/or from a direct interaction with intracellular signaling cascades involving, for example,

the phosphorylation of various proteins. These two types of actions will be considered in the following sections.

It should also be mentioned that in addition to sex steroids, other larger or more polar molecules such as the peptidergic or protein hormones (e.g., vasopressin/vasotocin, oxytocin, gonadotropin-releasing hormone, prolactin, and others) play a role in the control of male sexual behavior. These compounds exert their actions through binding to receptors located on the cell membrane that are coupled to the activation of a second intracellular messenger system (such as activation of adenylate cyclase leading to the synthesis of cyclic AMP). These other mechanisms of behavior control are, however, very diverse and their understanding is sometimes only partial. They also differ markedly between species and thus cannot be summarized within the scope of the present article. This presentation will thus be centered on sex steroid actions to illustrate general principles of behavioral neuroendocrinology.

Two key steps should be considered in the behavioral action of testosterone on its target neurons: its metabolism into other steroids and these metabolites binding to specific nuclear receptors.

Role of Testosterone Metabolism

When entering its target cells, testosterone undergoes important metabolic transformations. Two enzymes, aromatase and 5α-reductase, catalyze the transformation of testosterone into behaviorally relevant metabolites estradiol and 5α-dihydrotestosterone (5α-DHT), respectively. Additional enzymes inactivate the steroid (5β-reductase; **Figure 4**). Other reversible transformation can also transform

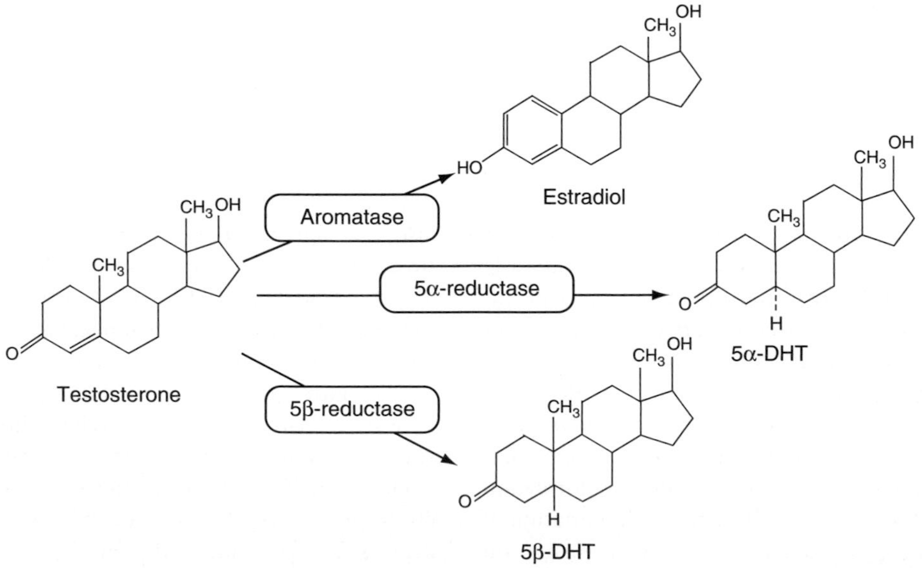

Figure 4 Primary transformations of testosterone into behaviorally active metabolites such as estradiol and 5α-dihydrotestosterone (5α-DHT) or into in behaviorally active compounds such as 5β-dihydrotestosterone (5β-DHT).

testosterone or its metabolites into more polar hydroxylated compounds that have generally a weaker behavioral activity although their behavioral significance deserves a more detailed investigation (e.g., transformation of testosterone into androstenedione by the 17β-hydroxysteroid dehydrogenase, transformation of the two DHTs into corresponding diols by the 3α- or 3β-hydroxysteroid dehydrogenase).

The ratio of active versus inactive testosterone metabolites that are produced in the brain and in peripheral structures can be affected by factors such as the sex, age, season, or hormonal condition of the subjects, providing a mechanism probably related to fine-tuned adjustments of the hormonal regulation of behavior. Depending on the species, 5α-DHT or estradiol alone is able to mimic most if not all effects of testosterone. Copulatory behavior can, for example, be activated by 5α-DHT alone in rabbits or guinea pigs, and by estradiol alone in the rat or in the Japanese quail. In most species, however, the full activation of male sexual behavior results from a

synergistic action of both estradiol and 5α-DHT. The relative role of these two steroids in this process varies between species, but it seems that both are usually involved.

Binding to Receptor

The neuroanatomical distribution of high-affinity binding sites for androgens, estrogens, and progestagens is remarkably consistent among vertebrate species. This distribution was originally mapped in the 1970s using in vivo autoradiographic techniques. More recently, this distribution was reanalyzed by immunocytochemical studies employing mono- or polyclonal antibodies against the steroid receptors molecules. These receptors have also been cloned and sequenced in many species and probes could therefore be synthesized and used to localize the corresponding mRNA by in situ hybridization. All these different experimental approaches have produced results that are in general in good agreement.

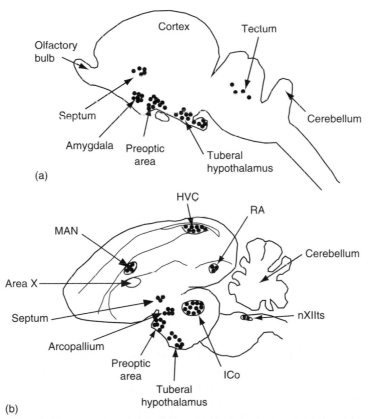

Figure 5 Steroid-binding sites in vertebrates in general (a) and in a specialized case, the songbird brain (b). In addition to the expression of receptors in the POA-hypothalamus, limbic system, and optic tectum, including the nucleus intercollicularis (ICo), songbirds express receptors for androgens, and in some case for estrogens, in specialized nuclei of the telencephalon that are implicated in song learning and production such as HVC (formerly High Vocal Center, now used as a proper name), RA (nucleus robustus arcopallialis), and MAN (magnocellular nucleus of the anterior nidopallium). In songbirds, androgen receptors have also been identified in the nucleus of the 12th nerve (XIIts) that innervates the synrinx. Balthazart J & Riters LV (2000) Ormoni e Comportamento. In "The biology of behavior" Bateson P and Alleva E (eds.) Trecani Publishers, Rome, pp. 85–97 translated in English in 2001 (Balthazart J & Riters LV (2001) Hormones and Behavior. In "Frontiers of Life" Baltimore D, Dubelcco R, Jacobs F and Levi-Montalcini R Eds., Academic Press, Orlando FL, pp. 95–108.

Cells with a dense expression of receptors for sex steroids are localized in the medial POA, the hypothalamus (anterior hypothalamic area, ventromedial nucleus, tuberal hypothalamus), telencephalic structures that are part of the limbic system (amygdala, lateral septum, bed nucleus of stria terminalis), and in specific parts of the mesencephalon (optic tectum). A schematic presentation of this distribution is shown in **Figure 5**. A similar distribution has been observed among a wide range of vertebrates ranging from fishes to mammals. Furthermore, the anatomical distribution of androgen-concentrating cells is in general similar to that of the estrogen-concentrating cells. However, differences have been observed in the intensity and the number of labeled cells, as well as in their precise distribution within a given nucleus.

Some cases of specialization have been observed in this distribution. A number of additional brain sites expressing sex steroid receptors have in particular been identified in vertebrate species that produce vocalizations in the context of reproduction. For example, in songbirds such as zebra finches (*Taeniopygia guttata*) and canaries (*S. canaria*), a network of neurochemically specialized brain nuclei controls both the learning and the production of song. Singing behavior is steroid sensitive and accordingly, most nuclei in the so-called song control system (located in the telencephalon as well as the mesencephalon and brainstem) contain androgen receptors. This presence of androgen receptors in telencephalic nuclei outside the limbic system is unusual among vertebrates and is functionally related to singing. Similar specializations are present in the midshipman fish (*P. notatus*) and several species of amphibians such as the African clawed frog (*Xenopus laevis*). In both species, the vocalizations are produced by a complex neuronal circuitry that includes a number of androgen-sensitive nuclei in brain regions that do not normally contain androgen receptors in other vertebrates.

Only a subset of the total number of binding sites for sex steroids in the brain have been implicated in the control of male sexual behavior. These critical sites have been identified by a combination of lesion experiments and of stereotaxic implantation of steroids directly in the brain of castrated subjects. In general, the medial part of the POA appears as a critical and sufficient site for the activation by steroids of male sexual behaviors. Castrated males that are sexually inactive will recover a rate of sexual activity that is often similar if not equivalent to the level seen in gonad intact sexually mature males when implanted with testosterone in this brain site (see section 'The Preoptic Area as a Key Site of Testosterone Action on Copulatory Behavior'). Additional sites are also implicated. For example, androgen action in the septum, bed nucleus of the stria terminalis, and amygdala modulate the expression of male sexual behavior even if the action of androgens in the POA alone is sufficient to activate this behavior in many cases.

The activation by testosterone or estradiol of these nuclear receptors modulates the transcription of a multitude of genes that encode for a variety of receptors (for neurotransmitters, for neuropeptides, for steroids themselves) and enzymes that control the synthesis or catabolism of neurotransmitters and neuropeptides, as well as protein (neuro)hormones themselves. These changes are confined to the anatomically specific sites that express steroid receptors and ultimately result in specific changes in neural activity and behavior.

The Emerging Role of Nongenomic Effects of Steroids

The steroid-induced behavioral changes described earlier usually require a time course extending from hours to days after the beginning of the exposure to the steroid. Such a time course can therefore explain changes in reproductive behavior that are observed over months during the annual cycle with animals alternating seasonally between a time of active reproduction and a period of sexual quiescence when no sexual behavior is usually observed.

Faster actions of steroids have also been identified suggesting that these hormones may also act via fundamentally different mechanisms. It has been noted for at least two decades that steroids are able to alter excitability of neurons in culture within seconds of their application. This is particularly the case for estradiol, the steroid produced in the brain by aromatization of testosterone that plays a key role in the activation of male sexual behavior. Besides their genomic actions, estrogens indeed exert effects that are too rapid (seconds to minutes) to be mediated through the activation of protein synthesis. Although purely cytoplasmic effects have also been described, nongenomic effects are generally initiated by steroids acting at the plasma membrane resulting in the activation of a wide variety of intracellular signaling pathways. The nature of the potential membrane estrogen receptors mediating these effects is still debated. It is now clear that multiple receptor systems are implicated and several of them have actually been identified. First, the well-characterized classical nuclear receptors for estrogens (estrogen receptor α and β) can associate with the cell membrane and generate intracellular signals through association with a G-protein-coupled receptor (GPCR). Additionally, novel membrane receptors such as GPR30 or ER-X have also been proposed as candidates for mediating membrane actions of estrogens. Finally, estrogens can also act as coagonists or allosteric modulators of GPCR or ion-gated channels/receptors. The intracellular signaling pathways activated by these receptors result in phosphorylations of enzymes or receptors leading, for example, to changes in enzymatic activities or receptor uncoupling from their effectors. It has therefore been hypothesized that these changes in cellular (neuronal) function could modify at the level of the organism specific aspects of male sexual behavior, and experimental evidence supporting this idea has recently been obtained in a number of model systems.

Evidence that estrogens acutely influence processes such as pain and also aggressive and sexual behaviors is indeed accumulating. It was demonstrated, for example, that a subcutaneous injection of estradiol stimulates mounts and anogenital investigations within 35 min in castrated rats (*Rattus norvegicus*), a latency too short to be compatible with an activation by genomic mechanisms. Subsequent studies demonstrated that a single injection of 17β-estradiol facilitates the expression of most aspects of male sexual behavior within 10–15 min in quail (*Cournix japonica*) and mice (*Mus musculus*). The existence of such rapid behavioral effects of estradiol seems to be an ancient feature in vertebrates since they are also observed in fishes. Injection of estradiol indeed modulates within minutes the production of courtship vocalization in the plainfin midshipman fish (*P. notatus*).

It is currently difficult to assess the overall significance of these rapid effects of steroids on behavior. Only a few examples have been identified and it is thus not possible to determine how widespread this type of effect will be. In several cases, the nongenomic effects of estrogens are best observed in animals pretreated with a suboptimal dose of testosterone or estradiol benzoate suggesting that these nongenomic actions require some steroid priming to occur. Why would these two types of the regulation of behavior by estrogens have evolved especially with latencies of effects that differ by several orders of magnitude? It is important to note that even during the reproductive season, males have to spend time with other activities than reproduction (search for food, hide from predators, etc.), and sexual behavior should thus not be expressed continuously. Alternations on a short-term scale between sexual activity and inactivity are usually considered to be under the control of neurotransmitter activity (e.g., dopaminergic, noradrenergic, or glutamatergic inputs to steroid-sensitive areas). The discovery of rapid actions of estrogens on reproductive behaviors provides, however, new insight into this question. One could argue that in its short-term context, estradiol displays most, if not all, functional characteristics of a neurotransmitter or at least a neuromodulator and thus can regulate short-term changes in behavior. Additional research on the functional significance and on the cellular mechanisms underlying such rapid effects of steroids is now required to evaluate their overall importance in the control of reproduction.

The Preoptic Area as a Key Site of Testosterone Action on Copulatory Behavior

As reviewed earlier, the neuroanatomical distribution of sex steroid receptors has guided many initial investigations on the neural circuit controlling male sexual behavior. There are still many gaps in our knowledge about this circuit and functional anatomical studies are based on only a limited number of species. However, lesion studies and hormone implant investigations have all clearly demonstrated the importance of the medial POA (mPOA), a brain region at the junction of the telencephalon and diencephalon. It is obviously a key site for the integration of information involved in the regulation of male sexual responding. Notably, lesions of the mPOA impair copulation in male rats and in a large number of other mammalian species as well as in all species of birds, reptiles, amphibians, and fishes that have been investigated.

The mPOA is clearly associated with the anterior hypothalamus based on functional considerations. It is a target of steroid hormone action and bidirectionally connected to a large number of brain regions. In particular, the mPOA receives, directly or indirectly, inputs from most, if not all, sensory modalities and is therefore ideally positioned to integrate information from the environment and adjust responses made by the organism to environmental and endocrine inputs (**Figure 6**).

It is also apparent from many studies that the POA sends a prominent projection to the periaqueductal gray (PAG is also referred to by its Latin name, substantia grisea centralis). This projection to the PAG represents a key link between the diencephalic center where endocrine stimuli and sensory inputs originating from the female are integrated with the spinal module controlling motor outputs reflected in sexual behavior itself.

In most species, the full complement of sensory inputs and motor outputs to the POA–PAG connection involved in the control of male sexual behavior still needs to be elucidated. In a number of selected cases such as rodents or Japanese quail (*C. japonica*), this circuitry has been investigated by multiple approaches including the analysis of the expression of immediate early gene expression to identify brain areas involved in sexual responses, tract-tracing studies of the afferent and efferent connections of the POA–PAG connection, and experimental lesions of the different nuclei identified by the previous approaches to confirm their implication in behavior control.

In this context, studies of the expression of the immediate early gene *c-fos* in particular have been helpful in birds and in mammalian species. In quail, for example, copulation induces the appearance of Fos-immunoreactive cells in the POA, the bed nucleus of the stria terminalis, the ventral mesopallium, parts of the arcopallium including the nucleus taeniae of the amygdala, and the mesencephalic nucleus intercollicularis. Fos induction was observed throughout the rostral to caudal extent of the preoptic region of male quail and in the rostral part of the hypothalamus to the level of the supraoptic decussation. It is unlikely that this Fos induction resulted from copulation-induced endocrine changes, because copulation did not affect

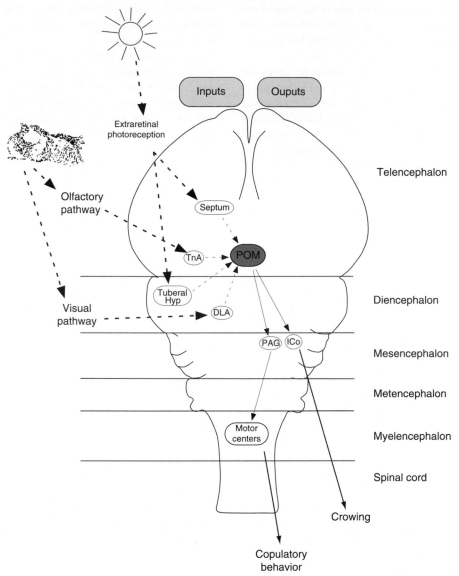

Figure 6 Schematic representation of the neural circuit mediating male sexual behavior based primarily on studies conducted with Japanese quail with special emphasis on the inputs and outputs of the medial preoptic nucleus (POM). The figure illustrates the putative visual and olfactory inputs to the circuit and the outputs to nuclei potentially mediating the expression of reproductive behaviors (copulation) and vocalizations (crowing). ICo: intercollicular nucleus; PAG: periaqueductal gray; TnA: nucleus taeniae of the amygdala; Tuberal Hyp: tuberal hypothalamus. Modified from Ball GF and Balthazart J (2009) Neuroendocrine regulation of reproductive behavior in birds. In: Pfaff DW, Arnold AP, Etgen AM, Fahrbach SE, and Rubin RT (eds.) *Hormones, Brain and Behavior*, pp. 855–895. San Diego, CA: Academic Press.

plasma levels of luteinizing hormone or of testosterone and a similar induction could be detected in castrated males whose plasma testosterone concentrations has been clamped to a stable high level by subcutaneous implantation of a capsule filled with testosterone. Rather, the Fos responses seem to be due to copulation-associated somatosensory inputs and, surprisingly, to olfactory stimuli originating from the female. This interpretation was recently confirmed in a study demonstrating that Fos induction is significantly decreased in birds rendered anosmic by occlusion of the nostrils and in which the cloacal region had been anesthetized before they could

interact with the female. Similar conclusions had been reached by experimental deafferentations in rats. This comparison is of interest, given that male birds do not have an intromittent organ and are not supposed to rely on olfactory stimuli to detect females. Additional work would therefore be in order to identify the nature of brain activation that mediates this Fos expression. Work in other vertebrate taxa would also be of interest to determine the generality of these broadly distributed changes in brain activity.

Taken together, these data indicate that a large number of brain areas are activated during the expression of male sexual behavior as revealed by an increased expression of immediate early genes and confirmed in some selected cases by an increased glucose accumulation following expression of the behavior. Tract-tracing studies show that these brain regions are connected, often bidirectionally, to the mPOA and therefore constitute a functional network where steroids are acting to activate the expression of male sexual behavior. The specific function of each node in this network has not always been identified, but information is available in some specific cases. For example, the amygdala in hamsters (*Mesocricetus auratus*) clearly serves as an integration site for the olfactory stimuli originating from the female and the endocrine stimuli produced by the gonads.

The Neuroendocrine Control of Appetitive Sexual Behavior

Descriptions of male sexual behavior in non-human animals have distinguished between two different phases: a highly variable sequence of behaviors that involves attracting and courting a female, the appetitive sexual behavior (ASB; a visible signal of sexual motivation), followed by the highly stereotyped copulatory sequence (consummatory sexual behavior, CSB). As mentioned previously, most of the work on the endocrine control of male sexual behavior has focused on the analysis of consummatory aspects of this behavior (intromission and ejaculation). It is only more recently that due consideration has been given to the analysis of the appetitive phases of this behavioral sequence.

Appetitive behaviors such as courtship displays are clearly necessary for and distinct from copulation itself, and though both are under the control of testosterone and its metabolites, the neural circuits that implement these different aspects of the behavior should by necessity be distinct to some degree. Investigations by Barrry Everitt and his colleagues in Cambridge, UK, were especially important in stimulating work on circuit specializations related to the control of different components of male sexual behavior. Their experiments indeed showed that if lesions to the mPOA in rats eliminate male-typical copulatory behavior, they have more limited or even no effects on some measures of sexual motivation. Rats with such lesions still pursue and attempt to mount females. They also perform learned instrumental responses (in operant conditioning paradigms) to gain access to females. In contrast, lesions to the basolateral amygdala inhibited the ability of males to acquire learned responses that are rewarded with access to females. Pharmacological studies also revealed selective effects on CSB of manipulations of the preoptic opioid system but left relatively intact measures of ASB. These observations lead to the idea of a double dissociation between brain areas mediating copulatory behavior on the one hand (the mPOA) and appetitive sexual behavior/sexual arousal/sexual motivation on the

other hand (amygdala, bed nucleus striae terminalis). This notion has since been investigated in a few other vertebrate species.

Sophisticated behavioral tests have now been designed to analyze and quantify in a reasonably independent manner these two aspects of sexual behavior. Studies on male ASB in a diverse set of species are particularly important from a clinical perspective. Patterns of male sexual performance are often stereotypic and can be species specific in nature. Generalizations from non-human animals to humans are sometimes difficult. In contrast, mechanisms underlying sexual motivation and ASB seem to be more widespread among vertebrates and could potentially be more easily transposed in humans.

In both rats and quail, the two species in which most research on this topic has been carried out, the expression of ASB is markedly inhibited if not completely suppressed by castration, and recovery is observed following treatment with exogenous testosterone. In both species, the action of testosterone on ASB also seems to be mediated by its aromatization into an estrogen. It was, for example, demonstrated that treatment of rats with the aromatase inhibitor Fadrozole[TM] markedly decreases a widely accepted measure of ASB, the number of level changes in a bilevel apparatus in which a male can freely pursue a female. Aromatase-knockout mice also exhibit major deficits in measures of partner preference and sexual motivation. Similarly, in Japanese quail, measures of ASB are enhanced by testosterone whose action is mediated by its aromatization into an estrogen. Thus, both ASB and copulation itself depend for their activation on similar if not identical endocrine stimuli. This is understandable from an ultimate causation point of view since natural selection should favor the control by similar hormones of behaviors that must by nature be expressed in sequence to ensure successful reproduction.

Interestingly, however, several studies have suggested dissociations between neural circuits underlying the expression of ASB and male copulatory behavior. In rats, the work of Everitt and colleagues mentioned before suggested that the mPOA, which obviously controls copulation, might not be implicated in the activation of ASB. However, there is also clear evidence coming from a variety of studies in different species that the mPOA plays some role in the control of these other appetitive aspects of sexual behavior. It has indeed been reported that lesions of the mPOA diminish the preference for a female partner in rats and ferrets (*Mustela putorius furo*), decrease pursuit of the female by male rats, and decrease anticipatory erections and anogenital investigations in marmosets (*Callithrix* sp.). Furthermore, in vivo dialysis experiments revealed that sexual interactions progressively increase the level of extracellular dopamine in the mPOA of male rats, and this increase begins as soon as the male is introduced to the female and initiates pursuits and anogenital investigations, that is, several minutes before the beginning of copulatory interactions (consummatory responses) sensu stricto. Pharmacological manipulations of the dopaminergic system in the mPOA additionally confirm its involvement in the control of appetitive sexual response. For example, microinjections of a dopaminergic antagonist within the POA decrease measures of sexual motivation such as the preference of a male for the female-baited arm in a maze.

Detailed studies in Japanese quail have also analyzed the neural circuit mediating ASB as reflected by measures of two of its components: the learned social proximity response and the rhythmic cloacal sphincter movements. During the 'learned social proximity response,' a male quail will stand for most of the day in front of a window that provides him with visual access to the female after he has been copulating with that female in the same arena. This is a robust, easily quantifiable, response reliably produced in the laboratory that provides a useful way to investigate the mechanisms regulating male ASB. This response is, however, learned only after the male performs copulatory behavior in the testing chamber and, as a consequence, cannot be studied completely independent of the occurrence of copulation. In contrast, the 'rhythmic cloacal sphincter movements' are produced in anticipation of copulation but do not require copulatory behavior to occur in order to be produced. These movements are greatly facilitated in males, including sexually naive males, by the simple view of a female. They produce by rhythmic movements of a sexually dimorphic striated cloacal sphincter muscle that is interdigitated with the proctodeal gland, a meringue-like foam that is transferred to females during copulation and enhances the probability of fertilization. These two responses are androgen dependent. They are not expressed in castrated males and are restored to rates observed in intact sexually mature males by treatments with exogenous testosterone either injected systemically or implanted directly in the mPOA. As observed for copulatory behavior, effects of testosterone on these two measures of ASB require aromatization of the androgen in the POA.

Experimental analysis of the neuroanatomical bases of these behaviors indicates a clear role for the mPOA in the control of ASB and even suggests an anatomical specificity within this region: the rostral part would be more specifically implicated in the control of ASB while copulatory behavior sensu stricto would rather be controlled by the posterior part of this brain region. This anatomical separation is supported by two independent types of experiments.

In one case, discrete electrolytic lesions aimed at the medial preoptic nucleus (POM) strongly inhibited, as expected, copulatory behavior but, in parallel, also decreased to various degrees the expression of the learned social proximity response. Closer inspection of the data revealed that behavioral effects of the lesions were closely related to their specific location within this nucleus. Lesions in the caudal POM, at the level of or just rostral to the anterior commissure, were associated with decreased expression of copulatory behavior, while slightly more rostral lesions were specifically associated with inhibition of the measure of ASB. In another experiment, POM lesions also completely abolished the expression of rhythmic cloacal sphincter movements induced in castrated males treated with exogenous testosterone by the view of a female, but the anatomical specificity of this latter effect could not be established due to the limited number of subjects available in the study.

In a second type of study, the analysis of the immediate early gene expression activated by the expression of ASB or by copulation revealed a differential activation of subregions of the POM by these two aspects of sexual behavior. In this study, castrated males treated with testosterone were allowed to interact freely with a female and express the full sequence of copulatory sexual behavior or only to express rhythmic cloacal sphincter movements in response to the visual presentation of a female. Quantification of the protein product of the immediate early gene *c-fos* in the brain of these subjects collected 90 min after this sexual experience demonstrated an increased *c-fos* expression throughout the rostrocaudal extent of the POM in males that had copulated, whereas the view of a female and expression of rhythmic cloacal sphincter movements induced an increased *c-fos* expression in the rostral POM only. These data thus provide additional support to the idea that the POM is implicated in the expression of both ASB and copulatory behavior but that there is a partial anatomical dissociation within this nucleus between subregions involved in the control of each aspect of male sexual behavior. The rostral and caudal POM seem to differentially control the expression of these two components of sexual behavior.

This anatomical specificity in the control of appetitive and consummatory sexual behavior by the POM might not be restricted to quail. In rats, as in quail, lesions of the caudal POA and anterior hypothalamus block the expression of copulatory behavior, but more rostral lesions in the POA had little or no effect in at least one experiment. In hamsters (*M. auratus*), pheromones alone are able to activate c-fos expression in neurons of the mPOA in the absence of copulatory interaction with females indicating that stimuli encountered during the appetitive phase of male sexual behavior are processed, at least in part, in the mPOA. Similarly, in a songbird, the house sparrow (*Passer domesticus*) female-directed song, an appetitive behavior that precedes copulation, relates positively to the induction of the immediate early genes *c-fos* and *zenk* in the rostral POM but not the caudal part of this nucleus.

Alternative Models of Reproduction in Vertebrates

The general overview presented earlier reflects mechanisms of behavior control that presumably apply to the vast majority of vertebrate species. These principles have often been tested on only a limited number of species of fishes, amphibians, reptiles, birds, and mammals, and information concerning the last two classes is by far more abundant than for fishes and other less studied tetrapod groups such as amphibians and reptiles. A number of species or families of vertebrates have, however, adopted unusual modes of reproduction or adjusted to different environmental conditions and the question arises as to whether these principles also apply for these species. This question has obviously not been answered in a general manner, but a few principles should be mentioned.

Tropical species do not experience the drastic changes in the duration of photoperiod that usually drive testicular physiology and thus reproductive cycles in species living in the temperate zone. They either reproduce during most of the year or have adapted to other factors that may limit reproductive success such as rainfall in arid environments. Based largely on studies in birds, these species often do not exhibit the pronounced seasonal rhythms in plasma testosterone concentrations that are seen in temperate zone animals. Plasma testosterone concentration is often low throughout the year and sometimes does not increase markedly at the beginning of the reproductive period. Although they may not be strictly seasonal in nature, reproduction in tropical species is cyclical and all species cease reproduction in association with molt in birds, for example. It has been suggested that the presence of reproductive behavior in association with low testosterone concentrations in the circulation was permitted by an increased sensitivity of the brain to these low levels of steroids. Available evidence suggests nevertheless that the action of testosterone on the mPOA and on the connected network of nuclei described before is still implicated in the activation of male sexual behavior in these species.

A large diversity of reproductive patterns is found in fishes including species that change sex during their life either as a function of age, of environmental factors or of the social situation. The change from male to female or female to male is also called 'sequential hermaphroditism' and is common in fishes, rare in amphibians and reptiles, and never reported to our knowledge in birds and mammals. Sex changes are, when documented, associated with huge variations in plasma concentrations of sex steroids and with a substantial remodeling of brain neurochemistry affecting variables such as the expression of various neuropeptides (vasotocin, gonadotropin-releasing hormone) or the intracellular metabolism of steroids. In all these species, testosterone or 11KT remains, however, the steroid responsible for the activation of male sexual behavior and the POA is a key area controlling reproduction. A few piscine species (deep-sea, some serraninae, or sea basses) have been identified that seem, based on examination of gonadal tissue, to exhibit simultaneous hermaphroditism (simultaneous

presence of testicular and ovarian tissue in the same subjects). Little or no information is however available on the endocrine control of reproductive behavior in such species.

Parthenogenetic reproduction, that is, asexual reproduction in which females can reproduce without fertilization by a male, is sometimes observed in invertebrates (aphids, some bees, and parasitic wasps) and very rare in vertebrates (occasionally described in some sharks and reptiles) but has been extensively documented in one species of whip-tailed lizard. In *Cnemidophorus uniparens*, all individuals are triploid females. They lay unfertilized eggs all of which develop into daughters. This contrasts to other lizards including species of the genus *Cnemidophorus*, where both sexes are present and male sexual behavior (mounting on the female from the rear and apposition of cloacal areas in a so-called doughnut posture) is activated like in other vertebrates by testosterone. Surprisingly, in *Cn. uniparens*, females are able to lay eggs that will produce offspring in the absence of sperm but they also display, over time, cycles of sexual activity during which they successively assume a female and then a male role. Depending on its endocrine state, one subject will either play the role played by the male in closely related species and mount a female that is about to lay eggs or will be mounted and will lay eggs. It has been demonstrated that the female-like receptive behavior is displayed just before ovulation when circulating estrogen concentrations are high, while the male-typical mounting behavior is displayed when plasma concentrations of progesterone are elevated. Accordingly, these two types of behaviors can be induced in the laboratory by injecting ovariectomized females either with estrogens or with progesterone. Interestingly, progesterone rather than testosterone is thus responsible for the activation of male-typical copulatory behavior in this species. The action of progesterone on behavior still takes place however in the POA and is associated with a marked increase in the expression of progesterone receptors. Although progesterone is usually considered as an inhibitory factor for the activation of male behavior, scattered studies in a variety of species including rodents indicate that it could in some circumstances have a facilitatory role and this notion should be investigated further. From a functional point of view, these pseudocopulations in lizards apparently play a significant role in reproduction: females that undergo mounting release more eggs and thus produce more offspring than females that do not engage in this behavioral interaction.

Finally, a few words should be added concerning species that adopt multiple reproductive phenotypes within a same sex, including large males that defend territories and smaller ('satellite') males that often display morphological and behavioral features of females (smaller size, absence of colorful displays) and usually steal copulations from the larger dominant male (sneakers). Such social systems have been observed in most, if not all, vertebrate classes. However, the hormonal bases of these alternative reproductive tactics have been studied in detail in only a few cases (the plainfin midshipman fish, *P. notatus*, or the tree lizard, *U. ornatus*). In these examples, sneaker males were shown to display reduced levels of androgenic steroids (testosterone or 11KT in fishes) as compared to dominant males. In contrast, a somewhat similar behavioral and morphological polymorphism in a shore bird, the ruff (*Philomachus pugnax*), does not seem to depend on a differential activation by testosterone but appears to be directly controlled by genetic autosomal factors. The ruff has two types of males: territorial males that defend small territories on leks that will be visited by females during the breeding season and satellite males (approximately 16% of the population) that do not defend such territories but stay in the vicinity and try to sneak copulations with females visiting the lek. These different behavioral strategies are associated with different patterns of plumage. Territorial males display dark (brown or black) long fluffy feathers in their neck (the ruff) and occipital 'head tufts,' whereas these feathers are lightly colored (white, creamy yellow) in satellites. Although plasma testosterone concentrations have not to our knowledge been compared in these two types of morphs, the steroid is unlikely to contribute to these behavioral and morphological differences between morphs because females that do not show these attributes spontaneously will display them with similar proportions as males if treated with a same dose of exogenous testosterone.

Conclusions

Research initiated in the middle of the nineteenth century by Adolph Bertold who identified the critical role of a secretion of the testis in the control of male sexual attributes in chicken has during the second half of the twentieth century produced a huge corpus of data that explain with reasonable detail and accuracy the (neuro)endocrine mechanisms controlling male reproductive behavior. A selection of species have been studied in all classes of vertebrates from fishes to mammals, and several general principles have emerged from this work. A number of broad questions remain, however. They concern the neural mechanisms underlying the differential sensitivity of male and female brains to the same endocrine stimuli, the interaction between genomic and nongenomic actions of steroids on the brain that support sexual behavior and also the detailed architecture of the neural circuits that control the appetitive and consummatory phases of this behavior and to what extent these circuits can be generalized to all vertebrates. From a comparative point of view, it would also be useful to determine to what extent these general principles have been modified in species that have adapted unusual reproductive strategies including hermaphroditism, parthenogenetic reproduction, and alternative mating strategies.

Acknowledgments

The preparation of this review and the experimental work from our laboratories was supported by grants from the NIMH (Grant number RO1 MH50388) to G.F. Ball and from the Belgian Fonds de la Recherche Fondamentale Collective (Grant number 2.4537.09) to J. Balthazart.

See also: Sex Change in Reef Fishes: Behavior and Physiology.

Further Reading

Adkins-Regan E (2005) *Hormones and Animal Social Behavior. Monographs in Behavior and Ecology.* Princeton and Oxford: Princeton University Press.

Ball GF and Balthazart J (2004) Hormonal regulation of brain circuits mediating male sexual behavior in birds. *Physiology & Behavior* 83: 329–346.

Ball GF and Balthazart J (2008) How useful is the appetitive and consummatory distinction for our understanding of the neuroendocrine control of sexual behavior? *Hormones and Behavior* 53: 307–311; author reply 315–318.

Balthazart J, Baillien M, Cornil CA, and Ball GF (2004) Preoptic aromatase modulates male sexual behavior: Slow and fast mechanisms of action. *Physiology & Behavior* 83: 247–270.

Balthazart J and Ball GF (2006) Is brain estradiol a hormone or a neurotransmitter? *Trends in Neurosciences* 29: 241–249.

Balthazart J and Ball GF (2007) Topography in the preoptic region: Differential regulation of appetitive and consummatory male sexual behaviors. *Frontiers in Neuroendocrinology* 28: 161–178.

Balthazart J and Hendrick JC (1976) Annual variation in reproductive behavior, testosterone, and plasma FSH levels in the Rouen duck, *Anas platyrhynchos. General and Comparative Endocrinology* 28: 171–183.

Becker JB, Berkley KJ, Geary N, Hampson E, Herman JP, and Young EA (2008) *Sex Differences in the Brain. From Genes to Behavior.* Oxford: Oxford University Press.

Becker JB, Breedlove SM, Crews D, and McCarthy MM (2002) *Behavioral Endocrinology.* Cambridge MA: MIT Press.

Berthold AA (1849) Transplantation der Hoden. *Archiv für Anatomie, Physiologie und wissenschaftliche Medicin* 16: 42–46.

Cornil CA, Ball GF, and Balthazart J (2006) Functional significance of the rapid regulation of brain estrogen action: Where do the estrogens come from? *Brain Research* 1126: 2–26.

Crews D (2005) Evolution of neuroendocrine mechanisms that regulate sexual behavior. *Trends in Endocrinology & Metabolism* 16: 354–361.

Everitt BJ (1995) Neuroendocrine mechanisms underlying appetitive and consummatory elements of masculine sexual behavior. In: Bancroft J (ed.) *The Pharmacology of Sexual Function and Dysfunction*, pp. 15–31. Amsterdam: Elsevier.

Goy RW and McEwen BS (1980) *Sexual Differentiation of the Brain.* Cambridge, MA: The MIT Press.

Hinde RA (1965) Interaction of internal and external factors in integration of canary reproduction. In: Beach FA (ed.) *Sex and Behavior*, pp. 381–415. New York: Wiley.

Lehrman DS (1965) Interaction between internal and external environments in the regulation of the reproductive cycle of the ring dove. In: Beach FA (ed.) *Sex and Behavior*, pp. 355–380. New York: Wiley.

Nelson RJ (2005) *An Introduction to Behavioral Endocrinology.* Sunderland, Massachusetts: Sinauer Associates.

Pfaff D, Arnold AP, Etgen AM, Fahrbach SE, and Rubin RT (2002) *Hormones, Brain and Behavior.* Amsterdam: Academic Press.

Wingfield JC and Moore MC (1987) Hormonal, social, and environmental factors in the reproductive biology of free-living male birds. In: Crews D (ed.) *Psychobiology of Reproductive Behavior: An Evolutionary Perspective*, pp. 148–175. Englewood Cliffs, NJ: Prentice-Hall.

Mammalian Female Sexual Behavior and Hormones

A. S. Kauffman, University of California, San Diego, La Jolla, CA, USA

Introduction

What is sexual behavior, and why (and how) do animals exhibit it? These are important questions which have been the basis of much scientific focus over the past century. Sexual behavior evolved to bring gametes together from two different individuals, thereby increasing diversity and variety in the genetic makeup of the offspring (in contrast to being a complete clone of the parent). Such genetic variety has been proposed to be adaptive in allowing animals to increase the diversity in their traits in an ever-changing natural environment. In mammals, sexual behavior often differs between males and females, reflecting differences in a multitude of ecological and physiological pressures, as well as anatomical distinctions between the sexes. This review will focus exclusively on sexual behavior of female mammals.

Historically, the empirical study of female sexual behavior was initiated well after that of male sexual behavior. Unlike male mating behavior, which was systematically studied as early as 1849 by Adolf Berthold, female reproductive behavior received little scientific attention prior to the twentieth century. The reason for this discrepancy is unclear, though it may reflect the fact that female reproduction, unlike that of males, occurs in cycles which significantly complicate its study. That is, female reproduction in mammals has specific phases which repeat throughout adulthood, beginning with courtship and mating, followed soon thereafter by ovulation and fertilization, leading to pregnancy and parturition, and culminating with lactation (and then repeating the entire cycle again). Unlike males, the reproductive status of a female mammal constantly changes, creating difficulties or complications in studying any one particular stage. Moreover, even when a female is not pregnant or lactating, her sexual behavior usually occurs only at specific stages of her estrous (or menstrual) cycle, further reducing the ease with which such behavior can be studied. In contrast, male mammals do not have cyclic restrictions regarding when they can show sexual behavior, nor do they exhibit reproductive stages of pregnancy or lactation, making them easier models to study.

In the first half of the twentieth century, researchers began to systematically study female sexual behavior in mammals, detailing its characteristics and components, as well as defining its hormonal and neural correlates (discussed in detail in later sections). Today, female reproductive behavior is studied within the fields of biopsychology, neuroscience, endocrinology, and animal behavior, and for some aspects we know more today about the neural mechanisms underlying female sexual behaviors than we do about male mating behaviors. Most information regarding mammalian female sexual behavior has been gleaned from laboratory studies involving rodents, primarily rats, guinea pigs, and hamsters, and more recently, mice (including transgenic and knockout mouse models). Thus, the majority of the information detailed in this chapter focuses on rodents, though additional discussion includes findings from non-human primates and, to a lesser extent, other mammalian species. Keep in mind that although there are many overlaps and similarities between different mammalian species in terms of female sexual behavior, there are sometimes differences as well, and not all mechanisms present in rodents can be commonly found in other species.

Cycles and Components of Female Sexual Behavior

Sexual behavior can be broadly defined as all behaviors necessary and sufficient to achieve fertilization of female gametes (ova) by male gametes (sperm). Female sexual behavior includes both copulatory and noncopulatory behaviors that are linked to sexual interaction. In other words, reproductive behavior includes not only copulation itself (i.e., penile insertion and deposition of sperm in the vagina) but also the various behaviors immediately preceding copulation, such as courtship displays or specialized vocalizations, which are crucial to promoting and achieving subsequent copulation. Likewise, sexual behavior could include postcopulatory behaviors which help ensure successful fertilization, such as postural adjustments which promote sperm transport to the egg, or mate-guarding behavior which occurs in some species. Despite this possibility, the study of female sexual behavior in mammals has predominantly focused on precopulatory and copulatory behaviors, with little attention given to postcopulatory aspects.

In contrast with males, most female mammals exhibit estrous cycles (menstrual cycles in humans and non-human primates). Early observations noted that for many species, female sexual behavior is restricted to a specific stage of the cycle known as 'behavioral estrus.' The word 'estrus' is derived from the Latin word 'oestrus,' which translates as frenzy (as well as horse-fly or gadfly). More loosely, oestrous translates to 'in a frenzied condition' or 'possessed by the gadfly,' and is commonly referred to as being 'in heat.' Females in estrus exhibit a range of typical behaviors that have long been recognized as relating to actively seeking, and initiating, copulation. In most species (humans being one exception), similar sexual behaviors are not typically observed in anestrous females (i.e., females not in behavioral estrus). For example, female rats (*Rattus norvegicus*) or mice (*Mus muscus*) housed with a sexually active male display mating behavior every fourth or fifth night, when she is said to be in estrus (i.e., in heat). If the male is unsuccessful in achieving fertilization, or has been experimentally vasectomized to prevent fertilization, the female continues to mate with him once every 4–5 days, but not on days in between. Hence, her estrous cycle is considered to be 4–5 days in length, with behavioral estrus comprising approximately a day or so, depending on the species. Similar observations were made long ago in other species, including Syrian hamsters (*Mesocricetus auratus*, 4–5 day cycles), guinea pigs (*Cavia porcellus*) and sheep (*Ovis aries*, 16-day cycles), and various canines (*Canis lupus*, variable duration; typically 7 months with estrus lasting approximately 10 days).

The specific timing of the occurrence of female sexual behavior, that is, behavioral estrus, is such that it is tightly coupled to the timing of ovulation. Estrus typically occurs just before ovulation (variable, but on the order of hours), ensuring that sperm will be readily available in the female's reproductive tract to fertilize the ovulated egg, thereby facilitating successful reproduction. For this reason, it is not surprising that a majority of mammalian species (though certainly not all) have linked the timing of female sexual behavior with the timing of ovulation. As will be discussed later, both ovulation and female sexual behavior are regulated primarily by changes in gonadal sex steroids (estradiol and progesterone), providing synchrony and concordance in the hormonal regulatory mechanisms underlying these two coupled phenomena. Some species, such as voles (Rodentia; subfamily Arvicolinae), musk shrews (*Suncus murinus*), and rabbits (Order Lagomorpha), do not have regularly occurring estrus cycles, but instead display 'induced' estrus which is caused by exposure to a conspecific male. Many mammals with induced estrus are solitary and hence restrict the triggering of their female mating behavior, as well as ovulation, to times when a male is present. For these species, females never come into heat unless exposed to a male (or certain male signals, such as pheromones and olfactory cues).

Although a general consensus emerged that female sexual behavior is elevated during behavioral estrus in most species, initial studies of female mating did not systematically parse out various sexual behaviors or components of behavioral estrus. This led to some confusion and discrepancies in the field, as different researchers defined sexual behaviors in different ways. In many cases, they failed to define aspects of sexual behavior at all. Because of this, a more detailed description of female

mating was eventually proposed by Beach, who suggested that female sexual behavior in mammals comprised three critical and separate components: *attractivity*, *proceptivity*, and *receptivity*.

Attractivity

Beach defined the first component of female sexual behavior, attractivity, as the stimulus value of the female to the male. Thus, attractivity is a relative measure of 'value' of a female to a given male (i.e., how attractive she is to him). This value is inferred by the observer based on the male's behavioral or physiological response to certain parameters of the female. Unlike the other components of female sexual behavior, attractivity cannot be determined without directly assessing the responses and actions of another animal other than the female, that is, the male, and therefore involves some additional variability based on inherent individual differences in the personal preferences of males (as described later). Attractivity can encompass both behavioral and nonbehavioral aspects. Nonbehavioral attributes include specific eye-catching body coloring or markings, or enticing body shape or morphology. Behavioral cues are varied and often species specific, and include typical courtship displays, such as alluring movements which the male finds attractive, enhanced presentation of certain appealing body features, or the active secretion of olfactory cues to which the male is attracted. Ear wiggles in female rats and mice are a common behavioral display that males find enticing. The adaptive value of attractivity includes bringing the male, a prospective mate, closer to the female, assisting males to identify the female's reproductive status and/or genital regions, and orienting the male's coital responses.

Attractivity in many species is typically highest during estrus. As we shall see in later sections of this article, we now know that many aspects of attractivity, like proceptivity and receptivity, are dramatically influenced by ovarian hormones (estradiol and progesterone). This serves to maximize a male's interest in the female around the time of ovulation (which is similarly induced by these same hormones). There are many examples of animals showing increased attractivity during behavioral estrus versus anestrus. For example, one of the earliest studies by Warner, in 1927, determined that male rats, trained to cross an electrified grid (which provides a mild shock) to gain access to a female, displayed more frequent crossing when the females were in estrus than not, suggesting the males found the females more attractive. Likewise, 'preference studies' in many species have assessed a female's attractiveness to males under different physiological conditions. Males tend to display preference for females in estrus rather than females who are anestrous. This has been convincingly shown for rodents and dogs, in which, given a choice between a female in estrus and another who is not, a male tends to spend significantly more time with the estrus female. This enhanced attractivity during estrus even extends to male preferences for the urine of estrus versus anestrous females (i.e., chemicals/odorants in the estrous female's urine are perceived as attractive to the male).

Increased attractivity during estrus has also been demonstrated for many non-human primate species, including chimpanzees (*Pan troglodytes*), baboons (*Papio ursinus*), Rhesus monkeys (*Macaca mulatta*), and pig-tailed macaques (*Macaca nemestrina*). Many of these monkey studies have used a PROX score to determine the amount of time a male spends in close proximity to a female; estrus females almost always have higher PROX scores than anestrous females, and hence, higher attractivity. Similar findings in primates have been observed with other related measures (such as the acceptance ratio, which is the proportion of female solicitations that elicit male sexual behavior).

Despite the pattern of increased sexual attractivity during estrus, which is induced by increased ovarian sex steroids at this time (discussed later), attractivity can sometimes depend on factors in addition to hormonal status. Even when several females are in behavioral estrus or under the same sex steroid conditions, certain females may promote a greater sexual response in males, suggesting that some nonhormonal element may also be involved in producing elevated female attractiveness. Likewise, males of many species often exhibit individual preferences for certain estrus females over others who are similarly in estrus. Such nonhormonal factors influencing a female's attractivity could include her age, whether she is sexually experienced or not, morphological features, or other unidentified 'X-factors' that one female possesses that another does not.

Proceptivity

Proceptivity, Beach's second component of female sexual behavior, is defined as the extent to which a female initiates mating; proceptive behaviors therefore reflect a female's willingness and motivation to mate (i.e., appetitive behaviors). This aspect is analogous to sex drive or libido. Thus, behaviors in which a female is sexually solicitous and initiates copulation (but not the act of copulation per se) are considered proceptive. Whereas attractivity reflects how much a male is attracted *to a female*, proceptivity reflects how much a female is attracted *to a male*. Importantly, the identification and study of proceptivity has contributed significantly to the understanding that female mating behavior is not simply a passive process whereby the female just waits for the male to copulate with her, but rather, females play an active role in initiating many aspects of sexual interaction and copulation. Functionally, proceptive behaviors serve to arouse the male and to facilitate, coordinate, and synchronize male and female behaviors and bodily adjustments necessary for the act of copulation. Proceptive behaviors may also play a role in mate seeking and identification, as well as mate selection. Supporting the importance of proceptivity in reproductive behavior, when female rats display proceptive behavior (i.e., actively seek out and promote sexual interaction), successful copulation occurs 90% of the time, whereas only 3% of male-initiated contacts result in successful copulation. Proceptive behaviors may increase the level of sexual excitement in the performing females, although this conjecture has not been systematically studied.

Different species exhibit different kinds and degrees of proceptive behavior. However, there are four general categories of proceptivity, as designated by Beach. The most common, almost universal among proceptive females, are affiliative behaviors, that is, female actions leading to the establishment and maintenance of proximity to a male. Affiliative behaviors are measured as the tendency of a female to approach and remain in the vicinity of the male. In monkey studies, this is characterized by the PROX score which was discussed earlier. For example, female rhesus monkeys do not typically display close proximity interactions with males but show increased approaches and spend more time sitting next to a male monkey right before mating. These higher PROX scores of the females translate to increased proceptivity.

A second general class of proceptive behaviors includes those that are sexually solicitous. Solicitation behaviors, also referred to as 'invitation' or 'presentation behaviors,' are varied and include the female assuming specific coital postures before physical contact with the male, performing specialized gestures (such as head bobbing or lip smacking in certain primates), presenting the female genitalia to the male, or making solicitous vocalizations to help initiate male contact and increase male arousal. A third general class of proceptivity is physical contact responses, in which the female initiates contact with the male. Examples include female rodents and canines investigating and touching the male's anogenital region or some primate species engaging in generalized grooming of the male. In sheep, females will often repeatedly nudge the ram with their heads, a common proceptive behavior preceding copulation. The last class of proceptive behavior encompasses alternating approaches and withdrawals of the female to the male. This is particularly common in rodent species ('hops and darts'), but also occurs in other mammalian orders. The behavior consists of the female approaching the male and then retreating if he follows her; when he stops following, she reapproaches and then again withdraws. Such a pattern is stimulating and enticing to males, and serves to increase males' sexual interest in the female, culminating in copulation. Approaches and withdrawals may also serve to orient the male so that he is in the best physical position relative to the female to achieve copulation.

As with attractivity (and receptivity, described later), females are most proceptive during behavioral estrus. Thus, females exhibit the highest levels of actively seeking out and initiating sexual interaction during the time when they are concurrently most attractive to potential mates. The synchrony of maximal attractivity and proceptivity is governed by ovarian sex steroids which peak at behavioral estrus and which also promote ovulation soon thereafter. There are numerous examples of proceptive behaviors peaking concurrent with behavioral estrus. In many non-human primates, female genital presentations to males increase with estrus, and female approaches to males are highest during midcycle (estrus). In baboons, lip smacking (a proceptive behavior) is most common during behavioral estrus. Interestingly, female rhesus monkeys trained to press a lever to gain access to a male exhibit significant increases in lever-pressing frequencies during midcycle. Similar motivational experiments in rats observed estrous females displaying a tenfold increase in the likelihood to cross an electrified grid in order to reach a male, or significantly more bar presses than anestrous females in order to gain access to a male. Numerous rodent preference studies involving Y-mazes (in which stimulus animals are confined to the arms of the Y, and an experimental animal is introduced with free access to all areas) have also documented that females given the choice of spending time with a stimulus male or another stimulus female spend dramatically more time next to the male when she is in estrus versus anestrus. Thus, proceptivity is increased when females are in behavioral estrus.

Although proceptivity is highest in estrus for most but not all species, additional factors can influence proceptive behaviors. Perhaps not surprisingly, the degree of female proceptivity depends, in part, on male attractiveness. Moreover, similar to a male's individual preferences in determining attractivity, personal preferences of the female can play a role in her proceptivity. Thus, females can sometimes display individual differences in what they find attractive and will therefore display more proceptive behavior to certain males over other males; in some cases, the males which are less attractive to one female may be more attractive to another. Such personal preferences have been documented in many species, including sheep, dogs, and non-human primates. The underlying basis for these personal preferences is not currently known.

Receptivity

'Sexual receptivity' is the most classic term associated with female sexual behavior. However, until Beach's 1976 categorization of female mating behaviors, researchers used the term 'receptivity' in different contexts and with different meanings, which often confused and complicated the study of female reproduction. Discrepancies in the literature were often due to differences in how receptivity was defined (or more often, how it remained undefined). Many classical studies of 'receptivity' included numerous proceptive behaviors along with the act of copulation itself, making it hard to tease apart specific aspects of behaviors and their underlying mechanisms.

Beach operationally defined receptivity in stimulus-response terms to include sexual behaviors exhibited by females in response to stimuli normally provided by conspecific males. Beach further defined receptive behaviors as those which comprise female reactions and responses necessary and sufficient to achieve penile insertion, that is, the act of copulation itself. This equates with a female's readiness to allow the copulatory act and often takes the form of species-specific postures adopted by the female during intercourse. For example, in rodents, receptive females usually display lordosis, a stereotyped behavior in which the female stands immobile, arches her back, and deflects her tail to the side, all in an effort to adopt a position that facilitates the male's penile insertion.

The experimental study of female receptivity typically includes quantifications of the receptive behavior and is usually expressed as a ratio of the male's attempts to copulate and his success in doing so. In rodents, this ratio is termed 'the lordosis quotient' (LQ); higher LQ values represent a greater frequency of the female displaying the receptive posture (lordosis) per mating attempt by the male. Similar ratios have been adopted for other species. In canines, the rejection coefficient is determined by dividing the total attempts by the male to mount the female by the number of times he is permitted to mount her and exhibit thrusting. In monkeys, similar scores are called 'the acceptance ratio' or 'the success ratio.'

In some cases, there is overlap of receptive behaviors with proceptive behaviors, or even attractivity (in fact, some experimental measures, such as the acceptance ratio, can also be used to determine attractivity or other components). In some rodents, a female actively initiates copulation (i.e., exhibits proceptivity) by exhibiting the lordosis posture, which is typically regarded as a receptive behavior. Thus, lordosis in this species is both proceptive and receptive. Given that the proceptive phase leads into the receptive phase, some overlap is not surprising, especially at the transition between the two phases. In the small insectivore the musk shrew (*Suncus murinus*), after an initially aggressive interaction with a courting male, a female shrew suddenly displays 'tail wagging' when she is ready to mate. This intriguing behavior involves the female shrew continually walking away from the male while rapidly flicking her tail back and forth; during this time, the female never stops to adopt an immobile posture, and males attempt to mount and copulate with her as she constantly walks and tail wags. In this case, tail wagging is an indicator of both proceptivity and receptivity.

As with attractivity and proceptivity, receptivity in most mammals with recurrent estrous cycles is highest during behavioral estrus, when the female is in heat. In many cases, receptive behavior is never displayed in anestrous states. For example, lordosis is never observed in normal female rodents that are not in behavioral estrus. Likewise, dogs and cats that are not in estrus usually avoid or actively discourage male-mounting attempts and will not permit penile insertion even if he does mount. Some non-human primates, but not all (e.g., chimpanzees), follow similar patterns, showing increased receptivity only during the phase of the cycle when ovulation is imminent. As we shall see in the next section, this is primarily due to elevated ovarian sex steroids which peak around behavioral estrus and serve to promote both enhanced receptivity and subsequent ovulation.

An Alternate Model of Classifying Female Mating

Beach's proposed classification of female sexual behavior has been used extensively since its inception in 1976, and it remains the most common method of categorizing female mating. Recently, another set of categories was suggested by Blaustein and Erskine. This new set of definitions highlights the active contribution of the female in the mating process. Blaustein and Erskine split female sexual behavior into three main components: copulatory, paracopulatory, and progestative behaviors.

Copulatory behaviors include those that result in successful transfer of sperm from the male to the female. This is similar to receptive behaviors as described by Beach and includes specific postures adopted during mating in order to promote penile insertion. Paracopulatory behaviors are similar to proceptive behaviors and include species-specific female behaviors which arouse and stimulate the male to mount her, such as approach and withdrawal cycles, vocalizations, and affiliative behaviors. Paracopulatory behaviors also include behavioral aspects of attractivity, such as ear wiggling, which attract the male and thereby help facilitate sexual interaction.

The third component, progestative behaviors, includes factors that increase the probability of reproductive success, such as the female becoming pregnant. Progestative behaviors include short-term behavioral adjustments in the timing of sexual intercourse, such as pacing behavior. Pacing is common in rodent species, notably rats and mice. In these species, females actively regulate the pace of the copulatory bout, serving as the principal controller in the rate and duration of sexual stimulation. When a female rodent paces intercourse, she periodically stops displaying lordosis, thereby stopping copulation, and avoids the male for short periods before reinitiating another lordosis bout. Thus, copulation proceeds as a cycle of lordosis (mounts with penile insertion) and nonlordosis

(no mounts), until it finally culminates in ejaculation. The duration and interval between successive mounting bouts is entirely determined by the female and can range from several seconds to several minutes, or more. Female pacing of copulation has been proposed to ensure that she receives a species-specific temporal pattern of vaginocervical stimulation which best optimizes the likelihood of ovulation, corpora lutea survival, and fertilization. Importantly, pacing and other progestative behaviors emphasize that the female's role in copulation is not passive, but rather she is an active participant, and in some cases, the principal regulator of the event. This may be adaptive, as females invest more time, energy, and resources in reproduction and offspring development than do males, and thus, it benefits females to actively regulate their mating efforts.

Hormonal Factors Modulating Female Sexual Behavior

As alluded to in previous sections, female sexual behavior is highly regulated by ovarian sex steroids, primarily estradiol (E2) and progesterone (P4). In fact, using Beach's classifications of female reproductive behavior, all three components (attractivity, proceptivity, receptivity) are affected by sex steroids. However, these components are not necessarily affected by sex steroids in the same way, and not always by the same hormones.

In the early 1900s, researchers first began to dissect the hormonal mechanisms underlying female sexual behavior. Following early work on males (beginning with Berthold), investigators began to assess the role of the ovaries in the control of female mating behavior. The results were consistent and conclusive: for most mammalian species, including rodents, canines, and sheep, removal of the ovaries drastically diminished, or eliminated, female sexual behavior. Because mating behavior stops after ovariectomy, researchers postulated that cycles in ovarian physiology and function influence cycles of sexual behavior. In 1917, Stockard and Papanicolo demonstrated that changes in guinea pig vaginal cytology were tightly correlated with changes in ovarian physiology and function, findings that were soon extended to rats, mice, and other mammalian species. This noninvasive technique allowed investigators to easily determine the stage of the ovarian cycle without removal of the ovaries, simply by observing the pattern of vaginal cell types taken easily from a vaginal swab. The vaginal cell types have since been used to designate four specific stages of the female's estrous cycle: diestrus I and II, proestrus, and estrus. It was later shown that these vaginal stages correspond to, and are induced by, changes in ovarian sex steroid secretion. It is now appreciated that, in general, estradiol is low in diestrus I, rising in diestrus II, high in proestrus, and low again during vaginal estrus

(**Figure** 1). Progesterone is essentially low throughout diestrus II and early proestrus, and is briefly elevated in late proestrus and again in early diestrus I after ovulation has occurred (**Figure** 1). Ovulation typically occurs at the end of proestrus, after estradiol has peaked (ovulation is itself induced by a preovulatory 'surge' in pituitary luteinizing hormone during mid-to-late proestrus under the governance of high estradiol levels) (**Figure** 1).

In 1922, before the discovery and identification of estrogens and progestins, Long and Evans correlated sexual behavior of female rats with stage of vaginal cytology. Female mating behavior (lordosis) was observed only when the vaginal smear was indicative of proestrus or early estrus. Thus, behavioral estrus (i.e., being in heat) is not necessarily equivalent with vaginal estrus, but actually occurs more frequently at the latter part of proestrus (extending into early vaginal estrus) (see **Figure** 1). Based on the results of Long and Evans and others showing that female mating was elevated primarily during proestrus, and that ovariectomy eliminated female sexual behavior in most species, researchers postulated that some ovarian substance (or substances) produced in the ovaries during proestrus was responsible for stimulating female mating. These substances were eventually shown to be estrogens and progestins.

Estradiol

Working on the hypothesis that ovarian products could promote female sexual behavior, early studies found that

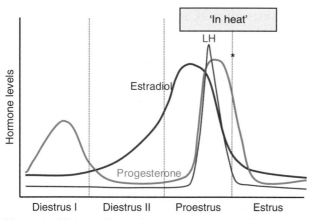

Figure 1   Diagram of reproductive hormone levels in blood across a typical rodent estrous cycle. Behavioral estrous, that is, when females are 'in heat,' occurs during the end of proestrus and early vaginal estrus. Ovulation, indicated by the blue asterisk, occurs in late proestrus/early estrus (species specific) and is generated by a surge in LH secretion which, in turn, is generated by rising estradiol levels. Note that hormone levels depicted are estimates rather than absolute values, and the maximal height of each line is not necessarily relative to the other hormones, just to itself. Thus, each line reflects each individual hormone's general pattern over the cycle. (Note also that there may be slight difference from one species to the next in terms of the temporal pattern for each hormone.)

injecting ovariectomized (OVX) rodents with ovarian chemical extracts (interestingly, isolated from other mammalian species, such as swine) caused significant increases in female mating behavior, measured as lordosis. However, the specific substance within the ovaries that was inducing behavior remained unknown. Soon thereafter, in 1929 and the early 1930s, Dosiy, Butenandt, and MacCorquodale isolated and identified the ovarian estrogens, of which 17-β-estradiol (E2) is the primary circulating form. As mentioned earlier, the secretion of E2 from the ovary was subsequently shown to increase slowly over the estrous cycle, peaking at proestrus and then declining rapidly thereafter; thus, the peak in E2 levels was shown to coincide with the peak in female behavioral estrus (**Figure** 1). To confirm that E2 played a role in driving female sexual behavior, early studies injected E2 into OVX female rodents and then assessed their sexual behavior. In most cases, E2 treatment substantially increased female reproductive behavior, findings which have since been repeated numerous times in many species, both rodents and nonrodents alike. However, in almost all cases, female sexual behavior was not maximal after E2 treatment (regardless of dose), and often a significant proportion of treated animals did not display any mating behavior. Thus, while E2 was capable of inducing some female sexual behavior, another factor appeared to be required as well. This additional factor was soon shown to be progesterone (P4), which is described in more detail in the following lines.

Although early studies focused either on receptivity (i.e., lordosis) or a combination or sexual behaviors (often lumped together as 'receptive behaviors'), later studies and post hoc reevaluation of earlier studies began to determine the effects of E2 on individual components of sexual behavior. It now seems clear that E2 has stimulatory effects for all main components of female reproductive behavior, though the mechanisms by which E2 exerts these effects may vary from one component to the next. Attractivity increases significantly in OVX females given E2 treatments. For example, the tendency of males to visit and affiliate with female rodents and dogs is robustly elevated after E2 treatment, and similar increases in male preferences are observed for urine taken from E2-treated females than from OVX females not given E2. In many non-human primates, male approaches and/or mounts to a female are higher in E2-treated OVX females than those not treated with E2, and male monkeys trained to press a lever for access to a female substantially increase their lever pressing when the female has been injected with E2, indicating that E2 increases female attractivity.

Like attractivity, proceptivity is also enhanced by E2. In some cases, proceptive behavior is only present when E2 is present, whereas in other cases, proceptive behaviors are exhibited with or without E2, but increase in magnitude and frequency when E2 is present. In general, most OVX females do not show proceptive behaviors and E2 replacement increases the display of proceptivity. For example, rats, hamsters, and mice that are OVX show little preference for males over females, and do not exhibit hops and darts in the presence of males. In contrast, OVX female rodents administered E2 spend much more time affiliating with males and even exhibit increased tendencies to traverse an electrified grid to gain access to a male. Like rodents, female dogs and sheep also show increased affiliative behaviors after E2 treatment compared with the OVX state. Similarly, in rhesus monkeys and other primates, the frequency of presentations of swollen genital regions to males increases with E2 treatment, as does the frequency with which a female rhesus monkey will press a lever to gain access to a male, indicating that E2 increases her proceptivity.

As mentioned earlier, receptivity is also enhanced by E2 treatment. Although some species, such as rabbits and rhesus monkeys, do not rely on E2 to display receptivity, OVX animals of most mammalian species, including rodents, cats, canines, pigs, cattle, and some primates (e.g., baboons), fail to display receptive behaviors until treated with E2. Even so, as observed by Long and Evans and others in the 1920s, E2 treatment alone increases but does not usually maximize the degree or magnitude of the receptive behavior. Moreover, in some cases, a good percentage of females (as high as 40% in some studies) do not show much receptive behavior after E2 treatment. Thus, E2 is stimulatory to receptive behavior, but in many cases, an additional factor is also needed to achieve maximal receptivity (i.e., normal levels observed in intact animals). Usually, this additional factor is P4.

Progesterone

The early finding that many OVX females treated with E2 did not display full receptivity suggested that some other factor, working in conjunction with E2, was also involved. Progesterone (P4) was discovered and isolated not long after E2, in 1934–1935 by Allen and others. Given that P4 is high during late proestrus at the same time that E2 levels are elevated and in synchrony with maximal estrous behavior, researchers began to test the notion that P4 was stimulatory to sexual behavior. This was indeed to shown to be the case. Initial studies by Young and others in the mid 1930s and 1940s established that whereas E2 treatment did not fully restore female sexual behavior in OVX guinea pigs, treatment with both E2 and P4 did maximally induce sexual behavior. Similar findings of P4's ability to stimulate female mating were also reported for female rats. However, early monkey studies, as well as other studies in rodents, reported that P4 decreased female sexual behavior in OVX E-treated females.

Although confusing at first, these contradictory findings regarding P4's effects on female receptivity were eventually reconciled by the discovery that P4 exerts biphasic effects on female mating. Elevated E2 during proestrus, followed by a rise in P4 in late proestrus, stimulates maximal induction of female sexual behavior in normal cycling females (typically studied in terms of lordosis). The rise in P4 is thought to be produced by the ovarian follicle, though recent studies also suggest the possibility of neurally derived P4 produced within the brain itself (from astrocytes). Afterward, in late estrus/ early diestrus, after ovulation has occurred, the corpora lutea secrete high levels P4; these high P4 levels during estrus/diestrus serve as an inhibitory signal to female sexual behavior. Unlike E2, which appears to be almost exclusively stimulatory to female sexual behavior, P4 has a dual role, providing both stimulatory and inhibitory effects on female mating, depending on the timing and dose of P4. P4 given alone to OVX females, in the absence of E2, does not stimulate female sexual behavior; thus, the stimulatory effects of P4 on female mating require E2 to be present (typically for 18–20 h or more beforehand). This observation may reflect, in part, the actions of E2 to upregulate the receptors for P4 in the brain. Interestingly, P4 given before or concurrently with E2 can actually inhibit or depress estrous behavior, a phenomenon called 'concurrent inhibition.' The P4 treatment appears to be mimicking the second rise in P4 during the vaginal estrus/diestrus stage, which as stated earlier, serves to terminate receptive behavior in E2-primed rodents. Although P4 (or several of its metabolites) can act initially to induce maximal female receptivity after prior E2 treatment, some species, such as rabbits, certain hamsters, and voles, do not require P4 at all to display high receptivity; in these species, E2 treatment alone is sufficient to induce maximal female receptivity.

In contrast to receptivity, there are far less data concerning the effects of P4 on attractivity or proceptivity, primarily because E2 alone often appears to induce high attractivity and proceptivity in the absence of P4. Moreover, additional treatment with P4 in conjunction with E2 does not usually further increase attractivity or proceptivity. However, as noted by Blaustein and Mani, there is some evidence that P4, particularly of adrenal origin, can increase certain proceptive behaviors in E2-treated rats, suggesting that P4 may be stimulatory for this behavioral component. The situation is complicated by the fact that in several species of non-human primates (rhesus monkeys and baboons), attractivity is sometimes reduced by P4 treatment, though the results are inconsistent. The role of P4 in these nonreceptivity components may be species specific or trait specific, and this issue requires more empirical testing before definitive conclusions can be made.

Although sex steroid signaling is important, if not essential, for female sexual behavior in most mammalian species, it should be noted that many other factors, such as age, experience, photoperiod, and various neuropeptides and neurotransmitters, have also been implicated in regulating (either stimulating or inhibiting) female mating behavior. Many of these additional factors modulate (or mediate) the main effects of sex steroids and often require the presence of at least E2 in order to affect female mating; others influence sexual behavior completely independently from E2 or P4. In-depth discussion of these numerous other factors is beyond the scope of this chapter. A brief condensed list of neuropeptides and neurotransmitters implicated in this system includes opioid peptides, neuropeptide Y, serotonin, oxytocin, vasopressin, cholecystokinin, dopamine, GnRH I and II, nitric oxide, corticotropin-releasing factor, and norepinephrine.

Receptor Mechanisms of Hormone Signaling for Female Sexual Behavior

By the mid-twentieth century, it was fairly well established that both E2 and P4 were important for female sexual behavior. However, the mechanism by which these hormones exerted their effects on female mating, and receptivity in particular, were unknown. Although it was assumed that E2 and P4 acted in the brain to elicit their effects, the site of actions and the molecular/cellular mechanisms were not yet studied. Not until the identification of the various receptor subtypes toward the end of the century, along with the recently harnessed ability of modern genetics to remove or alter select genes in mouse models, could such issues properly be addressed. Today, there is a wealth of information that has been gleaned from studies with transgenic and knockout mouse models; these, when combined with the use of specific agonist, antagonist, and other chemical/drug treatments, have begun to identify the molecular mechanisms underlying E2 and P4 signaling in relation to female sexual behavior.

The first estrogen receptor was identified by Jenson in the late 1950s. For approximately 40 years after, most estrogen effects were presumed to be mediated by this one receptor until the discovery in 1996 of a second estrogen receptor by Gustafsson and others. Upon this discovery, the first estrogen receptor was termed 'estrogen receptor α' (ERα), whereas the second receptor was termed 'ERβ.' Both ERα and ERβ are nuclear receptors that bind DNA and influence gene transcription. More recently, a membrane-associated ERα, as well as another membrane receptor (termed 'GPR30') capable of binding E2 and initiating intracellular signaling cascades, have been identified, but there is less information as of yet regarding their roles in female sexual behavior.

In the mid-1990s, investigators first used molecular genetic technology to knock out the functional ERα gene from mice. This allowed Rissman and others to test whether ERα, or some other pathway, was important for mediating E2's stimulatory effects on female sexual behavior, primarily lordosis. These researchers determined that intact female ERα knockout mice (termed 'ERαKos') had severely altered estrous cycles and diminished lordosis responses. When these knockouts and their wild-type littermates were tested for female receptivity following ovariectomy and sex steroid replacement, ERαKOs (unlike wild-type females) still failed to display any lordosis with a sexually experienced stimulus male (**Figure 2**). This indicated that female sexual behavior, in particular receptivity, is dependent on functional ERα signaling pathways. Perhaps not surprisingly then, intact ERαKO females were shown to be infertile (they also display impairments in ovulation). In addition, Rissman showed that P4 has no facilitory effect on lordosis in ERαKOs as it does in wild-type mice, perhaps reflecting the absence of available progesterone receptors, since E2 acts via ERα to upregulate most progesterone receptors in the brain (particularly, the hypothalamus). Interestingly, E-treated ERαKO females were still attractive to males, because males attempted to mount ERαKO females to a similar degree as wild-type females. Moreover, when given a choice to spend time with either wild-type or ERαKO females, males spent equivalent time between the two genotypes. Thus, attractivity, which is increased by E2, does not seem to require ERα in mice (though ERα may be sufficient to mediate attractivity in normal females).

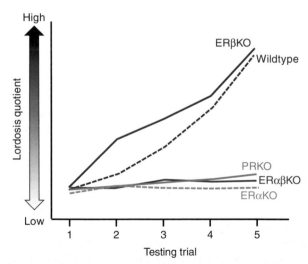

Figure 2 Composite diagram of general lordosis levels in female mice from five genotypes. Females were treated with E2 and tested multiple times for female sexual behavior. Note that normal female mice (wild types) usually require several trials of testing before they display high levels of receptive behavior. Female mice lacking ERβ display high levels of lordosis, similar to wild-type females. In contrast, females lacking either PR or ERα do not display significant lordosis, even after multiple testings. Like ERαKOs, females lacking both ERα and ERβ fail to display lordosis. This indicates that PR and ERα are each essential for proper female receptivity in mice.

Whether the stimulatory effects of E2 on attractivity are mediated by ERβ or GPR30 has not yet been determined.

Soon after the findings in ERα were published, additional studies were performed on female mice lacking a functional ERβ gene. When tested for sexual behavior, intact and E-treated female ERβKO mice showed lordosis to the same extent as wild-type females, suggesting that ERβ is not necessary for lordosis and sexual receptivity, at least in mice (**Figure 2**). Interestingly, some ERβKO females appeared to display *increased* sexual receptivity, at least during early testing trials; these and other data have led to the speculation that ERβ may be important during development for proper defeminization of the brain and behavior. Thus, in the absence of ERβ, ERβKO females may be hyperfeminized and therefore show higher levels of female sexual behavior. Supporting this possibility, male ERβKO mice appeared to be slightly feminized in that because they show higher levels of lordosis than wild-type males (but not as high as normal wild-type females). Despite this, after multiple trials of behavioral testing, ERβKO females and wild-type females show essentially equivalent receptivity levels, suggesting that ERβ (unlike ERα) is not critical for normal lordosis. ERβαKO females, lacking both receptor subtypes, do not display lordosis, even when hormone primed, likely reflecting the absence of ERα signaling (**Figure 2**).

In most tissues studied, including the brain, the progesterone receptor (PR) is induced by E2 via ERα signaling, and therefore, without E2, P4 cannot properly signal. This led to the postulate that some of the observed physiological and behavioral responses attributed to P4 signaling, including enhancement of sexual receptivity, might be due to the combined effects of P4 and E2, rather than just P4 alone. To clearly delineate the distinct roles of P4 and E2 in female sexual behavior, O'Malley and colleagues generated a novel mouse strain in which both forms of the PR were ablated using molecular gene-targeting techniques. Although male PR knockout (PRKO) mice were completely fertile, female PRKOs were shown to be infertile. When tested for sexual behavior, OVX PRKO females given E2 and P4 did not show lordosis, unlike their wild-type counterparts (**Figure 2**). This suggested that PR is essential for normal female receptivity. This conclusion was supported by findings that E2-primed female rats given neural infusions of PR antisense mRNA were incapable of displaying lordosis. Additional findings with PR subtype knockout mice soon determined that the PR-A receptor subtype, rather than PR-B, is likely the key PR subtype mediating P4's facilitory effects on receptivity in rodents. Interestingly, a membrane progesterone receptor was also recently identified, though like that of GPR30, the role of this G-protein-coupled receptor in female mating has not yet been sufficiently tested.

Interestingly, numerous classical studies showed that wild-type OVX mice and rats often display some receptivity, although not maximal, with just E2 treatment, suggesting that P4 is not required for minimal-to-moderate expression of lordosis. The fact that transgenic mice lacking functional PR show no lordosis at all, even with E2 treatment, required that earlier conclusions be modified. A more accurate statement is that PR signaling, but not necessarily P4 itself, is crucial for any display of female receptivity in rodents; in the absence of P4, some moderate lordosis can still be elicited by E2, provided neuronal PR is intact. Although this seems contradictory, recent evidence regarding nonligand activation of PR has shed some light on this issue.

In 1991, Power and others working in O'Malley's group made the fascinating discovery that nuclear PR can be activated in vitro by dopamine or dopamine agonists, in the complete absence of P4. Interestingly, the ability of dopamine to activate PR was not via direct binding of the steroid receptor. Rather, dopamine was shown to bind to its own membrane receptor (the D1 subtype of dopamine receptors) to initiate a second messenger intracellular cascade which subsequently activates nuclear PR. This indirect activation of PR via activation of D1 dopamine receptors is now referred to as 'ligand-independent activation' (or 'hormone-independent activation') and represents 'crosstalk' between the dopamine and PR systems. It has since been postulated that other factors can also activate PR and/or ER by ligand-independent activation, include GnRH, D5 dopaminergic agonists, phorbol esters, nitric oxide, prostaglandin E2, and cAMP.

Although Power determined that dopamine could activate PR in vitro, it was unclear if this process could occur in vivo, and in particular, in relation to female sexual behavior. This possibility was soon tested by Mani and others. Infusion of E2-primed female rats with dopamine agonists (for the D1 receptor) into the brain facilitated female receptivity in E2-primed females, similar to the effects of P4 treatment. Furthermore, dopamine's stimulation of female receptivity was prevented if the E2-primed rats were also treated with PR antagonists or antisense oligonucleotides directed at PR mRNA. This suggested that dopamine's stimulatory effects on lordosis were dependent on PR signaling (even if P4 was not present). This conjecture has since been supported by additional findings that dopamine agonists can stimulate receptivity in normal E2-treated wild-type females but not in E2-treated PRKO females. Thus, PR is essential for female receptivity, but P4 itself may not be required, provided other ligand-independent pathways still persist to activate PR pathways.

Prior to the generation and testing of ER and PR knockout mice, numerous sexual behavior studies had manipulated sex steroid receptor signaling capabilities through the use of agonist, antagonist, and antisense oligonucleotide treatments. The results of these studies generally agreed with later findings from knockout models. Thus, administration of antiestrogens (which bind both

ERα and ERβ) into the brain (or systemically) was highly effective in preventing lordosis in E2-primed female rodents of several species. Most recently, agonists selective for either ERα or ERβ have been used. When these agonists were delivered into the brains of female rats, only ERα agonists elicited lordosis (and proceptive behaviors), similar to E2 treatment. In contrast, an ERβ agonist was unable to induce lordosis. These findings support the conclusions of the knockout studies that ERα, but not ERβ, is involved in promoting female proceptive and receptive behaviors in rodents. Similar studies have been performed with progesterone signaling. Systemic or central (into the brain) treatment of female rats, guinea pigs, and mice with PR antagonists (e.g., RU486) prevented the facilitation of female sexual behavior by P4. Likewise, central infusions of antisense oligonucleotides directed against PR mRNA (which reduces synthesis of PR protein) inhibited P4's ability to facilitate female sexual behavior. The level of reduction in neuronal PR achieved with antisense infusions was approximately half that of normal levels in the brain, indicating that a certain amount of PR, above a given threshold, is necessary to achieve its facilitory effects.

Neuroanatomical Substrates Underlying the Control of Female Receptivity

The importance of E2 and P4 signaling in regulating female sexual behavior in most mammals is unquestioned. However, where are these hormones acting to achieve their effects on female mating? Given that the brain controls behavioral output, it was long assumed that sex steroids acted in the brain to induce sexual behavior. Although this topic has been studied in a number of species, including shrews, sheep, rabbits, and some nonhuman primates, the most data to date have been accumulated for rodents, in particular rats. Thus, this section and the next will focus primarily on the neuroanatomical circuits and hormonal sites of action mediating rodent female sexual behavior, in particular, lordosis. As the neuronal circuitry underlying lordosis has previously been extensively reviewed, this chapter will only serve to summarize the main findings.

Lordosis, a term derived from the curvature of the spine, is the most studied component of female sexual behavior. In the past 50 years, much work, spearheaded by Pfaff and others, has sought to identify the key brain regions involved in controlling this receptive behavior. As described earlier, a female displaying lordosis stands fairly immobile, arches her back, often raising her head and perineum up, and deflects her tail to the side, all with the goal of aiding penile insertion by the male (**Figure 3**). In the absence of lordosis, penile insertion and ejaculation are not possible. Lordosis has been shown to possess a

Figure 3 Picture (video capture) of lordosis in a female rat. Note the female's receptive posture and arched back, which allows the male to mount and copulate with her. Picture courtesy of Dr. Greg Fraley.

reflex component, and is elicited by a combination of hormonal priming (E2 and P4) and stimulation of mechanoreceptors in the periphery, specifically, tactile stimulation of the flank, rump, and perineum normally provided by the copulating male. Pressure responsive sensory neurons in the flank, rump, and perineum receive the tactile sensory input, which is then transmitted up the spinal cord to the brain. Interestingly, this peripheral aspect to lordosis is also regulated by sex steroids: the size of the receptive fields on the flanks is E2-dependent, and increases by as much as 30% when the female is in behavioral estrus. E2 therefore decreases the stimulus threshold required to activate the reflexive lordosis posture. Other sensory modalities, such as olfaction, vision, auditory input (vocalizations), or touch-pressure inputs from other skin areas, may, in some cases, modify the lordosis response, but none are sufficient or necessary for the lordosis reflex to occur.

Tactile information from the periphery enters the spinal cord and is relayed to the medullary reticular formation (MRF) in the hindbrain (a region known to mediate posture, movement, pain, and arousal). Lesions to this part of the pathway completely disrupt lordosis, indicating that this sensory input to the MRF is a necessary component of the lordosis mechanism. Importantly, the MRF forms the basis of the lordosis 'reflex arc,' as this region sends descending projections through the spinal cord to control motorneurons for back muscles that are critical for the lordosis posture. However, while lesion studies showed that destruction of parts of the MRF caused deficits in lordosis performance, midbrain lesions were also found to abolish lordosis. This indicated that the spinohindbrain (MRF) reflex arc is not sufficient by itself to mediate lordosis, and that input from the midbrain (or more anterior regions) is also essential. It was later shown that the midbrain central gray (also called 'periaquiductal gray,' PAG) receives some ascending spinal input. Lesions of the PAG also reduce lordosis, and electrical stimulation of just the POA was sufficient to promote lordosis. Although the PAG itself does not send direct descending projections down the spinal cord in rats, it does project to the MRF in the hindbrain. Thus, the same region (MRF) that receives sensory pressure signals from the flanks and rump also receives input from the PAG, placing the MRF in a key position to integrate both ascending and descending regulatory information for lordosis (see **Figure 4**). Although the MRF acts akin to a relay station for the PAG and spinal cord pathways, it is unclear if the same exact cells in the MRF receive both spinal and PAG input or if there are two separate neuronal populations.

In addition to the basic reflex circuit of which the MRF and PAG are critical components, other brain regions have also been shown to be critical for regulating lordosis. In particular, early studies determined that lesions of the ventromedial nucleus of the hypothalamus (VMN) dramatically diminish lordosis, as does destruction of efferent and afferent fibers of the VMN. Moreover, site-specific electrical stimulation of just the VMN in E2-primed females facilitates lordosis, indicating the importance of

this region in mediating the behavior. In regard to lordosis, efferent axonal fibers from the VMN travel caudally through the brain in both lateral and medial pathways, terminating in the PAG or nearby in the midbrain reticular formation (which also projects to the MRF), thereby connecting the hypothalamic portion of the circuit with the midbrain–hindbrain component (see **Figure 4**). Although the VMN has been shown to be the predominant forebrain nucleus controlling lordosis, over the years additional regions and nuclei have been added to the lordosis mechanism, including the preoptic area (POA, including the medial preoptic nucleus), the medial amygdala, the bed nucleus of the stria terminalis (BNST), and the arcuate nucleus. Many of these other sites serve to modulate the activity of the VMN, thereby modifying the degree, duration, and timing of the lordosis response. In fact, the VMN appears to be the key nodal point in the hypothalamus wherein multiple regulatory factors, such as metabolic, olfactory, and hormonal cues, converge and become integrated in order to control lordosis behavior.

Neuronal Targets of Hormonal Actions for Regulating Female Receptivity

Knowing the neuroanatomical regions involved in the lordosis circuit, one can ask where sex steroids, E2 and P4, act in the brain to facilitate lordosis, and are these the same brain regions comprising the lordosis circuitry? Much information regarding this question has been gathered. Most of the initial studies addressing the location of ERs were performed before the discovery of the second ER isoform (ERβ), and thus, were conducted under the assumption that there was only one ER. In hindsight, the results of some of these early studies obviously reflect the binding of both ERα and ERβ. To begin with, investigators in the 1960s and 1970s performed autoradiography studies with radiolabeled E2 in order to localize estrogen-binding neurons in the forebrain, midbrain, and hindbrain. Not surprisingly, these studies identified E2-binding sites (i.e., ER) in many places in brain, including regions comprising the lordosis circuitry. The highest density of E2 binding was observed in the forebrain and hypothalamus, including the POA, anterior hypothalamus, ARC, and VMN, as well as appreciable binding in the amygdala and PAG. Notably, the pattern of E2 binding in these regions is highly conserved among mammals (and indeed, among vertebrates), highlighting the evolutionarily conserved nature of many estrogen-regulated neuronal processes (including reproductive behavior).

The initial binding studies were unable to tease apart anatomical information regarding specific ER subtypes. Later studies determined that, while there is some overlap between ERα and ERβ, the two receptors have distinct expression patterns in the brain. ERα is most highly

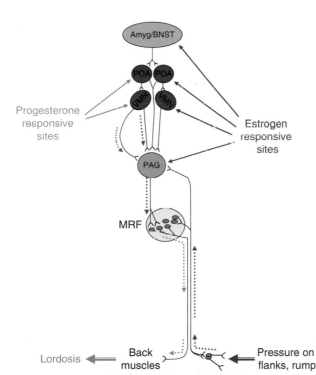

Figure 4 Schematic diagram of the neuronal circuits and sex-steroid-binding regions underlying lordosis in female rodents. Dotted arrows reflect the direction of neuronal signaling within the lordosis pathway. MRF: medullary reticulary formation; PAG: midbrain central gray (or periaquaductal gray); VMN: ventromedial nucleus of the hypothalamus; POA: preoptica area; AMYG/BNST: medial amygdala and the bed nucleus of the stria terminalis. Adapted with kind permission of Springer Science + Business Media from Pfaff DW (1980) *Estrogens and Brain Function: Neural Analysis of a Hormone-Controlled Mammalian Reproductive Behavior*, Fig. 13.1, p. 236. New York: Springer-Verlag.

expressed in the BNST, amygdala, POA, anteroventral periventricular nucleus (AVPV), arcuate nucleus, VMN, and PAG. Excluding the AVPV, most of these ERα regions have been directly implicated in regulating lordosis, as outlined in the previous section (see **Figure 4**), further supporting the contention that ERα is the critical estrogen receptor subtype that promotes lordosis. ERβ is abundant in many of the same regions as ERα, and is also present in additional regions such as the paraventricular and supraoptic nuclei, dentate gyrus, and cerebellum. Despite the high degree of regional overlap between the two receptor subtypes, a given region may participate in multiple behavioral and physiological processes, and thus, these two receptors may not be present in the same circuits (or cells) within each region. Whereas there are many cells in the amygdala, BNST, and AVPV that co-express both ER subtypes, there is much less coexpression in other nuclei and brain regions, suggesting the two ERs are likely mediating different processes in these sites. Within the VMN or PAG, for example, there may be subpopulations of neurons involved in the lordosis circuit which utilize ERα but not ERβ; this possibility remains to be directly tested.

An additional method of studying the site of hormone action in the brain has been to use site-specific infusions or implants or either E2 or E2 agonists/antagonists. Many studies in the 1970s and 1980s revealed that localized implants of E2 directly (and solely) into the VMN induce lordosis in OVX female rats, though not as frequent or robust as in intact estrous rats or rats treated systemically with E2 + P4 injections. However, there is an improved lordosis response with E2 implants given into VMN followed with a subcutaneous injection of P4, suggesting that much of the stimulatory actions of E2 on lordosis can be achieved simply via estrogenic action within the VMN. Conversely, anti-estrogens given directly into the VMN prevent lordosis in E2-primed rats, also indicating that the VMN represents a key, if not *the* key, site of E2 action for behavioral receptivity. It should be noted that compounds infused or implanted into a given brain region do not always stay localized, and may diffuse to other nearby areas. Thus, rather than concluding that the VMN by itself is the key site of E2 action, it may be more appropriate to conclude that the VMN and/or adjacent areas are involved in the estrogenic response. In addition, the VMN is not a uniform, homogenous nucleus, and contains several subregions; some subregions, such as the ventrolateral VMN, may play a bigger role in female receptivity than others, and implant or lesion studies need to consider this possibility when interpreting results.

Despite the important caveats mentioned above, when taken into consideration along with the numerous other techniques that have studied the VMN (lesion studies, tract tracing studies, electrical stimulation, ER expression, etc.), these localized implant studies further strengthen the argument that the VMN is indeed critical for mediating sexual receptivity in rodents. Other regions have also been implicated in mediating some of the effects of E2 on lordosis. In some studies, implants of E2 into the POA, where ERα is also expressed, also modulate lordosis and receptivity in female rodents. Intriguingly, E2 may have a dual inhibitory/stimulatory role in the POA. Electrical stimulation of the POA typically prevents lordosis from occurring, and POA lesions enhance lordosis, implying that the POA provides inhibitory input to lordosis circuits. Selective E2 treatment into the POA usually inhibits lordosis, especially during initial exposure, but may increase behavior after longer exposure. This may relate to the duration of E2 needed to achieve onset of lordosis behavior, which is typically 18–20 h. At present, the POA and other regions (such as the PAG) have not received the same level of attention as the VMN (in terms of the number of studies looking at hormonal action for lordosis).

Although it is now fairly clear where E2 acts in the rodent brain to elicit (or modify) lordosis behavior, it still remains to be determined what E2's specific actions are that result in the induction of lordosis. That is, what effect does E2 binding in the VMN, POA, or PAG elicit, and how does this relate to altered lordosis response? One possibility is that E2 directly stimulates the neuronal activity (i.e., action potentials and neuronal firing) of key neuronal populations, thereby increasing the signaling of these neurons to downstream parts of the circuit (culminating in the stimulation of motor neurons controlling the lordosis posture). In vitro electrophysiology studies found that E2 treatment onto VMN slice preparations did not by itself change neuronal resting potentials; however, E2 did increase the responsiveness of VMN neuronal firing to other neurotransmitters or to electrical stimulation. Thus, one role of E2 may be to bias neuronal responsiveness in the VMN (and perhaps elsewhere), making ER-sensitive neurons more likely to be excited and/or less prone to inhibition by other lordosis-modulating inputs (neurotransmitters, metabolic cues, pheromone/olfactory input, etc.). Supporting this possibility, E2 has been shown to be capable of changing the pattern and frequency of neuronal firing in vivo in the VMN region, though this could also be a direct effect of E2 on generating action potentials. Regardless, it appears that E2 can alter neuronal firing within key lordosis brain regions, though the mechanism by which E2 increases neuronal activity remains to be elucidated.

Another likely effect of E2, which is not mutually exclusive from its ability to promote neuronal firing, is to regulate DNA activity (mRNA and protein synthesis) in neurons in the lordosis circuit. Once E2 binds nuclear ER (in the case of lordosis, ERα), it can directly interact with DNA as a transcription factor. Thus, E2 can directly activate (or inhibit) gene transcription, thereby influencing protein synthesis. Indeed, investigators have observed increased mRNA and protein synthesis after E2 exposure in the VMN, and protein synthesis inhibitors given directly into the VMN reduce E2's stimulation of lordosis behavior. Thus, certain proteins need to be synthesized within the VMN (and perhaps other lordosis nuclei) following E2 exposure in order to achieve full lordosis. However, determining the identity of these critical proteins is complicated, as the transcription of many genes changes in response to E2. Many candidate genes (and hence, proteins) in the VMN, POA, ARC, and PAG have been shown to be modulated by E2, and over the years the list has grown. These proteins include various neurotransmitters and neuropeptides that may be involved in stimulating other downstream sites in the lordosis circuit (and whose synthesis is increased by E2), or neuropeptides which act to inhibit or diminish lordosis (and whose synthesis is correspondingly reduced by E2). Further complicating the issue, some neuropeptides are both downregulated and upregulated by E2, depending on the specific neuronal site or the temporal pattern of E2 signaling. Examples of E2-regulated neuropeptides which have been implicated in facilitating and/or diminishing lordosis include oxytocin, β-endorphin, enkephalin, neuropeptide Y, galanin, and cholecystokinin. Which of these neuropeptides is absolutely critical to female receptive behavior, and in what manner (i.e., what is their function for the lordosis process), is currently an active area of ongoing research.

In addition to neuropeptides, E2 also regulates the synthesis of several receptor proteins, which may be critical for receiving incoming neurotransmitter or hormone signals that regulate or time the display of lordosis. Perhaps the best characterized is the progesterone receptor (PR), which is present throughout the brain and robustly upregulated in many but not all brain regions, including key nuclei within the lordosis circuit. As discussed earlier, when PR signaling is reduced, either by interference with the receptor's binding (via an antagonist), genetic knockout of the PR gene, or infusions of PR mRNA antisense, females are hyposensitive or unresponsive to P4's ability to facilitate receptivity. In the absence of sufficient E2 (either a high enough dose or a long enough duration of exposure), PR availability is significantly diminished in several sites, including the POA and VMN, and P4 treatment is then far less effective in promoting lordosis. With elevated E2 (exogenously administered or during proestrus), PR is dramatically upregulated in the POA and VMN, correlating with elevated facilitation of lordosis by P4 (or ligand-independent dopamine signaling). It takes approximately 18–20 h of elevated E2 exposure to induce significant PR numbers, and this is often maximal after 24–26 h, corresponding with the temporal induction of female receptivity after 20 h of E2 treatment, and increased lordosis after even longer durations.

Supporting the role of PR in select lordosis nuclei, P4 implants exclusively into the VMN increase lordosis (and proceptivity). Conversely, PR antagonists (RU486) delivered into the VMN can diminish both lordosis and proceptivity, further supporting the contention that PR specifically within the VMN is critical for female sexual behaviors. Whether, in intact behaving females, this is due to P4 signaling, ligand-independent signaling, or a combination of both remains to be teased apart. Moreover, the exact role of P4 and PR is still being determined in terms of facilitating receptive behavior at the peak of behavioral estrus as well as terminating lordosis soon thereafter (the biphasic response); PR signaling thereby contributes to regulating both the onset and the offset of sexual behavior, though the mechanisms for this are unknown. The important refractory-inducing role of later P4 signaling may serve to 'reset' the lordosis circuit for the next period of receptivity. It has been proposed that one possible mechanism for this may be that extended P4 exposure eventually downregulates its own receptor, thereby preventing additional P4 from stimulating lordosis. Support for this possibility derives from pharmacological treatments that prevent degradation of neuronal PR and concurrently prevent P4 from inducing a refractory period in receptivity. Interestingly, several researchers have postulated that P4's both facilitory and refractory effects occur in the same brain regions (in and near the VMN), and perhaps even in the same neurons, though the latter has not yet been tested.

As alluded to earlier, PR signaling may play a role in not only receptivity, but also proceptivity. Whereas E2 implants into VMN stimulate some lordosis, often times little proceptive behaviors are observed with such treatment. However, P4 implants into the VMN (but not POA or PAG) of E2-primed OVX female rodents typically induce all proceptive behaviors (hopping, darting, ear wiggling, etc.), suggesting that PR signaling in the VMN is sufficient for rodent proceptivity. In support of this, VMN lesions decrease proceptive (and receptive) behaviors. Even so, one cannot rule out a role of the POA in proceptive behaviors, since lesions of the POA have been shown to reduce proceptive behavior (and increase receptivity). Thus, the POA may be a stimulatory and necessary component of proceptive behavior. Unlike the VMN and POA, the PAG has not been implicated in proceptivity, only receptivity. These findings indicate that there are both different and overlapping sites/circuits for receptive and proceptive behaviors, including the sites of hormone actions. However, compared to receptivity, the brain regions and hormonal actions underlying proceptivity have received far less attention and require more experimental investigation.

Generalizations to Other Species

It should be restressed that the vast majority of data for neuronal substrates and hormonal actions in the brain in regard to female sexual behavior have come from rodent models, and other mammalian species may not utilize the same brain pathways or hormonal mechanisms. This appears to be true for rabbits and some non-human primates, though there are currently far less data available for these nonrodent species to derive strong conclusions and detailed models. Whether or not humans exhibit similar or dissimilar hormonal mechanisms and sites of actions for sexual behavior compared to rodents is even more unclear. There are not many detailed studies addressing this issue, primarily given the ethical and logistical limitations in experimental design for studies with human subjects (e.g., lesions or neuronal implant studies are not simply possible). Moreover, there are often numerous confounding variables present in human studies of sexual behavior, thereby clouding the view of the underlying mechanisms and processes. For example, steroid hormones may have an impact on human sexual behavior, but confounding factors may also affect sexual actions in women, including stress or metabolic factors, social pressures, cultural influences, or religious practices. Moreover, most human studies do not (and often cannot) control for the man's role in sexual interactions; men may find women more or less attractive at certain times (or for various reasons) and therefore alter their involvement in sexual activity accordingly; this could have large unforeseen impacts on any experimental results concerning the female's sexual behavior patterns. Thus, it is difficult to empirically test female sexual behavior in humans, especially the hormonal and neuronal underpinnings.

Despite the limitations of human studies, some researchers have reported increased sexual activity and increased erotic thoughts in humans around the time of ovulation, corresponding to the time of peak estrogen levels, although this has not been consistently observed among all studies. ER and other sex steroid receptors have been identified in the human brain, including the hypothalamus and amygdala (similar to rodents), suggesting that sex steroids could act in a similar manner in rodents and humans. However, it is well recognized that humans, like some other non-human primates, do not limit sexual activity to a particular time of the menstrual cycle, and many females engage in sexual behavior at all stages of their cycle. Moreover, women can still engage in sex after ovariectomy (or menopause), further suggesting that sex hormones are not critical for human sexual behavior. Although sex often still occurs in these low sex steroid conditions, it may not occur as frequently, or with the same motivation, desire, or fulfillment, indicating the importance of looking at specific parameters (i.e., proceptivity, receptivity, etc.) of sexual behavior, rather than simply its presence or absence. Interestingly, while sex steroids may not strongly influence human female sexual receptivity, they may have a more noticeable role in inducing proceptivity (i.e., libido and sex drive) in women. In particular, androgens (such as testosterone and other metabolites) may be important for motivational aspects to engage in sexual activity in humans, in both men and women. Such androgens may come from ovaries and/or adrenals and that testosterone levels over a woman's cycle correlate with sexual desire much better than estrogen levels. Moreover, androgen treatments can significantly increase sexual behavior (particularly sexual desire and libido) in postmenopausal women or OVX women. Despite these findings, there is still not a clear consensus at present on the role of hormones, nor specific brain sites, in controlling sexual behavior in women. Hopefully, the next generation of research will enlighten us more on this intriguing and important issue.

See also: Female Sexual Behavior and Hormones in Non-Mammalian Vertebrates; Hormones and Behavior: Basic Concepts; Male Sexual Behavior and Hormones in Non-Mammalian Vertebrates; Mammalian Female Sexual Behavior and Hormones; Maternal Effects on Behavior; Neural Control of Sexual Behavior; Pair-Bonding, Mating Systems and Hormones; Parental Behavior and Hormones in Mammals; Parental Behavior and Hormones in Non-Mammalian Vertebrates; Reproductive Skew, Cooperative Breeding, and Eusociality in Vertebrates: Hormones; Seasonality: Hormones and Behavior; Sex Change in Reef Fishes: Behavior and Physiology; Sexual Behavior and Hormones in Male Mammals; Stress, Health and Social Behavior; Vertebrate Endocrine Disruption.

Further Reading

Beach FA (1976) Sexual attractivity, proceptivity, and receptivity in female mammals. *Hormones and Behavior* 7: 105–138.

Beyer C, Hoffman K, and Gonzalez-Flores O (2007) Neuroendocrine regulation of estrous behavior in the rabbit: Similarities and differences with the rat. *Hormones and Behavior* 52: 2–11.

Blaustein JD and Erskine MS (2002) Feminine sexual behavior: Cellular integration of hormonal and afferent information in the rodent forebrain. In: Pfaff DW, Arnold A, Etgen A, Fahrbach S, and Rubin RT (eds.) *Hormones, Brain, and Behavior*, pp. 139–214. New York: Academic Press.

Blaustein JD and Mani SK (2007) Feminine sexual behavior from neuroendocrine and molecular neurobiological perspectives. In: Blaustein JD (ed.) *Handbook of Neurochemistry and Molecular Neurobiology*, 3rd edn., pp. 95–150. New York: Springer.

Mani S (2001) Ligand independent activation of progestin receptors in sexual receptivity. *Hormones and Behavior* 40: 183–190.

Micevych P and Sinchak K (2007) The Neurochemistry of limbic-hypothalamic circuits regulating sexual receptivity. In: Blaustein JD (ed.) *Handbook of Neurochemistry and Molecular Neurobiology*, 3rd edn., pp 151–194. New York: Springer-Verlag.

Nelson RJ (2005) Female reproductive behavior. *An Introduction to Behavioral Endocrinology*, 3rd edn., pp. 319–386. Sunderland, MA: Sinauer Associates.

Pfaff DW (1980) *Estrogens and Brain Function: Neural Analysis of a Hormone-controlled Mammalian Reproductive Behavior.* New York: Springer-Verlag.

Pfaff DW, Sakuma Y, Kow LM, Lee AWL, and Easton A (2006) Hormonal, neural, and genomic mechanism for female reproductive behaviors: Motivational arousal. In: Knobil EK and Neill JD (eds.) *Physiology of Reproduction,* 3rd edn., pp. 1825–1920. New York: Academic Press.

Rissman EF, Wersinger SR, Fugger HN, and Foster TC (1999) Sex with knockout models: behavioral studies of estrogen receptor α. *Brain Research* 835: 80–90.

Wallen K and Zehr JL (2004) Hormones and history: The evolution and development of primate female sexuality. *Journal of Sex Research* 41: 101–112.

Mammalian Social Learning: Non-Primates

B. G. Galef, McMaster University, Hamilton, ON, Canada

Introduction

Preceding articles in this section demonstrate that social interactions of various kinds facilitate the acquisition of adaptive patterns of behavior by insects, fishes, and birds, and articles to follow will provide similar evidence of social learning in monkeys and apes. The nonprimate mammals (hereafter mammals), with which this article is concerned, are similar to animals with nervous systems both more and less complex in that interaction of the naïve with the knowledgeable often guides the behavior of the naive in adaptive directions.

Arbitrary Behaviors

Early work on social learning by mammals was concerned with the rapidity with which they learned arbitrary responses (such as pressing a lever to acquire food or stepping on a treadle to open a door) that were unrelated to their natural behavior. Results of such experiments were often discussed as demonstrating imitation, although today the same data would almost certainly be interpreted as demonstrating simpler types of social learning such as local or stimulus enhancement. For example, in a classic study conducted 40 years ago, kittens (*Felis catus*) were found to learn to press a lever to obtain food far more rapidly after watching their mother press the lever and get food than after watching a strange female cat do so. The more rapid learning by kittens that watched their mother was interpreted as showing that kittens imitated her behavior, although it can be explained more parsimoniously as showing only that kittens attend more closely to objects that their mother manipulates than to objects with which other adult female cats interact (i.e., as an instance of local enhancement).

More recently, Norway rats (*Rattus norvegicus*) that observed a rat pushing a joy stick either to left or right were reported to learn to press a joy stick in the same direction (left or right) as had their respective demonstrators, and it was suggested that the observer rats imitated their demonstrator's behavior. However, subsequent work showed that demonstrator rats left olfactory cues on the side of the joy stick that they had touched and that these cues influenced the behavior of other rats when they encountered it.

Perhaps the most striking instance of social learning of an arbitrary action by a mammal concerns golden hamsters (*Mesocricetus auratus*) that learned to use their teeth and forepaws to retrieve a piece of food dangling at the end of a short chain attached to a shelf. Three quarters of young hamsters whose mothers demonstrated food-retrieval behavior for them learned to pull the chain to obtain the food, while only a fifth of pups reared by a mother that did not exhibit retrieval learned the trick. Unfortunately, nothing is known of the behavioral mechanisms supporting this instance of social learning.

Natural Behaviors

Although, the early history of laboratory studies of social learning in mammals was largely concerned with the social learning of arbitrary responses, more recent work has focused almost entirely on the social influences on behaviors similar to those observed in free-living animals. In the following sections, representative experimental studies based on observations of the behavior of mammals living in natural circumstances in which subjects (1) chose appropriate substances to ingest, (2) overcame the defense of potential foods, (3) avoided predators, and (4) selected a mate are described.

Choosing Food

Much work on social learning in mammals has been concerned with learning how to forage successfully. Here, three examples of such social learning each of which depends upon quite different social learning processes are considered.

Rats avoiding poisons

In the 1950s, rodent-control operatives evaluated a method of rodent control that appeared to have considerable potential to reduce the cost of exterminating rodent ests. By placing permanent poison-bait stations in rat-infested areas, the rodent-control experts hoped to substantially reduce the expense of constantly replacing temporary baits.

The permanent bait stations had great initial success, with rats eating ample amounts of poison and dying in large numbers. However, later bait acceptance was very poor, and targeted rat colonies soon returned to their original sizes. The failure of permanent stations resulted from a few adult colony members surviving their first ingestion of bait and, as a result of suffering the ill effects associated with ingesting the poison, learning to avoid eating the bait. These knowledgeable survivors somehow dissuaded their young from even tasting the poisoned bait that the adults had learned not to eat.

Laboratory analyses of the transmission of food choices from adult rats to their young revealed that adults do not directly dissuade their young from eating poisonous substances. Rather, young rats are both strongly inclined to eat whatever foods adults of their colony are eating and extraordinarily reluctant to eat foods that they have not eaten before. Consequently, young rats eat foods that adults of their colony are eating, not foods that those adults are avoiding. Poison avoidance by the young is a byproduct of tendencies to avoid ingesting novel foods and to learn from others what foods to eat.

Starting before birth and extending throughout life, many different socially mediated experiences are involved in such social induction of young rats' food choices. For example, if a gestating female rat is fed garlic, garlic is subsequently detectable in her amniotic fluid, and following parturition, her young show an enhanced preference for the scent of garlic. When young rats begin to nurse, flavors incorporated into maternal milk reflect the flavors of foods that a lactating female is eating, and experience of these flavors in mother's milk causes weaning young to prefer foods their mother ate during the weeks that she was suckling them. Also, lactating rats are great hoarders of food, returning large quantities of food to the burrows where their young shelter. When an adult rat takes food from such a hoard, any young in its vicinity become intensely interested in the particular piece of food that the adult is holding. The young rats often try to steal the food that the adult is eating, and adults are surprisingly willing to give up food to juveniles. After a juvenile eats food taken from an adult, the juvenile shows an increased preference for that food that it does not show after taking the same food directly from the floor and eating it.

As young rats grow older and leave the nest site to feed in the larger world, they use visual cues to locate an adult rat at a distance from the nest entrance and approach and feed with that adult. Because approaching young tend to crawl up under an adult's belly and to begin to feed with their heads right under an adult's chin, adults can rather precisely direct young to foods that they are eating. And when an adult rat leaves a feeding site to return to its burrow, the adult deposits a scent trail that leads young rats seeking food to the same location where the adult has fed. Also, while feeding, adult rats deposit, both on and near foods, olfactory cues that are highly attractive to pups and cause them to prefer feeding sites and foods that adults have previously exploited.

In a number of mammalian species, in addition to Norway rats (mice, voles, European rabbits, Mongolian gerbils, golden and dwarf hamsters, bats, and dogs), a naïve animal (an observer) that interacts with another of its species that has recently eaten a food (a demonstrator) subsequently shows a substantial increase in its preference for whatever food its demonstrator ate. Exposure to a demonstrator rat can markedly increase the survival of rats in environments where ingesting the most palatable foods present does not lead to selection of a nutritionally adequate diet. For example, young rats placed in enclosures where they had continuous access to four different foods, three relatively palatable but low in protein and one relatively unpalatable but protein rich, lost weight, and would surely have died of protein deficiency. By contrast, pups that shared their enclosures with adult rats previously trained to eat the relatively unpalatable, protein-rich food grew at almost the same rate as pups offered just the protein-rich diet.

The relatively simple social learning mechanisms available to rats are also sufficient to support the sort of behavioral traditions that are common in our own species and present in other primates as well. All four members of each of several colonies of rats assigned to one condition were trained not to eat a pepper-flavored food and to eat a horseradish-flavored food, whereas all four members of each colony assigned to a second condition were taught the reverse. Following this training, each colony was offered a choice between pepper- and horseradish-flavored foods for $3\,h\,day^{-1}$, and each day immediately after a colony had been fed, one of its members was removed and replaced with a naïve rat. After 4 days, all members of original colonies had been replaced, and for 10 days thereafter, the individual in each colony that had been there longest was replaced with a naïve rat. Even after replacement of original colony members, large effects of the food preferences learned by original colony members were still evident (**Figure 1**). Similar transmission chains have also been found among colonies of rats trained to dig in sawdust for food.

Overcoming the Defenses of Prey

Pinecone stripping by roof rat

Roof rats (*Rattus rattus*) living in the pine forests of Israel and of Cyprus (places where no squirrels are present to compete for pine seeds), but not roof rats living elsewhere, subsist on a diet of pine seeds that they secure by stripping the scales from pinecones and eating the seeds that the scales protect. Laboratory studies of pinecone stripping by wild-caught rats revealed that to recover more energy from eating pine seeds than is expended in removing scales from pinecones, rats must take advantage of the architecture of pinecones, first stripping the scales from the base of a cone, and then removing the remaining scales in succession as they spiral around the cone to its apex (see **Figure 2**).

Less than 6% of rats captured outside pine forests and given pinecones to eat learn to open them efficiently.

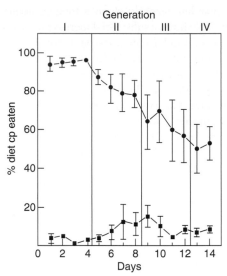

Figure 1 Amount of pepper-flavored diet (diet cp) eaten by members of colonies offered a choice between diet cp and horseradish-flavored diet (diet hr) and initially trained to eat either diet cp or diet hr. Galef BG, Jr. & Allen C (1995) A new model system for studying animal tradition. *Animal Behaviour* 50: 705–717. (Figure 5).

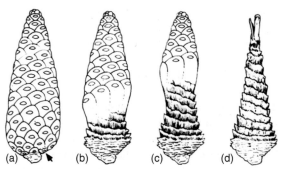

Figure 2 (a) Schematic diagram of pinecones being efficiently stripped of their scales in the efficient manner taking advantage of the architecture of the pinecone and (b) photographs of two efficiently pinecones that were attacked inefficiently (on left) and two that were attacked efficiently (on right). Terkel J (1996) Cultural transmission of feeding behavior in the black rat (Rattus rattus). In: CM Heyes & BG Galef, Jr. Eds. Social Learning in Animals the Roots of Culture. San Diego: Academic Press (Figure 5).

However, more than 90% of rats born to mothers that could not remove the scales from pine cones efficiently but reared by foster mothers that stripped pinecones in the presence of their foster young, learned the efficient method of removing scales from cones.

When a roof rat mother removes the scales from a cone, her young gather around her and attempt to snatch the pine seeds as she uncovers them. As the young mature, they snatch entire partially opened pinecones from their mother and then continue the stripping process that their mother started. Indeed, just providing young rats with

pinecones started properly by an adult rat or even by a human experimenter using a pair of pliers to remove scales from its base results in 70% of the young rats learning the efficient method of removing scales. Thus, a very simple sort of social learning enables young rats to learn a skill that enables them to survive in pine forests, a habitat that would otherwise be closed to them.

Similarly, juvenile red squirrels (*Tamiasciurus hudsonicus*) that have watched an experienced adult squirrel open hickory nuts open similar nuts at a substantially younger age and with greater efficiency than siblings lacking such experience.

Meerkats learning to eat scorpions

Meerkats (*Suricata suricatta*) are highly social animals that live in arid regions of southern Africa where they feed on a range of vertebrate and invertebrate prey, some of which, such as scorpions, are potentially dangerous. Young meerkat pups are initially incapable of foraging for themselves, and when from 30 to 90 days of age, are provisioned by adult group members that respond to begging calls pups emit when hungry. Adults typically kill or remove the sting of scorpions before they give them to very young meerkats. However, as the pups grow older and better able to handle intact, live scorpions, adults provide an increasing proportion of intact prey to pups. When human experimenters provisioned young meerkats in the field with live, scorpions with their stings removed, the pups' subsequent ability to handle such 'dangerous' prey without being either pseudostung by them or letting them escape increased markedly. Thus, adult meerkats' provisioning of their young facilitates their acquisition of an important skill.

Learning to Avoid Predators

Predator recognition and avoidance pose a challenge both to the young of many mammals and to scientists trying to understand how animals learn to avoid predators without any personal experience of the potentially disastrous consequences of direct contact with them. Although there have been far fewer studies of the role of social learning in the development of antipredator than of foraging behaviors, work on predator avoidance learning in birds, fish, and primates, together with that conducted in nonprimate mammals, suggests a potential solution to the problem. Such work is of some practical importance in that attempts to reintroduce captive-reared endangered species into natural habitat often fail because captive-reared animals released into the wild often respond inadequately to the approach of a predator.

Predator recognition in wallabies

Captive-reared Tamar wallabies (*Macropus eugenii*) were given the opportunity to observe either a demonstrator wallaby that had been previously trained to avoid a stuffed fox or a naïve demonstrator wallaby that was indifferent to foxes. Observer wallabies that had watched a fearful demonstrator interact with the stuffed fox showed significantly longer periods of vigilance in response to presentation of the fox than observer wallabies that had seen an indifferent demonstrator interact with the fox, and the response was specific to foxes and not shown to other stuffed animals.

In a conceptually similar study, juvenile captive-reared black-tailed prairie dogs (*Cynomys ludovicianus*) were exposed to various animals restrained behind a screen barrier: a ferret, a rattlesnake, a hawk, and a harmless rabbit. The prairie dogs were then given additional exposure to each stimulus animal either with or without an experienced adult demonstrator present. During this training, the alarm vocalizations and vigilance behavior of the juveniles closely matched that of their demonstrators, and following training, juveniles trained with an experienced adult were more wary of the three predatory animals than were juveniles that had experienced the predators without a demonstrator. Perhaps most interesting, when the prairie dogs were released back into the wild, those that had been exposed to predators in the presence of an experienced demonstrator had a significantly greater probability of surviving for 1 year than those prairie dogs lacking such training.

Learning to avoid biting flies

Blood-feeding biting flies are among the most common of mammalian predators, and their attacks elicit avoidance responses ranging from elephants manufacturing tools from branches for fly switching to self burying in mice. Deer mice (*Peromyscus maniculatus*) experiencing a single 30-min session of attack by biting flies and then exposed to flies that had been surgically deprived of the ability to bite buried themselves in the substrate, whereas mice without prior experience of biting flies did not. Most interesting mice that had no personal experience of biting flies but had witnessed another mouse under attack by biting flies, engaged in self burying when subsequently exposed to flies that were unable to bite.

Development of response to alarm calls

Adult Belding's ground squirrels (*Spermophilus beldingi*) that detect an avian predator such as a hawk or eagle whistle, and other adults respond to their whistles by running to the nearest burrow entrance. When adults detect a relatively slow-moving ground predator, they emit a trill to which other adults respond by standing on their hind legs and looking about.

Newly emerged young ground squirrels do not behave differently either to the two alarm calls of adults or to alarm calls and other sounds. Development of appropriate

responses to alarm calls of juvenile squirrels maintained in captivity without their dams was slower than that of captive young squirrels maintained with their dams, suggesting that interaction with dams exhibiting appropriate responses to alarm calls sped juvenile's learning of the appropriate responses.

Choosing a Mate

Rats and mice

Although most experiments on social influences on sexual behavior have been carried out in birds and fishes, a few studies suggest that in mammals as well, social interactions of various kinds can influence both the choice of a mate and sexual performance. Female Norway rats prefer as sex partners males that have recently copulated with other females, and female mice spend more time investigating urine collected from males exposed overnight to an estrous female than to urine from males exposed to a female not in estrous, although as yet, there is no evidence that this change in the attraction of female mice to male urine causes females to change their preferences for a partner.

Farm animals

Although strictly speaking a case of social influence that only suggests possible social learning, many species of farm animal (e.g., goats, cattle, horses, and pigs) exhibit enhanced sexual performance after viewing conspecifics copulating. For example, sexual performance of male sheep (*Ovis aries*) is enhanced following interaction with another male that has recently interacted with a ewe. It has been hypothesized that olfactory cues transferred from females to males during their period of interaction have a stimulating effect on other males.

Animals Inconvenient for Controlled Studies

There is an expectation that animals with a large brain are more likely to engage in complex sorts of learning, including social learning, than animals with a small brain. However, many large-brained mammals from elephants to whales have large bodies that make them inconvenient subjects for controlled, experimental studies. Despite the difficulty of providing conclusive evidence of social learning in such creatures, there is a growing body of evidence suggesting that many such animals may be sophisticated social learners.

Bottlenose Dolphins

In the wild, young dolphins (*Tursiops* sp.) and their mothers forage together for several years giving the young ample opportunity to learn complex foraging behaviors from their mothers. For example, while foraging in deep-water channels, some adult female dolphins carry marine sponges that are believed to be used to protect their noses while probing the sea floor to locate small, bottom-dwelling fish. At Shark Bay in Western Australia, the only study site where sponge carrying has been observed, the behavior occurs almost exclusively within a single maternal line, with most daughters (and a few sons) of sponge-carrying females adopting the habit. Although a genetic explanation of the pattern of sponge use at Shark Bay seems plausible, examination of several possible modes of genetic inheritance make it unlikely that a genetic propensity is responsible for the observed distribution of the behavior. Further, because only some of the many female dolphins that forage in deep-water channels use sponges while foraging there, it is unlikely that exposure to deep channels in itself results in sponge use.

Whales

There are numerous reports of behavior consistent with the view that many cetaceans (i.e., whales and dolphins) engage in social learning. For example, the rate of spread among humpback whales (*Megaptera novaeangliae*) in the Gulf of Maine of a novel foraging behavior, 'lobtail feeding' (in which the whales slam their tail flukes in the water before diving for prey), is consistent with social transmission of the behavior, although explanation in terms of individual learning in response to a change in prey availability is also possible. Similarly, although scattered reports of mother killer whales (*Orcinus orca*) 'teaching' their young to beach themselves to capture seals are consistent with the view that such behavior is socially learned, the reports do not offer strong support for that interpretation.

Elephants

The social knowledge possessed by the matriarch in a family of elephants (*Loxodonta africana*) influences the social behavior of other family members, reducing the probability that they will engage in unnecessary defensive behaviors when encountering familiar families that pose no threat. The older the family matriarch is, the better the family members are at discriminating vocalizations of familiar and unfamiliar individuals and responding appropriately to them. The age of a family matriarch predicts more than 30% of the variation among families in the number of young that they produce, suggesting that the social knowledge of older females has adaptive consequences for her kin. Although it has not been shown that

other family members learn from the matriarch which female's vocalizations to respond to and which to ignore and continue to respond appropriately in her absence, it seems probable that such a social transmission of social knowledge occurs.

Conclusion

Although the study of social learning in mammals is still in its infancy, many of the biologically important activities in which mammals engage have already been found to be modifiable by socially acquired information. In future, we can expect to see both more examples of behavior in free-living mammals that are likely to be a product of social learning and ever more convincing experiments leading to a deeper understanding of the ways in which social interactions improve acquisition of adaptive patterns of behavior.

See also: Apes: Social Learning; Imitation: Cognitive Implications.

Further Reading

Bilko A, Altbacker V, and Hudson R (1994) Transmission of food preference in the rabbit: The means of information transfer. *Physiology and Behavior* 56: 907–912.

Box HO and Gibson KR (eds.) (1999) *Symposia of the Zoological Society of London, Vol. 72: Mammalian Social Learning: Comparative and Ecological Perspectives*. Cambridge: Cambridge University Press.

Galef BG (2003) Traditional behaviors of brown and black rats (*R. norvegicus* and *R. rattus*). In: Perry S and Fragaszy D (eds.) *The Biology of Traditions: Models and Evidence*, pp. 159–186. Chicago: University of Chicago Press.

Galef BG (2007) Social learning by rodents. In: Wolff JO and Sherman PW (eds.) *Rodent Societies: An Ecological and Evolutionary Perspective*, pp. 207–215. Chicago: University of Chicago Press.

Griffin AS (2004) Social learning about predators: A review and prospectus. *Learning and Behavior* 32: 131–140.

Heyes CM and Galef BG (2004) Special issue: Social learning and imitation. *Learning and Behavior* 32(1): 1–140.

Janik VM and Slater PJB (2003) Traditions in mammalian and avian vocal communication. In: Fragaszy DM and Perry S (eds.) *The Biology of Traditions: Models and Evidence*, pp. 213–236. Cambridge: Cambridge University Press.

Mann J and Sargeant B (2003) Like mother like calf: The ontogeny of foraging traditions in the wild Indian Ocean bottlenose dolphin. In: Fragaszy DM and Perry S (eds.) *The Biology of Traditions: Models and Evidence*, pp. 236–266. Cambridge: Cambridge University Press.

Rendell L and Whitehead H (2001) Culture in whales and dolphins. *Behavioral and Brain Sciences* 24: 309–382.

Zentall TR (1988) Experimentally manipulated behavior in rats and pigeons. In: Zentall TR and Galef BG, Jr (eds.) *Social Learning: Psychological and Biological Perspectives*, pp. 191–206. Lawrence Erlbaum Associates: Hillsdale.

Maps and Compasses

P. J. Fraser, University of Aberdeen, Aberdeen, Scotland, UK

Introduction and Definitions

Much of our understanding of animal navigation has centered around Kramer's map and compass hypothesis, which proposes that an animal has to determine its position relative to a goal (map step) and then set out in a particular direction to reach its goal (compass step). A human using a map and compass is able to align the map with the true north, having allowed for the difference between this and the magnetic north read by the compass. He can then work out his current position from the features on the map and the visible landmarks and determine the angle and distance home. Many orientation behaviors and homing abilities in animals suggest that they can use a set of equivalent processes, using only their brains and sensory systems.

Magnetic Compass and Other Compasses

The common meaning of compass is an instrument with a magnetized needle used to steer ships. This is derived from an older meaning of a space limiting circle, which clearly relates to the compass card, the circular card with degree markings used to read the bearing. The compass has now come to mean an instrument that indicates an absolute direction relative to the magnetic field of the earth. Those involved in animal navigation have also used the term for nonmagnetic field-based direction indicators, so sun compasses, star compasses, wind compasses, moon compasses, and polarized light compasses have all been proposed. Additionally, because of the daily rotation of the earth, the use of celestial cues as a reference demands a time sense. Animals have to vary angles with respect to the sun's position in the sky, which changes at $15° h^{-1}$.

A compass points to the magnetic north, a point around 2000 km south of the axis of rotation of the earth; true north is where all longitude lines meet. The alignment between the magnetic north and the true north varies over the surface of the earth; the difference is called the declination and is usually less than 20° except at very high latitudes.

There is of course a line, the agonic line, where the magnetic north and the true north are in alignment. Declination is measured East or West of this line. Any small map of an area of the world will have its particular declination marked on it. Because declination can vary considerably, accurate general strategies for locomotion allow for declination and reference everything to the true north.

There is evidence that animals can also compensate for declination drift. Of course, in limited areas where there is little change in declination angle, it is less necessary to take this into account.

The earth acts like a large weak dipole magnet and at the magnetic poles, the direction of the lines of force are vertical, emerging from the antarctic magnetic pole and returning downwards at the arctic magnetic pole. At the magnetic equator, the lines of force are parallel to the surface of the earth. The lines of force thus have a dip angle or inclination at any point that varies between 0° and 90°. Lines of equal inclination, or isoclinics, are roughly parallel to lines of latitude. It is to be noted that inclination should not be confused with declination, which is the difference in alignment of pointers to magnetic North and true North.

The earth's magnetic field is changing by around 7 min per annum in a westward direction, although for most purposes it is regarded as a static field. Nevertheless, declination angles on maps need to be updated every 5 years. Over timescales of several hundred thousand years, polarity reversals of the dipole field may occur with the poles changing hemispheres.

The magnetic field originates in convection currents in the molten core of the earth and significant local variations can be produced by magnetized rocks in the earth's crust. These anomalies are known to affect animals. Yet another source of variation comes from electric currents in the ionosphere, leading to diurnal variation as the earth rotates around its axis. Magnetic storms associated with solar wind and solar flares and local disturbances due to lightening add to the variation. Intensity varies between 60 000 nT at the poles to around 30 000 nT near the magnetic equator. The regular spatial variations in inclination (which varies with latitude) and intensity provide important information to animals and have been proposed as a map component for some animals.

Map

A map is a representation of the surface of the earth or a part of it on a plane surface. Topographical maps show changes in level of the ground, landmarks, and features such as rivers in the valleys between hills. A map allows us to orient in an unfamiliar locality because of its representation of landmark features. Animals have the ability to use landmarks and have memory capabilities that allow them to recognize their home locality and familiar neighboring localities. Special cases where a single landmark close to a goal acts as a beacon are known. More controversial are examples where a rule-based grid rather than a particular learned topography is used, allowing extrapolation and hence orientation outside familiar territory and providing a basis for inheritance of the map information. In general, the evidence for compass senses is much stronger than the evidence for maps. Magnetic field maps, olfactory-based maps, and infrasound-based maps have been proposed and are supported with evidence.

Odometer

Odometers measure distance. Evidence is emerging from fitting ants with stilts or shortening their legs, to the effect that sometimes this information comes from counting steps or by utilizing the visual flow field in the case of flying bees. Landmarks are also needed to identify home.

Navigation in Animals

Over a fruitful half century of research into mechanisms of animal navigation used by terrestrial and aquatic animals in air, in water, and on land, our knowledge of sensory mechanisms involved in navigation and the ways in which they are used has increased and is still increasing. Agreement has not been reached regarding the detailed mechanisms involved in navigation. At present, there are split opinions regarding the role of olfactory-based maps and about the importance of radical pair-based magnetic field sensing mechanisms compared to magnetite-based mechanisms in birds and other animals.

In 1952, Griffin classified orientation related to homing into three categories. In the first category, independent of an external reference, animals search until they encounter landmarks by chance. His second category involves unidirectional orientation, following a single compass direction (e.g., sun compass) back home where landmarks could be recognized. This uses a single external reference. More complicated is his third category, which uses at least two independent external factors. This equated to human seafaring and was called 'true navigation.' Forty years later in 1992, Papi used 11 different categories, although arguably there could be overlap in some of his divisions. The second category together with an odometer can lead to accurate homing.

Animals make use of a remarkable number of passive and active transport mechanisms, and often optimize routes to save energy or reduce distance between points over the globe. For example, arctic waders which migrate between their breeding area in the high arctic tundra and their wintering areas in the Southern hemisphere are thought to approximate to a great circle route, which is the shortest possible. They could do this by using a sun compass without compensating for the time shift.

With migration speeds around one eighth of the peak flight speed and frequent fuelling stops, wild birds undergo an exercise different from that of homing pigeons returning quickly from relatively short distances. Insects and birds make use of prevailing winds and migrating plaice rise from the bottom when tidal currents move in an appropriate direction and hence utilize tidal transport. An interesting 'static migration' is where zooplankton in the ocean use depth sensors to maintain depth by swimming at up to 10 body lengths per second against upwelling or downwelling vertical currents, but drift passively with horizontal components of the current. In a featureless ocean, this allows them to optimize their filter feeding and accounts for their occurrence in dense patches. With our knowledge of the horizontal as well as the vertical currents inshore or around submerged features, the aggregation of plankton, which is important for predators further up the food chain, can be predicted with simple models.

Successful navigation need not require learning or experience of routes or parts of routes, but experience can count, greatly enhancing navigation ability in several ways. Often, inexperienced birds use different mechanisms when they gain experience. Clearly, some redundancy is present regarding the sensory modalities used by animals and anomalies and peculiarities in gradients of sensory cues used may lead to conflicts that can give disorientation or can be compensated using elegant mechanisms. Passerine birds migrating from areas near the magnetic north pole reset their reading of their magnetic compass to the North South direction of dawn or dusk polarized light.

The redundancy has led to the discovery of multiple mechanisms and, not surprisingly, to arguments over the relative importance of particular cues. There are still significant gaps in our knowledge of how magnetite and radical pair-based mechanisms transform their interaction with the magnetic field into sensory signals in the nervous system.

The pinnacle of navigation ability is so-called true navigation involving a map and compass mechanism, which may either involve a mosaic of local cues or for grid-based navigation or bicoordinate navigation, it involves two or more physical or chemical gradients.

Why Maps and Compasses for Animals?

Early attempts to explain the remarkable migration and homing ability of animals considered a variety of mechanisms including the use of celestial cues and path integration by which an animal senses where it is going during locomotion and then can compute where it is by integrating all the turns and knowing the velocities of straight line locomotion. Studies in the 1920s and 1930s showed that several species of wild birds such as swallows, starlings, and seabirds returned over the sea from unfamiliar release sites over 1000 km from their island home. Further work on petrels and shearwaters across the open ocean showed that landmarks were not necessary. While the returns were impressive, they did little to tell us about the mechanisms involved.

Most research work has concentrated on a small number of model animals with birds and homing pigeons in particular as especially well-worked examples. Different animals of course have access to different set of sensory cues so it is difficult to generalize about the importance of mechanisms. Some such as a littoral animal, the amphipod, *Talitrus*, use the sun, polarized light, the moon, spectral differences in sky radiance over sea and land, magnetic fields, substrate slope, and landscape features as cues to orient on their beach. The foraging desert ant, *Cataglyphis*, counts its steps to measure distance during path integration and uses polarized ultraviolet light sensed with the oriented microvilli of retinula cells in a small dorsal part of their compound eye as a compass cue, to return in a straight line to a small nest hole in a featureless flat desert. It can use wind if deprived of all other cues. The mole rat, which uses path integration and a polarity compass to sense magnetic fields underground, clearly does not have access to visual-based mechanisms. Many of these animals make frequent use of a compass such as the axis of polarized light or the magnetic field as an external reference to avoid the continual accumulation of errors, which is an inherent problem in path integration. There is evidence that harbor seals in an aquarium are able to identify single lodestars and indicate their azimuth. Hence they have the capability of steering by the stars when traveling at sea.

Sun Compass

For many animals and birds in particular, early ideas were based around the sun compass, and this learned mechanism, although more complicated because of its dependence on time cues, is preferred as long as the sun can be seen. We now know that it can be replaced by a magnetic compass without loss of navigation ability under overcast conditions. Due to the 24 h revolution of the earth, the position of the sun varies by 360/24, that is, $15° h^{-1}$. Therefore, using a sun compass requires compensation for this time shift, and this involves the internal clock of the animal. By resetting the internal clock using an artificial light regime for a few days, orientation can be predictably altered. For example, setting the clock back by 6 h causes a 90° shift in orientation of pigeons. Normal birds released north of their loft head south, whereas the 6 h slow clock shifted birds head west. This has been tested over a full range of directions and distances from less than 1.5 to 167 km with the same magnitude and direction of deviation. Starlings in a circular cage before migration show migratory restlessness and tend to orient toward their migratory goal when they can see the sun. If their view of the sun is blocked and replaced with a stationary light source, the birds shift their preferred orientation by $15° h^{-1}$ consistent with their use of a time compensated sun compass.

The idea that in addition to its daily east–west progression, the sun could provide north–south information in terms of its noon altitude in the sky, which is of course greater at the equator than near the poles, was formulated by Matthews and elegantly disproved using clock shifted birds.

Clearly, the direction home is established with reference to an external reference, the sun in this case, and this excludes the necessity of the use of navigational strategies not based on an external reference. Hence, pure inertial navigation and piloting including beaconing can be ruled out as strictly necessary for navigation in these birds. Note that navigation to food sources by ocean-going seabirds such as petrels is known to involve olfactory beaconing to find the chemical dimethyl sulphide released by phytoplankton in particular abundance in productive foraging areas.

Thus, birds either gain route-specific information by somehow recording the overall direction of an outward journey, including accounting for detours and reversing this to return, or use local site-specific information to work out the directional relationship to the goal. In fact, both types of mechanism are used.

Young pigeons displaced in a distorted magnetic field lose their ability to orient back to the loft, but are able to orient when displaced without interference with the magnetic field. Distorting the magnetic field only at the release site still did not disturb their ability to orient showing that the direction had been learned in the outward journey. When 3-months old or if given experience before this time, pigeons learn to use site-specific information and are then able to work out their directional orientation from a site-specific map, and this must be considered the dominant mechanism involved.

Early experimenters also considered that by using vision and memory, the route back, for example, for a homing pigeon, could be worked out from the landmarks passed on the way. Remarkably, pigeons wearing frosted contact lenses, which allowed some light through but prevented detailed object recognition, not only flew back in the right direction but also found their way to the vicinity of the loft where they then had difficulty locating the loft. Furthermore, pigeons anesthetized and taken to their release site still performed their homing perfectly well.

These experiments rule out a dependence on path integration mechanisms and also rule out the necessary use of detailed visual cues for the main return part of the homing journey. Note the apparent contradiction that the sun compass, which we have just stated is the preferred mechanism for navigation and which allowed development of map and compass ideas before the existence of magnetic field sensitivity was established for animals can also be regarded as unnecessary. Since many migrations take place at night, this is in retrospect less surprising and has led to our appreciation of the role of other senses such as magnetic field reception in true (map and compass) navigation.

Early experiments used vanishing bearings to determine the direction of return, where the departing bird is followed with binoculars until it becomes a vanishing tiny speck when the angle is taken, hence minimizing the effect of small-scale fluctuations in direction. Recent studies of birds that had been confined to a $100 \, m^3$ wire aviary with only a view of landmarks a few hundred meters from the aviary and then released for the first time with a miniature GPS data logger from a point 20 km away over a lake to make sure that they did not land immediately after release, showed that their homing ability was not different from pigeons allowed free flights and hence free access to landmarks. Once they got close, they had more difficulty than the control birds actually getting back to the loft, consistent with the frosted contact lens experiments.

These sorts of experiments demonstrate that the pigeons have a map mechanism that gives them information about the direction of displacement. Although the birds ended up near the loft, the detailed analysis of the tracks possible with the GPS technology showed that some birds in fact overshot the loft location and then turned to come back as if their direction was calculated from their map sense rather than using an odometer. Although there is evidence that flying bees can use the visual flow field as an odometer to measure distance, there does not seem to be evidence that birds similarly use this.

Animal Maps

A variety of factors could contribute to a map, including magnetic fields, gravitational cues, infrasound, odors, views of distant landmarks, and hydrostatic pressure. Gradients aligned approximately orthogonally allow bicoordinate position fixing necessary for true navigation. While gravitational maps based on variations in gravity over the surface of the earth are possible, there are difficulties regarding the sensitivity of known gravity receptors in animals.

Magnetic Maps

Magnetic maps using inclination and intensity were taken seriously when it was realized that there was a relationship between natural temporal variation in the magnetic field and homeward orientation in pigeons and that orientation was also disrupted in the vicinity of magnetic field anomalies.

More recently, Lohmann has shown that young loggerhead turtles, which swim out from their natal beach and remain for several years in the North Atlantic gyre, can distinguish different inclination angles. They also swim

appropriately with regard to the inclination angles to keep themselves in the gyre. Thus, exposure to an inclination angle found on the northern boundary of the gyre caused the turtles to swim south-southwest. The turtles could also distinguish different intensities. By replicating both the intensity and the inclination found at three separate points on the gyre and testing the orientation of hatchling turtles to these values, it was shown that the turtles behave as if they are using a bicoordinate map made up of isoclinics (lines of equal inclination) and isodynamics (lines of equal intensity). Similar magnetic maps are known in a crustacean, the spiny lobster. This sort of map based on inclination and intensity has been criticized because it is only near certain magnetic anomalies that the angles of the field parameters intersect at a sufficiently high angle to allow accurate positional information, and these anomalies are mobile over evolutionary time. An alternative map based on detectors allowing separate extraction of the direction and intensity signals would allow determination of the external field vector (magnitude and direction), and an array of 1000–1000 000 cells is estimated as giving sufficient signal-to-noise ratio. Another model involves the gradient in the intensity slope and variations in intensity of the main field of the earth. Orientation errors are symmetrical about the line of intensity slope through the loft. Note total field intensity is a scalar, not a vector, and can be measured at any point but measuring the direction of intensity slope requires movement by the animal over known spatial coordinates. This fits with the disturbances in orientation correlated with normal variance in the earth's magnetic field and erroneous orientation around anomalies.

Attached magnets will impair magnetite-based detection mechanisms and there is good laboratory evidence in sea turtles and honeybees that this can happen but less clear evidence from field experiments with homing pigeons.

Although bicoordinate maps offer a complete mechanism for homing, for animals returning along a coastline for example, a single parameter may be enough to locate a target. While some potential position indicators forming navigational maps such as the solar arc have been discounted, others such as the geomagnetic field, infrasound, or natural odor sources have been supported by reasonably convincing evidence.

The magnetic field can be described as a vector in 3-dimensional space at any point on the earth's surface. We know some of the detail of receptors required to interact with this magnetic field. It is also known that because the optimum magnetic to thermal energy ratios for determining direction and intensity are 2 and 6, it is likely that different structures will be involved in detecting these. For direction, only a small number of cells, perhaps minimally 6 for each direction in 3 planes, would be needed for a magnetic compass sense. This small group would be difficult to locate and although electrophysiological responses to intensity of magnetic fields are known in primary afferent cells in the trigeminal nerve of rainbow trout and a migrating bird, the bobolink, no directional responses have been found in this way. In honeybees from behavior, similar thresholds from 25 to 200 nT are also known.

Radical Pair Mechanisms

Early observations by Wiltschko that caged European robins changed their orientation when magnetic North was experimentally altered showed that orientation to a magnetic field was possible and hence demonstrated the use of a magnetic compass. Two types of compass are now known, a polarity compass which works like those built by humans with a magnetized needle using the polarity of the magnetic field, and an inclination compass. This inclination compass relies on whether the magnetic field lines run up or down. By testing animals in an altered magnetic field in which the vertical component is inverted, it is easy to distinguish between polarity compasses (found in spiny lobsters and rodents) and inclination compasses (found in birds, salamanders, and turtles). Animals with an inclination compass also show a dependence on the wavelength of light. Birds show magnetic orientation when illuminated with the blue-green end of the spectrum and are disoriented under yellow and red light. It is thought that a radical pair mechanism involving cryptochromes is responsible, and in an insect, *Drosophila*, the involvement of cryptochromes in magnetic field orientation has been convincingly demonstrated. In the robin, a single eye, the right eye is now considered the site of magnetic compass information. At present, although doubts have been raised about the necessary sensitivity of this sort of system, it is now well accepted that cryptochromes in the eyes of night flying migrants are an important component of their magnetic field orientation. Low levels of light are involved and this may explain the special role assumed by cryptochromes in the eye rather than those sited elsewhere in the body. The eye with transparent cornea and lens allows photoreceptive components unhindered access to low intensity levels of light. Furthermore, the eye and head of birds are stabilized by the vestibular and other balancing senses during flight and this may help in the long-term integration of magnetic field information necessary for this system to work. In addition, birds have superparamagnetic magnetite-based receptors in the upper beak, which are used for recording magnetic intensity. Despite the presence of single domain magnetite particles in the nasal area, when migratory silvereyes, *Zosterops*, were given a strong, brief magnetic pulse designed to reverse the polarity, there was a marked effect on their orientation behavior, showing that magnetite was involved. By applying local anesthetic to temporarily deactivate the upper beak receptors, this effect was abolished. It was thus shown that these are the crucial receptors involved in orientation.

Infrasound Maps

Pigeons have long been known to be extremely sensitive to infrasound, which is defined as sound frequencies below 20 Hz. They respond down to 0.05 Hz. Such sound waves are much less attenuated by the atmosphere than higher wavelength sounds. Variation in atmospheric temperature has a refractive effect on infrasound, bending sound waves upwards toward the upper atmosphere where a temperature inversion reflects them back toward the ground. Although this will also be affected by wind, the net effect for a bird flying close to the earth's surface is a varying pattern of encountering sound from a source such as wave interactions in the ocean or from the surface of large lakes. Magnetic storms also produce infrasound as do supersonic airplanes such as Concorde.

Hagstrum has analyzed disruptions to pigeon releases at sites of anomalies where the birds normally show random orientation. Interestingly on odd occasions, birds released at these sites do orientate well and this correlates with a change in the speed and direction of the winds in the upper troposphere nearby. This would be explained if pigeons use infrasound cues and it is this, rather than a magnetic anomaly, that is involved in disrupted orientation. Annual variation in homing ability could correlate with annual variation in the intensity of the atmospheric background or microbaroms, due to winter storms. Releases of pigeons at lakes and temperature inversions lead to poor initial orientation and again this is best explained by the birds using infrasound. A small number of pigeon races which incurred large losses of birds could have encountered infrasonic shock waves from Concorde supersonic aircraft. Results from removal of cochlea and lagena in the ear which abolish reception of infrasound on releases are not conclusive although perfectly good homing can clearly occur without infrasound detection.

Olfactory Maps

The role of olfaction in pigeon homing has proved controversial over the years. Nevertheless, there is good evidence that pigeons can derive information on their position relative to home from trace substances in the atmosphere. Section of the olfactory nerve gave decreased orientation. Sealing one nostril together with section of the contralateral olfactory nerve had the same effect, whereas the same treatment as a control, but with ipsilateral section of the olfactory nerve, did not affect orientation. This mechanism worked with inexperienced and experienced birds over a range of several hundred km. Similar results were obtained with temporary interference with olfaction, using zinc sulphate irrigation of the olfactory epithelia.

Filtering airborne substance from the air given to pigeons during transport to release sites together with local anesthetic sprayed into the nostrils of the birds before release interfered with orientation. Experiments changing the wind direction before release had an effect on the vanishing bearings of pigeons, suggesting that pigeons develop an olfactory map by associating olfactory activity with current wind directions. Spatial chemical gradients are known to exist in the atmosphere. Procellariform seabirds have particularly well developed olfactory systems and are known to find food sources by smell.

Objectors to the olfactory nerve interference experiments suggested that the operation could have interfered with magnetic field receptors, but the recent finding that the upper beak receptors are the main magnetite-based receptors in birds has negated that objection. Since adding or withholding odors during transport does alter the strength of orientation, it is difficult to argue that odor is irrelevant to the navigational map used. Recently, birds were provided with odorless air, ambient air, or artificially scented air during transport to a release point only 8 km from home. Surprisingly, the scented air birds oriented as well as the ambient air birds, with the odorless air birds largely disoriented. An idea now proposed is that the odors function as a primer – an 'olfactory wake-up call.' At longer distances, all the birds adopted accurate homeward bearings. The idea is that with the olfactory deprivation, the birds were not paying attention. The role of olfaction may hence be more complicated than a component of an independent map.

Other Maps

Perhaps also not all possible maps have been considered. In a recent study of migrating plaice fitted with tags logging temperature and pressure, it proved possible for the plaice on the bottom of the sea, to locate the position of each data fix from the unique pairings of temperature and depth, and hence plot the migration patterns of the plaice over time. Clearly, if we can do this, then the animal could use temperature and depth as a map component. Fish are well known to be able to sense depth using hydrostatic pressure as a proxy and fish with swim bladders with thresholds around 0.5 cm of water pressure are about ten times more sensitive than fish such as sharks or crustacea such as crabs in which the sensors are known to be the angular acceleration receptors in their vestibular systems.

See also: Amphibia: Orientation and Migration; Bat Migration; Bats: Orientation, Navigation and Homing; Behavioral Endocrinology of Migration; Bird Migration; Circadian and Circannual Rhythms and Hormones; Fish Migration; Insect Migration; Insect Navigation; Irruptive Migration; Magnetic Compasses in Insects; Magnetic Orientation in Migratory Songbirds; Migratory Connectivity; Pigeon Homing as a Model Case of Goal-Oriented Navigation; Pigeons; Sea Turtles: Navigation and Orientation; Spatial Memory; Spatial Orientation and Time: Methods; Vertical Migration of Aquatic Animals.

Further Reading

Able KP (1994) Magnetic orientation and magnetoreception in birds. *Progress in Neurobiology* 42: 449–473.

Fraser PJ, Cruickshank SF, Shelmerdine RL, and Smith LE (2008) Hydrostatic pressure receptors and depth usage in crustacea and fish. *Navigation: Journal of the Institute of Navigation* 55: 159–165.

Gagliardo A, Ioale P, Savini M, Lipp H, and Dell'Omo G (2007) Finding home: The final step of the pigeons' homing process studied with a GPS data logger. *The Journal of Experimental Biology* 210: 1132–1138.

Hagstrum JT (2000) Infrasound and the avian navigational map. *The Journal of Experimental Biology* 203: 1103–1111.

Jorge PE, Marques AE, and Philips JB (2009) Activational rather than navigational effects of odors on pigeon homing. *Current Biology* 19: 650–654.

Lohmann KJ, Lohmann CMF, and Putman NF (2007) Magnetic maps in animals: Nature's GPS. *The Journal of Experimental Biology* 210: 3697–3705.

Metcalfe JD, Hunter E, and Buckley AA (2006) The migratory behaviour of North Sea plaice: Currents, clocks and clues. *Marine and Freshwater Behaviour and Physiology* 39: 25–36.

Papi F (2006) Navigation of marine, freshwater and coastal animals: Concepts and current problems. *Marine and Freshwater Behaviour and Physiology* 39: 3–12.

Phillips JB (1996) Magnetic navigation. *Journal of Theoretical Biology* 180: 309–319.

Walker MM, Dennis TE, and Kirschvink JL (2002) The magnetic sense and its use in long-distance navigation by animals. *Current Opinion in Neurobiology* 12: 735–744.

Wallraff HG (2004) Avian olfactory navigation: Its empirical foundation and conceptual state. *Animal Behaviour* 67: 189–204.

Wiltschko R and Wiltschko W (2003) Avian navigation: From historical to modern concepts. *Animal Behaviour* 65: 257–272.

Wiltschko W, Munro U, Ford H, and Wiltschko R (2009) Avian orientation: The pulse effect is mediated by the magnetite receptors in the upper beak. *Proceedings of the Royal Society B* 276: 2227–2232.

Marine Invertebrates: Genetics of Colony Recognition

R. Grosberg and D. Plachetzki, University of California, Davis, CA, USA

Introduction

Many sessile, encrusting clonal and colonial marine animals – notably sponges, cnidarians, bryozoans, and colonial ascidians – exhibit a suite of life-history traits that promote intraspecific competition for space and the evolution of complex behaviors that mediate the outcomes of somatic interactions. These traits include the capacity for *indeterminate growth* and reproduction, often augmented by limited dispersal and, in some cases, kin-directed settlement behavior of their sexual and asexual *propagules*. In these taxa, intraspecific competitive interactions elicit behaviors ranging from no apparent response, through active cytotoxic rejection, to intergenotypic fusion of individuals and colonies. Among cnidarians and some bryozoans, incompatibility responses extend beyond simple rejection, often eliciting a complex suite of agonistic behaviors. In addition, developmental transitions in the expression of these complex fusion and rejection behaviors may also occur.

As in many social insects and vertebrates that modify the expression of their social behaviors according to the relatedness of *conspecifics*, a growing number of field and laboratory studies on colonial marine invertebrates show that neither intergenotypic rejection and aggression, nor fusion randomly occur with respect to the genotypes of interacting conspecifics. Instead, the initiation of agonistic behavior often depends on the relatedness of contestants: interactions between clonemates and close kin generally do not elicit cytotoxicity or aggression, whereas interactions between more distant relatives do. Likewise, *somatic fusion* usually occurs only between clonemates and close kin. Thus, precise allorecognition, the ability to distinguish self from conspecific nonself, once thought to be the hallmark of the vertebrate immune system, is phyletically broadly distributed, and appears to be a ubiquitous feature of all multicellular animals, along with fungi, myxobacteria, and myxomycetes.

Pioneering studies of invertebrate allorecognition, dating back a century to classic studies on the colonial ascidian *Botryllus schlosseri*, made it clear that specificity in the expression of intercolony fusion and rejection was heritable. Thus, allorecognition in these taxa provided an early model for the studies of behavioral genetics. The dependence of *somatic rejection*, aggression, and fusion on relatedness, together with discrimination reliabilities that often exceed 95%, implies that (1) these behaviors enhance individual and *inclusive fitness* by mediating responses with respect to the genetic identities of interactors; (2) genetically based recognition cues govern the expression of these behaviors; and (3) the diversity of these cues is built on unusually high levels of genetic variation.

In this way, several features of invertebrate allorecognition systems mirror several aspects of the major histocompatibility complex (MHC), a key element of the vertebrate adaptive immune system. For this reason, analogies between the vertebrate immune system and invertebrate allorecognition behavior inspired many early studies of invertebrate allorecognition, with the hope of discovering retained ancestral features of our own MHC. We now understand that most of the similarities between invertebrate allorecognition and vertebrate MHC systems are superficial and likely reflect convergent evolution, not common ancestry. Still, these parallels may reveal common selective forces that have led to the evolution of the diverse array of allorecognition systems, including the vertebrate MHC. For instance, the function of both allorecognition and the MHC relies on extremely high levels of genetic polymorphism that confer cue specificity. The use of highly *polymorphic*, genetically based phenotypic cues to regulate the expression of these social behaviors (including inbreeding avoidance) and immune function potentially imposes selection on the genes that produce these cues.

Understanding how natural selection influences the evolution of this exacting specificity and its underlying genetic diversity fundamentally requires an integrated analysis of both formal and molecular genetics of allorecognition. Formal genetic approaches generally involve breeding animals with different phenotypes, and then correlating the phenotypes of progeny with the presence or absence of genetic markers over successive crosses. In so doing, breeding studies can circumscribe how many loci are involved in the trait (i.e., how many distinct markers correlate with the allorecognition phenotype), the degree to which certain alleles are dominant over others in the expression of a given allorecognition phenotype, and the level of standing genetic variation that exists in a given population for an allorecognition phenotype (i.e., how many allorecognition classes, or *allotypes*, segregate in a population). Alternatively, studies on the molecular genetics of allorecognition can reveal the specific genes involved in these phenotypes and provide primary sequence-level resolution on the identity of loci involved in allorecognition. Such data make it possible to compare how alleles differ from each other, provide direct measures of how natural selection acts on individual loci, and reveal how specific regions or positions within these loci evolve. In this way, formal and molecular genetic approaches complement each other and together offer powerful tools for deciphering the evolutionary genetics of allorecognition.

To this end, two marine invertebrate model systems have emerged over the last two decades, one focusing on the colonial ascidian genus *Botryllus* (Phylum Chordata), and the other on colonial *hydrozoans* in the genus *Hydractinia* (Phylum Cnidaria). In both these cases, formal and molecular genetic approaches have begun to reveal key components of the allorecognition machinery. These components minimally include the genes encoding receptors and cues that determine behavioral responses to other conspecifics. Here, we summarize what is presently understood about the genetics of allorecognition in these two taxa, along with the far more limited data that presently exist for bryozoans and sponges. We also evaluate the functional significance of allorecognition in colonial marine invertebrates and consider how various forms of natural selection can explain what is presently understood of the genetics of allorecognition. Finally, we review the broad *macroevolutionary* patterns of allorecognition behavior in colonial invertebrates and consider the factors that may have contributed to their evolution.

Genetics of Allorecognition in *Botryllus schlosseri*

The formal and molecular genetics of allorecognition are better understood in the colonial ascidian *B. schlosseri* than in any other invertebrate system. Ascidians are soft-bodied invertebrate chordates (Phylum Chordata) that belong to the Subphylum Tunicata, a clade that diverged from its sister taxon, the Craniata (including the vertebrates), over 600 Ma. The life cycles of botryllid ascidians such as *B. schlosseri* offer many opportunities for allorecognition behavior to be expressed. First, the fertilized egg develops into a motile tadpole-like larval stage. This tadpole stage possesses all the diagnostic chordate features, including a notochord, a dorsal hollow nerve tube, and pharyngeal gill slits. The larvae swim for a few minutes to hours, and then attach to hard substrates, often dispersing so little that they settle in the vicinity of their kin, apparently using shared allorecognition alleles to detect their relatives. Once attached, the tadpole metamorphoses into a minute, founding oozooid. During the metamorphic transition to an attached phase, the juvenile *Botryllus* loses all its chordate features, except for its gill slits, which it uses for respiration and feeding. The oozooid then asexually buds off additional zooids (**Figure 1(a)**), which in turn bud still more zooids. Repeated cycles of asexual budding ultimately give rise to a modular colony of genetically identical zooids, each with its own set of ovaries and testes. The zooids lie embedded in a

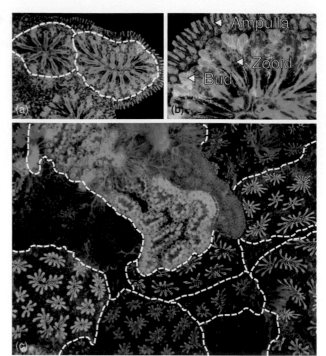

Figure 1 Colony structure and ecology of the colonial ascidian *Botryllus schlosseri*. (a) The adult zooids (the modular, cloned units that compose a colony) of *B. schlosseri* form star-shaped systems embedded in a gelatinous tunic. Numerous tiny saccular ampullae of the tunic's interzooidal blood-circulatory complex fringe the colony's complex blood vascular system. (b) A group of adult zooids, asexual buds (which sequentially and synchronously develop into new zooids), and peripheral finger-like projections of the blood-vascular system called 'ampullae' (sites of allorecognition). (c) Competition for spatial resources is fierce. Individual colonies of *B. schlosseri* (outlined in white dashed lines) surrounding a single colony of *Botrylloides leachi*, a species closely related to *Botryllus* (red dashed lines).

cellulose matrix (the tunic), interconnected by a ramifying and anastomosing blood vascular system (**Figure 1(b)**). Colony size has no intrinsic physiological or structural limit; consequently, *B. schlosseri* colonies can continually grow, often encountering themselves (self-recognition) or conspecifics (allorecognition) as they expand (**Figure 1(c)**). When colony edges meet, they interact via ampullae, finger-like projections of their vascular network (**Figure 2(a)**). The ampullae may either fuse, establishing blood flow between the colonies, or reject, a response accompanied by cytotoxic reactions and the formation of a barrier between incompatible colonies.

Extensive breeding studies, dating back to the early 1960s, show that the outcome of these interactions – fusion or rejection – depends on a single highly polymorphic allorecognition locus, now called *FuHC*. Alleles at *FuHC* are expressed codominantly in *Botryllus*. Upon contact, individuals that share one or both alleles at *FuHC* fuse (**Figure 2(b)**), whereas pairs of colonies that do not share an allele reject (**Figure 2(c)**). The DNA sequence of *FuHC*

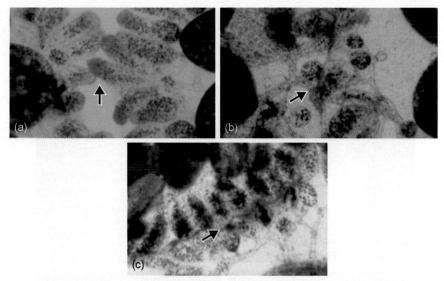

Figure 2   Allorecognition reactions in *Botryllus schlosseri*. (a) Initial contact between the ampullae of two colonies. (b) Vascular fusion between two colonies that share one or both alleles at their *FuHC* locus. (c) Cytotoxic rejection between two incompatible colonies that lack a shared *FuHC* allele. Arrows denote initial points of contact.

shows no obvious homology to any known vertebrate gene. Nevertheless, like other genetic systems mediating allorecognition, the *FuHC* locus displays an extreme level of allelic variation, with several studies yielding estimates of polymorphism in excess of 100 alleles, and heterozygosities that approach 1. These extraordinary levels of genetic polymorphism mean that individuals are only likely to share alleles with themselves (self) and close relatives; thus, this polymorphism permits individuals to discriminate kin relationships with far greater resolution than if fewer alleles were present in the population. Also, because only a single shared *FuHC* allele is required for fusion, individuals that are only related as kin, not clones, can fuse. Thus, genetic chimeras (single colonies composed of multiple genotypes) arise with appreciable frequency in *Botryllus*, but are only likely to form between close kin.

Other loci in addition to *FuHC* may be involved in histocompatibility and allorecognition responses in *Bortryllus*. The same genetic mapping and functional approaches that revealed the identity of the *FuHC* locus also hinted at another locus involved in allorecognition. This locus, called *fester*, is polymorphic, though to a lesser extent than *FuHC*. However, in addition to sequence polymorphisms, *fester* expresses a large number of unique mRNA splice products that yield a higher diversity of *fester* gene products in the population than would be expected from allelic diversity alone.

What is the evidence that *fester* functions with *FuHC* to mediate allorecognition behavior in *B. schlosseri*? First, in adult *B. schlosseri fester*, gene expression is restricted to the ampullae (the site of either fusion or rejection) and to a subset of blood cells thought to play an important role in allorecognition. Furthermore, *B. schlosseri* can express

allorecognition behavior as early as the tadpole larval phase, and both *fester* and *FuHC* share a common domain of gene expression in early tadpole and oozooid developmental stages. Most compelling is the fact that knocking down the expression of *fester* produces altered allorecognition phenotypes. It seems that *fester* is a receptor for *FuHC* gene products; however, its exact role in allorecognition is still uncertain.

Genetics of Allorecognition in *Hydractinia symbiolongicarpus*

The cnidarian genus *Hydractinia* encompasses a clade of marine, colonial hydrozoans, many of which inhabit the discarded shells of marine gastropods that are subsequently occupied by hermit crabs (**Figure 3**(c), inset). Like *Botryllus*, several species in this genus have relatively short generation times (on the order of weeks to months) and can be cultured and bred in the lab, making them ideal candidates for the genetic studies of allorecognition behavior. Colonies of *Hydractinia* are either male or female: males shed sperm into the water, and fertilized eggs develop into minute, wormlike larvae (planulae) while held on the female colony. When a crawling planula contacts a hermitted shell, it metamorphoses into a founder polyp, analogous to the oozooid of *Botryllus*. Through repeated episodes of asexual budding, a colony develops, which – in the absence of competitors – could expand to cover the entire shell. However, in many cases, multiple sexually produced planulae colonize a single shell, and the ensuing intraspecific competition for space may be fierce.

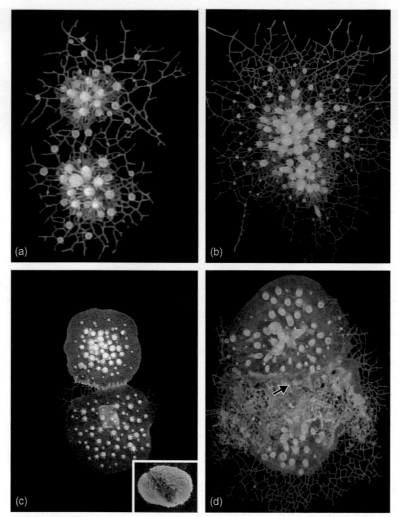

Figure 3   Fusion and rejection responses in *Hydractinia symbiolongicarpus*. (a and b) Fusion between compatible colonies, in this case, full siblings. (c and d) Aggressive rejection between incompatible colonies, accompanied by the production of nematocyst-laden hyperplastic stolons (arrow). (b and d) were taken 2 weeks after (a) and (c), respectively. Inset, a hermit-crab-occupied snail shell colonized by two incompatible *H. symbiolongicarpus* colonies, separated by a conspicuous zone of rejection.

As *Hydractinia* colonies grow, they extend tubelike stolons over the shell, from which specialized feeding, defensive, and reproductive polyps, emerge. The stolons themselves are extensions of the guts of each of the polyps and form a gastrovascular system that links the members of a colony. When a colony encounters itself as it grows around a shell, its stolons invariably fuse, preserving the integrity of self and functionally unifying the colony. When the stolons of genetically distinct colonies grow into contact, one of the three outcomes ensues: (1) fusion, forming a functionally and behaviorally integrated, but genetically chimeric individual (**Figure 2(a)** and **2(b)**); (2) aggressive rejection, accompanied by the induction of specialized organs of aggression, the hyperplastic stolons (**Figure 2(b)** and **2(c)**); or (3) transitory fusion, in which initial fusion is followed by varying degrees of rejection. As with *Botryllus*, the probability of fusion is closely tied to kinship: parents

and offspring invariably fuse, but full sibs usually fuse < 40% of the time, and more distantly, relatives are rarely compatible, and usually aggressively reject each other.

With aggressive rejection, closely apposed stolons begin to accumulate specialized *nematocytes*, the diagnostic stinging cells of cnidarians, to their tips, and become hyperplastic. Interestingly, the recruitment of nematocytes to form hyperplastic stolons begins before rejecting individuals actually touch, suggesting the action of a diffusible chemical cue that signals allotypic identity or disparity. By some unknown trigger, nematocytes from one of the hyperplastic stolons synchronously discharge, injuring, and sometimes, eventually killing an opponent. Alternatively, aggressive bouts can persist as standoffs for weeks or months with no clear winner.

Despite over half a century of research on allore-cognition in *Hydractinia*, the genetic basis of specificity is just beginning to be understood. Early accounts suggested that a single genetic locus with multiple codominant alleles controlled allorecognition specificity, as in *Botryllus*. However, subsequent genetic models and mating studies confirmed that multiple loci likely control allorecognition, at least in *Hydractinia symbiolongicarpus*. One recent study, using highly inbred lines, implicated two distinct genetic loci, *alr1* and *alr2*, that cosegregated with allorecognition phenotypes. Positional cloning of the genomic region that contained these markers showed that *alr2* is an immunoglobulin-like protein with both transmembrane and hypervariable amino acid sequence regions, making *alr2* a candidate allorecognition surface protein. If *alr2* is an allorecognition surface protein, the role that *alr1* plays in mediating allorecognition in *Hydractinia* remains to be determined.

In contrast to our understanding of the genetic basis of allorecognition in *Botryllus*, a correlation between specific polymorphisms at these loci and the expression of allorecognition phenotypes has not yet been fully demonstrated. Much of the confusion over this question relates to the fact that key experiments have yet to be conducted in the *Hydractinia* system. For instance, the power of our understanding of the genetics of allorecognition in *Botryllus* stems from experiments where wild-caught (noninbred) individuals were tested against lab strains that had been characterized at the *FuHC* locus. Importantly, both fusion and rejection phenotypes were observed in these allorecognition experiments allowing the sequences at *FuHC* to be correlated with the observed phenotypes. Similar experiments need to be conducted in *Hydractinia*. Furthermore, the types of functional assays that were pivotal to demonstrating that *FuHC* is an allorecognition locus in *Botryllus* have yet to be performed in *Hydractinia*.

Once candidate genes are identified by formal and molecular genetic analysis, the best standards of evidence linking these genes to specific allorecognition phenotypes are gain- and loss-of-function experiments. Here, the hypothesis that a specific gene is involved in the expression of a given phenotype (for instance, allorecognition) is tested by turning off the function of the specific gene and assessing any change that results in phenotype. Changes that do occur may then be 'rescued' by turning the gene of interest back on. Such experiments, commonly referred to as functional genomics, are often done using gene knockdown methodologies that include RNA interference (RNAi), morpholinos, and other methods. Irrespective of the method used, these techniques allow the hypothesis that a specific gene of interest is involved in an observed phenotype to be tested directly. Although the *alr1* and *alr2* loci are likely to be involved in *Hydractinia* allorecognition, functional genomics experiments are the crucial next steps toward understanding the genetics of allorecognition in this system.

Genetics of Other Allorecognition Systems

As we have seen, the life histories of many colonial marine invertebrates make allorecognition a critical aspect of their behavior and ecology. In addition to colonial ascidians and hydrozoans, many sponges, bryozoans, and anthozoan cnidarians such as anemones and corals are capable of precise allorecognition. Sponges and bryozoans tend to exhibit fusion-rejection behaviors like those of colonial ascidians, with rejection associated with cytoxicity, and the preservation of the genetic integrity of self, but not the induction of behaviors or structures that are overtly agonistic. On the other hand, as in *Hydractinia*, incompatibility in many anthozoan cnidarians is often accompanied by the production of aggressive structures, heavily armed with specialized nemaotocysts, that include modified tentacles (e.g., sweeper tenatcles and acrorhagi), extensions of the gut (e.g., mesenterial filaments), and entire polyps (e.g., dactylozooids).

While a great deal is known about the occurrence of allorecognition behaviors in sponges, corals and anemones, and bryozoans, the challenges of breeding most colonial marine invertebrates in the lab have left the formal and molecular genetics of allorecognition in taxa, other than *Botryllus* and *Hydractinia*, virtually unknown. Consequently, most studies of the relationship between fusion and rejection frequencies and relatedness involve various proxies for kinship, often distance between sources of experimentally grafted colonies. Because both sexually and asexually produced propagules in most colonial marine invertebrates have relatively limited dispersal potential, kinship should decline with the distance separating two individuals.

For example, in the Pacific sponge *Callyspongia* (Phylum Porifera), grafting experiments show that the likelihood of fusion between fragments increases as the distance between source colonies decreases. Similar patterns of fusion frequencies declining with distance are well documented in other sponges; however, it is often unknown whether compatible grafts are limited to clonal fragments, or whether kin can fuse as well. Allorecognition also occurs in colonial, encrusting bryozoans (suspension-feeding members of the *Lophotrochozoa*). In many bryozoans, sexually produced larvae are shed into the water column daily. In some species with nonfeeding larvae, settlement habitually occurs near the parental colony. And, as in *Botryllus*, the larva of at least one species of bryozoan seems to take relatedness into account when making their settlement decisions. Analyses of fusion and rejection in several bryozoans and many corals and sponges confirm the general pattern that clonemates are always compatible, and that compatibility declines with relatedness between allogeneic individuals.

Evolution of Allorecognition Systems

The specificity of fusion and rejection behaviors in all colonial marine invertebrates studied to date suggests that the loci controlling allorecognition specificity are extremely polymorphic. This raises two questions: (1) what is the source of the underlying genetic variation? and (2) how are such high levels of polymorphism maintained in natural populations? In terms of sources of variation, simple nucleotide substitutions may generate most of the observed allelic variation. But there are other ways by which hypervariable recognition cues could be generated, one likely candidate being the structural alteration of genetic loci themselves via *somatic recombination*. Examples of this polymorphism-generating mechanism include intraindividual genomic recombination, as is the case for the V(D)J system in the vertebrate adaptive immune system, and the generation of alternative splice products, as is the case for the *fester* locus in the *Botryllus* allorecognition system. These changes to the physical structural of genes or mRNA transcripts represent direct polymorphism-generating mechanisms, but examples of such mechanisms are rare.

It seems inescapable that some form of selection favoring rare alleles (negative frequency-dependent selection) drives the maintenance of the extreme levels of polymorphism inferred in most populations of marine invertebrates. The expression of behaviors such as fusion and rejection in colonial marine invertebrates is obviously functionally important, both in terms of the maintenance of the genetic integrity of self (and avoiding the costs of various forms of somatic and germ line parasitism that 'defector' genotypes may inflict on their fusion partners) and competition for space. By limiting fusion to clonemates and close kin, highly polymorphic allorecognition systems minimize the possibility of fusing with a parasitic genotype and maximize the inclusive fitness benefits of behaving altruistically toward a fusion partner. In cnidarians especially, by directing aggressive behavior away from clonemates and close kin, allorecognition systems reduce the inclusive fitness costs of harming self or a relative.

In *Botryllus*-like systems where the genetics of allorecognition determine whether intergenotypic contacts elicit fusion or passive rejection, simple population genetic models confirm our intuition that rare allorecognition alleles will be favored when the costs of intergenotypic fusion exceed the benefits. These costs include various forms of intraspecific parasitism, whereas the benefits of fusion include enhanced competitive ability or a greater range of environmental tolerance arising from increased genetic diversity in chimeric individuals. However, the situation is more complicated and daunting when aggression, rather than merely passive rejection, is an alternate outcome to fusion. The paradox arises because any new mutant, that by definition must be initially rare, will face nearly universal assault from more common allotypes. For this reason, it is hard to imagine how rare allotypes could increase in frequency and become established in a population. There are several ways that selection might circumvent this obstacle. For one, as we have seen, populations of colonial invertebrates are often not randomly distributed spatially with respect to relatedness among individuals: kin are far more likely to interact than would be expected if there were extensive dispersal of motile larval and asexual propagules. Consequently, a newly arising allotype might encounter kin with appreciable frequency, decreasing the risks of attack and at least allowing an initial increase in the frequency of a rare mutant. Alternatively, selection favoring rare alleles for some other phenotype, that is, mating and inbreeding avoidance as in mice and perhaps other mammals, or in disease resistance, could initially favor allotypes that could subsequently be employed as allorecognition markers.

Phylogenetic Distribution of Allorecognition in Colonial Invertebrates

Allorecognition systems are distributed widely across the animal tree of life. Sponges, bryozoans, ascidians, and colonial hydrozoans each occupy distinct branches on this tree, and have likely been on their own independent evolutionary pathways for well over half a billion years (**Figure 4**). The extremely divergent phylogenetic distribution of allorecognition and the absence of clearly shared genetic elements in the systems that have been explored thus far on the molecular level (e.g., the *FuHC* and *fester* genes in *Botryllus*, the *alr1* and *alr2* genes of *Hydractinia*, and the vertebrate MHC genes) suggest one of the two possibilities for the macroevolutionary distribution of allorecognition behaviors. Either they are so distantly related that the signature of common ancestry has been lost (the last common ancestor of sponges and ascidians may have existed as many as 1 billion years ago), or the different allorecognition systems have evolved independently in animals.

Most modern views of animal phylogenetic relationships place the sponges as the earliest branching metazoan lineage. Sponges possess several highly specialized cell types, but lack tissue-grade organization. If the diversity of animal allorecognition systems evolved from a common ancestor that predated sponges, it could have done so to mediate interactions between the cells of isogeneic and allogeneic individuals. Such interactions occur in modern sponges (not to mention myxobacteria, many fungi, red algae, and cellular slime molds) and can give rise to genotype-specific partitions of sponge cell types in chimeric individuals. Indeed, the ability to distinguish

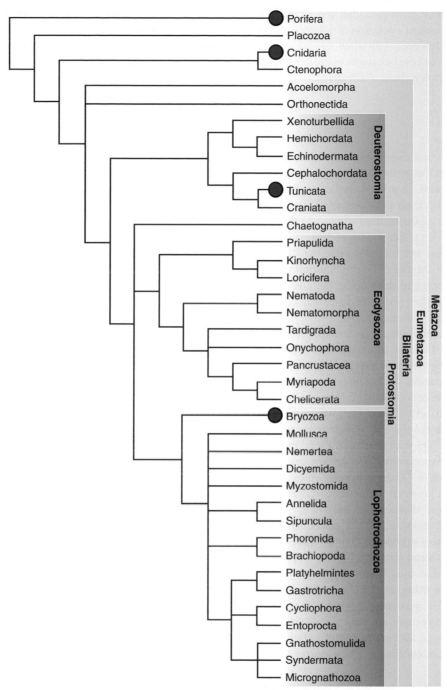

Figure 4 One recent view of animal phylogeny showing the distribution of allorecognition behavior (red circles). Much of the living phyletic diversity of animals can be found in the three major clades of bilaterian animals (deuterostomes, ecdysozoans, and lophotrochozoans). Earlier branching lineages include sponges (Porifera) and cnidarians. That allorecognition occur across such vast phylogenetic intervals may indicate the independent origins of these systems.

conspecific self from nonself extends well beyond the evolutionary history of animals and appears to be a basic attribute of multicellular life.

All allorecognition systems may have evolved several billion years ago from a common ancestor to the three major domains of life. However, it seems far more likely that allorecognition evolved multiple times in the deep history of multicellular life and occurred independently in the four major groups discussed here. If so, the evolution of allorecognition behavior in disparate lineages represents an incisive and fundamental example of convergent evolution of a behavioral phenotype. The organismal, behavioral, and ecological attributes shared among these taxa (e.g., indeterminate growth, asexual propagation, and limited dispersal) may have provided the adaptive substrate for the independent evolution of allorecognition in disparate lineages. However, only when we understand the genetics that underlie these behaviors in greater depth can we discriminate between these alternatives.

Conclusions

Allorecognition is a conditional behavior whose expression depends not only on the genotypes of both participants in an interaction, but also, as several studies on hydrozoans and corals suggest, on their developmental state. This, in and of itself, makes characterizing phenotypes and their underlying genotypes a major challenge. In addition, many of the organisms that exhibit allorecognition-dependent behaviors are long lived and difficult to culture in the lab, posing major challenges to developing the kind of broad taxonomic coverage that promises to reveal underlying functional and genetic patterns.

Nevertheless, several clear patterns do emerge. For instance, virtually all known allorecognition systems have very high specificity, which is presumably controlled by numerous variable genetic factors. In addition, despite the fact that such specificity could be controlled by genetic variation distributed across many loci, what we presently understand from the *Botryllus* and *Hydractinia* systems suggests that just a few loci with extensive allelic variation at each locus controls specificity. This pattern could reflect functional constraints on (1) the genes that confer specificity, (2) co-evolution between the genes that confer specificity (cues) and those that actually encode the receptors that facilitate recognition, or, (3) the genes that mediate how cues and receptors interact to yield specific behaviors such as somatic fusion and rejection. Finally, it appears that only partial genetic matching is necessary for two allotypes to be compatible; in other words, the available evidence suggests that self is recognized, rather than nonself. Whether this reflects recognition errors, or selection favoring the ability to distinguish not just self from nonself, but close from distant kin, continues to be a matter of considerable debate.

Regardless of whether individual or kin recognition is the primary selective factor favoring the evolution of genetic diversity in allorecognition systems, growing evidence from many groups of multicellular organisms confirms that the capacity to distinguish self from nonself may be a universal and essential feature of multicellular life. The genetic data currently available suggest that allorecognition evolved numerous times in the history of life, and was likely co-opted in many different ways to regulate the expression of traits such as agonistic behavior, mating preferences, and pathogen defense.

See also: Dictyostelium, the Social Amoeba; Kin Recognition and Genetics; Recognition Systems in the Social Insects; Social Insects: Behavioral Genetics.

Further Reading

Bancroft FW (1903) Variation and fusion in colonies of compound ascidians. *Proceedings of the California Academy of Sciences* 3: 137–186.

Buss LW (1982) Somatic-cell parasitism and the evolution of somatic tissue compatibility. *Proceedings of the National Academy of Sciences of the United States of America: Biological Sciences* 79: 5337–5341.

Buss LW (1987) *The Evolution of Individuality*. Princeton, NJ: Princeton University Press.

De Tomaso AW, Nyholm SV, Palmeri KJ, et al. (2005) Isolation and characterization of a protochordate histocompatibility locus. *Nature* 438: 454–459.

Fletcher DJC and Michener CD (1987) *Kin Recognition in Animals*. New York, NY: Wiley.

Grosberg RK (1988) The evolution of allorecognition specificity in clonal invertebrates. *Quarterly Review of Biology* 63: 377–412.

Grosberg RK, Levitan DR, and Cameron BB (1996) Evolutionary genetics of allorecognition in the colonial hydroid *Hydractinia symbiolongicarpus. Evolution* 50: 2221–2240.

Grosberg RK and Strathmann RR (2007) The evolution of multicellularity: A minor major transition? *Annual Review of Ecology Evolution and Systematics* 38: 621–654.

Hamilton WD (1964) Genetical evolution of social behaviour I and II. *Journal of Theoretical Biology* 7: 1–16 and 17–52.

Ivker FB (1972) Hierarchy of histo-incompatibility in *Hydractinia echinata. Biological Bulletin* 143: 162–174.

Jackson JBC, Buss LW, and Cook RE (eds.) (1985) *Population Biology and Evolution of Clonal Organisms*. New Haven, CT: Yale University Press.

Nicotra ML, Powell AE, Rosengarten RD, et al. (2009) A hypervariable invertebrate allodeterminant. *Current Biology* 19(7): 583–589.

Nyholm SV, Passegue E, Ludington WB, et al. (2006) *Fester*, a candidate allorecognition receptor from a primitive chordate. *Immunity* 25: 163–173.

Oka H and Watanabe W (1960) Colony-specificity in compound ascidians. *Bulletin of the Marine Biological Station at Asamushi* 10: 153–155.

Mate Choice and Learning

E. A. Hebets and L. Sullivan-Beckers, University of Nebraska, Lincoln, NE, USA

Introduction

While an individual's genetic framework is a major contributor in determining its eventual mate choice, the role of the environment in further influencing mating decisions has long been recognized. Animals gather information from the environment throughout life, and in some cases, may apply this information to increase their odds of obtaining a high-quality mate. In short, these individuals learn. Moreover, such learning can have a social component. 'Social learning' is a general term that describes any learning based on observing, interacting with, and/or imitating others in a social context. Social learning can transmit information vertically, generation to generation (e.g., parent to offspring) and/or horizontally, within a generation (as individual to individual). This form of information transfer is generally referred to as 'cultural transmission.' This entry will focus on social learning that relates to mate choice – mate-choice learning.

Mate-choice learning can be separated into two broad categories: learning based on personal experiences with others (referred to as 'private' or 'personal information') or learning that results from the observation of others (referred to as 'public information'). Learning from private experiences can occur at the juvenile or adult stage and may include encounters with conspecifics or heterospecifics, same sex or opposite-sex individuals (**Figure 1**). Mate-choice imprinting, for example, demonstrates how an early experience based on private information shapes subsequent mate choice. Conversely, public information refers to any information gained through the observations of other individual's experiences. An example of the use of public information is mate-choice copying, for example, when a female mimics the mating decision of another female in the population. Mate choice that is influenced by private information is sometimes termed 'independent mate choice,' whereas mate choice based on public information is 'nonindependent mate choice.'

Mate-choice learning, whether it is through the acquisition of private or public information, balances various costs and benefits. For example, the process of learning itself can be costly, a topic covered in depth elsewhere. Additionally, costs can come in the form of imprinting on the wrong species (which could lead to reduced fitness), or from copying another individual that has chosen poorly itself. Nonetheless, the prevalence of mate-choice learning across taxonomic groups suggests that there are significant benefits associated with mate-choice learning. For example,

the use of public information relieves an individual from personally gathering information and could minimize costs typically associated with mate assessment such as exposure to predators or decreased time devoted to other important activities such as foraging. Mate-choice learning more generally permits flexibility in mate choice, which could be extremely important in a changing environment. In the following text, examples of different forms of mate-choice learning will be provided and the state of research in this area summarized.

Private (Personal) Information

Juvenile Experience: Mate-Choice Imprinting

'Mate-choice imprinting' refers to the learning process, or processes, by which young individuals acquire sexual preferences based on their observation of adults. Several specific forms of imprinting exist and the general tenet was first described by Douglas Spalding in the nineteenth century as he recounted his observations of newly hatched chicks following random moving objects. Despite its early description, however, the notion of imprinting was not popularized until the 1930s by the pioneering work of the Nobel Prize winning Austrian ethologist, Konrad Lorenz. Similar to other forms of imprinting (e.g., filial imprinting), sexual imprinting, or mate-choice imprinting, typically takes place during a sensitive period early in life. Historically, it has most frequently been observed in species with parental care, where the young use the parent of the opposite sex as the model upon which they base their future mating preferences. This kind of early mate-choice imprinting is thought to function to ensure conspecific matings, enabling individuals to avoid presumably costly heterospecific matings. Nonetheless, it is now clear that mate-choice imprinting is not always restricted to an early sensitive period and that preferences often continue to be modified throughout development.

Crossfostering experiments are one of the primary means by which scientists study early mate-choice imprinting and such studies are most easily, and frequently, conducted with birds. In crossfostering experiments, offspring are raised by parents of either another phenotype (e.g., a different color morph) or another species and, subsequently, their adult mate choice is examined. Using crossfostering experiments, mate-choice imprinting has been demonstrated in numerous bird species including, but not confined to snow geese, zebra finches, Bengalese finches, great tits,

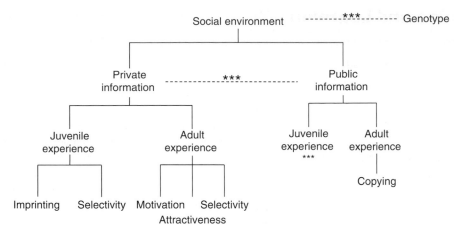

Figure 1 Mate-choice learning as influenced by social environment. This diagram depicts the various sources of social information that can impact mate-choice learning; the various life stages during which learning might be important; and some of the documented outcomes of mate-choice learning. Sections marked with '***' indicate topics for which research is lacking or nonexistent, and we suggest that these might be potentially fruitful areas for future focus.

blue tits, and red jungle fowl. In mammals, reciprocal cross-fostering of sheep and goats has demonstrated a role of maternal imprinting on subsequent sexual preferences. Similarly, in Lake Victoria fishes, females of some species appear to imprint on the phenotype of their mother. Crossfostering experiments supporting a process of mate-choice imprinting are prevalent, yet studies do exist for which such early experiences have not influenced adult mate choice – raising interesting questions about species-level differences in the potential for, and importance of, mate-choice learning.

Traditional examples of mate-choice imprinting, as outlined earlier, are often restricted to species in which young spend significant time with their parents, thus, enabling parental imprinting (either paternal or maternal). However, mate-choice learning may also be prevalent in species that lack parental care, yet still have significant exposure to other conspecifics. For example, female wolf spiders are known to choose to mate with mature males of a phenotype with which they had experience as a subadult. This type of imprinting is referred to as 'oblique imprinting' – imprinting on a nonparental adult. In another example of oblique imprinting, damselfly males alter their preference of female morphs based upon prior experience – males raised in the absence of females show no preference, while those raised with one female form subsequently exhibited a preference for females of that form. Planthoppers have also been shown to exhibit a learned preference for conspecifics. Finally, in humans for whom arranged marriages are the norm, the experiences young women have outside the traditional family environment, including exposure to outside media and participation in youth groups, influence their involvement in marriage arrangements.

In addition to empirical studies that utilize crossfostering or various early exposure techniques, numerous

mathematical models have been constructed to examine the various aspects of mate-choice imprinting. For example, population genetic models have been used to explore the evolution of different forms of imprinting. In these models, Tramm and Servedio compared the evolution of paternal, maternal, and oblique imprinting and found that paternal imprinting was the most likely to evolve. Their results suggest that the success of a particular imprinting strategy is most influenced by the group of individuals that are imprinted upon (termed the 'imprinting set').

Juvenile Experience: Mate Selectivity

Mate-choice imprinting involves juvenile individuals imprinting on, or learning, various characteristics of an adult model, whether the model is their mother, father, or another nonparental adult. Subsequently, these learned characteristics are incorporated into the individual's mate-choice criteria, and mating partners with similar characteristics are preferred. However, experience with conspecific adults may not always lead to a preference for individuals resembling a model. Sometimes, early experience may simply increase choosiness. Such effects of early experience have been documented in various animal taxa. For example, in both field crickets and wolf spiders, research has shown that early experience by females with courtship songs or displays can lead to increased selectivity for mates.

Adult Experience: Mate Selectivity

Effects of experience on mate choice need not be restricted to young or immature individuals. As adults, encounters with rivals and potential mates can also alter mating behaviors for both males and females. For example, in some spiders, fruit flies, crickets, and newts, naive females are less discriminating in mate choice than older and more experienced females. A female's threshold to accept a male can also change with successive encounters, both pre- and postmating. Presumably, as females gain experience with mates, they learn to distinguish among them. A significant literature exists on female search strategies (e.g., sequential search, best-of-*n*, and variable threshold), many of which implicitly assume learning.

Not only do adult females alter their mate choice based upon their personal experiences with mature males, but they may also alter their preferences based on personal information regarding their own attractiveness. In humans, for example, attractive females have stronger preferences for high-quality males than less attractive females, and in zebra finches, a female's self-perception has been shown to influence her mate choice. In nature, this self-assessment may or may not be learned, but theoretical models suggest that the perception of one's own attractiveness could develop through previous experiences with the opposite sex, resulting in increased choosiness following successful encounters and decreased choosiness following rejection by potential partners.

Thus far, we have been focusing mostly upon female mating preferences. However, males have also been shown to alter mating behaviors with experience. As males are rejected or accepted by females, they may become more or less sexually aggressive and/or more or less discriminating. Trinidad guppy males, for example, learned to direct courtship at conspecific females after 4 days of contact with conspecific and heterospecific females. In damselflies, males prefer females of a morph with which they have had previous experience. In *Drosophila*, a male's experience with a heterospecific female often leads to reduced future courtship effort toward heterospecific females. In wolf spiders, previous mate effects are known to shape a male's future mating success. Males that had experienced, but not mated with, a female were less likely to mate in the next encounter. However, if the male had mated with the previous female, it was more likely to mate with the next.

Public Information

Adult Experience: Mate-Choice Copying

In various taxa (although primarily in fish and birds), females observe and copy the mating decisions of conspecific females. In some cases, mate-choice copying leads to an increased preference for the male traits observed in the mated male. In other cases, females may prefer the actual male that was observed mating with another female. Mate-choice copying has the benefit of decreasing the investment in mate assessment that a female must make. The reliability, consistency, and agreement between sources of information available to a female may determine when a female

copies mating decisions and when she will forego mate-choice copying, relying instead on private assessment. In some species, when public and private information conflict, females base decisions on their own assessment, while in other species, females revert to mate-choice copying in such situations. In humans, mate-choice copying has been documented to depend on the quality of the model female observed with a potential mate. Additionally, in humans as well as other taxa, the degree to which females will copy mating decisions of others is influenced by sexual experience. In many cases, virgin females are more likely to copy mate-choice decisions than more sexually experienced females. Mate-choice copying has been documented in vertebrate (e.g., fish, birds, and mammals) and invertebrate (e.g., insects) species.

Mechanisms of Mate-Choice Learning

Identifying and describing the physiological mechanisms that underlie the relationship between learning and mate choice is a vast area of research. Here, some of the major findings of the field are summarized. The neurophysiology of early mate-choice imprinting in zebra finches has been extensively explored. Immediate-early genes (*c-fos* and ZENK) have been used to estimate neuronal activity and to identify activated brain regions with exposure to novel and previously experienced stimuli. Researchers have also investigated neuronal control of the length and timing of the sensitive period for sexual imprinting. In *Drosophila*, the neurosensory pathway that functions in the male and female brain to determine whether to attempt courtship with a potential mate based on previous experience, has been described. In mice, after investigating the volatile chemical signals present in female urine, males acquire more complex and extensive preferences for the odor of sexually receptive females. These male preferences correspond to changes in the piriform cortex of the brain, and knockout studies have demonstrated that the gene *Peg3* disrupts these effects of experience. Thus, in disparate taxonomic groups, significant information is available on the physiological mechanisms underlying mating choice learning, and this remains an active area of research.

Evolutionary Consequences of Mate-Choice Learning

One of the most intriguing and intellectually stimulating aspects of mate-choice learning is its potential to drive evolutionary change. Not surprisingly then, exploring the evolutionary consequences of mate-choice learning is an extremely active area of research, rich with theory and modeling. The most frequently discussed aspects of mate-choice learning involve its putative influence on such evolutionary processes as speciation, hybridization, and sexual selection.

Speciation and Mate-Choice Learning

It has frequently been suggested that mate-choice imprinting can facilitate reproductive isolation. Imprinting on one's parental phenotype, for example, leads to positive assortative mating, where similar phenotypes preferentially mate with each other. Any new phenotype, or novel trait, appearing in a population could rapidly lead to reproductive isolation via mate-choice imprinting, even if it is initially present at a low frequency. Empirical work with collared flycatchers (*Ficedula albicollis*) has provided support for such a mechanism, as the artificial introduction of a novel trait (a red stripe on a male's forehead) led to positive assortative mating – females having experienced males with a red stripe were more likely to pair with males possessing red stripes. The initial effects of such mate-choice imprinting could then be followed by disruptive selection. In fact, a recent mathematical model has demonstrated that reinforcement (enhancement of premating isolation) can occur via learned mating preferences. It is important to note, however, that the influence of mate-choice imprinting on evolutionary processes such as speciation depends implicitly upon the imprinting set, or the individuals used as models. For example, imprinting on a nonparental phenotype (oblique imprinting such as mate-choice copying) would likely inhibit population divergence. Nonetheless, the involvement of mate-choice imprinting on speciation and diversification has likely been important for numerous taxonomic groups and has been explicitly suggested to have played a role in the diversification of various birds (e.g., Galapagos finches; various brood parasites) as well as fishes (e.g., Lake Victoria cichlids).

The occurrence of interspecific brood parasitism raises unique questions with respect to the evolutionary implications of mate-choice imprinting. Consequently, a significant amount of research addresses the role of mate-choice imprinting on speciation and diversification in avian brood parasites. Interspecific brood parasites constitute approximately 1% of all bird species and are defined as those species for which adults do not care for their young, but instead deposit their eggs in the nests of other species, where the young are left to be raised by foster parents. Given the common occurrence of mate-choice imprinting in birds, an obvious question arises regarding how imprinting on a foster parent might influence subsequent reproductive success of the parasitic offspring. For example, if parasitic offspring imprint on visual aspects of their foster parent, their subsequent ability to find a conspecific mate could be severely compromised. However, imprinting on the song of the foster parent (which can be learned), for both males and females, could facilitate conspecific matings. Indeed, in whydahs and indigobirds (interspecific brood parasites in the genus *Vidua*), parasitic male offspring copy the song of their foster fathers. Parasitic female offspring also imprint on their foster father's song. This host imprinting ultimately enables parasitic offspring to find conspecific mates as adults. This process of host imprinting has been proposed as a mechanism promoting diversification in this group, as host shifts could readily lead to reproductive isolation. However, one could also imagine a scenario where mate-choice imprinting on a host could lead to hybridization. For example, if numerous species utilize the same host, the likelihood of parasitic individuals mating with a heterospecific brood parasite increases, and recent work has indicated that continued gene flow does exist between some host races.

Hybridization and Mate-Choice Learning

Although mate-choice imprinting often results in positive assortative mating, typically with conspecifics, the potential exists for misimprinting, or imprinting on the wrong species. Hybridization between species of Darwin's finches, for example, is known to occur and is thought to result from misimprinting. Additionally, crossfostering experiments conducted in the wild have demonstrated that some bird species will imprint on a foster parent of another species, resulting in heterospecific pairings.

Heterospecific matings could result in hybrid offspring and hybrid zones are not uncommon in nature. What role then, if any, does mate-choice imprinting play in hybrid zones? Using an artificial neural network, Brodin and Haas demonstrated that phenotypes of pure species are learned faster and better than those of hybrids, potentially leading to selection against hybrids. Further spatial simulations combined with empirical data on dispersal demonstrate that mate-choice imprinting can maintain a hybrid zone under natural conditions.

Sexual Selection and Mate-Choice Learning

In addition to its role in speciation and hybridization, mate-choice learning might also lead to the evolutionary change of specific traits within a species, especially traits that are sexually selected. For example, mate-choice imprinting can lead to sexual preferences for extreme phenotypes beyond which an individual has experienced, potentially driving trait elaboration. One mechanism by which this is possible is via peak shift – a consequence of discrimination learning of differentially reinforced stimuli (e.g., individuals are trained such that one stimulus is rewarded and the other is punished). Essentially, peak shift can lead to a preference for an exaggerated trait never previously experienced. For example, in an elegant study using zebra finches, ten Cate and colleagues raised males with the parents of artificially painted beaks (orange or red). In subsequent mating trials, they were able to show a shift in male beak color preference, with males directing more courtship to females at the extreme maternal end of the spectrum, despite the fact that this beak color was more extreme than seen in the model parent.

The above-mentioned example addresses the role of parental imprinting on trait evolution. However, oblique imprinting, or imprinting on a nonparental adult, also has intriguing potential regarding the evolution of secondary sexual traits. The cultural transmission of mating preferences, or passing on of mating preferences through nongenetic mechanisms, could lead to evolutionary changes in secondary sexual traits, or cultural inheritance. Cultural transmission refers to the process by which the phenotype of a species can change based upon information acquired during an individual's lifetime. Essentially, the cultural transmission of female preferences (via juvenile experience effects with nonparental adult conspecifics or via mate-choice copying) could drive the cultural inheritance of male secondary sexual traits. The details of such evolutionary change would depend explicitly on the form of imprinting and on the imprinting set.

Genotype-by-Environment Interactions and Mate-Choice Learning

Thus far, we have focused solely on various environmental effects on mate-choice learning, with no discussion of the underlying genetics. Yet, all organisms are influenced by both their genes and their environment. Much recent work has been directed explicitly at understanding the interactions between an individual's genotype and its environment. Genotype-by-environment interactions (GEIs) have become one of the major explanations regarding the maintenance of genetic variation in secondary sexual traits, despite putatively strong sexual selection that should diminish this variation. While most studies of GEIs have focused on male signaling traits, it seems equally likely that female preferences are influenced by GEIs. For example, a female's genotype may impact her likelihood and/or her ability to learn mating preferences. Such GEIs with respect to mate-choice learning would certainly influence the interactions between learned mate choice and the evolution of male secondary sexual traits. Future work exploring the interactions between genotypes and social environments will surely provide a rich source of new knowledge and insights regarding mate-choice learning and its role in evolutionary processes.

See also: Alex: A Study in Avian Cognition; Apes: Social Learning; Avian Social Learning; Behavioral Ecology and Sociobiology; Collective Intelligence; Costs of Learning; Cultural Inheritance of Signals; Culture; Decision-Making: Foraging; Fish Social Learning; Flexible Mate Choice; Imitation: Cognitive Implications; Insect Social Learning; Isolating Mechanisms and Speciation; Learning and Conservation; Mammalian Social Learning: Non-Primates; Mate Choice in Males and Females; Memory, Learning, Hormones and Behavior; Monkeys and Prosimians: Social Learning; Psychology of Animals; Sexual Selection and Speciation; Social Cognition and Theory of Mind; Social Information Use; Social Learning: Theory; Vocal Learning.

Further Reading

Bischof H-J and Rollenhagen A (1998) Behavioral and neurophysiological aspects of sexual imprinting in zebra finches. *Behavioral Brain Research* 98: 267–276.

Drullion D and Dubois F (2008) Mate-choice copying by female zebra finches, *Taeniopygia guttata*: What happens when model females provide inconsistent information? *Behavioral Ecology and Sociobiology* 63: 269–276.

Dukas R (2005) Learning affects mate choice in female fruit flies. *Behavioral Ecology* 16: 800–804.

Fawcett TW and Bleay C (2009) Previous experiences shape adaptive mate preferences. *Behavioral Ecology* 20: 68–78.

Hebets EA (2003) Subadult experience influences adult mate choice in an arthropod: Exposed female wolf spiders prefer males of a familiar phenotype. *Proceedings of the National Academy of Sciences of the United States of America* 100: 13390–13395.

Immelmann K (1972) Sexual selection and other long-term aspects of imprinting in birds and other species. *Advances in the Study of Behavior* 4: 147–174.

Irwin DE and Price T (1999) Sexual imprinting, learning and speciation. *Heredity* 82: 347–354.

Mery F, Varela SAM, Danchin E, et al. (2009) Public versus personal information for mate copying in an invertebrate. *Current Biology* 19: 730–734.

Ophir AG and Galef BG (2004) Sexual experience can affect use of public information in mate choice. *Animal Behavior* 68: 1221–1227.

Qvarnstrom A, Blomgren V, Wiley C, and Svedin N (2004) Female collared flycatchers learn to prefer males with an artificial novel ornament. *Behavioral Ecology* 15: 543–548.

Servedio MR, Saether SA, and Saetre GP (2009) Reinforcement and learning. *Evolutionary Ecology* 23: 109–123.

Sirot E (2001) Mate-choice copying by females: The advantages of a prudent strategy. *Journal of Evolutionary Biology* 14: 418–423.

Slagsvold T, Hansen BT, Johannessen LE, and Lifjeld JT (2002) Mate choice and imprinting in birds studied by cross-fostering in the wild. *Proceedings of the Royal Society of London Series B: Biological Sciences* 269: 1449–1455.

ten Cate C, Verzijden MN, and Etman E (2006) Sexual imprinting can induce sexual preferences for exaggerated parental traits. *Current Biology* 16: 1128–1132.

Tramm NA and Servedio MR (2008) Evolution of mate-choice imprinting: Competing strategies. *Evolution* 62: 1991–2003.

Mate Choice in Males and Females

I. Ahnesjö, Uppsala University, Uppsala, Sweden

Introduction

Choosing the right mate can be a potent evolutionary force resulting in the evolution of characters and behavior in the chosen sex, including many elaborate courtship displays and ornamental traits. Survival is a prerequisite for reproduction. Therefore, in an evolutionary sense, it is the individual differences in ability to contribute genes to the next generation, that is, reproductive success that is important and not survival per se. In many animals, there are two sexes and reproduction is primarily sexual. When the gametes of the two sexes differ in size, we use the size of gametes to define female and male: Females produce larger gametes, eggs, whereas males produce smaller gametes, sperm. This means that when reproducing, highest fitness will be achieved for an individual, within a sex, that is most successful in combining its genetical material with that of the other sex. Consequently, the ability to more successfully influence the interaction with individuals of the other sex becomes important, for instance, by being attractive. The problem is, though, how to decide which mates will contribute the genes and resources that will result in an offspring production that is relatively better than that of other individuals of the same sex in the same population at the same time.

In sexually reproducing animals, mate choice is a process by which individuals of one sex gain higher fitness by preferring to mate with some individuals to others. An individual that discriminates among encountered potential mates is called 'choosy.' When the opposite sex is choosy, it is beneficial for an individual to signal attractiveness to individuals of that sex, which will be able to assess the quality and compatibility of potential mates. Therefore, characters and behavior that affect attractivity may be selected by sexual selection and result in so-called secondary sexual characters (i.e., characters that provide reproductive rather than survival benefits). Many extravagant characters have been selected because they provide their bearers with reproductive advantages by being chosen as mates. These characters can be smells, sounds, behaviors, visual displays, morphological structures, etc. Competition is the unifying aspect of sexual selection: either as a process where individuals of one sex compete among themselves to become chosen as mating partners by the other sex (intersexual selection) or as a competitive process within a sex for access to mating partners of the opposite sex (intrasexual selection). Although, the concept of 'female choice and male–male competition' is

common, as a sort of generalized description of animal reproduction, it is not the one and only perspective. We now know that mate choice occurs in both sexes and the sex competing for access to mates may also be discriminate and perform mate choice. Mate choice and mating competition can occur in both sexes simultaneously. Yet, they may often vary dynamically in space and time, and one process may often predominate over the other in one or both sexes for a prolonged time. When reproducing, individuals first have to become sexually mature and then they may need to acquire specific resources to become ready to breed (such as food, territories, etc.). Once ready to mate, they have to choose among mates and/or compete for mates and once mated, there are postmating processes. For example, cryptic choice or parental progeny choice mechanisms may affect the outcome. Here, the mate choice refers to both female and male mate choice in the premating stage, that is, choice based on resources, ornaments acquired, and inherent mate qualities. The postmating stage may also involve mate choice in terms of cryptic mate choice or progeny choice, but this will not be covered in this chapter.

Historical Perspective

In 1871, Darwin hypothesized that female mate choice results in ornamental traits in males and saw many colorful male birds as typical examples of this. However, he was also aware of those females of many species that were conspicuously ornamented, and he viewed these as exceptional cases where males have been the selectors instead of being selected. Historically, the perspective of female choice and male–male competition resulting in sexual selection has been in focus. Why female mate choice seems more prevalent than male mate choice was not approached until about 100 years later by Williams and then by Trivers. They used Bateman's study on fruit flies, *Drosophila melanogaster*. In his experiment, males and females were allowed to mate freely and the variance in mating success among females was found to be much lower than among males. This sexual difference in variance in mating success and the observation that mating success continues to increase with the number of mating more steeply in males than in females (the Bateman gradient) are frequently used as indicators of sexual selection. Though later stochastic models have shown that these results could be an outcome of random mating, questioning them as main indicators of sexual selection. However, Bateman's adaptive framework, of female choice and indiscriminant male mating behavior, led on to the parental investment ideas of Triver's. He defined "parental investment as any investment by the parent in an individual offspring that increases the offspring's chance of surviving (and hence reproductive success) at the cost of the parent's ability to invest in other offspring," and he concluded that "what governs the operation of sexual selection is the relative parental investment of the sexes in their offspring." Recent research has, however, made clear that it may also be the other way around, that the operation of sexual selection may influence patterns of parental investment. Triver's ideas of parental investment and sexual selection still have great value, but parental investment is empirically hard to assess. Consequently, later population-related models for predicting which sex that predominates in mating competition have been put forward, circumventing the problem of estimating relative parental investment. The operational sex ratio (OSR) models consider the number of individuals of each sex that are ready to breed at any given time and place, and the sex in excess is predicted to predominantly compete for access to mates whereas the opposite sex could be more choosy. In a population, OSR biases can arise due to the sexual difference in potential reproductive rate (PRR) If individuals of one sex are slower in processing matings and take longer to become ready to breed again, this sex will have a lower PRR and is predicted to be choosier and less competitive. The processing of mating involves gamete production, parental care, etc. However, it is important to note that the realized reproductive rate will always be equal (each offspring has a mother and a father), but the potential (when unconstrained by mate availability) reproductive rate may differ between the sexes. Possibly, the sexual difference in PRR may reflect sexual differences in parental investment in many animals, but PRR is more empirically accessible and also includes time expenditures that may matter more than energy investments when predicting patterns of mating competition and mate choice in a population. It is though important to recognize that within a sex individuals can easily be both choosy and competitive and that choosiness may relate to the variation in the quality of potential mates, quite independently from mating competition. In many population models, such individual differences may be overlooked.

A classic example of female mate choice is provided by Malte Andersson's 1982 studies of long-tailed widowbirds. In this African bird, males have substantially elongated tails and females are attracted to mate with males having longer tails. Andersson experimentally demonstrated this by manipulating male tail length. Some males had their tails shortened, some males maintained the same tail length, and others were provided elongated tails (males

in all three treatments had their tails cut and glued). Females showed a clear mate choice for males displaying elongated tails. Another classic example comes from the peacock, where peahens show a clear preference for males having more eye-spots on their elongated feathers. Many more examples of both male and female mate choice are found in Andersson's book, *Sexual Selection*.

Mate Choice Evolution

Mate choice is usually costly (in terms of time, energy, risk) and has to be balanced by benefits (resulting in a net fitness gain) in order to be selected. Such benefits may be both direct and indirect. Direct benefits are when immediate effects on fitness occur such as provisioning of resources to offspring or improved fertilization success. Indirect benefits, on the other hand, enhance offspring fitness by increasing their viability or attractiveness through inheriting good or attractive genes. In nature, it is presumably a combination of both and of multiple kinds. However, costs and benefits of mate choice vary between populations, contexts, and over the season.

Direct Benefits

In animals where there are nuptial gifts, territories, parental care, or other resources provided, we can easily envision that choosy individuals will benefit by being able to assess these benefits directly or via one or several cues indicating the gain. In many organisms with indeterminate growth (fishes, reptiles, amphibians, many invertebrates), male choice for larger and more fecund females provides good examples of male mate choice for direct benefits. Examples are many but, for instance, in the broad-nosed pipefish, both more and larger eggs are gained for males choosing to mate with larger females (**Figure 1**). Male mate choice for more fecund females has also been documented in animals with determinate growth, such as insects and in the zebra finch. Female mate choices for direct benefits are also common: In many birds and fishes, female preferences for male territorial qualities or paternal abilities have been demonstrated. For instance, in the fifteen-spined stickleback, females prefer males that court more intensely and these males are also better at fanning the nest, which result in a higher hatching success. Similarly, in the sand goby, females prefer the most competent fathers and not the males that are most successful in male–male competitive interactions. Males providing good territories or oviposition sites are preferred by females in for instance, dragonflies, frogs, birds like pied flycatchers, dunnocks, red-winged blackbirds, and male pronghorn antelopes defending a good feeding territory attract more females. In many insects

(a)   (b)

Figure 1   (a) Two female broad-nosed pipefish in the front with zigzag patterns on their trunk. The male pipefish is in the back. Photo: A. Berglund. (b) A close-up of a male's brood pouch with almost fully developed embryos (eyes visible). While brooding for more than a month, the embryos are osmoregulated, oxygenated, protected, and provided some nutrients. Photo: O. Jennersten.

(crickets, butterflies, dance flies), males provide nuptial gifts as packages of additional resources and sperm (spermatophores), or prey gifts can be provided by males or females. Commonly, there are mate choice processes in operation for larger gifts, providing direct benefits. Other examples of male choosiness that may co-occur with females being choosy as well are when males are choosy for females that have not mated recently, or are close to conception. Similarly, in chimpanzees, males prefer older females as they have higher breeding success, as compared to younger ones. Important to bear in mind is that mate choice for direct benefits may be heritable, but does not require heritability as fitness gains may affect immediate resource situations.

Another way for mate choice evolution is to exploit an already naturally selected sense. An animal may already be sensitive to certain features (colors, smells, or sounds) that, for example, occur in their diet and are preferred as food items and an inherited ability to sense these features may confer fitness advantages. When such a preexisting sensory bias, in a nonmating context, is present it may also affect the evolution of mate choice preferences and result in mating biases. In some guppy fish species, a general attraction to orange food can largely explain differences between populations in female mating preferences for males with larger orange dots. However, predation is another factor that influences and limits the expression and preference for orange dots.

Indirect Benefits

Indirect benefits of mate choice may occur in combination with direct benefits. However, indirect benefits are more commonly considered when direct benefits are lacking or assumed to be of minor importance. The indirect benefits of mate choice require inheritance, such that the offspring inherit the genes that give the improved viability or attractiveness. A mating preference may select for traits indicating viability, for instance, when the quality of an individual's immune defense or ability to acquire specific nutrients are indicted via this trait. Such indicators may indicate 'good genes' in general, that will be inherited to the offspring, or 'handicaps' if an individual can obtain heritable resistance to diseases or parasites by choosing mates that indicate their ability to invest in larger (i.e., handicapping) displays as well as investing in a costly immune defense. For instance, starling males that sing more frequently also show a stronger immune response, and females choose to mate with males singing more frequently, which possibly results in more viable offspring. An additional form of indirect benefits of mate choice is the process known as 'the Fisherian run-away process.' It refers to when there is inheritance both for an attractive trait and the preference for it, which results in a self-reinforcing co-evolution of the trait and the preference (the trait may be an indicator trait or an arbitrary trait). The Fisher run-away process is theoretically well founded, but good empirical examples are less obvious. This is possibly because when a trait becomes extreme ('runs away') it becomes costly and may start functioning as a 'handicap' trait, since an individual has to be of good quality to afford it.

As long ago as 1972, Trivers suggested that the choice of a mate should favor a mate that is most compatible with the chooser in terms of producing adaptive gene combinations in the offspring. Consequently, males and females may choose a mate that either has particular genes that will result in a more successful combination, for instance, in terms of heterozygosity or major histocompatibility complexity (MHC – an important function of the vertebrate immune system). There are many examples of maternal and paternal allele combinations that affect offspring fitness; however, the extent to which this is used in mate choice is less documented. It should also be noted that there are distinctions between choice for good genes, compatible genes, and the conflicts that may occur between optimizing these choices under various circumstances. Thus, directional sexual selection may operate to generate an ornament that signals good genes. However, the individual that carries the ornament may not always be the most genetically compatible mate to all individuals. When choosing a genetically compatible mate, it may be important to optimize similarity or dissimilarity, as shown in three-spined sticklebacks where females optimize the MHC-complexity of their mate choice in relation to their own MHC profile.

Other Factors Affecting Mate Choice Evolution

A complicating factor when it comes to demonstrating benefits of mate choice is that there may always be maternal or paternal effects, that is when an individual choosing a mate also allocates resources in relation to the mate's attractiveness or quality. Such maternal or paternal effects may either be a differential allocation (i.e., when the chooser allocates additionally when mating to a preferred mate), or it may be compensatory (i.e., when the chooser compensates when mating to a mate they prefer less). In many animals, females are in control of egg numbers and quality and may thus adjust their egg allocation depending on mate quality. Similarly, allocation to care by both males and females may also depend on mate quality. Furthermore, the importance of sexual conflict between females and males is becoming increasingly clear. Selection for a trait in one sex may have a negative influence in the other sex, which can result in a sexually antagonistic co-evolution. Obviously, this may apply to mate choice evolution, as preferences and chosen traits may have different optima in the two sexes.

A Pipefish Example

In the broad-nosed pipefish (*Syngnathus typhle* L.), swimming in the eelgrass meadows of the sea, females usually display in groups and compete for access to male mating partners. Females display contrasted zigzags (ornaments) on their body trunk (**Figure 1**(a)), a pattern that attracts males, but intimidates other females. Males are also better able to assess female size using this pattern. In their mate choice, males prefer larger, more ornamented, and dominant females, and benefit by receiving larger eggs that will result in larger offspring of higher fitness (better survival to anemone predation and higher growth rate). However, female mate choice also selects for larger males, presumably as they are better care providers. If a female is constrained to mate with a smaller, less preferred male, she compensates by providing eggs with a higher protein concentration. Both male and female mate choice has been demonstrated by A. Berglund and coworkers, and both males and females produce newborn that survive predation better if mated to a preferred mate as compared to a less preferred mate (even when standardizing for offspring size differences). Although both male and female mate choice occur, only males (not females) copy the mate choice of other males. Similarly, mating competition for

access to mates can be prominent in both sexes, but the predominant female–female competition characterizes the mating pattern of this species in particular. Consequently, multiple sexually selective processes operate dynamically and the same cues are used in several contexts. Also postmating selective processes may occur, as brood reduction in the male pouch is common while brooding the embryos.

Mate Assessment

How mates are encountered, simultaneously or in sequence, will influence the opportunity for performing mate choice and assess mates. In nature, the options for simultaneous comparisons of large number of mates are usually relatively limited. However, lekking species are an example where several individuals of one sex can be assessed. Still, the interactions between competition and mate choice on the lek are complex and vary between animals. For example, lekking birds such as the ruff, black grouse, and manikins all differ in how competition and mate choice contribute to fitness. When mate assessment is more sequential, various assessment tactics can be employed. For example, a fixed threshold tactic can be used, mating with the first mate encountered that fills the minimum requirements, or sequential comparisons where the best of a number of possible mates is chosen. The assessors can then optimize costs and benefits of continued search or acceptance of the present mate. Mate assessment can also be done by copying the mate choice of others. This may be a good option when a mate searcher is unable to discriminate various mate qualities or if copying others reduces the search and discrimination costs. Mate choice copying has been demonstrated many times in fishes, for instance, in guppies, gobies, medakas, and pipefish (**Figure 1**). Mate assessment strategies have mostly been studied when animals are breeding, but mate qualities can also be assessed and evaluated during nonbreeding seasons.

The influence of mating competition on mate assessment can be both positive and negative. The choosing individuals can use competition within the other sex to assess mate qualities, and competition may also be incited for this purpose. However, competitive interactions may also hamper mate assessment and dictate mating in conflict with mate choice. On the other hand, mate choice can override status rankings from competitive interactions and be based on other abilities and characters. Consequently, the actual mate in the wild may not necessarily be a preferred mate; the actual mate can be a compromised or a constrained 'decision' resulting from other processes than mate choice.

Obviously, rarely is the choice of mate based on only one single signal or ornament. Instead, there are multiple cues, and each cue may also have multiple functions. Visual, olfactory, and sound cues can easily be combined and they can be used in different environments, contexts, and at different distances. Some cues may have evolved as species-recognition cues and others as mate-quality signals. Efficiency and selection of signals are also context dependent, and anthropogenic disturbances may affect mate assessments. For instance, color displays in fishes may be difficult to assess or be distorted in turbid (polluted, eutrophic) environments, and bird song may in a noisy environment be hampered as a mate attractor.

Contrasting Female and Male Mate Choice

Mate choice is important to the evolution of secondary sexual character in both sexes, and the process of mate choice and sexual selection works according to the same principles in both sexes. Though the operation of sexual selection does contrast somewhat between males and females, some general tendencies can be discussed. Females by definition invest more energetically into each gamete, and often they also provide parental care (at least in mammals and birds; however, less often in fishes where males are the main care provider). As a consequence, females often compete with other females for resources necessary for a successful reproduction, whereas males may compete for these females. However, that this always is a consequence of differences in gamete size should be challenged. Interesting comparisons among *Drosophila* fly species show that differences in relative gamete size do not necessarily predict sexual patterns of mating discrimination. In general, the relative intensity of mate choice and mating competition varies, and choice can develop in both sexes whenever there is variance in the quality of mates that may affect fitness. It is more rare with very costly (i.e., highly extravagant) secondary sexual characters in females, and possibly this is because resources traded to such character may have to pay fecundity costs which may somewhat constrain the development of the secondary sexual characters in females. Males, however, may pay survival costs that may constrain ornament development. Notable is that ornaments selected by mate choice occur in females and males in many kinds of animals.

Future Perspectives

In the near future, I envision a broader attention to mate choice studies, unbiased with regard to the sexes. We need more behavioral studies in many different kinds of animals and in various contexts. We will be able to explore the genomics of mate choice; today's and tomorrow's molecular and genomic tools make this possible. Revealing how mate choice operates and imposes selection on behavior, morphology, reproductive physiology, genes, and proteins in both sexes will be exciting. Selection experiments, cross-fostering experiments, and in particular a systematic approach to investigate fitness consequences of mate choice in both sexes, by comparing outcomes from preferred and less preferred (or random) mating, are important future avenues. This means approaching mate choice both on the individual and the population level. Mate choice evolution is context dependent and we need detailed studies on how mate choice interacts with mating competition, ecological circumstances, social circumstances, and the dynamic variation in time and space.

Acknowledgments

I am most grateful to P. Gowaty, M. Ah-King, and L. Kvarnemo for inspiring text comments.

See also: Bateman's Principles: Original Experiment and Modern Data For and Against; Compensation in Reproduction; Cryptic Female Choice; Differential Allocation; Flexible Mate Choice; Social Selection, Sexual Selection, and Sexual Conflict; Sperm Competition.

Further Reading

Ahnesjö I, Forsgren E, and Kvarnemo C (2008) Variation in sexual selection in fishes. In: Magnhagen C, Braithwaite VA, Forsgren E, and Kapoor BG (eds.) *Fish Behaviour*, pp. 303–335. Enfield, NH: Science Publishers.

Andersson M (1994) *Sexual Selection*. Princeton, NJ: Princeton University Press.

Andersson M and Simmons LW (2006) Sexual selection and mate choice. *Trends in Ecology and Evolution* 21: 296–302.

Berglund A and Rosenqvist G (2003) Sex role reversal in pipefish. *Advances in the Study of Behavior* 32: 131–167.

Gowaty PA, Steinichen R, and Anderson WW (2003) Indiscriminate females and choosy males: within- and between species variation in *Drosophila*. *Evolution* 57: 2037–2045.

Johansson BG and Jones TM (2007) The role of chemical communication in mate choice. *Biological Reviews* 82: 265–289.

Jones AG and Ratterman NL (2009) Mate choice and sexual selection: What have we learned since Darwin? *Proceedings of the National Academy of Sciences of United States of America* 106: 10001–10008.

Kokko H, Brooks R, Jennions MD, and Morley J (2003) The evolution of mate choice and mating biases. *Proceedings of the Royal Society, London Series B* 270: 653–664.

Snyder BF and Gowaty PA (2007) A reappraisal of Bateman's classic study of intrasexual selection. *Evolution* 61: 2457–2468.

Trivers RL (1972) Parental investment and sexual selection. In: Campbell B (ed.) *Sexual selection and the Descent of Man, 1871–1971* pp. 136–179. Chicago: Aldine.

Wong BM and Candolin U (2005) How is female mate choice affected by male competition? *Biological Reviews* 80: 559–571.

Maternal Effects on Behavior

H. Schwabl, Washington State University, Pullman, WA, USA
T. G. G. Groothuis, University of Groningen, Groningen, Netherlands

Brief History and Definitions

Organisms develop under the continuous interaction of genes and environmental factors. Parents not only transmit genes to their offspring but also influence their environment, which can profoundly affect offspring traits. Both mother and father can cause such environmental effects although maternal effects are often more pronounced than paternal effects, as in most species, mothers provide more care and in all species, mothers provision the eggs, influencing the embryo. Perhaps because of this, parental effects are commonly labeled as 'maternal effects.' Maternal effects are very widespread in the plant and animal kingdoms and can be mediated by diverse and multiple proximate means and pathways. Maternal reproductive decisions such as the timing of reproduction, choice of breeding location, and the number of siblings in a given propagule will, for example, indirectly influence the social environment or food availability for the offspring. The differential bestowment of the embryo, fetus, or newborn with bioactive supplements such as immune-active substances, antioxidants, growth factors, and hormones will affect the development and differentiation of physiological functions and is also known as 'prenatal' or 'developmental programming.' Quality and quantity of food provisioning will influence growth and health. Finally, maternal effects are also transmitted directly via behavior, by processes such as social facilitation, which may lead to cultural transmission of certain traits. In this article, we focus on maternal effects that are mediated by hormones, especially androgens since this proximate pathway has been studied most extensively in the ecologically relevant context.

Two classes of maternal effects are distinguished: the so-called indirect genetic effects and indirect environmental effects (**Figure 1**). The first refer to the situation in which the parental effect on the offspring depends on the genetic background of the parent. For example, the quality of food provisioning by the parent to the young may depend on the genes of that parent. In this case, the young would receive not only the genes for good food provisioning from their parent but also relatively high-quality food, increasing their survival and thereby strengthening the propagation of the relevant genes in the population.

Clearly, such indirect genetic effects can profoundly affect evolution. Indirect environmental effects refer to the situation in which the environment is 'translated' to the offspring by the parent. For example, parents reproducing

in environments with high food quality/quantity can provision their young with more, better, or different food, indirectly leading to offspring phenotypes different from those in lower-quality environments. Or, in anticipation of the offspring environment, mothers may directly modify offspring phenotype, based on how she is experiencing the environment, via the transmission to the offspring of certain signals that alter their development. The latter points to the possibility of maternal effects to flexibly adjust specific offspring traits to relevant environmental factors in which it develops and lives, a flexible adjustment that cannot be achieved by the transmission of genetic material alone. The concept of maternal effects being adaptive receives currently much attention. Finally, maternal effects on the environment of the offspring can interact with transmitted genetic information. Via several pathways, maternal signals or the maternally provided environment may induce changes in gene expression, for example, DNA silencing by methylation of some genes. This could not only have profound effects on first generation of offspring but also carry over to subsequent generations if such epigenetic marks are not erased during gametogenesis.

Historically, the term 'maternal effect' was used in quantitative genetics to account for variation in offspring phenotype that is not accounted for by additive genetic variance and the developmental environment. As a consequence, maternal effects were seen as undesired noise in artificial selection and breeding with no apparent function in selection and adaptation. Maternal effects have also been known to biomedical research for some time, and they were originally viewed as pathological perturbations of resilient genetic developmental programs by suboptimal maternal condition and health. Maternal effects that, for example, lead to alterations in sexual behavior within a sex were thought to derail or interfere with the cascade of development events that leads from genes via hormonal signals (acting as a developmental switch) to male or female phenotype. Those that lead to modifications in the stress responsiveness were thought to interfere with sex-specific pathways of the differentiation of this neuroendocrine system. This perspective is reflected in still-accepted terminology such as 'demasculinization' of males and 'masculinization' of females and entails a lack of appreciation for variation in phenotype within the sexes. This pathophysiological perspective is applied also to physiological systems that maintain organismal homeostasis such as body mass and energy regulation.

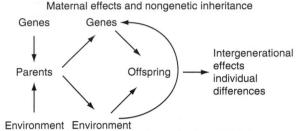

Maternal effects and nongenetic inheritance

Figure 1   Scheme depicting the principle of maternal or parental effects in which the mother or parent influences the environment of the developing offspring, potentially leading to intergenerational effects and individual differences in behavior. Indirect genetic effects are due to effects depending on the genes of the parents; indirect environmental effects are due to effects that depend on environmental effects on the parents. In some cases, the environment of the offspring, affected by the parents, may influence gene expression, such as DNA silencing by methylation. See text for details.

A paradigm shift took place in the early 1990s when maternal effects were proposed to be adaptations that evolved to enhance Darwinian fitness. This change in perspective was facilitated by two independent, but simultaneously occurring developments. Developmental biology joined forces with evolutionary biology and ecology, resulting in an appreciation for developmental plasticity; and molecular biology and genomics began exploring the regulation of gene expression rather than gene sequences. These new developments set the stage for a rapid proliferation of empirical studies of maternal effects from an evolutionary, ecological, and adaptationist perspective.

Maternal effects influence a wide array of offspring traits expressed in early life as well as adulthood. Affected traits range from growth rate, immune function and susceptibility to disease, morphological characters, and food, habitat, and mate preferences to behavioral strategies and tactics. This has been demonstrated in many plant and animal taxa, the latter ranging from mammals to insects. This article focuses on maternal effects on behavior although one has to appreciate that effects on behavior often go in tandem with effects on nonbehavioral traits and physiological functions.

Behavior is a strong force in evolution, no matter whether it promotes or slows down evolutionary change. Maternal effects, in particular those mediated by hormonal signaling between mother and embryo/fetus, cause variation in behavioral phenotype in many taxa. Consequently, maternal effects can generate multiple behavioral phenotypes within families, in populations, and among populations and therefore may be particularly important and strong forces in evolutionary processes and trajectories, for example the rapid adaptation to novel or changing environments. The evolutionary impact of maternal effects on behavior can only be fully appreciated when one also understands the physiological and developmental mechanisms by which they

result in the modification of a behavior or suites of behavioral traits, a topic addressed later.

Finally, the prevalence of maternal effects is relevant not only for those studying function and evolution of behavior, but also for those studying its genetics and differentiation. Such studies often make use of genetic selection lines and cross fostering design and generate measures of heritability by analyzing parent–offspring similarities. In particular, prenatal maternal effects can have a profound influence on the interpretation of results from such analyses and are difficult to account for without the implementation of specific statistical techniques or embryo transplantation (accounting for prenatal maternal effects).

Hormone-Mediated Maternal Effects

Commonly, hormones are mediators of maternal effects and influence offspring phenotype either indirectly or directly. The first is, for example, demonstrated by the finding that testosterone production in the young may be stimulated by frequent social interactions when offspring are raised in high density, an environment that can result from the choice of nest sites by the parents. Such early stimulation of testosterone production may cause long-lasting changes in the sensitivity to testosterone later in life, as has been demonstrated in juvenile black-headed gulls, *Larus ridibundus*, by T. Groothuis and colleagues. Even more intriguing and subtle are direct effects of prenatal exposure to maternal hormones. In many animal taxa, not only does the embryo and fetus produce its own hormones, but it is also exposed to those of the mother. A substantial body of research was devoted to the effects of maternal stress, resulting in elevated embryonic exposure to glucocorticoids (cortisol or corticosterone) with strong impact on stress sensitivity, sexual behavior, and cognitive functions of the offspring in later life. In humans, exposure to maternally transmitted drugs that mimic androgen action as a result of medication of the pregnant mother affects both morphology and play behavior of daughters. Such observations have strengthened the perspective of hormone-mediated effects being maladaptive. However, proliferation of research on hormone-mediated maternal effects from an adaptive perspective was spurred by a study of domesticated canaries (*Serinus canaria*) by H. Schwabl that demonstrated the presence of various maternal steroid hormones, in particular androgens such as testosterone, in the avian egg. Even more important was the observation that the eggs of a clutch of an individual female can vary systematically in the concentrations of these hormones. This within-propagule variation and the variation of hormone concentrations among the clutches of different females prompted numerous adaptive hypotheses of the function of hormone-mediated maternal effect in intra- and interfamily context.

The study of prenatal hormone-mediated maternal effects has greatly benefited from the inclusion of oviparous species as model systems and by now most of the ecologically relevant research is performed on birds and not on laboratory rodents. There are several reasons for this development. First, the accumulation of maternal hormones in the eggs of oviparous species, to which the embryo is then exposed, occurs during a relatively short period, ending at oviposition. Therefore, the concentrations of maternal hormones can be measured in the laid egg before the embryo's own hormone production starts. In mammals, in contrast, maternal hormone levels fluctuate during pregnancy and are difficult to measure without interfering with the mother (affecting her hormone production) and in separation from embryonic hormonal contributions. Moreover, the mammalian placenta serves as an endocrine interface between mother and fetus that metabolizes and converts hormones. Second, prenatal hormone exposure is easier to manipulate in oviparous than in viviparous species. Third, maternal steroids occur in substantial and variable concentrations in the eggs of fishes, reptiles, and amphibians providing ample opportunity for ecological and evolutionary studies. Their role is now studied best in birds, since they lay relatively large eggs and their ecology, reproductive strategies, and development are well known.

Many recent studies of adaptive maternal effects in birds focused on those mediated by the transmission of maternal hormones into the egg because, in contrast to other maternal effects such as egg size, hormonal signaling has unique features that may allow for the evolution of potent and specific maternal influences on the offspring. Hormones are chemical messengers of integrated neuroendocrine systems that induce specific changes in an organism's physiological state in response to or in preparation for environmental change; they regulate transitions between life-history states, maintain homeostasis, integrate multiple-component traits, and they regulate and influence development. In this way, hormones are signals rather than resources such as nutrients. Hormones cause their effects by binding to their receptors in or on specific target cells which, upon binding of the hormone, initiate the first step of signal transduction pathways to achieve changes in cell function and properties. The same hormone can have different effects in the sexes and influence both developing and adult organisms. A single hormone can have multiple targets and a single target cell can be affected by several hormones. Hormones influence the probability of a behavior to occur in a certain context and modify the differentiation of a behavior during development and/or its expression in adulthood. These properties of hormonal regulation render them potent signals for the communication from mother to developing offspring.

Classical studies of mechanisms of sex differences in behavior revealed that hormones can influence behavior in two ways which is known as the 'concept of organizational versus activational action.' First, a hormone can cause permanent behavioral differences (e.g., between the sexes) by irreversibly organizing the functions and properties of neural and muscular hardware during a critical sensitive phase of development. Second, a hormone, when secreted later in life, can transiently and reversibly activate a behavior. Much of the progress of research in underlying mechanisms of sex differences in behavior rests on this heuristically extremely useful concept but, as discussed later on, it now needs modification to accommodate variation in behavior within the sexes, maternal hormonal effects on behavior, and nonhormonal behavioral differentiation.

Abbreviated Review of Hormonal Maternal Effects on Behavior

Introduction

Prenatal hormone-mediated maternal effects have been extensively studied in mammals, especially rodents, in the contexts of stress physiology (pre- and perinatal exposure to cortisol and corticosterone) and sexual differentiation (exposure to androgens, see section 'Why Do Maternal Hormones Not Interfere with Sexual Differentiation?'). In an ecological/evolutionary framework, they are now most studied in birds and to some extent in lizards and fish species (mostly exposure to androgens). The results obtained with birds, but also with rodents and other mammals have been summarized in several comprehensive recent reviews (see Further Reading). Therefore, we provide only a compressed summary necessary for our subsequent discussion of fundamental and conceptual issues. We focus on androgens as these are the hormones studied the most in an ecologically relevant context.

Maternal hormonal effects go well beyond those on behavior and modifications of nonbehavioral traits have to be taken into account when discussing those on behavior. Affected nonbehavioral systems and functions include immune defense (mostly suppressing), growth and metabolism (mostly enhancing), the neuroendocrine stress response axis and the production of steroid hormones, and morphological structures such as sexual signals (enhancing). Maternal hormonal effects on behavior (as well as those on nonbehavioral traits) may be classified into those expressed in early life (relatively soon after birth or, in oviparous species, after hatching and the consumption of hormone-laced yolk and albumin in which maternal hormones are accumulating during egg production) and those expressed in late life (long after differential developmental exposure to maternal hormones). They may also be classified into those that are sex linked and those that occur in both sexes.

Birds

Examples for early life effects of maternal androgens in birds are earlier hatching and enhanced nestling begging behavior in altricial and semiprecocial birds and the behavior of hatchlings in novel environments in precocial birds. A consequence of differential begging behavior is variation among siblings in the amount of food they obtain from parents leading to differential growth trajectories. Functionally, these effects have been interpreted to reflect an adaptive maternal influence on sibling competition (hormonal favoritism).

Long-term effects, expressed in adult offspring life, of differential embryo exposure to maternal steroids were reported for nonreproductive juveniles and reproductively competent adults of several taxa. So far, they include modified behavioral responses to novel objects in domesticated zebra finches (*Taeniopygia guttata*), dispersal distance in free-living great tits (*Parus major*), and aggressive and sexual behavior in captive house sparrows (*Passer domesticus*) and black-headed gulls. The enhanced aggressive and sexual behavior of sparrows and gulls might be caused by specific developmental modifications of the neural structures that influence the probability to show aggressive or sexual behavior in response to certain cues and in a certain context. Given the fact that these behaviors are androgen dependent the maternal effects may be caused by upregulation of either androgen production or androgen sensitivity. However, some altered behaviors are unlikely to be under the control of androgens. The altered response of zebra finches to novelty could point to yet-to-be-identified organizational modifications of neural functions that underlie differences in coping style or personality. Alternatively, they could be nonspecific and general carry-over effects of differential growth in early life (see later) or be caused by neurochemical modifications that influence a wide array of behaviors such as the dopaminergic or serotonergic system that in turn influences for example, impulsivity, a trait related to personality too. For an understanding of the evolutionary consequences of maternal effects on behavior, it is now critical to identify the potential developmental hormone targets in the embryo. If brain areas of the neuroendocrine circuitry and pathways of aggressive and sexual behavior are targets (i.e., by expressing functional androgen receptors) and the properties of these targets in the adult offspring vary with early hormone exposure, we will have evidence that maternal hormones modify specific behavioral modules rather than having unspecific effects of a more general nature.

Sex-specific (linked) and sex-independent effects provide another type of results whose better understanding is critical to test or generate hypotheses about the role of maternal effects on behavior in evolution. Depending on the species, nestling begging, mass gain, structural growth, and immune function can be impacted in either both or only in one sex. Adult aggressive behavior was enhanced by prenatal testosterone exposure in both sexes of the house sparrow (in a nonreproductive as well as a reproductive context) while enhanced sexual (courtship) behavior, normally performed by the male sex only in this species, was restricted to males. In black-headed gulls, in contrast, both sexes perform the same aggressive and sexual displays and both displays were enhanced in both sexes by exposure to androgens in the egg.

A nonbehavioral example for a sex-linked effect comes from the dichromatic plumage of the house sparrow. Its male-specific plumage signal, the throat bib of black feathers, was enlarged by developmental testosterone exposure in males, but no such badge was induced by testosterone exposure in females that normally lack this trait. In contrast, in the sexually monochromatic black-headed gull, development of breeding plumage coloration was enhanced (or developed earlier) in both sexes. The molecular mechanisms of sex limitation and sex independency of maternal hormonal effects still wait to be studied, but three important messages can be extracted from these results. First, the degree of sexual dimorphism in the trait/species seems to play a role so that maternal steroids interact with both sex-limited and autosomal genes. This may not only be limited to traits that show sex-specific sensitivity to androgens such as aggression. For example, yolk androgens differentially affected growth in male and female chicks of several species. Second, in case of sex-independent effects, the evolution of a maternal hormonal effect for the benefits accrued by modifications of a certain trait in one offspring sex can be constrained by concomitant effects on the other sex. Third, maternal hormonal effects do not just shift the phenotype of one sex toward that of the other sex.

Related to the issue of sex-specific effects is the potential influence of the hormonal state of the mother on propagule sex ratio. For example, an increase in the maternal concentration of corticosterone during egg production decreases the proportion of sons in broods of the homing pigeon (*Columba livia domestica*), while an increase in testosterone concentrations increases the proportion of sons; similar findings have been reported for other bird species. Sex-specific mortality in response to prenatal hormone exposure is an unlikely explanation as there was no effect on embryo mortality. One possibility is that these hormones affect the segregation of the sex chromosomes (meiotic drive). In birds, the female is the heterogametic sex and is therefore potentially in some control of which sex to produce. Sex ratio may depend on female hormonal state as suggested by studies in the group of T. Groothuis. Sex determination obviously affects behavior, but a wider discussion of maternal effects on sex ratio is outside the scope of this contribution.

Adaptive explanations of hormone-mediated maternal effects have been inspired by the systematic variation in concentrations of maternal hormones in eggs resulting in differential exposure of the embryo among and within species. Most of the knowledge of this variation comes from studies of birds as well. The eggs of all bird species studied to date contain substantial amounts of androgens. Comparative analyses of altricial songbirds by the groups of H. Schwabl and T. E. Martin and others suggest that variation in androgen concentrations may be an adaptation to ecological conditions to modify development rate of the embryo and nestling in relation to time-dependent mortality by predation. The results indicate that high predation rate on eggs or chicks may be linked to androgen-mediated higher rate of development. This is consistent with results that manipulations of egg androgen concentrations can shorten the time from onset of incubation to hatching (i.e., the embryo period) in some species. Whether and how slow or fast development influences behavior and other traits remains to be shown.

Variation in egg hormone levels among females of the same species can be attributed to variation in environmental factors such as breeding density, food abundance, or paternal quality. These environmental factors may affect food provisioning by the parents as well as competition among unrelated individuals in early life (after fledging in altricial species and already before that time in semiprecocial species). This is consistent with results from experimental studies demonstrating effects of androgen manipulations in the egg on begging and aggressive behavior of chicks.

Variation of androgen levels among the eggs within a clutch (eggs are usually laid with an interval of one or several days) often reveals a systematic pattern. For example, yolk androgen concentrations increase with every successively laid egg of a clutch in many bird species. Later-laid eggs usually hatch later than those from earlier-laid eggs in the sequence (due to the onset of egg incubation by the parent(s) before a clutch is complete) that results in an age hierarchy among siblings. Consequently, chicks hatching from later-laid eggs have to compete with older nest mates. Greater embryonic exposure to androgens may mitigate their disadvantages. In some other species, in contrast, a late onset of incubation and lower androgen levels in later-laid eggs may work in tandem to result in an even more tilted competitive hierarchy in order to facilitate brood reduction. This hypothesis of hormonal favoritism formulated by H. Schwabl and D. Mock is also consistent with the effects of androgens on begging and hatching. To what extent the offspring may be able to modify the maternal signal or its response to it is addressed in section 'Who Is in Control: The Mother or the Offspring, None or Both?'.

Mammals

A classical case for studies of androgen-mediated maternal effects in mammals is the masculinization of the female spotted hyena (*Crocuta crocuta*). Like in some other mammals, female spotted hyenas possess a pseudopenis, seem to be relatively aggressive, and are dominant over males. In addition, the pups are born precocially and engage in sibling competition, expressed as overt aggression, soon after birth. It has been hypothesized that these unusual traits result from exposure of females to enhanced levels of androgens such as androstenedione circulating in the mother during pregnancy. Experimental evidence by C. Drea and colleagues suggests that the development of the pseudopenis is at least partly under direct genetic control, while its size depends partly on maternal androgens. The work of K. Holekamp and her colleagues provides some evidence for rank-related and androgen-mediated maternal effects on offspring aggressive behavior, but a causal relationship between exposure to maternal androgens and sibling competition, female aggression, and dominance is as yet not fully established.

In several mammalian species, for example, pigs and rodents, in which several fetuses develop next to each other along one of the two uterine horns, adjacent siblings hormonally influence each other's behavioral and physiological differentiation. This is best documented for mice and gerbils by the work of F. vom Saal, J. Vandenbergh, M. Clark, and J. Galef. They showed that for example females positioned between two brothers differ from females positioned between two sisters in multiple traits, including aggression, length of the estrous cycle, morphology, and the sex ratio they will produce themselves. Moreover, prenatal exposure to androgens produced by brothers induces a larger and more male-like urogenital distance in females. Although these effects are not directly caused by maternal hormones, they still can be considered hormone-mediated maternal effects since the position of a fetus relative to its brothers and sisters is a maternal trait. Moreover, females producing sex ratios skewed to sons expose their daughters to more testosterone via their sons.

Other Vertebrate Classes

Another intriguing example by which mothers indirectly influence early hormone exposure of their offspring is temperature-dependent sex determination, for example of reptiles. Mothers of these species bury their eggs and the location and depth of the nest in the substrate determines incubation temperature. Incubation temperature, in turn, determines the sex ratio of the propagule. Elegant work by D. Crews and others suggests that developmental temperature influences sex determination by temperature-dependent hormonal mechanisms such as the rate of conversion of testosterone to estradiol. Interestingly, reptile eggs contain also maternally deposited testosterone and estradiol suggesting that, in addition to nest site selection, maternal steroids may provide a direct pathway for the mother to affect the sex, and thereby the behavior, of her offspring.

Our brief review of hormone-mediated maternal effects emphasizes the abundance, complexity, and diversity of maternal effects on behavior and other offspring traits raising a series of fundamental questions about their mechanisms, functions, and evolution which we will address in the next section.

Fundamental Questions

Despite the current attention for hormone-mediated maternal effects by evolutionary and behavioral ecologists, several critical and fundamental issues are unresolved and underappreciated for further understanding. Their following discussion may provide readers with both an up-to-date impression of relevant critical questions and state-of-the-art current information.

What Are the Mechanisms Underlying Hormone-Mediated Maternal Effects?

For historical reasons, research in behavioral biology is somewhat dichotomized. On the one hand, behavioral and evolutionary ecology focuses on the function and evolution of behavior; on the other hand, behavioral physiology, neuroethology, and biomedical behavior research focuses on the proximate mechanisms of behavior. For example, behavioral ecology is concerned with how different behavioral phenotypes are related to ecology, how behavioral phenotypes affect fitness components, and why they coexist in a population. Neuroethology and neuroendocrinology, in contrast, are interested in how a certain environmental cue is translated into a behavioral response or how behavioral differences, for example between the sexes, develop. To move forward in our currently limited understanding of the role and scope of maternal effects on behavior in evolutionary processes, these complementary approaches and perspectives need to be integrated and combined with concepts of evolutionary developmental biology and life-history theory such as developmental plasticity and reactions norms.

The mechanisms by which maternal effects are mediated in the developing offspring are one of the keys to evaluate and test proposed adaptive hypothesis. Yet, these mechanisms are currently little understood hampering progress at the ultimate research front. Exceptions are very informative studies of a few laboratory model systems such as the impact of maternal parenting style on the differentiation of the stress response axis and adult behavior in laboratory rats. Hormone-mediated maternal effects, in particular those employing signaling by maternal steroid hormones in avian eggs, provide a suitable system for such mechanistic analyses in an ultimate context. With caveats pointed out later on, the classical concept of activational versus organizational hormone action provides us with a useful tool and framework to guide research into the mechanisms by which maternal steroids modify behavior.

Injections of androgens into avian eggs (most studies used testosterone, although the hormone cocktail in eggs includes other steroids, such as the androgenic testosterone precursor androstenedione, the testosterone metabolite and more potent androgen 5α-dihydrotestosterone, and the glucocorticoid corticosterone) influence multiple and diverse behavioral and nonbehavioral traits in early and in adult life of the offspring (see section 'Abbreviated Review'). What are the underlying mechanisms of these effects? A first step to address this question is the identification of the hormonal targets in the embryo. Such information is currently limited by the focus of previous research on sexually dimorphic structures and sex differences. For begging, for example, such targets could be neurons and/or myocytes of the neuromuscular module that underlies the begging reflex, for which there is already some evidence. However, they might also be components of systems and organs that regulate metabolism and growth (e.g., the liver and the growth-hormone pathway) if hormonal effects on begging are indirect. For influences on neonatal exploratory behavior, hormonal targets might be in sensory, perceptual, motivational, and neuroendocrine systems. Early effects of maternal testosterone on nestling phenotype also include nonbehavioral traits such as immune function, metabolism, and growth. Consequently, we might expect to identify immune organs/cells (e.g., bursa Fabricius and thymus), mitochondria, and/or components of the hypothalamus–pituitary–thyroid axis and hypothalamus–pituitary–somatic axis as targets as well. Delineating these developmental targets of maternal steroids is important to understand effects on certain traits and crucial to inform research focusing on ultimate adaptive functions. Such research will also provide critical information on physiological constraints resulting from effects on multiple traits (pleiotropy, see section 'How to Explain the Multiple Effects of Maternal Hormones?').

Long-term effects of maternal steroids on behavior, for example on the expression of sexual and aggressive behavior later in life, resemble those of organizational hormonal actions during sexual differentiation. Organizational actions are permanent and irreversible developmental modifications that are not a consequence of sex differences in adult hormone levels, although sex-specific hormone production profiles inducing sex-specific behavior profiles can themselves be consequences of early organizing effects. Given the fact that the yolk is entirely consumed a few days after hatching, altered chick behavior several weeks after hatching are long term as well and may be considered organizational too, although its reversibility is difficult to test given the disappearance of most juvenile behavior later in life. This illustrates that the dichotomy of organizational versus activational effects is somewhat artificial.

A large body of research of sexual differentiation identified central and peripheral structures to be hormonally organized by sex-specific differential hormone secretion in early life. The mechanisms include modifications in the capability/sensitivity to respond to hormonal or nonhormonal signals in adulthood due to sex differences in neuron number, circuitry, and/or hormone response pathways such as receptors and hormone-metabolizing enzymes. At a first glance, long-lasting maternal hormonal effects seem to involve similar mechanisms or may even be caused by the co-option of the existing pathways of sexual differentiation for maternal hormonal modifications. For example, in rodents, males have a larger urogenital distance and are more aggressive than females, and prenatal exposure to androgens produced by siblings induces a larger urogenital distance and enhances aggression in females. Likewise, in species in which females show signs of masculinized genitalia, females are often dominant over males. As described earlier, aggressive behavior is enhanced by yolk androgens in both sexes of adult house sparrows and black-headed gulls. At least in the house sparrow, the differences in adult aggression are not associated with differences in plasma levels of progesterone, testosterone, 5α-dihydrotestosterone, and 17β-estradiol providing strong evidence that differential exposure of the embryo to maternal testosterone permanently organizes neural pathways of aggression, that is, their sensitivity to hormones in adulthood, in both sexes.

However, maternal hormonal effects appear to also differ from classical hormonal organization of sex differences, unless there is sex-specific developmental exposure to maternal hormones, for which there is some evidence in birds. Organization of sex differences results from a lack of embryonic/fetal production of a certain hormone during a critical developmental phase in one sex, termed the 'default' sex (differentiation in the absence of the hormone). When this 'default' sex is experimentally exposed to the hormone during the critical phase, the behavioral phenotype is shifted toward the hormonally organized sex (in which the embryo does produce the hormone). The conventional terminology for such effects is 'masculinization' of females in mammals, where exposure to exogenous testosterone can induce male characteristics, or 'demasculinization' of males in birds, where exogenous estradiol can induce female characteristics (see also later). However, some traits, for example the badge of black feathers on the chin of male house sparrows which is absent in female house sparrows, can be increased by exogenous (maternal) testosterone in genetic males, but

cannot be induced at all in genetic females. This is inconsistent with the view that maternal androgens simply 'interfere' with sexual differentiation and 'masculinize' females. Rather, the observation points to an interaction between sex-limited genes and testosterone. Thus, on the one hand, maternal hormonal effects do resemble those of the classical vertebrate sexual differentiation model of brain and behavior. But, on the other hand, they hint at novel, yet-to-be-identified mechanisms and targets that result in individual variation of traits including behavior within sexes. Research is urgently needed to dissect these sex-linked and sex-independent mechanisms of long-term organizational actions of maternal hormones. Clearly, maternal hormones do not appear to just interfere with the basic hormonal mechanisms of sexual differentiation.

Why Do Maternal Hormones Not Interfere with Sexual Differentiation?

Natural levels of maternal androgens in the avian egg do not interfere with sexual differentiation as expected by the classical model of vertebrate sexual differentiation. Although this seems puzzling at first, there are simple explanations for this apparent paradox. First and sufficient is 'dosis facit venenum' – as noted in 1535 by Paracelsus. All experiments demonstrating a role of steroids (androgens and estrogens) in the sexual differentiation of brain and behavior of birds applied supraphysiological doses that often affected differentiation of the primordial gonad into an ovary or testes. Lower doses, still orders of magnitudes higher than those of maternal androgens and estrogens occurring naturally in avian eggs, were, however, ineffective to reverse the pathways guided (without the hormone) by genetic sex. Thus, the effects of variable concentrations of maternal androgens in the egg on many traits cannot be explained by interference with the basic processes of sexual differentiation with either females just being 'masculinized' or males being 'feminized' at the gonadal or secondary sexual trait level. Other explanations, such as different timing of hormone-regulated developmental processes, critical periods, metabolism and inactivation of maternal steroids by the embryo, different endogenous doses, and different hormones being involved, put forward to explain this apparent paradox might also apply but one of the most parsimonious explanations is dose.

Second, in birds, the steroid most closely associated with sexual differentiation of the gonad, the brain, and behavior is estradiol. Almost all experimental studies of maternal hormonal effects have been conducted with androgens, particularly with testosterone which can be converted to estradiol by the enzyme aromatase. Since male avian embryos do not seem to have high levels of aromatase activity, we do not expect testosterone to affect their sexual differentiation. Indeed, natural concentrations of estradiol in egg yolk are much lower than those of androgens, perhaps to avoid interference with estradiol-regulated components of sexual differentiation. In mammals, however, testosterone is the effective hormone for sexual differentiation (channeling phenotype into the male pathway), but again, experiments in mammalian sexual differentiation have used pharmacological and not physiological levels of testosterone.

Third, the last decades of research of sexual differentiation of brain and behavior experienced a paradigm shift changing focus from hormonal organization to direct effects of genes without hormones as intermediaries from gene to phenotype. This shift resulted among other findings from a failure to completely sex-reverse singing behavior and the sexually dimorphic neural song control system of songbirds by early steroid exposure. Indeed, evidence is rapidly accumulating that sexual differentiation of at least some sexually dimorphic behavioral traits and brain structures and functions does not require a hormonal link between gene(s) and trait differentiation. Moreover, both male and female genes and not only the absence of a certain gene in one sex and its presence in the other turn out to be important to determine sex differences in vertebrates. Consequently, the view of a default sex (no sex gene and no hormone in the homogamete) and an organized sex (sex gene and hormone in the heterogamete) became too narrow to understand the development of sex differences in adult traits including behavior. A re-evaluation of the gene-hormone-sexual differentiation hypothesis and relaxation of some of its stringent assumptions is warranted. A wider view of sexual differentiation mechanisms needs to accommodate classical and new results from sexual differentiation research as well as results of maternal hormonal effects. We propose that, rather than hormones guiding sex differences, they are modifiers of genetically regulated developmental pathways with maternal steroids acting to cause some of the within-sex, individual variation in traits (developmental plasticity). It is suggested that the cascade of events from gene(s) via hormones to phenotypes opens a window to the environment to induce variation in phenotypes. In comparison to classical sexual differentiation in which the hormone is seen as a developmental switch channeling differentiation into one of two pathways (the sexes), our view suggests hormones cause developmental modification along a continuum within these pathways. This hypothesis can be examined by identifying embryonic hormone targets and studying differential expression of sex-chromosomal and autosomal genes in response to physiologically, developmentally, and ecologically relevant doses of exogenous hormones.

From an evolutionary perspective, it might be the hormonal intermediary (providing a window to the environment) in the pathway from genes to phenotype that allowed for the evolution of adaptive maternal effects by

co-option of the basic hormonal pathway. It will be informative to investigate which traits include a hormonal link in their developmental differentiation and which do not. In this context, our next two fundamental questions – pleiotropy of maternal hormonal effects and indirect and direct effects – become relevant.

How to Explain the Multiple Effects of Maternal Hormones? Understanding Pleiotropic Actions, Integrated Phenotypes, Modularity of Traits, and Direct and Indirect Effects

One of the hallmark properties of hormones is their action on multiple targets. On the one hand, this allows for a coordination and integration of components that interact to produce a trait, but, on the other hand, this can cause tradeoff by antagonistic effects. For example, in the adult male vertebrate testosterone affects immune function, sexual and aggressive behavior, brain function, sperm maturation, metabolism, muscle mass, and sexual ornamentation. Similarly, the effects of maternal androgens in the avian egg on offspring phenotype are diverse (see section 'Abbreviated Review'). Are these multiple actions reflecting integrated maternal modification of offspring trait networks to enhance fitness? Or, are they reflecting constraints and tradeoffs by antagonistic pleiotropy that will set limits to the modification of a certain trait by the mother? A first step to differentiate between these alternatives is to identify the cellular targets of maternal hormones in the early embryo. For example, the presence of androgen receptors in certain brain structures such as the *Nucleus taenia*, the avian homolog of the mammalian amygdala and potentially involved in the regulation of aggressive behavior, and the *preoptic area*, involved in the regulation of sexual behavior, and in immune organs such as the *bursa Fabricius*, important in humoral immunity would suggest that these traits are affected as a suite and may not be modified independently by the mother. Combined with complementary studies of developmental hormone exposure on the function of target organs, such knowledge will lead to a better foundation for research concerned with adaptive functions. Both are currently not available. It also is important to consider dose–response relationships in such studies as these might differ among targets (traits) allowing for some flexibility.

Mechanisms likely limit maternal ability to adaptively modify offspring phenotype. The observed modifications by maternal steroids of multiple traits may result from an effect on a single system with indirect consequences on other systems and traits. For example, enhanced metabolism as a result of maternal testosterone might indirectly influence begging behavior, growth, overall activity levels, and immune function. Moreover, differential growth resulting from variation in begging performance could theoretically carry over into the adult phenotype with individuals in better condition during development becoming dominant over others and showing elevated levels of aggression. Therefore, the observed long-term effects of developmental exposure to maternal hormones on adult behavior could just be indirect with general consequences on the activation of begging or even developmental metabolism rather than representing organizational hormonal effects on specific behavioral modules. As a consequence, such a scenario would not require that a maternal hormone impacts the differentiation of certain brain structures underlying the expression of a behavior such as aggression in adulthood. As mentioned before, a wide array of behavioral effects may be caused by hormones acting on one specific property of the brain influencing many behavioral domains, for example, impulse control or sensitivity to environmental cues. From the work of L. Rogers and the group of T. Groothuis, there is some evidence that prenatal exposure to steroid hormones affect brain lateralization, and the latter is known to affect a whole suite of traits including perception, cognition, and motor behavior. Also here, the identification of the developmental targets of maternal hormones using molecular tools is essential to differentiate between these alternatives that have critical implications for consideration of maternal effects as adaptations.

Current research in behavioral biology shows great interest in correlated traits of behavior and physiology that seem to come as an integrated suite of diverse characters on which selection can act. Such suites of traits are now known as 'animal personalities,' 'coping styles,' 'behavioral syndromes,' or 'temperaments,' and have been demonstrated in many animal species. For example, the group of J. Koolhaas demonstrated that mice artificially selected for short or long attack latency differ in a wide array of other behavioral and physiological traits, such as adrenocortical stress response, androgen production, perception of environmental cues, nest building, and entrainment of the biological clock. Similar multiple trait differences were demonstrated by the group of P. Drent and T. Groothuis and their collaborators in great tits *P. major* selected for exploration strategy in novel environments. Are these phenotypic correlations the result of gene correlations (epistasis), gene pleiotropy, or pleiotropic effects of hormones during development? Quantitative genetic studies indicate large environmental, in fact maternal, components on trait variation such as sexual behavior in zebra finches *T. guttata* and personality in great tits. Recent research hints at an important role of maternal hormonal effects here.

Most exciting, the eggs of females of the great tit lines artificially selected for exploration strategy differ in their concentrations of maternal hormones. Japanese quail *Coturnix japonica* lines artificially selected for their behavioral stress responses differ in yolk corticosterone and those selected for social reinstatement behavior in yolk

androgen concentrations. Experimental manipulation of these yolk hormones affects ecologically relevant behavioral traits in the offspring, for example exploration of novel environments and adrenocortical stress response in domesticated Japanese quail, dispersal distance in free-living great tits, and social behavior in several species. Finally, personality affects fitness in natural environments. This raises two exciting and novel possibilities that have strong implications for evolutionary synthesis. First, selection for behavior coselects for a maternal effect mechanism (hormone exposure of the embryo). Second, adult behavior that was selected for is caused by differential developmental exposure to maternal hormones rather than by selection for certain behavior genes. Moreover, trait integration (functional correlation) in these phenotypes (e.g., behavioral style and physiological stress response) might result from the pleiotropic effects of maternal hormones on multiple systems during development rather than from genetic epistasis (correlated genes). Again, this hints at a modularity of maternal effects and the role of hormones in integrating module components.

Who Is in Control: The Mother or the Offspring, None, or Both?

The exposure of the embryo to maternal hormones has, as already mentioned, historically been considered a pathological epiphenomenon (i.e., insufficient protection of the embryo/fetus and leakage of maternal hormones, for example, across the placenta). With the current view of embryonic exposure to maternal hormones being an adaptive process and the finding that maternal hormone concentrations vary systematically in bird eggs, the pendulum has swung around and the predominant perspective in behavioral ecology is now that mothers 'control' of how much hormone the embryo is exposed to. Current evidence does, however, not justify this extreme view. We need fundamental physiological research to establish whether mothers have for example evolved a specific transfer mechanism that may regulate embryo hormone exposure independently from their own exposure to these hormones. While the absence of such a mechanism does not exclude adaptive effects of maternal hormones, it may generate physiological tradeoff between the effect of the hormones on the mother and those on her offspring.

Equally relevant here is the question to what extent the embryo is just a passive receiver of and responder to the hormonal signals of the mother, or whether it is able to modify its response in its own interest. This is especially important in the case of obvious parent–young conflict such as in the case of variation of maternal androgen concentrations among eggs of the same nest by which mothers may favor some siblings over others. Some have recently argued that maternal hormonal favoritism might not be evolutionarily stable because embryos might counter maternal 'manipulation.' For example, they might increase their sensitivity to the maternal hormone in brain areas that facilitate competitive behavior or increase their own production of the same hormone to make up for low levels of hormones provided by the mother. This argument is invalid for two reasons. First, it assumes that an embryo has information on the position in the laying sequence it is developing in, which is implausible. Second, it ignores that offspring countermeasures to maternal manipulation may have costs for the fitness of its siblings and the fitness of the mother, severely hampering this scenario. However, it is possible that embryos downregulate hormone sensitivity in certain organs such as those of the immune system to avoid concomitant immunosuppressive effects that might come with the beneficial effects of the maternal hormone, for example, on begging. Very recent studies in the groups of Groothuis (birds) and Bowden (turtles) suggest that embryos metabolize maternal steroid hormones already very early in development, but much more work is needed here before we will be able to decide in how far the offspring can play an active role in shaping its response to maternal hormones. Perhaps embryonic hormone conversion functions to enable the chick to regulate its exposure to the hormone, using the metabolites as a depot, avoiding detrimental effects of exposure to too high levels at vulnerable times.

How to Make Evolutionary Sense of Maternal Effects Expressed During Different Life-History Stages?

As already discussed, hormone-mediated effects are expressed both in early life and adulthood. We suggest that different selection scenarios favor actions on traits of these different life-history stages. Modifications expressed in early life-history stages such as those on begging are most likely shaped by intrafamily genetic conflict (parent offspring, siblings) over parental investment. Modifications becoming apparent in later stages, for example those on aggression and sexual behavior, in contrast, may be shaped by common interests of mother and offspring in competing for resources with nonrelated conspecifics. As explained in the previous section, in sibling competition, the interests of the mother are in conflict with those of the offspring, and therefore evolution of mechanisms in the offspring to counteract maternal hormonal manipulation might be expected based on theoretical considerations, but not occur because of constraint on their evolution. In competition with unrelated conspecifics, in contrast, the mother's and offspring's interests coincide and therefore co-evolution of maternal signaling mechanisms and offspring responses to the maternal signal as an integrated adaptive maternal effect can be expected. A comparison of intra-versus interfamily

variation in the intensity of the maternal signal (hormone concentrations) might hint at the relative importance of within- and among-family conflict in shaping these maternal effects. For example, larger variation among than within families in yolk testosterone concentrations might suggest that it is among-family conflict that drives the maternal effect. And consequently, one might predict that in such species offspring countermechanisms to maternal manipulation are unlikely to evolve. However, early and late effects may also be coupled, again begging for more mechanistic research.

Why Is Phenotypic Plasticity Relevant for Understanding Maternal Effects?

The causes of variation in existing phenotypes, the raw target units for selection, are still not well understood. Mutations cannot easily explain, for example, rapid adaptations to novel or changing environments. Maternal effects, in particular those that are environment based, could play an important role in the nongenetic induction of modified phenotypes that might be highly adaptive in changing or novel environments. This hypothesis puts a premium on phenotypic plasticity in two ways. First, developmental plasticity will determine in how far mothers can influence offspring phenotype both of early and later life-history stages. The concept of reaction norms that describes how variation of a certain environmental factor (e.g., the different concentrations of maternal hormones in the eggs of a clutch) modifies the phenotype produced from a certain genotype (family) might be useful here as, in physiological terms, reaction norms are dose–response relationships in different genotypes. Such studies, integrating tools and approaches of population genetics, developmental biology, and life-history theory are still very rare. Second, plasticity of the mother in her reaction to environmental cues will determine in how far she is able to modify the developmental environment of her offspring (e.g., hormone) to modify their phenotype in anticipation of a certain offspring environment. This requires integration of approaches of behavioral ecology, neuroethology, reproductive biology, and neuroendocrinology to study adult plasticity.

As addressed in the abbreviated review section, mothers, at least avian mothers, transfer different amounts of hormones to their embryos depending on a range of environmental factors. Recent research indicates that they also seem to be able to make such adjustments rather rapidly, within a few days. This suggests strong selection acting on maternal neuroendocrine systems that function to transduce environmental cues into variable exposure of the embryo to maternal hormones to achieve effects on offspring phenotype, often its behavior. Yet, we still know very little of how these egg hormone levels are regulated in response to cues experienced by the mother and how they reach the egg. These are two important pieces of mechanistic information whose lack severely hampers progress in understanding adaptive maternal effects in evolution.

Although comparative studies generated evidence that, for example, yolk androgen levels correlate with variation in selection pressures, for example, time-dependent offspring mortality by predation rates on eggs and nestlings, and that in turn embryo and nestling development rates correlate positively with androgen levels in eggs, we do not know if individual females can translate perceived probabilities of nest predation into differential deposition of growth-promoting hormones into eggs. Has the female brain and her neuroendocrine reproductive control system evolved under the selection of fitness consequences of maternal effects realized in the offspring?

Are Maternal Effects a Bet-Hedging Strategy?

Phenotypes are the targets of selection (natural and sexual) and the ability of females to vary the intensity or quality of the maternal effects can potentially allow them to 'create' multiple phenotypes within and among propagules. In birds, for example, this could be achieved by intraclutch and interclutch variation in yolk hormone concentrations. The existence of multiple phenotypes of siblings within a propagule might be particularly advantageous in unpredictably fluctuating environments, representing a form of bet hedging by the female that could ensure that at least some of the offspring fit an environment. Production of different offspring phenotypes among propagules produced over an extended reproductive period might also be advantageous when there are predictable short-term changes in ecological conditions. For example, adjusting the phenotype of offspring born later in the season by hormonally enhancing their propensity to aggressively compete for food resources as population density and the number of competing nonrelated juveniles are expected to increase could enhance reproductive success as well as survival of the offspring. Elevated yolk testosterone levels with progress of the breeding season and enhanced aggression of independent juvenile and adults in response to their exposure to high levels of testosterone in the eggs in some species such as the house sparrow lend support to this function.

Are Maternal Effects Epigenetic Transgenerational Effects?

As we have seen, maternal effects include modifications of development that carry through an organism's entire life span likely caused by permanent developmental influences on the expression of certain genes. Consequently, they can be viewed as epigenetic in the broad

sense of Waddington. Are they also epigenetic in the narrow sense that requires transmission of modifications in gene expression over several generations? Here, two sets of laboratory studies with rodents are instructive. The group of M. Meaney has demonstrated that style of maternal care epigenetically programs rat pups in the broad sense. Individuals raised by mothers that differ in licking and grooming of their pups differ in their adult adrenocortical stress response resulting from permanent, but reversible methylation of a promoter for the expression of the glucocorticoid receptor. Maternal parenting style also programs behavioral syndromes (modules) in both sexes, such as sexual behavior and behavior in open field tests that reflects motivational state in novel environments. Importantly, mothering style is affected by the maternal stress response system perpetuating the effect into the following generation. In this case, the epigenetic maternal effect is transgenerational, although transmission between generations is via maternal behavior rather than the transmission of epigenetic modifications of gene expression. Similarly, D. Crews, M. Skinner, and their collaborators found that in utero exposure of mouse fetuses to vinclozoline, a fungicide that interacts with the androgen receptor programs a suite of behaviors and traits of the adult offspring including their mate choice, sexual behavior, and anxiety-related behavior. Effects of in utero vinclozoline exposure also include a host of pathologies that are carried on, through the male germ line, for several generations. Thus, in this case, there is some evidence for the transmission of modification of gene expression across generations. Does such transgenerational transmission of nongenetic developmental modifications also hold for maternal effects that are mediated by naturally variable exposure to maternal hormones, such as testosterone in birds? This exciting possibility is in urgent need of investigation and has tremendous implications for the role of maternal effects in evolutionary trajectories.

From a functional evolutionary perspective, both narrow sense (transgenerational via modified gene expression) and broad sense (occurring only in the first generation) epigenetic maternal effects might differ dramatically in their consequences. Broad epigenetic effects will allow each individual mother (of subsequent generations) to gauge environmental conditions and adjust her offsprings' phenotypes, with the epigenetic marks being erased after each generation. Epigenetic maternal effects in the narrow sense, in contrast, would influence several generations of offspring limiting flexibility for individual mothers of successive generations to adjust offspring phenotype. It remains to be shown if maternal effects are transgenerational in nature, if so how they are transmitted and how long modifications last, which traits are affected, and how ancestral epigenetic marks are being erased.

Maternal Effects on Behavior as Evolutionary Force

Maternal effects on behavior may have a particularly strong impact on evolutionary processes, regardless of whether behavior drives evolution, as assumed traditionally, or whether it slows evolution, as proposed as a recent alternative hypothesis. After all, it is behavior that allows animals to compete and interact with conspecifics, explore and exploit novel environments, and respond to changes in their environment. Indeed, maternal hormonal effects often impinge on behavioral traits particularly those that regulate competitive and sexual behavior and responses to environmental change. Again, an instructive example is the aviary study by the group of H. Schwabl of house sparrows in which both males and females hatching from testosterone-injected eggs exhibited higher aggression and showed greater ability to obtain and defend a nest site. Although we presently do not know if such modifications of competitive behavior influence fitness in normal environments, it is likely that they do, for example when there is great competition for limited nesting sites. A related example comes from research by the group of T. Groothuis on black-headed gulls. Mothers breeding in colonies with high density produce eggs with higher concentrations of androgens. This enhances the chicks' ability to defend the territory when the parents are absent to forage. This is especially important at high breeding density because chicks from other broods may try to steal regurgitated parental food. A third instructive example is maternal hormonal effects on dispersal distance in great tits in relation to ectoparasite prevalence in the nest, as found by B. Tschirren, an observation suggesting potential implications for the co-evolution of host and parasites.

Such evolutionary considerations need, however, to be informed by knowledge of mechanisms. This can be illustrated with results from the house sparrow. Enhanced adult aggression by early testosterone exposure in the egg occurred without any differences in adult hormone levels, indicative of 'organization' of brain function in the classical neuroendocrine sense. By this mechanism, the aggressive potential of the offspring can be enhanced without increased adult testosterone levels possibly avoiding antagonistic pleiotropic effects such as immunosuppression and removing constraints that might impede evolutionary change via the maternal effect on behavior.

Is There Also a New Role for Behavioral Ecologists?

We have argued that more insight into the mechanism underlying hormone-mediated maternal effects is critical for understanding all four of Tinbergen's questions regarding these effects: their causation, function, evolution, and developmental plasticity. This requires input from for example neurobiology, endocrinology, embryology, and molecular genetics. There is an important role for behavioral ecology to play. First, the increasing appreciation for maternal effects as adaptations is not yet sufficiently supported by data demonstrating enhanced fitness consequences of such effects. Some data show positive effects on fitness-related processes such as early growth, but others show negative effects such as on immune function suggesting tradeoffs. Because of these potential tradeoffs, it is indispensable to obtain information on survival and reproductive success. Only very few studies have generated data on chick survival and as yet no study has demonstrated effects on reproductive success, let alone inclusive fitness. Second, if hormone-mediated maternal effects function to adjust offspring to specific environmental conditions, and if such effects generate tradeoffs in the offspring, their fitness benefits should depend on context, but context was hardly taken into account in studies, probably because it requires additional experimental groups and more complex experimental designs. Third, because of the potential of parent–offspring conflict, fitness consequences have to be analyzed separately for both the mother and the offspring. Thus, for the time being, the interpretation of mothers 'adjusting' their hormonal 'investment' in chicks by differentially 'allocating' hormones should be taken with caution. As long as we do not know whether mothers are able to actively modify hormones exposure of offspring and to what extent these hormones are costly for the mother or the young, such a perspective is unjustified.

See also: Development, Evolution and Behavior; Differential Allocation; Female Sexual Behavior and Hormones in Non-Mammalian Vertebrates; Neural Control of Sexual Behavior; Neurobiology, Endocrinology and Behavior; Sex Allocation, Sex Ratios and Reproduction; Spotted Hyenas.

Further Reading

Eising CM, Müller W, and Groothuis TGG (2006) Avian mothers create different phenotypes by hormone deposition in their eggs. *Biology Letters* 2: 20–22.

Gil D (2008) Hormones in avian eggs. Physiology, ecology and behavior. *Advances in the Study of Behavior* 38: 337–398.

Groothuis TGG, Carere C, Lipar J, Drent PJ, and Schwabl H (2008) Selection on personality in a songbird affects maternal hormone levels tuned to its effect on timing of reproduction. *Biology Letters* 4: 465–467.

Groothuis TGG, Müller W, Von Engelhardt N, Carere C, and Eising CM (2005) Maternal hormones as a tool to adjust offspring phenotype in avian species. *Neuroscience and Biobehavioral Reviews* 29: 329–352.

Groothuis TGG and Schwabl H (2008) Hormone-mediated maternal effects in birds: Mechanisms matter but what do we know of them? *Philosophical Transactions of the Royal Society B: Biological Sciences* 363: 1647–1661.

Groothuis TGG and von Engelhardt N (2005) Investigating maternal hormones in avian eggs: Measurement, manipulation and interpretation. *Annals of the New York Academy of Sciences* 1046: 168–180.

Martin TE and Schwabl H (2008) Variation in maternal effects and embryonic development rates among passerine species. *Philosophical Transactions of the Royal Society of London Series B: Biological Sciences* 363: 1663–1674.

Mousseau TA and Fox CW (eds.) (1998a) *Maternal Effects as Adaptations.* New York, NY: Oxford University Press.

Müller W, Lessells C, Korsten P, and Von Engelhardt N (2007) Manipulative signals in family conflict? On the function of maternal yolk hormones in birds. *The American Naturalist* 169: E84–E96.

Partecke J and Schwabl H (2008) Organizational effects of maternal testosterone on reproductive behavior of adult house sparrows. *Developmental Neurobiology.* Published Online: DOI:10.1002/dneu.20676.

Rhen T and Crews D (2002) Variation in reproductive behavior within a sex: Neural systems and endocrine activation. *Journal of Neuroendocrinology* 14: 517–532.

Schwabl H (1993) Yolk is a source of maternal testosterone for developing birds. *Proceedings of the National Academy of Sciences of the United States of America* 90: 11446–11450.

Schwabl H, Palacios MG, and Martin TM (2007) Selection for rapid embryo development correlates with embryo exposure to maternal androgens among passerine birds. *The American Naturalist* 170: 196–206.

Sockman KW, Sharp PJ, and Schwabl H (2006) Hormonal orchestration of avian reproductive effort: Toward an integration of the ultimate and proximate bases for flexibility in clutch size, incubation behaviour, and yolk-androgen deposition in birds. *Biological Reviews* 81: 629–666.

Weaver IC, Cervoni N, Champagne FA, D'Alessio AC, Sharma S, and Seckl JR (2004) Epigenetic programming by maternal behavior. *Nature Neuroscience* 7: 847–854.

Mating Interference Due to Introduction of Exotic Species

A. Valero, Instituto de Ecología, UNAM, México

Exotic Species

Exotic species have been intentionally introduced world-wide for economic purposes or for biological control. However, accidental introduction has been common too. In any case, a frequent outcome of introductions is the extinction of the exotic species. If, however, exotics are able to adapt to novel conditions, a variety of outcomes is expected. In some cases, exotic and native species may reach a stable equilibrium if resources are abundant and competitive interactions between the two are sufficient to maintain them close to equilibrium. In others, interactions between exotics and natives may result in competitive exclusion and displacement of the native species.

For example, Mauritius Island was once home to a variety of endemic geckos of the genus *Nactus*. After the introduction of the house gecko (native to southeast Asia), *Hemidactylus frenatus*, most populations of *Nactus* geckos were displaced to the surrounding islets, including the night gecko, *Nactus soniae*, a species that was described only last year. The ability of house geckos to aggressively occupy refugia already taken by endemic geckos was responsible for such displacement. In the only islet where house geckos coexist with the night gecko, the night gecko possesses a superior ability at gripping the powdery tuff rock substrate of this Mauritian islet.

Displacement from preferential sites does not always take place aggressively. Red-eared sliders, *Trachemys scripta elegans*, are present in the pet trade in virtually any location. In Europe, these popular turtles have been introduced into aquatic ecosystems where freshwater turtle diversity is lower than in the native habitat of the red-eared slider. Therefore, native species, such as the European pond turtle, *Emys orbicularis*, which are commonly the dominant species in a given pond, have gradually been displaced from preferred basking sites by red-eared sliders. In such cases, the native turtles simply avoid basking spots already chosen by the exotic species, with a subsequent reduction in heating efficiency that may negatively impact the other daily activities of *E. orbicularis*.

Indirect feeding competition is another mechanism responsible for the decrease of native populations after introductions of exotic species. In Hawaii, the introduced house gecko (*H. frenatus*) feeds on the same insects as the native gecko *Lepidodactylus lugubris*. However, the superior ability of *H. frenatus* to deplete insects rapidly, especially when these occur in clumps, mean reduced acquisition rates for *L. lugubris* and, in turn, imply reductions in body condition, fecundity, and survivorship of the native gecko.

Although competition may take place over feeding resources or space (refugia, sun-basking locations, or nesting places), it may promote other kinds of interspecific interactions that are also detrimental to the native species.

Mating Interference

Mating is a biologically universal process. Successful mating depends, in part, on the proper functioning of premating and postmating reproductive barriers and also on the successful fusion of gametes to produce an embryo. Postmating mechanisms take place during or after the release of gametes and include biochemical recognition (or rejection) of gametes, and premating mechanisms comprise the production of species-specific visual, chemical, or acoustic signals expressed during courtship rituals aimed at achieving mate recognition and/or successful release of gametes. For example, the acoustic signals produced by frogs of different species that live in sympatry have distinct acoustic properties used by females to orient to conspecific males. In some nocturnal moths, females produce pheromones that attract only conspecific males. Male courtship in several species is a set of species-specific movements that allow the display of particular visual signals (color or size of secondary sexual characters) that are attractive only to conspecific females. The male vermillion flycatcher (*Pyrocephalus rubinus*) performs a 'nuptial flight,' a series of complex movements in front of and at the back of the female that allow it to display in full its bright red color. Although there are several bird species that also display red plumage in the flycatcher's habitat, conspecific male courtship is needed for the female to accept copulations.

However, premating mechanisms are a relatively weak barrier and do not completely prevent interspecific mating from taking place. In addition, some environmental factors may hinder mate recognition: man-made or natural noise may prevent the transmission of acoustic signals; water turbidity may interfere with the reception of visual signals; or water pollution may make chemical mate recognition confusing. Even when the transmission of mate recognition signals is working appropriately, courtship persistence may have been favored in some organisms, causing some species to mate or attempt mating with biologically different, yet attractive species. Incomplete mate recognition leads to mating attempts, which amount to mating interference and may carry associated costs to one or both species.

Anthropogenic effects may promote mating interference: mate recognition is reduced due to chemical interference as a result of the disposal of chemical substances into rivers; female swordtail fish, *Xiphophorus birchmanii*, fail to show preference for either conspecific males or heterospecific *Xiphophorus malinche* males when they are tested in water from a river subject to sewage effluent and agricultural runoff, while they show preference for conspecific over heterospecific males in spring or tap water. Eutrophication of aquatic systems leads to incomplete mating isolation and/or hybridization that alters sexual selection; for example, eutrophication of Lake Victoria, the largest of the Great Lakes of Africa, has resulted in the loss of diversity of cichlid fishes; given that male coloration is costly to produce and visibility is greatly reduced under turbid water conditions, sexual selection has relaxed and male coloration has decreased. Increased anthropogenic noise interferes with the acoustic communication of some insects, amphibians, and birds: female treefrogs (*Hyla chrysocelis*) in Minnesota take longer to orient toward a speaker playing a male's song that is being broadcast along with traffic noise, than when the same stimulus is played back without masking noise. Introduction of exotic species due to accidental release, planned cultivation for commercial purposes, or biological control measures, can impact the survival of native species as in the case of introduced Trinidadian guppies (*Poecilia reticulata*) that sexually harass endemic fish of the Mexican family Goodeidae, like *Skiffia bilineata*.

Mating interference is not exclusively associated to anthropogenic activities that have mobilized species at a global level (accidental transport of continental species to islands on board ships; planned transport of exotic species to colonize habitats along with humans, or to be sold as pets or utility animals). It is also the result of natural dispersal. Natural dispersal may be spontaneous or gradual, depending on the conditions before dispersal: spontaneous dispersal occurs in association with drastic climatic phenomena such as hurricanes or floods. During these events, many plant and animal species travel hundreds or thousands of kilometers, eventually stopping at novel locations where they either perish, or are faced with completely new living conditions where mating interference may take place. Gradual dispersal is more frequent than spontaneous dispersal and occurs as part of the life cycle of many species. Mating interference occurs in several species regardless of phylogenetic relationships, that is, genetic incompatibility between species is not a guarantee against interference. If interference occurs between exotic and native species, it may be of concern for the conservation of endangered species.

Cases of Mating Interference in General

Mating interference in animals has been extensively documented and different types of mating interference

recognized, depending on the stage of mating at which interference takes place: before, during, or after mating.

Premating interference may take place if the signals used in mate attraction do not reach the individual that has to be attracted because heterospecific signals emitted simultaneously prevent this type of communication to be completed. A summer night in any humid forest is bathed with the choruses of many sympatric frogs calling out for a suitable mate. For any of these frogs, distinguishing the right mate is complicated given the similarity among simultaneous calls; thus, such auditory masking can lead individuals to mistake heterospecifics for mates. Chemical signals used in courtship may also get jammed by heterospecific signals and cause males of some butterfly species to be attracted to heterospecific females, as is the case in male butterflies of *Heliothis zea* exposed to a synthetic pheromone that mimics that of *Heliothis virescens*.

Sometimes, when heterospecific males resemble conspecifics, males may engage in rivalry, displaying territorial behaviors to chase away individuals they have mistakenly recognized as intruders. Individuals engaging in territorial interactions incur substantial costs because time and energy and nutrients are wasted. Thus, this also amounts to mating interference. An interesting case is that of the amberwing *Perithemis tenera*. Although dragonflies usually compete for resources with amberwings, males chase away horse flies of the genus *Tabanus* and butterflies like *Ancyloxypha numitor*, which resemble conspecifics, instead of the intruding dragonflies!

A third type of premating interference may arise when courtship is erroneously directed at heterospecifics. Note that misdirected courtship is different from signal jamming: while the latter prompts a modification of courting rituals in order to reach the target organism, or in some cases, it means that courtship activities are halted, the former does not prevent courtship from taking place. Misdirected courtship is the result of various factors; mate recognition traits among different species may overlap, or heterospecifics may resemble high-quality mates. In any of these cases, the courting individual will direct courtship effort toward the 'wrong' potential mate, and not toward a conspecific. In some islands of the Pacific, misdirected courtship occurs between invasive male *Hemidactylus frenatus* geckos and the larger females of the native *H. garnotii* geckos. Experiments with grasshoppers have also shown that male *Tetrix cepero* prefer to court the larger heterospecific *Tetrix subulata* females even when these reject them. Courted individuals may incur in costs due to wasted time and energy in rejecting approaches from unwanted males. In Mexico, males of the introduced Trinidadian guppy *P. reticulata* court conspecific females preferably, but also approach and attempt mating with females of the native *S. bilineata*, a distantly related fish whose appearance resembles that of a large *P. reticulata* female. There is a cost associated to such interactions: when females of this species spend their pregnancy with exotic males, their growth is slowed down in comparison to when conspecific males are around, suggesting that a compromise in resource assignment is established. In this matrotrophic species, the developing offspring may also suffer the consequences of reduced nutrient availability.

Finally, a fourth type of premating interference has to do with females erroneously choosing heterospecific males over conspecific males. Note that this is different from misdirected courtship in that the sex that exerts choice (in the majority of cases it will be the female), and not the one that courts, is responsible for the 'wrong' decision. As in the case of misdirected courtship, erroneous female choice may arise from an overlap in mate selection traits, from female insensitivity to differences in mate selection traits, or to sensory biases already present in females. Sensory biases are behavioral traits that evolved in contexts different to mating, but can, due to co-evolution of male–female mating traits, be adopted in mating rituals. For example, the freshwater fish *Ameca splendens* preys on insect larvae; male *A. splendens* show a terminal yellow band in the caudal fin that, when in movement, resembles a larvae. Because of a sensory bias, females are immediately attracted to such movements, facilitating female approach to the male, and thus, courtship.

Interference that takes place during mating may occur if males are indiscriminate about which female to mate with, and skip courtship behaviors; it may also take place if one sex is not able to reject forced copulations from heterospecific mates. In the spider mites *Panonychus mori* and *Panonychus citri*, male *P. citri*, unlike its congeneric *P. mori*, guard females indiscriminately, resulting in equal chances of copulating with conspecifics and heterospecifics. Female *P. mori* do not reject males. Heterospecific copulation is completed earlier than in intraspecific matings, and although a small proportion of heterospecific gametes are fertilized, all die during development or at the larval stage. Interspecifically, mated females of *P. mori* refuse subsequent males, which means that the operational sex ratio becomes biased toward males, a situation that could drive sexual selection. However, eggs of *P. citri* females can be fertilized successfully with conspecific sperm from subsequent males, and as a result, the reproductive rate of *P. citri* males is higher than that of *P. mori*. Taken together, the evidence suggests that interference is more intense for *P. mori* than for *P. citri*, and perhaps it is this that has driven some populations of *P. mori* to local extinction even in areas of their distribution with more favorable climatic conditions.

Finally, postmating interference takes place if postmating isolation mechanisms are poorly developed and fertilization of heterospecific gametes takes place with the subsequent production of an embryo. This phenomenon, which has been studied extensively, is called 'hybridization.' For a long time, biologists viewed hybridization

as a negative consequence of contact between different species, because it lowered the fitness of hybrids, as well as the fitness of the hybridizing individuals. However, recent evidence suggests that hybridization does not always have a negative impact on hybrids produced, and that in a few cases, it may even increase hybrid fitness. Such hybrid vigor depends on the phylogenetic relationships of the hybridizing species (whether they are closely or distantly related, and thus, genetically compatible), the amount of time they have spent in secondary contact (i.e., the amount of time needed for natural selection to act), and the life cycles of both species (new variants subject to selection are produced more rapidly in species with short life cycles).

Mating Interference During Biological Invasions

The presence of alien or exotic species in a given ecosystem poses a threat to the survival of native species when competition for food and space also occurs. In addition, exotic species may interfere with the reproduction of the native species at various stages of reproduction, depending on their phylogenetic relationships. Congeneric exotic species may impose costs due to hybridization, in addition to the costs mentioned here. More distantly related species that are genetically incompatible with native species may interfere with mate recognition processes, or female choice, and may even modify the reproductive schedules of native species.

For example, in Mexico, the introduced Trinidadian guppy (*P. reticulata*, Poeciliidae) males court, and attempt copulation with, females of the native *S. bilineata* (Goodeidae). Although male guppies do not preferentially court heterospecific females, they approach and court them regardless of how many conspecific females are present at any time. This suggests that mate recognition barriers are not sufficient to overcome the persistence of male guppies at mating attempts. A more detailed examination has revealed that females of both viviparous species are morphologically very similar; in fact, some heterospecific females look even more attractive to the guppy eye than conspecifics, given their large size, which in guppies, is a sign of fecundity the males have evolved to respond to.

The presence, at a crucial stage of reproduction, of the exotic species may be detrimental to the female or to the development of the offspring. Female *S. bilineata* that have completed their pregnancy in the presence of courting guppy males do not increase body weight or size as rapidly as females that completed their pregnancy in the presence of conspecific males. This could be a result of reduced feeding rates due to behavioral inhibition or to time wasted in the avoidance of male harassment. Such an impact on growth not only affects the female but also the developing offspring in this viviparous species.

Guppies show remarkable abilities at learning the identities of conspecifics and make adaptive decisions based on their association preferences with certain individuals. Although in the laboratory, male guppies seem to desist from approaching unresponsive heterospecific females after some time, they also learn to recognize novel heterospecific females and court them preferentially. Thus, under natural conditions, the renewal of the heterospecific population could lead to a constant rate of male guppy approaches to novel females, with negative consequences to their activities and growth.

Competitive interactions or mating interference between exotic and native species may lead to a variety of outcomes. The novel environmental conditions experienced by exotic species may facilitate the expression of morphological or behavioral traits that have positive consequences on their fitness. However, native species may suffer niche displacement and an increase in extinction probability if they cannot adapt to the novel environmental conditions that niche displacement imposes on them. Exotic species may also evolve in response to the new set of selective pressures encountered after dispersal, resulting in shifts that allow them to coexist in equilibrium with, or to be more successful at the expense of the native species.

In some cases, life-history strategies of endemic and native species may prevent or reverse these outcomes; for example, native species may be at an advantage over exotics if they reproduce continually throughout their lifetime (iteroparous) and the exotic species are semelparous (individuals that reproduce once or very few times), because they will produce more offspring per year and will increase their chances of survival. Species that reproduce earlier in the year will also have an advantage over those that reproduce later, and may even temporarily avoid the negative consequences of interspecific interactions, specifically those resulting from mating interference.

Differences in resistance to drastic temperature changes may also influence the outcome of interspecific interactions. Species that are introduced from a tropical to a temperate habitat may find themselves at a disadvantage with respect to native species because they will not be able to reproduce early in the year. However, if they reproduce rapidly and several times throughout the year, and enough time passes without any other drastic selective pressure acting, they will have increasing chances to adapt to temperate periods, invading a novel niche that may result in an increase in fitness and possibly, a new displacement event for native species.

Conservation and Management

Methods for the control of exotic species can be aimed at prevention, detection, and/or eradication (after invasion has progressed).

Preventive measures include the application of demographic models of population growth and expansion to potential invasive species that may enter an area continuously. These models, in turn, could be used to create regulations aimed at controlling the entry of exotics. Prevention can also be achieved through the identification of potential sources of exotic species introduction (e.g., movement of commercial ships through different areas, or transportation of goods that are known to be associated to other organisms). Detection of introduced species requires some knowledge of recent transport of exotic species carriers in order to make a search more effective.

Eradication measures are aimed at the complete removal of the introduced species. Methods include the use of chemical treatments, manual removal, or biocontrol. The success of eradication depends on several factors such as availability of resources to implement it, maintenance of a sustained eradication effort to prevent reinvasion, early detection of introduced species, resilience of species to eradication or even their ability to adapt to novel conditions imposed by eradication efforts (e.g., resistance to insecticides), and the time that has passed since the introduction took place. Examples of successful and unsuccessful eradication procedures are abundant in the literature. Eradication should be oriented to failure prevention, rather than to absolute success, but to achieve this, alternative strategies to counter potential failure should be thought in advance. A thorough knowledge of the ecology and behavior of the species for which eradication is being considered is the best help in planning strategies, making them efficient, and preventing failures.

See also: Anthropogenic Noise: Implications for Conservation; Male Ornaments and Habitat Deterioration.

Further Reading

Carroll SP and Fox CW (2008) *Conservation Biology. Evolution in Action*. New York, NY: Oxford University Press.

Elton CS (1958) *The Ecology of Invasions by Animals and Plants*. Chicago, IL: The University of Chicago Press.

Groning J and Hochkirch A (2008) Reproductive interference between animal species. *The Quarterly Review of Biology* 83: 257–282.

Mooney HA and Cleland EE (2001) The evolutionary impact of invasive species. *Proceedings of the National Academy of Sciences of the United States of America* 98: 5446–5451.

Sax DF, Stachowicz JJ, and Gaines SD (2005) *Species Invasions. Insights into Ecology, Evolution, and Biogeography*. Sunderland, MA: Sinauer Associates.

Mating Signals

H. C. Gerhardt, University of Missouri, Columbia, MO, USA

Introduction

Signals used to attract or stimulate prospective mates are representative of the most spectacular and bizarre behaviors observed in nature. Mating signals are especially important for understanding evolutionary theory, because they often have the dual function of attracting potential mates and warning or repelling rivals and because they are subject to multiple and often conflicting forms of selection. For example, conspicuous long-range signals that effectively attract mating partners are potentially costly in several ways. First, signals are often energetically costly to produce. Second, signal production often precludes feeding and other maintenance activities. Third, conspicuous signals often attract reproductive competitors, including sexual parasites such as satellite males, which do not signal but instead try to intercept prospective mates attracted to the signaler. Finally, as emphasized by Darwin, conspicuous signals may attract predators and parasitoids, and signaling behavior may reduce the time during which a signaler is vigilant for predators, thus lowering the signaler's chances of survival. One counter-measure that may avoid or mitigate the negative effects of illegitimate receivers is to switch to less conspicuous, 'courtship' signals once a prospective mate has been detected nearby.

Another theoretical consideration regarding the evolution of mating signals is the reciprocal selection operating on both signalers and receivers, which can constrain the degree of change in signals and in the criteria used to decode them. In general, prospective mates will be selected to attend to signals or properties of signals that are honest advertisements of the signaler's fitness or direct benefits that may be provided such as parental care. But mates that are too demanding risk not finding a mate at all unless the operational sex ratio (ratio of the numbers of reproductively active individuals of each gender) is highly skewed toward signalers.

Operational sex ratios can also help us understand sex-specific investment in signals. In general, females are the choosing gender because their potential reproductive output is less than that of males: they produce fewer, more costly gametes and are more often burdened with parental care. This theoretical viewpoint is supported by the fact that in most sexually dimorphic species, the male is extravagantly ornamented, produces conspicuous mate-attraction signals, or both. The situation is reversed in some species in which males provide most or all of the parental care.

Divergence in mating signals is a hallmark of speciation; indeed, many cryptic species, which are difficult to distinguish by external appearance, were first detected by differences in mating signals. Mating with a member of another species is probably the biggest mistake an individual can make, and so the direction of changes in mate-attraction signals may be limited by the properties of the mate-attraction signals of other species in the same community. As discussed below, this kind of selective pressure can also lead to differences in mate-attraction signals between populations within a species depending on the presence or absence of a closely related species with similar signals. Such a geographical pattern is termed 'reproductive character displacement.'

Approaches to the Study of Mating Signals

Physical Description

The starting point for describing an animal mating signal is to identify its physical properties and their typical range of values. One reason is that within each sensory modality, some animals produce signals that cannot be detected by humans, and even if detectable, differences between the sensory systems of the animal in question and humans are bound to result in differences in perception. Some common examples are the ultraviolet patterns of some butterflies and birds, the infrasonic sounds (frequencies below 30 Hz) produced by elephants, and of course, the ultrasonic signals (frequencies above 20 kHz) of many insects, rodents, and bats. Other animals communicate with vibrational signals that travel through plants or underground and with weakly electric signals that travel underwater. Special equipment is needed to detect and characterize these kinds of signals.

Each property or combination of properties of a signal has the potential to convey information about the signaler to a prospective mate or rival. Researchers must describe these properties objectively and assess their interrelationships and variability. Not only is the sensory world of each species unique and different from that of humans, but also the kinds of messages that will be important to each species will depend on its ecology and evolutionary history. Hence, field observations of the behavior of both the signaler and receiver after the production of a signal are necessary to establish that a signal serves to promote mating and which particular properties of such signals have the largest impact on mating decisions.

Patterns of Variation in Different Properties of Mating Signals

One generalization about mating signals is that some properties of a mating signal – sometimes termed 'static' – are highly stereotyped within an individual. Visual patterns and colors are familiar examples, but some patterns of movement in visual displays and the frequency spectrum or fine-scale temporal ('pulse') pattern of acoustic signals are also consistent enough to qualify (**Figure 1**). Some static properties also show low variability at the level of individuals (i.e., *between* individuals) within the same population, perhaps caused at least in part by stabilizing selection imposed by prospective mating partners because of the high costs of hybridization. A pattern in which low variability within and between individuals within populations is coupled with significant variation among populations across the species' range of distribution may reveal the potential for speciation. Other static properties that show relatively high variation between signalers in the same population may be useful for identifying individuals; if subject to sexual selection, such properties are likely to be under either stabilizing or directional selection by prospective mating partners.

Other properties of mating signals – sometimes termed 'dynamic' – vary considerably within individuals and may be useful for assessing an individual's physical or genetic fitness. Dynamic properties such as the duration or rate of displaying or calling are familiar examples that are often correlated with the energetic costs of signaling (**Figure 2**). In some species, within-individual variation is so great relative to between-individual variation that the current value of a dynamic property merely reflects the current condition of the signaler, such as whether it has been successful in foraging, rather than its inherent genetic quality. In other species, between-individual variation is sufficient to distinguish among individuals, and the current value of a dynamic property may indicate both current condition and genetic fitness. Dynamic properties are typically under moderate to strong directional selection, and prospective mates often strongly reject signalers whose display duration and rates are at the low end of the range of variation (**Figure 2**).

While particular properties of mating signals and even particular ranges of values of those properties strongly affect the preferences of prospective mates, these results rest largely on experiments in which the value of one property of a signal is varied while those of all or most

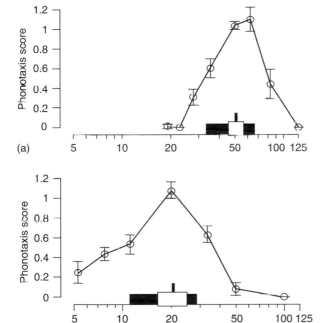

(a)

(b) Pulse rate (pulses/s)

Figure 1 Pulse-rate, a static acoustic property, and female response functions in two cryptic species of gray treefrogs: (a) *Hyla chrysoscelis*; (b) *H. versicolor*. These frogs can be distinguished only by differences in their advertisement (mating) calls and chromosome number. Bars on the X-axis show the mean pulse rate at 20 °C (vertical lines), the standard deviation (white boxes) and the total range of variation over all breeding temperatures (black boxes). Phonotaxis scores (open circles) show the response time of females to synthetic calls with different pulse rates relative to the response time to mean pulse rate for the species. Error bars are standard deviations. Note that females responded best to pulse rates at or close to the mean value and less well to lower and higher values. Reproduced from Bush SL, Gerhardt HC, and J. Schul. (2002) Pattern recognition and call preferences in treefrogs (Anura: Hylidae): a quantitative analysis using a no-choice paradigm. *Animal Behaviour* 63: 7–14.

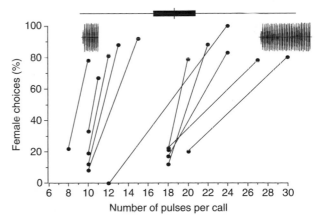

Number of pulses per call

Figure 2 The number of pulses per call (call duration), a dynamic acoustic property, and female choices in the gray treefrog (*Hyla versicolor*). The vertical line indicates the mean, the black boxes, the standard deviation, and the horizontal line, the range of variation among males. Oscillograms show synthetic calls at the two ends of the range of values tested. Points connected by lines show the percentages of females that chose each of two alternatives that differed only in the number of pulses per call. Note that in every test the majority of females choose the longer of the two calls. Reproduced from Gerhardt HC, Tanner SD, Corrigan CM, and Walton HC (2000) Female preference functions based on call duration in the gray treefrog (*Hyla versicolor*). *Behavioral Ecology* 11: 663–669.

other properties are held constant (**Figures 1** and **2**). On the one hand, this is a powerful way to demonstrate the relevance and form of selection (e.g., stabilizing or directional) by receivers. On the other hand, signals of different individuals differ in the values of several to many properties, and prospective mates must usually base their final mate-choice decisions on some kind of overall assessment of the signal. Recent experiments have begun to assess the effects and interactions of varying multiple properties of mating signals in crickets and frogs. The results of these studies are not readily comprehended because they rely on complex statistical analysis and modeling.

A more easily understood example involves two properties of the same signal, whose biological significance is well known and for which patterns of female preference have been estimated in isolation (**Figures 1** and **2**). The overall message is that the relative importance of these two properties to females differs depending on the biological community in which they exist. As discussed earlier, the pulse rate of the advertisement call is under stabilizing selection by female mate choice in both species; the difference in these properties is the only reliable indicator of species identity (**Figure 1**). The duration of the advertisement call is a dynamic property, which overlaps completely between the calls of the two species. This property indicates the energetic costs of calling and genetic quality of the signaler and is under strong directional selection by females of both species. When females of one of these treefrog species were offered a choice between long-duration calls with a pulse rate outside the range of variation (and more similar to that of the other species) and short-duration calls with a pulse rate typical of conspecific males in the same population, their choices depended on whether females were collected in populations where both species occurred (sympatry) or from populations where the other species was absent (allopatry). Females from sympatric areas nearly all preferred the short call with the correct pulse rate, whereas females from allopatric areas did not show a preference.

Signaling Environment

Another set of constraints on the production of mating signals involves the physical and biotic environments and the timing of signal production during the daily cycle. These factors can affect the choice of signal modality, the values of the physical properties of a signal, and the location from which signals are produced. For example, unless an animal can produce its own source of light, visual signals depend on light from the sun or moon. Depending on their sound frequency, acoustic signals propagate most effectively when an animal calls from an elevated position in most environments. Chemical signals are of limited use for long-range communication unless they can be broadcast into a wind or current, and

even then the chances of a signal's reaching prospective mates are uncertain. Many animal species ameliorate these constraints by producing signals in more than a single modality and by changing the values of particular properties according to both the physical and biological circumstances.

Major Sensory Modalities

In the next sections of this article, the main signaling modalities will be considered. Within each subsection, the following topics will be discussed: (1) The physical nature of each class of signals; (2) the advantages and disadvantages of long-range communication in particular environments; (3) energetic and other constraints on signal production; (4) modifications of signaling behavior that minimize exploitation by reproductive competitors and predators; and (5) how prospective mating partners evaluate variation in signals.

Chemical Signals

Chemical signals transmitted from a signaler to a potential mate are termed pheromones, a term based on a Greek word that means 'to carry excitement.' Pheromones can also mediate aggressive interactions between members of the same gender. Chemical signals were probably the first signals to evolve, since chemical messages are universal within and between cells; their efficient use outside of an animal requires the further evolution of structures to store and expel them into the environment.

Nearly all chemical signals are organic compounds. If transmitted by air, pheromones are limited in size to a molecular weight of about 300 and contain a maximum of about 20 carbon atoms. This is the upper limit on volatility in air; the lower limit is about five carbon atoms because smaller molecules are not only likely to be too volatile but they also limit the signal diversity. Much larger compounds can be used if pheromones are transmitted by water or direct contact between animals.

Chemical signals are transmitted directly to receivers, in which they directly cause changes in biochemical activity within single cells. These changes often take place in specialized chemosensory cells and result in the depolarization of cell membranes and the generation of nerve impulses in the same or adjacent cells. Some pheromones, particularly those used in mate attraction, have highly specific effects on narrowly 'tuned' chemoreceptors. One example is bombykol, a pheromone of the silkworm moth: a single isometric change in this molecule can render this substance ineffective. Because chemoreceptors are generally much less specific, the relative excitation of different receptor types is the key to signal identification, and the situation is even more complicated because even lower animals often produce pheromones that consist of blends of several to many different compounds. Cross-species attraction by pheromones and pheromone mixtures has often been documented in the laboratory, but mismatings are rare because different species produce sexual pheromones in different habitats, different times of the diurnal cycle, or both.

The limited range of most chemical signals is attributable to the fact that simple diffusion is inefficient and slow, and transmission via wind and water currents is uncertain (**Figure 3**). For example, strong winds result in the lateral dispersion of pheromone molecules, and even if the velocity is optimal, there is no guarantee that receptive mates will be located downwind. Another disadvantage of chemical signals is that if their effects are long-lasting on receivers, the signaler has little scope for

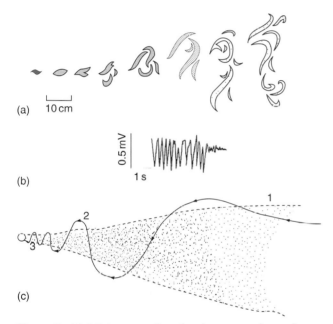

(a) 10 cm

0.5 mV

1 s

(b)

(c)

Figure 3 (a) Artists conception of a pheromone plume of a female moth. Adapted from Connor WE, Eisner T, Vander Meer RK, Guerrero A, Ghiringelli D, and Meinwald J (1980) Sex attractant of an arctiid moth (*Urthetheisa ornatus*): A pulsed chemical signal. *Behavioral Ecology and Sociobiology* 7: 55–63; Farkas SR and Shorey HH (1974) Mechanisms of orientation to a distant pheromone source. In: Birch MC (ed.) *Pheromones*, pp. 81–95. North Holland: Amsterdam. (b) Electrical response of the chemosensory organ of a male moth situated about 6 cm downwind from a scenting female. The rhythmic changes in neural activity reflect the temporal pulses of scenting by the female, but similar, more irregular activity would be observed further downwind even if pheromone production were not pulsed because of the scatter of molecules in the wind. (c) Hypothetic path of a moth locating a pheromone source by first flying upwind and then flying crosswind when losing the trail and reversing the crosswind direction when overshooting. Reproduced from Gerhardt HC (1983) Communication and the environment. In: Halliday T and Slater P (eds.) *Communication and Social Interactions*, pp. 82–113. Oxford: Blackwell Scientific Publications.

changing the message by changing to another signal. For example, a persistent alarm pheromone could cause inappropriate behavior that interrupts mating long after the predator that elicited the signal has left the area. A long-duration signal can, however, be advantageous in marking a large territory, both in terms of discouraging potential rivals and perhaps in attracting prospective mates. In a classic paper, Bossert and Wilson showed how variation in quantity of a pheromone released (Q) relative to the sensitivity (K) of potential recipients can affect the dynamics (extent and duration) of the 'active space,' the area in which a pheromone can be detected by a receiver, and the 'fade-out time,' the time at which the concentration of a pheromone remains above the receiver's detection threshold. Long fade-out times are useful for marking large territories, but make it difficult to change rapidly from one signal to another.

Visual Signals

Visual signals usually involve distinctive patterns of movement, which are usually termed displays. Static properties such as color patterns, patterns that contrast with their background, and elaborate body parts (**Figure 4**) are even more conspicuous when combined with particular movements that show off the pattern. The potential combinations of patterns and movements are endless, and multiple signals can be produced in rapid sequences. Flight displays and displays by lekking birds are especially spectacular in this regard. In general, some movement is required to elicit robust responses in visual receptors and interneurons, which are also sensitive to contrasts between dark and light objects.

To avoid unwanted detection by predators, animals can often hide conspicuous color patterns and remain still. Predation is also reduced because bright plumage and other sexual colors and patterns are usually developed only during the breeding season. Of course, some animals that are protected by toxic secretions or venom (and their mimics) often display gaudy colors and patterns and move about in open areas during daylight. Still other animals, such as some species of predatory fireflies, exploit the visual mate-attraction signals of other species. Indeed, a male responding to a flash that is exquisitely timed relative to the end of his own flash may find a receptive female of his own species or a large female of a predatory species (**Figure 5(a)**).

With the exception of animals such as fireflies and many marine animals that can produce their own light, visual signals depend on light from the sun or moon; hence most displaying is confined to daylight or bright moonlit nights. Even fireflies usually signal during a short time window in the evening when the spectrum of their flashes best contrast against the dominant wavelengths of the nighttime sky.

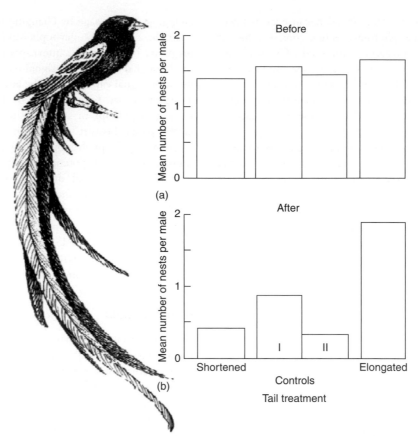

Figure 4 Tail length and mating success in long-tailed widow birds. Top histograms show males with about the same tail length attracted about the same number of females that nested in their territories. Bottom histograms show the change in mating success when the tails of males were elongated or shortened. Reproduced from Andersson M (1982) Female choice selects for extreme tail length in widowbirds. *Nature* 299: 818–820.

Visual signal transmission is also limited because light cannot travel through dense or opaque objects; generally there must be a clear line-of-sight between the signaler and its prospective mate. Thus, visual signals are less useful than chemical and auditory signals in densely vegetated habitats. On the positive side, this requirement also means that the source of the signal is easily located, which is often problematic in communication using other modalities. In open, well-lit habitats, visual displays can work rapidly over great distances.

Visual displays are usually 'honest' advertisements in that there are considerable costs to their production that can be easily assessed by prospective mates. First, in most species, displays tie up appendages needed for locomotion and feeding and displaying are often mutually incompatible activities (**Figure 6**). Second, repetitive motion can be energetically costly, and a signaler's relative condition (and perhaps genetic fitness) may be correlated with the rate and duration of its displays.

Acoustic Signals

An acoustic signal is a mechanical disturbance that propagates through air, water, or solids. Because so many

properties of sounds – frequency, temporal patterns, amplitude – can usually be varied rapidly and independently, the potential for acoustic mating signals to convey information to receivers is comparable to that of visual signals. Like visual signals, auditory signals can also be rapidly transmitted between signalers and receivers. Like chemical signals, sounds can move around objects between communicating animals if the wavelength of the sound is greater than the diameter of the object.

Acoustic signals work during both day and night. Moreover, acoustic signals can be designed to be nearly as easy to locate as visual signals or more difficult to locate when acoustically orienting predators or reproductive competitors are around. In general simple, one-component sounds of intermediate frequency and a slow onset and ending are harder for higher-vertebrate predators to locate than are broadband, noisy signals with sharp onsets or endings. The latter properties enhance the differences in time of arrival and amplitude at the two ears that allow sound localization. These properties are often characteristic of short-range courtship calls whose overall amplitude is usually much less than long-range calls.

In some communities, species can use different frequencies that reduce masking interference between

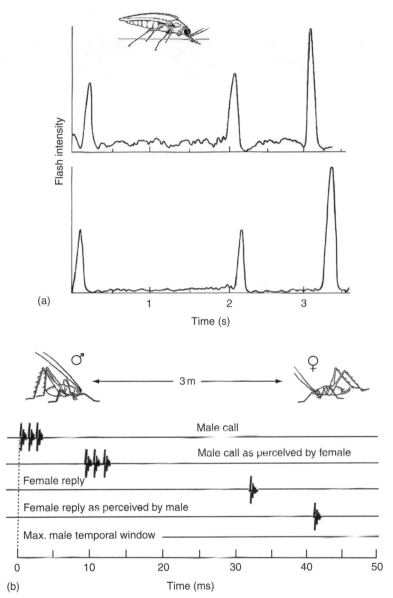

(a)

Time (s)

(b)

Time (ms)

Figure 5 Exchanges of mating signals and exploitation by a predator. (a) The first two flashes in both traces were produced by a male firefly (*Photinus macdermotti*). The third flash of higher intensity was produced by a receptive female of the same species (top trace) and by a predatory female (*Photuris versicolor*) mimicking the female in flash pattern (single flash) and timing (delay from the last flash of the male). (b) Production of a calling song by a pair of katydids (*Leptophyes punctatissima*). The more elaborate male call is answered after a species-specific delay by a simple pulse by the female. The male's acceptance time window is longer than in fireflies so that it can respond to females at various distances, taking into account the longer propagation time of acoustic rather than visual signals. Adapted from Andersson M (1982) Female choice selects for extreme tail length in widowbirds. *Nature* 299: 818–820; Gerhardt HC and Huber F (2002) *Acoustic Communication in Insects and Anurans: Common Problems and Diverse Solutions*. Chicago, IL: University of Chicago Press.

conspecific mating signals and those of other species. Signals with certain frequencies may also be inaudible to other species with which hybridization is potentially possible and may also be out of the hearing range of some predators.

The acoustic environment and an animal's ecology can also influence the design of acoustic mate-attraction signals. In general, sounds with relatively low frequencies propagate further than sounds with high frequencies,

especially near the ground. If an animal cannot produce low-frequency sounds, a common constraint in small animals, then they are likely to seek elevated calling sites on rocks or in trees.

In densely forested areas acoustic signals are likely to be distorted by reverberation. That is, when sounds with relatively high frequencies (and hence short wavelengths) hit trees and leaves, they are redirected and arrive later in time at a receiver's position than do sounds that are not so

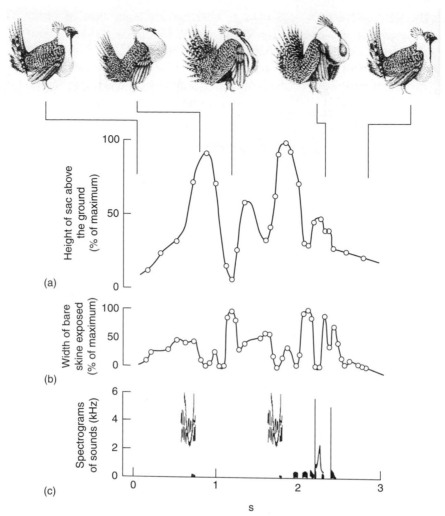

Figure 6 Visual and acoustic displays of the greater sage grouse. (a) Strutting male inflates his large esophageal sac by heaving it upwards and letting it fall twice; (b) This movement exposes bare patches of olive-colored skin on the chest; (c) Toward the end of the display, the male compresses the inflated sac and releases the air explosively to produce a pop and low-pitched coos. Reproduced from Wiley RH (1983) The evolution of communication: Information and manipulation. In: Halliday T and Slater P (eds.) *Communication and Social Interactions*, pp. 156–189. Oxford: Blackwell Scientific Publications.

impeded. This can reduce the coherence of the time properties of a signal, so that, for example, the silent intervals between rapidly repeated sound pulses are obscured. One solution, often adopted by forest-dwelling birds in the tropics, is to produce relatively low-frequency sounds of long duration; if pulsed, the rate of pulsing is distinctly lower than that of birds living in open areas.

In open areas, an acoustically signaling animal must contend with wind and rising pockets of air that interrupt a sound in a nearly random fashion. Animals living in such areas often deal with such amplitude fluctuations by producing rapidly pulsed signals, which will be interrupted only periodically. Thus, enough successive pulses will be transmitted undistorted to allow the prospective mate to recognize the signal.

For example, studies of the wide-ranging rufous-collared sparrow by Hanford and Lougheed in South America show that its song is much more likely to contain rapid trills in open areas and longer notes and slower trills in denser habitats where reverberations are a problem.

In most vertebrates, sound production uses the respiratory system and hence animals can move freely, engage in other activities, and signal all at the same time. In other animals such as many orthopteran insects (crickets, grasshoppers, and katydids), sounds are produced by scraping movements of legs or wings. Whatever the mechanisms, prolonged acoustic signaling can be energetically costly. Roaring in red deer and calling in some frogs can result in losses of more than 20% of the body weight just over the course of the breeding season. For this reason, prospective mates may pay special attention to the amount and rate of signaling because these properties of acoustic mating signals are correlated directly with energetic costs. Such a relationship means that these properties are likely to be 'honest indicators' of the signaler's genetic fitness, in addition to insuring that the signaler is not ill or heavily parasitized.

Dual Signalers

The rapid production and fade-out time of visual and acoustic signals make it possible to add another dimension to communication in species in which individuals of both genders signal: their timing. For example, relatively elaborate flash patterns produced by male fireflies are answered by simple flashes of receptive females within a species-specific time window. Similarly, in some species of katydids, females respond to the calling song of the male within a somewhat wider time window that allows for time delays when the two individuals are widely separated (**Figure 5(b)**). Although not strictly mating signals, many paired birds that live in dense habitats keep in acoustic contact by duetting, which sometimes results in what is often perceived by humans as a single complex note.

Multimodal Signaling

Some signals consist of combinations of acoustic, visual, and chemical elements. Lek-breeding birds such as sage grouse and prairie chickens are very vocal as they perform their dances (**Figure 6**), and bowerbirds not only vocalize and display but build, decorate, and defend special structures called bowers, to which females are attracted and within which mating takes place. Combinations of acoustic and visual signals are also common in some mammals, such as red deer. Many animals sequentially produce a variety of signals that differ in modality and range. Visual and acoustic signals are most likely to be used at a distance, and for the reasons discussed earlier, acoustic signals will also be favored at night and in dense habitats. Long-range signals should at least have properties that identify the signaler as a conspecific individual. More subtle information about the fitness and status of a signaler can be extracted at closer range from the same signal or other signals. For example, visual and chemical signals are useful at close range in most habitats, and chemical signals can also be used at night.

See also: Communication Networks; Flexible Mate Choice; Honest Signaling; Mate Choice in Males and Females; Multimodal Signaling; Sexual Selection and Speciation.

Further Reading

Andersson M (1994) *Sexual Selection*. Princeton, NJ: Princeton University Press.

Bentsen CL, Hunt J, Jennions MD, and Brooks R (2006) Complex multivariate sexual selection on male acoustic signaling in a wild population of *Teleogryllus commodus*. *American Naturalist* 167: E102–E116.

Bradbury JW and Vehrencamp SL (1998) *Principles of Animal Communication*. Sunderland, MA: Sinauer Associates, Inc.

Endler JA, Westcott DA, Madden JR, and Robson T (2005) Animal visual signals and the evolution of color patterns: Sensory processing illuminates signal evolution. *Evolution* 59: 1795–1818.

Gerhardt HC (1994) Reproductive character displacement of female mate choice in the grey treefrog *Hyla chrysoscelis*. *Animal Behaviour* 47: 959–969.

Gerhardt HC and Huber F (2002) *Acoustic Communication in Insects and Anurans: Common Problems and Diverse Solutions*. Chicago, IL: University of Chicago Press.

Handford P and Lougheed SC (1991) Variation in duration and frequency characters in the song of the Rufous-collared sparrow, *Zonotrichia capensis*, with respect to habitat, trill dialects and body size. *Condor* 93: 644–658.

Hauser MD (1966) *The Evolution of Communication*. Cambridge/London: MIT Press.

Marler PR and Slabberkoorn H (eds.) (2004) *Nature's Music: The Science of Birdsong*. San Diego/London: Elsevier Academic Press.

Ryan MJ and Rand AS (2003) Sexual selection in female perceptual space: How female tœngara frogs perceive and respond to complex population variation in acoustic mating signals. *Evolution* 57: 2608–2618.

Welch AM, Semlitsch RD, and Gerhardt HC (1998) Call duration as an indicator of genetic quality in male gray treefrogs. *Science* 280: 1928–1930.

Wilson EO and Bossert WH (1963) Chemical communication among animals. *Recent Progress in Hormone Research* 19: 673–716.

Measurement Error and Reliability

S. W. Margulis, Canisius College, Buffalo, NY, USA

Introduction

Think about the study of behavior and what do you envision? More likely than not, 'animal behaviorist' conjures up the image of a disheveled, khaki-clad individual with binoculars and a clipboard, sitting in the midst of a jungle, jotting down notes about the fascinating behaviors he or she sees amidst a large and complex group of mammals. The idea that behavioral observation is a subjective, casual endeavor is far from true. With the expansion of ethology in the 1930s, the idea that animals could be observed in natural settings steadily grew in scientific importance. As the field of ethology and behavioral ecology expanded, there came an explosion of research methods, conventions, and practices. While all of these may have been internally valid (i.e., provided quantitative, reliable measures for the particular study for which they were designed), it was difficult, if not impossible, to generalize to a larger population or compare across studies as a result of these methodological and analytical differences. Thus, external validity was compromised because of a lack of standardization and systematic data collection rules. In 1974, the seminal paper published by Jeanne Altmann provided a critical conceptual framework and operational guide for behavioral data collection and quantification. Virtually all observational data ascribe to one of the methods outlined in this paper. These methods were designed not only to provide some degree of standardization to the discipline, but also to reduce bias by structuring observations such that an observer's choice of which subject to watch and what behaviors to record was based on a priori decisions and statistically valid procedures.

As technology has changed, so too have data collection methods. The image of the field researcher with a clipboard has been replaced by the researcher with a PDA (personal data assistant) or other handheld device; live observation may be replaced by digital video recording followed by playback, and analytical methods have grown in complexity as computers have become routine. That being said, as with any type of scientific investigation, there can be sources of error. As in any science, understanding the nature or these errors is essential in order to proactively control for their effects methodologically, or to account for them statistically at the conclusion of the study. Here, the key issues in the measurement and control of inter- and intraobserver reliability in observational research, and the methods and strategies for understanding and controlling these sources of error are discussed.

Behavioral Methodology: A Brief Overview

While every observational research study has its own unique study design and methodology, virtually all studies use as their starting point one of several basic observation techniques. Although other articles in this series will go into the details, a broad brush overview of these methods is warranted here.

Behavioral data collection schemes are based on several key concepts: (1) what does one observe (a single individual or a group); (2) how does one record observations (continuously, or instantaneously), and (3) what behaviors does one observe (establishment of a clearly defined ethogram). By utilizing various combinations of these three conceptual ideas, a study can focus on particular aspects of individuals, groups, and behaviors. Each factor requires careful planning, testing, and training to minimize errors.

Error can be introduced at a number of junctures in a study. For example, if observers are inaccurate in their ability to identify individuals quickly and accurately, they may erroneously ascribe behaviors to the wrong individuals. An additional source of error can be introduced into the data recording scheme if observers fail to time behaviors accurately. Ethograms with incomplete or vague behavioral descriptions can lead to excess variability in how observers interpret behaviors and thus lead to missed or misidentified behaviors.

In general, most behavioral data collection schema involve one of two approaches: in some cases, a continuous recording approach, in which a single individual is observed (continuous, focal observation sensu Altmann), and the onset time of every behavior or behavior transition is recorded. Alternatively, an instantaneous approach is used, which may be applied to a single individual or to a group (point or scan sampling). In this case, the behavior in which an animal is engaged at a particular point in time, usually signaled by a stopwatch, is recorded. Each of the methods has its own set of advantages and disadvantages such that no single methodology is appropriate in all cases. Thus, one must be well versed in the particular strengths and pitfalls of each method in order to decide on the best fit for a particular study, and to recognize that each method has inherent sources of error that must be understood and addressed.

Which method to choose is based on the question that the researcher is trying to answer. An instantaneous or scan-sampling approach is most appropriate when behaviors of interest are defined as state behaviors; this method does not necessarily require that individuals be recognizable (though it is preferable). Detailed interactions are not readily quantified using this method (though it is possible to combine a scan approach with select, continuous observation for highly visible, key behaviors). A continuous, focal approach is often used when interactions are an important component of a study, and when both event and state behaviors are to be recorded. This method often requires more rigorous training before observers attain a sufficient level of comfort with the procedures. Other standard methods are also available to the researcher, but are not discussed here and will be covered fully in other articles.

The Importance of Ethogram Construction

The importance of a clearly defined and consistent ethogram is often overlooked. Recent efforts to develop some level of standardization of ethogram structure and terminology (e.g., EthoSource, described by Martins, and related ontologies described by Midford and colleagues; or SABO, as outlined by Catton) have made progress in this regard. However, there is still considerable variation in the structure, detail, and terminology of ethograms. Use of terms that may be synonymous in ethograms can lead to confusion, and lack of specificity of definitions can result in errors. In most ethograms, behaviors may be defined functionally, in which the presumed use of the behavior is implied, or operationally, in which no specific function is assigned and the description, or definition, provides details on the motor patterns associated with the performance of the behavior. Animal behaviorists often use functional definitions and assumptions; however, in some circumstances, it can be difficult for an observer to reliably identify behaviors functionally. Play and aggression – both functional categories – may involve similar motor patterns, and clear and precise operational definitions may be critical, particularly for novice observers or for a species that has not been well studied such that functionality cannot be satisfactorily ascribed.

The level of ethogram detail is another critical component of study design that influences observer accuracy. A hierarchically structured ethogram can facilitate ease of use, with more detailed, deeper levels of behavioral description used for studies that are narrowly focused or when highly experienced observers are available.

Sources of Error

The nature of behavioral research is fundamentally no different from any other branch of scientific inquiry. Sources of error can be introduced into any study at various levels. While they can be controlled for and minimized, it is impossible to eliminate them. Being aware of these sources and how they might bias data are fundamental to the conduct of good science. The three most common means by which error can be introduced into a behavioral study include observer error, equipment error, and computational error. It must be stressed that these sources of error are common to all scientific investigations and not unique to behavior. The role of the observer is perhaps more critical to behavioral observation than what may be the case for certain types of laboratory sciences, and will be the focus of the remainder of this article. Equipment and computational error are briefly touched upon in the next section.

Equipment and Computational Error

While behavioral data collection has advanced from paper and pencil check-sheets to, in the majority of studies, electronic data collection systems, data collection is nevertheless subject to recording errors. These may involve the failure of electronic devices (particularly in the field), transcription error, and coding error. Careful review and proofreading of all data can alleviate many of these problems. Use of computer-aided data collection tools does not negate the need to review entered data. Tapping an incorrect box on a PDA screen is no less likely than checking the wrong box on a paper check-sheet. There have been numerous times when I have proofread a dataset, confident that it was error-free, only to discover data entry errors.

Computational errors generally occur during the data analysis phase of a study; however, use of statistical packages minimizes the errors here, provided the user understands the assumptions and rationale of the statistical software being used. Statistical textbooks and software user guides are of course essential; however a number of recent works have emphasized statistical issues that are more common in behavioral studies, particularly those relating to small sample size, repeated measures, and generalizability (see e.g., Kuhar's or Plowman's more extensive treatment of this subject). Behavioral data analysis often involves one or more levels of data tabulation and summary before statistical analyses can be conducted. These may be done in an automated fashion using behavioral or statistical software, or it may be done by hand before data are entered into a computerized system. Again, double-checking and proofreading all such intermediate phases can minimize the probability of such calculation errors.

Observer Error

The role of the observer is critical to the successful collection of behavioral data, but observer error has the potential to be the most significant error component of behavior studies. However, clear guidelines and methods exist to ensure that such errors are minimized, acknowledged, and controlled. Because of the tremendous variability among observers, we must be cognizant of how to recognize, measure, and control for individual variation to assure a sound study design. Observers have the potential to introduce variation into behavioral investigations in several ways. First, the very presence of an observer may alter the behavior of the subjects. Second, observers may perceive events differently, based on their view of a particular situation or group (errors of apprehension). Third, observers may err because of lack of training or experience or because protocols and ethograms are unclear. Individual observers enter into an investigation with their own personal biases which may have the potential to influence the quality of their data collection as well. Finally, as already discussed, observers may record their observations incorrectly or may have difficulty utilizing equipment. All of these sources of observer error can be addressed and reduced via training and regular assessment of reliability and validity.

Observer Presence

The idea that an observer alters the behavior of the animals he or she observes has been debated for decades and leads to a conundrum: how can we observe natural behavior, if, by definition, we alter the behavior that we are observing simply by our presence? Use of video and remote recording devices is one way to address this concern; however, much observational data collection is – and will always be – done via live observation. Maintaining standard observational protocols holds the observer effect constant and while it may be that behavior is altered, it is in theory altered consistently across all subjects, thus enhancing internal validity. Long-term field studies have demonstrated that most animals can habituate to observer presence, suggesting that the observer's effect on the individuals that are observed may be relatively minor.

Errors of Apprehension

When two observers watch the same animal from different vantage points, differences in perspective may alter the extent to which they perceive a particular event. This is a problem primarily when observers' movements are constrained in how and where they are able to move in the area in which they are observing. This may be the case in a laboratory or captive situation in which animals may be out of view of the observer, or the observer's vantage point may prevent a clear view. In nature, observers' movements may be constrained by the activity of the animal they are watching or by other animals in the group. Ideally, simply changing one's physical position (when the observer is able to do so) to obtain the best possible view of an interaction can mitigate apprehension error. This can be a problem when conducting interobserver reliability tests (to be discussed later).

Observer Error and Bias

As already discussed, there is no single standard protocol for observing behavior. Although there are methodological standards, every study, every individual subject, and every study setting is unique. Thus, training observers is a time-consuming, tedious, but critical component of any investigation and will improve internal validity. It is only through rigorous training and ongoing monitoring and evaluation that one can maintain an acceptable level of interobserver agreement. Even an experienced researcher will require some time to become familiar with their subjects, and to ascertain the validity of their ethogram. Vague or equivocal definitions, for example, can lead to confusion among observers. Lack of experience with data recording systems can be a source of error, until observers have practiced sufficiently and are comfortable with the protocol, the layout of the datasheet, the codes used to record information, and so on. Novice observers often enter into observations with preconceived and oftentimes erroneous notions about behavior, and it may take some time and effort to move them from a subjective view of behavior in which they interpret and read meaning into behavioral patterns and events, to a more objective, consistent ability to record actions without assuming intent or meaning. Dissuading observers of their preconceptions is often the most challenging part of training observers.

Once this challenge of reducing observer bias is met, even a trained observer who has passed standardized reliability tests may diverge from that standard over time. Just as any process may need to be calibrated periodically, so must observer reliability to avoid observer 'drift' in recording of behavioral information. Regular review and repeated reliability testing can address this error.

Reliability and Validity

Reliability is an indicator of how repeatable one's results are, and is critical to maintaining accurate data collection. Unlike measuring weight or length for example, in which the potential exists for getting precisely the same measurement repeated times, it is highly improbable that an animal or a group of animals will perform exactly the same behaviors in the same way if measured multiple times. Careful data collection designs, however, can ensure consistency and standardization, which in turn improves repeatability. This is particularly important in long-term field studies, where data may be gathered by multiple observers over a period of years, or decades.

Training and adhering to a standard of accuracy and precision is critical.

The terms 'accuracy' and 'precision' may be considered synonyms in some disciplines (in fact, the thesaurus program of my word processor indicates that they are indeed synonymous), but in the case of behavioral observation, they are subtly but distinctly different. Accuracy refers to how close a recorded observation is to reality ('the truth'), whereas precision refers to how consistently an observer records the same behavior in the same way. Methodological differences sometimes necessitate a trade-off between accuracy and precision. A simple ethogram with clearly defined definitions may facilitate good precision among observers – for example, it is relatively simple to identify an animal as being active or inactive. However, there may be a loss of accuracy in that the behavioral categories may be too broad to adequately answer the study's main questions. In addition, precision may be used as an indicator of intraobserver reliability: that is, to what extent does an individual consistently observe behaviors in the same way? Accuracy is an important element of evaluating how good a study design is at collecting data to answer the question at hand: that is, to what extent do data reflect reality? How suitable is the chosen research design in answering the question that one has posed? Thus, the internal validity of an investigation is closely linked to the applicability of the methods chosen to answer the question posed. External validity is a measure of the generalizability of results to other study populations or species, as the case may be. This may be linked to the ethogram chosen and how broadly applicable it is. Reliability and validity are both essential measures that one must evaluate in terms of both inter- and intraobserver reliability.

Intra- and Interobserver Reliability

Intraobserver reliability can provide a measure of consistency and repeatability. Regular review of methods and ethogram, and reliability testing (to be described below) can provide a quantifiable measure of intraobserver reliability. Because it is common to use multiple observers for behavioral studies, either simultaneously (to maximize efficiency of data collection) or sequentially (to maintain ongoing, longitudinal investigations), it should come as no surprise that maintaining a high standard of interobserver reliability may be the most important aspect of ensuring accurate and precise data for behavioral investigation. Every observer comes into a study with his/her own set of biases and tendencies. Careful and rigorous training are essential to the conduct of behavioral studies. While there is no single training protocol for observers, convention necessitates extensive training on observation methods and animal identification, familiarity with the ethogram and data collection devices, and practice, either

supervised or unsupervised, until the observer feels a degree of comfort with the methods. It is at this point that formal interobserver reliability testing should be initiated. Interobserver reliability encompasses a number of statistical approaches that facilitate a comparison between observers: that is, how similar are the data collected by two researchers who observe the same individual at the same time? Theoretically, they should be identical, but in practice, this is rarely the case. Two individuals weighing the same standardized weight on a balance are unlikely to get exactly the same measure, but they should be quite close; similarly, two researchers observing the same individual at the same time may not record exactly the same sequence of behavior, but differences should be minimal and most importantly, they should be random. Often, the conduct of interobserver reliability tests can highlight weaknesses in the study design or protocols. If for example, an observer is consistently misscoring a particular behavior, it may be that the observer needs more training and practice; however, it may also be the case that the behavior is not adequately defined on the ethogram.

Techniques for Measuring Reliability

Most measures of inter- or intraobserver reliability utilize simultaneous observation of the same individual, or independent scoring of videotaped footage. In both cases, the goal is to have observers independently score samples of behavior that should be identical if there were no observer error or bias. When two observers conduct simultaneous, live observations, it is critical that they not communicate with each other as this could influence the outcome by violating assumptions of independence. This can be challenging. If, for example, one observer notices that a second observer is entering a behavior that the first observer may have missed, this could lead the first observer to rethink his/her data entry and add a behavior that he/she might otherwise have erroneously missed. Conversely, two observers are, by definition, viewing a situation from slightly different vantage points and therefore may not be able to see exactly the same sequence of behavior because of errors of apprehension. However, this does not necessarily imply that their data are not reliable, since they may have been unable to adjust their position.

When using live observation, the likelihood that only a small subset of possible behaviors will be seen is high. Should two observers be considered to have high reliability if they both correctly score a subject as sleeping for 20 consecutive scans? The use of videotaped sequences of behavior resolves a number of problems. First, observers are able to watch and score videotape individually and independently, without possible influence from other observers. Second, the researcher can utilize one or more segments of footage that encompass a greater range of behaviors on the ethogram, thus providing a more rigorous test of observer accuracy. Finally, all observers are able to view the sequence portrayed on the videotape from the same perspective.

Details of reliability measurements can be found in sources listed at the end of this article, and a particularly clear example of how to calculate the various reliability metrics can be found in Lehner's book; however, they are briefly described here.

Assessing Reliability via Concordance

A number of statistical methods exist for quantifying observer reliability, and all are based on a similar premise: to what extent do data collected by two individuals (or by one individual at different points in time) agree? In its simplest form, this may mean evaluating percent agreement. For example, consider an animal that is observed for 10 min, and the state behavior in which it is engaged is noted every minute on the minute (an instantaneous sampling approach). If two observers record data on the same individual for these 10 min, the 'agreement' between their datasets is easily calculated: How many of the 10 point observations are the same? If all are identical, then the agreement is 100%; if nine out the ten are identical, then agreement is 90%. A variation on this is the kappa coefficient, which corrects for chance agreement.

Kendall's coefficient of concordance can evaluate reliability evaluations with more than two observers; however, data must be converted to ranks to accommodate this nonparametric approach. Most behavioral studies look for agreement at or above 90% before an observer is considered to be 'reliable.' There is no hard and fast rule on this, however, so this value should be thought of as a guideline only. Most often, new observers are tested against a standard (the lead investigator, or main field assistant, for example).

Assessing Reliability via Correlation

Several statistical tools are available to measure correlations between nominal, ordinal, interval, and ratio data. The Phi coefficient measures correlation between nominal variables; for example, comparing the number of times two observers score a particular behavior. Similar standard statistical measures of correlation are appropriate for evaluating interobserver reliability. Spearman correlation is used for ordinal or ranked data, and Pearson correlation for interval or ratio data. Correlation coefficients range from 0 to 1, with higher values indicating better agreement. In general, a correlation coefficient > 0.7 is considered a strong correlation.

Maintaining Reliability and Consistency

The goal of behavioral research, as with any scientific endeavor, is to collect accurate, reliable data that allow the scientist to answer the question posed. The methodology chosen should fit the question at hand; it should be tested and modified to maximize its effectiveness, and its efficacy evaluated before finalizing data collection plans. It is imperative that observers be trained and their reliability – their accuracy, precision, repeatability, and validity – tested prior to utilizing their data, and regularly throughout the period of data collection.

See also: Ethograms, Activity Profiles and Energy Budgets; Experiment, Observation, and Modeling in the Lab and Field; Experimental Design: Basic Concepts.

Further Reading

Altmann, J (1974). Observational study of behavior: Sampling methods. *Behaviour* 49: 227–266.

Caro, TM, Roper, R, Young, M, and Dank, GR (1979). Inter-observer reliability. *Behaviour* 69: 303–315.

Catton, C, Dalton, R, Wilson, C, and Shotton, D (2003). SABO: A proposed standard animal behaviour ontology. www.bioimage.org/pub/SABO/SABO.

Kuhar, CW (2006). In the deep end: Pooling data and other statistical challenges of zoo and aquarium research. *Zoo Biology* 25: 339–352.

Lehner, PN (1996). *Handbook of Ethological Methods,* 2nd edn. Cambridge: Cambridge University Press.

Martin, P and Bateson, P (2007). *Measuring Behaviour: An Introductory Guide,* 3rd edn. Cambridge: Cambridge University Press.

Martins, EP (2004). EthoSource: Storing, sharing, and combining behavioral data. *Bioscience* 54: 886–887.

Midford, PE (2004). Ontologies for behavior. *Bioinformatics* 20: 3700–3701.

Paterson, JD (2001). *Primate Behavior: An Exercise Workbook.* Long Grove, IL: Waveland Press.

Ploger, BJ and Yasukawa, K (eds.) (2003). *Exploring Animal Behavior in Laboratory and Field.* New York, NY: Academic Press.

Plowman, AB (2008). BIAZA statistics guidelines: Toward a common application of statistical tests for zoo research. *Zoo Biology* 27: 226–233.

Stamp Dawkins, M (2007). *Observing Animal Behaviour.* New York, NY: Oxford University Press.

Memory, Learning, Hormones and Behavior

V. V. Pravosudov, University of Nevada, Reno, NV, USA

Introduction

Importance of Memory and Learning

Memory is a process of acquiring, retaining, and retrieving information about past experiences, while learning is a process of using these experiences to adaptively modify behavior. Learning is generally classified as nonassociative (e.g., habituation) or associative (classical and operant conditioning, active and passive avoidance).

Memory and learning often play critical roles in the survival and reproductive success of many animal species. Animals may need to remember numerous pieces of information related to locations of breeding sites/nests, shelters, food sources, territory boundaries, locations of neighbors as well as their identity, etc. While it appears that memory and learning abilities have fitness consequences in most animal species, in some, memory might be especially crucial for fitness. Food-caching animals, for example, store food when it is abundant and then use their caches during the periods of food scarcity (e.g., winter). Many food-caching species rely, at least in part, on memory to find previously stored food and failure to recover caches during energetically demanding times may result in death. Female parasitic cowbirds lay their eggs in the nests of host species. This strategy requires remembering locations of multiple nests in addition to the condition of these nests so that the eggs can be laid at the proper time to have the best chance of survival. Good memory in female parasitic cowbirds is thus crucial for their reproductive success and hence biological fitness. Polygynous male meadow voles that have large home ranges encompassing the home ranges of several females need to remember the locations of multiple potential mates.

There are numerous other examples of behaviors that are strongly dependent on memory and learning skills, but the main issue in all of them is that memory and learning are crucial for successful survival and reproduction. Any variation in memory and learning is likely to have strong fitness consequences. Understanding the causes of variation in memory and learning is thus of great importance to biologists. A great deal of work on memory and learning has been done in the biomedical field, which is mostly focused on how to maintain healthy cognitive functions in humans as well as what causes impairments in memory and learning. An evolutionary approach may also be useful to understand the variations between species as well as to reveal how selection pressures might have molded both behavior and the mechanisms associated with memory and learning.

Memory Types

Psychologists usually recognize several types of memory on the basis of the type of information processed and the area of the brain that is involved in the processing.

Spatial memory appears to be dependent on an area of the brain called the hippocampus, in mammals and birds. Spatial memory allows animals to remember information for location of shelters, food patches, nests, etc. One group of animals that relies heavily on spatial memory is food-caching species, both in mammals and birds. Some species of birds, for example, can store tens of thousands of individual food items throughout fairly large home ranges and then recover them, using spatial memory. Clark's nutcracker (*Nucifraga columbiana*) is the most well-known food-caching bird species and has been shown to use spatial memory to recover caches months after creating them. Some parids (tits and chickadees) are also known to store tens to hundreds of thousands of individual caches throughout autumn and sometimes spring. Even though there are debates about whether chickadees use spatial memory to recover their long-term caches (several months after making caches), they clearly do employ spatial memory on a relatively short-term basis (4–6 weeks), and their reliance on spatial memory appears to correlate with their reliance on cached food. In contrast to hippocampus-dependent spatial memory, memory for other cues, such as color, appears to be independent of the hippocampus.

Hippocampus-dependent declarative memory has been described mostly in humans, and it usually refers to memories about specific facts, such as learning facts from a book. Declarative memory is sometimes further subdivided into semantic (abstract knowledge that is not necessarily connected to specific time or place) and episodic (events or facts connected with specific time and location) memory. Episodic memory has been historically thought to be uniquely human, but recent research on food-caching birds and some mammals suggests that animals might be capable of episodic-like memory in which they can remember 'what,' 'when,' and 'where.' Western scrub-jays (*Aphelocoma californica*), for example, have been shown to remember what type of food they cache (perishable vs. nonperishable), when they cached it (so they can retrieve perishable food faster), and where they cached it. This type of memory may be important for many food-caching species that store both perishable and nonperishable food. Remembering location, food type, and timing of each cache allows these animals to successfully retrieve their caches when needed, rather than risking a chance to recover spoiled food.

In addition, memory can also be categorized into working or reference memory. Working memory is usually defined as a relatively short-term memory that holds and manipulates information on a temporary basis. This type of memory is usually associated with learning information that changes on a regular basis. For example, a chickadee that is recovering previously made food caches may use working memory to remember all sites that it has inspected prior to finding the correct cache location. Reference memory, on the other hand, is long-term and is usually concerned with associative or discrimination learning of more stable information by repetitive training. An animal may use reference memory to remember locations (trees, feeders) that contain food on a regular basis.

Hormones and Memory

While neural mechanisms mediate memory and learning in animals, hormones are well known to interact with these mechanisms of learning and hence have significant effects on cognitive processes. The memory process is usually subdivided into three phases: acquisition, consolidation and storage, and recall (or retrieval), and hormones might affect either one or all of these phases. Among the hormones that are known to impact on learning and memory, the most well studied are glucocorticoids, or 'stress' hormones produced by the adrenal glands, and gonadal steroids such as testosterone and estradiol.

Glucocorticoid Hormones

Glucocorticoid hormones have strong effects on memory and learning in most animals. Glucocorticoids are steroid hormones produced by the adrenal glands, and elevation in glucocorticoids is usually associated with stressful events. The most common glucocorticoid hormone in some mammals including humans is cortisol. Corticosterone, on the other hand, is the main glucocorticoid hormone in reptiles, birds, and other mammals (e.g., some rodents). Glucocorticoids (hereafter CORT) are essential hormones necessary for multiple physiological functions including regulation of metabolism and gluconeogenesis.

Stress and Glucocortocoid Hormones

Most animals are known to respond to a variety of ecologically relevant stressful conditions (food deprivation, social stress, predation risk, etc.) by significantly elevating CORT levels. The magnitude of CORT elevation may be related to the nature of the stressors. Hence, knowing the relationship between CORT and memory and learning is

important not only from the biomedical perspective but also from the evolutionary perspective. For example, changes in memory mediated by changing CORT levels may have significant impact on biological processes and fitness.

First, it is important to note that CORT is essential for the maintenance of memory and learning. Numerous experiments on both mammals and birds demonstrate that removal of adrenal glands, and hence that of circulating CORT, results in severe hippocampus-dependent memory and learning impairments. Providing exogenous CORT via implants or injections to restore physiologically normal CORT levels restores cognitive functions. Prefrontal cortex-dependent working memory also requires glucocorticoid hormones for normal functioning and is also negatively impacted by adrenalectomy and restored with CORT implants.

Naturally occurring ecological conditions that result in significant reductions in CORT levels below baseline are probably not relevant, but these data strongly suggest that presence of CORT is essential. On the other hand, elevation in CORT levels is a typical response to numerous ecologically relevant ecological perturbations and therefore, from both ecological and biomedical standpoints, it is important to understand the effects of elevated CORT levels (both short-term and long-term or chronic) on cognitive processes.

Short-Term Elevations in Glucocorticoid Hormones

Numerous environmental variables may trigger fairly short-term (minutes to hours) CORT increases. The presence of, or attack by a predator, temporary food shortages, rapid changes in weather patterns (e.g., snow storm), social interactions, etc. may all elicit a relatively short-term elevation in glucocorticoid hormones. Short-term CORT elevations have been reported to have both negative and positive effects on memory and learning. Whether the effect is positive or negative generally depends on the magnitude of CORT elevation and on timing of such elevation in relation to the memory phase (acquisition, consolidation, or retrieval).

Most effects of CORT elevation on memory and learning appear to follow an inverse U-shape relationship in which an initial increase in CORT enhances memory, and after a certain threshold is reached, any further increase usually results in memory impairments. Many experimental studies with rodents established that memory appears to be dependent on glucocorticoid hormones in a concentration-dependent fashion, and experimentally elevated CORT levels result either in improved memory performance caused by moderately elevated CORT levels or in impaired memory caused by strong CORT elevation. In domestic chicks, for example, experimentally induced moderate, but not strong, increases in CORT levels resulted in enhanced memory for a weak aversant in a passive avoidance task. Interestingly, when a strong aversive stimulus was used, both moderate and high CORT elevations impaired memory performance.

In many experimental studies, induced acute CORT elevation at the time of memory acquisition or immediately after training on a learning task results in improved memory performance, which suggests that elevated CORT enhances memory acquisition and consolidation. On the other hand, acute CORT increases prior to memory retrieval usually impair it. These are, however, only general patterns and the results of many studies have not conformed to these generalities.

For example, adding moderate doses of CORT shortly after training on a memory task generally enhances long-term memory in rats. Numerous studies suggest that CORT increased immediately following a stressful event can enhance memory of that event, which may be highly adaptive as the animal might be able to avoid such stressful events in future. Administration of CORT several hours after learning generally has no effect on memory of the trained event. In contrast, CORT elevation prior to memory retrieval seems to impair memory retrieval. Thus, short-term peaks in CORT appear to enhance memory encoding and consolidation when they occur either during or immediately after training on a memory task, but they tend to impair memory retrieval when applied directly prior to retrieval. It is not clear, however, whether this pattern is always consistent. In rodent studies, when memory testing was done shortly after memory acquisition that was associated with elevated CORT levels, memory retrieval was impaired if CORT remained elevated throughout both memory acquisition and recall. In humans, cortisol elevation associated with social stress results in impaired social memory (such as face recognition).

In many ecologically relevant conditions, however, short-term stressful events may occur unpredictably and memory retrieval may be essential for survival during such events. For example, food-caching animals store food when it is abundant and when energy budget favors storing food over eating it. In northern latitudes, naturally available food may be unpredictable because of changing weather and difficult to obtain; memory-based cache retrieval may be the only option to gather necessary energy reserves. During such times, CORT levels are usually elevated and, sometimes, even highly elevated for prolonged periods of time encompassing both food caching and cache retrieval. If CORT elevations were detrimental to memory retrieval, the ability of these animals to retrieve their caches in times of hardship would be compromised.

At least one study showed marginal improvement in spatial memory during memory-based food cache recovery in mountain chickadees (*Poecile gambeli*) with clinically high CORT levels induced five minutes prior to memory retrieval. On the other hand, applying acute-stress-like high CORT levels only prior to memory encoding (food caching) in these chickadees had no effect on memory-based cache retrieval. These results suggest that in food-caching chickadees, strongly elevated CORT prior to memory retrieval specifically improves memory retrieval. In nature, chickadees and other food-caching birds may frequently experience rapidly and unpredictably changing weather conditions, which may cause spikes in CORT levels, and elevated CORT appears to enhance spatial memory needed for successful cache retrieval. In addition, some studies of long-term CORT treatment in which elevated CORT was present during memory acquisition, consolidation, and recall found memory improvements suggesting no impairing effects of high CORT on memory retrieval. Thus, although many studies showed negative effects of elevated CORT on memory retrieval, it appears that such effects are not necessarily general and may depend on the evolutionary history of a particular species.

Long-Term, Chronic Elevations in Glucocorticoid Hormones

Most studies investigated the effects of chronic elevation in CORT on memory by one of the two main techniques. Some studies provided chronic CORT via long-term implants, whereas others used stressful conditions (social stress, bright light, etc.) rather than hormone implants. Such stressful conditions almost always result in elevated CORT levels, but it is possible that some other factors besides the hormones may directly cause changes in memory performance. This problem is usually addressed by experiments in which animals' adrenal glands are removed by effectively removing CORT from the system. Then, CORT implants are added simulating normal CORT levels (adrenalectomy without adding exogenous CORT results in severe memory impairments). When adrenalectomized animals are stressed, they cannot increase CORT levels and usually show no consequences of stressful environment on memory, suggesting that stress affects memory via elevated CORT.

Long-term, chronic (weeks or months) elevations in glucocorticoids have usually been associated with memory impairments, most often those that are hippocampus dependent. Numerous rodent studies have documented impairments in spatial memory performance following several weeks of repeated stress. Such impairments are usually accompanied by dendritic atrophy within the hippocampus, which appears to be reversible if normal conditions are restored. Dendritic morphology appears to be directly related to memory processes and therefore elevated CORT may affect memory by causing dendritic atrophy. However, it remains somewhat unclear whether only relatively high chronic elevations in glucocorticoids have severe negative impact on memory. Some studies have shown that moderate elevations might actually have positive or no effects on learning and memory.

In addition, numerous studies suggest that negative effects of long-term elevated CORT on memory may be age, sex, and/or time dependent.

The effects of long-term CORT elevation on memory and learning are somewhat controversial. In one study, experimental doubling of CORT plasma levels in rats for 30 days had no effect on spatial learning. Similar chronic CORT increase for 60 days, on the other hand, resulted in spatial memory impairments in middle-age rats, but had no effect on young rats. It has been suggested that chronic CORT elevation may negatively affect memory via increasing neuronal death and decreasing their number. Interestingly, the hippocampal neuron numbers were not affected even by the longest administration of CORT in either middle-aged or young rats. In a different experiment, chronic stress applied to male rats for 14 days (presumably increasing circulating levels of endogenous CORT) resulted in improved spatial memory performance, while memory performance tested after 21 days of chronic stress was impaired. In female rats, 21 days of chronic stress resulted in enhanced spatial memory performance but females rats tested after 28 days of chronic stress showed neither improvements nor impairments in spatial memory. Such differences in responsiveness to elevated CORT between males and females seem to be mediated by gonadal hormones (e.g., estrogen) in females. Interestingly, chronic stress also has a lesser impact on dendritic atrophy in female rats compared to males, which also suggests counter effect by estrogen. More work is needed to better understand the interactions between gonadal and stress hormones and how they affect memory and the brain.

Some studies, on the other hand, showed that long-term stress might have an enhancing effect on memory. Tree shrews (*Tupaia belangeri*) placed in conditions of chronic stress and with chronically elevated CORT levels for four weeks performed significantly better on a hippocampal-dependent spatial memory test compared to control shrews. There were no effects of elevated plasma CORT levels on hippocampus-independent memory task involving sites marked with local color cues (e.g., when a target site was clearly marked with a unique color and so spatial memory was not necessary to relocate it). Interestingly, hippocampal cell proliferation rates in tree shrews were negatively affected by chronic stress. Cell proliferation in the hippocampus is a component of neurogenesis (production of new neurons), which has been implicated in memory function. Reduction in neurogenesis might usually be considered bad for memory process, but the fact that tree shrews showed enhanced memory and reduced hippocampal cell proliferation at the same time suggests no negative effects of reduced cell proliferation rates on memory. All these results were in contrast to previous studies of tree shrews reporting negative effects of chronic stress on hippocampus-dependent but not hippocampus-independent memory.

Just as in the case of short-term CORT elevation, the effects of long-term CORT elevations appear to follow a dose-dependent inverted-U shape relationship with memory. Moderate CORT elevations may have memory-enhancing effects, while high stress-induced CORT levels are likely to produce detrimental effects on memory and the brain.

There are several examples of improved learning and memory performance associated with long-term moderately elevated CORT levels. In greylag goslings (*Anser anser*), higher levels of excreted CORT correlated with better learning performance. Food-caching parids (tits and chickadees) provide another example. As mentioned before, these birds store numerous food items during the winter when these birds experience energetically demanding adverse conditions that are characterized by poor food availability and predictability. Under such conditions, finding previously made food caches appears to be crucial for survival. Field studies suggested that these birds might have moderately elevated CORT levels for several months during the winter, likely caused by unpredictable foraging and adverse weather conditions. Such long-term CORT elevations may persist in addition to short-term CORT changes caused by quickly changing environmental conditions. In the laboratory, experimentally induced long-term (several months) unpredictable foraging conditions resulted in moderate but significant CORT elevations and in enhanced spatial memory performance in mountain chickadees. This type of CORT elevation also resulted in enhanced spatial performance in mountain chickadees, suggesting that adverse environmental conditions may enhance spatial memory via moderately elevated CORT levels. Interestingly, nonspatial memory performance was not affected by either unpredictable foraging conditions or by experimentally elevated CORT levels. These results are consistent with the study of tree shrews that showed improved spatial, but not nonspatial memory performance in animals with chronically and moderately elevated CORT levels.

If moderately elevated CORT levels enhance memory, then why would selection not favor constantly increased levels of these hormones? Because elevated CORT also comes with negative effects, for example, a compromised immune system, even moderate CORT elevations may represent a trade-off between improvements in learning and memory and negative effects on other physiological systems. Thus, CORT elevation would be favored only when benefits from positive effects, such as enhanced memory, outweigh negative effects. For example, if food-caching birds do not retrieve cached food during energetically challenging times during the winter, they will most likely die from starvation. Compromised immune system caused by elevated CORT, on the other hand, is unlikely to result in imminent death during such times.

Development, Glucocorticoids, and Memory

Conditions under which animals develop and grow appear to have lasting effects on learning and memory, among other things. Many young growing animals cannot obtain energy resources by themselves and may have to depend on their parents for such resources. If parents are unable to provide sufficient resources, developing young often respond by elevating glucocorticoid levels. Such effects are especially prominent in altricial birds that depend entirely on their parents' ability to collect and bring sufficient amounts of food. When developing chicks do not get enough food, CORT levels rise, which has been shown to increase begging rates. Increased begging rates signal parents that more food is needed. If despite increased begging, parents are not capable of providing more food, CORT elevation may become prolonged and appears to result in long-term cognitive impairments of learning abilities, which extend well into adulthood and likely have negative consequences for survival.

For example, western scrub-jays, *Aphelocoma californica*, that were food restricted during posthatching development were reported to have significantly higher circulating CORT following the start of food restrictions. After several months, they showed impaired hippocampus-dependent spatial memory performance compared to young fed ad libitum. Young that were kept on restricted diets during posthatching development also had smaller hippocampi with fewer neurons when they were about a year old, even though they had months of unlimited food since the time they became nutritionally independent. It is likely that the negative effects of food restrictions during a relatively short period of posthatching development on memory were mediated by high CORT. These negative effects were limited to hippocampus-dependent spatial memory, while hippocampus-independent memory for color did not seem to be affected. In zebra finches (*Taenopygia guttata*), experimentally elevated plasma CORT levels during posthatching development (as well as an independent food restriction) resulted in reduced song learning abilities in males, suggesting that restricted food might affect the brain and learning via high CORT.

In both red-legged (*Rissa brevirostric*) and black-legged (*R. trydactila*) kittiwakes, experimentally enhanced CORT levels mimicking those triggered by naturally occurring food restrictions caused impairments in associative learning. This work suggests that variation in food supply may have serious long-term consequences, as learning ability appears to be crucial for survival. In addition, prenatal exposure to corticosterone in domestic chicks (*Gallus gallus domesticus*) had a negative effect on the learning component of filial imprinting. Corticosterone-treated naïve chicks were not as good as control birds in identifying specific individuals they encountered after hatching. Many mammalian studies also suggested that the long-term effects of food restrictions during postnatal development are often limited to the hippocampus and hippocampus-dependent cognitive processes.

Some avian species (e.g., tufted puffins, *Fratercula cirrhata*) that experience high variance in foraging success, which translates into highly irregular food provisioning rates for the developing offspring, appear to have evolved a muted CORT response to food deprivations. Because the young of these species do not increase CORT when food deprived on a regular basis, they avoid long-term negative effects of high plasma CORT on cognitive function.

Mechanisms of Glucocorticoid Effects on Learning

CORT may have both positive and negative effects on learning and memory, and so it is important to understand the mechanisms of such effects. There are two types of corticosteroid receptors described for the neurons in the mammalian brain – mineralocorticoid (MR; high affinity) and glucocorticoid (GR; low affinity). Activation of these receptors has been reported to induce changes in synaptic plasticity in the hippocampus, which, in turn, is likely to affect memory and learning. MR receptors are usually activated by low blood levels of glucocorticoids. GR receptors, on the other hand, are usually only partially activated with low to moderate levels. When glucocorticoid levels start increasing, more GR receptors become activated until they become fully saturated at very high levels. The strength of neuronal synaptic contacts is high when MR and only a fraction of GR receptors are activated. Activation of all GR receptors impairs long-term potentiation (LTP, strength of and number of synaptic connections) and results in reduced neuronal firing, which potentially explains impaired memory performance in animals with highly elevated CORT levels. Hippocampal neurons contain both MR and GR receptors, while neurons in the rest of the brain seem to contain mainly GR. Thus, stress-induced memory changes appear to concern hippocampus-dependent memory processes, for example, spatial memory.

The inverted-U shape relationship between glucocorticoid concentration and hippocampal-dependent learning processes appears to stem, at least in part, from the relative activation of both MR and GR receptors. When the concentration of circulating glucocorticoids is very low, all GR receptors are unoccupied and many MR receptors are unoccupied as well, resulting in a low strength of neuronal synaptic connections and low neuron firing rates, which may result in memory and learning impairments. When the concentration of blood glucocorticoids is moderately increased, MR and a fraction of GR receptors become occupied resulting in enhanced LTP and memory and learning enhancement. When the circulating level of glucocorticoids increases yet further, all MR and GR receptors are saturated resulting in impaired LTP in the hippocampus, and reduced neuron firing and hence greater impairments in memory and learning. Experimental blocking of MR receptors in rats resulted in impaired long-term potentiation (LTP) in the hippocampus, whereas blocking GR receptors resulted in enhanced LTP confirming the opposite effects of GR and MR receptor activation on hippocampal synaptic plasticity. Interestingly, zebra finches selected for high corticosterone response to acute stress showed impaired spatial learning and had lowered MR mRNA expression in the hippocampus compared to control birds.

Long-term chronic elevation in glucocorticoids may result in neuronal death, reduction in dendritic tree and dendritic atrophy, and hence likely a reduction in synaptic connections, as well as decreased hippocampal neurogenesis rates. More recent studies reported no effects of chronic CORT elevations on the total number of hippocampal neurons, but the negative effects on dendritic branching, synaptic plasticity, and neurogenesis seem to be undisputed. It also appears that most of these negative effects are not necessarily permanent and can be reversed. Furthermore, it is likely that the magnitude of increased CORT level may be important. For example, experimentally induced, moderate, long-term elevation in CORT did not affect hippocampal cell proliferation rates in mountain chickadees. Likewise, in rats, social stress did not affect hippocampal cell proliferation rates while affecting new neuron survival. In tree shrews, on the other hand, chronic stress associated with chronic CORT elevation did result in lowered hippocampal cell proliferation rates, even though spatial memory was actually improved by the treatment in this particular study.

Finally, positive effects of higher CORT titers on memory and learning may also be potentially mediated by increased glucose levels triggered by higher CORT and elevated glucose has been shown to directly enhance memory and learning.

Epinephrine

Epinephrine or adrenalin is a classical 'fight or flight' hormone produced by the adrenal glands in response to stress. Epinephrine release is fast and relatively short lived, and it usually precedes elevation of glucocorticoid hormones. Epinephrine functions similar to glucocorticoid hormones but much more rapidly to boost glucose and oxygen supply to the muscles and the brain while restricting nonvital physiological functions in quick preparation for a stressful event.

Like glucocorticoid hormones, epinephrine seems to affect memory in an inverted-U shape relationship; moderate increases enhance while low and high concentrations seem to impair learning and memory. Similar to CORT, memory is enhanced only when epinephrine is present immediately following training. If epinephrine is administered before training or fairly late after training, learning is not affected. Memory-enhancing epinephrine effects are likely adaptive because remembering circumstances under which stressful event had occurred can help an animal avoid those circumstances in the future.

Since epinephrine cannot cross the blood–brain barrier, it has been hypothesized that elevated glucose levels may mediate its effects on memory, which may be similar to the effects of CORT that also results in higher glucose levels.

Gonadal Hormones

Testosterone

Testosterone is produced by gonads in males, by ovaries in females and, in smaller amounts, by the adrenal cortex in both sexes. Plasma testosterone levels are higher in sexually mature males in most species and its concentration varies during the breeding cycle. The pattern of circulating testosterone levels depends on multiple ecological and physiological factors.

Results of numerous studies investigating the effects of testosterone on cognition have been mixed suggesting that the relationship between testosterone and learning and memory is complex and depends on numerous factors.

One of the easiest methods to practically eliminate testosterone in males is castration. Early studies comparing castrated and intact male rodents concluded that testosterone has no effect on learning and memory. Later work, however, showed enhancing effects of testosterone on memory and learning by applying testosterone to castrated animals. Numerous studies also suggested that testosterone may be important for normal memory maintenance. Removing androgens by castration decreases hippocampal synaptic density in rats and in monkeys, indicating importance of these hormones for synaptic maintenance and hence hippocampus-dependent memory and learning. Provision of exogenous testosterone restores synaptic density, suggesting a link with hippocampal synaptic morphology. Male meadow voles (*Microtus pennsylvanicus*) with naturally higher testosterone levels have been reported to have larger hippocampi, suggesting a role in hippocampal enlargements, which, in turn, might be expected to be associated with enhancement of hippocampus-dependent memory. Another study with the same species, however, reported no differences in spatial memory performance between males with naturally high and naturally low testosterone levels. Castration in male rats also did not seem to impair acquisition in a spatial memory task, but did impair spatial working memory in one study. Removal of gonads in adult male rats resulted in impaired spatial memory acquisition, and such impairments were reversed by administration of exogenous testosterone. Some hippocampal-independent learning, however, does not seem to be affected by removing testosterone.

In male white-footed mice (*Peromyscus leucopus*), testosterone levels and spatial memory performance vary with photoperiod; during long days, testosterone levels were higher and memory performance was better compared to short, winter-like days. Administration of exogenous testosterone to castrated males on short photoperiod enhanced spatial memory performance, but no effects of castration or addition of exogenous testosterone to castrates were observed on spatial memory performance in mice maintained on long days. These results suggested that photoperiod somehow interacts with the effects of testosterone on spatial memory.

The interactions between testosterone and cognition also appear to be dependent on sex and age. In many animals, including humans, testosterone levels in males decline gradually with age. Such a decline is usually correlated with declines in memory and learning. Several studies also reported correlations between performance on memory tasks and testosterone levels in humans. Experimentally increased testosterone levels usually enhance spatial and working memory in older human males, but such effects are not found in young males.

Aromatase enzymes can convert testosterone into estradiol within the brain in cells of the hippocampus and amygdala. In mammals, both males and females appear to have receptors for androgens and estrogens in the hippocampus, amygdala, and prefrontal cortex, all brain areas involved in cognitive functions. Testosterone may potentially affect cognition either directly as androgen or indirectly as estrogen when converted into estradiol within the brain. Some studies, on the other hand, suggested that there are no testosterone receptors in the hippocampus while there are estrogen receptors. Because testosterone may affect cognition both as an androgen and as an estrogen, it remains unclear whether and when testosterone may directly affect memory and learning or whether most of the effects of testosterone on memory are via conversion of testosterone into estradiol. Careful studies using either aromatase inhibitors or providing exogenous estradiol or testosterone to gonadectomized animals may help to separate the effects of these two hormones on cognitive functions.

One study reported that when aromatase inhibitors were used to prevent conversion of testosterone into estradiol in the brain, extra testosterone improved spatial but not verbal memory in human males, suggesting the direct involvement of testosterone in spatial memory regulation. Improvements in verbal memory were observed only in the absence of aromatase inhibitors, suggesting a role of estradiol in verbal memory. In a different experiment, castrated male rats were given either testosterone or estradiol and rats treated with estradiol showed improvements in acquisition of a spatial memory task while rats treated with testosterone only improved working memory. These results suggest that testosterone may indeed affect memory directly and that testosterone and estradiol may affect different memory systems.

In castrated male zebra finches, implantation of estradiol resulted in enhanced spatial memory performance. Administration of testosterone implants also resulted in enhanced acquisition of spatial memory and in enlarged size of neurons in the rostral hippocampus, but administration of dihydrotestosterone, an androgen that cannot be converted into estradiol by aromatase, resulted in birds not learning the spatial task. The authors of this experiment concluded that testosterone affected the hippocampus and spatial memory via conversion into estrogen. A surprising result of that experiment was that castrated males with no implants did better on acquisition of a spatial memory task than males implanted with dihydrotestosterone. One possibility is that androgens not converted into estrogens by aromatase may actually be detrimental to spatial memory, but more research is needed to verify this claim.

Most studies investigating testosterone effects on memory used manipulated castrated animals, while not much has been done with hormonal manipulations in intact animals. Future studies should also aim to integrate the interplay between environmental variables, testosterone levels, and cognition in different species to better understand the biological trade-offs associated with elevated testosterone.

Estradiol

Estradiol is produced by ovaries in females and to a lesser extent by the adrenal cortex in both sexes, but it is also produced in the brain in both males and females by aromatization of testosterone. Estradiol levels are naturally elevated in sexually mature females in mammals, in addition to estrous cycle fluctuations. In birds, estradiol is elevated during breeding and is low during nonbreeding periods.

It has been well established that estradiol has a direct effect on the properties of the hippocampal neurons (see also earlier), which are likely to affect hippocampus-dependent learning. Higher levels of estrogens appear to be associated with greater dendritic spine density in the hippocampus of the rats and were reported to vary by as much as 30% during the estrous cycle. Rodent studies showed that providing exogenous estradiol results in increased number of dendritic spines in the hippocampal neurons, along with the increased number of synapses. Such changes in hippocampal neurons caused by elevated estrogen levels are likely the cause of enhanced hippocampus-dependent memory. Administration of exogenous estrogen to gonadectomized rats resulted in enhanced spatial memory performance in many experiments. In meadow voles, estradiol treatment of castrated males resulted in increased survival of hippocampal neurons and better spatial memory performance.

However, on the basis of its effects on hippocampal morphology, it could be expected that estradiol should always have an enhancing effect on hippocampus-dependent memory. It has been reported that estradiol might have both enhancing and impairing effects. First, it appears that the relationship between estradiol and memory follows an inverted-U shape relationship as observed for several other hormones (see earlier). Absence or very low plasma concentrations and very high levels are usually associated with impaired memory, while moderate concentrations are associated with enhanced memory, especially hippocampus-dependent memory. Second, it has been suggested that estradiol might improve some memory types (e.g., working memory) while impairing the others (e.g., reference memory), but support for this idea has been mixed. It has also been suggested that estrogens might specifically affect memory acquisition and consolidation, but not memory retrieval. Administration of estradiol prior to or immediately following training on a memory task seems to enhance memory performance in rodents, while application of estradiol after a period of time following training has no effect on memory.

The reported effects of estrogens on cognition also appear to vary depending on specific cognitive tasks. For example, acquisition of a spatial task in gonadectomized rats appears to be enhanced by providing exogenous estradiol, but once the animals acquired the task, estrogens did not seem to provide any more enhancements. Adding exogenous estradiol to ovariectomized rats also seems to enhance working memory on spatial memory tasks. Thus, estrogens appear to enhance recognition memory in ovariectomized rats allowing animals better discrimination between familiar and unfamiliar objects. Such effects may be adaptive and better discrimination memory especially during breeding period may enhance animals' fitness.

The relationship between estradiol and memory also appears to be age dependent. Experiments showed that spatial memory performance in older female rats correlates with blood estrogen levels; females with lower estrogen perform worse on spatial memory tasks than females with higher levels. In human postmenopausal females, estrogen supplement seems to have a positive effect on memory and learning. Estradiol replacement in rats with removed ovaries enhanced spatial memory and higher concentration of estradiol resulted in better memory performance. Simply removing ovaries, on the other hand, produced mixed results in rats. Young rats showed impaired memory after one day following ovariectomy, while memory performance of middle-aged rats was not affected by ovariectomy. In non-human primates, estrogen effects on memory appear to be especially pronounced in aged animals.

Several studies in rodents and primates reported correlational results showing that females with naturally higher estrogen levels performed worse on a spatial memory task compared to females with naturally lower levels. Interestingly, one study reported that female meadow voles with high estrogen levels had significantly larger hippocampi compared to low estrogen females. This result seems contradictory to the findings of negative effects of high estradiol levels on memory, as larger hippocampus and denser dendritic branching with more synapses, all of which are usually associated with elevated estrogen levels, have been linked to enhanced memory performance. Administration of a high dose of estradiol to ovariectomized female meadow voles, however, also resulted in impaired spatial memory confirming the hypothesis of the inverse-U shape relationship between estradiol and memory.

Although males maintain lower estrogen levels than females, no differences in the number of estrogen receptors in the hippocampus between males and females have been reported in rodents, again confirming the idea that in males testosterone affects the hippocampus via conversion into estrogen. Interestingly, adding exogenous estradiol to castrated males resulted in improved memory performance on some spatial memory compared to control castrated males.

Overall, numerous studies have suggested that estrogens are important for learning and memory processes and that the effects of estrogens on memory may depend on estrogen concentration, memory type, and age of animals. Most of the studies, however, used ovariectomized animals, while the effects of temporary elevations in estrogens on memory in intact animals remain less clear. Estrogen levels increase naturally in many animals during breeding, and thus it is important to understand how such elevations might affect fitness-related functions such as memory and learning. Some studies showed that animals with high estrogen levels perform worse on spatial memory tasks compared to individuals with lower estrogen levels. But these comparisons are correlative and it is possible that other differences exist between animals with high and low estrogen levels. Future studies should investigate the suggested inverse-U shape relationship between estradiol levels and memory. It is important to test the effects of temporary moderate elevations in estrogens on memory in intact animals. In some species, memory appears to be especially crucial for successful reproduction and understanding the relationship between estrogens and learning and memory in such species might be especially relevant. For example, female parasitic cowbirds need to maintain complex memories for locations of numerous host nests, content/stage of these nests, and when these nests were found. Enhanced memories appear to be crucial for the fitness of these birds and so it will be interesting to determine whether estrogens elevated during egg laying have an effect on memory in these birds.

Other Hormones

Progesterone

Progesterone is produced in gonads, adrenal cortex, and also in the brain in both males and females. Progesterone concentration is low in young and in old individuals (e.g., postmenopausal women), and it also fluctuates during estrous cycle in sexually mature females.

Reports of the effects of progesterone on memory appear to be conflicting. In some cases, elevated progesterone has been suggested to increase dendritic spine density in hippocampal neurons, and thus progesterone may potentially be beneficial for memory function. In many studies, the effects of progesterone on memory have been studied in conjunction with estradiol, and combination of these two hormones seems to be important for regulation of memory. Rodent studies, for example, showed a positive correlation between the concentration of both estradiol and progesterone the density of dendritic spines on hippocampal neurons. In ovariectomized rats, administration of progesterone enhanced the positive effect of estradiol on dendritic spine density in the hippocampal neurons.

Other studies, however, suggested that progesterone interferes and counteracts the enhancing effects of estradiol on dendritic spines. Some investigations, for example, showed negative effects of exogenous progesterone on spatial memory task in young rats. When applied together with estradiol, progesterone has often been reported to block the enhancing memory effects of estradiol in young female rats. On the other hand, progesterone when administered with estradiol to older rats improved their spatial memory. Progesterone also seems to counter the memory-enhancing influence of estradiol in a concentration-dependent fashion in which only large doses are effective.

Adrenocorticotropic Hormone

Adrenocorticotropic hormone (ACTH; produced by the pituitary gland, but also expressed centrally) appears to be important for memory maintenance, and administration of exogenous ACTH to hypophysectomized rats (with pituitary gland removed) seems to improve learning and memory. Experimentally elevating ACTH levels in intact animals, on the other hand, does not seem to produce any enhancements.

Some other hormones (such as insulin, vasopressin and oxytocin, thyroid hormones) have also been suggested to affect memory and learning, but their effects on cognitive functions have not been investigated as intensively as those of glucocorticoid and gonadal hormones.

See also: Fight or Flight Responses; Hormones and Behavior: Basic Concepts; Spatial Memory; Stress, Health and Social Behavior.

Further Reading

Bartolomucci A, de Biurrun G, Czeh B, van Kampen M, and Fuchs E (2002) Selective enhancement of spatial learning under chronic stress. *European Journal of Neuroscience* 15: 1863–1866.

Daniel JM (2006) Effects of oestrogen on cognition: What have we learned from basic research? *Journal of Neuroendocrinology* 18: 787–795.

De Kloet ER, Oitzl MS, and Joels M (1999) Stress and cognition: Are corticosteroids good or bad guys? *Trends in Cognitive Sciences* 22: 422–426.

Janowski JS (2006) Thinking with your gonads: Testosterone and cognition. *Trends in Cognitive Sciences* 10: 77–82.

Kitaysky AS, Kitaiskaia EV, Piatt JF, and Wingfield J (2006) A mechanistic link between chick diet and decline in seabirds? *Proceedings of the Royal Society B* 273: 445–450.

Luine VN (2008) Sex steroids and cognitive function. *Journal of Neuroendocrinology* 20: 866–872.

Lupien SJ and Lepage M (2001) Stress, memory, and the hippocampus: Can't live with it, can't live without it. *Behavioural Brain Research* 127: 137–158.

McEwen BS and Sapolsky RM (1995) Stress and cognitive function. *Current Opinion in Neurobiology* 5: 205–216.

McGaugh JL and Roozendal B (2002) Role of adrenal stress hormones in forming lasting memories. *Current Opinion in Neurobiology* 12: 205–210.

Pravosudov VV (2005) Corticosterone and memory in birds. In: Dawson A and Sharp P (eds.) *Functional Avian Endocrinology*, pp. 257–268. New Delhi, India: Narosa Publishing House.

Rosendal B (2002) Stress and memory: Opposing effects of glucocorticoids on memory consolidation and memory retrieval. *Neurobiology of Learning and Memory* 78: 578–595.

Sandi C, Rose SPR, Mileusnic R, and Lancashire C (1995) Corticosterone facilitates long-term memory formation via enhanced glycoprotein synthesis. *Neuroscience* 69: 1087–1093.

Mental Time Travel: Can Animals Recall the Past and Plan for the Future?

N. S. Clayton and A. Dickinson, University of Cambridge, Cambridge, UK

Introduction

In an influential paper that was published in 1997, Suddendorf and Corballis argued that we humans are unique among the animal kingdom in being able to mentally dissociate ourselves from the present. To do so, we travel backwards and forwards in the mind's eye to remember and reexperience specific events that happened in the past (*episodic memory*) and to anticipate and preexperience future scenarios (*future planning*). Although physical time travel remains a fictional conception, mental time travel is something we do for a living, and the fact that we spend so much of our time thinking about the past and the future led to Mark Twain's witty remark that "my life has been filled with many tragedies, most of which never occurred."

Mental time travel then has two components: a retrospective one in the form of episodic memory and a prospective one in the form of future planning. In formulating their *mental time travel hypothesis*, Suddendorf and Corballis were the first to suggest that episodic memory and future planning are intimately linked and can be viewed as two sides of the same coin so to speak. In fact, their proposal consisted of two claims. In addition to integrating the retrospective and prospective components of mental time travel, they also argued that such abilities were unique to humans and reflected a striking cognitive dichotomy between ourselves and other animals. The latter idea was not new, however, but rather an extension of what others have argued makes episodic memory special.

Indeed in his seminal studies of human memory, Tulving coined the term episodic memory in 1972 to refer to the recollection of specific personal happenings, a form of memory that he claimed was uniquely human and fundamentally distinct from semantic memory, the ability to acquire general factual knowledge about the world, which he argued we share with most, if not all, animals. Ever since he made this remember–know distinction, most cognitive psychologists and neuroscientists have assumed that episodic memory is special because of the experiential nature of these memories, namely that our episodic reminiscences are accompanied by a subjective awareness of currently reexperiencing an event that happened in the past, as opposed to just knowing that it happened. Of course we also have many instances of knowledge acquisition in which we do not remember the episode in which we acquired that information. For example, although most of us know when and where we were born, we do not remember the birth itself nor the episode in which we were told when our birthday is, and therefore such memories are classified as semantic as opposed to episodic.

Episodic and semantic memory, then, are thought to be marked by two separate states of awareness; episodic remembering requires an awareness of reliving the past events in the mind's eye and of mentally traveling back in one's own mind's eye to do so, whereas semantic knowing only involves an awareness of the acquired information without any need to travel mentally back in time to personally reexperience the past event. It is for this reason that in later writings, Tulving has argued that one of the cardinal features of episodic memory is that it operates in 'subjective time,' and he refers to the awareness of such subjective time as *chronesthesia*.

Language-based reports of episodic recall suggest that the retrieved experiences are not only explicitly located in the past but are also accompanied by the conscious experience of one's recollections, feeling that one is the author of the memory, or of traveling back not in any mind's eye but in *my* mind's eye, what Tulving called *autonoetic consciousness*. In other words, Tulving and others argue that episodic memory differs from semantic memory not only in being oriented to the past, but specifically in the past of the owner of that memory. So while some semantic knowledge, such as the birth date example described earlier, does involve a datable occurrence, these memories are fundamentally distinct from episodic memories because they do not require any mental time travel. As William James so aptly wrote "Memory requires more than the mere dating of a fact in the past. It must be dated in *my* past."

From a biological perspective, the characterization of episodic memory in terms of these two phenomenological properties of consciousness, namely autonoesis and chronesthesia, presents major problems for two reasons. The first is that positing a subjective state of awareness is difficult to integrate with evolutionary processes of natural and sexual selection, which operates on behavioral attributes such as reproductive success and survival rather than on mental states. The second is that this definition makes it impossible to test in nonverbal animals, in the absence any agreed behavioral markers of non-Linguistic consciousness. Adopting an ethological approach to comparative cognition necessitates two requirements. The first is that the memory needs to be characterized in terms of behaviorally defined properties as opposed to purely phenomenological ones, such as the types of information encoded. Indeed, we shall argue that the ability to remember what happened, where and how long ago is a critical behavioral criterion for episodic memory. The second requirement is the identification of an ethological context in which these memories would confer a selective advantage. Note that by doing so, we transform this debate about the human uniqueness of mental time travel into an empirical evaluation in non-Linguistic animals as opposed to restricting it to the realms of philosophical personal ponderings. But before doing so, let us return to the two claims made by the mental time travel hypothesis: (1) future planning and episodic memory are subserved by a common process, mental time travel, and (2) this process is uniquely human. We shall evaluate each of these claims in turn, and argue that there is good evidence to support the first claim, but that considerably more controversy surrounds the second component of Suddendorf and Corballis' thesis.

Evidence to support the first claim comes from a number of sources. First, studies of brain activity while engaged in either memory retrieval or future-oriented tasks identify a specific core network of regions in the brain of healthy human adults that support both episodic recollection and future planning. Moreover, there are patients such as DB and KC, who show specific impairments in episodic but not semantic memory, and these patients have similar deficits in episodic but not semantic forethought. Finally, studies of cognitive development in young children suggest that episodic memory and future planning both emerge at about the same age, and are not properly developed until children reach the age of about four.

Is Mental Time Travel Unique to Humans?

Regarding the second claim about the uniqueness of episodic memory and future planning, if we are to adopt an ethological approach of the form we outlined earlier, then the question becomes one of asking where in the natural world these two processes might intersect, in which species, and under what conditions. One classic candidate is the food-caching behavior of corvids, members of the crow family that include jays, magpies, and ravens as well as the crows. These large-brained, long-lived, and highly social birds hide food caches for future consumption, and rely on memory to recover their caches of hidden food at a later date, typically weeks if not months into the future. So clearly food-caching is a behavior that is oriented toward future needs. Indeed, the act of hiding food is without obvious immediate benefit and yields its return only when the bird comes to recover the caches it made. Given that these birds are dependent on finding a significant number of these caches for survival in the wild,

it seems likely that the selection pressure for an excellent memory for the caches would have been particularly strong, especially as they cache year round.

These birds also cache reliably in the laboratory, providing both ethological validity and experimental control. At issue, however, is whether or not these birds episodically remember the past and plan for the future. For these reasons, we shall now turn our attention to assessing the evidence as to whether or not these food-caching corvids can remember the past and plan for the future.

Episodic Memory

As we stated earlier, language-based reports of episodic recall in humans suggest that the retrieved experiences are not only explicitly located in the past but are also accompanied by the conscious experience of one's recollections. From a comparative perspective, the problem with this definition, however, is that in the absence of agreed non-Linguistic markers of consciousness, it is not clear how one could ever test whether animals are capable of episodic recollection. For how would one assess whether or not an animal can experience an awareness of the passing of time and of reexperiencing one's own memories while retrieving information about a specific past event.

Behavioral criteria for episodic memory
This dilemma can be resolved to some degree, however, by using Tulving's original definition of episodic memory, in which he identified episodic recall as the retrieval of information about 'where' a unique event occurred, 'what' happened during the episode, and 'when' it took place. The advantage of using this definition is that the simultaneous retrieval and integration of information about these three features of a single, unique experience may be demonstrated behaviorally in animals. Clayton and Dickinson termed this ability 'episodic-like memory' rather than episodic memory because we have no way of knowing whether or not this form of remembering is accompanied by the autonoetic and chronesthetic consciousness that accompanies human episodic recollections. Indeed, we have argued that the ability to remember the 'what-where-and-when' of unique past episodes is the hallmark of episodic memory that can be tested in animals.

Empirical tests of episodic-like memory
We focus our analysis on one particular species of food-caching corvid, the western scrub-jay, capitalizing on one feature of their ecology, namely, the fact that these birds cache perishable foods, such as worms, as well as nondegradable nuts, and as they do not eat rotten items, recovering perishable food is only valuable as long as the food is still fresh. In a classic experiment published

in 1998, we tested whether the jays could remember the 'what, where, and when' of specific caching events.

Although the birds had no cue predicting whether or not the worms had perished other than the passage of time that had elapsed between the time of caching and the time at which the birds could recover the caches they had hidden previously, the birds rapidly learned that highly preferred worms were fresh and still delicious when recovered 4 h after caching, whereas after 124 h, the worms had decayed and tasted unpleasant. Consequently, the birds avoided the wax worm caches after the longer retention interval and instead recovered exclusively peanuts, which never perish. Following experience with caching and recovering worms and peanuts after the short and long intervals, probe tests, in which the food was removed prior to recovery, showed that they relied on memory to do so rather than cues emanating directly form the food. Subsequent tests revealed that the jays could remember which perishable foods they have hidden where and how long ago, and irrespective of whether the foods decayed or ripened.

Since the initial studies, a number of other laboratories have also turned their attention to the question of whether or not animals have episodic-like memory. Using paradigms analogous to those employed with the jays, there is now good evidence that rats, mice, and magpies can remember the what-where-and-when and what-where-and-which of past events.

Forethought

If forethought, at least in the form of episodic future thinking, falls under the general umbrella of mental time travel and is the reason why episodic memory evolved in the first place as we suggested in the introduction, then we should expect to find a concomitant development of episodic memory and episodic future thinking. So if one accepts the evidence that the scrub-jays can episodically recall the past, at least in terms of the behavioral criteria, then these birds should also be capable of planning for the future. The topic is of course a controversial one, and indeed there is much debate about whether non-human animals are capable of forethought (see, e.g., the arguments of Suddendorf and Corballis, and the responses from my laboratory). For how does one test whether the jays' caching decisions are controlled by future planning?

Behavioral criteria for future planning

The first distinction that one must draw is between prospectively oriented behavior and future planning. Several anticipatory activities, including migration, hibernation, nest building, and food-caching, are clearly conducted for a future benefit as opposed to a current one, but they would not constitute a case of future planning unless one could demonstrate the flexibility underlying cognitive control, and thereby rule out simpler accounts in terms of behavior triggered by seasonal cues or previous reinforcement of the anticipatory act.

So the first issue to address is whether the caching behavior of the jays is sensitive to its consequences. To do so, once again we capitalized on the fact that the jays love to eat and cache fresh worms but that they do not eat them once they have degraded. We used a variant of the Clayton and Dickinson (1998) caching paradigm in which the jays were given fresh worms and nuts to cache before recovering them 2 days later. In contrast to the original experiments on episodic-like memory, in which the state of the worm caches varied with the retention interval, in the future-planning experiment, the worms were always degraded at recovery in order to investigate their choice of what to cache, as opposed to where to search at recovery. The objective of this experiment was to assess whether or not the birds could learn that even though the worms were fresh at the time of caching there was no point in caching them because they would always be degraded and therefore unpalatable at the time of recovery. The jays rapidly learned to stop caching the worms, even though they continued to eat the fresh worms at the time of caching, thereby demonstrating that caching is indeed selective to its consequences in the sense that the jays could learn what not to cache.

The Bischof–Köhler hypothesis

Suddendorf and Corballis have also argued that a critical feature of future planning is that the subject can take action in the present for a future motivational need, independent of the current motivation. Indeed, they argued that mental time travel provided a profound challenge to the motivational system in requiring the subject to suppress thoughts about one's current motivational state in order to allow one to imagine future needs, and to dissociate them from current desires.

To illustrate this distinction between current and future motivational states, consider the following example. A current desire for a croissant at breakfast may lead to an early morning trip to the local baker. Of course it will take some time to reach the market, and therefore the croissant will not be eaten now but in a few minutes time. But although the croissant will be eaten at a future time as opposed to the present, this behavior would not fulfill the Bischof–Köhler criterion because the action is governed by one's current motivational state. By contrast, going to the baker's shop in order to ensure that there are croissants for tomorrow's Sunday brunch would be an example of the future planning envisaged by the Bischof–Köhler hypothesis because this action would be performed for a future motivational need, independent of one's current needs.

This hypothesis was inspired by a comparative perspective, from reviewing the evidence for human and non-human primate cognition, and indeed it has led to a number of tests of whether animals can dissociate current from future motivational needs. In one study to address this issue, Naqshbandi and Roberts gave squirrel monkeys the opportunity to choose between eating four dates and eating just one date. Eating dates makes monkeys thirsty, but rather than asking the monkeys to chose between water and the dates, the experimenters manipulated the delay between the choice (one vs. four dates) and receiving water such that the monkeys received water after a shorter delay if they had chosen the one date rather than the four dates. The monkeys gradually reversed their natural preference for four dates, suggesting that they were anticipating their future thirst. However, because the monkeys received repeated trials in which they learnt the consequences of their choices, one can give a simple associative explanation in terms of reinforcement of the anticipatory act by avoidance of the induction of thirst.

More convincing evidence for a dissociation of current and future motivational states comes from a study by Correia, Dickinson, and Clayton on the food-caching scrub-jays. Like many other animals, when sated of one type of food, these birds prefer to eat and cache another type of food. Correia and colleagues capitalized on this specific satiety effect to test whether the birds would choose to cache the food they want now or the food they think they will want when they come to recover their caches in the future. In the critical group, the birds were sated on one of two foods that were both then made available for caching. Then, immediately prior to the recovery of these caches, they were sated on the other food. Consequently, the food that was valuable at recovery was the one that was less valuable at the time of caching. At the beginning of the experiment, the birds cached the food they desired at the time, but then rapidly switched to storing preferentially the food that was valuable at the time of recovery rather than the one they wanted to eat at the time of caching, suggesting that the jays can plan future actions on the basis of what they anticipate they will desire in the future as opposed to what they need now. So this study supports the notion that jays can dissociate future from current motivational needs, and therefore provides direct evidence to challenge the Bischof–Köhler hypothesis (for further discussion see our recent review in *Animal Behaviour*).

For the skeptic, however, this kind of task need not require prospective mental time travel because the scrub-jay does not need to imagine a future situation. Suppose that the act of recovering a particular food recalls the episode of caching that food. If the bird is hungry for that particular food, then recovering it will be rewarding and therefore this could directly reinforce the act of caching the food through the memory of doing so. The point is that such memory-mediated reinforcement does not require the bird to envisage future motivational states.

Tulving's spoon test

Tulving has argued that it is possible to test whether animals are capable of such episodic future thinking, and devised what he calls the 'spoon test,' which he argues is a 'future-based test of autonoetic consciousness that does not rely on and need not be expressed through language.' The test is based on an Estonian children's story tale, in which a young girl dreams about going to a birthday party. In the dream, all of her friends are eating a delicious chocolate mousse, which is her favorite pudding, but alas she cannot because she does not have a spoon with her, and no one is allowed to eat the pudding unless they have their own spoon. As soon as she gets home she finds a spoon in the kitchen, carry it up to her bedroom and hides – or caches – it under her pillow, in preparation for future birthday parties and even dreams of future birthday parties for that matter.

The point then is to use past experience to take action now for an imagined future event. To pass the spoon test, an animal must act analogously to the little girl carrying her own spoon to a new party, a spoon that has been obtained in another place and at another time. Is there any evidence that animals and young children can pass this spoon test? Although some animals, notably primates and corvids (namely the scrub-jays we discussed earlier), have been shown to take actions now based on their future consequences, most of these studies have not shown that an action can be selected with reference to future motivational states independent of current needs as discussed in the previous section.

Mulcahy and Call were the first to devise a spoon test for animals. In their study a variety of species of non-human apes were first taught to use a tool to obtain a food reward that would otherwise have been out of reach, before being given the opportunity to select a tool from the experimental room, which they could carry into the sleeping room for use the following morning. Although most of the subjects did choose the correct tool on some trials, the individual patterns of success for each subject was not consistent across subsequent trials, as one would expect if they had a true understanding of the task. Furthermore, the apes received a number of training trials, so reinforcement of the anticipatory act cannot be ruled out. A more convincing case of planning was provided by Osvath and Osvath. In a recent series of experiments, these authors demonstrated that when selecting a tool for use in the future, chimpanzees and orangutans can override immediate drives in favor of future needs.

One of the most striking examples of the spoon test in animals comes from recent studies of the food-caching scrub-jays. In the laboratory, work by Raby and colleagues showed that our jays can spontaneously plan for tomorrow's breakfast without reference to their current motivational state. The birds were given the opportunity to learn that they received either no food or a particular type of food, for breakfast in one compartment, while receiving a different type of food for breakfast in an alternative compartment. Having been confined to each compartment at breakfast time for an equal number of times, the birds were unexpectedly given the opportunity to cache food in both compartments one evening, at a time when there was plenty of food for them to eat and therefore no reason for them to be hungry. Given that the birds did not know which compartment they would find themselves in at breakfast tomorrow and on the assumption that they prefer a variety of foods for breakfast, we predicted that if they could plan for the future, then they should cache a particular food in the compartment in which they had not previously had it for breakfast.

This the birds did, suggesting that they could anticipate their future desires at breakfast time tomorrow when they would be hungry. Importantly, because the birds had not been given the opportunity to cache during training, we can in this experiment rule out an explanation in terms of mediated reinforcement of the anticipatory act. These findings led Shettleworth to argue that "two requirements for genuine future planning are that the behavior involved should be a novel action or combination of actions and that it should be appropriate to a motivational state other than the one the animal is in at that moment ... Raby et al. describe the first observations that unambiguously fulfill both requirements."

Although it seems clear that the scrub-jays and chimpanzees do pass the spoon test, at issue, however, is whether or not these tasks truly tap episodic future thinking. Indeed, we have argued that in the absence of language, there is no way of knowing whether the jays' ability to plan for future breakfasts reflects episodic future thinking, in which the jay projects itself into tomorrow morning's situation, or semantic future thinking, in which the jays acts prospectively but without personal mental time travel into the future. In the latter case, all that the subject has to do is to work out what has to be done to ensure that the implement is to hand, be it a spoon, some other tool, or a food-cache. In no sense does this task require the subject to imagine or project one's self into possible future episodes or scenarios. As Raby et al. have argued, however, what these studies do demonstrate is the capacity of animals to plan for a future motivational state that stretches over a timescale of at least tomorrow, thereby challenging the assumption that this ability to anticipate and act for future needs evolved only in the hominid lineage.

Concluding Remarks

The mental time travel hypothesis of Suddendorf and Corballis makes two claims. We have argued that the first claim that episodic memory and future planning are intimately linked and subserved by the same common process of mental time travel has good support. However, we challenge the second claim about human uniqueness. Indeed, we have argued that at least some animals, notably a few primates and corvids, are capable of recollecting the past and planning for the future. In the case of the scrub-jays, the functional account of caching appears to be reflected in the psychological processes underlying this behavior; by fulfilling the behavioral criteria we have outlined, they therefore show at least some elements of episodic memory and forethought. It also serves as a superb illustration of the integration of the retrospective and prospective components of mental time travel for there is no benefit to the animal of hiding food at the time of caching. The benefit occurs when recovering the caches at a future time, and to do so effectively, the jays must rely on their episodic-like memories of past caching events to know where to search for their hidden stashes of food.

See also: Intertemporal Choice; Time: What Animals Know.

Further Reading

Clayton NS, Bussey TJ, and Dickinson A (2003) Can animals recall the past and plan for the future? *Nature Reviews Neuroscience* 4: 685–691.

Clayton NS, Correia SPC, Raby CR, Alexis DM, Emery NJ, and Dickinson A (2008) In defense of animal foresight. *Animal Behaviour* 76: e1–e3.

Clayton NS and Dickinson A (1998) Episodic-like memory during cache recovery by scrub jays. *Nature* 395: 272–274.

Correia SPC, Alexis DM, Dickinson A, and Clayton NS (2007) Western scrub-jays (*Aphelocoma californica*) anticipate future needs independently of their current motivational state. *Current Biology* 17: 856–861.

James W (1890) *The Principles of Psychology*. New York: Holy.

Mulcahy NJ and Call J (2006) Apes save tools for future use. *Science* 312: 1038–1040.

Naqshbandi M and Roberts WA (2006) Anticipation of future events in squirrel monkeys (*Saimiri sciureus*) and rats (*Rattus norvegicus*): Tests of the Bischof–Kohler hypothesis. *Journal of Comparative Psychology* 120: 345–357.

Osvath M and Osvath H (2008) Chimpanzee (*Pan troglodytes*) and orang-utan (*Pongo abelii*) forethought: Self-control and pre-experience in the face of future tool use. *Animal Cognition* 11: 661–674.

Raby CR, Alexis DM, Dickinson A, and Clayton NS (2007) Planning for the future by western scrub-jays. *Nature* 445: 919–921.

Relevant Websites

http://www.psychol.cam.ac.uk/ccl/ – Department of Experimental Psychology: Research.

http://www.neuroscience.cam.ac.uk/directory/profile.php?nsclayton – Professor Nicky Clayton: Cambridge Neuroscience.

http://www.youtube.com/watch?v=y_MnwNyX0Ds – Bird Tango.

http://www.sciencemag.org/cgi/content/full/315/5815/1074.

Metacognition and Metamemory in Non-Human Animals

R. R. Hampton, Emory University, Atlanta, GA, USA

Introduction

Metacognition generally means thinking about thinking. Metacognition can allow one to monitor and adaptively control cognitive processing. For example, a human student might improve his/her grade by dedicating more of his/her study effort to the longest textbook chapters and the most difficult topics on an upcoming exam. He/she might restudy the definitions of terms he/she is less familiar with or finds that he/she forgot after a single study session. During the exam, he/she might skip questions whose answers he/she is unsure of, returning to them only after first answering questions about which he/she is confident. In each case, our student has monitored the difficulty faced in learning or performing and has controlled his/her behavior appropriately.

While most interest in metacognition is focused on such monitoring and control of one's own cognitive processes, metacognition can also refer to a general knowledge about how cognition works. For example, metacognitive knowledge refers to a variety of information characterizing cognition in general, such as knowing that forgetting happens over time, that one has to attend carefully to follow complex directions, and that some people are better at math than others. This article deals with metacognition as monitoring and controlling one's own cognitive functioning rather than knowing more generally how cognition works.

Approaching the Study of Metacognition in Non-Human Animals

Metacognition in humans is often regarded as being associated with consciousness and complex cognition. These characterizations raise concerns about the feasibility of studying metacognition in non-human species. But metacognition can be operationalized and studied with objectively observable behavior as will be described in this article. Studies of metacognition in non-human animals have focused on the ability of subjects to monitor and control their own cognitive states. In order to objectively determine whether such monitoring and control occurs, experiments have been designed with three critical features. First, the experimenter defines a primary behavior that can be scored for accuracy or efficiency such as performance in a test of matching-to-sample (MTS; this is a memory test in which subjects are required to select a recently experienced stimulus from among a set of

distracter stimuli). Next, the experimenter defines a secondary behavior that can be used to infer monitoring and control of the cognition underlying the primary behavior, such as the subjects avoiding difficult tests, or seeking additional information when they do not know the correct response to make. Finally, the experimental design must allow for an explicit assessment of whether the primary and secondary behaviors are correlated. For example, were the tests that the subjects avoided indeed ones on which they were likely to respond incorrectly? This correlation can be assessed most powerfully when the subjects' state of knowledge is experimentally manipulated and can therefore be confidently known. If subjects avoid memory tests for which they have never been shown the correct answer while taking tests for which the answer was recently presented, this would be consistent with metacognition.

Studies of Metacognition in Non-Human Animals

Non-human animals have demonstrated metacognition in a variety of experiments with the features described earlier. These experiments can be classified according to whether they required metacognition about perception or about memory. Monkeys, dolphins, pigeons, and rats have been shown to either decline difficult trials or make accurate posttrial confidence judgments in perceptual tests. Apes and monkeys have similarly performed in ways consistent with metacognition on memory tests, while pigeons are generally reported not to do so. It should be emphasized, however, that while species differences in metacognition would clearly be of interest, there is currently insufficient data available to reach any firm comparative conclusions. In the following section, a few representative types of test of non-human metacognition are described.

Avoiding Difficult Perceptual Tests

The first study of metacognition in a non-human species was published by David Smith and his colleagues and described the performance of a bottle-nosed dolphin (*Tursiops truncatus*) in an auditory psychophysical task. The dolphin was required to discriminate between tones of 2100 Hz and tones of any lower frequency (ranging from 1200 to 2099 Hz). It was initially trained to make this discrimination (the primary behavior) by responding to a left paddle following 2100 Hz tones and to a right paddle for any lower frequency tone. As expected, the dolphin's accuracy decreased as the tested frequency approached 2100 Hz (the dolphin was likely to respond to the left paddle when the frequency was close to 2100 Hz, treating these tones as if they were 2100 Hz tones).

After the dolphin had acquired this primary discrimination, a third paddle was introduced that allowed the dolphin to decline a given discrimination trial (the secondary behavior) in favor of an easy discrimination (a 1200 Hz tone). With these contingencies in place, the dolphin could maximize the rate of reward by performing the primary discrimination (choosing the left or right paddle) when the discrimination was easy, while selecting the third paddle when the discrimination was difficult. The dolphin's behavior generally conformed to these contingencies. It was unlikely to use the third paddle following low frequencies (the easiest trials) and was increasingly likely to use this 'decline test' paddle following frequencies near 2100 Hz (the most difficult trials). Later work by Smith and his colleagues showed that monkeys behaved the same way in an analogous psychophysical test in which the density of pixels in a visual display substituted for tones. Humans given a nearly identical test showed patterns of behavior very similar to those shown by the monkeys. It is interesting to note that the humans reported that they used the 'decline test' response only when they felt uncertain.

Confidence Judgments Following Tests

A retrospective gambling paradigm was developed by Herb Terrace and his colleagues to assess the ability of monkeys to accurately judge how likely their choices on trials they had just completed were to be correct. In this paradigm, monkeys rated their 'confidence' by wagering either a large or small number of video tokens on the accuracy of each test trial immediately after they completed it. The video tokens were secondary reinforcers that were periodically 'cashed out' for actual food when a sufficient number had accumulated. Critically, monkeys placed their wager after answering, but before receiving feedback about their accuracy. In this paradigm, metacognition predicts large wagers following easy tests (i.e., when monkeys are confident of their answer) and small wagers following difficult tests (i.e., when monkeys would be unsure of their answer). This in indeed how the monkeys performed in tests on which they were required to discriminate line lengths. These results suggest that they knew whether they had responded correctly despite the lack of feedback prior to placing their bet. Monkeys trained to make these confidence judgments immediately generalized the ability from perceptual tests to memory tests, showing that performance was

not restricted to a specific set of test stimuli or even a particular cognitive domain.

Avoiding Difficult Memory Tests

When subjects are presented with lists of items to remember (such as the list of salad dressings available with your order at a restaurant), it is typical for items early and late in the list to be remembered better than items in the middle of the list. Such serial position effects have been a staple of memory research in humans and non-humans. Work with monkeys took advantage of this predictable pattern of memory performance to assess whether monkeys showed metacognition for memory. Monkeys saw a list of four consecutive random dot polygon figures and their memory for individual polygons from the list was probed using a yes–no recognition test. Monkeys showed the expected serial position effect; their memory was better for the first and last items than for the middle items. Monkeys were then presented with a decline test response, concurrently with a probe polygon that may or may not have been from the studied list and a 'not there' response used to indicate that the polygon was not from the studied list. The monkeys declined tests of the middle list items more often than tests of the first and last list items, thus showing that use of the metacognitive response again correlated with accuracy in the primary memory test.

Seeking Information When Ignorant

Metacognition is shown when subjects collect additional information when ignorant and act without expending the effort to seek information when already informed. The first tests of this capacity were conducted with human children, chimpanzees (*Pan troglodytes*), and orangutans (*Pongo pygmaeus*). A modified version of this same test was subsequently used with rhesus monkeys and capuchin monkeys (**Figure 1**). Subjects were presented with a set of opaque tubes in which food was hidden. Subjects either witnessed the baiting (seen trials) or did not (unseen trials), and therefore were either informed or ignorant about the food's location on each trial. On the test, subjects could select a single tube and collect the reward, if they were correct. This test is an interesting assessment of metacognition because the subjects could bend over and look down the length of the tubes to locate the food before choosing (see **Figure 1**). Subjects demonstrate metacognition by collecting information when ignorant (unseen trials) and choosing immediately when informed (seen trials). Human children, chimpanzees, orangutans, and rhesus monkeys clearly showed this pattern of behavior, while the case for capuchin monkeys was less clear (some capuchins made this differentiation under at least some conditions). Pigeons tested in related conditions in which they were given an opportunity to study before taking memory tests did not learn to do so,

Figure 1 *Left*, a rhesus monkey, ignorant of the food's location (unseen trial), makes the effort to bend down and collect more information by looking through the ends of the opaque tubes before making a choice. *Right*, an informed monkey makes a choice without going to the effort of confirming the location of the food (seen trial). Such selective information seeking suggests that the monkey knows when he knows, and only seeks more information as needed.

and instead proceeded to the tests without the information needed to succeed.

Avoiding Upcoming Tests

A few studies have required subjects to make a metacognitive judgment *before* seeing the actual test. In one study, monkeys were initially trained to match to sample, and then the delay between the study and test phases was gradually lengthened until monkeys performed at an intermediate level between chance and perfection. A metacognitive response was then introduced at the end of the delay interval that allowed monkeys to accept the memory test and receive a favored reward if correct, or decline the memory test and receive a guaranteed, but less desirable, reward. On other trials, only the option to take the memory test was offered at the end of the delay (**Figure 2**). Monkeys were more accurate on trials on which they accepted the test than on trials on which they were required to take the test, demonstrating that they accepted tests when memory was relatively good and declined tests when memory was relatively poor. Use of the decline test response generalized to conditions in which memory was directly manipulated either by providing no sample to remember (monkeys overwhelming declined subsequent memory tests) or by increasing the delay interval (monkeys were more likely to decline tests after long than after short delay intervals). Rats were similarly shown to avoid an upcoming auditory duration classification when the signal to be classified was of ambiguous duration. Two similar studies in which pigeons could avoid upcoming memory tests did not find metacognitive performance.

Interpreting Metacognitive Performance

The performances of some non-human animals in the tests described earlier clearly meet the criteria for

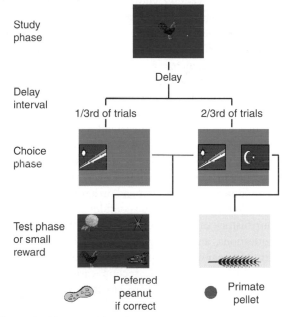

Figure 2 Metacognition about memory, or metamemory, in monkeys. Each panel depicts what monkeys saw on a touch-sensitive computer monitor at different stages in a trial.

metacognition. Subjects adaptively took easy tests and declined difficult tests. Animals judged past performance correctly, sought more information when needed, and predicted accuracy even before seeing the test. But behaving in a metacognitive way does not by itself specify what particular mechanism underlies the performance. Metacognition in humans is often associated with the conscious awareness of one's own cognitive states and is therefore presumed to reflect private monitoring of those states. But the evidence presented in this article proves neither that metacognitive performance is based on private monitoring of mental states, nor that if it were, those states would need to be conscious states. In the following section, some approaches to explaining metacognitive performance are described. It is likely that no one explanation is sufficient to account for all metacognition; rather, there is a diversity of ways in which metacognition can come about.

Private Versus Public Stimuli for Metacognition

It is useful to distinguish between private and public mechanisms for metacognition. Private mechanisms are those by which cognitive control is contingent on the privileged access the subject has to their own cognitive states. In the case of public mechanisms, adaptive cognitive control is based upon the use of publicly available information, such as the perceivable difficulty of a problem or the subject's reinforcement history with particular stimuli. Contrast the following two situations requiring a metacognitive judgment: (1) a colleague asks whether you remember the title of B. F. Skinner's first book and (2) a friend asks whether you can answer a question his 6 year old has about psychology. In the first case, you would surely check the contents of your memory and determine whether you can retrieve a memory of the book title. Your metacognitive judgment would therefore depend on your success or failure at privately retrieving the relevant explicit memory, a cognitive state to which you, as the one doing the remembering, have privileged access. In the second case, your friend has not even asked you to retrieve a specific memory. If you are an expert in Psychology, you might feel confident (probably correctly) that you can answer the question of a 6 year old. However, your confidence would not depend on a private evaluation of your memory. Instead, your confidence would depend on your history of expertise, your past ability to answer such questions, and your assessment of the intellectual capacity of 6 year olds – all publicly available information. It is significant that, in the second case, your friend's judgment about your ability to answer correctly would be about as accurate as your own. This would not be true if you were introspectively accessing a specific explicit memory, in which case you as the introspecting individual would have a distinct advantage over others in accurately estimating your knowledge. Thus, the observation of adaptive cognitive control should not be uncritically equated with private mechanisms. In the following section, several mechanisms for adaptive cognitive control are proposed that do not require access to private mental states.

Classes of Stimuli Sufficient for Metacognitive Control

Many cases of metacognition may be adequately accounted for by public mechanisms. Because we cannot obtain from non-humans the verbal reports that constitute part of the evidence for private introspective metacognition in humans, we can only infer private metacognition

in non-humans by excluding the likely public mechanisms. Below, four classes of mechanisms for metacognition are described. This list is representative rather than exhaustive.

Environmental cue associations

Some stimuli are more difficult to discriminate or remember than others, and some test conditions are more challenging than are others. Stimuli that are close together on a continuum are more difficult to discriminate than are those that are far apart. Highly similar images are difficult to identify in MTS tests. Memory tests after long delays are more difficult than those following short delays. Stimulus magnitude, image similarity, and delay interval are all types of publicly available information that indicate the difficulty of a particular test trial. Subjects performing tests with such stimuli might use the identity, magnitude, similarity, delay, or other publicly available information as a discriminative cue for declining tests or rating confidence. For example, if subjects have experienced low rates of reward with stimuli in a specific magnitude range, they could learn to avoid tests with all stimuli in that range. The probability that such can account for performance in a given paradigm is best assessed by generalization tests which determine whether or not performance is maintained across changes in the particular stimuli used and specific conditions of testing. If performance immediately generalizes to new test conditions or new stimuli, it is safe to conclude that metacognitive responding was not controlled by stimuli that were changed for the generalization test.

Behavioral cue associations

This account of metacognitive behavior is similar to environmental cue associations, with the exception that the discriminative stimuli controlling use of the metacognitive response are systematically generated by the subject in a way that correlates with accuracy in the primary task. For example, the subject may vacillate when it does not know the correct response on a given test. This vacillation itself does not necessarily represent metacognition by the subject that it does not know the answer, but can rather be an unmediated result of not knowing how to respond. It is common to see this sort of vacillation in monkeys taking MTS tests, for example, in which they look back and forth between the choice stimuli before choosing. It is also well known that response latency is often longer for incorrect than correct responses. Because vacillation and response latency correlate with accuracy, subjects could use these self-generated cues as discriminative stimuli for the metacognitive response, for example, by declining tests on which they experience a relatively long response latency. One way to assess whether behavioral cue associations account for metacognitive performance is to require subjects to make the secondary metacognitive judgment *before* they have seen the relevant primary test, and therefore before the test could have elicited vacillation or similar behavioral responses, as was done in some of the studies described earlier.

Response competition

In most reports of metacognition in non-human animals, subjects are confronted with the primary discrimination problem or memory test and the secondary metacognitive response option simultaneously. Because subjects can only make one response (a primary test response or a secondary decline test response, for example), simultaneous presentation puts these two behaviors in direct competition. As indicated earlier, animals are often slower to respond on error trials than on correct trials. On error trials with no prepotent primary test response, the probability that the subject will make the secondary metacognitive decline test response is greater, simply because no other competing response occurs immediately. On correct trials, when the inclination to make a primary test response is strong, it may dominate the tendency to decline the test or collect more information before responding. In all of the studies described earlier, the evidence for metacognition is that difficult primary test trials are declined or delayed (while more information is collected). Higher probabilities of the metacognitive response on difficult trials may therefore result from competition between primary choice responses and secondary metacognitive responses. As an example of how different behaviors can compete, consider a rat that has good knowledge of the location of food on a maze. Such a rat is likely to go directly to the baited locations and is consequently unlikely to explore other locations or engage in other behavior. Response competition can be ruled out as an account for metacognitive responding by presenting the secondary metacognitive response option either *before or after* the primary test, so that the two types of response do not compete directly.

Introspection

Metacognition could also be mediated by private introspective assessment of the subject's mental states. While introspection (i.e., the contemplation or perception of ones own mental states) might not necessarily require consciousness, it is closely allied with consciousness in humans. By the introspection account, the discriminative stimulus controlling a metacognitive response (e.g., declining to take a test) is the private experience of uncertainty or the weakness of memory. In the case of uncertainty, subjects are suggested to experience conscious (at least in humans) 'feelings of uncertainty' that differ from the experience of objective stimuli. In the case of memory, subjects are proposed to assess the strength

of their memory. The assessment of memory might be accomplished through several mechanisms that vary in sophistication from detecting whether a memory is present (while knowing nothing of the content of the memory) to attempting to retrieve the relevant memory and determining the success of that effort. Subjects use the decline response or other metacognitive response when memory is determined to be absent or weak. The important difference between this account and the preceding three is that the use of the metacognitive response is based on privileged introspective access to the subject's cognitive states, rather than on publicly available information or response competition. Due to the private nature of introspection, the conclusion that it accounts for metacognitive performance in non-humans can probably be reached only by ruling out other accounts.

Inferring Consciousness

While humans often describe metacognition as accompanied by conscious experience, it is difficult or impossible to specify the causal role that consciousness per se plays in metacognition. But the study of nonverbal species highlights the fact that the functional properties of cognitive systems, but not the phenomenological experiences associated with them, can be determined from behavioral experiments. Functional descriptions of cognitive systems can be applied equally well to human and non-human animals. In contrast, description of cognitive systems in terms of subjective experience and various conscious states creates a rift between the study of human and non-human cognition. Whether or not metacognition in other animals is associated with subjective conscious states like those experienced by humans, the growing literature on non-human metacognition demonstrates that the processes underlying metacognition can be effectively studied in non-human species. Metacognitive performance can be achieved through a variety of mechanisms, some of which may be entirely consistent with traditional views of non-human cognition and others that might call for a re-evaluation of the richness of comparative cognition.

See also: Intertemporal Choice; Mental Time Travel: Can Animals Recall the Past and Plan for the Future?; Time: What Animals Know.

Further Reading

Call J and Carpenter M (2001) Do apes and children know what they have seen? *Animal Cognition* 4: 207–220.

Flavell JH (1979) Metacognition and cognitive monitoring: A new area of cognitive-developmental inquiry. *American Psychologist* 34: 906–911.

Foote AL and Crystal JD (2007) Metacognition in the rat. *Current Biology* 17(6): 551–555.

Hampton RR (2001) Rhesus monkeys know when they remember. *Proceedings of the National Academy of Sciences of the United States of America* 98(9): 5359–5362.

Hampton RR (2005) Can Rhesus monkeys discriminate between remembering and forgetting? In: Terrace HS and Metcalfe J (eds.) *The Missing Link in Cognition: Origins of Self-reflective Consciousness*, pp. 272–295. New York, NY: Oxford University Press.

Hampton RR, Zivin A, and Murray EA (2004) Rhesus monkeys (*Macaca mulatta*) discriminate between knowing and not knowing and collect information as needed before acting. *Animal Cognition* 7: 239–254.

Kornell N (2009) Metacognition in humans and animals. *Current Directions in Psychological Science* 18(1): 11–15.

Kornell N, Son LK, and Terrace HS (2007) Transfer of metacognitive skills and hint seeking in monkeys. *Psychological Science* 18(1): 64–71.

Nelson TO (1996) Consciousness and metacognition. *American Psychologist* 51(2): 102–116.

Roberts WA, Feeney MC, McMillan N, MacPherson K, Musolino E, and Petter M (2009) Do pigeons (*Columba livia*) study for a test? *Journal of Experimental Psychology-Animal Behavior Processes* 35(2): 129–142.

Smith JD, Schull J, Strote J, McGee K, Egnor R, and Erb L (1995) The uncertain response in the bottle-nosed-dolphin (*Tursiops truncatus*). *Journal of Experimental Psychology-General* 124(4): 391–408.

Smith JD, Shields WE, Schull J, and Washburn DA (1997) The uncertain response in humans and animals. *Cognition* 62(1): 75–97.

Smith JD, Shields WE, and Washburn DA (2003) The comparative psychology of uncertainty monitoring and metacognition. *Behavioral and Brain Sciences* 26: 317–374.

Smith JD and Washburn DA (2005) Uncertainty monitoring and metacognition by animals. *Current Directions in Psychological Science* 14(1): 19–24.

Sutton JE and Shettleworth SJ (2008) Memory without awareness: Pigeons do not show metamemory in delayed matching to sample. *Journal of Experimental Psychology-Animal Behavior Processes* 34(2): 266–282.

Relevant Websites

http://www.psychology.emory.edu/lcpc/bailout.high.html – Video of a Monkey Performing a Metamemory Test.

Microevolution and Macroevolution in Behavior

J. E. Leonard, Hiwassee College, Madisonville, TN, USA
C. R. B. Boake, University of Tennessee, Knoxville, TN, USA

Introduction

To what extent do microevolutionary patterns and processes result in macroevolution? Widely accepted evolutionary theory posits that the accumulation of small genetic changes over time (microevolution) can lead to large-scale changes, including speciation (macroevolution). The vast majority of studies that address the link between micro- and macroevolution focus on morphology. By contrast, this topic has received relatively little attention from behavioral biologists. Few studies have reported genetic effects on behavior that could result in novel behavioral phenotypes. This article discusses recent work in behavioral genetics and the maintenance of behavioral polymorphisms within natural populations or species (the microevolution of behavior), and discusses methods and data that address major differences in behavior within a genus or higher taxon (macroevolution). Additionally, we suggest two major avenues of research that could be applied to the study of the relationship between micro- and macroevolutionary processes for behavior.

Microevolution and Behavior

Microevolution refers to the evolution of changes that are observed within a population or species, but also encompasses differences among populations within a species that have adapted to their respective local environments. Such changes have evolutionary consequences only if they have a genetic basis. The genetics of divergence can be evaluated by studying individuals in a common environment or by using formal breeding studies. Studies covering numerous taxonomic groups have identified intraspecific behavioral polymorphisms in natural populations, and we offer examples from four taxa.

Vertebrate Examples

Freshwater populations of the three-spined stickleback, *Gasterosteus aculeatus*, have two ecotypes, a benthic form and a limnetic form. Morphological differences between these ecotypes are directly related to feeding ecology: benthic forms have mouthparts suited to bottom-feeding and raking, whereas the mouth and jaw of limnetic forms are shaped for open water feeding. This close relationship between foraging behavior and feeding morphology within a species has a genetic basis, as demonstrated by rearing many generations of each morph in the laboratory. In addition, the ecotypes differ in mating behavior: members of the benthic populations are susceptible to egg cannibalism and males from these populations have a lower frequency of the prominent zigzag nest advertisement display than males from limnetic populations. The differences in courtship displays persist in laboratory populations, indicating that the differences are influenced by an underlying genetic component.

The probability of predation on *Taricha* newts by garter snakes (*Thamnophis sirtalis*) in the Western United States differs between populations as a result of the snakes' resistance to tetrodotoxin (TTX), the newt's defensive chemical (**Figure 1**). TTX is an extremely powerful neurotoxin. Resistance in snakes is due to a change in the sodium channels in nerves and muscles and is heritable. Individual snakes are capable of evaluating their own level of resistance and modifying their choice of prey accordingly, spitting out newts that are too toxic for them. Population analyses have shown that snakes are winning the arms race, with no populations having been found where the average level of snake resistance is lower than the average newt toxicity.

Invertebrate Examples

Populations of the parasitoid wasp *Nasonia longicornis* in Utah and in California differ in female mate choice and remating frequency. The differences are maintained in a common laboratory environment. The behavioral differences in this species are associated with different selection pressures imposed by competition for shared hosts with their widespread sister species *N. vitripennis*. Across the species' range, males of *N. vitripennis* stay on the host puparium after emerging and are highly aggressive; females emerge as virgins and mate on the puparium. Females of *N. longicornis* collected from California, where range overlap between *N. vitripennis* and *N. longicornis* is common, mate inside the host puparium and rarely remate. Females of *N. longicornis* collected from Utah, where the species rarely share hosts, mate outside the host puparium and remate frequently. For *N. longicornis*, the presence or absence of a congener has resulted in small microevolutionary differences that appear to be localized adaptations.

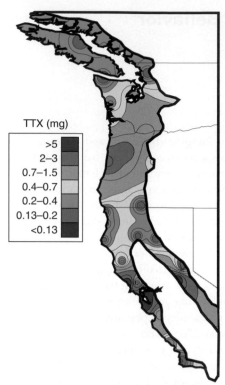

Figure 1 Geographic distribution of tetrodotoxin resistance in *Thamnophis* snakes in Western North America. The tolerance of snakes to various levels of TTX is color-coded, with a geographic range from Vancouver Island (in the north, at top) to the tip of Baja California. Reproduced from Hanifin CF, Brodie ED Jr, and Brodie ED III (2008) Phenotypic mismatches reveal escape from arms-race co-evolution. *PLoS Biology* 6: 471–482.

The North American spider *Anelosimus studiosus* shows variation in social behavior in which some populations contain up to 15% social spiders that live in aggregations, while other populations consist of solitary spiders. These patterns are associated with a suite of differences at the individual level that are maintained in a common laboratory environment and some of which are known to have a genetic basis. Individuals with the asocial phenotype are intolerant of conspecifics, aggressive toward prey, and kill superfluously, whereas social spiders aggregate, are less aggressive toward prey, less likely to kill superfluously, and generally less active. In populations with both phenotypes, the social phenotype provides a fitness advantage to females and a fitness disadvantage to males, suggesting that the polymorphism produces evolutionary conflict between the sexes.

These examples show that local environmental conditions can result in microevolutionary changes in behavior. We now examine macroevolutionary change, that is, how new behaviors evolve in novel environments and contribute to speciation.

Macroevolution and Behavior

Macroevolution results in phenotypic differences that can be observed at the species level or above. For example, an appendage or a new behavior for feeding that is not observed in related taxa might be seen as evidence for speciation or higher scale divergence. Changes in morphology are much easier to observe over evolutionary time than changes in behavior because morphological structures are preserved in the fossil record. The fossil record rarely preserves evidence of behavior, although we can make inferences about how a structure might have been used on the basis of what is known about current morphology. Because we cannot see how ancient organisms behaved, we must use a phylogenetic approach to studying macroevolution of behavior.

One of Nobel Prize-winning ethologist Niko Tinbergen's (1963) four aims of ethology is to understand the evolutionary history or origin of any behavior under study, which is essentially a phylogenetic question. Investigations into the phylogenetic relationship among species, genera, or families can provide information about behavioral evolution when an independently derived phylogeny is available. By placing traits into a phylogenetic framework, we can understand whether traits are homologous or homoplasious, which in turn can provide information on how behaviors have been modified through evolutionary time. Phylogenetic analyses can also be used to answer the question of whether a group of derived characters evolved together or independently.

Tracing the long-term evolution of behavior requires the availability of well-described traits for a group of related species and a reliable phylogeny for the group. Fortunately, an increasing number of published phylogenies is making this approach feasible for many taxa. However, the lability and complexity of behavior, coupled with poor information on its genetic basis, have deterred many systematists from using behavior as a phylogenetically informative character. Yet behavior can be as useful as morphology in constructing phylogenies. One such approach uses similarities in behavior as characters to infer relatedness, and another uses behavioral changes to infer evolutionary origins and subsequent history. A classic example of the first is from the Nobel Prize–winning ethologist Konrad Lorenz, who examined the origins and diversification of courtship behavior among three families of duck (Lorenz, 1971). He had demonstrated that the sequence of behaviors leading to successful mating is highly stylized and fixed for any particular species. In other words, these behaviors and the patterns themselves have a strong genetic basis. Lorenz found that a phylogenetic tree based on about 20 species and a mix of 48 morphological and behavioral traits allowed him to assess the degree of homology for many behavior patterns. In particular, he showed that the majority of courtship motor patterns were attributable to common descent rather than convergent evolution, and he identified branch points where novel behaviors evolved. The result was a comprehensive phylogenetic description of courtship evolution.

The second approach is to use studies of behavior in a phylogenetic context to uncover macroevolutionary trends. An example is the relationship between genetic variation and host plant affiliation in species of leaf beetle in the genus *Ophraella*. Larvae and adults vary between species in whether they eat host plants of congeneric beetles and in whether they survive on these other plants. The existence of genetic variation for feeding is an indicator of the ability of a beetle species to adapt to a new host plant. Out of more than 20 tests of different combinations of beetle and plant species, only a few cases revealed genetic variation in the beetles for larval or adult feeding behavior when not on their own hosts, and in most cases, the beetles died if they were not on their host species. The beetle species were more likely to show a feeding response if the tested plant was in the same tribe as their own host plant. The pattern of associations between beetles and plants shows a rough phylogenetic congruence, suggesting that the genetic variation in feeding behavior may have permitted host shifts in the past.

Another example is a study of the sexually selected traits of plumage and song in the New World Orioles (*Icterus*), a group that has great variety of patterns of song and plumage. Many tropical species have sexually monomorphic bright plumage, whereas temperate species tend to be dimorphic, with relatively dull females. A phylogeny that was developed with mitochondrial DNA showed that the dimorphism in the temperate species is a derived trait, with females having evolved a dull plumage. Without the phylogenetic analysis, the dimorphism would have been explained with the general hypothesis that sexual dimorphisms evolve through males becoming more showy. The same song is frequently given by both sexes, but because of the limitations of the available data, the analysis was restricted to the species level, where it was shown that song is highly evolutionarily changeable in these species. Many song characteristics are homoplasious, having been gained and lost repeatedly within the clade, and the similarity of components of songs between species is often due to convergent evolution.

Insights such as those seen earlier require both a reliable phylogeny and time-consuming analyses of the behavior of several species in a common environment. Phylogenies are often based on molecular traits, and the molecular and statistical tools for developing phylogenies are improving rapidly. Yet many species of great interest to behavioral biologists belong to groups for which reliable phylogenies do not exist. At present, macroevolutionary analyses of behavior are possible for a relatively small number of taxa and further progress on questions of behavioral macroevolution depends on progress in phylogenetic systematics.

Are Micro- and Macroevolution the Ends of a Continuum?

Investigating and comparing traits between closely related species allows us to address questions related to both micro- and macroevolution. Species differences in behavior inform questions of how genes encode behavior, the types of genetic changes needed to modify behavior, and how behaviors coevolve with other traits. The same studies can be used to place traits onto a phylogeny, allowing the identification of the nature and timing of major evolutionary transitions in behavior as well as novel functions for those traits. Our ability to address both micro- and macroevolutionary questions from similar data sets allows us to describe the relationships between these two broad fields.

The central issue is to determine the way in which small genetic changes within populations can lead to major transitions in phenotype that are involved in speciation. Two approaches show promise for meeting this challenge, one focusing on studies of related individuals, and the other taking a molecular perspective. The first approach starts by recognizing that traits do not evolve independently. Thus, macroevolutionary patterns can be detected by examining variation and covariation in multiple behavioral traits among populations. Understanding the co-evolution of multiple phenotypes requires data on the genetic variation and covariation for those phenotypes to uncover the ones that are the focus of adaptation, in other words applying quantitative genetics to behavioral evolution. A second approach, described in the following section, specifically addresses the way in which small changes in a DNA sequence produce major phenotypic changes by changing the timing of gene activity in closely related species. In both of the approaches, morphological traits have been the focus of most of the literature to date. Although studies that explicitly examine behavior are rare, many of the concepts and experimental procedures developed for morphological analyses can be applied to behavioral studies.

Microevolution to Macroevolution: Multivariate Analyses and the G-matrix

Evolution is a multivariate process. Selection affects a whole organism, not simply a single trait. Therefore, the relationships among different traits and their effects on fitness are key to understanding how a population responds to selection, and the types of genetic changes required to produce novel phenotypes. A concept and method developed to study multivariate evolution is potentially applicable to behavioral traits. This is the G-matrix, which was developed by breeders and which can be used to study inheritance in relation to evolution (see **Box 1**).

At present, the G-matrix for behavioral traits is a conceptual tool rather than an operational technique. Estimating genetic variances and covariances for behavior

---

Box 1   The Genetic Variance–Covariance Matrix, or G-matrix

The G-matrix is developed by displaying the same traits in columns and rows. The elements of the main diagonal of the matrix give values of additive genetic variances for individual traits while off-diagonal elements are additive genetic covariances between traits. This matrix thus describes the genetic relationship between multiple heritable traits. Equation [1], developed by Lande, provides a mathematical expression that allows us to understand how the G-matrix can be used to predict responses to selection when multiple traits are taken into consideration:

$$\Delta \bar{z} = G\beta \tag{1}$$

where $\Delta \bar{z}$ is a vector of the change in mean trait values across one generation for all traits measured, G is the matrix of additive genetic variances and covariances for the traits being studied, and β is the selection gradient, a vector that describes the direct force of selection on each trait. This equation allows us to predict the response of a population to selection when many traits are taken into consideration. For example, when two traits are negatively correlated genetically (have a negative genetic covariance), selection to increase one trait will decrease the other. Overall, then, the G-matrix permits computations of the rate and direction at which evolution will proceed.

---

traits is arduous, requiring large sample sizes from individuals within known pedigrees and application of sophisticated statistical algorithms. The great difficulty in applying statistical genetics to behavior is the enormous phenotypic variation in behavior. For behavioral traits, it is common for the same individual to show substantial differences across repeated observations of the same act. In other words, behavior often has a low repeatability.

Despite this limitation, the matrix can surely inform the study of behavioral evolution. Some quantitative genetic studies of morphology suggest applications to behavior. A morphological trait that is probably associated with behavior is throat coloration in the guppy *Poecilia reticulata*. Spots on male throats are important signals during courtship and hence the subject of sexual selection. Artificial selection for increasing the length of the spot on the throat decreased the width of the spot, indicating a genetic constraint on how big the spot can become. Thus, the morphological genetic variances and covariances that drive the evolution of spot size and shape must have correlated effects on mating behavior.

An important focus of current research is the temporal stability of the G-matrix. We do not yet know whether estimates of genetic variances and covariances made in a single generation can be used to extrapolate phenotypic changes over evolutionary time. This problem arises because the components of the matrix are affected by allele frequencies, and evolution results in changes in allele frequencies. The stability of the G-matrix can be tested by evaluating it in closely related species, or by manipulating it within a species. Closely related species are useful for tests if one assumes that stability within a species translates into similarity between close relatives. Thus, in a study of the genetic covariance structure for five morphological traits in each of three cricket species in the genus *Gryllus*, the G-matrix was consistent across the species, suggesting that it was also stable over time within species. Approaches that manipulate the G-matrix directly have given variable results. The G-matrix was found to be constant after a population bottleneck in

Drosophila melanogaster, but in another study, after laboratory lines of *D. melanogaster* had been inbred for two generations, the covariance structure changed in comparison with that of an outbred control population. Perhaps the G-matrix for certain suites of characters like morphological traits is more stable than for other traits like behaviors; this supposition has been the subject of considerable debate.

A recent promising approach to understanding the role of the G-matrix in behavioral evolution examines suites of behavioral traits within a relatively coherent group, such as those involved in competing for food or mates. Such groups of traits have been called behavioral syndromes. If these syndromes are also characterized by significant genetic correlations between the component traits, they could present an ideal opportunity to study behavioral evolution via G-matrix techniques. Overall, though, because so few studies have focused on genetic correlations for behavioral traits, the lability of genetic covariance structures for behavior remains an open question.

Gene Regulation and Macroevolution

Evolutionary behavioral genetics is a young science and only in the past 15 years has there been a strong effort to investigate genetic effects on behavior in natural populations. The majority of what we know about how genes influence behavior comes from laboratory organisms like *Drosophila melanogaster* and house mice, *Mus* spp. Although results from the laboratory are not necessarily applicable to natural populations, they do reveal mechanisms by which genetic change influences behavioral change. An important insight arising from such studies is that gene regulation can be a source of evolutionary novelty.

The study of gene regulation of behavior is a developing area of research that may be able to link micro- and macroevolutionary changes. Until recently, most evolutionary biologists investigated mutations within structural genes as the probable primary source of genetic variation, with macroevolutionary changes assumed to result from an accumulation of small genetic changes over time. The example mentioned earlier represents both a small change and a change in a structural gene: at least part of the populational variation in resistance to TTX in *Thamnophis sirtalis* is attributable to a mutation in a structural gene, one that codes for sodium channels in cell membranes. Recent studies of the role of gene regulation in evolutionary change could revolutionize the understanding of the genetic processes underlying micro- and macroevolution, though as of this writing such a statement is controversial. Gene regulation can take place through the action of either cis- or trans-acting factors. Trans-acting factors are coded by regions of DNA far away from the structural genes they influence, whereas cis-regulatory elements are close to the promoter regions. Recent studies investigating cis-regulatory elements suggest that small changes to the regulatory sequence and not the regulated gene can lead to large-scale changes in morphology. In Darwin's finches, beak shape is associated with the kind of food eaten and thus with feeding behavior. The large ground finch (*Geospiza magnirostris*) has a broader bill than the other finches, and this allows it to crush large seeds. The increased breadth is attributable to the action of a gene that regulates growth, Bmp4, which is expressed in the appropriate region of the developing beak at the appropriate developmental stage. The gene is not active in the same way as in other species in the clade that have narrower beaks. Cave fish (*Astyanax mexicanus*) have lost the ability to see; this has occurred several times within the species. The loss of eyesight is due to the activity of a regulatory gene, Hedgehog, which causes the developing eye to degenerate as a result of Hedgehog's stimulation of genes that affect eye morphology. Differences in social behavior between vole species appears to be due to changes in gene regulation: in the prairie vole (*Microtus ochrogaster*) pair bonds are long-lasting and males exhibit substantial parental behavior, whereas other closely related voles are solitary and do not have male parental care. The differences in male behavior appear to be due to the sensitivity of a brain region to the hormone vasopressin. Males of the monogamous prairie voles have a change in the promoter region of the vasopressor V_{1a} receptor gene, and when other species are transformed with this form of the receptor gene, they show increased affiliative behavior, similar to the affiliative behavior of the prairie voles. In some species, sufficient genetic tools are available to determine whether morphological evolution is attributable to changes in structural or regulatory genes; however, the application of such techniques to behavioral evolution is less common.

Challenges for Studying Micro- and Macroevolution in Behavior

Two major challenges need to be addressed in the near future. First, we know relatively little about the genetics underlying behavioral traits, that is, the field of behavioral genetics is still young. Second, studies in the laboratory may not always help us to understand evolution in natural populations. Recent morphological and behavioral genetic studies of laboratory organisms such as *Drosophila* species and mice have resulted in crucial insights such as the role of the G-matrix and the importance of gene regulation in evolution. Molecular genetic analyses are becoming feasible for a wider array of taxa and new species are becoming amenable to laboratory culturing. The honeybee is a species that is rapidly becoming a model species for understanding the genetic basis and evolution of sociality. Other species are also receiving attention in behavioral evolutionary genetics. For example, the burying beetle *Nicrophorus vespilloides* has been used to evaluate the effects of genetic correlations on the evolution of multiple mating, the genetic basis of sex pheromone production has been investigated in the moths *Helicoverpa armigera* and *H. assulta*, and the Argentine ant *Linepithema humile* is a new model organism for understanding genetic changes associated with invasiveness. With the rapid drop in the costs of sequencing genomes and studying gene expression, it is likely that using molecular tools to understand behavioral evolution will become common.

Laboratory studies of behavior genetics can inform our understanding of evolution in natural populations. Evolution results from a combination of direct and indirect natural selection, genetic drift, inheritance of individual traits, and genetic covariances between traits. Of these components, some are more easily estimated in natural populations and others generally need to be studied in captive populations. The interpretation of measures of natural selection and drift made in captivity is subject to many complications. In contrast, because of the needs for large sample sizes and known pedigrees, it may not be possible to measure the inheritance or genetic covariance of traits in natural populations. Genotype-by-environment interactions appear to be widespread and thus, the extrapolation of genetic measurements made in the laboratory to natural populations needs to be conducted with caution. For these reasons, behavior-genetic analyses are sometimes conducted on natural populations of animals such as vertebrates or social insects, where molecular genetic estimates of pedigrees are possible. Alternatively, such studies are conducted with laboratory populations of species that are known to maintain characteristic behavior in captivity (with careful replication of crucial elements of the habitat) such as stickleback fishes and burying beetles (**Figure 2**).

Macroevolutionary changes in behavior can often be observed in natural populations and in many cases, change little when species are studied in captivity. The fields of animal functional morphology and sensory biology, which are intimately related to behavior, make great use of

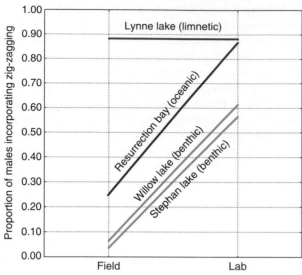

Figure 2 Differences between populations of stickleback fishes in a component of courtship behavior, the male zigzag display. Note that although values in the laboratory and natural populations differ, the rankings of the populations are the same. The limnetic population shows no plasticity for this trait, with both field and laboratory observations giving the same value. Reproduced from Shaw KA, Scotti ML, and Foster SA (2007) Ancestral plasticity and the evolutionary diversification of courtship behaviour in threespine stickleback. *Animal Behaviour* 73: 415–422, with permission from Elsevier.

laboratory studies, but sometimes also incorporate a field component to examine the traits under natural circumstances. As with microevolution, laboratory and field studies of behavioral macroevolution can be complementary. The laboratory biologist needs to be careful in designing the observations and in their interpretation. In analogy to the problem in physics of not being able to simultaneously know what a particle is and how it is moving, it is necessary to conduct a variety of observations and experiments to understand the evolution of behavior. We cannot let anxieties about the realism of a particular method paralyze us and restrict our work.

Conclusion

One of the core goals of evolutionary biology is to understand how major transitions in phenotypes arise over evolutionary time, the link between micro- and macroevolutionary processes. Our current understanding of these evolutionary processes, and the methods used to study them, should be applicable to behavioral traits. The major difficulty in applying the methods and

theory to behavior is that behavior is so variable, resulting in large standard errors for estimates. Much of our understanding of the genetics underlying behavior has come from major laboratory organisms like *Drosophila* sp. and *Mus* sp., but the advent of rapid and less expensive genomic technologies is allowing sophisticated studies of behavioral evolution to be conducted with many species. The principles developed with model organisms can be applied to species in more natural settings to understand truly the relationship among genetics, the environment, and evolutionary change.

See also: Development, Evolution and Behavior; Developmental Plasticity; Evolution and Phylogeny of Communication; Genes and Genomic Searches; Konrad Lorenz; Mating Signals; Niko Tinbergen; Phylogenetic Inference and the Evolution of Behavior; Sex and Social Evolution; Sexual Selection and Speciation; Threespine Stickleback.

Further Reading

Abzhanov A, Protas M, Grant BR, Grant PR, and Tabin CJ (2004) Bmp4 and morphological variation of beaks in Darwin's finches. *Science* 305: 1462–1465.

Hanifin CF, Brodie ED Jr, and Brodie, III ED (2008) Phenotypic mismatches reveal escape from arms-race coevolution. *PLoS Biology* 6: 471–482.

Hoekstra HE and Coyne JA (2007) The locus of evolution: Evo devo and the genetics of adaptation. *Evolution* 61: 995–1016.

Lande R (1976) Natural selection and random genetic drift in phenotypic evolution. *Evolution* 30: 313–334.

Lorenz K (1971) Comparative studies of the motor patterns of the Anatinae (1941). In: Martin R (trans.) *Studies in Animal and Human Behavior*, vol. 2, pp. 14–114. Cambridge: Harvard University.

McLennan DA and Mattern MY (2001) The phylogeny of the Gasterosteidae: Combining behavioral and morphological data sets. *Cladistics* 17: 11–27.

Price JJ, Friedman NR, and Omland KE (2007) Song and plumage evolution in the New World orioles (*Icterus*) show similar lability and convergence in patterns. *Evolution* 61: 850–863.

Prud'homme B, Gompel N, and Rokas A (2006) Repeated morphological evolution through cis-regulatory changes in a pleiotropic gene. *Nature* 440: 1050–1053.

Robinson GE and Ben-Shahar Y (2002) Social behavior and comparative genomics: New genes or new gene regulation? *Genes, Brain and Behavior* 1: 197–203.

Shaw KA, Scotti ML, and Foster SA (2007) Ancestral plasticity and the evolutionary diversification of courtship behaviour in threespine stickleback. *Animal Behaviour* 73: 415–422.

Tinbergen N (1963) On aims and methods of ethology. *Zeitschrift für Tierpsychologie* 20: 410–433.

Yamamoto Y, Stock DW, and Jeffery WR (2004) Hedgehog signalling controls eye degeneration in blind cavefish. *Nature* 431: 844–847.

Young LJ, Nilsen R, Waymire KG, MacGregor GR, and Insel TR (1999) Increased affiliative response to vasopressin in mice expressing the V_{1a} receptor from a monogamous vole. *Nature* 400: 766–768.

Migratory Connectivity

P. P. Marra and C. E. Studds, Smithsonian Migratory Bird Center, National Zoological Park, Washington, DC, USA
M. Webster, Cornell Lab of Ornithology, Ithaca, NY, USA

Published by Elsevier Ltd.

Introduction

Migration is a diverse behavior found in all animal taxa, and in its simplest form, is defined as a repeated seasonal movement to and from a breeding area. Animals exhibit several forms of migration, including seasonal migrations across latitudes, altitudinal migrations up and down mountains, migrations that overlap with key life-history stages such as molt, and migrations that can span multiple generations over space and time. This variation can occur within species or between species, but all forms of migration involve movement away from breeding areas and then a return. The geographic linking of individuals or populations between different stages of the annual cycle, including between breeding, migration, and winter stages, is known as *migratory connectivity*. In this article, we will discuss why understanding migratory connectivity is so critical from ecological, evolutionary, and conservation perspectives and also provide descriptions of the various approaches being used to track animals throughout the annual cycle. Most of our examples will be drawn from the bird literature because this is where the understanding is most advanced.

Nearctic–Neotropical migratory birds move north and south between breeding areas in North America and nonbreeding areas in the Caribbean, Middle America, and South America. More specifically, they spend approximately 3–4 months of the year on breeding areas at temperate latitudes. Most species then molt, build fat stores, and migrate south in August and September to a distant and ecologically different location, often in the tropics. It is here that they spend the majority of the annual cycle – 6–8 months. At the end of the stationary portion of the nonbreeding period, they once again build fat stores and leave on spring migration to return to breeding areas. Quantifying migratory connectivity is essential for understanding how events in one period of the annual cycle influence subsequent stages. Such interseasonal effects or 'seasonal interactions' are poorly understood within all bird migration systems, in large part because it has been difficult to determine how specific summer and winter populations, along with their stopover locations, are connected throughout the year.

A better understanding of migratory connectivity will allow researchers to follow populations or individuals throughout the annual cycle and thereby address questions regarding the ecological and evolutionary implications of seasonal interactions. The challenge in studying migratory connectivity is to understand not only the geographic connections among periods of the annual cycle but also how these connections influence the ecology, evolution, and conservation of migratory species. Here, we review the ecological and evolutionary considerations of understanding migratory connectivity as well as discuss the advances in marked animal approaches, genetic analyses, and stable isotope chemistry that now make it possible to gain some insights into the population origin of individual birds.

Why Study Connectivity?

Ecological Considerations

The fact that individuals spend time each year in two or more widely separated geographic areas has obvious but poorly studied consequences for population dynamics. The conditions and selective pressures at winter locations are likely to affect individual performance during the breeding season and vice versa. This simple fact has important implications for the ecology, evolution, and conservation of migratory birds. For example, factors and events on the wintering grounds (e.g., weather patterns and deforestation) may affect bird survival and, hence, subsequent recruitment on the breeding grounds. Similarly, differences in reproductive success in summer can lead to changes in the structure of the winter population. Consider an example from a long-distance migratory bird that breeds in Eastern North America. Seasonal interactions and carry-over effects are likely to be most pronounced if summer and winter populations are tightly linked (i.e., if individuals overwintering together in one location also breed near each other in a particular part of the breeding range (**Figure 1(a)** and 1(**b**))), but may be much less pronounced if population connections are weak (i.e., if individuals overwintering in one area spread out over a large geographic range for breeding (**Figure 1(c)**)).

Interestingly, if birds originating from three different breeding sites completely segregate on wintering areas, resulting in high migratory connectivity (**Figure 1(a)** and 1(**b**)), then one could theoretically differentiate between two alternate migration strategies. First, birds could use (1) *leapfrog migration* – where northern wintering populations breed in the southern portions of the breeding range, and southern wintering populations breed in the northern portions of the breeding range (**Figure 1(a)**). This would

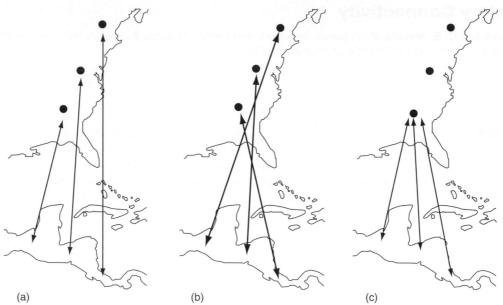

Figure 1   Illustrations of strong (a and b) and weak (c) migratory connectivity for a hypothetical population of Neotropical–Nearctic migratory bird. Strong connectivity may occur as either (a) leapfrog or (b) chain migration strategies as birds travel between tropical wintering and temperate breeding grounds.

result in differential migration distance between populations, where northern wintering populations migrate a comparatively short distance and southern winter populations migrate to northern latitudinal extremes. Alternatively, individuals may use (2) *chain migration* – where northern wintering populations breed in the northern portion of the breeding range and southern wintering populations breed in the southern portion of the breeding range (**Figure 1(b)**). Migration distance, therefore, is similar between populations. In either case, the result may have profound implications for life-history strategies as well as other aspects of the ecology and evolution of migratory species.

Similarly, large-scale climatic events can have effects on migratory populations throughout the year. For example, recent analyses show that global climate patterns can affect demographic rates (El Niño Southern Oscillation) and, ultimately, population dynamics (climate change). For migratory species, the magnitude of such effects on population dynamics will likely vary with the degree of connectivity between winter and summer populations. There is a need to determine this connectivity, along with detailed modeling efforts, in order to ascertain how focal breeding populations are affected by large and small-scale events affecting various wintering populations (and vice versa). Such information is essential for determining when in the annual cycle populations are most limited and what factors drive population dynamics.

Finally, the scope of natal dispersal will determine the extent to which events in one part of the breeding ground might affect the recruitment and gene flow to other parts.

Source/sink population models are a start for understanding these relationships, but more detailed spatial models that explicitly incorporate migration are required. Interestingly, natal dispersal on breeding areas may be determined in part by events on wintering areas. Using stable-hydrogen isotope ratios in the feathers of American redstarts (*Setophaga ruticilla*) captured as immature birds and again as adults, Colin Studds, Peter Marra, and Kurt Kyser, showed that habitat use during the first tropical nonbreeding season interacts with latitudinal gradients in spring phenology on the temperate breeding grounds to influence the distance traveled on migration and the direction of dispersal by first-year redstarts. Because natural selection acts on these animals throughout the annual cycle, as we gain a better understanding of migratory connectivity, more emphasis should be placed on studying the biology of these animals in the context of where they have come from and where they are going in the next phase of their life cycle.

Evolutionary Considerations

Individuals of migratory species experience two widely separated and ecologically different habitats during their lifetimes. Selective pressures are likely to vary between these summer and winter habitats, and this may affect the degree to which individuals are 'locally adapted' to either habitat. Models of nonmigratory species have shown that the effects of gene flow on local adaptation and niche breadth can be complicated by population dynamics and mating patterns, but few models have explicitly incorporated the effects of migration. Furthermore, for many species of migratory birds, males and females occupy different habitats on the wintering grounds (sexual habitat segregation) but not on the breeding grounds. Thus, males and females are subject to similar selection pressures in the summer, but different ones in the winter.

The degree of local adaptation is likely to be strongly affected by the strength of connectivity between summer and winter populations. For example, if summer and winter populations are tightly linked, it may be possible for particular populations to become 'well adapted' to both their summer and winter grounds. If, on the other hand, birds from a particular breeding population spread out over the entire winter range (and vice versa), we expect a somewhat poorer fit between the birds and their environment. Likewise, local adaptation on the breeding grounds may be hindered by strong connections among breeding populations via natal dispersal. This expectation is supported by the gene flow studies of nonmigratory species. To our knowledge, the effects of the microevolutionary consequences of migratory connectivity have not been explored in any migratory species, despite the fact that such effects should be more pronounced than in sedentary species.

How Can We Determine Migratory Connectivity?

Marked Animal Approaches

Capture–recapture methods developed within the last decade permit direct estimation of movement probabilities of individually marked animals across different locations. The methods involve the use of multistate models in which individual animals are categorized by the location at which they are recaptured or re-sighted. These models permit estimates of different detection probabilities for different locations and estimates of location-specific rates of survival and movement. Although suitable for many types of investigations, data on the return rates of marked individuals to both breeding and wintering grounds have not proven useful for understanding the connectivity of migratory bird populations. The problem is that it is all but impossible to re-sight or recapture the same individuals at multiple locations throughout the annual cycle. Satellite transmitters, unlike traditional markers such as leg bands, offer promise for understanding migratory connectivity. Unfortunately, they are quite expensive ($3500) and are limited to animals of large body size (>165 g), which excludes all passerine songbirds as well as smaller shorebirds and raptors. For larger birds, satellite transmitters allow the detailed collection of direct information on the movement patterns of individuals over large spatial areas.

Engineers at the British Antarctic Survey have recently developed a miniature and affordable daylight-level data recorder (geolocator) for tracking animals over long periods of time. These devices weigh as little as 0.8 g, and are rapidly becoming smaller and can be attached to birds by methods similar to long-standing VHF radio-transmitters used in radio-tracking songbirds. Geolocators take consistent readings of daylight timing for 1–2 years. Unlike radio-transmitters, the geolocators must be recovered from returning birds and archived data downloaded. The recovered data are then interpreted to determine the latitude and longitude of the individual bird twice per day for every day the logger was attached and exposed to suitable sunlight. These geolocators have returned accurate and detailed location information on large pelagic birds, and their utility on small migrating songbirds has recently been demonstrated with a single study of the Wood Thrush (*Hylocichla mustelina*) and Purple Martin (*Progne subis*) conducted by ornithologist Bridget Stutchbury. The use of geolocator tags for studies of migratory connectivity and seasonal interactions in small passerine songbirds may thus present an unparalleled opportunity to discover how distant breeding and nonbreeding areas connect and interact in space and time.

Molecular Genetic Approaches

Because *extrinsic markers*, such as the aforementioned tagging methods, require that the marked individuals be relocated at some point, some researchers have turned their attention to the use of *intrinsic markers* of population origin – that is, markers or indicators that come from the animal itself. One popular approach has been to use molecular genetic markers, because although only some birds have leg bands (or other extrinsic tags), they *all* have DNA. Genetic markers clearly hold considerable potential for the studies of migratory connectivity, but their use is complicated by a number of factors.

The basic logic of most genetic approaches is that, if certain genetic markers (e.g., alleles or haplotypes) are found, say, in one breeding population (X) but not another (Y), then finding those markers in a particular wintering population will indicate some level of connectivity between that wintering population and breeding population X. In some cases, it should also be possible to determine the degree or strength of that connectivity. For example, strong connectivity would be suggested if many individuals in the wintering population had the genetic marker from breeding population X.

This approach hinges on some level of genetic differentiation among breeding populations. Typically, markers will not be unique to particular populations, but instead might vary in frequency across populations. In this case, it is possible to calculate the probability that a wintering individual originated from one breeding population or another (or vice versa) – that individual has a high probability of originating from any population where its genetic markers are common, and a low probability of having come from populations where those markers are rare. Indeed, a number of sophisticated analytical methods ('assignment tests') have been devised to determine the probability (or likelihood) that an individual came from one population or another, including situations where the actual population of origin may not have been sampled. The strength of these probability calculations, and hence the ability to assign individuals and determine connectivity, depend on both the degree of genetic differentiation among populations (e.g., in the breeding range) and on the number of markers used. In the extreme case of complete differentiation, a particular genetic marker will be found in one population and not others, and hence a single genetic locus can indicate the population of origin. Typically, genetic differentiation among populations will not be so extreme, and a relatively large number of markers may be needed.

In recent years, a number of highly variable (polymorphic) genetic markers have been developed, thereby substantially increasing the likelihood of finding the genetic variation needed to assign individuals to populations and determine connectivity. Microsatellites are a particularly popular class of markers, owing to their typically high levels of polymorphism and ease of use. Moreover, the primary difficulty with using microsatellites – that they need to be developed for each species of interest – is becoming less of a limitation as new high-throughput genomic methodologies (e.g., 454 sequencing) become more widespread. Some studies have also used sequence data for specific loci that are variable enough to show polymorphism within or across populations, such as the highly variable mitochondrial 'control region.' Another potentially useful class of genetic markers that has been used only rarely for the studies of migratory animals is amplified fragment length polymorphism (AFLP). This method simultaneously surveys a large number of genetic loci and typically uncovers substantial variation, which can be used to differentiate among populations. Finally, recent technological advances in genomics have made it possible to scan the genome for single nucleotide polymorphisms (SNPs) which are common and distributed throughout the genome and therefore hold considerable potential for evolutionary and population genetic studies. To date, few studies have used SNPs to study migratory connectivity, though the potential power of these markers makes it likely they will be used in the very near future as costs decrease.

With the development of highly variable markers and sophisticated methods to analyze them, the principal difficulty with molecular genetic approaches to study connectivity is no longer technological, but rather biological. That is, for many organisms, genetic differentiation among populations is very low and may be insufficient for a robust assignment of individuals using genetic markers alone. This is not so much an issue of the markers, but rather the dispersal behavior of the organisms themselves, as high levels of gene flow will prevent or degrade genetic differentiation among populations. Finding low differentiation among populations is itself informative, as high levels of dispersal between populations suggest low levels of migratory connectivity. However, because it is really natal dispersal that affects genetic differentiation (i.e., how far, on average, an individual moves from where it was born to where it settles and breeds as an adult), high levels of natal dispersal can eliminate genetic differentiation among populations even if adults show very high levels of migratory connectivity, thereby making it difficult to assess the migratory connectivity of adults. Time is another factor affecting genetic differentiation among populations, as it takes some time for genetic differentiation to build up. Thus, migratory organisms that have recently expanded from a smaller population (e.g., since the last Pleistocene glacial maximum) may show limited genetic differentiation among populations.

Because of these factors, several recent studies of Nearctic–Neotropical migratory birds – which are thought to have high levels of natal dispersal and also have undergone recent population expansions – have found limited utility for genetic markers in disentangling migratory connectivity. However, these studies were able to use genetic markers to determine connectivity at broad geographic scales, and higher resolution (i.e., more variable) markers may allow the determination of connectivity at finer scales. In the end, genetic markers may be most useful to studies of migratory species with somewhat limited natal dispersal, and/or in combination with other types of markers.

Stable Isotope Approaches

Another technique that relies on intrinsic markers in biological tissues to trace the origin and movement of migratory animals is stable isotope analysis. Stable isotopes are nonradioactive forms of elements that have similar chemical properties but vary in their atomic mass because of differences in the number of neutrons. During geochemical and metabolic processes, the differences in mass cause separation among isotopes of the same element, a phenomenon known as *isotopic fractionation*. Approximately, two-thirds of the elements have more than one stable isotope, but isotopes of carbon (^{13}C), nitrogen (^{15}N), hydrogen (^{2}H or D), and sulfur (^{34}S), are among the most useful for studying migratory connectivity for two reasons. First, their patterns of isotopic fractionation are well understood and vary predictably across broad spatial scales. Second, their high natural abundance allows them to be present at detectable levels in biological tissues. Stable isotopes are analyzed using *isotope ratio mass spectrometry*, and sample results are expressed in δ units relative to a standard of known isotopic composition. For example, the results of carbon isotope analysis are calculated as: $\delta^{13}C = \{[(\delta^{13}C_{unk}/\delta^{12}C_{unk})/(\delta^{13}C_{std}/\delta^{12}C_{std})] \times 1000\}$. Some of the most informative research on migratory connectivity has involved multiple stable isotopes or used stable isotopes and genetic markers together, and we will highlight these studies.

Feathers are the most commonly used tissue in stable isotope investigations of migratory connectivity. Most species of migratory birds undergo a complete molt once each year between July and September on or near their breeding areas, and the isotopic signatures of foods eaten during this time become incorporated into feathers. Because isotopic signatures are mostly inert once stored in feather tissue, samples collected later during the year provide information about the geographic origin of birds during molt. Each of the aforementioned isotopes provides different potential clues about a bird's molt location. Stable-hydrogen isotopes in growing season precipitation vary strongly with latitude. Stable-carbon isotopes show a similar pattern due to broad-scale differences in plant water use efficiency and photosynthesis strategy. Finally, stable-sulfur isotopes differ between marine and terrestrial environments, making it possible to measure longitudinal origins of molt in species whose habitats extend to coastal regions.

In one of the earliest sets of studies using multiple stable isotopes, Richard Holmes, Page Chamberlain, Dustin Rubenstein, and colleagues, sampled feathers from Black-throated blue warblers at breeding sites from North Carolina to Michigan. As predicted, they found that δD and $\delta^{13}C$ values varied systematically with the latitude of the sampling location. Feathers collected from wintering populations in the Greater Antilles revealed considerable mixing of individuals from a variety of breeding populations, but also indicated strong regional connectivity between wintering and breeding populations. A greater proportion of individuals wintering in the western islands of the Greater Antilles originated from northern breeding populations, whereas those wintering on islands further east were from more southern breeding populations.

When examined alongside molecular genetic markers, analyses of multiple stable isotopes can potentially yield even more refined estimates of migratory connectivity. Jeff Kelly and colleagues analyzed δD, $\delta^{34}S$, and mitochondrial DNA (mtDNA) in the feathers of Swainson's thrush at 12 breeding sites throughout North America and 5 winter sites in Mexico, Central, and South America. Analyses of mtDNA indicated the existence of two haplotypes: inland and coastal. Patterns of δD were particularly useful at distinguishing birds from coastal sites, while variations in $\delta^{34}S$ were helpful in separating inland sites. Together, these data revealed that birds with coastal haplotypes migrated to more northern winter sites compared to inland ones, and that birds from northern winter sites

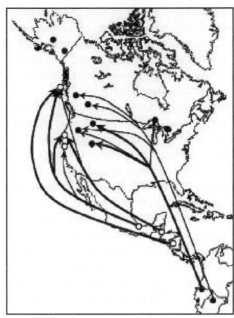

Figure 2 Map of predicted breeding sites of Swainson's thrushes sampled at five wintering sites. The weight of the arrows reflects the number of individual birds predicted to share that breeding origin. Heavy arrows indicate that 4–6 individuals share the origin; light arrows indicate that 1–3 individuals share that origin.

appeared to migrate shorter distances to southern breeding sites. This latter finding suggested that Swainson's thrushes engage in leapfrog rather than chain migration (**Figure 2**).

It is important to note that stable isotopes and molecular genetic markers have been used with great success in taxa other than birds. For example, Luciano Valenzuela and coworkers used $\delta^{13}C$ and $\delta^{15}N$ together with mtDNA to identify summer feeding areas in right whales and to understand the behavioral mechanism through which calves learn these locations. Furthermore, individual right whales that shared the same mitochondrial haplotype also had similar $\delta^{13}C$ and $\delta^{15}N$ signatures. This pattern suggested that individuals from each matrilineal lineage followed the same migratory route to summer feeding locations that they learned from their mothers during their first year of life.

An example using multiple stable isotopes with clear implications for conservation involves Monarch butterflies. The entire population of North American Monarch butterflies spends the winter at approximately ten winter sites in Mexico. Despite over 50 years of intensive study, it remained unknown whether the entire population mixed together at these winter sites or whether there was tighter connectivity between breeding and wintering populations. Len Wassenaar and Keith Hobson sampled δD and $\delta^{13}C$ in butterflies at their natal sites throughout North America and at 13 winter locations in Mexico.

Isotopic signatures indicated that individuals from the Midwestern United States were present at each of the winter sites sampled. However, butterflies with isotopic signatures indicative of more northern breeding areas were present at only two sites, making these locations strong candidates for protection.

Although the isotopic composition of several different tissues has proven to be useful for identifying regional and, potentially, even more localized populations of migratory species, there are several important caveats to this technique. Each isotope carries a unique set of assumptions, and it is necessary to understand these assumptions and to tailor experimental design accordingly. For example, despite its frequent use, the successful use of δD to unravel migratory connectivity depends on assigning individuals to the geographic location of molt by using a δD base map developed from 30-year running averaged values as a guide. Therefore, it is not only important to understand the natural history of the study species but also necessary to account for environmental, sampling, and analytical error. Bayesian statistical methods are quickly becoming recognized as an important tool in dealing with this bias because of their ability to incorporate prior information about the potential sources of error into models.

Future Considerations

Other intrinsic techniques have been attempted but with varying results. For example, populations of blood parasites within migratory species, such as malaria and bacteria, have been used in the studies of migratory connectivity, but these approaches have been met with mixed success. Recently, trace elements have also been explored for their utility in examining migratory connectivity. The growing number of studies showing differences in trace element concentrations among spatially discrete bird populations underscores the future potential of this technique. However, unlike the most often used stable isotopes, trace element signals are not known to change in continuous fashion across physical or environmental gradients. Thus, leveraging trace element data to advance our understanding of migratory connectivity will likely first require detailed mapping of these elements across the breeding and/or wintering ranges of migratory species.

For the time being, as far as smaller-bodied birds are concerned, geolocators, isotopes, and perhaps genetics, will be our best approach. With advances in analytical techniques, the research bottleneck has shifted from the lab to the field: although the isotopic and genetic tools are available, it remains difficult for a single researcher (or team) to collect samples from many hundreds or thousands of individual birds from across the range of a particular species. Hence, as banding studies increase throughout North and South America, Africa, and Asia,

there should be an organized and systematic feather-collection (as well as tissues from other taxa) initiative that will foster studies at scales of sampling intensity that are otherwise impossible to achieve.

In North America alone, approximately 1.2 million songbirds are banded each year. Yet, only in a few instances, are feathers being collected, and there is not yet any systematic effort within the ornithological community to collect and archive such samples. Clearly, this represents a lost opportunity for gaining valuable data. Not only can feathers be informative about extant patterns and processes, but the prospect of collections made over time offer the possibility of tracking temporal changes in breeding and/or wintering ranges of species. Such data would be important to evolutionary biologists interested in microevolutionary processes, population biologists investigating the causes for population declines, as well as conservation biologists concerned about the effects of climate change.

See also: Magnetic Orientation in Migratory Songbirds.

Further Reading

Bensch S and Akesson M (2005) Ten years of AFLP in ecology and evolution: Why so few animals? *Molecular Ecology* 14: 2899–2914.

Durrant KL, Marra PP, Fallon SM, et al. (2008) Parasite assemblages distinguish populations of a migratory passerine on its breeding grounds. *Journal of Zoology* 274: 318–326.

Haig SM, GrattoTrevor CL, Mullins TD, and Colwell MA (1997) Population identification of western hemisphere shorebirds throughout the annual cycle. *Molecular Ecology* 6: 413–427.

Kelly JF, Ruegg KC, and Smith TB (2005) Combining isotopic and genetic markers to identify breeding origins of migrant birds. *Ecological Applications* 15: 1487–1494.

Marra PP, Hobson KA, and Holmes RT (1998) Linking winter and summer events in a migratory bird by using stable-carbon isotopes. *Science* 282: 1884–1886.

Marra PP, Norris DR, Haig SM, Webster MS, and Royle JA (2006) Migratory connectivity. In: Crooks K and Sanjayan M (eds.) *Connectivity Conservation.* New York: Oxford University Press.

Poesel A, Nelson DA, Gibbs HL, and Olesik JW (2008) Use of trace element analysis of feathers as a tool to track fine-scale dispersal in birds. *Behavioral Ecology and Sociobiology* 63: 153–158.

Rubenstein DR, Chamberlain CP, Holmes RT, et al. (2002) Linking breeding and wintering ranges of a migratory songbird using stable isotopes. *Science* 295: 1062–1065.

Sillett TS, Holmes RT, and Sherry TW (2000) Impacts of a global climate cycle on population dynamics of a migratory songbird. *Science* 288: 2040–2042.

Studds CE, Kyser TK, and Marra PP (2008) Natal dispersal driven by environmental conditions interacting across the annual cycle of a migratory songbird. *Proceedings of the National Academy of Sciences of the United States of America* 105: 2929–2933.

Stutchbury BJM, Tarof SA, Done T, et al. (2009) Tracking long distance songbird migration by using geolocators. *Science* 323: 896–896.

Valenzuela LO, Sironi M, Rowntree VJ, and Seger J (2009) Isotopic and genetic evidence for culturally inherited site fidelity to feeding grounds in southern right whales (*Eubalaena australis*). *Molecular Ecology* 18: 782–791.

Wassenaar LI and Hobson KA (1998) Natal origins of migratory monarch butterflies at wintering colonies in Mexico: New isotopic evidence.

Proceedings of the National Academy of Sciences of the United States of America 95: 15436–15439.

Webster MS and Marra PP (2005) The importance of understanding migratory connectivity and seasonal interactions. In: Greenberg R and Marra PP (eds.) *Birds of Two Worlds: The Ecology and Evolution of Temperate–Tropical Migration Systems.* Baltimore, MD: Johns Hopkins University Press.

Webster MS, Marra PP, Haig SM, Bensch S, and Holmes RT (2002) Links between worlds: Unraveling migratory connectivity. *Trends in Ecology & Evolution* 17: 76–83.

Molt in Birds and Mammals: Hormones and Behavior

J. C. Wingfield, University of California, Davis, CA, USA
B. Silverin, University of Göteborg, Göteborg, Sweden

Introduction

During their lifetime, tetrapod vertebrates undergo several bouts of integument replacement, a phenomenon referred to as 'molt' – the periodic shedding, and replacement, of epidermal structures. This energy-demanding process is controlled by hormones and it involves not only replacement of feathers, hair, scales, etc., but also physiological events such as increased vascularization of feather/hair follicles, osteoporosis, changes in the rate of protein synthesis and overall metabolism, a shift in the heterophil/lymphocyte ratio, decrease in body fat, etc. Behaviors associated with the molting process included secretive habits, rhythmic movements to dislodge skin or abrade it, and changes in foraging behavior – sometimes to select nutrients essential for replacing skin. For these reasons, molt cycles are restricted to times of the year when trophic resources are sufficient to support replacement of the integument.

Molting systems have evolved great diversity in patterns and types and appear to be ubiquitous to vertebrates as well as many invertebrates (but only the former will be discussed here). Birds have particularly well-known molt cycles that appear to be highly variable. In those wintering in the temperate zone, and also for some species wintering in tropical areas, adults replace all their flight and body feathers immediately after breeding, and molt is finished before onset of migration. Juveniles follow the same time schedule, but can have more varying molting patterns. Many long-distance migrants do not molt before migration, but initiate and complete their entire molting cycles in the winter area (winter molt). Some tropical migrants have a seasonal split in the molting period (e.g., some feathers being molted in the breeding area and then completed in the winter area). A few long-distance migrants have two complete annual molting cycles: one immediately after breeding (before migration) and another in the winter quarters. Large birds, for example, eagles and albatrosses, molt only some feathers each year – serial molt and complete replacement of the integument may take longer than one year. Others, for example, some ducks and geese, molt all their wing feathers simultaneously. This speeds up the molting period but results in vulnerability to predators during molt. One way to reduce the risk of being taken by a predator during this period is to gather in huge flocks (molt migration) and aquatic birds gather out on open water. Thus, molting strategies vary considerably among birds.

At least among the Palearctic warblers, the summer molt seems to be the ancestral molt pattern, whereas winter molt appears to have evolved independently 7–10 times in this clade. The reason why winter molt evolved is unclear. Several hypotheses have been proposed, for example, if there is competition for winter territories, birds should migrate to these areas as soon as possible after breeding and postpone molt till after arrival in the wintering ground. Or, maybe winter molt was favored because it could speed up onset of the migratory flight to the breeding grounds in spring.

Why Molt?

Once the integument has developed in a mature individual, daily and seasonal routines such as foraging can result in accumulating wear and tear. Furthermore, social interactions including play behavior, sexual interactions, and aggression (especially fighting) can result in damage to the integument. In most species (e.g., many mammals including humans), the cornified outer layers and hair are replaced continuously. In fish, scales can be replaced as they are lost or worn. Indeed, the ability to replace damaged components of the integument is probably ubiquitous. However, many species also show periodic shedding of the integument and replacement that often involves specific changes in structure as well.

With time, bird feathers, for example, are worn out, affecting maneuverability in the air and flying performance as well as thermoregulatory capacity and coloration. Feathers, therefore, must be replaced at regular intervals (once or twice a year). As molt is very energetically expensive, it should not overlap with other energy-demanding events such as breeding and migration. The seasonal timing of molt is therefore of great importance, and it results from a trade-off between having a good integument quality during breeding or during the nonbreeding period. Examples of wear and tear are shown for a song bird in **Figures 1** and **2**. In this case, continuous replacement after loss or damage is limited and a complete periodic molt is essential.

Molting Processes

The vertebrate integument is diverse including dermal and epidermal structures adapted to aquatic environments

Figure 1 Why molt? Nuttall's white-crowned sparrows (*Zonotrichia leucophrys nuttalli*) on the left-hand side have fresh breeding plumage, while those on the right have extremely worn plumage at the end of the breeding season in July. Some feathers (e.g., on the head, right panels) may have been lost while fighting. Note also the faded coloration of feathers in the right-hand panels. Photos by J.C. Wingfield, taken on the Pacific coast of central California.

Figure 2 Close up of fresh plumage (left-hand panels) and worn plumage (right-hand panels) in Nuttall's white-crowned sparrows (*Zonotrichia leucophrys nuttalli*) on the Pacific coast of central California. Note the extreme wear on the right-hand panels and the dramatic fading of color. This worn plumage is completely replaced by molt. Photos by J.C. Wingfield.

(marine and freshwater) and terrestrial habitats (from humid and mesic environments to extreme aridity). In fishes, the integument usually produces scales of great diversity and also mucous. The development of a stratum corneum during embryogenesis first occurred in amphibians and is found in all tetrapod vertebrates. This multilayered structure involved programmed cell death as an impervious outer layer was produced. Such a skin structure also gave rise to scales, feathers, and hair in the amniote vertebrates. The structure allowed for periodic shedding of older and worn outer layers. Here we restrict discussion of molting to periodic replacement of epidermal structures in tetrapod vertebrates (with a focus on birds and mammals) because the various groups of fish, where scales etc. are replaced individually, apparently do not molt in the sense of shedding large portions of the integument, although fish do replace their integument and some species may be able to shed small portions at intervals.

In some species, the desquamation of epidermal derivatives may be continuous and in small fragments (e.g., dandruff in many mammals and birds), or cyclic in which a true molt occurs and the entire integument is replaced systematically. The speed of this process may vary from several months in some species to just a few weeks in others or even hours in the case of some reptiles. The process also involves distinct components of shedding and regeneration of new skin, hair, feathers, etc. in various combinations.

Amphibians shed the outermost layers of skin on a periodic basis – developmental and seasonal. The duration and the frequency of molts vary tremendously as a function of species and habitat (e.g., terrestrial versus aquatic). This shedding is accomplished by removal of fragments rather than the entire integument at once. Specific behavioral events are probably restricted to mild abrasion of the skin on environmental substrates to assist sloughing.

Reptiles, such as squamates, frequently shed the entire integument in one piece allowing new skin underneath to expand rapidly enabling further growth of the individual. These species show behavioral responses in terms of finding a refuge, muscular contractions to assist splitting of old skin, undulations and abrasion to aid sloughing – sometimes as a complete structure.

Birds usually show waves of feather shedding and replacement that progress in specific sequences over the entire body surface (**Figure 3** shows an example of sequential replacement of flight feathers). In a few species, feathers may be shed almost simultaneously resulting in flightlessness and the potential for severe challenges to thermoregulation. The skin is shed in small fragments. Behavioral components include molt migration or seeking a sheltered place (especially if flight is impaired), reduced aggression, and territoriality in some species.

In mammals, hair can be shed and replaced in waves on a periodic basis or continuously (as in humans). In others, such as cetaceans, skin is shed in fragments, probably

Figure 3 Feather replacement in the wing of a willow warbler, *Phylloscopus trochilus*. Old worn feathers are shed and new ones grow as pins between new feathers below, and old feathers above. This set sequence occurs each time the bird molts progressing from feathers shed in the midwing to feathers at the extremities. The pin feathers have a rich blood supply and gradually unfold into a mature feather as they grow out. Sequences (waves) of feather replacement occur elsewhere on the body and also in the hair replacement of mammals. Photo by B. Silverin.

continuously. Behavioral components include abrasion such as rubbing against substrate to hasten loss of old hair and skin.

Behavioral Effects

The physiological and morphological aspects of vertebrate molts have been described elsewhere, but behaviors associated with molting have received much less attention. In general, molt cycles have implications for expression of behavior such as those changes related to the molting process itself, and changes in integument structure that allow the individual to express behavioral traits at different seasons or in different habitats. These can be summarized as follows:

Dietary changes – to provide high sulfur-containing amino acids to produce keratin etc.
Sloughing behavior – to remove old skin, hair, or feathers; abrade scales.
Secretive behavior – such as seek a shelter while new skin cornifies, feathers and hair grow. This reduces possibility of damage to components of a developing new integument. Additionally, molt affects flight ability in birds and, thereby, the risk of being taken by a predator.
Changes in crypsis – seasonal change in pelage/plumage may allow renewal of integument to adapt to seasonal climatic changes. This could include development of white plumage in winter and cryptic pelage in summer (**Figure 4**).
Changes in insulation – seasonal changes may also include adjustment of insulation qualities of pelage

Figure 4 Examples of seasonal changes in pelage. Seasonal changes in plumage coloration in willow ptarmigan (*Lagopus lagopus*) on the North Slope of Alaska. Top left panel shows typical all white winter plumage, cryptic in snow but also highly insulated against the Arctic winter weather. Lower left panel shows summer plumage, more cryptic in the absence of snow. The top right panel shows a musk ox (*Ovibos*) on the North Slope of Alaska in full winter pelage with long outer guard hairs and thick insulating hair beneath. Lower right panel shows a musk ox in spring shedding large chunks of insulating hair. This will be replaced the following autumn. Photos by J.C. Wingfield.

(e.g., down or extra fur for the winter months) so that the organism can go about its daily routines despite major changes in weather (**Figure 4**).

Changes in "protective" structure – to allow individuals to take advantage of different environments at certain times of year. For example, some amphibians develop a more cornified skin during the terrestrial phase of their life cycle. Euryhaline fish undergo changes in their integument and degree of mucus production during movements between fresh and salt water (e.g., to cope with directions of salt and water transport in permeable components of the integument etc.).

Changes in nuptial structures – seasonal change in pelage may also allow development of nuptial appendages of the integument (e.g., antlers, plumes, etc.) and changes in color patterns for reproductive purposes – to attract a mate, defend a breeding territory (and paternity). Social interactions during molt (at least in birds) may have important effects on the coloration and patterns of plumage developed.

Hormone Mechanisms in Molt

Control systems are not well known but involve hypothalamo–pituitary secretions that in turn regulate

peripheral endocrine systems to orchestrate the development of a physiological and morphological state so that molting can actually begin, and also regulate the processes of specific loss and replacement of integument structures. These physiological and behavioral adjustments associated with molting are accompanied by changes in hormonal-secretion patterns. In birds, at the onset of postbreeding molt, sex-steroid levels as well as gonadotropins are basal. Elevated circulating testosterone, for example, may delay or even prevent molt. Furthermore, there is also increased and irreversible conversion of testosterone to biologically inactive metabolites in the anterior hypothalamus at the time of molt in great tits *Parus major*, essentially reducing effects of testosterone in target tissues and allowing the postnuptial (or prebasic) molt to begin. In contrast, in some squamate reptiles, molt may occur just prior to spring mating. In mammals, sex steroids such as testosterone enhance growth of some hair types and inhibit others. Inter-relationships of molt and other life history stages appear to be complex and deserve further study.

It is generally considered that the thyroid hormone thyroxine (T4) is responsible for the induction of molt. Most species studied to date, but not all, show a seasonal elevation of T4 during molt. Thyroidectomy in amphibians, reptiles, and birds delays molt although it may not eliminate it completely. However, even if molt does begin in thyroidectomized animals, the epidermal structures that develop are frequently malformed. Although triiodothyronine (T3) in some species shows a similar temporal change over seasons as T4, experiments indicate that T3 has rather little, if any, influence on the molting process. Injections of T4 into thyroidectomized animals tend to restore molt to its normal frequency and the structures developed appear normal. Thyroid hormones also appear to be important for development of thick skin, more impervious to water in newts moving from breeding ponds onto land.

The role of prolactin in molt is strongly suggested by some studies, but others have failed to relate high circulating levels of prolactin to molt. In birds, peaks of prolactin secretion are consistently associated with onset of molt. As this hormone is also associated with parental behavior, osmoregulation, metabolism, and growth, relationships to molt are complex and probably affected by season, life history stage, etc. In newts, prolactin is important for the development of smooth skin with rich mucous secretion prior to entry into breeding ponds. However, these effects are also dependent upon thyroid hormones. On the other hand, in mice, prolactin may inhibit onset of waves of hair replacement. Furthermore, disruption of prolactin receptor gene expression in mouse skin advanced onset of molting. Much more research is needed to understand the hormonal regulation of molt and associated behavior in general.

Molt and Environmental Perturbations

The molting process is critical for many species and disruption of molt can have serious consequences for survival, reproductive success, etc. Thus, the potential for environmental perturbations to affect molt has probably had an influence on the timing of molt to the most benign time of year, and the fact that many animals may seek a refuge while molting. Stress hormones such as glucocorticoids may play an important role during molt. Free-living birds show distinct seasonal variations in plasma levels of corticosterone with a nadir during prebasic molt when feathers are being replaced. The downregulation of the hypothalamo–pituitary–adrenal (HPA) axis appears to be mediated at different levels in different species of birds. The downregulation of the HPA axis during molt might enable the individual to avoid the protein catabolic effects of glucocorticoids such as corticosterone, and subsequent inhibition of feather growth and decreased quality of feathers (which would be important for over winter survival for example). Experimental elevation of plasma levels of corticosterone during molt decreases the rate of feather replacement. Because molt may dramatically reduce a bird's flight ability, high corticosterone levels during molt in turn may lower

survival rate. For example, free-living pied flycatchers, *Ficedula hypoleuca*, undergoing a simulated molt were depredated more frequently by sparrowhawks, *Accipiter nisus*, than were control birds.

On the other hand, downregulation of the HPA axis during molt and a reduced stress response might also have costs for the free-living animal because this results in reduced ability to cope maximally with unpredictable stressors, such as adverse weather conditions, human disturbance, etc. This may mean there is a trade-off between timing of molt to the optimal time for food supplies and a lowered ability to cope maximally with environmental stressors. To illustrate this trade-off further, a recent study showed that induced molt (by plucking feathers) in captive starlings, *Sturnus vulgaris*, resulted in only a moderate increase in their plasma levels of corticosterone when subjected to physical chronic stress (food restriction), psychological stressors, or daily disturbances (such as cage disturbances or loud music). The moderate rise in corticosterone did not slow down feather regrowth, nor did it affect feather quality as occurs in experiments using techniques that elevate corticosterone to much higher levels. Thus, evolution might have selected regulatory mechanisms that reduce the response to stress during molt so that the effects of high corticosterone levels that may interfere with the molting process can be avoided. Nonetheless, birds still retain some coping capacity in the face of stress.

Future Directions

Clearly, replacement of the integument is an essential component of the life cycle and represents not only a maintenance function aiding survival, but may also be important for attracting a mate and reproductive success. As such, patterns of molt and development/loss of epidermal structures are extremely diverse across vertebrate taxa. The role of behavior in molt cycles, both to aid the molting process and behavioral interactions that may influence molt, remains very poorly understood. Hormone control mechanisms have been well investigated in some species and suggest great diversity across taxa according to time of year and context. How hormones may influence molt behavior remains almost completely unknown in vertebrates. With the new research tools available today, the regulation of molt cycles and associated behavior is a rich research area awaiting exploration.

See also: Hormones and Behavior: Basic Concepts; Seasonality: Hormones and Behavior.

Further Reading

Adolf EF and Collins HH (1925) Molting in an amphibian, *Diemyctylus*. *Journal of Morphology and Physiology* 40: 575–592.

Alibardi L (2003) Adaptation to the land: the skin of reptiles in comparison to that of amphibians and endotherm amniotes. *Journal of Experimental Zoology* 298B: 12–41.

Bauwens D, Van Damme R, and Verheyen RF (1989) Synchronization of spring molting with the onset of mating behavior in male lizards. *Journal of Herpetology* 23: 89–91.

Craven AJ, Ormandy CJ, Robertson FG, et al. (2001) Prolactin signaling influences the timing mechanism of the hair follicle: Analysis of hair growth cycles in prolactin receptor knockout mice. *Endocrinology* 142: 2533–2539.

Dawson A (2006) Control of molt in birds: Association with prolactin and gonadal regression in starlings. *General and Comparative Endocrinology* 147: 314–322.

Dent JN, Ludeman A, and Forbes MS (1973) Relations of prolactin and thyroid hormone to molting, skin texture and cutaneous secretion in the red-spotted newt. *Journal of Experimental Zoology* 184: 369–382.

Ebling FJ (1976) Hair. *Journal of Investigative Dermatology* 67: 98–105.

Holmgren N and Hedenström A (1995) The scheduling of molt in migratory birds. *Evolutionary Ecology* 9: 354–368.

Jørgensen CB (1988) Nature of molting control in amphibians: Effects of cortisol implants in toads, *Bufo bufo*. *General and Comparative Endocrinology* 71: 29–35.

Kuenzel WJ (2003) Neurobiology of molt in avian species. *Poultry Science* 82: 981–991.

Ling JK (1970) Pelage and molting in wild mammals with special reference to aquatic forms. *Quarterly Review of Biology* 45: 16–54.

Ling JK (1972) Adaptive functions of vertebrate molting cycles. *American Zoologist* 12: 77–93.

McGraw KJ, Dale J, and Mackillop EA (2003) Social environment during molt and the expression of melanin-based plumage pigmentation in male house sparrows (*Passer domesticus*). *Behavioral Ecology and Sociobiology* 53: 116–122.

Nelson RJ (2005) *An Introduction to Behavioral Endocrinology*, 3rd edn. Sunderland, MA: Sinauer.

Nicholls TJ, Goldsmith AR, and Dawson A (1988) Photorefractoriness in birds and comparison with mammals. *Physiological Reviews* 68: 133–176.

Romero LM (2002) Seasonal changes in plasma glucocorticoid concentrations in free-living vertebrates. *General and Comparative Endocrinology* 128: 1–24.

Rougeot J, Allain D, and Martinet L (1984) Photoperiodic and hormonal control of seasonal coat changes in mammals with special reference to sheep and mink. *Acta Zoologica Fennica* 171: 13–18.

Silverin B and Deviche P (1991) Biochemical characterization and seasonal changes in the concentration of testosterone-metabolizing enzymes in the European great tit (*Parus major*) brain. *General and Comparative Endocrinology* 81: 146–159.

Slagsvold T and Dale S (1996) Disappearance of female pied flycatchers in relation to breeding stage and experimentally induced molt. *Ecology* 77: 461–471.

Stewart PD and MacDonald DW (1997) Age, sex and condition as predictors of molt and the efficacy of a novel fur-clip technique for individual marking of the European badger (*Meles meles*). *Journal of Zoology London* 241: 543–550.

Strochlic DE and Romero LM (2008) The effects of chronic psychological and physical stress on feather replacement in European starlings (*Sturnus vulgaris*). *Comparative Biochemistry and Physiology A* 149: 68–79.

Svensson E and Hedenström A (2008) A phylogenetic analysis of the evolution of moult strategies in Western Palearctic warblers (Aves: Sylvidae). *Biological Journal of the Linnean Society* 67: 263–276.

Swaddle JP, Williams EV, and Rayner JMV (1999) The effect of simulated flight feather molt on escape take-off performance in starlings. *Journal of Avian Biology* 30: 351–358.

Wilson FE (1997) Photoperiodism in American Tree sparrows. Role of the thyroid gland. In: Harvey S and Etches RJ (eds.) *Perspectives in Avian Endocrinology*, pp. 159–169. Bristol: J. Endocrinol.

Wingfield JC and Farner DS (1993) The endocrinology of wild species. In: Farner DS, King JR, and Parkes KC (eds.) *Avian Biology*, vol. 9, pp. 163–327. New York: Academic Press.

Wingfield JC and Silverin B (2009) Ecophysiological studies of hormone-behavior relations in birds. In: Pfaff DW, Arnold AP, Etgen AM, Fahrbach SE, and Rubin RT (eds.) *Hormones, Brain and Behavior*, 2nd edn, vol. 2, pp. 817–854. New York: Academic Press.

Monkeys and Prosimians: Social Learning

D. M. Fragaszy and J. Crast, University of Georgia, Athens, GA, USA

Introduction

In this chapter, we highlight examples of social influences on learning observed in prosimians and monkeys and consider the role of socially mediated learning in the biology of these animals. Learning is always the outcome of interacting physical, social, and individual factors and takes place over time. Thus, we cannot parse learning, either as a process or as an outcome, into portions that are socially influenced and portions that are not. Instead, we can document how social processes affect behavior relevant to the learning process, and we can seek evidence for social contributions to learning outcomes.

To begin, we provide some background on the taxonomic groups of interest in this chapter: monkeys and prosimians. Primates are a remarkably diverse order. Body size alone spans three orders of magnitude, from tiny prosimians weighing a few hundred grams to massive apes weighing more than 100 kg. Diet, morphology, mating systems, locomotor style, life history, and every other aspect of the biology of these animals is as diverse as body size, and this diversity is important when considering the contributions of the social context to learning in particular species.

Phylogeny of Prosimians and Monkeys

As Fleagle (1999) discusses in greater detail, the order Primates includes two suborders: Prosimii, prosimians, and Anthropoidea, monkeys, apes, and humans (see **Figure 1**). The two suborders have evolved separately for at least 55 Million years. Two infraorders are classified within Anthropoidea: the platyrrhines (New World monkeys) and catarrhines (Old World monkeys, apes, and humans). New and Old World monkeys diverged approximately 40 Millions of years ago (Mya), and apes and hominids (hominids include modern humans and their ancestors; superfamily Hominoidea) diverged from the Old World monkeys (superfamily Cercopithecoidea) approximately 20 Mya. Given their lengthy independent evolution, variation in the life histories, body sizes, social organizations, etc., within each suborder, infraorder, and superfamily in the order Primates is to be expected.

The suborder Prosimii includes the infraorders Lemuriformes (the lemurs of Madagascar), Lorisiformes (the lorises of Africa and Asia and the galagos of Africa), and Tarsiiformes (tarsiers of Southeast Asia). All prosimians live in tropical habitats in Africa and Asia and the vast majority are arboreal and nocturnal. Prosimians are sometimes referred to as 'living fossils' because they appear to have some physical similarities to ancestral primates of approximately 50 Mya. In general, prosimians rely to a greater extent than other primates on olfaction. Some are solitary foragers; others travel and forage in groups ranging from small family units to larger social groups of as many as 27 individuals. We know less about the lifestyles and behavior of prosimians than of monkeys.

In comparison with prosimians, species in the suborder Anthropoidea are characterized by a relatively larger brain for their body mass, diurnal lifestyle, and a greater reliance on vision than on olfaction. Anthropoid species generally exhibit greater manual dexterity than prosimians, and anthropoids are more likely to live in groups. New World monkeys (infraorder Platyrrhini) are arboreal and relatively small-bodied, ranging in size from approximately 100 g (the pygmy marmoset [*Cebuella pygmaea*]) up to 10 kg (the muriqui [*Brachyteles arachnoides*] and spider monkey [Genus *Ateles*]). Many genera live in small family groups; others live in medium-to-large social groups (as many as 50–60 individuals). Within the New World monkeys are the subfamilies Callitrichinae (marmosets and tamarins), Atelinae (muriquis, woolly, howler, and spider monkeys), Pitheciinae (titis, sakis, bearded sakis, and uakaris), Cebinae (squirrel monkeys and capuchins), and Aotinae (owl monkeys, the only nocturnal anthropoid). During the platyrrhine radiation in the Americas, the genera adapted to distinct niches, making a living in different parts of the forest canopy and resulting in great diversity in social organization, reproductive strategy, diet, and locomotor style.

Compared to New World monkeys, Old World monkeys (superfamily Cercopithecoidea) are mostly larger-bodied, ranging from around 1 kg to approximately 30 kg, and some are terrestrial. Old World monkeys include subfamilies Cercopithecinae (baboons, mandrills, drills, macaques, mangabeys, and guenons) and Colobinae (colobus monkeys and langurs), which differ particularly in their dietary adaptations (see Fleagle for details). Most cercopithecoid species live in large polygamous social groups with clear dominance hierarchies within and between matrilines (female kin groups), and some form multilevel societies during parts of the year. Of all primate superfamilies, cercopithecoids have the widest geographic range, greatest number of species, and form some of the largest groups and biomass

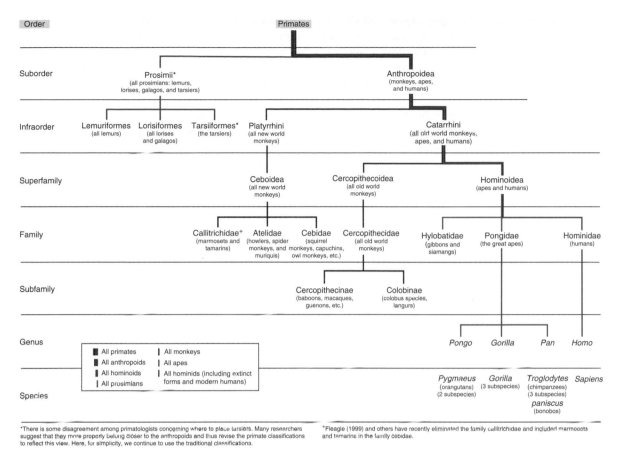

Figure 1 legend:

■ All primates	All monkeys
■ All anthropoids	All apes
■ All hominoids	All hominids (including extinct forms and modern humans)
All prosimians	

*There is some disagreement among primatologists concerning where to place tarsiers. Many researchers suggest that they more properly belong closer to the anthropoids and thus revise the primate classifications to reflect this view. Here, for simplicity, we continue to use the traditional classifications.

⁺Fleagle (1999) and others have recently eliminated the family callitrichidae and included marmosets and tamarins in the family cebidae.

Figure 1 Primate taxonomic classification. This abbreviated taxonomy illustrates how primates are grouped into increasingly specific categories. Only the more general categories are shown, except for the great apes and humans. Reproduced from Turnbaugh WA, Nelson H, Jurmain R, and Kilgore L (2002) *Understanding Physical Anthropology and Archaeology*, 8th edn., with permission from Wadsworth.

densities in the primate order. Despite this, Old World monkeys have less diversity in diet and social organization than New World monkeys.

Phylogeny and Socially Mediated Learning

This brief review of primate phylogeny suggests some reasons why we might expect socially mediated learning to vary across primate taxa. First, group demographics and social dynamics within groups define the social context, and thus influence socially mediated learning within a group. The number of groupmates, their age and sex, and the nature of social relationships within the group vary enormously across species, and may vary considerably within species as well. Groups of monkeys of the same species may live in smallish groups (4–7 individuals) or quite large groups (more than 40 individuals) depending on the local distribution of resources. Second, reliance on various sensory modalities (vision, olfaction, audition, and touch) in social interaction and in general activity varies across taxa. For example, species that are particularly attentive to smell (such as many prosimians) will be affected by

social partners in a way different from that of species that are highly reliant on vision. Third, motor patterns and action proclivities vary considerably across species. For example, leaf-eating monkeys are generally less likely to manipulate objects spontaneously than species that feed on seeds and nuts. Finally, the variability in behavioral ecology across species means that individuals of different species are interested in different kinds of activities, locations, objects, and events. For example, leaf-eating monkeys may be less likely to attend to sequences of actions during feeding than are seed- or nut-eating species; omnivorous species are less likely to attend to the odor of leaves eaten by another than are dietary specialists. Behavioral priorities and proclivities of each species constrain what an individual is likely to learn in the first place, and thus the role of social context in learning.

The Sources of Social Context

Social Organization

The social organization (i.e., the size, demographic composition, and spatiotemporal coordination of individuals within a group) and social relationships among individuals in a group provide the boundaries of the social context in which an individual can learn. As Coussi-Korbel and Fragaszy have proposed, conspecifics with which an individual has a long-term social relationship and that are frequently nearby are particularly important and enduring components of an individual's experience. In theory, the more closely individuals coordinate their activity in space and/ or time, the more likely an individual's activity is to influence the activity of others. Individuals of species in which social partners spend more of their time apart than together are likely to experience less direct social influences on learning specific actions than species that spend most of their time in the company of conspecifics. For example, adults of many nocturnal prosimians form sleeping groups during the day but travel and forage alone at night (e.g., dwarf and mouse lemurs [Genus *Cheirogaleus* and *Microcebus*, respectively] and some galagos [Genus *Galago*] and tarsiers [Genus *Tarsius*]). These animals are therefore not often in the company of others that might influence their behavior. However, all monkeys and some prosimian species, such as lemurs, sifakas, and indris, remain in cohesive groups and are near conspecifics virtually all the time. This intensely social lifestyle affects every aspect of experience through every sensory modality. Interactions with conspecifics structure where and how an individual budgets the time that it devotes to different activities (e.g., travel vs. feeding), and conspecifics also influence how an individual responds to events that occur nearby. For example, as Cheney and Seyfarth have shown, monkeys attend to overt signals made by others concerning objects, events, or locations of affective value (i.e., desirable or objectionable) such as a recruitment call to a food site or an alarm call to a predator, even if out of sight or some distance away from the other group members.

Social Relationships

Individuals are more likely to be near others with which they share a mutually affiliative relationship (e.g., dependent offspring with a parent). If social influences on learning are maximized when individuals are near one another, then a potential learner will be more influenced by those with which it shares positive affiliations than by others: a phenomenon Coussi-Korbel and Fragaszy have labeled Directed Social Learning. Over time, uneven social influences on learning across individuals within a group can lead to the generation of behavioral variations among subgroups. For example, young Japanese macaques living in Koshima, a small island in Japan, first began to wash sweet potatoes in the sea when these were provided for them on a sandy beach on the island. Initially, only juveniles washed potatoes. In subsequent years, the juveniles' older siblings and mothers started to wash potatoes. Older individuals adopted the behavior more slowly than

juveniles; adult males most slowly or not at all. If social influences contributed to the spread of the behavior, it did so unevenly across age and sex classes in accord with the predictions of the Directed Social Learning model. However, as Galef has indicated, a similar outcome could reflect accumulation of individual experience without any social influence, so we cannot definitively claim that social influence promoted the spread of the behavior. A similar caveat applies to several commonly cited examples of traditions in non-human primates. Observing the development of behaviors by new practitioners, with the requisite detail of social contexts and behavioral change over time, is necessary to make strong claims about the contributions of social context to learning a specific behavior. Such developmental studies are now underway with some species of monkeys.

Social influences within a group can be thought of as either vertical (across generations) or horizontal (within generations; among juveniles, for example). Vertical and horizontal social influences are common in primates. Vertical social influence is often discussed as promoting behavioral continuity between generations, while horizontal social influence is more likely to promote adaptive behavioral change; for example, in response to changing circumstances. Vertical social influence promotes continuity in commonplace and routine preferences and behaviors that young primates acquire gradually while traveling with adults, such as habitual travel routes and sleeping sites. Vertical social influence can also promote refinement of specific behaviors. For example, as Cheney and Seyfarth have shown, young vervet monkeys (*Cercopithecus aethiops*) gradually narrow the range of animals to which they give alarm vocalizations according to differential adult responsiveness to their calls. Adults respond to juveniles' calls in response to actual predators and ignore calls in response to benign animals.

Perry's studies of white-faced capuchins (*Cebus capucinus*) in Costa Rica provide examples of behaviors reflecting horizontal social influence. These monkeys sometimes develop idiosyncratic social behaviors ('games') that are played in pairs by close companions in a play context, but not between parent and offspring. One of the games identified by Perry and colleagues is the toy game, in which two monkeys take turns extracting an inanimate object, like a twig or leaf, from each other's mouth. In the toy game, one monkey holds the object tightly in its mouth without chewing it, and the other monkey attempts to pry open the first monkey's mouth and extract the item. Once retrieved, the monkeys then repeat the procedure or switch roles. Although initially one individual instigates a new game, eventually several different pairs in the group participate in the same game. Such behaviors are maintained by a particular social context and often disappear when that context disappears (e.g., when the key initiator of the behavior emigrates from the group).

For many species of primates, the most influential social partner from birth until independence is the mother, and in some species that share parental care, the mother and father (e.g., callitrichids, owl monkeys, and titi monkeys). Infants of most primate species are carried by the mother and thus are influenced by her activity as they travel together throughout the day. Even when able to travel independently, infants typically remain near their mother to nurse, rest, and feed, and this period of dependency is often considered important for skill learning by infants. Aye-ayes (*Daubentonia madagascariensis*), a nocturnal prosimian species, provide a striking example (see **Figure 2**). A significant part of the aye-aye's foraging activity involves extracting larvae from woody substrates, using a method called tap-foraging. In tap foraging, aye-ayes tap the substrate with a finger to locate a hollow cavity, gnaw the wood in the right place, and insert a specially adapted, long and skinny digit to probe the cavity and to extract the larva. Krakauer demonstrated several ways in which immature aye-ayes' proficiency in tap foraging is influenced by close proximity with their mother while she engages in the behavior. In general, the aye-aye mother allows her infant to remain nearby while she tap-forages. Over time, the infant begins to take over the site where the mother is working and extract the larvae itself. Infants of a naturally nontap-foraging mother attempted tap-foraging less often than other infants and never succeeded at extracting a larva.

Processes Mediating Learning in a Social Context

Facilitation and Enhancement

One common and powerful form of social influence on learning in primates is increased probability of performing a behavior when a conspecific is seen performing that behavior. Such socially facilitated behaviors are already in an individual's repertoire, for example, vocalizing or grooming. Another powerful social influence on behavior is increased interest in an object or in an area where another has recently been active or where others' previous activity has left artifacts (e.g., scents or physical alterations) (see **Figure 3**). Such increased interest in areas or objects where others have been active has been termed, respectively, local and stimulus enhancement; hereafter, enhancement. The bulk of empirical studies of social influences on learning in monkeys and prosimians have concerned these two phenomena.

Social facilitation is particularly common in primates in the context of feeding. For example, individuals are likely to begin eating, even if satiated, if nearby group members are eating. Social facilitation can lead to exposure to a new food item, or support exploratory activities that indirectly aid learning a foraging skill, as when young monkeys learn to locate hidden prey through repeated bouts of searching begun while or shortly after seeing others forage for hidden prey. This simple mechanism can support individuals developing the same dietary preferences as their groupmates, as individuals eating at the same time usually eat in the same place, and therefore often eat the same things. More generally, social facilitation results in temporal coordination of group activity.

Enhancement may occur through multiple senses and over an extended time period. For example, an individual's attention may be drawn to a foraging site through observation of another feeding, hearing the other's actions (such as breaking a stick), eating food items derived from

Figure 2 Mother and infant aye-aye foraging jointly. Aye-ayes (*Daubentonia madagascariensis*, lemurids) locate hidden prey by tapping on woody substrates. Infants begin to practice this technique at the same sites as their mothers. Photo by David Haring/Duke Lemur Center.

Figure 3 Infant Japanese macaques (*Macaca fuscata*, cercopithecines) attend closely to their mother's activity with stones. In groups of Japanese macaques provisioned with food, many individuals engage in stone-handling, and this behavior has been characterized as a tradition. Photo by Jean-Baptiste Leca/Primate Research Institute, Kyoto University and Iwatayama Monkey Park.

another's activity at the site, smelling another's mouth, and encountering artifacts (including scents) of past foraging activity, as well as through joint contact with materials another is handling. Any and all of these experiences increase the probability that an individual will investigate the site that another is exploiting or has exploited. Typically, young primates show strong interest in sites where others, especially adults, are foraging (see **Figure 4**). To the extent to which juveniles' proximity is tolerated by adults, young primates may approach and eat dropped food or even take bits of food from another's hand or mouth. However, even when young monkeys do not acquire food as a result of approaching, they are still intensely interested in sites where others forage.

Although most adult monkeys and prosimians do not overtly share food, enhancement of interest in foraging sites appears to be actively promoted in callitrichids. For example, Rapaport and Brown have found that adult golden lion tamarins (*Leontopithecus rosalia*; see **Figure 5**), which live in cohesive family groups that are led by a cooperatively breeding pair, emit food-offering vocalizations that draw their dependent offspring to a site containing live prey or large/tough-skinned fruit. Instead of taking the food for themselves, an adult waits until a juvenile reaches the site and allows the juvenile to extract the food item. This form of provisioning (or, as Rapaport and Brown refer to it, opportunity teaching) peaks around weaning (3–4 months) and continues untill infants are about a year old. Adults selectively provision infants with items that are difficult to process.

Callitrichids rely to varying degrees on extractive foraging for hidden foods, and participating in foraging with adults apparently helps youngsters learn to search in appropriate places and to perform appropriate actions. Research has shown more overt instances of adults actively providing social supports for youngsters learning to forage in callitrichids than in other monkeys, such as cercopithecines and colobines, which live in larger groups and show less shared parental care. Brown and Rapaport suggest that the degree of parental assistance in foraging seen in callitrichids is matched only by apes.

Motor Imitation

Motor imitation (i.e., performing a specific action after observing another perform the same action) is thought to contribute importantly to learning in humans. Currently, we have no evidence that prosimians or monkeys imitate novel actions spontaneously, as do humans. Nevertheless, recent experimental evidence indicates that marmosets and tamarins (callitrichids) will use the same part of the body to move an object that they have witnessed a conspecific use to solve a foraging problem. Currently, callitrichids provide the best evidence of imitation of familiar actions in monkeys. It is interesting that callitrichids have aptitude in this domain (as well as in opportunity teaching), whereas cercopithecoid monkeys do not, because callitrichids are phylogenetically more distant from hominids than cercopithecoid monkeys (see **Figure 1**), while true imitation is present in hominids.

Figure 4 Infant and juvenile bearded capuchins (*Cebus libidinosus,* cebids) watch an adult crack a palm nut using a stone hammer, a common behavior in many wild groups of this species. Young monkeys regularly attend closely to proficient crackers and collect bits of broken nut from sites where adults crack. This tolerant social context is thought to promote investigation of appropriate sites and materials by the youngsters, and thus to aid them in learning to crack nuts. Photo by Barth Wright/EthoCebus Project.

Figure 5 Adult golden lion tamarin (*Leontopithecus rosalia,* callitrichids), carrying twins. Parents in this species call their dependent offspring to places where a hidden food item can be procured, a phenomenon called 'opportunity teaching'. Photo by Jessie Cohen/National Zoo, Smithsonian Institution.

Learning a Decision Rule through Observation

Psychologists have long been interested in whether individuals can learn arbitrary decision rules from watching others select objects from a set. Typically, a subject observes a skilled partner and a short time later works on an identical problem. Monkeys have considerably greater success on this kind of task than in reproducing novel actions after watching others perform them. For example, Subiaul trained two rhesus macaques (*Macaca mulatta*, cercopithecines) to touch in fixed order each of four pictures appearing simultaneously on a touch-screen monitor. Each monkey trained alone and became an expert at a particular sequence of four pictures, and then each monkey learned the other monkeys' sequences as well as other new sequences. Both the monkeys more quickly learned the series that they had watched their social partner perform than series that they had not watched the other monkey perform. As each monkey was already skilled at touching pictures in a particular sequence, what each monkey learned from watching the other was the order in which to touch a new set of pictures. Subiaul labels this type of learning 'cognitive imitation,' because the observer adopts a rule demonstrated by another, rather than a particular action. Subiaul argues that monkeys can adopt novel decision rules, but not match novel actions, from watching others because matching novel actions depends upon 'derived neural specializations mediating the planning and coordination of fine and gross motor movements' that some hominids (see Whiten, this volume), but not monkeys and prosimians, possess.

Biological Significance of Socially Mediated Learning

Socially mediated learning probably serves biological functions in primates similar to those it serves in other taxa. Social partners provide a context for learning in non-human primates, both highlighting relevant features of the environment through enhancement and promoting behaviors that are generally appropriate for a particular place and time through social facilitation. In the short term, social mediation of learning reduces risk during the acquisition of useful skills and knowledge, and social mediation may be especially beneficial to acquiring certain foraging skills. Differentiated relationships with specific others produce a mosaic of learning opportunities across individuals within a group, thus promoting behavioral variation within a group.

Social mediation of learning can also have longer-term consequences when it results in traditions (i.e., relatively enduring behaviors acquired in part by socially mediated learning and practiced by at least two members of a group). Behavioral traditions hold strong interest for evolutionary biologists because traditions generate and maintain behavioral variation over time outside of, or perhaps even ahead of, changes in the genetics of a population. In this indirect manner, socially mediated learning contributes to evolution, and social learning becomes central to the contemporary debate about the relationship between traditions in non-human animals and the phenomenon of culture (for discussion see Perry, this volume, or Caldwell and Whiten, 2007).

Summary

Monkeys and prosimians have varied social lives, which influence how and what individuals learn. In general, monkeys and prosimians are interested in conspecifics and attend to what they are doing. The motivation to synchronize behavior with others (social facilitation) promotes behavioral coordination within a group. Interest in where another is acting (enhancement) draws attention to both places and objects. Such processes channel an individual's activity sufficiently that monkeys and prosimians tend to acquire preferences and behavioral patterns similar to those of their groupmates. Monkeys match the specific actions of others only in very limited circumstances. The influence of older on younger individuals promotes the maintenance of behaviors across generations (traditions), and enduring traditions may have an impact on natural selection.

See also: Apes: Social Learning; Culture; Imitation: Cognitive Implications.

Further Reading

Brown GR, Almond REA, and van Bergen Y (2004) Begging, stealing, and offering: Food transfer in non-human primates. *Advances in the Study of Behavior* 34: 265–295.

Caldwell CA and Whiten A (2007) Social learning in monkeys and apes: Cultural animals? In: Campbell C, Fuentes A, MacKinnon KC, Panger M, and Bearder SK (eds.) *Primates in Perspective.* New York, NY: Oxford University Press.

Cheney DL and Seyfarth RM (1990) *How Monkeys See the World: Inside the Mind of Another Species.* Chicago: University of Chicago Press.

Cheney DL, Seyfarth RM, Smuts BB, and Wrangham RW (1986) The study of primate societies. In: Smuts BB, Cheney DL, Seyfarth RM, Wrangham RW, and Struhsaker TT (eds.) *Primate Societies.* Chicago: The University of Chicago Press.

Coussi-Korbel S and Fragaszy DM (1995) On the relation between social dynamics and social learning. *Animal behaviour* 50: 1441–1453.

Fleagle JG (1999) *Primate Adaptation and Evolution*, 2nd ed. San Diego, CA: Academic Press.

Fragaszy DM and Perry S (2003) Towards a biology of traditions. In: Fragaszy DM and Perry S (eds.) *The Biology of Traditions.* Cambridge, UK: Cambridge University Press.

Fragaszy D and Visalberghi E (2004) Socially-biased learning in monkeys. *Learning and Behavior* 32: 24–35.

Galef BG Jr. (1992) The question of animal culture. *Human Nature* 3: 157–178.

Huffman MA (1996) Acquisition of innovative cultural behaviors in non-human primates: A case study of stone handling, a socially transmitted behavior in Japanese macaques. In: Heyes CM and Galef BG Jr. (eds.) *Social Learning in Animals: The Roots of Culture.* San Diego, CA: Academic Press, Inc.

Kappeler PM (1997) Determinants of primate social organization: Comparative evidence and new insights from Malagasy lemurs. *Biological Reviews* 72: 111–151.

Perry S, Baker M, Fedigan L, et al. (2003) Social conventions in wild white-faced capuchin monkeys – Evidence for traditions in a neotropical primate. *Current Anthropology* 44(2): 241–268.

Rapaport LG and Brown GR (2008) Social influences on foraging behavior in young non-human primates: Learning what, where and how to eat. *Evolutionary Anthropology* 17: 189–201.

Subiaul F (2007) The imitation faculty in monkeys: Evaluating its features, distribution and evolution. *Journal of Anthropological Sciences* 85: 35–62.

Relevant Websites

http://pin.primate.wisc.edu – Primate Info Net.

Monogamy and Extra-Pair Parentage

H. Hoi and M. Griggio, Konrad Lorenz Institute for Ethology, Vienna, Austria

Introduction

Social Monogamy

Monogamy is a mating system in which a single adult male and a single adult female mate. Such pair bonds may last for a single breeding attempt, a breeding season, or many breeding seasons as in some pair-living mammals and some geese and swans. An active termination of the bond and pairing with a new partner (mate switching) occurs in many species. In birds, in particular the subfamily of passerines, more than 93% of all species are socially monogamous. Social monogamy is often associated with biparental care; however, exceptions exist.

Genetic Monogamy

This is an exclusive mating relationship between a male and a female resulting in all offspring being genetically directly related to both partners. The use of molecular techniques revealed that many socially monogamous bird species obtain fertilizations outside their pair bond (called 'extra-pair fertilizations' (EPFs)), with frequencies of extra-pair young (EPY) reaching 70% in some. Less than 25% of all socially monogamous bird species so far studied practice true genetic monogamy, and true genetic monogamy occurs in fact in only 14% of surveyed passerine species whereas the remaining 86% species showed varying levels of extra-pair paternity (EPP). Thus, there is a discrepancy in the occurrence of social and genetic monogamy. Social bonds do not reliably predict genetic mating patterns. In fact, this discovery has led to a revision in terminology, such that species are now commonly classified depending on whether they are genetically or socially monogamous.

Extra-Pair Copulations

Extra-pair copulations (EPCs) that result in EPFs and ultimately EPY are responsible for the differences in social and genetic mating patterns. Many studies have demonstrated that such copulations outside the pair bond are an alternative reproductive strategy adopted by males to increase their reproductive success and adopted by females to obtain genetic benefits. The rate of EPP reflects the proportion of offspring, for example, within a nest or population, fathered by males other than the primary male (social mate).

EPCs and the Classical Mating Systems

EPCs have been reported in monogamous, polygynous, and polyandrous species, but are most common in monogamous mating systems. For example, in Southwestern Willow Flycatchers (*Empidonax traillii extimus*), polygynous and monogamous males engaged in EPFs. However, females socially paired with polygynous males are more likely to seek or accept EPC than females paired with socially monogamous males, which offsets overall higher apparent reproductive success of socially polygynous males. Two alternative hypotheses explain how EPP and social and genetic mating systems are interrelated in birds. The male trade-off hypothesis predicts that social polygyny increases EPFs because males concentrate on attracting additional social mates which prevents effective protection of females with whom they are already mated with (see section 'Variation due to time constraints: Mate guarding, parental care, or EPP?'). The second hypothesis is the female choice hypothesis, which states that social polygyny should decrease EPFs because a substantial proportion of females can pair with the male of their choice, and males can effectively guard each mate during her fertile period (see below). Dennis Hasselquist and Paul Sherman found that extra-pair chicks were twice as frequent in socially monogamous as in socially polygynous species (23% vs. 11%) and concluded that in socially polygynous species there is less incentive for females and males to pursue extra-pair mating and in contrast females very likely incur higher costs for sexual infidelity, for example, due to physical retaliation or reduction of paternal efforts than in socially monogamous species.

Why Is EPP the Alternative Mating Tactic for Monogamous Species?

In socially polygynous mating systems, theoretically, all females could pair with the best male available. Under the assumption that there is an overall best male that all females prefer, in a monogamous system only one female can be mated to the best, the second female to the second best male, and so on. Thus, females, as a consequence of later pairing, have to accept mating partners of lower quality. In such a situation, EPCs are one postmating strategy to increase offspring fitness. Quality variation in female mating partners alone may explain variation in

Table 1 Possible benefits and costs of EPCs for males and for females

	Benefits	Costs
Males & females	• Insurance against mate's infertility • Possible future mate acquisition • Production of genotypically better or diverse offspring	• Risk of acquiring sexually transmitted pathogens • Increased likelihood of divorce • Risk of predation
Females only	• Access to resources (e.g., parental care)	• Male retaliation • Risk of injury • Harassment from extra-pair males
Males only	• Increase the number of offspring	• Ejaculate production costs and sperm depletion • Increased risk of cuckoldry • Parental investment into nonrelative offspring

female extra-pair behavior, for example, why some but not all individuals engage in EPCs. This strategy may also allow females to choose a social mate for offspring care separate from choosing the genetic father (see below). Several benefits and costs are associated with EPCs as Patty Gowaty pointed out in 2006, but data so far do not support strong conclusions about the relative costs and benefits. Thus, we present a summary (**Table 1**) of costs and benefits of EPCs. Most of the costs and benefits are more or less the same for both sexes. The only big difference in terms of benefits is that males successfully performing EPCs can significantly increase offspring numbers, whereas females may gain from access to resources (access to food, nest sites, and paternal care). In terms of costs, females suffer from the aggression of guarding or retaliating males, whereas males mainly suffer because they waste reproductive energy (sperm, paternal care, time, etc.).

EPP and Female Choice: Genetic Benefits

We can divide the genetic benefits of EPP into two types: the 'good genes hypothesis,' with additive effect, and the 'compatible genes hypothesis,' with a nonadditive effect.

The Good Genes Hypothesis

This hypothesis predicts that males will be selected to signal (through, e.g., ornaments) their genetic quality and that females will prefer copulating with males carrying good genes. The point is that not all females can pair with the most preferred males because of the constraint of social monogamy. So, females paired to nonpreferred males might try to copulate with a better male, to obtain good genes for her offspring. Females might gain both the direct benefits (e.g., paternal care) from the social mate and the genetic benefits of the 'good genes' from the second male.

The 'Compatible Genes Hypothesis'

This hypothesis predicts that not all females choose the same genes, but rather, each of them prefers particular genes. Such differences in preference can be due to genetic incompatibility. In this way, genetic benefits to females are due to the interaction between maternal and paternal genomic contributions. So, this hypothesis predicts that females pursue EPCs to augment the chances of finding compatible partners who in turn, will confer to the EPY higher fitness than their paternal half-sibs. Many have attempted to investigate this question, but studies investigating these two hypotheses to explain EPP are largely inconclusive. The most common conclusion is that females are unlikely to indirectly benefit from having EPCs. Most of these studies are correlative, so open to a range of interpretations. A strong evidence for indirect benefits of female and male mate choice was found by one recent experimental investigation of *Drosophila pseudoobscura* demonstrating that females and males mating with partners they preferred had significantly greater offspring viability than subjects limited to reproduce with partners they did not prefer. Patty Gowaty and her colleagues then examined a variety of fitness components in similar experiments in insects, birds, and mammals and reported results similar to those reported for *D. pseudoobscura*.

Variation in EPP Within and Between Species

Variation in extra-pair behavior is, however, also influenced by many other factors. For example, in many populations, a varying number of individuals do not engage in EPCs, suggesting that there are costs or limited or absent benefits of EPC. There are many hypotheses to explain variation in EPFs among and within species (see below). The percentage of EPY in populations may range from 0 to more than 70%. In many songbird populations, the percentage of EPY is between 10% and 25%, suggesting that at least some individuals in a population benefit from EPCs. Among socially monogamous species, Reed Bunting (*Emberiza schoeniclus*) exhibit the highest rate of EPY. Indeed, it was found that 55% of all offspring were fathered by extra-pair males and 86% of broods contained at least one chick fathered outside the pair bond.

In cooperatively breeding Superb Fairy-wrens (*Malurus cyaneus*), 72% of offspring are fathered by males other than the putative father, and 95% of broods contained extra-pair offspring. Such high rates of EPY inspire questions about the adaptive function of EPP. However, several studies revealed very low levels of EPP, for example, below 5% of offspring. The EPP may also show seasonal fluctuations. For example, EPP increases in second broods of House Sparrows *Passer domesticus*.

Variation in the level of EPP exists in several species but has not received much attention, partly because there are only a few such long-term data sets. However, it can be helpful to explain EPP investigating variation of a single population between years and examine whether this is mediated by, for example, ecological variables.

Hypotheses Explaining Within and Between Species Variation in EPP

There are many hypotheses to explain variation in EPFs among and within species.

Variation in EPP is due to measuring error

DNA methods have been used to investigate the paternity of over 25 000 avian offspring, but only two studies combined have contributed over 12% to the total number, as reported in a recent review written by Simon Griffith and his colleagues, which indicates that the most studies are small, for example, more than 75% of all studies have less than 50 broods. Sample sizes range from 15 to 2013 offspring.

Variation in EPP due to the need of male parental care

When a female cannot rear young successfully without the help of a male, social monogamy is likely to become the reproductive strategy that best maximizes the fitness of both sexes. So, when male contributions to offspring survival are critical (biparental care), females may be constrained to social monogamy so that social monogamy may be the only option for males and females. Following the same reasoning, EPP may occur more frequently in situations where females are less constrained by the need for male parental care (this is one of the predictions of the Constrained Female Hypothesis, hypothesis proposed by Patty Gowaty) or in which they are able to compensate for reduced paternal care. Several studies investigated the possible relationship between paternity and paternal brood provisioning. One species that received some attention on this topic is the Reed Buntings (*Emberiza schoeniclus*). In the first study, Andrew Dixon and his collaborators found that males adjust their parental care (i.e., feeding effort) to the proportion of EPP in their nests. Subsequent studies failed to find similar results. Anyway, a recent study, conducted by Stefan Suter and his colleagues based on a large number of nests and in which many

hours of behavioral observations were performed, found similar results of Dixon's study: males adjusted parental care to the amount of EPP. Moreover, females compensated for low male parental effort, but the nestling mortality was higher in the nests with decreased male feeding effort. So, from these studies, we can conclude that for some species there is a cost for females when engaging in EPCs, and in some species this reduction in paternal care may increase offspring mortality.

Experimental evidence revealed a link between the need for paternal care and the incidence of EPP in the Serin (*Serinus serinus*). Maria Hoi-Leitner and her colleagues manipulated the abundance of food around the nest during the fertile phase of the female. The likelihood of EPP was significantly higher in territories with a high availability of food. They found a negative relation between environmental quality and paternity both in unmanipulated and manipulated habitats. Second, male parental assistance was related to food availability. A theory of coevolutionary selective pressures acting on males for the control of females' reproductive capacities and on females for resistance to males' efforts to control them proposes that social monogamy will often be genetic polyandry. This 'constrained female hypothesis' is in line with studies of sperm competition. However, it is Gowaty's assumption that it is female quality, or the quality of the environment where they live, that determines levels of EPP that is new and unique to this hypothesis. The female-constraint hypothesis proposes that if males retaliate with reduced parental care in response to low paternity certainty and females cannot compensate for the loss, that females will be less likely to seek EPCs. Results from Maria Hoi-Leitner and her collaborators, and other studies supported this hypothesis.

It must be noted that monogamy exists also without parental care. Fitness benefits through biparental care (father and mother collaborate in parental care) are thought to contribute to the evolution of monogamy. Anyway, it must be noted that social monogamy has evolved in the absence of biparental care in some mammals, coral reef fishes, reptiles, and amphibians. Two hypotheses for the evolution of social monogamy even without parental care were proposed: (i) the territorial cooperation hypothesis; (ii) the extended mate-guarding hypothesis. The first hypothesis is based on the fact that social behavior is often correlated to territoriality, and the majority of socially monogamous taxa are also territorial. This seems to suggest that individuals in pairs may benefit by sharing territorial defense. The second hypothesis suggests that selection for male mate guarding of females may play a role in the evolution of social monogamy. In particular, male mate guarding is predicted to evolve whenever the guarding sex benefits by limiting a mate access to other opposite-sex conspecifics, and it may lead to social monogamy if males are unable to monopolize more than one female at the same time.

Variation due to time constraints: mate guarding, parental care, or EPP?

Concerning the EPP topic, two hypotheses exist with the same name: the 'trade-off hypothesis.' These hypotheses suggest that males may be limited in their pursuit of extra-pair matings, because of constraints imposed by either caring for offspring (hereafter 'trade-off hypothesis for care') or paternity assurance (hereafter 'trade-off hypothesis for mate guarding').

The 'trade-off hypothesis for care' predicts a negative correlation between level of male contribution to parental care and frequency of EPFs. The incubation behavior in males may be more likely to limit pursuit of EPCs than other forms of parental care because time. Thus, males may face a trade-off between incubation of their offspring and seeking extra-pair mating opportunities.

The 'trade-off hypothesis for mate guarding' predicts a negative correlation between level of mate guarding and frequency of EPFs. In those species in which some males are polygynous, males are expected to face a trade-off between paternity assurance and acquisition of more than one mate. Polygynous males should be less efficient in guarding their mates than monogamous males, since they have to partition their time between two mates. Under this scenario, males face a trade-off between acquiring a second mate and defending their paternity. Tentative support for the so-called trade-off hypothesis has been found in a few bird species, in which socially polygynous males are cuckolded more frequently than monogamous males. The trade-off hypothesis predicts that (1) polygynous males are cuckolded more frequently than monogamous males, (2) mate guarding should be less intense in polygynous males than in monogamous males. Furthermore, if there is a trade-off between protecting paternity and looking for additional mates, males are expected to invest more time in guarding their mate when the probability of attracting a new mate is low. A study that supported the 'trade-off hypothesis' is the one conducted by Andrea Pilastro and his collaborators on Rock Sparrow (*Petronia petronia*), a facultative polygynous species. Overall, 32% of the chicks were not sired by the social father and about 57% of the broods contained at least one extra-pair young. Polygynous Rock Sparrow males allocated less time to guarding their mate during female's fertile period than monogamous males, and polygynous males were cuckolded more frequently than monogamous males (50.5 and 6.6% of the young, respectively). Reproductive success (number of young fledged/year) did not differ between monogamous and polygynous males once paternity was accounted for. These results indicate that mate guarding can be efficient in preventing cuckoldry, and that there is a trade-off between attracting an additional mate and protecting paternity, at least in the Rock Sparrow. It must be noted that other studies have shown that polygynous males do not lose paternity more often than monogamous males. This is probably because in some species polygamous males are better individuals. In other words, in some populations, polygynous males are usually the most preferred males.

Variation in EPP due to genetic variability

It has been frequently stated that genetic benefits influence the reproductive behavior of individual males and females. If females gain indirect benefits (e.g., good genes or genetic heterozygosity) when seeking extra-pair matings, one would expect that genetic variability among males in a population affects female extra-pair behavior as their benefits. Marion Petrie and Marc Lipsitch could show based on a game theoretic model that females should more likely mate with additional males if there was extensive additive genetic diversity among those mates with respect to fitness. Thus, if there is little genetic variation among males, females would not benefit from seeking EPFs and hence extra-pair behavior should be scarce. In a comparative study, Marion Petrie and her collaborators tested this 'genetic diversity' hypothesis at an inter as well as intraspecific level and found a significant positive relationship between EPF frequency and estimates of genetic variability (allozyme polymorphism); however, comparatively few, only 22%, of the variance in the variation of the EPP rate is explained but, for example, the level of sexual dichromatism, body size, and sample size already explained 85% of the variation in EPP. On the other hand, it is remarkable that such simple measures of genetic diversity can explain some of the proportion of variation in EPP among taxa. Comparisons among populations of species where males differ in levels of genetic variability would be needed for a more powerful test of this hypothesis.

Variation in EPP due to breeding density

Based on observations that EPCs are more common among colonially than among more dispersed nesting species, the prediction is that breeding density promotes EPCs because opportunities to pursue EPCs should be much greater for both sexes. Consequently, colonial species or species nesting at high densities are assumed also to have higher rates of EPP than species nesting at lower densities. In fact, some colonial or aggregate nesting species have high frequencies of EPFs.

There is some evidence for such a density effect when comparing the rate of EPP between individuals in the same population (within population comparison) but breeding at varying density situations. For example, a positive relationship between EPFs and nesting density was found in Bearded Reedlings (*Panurus biarmicus*), a species where some individuals within the population nest colonially and others nest solitarily. With the same approach, some other studies revealed a positive relationship between breeding density and EPP, but others did not. The best evidence within a single population level is provided by the study of Patty Gowaty and William Bridges where they experimentally manipulated breeding density by nest box placement to determine the effect on the incidence of EPFs, and they found a significantly higher EPF frequency in Eastern Bluebirds (*Sialia sialis*) breeding in nest boxes at high densities compared to areas with lower nest box density. A striking example for an intraspecific study but comparing different populations (between species comparison) which is in support of the density hypothesis comes from studies of Willow Warblers (*Phylloscopus trochilus*). A Norwegian population that had an EPF frequency of 33% had over twice the nesting density of a Swedish population, which reported no EPFs. On the interspecific level, there is not much evidence for an effect of breeding density on EPP in birds. In a comparative analysis involving 72 species, David Westneat and Paul Sherman found no relationship between nesting density and EPF frequency. Thus, nesting density may influence EPF frequency within populations of some species but does not appear to be a reliable predictor of whether a particular species will have extra-pair matings. In conclusion, breeding density appears to be to some extent important to explain differences in EPP between individuals in the same population and also explains possibly variation between different populations of the same species, but there is not much evidence for an effect of breeding density on EPP in birds. There are probably confounding factors, for example, differences in breeding synchrony and habitat (see section 'Variation in EPP due to variation in breeding synchrony') could have also been responsible for observed differences in EPFs.

Variation in EPP due to variation in breeding synchrony

It was proposed that breeding synchrony promotes EPFs. The logic behind this 'synchrony hypothesis' is that synchronous breeding allows females to more effectively compare potential extra-pair males that would be competing and displaying for EPCs at the same time. Breeding synchrony in this context refers to the proportion of females that are fertile at a given moment, and high synchrony refers to a situation where many females are fertile at the same time.

In contrast, the 'asynchrony hypothesis' suggests that asynchrony promotes EPFs and if males guard their mates, assuming that mate guarding constrains males from seeking EPF, asynchronous breeding allows them opportunities to seek EPCs when their own mates are no longer fertile.

A correlative study conducted by Herbert Hoi's group revealed a weak evidence for an effect of breeding synchrony on EPP in House Sparrows, but in an experimental study, the same researchers aimed to test whether an alteration of local breeding synchrony by means of

acceleration and postponement of egg laying could generate differences in the occurrence of EPP in House Sparrows (*Passer domesticus*), they found different results. Therefore, they swapped nest material between the nests of neighbors and found that higher occurrence of EPY within broods was associated with laying order. The latest broods within a local nesting aggregation contained significantly more EPY than those of earlier breeding pairs, but there was no clear evidence for breeding synchronization. There was only evidence for an interaction between laying order and breeding synchrony in that the latest broods within a nesting aggregation contained more EPY provided that females laid their eggs relatively synchronously. Thus, it was proposed that laying order and the time lag in egg laying among neighboring pairs may be important determinants for the occurrence of EPP too.

Variation in EPP due to combined effects of socioecological factors

In general, the two socioecological factors are thought to affect the degree of EPP either via influencing male control over females or female opportunities for EPCs. Local breeding density may also affect the information females have about the number and quality of potential sexual partners.

EPP is also influenced by other factors like (i) ecological parameters (e.g., food or nest predation which may influence the need of male paternal care), or (ii) the degree of sexual conflict. For instance, quality differences between pair members and possible extra-pair partners may influence whether females cooperate with their mate or not. Less attractive males may consequently invest more in mate guarding or other paternity guards to avoid paternity losses. Thus, extra-pair rate is not necessarily an adequate measure to identify the influence of socioecological factors on male- and female-mating strategies. Therefore, Anton Kristin and collaborators investigated the role of the two socioecological factors in relation to male investment in paternity assurance of Lesser Grey Shrikes. Male Shrikes perform a mixed strategy to ensure paternity. They copulate frequently, mainly after territorial intrusions by other males, and guard their mates throughout the whole fertile phase. They found that males seem to be constrained by the frequency of intrusions by neighboring males, and this risk is associated with laying synchrony. There is intense sperm competition and the risk of intrusions depends on the timing and overlap of breeding attempts, and males adjust their investment to paternity assurance accordingly. Neither breeding density per se nor breeding synchrony in terms of overlapping fertility of close neighbors (the usual measure in most studies) were related to the intensity of paternity guards. However, when including the breeding order in relation to neighboring nests as a second qualitative factor (timing of overlapping between the fertile phases of neighboring females), they detected that the intensity of paternity guards increased with an increase in breeding synchrony. Sperm competition and the effect of socioecological factors are not at all reflected when examining the level of EPP. Francisco Valera and his colleagues did not detect any case of EPP in this species, and male punishment of unfaithful females seems to be the main reason for the emergence of genetic monogamous system. Their results suggest three conclusions, one is that extra-behavior and tactics to avoid them (e.g., paternity assurance) may be more sensitive measures to investigate the importance of breeding density and synchrony than simple examining the final outcome (EPP rate). Second, independent of the time overlap between two females, the order of clutch initiation (breeding order) is important in such a way that pairs breeding later are more likely faced with a higher intruder frequency. The same result is also confirmed by the study performed by Herbert Hoi's group and reflected in EPP. Indeed, they found in their experiment on house sparrows that mainly the laying order among neighboring pairs is an important determinant for the occurrence of EPP. Third, it seems more reasonable to examine combined effects and the interaction of several socioecological factors including at least also breeding order. Studies conducted by Herbert Hoi's group, and Anton Kristin and collaborators found indication for an interaction between socioecological factors investigated. Such interactions may be due to females, for example, nest site choice as well as their ability to alter egg-laying patterns to either minimize synchrony in situations where they find themselves in dense breeding situations or increase synchrony. Thus, if the distance to the nearest neighbors affects the importance of synchronization, one could predict that the effect of synchronization might decrease with increasing internest distance, a prediction which should be examined in future studies.

Variation in EPP due to other ecological factors
Habitat visibility
One idea is that the level of EPP as well as the type of and investment in the paternity assurance tactic is influenced by habitat visibility. The basic assumption is that females are more able to escape male paternity guards in closed, that is, visually occluded, habitats and consequently the level of EPP should be higher and paternity assurance behaviors more frequent among species breeding in habitats with reduced visibility compared to those breeding in more open habitats, assuming that occurrence of paternity guards reflects an increased risk of cuckoldry.

In a comparative study, Donald Blomqvist and his colleagues found that species breeding in closed habitats had higher EPP rates than those breeding in more open habitats. Mate guarding was also more frequent in closed

habitats, but not high copulation rates. These relationships, however, were influenced strongly by taxonomic position, particularly by differences between passerines and non-passerines, implying that phylogeny and traits associated with it play an important role in explaining the occurrence of EPP and paternity guards. Such a comparison is also biased because the level of EPP depends not only on habitat visibility and female opportunities but also male intruders and the frequency of intrusions which may be influenced by other variables than habitat visibility (see above). In Lesser Grey Shrikes, Anton Kristin and collaborators found that despite the high visibility of females for their partners territorial intrusions have been observed to be very common and EPC attempts by intruding males occur and frequent within-pair copulations are used as a response to territorial intrusions. Similarly, Francisco Valera and his collaborators showed that male Shrikes also guard their females intensely throughout the breeding period in a very open habitat and adjust their mate-guarding behavior mainly to the occurrence of intrusions.

Thus, the question here is to what extent intruding males use the existence of a dense habitat to sneak into a territory to pursue an EPC undetected by the pair male.

Food availability
The role of food is already discussed in the section related to parental care. Some experimental studies revealed that manipulating food supply affects the level of EPP although with contrasting results. This may be due the different breeding situation (solitary vs. colonial) and food supplementation may benefit either females and enhance their EPP behavior or males by protect male paternity.

Predation pressure
There is no study as far as we are aware of investigating a possible role of variation in predation risk on EPP. However, it is likely that predator-free environments such as on some islands may affect female EPP strategies.

Parasite infestation
Parasites play an important role and are one of the driving forces in mate and probably extra-pair mate choice. However, there is no study we are aware of which has been investigating whether parasite loads of males and females may affect their own EPP behavior or whether parasite loads of potential EPP partners influence extra-pair mate choice.

Male Strategies in Relation to EPP

There are no particular behaviors males developed in response to increase the chances of EPCs we are aware of. However, to maximize their own fitness (reproductive success), males developed several tactics to prevent their mates from engaging in EPCs. To avoid having a mate engage in EPCs, and end up caring for another male's offspring, males may use different paternity guards.

These male tactics, including mainly mate guarding and frequent within-pair copulations, are already well reviewed for birds and also hold good for animal groups in the books of Tim Birkhead and Anders Møller. During mate guarding, males remain close to their fertile mates to prevent other males from seeking EPCs. Costs of mate guarding include time, energy, and opportunity costs. Frequent within-pair copulations are a strategy to increase the probability of fertilization success for a male and is very likely proportional to the relative number of sperm delivered to a particular female. Both mate guarding and frequent copulations are regarded as alternative compensatory paternity guards in the sense that, in general, the presence of one means the lack of the other and vice versa. As mate guarding is very time consuming, frequent copulations seem to be a logical alternative, for instance, in colonial seabirds, where one partner has to stay near the nest site to defend the nest during the feeding trips of the other. With many neighboring males, fertile females cannot be guarded properly and several authors have suggested that the risk of cuckoldry increases with colony size and density. For example, in raptors, the sperm competition intensity increases with breeding density, and males rely on frequent copulations to ensure paternity. Unless copulations are very costly for the male, there is no reason why males of 'mate-guarding species' should not also copulate frequently to increase paternity certainty.

As suggested by Tim Birkhead and Anders Møller, territoriality can be also seen as a paternity guard when it helps to keep away other males from the pair female to perform EPCs, or reduces the information pair females may get about the number and quality of potential sexual partners in the neighborhood. Moreover, postcopulatory mating plugs like in many mollusks, insects, mammals are a different strategy to prevent EPCs.

Creating a risky environment for unfaithful females where the risk may include any kind of retaliation, for example, reduction of paternal care or direct aggressive punishment is another possible strategy for males to guard their paternity. Finally, a proper choice of the breeding site as far as under male control may also play a role in determining the level of EPP (discussed earlier).

The investment into paternity guards, and consequently also the level of EPP, very much depends on the individual quality of the pair partners as well as the potential extra-pair candidate. In many species, the intensity of mate guarding is not fixed but varies between males, as found for example in Rock Sparrow by Matteo Griggio and his collaborators. Other than local socioecological conditions (see above), also individual characteristics, like phenotypic quality and age, may influence male ability or willingness to perform mate guarding. In literature, there are two opposite sets of findings. Some studies revealed that high-quality males guard their females more strongly than low-quality males. On the contrary, other studies found a negative relationship between mate guarding and male quality, like in Bluethroat, *Luscinia s. svecica*. In the first case, one possible explanation is that only high-quality males can afford spending time and energy on mate guarding. In the second case, an explanation could be that unattractive males guard more intensely because perceive a higher female infidelity (sexual conflict over fertilizations) or because those males have low success in obtaining EPCs (trade-off between mate guarding and perform EPCs).

Females Strategies in Relation to EPP

Studies and reviews on sperm competition intensively addressed this topic in detail (for review, see the books of Tim Birkhead and Anders Møller), and in principle females have pre- and postcopulatory tactics, including choice of the copulation partner, for example, via timing copulations accordingly. During copulation, females may have opportunities, for example, to avoid cloacal contacts. Postcopulatory tactics are what is generally summarized under cryptic female choice and may include active sperm selection and storage or sperm rejection. Another possibility is differential allocation into offspring (variation in maternal investment) in relation to quality of copulation partners. Females have been shown to invest differentially in eggs by either their sex or paternal phenotype. Inducing abortion like in many mice species may be seen as another postcopulatory strategy.

Precopulatory behaviors may also involve female behaviors to incite male–male competition by conspicuously advertising their fertile period. This includes behaviors where females apparently resist copulation attempts by other than the pair or dominant males. Such a 'resistance as a ploy' tactic is described in bearded tits, whereby females initiate several males simultaneously to chase her resulting in the fastest (best) male copulating with them. Opposite to conspicuously advertising the fertility, in many species females try to hide their fertile period. In birds, producing the biggest gametes in animal kingdom and hiding fertility are therefore more difficult, but there are still some species where females try to do so. It was experimentally demonstrated in penduline tits (*Remiz pendulinus*) that females try to hide their fertile period by hiding their eggs in the soft layer of the nest bottom which consequently enables females to mate with several males.

Concluding Remarks

The discovery of EPP via molecular tools is probably one of the most important empirical discoveries in avian mating systems over the last 30 years. Although there is still increasing interest in this topic as indicated by the number of published papers in last 5 years, there is not much advancement to detect and there is still no way to reliable predict whether a species may implement an extra-pair mating strategy or not. Some strong correlations have been identified, but many exceptions still exist. We think that more attention should be given to the behavioral interactions between actors involved in the EPP phenomenon (e.g., male, female, and extra mates). Experimental approaches in seminatural condition (e.g., big enclosures, where it is quite easy to follow behaviors of different players) seem to us a good direction to better understand the phenomenon of EPP. Our opinion is that, even if behavioral observations are time consuming, the behavioral approach is the best way to fully understand a behavioral phenomenon.

There is still a heavy bias toward species from temperate regions. Only a few tropical species are examined in relation to extra-pair behavior where most have rather low levels of EPP. For other taxa, for example, insects, reptiles, and mammals, there is only little information on EPP, probably because social monogamy is a rare mating system. However, to get a general picture, it would be important to include also other groups. Most research on the importance of various ecological factors on variation in EPP is still correlative but to understand whether there is a causal relationship between diverse ecological factors and EPP there is desperate need for experimental studies on the species level. Thus, the evolution of extra-pair mating systems remains an exciting field of research also because it relates to our own mating system and still an enigmatic field of research for evolutionary biologists.

See also: Differential Allocation; Mate Choice in Males and Females; Reproductive Success; Social Selection, Sexual Selection, and Sexual Conflict; Sperm Competition.

Further Reading

Anderson WW, Kim YK, and Gowaty PA (2007) Experimental constraints on female and male mate preferences in *Drosophila pseudoobscura* decrease offspring viability and reproductive success of breeding pairs. *Proceedings of the National Academy of Science* 104: 4484–4488.

Birkhead TR and Møller AP (1992) *Sperm Competition in Birds: Evolutionary Causes and Consequences*. London: Academic Press.

Birkhead TR and Møller AP (1998) *Sperm Competition and Sexual Selection*. London: Academic Press.

Blomqvist D, Hoi H, and Weinberger I (2006) To see or not to see: The role of habitat density on the occurrence of extra-pair paternity and paternity assurance behaviors. *Acta Zoologica Sinica* 52: 229–231.

Dixon A, Ross D, O'Malley SL, and Burke T (1994) Paternal investment inversely related to degree of extra-pair paternity in the Reed Bunting (*Eniberiza schoeniclus*). *Nature* 371: 698–700.

Gowaty PA (1996) Battles of the sexes and origins of monogamy. In: Black JM (ed.) *Partnerships in Birds: The Study of Monogamy*, pp. 21–52. Oxford: Oxford University Press.

Gowaty PA (2006) Beyond extra-pair paternity: Individual constraints, fitness components, and social mating systems. In: Lucas J and Simmons L (eds.) *Essays on Animal Behavior: Celebrating 50 years of Animal Behaviour*, pp. 221–256. Cambridge: Cambridge University.

Gowaty PA and Bridges WC (1991) Nestbox availability affects extra-pair fertilizations and conspecific nest parasitism in Eastern Bluebirds, *Sialia sialis*. *Animal Behaviour* 41: 661–675.

Griffith SC, Owens IPF, and Thuman KA (2002) Extra-pair paternity in birds: A review of interspecific variation and adaptive function. *Molecular Ecology* 11: 2195–2212.

Griggio M, Matessi G, and Pilastro A (2005) Should I stay or should I go? Female brood desertion and male counter-strategy in rock sparrows. *Behavioral Ecology* 16: 435–441.

Hasselquist D and Sherman PW (2001) Social mating systems and extra pair fertilizations in passerine birds. *Behavioral Ecology* 12: 457–466.

Hoi-Leitner M, Hoi H, Romero-Pujante M, and Valera F (1999) Female extra-pair behaviour and environmental quality in the serin (*Serinus serinus*): A test of the 'constrained female hypothesis'. *Proceedings of the Royal Society of London, Series B* 266: 1021–1026.

Johnsen A, Lifjeld JT, Rohde PA, Primmer CR, and Ellegren H (1998) Sexual conflict over fertilizations: Female bluethroats escape male paternity guards. *Behavioural Ecology & Sociobiology* 43: 401–408.

Kristin A, Hoi H, Valera F, and Hoi C (2008) The importance of breeding density and breeding synchrony for paternity assurance strategies in the lesser grey shrike. *Folia Zoologica* 57: 240–250.

Petrie M, Doums C, and Møller AP (1998) The degree of extra-pair paternity increases with genetic variability. *Proceedings of the National Academy of Sciences* 95: 390–9395.

Petrie M and Lipsitch M (1994) Avian polygyny is most likely in populations with high variability in heritable male fitness. *Proceedings of the Royal Society, London, Series B* 256: 275–280.

Pilastro A, Griggio M, Biddau L, and Mingozzi T (2002) Extrapair paternity as a cost of polygyny in the rock sparrow: Behavioural and genetic evidence of the 'trade-off' hypothesis. *Animal Behaviour* 63: 967–974.

Suter SM, Bielanska J, Rothlin-Spillmann S, et al. (2009) The cost of infidelity to female reed buntings. *Behavioral Ecology* 20: 601–608.

Václav R, Hoi H, and Blomqvist D (2003) Food supplementation affects extrapair paternity in House Sparrows (*Passer domesticus*). *Behavioral Ecology* 14: 730–735.

Valera F, Hoi H, and Krištín A (2003) Male shrikes punish unfaithful females. *Behavioral Ecology* 14: 403–408.

Westneat DF and Sherman PW (1997) Density and extra-pair fertilizations in birds: A comparative analysis. *Behavioral Ecology and Sociobiology* 41: 205–215.

Morality and Evolution

K. McAuliffe and M. Hauser, Harvard University, Cambridge, MA, USA

... of all the differences between man and the lower animals, the moral sense or conscience is by far the most important. This sense... has a rightful supremacy over every other principle of human action; it is summed up in that short but imperious word ought, so full of high significance. It is the most noble of all the attributes of man, leading him without a moment's hesitation to risk his life for that of a fellow-creature; or after due deliberation, impelled simply by the deep feeling of right or duty, to sacrifice it in some great cause.

Darwin C (1871) *The Descent of Man and Selection in Relation to Sex*, pp. 70–71. Princeton: Princeton University Press.

Introduction

In the *Descent of Man*, Darwin explored the evolutionary origins of our moral sense, and as the quotation above emphasizes, highlighted what we see as three essential points. First, to understand the origins of our sense of right and wrong, we must adopt a comparative perspective, drawing on studies of animals, to reveal what is uniquely human as opposed to what is shared across species. Second, understanding the moral sense is fundamentally a problem about the power of our conscience to guide what we ought to do. It is, in Darwin's terms, the highest of virtues, giving humans a sense of nobility, and fundamentally distinguishing them from other animals. Third, our sense of ought, of what should or could be done, can lead to either instinctive action ('without a moment's hesitation') or to a more contemplative stance ('after due deliberation') where we reflect upon particular principles of justice, and then based on this analysis, act in such a way that we support some great moral cause, often at personal cost ('sacrifice'). These three points target aspects of phylogeny and proximate cause, that is, the patterns of evolutionary change and the psychological mechanisms that either facilitate or constrain their appearance. Darwin also discussed the adaptive significance of morality, and in particular, the selective pressures that may have led to its appearance in our species. Characteristic of his thinking at the time, Darwin perceived a strong role for group-level pressure, such that individuals in groups acting in particularly altruistic ways would ultimately outcompete groups acting less cooperatively.

In this essay, we further explore the evolutionary origins of our moral sense, providing a synopsis of the current state of empirical play and the issues raised by the experiments and observations of animals. Like Darwin, we distinguish questions of proximate and ultimate cause, and specifically, separate out the issues of phylogeny, adaptation, and psychological mechanism. Like Darwin, we also distinguish between the psychological mechanisms that guide our intuitive and rather automatic sense of right or wrong, from those that underpin our contemplative reflection of what ought to be. From the perspective of psychological mechanisms and adaptive function, we can explore the building blocks of our capacity to decide what is right and what is wrong, and the conditions under which particular actions are permissible or forbidden. This perspective seeks an understanding of the core psychological mechanisms that enable organisms, both human and non-human, to decide what is fair, when harms are permissible, and when social contracts may be broken.

To this end, we review two distinct sets of literatures. The first explores studies that fall within the general area of behavioral economics, and in particular, the processes that guide cooperation, resource distribution, and a broad sense of fairness. We focus here on results that relate to some of the critical features of human cooperation and altruistic behavior, specifically, attention to inequities, reputation, punishment, and reciprocity. Second, we explore the mechanisms of action perception and production, and in particular, the extent to which animals distinguish intentional from accidental consequences, as well as the cues they use to decide goal-directed actions. Together, these processes comprise some of the fundamental building blocks that are evolutionarily ancient, appear early in human ontogeny, and ultimately lead to a full-fledged moral sense in healthy human adults. We then conclude with our current and personal sense of what makes human morality fundamentally different from what is observed in other animals, focusing specifically on how we evolved a brain that conceives of that imperious word ought.

Cooperation and Moral Judgment in Humans

A great deal of moral philosophy has been devoted to an exploration of our capacity to cooperate with others, sacrifice personal gains for the benefit of others, maintain social contracts, and appreciate that a sense of justice is premised on a sense of fairness. This rich tradition is generally aimed at understanding the guiding principles that appear to underpin not only our intuitive sense of cooperative action and distributive justice, but also the principles that ought to guide our decisions. Thus, for example, we observe in Rawls' thinking on justice a clear distinction between the intuitive principles that may guide our spontaneous judgments of fairness, and those that percolate up during a period of considered reflection, where we consciously divorce ourselves from the potentially powerful and biasing influences of in-versus out-group partiality (e.g., favoring kin over nonkin). The question of interest here is how human cooperation, including the psychological mechanisms that support it, as well as the selective pressures that led to its particular design features, evolved.

When biologists have discussed the evolution of cooperation, they have often focused on behavior and fitness consequences, without Taking into account psychological mechanisms that may be relevant or even required. Thus, Darwin puzzled over the possibility of altruism by asking how such a costly behavior could evolve, given that his theory of natural selection favored self-beneficial actions. This puzzle vanished when Hamilton, and later Williams, pointed the way to a different level of analysis, one that focused on genes as opposed to either individuals or groups. That is, we can explain why an animal engages in self-sacrifice for the benefit of another by the fact that the 'other' is a close genetic relative. As such, altruism evolves by benefiting genes shared in common. Where puzzles remain today is in explaining the evolution of cooperation among genetically unrelated individuals, and especially the kind of large-scale cooperation often observed among human societies.

A partial answer to these puzzles emerged, interestingly enough, when Trivers proposed his theory of reciprocal altruism, blending issues of adaptive function with psychological constraints. Specifically, and unlike Hamilton and Williams, Trivers' theory included not only a set of evolutionary conditions for the emergence of cooperation among unrelated individuals, but also a discussion of requisite psychological mechanisms including recognition of individuals, memory of past interactions, and strong emotional responses to defection, including moralistic aggression. Thus, Trivers' analysis paved the way for what has now become a major focus in the field: a consideration of both proximate and ultimate concerns related to the evolution of cooperation. More specifically, and as many have argued, cooperation among unrelated others can evolve if it takes place within stable groups with opportunities for repeated interactions (reciprocal altruism), or in groups where cooperative decisions can be based on reputation (indirect reciprocity).

However, these conditions demand consideration of the requisite psychological mechanisms, including recall of prior outcomes to evaluate reputation, quantification and tracking of the payoff matrices to evaluate fair distributions, and assessment of whether the resources were distributed intentionally or as a byproduct of otherwise selfish behavior.

While reciprocal altruism and indirect reciprocity can explain the evolution of cooperation in small social groups where individuals know each other, these mechanisms cannot account for the fact that modern humans often live in large groups of unrelated others, where reputation tracking is not possible, and where repeated and stable relationships are unreliable, thus making reciprocity difficult or impossible. In these situations, cooperation nonetheless evolves, demanding a different kind of account that can accommodate the fact that the optimal strategy is to defect and free ride on the contributions of others. Here, Boyd and Richerson pointed out that it appears that punishment evolved to crack down on the defector problem, and bring about stable cooperation. On an ultimate level, focused on evolutionary consequences, punishment is a behavior that reduces the fitness of a recipient at a temporary cost, but ultimate benefit to an actor (see the reading by Clutton-Brock and Parker for more on this topic). However, from a proximate perspective, punishment requires specific psychological mechanisms including the recruitment of motivating emotions (e.g., Trivers' moral outrage), the ability to determine when a norm has been violated, the assessment of just deserts, and a mechanism to distinguish whether the outcome (e.g., failure to cooperate) was intended or accidental.

Recent work on the evolution of fairness provides a good example of how proximate and ultimate concerns have come together. Specifically, using a combination of well-defined bargaining games from behavioral economics, together with rich psychological analysis and cross-cultural data, we can see that humans evolved a distinctive sense of fairness, one shared by all members of our species, but open to cross-cultural variation, and constrained by sensitivity to inequities and the ability to punish those who violate norms of distributive justice. Consider, as an example, the well studied, one-shot, anonymous Ultimatum Game. An experimenter first informs two individuals, a donor and a recipient, about the three rules of the game. Rule 1: the bank allocates a sum of money to the donor who has the option of allocating some proportion of this sum to the recipient. Rule 2: whatever amount the donor gives to the recipient, the recipient keeps, and the donor keeps the remainder. Rule 3: if the recipient chooses to reject the donor's offer, then neither donor nor recipient keeps any money. According to the rational economic model, the recipient should accept any offer from the donor as some money is surely better than no money. However, results from this game show that people across cultures consistently reject some offers, licensing the conclusion that humans approach this problem with a sense of what constitutes an unfair offer. Further, when subjects learn that the donor is a computer or random number generator, they are more likely to accept unfair offers. These results show that humans not only attend to the distribution of resources (i.e., outcomes), but also to the means by which resources are distributed. Returning to the ultimatum game, there is a fundamental (moral) difference between a human donor offering 1 out of 20 possible dollars to a recipient, and a computer program that uses a random number generator to offer $1 out of $20. Computers are not intentional agents, and thus, don't enter the moral domain. Regardless of how upset we are at the computer's offering, we can't hold them responsible. On the other hand, we do hold other humans responsible for their actions. Pushing the point further, if we forced a human to roll a die to pick the donation, and the die landed with a 1 facing up for $1, we would also not hold the human responsible for this outcome; this game functionally strips the human agent of his intentional control – of his free will. What is critical, then, is a combination of both the actual outcome and whether the agent was responsible for this decision.

Thus, a sense of fairness, together with an ability to discriminate between intentional and accidental actions, is critical to cooperation, and more generally, to our moral judgments about others' actions. Given the importance of these abilities, they raise the crucial question of whether these traits are unique to humans or whether they also play a role in governing behavior in non-human animals (hereafter animals).

Cooperation, Fairness and Action Perception in Animals

Cooperation in animals is taxonomically widespread, with numerous examples of individuals engaging in costly altruistic behavior for the benefit of others. In the early literature, most of the examples were consistent with the theory of kin selection and with optimization of inclusive fitness. That is, most altruistic acts of cooperation evolved to benefit close genetic relatives. Following on the heels of Trivers' conceptual work on reciprocity, however, several cases of reciprocal altruism emerged in the literature. These cases, as well as several more recent studies have, for the most part, been dismissed, either because of a failure to replicate, the weakness of the effects, alternative explanations (e.g., byproduct mutualism), or the highly artificial conditions under which the evidence has been obtained. At best, we argue, reciprocity is an uncommon form of social interaction among animals. We further argue, however, that consideration of both proximate and ultimate factors makes this conclusion unsurprising.

Specifically, there are at least two reasons why reciprocity might be rare among animals. First, the demographics of most animal populations may provide a sufficiently high density of kin to eliminate the pressure for nonkin based relationships. Despite these issues, we suggest that recent work on cooperation, including reciprocity, has provided new insights into some of the psychological mechanisms that are shared among human and non-human animals, and leads to one of our primary conclusions: though we share with other animals some of the core building blocks of morality, only one species – our species – has combined these core elements into a truly moral system that not only considers how we distinguish moral rights from wrongs, but also what ought to be the foundation for such decisions.

A Sense of Fairness in Animals?

Recent comparative work on primates and dogs has explored the problem of inequity aversion as an important component of the more general sense of fairness. One of the earliest treatments of this problem was Brosnan and de Waal's study of brown capuchin monkeys (*Cebus apella*). In this experiment, subjects that had been trained to trade tokens for food rewards watched a conspecific acquire and eat a high-value food item and then were given the opportunity to acquire and eat a lower-value food item. Subjects consistently refused to trade the token for the lower-value food, and this result was interpreted as evidence for inequity aversion and a sense of fairness. This experiment was heavily criticized because the authors could not rule out the effect of frustration as the driving force behind rejections. That is to say, subjects may have rejected unfair offers not because the other individual was getting a better deal, but because they were frustrated at not being able to obtain the higher-value food item that was right in front of them. Though subsequent experiments confirmed the validity of these critiques, Brosnan, de Waal, and their colleagues, have since replicated the original findings with relevant controls, and found that their results cannot be explained by frustration, or further extended to parallel findings with chimpanzees (*Pan troglodytes*). Adding to the comparative scope of this work, a recent experiment by Range and colleagues shows that domestic dogs are sensitive to inequities in reward distribution. In this experiment, subjects were given a command to perform an action (paw shake) in a social situation where one individual received a food reward for its performance while the other did not. Subjects that did not receive the reward were more reluctant to perform the action and displayed more stress behavior in the social condition compared to an asocial control.

In sum, although there is still much controversy surrounding the results on inequity aversion in animals, minimally, it appears that animals are sensitive to the distribution of rewards, in both social and nonsocial contexts, responding negatively when an outcome appears unfair.

Another approach to studying fairness in animals comes from a series of experiments investigating prosocial behavior, specifically, the tendency to help another in a situation where there are no personal gains, and little or no personal cost. In these experiments, subjects are given the option of acquiring a reward for themselves or for themselves as well as for another individual. The important difference between prosociality tasks and inequity aversion tasks is that a preference for equity is costless in the former (actors receive a payoff either way) and costly in the latter (actors receive nothing if they reject an unfair offer). Studies of common marmosets (*Callithrix jacchus*) and brown capuchin monkeys have shown that subjects consistently choose the prosocial option. Studies of chimpanzees, on the other hand, have shown that individuals are indifferent to the welfare of conspecifics and will choose indiscriminately between the two options. Interestingly, when chimpanzees are tested in a spontaneous altruism task, where subjects are given the opportunity to help another individual in the absence of a food reward, they do exhibit prosocial behavior. At present, it is not clear whether the difference in prosociality between these studies is due to the nature of the reward (i.e., food versus nonfood), or to specific details of the task demands.

Together, studies of inequity aversion and prosociality suggest that the ability to both detect and react to unfair outcomes, together with a preference for fair resource distributions, are not uniquely human traits. However, these experiments focus exclusively on outcomes and do not explore the means by which they are achieved. In the next section, we discuss experiments that tap into the psychological mechanisms involved in discriminating between intentional versus accidental actions.

Going Beyond Outcomes to Intentions and Goals

In this section we explore two questions: (1) Can animals draw inferences about an individual's intentions and goals, and (2) if so, does this capacity influence social behavior? Recent studies of rhesus monkeys on Cayo Santiago have explored their ability to use subtle details of an action sequence to draw inferences about the actor's goal. In the basic design, Wood and colleagues presented two potential food sources (overturned coconut shells) to a subject, acts on one, and then walks away, allowing the subject to selectively approach. Although coconuts are native to the island on which these animals live, rhesus cannot open the hard outer shells themselves, and therefore, only obtain the desired inner fruit when the coconuts open on their own or have been opened and discarded by a human. It, thus, logically follows that if subjects perceive the experimenter's action as goal-directed and potentially communicative,

then they should selectively approach the coconut contacted, as this maximizes the odds of obtaining food.

Results revealed that when the experimenter grasped the coconut with his hand, foot, or a precision grip involving the pointer finger and thumb, rhesus selectively approached this coconut over the other; in contrast, they approached the two coconuts at chance levels when the experimenter flopped the back of his hand on the coconut [accidental], touched or grasped the coconut with a tool, or grasped the coconut with his hand for balance while standing up. These results rule out low-level association accounts; many of these individuals have experience seeing humans perform goal-directed actions with tools, and yet they did not perceive tool-related actions as goal-directed, and none of these individuals have experience seeing humans perform grasping actions with their feet, and yet they did perceive foot-related actions as goal-directed. Further, these data show that when assessing the meaning of actions, rhesus are highly sensitive to the means used to achieve a goal – for example, perceiving a hand grasp action as goal-directed but a hand flop action as accidental, despite the fact that the experimenter's body position, eye gaze, and duration of contact with the coconut were identical across the two conditions. These studies, together with research on other species, suggest that non-human animals infer the meaning of an action by evaluating the actor's goals in relation to the environmental constraints on achieving such goals.

Given that some animal species are able to draw inferences about others' intentions and goals, we can ask whether this ability influences their social interactions. That is, are animals completely outcome-oriented or do they attend to the means by which outcomes are achieved? One of the first studies to explore this problem was Call and colleagues' experimental study of captive chimpanzees. In this study, a human experimenter faced a chimpanzee, seated on the opposite side of a Plexiglas partition. In each of several conditions, varying the nature of the experimenter's action, a grape was presented near the opening of the partition; the opening was large enough for the chimpanzee to reach and grab the grape. In one condition, the experimenter brought the grape within grasping distance, but then rapidly retracted it as soon as the subject reached. This action was defined as teasing. In a second, and highly parallel condition, the experimenter brought the grape forward, but then dropped it as soon as the subject reached. This action was defined as clumsiness. Chimpanzees showed much greater signs of frustration in the teasing than clumsy conditions, leaving the test chamber earlier, and acting aggressively toward the experimenter (i.e., banging on the Plexiglas partition). Thus, as Call and colleagues suggest, chimpanzees appear to make a distinction between unwilling (teasing) and unable (clumsy), thereby showing sensitivity to more than the mere outcome of an event, as in both of these cases, the subject failed to obtain the grape.

In the aforementioned study of reciprocity in tamarins, Hauser and colleagues showed that individuals were more likely to cooperate in situations in which a conspecfic's actions were truly altruistic, than when the same amount of food (outcome) was delivered as an accidental byproduct of an otherwise selfishly motivated action. Specifically, a game was set up such that the individual playing the actor-1 position was offered an opportunity to pull a tool, delivering one piece of food to self and three pieces to an unrelated partner. For the partner, or actor-2, pulling the tool resulted in no food for self, but two pieces for the partner (actor-1). A session was defined as 12 trials each for actor-1 and -2, alternating turns. If both actor-1 and -2 pulled on their respective turns, they would maximize the overall returns, with three pieces each, after an alternating round. This is what would be expected if actor-2 perceives actor-1's pull as altruistic, that is, motivated by the goal of giving food. In contrast, if actor-2 perceives actor-1's pull as selfish, with the three pieces obtained as a byproduct, then actor-2 should not pull. Results showed that actor-2 rarely pulled. This condition, combined with another showing that individuals will altruistically give food (i.e., paralleling the actor-2 position) when a partner altruistically reciprocates, reinforces the conclusion that tamarins attend to both the outcomes and the means by which they are obtained.

A final experiment adds to this literature by showing not only sensitivity to the means by which outcomes arise, but the agent responsible for such outcomes. Jensen and colleagues first presented one chimpanzee, A, with an apparatus involving a sliding tray full of food. Subject B was then introduced into an adjacent enclosure that contained a rope. By pulling on the rope, B moved the sliding tray away from A, thereby stealing A's food. However, subject A's enclosure also contained a rope that, if pulled, would collapse the sliding tray, thereby taking the food away from B. In conditions where B pulled the tray away from A, A frequently became agitated and collapsed the sliding tray. However, in a similar condition where, instead of B pulling the tray, the experimenter pushed the tray away from A to B, A rarely collapsed the tray. This result is interesting because it shows that chimpanzees not only respond to unfair outcomes, but that they distinguish between human and chimpanzee agents; a parallel set of findings was presented by Hauser and colleagues in their reciprocity study, showing that cooperation increased in the face of a unilateral tamarin cooperator, but not a game in which the payoffs remained the same, but a human cooperator delivered the rewards for another tamarin.

Conclusion

Research on inequity aversion, prosociality and action perception in non-human animals is still in its nascent stages, and there are currently several noticeable gaps in our understanding of these phenomena. First, while there is an emerging body of evidence suggesting that these capacities are present in some species of non-human primates and domestic dogs, little is known about its taxonomic distribution or the evolutionary pressures that may select for this capacity in different species. For example, given the increasing evidence that some food-caching jays are sensitive to where others are looking, what others have seen, and how such information guides cooperation and cheating, it would not be surprising to find that at least these birds, and possibly other animals, are sensitive to inequities, and to the distinction between means and outcomes. Second, it is unclear whether the capacities discussed here are specific to the social domain or are important in other domains as well. For example, do animals appreciate that some properties of artifacts such as tools are intentionally designed whereas others are simple byproducts of physical constraints or accidents? Third, the majority of studies that have investigated these capacities in animals have focused on captive individuals. Similar studies of wild populations will be crucial to understanding whether these abilities are expressed in nature or are only elicited under controlled laboratory conditions that often set up situations that would never arise in the wild. For example, though capuchin monkeys can work with a token economy and detect inequities, and though they can solve a joint action task, these situations never arise in the wild. It is, thus, essential to distinguish between the capacity to solve various cooperative tasks and the social and ecological pressures that might demand that such abilities be used to cooperate.

Despite these limitations, we believe the work on the foundations of morality reviewed here (and elsewhere; Hauser's 2006 book, *Moral Minds: How Nature Designed Our Universal Sense of Right and Wrong,* for a more extensive discussion) lead to at least two conclusions. First, we share with other animals several core psychological capacities that were most likely necessary for the evolution of our moral sense. Like other animals, a sense of justice as fairness is premised on an ability to take the perspective of another individual, show concern for others, detect inequities, and inhibit the temptation to feed self-interest. Though humans certainly show highly elaborated forms of these abilities, there are significant precursors in other species. Second, there are two ways in which human moral behavior, and the psychology that supports it, are unique. Animals show virtually no evidence of reciprocity and large-scale cooperation, and as far as we can tell, never engage in the problem of considering not only what is the moral state of play, but what could or should be – our sense of ought. Though a brief essay like this is not the place to develop these issues, we end with a speculative consideration. What allows humans to uniquely engage with the thought of the ought is our ability to combine different modular representations into new representations. Whereas other animals have evolved highly adaptive, modular, and informationally encapsulated domains of thought, targeted at single problems, humans evolved the capacity to create interfaces between these domains to create entirely new systems of thought. Thus, we alone can consider how we typically distribute resources, often based on matters of effort and need, step back from such norms, and consider a more enlightened perspective that not only considers matters of fairness but also individual welfare. And we can do this because we can prospectively evaluate the future, sideline current needs and temptations, and realize that progress is made by dissent as opposed to consent. What fuels the ought is the realization that we can, and often should, entertain a different moral landscape.

See also: Cooperation and Sociality; Empathetic Behavior; Mental Time Travel: Can Animals Recall the Past and Plan for the Future?; Punishment; Social Cognition and Theory of Mind.

Further Reading

Boyd R and Richerson PJ (1992) Punishment allows the evolution of cooperation (or anything else) in sizeable groups. *Ethology and Sociobiology* 13: 171–195.

Brosnan SF and de Waal FBM (2003) Monkeys reject unequal pay. *Nature* 425: 297–299.

Call J, Hare B, Carpenter M, and Tomasello M (2004) 'Unwilling' versus 'unable': Chimpanzees' understanding of human intentional action. *Developmental Science* 7: 488–498.

Clutton-Brock TH and Parker GA (1995) Punishment in animal societies. *Nature* 373: 209–216.

Dugatkin LA (1997) *Cooperation Among Animals*. Oxford: Oxford University Press.

Hamilton WD (1964) The genetical evolution of social behavior, I. *Journal of Theoretical Biology* 7: 1–16.

Hauser MD (2006) *Moral Minds: How Nature Designed Our Universal Sense of Right and Wrong*. New York: Ecco.

Hauser MD, Chen MK, Chen F, and Chuang E (2003) Give unto others: Genetically unrelated cotton-top tamarin monkeys preferentially give food to those who altruistically give food back. *Proceedings of the Royal Society, London, B* 270: 2363–2370.

Jensen K, Call J, and Tomasello M (2007) Chimpanzees are vengeful but not spiteful. *Proceedings of the National Academy of Sciences* 104: 13046–13050.

Range F, Horn L, Viranyi ZS, and Huber L (2009) Absence of reward induced aversion to inequity in dogs. *Proceedings of the National Academy of Sciences* 106: 340–345.

Rawls J (1971) *A Theory of Justice*. Cambridge: Harvard University Press.

Stevens JR and Hauser MD (2004) Why be nice? Psychological constraints on the evolution of cooperation. *Trends in Cognitive Sciences* 8: 60–65.

Trivers RL (1971) The evolution of reciprocal altruism. *The Quarterly Review of Biology* 46: 35–57.

Williams GC (1966) *Adaptation and Natural Selection*. Princeton, NJ: Princeton University Press.

Wood JN, Glynn DD, Phillips BC, and Hauser MD (2007) The perception of rational, goal-directed action in non-human primates. *Science* 317: 1402–1405.

Motivation and Signals

D. H. Owings, University of California, Davis, CA, USA

Introduction and Definitions

A discussion of signals requires consideration of communication, the broader context in which signals are used.
Most modern workers in animal communication recognize that two different roles underlie communication. The
role of signal production involves using signals to manage
the behavior of others, in part by exploiting their assessment systems. The role of assessment involves making
adaptive behavioral decisions by selectively attending to
the most reliable stimuli available, including both signals
and cues, for appraising individuals and situations. Notice
the linkage between these two roles. The definition of
each includes the other because each is targeted on the
other. All participants in communication play both signal
production and assessment roles, and it is the interplay
between these two individual activities that produces
the social process of communication. A signal is a trait
specialized for communication, that is, for managing the
behavior of others by working through their assessment
systems. A cue is any stimulus upon which assessment
can be based that has not been specialized for the communicative exploitation of assessment systems. The behavior of Belding's ground squirrels while contending with
mammalian predators illustrates the distinction between
signals and cues. When a female with young spots a mammalian predator, she is likely to produce a trill alarm call and
watch the predator vigilantly. Other squirrels not only
respond to her signal, the alarm call, but also use the direction of her gaze to locate the predator. The direction of the
caller's gaze is not a signal; it is a perceptual activity by
the caller to monitor the predator, but is opportunistically
exploited by other squirrels as a useful cue.

A discussion of motivational systems can be clarified
by considering their place among all classes of psychological systems that underlie behavior. These psychological
systems can be grouped into two general categories –
knowing and wanting. The mechanisms of knowing serve
the processes of information acquisition and processing,
also called perception and cognition, respectively. These
processes are structured by the mechanisms of wanting,
the motivational systems that focus an animal's efforts on
matters important for its proximate and ultimate success.
Animals are motivated to know about those matters that
are important to them and accomplish this with perceptual
and cognitive mechanisms that have been shaped to emphasize information crucial to their proximate and ultimate
success. So, the mechanisms of knowing and wanting

are intricately intertwined, representing two sides of the
same coin of behaving.

Many factors influence an organism's success, including its effectiveness in acquiring food, water, and oxygen,
reproducing, avoiding attack by predators and parasites,
maintaining its social status, and so forth. Animals have
different complex motivational systems associated with
each of these factors important to their success. These
systems specify pertinent physiological responses, relevant cues requiring attention, preferred outcomes of
behavior, the importance of these outcomes, and the activities most likely to generate them. In communication,
motivational systems are most clearly relevant to signal
production because they deal with what an individual's
behavior serves to accomplish proximately, and how hard
it is trying, but they are also relevant to assessment because
they focus an individual's attention on the features of the
environment most relevant to its current efforts.

The variety of motivational systems that comprise an
individual are linked to each other via a set of priorities.
Such a ranking with regard to urgency is important in
part because the demands addressed by different systems
vary in the immediacy of their impact on the well-being of
the individual. More immediate demands are given higher
priority. If you are suffocating, for example, you will die in a
few minutes unless you get air. So, work on all other demands
needs to be set aside until your oxygen need is met. But
this is an extreme example; animals can usually breathe
while meeting most other needs. Nevertheless, work on the
demands of different motivational systems can often be
incompatible, and this incompatibility also drives the need
to prioritize motivational systems. These priorities can produce motivational conflicts that must be resolved through
compromise. For example, Belding's ground squirrels must
focus their attention on the ground in order to forage, but
must raise their heads in order to monitor for predators.
Their resolution of this conflict between meeting nutritional
and antipredator needs varies depending on the individual's
nutritional state. The greater the nutritional deficit, the less
willing the squirrel is to shift its efforts from foraging to
antipredator vigilance when it detects an alarm trill.

Motivational Systems Drive and Direct Signal Production

When predators endanger the young of a species that
exhibits parental care, an activated parental motivational

489

system can drive and direct the production of anti-predator signals. This motivational effect can be observed when California ground squirrels deal with rattlesnakes. Rattlesnakes are an important source of predation on California ground squirrel pups, but not on adults. These squirrels have evolved the capacity to neutralize rattlesnake venom and this capacity is sufficient among adults to allow them to survive the injuries produced by rattlesnake bites. This neutralizing capacity provides adults with the option of confronting rattlesnakes, partly in defense of pups. Females with activated parental motivational systems (females with dependent young, aka maternal females) spend more time confronting rattlesnakes than nonmaternal females and males, neither of which contributes much to care of pups. When these squirrels deal with snakes, they invariably produce a visual signal, tail flagging, in which the fluffed tail is repeatedly waved from side to side as a means of managing the snake's behavior. The driving effects of the parental motivational system on signal production is revealed when maternal females engage in much more tail flagging than nonmaternal females and males.

But the motivational system that drives tail flagging also directs it, organizing that signaling activity in ways that are sensitive to the level of danger involved. Maternal squirrels tail flag more to large than small rattlesnakes, and more to snakes near than distant from their home burrows. Rattlesnake confrontation by adults can even be aggressive enough to induce the snake to rattle defensively at the squirrel. This sound incidentally includes cues about snake size and body temperature, both of which vary positively with the degree of risk that the confronting squirrel faces. (Larger and warmer snakes are more dangerous.) In playbacks of rattling sounds from rattlesnakes varying in their temperature and size, the tail flagging response by maternal females is stronger and more finely discriminating among these acoustic risk cues than the tail flagging of nonmaternal females and males.

Motivational Systems Focus Assessment

As noted earlier, motivational systems are relevant to assessment because they focus an individual's attention on the features of the environment most relevant to its current efforts. Why do animals not simply attend to everything important to them? In general, they do not because attention is a limited resource that can be allocated at each moment only to a fraction of all important matters. When, for example, an animal is seeking an object that is difficult to detect, this interferes with the detection of a second important object more than when the first object is easy to detect. Laboratory studies with blue jays illustrate this point. These birds were required

to detect a computer image of a mealworm in the center of a computer screen and a moth at the periphery of the screen (both representations of attractive food items). The experimenters varied the difficulty of detecting the center mealworm by changing the number of distractor stimuli presented with it. Jays were able to adjust to these changes and maintain their performance when the mealworm-detection task was more difficult. But these adjustments interfered with the detection of the peripheral moth stimulus. Performance in that task declined as the mealworm-detection task became more difficult, thus revealing limitations on the availability of attention for important tasks.

To understand the implications of such limitations for communication, imagine now that the focus of an animal's attention is on the courtship signals of a potential mate, and that discriminating between high- and low-quality mates is difficult because the differences in their signals are subtle. Imagine also that the important peripheral events are alarm signals evoked by an approaching predator. The problematic nature of such limitations is compounded by the fact that many animals work hard to be cryptic. This is true of individuals engaged in activities that conspecifics might contest, but is also true of the prey that predators hunt and the predators that need to avoid detection by vigilant prey.

Directing Attention and Amplifying Salience

Female Norway rats undergo striking motivational changes during pregnancy. Mothers who have just given birth are strongly attracted to pups, finding contact with them rewarding and expressing the full repertoire of maternal activities. In contrast, virgin females seem to find newborn pups noxious; they avoid pups and may even attack them. This remarkable shift from aversion to parental attraction is initiated a few days before the female gives birth, primarily by changes in ovarian steroid hormone secretion (progesterone drops and estrogen rises), but also depends on increased release of the peptide hormone prolactin and polypeptide hormone lactogen. These hormonal changes induce a suite of physiological and behavioral changes that enhance the salience of visual, auditory, and olfactory stimuli associated with rat pups, and even augment the capacity of pup-related stimuli to serve as incentives for maternal behavior. Not all of these stimuli are signals; some, such as the general sight of pups, are unspecialized cues. This illustrates an important point about assessment; it involves an active extraction of the information needed to guide behavior, whether or not the information source is specialized for communication. Nevertheless, communication and signals are involved. Maternal females are, for example, much more responsive than nonmaternal females to the ultrasonic vocalizations that pups emit when their body temperature drops as a result of separation from the nest and littermates. Mothers respond to these calls by searching for and retrieving pups, and returning them to the warmth of the nest.

The Interplay between Signal Production and Assessment

Maintaining and Fostering Motivational States through Repeated Signal Inputs

It is of some interest to note that Norway rat mothers are not completely in charge of their states of parental readiness. If a female prematurely loses her litter, her responsivity to playbacks of pup calls declines sooner than if she retains her litter to the time of weaning. In the normal course of infant–mother interactions, pups 'persuade' mothers to continue caring for them as the prenatal hormone effects dissipate during the first few postpartum days. Stimulation from pups induces the mother to release the peptide hormone oxytocin, and this maintains the mother's maternal motivation.

In fact, even virgin females exposed to pup contact for several days eventually begin to behave maternally under a sustained barrage of persuasive inputs from pups. Such slowly developing effects of repeated inputs illustrate an important point about communication. Signal production functions not only immediately, to trigger responses by stimulating already-active motivational systems, but also more gradually to foster activation of motivational systems. These more slowly developing effects of repeated signal inputs are called priming effects, in contrast with more immediate triggering effects. Processes involving such cumulative priming effects of repeated inputs are said to be tonic, and the associated communicative processes are said to reflect tonic communication. Tonic communication with its associated extended repetition of signals is a widespread phenomenon. Think, for example, of how often you have noticed a perched songbird singing in long bouts of repeated songs unbroken by any form of social interaction. Singing by songbirds provides a prime example of tonic communication.

Nevertheless, most studies of communication involve triggering rather than priming effects of signals. We need more studies of tonic communication in part because they will increase our understanding of the role of motivation in communication. Tonic effects are often mediated by hormonal changes that engender the broad thematic shifts in behavior that are the hallmark of transitions between motivational systems.

Maintaining and Fostering Motivational States through Repeated Signal Inputs to one's Self

We typically and usually correctly assume that signaling behavior is targeted at some individual other than the signaler itself. But that is not always the case, as research

with ring doves illustrates. Reproduction in ring doves involves a cascading series of priming effects in male and female. Males typically initiate courtship by bowing and cooing, which is followed by the male's cooing over prospective nest sites. The female gradually comes to join the male in cooing over the prospective nest site, and ultimately engages in a long stint of solo nest-cooing, before the two join forces in the construction of a nest. When nest-building reaches a threshold level, hormonal changes are triggered in the female that culminate in ovulation and copulation. What role do the female's coos play in this process? It seems reasonable to identify the male as the target of these vocalizations. However, muting the female in several different ways leaves the male's courtship activities relatively unchanged, but blocks the hormonal changes in the female leading to ovulation. And, playbacks of coos to the female restores those changes, especially when the vocalizations used are her own. Further playback studies indicate that the female is in fact the target of the male's calls, but these calls have their effects on the female by stimulating her to coo, which in turn induces her to ovulate through a process of vocal self-stimulation.

Structuring Signal Production to Capitalize on Motivational Features of Assessment Systems

Human adults speak to infants in ways that would be unusual and probably even offensive if directed at other adults. Compared with the choppy and rapid-fire patterns of normal conversation among adults, infant-directed speech is slower, has a higher pitch, and often contains smooth, exaggerated changes in pitch. Such patterns of intonation involve what are called the prosodic features, that is, 'melodies' of speech. When approving of something an infant has done, the exaggerated contours of pitch change involve a rise–fall pattern

In subtle contrast, the prominent pitch contours used to get an infant's attention end on an upswing

In more striking contrast, the melodies used to soothe an infant or disapprove of its behavior involve a much less pronounced variation in pitch, but differ from one another in patterns of change in amplitude (sound intensity). Sounds of soothing involve no abrupt shifts, changing slowly in amplitude and maintaining low amplitudes ('Thaaat's okaaay. Maaama's here.'). Sounds of disapproval, on the other hand, are short and sharp, onsetting abruptly ('**No! Uh uh! Don't do that!**').

Why do we use these different melodies to manage our infants' behavior in different ways? Playbacks to infants in which prosodic patterns are preserved but linguistic content is eliminated provide a simple answer. These are what work best. From birth, these various patterns tap into the infants' differing motivational reactions to different melodies; they are in effect different unconditioned stimuli for pleasing, alerting, soothing, and alarming the infant. It seems that evolution through natural selection, perhaps as well as the shaping effects of experience, has generated patterns of speech that can be used to manage the behavior of infants. These patterns are effective because they capitalize on the motivational components of infant assessment systems.

It is of interest to note that the motivational impact of infant-directed speech changes as the infant enters its second year. At this time, parents pair presentation of new objects (e.g., a teddy bear) with prosodic emphasis to draw the infant's attention to the verbal label for that object ('see the TEDDY BEAR?'). This fosters the development of the infant's vocabulary. Thus, the impact of infant-directed speech has more to do with the induction of motivational states in infants during their first year, but begins to contribute to language development in the second year.

There is evidence that the features of assessment systems exploited by infant-directed speech predated the production of these patterns evolutionarily, and so may have been sources of natural selection shaping the melodies of infant-directed speech. For example, the elevated pitch of infant-directed speech appears to be similar to a widespread vocal pattern among vertebrates, in which animals raise their voice pitch to be less threatening and drop pitch for a more threatening effect. Similarly, research with sheep-herding dogs indicates that additional conclusions about infant-directed speech also apply to non-humans. Individuals who use dogs as assistants in herding their sheep use short, rapidly repeated, broadband notes to stimulate movement by their dogs, and longer, continuous, narrowband notes to inhibit movement (these are similar to the attention-getting and soothing melodies of infant-directed speech). Experimental tests with domestic pups support the hypothesis that these two patterns of acoustic stimulation differ in their capacity to stimulate motor activity.

Future Directions

How Do Signals Acquire Their Salience to the Motivational Systems of Targets?

Adolescent male laboratory rats are very playful, and when they engage in bouts of rough-and-tumble play, they emit frequency-modulated (FM) ultrasonic vocalizations with an average sound frequency of about 50 kHz. As these males progress to adulthood, they become less playful and more aggressive. When interacting aggressively, they produce a different type of vocalization, a 22-kHz ultrasonic call. The FM 50-kHz and 22-kHz calls differ in their incentive value to listening rats. Rats are attracted to FM 50-kHz calls, performing operant responses both to produce playbacks of these calls and to gain access to a playful situation in which FM 50-kHz calls are emitted. In contrast, they avoid performance of responses that produce 22-kHz calls. These differences in incentive values are consistent with the hypothesis that the sound of the playful FM 50-kHz call has a positive motivational value that makes it a useful tool for attracting play partners, whereas the sound of the 22-kHz call has a negative motivational impact that facilitates its use to repel potential adversaries.

The area of signals and motivation would benefit immensely from additional research addressing the question of how signals such as the aforementioned rat vocalizations acquire such salience and incentive value. The general form of an answer to such questions is likely to appeal to the interplay between the roles of assessment and signal production. For example, there is evidence that the previously discussed ultrasonic retrieval call used by infant rats originated as a byproduct that was used as a cue of infant distress by vigilant mothers. When infants experience excessive cooling as a result of becoming separated from the nest, they use an abdominal compression maneuver to deal with the cardiovascular consequences of the excessive cooling. Such cooling increases blood viscosity and reduces cardiac functioning, thereby jeopardizing the infant by reducing blood circulation. The abdominal compression maneuver increases pressure in the abdominal cavity, thereby augmenting venous return of the blood to the heart. At the same time, this maneuver incidentally forces some air through the restricted larynx and so produces ultrasounds that in their original form were probably byproducts but also excellent cues that a pup was in jeopardy. The mother's proactive use of these cues could have then shaped this byproduct sound into a retrieval call.

But what are the actual processes that would generate such a scenario? Most often, researchers in animal communication cite evolution through natural selection as the causal processes. This is likely; the mother's use of the ultrasound cue could have been a source of selection favoring refinement of the sound into a signal. But this leaves unanswered proximate questions about the development and causation of the use of these sounds as cues and signals. Such proximate questions, especially when they deal with the roles of motivational systems in assessment and signal production, are important but have not yet received the attention they merit. The bit of data available for other signaling systems suggests that the type of interplay between assessment and signal production that clearly goes on evolutionarily can also play a proximate causal role in shaping these communicative roles. For example, young male brown-headed cowbirds need normal social interactions with adults of both sexes in order to develop the ability to integrate singing into an effective set of courtship maneuvers, to sing evocative songs, and to target females in their courtship-singing efforts. Some of the processes involved include the operant conditioning that psychologists have studied for many decades. Adult female brown-headed cowbirds, for example, respond preferentially to songs with particular properties and young males retain these properties in their developing songs while deleting other song characteristics.

A model of agonistic vocal communication among primates suggests another proximate route by which signals can become salient. Rather than involving independent discovery of cues by assessment systems, followed by modification of cues into signals, this scenario involves a higher-order mode of action of the signal-production role. Signalers augment the salience of their signals by pairing them with intrinsically evocative stimulation, such as attack. This involves another form of conditioning, classical rather than operant. Old-world monkeys, for example, emit individually distinctive calls, a candidate conditioned stimulus, while subjecting their adversaries to attack, an intrinsically noxious unconditioned stimulus. Such pairings have the potential to enhance the noxious motivational impact of the vocalizer's own calls on the conditioned individual, but may not have that effect on the same types of calls used by others, with their own distinct individuality in structure.

These findings indicate that a fruitful path for future research will involve a synthesis of the dominant evolutionary questions about signals and motivation with proximate questions about the roles played by fundamental motivational mechanisms in the causation and development of communicative behavior.

See also: Acoustic Signals; Aggression and Territoriality; Agonistic Signals; Alarm Calls in Birds and Mammals; Communication and Hormones; Communication Networks; Cultural Inheritance of Signals; Deception: Competition by Misleading Behavior; Electrical Signals; Food Signals; Honest Signaling; Interspecific Communication; Mating Signals; Parental Behavior and Hormones in Mammals; Parental Behavior and Hormones in Non-Mammalian Vertebrates; Parent–Offspring Signaling; Punishment; Signal Parasites; Visual Signals.

Further Reading

Berridge KC (2004) Motivation concepts in behavioral neuroscience. *Physiology and Behavior* 81: 179–209.

Blumberg MS and Sokoloff G (2001) Do infant rats cry? *Psychological Review* 108: 83–95.

Burgdorf J, Kroes RA, Moskal JR, et al. (2008) Ultrasonic vocalizations of rats (*Rattus norvegicus*) during mating, play, and aggression: Behavioral concomitants, relationship to reward, and self-administration of playback. *Journal of Comparative Psychology* 122: 357–367.

Cheng M-F (1992) For whom does the female dove coo? A case for the role of vocal self-stimulation. *Animal Behaviour* 43: 1035–1044.

Dukas R (2001) Behavioural and ecological consequences of limited attention. *Philosophical Transactions of the Royal Society of London B Biological Sciences* 357: 1539–1547.

Fernald A (1992) Human maternal vocalizations to infants as biologically relevant signals: An evolutionary perspective. In: Barkow JI I, Cosmides L, and Tooby J (eds.) *The Adapted Mind: Evolutionary Psychology and the Generation of Culture*, pp. 345–382. Oxford: Oxford University Press.

Owings DH and Morton ES (1998) *Animal Vocal Communication: A New Approach*. Cambridge: Cambridge University Press.

Owren MJ and Rendall D (1997) An affective-conditioning model of non-human primate vocal signaling. In: Owings DH, Beecher MD, and Thompson NS (eds.) *Perspectives in Ethology: Communication* vol. 12, pp. 299–346. New York: Plenum.

West MJ, King AP, and Freeberg TM (1997) Building a social agenda for the study of bird song. In: Snowdon CT and Hausberger M (eds.) *Social Influences on Vocal Development*, pp. 41–46. Cambridge: Cambridge University Press.

Young LJ and Insel TR (2002) Hormones and parental behavior. In: Becker JB, Breedlove SM, Crews D, and McCarthy MM (eds.) *Behavioral Endocrinology*, 2nd edn., pp. 331–370. Cambridge: MIT Press.

Multimodal Signaling

G. W. Uetz, University of Cincinnati, Cincinnati, OH, USA

Introduction – What Are Multimodal Signals?

As is clear from other sections of this encyclopedia, animals communicate in a wide variety of ways, both overt and subtle. A considerable amount of research has revealed that animals do not utilize channels of communication that are obvious to humans, for example, visual signals and auditory signals alone, but they also use a variety of other channels that are less visible or audible, including some that are not detectable by human senses. In addition to visual signals and acoustic signals, these communication modes may include chemical signals in the form of airborne sexual scents and territorial urine markings, seismic signals (i.e., vibrations sent through substrates such as plant stems, leaves, and soil), and electrical signals sent through water by electric fishes. While many animals are well known for their use of specific forms of communication – the songs of birds, crickets, and frogs are all examples of auditory or acoustic communication – a more detailed analysis has led to the realization that many animals are capable of producing signals in multiple sensory modes or channels. The use of multiple sensory modes or channels for communication is known as multimodal signaling, and it is turning out to be more common than originally thought.

Examples of Multimodal Signaling

Some of the earliest published observations of multimodal signaling were described in birds by ethologists, such as Niko Tinbergen, who characterized the combination of visual and acoustic signals as 'displays' (vocalizations accompanied by body postures and movements) used by gulls, pigeons, and domestic fowl. For birds, the most frequent modes used in multimodal signals are auditory and visual, and in many species the combination can be quite dramatic, as in this video of the North American Sage Grouse (**Figure 1**). For example, the familiar male North American Wild Turkey (*Meleagris gallopavo*) combines a visual signal (a 'fan' display of multicolored tail feathers) and an acoustic signal (the 'gobble' call) (**Figure 2**).

The primary context of multimodal signaling in birds is in courtship, although there are numerous examples of other social interactions, such as territory defense, antipredator behavior, and parent–offspring communication. Mammals also use multimodal signaling in a variety of contexts. Most people would recognize that the communication of aggressive threat by dogs is multimodal, since the acoustic signal of growling is usually accompanied by the visual signals of facial and body postures, such as bared teeth, raised fur, and ear position. It is also easy for us as humans to recognize multimodal signals in our relatives among the primates, perhaps because we are similar in many ways, and their often spectacular communication displays include a wide array of combinations of visual and acoustic signals. Primatologists like Marc Hauser and Sara Partan have cataloged natural facial expressions and vocalizations of rhesus macaques, and found that in many cases, vocalizations were accompanied by articulatory gestures and positions of the mouth and lips (**Figure 3**).

Prairie dogs and ground squirrels are well known for their antipredator alarm call vocalizations, and these acoustic signals are often accompanied by visual signals such as upright postures, jumping, and tail flicking. Recent studies by Aaron Rundus, Don Owings, and their colleagues have shown that alarm signals given by the California Ground Squirrel (*Spermophilus beecheyi*) in the presence of rattlesnakes are truly multimodal in that these mammals not only vocalize and flag their tails, but also emit heat signals from their tail that can be detected by the snakes' infrared vision. These signals deter predation by increasing the apparent size of the prey, making the snake more cautious in its approach (**Figure 4**).

Many examples of multimodal signaling can be seen in other vertebrates animals, including reptiles, amphibians, and fishes. For example, Emilia Martins and colleagues have shown that sagebrush lizards (*Sceloporus graciosus*) use 'head-bob' displays as visual signals in combination with chemical signals in territorial defense. While frogs and toads are best known for acoustic communication, Peter Narins, Mike Ryan, and their colleagues have demonstrated that the inflation of the vocal sac producing the sound also serves as a visual signal, and some responses can be elicited only when the two are combined in a multimodal signal (**Figure 5**). Likewise, several species of fishes known primarily for visual signaling behavior, including the Siamese Fighting fish *Betta splendens*, swordtails (*Xyphophorus* spp.), sticklebacks (*Gasterosteus* spp.), and numerous species of Lake Malawi cichlids, have also been shown to communicate using either chemical, tactile, and/or acoustic cues as part of multimodal signals.

Multimodal signals are less well studied among most invertebrates, but some of the best examples can be seen in insects, crustaceans, and spiders. Some insects, such as

Figure 1 Males of the Greater Sage Grouse (*Centrocercus urophasianus*) in the Great Plains of North American exhibit multimodal communication in courtship displays, using visual signals accompanied by acoustic 'booming' signals. Photo courtesy of Marc Dantzker, used with permission.

Figure 2 Males of the North American Wild Turkey (*Meleagris gallopavo*) exhibit multimodal signaling during a courtship display that combines visual and acoustic signals. Photo by Bob Schmitz Cornell Lab of Ornithology. From Cornell Lab of Ornithology Birds of North America, used with permission. On-line: http://bna. birds.cornell.edu/bna/species/022/galleries/photos/ BOS_070802_00143D_S/photo_popup_view.

Drosophila melanogaster and the Hawaiian *Drosophila* species, are well known for using visual, chemical, and vibratory signals during courtship and other communication contexts. Bert Holldobler has examined communication in a variety of ant species and demonstrated that not only do ants use a diverse array of chemical compounds but also that these signals are often combined with vibratory communication produced by stridulation, percussion, and tactile stimulation (**Figure 6**). Moreover, multimodal signaling in ants occurs in multiple contexts, including alarm signals, recruitment to food sources, and agonistic

behaviors used in colony defense. In decapod crustaceans, chemical signals are common, but Melissa Hughes, Paul Moore, and colleagues have found that the sexual and aggressive responses of snapping shrimp and crayfish often vary depending on whether visual and/or tactile information is present.

One group of invertebrates that has been extensively studied with respect to multimodal communication is the wolf spiders (Lycosidae), which often use visual and vibratory signals in their courtship behavior. Signal structure varies considerably within this family, as some species use only a single mode, while others incorporate both visual and vibratory components into their signals. Moreover, Jerry Rovner found that some species, for example, *Rabidosa rabida*, produce visual and vibratory signals separately at different times during precopulatory behavior, while others, for example, *Schizocosa ocreata*, use simultaneous vibratory and visual signals. With my colleagues and former students Gail Stratton, Sonja Scheffer, Will McClintock, Eileen Hebets, Matt Persons, Andy Roberts, Phil Taylor, Jeremy Gibson, and Dave Clark, I have conducted studies of multimodal communication in members of the genus *Schizocosa*, which includes several highly similar 'cryptic' species (the *S. ocreata* clade) that have apparently arisen from rapid speciation via behavioral isolation driven by sexual selection. Species within this clade are all similar in size, coloration, and genital characters, and females are hard to distinguish. Males, however, vary in foreleg decoration (partial pigmentation; full, dark pigmentation; tufts of bristles) and behavior. Male *Schizocosa* may use substratum-borne (seismic) signals, visual signals, or both in courtship behavior. The diversity of signal types used across species appears to serve as a premating isolation mechanism (**Figure 7**) (link to video of spider courtship w/sound). A number of examples from this research will be featured in the following sections.

Categories and Mechanisms of Multimodal Signals

Given the diversity of multimodal signals, questions naturally arise about what kinds of information they contain and whether a multimodal signal contains information that is the same or different from that in a unimodal signal. For example, ornithologists recognize bird species by both their songs and plumage patterns, but when birds perceive these signal components, what information do they obtain from them? Do the visual displays and sounds presented together in a multimodal signal by a singing bird 'say' essentially the same thing, for example, species identity? Alternatively, do they contain multiple kinds of information, for example, does song represent species identity, while plumage provides information on male condition or dominance status?

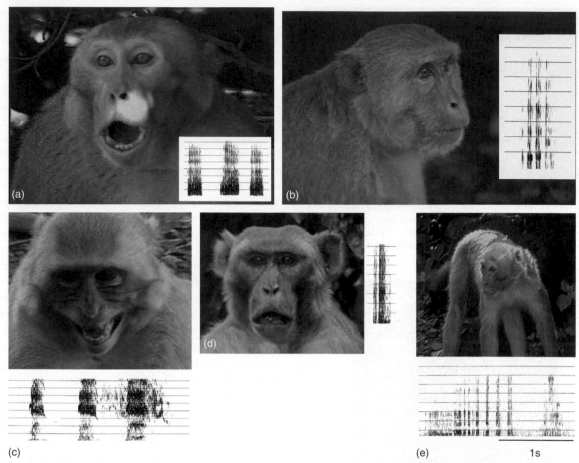

Figure 3 Images from videotape of vocalizing animals. Spectrograms were collected from videotape at the same frame as the picture, using a Kay Digital Sona-Graph model 7800. Horizontal bars mark 1–kHz intervals (1–8 kHz). (a) Adult male barking, with open mouth and ears back. (b) Adult female (845) giving a pant-threat vocalization. (c) Subadult female (X70) giving a broadband ('noisy') scream. (d) Adult male (C78) grunting. (e) Subadult male girneying and waving his tail. Although this girney contains primarily broadband components, girneys can also include narrow-band sounds (e.g., see spectrogram in Kalin et al., 1992). From Partan (2002) Single and multichannel signal composition: Facial expressions and vocalizations of rhesus macaques (*Macaca mullata*). Behaviour 139: 993–1027.

Classification of Multimodal Signals and Responses

Since the effectiveness of signaling is determined in large part by whether the sender receives an appropriate response from the receiver, much theory and research has focused on answering these questions from the recipient's point of view. Animal communication researchers Sara Partan and Peter Marler have categorized multimodal signals, on the basis of the information content of signals and the responses of receivers (**Figure 8**).

Redundant Signals

The first distinction that can be made between types of multimodal signals is between redundant and nonredundant signals. If individual components of a multimodal signal elicit the same response from a receiver when presented separately, it is likely that they contain the same or similar kinds of information about the sender (e.g., species identity), and are therefore redundant. Researchers often refer to these different components as 'back up' signals, as each one may suffice if another is obstructed by environmental constraints or noise. However, not all receiver responses to redundant multimodal signals are the same, as some may be equal to unimodal signals in their intensity (equivalence), while the intensity of others is increased (enhancement) (**Figure 9**).

Nonredundant Signals

Alternatively, signal elements in component modes may contain distinctly different kinds of information, and responses of receivers to each mode of these nonredundant signals may be entirely different (**Figure 10**). Researchers often refer to these kinds of multimodal

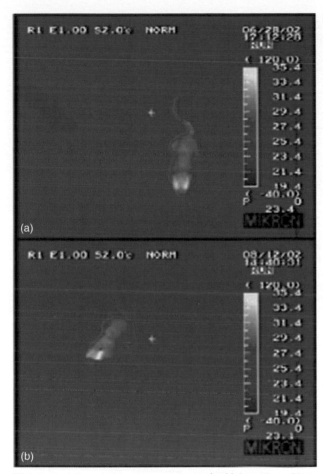

Figure 6   The 'tournament' behavior of the honey ant, *Myrmecocystus mimicus*. Many individuals from rival colonies engage in lateral displays, as between the two opponents here, vigorously antennating each others' bodies. These multimodal signals include visual, tactile/vibratory, and chemical components. From Holldobler B (1999) Multimodal signals in ant communication. *Journal of Comparative Physiology A* 184: 129–141. Photo by B. Holldobler; Copyright 1999 Springer, used with permission.

Figure 4   Infrared video frames of a squirrel interacting with a rattlesnake, showing heat emitted during tail flagging (a), and a gopher snake, with nonheat emitting tail flagging (b) during experimental trials. From Rundus AS, et al. (2007) *Proceedings of the National Academy of Science* 104: 14372–14376. Copyright 2007 National Academy of Science USA, used with permission.

Figure 7   Male *Schizocosa ocreata* wolf spiders signal to females with multimodal courtship displays that include visual signals (leg waving, tapping, conspicuous leg tufts) and seismic signals (substratum vibration produced by percussion and stridulation). Photo by George Uetz, used with permission.

Figure 5   The Tungara frog (*Physaelemeous pustulosus*), produces its characteristic acoustic signals by inflating the vocal sac, which also serves as a visual signal. Photo courtesy of Marc Dantzker, used with permission.

signals as containing 'multiple messages,' for example, while one mode provides information about species identity, the other signals male quality. When nonredundant sets of signal information are combined into a multimodal signal, there are a variety of different kinds of responses that receivers can show. One possibility, *signal independence*, occurs when the response to a multimodal signal includes the responses to each of its unimodal components.

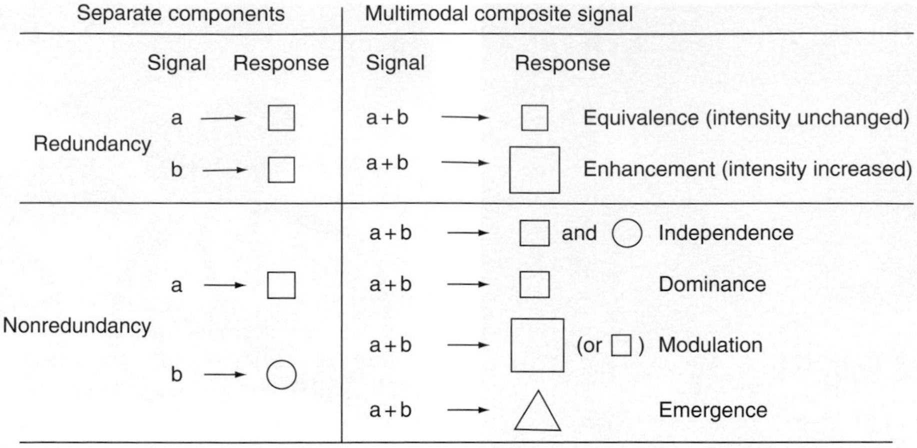

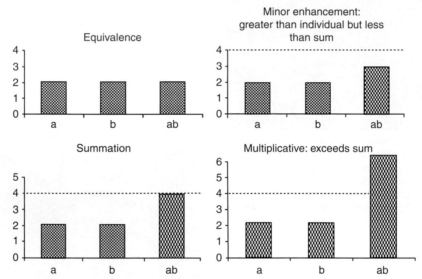

Figure 8 Classification of multimodal signals into categories. Each signal has components 'a' and 'b' (e.g., visual and acoustic), which can be redundant (upper section) or nonredundant (lower section). Geometric shapes symbolize responses of receivers to various signals when presented alone (left section), or when combined into a multimodal signal (right section). Size of geometric shapes indicates intensity of response, while different shapes indicate distinct responses. From Partan SR and Marler P (1999) Communication goes multimodal. *Science* 283: 1272–1273. Copyright 1999 American Association for the Advancement of Science, used with permission.

Figure 9 Idealized responses to redundant signal components 'a' and 'b,' as well as a multimodal signal, 'ab.' Responses can be equivalent (left) or enhanced in several ways (minor, summative, multiplicative). From Partan S (2004) Multisensory animal communication. In: Calvert G, Spence C, and Stein BE (eds.) *The Handbook of Multisensory Processes*, pp. 225–240. Cambridge, MA: MIT Press. Copyright 2004 MIT Press.

A second possibility is when the multimodal signal generates a response seen with only one of the component modes. Here the component generating the response is considered *dominant* in relation to other modes. In some cases, the presence of one mode may modulate the response to another, as when the visual motion of a courtship display increases responses to acoustic signals. Additionally, when signals in different modes are presented as multimodal signals, that signal may receive an entirely different type of response compared to the response elicited by any single component. This is called *emergence*. An example of this was found by Candy Rowe with

feeding studies of domestic chicks. Neither the color of food items nor the odor of toxic chemicals within them elicited a response, but when presented together, strong aversive behavior was seen.

Complexity in Multimodal Signal Components

While it is true that many animals create signals using two (or more) sensory modes, a closer look reveals that there may be even more to multimodal signals than simply the combining of communication modes. Whether or not the information contained by components of a multimodal

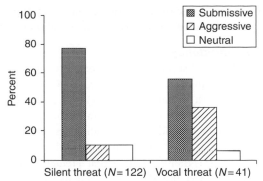

Figure 10 Proportions of responses of rhesus monkeys (submissive, aggressive, neutral) vary depending on whether a threat display is unimodal (silent – visual) or multimodal (visual accompanied by vocalizations). From Partan S (2004) Multisensory animal communication. In: Calvert G, Spence C, and Stein BE (eds.) *The Handbook of Multisensory Processes*, pp. 225–240. Cambridge, MA: MIT Press. Copyright 2004 MIT Press.

signal is redundant, multiple signals may still be present in each component mode. To return to a familiar example, the multimodal signal of male North American Wild Turkey combines a visual fan tail display of multicolored feathers and an accompanying acoustic 'gobble' signal. However, there are other visual signals provided by the size of the 'beard' (a set of modified hair-like feathers that protrude from the breast), spurs on the tarsi, and the coloration of the wattle or dewlap (a flap of skin below the neck), caruncle (skin flaps along the side of the head), and snood (a fleshy projection hanging over the beak) (**Figure 11**). In addition, the acoustic component not only includes the familiar 'gobble,' but also several other acoustic elements ('clucks,' 'putts,' 'purrs,' 'yelps,' 'cutts,' 'whines,' 'cackles,' and 'kee-kees') and a low-pitched drumming vibration. The various elements of this overall communication package have been shown to be associated with different aspects of the male's current physiological

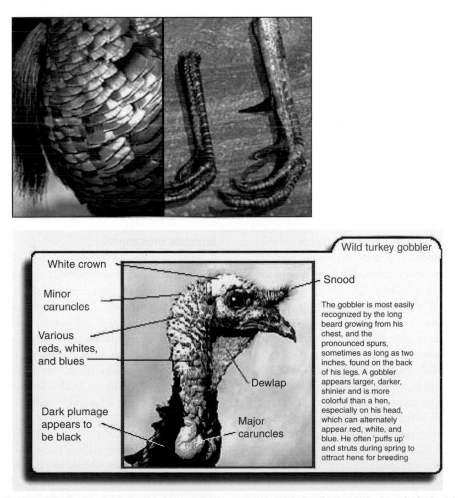

Figure 11 The American Wild Turkey exhibits complex, multicomponent visual signals that include a 'fan' of tail feathers, bright red caruncles on the side of the head, a beard of breast feathers, and a snood of flesh hanging over the beak, as well as multiple acoustic signals. Website – http://www.nwtf.org/all_about_turkeys/new_turkey_look.html

state (arousal, reproductive status), body condition or health, level of parasitic infestation, or dominance status. Moreover, these elements function in both male–male and male–female interaction contexts. Because the multiple individual elements that together make up multimodal signals consist of individual traits and behaviors controlled by different alleles, and function in different contexts, they may reflect different selection pressures and evolutionary pathways.

Function and Adaptive Value of Multimodal Signals

Courtship signals used by males have been studied more extensively than other signal types, and most researchers agree that they serve multiple functions in mate recognition and mate choice of females. Studies in many animal taxa suggest that male signals attract female attention, allow females to recognize their species, and also convey information about male dominance status, condition, or genetic quality. Multimodal signals may simultaneously serve some or all of these functions.

If One Sensory Channel Will Suffice, Why Use Multiple Channels?

Because the use of sensory channels varies among animal species, there is considerable diversity in signaling behavior. An important question to address then is why some species use multiple modes, whereas others use only a single mode. Some species may be limited to unimodal signaling by their phylogeny, while others may be constrained by their ecology. Entire groups of phylogenetically related animal taxa use the same mode of communication, suggesting common ancestry and evolution. For example, crickets and katydids use acoustic signals and have highly developed acoustic senses. Differences in song structure between species allow species recognition and mate discrimination. However, these insects are also nocturnal, which may preclude use of visual signals altogether.

One explanation for multimodal signaling is that the signal redundancy that is possible with multiple communication modes reduces mistaken identity at the species level or inaccurate assessment of suitability of a potential mate. A second possibility suggested by many researchers is that the use of multimodal signaling may compensate for variability in signal transmission under different sensory environments. Variability in signal transmission imposes constraints on signal detection, and creates the need for 'back up' signals. Another possibility is that different modes contain multiple messages and are useful in different contexts, for example, species recognition, territorial defense, male–male aggression, cooperation, male–female courtship, and mating. Finally, it is also possible that if senders can exploit multiple senses of the

receiver at the same time, the signal may be stronger and may result in increased receiver detection and learning.

Signal redundancy

Results of several studies have demonstrated redundancy in signal information. For example, Tim Birkhead has shown that both visual (beak color) and acoustic (song rate) courtship signals of male Zebra finches (*Taeniopygia guttata*) contain redundant information, as redder beaks and higher song rates signal better condition and higher mate quality. It is therefore likely that a female who is able to assess both visual and acoustic signals would receive more accurate information about male quality than with either mode alone. Redundant signaling may also prevent 'cheating,' as was demonstrated by Marlene Zuk, Dave Ligon, and Randy Thornhill in a study of red jungle fowl (*Gallus gallus*). They manipulated some male traits (color and size of comb) to exaggerate apparent quality, but females ignored those traits in favor of others when in conflict, suggesting that they were not fooled.

Different sensory environments

A subset of hypotheses regarding multimodal signaling suggest that redundant multimodal signals have evolved to compensate for environmental constraints on the efficacy of information transfer due to complex habitats and/or noise. A previous study with my colleagues Sonja Scheffer and Gail Stratton found that the vibratory component of male *Schizocosa ocreata* courtship was very quickly attenuated in complex leaf litter and could not be detected beyond the leaf upon which the male was courting. Nonetheless, females on leaves different from those of the males could still recognize males on the basis of visual cues alone. Likewise, Eileen Hebets found that male *S. retorsa* could successfully mate in the dark, as vibration signals alone were sufficient.

Multiple messages

A number of researchers have suggested that the individual components of multimodal signals each convey distinct information, often referred to as the 'multiple message hypothesis.' For example, studies of cardinal plumage, house finch coloration, and peacock displays have shown that different components of these complex signals may contain information on different aspects of male quality. Studies of the swordtail fish *Xiphophorus pygmaeus* by Mike Ryan, Molly Morris, and their colleagues suggest that each signal mode provides information about a different aspect of the sender. When visual signals are presented alone, females prefer heterospecific *X. cortezi* over conspecifics, presumably on the basis of size. However, when chemical cues are present, females more often choose their own species. Moreover, when an additional visual cue (vertical bars) is added, females almost always choose their own species, suggesting that multimodal signals are critical to ensuring correct species mating. These findings suggest that multimodal (chemical and visual) signals allow species identification, while visual cues alone may provide information on male quality.

Increased receiver detection and learning

Tim Guilford and Marian Stamp-Dawkins coined the term 'receiver psychology' to explain how the sensory capabilities of the intended receiver have shaped the evolution of signal design. A number of studies suggest that multimodal signals improve the detectability, discriminability, and memorability of signals. This can be critically important for insects that use antipredator strategies involving aposematic coloration to advertise unpalatable defensive chemicals, as the effectiveness of the strategy depends upon how well predators recognize and remember unpalatable prey. For example, Candy Rowe has examined the role of combined stimuli by studying the learning of food palatability in domestic chicks, and found that multimodal combinations of olfactory (chemical) and visual (color) stimuli increase the speed of avoidance learning and retention of learned aversions over unimodal stimuli. Dan Papaj and colleagues have examined multimodal cues and foraging in bumblebees, and found that bees make more effective decisions when multiple cues from flowers were present, and that multimodal flower 'signals' (odor and color) were associated with consistently higher accuracy in food finding.

Reliability of information

Even if multiple modes allow redundant (or 'back-up') cues for species recognition under environmental conditions that may occlude or constrain unimodal signals, communication in a single mode (e.g., visual) may be more important in another context (e.g., male–male conflict, mate choice). For some *Schizocosa* wolf spider species, individual channels (vibration or visual) may be more reliable as 'honest' indicators of condition as well as species identity. Gail Stratton, Sonja Scheffer, Will McClintock, Eileen Hebets, Jeremy Gibson, and I have found that variation in either visual male secondary sexual characters (presence/absence, size and symmetry of foreleg tufts) or seismic signals (stridulation, percussion) separately influence female receptivity (see **Figure 5** – *S. ocreata*). Even so, when females were able to perceive both visual and vibratory signals together, absence of tufts did not affect receptivity.

Potential Costs of Multimodal Signals

While most studies of multimodal signaling have focused on the benefits of multiple versus single modes, there may be costs as well; so cost–benefit trade-offs have likely influenced the evolution of multimodal signals. In their extensive review, Eileen Hebets and Dan Papaj have pointed out that for many animals, multimodal signaling is akin to 'multitasking,' for which energetic expenses would be expected to be greater than a unimodal signal. Alternatively, the energy expended producing a signal in one mode may reduce the animal's capacity to expend energy in another mode.

Multimodal signaling could incur other fitness costs in addition to potential energetic expense. Multimodal signals that are evolved to be conspicuous and easily recognized by intended receivers often inadvertently become 'public information,' and as a consequence, may be intercepted and exploited by other, unintended receivers (also known as 'illegitimate' or 'illicit' receivers). Usually, there are two categories of signal exploiters: (1) 'social eavesdroppers' such as conspecific males, which can potentially exploit the information content of intercepted signals (e.g., potential strength of rivals, location of females) and (2) 'interceptive eavesdroppers' or 'cue-readers,' such as predators or parasites, for which the signal itself reveals the location of potential prey/host but information content of the signal is unimportant.

A variety of animals are known to eavesdrop on the interactions of conspecifics, and thereby gain knowledge of the relative competitive ability of males or the mate choice preferences of females. Mathieu Amy and Gerard Leboucher studied eavesdropping in male domestic canaries, and found that they use both visual and acoustic cues to eavesdrop and responded differentially to winners and losers of agonistic interactions they had witnessed. Research with my colleagues Andy Roberts and Dave Clark indicates that male *S. ocreata* wolf spiders show evidence of eavesdropping, as they can discern the presence of another courting individual (**Figure 12**). However, eavesdropping on either visual, seismic, or multimodal signals from other males does not always result in social facilitation of courtship behaviors by the eavesdropper, but appears to depend on experience and density of courting males.

There are a number of examples of predatory and parasitic species using acoustic or visual cues to locate prey. Although predator detection of signals produced by prey or host species can be highly specific (e.g., parasitic flies of the genus *Ormia* attracted to mating calls of host cricket species), not all signal modalities are equally capable of exploitation (e.g., chemical signals are less detectable). In this case, multimodal communication may be a two-edged sword. There may be benefits of using redundant signals in different modes, in that they may deceive predators, but multimodal cues might make the sender more conspicuous. Using video/seismic playback, Andy Roberts, Phil Taylor, and I found that use of complex, multimodal courtship signals by *S. ocreata* increases the speed with which a common predator, the jumping spider *Phidippus clarus* (Salticidae), responds to courting males.

Our results indicated that the benefits of increased signaling efficacy of complex, multimodal signaling may be countered by increased predation risks.

Integration of Signaler Behavior and Receiver Sensory Perception

In order to understand the evolution of male courtship signals, it is necessary to assess the interaction of multiple sensory modes and signaling behaviors. Recent studies have attempted to integrate aspects of sensory biology and signal design theory with behavioral and ecological considerations to gain a better understanding of the evolution of courtship signals in mate choice. Some researchers have focused on what Tim Guilford and Marian Stamp-Dawkins call 'strategic design' (relating to the function of the signal, e.g., conveying male quality) or the fitness-related benefits of mate choice related to male signals (reproductive success, offspring survival). Others have addressed 'tactical design'

(relating to the effectiveness of the signal, e.g., getting female attention). The conclusion of much of their work is that the evolution of sensory systems, signal production, and behavioral responses are coupled, because when animals produce a signal to carry information in one modality, that behavior often results in additional cues that are perceived in other sensory modes, and thus a multimodal signal arises. As a consequence, multimodal signals are ultimately integrated messages. In a recent review, Eileen Hebets and Dan Papaj have suggested that understanding the evolution of complex multimodal signals requires a focus on proximate explanations (efficacy-driven hypotheses) for multimodal signals, as well as studying the interactions between signal modes (how the presence or absence of signals in one mode affects the perception of signals in other modes).

How to Study Multimodal Communication

To tease apart signal elements, and determine their effectiveness when presented alone and together, research on multimodal communication has involved several experimental techniques. One common technique involves isolating individual modes or channels of multimodal signals to observe responses of receivers to single versus combined modes. Cue isolation (single sensory modes) experiments allow manipulation of the sensory environment so that only one type of stimulus is received. For example, studies of wolf spiders by my colleagues Eileen Hebets, Phil Taylor, and Andrew Roberts and I have tested responses of females to unimodal cues from courting males (e.g., eliminating seismic signals on isolated substrates, eliminating visual cues with opaque barriers or darkness), and compared them to cue-combination experiments (selected multiple modes presented together) (**Figure 13**).

The use of video/audio digitization and playback (**Figure 14**) has created a number of opportunities for

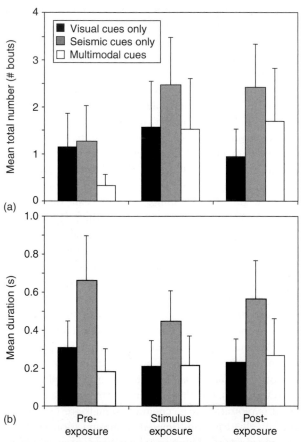

Figure 12 Responses of 'eavesdropping' male wolf spiders. (a) Mean total number (+SE) and (b) mean duration (s) (+SE) of bouts of jerky tap behavior during stimulus exposure periods for male *Schizocosa ocreata* exposed to live, courting and noncourting, male spider stimuli. From Roberts JA, Galbraith E, Milliser J, Taylor PW, and Uetz GW (2006) Absence of Social Facilitation of Courtship in the wolf spider, *Schizocosa ocreata* (Hentz) (Araneae: Lycosidae). *Acta Ethologica* 9: 71–77. Copyright 2006 Springer, used with permission.

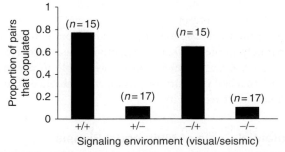

Figure 13 Example of cue isolation experiment from Hebets EA (2008) Seismic signal dominance in the multimodal courtship display of the wolf spider *Schizocosa stridulans*. From Stratton (1991) *Behavioral Ecology* 19: 1250–1257. Copyright 2008 Oxford University Press, used with permission.

the experimental study of multimodal signaling. Recent studies have used these techniques to manipulate sensory modes as well as various aspects of visual or acoustic/ vibratory signals. For example, Chris Evans and Peter Marler used digital video to create a virtual 'audience' (a video of another bird) as well as a virtual predator (hawk) flying above for a caged domestic chicken. They found that both visual and acoustic cues from the audience elicited more alarm calls in the presence of the hawk stimulus, but that multimodal cues increased the rate of alarm calling. Sarah Partan and colleagues have used audio playback alone, silent video playback, and both in combination to study the responses of female pigeons to male courtship behavior. They found that while audio playback elicited higher levels of female courtship 'cooing' responses than silent video, multisensory audio/video signals were more effective than either component presented alone in eliciting full female receptivity response.

My colleague Andrew Roberts and I have used video–audio playback in our studies of two wolf spiders, using digital recordings of male courtship to manipulate both the video and audio (seismic) components. We used cue-conflict (mixed conspecific/heterospecific components) experiments to identify which signals were important in species recognition. Female *Schizocosa ocreata* (a species with multimodal signals) responded most strongly when both video and audio playback from conspecific males were presented, and responded negatively to heterospecific signals. In contrast, *S. rovneri* (a species with unimodal seismic signaling) responded to audio/ vibration signals regardless of the video image (**Figure 15**).

Recent advances have enabled the use of robotic animals in experimental studies of animal communication, and a number of researchers have created them to mimic squirrels (Sarah Partan), birds (Gail Patricelli, Esteban Fernandez-Juricic), fish (Jens Krause), lizards (Emila Martins, Terry Ord, Chris Evans, Dave Clark,

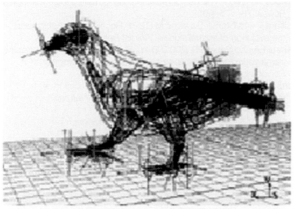

Figure 14 Playback of digital images – either rendered videos of real animals or 'wireframe' animations (below) – as visual stimuli alone or accompanied with acoustic cues, are gaining increasing use as experimental stimuli in the study of multimodal signaling behavior. See the 'Virtual Pigeon' website for more examples. http://psyc.queensu.ca/%7Efrostlab/neuro_eth.html; http://helios.hampshire.edu/~srpCS/Research.html

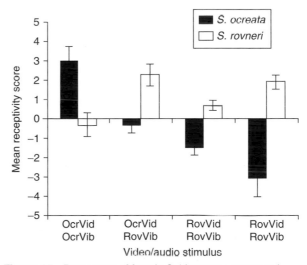

Figure 15 Responses of female *Schizocosa ocreata* and *S. rovneri* to video/audio playback of conspecific and mixed species visual/seismic signals. From Uetz GW and Roberts JA (2002) Multi-sensory cues and multi-modal communication in spiders: Insights from video/audio playback studies. *Brain Behavior and Evolution* 59: 222–230. Copyright 2002 Karger Press, Basel; used with permission.

Joe Macedonia), and frogs (Peter Narins, Mike Ryan). These studies have successfully demonstrated that animals recognize these simulacra as conspecifics, and respond accordingly, allowing experimental playback studies of multimodal and unimodal signals in the field.

Conclusion

Multimodal signaling in animals is a complex topic, and future researchers in animal behavior will find many research challenges. While a traditional (and highly useful) approach has been to study animal communication by partitioning modalities, this approach has tended to isolate disciplines and reduce scientific communication. Given the current explosion of integrative and interdisciplinary approaches to scientific questions in behavior, enabled by technological advances in neurobiology and genomics, opportunities for new research in multimodal communication seem limitless.

See also: Acoustic Signals; Electrical Signals; Mating Signals; Olfactory Signals; Robotics in the Study of Animal Behavior; Sound Production: Vertebrates; Vibrational Communication; Visual Signals.

Further Reading

Candolin U (2003) The use of multiple cues in mate choice. *Biological Review* 78: 575–595.

Guilford T and Stamp-Dawkins M (1991) Receiver psychology and the evolution of animal signals. *Animal Behaviour* 42: 1–14.

Hebets EA and Papaj DR (2005) Complex signal function: Developing a framework of testable hypotheses. *Behavioral Ecology and Sociobiology* 57: 197–214.

Partan SR and Marler P (2005) Issues in the classification of multimodal communication signals. *American Naturalist* 166: 231–245.

Rowe C (1999) Receiver psychology and the evolution of multicomponent signals. *Animal Behaviour* 58: 921–931.

Uetz GW and Roberts JA (2002) Multisensory cues and multimodal communication in spiders: Insights from video/audio playback studies. *Brain Behavior and Evolution* 9: 222–230.

Naked Mole Rats: Their Extraordinary Sensory World

T. J. Park, University of Illinois at Chicago, Chicago, IL, USA
G. R. Lewin, Max-Delbrück Center for Molecular Medicine, Berlin, Germany
R. Buffenstein, University of Texas Health Science Center at San Antonio, San Antonio, TX, USA

Naked Mole-Rat Ecology and Sociality

Naked mole-rats (Rodentia; Bathyergidae, *Heterocephalus glaber*) are hystricognath rodents, naturally found in the hot tropical regions of the horn of east Africa (Kenya, Ethiopia, and Somalia). The hystricognath suborder of rodents holds the records for both the largest-living rodent, the capybara (~50 kg), and the longest-living rodent, the naked mole-rat (~30 years). The naked mole-rat, as its name suggests, lacks a furry pelage. When it was first caught in 1842 by the famous naturalist Eduard Rüppell, he suspected he had a diseased specimen that had lost its hair and named it after its odd-shaped head (Hetero-different, cephalus – head) and smooth skin (glaber). Indeed, it was only after several specimens were collected that it became apparent that the naked mole-rat's weird unusual mammalian appearance resembling a saber-toothed sausage or a miniature walrus was normal.

Naked mole-rats lead a strictly subterranean existence. Earliest fossil records for the Bathyergid mole-rat family reveal that their ancestors have lived below ground since the early Miocene, more than 24 Ma. Not surprisingly, they have evolved a set of characteristics highly suited to life in dark, dank underground burrows: living in the dark, they have lost the ability to see beyond being able to tell light from dark, and they have very small eyes that they often do not even bother to open. As can be seen in **Figure 1**, naked mole-rats have a streamlined cylindrical shape, with no external ear pinnae, and the males also have internal testes; so it is extremely difficult to distinguish males from females within the colony. They also have very short limbs that enable them to run backwards and forwards with equal speed and use hairs that are very sensitive to touch (vibrissae) located on both their tail and face to detect objects in their path and sense their underground environment.

Naked mole-rats live in an extensive maze of underground tunnels and chambers up to 8 feet (2.5 m) beneath the soil surface. They dig these tunnels using their chisel-like, ever-growing incisor teeth that are actually situated outside the mouth (extra buccal), and large and powerful masseter jaw muscles. When digging with their teeth, their lips are actually closed, preventing soil from entering the mouth (the teeth grow through the lips). They kick the loosened excavated sand to the surface where the sand forms small volcano-shaped mounds, which are the only above ground signs that the mole-rats are living below ground. The main 'highways' underground are extended during the rainy season when soils become softer; during the rest of the time, the mole-rats rarely dig burrows, digging only the more superficial ones to find roots, tubers, and small onion-like bulbs to eat.

Because rainfall is infrequent, unpredictable, and patchy in these arid regions of northeast Africa, food resources have a patchy distribution, and consist of either clumps of small bulbs, corms, and occasionally, a single huge tuber that may weigh as much as 50 kg. These foods not only supply all the energy the mole-rats need, but crucially all their water too, for mole-rats do not naturally drink and have no access to fresh water underground. When naked mole-rats encounter large tubers, they carefully eat only the nonpoisonous inner components that are packed with nutrients and leave the epidermis intact. After feeding, they pack soil back into the holes made in the tuber, so that it continues to be healthy and can regrow and serve as a constant food supply for many years. Because animals cannot see where to forage and dig blindly hoping to come across food they can eat, it is an extremely energetically costly process. Living in large groups and dispersing widely to forage help this highly social species locate underground food resources and improve the chances of finding food and thus, foraging efficiency. Food when found is often carried back to the nest to be shared with smaller siblings and other members of the colony.

Like the eusocial insects (bees, ants, wasps, and termites), naked mole-rats exhibit complex social behavior;

Figure 1 Naked mole-rats have small eyes and small external ears but prominent whiskers and teeth.

animals live in large family groups of up to 300 individuals. The colony shows a marked reproductive division of labor, in that only a single reproductive female is solely responsible for the production of offspring. Several litters of different ages are present within the colony and most individuals will stay and work in the colony in which they were born throughout their long lives. The reproductive female retains her position as the sole breeder in the colony primarily through aggressive behavior and bullying. Any female in the colony has the potential to breed, but will remain in a prepubescent state until the breeding female dies or she is removed from the colony. In that situation, aspiring females will fight to death to establish dominance, and then the dominant will continue to breed throughout her long life. The breeding female may have as many as 30 pups (average 12 per litter) in a single litter, and produces up to four litters per year. The breeding female breeds with one to three males; all other non-breeding individuals in the colony have very low reproductive hormone levels. The nonbreeding animals within a colony assist with the direct care of the pups by huddling with them in the nest to keep them warm, and retrieving pups that wander from the nest, and they will also carry pups out of the nest in response to alarm calls. As pups are weaned, the nonbreeding animals practice another type of cooperative behavior, allocoprophagy. Allocoprophagy is where pups feed on the feces of other members of the colony members by begging for feces from adults. Consumption of fecal material from adults provides an inoculation of microfauna and flora that the pups will need to assist with the digestion of the high fiber content of their herbivorous diet.

Nonbreeding animals do most of the day-to-day maintenance of the colony. Besides foraging and carrying food back to a communal food cache, they cooperatively dig the burrow systems in digging chains or assembly lines, in which the animal at the front of the chain excavates the burrow and each of the animals lined up behind it sweeps dirt backwards until the last animal in the chain kicks the dirt out of a temporary opening at the surface. Similarly,

these individuals excavate out a latrine or toilet chamber and may maintain it by kicking in fresh soil, or by blocking it with soil when it is full.

Living in large groups underground, naked mole-rats are exposed to high levels of carbon dioxide and low levels of oxygen. The naked mole-rats show many adaptations to this underground air composition; animals have low metabolic rates to reduce oxygen consumption and their blood respiratory properties include a high oxygen affinity hemoglobin to extract as much oxygen as possible from the atmosphere; and pronounced acid base buffering capacity to neutralize an carbonic acid formed in blood from the dissolving of carbon dioxide.

Animals also maintain a low body temperature and pronounced tolerance of thermolability such that body temperature closely tracks that of the surrounding temperatures. These latter features help minimize overheating in an environment where the humidity is close to 100%, and this high humidity prevents the use of evaporative cooling. By having a body temperature similar to the environment, the naked mole-rat also does not need to spend large amounts of energy on thermoregulation like other mammals do. They rather rely upon the fact that temperatures in equatorial regions show little variation both daily and seasonally and that temperatures in underground burrows are high enough for them to function normally.

The subterranean habit also imposes constraints on other aspects of naked mole-rat physiology. Living in the dark with no cues to set circadian rhythm, most individuals have a free-running activity pattern and are active both day and night, sleeping for short periods of time several times a day. Furthermore, as light is essential for the formation of vitamin D in skin, it is not surprising that this species is naturally deficient in the principal circulating metabolite, 25-hydroxy vitamin D of this important hormone. They also do not acquire vitamin D through their diet as they do not consume animal products and are strictly vegetarian, eating only roots and tubers. Despite this naturally deficient vitamin D status, calcium metabolism is unaffected and calcium absorption from the diet is via a highly efficient, vitamin D-independent, passive process. These animals dump excess calcium that they have absorbed in their teeth and bones, and use these as calcium reservoirs should they need calcium for other functions. Not surprisingly, their bones and teeth have high calcium content and are extremely strong.

Somatosensory (Touch) Specialists

Because of their lightless environment, it is no surprise that many subterranean animals rely heavily on nonvisual sensory modalities to communicate and maneuver about in their complex tunnel systems. In fact, although there is wide variation across species, the visual systems of many subterranean animals are considered degenerate as far as their ability to detect objects is concerned. This is indeed the case for naked mole-rats. Naked mole-rats have very small eyes, and the region of brain cortex that usually receives inputs from the visual system has been taken over by somatosensory inputs. However, the visual system does retain the ability to detect changes in luminance, which can trigger escape responses.

Naked mole-rats also show a very poor ability to localize sounds using acoustic cues. Like many subterranean mammals, naked mole-rats predominantly produce and hear low frequency vocalizations (below 8000 Hz) which sound to us very much like bird cheeps (for sound localization, 'low' frequencies are considered to be those frequencies with wave lengths longer than the ear to ear distance, about 10 000 Hz and below for naked mole-rats). Low frequencies are actually well suited for auditory-vocal communication underground because these sounds travel along tunnels and through soil much better than high frequency sounds. However, for animals with small heads like naked mole-rats, low frequencies are not good for determining the directionality of a sound. The lack of external ear structures also hinders good sound localization. In contrast to subterranean mammals, small-headed surface dwelling mammals with well-developed external ears have good high frequency hearing which they use to accurately localize sounds.

It should be noted that even though naked mole-rats are poor sound localizers and have elevated auditory thresholds in general, their close range hearing is acutely sensitive and they use a sophisticated auditory-vocal communication system. Naked mole-rats utter at least 17 different vocalizations, they frequently call back and forth to one another, and they can use vocalizations to signal that they have located a food source and to identify animals and their position in the social hierarchy.

If naked mole-rats cannot use their visual or auditory systems to localize objects, then how might they determine the location of objects (other mole-rats, food, tunnel walls) in their burrows and orient themselves appropriately to those objects? Nonvisual and nonauditory cues that could be used for the basic tasks of spatial orientation, could include magnetic sensation, olfaction, somatosensation (touch), or any combination of these senses. However, a striking physical feature about naked mole-rats – a systematic array of sensitive sensory hairs on their bodies – suggests that touch might play a major role.

As mentioned earlier, the fine hairs that make up fur on most mammals are completely absent from naked mole-rats. However, these animals are not entirely hairless. In addition to having facial vibrissae (whiskers), they also display a grid-like pattern of sensory vibrissae across the entire body, which look very much like typical facial whiskers. A detailed anatomical analysis of these vibrissae

indicated that they are very large guard hairs. There are about 80 of these body vibrissae on a naked mole-rat's body, and the patterning of the hairs is fairly consistent from individual to individual. Compared to the guard hairs found on similarly sized animals (like the laboratory rat or common mole-rat, *Cryptomys hottentotus*), the guard hairs on the naked mole-rat are far fewer in number, and the hairs and follicles are much larger in size, being almost as stout as the whiskers found on the face. The body vibrissae and their locations on the body can be seen in the photographs and drawings in **Figure 2**.

Testing touch-guided orientation behavior showed that naked mole-rats are remarkably adept at orienting toward

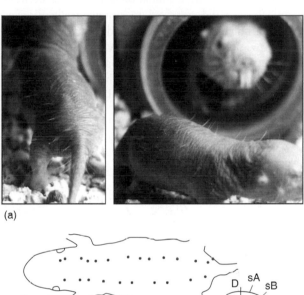

(a)

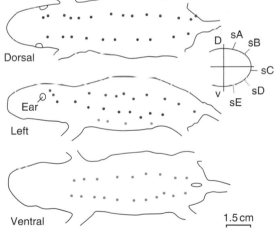

(b)

Figure 2 Naked mole-rats have rows of whisker-like sensory vibrissae on their bodies. (a) Body vibrissae as well as facial vibrissae (whiskers) are easily identified in these photographs. (b) Each naked mole-rat has ten rows of body vibrissae. The schematic indicates the placement of body vibrissae on one example animal. There are two rows of body vibrissae on the dorsal surface, two rows on the ventral surface, and three rows on each side (only the left side is shown here). **Figure 2(b)** reprinted from Crish SD, Rice FL, Park TJ, and Comer CM (2003) Somatosensory organization and behavior in naked mole-rats I: Vibrissa-like body hairs comprise a sensory array that mediates orientation to tactile stimuli. *Brain, Behavior and Evolution* 62: 141–151, with permission from S. Karger AG, Basel.

the point of contact with even a single body vibrissa. Orientation was tested by manually deflecting single vibrissa, or by vibrating one or more vibrissae via an electromagnetic field, and recording behavioral responses on video tape. Deflection of a single vibrissa triggered a pronounced orientation of the snout toward the point of stimulation, and the ability to evoke such an orientation response was extremely reliable (95% of trials). The drawings presented in **Figure 3** are reconstructions from video-taped test trials and they illustrate the type of responses recorded. Note that the angular orientation of the head axis at the completion of responses, relative to the original body axis, increased systematically as the stimulus position varied along the body. Stimulation of the skin in-between body vibrissae did not lead to systematically directed orienting which indicates that the body vibrissae represent specialized points of touch sensitivity, and with a function that – at least for some stimuli – parallels that of facial whiskers. Indeed, physiologically, the sensory hairs are quite unlike those found on the bodies of other rodents.

The sensory nerve fibers that innervate such hairs have two unusual features; first the hair has directional sensitivity so that moving the hair towards the animal's head produces optimal discharge. Second, responses from hair movement are slowly-adapting so that sensory neurons continue to fire action potentials during static displacement of the hair. Both of these features have been described for nerves on the facial whiskers of other species.

An interesting observation made during these experiments was that stimulation of vibrissae on the tail also triggered orientation responses that brought the snout near the point of stimulation. However, orientation to tail vibrissae involved a mixed repertoire of response patterns. About half of the time the animal rotated nearly 180° so that the snout was pointing toward the original position of the tail. However, the other half of the time, instead of rotating the head and torso toward the point of stimulation, the animal stepped straight backwards, stopping at a position that placed the snout in close proximity to the original position of the tail. Hence, two very different motor

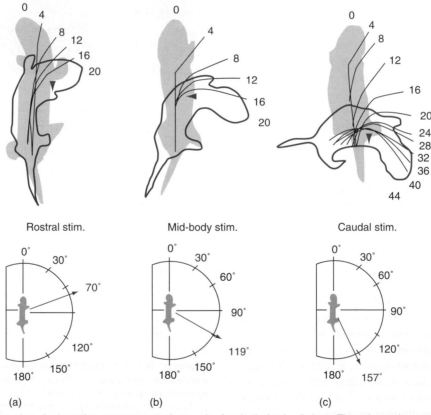

Figure 3   Three examples of orientation responses to the touch of a single-body vibrissa. The schematics were reconstructed from video tapes. In (a), a vibrissa near the shoulder (rostral) was stimulated, in (b), a vibrissa located along the mid-body was stimulated, and in (c), a vibrissa near the hip (caudal) was stimulated. In each schematic, the gray silhouette indicates the position of the body and head just prior to stimulation. The blue arrowhead indicates the location of the vibrissa that was stimulated. The lines labeled 0, 4, 8, etc. indicate the body and head axes at every fourth video frame after stimulation. The red outline indicates the position of the body and head at the completion of the orientation movement. The red numbers below indicate the head angle at completion relative to the head angle at the moment of stimulation. Reprinted from Crish SD, Rice FL, Park TJ, and Comer CM (2003) Somatosensory organization and behavior in naked mole-rats I: Vibrissa-like body hairs comprise a sensory array that mediates orientation to tactile stimuli. *Brain, Behavior and Evolution* 62: 141–151, with permission from S. Karger AG, Basel.

patterns could be initiated to accomplish the same goal of bringing the snout to bear on the point of stimulation. In their tunnels, naked mole-rats, like other mole-rat species, spend a considerable amount of time walking and running backwards, so their orientation behavior is well adapted in topography to their 'tubular world.'

The results presented above demonstrate that naked mole-rats can use their sparse array of body vibrissa to accurately orient toward discrete points of contact. It is yet to be determined if the degree of accuracy and reliability they display are typical of other mammals (certainly, humans can be keenly aware of even a small deflection of a hair on the back of the hand). Perhaps the most remarkable thing about the naked mole-rats' orientation behavior to vibrissa deflection is its utility as a model system for studying nonvisual sensory-motor integration since the number of sensors (vibrissae) is tractable, they are spaced widely apart, and the behavioral output is robust. For example, stimulation of two hairs on the same side of the body generates an averaged orientation response, as illustrated in **Figure 4**, similar to what is seen for visually and acoustically guided orientation behavior in other mammals. These data represent the first demonstration of this phenomenon for the somatosensory system. Interestingly, when two hairs on opposite sides of the body were simultaneously stimulated, animals oriented to either one side or the other (not averaging).

The body vibrissae of naked mole-rats may function in a variety of behaviors in nature. For example, they have been shown to play a role in learning a maze task. They may also play a role in functions, such as sensing changes in air flow, that are yet to be tested.

Insensitivity to Chemical Irritants

The body vibrissae of the naked mole-rat act as very sensitive touch detectors. In contrast, another aspect of this species' somatic sensory system is extremely insensitive: its response to chemical irritants. Both the skin and the upper respiratory tract of naked mole-rats are completely insensitive to specific irritants, including acid, ammonia, and capsaicin (the spicy 'hot' compound found in chili peppers). At least one aspect of this insensitivity appears to be quite adaptive, since naked mole-rats are exposed to acidic conditions every day in their burrows.

As mentioned earlier in this encyclopedia entry, naked mole-rats have a very unusual combination of ecological and social characteristics. They are fully subterranean, extremely social, and they live in colonies with many individuals. In other words, naked mole-rats live in large numbers in very tight quarters and in very poorly ventilated spaces where respiration depletes oxygen and increases carbon dioxide (CO_2) to extremely high levels. Usually,

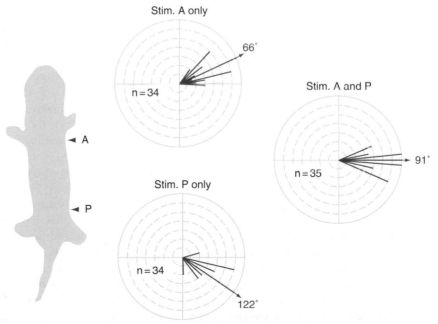

Figure 4 When two body vibrissae on the same side of the body are stimulated at the same time, the nervous system computes an averaged output. The three circular graphs show the distributions and average turning angles from the stimulation of an anterior vibrissa near the shoulder (Stim. A only), a posterior vibrissa near the hip (Stim. P only), or both simultaneously (Stim. A & P). Data represent 34 or 35 repetitions for each stimulus configuration. Note that orientation movements to simultaneous stimulation brings the snout to a position half way between the two points of stimulation. Reprinted from Crish SD, Dengler-Crish CM, and Comer CM (2006) Population coding strategies and involvement of the superior colliculus in the tactile orienting behavior of naked mole-rats. *Neuroscience* 139: 1461–1466, with permission from Elsevier.

these conditions would challenge an animal's ability to extract oxygen from the air and to maintain an appropriate acid–base balance. Also mentioned earlier, naked mole-rats have adaptive mechanisms to help deal with these challenges, including hemoglobin with a very high affinity for O_2, and blood with a high capacity to buffer CO_2.

Breathing high levels of CO_2 not only challenges the body's ability to maintain an appropriate acid–base balance, it also induces pain in the eyes and nose due to the formation of acid on the surface of those tissues as they come into contact with CO_2. To give some perspective, the concentration of CO_2 in room air is about 0.03%, and CO_2 levels in naked mole-rat tunnels is about 2%. Higher concentrations ($\sim$5%) have been measured in their nest chambers in the laboratory, and it is likely that concentrations are much higher in their natural nest chambers in Africa.

Recent experiments have shown that naked mole-rats have sensory adaptations that make them insensitive to stimulation of the nerve fibers that normally respond to high levels of CO_2, and other specific chemical irritants such as capsaicin and ammonia. The nerve fibers that respond to these irritants belong to a class of fibers called C-fibers. They are small in diameter, unmyelinated, and they release neuropeptides, notably Substance P and calcitonin gene-related peptide, onto their targets in the central nervous system. These are the nerve fibers that convey the stinging, burning sensation we experience when sniffing ammonia fumes or rubbing our eyes after handling hot chili peppers, and the neuropeptides they release are thought to be critical in signaling the unpleasant quality of irritants.

Remarkably, naked mole-rats naturally lack these neuropeptides from the C-fibers innervating their eyes and nose (and skin, described below). Behaviorally, the animals show no signs of irritation or discomfort from applying capsaicin solution to their nostrils, whereas in mice, capsaicin induces vigorous rubbing of the nose. Naked mole-rats also fail to avoid strong ammonia fumes. When placed in an arena with sponges that are saturated with ammonia or water, naked mole-rats spend as much time in close proximity to the ammonia as they do to the water. Rats and mice tested in the same procedure enthusiastically avoid the ammonia. These data, and a schematic of the testing arena, are shown in **Figure 5**. Interestingly, naked mole-rats do show an aversion to a different irritant, nicotine fumes, which act on a population of sensory fibers that are distinct from classical C-fibers.

The remarkable insensitivity that naked mole-rats display is not limited to their eyes and nose, but it extends to their skin as well. Naked mole-rats show no response to capsaicin solution or acidic saline (the strength of lemon juice) injected into the skin of the foot. The same irritants cause rubbing and scratching at the injection site in humans and vigorous licking in rats and mice. Responses of naked mole-rats and mice to capsaicin and acidic saline

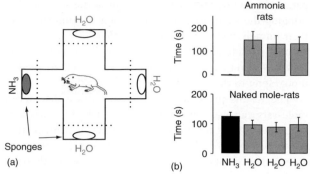

Figure 5 Naked mole-rats do not avoid ammonia fumes (considered to be a chemical irritant). (a) Shows a schematic of the testing arena used to measure avoidance. The arena included sponges saturated in ammonia (NH_3) or water (H_2O). Animals were free to move about while the time they spent near each sponge was recorded. The total testing duration was 20 min for each animal. (b) Shows the total amount of time spent near each sponge for laboratory rats and naked mole-rats. The time spent near the ammonia-soaked sponge is indicated by black bars labeled NH_3, while the time spent near each of the water-soaked sponges is indicated by grey bars labeled H_2O. Note that laboratory rats spent virtually no time near the ammonia sponge, while naked mole-rats spent as much time near the ammonia sponge as they did near the water sponges. Redrawn from a figure in LaVinka PC, Brand A, Landau VJ, Wirtshafter D, and Park TJ (2009) Extreme tolerance to ammonia fumes in African naked mole-rats: Animals that naturally lack neuropeptides from trigeminal chemosensory nerve fibers. *Journal of Comparative Physiology A, Neuroethology, Sensory, Neural, and Behavioral Physiology* 195(5): 419–427.

are shown in **Figure 6**. It is noteworthy that naked mole-rats do not have a complete loss of neuropeptides as their viscera appear to retain the full complement.

Surprisingly, physiological studies revealed that naked mole-rat C-fibers themselves respond to capsaicin. This finding suggested that the lack of neuropeptides, which would normally be released onto spinal neurons, acted to 'disconnect' the C-fibers from the central nervous system, preventing the brain from sensing irritation. To test this hypothesis, one of the missing neuropeptides was introduced into the spinal circuitry using two techniques. The first was gene therapy, where Substance P was introduced into the nerve fibers of the foot. This was done by applying a transgenic virus engineered to carry the DNA for Substance P. The second technique involved directly infusing Substance P into the spinal cord at the level where nerve fibers from the foot enter. In both cases, the introduction of Substance P caused naked mole-rats to respond behaviorally to capsaicin: after treatment, the animals licked at the injection site similarly to rats and mice.

Physiological studies also revealed another surprise. In contrast to their response to capsaicin, C-fibers in naked mole-rats were completely unresponsive to acidic saline. Consistent with this finding, introducing Substance P had no effect on acid insensitivity behavior – the mole-rats remain impervious to acidic saline injection.

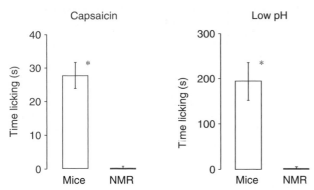

Figure 6 Naked mole-rats are immune to chemical irritants injected into the skin. A small amount of capsaicin solution (similar to chili pepper juice) or acidic saline (similar to lemon juice) was injected into the skin of one paw. Mice responded to both of these chemical irritants by licking the paw. Naked mole-rats showed virtually no response. The bars indicate the average amount of time spent licking. Reproduced from Park TJ, Lu Y, Jüttner R, et al. (2008) Selective inflammatory pain insensitivity in the African naked mole-rat (*Heterocephalus glaber*). PLoS Biology 6(1): e13.

Anatomical studies revealed yet another anomaly about the C-fibers of naked mole-rats. C-fibers in naked mole-rats have an unusual pattern of connectivity in the spinal cord. Almost half of the cells in the deep dorsal horn of the spinal cord receive direct (monosynaptic) connections from capsaicin-sensitive C-fibers, whereas in other species, almost all capsaicin-sensitive C-fibers terminate in the superficial dorsal horn. The significance of this unusual connection pattern is not clear, but it suggests that whatever signals might be conveyed from the C-fibers might not follow the usual irritant pathways once they reach the spinal cord.

Taken together, it appears that the C-fiber system of naked mole-rats has multiple mechanisms that make the system insensitive to specific irritants. Again, the working hypothesis is that these mechanisms are adaptations that have evolved for living under high CO_2, and therefore acidic, conditions that would otherwise cause chronic irritation of the eyes and nose. It is unclear if there are adaptive advantages to insensitivity in the skin. The extension of insensitivity to the skin may be an epiphenomenon. The trigeminal C-fibers that innervate the eyes and nose have a similar physiology and are developmentally orthologous to C-fiber sensory afferents in the dorsal root ganglia, which innervate the rest of the body. It is only speculation at this time, but it may be that adaptive changes in trigeminal C-fibers are necessarily reiterated in dorsal root ganglia C-fibers.

An Infantile Vomeronasal Organ

Studies on the neural structures of the nasal cavity revealed yet another sensory system anomaly in naked

mole-rats which may well be related to this species' lack of neuropeptides. The anomaly concerns the vomeronasal organ, a structure located at the base of the nasal cavity, separate from the olfactory area. In many mammals, and especially in rodents, the vomeronasal organ responds to pheromones and plays a critical role in mediating social and sexual behaviors. Contrary to what one might expect of a blind rodent, the vomeronasal organ in naked mole-rats is extremely small at birth, and it shows no postnatal growth, a characteristic that is unique among all rodents studied to date.

The unusual anatomical features of the naked mole-rats' vomeronasal organ suggest reduced or absent functionality, and this notion is supported by studies on sexual suppression, a function usually associated with the vomeronasal organ. Even though sexual suppression is extremely widespread among naked mole-rats (most animals remain nonreproductive throughout their extraordinarily long life), pheromones do not play a role in the suppression.

It may be that the apparent degenerate nature of the naked mole-rats' vomeronasal organ is related to this species' lack of neuropeptides, and the connection is quite interesting. The vomeronasal organ is a tubular structure encased in a fairly rigid capsule and it is equipped with a pump that acts to bring molecules into the organ where they interact with vomeronasal sensory cells. This differs from the olfactory epithelium which interacts with odorant molecules that are drawn in via respirations. The driving force for the vomeronasal pump is alternating dilations and constrictions of blood vessels and the dilations are triggered by Substance P which is released from trigeminal C-fibers (Substance P can be released both centrally, where it acts as a neurotransmitter, and peripherally, where it acts as a vasodilator). For now, it is still a theory, but it seems likely that the naked mole-rat's loss of neuropeptides may have disabled not only irritant detectors, but the vomeronasal pump as well, leading to degeneration of that organ.

A Sensory World Beyond Imagination?

Pitch dark, dank, fetid, fecal, foul-smelling, and seething with cool-bodied, lithe creatures equipped with razor-sharp, hyper-mobile teeth. The sensory world of the naked mole-rat does not seem all that appealing to sun-loving mammals like us. Yet, the fact that naked mole-rats are eusocial, a system reminiscent of brutal feudal hierarchies, should not tempt us to anthropomorphize their sensory world. Naked mole-rats are an ancient and highly successful species that are supremely adapted to, what we consider an unimaginable environment. It is clear that the sensory adaptations observed in these animals are all pieces in a larger puzzle, a puzzle that may be key to explaining their success in extremity. The sensory hairs along the body of naked mole-rats are superbly equipped to provide information about speed, direction, and passing of con-specifics in the narrow and complex passageways that are its natural home. The arrangement and physiology of the sensory hairs in naked-mole rats appears to be unique amongst mammals, and there is still a lot to be learned about how this sensory information is represented and used in this species. Crowded dank chambers have probably also necessitated the evolution of a differently tuned nociceptive system in naked mole-rats. The high CO_2 levels found in naked mole-rat burrows are normally not tolerated by other mammalian species and are, on the contrary, damaging and dangerous, a thing to be avoided. This is the core concept of nociception, avoid what causes damage, first proposed by Sherrington at the beginning of the last century. Avoidance of con-specifics (CO_2 pumps), would spell certain extinction. It is not only the chemical composition of the naked mole-rat burrow that provokes sensory adaptation, it is also the temperature. Naked-mole rats are poikilothermic, but that makes the temperature all the more relevant as a sensory parameter because the preferred body temperature is $32\,^{\circ}C$, around $5\,^{\circ}C$ below most other mammals. Thus, thermosensation in the naked mole rat must have a completely different set-point from other mammals, the sensory basis for this adaptation is at present not understood. Finally, to come a full circle, it is clear that the social behavior of the naked-mole is an essential contributor to its ecological success. Amongst all rodents, olfaction is hugely important for social interaction. Also, here, intriguing new evidence suggests that naked mole rats possess a vomeronasal organ that is distinct from other rodents, but how this might relate to olfactory-driven social interaction is, at present, unclear.

See also: Helpers and Reproductive Behavior in Birds and Mammals; Sex Allocation, Sex Ratios and Reproduction.

Further Reading

Bennett NC and Faulkes CG (2000) *African Mole-Rats: Ecology and Eusociality.* Cambridge: Cambridge University Press.

Buffenstein R (1996) Ecophysiological responses to a subterranean habitat: A Bathyergid perspective. *Mammalia* 60: 591–605.

Buffenstein R (2005) The naked mole-rat: A new long-living model for human aging research. *The Journals of Gerontology Series A: Biological Sciences and Medical Sciences* 60(11): 1369–1377.

Catania KC and Henry EC (2006) Touching on somatosensory specializations in mammals. *Current Opinion in Neurobiology* 16(4): 467–473.

Crish SD, Dengler-Crish CM, and Comer CM (2006) Population coding strategies and involvement of the superior colliculus in the tactile orienting behavior of naked mole-rats. *Neuroscience* 139(4): 1461–1466.

Crish SD, Rice FL, Park TJ, and Comer CM (2003) Somatosensory organization and behavior in naked mole-rats I: Vibrissa-like body hairs comprise a sensory array that mediates orientation to tactile stimuli. *Brain, Behavior and Evolution* 62(3): 141–151.

Hetling JR, Baig-Silva MS, Comer CM, et al. (2005) Features of visual function in the naked mole-rat *Heterocephalus glaber*. *Journal of Comparative Physiology A* 191(4): 317–330.

Jarvis JUM (1981) Eusociality in a mammal: Cooperative breeding in naked mole-rat colonies. *Science* 212: 571–573.

Larson J and Park TJ (2009) Extreme hypoxia tolerance of naked mole-rat brain. *Neuroreport* 20(18): 1634–1637.

LaVinka PC, Brand A, Landau VJ, Wirtshafter D, Park TJ (2009) Extreme tolerance to ammonia fumes in African naked mole-rats: Animals that naturally lack neuropeptides from trigeminal chemosensory nerve fibers. *Journal of Comparative Physiology A* 195(5): 419–427.

Park TJ, Comer C, Carol A, Lu Y, Hong HS, and Rice FL (2003) Somatosensory organization and behavior in naked mole-rats: II. Peripheral structures, innervation, and selective lack of neuropeptides associated with thermoregulation and pain. *The Journal of Comparative Neurology* 465(1): 104–120.

Park TJ, Lu Y, Jüttner R, et al. (2008) Selective inflammatory pain insensitivity in the African naked mole-rat (*Heterocephalus glaber*). *PLoS Biology* 6(1): e13.

Riccio AP and Goldman BD (2000) Circadian rhythms of locomotor activity in naked mole-rats (*Heterocephalus glaber*). *Physiology & Behavior* 71(1–2): 1–13.

Sherman PW, Jarvis JUM, and Alexander RD (eds.) (1991) *The Biology of the Naked Mole-Rat.* Princeton, NJ: Princeton University Press.

Smith TD, Bhatnagar KP, Dennis JC, Morrison EE, and Park TJ (2007) Growth-deficient vomeronasal organs in the naked mole-rat (*Heterocephalus glaber*). *Brain Research* 1132(1): 78–83.

Yosida S, Kobayasi KI, Ikebuchi M, Ozaki R, and Okanoya K (2007) Antiphonal vocalization of a subterranean rodent, the naked mole-rat (*Heterocephalus glaber*). *Ethology* 113: 703–710.

Nasonia Wasp Behavior Genetics

R. Watt, University of Edinburgh, Edinburgh, Scotland, UK
D. M. Shuker, University of St. Andrews, St. Andrews, Fife, Scotland, UK

Introduction

Nasonia is a genus of parasitoid wasp that attacks the pupae of many large fly species (across families such as the Muscidae, Sarcophagidae, and Calliphoridae; **Figure 1(a)** and **1(b)**). As a parasitoid, *Nasonia* kill the pupae they attack, being as much predatory as they are parasitic (Godfray, 1994). In common with many other parasitoids, *Nasonia* can influence the population density of the host species they parasitize and may act as biological control agents. Also known as 'jewel wasps,' there are four species in the *Nasonia* genus. By far the best known is *Nasonia vitripennis*, which is distributed across the whole of the northern Palearctic region, being the only *Nasonia* species so far found in Europe and Asia. *N. vitripennis* co-occurs with the other three species in North America. *N. longicornis* is found predominantly in the west of North America, with *N. giraulti* and *N. oneida* occurring in the eastern United States. Recent data suggest that range margins may be changing however, and the very recent discovery of *N. oneida* suggests that further species may await discovery in both America and across Europe and Asia. The four species are reproductively isolated from each other both prezygotically by various behaviors associated with mating (discussed later) and also postzygotically due to nuclear–cytoplasmic incompatibilities associated with the endosymbiotic bacteria *Wolbachia*. Bidirectional cytoplasmic incompatibilities between different *Wolbachia* strains mean that F1 hybrids usually fail to develop, although antibiotic curing of the different wasp species of *Wolbachia* facilitates hybrid formation (albeit with some more 'conventional' loss of hybrid fitness). The ability to make these crosses has played an important role in the success of many *Nasonia* genetics projects, as differences between the species are usually more pronounced (and so easier to resolve) than differences among individuals within a species.

Another factor in the success of *Nasonia* genetic studies is haplodiploidy (**Figure 1(c)**). As with all Hymenoptera (bees, ants, and wasps), *Nasonia* are haplodiploid. This means that females are diploid with both maternal and paternal chromosomes, developing conventionally from fertilized eggs. Males on the other hand are haploid, developing parthenogenetically from unfertilized eggs. Males therefore only contain maternal chromosomes. Haplodiploidy combines many of the advantages of haploid genetic analysis (no effects of dominance for example), with an

organism with complex behavior and ecology. Moreover, haplodiploidy facilitates powerful quantitative genetic and crossing designs (**Figure 2**), which can be important given the low heritability of many behavioral traits.

The study of the behavior and genetics of *Nasonia* wasps has a long history. As a genetic study organism, *Nasonia* played an important role in early studies of mutation (**Figure 1(d)**), but it is perhaps best known for its behavior, in particular its sex ratio behavior and for the presence of sex ratio distorting endosymbionts. More generally though, as a parasitoid wasp, *Nasonia* displays a broad array of behaviors, spanning host location and host choice, through to the reproductive allocation decisions that females need to make, and their elaborate courtship behavior. Our understanding of the genetic basis of many of these behaviors is only just beginning to take shape, but thanks to the *Nasonia* Genome Project the arrival of the full genome sequences of three of the four *Nasonia* species (*vitripennis*, *giraulti*, and *longicornis*) has resulted in a rich new source of genetic and genomic information. We will begin with an introduction to the field of insect behavior genetics and then review the behavior genetics of *Nasonia*, taking the life cycle of the wasp as our guide.

Insect Behavior Genetics

Behavior genetics seeks to characterize the genes and genetic networks that control or influence an animal's behavior. As such, behavior geneticists have to integrate whole organism phenotypes (the behavior or behaviors of interest) with increasing levels of genotypic detail, including the action of individual molecules. Since behavior is, at its most basic, a motor response to some aspect of the environment, behavioral genes include those associated with the development and action of the nervous system, including the associated sensory systems, as well aspects of physiology and cellular metabolism. Moreover, since behavior is a key intermediary between an organism and its environment, the ecological context of behavior is also crucial for understanding how genetics shapes behavioral variation and therefore influences evolution. All this means that behavior genetics is at the center of an integrated approach to understanding animal behavior, as envisaged by Tinbergen in his 'four questions.' Most obviously, genetics can tell us a lot about the mechanistic basis of behavior (what neural and physiological systems

Figure 1 Some aspects of *Nasonia* biology. (a) Blowfly pupae are the host for this parasitoid wasp. (b) A female *Nasonia* laying eggs on a host: females choose hosts depending on the host species, its size, and whether or not other females have already parasitized the host. (c) As with all Hymenoptera, *Nasonia* are haplodiploid, with females being diploid (2n), carrying chromosomes from both their mother (pink arrow) and father (blue arrow), while males are haploid (n), carrying chromosomes only from their mother (pink arrow). Inset: *Nasonia* are gregarious, with multiple wasp pupae developing on one fly pupa, within the host puparium. (d) There are a number of visible mutant markers available, derived from early studies of mutation in *Nasonia*; shown here is the red-eye mutant *STDR*, often used in studies of sex ratio in *Nasonia*. Photographs by David Shuker and Stuart West. Line drawings from Whiting AR (1967) The biology of the parasitic wasp *Mormoniella vitripennis*. *Quarterly Review of Biology* 42: 333–406, used with permission from the University of Chicago Press.

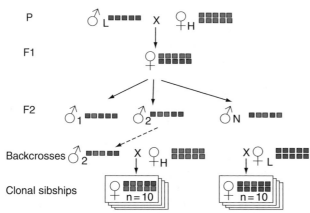

Figure 2 Haplodiploid genetics allow powerful breeding designs for genetic analysis. Here is illustrated a 'clonal-sibship' design for a QTL study. Parents from a 'high' and 'low' line (for a theoretical trait, such as body size) are first crossed. F1 daughters are collected as virgins and given hosts to parasitize, with the (all) male recombinant F2 offspring collected. These F2 males can then be backcrossed to high and low parental line females for the screening of the phenotypes of the recombinant male genotypes in both high and low genetic backgrounds (see Velthuis et al., 2005, for an example of this approach).

influence behavior), but genes can also tell us about how behavior develops, its evolutionary past (through the phylogenetic signal carried by genes), and its evolutionary present. Despite this integrative framework, behavior

geneticists tend to either focus on mechanistic, 'bottom-up' approaches to behavior, or on describing patterns of behavioral variation in populations, and from that inferring something about the genetics of behavior using quantitative genetic techniques in a more 'top-down' approach.

Insect behavior genetics, at least from a bottom-up perspective, has been dominated by study of the fruitfly *Drosophila melanogaster*. Considerable progress has been made in indentifying and characterizing the genetic pathways influencing neural patterning (including famous genes such as *fruitless*), neurohormones and their receptors, neurotransmitters, and other important cellular signaling cascades. In terms of top-down approaches, while *Drosophila* has again been popular, many more species have been studied, not least because of the array of behaviors available across species and the evolutionary (as opposed to mechanistic) puzzles they represent. However, most species are without the access to the molecular resources available in *Drosophila* necessary to link molecular processes with the biological variation that evolution acts upon. One of the major challenges currently facing insect behavior genetics is to link bottom-up and top-down approaches, and species with both rich behavioral repertoires and well-characterized genetics may be important in the coming years in facilitating this link up. *Nasonia* represents one such species.

Nasonia Behavior Genetics

Oviposition Decisions

The *Nasonia* life cycle starts with oviposition (egg laying) by females on a host. The parasitoid lifestyle means that the hosts are other insects, in this case being the pupae of large-bodied Diptera. *Nasonia* are gregarious ectoparasitoids, meaning that several eggs are laid on a host and that after hatching the wasp larvae attach to the host and consume it from the outside in. Females drill through the puparium wall that surrounds the fly pupa proper, laying eggs in the airspace between the puparium and the fly pupa. When the eggs hatch, the first instar larvae attach themselves to the host, ingesting host fluids. The number of wasp larvae developing on a host depends on the species of host (less than 10 on a host like the house fly *Musca domestica*, to more than 60 on large calliphorid blowfly pupae) and also on aspects of host quality (size, the presence of other eggs, and whether or not an adult has already fed on the host – so-called host feeding). The feeding of the wasp larvae during their development destroys the host pupa.

Female *Nasonia* face several important reproductive decisions when they encounter a host. First, females may not actually lay eggs immediately, but instead drill through the puparium wall and into the pupa and then feed on the host fluids that escape. Nothing is known about the genetic basis of host feeding, but resource stress increases host feeding (either as a result of intense larval competition during development, or prolonged time away from food sources). Next, females have to decide if the host is suitable for oviposition. Females explore hosts prior to oviposition, presumably obtaining information via their antennae and other sensory apparatus, including the ovipositor if a drill hole from a previous female is located. Data from the four species suggest there are species differences in host preference (with *N. vitripennis* appearing the most general of the four), which presumably are genetic in origin. Moreover, recent work has identified a chromosome region associated with host preference using crosses between *N. vitripennis* and *N. giraulti*. *N. giraulti* has a much stronger preference for *Protocalliphora* hosts than *N. vitripennis*, and crosses identified a 16-Mb region of chromosome 4 associated with this preference. Recent gene expression studies have also revealed a complex transcriptome activated during oviposition. It is currently not known if rearing environment also influences female host preference later in life, as seen in some herbivorous insect species.

Once a female has decided whether or not to oviposit on a host, the next decision is how many eggs to lay. Experiments have shown that females use a variety of cues to determine clutch size (including the presence and number of other eggs, whether those eggs are from conspecifics or heterospecifics, and the presence of venom from another female), but the presence of a drill hole by itself is not sufficient to influence clutch size. The genetic basis of clutch size has recently been explored with a quantitative trait locus (QTL) study in *N. vitripennis*. Unsurprisingly, the clutch size trait appears to be polygenic, with a clear QTL on chromosome 1, and further, weaker QTL on chromosomes 2, 4, and 5. The extent to which this variation is associated with genes identified by the gene expression study mentioned earlier (i.e., linking molecular patterns of variation with population level variation) remains to be explored.

One of the best-characterized behaviors in *Nasonia* is the sex ratio behavior of females. The population biology of the wasp (in particular, the small mating groups formed by offspring that share a patch of one or a few hosts) means that interactions among related offspring are common, including competition between related males (typically brothers) for mates. This localized competition for mates among sons favors females that limit this competition, leading to selection for female-biased sex ratios that vary with the number of females ('foundresses') laying eggs on a patch of hosts. William Hamilton suggested in 1967 that local mate competition (LMC) should influence female sex allocation, and much theory and experiment has followed, no more so than on *Nasonia*. Pioneering work by Jack Werren confirmed that LMC is an appropriate framework for understanding *Nasonia* sex ratios and confirmed the predicted patterns. Work since then has refined the general LMC models and identified how females estimate the likely level of LMC their offspring will face.

For all this theoretical and empirical work, much less is known about the genetic basis of sex ratio, but work is progressing. A series of heritability studies in the 1980s and 1990s showed that there is heritable variation in sex ratio in *N. vitripennis* (albeit quite low, around 10–15% or so), and that populations respond to artificial selection on sex ratio. More recently, a mutation accumulation study in *N. vitripennis* has been the first to show mutational variation in sex ratio in any species (shown by Pannebakker and colleagues to have a mutational heritability of about 0.001–0.002). That study again estimated the heritability of sex ratio (and again it was low) and then used our understanding of selection on sex ratios to estimate the strength of stabilizing selection on sex ratio. This selection against sex ratio mutants, combined with the rate of mutation revealed by the mutation accumulation study, allowed a prediction for the level of genetic variation expected in a population. This estimate suggested there should be more additive genetic variation (i.e., a higher heritability) than has so far been uncovered in studies of *Nasonia*, suggesting that other sources of selection may be acting against sex ratio mutants; in other words, sex ratio genes should be pleiotropic to some extent, thereby influencing other fitness-related traits.

This suggestion has been tested in the QTL study considered earlier. Not only did this study consider clutch size, it also considered sex ratio. If sex ratio genes are pleiotropic and also act as clutch size genes, they should co-occur in QTL studies. The researchers found a significant QTL influencing sex ratio, but this time it was on chromosome 2. However, weaker sex ratio QTL were also found on chromosomes 3 and 5, with some overlap to the weaker clutch size QTL. While these data only suggest that perhaps some of the same genes influence clutch size and sex ratio, not least since each QTL encompasses a genome region containing potentially many hundreds of genes, promising chromosome locations for further study have nonetheless been identified.

Although it seems as though all four *Nasonia* species vary their sex ratios in line with LMC theory, sex ratios do differ between them as the extent of LMC experienced by broods also varies. This is mostly due to differences in 'within-host mating' (the extent to which eclosing adults mate inside the remains of the puparium before 'emerging' to the outside world). For example, *N. giraulti* has the highest rate of within-host mating, leading to the most extreme LMC and hence the most female-biased single foundress sex ratios. Again, these species differences are likely genetic in origin, and new results from interspecific crosses have indicated that genes in a region of chromosome 5 are associated with this species difference. An exciting possibility is that this region is associated with the same genes as the QTL identified in the intraspecific study in *N. vitripennis* considered earlier.

Eggs are not the only thing produced by females as part of oviposition. Prior to egg laying, females sting the host pupa releasing venom. Venoms of parasitic wasps can have many effects on the host, and in the case of *Nasonia*, it is known that developmental arrest of the fly pupa, suppression of the host immune system, and an increase of lipid levels in the hemolymph result from envenomation. Venom is therefore clearly important in preparing a host for larval consumption and shaping the larval environment. Work is still on going to identify the proteins involved in aspects of host transformation, but it has recently been shown that a protein between 67 and 70 kDa in size may be responsible for host developmental arrest and death. Moreover, bioinformatic approaches (both genomic and proteomic) as part of the *Nasonia* Genome Project have identified 79 genes associated with the venom, approximately half of which have not been ascribed venom function in other species.

Larval Behavior

Compared to adult behavior, much less is known about larval behavior. Larvae undergo four instar molts before pupating. Circumstantial evidence from numerous studies has shown that larval density (assumed to be a correlate of larval competition) influences adult wasp size, fecundity, and energy reserves. As such, ovipositing females appear

to avoid laying eggs on hosts already containing many eggs that are likely to be highly competitive environments for their offspring. In *N. vitripennis*, male larvae develop faster, and there is evidence for asymmetric larval competition (whereby the two sexes represent unequal competitors for each other). However, the genetics of larval competition are completely unknown and represent a topic ripe for study, especially as theory predicts parent–offspring conflict over traits such as larval competition, which might lead to patterns of genomic imprinting influencing which genes are activated during larval development. Venom may also be an important mediator of larval behavior, influencing as it does how good a resource a host represents to the developing larvae (e.g., by debilitating host immune responses). The interactions between venom and larval behavior require further study however.

One interesting aspect of larval development has received more scrutiny however, namely, larval diapause. This developmental arrest of third instar larvae acts as an overwintering stage for larvae developing at the end of the temperate summer. Pioneering work by Saunders in the 1960s showed that this developmental switch is not controlled by the individual larvae however, but rather by the mother in response to changes in photoperiod (day length) and also to some extent to changes in temperature and host availability. Females therefore change some aspect of egg physiology to induce diapause, and resources now available from the *Nasonia* Genome Project mean that the molecular mechanisms controlling diapause induction are being teased apart, with the exciting prospect of a full molecular explanation of an important maternal effect.

Adult Emergence and Mating

N. vitripennis is protandrous, with males emerging first from the host puparium in order to mate with females that will emerge soon after. As mentioned earlier, this is not true for all *Nasonia* species, with *N. giraulti* showing the greatest extent of within-host mating and *N. longicornis* showing intermediate behavior. The genetic basis for this species-specific difference is currently under scrutiny. This difference in within-host mating is also reflected in differences in mating site preference. In *N. vitripennis*, males prefer to wait outside emergence holes in the puparium for females to emerge, competing to hold these small mating territories unless high male density makes defense impractical. These mate site preferences seem less important for the other two species so far studied, perhaps due to the greater likelihood of within-host mating. Certainly, males in *N. vitripennis* mark their mating territory following successful mating with the recently identified pheromones (4*R*,5*R*)- and (4*R*,5*S*)-5-hydroxy-4-decanolide (HDL) and 4-methylquinazoline (4-MeQ); these pheromones are very attractive to virgin females, such that males typically remain where they have been successful at obtaining mates. The species' differences in mating site are hypothesized to have evolved to reduce interspecific mating in areas of sympatry (i.e., East and West North America) and perhaps represent one of several mechanisms to facilitate prezygotic reproductive isolation.

Courtship and Copulation Behavior

Courtship begins with the male finding a female, usually as she is emerging from the host puparium. Female contact pheromones are an important part of a male recognizing a female. Courtship in *Nasonia* is an intricate display that induces female receptivity. Once a male has located a female, he will mount her back and position himself above her head. Once in this position, the male will begin a series of 'head nods' which always start with the extrusion of the mouthparts. This behavior has been shown to coincide with the release of an as yet unidentified pheromone from mandibular glands that is important for female receptivity. After a species-specific series of head nods, the female will indicate her receptivity by lowering her antenna and extruding her genitalia, physical cues which then induce the male to back up and attempt copulation. Once copulation is complete, the male will return to his original position above the female's head and engage in a second bout of (postcopulatory) courtship. This second courtship sequence is usually shorter and is again terminated by the female lowering her antennae. This signal leads to the male dismounting and is associated with females becoming almost completely unreceptive, both to further males and their pheromones.

The genetic basis of courtship has been of great interest given its possible role in the reproductive isolation of the different species in North America. A within-species study in *N. vitripennis* identified low heritability of both courtship duration and copulation duration (significant in the latter case), although for both traits dam effects were significant, suggesting nonadditive genetic effects. Use of interspecific hybrids has confirmed a genetic basis for the differences in head-nod series, but they also revealed a so-called grandfather effect, such that the species of the maternal grandfather of the male performing the courtship influenced the behavior observed (it should be remembered that in haplodiploid species, males do not have fathers, only mothers, but they will have maternal grandfathers). Similarly, an intraspecific study in *N. vitripennis* suggested both heritable variation in courtship behavior and at least a weak grandfather effect within-species. The mechanistic basis for these transgenerational genetic effects is not known, but an obvious possibility is some form of genomic imprinting, such that paternal chromosomes manage to influence expression of a trait when passed through to grand-offspring. One of the most exciting discoveries of the *Nasonia* Genome Project has been a full DNA methylation toolkit, with all three families

of DNA methyl-transferase genes (*Dnmt-1*, *Dnmt-2*, and *Dnmt-3*) present. DNA methylation is one the best-known mechanisms of genomic imprinting, with methylated cytosine residues influencing how a gene is expressed (both up- and downregulation is possible). There is also now direct experimental evidence from *Nasonia* for DNA methylation (21 of 42 randomly chosen genes showed patterns consistent with DNA methylation), so genomic imprinting as a mechanism by which parents influence offspring behavior is a very real possibility.

Since female *Nasonia* are assumed to disperse from the natal patch almost immediately after mating, it has been thought that females will have little opportunity to remate and will thus remain monandrous in the wild. Evidence from population genetic studies of *Nasonia vitripennis* however suggests that multiply mated (polyandrous) females do exist in natural populations, albeit at a low frequency. Moreover, there is genetic variation in *N. vitripennis* for polyandry, and laboratory-maintained cultures of wasps do appear to evolve a greater degree of polyandry as a result of artificial selection arising from some as yet unidentified aspect of lab rearing. Interestingly, although female polyandry is heritable in *N. vitripennis* (i.e., there is significant additive genetic variation in remating), there is also evidence for nonadditive effects influencing female remating rate, even after controlling for common-environment effects. These residual nonadditive maternal effects could be genetic in origin, again raising the possibility that genomic imprinting may be influencing aspects of *Nasonia* mating behavior. Understanding the genetic basis of female receptivity is now being considered in terms of which genes are activated during and after mating, but one interesting fact has already emerged thanks to the *Nasonia* Genome Project. None of the *Nasonia* species appear to have either the sex peptide gene, or the sex peptide receptor. The sex peptide gene encodes one of a large array of male accessory gland proteins (ACPs) in *Drosophila*, and is associated with male effects on many aspects of female reproductive physiology after mating, including shutting down female receptivity. As such, sex peptide and other ACPs have been the targets of a great deal of interest in *Drosophila*, especially in terms of sexual conflict over the control of reproduction. However, the absence of sex peptide and its receptor in *Nasonia* suggests that different mechanisms may be involved, perhaps a key lesson for insect behavioral geneticists in the coming years.

There has been rather little work on sexual selection in *Nasonia*, barring some early attempts to explore 'rare-male effects' on female mate preferences, some descriptions of patterns of sperm precedence in multiply mated females (first male sperm precedence appears the most common pattern), and the descriptions of male territoriality mentioned above. However, female mate preferences in terms of conversus heterospecifics have received attention, and indeed females exhibit preferences for their own species as expected. A QTL study using two strains of *N. longicornis* that varied in their willingness to mate with *N. vitripennis* identified three major QTL influencing female willingness to mate (**Figure 3**), but more remains to be done to explore the genetics of mate preference within and between *Nasonia* species.

Dispersal Behavior and Host Location

After mating, females disperse from the natal patch in order to look for new hosts. Little is known directly about this behavior, although population genetic studies of wild populations suggest that female dispersal is sufficient to limit population substructuring and levels of inbreeding to that occurring as a result of within-patch mating. In all species of *Nasonia*, females disperse from the natal patch. However, the extent to which males also disperse from the natal patch varies between the species. *N. vitripennis* males are brachypterous (short winged) and cannot fly. Any dispersal they undertake is done via walking away from the patch, and it is not known whether these males ever reach other patches and mate with females there. *N. giraulti* males on the other hand are fully winged and can disperse away from the host puparium (*N. longicornis* male wings are intermediate in size, and they have some ability to fly). QTL analysis by Gadau and colleagues of interspecific hybrids has shown that these differences between the species in wing morphology (including size and shape) are associated with 11 regions in the genome. That study also indentified epistatic interactions between the QTL influencing wing size and also wing bristle density. Intriguingly, very recent work was suggested that regulatory sequences associated with the gene *doublesex*, better known for its well-conserved role in the sex-determination cascade of many organisms, including *Nasonia*, is the basis for one of the wing size QTL.

Once they start to disperse, host location is primarily determined by olfaction. Females have, through a series of olfaction tests, been shown to be attracted to rotting meat and specifically rotting meat on which maggots of the host species have fed, indicating that the larvae produce a chemical signal while feeding. Females appear to be able to remember smells associated with a host they have parasitized, which may help them to locate hosts at different patches. The genetic basis of these behaviors remains unexplored though.

The Molecular Control of Behavior

We will end our consideration of the behavior genetics of *Nasonia* with a brief discussion of some of the latest

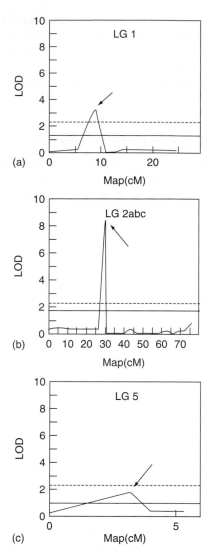

Figure 3 Results of a quantitative trait loci (QTL) analysis for female mate discrimination in *Nasonia longicornis*. The arrows represent QTL for willingness to mate with *Nasonia vitripennis*. (a), (b), and (c) are the results for linkage groups 1, 2, and 5, respectively. LOD signifies the 'likelihood-of-odds' score, used to calculate the effect-size and significance of a putative QTL. The position along each linkage group is denoted on the X-axis in centimorgans (cM). The dotted horizontal lines represent genome-wide significance thresholds and the solid horizontal lines represent linkage-group-wide significance thresholds (both calculated via permutation tests; $P < 0.05$). The QTL identified on linkage groups 1 and 2 have greater statistical support than the QTL found on linkage group 5 (this latter QTL was significant in an alternative genetic background: for further details of this study see Velthuis et al., 2005). Recent mapping work has coordinated the linkage groups from studies such as this to the five chromosomes of *Nasonia*. Adapted from Velthuis B-J, Yang W, van Opijnen T, and Werreh JH (2005) Genetics of female mate discrimination of heterospecific males in *Nasonia* (Hymenoptera, Pteromalidae). *Animal Behaviour* 69: 1107–1120, with permission from Elsevier.

findings from the *Nasonia* Genome Project in terms of the molecular control of behavior, to give a flavor of possible new avenues of research. Although *Nasonia*-specific pathways have yet to be elucidated, bioinformatic analysis of the *Nasonia* genome has revealed a number of important features.

First, from our review of *Nasonia* behavior earlier, olfaction is clearly an important aspect of behavior, especially in females. *Nasonia* has a large number of olfactory-binding proteins (OBPs), with ~90 annotated in the genome. In addition, *Nasonia* has a sizeable number of olfactory receptor proteins (around 300) and also gustatory receptor proteins (58 annotated genes). This overall repertoire is larger than both that of *Drosophila melanogaster* and the honeybee *Apis mellifera*, being more similar to that of the beetle *Tribolium castaneum*. Clearly, both smell and taste are important sensory modalities for *Nasonia* as it interacts with its environment and may be key to understanding many aspects of its behavior. Second, *Nasonia* has a suite of neurohormones and their associated G-protein-coupled receptors thought to control behavioral responses via their interaction with neural and cellular signaling pathways. In addition, the presence of neurotransmitter receptors such as cys-loop ligand-gated ion channels has been confirmed. While these findings are in some ways not especially surprising, they do provide a starting point for a more detailed molecular genetic analysis of the control of behavior in *Nasonia*. Finally, along with a number of other insects, *Nasonia* also boasts the recently discovered oxytocin/vasopressin-like protein inotocin and its receptor. Oxytocin and vasopressin are well-known molecules from vertebrates, with various cell signaling roles that are often associated with reproductive behaviors. Although the function of inotocin in insects is not yet known, it may well prove to be important in a number of *Nasonia* behaviors.

See also: *Drosophila* Behavior Genetics; Genes and Genomic Searches; Konrad Lorenz; Parasitoid Wasps: Neuroethology; Parasitoids; Sex Allocation, Sex Ratios and Reproduction.

Further Reading

Boake CRB, Arnold SJ, Breden F, et al. (2002) Genetic tools for studying adaptation and the evolution of behavior. *American Naturalist* 160: S143–S159.

Gadau J, Page RE, and Werren JH (2002) The genetic basis of the interspecific differences in wing size in *Nasonia* (Hymenoptera; Pteromalidae): Major quantitative trait loci and epistasis. *Genetics* 161: 673–684.

Godfray HCJ (1994) *Parasitoids. Behavioural and Evolutionary Ecology.* Princeton, NJ: Princeton University Press.

Hamilton WD (1967) Extraordinary sex ratios. *Science* 156: 477–488.

Ivens ABF, Shuker DM, Beukeboom LW, and Pen IR (2009) Host acceptance and sex allocation of *Nasonia* wasps in response to conspecifics and heterospecifics. *Proceedings of the Royal Society, Series B* 276: 3663–3669.

King BH and Skinner SW (1991) Proximate mechanisms of the sex ratio and clutch size responses of the wasp *Nasonia vitripennis* to parasitized hosts. *Animal Behaviour* 42: 23–32.

Pannebakker BA, Halligan DL, Reynolds KT, et al. (2008) Effects of spontaneous mutation accumulation on sex ratio traits. *Evolution* 62: 1921–1935.

Pultz MA and Leaf DS (2003) The jewel wasp *Nasonia*: Querying the genome with haplo-diploid genetics. *Genesis* 35: 185–191.

Shuker DM, Phillimore AJ, Burton-Chellew MN, Hodge SE, and West SA (2007) The quantitative genetic basis of polyandry in the parasitoid wasp, *Nasonia vitripennis*. *Heredity* 98: 69–73.

Steiner S and Ruther J (2009) Mechanism and behavioral context of male sex pheromone release in *Nasonia vitripennis*. *Journal of Chemical Ecology* 35: 416–421.

The *Nasonia* Genome Working Group (2010). Functional and evolutionary insights from the genomes of three parasitoid *Nasonia* species. *Science* 327: 343–348.

Van den Assem J and Werren JH (1994) A comparison of the courtship and mating behavior of 3 species of *Nasonia* (Hymenoptera, Pteromalidae). *Journal of Insect Behavior* 7: 53–66.

Velthuis B-J, Yang W, van Opijnen T, and Werreh JH (2005) Genetics of female mate discrimination of heterospecific males in *Nasonia* (Hymenoptera, Pteromalidae). *Animal Behaviour* 69: 1107–1120.

Werren JH (1980) Sex ratio adaptations to local mate competition in a parasitic wasp. *Science* 208: 1157–1159.

Werren JH (1983) Sex ratio evolution under local mate competition in a parasitic wasp. *Evolution* 37: 116–124.

Whiting AR (1967) The biology of the parasitic wasp *Mormoniella vitripennis*. *Quarterly Review of Biology* 42: 333–406.

Nematode Learning and Memory: Neuroethology

C. H. Lin and C. H. Rankin, University of British Columbia, Vancouver, BC, Canada

The Nematode *Caenorhabditis elegans*

As 1 mm long free-living, soil-dwelling nematode (roundworm) that feed on bacteria in the soil and on rotting vegetation, *C. elegans* (**Figure 1**) face many challenges in their dynamic environment. During their short life span (15–30 days) and rapid reproductive life cycle (2–3 days), *C. elegans* must quickly learn about their environment to successfully grow and reproduce. In the laboratory, they feed on a benign strain of *Escherichia coli* and navigate across the surface of agar in sinusoidal waves very much like snakes. Many elements of *C. elegans'* behavior have been well characterized and are easy to quantify.

One example is the *C. elegans'* escape from a mechanosensory touch by moving forward or backward away from the origin of the stimulation. This escape response can be quantified as the distance of travel resulting from a mechanosensory stimulus and/or the frequency of responses to the stimuli. In addition, *C. elegans* can thermotax toward or away from temperatures with high precision. The thermotactic behavior can be quantified as the percentage of worms that migrate to a specific temperature within a temperature gradient over a given period of time. Moreover, *C. elegans* can sense and chemotax toward or away from a variety of both soluble and volatile compounds, and the chemotactic behavior can be quantified in the same manner as the thermotactic behavior. These and other behaviors are mediated by the 302 neurons that make up the nervous system of a hermaphrodite *C. elegans*. Although its nervous system is simple, the *C. elegans'* behavior demonstrates a high degree of plasticity; *C. elegans* can show both nonassociative and associative learning in mechanosensory, thermosensory, and chemosensory modalities. The wealth of neurobiological, genetic, and developmental information available for *C. elegans* makes it an ideal model on which to study the neurological and molecular basis of learning and memory.

Not only is the behavior of *C. elegans* well characterized, but *C. elegans* researchers also have the advantage of a fully mapped neuronal wiring diagram. The 302 neurons in the hermaphroditic form of *C. elegans* are connected with ~5000 chemical synapses, 600 gap junctions, and 2000 neuromuscular junctions. To put this into context, the number of chemical synapses in an entire *C. elegans* is equivalent to the number of synapses made by a single hippocampal pyramidal neuron in mammals. In addition, *C. elegans* was the first multicellular animal to have its genome fully sequenced; it contains ~20 000 genes, of which ~5000 are homologous to human genes. The continually expanding mutant library is massive: currently ~6000 genes. These mutant and transgenic strains can be conveniently stored in −80 °C freezers because this animal can survive at extreme temperatures. Furthermore, *C. elegans* is transparent, which allows *in vivo* observation of protein localization in intact living animals, using transgenic strains that express genetically engineered proteins tagged with markers, such as green fluorescent protein (GFP; this protein was originally isolated from jellyfish and can be attached to genes encoding for specific proteins to make those proteins visible under fluorescent light). Another application of the fluorescence technology in this transparent animal is called fluorescence resonance energy transfer (FRET), which is used to monitor calcium activity *in vivo*, using a genetically encoded calcium sensor such as Cameleon that alters the wavelength of fluorescence emitted upon calcium binding. This transparency also means that specific neurons can be killed using a laser microbeam in an intact living animal in which the behavior can be studied in the absence of that neuron. Laser ablation studies have led to the identification of neuronal circuits responsible for many types of behaviors in *C. elegans*. Subsequently, the known circuitry provides a framework in which the molecular basis of behavioral plasticity can be explored in *C. elegans*.

Mechanosensory Learning and Memory

Nonassociative Learning for Mechanosensory Stimuli

Habituation, a form of nonassociative learning, is defined as a gradual decrease in responding to repeated irrelevant stimuli. To study habituation to mechanosensory stimuli in *C. elegans*, taps of a constant force were delivered to the side of the Petri-dish filled with the agar growth medium in which the worms had been cultivated. These taps stimulate the entire body of the worms, which move backwards in response. This is termed the 'tap-withdrawal response.' As the stimulus is repeated, worms learn to ignore it and respond less and less to the tap (**Figure 2**). Similar to other organisms, repeated mechanical stimuli delivered at shorter interstimulus intervals (ISIs; 10 s) result in faster habituation than stimuli delivered at longer ISIs (60 s). Shorter ISIs produce a lower asymptotic responding level than do longer ISIs, and worms recover more rapidly from the habituation elicited by shorter ISIs than by longer ISIs. Depending on the ISI of training, this short-term habituation to taps can last as long as 1 h.

The neuronal circuit for the tap-withdrawal response was mapped by laser ablation experiments. The tap stimuli are primarily sensed by the mechanosensory neurons AVM, ALM, PVD, and PLM (**Figure 3**). These sensory

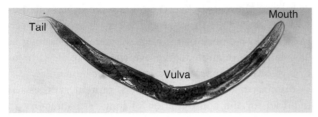

Figure 1 Microscopic image (10X) of a nematode *C. elegans*. This 1 mm long soil-dwelling nematode eats bacteria and lives 15–30 days. The mouth of the worm is on the right and the tail on the left. The vulva is the lip-like structure in the middle.

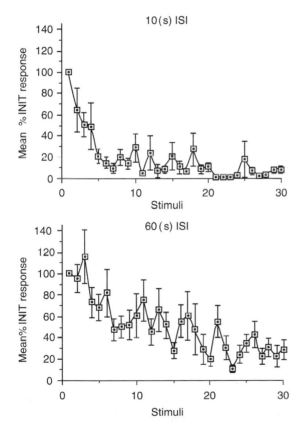

Figure 2 Habituation to mechanosensory stimuli, a form of nonassociative learning in *C. elegans*. When the mechanosensory stimuli (taps to the Petri plate holding the worm on agar surface) were presented repeatedly at 10 s ISI or 60 s ISI *C. elegans*, their responses (magnitude normalized to initial response) to taps became smaller and smaller. The rate of decrease was faster when the taps were presented at 10 s ISI than at 60 s ISI. Adapted from Rankin CH and Broster BS (1992) Factors affecting habituation and recovery from habituation in the nematode *Caenorhabditis elegans*. *Behavioral Neuroscience* 106(2): 239–249, with permission from American Psychological Association.

neurons synapse onto the command interneurons AVD, AVA, AVB, and PVC, which in turn drive a pool of motor neurons that mediate forward and backward movement. The response is produced by activity in gap junctions and modulated by activity at chemical synapses. A number of behavioral findings suggested the hypothesis that the site of habituation was the chemical synapses between the sensory neurons and the command interneurons. These findings were supported by the physiological finding using the genetically encoded calcium sensor Cameleon that during habituation to taps, the calcium currents in the mechanosensory neurons gradually decreased with repeated mechanical stimulation at 10 s ISI. In neurons, decreases in calcium influx often correlate with decreased release of neurotransmitter, and thus decreased behavioral responses. Once the neuronal circuit underlying tap habituation was determined, a candidate gene approach was used to study genes expressed in either the sensory neurons or the interneurons in the tap-withdrawal circuit (Ardiel and Rankin, 2008).

Glutamate neurotransmission is thought to play an important role in learning and memory in mammals. Therefore, the first candidate gene to be studied was

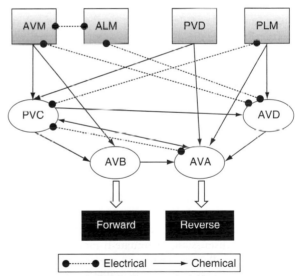

Figure 3 The tap-withdrawal circuit. Mechanosensory neurons (AVM, ALM, PVD, and PLM) are shaded boxes and interneurons (PVD, AVD, AVB, and AVA) are open ovals. AVB and AVA connect to motor neurons that eventually drive forward or reverse movement, respectively. Solid lines are chemical (directional) connections with arrow pointing towards the postsynaptic neuron (double arrow indicates that both neurons send chemical connections towards each other). The dashed lines indicate electrical (nondirectional) connections. AVB and AVA then drive pools of motor neurons to produce forward and backward movement. Modified from Wicks SR and Rankin CH (1995) The integration of antagonistic reflexes revealed by laser ablation of identified neurons determines habituation kinetics of the *Caenorhabditis elegans* tap withdrawal response. *Journal of Comparative Physiology A* 179(5): 675–685.

eat-4, a vesicular glutamate transporter expressed in the mechanosensory neurons. EAT-4 is homologous to a mammalian vesicular glutamate transporter 1 (VGLUT1) whose expression level has a large impact on the efficacy of glutamatergic transmission. *C. elegans* mutants containing lack-of-function *eat-4* alleles are defective in glutamatergic transmission. These mutants habituated more rapidly than wild-type worms to repeated tap stimuli presented at both 10 and 60 s ISIs. Despite the low glutamate levels in these mutants, *eat-4* mutants recovered faster after habituation to stimuli delivered at a 10 s ISI than a 60 s ISI, which suggests that the mechanisms of habituation are still intact. The behavioral data of *eat-4* mutants lead to the hypothesis that glutamatergic transmission plays a key role in the kinetics of habituation to taps, but not in the ability of worms to habituate to repeated mechanosensory stimuli.

Another candidate gene, a dopamine receptor homolog, *dop-1*, is also expressed in the mechanosensory neurons. Worms with a mutation in *dop-1* habituated more rapidly compared to wild-type. Interestingly, this more rapid habituation in *dop-1* mutants occurred only (1) on food but not in the absence of food, (2) in the frequency but not the magnitude of responses, and (3) at a 10 s ISI but not at a 60 s ISI. This is the first genetic evidence supporting the hypothesis that habituation to short and long ISIs are mediated by different mechanisms and that there is a food-dependency effect on the rate of habituation that involves the dopamine system. Additionally, the observation that the *dop-1* mutation selectively affected the frequency but not the magnitude of reversal suggests that the habituation of the response frequency and response to repeated stimuli may be mediated through different mechanisms. Further investigation into the role of dopaminergic signaling in habituation demonstrated that many other mutants with deficient dopaminergic transmission showed the same phenotype as *dop-1*. Conversely, a mutant strain hypothesized to have higher levels of dopamine at synapses showed slower habituation; a mutation in a dopamine reuptake transporter, *dat-1*, hypothesized to lead to stronger and more sustained dopamine transmission, habituated more slowly than wild-type. Parallel to what was seen at the behavior level, the decrement of calcium influx in the mechanosensory neurons after repeated stimulation measured using Cameleon was more rapid in *dop-1* worms than in wild-type worms. These findings supported the hypotheses that dopamine alters the rate of habituation by changing the excitability of the mechanosensory neurons mediated by calcium influxes.

Long-Term Memory for Mechanosensory Habituation

C. elegans is capable of forming long-term memory for habituation to tap stimuli that lasts at least 48 h after the training session. Similar to findings in other species, long-term memory for training is better retained if small blocks of stimuli are presented several times (spaced/distributed training) rather than all at once in a single large block of stimuli (massed training). When a single block of 60 taps was presented at a 60 s ISI, it led to an intermediate memory 12 h later, but did not lead to long-term memory 24 h later. However, when the 60 taps were presented as 3 blocks of 20 stimuli, separated by a 1 h break between each block, memory of training was retained as long as 48 h later. In relation to the average life span, 48 h for *C. elegans* would approximate 10 years for humans!

Similar to other animal models, protein synthesis is required for the consolidation of long-term memory in *C. elegans*. When heat shock, which disrupts protein synthesis, was administered during the 1 h resting period between training blocks, animals failed to form long-term memory. Heat shock was also used to demonstrate memory reconsolidation blockade in *C. elegans*. In mice when a memory for fear conditioning is recalled, that memory is transformed into an unstable liable state and then must be restored (reconsolidated) to be retained for future use. If the reconsolidation process after memory retrieval is disrupted, for example, by blocking protein synthesis, the memory will be lost. When *C. elegans* was given long-term habituation training, followed 24 h later by memory recall (10 taps) and then heat shock, the memory was lost. These characteristics of memory in *C. elegans* suggested that basic memory encoding and retrieval mechanisms in *C. elegans* may represent processes that are highly conserved throughout evolution.

Glutamatergic transmission also plays a central role in long-term memory formation in *C. elegans*. *C. elegans* with a mutation in the vesicular glutamate transporter, *eat-4*, did not show long-term memory for habituation. The same was true for *C. elegans* with mutations in *glr-1*, a homolog of the mammalian AMPA-type glutamate receptor subunit GluR-1, which has been implicated in learning and memory formation in mammals. Administration of the AMPA/ Kainate receptor antagonist DNQX during training in *C. elegans* blocked long-term memory for habituation to taps. These findings confirm the importance of the glutamate receptor subunit *glr-1* in long-term memory for habituation to taps. Furthermore, when genetically engineered *C. elegans* expressing *glr-1* tagged with GFP were used to visualize *glr-1* expression levels *in vivo*, GLR-1::GFP expression was found to be downregulated in the ventral nerve cord of the trained worms compared to naive worms. Interestingly, in the reconsolidation blockade paradigm when memory reconsolidation after a reminder was blocked by heat shock, GLR-1::GFP expression levels were upregulated and reset to control levels. Over a large number of studies, this downregulation of GLR-1::GFP in the ventral cord is a consistent correlate of long-term memory for habituation.

In conclusion, glutamate transmission plays a major role in learning and memory in *C. elegans*, as it does in mammals.

Mechanosensory Associative Learning

The association of environmental cues with an experience (i.e., habituation) is a particular type of associative learning called context conditioning. *C. elegans* that were habituated to taps in the presence of a chemosensory cue (sodium acetate) showed greater memory for habituation when tested in the presence of that cue an hour later than they would have if training and testing had occurred in different environments. This association was sensitive to latent inhibition: when worms were exposed to sodium acetate for an hour before habituation training, they did not show enhanced memory of the training. This association was also sensitive to extinction: worms exposed to sodium acetate during the 1 h break between training and testing also failed to show enhanced memory. The enhanced memory occurred only when the sodium acetate cue was present during both training and testing. These findings indicate that *C. elegans* can demonstrate associative learning along with habituation to mechanosensory stimuli.

Chemosensory and Thermosensory Associative Learning

C. elegans are also capable of associating the presence or absence of food with environmental cues. These cues include temperature, smell, and taste. If *C. elegans* are given two chemosensory cues and one is paired with food while the other one is not, then in a choice test, the worms will choose the cue that had been previously associated with food. An originally attractive cue can also become less attractive if it is paired with starvation or an aversive stimulus. In much the same way, when *C. elegans* are placed on a thermal gradient, they will migrate to the temperature where they were raised with an abundance of food and avoid a temperature that was paired with starvation.

Thermosensory Associative Learning

When *C. elegans* are cultivated with food at a specific temperature and then challenged with a temperature gradient (from 15 to 25 °C with no food present), they will navigate toward the cultivation temperature and stay at that temperature for 2–4 h (**Figure 4**). This behavior is called 'isothermal tracking.' When the worms were subjected to starvation for 3 h at a particular temperature, they learned to avoid this particular temperature on a temperature gradient. This association of food abundance and a specific temperature could be reversed rapidly. The temperature preference could be changed by cultivating the *C. elegans* at a different temperature for 2–4 h. When tested, the behavior of the worms indicated that they had learned to associate the new temperature with food. Furthermore, long-term memory of the thermal association with food appeared to last as long as 48 h after the training session. These studies demonstrated that *C. elegans* can retain both short- and long-term memory for the association between food and a specific temperature, and that this memory can be rapidly modified upon the presentation of a new temperature paired with food.

The major thermosensors in *C. elegans* are the AFD and AWC sensory neurons (**Figure 5**), located at the tip of the worms' nose. The AFD and AWC neurons innervate a pool of interneurons (AIB, AIY, AIZ, RIA, RIB, RIM) that integrate inputs from multiple sensory modalities (i.e., temperature and the smell of food) and send signals to command interneurons (AVE, AVA, and AVB) that dictate the direction of subsequent movement. The site of plasticity in the thermotaxis circuits was hypothesized to be located at the interneuron level. The AIY (cryophilic) and AIZ (thermophilic) interneurons of the circuit both feed onto the RIA interneurons to mediate movement to higher or lower temperatures.

ncs-1 encodes a neuron-specific calcium sensor that is expressed in the AIY interneurons, and was shown to be important in the plasticity of isothermal tracking behavior. An *ncs-1* knockout mutant failed to perform isothermal tracking to the temperature (20 °C) previously paired with

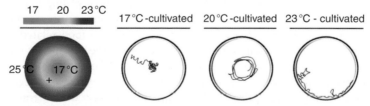

Figure 4 Isothermal tracking. The left-most panel illustrates the thermal gradient imposed on an agar plate and used to challenge worms after temperature-food conditioning. The center of the plate was 17 °C, the outermost ring was 23 °C, and somewhere in the middle was 20 °C. The black lines in the round agar plates indicate the tracks of worm movement. Worms that were cultivated with food at 17 °C migrated to the center (which was 17 °C). Worms cultivated at 20 °C displayed 'isothermal tracking' forming a circular ring where the agar was 20 °C. Worms cultivated on 23 °C formed a track at the outer edges of the agar where the 23 °C was. Modified from Mori I, Sasakura H, and Kuhara A (2007) Worm thermotaxis: A model system for analyzing thermosensation and neural plasticity. *Current Opinion in Neurobiology* 17(6): 712–719 with permission.

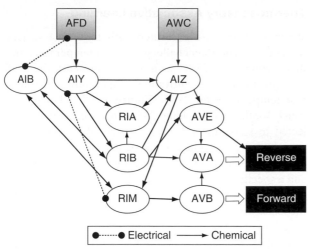

Figure 5 Thermosensory neural circuit. Sensory neurons (AFD and AWC) are shaded boxes and interneurons (AIB, AIY, AIZ, RIA, RIB, RIM, AVE, AVA, and AVB) are open ovals. Solid lines are chemical (directional) connections with arrow pointed towards the postsynaptic neuron (double arrow indicates that both neurons send chemical connections towards each other). AVB and AVA drive pools of motor neurons to produce forward and backward movement. The dashed lines indicate electrical (nondirectional) connections. Modified from Hobert O (2003) Behavioral plasticity in *C. elegans*: Paradigms, circuits, genes. *Journal of Neurobiology* 54(1): 203–223 and Kuhara A, Okumura M, Kimata T, et al. (2008) Temperature sensing by an olfactory neuron in a circuit controlling behavior of *C. elegans*. *Science* 320(5877): 803–807.

food. Conversely, overexpression of NCS-1 increased the rate of thermal learning acquisition and decreased the rate of extinction of this learned behavior. In addition, *C. elegans* mutants deficient in *ncs-1* showed less isothermal tracking behavior for a period of time after training, which suggests that *ncs-1* may play a role in the long-term memory of thermosensory learning. Another molecule studied appeared to play a role specifically in the cryophilic circuitry (AIZ-RIA). Calcineurin (*tax-6*) mutants showed defective thermotactic learning particularly at a lower temperature (17 °C) rather than at a higher temperature (23 °C). Cultivation at 17 °C paired with starvation inhibited the cryophilic AIZ-RIA circuitry by suppressing the temperature-evoked changes in calcium concentration (measured using Cameleon) through calcineurin (*tax-6*).

The involvement of neuroendocrine signaling in thermotactic learning was uncovered using a genetic screen for food-associated thermotactic plasticity. A genetic screen for defects in avoiding a temperature paired with starvation isolated a cohort of *aho* (abnormal hunger orientation) mutants. One of these mutations, *aho-2*, was found to be the same as an allele of *ins-1*, a human insulin homolog. Normally, when wild-type *C. elegans* are presented with a temperature previously associated with starvation, the calcium currents in the AIY interneuron would be lower than in naive animals. However, *ins-1* mutants failed to display decreased calcium currents in AIY when presented with a temperature previously associated with starvation. *ins-1* is an antagonist to the insulin receptor, *daf-2*, and the downstream target of *daf-2* is a PI3-kinase, *age-1*. A deficiency in AGE-1 (a central component of the *C. elegans* insulin-like signaling pathway) causes the animals to avoid the cultivation temperature associated with starvation much earlier than wild-type animals. The more rapid thermotactic learning in *age-1* mutants can be restored to wild-type level by expressing the *age-1* gene in any of the AIY, AIZ, or RIA interneurons. HEN-1, a secreted protein that is expressed in the AIY interneurons, also demonstrates a role in thermotactic learning; *hen-1* mutants continued to favor a temperature even though that temperature was paired with starvation. Together, this evidence suggests that calcium levels in AIY and AIZ and their communication to RIA interneurons are important sites of the plasticity, and that the plasticity is modulated by the insulin, the pathway, and LDL-like secretory proteins.

Chemical Associative Learning

A number of different chemosensory associative learning paradigms have been developed for *C. elegans*. For example, when two different chemoattractants were positioned at the opposite ends on the agar growth medium, *C. elegans* dispersed evenly between the two attractants; however, if one of the two cues was previously associated with food, and the other with starvation, worms preferentially migrated to the cue previously associated with food. If a naturally attractive cue (NaCl) was presented to the worm under starvation conditions, chemotaxis to that cue (NaCl) would be suppressed. This suppression occurred gradually over a period of time (3–4 h) and could be rapidly reversed (i.e., within 10 min) if animals were fed in the presence of the cue or if starved without the cue (extinction). Furthermore, an attractive odor such as diacetyl became aversive if paired with an aversive taste such as acetic acid. Taken together, these findings demonstrate that *C. elegans* can modify natural responses to chemical cues according to recent experiences.

The neurotransmitter serotonin plays an important role in many food-related behaviors in *C. elegans*. In *C. elegans*, the absence of serotonin appears to encode starvation. When an attractive chemical cue was paired with starvation, *C. elegans* became less attracted to the cue; this attractiveness was measured by the chemotaxis index (the proportion of worms that migrated to the source of chemical gradient within a given period of time). When a chemical cue was paired with starvation in the presence of exogenous serotonin, the reduction in chemotaxis to the cue was blocked. Moreover, serotonin has been implicated in a pathogenic bacteria aversion learning paradigm:

C. elegans that were fed a pathogenic bacteria strain and then given a choice test between the pathogenic and a benign strain of bacteria showed an aversion to the pathogenic strain. However, *mod-1* mutants missing a serotonin receptor on the interneurons of the chemotaxis circuit displayed less aversion than did wild-type worms. On the basis of these observations, serotonin has been hypothesized to play a role in both pathogenic aversion learning and in food-related behaviors.

Glutamatergic transmission, especially using AMPA and NMDA-type glutamate receptors, is important in several forms of learning and memory in mammals. The AMPA-type glutamate receptor subunit, *glr-1*, which is critical to long-term memory for mechanosensory habituation, was also shown to be critical to a form of olfactory associative learning. NMDA-type glutamate receptor subunits *nmr-1* and *nmr-2* were shown to be important for the memory retention of salt-starvation conditioning. Neuron-specific rescue of NMR-1 and NMR-2 in *nmr-1* and *nmr-2* mutant animals demonstrated that *nmr-1* and *nmr-2* functioned in the RIM interneurons to modulate chemotaxis learning and memory. A mutation in a glutamate-gated chloride channel homolog, *avr-15*, reduced starvation-induced gustatory learning (Ye et al., 2006). In addition, *casy-1*, a homolog of a gene shown to play a role in human memory, calsyntenin 2, was shown to disrupt associative learning in multiple modalities (gustatory, olfactory, and thermosensory) in *C. elegans*. The close functional relatedness of *casy-1* with its human homolog was demonstrated when the expression of human calsyntenin 2 gene in *C. elegans* rescued the behavioral defects of *casy-1* mutants. Interestingly, rescuing *casy-1* only in *glr-1* positive neurons rescued olfactory learning. In contrast, *casy-1* rescue in *glr-1* positive neurons did not rescue gustatory associative learning; this supported the hypothesis that although many forms of learning appear to require *glr-1*, salt (gustatory) conditioning is not *glr-1* dependent.

The Future of Learning and Memory in *C. elegans*

Since the publication of the first report that *C. elegans* could learn, learning and memory paradigms in *C. elegans* have been expanding. With the development of new paradigms, new learning abilities are being uncovered. For example, a recent study reported that *C. elegans* could learn an environmental oxygen level that was paired with food and another study suggested that *C. elegans* could learn mazes.

C. elegans has also been used to study learning and memory across the life span. For example, short-term habituation was shown to be similar across development. Additionally, the memory for mechanosensory habituation

training presented to 1-day-old juvenile worms could be retrieved in 5-day-old adults. At the other end of the life span, researchers have begun to investigate aging-dependent effects on learning and memory in the *C. elegans*. Similar to mammalian systems, not all learning and memory functions decline with age. Habituation increased in old animals and spontaneous recovery slowed with age. Examination of thermotaxis learning behavior across adult life indicated that the learning behavior declined between mid- and late-reproductive ages.

The powerful genetic, neurobiological, and developmental tools available for *C. elegans* make this organism a unique model to investigate the underlying mechanisms for learning and memory. From the first discovery of the learning ability of *C. elegans* in 1990 until today, neuronal circuitries, genes, and molecular mechanisms for learning and memory behaviors in *C. elegans* have been identified. This not only highlights the efficacy of the *C. elegans* as an animal model to dissect the underlying mechanisms for learning and memory, but also demonstrates that a number of mechanisms of plasticity found in *C. elegans* are highly conserved across evolution and may represent traits critical to survival throughout the animal kingdom.

See also: Non-Elemental Learning in Invertebrates.

Further Reading

Aamodt E (2006) *The Neurobiology of C. elegans*. New York, NY: Academic Press.

Ardiel EL and Rankin CH (2008) Behavioral plasticity in the *C. elegans* mechanosensory circuit. *Journal of Neurogenetics* 22(3): 239–255.

Bargmann CI (October 25, 2006) Chemosensation in *C. elegans*. In: Jorgensen E and The *C. elegans* Research Community (eds.) *Wormbook*. http://www.wormbook.org: doi/10.1895/ wormbook.1.123.1.

Chalfie M, Sulston JE, White JG, Southgate E, Thomson JN, and Brenner S (1985) The neural circuit for touch sensitivity in *Caenorhabditis elegans*. *Journal of Neuroscience* 5(4): 956–964.

de Bono M and Maricq AV (2005) Neuronal substrates of complex behaviors in *C. elegans*. *Annual Review of Neuroscience* 28: 451–501.

Giles AC and Rankin CH (2008) Behavioral and genetic characterization of habituation using *Caenorhabditis elegans*. *Neurobiology of Learning and Memory* 92(2): 139–146.

Giles AC, Rose JK, and Rankin CH (2006) Investigations of learning and memory in *Caenorhabditis elegans*. *International Review of Neurobiology* 69: 37–71.

Hobert O (2003) Behavioral plasticity in *C. elegans*: Paradigms, circuits, genes. *Journal of Neurobiology* 54(1): 203–223.

Hodgkin J, Horvitz HR, Jasny BR, and Kimble J (1998) *C. elegans*: Sequence to biology. *Science* 282(5396): 2011.

Mori I (1999) Genetics of chemotaxis and thermotaxis in the nematode *Caenorhabditis elegans*. *Annual Review of Genetics* 33: 399–422.

Mori I, Sasakura H, and Kuhara A (2007) Worm thermotaxis: A model system for analyzing thermosensation and neural plasticity. *Current Opinion in Neurobiology* 17(6): 712–719.

Murakami S (2007) *Caenorhabditis elegans* as a model system to study aging of learning and memory. *Molecular Neurobiology* 35(1): 85–94.

Nader K, Schafe GE, and Le Doux JE (2000) Fear memories require protein synthesis in the amygdala for reconsolidation after retrieval. *Nature* 406(6797): 722–726.

White JG, Southgate E, Thomson JN, and Brenner S (1986) The structure of the nervous system of the nematode *Caenorhabditis elegans*. *Philosophical Transactions of the Royal Society of London Series B* 314(1165): 1–340.

Ye HY, Ye BP, and Wang DY (2006) Learning and learning choice in the nematode *Caenorhabditis elegans*. *Neuroscience Bulletin* 22(6): 355–360.

Relevant Websites

www.wormbase.org – Worm Base.

www.wormbook.org – The online review of *C. elegans* Biology. Worm Book.

www.wormatlas.org – Worm Atlas.

Nervous System: Evolution in Relation to Behavior

J. E. Niven, University of Cambridge, Cambridge, UK; Smithsonian Tropical Research Institute, Panamá, República de Panamá

Introduction

The nervous system occupies a unique position at the interface between the genome and behavior and is the product of interactions between the genetically defined developmental program and the environment. The output of the nervous system and the effectors (e.g., muscles and secretory organs) it controls is behavior and, therefore, evolutionary change within the nervous system is the basis for the evolution of behavior. The remarkable behavioral diversity of animals has required changes in the morphology or physiology of the nervous system or sometimes both. Identifying these changes and showing how they affect behavior is essential for understanding the evolution of the nervous system.

All nervous systems are composed of broadly similar molecular components – voltage-gated ion channels, G-proteins, pumps, transporters, neurotransmitters, etc. Combinations of these components are found in nervous systems throughout the animal kingdom. The molecular components are found in neurons that are themselves elements of neural circuits. The structural motifs of these circuits formed by connections between neurons, such as lateral inhibition or feed-forward excitation, are similarly found throughout animal nervous systems, and many were originally described in invertebrate nervous systems. Thus, the challenge is to explain how, from similar building blocks and circuit motifs, nervous systems capable of generating such different behavioral output as a human and a nematode worm have evolved.

We can consider the evolution of the nervous system at many levels of organization from whole brains and brain regions to neural circuits, single neurons, and molecules. These different levels are all interconnected; molecules influence the input–output relationships of individual neurons altering the final output of neural circuits, local circuits form brain regions, and these regions connect to form whole brains. Where homologous elements can be identified, these different levels of organization can be compared among species with known phylogenetic relationships. Over larger phylogenetic scales, however, differences in the development of the nervous system, the number of neurons, and their anatomy mean homologous elements often cannot be identified. Nervous systems also differ substantially in the extent to which neurons are gathered into a central mass (centralization), and this mass concentrated towards the animals' anterior (cephalization).

Differences in the size and structure of nervous systems and their behavioral output pose a substantial challenge for understanding their evolution. Comparative studies of nervous systems are limited by the inability to find measures that permit meaningful comparisons among species over small and large phylogenetic scales. For example, the human nervous system is highly cephalized and centralized. With 3 billion neurons, the human brain is difficult to compare with the nematode *C. elegans*, which contains just 300 neurons. Yet, both of these nervous systems are the result of selective pressures, which affect all levels of organization, and ultimately should be encompassed by any framework hoping to address the evolution of the nervous system.

The aim of this article is to emphasize how consideration of adaptation, phylogeny, development, and mechanism can contribute to our understanding of the evolution of the nervous system and the behavior it generates. A comprehensive discussion of the evolution of the nervous system is not possible within a single article. Instead, principles will be drawn from different nervous systems (e.g., birds, mammals, fish, insects) at all levels of organization (whole brains, brain regions, neural circuits, neurons). Although relevant, a detailed discussion of the generation of behavior by neural circuits is beyond the scope of this article but is dealt with in other articles of the encyclopedia.

Evolutionary Changes in Nervous Systems for Improving Behavioral Performance

Nervous systems are under selective pressure to produce adaptive behavior in fluctuating, noisy environments. Adaptive behavior requires that sensory information is acquired and processed accurately, allowing decision making and motor planning to be adjusted to prevailing environmental conditions. Sensory information is interpreted in the context of memories that have been formed in response to previous experience. Behavioral performance may be improved by greater accuracy of information acquisition and processing in sensory systems, the formation of more accurate memories, improved decision making, or greater precision of motor control.

Selection to improve behavioral performance would be expected to produce nervous systems that are adapted to an animal's environment. The adaptation of components within nervous systems to the environment has been demonstrated in numerous studies. The majority of these studies have focused on sensory systems because the information that they encode is most easily determined. Both the sensitivity and selectivity of sensory receptors from several modalities including vision, olfaction, and audition have been shown to be adapted to the abiotic and biotic environment. Indeed, even the combination of sensory systems that an animal possesses is adapted for environmental conditions (**Figure 1**). Sensory systems may be adapted to abiotic factors, such as the intensity and spectral composition of light, but also to signals generated by predators, prey, and conspecifics. Adaptation may involve changes at all levels of the nervous system from molecular components, such as receptors or ion channels, to the size, number or connectivity of neurons.

Within motor systems, increases in the number of motor neurons controlling a particular muscle and in the specificity of their recruitment would be expected to produce more precise control of muscles. Imprecision in the motor system may be due to noise or to changes in the intended movement, making it difficult to quantify the precision of the motor system. There are little data, except in humans, that specifically address these relationships. The organizations of the motor systems controlling local limb movements are remarkably similar in vertebrates and insects, despite vertebrates typically having far greater numbers of motor neurons innervating their muscles. However, the impact of these differences in motor neuron numbers upon behavioral performance is difficult to interpret because of biomechanical differences between endo- and exoskeletons.

Improving the extraction and processing of information from the environment or the precision of motor control usually requires increases in the size and/or number of neurons. For basic biophysical reasons, larger

neurons can support faster signals, allowing them to encode information at higher frequencies (**Figure 2**). This explains the large size of neurons responsible for generating escape behaviors in which the speed of signaling is often vital for survival, such as the giant fibers in squid, crayfish and cockroach. Greater numbers of sensory receptors improve the resolution of information extracted from the environment. Within the central nervous system, increased numbers of neurons may increase the amount of processing of sensory information, while increased numbers of motor neurons may also improve the precision of motor control. Many species, relying primarily on one sensory modality, have relatively greater numbers of receptors and interneurons for detecting and processing information from that modality than closely related species relying primarily upon a different modality. For example, the visual cortical regions of the brain in the African hedgehog, which lives above ground, are well-developed in comparison to those of the Star-nosed mole, which is subterranean and has an enlarged cortical somatosensory representation (**Figure 3**). Likewise, in those species that rely heavily upon particular forms of memory, such as spatial memory, brain regions involved in storing those memories are relatively enlarged compared to those species that do not.

Energetic Costs of Nervous Systems

Selective pressures on behavior, which is the final output of the nervous system, will affect all levels of organization

Figure 1 The combination of sensory systems an animal possesses is adapted for prevailing environmental conditions. Cave populations of *Astyanax mexicanus* that have been isolated for ~1 Ma show eye loss. The photograph shows one eyeless cave fish (foreground) and two fish from closely related surface-dwelling populations. Courtesy of R. Borowsky. Reproduced from Niven JE (2008) *Current Biology* 18: R27–R29.

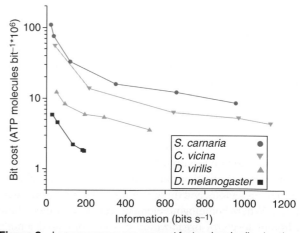

Figure 2 Larger neurons can support faster signals allowing them to encode more information. However, a plot of the information rates (bits s^{-1}) versus the energy efficiency of information transmission (ATP molecules bit^{-1}) shows a trade-off between energy efficiency and information coding in insect photoreceptors. Data from four fly species (smallest to largest): *Drosophila melanogaster*, *D. virilis*, *Calliphora vicina*, and *Sarcophaga carnaria*. Larger photoreceptors can transmit higher rates of information but are less energy efficient. Modified from Niven JE, Anderson JC, and Laughlin SB (2007) *PLoS Biology* 5: 28–40.

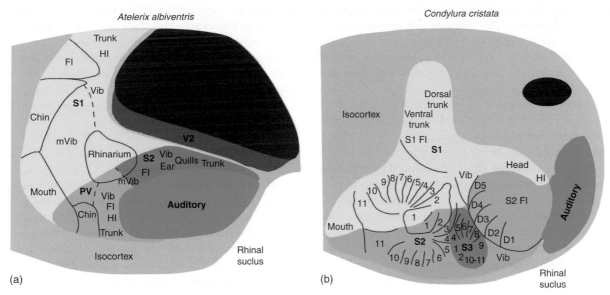

Figure 3 Species relying primarily on one sensory modality have relatively enlarged brain regions for processing information from that modality than closely related species relying primarily upon a different modality. A reduction in the size of visual cortical regions and an expansion in cortical regions associated with mechanosensory processing are associated with subterranean living. (a) The African hedgehog *Atelerix albiventris* lives above ground and has well developed visual (V) and auditory processing. (b) The star-nosed mole *Condylura cristata* is subterranean and has reduced visual (V) representation and an enlarged somatosensory (S) representation. Reproduced from Niven JE and Laughlin SB (2008) Energy limitation as a selective pressure on the evolution of sensory systems. *Journal of Experimental Biology* 211: 1792–1804 [after Catania (2005)].

within the nervous system. Many elements within the nervous system, including sensory and motor systems, contribute to numerous behaviors and, therefore, are subject to selective pressures on all these varied behaviors. Unopposed, these selective pressures would be expected to produce elaboration and expansion of the nervous system leading to more accurate behaviors. However, building, maintaining, carrying, and using a nervous system all have associated costs, which may be substantial. The human brain, for example, consumes ~20% of the resting metabolic rate while in flies, 8% of the metabolic rate is consumed by the retina alone. Thus, the direct and indirect energy consumption of the nervous system is a cost that opposes selective pressures to enlarge nervous systems and improve behavior.

The major source of energy consumption within the adult nervous system is electrical signaling, which is itself mediated by sodium (Na^+) and potassium (K^+) ion flow through voltage-gated ion channels along neurons. A signaling event reverses the polarity of these ions' concentrations across the neuron cell membrane, and that polarity must be restored by the Na^+/K^+ pump. The active transport mechanism of the pump consumes many ATP molecules, and additional processes within nervous systems consume energy including neurotransmitter production, vesicle loading, and transmitter recycling. The relationship between neural activity and energy consumption means that increased neural activity incurs greater energetic costs. Larger neurons or those

that support faster signals also consume more energy both at rest and during signaling.

The relationship between neural signaling and energy consumption has been quantified in fly photoreceptors (**Figure 2**). The information rate (measured in bits s^{-1} and combining both the speed and reliability of signaling) of a photoreceptor is dependent upon its size; larger photoreceptors encode more information than their smaller counterparts. Likewise, larger photoreceptors consume more energy than small photoreceptors but the energy costs increase out of proportion with information coding. Therefore, although photoreceptors become more efficient as they encode more information, larger photoreceptors are always less efficient, consuming more energy per unit of information. Thus, each additional unit of information that a photoreceptor can encode causes a drop in energy efficiency, which strongly penalizes any excess capacity and promotes the reduction of information coding to a functional minimum.

Within the nervous system, both morphology and information coding have evolved to reduce energy expenditure. The morphology of neurons within both vertebrate and invertebrate nervous systems minimize the total wiring length of connections (axons and dendrites). Reducing the total length of axons and dendrites should reduce the energetic cost of their construction and the energetic cost of transmitting signals. Computer simulations show that the positions of and connections among neurons in the *C. elegans* nervous system are close to the arrangement that would minimize the total length of axons and dendrites. This suggests that arrangement of neurons in the *C. elegans* nervous system has evolved to reduce the energetic cost of wiring. However, not all aspects of neural morphology conform to structures that minimize wiring length, including long range connections in *C. elegans* and the positions of ganglia within the insect nerve cord, reflecting the trade-offs between energy minimization and behavioral performance.

Many neural circuits use coding schemes that reduce the energy consumption of information coding including sparse coding, redundancy reduction, and predictive coding. These schemes have been characterized in the sensory systems of both vertebrate and invertebrates. One strategy is to reduce the amount of redundant information encoded at the periphery that is transmitted to central brain regions. Removing redundant information reduces the amount of energy consumed by encoding information because fewer electrical signals are transmitted to central brain regions. Along with reducing energy costs, redundancy reduction also makes the information processing easier. The energetic cost of information processing can also be reduced by encoding information as analog signals rather than action potentials. Information transmission using analog signals does not involve the large influxes of Na^+ ions that occur during action potentials and so does not incur the high energetic costs of extruding these ions. However, analog signals are restricted to short distances because the electrical signals degrade. Short neurons at the periphery of the vertebrate visual, auditory, and gustatory systems as well as neurons in invertebrate visual and motor systems lack action potentials and encode information as analog signals. Other strategies, such as sparse coding, make information coding more efficient by increasing the information content of each action potential.

The occurrence of energy saving strategies for information coding and cost minimizing neural architecture suggest that the nervous system has been under selective pressure to reduce energy costs. Components within the nervous system that deviate from structures or coding strategies that minimize energy consumption emphasize that costs can only be reduced to a functional minimum imposed by behavior.

Energetic costs have also been suggested to influence the evolution of entire brains. As mentioned earlier, the large relative brain volume of primate brains incurs a high energetic cost. The ability to support this high energetic cost has been hypothesized to be due to a reduction in the volume and energetic cost of other organs, such as the gut. Primate relative brain volume increases as gut volume decreases, suggesting a trade-off between these two organs that are energetically expensive to maintain. During evolution primates gaining access to high energy, more easily digestible food would be able to reduce their gut volume

and increase their brain volume, though alternative scenarios can also explain this trade-off. Trade-offs between brain volume and other expensive tissues, such as reproductive tissues, have been suggested in birds and bats. This interplay between behavior, physiology, and morphology emphasizes the difficulty of producing general evolutionary scenarios to explain brain evolution.

Constraints

Physical and developmental constraints may influence the extent to which selection for improved behavioral performance and reduced energetic costs can affect the evolution of the nervous system. Physical constraints on the nervous system include the amount of space that the nervous system can occupy and the minimum diameter of axons. Space may be a constraint in small insects such as the larvae of the beetle *Ptinella tenella* in which the volume of the head capsule is so small that the brain is pushed into the thorax. In larger animals, however, brain cases exceed brain volume suggesting that space is not constraining. There is also a physical limit upon the minimum diameter of axons that can transmit information imposed by the properties of voltage-gated ion channels (**Figure 4**). This limit corresponds to the smallest axons

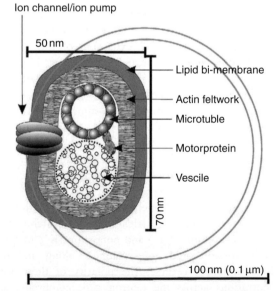

Figure 4   A physical limit upon minimum axon diameter imposed by the properties of voltage-gated ion channels. The minimum possible diameter of axons is set by mitochondria and other intracellular molecular components. The minimum diameter imposed by these components smaller than the limit imposed by voltage-gated sodium channels. The minimum diameter of axons in vertebrate and invertebrate nervous systems is close to the limit imposed by voltage-gated sodium channels. Reproduced from Faisal AA, White JA, and Laughlin SB (2005) *Current Biology* 15: 1143–1149.

observed in the adult nervous systems of both vertebrates and invertebrates. In very small insects, this constraint on minimum axon diameter may be particularly important because of the limited space the nervous system can occupy. Therefore, space and minimum axon diameter limit expansion and miniaturization of the nervous system, respectively.

In the absence of other constraints, selective pressures would be expected to produce independent changes in components of the nervous system producing mosaic evolution. Developmental mechanisms or the function of neural circuits, however, may impose constraints on the extent of independent change producing concerted evolution. In environments where particular sensory modalities are absent (e.g., light in caves or subterranean environments), sense organs detecting those modalities are reduced or absent strongly supporting the mosaic evolution of the nervous system. However, both mosaic and concerted evolution have been claimed to account for changes in the relative volumes of brain regions in mammals. Selection for changes in the volume of one brain region may influence the volume of other brain regions through axonal connections between them. Within the insect ventral nerve cord, individual ganglia may change in volume but also influence neighboring ganglia through axonal connections with other ganglia and the brain. Clearly, the connectivity between regions of the nervous system may constrain the influence of selection upon individual brain regions. Thus, mosaic and concerted changes may both occur to some extent within the nervous system.

The Evolutionary Significance of Brain Size

One of the most striking differences between nervous systems is their size (measured as weight or volume) – the brain of a blue whale weighs up to 9 kg while that of a locust weighs less than a gram (**Figure 5**). Differences in size correspond to differences in the total number of neurons within the brain and the size of individual neurons, larger brains generally having more numerous and larger neurons. Yet, the consequences of increased or reduced size of the nervous system, and especially the brain, for behavior remain unclear: honeybees have highly sophisticated behavior despite their small brains. Indeed, it is difficult to conceive of a single scale that would allow the behaviors of members of phyla as different as the Cnidaria, Arthropoda, and Chordata to be compared quantitatively. Nevertheless, an explanation is necessary for the differences between the largest and smallest brains; large nervous systems incur substantial energetic costs and should be strongly selected against unless they provide behavioral benefits. Yet it is not clear what behavioral benefits accrue to brains of large absolute size, for example those of whales, versus to brains of smaller absolute size including those of humans.

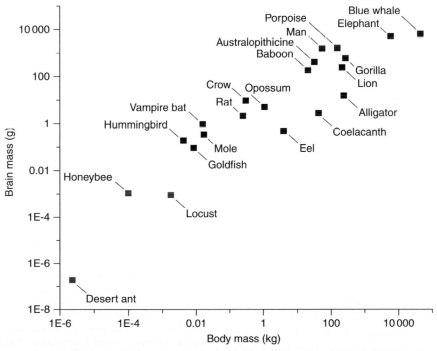

Figure 5 Differences among species in absolute brain volume. Brain volume increases with body mass over several orders of magnitude. Insects brains (red) are smaller than vertebrate brains (black), but this does not preclude them generating sophisticated behavior. Reproduced from Chittka L and Niven JE (2009) Are bigger brains better? *Current Biology* 19: R995–R1008.

One suggestion has been that it is the relative and not the absolute size of brains that is related to their behavioral output. Relative brain size takes into account the enormous differences in body mass among species. Numerous comparisons among vertebrates, and to a lesser extent among insects, have been made using this measurement. In both vertebrates and insects, brain size scales with body mass but individual species may have relatively larger or smaller brains. Within the vertebrates, humans have the largest relative brain size leading to the suggestion that it is the relative, rather than the absolute size of the brain, which affects behavioral output. This hypothesis provides an explanation for large mammals, such as whales, having larger absolute brain size than humans but not more complex behavior. There are problems, however, with using relative brain size as a measure of overall behavioral output. Relative brain size is a post hoc measure with no theoretical basis, chosen solely because it is a measure that ranks humans higher than other mammals. Additionally, the relative brain sizes of insects often far exceed that of mammals or birds suggesting that this measure does not even meet the criterion of ranking humans most highly.

The computational capacity of the nervous system is mainly determined by its absolute size and specifically by the number of neurons it contains, their size, and the number of connections between them. As nervous systems, including brains, become larger, the distances over which signals must be transmitted increases. Because conduction velocity is proportional to axon diameter, axon diameter must increase in larger brains to preserve speed. Thus, the nervous systems of larger animals contain larger diameter axons. Additionally, connectivity must be maintained between distinct regions and so, as brains become larger and the distances among brain regions increase, the number of long-distance connections also increases. In mammalian brains, long-distance axons increase out of proportion with the size of the brain. Thus, increases in brain size do not always increase the numbers of computational elements.

The number of computations that the brain can support is also affected by its energy consumption. The energy available for computation and for supporting the brain at rest is dependent upon the specific metabolic rate of neural tissue. Brains may have different specific metabolic rates and, therefore, support different numbers of computations. Relatively large brains from small animals may support more computations than similarly sized brains from larger animals. Thus, both absolute and relative brain size may influence the number of computations the brain can support and therefore its behavioral output.

One problem that must temper any attempt to relate absolute or relative brain size to behavioral output is that individuals within a single species may have remarkably different behavioral capabilities but similar brain volumes. In humans, for example, performance on IQ tests differs substantially among individuals, demonstrating that brain size or neuron number is not sufficient to explain behavioral differences. This emphasizes the need to understand the function of nervous systems, including brains, at the level of neural circuits rather than in terms of size.

Unlike total brain size, the size of particular central brain regions is positively correlated with performance in specific behavioral tasks in which those brain regions are thought to be involved (**Figure 6**). The implicit assumptions of these studies are that the size of a particular brain region is related to neural processing and energy consumption and that the role of a particular brain region does not differ between the animals being compared. While these assumptions may be reasonable in closely related species as the phylogenetic distance between the species being compared increases, such assumptions become difficult to support. Moreover, a further assumption is often that a particular brain region is involved in a single behavior but typically the regions being compared often contain millions of neurons forming numerous neural circuits, making this unlikely. Nevertheless, the correlations found in these studies suggest that with increased knowledge of central neural circuits, it may eventually be possible to move beyond the peripheral nervous system to quantify neural processing and energetic costs in central brain regions and relate these to behavioral performance.

Future Directions

Although considerable progress has been made in understanding the evolution of nervous systems, many aspects

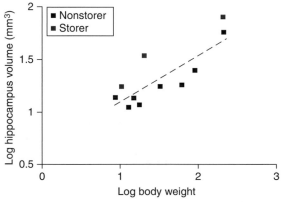

Figure 6 The volume of central brain regions is positively correlated with behavioral performance. Birds that store food have a relatively larger hippocampal volume than those that do not. Hippocampal volume is implicated in behavioral performance on spatial memory tasks, suggesting that birds that store food have a larger hippocampus because they retain spatial memories of their food stores. Modified from Krebs JR, Sherry DF, Healy SD, Perry VH, and Vaccarino AL (1989) *Proceedings of the National Academy of Sciences of the United States of America* 86: 1388–1392.

remain unclear including the extent to which neural components can change independently and the extent these changes influence behavior. Recent evidence suggests that the trade-offs between the information processing necessary for generating adaptive behavior and the energetic costs of that information processing have shaped the evolution of nervous system. Shifts in animals' environments can alter the balance between this trade-off, affecting almost any aspect of the nervous system. This trade-off may have influenced the evolution of nervous systems as diverse as those of fruit flies, cave fish, and the recently discovered hominin, *Homo floresiensis*. Trade-offs between energy consumption and behavioral performance can influence the properties of single neurons, sense organs, brain regions, or even entire brains. Yet, the extent to which changes in specific neural components can occur independently remains controversial. Even when selective pressures act on a specific neuron, neural circuit, or population of neurons, changes must be integrated into the remainder of the nervous system to produce adaptive behavior. Thus, evolutionary changes in one region of the nervous system are not entirely independent of other regions but the extent, from mosaic (independent) to concerted (dependent), remains unclear. Resolving this will require a better understanding of the evolution and development of the nervous system. Increases or decreases in the overall size of the nervous system may produce changes in the numbers and/or size of neurons, but these do not easily explain the differences in behavioral performance. Larger brains may not only contain greater numbers of neurons but also novel circuit elements and brain regions capable of additional serial and/or parallel processing of information. Additionally, it is not only the number of neurons and connections that influence the behavioral output of the brain, but novel neural pathways and connections between may also have dramatic impacts upon behavior. New neural connections can allow novel associations to be formed that were previously not possible. Thus, it is not only the size of brains but also the circuits within them that affect behavior.

The trade-off between behavioral performance and energetic cost produces evolutionary changes in neural components but the extent to which changes in single neurons and neural circuits produce changes in behavior is unknown. Changes in one component of the nervous system may not affect behavior because there is extensive plasticity within nervous systems that is capable of buffering change. Evidence also suggests that neural circuits can produce similar outputs in many different ways, such as different sets of ion channels and strengths of connections between neurons. These different outputs are likely to incur different energetic costs and, therefore, selection for minimum energetic costs will favor certain circuit configurations although they produce similar behavioral outcomes. Thus, changes in the properties of neurons and the strength of connections between them can affect both behavioral output and the energetic costs of generating behavior.

See also: Adaptive Landscapes and Optimality; Costs of Learning; Development, Evolution and Behavior; Levels of Selection; Naked Mole Rats: Their Extraordinary Sensory World; Non-Elemental Learning in Invertebrates; Predator Evasion; Problem-Solving in Tool-Using and Non-Tool-Using Animals; Spatial Memory.

Further Reading

Bullock TH and Horridge GA (1965) *Structure and Function in the Nervous System of Invertebrates*. San Francisco, CA: W.H. Freeman & Company.

Catania KC (2005) Evolution of sensory specializations in insectivores. *Anatomical Record Part A: Discoveries in Molecular, Cellular and Evolutionary Biology*, 287: 1038–1050.

Cherniak C (1995) Neural component placement. *Trends in Neurosciences* 18: 522–527.

Chittka L and Niven JE (2009) Are bigger brains better? *Current Biology* 19: R995–R1008.

Healy SD and Rowe C (2007) A critique of comparative studies of brain size. *Proceedings of the Royal Society B* 274: 453–464.

Kaas JH and Collins CE (2001) The organisation of sensory cortex. *Current Opinion in Neurobiology* 11: 498–504.

Laughlin SB and Sejnowski TJ (2003) Communication in neuronal networks. *Science* 301: 1870–1874.

Niven JE and Laughlin SB (2008) Energy limitation as a selective pressure on the evolution of sensory systems. *Journal of Experimental Biology* 211: 1792–1804.

Roth G and Dicke U (2005) Evolution of the brain and intelligence. *Trends in Cognitive Sciences* 9: 250–257.

Striedter G (2005) *Principles of Brain Evolution*. Sunderland: Sinauer.

Nest Site Choice in Social Insects

S. C. Pratt, Arizona State University, Tempe, AZ, USA

Introduction

Social insects are famous for their elaborate nest architecture; less well-known is their skill at moving from one nest site to another. Some, like army ants, move so often that they make no permanent structure, bivouacking instead in simple natural shelters. Others, like honeybees and polybiine wasps, build elaborate nests, but emigrate to new homes during colony reproduction. Still others, like ants of the genus *Temnothorax*, are often forced to move because of the fragility of their nests. House-moving is one of the most challenging tasks a colony faces. Its future success depends on finding a home that offers the right physical environment, protection from enemies, and access to resources. At the same time, choosiness must be balanced with speed, to minimize exposure to a hostile environment, and to prevent delays in growth and reproduction. In most cases, consensus must be reached among hundreds or thousands of individuals, lest the colony should divide among multiple sites to the detriment of all. Finally, all of this must be achieved without well-informed leaders or central control. Instead, the work of selecting and moving to a home is distributed across a population of workers, each informed about only a limited number of options, and influencing only a portion of its nestmates. Social insects have evolved impressively sophisticated solutions to these challenges, making nest site selection a leading model system of the collective intelligence of animal groups. This article reviews what has been learned about the two best-studied groups: *Temnothorax* ants and the honeybee *Apis mellifera*.

Nest Site Choice by *Temnothorax* Ants

Temnothorax are adept house-movers, an ability that is likely related to the fragility of their nests. The best-studied species, *T. albipennis* and *T. curvispinosus*, typically live in rock crevices or hollow nuts. In the laboratory, where most studies have been carried out, they thrive in artificial cavities made from a perforated slat sandwiched between glass slides. Emigrations can be induced by removing the roof slide and providing an intact nest nearby. Over the next few hours, the colony safely relocates to its new home. This process is best understood by considering the simple case when only one site is available, before turning to the more complex problem of deciding between sites.

Organization of Colony Migration

Emigrations are organized by a minority of active scouts, roughly one-third of the colony's workers. Each of these scouts sets out from the damaged nest to find a new home, thoroughly inspecting any candidate that she finds. If it passes muster, she returns to the old nest to inform other scouts of its location. She uses a behavior called tandem running, in which she attracts a single recruit to follow her toward the new site (**Figure 1(a)**). Their progress is slow and halting, as the leader must pause frequently to allow her follower to catch up. The pair often lose contact for good before reaching the site, but even these broken tandems recruit ants, because the orphaned follower enjoys a higher chance than a naive searcher of finding the target.

Tandem followers make their own assessment of the site and may also begin to recruit. The resulting positive feedback increases the site's population until it reaches a critical level and triggers a dramatic change in behavior. Scouts cease tandem runs from the old nest, and instead begin to carry nestmates, one at a time, to the new site (**Figure 1(b)**). These transports are roughly three times faster than tandem runs, and population growth accelerates sharply. Over the next few hours, the entire colony is brought to its new home.

Emigration is thus divided into two phases. In the first, discoverers use tandem runs to bring fellow scouts to their find. In the second, the assembled corps of scouts transports the bulk of the colony. Transported ants are generally not scouts, but members of the colony's passive majority, including brood items and queens. This change in targets may explain the difference in recruitment methods. Speedy transports are better for efficient movement of a large number of nestmates, but a scout needs more than quick transit. She must also learn visual landmarks that mark the route, so that she can later navigate independently. A tandem follower is better positioned to learn than a transported ant, because she adopts the same posture she will later use when recruiting on her own.

How does a scout decide when to switch from the first phase to the second? After completing a tandem run, she assesses the population at the new site, apparently through her rate of physical encounters with other ants. Once this population attains a threshold level, or quorum, she switches to transport (**Figure 1(c)**). Quorum-sensing is a logical way for scouts to tell when they have assembled enough transporters. It can also save them from

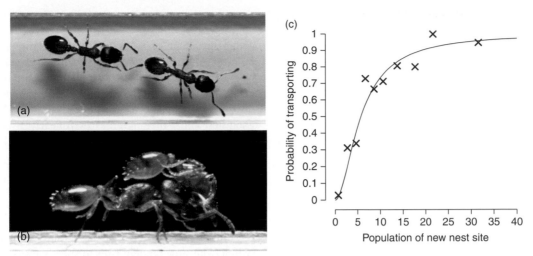

Figure 1 Recruitment behavior used in emigration by *Temnothorax* colonies. (a) Tandem run, in which a single ant is slowly led to a candidate site. (b) Social transport, in which a single nestmate is rapidly carried to the new site. (c) Quorum rule for switching from tandem runs to transports: Crosses show the proportion of ants deciding to transport, rather than lead a tandem run, as a function of the population of the site being recruited to. Line shows a nonlinear function fit to these data.

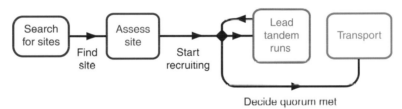

Figure 2 Summary of the decision algorithm used by scout ants during collective nest site choice.

unnecessary recruitment to a site that is near the old nest and easy to find, such that independent discoveries bring an adequate corps of transporters.

There is, however, another dimension to quorum attainment: it marks the last in a series of increasing levels of commitment to a site. A scout enters the first level when she decides to search for a new nest, spurred by the inadequacy of her current home. The second level begins when she finds a candidate and assesses its quality. If she judges it good enough, she advances to the third level in which she recruits fellow scouts to evaluate the site. The final level comes only when quorum attainment indicates that these others have confirmed her judgment by continuing to visit or recruit to the site. From that point on, she pays no further attention to population, and will continue to transport even if the site is experimentally emptied of nestmates.

This series of steps constitutes a decision algorithm that guides scout behavior (**Figure 2**). The algorithm clarifies two otherwise puzzling observations. First, a scout that has found the new site but not yet sensed a quorum will sometimes retrieve isolated brood items. She carries these not to the safety of the intact new site, but to the destroyed old nest. Second, after sensing quorum attainment, many scouts lead 'reverse' tandem runs from the new nest back to the old. Both behaviors make sense if we assume that recruitment behavior is described by two simple rules: tandem runs are only led away from home to a place where work needs to be done, and transports are made only toward home to repatriate lost or misplaced nestmates. Before a scout senses a quorum at a new site, the old nest is still her home, despite being heavily damaged. She transports lost ants there, and she leads tandem runs away from there to summon help in assessing a candidate site. Her allegiance switches to the new site only when it attains a quorum. From then on, she transports ants only to her new home and she leads tandem runs away from there to summon help in retrieving misplaced ants.

Collective Decision-Making Among Nest Sites

In most cases, colonies confront many candidate homes and must decide among them. Laboratory experiments show that colonies have strong preferences and are adept at choosing a favored site among a group of inferior competitors. They care about many site attributes, but give particular weight to having an intermediate cavity size and a small entrance. These features likely contribute to nest defense, the accommodation of future growth, and the regulation of internal nest environment. Ants also strongly favor a dark interior, perhaps as an indirect cue of nest wall integrity. Context matters as well, and ants avoid sites that are too close to competing colonies or that contain corpses of conspecific ants. Colonies integrate all of these attributes when assessing sites, weighting them according to importance.

Nest site choice is a challenging task, with inherent tradeoffs between decision speed, accuracy, and unanimity. A colony can improve its chances of finding the best site by evaluating many candidates, but this will take time and make it harder to winnow alternatives to a single choice. Coordination is also challenging, as scores or hundreds of ants must achieve consensus without any single ant learning about all sites, choosing the best and directing others to go there. Instead, the decision is shared by the population of active scout ants, each knowing only a subset of the options. In essence, the decision results from a competition among recruitment efforts at different sites, driven by two key components of the behavioral algorithm described earlier: quality-dependent recruitment initiation and the quorum rule.

Quality-dependent recruitment initiation

When a scout finds a site, she typically does not recruit to it right away, but first makes several visits in between trips to the old nest or further search of the surrounding landscape. The interval until the start of recruitment can be quite long, but it will be longer, on average, for worse than for better sites. That is, each scout conditions her probability of starting to recruit on her assessment of site quality. This effect is amplified by the positive feedback inherent in recruitment, because the scouts brought to a site will themselves initiate recruitment at a quality-dependent rate. This leads to faster population growth at a better than a worse site, driving the colony toward selection of the better site.

From the point of view of an ant that has found a mediocre site, this rule amounts to an investment of time to improve the colony's chances of finding a better option. The balance of exploitation versus exploration is a fundamental problem for any animal engaged in search, whether for a nest site, a mate, or a food source. If options are encountered sequentially, the animal must decide whether to settle for its current discovery or to search for a better one. A scout that delays recruitment to a site is essentially opting for further search. There is an interesting difference between her behavior and that of a solitary animal: her delay in recruiting buys time not only for her own search efforts, but also for those of her nestmates. Thus, she enhances

the colony's search effort, even if she herself never sees another site.

An advantage of this rule is that it allows a colony to hold out for an ideal site, but to settle eventually for the best that can be found. If only a mediocre site is available, ants will recruit to it, although it will take them longer to do so. As a result, colonies offered a choice between a good nest and a mediocre nest will nearly always choose the good one, but the same colonies offered a choice between a mediocre site and a still worse one will nearly always choose the mediocre one.

Quorum rule

The ants' quorum rule amplifies the quality-dependent recruitment effect. Once a site attains a quorum, the switch from slow tandem runs to speedy transports accelerates population growth. On average, a better site will experience this acceleration sooner, allowing it to expand its lead over inferior competitors. The quorum rule favors better nests by imposing an extra level of scrutiny. Each scout relies not only on her direct assessment of a site, but also on an indirect cue about the judgments of other ants. She fully commits only if some minimum number vote with their feet by spending time at the site. This rule can filter out errors by a small number of ants that start recruiting immediately to a site that is not very good.

This description of nest-site choice is somewhat idealized. Colonies may split between sites or even move into an inferior candidate, especially when moving rapidly under duress or when an inferior site happens to be very close to their current home. When this happens, the colony must launch a second emigration from the inferior to a better site. These multistage migrations are suboptimal outcomes, given the likely dangers of exposure during transport and the risk that the colony never reunites. The ants' decision algorithm does not eliminate these dangers, but it minimizes them by reducing the likelihood of splitting between sites.

It is tempting to divide emigrations into an early deliberative phase and a later implementation phase, with the boundary marked by quorum attainment. There is some value in this distinction, but these functions are really not so well separated. Decision-making continues after quorum attainment, most obviously when a temporary split must be resolved by secondary emigrations. At a more basic level, individual ants do not cease to assess a site's quality just because it has attained a quorum. Even those scouts that arrive at a nest after it has grown quite populous still condition their recruitment on its intrinsic quality. Scouts always consider both their own direct assessment of a nest and the 'votes' of their nestmates.

Speed/accuracy tradeoff

A crisis caused by nest destruction is not the only occasion for house-moving. A colony inhabiting an adequate nest will emigrate if a better site becomes available. In these unforced emigrations, colonies take far longer to finish the move, but their performance is much better, with less splitting between sites. This difference illustrates a fundamental tradeoff between speed and accuracy that is faced by all decision-makers. *Temnothorax* colonies have the ability to shift their stress from one to the other, sacrificing accuracy for speed when pressed to end their dangerous exposure, but investing time for a better result when urgency is less.

Interestingly, colonies use the same behavioral algorithm regardless of urgency, but they tune it for each setting. In a crisis, each active ant moves more rapidly through the algorithm's increasing levels of commitment to a site. The most striking change is their higher rate of recruitment initiation to a candidate site, and models suggest that this has a large effect on the speed/accuracy tradeoff. By delaying recruitment longer in less urgent circumstances, ants invest more in search, at the cost of taking longer to complete the move. More search effort improves chances of finding the best site, but the colony also gains in discriminatory power. When latencies are long, so are the differences between those at better versus worse sites. Greater latency differences mean greater differences in population growth, and thus a greater likelihood that a better site outstrips lesser ones to become the colony's choice.

Individual comparison

In the process described earlier, comparison among sites is an emergent property of the whole colony, not an activity of well-informed individuals. This does not mean, however, that individuals lack this capacity. Scouts almost certainly compare candidate sites with their current home, as indicated by their unwillingness to abandon an adequate nest unless they find a significantly better one. Simulations suggest that this ability is needed for a colony to settle stably in a site, rather than constantly initiating new emigrations. Whether a single ant can also pick the better of two candidate sites is less certain. Ants may simply forget about a nest if they leave it without recruiting, or they may retain a memory of it that causes them to ignore any subsequent finds of lower quality. Such comparisons are potentially quite important, given that a quarter or more of active ants are seen to visit multiple sites, at least in small laboratory arenas. Even without direct comparisons, emigrations may be strongly influenced by these ants, because of the opportunities created for better sites to divert potential recruiters from lesser ones.

Individual comparison is also relevant to rational decision-making, which requires that options be consistently ranked according to intrinsic fitness value, and not by comparison to available alternatives. Irrationality is commonly seen when decision-makers are faced with options that vary in multiple attributes, such that none is clearly

superior. Some strategies for resolving these difficult choices involve direct comparisons among options and can lead to irrational outcomes such as intransitivity or preference reversals. Faced with one such context, *Temnothorax curvispinosus* colonies behave quite rationally, possibly as a result of their highly distributed mode of decision-making, in which most ants lack the opportunity to make direct comparisons.

Nest Site Choice by HoneyBees

The house-hunting behavior of honeybees has many similarities to that of *Temnothorax*, but also many revealing differences. Like the ants, honeybees are cavity nesters, at least in the temperate zone, where house-hunting has been best studied. Colonies show strong site preferences based on multiple criteria, including cavity volume, entrance size, and entrance location. Honeybees sometimes abandon a nest site and move to a new one, typically when foraging conditions deteriorate, but house-hunting most often occurs during colony reproduction. A colony's queen, along with about one-third of its workers, bequeath their nest to a new daughter queen and the remaining workers. The departing bees settle as a compact swarm on a tree branch or similar site. From this bivouac, the bees spend up to several days scouring the countryside for candidate sites, deliberating among them and choosing one as their new home.

Collective Decision-Making

Like *Temnothorax*, bees rely on a competition among recruitment efforts at different sites, carried out by a minority of nest site scouts. These scouts, numbering only a few hundred of the swarm's several thousand bees, travel up to several kilometers from the bivouac. Upon finding a candidate home, typically a tree hole or similar cavity, a scout inspects it closely. If its quality is sufficient, she returns to the swarm and uses waggle dance communication to inform other bees of its distance and direction. Her dancing also encodes the quality of the site, principally as the number of dance circuits she completes during her stay at the swarm. The more circuits, the more opportunity for followers to read the dance, and so the more new bees show up at the site. The recruits themselves may join in advertising the site, also tuning their number of dance circuits to site quality. The result is a positive feedback cascade that swells the number of scouts visiting the site, but at a rate that depends on site quality.

The swarm's corps of scouts typically find many possible homes, and dances are present for several candidates at the same time. How does the group settle on a single one? It was once thought that the decision was made on the swarm's dance floor, on the basis of the typical course

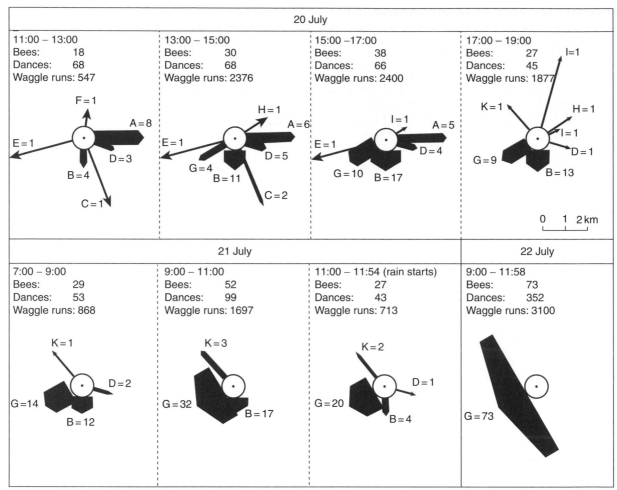

Figure 3   Summary of a honeybee swarm's decision process over 3 days. Each panel shows the number of dancers, dances, and waggle runs during a 1–3 h interval. The circle represents the swarm, and each arrow represents the distance and direction of a candidate nest site. The thickness of the arrow correlates with the number of bees advertising that site in the interval, also given by the number next to each site's letter designation. The swarm considered a total of 11 sites, but with no clear leader until the second half of the process, when site G gradually gained support and became the target of all the dances. Adapted from Seeley TD and Buhrman SC (1999) Group decision making in swarms of honeybees. *Behavioral Ecology and Sociobiology* 45: 19–31, with permission from Springer.

of events there: the number of advertised sites diminishes over several days until only one remains, and the swarm lifts off and flies to this site (**Figure 3**). It now appears, however, that a dance consensus is not the trigger that tells the bees that a choice has been made. Instead, like *Temnothorax*, each scout monitors her candidate site to determine when its population has reached a quorum. Upon sensing this, she returns to the swarm and pushes her way through it, delivering a brief vibrational signal called piping to scores or hundreds of bees. Piping stimulates recipients to warm up for flight by shivering their wing muscles. Within an hour, the flight-ready bees are prompted to lift off by buzz-runners, who break up the cluster of bees by burrowing rapidly through it. Interestingly, a similar combination of piping and buzz runs is also used earlier in emigration, to instigate the swarm's initial departure from its natal nest.

Once aloft, the diffuse but cohesive group flies directly for the new site. This is an impressive feat of collective orientation in which thousands of bees, 95% of them ignorant of their destination, travel up to several thousand meters to a pinpoint goal. An early hypothesis held that the bees are guided by pheromones released from the Nasonov glands of informed scouts. This does not appear to be the case, since sealing shut the glands of all swarm members does not interfere with normal orientation. Experiments and models better support the 'streaker bee' hypothesis, which holds that knowledgeable scouts point the way by flying through the swarm at high velocity in the direction of the target site.

Quorum Sensing, Attrition, and Consensus

As in the ants, quorum sensing amplifies a difference among sites created by quality-dependent recruitment effectiveness. Better sites experience faster population growth and so are more likely to reach a quorum and trigger lift-off. For the bees, however, quorum attainment is a much clearer watershed than it is for the ants. It marks the shift from a deliberative period lasting several days to an implementation period that may take only 1 h. This sharper distinction facilitates consensus on a single site by reducing the time window for a second site to reach a quorum. Indeed, bee swarms do not tolerate splitting between sites. If there is disagreement among scouts when the swarm lifts off, it soon resettles and continues to deliberate. This difference from the ants may be rooted in a greater cost of splitting for bees. Division of the swarm leaves one portion queenless and doomed to early extinction, as the workers cannot lay the fertilized eggs necessary to rear a new queen. *Temnothorax* colonies have brood from which new reproductives can be reared, and some colonies have multiple queens.

The importance of consensus for the bees is also suggested by another distinctive feature of their decision-making: dance attrition. Unlike *Temnothorax* recruiters, each honeybee dancer eventually ceases advertising a site, even before the swarm has reached a decision (**Figure 4**). This applies even to dancers for an excellent site, although it takes longer for their activity to decline from its high initial levels. An important effect of attrition is to slow population growth at each advertised site. Overly effective dancing poses the danger that more than one site will reach a quorum at the same time. This means either that the colony remains deadlocked or that it splits with disastrous consequences. By moderating recruitment strength, attrition lengthens the intervals between quorum attainment at different sites. It also fosters the achievement of a dance consensus. Although this consensus does not trigger the swarm's decision, it typically coincides with it, and it may help to avoid abortive lift-offs.

Another interesting difference from the ants is the lesser role for comparison or switching among sites by individuals. Given the importance of unanimity to the bees and their reliance on a centralized advertising location, it might be expected that scouts commonly follow one another's dances and determine for themselves which advertised site is best. Although such comparisons may occur, they appear not to be an important component of the swarm's decision. Very few scouts visit more than one site, and experimental suppression of comparison does not hinder the swarm's ability to settle on a single site.

Conclusion

A striking similarity between honeybee and ant emigration is the central role of quorum-sensing in coordinating

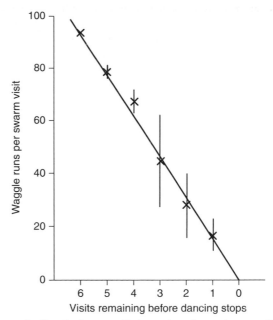

Figure 4 Scouts decrease their number of waggle runs with each successive visit to the swarm, eventually ceasing to dance altogether. This applies regardless of site quality, but bees advertising better sites start with a larger number of dance circuits, and so persist longer at dancing then bees advertising worse sites. Adapted from Seeley TD (2003) Consensus building during nest-site selection in honeybee swarms: The expiration of dissent. *Behavioral Ecology and Sociobiology* 53: 417–424, with permission from Springer.

behavior. The quorum rule provides a solution to a general dilemma faced by social organisms that must reach consensus decisions. On the one hand, they can benefit from the 'wisdom of crowds' if they filter out individual errors by taking into account the independent judgments of many individuals. On the other hand, if individuals are too independent, the group will have difficulty reaching consensus on a single option. The quorum rule offers a compromise between these demands: social influences are weak when exerted by only a few individuals, but their impact grows sharply once the numbers advocating an option surpass a threshold. This strategy is not exclusive to ants and bees. Many social animals, including fish, birds, and arthropods, use analogous threshold rules to optimize the integration of personal and social information.

The tradeoff between speed and accuracy is another general decision-making issue that emerges in both ants and bees. Effective discrimination among options improves with information, but gathering information requires an investment of time. Both ants and bees adopt strategies that markedly slow their decision-making, but make it more accurate. At least for ants, these measures can be adjusted to accelerate emigration at the cost of accuracy in urgent conditions. Individual decision makers face a fundamentally similar tradeoff and also have means to adaptively tune their behavior according to context. Their choices emerge from a neural network rather than a social one, but both systems address the same challenge and may use similar strategies. Thus, the future study of both individual and collective intelligence may benefit from seeking evidence of common solutions.

See also: Collective Intelligence; Communication Networks; Consensus Decisions; Decision-Making: Foraging; Distributed Cognition; Group Movement; Honeybees; Insect Social Learning; Rational Choice Behavior: Definitions and Evidence; Social Information Use.

Further Reading

Conradt L and List C (2009) Theme issue: Group decision making in humans and animals. *Philosophical Transactions of the Royal Society B* 364: 719–852.

Conradt L and Roper T (2005) Consensus decision making in animals. *Trends in Ecology & Evolution* 20: 449–456.

Franks NR (2008) Convergent evolution, serendipity, and intelligence for the simple minded. In: Morris SC (ed.) *The Deep Structure of Biology*, pp. 111–127. West Conshohocken, PA: Templeton Foundation Press.

Franks NR, Mallon EB, Bray HE, Hamilton MJ, and Mischler TC (2003) Strategies for choosing between alternatives with different attributes: Exemplified by house-hunting ants. *Animal Behaviour* 65: 215–223.

Franks NR, Pratt SC, Mallon EB, Britton NF, and Sumpter DJT (2002) Information flow, opinion polling and collective intelligence in house-hunting social insects. *Philosophical Transactions of the Royal Society of London B Biological Sciences* 357: 1567–1583.

Lindauer M (1955) Schwarmbienen auf Wohnungssuche. *Zeitschrift für vergleichende Physiologie* 37: 263–324.

Pratt SC (2009) Insect societies as models for collective decision making. In: Gadau J and Fewell JH (eds.) *Organization of Insect Societies*, pp. 503–524. Cambridge, MA: Harvard University Press.

Pratt SC, Mallon EB, Sumpter DJT, and Franks NR (2002) Quorum sensing, recruitment, and collective decision-making during colony emigration by the ant *Leptothorax albipennis*. *Behavioral Ecology and Sociobiology* 52: 117–127.

Pratt SC and Sumpter DJT (2006) A tunable algorithm for collective decision-making. *Proceedings of the National Academy of Sciences of the United States of America* 103: 15906–15910.

Seeley TD (2003) Consensus building during nest-site selection in honey bee swarms: The expiration of dissent. *Behavioral Ecology and Sociobiology* 53: 417–424.

Seeley TD and Buhrman SC (1999) Group decision making in swarms of honey bees. *Behavioral Ecology and Sociobiology* 45: 19–31.

Seeley TD and Morse RA (1978) Nest site selection by the honey bee, *Apis mellifera*. *Insectes Sociaux* 25: 323–337.

Seeley TD, Visscher PK, and Passino KM (2006) Group decision making in honey bee swarms. *American Scientist* 94: 220–229.

Sumpter DJT (2010) *Collective Animal Behavior*. Princeton, NJ: Princeton University Press.

Visscher PK (2007) Group decision making in nest-site selection among social insects. *Annual Review of Entomology* 52: 255–275.

Neural Control of Sexual Behavior

D. Crews, University of Texas, Austin, TX, USA

Introduction

The brain is a sexual organ, which like the gonad, is initially bipotential, differentiating into one of two types. More than a century of scientific research has established that the brain is the mediator and regulator of all aspects of reproduction. In this article, I trace the evolution of ideas related to brain organization and the control of sexual behavior. The question guiding investigators underwent a major paradigm shift 50 years ago: from the original emphasis on the bisexual nature of the brain to how the brain happens to differ in males and females. This may not seem to be an important distinction, but considering that, in the first instance, the emphasis is on the similarity of the sexes while in the second and current perspective, importance is placed on the differences between the sexes, there definitely is a shift in direction. In both psychology and biology, it is commonplace for investigators to not so much solve problems as to create new questions, without resolving the original question with the advent of newer techniques; in this instance, effort toward understanding the brain's inherent bisexuality was deflected to understanding the organ's sexual differentiation. Recently, a new paradigm has been introduced, which may reunite researchers as they address the two questions.

100 Years Ago

In the late 1800s, the focal question was why one sex would behave like the opposite sex, a phenomenon noticed more commonly in some species. The late 1800s and early 1900s marked the beginning of the realization that reproduction and sexuality differed in origin and consequence. In particular, Richard von Krafft-Ebing and Sigmund Freud speculated on the bisexual nature of the brain. It was during this period that 'bisexual' came to mean 'bipotential,' meaning that the same anlagen (the rudimentary beginnings of an organ, usually in the embryo) would give rise to one of two states, rather than the same structure housing two distinct states. Some time later, researchers such as Eugen Steinach (Austria) and Calvin Stone (USA) demonstrated that the interstitial (Leydig) cells (and not the Sertoli cells) of the testes produced the hormones (initially called 'incretions') responsible for seasonal as well as pubertal growth of secondary sex characters. These and other researchers (e.g., Carl Moore) suggested that hormones cause an 'eroticization'

of the central nervous system, though there was considerable debate as to whether they were acting generally or at specific sites. What was resolved by 1940 was that hormones changed the individual's sensitivity to specific stimuli (e.g., tactile, visual, and odor cues). It was also accepted that while males and females exhibited characteristic behaviors, they had the capacity to exhibit the behavior of the opposite sex. Indeed, Frank Beach in his compendium *Hormones and Behavior* devoted its second chapter ('Reversal or Bisexuality of Mating Behavior') to this common observation. Like others before him, he stressed that such heterotypical behaviors were exhibited alternately, never coincidentally, and were elicited by the stimulus context, and not by specific hormones. It is important to note here that early ethologists such as Tinbergen also emphasized the role of tonic inhibition in switching between behaviors.

> The general impression that one gains from a survey of the literature tends to throw some doubt on any concept of sex reversal which depends upon complete sex-specificity both of the behavioral mechanisms and of the gonadal hormones A somewhat more reasonable hypothesis would seem to be that in many if not all vertebrate species both males and females are equipped by nature to perform at least some of the elements in the overt mating pattern of the opposite sex.' (Beach, 1948, p. 69)

50 Years Ago

In 1959, a single publication by William C. Young and his colleagues changed the paradigm of behavioral endocrinology so much so there has been little work on bisexuality of the brain since that time; this seminal study set the trajectory of research on the neuroendocrinology of sexual behavior to the present day (this review is cited as Phoenix et al., in the readings at the end of this article). Indeed, for the past 50 years, almost all research in this area has focused on why males (or females) behave the way they do.

Drawing the analogy with the differential development of the accessory sex structures during embryogenesis as described by Alfred Jost a few years previously, Young and colleagues suggested that a similar dual anatomy exists in the brain, proposing that just as the early hormonal environment determined the fate of the ducts that transport eggs (Müllerian ducts) or sperm (Wolffian), these hormones also acted on the developing brain, specifically on the neural circuits subserving female- and male-typical sexual behaviors. In addition to its embryological foundation, the new perspective also built on the foundation laid earlier demonstrating that sexual behaviors were not simply dictated by sex steroid hormones, but reflected mechanisms intrinsic to the state of the brain itself. Although it was recognized that in some way the hormones were acting on the brain, the mechanism of this action was a mystery. It should be pointed out, however, that this new perspective in itself did not explain (nor did it seek to) the observation that individuals of either sex retain the capacity to, and commonly display, the behaviors typical of the opposite sex. The Organizational/Activational concept of Young and colleagues was further refined a few years later by Richard Whalen with the concept that the development of sex-typical behaviors resulted from two independent processes, namely, masculinization–demasculinization and feminization–defeminization (see section 'Are There Dual Circuits or a Single Circuit with Alternative Outputs?'). Put simply, 1959 marked a time when the salient question transitioned from 'why do males and females sometimes behave as the opposite sex,' to 'why do males behave like males and females like females.'

The Origin of Sexual Behavior

Before proceeding further, it is first necessary to raise the issue of the origin of sex itself. I am not referring to the evolution of sexual reproduction, or even why the preponderance of life forms exhibit two sexes. Instead of asking why sex evolved, it might be informative to ask who came first, male or female. The scientific view is that the 'female' was the first sex. (I would like to avoid the semantics for a moment as male and female are defined in terms of the opposite sex.)

There is little question among researchers that the first organisms simply cloned themselves. In each new generation, the complete genetic material of the parent and the siblings was identical. This same process occurs today in organisms that reproduce by parthenogenesis. In the process of evolution, the gametes were initially uniform in size (isogamy); but with time, they became different in size (anisogamy) and contained only one half of the genetic material that produced a new individual when the complementary types were fused (fertilization). This suggests – and evidence supports it – the supposition then that the first 'sex' was an egg producer. Put simply, what is called 'female' today was in fact the ancestral sex with males (sperm producer) relatively late entrants in the game of life.

Originally then, the brain was required only to coordinate and stimulate the production of eggs. With the development of two types of gametes came the need for behaviors that would be complementary, thereby ensuring fertilization. If one considers that the first sex was female, and males were derived much later in evolution, it stands to reason that behavior associated with ovulation (i.e., female-like receptivity) is the ancestral state and behavior associated with the delivery of sperm (i.e., male-like mounting) is a derived state. This more recent origin may account for 'male sexual behaviors' to be more plastic than are 'female sexual behaviors.'

Switching Between the Sex Roles

Early in development (the when and how varies between species), genes and hormones interact to organize the functional neuroanatomy such that later, as adults, males and females will exhibit complementary behaviors necessary for successful reproduction. This concept was originally built on an analogy with the sexual differentiation of the genital tract, and its characteristics were (i) completion during a limited sensitive window of embryonic development or shortly after birth, (ii) irreversibility, and (iii) the presumed existence of separate neural structures mediating male and female sexual behaviors. Although these particulars have been modified since to account for species differences, extensive research with rodents revealed a male-specific testosterone surge toward the end of in utero development, enabling later expression of male behavior (masculinization) and disabling later expression of female behavior (defeminization).

In formulating the Organizational/Activational Concept, Young and colleagues did not ignore, but did give rather short shrift to the observations available at the time that sexual behaviors characteristic of the opposite sex were displayed by individuals of most species studied, and particularly common in some. In view of this inherent and persistent bisexuality of the vertebrate brain, I believe that the analogy to the dual duct system was unfortunate and misleading. Rather, a more accurate perspective is to consider the network of limbic and hypothalamic nuclei involved in the control of sexual behavior to be a single entity, whose entirety is organized in a male- or female-typical way. How the implications of this perspective for the way research is conducted differ from those of a model based on the independent existence of separate 'centers' for male and female behavior in several ways is discussed further below.

Activation and Deactivation, Inhibition, and Disinhibition

A second kind of plasticity, observed in adulthood, is the activation and deactivation of behavior. Females display receptive behavior during the periovulatory phase of the ovarian cycle when estrogen levels are high, and at other times reject courting males. Males display mounting and other copulatory behaviors toward receptive females throughout the breeding season when androgen levels are high, and at other times show no particular interest in females. Activation of copulatory behavior appears to depend on gonadal sex steroids, being eliminated by gonadectomy and activated by exogenous testosterone (males) or estrogen and progesterone (females). However, gonadectomy followed by administration of the sex steroid typical of the opposite sex generally is not thought to activate the behavior typical of the opposite sex, a failure that is attributed to the permanent effects of developmental organization.

If one accepts this Developmental Organization followed by Adult Activation paradigm, one tends to view sex differences in brain structure as likely candidates for being involved in the display of male-typical behavior by males and female-typical behavior by females. Experimentally one asks how these differences arise during development, and then how the sexually dimorphic circuits are activated in adulthood. This perspective is little changed in recent years as the use of genetically modified mice has entered mainstream research on hormone-brain-behavior research.

Are There Dual Circuits or a Single Circuit with Alternative Outputs?

An influential conceptualization of how the brain might differentiate in males versus females was the Orthogonal Model of Richard Whalen (**Figure 1**, top panel). Summarizing the evidence to date, Whalen concluded that sexuality was not a one-dimensional or linear continuum, with masculine and feminine at opposite ends as originally

proposed by the early philosophers. Rather, Whalen suggested that sexuality comprises two distinct dimensions, one signifying the degree of masculinization and the other the degree of feminization. In the process of organization, these were affected differently to result in individuals typically displaying behaviors consistent with their gonadal sex.

The model was believed (and continues to be so by many) to reflect brain differentiation, along with an explicit identification of particular brain areas corresponding to masculine and feminine tendencies (e.g., see the work of McEwen listed in the readings at the end of this article). Early studies established that the medial preoptic area (mPOA) was the final integrative area necessary for the display of the male-typical mounting behaviors with the ventromedial nucleus of the hypothalamus (VMN) playing the comparable role in female-typical sexual receptivity (**Figure 1**, bottom panel). A common assumption was that when applied to the brain, the Orthogonal Model suggested that these particular nuclei were differentially influenced by early hormonal milieus and represented by two (dual) circuits, a view that is consistent with the canonical Organization-Activation paradigm outlined earlier. However, unlike the definitive work on the song system in birds, and despite the abundant work on sex differences on morphological and neurochemical aspects of mPOA and VMN in the mammalian brain, there is remarkably little evidence that the recorded differences are more than correlates of observed sexually differentiated behaviors. As Södersten put it, "the search for morphological sex differences in adult rat brains that are caused by the 'organizing effect of perinatal androgen' and that can be related to sex differences in behavior has not been fruitful and may continue unrewarded."

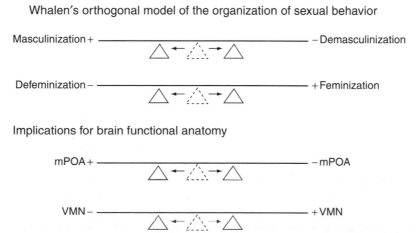

Whalen's orthogonal model of the organization of sexual behavior

Masculinization + ———————————————— – Demasculinization

Defeminization – ———————————————— + Feminization

Implications for brain functional anatomy

mPOA + ———————————————— – mPOA

VMN – ———————————————— + VMN

Figure 1   Whalen's (1974) Orthogonal Model for the Differentiation of Sexual Behavior (top panel). Masculinization and feminization are considered separate neuroendocrine organizational processes, such that during development in males masculine traits are enhanced and feminine traits are suppressed (=defeminization); the complement is postulated to occur in females. Extensive research indicated the final integrative area for mounting is the medial preoptic area (mPOA) and for receptivity the ventromedial hypothalamus (VMN). Thus, when the Orthogonal Model was applied to the brain (bottom panel), a parallel process of enhancement and suppression was believed to occur in the mPOA and VMN.

It is obvious that sexual behavior is the result of many brain nuclei acting in concert in addition to the external stimuli and the hormonal history of the participating individuals. An attractive formulation of this body of work is found in Sarah Newman's concept of a Social Behavior Network that underlies sexual behavior (**Figure 2**). By shifting the focus of study from single nuclei (nodes) in isolation to integrated networks, Newman predicted this would lead to new insights into brain-behavior relationships. Importantly, Newman focused on sex differences and did not consider the application of this model to address the question of the possible interactions within the network when animals display heterotypical sexual behaviors. Using this as a platform, I suggest that the sex-typical differences in behavior are the result of how the network activity varies as a result of the reciprocally inhibitory interaction of two root nodes (mPOA and VMN).

The Dual Circuits Model emphasizes how hormones act on dual neural circuits, one subserving male-typical mounting and the other female-typical receptivity, with each viewed as operating relatively independently of one another. Traditionally, research supporting this model is exemplified by study of a single sex with the dependent variable being the sex-typical behavior; c.f., mounting in male individuals, receptivity in female individuals. This has led to the development of models of the neural circuit of lordosis in female or mounting in males, but in isolation of its complement (**Figure 3**). On the other hand, the Common Network Model emphasizes how hormones act on a single neural network resulting in two mutually exclusive outputs. This model reflects the increasing appreciation of how brain nuclei are networked by neurochemical and molecular interactions and how these neural systems are fundamental (in an evolutionary sense), particularly when the brain must alternate between mutually exclusive behavioral outputs. The Common Network Model suggests then that sex-typical behavioral phenotypes are mirrored by specific neurotransmitter and molecular phenotypes in two functionally associated nuclei (as 'root nodes' of a larger network of nuclei). The whiptail lizard is instructive because it enables deconstructing the confounding properties of genotype-, sex hormone-, and developmental-specificity inherent in conventional mammalian model systems.

Reciprocal Inhibition Between the POA and the VMN

Certainly there is ample evidence that the mPOA and VMN are crucially involved in the control of male- and female-typical sex behaviors, respectively. What is less part of the current orthodoxy is the possibility that the two centers work in concert, albeit in a mutually antagonistic manner. However, the involvement of each brain

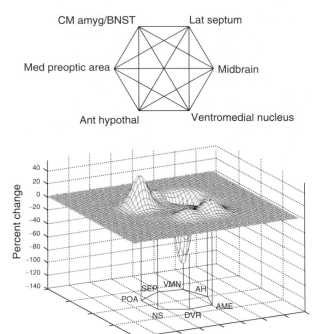

Figure 2 Newman's Social Behavior Network. Top panel illustrates how a limbic neural network consists of specific nuclei that are both hormone sensitive and reciprocally interconnected. The network of brain nuclei is similar in both sexes, but Newman proposed that the activity of the network is different in males and females when they display sex-typical (homotypical) behaviors. She did not speculate on the patterns of activity that may be reflected during the display of heterotypical sexual behaviors. Bottom panel depicts such a network as indicated by the pattern of metabolic activity (as measured by cytochrome oxidase histochemistry) in identified nuclei. The peaks and valleys indicate the *differences* in average abundance in each nucleus in sexually experienced male and female leopard geckos from the same incubation temperature; peaks indicate males greater than females and valleys indicate females greater than males. Geckos exhibit temperature-dependent sex determination and lack sex chromosomes, so differences are due to endocrine history and not genotype. Note the sex difference, particularly in the relationship between the preoptic area (POA) and the ventromedial nucleus of the hypothalamus (VMN). AH – anterior hypothalamus; AME – medial amygdala; DVR – dorsal ventricular ridge; NS – nucleus sphericus, homolog of the mediobasal amygdala; SEP – septum.

area in behaviors typical of the 'other sex' is not lacking. Two examples are that implantation of testosterone into the VMN restores sexual motivation, but not copulatory behavior itself, in castrated male rats; administration of either androgen receptor antagonists or microlesions within the dorsomedial VMN impairs sexual motivation and copulatory behavior in male rats. Further, multiple lines of evidence indicate the mPOA and VMN are functionally related in an opposing fashion; the mPOA projects to, and receives, projections from the VMN. The mPOA and VMN also have opposing roles in the control of autonomic function and female reproductive behavior characterized by Pfaff and colleagues: 'net effect of the outputs

Dual circuit

Common network

Male circuit

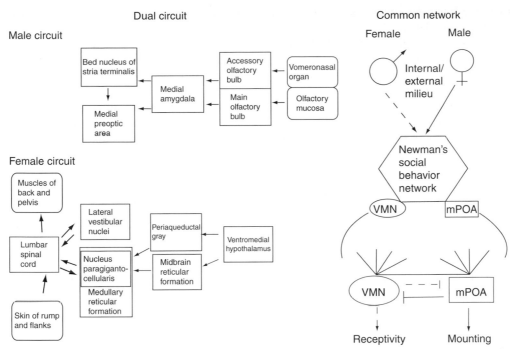

Figure 3 Two contrasting models of the neural mechanisms underlying sexual behavior. The Dual Circuits Model (left panel) suggests separate neural circuits (circles and squares represent brain nuclei/areas) ending with the medial preoptic area (mPOA) or ventromedial nucleus of the hypothalamus (VMN) as final integrative areas for male- and female-typical behaviors. The Common Network Model (right panel) posits that a common network of nuclei (Newman's Social Behavior Network) is involved in sexual behavior of both males and females, and it is the reciprocal interaction between two root nodes of this network (mPOA and VMN) as well as the hormonal history of the individual and the nature of the stimulus context that determines the type of behavior exhibited. Neuroanatomical pathways mediating copulatory behavior. In the Dual Circuits Model, the brain nuclei known to be important in the expression of the various behaviors are depicted as square boxes, and the projections between them (of which some are excitatory and some inhibitory) are shown as arrows. Sensory and effector organs are shown as rounded boxes. Brain nuclei critical for hormonal control over behavior are shown as bold boxes. Top portion depicts the Male Circuit involved in processing sexually relevant chemosensory information has been well studied in the male rodent, and involves the pathway from the main and accessory olfactory bulbs to the amygdala, particularly the medial division, and then via the bed nucleus of the stria terminalis and an alternative route via the ventral amygdalofugal pathway to the medial preoptic area. Exactly what happens to this information once it reaches the medial preoptic area is anyone's guess, but is presumed to result in the decision to attempt to mount. Once the mount is established, events in both the female and male involve reflex arcs mediated by well characterized neural circuits, shown by bold arrows. Bottom portion depicts the Female Circuit underlying lordosis in the female rat and involves the transfer of sensory information from the male's mounting and thrusting movements to the lumbar spinal cord, when it ascends to the brainstem motor nuclei responsible for integrating the muscular motor pattern of lordosis. Descending control over this reflex arc is exerted by the ventromedial hypothalamus via the periaqueductal gray and the midbrain reticular formation.

from the preoptic region is to reduce feminine-typical behavior and to increase male-typical behavior.' Neuronal activity increases in the VMH during sexual receptivity in the female rat, and is reduced when there is increased activity in the mPOA. Effects of excitatory and inhibitory amino acid neurotransmitters are opposite in the VMN and mPOA. Specifically, in the VMN, GABA is facilitatory, while NMDA is inhibitory, to lordosis, and in the POA, GABA inhibits and NMDA facilitates, lordosis in hormone-primed female. So, ought the two centers to be considered as two independent neuroanatomical anatomical units, one of which will be chosen by developmental events, to determine the sexual phenotype of the animal, while the other languishes? This model should be rejected in favor of a mutual inhibition model in which the centers work actively together antagonistically in both

sexes, in a way more analogous to the interactions of two political parties, the balance of whose power determines their joint decisions.

Looking to Nature

The central issue in science is the support (or lack thereof) of the hypothesis. In this particular case, however, there are inherent and seemingly insurmountable obstacles to proving whether the there are dual neural circuits in each individual, with one predominant in one sex and the another in the opposite sex, or whether there is a single sex behavior circuit that is modulated to produce one of two outputs. This hypothesis cannot be addressed in mammals and any other species having heritable sex chromosomes (e.g., XX females and XY males). Not only do the sexes differ in an elemental gene, but males and females develop and age in entirely different endocrine milieus and, as a consequence, have different life history experiences.

Fortunately, we can look to nature for the necessary evidence. Hermaphroditic species come in several varieties. Simultaneous hermaphrodites are species in which each individual produces both sperm and eggs, but curiously, never at the same time. When breeding, one individual will assume the 'male' role and shed sperm and its partner the 'female' role and shed eggs. In the next spawning, even the roles are completely reversed. In sequentially hermaphroditic species, the individual begins as one sex, but transforms into the opposite sex if the appropriate social events present themselves. Clearly, in both instances, the brain of each individual is bisexual in its organization and performance. In the only experiment that has been done to date, Leo Demski demonstrated that stimulating one brain area of the sea bass, a simultaneous hermaphrodite, would cause sperm release while stimulation in another brain area resulted in egg release.

But what about the 'higher' vertebrates, that is the reptiles, birds, and mammals that constitute the amniote vertebrates? Particularly revealing insights into the relationship between the sexual dimorphism of the brain (or lack of) and the display of sex-typical behaviors are afforded by my work on parthenogenetic whiptails of the genus *Cnemidophorus*. Some species of the genus are gonochoristic with male and female individuals that behave in a sexually dimorphic manner (i.e., males mount receptive females), while some species are parthenogenetic, all individuals being morphologically female and reproducing clonally. In the gonochoristic species, the brain is sexually differentiated in a typical vertebrate pattern.

For example, in *C. inornatus*, males mount while females do not, and male mounting is dependent on androgens acting on the mPOA. Female *C. inornatus* do not mount, but exhibit receptivity dependent upon estrogen acting at the level of the VMN. Individuals of the parthenogenetic species engage, at different times, in behaviors that physically are identical to both the male- and female-typical behaviors of their sexual congeners (albeit with the exception of intromission and insemination, hence called 'pseudosexual behavior'). When pairs of animals are observed displaying these complementary behaviors, there is a tight relationship between the behavior displayed and the ovarian state of the animal (i.e., the individual mounting and displaying other male-like copulatory behavior (pseudocopulation)) is generally postovulatory and has elevated progesterone levels, while the receptive individual is preovulatory, having high estrogen levels. Any given individual will thus display both behaviors at different points in the ovarian cycle.

Hormonal and neuroanatomical correlates of the two kinds of behavior in these animals parallel those observed in males and females of more commonly studied vertebrates. Examination of the mPOA and the VMN of the parthenogenetic lizards indicates that these nuclei do not change in cell size or number during these different behavioral phases, nor do these parameters respond to exogenous hormone treatment. However, they do differ in metabolic activity in predictable ways (**Figure 4**): Rand and Crews showed that during the male-like pseudocopulatory behavior metabolic activity as measured by 2-deoxyglucose uptake (2DG) is high in the mPOA but below baseline in the VMH (indicating suppression of activity); during female-like pseudoreceptive behavior, the opposite occurs, with 2DG suppressed in the POA and enhanced in the VMN. Intracranial implantation of

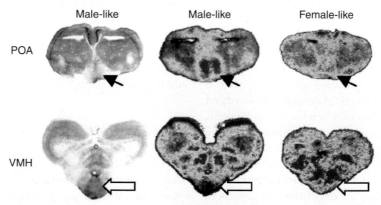

Figure 4 Metabolic activity during pseudosexual behavior in the unisexual lizard. Brains in two individual lizards engaged in a pseudocopulation. In the left column are light micrographs of brain sections at the level of the medial preoptic area (mPOA) (top row) and the ventromedial nucleus of the hypothalamus (VMH) (bottom row). Other columns are pseudocolor images where red denotes maximum accumulation of 2DG and green the lowest accumulation. Middle column is the brain of the individual exhibiting male-like pseudosexual behavior (same brain sections as on left), while the right column is the brain of the lizard exhibiting female-like pseudosexual behavior. Rand MS and Crews D (1994) The bisexual brain: Sex behavior differences and sex differences in parthenogenetic and sexual lizards. *Brain Research* 663: 163–167.

androgen (and progesterone) into the mPOA of both male *C. inornatus* and *C. uniparens* elicits mounting behavior, but fails to elicit either mounting or receptive behavior when placed in the VMN. On the other hand, while implantation of estrogen into the VMH elicits receptive behavior in female *C. inornatus* and in *C. uniparens*, it fails to do so in male *C. inornatus*, suggesting that the brains of these animals are not bisexual, but rather that either sex is capable of expressing male-typical behavior.

Both the mPOA and the VMN are dimorphic in size, with the mPOA being larger, and the VMN smaller, in sexually active male *C. inornatus* than in females or in the descendant parthenogenetic species. Castration of male *C. inornatus* causes the mPOA to decrease, and the VMN to increase to female size; androgen replacement restores the sex difference. The overall change in nuclear volume is paralleled in soma size of individual neurons in both areas, suggesting that the size of these neurons changes to reflect their functional activity. However, once again, this sex difference appears to be a correlate, rather than a necessary substrate of the expression of male-typical behavior, since *C. uniparens* exhibiting male-like pseudo-copulatory behavior (either as intact postovulatory or ovariectomized, testosterone-treated animals) do not show an increase in regional or somal area of the mPOA. Dias and Crews found that differences in the mPOA thus observed between the parthenogens displaying male- and female-typical behaviors have also been subtle at the levels of gene expression and neurotransmitter levels. The parthenogenetic whiptails thus oblige us, while continuing to accept the existence of sex differences in brain morphology, to consider the possibility that such developmentally long-term differences in morphology are less important in determining the behavior exhibited than is the short-term activity of the brain, which is determined by external stimuli as well as by immediate physiological state. However, in these animals, as in others studied, this sexual phenotype-determining 'activity' can be profitably studied by focusing on the interaction between the mPOA and the VMN.

Conclusions

Beach only reluctantly accepted the idea of sexually dimorphic central structures, not because he was stubborn, but rather because he was hesitant to concede that such organizational actions might be the mechanism underlying activational gating of the bisexual brain.

> . . . the specificity of the mating patterns for the two sexes, although probably inherited, is not rigidly dictated by the innately organized substratum. Although there may be a strong preference for the normal copulatory response it is obvious that in a few individuals at least, there exists

the innate organization essential to the mediation of the mating pattern of either sex. The presence or absence of such duplicative arrangement within all individuals is a matter for speculation. It is obvious, however, that the mating behavior to be displayed by a member of either sex may in part or (in the cases reported), entirely predetermined by the behavior of the partner. (Beach, 1938, p. 324)

Beach thus delineated four essential points: first, that both male and female individuals are capable of displaying the sexual behaviors of the opposite sex; second, that the brain must have the neural circuitry sufficient to support these opposite behaviors although third, each sex is predisposed to exhibit the behavior consistent with its sex; and fourth, that the stimulus animal is essential in eliciting the complementary behavior. If one accepts Beach's conclusions, one expects that male- and female-typical copulatory behaviors are mediated by brain structures that are present and (at least latently) fully functional in both sexes, that is, not sexually dimorphic. Experimentally one is then forced to examine how males and females can behave differently, and what, if not to mediate sex-typical copulatory behavior, are the functions of the observed sexual dimorphisms in brain structure. I propose that the neural mechanisms mediating both male and female copulatory behavior are under tonic inhibition from a range of sources, and that activation constitutes relief from some of these inhibitory inputs. Major sexual dimorphisms in brain structure are seen as mostly sex-specific sources of additional inhibition so that, for example, the large mPOA typical of males is responsible not for mediating male-typical copulatory behavior, but for allowing a more sophisticated pattern of inhibition. Sex differences, in other words, should not be seen as sex differences, but as male- and female-typical features that enable males and females to do better the things they do, rather than enabling them to do something that the other sex cannot. Either sex, the evidence shows, is intrinsically capable of doing either thing.

See also: Animal Behavior: Antiquity to the Sixteenth Century; Animal Behavior: The Seventeenth to the Twentieth Centuries; Comparative Animal Behavior – 1920–1973; Development, Evolution and Behavior; Endocrinology and Behavior: Methods; Ethology in Europe; Female Sexual Behavior and Hormones in Non-Mammalian Vertebrates; Future of Animal Behavior: Predicting Trends; Integration of Proximate and Ultimate Causes; Male Sexual Behavior and Hormones in Non-Mammalian Vertebrates; Mammalian Female Sexual Behavior and Hormones; Mate Choice in Males and Females; Mating Signals; Nervous System: Evolution in Relation to Behavior; Neurobiology, Endocrinology and Behavior; Neuroethology: Methods; Pair-Bonding, Mating Systems and Hormones; Psychology of Animals; Reproductive

Skew, Cooperative Breeding, and Eusociality in Vertebrates: Hormones; Sexual Behavior and Hormones in Male Mammals; Sexual Selection and Speciation.

Further Reading

Beach FA (1938) Sex reversals in the mating pattern of the rat. *Journal of Genetic Psychology* 53: 329–334.

Beach FA (1948) *Hormones and Behavior*. New York: Paul B. Hoeber.

Beach FA (1971) Hormonal factors controlling the differentiation, development, and display of copulatory behavior in the ramstergig and related species. In: Tobach E, Aronson LR, and Shaw E (eds.) *The Biopsychology of Development*, pp. 249–296. New York: Academic Press.

Crews D (2005) Evolution of neuroendocrine mechanisms that regulate sexual behavior. *Trends in Endocrinology and Metabolism* 16: 351–361.

Crews D and Moore MC (1986) Evolution of mechanisms controlling mating behavior. *Science* 231: 121–125.

DeVries GJ and Simerly RB (2002) Anatomy, development, and function of sexually dimorphic neural circuits in the mammalian brain. In: Pfaff DW, Arnold AP, Etgen AM, Fahrbach SE, and Rubin RT (eds.) *Hormones, Brain and Behavior* vol. 1, pp. 137–191. San Diego, CA: Academic Press.

Dias BG and Crews D (2008) Regulation of pseudosexual behavior in the parthenogenetic whiptail lizard, *Cnemidophorus uniparens*. *Endocrinology* 149: 4622–4631.

McEwen BS (1981) Neural gonadal steroid actions. *Science* 211: 1303–1311.

Newman SW (1999) The medial extended amygdala in male reproductive behavior: A node in the mammalian social behavior network. *Annals of the New York Academy of Sciences* 877: 242–257.

Pfaff DW, Schwartz-Giblin S, McCarthy MM, and Kow LM (1994) Cellular and molecular mechanisms of female reproductive behaviors. In: Knobil E and Neil J (eds.) *The Physiology of Reproduction,* 2nd edn., pp. 107–220. New York: Raven Press.

Phoenix CH, Goy RW, Gerall AA, and Young WC (1959) Organizing action of prenatally administered testosterone propionate on the tissues mediating mating behavior in the female guinea pig. *Endocrinology* 65: 369–381.

Rand MS and Crews D (1994) The bisexual brain: Sex behavior differences and sex differences in parthenogenetic and sexual lizards. *Brain Research* 665: 163–167.

Sengoopta C (2006) *The Most Secret Quintessence of Life: Sex, Glands, and Hormones, 1850–1950*. Chicago: University of Chicago Press.

Södersten P (1984) Sexual differentiation: Do males differ from females in behavioral sensitivity to gonadal hormones? *Progress in Brain Research* 61: 257–270.

Södersten P (1987) How different are male and female brains? *Trends in Neuroscience* 10: 197–198.

Tinbergen N (1951) *The Study of Instinct*. Oxford: Clarendon Press.

Wallen K and Baum MJ (2002) Masculinization and defeminization in altricial and precocial mammals: Comparative aspects of steroid hormone action. In: Pfaff DW, Arnold AP, Etgen AM, Fahrbach SE, and Rubin RT (eds.) *Hormones, Brain and Behavior,* vol. 4, pp. 385–423. San Diego, CA: Academic Press.

Whalen RE (1974) Sexual differentiation: Models, methods, and mechanisms. In: Friedman RC, Richart RM, and Van de Wiele RL (eds.) *Sex Differences in Behavior*, pp. 467–481. New York: Wiley.

Neurobiology, Endocrinology and Behavior

E. Adkins-Regan, Cornell University, Ithaca, NY, USA
C. S. Carter, University of Illinois at Chicago, Chicago, IL, USA

Introduction

Two types of mechanisms, neural and hormonal, have been prominent in the history of research directed at uncovering the proximate physiological causes of animal behavior. During the first part of this history, the nervous and endocrine systems were envisioned as separate systems and were studied by somewhat different research communities. As a result, research on physiological mechanisms of animal behavior has tended to develop along two somewhat separate and parallel tracks. These dual origins are reflected in the organization of this survey. Beginning in the twentieth century, several discoveries led to the realization that the nervous and endocrine systems are physiologically integrated to a highly significant extent, which is of great importance for animal behavior. Nerve cells can synthesize and secrete hormones; the behavioral effects of hormones are mediated by their actions on neurons, and the brain regulates the endocrine axes so that hormone levels related to behavior are responsive to both social and physical environments.

Origins of Behavioral Neurobiology: Sensory, Motor, and Motivational Systems in Comparative Perspective

The history of the study of the neural mechanisms of animal behavior is largely the history of neuroscience in a more general sense. The overarching motivation of the pioneers was often a desire to understand human minds and brains. However, because of the impossibility of doing experimental work with humans, investigations of animals have long played a significant role. The oldest and deepest scientific roots of the field are comparative neuroanatomy and comparative physiology. Then in the twentieth century, developments in ethology led to the rise of neuroethology (also called behavioral neurobiology), which emphasized naturally occurring behavior in nondomesticated animals, while developments in psychology produced the subfield of physiological psychology, with its emphasis on learning, memory, and motivation in domesticated laboratory animals.

Comparative Neuroanatomy

Writing in the 1600s, René Descartes emphasized that the brain and nervous system are responsible for behavior.

In subsequent centuries, many scientists examined and described the structure of the brain and nervous system in an array of animal species. A common theme was to note what seemed to be marked differences in the organization of the brain, especially the forebrain, and in the relative sizes of structures and brain divisions, and to speculate about their relationship to behavior and intelligence.

With the publication of Charles Darwin's theory of evolution by natural selection, these species differences in brain structure and size began to be interpreted in an evolutionary framework. Until the middle of the 1900s, the dominant view had been that the brains (especially forebrains) of different vertebrate classes (as represented by a small number of species from each of the largest classes) were fundamentally different in organization, that they formed an evolutionary series progressing toward the human brain, and that this series paralleled an increase in intelligence and behavioral complexity culminating in apes and humans.

A set of scientific developments beginning in the 1950s and 1960s then led to a substantial revision of these ideas: a veritable intellectual revolution in comparative neuroanatomy. New methods for tracing the neural connectivity between brain regions revealed that the structure of different vertebrate brains had in fact been highly conserved over evolutionary time, with the same basic ground plan from fish to mammals. The consequences of this revolution are still being felt, for example, in recent efforts to rethink and rename the structures of the avian brain. Another key development was the realization that phylogenetic relationships are tree-like, rather than ladder-like. As this more modern view of phylogeny was absorbed into comparative neuroanatomy and comparative animal behavior, efforts were made to expunge the remnants of teleological thinking (evolution as a guided progression toward human superiority) from the field as well. Tree thinking, along with improved methods for taking appropriate account of body size in the comparative study of brain size, led to the realization that large brains and large forebrains had evolved several times independently in vertebrates, and that mammals do not have larger brains than all other vertebrates when corrected for body size. This repeated convergent evolution of large brains then allowed researchers to more rigorously test hypotheses about the ecological or behavioral characters that are associated with large brains (long-distance migration? predatory foraging? group living and social life?). Such characters provide clues to the selective pressures for increases in brain size: a line of research that continues to be active. New methods for determining phylogenetic trees from molecular information and for statistically analyzing comparative data in a phylogenetic framework have increased the power and objectivity of the comparative approach to such hypothesis testing.

The popular world, and the neuroscientific world as well, have been slow to absorb this revolution, however. One still sees the words 'lower' and 'higher' applied to animal species. The incorrect assumption that nonmammals lack the forebrain structures of intelligent learning and therefore are largely 'instinctive' creatures is still widespread.

Comparative Physiology

Among his many other intellectual pursuits, Descartes was interested in whether and how (through what bodily processes) behaving animals were different from mechanical toys and automata. It was Descartes who developed the concept of the spinal reflex and proposed specific neural pathways for reflexive actions such as withdrawing a limb from fire. It was not until the late 1800s, however, that it was discovered (by Santiago Ramón y Cajal) that the nervous system consists of cells (neurons): an achievement for which he shared a Nobel Prize with Camillo Golgi in 1906. The studies of another Nobel Prize laureate, Charles Sherrington (**Figure 1**) (who originated the term synapse and won the Nobel Prize in 1932), were the beginning of a long and productive line of research on the nature of reflexes and their underlying neuronal activity. Among other discoveries, it was found that a rich array of biologically significant reflexes and even more complex actions still occurred in spinal and decerebrate preparations. The reflex concept remained an essential part of the neurobiology of behavior for several decades. It is still the case that some behaviors important for survival (e.g., coughing or withdrawing a limb from a sharp object) are best thought of as reflexes, along with all the stretch reflexes that posture and locomotion require. Spinal reflexes of mammals were found to include sexual reflexes such as ejaculation or estrous postures, raising questions about whether hormones act at the level of the spinal cord and the brain: one of many signs of the bridge forming between research on neural and hormonal mechanisms.

In 1786, Luigi Galvani discovered that muscle twitches and nerve function have an electrical basis, and in 1870, Fritsch and Hitzig found that weak electrical stimulation of the dog cortex produced muscle movements on the opposite side of the body. In subsequent decades, the neurophysiological approach to behavioral mechanisms flourished as parallel advances occurred in the apparatus for recording from and stimulating single and multiple

Figure 1   Sir Charles Sherrington. © The Nobel Foundation 2009.

neurons (e.g., amplifiers and oscilloscopes) and in the animal preparations themselves (e.g., J. Z. Young's discovery of the giant motor axon of squid). Especially exciting for those with a keen interest in animal behavior was the development of methods for stimulating or recording from the brains of freely moving animals. The studies of Walter Hess (**Figure 2**) (a 1949 Nobel Prize winner) in cats showed that a variety of normal appearing actions occurred following brain stimulation, including going to sleep, an early sign of the role of brain activity in this biologically important behavior.

Although not as technically sophisticated as neurophysiology, the use of ablations or lesions of specific brain regions has long been an important tool for testing hypotheses about the causal relation between the function of a region and the expression of a behavior. Marie Jean Pierre Flourens originated this experimental approach to the study of the brain in the 1820s, establishing through a systematic program of circumscribed ablations in rabbits and pigeons that damage to different brain divisions has different effects on behavior. For example, an ablation in a deep cerebellar layer produced locomotor deficits in pigeons, whereas an ablation of a part of the midbrain (in what then came to be called the optic lobe) caused blindness. Lesions are still a valuable stage of a brain and behavior research program, and technical improvements now permit lesions that are small, neurochemical (affecting only a subset of neurons such as dopaminergic neurons), or even reversible (e.g., temporary inactivation with lidocaine).

Figure 2 Walter Hess. © The Nobel Foundation 2009.

Figure 3 Theodore Bullock. With permission from the Scripps Institution of Oceanography, University of California – San Diego.

Neuroethology (Behavioral Neurobiology)

The work of ethologists on mechanisms of behavior tended to focus on nonphysiological mechanisms such as sensory cues, and much of what was referred to as behavioral physiology by this community did not actually go inside the animal, but instead used its responses to external cues to conceptualize internal processes. As recently as 1966, Peter Marler and William J. Hamilton's textbook *Mechanisms of Animal Behavior* contained rather little information about any neural mechanisms. Concepts such as releasing stimulus or hunger drive, and models such as Tinbergen's hierarchical model of instinct, were clearly meant to reflect some kind of neural processes, but explicit links to those were seldom proposed. A few researchers began to explore those links, however. For example, Erich von Holst and Ursula von St. Paul electrically stimulated the brains of freely moving chickens, producing behavior such as vocalization, grooming, feeding, and aggressive attack. These investigations were explicitly aimed at understanding the neural basis of drive. Jerram Brown and Robert Hunsperger used a similar method to study the neural basis of aggression in cats and applied the term neuroethology to such research.

Subsequent years saw the flowering of a very active research interest in the study of the neural mechanisms for the ecologically relevant adaptive behavior of nondomesticated animals such as insects (crickets, locusts, cockroaches), toads, and bats. This particular marriage of comparative physiology with animal behavior is what is sometimes meant today by the terms neuroethology or

behavioral neurobiology. The emphasis has been on sensory processes and motor output, and a number of these lines of research have become classics of animal behavior.

The neurophysiological approach has been prominent, and Theodore Bullock (**Figure 3**) in particular did much to ensure that neurophysiology would be comparative, directed at a diverse array of animals.

With respect to sensory processes, neuroethologists discovered and analyzed several previously unknown sensory systems, such as acoustic reception by moths and electroreception by weakly electric fish (the latter by Bullock, who also found the infrared receptors of pit vipers). They explored animals' abilities to detect stimuli out of the range of human detectability, for example, ultrasonic hearing by bats, ultraviolet wavelength vision by birds, and the exceptional binaural ability of owls when locating small prey by sound. These discoveries have reinforced the ethologists' insight that understanding an animal's umwelt is critical to an understanding of its behavior.

The classic neuroethological studies of motor processes produced several key concepts and discoveries about how the nervous system works. The importance of inhibition as well as excitation became apparent. Actions that have to occur very rapidly for the animal to survive (e.g., escape from a predator) could be triggered by the activity of a very small number of command neurons. Even in a vertebrate (a fish), the escape response was found to be triggered only by two very large cells, the Mauthner cells. Studies by Donald Wilson of locust flight revealed the existence of a central pattern generator (also called oscillator or pacemaker), the neural elements that produce rhythmic firing leading to rhythmic muscle movement even in the absence of stimulus inputs or any feedback from the periphery. It had been known since at least 1914 (through studies by T. Graham-Brown) that no proprioceptive input was needed for a dog's locomotor or scratch reflexes. Such central pattern generators might need an initial stimulus to get them going, and their exact frequency might be modulated by external stimuli, but the basic rhythmic pattern was clearly organized centrally, rather than resulting from a chain or sequence of stimulus–response reflexes. In a similar vein, the existence of endogenous circadian clocks became convincingly established, and so-called master clocks were then localized in the nervous system of an insect (in the subesophageal ganglion of a cockroach) in the late 1950s by Janet Harker and subsequently in the diencephalic suprachiasmatic nuclei of rats in the 1970s by Robert Moore and Victor Eichler and by Friedrich Stephan and Irving Zucker. Another set of key motor system concepts, developed by von Holst and Horst Mittelstaedt, were efference copy and reafference. Motor command signals are copied to another region of the nervous system where they can be compared to sensory feedback resulting from the motor performance. Processes of this kind allow the animal to tell the difference between active and passive movement (e.g., between moving vs. being windblown) and to avoid interference between sensory cues from the individual's own emissions versus echoes or emissions from other individuals (as in bats using ultrasound to catch insects). All these concepts have proven to be of enduring value in understanding how nervous systems produce adaptive behavior.

Physiological Psychology

The science of psychology has long sought to understand the behavior of all animals, not just humans. Early generations of comparative psychologists studied a highly diverse array of animals, including microbes, invertebrates, and vertebrates from all the larger classes. Physiological psychologists took the understanding of the neural and other physiological bases of behavior as their mission. Early on, and continuing up to the present, there was great interest in using the lesion method to study learning and memory. An important article was the research of Karl Lashley on cortical lesions and memory for learned tasks in rats – his search for the engram. He found that task memory did not seem to be located in any particular place in the cortex. Instead, how much cortex was damaged predicted whether and how much memory was lost. At the time this may have seemed like a failure to find the engram, but subsequent decades have revealed a great truth in his findings: the cortex works in a distributed manner. The secret of learning and memory is now thought to lie in part in the structural and functional

plasticity of neurons and their connections: a concept originated by Donald Hebb in the 1940s.

The 1950s and 1960s were a time of great interest in the hypothalamus and its role in motivated behavior of basic survival significance such as hunger, drinking, and regulation of temperature. Pictures of obese rats with lesions of the ventromedial hypothalamus are still compelling textbook images. Such research has taken on new significance recently with the occurrence of a pandemic of obesity in humans.

In recent decades, brain-oriented physiological psychology has become known as behavioral neuroscience. Additional brain regions such as the amygdala and prefrontal cortex have been thoroughly explored in relation to behavior. Their roles have been established in emotional responses such as fear (amygdala) and in cognitive functions such as switching problem solving strategies (prefrontal cortex). One of the most notable developments in the science of animal behavior has been a convergence of interest between behavioral neuroscientists and neurobiologists in neural mechanisms for use of space and memory for spatial locations. This line of research has produced insights into the role of hippocampal neurons in performance of rats in mazes, in memory for locations of stored food items in scatter hoarding birds, and in homing by pigeons.

Origins of Behavioral Neuroendocrinology: Social and Reproductive Behaviors

In the last quarter of the twentieth century, the biologist E. O. Wilson argued for a 'new synthesis' or 'consilience' ('jumping together of knowledge by the linking of facts and fact-based theory across disciplines'). The wisdom of this approach is especially relevant to the behavioral neuroendocrinology of social and reproductive behaviors. Here we will highlight a few of the milestones that allowed this truly integrative field of science to emerge at the intersection of disciplines such as agriculture, anatomy, biochemistry, ethology, molecular biology, physiology, and psychology.

Endocrinology

Awareness that endocrine systems played a role in behavior predates recorded history and was documented by Aristotle (*ca* 350 BC). Anatomical and behavioral changes associated with puberty and the external location of the testes probably provided ancient humans with their first knowledge of the importance of endocrine organs. Castration as a method for inhibiting the sexual behavior of male humans or as a punishment is ancient, and testes were consumed in the search for power and virility.

However, some of the earliest ideas regarding the role of the gonads in behavior were incorrect. Testicular hormones are not water soluble and beyond their nutritional benefits, ingested testes were unlikely to directly affect behavior.

With archaic roots in Chinese medicine and alchemy, modern chemistry is dated to the eighteenth century. During that period, pioneers such as Joseph Priestly, Carl Scheele, and Antoine Lavoisier documented the first extensive list of elements, including oxygen and hydrogen.

The idea that behaviorally active chemicals (hormones) are secreted by endocrine tissue into the blood stream and that they act on target tissues including the nervous system to influence behavior is comparatively modern. The first modern evidence of neurohormones is attributed to Otto Loewi in 1921. Loewi demonstrated that secretions from the vagus nerve ('vagusstuff') are capable of affecting heart rate. 'Vagusstuff' was later identified as acetylcholine and norepinephrine. Loewi shared the Nobel Prize in 1936 with Henry Dale; Loewi and Dale are sometimes referred to as the 'fathers' of neuroscience.

The formal concept of a 'hormone' was described in 1905 by Ernest Starling and William Bayliss. Dale had demonstrated in the early 1900s that pituitary gland extracts (later found to contain oxytocin) could be used to induce labor, first in domestic animals and shortly thereafter in humans.

The role in endocrinology of secretions of the central nervous system can be traced to Ernst Scharrer. In 1928, Scharrer had identified the largest cells in the hypothalamus, calling these the 'magnocellular neurons.' In collaboration with his wife Berta, he also articulated the concept of neurosecretion. However, the behaviorally active chemicals secreted by the magnocellular neurons were not identified until Vincent du Vigneaud synthesized oxytocin in 1953 and vasopressin in 1954. Du Vigneaud received the 1955 Nobel Prize in Chemistry for the 'first synthesis of a polypeptide hormone.' His Nobel lecture titled 'A Trail of Sulfa Research: From Insulin to Oxytocin' set the stage for the understanding that physiologically active hormones were produced not just in the pituitary or peripheral endocrine organs, but also in the nervous system.

Hormones, Behavior, and Neuroendocrine Systems

The classic tools of endocrinology arose in other disciplines, but rapidly spread to the study of behavior. Naturally occurring changes in behavior associated with maturation and naturally occurring pathologies were the source of many basic findings. Accidental lesions or tumors of the nervous system or endocrine abnormalities led to the earliest medical awareness of relationships between neuroendocrine systems and behavior.

The first experimental endocrine study is usually credited to A. A. Berthold. In 1849, Berthold described changes associated with removal of the testes and their reimplantation in roosters. In 1889, shortly after the invention of the hypodermic needle, an aging biologist, C.E. Brown-Sequard, injected himself with aqueous testicular extracts. Although likely the result of a placebo effect, Brown-Sequard's enthusiastic reports of renewed strength and vigor, published in the respected medical journal *Lancet*, launched the 'monkey gland' era. In the decades that followed, a Viennese physiologist Eugen Steinach initiated a widely publicized series of surgical manipulations aimed at boosting endogenous hormone production and thus revitalizing aging males. The 'Steinach Operation' was basically a vasectomy and probably primarily based on the power of suggestion, but it attracted celebrity followers such as the poet W. B. Yeats and Sigmund Freud. Taken together, work in this period generated intense interest in the behavioral effects of 'internal secretions.'

Although awareness of the effects of steroids is ancient, steroid chemistry exploded only between the 1920s and 1930s with the identification and synthesis of gonadal and adrenal hormones, including testosterone, estrogen, progesterone, and glucocorticoids. Putting specific steroids into their behavioral context also began in the first half of the twentieth century. For example, documentation of the rodent estrous cycle and early evidence for a role for ovarian secretions in the induction of behavioral estrus were provided in guinea pigs by Charles Stockard and George Papanicolau in 1917 and in mice by Edgar Allen and Edward Doisy in 1923. (The 'pap' smear was later developed based on knowledge gained from these studies).

Initially, measurements of hormones relied on bioassays. For example, Allen and Doisy in 1923 described the use of the immature rodent uterus as a bioassay for estrogen. More advanced methods for measuring hormones, initially based primarily on radioimmunoassay, became available through the work of Rosalyn Yalow and Solomon Berson in the 1950s and 1960s, for which Yalow received a Nobel Prize in 1977. Availability of quantitative hormone assays led to a flurry of studies correlating the release of gonadal steroids with reproductive behaviors.

Once synthetic steroids were available, it was possible to show that estrogen, often in combination with progesterone, could induce female proceptivity and receptivity. Parallel studies in males focused on testicular hormones, including testosterone, and tended to concentrate on mounting behavior, considered an 'appetitive behavior,' or the capacity to show an erection and an ejaculatory response, sometimes called a 'consummatory behavior.' Much of this research originated from psychologists and anatomists, including Calvin Stone, Frank Beach, William C. Young, Daniel Lehrman, and their colleagues or students.

However, when gonadal steroids were injected, the behavioral effects tended to require hours or even days to be seen. This left open the important possibility that other chemicals, perhaps indirectly affected by steroids, could influence behavior. One of these was a small decapeptide, luteinizing-hormone releasing hormone (LHRH). The first hypothalamic releasing hormones had been identified independently in 1969 by Roger Guillemin and Andrew Schally (earning for both a share of the 1977 Nobel Prize, with Yalow). LHRH was synthesized in the hypothalamus, regulated gonadal functions of the anterior pituitary, and thus coordinated various reproductive processes including gamete production and behavior in both sexes. In 1971, two investigators (Robert Moss and Donald Pfaff) independently demonstrated that LHRH was capable of facilitating mating in female rats. These findings marked the beginning of contemporary approaches to 'neuroendocrinology.'

Classical approaches to mapping neuroendocrine systems include ablation of brain areas and removing tissues in which a particular compound is synthesized or where receptors are concentrated. For example, aspiration of large segments of the cortex did not prevent the expression of maternal behavior or female sexual responses, but did interfere with male sexual behavior. In contemporary behavioral endocrinology, chemicals are typically manipulated by biochemical or molecular methods, either enhancing or preventing the effects of a particular compound. In addition, new methods for mapping hormone receptors have proliferated since the 1960s. Taken together, these strategies have allowed of the analysis of underlying neural substrates and circuits for various complex behaviors including those necessary for species-typical reproductive behaviors.

Sexual Differentiation of Behavior

Differences between males and females, as well as the processes associated with sexual differentiation, have been a long-standing theme in this field. In 1916, Frank Lillie described in genetic females the development of male-like anatomical changes, known as 'free-martinism,' in females that had cohabitated in utero with a male sibling. This observation implicated testicular hormones in phallic development and led in time to a detailed analysis of the biology of sexual differentiation. Cross-sexual testicular transplants or gonadectomies by Steinach and others supported the hypothesis that gonadal secretions could affect anatomy and sexual behavior in later life. Experiments involving injections of testosterone in early life in guinea pigs, published in 1959 by Charles Phoenix, Robert Goy, Arnold Gerall, and William Young, were particularly influential in identifying organizational, developmental effects of hormones – in contrast to activational, short-term effects more commonly seen in adulthood. (It is now known that the same molecules can have both organizational and activational consequences.)

The very notion of sex differences in the nervous system remained a source of controversy for much of the twentieth century, although clear evidence for sex differences in the structure of the brain and spinal cord was available in the 1960s and 1970s. Research on sex differences focused initially on steroid-regulated processes, but recent evidence suggests that at least some sex differences in brain and behavior may be steroid-independent. Steroid-independent sexual differentiation is more apparent in nonmammalian vertebrates. For example, temperature-dependent sexual differentiation is well documented in reptiles.

Parental and Pairing Behavior

One of the clearest activational effects attributed to hormones is female parental behavior. In mammals, because of its association with birth, maternal behavior was logically linked to the endocrine changes of pregnancy and parturition. Howard Moltz, Jay Rosenblatt, and many others conducted studies mimicking the endocrine changes preceding birth. These studies implicated estrogen and progesterone (withdrawal), as well as the anterior pituitary hormone prolactin, in maternal behavior. However, even after treatment with these hormones, most reproductively naive animals still required days prior to the onset of positive reactions to infants.

Oxytocin as a candidate for the rapid induction of maternal responsiveness was initially rejected; elimination of oxytocin as a factor in maternal behavior was based on the finding that females with the pituitary gland removed (thought to be the primary source of oxytocin) remained capable of expressing maternal behavior. However, in the 1970s, it was shown that when the blood supply from maternal animals was transfused into reproductively naïve females there was an almost instant onset of maternal reactions in the naïve animals. Clearly, something was missing in the understanding of the biochemical 'cocktail' for maternal behavior. Finally, in 1979, Cort Pedersen and Arthur Prange injected oxytocin directly into the nervous system and saw a quick onset of maternal behavior in estrogen-primed, naïve females; they were also able to block maternal responses with an oxytocin antagonist, providing compelling evidence for a direct role for oxytocin in this behavior.

In studies of maternal behavior in sheep conducted in the early 1980s, Barry Keverne and his colleagues also proved that oxytocin was involved in the formation of the mother–infant bond. Oxytocin was later shown to be released within the nervous system, confirming the fact that oxytocin could affect behavior even in the absence of its release from the pituitary gland. Oxytocin has since been implicated in the downregulation of anxiety and fear, while vasopressin and the functionally related peptide, corticotropin releasing hormone (CRH), typically have opposing effects on these processes. Generalized emotional effects of neuropeptides, mediated in part by effects on sensory systems and central and autonomic effects, probably allow mammalian females to respond appropriately to their newborn from the moment of birth. Several other social and reproductive functions have been attributed to neuropeptides. For example, studies of socially monogamous species, such as prairie voles, have revealed that both oxytocin and vasopressin are involved in pair bond formation, possibly in both sexes. However, oxytocin, which is estrogen-dependent, may play a particularly important role in female behavior, although it is also involved in male sociality. Vasopressin, which is androgen-dependent, appears to facilitate the more active behaviors, including mate defense and territorial behaviors, which may be especially critical in males.

In 1978, Carol Diakow showed that vasotocin, an evolutionary precursor to oxytocin and vasopressin, played a role in amphibian mating behavior. Vasotocin had previously been shown to be important in egg-laying. The gene for neuropeptides related to oxytocin predates the split between invertebrates and vertebrates and it is likely that these ancient molecules have been co-opted for various 'modern' functions during the course of evolution.

The Molecular Era

Methodologies arising from molecular biology are now revolutionizing our understanding of the role of specific hormones and their receptors in behavior. For example, research in 'knock-out' mice made mutant for the gene for oxytocin, the oxytocin receptor, or the vasopressin (V1a) receptor suggests that both oxytocin and vasopressin are important for selective, social recognition learning. However, it is interesting that mice with these genetic deficits are not asocial, can still give birth, and remain capable of maternal behavior. Taken together, and in the context of studies of pair bond formation, these findings suggest that in mammals, both oxytocin and vasopressin are necessary for the development of selective social interactions. These molecules, along with many others, work as components of a highly integrated and often sexually dimorphic neural circuitry for social behavior. Molecular methods have also been used to demonstrate that differences in the expression of the genes for neuropeptide receptor are correlated with species- and individual-differences in patterns of sociality. By over-expressing certain genes, it is possible to create animals capable of showing behavioral patterns that are not usually seen in their species. For example, increasing availability of the V1a receptor in specific brain regions produces males capable of forming pair bonds, even in species, such as montane voles, for which this is atypical.

Studies of mice that lack the gene for specific steroid receptors are also providing a new understanding of the behavioral effects of compounds such as estrogen, progesterone, and androgen. This research is complicated by interactions among different hormones and the presence of various subtypes of steroid receptors. However, such work has important translational implications because of the many medical manipulations of hormones, including widely used hormone replacement therapies and contraceptives.

Recent Years: Integration and Discovery

The last few decades have seen increased integration between research on neural and hormonal mechanisms of animal behavior, as well as increased scientific integration with other subfields, for example, molecular biology, as just illustrated in the previous section. Researchers now discover social influences on the expression of genes, or use the expression of immediate early genes to identify regions of neural activity as a substitute for the brain imaging that is not yet possible in freely moving animals. Although neuroethology and behavioral neuroendocrinology have always included an evolutionary perspective, the connection to evolutionary biology continues to produce new insights, for example, into the roles of co-evolution and sexual selection in shaping some neural and hormonal mechanisms. New approaches such as computational or network modeling of brain activity related to animal behavior are occurring through links to fields such as computer science that were not previously connected to animal behavior.

The types of behavior that have been studied physiologically show both change and continuity. There has been increased interest in animal cognition, social learning, and social relationships. At the same time, new discoveries of sensory and motor mechanisms and systems have continued to be made, for example, magnetic field detection by sea turtles and birds (leading to a search for the elusive receptors), the accessory olfactory system and its role in social behavior, and the remarkable neural system in the telencephalon of songbirds that is responsible for the perception, learning, and production of song. These last two are hormone regulated and provide excellent examples of the historical trend toward viewing neural and hormonal systems as interconnected.

See also: Acoustic Communication in Insects: Neuroethology; Aggression and Territoriality; Aquatic Invertebrate Endocrine Disruption; Bat Neuroethology; Behavioral Endocrinology of Migration; Circadian and Circannual Rhythms and Hormones; Communication and Hormones; Conservation Behavior and Endocrinology; Crabs and Their Visual World; Experimental Approaches to Hormones and Behavior: Invertebrates; Female Sexual Behavior and Hormones in Non-Mammalian Vertebrates; Field Techniques in Hormones and Behavior; Food Intake: Behavioral Endocrinology; Hibernation, Daily

Torpor and Estivation in Mammals and Birds: Behavioral Aspects; Hormones and Behavior: Basic Concepts; Immune Systems and Sickness Behavior; Insect Flight and Walking: Neuroethological Basis; Invertebrate Hormones and Behavior; Leech Behavioral Choice: Neuroethology; Male Sexual Behavior and Hormones in Non-Mammalian Vertebrates; Mammalian Female Sexual Behavior and Hormones; Maternal Effects on Behavior; Memory, Learning, Hormones and Behavior; Molt in Birds and Mammals: Hormones and Behavior; Naked Mole Rats: Their Extraordinary Sensory World; Nematode Learning and Memory: Neuroethology; Nervous System: Evolution in Relation to Behavior; Neural Control of Sexual Behavior; Neuroethology: What is it?; Pair-Bonding, Mating Systems and Hormones; Parasitoid Wasps: Neuroethology; Parental Behavior and Hormones in Mammals; Parental Behavior and Hormones in Non-Mammalian Vertebrates; Predator Evasion; Reproductive Skew, Cooperative Breeding, and Eusociality in Vertebrates: Hormones; Sex Change in Reef Fishes: Behavior and Physiology; Sexual Behavior and Hormones in Male Mammals; Sleep and Hormones; Sociogenomics; Sound Localization: Neuroethology; Stress, Health and Social Behavior; Tadpole Behavior and Metamorphosis; Vertebrate Endocrine Disruption; Vocal–Acoustic Communication in Fishes: Neuroethology; Water and Salt Intake in Vertebrates: Endocrine and Behavioral Regulation; Wintering Strategies.

Further Reading

Carter CS (ed.) (1974) Hormones and sexual behavior. In: Schein MW and Stroudsburg PA (Ser. eds.) *Benchmark Papers in Animal Behavior.* Stroudsburg, PA: Dowden, Hutchinson & Ross.

Carter CS and Getz LL (1993) Monogamy and the prairie vole. *Scientific American* 268: 100–106.

Ewert JP (1980) *Neuroethology: An Introduction to the Neurophysiological Foundations of Behavior.* Berlin: Springer-Verlag.

Hebb DO (1949) *The Organization of Behavior; A Neuropsychological Theory.* New York, NY: Wiley.

Hodos W and Campbell CBG (1969) *Scala naturae*: Why there is no theory in comparative psychology. *Psychological Review* 76: 337–350.

Lashley KS (1963) *Brain Mechanisms and Intelligence: A Quantitative Study of Injuries to the Brain; with a New Introduction by D.O. Hebb.* New York, NY: Dover.

Marler P and Hamilton WJ III (1966) *Mechanisms of Animal Behavior.* New York, NY: John Wiley & Sons.

Pfaff D, Arnold AP, Etgen AM, Fahrbach SE, and Rubin RT (eds.) (2009) *Hormones, Brain and Behavior,* 2nd edn. Amsterdam: Academic Press (Elsevier).

Roeder KD (1967) *Nerve Cells and Insect Behavior.* Cambridge, MA: Harvard University Press.

Sengoopta C (2003) 'Dr. Steinach coming to make old young!': Sex glands, vasectomy and the quest for rejuvenation in the roaring twenties. *Endeavor* 27: 22–126.

Stellar E (1954) The physiology of motivation. *Psychological Review* 61: 5–22.

Striedter G (2005) *Principles of Brain Evolution.* Sunderland, MA: Sinauer.

Young LJ and Hammock EA (2007) On switches and knobs, microsatellites and monogamy. *Trends in Genetics* 23: 209–212.

Zupanc GKH (2004) *Behavioral Neurobiology: An Integrative Approach.* Oxford, UK: Oxford University Press.

Neuroethology: Methods

S. S. Burmeister, University of North Carolina, Chapel Hill, NC, USA

Introduction

As articulated by Tinbergen in his 'four questions,' a complete understanding of behavior requires an understanding of it at multiple levels of analysis. Neuroethology represents the effort to understand the neurobiology of behavior, what Tinbergen called the causation of behavior. Neuroethologists typically work on a variety of animals, using the natural talents of particular organisms to investigate the basic principles of neurobiology. For example, to understand how the auditory system decodes the location of a sound source, neuroethologists turn to animals that are well adapted to locate sound, such as the barn owl, which relies on acoustic cues to find prey when hunting at night. Using this approach, neuroethology has been very successful in uncovering basic principles of neurobiology. But why should the behavioral ecologist be interested in the findings of the neuroethologist? Understanding the neural mechanisms of behavior not only gives us a more complete understanding of behavior, but can also inform our perspective on behavioral evolution by determining the sensory, cognitive, or motor constraints on the evolution of behavior.

Because neuroethologists are interested in the mechanisms of natural behavior (as opposed to clinically relevant behavior), they address questions that are relevant to the natural history of the organism under study. For example, when neuroethologists ask '*how do animals perceive the world?*' they use behaviorally relevant stimuli to, for example, determine how the toad's visual system discriminates prey from predator. When neuroethologists ask '*how is motor output generated?*' they investigate behaviors that are intimately tied to natural history, such as flight in locusts. Finally, neuroethologists are interested in the plasticity of these mechanisms across different time scales, as plasticity is a major source of individual variation. In temperate breeding songbirds, for example, the neural circuit controlling song may vary dramatically across the year, which helps explain why males are more vociferous in the spring. Even more broadly, experiences of all kinds are encoded by the nervous systems to shape future behavior, as has been elegantly demonstrated in the sea slug, *Aplysia californica.* In some cases, neuroethologists put these questions into an evolutionary perspective in order to better understand the evolution of behavior and its mechanisms. Doing so allows neuroethologists to address the question, *why do individuals or species differ in their behavior?* For example, why are prairie voles monogamous when montane voles are not? These are just some examples of the classic models in neuroethology. In many of these cases, neuroethologists used electrophysiological recordings, electrical stimulation, and lesions to determine the causal relationship between neural activity and behavior. These techniques are still invaluable to neuroethological studies, but they have been augmented in recent years by advances in molecular biology and computational biology.

Advances in Molecular Biology

Technical advances in molecular biology have influenced all aspects of biological research, and neuroethology is no exception. The molecular neuroethologist is typically interested in the genes that are expressed in the nervous system either during development or in adulthood. Understanding genetic differences among individuals or species is an important way of addressing the question, *why do individuals or species differ in their behavior?* In addition, measuring changes in gene expression that are associated with behavior is an important approach to understanding all aspects of the neurobiology of behavior. Both approaches depend, at some point, on knowledge of the relevant gene sequences. For neuroethologists, this can be a challenge. However, recent advances in sequencing technology have made these types of data more accessible than before.

High-Throughput Sequencing

In spite of the explosion of genome sequencing (www.genome.gov), scientists have successfully sequenced the genome of only a fraction of the species under study. Government agencies and institutes typically choose species because they have small genomes and because many scientists have chosen to work on them to answer a particular class of questions. These selection criteria systematically exclude many of the species of interest to neuroethologists, as neuroethologists select their study organisms for very different, often idiosyncratic, reasons (see Introduction). Recent advances in sequencing technology, however, have made large-scale sequencing efforts much more affordable. Thus, it is now feasible for a single laboratory to sequence a genomic or cDNA library of their study organism.

Genomic libraries and cDNA libraries provide different types of information. A cDNA library is a collection of cloned DNA sequences that are complementary to the mRNA that was extracted from an organism or tissue (the 'c' in cDNA stands for 'complementary'). Thus, the cDNA library represents a so-called transcriptome – that is, a collection of transcribed, or expressed, genes. As such, transcriptomes are tissue and state specific; for example, the transcriptome of a singing bird's brain would be different from the transcriptome of a quiet bird. A database of known cDNA sequences is an invaluable tool in studies that manipulate or measure gene expression. In contrast, a genomic library is a collection of cloned pieces of an organism's genome. As such, a genomic library is organisms-specific and does not vary with the tissue or state of the animal. A genomic library provides information about gene sequences, including regulatory regions that determine when and where a gene is expressed. Genomic libraries are useful tools when a neuroethologist is interested in species differences in gene sequences.

Comparative Gene Analysis

Many neuroethologists are interested in the evolution of behavior, and a powerful way to address this goal is to compare behaviorally relevant genes among closely related species that differ in behavior, or among distantly related species that are convergent in behavior. Neuroethologists can determine gene sequences from a genomic library, or by using PCR. In some cases, selection has acted on the coding region of genes, resulting in changes in the structure and function of proteins. For example, comparative analysis of coding sequences demonstrated that the independent evolution of electric communication signals in two lineages of fish has been accompanied by convergent evolution in the structure of the sodium channels that are important in producing the electric signal. In other cases, selection has acted on the regulatory regions of genes, modifying their expression patterns. For example, differences in the regulatory region of the vasopressin receptor gene results in distinct distribution of the receptor in the monogamous prairie vole compared to the polygamous montane vole, although the receptor itself is identical.

Gene Expression Analysis

One important way we understand brain–behavior relationships is to understand how variation in gene expression relates to variation in behavior within individuals, among individuals, and among species. Although two individuals (or species) may have similar genes, variation in the expression of those genes can have profound impacts on the function of neurons and, therefore, on behavior. Relating gene expression patterns to behavior is facilitated by an understanding of the underlying gene sequences.

A variety of tools are available for quantifying gene expression. Two of the most versatile and widely used are microarray analysis and quantitative reverse transcription PCR (RT-PCR), sometimes called real time RT-PCR. RT-PCR uses primers (short sequences of DNA) to amplify a target sequence from a cDNA pool that represents the genes that were transcribed in the tissue sample. RT-PCR is very sensitive and can detect the presence of very low abundance gene transcripts because the target DNA sequence increases exponentially during the PCR reaction, meaning that it doubles with every cycle. Variation in the number of transcripts in the sample will produce variation in the time at which amplification is detectable; greater numbers of initial transcripts will result in earlier amplification. For example, a tenfold difference in transcript concentration will result in a difference of three cycles in the RT-PCR reaction. Thus, in quantitative RT-PCR, the main interest is the initial cycle number that results in detectable amplification. In order to detect that initial cycle (usually called the cycle threshold), dyes that increase in fluorescence proportionally with the amount of the target DNA are added to the reaction. To use quantitative RT-PCR to measure changes in gene expression, one needs to know the sequence of the gene of interest. In addition, because this is a 'candidate gene' approach, meaning that the researcher has a specific hypothesis about a change in expression of a particular gene, it is typically used to detect changes in expression for a small number of known genes. A complementary approach to measuring changes in gene expression is to use a microarray, which is particularly suited to gene discovery.

In a microarray, or gene chip, tiny spots of DNA are attached to a solid surface, typically a glass slide, in an array. To detect differences in gene expression with a microarray, one hybridizes the array with cDNA synthesized from RNA extracted from experimental tissue samples (e.g., individuals before and after displaying a particular behavior). Before hybridization, one labels the two contrasting samples with different fluorescent dyes and then mixes them in equal volumes. This mixture is then incubated with the array where the two samples compete for hybridization with the DNA spots deposited on the slide. For example, if sample A, labeled with a red dye, has more of a particular transcript than sample B, labeled with a green dye, then the result of the hybridization will be more red dye associated with that transcript on the array. In many cases, these DNA sequences have been previously characterized and they might have been selected from a larger set for a particular experiment. But neuroethologists working on unusual animals may not have this luxury. In such cases, the DNA sequences on the array do not have to be previously characterized. If strong differences in hybridization are found with an array, the experimenter can then return to the identified

DNA sequences to determine their identity and characterize them more thoroughly. In either case, the combination of high-throughput sequencing of a cDNA library and a microarray is a powerful way to discover new genes that are associated with particular behaviors.

Manipulating Gene Expression

Quantitative RT-PCR and microarrays are useful for associating changes in gene expression with changes in behavior. However, because the expression of behavior, itself, can cause changes in gene expression in the behaving animal, it is important to go beyond correlations when testing hypotheses about the causal relationship between genes and behavior. Traditionally, this was accomplished by 'knocking out' a gene in a laboratory mouse. In this approach, a particular gene is silenced in a clonal line of mice. This continues to be an important tool for neuroscientists, but it does not lend itself to investigations of brain–behavior relationships in natural populations of animals. A number of novel approaches now allow neuroethologists to manipulate the presence or absence of genes without the genetic tools of the traditional laboratory mouse model. Two powerful ways to manipulate the expression of genes include the use of viral vectors to introduce novel genes and RNAi to silence the expression of native genes. A key advantage to both techniques is that they can be site-specific. That is, unlike whole-organism knock-outs, these approaches manipulate gene expression of specific brain regions while leaving others intact. This is a key innovation given the complexity of neural tissue because a particular gene product may regulate different behaviors depending on the neural circuit where it is expressed.

If a neuroethologist finds that expression of a particular gene is associated with a particular behavior, he or she may want to silence the gene to determine whether it is a causal factor in the behavior. A relatively simple way of silencing genes in vivo is to capitalize on a cell's RNA regulatory machinery. The RNA interference (RNAi) pathway regulates the likelihood that an RNA molecule will be translated into protein. One way to activate the RNAi pathway is to inject double-stranded RNA corresponding to the target gene into a particular brain region. The double-stranded RNA recruits the RNA regulatory machinery, resulting in the degradation of the target mRNAs, thus leading to gene silencing. Alternatively, a neuroethologist may want to introduce a novel gene, increase expression of a gene, or introduce a native gene into a novel location. One way of doing so is to use viral vectors that insert a gene of interest into the cells of a target brain region. Such viral vectors are engineered to carry the candidate gene and they capitalize on the ability of viruses to deliver their genetic material into foreign cells. This is a particularly useful tool when a neuroethologist wants to

test a hypothesis related to behavioral variation between closely related species. In an elegant example of this approach, neuroethologists used a viral vector to introduce a prairie vole gene for the vasopressin receptor into the brains of montane voles to demonstrate that the receptor's unique distribution in prairie voles is an important cause of species-differences in social behavior.

Novel Approaches to Identifying Neural Circuits

The central nervous system is one of the most complex of all organs and understanding its structure is a fundamental endeavor in neurobiology. This is of particular interest to neuroethologists since nervous system structure can vary dramatically among species and such variation is an important cause of species differences in behavior. A major goal of neuroanatomical studies is to determine how individual brain regions are interconnected. In addition, studies that investigate the function of neural circuits during specific behaviors are an important complement to studies of nervous system structure, as brain regions or neural circuits can contribute differentially to different behaviors.

Novel Neural Tracers

Most of what we know about the neural connections among brain regions comes from studies that used lesions, which can identify connections by subsequent degeneration of axonal fibers, or the introduction of tracers. A tracer can be any substance that is taken up by one part of a neuron and transported to another part. Some substances are transported retrogradely (from axon to cell body), anterogradely (from dendrites or cell body to axon), or both. Tracers vary in their effectiveness depending, in part, on how efficiently neurons take them up, and on the method used to visualize them.

There are two major constraints to traditional neural tracers. First, typically only large, robust connections can be identified with discrete injections. Second, only a single set of connections can be identified in a single animal. The application of self-amplifying, transneuronal tracers, such as pseudorabies virus, can potentially solve both these problems. In tracing studies, an attenuated form of the virus is typically used. The pseudorabies virus is taken up by axonal terminals and transported to the cell body where it is replicated. The replication of the virus effectively amplifies the signal and, ideally, results in all neural connections being identified with similar probabilities. Once replicated by the cell, the virus is distributed throughout the dendrites where, importantly, it crosses synapses to subsequently infect connected cells. This process is repeated and, with enough time, should identify the entire neural network connected, ultimately, to the brain region where the initial injection of virus was made. The pseudorabies virus can be made cell type-specific with genetic modifications that make its expression conditional on the type of neuron within which it is expressed. For example, neuroethologists used this approach to map the afferent network of neurons expressing GnRH, a releasing hormone that regulates reproductive physiology through its action on the pituitary. One constraint on the use of the pseudorabies virus in tracing studies is that it must be infectious in the animal being studied.

Functional Activity Mapping

An important complement to neural tracing studies, which reveal the structural connections among brain regions, are studies that investigate the activity of brain regions during the expression of a behavior or in response to behaviorally relevant stimuli. Most of what we know about nervous system function comes from studies that record electrical activity of neurons in anesthetized or restrained animals. A disadvantage of electrophysiology is that it requires the implantation of electrodes, which constrains the types of behaviors an animal can engage in during recording. In addition, researchers are typically able to study only one or a small number of brain regions at any given time. An alternative to electrophysiology is functional activity mapping, which uses markers that are correlated with neural activity to analyze the pattern of neural activity in multiple brain regions simultaneously. A range of markers are available, including endogenous metabolic markers, such as the mitochondrial enzyme cytochrome oxidase, exogenous metabolic markers, such as radioactively labeled glucose, and endogenous changes in the expression of genes or protein.

Recently, there has been an explosion in studies that utilize changes in expression of genes or proteins that are correlated with neural activity. These studies capitalize on the so-called immediate-early gene response of neurons. The immediate-early gene response was first described when researchers discovered that, in response to external stimulation, cells launch a rapid increase in gene expression that is followed by a wave of protein expression. Further studies showed that expression of immediate-early genes is controlled by cellular signaling cascades that are, in turn, regulated by changes in membrane depolarization. Thus, when neuroethologists detect an increase in immediate-early gene expression (or their protein products), they infer that those cells or brain regions have been recently activated. This approach has proved to be very fruitful, particularly because one can measure changes in neural activity of many brain regions at the same time. In addition, this technique is sometimes more readily adaptable to a variety of organisms than is electrophysiology. However, a major constraint on the interpretation of these studies stems from

the time course of changes in immediate-early gene expression. For example, changes in gene expression can typically be detected within minutes and peak levels of gene expression are detected 30 min to 1 h after stimulus onset. Thus, functional activity mapping is very useful for identifying functional attributes of individual brain regions, but is not typically suitable for identifying the underlying neural code.

Advances in Electrophysiology

Neurons communicate with electrical signals, so it should be no surprise that electrophysiology is one of the most important tools of neuroethologists. Electrophysiology is a versatile tool and is favored by many neuroethologists because of its temporal precision. Given the constraints of electrophysiogical recording, most in vivo studies use anesthesia or paralytic chemicals to restrain the animal during recording. In addition to the obvious disadvantage to neuroethologists of working on an animal that cannot move, the use of anesthesia or chemical restraint may pose a more fundamental problem, as neural activity sometimes varies according to the animal's state. For example, parts of the songbird pallium show robust auditory responses to song when the bird is asleep, but no responses when the animal is awake. In addition, in most electrophysiology studies, neuroethologists record from one neuron at a time. Since many functions of the nervous system are likely carried out by ensembles of neurons, rather than individual neurons, this approach may fail to reveal the underlying coding mechanisms. Recent advances in electrophysiology now allow neuroethologists to record from awake, freely behaving animals. The head-mounted equipment is much smaller than before, allowing researchers to work on a wider variety of animals, and the use of commutators or telemetry systems also allows the animal to move about during recording. In addition, the use of multielectrode arrays has facilitated the analysis of networks of neurons. The ability to record from multielectrode arrays has been enabled by advances in electronic technology as well as in the computational methods required to analyze the larger and more complex data sets. Finally, electrophysiology studies have been advanced conceptually by the integration of information theory, a field of mathematics that, when applied to electrophysiology data, can quantify how much information is encoded by a neuron's response.

Advances in Computational Biology

Computational biology represents the integration of computer science, mathematical modeling, and statistics to solve biological problems. As such, its reach is substantial and highly diverse. Within the field of neuroethology, computational approaches have had an impact in molecular biology, particularly in large-scale sequence analysis and microarray analysis, and in electrophysiology, where computational approaches have facilitated analysis of multielectrode recordings. In addition to advancing these fields, computational biology has also generated new fields of special significance to neuroethologists, namely, computational neuroethology and artificial neural network modeling.

Artificial neural networks are mathematical models that simulate biological neural networks, or nervous systems. They can be used to explore the relationship between nervous systems and behavior and to explore how nervous systems can constrain the evolution of behavior. Generally, neural network modeling is motivated by theory; neural network models are highly simplified and are meant to provide general models that can be used to test ideas. In a neural network, the essential element is a 'node,' conceptually akin to a neuron. Nodes have states and are interconnected with other nodes. The pattern of connections among nodes is referred to as the network architecture. Sometimes, the neural network can change as the result of experience. For example, input nodes may respond to a 'stimulus' and relay this information to output nodes that produce some response or 'behavior.' If the response does not match some standard, the model can specify changes to the nodes and/or their connections that may improve the output.

Computational neuroethology is a related approach to modeling the neural basis of behavior. Like neural network modeling, computational neuroethologists create mathematical models of biological nervous systems to simulate animal behavior. Computational neuroethology emphasizes the interaction of the simulated animal with its environment. Thus, when a simulated animal produces a behavioral response to a stimulus, that behavior changes the nature of the animal's stimulus environment, resulting in the so-called action-perception cycle. In computational neuroethology, the simulated animals may be computer simulations or robotic simulations, and they are generally inspired by specific species and the natural problems they face, such as an insect foraging for food.

See also: Neurobiology, Endocrinology and Behavior; Neuroethology: What is it?; Robot Behavior; Sociogenomics.

Further Reading

Beer RD (1990) *Intelligence as Adaptive Behavior*. San Diego, CA: Academic Press.

Borst A and Theunissen FE (1999) Information theory and neural coding. *Nature Neuroscience* 2(11): 947–957.

Enquist M and Ghirlanda S (2005) *Neural Networks & Animal Behavior*. Princeton, NJ: Princeton University Press.

Ewert JP (1974) The neural basis of visually guided behavior. *Scientific American* 230(3): 34–42.

Hawkins RD, Kandel ER, and Bailey CH (2006) Molecular mechanisms of memory storage in *Aplysia*. *Biological Bulletin* 210(3): 174–191.

Knudsen EI and Konishi M (1978) A neural map of auditory space in the owl. *Science* 200(4343): 795–797.

Lim MM, Wang Z, Olazabal DE, Ren X, Terwilliger EF, and Young LJ (2004) Enhanced partner preference in a promiscuous species by manipulating the expression of a single gene. *Nature* 429(6993): 754–757.

Nick TA and Konishi M (2001) Dynamic control of auditory activity during sleep: Correlation between song response and EEG. *Proceedings of the National Academy of Sciences of the United States of America* 98(24): 14012–14016.

Nottebohm F (1981) A brain for all seasons: Cyclical anatomical changes in song control nuclei of the canary brain. *Science* 214 (1527): 1368–1370.

Robertson RM (1986) Neuronal circuits controlling flight in the locust: Central generation of the rhythm. *Trends in Neurosciences* 9: 278–280.

Tinbergen N (1963) On aims and methods in ethology. *Zeitschrift fur Tierpsychologie* 20: 410–433.

Yoon H, Enquist LW, and Dulac C (2005) Olfactory inputs to hypothalamic neurons controlling reproduction and fertility. *Cell* 123(4): 669–682.

Young LJ, Wang Z, and Insel TR (1990) Neuroendocrine bases of monogamy. *Trends in Neurosciences* 21(2): 71–75.

Zakon HH, Lu Y, Zwickl DJ, and Hillis DM (2006) Sodium channel genes and the evolution of diversity in communication signals of electric fishes: Convergent molecular evolution. *Proceedings of the National Academy of Sciences of the United States of America* 103(10): 3675–3680.

Neuroethology: What is it?

M. Konishi, California Institute of Technology, Pasadena, CA, USA

Neuroethology and Central Pattern Generators

The starting point of a neuroethological study is the choice of a behavior, be it a spontaneous movement or a response to a sensory stimulus. Rhythmic movements such as walking, flying, and swimming have been favorite subjects in this field, partly because these movements can be easily induced, described, and measured. The source and the control of rhythms have been the major topics. The term 'central pattern generator (CPG)' means that the source of the pattern is not peripheral but in the central nervous system. A CPG should continue to produce its patterned output after the removal of all pace-making sensory inputs to it. The late D. M. Wilson was one of the early investigators who combined behavioral and neurophysiological methods to prove the existence of a CPG. He showed that the locust could maintain the normal pattern of sequential flapping of its wings after the removal of all the peripheral sensory nerves that could send timing information to the presumed CPG for flight in the thoracic ganglia. However, other researchers later reported that the locust could use certain sense organs on its head to detect the rhythmic movement of the air during flight. Thus, the identification of potential sources of sensory feedback for CPGs is not as simple as it sounds.

Nevertheless, neural circuits that generate rhythmic movements are known in many systems and animals. One of the most extensively studied systems is the stomatogastric ganglion of lobsters. This ganglion contains about 30 neurons and generates many different patterns of discharges in the participating neurons according to the roles that different parts of the stomach play as food moves by rhythmic as well as sequential contractions of the muscles involved. The coordination of even this small number of neurons involves not only electrical signals but also a large number of chemical signals called 'neuromodulators.' The use of CPGs for motor coordination is restricted to neither simple systems nor simple animals. A group of birds known as 'suboscine songbirds' develop normal songs of their species even when the birds cannot hear their own voices, that is, without auditory feedback. Domestic chickens use some 12 different vocalizations for social communication. All these calls develop in deaf chickens. All these avian groups must have CPGs for vocalizations, although no attempts have been made to look for them.

The Role of Sensory Feedback in Motor Coordination

Despite the early emphasis on central pattern generators, the role of sensory feedback in the control of movements has also been extensively studied. A simple example of sensory feedback is hearing one's own voice in speaking. If a person hears his own voice returning with a certain delay, it is hard to speak normally. Obviously, hearing one's own voice is essential for learning languages, because the speaker must decide whether he is pronouncing words properly or not. The importance of auditory feedback for vocalization has also been shown in animals. A large group of birds called 'oscine songbirds' learn songs. Young birds listen to their own father or other male adults sing and remember the song until they can sing in adulthood. As young birds become gradually mature, they try to match their own voices with the memorized song. If these birds cannot hear their own voices, they fail to reproduce the memorized song. Thus, auditory feedback is essential for song learning. Auditory feedback is also necessary for the maintenance of adult song. If a songbird hears its own song with a delay, the bird starts making errors in both composing and sequencing of different parts of his song.

The ability to control the voice by auditory feedback is a prerequisite for vocal learning. The CPG of vocal learners must be modifiable such that they can adjust the variables in the CPG in order to learn. However, the CPG loses its meaning if it is infinitely variable. Oscine songbirds solve this problem by using a different method of encoding song. They use a sensory 'template' to which auditory feedback must be matched. The template is not a part of the CPG for song but a separate reference to which song must be compared. A bird hears its own song and compares it with the template in the brain. If the song differs from the template, the bird changes it until it matches the template. The removal of auditory feedback deprives the bird of the means to compare its song and the template. The discovery of the brain pathways for the control of song opened the possibility to identify the site and mechanisms of this comparison.

Nottebohm and his associates discovered that lesion of a specific area in the male canary's brain caused dramatic changes in its song. Using anatomical tracing methods, they identified several discrete areas, which are now collectively called 'the song system.' The area is now called 'high vocal center' (HVC) and contains three different classes of neurons: the first group includes local inter-neurons; the second group includes neurons that project to RA, which is on the motor pathway to the hindbrain area that innervates the muscles for the control of vocalizations; and the third group projects to an area called X in the so-called anterior forebrain pathway (AFP), which is thought to be involved in the feedback control of song, because lesions of this pathway affect song development in young birds but not in adult birds.

Another interesting feature of the song system is the selective response of its neurons to the individual bird's own song. The template matching mentioned earlier may take place in the AFP. RA receives inputs from both the AFP and the HVC. RA neurons in sleeping zebra finches fire series of impulses that closely resemble those which occur during singing. Furthermore, when young zebra finches hear a tutor song, the discharge pattern of their RA neurons changes during the following night of sleep. These tutor-song-induced discharge patterns do not occur if auditory feedback is disrupted. These findings are not only consistent with the template theory of song learning but also with the idea of consolidation of memories during sleep in humans.

The song system is unique to oscine songbirds, although the brains of other vocal learners, such as parrots and some hummingbirds, seem to contain areas that are homologous or analogous to those of oscine songbird. The song system is present in both genders in species in which both sexes sing. The song system's nuclei such as HVC and RA are smaller in females than in males in species in which females are known to sing only occasionally. The song system is absent in females of species in which only the male sings as in the zebra finch. However, if one examines the brain of a young female zebra finch at 20 days of age, one can easily identify RA, which is not much smaller than the RA in a male of the same age. The absence of RA in the adult female is due to programmed cell death. We know today how the gender difference in RA emerges during the first 40 days of life in zebra finches. The neurons that are destined to become RA cells are born on the sixth day of incubation in both sexes. These cells are of equal size between the sexes at the time of hatching. The RA cells in the female gradually become smaller and die during the first month after hatching. This phenomenon is called 'programmed cell death.' It is, however, possible to stop the program by giving female chicks a small amount of estrogen, the female hormone! Estrogen and testosterone are similar to each other in their molecular structures and an enzyme named 'aromatase' can convert estrogen to testosterone. The avian brain can produce this enzyme. The female zebra finches treated with estrogen to save RA cells from death sing in adulthood, although their external appearance remains feminine.

The Mechanisms of Signal Processing

The processing of biologically important stimuli or signals is also an important subject in neuroethology. The classic Lorenz–Tinbergen model contained an 'Innate Releasing Mechanism,' which included a central gate that could be opened only by a specific releaser. Daniel Lehrman challenged this interpretation of behavior by pointing out that peripheral sense organs themselves might be tuned to such stimuli. The sex attractant of the silkworm moth is a good example here, because it binds selectively to special molecular receptors in the antenna.

There are many similar examples in other sensory systems and animals. An auditory organ called the 'Johnston's organ' occurs at the base of the insect antenna. As air molecules hit the antenna, it oscillates, causing the Johnston's organ to respond with nerve impulses. Experiments show that the mechanical response of the Johnston's organ of male mosquitoes is tuned to the frequency range of the flight noise of female mosquitoes of the same species. Insects such as noctuid moths that are preyed upon by bats have ears that are sensitive to the high frequency of the bat's echolocation sounds. These ears are served by only two auditory nerve fibers, which transmit signals to the central nervous system. The late Kenneth D. Roeder, one of the forefathers of neuroethology, carried out beautiful nightly field experiments in which he broadcast high-frequency sounds resembling the bat's echolocation signals as noctuid moths approached a loudspeaker at the tip of a long post. Moths showed two different evasive strategies, flying away when the sound source was far and diving down when the source was near.

Of course, not all examples of signal detection can be accounted for by the specialization of sense organs. Also, the central mechanisms of signal detection are hard to find, because one cannot predict the methods of signal representation in the successive stages of processing. However, this goal was achieved in the sensory pathways for sound localization in barn owls and in the sensory pathways for jamming avoidance response in electric fish *Eigenmannia*. An owl hears the same sound at the same time in its two ears if the sound source is located directly in front. If the source is now moved to the right by so many degrees, then the owl's right ear receives the sound earlier than the left ear. In reality, the owl does not use differences in the arrival times of the first sound wave but uses differences in the phase angle of the sounds between the two ears. Nevertheless, 'time' may be used in place of 'phase' in nontechnical literatures. The owl uses this time difference, called 'the interaural time difference' (ITD), for source localization in the horizontal plane (i.e., right-left direction). The owl uses the interaural level difference (ILD) for the vertical direction. The use of the intensity cue for the vertical direction is due to an asymmetry in the owl's external ears. Although the left and the right ear canals in the skull are identical, the skin flaps that cover them are asymmetrical. The left skin flap is located higher on the face than the right skin flap. This arrangement makes the left ear more sensitive to sounds coming from below and the right ear more sensitive to sounds coming from above. The owl's midbrain auditory area contains neurons that respond to noises coming from a particular direction in space. These neurons are arranged to form a map of auditory space. These 'space-specific' neurons respond only to particular combinations of ITD and ILD. These findings led to systematic surveys of the lower stages of signal processing leading to the map. The owl's brain processes ITD and ILD in two separate pathways starting from the first brain auditory stations, the cochlear nuclei. Processing of both ITD and ILD also takes place in separate frequency bands until these bands converge in the external nucleus where the map of space is found. The map of auditory space projects to the map of visual space in the optic tectum in which neurons respond to both auditory and visual stimuli coming from the same direction.

Combinations of behavioral and neural studies also greatly facilitated the work on the neural mechanisms of jamming avoidance response in electric fish *Eigenmannia*. This group of fish produces electrical signals to detect nearby objects including other fish. The waveform of these signals resembles sine waves. When two *Eigenmannia* fish detect each other, they change the frequency of their signals so as to avoid jamming each other. The late Walter Heiligenberg and his associates studied the jamming avoidance response (JAR) and its neural mechanisms in great detail. In contrast to the owl work mentioned earlier, the investigators started from the lowest stage of signal processing, the electrosensory cells in the skin. Heiligenberg developed a system in which he could record single neurons from the brain of a fish that was actively performing jamming avoidance responses. The electrosensory system also consists of time and amplitude pathways. Each pathway includes several stages in which information necessary for JAR is detected and encoded. The highest station contains neurons that fire when the fish's own frequency is higher than the other fish's frequency and neurons that fire when the fish own frequency is lower. These and other neurons form a brain network that controls the discharge of the electric organs near the tail.

Although owls and electric fish are very different, their algorithms of signal processing are similar; these include stepwise detection and encoding of signals in the ascending sensory pathways, separation of stimulus variables such as time and amplitude, single neuron representations of complex stimuli at the top of the hierarchy, and their connections to the motor control center of the behavior involved. These similarities do not occur by chance but reflect certain principles that underlie signal processing

by sensory systems. As more sensory systems are studied with reference to natural behaviors, general rules are likely to emerge.

Representation of Complex Stimuli by Single Neurons

The preceding section discussed how neurons selective for biologically important stimuli acquire their response properties through multiple stages of processing. How far can this process be extended? The presence of neurons selective for faces in the monkey and human brain has been known for years. Progress is also being made in explaining how such selectivity develops in successive stages of signal processing. However, there is still some skepticism about the presence of such neurons. One reason is the belief that single neurons cannot represent such complex stimuli. The second reason states that these neurons are responding to simpler aspects of faces, but this possibility has not been adequately excluded. Single face-selective neurons obviously do not carry out all stages of signal processing in face discrimination themselves. Like the space-specific neurons of the owl, the face neurons represent the results of all processes leading to them. In the primate visual system, the successive stages of signal processing go from the primary visual cortex (V1) to the middle temporal area (V5) and to the inferotemporal cortex (IT), which is the first site with face neurons in this pathway. IT neurons convey their face selectivity to the medial temporal lobe, parahippocampal gyrus, perirhinal cortex, entorhinal cortex, hippocampus, and amygdala. This example of face-selective neurons shows that the neuroethological approach is not limited to lower organisms. The existence of single neurons selective for complex biologically significant stimuli indicates how central sensory systems are organized. There is no contradiction between the general assumption that multiple neurons are involved in sensory coding and the fact that there are single neurons that are selective for complex biologically significant stimuli. The face neuron represents the results of all computations leading to it. The owl's space-specific neuron and the *Eigenmannia*'s sign selective neuron are the products of successive stages of computations.

Summary

The primary goal of neuroethology is to study the neural mechanisms of natural behavior. Stereotyped movements in animals led to the concept of central pattern generators, which are now known and well studied in many species. Control of movements by sensory feedback has also been shown in many systems and animals. This control also provides possibilities for learning of motor skills as in birdsong and speech development. Processing of sensory signals has also been an important subject in neuroethology. Contrary to the idea of early ethologists who thought of mechanisms within the central nervous system, peripheral sensory cells and organs may be highly selective for biologically important stimuli. Neuroethological approaches have also revealed the nature of neural organizations in which successive stages of signal processing are carried out in animal groups such as the barn owl and *Eigenmannia*. These studies also lend support to the idea that tapping a high-order neuron at a time allows one to deduce the computational organization of the system in which the neuron occurs. Although the present review does not cover this subject, the use of molecular genetical methods to study behaviors has greatly advanced in animal groups such as nematodes, the fruit fly *Drosophila*, and zebra fish. Combination of this and neuroetholgical approaches is likely to yield a new level of understanding natural behavior.

See also: Neuroethology: Methods.

Further Reading

Heiligenberg WF (1991) *Neural Nets in Electric Fish.* Cambridge, MA: The MIT Press.

Konishi M (1963) The role of auditory feedback in the vocal behavior of the domestic fowl. *Zeitschrift für Tierpsychologie* 20: 349–367.

Konishi M (1965) The role of auditory feedback in the control of vocalization in the white-crowned sparrow. *Zeitschrift für Tierpsychologie* 22: 770–783.

Kroodsma DE and Konishi M (1991) A suboscine bird (Eastern phoebe, *Sayornis phoebe*) develops normal song without auditory feedback. *Animal Behaviour* 42: 477–487.

Leonardo A and Konishi M (1999) Decrystallization of adult birdsong by perturbation of auditory feedback. *Nature* 399: 466–470.

Marder E, Bucher D, Schultz DJ, and Taylor AL (2005) Invertebrate central pattern generation moves along. *Current Biology* 15: 685–699.

Nottebohm F, Stokes TM, and Leonard CM (1976) Central control of song in the canary, *Serinus canarius. J. Comp. Neurol.* 165: 457–486.

Roeder KD (1963) *Nerve Cells and Insect Behavior.* Cambridge, MA: Harvard University Press.

Tinbergen N (1951) *The study of Instinct.* Oxford: Claredon Press.

Tsao D (2006) A dedicated system of processing faces. *Science* 314: 72–73.

Wilson DM (1961) The central nervous control of flight in a locust. *Journal of Experimental Biology* 38: 471–490.

Ziegler HP and Marlor P (2008) *Neuroscience of Birdsong.* Cambridge University Press.

Non-Elemental Learning in Invertebrates

M. Giurfa and A. Avarguès-Weber, CNRS, Université de Toulouse, Toulouse, France; Centre de Recherches sur la Cognition Animale, Toulouse, France
R. Menzel, Freie Universität Berlin, Berlin, Germany

Introduction

Cognitive science provides a fresh look at animal behavior, and its merge with neuroscience overcomes the conceptual limitations of traditional experimental psychology and ethology. Despite the multitude of approaches in cognitive neuroscience and the respective attempts to define these approaches, a general definition for the term 'cognition' remains elusive probably because the key terms are understood differently depending on the conceptual traditions to which the scientists relate themselves, the behaviors in question, and the considered complexity of the neural substrates underlying them. A key term is 'representation,' the understanding that the brain is actively involved in perceiving the world and creating motor patterns by recruiting memories, expecting outcomes, and making decisions between neural instantiations of behavioral options. Gaining information by learning and by storing it in multiple forms of memory, as a fundamental form of representation, is an essential and most likely a basic property of any neural system of some complexity. Here, we shall focus on *nonelemental forms of associative learning*, that is, on learning forms in which simple, unambiguous links between specific events in an animal's environment cannot account for experience-dependent changes in behavior, and which require operations on remote and recent memories. In this respect, *nonelemental associative learning* transcends *elemental forms of associative learning*, in which animals learn univocal connections between specific events in their environment. In particular, we shall ask whether animals with small brains like molluscs and insects are capable of performing *nonelemental associative learning*.

Elemental Forms of Associative Learning in Invertebrates

Associative learning allows extracting the logical structure of the world by evaluating the sequential order of events. Two major forms of associative learning are usually recognized: in *classical conditioning*, animals learn to associate an originally neutral stimulus (conditioned stimulus (CS)) with a biologically relevant stimulus (unconditioned stimulus (US)); in *operant conditioning*, they learn to associate their own behavior with a reinforcer and relate this connection to the context conditions of the environment.

In their most simple version, both learning forms rely on the establishment of associative links connecting two (or more) specific and unambiguous events in the animal's world. For instance, in *absolute classical conditioning* (A+), a direct link between an event (A) and reinforcement (+) is learned, while in *differential classical conditioning* (A+ vs. B−), simple, unambiguous links between A and reinforcement and between B and the absence of reinforcement are simultaneously learned.

Multiple cases of these simple learning forms have been described for invertebrates. For instance, in the honeybee *Apis mellifera*, olfactory conditioning of the proboscis extension response (PER) has been repeatedly used for the study of *elemental classical conditioning* and its neural substrates. Individually harnessed hungry bees that do not respond to an odor presentation with an extension of their proboscis do so when their antennae are stimulated with sucrose solution (the US). If the odor (the CS) is forward paired with sugar, the bees learn an association between odor and sugar reward so that they exhibit conditioned PER to future presentations of the odor alone (**Figure 1**). An example of *elemental operant conditioning* is provided by the aquatic mollusc *Lymnaea stagnalis*, which can be trained to suppress the opening of its pneumostome, a small respiratory orifice, when the animal surfaces and attempts to breathe. This is achieved by an aversive and repeated mechanical stimulation of the pneumostome, which determines that the mollusc learns to reduce its attempts to open the pneumostome as training progresses. In both examples, the neural networks mediating associative learning are relatively simple and well studied, thus underlining the advantages of invertebrates as model systems for the understanding the neural mechanisms of simple forms of learning.

Nonelemental Forms of Associative Learning in Invertebrates

In the higher-order forms of learning on which we focus here, simple links connecting specific events are generally not useful because ambiguity characterizes the events under consideration. For instance, in the discrimination termed *negative patterning discrimination*, an animal has to learn to differentiate a nonreinforced binary compound AB− from its reinforced elements (A+, B+). This situation is particularly challenging as each element A and B

Figure 1 Olfactory conditioning of the proboscis extension reflex. (a) An individual bee is immobilized in a metal tube so that only the antennae and mouth parts (the proboscis) are free to move. The bee is set in front of an odorant stimulation setup which is controlled by a computer and which sends a constant flow of clean air to the bee. The air flow can be rerouted through cartridges presenting chemicals used for olfactory stimulation (conditioned stimuli or CS). A toothpick soaked in sucrose solution (unconditioned stimulus or US) is delivered to the antennae and the proboscis. In this appetitive classical conditioning, the bee learns to associate odorants (CS) and sucrose solution (US). (b) The proboscis extension reflex of the honeybee. Bees exhibit this reflex when their antennae are touched with sucrose solution (US). After successful conditioning, bees extend the proboscis to the odorant (CS) which predicts the US.

appears as often reinforced as nonreinforced. Relying on elemental links between A (or B) and reinforcement (or absence of reinforcement) is useless to solve this problem. Another example of nonelemental learning is the so-called *biconditional discrimination* where the subject learns to respond to the compounds AB and CD and not to the compounds AC and BD (AB+, CD+, AC−, BD−). As in negative patterning, each element, A, B, C, and D appears reinforced as often as nonreinforced so that it is impossible to rely only on the associative strength of a single element to solve the task. In both examples, animals have to suppress linear processing of compounds and learn that a compound is an entity different from its components.

A second form of nonelemental learning is *contextual learning*, in which animals learn to produce adaptive responses that can be linked to a specific context. They learn that, given a certain stimulus or condition, a particular response is appropriate whereas, given a different stimulus or condition, the same response is no longer appropriate. This form of learning, usually referred to as conditional learning or occasion setting, cannot be viewed as elemental learning because a given stimulus may or not be predictive of a certain outcome, depending on the particular environment.

A third form of nonelemental rule is provided by *rule learning* in which animals respond to novel stimuli that they have never encountered before or can generate novel responses that are adaptive given the context in which they are produced. In doing this, animals exhibit a positive transfer of learning, a capacity that cannot be referred to as an elemental learning because the responses are aimed at stimuli that do not predict a specific outcome per se based on the animals' past experience.

One of the first works adopting a nonelemental learning perspective in invertebrates was performed on lobsters. These animals normally exhibit exploratory behavior when placed in an aquarium. They can be

aversively conditioned to stop searching by pairing an olfactory stimulus delivered in water with a mechanosensory disturbance produced by the experimenter. Lobsters were trained in this way with an olfactory compound AX reinforced by the aversive mechanosensory stimulation (AX+). Conditioning was either absolute (AX+) or differential, when a second compound AY (AX+ vs. AY−) was used. After absolute conditioning, lobsters inhibited their search behavior when presented with AX as expected, but searched when presented with A, X, or with a novel odor Y. This result is consistent with learning the compound AX as an entity different from its components A and X, as proposed by the configural theory (Pearce, 1994). After differential conditioning, lobsters again inhibited their searching behavior when presented with AX but not with AY. Interestingly, they also inhibited search when presented with the element X but not with the element Y. A was not useful as it was common to the reinforced and the nonreinforced compounds AX+ and AY−, respectively. In this case, lobsters seemed to have learned the compounds AX and AY in elemental terms, thus being able to fully generalize their respective responses to X and Y. This work shows that depending on the conditioning protocol, lobsters treat and learn an olfactory compound differently so that either elemental or nonelemental associations with the negative mechanosensory reinforcer are built.

In honeybees, several studies have addressed the issue of elemental versus nonelemental learning, using visual conditioning of free-flying animals or olfactory PER conditioning of harnessed animals. In the first protocol, bees flying between the hive and a feeding site are trained to discriminate different kinds of visual targets (colors, shapes, motion cues, etc.) at the food source. Correct choices are rewarded with a drop of sucrose solution. In the second protocol, described earlier, harnessed bees learn a Pavlovian association between odor and the sucrose reward. In both experimental protocols, bees were shown to solve a biconditional discrimination (AB+, CD+, AC−, BD−). In the visual modality, free-flying bees had to discriminate complex patterns that were arranged to fulfill the principles of this discrimination problem. In the olfactory modality, olfactory compounds were used and bees learned to respond appropriately to each compound, independently of the ambiguity inherent to the components.

Bees also proved to be able to solve a negative patterning discrimination (A+, B+, AB−) in the olfactory domain. It was shown that in situations in which ambiguity is created at the level of the odorants integrating a compound, olfactory processing is consistent with the *unique cue theory*, a form of processing in which animals detect to some extent the presence of the components in the compound but in which they also assign a unique identity to the compound (the unique cue), resulting from the interaction between its components.

Neural Bases of Nonelemental Learning in Invertebrates

The interest in nonelemental olfactory learning protocols in insects relates to the possibility of correlating the behavior with the plasticity of the underlying neural circuits. The olfactory circuit is relatively well known. In the case of honeybees (**Figure 2**), peripheral processing of odor molecules occurs at ~60 000 olfactory receptor neurons (ORNs) and in 160 glomeruli of the antennal lobe (AL). ORNs and glomeruli in the AL have broad, overlapping and combinatorial responses to a range of odors. Processed olfactory information is conveyed by ~800 projection neurons (PNs) to higher-order brain centers

(mushroom bodies (MBs) or lateral protocerebrum). MBs are particularly interesting from the perspective of nonelemental learning since they receive segregated information of different sensory modalities (visual, olfactory, mechanosensory) and provide multimodal output that reflects the integration of information between modalities at the level of the neurons that constitute them, the Kenyon cells and the mushroom body output neurons.

In honeybees, bilateral olfactory input to both antennae is required to solve a negative patterning discrimination. Given that the olfactory circuit remains practically unconnected between hemispheres until the MBs, this result suggests that the reading of a unique cue, arising from odorant interactions within the mixture, occurs upstream the ALs, that is, at the level of the MBs. Mushroom body-ablated honeybees were used to determine whether these structures are necessary to solve nonelemental olfactory discriminations. Bees were conditioned in a side-specific discrimination so that when odorants were delivered to one antenna, the contingency was A+ versus B−, while it was reversed (A− vs. B+) when they were delivered to other antenna. Bees without lesions could solve this nonelemental problem (each odor is as often rewarded as nonrewarded), while bees with unilateral lesions of the MBs were impaired in this problem solving but not in elemental discriminations. It was therefore proposed that MBs are required for solving nonelemental discriminations.

It thus appears that at least lobsters and honeybees are capable of nonelemental learning in the strict sense and that in insects, MBs are involved in such kind of problem solving. Such forms of learning are highly dependent on the way in which animals are trained, the number of trials, and on the similarity between elements in a compound.

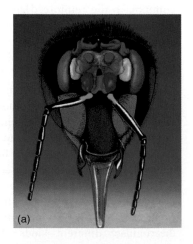 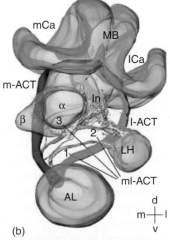

Figure 2   The basic organization of the honeybee olfactory system. (a) Frontal view of the brain with the main olfactory centers; (b) Three-dimensional reconstruction of the olfactory circuit based on confocal microscopy; AL: antennal lobe; LH: lateral horn; MB: mushroom body; m-ACT: medial antenno-cerebral tract; l-ACT: lateral antenno-cerebral tract; mCa: medial calyx; lCa: lateral calyx; alpha and beta: alpha and beta lobes of the mushroom body. Courtesy of Wolfgang Roessler.

Further research should ask whether other invertebrates particularly *Drosophila* solve nonlinear discrimination problems. Neurogenetic tools available in this insect could be a most useful tool for identifying in a more precise way the neuronal circuits involved in nonlinear discriminations.

Contextual Learning in Invertebrates

Contextual learning can be subsumed in the so-called occasion setting problem. In this problem, a given stimulus, the occasion setter, informs the animal about the outcome of its choice (for instance, given stimulus C, the occasion setter, the animal has to choose A and not B because the former but not the latter is rewarded). This basic form of conditional learning admits of different variants, depending on the number of occasion setters and discriminations involved, which have received different names. For instance, another form of occasion setting involving two occasion setters is the so-called *transwitching problem*. In this problem, an animal is trained differentially with two stimuli, A and B, and with two different occasion setters C1 and C2. When C1 is available, stimulus A is reinforced while stimulus B is not (A+ vs. B−), while it is the opposite (A− vs. B+) with C2. This problem does not admit lineal solutions as each element (A, B) and each occasion setter (C1, C2) appear equally as often connected with reinforcement as with absence of reinforcement. Focusing on A or B alone does not allow solving the problem. Animals have, therefore, to learn that C1 and C2 define the valid contingency.

In the mollusc *Aplysia californica*, exposure to two different contexts (a smooth, round bowl containing lemon-flavored seawater and a rectangular chamber with a ridged surface containing unscented seawater that was gently vibrated by an aerator located in one corner) and experiencing a series of moderate electric shocks (US) in one of these two contexts lead to the establishment of an association between the context and the shock. The context alone elicited a defensive reaction which was exclusive for the reinforced context.

Crickets *Gryllus bimaculatus* and cockroaches *Periplaneta americana* also exhibit contextual learning as they solve a typical version of the transwitching problem. Both crickets and cockroaches associate one odorant with water reward (appetitive US) and another odorant with saline solution (aversive US) under illumination, and learn the reversed contingency in the dark. Thus, the visual context affected the learning performance only when crickets were requested to use it to disambiguate the meaning of stimuli and to predict the nature of reinforcement.

Bumblebees *Bombus terrestris* have also been trained in a transwitching problem to choose a 45° grating and to avoid a 135° grating to reach a feeder, and to do the opposite to reach their nest. They also learn that an annular or a radial disc must be chosen, depending on the disc's association with a 45° or a 135° grating either at the feeder or at the nest entrance: in one context (the nest), access was allowed by the combinations 45° + radial disc and 135° + annular disc, but not by the combinations 45° + annular disc and 135° + radial disc; at the feeder, the opposite was true. In both cases, the potentially competing visuomotor associations were insulated from each other because they were set in different contexts. Comparable behavior was found in honeybees, where distinct odors or times of the day were the occasion setters for a given flight vector or rewarded color. Further examples for contextual learning could be provided but they would be redundant for the main conclusion of this section: invertebrates are capable of different forms of conditional learning. Despite this cumulative body of evidences, the nature of the associations underlying this kind of learning and the neural substrates underlying this form of learning remain unclear.

Studies of decision making in the fruit fly *Drosophila melanogaster* indicate that MBs are of fundamental importance for this behavior. In this case, an individual fly suspended at a torque meter from a copper wire glued to its thorax beats its wings when hanging in the middle of a cylindrical arena displaying a visual panorama with identifiable landmarks (**Figure 3**). An unpleasant heat-beam is focused on the fly's thorax and switched on whenever the insect fly toward a given landmark on the cylinder. The fly controls the reinforcer delivery as its flight maneuvers determine the on/off switching of the heat beam if the appropriate flight directions (i.e., landmarks) are chosen. In studies of decision-making in *Drosophila*, flies were conditioned to choose one of two flight paths in response to color and shape cues; after the training, they were tested with contradictory cues. Normal flies made a discrete choice that switched from one alternative to another as the relative salience of color and shape cues gradually changed, but this ability was greatly diminished in mutant flies with miniature MBs or with hydroxyurea ablation of MBs. Although this protocol does not provide a formalized nonlinear discrimination problem such as those presented earlier (e.g., negative patterning), it has the merit of moving from the traditional elemental learning protocols applied so far in *Drosophila* to a more sophisticated problem in which the cognitive richness of fly behavior could be revealed and related to the MBs. Furthermore, it was shown that salience-dependent choice behavior consists of early and late phases; the former requires activation of the dopaminergic system and MBs, whereas the latter is independent of these activities. Immunohistological analysis showed that MBs are densely innervated by dopaminergic axons, thus suggesting that the circuit from the dopamine system to MBs is crucial for choice behavior in *Drosophila*.

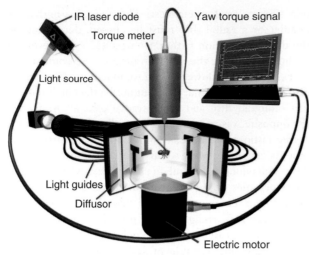

Figure 3 The flight simulator used for visual conditioning of a tethered fruit fly. A *Drosophila* is flying stationarily in a cylindrical arena homogeneously illuminated from behind. The fly's tendency to perform left or right turns (yaw torque) is measured continuously and fed into the computer. In closed-loop, the computer controls arena rotation. On the screen four 'landmarks,' two Ts and two inverted Ts, are displayed in order to provide a referential frame for flight direction choice. The illumination of the arena can be changed using color filters. A heat beam focused on the fly's thorax is used as an aversive reinforcer. The reinforcer is switched on whenever the fly flies towards a prohibitive direction. The fly controls therefore reinforcer delivery by means of its flight direction so that operant conditioning mediates the performance observed. However, classical associations between landmarks and reinforcer (or its absence) can also be established in this protocol. Courtesy of B. Brembs.

Positive Transfer in Rule Learning by Invertebrates

Nonelemental associative learning also underlies problem solving in which animals respond to novel stimuli that they have never encountered before or can generate novel responses that are adaptive given the context in which they are produced. In doing this, the animals exhibit a positive transfer of learning, a capacity that cannot be referred to as an elemental learning because the responses are aimed at stimuli that do not predict a specific outcome per se based on the animals' past experience.

A typical example of rule learning is the acquisition of the sameness or difference principle. These rules are demonstrated through the protocols of delayed matching to sample (DMTS) and delayed nonmatching to sample (DNMTS), respectively. In DMTS, animals are presented with a sample and then with a set of stimuli, one of which is identical to the sample and which is reinforced. Since the sample is regularly changed, animals must learn the sameness rule, that is, *'always choose what is shown to you (the sample), independent of what else is shown to you.'* In DNMTS,

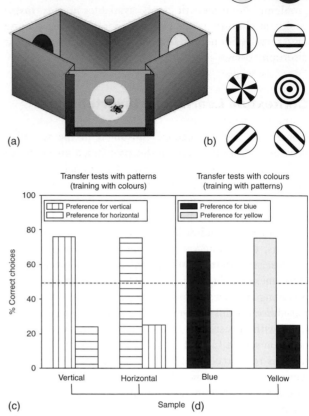

Figure 4 Rule learning in honeybees. Honeybees trained to collect sugar solution in a Y-maze (a) on a series of different patterns (b) learn a rule of sameness. Learning and transfer performance of bees in a delayed matching-to-sample task in which they were trained to colors (Experiment 1) or to black-and-white, vertical and horizontal gratings (Experiment 2). (c, d) Transfer tests with novel stimuli. (c) In Experiment 1, bees trained on the colors were tested on the gratings. (d) In Experiment 2, bees trained on the gratings were tested on the colors. In both cases bees chose the novel stimuli corresponding to the sample although they had no experience with such test stimuli. *n* denotes number of choices evaluated. Reproduced from Giurfa M, Zhang SW, Jenett A, Menzel R, and Srinivasan M (2001) The concepts of sameness and difference in an insect. *Nature* 410: 930–933.

the animal has to learn the opposite. Honeybees foraging in a Y-maze learn both rules. Bees were trained in a DMTS problem in which they were presented with a changing nonrewarded sample (i.e., one of two different color disks or one of two different black-and-white gratings, vertical or horizontal) at the entrance of a maze (**Figure 4**). The bees were rewarded only if they chose the stimulus identical to the sample once within the maze. Bees trained with colors and presented in transfer tests with black-and-white gratings that they have not experienced before solved the problem and chose the grating identical to the sample at the entrance of the maze. Similarly, bees trained with the gratings and tested with colors in transfer tests also solved the problem and chose the novel color corresponding to that of the sample grating at the maze entrance. Transfer was not limited to different kinds of modalities (pattern vs. color) within the visual domain, but could also operate between drastically different domains such as olfaction and vision. Furthermore, bees also mastered a DNMTS task, thus showing that they also learn a rule of difference between stimuli. These results document that bees learn rules relating stimuli in their environment. The capacity of honeybees to solve a DMTS task has recently been verified and studied with respect to the working memory underlying it. It was found that the working memory for the sample underlying the solving of DMTS is around 5 s and thus coincides with the duration of other visual and olfactory short-term memories characterized in simpler forms of associative learning in honeybees (Menzel, 1999). Moreover, bees trained in a DMTS task can learn to pay attention to one of two different samples presented successively in a flight tunnel (either to the first or to the second) and can transfer the learning of this sequence weight to novel samples.

The neural basis of rule extraction has not been addressed yet in invertebrates. The potentials offered by *Drosophila* with respect to molecular genetics and by the bee with respect to the recording of neural correlates will certainly be used in the near future to establish closer links to the neural substrates.

Conclusion

Here we focused on a particular basic cognitive faculty that relates to the ability of animals to process sequences of associative connections such that structures of interrelatedness are derived which are not housed in the elemental associations. In some cases, rules are learned and applied across sensory modalities, in others temporal relations are acquired. Learning under natural conditions will be much richer than implied here because bees, for example, are known to navigate along novel routes according to the expected outcome of the navigational choices, and *Drosophila* decides between flight goals by integrating multiple stimulus conditions. These and the examples discussed here require brain functions best conceptualized as representations, since the relations established during learning cannot reside in basic cellular modules of associative connections as they were so successfully studied in invertebrates. Rather they must be represented in network properties composed of multiple cellular association modules which incorporate new information into already stored information by some self-organization process, retrieve appropriate information from remote stores, and allow decisions to be made according to the current conditions, the internal status of

the animal, and the evaluated expected outcomes. Hints for memory processing during both memory storage and retrieval come from multiple observations. For example, consolidation of earlier forms of memory into later and stable forms changes the content of the memory and is accompanied by transfer between structures, for example, between the gamma lobe and the alpha/beta lobe neurons in the mushroom body of *Drosophila*. Memory retrieval initiates processes described as reconsolidation, and decisions between simultaneously activated memories are being made without access to stimuli according to the expected outcome. In this respect, memory processing during storage and retrieval in invertebrates resembles basic features described for mammals and humans, and it is conceivable that analog network processes may be responsible despite the large differences in the structure and functional organization between for example, insect and mammalian brains. How are we to discover these processes? A fundamental requirement for any experimental approach is that the working of the neural nets is monitored at the level of multiple but single neurons under conditions in which the animal learns, retrieves, and processes memory. Ideally, these neurons should be identifiable anatomically, aiming to establish a close relationship between structure and function. These strict requirements are not met by any animal although recent advances in optical and electrical recordings from neurons in the *Drosophila* and the bee brain come close.

Two streams of new developments have to meet in an attempt to take advantage of invertebrates as models for a cognitive neuroscience approach, a conceptual shift in addressing the phenomena of learning and memory, and a major methodological advance. Methodological advances are already on the verge. Calcium and voltage sensitive dyes as well as light driven dyes for controlling neural excitation can be expressed in defined neurons of the *Drosophila* brain, while recordings from multiple neurons in the bee brain can be performed for several days when the animals learn and perform. A more important achievement will be the conceptual shift, which relates to the necessity to include invertebrates into the cognitive view of behavior. It is the combination of stereotypical and highly flexible behavior of invertebrates that makes them such attractive study objects for a cognitive approach. Evidence presented and discussed in this article aims at promoting this cognitive framework to understand invertebrate behavior.

See also: Categories and Concepts: Language-Related Competences in Non-Linguistic Species; Crabs and Their Visual World; *Drosophila* Behavior Genetics; Insect Social Learning; Metacognition and Metamemory in Non-Human Animals; Nematode Learning and Memory: Neuroethology; Taste: Invertebrates; Vision: Invertebrates.

Further Reading

Davis R (2005) Olfactory memory formation in *Drosophila*: From molecular to systems neuroscience. *Annual Review of Neuroscience* 28: 275–302.

Giurfa M (2007) Behavioral and neural analysis of associative learning in the honeybee: A taste from the magic well. *Journal of Comparative Physiology A* 193: 801–824.

Giurfa M, Zhang SW, Jenett A, Menzel R, and Srinivasan M (2001) The concepts of sameness and difference in an insect. *Nature* 410: 930–933.

Greenspan RJ and van Swinderen B (2004) Cognitive consonance: Complex brain functions in the fruit fly and its relatives. *Trends in Neurosciences* 27: 707–711.

Guo J and Guo A (2005) Crossmodal interactions between olfactory and visual learning in *Drosophila*. *Science* 309: 307–310.

Heisenberg M (2003) Mushroom body memoir: From maps to models. *Nature Reviews Neuroscience* 4: 266–275.

Lachnit H, Giurfa M, and Menzel R (2004) Odor processing in honeybees: Is the whole equal to, more than, or different from the sum of its parts? In: Slater PJG (ed.) *Advances in the Study of Behavior*, vol. 34, pp. 241–264. San Diego, CA: Elsevier.

Livermore A, Hutson M, Ngo V, Hadjisimos R, and Derby CD (1997) Elemental and configural learning and the perception of odorant mixtures by the spiny lobster *Panulirus argus*. *Physiology & Behavior* 62: 169–174.

Menzel R (1999) Memory dynamics in the honeybee. *Journal of Comparative Physiology A* 185: 323–340.

Menzel R, Brembs B, and Giurfa M (2006) Cognition in invertebrates. In: Strausfeld NJ and Bullock TH (eds.) *The Evolution of Nervous Systems. Vol II: Evolution of Nervous Systems in Invertebrates*, pp. 403–422. London: Elsevier Life Sciences.

North G and Greenspan R (2007) *Invertebrate Neurobiology*, pp. 665. New York, NY: CSHL Press.

Pearce JM (1994) Similarity and discrimination: A selective review and a connectionist model. *Psychological Review* 101: 587–607.

Swinderen BV (2005) The remote roots of consciousness in fruit-fly selective attention? *BioEssays* 27: 321–330.

Zhang K, Guo JZ, Peng Y, Xi W, and Guo A (2007) Dopamine-mushroom body circuit regulates saliency-based decision-making in *Drosophila*. *Science* 316: 1901–1914.

Zhang SW and Srinivasan MV (2004) Exploration of cognitive capacity in honeybees. In: Prete FR (ed.) *Complex Worlds from Simpler Nervous Systems*, pp. 41–74. Cambridge: MIT Press.

Norway Rats

B. G. Galef, McMaster University, Hamilton, ON, Canada

The Rise of Rats

Not all that many years ago, before the ethological approach to the study of animal behavior came to the fore early in the 1970s, comparative and physiological psychologists conducted most behavioral research with animals. The intent of these investigators was quite different from that of many of today's scientists with an interest in animal behavior. For the most part, comparative and physiological psychologists were interested in exploring general laws of behavior, particularly those laws governing the formation of the associations that underlie learning.

Because the focus of research was on the discovery of general laws believed to be applicable to any species in any situation, convenience rather than theory or ecological considerations determined the choice of a species to study. And, one species proved more convenient than any other. Indeed, in the 1930s and 1940s, more than 60% of all papers published in two of the leading animal-behavior journals of the time (the *Journal of Comparative and Physiological Psychology* and the *Journal of Animal Behavior*) were concerned with the behavior of a single organism, the Norway rat (*Rattus norvegicus*).

Given the focus on discovery of general behavioral principles that might apply to all species, humans included, the Norway rat was not an altogether bad choice. Rats are members of the order Rodentia, the mammalian order with by far the greatest number of species (more than 2000), and of the genus *Rattus* (with more than 50 species), which is the most species rich of the murid family of rodents, the Muridae, that includes mice, gerbils, and hamsters.

Rats are about the average size for a mammal (reports of Norway rats as big as cats are considerable exaggerations; a very large adult male rat weighs 500–600 g, making it a very small cat indeed) and a convenient size for laboratory work. Rats are neither so large as to make their maintenance in large numbers impractical nor so small as to make direct observation of their behavior difficult or surgery on them particularly demanding. Domesticated rats are easy to produce. They become sexually mature at 3 months of age, have a dozen or more pups in a litter, can produce a litter every 21 days, and breed all year round. Further, the ability of rats to thrive on relatively low-protein diets makes them inexpensive to feed. Unlike their sometimes-vicious and very timid wild forebears, rats of domesticated strains are easy to handle and will go about their business undisturbed even when nearby humans are watching their every move.

Important advantages accrued from having, quite literally, hundreds of researchers working on various aspects of the behavior and physiology of a single species. Techniques developed, for example, in studies of rat learning were of use to scientists studying the rat nervous system, and students of learning in rats could benefit from information concerning rats' sensory systems. Further, the adequacy of the methods used in an experiment could be readily evaluated by others working with the same animal, so potentially important findings could be replicated (or not) almost immediately.

The Fall of Rats

The decline in the dominance of Norway rats as subjects in behavioral research had a number of causes. Foremost among these was the mid-twentieth century increased interest in studying behavior among of a group of European biologists who called themselves ethologists. Ethological investigation of a species started with the construction of an ethogram, a complete description of the behavioral repertoire of a species while in its natural habitat.

Norway rats are, unfortunately, most active in the dark and underground and therefore are difficult to observe in the wild. Even worse, humans have inadvertently transported Norway rats around the world, making identification of their place of origin, their natural habitat, all but impossible. Today, most free-living Norway rats live in man-made structures, feed on human refuse or crops, and because of their close association with humans, are protected from many potential predators. Their current habitat is hardly natural.

Ethologists were particularly interested in interactions between animals and the environments in which they evolved. For example, the last of Tinbergen's four questions defining the field of ethology, and the question that was to serve as a focus of research in ethology's descendant field, behavioral ecology, concerned the functions of behavior (i.e., the ways in which behavior increased survival and reproductive success in natural circumstances). The behavior of Norway rats in their natural habitat, wherever it may be, is simple not available for such studies.

Ethologists focused their research not on individually learned behaviors but on instincts, species-typical patterns of behaviors that ethologists believed reflected directly the action of natural selection on the genetic substrate of behavior. One of the more appealing characteristics of rats to comparative psychologists was that rats did not seem to show elaborate, heritable, species-typical behaviors that could interfere with the discovery of general principles of behavior.

And, there were other problems as well. For example, the domesticated rats that were subject to so much attention from comparative psychologists had undergone several hundred generations of breeding in captivity. Exposure to artificial selection by humans breeding rats for tameness and for fertility in captivity ensured that the genetic substrate of domestic rats was not that of their wild forebears. A century or more of such artificial selection, led some ethologists to assert that the behavior of laboratory rats was not 'natural,' and consequently, not worth studying.

As a practical matter, increasingly stringent regulations governing the breeding and maintenance of laboratory animals, made work with rats ever more expensive and reduced the number of laboratories that could afford to use rats in experiments. And invention of procedures for generating knockout strains of mice preceded the more-difficult development of knockout rat strains by 14 years. The importance of knockouts for exploring the genetic substrate of behavior made mice the species of choice for many scientists interested in studying the mechanisms of mammalian behavior, even if using mice as subjects meant repeating behavioral studies previously performed with rats. In sum, although work on the behavior of Norway rats and its mechanistic substrate continues today, the dominant position of the species in studies of animal behavior is over.

What Is the Origin of Laboratory Rats?

Norway rats, the forebears of all laboratory rat strains, are generally assumed to have originated somewhere in Asia, possibly in Northern China, although the long association of Norway rats with humans makes the species' point of origin difficult to determine. It is known that sometime in the mid-eighteenth century, Norway rats spread through Europe (the appellation Norway rat is believed to derive from the false, eighteenth-century presumption that the first of the species arrived in the United Kingdom on lumber ships coming from Norway, although there were probably no *Rattus norvegicus* in Norway when the species first invaded Britain). The black or roof rat (*Rattus rattus*), not the Norway rat, had been common in Europe before the larger and more aggressive Norway rat arrived in the eighteenth century and displaced them. Consequently, the black rat, not the Norway rat, was the principle vector in the recurring bouts of bubonic plague that killed 25% or more of the human population of Europe during the latter half of the fourteenth century. Norway rats arrived on the east coast of North America shortly after their arrival in Europe, and spread with the gold rush to California

in 1849. Today, Norway rats are to be found on every continent but Antarctica, and from Alaska ($64°$ N) to South Georgia Island ($55°$ S).

However, rats do not thrive in areas with continental climates, such as Alberta, Canada, and northern Montana, USA, where winters are long and cold and human habitations are relatively sparse. Indeed, Norway rats are most successful in temperate climates. In the tropics, they are largely replaced by other species of their genus, for example, the more arboreal, lighter and longer-tailed *Rattus rattus* and *Rattus exulans* (the Polynesian rat).

Norway rats, like humans, are great generalists and are able to thrive in a broad range of environments. Rats have been seen catching fingerling trout in hatcheries, diving in rivers to feed on mollusks, and catching and killing wild ducks and geese. Wherever rats are introduced onto islands by human visitors (or when they swim or float to islands from distant shores), they become a threat to the survival of any ground-nesting birds found there and have been implicated in the extinction or endangerment of numerous bird species.

Norway rats are the forebears of all domesticated rat strains whether albino, hooded, black or agouti colored. When, where, and how Norway rats were first domesticated is not known. In the nineteenth century, wild Norway rats served as prey in the brutal sport of rat baiting in which a dog was placed in a pit with large numbers of rats and bettors wagered on how many rats the dog would kill in a specified period. One story is that when rare albino wild Norway rats were trapped in the course of securing the large numbers of rats needed for rat baiting, the albinos were displayed in cages outside betting establishments, and that these albino rats were the ancestors of at least some of today's domesticated strains. Whatever their source, domesticated albino rats were first used in the laboratory in 1895 at Clark University in Worcester, Massachusetts, in studies of nutrition. Five years later, they were subjects in Willard Small's studies, also at Clark, of the behavior of rats in mazes. The decades of subsequent research on the behavior of Norway rats and its neural substrate have led to publication of many tens of thousands of research articles. Obviously, it is not possible to thoroughly review so vast a literature here. In the following sections, I describe a few of the many areas in which studies of Norway rats have played an important role, and mention a scattering of findings that I find either intriguing or amusing. A great deal more information concerning rats, both wild and domesticated, is available in the 'Further Reading.'

Regulatory Systems

Some of the earliest studies of rat behavior were concerned with the role of behavior in maintaining the internal environment of animals within the boundaries compatible with life. The ability of rats to select items for ingestion, to regulate their intake so as to neither lose nor gain appreciable amounts of weight, and to maintain a relatively constant body temperature each has an extensive literature.

Selecting Foods

Results of experiments conducted in the 1940s and 1950s were interpreted by many as demonstrating that rats that had been deprived of a specific nutrient (e.g., thiamine) could select the food containing thiamine from among a cafeteria of foods only one of which contained the needed vitamin. The results of these reports are responsible for the belief, widespread even today, that your body will lead you to seek out whatever foods you need to eat to remain healthy or to regain health should you become deficient in some nutrient. Unfortunately, the interpretation of this early research has proved exaggerated. Although rats that need salt can identify salt in a food or fluid, and thirsty rats will seek water, rats fail miserably in selecting appropriate foods when in need of almost any other of the dozens substances (vitamins and minerals) needed for health.

Controlling Body Temperature

Like other mammals, rats use evaporative cooling to avoid heat stress. However, unlike humans and horses, rats do not sweat. Instead, overheated rats spread saliva on the unfurred areas of their bodies (as do elephants). The rat's naked tail (which some people find repulsive) serves as a particularly effective window through which to release heat. Consequently, rats that have had their tails surgically removed have a markedly reduced ability to remain cool when exposed to elevated environmental temperatures.

Reproductive Behavior

Every aspect of reproduction from selection of a mate to weaning of young has been studied in Norway rats. There are, for example, extensive and detailed studies of: (1) the cues that male rats use to determine if a female is in the receptive phase of her estrous cycle, (2) patterns of copulation and their effects on the rewards both male and female rats garner from engaging in sexual activity and the impregnation of females, (3) the behavior and sensory experiences of fetal rats and effects of intrauterine experiences on postnatal behavior, (4) the nest building that females engage in before parturition, (5) behaviors during parturition when dams lick their pups, gather them in the nest and assume a nursing posture over them, (6) the behavior of young while both seeking their mothers nipples and nursing and when the mother is absent from the nest, (7) mother's behavior toward her

maturing young: her retrieval of pups that stray from the nest, the gradual reduction in time she spends in contact with her offspring and changes in maternal delivery of milk as her young mature, (8) the increased aggression of mothers with young, and (9) the changes in pups behavior as they develop from exothermic, blind, deaf and hairless eraser-sized newborns to independent juveniles.

Perhaps surprisingly, the seemingly helpless blind and hairless pups huddled together in a nest can behaviorally regulate their temperature, spreading apart and increasing the surface area of the huddle to increase heat loss when the environmental temperature is high and forming a tight ball with a small surface area when the environmental temperature is low. Equally surprising is the impact of prenatal life on later behavior. As first discovered in Norway rats, whether a fetal mammal is located in its mother's uterus between two brothers or two sisters (i.e., its intrauterine position) profoundly influences the amount of testosterone to which it is exposed before birth. As was subsequently established in studies with mice and gerbils, while an adult, much of an animal's hormonally influenced reproductive behavior is modified by intrauterine exposure to testosterone.

Social Behavior

Free-living wild rats are intensely social beings that live in colonies consisting of from a few to several hundred individuals. Colony members share a burrow system and paths through the environment that they defend against intrusions by unfamiliar conspecifics.

Communication

Such social life requires communication, and Norway rats communicate in a variety of interesting ways. They produce olfactory cues that allow both individual identification and guide movement through the environment. They vocalize during social and sexual interactions and in response to the presence of potential predators, and much of their vocalization is ultrasonic (i.e., in a frequency range too high for humans to hear).

Aggression

Books have been written about the aggressive behavior of rats, describing the stimuli that elicit, direct, and terminate aggressive interactions, the neural and endocrine substrates of aggression, and rats' postures and movements while interacting aggressively or stealing food from one another. Intruders into the territory of a colony of wild rats are vigorously attacked, and even brief attacks on intruders that do not produce any detectable wounds can have fatal consequences, though the causes of such death are not well understood.

Predation

Laboratory rats' predatory behavior toward mice and the response of rats to cats and other potential predators were also studied for many years. However, ethical concerns have largely ended experiments involving either staged aggressive encounters between mammals or between predators and potential mammalian prey.

Social Learning

Despite their aggressiveness toward strangers, members of established colonies of rats form stable dominance hierarchies and live relatively amicably, sleeping together, grooming one another, and following each other through the environment. Life in socially tolerant groups provides rats with opportunities to both observe and learn from the behavior of others of their species. More than 50 papers have been published concerned with the finding that after a naïve 'observer' rat interacts for a few minutes with a 'demonstrator' rat that has recently eaten a distinctively flavored food, the observer rat shows a substantial increase in its liking for whatever food its demonstrator ate.

Rats have also been shown to learn to dig for buried food by watching other rats do so. After learning socially either to eat a particular food or to dig for food, an observer rat can act as a demonstrator for new, naïve observers, and such chaining can be sustained for several 'generations' thus producing rat 'traditions.'

Sensory Systems

Rats are sensitive to a broad range of stimuli; they see, hear, taste, smell, and respond to touch. Each of the rat's sensory system has been fully explored, and each has its own voluminous literature.

Vision

In nature, Norway rats are most active at dusk and dawn, and possibly as a result, they are less dependent on vision than other well-studied mammals such as cats and ferrets. The visual, acuity, even of wild rats is quite poor, and domesticated rats have about half the visual acuity of their wild brethren with albino strains of rat suffering particularly from poor vision. All strains of Norway rat lack both color vision and a fovea, and their visual cortex is less clearly functionally differentiated than that of some other mammals.

Olfaction

Rats have a keen sense of smell, and throughout life, depend heavily on olfactory cues in their day-to-day functioning. Prenatal exposure to odorants can have lasting effects on rats postnatal behavior. Infants quickly learn to identify the odor of their mother and home nest and use olfactory cues to find their mother's nipples to nurse. Adult rats deposit scent marks in the environment that allow others to identify their age, sex, reproductive state, and dominance status. Most impressive, the reproductive behavior of female rats can be markedly affected by olfactory cues; exposure to strange males both accelerates the age of onset of puberty and the regularity and timing of estrous cycles.

Taste

The taste perceptions of rats and humans are surprisingly similar. Members of the two phylogenetically quite distant mammals almost always find the same flavors attractive or repulsive. Consequently, Norway rats have served as models for understanding human taste perception. Like humans, rats display different facial expressions when experiencing pleasant and unpleasant flavors. However, there is no evidence that the disgust faces of rats dissuade other rats from eating the food that a grimacing animal has found distasteful. Also like humans, rats find it particularly easy to associate experience of an unfamiliar flavor, but not an unfamiliar noise or visual cue, with later gastrointestinal upset. After a single experience, both rats and humans learn to avoid a novel flavor to which they were exposed hours before becoming ill.

Hearing

Relative to humans, hearing in rats is shifted toward higher frequencies, and rats can detect sounds with frequencies as high as 80 kHz. Rats produce a number of auditory signals both audible to humans and in the ultrasonic range. Relatively little work has been done on audible rat vocalizations, possibly because of their great variability. However, rats' more stereotyped ultrasonic vocalizations and the responses to them have received considerable attention.

Ultrasonic calls (40–50 kHz) are emitted by infant rats when they cool. Adult rats emit a 22 kHz 'long-call' in aversive situations (e.g., after losing a fight or detecting a cat) and, perhaps surprisingly, after ejaculating. Rats also emit a 50 kHz 'chirp' that may be associated with pleasant events (e.g., playing or being tickled) that has been described as a form of laughter, although it also occurs in some unpleasant circumstances, for example, during aggression or in response to some types of pain.

All these ultrasonic vocalizations can affect the behavior of rats that hear them. Mother rats are attracted to the ultrasonic vocalizations of pups, and there is some evidence that exposure to 22 kHz long calls increases the wariness of rats hearing them. Although rats can use their ultrasonic vocalizations to detect objects at a distance, they are far less sophisticated in their use of ultrasound for echolocation than are bats.

Somatosensation

The sense of touch plays in important role in rats' movements about their environment. Rats are 'thigmotactic'; they are biased toward remaining in physical contact with vertical surfaces, presumably to protect against predation. Rats' vibrissae, the whiskers around their noses, are extremely sensitive to tactile stimuli, and have been compared with human fingertips. Rats can move their vibrissae independently across surfaces, allowing them to discriminate among objects of different size, texture and shape.

Learning and Cognition

In the decades when rats were the predominant species in behavioral studies, they most often served as subjects in investigations of learning. At first, such studies took place in complex mazes with many choice points that were believed to mirror the complex burrow systems in which wild rats live. When behavior in such complex environments proved intractable to analysis, experimenters shifted to simple T-mazes with only a single choice point. Finally, rats were studied in highly automated Skinner boxes, where subjects were rewarded for pressing levers with food delivered on various schedules.

Most recently, studies of the ability of rats to solve cognitively demanding tasks have been in vogue. In the Morris water maze, rats are placed, on successive trials, in random locations in a small circular tank filled with water. To escape from the water, which the rats find mildly aversive, they have to learn the location of a platform hidden just beneath the water's surface. In different version of the task, the rats can use visual cues in the surrounding room, a beacon directly indicating the location of the platform, or information concerning the distance of the platform from the wall of the test chamber to find it. Solution of the task using cues outside the pool itself can require the rat to form a 'cognitive map' of the relationship between cues in the room and the location of the platform.

Perhaps the most challenging task with which rats have been presented is the multiarm maze. Here, as the name of the apparatus implies, a rat is placed on the central platform of a maze with several arms (8 is the most common number) and a small piece of food is placed at the end of each arm farthest from the central platform. The rat is free to explore the maze until it has recovered food from the end of each arm. Greatest efficiency requires that a rat enter each arm of the maze only once, a performance that requires the subject to remember which arms it has previously entered.

Rats are extraordinarily good at this task, and make relatively few errors, rarely reentering a previously visited arm of the maze.

Conclusion

In this brief article, I have just begun to scratch the surface of research on the behavior of rats. Many topics that have been the focus of extensive research have not even been mentioned, among them: play, circadian rhythms in activity, motivation, schedules of reinforcement, memory, the results of domestication, maternal effects on development, postures, locomotion, grooming, exploratory behavior, response to pain, or rats as model systems to study human behavioral disorders such as anxiety or obsessive–compulsive disorder. This list could be lengthened considerably without difficulty.

'Further Reading' provides both greater depth and greater breadth of coverage of the role of rats in both science and everyday life than this brief article. Barnett's *The Rat: A Study in Behavior*, although it is somewhat dated, provides classic descriptions of the behavioral repertoires of wild rats and discussion of some laboratory work with domesticated rats. Meehan's *Rats and Mice: Their Biology and Control* provides an introduction to the extensive literature on the control of pest populations of rats. Both Telle's and Calhoun's classic articles are difficult to find today, but provide some of the best descriptions available of the social behavior of wild rats. Munn's *Handbook of Psychological Research on the Rat* provides a summary of research with laboratory rats in its heyday. Other recommendations provide introductions to specific topics covered here. Most important among these is Wishaw and Kolb's recent, 500-page edited volume *The Behavior of the Laboratory Rat*. It provides a compact and up-to-date summary of much of the work on domesticated rats.

Several publications about Norway rats intended for lay audiences appeared during the last decade. S. A. Barnett's '*The Story of Rats*' and Lore & Flannelly's '*Rat societies*' are both trustworthy.

See also: Comparative Animal Behavior – 1920–1973; Food Intake: Behavioral Endocrinology; Hearing: Vertebrates; Hormones and Behavior: Basic Concepts; Mammalian Female Sexual Behavior and Hormones; Mammalian Social Learning: Non-Primates; Memory, Learning, Hormones and Behavior; Neural Control of Sexual Behavior; Ontogenetic Effects of Captive Breeding; Parental Behavior and Hormones in Mammals; Psychology of Animals; Sexual Behavior and Hormones in Male Mammals; Smell: Vertebrates; Social Information Use; Vision: Vertebrates; Water and Salt Intake in Vertebrates: Endocrine and Behavioral Regulation.

Further Reading

Barnett SA (1975) *The Rat: A Study in Behavior.* Chicago, IL: University of Chicago Press.

Barnett SA (2001) *The Story of Rats.* Crow's Nest, New South Wales: Allen & Unwin.

Burn CC (2008) What is it like to be a rat? Sensory perception and its implications for experimental design and rat welfare. *Applied Animal Behaviour Science* 122: 1–32.

Calhoun JB (1963) *The Ecology and Sociology of the Norway Rat.* Public Health Service Publication no. 1008. Bethesda, MD: U.S. Department of Health.

Galef BG, Jr (1991) A contrarian view of the wisdom of the body as it relates to food selection. *Psychological Review* 98: 218–224.

Lore R and Flannelly K (1977) Rat societies. *Scientific American* 236: 106–115.

Meehan AP (1984) *Rats and Mice: Their Biology and Control.* East Grinstead, UK: Rentokil Ltd.

Munn NL (ed.) (1950) *Handbook of Psychological Research on the Rat: An Introduction to Animal Psychology.* Cambridge, Massachusetts: Riverside Press.

Shair HN, Barr GA and Hofer MA (eds.) (1991) *Developmental Psychobiology: New Methods and Changing Concepts.* Oxford: Oxford University Press.

Shettleworth SJ (1998) *Cognition, Evolution and Behavior.* Oxford: Oxford University Press.

Stricker EM (ed.) (1990) *Neurobiology of Food and Fluid Intake.* New York: Plenum Press.

Telle HJ (1966) Beitrag zur Kenntnis der Verhaltensweise von Ratten, vergleichend dargestellt bei Rattus rattus und Rattus norvegicus. *Zeitschrift fur Angewandte Zoologie* 53: 129–196.

Whishaw IQ and Kolb B (eds.) (2005) *The Behavior of the Laboratory Rat: A Handbook with Tests.* Oxford: Oxford University Press.

Octopus

J. A. Mather, University of Lethbridge, Lethbridge, AB, Canada

Introduction to the Octopus

Like *Aplysia*, but unlike most of the animals featured in this section, octopuses are in the mollusca, a large diverse phylum that is mostly marine. Octopuses are in the class cephalopoda within the molluscs and are very different from clams, slugs, and snails. The biggest difference is the loss of the protective shell, but this loss has probably led to the evolution of the complex display system, the flexible arms, the acute sensory systems, and the large brain, all of which characterize them and make their behavior so interesting. The name cephalopod means 'head foot' and it is used because, unlike the vertebrate arrangement, the body is at the posterior, the head at the center, and the arms (not feet) at the anterior end of the animal. Cuttlefish (not fish!) and true squid are in the coleoid cephalopod subclass (including all but the genus *Nautilus*) with the octopuses.

When we say the octopus, we do not really mean one species. The genus *Octopus* has about 100 species, and while the behavior of most of them remains completely unstudied, new ones are being described every year. It is better to think of the family Octopodidae as octopuses, which allows us to include the giant Pacific octopus (GPO), which has been recategorized as *Enteroctopus dofleini* recently, the poisonous blue-ringed octopus *Haplochlaena maculosa*, and the deep-water genus *Bathypolypus*. The octopuses look pretty much alike, with eight flexible arms, and the body enclosed in a sac-like mantle, though they range from the 15 g *joubini* to the 25 kg *dofleini* in size. Almost all of them live on or close to the sea bottom and none lives long, from 6 months to 3–4 years for the GPO, raising the question of why they have developed their high intelligence when they do not live long to use it (**Figure 1**).

If any one species could be described as 'the' octopus, it would be *O. vulgaris* from the Mediterranean. It was originally described by Aristotle in 330 BC. Its range was described as so huge that taxonomists began to suspect it represented more than one species, and reassessment of the variants showed that this was true. The trouble is that taxonomists are still not sure which subgroup of animals is the true *Octopus vulgaris* and which ones are different species in a similar-looking subgroup. This makes it difficult for researchers to know if they are describing species-typical behavior.

One way in which octopuses are special is that their movement is not limited by a fixed skeleton. Instead, their skeleton is provided by a muscular hydrostat and movement, particularly of the flexible arms, theoretically has an unlimited number of degrees of freedom. There is a large set of different muscles in the arm – longitudinal, transverse, circular, and oblique – and a differential contraction of some of them sets up a skeleton against which the others can contract. Octopuses can also bend or twist these arms anywhere along their length – they are even reputed to be able to tie a knot in the arm and run it down its length and off the tip. This complexity of possible actions is probably why 3/5 of the neurons in the octopus's nervous system are outside the brain. However, video analysis has shown that there are patterns of muscle activation along the arm that control very stereotyped reaching movements. Behind the complexity are combinations of much more simple units. A catalog of the octopus arm movements by Mather described a wider but still limited set of actions that the octopus arms can undertake. For instance, in Webover, an octopus extends and spreads the arms, splays them, and pulls the web between them down to form a parachute-like spread over part of the environment, often a rock under which a hapless crab is hiding.

An important feature of the octopus's nervous system is the very complex skin display system. Each of the many chromatophores contains red, yellow, or brown pigment in an elastic sac. Sacs can be pulled out by a group of muscles that are connected to nerves from the brain, and when they do this, the color is displayed. Below this is a layer of leucophores and irridophores which reflect ambient or blue-green light, so when the sacs are retraced, these colors show. Skin colors can match the background

Figure 1   This giant Pacific octopus (*Enteroctopus dofleini*) shows the many flexible arms and the lateral lens-type eyes that are characteristic of the group.

in pattern and intensity, and the appearance is enhanced by extension or retraction of skin papillae and postures of those moveable arms. The result is a system which can mimic the appearance of many background features and change in less than 30 ms and 1 cm square area. Octopuses, being solitary, do not appear to use this system much to display to conspecifics and in fact do not have color vision. Instead, their appearance seems tuned to the visual system of the receivers, the bony fishes with whom coleoid cephalopods evolved in competition. The anatomy and physiology of this system is so complex that even a template to describe it is difficult to form, as the octopus builds its appearance from postural, chromatic, and textural components.

The unusual movement, the dazzling skin displays, and the molluscan history also combine with an unusual life history. Octopuses are mostly semelparous, which means that they leave reproduction until the end of the lifespan. After that, males die and females hide in shelter and tend and defend their eggs, up to tens of thousands and often only the size of a rice grain. After the eggs hatch, the mother dies, and the newborns float off in the open sea. They look somewhat octopus-like, with the right number of arms but a large body and with few of the chromatophores that are so important in adults. They feed on tiny crustaceans and fish and are eaten by fish; they may stay in the open ocean for months but eventually settle to the sea floor. In this transition, they grow longer arms, develop the network of chromatophores that give them their appearance, and become the subadult octopuses we know.

A Historical View of Learning Research

Years of research have linked octopuses to the study of learning. Work began at the Stazione Zoologica in Naples

in the 1950s, mainly on the initiative of the British researcher JZ Young, and flourished through the 1970s. *O. vulgaris* was an ideal animal for learning studies. It was big enough to handle, was resilient in recovery, adapted easily to being kept in captivity, and had a strong exploratory drive that could be used by researchers. Octopuses sheltered in a home of several bricks and readily emerged to investigate whatever stimuli were presented, and their acute vision (the eyes are a classic example of convergent evolution with the vertebrate eye) meant that they could readily be tested with visual stimuli. Paired figures were presented to them at the opposite end of their tank from the home (simultaneous discrimination) and octopuses were rewarded with food and punished with a small electric shock.

Young investigated the anatomy of the octopus brain, which was the result of major centralization of the molluscan system of several pairs of ganglia. In parallel to this, Wells and his associates began to look especially at the effect of brain lesions on learning. They found a visual learning center in the vertical lobe. Stimulation of this area did not result in any behavior, and the removal only affected octopuses' visual memories; they could no longer learn a discrimination or repeat one that they had learned. With further investigation, Wells also isolated another brain region, the subfrontal lobe, which seemed essential for tactile memory. After removal of that area, the octopuses could no longer discriminate between rough and smooth-surfaced cylinders.

As the brain of the octopus is bilaterally symmetrical and octopuses usually view visual targets monocularly (with little frontal visual field overlap), researchers could look at information storage. Octopuses could be trained with one eye but tested with the stimulus placed in the field of view of the other. The stored information did not transfer to the other side of the brain immediately but had done so within a day. If the vertical lobe of the favored side was removed, the octopus did not learn the discrimination. If the connections between the two brain halves were cut before the learning, no transfer took place, whereas if it was cut after the learning, it had done so.

In parallel with these studies, Sutherland used the octopus' ability to discriminate visual stimuli to search for the rules by which they encoded visual shapes. Initially, he found that they easily discriminated a vertical rectangle from a horizontal one but were much worse at discriminating a pair of oblique ones (this is also true for vertebrates) and theorized that the octopus discriminated patterns by their vertical and horizontal extents. However, it could also discriminate a square from a circle and a W-shaped figure from a V-shaped one, perhaps by the presence of visual angles or the ratio of edge to area. Sutherland advanced several different models of shape discrimination but octopuses were able to learn to discriminate complex shapes that did not follow any simple assessment rules. In the end, it became apparent that (like vertebrates) octopuses were able to attend to different dimensions of the figure and to choose which ones were important for the discrimination.

These studies led to an assessment of what might be called simple concept formation. When octopuses were given the two different cues of brightness and shape orientation for a discrimination, they learned faster than those that were given only one. Further testing showed that some relied on one cue and some on the other; when a separate group was trained to use one cue and then switch to another, they took longer to learn. If they were trained by finer and finer distinctions on an orientation discrimination that was initially too difficult, octopuses could learn it. They were also able to learn to respond to switches of the positive and negative stimuli in six successive reversals – if the criterion for correct choices was 70% (less than usual but appropriate for a win-switch forager).

This emphasis on lab investigation of learning flourished between 1955 and 1970, but dwindled thereafter. A downturn in funding was part of the cause, but further investigation into the control of octopus learning needed electrophysiological techniques that were not then available. There must also be a balance between lab and field investigations, and little or nothing was known of octopus behavior in the field. Ironically, Mather's field observations of octopuses in Bermuda suggested that some of the testing did not use an appropriate stimulus situation. In the field, octopuses foraged by moving over the sea bottom and feeling amongst and under the rocks, digging arms into the sand, and snaking them along algae to locate prey. Presenting paired visual stimuli for a food reward did not mimic a natural situation as prey was usually in hiding. Instead, octopuses used their vision to orient in the shallow waters, to find likely locations for prey, and return to their sheltering home.

Other Approaches to Octopus Behavior

Through these years, there was a low countercurrent of field investigation of octopuses that was more ecological than behavioral. Octopuses are top predators in many ecosystems: how are their populations regulated, what guides their prey choices, and how do they avoid predation? The availability of sheltering homes may limit habitat choice. One study found that discarded beer bottles may extend the range of *Octopus rubescens* by providing such shelter, and fishers have taken advantage of this necessity in the Mediterranean by providing sheltering pots as traps. Most octopuses choose a wide variety of crustacean and molluscan prey. Consumption is affected by prey availability but lab selection does not always match that in the field, commonly measured by sampling the hard-shelled remains of prey left in a midden outside the octopus home. Octopuses may be specializing generalist, with the species taking a wide variety of prey and yet

individuals specializing, perhaps fueled by learning. On the other hand, the soft octopus body makes them easy prey for many fish species and marine mammals, and their movement and distribution may be limited by predator pressure.

The life history of octopuses has meant that studies on the development of behavior are few. There are three major transitions in the life of octopuses. The first is hatching, assisted by a hatching gland on the posterior tip of the mantle that dissolves the membrane of the egg. Hatchlings are immediately positively phototactic and negatively geotactic, so they move fairly quickly to near the water surface, where they are carried away by ocean currents. The paralarvae are nearly transparent slow swimmers and are swept by these currents, though they can use bursts of swimming to evade their (mostly) fish predators or to catch their mostly crustacean larvae prey. After some weeks or months in the plankton, octopuses undergo a second transition, settling from to water column to the sea bottom. During this transition, the arms grow a great deal and the buccal lobes of the brain that control them also grow, the chromatophores develop from a few to many, thickly covering the body surface, and the octopus preferentially seeks shelter and darkness. During the long subadult period, octopuses are voracious predators and have a very high conversion efficiency of food to body weight of 50%. Toward the end of the life cycle, they undergo a third major transition. The optic gland matures, suppression of growth of reproductive tissues is removed, and the body metabolizes protein from the muscles to form eggs or sperm. Males appear to become more active and seek out others with whom to mate (by passage of a large spermatophore that contains hundreds of sperm). Boal's review of mating strategies notes that only *Abdopus* has mate guarding and competition for females; octopuses are generally solitary. The digestive gland shrivels and octopuses reduce their food intake, which also serves to protect the eggs while females guard them for the last weeks of their lives.

Studies of behavior development have been done not on octopuses but on cuttlefish, which hatch from their eggs as miniature versions of the adult. Hatchlings have a narrow visual search image for preferred prey, resembling Mysid crustaceans, and find it difficult to learn not to strike at one with the extended tentacles when it is enclosed in a glass test tube. Over the first 6 weeks of life, they gradually expand their prey preference and learning capacity, and the growth of the vertical lobe of the brain parallels the learning growth. But, like vertebrates, cuttlefish can learn to modify these preferences with early exposure. Dickel and colleagues showed that the presence of crabs immediately after and even before hatching, modified their prey preference in a process reminiscent of mammalian early imprinting. No one knows whether similar learning occurs in octopus paralarvae.

We may have to similarly turn to other cephalopods to understand the full range of the skin display system. Much of the information about the system is descriptive, despite Packard's lifelong study of its physiology and structure. On a behavioral level, octopuses can produce counter-shading of dark above and paler below. They can make a deceptive resemblance to features of the environment or other noxious animals such as poisonous sea snakes. They can put on disruptive coloration that breaks up the pattern of the body (see the bars that extend out from the linear pupils of the eyes and break up its round outline). When threatened, they can darken the skin around the eyes and at the edge of the extended arms and web, looking both larger and threatening in a deimatic display. Octopuses often change these patterns many times and unpredictably, also extruding ink that acts as a visual screen. But they seldom perform these displays to conspecifics, and the true squid, which have a dazzling repertoire of such patterns as the Zebra, the Saddle, and the Stripe, may be a better group in which to investigate the meaning of such a system and whether it could be considered a language.

Modern Research on Cognition and Neural Control

As modern technology combines with a return to studying adaptive behavior in field and laboratory, there is a reemphasis on the octopus's cognitive capacity and its neural base. Williamson and Chrachri review the four most interesting neural networks for study in the cephalopods. One is the giant fiber escape system of the squid, with nerves of such large diameter that they have been used historically for neurophysiological study. Another is the hierarchical control of the chromatophore system described above, starting with the optic lobe behind the eyes and extending to the mantle motorneurones. A third is the statocyst-based balance system, an effective parallel of the one in vertebrates. A fourth is the visual system, particularly the optic lobe and its structural parallel with the retina of vertebrates. The last is the memory systems, the two are not as completely separated as earlier researchers believed and visual and tactile memories may interact. A separate analysis of the neural basis of behavior is the approach of Hochner and colleagues, investigating the electrophysiology of the neurons and field potentials from different areas of the brain.

In parallel to this are investigations of the learning and cognitive capacity of the octopus from new directions. For one, we have looked beyond the population to the individual, as octopuses are very different from one another. Mather began investigating octopus personalities in the 1990s and Sinn has carried on the work with *Euprymna* squid. Sorting behaviors in common situations by factor analysis has resulted in three dimensions, Activity, Reactivity, and Avoidance. These dimensions are again a parallel to those found for many vertebrates. Sinn's squid work suggests that such individual differences, sometimes called behavioral syndromes, have clear developmental patterns and also convey adaptive advantages. Such differences must, as in vertebrates, adapt the individual well to the complex changing near-short tropical marine environment and allow it to select appropriate micro-habitats.

Our understanding of the adaptive capacity of octopus learning has led in new directions. Boal's test of their spatial learning, based on the capacity Mather found in the field, may be a more ecologically appropriate paradigm for learning. Fiorito has studied problem-solving in octopuses, testing their ability to take a lid off a glass jar to gain access to the crab hiding inside. Octopuses can also problem-solve in gaining access to a captured clam. They can manipulate the clam while it is held in the arm web out of sight and use three techniques, pulling apart, chipping with the beak, and drilling a hole in the shell and injecting a venomous neurotoxin. Each is done with the appropriate orientation, and both the Boal work and this study suggest that the octopuses, despite Wells' earlier assumptions, have an understanding of where they are in space and of where appropriate body parts are situated. There must be limitations of the feedback about arm positions and actions, as so much of the nervous system is not in the brain, but Grasso's recent work on arm actions suggests that their control is very sophisticated.

Two areas suggest abilities outside what one might expect given the life history of the octopus. One is the suggestion of observational learning by Fiorito, who allowed one octopus to learn a visual discrimination by observing another make the correct choices. Such an ability, which is not adaptive for a solitary animal, may be an offshoot of their general drive to investigate and learn by observation. Another offshoot of this drive to investigate is the appearance of play behavior in octopuses. Given a floating pill bottle, two of the six octopuses repeatedly aimed water jets at it with their funnel, driving it to the opposite end of the tank where it was returned by the intake current. This behavior was observed at a lesser level in the manipulation of plastic toys by octopus arms. Play is now known in many vertebrates, possibly with the long-term gain of practice by sheltered young for the adult social world and for long-term survival. In solitary octopuses, it cannot serve these functions. Instead, these results suggest that observational learning and play are the results of a big brain, a complex sensory system, and an exploratory drive, whatever animal they are observed in.

The above account demonstrates that the octopus is a particularly good model for understanding learning and cognition. Its learning and memory are not based on the same nervous system as vertebrates, as its phylogenetic derivation is very different. In addition, it lives differently as it is marine, has a semelparous life history, major life transition from a pelagic paralarva to subadults, and a complex but completely different brain organization. On the one hand, recent research has suggested that it has a wide-ranging capacity for learning and many of the cognitive abilities such as personalities, play, observational learning, and problem solving that we tend to associate with higher vertebrates. Octopuses may even have a simple form of consciousness. On the other, it has complex brain circuitry that parallels that of vertebrates, as well as complex neurophysiological and neurochemical functions. The combination of behavioral and neurophysiological information should ensure that the octopus remains an alternative model for the study of learning, memory, and cognition.

See also: Playbacks in Behavioral Experiments; Visual Signals.

Further Reading

Boal JG (2006) Social recognition: A top down view of cephalopod behaviour. *Vie et Milieu* 56: 69–79.

Borelli L and Fiorito G (2008) Behavioral analysis of learning and memory in cephalopods. In: Menzel R and Byrne JH (eds.) *Learning and Memory: A Comprehensive Reference, Vol. 1, Learning Theory and Behavior*, pp. 605–627. New York, NY: Elsevier.

Dickel L, Darmaillacq AS, Poirier R, Agin V, Bellanger C, and Chichery R (2006) Behavioural and neural maturation in the cuttlefish *Sepia officinalis*. *Vie et Milieu* 56: 89–95.

Hanlon RT and Messenger JB (1996) *Cephalopod Behaviour*. Cambridge, UK: Cambridge University Press.

Hochner B, Shomrat T, and Fiorito G (2006) The octopus: A model for a comparative analysis of the evolution of learning and memory. *Biological Bulletin* 210: 308–317.

Mather JA (2004) Cephalopod skin displays: From concealment to communication. In: Oller DK and Griebel U (eds.) *Evolution of Communication Systems: A Comparative Approach*, pp. 193–214. Cambridge, MA: MIT Press.

Mather JA (2008) Cephalopod consciousness: Behavioral evidence. *Consciousness and Cognition* 17: 37–48.

Messenger JB (2001) Cephalopod chromatophores: Neurobiology and natural history. *Biological Reviews* 76: 473–528.

Nixon M and Young JZ (2003) *The Brains and Lives of Cephalopods*. Oxford, UK: Oxford University Press.

Scheel D, Lauster A, and Vincent TLS (2007) Habitat ecology of *Enteroctopus dofleini* from middens and live prey surveys in Prince William Sound, Alaska. In: Landsman NH (ed.) *Cephalopods Past and Present: New Insights and Perspectives*, pp. 434–458. New York, NY: Springer.

Villanueva R and Norman M (2008) Biology of the planktonic stages of benthic octopuses. *Oceanography and Marine Biology: An Annual Review* 46: 105–202.

Warnke K, Soller R, Blohm S, and Saint-Paul U (2004) A new look at geographic and phylogenetic relationships within the species group surrounding *Octopus vulgaris* (Mollusca, Cephalopoda): Indications of very wide distribution from mitochondrial DNA sequences. *Journal of Zoological Systematics and Evolutionary Research* 42: 306–312.

Wells MJ (1978) *Octopus: Physiology and Behavior of an Advanced Invertebrate*. London: Chapman & Hall.

Williamson R and Chrachri A (2004) Cephalopod neural networks. *Neurosignals* 13: 87–98.

Relevant Websites

Cophbase.utmb.edu – Formal scientific information about cephs.
Thecephalopodpage.org – More general and informal information.

Olfactory Signals

M. D. Ginzel, Purdue University, West Lafayette, IN, USA

Introduction

Communication involves the transfer of information via a common system of signals. These signals can be sent along visual, auditory, chemical, tactile, and even electrical channels. Chemical communication is a widespread phenomenon among animals, ranging from unicellular prokaryotes to humans. The olfactory systems of these organisms are capable of detecting both general odorants derived from food or the environment and semiochemicals that influence the interactions between organisms. Semiochemicals can be further classified into pheromones, allomones, kairomones, and synomones based on the nature of the interactions they mediate.

Types of Semiochemicals

A *pheromone* is an externally secreted signal that sends meaningful information to members of the same species. The first pheromone was identified in 1959 from the common silk moth *Bombyx mori*. Like many other nocturnal moths, virgin females produce this volatile signal from eversible glands on the tip of their abdomens. After more than 20 years of research, which included extracting abdominal tips of over 250 000 female moths, Butenandt and others identified the active component of silk moth pheromone as bombykol.

Pheromones act as either releasers or primers based on their mode of action. A releaser pheromone elicits an immediate change in the behavior of the receiver, while a primer causes a less rapid and longer term physiological change in the receiver. Interestingly, a single chemical signal can perform both releaser and primer functions, depending on the context in which it is sent. For example, the pheromone of a queen honeybee (trans-9-keto-2-decenoic acid) acts as a primer by inhibiting the ovarian development of female workers in the colony and preventing the rearing of additional queens. This pheromone is picked up by a retinue of females that groom the queen and is transferred throughout the colony as workers feed each other by tropholaxis. Virgin queens also release this same compound while on nuptial flights, where it performs a releaser function – acting as a sex attractant by calling in males for mating.

Allelochemicals are interspecific signals that affect the growth, health, behavior, or population biology of the receiver. These signals can be further categorized as allomones, kairomones, or synomones. An *allomone* elicits a behavioral or a physiological response in the receiver that results in an adaptive advantage to the senders. These are often defensive compounds that act as repellents or feeding deterrents. For example, green lacewings in the genus *Chrysopa* produce skatole-rich defensive secretions that repel invertebrate predators. In some cases, allomones can also come in the form of chemical mimicry. For example, the bola spider mimics the sex pheromone of a noctuid moth. After luring a moth into range, the spider captures its prey with a silken bolas.

Kairomones, on the other hand, are chemical signals that benefit the receiver rather than the emitter. Phytophagous insects often use kairomones to locate appropriate host plants. The western pine beetle, a bark beetle, responds more readily to aggregation pheromones when they are accompanied by a terpene called myrcene which is released from its host, weakened ponderosa pine trees, *Pinus ponderosa*.

Finally, *synomones* mediate mutualistic interactions that benefit both sender and receiver. Floral scents which attract pollinators are an example of synomones, wherein the pollinator receives food in the form of nectar or pollen and the plant is, in turn, pollinated. Plant odors are generally complex mixtures, though only a few compounds may mediate behavior. These odors are often species specific blends of secondary compounds.

Not all chemical signals fit neatly, however, into the categories described earlier. Some signals originate from abiotic factors. For example, fermentation products attract parasitic wasps to decaying fruit inhabited by fruit flies that the wasps parasitize. In fact, the very process of classifying the response of organisms to chemical signals may color or limit our interpretation of them.

Structure of Olfactory Signals

Chemical signaling is the dominant means of communication within and among species. These signals are largely composed of secondary metabolites – compounds not involved in primary physical processes. Chemical signals are structurally diverse with properties that tend to vary with the medium through which they are propagated. Nonetheless, some common features unite the sex pheromones of terrestrial and aquatic organisms. Air-borne pheromones are volatile and of low molecular weight, allowing them to diffuse rapidly. These signals are relatively simple organic compounds often composed of a basic hydrocarbon structure to which functional moieties are attached. Most airborne compounds range in length from 5 to 20 carbon atoms and have molecular weights from 80 to 350 amu. Moreover, many insect pheromones are composed of a blend of structurally related compounds. Signals involved in processes requiring a high degree of specificity (e.g., sex pheromone) usually have a higher molecular weight, allowing for a greater diversity of configurations and more specificity. In the aqueous environment where diffusion rates depend on solubility rather than volatility, chemical signals are often soluble compounds similar in size to those of terrestrial organisms or large polar proteins. Many marine organisms, for example, employ polypeptides as pheromones.

Advantages and Disadvantages of Chemical Signaling

There are unique advantages and limitations to chemical signaling when compared to other channels of communication. For example, chemical communication is independent of light, and signals can be transmitted during both day and night. Unlike visual signals, and to some extent auditory signals, chemicals can travel around obstacles quite easily. They also persist in time, an advantage over auditory signals. Persistence, however, can be a liability if the sender needs to augment the signal or quickly advertise a change in status. Nevertheless, chemical signals are energetically rather efficient to produce and also have a broad transmission range – anywhere from contact pheromones detectible only on the surface of an animal to sex pheromones that are effective over distances of several kilometers. There are a number of drawbacks related to chemical communication, however. Chemicals are often borne on the wind or carried by currents of water and, as a result, delivery can be quite slow and the accuracy by which the signal is delivered diminishes. Also, it is very difficult to modify the frequency or amplitude of a chemical signal once it is released, which may make it difficult for a receiver to localize the source of a distant signal.

While chemical signals are known from a wide variety of organisms, they are particularly well characterized in insects and vertebrates. For this reason, I will focus on olfactory signals in these two taxa.

Insect Pheromones

Sex Pheromones

Sex pheromones are arguably the best studied of semiochemicals, having been identified from nearly all orders of insects. These signals can be produced by either sex and advertise the identity and status of the sender for the purpose of attracting a mate. Much of the interest in these signals is fueled by the notion of exploiting them as

a means of pest management. A greater understanding of the chemically mediated mating behavior of insects will likely lead to new methods for the monitoring, mating disruption, and mass trapping of these important pests. For example, the synthetic sex pheromone of the codling moth, *Laspeyresia pomonella*, was first used to monitor populations 30 years ago and later used in mass trapping and mating disruption efforts.

Aggregation Pheromones

Aggregation pheromones attract conspecifics of both sexes and are particularly common among insects and other arthropods that exploit food sources that are patchy in distribution and sporadically available. These pheromones mediate the formation of a group of individuals for the purpose of mating, overwhelming predators, or overcoming host resistance by mass attack. It has been suggested that aggregation pheromones arose from sex pheromones when members of the producing sex opportunistically responded to sex pheromones. Among beetles that infest stored products, these pheromones are commonly produced by males. For example, males of the red flour beetle, *Tribolium castaneum*, produce 4,8-dimethyldecanal as an aggregation pheromone. Moreover, the aggregation pheromone of the pea and bean weevil, *Sitonia lineatus*, has been used to trap in mass this widely distributed pest of legumes. Perhaps the most widely studied aggregation pheromones are those of the conifer-attacking bark beetles. For example, in the mountain pine beetle, *Dendroctonus ponderosae*, the production of the aggregation pheromone component exobrevicomin is induced when a pioneering female feeds on host phloem while attacking a tree. If a sufficient number of beetles respond to this chemical signal, host defenses can be overcome and the tree colonized. Another remarkable example of aggregation pheromones are those of the gregarious locust. Aromatic hydrocarbons structure the formation of great swarms of the desert locust, *Schistocerca gregaria*, which can cover $1200\,km^2$ and contain as many as 80 million individuals.

Aphrodisiac Pheromones

Aphrodisiac pheromones are commonly produced by males of Lepidoptera and act as close-range sex pheromones to mediate courtship behaviors. In comparison with most sex pheromones, these signals are often perceived through contact alone, produced in staggeringly large quantities, and are by and large more structurally diverse. Moreover, many of these compounds are often derived from diet. For example, male danaid butterflies release a pyrolizidine alkaloid called 'danaidone' from large eversible brush-like 'hair pencils.' When a male butterfly overtakes a female in flight, he dusts her with particles containing the pheromone. These compounds induce the female to land. The male continues to hover

over the female, dusting her with pheromone, until finally he too lands nearby and mates with her. Interestingly, if the 'hair pencils' are extirpated, males are unsuccessful at attracting a mate. Another example of close-range sex pheromones are cuticular hydrocarbons that act as contact pheromones that mediate mate recognition. In some beetles, such as longhorned beetles, these pheromones can be single components or blends of branched or straight-chained saturated and unsaturated hydrocarbons. Males are unable to recognize females, even those within a few millimeters, until physically contacting them with their antennae. In fact, some males even attempt to mate with glass rods or dummies that have been treated with solvent extracts of the female cuticle containing the pheromone. In flies, these pheromones are often methylated hydrocarbons with chain length greater than 25 carbons and can function both as contact pheromones and volatile attractants that function over very short distances. In some species of flies, such as those belonging to the genus *Glossina*, males transfer cuticular hydrocarbons to females while mating, and these compounds act as abstinons – inhibiting further courtship by other males.

Alarm Pheromones

Alarm pheromones are usually volatile compounds that are released by either clonal or social insects in response to a disturbance. These signals can be monoterpenes, sesquiterpenes, or short-chain aliphatic hydrocarbons. They often comprise mixtures of compounds and are less specific than other types of pheromones. In response to alarm pheromones, nonsocial insects usually disperse. For example, aphids fall from their host plant. Social insects, on the other hand, often respond aggressively to alarm pheromones, which are often blends of compounds, with each component eliciting a different response in the receiver. For example, worker leaf-cutting ants in the genus *Atta* raise their heads and open their jaws upon perceiving hexanal – the most volatile and rapidly spreading component of the alarm pheromone. Other components, such as hexanol, are less volatile, spread more slowly, and attract other ants to the site of release. Finally, other components of the blend elicit aggressive behaviors such as biting in ants that are in proximity to the release point. In social insects, these signals are often produced by mandibular glands, although other glands can also be associated with their release. For example, guard bees in the genus *Apis* mark intruders to the hive with mandibular gland secretions that stimulate aggressive behavior in other bees, while the glands associated with the sting release another alarm pheromone.

Epideictic Pheromones

Nonsocial insects commonly release chemical signals that mark their eggs or pattern the spacing of populations.

Many of these compounds are the result of competition for limited resources, and chemical signals that indirectly affect population density are commonly referred to as 'epideitic pheromones.' The most thoroughly studied examples of epidiectic pheromones are those of stored grain pests. For example, larvae of the flour moth produce a pheromone from their mandibular glands. The pheromone, targeted to larvae of these same species, causes increased wandering, delayed pupation, and smaller pupae. These smaller pupae ultimately result in smaller adults that lay fewer eggs and thereby lessen density-dependent mortality factors. Oviposition deterrents are also quite common among parasitic Hymenoptera and inhibit oviposition on the same host, thereby reducing competition.

Trail Pheromones

Social insects also commonly use pheromones to mark feeding or nest sites and trails. Trail pheromones are produced primarily by social insects that forage on the ground, such as termites and ants, although a few nonsocial insects also use them. Termites, for example, continually produce trail pheromone from abdominal glands and deposit a drop each time the abdomen touches the substrate. In this way, insects lay down trail pheromones as they walk and move, and the persistence of well-used trails is reinforced by those who follow. Tent caterpillars even overmark their original paths to a food supply by pressing the terminal abdominal segment along the trail as they return to their nest. Chemical trails can even be followed by flying insects. For example, stingless bees in the genus *Trigona* mark a food source with a pheromone composed mostly of citral and continue to lay down pheromone on the way back to the nest.

Pheromones also play important roles in maintaining the colony structure of social insects. For example, pheromones regulate many aspects of colony life in honeybees. In fact, honeybees use at least 36 different pheromones components, secreted by 15 different glands. Pheromones are involved in such diverse behaviors as foraging, trail marking, colony defense, nestmate recognition, colony fission, swarming, and mating.

Vertebrate Pheromones

The pheromones of vertebrates commonly exist as complex mixtures rather than as single compounds or the simple mixtures characteristic of many arthropod and invertebrate pheromones. Signal specificity of these pheromones is often achieved by varying the proportions of individual components of the mixture. For example, the preorbital and pedal gland secretions of antelopes contain as many as 50 individual compounds. Among vertebrates, a sense of smell is well developed in mammals, especially carnivores and ungulates, where the recognition of individuals is important in maintaining dominance hierarchies, defending territories and in providing parental care. These messages often arise from odor-producing glands in the skin but can also be found in feces, urine, vaginal secretions, saliva, and even expired air. Glands that release odors in mammals are often associated with hairs that may serve to further disperse the signal.

Interestingly, both primer and releaser pheromones structure the highly social lives of mammals. Primer pheromones appear to be most closely associated with urine. For example, the estrous cycle of the laboratory house mouse *Mus musculus* can be suppressed by the presence of another female, but also accelerated by odors present in the urine of males. Releaser pheromones are more common among mammals. For example, female boars display receptive behavior in response to androgen-derived primer pheromones found in the saliva of males. Moreover, 2-methylbut-2-enal is produced in rabbit milk and mediates nipple searching behavior in pups.

The semiochemistry of amphibians and reptiles remains to be studied intensely. Nevertheless, there is evidence of sex pheromone use by these groups. For example, the male newt *Cynops pyrrhogaster* attracts conspecific females by the decapeptide sodefrin which is released into the water from an abdominal gland. Males of the magnificent tree frog, *Litoria splendida*, release a peptide sex pheromone that attracts local females. Moreover, toad tadpoles respond to alarm pheromones released from injured kin by leaving the large conspicuous shoal and swimming to deeper water. Alarm pheromones have also been identified in fish, including carp and minnows. In fact, these signals were among the first vertebrate pheromones to be recognized. The alarm pheromones of minnows, and other fish in the subfamily Leuciscinae, are held within specialized 'club cells' within the skin that are released only when the skin is damaged. The homing behavior of salmon also appears to be mediated by pheromones and kairomones. Apparently, these fish return to their natal streams by orienting to distinctive odors that were imprinted as juveniles. There is also evidence of primer pheromones in fish that induce ovulation. These pheromones are responsible for the maturation of oocytes and raising the volume of milt. Releaser pheromones also act as sex attractants and mediate spawning behavior and the release of gametes.

Perception and Interpretation of Olfactory Signals

How are odorant molecules like pheromones and host odors converted into signals that travel to the brain and trigger a behavioral response in the organism? In this regard, there are striking similarities between olfaction in vertebrates and invertebrates at both the cellular and the organ levels. All systems are composed of olfactory receptors cells. These receptors are polarized neurons that are exposed to the outside world on one end, where they are specialized for chemical detection, while the other end extends to the brain and is specialized for signaling. After an odorant molecule binds to the odorant receptor protein in the cell membrane, the protein undergoes a conformational change and an intracellular cascade of secondary messengers is set into course. This signal cascade causes hyperpolarization of the cell membrane which ultimately results in the transmission of an action potential that conveys information to the brain. Specifically, the binding of an odorant to a specific receptor protein activates a G-protein in the cell membrane. Interestingly, G-proteins are involved in other processes that incorporate cellular receptors (e.g., hormone reception and vision), and all share amino acid sequences including seven-transmembrane domain regions that crisscross the membrane. In the nerve cell, GTP interacts with adenylate cyclase to produce cyclic adenosine monophosphate (cAMP), which serves as a second messenger to open a cation channel permitting an influx of ions. This flush of ions into the nerve cell, in turn, causes a depolarization and the nerve then fires. There is new evidence, however, that in insects signal transduction may be independent of G-protein-coupled secondary messengers and rather the receptor neurons themselves act as cation channels. Nevertheless, these systems are extremely sensitive and most animals are able to detect a host of compounds including completely novel odorants. This astonishing ability is partly due to the diversity of receptor types and the broad yet overlapping specificities of the receptors.

Insects have a highly tuned olfactory system and can detect vanishingly small amounts of pheromones in the environment. A male moth may rapidly respond over a distance of 100 m to as little as 200 molecules of pheromone released by a calling female. It has been estimated that 50% of odorant receptor cells on the antennae of the male silk moth are tuned to respond to female sex pheromone. There is often strong sexual dimorphism with regard to antennal morphology. For example, the antennae of male silk moths possess gross anatomical and fine structural morphology that serves to increase the reception rate of chemical signals. Surface area is enhanced by comb-like structures that project from a central shaft of the antennae, and each of these projections is covered by odorant receptors. In fact, the antennae essentially act as a molecular sieve composed of more than 100 000 individual sensillae through which air passes. Odorant molecules are first caught by sensory hairs that are perforated by many pores and then pass through these openings on the sensillum. The space within a sensillum is filled with hydrophobic lymph, similar to the mucus of vertebrates, and hydrophobic pheromone molecules are carried through the lymph by specialized odorant binding proteins to odorant receptor neurons. The axons of these neurons terminate in the antennal lobes of the brain where they synapse with other neurons at glomeruli. The antennal lobes are equipped with excitatory projection neurons that further send their axon terminals to a portion of the brain called 'the protocerebrum' for higher processing.

In vertebrates, membranes of chemosensory cells, called 'olfactory epithelia,' are bathed in mucus and often modified to increase surface area. There can be millions of olfactory receptors cells. These olfactory cells also tend to be concentrated at the anterior end of an organism and in many vertebrates are often located on the roof of the mouth. The importance of olfaction in the lives of organisms is often reflected in the number of olfactory receptor cells they possess; some dogs, for example, are endowed with approximately one billion receptor cells, while humans have only 10–40 million. The axons of these cells extend directly to the brain, where they connect with interneurons in spheres of nerve tissue called 'glomeruli.' Interestingly, the axons of receptor cells that respond to a specific odorant or related molecule converge on an individual glomerulus, bringing together information from large numbers of neurons. For example, each glomerulus of a rabbit is composed of approximately 25 000 receptor cells.

There are two dominant olfactory systems in vertebrates. The common or main olfactory system senses the environment and is used to find food, detect predators and prey, and mark territories. However, some amphibians, reptiles, and mammals perceive chemical signals, particularly those involved in mate attraction, courtship, parental care, and aggression, using a specialized structure called the 'vomeronasal organ' (VNO) located on the roof of the mouth or between the nasal cavity and mouth. This secondary or accessory system is separate from other chemosensory organs and the neural wiring innervates brain regions other than the main olfactory system, particularly the hypothalmic-pituitary axis – a region of the brain important in hormonal regulation. The VNO is specialized to detect nonvolatile pheromones. Snakes use their tongues to deliver compounds to the VNO, while in mammals, many pheromone signals are in urine or specialized secretions, and the receiver must lick or touch its nose to the chemical for it to be perceived. Mammals also show a characteristic grimace, called the 'flehmen,' where the head is raised and lips curled after making contact with pheromones. This response helps transfer the compounds to the VNO.

Conclusion

Olfactory signals mediate critical processes in the lives of animals, from finding a mate to avoiding danger.

Nevertheless, pheromones often work in concert with other communication channels to form composite signals. Signals from parallel sensory channels may provide redundancy, ensuring that a signal gets through to the receiver. For example, black-tailed deer transmit an odor alarm signal along with sound and visual signals. Also, signals from additional sensory channels (e.g., auditory or visual) may modulate the intensity of the message or may even be necessary for the message to be received altogether. Ants in the genus *Novomessor* enhance chemically mediated recruitment of nestmates to a food source by adding stridulatory vibration signals. Moreover, pheromones alone are not sufficient to mediate mating behavior in the fruit fly, *Drosophilia melanogaster*. Visual, acoustic, olfactory, and tactile signals are used to stimulate receptivity in females and copulatory behavior in males. A greater understanding of the genetic and neural architecture underlying communication will undoubtedly shed light on the evolution of these multimodal and other olfactory signals.

See also: Alarm Calls in Birds and Mammals; Communication Networks; Interspecific Communication; Mating Signals; Smell: Vertebrates.

Further Reading

Andersson M (1994) *Sexual Selections.* Princeton, NJ: Princeton University Press.

Brown RE (1994) *An Introduction to Neuroendocrinology.* Cambridge: Cambridge University Press.

Finger TE, Silver WL, and Restrepo D (2000) *The Neurobiology of Taste and Smell,* 2nd edn. New York: Wiley-Liss.

Gosling LS and Roberts SC (2001) Scent making by male mammals: Cheat-proof signals to competitors and mates. *Advances in the Study of Behavior* 30: 169–217.

Greenfield MD (2002) *Signalers and Receivers: Mechanisms and Evolution of Arthropod Communication.* New York: Oxford University Press.

Hardie J and Mink AK (ed.) *Pheromones of Non-Lepidopteran Insects Associated with Agricultural Plants.* Wallingford, UK: CAB International.

Nelson RJ (1995) *An Introduction to Behavioral Endocrinology.* Sunderland, MA: Sinauer Associates.

Nijhout HF (1994) *Insect Hormones.* Princeton, NJ: Princeton University Press.

Smith RJF (1999) What good is smelly stuff in the skin? Cross function and cross taxa effects in fish 'alarm substances'. In: Johnston RE, Müller-Schwarze D, and Sorenson PW (eds.) *Advances in Chemical Signals in Vertebrates,* pp. 475–488. New York: Kluwer Academic/Plenum Press.

Touhara K and Vosshall LB (2009) Sensing odorants and pheromones with chemosensory receptors. *Annual Review of Physiology* 71: 307–332.

Traniello JFA and Robson SK (1995) Trail and territorial communication in insects. In: Cardé RT and Bell WJ (eds.) *Chemical Ecology of Insects,* 2nd edn, pp. 241–286. London: Chapman and Hall.

Wyatt TD (2003) *Pheromones and Animal Behavior: Communication by Smell and Taste.* New York: Cambridge University Press.

Ontogenetic Effects of Captive Breeding

J. L. Kelley, University of Western Australia, Crawley, WA, Australia
C. M. Garcia, Instituto de Ecología, UNAM, México

An Introduction to Captive Breeding

Animal populations worldwide are becoming increasingly endangered due to habitat loss, climate change, unsustainable harvesting practices, and impacts from invasive and pathogenic organisms. Human intervention is often necessary to maintain viable populations in the wild and to prevent species from becoming extinct. One approach to prevent extinction is to mange vulnerable or endangered species in situ (in their natural habitat) and attempt to control the factor(s) that is(are) causing the species' decline, for example, through habitat restoration and/or the eradication of nonnative predators. If in situ intervention is not possible, however, individuals or populations may be removed from the wild and bred in captivity until the problem(s) causing the decline can be resolved, and there are sufficient numbers of animals available for release into the wild. This form of ex situ conservation (or ex situ preservation) is called *captive breeding.*

The primary aim of captive breeding is to maintain viable populations in captivity that can later be used to augment or reestablish populations in the wild via *reintroduction.* Other important goals of captive breeding include providing a 'gene bank' for long-term species preservation and producing animals as subjects for conservation research projects. Captive-bred animals are also often required as exhibits for zoos and aquaria to avoid removing additional animals from the wild. Importantly, captive breeding programs play a crucial role in raising public awareness of conservation issues, providing support and opportunities for fundraising. It should be noted that captive breeding programs are not always undertaken for species under imminent threat, but may be implemented for species that are predicted to face population declines in the future. For example, an 'insurance population' of Tasmanian devils has been set up on mainland Australia as wild populations are under threat from Devil Facial Tumour Disease.

There was a great deal of interest in the use of captive breeding as a recovery tool for endangered species during the 1980s and 1990s. This culminated in the suggestion that zoos, aquaria, and other breeding facilities could effectively operate as 'arks,' in which species could be preserved over long periods of time before eventually being reintroduced into the wild. Despite this early enthusiasm, however, it is now generally recognized that captive breeding may not be suitable for all species.

Indeed, it has proved difficult to achieve self-sustaining captive populations of animals such as whooping cranes, giant pandas, and northern white rhinos due to a shortage of reproductively active animals, high mortality, and poor reproductive success. On the other hand, some animals appear well suited to captive breeding programs. An analysis of the relationship between body mass and the intrinsic rate of population increase in captive populations of threatened animals revealed that species that are most likely to have high population growth rates are those with small body mass such as black-footed ferrets (*Mustela nigripes*), pink pigeons (*Columba mayeri*), and golden lion tamarins (*Leontopithecus rosalia*).

If an endangered species is successfully bred in captivity, it does not necessarily mean that its reintroduction into the wild will be successful. Only a small number of captive breeding and reintroduction programs have been considered successful, that is, where released animals can potentially form self-sustaining populations in the wild. Captive breeding and reintroduction is therefore often considered as a last resort strategy, particularly since this form of management can also be very expensive and time consuming. Nonetheless, several species such as the Arabian oryx and the Californian condor have been brought back from the brink of extinction through such schemes. Successful reintroduction schemes require expertise from a variety of groups including biologists, policy makers, sociologists, and organizational consultants, and problems may emerge if communication among these parties is poor.

The limited success of reintroduction programs is also frequently attributed to behavioral problems associated with captive breeding; captive-bred animals often show deficiencies associated with foraging skills, predator avoidance, and social interactions. These behavioral deficiencies can arise through genetic adaptation to the captive environment (domestication) and/or as a result of developmental (ontogenetic) effects. Ontogenetic effects may be of particular interest to breeding managers as their effects on animal behavior may be more easily reversed (e.g., through environmental enrichment and learning) than those resulting from unintended selection. In another category falls mating behavior, in particular mate choice. This is the consequence of adaptive preferences that make animals choosy when it comes to mate. In most species, females are unlikely to accept any potential partner; thus, mate choice limits the success of captive breeding and introduction programs when partner availability is limited, as is often the case.

Here, we first consider how genetic and ontogenetic processes operate in captivity and how these might influence the behavior of captive animals. We then examine the environmental factors that may contribute to behavioral deficiencies in captive animals, focusing on the availability of space and food, the reduction or loss of predation risk, and the social environment. We also discuss the relationship between the developmental environment and abnormal repetitive behaviors (or stereotypic behaviors) that may be observed in captive animals. Finally, we explore the ways in which breeding managers can modify the behavior of captive animals by providing environmental enrichment and opportunities for learning.

Effects of Captivity on Animal Behavior

The behavior of wild and captive animals is often markedly different, and there are numerous reports of behavioral differences being observed between wild and captive-reared or captive-bred animals. Behaviors are not usually 'lost' when animals are bred in captivity; rather, there tend to be quantitative differences in the behavioral repertoire of wild versus captive animals. Behavioral differences between wild and captive-bred animals can arise as a result of genetic changes occurring in captivity (through selection), through differences in the environment experienced during development, or a combination of both these factors.

Genetic Effects of Captive Breeding

The process by which animals become adapted to captivity as a result of genetic changes occurring over generations is referred to as *domestication*. The selective processes that result in domestication can either be artificial or natural. *Artificial selection* occurs when individuals with particular characteristics are mated in order to produce offspring with preferred traits. For example, for thousands of years, humans have selectively bred animals to produce desirable traits in modern livestock such as large muscle mass in sheep and pigs. On the other hand, *natural selection* in captivity promotes traits that are advantageous in the captive environment. Both types of selection build on the genetic effects of inbreeding depression and random genetic drift which result from the commonly small population size of captive-bred stocks.

Genetic adaptation to captivity has been demonstrated in insects, amphibians, fish, and mammals and is often associated with increased levels of fecundity. For example, levels of fecundity in large white butterflies (*Pieris brassicae*) that had been maintained in captivity for 100–150 generations were around 13 times higher than levels observed in wild populations. However, traits that are advantageous and selected for in captivity are often highly maladaptive

if captive-bred animals are introduced into the wild. For example, hatchery-reared salmonid fish have lower fitness in natural environments than wild fish, with declines in fitness sometimes being observed after just 1–2 generations of hatchery rearing.

Behaviors that are favored by natural selection in wild animals (e.g., predator avoidance) tend to lose their adaptive significance in captivity resulting in what is often referred to as *relaxed selection* on these traits. Isolation of prey from predators can cause some components of antipredator behavior to be lost rapidly (over the course of a few generations), while others may persist. For example, tammar wallabies (*Macropus eugenii*) inhabiting Kangaroo Island, South Australia (which has been isolated from mammalian predators for around 9500 years), retain a fear response toward visual, but not acoustic cues from mammalian predators. Loss of antipredator behaviors has also been shown to occur in captive breeding programs: antipredator responses of the threatened Mallorcan midwife toads (*Alytes muletensis*) were found to diminish after around 9–12 generations of captive breeding.

Ontogenetic Effects of Captive Breeding

Differences in the behavior of wild and captive animals can also arise as a result of ontogenetic factors, which are those that take place during the animals' development. Ontogenetic differences occur because many species are *phenotypically plastic* meaning that their appearance and/or behavioral traits exhibit different responses depending on the environment that they encounter during development. Learning is a form of phenotypic plasticity as it allows animals the chance to modify their behavioral skills in response to specific stimuli in the environment. Learning is not restricted to mammals but occurs in most animal groups, as Darwin showed in his early study of earthworm behavior. Yet there is a disproportionate representation of vertebrates, and in particular of fish, birds, and mammals in the literature on ontogenetic effects of captive breeding. This may be a result of human biases toward certain groups of organisms, and/or of a widespread misconception than amphibians and reptiles – and invertebrates in general – are not capable of much learning.

Learning can occur at the level of the individual through trial and error (*individual learning*), or it may arise through observing and/or interacting with others (*social learning*). Some animals may have a repertoire of complex learned behaviors such as parental care, sexual behavior, social interactions, and foraging skills, and development of these skills may be constrained if the captive environment does not provide sufficient learning or training opportunities in the form of realistic behavioral (e.g., foraging) tasks. Opportunities for social learning may also be limited if the structure of the social group does not allow information to be transmitted between particular individuals, such as from parent to offspring. Behaviors that are learned and socially transmitted to other group members have the potential to be lost much faster than genetic diversity (i.e., within a single generation) and are therefore of major concern to captive breeding managers.

Environmental Factors Affecting Behavioral Development

The environment that an animal experiences in the wild may differ greatly to that which it encounters in captivity. Captive animals typically live in a confined space and those kept indoors may be subjected to artificial lighting conditions (which may result in abnormal circadian rhythms), different temperature regimes, and are generally not exposed to the variation in physical conditions experienced by wild animals. In contrast, animals kept in outdoor enclosures may experience greater variation in environmental conditions than their wild counterparts if the enclosure does not provide a range of habitats (e.g., shelters, tunnels, and exposed rocks) to allow behavioral buffering through the use of such refuges. The time budgets of wild and captive animals can also differ dramatically. For example, captive black and white ruffed lemurs (*Varecia variegata*) spend more time self-grooming and less time feeding than lemurs in the wild. The social environment is also very different in the wild, with group organization likely being more dynamic in terms of group size, sex ratio, age structure, and social hierarchy than that occurring in captivity. The environment experienced during ontogeny can also result in captive and wild animals displaying different responses to stress. However, the specific cause of behavioral deficiencies observed in captive-born individuals is often not known and likely attributable to a combination of factors, including genetic effects. Major differences between wild and captive environments and their likely affects on behavioral development are discussed below.

Space

The amount of space used by animals in the wild largely depends on the spatial distribution of resources such as food, water, and shelter. Some wild animals are highly territorial while others are dispersed or migratory with space requirements varying on an annual or seasonal basis. In contrast, the amount of space available for captive animals is almost always reduced such that animals live at higher densities than they would otherwise experience in nature. Although the basic nutritional and housing requirements of animals may be met, the environment must also provide for the animals' physiological and psychological

welfare; for example, there may be the basic need to perform species-specific behaviors such as exploration.

As larger population sizes are generally preferable for maintaining viable populations in captivity (e.g., through avoiding inbreeding), there is a conflict between the size of the captive population that can be housed and the amount of space available for each individual. Many zoos overcome this problem by incorporating animals from different institutions into their captive breeding programs, hopefully increasing the size of the breeding population if translocated individuals are acceptable breeding partners to the local animals (see section 'Sexual Behavior'). The amount of space provided for an animal in captivity must allow for the appropriate expression of the 'normal' behavioral repertoire. Animals with insufficient space often display stereotypical behaviors such as pacing. Crowding is often associated with high levels of aggression and reduced reproduction in rodents, and increasing cages sizes has been shown to alleviate these effects. The amount of space available to an animal group can also directly influence its social structure. For example, house mouse populations switch from territorial societies to dominance hierarchies if space is limiting.

Food Availability

Wild animals spend a large proportion of their time foraging, including deciding where to forage, how long to spend foraging, and which food items are likely to be most profitable. However, animals in captivity are typically presented with their food at the same time and place every day. As a consequence, captive animals spend much less time and energy foraging than wild animals. Captive animals therefore do not need to trade foraging behavior for other important activities such as reproduction and predator avoidance. However, the extra time created in their activity budgets can be detrimental to their welfare, and a number of studies have shown that captive animals benefit when provided with foraging activities that are more time consuming. For example, captive common marmosets (*Callithrix jacchus*) presented with hidden food showed increased foraging and movement activity and reduced scratching and grooming compared to marmosets whose food was presented in a bowl. Indeed, it is common practice in modern zoos to allow animals the opportunity to forage for their food by presenting it at various locations around the enclosure or by presenting it in such a way that it takes animals time to obtain it (e.g., presentation boxes, foraging puzzles).

Predation

Most animals are subjected to predation risk and must spend a considerable portion of their time being vigilant toward potential risks and responding to threats. In contrast, animals in captivity are rarely exposed to predation threats or any of the cues that are associated with increased risk (unless they receive training as part of their management program, see section 'Predator Training'). It was previously thought that animals would develop their full repertoire of antipredator skills without requiring any experience. However, antipredator behavior in animals is often modified as a result of experience with predators or predator-related cues (e.g., odors), and/or as a result of observing the antipredator responses of others. Animals born in captivity therefore may not have the opportunity to learn how to respond appropriately to a predation threat. As a result, the antipredator behavior of captive-reared animals can be impoverished when compared to the skills of wild-caught animals. Indeed, high predation on released animals has been reported as a major cause of failure in many reintroduction schemes.

Social Environment

Wild animal populations are hugely variable in social structure ranging from species with largely solitary lifestyles to those with highly complex hierarchical societies. However, the social environment in captivity may depend on the availability of animals or enclosure space with social organization typically being more uniform (e.g., a few adults of each sex) than that observed in the wild. Animals often fail to breed when housed in an inappropriate social environment, and mistakes are often made because of a lack of information about the social structure of populations of endangered species. Social isolation in an otherwise gregarious species can cause high levels of stress, poor health, and the development of abnormal repetitive behaviors. Animal keepers often obtain information about the social requirements of a particular species through simple field research or experiments in captivity where they are allowed to chose their social partners.

Parent–offspring interactions

Captive-born young may either be allowed to remain in the enclosure with their parents or, if necessary, may be removed for hand rearing. Hand rearing is often used for primates because of high rates of maternal neglect in captive populations. Indeed, poor social environment is often blamed for poor parenting skills in captive populations, possibly because females are unable to learn these skills from other females in the population. In captive breeding programs for birds such as whooping cranes (*Grus americana*), eggs are often removed to induce the production of larger brood sizes than would normally be produced. This is done to maximize egg production to provide a large number of individuals for reintroduction. As the parents cannot care for all the young, some of them may be hand reared or cross-fostered by either different parents or

members of a closely related species. This can subsequently create problems associated with species recognition, song learning, habituation to humans, and the development of antipredator skills. For example, postrelease monitoring of captive-born whooping cranes found that those that were raised by their parents were more vigilant and formed larger groups (a potential antipredator strategy) than those which were hand reared. Reproductive suppression has been reported in animals such as rhinoceros, naked mole rats, marmosets, and tamarins when young adults are forced to remain in their family groupings. In other species, however, the presence of 'helper' family members may be required for successful reproduction. Female monkeys that are isolated from their mothers at an early age may be more likely to reject their own offspring, and there are also reports of females pacing instead of caring for their young.

Sexual behavior

Many species require particular conditions in order to breed and the key to success is to ensure that these are met, where possible, by the captive environment. One problem with identifying these conditions in endangered species is that their natural history is often not well known. In these cases, animal breeders may use a closely related species for which more information is available, and it may take several years for animal managers to find the conditions required for successful breeding.

Captive breeding facilities often provide limited opportunities for mate choice, either because few sexually mature adults are available or because management practices require matings between particular individuals (e.g., through the use of studbooks) in order to enhance genetic diversity. Such animals may refuse to mate or these pairings may result in low or no reproductive success, if for example the individuals selected are genetically incompatible. However, knowledge of the mating preferences of captive species can be used by breeding managers to facilitate pairings between particular individuals. Mating preferences can be determined by early social interactions and animals display preferences based on either familiar or novel phenotypes that were encountered at an early stage of development. For instance, experimentally released mallards may remain in compact flocks which do not mix with wild birds, thus reducing the opportunity for interbreeding (and baffling attempts to perform genetic rescue of endangered populations). Mating patterns in captivity can also be assessed using genetic parenting. This can be particularly useful for animals kept in groups, as it may reveal that only a few males are siring most of the offspring produced, with the consequent negative effects of reducing the effective (genetic) population size of the captive population.

Abnormal Repetitive Behaviors (ARBs) and Stress in Captive Animals

Abnormal repetitive behaviors (ARBs) also referred to as 'stereotypical behaviors' are defined as repetitive, unvarying, and apparently functionless behavior patterns. These are commonly observed in captive animals held in zoos and breeding institutions but are rarely observed in nature. It has been estimated, for example, that 80% of giraffes, 69% of gorillas, and 43% of elephants in captivity display these behaviors, and that up to 10 000 animals are affected worldwide. Examples of common ARBs include pacing, rocking, overgrooming, self-harm (e.g., biting), and eating inedible objects. ARBs may be performed in response to cues in the captive environment and be driven by frustration, fear, or discomfort ('frustration-induced stereotypic behavior'). These sorts of ARBs can develop from attempts to escape; examples include bar chewing in rodents and pacing in wombats and leopards. ARBs are also thought to be linked to abnormalities of the central nervous system caused by sustained stress and/or a suboptimal environment experienced during early development ('malfunction-induced stereotypical behavior'). Impoverished rearing environments are known to impair brain development in rodents, and the observation that ARBs are more often observed in captive-born animals than in wild-born animals subsequently kept in captivity provides some evidence for underlying pathological problems. Regardless of the underlying causes of ARBs, they are an important consideration in terms of animal welfare, public education, and for the conservation of 'normal' behavior patterns.

It has been suggested that performing repetitive behaviors may be a method that allows animals to 'cope' with the stresses associated with captivity. Indeed, the proportion of time spent performing ARBs tends to increase when individuals are under increased stress (which is why Broom suggested that stereotypies should be viewed with alarm if they take up more than 10% of an animal's time), for example, when zoo visitor numbers are high and there is more human contact. Animals in the wild have a suite of behaviors allowing them to respond to stresses such as predator encounters or aggressive interactions with conspecifics, but animals in captivity may face conditions (e.g., human interactions) that they are less equipped to deal with. Although ARBs tend to be linked with poor welfare, animals in these environments that perform ARBs often fare better (e.g., reduced response to stress) than those that do not display them, providing some evidence for a 'coping effect.' Some types and/or levels of stress may actually be beneficial for captive animals and, if experienced during early development, may allow animals to more effectively deal with stressful situations encountered in later life.

Modifying the Behavior of Captive Animals

The realization that the captive environment often does not allow for the appropriate expression and development of behavior has led to improvements to the physical environment, through environmental enrichment. This not only improves the general welfare of captive animals (e.g., through stress reduction), but also ensures that where possible, a full repertoire of behaviors is expressed. Where learning is required for the development of a species' behavior, there have been attempts to either provide the relevant learning experiences and/or to rear animals in seminatural conditions so that they might be exposed to the appropriate stimuli. These can also be regarded as forms of environmental enrichment. These different approaches and their success at reversing the behavioral effects of captivity are discussed below.

Environmental Enrichment

Environmental *enrichment* refers to modifications that act to enhance the level of physical and social stimulation provided by the captive environment (Würbel et al., 1998). The method is commonly employed by zoos and is beneficial both for an animal's behavior and its physiology, primarily through stress reduction. Chronic stress is associated with reduced immune responses, poor growth, and poor reproductive success and may be responsible for the ARBs that are sometimes observed in captive animals. Numerous studies in the zoo literature document the positive effects of environmental enrichment (e.g., adding climbing structures, hiding places, and foraging tasks) on animal health and behavior. Besides increasing the well-being of captive animals, environmental enrichment ensures that an animal's behavior as well as its genes is conserved. Thus, enrichment can increase the behavioral repertoire displayed by a species when in captivity and potentially enhance the success of a captive breeding and reintroduction program.

Although environmental enrichment is the most common method used for reducing the prevalence of ARBs, it is only successful about 50% of the time. It is possible that the enrichments provided are insufficient to promote 'normal' behavioral patterns and/or that the ARBs were acquired during early development and are more difficult to lose. Other suggestions for reducing the prevalence of ARBs in captive animals include genetic selection, the use of drugs, and encouraging animals not to perform ARBs through either positive (rewards) or negative reinforcement (mild punishments or prevention – such as obstructing pacing routes). However, enrichment is the simplest and most commonly performed method and any alternative/complementary strategy should, if possible, be based on deeper understanding of the species' natural history.

One method of using environmental enrichment to increase the chances of reintroduction success is to rear captive-bred animals in seminatural environments. Studies have shown that the behavioral skills and postrelease survival of animals reared in seminatural environments are enhanced in comparison to those reared in standard enclosures. For example, black-footed ferrets (*Mustela nigripes*) reared in enriched environments and provided with live prey are more efficient hunters than those reared in standard pens. Another method for increasing postrelease survival is to initiate a 'soft release,' where animals are acclimatized shortly before release into the wild, or they are released but provided with food and shelter for a while as in the traditional falconry's method of 'hacking back.' Individuals may be held in enclosures positioned at the release site, allowing them to acclimatize to natural conditions or they may be allowed to roam over a relatively large area prior to release. For example, the reintroduction of golden lion tamarins was more successful when they were freed into a remote area of the zoo (containing foraging and climbing tasks) for 2 months before being transferred to their natural habitat in Brazil.

Predator Training

Species whose antipredator skills are most likely to be affected by captive rearing are those whose behaviors are acquired as a result of learning. Although these animals may be considered less suitable candidates for release than those whose antipredator skills are 'hard wired,' these skills can potentially be reacquired if the correct stimuli are provided. The success of training may depend on when it is performed; many animals have 'sensitive periods' or developmental stages during which particular behaviors are more readily acquired. A species' sensitive periods will dictate whether attempts to 'train' individuals to recognize and respond to predators should be performed when animals are at an early stage of development or just prior to release, when they are adults.

Experiments with mammals, birds, and fish have shown that individuals can learn to recognize novel predators relatively rapidly, often after just a few exposures. Individuals that observe conspecifics responding fearfully toward a novel predator retain a fear response toward that predator during future encounters. Learning can also occur on the basis of auditory and chemical cues. Model predators are often used for training purposes as it avoids the ethical issues associated with live predators. For example, predator models have been used to enhance predator avoidance in Siberian polecats (*Mustela eversmanni*); polecats reduced their escape times after just

one exposure to a model badger presented in conjunction with a mild aversive stimulus.

Individuals readily habituate if overexposed to model predators, or if their detection is not associated with fear. For example, predation by red foxes is the main factor influencing the survival of reintroduced juvenile houbara bustards in Saudi Arabia. Both live and model fox predators were used in an attempt to train captive-bred houbara bustards to recognize predators. Training with the live predator and not the model increased the survival probability of released animals, possibly because the birds had become habituated to the model as its presence was not associated with a negative experience. The level of negative experience required to elicit learning is an interesting topic of research, both in the context of designing simulated predator encounters and also in relation to animal welfare.

Summary

In summary, captive breeding can have a host of effects of various seriousness on the animals' behavior, which can limit the success of captive breeding and of reintroduction programs. Both genetic and ontogenetic effects can cause behavioral differences to arise between wild and captive animals. The ontogenetic effects of captive breeding may be reversed more readily than the genetic effects if the relevant environmental stimuli and learning opportunities can be identified and provided.

Acknowledgments

We are extremely grateful to Dr Helen Robertson of Perth Zoo for sharing her insights into captive breeding management and for providing comments on this article. J.L. Kelley acknowledges funding from the University of Western Australia.

See also: Learning and Conservation.

Further Reading

Araki H, Berejikian BA, Ford MJ, and Blouin MS (2008) Fitness of hatchery-reared salmonids in the wild. *Evolutionary Applications* 1: 342–355.

Balmford A, Mace GM, and Leader-Williams N (1996) Designing the ark: Setting priorities for captive breeding. *Conservation Biology* 10: 719–727.

Beck BB, Rapaport LG, Stanley Price MR, and Wilson AC (1994) Reintroduction of captive-born animals. In: Olney PJS, Mace GM, and Feistner ATC (eds.) *Creative Conservation: Interactive Management of Wild and Captive Animals*. London: Chapman & Hall.

Bjonet SJ, Price IR, and McGreevy PD (2006) Food distribution effects on the behaviour of captive common marmosets, *Callithrix jacchus*. *Animal Welfare* 15: 131–140.

Blumstein DT (2002) Moving to suburbia: Ontogenetic and evolutionary consequences of life on predator-free islands. *Journal of Biogeography* 29: 685–692.

Broom DM (1991) Animal Welfare: Concepts and measurement. *Journal of Animal Science* 69: 4167–4175.

Calisi RM and Bentley GE (2009) Lab and field experiments: Are they the same animal? *Hormones and Behavior* 56: 1–10.

Carlstead K, Mellen J, and Kleiman DG (1999) Black rhinoceros (*Diceros bicornis*) in US zoos. I: Individual behavior profiles and their relationship to breeding success. *Zoo Biology* 18: 17–34.

Cheng KM, Shoffners RN, Phillips RE, and Lee FB (1978) Mate preference in wild and domesticated (Game farm) mallards (*Anas platyrhynchos*). I: Initial preference. *Animal Behaviour* 26: 996–1003.

Clark TW, Reading RP, and Clarke AL (eds.) (1994) *Endangered Species Recovery: Finding the Lessons, Improving the Process*. Washington, DC: Island Press.

Cyr NE and Romero LM (2008) Fecal glucocorticoid metabolites of experimentally stressed captive and free living starlings: Implications for conservation research. *General and Comparative Endocrinology* 158: 20–28.

Frankham R (2008) Genetic adaptation to captivity in species conservation programs. *Molecular Ecology* 17: 325–333.

Griffith B, Scott JM, Carpenter JW, and Reed C (1989) Translocation as a species conservation tool: Status and strategy. *Science* 245: 477–480.

Hogan LA and Tribe A (2007) Prevalence and cause of stereotypic behavior in common wombats (*Vombatus ursinus*) residing in Australian zoos. *Applied Animal Behaviour Science* 105: 180–191.

Jensen P (ed.) (2002) *The Ethology of Domestic Animals*. Wallingford: CABI Publishing.

Kerridge FJ (2005) Environmental enrichment to address behavioural differences between wild and captive black-and-white ruffed lemurs (*Varecia variegata*). *American Journal of Primatology* 66: 71–84.

Kraaijeveld-Smit F, Griffiths RA, Moore RD, and Beebee TJC (2006) Captive breeding and the fitness of reintroduced species: A test of the responses to predators in a threatened amphibian. *Journal of Applied Animal Ecology* 43: 360–365.

Kreger MD, Hatfield JS, Estevez I, Gee GF, and Clugston DA (2005) The effects of captive rearing on the behaviour of newly-released whooping cranes. *Applied Animal Behaviour Science* 93: 165–178.

Latham NR and Mason GJ (2008) Maternal deprivation and the development of stereotypic behavior. *Applied Animal Behaviour Science* 110: 84–108.

Lewis OT and Thomas CD (2001) Adaptations to captivity in the butterfly *Pieris brassicae* (L.) and the implications for ex situ conservation. *Journal of Insect Conservation* 5: 55–63.

Mason G, Clubb R, Latham N, and Vickery S (2007) Why and how should we use environmental enrichment to tackle stereotypic behaviour? *Applied Animal Behaviour Science* 102: 163–188.

Mason GJ (1991) Stereotypies: A critical review. *Animal Behaviour* 41: 1015–1037.

Miller B, Biggins D, Wemmer C, et al. (1990) Development of survival skills in captive-raised Siberian polecats (*Mustela eversmanni*). II: Predator avoidance. *Journal of Ethology* 8: 95–104.

Russell A (1970) Efects of maternal deprivation treatments in the rat. *Animal Behaviour* 18: 700–702.

Shier DM and Owings DH (2006) Effects of predator training on behavior and post-release survival of captive prairie dogs (*Cynomys ludovicianus*). *Biological Conservation* 132: 126–135.

Snyder NFR, Derrickson SR, Beissinger SR, et al. (1996) Limitations of captive breeding in endangered species recovery. *Conservation Biology* 10: 338–348.

Vargas A and Anderson SH (1999) Effects of experience and cage enrichment on predatory skills of black-footed ferrets (*Mustela nigripes*). *Journal of Mammalogy* 80: 263–269.

Ward C, Bauer EB, and Smuts BB (2008) Partner preferences and asymmetries in social play among domestic dog, *Canis lupus familiaris*, littermates. *Animal Behaviour* 76: 1187–1199.

Wemelsfelder F (1993) The concept of animal boredom and its relationships to stereotyped behaviour. In: Lawrence AB and Rushen J (eds.) *Stereotypic Animal Behaviour: Fundamentals and Applications to Welfare*, pp. 65–96. Wallingford: CABI.

Wielebnowski N (1998) Contributions of behavioural studies to captive management and breeding of rare and endangered mammals. In: Caro T (ed.) *Behavioral Ecology and Conservation Biology*. New York: Oxford University Press.

Wingfield JC, Hegner RE, Dufty AM, and Ball GF (1990) The 'Challenge Hypothesis': Theoretical implications for patterns of testosterone secretion, mating systems, and breeding strategies. *The American Naturalist* 136: 829–846.

Würbel H, Chapman R, and Rutland C (1998) Effect of feed and environmental enrichment on development of stereotypic wire-gnawing in laboratory mice. *Applied Animal Behaviour Science* 60: 69–81.

Young RT (2003) *Environmental Enrichment for Captive Animals*. Great Britain: Blackwell Publishing.

Relevant Websites

http://www.cbsg.org – The Captive Breeding Specialist Group.

http://www.iucn.org – The International Union for the Conservation of Nature.

http://www.waza.org – World Association of Zoos and Aquaria.

Optimal Foraging and Plant–Pollinator Co-Evolution

G. H. Pyke, Australian Museum, Sydney, NSW, Australia; Macquarie University, North Ryde, NSW, Australia

General Concepts of Co-evolution

Oaks protect their leaves from herbivores with toxic tannins. The caterpillars that eat oak leaves have digestive mechanisms that detoxify tannins. The oak defense and the caterpillar response provide an example of co-evolution. Co-evolution occurs when one species evolves in response to evolutionary changes in another, the result being an evolutionary feedback involving two or more species. Co-evolution is an important and ubiquitous process in nature that can occur any time two evolving populations interact through evolutionary time. As the oak-caterpillar example illustrates, many species interact via foraging behavior. These interactions may be predator–prey relationships or interactions between animals that compete for similar food resources, and it follows that these interactions set the stage for foraging behavior to coevolve with the attributes of competitors and prey.

To take a more behavioral example, consider a small herbivorous mammal foraging in a meadow. Rocks and shrubs in the meadow offer safety from predators, but they are far from the mammal's food which occurs out in the open away from shelter. In a system like this, we expect that natural selection will favor a pattern of activity in which our small mammal allocates some time every day to feeding in the open and some time to sheltering behind rocks. Yet, our small mammal's predator evolves too. Changes in the prey animal's activity pattern will generate selection on the predator's activity pattern. Co-evolution between the two species would thus lead to patterns of allocation of foraging time for each species with distance from cover.

Big predators eat big prey items. And, coevolutionary pressures offer a partial explanation of this phenomenon. On the predator side of the equation, predator and prey body sizes influence the predator's encounter rate with prey, the likelihood that the predator can successfully capture a prey item, the time it takes to eat a prey item, and the metabolic costs of these activities. On the prey side of things, predator and prey body sizes influence the prey's rate of food intake, risk of being eaten by the predator, and metabolic costs. Larger prey species commonly experience less predation, and this selects for larger prey bodies. Large predators can capture larger prey, so selection will often favor larger predators in response to larger prey. Co-evolution can, as this example shows, contribute patterns that occur at higher levels of biological organization such as between two trophic levels (predators and prey) such as a correlation between the body sizes of predator and prey species.

Theories of co-evolution seek to explain and predict the outcomes of co-evolution. As this article explains, it can be difficult to determine how coevolutionary processes work. The remainder of this article focuses on co-evolution between plants and their pollinators. This topic demonstrates both the potential and the difficulties associated with developing theories of co-evolution.

Plants and Their Pollinators

Investigators have recognized co-evolution between plants and their pollinators since Darwin first put forward his theory of evolution. For example, pollination biologists have traditionally identified 'pollination syndromes' that relate flower characteristics to pollinator characteristics: hummingbird pollinated flowers tend to be red and tubular, for example, while bee pollinated flowers tend to be blue or yellow and open. At a somewhat finer scale, we often find that long-tongued (or long-proboscized to be technically correct) bees visit flowers with long corollas (**Figure 1**) and that flowers visited by larger pollinators (birds, bats, etc.) tend to offer larger amounts of nectar.

Co-evolution must, in particular, occur between the foraging behavior of flower-visiting animals and plant traits such as nectar production, nectar concentration, and flower morphology. Consider, for example, a plant that increases its nectar production. This simple change can influence pollinator behavior in several ways. A pollinator may spend more time at each flower, simply because it takes more time to lap up more nectar. Or, it may visit more flowers per plant because one 'rich' flower indicates the presence of other rich flowers.

Many explanations of pollination syndromes are unsatisfying because they make unrealistic assumptions about the process of natural selection. To argue, for example, that a plant species evolved a long corolla to exclude pollinators with a short proboscis requires the following sequence of events: corolla length increases across the entire plant species; this reduces the ability of pollinators with a short proboscis to collect nectar; these pollinators switch away from this plant species to others; and this somehow benefits the plant species (with the newly elongated corolla). This reasoning incorrectly implies that selection acts at the species level and fails to consider the consequences of changes in corolla length for an individual plant.

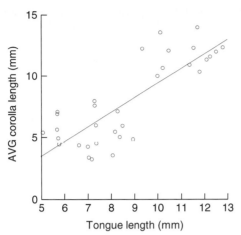

Figure 1 Average corolla length for flowers visited by each bumblebee species and caste versus bumblebee tongue length (Observations made during transect surveys of bumblebees and visited flowers during summer 1974 near the Rocky Mountain Biological Laboratory, Colorado; Pyke, unpublished).

Some explanations of pollination syndromes are also unsatisfying because they imply that an evolutionary change in one species occurs because it benefits another species. To say, for example, that plants pollinated by larger animals provide larger nectar rewards because their pollinators need more food wrongly assumes that the plants evolve to serve the needs of their pollinators. Instead, we need to explain this correlation by considering how changes in nectar production affect individual plants.

Optimal Behavior for Plants and Pollinators

An approach from evolutionary game theory helps to overcome some of these problems. Instead of thinking about species-level benefits, we consider the evolutionarily stable state (ESS) for a frequency-dependent trait, like corolla length or nectar production. An evolutionarily stable corolla length, for example, would be a corolla length such that if all plants in a population adopt the ESS corolla length, no mutant individual adopting a longer or short length can do better. In practice, this is a form of optimization, because at the ESS a given individual's corolla length should be the optimal evolutionary 'reply' to the actions (corolla lengths) of the other individuals in the population. This means that we can use results from optimal foraging theory when we analyze the co-evolution of pollinator foraging behavior and floral traits.

Applying optimal foraging theory to the behavior of nectar-feeding animals has been a relatively straightforward and much used approach. For example, worker bumblebees often restrict their foraging to nectar collection, they are not looking for mates or defending territories, they

experience virtually no risk of predation while they forage, and their colony's survival and growth of depends on their nectar-collecting efforts. Hence, for nectar feeders like bumblebees, we can reasonably apply foraging theory's premise that behavior will maximize the net rate of energy intake. Using this premise, we can consider a wide variety of foraging behaviors including which plants nectar feeders visit, how they move within and between plants, the numbers of flowers they probe per visit to a plant, and the spatial distribution of individuals across different flower patches. Consider, for example, how one might apply the classical foraging problem of patch departure to nectar feeders. We can readily observe rates of nectar consumption and travel times within and between plants. Moreover, we can find the statistical relationship between the nectar obtained at one flower and the forager's expectation of rewards at subsequent flowers on the same plant by recording the correlations between the nectar volumes of flowers on the same plant. Observations indicate that the amount of nectar obtained at a flower influences a nectar feeder's decision to probe more flowers on same plant. With this information in hand, we can determine the optimal departure rule.

Indeed, applications of optimal foraging models to nectar feeders have experienced considerable success, typically finding at least qualitative agreement with theoretical expectations. The optimality approach leads us to expect, for example, that an individual encountering a high-nectar flower should be more likely to probe another flower on the same plant, because flowers on the same plant tend to have similar amounts of nectar. In addition, we expect an increased tendency to move to a nearby plant, because neighboring plants tend to have similar nectar levels. Observations support these qualitative predictions. The quantitative predictions of optimal foraging models also have an impressive record, even though it is not quite as impressive as record for qualitative predictions. One model has, for example, accurately predicted the average number of flowers that a nectar-feeding animal probes per plant visit.

The plant side of the coevolutionary problems presents a more difficult challenge. A model of optimal nectar production, for example, must consider a fairly complex sequence of questions: how do changes in nectar production affect pollinator behavior, how does pollinator behavior affects pollen transfer from one plant to another, how does pollen transfer affects plant fitness through both male and female function, and what does it cost a plant to produce nectar. In the next paragraph, I review how one might construct a model that incorporates this information.

To begin, we would consider a (mutant) plant with a slightly higher rate of nectar production and hence a slightly higher amount of nectar per flower. A nectar-feeding pollinator that visits such a plant would respond to the higher nectar volumes by visiting more flowers than average on the plant. As a result, the pollinator will deposit more pollen (from previously visited plants) on our mutant's stigmas, and it will also collect more of the mutant's pollen that it will ultimately deposit on the stigmas of other plants. Changes in pollinator behavior could also affect pollen flow within the subject plant (i.e., from one of our mutant's flowers to another on the same plant). Taken together, all these changes in pollen flow can, at least in theory, affect the mutant's fitness. For example, receiving more pollen from other plants could increase seed production (in cases where pollen is limiting). Alternatively, if the plant already receives enough pollen for maximum seed production, more pollen may allow the mutant to choose its mates (i.e., selectively use the pollen it receives) and produce higher quality seeds. In theory, then, a higher rate of nectar production should increase our mutant's fitness by enhancing the quantity and quality of its seeds and by increasing the amount of pollen transported to other plants. However, because it costs a plant energy and other resources to produce nectar, a higher rate of nectar production could also reduce the mutant's fitness. Assuming that nectar production is at an evolutionary equilibrium, we would predict that our mutant's increased nectar production will actually decrease its fitness, because departures from the population norm should not payoff at the evolutionary equilibrium. This suggests an optimization approach, because equilibrium nectar production should be at a local peak.

While the recipe outlined seems straightforward, using it to build and test a model of nectar production presents significant practical difficulties. Determining the source and the ultimate destination of transported pollen grains is technically challenging and labor intensive; and it follows that determining how pollinator behavior influences pollen flow just compounds this difficulty. Assessing how variation of pollen flow affects seed quantity is experimentally difficult and time consuming. Assessing the consequences of variation of pollen flow for seed quality requires even more time because seed quality varies along many dimensions (e.g., likelihood of germination, rate of growth, subsequent reproduction) and measuring these dimensions requires sowing seeds and growing them to maturity. Determining the costs of nectar production is also difficult because it is difficult and time consuming to measure plant-level differences in nectar production and difficult to translate such differences into differences in fitness. In addition, we need large sample sizes to separate variation in nectar production from other plant traits because it covaries with other plant traits.

Given these complexities and associated difficulties, it is not surprising that efforts to develop and test optimality models of co-evolution have moved slowly. Moreover, the conceptual and practical difficulties we face in the analysis of pollination reflect more general problems that plague the analysis of co-evolution generally. The next paragraph discusses one study designed to consider pollinator-plant co-evolution, and it illustrates the difficulties and potential of this approach.

Hummingbirds and Scarlet Gilia: A Worked Example

In the rocky mountains of Colorado, hummingbirds collect nectar from the flowers of scarlet gilia (*Ipomopsis aggregata*, see **Figure 2**). In the late 1970s, I studied this system with the goal of explaining both hummingbird foraging behavior and the level of nectar production per flower. As expected, measurements show that flowers on a single plant offer similar nectar rewards (see **Figure 3**). So one might expect that hummingbirds will probe a second flower, if they discover a rich flower, since this means that other flowers on the plant are likely to be similarly rich. I expressed this idea mathematically by assuming that birds use a threshold: visiting a second flower on a given plant if they visit a flower that offers more than this threshold and leaving the plant to find another if they obtain less than the threshold. Using optimality reasoning, I determined the threshold that maximized the birds' net rate of energy intake. Observations of hummingbird departures from plants supported this threshold model. I then used this model to ask how changes in nectar reward would change hummingbird behavior (**Figure 4**).

Figure 2   Scarlet gilia (*Ipomopsis aggregata*).

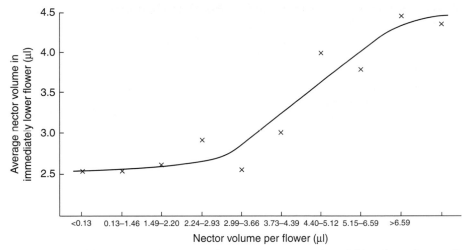

Figure 3 Average nectar volume (μl) in the immediately lower flower versus nectar volume (μl) in a flower (grouped into intervals), with line fitted by eye. Observations on *Ipomopsis aggregata* in vicinity of Rocky Mountains Biological Laboratory, Colorado. Reproduced from Pyke GH (1978) Optimal foraging in hummingbirds: Testing the marginal value theorem. *American Zoologist* 18: 627–640.

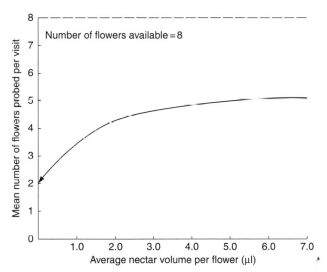

Figure 4 Predicted average number of flowers probed per hummingbird visit to an *Ipomopsis aggretata* plant versus average nectar volume per flower. Reproduced from Pyke GH (1981) Optimal nectar production in a hummingbird-pollinated plant. *Theoretical Population Biology* 20: 326–343.

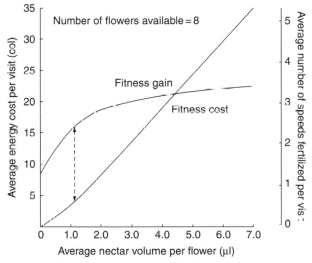

Figure 5 Average energy cost (cal) and average number of seeds fertilized (i.e., fitness gain) per hummingbird visit to an *Ipomopsis aggregata* plant. The scales for each correspond on the basis of the average energy content per seed. Reproduced from Pyke GH (1981) Optimal nectar production in a hummingbird-pollinated plant. *Theoretical Population Biology* 20: 326–343.

This is first step in considering the plant side of the coevolutionary problem, because one could use this information to draw inferences about the fitness of a mutant plant that offers more or less nectar than average. The second step is to ask how changes in hummingbird behavior influence pollen flow. To assess this, I used a stuffed hummingbird to simulate the process of probing a sequence of flowers and measured seed production in the experimentally probed flowers. This technique showed that the number of flowers probed per visit did not affect seed production (i.e., female reproduction), apparently because the plants already received more than enough pollen to fertilize their seeds. Yet, the same technique showed that more flower visits did increase male reproduction. A plant that experienced more flower visits, transferred more pollen to other plants, and 'fathered' more seeds. This suggests that a mutant offering more nectar would primarily benefit via enhanced male function. It might be costly to produce more nectar, and this presents a logical problem because we must express costs of nectar production and the fitness value of 'fathering' more seeds in the same currency. To solve this problem, I expressed both the costs and the benefits in terms of energy (see **Figure 5**). Using this approach, I calculated that the optimal average nectar volume (per flower) for a plant is about 1.1–1.2 μl. The observed nectar volume per flower was consistent with (i.e., not significantly different from) this. And it follows that the observed nectar production (per flower per day) also agreed with predictions. Hence, both the observed foraging behavior of the hummingbirds and the observed nectar production per flower were consistent with an evolutionarily stable strategy.

See also: Co-Evolution of Predators and Prey; Cost-Benefit Analysis; Game Theory; Optimal Foraging Theory: Introduction; Patch Exploitation.

Further Reading

Harder LD and Johnson SD (2009) Darwin's beautiful contrivances: Evolutionary and functional evidence for floral adaptation. *New Phytologist* 183: 530–545.

McGill BJ and Brown JS (2007) Evolutionary game theory and adaptive dynamics of continuous traits. *Annual Review of Ecology Evolution and Systematics* 38: 403–435.

Pyke GH (1978) Optimal foraging in bumble bees and co-evolution with their plants. *Oecologia* 36: 281–294.

Pyke GH (1978) Optimal foraging in hummingbirds: Testing the marginal value theorem. *American Zoologist* 18: 627–640.

Pyke GH (1980) Optimal foraging in nectar-feeding birds and coevolution with their plants. In: Kamil AC and Sargent SD (eds.) *Foraging Behaviour.* New York, NY: Garland Press.

Pyke GH (1981) Optimal nectar production in a hummingbird pollinated plant. *Theoretical Population Biology* 20: 326–343.

Optimal Foraging Theory: Introduction

G. H. Pyke, Australian Museum, Sydney, NSW, Australia; Macquarie University, North Ryde, NSW, Australia

Introduction

Foraging, which is the process by which animals obtain food, is a fundamental activity for animals. Animals require food to sustain their metabolism, provide energy for a wide range of activities, and support reproduction. In some situations, foraging occupies a high proportion of available time, and since animals often cannot do two things at once, increasing the time spent on foraging may reduce the time available for other activities such as mating, resource defense, and predator avoidance. Optimal foraging theory is an approach to the study of foraging behavior that uses the techniques of mathematical optimization to make predictions about this critical aspect of animal behavior.

Optimal Foraging: The Classic Models

Consider a hummingbird drinking nectar from flowers. When our hummingbird arrives at a fresh flower it obtains nectar quickly, but as it spends more time the nectar becomes harder to obtain because the hummingbird has depleted the supply. Most food patches work this way. Fresh patches provide food quickly, but the rate of intake declines as the forager depletes the patch. This simple observation presents a foraging problem, because it takes time and energy to move to a fresh patch. How long should the forager spend exploiting a patch before it moves to a fresh one? This is one of the classical problems of foraging theory. **Figure 1** shows the idea of patch depletion graphically: the amount of energy extracted from the patch increases with the time an animal spends in the patch, but the instantaneous rate of food gain (given by the slope of this function) declines steadily; so this *gain function* increases but bends down. Now, it takes T units of time for the animal to travel from one patch to another; and t is the time the animal spends in each patch. The classic patch model finds the patch time, t, which gives the highest rate of food intake. **Figure 1** shows how we can find this 'optimal patch time' graphically. The slope of the line that connects the point $-T$ on the x-axis (which is the time axis) to the point $(t, e[t])$ on the gain function gives the rate of energy intake an animal can expect if it spends time t. A bit of reflection shows that the highest slope (and hence the maximal intake rate) occurs at time t_{opt1} when the line is tangent to the gain function. This simple graphical approach predicts that foragers should stay longer when it takes longer to travel to fresh patches. Compare t_{opt1} and

t_{opt2}, which correspond to short (T) and long $(4T)$ travel times in **Figure 1**. A number of early empirical studies support this prediction qualitatively. The theoretical and empirical results of patch exploitation are reviewed elsewhere in this encyclopedia.

We can view foraging behavior as the outcomes of a set of decisions. As described earlier, an animal can decide whether to stay in a patch or leave it. Foraging animals make many other types of decisions, of course. For example, they decide what types of food to eat; and they decide where and when to search for food. These decisions result in the foraging behavior that we observe.

We can understand these behavioral decisions if we can explain and predict them in terms of underlying processes, and our understanding will be greater the more quantitative (rather than qualitative) we can be in matching predictions and observations. We might, for example, assume that an animal can determine its average energy yields and its handling times associated with consuming various potential food types when encountered, as well as the average time it spends between successive food items. Then, based on our own measurements of these variables, we could predict which food types a forager should include in its diet. The extent to which our observations match our predictions would indicate how well we understand the forager's dietary decisions. Optimal foraging theory seeks to understand foraging behavior in this way.

At the most fundamental level, optimal foraging theory assumes that foraging decisions have evolved, and, consequently that the fitness associated with the foraging behavior of an individual animal has been maximized; hence, the underlying processes are 'optimal.' We can therefore use the mathematical machinery of optimization to critically formulate our predictions about foraging behavior. To apply this logic, optimal foraging models must first describe the foraging process mathematically. In the patch model described earlier, for example, our description includes the travel time, the gain function, and the assumption that the forager can choose a range of patch residence times. The next step in the process is to calculate how changes in patch residence time affect the forager's evolutionary fitness.

Ideally, we would find the fitness, measured in terms of offspring production, associated with a given patch residence time, but this is quite difficult in practice. Instead, foraging models maximize a currency that acts as a proxy for fitness. Classical foraging models (like the patch exploitation model outlined earlier) use the currency of maximization of long-term rate of net energy gain.

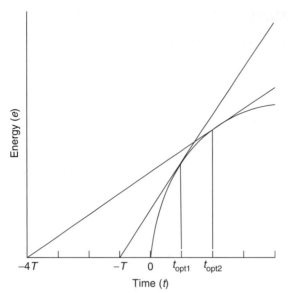

Figure 1 Graphical representation of optimal departure rule.

Rate maximization has a prominent place in foraging theory, but investigators also use other currencies, such as minimizing the probability of starvation, or mortality rate. To take one example, a model of herbivore foraging might include food contents other than energy – such as nutrients and toxins – in the currency of maximization because plant tissues vary dramatically in quality and constituents.

Optimal foraging models can and do take many forms. Models can differ in the behavioral decision they consider (patch use, prey choice, habitat use), and they differ in how they model the environment (e.g., sequential encounter with resources vs. simultaneous encounter) and in which currency they maximize (e.g., rate of net energy intake vs. probability of survival). Notwithstanding this diversity, we recognize a classic set of foraging models that are important because they serve as the starting point for further development. These classical models consider two basic decision problems (patch use and prey choice) using the currency of rate maximizing and assuming that the forager encounters resources (prey or patches) sequentially. These classic models recognize that, in deciding to do something, an animal may forgo other choices and, in this sense, a forager will typically pay an 'opportunity cost' when it chooses one action instead of another. The idea of opportunity cost is a central feature of many optimal foraging models.

Beyond the Classic Models

Investigators have extended and improved the classical models in many ways. For example, a fairly large family of models considers tradeoffs between foraging and other aspects of behavior. The best location for foraging might,

for example, be the worst location in terms of the risk of predation. Tradeoffs between foraging and predation risk have been the focus of many recent theoretical and empirical studies. To take another example, the classical models assume that the forager's behavior is tuned to environmental conditions as if it has perfect information about the properties of the environment such as prey quality or encounter rates. Realistically, however, variables like these will often change, and a forager will need to adjust its behavior in response to these changes. Several models have considered the problems of 'incompletely informed foragers.' Commonly, these models make assumptions about how the environment varies, and consider how experience and information acquisition should influence foraging decisions. This approach, therefore, provides an important bridge to other aspects of animal behavior such as learning, cognition, and decision making.

Another important trend is the development of so-called dynamic foraging models. In the classical models, we imagine that the animal adopts, for example, a fixed patch residence time that represents the single best choice. Dynamic optimization models suppose, instead, that the best patch residence might change as the animal's state (e.g., it's hunger) changes. Instead of predicting a single optimal choice, dynamic models predict an optimal decision trajectory that predicts how decisions might change over the course of a day, and how this change covaries with a state variable like hunger.

A fascinating recent development is the extension of the optimal foraging approach to phenomena outside the realm of animal feeding behavior. Engineers, computer scientists, sociologists, anthropologists, psychologists, and neuroscientists have all adapted foraging models for their purposes. To give some specific examples, investigators have adapted foraging models to consider criminal behavior (how long a burglar remains in a house collecting things to steal before he/she moves on to another house), military search strategies, and how human computer users distribute their time among various web sites.

Growth and Prognosis

The optimal foraging approach has also grown enormously in terms of numbers of publications and it continues to grow (see **Figure 2**). Beginning in the mid-1960s, the annual number of publications considering foraging theory grew exponentially, especially during the late 1970s. Since then, this annual rate has grown steadily but much more slowly. Unlike many other areas of research, optimal foraging theory has not yet begun to show a decline in publication rate.

Optimal foraging theory has survived a number of criticisms and passed all the reasonable tests that one could apply to any theoretical approach. Some have

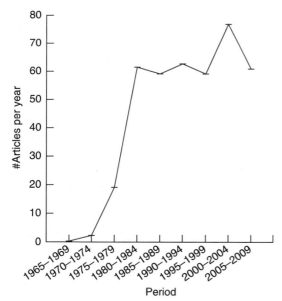

Figure 2 Number of scientific articles relating to *Optimal Foraging Theory* published per year versus 5-year period.

criticized it for being on overly simplistic and unrealistic; but most significant conceptual paradigms develop iteratively, improving assumptions and refining models as new data comes to light. Some critics argue natural selection has not had enough time to optimize foraging behavior. For others, the premise of optimization is valuable and justified, because behavior can evolve relatively rapidly, and because it has manifestly improved our understanding of animal foraging behavior. Other critics point to quantitative disagreements between expectations and observations and

pronounce the theory dead. In contrast, proponents point to consistent qualitative agreement and reasonable (but more modest) quantitative agreements with the theory.

Investigators have used ideas from optimal foraging theory in several other areas of biology. Ecologists have, for example, used the theory to predict (1) how food density affects consumer behavior (via the so-called functional response), (2) population dynamics of foraging animals, and (3) species coexistence. It has also had a major impact on the area of psychology through its involvement with issues such as learning, memory, and decision making. Optimal foraging theory has therefore demonstrated its usefulness and emerged as a strong theory of behavior and ecology.

See also: Cost-Benefit Analysis; Ecology of Fear; Habitat Selection; Optimal Foraging and Plant–Pollinator Co-Evolution; Patch Exploitation.

Further Reading

Pyke GH (1984) Optimal foraging theory: A critical review. *Annual Review of Ecology and Systematics* 15: 523–575.

Pyke GH, Pulliam HR, and Charnov EL (1977) Optimal foraging: A selective review of theory and tests. *Quarterly Review of Biology* 52: 137–154.

Sih A and Christensen B (2001) Optimal diet theory: When does it work, and when and why does it fail? *Animal Behaviour* 61: 379–390.

Stephens DW, Brown JS, and Ydenberg RC (eds.) (2007) *Foraging Behavior and Ecology.* Chicago & London: University of Chicago Press.

Stephens DW and Krebs JR (1986) *Foraging Theory.* Princeton, NJ: Princeton University Press.

Orthopteran Behavioral Genetics

Y. M. Parsons, La Trobe University, Bundoora, VIC, Australia

Introduction

The orthopteran order of insects includes the grasshoppers, crickets, and their relatives. The order comprises two suborders: Caelifera (grasshoppers, locusts, and mole crickets) and Ensifera (true crickets, katydids, and bush crickets). Members belonging to Caelifera have short antennae and abdominal tympanal organs (ears), whereas antennae of the Ensifera reach at least to their abdomen and their ears are located in their fore tibia (front legs). Orthopteran insects have a worldwide distribution and can generally be recognized by their ability to produce sound. The ability to sing is a fascinating feature of orthopteran insects and has formed the basis of a great deal of behavioral research. Orthopteran insects also display other behaviors such as those involved in defense, camouflage, and temperature control, but the most studied behaviors have been those associated with singing and, to a lesser extent, swarming: singing due to its relevance as a model system of stereotypic behavior, courtship behavior, and reproductive isolation; and swarming due to its relevance as an extreme example of phenotypic plasticity and because of the economic damage caused by locust plagues. Both cricket singing and locust swarming represent complex phenomena that are often difficult to analyze, but the study of these traits in orthoptera has led to a detailed understanding of the underlying mechanics and neurophysiology that has enabled targeted genetic analysis. Behavioral variation is driven by variation at both the genetic and the environmental level, and elucidating the contribution of these to acoustic and swarming behavior is the focus of this article.

Genetics of Acoustic Behavior

For me, the sound of crickets is inextricably linked with long summer evenings. Crickets can and do sing at other times of the day and in other seasons, but for those of us living in urban areas noise levels often have to drop appreciably before the chirping becomes noticeable. Why do crickets and grasshoppers sing with such monotonous regularity? Do they have ears? Is singing a learned behavior? These and similar questions have stimulated a remarkable number of investigations over many years and valuable insight into the neuroethology of orthopterans and neurobiology in general. Cricket song, in particular, has been extensively studied as a model system of highly stereotypical repetitive behavior led

by Franz Huber, the founding father of cricket neurobiology. Such rhythmic behaviors are found in many organismic processes (digestion, heartbeat, respiration, locomotion, etc.), and studies of simple model systems have led to an understanding of the principles underlying rhythmic motor pattern generation.

The song of both crickets and grasshoppers is a sequence of sound pulses generated by rubbing specialized structures (e.g., stridulatory pegs). There are a number of variations, but in many crickets, these elements (toothed file and scraper) are located on the forewings, while most grasshoppers have stridulatory pegs on the hind legs that they rub over their forewing or against their abdomen. Stridulation is a behavior similar to respiration and flight, with repetitive contracture of antagonistic and synergistic muscles under the control of a small network of neurons (rhythmic motor generators) that coordinate the required contraction and relaxation of opposing groups of muscles. An interesting aspect of these oscillators is that once triggered the rhythm will continue without sensory input. Sensory feedback, although not necessary for basic rhythmic output, is involved in cueing the control center that triggers the pattern generator. In the cricket, the song pattern generator has been localized to the thoracic ganglia and the control center higher up in the neural circuitry.

Cricket song represents a complex communication system whereby males and females identify members of their own species, females locate potential mates, and males advertise their presence to ward off potential competitors. The repertoire of cricket songs was classified by Alexander in 1962 into three main types: calling, courtship, and aggressive. Calling song serves both to attract females and advertise a male's presence (males call from a stationary location and females approach). Courtship song is elicited after males and females have come together, and facilitates copulation. Males sing an aggressive song during territorial fights.

The songs of crickets and grasshoppers have also been extensively studied due to their potential importance in species formation. Many song patterns are species specific, especially where different species occur in the same locality. Females recognize the song of their conspecifics, and hence the song represents an important premating reproductive isolating barrier. Variation in song is therefore linked to the speciation process itself. In grasshoppers, naturally occurring hybrids exist where variation in song structure and morphology has provided insight into the evolutionary processes underlying the formation, extent, and duration of hybrid zones. In crickets, investigation of male mating song and female preference variation in closely related species has provided information on the evolutionary genetics of mating behavior.

How Much of Acoustic Behavior Is Under Genetic Control?

The song of crickets can be affected by the environment, particularly temperature, whereby the rate of calling varies in a linear fashion relative to the ambient temperature. However, song is stereotypic and not a learned behavior; in other words, under genetic control. Bentley and Hoy showed this to be the case by crossing different species with distinctive song patterns and experimentally changing the environmental and genetic input of the hybrids produces. They observed that hybrids always produced song with characteristics intermediate to that of the parents (e.g., the number of pulses/trill and the number of trills/phase; **Figure 1**) regardless of the environmental changes. Not all song characters observed display this pattern of inheritance. Indeed, some song characters did not vary between parents or hybrids, whereas some hybrid characters were more similar to that of the male of the maternal parent (only male crickets sing) indicating the involvement of X-linked genes (female Orthoptera are XX whereas males have a single X).

The pattern of inheritance of intermediate characters in hybrids indicated a polygenic system with the involvement of a number of genes. This has been illustrated in a number of crossing experiments including those involving the Hawaiian cricket *Laupala* by Shaw in 1996. The male courtship song of these small, flightless crickets is important in species identification and is believed to be central

in maintaining reproductive isolation between sympatric species. Shaw and colleagues have studied the molecular and behavioral aspects of the group and found that these crickets are undergoing an extremely rapid rate of speciation, very similar to that of the African cichlid fish. The *Laupala* cricket song has a very simple structure and one feature, the pulse rate, distinguishes different species (**Figure 2**). Shaw crossed two closely related species endemic to the big island of Hawaii, *Laupala kohalensis* and *Laupala paranigra*, to produce F_1 and F_2 hybrid generations and analyzed the inheritance of the pulse rate (**Figure 3**). The intermediate nature of the pulse rate of the F_1 generation as well as the wider variation in pulse rates observed in the F_2 generation support a polygenic model. If this characteristic was controlled by only one gene, the pattern would be similar to the results Mendel obtained in his famous pea experiments where the F_1 offspring all resembled just one of the parents, and only forms resembling one or other parent (i.e., no intermediates) were recovered in the F_2 generation. When one considers that singing behavior requires coordination of various morphological features as well as neuronal input, it is not surprising that a number of genes are involved. This is not to say that all the genes involved vary in all species, but some of them do and these have no doubt been important in leading to the difference we observe between species and hence, speciation itself. Natural selection as well as sexual selection has been implicated in the evolution of genetic variation underlying acoustic behavior – selection can act on morphological as well as the behavioral features and hence we have the immense variation we hear.

How Much Genetic Variation in Calling Song Is Present?

Male-calling song is an important feature of mate recognition in crickets and grasshoppers and, as such, there is considerable variation between species and significant species level differences have been identified in various taxa. Indeed, cryptic species complexes can be distinguished based on species specific calling songs as in *Laupala* and other cricket and grasshopper species as well. This suggests then that the level of genetic variation between species is quite considerable. Evidence indicates that song characters are broadly polygenic and that genes on the sex chromosomes are involved. Studies in both crickets and grasshoppers have indicated a polygenic mode of inheritance, and heritability estimates for calling-song parameters indicate that heritable genetic variation is reasonably substantial.

Female Preference

Breeding experiments have demonstrated that female preference is also under genetic control and that similar

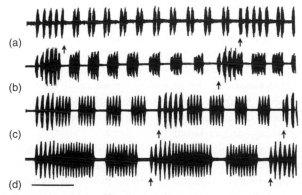

Figure 1 Calling-song pulse patterns of *Teleogryllus* species and F1 hybrids. (a) *T. oceanicus*; (b) *T. oceanicus* ♀ × *T. commodus* ♂ F1 hybrid; (c) *T. commodus* ♀ × *T. oceanics* ♂ F1 hybrid; and (d) *T. commodus*. Each trace starts at the beginning of a phrase and arrows indicate the beginning of subsequent phases. Reproduced from Bentley DR and Hoy RR (1972) Genetic control of the neuronal network generating cricket (*Teleogryllus gryllus*) song patterns. *Animal Behavior* 20: 478–492, with permission from Elsevier.

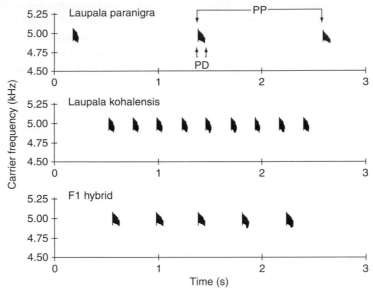

Figure 2 Sonograms of the male-calling song of *Laupala paranigra*, *L. kohalensis*, and an F1 hybrid. PP, pulse period; PD, pulse duration. Reproduced from Shaw KL (1996) Polygenic inheritance of a behavioral phenotype: Interspecific genetics of song in the Hawaiian cricket genus *Laupala*. *Evolution* 50: 256–266.

genes could be involved. Early studies demonstrated that hybrid females preferred the song of hybrid males over either parents suggesting the involvement of pleiotropic gene(s) affecting both song and preference. Association between male-calling song and female preference at the genetic level was thought to be necessary to ensure that they remain synchronous across evolutionary change. However, more recent studies have shown that this may not be the case.

Identifying Genes Underlying Acoustic Behavior

Despite the wealth of information available on the mechanics and neurophysiology of cricket mating behavior, little is known about the underlying molecular genetics. Crickets proved an ideal model for investigating neurophysiological aspects and crossing studies demonstrated the polygenic nature of cricket song and the potential involvement of the X chromosome, but identifying the genes responsible has proven more difficult. *Drosophila* is one of the most commonly used models for genetic studies due to the combination of small genome size, tractable chromosome number, short generation time, and ease of laboratory rearing. Generation time in crickets, on the other hand, varies from 6 weeks to several months, while genome size is considerably larger and chromosome numbers also vary considerably. Progress has been made, however, by following a quantitative trait loci (QTL) mapping approach in the Hawaiian cricket. The simple song structure in males together with the rapid rate of speciation observed makes *Laupala* an ideal system for studying the genetic basis of acoustic behavior. As this behavior is central to conferring reproductive isolation,

it provides a model for identifying genes involved in speciation, the holy grail of evolutionary biology.

Quantitative traits are characteristics that are measured, such as height and weight, that generally follow a bell-shaped distribution when measured in a population such as that seen in *Laupala* populations (**Figure 3**). Variation between individuals in these traits (phenotypic variation) is due to genetic variation in the genes that are involved in specifying the trait. QTL represent the genomic regions that contain one or more of these genes and can be mapped through crossing individuals that vary in the trait of interest and analyzing their offspring for association between specific gene markers and phenotypic variation. To identify QTL at the molecular level, two approaches are possible. The first involves the development of a linkage map using DNA markers and computer-based analysis to identify the number and location of genomic regions involved in phenotypic variation. The second is a random or 'candidate' approach. As the *Laupala* genome is relatively unknown, both approaches are appropriate to search for QTL effecting mating-song variation.

Mapping studies in replicate populations have identified eight QTL in *Laupala* that account for more than half of the genetic variation underlying pulse rate variation. Other QTL that were not identified would account for the remaining genetic variation. QTL were mapped to the X and autosomal linkage groups that correspond to six of the eight chromosomes in *Laupala*. The results are consistent with a model whereby pulse rate phenotype has diverged through substitution of alleles of small-to-moderate effect in many genes under direction selection. The results of these studies have helped to shed light on

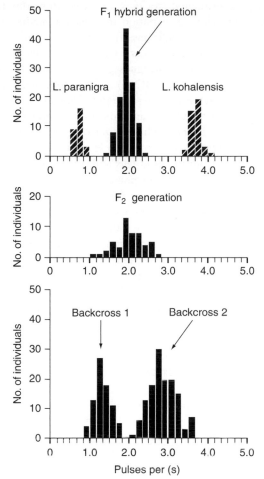

Figure 3 Histogram of pulse rates from *Laupala paranigra* and *L. kohalensis* parental populations F1, F2, and backcross hybrid populations. Reproduced from Shaw KL (1996) Polygenic inheritance of a behavioral phenotype: Interspecific genetics of song in the Hawaiian cricket genus *Laupala*. *Evolution* 50: 256–266.

the evolutionary mechanisms involved in reproductive isolation and the speciation process. They confirm the role of sex chromosomes, indicate that male song has evolved via directional selection, and suggest that the substitution of alleles of moderate effect in genes underlying behavioral variation can lead to species divergence.

Candidate Gene Approach

A 'candidate' gene is one for which the function suggests it may play a role in the variation associated with a particular trait. Candidate genes for cricket singing include those that function in the signaling mechanism of the neuron, the basic component of behavior. All animal behavior is ultimately the result of information transfer between and within nerve cells that occurs via electrical signals generated by the flow of inorganic ions across the cell

Table 1 Voltage-gated ion-channel genes

Drosophila locus	Gene	Homologs isolated
Cacophony	Calcium-channel a1 subunit	Mammal, nematode, mollusk, rodent
Slowpoke	Calcium-activated K⁺ channel	Mammal, nematode, turtle, rodent
Seizure	Potassium channel	Mammal, rodent
Paralytic	Sodium channel	Mammal, rodent, mollusk, cnidaria
Shaker	Potassium channel	Mammal, rodent, mollusk, crustacean

membrane. Ion flow is controlled by ion channels, a ubiquitous class of proteins that span the cell membrane forming pores through which ions flow. These proteins play critical roles in the propagation of action potentials and modulation of membrane potential, and mutations in ion-channel genes are associated with neurological defects, such as ataxia, cardiac arrhythmia, migraine, epilepsy, and myotonia. Gene family members share sequence, structural and functional properties and display significant evolutionary conservation making them ideal genes to investigate in diverse taxa such as *Laupala* for which genome sequence is not available (**Table 1**).

Evidence implicating ion-channel genes in mating-song variation in *Laupala* comes from *Drosophila* research where courtship song aberrations were found to be due to mutations in two ion-channel genes, *cacophony* and *slowpoke*. The *cacophony* gene, in particular, is a very strong candidate for effecting variation in *Laupala* cricket song for four reasons: (1) the song of *cacophony* males exhibited differences in song parameters that have been observed among different *Drosophila* species suggesting a role in naturally occurring variation; (2) the gene has been mapped in *Drosophila* to the X-chromosome and the X-chromosome has been implicated in pulse rate song variation in *Laupala*; (3) the mutations are not physiologically deleterious suggesting that genetic variation does not significantly reduce the fitness of the individual and (4) *cacophony* mutants exhibited longer interpulse intervals (analogous to pulse period, the inverse of pulse rate in *Laupala*) compared to wild-type flies. Using sequence information from other organisms, coding and noncoding regions of the putative *Laupala* homolog of *cacophony* has been isolated and sequenced. No species-specific variation at the molecular level has yet been found, however, which suggests that either more *cacophony* gene regions need to be investigated or that this gene is not a candidate for controlling phenotypic variation in pulse rate in *Laupala* crickets. Given the polygenic nature of variation in cricket song, however, there are certainly many more potential candidate genes to investigate.

Gregarious Behavior in Locusts

Why do reasonably well-behaved solitary grasshoppers change into gregarious voracious migratory swarms? This spectacular phenomenon, peculiar to a group of grasshoppers known as 'locusts,' has been the subject of intense research for many decades. When locusts are subject to increased population density, they undergo an extreme phase shift in behavior, coloration, morphology, development, and endocrine physiology that can result in sweeping plagues causing widespread devastation. Density-dependent phase transition occurs in other insects including moths, beetles, and aphids, but locusts alone display the ability to gregarize, changing from the low-density solitary to high-density gregarious phase. Indeed the terms 'gregaria' and 'solitaria' were coined by Uvarov in 1921 to describe the two extreme phases of *Locusta migratoria*.

Phase transition in locusts involves change in a suite of characters; however, the key aspect is the transition from the normal solitary to the gregarious behavior that is a prelude to other observable changes. This phenotypic change in behavior occurs relatively quickly, within only a few hours of increased population density, in contrast to the other changes, and it has recently been shown that increased levels of serotonin effect the behavioral shift. This work was conducted by Simpson and colleagues who have gained considerable insight into the stimuli and neurophysiological and ecological mechanisms involved in phase transition. Dense populations can develop as a result of environmental and biotic factors: seasonal winds lead to convergence of solitary individuals; restricted local vegetation encourages aggregation; precipitation and increased food encourage larger population sizes. Chemical communication contributes to maintenance of high local population density which ultimately triggers phase transition. Swarming occurs when the locusts maintain density during migratory flight. Migration is associated with reduced food quality and/or quantity, and hence the response to increased density reflects food resource availability.

Phase transition represents a radical example of phenotypic plasticity where a given genotype can produce different phenotypes in different environments. Transition therefore represents an environmental polymorphism rather than a genetic polymorphism, and full sib individuals with similar genetic makeup will develop solitary or gregarious behavior depending on rearing density. This does not mean that there is no genetic component, however. Plasticity is a function of the genotype and the co-coordinated changes are effected through selective expression of genes from within a constant genotype. This has been demonstrated quite convincingly by Kang and colleagues who conducted a large-scale gene expression analysis of phase change in a gregarious population of migratory locusts from China. Both gregarious and

solitary experimental stocks were raised from this field population, with the variation in behavior mediated by different rearing conditions: the gregarious culture at high density and the solitary culture as individuals in isolation. Kang and colleagues generated gene libraries from messenger RNA (mRNA) isolated from head, hind leg, and midgut tissue from representatives of both gregarious and solitary cultures and compared expression profiles between the two. They found over 500 genes that differed statistically in expression levels between the two phases; most from the hind legs and midgut were down-regulated in the gregarious phase, whereas several in the head were upregulated. The function of many of these expressed genes was determined and included genes, for example, that function in cell growth, carbohydrate metabolism, and neuromodulation providing some exciting insight on the molecular mechanisms of phase change and a huge repertoire of genes to investigate further. The majority of genes discovered in the study could not be assigned a function, however, due to lack of sequence homology to known genes in other insects. The order Orthoptera is an ancient lineage, having diverged over 300 Ma, so this is not surprising, but nonetheless highlights the need for similar large-scale genetic studies in other orthopteran taxa.

Grasshoppers that undergo phase transition represent only a small number of grasshopper species, and even within this small number of species some populations are less likely to form aggregations than others. This represents variation in the level of plasticity at the population level, another aspect of phase transition that has been investigated at the genetic level. In *L. migratoria*, populations are identified historically as 'outbreaking' or 'nonoutbreaking,' and laboratory experiments in a common environment were conducted by Chapuis and colleagues to compare parental effects on the expression of phase transition in offspring. They observed larger phase changes in offspring from the historically outbreaking population and concluded that this was due to genetic variation in the expression of parentally inherited gregarization. This indicates that although phase change itself is subject to environmental cues, genes control the level of phase change expression and evolution has resulted in differences between populations. Furthermore, their work may have important implications in assessing the dynamics of plague outbreaks and pest management control.

Concluding Remarks

The variation we observe in acoustic behavior in crickets is largely controlled by variation at the genetic level, whereas the phenotypic phase change that occurs in locusts is largely driven by the environment. Genetic variation is seemingly more important in specifying phenotypic variation in acoustic behavior than it is in behavioral phase change, but genes are nonetheless the ultimate basis of the differences we observe. As with all observable characteristics, both genetic and environmental factors play a role, and information on the contribution of each is critical to understanding the evolution of complex behavior.

Acknowledgments

The author thanks the two anonymous reviewers who provided valuable comments on an earlier version of this review. Research investigating candidate genes for acoustic behavior in Hawaiian crickets was supported by the Radcliffe Institute for Advanced Study, Harvard University.

See also: *Drosophila* Behavior Genetics; Genes and Genomic Searches.

Further Reading

Alexander RD (1962) Evolutionary change in cricket acoustical communication. *Evolution* 16: 443–467.

Anstey ML, Rogers SM, Ott SR, Burrows M, and Simpson SJ (2009) Serotonin mediates behavioral gregarization underlying swarm formation in desert locusts. *Science* 323: 627–630.

Applebaum SW and Heifetz Y (1999) Density-dependent physiological phase in insects. *Annual Review of Entomology* 44: 317–341.

Bentley DR and Hoy RR (1972) Genetic control of the neuronal network generating cricket (*Teleogryllus gryllus*) song patterns. *Animal Behavior* 20: 478–492.

Butlin RK and Hewitt GM (1988) Genetics of behavioral and morphological differences between parapatric subspecies of *Chorthippus parallelus* (Orthoptera: Acrididae). *Biological Journal of the Linnean Society* 33: 235–246.

Chapuis M-P, Estoup A, Auge-Sabatier A, Foucart A, Lecoq M, and Michalakis Y (2008) Genetic variation for parental effects on the propensity to gregarise in *Locusta migratoria*. *BMC Evolutionary Biology* 8: 37.

Ewing AW (1989) *Arthropod Bioacoustics: Neurobiology and Behavior*. Ithaca, NY: Cornell University Press.

Hoy RR (1974) Genetic control of acoustic behavior in crickets. *American Zoologist* 14: 1067–1080.

Huber F, Moore TE, and Loher W (1991) *Cricket Behavior and Neurobiology*. Ithaca, NY: Cornell University Press.

Kang L, Chen XY, Zhou Y, et al. (2004) The analysis of large-scale gene expression correlated to the phase changes of the migratory locust. *Proceedings of the National Academy of Science USA* 101: 17611–17615.

Parsons YM and Shaw KL (2002) Mapping unexplored genomes: A genetic linkage map of the Hawaiian cricket *Laupala*. *Genetics* 162: 1275–1282.

Shaw KL (1996) Polygenic inheritance of a behavioral phenotype: Interspecific genetics of song in the Hawaiian cricket genus *Laupala*. *Evolution* 50: 256–266.

Shaw KL and Parsons YM (2002) Divergence of mate recognition and its consequences for genetic architectures of speciation. *American Naturalist* 159: S61–S75.

Shaw KL, Parsons YM, and Lesnick SC (2007) QTL analysis of a rapidly evolving speciation phenotype in the Hawaiian cricket *Laupala*. *Molecular Ecology* 16: 2879–2892.

Simpson SJ and Sword GA (2008) Locusts. *Current Biology* 18: R364–R366.

Walker TJ (1962) Factors responsible for intraspecific variation in the calling songs of crickets. *Evolution* 16: 407–428.

P

Pair-Bonding, Mating Systems and Hormones

W. Goymann, Max Planck Institute for Ornithology, Seewiesen, Germany

Introduction

Reproductive relationships between individuals can be described as social mating systems, which are often related to parental care patterns. A broad classification of social mating systems includes: (1) Monogamy, in which one male and one female express a partner preference, which leads to the formation and maintenance of either a temporary or a permanent pair bond. In this case, neither sex monopolizes additional members of the opposite sex. (2) Polygyny, in which individual males control or gain access to multiple females. (3) Polyandry, in which individual females control or gain access to multiple males. In the latter two cases, partner preference, pair-bond formation, and maintenance are relaxed or at least biased toward one sex. Finally, there is (4) promiscuity, in which neither females nor males control access to members of the opposite sex and each individual may mate with multiple partners. In promiscuous species, pair bond formation is absent.

After the development of genetic parentage analyses in the 1980s, it soon became clear that these social mating systems are built upon genetic mating systems with various degrees of extra-pair fertilizations. As a consequence, many more females produce offspring with multiple males than previously suspected. In other words, genetic mating systems describe who produces how many offspring with whom and with how many others. Genetic mating systems can also explain alternative mating tactics, such as those of satellite or sneaker males, who attempt to fertilize females visiting the territories of males.

Why should the circulating concentration of reproductive hormones be related to a particular mating system? Several comparative studies have shown that testis size is related to the social mating system or sperm competition across vertebrate and nonvertebrate species, suggesting that high levels of testosterone – the major steroid involved in male vertebrate reproduction – are associated with high levels of sperm competition. But large testes in

vertebrates do not necessarily and/or continuously produce large amounts of testosterone, although a recent comparative study suggests a slightly positive relationship between testis size and seasonal testosterone peaks in birds. What is the evidence for a relationship between mating systems and testosterone?

The Challenge Hypothesis and Comparative Evidence Relating Testosterone and Mating Systems

When discussing mating systems and testosterone, it is important to distinguish between circulating testosterone concentrations and androgen responsiveness, which is the ability of an individual to increase testosterone secretion from breeding baseline levels to the physiological maximum. A major step in understanding the relationship between hormones, mating systems and behavior was achieved through the Challenge Hypothesis by John Wingfield and colleagues. This hypothesis represents one of the first formalized attempts to explain the huge variation in plasma testosterone levels and focuses on the androgen responsiveness rather than absolute testosterone concentrations.

In principle, the Challenge Hypothesis is based on the observation that there is a bidirectional relationship between hormones and behavior: hormones influence behavior and behavior feeds back on hormone secretions. The Challenge Hypothesis assumes that elevations of circulating androgens above a certain level are for the most part associated with temporal variations in aggressive and, to a lesser extent, sexual behavior, rather than with basal reproductive physiology. Many studies have demonstrated rapid effects of social interactions on plasma concentrations of androgens (i.e., testosterone) in a wide array of vertebrate taxa, such as fish, amphibians, reptiles, birds, and mammals including humans. In line with these data, seasonally breeding birds with a high degree of male–male competition have high plasma androgen concentrations during periods of social instability and/or when females are receptive. In contrast, high concentrations of circulating androgens are virtually absent in species that do not compete for territories or mates.

More precisely, the Challenge Hypothesis postulates three (idealized) levels at which testosterone or other androgens are present in the circulation (**Figure 1**): First, there is a constitutive homoeostatic 'Level A' which represents the basal secretory activity of the Leydig cells during the nonbreeding season. This level is presumed to maintain feedback regulation of both gonadotropin-releasing hormone (GnRH) from the hypothalamus and gonadotropin release from the pituitary gland. Second, there is a regulated (periodic) breeding season baseline 'Level B,' which represents the constitutive secretory activity stimulated by environmental cues such as day length (**Figure 1**). Level B is considered sufficient for spermatogenesis, the development of secondary sexual characters and accessory organs, and the expression of reproductive behaviors. Levels A and B can be considered as the basic levels at which hormones influence behavior, morphology, and physiology. Finally, there is a maximum response 'Level C' that is considered to be achieved through social stimulation from competing males or via interactions with receptive females. Thus, Level C is induced by behavior feeding back on the secretion of the hormone. The increase of testosterone to Level C can be short or long

in duration, and small or great in magnitude. In contrast to the increase from Level A to Level B, which periodically occurs at the onset of the breeding season, the increase from Level B to Level C is considered facultative and may be expressed throughout all phases of the breeding life-cycle stage (**Figure 1**).

The three levels of testosterone release represent the first important cornerstone of the Challenge Hypothesis. The second important ingredient of this hypothesis is the observation that high levels of testosterone interfere with male parental care or result in other costs that should be avoided. But what do these idealized levels A, B, and C of testosterone and the interference of testosterone with paternal care imply for the relationship between testosterone and mating systems? The Challenge Hypothesis states that temporal patterns of plasma testosterone are the result of a trade-off between the degree to which male parental care is necessary for reproductive success and the

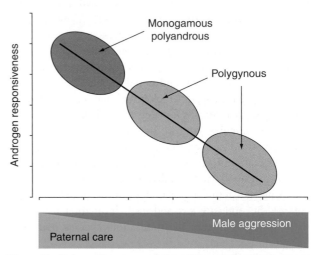

Figure 2 Schematic representation of the relationship between androgen responsiveness and mating systems in birds, as originally formulated in the Challenge Hypothesis: on the x-axis, species are listed according to a decrease in paternal care from left to right and an increase in the importance of male–male aggression from left to right. The y-axis represents the androgen responsiveness, that is, the magnitude of the difference between Level B and Level C testosterone concentrations. Socially monogamous and polyandrous species, in which males provide substantial amounts of paternal care, show a high level of androgen responsiveness. This means that they express low levels of testosterone (Level B) most of the time, but are capable of quickly increasing their testosterone concentrations to maximum (Level C) during brief periods of male–male encounters. In contrast, polygynous males that do not participate in parental care and frequently interact with competing males express high levels of testosterone close to Level C throughout the breeding season, thus showing a low level of androgen responsiveness. Polygynous males that provide paternal care show an intermediate pattern. Redrawn from Wingfield JC, Hegner RE, Dufty AM, and Ball GF (1990) The ''challenge hypothesis'': Theoretical implications for patterns of testosterone secretion, mating systems, and breeding strategies. *American Naturalist* 136: 829–846.

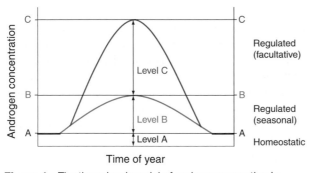

Figure 1 The three-level model of androgen secretion in seasonally breeding male birds: Level A represents the nonbreeding androgen baseline required for feedback regulation of GnRH and gonadotropin release. Level B represents the androgen baseline during breeding induced by environmental cues such as the increase in day length. Level B is sufficient for spermatogenesis and for the expression of secondary sexual characters and reproductive behaviors. Level C represents the physiological testosterone maximum that – in the original Challenge Hypothesis – can be achieved during interactions with other males or receptive females. The increase from Level A to Level B occurs seasonally at the onset of the breeding season, while the increase from Level B to Level C is facultative, that is, only triggered by social stimulation or challenge during the breeding season. Redrawn from Wingfield JC, Hegner RE, Dufty AM, and Ball GF (1990) The ''challenge hypothesis'': Theoretical implications for patterns of testosterone secretion, mating systems, and breeding strategies. *American Naturalist* 136: 829–846.

necessity or benefit of expressing aggressive behavior (**Figure** 2). The balance between costs and benefits of the effects of permanent high levels of testosterone is assumed to differ between monogamous and polygynous species. Socially monogamous species with a high degree of male parental care are predicted to show an increase in androgens from Level B to Level C (the androgen responsiveness) only during periods of territory establishment, during acute male-male challenges, or when females are fertile, so that paternal care is not compromised or other costs of permanent high levels of testosterone are avoided. In other words, these species express a large androgen responsiveness, because they show a strong rise in testosterone when challenged (**Figure** 2).

The same is true for classically polyandrous species in which males provide exclusive parental care. In contrast, androgen levels in polygynous species with little or no paternal care should be close to the maximum Level C throughout the breeding season because males continuously and intensely interact with each other and with receptive females. Hence, in such a competitive environment, the benefits of permanent high levels of testosterone may outweigh the associated costs. In other words, polygynous species express a low level of androgen responsiveness, because their testosterone levels are high throughout the breeding season. Polygynous species in which males contribute to parental duties at the nest, however, should show an intermediate level of androgen responsiveness between those two extremes. These predictions were confirmed in interspecific comparisons of seasonal androgen patterns in birds and fish, confirming the existence of a relationship between mating system and androgen responsiveness in these taxa. The data suggest that socially monogamous species show a larger difference between Level B and Level C testosterone concentrations than polygynous species. This difference seems to be based mainly on lower Level B values in socially monogamous than polygynous species rather than on variation of Level C testosterone concentrations (but see below). In these comparative studies, the influence of paternal care on the androgen responsiveness persisted only in passerine birds: in passerines, different androgen responsiveness probably evolved in passerines in response to changes in the male's paternal contribution during the incubation phase. Thus, it is possible that other costs (unrelated to paternal care) associated with high levels of testosterone drive this relationship between mating system and androgen responsiveness in other species. One direct and potentially costly effect of testosterone may be that permanently high levels may enhance the likelihood of escalating conflict behavior. This may lead to inappropriate expression of aggressive or risk-taking behavior and hence increase the risk of injury or predation.

Other studies looked at maximum testosterone concentrations rather than the androgen responsiveness. These studies found that polygynous birds appear to have higher testosterone peaks than socially monogamous birds. A better predictor, however, was the rate of extra-pair behavior: bird species with higher rates of extra-pair paternity expressed higher peak levels of testosterone. This fits with another observation in vertebrates, that – independent of mating system – high testosterone levels were strongly associated with high frequencies of sexual behavior, with the largest effect observed in nonpaternal vertebrates, in particular mammals. As more and more data become available on genetic mating systems and the degree of within-pair and extra-pair paternity, one may base future comparative analyses of testosterone concentrations or the androgen responsiveness on genetic rather than social mating systems. To do this properly, a combination of testosterone concentrations and genetic paternity data of individuals from the same study population would be needed. Thus, combined efforts of behavioral ecologists and behavioral endocrinologists to obtain such kind of data would certainly be helpful.

Recent findings in birds suggest that some refinements of the Challenge Hypothesis or its interpretation may be necessary in the future. First, a number of recent studies showed that testosterone does not universally suppress paternal behavior in all species. According to the Essential Paternal Care Hypothesis, males of some passerine bird species may have become insensitive to the suppressive effect of testosterone on paternal behavior. This behavioral insensitivity should mainly occur in species in which the survival of the offspring is severely hampered if the male stops feeding the young.

A second issue is related to a certain methodological discrepancy between the predictions and the support of the Challenge Hypothesis. The predictions of the Challenge Hypothesis, namely, the increase from Level B to Level C testosterone concentrations, relate to a situational increase of testosterone mainly during agonistic interactions between males, that is, a behavioral effect on testosterone secretion during social challenges. But support for the Challenge Hypothesis is mainly based on seasonal patterns of testosterone, that is, Level C concentrations of testosterone measured during the periods of territory establishment or mate guarding and Level B concentrations during the rest of the breeding season, assuming that the higher levels during territory establishment and mate guarding are the results of these social interactions. Unfortunately, quite a substantial number of species do not show the expected increase in testosterone from Level B to Level C during experimental inductions of situational male–male interactions. This discrepancy led to the distinction between the seasonal androgen response (R_{season}, based on seasonal profiles of Level B and Level C testosterone concentrations) and the androgen responsiveness to male–male interactions ($R_{male–male}$, based on testosterone concentrations measured during experimental inductions of territorial conflicts between males compared to those during control situations). Data based on R_{season} broadly support the predictions of the Challenge Hypothesis, but data based on situational $R_{male-male}$ are ambiguous. Future work has to show whether the lack of androgen responsiveness to male–male challenges ($R_{male-male}$) in many species and the discrepancy between R_{season} and $R_{male-male}$ is related to specific ecological factors, which then need to be incorporated into refinements of the Challenge Hypothesis. Alternatively, the seasonal androgen response (R_{season}) may not entirely be caused by social stimulation from competing males or receptive females. If so, additional and potentially intrinsic factors need to be considered and also implemented in refinements of the Challenge Hypothesis.

In summary, there is substantial evidence that androgen responsiveness based on seasonal testosterone profiles (R_{season}) is related to mating systems: males of socially monogamous species are more likely to show large androgen responses, whereas males of polygynous species are more likely to express small androgen responses. Very likely, the disparity between males of monogamous and polygynous species is founded on differences in the benefits and costs of the effect of persistent high levels of testosterone. Unfortunately, and in contrast to males, we know very little about the influence of sex steroids on mating decisions or strategies in females: greylag geese (*Anser anser*) are socially monogamous and form long-term pair bonds: pairs with a higher synchrony of each other's seasonal testosterone profile (the so-called testosterone compatibility) produce larger clutches, have heavier eggs and a higher life-time reproductive output than less synchronous pairs. Thus, in monogamous geese, pair-bond quality is not predicted by just measuring a physiological parameter in one partner. The physiology of both partners is needed to predict the quality of the pair bond. So far, it is unknown whether a high level of testosterone compatibility is the cause or a consequence of pair synchronization in geese.

Experimental Evidence for a Role of Testosterone as a Proximate Regulator of Mating Systems

So far, we looked at comparative data to investigate whether there is a relationship between testosterone concentrations and mating system. But is there experimental evidence that testosterone is a proximate factor for the expression of different mating strategies so that mating systems can be manipulated using testosterone treatment? Once again, most investigations have been done in birds. Testosterone implants increased the likelihood of males to become polygynous or show extra-pair behavior in song sparrows (*Melospiza melodia*), white-crowned sparrows (*Zonotrichia leucophrys*), dark-eyed juncos (*Junco hyemalis*),

starlings (*Sturnus vulgaris*), mallards (*Anas platyrhynchos*), and red grouse (*Lagopus lagopus*), but not in pied flycatchers (*Ficedula hypoleuca*), house finches (*Carpodacus mexicanus*), spotless starlings (*Sturnus unicolor*), red-winged blackbirds (*Agelaius phoeniceus*), and blue tits (*Cyanistes caeruleus*), although male blue tits with testosterone implants showed a greater interest in interacting with females other than the one they were paired with. A potential confound for studies that investigate the effect of testosterone on the genetic mating system is that treatment with testosterone may lead to a shutdown of the internal production of testosterone and sperm. This is a pharmacological effect that should be kept in mind when comparing the within- and extra-pair fertilization success of testosterone-treated males with controls.

Additional evidence for an effect of testosterone in the proximate control of mating strategies comes from male side-blotched lizards (*Uta stansburiana*) that exist in several color morphs related to mating tactics. Adult males with an orange throat have high levels of testosterone, are highly aggressive, and defend large territories that overlap with the territories of multiple females. Adult males with a blue throat have intermediate levels of testosterone, are less aggressive, and defend small territories overlapping with those of few females. Adult males with a yellow throat express low levels of testosterone, do not defend territories but mimic females, and sneak copulations. Throat color has a genetic basis and is influenced by hormone levels during development. But blue- and yellow-throated males implanted with testosterone during adulthood start to defend territories which are as large as those of orange-throated males.

Whereas testosterone seems to activate mating strategies in adult side-blotched lizards, organizational effects of testosterone play a major role in a closely related species, the tree lizard *Urosaurus ornatus*. Also, tree lizards come in several color morphs: males with an orange-blue dewlap are aggressive and defend territories, whereas males with an orange dewlap are nonaggressive and do not establish territories. Males with high levels of testosterone and progesterone during ontogeny develop into the orange-blue morph which become territorial and defend the home range of several females. Males with low levels of testosterone and progesterone during development turn into the orange morph, which follow a sneaker or nomadic strategy. Adult circulating levels of testosterone do not differ between orange-blue and orange tree lizards and – unlike in side-blotched lizards – mating strategies are fixed and cannot be manipulated via testosterone implants. Thus, the major hormonal effects on mating strategies seem to occur at different times during the life history of tree- and side-blotched lizards. Organizational effects prevail in tree lizards, whereas side-blotched lizards remain plastic and mating strategies can be manipulated with hormones during adulthood. In summary, these data suggest that testosterone may facilitate polygynous mating strategies in some species, but since this is not universally the case, testosterone alone cannot explain mating decisions.

An important factor that has been largely neglected in the discussion of proximate factors for the control of mating strategies is the female part: it always takes two to tango and it takes even more to become polygamous or promiscuous. Thus, treatment with testosterone may increase the propensity of males to seek additional mates, but this does not mean that females of all species, and under all circumstances, are ready to accept these offers. There is a large body of ecological literature discussing factors that might influence the decision of females to become the secondary mate of a polygynous male. Such factors include for example male quality, territory quality, availability of unpaired males, etc. But we know very little about the physiology of such females. Potential hormonal factors that lead to or prevent the formation of a pair bond between females and males are discussed in the next section.

The Role of Oxytocin and Arginine-Vasopressin in Pair Bonding and Mating Systems: Voles as a Model

A highly instructive model for the study of the role of hormones in mating systems have been voles – small arvicoline rodent species that resemble mice with a shorter hairy tail, a rounder head, and stouter body. The prairie vole (*Microtus ochrogaster*) is a socially monogamous species found in grasslands throughout North America. A female and a male form a life-time pair bond and maintain a common nest and territory, which they defend against other voles. They live in communal family groups consisting of the breeder pair and their offspring. A large proportion of the offspring does not leave their family and serve as 'helpers at the nest.' When prairie vole pups are isolated from their families for just 5 min, they emit distress calls and their plasma corticosterone levels – a hormone released during stress – increase.

In contrast, montane voles (*M. montanus*) or meadow voles (*M. pennsylvanicus*) are much less social: males and females have separate territories and nests, and they only meet for mating. They are highly promiscuous and parental care is provided by the female only. When montane vole pups are socially isolated, they do not emit distress vocalizations and their plasma corticosterone levels remain low.

Which hormones are involved in the regulation of these differences in the life history of monogamous and promiscuous voles? As outlined earlier, testosterone would be an obvious candidate and indeed, plasma testosterone levels of monogamous prairie voles are far lower

than those of promiscuous voles. But testosterone treatment does not render male prairie voles polygynous, nor does castration of polygynous meadow or montane voles turn them monogamous. Thus, testosterone does not appear to play a direct role in the modulation of mating preferences in voles. Unfortunately, field data reflecting the seasonal pattern of testosterone and the influence of social stimuli on testosterone levels in monogamous and promiscuous voles are not available yet.

In contrast to gonadal steroids such as testosterone, the adrenocortical 'stress hormone' corticosterone seems to have an effect that apparently differs between the sexes. Injections of corticosterone, or stress leading to an increase in circulating corticosterone levels, facilitate pair bonding in monogamous male prairie voles, but the same treatment inhibits pair bonding in females of this species. It is unknown whether treatment with corticosterone would induce partner preference in promiscuous vole species.

The most important modulators of pair bonding and social behavior in voles appear to be oxytocin and arginine-vasopressin (or arginine-vasotocin in nonmammalian vertebrates). These peptides belong to a very ancient hormone family present in all vertebrates. In mammals, oxytocin plays an important role during parturition and lactation, as it stimulates contraction of the uterus and initiates the milk let-down from the mammary glands. Vasopressin reduces urinary water loss as a result of increased osmotic reabsorption of water from the kidney tubules. But, centrally, both hormones also play an important role in the regulation of affiliative behavior, that is, oxytocin is involved in the mother–infant bonding, grooming, and sexual behavior. Vasopressin has been implicated in male social behaviors including aggression, territorial behavior, and courtship. Both neuropeptides are also involved in the neural processing of sensory cues involved in social learning. For example, oxytocin knockout mice cannot recognize individuals to which they have been previously exposed. So, what is the role of oxytocin and arginine-vasopressin with respect to mating systems in voles?

Monogamous prairie voles typically form a pair bond as a consequence of intense mating during a 24-h period. Because mating results in central oxytocin release in mammals, it is likely that intense mating in prairie voles stimulates oxytocin release and facilitates the social pair bond of a female prairie vole to her mate. Indeed, oxytocin injections into the ventricle of the brain of unmated female prairie voles facilitate pair bonding. In male prairie voles, arginine-vasopressin rather than oxytocin mediates this effect: arginine-vasopressin is released during mating and injection of arginine-vasopressin facilitates pair bonding in unmated males. In addition to its effect on pair-bonding behavior, administration of arginine-vasopressin also stimulates paternal care and increases aggression toward strangers. Interestingly, injection of arginine-vasopressin into promiscuous male montane voles did not increase aggression toward strangers, but increased self-grooming behavior, suggesting that the differences between monogamous and promiscuous voles are not related to the release of the peptides, but lie further downstream in how these signals are processed in the brain.

The prairie vole brain expresses high levels of oxytocin receptors and arginine-vasopressin receptors (of the V1aR subtype) in brain regions that are involved in the dopamine reward and reinforcement circuits, brain areas that are involved in becoming addicted to various substances. The release of oxytocin and arginine-vasopressin upon mating in prairie voles activates the dopaminergic reward pathway in the brain. As a consequence, mating with a particular partner is reinforcing, rewarding, and presumably hedonic in monogamous prairie voles. These reinforcing effects may become coupled with the identity of the mate, resulting in conditioned partner preference. These results are supported by the finding that blocking dopamine receptors in these reward areas prevent the formation of partner preferences in monogamous prairie voles.

In contrast to prairie voles, promiscuous montane voles, which do not form partner preferences, have only few oxytocin and arginine-vasopressin receptors in these brain reward regions, suggesting that mating with a particular partner is not exceptionally rewarding for montane voles, preventing the formation of a partner preference. Also, the comparison of the brains of several vole species suggests that the distribution of oxytocin and arginine-vasopressin receptors differs between monogamous and promiscuous species: monogamous pine voles (*M. pinetorum*) are similar to monogamous prairie voles, whereas promiscuous meadow voles resemble promiscuous montane voles.

In a highly sophisticated experiment, viral vectors were used to overexpress the *Avpr1a* gene (the gene that encodes the arginine-vasopressin receptor V1aR) in the reward circuit of male promiscuous meadow voles. Unlike control males, these transgenic males showed an increased partner preference toward a female they were cohabitated with. Again, if dopamine receptors were blocked in the reward areas of these transgenic animals, the formation of a partner preference was prevented. These results suggest that mutations that alter the expression of a single gene (in this case the receptor for arginine-vasopressin) in a specific region of the brain can have a remarkable impact on complex social behaviors such as pair bonding and could potentially transform a promiscuous vole into a monogamous one (or vice versa).

In a similar experiment with female voles, viral transfection of the gene for the oxytocin receptor into the reward circuit of the brain led to accelerated partner preference in female prairie voles, but was not sufficient to induce partner preference in promiscuous meadow voles. These data suggest that differences in the expression of oxytocin receptors in the reward circuit of female prairie voles may contribute to natural variation in partner preference behavior, but unlike the situation in males, it is not sufficient to induce partner preferences in promiscuous species.

How did the different patterns of expression of oxytocin and arginine-vasopressin receptors evolve in monogamous and promiscuous vole species? Most is known about the *Avpr1a* gene coding the receptor for arginine-vasopressin. This gene is highly homologous in prairie voles and montane voles, but upstream from the transcription start site, the prairie vole gene contains a highly repetitive sequence, a microsatellite. In montane voles, this repetitive sequence is much shorter. Transgenic mice with the prairie vole *Avpr1a* gene (including this microsatellite sequence) express the gene in brain regions similar to those found in prairie voles and, unlike control mice, show a strong partner preference. Thus, the change in the microsatellite region of the *Avpr1a* gene may have been the molecular event inducing a change in the expression of the arginine-vasopressin receptor in the reward circuit of the vole brain. This change resulted in the biological potential to develop a conditioned partner preference.

Implications for Pair-Bonding Behavior or Mating Systems in Humans

What do these data on pair-bond formation in voles implement for pair-bonding behavior in humans? It is unknown whether there is a common physiological mechanism for pair-bonding behavior in voles and humans. Similar to voles, plasma oxytocin levels are elevated during orgasm in women and plasma arginine-vasopressin levels are elevated in men when they are sexually aroused. Furthermore, when humans view photographs of people with whom they are 'truly, deeply, and madly in love,' their brain activity patterns resemble those observed after cocaine or opioid infusions: they show strong activations of brain areas involved in the dopamine reward and reinforcement circuits, areas rich in oxytocin and arginine-vasopressin and their receptors, suggesting that there is some truth in the saying that 'love is an addiction.'

A recent study also links the findings of the *Avpr1* gene encoding the arginine-vasopressin receptor in voles and a polymorphism of the equivalent AVPR1A gene in humans. Men with a particular variant of the AVPR1A gene are more likely to remain unmarried. And when they get married, they are more likely to report a recent crisis in their marriage. Also, their spouses are more likely to express dissatisfaction in their relationships than spouses of men with a different variant of the gene. These data suggest that the gene for the receptor being involved in differences in pair-bond formation in voles is probably of some relevance also in humans.

According to the saga of 'Asterix and Obelix,' Getafix, the Gaul druid, was famous for brewing magic potions more than 2000 years ago, among them the famous potion giving its drinker superhuman strength. To my knowledge, a 'love potion' was not on the druid's portfolio of drinks, but, given the recent advances in understanding the biology of pair-bonding behavior, we may not be too far from such a potion now.

Acknowledgments

Discussions with and constructive criticism from Silke Kipper, Nicole Geberzahn, Barbara Helm, and Katharina Hirschenhauser helped to improve previous versions of this contribution.

See also: Aggression and Territoriality; Mate Choice and Learning; Monogamy and Extra-Pair Parentage; Parental Behavior and Hormones in Mammals; Parental Behavior and Hormones in Non-Mammalian Vertebrates; Reproductive Skew, Cooperative Breeding, and Eusociality in Vertebrates: Hormones; Seasonality: Hormones and Behavior; Sex Change in Reef Fishes: Behavior and Physiology; Sperm Competition.

Further Reading

Adkins-Regan E (2005) *Hormones and Social Behavior*. Princeton, NJ: Princeton University Press.

Bartels A and Zeki S (2000) The neural basis of romantic love. *Neuroreport* 11: 3829–3834.

Carter CS, DeVries AC, Taymans SE, Roberts RL, Williams JR, and Gotz LL (1997) Peptides, steroids, and pair-bonding. *Annals of the New York Academy of Sciences* 807: 260–272.

Garamszegi LZ, Eens M, Hurtrez-Bousses S, and Møller AP (2005) Testosterone, testes size, and mating success in birds: A comparative study. *Hormones and Behavior* 47: 389–409.

Goymann W (2009) Social modulation of androgens in male birds. *General and Comparative Endocrinology* 163: 149–157.

Hirschenhauser K, Möstl E, and Kotrschal K (1999) Within-pair testosterone covariation and reproductive output in greylag geese (*Anser anser*). *Ibis* 141: 577–586.

Hirschenhauser K and Oliveira RF (2006) Social modulation of androgens in male vertebrates: Meta-analyses of the challenge hypothesis. *Animal Behaviour* 71: 265–277.

Hirschenhauser K, Winkler H, and Oliveira RF (2003) Comparative analysis of male androgen responsiveness to social environment in birds: The effects of mating system and paternal incubation. *Hormones and Behavior* 43: 508–519.

Lynn SE (2008) Behavioral insensitivity to testosterone: Why and how does testosterone alter paternal and aggressive behavior in some avian species but not others? *General and Comparative Endocrinology* 157: 233–240.

Moore MC, Hews DK, and Knapp R (1998) Hormonal control and evolution of alternative male phenotypes: Generalizations of models for sexual differentiation. *American Zoologist* 38: 133–151.

Oliveira RF (2004) Social modulation of androgens in vertebrates: Mechanisms and function. *Advances in the Study of Behavior* 34: 165–239.

Ross HE, Freeman SM, Spiegel LL, Ren X, Terwilliger EF, and Young LJ (2009) Variation in oxytocin receptor density in the nucleus accumbens has differential effects on affiliative behaviors in monogamous and polygamous voles. *Journal of Neuroscience* 29: 1312–1318.

Sinervo B, Miles DB, Frankino WA, Klukowski M, and DeNardo DF (2000) Testosterone, endurance, and Darwinian fitness: Natural and sexual selection on the physiological bases of alternative male behaviors in side-blotched lizards. *Hormones and Behavior* 38: 222–233.

Walum H, Westberg L, Henningsson S, et al. (2008) Genetic variation in the vasopressin receptor 1a gene (AVPR1A) associates with pair-bonding behavior in humans. *Proceedings of the National Academy of Sciences* 105: 14153–14156.

Wingfield JC, Hegner RE, Dufty AM, and Ball GF (1990) The "challenge hypothesis": Theoretical implications for patterns of testosterone secretion, mating systems, and breeding strategies. *American Naturalist* 136: 829–846.

Wingfield JC, Moore IT, Goymann W, Wacker D, and Sperry T (2006) Contexts and ethology of vertebrate aggression: Implications for the evolution of hormone-behavior interactions. In: Nelson R (Ed.) *Biology of Aggression*, pp. 179–210. New York: Oxford University Press.

Young LJ, Nilsen R, Waymire KG, MacGregor GR, and Insel TR (1999) Increased affiliative response to vasopressin in mice expressing the V1a receptor from a monogamous vole. *Nature* 400: 766–768.

Young LJ and Wang Z (2004) The neurobiology of pair-bonding. *Nature Neuroscience* 7: 1048–1054.

Parasite-Induced Behavioral Change: Mechanisms

M.-J. Perrot-Minnot and F. Cézilly, Université de Bourgogne, Dijon, France

Parasites can manipulate the behavior of their hosts to their own benefit – this is what evolutionary parasitology studies tell us. But let us go a step further and take up the challenge raised by these manipulative parasites messing with the brains of their hosts and giving our own brains a serious puzzle. How can a so-called 'simple' (not 'regressed') parasite hijack the behavior of its host, which in some instances might be a so-called 'higher' vertebrate? Is there anything like a 'manipulative molecule' secreted by the parasite to directly target its host's CNS and specifically modulate the behaviors affecting transmission success? Or does manipulation come as a fortuitous side-effect of the infection on the host immune system and metabolism?

Despite the growing number of studies reporting on behavioral manipulation by parasites, the proximate mechanisms underlying this phenomenon have been investigated in only a few of them (**Table 1**), and no mechanism has been completely elucidated. This article aims at reviewing these few cases, where the mechanisms of parasite manipulation have been investigated. However, we use here a broader approach, looking for a causal connection between altered host behavior and the modulation of gene expression in both the host and the parasite. The analysis is focused on the mechanisms underlying changes in behavior that increase parasite transmission success (strictly speaking, parasite-induced behavioral manipulation). Mechanisms associated with disease-related behavioral disorders, such as immune-generated alteration of the CNS, will not be addressed here.

The phenomenon of parasite manipulation can be fully understood only if the demonstration of a fitness gain for the parasite (ultimate cause) is coupled to the identification of the mechanisms underlying the observed behavioral changes (proximate causes). Understanding proximate causes of manipulation will contribute to our evolutionary analysis in two ways: (1) it will help evaluate the costs a parasite pays to invest in manipulation and whether these costs are shared with investment in parasite survival (i.e., defense against the immune system) and (2) it will reveal how complex and specific the manipulation process is. These two criteria are currently acknowledged as important in assessing the adaptive significance of manipulation.

The mechanisms involved in parasite manipulation have been explored since the pioneering work on rodents infected with *Toxoplasma gondii* and on the amphipod *Gammarus lacustris* infected with the acanthocephalan bird parasite *Polymorphus paradoxus*. Since then, several studies have attempted to identify the changes in host neurophysiology or gene expression associated with parasite manipulation (**Table 1**). The expected complexity of the interactive network connecting a host and its manipulative parasite comes from the modulatory connections between the neuronal, hormonal, and immune systems of the host. The investigation of proximate mechanisms therefore relies on an integrative approach combining behavioral ecology, neurophysiology, pharmacology, molecular biology, and biochemistry.

From Phenotypic Behavioral Changes to Altered Gene Expression

Changes in host behavior following infection are not necessarily profitable to the parasite. They may actually benefit the host through compensating for the effect of infection or getting rid of the parasite. Alternatively, they can be pathological side-effects, with no benefits for the host or for the parasite. When beneficial to the parasite, changes in host behavior can result from a combination of direct and indirect effects of a parasite on its host's CNS. The most likely indirect effect relies on the connection between the neuronal and immune systems: the host's immunological response to infection can be involved in changing the host's behavior into a behavior that favors parasitic transmission. Therefore, the methods used to investigate mechanisms of parasite-induced behavioral changes must not only identify the biochemical or physiological changes in manipulated hosts, but also demonstrate that these changes are indeed the proximate cause of behavioral manipulation.

More precisely, we have to identify the following:

1. The functional connection between an altered behavior and the corresponding genes expressed in the host.
2. The parasite's biochemical signal (in the excretory/secretory (E/S) parasite products) targeting host's genes, whether it corresponds to 'manipulative molecules,' or molecules with a broader spectrum, including physiological targets.
3. The causal link between some of those genes and the direct target of the parasite's E/S products.

To that end, several complementary approaches are possible:

– The exploration of specific neurophysiological pathways by means of ethopharmacology and techniques used on candidate proteins (immunohistochemistry or

Table 1 Review of studies attempting to identify the changes in host's neurophysiology or proteome associated with parasite manipulation in invertebrates (for a review on vertebrates, see Klein 2003)

Host	Parasite	Manipulated behavior	Method of investigation	References
Grasshopper, *Meconema thalassinum* (Orthopteran insect)	Hairworm, *Spinochordodes tellinii* (Nematomorph)	Seeking water and jumping into it	Brain proteome (host) and Parasite proteome	1
Cricket, *Nemobius sylvestris* (Orthopteran insect)	Hairworm, *Paragordius tricuspidatus* (Nematomorph)	Seeking water and jumping into it	Brain proteome (host) and Parasite proteome	2
Mosquitoe *Anopheles gambiae* (Dipteran insect)	*Plasmodium berghei* (Protozoa: Apicomplexa)	Increased biting rate	Brain proteome (host)	3
Tsetse fly *Glossina palpalis gambiensis* (Dipteran insect)	*Trypanosoma brucei* (Protozoa: Sarcomastigophora)	Increased probing, Extended engorging duration	Brain proteome (host)	4
Amphipod *Gammarus insensibilis* (Crustacea)	Flatworm *Microphallus papillorobustus* (Platyhelminthes: Trematoda)	Negative geotaxis: water surface	Brain proteome (host)	5
		Positive phototactism	Immunocytochemistry on the brain (5-HT)	6
Amphipod *Gammarus pulex* (Crustacea)	Thorny-headed worm *Polymorphus minutus* (Acanthocephala: Polymorphidae)	Negative geotaxis: water surface	Brain proteome (host)	5
Amphipod *Gammarus pulex, Gammarus roeseli* (Crustacea)	Thorny-headed worm *Pomphorhynchus laevis, Pomphorhynchus tereticollis* (Acanthocephala: Pomphorhynchldae)	Positive phototactism, Increased drifting behavior and activity	Ethopharmacology (phototactism), Immunocytochemistry on the brain (5-HT)	7
		Attraction to chemical cues from fish predator		
Amphipod *Gammarus lacustris* (Crustacea)	*Polymorphis paradoxus* (Acanthocephala: Polymorphidae)	Clinging behavior Positive phototactism	Ethopharmacology (clinging), Immunocytochemistry on the nerve cord (5-HT)	8 9

[1] Biron DG, Marché L, Ponton F, Loxdale HD, Galéotti N, Renault L, Joly C, and Thomas F (2005) Behavioural manipulation in a grasshopper harbouring hairworm: a proteomics approach. *Proceedings of the Royal Society B* 272: 2117–2126.

[2] Biron DG, Ponton F, Marché L, Galeotti N, Renault L, Demey-Thomas E, Poncet J, Brown SP, Jouin P, and Thomas F (2006) 'Suicide' of crickets harbouring hairworms: a proteomics investigation. *Insect Molecular Biology*: 15: 731–742.

[3] Lefevre T, Thomas F, Schwartz A, Levashina E, Blandin S, Brizard J-P, Le Bourligu L, Demettre E, Renaud F, and Biron DG (2007a) Malaria Plasmodium agent induces alteration in the head proteome of their Anopheles mosquito host. *Proteomics* 7: 1908–1915.

[4] Lefèvre T, Thomas F, Ravel S, Patrel D, Renault L, Le Bourligu L, Cuny G, and Biron DG (2007b) *Trypanosoma brucei brucei* induces alteration in the head proteome of the tsetse fly vector *Glossina palpalis gambiensis*. *Insect Molecular Biology* 16: 651–660.

[5] Ponton F, Lefevre T, Lebarbenchon C, Thomas F, Loxdale H, Marche Renault L, Perrot-Minnot M-J, and Biron D (2006) Behavioural manipulation in gammarids harbouring trematodes and acanthocephalans: a comparative study of the proximate factors using proteomics. *Proceedings of the Royal Society B: Biological Sciences* 273: 2869–2877.

[6] Helluy S and Thomas F (2003) Effects of *Microphallus papillorobustus* (Plathyhelminthes: Trematoda) on serotonergic immunoreactivity and neuronal architecture in the brain of *Gammarus insensibilis* (Crustacea: Amphipoda). *Proceedings of the Royal Society B* 270: 563–568.

[7] Tain L, Perrot-Minnot M-J, and Cézilly F (2006) Altered host behaviour and brain serotonergic activity caused by acanthocephalans: evidence for specificity. *Proceedings of the Royal Society B* 273: 3039–3045.
Tain L, Perrot-Minnot M-J, and Cézilly F (2007) Differential influence of *Pomphorhynchus laevis* (Acanthocephala) on brain serotonergic activity in two congeneric host species *Biology Letters* 3: 68–71.

[8] Helluy S and Holmes JC (1990) Serotonin, octopamine, and the clinging behavior induced by the parasite *Polymorphus paradoxus* (Acanthocephala) in *Gammarus lacustris* (Crustacea). *Canadian Journal of Zoology* 68: 1214–1220.

[9] Maynard BJ, DeMartini L, and Wright WG (1996) *Gammarus lacustris* harboring *Polymorphus paradoxus* show altered patterns of serotonin-like immunoreactivity. *Journal of Parasitology* 82: 663–666.

immunocytochemistry, HPLC-ED, etc.). This approach can establish a functional link between a neuromodulatory pathway and the observed altered behavior, without establishing how the parasite directly hijacks the neurophysiology of its host.

- The differential screening of the host proteome or transcriptome between infected-manipulated individuals and nonmanipulated ones (uninfected and infected), to reveal proteins or mRNA associated with a manipulated phenotype (as the cause or the consequence of altered behavior and physiology).

- The proteomic analysis of parasite's E/S products followed by the identification of the biological fractions modulating host's behavior. Proteomic tools applied to the analysis of E/S products screen for molecules released by a manipulative parasite that could trigger the observed phenotypic changes.

The first two approaches must compare infected-manipulated individuals with nonmanipulated ones (infected by a nonmanipulative stage of parasite and uninfected), to specifically identify neurophysiological or biochemical changes associated with manipulation. Still, will these pathways or molecules in the host's repertoire be the direct target of parasite? The third approach is thus necessary to identify the initial molecular dialog setting up behavioral manipulation.

Several inferences emerge from the astonishing fact that parasites increase their own transmission success by taking control of their host's behavior. (1) The molecular cross-talk between a host and a parasite that results in fine-tuned phenotypic alterations is probably complex and intimate. (2) A parasite manipulating the behavioral flexibility of its host so that it performs the appropriate behavior likely uses either molecular mimicry (biochemical evolutionary convergence) or highly conserved molecules (phylogenetic inertia). The parasite may thereby control some of its host modulatory pathways by usurping signaling processes. (3) Changes in host behavior are often a mix of direct and indirect effects, and it may prove difficult to differentiate between the two. Investigations of proximate mechanisms involved in parasite manipulation must keep these inferences in mind.

Multidimensionality and Mechanisms of Parasite Manipulation

The capacity of a parasite to manipulate several behavioral and physiological traits together has been largely ignored in most empirical studies so far, although a review of studies on the same host–parasite systems shows that manipulative parasites generally modify more than a single dimension in the phenotype of their hosts. For instance, the acanthocephalan fish parasite *Pomphorhynchus laevis* reverses the photophobic behavior

of its host *Gammarus pulex* and its antipredatory behavior in reaction to olfactory cues, and increases its activity and its drifting behavior. Several physiological changes have been reported as well in *G. pulex* infected with *P. laevis*, such as increased hemolymph protein titers (in particular haemocyanin), reduced O_2 consumption, increased glycogen content, fecundity reduction, and immunosuppression. In wild rats infected with the protozoan *T. gondii*, changes in activity and in motivational level in various contexts have been reported. *T. gondii*-infected rats were found to be significantly less neophobic toward food-related novel stimuli. In outdoor captive environment, they were more likely to be trapped than their uninfected conspecifics, and their propensity to approach a mildly fear-inducing object was higher than that of uninfected rats (reduced neophobia). Alteration of innate behavior (such as neophobia) extends to the reversal of antipredatory behavior from a strong aversion to a preference for cat-treated areas in infected rats. This 'fatal attraction' is expected to increase the chances of transmission of *T. gondii* to its feline definitive host. Such multidimensionality of manipulation makes sense from an ecological and evolutionary point of view: having the 'vehicle' host reaching the right place at the right time (through being predated by, or stinging, or biting the next host species in the cycle) probably involves several behaviors related to environmental sensing and microhabitat choice. In parasites with a direct life cycle, transmission by contact or wounding can be increased by modulating a number of social behaviors, such as aggression and exploration. Several cue-oriented behaviors are generally altered in infected invertebrates (among phototaxis, chemotaxis, rheotaxis or wind-evoked behavior, geotaxis, etc.) that together contribute to increased transmission success of the manipulative parasite.

Are these multiple dimensions of a manipulated phenotype functionally independent? Or do the proximate mechanisms of manipulation have 'pleiotropic effects'? The best argument supporting the hypothesis of 'pleiotropic effects' lies in the functional connection between host's neuronal, immunological, and endocrine/metabolic systems, be the host an invertebrate or a vertebrate. Because the very first conditions for a parasite to develop are to successfully establish in a host and exploit its energy reserves, some mechanisms must exist that allow the parasite to interact with its host's physiology, especially the host's immunity. As pointed out by several authors, the evolutionary transition leading to parasite manipulation may simply consist in an extension of the effect of the parasite on the immune system of its host to its neuronal system. Targeting diverse and flexible neuromodulatory pathways to induce adaptive behavioral change in its host would thereby be a small evolutionary step. The understanding of proximate mechanisms of parasite manipulation allows us to test this evolutionary and functional scenario.

Investigating Host's Neuromodulatory Pathways

Biogenic amines (serotonin, dopamine, octopamine among others) and other chemical signals such as neuropeptides or the gas nitric oxide (NO) play a neuromodulatory role in numerous sensory, motor, and endocrine functions, in both invertebrates and vertebrates. They modulate the behavioral or physiological responses of an organism to external information, according to its internal status. By 'manipulating' these neuromodulatory pathways in its host's CNS, a parasite could adjust its host's behavioral response to reflect the parasite's own interest. The E/S products of the parasite would thus be akin to the venom of several predators or parasitoid wasps manipulating the monoaminergic system of their hosts to improve prey handling and use.

Several studies have shown a major role of biogenic amines and neuropeptides in the physiological and behavioral alterations induced by parasites (**Table 2**). The 'candidate neuromodulatory pathway' approach to parasite manipulation targets simple tropisms or cue-oriented behaviors such as phototaxis, geotaxis, chimiotaxis, thermal gradient sensitivity (in biting or sucking vectors of warm blood animals), and reflectance (**Table 2**). In vertebrates, several viruses and protists increase their hosts' exploratory behavior or aggression, two behaviors suspected to enhance parasite transmission either by predation or by conspecific wounding/contact respectively. These behavioral effects have been related to changes in concentrations or receptor binding of amines (dopamine, serotonin) or opioids.

However, few studies have combined ethopharmacological analysis to biochemical techniques (immunohisto- or cytochemistry, western blot, and ELISA) to establish or invalidate the involvement of a neuromodulator in parasite manipulation of behavior. One pioneer ethopharmacological study investigated the role of several neuromodulators in the behavioral alterations induced by *P. paradoxus* in its intermediate host *G. lacustris*. Uninfected individuals injected with serotonin responded to mechanical stimulation by skimming to the water surface until clinging to floating material and exhibited positive phototaxis, two behavioral mimics of amphipods infected with this parasite of dabbling ducks. Immunocytochemistry on the nerve cord of amphipods infected with *P. paradoxus* revealed an increase in the number of varicosities exhibiting serotonin-like immunoreactivity in the third thoracic ganglion. Serotonin was altered either in the amount or in the number of local storage and release sites along neural fibers in *P. Paradoxus*-infected amphipods, but not in *G. lacustris* infected with *Polymorphus marilis*, a parasite of diving ducks inducing positive phototactism but no escape response.

Exogenously supplied serotonin can mimic the effect of parasitism in other amphipod-acanthocephalan systems:

G. pulex – P. laevis and *Pomphorhynchus tereticollis*. Injection of serotonin in uninfected *G. pulex* reversed their reaction to light, mimicking the positive phototactism of gammarids infected with these two fish parasites. The serotonergic activity in the brains of infected-manipulated gammarids was significantly increased, compared to that of four controls: uninfected *G. pulex*, *G. pulex* infected with *P. tereticollis* but not manipulated, *G. pulex* infected with the bird acanthocephalan *Polymorphus minutus* (not altering phototaxis), and a nonmanipulated sympatric amphipod species *Gammarus roeseli*, infected with *P. laevis* (**Figure 1**).

In *Gammarus insensibilis* infected by the cerebral trematode *Microphallus papillorobustus*, immunocytochemistry on brain has revealed the degeneration of discrete sets of serotonergic neurons: immunoreactivity to serotonin (5-hydroxytryptamine or 5-HT) was decreased in the optic neuropils but increased in the olfactory lobes. This imbalance in brain serotonergic activity is suspected to contribute to the behavioral alterations reported in this brackishwater amphipod species, in particular, positive geotactism and attraction to light. In vertebrates, several viral and protozoan parasites infecting the CNS of their rodent hosts alter neurochemical pathways in the brain. In the brains of infected mice and rats for instance, rabies virus decreases 5-HT and GABA neurotransmission, and *T. gondii* increases the concentration of dopamine and decreases the concentration of norepinephrine. These changes in neuromodulatory pathways may be linked to elevated aggression exhibited by infected rodents (and exploratory and fearless behavior in the case of *T. gondii* infected rats). These behavioral alterations presumably enhance the transmission of rabies virus by increased conspecific biting and of *T. gondii* by increased predation rate.

Although the exploration of these neurophysiological changes can provide evidence that a neuromodulator plays a key role in one or few behavioral dimensions of parasite manipulation, it also has several limitations. First, the neuromodulatory and signaling network is complex: several neuropeptides or amines may act together to modulate a given behavior, while a single neuromodulator may regulate several behaviors. If this may fit well with the multiple dimensions of parasite-induced alteration on host's phenotype, it makes the full understanding of the underlying neurophysiological process difficult. Second, showing a change in brain CNS does not establish a causal connection with the manipulative process.

Screening the Host's Proteome and Transcriptome

In the few host–parasite systems to which it has been applied, the proteomic approach appeared sensitive enough to detect proteome differences between infected and noninfected hosts that can be attributed to the manipulative syndrome.

Table 2 Review of studies suggesting the involvement of certain neuropeptides or biogenic amines in parasite-induced alteration of invertebrate hosts' behavior (used as intermediate hosts by trophically transmitted parasites or as food store and shelter by parasitoid larvae)

Host species	Parasite species	Technics	Results	References
Amphipod *Gammarus lacustris*	*Polymorphus paradoxus* (Acanthocephalan)	– Injection: serotonin, dopamine, octopamine, norepinephrine	Only 5-HT injection mimics clinging behavior	1
		– Immunocytochemistry (anti-5-HT) on nerve cord	– Increased 5-HT immunoreactivity in the third thoracic ganglion (increase in varicosities)	2
Amphipod *Gammarus insensibilis*	*Microphallus papillorobustus* (Trematode)	Immunocytochemistry (anti-5-HT) on the brain	Serotonergic activity depressed in specific areas of the brain	3
Amphipod *Gammarus pulex*	*P. laevis, P. tereticollis* (Acanthocephalan)	– Immunocytochesmitry (anti-5-HT)	– Increase in brain 5-HT immunoreactivity in infected amphipods (correlates to their degree of manipulation of phototactism)	4
		– Injection of serotonin	– Injection of 5-HT to uninfected animals mimics positive phototactism of infected ones	
Crab *Macrophthalmus hirtipes*	*Maritrema* (Trematode) *Profilicollis* (Acanthocephalan)	HPLC-ED on brain extracts	Increase in 5-HT content in the brain of crabs coinfected with both parasite	5
Crab *Hemigrapsus crenulatus*	*Profilicollis antarcticus* (Acanthocephalan)	HPLC-ED on hemolymph extracts	Increase in hemolymph dopamine content, but not serotonin, in infected crabs	6
Moth *Manduca sexta*	*Cotesia congregata* (Hymenoptera Braconidae)	– HPLC-ED on haemolymph extracts	– Increased octopamine content of the brain, thoracic ganglia, and abdominal ganglia	7
		– Injection of octopamine or blood from postemergence parasitized larvae	– Injection mimics the decreased peristaltic activity in the foregut (related to decreased feeding)	8

Periplaneta americana (and other cockroaches)	Ampulex compressa (Hymenoptera, Sphecidae)	– Injection (dopamine) – GC-MS and HPLC-ED on venom – Immunocytochemistry (anti-dopamine) – Electrophysiology	– Injection of dopamine mimics venom-induced grooming – Pharmacological depletion of monoamines mimics venom-induced nonparalytic hypokinesia and reduced-escape response – Dopamine present in the venom – Decrease in octopamine neurons activity in the thorax, modulated by input from descending neurons from the brain, themselves modulated by venom injection	9

[1]Helluy S and Holmes JC (1990) Serotonin, octopamine, and the clinging behavior induced by the parasite Polymorphus paradoxus (Acanthocephala) in Gammarus lacustris (Crustacea). Canadian Journal of Zoology 68: 1214–1220.

[2]Maynard BJ, DeMartini L, and Wright WG (1996) Gammarus lacustris harboring Polymorphus paradoxus show altered patterns of serotonin-like immunoreactivity. Journal of Parasitology 82: 663–666.

[3]Helluy S and Thomas F (2003) Effects of Microphallus papillorobustus (Platyhelminthes: Trematoda) on serotonergic immunoreactivity and neuronal architecture in the brain of Gammarus insensibilis (Crustacea: Amphipoda). Proceedings of the Royal Society B 270: 563–568.

[4]Tain L, Perrot-Minnot M-J, and Cézilly F (2006) Altered host behaviour and brain serotonergic activity caused by acanthocephalans: evidence for specificity. Proceedings of the Royal Society B 273: 3039–3045.

[5]Poulin R, Nichol K, and Latham ADM (2003) Host sharing and host manipulation by larval helminths in shore crabs: cooperation or conflict? International Journal for Parasitology 33: 425–433.

[6]Rojas JM and Ojeda FP (2005) Altered dopamine levels induced by the parasite Profilicollis antarcticus on its intermediate host, the crab Hemigrapsus crenulatus. Biological Research 38: 259–266.

[7]Adamo SA and Shoemacker KL (2000) Effects of parasitism on the octopamine content of the central nervous system of Manduca sexta: a possible mechanism underlying host behavioural change. Canadian Journal of Zoology 78: 1580–1587.

[8]Miles CI and Booker R (2000) Octopamine mimics the effects of parasitism on the foregut of the tobacco hornworm Manduca sexta. Journal of Experimental Biology 203: 1689–1700.

[9]Several references in.

Wiesel-Eichler A and Libersat F (2004) Venom effects on monoaminergic systems. Journal of Comparative Physiology A 190: 683–690.

Libersat F and Pflueger HJ (2004) Monoamines and the orchestration of behavior. Bioscience 54: 17–25.

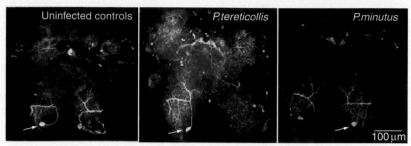

Figure 1 5-HT immunoreactivity (yellow) within the brains of uninfected *Gammarus pulex*, *P. tereticollis*-infected *G. pulex*, and *P. minutus*-infected *G. pulex*. Arrows show position of tritocerebrum giant neuron (TGN) cell body. No differences in brain anatomy from infected and uninfected individuals were observed. Bar shows 100 mm.

Indeed, the use of comparative screening of whole proteome or transcriptome between infected hosts and uninfected hosts appears a powerful means to cope with the predicted complexity of proximate mechanisms involved in parasite manipulation, if several conditions are met (e.g., the quality of controls run, the access to database allowing protein identification, and other limitations listed here). Proteins or transcripts differentially produced and specifically associated with the manipulative process can be identified, if one compares manipulated hosts with uninfected and infected nonmanipulated hosts. The analysis of infected nonmanipulated hosts (i.e., usually containing a developmental stage of the parasite not infective to the next host) is an important control to run, to distinguish the proteins or transcripts specifically associated with the manipulative process from the ones produced in response to infection. Similarly, noninfected hosts exposed to the same environmental conditions as infected manipulated ones should be analyzed (in addition to noninfected ones in their natural environment) to distinguish the proteins or transcripts specifically associated with the manipulative process from the ones produced in response to the environmental changes associated with manipulation (for instance, living at the surface instead of the bottom of a body of water). The differences in brain proteome between infected-manipulated hosts and controls are either in the presence/absence, the quantity, or the posttraductional processing of certain proteins.

From the studies reviewed (**Table 3**) it seems that the alteration of the CNS is a common feature in the proteome of infected manipulated animals. In addition, key metabolic pathways are often perturbed, as well as proteins involved in cellular stress (HSP, other chaperones), immunomodulation, or oxidative damage. Alteration in energy metabolism in the brains of infected blood-feeding insects can be interpreted as a parasite strategy to manipulate vector-feeding behavior by inducing a nutritional stress. Interestingly, several proteins putatively involved in similar behavioral modifications in different host–parasite systems belong to the same family. For instance, differential expression of proteins from the family (CRAL–TRIO) implicated in the vision process has been found in the brain of the wood cricket *N. sylvestris*

infected by the manipulative nematomorph *P. tricuspidatus* and in the brain of *G. insensibilis* infected with the bird trematode *M. papillorobustus* (**Table 3**). Such pattern is suggestive of a limited 'jeu des possibles,' with molecular convergence or conserved proximate mechanisms being the only way to alter the behavior of phylogenetically distant hosts such as an amphipod and an insect.

However, despite their power in investigating the molecular basis of parasite manipulation, proteomic studies have several limitations:

– Some peptides might be undetected because of their low concentration, specific pI, or small size (for instance, neuropeptides).
– Most proteomic studies have focused on differential expression of proteins; however posttranslational modifications of proteins might be involved as well in the modulation of the host's phenotype induced by parasites, and their importance in interpretating transcriptional data has been emphasized by some authors.
– Protein identification becomes problematical in nonmodel organisms: it relies on cross-species protein identification, and therefore, on either highly conserved proteins or proteins known from organisms closely related to the host. For instance, the power of proteomic studies on *Gammarus* amphipods in revealing proteins specifically linked to the manipulative process was limited by the impossibility of identifying 27 out of 72 proteins spots differentially present or absent from the brain of *G. insensibilis* manipulated by *M. papillorobustus* and 60 out of 68 proteins spots differentially present or absent from the brain of *G. pulex* manipulated by *P. minutus*.

Microarrays provide an alternative to proteomics but have been used so far in a limited number of host–parasite systems (rodents, salmons, mosquitoes, and bees infected with Apicomplexa protists, helminths, or mites). Most studies have performed transcriptional profiling of infected and uninfected individuals or resistant and susceptible strains, to identify the immune and metabolic response to infection. In parasites, microarray profiles coupled to other techniques in functional genomics have

Table 3   A review of studies screening the proteomes of infected manipulated and noninfected individuals in insects and amphipods. The role of proteins differentially produced has been putatively ascribed to the functioning of the CNS, to metabolic pathways, or to other functions, including immunity and stress response

Host–parasite system	Differential changes in the brain proteome of infected animals		
	Functioning of the CNS	Metabolism	Others
Tsetse flies with mature infection of *Trypanosoma gondii*	Dopa decarboxylase (−) (synthesis of dopamine and serotonin)	Glycolytic enzymes	Pheromone and odorant-biding protein (+)
			Molecular chaperone (stress-activated proteins) Signaling,...
Anopheles gambiae (mosquitoe) with mature infection of *Plasmodium berghei* (sporozoite stage in salivary glands)	Tropomyosin (+up)	Metabolic enzymes involved in the production of ATP[b] or in glucose oxidation pathway (+up)	Molecular chaperone (HSP20, stress protein)
	Calmodulin (+up) (activation of NO synthase)[a]		Several other proteins involved in cell cycle, signaling, etc....
Meconema thalassinum (grasshopper) during manipulation by *Spinochordodes tellinii* (nematomorph)	Six protein families involved in the development of the CNS (+) (such as Actins, ATPase, Wnt): two in the release of neurotransmitters, one in apoptosis[c]	?	
Nemobius sylvestris (cricket) during manipulation by *Paragordius tricuspidatus* (nematomorph)	Five protein families involved directly and/or indirectly in the development of the CNS (such as Actins, ATPase, Wnt)	Glucose metabolism	Glutathion S-transferase (oxidative stress)
	Two proteins implied in the vision process (including CRAL−TRIO)		Nonspecific stress Signal transduction (serine/threonine protein phosphatase)
Gammarus insensibilis infected with *Microphallus papillorobustus* (Trematode)	Arginine-kinase (−) (regulating factor in NO synthesis)[a]	?	Proteins implied in immunity defenses (PBP–GOBP (pheromone binding protein) (+) and ATP-gua_Ptrans: (−))
	Aromatic-L-amino acid decarboxylase (−) (synthesis of serotonin)		
	Two proteins implied in the vision process (including CRAL_TRIO)		
Gammarus pulex infected with *Polymorphus minutus* (Acanthocephalan)	Arginine-kinase (+) (regulating factor in NO synthesis)[a]	?	Proteins implied in immunity defences (prophenoloxidase: (+); MAM, sushi and ATP-gua_Ptrans: (+))
	Tropomyosin (+) (interacts with the development and plasticity of the CNS)		

[a]NO is a neuromodulator involved in memory, neuronal development as well as immune defense.
[b]ATP can also play a role as neuromodulator.
[c]Inducing apoptosis in the brain can alter the chemical signals in the brain (inflammatory immune responses), as reported in vertebrates infected with certain viruses or protozoans.
(−): Not detected in the brain proteome of infected individuals compared to uninfected ones.
(+): Detected in the brain proteome of infected individuals but absent from the one of uninfected individuals.
(−do) Downregulation.
(+up) Upregulation.

been used to discover new drug targets, or to understand the genetic basis of drug resistance. To our knowledge, the method has not yet been used to screen for 'manipulative molecules,' possibly because of the lack of genomic data on the historically best model systems of behavioral manipulation (acanthocephalans or trematodes and their arthropod intermediate hosts). This could, however, be done on mosquitoes/*Plasmodium* (with altered behavior driven by olfactory cues), tsetse flies and *Trypanosoma*, and bee/*Varroa* systems. Indeed, in the later model, a recent microarray-based analysis reveals that resistant and susceptible strains of bee to *Varroa* differ more on olfaction-related genes than on immune genes.

The Search for the 'Molecules of Manipulation' in E/S Products of Manipulative Parasites

Differential screening of infected-manipulated hosts and noninfected or infected and nonmanipulated hosts will not reveal which molecules are actually released by the parasite to induce changes in the proteome or transcriptome. Even if proteomic studies can reveal a protein specifically produced by a parasite during the manipulation process, such as the mimetic protein Wnt by hairworms infecting crickets, the demonstration of its role as a 'manipulative molecule' requires that its excretion in the host's hemolymph is established. Therefore, proteomic analysis of E/S products from manipulative parasites is necessary to identify the parasitic molecules initiating the manipulation process. The separation of biological fractions in the 'secretome' allows their biochemical identification and must be followed by the biological testing in vivo of their functional role (in the manipulative process).

Such analysis is currently limited to parasites that can be maintained or cultivated in vitro. The proteomic identification of E/S parasite products also relies on ongoing sequencing (genomic and EST) projects. So far, it has been possible in several nematodes and in the trematode *Schistosoma mansoni*, in the prospect of designing vaccines or drugs. Given these limitations, no such analysis has yet been done to identify the E/S products involved in behavioral manipulation. Studies analyzing E/S products from worms have revealed the release of proteins involved in a diversity of functions, such as stress response proteins/chaperones, antioxidant enzymes, energy metabolism and structural/cytoskeletal proteins, immune evasion, protease inhibitors, and lipid binding. No doubt the proteomic analysis of 'secretome' is a promising route to the discovery of the 'manipulative molecules' (either directly or indirectly targeting the host's CNS).

Conclusion

The most tricky aspect in our quest for the mechanistic basis of parasite-induced behavioral alterations is a causality problem akin, at first sight, to 'the chicken or the egg' dilemma. Whenever one spots a change associated with the manipulative process/the manipulated phenotype (be a neurophysiological change, a gene or a protein differentially expressed or produced), is it the cause or a side-effect of parasite manipulation? The reviewed studies addressing the mechanisms underlying parasite manipulation reveal that the alteration of host's CNS is a common feature of manipulated hosts. Given the diversity of host–parasite systems in which behavioral alterations have been reported, further studies should help in answering several key questions: (1) Have the same constraints on parasitic transmission led to a similar solution (evolutionary convergence of behavioral manipulation)? (2) Are similar behavioral alterations induced by the same biochemical tools (molecular convergence)? For instance, geotaxis is altered in several host–parasite systems. Ants parasitized with parasitic fungus of the genus *Cordyceps* and ants infected with the liver fluke *Dicrocelium*, as well as the caterpillar *Mamestra brassicae* infected with *M. brassicae* nuclear polyhedrosis virus, climb to the top of a plant. Several amphipods infected with various acanthocephalan species (*G. pulex/P. minutes*, *G. lacustris/P. paradoxus*, and *G. insensibilis/M. papillorobustus*) swim at the water surface. Chemosensing and possibly learning are also altered in several host–parasite systems. Changes in olfactory perception and/or learning have been reported in various host–parasite systems as different as rats infected with *T. gondii*, amphipods infected with acanthocephalans, or dipteran vectors carrying protozoans. Is molecular convergence at the heart of these similar changes? The understanding of the mechanisms underlying parasite manipulation will likely reveal how parasites play tricks on their hosts, guiding them round more or less directly, by using molecular mimics or conserved molecules that affect the host's flexible neuromodulatory network.

See also: Defensive Avoidance; Evolution of Parasite-Induced Behavioral Alterations; Immune Systems and Sickness Behavior; Intermediate Host Behavior; Nematode Learning and Memory: Neuroethology; Neuroethology: Methods; Neuroethology: What is it?; Non-Elemental Learning in Invertebrates; Parasite-Modified Vector Behavior; Parasitoid Wasps: Neuroethology; Propagule Behavior and Parasite Transmission.

Further Reading

Adamo SA (2002) Modulating the modulators: Parasites, neuromodulators and host behavioral change. *Brain, Behavior and Evolution* 60: 370–377.

Biron DG, Moura H, Marche L, Hughes AL, and Thomas F (2005) Towards a new conceptual approach to 'parasitoproteomics'. *Trends in Parasitology* 21: 163–168.

Klein S (2003) Parasite manipulation of the proximate mechanisms that mediate social behavior in vertebrates. *Physiology & Behavior* 79: 441–449.

Lefèvre T, Koella J, Renaud F, Hurd H, Biron DG, and Thomas F (2006) New prospects for research on manipulation of insect vectors by pathogens. *PloS Pathogens* 2: 633–636.

Libersat F, Delago A, and Gal R (2009) Manipulation of host behavior by parasitic insects and insect parasites. *Annual Review of Entomology* 54: 189–207.

Tain L, Perrot-Minnot M-J, and Cézilly F (2006) Altered host behaviour and brain serotonergic activity caused by acanthocephalans: Evidence for specificity. *Proceedings of the Royal Society of London, B* 273: 3039–3045.

Thomas F, Adamo S, and Moore J (2005) Parasitic manipulation: Where are we and where should we go? *Behavioural Processes* 68: 185–199.

Parasite-Modified Vector Behavior

H. Hurd, Keele University, Staffordshire, UK

Introduction

Hematophagy is a feeding behavior that has been adopted by many invertebrates, including some insects and arachnids. It is thought to have evolved independently in many different lineages and is often accompanied by specific adaptations of mouthparts for biting and sucking. Blood-feeding arthropods can be regarded as ectoparasites that feed on the blood of their hosts when they are in temporary contact with them (such as mosquitoes, sandflies, tsetse flies, blackflies, tabanids, and blood-feeding bugs), in permanent contact (such as lice, sheep ked, and tungid flies), or in periodic contact (such as fleas and ticks). Many feed on blood only during one particular phase (often the adult female), while others feed on blood throughout their life cycle.

There are two major strategies associated with the acquisition of a blood meal, namely pool feeding, where biting mouthparts cut open superficial blood vessels and feeding occurs on blood that leaks out to the surface of the host; and piercing and sucking, where specialized mouthparts are inserted into the skin and blood vessels are severed or cannulated. During these feeding events, organisms present in the host blood or superficial skin layers may be imbibed inadvertently. Many of these parasitic organisms continue their life cycles within the arthropod and are transmitted back to the vertebrate host during a future blood-feeding episode. Hematophagous arthropods thus act as vectors of parasites and pathogen, including many of medical and veterinary importance. Because of this, they are indirectly responsible for tremendous suffering and economic loss to mankind.

Vector-Transmitted Parasites and Pathogens

Although insects were confirmed as vectors of parasitic diseases only at the end of the nineteenth century, travelers and explorers had already recorded the belief by tribes in Africa and South America that febrile illnesses were associated with mosquitoes and sleeping sickness with tsetse flies. We are now aware of a multitude of vector-borne disease caused by parasitic organisms ranging in size from arboviruses to helminths. Many diseases disproportionately affecting poor and marginalized populations are transmitted by insects. These include malaria, caused by a protozoan parasite of the genus *Plasmodium;* African trypanosomiasis or sleeping sickness, caused by the flagellates *Trypaonsoma brucei rhodesiense* and *T. b. gambiense;* Chagas' disease, caused by *Trypanosoma cruzi;* leishmaniasis caused by *Leishmania* spp.; lymphatic filariasis caused by *Wuchereria bancrofti* or *Brugia malayi;* river blindness caused by *Onchocerca volvulus,* and dengue fever, caused by DEN viruses. In addition, many wild and domesticated animals are afflicted by diseases caused by vector-borne parasites and pathogens. In the majority of cases, these organisms continue to develop within their vector, either just passing through developmental stages or undergoing sexual and/or asexual reproduction. In this process, complex biochemical, physiological, and behavioral interactions occur between the parasite and the vector that affect the fitness of the vector and the transmission prospects of the parasite.

Vector-borne parasites can only complete their life cycles once they have matured and are located in a site in the vector that gives them access to their next vertebrate host. This contact will occur during a blood-feeding episode. Some parasites, such as the sporozoite stages of *Plasmodium,* migrate to the salivary glands and a few are transferred to the vertebrate in the saliva that is injected into the wound made by the probing mosquito. Others, such as *Leishmania,* migrate toward the front of the gut and are regurgitated during feeding. Some trypanosomes are transmitted via the proboscis, whereas, in contrast, other trypanosomes are defecated with the feces close to the bite site and enter the wound from the surrounding skin.

To complete its life cycle, a vector-borne parasite or pathogen thus depends upon its vector making a minimum of two contacts with its vertebrate host, one to transmit the infectious organism to the vector and one to return it to a vertebrate host. However, in both the cases, successful establishment of an infection is by no means guaranteed as the host and vector will use their immune systems and other mechanisms to repel the invaders. We can therefore predict that selection will be operating in favor of those parasites that have evolved mechanisms to enhance their prospects of transmission. Evidence suggests that one such mechanism operates via manipulation of the vector's feeding behavior.

Vector Feeding Behavior

Examples of parasites that appear to manipulate the behavior of their host, such that their chances of transmission to the next host are enhanced, are well documented. Amazing examples of trophically transmitted parasites that cause their host to be eaten by their next host in preference to uninfected conspecifics are beloved by Parasitologists. Hosts of parasites transmitted in this way would not normally be seeking contact with the parasite's next (definitive) host, and indeed, if uninfected, they would be avoiding predation. Parasite transmission via blood feeding is, in contrast, very different in that it is in the interest of both the vector and the parasite that contact is made with the parasite's next host. Host manipulation that causes this to occur seems, on the face of it, unnecessary. However, blood feeding is a risky strategy, as vertebrates try to defend themselves against blood feeders. Vector decisions concerning feeding will therefore tend to minimize these risks by limiting host contact. In contrast, parasite transmission success will increase when more contact is made and when more hosts are contacted. Vector fitness will be optimized when the maximum amount of blood can be taken, with minimum risk to the insect. Once a parasite has reached a mature, infective, stage, parasite fitness will be enhanced, the more host contacts are made, regardless of the risk.

Insect blood-feeding behavior can be broken down into a sequence of events, namely; appetitive search, attraction to the host, landing, probing, and feeding. Many of these events have been shown to be altered by vector borne infectious agents, although probing behavior has been studied the most. Although the mechanisms underlying these changes differ, the resulting consequences are similar, namely, that more host contact occurs and parasite transmission is likely to be enhanced.

Feeding Persistence in the Face of Host Defensive Reactions

In a natural situation, hematophagy is often interrupted by host defensive responses. As this can result in death, insects are wary feeders and desist if they are aware that they will be interrupted. Feeding persistence is the repeated attempt to feed, despite these interruptions.

If a parasite is still in an immature stage, and not yet infective to the next vertebrate host, then vector contact with defensive hosts will not be beneficial, as it poses a major mortality risk to the vector, and hence to the parasite. The vector, however, always gains from host contact, especially in the case of females, because a blood meal will result in the production of a batch of eggs. So there is a conflict of interest between parasite and vector when the parasite is immature. A rodent malaria (*Plasmodium yoelii nigeriensis*)/mosquito (*Anopheles gambiae*) model has been used to investigate this conflict. Mosquito feeding was repeatedly interrupted before blood was imbibed and it

was shown that, while the parasites are developing as oocysts on the midgut wall and not yet infective, feeding persistence is significantly reduced in comparison with uninfected mosquitoes. This suggests that the malaria parasites had altered the mosquito's feeding behavior in such a way that fewer risks were taken and the vector was more likely to survive long enough for the parasite to mature. Interestingly, another study showed that the proportion of infected mosquitoes that fed from a mouse that was anesthetized and therefore unable to engage in defensive behavior, actually rose if they were infected with oocysts of another rodent malaria species, *Plasmodium chabaudi*. It would appear that, when oocysts are present, host contact is reduced only if it is risky, but there is more likelihood that the vector will obtain a nutritious blood meal if it is safe to do so. In contrast, the former study also showed that when the mature-stage parasites, the sporozoites, were present in the salivary glands, feeding persistence was significantly greater than that of uninfected mosquitoes. This remarkable reversal of effects could be predicted to benefit the parasite at both the stages of its development as the vector will engage in less risky host contact when contact does not benefit the parasite and more contact when the parasite could be transmitted.

Protozoan parasites of the genus *Leishmania* are the causative agents of different forms of leishmaniasis in mammalian hosts. In an experiment performed to examine feeding persistence, sand flies infected with infective stages were allowed to feed on an anesthetized mouse but feeding was artificially interrupted by brushing the antennae of the flies. The more feeding persistence, the greater the number of infective stages that were present per fly. In the case of both the malaria parasites and *Leishmania* parasites, only the infective stage of the parasite caused this behavioral change.

Experiments that monitor animal behavior in the laboratory should always be interpreted with care as activity may not be the same as that occurring in the wild. Studies of vector transmission are much more difficult to perform in a natural setting in the field than in the laboratory, not least because of ethical issues surrounding the experiments with human subjects. However, it has been demonstrated that mosquitoes infected with sporozoites of the human malaria *Plasmodium falciparum*, engorge more fully than uninfected mosquitoes in a field situation. This suggests that more tenacious feeding behavior also occurs in naturally infected mosquitoes, although we must bear in mind that another interpretation is that more tenacious mosquitoes are more likely to become infected.

Biting and Probing Behavior

Once hematophagous insects have landed on a host, they must probe the skin to find a blood source. During this probing phase, parasites may be deposited in, or on, the skin. Probing will cease if no blood vessel is pierced, or if host defensive responses endanger the vector. In addition, once the vector begins imbibing blood, feeding may be discontinued if the blood flow dries up or, again, if the vector is interrupted. Persistent insect will then attempt to feed again by probing in a different place or by moving to a new host. Any of these interruptions could result in increased parasite transmission as a new probing or feeding attempt will deposit additional parasites in the new wound site. Thus, the parasite will be advantaged if the vector needs several attempts to obtain blood, or takes multiple, small, blood meals. Several parasites have been shown to change the biting behavior of their vectors such that they probe more or take multiple meals and a few studies have demonstrated that this is likely to enhance transmission prospects for the parasite. Four different examples are given below.

The rat flea acts as a vector for *Yersinia pestis*, the plague bacillus responsible for the Black Death of 1348 and subsequent plague epidemics. The bacilli multiply in a mass of coagulated blood that blocks the flea's foregut. When the flea attempts to feed, the blockage causes a regurgitation of blood and bacilli into the mammalian host, thereby transmitting *Y. pestis* to the vertebrate host. The formation of the blood mass is caused by the product of a gene that is encoded on the plasmid of the bacillus. Difficulties experienced when trying to take blood up into a blocked gut cause the flea to make multiple attempts to bite. The bacilli pathogens themselves have thus been shown to directly cause the change in the vector's biting behavior. This will increase the likelihood that transmission to several hosts will occur and thus increase the fitness of the pathogen.

Blockage of the vector's gut and subsequent regurgitation of parasites also account for the transmission of the protozoan *Leishmania* from the sand fly back to the vertebrate. The infective stage of the parasite is located within a gel-like material that fills the lumen of the valves in the foregut that help to pump up the blood meal. This gel causes a blockage, which inhibits feeding. A major component of the gel is a filamentous substance that is produced by the parasites. It forms the 3D matrix of the plug that blocks the gut and keeps the valve open. By restricting blood flow, the plug facilitates regurgitation, prolongs feeding time, and causes flies to bite more. A comparison of the behavior of infected and uninfected sand flies feeding on an anesthetized mouse demonstrated that flies harboring infective stages take longer to feed but do not probe more often. As it is additional probing that creates more chances of transmission, this observation stresses the importance of host defensive behavior (not operating in an anesthetized mouse) in initiating vector behavior that enhances transmission of *Leishmania* parasites, rather than a blocked gut. It has been suggested that

the presence of the gel may inhibit the functioning of the mechanoreceptors in the foregut that detect blood flow, thereby altering the hunger state or increasing the threshold blood volume at which feeding is deceased. In common with *Y. pestis*, it would seem that *Leishmania* produce a molecule that is manipulating the biting behaviors of their vector.

Mosquitoes infected with malaria parasites also display increased probing behavior when they are infected with the infective stage of the parasite, the sporozoite. Here too, this change in vector biting behavior would appear to enhance transmission prospects, but there is no evidence as yet that this is the result of the action of a parasite molecule. In the mosquito, *Plasmodium* parasites divide asexually in the oocyst phase, which is located on the gut wall. A proportion of the thousands of sporozoites that are produced in the oocyst invade the distal lobes of mosquito salivary glands. Sporozoites are released into the vertebrate host with the saliva that is discharged during probing and feeding.

The distal lobes of the salivary glands of female mosquitoes produce an enzyme, apyrase, which inhibits the action of the platelet-aggregation factor. When apyrase is released into the vicinity of the wound made by the mosquito's stylets, the enzyme inhibits the action of platelet-aggregation factor and thus platelets are not recruited to plug the wound in the lanced blood vessel. Instead, a pool of blood will form in the surrounding tissue and the mosquito can feed from this pool. Feeding time has been related to the quantity of apyrase in the glands of different anopheline mosquitoes. Mosquitoes in which the apyrase gene had been knocked down take twice as long probing before obtaining a blood meal. The presence of sporozoites has a similar effect to the gene knock down. A fourfold decrease in the activity of salivary apyrase has been detected in malaria-infected mosquitoes, even though the volume of saliva produced is constant. Several studies have shown that probing behavior and blood location time increases in sporozoite-infected mosquitoes. We could thus predict that the inactivation of apyrase and consequent increase in probing behavior will increase the number of hosts contacted by infected mosquitoes. This prediction was upheld in a study that showed that a mosquito infected with avian malaria took multiple blood meals on different hosts. In addition, markers were used to type the human blood meals taken by field-caught mosquitoes and match them with the individuals they had fed upon. Significantly more malaria-infected mosquitoes were shown to have fed on more than one person during the night, compared with uninfected ones. Clearly, the increase in host-seeking and probing behavior that occurs when mosquitoes are infected with sporozoites is likely to increase the chances for parasite transmission.

A final example is that of a kissing bug infected with the protozoan, *T. cruzi*, which causes Chagas disease.

The parasite multiplies in the digestive tract of the bug; forming an infectious form that is excreted with the feces and invades the definitive vertebrate host via the mucous membranes. Detection of, and orientation to, a potential host is almost twice as fast as that in uninfected kissing bugs, the biting rate of the bug is significantly higher and defecation occurs 3 min earlier. Unlike the previous examples, neither salivary gland pathology nor blockage of the gut can explain the increase in the biting rate and it has been suggested that infection may lead to depletion of trace nutrients that leads the infected kissing bugs to be more eager to bite. Can this be regarded as true manipulation or is the vector's feeding behavior a by-product of the infection?

Is the Parasite Manipulating Vector-Feeding Behavior?

In the 1980s, Richard Dawkins discussed the idea of the extended phenotype in his book of that name. The concept that products of the genes of an organism can influence factors outside of that organism has been applied to parasites and developed into the 'manipulation hypothesis.' This states that changes in host behavior that occur following infection are the result of parasite adaptations that favor transmission. Alternative hypotheses that explain these behavioral changes are that they are secondary outcomes of the pathological effects of the infection, such as resource depletion, or that they are host adaptations that compensate for the effects of the parasites. Finally, it has been suggested that host behavioral changes may be coevolved traits and that there may be fitness benefits for the host as well as the parasite. It has been suggested that several criteria should be met before behavioral changes can be regarded as manipulations. One of these is that there should be obvious fitness benefits to the parasite. Clearly some of the examples of parasite-induced alterations in vector behavior outlined earlier would be predicted to enhance the transmission prospects and thus would confer benefits upon the parasite. This has been confirmed for mosquitoes with sporozoite infections and also for *Leishmania*-infected sandflies as both the vectors have been shown to feed on multiple hosts when infected. Finally, the discovery that both plague bacilli and *Leishmania* parasites produce molecules that manipulate the blood-feeding behavior of their vectors supports the manipulation hypothesis. Despite these examples, a great deal of caution must be taken before categorizing the majority of changes in the vector blood-feeding behavior as manipulations and many more investigations of underlying mechanisms need to be performed before we can do this with any degree of assurance.

See also: Ectoparasite Behavior; Evolution of Parasite-Induced Behavioral Alterations; Intermediate Host Behavior; Parasite-Induced Behavioral Change: Mechanisms; Propagule Behavior and Parasite Transmission.

Further Reading

Dawkins R (1982) *The Extended Phenotype: The Long Reach of the Gene.* Oxford: Oxford University Press.

Hurd H (2003) Manipulation of medically important insect vectors by their parasites. *Annual Review of Entomology* 48: 141–161.

Koella JC (2005) Malaria as a manipulator. *Behavioral Processes* 68: 271–273.

Lefèvre T, Roche B, Poulin R, et al. (2008) Exploiting host compensatory responses: The 'must' of manipulation? *Trends in Parasitology* 24: 435–439.

Lehane MJ (2005) *The Biology of Blood-Sucking Insects.* Cambridge: Cambridge University Press.

Marquardt W (ed.) (2004) *Biology of Disease Vectors,* 2nd edn. San Diego, CA: Elsevier Academic Press.

Moore J (2002) *Parasites and the Behavior of Animals.* New York: Oxford University Press Inc.

Mullen G and Durden L (2002) *Medical and Veterinary Entomology.* San Diego, CA: Academic Press.

Poulin R (2007) *Evolutionary Ecology of Parasites*, 2nd edn. Princetown: Princetown University Press.

Rogers ME and Bates PA (2007) *Leishmania* manipulation of sand fly feeding behaviour results in enhanced transmission. *Plos Pathogens* 3: 818–825.

Parasites and Insects: Aspects of Social Behavior

M. J. F. Brown, Royal Holloway University of London, Egham, UK

Introduction

Parasites are ubiquitous. Probably 50% of all animal species are parasites, and every free-living animal is likely to be parasitized at some point during its life cycle. A parasite can be broadly defined as an organism that lives in or on another organism, the host, from which it derives resources, resulting in a reduction in fitness of the host and an increase in its fitness. This definition includes viruses, bacteria, microparasites, fungi, and macroparasites. It has been known for a long time that parasites can, and do, alter the behavior of their hosts. While many insect hosts are solitary animals, they nearly all engage in some form of social behavior, for example, mating behavior or competitive interactions. However, the social insects – ants, bees, termites, and wasps – have taken social behavior to extremes, evolving arguably the largest and most complex social systems on the planet, bar *Homo sapiens*. The success of these societies has made social insects the dominant form of terrestrial life. However, only recently have the interactions between parasites and insect social behavior been put under the spotlight. Recent discoveries suggest that the impacts of parasites on social behavior in insects have been profound, and considerably more diverse than were initially expected. In this article, I will start by examining the relationship between parasitism and social behaviors seen in both solitary and social insects, before examining the more complex effects of parasites within insect societies.

Mating Behavior and Parasites

In many solitary species, mating may be the only time when individuals come together and exhibit social behavior. The function of this behavior is to maximize fitness of both male and female parties. Parasites might influence mating behavior in at least four distinct ways.

First, and most prosaically, parasitized individuals may be less capable of successful mating behavior because of the impact of parasitism. For example, male fruit flies (*Drosophila nigrospiracula*) infested with a mite (*Marrocheles subbadius*) are less capable of mating with female flies, and male dragonflies (*Libellula pulchella*) infected with a gregarine are less capable of gaining a territory and thus have lower mating success.

Second, parasite-induced reductions in host fitness may have an impact on mate quality. If insects can detect reduced fitness in prospective mates, parasites should affect the likelihood that their host will be chosen as a mating partner. In contrast to the first mechanism, there is little if any good evidence that mate choice behavior is driven by parasites. One possible example is seen in midges (*Paratrichocladius rufiventris*) parasitized by ectoparasitic mites (*Unionicola ypsilophora*), which have higher mating success. However, this effect appears to be driven by the presence of the mite being the signal for the absence of a more virulent, castrating nematode worm (probably *Gastromermis rosea*). Despite this disparity in evidence, these first two mechanisms overlap considerably.

Third, and most interestingly, some parasites are sexually transmitted diseases (STDs). Consequently, mating behavior should have been selected to minimize the potential of acquiring an STD. In contrast, parasites that are sexually transmitted should have been selected to manipulate their host and increase the number and frequency of sexual contacts. There is little or no evidence to support the former hypothesis, and while several studies have provided evidence supporting the latter effect (e.g., male milkweed leaf beetles (*Labidomera clivicollis*) are more successful in obtaining matings when infected by the subelytral mite (*Chrysomelobia labdomerae*)), such data can also be interpreted as host adaptations to increase their fitness against the background of parasite virulence. For example, life-history theory predicts that if an individual's lifespan is going to be curtailed, one response is to increase the energy allocated to mating opportunities.

A fourth way in which parasites might affect mating behavior is selection on mating frequency. The Tangled Bank theory proposes that sexual reproduction results in an array of offspring that vary in their ability to cope with environmental challenges. Parasites are a key feature of every organism's environment. Mating with multiple individuals enables individuals to produce more genetically heterogeneous offspring, increasing the likelihood that some of them may escape parasitism. The evolution of polyandry through natural selection by parasites has been particularly closely examined in social insects (mainly the bees and ants). This is because polyandry also has the effect of reducing relatedness within colonies, decreasing the genetic benefit to individual workers of raising their mother's brood (see section 'Evolution of Sociality' below for a more detailed explanation of why relatedness matters within insect societies). Comparative analyses, theoretical studies, and experiments manipulating polyandry in bumble bees and honeybees all provide evidence that parasites might have selected for the evolution of polyandry. However, counter studies and alternative hypotheses for polyandry exist, and the question remains open.

Evolution of Sociality

Standard explanations for the evolution of sociality are based upon Hamilton's theory of kin selection. This states that apparently altruistic behavior, in this case living in a group and helping to raise another individual's offspring, can evolve if the benefits of the behavior multiplied by the relatedness of the recipient to the actor are greater than the costs to the actor. While most studies have examined the importance of relatedness, there is an increasing recognition that the costs and benefits require equal study. Parasites might affect this balance if they change the probability of being able to produce offspring, that is, if they decrease direct fitness. This is, in fact, a general result of parasitism, which must, by definition, reduce the Darwinian fitness of hosts. Consequently, whether offspring stay at home to help rear their siblings or whether individuals join a social group and forego their own solitary reproduction may be significantly affected by parasites. If parasites reduce the ability of an animal to produce its own offspring, then selection may act to favor parasitized individuals that redirect their energy to help relatives that are able to reproduce, resulting in an increase in both inclusive fitness and social behavior. Thus, the origin of insect societies may lie in parasitism. Like many of the hypotheses or ideas discussed in this article, there is a general lack of tests of this hypothesis.

Division of Labor

Insect societies have been likened to factories, where not only are many tasks done in parallel, but also individuals are specialized on one task or a small subset of tasks. While extreme task specialization does exist, most social insect workers are totipotent (in terms of behavioral repertoire) and switching between tasks is common at both short (minutes) and long (days) time scales. Nevertheless, at any point in time tasks are distributed across the individuals within a colony. This division of labor has been suggested to explain the dominance of social insects in terrestrial habitats. Division of labor was initially thought to result from natural selection operating on colonies to maximize efficiency, and thus, by extension, reproductive fitness. While this explanation has intuitive appeal, convincing evidence to support it is sparse. In contrast, Paul and Regula Schmid-Hempel suggested that the way labor is divided and colonies are organized might have been selected for by parasites. It has been suggested that in

many social insect species, there is a general temporal pattern in the organization of workers, with young workers living inside the nest and older workers acting as foragers outside the nest. The queen, the most valuable member of the colony, usually resides with the younger workers and brood. The Schmid-Hempels argued that parasites generally come from outside the colony. The internal-to-external drift of workers as they age would thus serve to counteract the spread of a parasite from foragers to internal and younger nest workers, and ultimately to the queen (**Figure 1**). Similarly, spatial partitioning of workers would reduce the spread of disease within a colony.

More recent work with a trypanosome parasite (*Crithidia bombi*; **Figure 2**) and bumble bee (*Bombus impatiens*) colonies has shown that the prerequisite for this hypothesis – that transmission occurs through contact networks – is true. Similarly, work in leaf-cutting ants (*Atta* spp. and *Acromyrmex* spp.) has demonstrated that ants that work with pathogen-rich material (waste from the fungus garden and dead nestmates) have reduced contact rates with other groups of workers. Presumably, this compartmentalization limits the flow of parasites and toxic waste back in to the colony. However, whether systems of division of labor have indeed

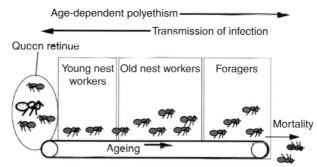

Figure 1 The conveyor belt model showing the potential interaction between division of labor and parasite epidemiology within a colony of social insects. Modified from Schmid-Hempel P (1998) *Parasites in Social Insects*. Princeton, NJ: Princeton University Press.

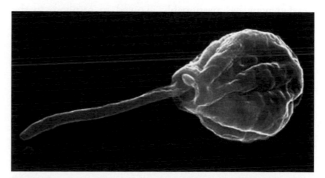

Figure 2 A scanning electron micrograph of *Crithidia bombi*, a prevalent parasite of bumble bees (*Bombus* spp.). This trypanosome and its bumble bee hosts have become a model system for the study of parasites in social insects. Image courtesy of Professor Paul Schmid-Hempel, ETH-Zürich, Institute of Integrative Biology, Switzerland.

been selected by parasites, or simply provide fortuitous protection, remains to be tested.

In addition to selecting for the structure of division of labor, parasites may change how workers behave within a given system. For example, honeybees (*Apis mellifera*) infected by a microsporidian (*Nosema apis*) become foragers more quickly and are less likely to exchange food via trophallaxis with the queen or young workers. However, in most such cases, whether these behavioral changes are an adaptation of the host to minimize spread of the disease, manipulation by the parasite to maximize its transmission, or nonadaptive side effects of parasitism remains unknown. Nevertheless, such changes have knock-on effects for the distribution of other workers across tasks and thus will effect the functioning of the colony as a whole.

Social Immunity

Social immunity can be broadly defined as group or social behaviors that lead to an increased protection – at the colony level – against parasites. By definition, social immunity is a social behavior selected for by parasites.

The earliest example of social immunity was the discovery of *necrophoresis* and undertakers in ants and honeybees. In honeybees and, to a lesser extent, some ant species, dead workers are removed from the colony by a subset of workers that appear to be task specialists. In honeybees, these individuals are known as undertakers, whereas in ants, this task can be conducted by a broader group of less specialized workers. This process is known as necrophoresis. Originally suggested by E. O. Wilson and colleagues to be elicited by the presence of oleic acid – which increases in concentration as insects decay – later work by Deborah M. Gordon demonstrated that ant responses to this chemical are context dependent. The most recent study suggests that undertaker ants respond not to the products of death, but to the absence of chemical products indicating life. Regardless of the mechanism, this undertaking behavior has presumably evolved in response to the threat that decomposing corpses pose to the colony, through the production of parasite transmission stages (if the individual was killed by a parasite that requires host-death for transmission, as many fungi do), colonization by opportunistic pathogens, or the release of generally toxic products.

More recent discoveries in termites have inspired an increasing focus on the study of social immunity. Dampwood termites (*Zootermopsis angusticollis*) suffer high parasite pressure from fungal pathogens. Rebeca Rosengaus and colleagues showed that individuals respond to contact with spores by rapidly vibrating themselves against the substrate. Nestmates that detect these vibrations flee from the location of the spores. Thus, this social behavior acts to minimize infection.

A second and very different behavior that minimizes infection is the collection of resin by ants, and resin or

propolis by honeybees. Wood ants (*Formica paralugubris*) forage for resin, a nonfood stuff. Resin protects colonies against the proliferation of detrimental microorganisms by inhibiting the growth of bacteria and fungi. Similarly, honeybees (*A. mellifera*) collect resin, or propolis. Foraging is a social task in ants, bees, termites, and wasps because collected items are brought back to the colony and distributed among colony members. Consequently, collection of nonfood items with prophylactic effects is an example of a social behavior that most likely is selected for at the colony level, and which appears to have evolved convergently in two very different social species.

Once infection has occurred, social immunity can act to minimize the spread of the parasite. This can be achieved by social exclusion of infected individuals or, as has been more recently discovered, acquiring immunity through deliberately interacting with infected animals. Sylvia Cremer and colleagues showed that ants (*Lasius neglectus*) raise their contact rates with nestmates that have been exposed to a fungal parasite. When treated with the same parasite later, these individuals have higher resistance. This immune priming may well be a widespread mechanism of social immunity across the social Hymenoptera.

A much more dramatic method of control is social behavioral fever. Behavioral fever, the induction of higher body temperatures through behavioral means, is used by individual locusts to control the growth of pathogenic fungal parasites. Honeybees have evolved a social form of behavioral fever in which together a large group of bees generate high temperatures through synchronized muscle vibration within the hive. This behavior was first seen as a way to increase the rate of brood development, and then as a defense against predators; Asian honeybees (*Apis cerana*) use it to 'bake' to death predatory hornets that invade their hives. More recently, Philip Starks and colleagues demonstrated that in the European honeybee, *A. mellifera*, behavioral fever can kill a lethal fungal parasite (*Ascosphaera apis*). The presence of this parasite is either sensed by adult bees prior to visible symptoms in infected brood, or is communicated to the adults by infected brood, which would indicate an even more complex social behavior.

Working for the Colony and Adaptive Suicide

Unlike most solitary animals, social insects work mostly to increase their inclusive fitness, rather than their direct fitness. This means that the work they do increases the production of new individuals, and particularly new queens and males, by the colony. Consequently, parasites may have selected for novel behaviors in social systems. One such behavior is seen in bumble bees that are parasitized by conopid fly parasitoids. These flies oviposit into the abdomen of foraging bees. Rather than going back to their colonies at the end of the foraging day, these infected bumble bees stay outside their nest at night. This serves to reduce their body temperature and prolong the development time of the fly and thus their own lifespan. All infected bees die eventually, and they only rarely reproduce themselves, so this is an example of a behavior that has been selected at the colony level to maximize the amount of work a bumble bee can provide to its colony.

While workers rarely reproduce, in many social Hymenoptera, they have the potential to lay haploid eggs that can develop into males, giving the workers direct fitness through their sons. Parasites that incur a physiological cost may prevent workers from developing their ovaries, and thus act to maintain social cohesion within a society and prevent selfish behavior. One such parasite is the trypanosome *C. bombi*, which is a common parasite of numerous bumble bee species worldwide. Infection by this parasite reduces ovarian development, and thus workers carry on working for the colony and therefore aid the queen in her reproduction, rather than competing with each other for the opportunity to lay male eggs. This should be advantageous both to the colony as a whole, as it should maximize colony productivity, and to the parasite, which needs new queens to carry it through into the next generation.

A second kind of colony-level behavior is that of adaptive suicide. The essential idea is that workers that become infected leave the colony to die elsewhere. Such behavior could be advantageous to the colony, if it would prevent the spread of parasites and disease to other workers. Consequently, it would increase the inclusive fitness of the worker committing suicide. Such adaptive suicide is not predicted to occur in solitary animals, as there would be no selective benefit. The transition of parasitized individuals to external work, such as foraging, which has a high associated mortality risk, has been interpreted as an example of adaptive suicide. Similarly, the loss of fleeing behavior in parasitized aphids when approached by coccinellid predators has been interpreted as providing protection to clone mates. However, there remains little or no convincing evidence that adaptive suicide has evolved within insect societies.

Conclusion

The effects of parasites on social behavior appear to be many and varied. However, most of our evidence for these effects is from a very small number of study systems. While some of the effects seen in these model systems are likely to be widespread across the interactions between parasites and social groups, others are likely to be specific to a given host–parasite interaction. Much further work is needed to determine which is which, and to discover if these effects follow general rules. Such studies across diverse host taxa and parasite groups will undoubtedly discover new and unimagined impacts of parasites on social behavior in insects.

See also: Avoidance of Parasites; Disease Transmission and Networks; Group Living; Parasite-Induced Behavioral Change: Mechanisms; Social Behavior and Parasites.

Further Reading

Brown MJF and Schmid-Hempel P (2003) The evolution of female multiple mating in social hymenoptera. *Evolution* 57: 2067–2081.

Chapuisat M, Oppliger A, Magliano P, and Christe P (2007) Wood ants use resin to protect themselves against pathogens. *Proceedings of the Royal Society B* 274: 2013–2017.

Choe D-H, Millar JG, and Rust MK (2009) Chemical signals associated with life inhibit necrophoresis in Argentine ants. *Proceedings of the National Academy of Sciences of the United States of America* 106: 8251–8255.

Knell RJ and Webberloy KM (2004) Sexually transmitted diseases of insects: Distribution, evolution, ecology and host behavior. *Biological Reviews* 79: 557–581.

McAllister MK and Roitberg BD (1987) Adaptive suicidal-behavior in pea aphids. *Nature* 328: 797–799.

Müller CB and Schmid-Hempel P (1993) The exploitation of cold temperature as defense against parasitoids in bumblebees. *Nature* 363: 65–67.

Rosengaus RB, Jordan C, Lefebvre ML, and Traniello JFA (1999) Pathogen alarm behavior in a termite: A new form of communication in social insects. *Naturwissenschaften* 86: 544–548.

Schmid-Hempel P (1998) *Parasites in Social Insects.* Princeton, NJ: Princeton University Press.

Schmid-Hempel P and Schmid-Hempel R (1993) Transmission of a pathogen in *Bombus terrestris*, with a note on division of labor in social insects. *Behavioral Ecology and Sociobiology* 33: 319–327.

Shykoff JA and Schmid-Hempel P (1991) Parasites delay worker reproduction in bumblebees – consequences for eusociality. *Behavioral Ecology* 2: 242–248.

Starks PT, Blackie CA, and Seeley TD (2000) Fever in honeybee colonies. *Naturwissenschaften* 87: 229–231.

Ugelvig LV and Cremer S (2007) Social prophylaxis: Group interaction promotes collective immunity in ant colonies. *Current Biology* 17: 1967–1971.

Visscher PK (1983) The honey bee way of death: Necrophoric behaviour in *Apis mellifera* colonies. *Animal Behaviour* 31: 1070–1076.

Parasites and Sexual Selection

A. Jacobs and M. Zuk, University of California, Riverside, CA, USA

Introduction

At first glance, parasites and sexual signaling might seem like an odd combination. For a long time, the study of parasites was relegated to the field of parasitology, most other scientists overlooking them or dismissing them as unimportant. However, in more recent years, a broader range of scientists have begun to appreciate that parasites can impose a heavy cost on their hosts in terms of energy and may affect many aspects of their hosts' life history and ecology, playing a role in everything from predator–prey interactions to host reproduction and sexual selection.

Upon consideration, it makes sense that animals looking for mates would do well to avoid heavily parasitized individuals. Some parasites can be easily transmitted during the close contact required for mating, making infected mates a risky choice. Also, in species in which both parents care for the offspring, such as most North American songbirds, a mate with parasites may have less energy to devote to raising young, thus lowering the number of offspring raised and the fitness of that individual's mate. These ideas are known as the *transmission avoidance hypothesis* and the *good parent hypothesis*, respectively.

Both of these hypotheses rely on individuals gaining direct benefits from their choice of nonparasitized mates. However, it is also conceivable that choosing parasite-free mates could provide indirect benefits as well in the form of good genes that would help offspring survive better. If parasite resistance is heritable, then that resistance would be passed on to any offspring, giving them higher fitness as well. This idea was put forward in 1982 by William D. Hamilton and Marlene Zuk and thus came to be known as the *Hamilton–Zuk hypothesis*. This hypothesis proposed that many of the flashy signals males use to attract mates evolved as a way of signaling parasite resistance. The healthiest males would develop the most elaborate sexual signals. These signals could be anything from the chirp of a cricket to the long, elaborate tail of a male widow bird. Females would choose to mate with males on the basis of these cues and by doing so gain resistance genes for their offspring.

One of the reasons why the Hamilton–Zuk hypothesis gained so much attention was that it offered a possible solution to a problem that had puzzled evolutionary biologists for some time. In some species, females raise offspring on their own with no input from males except the sperm needed to fertilize their eggs. The species include lekking species such as the sage grouse, in which males

display in a large area known as a lek. Females come to the lek, survey the males assembled there, and mate with one of them. These females are often picky about which mate they choose, and the preferred males in the population often end up fathering most of the offspring. However, after several generations of such strong selection, one would expect that the genes behind the trait preferred by females would be driven to fixation. Thus, all males would be fairly genetically similar, and females would not gain any genetic benefits for their offspring. Why then do females in such species remain choosy? This is a problem known as the lek paradox.

The Hamilton–Zuk hypothesis offers a solution to the lek paradox. Because parasites and diseases are constantly coevolving with their hosts, the fitness of a given genotype is subject to change. Parasites often evolve to exploit whichever host genotypes are most common. Thus, a genotype that confers resistance now may not have any advantage against parasites a few generations later when they have evolved to exploit it. Since the 'fittest' genotype is constantly changing, females are not selecting for any particular genotype generation after generation, thus maintaining genetic variability in the population.

The Hamilton–Zuk hypothesis allows scientists to make various predictions about the relationship between bright signals and parasites, both within and between species. Within a species, the hypothesis states that less parasitized males should display more extravagant secondary sexual signals and that females should prefer to mate with these males. The hypothesis also predicts that when looking at many different species, those with the most parasites should have evolved the flashiest signals. Both predictions have been examined, although most scientists have focused on the relationship between parasites and sexual signals within a single species.

Examples of the Role of Parasites in Animal Signaling

Jungle Fowl

Jungle fowl are the ancestors of domestic chickens, and in their native Asia, they live in groups consisting of a dominant rooster, one or more subordinate roosters, hens, and their chicks. The males are elaborately ornamented with colorful feathers and a bright red fleshy comb, and both male competition and female choice of particular males have been well established in this species. Jungle fowl adapt readily to captivity, and because of their similarity to domestic poultry, they have been widely used in studies of the effect of parasites on sexual selection.

Of all the secondary sexual characteristics male jungle fowl possess, the most important in mate preference seem to be the size and color of the comb and the color of the male's eye, which ranges from pale yellow to bright orange. The color and length of the ornamental feathers do not seem to matter as much to females in selecting a mate, and although larger males are more likely to win fights with other roosters, the hens do not seem to prefer larger males. The traits used by females are particularly interesting because they are condition dependent; in other words, they can change relatively readily depending on the physical condition of their bearer. Comb size and color can change in a matter of days, but feathers molt only once a year and are fixed after that. Condition-dependent traits are thought to be important in sexual selection because they allow females to assess male quality in a fine-tuned manner.

When a group of jungle fowl chicks were given infections of a common intestinal roundworm parasite, they grew more slowly and had less well-developed secondary sexual characteristics than a control uninfected group of birds. Eventually, the infected males were able to 'catch up' to the uninfected roosters in body size, but the parasites seemed to exert a long-lasting effect on precisely those traits that the females used in mate choice, and the hens were much less likely to mate with a parasitized male. This result is important, because it suggests that parasites disproportionately diminish ornamental characters, which means that those characters are good indicators of freedom from parasites, as the Hamilton–Zuk hypothesis suggests. Hens were not simply avoiding the overall most sickly-looking roosters, but were selectively paying attention to the handful of traits affected by the parasites.

Barn Swallows

Barn swallows are a common sight over fields in North America and Europe. These fork-tailed birds are socially monogamous, and the tails of the males serve as a secondary sexual character. Anders Pape Møller, who works at the Pierre and Marie Curie University in Paris, has conducted extensive work on sexual selection in barn swallows and what factors determine how many offspring a particular male swallow has. Through observations as well as manipulating experiments, he found that females tended to focus on male tail length in particular when choosing a mate, so that males with longer tails are more likely to mate earlier, giving them more time to raise more offspring. Also, these males were more successful at gaining extra-pair copulations, or mating with females already paired to another male. Both of these factors increased the

number of offspring a male could have, and thus a male's reproductive success was directly correlated with his tail length.

How did parasites fit into this picture? After his initial work on the tails, Møller set out to answer that very question. He examined the birds and found a blood-sucking mite that seemed to be important to the barn swallows' fitness. If a barn swallow carried enough mites, it became anemic and emaciated from constant blood loss. This debilitating parasite seemed to play an important role in sexual selection as well. When Møller examined the effects of parasites on males, he found that males with fewer mites tended to have longer tails, making them more attractive to females and increasing their reproductive success.

It is not enough to simply say that females avoid parasitized males, however. Mites are ectoparasites that are easily transmitted between individuals and male barn swallows help raise the nestlings. Thus, female preference in this species could be explained by the good-parent hypothesis, the transmission avoidance hypothesis, or the Hamilton–Zuk hypothesis. To test which of these models best explained the female preference in this case, Møller also looked at the heritability of mite resistance. If females were choosing males because they had the genetic ability to resist the mites, as predicted by the Hamilton–Zuk hypothesis, then their offspring should also have fewer mites. However, there was the possibility that nestlings might resemble their fathers in terms of mite numbers simply because their fathers shared mites with them when they went to the nest (a variant of the transmission avoidance hypothesis). To control for this, Møller performed a cross-fostering experiment in which he moved very young nestlings between nests and observed them. He found that nestlings whose fathers had fewer mites also had fewer mites regardless of whether they were raised in their parents' nest or another nest, implying that there was some genetic component to resistance that passed from the father to the offspring. This heritability of resistance is one of the crucial predictions of the Hamilton–Zuk hypothesis that sets it apart from other explanations for why females might avoid parasitized males. Females in this species appear to be choosing males on the basis at least in part of their ability to resist parasites, just as the hypothesis predicts.

Soay Sheep

Soay sheep are the descendants of domestic sheep that were once kept on the islands of St. Kilda off the coast of Scotland. The human inhabitants of these islands have since left, but the sheep remain, living and breeding in the wild. Their presence on the island provides scientists with an excellent study system, since the sheep live without predators or human interference. Long-term studies continue to be conducted on the sheep, with variables such as population density, breeding success, and parasite load having been measured for many years.

Soay sheep have a polygynous mating system in which males mate with multiple females. Dominant males will locate females in oestrus and accompany them, chasing off other males who approach. Dominance, in this case, is usually linked with the size of the horns, which are a secondary sexual character. Two types of horn exist in the male population: normal males, and scurred males, which have smaller horns. These scurred males are often unsuccessful at defending females and must rely on sneaking copulations when the dominant male is elsewhere. Females also seem to prefer dominant males with larger horns, although this may simply be because of such males protecting them from harassment by other males.

Scurring is apparently a hereditary trait, but horn size can be influenced by parasites. Of the parasites that infect the sheep, most seem to be fairly benign, but one, a nematode called *Teladorsagia circumcincta*, can cause symptoms such as lack of appetite, weight loss, and diarrhea. Sheep infected with these worms are more likely to die during the winter, particularly in years when food is scarce. In addition to these symptoms, males that had higher fecal egg counts (meaning that they were infected by a greater number of worms) tended to have smaller horns. They also spent less of their time pursuing females or engaged in sexual activities. Fecal egg count also showed some heritability, implying that whatever allowed some sheep to avoid heavy parasite loads was at least partly genetic: a prediction consistent with the Hamilton–Zuk hypothesis. These results do not demonstrate a conclusive link between parasites and mating success in this species, but they do suggest a connection between the two things.

Sex, Parasites, and the Immune System

When scientists initially set out to test the Hamilton–Zuk hypothesis, many of them chose to do so by studying the direct effects of a single parasite species on mating success. However, sometimes problems would arise when it came to picking a proper parasite. Parasites are often present only in certain locations or at certain times, so their impact on the host might not be readily apparent. In addition, not all parasites are equal when it comes to the evolution of male sexual signals. A highly virulent parasite or pathogen that sweeps through a population quickly would be unlikely to cause the kind of selection pressure necessary for males to evolve flashy signals. Such diseases usually leave their hosts either dead or with lasting immunity; thus any male or female encountered would almost certainly be uninfected and there would be no need for a special signal to convey that information. Parasites that cause little or no damage would also be unable to drive

the evolution of secondary sexual signals. Unless the parasite causes some harm to its host, females choosing elaborate males would gain no benefit in terms of fitness for their offspring. Under such conditions, female choice, which can be costly to the female, would not be maintained.

How then does one decide which parasite to choose when testing the Hamilton–Zuk hypothesis? Many scientists have chosen to bypass the issue entirely by instead looking at immunity. An animal that is able to mount a stronger immune response can presumably fend off parasitic infections more easily, giving scientists a proxy of overall parasite resistance. Measuring immunity also allows scientists to get around the problem of measuring multiple parasite species. Animals are almost always infected with more than one kind of parasite, but it is often difficult to measure all the parasites involved and their relative importance. The immune system, however, must be able to deal with all of them to some degree, and so provides a way to estimate the overall importance of parasites to a given species.

Measuring immunity also opens up a whole new field to those studying sexual selection: the field of immunoecology. It takes energy and resources to mount an efficient immune response, both of which may be limited. Thus, maintaining immunity may require trade-offs with other functions, such as reproduction. These trade-offs can influence an animal's life history strategy, including things such as how many offspring to have and whether to have them all at once or to spread them out over the course of several breeding seasons.

If immunity can be used as a measurement of parasite resistance, then choosy individuals would benefit from being able to gauge whether their potential partner can mount a strong immune response. The same secondary sexual signals that females use to judge whether a male is parasitized or not should also reveal information about his general immune status. Thus, measuring immunity in place of parasites is a useful tool when examining how parasites influence sexual signaling. It is particularly useful in animals such as invertebrates, whose parasites are not as well known.

Invertebrate Examples

When one thinks of secondary sexual characters, often the first things to come to mind are the elaborate and colorful feathers of a bird like a peacock or the croaking call of a male frog. Yet traits such as cricket song have also evolved through sexual selection. Invertebrates are sometimes neglected in studies of how parasites influence sex and sexual selection, and yet in many ways they are an excellent system in which to test various hypotheses. Their short generation times and the fact that scientists can easily keep them in the lab make them amenable to such tests.

Crickets

Most people have heard the chirping of crickets on a warm summer's night. What fewer people realize is that this song is the product of male crickets only and that they use it to attract females. A female cricket will listen to and assess the song of a male and then move toward the source of the song she prefers: a process known as phonotaxis. Once the female reaches the male, he will continue to court her using a softer song known as courtship song, and the female will decide whether or not to mate with him.

Calling song is the first cue a female uses to assess a male, and females find the calling songs of some males more attractive than others. Specifically, females find elements of the song such as long trills or fast chirps particularly attractive. Which elements females prefer often depends on species, but presumably females hone in on those parts of the song because they convey some information about the male. This idea was tested by Leigh Simmons, who works at the University of Western Australia, and his colleagues. To see whether any components of the song told the females about male immunity, they first recorded males singing in the field, then captured them and brought them into the lab to test their immune response. They measured the ability of the males to encapsulate a piece of nylon thread. Encapsulation is an immune response often used by insects to defend against parasitoids or other foreign objects that get into the body. What they found was that certain elements of the song were indeed correlated with the male's ability to encapsulate a foreign object. Thus, a female can use this information when choosing which male she wants to mate with.

How might this information benefit the female? Given that crickets do not have parental care, the good parent hypothesis does not explain this preference. Whether males with stronger immunity succeed in passing that trait on to their offspring is not known in the species, although it is certainly possible. Thus, females may benefit from preferring some male songs to others either by avoiding becoming parasitized themselves or by gaining good genes for their offspring.

Bedbugs

Up until now, most of the examples discussed have dealt with females discriminating against infected males and how this may have driven the evolution of male sexual characters. However, parasites and pathogens are ubiquitous, and males are not the only ones they infect. In bedbugs in particular, females have been the ones under pressure to evolve a special character related to pathogens. This is due to the rather odd mating habits used by this species. Males have a special, needle-like copulation organ known as a paramere, which they use to pierce the female's body and inject their sperm into her abdomen, despite the fact that the females have a fully functional genital tract. This process, which is called traumatic insemination, has dire consequences for the female. Such mating practices reduce female lifespan, and this reduced lifespan results in lower egg production, or lower fitness for the female.

In response to this unusual mating situation, female bedbugs have evolved a special organ at the site where males insert their parameres. This organ is called the spermalege. Scientists have proposed many hypotheses about the actual function of the spermalege, including one that suggests that it helps to prevent infection of the puncture wound caused by the male. To test this idea, they used needles to mimic the male's paramere and punctured the bodies of female bedbugs either at the spermalege or at a different site on the abdomen. The needles used were either sterile or dipped in a bacterial solution. For females pierced with a bacteria-laden needle, the site of the injection made a huge difference. If the needle went into her spermalege, she was much more likely to survive longer than if it pierced a random site on her abdomen. These results suggest that pathogens have helped to drive the evolution of an entirely new organ in this species.

Sex Hormones, Signaling, and the Immunocompetence Handicap Hypothesis

Signals as diverse as a cricket's song and a bird's tail all seem to convey some information about parasites and immunity. However, how those signals can give accurate information is a subject much debated by scientists. What are the mechanisms that link the length of a male barn swallow's tail to his ability to resist parasites? One possible clue comes from the hormones needed to develop such a signal in the first place. In vertebrates, most male sexual signals develop under the influence of testosterone. However, in addition to the role it plays in the development of male signals, testosterone also seems to depress the immune system. This observation led to a proposed mechanism for how sexual traits are linked to immunity.

In 1992, Ivar Folstad and Andrew Karter proposed the Immunocompetence Handicap Hypothesis. According to this hypothesis, signals that indicate a male's resistance to parasites are kept honest – in other words, males cannot produce the elaborate signal unless they also possess the genes for resisting parasites – by the relationship between immune suppression and testosterone. If testosterone is needed for the production of secondary sexual characteristics, like combs on roosters, that males need to get mates, but if at the same time it also increases the males' vulnerability to disease, then only those males of particularly high quality would be able to maintain their showy ornaments despite the challenge to their immune systems. Females therefore have a way to detect the real studs, because a male who cheats by producing a long tail or big comb, but is not of high quality, will be unable to pay the price of compromised immunity and will succumb to illness.

Testing this hypothesis has been very challenging. The interactions between hormones and the immune system are complex, and so it is difficult to make falsifiable predictions about the way ornaments or mate choice should react to changes in either. Some recent work by Deborah Duffy and Greg Ball from Indiana University was promising, however. The scientists used starling song as their ornamental signal. As in many birds, male starlings mainly sing during the breeding season, and the part of the brain controlling song is under the influence of testosterone. Duffy and Ball recorded the songs of 16 male starlings that were placed in an aviary with a receptive female. Then they noted how many bouts of continuous song the males produced, assuming that a more vigorous and sustained song is the equivalent of a longer tail or brighter color. It turned out that the birds with more vigorous songs also had more robust immunity, supporting the idea that females can gauge resistance to disease using male signals as a cue.

Current Issues and Future Directions

Despite the fact that parasites and sexual selection seem like two disparate areas, much evidence now suggests that the two are indeed connected. Since the publication of the Hamilton–Zuk hypothesis over 20 years ago, a considerable amount of work has been done linking parasites to sexual signaling in animals. However, most of the studies done to date have focused on one species of host at a time, or looked only at a single parasite in relation to sexual features. This raises the question of how broadly the results of such studies can be applied. While parasites play an important role, they are likely only one part of the picture of how sexual signals evolved. There are many hypotheses about what females look for when they pick a mate. For example, males may carry 'good genes' that do not relate to parasite resistance, or females may choose a male on the basis of the quality of his territory. One possible way to examine how broadly the Hamilton–Zuk hypothesis applies across species is to use meta-analyses, which compile data from many different studies of multiple species. Such analyses may prove useful in the future for examining larger evolutionary questions.

In all probability, no one explanation can explain the evolution of all sexual traits, and the relative importance of parasites in this process has not yet been determined. In addition, how factors such as parasites affect the degree to which secondary sexual traits are developed is also not well understood. Why are some species bright and flashy while others are dull and drab, even if they have the same

mating system? How have parasites shaped the mating systems themselves, which determine how much selection pressure an animal faces because of sexual selection? Whether evolution has shaped animals to be monogamous, polygynous, or somewhere in between may be related to the risk of acquiring parasites from potential mates. Sexually transmitted diseases are thought to be particularly important in this regard. However, while theoretical models indicate that sexually transmitted diseases may change the optimal number of mates for an animal, actual data supporting the idea are still lacking.

Another underexplored area has to do with differences in rates of parasitism between the sexes. In most species, males seem to be more prone to infection by parasites than females. The reasons for this are not well understood. Sexual selection is thought to be responsible for many of the differences between males and females, and it may help to explain this one as well. For instance, males may be subject to higher parasite loads because they must use more of their energy than females in competing with rivals and attaining mates or because territory defense and fights with other males may expose them to more parasites. Such explanations require further testing, however, before they can be accepted. In short, the study of parasites and their impact on host sexual behavior is still a burgeoning field with great potential for future work.

See also: Avoidance of Parasites; Beyond Fever: Comparative Perspectives on Sickness Behavior; Conservation, Behavior, Parasites and Invasive Species; Ectoparasite Behavior; Evolution of Parasite-Induced Behavioral Alterations; Flexible Mate Choice; Immune Systems and Sickness Behavior; Intermediate Host Behavior; Mating Signals; Parasite-Induced Behavioral Change: Mechanisms; Parasite-Modified Vector Behavior; Parasites and Sexual Selection; Propagule Behavior and Parasite Transmission; Reproductive Behavior and Parasites: Invertebrates; Reproductive Behavior and Parasites: Vertebrates; Reproductive Success; Self-Medication: Passive Prevention and Active Treatment; Social Behavior and Parasites; Social Selection, Sexual Selection, and Sexual Conflict; Stress, Health and Social Behavior.

Further Reading

Altizer S, Nunn CL, Thrall PH, et al. (2003) Social organization and parasite risk in mammals: Integrating theory and empirical studies. *Annual Review of Ecology Evolution and Systematics* 34: 517–547.

Folstad I and Karter AJ (1992) Parasites, bright males, and the immunocompetence handicap hypothesis. *American Naturalist* 139: 603–622.

Hamilton WD and Zuk M (1982) Heritable true fitness and bright birds: A role for parasites? *Science* 218: 384–387.

Møller AP (1994) *Sexual Selection and the Barn Swallow*. Oxford: Oxford University Press.

Møller AP, Christe P, and Lux E (1999) Parasitism, host immune function, and sexual selection. *The Quarterly Review of Biology* 74: 3–20.

Sheldon BC and Verhulst S (1996) Ecological immunology: Costly parasite defenses and trade-offs in evolutionary ecology. *Trends in Ecology and Evolution* 11: 317–321.

Wilson K, Grenfell BT, Pilkington JG, Boyd HEG, and Gulland FMD (2004) Parasites and their impact. In: Clutton-Brock T and Pemberton J (eds.) *Soay sheep: Dynamics and Selection in an Island Population*, pp. 113–165. New York: Cambridge University Press.

Zuk M (2007) *Riddled with Life: Friendly Worms, Ladybug Sex, and the Parasites that Make Us Who We Are*. Orlando: Harcourt, Inc.

Parasitoid Wasps: Neuroethology

F. Libersat, Institut de Neurobiologie de la Méditerranée, Parc Scientifique de Luminy, Marseille, France

Introduction

Predators as diverse as snakes, scorpions, spiders, insects, and snails manufacture venoms to incapacitate their prey. Most venoms contain a cocktail of neurotoxins and each neurotoxin is designed to target-specific receptors in the nervous and muscular systems. Most neurotoxins act peripherally and interfere with the ability of the prey's nervous system to generate muscle contraction or relaxation, resulting in immobilization and often death of the prey to be consumed immediately. However, in a few species of predatory wasps, venoms appear to act centrally to induce various behaviors. These venomous wasps use mostly other insects or spiders as food supply for their offspring. Most parasitoid wasps eat only nectar from flowers and other small insects, but as larvae they eat something totally different. Many of these wasps paralyze their prey and then lay one or more eggs in or on the host, which serves as a food source for the hatching larvae. In a few instances, the parasitoid wasp often manipulates the host's behavior in a manner that is beneficial to and facilitates the growth and development of its offspring. Although the alteration of host behavior by parasitoids is a widespread phenomenon, the underlying neuronal mechanisms are only now beginning to be deciphered. As of today, only a few behavioral alterations can be unambiguously linked to alterations in the central nervous system (CNS).

The direct manipulation of the host nervous system and behavior may take several forms. In some instances, the venom is purely paralytic, affecting either the peripheral or CNS to induce partial or total paralysis, which can be transient (seconds to minutes) or long-term (hours to days). In other instances, the venom might affect behavioral subroutines to produce finer manipulations of the host behavior. In this article, I will discuss selected case studies where the neural mechanisms underlying host manipulation by parasitoid wasps have been identified. I will then focus on one case study where a wasp hijacks the brain of its host to control its motivation to perform specific behaviors.

Most ectoparasitoid wasps incapacitate their prey and then drag it to a burrow or a nest. In this protected nest, the wasp lays its egg on the prey and seals the burrow with the inert prey inside. When the larva later hatches, it feeds on the host, ultimately killing it, and pupates in the nest, sheltered from predators that could harm the cocoon. The hunting and host-manipulation strategies of these wasps are diverse and, at least to some extent, depend on the host natural behavior. Hunters of relatively small or harmless prey usually inflict a single or double sting to the prey item. This typically results in deep paralysis by affecting, for example, the peripheral nervous system (i.e., the neuromuscular junction: synapse between the motoneurone terminals and the muscle). In those species of wasps where the paralyzing venom is injected into the hemolymph of the prey, as in the beewolf (the Egyptian digger wasp *Philanthus triangulum*), the venom has been shown to affect the peripheral nervous system (**Figure 1**). *P. triangulum* feeds its larvae almost exclusively with Honeybees (*Apis mellifera*). The beewolf paralyzes bees by stinging them on the ventral side of the thorax through the membrane between the first and second segments. These wasps are sufficiently strong to airborne cargo the prey item back to the nest (**Figure 1(a)**). After provisioning the nest with a few bees, the wasp lays an egg in it and seals it. The venom of the beewolf contains potent neurotoxins known as philanthotoxins, which evoke neuromuscular paralysis in the bee prey. Such philanthotoxins interfere presynaptically and postsynaptically with glutamatergic synaptic transmission (**Figure 1(b)**). Because glutamate is the neurotransmitter at the insect neuromuscular junction, philanthotoxins in the venom block the neuromuscular transmission to induce flaccid paralysis of the prey. One potent component of the *Philanthus* venom is δ-philanthotoxin, which blocks open ionotropic glutamate receptors in the insect neuromuscular junction (**Figure 1(b)**). Paradoxically, the very same δ-philanthotoxin blocks glutamate uptake (it interferes with the glutamate transporter) at the insect neuromuscular junctions thereby, prolonging the presence of glutamate at the neuromuscular junction (**Figure 1(b)**). This venom-induced hyperexcitation of muscle contraction is presumably responsible for the initial tremor, which immobilizes the prey until flaccid paralysis begins. Hyperexcitation preceding flaccid paralysis is a common venom strategy seen in several types of venomous animals, such as octopus, spiders, coelenterates, and some cone snails where the hyperexcitation is produced by different classes of substances. Apparently, the hyperexcitation immediately immobilizes the prey, so that it cannot get out of reach of the predator, until the slower acting flaccid paralysis begins. The wasp paralyzes several bees and drags them into a concealed burrow. It then lays an egg on one of the bees, seals the burrow, and leaves. The hatching larva is, thus, provided with a large, paralyzed food supply to feed on until pupation.

On the other hand, wasps, which hunt on large prey such as tarantula spiders, face a much more considerable

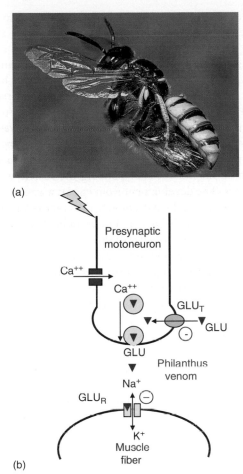

(a)

(b)

Figure 1 (a) A photograph of an air-borne *Philanthus* wasp carrying its bee prey back to the nest. (b) Schematic representation of the insect neuromuscular junction where *Philanthus* venom affects glutamatergic synaptic transmission. Calcium (Ca++) ions move in when an action potential (blue arrow) reaches the motoneuron terminal and facilitates the vesicular release of glutamate (GLU). One potent component of the *Philanthus* venom (δ-philanthotoxin) blocks open ionotropic glutamate receptors (GLU$_R$) and glutamate uptake (GLU$_T$) to induce muscular paralysis.

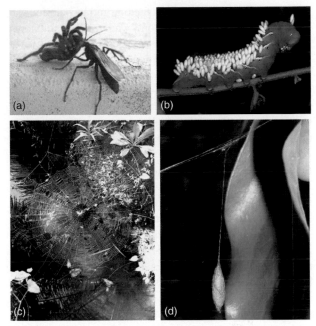

Figure 2 (a) The spider wasp, *Tachypompilus ignitus*, dragging an immobilized *Palystes* spider to her nest. (b) The tomato hornworm, *Manduca*, parasitized by the solitary braconid endoparasitoid wasp *Cotesia* pupae. (c) The normal web of the orb-weaving spider *P. argyra*. (d) The cocoon web of a spider parasitized by the Ichneumonid wasp and wasp cocoon from above.

danger (**Figure 2(a)**). The tarantula-hawk (*Pepsis*) is the fearsome enemies of spiders. These wasps usually first disarm the spider of its most powerful weapon, the fangs, with multiple stings into the cephalo-thorax but sometimes directly in the mouth. After this stinging sequence, the spider is totally paralyzed, which allows the wasp to drag the spider back to the nest, walking backwards facing its formidable opponent. Once the host is concealed, the wasp lays a single egg on the abdomen of the spider and seals the entrance to the nest. Depending on the species, the spider would completely or nearly completely recover from paralysis within a few hours to 2 months. If the tarantula survives what usually happens next, it can revive and continue living a normal life. But another fate awaits the spider as the larva hatches from the

egg after 2 days and feeds on the entombed spider for 5–7 days. The satiated larva then pupates inside the nest, safe from predators.

As we shall see in this article, some parasitoid wasps alter the behavior of their host to the finest degree. The unique effects of such wasp's venom on prey behavior suggest that the venom targets the prey's CNS. A remarkable example of such manipulation is that of the braconid parasitoid wasp (*Glyptapanteles* sp.) that induces a caterpillar (*Thyrinteina leucocerae*) to behave as a bodyguard of its offspring. After parasitoid larvae exit from the host to pupate, the host remains alive but displays stunning modifications in its behavior: it stops feeding and remains close to the parasitoid pupae to defend these against predators with violent head swings. The parasitized caterpillar dies soon after while unparasitized caterpillars do not show any of these behavioral changes. In another example of host manipulation, the wasp *Cotesia congregata* and its host, the tobacco hornworm *Manduca sexta*, we have some information about the underlying mechanisms of manipulation. The female wasp injects a mixture consisting of venom, polydnavirus, and wasp eggs into its caterpillar host (**Figure 2(a)**). The wasp larvae hatch and develop inside the host's hemocoel, exit through the cuticle, and spin a cocoon which stays attached to the host. One day before exiting the host, host feeding and spontaneous locomotion decline. The host remains in this torpor until death. The decline in host feeding and locomotion can be induced by wasp larvae alone which, by an unknown chain of events, target the subesophageal ganglion (SEG) of the host to induce neural inhibition of locomotion. Furthermore, the change in host behavior is accompanied with an elevation of CNS octopamine (OA), a neuromodulator, suggesting that alterations in the functioning of the octopaminergic system may play a role in depressing host feeding or locomotion. But, the most exquisite alteration of behavior ever attributed to a parasitoid wasp is probably the Ichneumonid wasp *Hymenoepimecis*'s manipulation of its spider host. In this exceptional example of host behavioral manipulation, the parasitoid wasp takes advantage of the natural behavior of web waving of its prey to provide the larva with a shelter. Instead of paralyzing and then burrowing into the host, this wasp literately coerces the host to build the shelter for its future larva. The wasp stings its spider host, *Plesiometa argyra* (Araneidae), evoking a total, but transient, paralysis during which the wasp lays its egg on the paralyzed spider and flies away. Soon after the sting, the spider recovers to resume apparently normal activity. It builds normal orb webs to catch prey (**Figure 2(c)**), while the wasp's egg hatches and the larva grows by feeding on the spider's hemolymph. The larva feeds for about 2 weeks and just before it kills the spider, a dramatic behavioral change occurs in the spider. The prey, driven by an unknown mechanism, starts weaving a unique web with a design that seems tailored to fit the needs of the larva for its next stage in development, the metamorphosis. The new web is very different from the normal orb-shaped web of *P. argyra*, and is designed to support the larva's cocoon suspended in the air, rather than lying on the ground (**Figure 2(d)**). In this safe net, the wasp larva consumes the spider, ultimately killing it, and then pupates in the suspended net. Interestingly, if the wasp larva is removed just prior to the execution of the death sentence, the spider continues to build the specialized cocoon web. Hence, the changes in the spider's behavior must be induced chemically rather than by direct physical interference of the wasp larva. The wasp larva must secrete chemicals to manipulate the spider's nervous system to cause the execution of only one subroutine of the full orb web construction program while repressing all other routines. The nature of the chemicals involved in this extreme alteration of the spider's behavior, is unfortunately unknown.

Sphecid wasps often hunt large and potentially harmful orthopteroids (crickets, katydids, grasshoppers, etc.). They usually sting their prey to evoke total transient paralysis, although in some instances, a more specific manipulation takes place. One example of total transient paralysis of the host can be found in the *Larra* – mole cricket system. Mole crickets spend most of their time in a burrow. A larrine wasp (e.g., *L. anathema*) in a hunting mood penetrates the underground refuge of the cricket and attacks it. The frightened cricket may emerge in panic

from its burrow pursued by the wasp. The wasp then wrestles with its prey to finally inflict multiple stings, mainly in the thoracic region. The stings induce a total transient paralysis of the legs, lasting just a few minutes. The wasp performs host feeding, sucking some hemolymph before laying a single egg between the first and second pairs of legs of the inert cricket. The wasp then leaves the cricket which fully recovers from paralysis and burrows back into the ground, apparently resuming normal activity. The egg soon hatches and the larva starts feeding on the cricket after piercing the cuticle with its mandibles. The development from egg laying to pupation lasts between 2 weeks and a month, during which the mole cricket appears to behave quite normally, demonstrating complete recovery from paralysis.

An example for central paralysis can be found in the Palearctic Larrine digger wasp *Liris niger* which hunts crickets as food supply for its brood. To transport the cricket to a burrow and lay an egg on its cuticle, the wasp incapacitates the prey with four stings, which are applied near, or perhaps inside, the CNS. First, the wasp disarms the cricket's most powerful weapons, the metathoracic kicking legs, by injecting venom presumably into the metathoracic ganglion. This sting paralyzes the metathoracic legs for several minutes. Successively, the wasp injects venom into the two other thoracic ganglia, transiently paralyzing the legs associated with these ganglia and rendering the stung cricket lying helplessly on its back for several minutes. Last, the wasp stings into the neck, probably directly into, or in the vicinity of, the subesophageal ganglion. This last sting is responsible for the next phase of envenomation, a long-lasting hypokinetic state. The wasp drags the paralyzed cricket to a burrow, glues an egg between its fore and the middle legs, and seals the burrow with soil particles or pebbles. After the burrow has been sealed, the cricket fully recovers from its paralysis and can maintain posture and even walk. However, at this time, a different story unfolds, as the stung cricket never attempts to escape the burrow; rather it stays motionless, although not paralyzed, in its tomb. The wasp larva, after hatching from the egg, feeds on the lethargic cricket and then pupates. If the cricket is experimentally removed from the burrow, no spontaneous and only little evoked activity can be observed in the stung cricket until it dies, probably due to lack of feeding. Thus, *Liris* venom induces not only total transient paralysis but also a partial irreversible paralysis which renders the cricket prey submissive in its future grave. It has been suggested that the latter effect of *Liris* venom is a result of the neck-sting, which is, for comparison, not typical for mole cricket-hunting *Larra* and does not evoke such long-term effects.

The short-term paralysis of the cricket legs has been thoroughly investigated in this *Liris*-cricket system. The venom's effect on the CNS of crickets has been studied in dissected preparations in which venom was manually applied to thoracic or abdominal ganglia by means of manipulating the wasp to sting directly into the ganglion or by pressure-injecting sampled venom into the ganglion. All experiments with dissected preparations demonstrate two pronounced phases of envenomation. First, the sting typically evokes a short (15–35 s) tonic discharge of the motoneurons innervating the legs. This discharge is most likely responsible for the convulsions of the cricket's limbs, which is the first venom effect observed immediately after the sting. The cellular mechanism by which the motoneurons' discharge rate increases, is not yet fully understood, but it is either due to the presence of an excitatory agonist (e.g., ACh receptors) in the venom or due to the low pH of the venom. The short tonic discharge of the legs' motoneurons then completely disappears, marking the onset of the second effect of the *Liris* venom: total transient paralysis of the legs. This paralysis, lasting from 4 to 30 min, is characterized by a complete absence of spontaneous or evoked activity in the affected neurons. Then, after the total paralysis phase is over, responses of the leg muscles and motoneurons to sensory stimuli recover. However, behaviorally, the prey fails to initiate locomotion, which underlies the beginning of the third, hypokinetic, and irreversible phase of envenomation, at the end of which the cricket dies. The venom's paralytic effects are restricted to the stung ganglion, indicating that the venom affects central (rather than peripheral) targets. For instance, excitation of leg sensory receptors of stung crickets evokes afferent sensory potentials that reach the stung ganglion but fail to engage a motor reflex in that ganglion, demonstrating that the venom's effect is restricted to the stung ganglion. Various physiological experiments have uncovered at least three types of effects in the CNS. First, the venom prevents the generation and propagation of action potentials in the affected neurons, presumably by interfering with voltage-dependant inward sodium currents. Second, the venom decreases central synaptic transmission, the underlying mechanism of which is not yet fully understood. Third, the venom increases leak currents in central neurons and consequently their excitability.

A Case Study in the Neural Mechanisms of Host Manipulation: *Ampulex compressa* and its Prey, the Cockroach *Periplaneta americana*

Ampulex Hunting Strategy and Offspring Development

The best understood manipulation of host nervous system and behavior is the case of the Sphecid cockroach-hunter *A. compressa*. After grabbing its cockroach prey (usually *P. americana*) at the pronotum or the base of the wing, the wasp inflicts a first sting into the thorax. This sting

renders the prothoracic legs transiently (1–2 min) paralyzed and presumably facilitates the second sting into the neck, which is much more precise and time-consuming (**Figure 3(a)**). After the neck-sting is complete, the wasp leaves the cockroach for about 30 min and searches for a burrow. During this period, the cockroach is far from being paralyzed but grooms frenetically for about 20 min. When this period is almost over, the wasp returns to the cockroach and pushes him around with its mandibles as if to evaluate the success of the sting. This is when another effect of the venom begins to take place, as the cockroach becomes a submissive 'zombie' capable of performing, but not initiating, locomotion. The wasp cuts the cockroach's antennae with the mandibles and sucks up hemolymph from the cut end. It then grabs one of the cockroach's antennal stumps and leads the host to the preselected burrow for oviposition, walking backwards facing the prey. The stung cockroach follows the wasp in a docile manner, like a dog on a leash, all the way to the burrow. Then, the wasp lays an egg and affixes it on

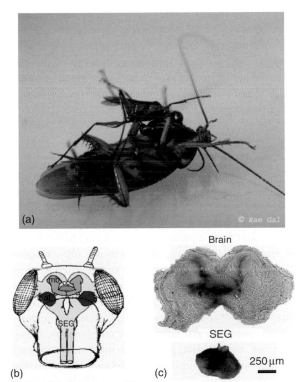

Figure 3 (a) The wasp *Ampulex compressa* stings a cockroach *Periplaneta americana* in the head. (b) A schematic representation of a dorsal view of a cockroach head shows the relative positions of the head ganglia in the head capsule. The brain and SEG are shown in yellow. The major structures of the brain include the central complex (cc, red), the mushroom bodies (mb, green), and the antennal lobes (al, blue). (c) Two sections of representative head ganglia (brain and SEG) preparations of a cockroach stung by a radiolabeled wasp. Radiolabeled venom is located posterior to the central complex and around the mushroom bodies of the brain and in the center of the SEG.

the cuticle of the coxal segment of the middle cockroach's leg. Having its egg glued on the live food source, the wasp exits the burrow and blocks the entrance with small pebbles collected nearby, sealing the lethargic host inside. The larva hatches within 2–3 days and perforates the cuticle of the cockroach's coxa to feed on hemolymph for the next few days. About 5 days after the egg was laid, the larva moves to the thoracic–coxal junction of the metathoracic leg and bites a large hole along the soft cuticular joint, through which it then penetrates the cockroach. The larva feeds on the internal organs of the cockroach until, 2 days after entering the host, it occupies the entire cockroach's abdominal cavity. Pupation occurs inside the cockroach's abdomen, roughly 8 days after the egg was laid.

The two stings by *A. compressa* induce a total transient paralysis of the front legs followed by grooming behavior and then, by a long-term hypokinesia of the cockroach's prey. In this state, the cockroach remains alive but immobile and unresponsive, and serves to nourish the wasp larva. The long-lasting lethargic state occurs when the venom is injected into the head but not when it is injected only into the thorax. Under laboratory conditions, and if not parasitized by the wasp larva, cockroaches gradually recover from this lethargic state within 1 or 2 weeks, demonstrating a partial long-term paralysis of the cockroach. In nature, cockroaches probably rarely reach recovery as the *A. compressa* larva consumes them before the end of this convalescent time.

Where Is the Venom Injected?

For more than a century, there has been a controversy over whether some parasitoid wasps deliver their venom by stinging directly into the CNS. In 1879, the French entomologist Jean Henri Fabre, who observed that specific wasps sting in a pattern corresponding to the location and arrangement of nerve centers in the prey, suggested that the wasp stings directly into target ganglia. Others challenged Fabre's idea and claimed that the wasp stings in the vicinity of, but not inside, the ganglion. In fact, solitary wasps' venoms usually consist of a cocktail of proteins, peptides, and subpeptidic components, some of which are very unlikely to cross the thick and rather selective sheath (the insects' blood–brain barrier) around the nervous ganglia. Thus, it is most unlikely that neurotoxins in the venom make their way into the CNS by simple diffusion from the hemolymph. It was, therefore, suggested that some wasps use a common strategy of 'drug delivery,' injecting venom directly into a specific ganglion of the CNS of the prey.

The unique effects of *Ampulex*'s venom on prey behavior and the site of venom injection both suggest that the venom targets the prey's CNS. Until recently, the mechanism by which behavior-modifying compounds in the venom reach the CNS, given the protective ganglionic sheath, was unknown. The *Ampulex* stinger, which is about 2.5 mm in length, is certainly long enough to reach the cerebral ganglia that lie 1–2 mm deep in the head capsule. But to obtain a direct proof of the central injection of the venom, we produced so-called 'hot' wasps by injecting them with a mixture of C^{14} radiolabeled amino acids which were incorporated into the venom. In cockroaches stung by 'hot' wasps, most of the radioactive signals were found in the thoracic ganglion and inside the two head ganglia: the supra and the subesophageal ganglia (**Figure 3**). Only a small amount of radioactivity was detected in the surrounding, nonneuronal tissue of the head and thorax. A high concentration of radioactive signal was localized to the central part of the supraesophageal ganglion (posterior to the central complex and around the mushroom bodies) and around the midline of the subesophageal ganglion. The precise anatomical targeting of the wasp stinger through the body wall and ganglionic sheath and into specific areas of the brain, is akin to the most advanced stereotactic delivery of drugs. Sensory structures located on the stinger might be responsible for mediating nervous-tissue recognition inside the head capsule to allow such precise venom injection inside the head ganglia. These experiments represent, to date, the only unequivocal demonstration that a wasp injects venom directly into the CNS of its prey, consistent with Fabre's ideas. *A. compressa* is almost certainly not the only wasp which injects venom in its prey CNS, although the use of such method of drug delivery remains to be proven in other wasp species.

Wasp Venom Induces Transient Paralysis of the Front Legs

The *Ampulex* venom, similar to the *Liris* venom, is a complex cocktail of proteins, peptides, and subpeptidic components. Of this cocktail, only low molecular weight fractions seem to be responsible for the short-lived front legs paralysis. Electrophysiological studies on the *Ampulex* venom have demonstrated that it dramatically affects central cholinergic synaptic transmission. For instance, injection of venom to the cockroach's last abdominal ganglion eliminates synaptically evoked action potentials in the postsynaptic giant interneuron (GI) (**Figure 4(a)**). Likewise, venom injections block the postsynaptic potentials evoked by exogenous cholinergic agents at the same synapse.

To identify the venom components responsible for the total transient paralysis of the front legs, fractions of the venom (based on different molecular weights) were applied to neurons and the responses quantified. The fractions that reduced neuronal activity caused a synaptic block in central synapses. Biochemical screening of the active fractions revealed that the venom contains high levels of the inhibitory neurotransmitter GABA, and

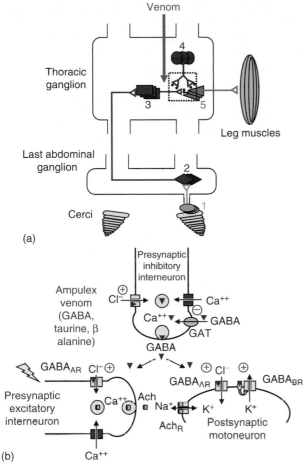

Figure 4 A schematic and simplified drawing of a cockroach's nervous system depicting the circuitry that controls escape behavior and leg movements. For escape, sensory mechanoreceptors of the cerci (1) recruit ascending GIs located in the last abdominal ganglion (2); these GIs converge onto the thoracic interneurons in the thoracic ganglion (3). The thoracic interneurons excite the leg motoneurons (5). Local inhibitory interneurons controls the activity of motoneurons involved in leg movements. The doted rectangle shows the area enlarged in (b). (b) A model of a wasp venom-induced short-term leg paralysis. The natural release of GABA at a synapse causes the opening of the chloride channels in the postsynaptic membrane. When GABA-gated chloride channels (GABA$_{AR}$) are opened by venom components, the resulting GABA current shunts the depolarization generated by the release and binding of acetylcholine (Ach) from the presynaptic excitatory interneuron when an action potential reaches the terminal (blue arrow). GABA can also act on the presynaptic terminals of the cholinergic interneuron by decreasing the synaptic release. Because taurine and β-alanine suppress the reuptake of GABA (GABA transporter: GAT) from the synaptic cleft, they may contribute to a prolongation of chloride channel open times. Together, these effects result in a failure of action potential generation in the postsynaptic motoneuron.

a GABA receptor agonist β-alanine. Another component in these fractions was identified as taurine, which is known to impair the reuptake of GABA by the GABA transporter from the synaptic cleft (**Figure 4(b)**). These constituents

mimic the transient action of whole venom, synergistically causing a total transient block of synaptic transmission at the cercal-giant synapse through GABA inhibition. Patch-clamp recordings from isolated thoracic motoneurons of the cockroach demonstrate that the *Ampulex* venom induces picrotoxin-sensitive currents, further implicating venom action on cockroach GABA receptors. The natural release or artificial injection of GABA at a synapse causes the opening of chloride channels in the postsynaptic membrane. When GABA-gated chloride channels are opened by venom components, if the sodium channels in the postsynaptic membrane are also opened by a synaptic release of ACh, then for each sodium ion entering the cell, a chloride ion will accompany it. The simultaneous entry of a negative ion and positive ion will produce no change in the membrane potential. The utilization of venom cocktails containing multiple toxins with distinct but joint pharmacological actions has been described previously in venoms of spiders and marine cone snails. To conclude, for the total transient paralysis, the study of *Ampulex* venom has demonstrated a novel strategy for venom-induced synaptic block through chloride channel activation.

Wasp Venom Initiates Prey Grooming Behavior

After the total paralysis effect of the venom is over, the *Ampulex* venom evokes a stereotyped, though excessive, uninterrupted grooming behavior in the stung cockroach. The venom-induced grooming is similar in all respects to normal grooming and involves the coordinated movements of different appendages. The grooming behavior is evoked only if venom is injected into the head, and cannot be accounted for by the stress of the attack, the contact with the wasp, a mechanical irritation or venom injection into a location other than the head. Thus, the *Ampulex* venom appears to engage a central neuronal circuit in the head ganglia responsible for grooming. Experimental manipulation of the monoaminergic system in unstung cockroaches affects grooming behavior. For example, a single injection into the subesophageal ganglion of the alkaloid reserpine, which transiently elevates the concentration of all monoamines in central synapses, induces excessive grooming similar to venom-induced grooming. The specific cause for this is probably an elevation in the levels of the monoamine dopamine (DA), since an injection of DA or DA receptor agonists similarly induced excessive grooming. Moreover, the injection of a DA-receptor antagonist (Flupenthixol) prior to a wasp sting markedly reduced venom-induced grooming. Thus, grooming behavior could result from the existence of a DA (or DA-like) component in the venom, or from a venom component that would activate a DA-releasing mechanism inside the cockroach head ganglia. A gas chromatography–mass spectrometry study has identified a DA-like substance in the venom. This substance might be responsible for a direct stimulation, via DA receptors, of grooming–releasing circuits within the head ganglia.

One can only speculate regarding the adaptive significance of the grooming phase of envenomation. First, it is possible that grooming is merely a side effect of the venom. For example, the DA-like substance in the venom could be involved in inducing the long-term hypokinetic state, in which case, grooming has no adaptive value to the wasp. Second, and more provocative, is the possibility that excessive grooming cleans off the ectoparasites such as bacteria and fungi on the cockroach's exoskeleton which are potentially harmful to the developing wasp's larva. Interestingly, beewolf females (**Figure 1(a)**) cultivate the *Streptomyces* bacteria in specialized antennal glands and smear them on the ceiling of the brood cell prior to oviposition. The bacteria enhance the survival probability of the larva as they are taken up by the larva later to protect the cocoon from fungal infestation.

Wasp Venom Controls Cockroach Motivation to Walk and Escape

The third phase of cockroach envenomation by *A. compressa* is probably the most interesting in terms of host behavioral manipulation, because a stung cockroach becomes a submissive 'robot,' and fails to initiate spontaneous or evoked locomotion. It is almost certainly of adaptive value to the wasp, since it enables resistance-free host feeding, transportation to the burrow, and oviposition. The venom-induced hypokinesia persists, if the egg is removed experimentally, for at least a week, after which the cockroach resumes a normal activity. In nature, however, the cockroach meets its inevitable fate about a week later.

The long-term hypokinesia is induced only if *A. compressa* stings the cockroach in the head ganglia. Hence, the inability of stung cockroaches to start walking cannot be accounted for by a direct effect of the venom on locomotory centers in the thoracic ganglia of insects. We propose that, unlike most paralyzing venoms, *Ampulex*'s venom affects the 'motivation' of its host to initiate movement, rather than the motor centers. Indeed, the wasp injects its venom directly into the subesophageal ganglion and into the central complex and mushroom bodies in the supraesophageal ganglion of the cockroach, all considered 'higher' neuronal centers modulating the initiation of movement. We investigated whether the venom-induced hypokinesia is a result of an overall decrease in arousal or, alternatively, a specific decrease in the drive to initiate or maintain walking. We found that the venom specifically increased thresholds for the initiation of walking-related behaviors and, once such behaviors were initiated, affected the maintenance of walking. Nevertheless, we show that the thoracic walking pattern generator itself appears to be intact. Thus, the venom, rather than decreasing the overall arousal,

manipulates neuronal centers within the cerebral ganglia that are specifically involved in the initiation and maintenance of walking. Furthermore, stung hypokinetic cockroaches show no deficits in spontaneous or provoked grooming, righting behavior, or the ability to fly in a wind tunnel. Hence, the head sting affects specific subsets of motor behaviors, rather than affecting behavior in general. How this comes about is not completely worked out, but we have uncovered some important pieces of the puzzle.

The hypokinetic state is characterized by very little spontaneous or provoked activity; an important hallmark of this hypokinetic state is the inability of stung cockroaches to produce normal escape responses. Wind stimuli directed at the cerci, which normally produce strong escape responses, are no longer effective in stung cockroaches. Normally, wind-sensitive hairs on the cerci detect the minute air movements produced by a predator's strike and excite GIs in the terminal abdominal ganglion (TAG) to mediate escape running behavior (**Figure 5**). The GIs activate various thoracic interneurons in the thoracic locomotory centers, which, in turn, excite various local interneurons or motoneurons associated with escape running. In addition, escape running can be triggered by tactile stimuli applied the antennae that recruit GIs descending from the head ganglia to the thorax. Tactile and wind information is carried by two distinct populations of interneurons, each located at the far and opposite ends of the nervous ganglionic chain, to converge on the same thoracic premotor circuitry which controls similar escape leg movements. Studies on stung cockroaches show that the sting affects neither the response of the sensory neurons and associated ascending GIs nor that of the descending interneurons. Moreover, thoracic interneurons receive comparable synaptic drive from the GIs in control and stung animals. Thus, the ultimate effect of the venom injected into the head ganglia must take place at the connection between the thoracic interneurons and specific motorneurons.

Unlike normal cockroaches, which use both fast and slow motoneurons for producing rapid escape movements, stung cockroaches activate only slow motoneurons, which are also important to maintain posture, and do not produce rapid movements. This lack of response of fast motoneurons appears to be due to a reduction in the synaptic drive they receive from premotor interneurons. Such reduction could be due to a modulation of a particular neuromodulatory system that controls a specific subset of behaviors. In this case, the venom would chemically manipulate specific pathways in the head ganglia which themselves regulate neuromodulatory systems involved in the initiation and/or execution of movement (**Figure 5**). Monoaminergic systems are again probable candidates, as alterations in these systems are known to affect specific subsets of behaviors. For instance, depletion of the synaptic content of monoaminergic neurons, and especially of dopaminergic or

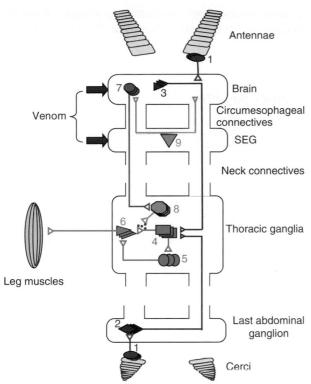

Antennae

Brain

Circumesophageal
connectives

SEG

Neck connectives

Thoracic ganglia

Leg muscles

Last abdominal
ganglion

Cerci

Venom

Figure 5 A current model of the neurophysiological events leading to venom-induced hypokinesia in cockroaches stung by *Ampulex compressa*. A schematic and simplified drawing of a cockroach nervous system depicting circuitries that affect walking-related behaviors. The walking pattern generator that orchestrates leg movements is located in the thorax. It consists of motor neurons innervating leg muscles (6), sensory neurons associated with sensory structures on the legs (not shown) and thoracic interneurons (TIAs; 4), which synapse onto the motor neurons directly and indirectly via local interneurons (5). The TIAs receive inputs from several interneurons. For example, sensory neurons in the antennae or cerci (1) recruit ascending (2) or descending (3) GIs, which converge directly onto the TIAs to ultimately evoke escape responses. In addition, neurons of the pattern generator receive input from thoracic neuromodulatory cells (8). One example of these is the DUM neurons, which secrete OA and modulate the efficacy of premotor-to-motor (4-to-6) synapses. The neuromodulatory cells, in turn, receive tonic input through interneurons descending from the brain (7) and SEG (not shown). This tonic input affects the probability of the occurrence of specific motor behaviors by modulating the different thoracic pattern generators. The wasp *A. compressa* injects its venom cocktail directly into both cerebral ganglia to modulate some specific yet unidentified cerebral circuitries. Our current hypothesis states that in the SEG, the venom suppresses the activity of brain-projecting DUM neurons (9), which control the activity of brain-descending interneurons (7) that modulate, either directly (not shown) or indirectly via the neuromodulatory cells (8), the walking pattern generator. Hence, the venom injected into the cerebral ganglia decreases the overall excitatory input to the thoracic walking pattern generator. As a result, walking-related behaviors are specifically inhibited, and the stimuli to the antennae or cerci fail to evoke normal escape responses.

octopaminergic neurons, induces impairment in the ability of cockroaches and crickets to generate escape behavior. The activity of octopaminergic neurons, known to modulate the excitability of specific thoracic premotor neurons in the cockroach, is compromised in stung cockroaches. The alteration in the activity of OA neurons could be part of the mechanism by which the wasp induces a change in the excitability of thoracic premotor circuitries.

The wasp injects its venom directly into the SEG and in and around the central complex in the brain. We have shown that, in stung cockroaches, the focal injection of a potent OA receptor agonist around the central complex area partially restores walking. Conversely, in controls, the focal injection of a selective OA receptor antagonist into the same area reduces walking. However, it appears that the relevant neurons that modulate walking reside in the SEG and send axons to innervate the motor centers, such as the central complex. Within this group of ascending interneurons, at least three OA ascending SEG neurons provide dense innervation in the central portion of the brain. Thus, the SEG sting might be affecting the activity of SEG octopaminergic ascending neurons to reduce OA levels in the walking centers of the brain (**Figure 5**).

To conclude, we propose that venom injection into the head ganglia selectively depresses the initiation and maintenance of walking by modifying the release of OA as a neuromodulator in restricted regions of the cockroach's brain. Then, it is likely that *A. compressa* alters some yet unidentified descending pathways in the cockroach head ganglia which affect, at the least, OA secretion from thoracic dorsal unpaired median (DUM) neurons (**Figure 5**). The latter are known to dramatically affect locomotion, which can explain the long-lasting hypokinetic state induced in stung cockroaches.

Conclusions

In this article, I introduce the reader to the astonishing world of parasitoid wasps and their insect hosts. There are several reasons to be interested in these wasps. First, they are increasingly used as a biological control of crop pests to preserve the environment. But, the most relevant reason is that these wasps are considerably better than we are at manipulating the neurochemistry of their prey with specific neurotoxins. Thus, for those interested in the neural mechanisms of animal behavior, parasitoid wasps have evolved, through years of co-evolution with their prey, a better 'understanding' of the neuromodulatory systems of their insect hosts than insect neuroethologists. Moreover, neurotoxins are invaluable as tools to reveal the physiological mechanisms underlying nervous system functions. Because neurotoxins are the outcome of one animal's evolutionary strategy to incapacitate another, they are usually highly effective and specific. Chemical engineers can generate hundreds of neurotoxins in their labs, but these products are random and often useless, whereas any natural neurotoxin has already passed the ultimate screening test, over millions of years of co-evolution. As such, wasp neurotoxins may provide us with new highly specific pharmacological tools to investigate cell and network function. Although, the alteration of host behavior by parasitoids is a widespread phenomenon, the underlying mechanisms are beginning to be revealed only now. I have focused here on the neuronal mechanisms by which parasitoid wasps manipulate the behavior of other insects using chemical warfare. In a case study, I have surveyed the unique venom effect of an unusual predator, the parasitoid wasp *A. compressa*. *Ampulex* does not kill its prey but instead performs a delicate brain surgery to take away the 'free will' of its prey to initiate locomotion. Much work remains to be done until we know the exact neuro-chemical cascade taking place in the host's CNS to alter its behavior. Given the breath of such investigations, a multidisciplinary approach, combining molecular techniques with cellular electrophysiology and behavior analysis, is essential. It is my great hope that such host–parasite interactions will stimulate the curiosity of young and talented minds to investigate the neuronal basis of parasite-induced alterations of host behavior, with the goal of increasing our understanding of the neurobiology of the initiation of behaviors and the neural mechanisms underlying changes in responsiveness, which are prime questions in the study of arousal and motivation.

See also: Evolution of Parasite-Induced Behavioral Alterations; Experimental Approaches to Hormones and Behavior: Invertebrates; Neuroethology: Methods; Parasite-Modified Vector Behavior; Parasitoids; Predator Evasion.

Further Reading

Adamo SA (2002) Modulating the modulators: Parasites, neuromodulators and host behavioral change. *Brain Behaviour and Evolution* 60(6): 370–377.

Eberhard WG (2000) Spider manipulation by a wasp larva. *Nature* 406 (6793): 255–256.

Fabre JH (1879) *Souvenirs Entomologiques* (1945 ed.), vol. 1, pp. 108–112. Paris: Delagrave.

Gal R and Libersat F (2008) A parasitoid wasp manipulates the drive for walking of its cockroach prey. *Current Biology* 18(12): 877–882.

Gnatzy W (2001) Digger wasp vs. cricket: (Neuro-) biology of a predator–prey-interaction. *Zoology* 103: 125–139.

Grosman AH, Janssen A, de Brito EF, et al. (2008) Parasitoid increases survival of its pupae by inducing hosts to fight predators. *PLoS ONE* 3(6): e2276.

Libersat F (2003) Wasp uses venom cocktail to manipulate the behavior of its cockroach prey. *Journal of Comparative Physiology A* 189: 497–508.

Libersat F, Delago A, and Gal R (2009) Manipulation of host behavior by parasitic insects and insect parasites. *Annual Review of Entomology* 54: 189–207.

Libersat F and Gal R (2007) Neuro-manipulation of hosts by parasitoid wasps. In: Yoder J and Rivers D (eds.) *Recent Advances in the Biochemistry, Toxicity and Mode of Action of Parasitic Wasp Venoms*, pp. 96–114. Kerala, India: Research Signpost.

Libersat F and Pflueger HJ (2004) Monoamines and the orchestration of behavior. *Bioscience* 54(1): 17–25.

O'Neill KM (2001) *Solitary Wasps: Behavior and Natural History*, p. 58. Ithaca and London: Comstock Pub., Cornell University.

Piek T (1990) Neurotoxins from venoms of the Hymenoptera – twenty-five years of research in Amsterdam. *Comparative Biochemistry and Physiology Part C* 96: 223–233.

Quicke DLJ (1997) *Parasitic Wasps*. London: Chapman and Hall.

Zimmer C (2000) *Parasite Rex: Inside the Bizarre World of Nature's Most Dangerous Creatures*. NY, USA: Free Press/Simon & Schuster.

Parasitoids

L. M. Henry, University of Oxford, Oxford, UK
B. D. Roitberg, Simon Fraser University, Burnaby, BC, Canada

Introduction and Definitions

Over half of the organisms on earth live parasitic lifestyles. The fact that almost all organisms are attacked by at least one, if not many, different parasites is a testament to the success of this feeding strategy. The intimate association between host organisms and the parasites that attack them cultivates coevolutionary dynamics where host defenses and parasite counterdefenses frequently ensue. These arms races have given rise to an astonishing diversity of morphological, physiological, and behavioral adaptations by hosts to avoid parasitism. The focus of this article will be the ecology and evolution of behavioral adaptations to avoid parasitism, with a particular focus on the often-elaborate defenses and counterdefenses of hosts and insect parasitoids.

Before addressing antiparasitoid host behavior, it is important that we first explain what we mean by parasitoids as well as the unique threat that they present to their hosts. A generally accepted definition of a parasite can be found in *Webster's International Dictionary*: "An organism living in or on another living organism, obtaining in part or all of its organic nutriment, commonly exhibiting some degree of adaptive structural modification, and causing some degree of real damage to its host." Parasites can be further categorized based on their interactions with hosts and on their life cycle, with an important distinction separating parasites from parasitoids. The defining difference between parasites and parasitoids is in the fate of their hosts. Parasitoids are any organisms whose larvae develop in or on another organism resulting in the death of the host, whereas most parasites do not obligately kill their hosts. Parasitoids can be further characterized by their life-history strategies. Idiobiont parasitoids are those that prevent further development of their host by immediately killing or permanently paralyzing the host before oviposition, whereas koinobiont parasitoids allow their hosts to continue to grow and develop after parasitization, which typically involves the parasitoid larvae living within an active, mobile host. (As we will see later, the koinobiont lifestyle may preadapt such parasitoids for exploiting or usurping behaviors typically expressed by healthy, active hosts.) Ectoparasitoids and endoparasitoids differ in that the latter develop inside their hosts, while the former feed outside the host body by adhering to the host. Parasitoid lifestyles occur in four orders of insects, including Hymenoptera (bees, wasps, and ants), Diptera (flies), Strepsiptera (twisted-wing parasites), and Coleoptera (beetles); however, the vast majority of parasitoid species are in the Hymenoptera. Within parasitoid families, there is evidence of extensive adaptive radiation, as demonstrated by the vast number of species in the major parasitoid clades. In fact, parasitoid wasps account for 20% of all known insect species and the Techinidae, which is a family of parasitoid flies, is one of the most speciose of all the Diptera.

Parasitoids have a particularly intimate relationship with hosts because a single host harbors the parasitoid's offspring until maturity. This is one of the major differentiating characteristics that separates parasites and parasitoids from predators in that parasites gain the majority of their nutrients developing from a single-living organism, whereas predators gain nutrients from many individual prey. The fact that hosts are required for parasitoids to complete their reproductive cycle is a driving force in parasitoid evolution and is thought to be one of the main reasons why parasitoids are often highly host specialized and possess unique adaptations for overcoming host defenses. Conversely, the variation in ecological interactions between host insects and the suites of parasitoids that attack them has contributed to the evolution and diversification of antiparasitoid defenses. One theoretical consideration on the evolutionary trajectories of aggressive and defensive traits in host–parasitoid associations is that parasitoids have a choice when selecting host species, whereas host insects may be attacked by many different types of parasitoids (and predators) at once. Therefore, specialization is typically advantageous in parasitoids as this allows them to coevolve with a single host species, whereas a general defensive strategy is often more effective in hosts as this protects against multiple natural enemies (cf., horizontal vs. vertical resistance in plants).

Throughout this article, we will approach the topic of antiparasitoid behavioral defenses by embedding theory with experimental evidence and examples from natural systems. To aid in understanding the variation in behavioral defenses, we categorize the different strategies based on characteristics of the host. Thus, two broad categories of defenses are defined: (1) 'preventative' traits that provide protection by reducing signals that parasitoids use to locate hosts, and (2) 'interactive' traits that are direct defensive interactions and provide protection against parasitoid attack once the parasitoid and host come into contact. Each broad category will then be subdivided into several subsections. Preventative traits may include forms of circumvention such as hiding, specialized feeding, defensive structures, and grouping effects. Interactive traits will be discussed at different levels of selection, including defenses that occur before parasitoid oviposition (egg laying) such as those defenses that are local (e.g., kicking, spitting) and global (e.g., alarm pheromone), as well as those that occur after parasitoid oviposition (e.g., adaptive suicide).

Preventative Traits

One of the ecological distinctions in any predator–prey interaction is defined by the cues predators use to locate prey. A major distinction between the foraging of predators and parasitoids is the use of chemical cues to locate prey items. Although predators use scent when foraging, it is typically in combination with a variety of other sensory systems. Insect parasitoids, on the other hand, rely almost exclusively on chemical cues to focus their search efforts to the most profitable environments and then use chemical, tactile, and visual cues to locate hosts at close range. Chemical cues that reveal a host's presence can be released by plants through herbivore feeding, defecation, egg laying, mating pheromones, exuviae, or any other chemical signals that are produced by the host interacting with its environment. Long-distance cues are typically general signals produced in large quantities, such as green leaf volatiles that are produced whenever a plant is damaged. These general signals can be thought of as having high detectability but low reliability in that they typically indicate the presence of a feeding animal but are not reliable indicators of a particular host species. Short-range cues, such as frass, exuviae, or mating pheromones, are highly reliable indicators of a specific host's presence; however, these signals are produced by the host themselves and are therefore under direct selective pressure to become inconspicuous to natural enemies. This evolutionary battle of hide and seek has resulted in highly sensitized olfactory systems in parasitoids and an array of elaborate means of avoiding detection in hosts.

Hiding and Avoidance in Space and Time

Almost all insects display some form of hiding or avoidance as the first line of defense against parasitoids. Hiding differs from avoidance in that avoidance is the use of a habitat that has few associated natural enemies (i.e., enemy-free space), whereas hiding refers to remaining inconspicuous in the presence of natural enemies. Enemy-free space is thought to be an important mechanism involved in both population dynamics and evolutionary responses in insects, such as in niche partitioning wherein prey or host organisms expand to new resources as a means of escaping predation and competition. For example, the shift from hawthorn to apple by the apple maggot fly, *Rhagoletis pomonella*, has facilitated escape from parasitism because the flies can feed much deeper in the fruits of apples thus avoiding parasitoid ovipositor. Enemy-free space may also be achieved by the timing of exposure to natural enemies, for instance via host diapause, thus creating a temporal refuge from parasitoids.

A common and effective antiparasitoid strategy employed by many larvae is to mine under the surface of host plants (e.g., cherry bark tortrix), thereby concealing themselves and reducing exposure to natural enemies. Such larvae maintain their position at considerable distance from their tunnel entrance making it difficult for parasitoids (i) to assess whether the tunnel is occupied and (ii) to reach hosts. Thus, parasitoids may be able to detect the herbivore-induced plant volatiles through their highly sensitized chemosensory system, but the host insect still remains elusive, as it is then hiding somewhere within the plant. Although this strategy eliminates external visual exposure, parasitoids have evolved several mechanisms for locating and accessing concealed hosts, such as an acute sensitization to vibrations generated by larval feeding or the fascinating behavior of vibration sounding. Vibration sounding is a form of echolocation through solid medium that several ichneumonid parasitoids (e.g., tribe: Cryptinae) use to locate concealed larvae or pupae by drumming the substrate with their antennae. Once located, many parasitoids of concealed hosts have highly specialized ovipositors designed to probe through wood, stems, or soil to lay eggs in hosts. As discussed earlier, some parasitoids have usurped the behavior of their hosts; here, the host defense of concealing itself to avoid parasitism is exploited by the parasitoid in that once the host is parasitized the shelter provides protection for the parasitoid's larvae by reducing attacks from their own parasitoids (hyperparasitoids).

Insect feeding (see later) and defecation are the most common source of chemical cues used by parasitoids to locate hosts. Frass volatiles, which can attract parasitic tachinid flies at a distance, are used to distinguish host from nonhost larva in Hymenoptera and can result in area-restricted searching or larviposition behavior in parasitoids. Thus, some insects have evolved strategies that allow them to distance their fecal material from themselves, such as frass ejection. This behavior is displayed by most species of skipper caterpillars (Hesperiidae), which can shoot a ballistic frass pellets as far as 1 m in distance. This behavior is not exclusive to the hesperiids, with geometrid and noctuid larvae adopting similar strategies. Although projectile frass ejection is unusual, many insect species, especially those in shelters, remove or defecate away from their local area, which has been shown to greatly reduce the incidence of parasitoid attack in Lepidopteran larvae (e.g., *Epargyreus clarts*). When moth outbreaks occur, their frass can rain down (away) from trees at rates approaching that of raindrops from thunderstorms.

Specialized Feeding

Hosts may be more detectable than they would be otherwise while exploiting their own hosts or prey. For example, injured plants may release specific odors after being fed upon, and these odors can act as cues to host-seeking parasitoids. Thus, there should be selection for hosts to minimize any cues that would improve natural enemy search efficiency. With this in mind, studies have showed that palatable caterpillars express feeding behavior that minimizes visual feeding damage to reduce predation from birds. Of course, birds are visual predators whereas parasitoids are primarily olfactory foragers, so the question is: Do herbivores reduce their chemical apparency via specialized feeding? There is no conclusive evidence for this to date; however, caterpillars rarely concentrate their feeding on a few leaves, which should make such individuals harder to find. Along that same line of reasoning, it has been suggested that herbivores may feed or oviposit (see earlier) in a manner that provides minimum information to their parasitoids, thereby reducing the efficacy of these natural enemies. Parasitoids make foraging decisions based upon host distributions; a random distribution provides little information on the expected whereabouts of hosts and is difficult to exploit. The jury is still out on this one, though research on leafminer-parasitoid interactions suggests that these two organisms may be involved in a 'princess-monster' game, wherein mine geometry impacts probability of escape (i.e., mine geometry evolves in direct opposition to the search patterns of the parasitoid).

A feeding preference by hosts that causes them to be found in enemy-free space as discussed earlier is a defensive form of feeding specialization. For most insects, feeding preference is often based upon secondary plant compounds, and in some species these compounds are sequestered in special locations (e.g., exoskeletal tissues in monarch butterflies), and these can reduce the vulnerability of herbivores to either parasitoid attack or parasitism. So, insect hosts might choose to feed on hosts because those secondary plant chemicals provide prophylactic protection against parasitoids; however, we are not aware of any demonstration of self-medication in any parasitized hosts (see postoviposition behavior later). Similarly, as already discussed, it has been known for some time that feeding internally on plants provides better protection against parasitoids than external feeding.

Grouping Effects

From the perspective of the individual, living within a group can be particularly effective as a defense against parasitoids (especially visual parasitoids) for two reasons. First, there is now ample evidence that natural enemies can be confused by the individuals within a group moving en masse when confronted by threats. Second, living in a group can be beneficial if there is a dilution effect (i.e., the per capita probability of parasitism declines as the size of the group increases). Unfortunately for the host, larger groups are also easier to detect. So, in addition to intraspecific competition for resources, living in a group has both costs and benefits for the individual that can vary with group size. When evaluating this tradeoff problem, evolutionary biologists generally agree that the approach of join-or-leave quandary per individual is preferable to focusing on optimal group size issue simply because group selection is generally a weaker force than individual selection, all else being equal.

Defensive Structures

Many insect larvae, in particular beetles, cover themselves with fecal material in the form of coatings, cases, or shields. These coverings may function as antiparasitoid defenses either as simple mechanical protection or via secondary defense compounds that they acquire from the plants that they feed on. In some cases, these larval products may be carried over into the pupal stage. For similar reasons, adult insects may also use excreta to cover their eggs. It is important to keep in mind, however, that fecal materials will release odors and thus may make producers more detectable than those that dispose of such materials.

Some caterpillars live in groups inside tents with all the benefits and costs noted in the group-living section. Currently, there is not much evidence that the shelters provide protection from parasitoids over and above the group-living benefits, but some analogous systems involving leafhoppers and mutualist ants suggest that leafhoppers in ant-built shelters were less likely to be attacked by parasitoids than those outside shelters.

Interactive Traits

Preoviposition

When selecting a suitable host, parasitoids, like many predatory animals, tend to abide by the principles of optimal foraging theory. Optimal foraging assumes that natural selection has resulted in foraging behavior that maximizes fitness, while taking into account the dependence of energy intake rate on the forager's ability to detect, capture, and handle each prey item. Parasitoids selecting hosts optimally, therefore, make decisions that maximize the net rate of fitness gained based on the profitability of each host, where profitability can be defined as the fitness of the offspring produced per unit time spent handling the host. This theory is exemplified by the life stages and size of hosts that parasitoids typically attack. The vast majority of parasitoids attack the eggs, early larvae, or pupae of their hosts; far fewer attack the highly mobile, often heavily sclerotized, adult insects as such stages are difficult to subdue and penetrate and thus less profitable. Euphorinae parasitoids are one of the few groups that have invaded the niche of adult insect hosts. Euphorins and some species in the family Tachinidae have been observed utilizing elaborate oviposition tactics to overcome adult defenses, such as ovipositing through very small openings in beetle armor, into the mouth of flea beetles, and even into the abdomen of chrysomelid beetles in the brief moment the beetles raise their elytra to fly away. Attacking the early larval stages of caterpillars and aphids (early instars) is a common strategy exhibited by most parasitoids of hosts that develop aggressive defensive behaviors as they mature, even though larger hosts universally yield larger parasitoid offspring of greater fitness. In wasps that attack spiders, aphids, and caterpillars, host size and aggressiveness have also been correlated with the range of acceptable host species and host size classes a parasitoid can successfully attack. Furthermore, this relationship can be a function of parasitoid size, with larger wasps able to maintain a broader range of larger host species or classes.

The physical defenses (i.e., morphological and behavioral) of hosts present formidable barriers to egg laying in parasitoids that have resulted in the evolution of very unusual and diverse oviposition strategies; note the distinct differences in the strategies that have evolved in the Hymenoptera versus the Diptera. As a general rule, Hymenopteran parasitoids must contact hosts with their ovipositor in order to lay an egg, thereby directly exposing them to physical defenses of their hosts, whereas only around half of the Dipterans have adopted this strategy. The remainder of Dipteran parasitoids have adopted methods of oviposition that do not require contact with the host, thereby circumventing the risks associated with physical contact. These strategies typically involve laying eggs away from hosts with the larvae hatching soon after the eggs are laid; they then either wait for passing hosts to attach onto (e.g., many Tachininae) or actively search for hosts (e.g., Dexiini). One of the most interesting oviposition strategies is the use of 'microtype' eggs, in which the parasitoid lays an egg on a host-consumed substrate (e.g., plant material) with the intent of having the egg ingested by the host insect. Once ingested the larvae hatches and tunnels into the host hemocoel.

Escape Behavior

High mobility is one of the most effective methods of avoiding parasitism; however, not all insects can move quickly, if at all. Those that cannot run or walk away, such as pupae and many larvae, often rely on violent wiggling to deter parasitoids. Other widely employed methods of escaping parasitoid attack are dropping or jumping, which are thought to be some of the most effective strategies and are used by many hosts as this removes the host from the local environment. However, this strategy also comes with

a tradeoff to the host insect, in that leaving the plant may result in death from desiccation or starvation. Natural selection clearly acts on such tradeoffs. For example, older aphids are more likely to drop from plants in response to parasitoids, but they can more readily relocate to new host plants than their younger colony mates. Similarly, caterpillars (e.g., *Plathypena scabra*) mitigate this risk by attaching a thread to the substrate prior to dropping, which is then used to climb back onto the plant once the threat has passed. However, parasitoids have been observed counteracting this adaptation by using the suspension thread as a source to locate the caterpillar that is attached to the other end by either sliding down the thread (e.g., *Diolcogaster factosa*) or actually reeling the caterpillar back up to the awaiting parasitoid. Combined with an alarm pheromone (see later), which reveals the presence of a parasitoid to your kin or neighboring community of insects, escape behaviors can be a very effective means of avoiding parasitism.

Studies have shown that escape behavior induced by alarm pheromones in aphids can reduce parasitism by almost half. In addition to providing the alarm call via diffusion in the air, some aphids dab the alarm pheromone onto the parasitoid thus providing directionality to the threat as the parasitoid forages. Defensive alarm pheromones that warn kin of attacking parasitoids, as in aphids that reproduce clonally, have most likely evolved through the benefits of increasing fitness in related individuals (i.e., kin selection). To counteract escape behaviors in aphids, parasitoids have evolved a number of different oviposition strategies, such as a fast sting in *Aphidius* parasitoids to oviposit in as many aphids as possible prior to escape or a stealthy sting, in *Aphelindae* parasitoids so as not to disrupt the aphid patch.

Aggressive Behavior

Aggressive behaviors are physical attacks to avoid parasitism such as kicking, spitting, thrashing, or biting. As mentioned previously, larger hosts are often more effective at physical combat, and therefore hosts may develop different defensive strategies as they mature. Small larvae and early instar aphids often resort to escape behaviors or alarm pheromones (or have no defenses at all), whereas later instar larvae and adult aphids often employ aggressive defensive behaviors.

Aphids raise their body and strike with kicks, antennae, or wings to knock parasitoids away, or some species vigorously wiggle as a group. Caterpillars are renowned for their spitting and regurgitant (e.g., pyralid, noctuid, and ericace larvae), which serve to entangle parasitoids, causing extensive grooming, paralysis, or sometimes death. In cases where parasitoids are attacking carnivorous hosts, the host may pose a considerable risk of death to the parasitoids. Pompilid wasps that attack spiders, and parasitoids of carnivorous moth larvae in the family Geometridae, have been observed actually eating their parasitoids. Cyphomyrmex fungus-growing ants have been reported to grasp, bite, and kill some species of diapriine wasps. Caterpillar bites have been known to sever parasitoid appendages, such as legs and ovipositors, which can maim or kill the parasitoid (e.g., *Cotesia* sp.). Head jerking and flicking is common in caterpillars, which can knock parasitoids away or deliver defensive oral or anal secretions. Studies have shown that parasitoids take a substantially longer time to handle and oviposit in aphids and caterpillars that exhibit aggressive defensive behaviors, thus demonstrating the effectiveness of these forms of defensive behaviors. Some insects are even willing to risk injury to defend their eggs, thus exhibiting maternal behavior: Pentatomid bugs aggressively protect their eggs from attacking parasitoids. Brood guarding has also been reported in parasitoids (e.g., *Goniozus nephantidis*) and secondary hyperparasitoids (e.g., *Trichomalopsis apanteloctena*), which defend the host insect until their larvae pupates, thus guarding against hyper or secondary parasitoid attack.

Chemical Defenses

Alarm pheromones and noxious secretions are common antiparasitoid defenses in aphids. Once an individual is disrupted by a parasitoid's attack, a pheromone is released from the cornicles that can trigger an escape or aggressive response in neighboring aphids.

Defensive secretions are typically more useful against predators than parasitoids, as parasitoids tend to be more specialized and have evolved behaviors to overcome defensive secretions (e.g., quick or stealthy oviposition strategies). A tachinid parasitoid, *Sturmia inconspicua*, will actually wait near its host *Diprion pini* until the defensive oral secretions subside, and then quickly move to oviposit. Secretions can be irritants or even poisons resulting in the death of the parasitoid. In a relatively small number of cases, insects use their waste products as a defensive compound. Cicadas and spittlebugs are reported to discharge anal fluids toward approaching natural enemies that act as potent chemical defenses.

A mutualistic use of secretions can be found in the production of nourishing compounds that attract beneficial organisms as bodyguards. A range of hemipteran taxa, including aphids, membracids, and coccids, all produce secretions that attract ants. The ants receive a food source from the secretions, and in return some ant species (e.g., *Formica* ants) defend the hemipterans from attacking parasitoids and predators.

Postoviposition Defenses

Changes in host behavior, postattack, can be difficult to evaluate because such changes could be due to control of behavior by the host, control of behavior by the parasitoid, or an interaction between the two participants. With that in

mind, there is some evidence for postoviposition behavior that benefits the host. For example, at least one species of aphid has been shown to choose high-danger behaviors after parasitization, whereas when healthy, only the safest behaviors are expressed. The interpretation here is that death to the individual host from these dangerous choices will also lead to death of the parasitoid and negation of parasitoid threat to other colony members. Since aphids are parthenogenetic during the summer months, the killing of the parasitoid by its terminally ill host will greatly benefit that aphid's genetically identical sisters and nieces and thus the inclusive fitness of the victim.

Several different hosts have been shown to relocate following oviposition by their parasitoids, including caterpillars and aphids, and, in a few rare cases, larvae have escaped their parasitoids while parasitoids cocoons were being spun. Except for the last case, none of these responses have been shown to benefit the host, but here interpretation becomes a bit tricky because parasitoids can become hosts for hyperparasitoids. Scientists have suggested that the parasitoid has usurped the behavior of its host to reduce the threat that it faces from its own parasitoid, the hyperparasitoid. If correct, this is in fact, antiparasitoid behavior at a higher trophic level than normally considered.

Similarly, some hosts that have been attacked by microparasites (e.g., viruses, fungi) will change microhabitats to create a kind of behavioral fever that is highly detrimental to the pathogen. Nothing similar has been demonstrated for any parasitoid–host system likely because the tolerance range between hosts and their parasitoids greatly overlap. Finally, although not normally considered a behavioral response, many hosts mount an immunological response to oviposition from parasitoids primarily through encapsulation of parasitoid eggs.

Conclusions

Due to the constraints on any review article, we have been limited to an overview of some of the most common and interesting defensive interactions between parasitoids and their hosts. Many interesting and active areas of research presented in this review have only been touched upon or not mentioned at all. Some of the notable omissions include the removal of ectoparasitoids and counteradaptations in parasitoids, parasitoid learning to circumvent host defenses, and the fascinating interplay of behavioral (and chemical) camouflage and mimicry between parasitoids and hosts. A more detailed explanation of the theory and biology of antiparasitoid defenses can be explored in the reference articles listed later.

The ecological interactions of insects and the parasitoids that attack them are a potent selective force that influences evolutionary directionality and has the potential to shape the structure of communities. It has been suggested that the dynamic arms race of defenses and counterdefenses (i.e., Red Queen's Hypothesis) can lead to a rapid burst of evolutionary change in both hosts and parasitoids, which facilitates diversification. Given that every defensive improvement by a host results in a selective advantage that must be overcome in order for parasitoids to reproduce, it is not surprising that specialization is such a common feature in most parasitoid species, and, as a result, the parasitoid guild has some of the highest levels of species diversity in the class Insecta. One of the major differentiating features between defenses in hosts and counterdefenses in parasitoids is that host insects must contend with a plethora of selective pressures from predators and parasitoids, while parasitoids can evolve to exploit a single host. Thus, the development and expression of morphological and behavioral defenses in any insect depends on the frequency of attack from parasitoids and predators throughout their coevolutionary history.

See also: Beyond Fever: Comparative Perspectives on Sickness Behavior; Co-Evolution of Predators and Prey; Defensive Chemicals; Group Living; Optimal Foraging Theory: Introduction; Risk-Taking in Self-Defense.

Further Reading

Cocroft RB (2001) Vibrational communication and the ecology of group-living, herbivorous insects. *American Zoologist* 41: 1215–1221.

Djemai I, Meyhöfer R, and Casas J (2000) Geometrical games between a host and a parasitoid. *The American Naturalist* 156: 257–265.

Gentry GL and Dyer LA (2002) On the conditional nature of neotropical caterpillar defenses against their enemies. *Ecology* 83: 3108–3119.

Godfray HCJ (1994) *Parasitoids: Behavioral and Evolutionary Ecology.* Princeton, NJ: Princeton University Press.

Godfray HCJ and Shimada M (1999) Parasitoids as model organisms for ecologists. *Researches on Population Ecology* 41: 3–10.

Gross P (1993) Insect behavioral and morphological defenses against parasitoids. *Annual Review of Entomology* 38: 251–273.

Heinrich B and Collins SL (1983) Caterpillar leaf damage and the game of hide-and-seek with birds. *Ecology* 64: 592–602.

Henry LM, Roitberg BD, and Gillespie DR (2006) Covariance of phenotypically plastic traits induces an adaptive shift in host selection behaviour. *Proceedings of the Royal Society: Series B* 273: 2893–2899.

Holt RD and Lawton JH (1993) Apparent competition and enemy-free space in insect host–parasitoid communities. *The American Naturalist* 142: 623–645.

Kawecki TJ (1998) Red queen meets Santa Rosalia: Arms races and the evolution of host specialization in organisms with parasitic lifestyles. *The American Naturalist* 152: 635–651.

Stireman JO and O'Hara JE (2006) Tachinidae: Evolution, behavior and ecology. *Annual Review of Entomology* 51: 525–555.

Tanaka S and Ohsaki N (2006) Behavioral manipulation of host caterpillars by the primary parasitoid wasp *Cotesia glomerata* (L.) to construct defensive webs against hyperparasitism. *Ecological Research* 21: 570–577.

Thompson JN (1986) Oviposition behaviour and searching efficiency in a natural population of a braconid parasitoid. *Journal of Animal Ecology* 55: 351–360.

Vet LEM and Dicke M (1992) The ecology of infochemical use by natural enemies in a tritrophic context. *Annual Review of Entomology* 37: 141–172.

Vinson SB and Iwantsch GF (1980) Host suitability for insect parasitoids. *Annual Review of Entomology* 25: 397–419.

Weiss MR (2006) Defecation behavior and ecology of insects. *Annual Review of Entomology* 51: 635–661.

Parental Behavior and Hormones in Mammals

K. E. Wynne-Edwards, University of Calgary, Calgary, AB, Canada

Universality of Mammalian Maternal Care

Mammals are distinguished from all other vertebrates by the inescapable need for maternal care in the form of lactation. Mammary glands are a distinguishing feature of the mammals, and all offspring are suckled, to a greater or lesser extent, on their mother's milk. In some cases, such as in rabbits, the milk transfer from mother to offspring can be for as short as a few minutes every day, while in other species, such as humans, fetal demands for milk meals are considerably more frequent and the milk is more dilute. When added to the intimate proximity of mother and fetus during internal gestation (pregnancy), this enhances the initial investment bias that eggs are more costly than sperm, 'traps' mammalian mothers into proximity to their offspring, and is responsible for the general bias that mammalian mothers provide considerably more parental care than mammalian fathers.

Of course, mammalian offspring need more parental care than just having their early nutritional needs met. They also need shelter, defense, grooming, and opportunities to learn from example and experience. In principle, all the nonlactational parental care needs could be met by male parents (**Figure 1**), or by other members of the social group. However, although scientists have learned a great deal from those types of care by nonmothers, mammalian mothers rarely escape responsibility for all these aspects of parental care.

Theoretically, it might be possible for female mammals to be born 'parental', and to shower care upon all offspring they meet in response to 'infantile' stimuli from those needy individuals. In practice, the evolutionary costs of providing parental care to an unrelated individual are measured in terms of decreased resources available for individual reproductive success, and this kind of indiscriminate caregiving would be advantageously exploited by others for their own benefit. Instead, there are adaptive advantages to the other side of the caregiver behavioral strategy, infanticidal behavior. Most female mammals are not indifferent to offspring other than their own. Instead, they tend to be specifically aggressive toward unprotected unrelated offspring and will attack, injure, and often kill them if they are undefended. Caregiving behavior by individuals other than the mother, such as fathers, close kin, and extended social groups, is equally sensitive to exploitation if care is given indiscriminately, and the animals may benefit, evolutionarily, from aggression toward unrelated or unfamiliar young. Thus, mammalian

parental behavior requires the appropriate expression of caregiving behaviors, at the appropriate times, toward appropriate individuals, and simultaneously requires the inhibition of incompatible behaviors that are adaptive at other phases of the life span.

Pregnancy as Preparation for Motherhood

The appropriate time for the expression of maternal behavior is immediately before birth. In many species, there is essential behavior before birth, including the choice of a secluded location for birth and the building of an appropriate nest or 'nursery.' At the time of parturition (=birth), there is no room for maternal 'mistakes' in the types of behavior she expresses toward her offspring. If she injures them, there is no adaptive value in becoming parental a few hours too late. Hormonal changes associated with the last stages of pregnancy, including the hormonal changes that precede and accompany parturition, are prime candidates for roles as the physiological signals involved in the transition from indifferent, or aggressive, behavior to parental behavior. Estrogen is well established as a key steroid hormone involved in the initiation of maternal behavior.

Estrogen and Maternal Behavior

Estrogen is the final step on the steroid biosynthesis pathway, with progesterone, followed by testosterone, as its precursors. In general, the concentration of estrogen is lower than that of other steroids in the circulation, with estrogen concentrations often measured in picogram per milliliter (10^{-12} g ml^{-1}) and its precursor steroids typically measured in nanogram per milliliter (10^{-9} g ml^{-1}). In female mammals, estradiol 17β is the major estrogen synthesized by the ovarian preovulatory follicle in the days and hours leading up to ovulation and mating. Its concentration is then typically reduced as the follicular tissue redifferentiates into the corpus luteum of pregnancy. Mammalian corpora lutea synthesize progesterone during early pregnancy, can persist throughout pregnancy, and also synthesize estrogen. However, in later pregnancy, steroid biosynthesis by the placenta often takes on a more important steroidogenic role, including estrogen synthesis. Steroid hormones are lipophilic and readily pass through cell membranes. Thus, the third trimester of mammalian pregnancy is typically characterized by the

Figure 1 Male Djungarian hamster retrieving young pups experimentally displaced from the nest area. Credit: Katherine Wynne-Edwards.

highest concentrations of estrogen seen by a female in her lifetime. Birth then initiates an extraordinary transition in estrogen concentration, with delivery of the placenta and rapid withdrawal of the steroid biosynthesis from that source.

As early as the mid-1960s, studies in the laboratory rat demonstrated that premature delivery of pups by Caesarian section resulted in acceleration of the onset of maternal retrieval, huddling, and grooming of pups. Within a decade, Jay Rosenblatt of Rutgers University had begun the first studies of a long and productive career studying maternal behavior in the laboratory rat. Steroid hormones were soon implicated in this process as hysterectomy (excision of both uterine horns) reduced the latency to express maternal behavior, but only if the ovaries remained intact. Exogenous estrogen could rescue parental behavior during short latencies following hysterectomy, even when the ovaries were removed. However, the relationships were not simple.

In a process known as 'concaveation,' that is a specialized subset of sensitization, virgin female rats could be induced to show a full spectrum of 'maternal' behavior after days of repeated exposure to pups. This process involved a decrease in fear and neophobia reactions, followed by a gradual increase in contact with those pups. However, it was not accompanied by hormonal changes such as those experienced by pregnant females before birth. Virgin rats with neither uteri nor ovaries responded to estrogen treatment with reduced latencies to express maternal behavior toward a stimulus pup, but the treatment was not effective if their uteri were intact.

Male rats also proved to be an exceptional model in which early developmental manipulations of hormonal experiences could be differentiated from the effects of lifelong exposure to ovarian sex steroids. Male rats that were castrated soon after birth, in the neonatal period when brain differentiation is not complete, experienced 'feminization' of their brain. In adulthood, they could be exposed to hormonal regimes that mimic late pregnancy and would respond with 'maternal' behavior. As was seen in females, estrogen acting centrally at the medial preoptic area of the hypothalamus induced 'maternal' behavior, whereas lesions to the same area prevented 'maternal' behavior by estrogen-primed male rats.

Over time, researchers have effectively used each new technology, from steroid implants targeted to specific hypothalamic nuclei, through immediate early gene expression after pup exposure, estrogen receptor distribution maps, and gene knockout mouse models, to identify the maternal estrogen-sensitive circuitry of the female rat and its role in the expression of maternal behavior. The emergent consensus has described an essential maternal hypothalamic circuit requiring the medial preoptic area, bed nucleus of the stria terminalis, and medial amygdala. As birth approaches, female rats and mice have increased estrogen receptor alpha expression that enhances sensitivity to estrogen in those brain regions. After birth, these hypothalamic regions remain important in the expression of appropriate maternal behavior, receiving input from other brain regions and gaining sensitivity to a broad diversity of other, biologically relevant, hormones.

Prolactin and Maternal Behavior

Prolactin is a peptide hormone that was named for its role in an essential form of mammalian parental behavior, lactation. However, it is also involved in many other aspects of maternal physiology, ranging from maternal recognition of pregnancy in rats and mice to sexual receptivity. Prolactin is also implicated in adult neurogenesis within the social brain and is selectively transported into the brain through the choroid plexus. Given these roles around the time of parturition, prolactin is ideally placed to be involved in the induction and maintenance of maternal behavior.

Prolactin increases before birth are involved in the maternal behavior of rodents although there are species differences in whether the role is restricted to the onset of maternal behavior or both the onset and maintenance of maternal behavior. The medial preoptic area of the hypothalamus is the critical site for prolactin induction of maternal behavior, and mice with a knockout mutation of the prolactin receptor show deficits in pup retrieval as a critical component of rodent maternal behavior. However, just as it is difficult to manipulate estrogen levels in pregnant rats without terminating the pregnancy, it is difficult to manipulate prolactin without altering lactational performance. As a result, studies are limited.

Another challenge is the unique relationship between the neurotransmitter, dopamine, and prolactin. Unlike other peptide hormones synthesized by the anterior pituitary, prolactin release in mammals is not stimulated by releasing hormones from the hypothalamus. Instead, prolactin release is tonically (=continuously) inhibited by dopamine released from the synaptic terminals of neurons in the hypothalamus. Thus, pulsatile release of prolactin into the peripheral circulation is dependent upon the withdrawal of dopamine inhibition rather than an increase in a stimulatory hormone. Pharmaceutical agents used to suppress prolactin in parental behavior research are actually dopamine receptor agonists. Since dopamine is central to brain pathways for reward, these pharmaceutical treatments can have broad effects that are not specific to prolactin.

Glucocorticoids and Maternal Behavior

There has been relatively little research looking at the relationship between adrenal glucocorticoid hormones and maternal behavior onset and maintenance. However, there is typically an increase in glucocorticoids soon before birth in both mother and fetus, and labor results in increased glucocorticoid concentrations in the shared maternal–fetal circulation. Although commonly referred to as 'a stress response,' glucocorticoids are secreted at diverse points in the life span when focused attention to salient environmental information, such as the approach of a predator or the identity of one's partner during sex, is biologically adaptive. Thus, cortisol and corticosterone, depending on the species, increases around the birth have appropriate timing to affect the salience of cues from the infant for the mother. There is also clear evidence that maternal stress during pregnancy, or fetal stress as a result of poor maternal care during infancy, results in epigenetic modification of the juvenile and adult stress responses of the affected individuals. These epigenetic changes are then reflected in the transgenerational transmission of maternal behavior styles.

Oxytocin and Vasopressin and Maternal Behavior

Oxytocin is an important hormone in the establishment of pair bonds between conspecific individuals, and parent–offspring social affiliation and bonding shares many traits with pair bonding. Both are social affiliation bonds that result in tolerance for physical proximity and altered aggressive reactions to intruders. Oxytocin is released during labor and parturition, and works with estrogen and progesterone to prime females to be maternal immediately. A direct role in maternal behavior is confirmed by the ability of oxytocin to enhance pup responsiveness when infused into key nuclei in the maternal behavior circuit after estrogen priming. Like oxytocin, vasopressin is secreted before and after birth, and infusions of vasopressin or antagonism of the vasopressin V1a receptor into the medial preoptic area enhance or reduce maternal behavior latencies. Also like oxytocin, these effects depend on the prior priming actions of estrogen.

Insights from Sheep

The ewe has also been studied extensively and highlights both similarities and differences from the maternal rodent brain. Their larger size has facilitated studies using local anesthesia to interrupt the components of the neural pathways leading to maternal behavior as well as microdialysis approaches to alter hormone exposures at specific brain sites. Unlike altricial rat pups that are naked, blind, and completely dependent on care for thermoregulation as well as nutrition, a lamb is precocial, independently mobile, and must maintain contact with its mother in a complex social herd. As farmers have long known, ewes exhibit a strong selectivity, in the form of exclusive access to nursing, toward their own lamb after birth that makes fostering an orphaned lamb onto a new mother challenging. As a result, it has been possible to clearly distinguish between maternal responsiveness toward any lamb and maternal selectivity for her own lamb.

Estrogen and Progesterone

The onset of maternal responsiveness is under the combined influence of elevated estrogen and a drop in progesterone before birth, with rapid declines in both steroids after the placenta is expelled. However, this effect is priming rather than sufficient for maternal behavior, as physiological doses of the steroids alone cannot elicit maternal behavior in the ewe without vaginal–cervical stimulation. Immediately postpartum, maternal attraction to amniotic fluid results in licking and cleaning of the neonate, although amniotic fluid is not attractive to females at other phases of the life span, and there are distinct auditory 'bleats' that enhance individual recognition. This attraction involves specific hypothalamic nuclei, including the preoptic area, bed nucleus of the stria terminalis, and the paraventricular nucleus. Within 2 h of birth, the lamb suckles its first meal of colostrum, and maternal selectivity is so firmly established that the maternal process can be considered an adult version of the immediate posthatch imprinting of some avian species on the first individual or animal they see. The neurobiology of this individual recognition pathway involves the main olfactory bulb and the amygdala. Within a few additional hours, the mother is also capable of recognizing her lamb from a distance through a combination of auditory and visual cues, and that recognition is achieved even if the olfactory stimuli necessary for nursing exclusivity are denied to the mother.

Oxytocin

Not unexpectedly, given its roles in social bonding and in the contractions of labor, oxytocin is also active at many points in the maternal selectivity pathway over the same time interval. As the lamb is expelled, oxytocin is released into the peripheral circulation from oxytocin synthesis in the paraventricular and supraoptic nuclei of the hypothalamus. Other sources of synthesis, plus increased receptor expression, also emerge along the maternal behavior pathway that includes the preoptic area and the bed nucleus of the stria terminalis.

Insights from Naturally Paternal Males

A major challenge associated with studying the hormonal basis for maternal behavior is that the hormones of interest are also involved in the physiological management of the pregnancy, the birth, and the establishment of lactation. This means that it is difficult to manipulate the social and behavioral aspects of maternal behavior without also altering pregnancy, birth, and lactational outcomes at the same time. Naturally, paternal male mammals avoid this problem because they are neither pregnant, nor lactating, as they become parents.

A second challenge is the severe adaptive consequence of failure to exhibit maternal behavior. The offspring die, and those maternal genes are not represented in the next generation. As a result, female mammals are likely to have multiple, redundant, pathways to ensure that maternal behavior occurs when needed and which might compensate for interventions that experimentally alter a single hormonal pathway. Natural mammalian paternal care, on the other hand, is relatively rare, and the trait is not always shared across closely related species. This suggests that there have been multiple independent evolutionary origins for paternal behavior. In each case, a parsimonious route

to the expression of paternal behavior would be the activation of preexisting maternal neural and endocrine circuits. Thus, each species is likely to have 'activated' one preexisting maternal neuroendocrine circuit involved in paternal and maternal behavior, and that circuit should be amenable to experimental 'dissection.'

Finally, naturally paternal fathers lack the mother's absolute security in her parentage. Her egg has not left her body from fertilization until birth, whereas his paternity depends on her mate fidelity, and that often involves sperm competition between males. For this reason, the provision of direct paternal care is likely to be facultative, and influenced by competing opportunities to attract new mates as well as his 'confidence' in his paternity. This facultative component of the expression of paternal behavior is expected to delay and focus the male's behavioral transition to the window of time during which his young are born. Female mammals are likely to undergo multiple developmental priming events for maternal sensitivity, whereas males are most likely to be sensitive to immediate environmental cues that are, therefore, amenable to experimental intervention.

Thus, paternal males represent a natural biological model system in which parental care is dissociated from the physiology of pregnancy, birth, and lactation, homologous to a maternal neurobehavioral circuit, and activated at a specific time. For these reasons, as well as a basic scientific interest in how males become 'Dads,' there has been considerable interest in the hormonal control of mammalian paternal behavior.

Prolactin and Paternal Behavior

The correlation between prolactin and naturally occurring paternal behavior is exceptionally strong. Positive associations between prolactin and the expression of natural parental care by mammals have been established for biparental rodent species, including the California mouse (*Peromyscus californicus*), the Mongolian gerbil (*Meriones unguiculatus*), the Djungarian hamster (*Phodopus campbelli*), and the striped mouse (*Rhabdomys pumilio*). Similar associations have also been reported in non-human primate species that have twins, and need the males to carry offspring, including the Common marmoset (*Callithrix j. jacchus*) and the Cotton-top tamarin (*Saguinus oedipus*). Even in human fathers, positive associations have been documented between prolactin concentration and responses when listening to recorded infant cries, self-report of two or more symptoms of pregnancy, and interactions with their children. However, experimental tests of this relationship in male Djungarian hamsters demonstrated that the association between prolactin and paternal behavior was not causal. Despite an increase in prolactin receptor mRNA in the choroid plexus of males after the birth of their litter, and an increase in serum prolactin concentration when their pups were 5-days old, pharmacological manipulation of prolactin concentration failed to reduce paternal behavior.

Testosterone and Paternal Behavior

Unlike prolactin, the relationship between testosterone and mammalian paternal behavior is generally seen as 'getting out of the way' to allow males to resist distraction from competing demands such as sexual opportunities, rather than facilitating or inhibiting the behavior. For example, a tradeoff between parental behavior and sexual behavior results in different adult male behavioral trajectories in Mongolian gerbils exposed to different uterine steroid environments as a result of secretions from their neighboring siblings. However, castration, which removes testicular testosterone synthesis, does not always affect direct parental behavior toward pups. In some species where castration does reduce paternal behavior, those changes turn out to be the result of reduced estrogen. Since estrogen is synthesized from testosterone, surgical depletion of the testosterone pool through castration also reduces estrogen synthesis. For the same reason, pharmacological supplementation of testosterone can show behavioral effects as a result of aromatization of that testosterone to estrogen.

Estrogen and Paternal Behavior

Given the strong, integrated, multilevel evidence for third trimester estrogen in the initiation of mammalian maternal behavior, it is not surprising that researchers have looked for evidence that estrogen is involved in the expression of parental behavior in species where direct paternal care occurs naturally. In the male California mouse, estrogen facilitates paternal contact with his pups, and there is evidence that the enzyme responsible for estrogen synthesis from testosterone is expressed in a key hypothalamic nucleus for maternal behavior, the medial preoptic area. In contrast, the male Djungarian hamster has surprisingly high levels of estrogen in its circulation, but there is no evidence of estrogen priming before the male becomes a father, or that reducing estrogen concentration through castration, or reducing estrogen exposure through the inhibition of estrogen synthesis, results in any deficits in paternal behavior. Results are equally mixed in non-human primates. Estrogen concentration increases as birth approaches in experienced Cotton-top tamarin, fathers. However, the associations are opposite in the Black Tufted-Ear marmoset (*Callithrix kuhlii*), with estrogen concentration highest at stages of the life span when attention to young is lowest, and an inverse association between individual estrogen concentration and parental care provided for their young. In men, there is evidence of

estrogen changes associated with being in a stable relationship and with fatherhood, but the biological samples were saliva, in which concentrations of estrogen are very low and difficult to quantify.

Progesterone and Paternal Behavior

In female mammals, the role of progesterone in maternal behavior is challenging to study because manipulations result in pregnancy failure. However, there has been strong evidence both for and against a role for progesterone in male parental behavior. Male progesterone receptor knockout mice and wild-type mice with pharmaceutically antagonized progesterone receptors show enhanced paternal behavior and reduced infant-directed aggression, suggesting an inhibitory role for progesterone in paternal behavior. In contrast, the same receptor antagonism treatment in male Djungarian hamsters does not increase the responses of naïve males to pups or decrease their infant-directed aggression.

Glucocorticoids and Paternal Behavior

As is true for research into the hormonal basis for maternal care, there is relatively little research on cortisol and corticosterone effects in mammalian fathers. However, given the diversity of contradictory evidence surrounding the roles for each of the sex steroids in the activation of a paternal behavior, glucocorticoid dynamics that alter the salience of pup and maternal stimuli are likely to be important. In men becoming fathers, in experienced Cotton-top tamarin fathers, and in Djungarian hamster males, there are increased glucocorticoid concentrations before birth and decreased glucocorticoid concentrations associated with a stable pair-bonded relationship.

Oxytocin and Vasopressin and Paternal Behavior

Expectant California mouse fathers have elevated plasma oxytocin concentrations relative to new fathers and naïve males, but currently there is little available data relative to paternal behavior and comparison between different species. Like females, biparental prairie voles (*Microtus ochrogaster*) respond to the infusions of arginine vasopressin with increased contact with pups and vasopressin V1a receptor antagonism with decreases in pup contact. Interestingly, these effects have also been shown to affect the incidence and extent of paternal behavior in species such as the meadow vole (*Microtus pennsylvanicus*) that express paternal behavior facultatively, depending on social and environmental conditions. There is also evidence that higher vasopressin synthesis levels within the bed nucleus of the stria terminalis are associated with more extensive paternal behavior in California mice.

Hormonal Homology Between Maternal and Paternal Behavior

Evidence to date suggests that there is considerable diversity between species in the roles that different hormones play in the activation of natural paternal behavior. In the male Djungarian hamster that acts as a midwife to deliver his offspring and is highly attentive to displaced pups, experimental tests of causal relationships between parental behavior and estrogen, testosterone, prolactin, and progesterone, have all been negative. Paternal behavior in that species is so essential for survival that there might be redundant, or hormone-independent, mechanisms of paternal behavior activation. On the other hand, few studies have looked at the developmental effects of hormones on these males. It is possible that there are organizational changes during development, whether in the uterus, or after birth, that sensitize the paternal male brain to multiple, appropriate cues for the activation of parental behavior and selective inhibition of pup-directed aggression. Future research testing the inference that correlation between a hormone and parental behavior is causal, in diverse species, will be needed to establish how common, or how divergent, those neuroendocrine circuits are.

It is also possible that male mammals depend on tracking signals from the pregnant female to time their transition to parental care. However, evidence to date does not support this hypothesis. Male Djungarian hamsters, separated from the female 2 h after mating and housed in a separate room with independent air handling, are welcomed back into the female's nest a few hours before birth and participate as midwives in birth. This hypothesis has also been explicitly tested through parallel salivary samples from men and women expecting their first child, which revealed no evidence that the man's hormones tracked those of his female partner.

Placentophagia as a Source of Hormones

Another potential source of steroid hormones for new parents is through placentophagia. Except for aquatic mammals, camels, and some primates, essentially all parturient mammals consume placenta (afterbirth) during the few hours surrounding the birth but shun it at all other stages of the life span. It is generally assumed that placentophagia functions to clean the nest area and avoid attracting predators. However, the female is typically so focused on eating the placenta that she prefers it over the pups. The complete sex steroid biosynthesis pathway, plus compounds capable of activating the endogenous opiate receptors, are present in placenta. This makes placenta a potentially potent hormonal 'hit' for a new mother or a new father. Placentophagia is not required for the normal expression of maternal behavior. However, there might be

modulatory effects associated with placentophagia drawing the female to the pups and exposing herself to stimuli from them while licking. Placentophagia might also have a significant impact on males and older offspring that are present during birth. In highly parental Djungarian dwarf hamsters that act as midwives during the birth of their pups, placentophagia is common, whereas, the closely related Siberian dwarf hamster (*Phodopus sungorus*), which is not paternal, refuses the placenta (**Figure 2**). Juvenile Djungarian hamsters also assist in the birth if they are present and that interaction includes placentophagia. However, the impact of placentophagia on male parental behavior is not known, and males of many highly parental mammalian species are excluded from the nest area during birth, which would preclude placentophagia.

Parental Behavior as an Addictive Reward

It is possible that parental behavior was the original 'adaptive' value for the neural pathways leading to addiction and reward in the brain. The challenges for a new parent are extraordinary. First, there must be a halt to the behaviors directed toward infants that are unrelated. These include an essential neophobia, aggression, infanticidal attack, and neglect. On the other side, there is an exquisitely timed transition into selective aggression toward adults that would have been possible sexual partners a few days ago but are now threats to the infant. At the same time, there is the initiation of an active set of licking and retrieval behaviors. Physiologically, there are shifts in metabolic rate that increase the overall metabolic demand. Behaviorally, there is a substantial readjustment of the time budget to build, maintain, and thermoregulate the nest environment that typically prioritizes the viability of

Figure 2   Male Djungarian hamster, that has never been a father, eagerly eating fresh placenta although most mammals, except parturient females, refuse placenta. Credit: Jennifer Gregg and Katherine Wynne-Edwards.

the infants over personal homeostasis and metabolic demands. Each of these behaviors then needs to be repeated, as often as necessary, 24 h per day. Such sudden devotion to a repetitive subset of previous behavioral diversity has all the hallmarks of addiction. At the level of the brain, parents need to become addicted to caring for their offspring.

Conclusions

Mammalian maternal behavior is under hormonal control. Female mammals are primed by the estrogen of late pregnancy to be sensitive to subsequent inputs, ranging from the vaginal–cervical stimulation of parturition, to the chemical and auditory characteristics of the neonates. Other steroid hormones, peptide hormones, and reward/reinforcement neural circuit modifications, also make important contributions to the timing and activation of specific components of appropriately timing and directing maternal care. Mammalian fathers, on the other hand, still have to ensure that they do not injure their offspring, but rarely make the same behavioral transition to the provision of direct retrieval, feeding, and care that mothers make. Nevertheless, those males that do become behavioral fathers challenge our understanding of how the brain forms strong social bonds, and have much to offer as animal models of the role of hormones in that process.

See also: Cooperation and Sociality; Infanticide; Maternal Effects on Behavior; Monogamy and Extra-Pair Parentage; Parental Behavior and Hormones in Non-Mammalian Vertebrates; Parent–Offspring Signaling; Sex Allocation, Sex Ratios and Reproduction.

Further Reading

Bridges RS (2008) *Neuroendocrinology of the Parental Brain*. Burlington, MA: Academic Press.
Carter CS and Keverne EB (2002) The neuroendocrinology of social attachment and love. In: Pfaff D (ed.) *Hormones, Brain and Behavior*, vol. 4, pp. 299–337. San Diego: Academic Press.
Francis D, Diorio J, Liu D, and Meaney MJ (1999) Nongenomic transmission across generations of maternal behavior and stress responses in the rat. *Science* 286: 1155–1158.
Hrdy SB (2000) *Mother Nature*. New York, NY: Pantheon Books.
Kristal MB (1980) Placentophagia: A biobehavioral enigma. *Neuroscience and Biobehavioral Reviews* 4: 141–150.
Lévy F and Keller M (2008) Neurobiology of maternal behavior in sheep. *Advances in the Study of Behavior* 38: 399–428.
Lim MM, Wang Z, Olazabal DE, Ren X, Terwilliger EF, and Young LJ (2004) Enhanced partner preference in a promiscuous species by manipulating the expression of a single gene. *Nature* 429: 754–757.
Nelson RJ (2005) *An Introduction to Behavioral Endocrinology*, 3rd edn. Stamford, CT: Sinauer Associates.
Numan M (2006) Hypothalamic neural circuits regulating maternal responsiveness toward infants. *Behavioral and Cognitive Neuroscience Reviews* 5: 163–190.
Rosenblatt JS (1967) Nonhormonal basis of maternal behavior in the rat. *Science* 156: 1512–1514.
Wynne-Edwards KE (2003) From dwarf hamster to daddy: The intersection of ecology, evolution, and physiology that produces paternal behavior. *Advances in the Study of Behavior* 32: 207–261.
Wynne-Edwards KE and Reburn CJ (2000) Behavioural endocrinology of mammalian fatherhood. *Trends in Ecology & Evolution* 15: 464–468.
Wynne-Edwards KE and Timonin ME (2007) Parental care in rodents: Weakening support for hormonal regulation of the transition to behavioral fatherhood in rodent animal models of biparental care. *Hormones and Behavior* 52: 114–121.

Parental Behavior and Hormones in Non-Mammalian Vertebrates

J. D. Buntin, University of Wisconsin-Milwaukee, Milwaukee, WI, USA

Introduction

Parental behavior patterns are remarkably diverse among the nonmammalian vertebrates, and although hormones are likely to regulate the expression of parental behavior in most if not all cases, the hormonal mechanisms involved have been investigated experimentally in less than 1% of extant species. Despite our overall ignorance of the diversity of hormonal mechanisms that influence parental behavior expression, a survey of the limited evidence available suggests a few broad themes in the relationship between hormones and parental behavior that are likely to apply to a variety of vertebrate taxa. Because these themes frame the discussion that follows, they are important to consider at the outset. First, hormones rarely work in isolation to influence parental behavior. More commonly, they work together in permissive, additive, or synergistic fashion to facilitate these changes. Second, the relationship between hormones and parental behavior is frequently bidirectional. As a result, the performance of parental behavior can generate cues that alter the secretion of the hormones that control its expression. Finally, hormones do not 'switch on' neural circuits that are responsible for parental behavior expression. Instead, they alter the sensitivity of specific brain circuits to cues from nests, eggs, and young, thereby lowering the threshold for parental behavior to be displayed. Prior experience in parenting has similar effects in many species, and may in some instances be sufficient to elicit parental behavior in the absence of any hormonal stimulation.

Hormones and Parental Behavior in Fishes

Parental care is apparently absent in the jawless fishes and cartilaginous fishes, but the bony fishes (teleosts) exhibit the same diverse array of parental care patterns as seen in birds, including brood parasitism, in which eggs are deposited in the nests of a host species. Unlike birds, however, exclusive parental care by the male is the most common pattern and is seen in over 50% of teleost families. It is also the pattern that has been studied most extensively by behavioral endocrinologists. Many teleost species also differ from avian species in exhibiting striking polymorphisms in male size and mating strategy, with only the large nesting males exhibiting parental behavior. The most common components of male parental care are nest building or substrate clearing, aeration of eggs by fanning, and guarding of eggs from brood predators, although other adaptations, such as mouthbrooding, are also displayed in some species. Parental care in most male teleosts ends at the time of hatching, although continued care of the newly hatched fry occurs in some species.

Male teleosts secrete significant amounts of testosterone and 11-ketotestosterone (11KT), an androgen with strong bioactivity that, unlike testosterone, cannot be converted to estrogen. Plasma concentrations of one or both of these hormones are elevated during the prespawning period when nest building or nest excavation occurs, but their role in nest-building behavior is uncertain. Three-spined sticklebacks (*Gasterosteus aculeatus*) have been the most intensively studied in this context, and there is general agreement that male nest building is suppressed by castration and restored by 11KT or its precursor, 11-ketoandrostenedione (11KA) in this species. However, castration has no effect on nest building in several other species. In male bluegill (*Lepomis macrochirus*) and pumpkinseed sunfish (*Lepomis gibbosus*), antiandrogen treatment suppresses nest building and other male-typical reproductive behaviors. However, while treatment with mammalian gonadotropins restores nest-building behavior in pumpkinseed sunfish (presumably by increasing gonadal androgen secretion), testosterone or 11KT pellets do not induce early nest-building behavior in untreated male bluegill. It is possible that some of these inconsistencies reflect the participation of other hormones that act permissively or synergistically with androgens to promote this behavior. One or more of the three gonadotropin-releasing hormone (GnRH) neuropeptides in the teleost brain are promising candidates based on recent evidence that nest building is reduced in Nile tilapia (*Oreochromis niloticus*) when GnRH-3 action is inhibited by intracerebroventricular injections of antiserum generated against this peptide.

In many teleost species, plasma levels of androgens decline after spawning when parental behavior toward eggs or young is exhibited, but this relationship is unlikely to be a causal one. In males of some species, such as the plainfin midshipman (*Porichthys notatus*), plasma 11KT levels remain elevated well into the egg stage of parental care, and in the bluebanded goby (*Lythropnus dalli*), levels are actually highest during this parental stage. In other species, such as bluegill and black-chinned tilapia (*Sarotherodon melanotheron*), androgen levels are low during the egg stage but increase after hatching while the males are caring for embryos or fry. This suggests that elevated androgen levels do not interfere with parental care expression, and in support of this interpretation, there is no convincing evidence from testosterone or 11KT treatment studies that androgens suppress parental fanning or brood defense.

As in birds and mammals, there is evidence that prolactin (PRL) stimulates parental behavior in some teleost fish. Both mammalian and fish PRL preparations increase parental fanning behavior in male three-spined sticklebacks, and mammalian PRL preparations facilitate parental fanning behavior in a Mediterranean wrasse (*Symphodus ocellatus*) and several species of cichlid fishes. Conversely, parental fanning behavior is inhibited in pumpkinseed sunfish and convict cichlids (*Amatitlania nigrofasciata*) following the administration of drugs that enhance the activity of dopamine, a PRL secretion inhibitor. In contrast to these findings, PRL reportedly failed to stimulate parental behavior in male breeders and immature helpers in the cooperatively breeding daffodil cichlid (*Neolamprologus pulcher*). Whether this reflects species differences in the hormonal control of parental care is difficult to assess, since studies in cichlids indicate that the dose–response relationship between mammalian PRL and parental behavior can be complex, with only the middle range of tested doses being effective. In addition, PRL-induced parental care in some species depends upon, or is enhanced by, the coadministration of testosterone, progesterone, or pituitary gonadotropins.

Hormones and Parental Behavior in Amphibians and Reptiles

Some components of parental care are seen in ~10% of frog species and in the majority of salamander species, but with the exception of one study in a neotropical frog, the effects of hormone manipulations on parental behavior expression have not been investigated. In the Puerto Rican frog (*Eleutherodactylus coqui*), the high plasma levels of androgens that are seen prior to mating drop precipitously after the male acquires a clutch of eggs and begins to exhibit strong nest attachment, egg brooding, and decreased egg cannibalism. Nevertheless, the implantation of testosterone pellets has no disruptive effects on any parental care behavior in males that are already brooding eggs. Accordingly, the hormonal mechanisms involved in parental behavior expression, if any, remain unknown.

Parental behavior in many species of reptiles consists only of nest excavation and burying of eggs, with no guarding of the nest site after egg laying. However, more elaborate or specialized forms of parental behavior have been reported, including parental defense of eggs in squamates (lizards and snakes), incubation of eggs through shivering

thermogenesis in pythons, and care of both eggs and young in crocodiles. As in the amphibians, however, an understanding of the hormonal basis of these behaviors, if any, must await the completion of experimental studies involving hormone manipulations.

Hormones and Parental Behavior in Birds

Parental care patterns in birds range from brood parasitism, in which eggs are laid in the nests of a host species and no parental behavior is displayed, to intensive and prolonged incubation of eggs and care of young by both parents. Over 99% of the ~9000 species of birds show some form of parental behavior, and unlike other vertebrates, over 90% exhibit varying degrees of biparental care. Over 70% of avian species rear altricial young, which are helpless at birth and dependent on their parents to provide warmth, protection, and food for an extended period. The remaining 30% rear precocial young which are able to leave the nest, follow their parents, and forage for food on their own soon after hatching. In these species, parental care after hatching consists of brooding the chicks to keep them warm, defense of the chicks, and leading them to food sources or shelter.

Nesting Behavior and Onset of Incubation

Nest site attachment, nest building, and onset of incubation represent a continuum of activities in birds that are largely controlled by the same suite of hormones. In species in which the female incubates, the early stages of nest site occupation and nest building are stimulated by the rising levels of estrogen and progesterone from the developing ovarian follicles. However, based on the few species that have been studied in detail, the hormonal requirements for later stages of nesting and the onset of incubation could vary depending on how many eggs are laid and when functional incubation begins during the egg-laying period.

In females of species with large clutches that lay one or more eggs before full incubation begins, the hormonal events associated with follicle development, ovulation, and egg laying are necessary, but not sufficient, for the onset of incubation. In these birds, incubation induction requires the priming action of one or more gonadal steroids, followed by an increase in plasma levels of PRL. In budgerigars (*Melopsittacus undulatus*), which show this pattern, the optimal hormonal treatment combination for stimulating incubation onset is estrogen and PRL. In chickens (*Gallus domesticus*) and turkeys (*Meleagris gallopavo*), similarly, exogenous PRL administration is effective in stimulating incubation onset, but only in females that have been exposed to estrogen and progesterone. Not surprisingly, there is a strong correlation between rising titers of PRL in plasma and time spent in the nest as full incubation develops in these birds. A particularly convincing demonstration of PRL's importance in sitting onset in chickens and turkeys is provided by evidence that incubation onset is prevented by active immunization against PRL or vasoactive intestinal peptide (VIP), a brain neuropeptide that stimulates PRL release.

In females of species that lay smaller clutches and begin to sit as soon as the first egg is laid, the hormonal requirements for incubation onset may be the same as those required for egg laying. The best studied species that fits this description is the ring dove (*Streptopelia risoria*). In this species, the induction of sitting behavior is a natural extension of earlier nesting behavior and is stimulated by the same gonadal steroids that support oviduct development and ovulation. Unlike galliform species such as chickens and turkeys, PRL plays little or no role in the onset of incubation behavior in ring doves. In fact, plasma PRL levels do not begin to rise until several days after incubation begins in this species, and progesterone injections are more effective than PRL injections in inducing incubation onset in nonbreeding doves of both sexes.

Less is known about hormonal mechanisms responsible for the onset of incubation behavior in male birds that share incubation duties with their mates or incubate without the breeding partner's assistance. In male ring doves, which sit on the nest for over 6 h each day, progesterone is effective in inducing incubation behavior in nonbreeding birds, provided that testosterone or estradiol is also present. Nevertheless, incubation can be expressed in male ring doves in the absence of adrenal or gonadal sources of steroid hormones, provided that they receive adequate social and environmental signals from the mate and nest. In addition, male ring dove, unlike females, do not show changes in plasma progesterone levels with onset of incubation. Although this raises questions about the importance of plasma progesterone in natural incubation induction in males, it does not rule out progesterone involvement, since Lea and his collaborators have shown that progesterone can also be synthesized locally in the ring dove brain. Progesterone receptors in the dove brain also show marked fluctuations during the breeding cycle that could influence progesterone sensitivity to constant levels of the hormone. Progesterone receptors are plentiful in the preoptic region of the dove brain and could be particularly important for incubation behavior expression, since progesterone implants in this region are effective in inducing incubation behavior in nonbreeding male doves.

Apart from the demonstration that PRL is less effective than progesterone in inducing incubation behavior in nonbreeding male ring doves, the importance of PRL in incubation onset in male birds has yet to be experimentally tested in any species. Descriptive studies, however, indicate considerable interspecies variation in the pattern of changes in plasma PRL levels in males sampled during different breeding stages. Interpretation of these changes is complicated by the fact that in many temperate zone songbirds, PRL secretion in males is stimulated by increasing day length, which coincides with the breeding season. As a result, elevated PRL levels in males may have little to do with parental behavior expression. A particularly dramatic example of this dissociation is seen in male brown-headed cowbirds (*Molothrus ater*), which exhibit no parental care at all, but nevertheless show robust elevations of plasma PRL during the breeding season.

Maintenance of Incubation

There is strong circumstantial evidence for PRL involvement in incubation maintenance in many species of birds. Plasma PRL concentrations increase during incubation in virtually all species of birds that have been studied, although there is interspecies variation in the timing of this increase in relation to egg laying. In biparental species in which both breeding partners contribute to incubation, PRL levels in blood tend to be higher during incubation than during earlier breeding stages in both sexes, and in species in which only one breeding partner incubates, PRL levels tend to be substantially higher in the incubating sex. Studies in canaries (*Serinus canaria*), ring doves, ducks, chickens, and turkeys indicate that PRL levels decrease when incubation behavior is interrupted by egg removal or by anesthetization or denervation of the brood patch, which transfers heat from the parent's body to the eggs. In addition, PRL levels rebound when nests and eggs are returned after an absence of 24–72 h. This strongly suggests that PRL is secreted in response to tactile cues from the eggs or from the act of sitting itself, but it is clear that other factors also contribute to PRL secretion at this stage. Social stimuli are capable of sustaining elevated PRL levels in incubating doves during sitting recesses of several hours, since visual and auditory cues from incubating breeding partners have been shown to maintain high plasma PRL levels in male ring doves that are deprived of opportunities to sit on eggs for several days. In pelagic seabirds such as albatrosses and penguins, however, breeding birds must travel from the nesting site to remote oceanic sites to forage for food. As a result, nest recesses of several days to several months are common during the incubation and chick-rearing periods, and PRL levels typically remain elevated during these lengthy absences without the benefit of tactile contact with eggs or social stimulation from their mates. Many of these species also show no decrease in plasma PRL levels when nesting failure occurs, which suggests that PRL secretion is maintained at high levels through the incubation and brooding periods by an endogenous mechanism that is buffered from social and environmental influences.

Although the effects of PRL injections on incubation maintenance have only been conducted in two species, they provide convincing evidence for PRL involvement in sustaining this behavior. In bantam hens (*G. domesticus*), the decrease in interest in sitting on eggs that normally occurs within 72 h of nest removal is prevented by PRL injections during the nest deprivation period. Furthermore, passive immunization against the PRL-releasing neuropeptide VIP induces nest abandonment, while coadministration of PRL maintains sitting activity. In ring doves with previous breeding experience, however, passive immunization against VIP does not disrupt incubation behavior. Whether this is because experienced birds are less dependent on hormonal stimuli for maintaining incubation than are inexperienced birds is unclear. Regardless of whether PRL is necessary for incubation maintenance, it is clear that it strongly facilitates this activity in ring doves (**Figure 1**). PRL injections extend the incubation period of ring dove pairs sitting on infertile eggs and maintain interest in sitting on eggs in both sexes when birds are deprived of exposure to their mates, nests, and eggs for 10 days. Subcutaneous administration of PRL is more effective than intracerebroventricular (ICV) injections in maintaining incubation readiness in nest-deprived doves, which suggests that PRL may be acting in part on peripheral target organs to sustain

this motivational state. These studies, together with the hormone assay studies described earlier, suggest that PRL secretion and sitting activity are mutually reinforcing in these and probably other avian species.

Care of Young

In general, PRL levels in blood tend to drop sharply after hatching in chickens, turkeys, ducks, and geese that rear precocial young. Studies in chickens suggest that this decline may be triggered by the disruptive effects of the chicks and their movements on sitting activity of the hen. Although levels decrease from those present during incubation, they remain higher than those seen in hens without eggs or young, which suggests that stimuli from the chicks, while having disruptive effects, may also retard this decline. In turkeys, pheasants, and chickens, PRL injections facilitate parental responses toward chicks. Similarly, parental behavior expression is strongly correlated with PRL levels after hatching in mallard duck hens (*Anas platyrhynchos*). These observations suggest that PRL may facilitate responsiveness toward young even though plasma levels are declining at this stage. On the other

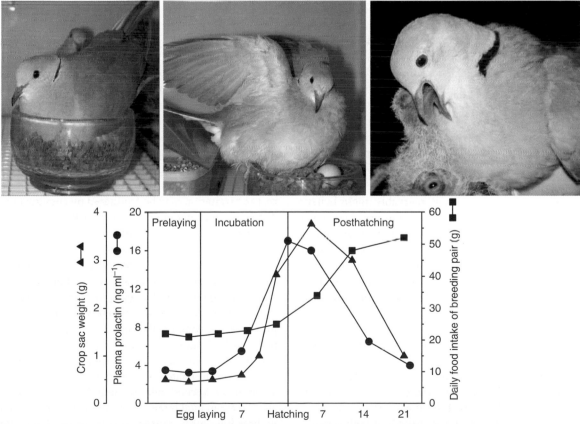

Figure 1 Prolactin and parental behavior in ring doves (*Streptopelia risoria*). Rising plasma levels of prolactin during the incubation and posthatching periods of the breeding cycle maintain incubation behavior, facilitate nest defense, stimulate crop sac development and crop milk formation, increase parental foraging behavior, and promote parental regurgitation behavior to transfer crop milk and seeds to the young in ring dove parents of both sexes.

hand, full maternal behavior can be induced in nonlaying hens in the absence of changes in PRL secretion simply by confining them with chicks, which suggests that cues from the chicks may promote parental behavior even in the absence of hormonal stimulation.

Plasma PRL levels tend to remain elevated for a longer period after hatching in species that rear altricial young. In female pied flycatchers (*Ficedula hypoleuca*) and ring doves of both sexes, PRL secretion is stimulated by signals from young or cues generated by parent–young interactions such as brooding or parental feeding. Replacing the growing young with newly hatched nestlings can prolong this period of elevated PRL levels in female pied flycatchers, but not in doves. However, PRL levels in both species eventually decline to baseline levels regardless of the type of stimulation provided, which suggests that PRL secretion is also controlled by endogenous timing mechanisms.

During the posthatching period, as during incubation, PRL secretion and parental behavior expression are likely to be mutually reinforcing in many avian species. PRL levels in blood are elevated in parents rearing altricial young and in nonbreeding 'helpers at the nest' that assist breeders in parental care in communal breeding species. However, the question of whether these results reflect PRL-induced provisioning behavior or provisioning-induced PRL secretion cannot be determined from correlational studies. Relatively few studies have directly examined the effects of PRL on parental responses toward young. Particularly compelling evidence for PRL involvement in parental behavior expression in wild birds comes from Badyaev and Duckworth's studies of a polymorphic population of male house finches (*Carpodacus mexicanus*) in which parental investment is strongly correlated with plumage color. Specifically, males with dull yellow plumage have high plasma PRL levels and provision their incubating mates and nestlings frequently, while males with bright red plumage have low PRL levels and spend little time on these activities. Remarkably, these behavioral phenotypes can be reversed by appropriate hormone manipulations. Specifically, red plumaged males show a threefold to fourfold increase in parental provisioning when given the PRL-releasing neuropeptide VIP, while parental provisioning ceases entirely in yellow plumaged males given bromocriptine, a dopamine agonist and PRL secretion inhibitor (**Figure 2**).

The strongest evidence for a direct effect of PRL on avian parental behavior comes from studies in captive ring doves. These birds are unusual in their ability to manufacture food for their young in the form of crop milk, which is produced by an esophageal pouch called 'the crop sac' (**Figure 1**). Rising levels of PRL during the second half of incubation stimulate crop sac growth in both parents, which leads to crop milk production. At hatching, PRL stimulates parental regurgitation, which is necessary to transfer the crop milk to the young squabs (**Figure 1**).

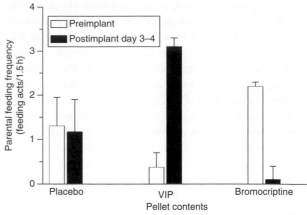

Figure 2   Evidence for a role of prolactin in parental provisioning of young in male house finches (*Carpodacus mexicanus*). Subcutaneous implantation of pellets containing vasoactive intestinal peptide (VIP), which stimulate prolactin release, caused a marked increase in parental provisioning in males with bright red plumage, which typically exhibit little parental provisioning of young. Subcutaneous implantation of pellets containing bromocriptine, which inhibit prolactin release, suppressed parental provisioning in males with dull yellow plumage, which typically show high levels of parental provisioning. Adapted with permission from Badyaev AV and Duckworth RA (2005) Evolution of plasticity in hormonally integrated parental tactics. In: Dawson A and Sharp PJ (eds.) *Functional Avian Endocrinology*, pp. 375–386. New Delhi: Narosa Publishing House.

PRL appears to act in two ways to stimulate this behavior. A direct action of PRL on target cells in the brain is likely, since ICV injections of PRL stimulate parental feeding invitations and regurgitation movements in nonbreeding doves with undeveloped crop sacs. Nevertheless, regurgitation feeding responses are more frequent in nonbreeding doves given subcutaneous injections of PRL, which stimulates crop sac development and crop milk formation. Because anesthetization of the crop sac reduces PRL's ability to stimulate regurgitation, these findings suggest that regurgitation may be facilitated by cues generated by distension and engorgement of the PRL-stimulated crop sac. Previous parental experience strongly influences these relationships, since doves with previous breeding experience show much greater parental responses to PRL than doves without previous breeding experience.

In addition to stimulating regurgitation behavior, PRL may initiate the increased foraging behavior needed to provision the young with seeds as they grow older. A link between PRL and parental hyperphagia is suggested by two lines of evidence. First, the increase in food intake begins soon after hatching when peak levels of plasma PRL are attained. Second, nonbreeding doves given PRL show an increase in food intake of similar magnitude to that seen in breeding birds raising young. Intracranial injection studies indicate that PRL acts directly on the brain to elevate food intake, and does so in part by promoting the synthesis of the appetite-stimulating neuropeptides

Neuropeptide Y (NPY) and Agouti-related peptide (AgRP) in the hypothalamus. Dove parents enter a negative energy state as they continue to provision their young, which may be responsible for the modest elevation in baseline corticosterone that occurs during the posthatching period. Because corticosterone also increases NPY, AgRP, and foraging activity in doves, these elevated levels may sustain parental hyperphagia after the second week of the posthatching period when PRL levels begin to decline.

Gonadal Hormones and Parental Behavior

While elevated levels of gonadal steroids stimulate nesting behavior and incubation onset, either by acting alone or in combination with PRL, they can have inhibitory effects on parental behavior at later breeding stages. Gonadal steroid levels in both sexes decrease markedly after egg laying in birds that participate in parental care and usually remain at low levels throughout the posthatching period. In part, this decrease is due to suppressive effects of high PRL levels on gonadotropin and gonadal steroid secretion. Breeding females of most species show little or no disruption of incubation behavior or nestling care when this decline is prevented by exogenous hormone administration, but males typically show serious deficits in parental responsiveness. In male spotted sandpipers (*Actitis macularia*), which are solely responsible for incubation, the antiandrogenic drug flutamide advances the onset of incubation when given during the egg-laying period, and the implantation of testosterone capsules after egg laying disrupts ongoing incubation behavior. Similar disruptive effects of testosterone treatment on male incubation behavior have been reported in one gull species and in three other songbirds that share incubation duties with their female partners; however, the fact that male ring doves and pigeons (*Columba livia*) are resistant to these effects indicates that this response is not universal. Testosterone implants also suppress parental provisioning of young in many male songbirds by increasing sexual and aggressive behavior, which is incompatible with the display of parental responses. Nevertheless, in a small number of species, including two species of longspurs (*Calcarius* sp.), males are apparently insensitive to the disruptive effects of testosterone on parental feeding of young. The evolution of behavioral insensitivity to androgens would appear to be adaptive here, since parental provisioning of young by the male parent is critical to nestling survival in these species.

Corticosterone and Parental Behavior

Activation of the hypothalamo–pituitary–adrenal (HPA) axis is an adaptive response to inclement weather, reduced food availability, social instability, and other stressors in birds. In breeding birds, the increased secretion of corticosterone that results from HPA activation is frequently associated with behaviors that divert effort away from reproduction (e.g., nest abandonment, dispersal) and toward immediate survival (e.g., increased food consumption). Field studies conducted in the 1980s with breeding pied flycatchers and song sparrows (*Melospiza melodia*) indicated that exogenous corticosterone treatment induces this same behavioral shift, thereby lending support to the idea that elevated titers of plasma corticosterone are incompatible with parental care. However, treated birds in these experiments achieved levels of corticosterone in plasma that were in the pharmacological range, thus complicating the interpretation of the results. Subsequent studies have dramatically revised the view of corticosterone's involvement in parental behavior regulation by demonstrating that corticosterone can either facilitate or inhibit parental behavior depending on the plasma levels achieved. Studies in pied flycatchers, for example, have shown that the frequency with which nestlings are fed increases when the parents' levels of plasma corticosterone are slightly elevated, decreases with somewhat larger elevations, and ceases altogether with large elevations that cause nest abandonment.

While there are many situations in which it would be advantageous for stress-induced HPA activation to suppress reproduction until conditions are more auspicious for success, there are situations in which tradeoffs between reproductive success and survival are such that it would be more adaptive to maintain the breeding effort despite stressful conditions. For example, pied flycatchers that are forced to rear young alone after loss of their mates increase their parental provisioning rates despite the fact that baseline corticosterone levels are elevated and corticosterone responses to capture and handling stress are enhanced. Similarly, willow warblers (*Phylloscopus trochilus*) rearing one brood during the brief and unpredictable breeding seasons of the subarctic forests of northern Sweden have lower corticosterone responses to stress than willow warblers breeding at lower latitudes under more temperate conditions where the longer breeding seasons permit the rearing of two or more broods. These two examples also illustrate the diversity of mechanisms that could underlie stress resistance. In the pied flycatcher example, resistance could result from decreased sensitivity of the reproductive system to glucocorticoids, while in the northern population of willow warblers, decreased stress sensitivity of the HPA axis itself could be involved.

In addition to diversity in the patterns of corticosterone secretion and responsiveness, there is complexity in the mechanisms by which corticosterone acts on its target cells. Recent studies in house sparrows (*Passer domesticus*) indicate that corticosterone can interact with two different cytoplasmic receptors as well as a third receptor on the plasma membrane. Very little information is available on the receptors that mediate the behavioral effects of corticosterone on parental behavior, but in the ring dove, corticosterone has been shown to increase food intake by binding to one of the cytoplasmic receptors in corticosterone-sensitive neurons in the brain. Since baseline levels of plasma corticosterone increase in parent doves after their young hatch, this action of corticosterone could contribute to the hyperphagia shown by parents that are provisioning their young.

The fact that most of the avian corticosterone that circulates in blood is bound to corticosteroid-binding globulin (CBG) also complicates the picture. The binding of corticosterone to CBG may decrease the biological activity of corticosterone by preventing access of the hormone to the two cytoplasmic receptors in target cells. Because CBG levels in blood can and do change independent of changes in corticosterone secretion, significant changes in free plasma corticosterone levels can occur across the breeding cycle in some species, such as the European starling (*Sturnus vulgaris*), without any detectable changes in total plasma corticosterone levels.

Based on the variety of ways in which corticosterone can interact with its target cell receptors, and the fact that bioactive corticosterone can be regulated by CBG levels, it is not surprising that the effects of corticosterone on parenting can be very complex. A particularly illuminating example comes from a study by Love and colleagues, who examined changes in baseline and stress-induced corticosterone levels in female European starlings sampled during laying, incubation, and chick rearing. While stress-induced changes in free corticosterone levels and baseline levels of total corticosterone did not vary across breeding stages, a significant decrease in CBG was observed in females rearing young, which resulted in higher levels of free corticosterone at this stage than during the incubation or laying periods. Females that abandoned their nests had higher levels of free corticosterone at all three breeding stages than females that did not, which supports the idea that large elevations in free corticosterone suppress parental care. However, among those birds that did not abandon their nests, chick-rearing females had higher levels of free corticosterone than laying and incubating females. These results, and similar findings in other birds (e.g., male Northern mockingbirds (*Mimus polyglottos*), ring doves, pied flycatchers, wandering albatross (*Diomedea exulans*), Adélie penguins (*Pygoscelis adeliae*)) indicate that moderate elevations in baseline levels of corticosterone during the parental period are not necessarily incompatible with parental behavior and may in fact be beneficial by stimulating parental foraging activity to meet the energetic demands of provisioning the young. This idea also receives support from hormone administration studies, which demonstrate that corticosterone acts either directly or permissively to increase locomotor activity and foraging in many species. Notably, free corticosterone levels of chick-rearing

female starlings that did not abandon their nests in the Love et al. study were no different from those of laying or incubating females that did abandon their nests. This suggests that corticosterone sensitivity can also change across the breeding cycle.

Given the fact that corticosterone can increase foraging and stimulate metabolic processes such as lipogenesis and gluconeogenesis for energy mobilization, it is not surprising that corticosterone secretion and sensitivity to corticosterone varies with energy stores and body condition. However, the effects of body condition may differ between species or between populations of the same species based on food availability and the risks and benefits of abandoning the breeding effort in a particular environment. For example, common murre parents (*Uria aalge*) raising young in years in which food is relatively scarce have higher baseline levels of corticosterone than in years in which food is abundant. Furthermore, parents that feed their young at higher than average rates under these conditions have higher baseline plasma corticosterone levels and lose less weight than parents with lower parental feeding rates. In this case, it would appear that birds in good condition are better able to mount a robust corticosterone response to food shortage, which in turn promotes increased foraging activity so that both they and their chicks are adequately provisioned. In contrast to this pattern, male Black-legged kittiwake parents (*Rissa tridactyla*) in good condition have been reported to lose weight during the chick-rearing period, while those in poor condition gain weight. In this instance, it would appear that males in these two energy states were adopting different strategies, with males in good condition foraging to provision young at their own energetic expense and males in poor condition foraging to replete their own energy stores at the expense of their young.

Corticosterone–PRL Relationships in Parental Behavior

How PRL and corticosterone interact to regulate parental behavior in birds is not well understood, but it is clear that there are important interactions between the two that could potentially influence reproductive success. In some species, or in some situations, PRL and corticosterone may have additive or synergistic effects on parental care, while in others, the effects could be antagonistic. In female ring dove parents, both plasma PRL and plasma corticosterone levels are elevated during the posthatching period, and PRL is capable of increasing circulating corticosterone levels by increasing HPA axis activity at the level of the brain. Since both corticosterone and PRL increase food intake, these findings suggest that PRL-induced elevations in plasma corticosterone could be one mechanism by which PRL stimulates the hyperphagia that is necessary for parents to provision their young. On the other hand,

PRL and corticosterone often exhibit divergent patterns of secretion in response to stress. When subjected to capture and handling stress, incubating snow petrels (*Pagodroma nivea*) and chick-rearing Black-legged kittiwakes show an increase in plasma corticosterone and a decrease in plasma PRL. Nevertheless, parent kittiwakes with chicks show a less pronounced suppression of PRL during stress than failed breeders that lost their eggs. In snow petrels, similarly, the likelihood of egg abandonment is higher in younger breeders than older breeders, and this in turn is correlated with the fact that the older birds maintain higher levels of circulating PRL during stress than younger birds. These findings suggest that the magnitude of the PRL response to stress may be important in determining whether stress-induced increases in corticosterone result in abandonment of the breeding effort and redirection of energy toward self-maintenance activities.

See also: Communication and Hormones; Food Intake: Behavioral Endocrinology; Hormones and Behavior: Basic Concepts; Parental Behavior and Hormones in Mammals; Parent–Offspring Signaling.

Further Reading

Angelier F and Chastel O (2009) Stress, prolactin, and parental investments in birds: A review. *General and Comparative Endocrinology* 163: 142–148.

Adkins-Regan E (2005) *Hormones and Animal Social Behavior*, pp. 411. Princeton, NJ: Princeton University Press.

Badyaev AV and Duckworth RA (2005) Evolution of plasticity in hormonally integrated parental tactics. In: Dawson A and Sharp PJ (eds.) *Functional Avian Endocrinology*, pp. 375–386. New Delhi: Narosa Publishing House.

Buntin JD (1996) Neural and hormonal control of parental behavior in birds. *Advances in the Study of Behavior* 25: 161–213.

Buntin JD, Strader AD, and Ramakrishnan S (2008) The energetics of parenting in an avian model: Hormonal and neurochemical regulation of parental provisioning in doves. In: Bridges RS (ed.) *Neurobiology of the Parental Brain*, pp. 269–291. New York, NY: Elsevier.

Love OP, Breuner CW, Vézina F, and Williams TD (2004) Mediation of a corticosterone-induced reproductive conflict. *Hormones and Behavior* 46: 59–65.

Lynn SE (2008) Behavioral insensitivity to testosterone: Why and how does testosterone alter paternal and aggressive behavior in some avian species but not others? *General and Comparative Endocrinology* 157: 233–240.

Oliveira R and Gonçalves DM (2008) Hormones and social behaviour of teleost fish. In: Magnhagen C, Braithwaite VA, Forsgren E, and Kapoor BG (eds.) *Fish Behaviour*, pp. 61–150. Enfield, NH: Science Publishers.

Townsend DS, Palmer B, and Guillette LJ, Jr (1991) The lack of influence of exogenous testosterone on male parental behavior in a neotropical frog (Eleutherodactylus): A field experiment. *Hormones and Behavior* 25: 313–322.

Wang Q and Buntin JD (1999) The roles of stimuli from young, previous breeding experience, and prolactin in regulating parental behavior in ring doves (*Streptopelia risoria*). *Hormones and Behavior* 35: 241–253.

Wingfield JC and Sapolsky RM (2003) Reproduction and resistance to stress: When and how. *Journal of Neuroendocrinology* 15: 711–724.

Parent–Offspring Signaling

A. G. Horn and M. L. Leonard, Dalhousie University, Halifax, NS, Canada

Introduction

In species with parental care, signaling between parents and offspring is key to regulating the nature, amount, and timing of that care. Here, we first show how this signaling can take a wide variety of forms, well beyond familiar examples such as the begging of nestlings or the crying of infants. We will then explain why parent–offspring signaling is of particular interest, not only because it regulates the delivery of care during such a critical phase of an animal's life history, but also because it contributes to our understanding of behavioral development and the evolution of animal signaling.

Parent–Offspring Signals Are Diverse, but Perform a Few Main Functions

To illustrate the wide variety of parent–offspring signals across the taxa, here we describe several examples from insects and vertebrates, arranged by the main functions they serve. The survey is not meant to be comprehensive, but is a selection of particularly interesting or well-studied examples. It will show that parent–offspring signaling regulates virtually every aspect of parental care and offspring development, often starting well before hatching or birth and sometimes continuing well beyond the dependent period. The signals use virtually every modality, although certain modalities are more prevalent in particular taxa. Olfactory signals, for example, are especially prevalent in insects and mammals, whereas acoustic and visual signals are more common in birds.

Signaling for Prenatal Care

Parent–offspring signaling starts at the earliest stages of parental care, when it often seems more akin to physiological regulation than signaling. In some live bearing amphibians, for example, fetuses scrape the mother's uterine wall with their teeth to stimulate production of the specialized tissue that the fetuses feed on. In mammals, fetuses produce hormones that perform a similar function, stimulating an increase in the placental resources available to the fetus. In several taxa of birds, including gulls and waterfowl, embryos give a call when eggs are cooled that, in turn, stimulates incubation by adults. Similar calls synchronize hatching among nest mates and likely stimulate parents to shift from incubation to feeding young. Such calls are also found in crocodilians, which do not incubate their eggs directly like birds, but instead do so by leaving them in heaps of rotting vegetation. Calls of the young crocodilians just before hatching stimulate the mother to open the nest so the young can scramble out.

Assembling or Contacting the Young

Once offspring are hatched or born, patterns of parental care vary, depending on whether young feed on their own (precocial species) or are fed by the parents (altricial species). In precocial species, young can wander widely, so a variety of signals serve to keep and restore contact with parents. When precocial chicks, such as chickens or ducks, are separated from the hen, they give isolation cheeps and the hen gives assembly clucks. Similarly, fawns that have hidden from a predator will call to reveal their location to the doe after the predator has passed. Even some altricial species occasionally require such signals. Rodent pups that become isolated outside the nest, for example, give ultrasonic vocalizations that stimulate and guide retrieval by the mother. For all these examples, gradations in the call structure or rate can signal gradations in the distress of both the offspring and the parent.

Maintaining and restoring contact effectively may be particularly challenging in species, such as colonial species, in which mixing of different families is likely. Such situations may require signals that differ among broods or litters in ways that allow parents and/or young to discriminate the signals of their own family. Particularly dramatic examples of such coding are found in penguins and seals, in which parents returning from foraging trips must find their own offspring in a colony or creche of thousands of youngsters. In many mammals, including rodents and ungulates, mothers frequently nuzzle and lick their young, partly for nonsignaling purposes such as grooming and stimulating defecation, but probably also to pick up olfactory and gustatory cues that help them to discriminate their own young.

Warning and Predator Defense

The assembly or contact calls just discussed are often used after the young have scattered or hidden because of a predator's approach. Warning of such approaches in the first place is an important function of many parent–offspring signals. Birds and mammals, in particular, have a variety of calls that warn their offspring of the presence of predators and that may cause offspring to fall silent, crouch, or hide. In cichlid fish, parents jerk their body or pelvic fins in response to potential predators, causing young to cluster around their parent or against the substrate.

Offspring may also signal when they detect a predator, to solicit help from parents. In precocial birds and mammals, when attack is imminent and hiding is futile, offspring may give distress calls that attract the parent's help; altricial nestlings and pups may give similar calls when handled. A particularly elaborate antipredator signal is shown by membracid treehopper nymphs. These nymphs live in large groups at exposed positions on host plants, where they gradually develop into adults. When predatory insects or spiders approach, the nymphs send a vibrational signal that is coordinated in a wave across the group. Adult females, which linger near their nymphs through the developmental period, respond to the vibrations by rushing the predator, waving their wings, and kicking their hind feet to repel it.

Provisioning Food

In most species with parental care, the main form of care is feeding the young, so not surprisingly these food solicitation signals, or begging displays, are the best-studied parent–offspring signals. A few examples from insects will illustrate their diversity. Even just among the colonial Hymenoptera, larvae stimulate feeding from workers (the deliverers of parental care) with a wide array of signals, from swaying of the head and jaws in certain species of ant, to scraping mandibles against nest walls in hornets, to chemical signaling (pheromones) throughout the taxon. Larval burrower bugs release volatile chemicals that stimulate provisioning by female parents, releasing more of the chemicals the more they are food deprived. Similarly, burying beetle grubs stimulate parents to feed them carrion by raising their heads in the air and tapping the parent's mandibles, doing so more vigorously the hungrier they are, thus stimulating higher provisioning rates by parents.

To appreciate the variety of information conveyed by begging signals, we turn to birds and mammals, which are particularly well studied. Altricial nestling birds are dependent on their parents for food and warmth, and nestlings beg for food by raising their heads high and opening their mouths to expose brightly colored gapes. They may also call, wave their wings, and stretch their bodies toward the parent. The display is more intense the more hungry the nestlings are, and some components of the display may actually carry information on the body temperature, immunocompetence, and long-term condition (mass relative to size) of the nestling. Parents, in turn, often call quietly when they arrive at the nest to stimulate begging, especially when nestlings are too young to be capable of detecting their arrival otherwise.

Analogous signals occur in mammals, although of course here they are related to nursing. Acoustic solicitation signals include the calls of infant seals that bring their mothers ashore to nurse and the squeals of piglets that signal their need for nursing. Tactile signals, including suckling and kneading, such as the familiar kneading of kittens, also stimulate lactation, as do various odors emanating from the young. Mothers, in turn, give various signals, such as the grunting in sows and calling and head bobbing in antelopes, that gather their young to be nursed.

Manipulating Host Parents

The final category of parent–offspring signaling, its exploitation by parasitic species, vividly illustrates the manipulative power that begging can have over parents. The best-known example, of course, is the begging of the cuckoo chick, which is often several times larger than the host parent that is feeding it. However, brood parasitism is not limited to cuckoos. Several species of birds, including certain cowbirds, viduine finches, honeyguides, and ducks, lay their eggs in the nests of other species, leaving those host species to raise the foreign parasite. In cowbirds and cuckoos, nestlings signal more intensely and effectively than host nestlings, and may even mimic the begging calls of their host's nestlings; in parasitic viduine finches, visual signals in the gapes of nestlings mimic those of their hosts.

Several insect species are also highly specialized brood parasites. For example, to extract care from their ant hosts, lycaenid butterfly caterpillars mimic larval ant pheromones, while staphylinid beetle larvae tap like ants on their hosts' mandibles. Brood parasitism has proved to be a powerful tool in exploring the theoretical aspects of parent–offspring signaling, because it represents such an extreme case of manipulation by 'offspring.'

Parent–Offspring Signaling Changes as Young Develop

Parent–offspring signaling changes dramatically as offspring develop, especially in birds and mammals. Indeed, this dynamic nature of the signaling system is one of its most fascinating aspects, one not found in most other animal signaling systems. The changes arise not just because of maturation in the perceptual and motor abilities of young, but also because the young learn to fine tune their responses, sometimes very quickly, to what is most effective at getting care from the parent.

Songbird nestlings illustrate these changes especially dramatically. They hatch blind, nearly deaf, and unable to do much beyond raising their heads and opening their mouths. Gradually through the nestling period, they begin to see shadows and then, once their eyes open, distinguish shapes. Their hearing improves, expanding its sensitivity into higher frequencies, and they get better at controlling their movements, orienting their begging response toward the approaching parent and ducking when the nest is shaken or, late in the nestling period, when the parent gives an alarm call.

The parental cues that are effective in stimulating begging shift in step with these changes. Early on, many species of songbirds give a loud, low-frequency call when arriving at the nest with food to stimulate nestlings to beg. Gradually, young respond to other signs of the parent's arrival, such as the thud or vibration made as it lands on the nest, its sudden shadow, and eventually, its shape. Eventually, they respond to the parent's alarm calls outside the nest by ceasing begging and scrunching down in the nest material.

As young birds transition into adulthood, the signals of the juvenile period may disappear or may be incorporated into adult behavior. For example, begging calls resurface in adult females in species in which males feed females as part of courtship or interactions during laying. In some species, they diversify early in the fledging period into the adult call repertoire or are directly incorporated into adult male song.

Carryover of parent–offspring signaling into adult life is perhaps most dramatically illustrated by imprinting. Here, the cues that offspring use to recognize and respond to their parents shape the cues that they use as adults to choose mates and other social companions. The carryover to adulthood can also be subtler, however. For example, great tits (a songbird) that beg more vigorously have more exploratory personalities as adults. Similarly, parent–offspring interactions during the juvenile period in mammals have long been seen as mechanisms for socialization. For example, primate parents that aggressively quash the more extreme demands of their near-independent young might be seen as teaching them not to be selfish so that they can fit in to the more equitable environment of the social group.

The evolutionary implications of such learning and development are intriguing. They fall under the broader class of so-called maternal effects, nongenetic influences on offspring phenotype that may direct the course of subsequent evolution. Imprinting is the clearest example, but the subtler effects just referred to may also be important, and more ubiquitous. Developmental stress in nestlings has been shown to affect their song repertoire size and other correlates of reproductive success in adulthood; it is a short step to imagine that the dynamics of signaling in the nest play a key role in communication later in life.

Evolutionary Questions Raised by Parent–Offspring Signaling

All these varied forms of parent–offspring signaling have one main function: the regulation of parental care. One might legitimately ask, however, why parental care should require regulation through signaling. Why do not parents simply give offspring what they need? After all, one would think that any parent that does not care sufficiently for its own offspring would be swiftly eliminated by natural selection. In fact, however, it may be adaptive for parents to withhold care that offspring demand, and intensive signaling might be one result of this parent–offspring conflict. The concept of parent–offspring conflict has launched a large theoretical literature on the evolution of parent–offspring signaling.

Parent–Offspring Conflict

Parent–offspring conflict is the disparity between the level of care that natural selection favors for parents and the level that it favors for offspring. To understand this disparity, one must see parental care in cost/benefit terms, like any other behavior. Any effort a parent spends on a given offspring entails the cost of withholding that care from another offspring or decreasing the chance that the parent can produce future offspring. For any given offspring, however, that cost is less important than it is for the parent. Specifically, while the parent is equally related to all of its offspring, the offspring is only half as related to its siblings as it is to itself. Thus, a parent gains equal amounts of fitness through any given offspring, while a given offspring gains twice as much fitness through its own sons and daughters than through its nieces and nephews. Since the costs of parental care are half as important to the offspring as they are to the parent, offspring are selected to get twice as much care from the parents as the parents are selected to give.

This parent–offspring conflict could affect the amount and timing of care. When a parent divides resources among its current offspring, as when a parent bird feeds its nestlings, it should divide the amount of food relatively evenly, but each offspring should demand twice as much food as its even share. Similarly, when a parent curtails care on one set of offspring so it can raise a second, as when a mother deer stops nursing to save energy for its next fawn, the offspring may resist by trying to prolong the period of care. In both situations, offspring might be selected to signal vigorously to parents in an attempt to coax extra resources from them.

Indeed, at first glance, these ideas fit quite well with many patterns of parent–offspring signaling. Nestling birds, for example, noisily beg and jostle each other when the parent arrives at the nest with food, as if they are competing to take food away from their siblings. Similarly, many young mammals, including young humans, often appear to resist vigorously when parents try to withhold care at the end of the dependent period. Nonetheless, such behavioral conflict between parents and offspring need not reflect an underlying genetic conflict. Weaning conflict in mammals, for example, might occur simply because offspring are not privy to the schedule of parent's activities or to the resources available to them. Thus, the amount and timing of care might indeed have to be negotiated, but not because parents and offspring disagree over the ultimate outcome.

Instead of looking for behavioral conflict, then, a proper test of the theory of parent–offspring conflict requires measures of the fitness consequences of different levels of care for both parents and offspring. These data are hard to get, so, not surprisingly, parent–offspring conflict has proved to be notoriously hard to test. Most work has instead assumed that the conflict exists, that is, that parents and offspring disagree over the level of care, and has turned to studying the signals with which parents and offspring settle on the level of care.

Models of Signaling

The models of parent–offspring conflict we have reviewed so far assume that offspring use some sort of signal, such as begging, to influence parental care. Given that these signals might be used to manipulate parents to give more care than they otherwise would, it seems reasonable to ask why parents respond to the signals in the first place.

One reason is that both parents and offspring have a shared interest in adjusting care to the offspring's needs, even though they may not agree on the precise level of those needs. To the extent that they do agree, both parties would benefit from a signal that correlates with aspects of an offspring's need that would otherwise be hard for a parent to perceive. For example, a parent can easily see how large a nestling is and can adjust its level of care accordingly without any signaling. A parent cannot, however, directly detect how much nutrients are in the nestling's bloodstream; for this information, some kind of signal is required.

Because of parent–offspring conflict, however, offspring might gain at the expense of parents by exaggerating their needs, so what keeps these signals honest? Numerous models have explored this question. The best-known answer is that, if begging is costly for offspring, then a reliable relationship between signaling and need can be evolutionarily stable, because any gains an offspring might achieve by exaggerating its needs are offset by the costs of the added begging. The result is a one-to-one relationship between the level of begging and the needs of offspring.

Other models suggest that begging might reliably signal offspring needs without being costly. For example, if begging is not a continuously varying signal, but an either/or signal (as might be the case, e.g., in very young nestlings that can only shoot their necks up and gape or not beg at all), then begging need not be costly to be evolutionarily stable, because it pays both parents and offspring for the latter to beg when they have passed a threshold of need. Alternatively, exaggeration might be kept in check by repeated interactions between parents and offspring. For example, if a parent bird senses that a nestling is exaggerating its needs, the next time the parent visits the nest it might be less inclined to feed that nesting, whatever the nestling's level of signaling, in effect, punishing the nestling for its bad behavior. The costliness of begging might also be affected by errors in how parents perceive begging levels or in how offspring assess their own needs, by trade-offs in effort between the parents, and by interactions among the siblings.

Empirical evidence that might help to support any one of these signaling models over the others is mixed. Many studies have confirmed that parents provision according to begging levels and offspring beg according to their needs. A key sticking point, however, is the cost of begging. Metabolic studies and (less conclusively) studies of nestling growth suggest the energetic costs of begging are quite low, but field experiments suggest that the costs of attracting predators with noisy begging can be high. To be properly assessed, however, the costs and the benefits of begging must be related to each other in the same fitness currency, which is difficult.

Not only are signaling models hard to distinguish empirically from one another, but they are also hard to distinguish from another set of models that do not involve signaling at all. The latter models treat begging not as a signal that coaxes care from the parent, but simply as a struggle among siblings to be the closest to their parents. Here, instead of offspring signaling how much they need and parents allocating resources accordingly, parents simply reward whichever offspring wins the scramble to get near them. Thus, the chief issue distinguishing the two sorts of models is whether offspring or parents control the allocation of resources, a distinction that has proved surprisingly hard to test.

One important criticism of all these models is that they overemphasize the competitive aspects of parent–offspring signaling. In fact, although offspring beg partly in order to attract feedings toward themselves, they also beg to make the parent provision the brood as a whole. Thus, the benefits of begging are partly shared among the whole brood, and so are the costs; nestlings in a nest are the proverbial eggs in one basket, so if any of them makes too much noise, they may all go down together. The interplay between how parent–offspring signaling functions at the individual and the whole brood level is one of the most interesting areas of current research.

Promising Research Areas

Having started with a strong focus on parent–offspring conflict and reliable signaling, the study of parent–offspring signaling, especially begging, has branched into several promising research areas. The dynamics of signaling within families, for example, how male and female parents mutually adjust their efforts, and how nestlings trade off competitive versus cooperative begging, is attracting particular interest. Another growing area will likely be the study of how parent–offspring signaling shapes the behaviors that offspring eventually display as adults, and the evolutionary implications of these effects (see 'Development' above). Of these effects, it is perhaps the nongenetic, maternal effects on offspring phenotype, such as recently discovered effects of maternal hormones on begging levels, that will have a particularly strong impact on how we think about the evolution of behavior. Finally, the study of the evolution of begging signals has been dominated by work on nestling birds, especially by songbirds. A taxonomically broader approach is likely to yield new insights, as already illustrated by emerging research on taxa as diverse as burying beetles and meerkats.

See also: Barn Swallows: Sexual and Social Behavior; Development, Evolution and Behavior; Differential Allocation; Honest Signaling; Parental Behavior and Hormones in Mammals; Parental Behavior and Hormones in Non-Mammalian Vertebrates; Signal Parasites.

Further Reading

Hinde CA and Kilner RM (2007) Negotiations within the family over the supply of parental care. *Proceedings of the Royal Society (London) B* 274: 53–60.

Hudson R and Trillmich F (2008) Sibling competition and cooperation in mammals: Challenges, developments and prospects. *Behavioral Ecology and Sociobiology* 62: 299–307.

Kilner RM and Hinde CA (2008) Information warfare and parent-offspring conflict. *Advances in the Study of Behavior* 38: 283–336.

Mas F and Kölliker M (2008) Maternal care and offspring begging in social insects: Chemical signaling, hormonal regulation and evolution. *Animal Behaviour* 76: 1121–1131.

Wright J and Leonard ML (eds.) (2002) *The Evolution of Nestling Begging: Competition, Cooperation and Communication.* Dordrecht, The Netherlands: Kluwer Academic Press.

Parmecium Behavioral Genetics

J. L. Van Houten, M. Valentine, and J. Yano, University of Vermont, Burlington, VT, USA

What Is a *Parmecium* and What Are Cilia?

Parmecium species have been useful model organisms for the studies of swimming behavior for over 100 years. Perhaps the most influential monograph on the behavior of ciliates, including *Parmecium*, was Jennings' 'Behavior of the Lower Organisms' (1906) in which he described his careful observations of the manner in which these cells respond to many kinds of stimuli by changing their swimming behavior. *Parmecium* cells behave in response to environmental stimuli by changing the way they beat their cilia, the organelles that propel them through the pond water that is their home. These cells can efficiently relocate to areas rich in food or escape noxious areas by manipulating just two aspects of swimming: the speed at which they swim, by beating their cilia and the frequency of their turns, by changing their ciliary power stroke. Their motor responses to stimuli are key to their sensory behavior.

Parmecium is a eukaryotic protist in the alveolatea superphylum and ciliophora phylum. It is a holotrichous ciliate, which means that it is a single-cell organism covered in cilia (**Figure 1**). On a *Parmecium* cell, there are about a thousand or more cilia that are slender organelles protruding from the surface of the cell. The cilia are about 10-μm long compared to the cells that range from 150- to 250-μm long. The cells can be seen with the naked eye, but the cilia cannot. Cilia are fitted with a central structure, the axoneme, and motor proteins called dyneins. They are covered with a membrane that is contiguous with the cell body membrane. If the dyneins are supplied with ATP and Mg^{2+}, the cilium moves at 10–20 Hz in a power stroke toward the rear of the cell and returns to the starting position in a lazier return stroke. This cycle of power and return stroke makes the cell swim forward, but the cells can also make abrupt turns by transiently changing the stroke toward the rear of the cell into the lazier stroke and the return stroke into a stronger stroke. This is the abrupt turn or 'avoiding reaction' described by Jennings. Since ciliates swim by beating their cilia in ways that move them forward or allow them to turn, the control of the ciliary beat is at the heart of understanding how these animals behave in response to environmental stimuli such as heat, ions, organic chemicals, gravity, water currents, or touch.

Parmecium has the properties of a neuron, and it is these properties that inspired the nickname of the swimming neuron. The cells are amenable to electrophysiological, biochemical as well as transmission and modern molecular genetic approaches, which make them appealing as a model organism, especially for the genetic dissections of cilia formation and function, ion channel function, pattern formation, DNA rearrangements, chromosome structure, secretion, and more. Here, we focus on their neuronal properties that drive behavior. *Parmecium* has many ion channels, and its membrane potential rests at about −45 mV in buffers that simulate pond water. The membrane's electrical properties that make it so much like a neuron are key to understanding the control of its cilia. The connection is simple on one level: ions that enter or leave the cell through ion channels manipulate the beating of the cilia. Calcium is the most important player, with K^+, Na^+, and Mg^{2+} also playing important roles. As long as calcium remains low (<100 nm) inside the cilia, the cilia beat to send the cell forward through the pond water in a helical path (**Figure 2**). If the cell depolarizes sufficiently to cause the opening of the voltage-gated calcium channels that are exclusively on the cilia, ciliary calcium rises, the stroke changes, and cells swim backward as long as calcium remains high in the cilium. Channels quickly close, the calcium is sequestered or removed, and the cell resumes its forward swimming in a new direction. An observer would see this as an abrupt turn of the cell because the backward swimming components are short lived. The cell does follow its old path when it starts swimming again, because the cell whirls in place as the backward swimming ends, and the cell shape is asymmetrical in part due to its oral groove being full of beating cilia that contribute to the cell's helical movement. Almost all new directions are equally favored, except for a 180° turn, which is underrepresented.

Combining Physiology, Genetics, and Behavior

There is a large body of work on the genetics of *Parmecium*, beginning with Mendelian and nonMendelian genetics, studied in great depth by Tracy M. Sonneborn and his students. *Parmecium* species presented an opportunity to use crosses and even grafts of paramecia to study very complex patterns of inheritance of mating type and surface antigens that had interesting patterns of inheritance. It is possible to mutagenize *Parmecium*, search for mutants, and analyze their mode of inheritance. The applicability of these genetic manipulations and Sonneborn's groundwork made *Parmecium* an attractive model organism by the mid-twentieth century.

Overlapping in time with the development of genetics of *Parmecium* were physiological studies that elucidated the basis of the membrane potential and ion currents that led to the nickname of swimming neuron. In Kaneko, Naitoh, Eckert, and others, used the relatively new technologies of electrophysiology to make the connection between ionic conductances and the beating patterns of cilia on *Parmecium*. They brought to light the basis of the graded calcium action potential that causes the transient influx of calcium into the cilia where the calcium changes the sliding of mictorubules of the axoneme, the structural heart of the cilium, and the bending of the cilia during the power and return strokes. The voltage-gated calcium channels reside only on the cilia, and when environmental stimuli cause sufficient depolarization to open these channels, the calcium that enters causes a change in ciliary

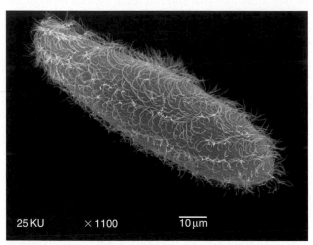

Figure 1 *Parmecium tetraurelia* cell. Scanning electron microscope image of a wild type cell from the dorsal side.

beat. As long as calcium remains high in the cilium, the change in beat will be sustained, but normally, this is very transient and lasts just a fraction of a second. As the calcium levels are reduced, the cell whirls in place, and eventually swims forward again. This is the avoiding reaction described by Jennings so long ago. The avoiding reaction can be elicited by high ionic concentrations, heat, repellents, or touching the anterior of the cell.

There are other ion channels that modify swimming behavior. There are many potassium conductances in the cell, and when the cells encounter attractants, lower ionic strength, or are tapped on the posterior, they hyperpolarize due to the opening of potassium channels. This hyperpolarization causes fast forward swimming of the cell.

Behavioral Genetics

This background of physiology, genetics, and behavioral observations, set the stage for Ching Kung to go about selecting for the behavioral mutants of *Parmecium*. Kung and J. Preer, a student of Sonneborn, chose to use *P. tetraurelia* because this species had the process of autogamy through which all loci are rendered homozygous at all loci with one simple manipulation in the laboratory (**Figure 3**). Autogamy facilitates the recognition of recessive mutations, speeds up crosses, and makes it unnecessary to do repeated back-crosses to generate mutants with identical genetic backgrounds. Kung used the established methods for mutagenesis and genetic analysis, but designed selections that would yield mutants that did not produce normal avoiding reactions. He trapped wild type cells in fits of avoiding reactions at the bottom of tubes filled with solutions high in Na$^+$, for example. Mutants unaffected by the Na$^+$ could swim upward following a

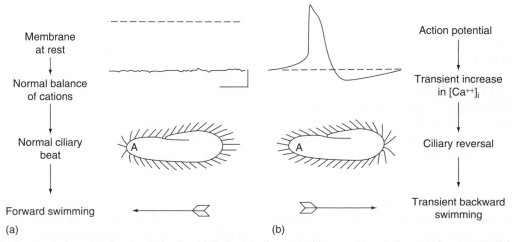

Figure 2 Calcium basis for cell swimming behavior. (a) Cells swim forward with a negative resting membrane potential; cilia beat toward the rear of the cell. (b) When an action potential is elicited through a depolarization (note dashed line is 0 mV potential), calcium rises inside the cilia, which change their power stroke toward the anterior and the cell transiently swims backward. From Kung C, Chang S-Y, Satow Y, Van Houten J, and Hansma H (1975) Genetic dissection of behavior in *Parmecium*. *Science* 188: 898–904, reprinted with permission from AAAS.

normal behavior of negative geotaxis to the top of the tube where they were collected. **Figure 4** shows how wild type cells generate trains of action potentials in depolarizing solutions with Ba^{2+} and jerk around in place without making much progress. In contrast, pawn cells depolarize but do not generate the action potentials. They swim slowly forward away from where they were first put in the pool of Ba^{2+}. Later, Kung selected for cells that had too robust an

avoiding reaction. The first mutants were called 'Pawns' because, as the chess piece, they could not back up. Crosses demonstrated that there were three pawn complementation groups or genes (*pwA*, *pwB*, *pwC*). The mutants with prolonged avoiding reactions were called Paranoiacs because they backed away for long periods of time as though persecuted. These mutants were generally single-site Mendelian mutants. Pawns were recessive and Paranoiacs dominant or semi-dominant. Kung identified several gene loci responsible for each of these phenotypes and eventually identified Dancer, *Dn*, a mutant that increases the calcium conductance by slowing the calcium channel inactivation, causing the cell to repeatedly turn. **Figure 5** shows a variety of the early mutants and their swimming tracks.

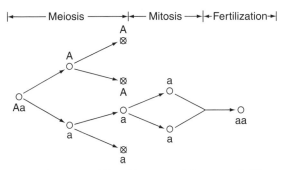

Figure 3 Autogamy. When *P. tetraurelia* become slightly starved, they will become mating reactive and reorganize their nuclei. As shown here, the nucleus undergoes the usual two meiotic divisions, but only one haploid nucleus survives. This one divides again by mitosis. If there is a cell of complementary mating type, it will receive one of these haploid nuclei. If there is no complementary cell, the two identical haploid nuclei will fuse, making a diploid nucleus that is homozygous at all loci. From Kung C, Chang S-Y, Satow Y, Van Houten J, and Hansma H (1975) Genetic dissection of behavior in *Parmecium*. *Science* 188: 898–904, reprinted with permission from AAAS.

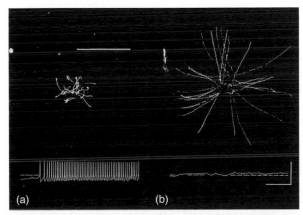

Figure 4 Cells swimming in Ba^{2+}. (All cells are introduced into the Ba^{2+} solution in the middle of the image. Each trace is a track of a cell swimming in the Ba^{2+} solution.). (a) Wild type cells in $BaCl_2$ buffers show repeated action potentials, each one of which cause a transient turn in the swimming path. The repeated turns are shown by the long-exposure micrograph of cells swimming. The tracks appear white because dark-field microscopy was used. (b) Pawn B mutants in $BaCl_2$ buffers show depolarization but no action potentials. As a result, the cells swim slowly out from the center of the picture where they were first placed in the $BaCl_2$ solution. From Kung C, Chang S-Y, Satow Y, Van Houten J, and Hansma H (1975) Genetic dissection of behavior in *Parmecium*. *Science* 188: 898–904, reprinted with permission from AAAS.

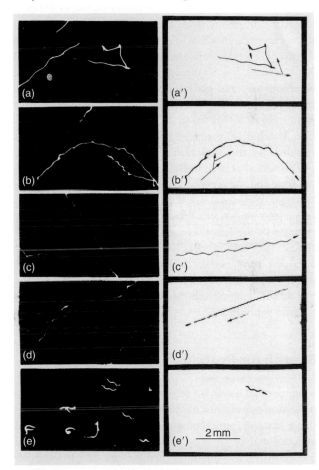

Figure 5 Variety of mutants. Cells were recorded in dark-field long-exposure micrographs while swimming in culture medium, which is rich in Na^+. The left column shows the long-exposure micrographs; the right column shows line drawings of the paths with arrows to indicate the direction of movement. (a) Wild type show paths with occasional turns from action potentials. (b) Fast-2 (*cam[11]*) mutant showing fast smooth swimming with no full turns from action potentials. (c) Pawn B mutant showing smooth swimming with no turns. (d) Paranoiac (*cam[3]*) showing the tight helix which is indicative of backward. Note the very long backward swimming. (e) Sluggish mutant hardly moving. From Kung C, Chang S-Y, Satow Y, Van Houten J, and Hansma H (1975) Genetic dissection of behavior in *Parmecium*. *Science* 188: 898–904, reprinted with permission from AAAS.

In analyses of these mutants, Kung found that the Pawns lacked the conductance of the voltage-gated calcium channel and that the Paranoiacs lacked a calcium-activated potassium conductance that normally truncates the action potential so that the cell usually goes backward for only a short period of time. Using a method of making 'models,' that is, cells treated with nonionic detergent so that components in the extracellular space can reach the inside of the cell and cilia, Kung showed that the Pawn mutants had functional axonemes. Adding Mg^{2+} and ATP causes models to move forward; adding calcium with Mg^{2+} and ATP reverses the swimming direction. Pawn models were able to reverse their swimming direction, but models of other mutants such as Atalanta, which was thought to have some defects in the axoneme, could not.

Over time, many more kinds of mutants were isolated using clever methods of selection. These mutants showed disrupted calcium-activated Na^+ conductances, reduced K^+ currents, or abnormal hyperpolarization-activated conductances, and had interesting names such as Pantaphobiacs, eccentrics, and restless. Eventually, almost every conductance identified in *Parmecium* had a mutant associated with it. This rich zoo of mutants was a tremendous resource for the dissection of other swimming behaviors of *Parmecium*, as seen in Bell et al. (2007).

Kung and workers had begun a 'genetic dissection' of the behavior of *Parmecium* in the 1970s, preceding the molecular genetic innovations that would revolutionize *Parmecium* genetics in the late twentieth and early twenty-first centuries. The insights that Kung brought with his genetic approach, combined with the electrophysiological analyses of mutants, were truly groundbreaking. The 1987 review in Further Reading summarizes Kung's progress since his seminal paper in 1975.

Search for Proteins

Kung pointed out in his seminal 1975 paper on his genetic dissection of behavior of *Parmecium* (see Further Reading) that *P. tetraurelia* presented advantages such as well-developed genetics and physiology, easily monitored behavior, and biochemistry. Since paramecia divide by fission and grow in high densities in the laboratory, they are amenable to searches for proteins that are the products of the interesting genes that govern behavior. The search for the protein products of pawns and Paranoiac genes was difficult and generally not successful in the early years. Comparison of two-dimensional gels from wild type and mutants did not yield results, and the sensitive mass spectrometers that are so helpful in protein identification today were not available in the 1970s–1980s. However, another convenient aspect of *Parmecium* came to the rescue. It is possible to inject protein or cell fractions into the cell. Through microinjection, one can assay whether the microinjected material 'cures' a mutant phenotype, implying that the protein product of the wild type version of the gene has now restored the wild type phenotype. Using this approach, Kung's group found that a single molecule, calmodulin, could cure many different mutants, Pantaphobiacs, Paranoiacs, and fast mutants. This important set of results made sense of the early genetic data that put fast mutants that do not turn and move very fast in Na^+ solutions in the same complementation group as Paranoiacs that swim backward for a long time in Na^+ solutions.

The study of *P. tetraurelia* calmodulin demonstrates an extraordinary dissection of the calmodulin molecule. First, the mutants are viable even though a molecule as important as calmodulin is changed in each. Second, the phenotypes are strong and easily identified. That is, swimming behavior tremendously magnifies subtle genetic changes so that selection for altered swimming behavior provides very effective selection strategies for clear phenotypes. Last, we have a better appreciation that calmodulin can interact not only with multiple ion channels but also with different classes of channels through each lobe of the calmodulin molecule. These outcomes would not have been possible with any other model organism.

To summarize the findings, mutations in the C-terminal (carboxyl) lobe of calmodulin cause the molecule to fail to properly activate the Ca^{2+}-dependent K^+ conductance that is responsible for returning membrane potential to basal levels after the action potential. There is a faster voltage-activated K^+ conductance that also works to short circuit and end the action potential, but this voltage-dependent conductance is rapidly activated and inactivated. If depolarization is prolonged, this conductance will be insufficient to repolarize the cell, and the slower to activate Ca^{2+}-dependent K^+ conductance will be responsible for returning the membrane potential to basal levels. This Ca^{2+}-dependent K^+ conductance requires calcium/calmodulin for its activation. Mutants, once called Pantaphobiacs and Paranoiacs, are now renamed *cam* mutants and they have very prolonged backward swimming in depolarizing solutions, such as those high in Na^+.

Mutations in the N-terminal lobe of the calmodulin gene produce entirely different phenotypes. These mutations cause the loss of a Ca^{2+}-dependent Na^+ current. This current requires the binding of calcium/calmodulin to the Na^+ channel as shown in patch clamping. These mutants have the common phenotype of no backward swimming, and fast swimming in high Na^+ buffers. The two have been renamed *cam* mutants, such as *cam¹¹*, that formerly was *fast-2*.

A close scrutiny of the large collection of *cam* mutants has uncovered a more complex picture with mating and growth impairments for some alleles. Each allele has a signature set of defects, including those in various ion conductances, in addition to the Ca^{2+}-activated Na^+ or K^+ conductances described earlier. Some *cam* mutants (formerly Paranoiacs) have an increased Ca^{2+}-activated Na^+ conductance, in addition to a reduced Ca^{2+}-activated K^+ conductance, which reinforces their long backward swimming in Na^+. Other *cam* mutants (formerly Pantophobiacs) have lost the hyperpolarization-activated Ca^{2+}-dependent K^+ conductance. In short, the mutations in the calmodulin gene can be grouped into those that cannot activate the Ca^{2+}-dependent K^+ channel or the Ca^{2+}-dependent Na^+ channel, but individual mutants have other changes in channel activation that lead to more complex behavioral phenotypes.

The identification of the gene products of the Pawn genes followed the same path as the identification of calmodulin as the gene product of the Pantophobiac genes. However, microinjection of fractionated cytoplasm did not provide a clear answer about the nature of Pawn genes, except to say that the gene products of the *pwA* and *pwB* genes were cytoplasmic or possibly endoplasmic reticulum-associated and probably function in trafficking of proteins like the calcium channel to the cilia.

Groups in Japan, some in collaboration with Kung, selected the behavioral mutants of another species, *P. caudatum*, which does not have the process of autogamy. The *cnr* (caudatum nonreversal) mutants fell into four complementation groups. Even though the *P. caudatum* *cnr* and *P. tetraurelia* *pawn* mutants could not be crossed to test for complementation, they could exchange cytoplasm by microinjection. While wild type *P. caudatum* cytoplasm cured the *cnr* mutants, as the wild type *P. tetraurelia* cytoplasm cured the *pw* mutants, no cytoplasm from *cnr* could cure any of the three *pw* mutants and vice versa. Kung concluded that there are at least seven gene products that are necessary for the functioning of the voltage-gated calcium channel.

The genes for both *pwA* and *pwB* have been cloned through a series of microinjection of smaller and smaller fractions of the *P. tetraurelia* genome. The genes look nothing like those coding for the subunits of a calcium channel, and the mystery still remains about how the *pw* gene products function to gate and otherwise control the function of the channels that are crucial to the cells' swimming.

Expanding the Genetic Dissection to Other Behaviors

We have focused on the swimming behavior mutants of Kung and workers in this description of the behavioral genetics of *Parmecium*, but there are other laboratories that have selected for mutants that cannot respond normally to environmental stimuli. None of these mutants have mapped back to the calmodulin gene. There are mutants that cannot respond normally to chemical stimuli by

changing their swimming to be attracted or repelled. There are mutants that fail to adapt in high K^+ or that fail to go backward in Mg^{2+}. There are even mutants that cause the Ca^{2+}-dependent K^+ to activate early in a depolarization, short circuiting the action potential. These are the TEA-insensitive mutants and, while they affect the same conductance as do the mutants with changes in the N terminal of calmodulin, the TEA-insensitive mutants are calmodulin mutants.

Modern *Parmecium* Genetics

Since the enormous effort to clone the *pw* genes by microinjection and curing, the *P. tetraurelia* genome has been sequenced and annotated http://www.genoscope.cns.fr/spip/Parmecium-tetraurelia-whole.html. The genome size appears to be about 35 000 genes, which have few to no introns and small introns when they are present. The ability to search the genome database has spurred enormous progress in the identification of genes and made it less crucial that there be clever screens to identify useful mutants. The genes for critical proteins that are necessary for homeostasis have been generally conserved well enough to be found through homology cloning, for example, plasma membrane calcium and SERCA pumps. But membrane proteins and signal transduction components have been generally less well conserved and the annotated genome is crucial in identifying the *P. tetraurelia* orthologs. Now, receptors for attractants and other proteins involved in signaling and behavior are coming to light.

The challenge of the annotated genomic sequences is that the genome has duplicated three times. Duplications have provided a large number of paralogs and homologs that need to be sorted through to determine whether these similar sequences code for proteins of similar and redundant functions, or have been assigned to different purposes in the cell. The advent of effective protocols for feeding RNA interference (RNAi) has made it possible to sort through these similar sequences and determine whether a 'phenotype' can be assigned to the reduction of the expression of a gene sequence. Now, it is possible to assign a function to a gene without having to generate a mutant.

Technologies for expression have also matured, and now it is possible to follow the tagged protein of interest in the cell. It is possible to inject plasmids or otherwise incorporate them into the cell and, until the cell goes through autogamy, retain the expression of the plasmid for study. In this way, it is possible to cure the existing mutants with an injection of plasmids and not use complex cell or nucleic acid fractions. When there are no existing mutants, it is possible to use the feeding RNAi to determine a phenotype and even combine RNAi feedings to look for synergies or additivity.

Summary

In summary, *Parmecium* behavioral genetics have come a very long way from Jennings' observations of swimming behavior. The genetic dissection approach taken by Kung rested upon two foundations. The first was built by Eckert, Kaneko, Naitoh, and others who used electrophysiology to understand the physiological correlates of the ciliary direction of beat, and the basis of the mechanostimulation from touching the front or back of the cell. The second foundation was the deep understanding of *Parmecium* genetics from the work of Sonneborn and Preer. Kung creatively demonstrated that a complex behavior can be dissected down to its root cause. He outpaced the technology that was needed to identify genes and gene products. Nonetheless, the exquisite dissection of the calmodulin molecule and the elucidation of its interaction with channels showed that painstaking microinjection and fractionation could result in the identification of a gene product in *P. tetraurelia*. Kung's combination of physiology, behavior, genetics, and biochemistry, paid off, as he predicted in his 1975 paper. We are now in the postgenomic era and groups are using bioinformatics and proteomics to understand the proteins of the cilia that govern behavior and finding elusive membrane and signal components that are crucial to *Parmecium* behavior at a very fast pace. However, we are standing upon the foundation laid by the geneticists, physiologists, and biochemists who went before.

Acknowledgments

We thank the NIH NIDCD for support of our work, the Vermont Genetics Network and Neuroscience COBRE at the University of Vermont for use of facilities, the colleagues of the *Parmecium* community, and especially the European Group of Research (GDRE) 'Parmecium Genomics' created by CNRS in 2002 that made the annotated genome possible, and the reviewers for their insightful comments.

See also: Electrical Signals; Evolution: Fundamentals; Smell: Vertebrates.

Further Reading

Arnaiz O, Cain S, Cohen J, and Sperling L (2007) *Parmecium* DB: A community resource that integrates the *Paramecium tetraurelia* genome sequence with genetic data. *Nucleic Acids Research* 35: Database issue D439–D444. website: http://parmecium.cgm.cnrs-gif.fr/.

Bell WE, Preston RR, Yano J, and Van Houten J (2007) Genetic dissection of chemosensory conductances in *Parmecium*. *The Journal of Experimental Biology* 210: 357–365.

Eckert R (1972) Bioelectric control of ciliary activity. *Science* 176: 473–481.

Jennings HS (1906) *Behavior of the Lower Organisms*. Bloomington, IN: Indiana University Press.

Kung C, Chang S-Y, Satow Y, Van Houten J, and Hansma H (1975) Genetic dissection of behavior in *Parmecium*. *Science* 188: 898–904.

Machemer H (1989) Cellular behaviour modulated by ions: Electrophysiological implications. *The Journal of Protozoology* 36: 463–487.

Preston RR, Kink JA, Hinrichsen RD, Saimi Y, and Kung C (1991) Calmodulin mutants and Ca^{2+}-dependent channels in *Parmecium*. *Annual Review of Physiology* 53: 309–319.

Saimi Y and Kung C (1987) Behavioral genetics of *Parmecium*. *Annual Review in Genetics* 21: 47–65.

Takahashi M (1988) Behavioral genetics in *P. caudatum*. In: Görtz H-D (ed.) *Parmecium*, pp. 271–282. New York, NY: Springer-Verlag.

Van Houten J (1994) Chemoreception in microorganisms: Trends for neuroscience? *Trends in Neurosciences* 17: 62–71.

Van Houten JL (1979) Membrane potential changes during chemokinesis in *Parmecium*. *Science* 204: 1100–1103.

Patch Exploitation

P. Nonacs, University of California, Los Angeles, CA, USA

Introduction

Foraging requires a set of hierarchical decisions. Where should the animal forage? Once it decides where, what set of potential prey or food types should be in its diet? And, if we define the 'where' as a patch, when should it leave that patch to go look for food elsewhere? What the forager does from the point it enters a patch to when it leaves is its 'patch exploitation' behavior. Identifying the behavioral rules for patch exploitation is critical to understanding much about foraging behavior in general. Knowing why animals leave patches leads to important insights on how entire habitats are used, how populations are spatially distributed, and how species may interact and coexist.

So let us consider how you might exploit your favorite restaurant as a patch. You are foraging on all your favorite dishes, prepared to perfection. Assume that the restaurant's food supplies replenish faster than you can eat and the check never arrives. In the parlance of foraging, we call this a 'nondepleting patch' (**Figure 1**). Although one might eventually leave the restaurant for reasons unconnected to eating, optimal foraging theory predicts one would never leave this patch expecting to find better food at a lower cost elsewhere. A number of organisms, such as barnacle larvae looking to settle on a rock, may indeed make only one patch choice decision in their entire lives. Thereafter, these animals treat the world like a single nondepleting patch that will never be abandoned.

Most animals, however, are not barnacle-like. They move from place to place often. And most patches, in fact, deplete (**Figure 1**). In the restaurant analogy, suppose the kitchen runs out of your favorite foods and starts serving less palatable entries, or that lots of other customers arrive causing the service to deteriorate and increasing the time it takes for food to arrive, or that they start asking you to pay the bill. If one or all these things happen, you may decide to quit this particular establishment and go somewhere else to eat. In terms of foraging jargon, the net rate of energy intake (calculated as the value of the food minus its cost, and divided by the time it takes to get and eat it) may be better at another restaurant. Graphically, the difference between a nondepleting and depleting patch is that when a forager is in a nondepleting situation, its total net energy intake goes up linearly (**Figure 1**). If it spends twice as long in a patch, it gets twice as much food. For a depleting patch, the longer an animal spends in a patch, the less food it will collect per unit of time.

Now let us consider a more natural situation: a bird gleaning insects in a bush. As the bird hunts successfully, it is emptying the bush of edible prey. At what point does it benefit the bird to quit searching a particular bush and fly to another one? A number of variables could be involved in this decision. First, what is the current capture rate of insects in the occupied bush? Second, is there another bush off in the distance with even more bugs on it? Third, how much does it cost the bird in time and energy to move? Note that all these variables deal only with energetic, food-related issues. We will add in other factors, such as predation risk later.

One solution for the bird is to leave the bush when it is empty and no insects are left. Practically, however, this solution is almost always unworkable. One could never be 100% certain that all the insects are gone or that another has not just arrived. Furthermore, the bird might need to search for a long, long time without any food reward to confidently conclude that patch is empty. Certainly, it could be doing better than that! Indeed, the mathematical solution for the best time to leave a depleting patch was derived in the mid 1970s by Eric Charnov. This solution is known as the 'Marginal Value Theorem' (often seen as abbreviated to the MVT), and it is probably the single most predictive and useful equation in the history of the behavioral ecology of foraging.

The Marginal Value Theorem

We start with the realization that the value of a depleting patch (let's call it patch type i) is a decelerating function of energy gained per time (**Figure 2**). Mathematically, we can represent the ever-changing gain function curve for i as depending on the amount of time spent in that patch, or $g_i(t_i)$. Each patch type has a specific encounter rate (λ_i). For each patch type, there is also an optimal length of time the animal should spend foraging. This is known as the 'patch residence time' (t_i). Summing across n patch types, the net rate of intake for a forager is:

$$R = [\lambda_1 g_1(t_1) + \lambda_2 g_2(t_2) + \cdots + \lambda_n g_n(t_n) - s]/$$
$$(1 + \lambda_1 t_1 + \lambda_2 t_2 + \cdots + \lambda_n t_n) \qquad [1]$$

For the forager, there exists a set of t values that will maximize R. Rather than solving all of them simultaneously, we simplify this problem by assuming that the animal spends the optimal amount of time in patch two

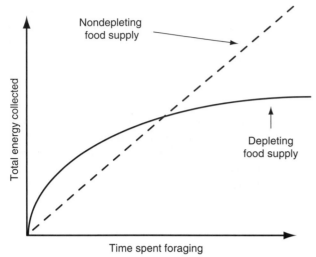

Figure 1 The relationship between time spent foraging and the total amount of energy collected. The slopes of the lines give the encounter rates. If the patch approximates an infinite food supply (i.e., the foragers never seriously deplete the amount of food), then the amount of food collected continues to increase in a linear manner (dashed line). If foraging depletes the patch, then a diminishing return function is expected and encounter rate decreases with time spent in the patch (solid line).

through n, and solve for the more tractable question of how long it should spend foraging in patch type 1. Therefore, we can then gather all the nonpatch one terms into two constants, c and k. This gives:

$$R = [\lambda_1 g_1(t_1) + k]/(c + \lambda_1 t_1) \qquad [2]$$

From basic calculus, we know that R is maximized with respect to time, with the t_1 value for which the first derivative (dR/dt) equals zero. This results when:

$$g'_1(t_1) = [\lambda_1 g_1(t_1) + k]/(c + \lambda_1 t_1) \qquad [3]$$

where $g'_1(t_1)$ is the instantaneous net gain in energy in patch type 1. The right-hand side of eqn [3] is an approximation of the average net rate of energy intake across the encounter-rate weighted average of all the patch types (plus search and travel costs). So intuitively, eqn [3] predicts that a forager should stay in a patch only as long as it is doing better at the moment than it would do, on average, by picking up and going elsewhere.

MVT Predictions: Patch Residence

The effect of catching prey on the forager's net intake rate in a depleting patch is a declining rate at which net energy accumulates (**Figure 1**). It takes the foraging animal longer and longer to collect equivalent amounts of energy. As the gain per unit effort decreases, it will become to the forager's advantage to leave the patch and find a richer one. How long the animal actually spends in a patch is called the 'patch residence time.' The optimal solution

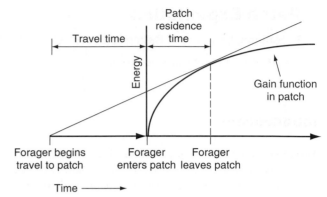

Figure 2 The optimal patch residence time when foragers travel between patches and patches decline in quality due to foragers' actions. The maximum rate of energy intake can be found by drawing the line with the highest slope that intersects the gain function (i.e., the tangent). The intersection predicts the optimal patch residence time.

can be shown graphically (**Figure 2**). Time runs both before and after a forager encounters a patch because there must be a travel time between patches. Patch residence times are given by the family of lines that initiate from where travel commences and intersect the gain curve. The optimal patch residence time is defined by the line with the greatest slope, which is the one line that is tangential to the gain curve.

In economic terms, the MVT predicts that a patch's immediate utility (i.e., its marginal value) to a forager depends on its value relative to all other options. Similar to a prey choice model, there is an all or none predictive equivalent: foragers should avoid patches whose return rates are expected to be below the overall environment's average. However, in better-than-average patches, animals should forage until the patch declines to the habitat's average. Thus, patch residence times should be variable in relation to the patch's quality at initial encounter. For slightly above-average patches, a short patch residence time is expected. For very good patches, long residence times are predicted.

The distance a forager needs to travel between patches should also affect patch residence time. One can see in **Figure 2** that if the travel time is increased, the tangent to the gain curve will move further to the right. Thus, the predicted time in the patch will increase. Similarly, reducing travel time will decrease the optimal patch residence time.

MVT Predictions: Prey Density

Unfortunately, for many foraging situations, the experimenter neither knows the overall average of the environment, nor is able to measure patch residence times without disturbing the forager. However, if we assume that the animal has an estimate of overall environment quality, the number of prey it leaves behind in the patch may allow us to measure the animal's estimation of its environment. Students of foraging call the number of prey left in a patch after a foraging bout the 'giving-up density' (GUD). Because an animal's foraging rate will decline as prey become scarcer in a patch, the MVT predicts that it should leave every above-average patch with approximately the same number of prey items still in it. Therefore, the GUD should approximate the average return rate of the habitat. Indeed, most experimental tests of the MVT measure prey densities before and after patch visits and then compare these values across patches.

The utility of measuring GUD applies to a wide variety of foraging problems beyond movement in and out of discrete habitat patches. For example, should a spider attempt to remove every drop of fluid from a fly before returning to hunting? Should a bee remove all the nectar from a flower before traveling to the next inflorescence?

MVT Predictions: Decision Rules

Despite its elegance, it is inconceivable that the MVT could perfectly predict quantitative animal behavior. There is a twofold problem. First, prey items come in discrete packages. A predator either has a prey item in its grasp or it does not. Therefore, the instantaneous gain in a patch is either very high (when the predator is feeding) or zero (in between prey captures). Thus, a literal interpretation of the MVT predicts that the forager should leave a patch whenever it is not handling a food item. Realistically, foragers should use a time-weighted average estimation of patch quality. It is unlikely, however, that any scheme averaging across varying times to capture of prey items could consistently identify the precise point where the instantaneous gain crosses the threshold for patch abandonment. Second, the forager must have an accurate estimate of the overall environment's quality. Again, it is difficult to imagine how in a changing world a forager could have direct knowledge of all the patches in its environment.

Given that foragers are unlikely to solve the MVT directly, can foragers reasonably approximate the optimal solutions? Indeed, simpler 'rules of thumb' come very close to optimal performance under a range of conditions. Three such rules have been proposed for how foragers decide to leave patches.

The first is a fixed time rule (**Figure 3(a)**) where an animal spends a set length of time in each patch and then moves on, regardless of how successful it has been. A second simple rule is to leave after a fixed number of prey captures (**Figure 3(b)**). Both fixed time and number rules can work very well when patches do not vary much in what they contain. However, if there is a high variability in patch quality, these simple rules can have serious

drawbacks. A fixed time rule can cause foragers to leave too quickly from very good patches and a fixed number rule can trap foragers for long periods of time in rather poor patches.

A more sophisticated rule is to for the forager to leave if a fixed time has passed since its last prey encounter (**Figure 3(c)**). These giving-up time (GUT) models have several variants. The simplest is a constant time rule: the animal leaves whenever a fixed period of time has passed without a prey capture. Whenever a capture is made, the forager's 'clock' is reset to its original value and begins counting down again. If patches are likely to be depleted rapidly due to the foragers' activities, it may be more advantageous for the animal to have a constantly decreasing fixed time interval. A prey capture resets the clock, but to new, shorter time. Conversely, if good patches are rare but not rapidly depleting, then it may be more advantageous for the fixed time to increase with every capture. Thus, a short run of bad luck would not cause a forager to abandon a still relatively rich patch. Foragers that use versions of GUT decision-making processes can be said to be Bayesian in their behavior. In other words, such foragers are continually updating or reevaluating their expectations of environmental quality. A review by Tom Valone suggests Bayesian abilities and behavior based on learning and updating one's expectations about the environment is widely distributed across animal species.

Testing the MVT: Patch Residence Times

Whichever rule of thumb an animal may use, a consistent prediction of the MVT is that the rule should be adjusted as the travel time between patches changes. For example, an increase in travel time would mean that the average quality of the habitat has declined (even if individual patches remain unchanged).

Bernie Roitberg tested this prediction in the foraging behavior of tephritid fruit flies on hawthorne fruits. Female flies lay eggs on the mature fruit, which they use vision to locate. In the experiment, Roitberg released the females on hawthorne trees that had a fixed number of fruit. Surrounding the test tree, the authors placed more trees that were either 1.6 or 3.2 m distant. In a third treatment, there were no other trees within sight for the flies. Thus, the experiment gave females access to patch of known and constant quality, but with visibly different travel distances to the next patch. The MVT predicted that: (1) the females would stay longer in a tree as their perception of the distance to the next patch increased, and (2) the females would spend this time searching more leaves, even though their success rate per time spent searching would decline. Both predictions were supported (**Table 1**): GUT and the number of leaves visited increased.

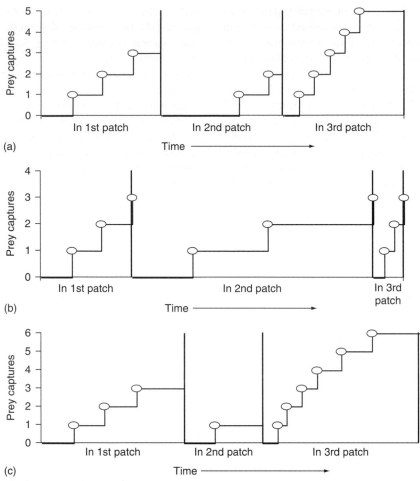

Figure 3 Rules of thumb for leaving patches. With a fixed time rule (a) the forager stays for a set length of time in each patch. Thus, the number of prey eaten may vary across patches. With a fixed number of captures rule (b) the forager stays in each patch until it captures a set number of prey items (three in this example). Thus, time spent in each patch may vary. With a GUT rule (c) the forager leaves when it fails to encounter a prey item for a given period of time. In this case, both the number of prey captures and the patch residence time may vary across patches.

Table 1 Tephritid fruit fly foraging behavior as a function of host density

	1.6 m	*3.2 m*	∞
GUT (min)	16.4	22.1	32.0
No. of leaves visited	59.3	89.3	103.4

Flies were released into three types of habitats that differed by having the trees spaced 1.6, 3.2 m, or out of the range of the fly's eyesight (∞). The MVT predicts that as flies perceive greater travel times between patches, they stay longer in patches (with longer GUTs) and search more thoroughly. The results support both predictions.

Testing the MVT: GUDs

As noted before, it is often difficult to measure the behavior of an animal as it occurs without your presence or equipment altering that behavior. A bird will probably stop hunting for insects if a large mammal is intently peering at it from a few feet away! After an animal departs a patch, however, it may be relatively simple to measure what it has experienced. Thus, after observing from afar that a gleaning bird has left a bush, we can measure all the insects it failed to catch and use this as an estimate of the bird's GUD. The measurement of GUD is one of the main methodologies in behavioral ecology for inferring patch exploitation decisions.

Burt Kotler used GUD to measure how two species of gerbils (genus *Gerbillus*) exploited patches of food. He set out trays of sand with seeds mixed in. The gerbils had to dig to find the seeds. After a night of foraging, Kotler collected the trays and counted the seeds left behind (i.e., the GUD). In one set of trials, the trays were the only food patch available. In a second set of trials, he provided 'free' food at same time he set out the sand trays. This free food was piles of seeds without the sand. Such easily collected food made the sand-tray patches relatively less valuable: Why dig when you do not have to?

In the context of the MVT, free food increased the overall quality of the habitat. Supporting the MVT prediction, both species of gerbils left significantly more seeds uncollected in the sandy trays (**Figure 4**). Interestingly, in the absence of free food, one species (*G. allenbyi*) had a significantly higher GUD than the other species (*G. pyramidum*). This species-level difference indicates that something more than food intake rate affected behavior. Kotler also found significantly higher GUDs in trays in open locations rather than under bushes. Small nocturnal rodents tend to view open locations suspiciously because open locations make them more susceptible to attack from owls. Thus, differences in GUD in this case also measured species differences in responding to predation risk. The smaller species, *G. allenbyi*, placed a higher premium to foraging in safer areas.

Testing Patch Departure Rules

Although there are situations in which fixed time or number departure rules appear to fit observed animal behavior, the majority of studies find patch departure decisions to be most consistent with some variant of a GUT rule. A variety of species facing a diversity of patch exploitation problems clearly modulate how long they stay in patches relative to the prey-capture rates they experience. For example, Lefebvre and colleagues found that bumblebee's decisions on when to leave a patch of

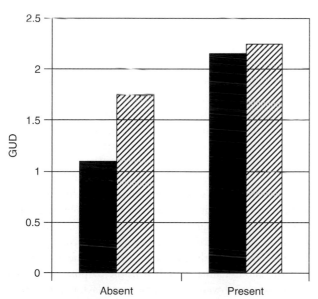

Figure 4 The mean GUDs for two species of foraging gerbils. GUD (as grams of uneaten seeds) is measured in sand trays containing seeds under two conditions: when piles of seeds not buried in sand are either absent or present. Both species forage similarly, with higher GUDs, when free seeds are present. In the absence of free seeds, however, the larger species, *G. pyramidum* (solid bars), took significantly more seeds from the sand trays than did the smaller species, *G. allenbyi* (striped bars).

flowers can depend on a multitude of experiences. Of particular importance is the number of rewarding and nonrewarding flowers encountered in a foraging bout. Eric Wajnberg reviewed an extensive set of studies on parasitoid wasps searching for hosts within patches. Wasp behavior across species supported a variety of potential decision rules, and GUT rules were commonly used. However, since the majority of studies were under laboratory conditions, fieldwork under more complicated environmental conditions and the presence of competitors is greatly needed.

Extending the MVT

The species-level differences across gerbils reveal a key limitation to the MVT. Clearly, if all gerbils cared about was maximizing net energy intake rate, there should have been little to no difference across the two species in the GUDs because they experienced very similar environments and foraging problems. Similarly, Peter Nonacs surveyed 26 different studies across a wide range of taxonomic groups that tested MVT predictions. Qualitatively, MVT predictions were strongly supported across the majority of studies: patch residence time tended to increase or decrease in the appropriate direction as the environment was changed. Quantitatively, however, only 3 of the 26 studies made accurate predictions. In 19 of the 23 misses, animals stayed longer in patches than predicted. Random errors could fit with the MVT because of the aforementioned problems in accurately estimating intake rates. A significant bias in one direction, however, suggests that a systematic error is present. Perhaps this error results from animals doing more than just foraging at any given time? There are several methodologies that try to model foraging behavior as a tradeoff between feeding and other activities. I will discuss two of these here.

Gilliam's Rule: Minimize the Ratio of Mortality Risk to Foraging Gain

In the mid-1980s, Jim Gilliam cleverly employed the same mathematical techniques used by rocket scientists to predict optimal trajectories for missiles to predict the behavior of animals simultaneously trying to maximize growth rate and minimize predation risk. The mathematics of analytical solutions to targeting missiles or describing tradeoffs between foraging, avoiding predators, and reproducing is very complex and beyond the scope of this study. However, if the animal is either too young to reproduce or the behavior is being considered in a time when there is no reproduction happening, then an attractively simple prediction falls out as regards the tradeoff between foraging gain and predation risk. An animal can maximize its fitness by minimizing the ratio of mortality risk (μ) over growth rate (g). This rule of minimize μ/g appears to be particularly useful in predicting patch choice and usage decisions because a comparison of two patches (μ_1/g_1 vs. μ_2/g_2) can be rewritten as μ_1/μ_2 versus g_1/g_2. This gives an effectively simple rule of thumb in patch choice. If one patch is twice as risky as another, it has to provide a growth rate that is double or better for a forager to prefer it.

Gilliam tested the μ/g rule with juvenile Creek Chubs (*Semotilus atromaculatus*) in an experimental apparatus consisting of three linear patches. In the patches on both ends were food dishes and adult fish. The adults are predatory to the point where they will cannibalize small, juvenile chubs. On one side was one adult, and in the other there were either two or three adults. The center patch was a safe refuge for the juveniles in that screening kept the adults from entering it. However, the refuge contained no food and the juveniles would eventually have to risk entering one of the food patches.

Predation risk was real and measurable: juveniles were on occasion eaten by adults. Two adults resulted in twice the risk of one and three were three times the risk. Foraging rate (f) was measured rather than growth rate, but they are obviously intimately connected. With equal foraging rates on both sides of the refuge, the juveniles should bias their foraging toward the less risky side. However, the μ/f rule predicts that by simply improving the foraging rate enough in the riskier patch, this preference could be reversed. Furthermore, since both μ and f were known, the point at which the shift should occur could be quantitatively predicted. Indeed, as the density of prey was increased in the riskier patch, the shift occurred where the rule predicted it would (**Figure 5**).

Nevertheless, μ/g models have a significant limitation in that time is not implicitly considered, except as instantaneous growth rate. If, for example, an animal has to gain a certain size or pass a certain developmental stage by a fixed date, the μ/g predictions can be grossly inaccurate.

State-Dependent Models

Gilliam's rule is an example of a solution to a simplified state-dependent model. The MVT model is static in that it predicts the behavior of the average animal responding to the average state of the environment. In a static model, the animal's physical state never changes. However, animals are dynamic in that their physiological states and needs and behavioral motivations are continually changing. Animals come in all ages, sizes, and conditions. Some are hungry, others are well fed. Some have offspring to feed, others are foraging only for themselves. Moreover, the environment is dynamic, too. There are good and bad days and changes over time. The mathematical techniques that give the μ/g rule produce predictions that are analytical solutions for unique and optimal behaviors for a given set of

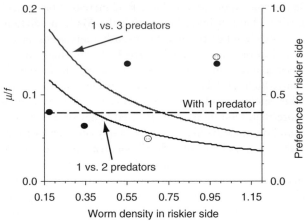

Figure 5 A test of Gilliam's rule with creek chubs. The fish had a choice to forage in two patches for worms. Both patches had predation risk, but the risk (μ) was not equal. One patch always had one predator and the other had either two (black line and points) or three predators (red line and points). Foraging rate was changed by adding more worms only in the riskier side. Thus, μ/f (measured on the left-hand y-axis) was always a constant for the safer patch (dashed line). However, μ/f for riskier side was greater at low worm density and the ratio declined as worms were added. Thus, the point where the solid lines cross the dashed line is where the minimize rule predicts the worm density at which fish should begin to value the riskier side more than the safer side. Fish decisions are the points plotted as the proportion of time spent foraging in the riskier patch (measured relative to the right-hand y-axis). For both experiments, the fish always preferentially forage in patch with the smallest μ/f. This means they switch from preferring the safer patch to preferring the riskier patch as the foraging rate there increases. Because of its higher μ, this switch occurs only at higher worm densities when the risky patch contains three rather than two predators.

environmental conditions. However, the mathematics for even simple problems (like the predation-risk/food-gain tradeoff) can be dauntingly complex. More complex problems rapidly produce unsolvable equations.

A contrasting approach is solving tradeoff problems numerically. If we know or can estimate the consequences of any decision or set of decisions, then we can let a computer calculate the fitness for all possible outcomes and from that set of outcomes, choose the behavior or sets of behaviors that give the highest fitness. This approach has many names, but most often is known as 'stochastic dynamic optimization.'

One example in the difference between the predictions of static and dynamic models is in the oviposition behavior of parasitoid wasps. A female wasp lays her eggs on a host. The size and the survivorship of her offspring depend on the size of the host and the number of eggs she lays on it. Therefore, for each size of host, Eric Charnov predicted the optimal number of oviposited eggs that maximizes the parasitoid wasp's reproductive success. Although this is not a classical foraging problem, wasp behavior can be easily related to the MVT. At what point does the marginal value of laying another egg in the current host drop below the value that egg would have if placed in the next found host? However, when the model was tested, the actual clutch size laid by *Nasonia vitripennis* females only partially supported the predictions. Females rarely exceeded the predicted optima, but quite often laid far fewer eggs than predicted.

Marc Mangel and Colin Clark derived a dynamic model for the same situation, including two real-world complications: a finite number of eggs that could be produced by a female wasp per time period and time effects (each foraging session is ended by onset of night-time, the wasps grow older, and winter approaches). The dynamic model was similar to the static model in predicting the maximum number of eggs laid, but the dynamic model's predicted distribution of clutch sizes was far more similar to the observed distribution of clutch sizes than the predictions resulting from the static model (**Figure 6**).

After noting the consistent bias in quantitative tests of the MVT, Peter Nonacs created a dynamic model of patch exploitation where individuals maximize fitness as a balance of foraging success, predation avoidance, mating opportunity, and parenting behavior. The results were encouraging in that they almost uniformly predicted patch

residence times that were longer than those from the MVT alone. Hence, they could account for the consistent deviations from MVT predictions as being caused by state-dependent behavior simultaneously reacting to multiple environmental variables.

Conclusion

There is abundant evidence that animals exploit patches not only relative to what they find in patches, but also relative to what they expect to find elsewhere. The best single predictive model for these decisions is the MVT. Through the predictions of the MVT, experimentalists have developed two extremely useful metrics for studying animal behavior. These are patch residence time and GUDs. Both allow us to explore how animals value patches not only relative to food, but to other factors such as predation risk. Even obvious limitations in the MVT have advanced our understanding. Decision rules based on simple measures, such as time from last prey capture, can reasonably approximate optimal solutions and do appear to match animal behavior. Finally, systematic errors in MVT predictions have led to creating more sophisticated and accurate models that incorporate predation risk and the animal's current physiological state.

The future of studying foraging behavior will surely often include the approach of determining the 'marginal values' of outcomes. Coupled with the techniques of dynamic optimization, models will be more ambitious in connecting life history considerations with foraging behavior. Thus, the ideas initially developed for how best to exploit a patch will continue to be have widely influential throughout behavioral ecology.

See also: Optimal Foraging Theory: Introduction; Trade-Offs in Anti-Predator Behavior; Wintering Strategies: Moult and Behavior.

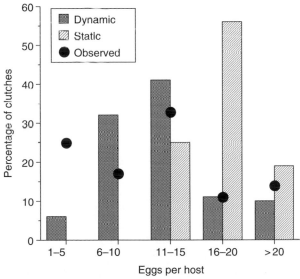

Figure 6 The predicted and actual behavior of parasitoid wasps encountering prey of four different size ranges. Upon encountering a suitable host, a wasp can lay from 1 to over 40 eggs on the host. The gray bars are the percentages of optimal clutch sizes (=eggs per host) predicted by a dynamic model that takes into consideration time of day and season, and different possible states and previous foraging experiences across wasps. The striped bars are the percentages predicted by a static model for each prey size that assumes a constant behavioral and physiological state for the wasps. The black dots are the observed percentages of clutch sizes for ovipositions on 36 hosts. The dynamic model fits the observed data better by predicting a range of small clutch sizes.

Further Reading

Charnov EL (1976) Optimal foraging: The marginal value theorem. *Theoretical Population Biology* 9: 129–136.

Charnov EL and Skinner SW (1984) Evolution of host selection and clutch size in parasitoid wasps. *Florida Entomologist* 67: 5–21.

Clark CW and Mangel M (2000) *Dynamic State Variable Models in Ecology.* Oxford: Oxford University Press.

Gilliam JF and Fraser DF (1987) Habitat selection under predation hazard: Test of a model with foraging minnows. *Ecology* 68: 1856–1862.

Kotler BP (1997) Patch use by gerbils in a risky environment: Manipulating food and safety to test four models. *Oikos* 78: 274–282.

Lefebvre D, Pierre J, Outreman Y, et al. (2007) Patch departure rules in bumblebees: Evidence of a decremental motivational mechanism. *Behavioral Ecology and Sociobiology* 61: 1707–1715.

Mangel M and Clark CW (1988) *Dynamic Modeling in Behavioral Ecology.* Princeton, NJ: Princeton University Press.

Nonacs P (2001) State dependent patch use and the Marginal Value Theorem. *Behavioral Ecology* 12: 71–83.

Roitberg BD and Prokopy RJ (1982) Influence of intertree distance on foraging behaviour of *Rhagoletis pomonella* in the field. *Ecological Entomology* 7: 437–442.

Valone TJ (2006) Are animals capable of Bayesian updating? An empirical review. *Oikos* 112: 252–259.

Wajnberg E (2006) Time allocation strategies in insect parasitoids: From ultimate predictions to proximate behavioral mechanisms. *Behavioral Ecology and Sociobiology* 60: 589–611.

Pets: Behavior and Welfare Assessment

B. L. Sherman, North Carolina State University, Raleigh, NC, USA

Introduction

Over the last 20 years, there has been a great deal of discussion and debate among animal scientists and veterinarians about the role of behavioral measures to assess the welfare of animals, including pets. Historically, those who argue the merits of objective scientific measurements to assess welfare and those who value assessment of affective states and behavior have been polarized. However, welfare scientists and veterinarians have recognized that objective measures of physiological states cannot fully describe the emotional state of an animal, in spite of the fact that affective measures are subject to problems of anthropomorphism and interpretation bias. In addition, behavioral observations in the context of an animal's environment provide critical information about welfare. The current comprehensive view, elucidated by Fraser, is that animal welfare is best assessed on the basis of three distinct but overlapping domains: body, mind, and nature. To varying degrees, the assessment of each component involves behavioral measures. The goal is the welfare of pets, to prevent their suffering and to insure their well-being, a state often described by pet owners as happiness.

The 'body' component encompasses the physical measures of basic health and physiological functioning, including measures of general body condition, health measures including heart rate, respiratory rate, body condition, and disease states. Physiological measures such as plasma glucocorticoids may be used to asses the function of the hypothalamic–pituitary axis (HPA) as indices of fear and pain. Values that reflect maintenance behaviors such as eating, drinking, and sleeping are included.

The 'mind' component encompasses psychological evaluation of an animal's affective state, motivation, and subjective experience. Such assessments are based on observations of an animal's behavior, which allow us to assess its motivation and emotional state. This approach utilizes the extensive literature on the cognitive abilities of animals.

The 'nature' component encompasses social functioning, the animal's ability to express natural behaviors in an appropriate environment, including normal development, behavior, and temperament. Information on the animal's actual living conditions compared to a natural environment is included in this interpretation. In some cases, a restricted environment, even if considered acceptable welfare, does not allow the expression of normal behaviors, such as reproductive and territorial behaviors or the full duration of maternal behavior.

Welfare must be considered on an individual basis and on the basis of the animal's species, age, history, individual temperament, physical condition, and circumstances under which data were collected. For example, for one dog, daily vigorous physical exercise may be an important component of its welfare. In the case of another dog, vigorous physical exercise might be eschewed by the animal and be considered abusive when forced. However, most would agree that some physical exercise is a requirement of welfare for dogs.

Current Uses of Behavior to Determine Welfare Status

The following schema use behavioral measures to determine welfare status and may be applied to pets. These include quality of life, behavioral assessment, welfare illustrator grid, and free-choice profiling. A discussion of the use of behavior to assess pets in clinical practices follows.

Quality of Life

The concept of 'quality of life' (QOL) has been applied to human patients with incurable diseases in order to make health care decisions attending to both the perspective of the health care provider and the patient's enjoyment of life. This approach is in contrast to traditional objective measures of medical treatment for human patients, which may fail to consider qualitative, affective factors. A focus on QOL emphasizes the patient's well-being and is a conscious mental experience. Thus, as with welfare, QOL is an abstract idea that cannot be measured directly, but its assessment includes the three broad domains, described earlier, of physical, psychological, and social functioning.

QOL varies along a continuum of feelings ranging from pleasant to unpleasant, where points might be considered poor, adequate, good or excellent. Individual preferences and needs influence QOL decisions. For example, illness or disability may or may not affect QOL, depending on the individual affected and a number of factors that are individualized. Thus, QOL is an individual assessment, suitable for the assessment of individual pets. The QOL concept has been applied to consideration of humane care of animals and the assessment of animal welfare, particularly pets (**Table 1**). More specifically, the QOL assessment has been applied to pet

Table 1 General concepts used to assess quality of life (QOL) of pets

Concept	Description	Low QOL	High QOL	Measures
Comfort and discomfort	Continuum from comfort to suffering	Discomfort, pain, distress	Comfort, absence of pain	Behavioral observation and physiological measures
Pleasure states	Opportunities for pleasure	No or infrequent states of pleasure, mistreatment, abuse	Continuous or frequent states of pleasure	Behavioral measures
Physical needs	Requirements for normal function (food, water, shelter)	Few needs met	All needs met	Physical observation, physical and behavioral measures
Environmental control	Ability to exert control over circumstances	Absence of control over situations, learned helplessness	Control over situations	Behavioral measures
Social relationships (appropriate to species)	Social emotions	No important social relationships	Social bonds	Behavioral measures
Health	Biological functioning; ranges from good health to disease	Poor health	Good health	Physical evaluation, behavioral measures
Stress	Measures the ability to cope effectively with challenging and aversive stimuli	High stress (chronic vs. acute)	Low stress	Physical, physiological behavioral measures
Fear/anxiety	Emotional responses that activate stress response and defensive or avoidance responses	Constantly/frequently experiencing fear/anxiety	Rarely (if ever) experience fear/anxiety	Behavioral observations and measures, physiological measures

Adapted from McMillan FD (2000) Quality of life in animals. *Journal of the American Veterinary Medical Association* 116: 1904–1910.

dogs and kennel dogs by Hewson and associates, integrating the three elements of the animal's health, its affect, and how it lives its life in its behavioral interactions with its environment. Odendaal and Meintjes correlated affiliative behavior with physiological measures, including heart rate, circulating catecholamines, circulating and urinary cortisol, and immune status. The concept of affect is critically important, since it is the influence of all other factors on affect that determines QOL. Behavioral measures include abnormal behavior (stereotypies and repetitive behavior, self-mutilation, coprophagy), frustration behaviors (chewing, vocalizing), and conflict behaviors (body-shaking, paw-lifting, and a lowered fearful posture), although these may vary depending on whether or not the stressor is acute or chronic. This approach emphasizes the whole animal and attempts to integrate all elements, including individual differences in temperament and experience.

Proxy assessment

Among humans, QOL is measured by self questionnaires in combination with health assessments. Since QOL refers to a conscious mental experience that pets cannot verbalize, how can we, as members of another species, assess their QOL? The tools used to assess QOL of humans with communicative and cognitive limitations serve as a model for the evaluation of pet QOL.

What is needed is an instrument for proxy assessment of QOL. Hewson and associates have integrated objective as well as proxy approaches to define the construct that is QOL. With regard to physical health as a component of QOL, the role of proxy health informant may be assumed by the pet's veterinarian, who can measure heart rate, respiratory rate, weight, and body condition, and conduct specific health assessment tests. The contribution of disease-specific clinical signs to overall QOL may then be combined with the contributions of other aspects.

With regard to affect and behavior, the role of proxy informant may be assumed by the pet's owner, who is familiar with the animal's personality, behavior, daily routine, and environment.

However, this process suffers from anthropomorphism, possible inadequate knowledge of species-typical behavioral signals, and the fact that the owner is not unbiased with regard to interpretation of the pet's behavior as reviewed by Hewson and associates. These problems can be attenuated by defining terms carefully, avoiding redundant terms, not relying on global questions alone to assess QOL, taking into account each animal's preferences in the most objective way possible, and being rigorous in all aspects of study design and analysis.

QOL approaches have been used in the evaluation of the effect of chronic pain on health-related QOL in dogs by Weisman–Orr and Yazbek and Fantoni. These have evaluated physiological and immunological responses as well as behavioral responses on the part of the animal. Thus, pain impacts QOL because of its associated feelings of discomfort. The magnitude of this effect of pain on QOL depends on many factors, such as the presence of familiar persons, the individual pet's pain threshold, its prior experiences, and other environmental factors.

Another study, by Freeman and associates, assessed the effect of cardiac disease on health-related QOL in dogs. One study developed and evaluated 'a questionnaire for assessing health-related QOL in dogs with cardiac diseases' named the FETCH (Functional Evaluation of Cardiac Health) questionnaire. The questionnaire's criterion validity was established by correlating the FETCH scores with the International Small Animal Cardiac Health Council's classification of disease severity. The questionnaire evaluated how each dog's cardiac disease impacted the dog's comfort and sociability as evaluated by the owner.

These approaches utilize standardized classifications of pain and cardiac disease, respectively, paired with validated questionnaires completed by the dogs' owners. Although details of the methodology have been criticized, these studies are major contributions in our appreciation of the effects of disease processes on QOL of pet animals. Such QOL approaches can be used for the assessment of welfare and to make the best clinical decisions about future treatments.

The QOL concept has been applied in a small study ($N = 27$) by Mullan and Main to evaluate the QOL of pet dogs visiting a veterinary practice on two sequential days. A dual instrument was developed and applied to QOL assessment in pet dogs. Owners were asked to provide biographical information regarding their dog, assess the resources provided to their dog (comfort, exercise, diet, mental stimulation, and companionship), and score behavioral and medical signs on a visual analog scare. The questionnaire was found to be repeatable, feasible, and to have good internal consistency and validity. The authors concluded that the evaluation could be used to assess welfare among pet animals in veterinary practice. In addition, the evaluation serves as an opportunity for veterinarians to advise clients to improve the pet's QOL. The very process of involving the owner in the process of assessing QOL is to increase the owner's awareness of welfare concerns to the benefit of the animal.

Behavioral Observations

As stated by Marian Stamp Dawkins, 'The most obvious, least intrusive, and potentially most powerful [indicator of welfare] is the animal's behavior.' Behavior reflects how pet animals interact with their environment and thus provides information about their needs, preferences, and internal states. Behavior can reveal when pets are ill, in pain, frightened, anxious, frustrated, or relaxed.

Pet behavioral assessments may be used to make management or treatment decisions in order to enhance welfare. For example, in order to asses the effect of pain in postoperative cats, the behavior of animals given postoperative analgesics was compared with those not given analgesics. Observations of cats' postures and movements, including general posture, position of the tail and ears, and degree of opening of the eyes and activity were measured. Kessler and Turner used this information to develop a scheme for evaluating the emotional state of cats, including how cats adapt to a new environment.

Normal behaviors

Knowledge of species-typical behavior and the range of individual behavior is a prerequisite of using behavior to assess welfare. Direct observations of activity, posture and body position (**Figure 1**), tail and ear carriage, facial expression, actions, piloerection (**Figure 2**), hypersalivation (**Figure 3**), urinations, and defecations can reveal information about the animal's internal state and motivation. Pleasurable behaviors, including play, grooming, scent exploration, and certain affiliative social behaviors and vocalizations, contribute in a positive way to an animal's welfare. For example, purring by kittens may be a

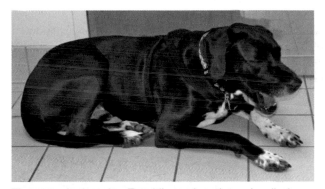

Figure 1 Anxious dog. Trembling and panting, a dog displays a posture consistent with tonic immobility.

Figure 2 Anxious cat. Cat displays piloerection along its dorsum and a dorsiflexed spine.

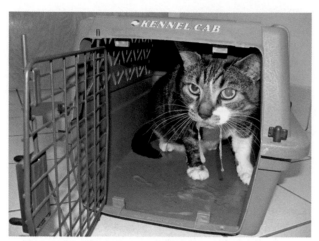

Figure 3 Confinement distress. A cat, unaccustomed to a carrier, displays anxiety-associated hypersalivation.

contentment signal that promotes positive social interactions. The play bow in dogs signals exuberant social play.

Anomalous behaviors

Abnormal behavior may be an expression of welfare concerns in pet animals. It suggests illness or that an animal's environment is inadequate in some way. For example, changes in behavior, such as increased water consumption, may be the first sign of illness observed in a pet animal. Abnormal behavior may indicate that welfare is compromised and that the environment lacks some important element needed for the normal functioning of the animal. Abnormal behavior may also indicate a negative affective state, such as frustration at not being able to act in a way that the animal is motivated to act. Self-injury, redirected behavior, displacement behavior, and vacuum activities may be observed as indicators of poor welfare.

Stereotypies, repetitive behaviors without apparent function, may indicate that the animal's needs are insufficiently met and the pet's welfare needs to be assessed. For example, a dog that exhibits stereotypic circling in the absence of an underlying medical etiology may suffer insufficient exercise and social interactions. However, stereotypies may have beneficial effects, such as increasing endorphins or preventing aggression. Neurochemical etiologies may underlie some repetitive behaviors, called compulsive disorders, where welfare is not suspect and pets respond positively to serotonergic medications.

Other Constructs

Welfare illustrator grid

Another method of assessing welfare using behavioral measures utilizes a welfare illustrator grid to increase objectivity as described by Wolfensohn and Honess. The grid technique may be used on a case-by-case basis to assess the welfare of pet animals over their lifetime or to compare the relative welfare states of a number of animals in a household at one point in time. In a typical case, there are four axes, each with an ordinal scale meant to indicate a level of welfare rather than an absolute measure. One of these axes is behavior, indicating the animal's deviation from a normal behavioral repertoire. Factors might include time budgets, social interactions, incidence of fighting, etc. Another axis is a clinical axis, which takes in to account the assessment of the clinical condition of each animal. Parameters might include heart rate, weight, body condition score, hormone assays. A third axis, causation, gives a score for the cause of the suffering, with a high score being intentional and a low score being inevitable or unpreventable. A fourth axis, duration, reflects the time span of the incident being evaluated in proportion to the actual time span of the animal. When a pet animal is evaluated and the results plotted, the result is a cognitive matrix of elements and constructs that can be explored in both a qualitative and a quantitative manner. The overall purpose is to heighten the perception of welfare along several axes, including behavior. The outcome of this approach is to illustrate for individuals responsible for animal welfare to improve their objectivity and consistency of action, with resulting benefit to the animal.

Free choice profiling

Another method utilized by Aerts and associates, and by which behavior may be used to assess pet welfare is free choice profiling, in which observers develop their own individual terminologies to describe the behavior of their subjects when observing the behavior of an animal as it interacts with its environment, including the observer. Then, each of the observers' terminologies, such as 'confident' or 'playful,' is applied to a linear analog scale, and the observers use that scale as a quantitative measurement tool for subsequent observations. This form of assessment is used widely in the study of animal temperament and personality. Although a seemingly anthropomorphic approach, naïve observers were very consistent in rating the behavioral expression of the same confined pigs on different occasions, and there was a high degree of agreement between observers. However, strong agreement between observers does not indicate that a judgment is correct. These sorts of overall evaluations may be more indicative of animals' internal state than specific measures that can be more precisely quantified. This sort of scheme may be used to quantify affective states such as pain, fear, and contentment.

As Fraser point out, such assessments have limitations, but they may be used to assess welfare in pet animals as well, with observers (pet owners) developing their own terms, then applying these terms to assess behavioral expression of their pet. Such an approach might permit an assessment of the affective state of the 'whole animal' in a manner that might increase assessment compliance by owners when treating clinical behavior problems.

Assessment of the Welfare of Pets in Clinical Practice

In clinical ethology or veterinary practice, how is it possible to assess the welfare of pets? The methodologies that have been developed for QOL and other assessments may require more time, resources, and owner commitment than are available. Yet, there is a need to increase awareness of pet animal welfare among owners and veterinarians. Behavior problems that impact on the human–animal bond are important welfare concerns. A pragmatic approach has been suggested by Dawkins, who posits two (and only two) basic questions with regard to the pragmatic assessment of the welfare of domestic animals: 'Are the animals healthy'? and 'Do the animals have what they want'? The response to the former query is in the domain of the veterinarian in consultation with the animal's owner, who can report on behavioral signs related to health, such as eating and drinking habits. When identified, health problems that are identified may then be treated. A QOL assessment may be used to determine the effects of chronic illness on the welfare of pets.

The response to the latter query, 'Do animals have what they want?,' addresses the mental and behavioral aspects of animal welfare. This question can be answered in a number of ways. One method, developed by Mullan and Main, is a QOL assessment that may be conducted in a companion animal practice. Another method is to specifically identify behavior problems or abnormal behaviors that signal reduced welfare. A simple questionnaire may be given to owners to fill out when waiting for veterinary appointments to identify common behavior problems in dogs (**Table 2**) and cats (**Table 3**), and modified for other species. When identified, behavior problems may then be treated by a veterinary behaviorist or clinical ethologist working with the veterinarian. Treatment of behavior problems can improve welfare.

A quantitative approach may be used to assess pet welfare, using an ethogram or catalog of the animal's

Table 2 Behavior checklist to identify behavior problems in pet dogs

Canine behavior checklist *Your Name:* _____ *Name of pet:* _____		*Date:*_____	
Does your dog or puppy?	**Yes**	**Occasionally or rarely**	**No**
1. Urinate (pee) in the house when you are at home?			
2. Defecate (poop) in the house when you are at home?			
3. Destroy/chew/claw objects in your home when you are at home?			
4. Urinate, defecate, salivate, or exhibit destructiveness when LEFT ALONE at home (in or out of a crate)?			
5. Whine, bark, or howl excessively?			
6. Dig excessively?			
7. Tremble, pace, or whine at the time of thunderstorms or other loud noises?			
8. Repetitively or excessively lick, stare at objects, pace, circle, or chase its tail?			
9. Wake you at night?			
10. Seem hyperactive or excessively excitable?			
11. Stare at, growl, bark, snap, or bite *family members* (including children)?			
12. Avoid new people or appear reluctant to travel or go into new places?			
13. Stare at, growl, bark, snap, or bite *nonfamily members* (visitors, strangers, children, veterinary team members)?			
14. Stare at, growl, bark, snap, or bite at other dogs (familiar or unfamiliar?)			
15. Chase or attack cats?			
16. Jump up on humans in greeting?			
17. Pull or lunge when walked on a leash?			
18. Have difficulty with basic commands (sit, stay, come)?			
19. Eat feces (stool)?			
20. Other: _____			

Table 3   Behavior checklist to identify behavior problems in pet cats

Feline behavior checklist   *Date:_____*
Your Name: _____
Name of pet: _____

Does your cat or kitten?	**Yes**	**Occasionally or rarely**	**No**
1. Urinate (pee) or urine mark in the house, outside the litter box?			
2. Defecate (poop) in the house, outside the litter box?			
3. Scratch/claw objects in your home?			
4. Vocalize (cry, mew, or meow) excessively?			
5. Wake you at night?			
6. Repetitively or excessively lick itself (overgroom), resulting in hair loss?			
7. Fail to adequately groom itself (undergroom)?			
8. *Repetitively or excessively* pace, circle, or chase its tail?			
9. Exhibit rippling skin on its back (dorsum)?			
10. Seem hyperactive or excessively excitable?			
11. Growl, chase, bite, or claw *family members*?			
12. Growl, chase, bite, or claw *nonfamily members* (visitors, strangers, veterinary team members)?			
13. Chase or displace dog(s) in the home?			
14. Chase or attack other cats in the home?			
15. Often retreat under the bed or up on a high perch?			
16. Exhibit any problems related to eating or drinking?			
17. Other: _____			

behaviors and comparing these under several conditions. The behavioral repertoire of a pet may be evaluated to determine the amount of time spent in various behaviors, such as play, resting, eating, grooming, or locomoting under two different conditions. For example, a dog's behavior when the owner is present and when he/she is not may be compared (via video recording) to determine whether the dog suffers from separation anxiety, a clinical state of reduced welfare characterized by excessive motor activities, distress vocalizations, destructiveness, and other negative behaviors not observed in unaffected animals.

Another method that utilizes behavior to assess what animals want is direct measurement of choice and preference as described by Fraser. This may be formalized with choice tests and the use of demand analysis. Modifications of this approach may be utilized in veterinary practice, to observe the behavior of pets in various locations and on different substrates, when in proximity to other animals and when approached by people. Then, this information may be used to devise handling methods that reduce fear and anxiety and ease management with each subsequent visit. Similarly, owners may be trained to note conditions under which their pets seem anxious or fearful or display defensive aggression or escape/avoidance behaviors. Then pets may be offered preferred locations or situations

through environmental management strategies, such as safe places when unfamiliar persons come to visit or private locations when being fed. Subsequently, the pet may be systematically desensitized to locations or situations that previously elicited fear or anxiety. In addition, as Ladewig proposed, veterinary practices should implement socialization and training programs for young and adopted pets to prevent behavior problems to reduce the incidence of behavior problems and resulting reduced welfare in pet animals.

Unique Problems in Assessing Pet Welfare Using Behavior

Assessment of Quality of Life

Although many attempts have been made to use scientific information, including behavior, to assess welfare, pragmatic and value-based decisions underlie these attempts. Certain types of behavioral information are subject to these biases because they may be measured more easily or accurately or with less apparent individual variability, or may be more prominent. For example, pet dogs who, when left alone exhibit obvious destructive behavior at the door of the owner's egress, are much more likely to be diagnosed ... and treated than dogs that exhibit immobility or catatonia as a manifestation of separation anxiety. The welfare of these dogs will remain compromised in spite of the owner's interest to the contrary. As human observers, behavioral information in many olfactory communication signals is beyond our detection. Thus, we cannot appreciate a cat that eschews use of its litter pan because it is appropriately deterred by olfactory signals produced by another cat in the household. In other cases, decisions, from our anthropocentric and culture-specific view, invoke our assumptions and opinions about what constitutes a good life for an animal. For example, cats confined to homes suffer less traumatic injury and exposure to infectious diseases and consequently, on average, live longer, healthier lives than unconfined cats. However, confined cats are unable to express a number of species-typical behaviors, including live prey acquisition. Underlying assumptions are retained when we combine different behavioral measures into an overall evaluation of QOL.

Utilizing behavioral information obtained from the owner in QOL assessments has been justifiably criticized as being biased by the anthropomorphism that characterizes owner–pet relationships. Anthropomorphism is pervasive and difficult to avoid, particularly when owners attribute complex human emotions and social motivations, such as jealousy, spite, and love, to their pets and value them for those qualities. Such attribution can lead to diminished welfare; for example, an owner may punish its pet retrospectively, erroneously attributing submissive behavior to the human emotion of guilt. Owners may over- or underestimate their pets' perceptive abilities or ability to understand human language. In addition, many pet owners have an inadequate understanding of communication signals given by their pet and consequently misunderstand the pet's perception of its world. For example, a dog that growls may be interpreted as dominant by its owner in spite of the fact that its visual communication signals indicate fear. The dual problems of anthropomorphism and misinterpretation of communication signals bias the owner as reporter with regard to the animal's welfare and may result in the pet being treated by the owner in a manner considered conducive but actually detrimental to its welfare, failing to meet its species-typical behavioral needs.

Areas Where More Research Is Needed

There is a paucity of data on several topics relating to the behavior and welfare of pets.

Behavioral Correlates of Disease Processes

Pioneering studies correlate behavior signs with pain, canine cardiac disease, and feline interstitial cystitis.

This approach is needed for other common diseases and conditions, particularly since owners often present ill pets to veterinarians on the basis of behavioral signs. For example, behavioral concomitants of endocrine diseases such as diabetes, hypothyroidism in dogs, and hyperthyroidism in cats will improve our understanding of behavioral correlates of disease processes.

Improved Understanding and Management of Fears and Anxieties

Fear and anxiety states in pet animals have important welfare implications. There is a need for improved understanding of the pathophysiology and heritability of these conditions. Artificial selection of pets, particularly dogs, appears to have resulted in an increase in fearfulness and fear-motivated aggression in certain breeds.

Using fear and intimidation to manage pets is pervasive and results in conditioned responses that perpetuate fearful responses and fear-aggression. Additional research is needed to better understand the physiological changes that underlie acute and chronic stress responses when forceful methods are used to manage pets. These findings may then be compared to management techniques that establish positive conditioned responses and enhance welfare. In addition, the differential behavioral response of pets to familiar and unfamiliar individuals, both conspecific and heterospecific, warrants further research.

See also: Disease, Behavior and Welfare; Domestic Dogs; Welfare of Animals: Behavior as a Basis for Decisions; Welfare of Animals: Introduction.

Further Reading

Aerts S, Lips D, Spencer S, Decuypere E, and DeTavernier J (2006) A new framework for the assessment of animal welfare: Integrating existing knowledge from a practical ethics perspective. *Journal of Agricultural and Environmental Ethics* 19: 67–76.
Beerda B, Schilder MBH, Van Hooff JARAM, deVries HW, and Mol JA (1998) Behavioural, saliva cortisol and heart rate responses to different types of stimuli in dogs. *Applied Animal Behaviour Science* 58: 365–381.
Dawkins M (2004) Using behaviour to assess animal welfare. *Animal Welfare* 13: S3–S7.
Fraser D (2008) *Understanding Animal Welfare: The Science and Its Cultural Context.* Ames, IA: Wiley-Blackwell.
Freeman LM, Rush JE, Farabaugh AE, and Must A (2005) Development and evaluation of a questionnaire for assessing health-related quality of life in dogs with cardiac disease. *Journal of the American Veterinary Medical Association* 226: 1864–1868.
Hewson CJ, Hiby EF, and Bradshaw JWS (2007) Assessing quality of life in companion and kenneled dogs: A critical review. *Animal Welfare* 16(S): 89–95.
Kessler MR and Turner DC (1997) Stress and adaptation of cats (*Felis silvestris catus*) housed singly, in pairs and in groups in boarding catteries. *Animal Welfare* 6: 243–254.
Ladewig J (2005) Of mice and men: Improved welfare through clinical ethology. *Applied Animal Behavior Science* 92: 183–192.
McMillan FD (2000) Quality of life in animals. *Journal of the American Veterinary Medical Association* 116: 1904–1910.

Mench JA and Mason GJ (1997) Behaviour. In: Appleby MC and Hughes BO (eds.) *Animal Welfare*, pp. 127–141. Oxon, UK: CAB International.

Mullan S and Main D (2007) Preliminary evaluation of a quality-of-life screening programme for pet dogs. *Journal of Small Animal Practice* 48: 314–322.

Odendaal JSJ and Meintjes RA (2003) Neurophysiological correlates of affiliative behaviour between humans and dogs. *Veterinary Journal* 165: 296–301.

Rochlitz I (2005) *The Welfare of Cats*. Dordrecht, the Netherlands: Springer.

Sherman BL and Mills DS (2008) Canine anxieties and phobias: An update on separation anxiety and noise aversions. *Veterinary Clinics of North America: Small Animal Practice* 38: 1081–1106.

Stafford K (2006) *The Welfare of Dogs*. Dordrecht, the Netherlands: Springer.

Taylor KD and Mills DS (2007) Is quality of life a useful concept for companion animals? *Animal Welfare* 16(S): 55–65.

Wemelsfelder F, Hunter TFA, Mendl MT, and Lawrence AB (2001) Assessing the 'whole animal': A free choice profiling approach. *Animal Behavior* 62: 209–220.

Wiseman-Orr ML, Scott EM, Reid J, and Nolan AM (2006) Validation of a structured questionnaire as an instrument to measure chronic pain in dogs on the basis of effects on health-related quality of life. *American Journal of Veterinary Research* 67: 1826–1836.

Wojciechowska JI and Hewson CJ (2005) Quality-of-life assessment in pet dogs. *Journal of the American Veterinary Medical Association* 226: 722–728.

Wolfensohn S and Honess P (2007) Laboratory animal, pet animal, farm animal, wild animal: Which gets the best deal? *Animal Welfare* 16(S): 117–123.

Yazbek KV and Fantoni DT (2005) Validity of a health-related quality of life scale for dogs with signs of pain secondary to cancer. *Journal of the American Veterinary Medical Association* 226: 1354–1358.

Pheidole: Sociobiology of a Highly Diverse Genus

J. Traniello, Boston University, Boston, MA, USA

Overview of the Ant Genus *Pheidole*

The genus *Pheidole* stands apart even when compared to the remarkable radiation of the ants (one insect family, the Formicidae) and the species richness of all animals and plants. Comprising more than 1100 described species, today *Pheidole* has the status of being the most diverse ant genus in the world. Because *Pheidole* is so extraordinarily rich in species, the genus has been called 'hyperdiverse.' Colonies can be abundant and ecologically dominant, particularly in the New World tropics. Nests in soil, decayed wood, and twigs are often readily collected and colonies are relatively easy to culture, enabling detailed mechanistic analyses of sociality that complement and extend field research. *Pheidole* has thus emerged as an important model to understand proximate and ultimate causes of social behavior and group structure.

Pheidole colonies generally have a single queen and are composed of morphologically distinct sterile workers of two different sizes (subcastes), minors and majors (**Figure 1**). This caste system is called complete dimorphism. The proportions of minors and majors in colonies can vary within and between species. The age of workers, whose bodies darken with maturation, can be estimated from the pigmentation of the cuticle (**Figure 2**). The size and age distribution of workers in colonies can thus be assessed and these two significant axes of the organization of division of labor readily studied. This demographic modularity also enables the experimental disassembly and restructuring of colonies to examine how *Pheidole* societies function and maximize reproductive success. These characteristics have facilitated the use of species of *Pheidole* as models for a wide variety of research in behavioral ecology, sociobiology, development, and neuroethology.

Workers are small, although the 'supermajor' subcaste of a few species may be roughly 4 mm in head width. Minor workers are uniform and unremarkable in form, but majors are striking: they have disproportionately large and sometimes bizarrely structured heads, typically adapted for combat and/or food processing. Majors are often called 'soldiers' because of their defensive function and usually have a small repertoire of behaviors, although in some species they may perform a broader range of tasks. Minor workers, in contrast, nurse brood (eggs, larvae, and pupae), construct and maintain the nest, and forage. These size and form characteristics are the foundation of the behavioral specializations of minors and majors, leading to low overlap between subcastes in task

performance. Because task specializations are considered important to the efficiency of colony operations, *Pheidole* has prominently served as a model to understand how labor is divided among workers and how natural selection has favored the design of an insect society.

Molecular analyses of *Pheidole* indicate a monophyletic origin in the New World, ~58–61 Ma. Diet ranges from specialized predation (e.g., on orabatid mites or termites), to scavenging and granivory. Current research on tropical species is providing a fascinating glimpse of the structure of *Pheidole* communities. Amy Mertl's recent examination of 59 ground-foraging species in Amazonian Ecuador showed that *Pheidole* vary widely in abundance, nest type, flood tolerance, and foraging range. Major workers differ in their involvement in foraging and ability to provide brood care when needed. A preliminary molecular phylogeny of these Amazonian *Pheidole* suggests large genetic distances between species and long periods of independent evolution. Strong interspecific variation in behavior and ecology suggests the presence of distinct groups of species that appear to segregate on the basis of nest site usage and/or tolerance to flooding disturbance.

In other ecological studies, *P. megacephala* has served as a model of invasive species biology. Desert *Pheidole* have been used to examine intercolony aggressive interactions and antipredatory behavior. *Pheidole* species have also provided excellent examples of the influence of parasites on competitive interactions.

Pheidole as a Model of the Evolution and Ecology of Caste

Because of its completely dimorphic worker caste, pioneering studies of physical caste evolution and the role of worker size variation (polymorphism) in ecological interaction have featured *Pheidole*. Species of *Pheidole* also provide paradigms for the analysis of age-related task performance (called temporal polyethism, or age-related division of labor). Additional path-breaking research has concerned how colony defense is organized through communication between minors and majors, how worker subcastes are determined physiologically, and why colonies have certain proportions of workers in each subcaste. Indeed, the genus *Pheidole* has provided an excellent system to study the adaptive nature of physical caste, caste differentiation and the evolution of development, worker

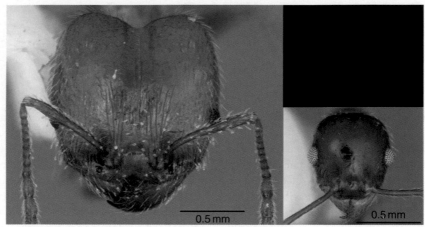

Figure 1   *P. dentata* major (left) and minor (right) worker.

Minor worker age (days)

Figure 2   Age-related changes in cuticular pigmentation in *P. dentata* minor workers.

behavioral ontogeny, colony demography, and the plasticity of social behavior and colony structure.

The Evolution of Major Worker Task Specialization

One trait thought to have been a key to the remarkable success of *Pheidole* is the major worker subcaste. Major workers provide colony defense though their morphological adaptations. They occupy locations within the nest that may require defense, and minors can mobilize majors to areas outside the nest where their ability in combat is required. Their defensive specializations make them efficacious in attacking competitors that threaten to usurp food sources or invade the nest. Depending upon the colony size and nest type, majors may provide in situ defense within the nest by occupying strategic positions, such as around nest entrances. Their specialization can enhance colony survival, but their flexibility to perform a broad range of task could be constrained by their morphology (i.e., their large, defense-adapted heads). Most of the differences between minor and major worker morphology are due to body size. The relatively massive

heads of majors show strong variation among species, perhaps because of the specificity of their colony functions.

Minors tend to most operations required for day-to-day colony functioning, but majors may serve as an 'emergency standby caste' and thereby compensate if there is a reduction in minors, such as might happen in a predation event by ants, other invertebrates, or reptiles, amphibians, and mammals, by absorbing their tasks. This indicates that majors can assess colony needs and respond accordingly by changing their behavior. Laboratory studies that model predatory losses have shown that if enough minors are removed from a colony, majors begin to perform tasks such as brood care that normally are outside of their repertoire. In some species of *Pheidole*, majors nurse brood, but seem to be incompetent at the task. This causes the minor workers remaining in the colony to increase their nursing, seemingly to compensate for the ineptitude of majors. Yet majors in other species of *Pheidole* appear to be as capable as minors at performing brood care.

These data suggest that the specialized morphology of *Pheidole* majors may compromise their ability to provide compensatory labor in the event of colony need, limiting them to defensive or trophic functions. In a comparative study in twig-nesting *Pheidole* species in Amazonian Ecuador, Amy Mertl studied the relationship between minor and major worker morphology, colony demography (the proportion of majors in a colony), and major worker ability to provide brood care, by quantifying nursing by majors in natural colonies and in colonies from which minors were experimentally removed. Across species, majors performed significantly less brood care than minors in intact colonies, but in subcolonies lacking minors, majors did not differ from minors in nursing acts after 1 week of contact with immatures. Apparently, majors perceived that immatures were not appropriately attended to during this time and upregulated their nursing. Brood nursed by majors, however, had lower survival than brood tended by minors, although rates of brood growth did not vary between subcastes.

Significant variation among these Amazonian species in brood care by major workers did not correlate with significant differences in brood growth or survival. Additionally, there was no significant association between the degree of major worker morphometric specialization and rates of nursing, growth or survival of brood, and the proportion of majors in a colony. Therefore, major worker morphology and specialization generally reduced the efficacy of brood care, but the extent of specialization in form did not further compromise their nursing ability. This lack of correlation may be due to the nature of predation on twig-nesting species: attacks may have either an extreme impact leading to colony death or negligible consequences on survival and fitness as colonies fortify defense at the nest entrance, secure their safety, and wait for danger to pass. In the former scenario, brood are killed along with the entire colony population, and in the latter case, colonies survive intact. Under both conditions, selection may be inadequate to favor major worker task flexibility.

Pheidole Defense: Social Organization, Colony Demography, and Plasticity

The striking morphological differences between *Pheidole* minor and major workers clearly reflect task specialization. Major worker head shape and the size and morphology of the mandibles are products of selection for combat, providing weaponry for their primary defensive role. The defensive behavior of *Pheidole* has also been examined from the perspective of the specificity of response to intruders and systems of communication between minors and majors responsible for organizing colony protection. Studies have also considered how patterns of investment in majors may adaptively change with geography, threat of competition, food availability, and colony life cycle.

The Organization of Colony Defense

P. dentata provides a model to examine the ecology of defense and the mechanisms of alarm communication that underscore effective colony-level responses to competition and predation. Over its wide geographic range, this species is sympatric with other ants, some of which are highly significant competitors. Native fire ants, *Solenopsis geminata*, and the imported fire ant *S. invicta*, compete with *P. dentata* and may raid neighboring colonies. In response to pressure from fire ants, *P. dentata* has evolved 'enemy specificity' in its alarm/recruitment communication. When minor workers encounter fire ants, they do not directly engage them in combat as they do other ants, but return home, laying a chemical trail. Within the nest, minors direct motor displays toward majors, raising their level of excitation and responsiveness. Together with the perception of the odor of fire ants on the body of minors, majors are stimulated to depart the nest and follow recruitment trails to attack the intruders. This alarm/recruitment response is specific to ants of the genus *Solenopsis*.

Additional studies of defensive responses suggest that *P. dentata* may have a degree of flexibility that allows colonies to adapt in ecological time to local threats. When laboratory colonies were subjected to repeat assaults from pavement ants, *Tetramorium caespitum*, major workers were eventually recruited. Colonies apparently 'learn' to recognize and respond to a novel threat.

Investigations of *Pheidole* have also increased our understanding of the antipredatory behavior of social insects. For example, to cope with the army ant *Neivamyrmex nigrescens*, *P. desertorum* in the US southwest frequently emigrates, moving among nest sites, and thus playing a shell game to decrease the risk of predation. Colonies of *P. desertorum* and *P. hyatti* prepare for defensive action with an 'alert phase' during which workers mass together around the nest entrance. In addition to directly combating army ants, major workers of both species participate in nest emigration by moving brood.

Ecological interactions between *Pheidole* and other ants can be impacted by the way in which *Pheidole* majors participate in defense. Parasitoid flies in the family Phoridae oviposit preferentially on *P. dentata* major workers when they are foraging together with minors. These flies are sometimes called 'decapitating flies' because after the egg hatches, the larva burrows into the ant's head and consumes the muscles and nervous tissue in the head capsule. Majors respond to attacks by these flies by absconding, rendering colony defense less efficacious, lowering competitive ability, and thus altering the outcome of interactions with *Solenopsis texana*. Ant-decapitating flies also influence the seasonality of foraging in the desert-dwelling *P. titanus*.

Adaptive Demography and Colony Plasticity in *Pheidole*

Pheidole colonies, like colonies of polymorphic ants in general, are thought to have proportions of worker sub-castes that have been determined by natural selection to optimize fitness. The proportions of minors and majors in *Pheidole* colonies may thus represent adaptation to local environments. Tests of this theory involve excavating colonies in the field and determining the frequency distribution of minors and majors. If adaptive demography theory is correct, then colonies inhabiting environments where predation or competition is more intense should have higher proportions of majors. The results of some studies support this prediction. For example, colonies of *P. morrisi* in Florida have more majors in relation to colony size than colonies in North Carolina or New York. One reason for this difference in colony structure could be that fire ants, which are potent competitors that may attack and destroy colonies, overlap with *P. morrisi* in Florida but not in more northern regions. The greater number of majors can provide more effective defense against fire ant attacks. Comparative studies of the ecology of sub-caste distribution patterns in *P. dentata*, however, have not provided support for adaptive demography theory, but the geographic range over which colonies were examined was much lower than that for *P. morrisi*.

In addition to adapting genetically to local variation in competition by evolutionary changes in the number of majors, *Pheidole* colonies can also facultatively increase production of majors in the short term if exposed to the odors of competitors. Colonies of the European *P. pallidula* will raise more majors when they have contact with alien conspecific workers. However, not all species of *Pheidole* appear to have equivalent plasticity in adjusting subcaste ratios. *P. dentata* laboratory colonies reared with fire ants nearby did not change their proportion of majors as they reared brood through several cycles. Additional research is needed to determine why species differ in their ability to adjust subcaste proportions.

Minor Worker Age and Division of Labor

Worker size- and age-related task specializations are thought to be critically important to the efficiency of labor and hence colony productivity. Age-related task performance, also called behavioral development or temporal polyethism, has often been studied in this respect and appears to be a conserved trait in ants and other social insects. *P. dentata* has served as the primary model in the study of temporal polyethism in ants, and social insects in general.

There has been controversy concerning the relationship of division of labor to developmental processes or changes in patterns of task allocation among workers that emerge independently of the biological properties of individuals. Some theories of polyethism do not require or consider age-related physiological development, suggesting instead that labor dynamics are self-organizing and worker age is only a correlate of behavior. Nevertheless, a variety of studies, many of *Phediole*, indicate that division of labor is species-typical, with hormonal, neurobiological, and other physiological changes having causal links to polyethism. Genetic mechanisms of polyethism are also well documented. Studies of *Pheidole* have yielded significant insights into how age, on-task experience, and flexibility in task performance, all contribute to the colony-level division of labor.

P. dentata minors transition from queen attendance and brood care, through other within-nest tasks, to foraging and other activities outside the nest during the first 16 days of adult life, forming nonoverlapping temporal castes that specialize on different roles. E. O. Wilson, who pioneered research on ant polyethism, called this model of age-related task performance *temporal caste discretization*, and hypothesized that it produced an efficient division of labor to maximize colony fitness.

The Repertoire Expansion Model

Recent studies show that *P. dentata* minors expand their repertoires from 5 to 17 tasks as they age, rather than shift between nonoverlapping sets of tasks as predicted by the temporal caste discretization model. Tasks typically performed by young minors (newly enclosed to 2–3 days in age), including brood care and queen attendance, are thus retained in the repertoire of older minors, rather than eliminated with increasing age. This pattern of behavioral development is termed *repertoire expansion*. The repertoire expansion model (REM) is supported by studies demonstrating that older minor workers of *P. dentata* perform both brood-care and nonbrood-care tasks with a high frequency. Moreover, only older minors increase nursing as the demand for brood care increases. The REM also predicts that minor worker behavior does not shift among spatially associated sets of behaviors that minimally overlap. Instead, because repertoire size expands with age, older workers may contribute significantly more to brood care through greater efficiency at this task suite than young minors.

The REM can be tested, for example, by determining whether the breadth of responsiveness to task-related olfactory stimuli increases with age. According to the REM, young workers should be capable of perceiving odor cues associated with nursing, whereas older minors should be able to detect and respond to odors associated with both brood and foraging. In an assay in which young and old minor workers could chose between orienting toward odors of brood or prey odors, old minors moved toward both brood and food odors, but young minors responded only to cues emanating from brood. These stimuli mediate responses to tasks associated with brood care and foraging, respectively. These differences in olfactory responsiveness demonstrate age-related variation in sensory abilities consistent with the predictions of the REM.

How well do young and old minor workers care for brood?

Young and old minor workers not only respond differentially to task-associated odors; they also differ significantly in the efficiency of performance of the principle task of nursing, which is dependent on the detection of cues related to brood. To examine temporal polyethism in *P. dentata*, Mario Muscedere and Tara Willey assessed the ability of young and old minors to care for the queen and developing brood in single age-cohort subcolonies. Larvae reared by old workers gained significantly more mass than those reared by young workers and did not differ significantly in mass from larvae reared by groups of workers of intermediate age. Additionally, old minors were more responsive to brood than young minors: old minors approached and cared for brood more frequently than their young siblings. Old minors also retrieved brood and assembled them in groups more rapidly than young minors. Furthermore, old minors engaged in more queen-directed behaviors in spite of the fact that young minors tended to be in closer proximity to the queen.

Young minors and older minors differ significantly in their efficiency at performing brood-care, so the performance of nursing by mature minors is not simply compensatory and due to behavioral flexibility of an age class that is typically specialized on outside-nest tasks. These results do not support the characterization of young workers as a discrete temporal caste specialized to respond to queen and brood needs. Although brood and queen care are performed earlier in development than foraging, older minors retain the ability to perform these tasks and do so with high efficiency. Histological studies show that young minors have poorly developed mandibular muscles, apparently limiting their ability to work. This suggests that young minor workers are developmentally immature and physically unable to perform many tasks. Minors acquire task proficiency and improve their efficiency at brood care with increasing age, rather than specializing on foraging and other outside-nest tasks later in adulthood to the exclusion of inside-nest tasks. Accordingly, the traditional view that young minors specialize on queen attendance and brood care may simply be based on observations that these are the first tasks performed by minors following eclosion, rather than tasks performed with the high efficiency expected of an age cohort that are nursing specialists.

The Neurobiology of Division of Labor

Theories of caste evolution and division of labor can be tested by examining the neural properties of workers of different subcaste and age, thus providing new insights into the mechanisms associated with social organization. Structural and neurochemical changes in areas of the brain that process sensory information associated with inside-nest tasks like brood care (occurring in darkness) and external tasks like foraging (performed in light) are likely to accompany temporal polyethism. The neural basis of task performance has been most extensively studied in honeybees; similar neurobiological techniques can also be applied to understand behavioral development in the worker caste of adult ants. Task transitions should reflect neuroanatomical and neurochemical change. For example, labor required within the nest does not require vision, which becomes significant as ants eventually depart from the nest to navigate foraging routes and patrol territory.

Pheidole provides an excellent model to analyze the neurobiology of division of labor. In *P. dentata*, changes in the volume of brain regions and modifications to neural connections accompany repertoire expansion in minor workers. Marc Seid's examination of the ultrastructure of one region of the mushroom body, a brain center for complex multisensory integration, learning and memory, showed that individual presynaptic boutons enlarge and acquire more synapses and vesicles as minors age. Older *P. dentata* minor workers, which both forage and nurse, have enlarged presynaptic boutons and more synapses and vesicles per axonal bouton than young minors. The total number of boutons decreases while the size and vesicle content of remaining boutons, number of synapses per bouton, and average size of postsynaptic elements all increase with age. These results indicate expanded and enhanced efficacy at synaptic connections important in processing sensory input, and the loss of other connections as minors age and increase their task breadth and efficiency. This synaptic pruning could be a mechanism of age-related processing of task-associated sensory information, and thus a neuroanatomical basis for behavioral development and repertoire expansion.

Brain Chemistry and Division of Labor

Biogenic amines such as serotonin, dopamine, and octopamine are present in the insect brain and commonly regulate different aspects of behavior, including aggression, olfactory sensitivity, and learning. Because olfaction and aggressive behavior are associated with tasks such as defense and foraging, neurotransmitters may cause subcaste and age-related differences in division of labor. Although brains of *P. dentata* minors are miniscule (roughly $0.00125 \, \text{mm}^3$, or about one-hundredth of the

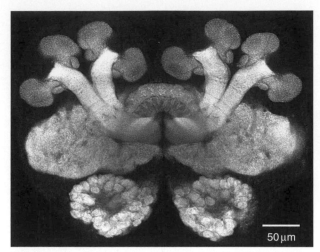

Figure 3 Confocal microscope scan of the brain of a *P. morrisi* minor worker.

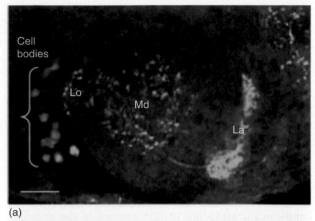

(a)

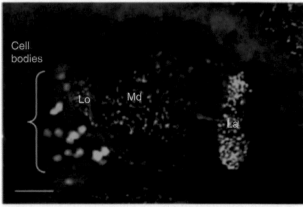

(b)

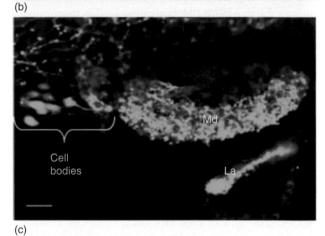

(c)

Figure 4 Serotonergic neurons in the optic lobe of the brain of a young (a) and old (b) minor worker, and an old major worker (c) La, lamella; Md, medulla; Lo, lobula. Scale bar is 20 µm.

size of a honeybee brain; **Figure 3**), it is possible to individually measure their amine levels. In *P. dentata*, titres of serotonin and dopamine (but not octopamine) increase significantly with age, the greatest increase in serotonin being coincident with the onset of outside-nest tasks like foraging. Serotonin could therefore activate foraging and the perception of its associated cues and signals such as prey odors, trail pheromones, and other orientation stimuli, as well as the aggressive actions associated with prey capture and territory defense. These age-associated changes in biogenic amine levels suggest their involvement in the neuromodulation of minor worker behavioral ontogeny and temporal polyethism.

There are also age- and subcaste-related patterns of serotonin immunoreactivity in the optic lobes of the brains of both minor and major workers of *P. dentata*. Serotonergic nerve cell bodies in the optic lobes increase significantly in number as major and minor workers mature (**Figure 4**). Old majors have greater numbers of serotonergic cell bodies than minors of a similar age. This age-related increase in serotonergic neurons, as well as the presence of diffuse serotonin networks in the mushroom bodies, antennal lobes, and central complex, occurs concomitantly with an increase in the size of worker task repertoires. Serotonin thus appears to be associated with the development of the visual system, enabling the detection of task-related stimuli outside the nest, thus playing a significant role in worker behavioral development and colony-wide division of labor. Serotonin and/or other amines may be involved in navigation, which is required during foraging.

Pharmacological manipulations of serotonin titers in the brain demonstrate its modulatory role in minor worker behavioral development. *P. dentata* minors have been treated with either serotonin precursors or antagonists. Following the oral administration of these compounds, trail-following assays have been conducted to

determine the effect of experimentally elevated brain serotonin on foraging behavior. Dietary administration of the serotonin precursor 5-hydroxytryptophan (5HTP) increased brain serotonin levels in treated *P. dentata* minors, which followed artificial trails for significantly longer distances than control minors. Minor workers fed the serotonin agonist α-methyltryptophan (AMTP), which inhibits tryptophan-5-hydroxylase (the rate-limiting enzyme of the serotonin synthesis pathway) and thus lowers brain serotonin, showed lower responses to trail pheromone. These results indicate that serotonin has neuromodulatory effects on the integration of chemical information related to foraging behavior.

Interspecific Variation in Behavior, Social Organization, and Brain Structure

Pheidole species differ widely in diet and vary in worker size, proportion of majors in colonies, and subcaste specialization and plasticity. This adaptive variation is reflected in interspecific differences in the structure of the brain. For example, *Pheidole morrisi* has mature large colonies of 5000–10 000 workers with a high proportion of majors that attend to a relatively broad range of tasks. Both minors and majors of *P. morrisi* forage over long distances, and majors can transport brood and will nurse if needed. Minors and majors are highly aggressive and this species is dominant in its community. In sharp contrast, minors and majors of *P. pilifera* form small colonies of 200–400 timid workers that have small foraging ranges and feign death when disturbed. Majors are passive, few in number, and do not nurse. Rarely seen above ground, majors are typically found in seed-filled nest chambers and are primarily millers. Minor and major workers of these and other *Pheidole* species, after correcting for brain size, differ significantly in the volume of brain regions such as the mushroom bodies, which likely play a neural role in the organization of division of labor. The ranks of residual volumes of brain compartments of majors parallel the ranks of their task diversity. These species also differ significantly in biogenic amine levels in the brain.

Summary

The hyperdiverse genus *Pheidole* provides outstanding opportunities to examine the relationship between adaptation, evolutionary success, social organization, and its mechanistic basis. Neuroethological studies of *Pheidole* species may help identify the key innovations in social structure associated with the extraordinary diversity of this genus. Proximate analyses of colony organization can be coupled with ecological studies to understand adaptive radiation.

Effective and efficient division of labor is significant to colony survival and reproductive success in social insects. *Pheidole* has been a principal system for the analysis of division of labor in an insect society because the importance of size and age can be studied simultaneously and colony structure can be experimentally altered to determine the importance of demography and explore behavioral plasticity. Recent studies have begun to examine the relationship of the neuroanatomy and neurochemistry of the brain to age- and size-related task performance,

providing new and detailed examinations of the role of brain structure and neurotransmitters in subcaste task specializations and age-related division of labor. Integrative research bridging brain, social behavior, and ecology will advance our understanding of how ant colonies are organized socially and our knowledge of the role of social organization in the evolution of the extreme diversity found in this genus. *Pheidole* can thus serve as a model system for comparative studies of other evolutionarily dominant clades.

See also: Ant, Bee and Wasp Social Evolution; Caste Determination in Arthropods; Division of Labor; Neurobiology, Endocrinology and Behavior; Neuroethology: Methods; Neuroethology: What is it?.

Further Reading

Beshers SN and Fewell JH (2001) Models of division of labor in social insects. *Annual Review of Entomology* 46: 413–440.

Brown JJ and Traniello JFA (1998) Regulation of brood-care behavior in the dimorphic castes of the ant *Pheidole morrisi* (Hymenoptera: Formicidae): Effects of caste ratio, colony size, and colony needs. *Journal of Insect Behavior* 11: 209–219.

Calabi P and Traniello JFA (1989) Social organization in the ant *Pheidole dentata*: Physical and temporal caste ratios lack ecological correlates. *Behavioral Ecology and Sociobiology* 24: 69–78.

Feener DH, Jr (1988) Effect of parasites on foraging and defense behavior in a termitophagous ant, *Pheidole titanis* Wheeler (Hymenoptera: Formicidae). *Behavioral Ecology and Sociobiology* 22: 421–427.

Hölldobler B and Wilson EO (1990) *The Ants*. Cambridge, MA: Harvard University Press.

Kaspari M and Byrne MM (1995) Caste allocation in litter *Pheidole*: Lessons from plant defense theory. *Behavioral Ecology and Sociobiology* 37: 255–263.

McGlynn TP and Owen JP (2002) Food supplementation alters caste allocation in a natural population of *Pheidole flavens*, a dimorphic leaf-litter dwelling ant. *Insectes Sociaux* 49: 8–14.

Mertl AL and Traniello JFA (2009) Behavioral evolution in the major worker subcaste of twig-nesting *Pheidole* (Hymenoptera: Formicidae): Does morphological specialization influence task plasticity? *Behavioral Ecology and Sociobiology* 63: 1411–1426.

Moreau CS (2008) Unraveling the evolutionary history of the 'hyperdiverse' ant genus *Pheidole* (Hymenoptera: Formicidae). *Molecular Phylogenetics and Evolution* 48: 224–239.

Muscedere ML, Wiley T, and Traniello JFA (2009) Age and task efficiency in the ant *Pheidole dentata*: Young minor workers are not specialist nurses. *Animal Behaviour* 77: 911–918.

Passera L, Roncin E, Kaufmann B, and Keller L (1996) Increased soldier production in ant colonies exposed to intraspecific competition. *Nature* 379: 630–631.

Pie MR and Traniello JFA (2007) Modularity, morphological integration and caste evolution in the ant genus *Pheidole*. *Journal of Zoology* 271: 99–109.

Seid M, Harris K, and Traniello JFA (2005) Age-related changes in the number and structure of synapses in the lip region of the ant *Pheidole dentata*. *Journal of Comparative Neurology* 488: 269–277.

Seid M and Traniello JFA (2005) Age-related changes in biogenic amines in individual brains of the ant *Pheidole dentata*. *Naturwissenschaften* 92: 198–201.

Seid M and Traniello JFA (2006) Age-related repertoire expansion and division of labor in *Pheidole dentata* (Hymenoptera: Formicidae): A new perspective on temporal polyethism and behavioral plasticity in ants. *Behavioral Ecology and Sociobiology* 60: 631–644.

Wheeler DE (1991) The developmental basis of worker caste polymorphism in ants. *American Naturalist* 138: 1218–1238.

Wilson EO (1976a) Behavioral discretization and the number of castes in an ant species. *Behavioral Ecology and Sociobiology* 1: 141–154.

Wilson EO (1976b) The organization of colony defense in the ant *Pheidole dentata*. *Behavioral Ecology and Sociobiology* 1: 63–81.

Wilson EO (1984) The relation between caste ratios and division of labor in the ant genus *Pheidole* (Hymenoptera: Formicidae). *Behavioral Ecology and Sociobiology* 16: 89–98.

Wilson EO (2003) *Pheidole in the New World: A Dominant, Hyperdiverse Ant Genus.* Cambridge, MA: Harvard University Press.

Yang AS, Martin CH, and Nijhout FH (2004) Geographic variation of caste structure among ant populations. *Current Biology* 14: 514–519.

Relevant Websites

www.myrmecos.net/myrmicinae/pheidole – Myrmecos.net.

www.antweb.org – AntWeb.

www.discoverlife.org – Discover Life.

www.eol.org – Encyclopedia of Life.

Phylogenetic Inference and the Evolution of Behavior

K. M. Pickett, University of Vermont, Burlington, VT, USA

Introduction

The use of behaviors in phylogenetic investigations has a long history. As with many evolutionary matters, Darwin can be cited as having been one of its promoters, for example, in his studies of the unfortunately named 'slave-making' ants. But others preceded Darwin. For example, in 1854, Henri de Saussure published a systematic hypothesis of the relationships of family Vespidae based on nearly as many nest architectural attributes as morphological; despite de Saussure's lack of computational tools, the details of his proposed phylogeny are strikingly similar to modern treatments based upon architecture, morphology, and genetic data. Brooks and McLennan provide an excellent historical review. The nineteenth-century ethologist C. O. Whitman commented, "instincts and organs are to be studied from the common viewpoint of phyletic descent." Early studies of birds, caddisflies, spiders, and social insects – especially due to the legendary systematist W. M. Wheeler – firmly positioned behaviors as important traits for phylogenetic analysis. Thereafter, the halcyon era of ethology, inaugurated by Nobel laureates Konrad Lorenz and Niko Tinbergen, sought to fuse behavior and taxonomy into a cohesive whole.

Since the work of these early ethologists, a revolution in taxonomy has taken place, complete with new tools and philosophy, making the practice of systematics much more explicit, testable, and therefore scientific. Beginning in the 1960s, this movement, often called the Cladistic Revolution, firmly established scientific principles set forth by fly taxonomist Willi Hennig. As a result, the power of systematic inquiry was greatly strengthened, and so were investigations of behavioral evolution. A brief list of studies employing modern cladistic methods to elucidate the phylogeny of behavior includes swordtail fish sexual selection, architecture in orb-weaving spiders, courtship behaviors in fruitflies, display behavior in birds, water mite sexual selection and display, termite caste evolution and behavior, and wasp social behavior.

After the Cladistic Revolution, the use of behavior in phylogenetics became controversial, but today the majority of practicing systematists readily exploit this rich source of character information. Nevertheless, some questions remain. To answer those questions, the general structure of modern phylogenetics, historical objections to the use of behavior in phylogenetics, and the responses to those objections are discussed here. Then, general benefits of behavioral characters are discussed. Finally, some advances in behavioral phylogenetic techniques are reviewed.

General Principles

Before treating the use of behavioral characters themselves, some of the underpinnings of phylogenetics are first reviewed, since objections to the use of behavioral characters often stem from a misunderstanding of the goals of phylogenetics. When viewed from first principles, the use of behavioral characters is not only substantiated but desired as well.

What Is Taxonomy?

The field of study often referred to today as *phylogenetics* is a subdiscipline of a field of study known as *systematics*. Systematics, in turn, is a subdiscipline of one of the most ancient fields of science: *taxonomy.* Some readers may be surprised by this characterization, as many today erroneously believe the relationship to be the reverse of what has just been described – namely that taxonomy (as they understand it) is a subdiscipline of systematics. This is in part due to the marginalization of taxonomy by population geneticists in the mid-1900s. Nonetheless, taxonomy is the field of biology interested in deciphering the patterns and relationships of taxa and the naming of taxa. The latter of those two is known as *nomenclature.* The former – deciphering the relationships of taxa – is systematics.

What Constitutes the Phylogenetic System?

Prior to the work of Hennig, the field of systematics was often characterized by careful study of specimens by experts in the field. That study, however, was almost never codified into a form amenable to scientific test. Prior to Hennig, systematists simply asserted the shape of phylogenies and pronounced which characters were homologies, without any formal procedure or test that other investigators could scrutinize.

In 1966, Hennig proposed a method, which he called *Phylogenetic Systematics*, that allows of the use of *characters* in such a way that the investigator can determine if the characters are similar due to ancestry or similar for other reasons. That is, phylogenetic systematics allows the investigator to distinguish *homology* from *homoplasy*. This is the fundamental feature of Hennig's system, and it is also one of the fundamental pursuits of evolutionary biology. In phylogenetics, homologies are identified by the *test of congruence*, in which characters are analyzed simultaneously with other characters (which may suggest different patterns), and the preponderance of character evidence is revealed, thereby identifying homologies. In other words, as any single character will tell whatever story it tells, the only way to test its story is to allow other characters to contend with it; homoplasy will not generate a single signal, given enough data, but the actual homologies will join to show a single pattern or signal. In this way, homologies reinforce one another and drive the structure of the phylogeny, whereas homoplasies do not. Prior to Hennig, investigators simply asserted which characters they believed to be homologous or homoplasious, but modern phylogenetics requires the construction of character matrices and explicit tests of the hypotheses of homology asserted in those matrices. In short, Hennig made systematics scientific.

Phylogenetic systematics is very flexible and permits the investigator wide latitude in selecting sources of evolutionary evidence. Originally, those sources were morphological and (frequently) behavioral. Today, character data include amino acid residues and nucleotide identity, among other biochemical sources. What makes phylogenetic systematics scientific is the manner in which characters are treated, not the source of the character data.

What Is a Character?

Phylogenetics is dependent on the use of character information to discover phylogeny. A working definition of a character is both philosophically and pragmatically necessary. When the term *character* is used herein, the definition offered by Freudenstein and co-authors in 2003 will be followed: a character is a biologically transmitted attribute of a species. This definition may seem simple, but it provides quite a bit of circumscription that will guide the identification of attributes that constitute valid characters and those that do not.

First, the definition indicates one of the most basic components of a character: it must be 'biologically transmitted'; that is, it must be heritable. Heritability can be difficult to assess directly under some circumstances; as with all propositions in science, the proposition of heritability is sometimes a tentative hypothesis.

The notion of biological transmission also stipulates why some features are not valid characters. For example, the geographic placement of an organism is not biologically transmitted from one generation to the other, even though it might be argued that in some nonbiological way, organisms inherit their locality from their parents. Much has been

written about the use of geographic location in phylogenetics, and while location may be interpretable via phylogenetics for investigations of biogeography, locations are not valid characters, because they are not transmitted biologically. The criterion of heritability via biological transmission does not necessarily imply genetic transmission alone, however. Provided the attribute is biological, and is transmitted from one organism to another via a biological process, the criterion is met. In another section, circumstances in which behavioral traits may be valid characters, despite their not being genetically encoded, are elaborated. Indeed, the involvement of behavioral characters permits the incorporation of novel sources of phylogenetic information.

Next, the working definition states that the attribute is that of a species. This taxonomic level is important, as it is the lowest level of the taxonomic hierarchy in which phylogenetics is involved. Limiting the definition to those traits transmitted by species means that within-species, population-level variants are not characters in the phylogenetic sense. Character formation occurs only at the level of the species and above. As a result, characters are fixed in a species (though they may be polymorphic at higher levels). The fixation of characters at the level of the species is critical to the concept of *synapomorphy*.

As reviewed earlier, systematists used behavioral data to inform phylogeny long before the development of phylogenetic systematics. Thus, it is somewhat surprising that since then, a variety of objections have been raised suggesting that behavioral attributes are not suitable as characters. As I will show below, behavior does not pose special problems relating to the character concept. I address some of the most common objections below, and respond to each.

Objections to the Use of Behavioral Characters in Phylogeny

Nearly 50 years after the cladistic revolution made systematics a hypothesis-driven, explicit, testable and repeatable science, recent treatments still object to the use of behaviors as characters.

Objections to the use of behavioral characters in phylogenetic analysis have, for the most part, relied on one or more of the following arguments: (1) the characters do not reflect the phylogeny of the taxa, but merely the evolution of behavior itself; (2) the use of characters of interest renders any deductions about the behavior circular; (3) behavioral characters are more prone to local adaptation and are therefore too plastic to be informative; (4) behavioral character delimitation is more likely to result in the treatment of nonindependent attributes as independent; or (5) behavioral characters are often not heritable. While these potential problems should be considered carefully by investigators proposing behavioral homology, in most cases it can be shown that these objections are of no consequence to behavioral phylogenetics. Further, close examination reveals that most of the arguments can be equally applied to other character types.

Phylogeny of Characters Versus Phylogeny of Taxa

Some have objected that the use of behavioral characters in phylogeny produces only a phylogeny of the behaviors, which may not be congruent with the true (and unknown) taxon phylogeny. This was an early objection to the use of molecular characters in phylogeny, now in such common use. It is relevant to reiterate here that two of the fathers of the field of ethology, Konrad Lorenz and Niko Tinbergen asserted that behavior does not evolve independently of phylogeny.

Indeed, all phylogenies are the result of character analysis. Thus any conclusions about the relationships of taxa are always based on character phylogenies, whether those characters are behavior, DNA sequences, or morphological traits. The power of the phylogenetic system is that additional independent characters can be folded into a phylogenetic analysis to further test the robustness of a given (character) phylogeny. The best phylogenetic approximation is the one based on the most character evidence, regardless of what those characters are. Thus, a priori exclusion of certain types of data detracts from the goal of obtaining a reasonable phylogenetic estimate.

Putative Circularity

The argument of phylogenetic circularity asserts that the character unduly influences the chosen tree shape, thereby making evolutionary deductions circular. For example, in a recent book, James H. Hunt writes: "To use [social behaviors] as evidence of common ancestry for taxa categorized as 'eusocial' constitutes a fallacy of affirming the consequent."

The circularity argument asserts that characters used in a phylogeny are off limits to evolutionary interpretation on that tree. Yet scientists are motivated to include characters for just that reason. Biologists interested in the evolution of a gene must use information from that gene's DNA sequence. Similarly, we must use morphological data to infer how morphologies evolved. In short, phylogeneticists *must* use the very characters of interest because those reveal their history better than other characters. The objection to inclusion of behavioral traits is particularly troublesome, as some morphological characters were originally codified with explicit reference to behavior. There is no logical reason to exclude any particular kind of character *a priori*. All potential characters are candidates for phylogenetic inference, and choosing from them requires consideration of information on stability, heritability, and fixity in taxa, as outlined in one of the following

sections. At the outset, any valid character is as good as any other, and when all characters are analyzed together, congruence allows us to infer the phylogenetic pattern.

Putatively Increased Plasticity

A major objection is that behavior is uniquely plastic; that is, it evolves rapidly and repeatedly in separate lineages. In other words, behavior is more likely than other character types to be homoplasious. Yet rigorous assessment shows that homoplasy and lack of signal in behavioral characters often derive from poor character delimitation (e.g., grouping many individual characters into suites of composite characters). Studies that compare *consistency indices* of behavioral to morphological characters have found that behavior is no more plastic than morphology. We must regard the charge of behavioral plasticity as an assertion unsupported by rigorous analysis.

Questions about Independence

Delimitation of behavioral attributes lends itself to lumping characters into reified classes that are actually composed of many independent characters. Yet this practice is hardly unique to behavioral phylogenetics – such questions plague morphology and molecular data as well. Indeed, the still young field of molecular phylogenetics must disentangle genetic functional constraints, selection, secondary structure, and reading frames, all of which can cause nucleotides within and across loci to violate the key assumption of independence. In fact, the functional and developmental complexities of molecules and morphology can place those character types at a disadvantage relative to behavior. A recent study shows elegantly how decades of careful behavioral observation can be employed to create a large matrix of independent behavioral characters. In this work, Fernando Noll identified and employed 42 independent, heritable behavioral characters to study bee phylogeny, thereby disarticulating the complex character 'social behavior.' Bypassing problems of defining and discretizing the compound character of 'sociality,' Noll simply coded observed variation in clearly defined and independent characters. The result was a robust phylogeny by which the evolution of sociality could be inferred for that group.

Questions about Heritability

The question of heritability is often raised for behavioral attributes, especially in the field of sociobiology, and most especially in reference to primate behavior, including humans. Advocates of a molecular-only approach to phylogenetics often point out that molecular data are known to be heritable without any question, whereas morphology and behavior can be altered by norms of reaction or learning. In fact, all character types are prone to errors associated with heritability. Molecular sequences used in a phylogeny can, in fact, represent laboratory contamination. Attributing contaminant sequences to loci from which they did not come clearly violates the heritability criterion. For behavior, resolving the 'norm of reaction' problem only requires that behavioral characters be stereotypical of a taxon.

Behaviors affected by learning pose special challenges for phylogenetic analysis. Prior to elucidation of the rules governing inheritance, behavioral traits were widely used. However, after Watson and Crick's landmark paper, nongenetic mechanisms of transmitting biological information from generation to generation, or taxon to taxon, became suspect. We now know that genetic material is not the sole reservoir of genealogical information and that several classes of biologically transmitted attributes of species are not encoded in the genome. Consider the classic example of bird song. Offspring learn from parents many varieties of bird song, and evidence shows that the elements of the songs themselves and the persistent differences in dialects of such songs are not transmitted genetically. Yet bird species can be distinguished by their songs, making song characters valuable for phylogenetic studies. Nongenetic transmission satisfies assumptions of phylogenetic analysis, and investigators can look for congruence with other characters to reveal those that are unreliable.

The Use of Behavioral Characters in Phylogenetic Analysis

Because of what is described here as character chauvinism – that is, the a priori preference of one character type over others – most treatments of behavioral characters have focused on the debates outlined earlier. However, in these discussions, the genuine benefits derived from using behavioral characters are rarely featured. My own research has shown time and again that the best phylogenies result from combined analyses of molecular, morphological, and behavioral characters. In these studies, resampling support and concordance with traditional taxonomic expectations are enhanced by the addition of behavioral characters. All character types (behavior, morphology, molecules) have liabilities, and an expanded approach provides results that are more robust than those achieved by excluding characters a priori. Furthermore, behaviors have some features that make them superior to DNA sequences for certain kinds of analyses.

Age and Vetting of Characters

One good reason to involve behavioral (or morphological) characters in phylogenetic analysis is the long history of

their use. Researchers across nearly 200 years of published systematic work have been using behavioral attributes. In these studies, homology assessments have been vetted, tested, altered, and reworked for these suites of characters. By contrast, molecular data lack a lineage of such vetting. Indeed, the era of molecular sequence data is younger than many who are gathering the data. Scientists have barely scratched the surface regarding which loci are most useful for phylogenetic analysis or how to treat them. As a result, behavioral characters comprise an especially firm foundation or phylogenetic analysis.

Rates of Evolution

The use of multiple kinds of evidence increases the stability and support of the phylogenetic result for two reasons. First, more data are better in general. Second, different kinds of characters evolve at different rates, and so inform different parts of the tree. For example, certain genetic loci may evolve much more rapidly than behaviors. Consider sequence variation in a gene like *cytb*, which differs among populations, versus the expression of the trait 'swims away from moving, overhead shadows.' Slower evolving characters support 'deeper' (i.e., older) nodes on the tree, whereas faster characters support more recent divergences. Clearly then, including characters like behavior can inform divergences for which other characters are not useful.

Levels of Analysis

Morphological and behavioral characters that a phylogeneticist chooses to code and include in a matrix result after examination of many species, genera, and higher-level taxa. Such investigation, which underlies phenotypic character descriptions, means that the discovery of synapomorphy can be quite reliable. That is, deciding which characters are fixed for the taxa under analysis is straightforward for morphology and behavior, in general. By contrast, molecular studies use single individuals to represent large assemblages of taxa. This practice has at least two unfortunate consequences avoided by phenotypic coding. First, single strings of nucleotides may contain both fixed characters and nucleotide sites that vary within populations, thereby violating an important assumption. Second, the exclusive use of sequence data tends to produce trees that are very skeletal, often with long branches. Inclusion of nonmolecular characters can resolve these difficulties.

Homology Assessment

Homology assessment in behavior is conducted by direct observation of multiple (often hundreds) of individuals. Homology assessment within loci (i.e., nucleotide positional

at this site (and at subsequent sites crossed later while flying) with memorized corresponding signals previously perceived at home and to deduce from this comparison an estimated bearing that most likely might guide them back to their loft. It has been shown that this bearing is determined as an angle to a directional reference that is available at home as well as everywhere abroad. To the extent to which utilized signals have been identified so far, they shall now be inspected.

Environmental Signals Involved in the Process of Home-Finding

The sun

In the early 1950s, Gustav Kramer detected the sun as a signal from which birds can derive compass directions. As the sun's position changes in the course of a day, its usability as a spatial direction marker requires a clock and some knowledge about the temporal progress of change. Like almost all animals, pigeons have an internal (circadian) clock that can easily be shifted by exposing the birds for several days to an artificial light–dark regime being out of phase of the natural day–night cycle. Depending on geographic latitude, season and time of day, and thus of the sun's current position along its orbit, a 6-h forward-shift causes, at median latitudes in summer, an error of the sun compass of about 100–130° counterclockwise. If correspondingly clock-shifted pigeons with little homing experience are released at unfamiliar sites, they deviate by approximately this amount from control birds coming from a light regime in phase with the natural one (**Figure 2**).

Figure 2 and many corresponding results make clear that pigeons during homing use the sun's azimuth (i.e., its vertical projection to the horizon) as a compass cue. If the bearings of the clock-shifted birds are rotated clockwise by the angular difference between expected and observed sun azimuth ($S_{T+6} - S_T$), they are statistically no longer distinguishable from the bearings of the controls. The outcomes of this kind of experiment corroborated Kramer's (metaphoric) map-and-compass concept. The clock-shifted pigeons obviously knew their position relative to the home site (e.g., north or south) on some sort of map, but their compass was rotated more than 90° counterclockwise and thus misled them toward a false direction. One part of the homing mechanism, its compass component, could from the late 1950s onward be seen as identified, but the nature of the other part, the map component, remained enigmatic.

The geomagnetic field

Everybody knows that a magnetic compass can be used everywhere and a widespread feeling regards the magnetic field of the earth as a suitable tool to cope with any problem of navigation around the globe. It is not surprising, therefore, that magnetism is the oldest and

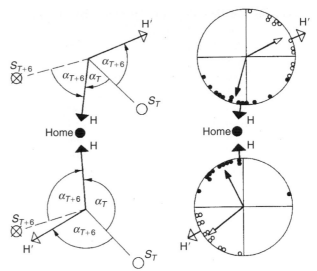

Figure 2 In two groups of pigeons, the circadian clocks were shifted forward by keeping the birds in an artificial light-dark regime advanced against the natural day by 6 h. The clock-shifted birds (○) were released, in alternation with unshifted controls (●), at two sites about 30 km north and south of home. The diagrams at the left show theoretically expected flight directions on the assumption of perfect position determination and use of a sun-azimuth compass. If the birds, at time T (here 10.00 a.m.), select an angle α_T relative to the sun S_T, they fly in the direction toward home H. At time T, the shifted clock of the experimental birds shows $T + 6$ h. At that time (here 4.00 p.m.), the sun would be at position S_{T+6}; H would be reached by selecting an angle α_{T+6}. By keeping this angle relative to the actually visible sun S_T, the birds would achieve the course H'. Under the given conditions (summer at latitude ~50°N), H' is ~120° left from H (like S_T from $S_{T+6} = \Lambda_{\text{azimuth}}$). Right-hand diagrams show actually observed vanishing bearings with their mean vectors (maximum possible length = all birds in one direction = radius). The experimental pigeons were equally well oriented toward H' as the controls toward H. Reproduced from Wallraff HG (1988) Navigation mit Duftkarte und Sonnenkompass: Das Heimfindevermögen der Brieftauben. *Naturwissenschaften* 75: 380–392.

still vivid candidate as a possible physical basis of pigeon homing. In fact, as many other birds, pigeons can perceive the magnetic field and they actually have a magnetic compass to which they resort if the sun is obscured by clouds (**Figure 3**).

Little experienced pigeons clock-shifted for the first time and released in reliably unfamiliar areas follow their sun-azimuth compass completely (cf. **Figure 2**) without showing any irritation from the discrepancy between signals from clock and sun on the one hand and from geomagnetism on the other hand. Deflections caused by clock-shift tend to be smaller, however, in older and more experienced pigeons, particularly if they are more or less familiar with the area and/or had been repeatedly released under clock-shift before. In such cases, attached magnets making magnetic signals unusable result in larger deflections and thus show that magnetic compass information is not definitely ignored.

Figure 3 shows not only an effect of attached magnets, but also a noneffect. Under clear skies, the magnets did not affect homeward orientation. Also, bisecting the trigeminal nerve branch connecting to a recently detected magnetoreceptor based on specifically arranged nanocrystals of magnetite in the pigeons' upper beak, being able to detect changes in magnetic intensity and suspected to act as a map receptor, does not impair their homing capabilities (**Figure 4**, (b) vs. (a)). Like the sun, the magnetic field obviously provides only directional, no positional, information; it is not a basis of the pigeons' map. The geomagnetic field is more suited as a directional reference than as an indicator of position. In principle, magnetic total intensity as well as inclination can indicate geographic latitude, but there is no obvious magnetic indicator of longitude. Even the latitudinal gradients are very gentle with slopes of less than 1% of total intensity and less than 1° difference in inclination per 100 km. Landmark-independent pigeon homing functions down to 30 km distance and less, and the birds would have to compare currently measured values with memorized home values.

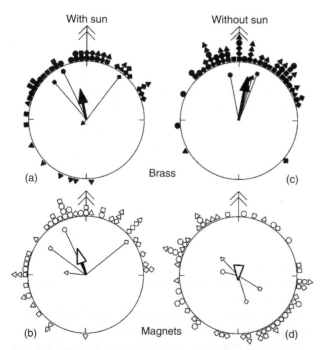

Figure 3 Vanishing bearings of pigeons under sun (a,b) and under overcast (c,d), respectively, with magnets (b,d) or brass bars (a,c), respectively, attached to their heads and wings. (a,b) Simultaneous releases of inexperienced experimental and control birds at four symmetrical sites 20–27 km from their loft in central Italy. (c,d) The same birds, after having completed flights from all four sites used in (a) and (b), now released at four unfamiliar sites at 40–43 km distance from home. Modified from Ioalé P (1984) Magnets and pigeon orientation. *Monitore Zoologico Italiano (NS)* 18: 347–358.

It is unclear to what purpose the pigeons might have the magnetoreceptor in their beak, if not in a map context. Stimuli for compass orientation are, according to current opinion, not transmitted by this receptor either, but by light-dependent processes in the birds' retina.

Neither the map nor the compass component of navigation appears to be involved in responses of pigeons to artificial magnetic fields that have been observed only during the first minutes after release. Lack of initial homeward orientation could be induced by exposure to irregularly oscillating magnetic fields, to a near-zero field, or to a very strong field over some time before release. In all these cases, the disturbance was transitory and ceased within several minutes. As the experimentally treated pigeons returned to their loft as fast as the untreated control birds, a mechanism being essential for homefinding was obviously not affected. Magnetic oscillations interfere with the birds' antistress system and thus with their instantaneous motivation to steer homeward when released from a cage. Application of a tranquillizer could compensate for the disturbance.

Atmospheric chemosignals

For over more than two decades of intense research on pigeon homing, it remained enigmatic from what kind of environmental signals displaced birds might deduce which compass direction they should fly in order to fly back to their home from an unfamiliar area. Yet it became clear what they do not use for this purpose: path integration, the sun, the geomagnetic field, visual cues, the Coriolis force, and infrasounds. The breakthrough came in 1971 when Floriano Papi and his colleagues released ten pigeons with sectioned olfactory nerves. After subsequent years of ongoing skepticism, but also of accumulating experimental evidence, there can hardly remain any doubt that pigeons are able to deduce the particular compass direction they should fly from atmospheric trace gases perceived by the sense of smell. The most important experimental findings shall briefly be specified.

- Pigeons with sectioned olfactory nerves, first-time displaced from home and released far away, failed to orient their routes homeward while often covering considerably long distances (**Figure 5**). Other such pigeons, pretrained over shorter distances and released together with those whose magnetoreceptor nerves were cut, departed completely disoriented; a few returned slowly over 55–60 km, none over 80–105 km (**Figure 4**, (c) vs. (a) and (b)).

- Homing behavior of pigeons with a unilaterally sectioned olfactory nerve combined with unilateral plugging of a nostril depended on the kind of combination. If the two treatments were applied ipsilaterally, the birds flew homeward oriented as usual, but they failed to do so if contralaterally impaired. Thus, lacking homeward

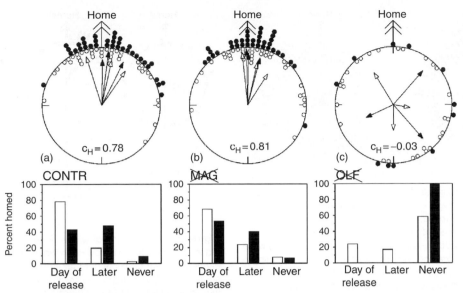

Figure 4 Vanishing bearings (top) and homing performances (bottom) of (a) untreated control pigeons and of pigeons with (b) sectioned branches of trigeminal nerves connecting to magnetoreceptors in the upper beak or (c) sectioned olfactory nerves. Open symbols and bars refer to three release sites at 55–60 km from home, filled symbols and bars to subsequent three releases of the returned birds at 79–107 km. Peripheral dots show individual bearings, arrows resulting mean vectors per release site; c_H = mean homeward (=upward) component of every six release-site vectors (possible range +1 to −1 = radius of circle). Data from Gagliardo A, Ioalè P, Savini M, and Wild M (2008) Navigational abilities of homing pigeons deprived of olfactory or trigeminally mediated magnetic information when young. *Journal of Experimental Biology* 211: 2046–2051.

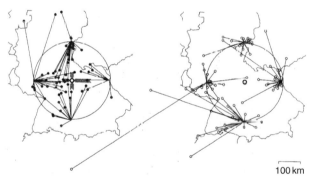

Figure 5 Recovery sites of inexperienced (=previously not yet displaced) homing pigeons, connected by lines with four release sites at 180 km distance from their home loft near Würzburg in northern Bavaria (=center). Normal control birds (left) and anosmic birds with bilaterally sectioned olfactory nerves (right). Arrowheads near the center symbolize returned pigeons (not existing in anosmic birds of which similar numbers were released). Modified from Wallraff HG (1980) Olfaction and homing in pigeons: Nerve-section experiments, critique, hypotheses. *Journal of Comparative Physiology* 139: 209–224.

orientation was not caused by a nonspecific influence of one or the other invasive manipulation per se.

- With an ablated piriform cortex that processes signals transduced from the olfactory bulb, pigeons failed to orient homeward in a similar way as pigeons with sectioned olfactory nerves.

- Pigeons were kept during transport and during waiting at the release site in an airtight container ventilated with air sucked through a charcoal filter. Upon release with locally anesthetized nasal cavities, they failed to orient their initial courses homeward. Alternately released control pigeons that could smell natural air before release, but whose nasal cavities were equally anesthetized, departed predominantly homeward oriented (**Figure 6**).

- Filters retaining aerosol particles, but being pervious to molecules in the gas phase, were ineffective.

- Pigeons were allowed to smell natural air for a period of 3 h at a site distant from home and were subsequently kept in a container ventilated with charcoal-filtered air. Afterward released under nasal anesthesia at another site in the opposite direction, they departed predominantly away from home. Control birds that had been allowed to smell natural air synchronously with the others, but at the site of release itself, also flying under nasal anesthesia, departed predominantly homeward. Homing speeds of the two groups were accordingly different.

- Gaining sufficient site-specific information from ambient air appears to require some time. A few minutes of smelling natural air during initial flight were too short, whereas an hour or more before departure resulted in better orientation even with impeded smelling during flight.

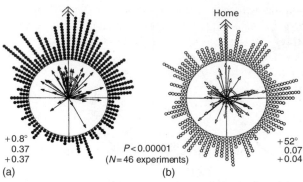

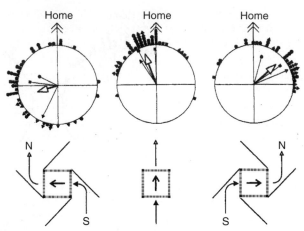

Figure 6 Air filtration. Until shortly before release, the pigeons were sitting in an airtight container ventilated with (a) natural environmental air or (b) air sucked through a charcoal filter. Immediately after removal from the container, the olfactory epithelia of each pigeon (experimentals as well as controls) were anesthetized by a spray of xylocain. Diagrams show vanishing bearings of 46 experiments, pooled with respect to angular deviations from homeward. The releases were conducted pairwise from opposing directions and similar distances with birds of equal or similar experience. The arrows symbolize the mean vectors per release (maximum length = 1 = radius). The overall mean (in numbers: direction, length, and homeward component of vector) from 46 single-release vectors corresponds to the center of the small ellipses which indicate the 95% and 99% confidence intervals. P refers to differences between homeward components per release. Data from Wallraff HG and Foà A (1981) Pigeon navigation: Charcoal filter removes relevant information from environmental air. *Behavioral Ecology and Sociobiology* 9: 67–77, and additional equivalent experiments.

Figure 7 Manipulation of winds. At the home site, some of the pigeons were living in an aviary in which the wind was deflected either to the left or to the right (examples showing wind from south). The diagrams (home upward) show corresponding vanishing bearings at three different sites as distinguished by symbols. Arrows indicate mean vectors calculated from three single-release vectors. Data from Baldaccini NE, Benvenuti S, Fiaschi V, and Papi F (1975) Pigeon navigation: Effects of wind deflection at home cage on homing behavior. *Journal of Comparative Physiology* 99: 177–186.

- Near-ground-level air in an open landscape appears to contain better positional information than air in a forest or amongst other vegetation allowing little air exchange.

In conclusion, homeward orientation and successful homing from unfamiliar distant sites require unobstructed and sustained olfactory contact to the local open-field air.

The wind

Some years before olfaction came into focus of research, it was found that pigeons coming from an aviary in a pit below ground level or from an aviary surrounded by a wall made of wood or glass failed to orient homeward, whereas visual shieldings allowing airflow did not impair homeward orientation. Later, when seen in the context of olfaction, related experiments became more specific. Pigeons were confined in aviaries in which winds were deflected or reversed; their initial flight courses upon release at distant sites were correspondingly deflected (**Figure 7**) or reversed.

These results suggested that pigeons at their home site associate varying olfactory inputs with concurrently varying wind directions and make use of these correlations at distant sites for appropriate spatial interpretation of olfactory input gained there. Specific experiments could

even show that pigeons actually correlate deflected winds with concomitant olfactory stimuli and artificial winds with artificial odors.

Sun, Magnetism, Odors, and Wind: How Are They Interrelated?

Four environmental parameters have been identified to be used by pigeons for home-finding. Functionally, the first two can be united to one, to a compass. Why pigeons and other birds have two compasses, is not clear. A fully operational magnetic compass, once existing at all, would suffice. Its handling appears much easier than that of a sun compass, as it is independent of cloudiness, time of day, season and largely of geographic latitude. Nevertheless, if it is visible, pigeons clearly prefer the sun to determine compass directions (cf. **Figures 2** and **3**).

If winds had been shielded during the long-term habituation at the home site or odors are excluded during and after displacement, pigeons are obviously unable to select the correct compass bearing leading toward home. These results forced us to think about the feasibility that trace gases dispersed in the ambient air, together with winds experienced at home and a compass, might constitute a navigation mechanism operating over hundreds of kilometers. Thus, in spite of intuitive disbelief, it was necessary to think about potential large-scale regularities in the chemical atmosphere that allow directionally distinctive extrapolation from the home site to never experienced areas far away.

Extrapolations could be possible if (1) proportions among concentrations of a number of atmospheric trace gases would regularly change over longer distances, differently in different directions, and if (2) the birds could determine, at home, in what directions what proportional changes can be expected (in terms of classical rules of olfaction, changes of ratios would result in changes of perceived odor quality). The first premise is not as unrealistic as it might seem at first glance. Gas-chromatographic analyses of air samples have shown that, within an area covering 400 km in diameter, the ratios of a number of volatile hydrocarbons actually did have spatial gradients in compound-specific directions (examples in **Figure 8**). The gradients are noisy, but statistically highly significant. The second premise implies that these directions should be correlated with wind directions at home under which relevant ratio spectra vary correspondingly, so that memorized odor/wind relations can provide a template for predicted odor/space relations. Also this premise has gained some empirical support, albeit in the only so far available data set the corresponding directions (ratios/wind vs. ratios/space) are, on average, systematically shifted against each other by about 50°. It is not yet clear whether this discrepancy is generally typical or merely an accidental outcome within a data sample of limited size. Even within this sample, the two bearings are similar in a number of compounds, though not in the majority.

Using the atmospheric data shown in **Figure 8** and proposing that model pigeons roughly know the directions of the ratio gradients, it was possible to simulate

homing within a radius of 200 km. Model calculations revealed levels of homeward orientation that were at least as good as those obtained with real pigeons. Further calculations including more of the captured volatile compounds support the hypothesis that, by selecting and optimized weighting of the most appropriate trace gases, evolution could have created a navigation system based on chemical ratio spectra varying in space as well as with wind direction with sufficient regularity.

It must be emphasized that most of the trace gases measured in the ambient air were anthropogenic hydrocarbons and thus could not have been the basis of phylogenetic development. They can, however, serve as substitutes of the unknown substances that birds may actually use for navigation. All volatile trace compounds in the atmosphere follow the same basic rules of emission, diffusion, turbulent mixing, advective transport with winds, chemical conversion, etc. so that ratio gradients observed in some example sets of compounds can be expected to exist in other combinations in a similar manner.

It should further be emphasized that the hypothetical mechanism of olfactory navigation described earlier reflects merely an attempt to show that home-finding on the basis of atmospheric chemosignals is at all feasible. It does not imply that the mechanism applied by the birds necessarily follows the algorithm of the model. Moreover, even if it should turn out that the model is completely wrong (no use of ratio gradients, etc.), the issue of olfactory navigation would persist as an empirical outcome and thus would demand an alternative explanation.

Do Pigeons Use an Olfactory Map?

At first sight, the pigeons behave as if they had an olfactory map which has been directionally adjusted, at home, to the compass scale. False adjustment can be created experimentally by rotating either the map as related to the compass (deflector aviary) or the sun compass as related to the map (clock shift). With respect to cognitive skills, however, it may be misleading to say that pigeons have an olfactory map. While living permanently at one site, perhaps in an aviary, the birds cannot develop an analog of what we call a map, that is, a two-dimensional representation of spatial structures in an area ranging over hundreds of kilometers. If they correlate, at home, varying olfactory compositions with varying wind directions and later compare, at a strange site, the current composition with those remembered, they might fly that compass direction toward which the wind blew while the memorized olfactory spectrum was, in some respect, most similar to the currently perceived spectrum. The pigeons cannot and need not know anything about spatial configurations, but natural atmospheric conditions must include structures including far-reaching regularities which we humans could draw as a map. Only by using a loose

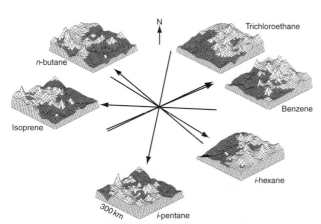

Figure 8 Relief maps showing standardized ratios of six atmospheric hydrocarbons, each as a portion of the sum of all six, in a radius of 150–200 km around Würzburg in Germany (=center of each map). Areas in which the relative abundance of the respective compound is above its overall mean are light gray, those with values below average are dark. The maps (each covering the same area) are arranged in the directions of the mean upward slopes of the respective compound-specific gradients as indicated by the central arrows. Modified from Wallraff HG and Andreae MO (2000) Spatial gradients in ratios of atmospheric trace gases: A study stimulated by experiments on bird navigation. *Tellus* 52B: 1138–1157.

terminology and considering merely the functional appearance, we may speak of a hypothetical olfactory gradient map or grid map, aligned in a compass scale.

Homing Within a Familiar Area

General Familiarity with a Larger Area

Distances flown by pigeons during voluntary flights at home are usually quite small, hardly more than 1 or 2 km around the loft. A larger area can be made more or less familiar, however, by releasing pigeons from various sites. The degree and range of familiarity depends on the density of a grid of release sites, on the number of releases within a given area, and on their distances from home. If pigeons are released at a site at which they had been released before or, at least, at a site with a number of previously experienced release sites around, olfactory deprivation has no or little effect (**Figure 9**). Thus, in a familiar area, olfactory inputs are largely redundant for home-finding. Olfaction-independent homing within a familiar area is obviously based on remembered visual-landscape cues which are processed and memorized primarily in the hippocampus. Visual-landmark orientation is an ability that pigeons share with many other animals of various taxa, but not in all other animals is its closer

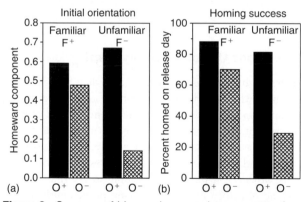

Figure 9 Summary of (a) mean homeward components of vanishing bearings (theoretically variable between +1 and −1) and (b) homing performances of four types of pigeons released at two sites about 55 km from home in opposing directions (two releases per site, $n = 44$–48 birds per type). Half of the pigeons had participated in training releases around – and at least 10 km from – either the one or the other test site. F^+ = familiar, F^- = unfamiliar with the surrounding area; O^+ = always unimpaired olfactory access to ambient air; O^- = confined to charcoal-filtered air previous to release and largely anosmic by nasal anesthesia from few minutes before release onward. Note that the F^-O^- pigeons recovered from anesthesia after a while; those that returned during daytime arrived significantly later than the other groups. Data from Wallraff HG and Neumann M (1989) Contribution of olfactory navigation and nonolfactory pilotage to pigeon homing. *Behavioral Ecology and Sociobiology* 25: 293–302.

investigation complicated by a second mechanism that can be used in parallel. With unimpeded smelling, contribution of vision and olfaction to the birds' behavior cannot be separated.

It is unknown in what way the pigeons deduce what kind of information from the landscape over which they fly if they are not specifically familiarized with a particular release site and subsequently crossed areas. The birds whose performances are shown in **Figure 9** had not been released previously at a site closer than 10 km from the current release site, but from a number of somewhat more distant sites around. Some pigeons may have flown over a more adjacent area, but generally they could not replicate from the beginning a former route home by following a sequence of individually learned landmarks. It seems more likely that they brought the aerial panorama underneath and around into coincidence with a remembered global image they had in mind. With their wide bimonocular visual field, pigeons can overlook a large area, may conserve its image, and later on try to match it with the current panoramic view. Owing to problems of parallax from different viewpoints, such matching may be imperfect so that the pigeons guided by landscape features alone were not better in initial orientation and homing speed than those guided solely by olfactory signals (compare F^+O^- with F^-O^+ in **Figure 9**).

Unlike atmospheric chemosignals, the visual landscape provides more than rough information on the compass direction toward home. It constitutes a spatial pattern ranging over an extended area around and implies in itself a directional guideline for piloting toward home without use of a compass. Including the everywhere-available sun in the topographical pattern might nevertheless be helpful. What, however, happens if the sun does not fit with the landscape in the expected direction? When released with their clock shifted by 6 h, anosmic pigeons behave much more variable and inconsistent than pigeons able to smell in an unfamiliar area (cf. **Figure 2**). They tend to depart somewhere between true home and home as expected according to the sun's position. Thus, the birds take notice of the sun also while visually piloting; noncoincidence between the sun's position and the landscape leads to conflicting responses. Probably depending on varying concomitant circumstances (structure of the landscape, degree of familiarity, effect of directional training, etc.), one or the other sort of signal may more or less dominate the directional choices.

Specifically Exercised Homing Routes

If pigeons are not only roughly familiar with a larger area around the release site, but have been consecutively released at one particular site, they often end up with an individually stereotyped homing route (**Figure 10**). This route seems to follow a learned chain of landmarks which

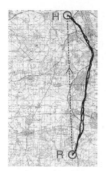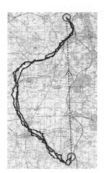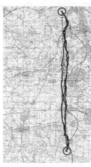

Figure 10 GPS-tracked homing routes of three individual pigeons during the final 3 of 20 training flights from the same release site (R) 25 km south of the home site (H). Adapted from Biro D, Meade J, and Guilford T (2006) Route recapitulation and route loyalty in homing pigeons: Pilotage from 25 km? *Journal of Navigation* 59: 43–53.

are usually not exactly on the shortest way home. In some cases, it appeared obvious that the birds were attracted by outstanding landmarks. So, pigeons flew clearly along a highway that deviates, though not very much, from the direction toward home; then, where continued road-following would have led too far away, at a conspicuous junction, they corrected the course homeward. If pigeons trained to home from a given site were released at a novel site 1–3 km sideways of the established route, they did not fly a novel route home, but were attracted back to the old accustomed path. Such behavior in detail is somewhat variable (e.g., route recapitulation or road-following not everywhere equally pronounced) and appears to be influenced by regional topographical peculiarities. As related experiments with GPS-tracking have so far not been conducted with anosmic pigeons, it is not yet clear whether they reflect exclusively visual orientation.

Do Pigeons Use a Visual-Landscape Map?

In the case of visually guided homing over a familiar landscape (with a spatial configuration the birds had memorized during earlier homing flights), the term map fits better than in the case of olfaction-based homing. Here, it may be appropriate to say that pigeons use a familiar-area map, a visual-landscape map, a topographical map, a mosaic map or (probably) a cognitive map, as here the birds are thought to operate with a neural spatial representation. At least when released at a site where they had not yet started previously (cf. **Figure 9**), the fairly well-oriented anosmic pigeons must have referred to some kind of vision-based two-dimensional map, achieved earlier while flying over adjacent parts of the area, and could not sequentially follow a chain of remembered landmarks. With sufficient knowledge of the landscape, additional use of a compass would not be essential. However, it may be helpful also for a pilot. Actually, the pigeons orient

their courses by paying attention to both the landscape and the sun.

In the case of route recapitulation during repeated homing from the same site, the term 'map' appears less clearly appropriate. Depending on the birds' more general or more specific experience, no explicit distinction can be made between map use and point-by-point path replication. Even while flying stereotyped routes, the pigeons obviously pay attention to a broader corridor of landmarks within which they fly back to their route when released a few kilometers away from it.

Conclusions and Perspectives

The homing behavior of pigeons is a simple output (flight in a particular direction) of a variety of inputs which are usually not completely under the experimenter's control. In part, they depend on the individual bird's life history (preceding homing flights). Also, they are not all directly involved in the process of navigation and they are subject to stochastic noise. There are, however, five classes of environmental signals that have been identified as sources of information used for home-finding. Pigeons deduce directional (compass) information from (1) the sun's position as well as from (2) the geomagnetic field. They deduce positional information from (3) the visual landscape as far as they could explore its configurations during earlier flights over a given area. Without such explorations, that is, also in unfamiliar areas, they deduce positional information from (4) trace gases dispersed in the atmosphere perceived by olfaction, provided that they had been living at home exposed to the free airspace with (5) undisturbed winds.

Within each of these classes of cues, more or less decisive details have not yet been clarified. Least problematic is the sun compass, albeit fine tuning to the sun's path in dependence on latitude, on time of day and season could still deserve closer attention. Magnetoreception is currently under vivid investigation. Research on visual piloting has recently gained considerable upswing by modern GPS technology allowing precise recording of homing routes. It is, nevertheless, difficult to analyze the methods by which the pigeons exploit their wide aerial view over the landscape, because its configurations cannot be manipulated experimentally like single landmarks in the laboratory.

By far the most challenging problem is olfactory navigation. Empirical research leaves hardly any doubt that it works, but very little is known about the environmental conditions that make its functioning possible. Currently, more revealing than further investigations of pigeon homing would be closer investigations of the atmosphere in order to clarify which spatial configurations it places at the birds' disposal. So far, it is merely a working hypothesis that they exploit gradients of ratios among a number of particular trace gases. Neither the actually involved chemical compounds are identified nor are their origins and spatiotemporal distributions in the airspace known. At least, however, an encouraging pilot study has revealed that ratio gradients do exist in the atmosphere that potentially could be employed for home-finding within a radius of 200 km around one particular geographic position. Further progress can only be achieved by an interdisciplinary approach on a broader basis that links avian navigation, olfactory signal processing and, most urgently, atmospheric chemistry connected with meteorology.

See also: Insect Navigation; Magnetic Compasses in Insects; Magnetic Orientation in Migratory Songbirds; Maps and Compasses.

Further Reading

Bingman VP, Riters VR, Strasser R, and Gagliardo A (1998) Neuroethology of avian navigation. In: Balda RP, Pepperberg IM, and Kamil AC (eds.) *Animal Cognition in Nature*, pp. 201–226. London: Academic Press.

Holland RA (2003) The role of visual landmarks in the avian familiar area map. *Journal of Experimental Biology* 206: 1773–1778.

Kramer G (1961) Long-distance orientation. In: Marshall AJ (ed.) *Comparative Physiology of Birds,* vol. 2, pp. 341–371.

Papi F (1991) Olfactory navigation. In: Berthold P (ed.) *Orientation in Birds*, pp. 52–85. Basel: Birkhäuser.

Papi F (ed.) (1992) *Animal Homing.* London: Chapman and Hall.

Wallraff HG (2004) Avian olfactory navigation: Its empirical foundation and conceptual state. *Animal Behaviour* 67: 189–204.

Wallraff HG (2005) *Avian Navigation: Pigeon Homing as a Paradigm.* Berlin: Springer.

Wiltschko R and Wiltschko W (2003) Avian navigation: From historical to modern concepts. *Animal Behaviour* 65: 257–272.

Pigeons

C. A. Stern and J. L. Dickinson, Cornell University, Ithaca, NY, USA

Introduction

Origins and Range

Rock pigeons (Columbidae: *Columba livia*) are widespread, common, and apparent; their combined status as domestic, feral, and wild birds gives them a distinct familiarity to people all over the world. Their behavior is easy to observe, so much so that citizen scientists have been recruited to study pigeon courtship behavior as a function of plumage through The Cornell Lab of Ornithology's Project PigeonWatch. Pigeons have served as a focus for a wide range of evolutionary and behavioral studies, including prominent recent studies of social and breeding behavior, mating systems, sex ratio, and navigation (pigeon navigation is dealt with elsewhere in this encyclopedia), that go back to the work of Charles Darwin. Recently, pigeons were even fitted with small sensor backpacks to monitor air pollution at altitudes at which sampling by ordinary means is difficult. Ubiquitous 'city rats,' pigeons continue to provide new insights into behavioral ecology and animal behavior.

Feral and domestic pigeons are found worldwide but, although common in the Americas, are not native there. In this article, we use the term 'feral pigeon' to refer to free-living birds that are descended from escaped or released domestic pigeons (in some cases, these domestics may have interbred with synanthropic rock pigeons). The term 'domestic pigeon' refers to those individuals, including all the various artificially selected breeds, that live in captivity, and the term 'rock pigeon' designates wild birds living in their indigenous range that are descended from wild, free-living ancestors.

At their maximum natural range, pigeons were distributed throughout Europe, from a northern limit in Great Britain and southern Scandinavia, southward into North Africa, and east into India and Mongolia. Their historical range is difficult to delineate precisely, for pigeons have a long history of both domestication and escape from domestication. Some wild rock pigeon populations have become casualties of the worldwide distribution of feral pigeons, which has caused the extinction of native populations through inter-breeding.

Domestication of rock pigeons began as long as 5000 years ago; existing domestic pigeons are the product of up to 5000 generations of artificial selection by humans. However, the association of rock pigeons with humans may have begun even earlier, when wild rock pigeons colonized human settlements in the eastern Mediterranean region.

The modern feral pigeon, which has been studied extensively by Richard Johnston and his students at the University of Kansas, derives part of its mongrel pedigree from these synanthropic rock pigeons, which later bred with escaped domestics to produce individuals of mixed ancestry.

Artificial selection by humans has resulted in a wide range of domestic pigeon breeds, with distinctive traits ranging from holding the tail erect (the fantail) to navigational abilities (the homing pigeon). Charles Darwin used domestic pigeons as an example of the power of selection to create diverse descendent types from a common ancestor, asserting in his 1859 masterpiece, *On the Origin of Species*, 'Great as are the differences between the breeds of the pigeon, I am fully convinced that ... all are descended from the rock-pigeon (*Columba livia* [sic]).' The release and escape of domestic pigeons into the wild provided a natural experiment for Darwin: Which of the artificially selected traits persist in feral populations under natural selection? By crossing different breeds of his own domestic pigeons, Darwin produced an individual identical in appearance to a wild rock pigeon, which he cited as an example of the 'principle of reversion to ancestral characters.' When domestic animals possess all of the genetic variation that was present before domestication, and assuming that the selective pressures remain unchanged, we would expect a population to revert to wild type appearance within 12–20 generations. Feral pigeons have not reverted and differ from natural populations of rock pigeons in several respects, including variability of plumage color and patterning, as well as aspects of morphology, such as tarsus length.

Why have feral pigeons not evolved to uniformly resemble the ancestral rock pigeon, losing the heterogeneity remaining from their artificially selected past?

Plumage variation may be maintained in feral populations because of positive or negative assortative mating. Alternatively, possessing nonancestral plumage types may confer benefits in urban environments, and these benefits may allow domesticated plumage patterns to persist. There is little evidence of natural selection favoring reversion to the wild type character state, but a study in Spain suggests selection against the longer tarsus of domestic pigeons in some feral populations, resulting in evolution towards a morphology more similar to that of the wild rock pigeon, but only for this one particular trait.

Plumage variation may be maintained because of release from natural selection in urban environments, which are likely characterized by predator scarcity and super-abundant food. Alberto Palleroni and colleagues found that, in a Californian population, feral pigeons with 'wild' variant coloration appear to have an advantage against peregrine falcons (*Falco perigrinus*). Peregrine falcon conservation efforts have resulted in successful reintroductions to North American cities, and the next 20 years may offer a replicated series of natural experiments to test the hypothesis that pigeon color variation is maintained because of release from predation. With domestic pigeons continuing to escape and interbreed with feral populations, such evolution may be slow, but the process represents an interesting opportunity to evaluate the different selective pressures exerted upon feral pigeons compared to their wild rock pigeon ancestors.

Mating Behavior

Courtship and Pair-Bonding

Feral pigeon courtship behavior has been studied extensively by Richard Johnston and colleagues. In the wild, mate choice takes place within courtship displays. Before courtship, males establish territories and defend them against other males. Each territory includes an existing nest or nest site; preferred nest sites of feral pigeons resemble the rocky cliffs with caves and ledges used by rock pigeons. An unpaired male begins to sing from the nest site, attracting females to his territory. When a female appears, the male approaches her and after determining that she is not another male by testing her aggressiveness, he performs a 'bow-coo' display in which he struts around the female while bowing, fanning his tail, and making sounds like 'coo-cuk-cuk-cuk-coo' (**Figure 1**). A receptive female allows the male to approach more closely and touch her head and neck feathers with his bill (heteropreening). This sequence of behaviors may be repeated multiple times over the course of hours or a few days. For a diagrammatic explanation of pigeon courtship, see **Figure 1** or the Project PigeonWatch web site at The Cornell Lab of Ornithology (see Relevant Websites).

Pair-bonding is initiated by the female begging for food from the male, which he eventually provides in the form of crop milk (a nutritious fluid secreted in the crop) or a seed. Once the pair is well-bonded, approximately a week after they begin courting, they copulate after most courtship feedings. The pair continues to copulate regularly throughout the nest-building period, and until the beginning of egg laying.

During the breeding season, the pair bond is maintained by periodic repetition of the courtship behaviors, including heteropreening and singing, as well as by the joint duties of rearing young. Other coordinated behaviors, such as territory defense and joint roosting, occur throughout the year. In feral pigeons, pairs usually endure until one pair member dies, although divorce and repairing does occur.

Mate Guarding and Extra-Pair Copulations

In feral pigeons, just as in many other species of birds, pair-bonded females engage in extra-pair copulations with males other than their social mate. Males guard their fertile mates intensely by following the female closely on the ground and in flight, pecking the back of her head, and harrying her until she moves further away from other pigeons. Males usually only guard when other pigeons are nearby, and such guarding probably prevents extra-pair copulations. In addition to preventing extra-pair copulations, which is certainly in the male's interest, guarding males may also protect females from costly harassment by extra-pair males. Claire Lovell-Mansbridge and Tim Birkhead suggested that females trade pair copulations for protection from harassment by extra-pair males. When females solicit more copulations from their partners, their partners invest more in guarding. Females do not approach extra-pair males and mate guarding does appear to reduce harassment.

Rates of extra-pair copulations appear to be moderate to very low in both feral and domestic pigeons, depending upon the population. The type of nest provided to domestic pigeons may influence the extra-pair copulation rate, with estimates of 2% and 17% of offspring resulting from extra-pair copulations in closed shelf and open-front nest types respectively. We do not yet know what makes males successful at extra-pair fertilization, because although polymorphic DNA microsatellites for pigeons have been identified by Traxler and colleagues, to our knowledge no researchers have yet used these to assign paternity to extra-pair males.

Sexual Competition Strategies

When the operational sex ratio (ratio of reproductively available males to available females) is male-biased, unpaired feral pigeon males sometimes induce breeding failure by damaging eggs. This evens the playing field and potentially causes divorce of the breeding pair. In feral pigeons, however, there is no convincing evidence that divorce rates increase following breeding failure. Among domestic pigeons, the duration of the pair bond is a more important predictor of pair bond stability than is breeding success.

Feral pigeon males may also attack copulating pairs, pushing the male off the female before pecking the female on her head. Attacking males then either perform the bowing display, a display given by males in breeding condition when females are present, or more rarely, mount and copulate with the female. In 1983, Derek Goodwin suggested that reproductively active males attack copulating pairs because selection favors a response to any copulation as if it is an extra-pair copulation with their own mate. If disrupting a copulation is not very costly, males benefit by attacking the copulating pair first and checking the identity of the mounted female later.

Courtship behaviors
Illustrations by julie zickefoose

	Bowing: a male puffs out his neck feathers, lowers his head and turns around in circles
Tail-dragging: a male spreads his tail and drags it while he runs after a female	
	Driving: one pigeon runs closely behind another
Billing: when a female puts her beak (or bill) inside the male's beak during courtship	
	Mating: when mating, a male stands on top of a female
Clapping: after mating, a male pigeon may make a display flight. In this display, he 'claps' his wings twice behind his back	

Figure 1 Pigeon courtship behaviors. Illustrations show bowing, tail-dragging, driving (mate guarding), billing, mating (copulation), and clapping. Reproduced from The Cornell Lab of Ornithology's Project PigeonWatch.

Mate Choice and Plumage Polymorphisms

One outstanding characteristic of feral pigeons is their extraordinary variation in plumage and mating behaviors. In 1981, Nancy Burley, a pioneer in devising methods for testing female mate choice in birds, assessed the mate choice of pigeons, using an experimental design that tested preferences for multiple traits. Although feral pigeon females are more selective in mate choice than are males, both sexes use multiple characteristics in selecting a mate (**Figure 2**). These include age, dominance status, reproductive experience, size, and feather condition. Results from Nancy Burley's experiments, along with several other investigations, indicate that feral pigeons prefer mates that are young, dominant, experienced, large, and have a low feather parasite load.

Feral pigeon mate preferences for plumage color and pattern are much more difficult to summarize neatly, as there is a large range of plumage polymorphisms present in most feral populations. While the plumage of domestic pigeon breeds is standardized and well-described, feral pigeon plumage varies in both color and pattern. Pigeon plumage is categorized using two parameters: color (blue, ash red, brown, or white) and melanism (plumage darkness, which causes patterning). From lightest to darkest, the commonly recognized patterns of melanism are bar, checker, T-pattern, and spread (**Figure 3**). The ancestral rock pigeon has blue bar plumage, and this plumage type remains common among feral pigeons.

Given the high degree of plumage polymorphism in feral pigeons, is plumage an important factor in their

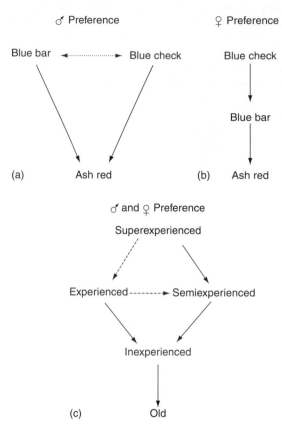

Figure 2 Mate preferences of male and female pigeons in mate choice experiments. (a) Shows the plumage type preferences of male pigeons; (b) shows the plumage type preferences of female pigeons; (c) shows preferences of both males and females with respect to mate age and breeding experience. The dashed lines indicate preferences that were not directly tested in the experiments. Reproduced from Burley N (1981) Mate choice by multiple criteria in a monogamous species. *The American Naturalist* 117: 515–528.

choice of mates? In mate choice experiments, both male and female feral pigeons show nonrandom preferences for plumage color and pattern, and data from several feral populations show nonrandom distributions of plumage color and pattern in mated pairs.

What type of nonrandom mating occurs within feral pigeon populations? Is there absolute preference for a single plumage type, or do individuals prefer mates that are either similar to or different from themselves? The results are mixed: feral pigeons seem to vary locally in their plumage preferences, and there is evidence from different populations for both negative assortative mating and positive assortative mating.

Among wild rock pigeons in Europe, plumage is monomorphic: all individuals have the blue bar plumage. In terms of characteristics such as age and breeding experience, mate preferences within rock pigeon populations probably resemble those found in feral pigeon populations, but plumage pattern may not be variable enough to

play a large role in mate choice. Little is known about the preferences of domestic pigeons, since the humans who breed them often constrain their mate choice to individuals of the same breed or type.

An interesting question recently raised in the literature is whether plumage morphs are indicative of polymorphic mating strategies within feral pigeon populations. Melanic males, whose plumage pattern is darker than bar, are more likely to retain gonadal function during the winter: their gonads fail to regress, making these males physiologically capable of breeding year-round. This lack of regression may be accompanied by decreased deposition of winter fat, which could explain why melanics have higher over-winter mortality rates than blue bars in rural habitats. However, in urban habitats, melanics enjoy lower over-winter mortality rates than blue bars, probably because of continuous food availability in urban areas. Blue bar morphs tend to have higher summertime breeding success than do melanics. The blue bars' strategy may be to concentrate breeding effort in a single season, producing fledglings with high rates of success, while the melanics' strategy may be to exert continuous but lower-level breeding effort year-round. These strategies could have equal overall success or vary in effectiveness depending upon habitat type or other environmental factors. A rigorous comparison of these strategies across urban and rural habitats would be an exciting future research project.

In a Californian feral population, Palleroni and colleagues discovered that individuals with blue bar plumage are significantly less likely to suffer capture by peregrine falcons (*Falco peregrinus*). This indicates that predation pressure confers a survival advantage to birds with the ancestral phenotype. If females prefer other phenotypes over the blue bar phenotype, natural selection may oppose sexual selection in populations under high peregrine falcon predation pressure. Assessing the relative strengths of these potentially conflicting selection pressures across different feral pigeon populations presents another interesting avenue for future investigation.

Breeding Behavior

Biparental Care

Pigeons breed in monogamous, heterosexual pairs, although there are occasional reports of joint laying, intraspecific brood parasitism (egg dumping), and male–male or female–female pairings. The female begins laying 2–3 days after nest completion, and then usually lays a second and final egg 1–3 days after the first. Although the mother incubates the eggs overnight, both parents incubate during the day, alternating to contribute roughly equal amounts of daylight incubation time. The eggs hatch about 18 days after the onset of incubation and nestlings are initially fed regurgitated cropmilk, a nutritious substance secreted

Figure 3 This poster displays some of the more common feral pigeon color morphs, including the ancestral blue-bar plumage type. Reproduced from The Cornell Lab of Ornithology's Project PigeonWatch.

in the crops of both parents. After about a week, as nestlings mature, parents gradually substitute seeds for crop-milk. Nestlings beg loudly for food, emitting calls of increasing amplitude as they grow.

Nestlings begin to walk away from the nest to forage independently when they are about 20 days old. However, they are not capable of flight until at least 25 days of age, reaching independence at 30–45 days. They are dependent upon their parents for food until about 7 weeks old, and afterward they continue to follow their parents to foraging sites, where they learn how to forage by imitating other pigeons' feeding behavior.

Sex Ratio Manipulation

R.A. Fisher correctly pointed out in 1930 that population sex ratios should reflect equal investment in sons and daughters. However, in 1973 Robert Trivers and Dan Willard added the insight that individual females may still benefit from facultatively adjusting the sex ratio of their young if one sex has a better prospect of reproductive success in a given maternal or social environment. For example, if females in poor condition produce low-quality sons and competition for access to females is intense, then poor-condition females may be selected to bias their

offspring production toward daughters, the reproductive success of which varies less as a function of condition. These females have a better chance of producing grand-children through daughters than through sons, which are likely to be out-competed by higher quality males.

Experimentally altering the mother's internal state (e.g., hormones) or external environment (e.g., food availability) provides a test of the ability of mothers to manipulate primary offspring sex ratio (sex ratio at conception), as well as the conditions under which it is beneficial for them to do so. Two recent experiments using domestic pigeons have produced interesting results. Using homing pigeons, Vivian Goerlich and colleagues found that females given testosterone-elevating implants produce a higher proportion of sons compared to sham-control females, given neutral implants, but this bias only affected first-laid eggs. In a different experiment, Thomas Pike found that females that were food-deprived subsequently produced a higher proportion of daughters in both the first- and the second-laid eggs compared with pigeons in an ad libitum food control. These results suggest that female pigeons are indeed capable of manipulating the primary sex ratio of their offspring in response to internal or external environmental changes.

Do female pigeons manipulate offspring primary sex ratio in the wild? Research at Groningen in the Netherlands suggests that pigeons manipulate the primary sex ratio in response to season; however, these results are as yet unpublished, and the importance of sex ratio manipulation in natural feral pigeon or rock pigeon populations remains unaddressed. Little is known about the mechanisms governing sex ratio adjustment in birds generally; pigeon studies could make a valuable empirical contribution to this area of research.

Brood Reduction

Brood reduction is a life history strategy that enables parents to reallocate their resources to fewer offspring during times of food stress. In pigeons, the two eggs in a clutch are laid ~40 h apart and usually hatch ~24 h apart. This hatching asynchrony gives the first-hatched chick a perceptible growth advantage over the second and, as the degree of hatching asynchrony increases, the imbalance in competitive abilities of the two chicks increases, resulting in increased chances of mortality for the second chick. The difference in size between the chicks caused by hatch order is further exacerbated by sexual dimorphism: the first-laid egg is more likely to be male, males are larger than females, and this produces an even greater difference between the first and second chick. Studies in Russia revealed that nestlings from first-laid eggs are more likely to fledge than are nestlings from second-laid eggs, in both summer and winter clutches. What do females gain by sacrificing their second chick's chances of survival?

While parents could avoid hatching asynchrony by beginning incubation only when both eggs have been laid, they could also simply feed the second-hatched chick more to help it catch up. Laying a second egg that hatches with a severe competitive disadvantage could be an adaptive strategy for fitting brood size to local food abundance. If food is scarce, the larger nestling can easily out-compete its younger sibling, leaving the parents capable of raising one nestling successfully rather than suffering total brood loss. If food is abundant, the smaller chick will receive enough food to catch up to its older sibling and the parents will benefit by producing two offspring instead of one. Evidence suggests that second chicks often grow faster than first chicks, supporting the idea that second chicks are able to catch up quickly under conditions of food abundance.

Dispersal and Recruitment

Natal dispersal, where young leave the area where they were born and move elsewhere to breed, is a key life history trait that is poorly understood in rock pigeons. In a Polish population of feral pigeons studied by Tomasz Hetmański and colleagues, young of both sexes were equally likely to disperse from their natal colonies and were particularly likely to disperse when their natal colonies contained a high density of breeding pairs. When the young pigeons dispersed from a high-density natal colony to a lower-density colony, they enjoyed increased breeding success. Thus, a crowded natal colony probably indicates low future probability of breeding near home, favoring dispersal to a colony with a lower density of breeding pairs.

Dispersal of young feral pigeons may occur in multiple phases: in a Russian population, 70% of 1-year-old immigrants to a colony moved to a different colony after 1 year. The choice of a breeding colony may have such a strong effect on a young pigeon's lifetime reproductive success that it is willing to make a corrective move if the first choice appears to be unsuitable. Once a young bird has made its final choice, however, it is unlikely to reverse it: adult pigeons show strong breeding-site fidelity in Poland and defend breeding territories year-round in some areas, often roosting in the nest at night when they are not breeding.

Social Behavior

Flocking and Group Foraging

Many species of birds live for part of the year in flocks – small social groups that form for a variety of functions, including feeding, commuting to feeding sites, roosting, and avoiding predators. Feral pigeon flocks may be comprised of individuals from the same breeding colony, or of individuals from different colonies that aggregate at a feeding site or that come together for defense against predators. When flying to distant feeding grounds, feral pigeons almost always travel in flocks, especially when the likelihood of encountering a predator is high, and pigeons respond to the predators by flocking tightly together and performing evasive flight maneuvers.

Although they tend to form smaller groups, wild rock pigeons also perform most of their important activities in flocks and appear to be more cohesive in predator avoidance maneuvers than are feral pigeon flocks, which sometimes splinter, isolating the more vulnerable individuals.

Within feeding flocks, dominant feral pigeon individuals occupy favorable central locations, forcing more subordinate birds to the periphery of the group. In studies of feral flocks in the United Kingdom, Ronald Murton and colleagues found that social hierarchies are generally stable across both feeding and roosting flocks, with dominant birds retaining advantageous positions. Subordinate birds suffer from lighter body weights than dominants, but this poorer condition could be either the cause or an effect of their subordinate status.

Individual pigeons within flocks vary in foraging ability, foraging behavior, and dominance rank. Flock-mates may identify and follow successful foragers, learning how to employ novel foraging methods by observing them. Such opportunities for social learning may be a premier benefit to joining feeding flocks, especially for young, inexperienced birds.

Coloniality

While most feral pigeons nest in colonies, single nests are not uncommon, and neither feral pigeons nor wild rock pigeons seem to be obligate colonial nesters. Feral pigeon colonies usually form near food resources, although some urban colonies migrate daily to exploit rich food resources outside the city. Abundant, accessible food may be a prerequisite for colony formation and persistence, and solitary nesting may be the result of food stress.

Feral pigeon colonies consist of different classes of residents, including a core of reproductive individuals that breed in the colony for four or more years, individuals that breed in the colony for 2–3 years, transient individuals that remain for less than 6 months and may or may not attempt breeding, and nonbreeding individuals. As this variable colony structure suggests, interactions and migration events between neighboring colonies are common; males may seek extra-pair copulations with females in nearby colonies, and individuals may immigrate to favorably placed colonies for the winter.

Pigeons nesting in colonies engage in cooperative defense of nestlings against predators, and the effectiveness of such defense compared to the defense capabilities of a solitary pair may represent one of the most important benefits to individuals of nesting colonially. Colony members may also benefit from sharing information about food resources and from foraging in flocks that are comprised of fellow colony members.

Individual Recognition

Feral pigeon parents identify their nestlings by location until their offspring fledge, at which point parents become capable of recognizing their young individually. With training, some domestic pigeons visually discriminate male from female pigeons, using photographs. This finding is surprising because wild feral pigeons do not seem to be able to tell male from female on sight alone. Instead, feral pigeons use behavioral tests, such as the courting male's aggressive challenge (which another male would respond to aggressively while a female would not) to determine whether an unfamiliar pigeon is a rival or a potential mate. Given that training can result in better discrimination of the sexes by domestic pigeons, it is surprising that feral pigeons do not exhibit this same discriminatory ability.

Navigation

One of the traits for which pigeons are most famous is their remarkable ability to navigate accurately over long distances. A thorough discussion of this aspect of pigeon behavior is presented elsewhere in this encyclopedia. Wild rock pigeons and feral pigeons seem to have roughly comparable navigational abilities, although both home less accurately than do trained homing pigeons: in one comparative experiment, feral pigeons proved to be 20–30% less adept at homing than were homing pigeons. Gaia Dell'Ariccia and colleagues in Zurich recently showed that homing pigeons home significantly more effectively in small flocks than they do individually, suggesting an important social dimension to navigational behavior that has been largely ignored in previous studies.

Invasive Behavior

Feral pigeons are distributed worldwide, inhabiting such diverse locations as Ghana, Sri Lanka, Japan, and Australia. Although humans often transport domestic pigeons to locations outside the rock pigeon's native range, escaped or released domestic pigeons have been extraordinarily successful, both in colonizing new areas and persisting in urban environments. What behavioral traits make recently domestic pigeons so successful at establishing feral populations in a wide variety of locations?

Feral pigeons possess several traits, including omnivory and sociability, that promote successful establishment in novel environments. They are quick to investigate and

sample new food resources, and consume a variety of food types. Feral pigeons' history of domestication and artificial selection may be the cause of their reduced fear of novel foods and of human proximity, compared to synanthropic doves with wild ancestors. However, feral pigeons do not appear to possess characteristics associated with competitive displacement of native species, such as interspecific aggression. Consistent with this observation, feral pigeon populations are usually concentrated in human-modified habitats, particularly cities, although they sometimes spread into the surrounding countryside. Feral pigeons are primarily urban birds and do not usually threaten the persistence of native species in the rural areas around the cities in which they establish populations.

Conclusion

Pigeons are excellent study subjects for answering a variety of questions in animal behavior and behavioral ecology. The complex social interactions that take place within the flocks and colonies in which pigeons perform most of their important activities are still poorly understood and constitute a fascinating topic for future study. Empirical studies of pigeons may help test predictions in important areas of animal behavior such as sex ratio theory and intrasexual competition, and from this perspective, they have been underutilized. Recent attention focused on urban ecology will likely increase the interest of scientists in using the pigeon as a model organism for tackling large, evolutionary, and behavioral questions. The evolutionary pressures affecting abilities such as visual discrimination of the sexes and of individuals remain mysterious, as do the potential roles of learning and experience in facilitating such discriminations in wild birds. The outcome of the large-scale natural experiment currently ongoing with increased peregrine falcon predation pressure on polymorphic feral populations remains to be determined, and presents an exciting opportunity to watch the opposing forces of sexual selection and natural selection in real-time action. There are few organisms that are as common, easy to observe, easy to raise, and interesting with respect to the broad range of questions that can be addressed with experimental studies in the lab and field as is the pigeon.

See also: Pigeon Homing as a Model Case of Goal-Oriented Navigation.

Further Reading

Burley N (1981) Mate choice by multiple criteria in a monogamous species. *The American Naturalist* 117: 515–528.

Dell'Aricca G, Dell'Omo G, Wolfer DP, and Lipp H-P (2008) Flock flying improves pigeons' homing: GPS track analysis of individual flyers versus small groups. *Animal Behaviour* 76: 1165–1172.

Goerlich VC, Dijkstra C, Schaafsma SM, and Groothuis TGG (2009) Testosterone has a long-term effect on primary sex ratio of first eggs in pigeons – In search of a mechanism. *General and Comparative Endocrinology* 163: 184–192.

Goodwin D (1983) Behaviour. In: Abs M (ed.) *Physiology and Behaviour of the Pigeon*, pp. 285–308. London: Academic Press.

Hetmański T (2007) Dispersion asymmetry within a feral pigeon *Columba livia* population. *Acta Ornithologica* 42: 23–31.

Johnston RF and Janiga M (1995) *Feral Pigeons.* New York, NY: Oxford University Press.

Lovell-Mansbridge CL and Birkhead TR (1998) Do female pigeons trade pair copulations for protection? *Animal Behaviour* 56: 235–241.

Marchesan M (2002) Operational sex ratio and breeding strategies in the feral pigeon *Columba livia*. *Ardea* 90: 249–257.

Murton RK, Coombs CFB, and Thearle RJP (1972) Ecological studies of the feral pigeon *Columba livia* var. II. Flock behavior and social organization. *Journal of Applied Ecology* 9: 875–889.

Nakamura T, Ito M, Croft DB, and Westbrook RF (2006) Domestic pigeons (*Columba livia*) discriminate between photographs of male and female pigeons. *Learning & Behavior* 4: 327–339.

Palleroni A, Miller CT, Hauser M, and Marler P (2005) Predation: Prey plumage adaptation against falcon attack. *Nature* 434: 973–974.

Pike TW (2005) Sex ratio manipulation in response to maternal condition in pigeons: Evidence for pre-ovulatory follicle selection. *Behavioral Ecology and Sociobiology* 58: 407–413.

Sol D (2008) Artificial selection, naturalization, and fitness: Darwin's pigeons revisited. *Biological Journal of the Linnean Society* 93: 657–665.

Traxler B, Brem G, Müller M, and Achmann R (2000) Polymorphic DNA microsatellites in the domestic pigeon, *Columba livia* var. *domestica*. *Molecular Ecology* 9: 366–368.

Wosegien A (1997) Experiments on pair bond stability in domestic pigeons (*Columba livia domestica*). *Behaviour* 134: 275–297.

Relevant Websites

www.pigeonwatch.org – The Cornell Lab of Ornithology's Project PigeonWatch.

http://www.birds.cornell.edu/pigeonwatch/GettingStarted/courtship-behaviors – Illustrations of pigeon courtship behaviors.

Pigs: Behavior and Welfare Assessment

J. Deen, University of Minnesota, St. Paul, MN, USA

Current Uses of Behavior to Determine Welfare Status

Introduction

The welfare of farmed animals has received increased attention in the recent times. Compromised welfare leads to reduction in productivity. Additionally, market forces can be negative when the welfare is shown to be poor. Thus, compromised animal welfare affects the economic sustainability of animal agriculture in more than one way. However, from the producers' perspective, there is considerable difficulty in demonstrating that adequate welfare has been ensured. The major reason for this is the lack of objective and scientific tools to evaluate animal welfare. Without objective quantification, a compromise in welfare cannot be approved nor can the efficacy of corrective measures assessed. Animal welfare involves several factors, many of which are influenced by factors other than threats to welfare. Nevertheless, behavioral changes are the first and obvious indicators of compromised welfare in animals, though the identification of the same requires thorough knowledge about the normal behavior of the species concerned. The aim of this article is to explain the importance of behavior in swine welfare assessment.

Importance of Behavior in Animal Welfare Assessment

Animal welfare is a multidisciplinary concept. Behavioral, physiological, and hormonal indicators are employed to assess the welfare of animals. Among these disciplines, the study of animal behavior is an important component. A major reason for the prominence of behavior as a tool in welfare assessment is that changes in behavior patterns are among the most rapid and visible responses of an animal to changes in its environment. Behavior is the way in which the animal expresses its basic needs, deficits, and happiness in an environment. The study of behavior (ethology) involves not only what an animal does but also when, how, why, and where the behavior occurred. Understanding animal behavior is the best way to know whether we are keeping the animals appropriately. The behavior of an animal is influenced by both the structure and function of the animal as well as by the interaction of the animal with the external environment.

All categories of pigs under farm conditions are exposed to different types of challenges to their welfare. These challenges vary depending on different aspects of farming such as routine management, housing, slaughter, and transport. The major welfare concern associated with housing is close confinement in barren environments. The significance of animal welfare is associated with an animal's adaptive response to stress and its impact on biological functioning. The first response to stress is an alteration in behavior in order to cope with the stressful environment. This may or may not be accompanied by changes in physiological parameters such as respiration and/or heart rate to support the coping efforts. In many instances, it is possible to objectively assess the changes in physiological responses. However, the difficulty is in using the physiological changes alone to identify or quantify the compromise in welfare as many of the physiological parameters are influenced by factors not necessarily threats to the welfare. Therefore, a system evaluation, using a multidisciplinary approach, has been suggested to be the ideal way to assess animal welfare.

By nature, each species has its own unique behavioral repertoire (the ethogram) to cope with problems in its natural environment. Scientific studies in farm animals under seminatural conditions have indicated only meager changes in their behavioral repertoire despite domestication. This is the major reason for the widespread consensus that animals should be able perform their natural behavior in order for their welfare to be at a high level. Thus, animal welfare is often defined in terms of natural living or normal functioning of behavioral systems. In other words, successful coping means both physiological health and the ability to perform normal behavior. Therefore, the occurrence of normal behavior is viewed as an indicator of welfare in farm animals.

Despite the admittance of behavior as an important indicator of animal welfare, there has been a serious dearth of scientific studies on different aspects of animal behavior. This was a serious limitation as a thorough knowledge about the normal behavior of a species is a prerequisite to understand behavior pathologies. This fact was well understood by the formal official movement to ensure animal welfare, the Brambell committee of the United Kingdom. This committee reported that scientific research on animal behavior and related fields is essential to answer many questions that arose over the welfare of captive/farmed animals. The committee report proposed how behavioral research could be used to detect pain and discomfort, understand the cognitive powers of animals, and identify motivations that are thwarted in captivity by the use of preference tests. The committee stimulated several animal behavior studies in the United Kingdom and other parts of the world.

In addition to the obvious use in animal welfare assessment, knowledge of animal behavior enables human caretakers to solve practical problems of animal housing, management, handling, transport, and health. It also contributes to the science of animal welfare and permits the humans to better use the skills and abilities of animals. Knowing the reasons for abnormal behavior, ethologists can offer ways to prevent or mitigate the problem. Further, behavior principles will help to better design animal accommodations thereby making the environments function better and improve the welfare of animals that live in such facilities. A mere knowledge of the behavior repertoire is not enough to assess the welfare of an animal. It is equally crucial to know whether animal welfare is at risk if the behavior observed differs from that in the natural environment for each species and farming system. This is achieved by animal motivation studies and by studies exploring the effect of environmental structures on animal behavior.

In many instances, physiological and/or behavioral responses essential for survival under natural conditions become part of the biology of the species. Such responses persist despite a need for them under domestication. However, from the animal's perspective, this is a true behavioral response to ensure survival and therefore not being able to perform such a behavior can be a compromise in welfare. An ideal example of such a behavior is altered nest building in periparturient sows despite the need for a nest for piglet survival. Although abnormal behaviors are the common indicators of a compromised welfare, there are positive behavior indicators as well suggestive of good welfare such as cheerful vocalizations and facial expressions in some species. Thus, well being comprises both the absence of negative effects and the presence of positive effects.

Motivation and Behavior

Behavior is a reflection of the inner state (motivation) of the animal, the brain process behind the overt behavior. Motivation also dictates when behavior is to be exhibited. Motivation studies are thus essential to understand the meaning of the behavior observed. Many altered/abnormal behaviors in animals suggestive of compromised welfare have been linked to motivational problems. Causal factors are the inputs to the decision-making center which are actually the interpretations of different internal and external states of the body like fear or hunger. The motivational state of an animal represents the combination of the levels of all relevant causal factors.

Motivation may arise from internal or external source. For example, group-housed sows prefer to eat simultaneously. That means when one sow sees another sow eating, she also is motivated to eat even if she is not hungry. Here, the source of motivation is external. Similarly, nest building is another behavior expressed by sows prior to farrowing. Even if there is no bedding material in the vicinity, sow will follow the same body movements as if she is building a nest. Here, the motivation is internal. There is no consensus regarding the strength of internal and external motivations. The causal components of a motivational state are difficult to quantify although behavior observation permits us to understand the motivational state of the animal. The motivational state and its strength in animals are studied using preference tests. Preference tests are generally performed in one of the two forms, to identify the motivation or to assess the strength of motivation. In the former, the animal is exposed to two conditions suitable to perform two different behaviors and is permitted to select one depending on its motivational state at that time. In the latter, the animal is given a condition suitable for a particular behavior and required to perform some work to access the condition. Depending on the amount of work performed, the strength of motivation is evaluated.

Behavioral Patterns in Swine

The changes in housing and management patterns in swine farming have been rapid with respect to both evolutionary time and the domestication history of the pig. Thus, there is every chance to have problems in meeting the behavioral needs of pigs. This becomes a serious issue given the fact that the behavioral differences between wild pigs and domestic pigs are more quantitative than qualitative in nature.

Social Behavior

Pigs in nature live in large social groups. The members within a group exhibit a fixed pattern of relationship between them and with their environment. Social behavior of pigs within a group comprises both agonistic and nonagonistic behaviors. Physical (size, age and sex composition of the group, degree of relationship between members) and social structures of the group as well as group cohesion (length of association between members) have effects on social behavior. Attempts to use the same resource at the same time by more than one individual in the group result in competition which may not necessarily lead to physical conflict. Hierarchy in the social group means that the dominant member is able to restrict the movements of subordinates and their access to resources.

The key factor in a social group is communication. Vocal signals are the most important means of communication in pigs. The common vocalizations in pigs are the following.

- *Grunt*: lasts for < 1 s and produced in response to familiar voices or during activities such as rooting.
- *Bark*: produced when pigs are startled.
- *Squeal*: intense vocalization produced when aroused.
- *Scream*: vocalization produced by a pig when hurt.

Olfactory signals are predominant in recognizing a new member in the group. Experiments using blindfolded pigs have indicated that olfactory stimuli are more important than visual or hearing senses in forming a dominance hierarchy among group members. Although vision plays a less significant role in social behavior, pigs have color vision, a panoramic range of about 310°, and binocular vision of 35–50°.

Dominance hierarchy in social groups of pigs

When unfamiliar pigs are mixed together for the first time, they compete with others, characterized by aggressive interactions to establish a dominance hierarchy. Generally, a new hierarchy is established within 24 h of mixing, though the level of aggression drops considerably after 1 h of mixing. The time to establish a hierarchy is longer in older pigs. Once a hierarchy is established, the group may become stable. Hierarchy is not permanent; when a new pig enters or leaves the group, the hierarchy is reorganized. Both sex and body weight are predictors of success in achieving dominance. Female pigs are aggressive for longer duration than castrated males. Although boars are generally dominant over sows, in a group of castrated male pigs and sows, the males may not be dominant. Breed difference in aggression has also been noted, with the Yorkshire more aggressive than Berkshire breed. A breed x sex interaction in the extent of aggressiveness has also been observed, with Hampshire male dominance over females exceeding that of Duroc male pigs. Once a hierarchy is established in a group of pigs, it is maintained by the subordinate pigs in the group avoiding conflicts. Recognition between members in the group and the memory of the social encounters are thus vital factors in maintaining social stability; pigs can recognize and remember 20–30 of their group mates. During the process of hierarchy establishment, dominance is established between each pair of pigs, with aggression usually expressed by the dominant animal. However, once hierarchy is established, the same animal may not appear to be very aggressive, as the relationships are understood by other group members. Following hierarchy formation, overt aggression is replaced by threat (sharp loud grunt and feint with the snout by the dominant pig). Thus, a stable social group of pigs is characterized by a high level of group bonding with a minimal level of aggression.

Aggression is an important social behavior in pigs even after the establishment of a social hierarchy since there will be competition for some resources at all times (see section 'Social Behavior'). The most common aggressive behavior is thrusting the head upward or sideways against the head or body of the opponent pig. The fighting pigs strut shoulder to shoulder while attempting to bite each other, usually at the neck, ears, and shoulder region while in a parallel posture. Another common aggressive behavior is levering (snout is put under the body of the opponent from behind). The losing pig in a fight will exhibit the characteristic submissive gesture (twisting the head away from the opponent) and will be usually chased by the winner. The various components of agonistic interactions in pigs, as developed by Jensen and Wood-Gush, are given in **Table 1**.

In many swine-breeding herds, a dynamic grouping system (sows are removed and new sows are added to the group at frequent intervals) is followed to make best use of resources. This results in disruption of the social hierarchy and consequent aggression. When sufficient space is available in such situations, subordinate pigs could inhibit aggression by moving away from the dominant ones. However, if there is physical limitation in performing submissive responses, the aggression remains unresolved and persists at a high level. The problem is exacerbated when resources such as feed are limited. This can be a serious challenge to the welfare of pigs. Similarly, in stall housing systems, the pigs are unable to exhibit submissive behaviors resulting in unresolved aggression. Both the quantity and the quality of space are important as a resource given that in natural surroundings groups of pigs occupy vast areas and forage over long distances daily.

Characteristic association between individuals within a group is another feature of social behavior. Certain

Table 1 Agonistic interaction patterns in swine

Agonistic behavior components	Description
Inverse parallel pressing	Pressing of shoulders against each other, facing opposite directions
Parallel pressing	Pressing of shoulders against each other, facing same direction
Head-to-body knock	Hitting with the snout against the body of the receiver
Head-to-head knock	Hitting with the snout against the head of the receiver
Nose to nose	The nose approaches the snout or the head of the receiver
Nose to body	The nose approaches the body of the receiver
Anal-genital nosing	The nose approaches the anogenital area of the receiver
Head tilt	The head is lowered and turned away from another animal
Aiming	An upward-directed thrust of the snout, slightly directed at the receiver, from a distance of 2–3 m
Retreat	Takes several steps away from the other animal

individual members within a group will have closer relationships (affiliation). These pairs evolve by themselves and support each other for their benefit. Strong social relationship between and within sexes and age groups has also been observed. Allogrooming (in pigs – nosing another pig) is another social behavior observed in pigs which is considered to reinforce social bonds within groups and reduce tension between group mates.

Feeding and Drinking Behavior

Pigs show characteristic behavior during feeding. Feeding behavior is affected by diurnal rhythms and social factors. Brain inputs such as visual input or input from taste receptors can also influence feeding behavior. Drinking is closely associated with feeding; social factors, diet composition and texture, and environmental temperature strongly influence drinking behavior as well. Social facilitation (tendency for animals to join in an activity) is another factor influencing feeding behavior. Pigs in general prefer to eat simultaneously. This means that a pig fed ad libitum will be motivated eat again if it sees another pig eating. Thus, social facilitation can influence both the frequency of feeding and the quantity consumed. In facilities like the electronic feeding system (ESF), sows are not able to eat simultaneously. This blocks social facilitation and is suggested to cause behavioral and welfare problems. Often, waiting to enter the ESF appears to frustrate the sows, making them aggressive and causing injuries and compromised welfare. The competitive order for feeding is not necessarily linear and varies considerably in different situations. The social hierarchy dictates the feeding order to a great extent, with the dominant sow entering the feeder first. It is also possible that a dominant sow may enter the station repeatedly even after eating their allotted feed, thus restricting the access of submissive sows. Rooting is another important component of feeding behavior in pigs. The snout of the pig is a highly developed sense organ and olfaction plays an important role in feeding behavior. Rooting involves moving the feed material or straw or mud using the snout.

Pigs also show distinctive defecation behavior. Pigs prefer to defecate in an area away from the place where they feed or rest.

Nonlocomotory Postural Behaviors

Stretching, standing, sitting, lying down, and getting up are some of the nonlocomotory behaviors exhibited by pigs. The definitions of various nonlocomotory behaviors are provided as follows.

- *Standing*: an upright position on extended legs while remaining stationary.

- *Sitting*: the posture in which most of the body weight and the posterior part of the trunk are in contact and supported by the ground.
- *Lying*: the posture in which the side or ventral part of the body is in contact with the ground.

Specific sequences of movements are involved in assuming each posture with a pause of few seconds between phases. Animal needs a certain amount of space for these postures. The first step in preparation for a lying down posture is adopting a stable standing posture. This is followed by the sow dropping down onto the knees and adopting a half-kneeling position. The sow then slides one knee under the body and rests on one knee and one shoulder. This is followed by gradual rotation of the upper part of the body to rest the shoulder and side of the head on the floor. Then, the sow rotates the front half of the body so that the front part becomes almost recumbent on the floor. Subsequently, the sow drops the hind quarters and adopts a sternal or lateral recumbency, in which the ventral surface or the side of the body is supported by the ground, respectively. Similarly, when standing up, the sow first raises up on the front knees to a position similar to sitting, followed by lifting the hind quarters. A sow may lie down vertically (on the belly), laterally (on the side); she may fall down on the side or may lie down by leaning on a wall. The lying down process can thus significantly affect the chances of piglet crushing since most crushing occurs when the sow lies down.

- *Stretching*: Stretching is generally performed after a period of rest or after a period when the limbs are folded. Stretching helps the animal to keep its joints and muscles in a functional state.

Sideways, backward, and forward movements are associated with nonlocomotory postural behavioral changes. These impose a dynamic space requirement, which is greater than the static space requirements associated with standing or lying stationary. For instance, while lying down or getting up, the animal makes forward and backward movements. During these postural changes, the animal moves its center of gravity and uses its weight as a counterbalance for rising or as a direct pull when lying down. The movement of center of gravity is achieved by the forward and backward movements of the body and this requires additional space.

Locomotion and Gait

Locomotion is an important component of an animal's activities. Locomotion enables an animal to respond in space and time to meet its different needs. Locomotion disorders have a major impact on the welfare status of animals. In locomotion, limbs act in definite synchronized patterns which are called 'gaits.' Gait may be symmetrical (limbs on one side repeat those of the other side, but half a stride later) or asymmetrical (limbs from one side do not repeat those of the other). Walk, pace, and trot in horses are symmetrical gaits, whereas canter and gallop in horse are examples of asymmetrical gaits. The full cycle of leg movement during the phases of support, propulsion, and movement of the body through the air is called 'stride'; stride length is the distance between successive imprints of the same foot. Various aspects of locomotory behavior are important in diagnosing conditions like lameness. For instance, in lame animals, the stride length will be affected. Lame animals will adopt peculiar body movements to adjust the center of gravity.

Spacing Behavior

There are behavioral parameters associated with space also, which are important especially in social animals such as pigs. Individual space depends on the body dimensions of the animal in a particular posture and individual space moves with the animal. The static area used by an animal refers to its home range and territory. The common terms related to spacing behavior are given as follows.

- *Flight distance*: the radius of space within which the animal will not voluntarily permit the intrusion of man or other animals. The animal may respond by startle, alarm, fight-or-flight display, and vocalizations if the flight distance is violated.
- *Territory*: an area defended by fighting or by demarcation that deters other animals from entering the area. There may not be a permanent territory for an animal.
- *Individual distance*: the minimum distance between an animal and other members in the group; each animal prevents other animals from entering this space. Individual distance includes the physical space (space for its basic movements of lying, rising, standing, stretching, and scratching). The space will be larger in the head region to accommodate the head movements during ingestion, grooming, and gesturing and it includes both vertical and horizontal distances. Individual space protects the animal from body damage due to contact and reduces competition and interference while feeding. Individual distance also provides sufficient space for escape when needed and for separation from diseased animals.

Animals have quantitative and qualitative needs in relation to space. The former are related to space occupation, social distance, flight distance, and actual territory, whereas qualitative needs are related to activities such as eating, body care, exploration, kinetics, and social behavior which are space dependent. The quality of space includes the presence of barriers to avoid visual contact with others and concealment locations to avoid aggression from other animals.

- *Social limit*: the maximum distance any animal moves away from the group. There is always a balance between individual distance and social limit in stable social groups.

The space occupied by an animal is a function of its body weight and its need for various activities. The relationship between space occupied and weight is a constant ('k') multiplied by the body weight to the power of 0.67. This constant depends on the extent to which the body shape of the animal deviates from a sphere and the activities that are important for the animal to perform. The values of 'k' are different for lateral recumbency and sternal recumbency as the space required differs. Sharing of space occurs when animals are housed in groups since all animals are not performing the same postural behavior at the same time. Generally a 'k' value of 0.034 is considered sufficient in group housing systems. Crowding in a group means that the movement of members in the group is restricted by the physical presence of other group members; this increases the chance for violating individual space, which may affect the fitness of the group members especially when resources are limited.

Exploratory Behavior

Exploration is another important component of swine behavior. Animals are strongly motivated to explore when they are in a new environment. Exploration subsides as the environment becomes familiar and it resumes when there is a change in the environment. Fear associated with novelty is a major reason for exploratory behavior; a novel situation causes the animal some degree of fear. Exploration also helps in preparation for the future such as escape from predator, escape from inclement weather, and ensuring food security. A high degree of awareness is needed to evaluate the situation and to prepare for the future. Given the link between fear and exploration, welfare scientists have suggested that estimating the potential of a new environment or a novel object to induce fear can be used to evaluate the welfare status associated with the housing system. Exploration may be inquisitive (the animal looks for a change) or inspective (the animal responds to a change). The former is considered to be a pleasurable activity for the animal and indicates no immediate needs, whereas the latter represents the behavioral outcome of an interaction between fear and curiosity. In any case, interpretation of this behavior remains the subject of some debate.

Sexual Behavior

Both sexes of swine exhibit characteristic sexual behaviors. Standing still when pressed on the back (immobility) which is facilitated by the boar odor is a characteristic sign of female sexual behavior in pigs. While in estrus, the female pig may become restless at night. Some breeds also show erect ears (ears held stiffly close to the head and turned upward and backward). Female pigs in estrus may also urinate frequently and may produce a soft rhythmic grunt. The female pig will also approach the boar and sniff the boar's head and genitals. The female pig in estrus will try to mount other estrus females. This behavior eventually transforms into searching behavior for the boar, a behavior that is successfully used to detect estrus in sows using electronic monitoring system. Female pigs are able to identify intact males by the strong boar odor produced by androgen metabolites present in both the saliva and the preputial secretions of boars.

Approaching a female pig and nosing her sides, flanks, and vulva is a major sexual behavior in boars. Boars also make a characteristic series (6–$8\,s^{-1}$) of soft guttural grunts (courting song). The boar urinates rhythmically and the pheromones in the urine further stimulate the standing behavior of the female pig. The boar will also foam at the mouth and move its jaw from side to side. Boars do not exhibit flehmen (lip curling; turning up of the upper lip in response to an odor). Instead, boars gape when they encounter sow urine. The major sexual arousing factor in boar is suggested to be the female pig's willingness to stand still and not the olfactory signals.

Prepartum Behaviors

During the preparturient period (ranging from late gestation to the beginning of the first stage of labor), sows exhibit different behaviors when in the natural surroundings. However, not all of them may be visible in the confined systems. Nest building (gathering nest materials such as grass or sticks and arranging it by rooting and nosing), which starts between 3 days and 24 h before farrowing, is the most notable prepartum behavior. The duration of nest building may vary among sows. During nest building, the sow uses her forelimbs to move the bedding, known as 'pawing.' Sows may show nest-building behavior even if there are no materials available to them. Sows will stop all these activities as time advances and adopt a lateral recumbency before farrowing. This is followed by a period of restlessness and frequent change of position. The frequency of postural changes increases before the sow adopts a final lying posture. Piglets are farrowed every 15–30 min and the sow will remain in lateral recumbency until farrowing is completed. Sows remain mostly inactive during the first 48 h after farrowing which helps to minimize piglet crushing and to facilitate establishment of a teat order.

Postpartum Behavior

Postpartum behavior by the sow forms part of the strategy to protect the piglets. Like bovines, sows also eat placenta (placentophagia) which may help to recycle nutrients and to minimize chances of predation by removing odors. Sows generally do not lick or groom their piglets, but may nose the piglets. The sow also produces repeated short grunts to invite the piglets to feed on colostrum. She appears very nervous at this time and is easily disturbed by the presence of an intruder; this may lead to movements by the sow causing piglet crushing. Sows are very defensive of the piglets and may attack with barks and open mouth if approached.

Piglets will stand on their feet within minutes after birth and will start searching for a teat, facilitated by the grunts of the sow. The firstborn piglet uses thermal, tactile, and olfactory cues to locate the teat and may be slow to find teat. However, subsequent piglets respond to the voices of their littermates and quickly locate the udder. The sow will remain stationary in lateral recumbency and may change position to help the piglet locate a teat. Sensory inputs such as vocalization, odors from mammary and birth fluids, and hair patterns of the sow are suggested to help the piglets to locate the teat.

Colostrum is available continuously for the first few hours; 10 h after farrowing, milk let down becomes synchronized and periodic. Letting down occurs once in every 50–60 min and each letting down lasts for 10–25 s. The piglets learn the sow's call and will be ready at the teat for milk. The interval between letting down is longer at night than during the day. Sows generally do not respond to feel or sight of a piglet under them though they respond to loud squeals. Sows are reported to be most responsive to piglet squeals on the first 2 days postpartum.

New-Born Piglet Behavior

The most import aspect of neonatal behavior is the huddling together of piglets to conserve heat since their thermoregulatory mechanism is poorly developed. The first stage in the suckling process is marked by the vigorous, rhythmic up-and-down movements around one segment of udder by the piglet using its snout for about a minute. This is followed by slow suckling with the tongue wrapped around the teat for about 20 s. Then, the slow suckling is followed by rapid mouth movements as milk starts to flow for about 10–20 s. During this phase, the piglets' ears are flattened, and their head moves along with suckling movements. After 10–20 s of milk flow, the piglets may show slow mouth movement or move to another teat for more milk. After nursing, the piglets leave the udder or fall asleep in position. The sow stops rhythmic grunts once the piglets stop suckling.

Piglets establish a teat order within the first day of their life and will return to the same teat during the entire lactation. Piglets compete for the most productive teat regardless of its position in the udder. Piglets show their characteristic defecation behavior within 4 days of birth. Piglets sleep on average 10.5 h day^{-1} during the first 5 weeks. In nature, sows and their piglets stay together for several weeks. Piglets start showing play behavior (tossing and waving the head, spinning around) when they are 2-days old and will reach a peak between 2 and 6 weeks of age. The play behavior facilitates socialization with group members.

Abnormal Behavior

Abnormal behaviors are generally considered indicators of compromised welfare in animals. Abnormal behavior is defined as a clear deviation from the normal behavior expressed by majority of the members of the species when they are allowed to perform their full range of behaviors. Abnormal behavior may be harmful to animal itself and/or to other members in the group. Provision of manipulable substances and thereby avoiding barren environments will help reduce most of the abnormal behaviors. Common abnormal behaviors in pigs are described as follows.

Stereotypy

Stereotypy is a repetition of sequence of movements without any apparent reason. The repetition may be regular or irregular and may be short or long. Stereotypies are linked to the dopamine release systems in the brain that control body movements. Stereotypies are generally exhibited when the animal has no control over its environment, causing frustration. The argument that stereotypy is part of an animal's coping mechanism has not been widely accepted. Even if it is so, the indication is that welfare is compromised. Sham chewing (pig imitates chewing movements even when there is no food to chew, characterized by chewing, mouth gaping, and frothing and foaming of saliva), bar biting (animal opens and closes its mouth around a bar/metal piping of the stall, engaging the tongue and teeth with surface and performing chewing movements) are common stereotypies noticed in pigs especially in stall-housed pigs. Continuous repeated pressing of the drinker by some sows is also considered to be a stereotypy. Provision of straw or other manipulable material has shown to reduce most of the stereotypic behaviors.

Tail biting

Tail biting is an abnormal behavior that occurs mainly in grow-finish pigs and that is directed to group mates. Crowding and consequent inadequacy of feeder and waterer spaces in group housing is associated with tail biting. Environmental factors such as high temperature, humidity, and noise may exacerbate the situation. However, lack of oral stimulation is suggested to be the major reason for this abnormal behavior. The incidence of tail biting is increased among pigs housed without bedding on slatted floors and ones that are fed automatically. Hunger also predisposes pigs to tail biting. The bleeding from tail biting stimulates further tail biting. Provision of rooting materials has been shown to reduce tail biting. Docking may reduce tail biting though it may lead to another abnormal behavior, ear chewing.

Vulva biting

This is very common in ESF systems where the hungry sows are forced to wait to access the feeder. As in tail biting, the bleeding from the wound stimulates further biting.

Anal massage

Anal massage is an abnormal behavior noticed in growing pigs, especially in docked pigs. Providing objects to chew and root will help to minimize this problem.

Belly nosing

It is the up-and-down movement of the snout and the top of the nose on the belly of other pigs, on the soft tissue between their hind legs and between their forelegs. This is common in early weaned piglets, especially those that continue to show teat seeking behavior, which gets directed toward other piglets.

Cannibalism

This is an abnormal behavior mostly seen in sows with first litters where the sow may kill and eat her own piglets. In its mildest form, the hyper-reactive sow may accidentally crush and kill the piglets which are then eaten partially or fully. In severe forms, the sow tries actively to avoid piglets and approaching piglets are attacked, killed, and eaten fully or partially. Once started, cannibalism usually stops only after the death of the entire litter. Cannibalism is associated with hyperexcitability of the sow following farrowing, due to the novelty of the farrowing environment. Providing bedding before farrowing and allowing sows to build nests may reduce cannibalism.

Pigs are highly motivated to explore their environment and spend 75% of their active time in foraging-related behavior. Provision of concentrates in a single location as in confined systems requires the animal to satisfy all its food intake needs within a short time, while not satisfying the motivation to explore and forage. In the barren environment, these behavioral needs are redirected at unsuitable targets such as pen mates. Lack of substrates is the major reason for abnormal behaviors such as tail biting and ear chewing. Studies have suggested that even the modern swine breeds retain the behavioral features of wild boars. Most of these natural behaviors are controlled by internal factors, and therefore swine welfare is compromised by a lack of opportunities to perform the behaviors that they are strongly motivated to perform.

Sickness Behavior

Sickness behavior is defined as the expression of the adaptive reorganization of the priorities of the host during an infectious period. The major feature of sickness behavior is a general decrease in activity (immobility, sleepiness, reduced food and water intake). Sickness behavior may be interrupted in response to important strong stimuli (e.g., a sick sow may respond to a newborn piglet). The reduction in activity helps the animal conserve energy, and spend more time resting and sleeping. However, reduction in activity may not occur in all cases. A sick animal will try to move away from its group. This helps the animal to be away from the disturbances of its group mates and minimizes the chances of transmission of disease. It is argued that the behaviors shown by sick animals are part of a strategy to fight disease. Behaviors such as play, grooming, sexual behavior will be reduced in illness, as animals divert resources for maintaining body temperature and combating pathogens.

Using Behavioral Principles to Improve Pig Welfare

Pigs prefer to live in large static groups with plenty of space and opportunities to perform many of their natural motivations. However, it is a practical and economic challenge to offer these facilities under commercial farming. For instance, space is a limiting factor in farms. Similarly, dynamic grouping with frequent mixing of pigs may be required to ensure better facility use. However, a knowledge of swine behavior can be successfully utilized to mitigate the threats to the welfare of domestic swine. Aggression is unavoidable when unfamiliar pigs are mixed together. However, the extent of aggression is determined by the method of feeding, and the amount and quality of space. Similarly, aggression at mixing can be influenced by design of the system and by management techniques. For instance, allowing small subgroup formation has been indicated to reduce aggression at mixing. The shape of the pen is also important in determining the level of aggression, with higher aggression in circular pens, and lower aggression in pens with a solid barrier. Aggressive interactions can be reduced over the short term by mixing pigs after sunset, administering some pharmacologic compounds, and the presence of boar at the time of mixing. Space availability is another issue as pigs in groups require space to show submissive behavior (e.g., running away). Providing the space necessary to perform this behavior will ensure quick establishment of social hierarchy and stability in the group. It has been suggested that the psychological stress associated with nonresolvable aggressive interactions among stall-housed sows is a chronic stressor.

Unique Problems in Assessing Welfare by Behavior

Obviously, welfare is a multifactorial concept. Therefore, behavior alone cannot be an adequate indicator of welfare in animals. However, given that behavior change is the first overt and perhaps the major indicator of compromised welfare, we are justified in using it in animal welfare assessment. At the same time, it is important to remember that we as humans are ascribing meanings to the behaviors of another species, and to that extent our interpretations can be erroneous.

Often, behavioral pathologies are used as indicators of compromised welfare. However, it is important to understand the positive behaviors in animals and their link to welfare. Many times, stress physiology is used to assess the existence of a positive state or at least the absence of a negative emotional state in animals. However, physiological parameters such as heart rate, respiration rate, and cortisol levels are influenced by factors not adversely affecting the welfare of the animals. For instance, sexual excitement can increase the heart rate and respiration rate without any adverse effect on the welfare of the animals. Individual differences, developmental changes, and diurnal variations in measures can also affect the interpretation of the final result. Thus, subjective assessment is also frequently needed for successful welfare assessment.

Nonetheless, the extent to which a human observer can accurately assess the mental feeling of the animal is debatable. This is a serious limitation in using behavioral indicators for welfare assessment. Motivational studies evaluating the animal's 'willingness to work' have been suggested as useful in assessing the state of the animal. Even then the correct interpretation demands appropriate experimental designs and sophisticated operant equipment. Even though preference tests are used in housing system studies, they are not perfect. The main limitation is that the results are unique to the testing situation. The behavioral choice of the animal may vary with the duration of exposure. Above all, it may be difficult to establish a scientific rationale for the choices made by the animal because those choices may not always be the best for the subject.

The multifactorial nature of animal welfare also raises the issue of how to weight those factors. Often, a higher importance is given to factors which humans value had they been exposed to such a welfare threat. It is also possible that different indicators may provide conflicting interpretations. For instance, physiological and behavioral indicators provide different assessment of the pain associated with castration. Similarly, behavioral and health indicators have come up with different conclusions regarding the welfare implications of providing bedding for pigs.

Another difficult issue, even when positive behaviors are used to assess the welfare status, is the level of such behaviors. For example, allogrooming is generally considered as a positive behavior in social groups of pigs. Nevertheless, excessive levels of allogrooming activity may not necessarily reflect positive states of the animal as they are likely to disturb the receiver. Obviously, there are no objective indicators to assess this and no gold standard has been established. Similarly, the ability of the animal to perform various postural behaviors is critical with respect to its welfare. But there are no scientifically valid numbers available regarding the duration or frequency of postural behaviors, leaving the assessor to resort to subjective judgments.

Areas Where More Research Is Needed

Obviously, all the limitations mentioned earlier warrant further research. However, there are promising areas that are expected to advance human knowledge about animal behavior. These have the potential to improve the welfare of domestic animals. One little-researched area is sickness behavior and its value as a disease indicator. Similarly, studies are needed to understand how to assess the positive experiences in animals that are suggested to be important components of good welfare. This issue is crucial because well being is not merely the absence of negative effects, but instead, predominantly the presence of positive effects. Affiliation (closer relationship between certain pairs of individuals) is a type of social behavior exhibited by many farm animal species. Affiliative behavior has been suggested to create a 'positive mood' in animals. However, the number of studies focusing on this behavior is considerably less than those involving social competition. Our efforts to focus on negative behaviors will only enable us to help the animal meet its needs and thereby avoid suffering. Research is also needed to understand how to use positive experiences to a better quality of life to the animals and to improve their health.

See also: Beyond Fever: Comparative Perspectives on Sickness Behavior; Self-Medication: Passive Prevention and Active Treatment; Welfare of Animals: Introduction.

Play

G. M. Burghardt, University of Tennessee, Knoxville, TN, USA

Introduction and History

Play has been recognized in non-human animals for many centuries, but the study of animal play, like much of animal behavior, really did not develop until after the writings of Charles Darwin and the rise of natural history, comparative psychology, and, in the early twentieth century, ethology. It is useful at this point to mention that play in animals is generally categorized as locomotor play (jumping, leaping, twisting, swinging, and running), object play (biting, mouthing, and manipulating), and social play (chasing, wrestling, and mounting). These are not completely independent as all three can occur at the same time when, for example, two dogs chase after a stick, both grab it, and then proceed to engage in a tug-of-war for possession. The need for a more precise definition of play will soon become apparent, however.

Although some early writers on animal behavior, including Charles Darwin, claimed that play occurred in a wide variety of animals, even crabs, ants, and fish, these were based on anecdotal evidence in the days before film and video documentation. With the rise of behaviorism and the disparaging of anthropomorphism, play soon became identified as a phenomenon largely limited to humans and other 'intelligent' mammals such as monkeys, apes, dogs, and cats. Furthermore, by early in the twentieth century, play was viewed by many, especially educators, as existing in order to help animals (and children) learn how to survive in adulthood. Indeed, Karl Groos, in a classic book published in 1898, put forth the enduring view that play is a necessary means for animals to develop and perfect their instinctive behavior (finding food, fighting conspecifics, repulsing predators, courting and mating, building nests, etc.). This soon became the major, but far from universal, theoretical assumption, although with many variants. Associated with this perspective was the position that the benefits of play are delayed until adulthood and that is why play appears to have no function when it appears in young animals. About 50 years later, the view that play also may be important when it occurs, such as in stimulating activity that provides immediate general neuromuscular and physiological benefits for performers outside of, and in addition to, any long-term benefits, was developed. The seminal 1981 book by Fagen reflected the dominant theme of the functional benefits of play, although when written, the empirical evidence for any benefits of play was scant indeed, even in the human child and education literature. This literature promulgated an optimistic positive role for play and largely ignored the costs. Although play may appear to be fun or enjoyable, its real importance lies elsewhere.

There were some alternatives to this 'play as practice for the future' view, including the surplus energy theory of Schiller and Spencer, the recapitulation theory of G. Stanley Hall, and the psychodynamic anxiety reduction approach of the Freudians, such as Winnicott, but the acceptance of the claim that play is linked to intelligence, large brains, and prolonged parental care seemed to support the practice and delayed benefits views. For example, the seminal studies on the development of intelligence and the resulting stage theories of Piaget found play part of the earliest 'circular' reflex-like reactions of infants, but higher order 'pretend' or pretense play did not occur until much later in childhood and most likely beyond the capabilities of other animals. However, even applying Piagetian methods, such pretend play has now been documented in some great apes. Nevertheless, such views hardened the conclusion that 'true' play was most common, if not found exclusively, in 'higher' mammals. In other species, play-like behavior was largely dismissed as misidentified or misfiring 'instincts' or their developmental precursors. Even the acceptance of play in birds was suspect in authoritative writings into the 1980s. And until quite recently, virtually all authorities dismissed reports of play in ectothermic (cold-blooded) vertebrates and all invertebrates. Nevertheless, while many play researchers accepted that play may have immediate benefits for 'higher' animals, not just benefits delayed until adulthood and *serious* tasks in life, the problem remained that none of these benefits, immediate or delayed, had been empirically confirmed with careful experimentation in either human or non-human animals of any species, as documented in papers by Martin and Caro and Peter Smith. In fact, virtually all tests of the putative benefits of play either failed or had serious methodological flaws. In this situation, it was perhaps understandable that outside the fraternity of play researchers, few found reason to consider play a serious topic for research. The reason turns out to be that scientists were searching for the obvious connections. The meaning of play lies elsewhere than in a stark utilitarianism or obvious appearing functionalism (play fighting leads to better fighting as adults, play with prey-like objects in kittens leads to better hunting ability in cats, play with dolls leads to better mothering in children, etc.).

Recognizing Play

But there is a more fundamental problem. A careful analysis of the literature of play in animals actually reveals that until recently there were no clear criteria for identifying play in them other than an uncritical anthropomorphic extrapolation from human play and 'obvious' animal play coupled with post hoc definitions. In other words, from apparently unambiguous examples of play in monkeys, dogs, rats, horses, and other mammals, definitions were formulated that were full of words such as 'may' or 'might,' or were restricted to only one kind of play, such as social play. They also included behavior most would not call play, such as exploration or behavioral stereotypies.

Let us elaborate on this important point. We might all agree on what is play in a dog or a monkey, but we typically do so by identifying the behavior and its underlying emotion with our own assumed feelings when performing a similar behavior. Thus, we develop the understandable view that play should be pleasurable or fun. Perhaps we can assess a dog having fun by observing its leaps, ears, tail position, and vocalizations. But what about turtles, fish, frogs, lizards, and snakes? If we want to determine how ancient and basal play is in vertebrate evolution, it is imperative to find out if play is, like endothermy or enlarged frontal lobes, a recent evolutionary innovation, as championed by some writers, or if it also occurs in much older, more 'primitive' animals. Below are five criteria which, when all are met, quite confidently permit a behavior to be characterized as play. Since using this approach has resulted in play being claimed to exist in animals where it had not been seriously entertained by scientists, this definitional treatment needs some details.

The *first criterion* for recognizing play is that performance of the behavior is not fully functional in the form or context in which it is expressed; that is, it includes elements, or is directed toward stimuli, that do not contribute to current survival. The critical term is 'not completely functional,' instead of 'purposeless,' nonadaptive, or having a 'delayed benefit.' This distinction recognizes that play may have an important current utility while not being focused directly on survival such as eating or fighting. The *second criterion* is that the behavior is spontaneous, voluntary, intentional, pleasurable, rewarding, or autotelic (i.e., 'done for its own sake'). Here, only one term of these often overlapping concepts need apply. Note that this criterion accommodates any subjective or emotional concomitants of play (having fun, enjoyable), but does not make them essential for recognizing play. The *third criterion* is that the behavior differs from the 'serious' performance of ethotypic behavior in at least one respect: incomplete (generally through inhibited or dropped final elements), exaggerated, awkward, precocious, or involves

play signals, role reversals, or other behavior patterns with modified form, sequencing or targeting. This criterion acknowledges, but does not require, that play may be found predominantly in juveniles in many species. The *fourth criterion* is that putatively playful actions be repeatedly observed during at least a portion of an animal's life. Repetition is found in all play and games in both human and non-human animals. This criterion also distinguishes transient responses to novel stimuli or environments from the play actions that may follow such initial exploratory behavior. The *fifth criterion* is that the behavior is initiated only when an animal is adequately fed, healthy, and free from intense stress (e.g., predator threat, harsh microclimate, crowding, and social instability), or intense competing motivations (e.g., feeding, mating, resource competition, and nest building). This contextual criterion appears essential for the occurrence of play, as it has been repeatedly shown that play is one of the first types of behavior to drop out when animals, including children, are hungry, threatened, mistreated, or exposed to nasty weather.

The five criteria can be summarized in one sentence: 'Play is repeated incompletely functional behavior differing from other versions in form, context, or developmental stage, and is initiated voluntarily when the animal is in a relaxed or low-stress setting.' The term 'initiated voluntarily' could refer to pleasure, fun, excitement, rewards, or other emotional attributes. However, these attributes are not explicitly included since they may be hard to ascertain in animals less likely to be viewed as similar to humans (e.g., turtles or fish).

The Taxonomic Diversity of Play

With this set of criteria as a tool, we can see that much of our behavior from gourmet cooking to biking to watching sports can be viewed as play, but also that the same behavior may sometimes be work, punishment, or fulfilling an obligation. More importantly, applying the criteria allows us to see play in the behavior of many animals other than birds and placental mammals, as extensively documented in Burghardt (2005). Here are some examples.

Many marsupials such as kangaroos, wallabies, Tasmanian devils, and wombats are playful, though as a group, they are nowhere near the richness of playfulness one sees in dogs, monkeys, and otters. Even the egg-laying monotreme, the duck-billed platypus, seems to play. In fact, data strongly suggest that some animals from many other groups, including fishes, insects, molluscs, and reptiles can, and do, play. Since this may seem a rather bold, if not unsettling, claim, here are a few examples. The Komodo dragon is the largest lizard in the world and is a deadly carnivore, capable of hunting and eating deer and water buffalo. A lizard observed in captivity at the National Zoo in Washington, DC, when several years old, explored objects such as old shoes, small boxes, and even soft drink cans. It grabbed and shook them like a dog with a slipper. Like the dog, the lizard would not try to eat the object. She also engaged in tug-of-war games with her favored keeper, removed handkerchief or notebooks from a keeper's pocket, and tried to run away with them. A large adult aquatic turtle at the same zoo, given a basketball, repeatedly and for years, banged it around his tank. A large Nile crocodile frequently chased and attacked a large ball attached to a rope thrown around and pulled by a keeper outside his large naturalistic enclosure. Great white sharks have done similar things. Several species of fish have been known to push around balls and balance them on their snouts. A cichlid fish may engage in behavior with a larger less agile fish that looks, objectively, like something we would term 'teasing' when seen in kids or even a dog. Play fighting is the best-studied type of play in animals and is the focus of literally hundreds of studies in laboratory rats. Comparable recent documented observations of play-like agonistic social interactions have been made of dart poison frogs, young turtles, some fishes, and even in wasps. Octopi have been documented performing complex manipulations with Lego blocks and using their water jet abilities to repeatedly 'bounce' floating balls. Honeybees engage in practice take-off behavior before their first successful flights. Freshwater stingrays are so attracted to balls that sink to the substrate that two will engage in a 'game' of keep-away to control access to them.

Some Processes Underlying Play

As diverse as the above examples are, it is important to note that most species in many taxonomic groups have not been recorded as playing, nor do they all play in the same way or to the same extent. Social, locomotor, and object play are all very common, but not universal, in mammals and birds, and some species that engage in one, do not engage in others. Other types of play typically found in humans, such as construction (building) play, social-dramatic play, language play, pretense, and games with implicit or explicit 'rules', are a bit more complicated, but rudimentary versions of all may be found in non-human animals as well. Even social play in animals can be extraordinarily complex. When animals play, especially when agonistic behavior patterns are employed involving hits, bites, attacks, leaps, pins, and so forth, it is important that both participants (and there can be more) understand that they are engaging in play and not real fights. Thus, play signals are often employed to signal 'let's play' and these can be postures, such as the play bow in dogs, play faces in primates, and even play odors in rodents, among many examples.

Such signals are only a first step in play. To maintain a playful encounter, both partners need to intermittently convey that they are still playing, even as the play fighting becomes more intense. Thus, play signals can become *metacommunication signals*, signals about other communicative interactions. Thus, play fights in dogs are periodically punctuated with play bows. Most social play involves competition, including pinning in rats, tug-of-war in dogs, and king of the hill in goats. Whereas in serious fights and competitive encounters, the goal appears to be to defeat the opponent as quickly as possible and gain the resource in contention, this is not the dynamic underlying play. In social play, the goal seems to be to continue the encounter as long as possible and have partners willing to engage in an interaction again in the future. Thus, while in serious fights, the larger, stronger, or more skillful partner wins and defeats the other one, such behavior largely defeats the reason animals engage in socially playful interactions. Consequently, in many species, we observe a phenomenon termed *self-handicapping*. This refers to the fact that in play interactions, just as when adults play with children, always defeating your opponent is not the best way to ensure future interactions. So, what is found in many social play interactions is turn-taking and an alternation of who wins and this often involves the better player not competing up to his or her ability in order to maintain the interaction or the probability of future ones. This may be one of the most important aspects of play, as some recent empirical and modeling work indicates that through social play, animals learn such things as fairness and how to interact in socially useful, rather than destructive, ways. Such skills and knowledge may also help individuals in courtship and other important endeavors.

From an evolutionary perspective, play has originated numerous times in animals throughout evolutionary history and has altered course in many ways, even in the most playful mammals. Thus, adult play in monkeys can differ in type and amount dramatically even in closely related species. Furthermore, sex differences in the amount or type of play are pronounced in many species. Hormones underlie such differences in many cases, but such hormonal bases are themselves related to evolutionary history and behavioral ecology, including mating systems, foraging and fighting modes, type of predators and other dangers, amount and extent of parental care and protection, and so on and so forth. Thus, a satisfactory play ecology needs to recognize that play taps into ancient behavioral systems that manifest themselves in many species.

Development and the Adaptive Function of Play

Play is also a developmental phenomenon and different kinds of play may have different developmental trajectories even in the same species, such as locomotor, object, and social play in domestic cats. Furthermore, different features of play may have their own developmental pacing; this has been most well-worked out in rat play fighting. Broad comparative analyses indicate that play in young animals is related to the relative length of the juvenile period, and to this extent, the ideas of Groos have some support. On the other hand, there are many exceptions. We now have quite a few descriptive studies on the development of play but, as mentioned above, few empirical demonstrations of the function of play. These need to be experimental to be more than just suggestive. Also, attempts to deprive animals of play almost invariably affect other aspects of behavior than play itself, such as social or object interactions of a nonplayful sort. It has, thus, been hard to go from suggestive and plausible to documented and proved. The work on rats, as reported by Pellis and Pellis, is the most sophisticated experimentally and shows that one needs to examine the execution of the specific movements involved in play, whether refinements in them cross over into improvements in 'important' tasks such as foraging or courtship behavior, and what specific kinds of deprivation have what consequences. The study of the functions of play, so much the focus of research, is only recently showing progress.

The Brain and Play

While play has now been identified in some invertebrate and ectothermic vertebrate animals, most of our knowledge concerning the neural bases of play is based on research in mammals, especially laboratory rats. During the 5–8 weeks after birth, postweaning rats are extremely playful in terms of engaging in play fighting. Until recently, most inferences on the role of the brain in play had to do with looking for correlations between playfulness and overall brain size across species in comparative studies of mammals, both placentals and marsupials. There is a roughly positive correlation, but this relationship largely disappears when phylogenetic correlations and finer within taxa (e.g., primates and canids) are looked at in some detail. What about the telencephalon or cerebral cortex, the more 'advanced' or later evolving brain structures. Data generally suggest that the cortex is not necessary for the basic motivation to play and the movements involved, at least in rats. However, the cortex does allow the animal to play much more flexibly. Indeed, the motor cortex is implicated in developmental changes in the defensive tactics used by rats but does not seem to be involved in changes in the frequency of instigating attacks in play. There are also relationships between the areas of the brain such as the hypothalamus and amygdala, and social play in primates. In fact, many parts of the brain contribute to play, understandable when the diversity of play is considered. Thus, the cerebellum (posture and rapid movement coordination), limbic system (emotion), and other areas are involved. But to be more specific on how play is neurologically organized is largely impossible at this point, except for the laboratory rat.

Experimental research using a variety of techniques including ablation, imaging, and pharmacologic studies is showing that one can measure both those areas of the brain involved in various types of play and ways of manipulating play. Pellis and Pellis (2009) present an in-depth treatment of play and its neurological bases in mammals. The role of specific areas of the developing rat brain in social play is being disentangled via ingenious experiments. It thus seems that play does have important consequences, but play itself needs to be carefully dissected before convincing and specific conclusions can be discovered. Comparative work is also urgently needed, and parallel studies in several species, including playful birds, would be most useful.

The Origins and Evolution of Play

The importance of play and its role in an animal's life and development may, as we have seen, differ greatly, even at the simple level of its causal mechanisms and developmental consequences. Such differences can even occur in the same species. The mechanisms and consequences of play can be categorized into three groups, though of course, in reality, a continuum most likely exists. Thus, there is *primary process play* that is somewhat atavistic and due to boredom, low behavioral thresholds, immature behavior, excess metabolic energy, and other factors with no necessary long-term effects, good or ill. In contrast, *secondary process play* helps maintain the condition of the animal physiologically, behaviorally, and perceptually. For example, physical exercise may be necessary for maintaining cardiovascular functioning and body flexibility, and mental games may aid in slowing the effects of senile dementia. Finally, there is *tertiary process play* that may be crucial for reaching developmental milestones, cognitive accomplishments, social skills, and physical abilities. It is not currently known which examples of play in human or non-human animals fall under which rubric and at which times in life. Neither are the specific consequences known for different kinds of play, with the notable exception of social play in rats. Even in rats, each of the movements may contain primary, secondary, and tertiary elements. Do play fighting and competitive games foster war and aggression or a sense of fairness and the necessity of rule following? Such questions may not be easy to resolve, but the field needs to keep an open mind on them and help provide answers and not accept assertions that fit our respective ideologies.

Rather than search for the 'true' or 'real' meaning of play, as if it is a unitary phenomenon, the conceptual framework outlined above encourages examination of factors in both the environment and the organism that facilitate the performance of play. Some kinds of play are more individually or socially adaptive than others. Thus, we must not forget that hazing, bullying, animal cruelty, gambling, risk taking, compulsions, and addictions of many kinds can have their origins in play. Ironically, it was work on reptiles, which did not seem to play much if at all, that led to the ideas underlying Burghardt's Surplus Resource Theory. Reptiles, lacking parental care, must largely survive on their own with little parental care providing them with nutrition, protection, and time to engage in behavior not directed toward immediate ends. Furthermore, reptiles have a physiology that is generally not conducive to the sustained vigorous behavior often seen in play. Reptiles also do not possess the rich behavioral repertoire of limb, body, facial, and other body parts of many mammals and usually operate on slower time scales, having a metabolic rate averaging only 10% that of the typical mammal. Still, as noted above, some reptiles do play, as do fish and other 'lower' animals. By examining those animals that do and do not play, and also looking at the great extent of play diversity in mammals and birds, one sees that several major groups of factors underlie the surplus resources that allow animals to play and to engage in behavior not totally needed for current survival demands.

Some organismal factors that facilitate play are good health, a physiology conducive to vigorous and sustained activity, and a diet that can sustain such behavior. Developmental factors, such as having parental care, allowing the animal to explore and play in relative safety, and sufficient time to do these, are also important, as well as possessing a rich repertoire of instinctive and motivational resources. Ecological factors, such as weather, potentially dangerous environments (trees, water, and predators),

and foraging styles, along with social factors, such as the type and number of potential play partners and social openness/rigidity, affect play in other species and certainly do so in people. Individual differences in play propensity and skills are found in human and non-human alike. Such differences provide the raw variation needed for natural selection, including sexual selection, to operate its transformative magic. Evolutionary and ecological considerations thus help explain why some species play and other less so, or not at all, as well as variation within the same species in play. We are just entering into a time of exciting and innovative research on play in all its manifestations and complexities.

See also: Cognitive Development in Chimpanzees; Domestic Dogs; Innovation in Animals.

Further Reading

Bekoff M and Byers JA (eds.) (1998) *Animal Play: Evolutionary, Comparative, and Ecological Perspectives.* Cambridge, UK: Cambridge University Press.

Burghardt GM (2005) *The Genesis of Animal Play: Testing the Limits.* Cambridge, MA: MIT Press.

Fagen R (1981) *Animal Play Behavior.* New York: Oxford University Press.

Groos K (1898) *The Play of Animals.* New York: Appleton.

Martin P and Caro T (1985) On the function of play and its role in comparative development. *Advances in the Study of Animal Behavior* 15: 59–103.

Pellegrini AD and Smith PK (1998) Physically active play: The nature and function of a neglected aspect of play. *Child Development* 69: 577–588.

Pellegrini AD and Smith PK (eds.) (2005) *The Nature of Play: Great Apes and Humans.* New York: Guilford Press.

Pellis SM and Iwaniuk AN (2000) Adult-adult play in primates: Comparative analyses of its origin, distribution, and evolution. *Ethology* 106: 1083–1104.

Pellis S and Pellis V (2009) *The Playful Brain: Venturing to the Limits of Neuroscience.* Oxford, UK: Oneworld Press.

Sutton-Smith B (1997) *The Ambiguity of Play.* Cambridge, MA: Harvard University Press.

Playbacks in Behavioral Experiments

G. G. Rosenthal, Texas A&M University, College Station, TX, USA

Introduction

Defined in an influential volume by Peter McGregor as 'the technique of rebroadcasting natural or synthetic signals to animals and observing their response,' playback is to animal behavior what the polymerase chain reaction is to molecular biology. From neuroethology, to behavior genetics, to animal cognition, it is difficult to conceive of any area of animal behavior where playback experiments have not made a major contribution. Playbacks provide an analytical approach to studying how animals respond to stimuli, where an experimenter can quantitatively manipulate some signal components while holding others constant.

Hunters and herders have long used artificial stimuli to manipulate animal behavior, and the mirrors, painted models, and chemical swabs used since the early days of ethology could well be considered playback stimuli. The term 'playback,' however, is generally applied to the electronic presentation of temporally patterned stimuli applied in an experimental setting. Hundreds of studies across numerous taxonomic categories have used playback of visual and acoustic cues, while a smaller number have presented electrical and vibrational stimuli. While visual, vibrational, and electrical playbacks are mainly used in laboratory settings to study perception, cognition, and communication, acoustic playbacks are frequently performed in the field, where they can be used to census individuals, lure them to capture, or quantify territory size.

Acoustic Playback

Playback of acoustic signals dates back to the end of the nineteenth century, when the newly invented gramophone was used for audio playback of conspecific signals to rhesus macaques; subjects would thrust their arms into the gramophone's horn in search of the other monkey. By the mid-twentieth century, inexpensive and portable equipment for recording, analysis, and broadcast made sound playback broadly accessible to researchers. Sound playback is ubiquitous in studies of vocal communication in birds and anurans, and has also been widely used in fish, mammals, and insects. Most studies have focused on signaling in terrestrial environments, but there is a growing body of work on acoustic playback to aquatic animals.

Stimulus Preparation

The most straightforward type of stimulus in an acoustic playback is simply a recording of a natural vocalization. Numerous experiments have compared responses of subjects to conspecific versus heterospecific calls, local versus foreign song dialects, or vocalizations of familiar versus unfamiliar individuals. Playback of unmanipulated recordings can also be used to obtain demographic information, for example by transect counts of the number of males that respond to conspecific vocalizations. Playback of recorded sounds can also be used, in the absence of a receiver, to measure attenuation and degradation of signals. This is typically done by recording sound at a series of standard distances from a speaker. The experimenter can compare how different vocalizations are affected within the same environment, or alternatively compare attenuation and degradation of the same signal across environments.

Natural recordings are often edited before presentation. Until the late twentieth century, natural sounds were recorded on magnetic tape, a variety of analog electronic devices used for filtering, and signal components repeated, excised, or rearranged via manipulations of audiotape. Contemporary editing is done on digitized sounds, using sound-editing software such as ProTools or Signal. At a minimum, editing involves application of band-pass frequency filters to minimize background noise and remove extraneous sounds. Different components of a signal can be 'cut and pasted,' for example to increase or decrease the interpulse interval in a repeated call, or to evaluate the effect of song syntax (the structure of distinct vocal elements, or 'syllables') on receiver response. Harmonic components of a signal can be removed, amplified, or attenuated, and signal components can be selectively accelerated or decelerated. These kinds of manipulations have proved invaluable in analyzing how receivers attend to signals. For example, in the *túngara frog* (see entry) *Physalaemus pustulosus*, males produce a two-component sexual advertisement call comprising a tonal frequency sweep (a 'whine') and one or more broad-band, high-energy harmonic 'chucks.' By adding and removing whines and chucks, researchers were able to determine that the whine is both necessary and sufficient to elicit a female response; females respond positively to whines alone, but fail to attend to isolated chucks. However, adding chucks to a call and increasing chuck number both increased the attractiveness of this compound signal.

Synthetic stimuli, where acoustic signals are generated based on specified parameters, offer the most control and flexibility over stimulus design. Sound synthesis allows experimenters to independently decouple specific variables and create hypothetical, mathematically specified stimuli that are nonexistent in nature. Although analog synthesizers are still widely used by musicians, sound synthesis in animal behavior is now performed digitally using a variety of software packages. There are two major classes of approaches to sound synthesis. Tonal synthesis represents sounds as sums of sinusoidal functions varying in frequency, amplitude, and phase. Sinusoidal functions can be convoluted with any number of mathematical functions to produce, for example, signals that ramp up in amplitude over time or vary in pulse repetition rate. The parameters in tonal synthesis are all based on the physical properties of the sound itself, and are independent of the signaler.

By contrast, physically based synthesis, which is less widely used in animal communication studies, reconstructs sounds based on a model of the sound production system; for example, linguists make extensive use of models of the human vocal production apparatus in generating speech sounds for playback studies.

Synthesized sounds are widely used in neurophysiological studies, where they can be used to determine neural responses to specific acoustic parameters. 'Morphing' one signal into another permits investigation of categorical perception. For example, acoustic intergrades between the 'ba' and 'pa' phonemes in human speech are always perceived by subjects as one or the other. A particularly powerful application of sound synthesis involves the ability to generate entirely novel stimuli. Several studies of túngara frogs, mentioned earlier, have assayed female mating preferences for the inferred calls of ancestral taxa, which are synthesized using acoustic parameters inferred by phylogenetic reconstruction.

Stimulus Presentation

The output device for audio playback is typically a commercially available loudspeaker, including underwater speakers for aquatic systems. The most important consideration is that the frequency–response curve of the speaker be relatively flat over the range of frequencies and sound pressure levels being played back. For stimuli outside the range of human hearing (like infrasonic or near-infrasonic calls in elephants), speakers have to be specially modified. Ultrasonic playback (e.g., to bats or mice) requires specially designed electrostatic speakers.

Playback of acoustic signals is typically sequential, with stimulus order varied and interstimulus intervals designed so as to minimize order effects. Simultaneous-choice experiments, where calls are paired antiphonally on opposite sides of an arena, is particularly common in laboratory experiments on frogs and insects. Studies of acoustic localization or interference can have multiple sounds playing out of multiple audio channels into an array of speakers.

Depending on the receiver, a variety of assays are used to measure receiver response. For examples, males typically respond vocally to acoustic signals of other males, and these responses can often be quantified automatically with the appropriate software. In some species, receivers will exhibit specific postural changes in response to an acoustic signal (e.g., a 'copulation solicitation display' involving raising of the tail in female birds). More general response measures include phonotaxis (approach to a speaker), habituation/dishabituation approaches (particularly useful in psychophysical assays of just meaningful differences), and changes in locomotor activities (e.g., number of perch changes in birds).

Electrical and Vibrational Playback

Electrical signals are produced by specialized electric organs in gymnotiform and mormyriform fishes, and substrate-borne vibrational, or seismic, signals have been most extensively studied in hemipteran insects. Despite vast differences in signal production and transmission, these can be parsed into frequency, temporal, and amplitude spectra just as acoustic signals can. For electrical playback, recorded or synthetic signals are transmitted via an amplifier to paired electrodes, often at either end of a plastic pipe that serves as a shelter. Electrical signals emitted in response to stimuli can then be recorded by the experimenter. For vibrational signals, the amplifier is connected to an electromagnet which vibrates the substrate, typically a plant, and vibrational responses are recorded with an accelerometer. Playback of white noise allows the experimenter to determine the response function of the substrate, and accelerometer recordings of playback stimuli provide an assessment of signal fidelity.

Playback of Visual Stimuli

Presentation of moving images to research animals dates back to the 1960s. As with acoustic playback, rhesus macaques provided proof of concept of visual playback, in this case by attending preferentially to ciné stimuli over stills or mirror images. Most studies of visual playback have focused on presenting subjects with moving images of other animals (conspecifics, closely related heterospecifics, predators, or prey) performing behaviors of interest to the experimenter.

Stimulus Preparation

The ability to manipulate visual stimuli has gone hand in hand with advances in technology available to researchers; one early study, for example, evaluated the importance of temporal structure in *Anolis* displays by comparing the aggressive response of subjects to films played forwards and backwards. Analog video-editing techniques like chroma-keying (colloquially known as 'green-screening,' where a selected color range can be changed or overlain with another video stimulus) allowed experimenters to standardize background features and isolate specific behaviors. Such approaches allowed of a range of tests on how visual cues in isolation elicit a particular behavioral response. For example, roosters produce 'ground alarm calls' in response to a video of a ground predator on an adjacent monitor, and 'aerial alarm calls' in response to a video of a looming hawk on an overhead monitor.

The ability to digitize video represented an important methodological advance. Numerous studies in the 1990s applied frame-by-frame manipulation of video sequences. This is a tedious process in which a video sequence is digitized, individual frames are imported into an image-editing program, and each frame is individually altered. Numerous studies used this approach to independently decouple stimulus behavior and morphology.

At 30 frames per second for the NTSC analog video standard used in most of the Western Hemisphere and 25 frames per second for the PAL standard used in most other countries, this represents a time-intensive process even for brief sequences; and each procedure results in only a single-parameter manipulation in a single exemplar. Frame-manipulated video is prone to producing a number of artifacts, including spatial discontinuities and aliasing effects (visual distortions caused by under-sampling) between a manipulated trait and background features. Moreover, it is difficult to manipulate two-dimensional projections of animals performing behaviors in three dimensions. Despite these concerns, frame manipulation can preserve much of the spatiotemporal complexity of an original video sequence while allowing broad flexibility in morphological manipulations. It is particularly appropriate for creating stimuli of short duration so that motion is confined to the plane of the screen.

Most contemporary visual-playback studies use some form of synthetic computer animation: a familiar feature of popular films and television shows. Like synthetic acoustic, electrical, and vibrational signals, synthetic animations are mathematical descriptions of a set of features chosen by the experimenter. Visual signals are fundamentally distinct, however, in that while these other signals involve energy generated by the signaler, visual signals typically involve the manipulation of incident light by the receiver, for example sunlight reflecting off feathers or skin. Further, visual perception depends critically on stimulus contrast with background elements. Synthetic acoustic and other generated signals are typically presented in isolation, or occasionally coupled with masking noise or interfering cues. With synthetic animation, the experimenter needs to generate an entire visual scene, specifying the color, intensity, and spatiotemporal distribution of both the light regime and the visual background.

Like sound synthesis, synthetic animations are parameter-based. This makes it possible to measure features of interest, like the relative size of morphological ornaments, on animals and their habitats (like the temporal frequency distribution of moving background vegetation) and then apply these to a synthetic model. This also allows the experimenter to generate morphologies and behaviors outside the range of natural variation; for example, chimeric individuals bearing traits of more than one species, or inferred ancestral states.

A very large number of parameters is needed to specify even a simple visual scene; for example, the appearance of a single point on an animal's skin depends on how the color, brightness, opacity, and shininess of the point interact with light as a function of intensity, wavelength, and incident angle, all of which in turn vary over space and time depending on the animal's orientation, position, and posture relative to light sources, other objects in the scene, and the receiver. Since it is unfeasible for experimenters to collect quantitative data on every parameter of a scene, some parameters are often fixed to arbitrary values, while others are based on individual exemplars. For example, body patterns are often based on digital photos of a single individual. Complex motor patterns, meanwhile, are often derived by superimposing a synthetic model over video footage of a live animal. This technique, called rotoscoping, dates back to early twentieth century ciné animations.

Stimulus Presentation

Visual stimuli are generally presented on cathode-ray-tube (CRT) video or computer monitors. Flat-screen monitors, which provide a limited viewing angle, have proved to be less generally appropriate. Color fidelity, spatial and temporal resolution, and the lack of depth cues can pose problems in interpreting responses to video stimuli; these issues are discussed in more detail below.

By contrast with acoustic playback experiments, where sounds are routinely broadcast to animals in the field, there is only one published study of video playback in the field, where a monitor placed in *Anolis* lizard territories elicited stereotypical responses from males and females. Video playback in the field is problematic, since ambient light tends to make it difficult to detect images on a video monitor, and since detection of a video is contingent on being in the line of sight of the monitor. *Robots* (see entry) may be more appropriate for field playback studies.

In the laboratory, video stimuli are typically presented adjacent to an arena, cage, or aquarium that restricts the animal from moving behind the monitor, thus increasing the likelihood that visual stimuli are detected. As with audio playback, presentation may be sequential, with stimuli presented in succession, or simultaneously, with stimuli usually presented on opposite sides of the arena. Response is assayed by the performance of specific behaviors directed at a particular stimulus; for example, *Anolis* lizards perform a characteristic 'head-bob' display in synchrony with a simulated intruder on video, while chickens produce a 'ground alarm call' when confronted with a video of a raccoon. Many video playback experiments, however, rely on simple proximity measures as an assay of preference. By itself, proximity does not provide information about the behavioral context in which an animal is responding, raising concerns that animals may be attending to artifacts in stimulus representation (see section Signal Fidelity).

Multimodal Playbacks

In almost all communication systems, receivers are likely to attend to multiple sensory modalities. Given the large number of playback studies in each individual modality, it is surprising that relatively few studies have used a multimodal approach. In both pigeons and túngara frogs, female receivers respond more strongly to a combination of visual and acoustic cues than to either cue alone. In the frogs, a combination of synthetic and edited-video stimuli was used to show that females attend specifically to the form and inflation pattern of the male vocal sac – a moving rectangle of the same size was no more effective than a blank screen at eliciting phonotaxis. A similar result is obtained by combining visual cues with substrate-borne vibration playback. For both wolf spiders and their predators, a combination of video and vibrational cues is more likely to elicit a response than either cue alone.

Interactive Playbacks

In nature, communication is an interactive process. Signalers dynamically adjust signal output, signal type, and signal parameters, depending on changes in orientation, position, and behavior in both intended receivers and eavesdroppers (e.g., predators or sexual rivals). Interactive playback attempts to mimic this property of senders: signal presentation is determined by receiver behavior. Typically, the experimenter specifies a set of rules (e.g., matching or escalating calls emitted by the subject in an aggressive display). Interactive playbacks are often manual, where the experimenter identifies a subject behavior and plays a signal in response. Since one of the benefits of playback is that subjects are presented with

consistent, repeatable sets of stimuli, interactivity in playback experiments may not always be desirable, particularly if experimenter subjectivity or error is an important factor. A potentially more rigorous approach is to have real-time signal-processing algorithms automatically determine the interaction. This approach has been used successfully with acoustic and electrical signals; more recently, real-time tracking of subject behavior has been used akin to a video-game controller, determining the behavior of an animated stimulus on screen.

Potential Hazards of Playback Techniques

Playbacks offer a degree of control and precision that is unavailable from observational studies or direct manipulation of live exemplars; by their very nature, therefore, they are prone to a number of potential pitfalls that may limit the external validity of experimental results. As noted in the preceding section, the appropriateness of interactivity is a matter of some debate: a signal that is not contingent on subject behavior may elicit artifactual responses, while one that is interactive may make it more difficult to compare responses across trials. Two main issues have been the focus of attention: pseudoreplication, whereby playbacks fail to adequately sample natural signal variation, and signal fidelity, whereby playbacks fail to represent signals appropriately.

Pseudoreplication

With regard to playback experiments, pseudoreplication was defined by McGregor and colleagues as 'the use of an *n* (sample size) in a statistical test that is not appropriate to the hypothesis being tested.' Many early acoustic playback experiments would use, for example, a single recorded exemplar per species when studying conspecific recognition in a territorial context. Without adequately sampling responses to multiple exemplars, it is impossible to discern whether differences in response are due to differences in stimulus classes or due to idiosyncratic differences among individuals. This problem can be addressed by using multiple natural exemplars and performing appropriate statistical analyses. Synthetic stimuli, which are generated from specified parameters, offer the opportunity to eliminate this idiosyncratic variation. Parameters can be modeled on data sampled from multiple natural signals. Even synthetic stimuli, however, leave open the possibility that responses depend on interactions between a manipulated parameter and a parameter that is arbitrarily fixed. For example, the attractiveness of a repeated acoustic mating signal might depend on an interaction between pulse rate and dominant frequency. Merely holding dominant frequency constant and varying pulse rate would provide an incomplete picture of how sexual selection acts on the signal. This problem is particularly difficult with video animations, where the number of possible parameters is very large; in practice, many animations use sample data for a tiny fraction of model parameters (typically morphological traits), and behavior and texture are modeled after single exemplars.

Signal Fidelity

Signal fidelity has garnered particular attention for visual playback studies, but is nevertheless an important concern in other modalities. This is particularly the case for sound in aquatic systems, where signals in small aquaria are distorted by reverberation and resonance. In the field, the directionality of sound and the transmission of sound through acoustic microenvironments may also alter acoustic signals in artifactual ways.

Video playback involves representing a three-dimensional signal on a two-dimensional surface, breaking a continuous visual stimulus into discrete, pixilated still frames at temporal intervals on the order of 33 ms, and collapsing spectral radiance and irradiance functions into red, green, and blue outputs on a monitor. In the absence of real depth cues, a large object far away subtends the same visual angle as a small object close up. Occlusion cues (static objects at varying apparent distances from the foreground) can provide depth information in a two-dimensional image.

The standard refresh rate of most video monitors (25–30 Hz) is just above the flicker-fusion threshold for humans. Many animals, particularly birds, have higher flicker-fusion frequencies. Depending on monitor type and the species being tested, subjects may perceive a series of static 'slides' as opposed to a continuous image. Newer computer monitors can refresh over 120 Hz, which is suitable for most species.

Color fidelity is perhaps the most intractable problem with video outputs for playback to non-human animals. The red, green, and blue phosphors or pixels of television or computer monitors are tuned to match the sensitivity of human red, green, and blue cone photoreceptors. By differentially stimulating each photoreceptor class, a video image is able to represent a wide range of illusory colors, or metamers. The correct perception of these metamers, however, is contingent on matching the three output classes to the sensitivity of receivers. Spectral tuning varies widely among and even within species. Researchers have developed a methodology for adjusting monitor color output to match sensitivity of known photoreceptor classes in a study species; problems arise, however, when testing animals with more than three color receptors. Colors can, however, be simulated by carefully selected color filters over the output screen. A harder problem is posed by the many highly visual animals, including many birds, fishes, and arthropods, that perceive color into the ultraviolet. Since video monitors do not emit directed ultraviolet light, current technology lacks a way to represent UV signal components in an experimental setting.

Despite these caveats, playback across modalities has been an indispensable tool for understanding communication systems. Numerous studies have quantitatively compared responses with playback and live stimuli; when transmission among live animals is prevented in modalities other than the one being played back, playback stimuli are typically as effective as live animals in eliciting responses. Moreover, there are few, if any, studies where predictions made by playback experiments have been directly refuted by observational work or by complementary physiological or molecular measures.

Conclusion

Playback techniques have grown hand in hand with available technology. Acoustic, electrical, and vibrational playback collectively represent a mature technique for experimentally manipulating emitted signals. Visual signals are both more complex and more contingent on the environment in which they are produced and perceived, but ongoing advances in image acquisition, analysis, and presentation continue to expand the scope of questions that can be addressed using video playback.

See also: Acoustic Signals; Electrical Signals; Experimental Design: Basic Concepts; Olfactory Signals; Robotics in the Study of Animal Behavior; Vibrational Communication; Visual Signals.

Further Reading

Hopp SL, Owren MJ, and Evans CS (eds.). *Animal Acoustic Communication: Sound Analysis and Research Methods*. Berlin: Springer-Verlag.

Kroodsma DE, Byers BE, Goodale E, Johnson S, and Liu WC (2001) Pseudoreplication in playback experiments, revisited a decade later. *Animal Behaviour* 61: 1029–1033.

McGregor PK (ed.) (1992) *Playback and Studies of Animal Communication.* New York, NY: Plenum Press.

Oliveira RF, McGregor PK, Schlupp I, and Rosenthal GG (eds.) (2000) *Special Issue: Video Playback Techniques in Behavioural Research. Acta Ethologica.* 3, s. 1.

Poultry: Behavior and Welfare Assessment

J. A. Linares, Texas A&M University, Gonzales, TX, USA
M. Martin, North Carolina State University, Raleigh, NC, USA

Introduction

Poultry behavior is a useful tool in the assessment of the welfare of poultry. Historically, evaluation of production parameters has been an effective way to assess poultry welfare. In general, a healthy bird will produce better than one that is sick or stressed. Although historic poultry behavioral research has helped us develop management practices to maximize production, we cannot adequately assess welfare using production parameters alone. Evaluation of production parameters alone is reactive rather than proactive, mostly considers lost production rather than potential production, and does not take into consideration the birds' needs based upon the natural state. Therefore, an understanding of poultry behavior and its continued application should be an integral part of the assessment of the welfare of poultry. The use of behavior in the assessment of poultry welfare is advantageous because it serves as an early indicator of problems, can be performed without specialized equipment, is instantaneous and may lead to increased unrealized production. The assessment of the welfare of poultry occurs at the confluence of various disciplines such as ethology, ethics, genetics, poultry science, and veterinary medicine. The proper assessment of poultry behavior and welfare requires an interdisciplinary and integrative approach. Behavioral assessments are beneficial during routine flock evaluations or when conducting poultry welfare audits.

This article is designed to be an introduction to practical items in the use of behavior in the assessment of the welfare of poultry and to serve as a guiding tool in the pursuit of further knowledge and experience. The article is written from the veterinary perspective based on a review of existing poultry behavior literature and our experience with poultry medicine, diagnostics, and commercial poultry production. Discussion will be centered mainly on chickens as they are the most abundant and the most studied poultry species.

Uses of Behavior to Assess the Welfare of Poultry

Poultry behavior is a good indicator of welfare and the performance of some behaviors is important to the welfare of poultry. Poultry behaviors include social behaviors, such as breeding behaviors, competitive behaviors and aggression to determine social order; individual behaviors such as broodiness, roosting, pecking, foraging, ground scratching, preening, dust bathing, head shaking, head scratching, feather ruffling, beak wiping, wing-leg stretching and wing flapping; and fear, distress, or frustration responses, such as flightiness, displacement preening or pecking, pacing, and aggression. The presence, absence, frequency, and intensity of a behavior are the parameters of its objective assessment.

To define and organize this section, we found it useful to employ four central components of animal welfare: health, natural living, mental states, and biological function. A discussion on the practical aspects of the use of behavior to assess poultry welfare follows. These components will overlap in flock assessment.

Health

An understanding of poultry behavior is an integral part in the assessment of their health status. Assessment of the vocalizations of an undisturbed flock to determine health status should be done before and after establishing visual contact. While active, healthy birds make certain vocalizations. Over time, one develops an ear for the sounds of an undisturbed healthy flock. A serious deviation from normal during daylight hours would be relative silence, or excessive vocalization. Infectious Coryza is a highly contagious chicken disease that is known to cause a virtual silence in an infected flock. Young turkeys with severe enteritis vocalize incessantly with high-pitched calls. One could quite commonly hear coughing and sneezing in a flock with severe respiratory disease. However, mild respiratory signs within a flock, such as with a vaccine reaction, may not be heard easily if drowned out by other vocalizations. As an example of practical use of poultry behavior to facilitate the assessment of flock health, one can utilize normal fear responses. Flocks become momentarily silent in response to a brief high-pitched whistle or loud hand clapping. Before vocalization resumes, one can hear birds with mild respiratory disease that cannot stop making abnormal respiratory sounds. This enables the evaluator to hear low-volume respiratory sounds.

On visual assessment, the first observations relate to appearance, posture, and locomotion. While active, healthy poultry stand holding their head relatively high, wings folded close to their body, and legs extended directly under their body (**Figure 1**). Preening, the act

Figure 1 Normal bird posture. Standing with head relatively high, wings folded close to body, and legs extended directly under their body.

Figure 2 Sick turkey. Note crouched posture with head drawn close to the body, eyes closed, feathers ruffled and soiled.

Figure 3 Birds are clearing a path while people are walking through a house. If the path does not fill in a short period of time, birds may be ill or reluctant to walk.

of smoothing or cleaning the feathers with the beak, is common. Sick poultry tend to be in a crouched position with head drawn close to the body, eyes closed, the feathers ruffled and often soiled (**Figure 2**). Severely diseased and depressed birds may be unresponsive to external stimuli. Healthy flocks are responsive to stimuli as part of their fear responses, such as reacting to a foreign presence, noises, bright lights, and sudden movements. As one approaches a healthy flock, birds may walk, run, or fly away. Normal avoidance responses can be used to assess flock health status. When walking among a healthy poultry flock in floor-type housing, the birds will typically move away clearing a path in front but they also tend to fill the void left behind relatively quickly (**Figure 3**).

Poultry express varying degrees of fear responses depending on the species, breed, environment, housing type, previous exposure to stimuli, and health status. Certain species and breeds, such as domestic turkeys and floor-raised brown egg layers are relatively docile and will approach people after a brief period of avoidance. Decreased avoidance, failure to fill the void left behind on one's path, or a failure to approach could be an indication that the birds are reluctant to move and could possibly be severely ill. However, white leghorn chickens and some meat chicken lines tend to be more flighty, and sudden movements, a flash of light or a loud noise could cause a commotion whether the birds are in cages or on the floor. In floor-raised birds, flightiness may also help assess management, as excessive flying and running beyond what is

anticipated for the breed could indicate that the birds are less familiar with human presence and may not be evaluated frequently (**Figure 4**). Excessive flightiness may also predispose birds to trauma.

Specific systemic, musculoskeletal, or nervous system diseases may also alter normal stance and locomotion. For example, birds with spinal lesions may sit back on their hocks and move backward for avoidance rather than forward. These birds may range from bright, alert, and responsive to severely depressed, based upon the severity of the lesion and ability to access feed and water. Also, one could observe a decrease in male strutting and mating behaviors in a breeding flock with a high incidence of lameness due to conditions such as bumblefoot.

The degree of preening and cleanliness is relative to the environment. Commercial poultry usually have white feathers, making cleanliness easy to assess. Poultry housed in a dirty environment may not be able to stay clean and may be at greater risk of disease because of the opportunity for pathogens to multiply in dirty environments.

Figure 4 Excessive flightiness can predispose birds to trauma and may be an indication that birds are not being inspected frequently.

Figure 5 Increased mucus production. Observe clear mucus from the nostrils.

Figure 6 Birds with upper respiratory infections may wipe their beaks on their shoulders, which will give the birds a brown collar appearance over time. Notice brown discoloration of the feathers at the shoulders and base of the neck of the turkey in the center.

Deviation from normal could be related to disease. Birds with an upper respiratory infection, including conjunctivitis, produce excess mucus (**Figure 5**). Normal behaviors such as beak wiping on the shoulders and head scratching are used to keep their eyelids and nostrils clean. Over time dust and feed particles will cling to the excess mucus and feathers may become matted with dry, crusting exudate. Therefore, a brownish collar appearance along the shoulders may be the first observation on a flock with an upper respiratory infection or facial lesions (**Figure 6**). Similar observations could be made in chickens with corneal ulcers and reactive conjunctivitis due to excessive ammonia levels.

Pecking trauma could result in fresh or dried blood on the feathers, feather loss, abrasions, and lacerations. The location of the skin trauma could be indicative of the cause. Head and facial trauma in roosters are usually associated with fighting or male aggression. Poultry will often peck at other birds that look or act differently. An example of this behavior is neck and tail trauma in growing turkeys. Feather loss and trauma to the nape and back of hens is associated with excessive male mating behavior. This type of trauma may be severe and lead to increased hen mortality. Excessive breeding of hens may be due to male aggression or if the hen has neurological or musculoskeletal signs that resemble a breeding receptive posture or lordosis. Trauma to the toes may be associated with toe pecking in young poultry. Vent pecking in laying or breeder hens may occur for several reasons. After a bird lays an egg, they can prolapse part of the cloaca. This can be accentuated by obesity or underlying disease. Although the cloaca will invert back within the bird, other birds may become attracted to it and peck at it. This behavior may become habitual and spread within the flock and lead to cannibalism. In large flocks, excessive vent pecking could be first spotted by the observation of blood on the eggshells.

Natural Living

In order to appropriately understand certain behaviors, people often look at the behavior of ancestral poultry. For example, the need for dust bathing is often researched using jungle fowl, which are considered an ancestral chicken lineage. However, domestication over thousands of years and selective breeding over the last several decades make this comparison difficult. Now there is great variability in the presence or absence of certain natural behaviors. The domesticated birds' natural behavior differs depending on breed, genetic selection, and use. For instance, layer chickens have been selected for increasing egg production. If layer chickens start exhibiting broodiness, they will stop laying eggs. This is contrary to the goal of egg production. Therefore, genetic selection has, in part, generated lines that are less likely to become broody. Genetic selection of certain traits like egg production and decreased broodiness may be genetically linked to less desirable traits such as increased aggressive behaviors. Aggressive behaviors would be more similar to that of the ancestral jungle fowl in contrast to meat-type birds or broilers, where the genetic selection for egg production and decreased broodiness is not as strong. Because the selection parameters that are chosen for our domestic poultry breeds can be complex, it is critical for us to understand to the best of our ability the specific innate behaviors in the natural state of living for our poultry species and breeds. Behavioral insight can greatly improve poultry welfare by rapidly identifying problems when there are deviations from natural behavior. Future problems can be minimized by troubleshooting management conditions that may limit natural behavior, thus leading to future husbandry improvements for the health and welfare of the birds and to improved productivity.

Common poultry species such as chickens, turkeys, ducks, and geese are innately social animals. The natural social structure within a population of poultry leads to competition for resources, breeding behaviors, parental care, aggression and submission behaviors, pecking orders or hierarchical ranking, and synchronized behaviors. Poultry have a natural predilection for competition whether it is for mating, feed, nest areas, or preferred locations. Competitive exchanges are worked out through aggressive and submissive behaviors that occur naturally within a flock. Aggressive behaviors are more prominent in males and include posturing, pecking, and scratching. Posturing in male turkeys may start as early as 1 week of age and include fluffing and holding wings down to appear larger (**Figure 7**). Submissive birds attempt to hide behind other birds, in corners, under feed lines, roosts, or nest boxes.

Not all social pecking comes from aggressive responses from higher ranking birds. Typically pecks to the head and neck area are more aggression related. However, pecking at the vent or other areas can be for reasons other than aggression. For instance, vent pecking may be associated with birds pecking at a partially prolapsed vent

Figure 7   Posturing in male turkeys including fluffing of feathers and dropping wings down toward toes may start as early as 1 week of age.

after egg laying because it looks different, with redirected feeding behavior, or with nutritional deficiencies. Hierarchical status is not always linear as one bird may be dominant over the second but not necessarily over the third even if the second bird is dominant over the third. These pecking orders may become stable in smaller flocks; however, with larger flocks such as in commercial setting, these rankings will likely be constantly in flux. This may predispose flocks to more aggressive interactions as the hierarchical rankings change. Adequate access to resources such as feed and water can help minimize aggressive behavior.

Natural breeding behaviors are also part of the social structure of poultry species as well as individual behaviors such as broodiness. Female birds receptive to breeding will go into squatting position with the wings slighting out from the body and will not move. This is called lordosis. Hens, especially as they start becoming reproductively active, will be less discriminating of potential mates and may even go into lordosis when approached by human workers. As males become reproductively active, they have greater aggression and competition behaviors. Sometimes these aggressive behaviors can be directed at females if they are not reproductively active yet. In a comingled breeding operation, timing of reproductive activity between the males and females is critical. The timing of initiating reproductive activities is modified by lighting and feeding programs and is beyond the scope of this document.

The concentration of males to females within a comingled population also can decrease aggression while increasing fertility. In commercial chickens between 1–8 and 1–12 male to female ratio is desired with 1–10 optimal. Too few males may lead to some female birds within a population to start exhibiting hormonal changes to take on male characteristics such as crowing and aggression. Females with ovarian tumors may also develop male characteristics. In commercial broiler breeder houses, where breeder birds are typically comingled, management practices have moved toward an environment that allows some separation of breeding, nesting, male and female areas. This is designed to decrease aggressive behavior and maximize egg production and use of nest boxes. A floor area of litter is usually where breeding takes place, as the males have better coordination for breeding on the ground. Male feeders that are elevated to good height for the male birds are placed in the floor area and females are usually excluded by competition. Nest boxes and female feeders are usually placed on elevated slats. This provides the hens an area away from breeding males. Female feeders are constructed so that the males cannot access the feed well because of their larger heads. The slat area also provides subordinate birds a place to hide from aggressive birds.

Egg production for either food consumption or breeding purposes relies on continual egg laying. As such, genetic selection for increased egg production and decreased broodiness has been established as part of our poultry's domestication. Broodiness is the natural behavior in which birds stop laying and sit on eggs to incubate them. Often hens become more defensive of the nest when broody. In addition to genetic selection, there are environmental ways to reduce the chances of broodiness naturally. One is to make sure that you have even lighting, as areas that are perpetually dark may predispose birds to broodiness. Another is frequent collection or removal of eggs from the nest boxes or cages, as birds are more likely to become broody if they have a clutch of eggs on which to sit (**Figure 8**). Most commercial systems have automated egg removal from nest boxes and cages. Domestication and genetic selection of layer chickens has greatly minimized the natural broody behaviors, so much so that it is sometime difficult to get the birds to lay eggs in nest boxes in cage-free environments.

Even in chicken genetic lines without as much commercial selective pressure, there is great variability in broody and parental behavior. For instance, Leghorn strains historically are difficult to get to sit on their eggs and have maternal instincts toward hatched chicks whereas Silkies are often used to incubate and hatch other birds' eggs in mixed poultry environments because of their naturally broody and maternal tendencies.

Other behaviors exist in poultry, including roosting, pecking, foraging, ground scratching, preening, dust bathing, head shaking, head scratching, feather ruffling, beak wiping, wing-leg stretching, and wing flapping. These behaviors may be learned or innate and seen in individuals within a flock. The assessment of poultry welfare based on these behaviors is difficult because there may not be a discernable benefit or detriment associated with having a management system that does not support expression of these behaviors.

Figure 8 Hen becoming broody after being allowed to sit on eggs that were not collected.

Mental States

The evaluation of animal welfare requires the assessment of animals' mental states. Certain poultry behaviors have a strong motivation and poultry welfare may be negatively affected if the birds are not allowed to express these behaviors. Frustration, fear, and pain are well-documented negative mental states of poultry. Responses to frustration include increased aggression, displacement preening, and stereotyped pacing. Hens in battery cages, particularly light hybrid strains, show frustration due to the inability to perform nesting behavior. Frustration is demonstrated by stereotyped pacing and increased aggression during the hour before they lay an egg.

Increased aggression is observed when poultry have to compete for limited feeding space or when attempts to eat are frustrated. For example, broiler breeders are placed on feed restriction programs to prevent obesity, reproductive disorders, and other health problems. Feeding smaller amounts of feed everyday increases the competition for the feed, so that the dominant birds eat the majority of the feed leaving little or no feed for the more subordinate birds in the population. Therefore, skip-a-day and similar feeding programs have been put in place to accommodate this aggressive behavior so that there is more food on less days allowing the dominant birds to eat their fill and still have plenty of food available for the more submissive birds. Still, increased aggression among cockerels is observed on mornings when feed is absent. Frenzied behavior may also be seen on feed days in a restriction program and may cause piling of birds and trauma; however, this behavior may be modified by applying our understanding of the bird's mental state and responding to it. Signal light programs have been successfully used to decrease aggressive and frenzied behavior at the time of feeding especially if started when birds are first exposed to a restriction program. Signal light programs are implemented by turning on a single dim or tinted signal light 10–15 min prior to running the feed line. After the feed line is full, the house lights would be turned on. Birds become trained to line up at the feed lines prior to food being available without the frenzied activity. Signal light programs are best implemented as soon as the feed restriction program starts and should be performed at the same time every feeding day.

Postural changes such as crouching with eyes closed and vocalizations such as those turkeys with enteritis make (both described previously) may be signs of chronic and acute pain respectively. Regarding fear responses, alertness and avoidance are normal responses and signs of good welfare. These birds are prey species by nature, and fear and aversion may prevent a bird from entering a situation that could cause pain and distress. Extreme nervousness and hysteria are indications of poor welfare. Extreme nervousness and hysteria may be associated with high population densities in floor-type housing and large group cages, genetic predisposition or exposure to noise, sudden movement, or unfamiliar situations. Multiple risk factors such as dense floor-raised white leghorn chickens exposed to new people moving rapidly increase the chances of hysteria within a flock. Signs of hysteria include wild flying, vocalizing, and attempts to hide. Hysteria may result in skin trauma from scratches, spontaneous fractures, decreased feed consumption, decreased egg production, and reproductive disorders such as internal laying and peritonitis secondary to abdominal trauma.

Biological Function

Some poultry behaviors are innate and related to the general metabolism of the bird (e.g., regulation of body temperature). Management systems should be in place to optimize environmental temperature so that the birds do not have to exert excessive energy toward maintaining optimal body temperature. However, because the flock is made up of many individuals and each individual within the flock may have a slightly different comfort zone for temperature, management systems often work as a best guess for optimal house temperature. Also, the bird's ability to maintain body temperature may be altered by stressors such as accessibility of feed and water, competition and aggression within a flock, exposure to humid or wet conditions, drafty or arid conditions, excessive dust, exposure to pathogens, and disease status. Therefore, textbook guidelines for housing temperatures should only be used as guidelines, and great attention should be paid to the birds' behavior to help determine the best temperature and management conditions for the flock as a whole. Special consideration should be given to chickens during their first 2 weeks of age as they have difficulty in regulating body temperature until they are older. Feeders and waterers should be evenly distributed over a gradient of temperatures for birds this age so that birds that are too warm have access to feed and water equivalent to that of birds that are too cold. A flock of birds that are comfortable should be evenly spread over the space provided for them (**Figure 9**).

Birds exhibit several behavior changes when they become overheated. Primarily, when birds are too hot, they will try to avoid the source of the heat unless they are debilitated in some way. Areas under brooder pans or near the heat source will be vacant except for potentially a few birds (**Figure 10**). Birds can end up in the corners of pens or houses and may pile, which could lead to suffocation and lack of access to food and water. Birds will open mouth breathe to cool themselves and may also hold their wings away from their bodies or down near their feet. Birds that are too hot will also eat less feed and may consume excessive amounts of water leading to watery droppings or flushing.

Poultry can also respond behaviorally to conditions that are too cold. Birds will likely huddle together or near a heat source if provided. Birds avoiding drafty conditions may huddle in corners of the house or pen. Cold birds will fluff up, which is a behavior that traps air between the body and the feathers. The trapped air will then be warmed by the body of the bird acting as a cushion between the bird and the cold conditions. If conditions are too cold, birds may be reluctant to seek out feeders and waterers.

Unique Problems in Assessing Welfare via Behavior

A unique problem in assessing poultry welfare via behavior is its inherent complexity. Poultry behavior is affected by variables such as species, breed, sex, age, hormone levels, neurobiology, health status, environment, and management practices. Therefore, the assessment of

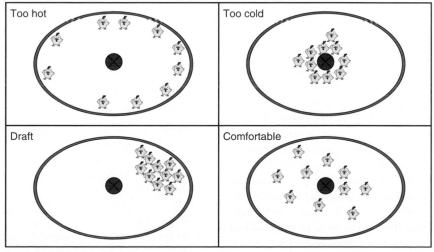

Figure 9   Diagram of how birds respond to environmental temperature. Red circle in the middle of each ring represents the heat source.

Figure 10 Birds are too hot near the brooder pans. Note how no birds are present directly under the hovering pans.

poultry welfare via behavior requires a broad knowledge base or interdisciplinary approach. The outcome of any assessment depends on the knowledge and experience of the observer and many observers may not be well versed in ethology, poultry science, and poultry medicine. An important consideration is that behavior occurs at the individual and flock levels. It is easier to assess individual behavior in small flocks up to 100. In larger flocks, individual behaviors may be overlooked. Another unique problem is the assessment of the mental state of poultry. A central question in the assessment of poultry welfare is whether a bird is in distress when it is prevented from performing a normal behavior. The answer requires an understanding of the causation and function of the behavior plus a determination of the strength of the bird's motivation to perform the behavior. The complexity of this task may require a well designed and analyzed behavioral study, which would be beyond the scope of a site visit or animal welfare audit.

Animal welfare is also complex. The challenge of assessing poultry welfare via behavior is that poultry welfare cannot be assessed by behavior alone. There may be times when restriction of a natural behavior is imposed to improve an animal's welfare in another way. Poultry confinement may limit some natural behaviors but improves the welfare of birds by providing a controlled environment, limiting the exposure to the elements, predation and potentially devastating diseases such as Avian Influenza. Battery cage housing systems keep the hens from nesting and dust bathing but provide the hens a controlled environment, ready access to feed and water, decreased external and intestinal parasitism and decreased traumatic pecking and subsequent cannibalism. Egg laying hens in free-range and aviary (cage-free) systems have the opportunity to express most behaviors but have higher mortality rates from bacterial diseases, parasitism, trauma, and cannibalism. Most commercial

poultry strains are highly productive and bred for specific purposes such as egg production or meat production. These birds are bred to perform under exacting housing, nutrition, and preventive medicine requirements. Some breeding programs select robust characteristics in lieu of maximizing productivity to develop birds that are better able to adapt to rustic, antibiotic-free, or free-range management conditions.

Areas Where More Research Is Needed

Although behavior in poultry has been studied for years, there are numerous opportunities for future research. Development of research models that more objectively assess pain and distress is crucial for poultry behavior research. Very little work has been done on determining the positive mental states of birds and the welfare value of retaining individual behaviors such as dust bathing and perching. Pecking conditions that were mentioned in this text may be due to boredom in intensive commercial environments and should be evaluated. Advances in these areas could lead not only to better welfare, but decrease health problems associated with aggressive and displacement behaviors and overall lead to better productivity of the birds.

See also: Group Living; Molt in Birds and Mammals: Hormones and Behavior; Stress, Health and Social Behavior; Trade-Offs in Anti-Predator Behavior.

Further Reading

Appleby MC, Mench JA, and Hughes BO (2004) *Poultry Behaviour and Welfare*. Cambridge, MA: CABI Publishing.
Craig JV (1992) Measuring social behavior in poultry. Symposium: Quantifying the Behavior of Poultry. *Poultry Science* 71: 650–657.
Craig JV and Swanson JC (1994) Review: Welfare perspectives on hens kept for egg production. *Poultry Science* 73: 921–938.
Dickson JG (ed.) (1992) *The Wild Turkey – Biology and Management*. Mechanicsburg, PA: Stackpole Books.
Duncan IJH (1991) Measuring preferences and the strength of preferences. *Poultry Science* 71: 658–663.
Duncan IJH (1998) Behavior and behavioral needs. First North American Symposium on Poultry Welfare. *Poultry Science* 77: 1766–1772.
Duncan IJH and Mench JA (1993) Behavior as an indicator of welfare in various systems. In: Savory CJ and Hughes BO (eds.) *Proceedings of the Fourth European Symposium on Poultry Welfare* Edinburgh, Scotland, UK, 18–21 September, pp. 69–80. Potters Bar, England: Universities Federation for Animal Welfare.
Hansen RS (1976) Nervousness and hysteria in mature female chickens. *Poultry Science* 55: 531–543.
Hewitt OH (ed.) (1967) *The Wild Turkey and Its Management*. Washington, DC: The Wildlife Society.
Lehner PN (1991) Sampling methods in behavior research. *Poultry Science* 71: 643–649.
Mauldin JM (1992) Applications of behavior to poultry management. Symposium: Quantifying the Behavior of Poultry. *Poultry Science* 71: 634–642.
Mench JA (1992) Introduction: Applied ethology and poultry science. Symposium: Quantifying the Behavior of Poultry. *Poultry Science* 71: 631–633.

Predator Avoidance: Mechanisms

R. G. Coss, University of California, Davis, CA, USA

The Relationship of Mechanistic Science and Higher-Order Systems

This discussion focuses on the antecedent causes of anti-predator behavior from a mechanistic perspective. The antipredator behavior of different organisms will also be described from the perspective of different levels of organization and time scales of change. As typically used in biology and psychology, the term *mechanism* characterizes a world-view perspective that maintains that reality is best represented as stable and fixed. Any change in this stability results from antecedent causes that are specifiable. This perspective has its origins in the atomist tradition in pre-Socratic Greek philosophy and later refined into the tradition of formal hypothesis testing in the early seventeenth century by Francis Bacon. During the late nineteenth century, British and American empiricists, like Thomas Chamberlin, promoted the testing of multiple hypotheses using eliminative induction to identify antecedent causes and consequences of discontinuity. In short, the empiricist–positivist–inductivist perspective of modern analytic philosophers attempts to explain change in the stability and continuity of the object of inquiry by antecedent causal sequences that can be decomposed into simpler elements at lower levels of organization. When trait stability over evolutionary time is considered as primary and change requires explanation, it must be noted that the ubiquitous, mechanistic term 'selection pressure' does not characterize the differential filtering process of natural selection that shapes evolutionary change. The less common expression 'sources of natural selection' will be used instead to describe the historical circumstances in which prey species failed to detect and recognize predators as threats as well as failed to engage in appropriate antipredator behavior. With the decline or absence of these sources of selection, the ability to recognize predators remains intact in some species for long periods, leading to the enquiry of the mechanisms of trait stability or slow reorganization.

As an extension of mechanistic science, control-systems theory incorporates higher levels of organization with multiple goal-directed reference signals and error-correcting feedback loops that regulate behavior over time. Nevertheless, while causal relationships are easiest to describe at the lowest level of organization with one-way causation, higher levels of organization involve the complex integration of regulatory elements with two-way causation that are more difficult to describe as having antecedent causes.

The highest levels of organization in a system can be difficult or impossible to characterize in this manner, especially if a large number of higher-order relationships exhibit emergent properties that preclude decomposition into simpler elements. As a result, empirical research in antipredator behavior consists typically of assessing causal relationships of perception and action at lower levels of organization.

One interesting facet of control-systems theory relevant to assessing proximate time scales of change in antipredator behavior is that simpler elements at low levels of organization react much more quickly to causal sequences than elements at higher levels of organization that incorporate much more generality in various inputs. As will be discussed, prey can engage in periodic or sustained vigilant behavior while foraging with the goal of evaluating their state of vulnerability based on various microhabitat qualities that include nearness to refuge and the ability to detect predators. Such tonic 'nonconsumptive effects' such as prolonged wariness when predators are first detected and then disappear from view operate at the highest level of the organism-environment relationship or cognitive level. Although predators are more likely to be detected by prey with tonic states of vigilance, this redirection of attention has energetic costs that impact prey health, growth, and fecundity, especially when vigilance is directed at detecting predators that use sit-and-wait ambush tactics. Experimental manipulations of this phenomenon in natural settings typically require the presentation of predator models to engender antipredator behavior.

Affordances

James J. Gibson developed the ecologically relevant perceptual theory positing that perceivers extract information from their environment as fluid 'space–time events' that do not require partitioning into a succession of immediate time steps that are reconstructed by the perceiver into a single event. Perception is thus an ongoing activity that involves the detection by the perceiver of the *invariant* features of the environment that can include an invariant pattern over time. The ecological perception approach asks how an animal knows what to do in complex settings with enormous amounts of information and what information needs to be ignored. This view readily acknowledges that natural selection can shape the active properties and selectivity of information gathering, especially habitat features that are useful. Such selectivity permits the recognition of predators, including how they might constitute a particular kind of threat and how the physical properties of the environment might be used differently to evade predation. In some contexts, antipredator behavior includes assessment of the temporal properties of predators and, for knowledge that a predator is nearby but out of view, expectations of the likelihood of it remaining in the area.

Animals with well-developed senses are active perceivers of structural invariants that include environmental features that engender particular kinds of action. Any information perceived, however, must be considered in the context of what it offers or affords the perceiver. From the perspective of perceivers with excellent climbing ability, a large steep boulder and tall tree share the same affordance of an elevated perch for evading terrestrial predators. Looking out is another affordance of a high perch, permitting predator detection and monitoring. Conversely, such predators would likely perceive the same structures as not affording accessibility to prey and indication that the predator was detected if alarm calls were emitted. Trees, however, would afford relative protection from avian predators attacking on the wing as would bushes, shrubs, and thickets that preclude prey seizure; both California ground squirrels and vervet monkeys recognize this affordance and readily dash to nearby bushes after spotting hawks and eagles. Both boulders and trees might afford access to adjacent structures if the predator can climb, and for trees with wide crowns, the low weight-bearing properties of thin branches can afford protection. During daytime attacks, baboons can evade heavier-bodied predators with facile climbing ability, such as leopards and pythons, by seeking refuge on thin branches near the edge of the crown, and macaques and langurs choose the crown edge as primary sleeping sites. In cold habitats, tall conifers are selected by colobine monkeys as sleeping sites to avoid leopards and for thermoregulation.

Escape to burrows affords rodents immediate refuge from nonburrowing predators, but burrows are also places of danger for digging predators and especially venomous snakes that use burrows for thermoregulation and ambushing. Rock squirrels and California ground squirrels recognize this difference in burrow affordance based on predator type, because they become wary of burrow entrances in a sustained or tonic manner after engaging rattlesnakes above ground that are no longer in view.

Other examples include perception of distance to refuge by both diurnal and nocturnal rodents that prefer traveling next to structures that occlude overhead detection. Well-studied laboratory open-field experiments illustrate the aversive arousal of rats and mice exposed in the center of an arena. Their quick scurrying to adjacent walls suggests that they immediately sense their vulnerability to overhead

threats. Although not as well documented, animal trails afford guidance, ease of travel to known resources, and the opportunity to detect partially concealed predators along the way. In the latter context, trails afford familiar routes with landmarks, patterns of vegetation, and engender motor learning along the route that facilitates escape. For diurnal mammals with widely spaced eyes, repeated travel without aversive consequences engenders habituation, enabling the rapid detection of partially concealed ambush predators. Trails with a smooth, well-trodden substrate also allow less energy expenditure during transit, and species as diverse as rodents and ungulates move easily and fluidly without appearing disabled to predators. In particular, elephant shrews recognize the affordances of well-maintained trails, spending up to 25% of their time on trail maintenance.

Predator Recognition

The recognition of predators by prey is paramount for survival, and, in some species, natural selection has engendered the evolution of well-integrated perceptual capabilities coupled with higher-order inferences of how predators hunt in specific settings. Depending on the prey species, predator recognition involves the use of different sensory modalities, some of which detect gradients in intensity and conspicuousness of predator features that are perceived in both static and dynamic contexts. For example, predator-experienced moose exhibit a heightened wariness to the odors and playback vocalizations of familiar and novel predators in contrast with predator-naïve moose, suggesting that predator exposure engenders a broad 'climate of fear' affecting moose readiness to respond appropriately.

Natural selection can also promote selectivity in how prey learn about predators. In historical situations in which prey encountered predators in highly variable contexts without predictable properties, natural selection operated on prey success in learning to recognize specific morphological characteristics of predators in different settings. In circumstances where stealthy sit-and-wait predators attack quickly during the day or at night, minimizing the ability of prey to assess the predator's physical appearance, prey that escape learn to be wary of specific locations where they were previously attacked. Learning in both contexts can occur rapidly with several aversive experiences. In laboratory studies using contextual fear conditioning, laboratory rats can exhibit one-trial learning by freezing in an experimental apparatus in which they received painful electric shocks the previous day. Such rapid associative learning of a specific spatial location as dangerous is analogous in nature to an animal escaping a painful bite by an unseen predator in a specific setting and later avoiding that area. This evolutionarily 'prepared' associative learning of the predictors of dangerous circumstances is contrasted by much slower 'unprepared' learning, requiring many more trials to make less urgent or ecologically relevant associations. In low-visibility habitats used by stealthy predators for hunting during the day, or less stealthy predators hunting in dark burrows and at night, some prey have evolved sensory specializations for predator recognition that involve the assessment of the direction and amplitude of specific sounds and detection of predator scent which is diffused and provides much less information on predator location.

Olfactory Predator-Recognition Cues

Marine invertebrates are sensitive to predator odors and evidence of predatory activity, such as the alarm odors of crushed conspecifics that is evident in 25 genera of gastropods. Snails can distinguish the odors of predatory and nonpredatory crustaceans as well as food and refuge. In particular, crabs that eat snails are especially provocative, as characterized by the ability of the gastropod mollusc *Littoraria* to distinguish the odor of their blue crab predator from fiddler crab and grass shrimp odors. Sea urchins respond to large sea stars within 5–10 cm upstream by moving their spines defensively with gaping pedicellariae. Odor discrimination can occur early in development when other sensory modalities are undeveloped. Among aquatic larvae, western toad tadpoles from Oregon can distinguish the odors of predatory garter snakes, backswimmers, and giant waterbugs from those of roughskin newts and rainbow trout that treat these tadpoles as unpalatable.

Age-related changes in antipredator behavior to odor cues are evident in terrestrial vertebrates, notably the broader predatory threat to smaller, more vulnerable juveniles. In a choice test of refuge scented with sympatric and allopatric invertebrate- and snake-predator odors, juvenile Australian scincid lizards avoided refuges scented by predator odors, especially the venomous funnelweb spider, whereas adults failed to distinguish the odors of predators from nonpredators.

Because the chemical attributes of predator recognition can be examined at a low level of organization involving neural pathways, and because prey species can be examined in laboratory settings in either animate or anesthetized states, research on predator-odor recognition has engendered considerable understanding of the integration of olfactory neurophysiology that mediates antipredator behavior. Integrative research has emphasized study of the provocative effects of the predator odor, trimethylthiazoline (TMT) originally isolated from fox feces. Laboratory rats and mice exhibit innate aversion to this odor illustrated by freezing and withdrawing. The causal sequences of TMT recognition begin with its activation of odor receptors exclusively in the D-domain glomeruli in the olfactory bulb that transmit information to the olfactory cortex for odor identification. The olfactory cortex projects to many forebrain areas, including the orbital frontal cortex (OFC), which regulates arousal, but the critical circuit for engendering a rapid antipredator response is the olfactory cortex activation of the medial division of the bed nucleus of the stria terminalis (BST). BST activation simultaneously triggers the release of stress hormones via the hypothalamus and pituitary gland and rapid freezing mediated by its direct projection to the midbrain periaqueductal gray (PAG). Although the basolateral complex of the amygdala interacts with the BST, and there is considerable evidence that the amygdala plays an essential role in the production of learned emotional memories, the amygdala plays no substantive role in innate TMT recognition and avoidance behavior. Research using mutant mice has shown that the mouse olfactory bulb has two functional modules, one of which participates in the process of associative learning while the other incorporates specialized neural organization for detecting mammalian predators essential for survival.

Acoustical Predator-Recognition

This facet of antipredator research emphasizes simulations of predator presence to prey using playbacks of predator vocalizations and predator-generated sounds. If the circumstances of acoustic predator recognition enhancing fitness were consistent for a long enough period during evolution, then neural specialization in auditory processing of acoustic structure might be expected to have evolved. If, in the developmental time frame, predator sounds were heard consistently in similar circumstances, but inconsistently in the evolutionary time frame, then learning would be expected to play an important role in acoustical predator recognition. Some facets of sound processes might involve innate perceptual biases that could facilitate learning. For example, captive-born cotton-top tamarins are not responsive to playbacks of predator vocalizations as would be predicted if acoustic predator recognition were innate; nevertheless, these monkeys do respond to low-frequency, noisy sounds that characterize larger body size and potential aggressive threats.

The manner in which bonnet macaques in southern India learn to ignore irrelevant sounds but react strongly to alarm calls and predator vocalizations is illustrative of the coupling of two learning processes, conditioned inhibition and Pavlovian conditioning. Conditioned inhibition involves the suppression of attention (i.e., selective habituation) to irrelevant sounds after repeated exposure without emotionally laden consequences, such as fearful running to trees or watching others run to avoid predators. Pavlovian (classical) conditioning results when salient, but initially irrelevant sounds act as predictors of emotionally laden consequences. For example, bonnet macaque infants transported ventrally by their mother on the ground respond to a broad range of loud noises (e.g., conspecific and heterospecific alarm calls) by rapidly clinging harder to their mother to prevent dislodging as she runs to trees and jumps from branch to branch as she climbs. With the exception of alarm calls and predator vocalizations, juveniles begin to ignore ecologically irrelevant sounds by actively observing adult inattention. In particular, watching others run is highly contagious and engenders a fearful emotional state that acts as an unconditioned stimulus maintaining the provocative properties of an ecologically important sound, the conditioned stimulus. In experimental study, playbacks of tiger growls elicited flight in only younger monkeys from Bangalore city where tigers are absent, a property reflecting their sensitivity to loud noises, whereas all monkeys fled in forest troops where tigers are present. Playbacks of eagle calls and leopard growls engender antipredator behavior in Diana monkeys from West Africa. When the leopard-initiated alarm calls of crested guinea fowl are paired with leopard growls in a higher-order Pavlovian conditioning experiment, Diana monkeys treated the guinea-fowl alarm calls as if a leopard were present. This result demonstrates in a natural setting how second-order Pavlovian conditioning of heterospecific alarm calls associated with first-order conditioning of leopard growls would be useful for detecting leopards prior to troop members spotting them.

The ability to selectively winnow out irrelevant sounds while retaining sensitivity to relevant predator sounds via selective habituation is evident in harbor seals in the northwest Pacific. Playbacks of the vocalizations of transient mammal-eating killer whales and unfamiliar fish-eating killer whales engendered strong submerging antipredator responses, but there was little submerging to the familiar calls of local fish-eating killer whales that posed no danger.

Novel predator vocalizations that share acoustic properties with sympatric predators can also be evocative due the perceptual process of stimulus generalization. Research on acoustic owl-predator recognition examined the antipredator behavior of migrant and resident tropical birds on the Yucatan Peninsula, Mexico. During playbacks of eastern screech-owl and ferruginous pygmy-owl vocalizations, migratory passerines responded to only familiar screech-owl vocalizations. However, both migrant and resident birds responded to pygmy-owl vocalizations that share acoustical properties with eastern screech-owl vocalizations, allowing migrants to generalize the familiar predator vocalization to the unfamiliar one. Similarly, yellow-bellied marmots in the Rocky Mountains, Colorado, respond to familiar coyote vocalizations and generalize this predator recognition to unfamiliar, but longer duration wolf vocalizations that share similar acoustical properties.

In contrast with the processes of sound generalization, the distinct sound differences of defensive rattling and hissing by rattlesnakes and gopher snakes are used by California ground squirrels to distinguish these predators when squirrels confront these snakes at close proximity, flagging their tails from side-to-side and throwing loose substrate at these snakes with their forepaws. The rattling sound also leaks cues to a rattlesnake's body temperature and vulnerability because cooler rattlesnakes strike with lower velocity, reduced accuracy, and more hesitation. Smaller rattlesnakes rattle with lower amplitude and emphasize higher sound frequencies than larger, more dangerous rattlesnakes, and ground squirrels become less cautious when they hear playbacks of the rattling sounds of more vulnerable smaller and cooler rattlesnakes.

Visual Predator Recognition

The ability of prey to detect and recognize predators visually during the daytime during which the predator's body is immediately distinct from the background as a recognizable Gestalt (unified pattern) can involve the complex perceptual integration of predator features, such as body shape, coloration, texture, and movement. Sit-and-wait ambush predators, such as felids, counter the predator-detection ability of visually adept prey by remaining still, crouching with flattened ears to diminish projecting contours, hiding in vegetation that disrupts body contours, and evolving pattern-blending camouflage. Hunting at night is another way to circumvent visual predator recognition.

Because depth perception is essential for ambushing, predators that rely on vision must face their prey, usually exposing both eyes to prey as a detectable schema. Pike and bass also face their prey using vegetation as cover before striking. Because ambush predators use both eyes to monitor prey, the schema of two facing eyes has been available as a consistent predator-recognition cue during the phylogeny of numerous vertebrate prey species. Research has shown that, during early development, paradise fish larva and jewel fish fry become alarmed by eye-like patterns consisting of two dark spots in the horizontal plane. Two horizontal light-emitting diodes simulating the moonlit eye shine of nocturnal carnivores augment fear in wild house mice associated with foot shock. Similarly, the horizontal arrangement of two concentric circles appears dangerous to a variety of primates, ranging from mouse lemurs to humans.

Social primates in general are excellent detectors of the direction of gaze of nearby conspecifics either as a potential threat or signal of appeasement. Such ability translates well for determining whether predators appear interested. From a theory of mind framework, this urgent inference that another agent is interested in the perceiver based on its two facing eyes is evidence of a second-order intentional system in which the perceiver has beliefs and desires about the beliefs and desires of others.

As evidence of the speed of assessing predatory interest by bonnet macaques, leopard-experienced forest and leopard-inexperienced urban monkeys start running toward nearby trees with latencies of 200–300 ms after detecting a realistic-looking model leopard with a spotted yellow coat appearing to stare at them in the open. These monkeys react similarly to only a forequarter view of the leopard model, illustrating the threatening appearance of the leopard's facing orientation.

A number of studies of human and non-human primates have documented rapid neural activity during face processing using brain scanning and electrophysiological recordings that provides insight into the rapid assessment of predatory threats. Perception of two facing eyes engenders a cascade of neural activity in the occipitotemporal cortex encompassing pattern recognition that peaks around 170 ms with near simultaneous activation of fearful emotions via interactions of the basolateral amygdala, OFC, and PAG that rapidly initiates flight. Longer cognitive assessment in the 400 ms time frame involves the recruitment of the parietal, medial, and lateral prefrontal cortices that regulate more deliberate action. Wild bonnet macaques respond more slowly to the presentation of a dark morph without the spotted yellow coat that acts as a leopard-recognition cue, and some monkeys continued to monitor the model without fleeing. Presentation of upside-down spotted yellow or dark leopard models also reduces the flight response either by disrupting the leopard's shape or by providing contextual information of a nonhunting cat resting on its back. Despite this postural difference, the spotted yellow coat of an upside-down leopard is still provocative, possibly because spotted patterns activate texture processing via dot-pattern selective neurons in macaque inferotemporal cortex and primates with trichromatic vision are especially sensitive to yellow. The responses of captive-born West Indian green monkeys, sooty mangabeys, pigtailed macaques, and rhesus macaques to leopard models suggest that these primates have also evolved the ability to recognize leopards as predatory threats.

Predatory snakes are a threat to variety of rodents with large venomous snakes capable of eating young or small monkeys. Whereas venomous snakes constitute major dangers during unexpected encounters, larger pythons and boas constitute more systemic predatory threats and can be dealt with effectively if detected early; thus, large snakes are provocative to juvenile and adult monkeys. An experimental study using snake models revealed that the Indian python was the only model snake among a series of smaller model venomous and nonvenomous snakes that engendered alarm calling by wild bonnet macaques.

One major cue for identifying partially occluded snakes in leaf litter is the crosshatch scale pattern. In experimental presentations, captive-born rhesus macaques are more vigilant toward a snake model with a crosshatch pattern than one without. Recognition of a snake-scale pattern is clearly innate in California ground squirrels as evidenced by their precocious ability to recognize a gopher snake and a textured strip resembling a gopher snake the first day pups use vision to navigate. The innate properties of snake recognition are also evident in the fast reaction times for expressing protective behavior. For example, in humans and wild bonnet macaques, unexpected detection of a nearby snake engenders immediate freezing, startling, or jumping back with latencies as brief as 200 ms. Such a fast reaction precludes higher-order cognition as apparent from interviews of experienced herpetologists surprised by snakes; they mentioned that they became consciously aware of the snakes mostly after the event. The visual processes underlying the innate aspects of snake recognition involve the same brain loci used for face recognition, but with different neural specializations useful for processing highly periodic patterns characterizing crosshatched snake scales, repetitive bands, and blotches. Research on cats has shown that the first phase of integrating moving gratings resembling snake scales into a global coherent pattern involves higher-order motion analysis by area V1 in the primary visual cortex, area MT in macaque visual cortex, and cortico-thalamo-cortical loops involving the pulvinar and superior colliculus. Also, to facilitate rapid responses to snakes to avoid envenomation or being seized by pythons, the OFC inhibiting physiological arousal is less activated during snake perception, allowing snake recognition by the occipitotemporal cortex and concomitant amygdala activation to drive freezing and jumping responses by the PAG. As evidence for elevated involvement of the amygdala and reduced involvement of the OFC during interactions with snakes, rhesus macaques with bilateral lesions to the OFC showed initial retention of snake fear whereas bilateral lesions of the amygdala eliminated emotional expressions of fearfulness. Similarly, recognition that a rattlesnake is dangerous by snake-naïve rock squirrels during a staged rattlesnake encounter is not disrupted by bilateral OFC lesions, but this removal of OFC inhibition of the amygdala increases sympathetic nervous system arousal substantially.

With evidence of ecologically relevant biases for detecting and recognizing snakes, learning can play an important role in fear augmentation or reduction. Watching an experienced rhesus macaque observing a snake fearfully on video enhances selective learning in captive-born monkeys that a snake is dangerous, while watching a similarly fearful response to flowers on video does not enhance fear of flowers. With this result, it seems reasonable to argue that, with numerous snake encounters in the wild, fear should increase with age. With the exception of pythons, which are provocative to all age classes, and unexpected encounters with snakes, adult bonnet macaques are less excited by snakes than juveniles; adults do remain vigilant when they forage near snakes, a property consistent with processes of selective habituation when snakes do pose a direct threat.

The Effects of Relaxed Natural Selection on Antipredator Behavior

Unlike the emphasis of evolutionary biologists on explaining how natural selection shapes evolutionary change, studies of relaxed selection examine trait disintegration or trait persistence as relics when the original sources of selection are diminished or no longer present. The study of antipredator behavior offers unique circumstances for investigating the effects of relaxed selection because the presence or absence of predators and their relative density and prey preferences can be quantified. Moreover, the survival functions of different antipredator behaviors in which failure to respond appropriately immediately impacts fitness are more easily interpreted than many other behavioral traits that affect fitness in a more progressive manner.

An important property of studying the effects of relaxed selection on trait stability is estimating the time scale for changes in the sources of selection from predators. There are several ways to estimate the time frame for the divergence of predator selected- and relax-selected populations. If the predator is exothermic, such as snakes, historical fluctuations in temperature can be estimated from the fossil pollen of temperature-sensitive trees and from sea-surface temperatures derived from temperature-sensitive foraminifera in ocean-core samples. Rising sea-level and the emergence of islands separating prey populations from mainland populations can provide relatively sharp demarcations in gene flow, as can geological events leading to the formation of other barriers to gene flow, such as large rivers. The presence of predators and prey in the same fossil assemblages provides clear evidence of historical sympatry, and inferences can be made of possible predator–prey sympatry from fossils appearing in adjacent assemblages of the similar ages. Together, these indices can be linked to genetic distances calibrated to time, showing population divergences in regions with different predator densities. Finally, coarse estimates of time scales of relaxed selection can be made from phylogenetic analyses of species with and without specific patterns of antipredator behavior.

Constraints on Trait Stability and Evolutionary Plasticity

There are three constraints at different levels of organization that potentially buffer antipredator behavior from undergoing rapid disintegration under relaxed selection: (1) the shared functionality of coping behaviors to mitigate predation at higher levels of organization, such as elevated vigilance, aggressive mobbing, or flight behavior, (2) neuronal organization at lower levels of organization engaged in the multifunctional processing of perceptual features useful for recognizing different predators, food

resources, and variegated habitat features, and (3) the number of predator-recognition cues integrated into a coherent Gestalt.

The first constraint of shared functionality in behavioral expression is evident in the antipredator behavior of cat-sized tammar wallabies on Kangaroo island, South Australia, that have been isolated from terrestrial mainland predators, but not large wedge-tailed eagles, for an estimated 9500 years. Presentations of mammalian predator models to inexperienced wallabies engendered heightened vigilance and antipredator behavior typical of mainland wallabies. This finding of visual predator recognition led to the multipredator hypothesis, positing that antipredator behavior might be buffered from rapid disintegration under relaxed selection from one class of predators if the sources of selection from any extant predators have properties that maintain similar patterns of coping.

With a focus on texture-based predator recognition, the second constraint on evolutionary change at lower levels of organization also involves intertwining multifunctional systems. As discussed earlier, the periodic pattern-processing aspects of rodent, cat, and primate V1 neurons in visual cortex are essential for distinguishing important visual features. Ground squirrels likely employ the same pattern-processing ability to distinguish periodic snake scales and grass seed heads from the complex backdrop of irregular detritus and leaf litter. Similarly, dot-pattern selective neurons in the primate inferotemporal cortex that receive input from the visual cortex are attuned to less regularity but have properties theoretically proposed to facilitate pattern segmentation essential for distinguishing spots and rounded patterns. Despite the enormous amount to pleiotropy at the genetic level playing a role in trait stability, any mutations that disrupt the developmental integrity of neural circuitry specialized for these critical visual pattern-processing abilities would thus be scrubbed from the gene pool.

The question remains as to whether under relaxed selection highly conserved neural processes mediating lower levels of visual information processing with broad functionality can protect more specialized neurological processes subserving higher-level pattern recognition from undergoing rapid disintegration. Although such purifying selection accounts for the ancient continuity of visual-texture processing by diverse mammalian species living in different habitats, further specialized neural circuitry clumps visual information, so it pops out to the perceiver as a meaningful Gestalt.

Recent single-unit recordings of humans engaged in visual-recognition tasks showed that only a fraction of neurons in the medial temporal cortex receiving visual input from the inferotemporal cortex were active out of the hundreds presumably addressed by this memory system and among the millions of neurons activated by a typical stimulus. If applicable to the neural circuitry underlying innate predator recognition, such sparsely distributed coding by only a few million neurons could explain how selection is less encumbered in modifying variation in the pattern of interneural connectivity among only a small proportion of units within a bank of many units. As a result of the much greater historical entrenchment of the neural organization for visual-texture processing than for pattern recognition, this latter specialization in information processing is much more plastic in the evolutionary time frame. Although tentative from only a few experimental studies, the robustness of predator recognition under relaxed selection appears to be associated with the number of distinctive predator features integrated perceptually into a meaningful whole. For example, the Spanish subspecies of pied flycatchers relies on two recognition cues for identifying their redback shrike predator: the black band that masked the shrike's eyes and its passerine-like body shape. Relaxed selection from shrike predation in the 1000 year time frame is sufficient to induce loss of recognition of a model shrike in nearly all pied flycatchers examined, compared with their ability to recognize more visually complex owl models which is facilitated by the owls' large facing eyes.

West Indian green monkeys imported from West Africa to Barbados Island in the sixteenth century will gather and alarm call vigorously at a leopard model and even alarm call at a yellow disk with leopard spots. In the longer time frame of an estimated half million years of relaxed selection from jaguars, black-tailed deer in Northern California appear to have lost their ability to recognize a model jaguar as dangerous because the spotted coat has seemingly regained its original obliterative shading function, disrupting jaguar shape. On the other hand, black-tailed deer do recognize a puma model as dangerous and generalize recognition of a large felid shape to a novel tiger model that has less disruptive stripes.

While intuitively simple in their linear configuration, snakes do constitute complex configurations that challenge the recognition process because their appearance can change substantially from coiled to moving in a sinusoidal fashion. Motionless snakes, especially their heads, need to be detected quickly, and coiling disrupts repetitive patterns, leaving the finer-grain snake scales as the only consistent snake-recognition cue. Nevertheless, rattlesnakes and gopher snakes are distinguished in experimental presentations by wild California ground squirrels experiencing relaxed selection for an estimated 300 000 years, a time frame in which their serum-based resistance to rattlesnake venom has dissipated completely (**Figure 1**).

While the snake-recognition system persists under prolonged relaxed selection, changes in the cohesiveness of antisnake behaviors are evident, and all populations studied show an elevation in sympathetic nervous system arousal. Based on human studies, high states of arousal can compromise behavioral expression, leading

Figure 1 Evolutionary persistence and loss of snake recognition by ground squirrels. Photographs are taken through a one-way mirrored window during experimental study revealing behavioral differences among California ground squirrels (*Spermophilus beecheyi*) that currently or historically encountered rattlesnake and gopher snake predators and Arctic ground squirrels (*Spermophilus parryii*) whose ancestors evolved for the past 3 My in snake-free central Alaska. (a) California ground squirrel from the snake-abundant Folsom Lake area harassing a caged rattlesnake by tail flagging vigorously. (b, c) California ground squirrels from the Lake Tahoe basin and Mt. Shasta are shown preparing to harass the rattlesnake by throwing substrate. Tahoe basin and Mt. Shasta squirrels have experienced relaxed selection from snakes for estimated times of 70 000 and 300 000 years, respectively. (d) Arctic ground squirrel exhibiting a vulnerable standing posture indicative of loss of snake recognition. Folsom Lake squirrels exhibit the highest serum-based resistance to rattlesnake venom recorded in this species. Lake Tahoe squirrels show an intermediate loss of venom resistance and squirrels from Mt. Shasta show the same lack of venom resistance as Arctic ground squirrels.

to recklessness. As discussed previously with rock squirrels engaging a rattlesnake in staged encounters, lesions of the OFC disinhibit amygdala activity that elevates arousal markedly. In a similar manner, elevated arousal while dealing with snakes in relax-selected California ground squirrel populations might reflect neural reorganization of the OFC due to genetic drift in combination with unrelated sources of selection. From this insight, natural selection from snakes for thousands of years has apparently dampened physiological arousal during snake encounters, which might explain the calmer, more coordinated pattern of antisnake behavior by these squirrels. Although biophysical evidence from rats indicates that the expense of generating neuronal action potentials is high, the persistence of antisnake behavior under relaxed selection argues for sparse coding for the underlying neural circuitry that engenders a low metabolic burden.

A final point should be made about higher-order predator-recognition processes under relaxed selection that include how prey anticipate the motivational states of predators, what actions they might take, and the kinds of habitat features and circumstances useful for evading predation. In experimental presentations, a predator can be recognized visually outside its natural surroundings by inexperienced prey. Nevertheless, it is reasonable to argue that predator recognition evolved to operate most effectively when the predator is embedded in the appropriate context of its typical surroundings. Prolonged relaxed selection is likely to compromise the habitat-related contextual aspects before predator-recognition fails completely, in part, because the integration of predator appearance and background habitat has been historically much more variable than the continuity of predator-recognition cues of shape, texture, sounds, and odor. Future study of relaxed selection should address whether species that continue to recognize predators robustly in appropriate habitats maintain this ability in experimental contexts altered to reflect historical circumstances no longer present.

See also: Conservation and Anti-Predator Behavior; Ecology of Fear; Predator's Perspective on Predator–Prey Interactions; Risk Allocation in Anti-Predator Behavior.

Further Reading

Berger J, Swenson JE, and Persson I-L (2001) Recolonizing carnivores and naïve prey: Conservation lessons from Pleistocene extinctions. *Science* 291: 1036–1039.

Blumstein DT (2006) The multipredator hypothesis and the evolutionary persistence of antipredator behavior. *Ethology* 112: 209–217.

Blumstein DT, Cooley L, Winternitz J, and Daniel JC (2008) Do yellow-bellied marmots respond to predator vocalizations? *Behavioral Ecology and Sociobiology* 62: 457–468.

Caro T (2005) *Antipredator Defenses in Birds and Mammals.* Chicago, IL: The University of Chicago Press, Chicago.

Coss RG (1999) Effects of relaxed natural selection on the evolution of behavior. In: Foster SA and Endler JA (eds.) *Geographic Variation in Behavior: Perspectives on Evolutionary Mechanisms*, pp. 180–208. Oxford: Oxford University Press.

Coss RG and Ramakrishnan U (2000) Perceptual aspects of leopard recognition by wild bonnet macaques (*Macaca radiata*). *Behaviour* 137: 315–335.

Curio E (1993) Proximate and developmental aspects of antipredator behavior. *Advances in the Study of Behavior* 22: 135–238.

Deecke VB, Slater PJB, and Ford JKB (2002) Selective habituation shapes acoustic predator recognition in harbour seals. *Nature* 420: 171–173.

Kats LB and Dill LM (1998) The scent of death: Chemosensory assessment of predation risk by prey animals. *Écoscience* 5: 361–394.

Lahti DC, Johnson NA, Ajie BC, et al. (2009) Relaxed selection in the wild: Contexts and consequences. *Trends in Ecology & Evolution* 24: 487–496. Doi:10.1016/j.tree.2009.03.010.

Lennie P (2003) The cost of cortical computation. *Current Biology* 13: 493–497.

Lima SL and Dill LM (1990) Behavioral decisions made under the risk of predation: A review and prospectus. *Canadian Journal of Zoology* 68: 619–640.

Michaels CF and Carello C (1981) *Direct Perception.* Englewood Cliffs, NJ: Prentice-Hall, Inc.

Overton WF and Reese HW (1981) Conceptual prerequisites for an understanding of stability-change and continuity-discontinuity. *International Journal of Behavioral Development* 4: 99–123.

Owings DH and Coss RG (2008) Hunting California ground squirrels: Constraints and opportunities for northern Pacific rattlesnakes. In: Hayes WK, Beaman KR, Cardwell MD, and Bush SP (eds.) *The Biology of Rattlesnakes*, pp. 155–168. Loma Linda, CA: Loma Linda University Press.

Powers WT (1973) Feedback: Beyond behaviorism. *Science* 179: 351–356.

Preisser EL, Orrock JL, and Schmitz OJ (2007) Predator hunting mode and habitat domain alter nonconsumptive effects in predator–prey interactions. *Ecology* 88: 2744–2751.

Quiroga RQ, Kreiman G, Koch C, and Fried I (2008) Sparse but not 'Grandmother-cell' coding in the medial temporal lobe. *Trends in Cognitive Sciences* 12: 87–91.

Stankowich T and Coss RG (2007) The re-emergence of felid camouflage with the decay of predator recognition in deer under relaxed selection. *Proceedings of the Royal Society, Series B* 274: 175–182.

Zuberbühler K (2007) Causal cognition in a non-human primate: Field playback experiments with Diana monkeys. *Cognition* 76: 195–207.

Predator Evasion

D. D. Yager, University of Maryland, College Park, MD, USA

Introduction

With relatively few exceptions, all animals must worry about becoming a meal for another animal. Thus, survival long enough to reproduce successfully requires defensive strategies. The dominant strategy for any animal is determined by its behavior and ecology. Relatively sedentary animals often rely on camouflage to avoid detection by predators. Others may advertise their distastefulness or formidable armament with bright colors and displays. Some creatures that must move around frequently for foraging, dispersal, or finding mates minimize their risk by avoiding the times and/or places most used by their primary predators. Others stay in large groups that can provide maximal vigilance and divert attention from individuals. From the predator's viewpoint, survival and reproduction require preventing or circumventing the defenses of their prey. The conflicting interests of predator and prey can lead to an evolutionary one-upmanship often likened to an arms race.

When other defenses fail and the predator is closing in, the best, and often only remaining strategy to avoid being captured is strong, active response. This means getting out of the predator's capture range and/or making continued capture attempts too time consuming or difficult to be worth the predator's effort. Effective escape typically has two components: timely detection of the threat and rapid evasive or deterrent behavior.

Basic Characteristics of Escape Behaviors

Escape systems have provided neuroethologists especially good opportunities to study the neural control of behavior, from the processing of sensory information (detection) to the coordination of the evasive behaviors. They have the advantages of easily elicited, robust, stereotyped behaviors combined with neural circuits that are relatively simple (emphasis on 'relatively') because they need to be fast and foolproof. Among the classic model escape systems are the crayfish tail flip, the fast start elicited by the Mauthner cells in fish, and the visually triggered escape in fruit flies and locusts. This article will discuss the neuroethology of effective escape using two model systems from the insect world: escape based on wind detection (cockroach, cricket) and based on sounds produced by predators (moth, praying mantis, several others).

It will be useful first to summarize the basic requirements for a successful escape system.

Optimally timed: In many cases, evasion should start as soon as possible after detection. Sometimes, however, strategic timing can enhance escape by maximally disrupting the attack sequence or startling the predator.

Fast: Once initiated, the response should be sudden and proceed rapidly.

Innate: The escape system should work the first time it is required. Death precludes learning.

Resistant to false negatives: Similarly, failure to detect or respond to an attacking predator is not acceptable.

Resistant to false positives: Executing an escape when there is not actually a threat wastes energy. More importantly, it disrupts important behaviors such as foraging or searching for mates. Habituation often helps limit false-positive responses to repeated stimulation.

Matched to predator(s): An effective escape system should be tailored to the strengths and weaknesses in the hunting styles of the most common predators. The level of risk posed by the predator influences the strength and/or nature of the response.

Translational/rotational: Evasion requires movement. This may be subtle such as shifting position to blend in better with the background or to hide. More often, it involves translation from one place to another, sometimes preceded by a rotation of the body. The movement may be in a particular direction relative to the attack, or may be random.

Integrated with the animal's biology: There are situations in which an escape system would not be needed and might even be counterproductive. Activation of escape systems is often context specific and may also be modulated by the reproductive state or social status of the animal.

Wind-Mediated Escape

When a predator moves, it produces a variety of stimuli that might be used by prey as a warning. An active nocturnal animal with poor vision and minimal hearing such as the cockroach *Periplaneta americana* could profitably use the substrate vibrations created by the predator's approach, for instance, and most insects have vibration sensors in their legs (subgenual organs) that are exquisitely sensitive. A careful or a stationary predator avoids producing vibration, but cannot avoid creating air currents (wind) as it attacks the cockroach. Such a stimulus indicates immediate, maximal danger and should reliably elicit a fast, strong evasive response.

The Evasive Behavior

Cockroaches are very good at avoiding capture and death, be it by a predator or a rolled-up newspaper. Kenneth Roeder first asked how they do this in the 1940s. Beginning in the mid-1970s, Jeffrey Camhi and colleagues, including Roy Ritzmann and Christopher Comer, have pursued the question in great depth. This work has been paralleled by very detailed studies of the comparable escape system in crickets in the labs of John Palka, Roderick Murphey, Gwen Jacobs, John Miller, and others.

Camhi and colleagues introduced a cockroach into a circular arena where a cockroach predator, the large toad *Bufo marinus*, was waiting. By filming the interactions with an overhead high-speed camera, they could document the evasive movements of the cockroach and their effectiveness. When the cockroach entered the toad's field of view, the toad lunged, protruding its sticky tongue in an attempt to capture the cockroach. It was successful only 45% of the time (although one might say that predator and prey were equally matched). The cockroach's response to the attack had two stages: an abrupt turn away from the toad that started well before the tongue came out, and then 50–70 ms later, running away (**Figure 1**). Two antennae-like appendages called 'cerci' near the tip of the abdomen carry long, thin hairs that are sensitive wind detectors (**Figure 2**). Immobilizing those filiform sensilla with glue increased the toad's success rate to 92%. Experiments using a precisely controlled wind source instead of a toad elicited the same evasive response confirming that wind was the necessary and sufficient trigger. Even newly hatched cockroach nymphs perform the escape competently.

Studies using high-speed videotaping of the cockroach evasive turn coupled with measurements of the forces generated by each leg have defined the sequence of movements leading the evasive turns. The metathoracic legs (third pair) provide power for the turn, but little directional force. The mesothoracic legs have the greatest effect on the direction of the turn. Segments of all the individual legs also have a division of labor. Force generated at the first leg joint creates predominantly forward or backward movement and force at the more distal joints pushes toward the opposite side. The precise coordination of the legs differs in specific detail when the threat comes from the front, the back, or the side.

The evasive maneuver is very fast. A stationary cockroach's turn away from the toad or a controlled wind source starts on average only 54–58 ms after the wind reaches the cerci, and the initial turn is completed 20–30 ms later. The average behavioral latency depends on the activity of the animal when stimulated and can be as short as 14 ms during slow walking. The entire attack from the first movement of the toad to the end of the tongue's forward movement lasts 200–250 ms. Latencies in these ranges – 40–80 ms for first movement, 100–250 ms for full execution – are typical of many insect escape systems.

False-negative responses are very unlikely based on the extreme sensitivity of the cercal wind detection system. Wind velocities as low as 3 mm s^{-1} elicit some behavioral response, and the threshold for an evasive turn is only 12 mm s^{-1} (0.007 mph), well below the wind velocities produced by a lunging toad.

Because the system is so sensitive, false positives could pose a severe problem. Environmental air currents caused by small thermal gradients are typically more than an order of magnitude above the cockroach's behavioral threshold as are air currents the insect itself creates when it moves. The characteristic of attack wind that is normally absent in background wind is high acceleration. Thus, cockroaches minimize false positives by requiring both a minimum wind velocity and an acceleration well above normal background levels to trigger the escape system. This also constitutes an example of 'feature detection,' in which a distinctive, but small subset of salient attributes is sufficient to recognize something. In other words, the cockroach does not need to know whether the toad is green or brown or whether the rolled-up newspaper is the *Post* or the *Times*, only that wind above threshold for both velocity and acceleration means immediate danger.

The initial pivoting turn away from the attack is crucial to the cockroach's success. In theory, there are two basic strategies for the direction of an escape maneuver. Going directly away from the threat makes sense geometrically, but it is a predictable response that even a relatively dim predator could learn to anticipate. Moving randomly eliminates that problem, but also means that sometimes the prey goes directly toward the predator. Nonetheless, a turn and run almost directly away from the attack emerged in the early studies as the strategy used by the cockroaches. The data from those experiments were puzzling, however, because of the very high variation in turn angle. A recent study employing a different analysis technique suggests that *Periplaneta* actually uses a hybrid strategy. Its direction of turn is almost always 90–180° away from the attack. Within that range, however, each individual has 3–4 preferred directions that it uses randomly. This suggests a full 360° divided into ~16 headings or a response 'resolution' in the range of 22–23°.

Neural Control

There are four layers of neurons controlling escape: wind afferents (detection), giant interneurons and thoracic interneurons (processing), and motor neurons (output, behavior).

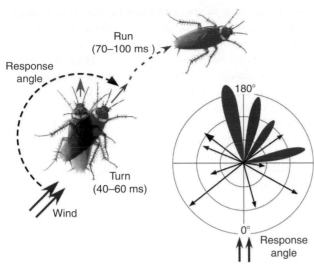

Figure 1 The basic escape response to a wind puff comprises a rapid turn followed by a run. The polar plot shows predicted response angles for three strategies of choosing an escape direction. The width of each response shape indicates variability and the length shows number of responses out of a large series. The black arrows indicate turns in random directions. A turn directly away from the threat is shown in red. An intermediate strategy (blue) is to turn generally away from the threat, but to choose randomly among 3–4 preferred directions within that range.

Detection

In *Periplaneta*, each cercus has ~200 long, slender filiform hairs on its ventral surface (**Figure 2**). Each hair sits on flexible cuticle in a socket (hence the common term 'socketed hair'). When wind moves the hair shaft, the base also moves causing stretch in the dendrite of a single bipolar afferent neuron, which in turn triggers action potentials ('spikes'). Filiform sensilla are the most sensitive biological sensors known. Mechanical energy as low as 4×10^{-21} W s acting on a cricket filiform hair generates an action potential in the afferent neuron. This is 100 times below the threshold of the most sensitive photoreceptive cell. Invertebrate filiform sensilla have recently inspired MEMS (micro-electromechanical systems) analogs in silicon for use as ultrasensitive flow sensors and as improvements for cochlear implants to assist the hearing impaired.

The socketed hairs on the cockroach cerci form a map of wind direction (an anemotopic map). Each hair has a preferred stimulus direction determined by the attachment site of the dendrite on the inner wall of the hair and by the physical geometry of the socket itself. Wind in the preferred direction increases the rate of afferent firing. Further, the sensilla form an organized array on the cercus. The largest hairs form nine columns along the length of the cercus each containing only hairs with the same

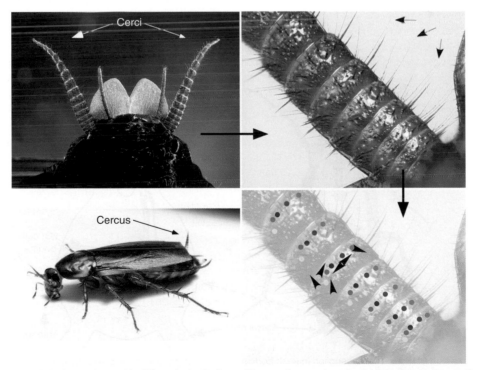

Figure 2 The cerci with their very long, thin filiform hairs (indicated by small arrows) protrude from the abdomen tip. The wing tips have been removed for a clearer view of the cerci. The base of a hair is visible as a domed socket (colored dots). Each hair in the row on every cercal segment has a preferred stimulus (wind) angle shown by the arrowheads. The result is columns along the length of the cercus that code a particular wind direction and form an 'anemotopic map.'

preferred direction. Thus, a wind will cause a specific pattern of afferent activity depending on its direction. The left and the right cerci point in different directions, which taken alone suggests that the anemotopic map could have a resolution in the range of 20° (360°/18 columns).

Comer and colleagues have discovered that wind stimulation of the antennae can also elicit brisk evasive turns. It certainly would be advantageous to have both rear- and front-facing sensors to detect predators. Antennal-based evasion thresholds are quite high, and the wind may actually be tapping in to circuitry that responds primarily to touch.

Processing

Speed is of the essence, and it is typical for escape systems to use relatively few interneurons with high conduction velocities and simple processing, that is, few synapses (**Figure 3**). The cockroach's head (containing the brain and subesophageal ganglion) is not necessary for directional escape turns, which saves on transmission and processing time. The turn does not rely on sensory feedback for control, that is, it is 'open-loop' versus a slower 'closed-loop' system that uses feedback. The ~400 afferent neurons synapse with seven bilaterally symmetrical pairs of 'giant interneurons' (GIs) organized into a dorsal group and a ventral group (dGIs and vGIs). The axon diameters are 25–30 μm for the three dGIs and ~60 μm for the three largest vGIs, whereas the diameters for the rest of the axons in the abdominal connectives are predominantly 0.5–10 μm. Diameter is the primary determinate of conduction velocity for axons without myelin, absent in most invertebrates, and the presence of large-diameter axons often indicates a high-speed processing system, probably for escape (although not all escape systems have GIs). Several lines of evidence indicate that the largest and fastest GIs, the vGIs GI1, GI2, and GI3, control the first stage of the evasive response. The dGIs are more involved in the second stage of evasion and in behaviors during flight.

The GIs have their cell bodies and dendrites in the last abdominal ganglion (A6). The vGIs receive only wind information, and the connections are monosynaptic. Action potentials in these interneurons leave the A6 ganglion ~2.5 ms after a strong wind stimulus strikes the cercus and reach the thorax 2–3 ms after that, thanks to the high conduction velocities (5–7 m s^{-1}). Even for the shortest behavioral latencies (11–14 ms), this leaves ≥6 ms to activate the motor neurons and initiate movement. Selective ablation and stimulation experiments support the necessity and sufficiency of activity in the vGIs as a group for the evasive turn triggered by low-intensity wind. However, higher wind levels, especially from the front, can trigger the same evasive movements via antennal stimulation.

The vGIs synapse with a population of ~100 interneurons (the TI$_A$s) in the thoracic ganglia. The TI$_A$s control the leg motor neurons, directly and indirectly through other interneurons.

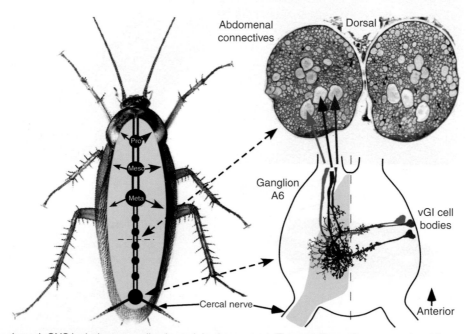

Figure 3 The cockroach CNS includes a ganglion in each body segment. The ganglia receive sensory input from and produce motor commands for their own segment. The three thoracic ganglia (metathoracic, mesothoracic, and prothoracic) control leg movement. Bundles of axons (connectives) provide communication among ganglia. The afferent neurons from the filiform hairs travel in the cercal nerve to the ganglion to an area (gray shading) where they synapse with interneurons including the GIs. The figure shows GI1, GI2, and GI3.

Directionality

A central neuroethological question is how the circuitry in the CNS uses wind direction information to orchestrate activity in the leg muscles to produce turn in the appropriate direction.

The initial experiments with cockroaches presented a puzzle. On one hand, the overall design of the CNS processing system is typical for creating directional behavior. The central strategy is to keep incoming information in parallel, but separate left and right channels. In the cockroach wind system, the cercal afferents do not cross the midline, and the vGIs get their input almost exclusively from the ipsilateral cercus. The six vGIs are not connected to one another. The stage is set for comparisons, in the thorax, of excitation strength and/or timing on left and right to determine the direction of the wind and the appropriate response. On the other hand, the anemotopic map created by the ~18 cercal columns converges on just six interneurons relevant to the turn, vGIs 1–3 on both sides. Intracellular recordings of vGI responses to wind from different directions revealed only a crude directionality: wind from the entire frontal region stimulates GI1, broadly from the side for GI3, and from the rear quadrant for GI2 (the contralateral vGIs have the mirror-image pattern). Selective ablation of vGIs individually and in combinations using injected enzyme further suggested a relatively low-resolution directional system.

Three contrasting ideas emerged: 'winner takes all' in which the side with the greatest number of spikes or shortest latency solely determines the turn direction; 'steering wheel' in which the result is a balance of the summed activity on the two sides; and 'population code' in which the activity of all the vGIs contribute equally to the final turn direction. Computational models based on and refined by experimental data show that a variant of the population code model, called the 'Darwinian population code,' best explains the turn directions.

Sound-Mediated Escape

Animals with sensitive hearing can take advantage of sounds produced by predators. (We deal here only with the pressure component of sound detected by ears using an eardrum (tympanum). The displacement component of sound behaves in much the same way as wind.) Tympanate hearing confers at least two major advantages: (1) sound can travel considerable distances before it becomes too faint to hear. A sensitive ear can detect an approaching predator at least several meters before it arrives. This translates to at least 0.5–2.0 s even for a fast-moving predator, much longer than needed for an effective escape; (2) it can detect frequencies above 1–2 kHz, whereas wind and vibration are most useful below that, often at <300 Hz.

Systems for avoiding capture by echolocating bats provide some of the best examples of sound-based escape systems in insects. In the 1950s and 1960s, Kenneth Roeder along with Asher Treat, working with noctuid moths, published classic studies first documenting such evasion strategies. Insectivorous bats are superb predators with highly effective echolocation (ultrasonic sonar). The echolocation is also their Achilles' heel, from the prey's perspective. The ultrasonic cries used during hunting must be very intense to insure adequate returning echo strength, but this also makes them excellent early warning signals for tympanate insects. Further, the prey have two cues to tell them exactly how much trouble they are in: increasing signal strength as the predator approaches; and the stereotyped sequence of increasing pulse repetition rates used by the bats during an attack. Bat–moth interactions have proven invaluable for our understanding of the sensory ecology of escape thanks to the continuing work of Fullard, Surlykke, and many colleagues. Comparable systems have been studied in crickets, locusts, katydids, green lacewings, scarab and tiger beetles, and praying mantids, among others, attesting to the intense predation pressure exerted on nocturnally active insects by hunting bats. Despite the diversity, complexity, and elegance evident in these escape systems, they follow the same fundamental principles detailed here for wind-triggered escape. Praying mantises provide an especially interesting case study.

The Evasive Behavior

Because they often fly at night, because they live in the same habitats world-wide as many insectivorous bat species, because they fly relatively slowly, and because they are palatable prey, mantises are highly vulnerable to attack by bats. Nonetheless, as long as they can hear, behavioral studies show that they evade capture in 76% of bat attacks. When the hearing is impaired by filling the ear with water or Vaseline, successful evasion drops to 34%.

Faced with a strong threat, a flying praying mantis dives toward the ground under full power, often with a spiraling trajectory (**Figure 4**). Strobe photography studies of free-flying mantids responding to artificial bat cries show that flight direction starts to change 125–230 ms after the stimulus. The first stage is an upturn in flight path with a reduction in flight speed. Then, the mantis rolls off to one side, which initiates the dive. The direction of dive is random. Such abrupt, unexpected, unpredictable changes in speed and direction pose a significant problem for bats, given their much greater momentum. In videotapes of flight room encounters and in field observations, the bat chased the mantis as it dove, but broke off the dive without capture in ~35–40% of attacks. Other times (10–15% of trials) the bat began the attack, but aborted before committing to a dive. Many times

(∼50%), the mantis dove out of the bat's detection range before an attack actually began.

Mantis hearing sensitivity is sufficient to detect the echolocation cries of most bats at >15 m (free-flight behavioral threshold at 10 m), whereas the bat must be within 6–8 m to detect mantis-size insects. False negatives are most likely when the bat suddenly turns toward the mantis at close range, so the insect's detection of the directional sonar beam is too late for effective evasion. This scenario accounted for many of the capture successes in the behavioral studies. False positives must be very rare simply because bats are by far the most common aerial source of ultrasound at night.

The power dive comes from the mantis' fast, complex response to ultrasound that involves all parts of its body. The front legs fully extend, the head rolls to one side (random), the abdomen curls up to a right angle with the body, and wing beat rate and excursion change. The response starts as early as 30 ms after stimulus onset (mean = 72 ms) with the foreleg movement first, followed in sequence by the head, wings, and abdomen. It reaches its maximum within 150–300 ms. Aerodynamically, the foreleg extension shifts the center of mass forward, but initially the greater effect is most probably the upward abdomen curl. This would create increased drag, quickly slowing the mantis and causing upward angle of flight preceding a roll one side or the other, randomly.

Mantis evasive behavior most resembles that of moths. Both have a powerful, unpredictable response to imminent threat, and both use a less intense response to lower threat-level stimuli, although it is directional in moths. In contrast, green lacewings and tettigoniids drop passively toward the ground. Cricket and locust responses are highly directional. Arctiid moths and tiger beetles take a highly effective, but entirely different approach by producing intense ultrasonic clicks when a hunting bat approaches. In some species, the clicks are a powerful warning of distastefulness (aposematism), which deters the bat. In many cases, the bat is startled and the attack disrupted. In yet other cases, very rapid, prolonged ultrasonic clicking interferes with the bat's sonar system causing an aborted or missed attack (jamming). The clicking defense works without evasive maneuvers in arctiid moths, so the insect can continue about its business unscathed and uninterrupted.

Just as cockroach behavioral threshold to wind varies depending on its ongoing activity, so is sound-triggered evasion integrated with the overall biology of the animal. In mantids, the evasive response is highly context specific, only occurring when the flight muscles are active. Ultrasonic pulses in any other situation elicit no change in behavior at all. In crickets and moths, reproductive demands can override predator evasion. In the former, sufficiently intense lower frequency (3–5 kHz) mating calls can inhibit activity of the neuron that triggers evasion (two-tone inhibition). In the latter, studies have

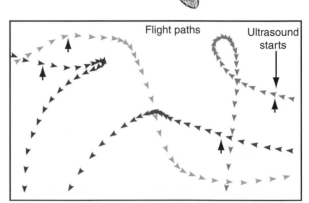

Figure 4 A mantis suspended in the air by a tethering wire flies normally with a streamlined posture. Ultrasonic pulses like those produced by a hunting bat elicit a complex evasive behavior after a very short latency which differs somewhat for the various components of response (means measured from video recordings at 1000 frames per second). In free-flight recorded under strobe lighting (22 ms between flashes; individual arrowheads), the evasion triggered by ultrasound comprises a dive, often with a strong turning component (spiral) during which the mantis continues flapping its wings. The response usually ends with the mantis landing, but sometimes it skims just above the ground, and then resumes flight. Latencies to the change in flight path typically range from 120 to 250 ms.

assessed behaviors when a flying male simultaneously receives pheromonal signals from a receptive female and ultrasonic signals from an attacking bat. The results show a clear balance point at which the potential for successful mating outweighs the threat posed by the bat, and the evasive response is suppressed.

Neural Control

As with other escape systems, the goal for flying mantids is a fast, but strategically timed movement out of danger (with disruption of the predator's attack, if possible). This insect has opted for randomness of evasion ('evitability' and 'Protean response' are terms used in reference to this strategy), which makes sense in the context of flight. It also implies less CNS processing. There are four layers of neural elements: afferent input, interneurons for processing and for ascending/descending transmission, and motor neurons. For insect sound-mediated escape in general, studies on the input side dominate, and less is known about processing above the neck and descending control.

Detection

The males in ~80% of praying mantis species have sensitive ultrasonic hearing (females in many species have secondarily lost their hearing, that is, there is strong auditory sexual dimorphism). Mantises have just one ear that is located in the ventral midline between the third pair of legs. In most species, the auditory system is most sensitive to 30–60 kHz sound (matching the range of dominant frequencies used by the most common insectivorous bats worldwide) with lowest mean thresholds of 40–60 dB SPL (15–30 dB SPL in some cases) (**Figure 5**). A few Old World species have their best sensitivity as high as 130 kHz, which could reflect vulnerability to predation by hipposiderid and rhinolophid bats that use echolocation cries in that range.

Although uniquely situated, the mantis ear functions on the same basic principle as in all animals with sensitive, high-frequency hearing. There are tympana that, in this case, sit in the walls of a deep auditory chamber. Sound is effective in causing the tympana to vibrate because there is air on both sides of the tympanal membranes, thanks to large air sacs, analogous to the vertebrate middle ear. Tympanal organs containing sensory receptors (chordotonal sensilla) similar to those in other insect ears detect the vibration and report via a tympanal nerve to the CNS. The bioacoustics of the mantis ear are complex because (1) the chamber itself affects tympanal vibration, selectively increasing the sensitivity by 10–12 dB at 40–70 kHz for the species tested; and (2) the single ear has four vibrating membranes that differ in size, thickness, and location within the chamber.

As expected, behavioral and neurophysiological tests confirm that the mantis auditory system is nondirectional.

Processing

A tympanal nerve on each side carries 40–50 afferents into the metathoracic ganglion where they synapse ipsilaterally with a population of at least six auditory interneurons (each of which has a mirror-image twin on the other side of the ganglion). The best-studied auditory interneuron, 501-T3, has an ascending axon that is among the 2–3 largest in the thoracic connectives with an appropriately high conduction velocity (4–$5\,\mathrm{m\,s^{-1}}$), and thus qualifies as a 'giant interneuron.' The two 501-T3s have a high degree of spike-by-spike synchronous firing arising from the similar activity in the right and left tympanal nerves and also from coordinating circuitry in the CNS. The result is essentially a single, powerful signal going to processing centers in the head.

The auditory interneuron 501-T3 has a distinctive phasic-tonic firing pattern shaped by inhibitory interactions. The firing rate in the initial burst of 3–6 action potentials is extremely high, 600–$800\,\mathrm{s^{-1}}$. An initial high-rate burst has emerged as a common feature of neural responses in vertebrate and invertebrate CNS circuits controlling time-critical behaviors. In cricket and probably moth ultrasound-triggered bat evasion, the firing rate in the burst must exceed a threshold ($220\,\mathrm{s^{-1}}$ and 350–$400\,\mathrm{s^{-1}}$, respectively) to elicit the behavior. The rate threshold minimizes false-positive responses and is a neural component of feature detection.

Headless mantids will fly readily and well, but do not perform any component of the evasive response to ultrasound. However, processing in the head, which contains the brain and the suboesophageal ganglion, must be very limited. Action potentials from 501-T3 and 401-T3 reach the head *ca.* 15 ms after stimulus onset, and descending auditory responses in at least three large axons leave the head only 4–5 ms later. Thus, the brain may play a permissive role (yes or no, depending on context and input from other sensory modalities), rather than exert fine control over the behavior itself. Involving the cephalic ganglia slows the response to bat attack, and there are other ultrasound-triggered defensive systems (in arctiid moths, for instance) that act more like segmental reflexes with no contribution from the head.

The specific role of the auditory interneuron 501-T3 has been investigated with combined behavioral and neurophysiological experiments. Extracellular recording with implanted electrodes during attack by a flying bat showed that this GI accurately tracked the bat cries up until ~250 ms prior to capture, but then ceased firing entirely. That is the point at which the evasive response begins. The key triggering stimulus feature for the full dive seems to be a bat echolocation pulse repetition rate exceeding $\sim55\,\mathrm{s^{-1}}$. In the typical attack sequence, this occurs at the transition from approach phase (chase) to the final capture maneuver. Thus, 501-T3 most probably provides the trigger signal for the evasive response and then drops out of the action,

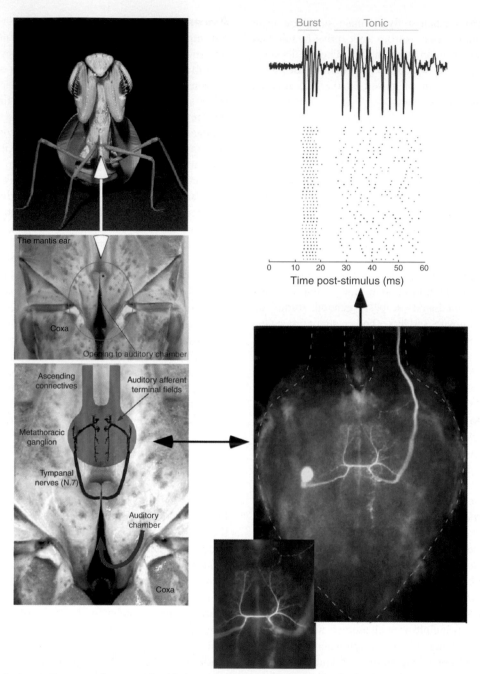

Figure 5   The single mantis ear can be recognized between the metathoracic legs by the knobs at its anterior end and the slit-like opening to the auditory chamber. Auditory afferents from the ear synapse with interneurons in the metathoracic ganglion, the best studied of which is 501-T3 shown here filled with lucifer yellow dye. The ~20-μm axon travels to the head. The firing pattern produced by 501-T3 in response to ultrasound is distinctive and consistent as shown by the raster plot from extracellular recording in an ascending connective. After a latency of 9–12 ms, there is a short, high-rate burst of action potentials followed by a pause and then a variable number of spikes that tracks the duration of the stimulus (50 ms in this case).

possibly to minimize habituation. This neuron is undoubt- edly necessary, but sufficiency has not been shown. Other interneurons such as 401-T3 may also play important roles. Proving both necessity and sufficiency of a single auditory interneuron for an evasive behavior is very difficult, and has so far only been accomplished for Int-1 in the cricket's ultrasound-triggered response.

Evolution

Underlying the evolution of adaptive behaviors are altera- tions in the CNS, often accompanied by anatomical changes in the body. None of these are likely to arise de novo, but are modifications of existing circuits and struc- tures. Uncovering the evolution of neuroethological systems must rely on comparisons of corresponding components and behaviors among existing species because behavior and neurons are rarely preserved as fossils. The comparative approach can be very powerful when combined with good information about the evolution (phylogeny) of a group of animals.

The auditory neuroethology of mantises provides an especially good example. There are extensive data across almost all of the >400 mantis genera on the various anatomical characteristics of the ear, on neurophysiological responses to ultrasound, on sound-triggered escape behaviors, and on the phylogeny derived from molecular genetics. We also have detailed information about the structures and sensory organs in mantis' closest relatives (cockroaches – all earless) that are precursors to the cyclopean ear.

Modern mantises first appeared about 150 Ma, and they were earless. The mantis ear evolved just once, about 35 My later, and must then have looked much as it does today. It arose from a pair of sensory structures that monitored the positions and movement of the legs. The structures fused in the middle of the body forming the auditory chamber which improved hearing. More importantly, the midline is the only place in the area where the ear could be protected from deformation by leg movements. Why the mantis ear evolved between the metathoracic legs remains a mystery.

Many mantis lineages have independently become sexually dimorphic: males hear normally, but females are earless and deaf (and flightless). This transformation must be 'easy' in the evolutionary sense. A simple change in developmental timing so that females retain juvenile characteristics is a probable mechanism.

Most remarkably, a few mantis lineages have evolved a second ear. It is located in the ventral midline of the mesothorax between the middle pair of legs and built from the same components as the metathoracic ear, that is, it is a serial homolog. The mesothoracic auditory system detects only frequencies of 2–4 kHz and operates independently from the ultrasound-sensitive ear. Its function is not yet known.

As would be expected, mantis' sound-triggered behaviors and the underlying neural systems evolved at about the same time as the ear. Modern bats, however, did not appear until ~60 Ma. This makes the important point that the current function of a system need not be the same as its original function. We do not know what sounds those early mantises listened to, nor from what predators they might have escaped, nor even if predator avoidance was the primary function of the earliest auditory system.

Conclusions

Wind-mediated escape and sound-mediated escape have become models for the study of both escape behavior and the neural machinery underlying it. Predator evasion using other sensory modalities (vision and substrate vibration, for instance) follows the same principles. The intense evolutionary pressure on both prey and predator has yielded sensors of astounding sensitivity and selectivity, neural circuits optimized for speed and reliability, and behaviors that either frustrate or feed, depending on your point of view.

See also: Acoustic Communication in Insects: Neuroethology; Bat Neuroethology; Cockroaches; Defensive Avoidance; Empirical Studies of Predator and Prey Behavior; Hearing: Insects; Predator Avoidance: Mechanisms; Sound Localization: Neuroethology.

Further Reading

Barth FG (2000) How to catch the wind: Spider hairs specialized for sensing the movement of air. *Naturwissenschaften* 87: 51–58.

Comer C and Leung V (2004) The vigilance of the hunted: Mechanosensory-visual integration in insect prey. In: Prete FR (ed.) *Complex Worlds from Simpler Systems*, pp. 313–334. Cambridge: MIT Press.

Domenici P, Booth D, Blagburn J, and Bacon J (2009) Cockroaches keep predators guessing by using preferred escape trajectories. *Current Biology* 18: 1792–1796.

Eaton RC (ed.) (1984) *Neural Mechanisms of Startle Behavior*. New York, NY: Plenum Press.

Edwards DH, Heitler WJ, and Krasne FB (1999) Fifty years of command neurons: The neurobiology of escape behavior in the crayfish. *Trends in Neurosciences* 22: 153–161.

Fullard JH (1998) The sensory coevolution of moths and bats. In: Hoy RR, Popper AN, and Fay RR (eds.) *Comparative Hearing: Insects*, pp. 279–326. Heidelberg: Springer Verlag.

Jacobs GA, Miller JP, and Aldworth Z (2008) Computational mechanisms of computational processing in the cricket. *Journal of Experimental Biology* 211: 1819–1828.

Ritzmann RE (1993) The neural organization of cockroach escape and its role in context-dependent behavior. In: Beer RD, Ritzmann RE, and McKenna T (eds.) *Biological Neural Networks in Invertebrate Neuroethology and Robotics*, pp. 113–138. New York and London: Academic Press.

Roeder KD (1998) *Nerve Cells and Insect Behavior*, revised edn. Cambridge: Harvard University Press.

Triblehorn JD, Ghose K, Bohn K, Moss CF, and Yager DD (2008) Free-flight encounters between praying mantids (*Parasphendale agrionina*) and bats (*Eptesicus fuscus*). *Journal of Experimental Biology* 211: 555–562.

Triblehorn JD and Yager DD (2002) Implanted electrode recordings from a praying mantis auditory interneuron during flying bat attacks. *Journal of Experimental Biology* 205: 307–320.

Triblehorn JD and Yager DD (2005) Acoustical interactions between insects and bats: A model for the interplay of neural and ecological specializations. In: Barbosa P and Castellanos I (eds.) *Ecology of Predator–Prey Interactions*, pp. 77–102. Oxford: Oxford University Press.

Yager DD (1999) Hearing. In: Prete FR, Wells H, Wells PH, and Hurd LE (eds.) *The Praying Mantids*, pp. 93–113. Baltimore, MD: Johns Hopkins Press.

Yager DD and Svenson GJ (2008) Patterns of praying mantis auditory system evolution based on morphological, molecular, neurophysiological, and behavioral data. *Biological Journal of the Linnean Society* 94: 541–568.

Predator's Perspective on Predator–Prey Interactions

S. Creel, Montana State University, Bozeman, MT, USA

Predation is an interaction between individuals of two species. Given this, it is somewhat surprising that studies of predation often focus data collection almost exclusively on one of the two species. In research on antipredator behavior, the most common approach has been to focus observations on the prey, recording their responses to real or simulated encounters with predators. This approach has been productive and has revealed a great deal about the behavioral and physiological responses of prey to immediate risk. Despite the ubiquity, utility, and tractability of this approach, it is important to keep in mind that the behavior of prey only constitutes half of a predator–prey interaction. Prey-based data collection tends to promote a style of analysis that treats predation risk as a fixed environmental condition to which prey respond dynamically. However, predators respond dynamically to spatial and temporal variation in the distribution, abundance, and behavior of their prey, just as prey respond to spatial and temporal variation in predation risk. Recognizing that predation is a 'game' with players simultaneously in motion on both sides of the board helps put the antipredator behavior of prey in context. Ideally, research on predator–prey interactions should combine prey-based and predator-based data collection, but relatively few studies have adopted this approach. However, studies that center data collection on the predator are reasonably common, and this approach has yielded some important insights about antipredator behavior, including the following:

- Identifying the stages of predation from the perspective of the predator makes it clear that the probability of making a kill once an attack has begun is only the final stage of the predation sequence. Much research on antipredator behavior focuses on this final stage, but broader approaches that consider the ways that prey can manipulate the probability of being encountered or attacked yield a more complete understanding of the costs and benefits of antipredator response.

- Taking this broader view of the predation sequence also promotes analysis that considers both behavioral and ecological responses by prey. Changes in behavior (such as an increase in vigilance) are important from the perspective of the prey species, but ecological responses (such as a shift in habitat selection) are probably more important from the perspective of understanding the consequences of predation for community and ecosystem structure and function. As we seek to understand the consequences of predator loss (and reintroduction), this perspective will be important.

- Considering the interaction between predators and prey from both perspectives encourages analysis that considers the effects on antipredator behavior on predator–prey dynamics. Predators affect prey numbers through direct killing and through the costs of antipredator responses, which are known as 'risk effects' or 'nonconsumptive' effects. Recent research shows that risk effects can constitute a large proportion (even the majority) of a predator's limiting effect on prey numbers, but we do not yet have a clear understanding of the general importance of risk effects, the mechanisms that mediate them, or the factors that affect their magnitude.

Methods of Studying Predation

Predator-focused research on predation is generally conducted in two ways. Direct observation of predators has the advantage of providing information on the outcome (rates of predation and patterns of prey selection) and the processes that produce this outcome. Direct observation is tractable for many carnivores, but predators occupy positions high in the food web, and consequently live at much lower population densities than their prey. Many vertebrate predators are shy, nocturnal, or solitary (approximately 85% of the species in the order Carnivora are solitary), which further complicates direct observation and often requires radiotelemetry to obtain representative data. Indirect studies are usually based on identifying the remains of prey in carnivore scats. Such studies provide useful data on patterns of prey selection and relative rates of predation, but do not yield absolute rates of predation, and do not give information on the processes that produce the observed pattern of predation. However, it can be difficult to directly observe a large sample of predation events, and indirect measures of the relative importance of prey are a useful way to broaden the scope of sampling. Direct observation and scat analysis can be combined to take advantage of the strengths of each method.

The Predation Sequence, Prey Selection, and Antipredator Behavior

Predators rarely take prey in proportion to their availability. The species, age, and sex of prey (among other attributes) affect the likelihood of being taken by predators. It seems intuitively obvious that individuals of a type that is killed by predators often would be under stronger selection to engage in antipredator behavior, but this is not necessarily correct: a prey type might be killed rarely because it is rarely targeted, or because it engages in effective antipredator behavior. In other words, predation rates and antipredator behavior are linked by a feedback loop, and inferences about the strength of predation pressure must take this feedback into account. For example, it is possible that one species is directly killed by a predator half as often as another prey species, but the less-frequently killed species invests more time and energy in antipredator behavior, so that the total effect of predation is the same for both prey species.

Studies of the predation sequence can reveal the processes by which a prey species becomes a large or small proportion of a predator's diet, and thus help resolve the total effect of a predator's presence on prey dynamics. Many studies have compared the diet of a carnivore with the relative abundance of each prey species. Such data describe the end-point of prey selection, but do not reveal whether prey selection is affected by nonrandom prey encounter rates, nonrandom decisions to hunt once prey are encountered, nonrandom success once a hunt has begun, or dilution of individual risk within a group of prey. Taking the view that prey selection is the outcome of a process with several stages reveals the complexities of predator–prey dynamics. Predators might focus disproportionately on given prey types at each of these stages, antipredator behavior might be employed with varying effort at each stage, and the mechanisms that affect one stage may complement or oppose the effectiveness of antipredator behavior at other stages. For example, a larger group promotes dilution of risk, but might also make the group easier to detect, or more likely to be attacked if it is detected (or less likely: even the sign of some effects is difficult to predict using logic alone, and empirical data are required to resolve these issues).

Formally, relative risk, or an individual's risk of predation given the abundance of its type, is the product of four conditional probabilities, each representing a stage of the predation sequence:

Encounter = probability of being encountered, given availability
Attack = probability of being attacked, given encounter
Kill = probability of a kill occurring, given attack
Dilution = probability of being the individual selected, given a kill

Encounter

Relatively few studies have examined whether predators encounter their prey in proportion to their abundance or in some nonrandom manner. While data are sparse, most studies suggest that encounter rates are not random across prey species or group sizes. Habitat selection and active searching by predators could produce nonrandom patterns of prey encounter. Prey might reduce the likelihood of encounter by morphological crypsis, behavioral crypsis (e.g., snails show lowered activity rates in response to predator cues in water; many prey species freeze when a threat is nearby), active avoidance (e.g., duikers move to the opposite side of a bush when a threat is detected), reductions in group size (e.g., female elk disaggregate in response to the local presence of wolves), or shifts in habitat selection (e.g., bottlenose dolphins move to safer waters in response to the presence of tiger sharks).

Attack

Few studies have directly examined the factors that influence the decision to hunt, once prey have been detected, but limited data suggest that attack probabilities are not random. Attack abatement strategies are seen in many prey species and generally fall into two categories. Some behaviors dissuade attack by sending an honest signal to the predator that a prey animal is in good condition. If the signal is honest, the predator benefits by avoiding the costs of a hunt that is more likely than average to fail, and the prey benefits by avoiding higher cost antipredator behaviors such as full speed flight. A good example is stotting by Thomson's gazelles in response to cheetahs (stotting is a difficult and energetically costly display, in which gazelles repeatedly throw themselves as high as possible in a series of abrupt, four-footed bounds). The second category of attack abatement consists of behaviors that inform the predator that it has been detected (e.g., alert postures, alarm calls). For predators that rely upon stealth (e.g., most stalkers), attacks are likely to fail if the prey are aware of the predator's approach while it is still too far for a final rush. Examples include woodpigeons preyed on by hawks and antelopes preyed on by leopards.

Kill

Mechanisms that affect the success of a direct attack are probably the most heavily studied aspect of antipredator behavior, because this is the stage that is most easily examined by studies that focus data collection on the prey. This stage is also considered often by studies that focus data collection on the predator, which often use the term 'hunting success' to describe the ratio of kills to hunts, or the probability of making a kill once a hunt is initiated. Quantitative comparisons of hunting success across species or studies are notoriously difficult, because differences among species (and habitats, group size, etc.) in the behavioral details of a hunt can make it difficult to define when a hunt occurs in a manner that is comparable across studies but also appropriate to the specific predator-prey pair that is being studied. One clear-cut example of this problem is that coursers pursue their prey openly, and making prey flee at the outset of a hunt often allows weaker prey to be identified. Stalkers, on the other hand, are more likely to succeed if they can launch a final rush without having alerted their prey. Because the fundamental methods of capturing prey differ for stalkers and coursers, the criteria used to operationally define the initiation of a hunt also differ. Thus, statements such as 'African wild dogs succeed in 44% of their hunts' must be taken in context.

Nonetheless, we know a good deal about the factors that affect hunting success. Aspects of antipredator behavior that can reduce the probability of a kill occurring once a hunt has begun include active defense (e.g., adult ungulates charging predators that attack juveniles), collective defense (e.g., ring formations in wildebeest or musk ox), coordinated flight (e.g., in fish schools), or predator confusion. It should be noted that the confusion produced by scattering groups can work against prey as well as the predator; predator confusion may be more effective for fish and birds, operating in three dimensions, than for species like mammals that operate in two dimensions and are thus more likely to intersect paths when moving quickly and erratically.

Dilution

Individual prey can reduce their immediate risk of predation by sharing risk with group mates, provided that predators cannot kill a large subset of a prey group and the detectability of groups by predators does not increase too substantially with group size. Empirical studies generally suggest that the benefit of risk dilution is a strong factor in reducing the per capita risk of death, when risk is assessed on a per-attack basis. Thus, dilution can provide selection pressure in favor of grouping, even if grouping does not yield antipredator benefits through other mechanisms that reduce the total rate of predation. If dilution of risk is the only benefit of grouping (on a per attack basis), then the time scale over which prey minimize the risk of predation becomes important.

Risk Effects: The Costs of Antipredator Behavior and the Total Limiting Effect of Predators on Prey

A large body of research has identified and quantified the behavioral responses of prey to predators. A large body of theoretical and empirical research has considered the manner in which direct predation can limit prey populations. Recent research has combined these approaches to consider the possibility that predators can limit prey populations in part through the costs of antipredator

responses. These costs are often termed 'nonconsumptive effects' or 'risk effects'. Experimental studies with invertebrates have shown that risk effects can comprise a large part of the limiting effect of predators on prey populations. For example, spiders rendered incapable of killing grasshopper prey (by gluing the spiders' mouthparts shut) induce typical antipredator behavior by the grasshoppers, and reduce grasshopper populations by an amount that is comparable to unmanipulated, lethal spiders. As a second example, the presence of the invertebrate freshwater predator *Bythotrephes* causes its *Daphnia* prey to migrate to colder and darker water to reduce predation risk. While Daphnia benefit from reduced predation risk in such conditions, they also grow more slowly and produce fewer eggs. This cost of responding to predation risk reduces the population growth rate of *Daphnia* by a greater amount than the effect of direct predation itself. Several studies suggest that risk effects can also be important for the dynamics of vertebrates in the wild, though we still have little information on this subject. For example, elk adjust their behavior, habitat selection, and diet in response to the presence of wolves, and these responses are associated with substantial decreases in food intake, progesterone concentration (progesterone is responsible for maintenance of pregnancy), and calf production. Such results suggest that the costs of antipredator responses can be an important part of the total effect of predators on prey dynamics. In turn, this result suggests that the effects of predator loss or reintroduction will often be more complicated than the result that would be predicted solely on the basis of direct predation rates.

See also: Defensive Avoidance; Group Living; Risk-Taking in Self-Defense.

Further Reading

Caro T (2005) *Antipredator Defenses in Birds and Mammals.* Chicago: University of Chicago Press.

Creel S and Christianson D (2008) Relationships between direct predation and risk effects. *Trends in Ecology & Evolution* 23: 194–201.

Creel S, Christianson D, Liley S, and Winnie JA (2007) Predation risk affects reproductive physiology and demography of elk. *Science* 315: 960.

Elgar MA (1989) Predator vigilance and group size in mammals and birds: A critical review of the empirical evidence. *Biological Reviews of the Cambridge Philosophical Society* 64: 13–33.

Fitzgibbon CD (1990) Mixed-species grouping in Thomson's and Grant's gazelles: The antipredator benefits. *Animal Behaviour* 39: 1116–1126.

Krasue J and Godin JG (1995) Predator preferences for attacking particular prey group sizes: Consequences for predator hunting success and prey predation risk. *Animal Behaviour* 50: 465–473.

Lima SL (1998) Nonlethal effects in the ecology of predator–prey interactions. *Bioscience* 48: 25–34.

Lima SL (2002) Putting predators back into behavioral predator–prey interactions. *Trends in Ecology and Evolution* 17: 70–75.

Lima SL and Dill LM (1990) Behavioral decisions made under the risk of predation: A review and prospectus. *Canadian Journal of Zoology* 68: 619–640.

Preisser EL, Bolnick DI, and Blumstein DT (2005) Scared to death? The effects of intimidation and consumption in predator prey interactions. *Ecology* 86: 501–509.

Schmitz OJ (1998) Direct and indirect effects of predation and predation risk in old-field interaction webs. *The American Naturalist* 151: 327–342.

Schmitz OJ, Grabowski JH, Peckarsky BL, Preisser EL, Trussell GC, and Vonesh JR (2008) From individuals to ecosystem function: Toward an integration of evolutionary and ecosystem ecology. *Ecology* 89: 2436–2445.

Werner EE, Gilliam JF, Hall DJ, and Mittelbach GG (1983) An experimental test of the effects of predation risk on habitat use in fish. *Ecology* 64: 1540–1548.

Problem-Solving in Tool-Using and Non-Tool-Using Animals

A. M. Seed and J. Call, Max Planck Institute for Evolutionary Anthropology, Leipzig, Germany

The majority of mobile animals need to locate their food in space (and in some cases, time). In addition, some kinds of food need to be extracted or processed before they can be consumed, and the pressure to exploit these resources, through extractive foraging or tool-use, has been hypothesized to have selected for advanced cognitive abilities. This article will focus on physical problem-solving in mammals and birds and review the evidence for the cognition underpinning it. A recurring question is whether animals that solve complex problems in their environment differ from those that do not, in terms of what, or how, they learn. Early research on the ability of animals to solve physical problems focused on tool-users, but these are not the only species to exploit embedded or otherwise defended resources, and so our review will encompass evidence from tool-using and nontool-using animals alike.

When Does a Problem Become a Problem?

Animals face many problems in their lives; they need to find food, avoid predators, reproduce and, in some cases, care for their offspring. This article is concerned with the problems animals face in the course of accessing and processing food. Some species never face these sorts of physical problems in their natural habitat; their food can be processed directly through morphological adaptations. However, some species take advantage of embedded or otherwise defended foods, despite the fact that evolution has not equipped them to process such resources directly. For example, although rodents, such as agoutis, can gnaw through tough nuts to get access to their kernels, other species must use tools to smash them open (such as chimpanzees and capuchin monkeys), or drop them from a height onto a hard surface (a tactic employed by several corvid species). Similarly, the aye-aye, a species of prosimian, native to Madagascar, has an elongated middle digit which it uses to probe tree holes for insect larvae. However, species without this adaptation, such as chimpanzees, woodpecker finches, and New Caledonian crows use stick tools to fish for invertebrates instead. Physical problems then, such as extracting a larva from a tree or opening a nut, are not simply a feature of the environment, but are modulated by the nature of the animal encountering them. Kohler, one of the first to research

the physical problem-solving abilities of animals, described a problem as follows:

> Something is to be achieved with regard to … a situation; but, as the situation is given, it cannot be achieved. How must we change the situation so that the difficulties disappear and the problem is solved? (Kohler, 1969, p. 134)

One of the most conspicuous candidates for problem-solving is tool-use, when an animal can be seen to reach beyond the limitations of its own body to gain an otherwise inaccessible resource. Animals in the wild also solve problems through direct manipulation of the environment; this can be identified when an expert in a particular species witnesses an individual encountering a new problem and finding a solution to it outside of the usual behavioral repertoire of that species. However, some behavioral solutions have a large genetically determined component, and are therefore, comparable to a morphological adaptation. Further study is needed to identify behavior as novel, and because of the interplay between genes and the environment, this may not be an easy task. For example, chimpanzees and orangutans are known to use tools customarily in the wild, but gorillas and bonobos do not. However, in captivity, all four species of great apes use tools. Nevertheless, regardless of whether tool-using per se is genetically predetermined in some or all of the ape species, the uses to which they are put are very variable and do not seem to be genetically fixed. Some of the difficulties in defining innovation and problem-solving can be circumvented by presenting animals with novel problems in the laboratory that prevent them from using morphological or behavioral adaptations; their problem-solving abilities can then be studied, although this approach is not without its caveats. This article will address four questions about problem-solving in animals. What selective pressures favor its evolution? What sort of problems can animals solve, and what makes one species better at solving problems than another? Finally, what cognitive mechanisms underpin problem-solving in different animal species?

The Evolution of Problem-Solving

Problem-solving is costly, for two main reasons. First, interacting with new objects leads to risks, both of actual physical harm (from resources such as poisonous caterpillars and nettles), and from the prospect of investing valuable foraging time into a nonprofitable venture (some problems may prove insoluble). In environments where these risks are large and the incentives small (because of the direct availability of reliable resources) it will be more adaptive to be cautious and conservative. Second, in both birds and primates, those species for which higher rates of innovation and tool-use are reported have relatively enlarged areas of the brain involved in cognitive processes such as inhibition and rule-learning (the mesopallium and nidopallium in birds and the isocortex and striatum in primates). Neural tissue is expensive in terms of energy demands (requiring around nine times more energy than the average requirements of body tissue), and animals maintaining a surplus would be selected against. Given the costs of problem-solving, why would it evolve?

One theory involves direct selection for problem-solving itself, in environments where the benefits of accessing defended food outweigh the costs of the risk of injury and a large brain. Candidates for the important features of these environments are: low availability of directly accessible foods; high availability of defended foods; availability of defended foods with high nutritional quality; and unpredictability (either through seasonal variation or rapid environmental change such as occurs in urban environments). Of course, these features are not mutually exclusive; need and opportunity can act in combination, and different features may have been acting during the evolution of different problem-solving species. Evidence for the importance of different environmental features comes from two sources. The first is the distribution of problem solving behavior across populations of a single species occupying different habitats. For example, woodpecker finches use tools to probe tree holes for invertebrates more frequently in arid habitats in the dry season, where the availability of surface prey is low and the availability and nutritional content of embedded prey is high, compared both to the wet season and to other habitats, where the opposite is true (Tebbich et al., 2002). Similarly, capuchin monkeys use stones to dig for high-quality tubers in arid environments with low availability of easily accessible foods, but have not been reported to do so in habitats rich in other food sources (Moura and Lee, 2004). Another source of evidence comes from the survival rates of different species of bird during introduction events to new environments. Large-brained species show higher survival rates, through higher rates of innovation. Such studies give clues as to the selective pressures that originally favored the evolution of problem-solving and the associated neural structures.

Alternatively, selection in another domain may have resulted in neural adaptations that, as a by-product, also allowed animals to solve physical problems. There are several hypotheses for the evolutionary pressures favoring increased brain size and intelligence. Some theories cite ecological factors other than those mentioned above, such

as a patchy distribution of food in time and space. Other theories emphasize the challenges and opportunities stemming from social living, such as increased competition; cooperating to secure resources unavailable to individuals; and learning from others. Evidence supporting the Social Brain hypothesis and its variants comes from positive correlations between relative brain size and group size in mammals. Social system is also related to brain size in birds, with cooperative breeders and long-term monogamous species having the largest brains. Amici et al. (2008) report evidence for a link between the social environment and adaptations that impact on problem-solving. They found from a comparative study of inhibitory skills among seven species of primate that the degree of fission-fusion dynamics (splitting and merging into subgroups) predicted success on tasks in which a prepotent response, such as reaching directly for food, had to be inhibited. Inhibition is certainly a cognitive skill that would benefit a species with high fission–fusion dynamics (if group mates are not seen for some time, relations among them can change), but that also would be important for problem-solving, which requires the repression of an ineffective direct approach.

Whether direct or indirect selection is responsible (and the two alternatives are not mutually exclusive), comparing animal problem-solving in the wild certainly reveals differences between species, with some groups of animals emerging as the most frequent innovators and tool-users; the great apes amongst primates (such an analysis has not been done with other mammals), and the corvids amongst birds. However, are these differences ascribable to a property of the animals themselves or the habitats in which they are found? Of course, the relatively large forebrains of these groups suggest that it is the former, but it could be the case that infrequent tool-users and innovators lack the incentive, rather than the ability, to solve problems. To comprehensively address whether there are differences between species in the types of problems they can solve, comparative tests under controlled conditions are needed.

What Sorts of Physical Problems Can Animals Solve?

Observations of animal problem-solving in nature have inspired experiments to try and uncover what problems animals can solve under controlled conditions, and to characterize the cognitive processes involved. Most of this work has been conducted with chimpanzees, inspired by the striking observations of tool use and manufacture by this species in the wild and its close relationship to humans. However, other species have been incorporated, mostly from those highly innovative groups, the corvids and primates. A few studies have also been conducted on

Table 1 The main physical problem-solving studies conducted in the laboratory on more than one species (species are shown in the legend)

Property	Task	Apes					Primates								Mammals			Corvids			Birds		
		1	2	3	4	5	1	2	3	4	5	6	7	8	1	2	3	1	2	3	1	2	3
Shape																							
Form	1. Bundled sticks	Y	Y	Y	–	–	Y	–	–	–	–	–	–	–	–	–	–	Y	–	–	–	–	–
	2. Unbend wire tool	Y	–	–	–	–	Y	–	–	–	–	–	–	–	–	–	–	Y	–	–	–	–	–
	3. Modify tool	Y	Y	Y	–	–	Y	–	–	–	–	–	–	–	–	–	–	Y	–	–	Y	–	–
Length	4. Tool length choice	Y	Y	Y	Y	–	Y	–	–	–	–	–	–	–	–	–	–	Y	–	–	Y	–	–
	5. Assemble tool	Y	Y	Y	–	–	Y	–	–	–	–	–	–	–	–	–	–	–	–	–	–	–	–
	6. Meta-tool	Y	–	Y	Y	–	Y	Y	–	–	–	Y	–	–	–	–	–	Y	–	–	–	–	–
Width	7. Tool diameter	–	–	–	–	–	Y	–	–	–	–	–	–	–	–	–	–	Y	–	–	–	–	–
Contact																							
Connection	8. Inverted rake	N	–	–	–	–	Y	Y	–	–	Y	N	Y	–	Y	–	Y	–	–	–	–	–	–
	9. String pull	Y	Y	Y	Y	Y	Y	Y	Y	Y	–	Y	Y	Y	Y	Y	Y	Y	Y	Y	Y	Y	Y
	10. Parallel	Y	Y	Y	Y	Y	Y	Y	Y	Y	–	Y	Y	Y	Y	Y	–	Y	Y	Y	Y	Y	Y
	11. Crossed	Y	Y	Y	Y	–	Y	Y	Y	N	–	–	Y	–	N	–	–	–	–	Y	–	–	Y
	12. Broken	Y	Y	Y	Y	–	Y	–	–	–	–	–	–	–	–	Y	–	–	–	–	–	–	–
	13. Contact tube	–	Y	–	–	Y	–	–	–	–	–	–	–	–	–	Y	–	–	Y	–	–	Y	–
	14. Cloth	Y	Y	Y	Y	Y	Y	Y	–	–	Y	Y	–	–	–	Y	–	Y	–	–	–	Y	–
Support																							
Material																							
Rigidity	15. Flimsy tool	Y	–	Y	–	–	–	–	–	–	Y	Y	–	–	–	–	–	N	–	–	Y	–	–
	16. Trap tube	Y	Y	Y	N	–	Y	–	–	–	–	–	–	–	–	Y	–	Y	Y	Y	Y	–	–
Continuity/solidity	17. Two-trap task	Y	–	–	–	–	Y	–	–	–	Y	N	–	–	–	–	–	Y	Y	Y	Y	–	–
	18. Trap table	Y	–	–	–	Y	Y	–	–	–	Y	–	–	–	–	–	–	Y	–	–	–	–	–
	19. Obstacles	–	Y	Y	Y	–	Y	Y	–	–	–	–	–	–	–	–	–	–	–	–	–	–	–
	20. Boards	–	–	–	–	–	–	–	–	–	–	–	–	–	N	–	–	–	–	–	–	–	–
Weight	21. Discrimination	Y	Y	–	–	–	Y	Y	–	Y	–	–	–	Y	–	Y	–	–	–	–	–	–	–

Y indicates that at least one individual solved at least one configuration of the problem.

N indicates that no individuals solved the problem. A dash indicates that the species has not been tested. Species shaded in gray are habitual tool-users in the wild.

A (Apes):1 – Chimpanzee; 2 – Bonobo; 3 – Orangutan; 4 – Gorilla; 5 – Gibbon. P (Primates – other); 1 – Capuchin; 2 – Macaque (various); 3 – Spider monkey; 4 – Squirrel monkey; 5 – Vervet monkey; 6 – Callatrichid; 7 – Baboon; 8 – Long-tailed lemur. M (Mammals – other); 1 – Dogs; 2 – Elephants; 3 – Degus. C (Corvids): 1 – New Caledonian Crow; 2 – Rook; 3 – Raven. B (Birds – other): 1 – Woodpecker finch; 2 – Parrot; 3 – Pigeon.

other mammals and birds. **Table 1** summarizes the results of the main studies on which more than one species has been tested, organized in terms of the physical principle involved. In each case, the problem is the retrieval of a food reward, to which direct access has been blocked. The subject either needs to modify the situation (e.g., creating a long stick tool by putting three smaller ones together), or make a choice between two or more options (e.g., pulling an effective tool rather than an ineffective one, which might be long, complete, rigid or connected to food, rather than short, broken, floppy, or unconnected to food). The problems and their solutions are shown pictorially in **Figure 1**. For a detailed description of the results of most of these studies, see Tomasello and Call (1997). From this overview, it is clear that animals are able to solve problems of a great many varieties involving the physical principles of shape, connectedness, and material properties (such as rigidity, continuity, and weight), including some that were thought to be the unique preserve of humans, such as meta-tool use.

However, **Table 1** merely indicates which species were able to find the solution to the problem over the course of an experiment, and it must be emphasized that species differ considerably in other performance measures, such as the amount of experience they need to solve a problem, and the proportion of successful individuals. The broken string paradigm has been used to test several species, and a more detailed analysis of this problem can illustrate how performance can differ. When presented with two objects attached to food rewards (strings, or other objects such as strips of cloth or prepositioned tools), one intact and one with a clear break in the middle, great apes, vervet monkeys, cotton-top tamarins, elephants, and pigeons, are able to pull the connected, continuous object to bring the food within reach (**Table 1**; **Figure 1(12)**). However, whilst some species performed significantly above chance, right from the start of the experiment, in at least some configurations (great apes, vervet monkeys, and elephants), pigeons and cotton-top tamarins required extensive training (they required over a hundred trials before the correct solution was learned).

It is tempting to infer that the species that solved the problem spontaneously used a qualitatively different cognitive mechanism, involving an appreciation of the principle of connectedness such as would underpin an adult human's behavior. However, an animal's performance depends on a number of both cognitive and noncognitive processes. For example, those animals that take longer to solve a particular task, or even fail it completely, may be less motorically dexterous, less motivated, more easily distracted, find the task at hand harder to perceive, or find irrelevant features of the task more attention-grabbing, compared to the species that solve it quickly. Even the same individuals can perform very differently on two tests supposedly probing the same ability. For example,

although the great apes tested by Herrmann et al. (2008) on the broken string problem were able to solve it spontaneously when the material involved was string or cloth, they performed at chance (in the 6 trials given) when the objects were two wooden canes, prepositioned around the food rewards. Similarly, Girndt et al. (2008) tested a group of chimpanzees on the trap table test (in which food should be raked in over a solid surface rather than one with a trap in it, **Figure 1(18)**). When tested with two prepositioned tools, they failed to find the solution in 20 trials, but they passed when given one tool and required to choose which reward to rake towards themselves. These small differences in task presentation had such a large effect on performance that had only one configuration been given, a very misleading picture of the animal's abilities would have emerged. Given that many species comparisons rely on comparing studies done using not only different materials but also completely different research setups, it is clear that caution must be exercised. Importantly, even an identical setup may not be equivalent for different species. Ideally, results of many tests probing the same ability in different ways should be employed to build confidence in the interpretation, a process referred to as 'triangulation.'

Although several processes (both cognitive and non-cognitive) go into the make-up of a successful problem-solver, comparing performance measures across a range of problems can provide us with an idea of different species' proclivity to solve new problems. Fine-grained analyses of performance data provide little evidence that apes outperform monkeys in problems involving space and objects, despite hypotheses to the contrary (Call, 2000). Furthermore, animals that customarily use tools in the wild do not systematically outperform nontool-users on the tasks in which both have been tested. Interestingly, on the tests conducted so far, corvids have performed comparably to primates, suggesting that there has been a convergent evolution of problem-solving skills in these two distantly related groups (Seed et al., 2009). However, to study the cognition underpinning problem-solving we need to go beyond absolute performance differences, and look at the results of tests that aim to isolate the cognitive processes from each other (e.g., perception, inhibition, learning, memory), and from the noncognitive ones (e.g., motor skills, motivation, attention, and temperament).

What Cognitive Mechanisms Underpin Animal Problem-Solving?

When animals solve problems, such as the broken string task described above, do they use a knowledge of object properties (only materials that are continuous and connected to the reward are worth pulling), or have they simply associated certain perceptual cues with rewarding

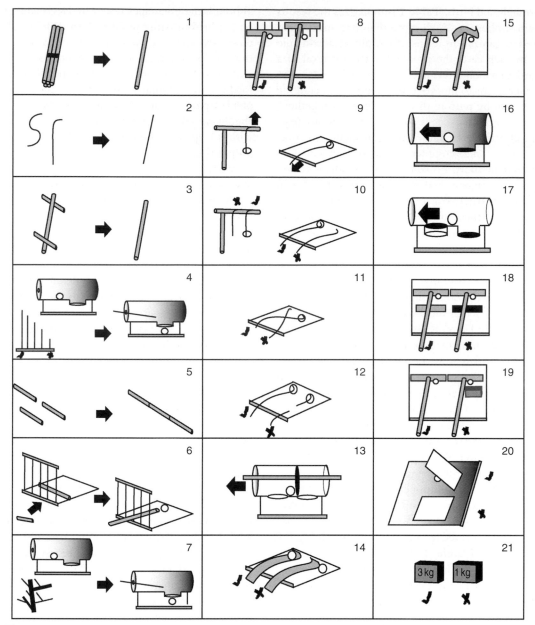

Figure 1   The 21 problems as numbered in **Table 1**. The arrow shows the transformation from one state to another or the direction in which the food should be moved for the problem to be solved.

outcomes? The question of whether animals solve problems through sophisticated cognitive processes, such as mental representation and reasoning, is a century old debate that is still on-going. At one pole of the debate is the view expounded by Thorndike, an American psychologist working at the beginning of the twentieth century. Thorndike presented animals with puzzle boxes that they had to learn to open, either to gain their freedom or gain access to food. From these studies, he concluded that whilst the animals were able to learn through trial-and-error to associate perceptual features of the stimuli with behavioral responses, repeating those combinations that led to reward and ceasing those that did not, there was no

evidence for any human-like reasoning. Kohler, a contemporary of Thorndike, objected to this conclusion on the grounds that the methodology employed did not allow the animals to engage in their powers of reasoning, because the causal mechanisms controlling the boxes' release-mechanisms were made too obscure. In his studies with chimpanzees, he made food inaccessible, but the means to reach it clearly visible (e.g., putting food too high for the chimpanzees to reach and leaving sticks to knock the food down, or boxes that could be stacked). From the rapid success of some of his subjects, he concluded that chimpanzees used insight, and not blind trial-and-error, to solve problems. Importantly, both accounts contain a hypothesis about two facets of the cognitive process, namely what animals see in a problem (shallow perceptual features or more abstract representations of causally relevant structural features) and how animals solve problems (associative learning or more complex 'human-like' processes, such as reasoning and insight). These two facets will be discussed in turn.

What Do Animals See in a Problem?

Objects in the environment have physical properties that dictate the possible ways in which they can interact with one another (e.g., solid objects cannot pass through one another). These properties, such as solidity, continuity, weight, and rigidity can be directly sensed, but the principles themselves can also be represented at a deeper level of abstraction (where more 'abstract' means that the information is not equivalent or reducible to concrete, analogue sensory input, but rather has undergone further processing in which meaning is extracted). This would enable them to be used to make predictions in novel situations that do not share any surface perceptual features with those in which they were first encountered. This is a critical feature of flexible problem-solving in humans, but in the absence of verbal report, when an animal transfers its solution to a new context, it is often impossible to know if it has inferred anything about causal properties, or if its behavior can be explained by a combination of past associative learning and generalization based on surface-level perceptual characteristics. Controlled experimentation that pits one account against another is needed.

Once paradigms have been found that subjects can at least learn to solve, different features can be varied systematically in transfer tasks, to ascertain which of the possible features the subjects used to solve the original problem. For example, in the broken string task, cotton-top tamarins, having learned the original discrimination (pull the unbroken rather than the broken cloth), were given transfer tasks that varied functionally irrelevant features, such as the cloth's color, shape, texture, and the shape and size of the gap. The tamarins readily transferred their solution across the majority of these changes, suggesting that they had used functionally relevant properties to solve the original discrimination (Hauser et al., 1999). In contrast, pigeons that had learned to solve one version of the broken string task failed to transfer to a new version in which the shape and color of the material was changed, suggesting that they had relied on perceptual cues to solve the original task (Schmidt and Cook, 2006).

However, even the transfer tasks solved by the tamarins could be solved by learning to avoid the gap as a perceptual feature, and then generalizing along this parameter. Animals certainly do rely on perceptual cues to solve some problems. Povinelli (2000) conducted a

series of experiments with chimpanzees, including several of the tasks depicted in **Figure 1(8, 12, 14–16, 18)**. The results of many of these studies were commensurate with the idea that the chimpanzees had used a perceptually based rule, rather than one that encompassed an abstract notion of object properties. For example, in the trap tube task, in which the subject needs to push a piece of food out of a horizontal tube away from a trap (**Figure 1(16)**), the one subject that learned to do so continued to use this strategy even when the tube was inverted, and therefore nonfunctional. It seemed therefore that she had treated the trap as a perceptual cue but had not encoded its functional significance. But does the use of perceptual information in one task mean that structural information cannot be encoded by members of this species? It must be emphasized that the two sorts of knowledge are not mutually exclusive, and indeed human researchers have argued for the existence of two cognitive processes working in parallel during problem-solving. Furthermore, the many factors affecting performance mentioned at the end of the last section also impact the results of transfer tests, and so the results of one study (especially, if they are negative) need to be interpreted cautiously. Seed et al. (2009) found that eight chimpanzees solved a version of the trap problem that did not require them to use a tool. The performance of these subjects was then compared with naïve subjects on a perceptually distinct transfer test made of new materials. Chimpanzees without experience on the first problem performed poorly on this task (only one subject was successful), but all of the experienced subjects in a group tested without a tool solved the new test in very few trials, suggesting that, in contrast to the results of the original trap tube study, they had encoded information about the functional properties of the objects involved in the initial testing phase. However, another group of experienced subjects tested with a tool required many more trials, and only half of the subjects were successful, revealing the critical importance of the manner of task presentation.

Seed et al. (2006) made some changes to the trap tube task in a study of rooks, and aimed to pit a perceptual and structural account directly against one another. Eight birds were tested on a version that featured two 'traps' along a horizontal tube (**Figure 1(17)**). One of the traps was sealed with a black disc at the bottom and would trap the reward if the rooks pulled the food over it. The other was nonfunctional; in Design A, it had a black disc at the top, which the food could pass across; in Design B, it had no black disc, so the food could fall through it. Once the birds had solved these tasks, they were given two transfer tasks, each featuring both previously nonfunctional traps (pass-across and fall-through). In Design C, both ends of the tube were blocked with bungs, and in Design D, the tube was lowered to the surface of the testing shelf. Crucially, therefore, both tasks featured the same familiar cue, but each required the opposite response to it (pull away from the black disc in Task C, pull towards it in Task D). One of the seven rooks was able to solve these transfers, suggesting that she did not simply use the appearance of the functional trap as an arbitrary, surface-level cue. Seed et al. (2009) recently conducted a similar experiment with chimpanzees. All of the eight chimpanzees tested learned to avoid the trap. Furthermore, one chimpanzee passed both designs C and D. Like the successful rook, this chimpanzee could not have been using a rule based on an arbitrary perceptual cue to solve the task.

Hanus and Call (2008) also pitted perceptual and abstract representational accounts against one another in an experiment designed to probe what chimpanzees know about weight. Chimpanzees were shown a seesaw balance, with an empty cup at either end. The experimenter surreptitiously placed the banana in one of the cups behind a screen. The chimpanzees saw the balance tip, and strikingly, they chose the lower cup significantly more often than the higher cup from the first trial. Perhaps most interestingly, they chose at chance when the experimenter's action tipped the balance, revealing that the shallow-level perceptual information, namely the downward movement of the lower cup, was not sufficient to elicit the chimpanzees' choice. Even subjects tested first on the casual condition that had been choosing the lower cup reverted to random responding when the weight of the banana was no longer the cause of the movement.

From these and other experiments, it seems that some animals do form abstract representations of some object properties. However, this assertion is still controversial, and a number of questions arise: which species; which specific properties; how do the abilities develop; and how (if at all) do they differ from those of humans?

How Do Animals Translate Information and Knowledge into Action?

Associative learning as put forward by Thorndike to explain the performance of mammals in his puzzle boxes is based purely on covariation, and considerable trial-and-error is needed before the emergence of a correct solution. Experiments in the laboratory have shown that animals are adept at learning to associate a response to a given cue if doing so reliably leads to a certain outcome (e.g., pressing a lever or pecking a key when a light is turned on, to gain access to food). Putting aside the question of the level of abstraction at which the learned knowledge is represented (perceptual or structural), is this the only mechanism available for animals to discover the solution to a problem? In the real world, events are not random and arbitrary, but are instead underpinned by predictable causal structures (e.g., heat causes water to evaporate, and not the other way round). Adult humans are able to infer the solution to a completely novel problem

without interacting with it at all, drawing on their knowledge of causality, logic, and other applicable concepts. Processes such as inference are little understood and lack rigid definitions; they are simply referred to as 'nonassociative processes.' What evidence is there for inference in animals?

Call (2007) tested great apes on a task in which subjects needed to locate a food reward based on the inclination of two boards lain on a table. One board had a food reward underneath it, and was therefore at an incline, whilst the other was flat (**Figure 1(20)**). All great apes were able to locate the reward at above chance levels. They did not show a preference for the inclined board when the cause of the incline was a wedge, despite being rewarded if they did so. It therefore seems that apes were able to use their knowledge of solid objects and the effect they have on one another to infer the location of food, because a learned preference for inclined boards could not explain the results. Similarly, they were also able to infer the presence of food when two cups were shaken and only one made a rattling sound, but did not choose a tapped cup in another condition even though this choice was rewarded (Call, 2004). In the experiment by Hanus and Call (2008) described earlier, chimpanzees inferred the location of food on the basis of its weight, choosing the lower of two cups on a see-saw balance when the baiting of one of the cups was the cause, but not when the experimenter moved the balance by hand. Call (2004) also found evidence for inference by exclusion in apes. Subjects were shown two cups, and the hiding of a piece of food in one of them behind a screen. Apes were then shown the empty cup, or in another condition, the empty cup was shaken and produced no rattling sound. The apes chose the untouched, baited cup significantly above chance levels.

Conclusion

Questions concerning the cognitive processes involved in animal problem-solving are still hotly debated, but they have relevance for fundamental questions concerning thinking in nonverbal animals, and the evolution of the human mind. The debate has moved on from a stark dichotomy between simple associative learning and more complex processes, and most modern researchers acknowledge the interplay between inherited predispositions, learning, and reasoning. Indeed, even Kohler, the great exponent of insight in animals, recognized that '... some previous learning is often needed not only for the solution of a problem but also for its understanding as a problem' (Kohler, 1969, p. 135). Physical problem-solving provides researchers with an excellent window onto these questions, because unlike the social domain, objects and events in the physical world can be tightly controlled by experimenters. However, because so many variables impact on performance on these tasks, experiments must be carefully designed and interpreted with caution, especially when comparing species. Evidence so far suggests that for large-brained, innovative species such as primates and corvids, associative learning of perceptual rules is not the only process available for solving problems. However, further work is needed to characterize these nonassociative processes in more detail, and the precise role they play in animal problem-solving. We have taken some steps towards an understanding of animals' knowledge of object properties, and their inferential abilities, but we know little about how these two facets interact. A still greater challenge is to explore the distribution of these cognitive skills amongst animals, to address the question of the evolutionary pressures selecting for problem-solving skills.

See also: Animal Arithmetic; Apes: Social Learning; Categories and Concepts: Language-Related Competences in Non-Linguistic Species; Costs of Learning; Development, Evolution and Behavior; Foraging Modes; Innovation in Animals; Intertemporal Choice.

Further Reading

Amici F, Aureli F, and Call J (2008) Fission-fusion dynamics, behavioral flexibility, and inhibitory control in primates. *Current Biology* 18(18): 1415–1419.

Beck B (1980) *Animal Tool Behavior: The Use and Manufacture of Tools by Animals.* New York: Garland.

Bitterman ME (1965) Phyletic differences in learning. *American Psychologist* 20: 396–410.

Blaisdell AP (2008) Cognitive dimension of operant learning. In: Roediger HL, III (ed.) *Cognitive Psychology of Memory. Vol. 1 of Learning and Memory: A Comprehensive Reference,* 4 vols., pp. 173–195. (J. Byrne Editor). Oxford: Elsevier.

Call J (2000) Representing space and objects in monkeys and apes. *Cognitive Science* 24(3): 397–422.

Call J (2004) Inferences about the location of food in the great apes (*Pan paniscus, Pan troglodytes, Gorilla gorilla,* and *Pongo pygmaeus*). *Journal of Comparative Psychology* 118(2): 232–241.

Call J (2007) Apes know that hidden objects can affect the orientation of other objects. *Cognition* 105(1): 1–25.

Call J and Tomasello M (2005) Reasoning and thinking in non-human primates. In: Holyoak KJ and Morrison RG (eds.) *The Cambridge Handbook of Thinking and Reasoning,* pp. 607–632. New York, NY: Cambridge University Press.

Fujita K, Kuroshima H, and Asai S (2003) How do tufted capuchin monkeys (*Cebus apella*) understand causality involved in tool use? *Journal of Experimental Psychology Animal Behavior Processes* 29: 233–242.

Gibson KR (1986) Cognition, brain size and the extraction of embedded food resources. In: Else JG and Lee PC (eds.) *Primate Ontogeny, Cognition and Social Behaviour,* pp. 205–218. Cambridge: Cambridge University Press.

Girndt A, Meier T, and Call J (2008) Task constraints mask great apes' ability to solve the trap-table task. *Journal of Experimental Psychology Animal: Behavior Processes* 34(1): 54–62.

Hanus D and Call J (2008) Chimpanzees infer the location of a reward on the basis of the effect of its weight. *Current Biology* 18(9): R370–R372.

Hauser MD (1999) perseveration, inhibition and the prefrontal cortex: A new look. *Current Opinion in Neurobiology* 9: 214–222.

Hauser MD (2003) To innovate or not to innovate? That is the question. In: Reader SM and Laland KN (eds.) *Animal Innovation,* pp. 329–338. Oxford: Oxford University Press.

Hauser MD, Kralik J, and Botto-Mahan C (1999) Problem solving and functional design features: Experiments on cotton-top Tamarins, *Saguinus oedipus oedipus. Animal Behaviour* 57(3): 565–582.

Herrmann E, Wobber V, and Call J (2008) Great apes' (*Pan troglodytes, Pan paniscus, Gorilla gorilla, Pongo pygmaeus*) understanding of tool functional properties after limited experience. *Journal of Comparative Psychology* 122(2): 220–230.

Heyes CM and Huber L (2000) *The Evolution of Cognition.* Cambridge, MA: MIT press.

Hurley S and Nudds M (2006) *Rational Animals.* Oxford: Oxford University press.

Kohler W (1969) *The Task of Gestalt Psychology.* Princeton, NJ: Princeton University Press.

Mackintosh NJ, Wilson B, and Boakes RA (1985) Differences in mechanisms of intelligence among vertebrates. *Philosophical Transactions of the Royal Society of London B Biological Sciences* 308: 53–65.

Moura ACdA and Lee PC (2004) Capuchin stone tool use in caatinga dry forest. *Science* 306: 1909.

Penn DC, Holyoak KJ, and Povinelli DJ (2008) Darwin's mistake: Explaining the discontinuity between human and non-human minds. *Behavioral* and Brain Sciences 31: 109–130.

Povinelli DJ (2000) *Folk Physics for Apes: The Chimpanzee's Theory of How the World Works.* Oxford: Oxford University Press.

Reader SM and Laland KN (2002) Social intelligence, innovation and enhanced brain size in primates. *Proceedings of the National Academy of Sciences USA* 99: 4436–4441.

Roberts WA (1998) *Principles of Animal Cognition.* Boston: McGraw-Hill.

Schmidt GF and Cook RG (2006) Mind the gap: Means-end discrimination by pigeons. *Animal Behaviour* 71(3): 599–608.

Seed AM and Call J (2009) Causal knowledge for events and objects in animals. In: Watanabe S, Blaisdell AP, Huber L, and Young A (eds.) *Rational Animals, Irrational Humans,* pp. 173–187. Tokyo: Keio University press.

Seed AM, Call J, Emery NJ, and Clayton NS (2009) Chimpanzees solve the trap problem when the confound of tool use is removed. *Journal of Experimental Psychology: Animal Behavior Processes* 35(1): 23–34.

Seed A, Emery N, and Clayton N (2009) Intelligence in corvids and apes: A case of convergent evolution. *Ethology* 115: 401–420.

Seed AM, Tebbich S, Emery NJ, and Clayton NS (2006) Investigating physical cognition in rooks, *Corvus frugilegus. Current Biology* 16(7): 697–701.

Shettleworth SJ (1998) Cognition, evolution, and the study of behavior. In: Shettleworth SJ (ed.) *Cognition, Evolution, and Behavior,* pp. 3–48. New York, NY: Oxford University Press.

Sol D, Timmermans S, and Lefebvre L (2002) Behavioural flexibility and invasion success in birds. *Animal Behavior* 63: 495–502.

Tebbich S, Taborsky M, Fessel B, and Dvorak M (2002) The ecology of tool-use in the woodpecker finch (*Cactospiza pallida*). *Ecology Letters* 5: 656–664.

Tomasello M and Call J (1997) *Primate Cognition.* New York: Oxford University Press.

Visalberghi E and Fragaszy DM (2006) What is challenging about tool use? The capuchin's perspective. In: Wasserman EA and Zentall TR (eds.) *Comparative Cognition: Experimental Explorations of Animal Intelligence,* pp. 529–552. New York, NY: Oxford University Press.

Visalberghi E and Limongelli L (1994) Lack of comprehension of cause-effect relations in tool-using capuchin monkeys (*Cebus apella*). *Journal of Comparative Psychology* 108: 15–22.

Propagule Behavior and Parasite Transmission

P. H. L. Lamberton, A. J. Norton, and J. P. Webster, Imperial College Faculty of Medicine, London, UK

Introduction

Propagules – a general term relating to all forms, such as eggs and larvae, which serve in dissemination and transmission between habitats and hosts – are one of the major parameters of success of living organisms. If in a population certain individuals have a greater capacity to disperse than others, then their or their offspring's chances of survival, and subsequent fitness, may be increased, either by an increase in the probability of finding a suitable habitat, by the conquering of new areas, and/or by the reduction of intraspecific competition. However, propagules of living organisms only serve their function if, either actively or passively, they end up in a favorable habitat/host.

Parasites that are not directly transmitted or vector-borne require some method of movement between their hosts, and it is the propagules of these indirectly transmitted parasites that enable the life cycles to be maintained. The majority of parasitic species spend at least one portion of their life cycles as 'free-living propagule' stages. The mobility of these propagules varies along a scale with some parasite propagules appearing not to require mobility or associated behaviors for successful transmission. For example, some parasite propagules, such as the highly resistant and ubiquitous *Ascaris* (roundworm) eggs, are released with the feces of an infected host, and subsequently simply rely on a mobile new host to make soil-transmitted contact with the eggs and their subsequent larvae for onward transmission. However, in general, in parasite species where the reproductive stages are relatively immobile, due to the lack of host mobility or habit (where they return to the same parasite-shedding location), the more likely it is that some elaborate means of dispersion for its propagules has been selected. For example, fixed aquatic animals ensure the dispersion of their gametes or planktonic larvae by developing various appendages, such as bristles, blades, or umbrellas, which allow them to be carried away by the wind or to drift with the currents. Associating oneself durably to a mobile organism amounts to selecting an equally efficient mechanism of dispersion. All parasites in the digestive tract of a mammal, for example, have their eggs or larvae dispersed more or less constantly in this manner as the living host 'vehicle,' as Claude Combes puts it, travels. Similarly, all bird parasites fly with their hosts and all fish parasites swim with theirs. In contrast, in insects parasitic of other insects and crustaceans from the monstrillid group, only the larvae are parasitic, since dispersion is ensured by mobile adults.

Propagule dispersion is ensured by a range of different processes. Furthermore, parasites that infect multiple hosts are impressive in the complexity of the cyclical changes of milieu that they must undergo to accomplish dispersion through the exterior environment, infection of successive and often dramatically contrasting host species, and transfers between different sites within hosts, for the life-cycle to be continued and their genes transmitted. In each of such parasites, the same genome must be capable of building a series of six or seven 'successive organisms,' each different from the others and each able to discover and exploit different habitats.

The Complexity of Parasite Behavior

Larval propagules are often nonfeeding and obtain their energy through stored glycogen. As their energy stores deplete with time and activity, both swimming capabilities and infectivity decrease. This provides strong constraints and selective pressures for the parasite propagules to locate and penetrate a suitable host rapidly postemergence. The rate of glycogen depletion, survival and infectivity of these propagules can be influenced by environmental, genotypic, and phenotypic factors.

Parasites appear to have thereby evolved a broad range of complex species-specific and stage-specific phenotypic behavioral traits to aid the transmission of their propagules, often including trade-offs between quantity and quality, with some species investing in sophisticated behaviors, while others produce multitudes of behaviorally less complex propagules. The degree of separation between these two extremes depends in part on the behavioral and ecological characteristics of the subsequent host, including host's environmental niche, behavior, density, and distribution, all of which influence the chance of a propagule encountering a host and therefore whether it is more advantageous to conserve energy and wait for host stimuli or to keep searching.

Transmission-enhancing behavioral traits can either be host independent (where they increase the chance of the parasite entering and remaining in the host's natural niche at a time when the host is likely to be within this environment) or host dependent and involve attraction to a host. Such behavioral traits can also involve either taxis or kinesis depending on whether they result in a direct movement toward or away from a stimulant (taxis), or involve alterations in behaviors such as an increase or a decrease in turning rate, thereby increasing the chance of a propagule remaining within a given area/environment (kinesis). These behaviors can also be innate and/or involve sensory perception. They are however rarely learned behaviors, as parasite propagules tend to penetrate the first host they come into contact with, although this can also depend on the maturity of the propagule. The role of parasite propagule behaviors in achieving and enhancing transmission will be discussed within this review in relation to several parasitic species, although focusing in particular on various species of trematode as model systems.

A Brief History of Research into Propagule Behavior

Knowledge of the range of parasite propagule behavior can provide insights into the ecology of such parasites and could potentially have strong implications for the success of parasite control programs and our general understanding of the diseases they cause. Although the study of parasite behavior has lagged far behind other fields of behavioral study, some of the earliest reports date back to 1883, with studies on *Fasciola hepatica* miracidia (a larval propagule) and their attraction to lymnaeid snails by A.P. Thomas. Similar host-finding behavioral studies seldom reappeared until the mid-1930s, often expanding from entomology investigations on parasitic insects. Most early studies were performed on trematode miracidia and cercariae, due to their medical importance, and such investigations still dominate parasite behavior research today. Behavioral strategies for location of a suitable host, food, environment, or mate have, however, now been demonstrated in many different genera with a wide range of mechanisms and inter- and intraspecific variation.

This review will focus primarily, but not exclusively, on trematodes, in particular schistosomes. Schistosomes, of the genus *Schistosoma*, are the causative agent of schistosomiasis a parasitic disease second only to malaria in terms of its socioeconomic and public health importance. They are platyhelminth macroparasites with an indirect life-cycle involving two propagule stages; miracidia and cercariae.

Host location by free-swimming parasite propagules is thought to consist of five essential phases: hatching/release, dispersal, microhabitat selection, orientation to the host, and penetration. We will later consider each of these phases for trematode miracidia and cercariae in turn. These two parasite propagules have been chosen as they contrast greatly in the type of host they infect, and therefore the type of behaviors that have evolved to maximize transmission. Trematoda miracidia are mainly infective to molluscan intermediate hosts which are slow

moving and aquatic or semiaquatic, while cercariae in contrast mainly infect the definitive mammalian hosts which are nonaquatic and warm blooded with sporadic presence in the environment that the propagules are released into. The specific behavioral, biological, and environmental characteristics of these hosts are strongly reflected in their infecting propagules' behavior, and therefore these two more specific examples of trematoda miracidia and cercariae provide excellent opportunities to demonstrate the wide variety of characteristics that parasite propagules of wider genera are capable of.

Miracidia

Hatching of Miracidial Propagule Stages

The first step in transmission of a miracidial propagule from the definitive host is hatching, and egg production can be rhythmic to increase the chances of coming into contact with a suitable environment and the next host. For example, *Schistosoma haematobium* egg excretion peaks in the urine of children around midday, when they are more likely to be near a suitable hatching site. The hatching process of the parasite itself further requires certain environmental conditions, paralleling those preferred by the molluscan host, and is prevented from occurring within the definitive host through high osmotic pressure. The shift from high to low osmotic pressure, light, and temperature are all strong triggers for the hatching of propagule stages, for example schistosomes, *Fasciola hepatica*, and *Echinostoma trivolvis* eggs each reflecting their subsequent host habitat preference.

Miracidial Dispersal and Host Attraction

Due to their finite glycogen reserves, miracidial activity and infectivity are ultimately both age- and temperature-dependent. Posthatching, miracidia either demonstrate dispersal behavior or attraction toward host-space and time locations. Dispersal behaviors tend to be host-independent innate behaviors that act to spread the propagules out, reducing the chance of high intensity infections and their associated density-dependent effects such as competition and increased host mortality. In some species, miracidia appear oblivious to their snail hosts for up to three hours posthatching, which initially led researchers, such as Chernin and Dunavan in 1962 and La Rue in 1951, to believe that miracidia were not capable of host-specific recognition.

This initial dispersal phase however can also be controlled by environmental stimuli matched to the species-specific substrate preferences of their intermediate hosts, indirectly increasing the likelihood, through sharing the same environment, of propagule contact with their appropriate host. *E. trivolvis* miracidia for example show strong negative geotaxic and positive phototaxic responses which act together to increase the numbers of miracidia in the upper half of the water. *Helisoma trivolvis*, the snail host for these *E. trivolvis* miracidia, tend to feed at the surface of the water, and therefore both these responses would act to increase the chance of contact with the subsequent host. The miracidia of *F. hepatica* are also strongly photopositive and their snail hosts are usually found near the surface, at the edges of the ponds, whereas the miracidia of the eye flukes, *Philophthalmus lucknowensis* and *P. gralli*, are geopositive and photonegative in accordance to their bottom-dwelling snail hosts, *Melanoides tuberculata* and *Tarebria granifer*, respectively. Likewise, *S. mansoni* and *S. haematobium* miracidia show geonegative and geopositive responses and photopositive and photonegative responses, respectively, directing them toward their contrasting *Biomphalaria glabrata* and *Bulinus globosus* snail host environments. In these instances, combinations of simple taxis behavior can greatly influence the chances of contact with the necessary subsequent host. Even though in these cases the individual host species are all snails, their specific environments vary and the parasites behavior has evolved to exploit that.

As water temperature and water velocity can also affect snail distributions, it is therefore not surprising that miracidia have evolved adaptations in their geo- and photoresponses with turbulence and temperature changes demonstrating the highly complex interactions between various stimuli and propagule behavior continually working in a balance to increase the chance of contact with the next host and continuation of the parasite life-cycle.

Miracidial Attraction to Host in Response to Host Stimuli

Propagules can further increase their chances of transmission by being directly attracted to host cues, with species-specific mechanisms of host-finding and -recognition, often evolved to minimize contact with nonsuitable hosts. *E. trivolvis* demonstrate klinokinesis as a positive reaction to snail-conditioned water, turning in tighter circles, causing them to remain within the vicinity of the chemicals associated with their *H. trivolvis* snail host. Both *S. mansoni* and *S. haematobium* demonstrate similar behaviors after the initial dispersal stage, with an increase in rate of change of direction in increasing gradients of snail-conditioned water and a 'turn-back response in decreasing gradients.' *S. japonicum* demonstrate even more sophisticated traits, swimming directly along a chemical gradient toward their intermediate snail host *Oncomelania hupensis*. This may have evolved as a result of *S. japonicum's* snail host being semiaquatic and therefore potentially more difficult to locate than *S. mansoni* and *S. haematobium's* fully aquatic snail hosts.

Intraspecific variation in sophistication of behavioral traits also occurs with *S. mansoni* parasites from Egypt capable of differentiating between sympatric snail species, while a Brazilian strain did not. These differences may be due to a parasite bottleneck occurring when transported to South America during the slave trade, with those parasites which were more generic in their attempts to penetrate nonsympatric snails more likely to succeed in finding a new host than those which retained their sympatric specificity.

Miracidial Attachment and Penetration

S. japonicum miracidia are attracted to host and nonhost species in their swimming behavior; however, once they encounter their intermediate host *O. hupensis*, or agar-containing *O. hupensis* chemicals, they perform host-specific responses of 'contact with return,' 'repeated investigation,' and 'attachment' which they do not exhibit when encountering the dead-end hosts *B. glabrata*. These results not only demonstrate the ability of host penetration to be species-specific, but they also indicate that the miracidia respond to different signals when they approach a host compared to when they attach and penetrate. As you can see, parasite behavior is much more complex than one might assume at first glance; however it is also much more varied. *S. mansoni* and *S. haematobium* miracidia are also capable of 'repeated investigation' followed by attachment and penetration attempts, but the chemical host cues are very similar to the macromolecule stimuli for their host seeking. These responses to complex macromolecules may enable parasites to locate their hosts in aquatic environments, but without wasting energy on behaviors triggered by smaller molecules which could be more diffused and less localized around a snail host. These simple variations in what stage of the host location and penetration process is species-specific illustrate the extensive variation between even very closely related parasite species, with some demonstrating more complex host space location, or host attraction and others demonstrating more complex behavioral traits and specificity at the point of attachment and penetration.

Much research into propagule behavior, including that discussed here, focuses on kinesis and taxis in response to a chemical, physical, and/or biological stimulus. Some however exhibit a wider repertoire of swimming behaviors, with those of *S. mansoni* recently elucidated for the first time by the authors of this review. These more specific behaviors have been observed to vary with parasite genotype and selective pressures, such as *in vivo* chemotherapy. It has been speculated that drug resistance may be associated with biological costs or life-history trade-offs and that such costs could occur within these vital dispersal, host-seeking and/or penetrating stages. *S. mansoni* miracidia with a reduced sensitivity to with praziquantel, the current drug of choice, have been observed to carry out significantly less straight-line swimming than susceptible miracidia in a snail-free environment, with these differences enhanced under praziquantel pressure. It was suggested that this relative reduction in straight swimming could confer a lower propensity for dispersal which could act to reduce the rate of spread of resistance within natural foci, particularly where snail numbers were low due to wide-scale mollusciciding, such as in Egypt.

The concept of different behavioral repertoires with intraspecific variation becomes mind boggling when one considers the range between contrasting larval stages, such as miracidial and cercarial trematode propagules, exposed to contrasting selective pressures and therefore differing in certain key predictable behavioral aspects.

Cercariae

Cercarial Emergence

Although host finding by cercariae is strategically similar to that of miracidia, with a dispersal phase, movement into a suitable host habitat consisting of energy efficient searching, orientation toward the host, attachment and penetration, the stimuli and specific behavior responses themselves are often very different reflecting their different hosts.

Cercariae demonstrate peaks in emergence from snails which maximize their numbers within the 'host-time' with peak release occurring just before and during the time when their definitive hosts are most likely to be within the water. These peaks can be induced through various physiochemical factors, such as light, temperature, mechanical disturbance of the snail, humidity, and pH. Cercarial production peaks vary between schistosome species, with schistosomes of ungulates such as *S. bovis, S. curassoni*, and *S. leiperi* being released from the snails during the early morning, human schistosomes *S. mansoni* and *S. haematobium* cercariae emerging around midday, and the rodent schistosome *S. rodhaini*'s emergence at dusk, each coinciding with water contact periods of their respective definitive host species.

Even more interesting is that *S. mansoni* emergence also depends on which host species it is mainly cycling through, as rodents can also act as a definitive host in some locations. In Guadeloupe, peaks have been observed at night in rodent-dense 'sylvatic' foci, and during the morning in more 'urbanized' foci where human are responsible for maintaining transmission, and even potentially a midpoint tripeak in 'swampy' foci where both humans and rodents maintain transmission. Recent studies on *S. japonicum* have also demonstrated that although it is capable of infecting up to 40 different definitive host species, peaks in cercarial emergence support population genetics findings of different hosts being the keystone species for transmission in different areas. For example,

in the marshlands in mainland China, where cattle have been shown to be the most important species for maintaining parasite life-cycles, cercarial peaks have been observed to occur in the morning, while in the hilly regions, where rodents are more important in the transmission, cercarial emergence peaks in the evening, thereby providing high numbers of viable cercariae at night time when the rodents may be coming into contact with the water.

The transmission of cercarial propagules, however, becomes more complicated when two parasite species share the same host. The chronobiology of both laboratory-bred hybrids of *S. mansoni* and *S. rodhaini* as well as intraspecific hybrids of two strains of *S. mansoni* with different shedding peaks from Guadeloupe has shown chronobiology to be genetically determined with peak emergence at both times relating to each parental species.

The presence of the next host need not be the only selective pressure for peaks in cercarial emergence, as the two parasites *Protorometra edneyi* and *Protorometra macrostoma* emerge during the day or night, respectively, even though they both infect the same host, *Goniobasis semicarinata*, a day feeder. It is argued that predation by diurnal feeding nonhosts of *P. macrostoma* might select against diurnal emergence.

In addition to these periodicities in cercarial emergence, investigations by Helen McCarthy in the early 2000s on microphillid trematodes have demonstrated trade-offs between the complexity and numbers of cercariae released depending on the mobility and the predictability of the next host. *Microphallus similis* produced few, but large cercariae without periodicity, each strong swimmers with greater host-seeking behavior capable of locating a mobile less predictable crab host, while *Maritrema arenaria* periodically produced many smaller cercariae, almost twice as much in volume as *M. similis*, which remained suspended and survived longer, increasing the chance of contact with their sessile barnacle hosts.

Cercarial Dispersal

As with miracidia, emergence from the snail into the environment is thought to be followed by innate swimming behaviors which initially facilitate dispersal into their host's habitat. Many species, such as *S. haematobium*, *E. caproni, Diplostomum spathaceum*, and *Cryptoctyle lingua*, perform highly active swimming until they reach a suitable habitat in which they then slow down, and as with miracidia they tend to be nonresponsive to a suitable host during this early phase. While such nonresponsiveness in miracidia may act to reduce density-dependent effects within snails, a stronger selective pressure at the cercarial stage could be to reduce the chance of inbreeding and loss of genetic diversity, should a high number of asexually produced cercariae all penetrate the same host causing extreme levels of genetic aggregation.

During these dispersal and host-space orientation phases, cercariae appear most susceptible to light and gravitational stimuli. Claude Combes and colleagues published an extensive study in 1994 on a Mediterranean lagoon, which demonstrated these dispersal responses and how they are tapered toward the subsequent host stage. They argued that to maximize cercarial success, there would be a trade-off between quantity and quality of cercariae, and investigated how the 'quality' of cercariae can alter depending on sophisticated behavioral mechanisms which favor transmission. They reinforced the ideas of host-independent behaviors and host-dependent responses.

The range in behavioral traits of trematode cercariae is extremely large. While some species, often with only limited development of tails, crawl along the substrate to disperse, the majority perform independent alternations of short swimming periods with long floating or sinking periods, termed the 'active' and 'passive' phase, respectively. Even the mechanisms of this swimming can vary, with some controlled by a pacemaker in the tail (e.g., *P. macrostoma*), while others are controlled by the brain (e.g., *C. lingua* and *Transversptrema patialense*). Cercariae are assumed to be capable of perceiving and responding to stimuli from the environment, because they become orientated toward their host space. Phototactic behavior has been reported for many cercariae with pigmented photoreceptors, such as *Posthodiplostomum cuticola* and *Trichobilharzia ocellata*. Photokinetic swimming behaviors have been observed in cercariae with unpigmented photoreceptors, such as *S. mansoni*, where swimming frequencies increase with irradiance. Cercariae are also strongly affected by gravity, with interactions between geo- and phototaxis maintaining the cercariae at optimum levels for host contact. For example, *Bunodera mediovitellata* are released from their bottom-dwelling intermediate host, a clam, and although they actively swim and demonstrate photopositive responses, their geopositive response is dominant and they do not rise more than 3–5 mm off the bottom, keeping them within their bottom-dwelling host-space, but moving them toward the edge of water bodies in the more illuminated areas. Bartoli and Combes published research in 1986 on various cercariae taxis responses, such as *Lepocreadium pegorchis* which have strong geopositive responses but are weak swimmers, remaining low where they can be inhaled by their bivalve host. *Cardiocephalus longicollis* demonstrate strong responses to light and gravity and concentrate themselves approximately halfway between the surface and the bottom, where they are inhaled though the gills of fish for transmission. *Renicola lari* are strongly photopositive, which takes them to the surface, where their subsequent host preferentially feeds. *S. mansoni* cercariae also demonstrate strong photopositive and geonegative responses taking them toward the surface of the water where they perform periodic resting behavior to reduce unnecessary energy use and predation while being located near definitive mammalian hosts contact points.

Helen McCarthy and colleagues have also shown that propagule behaviors can even vary between day and night. For example, *M. similis* cercariae, which infect shore crabs, are active and remain well distributed throughout turbulent water in the dark when the crabs are most active and feeding. But in light conditions, cercariae swim downward and move horizontally into the dark, increasing the chance of encountering the crabs hiding under rocks during the day. In addition, cercariae have been shown to be capable of recognizing an already-infected host, preventing the detrimental effects of superparasitism and thus enhancing the success of transmission. For example, Nolf and Cort showed in the 1930s that cercariae of *Cotylurus flabelliformis* swim away from rather than penetrating already-infected snails.

As with miracidia, cercariae not only exhibit kinesis and taxis, but they also have a wide repertoire of swimming behaviors which to date are little understood. These different behaviors were first fully described for *S. mansoni* cercariae by Alice Norton. External pressures (intraspecific competition and host-defense responses) appear to reduce activity and therefore influence the range of behaviors displayed, presumably through their effect on the cercarial glycogen stores. These behavioral changes were also linked to a reduced infectivity of these parasite lines to the laboratory mouse host. Interspecific competition between *S. mansoni* and *S. rodhaini* within the *B. glabrata* snail hosts also influenced their miracidial propagule's behavioral repertoire, reducing activity.

Cercarial Attraction to Host in Response to Host Stimuli

Cercariae tend to have more stimuli specific responses than miracidia which makes sense when you consider that the definitive host is a larger, warm-blooded organism which is not permanently aquatic, providing mechanical stimuli such as water turbulence, shadows, and temperature that cercariae can respond to. Haas and colleagues observed that *T. ocellata*, which tend to be photopositive and geonegative while dispersing, demonstrated downward bursts of swimming away from light in response to shadows, propelling them down increasing the chance of contact with the feet of their duck host. Even bottom-dwelling cercariae with poorly developed tails and limited swimming capabilities respond to shadows, with Combes and colleagues observing *Cainocreadium labacis* cercariae standing up on their reduced tails in any water currents, ready to attach to a passing benthic fish.

Although many different responses to stimuli have been observed, they all tend to demonstrate repeatable and predictable patterns, so much so that Wilfred Haas has defined simple mathematical models that can accurately mimic activity levels of *D. spathaceum*. Some of behavioral traits described earlier are so innate, in fact, that the cercaria's tail of *P. macrostoma*, continues to perform even when the head has been removed.

Cercarial Attachment and Penetration Stimuli

Most cercariae infect mobile mammalian definitive hosts; chemoattraction is probably relatively less important than environment physical cues, as the hosts do not stay in one place long enough for such mechanisms to be efficient. Some cercariae which infect a further molluscan host stage do however orientate themselves toward their hosts, such as *E. trivolvis*, *E. revolutum*, *E. echinatum*, and *Gorgodera amplicava*.

Summary Discussion

In this review, we have focused on aquatic free-living larval stages, primarily of trematode parasites; however, it must be emphasized that many other parasite propagules demonstrate a range of behavioral traits to enhance transmission, which, for constraints of space, are beyond the scope of this current review. Nevertheless, across the diverse range of multiple host life-cycles which parasites employ, selection on life-history characteristics is complex and ultimately driven by transmission success. As the energy sources in many propagule stages are limited, allocation of resources must be partitioned between host seeking and longevity, with resources remaining for establishment and development even after successful location and penetration of the host. Trade-offs in such allocations will evolve depending on the particular hosts they need to infect. Maximum resource use in host-seeking behavior will hasten the parasite propagule's demise, but such trade-offs not only occur within swimming behaviors such as active and passive phases, but also for the cercarial propagules within the asexual stage of reproduction in the snail. If the subsequent hosts are abundant and widespread, then a maximum rate of propagule production may be optimal. If however the next hosts are variably and unpredictably available, then more energy will be required for host location, in these instances fewer propagules but with greater energy reserves, temporal variation, and extended periods of release at a less virulent production level, enabling the host to stay alive longer and parasite production to continue all increasing the chances of completing a life-cycle.

This article has highlighted the scope of some of the work which has been performed to date in this field as well as revealing the large amount which remains to be investigated. Future research in this field needs to take advantage of more advanced technology such as fine scale automated video-capture techniques in order to elucidate many important issues related to these, often medically and/or veterinarily, important parasites. Indeed parasite propagule stages, being so often easily available and abundant, are very useful in investigating several fundamental parasite traits such as drug resistance and therefore treatment outcomes, and we look forward to more behavioral research on these topics in the future. Today, we know many of the modes of transmission used by the various groups of parasites; however, we still have major gaps at the level of the species simply because the experimental transmission of biological cycles is often difficult to perform under controlled laboratory settings. Most research to date has really only skimmed the surface of what is actually happening, with many exciting avenues opening up as interest increases in this relatively understudied arm of behavioral ecology.

See also: Avoidance of Parasites; Beyond Fever: Comparative Perspectives on Sickness Behavior; Conservation, Behavior, Parasites and Invasive Species; Ectoparasite Behavior; Evolution of Parasite-Induced Behavioral Alterations; Intermediate Host Behavior; Parasite-Induced Behavioral Change: Mechanisms; Parasite-Modified Vector Behavior; Parasites and Sexual Selection; Propagule Behavior and Parasite Transmission; Reproductive Behavior and Parasites: Invertebrates; Reproductive Behavior and Parasites: Vertebrates; Self-Medication: Passive Prevention and Active Treatment; Social Behavior and Parasites.

Further Reading

Combes C (2001) *Parasitism. The Ecology and Evolution of Intimate Interactions.* Translated by Isaure de Buron and V. A. Connors. Chicago/London: The University of Chicago Press.

Combes C, Fournier A, Moné H, and Théron A (1994) Behaviors in trematode cercariae that enhance parasite transmission: Patterns and processes. *Parasitology* 109: S3–S13.

Haas W (1992) Physiological analysis of cercarial behavior. *Journal of Parasitology* 78: 243–255.

Lu D-B, Wang T-P, Rudge JW, Donnelly CA, Fang G-R, and Webster JP (2009) Evolution in a multi-host parasite?: Chronobiological circadian rhythm and population genetics of *Schistosoma japonicum* cercariae indicates contrasting definitive host reservoirs by habitat. *International Journal of Parasitology.* E-pub ahead of print.

McCarthy HO, Fitzpatrick S, and Irwin SWB (2002) Life history and life cycles: production and behavior of trematode cercariae in relation to host exploitation and next-host characteristics. *Journal of Parasitology* 88(5): 910–918.

Moore J (2002) *Parasites and the Behavior of Animals. Oxford Series in Ecology and Evolution*, p. 338. Oxford: Oxford University Press.

Salt G (1935) Experimental studies in insect parasitism. III. Host selection. *Proceedings of the Royal Society of London. Series B Biological Sciences* 117: 423–435.

Sukhdeo MVK and Sukhdeo SC (2004) Trematode behaviors and the perceptual worlds of parasites. *Canadian Journal of Zoology* 82: 292–315.

Psychology of Animals

D. A. Dewsbury, University of Florida, Gainesville, FL, USA

Animal Psychology

Psychology is usually defined as the study of mind and behavior; often such definitions specify *human* behavior. From its historical beginnings as a separate discipline in the late nineteenth century, however, some psychologists have been interested in the study of the minds and behavior of non-human animals. This has often, but not always, been to shed light on humans. Students of animal psychology have grappled with the problem of studying animals in a discipline devoted primarily to humans throughout the field's history. This has led to a variety of approaches that have often been labeled 'comparative psychology.' Because the term *comparative psychology* has been defined in many different ways, I prefer to treat this field as *animal psychology.*

Six Approaches to Animal Psychology

I will differentiate six varieties of animal psychology. They are not exhaustive categories; surely, they overlap. Further, many individuals used more than one approach, especially during the early part of the history. However, I think they are useful in differentiating the many faces of animal psychology.

Zoological psychology lies at the boundary between psychology and zoology. The term has a long history, having been used, for example, by Linus Kline in 1904. The approach is animal-centered in that the focus is primarily on studying the life of the animal rather than on asking arbitrary questions in a so-called animal model. The emphasis is often on the natural behavioral repertoire of the animal rather than on training the animal to engage in some arbitrary task.

Although virtually all research with animals depends on observations of behavior, *behavioristic psychology* is generally process-oriented. The goal is to use the animal in an effort to understand a process of interest. Such processes include the mechanisms of learning, the prediction and control of learned behavior, and motivational processes.

The first two approaches deal only with behavior, while *physiological psychology* is concerned with mechanisms internal to the animal that affect and are affected by behavior. Included are studies of the nervous system, endocrine function, and other internal processes.

Developmental psychology is focused on the changes in behavior as the animal matures. This has been an interest

in psychology almost throughout its history. Some animal behaviorists have shown little interest in development, while others emphasized the approach. Alternative views on the roles of genes and environment in behavior have been the cause of much controversy in the field.

Cognitive psychology is technically the study of the mind's function, including perception, attention, memory, imagery, and decision making. However, the primary focus has been upon the so-called higher processes. These are mechanisms of such phenomena as thinking and problem solving that appear necessary to explain aspects of behavior patterns that appear more complex than those that reflect basic principles of learning.

In *mentalistic psychology*, the scientist attempts to understand the mental life of the animal. One tries to infer in the animal mind the kind of conscious experiences that are exemplified in human experience. One can know the conscious processes of other humans only by extrapolating to them from our own experiences. In this approach, the extrapolation is taken one step further. Although the line between cognitive and mentalistic approaches may seem to be a fine one, it is important to make this distinction. One can investigate the complex mechanisms that sometimes appear to affect thought and behavior without dealing with the difficult problems of mentalistic inferences.

It should be noted that all the six of these approaches have been present throughout the history of comparative psychology to some degree but their emphasis has ebbed and flowed throughout that history. Various combinations of the approaches have often been labeled 'comparative psychology.' As is appropriate for this volume, I will not try to provide a complete treatment of the physiological and behavioristic approaches. Rather, I emphasize those approaches that had the most direct effects on the broad field of animal behavior.

Predecessors of Animal Psychology

The beginning of scientific psychology as a discipline is generally dated in 1879, with the founding of Wilhelm Wundt's laboratory at the University of Leipzig. However, interest in psychological phenomena dates back well before that time. Interest in animal psychology can be seen in prehistoric cave paintings, in the work of philosophers such as Aristotle, Michel de Montaigne, and René Descartes, in the writings of early naturalists, and in those interested in applied animal behavior such as farmers and falconers.

Impetus for a serious study in comparative psychology came from Charles Darwin. His writings are suffused with behavioral observations. Such books as the *Origin of Species*, *The Expression of the Emotions in Man and Animals*, and *The Descent of Man and Selection in Relation to Sex* and his 1877 "*A Biographical Sketch of an Infant*" were especially important for psychology. Other British scientists who were important forerunners included George John Romanes, C. Lloyd Morgan, Sir John Lubbock, Douglas A. Spalding, and L. T. Hobhouse. Their work is described elsewhere in this volume. However, it is important to note that, although Romanes was a keen observer of, and very capable experimenter on, behavior, in some of his writings, he was prone to anthropomorphic interpretations based on anecdotal data. He was criticized by Morgan. This difference set the occasion for some of the controversies that would characterize animal psychology in the twentieth century. Overall, these authors represented antecedents of the zoological, behavioristic, cognitive, and mentalistic themes to varying degrees.

1879 to World War I

In his 1863 *Lectures on the Human and Animal Psychology*, and later editions thereof, Wundt followed Darwin in arguing that the elements and general laws that hold for animals are the same as those of humans. Wundt was especially interested in the mental processes that characterize conscious experience. He had already seen two different approaches that can be mapped into the definitions noted earlier: one in which animals were studied for their own sake and the other in which they were studied to enlighten human behavior.

At least from a North American perspective, activity in animal psychology soon moved to North America. Some notable research continued elsewhere, however. Perhaps most notable is the study of Clever Hans, a horse that was supposedly capable of remarkable feats of arithmetic. This claim was eventually debunked in Oskar Pfungst's *Clever Hans* (*The Horse of Mr. Van Osten*), which appeared in German in 1907 and in English in 1911. It was found that Hans appeared to base his performance on inadvertent cues from humans rather than feats of arithmetic. Other horses were also studied with similar results. We still use the term 'Clever Hans effect' to refer to the susceptibility of subjects to inadvertent cues.

Some of the most famous studies in the history of the field were conducted by German psychologist Wolfgang Köhler on the island of Tenerife during World War I. Köhler studied problem solving in chimpanzees that either had to move, climb, or sometimes stack boxes in order to reach a banana suspended out of reach above them. He also used various stick problems, such as when the chimpanzees had to join together two short sticks to form a single stick long enough to reach a food item outside of their cages. Köhler believed that the animals demonstrated insight in solving the problems, thus favoring a cognitive, over a behavioristic, interpretation.

Animal psychology flourished in the United States during this period. William James, arguably the founding American psychologist, wrote his 1890 *Principles of Psychology*, a work, more than any other, that led to the spread of psychology in North America. In the book, James, who was quite conversant with the animal work in England and elsewhere, adopted an evolutionary perspective. Between 1890 and World War I, journals were founded, laboratories were established, the American Psychological Association was founded (1892), and departments began to emerge. Animal psychology spread as the discipline of psychology was secured.

Several universities were especially important. The key person at James's Harvard was Robert M. Yerkes, who completed his Ph.D. there in 1902. Yerkes began his career, studying a great variety of species, including earthworms, frogs, and crabs. His book *The Dancing Mouse* (1907), an analysis of mutant strains of house mice, was one of the earliest behavior-genetic studies. This work was necessarily of a behavioristic orientation, though Yerkes also retained a strong cognitive interest that would become more dominant in his later work.

G. Stanley Hall, James's first Ph.D. student in psychology, assumed the presidency of nearby Clark University in 1898. Important early animal research was conducted in Hall's psychology department, especially by two graduate students. Like Yerkes, Linus Kline studies utilized a wide range of species, including vorticella, wasps, chicks, and rats. He was interested in both naturally occurring behavior and learning, but with a zoological psychology orientation. Kline believed that behavior could only be effectively studied in the context of the natural conditions under which the species lived. His colleague, Willard S. Small, conducted the first study of maze learning in rats. The task was designed to simulate the runways in which Small had observed rats to live under his father's cabin in Virginia. Small also studied the development of behavior in rats.

The program at Columbia University was led by Edward L. Thorndike, who was a student of James at Harvard before going to Columbia for his Ph.D. Thorndike is best known for his studies of cats in puzzle boxes. Cats were enclosed in wooden crates of various sorts that required operation of some kind of device for them to except and get food. The cats learned the tasks but Thorndike was interested in how they did so. He believed that cats solved the puzzles not with insight, but with a simpler process of trial and error. Thorndike thus followed in Morgan's tradition and adopted a behaviorist, rather than a cognitive, interpretation. Thorndike's conclusions were challenged by Canadian T. Wesley Mills, who argued that the task Thorndike used determined the conclusions in that there was no room for insight in the puzzle boxes.

John Broadus Watson completed his Ph.D. in 1903 at the University of Chicago and remained on the faculty until 1908 when he moved the Johns Hopkins. The dissertation *Animal Education* typifies the developmental and physiological threads, as Watson tried to correlate the development of rats' nervous systems with that of their learning abilities. It was Watson who deserves credit for developing the behavioristic approach into a full-blown behaviorism, as with his 1914 book, *Behavior: An Introduction to Comparative Psychology*. Watson believed that inferences about cognitive and conscious processes had no role in a science of psychology. He would gradually develop an environmentalist approach in which instinctive processes played little role. Less well known were his prewar studies of noddy and sooty terns on Bird Key in South Florida. He studied both homing and naturally occurring behavioral patterns in these birds. These field studies of the natural behavior of two closely related species of birds demonstrate the range of interest of comparative psychologists and predate the similar approaches of the later European ethologists.

The first Ph.D. recipient from the laboratory of Edward B. Titchener at Cornell University was Margaret Floy Washburn, probably the first woman with a Ph.D. in psychology. Washburn is best know in animal psychology for her 1908 textbook, *The Animal Mind*, which, in its four editions, would be the standard textbook in the field for the next quarter century. Although Washburn was well aware of the difficulty of inferring conscious processes from the behavior of animals, she followed in Titchener's footsteps and adopted a mentalistic perspective.

Karl S. Lashley was originally trained as zoologist, but he spent most of his career working in psychology. He completed the Ph.D. at Johns Hopkins, where he was strongly influenced by Watson, in 1914. He was a strong advocate of a physiological approach to the study of behavior and was well known for work using ablations and other techniques on the cerebral cortices of rats and monkeys. He also had a strong approach in zoological psychology making observations on, and writing about, the natural behavior of animals. He worked one summer with Watson at Bird Key and later was the mentor for some important animal psychologists.

This is not a complete list of the centers or the important animal psychologists and events of this period. However, it should be clear that the field was spreading and that most of the varying approaches were already developed as in the zoological approach of Kline, the behavioristic approaches of Thorndike and Watson, the cognitive approaches of Köhler and Yerkes, the physiological approach of Lashley, and the mentalistic approach of Washburn.

Many of these psychologists had students working with them and then ready to go out to spread the field; animal psychology seemed ready to explode. The problem was that there were few jobs in the field. Psychology was expanding largely because of its applicability to human behavior – especially in the context of education. In order to find employment, many of these psychologists switched interest to other parts of psychology – especially educational psychology. Then many were called to serve in the war. After the war, there were few animal psychologists ready to redevelop the discipline.

Between World Wars I and II

Although zoological psychology developed in Germany under the influence of Otto Koehler and others, a more psychological orientation, advocated by such scientists as Mathilde Hertz, never gained institutional status. Comparative perspectives were rebuilt gradually in North America.

The early journals in the field were the *Journal of Comparative Neurology and Psychology* (1904–1910), *Journal of Animal Behavior* (1911–1917), and *Psychobiology* (1917–1920). Their successor, *Journal of Comparative Psychology*, was founded as the prime journal in the field in 1921. During its first few years, there were many studies of human behavior included, probably because of a dearth of comparative studies.

A few animal psychologists entered the field during the 1920s but it was during the 1930s that the core of psychologists who would develop the field for the next 40 years or more entered the field. One of the few major products of the 1920s was Calvin P. Stone, who earned his Ph.D. with Lashley at the University of Minnesota in 1921. Stone spent most of his career at Stanford University, where he adopted an eclectic approach. Stone contributed to the study of learning and development, as with his work on the ontogeny of sexual behavior in rats. He also conducted physiological studies and displayed a concern for the natural behavior of animals characteristic of the zoological approach. He was a long-time editor of the *Journal of Comparative Psychology*.

The 'nature–nurture controversy,' the debate over the relative roles of genetics and environment in the development of behavior, has often been a main issue separating behavioral scientists; it flared up during the 1920s. Zing-Yang Kuo, a Chinese psychologist with a Ph.D. completed at the University of California, Berkeley in 1923, wrote articles suggesting that the concept of instinct had no role in psychology. He emphasized the importance of prenatal development. Later, he softened his position to suggest that nature and nurture were so intertwined that they could not be separated. Others argued for the value of the concept of the innate. Leonard Carmichael, a 1924 Harvard Ph.D., raised frog and salamander embryos in an anesthetic that eliminated movement and found that when returned to a normal environment the coordination and vigor of these animals was the same as in animals reared under control conditions, suggesting some innate organization. Nevertheless, the tide was turning in favor of interpretations based on nurture over nature.

The most visible approach during the interwar interval was that of behaviorism. Although Watson would be forced to leave academic life soon after the war, he was still able to popularize his behavioristic-environmentalistic approach broadly. The epitome of this approach was reached with B. F. Skinner's version of radical behaviorism, a 1931 Harvard Ph.D. Skinner focused on the modification of behavior in species such as rats, pigeons, and humans. He developed the principles of reinforcement originated by the British scientists and refined by Thorndike into his own approach. Skinner believed that the goals of his field were the prediction and control of behavior and he believed that physiological analyses contributed very little to those goals. Skinner was a very effective communicator who became well known outside of the field and recruited many students and followers. The approaches of Watson and Skinner became so well known that behaviorism was sometimes regarded as coincident with the field of animal psychology itself. That was not the case.

A balanced presentation of animal psychology during this period would note the overwhelming role of learning theory in 'neo-behaviorism.' Dominated by such psychologists as Clark L. Hull, Edward C. Tolman, and Edwin R. Guthrie, the field of learning theory was as prominent in psychology as virtually other approaches of the time. During much of this period, these theorists, not Watson or Skinner, prevailed.

In 1930, Robert Yerkes founded the facility that became the Yerkes Laboratories of Primate Biology in Orange Park, Florida. Yerkes brought to the facility the largest collection of captive chimpanzees for research in the world. He developed methods for their care, maintenance, and breeding and both he and his staff conducted many research studies, including those of learning, development, and social behavior. Yerkes adopted a cognitive approach to behavior, believing that the chimpanzees used higher mental processes in solving some of the complex tasks with which they were faced. Indeed, Yerkes sometimes even adopted a mentalistic perspective. In 1942, he was succeeded as director of the laboratories by Lashley.

Many of the leaders of the rebuilt field completed graduate study during the 1930s. C. Ray Carpenter completed his graduate work in Stone's Stanford department in 1932. Upon becoming associated with the Yerkes Laboratories and thereafter, Carpenter undertook some

of the most important early field studies of non-human primates around the world. He provided complete catalogs of the social, reproductive, and individual behavioral patterns of primates. Carpenter thus was a psychologist conducting field studies in a zoological psychology context, thus challenging the stereotype of animal psychologists as students only of learning in rats. He is recognized as the grandfather of modern primate field studies.

Harry F. Harlow was another Stanford Ph.D. (1930). Although he conducted important studies of learning in primates including some of its neural bases, he is best known for his developmental studies of rhesus monkeys. He uncovered important roles for social contact between the infant and its parents and siblings in the development of normal naturally occurring and learned behavior.

Lashley served as pre- or postdoctoral mentor for a cluster for budding animal psychologists. Norman R. F. Maier believed that there is a clear difference between learning and reasoning in rats. He designed experiments to show the different roles of these hypothesized processes. In some of these, rats would be given experience with several parts of an apparatus, such as a room with several connected tables, then shown a food incentive, and then challenged to locate the food. The rats behaved in ways suggesting their capacity to reason. As with Yerkes' approach, this work illustrates how cognitive perspectives were not completely dead during the era of learning theory and behaviorism.

Theodore C. Schneirla, like Maier a Michigan Ph.D., who went to work with Lashley, also became prominent in the field. He worked at the American Museum of Natural History in New York and at New York University. He was known most for his research in the natural behavior and learning of ants and for his theoretical approaches. Although his writing style was turgid, he was charismatic enough to attract students and colleagues to a group and approach I have called the *New York Epigeneticists*. Epigenetic approaches emphasize the dynamic nature of the interaction of genes and environment during development. Today, most psychologists have adopted such a position. The Schneirla group came to emphasize environmental factors over genetic influences during epigenesis. This embroiled them in various controversies, as will be seen in a following section. There was a reliance on the 'levels' theory according to which there are different levels of development in different species and it is difficult to generalize across levels. Maier and Schneirla collaborated on their 1935 *Principles of Animal Psychology*, one of several important textbooks published in the 1930s.

Donald O. Hebb worked with Lashley at Chicago, Harvard, and the Yerkes Laboratories. He was remarkable in his ability to integrate all the approaches. In his dissertation, he showed that some aspects of visual function in rats appear to be innately organized, yet he later cautioned against the nature–nurture dichotomy. He conducted physiological studies as well as research on the behavior and emotion in chimpanzees and the courtship of dolphins in the zoological perspective while in Orange Park. He focused on problems of the mind and thus represented the cognitive approach. His best-known contribution was his 1949 *The Organization of Behavior*, a book that combined neural and behavioral perspectives in relation to intelligence, emotion, learning, memory, and perception.

Another student of Lashley was Frank A. Beach, who has been called the 'conscience of comparative psychology.' Beach studied a variety of behavioral patterns but mainly reproductive behavior. His classic 1950 article, 'The Snark Was a Boojum,' was an indictment of the apparent narrowing of both species and range of problems studied in comparative psychology. It was influential despite some flaws in the analysis. Beach lamented the emphasis of some psychologists on the study of learning at the cost of understanding the natural behavioral patterns of the animal. Beach conducted numerous developmental studies and, like some of his colleagues, questioned the utility of the nature–nurture dichotomy. He was a founder of the field of behavioral endocrinology and conducted many studies of the neural bases of behavior. He thus represented the best of zoological and physiological perspectives.

Animal psychology between the wars is sometimes portrayed as dominated by behaviorism and the study of learning in rats. However, it should be apparent that many other threads were present in the fabric of animal psychology, even if they were less prominent. All the diverse threads I have delineated, and others, were intertwined during this pivotal period. This group of psychologists from the 1930s would provide the core for postwar animal psychology.

World War II to the Present

After the war, the six approaches delineated here became more differentiated as the psychology became more specialized and fragmented. With the shift in primary research funding from private foundations to the Federal government, there was a substantial growth of activity in all fields of psychology and this fostered specialization.

As physiological research grew and then mushroomed, the physiological approach came into significant prominence and diverged away from other orientations. The Society for Neuroscience was founded in 1969. In 1985, membership reached 10 000; by 2002, it reached 30 000. Although not all neuroscientists were interested in animal behavior, the growth of this approach pulled activity away from the other approaches. In 1947, given the growth of physiological research, the *Journal of Comparative Psychology* was renamed the *Journal of Comparative and Physiological Psychology*; Stone continued as editor. By 1983,

physiological work had grown so that it threatened to overwhelm the reset of the material. It was decided to split the journal, reestablishing the *Journal of Comparative Psychology* and creating a new journal, *Behavioral Neuroscience*. Some cross-approach work remained as in the work in behavioral endocrinology of Beach and others and with the growth of a field of neuroethology. In recent years, this interaction appears to be again increasing.

Since World War II, behavioristic approaches have continued. The theories of Hull, Tolman, and Guthrie gradually receded, though some of those trained in that approach became prominent in various fields of psychology. The radical behaviorism of Skinner, by contrast, was in the ascent as the combination of basic animal research and the application of Skinnerian principles to human behavior grew. Although it is sometimes written that behaviorism is dead, in fact radical behaviorism lives on and flourishes. The field of behavior analysis, its current title, with its own journals and meetings, though it is somewhat isolated from much of the rest of psychology.

Developmental research increased along with the other fields. The journal *Developmental Psychobiology* was founded in 1968. Controversy over the nature–nurture issue flared up for a period and has been resolved with general adoption of an epigenetic view that there is a continuous, dynamic interaction between genes and environment in the ontogeny of behavior. With recent advances in molecular biology and genotype mapping, the role of genes in influencing, not determining, behavior has become more widely accepted.

The New York epigeneticists clustered around Schneirla and his developmental approach. New recruits to his approach, such as Ethel Tobach, Daniel Lehrman, Jay Rosenblatt, and Gary Greenberg, carried the banner after Schneirla. The core principles of the method of levels and resistance to conclusions about genetic influences, often labeled 'genetic determinism,' are still defended within this group. There is some indication that humane concerns, fears that genetic determinism could lead to racism, sexism, and other discrimination, played a role in this approach.

Before and after World War II, European ethology, led by Konrad Lorenz and Niko Tinbergen, reached maturity. The approach presented an apparent contrast with many of the approaches in animal psychology. This new 'objective study of animal behavior' was presented as based in Europe rather than North America, practiced by zoologists rather than psychologists, conducted in the field rather than the laboratory, studying birds, fish, and insects rather than mammals such as rats, studying instinct rather than learning, and relying on observation rather than controlled experimentation. As should be apparent by now, there is a grain of truth in this characterization of ethology and animal psychology but there are many exceptions on both sides. Ethology was presented as a

David fighting the Goliath of comparative psychology. The ethological approach proved highly appealing, culminating in the awarding of the Nobel Prize in Physiology of Medicine to Lorenz and Tinbergen in 1973. Some zoologically oriented animal psychologists went to study and work in the laboratories of European ethologists.

The increased prominence of ethology led to one of the most hotly contested battles in the history of animal psychology. The key event was the publication of Daniel Lehrman's critique of Lorenz's instinct theory in 1953. The nature–nurture issue was central. Some psychologists, especially the New York epigeneticists, viewed ethologists as ignoring the role of the environment in development, while ethologists viewed psychologists as being ignorant of natural behavior and obsessed with antiinstinct rhetoric. Eventually, the disputes were settled. Led by Lehrman, Beach, and others, the psychologists became convinced of the value of the ethological approach and the ethologists came to see the benefits of controlled research of the psychologists. Today some differences in emphasis remain but it is sometimes difficult to tell the difference between an ethologist and a zoologically oriented animal psychologist. Both fields were changed by their interactions.

The zoological orientation came into prominence during the 1960s and 1970s. Following Beach's snark article, there were many reevaluations of comparative psychology with respect to its adherence to basic biological principles. Numerous psychologists again conducted their research with an eye toward natural behavior and the natural environment.

The next big event was the appearance of sociobiology, most dramatically represented in Edward O. Wilson's 1975 book with that title. The core principles of 1970s sociobiology, an evolutionary approach, were the theories that natural selection works at the level of the individual or gene, not the population or species and that the representation of one's genes in future generations could be achieved by facilitating the reproductive success of close relatives. This approach too proved controversial with accusations of genetic determinism. Wilson's final chapter applied the principles to human behavior and proved especially contentious. The approach was largely stripped of the controversial applications to humans and repackaged as behavioral ecology. This proved to be a highly successful endeavor as many diverse data appeared explicable in its context and new areas of research, such as that on kin selection, developed. Behavioral ecology, rather than ethology, came to dominate the study of behavior among zoologists. Zoological psychologists received yet another shot in the arm from this approach and many applied these principles to the systems they were studying. With the impetus from core ethology and behavioral ecology and with new sophistication concerning principles of evolution, the zoological approach thrived. In 1975,

Wilson predicted a shrinkage of ethology, comparative psychology, and physiological psychology with a growth of sociobiology and integrative neurophysiology. Animal psychology lives on, but there are some signs that, after initial invigoration, the zoological approach is decreasing as other orientations have come to the fore, as will soon be discussed.

In the mean time, another transformation of sociobiology occurred. In the application of these principles to human behavior, sociobiologists sought to understand the adaptive significance of current human behavior. Soon the emphasis shifted to the view that human behavior is the product of mechanisms that evolved early in human history, possibly in the Pleistocene epoch, and may not be adaptive in the present environment. This new orientation was labeled 'evolutionary psychology.'

Yet another influence on animal psychology was the so-called cognitive revolution. It developed in psychology and other disciplines from the 1950s, to the 1970s as an alternative to behaviorism and a return to psychology as the study of the mind. As we have seen, cognitive approaches never completely disappeared in animal psychology. Among the core principles in this incarnation were a belief in core mental mechanisms that appeared universal and with strong genetic base and a grounding of mental events in physical bases. Noam Chomsky's theory of linguistics, the impact of Jean Piaget's developmental theories, the research of Jerome Bruner and George Miller, the increase in computer analogies, and other events interacted to produce a greatly revitalized cognitive approach in human psychology. As would be expected, the approach spread to animal research as well. More and more results were given a cognitive interpretation. New interpretations and topics, such as representation, expectancy, concept learning, attention, timing, and memory, came to the fore. Some of the research that had been conducted on animal learning, both within and outside of the radical behavioristic environment, was reinterpreted and repackaged so that it emerged as 'animal cognition.' Of course, this work built on earlier approaches, but was highly successful in gaining representation in animal psychology. A pivotal event may have been the appointment of a new editor for the *Journal of Comparative Psychology* in 1989. With representatives of other approaches, including the zoological and developmental, Gordon G. Gallup, Jr., a strong advocate of a cognitive-mentalistic approach, was chosen as the new editor. The representation of cognitive studies increased at this time and has continued even as editors who emphasize different approaches have succeeded Gallup. It is difficult to determine whether Gallup's selection was a cause or effect of the increased representation of cognitive approaches in the journal.

A return of a mentalistic approach also occurred during this time. With his 1976 *The Question of Animal Awareness* and subsequent works, zoologist Donald R. Griffin reintroduced issues of consciousness and related processes to animal psychology; the approach became known as cognitive ethology. The claim was that behaviorism had inhibited study of the animal mind; animals have intentions, beliefs, and self-awareness. It may be difficult to see these processes but we should start by reinterpreting behavior in this context in order to attain a more complete understanding of animal mind and life. These proposals set off yet another controversy. Of course, we can never really know the consciousness of another human or non-human. Although cognitive ethology lives on, most students of animal cognition and related fields seem to reject its approach, believing that there are no documented cases of animal behavior that necessitate inferences of conscious experience. Nevertheless, Griffin and his associates succeeded in moving animal psychology in a more cognitive direction and discussions of animal minds became more acceptable.

Conclusion

Animal psychology and its history are sometimes presented as if they are coincident with a behavioristic approach. By contrast, the history and current state of animal psychology as I view them are product of the complex interplay of different approaches within animal psychology and the effects of events from outside. Its growth and development have been via dynamic interaction via processes somewhat reminiscent of the dynamic interaction of epigenetic development within the individual organism.

One must be concerned about the future. Animal laboratories are expensive and some universities appear to be phasing them out to make room for less expensive research. Pressures from animal rights extremists have made compliance with increased regulation more expensive and made departments think twice about investing in animal research.

In the space allotted, I have been able to only sketch some of these approaches and events. I could mention only a few of the significant individuals influencing this field. With this brief overview, the reader is unlikely to remember the names and events I have summarized. I do hope, however, that the reader will understand the complexity of animal psychology and be wary of simplistic presentations. If I can achieve that, I will have accomplished something.

See also: Agonistic Signals; Animal Behavior: Antiquity to the Sixteenth Century; Animal Behavior: The Seventeenth to the Twentieth Centuries; Apes: Social Learning; Behavioral Ecology and Sociobiology; Categories and Concepts: Language-Related Competences in Non-Linguistic Species; Chimpanzees; Communication: An Overview; Comparative Animal Behavior – 1920–1973; Costs of Learning; Darwin and Animal Behavior; Development, Evolution and Behavior; Dominance Relationships, Dominance Hierarchies and Rankings; Endocrinology and Behavior: Methods; Ethology in Europe; Experiment, Observation, and Modeling in the Lab and Field; Experimental Design: Basic Concepts; Future of Animal Behavior: Predicting Trends; Hormones and Behavior: Basic Concepts; Imitation: Cognitive Implications; Integration of Proximate and Ultimate Causes; Konrad Lorenz; Levels of Selection; Mate Choice and Learning; Mate Choice in Males and Females; Neurobiology, Endocrinology and Behavior; Niko Tinbergen; Norway Rats; Parental Behavior and Hormones in Mammals; Pigeons; Play; Reproductive Success; Sentience; Social Cognition and Theory of Mind; Social Learning: Theory; Spatial Memory; Welfare of Animals: Introduction.

Further Reading

Beach FA (1950) The snark was a boojum. *American Psychologist* 5: 115–124.

Boakes R (1984) *From Darwin to Behaviourism*. Cambridge: Cambridge University Press.

Darwin C (1871) *The Descent of Man and Selection in Relation to Sex*. London: John Murray.

Dewsbury DA (1984) *Comparative Psychology in the Twentieth Century*. Stroudsburg, PA: Hutchinson Ross.

Dewsbury DA (1992) Triumph and tribulation in the history of American comparative psychology. *Journal of Comparative Psychology* 106: 3–19.

Dewsbury DA (2003) Comparative psychology. In: Weiner IB (ed.) *Handbook of Psychology*, vol. 1, pp. 67–84. New York, NY: Wiley.

Dewsbury DA (2006) *Monkey Farm: A History of the Yerkes Laboratories of Primate Biology, Orange Park, Florida, 1939–1965*. Lewisburg, PA: Bucknell University Press.

Griffin DR (1976) *The Question of Animal Awareness*. New York, NY: Rockefeller University Press.

James W (1890) *Principles of Psychology*, vol. 2. New York, NY: Holt.

Lehrman DS (1953) A critique of Konrad Lorenz's theory of instinctive behavior. *Quarterly Review of Biology* 28: 337–363.

Morgan CL (1894) *An Introduction to Comparative Psychology*. London: Walter Scott.

O'Donnell JM (1985) *The Origins of Behaviorism: Animal Psychology, 1870–1920*. New York, NY: New York University Press.

Pfungst O (1908/1965) *Clever Hans: The Horse of Mr. Von Osten*. New York, NY: Holt, New York (Reprint: New York: Holt Rinehart, and Winston (Ed. R. Rosenthal).

Richards RJ (1987) *Darwin and the Emergence of Evolutionary Theories of Mind and Behavior*. Chicago, IL: University of Chicago Press.

Roitblat HL (1987) *Introduction to Comparative Cognition*. New York, NY: Freeman.

Watson JB (1914) *Behavior: An Introduction to Comparative Psychology*. New York, NY: Holt.

Relevant Websites

http://www.apa.org/divisions/div6/ – American Psychological
    Association (APA) Division of Behavioral Neuroscience and
    Comparative Psychology.
http://www.historyofpsych.org/ – APA's Society for the History of
    Psychology.

http://www.animalbehavior.org/ – Animal Behavior Society.
http://comparativecognition.org/ – Comparative Cognition Society.
http://www3.uakron.edu/ahap/ – Archives of the History of American
    Psychology.
http://psychclassics.yorku.ca/ – Full-Text Important Articles in the
    History of Psychology.
http://darwin-online.org.uk/ – Works of Chares Darwin.

Punishment

K. Jensen, Queen Mary University of London, London, UK
M. Tomasello, Max Planck Institute for Evolutionary Anthropology, Leipzig, Germany

What Is Punishment?

In everyday parlance, punishment occurs when one individual performs an act with the goal of discouraging another individual from engaging in a particular behavior. This may involve doing something aversive to the offender or withholding something of value. For example, owners might punish their dogs for chewing on the furniture by swatting them on the rump, and parents might punish their toddlers for drawing on the walls by withholding dessert. This commonsense view of punishment involves intent – the punisher has the goal of modifying the behavior of another individual. Importantly, it involves a change in behavior in the future; it is not simply reactive or malicious.

This future orientation is consistent with the biological view of punishment in which an individual reacts to harmful behavior by reducing the fitness of the instigator and thereby decreases the likelihood of future harm. Inflicting harm on another at a personal fitness cost with no direct fitness benefit (biological spite) is not likely evolve, except under very restrictive conditions leading to indirect fitness benefits (kin selection). However, by being reciprocal – inflicting a harm for a harm done – the punisher gains net fitness benefits in the future. Biologists are agnostic about the cognitive mechanisms of the punisher; intentions are not required.

There are two schools of thought on punishment in psychology. For learning theorists, punishment is any stimulus that causes a decrease in an organism's behavior. In this vein, a rose thorn is a punisher because it decreases the likelihood that the pricked individual will pick roses in the future. Roses, of course, do not have goals and are cognitively uninteresting, nor does the flower picker need any understanding of intentions. Social psychologists, on the other hand, restrict punishment to cases where impartial, outside observers mete out corrections on the basis of normative principles. This requires sophisticated cognitive abilities, as well as social norms and rational, rather than emotional, motives. The social psychological view not only restricts punishment to humans, but also it does not encompass all – nor arguably much of – human punishment in the everyday sense of the word. As for other animals, it is likely that their punishment lies along a continuum between simple reflexive responses to harmful events and intentional, impartial, norm enforcement.

Punishment Shapes the Social Environment

Punishment is a powerful means by which an individual can shape its social environment. At the simplest level, punishment achieves selfish benefits. Punishment is distinguished from mere aggression and avoidance by the delay in benefits. Aggression and avoidance, though potentially costly, provide immediate benefits. For instance, a large male that sexually harasses a female leaves behind offspring, and subordinate animals begging from dominant individuals will receive scraps of food. From an evolutionary perspective, the delay in benefits involved in punishment results in both the punisher and the target suffering costs, making the behavior – at least at the time that it is performed – a form of biological spite. For example, male hamadryas baboons (*Papio hamadryas*) will threaten or attack females that stray from the harem, and dominant male chimpanzees (*Pan troglodytes*) will attack rivals as well as supporters of rivals. Ultimately, for punishment to evolve, it must eventually benefit the actor, either directly or indirectly (i.e., through kin). As with positive reciprocity (often called reciprocal altruism), negative reciprocity is costly at the time that it is performed but yields benefits in the future. Because of delayed benefits, punishment is 'return benefits spite.' The functions of punishment in animal societies include achieving and maintaining social dominance, deterring cheaters and parasites, as well as disciplining and coercing offspring and sexual partners. In all these cases, the punisher receives delayed, direct fitness benefits from its actions.

Punishment and Cooperation

Retaliation against personally harmful behaviors (negative reciprocity) is a particularly interesting form of punishment. While the punisher still benefits from its actions, other individuals in the group might also benefit from changes in the target's behavior. For instance, a juvenile that is attacked when it tries to take food from one adult might learn to not take food from others. The role of punishment in maintaining cooperation is of great interest, because the benefits to the punisher are not so transparent and the consequences are far reaching. The reasons why animals cooperate – particularly when they make costly sacrifices for the benefit of others – is one of the longest-standing and most important questions in evolutionary biology. Individuals risk being exploited by group members: free-riders reap the benefits of cooperation without sharing the costs and can cause the collapse of cooperation. Punishment can be an important force for the maintenance of cooperation.

In animal societies, one context in which punishment would appear important for maintaining cooperation is in cooperative breeders. In cooperative breeders, breeding individuals will coerce their offspring into forfeiting reproduction to aid in the care of the dependent offspring. In some cases such as social insects, the nonreproductive workers will punish egg-laying cheats (biologists refer to this as policing since the punishers do not benefit directly). Cooperative breeding is a fascinating, but limited form of cooperation; punishers either directly benefit or provide benefits to kin (especially true in the social insects), and helpers remain because the costs of being exploited are less than the risks of leaving. Surprisingly, despite temptations to free-ride, examples of punishment of helpers that fail to work are rare. Punishment in cooperative breeders functions as a coercive strategy.

Outside of cooperative breeding, examples of punishment to maintain cooperation in non-humans are few, and these are equivocal. In one field experiment designed to test for tit-for-tat reciprocity, African lions (*Panthera leo*) did not punish laggards that failed to reciprocate the potentially hazardous behavior of approaching a perceived threat. In chimpanzees, there was one observation in captivity of a male attacking a group member who failed to provide support in a conflict, the inference being that the attacker punished the recipient for uncooperative behavior. Alternatively, in a systematic observational study of reciprocity and aggression in chimpanzees, failure to reciprocate grooming or support did not lead to punishment. In the only experiment designed specifically to probe punishment of noncooperative behavior in chimpanzees, captive individuals had their food stolen from them by a conspecific. They then punished the theft by collapsing the table on which the food sat, thereby preventing the thief from eating. Chimpanzees were punitive in that they were negatively reciprocal (vengeful). It is not clear that they were punishing a noncooperative behavior (theft) for the purpose of maintaining a cooperative relationship, however. In fact, theft increased over time while punishment declined, suggesting that punishment failed to deter noncooperative behavior.

Altruistic Punishment

A special form of punishment has been described recently in the experimental economics literature. Altruistic punishment is the punishment of free-riders without a return in benefits for the punisher. Because the punisher does not directly benefit from an increase in cooperation, altruistic

punishment has been claimed to pose a potential challenge to natural selection at the individual level. Theoretical models and experimental economic studies conducted on humans have shown that in the absence of punishment, the level of cooperation in groups declines with repeated interactions. In fact, the standard prediction in economic theory is that in one-shot encounters, pairs or groups of players should not cooperate at all: free-riders fare better in a population of cooperators by receiving the benefits of cooperation without the costs. As a result, cooperation is driven to extinction. However, if individuals are allowed to punish others – even if this imposes an additional cost on the punishers – cooperation can be maintained as a stable strategy. Punishment is more effective in maintaining cooperation than is direct (positive) reciprocity because the individual costs of punishment decline as the number of free-riders declines, whereas the cost of cooperation rises as the number of cooperators increases. As well, the threat of punishment can be a sufficient deterrent to cheating. The nature of altruistic punishment remains contentious. The reason for the debate is that the experiments that elicit altruistic punishment are artificial, and there are some questions whether the participants truly play the games as if they are anonymous and one-shot. However, studies outside of economics, as well as examples of people intervening on behalf of others, lend credence to the notion of altruistic punishment.

While there have been many studies of altruistic punishment in humans, the sole experimental attempt to find something like altruistic punishment in animals is an adaptation of the widely used economic experiment, the ultimatum game. The ultimatum game involves a division of a resource between two players. The first player proposes a division which the second player can accept. But if the second player rejects it, both get nothing; the responder pays a cost to punish the proposer for his offer. This experiment has been run hundreds of times in many human cultures, and contrary to economic models of rational self-interest, people routinely pay this punishment cost. In a reduced form ultimatum game played by chimpanzees, the apes behaved like self-interested maximizers, conforming to the predictions of standard economic theory. Proposers did not choose equal divisions, and responders did not reject any nonzero offer; that is, chimpanzees did not pay a cost to prevent a conspecific from getting more of a resource. This finding is consistent with the punishment experiment described earlier in which they did not react especially to the outcome of another individual eating as long as it was not stolen.

Third-Party Punishment

Clear evidence for altruistic punishment would require impartial, third-party punishment, where the punisher has no direct stake in the matter. (Previous examples were personal, or second-party, punishment.) As stated earlier, third-party intervention is an essential condition for punishment according to social psychologists. Humans do engage in third-party punishment, for instance, by policing all kinds of social norms that are violated – from smoking in restaurants to harming children. Some primates will intervene in disputes and some social insects will destroy eggs not laid by the queen. Biologists use the term policing for this kind of behavior, though it is fundamentally different from the human case. The examples of policing in animal societies are typically executed by dominant individuals who have their own interests at stake, and there is no evidence for punishment by disinterested individuals following the disruption created by a conflict in a group. Experimental evidence for third-party punishment in animals is needed to help resolve the question of altruistic punishment in non-humans.

Norm Enforcement

A special case of third-party punishment is norm enforcement. In this case, the simple violation of social norms, such as a teenager wearing torn jeans and a T-shirt to a wedding, will evoke punishment from others (such as expulsion from the wedding or disparaging gossip). The 'altruism' refers to the fact that the punisher's behavior benefits not the offender, but others in the group (e.g., by maintaining certain behavioral traditions as markers of group identity). There may be indirect benefits to the punisher, including a reputational benefit for being an altruistic punisher (attracting more cooperators and deterring cheats). But both third-party punishment and norm enforcement may also benefit the punisher indirectly by benefiting the group, as described earlier. Punishment can maintain any behavior – such as conformity to styles of dress – and not just cooperation, though the selective pressures favoring cooperation are more obvious than for wedding attire.

There have been a few claims for norm enforcement, or punishment of rule violation, in animals other than humans. One example was suggested in rhesus macaques (*Macaca mulatta*): higher ranking individuals aggressed against lower ranking conspecifics that failed to give food calls when the latter found food. The interpretation was that noncallers were punished for deception. However, a plausible alternative was provided in a study on white-faced capuchins (*Cebus capucinus*): individuals called to signal ownership of food resources and those that failed to call were more likely to enter into conflict with higher ranking individuals, since ownership was not clear. An observation in captive chimpanzees saw two individuals that failed to come into the enclosure at night in a timely manner (and hence delayed the feeding of the group); they were attacked the next morning by the others when reunited. And an example from the field reported a violent attack by eight individuals on single male who had failed to conform to a social rule such as not exhibiting species-typical submissive behavior. It is not clear, however, that these acts of aggression were really cases of punishment to enforce specific behaviors or were cases of redirected aggression, dominance behavior, or something else. Experiments modeled on such observations will be needed before we can attribute rule enforcement to non-human animals.

Social Preferences and Punitive Motives

Given the role of punishment in societies, it is important to know why individuals punish others, namely their intentions. As an example for why intentions are important, consider the difference between first-degree murder (intentional killing) and involuntary manslaughter (unintentional killing). From the victim's perspective, the outcome is the same, but psychologically (and legally), these are very different acts. Of particular importance for punishment are social preferences, namely whether the punisher has the intention or motivation of causing harm in other individuals.

Self-Regarding Preferences

Self-regarding preferences are nonsocial – the goals involved are purely personal with no regard for the consequences for others. Any effects on other individuals are byproducts. With regard to punishment, any change in future behavior that subsequently benefits the punisher is fortuitous. The goal of the individual is only that the offender immediately refrain from harming it, or that it remove itself from the harmful situation. Effects on the target of punishment – and on others that may be affected (e.g., through a decrease in uncooperativeness) – do not motivate the punishers choices. In other words, the individual is indifferent to the consequences of its actions apart from immediate, personal outcomes. Much aggression and harm avoidance may be seen in this light and would not be seen as punishment in the commonsense use of the term: the aggressor's motivation is not that the offender learn to refrain from doing something in the long term, but only that it stop it now. However, in terms of delayed costs and benefits for both the punisher and the target, unintended punishment may still have the same effects as intended punishment.

The cognitive processes required for self-regarding behaviors such as aggression and avoidance are minimal. In social insects, for example, an aggressive response is triggered by a biochemical cue. It is a fixed, evolved response and is cognitively not much more interesting than a rose thorn. The cognitively relevant cases are those that are intentional and flexible. Retaliatory punishment is an aggressive response to an aversive stimulus.

Shunning need only be explained by avoidance. However, the behaviors in many species are not inflexible. For instance, a low-ranking baboon will not behave aggressively to a dominant individual, or even the kin of the dominant. It may have learned this lesson through past experience (i.e., losing violent conflicts). The motivations for punishment are self-regarding – the punishing animal need only have as its goal the reduction of annoyances in its social environment. Consideration for how this affects the target of the punishment is not required.

The simplest mechanism for dealing with noncooperators would be to stop cooperating after being exploited; however, indiscriminate shunning punishes cooperators as well and would cause cooperation to fall apart. Cognitive mechanisms that would allow targeted avoidance of noncooperators should include individual recognition and perhaps a memory for specific events (episodic-like memory). Emotions such as anger are likely important in mediating punishment. In the punishment experiment with chimpanzees described earlier, when chimpanzees exhibited aggressive displays and tantrums – indicators of anger – they were more likely to collapse the food table. While likely important for motivating immediate reactions to undesirable events, it would be valuable to find evidence for planning punishment after the eliciting event (e.g., cold-blooded revenge). At present, there appears to be no evidence for delayed retributive punishment in non-human animals.

Social (Other-Regarding) Preferences

Other-regarding preferences are social. These are behaviors that are motivated by a concern for the welfare of others. Even though the outcome of punishment may benefit the punisher, the motivation to punish need not be self-regarding; the goal might be the effect it has on the behavior or the psychological state of the target, with any personal benefits arising as unintended byproducts. In addition to having the goal of decreasing harmful behavior directed at the self, the punishing individual may have as goal a change in the noncooperative behavior of the target, or may have the goal of maintaining a cooperative social environment. However, selfish benefits make it hard to rule out self-regarding motives, and it is difficult to establish other-regarding punitive motives. Furthermore, any individual action can arise from multiple motivations. For example, someone might scold a youth for crossing the street against a red light so as to protect him from potential harm, as well as to uphold the norms of society, to set an example for children who are present, to impress upon one's peers that one is an upstanding member of society, to relieve the moral outrage he feels, to experience pleasure in causing embarrassment in the youth, and to be the first to cross the street when the light is green so as to get the best seat on the waiting tram. The important point

is not to elucidate the complex suite of motivations for every action, but to determine whether certain motives even exist.

Antisocial preferences

Punishment can be motivated by a concern for the negative welfare (suffering) of others. This is not the same thing as self-regarding punishment, which is neutral to the consequences of others; it has as its primary goal that the target suffer. Antisocial (negative other-regarding) preferences such as spite and schadenfreude (pleasure in the misfortunes of others) can motivate punishment. For example, humans (at least males) experience pleasure in seeing an individual who previously cheated them receive a painful stimulus, yet will exhibit empathy for a cooperator in pain.

An important aspect of human cooperation may be a sensitivity to fairness. Aversion to inequity can motivate people to correct unfairness. People are ultracompetitive in that they compare their gains and losses to the gains and losses of others. They are sensitive to – and reciprocate strongly against – personally unfair outcomes as well as unfair intentions. When faced with personal unfairness in experiments such as the ultimatum game, people report feelings of anger and they show appropriate facial expressions and physiological responses such as an increased heart rate. This 'wounded pride' can motivate punishment in the absence of anticipated rewards such as future reciprocity or reputation. Spite (in the typical, psychological sense) has the suffering of the target as the ultimate goal and is not a means to an end. Altruistic punishment may therefore be a byproduct of spiteful (antisocial) punishment; the punisher inflicts harm on another individual for the sake of causing harm, rather than out of altruistic motives for others.

There is considerable debate about whether other animals have a sensitivity to fairness, particularly disadvantageous inequity aversion. Several experiments have suggested that some non-human primates and even dogs (*Canis familiaris*) are sensitive to inequity. However, several other experiments dispute these findings. In none of these studies can the animals inflict harm on others in response to inequity; they can only react to the experimenter. Two studies in chimpanzees have attempted to address antisocial punishment directly. In the punishment study described earlier, chimpanzees did not react spitefully by 'punishing' unfair outcomes when an experimenter pulled their food away and gave it to a conspecific. Nor, in the mini-ultimatum game did responders react in any way to unfair outcomes. Taken together, these studies suggest that harming others for the sake of causing others to experience loss might be unique to humans. Counterintuitively, such antisocial preferences might lead to altruistic outcomes; altruistic punishment is not necessarily motivated by altruism, and instead may be driven by spite.

Prosocial preferences

Punishment need not be self-regarding nor negatively other-regarding. One might be motivated by the positive outcomes for others. Empathy is the standard for prosocial preferences. Many parents, for instance, will tell their children that they are punishing them 'for their own good' and that it hurts them more than it hurts the child. The punishment is intended to produce delayed benefits for the target through the imposition of immediate costs; any benefits or costs to the punisher are unintended. Parental discipline could qualify as being prosocially motivated. Examples of this are intentional teaching (discipline) and reform. For example, humans will punish their child if it engages in a potentially risky activity to teach it to not do this; they will reform it for having done something that is harmful to others to prevent that from happening in the future. The majority of examples of discipline in animals are inconclusive, however, since the parents punish behavior that is personally aversive; the punishment therefore provides immediate, direct benefits to the parents. One potential exception is a field observation in chimpanzees of a mother aggressing against her infant that was about to eat poisonous leaves, though single observations are always inconclusive.

It is not yet clear what motivates altruistic, third-party or normative punishment in humans. It is possible that people have altruistic motives, that they have as a goal the modifying of the behavior of defectors so that others will benefit. They can be motivated by advantageous inequity aversion and empathy. It is also possible that they are indifferent to the effects of the punishment on both themselves and the target, but are motivated to achieve a particular social effect, such as a sense of justice. Punishment is used as a means to an end, and a punitive sentiment is a motive to see a noncooperator harmed for a cooperative end (and not an end in itself as in the case of spite). An altruistic punitive sentiment is a motive to harm the target for the target's own good, or for the benefit of others. Altruistic punitive motives can also be for the benefit of individuals other than the target, such as preventing free-riders from exploiting others in the future. The important point is that the benefit of the target or of other individuals is the final goal of altruistic punishment, and that it is not a self-serving means to an end.

Other Cognitive Mechanisms

As alluded to earlier, emotions are likely important components for punishment. Anger, in particular, appears to be a feature of all forms of punishment in humans and is likely to be prevalent in other animals. The circumstances that motivate anger are likely to vary; moralistic outrage, for instance, will only apply to altruistic, normative, or third-party punishment, since the emotion is in response to rule violations and not personal offenses. Other cognitive features that might be important for punishment in humans – and potentially their closest living relatives – are perspective-taking and intention reading. Chimpanzees, but not capuchin monkeys, have been shown to take the visual perspective of conspecifics in a competitive situation. Such visual perspective-taking allows them to know what another individual sees, has seen, and presumably knows. Chimpanzees, as well as capuchin monkeys, also recognize intent in others and are less likely to remain engaged in an interaction with a human who is unwilling to give them food than one who attempts to give them food but is unable to do so. However, there is as yet no evidence for an understanding of beliefs and desires (theory of mind) in non-human animals. Secondary (social) emotions such as satisfaction at revenge. pride in seeing justice done, guilt in failing to punish, and forgiveness when the punishment has been effective may also be important for punishment, at least in humans.

Conclusion

Punishment is an important way in which organisms can control or manipulate their social environments. Most often, it provides direct benefits to the punisher in that the offender stops engaging in the harmful behavior – ideally both at the moment and in the long term. In some cases, the direct benefits are great enough that the punisher will suffer some costs to bring it about. In addition, by punishing free-riders, others can benefit, and this can stabilize cooperation. Altruistic punishment, third-party punishment, and norm enforcement are special forms of punishment that may allow for the development of cooperation of unrelated individuals on a large scale seen only in humans. The cognitive mechanisms of particular importance to punishment are other-regarding preferences. Cooperative outcomes can result from purely self-regarding motives. The evolution of complex, stratified societies are likely to require other-regarding preferences. Antisocial preferences such as spite and prosocial preferences such as empathy can facilitate cooperation by being especially sensitive to cheaters and other norm violators.

See also: Conflict Resolution; Emotion and Social Cognition in Primates; Empathetic Behavior; Morality and Evolution; Social Cognition and Theory of Mind.

Further Reading

Boyd R and Richerson P (1992) Punishment allows the evolution of cooperation (or anything else) in sizable groups. *Ethology and Sociobiology* 13: 171–195.

Clutton-Brock TH and Parker GA (1995) Punishment in animal societies. *Nature* 373: 209–216.

de Waal FBM and Luttrell LM (1988) Mechanisms of social reciprocity in three primate species: Symmetrical relationship characteristics or cognition? *Ethology and Sociobiology* 9: 101–118.

Fehr E and Fischbacher U (2003) The nature of human altruism. *Nature* 425: 785–791.

Fehr E and Gächter S (2002) Altruistic punishment in humans. *Nature* 415: 137–140.

Gächter S and Hermmann B (2009) Reciprocity, culture and human cooperation: Previous insights and a new cross-cultural experiment. *Philosophical Transactions of the Royal Society B* 364: 791–806.

Gardner A and West SA (2004) Cooperation and punishment, especially in humans. *American Naturalist* 164: 753–764.

Gros-Louis J (2004) The function of food-associated calls in white-faced capuchin monkeys, *Cebus capucinus*, from the perspective of the signaller. *Animal Behaviour* 67: 431–440.

Hauser MD (1992) Costs of deception: Cheaters are punished in rhesus monkeys (*Macaca mulatta*). *Proceedings of the National Academy of Sciences of the United States of America* 89: 12137–12139.

Jensen K, Call J, and Tomasello M (2007a) Chimpanzees are rational maximizers in an ultimatum game. *Science* 318: 107–109.

Jensen K, Call J, and Tomasello M (2007b) Chimpanzees are vengeful but not spiteful. *Proceedings of the National Academy of Sciences of the United States of America* 104: 13046–13050.

Seymour B, Singer T, and Dolan R (2007) The neurobiology of punishment. *Nature Reviews Neuroscience* 8: 300–311.

Sigmund K (2007) Punish or perish? Retaliation and collaboration among humans. *Trends in Ecology & Evolution* 22: 593–600.

Silk J (2008) Social preferences in primates. In: Glimcher PW, Camerer CF, Fehr E, and Poldrack RA (eds.) *Neuroeconomics: Decision Making and the Brain*, pp. 269–284. London, UK: Elsevier.

West SA, Griffin AS, and Gardner A (2007) Social semantics: Altruism, cooperation, mutualism, strong reciprocity and group selection. *Journal of Evolutionary Biology* 20: 415–432.

Glossary

11-ketotestosterone (11KT)   Potent androgenic steroid hormone in teleost fishes that is analogous to dihydrotestosterone in tetrapod vertebrates in terms of inducing the development of secondary sexual characters often associated with territoriality and courtship in large males.

Abiotic   Nonliving.

Absconding   In honeybees, an absconding colony leaves its nest and searches for a new nest site. Absconding in response to low food availability, parasites, or predation is more common in 'African' strains of *Apis mellifera* than in 'European' strains.

Absolute sensitivity   The lowest amount of light that can be perceived by an animal.

Acanthocephalan   A phylum of parasitic worms known as 'acanthocephalans,' 'thorny-headed worms,' or 'spiny-headed worms,' characterized by the presence of an evertable proboscis, armed with spines, which it uses to pierce and hold the gut wall of its host. Acanthocephalans typically have complex life cycles, involving a number of hosts, including invertebrates, fishes, amphibians, birds, and mammals. About 1150 species have been described.

Accessory gland   A gland associated with reproductive organs of either males or females and producing substances accompanying the sperms or eggs.

Accommodation   An optical adjustment made by the eye to focus an object at a given distance.

Acoustic startle response   The behavioral and/or physiological response of an individual to an unexpected acoustic stimulus such as the sound of a nearby predator.

Action component   Any behavior elicited on the part of an evaluator by the act of recognition.

Action-oriented representations   The idea that internal representations should describe the external world by depicting it in terms of the possible actions an animal can take.

Activational effects   A change in behavior and/or physiology that occurs in response to a hormonal signal and that disappears once the influence of the hormonal signal ends.

Active electrolocation   The ability of weakly electric fish to detect objects and orient in their environment based on their electric sense. For active electrolocation, fish generate a carrier signal (EOD), which is modulated in amplitude and phase by the environment, resulting in the projection of a modulated signal onto their electrosensory skin surface. By sampling, the thus projected electric image fish can gain information about the properties and the location of nearby objects.

Active space   The area in which a signal (or cue) can be detected from the source.

Actual conflict   Observed conflict over reproduction in a social group; actual conflict can be much lower than potential conflict.

Acute phase response   A rapid, systemic, innate immune response that includes heterothermy (fever or hypothermia), production of proinflammatory cytokines, synthesis of defensive and other immune regulatory proteins, and sickness behaviors.

Ad libitum sampling   Noting whenever something of interest occurs.

Adaptation   (1). At the level of evolution, a process, driven by natural selection, whereby species or populations become better suited to the environment. It occurs over generations and results in an increase in those genes that allow individuals in a population to better survive and reproduce in an environment. (2). At the individual level, the use of regulatory systems, with their behavioral and physiological components, in order to allow an individual to cope with its environmental conditions.

Adaptive demography   The composition of eusocial insect workers within a colony so that different sizes and/or ages enhance the efficiency of colony operations and fitness.

Adaptive flexibility   The ability of individuals to adjust behavior or physiology as ecological or social conditions change in ways that enhance their fitness.

Adaptive radiation   Evolutionary diversification of a lineage into multiple species or differentiated populations (radiation), in which natural selection in novel environments has played a prominent role (adaptive).

Adaptive response   Refers to flexible behavior that an individual uses to adjust to another type of behavior or situation. Adaptively flexible behavior allows an individual to enhance its reproductive success or survival.

Adaptive suicide   Individual mortality that enhances inclusive fitness by benefiting relatives.

Additive character optimization   A type of character coding that applies differential costs for transformations across character-states arranged in leaner order. For example, if character-states {0,1,2} are observed, and 1 is assessed to be of intermediate similarity, additive coding can be employed to apply this conclusion. Thus, transformations from 0 to 1 and from 1 to 2 would cost the same, but the cost of transforming 0 directly to 2 would be equal to the cost of transforming from 0 to 1 plus the cost of transforming from 1 to 2 (hence, additive).

Additive genetic variance   The genetic variance of a quantitative character associated with the average effect of substituting one allele for another. Additive genetic effects are the only strictly heritable genetic effects.

Adipose tissue   Tissues that serve as the principal storage sites for body fat.

Adrenal glands   Endocrine glands located on the kidneys, which play a role in water and electrolyte balance.

Adrenocorticotropic hormone (ACTH)   Small polypeptide hormone derived from a larger precursor (proopiomelanocortin) produced by the anterior pituitary gland that stimulates the adrenal cortex (inter-renal glands in nonmammalian species) to produce corticosteroids (primarily glucocorticoids).

Aestivation   Spending the summer in a dormant stage. It occurs in crustaceans, snails, amphibians, reptiles, and lungfishes.

Affect   Subjective feelings.

Affective states   Emotional state, that is, feelings.

Affiliative social relationship   Strong association between individuals, usually manifested by high rates of physical proximity to one another and nonaggressive social interactions.

Affordance learning   A form of observational learning in which the crucial information an observer acquires is about properties of objects manipulated by the model and the opportunities they 'afford,' the observer then exploits this information rather than imitating the model's actions.

'African' honeybee   Bees derived from *A. mellifera* ecotypes that evolved in Africa were introduced into Brazil in the 1950s. Fiercely defensive of their nest, these bees have caused public health problems, due to the dangers of massive stinging, and agricultural management problems as their range has increased to cover much of South America, all of Central America, Mexico, California, Arizona, Texas, and parts of the southern United States.

Age polyethism   A mechanism for division of labor in which individuals within a social group specialize in different tasks at different developmental stages or different ages.

Aggression   Overt, complex, social behavior with the intention of inflicting damage or status change upon another individual.

Aggression against females   A category of male aggression. Some aggression against females may be mistaken for forced copulation.

Agonism (adj., agonistic)   Aggressive behavior including responses to aggression such as flight and submission. Conflict resolution through a series of aggressive or submissive signals.

Alarm call   A chemical, auditory, or visual signal emitted in the presence of a predator that may serve one or more functions, including advertisement of perception to the predator, advertisement of signaler quality, and warning conspecifics.

Alarm pheromone   Pheromones released in response to threats. In honeybees, the alarm pheromones are associated with the sting.

Alarm reaction   A behavior induced by chemical stimuli that tend to bring the animal in a position where it is less exposed to predation.

Alarm substance   Substance(s) in the skin of fishes that induce alarm reactions.

Allele   One of several alternative forms (nucleotide sequences) of a gene.

Allelochemical   A chemical signal produced by an organism that influences the behavior or physiology of an organism of a different species.

Alliance   A close social bond between two or more adult individuals. Alliances often support each other during conflict and are more likely to share resources with each other than with other animals.

Allochthonous   Originating elsewhere; not native to a place.

Allometric   Describing the relationship between the size of an organism and the proportional size of its parts.

Allomone   A chemical produced by individuals of a species used in communication with other species; typically used in defense against predators, etc.

Alloparental care (Alloparenting, alloparents)   Care for infants and juveniles that mimics and substitutes for the parental behavior of a parent. Typically, the caregivers are kin and the social group is cohesive and related.

Allopatric (n. Allopatry)   Geographically separated; for example, populations on different islands with little or no movement between islands. Allopatric speciation is the development of isolating mechanisms while incipient species are separated by a geographic barrier.

Allostatic load   The cumulative wear and tear and energetic demand of daily and annual routines. Allostatic load can also include increased demands of poor habitat, injury and infection, human disturbance and life history stages, such as breeding, migration, etc.

Allostasis   An elaboration on the concept of homeostasis, where there is an emphasis on the fact that (1) what counts as an ideal physiological measure can change over time, and (2) numerous physiological systems may become activated in the body's attempt to solve a challenge to its equilibrium.

Allotype The allorecognition phenotype of an individual. An allotype is the composite of an individual's allorecognition genes, or the expressed gene products of allorecognition loci that confer cue specificity.

Alternative reproductive tactics (ARTs) Discontinuous variation in mating behavior within one sex of a species, often associated with morphological variation.

Alternative splicing Different exons of an RNA transcript from a single gene are spliced together to produce different mRNA transcripts and thus different proteins.

Altricial Relatively immobile young (usually birds or mammals) depend on parents for food and warmth.

Altruism A behavior that is costly to the performer's fitness, but beneficial to others (evolutionary biology); helping behavior resulting from selfless concern for the positive well-being of other individuals (social science).

Altruistic punishment The costly infliction of harm on another individual or group that produces net benefits for all the individuals in the group (social science).

Alzheimer's disease Neural disease accompanied by cognitive dementia and occurrence of plaques and tangles in the brain.

Ammocoete Premetamorphic larva of a lamprey.

Amoebic dysentery (or amoebiasis) An infection of the intestine caused by *Entamoeba histolytica*, a unicellular protozoan parasite, which causes severe diarrhea. Infection occurs by consuming food or water contaminated with amoeba cysts.

Amplified fragment length polymorphism (AFLP) Genetic variation found by cutting DNA strands with restriction enzymes and amplifying the resulting segments by using PCR (polymerase chain reaction).

Amplitude Sound intensity, as determined by the magnitude of vibration by a sound-producing object. This physical property of sound is the primary determinant of our psychological experience of the loudness of sound.

Amplitude modulation Changes in the amplitude of a sound over time. The process of modulation produces extra frequencies in the sound, called sidebands.

Amygdala A brain region that (among other functions) plays a critical role in fear, anxiety, and aggression.

Anadromy Migratory pattern of fish that hatch and develop in fresh water and then migrate to saltwater for adult development to return to fresh water and breed.

Analogy (Analogical reasoning) (In the field of *logic*) A form of reasoning in which one thing can be inferred (see *inference*) as similar to another thing in certain respects, on the basis of the known similarity between the two things in other respects.

Anautogeny The adult female ectoparasite requires that protein be ingested, often in the form of a blood meal, in order to mature her eggs.

Androgen (pl. androgens) A steroid hormone with 19 carbon atoms, so named because of their *andros* (male)-generating effects. Examples include testosterone, androstenedione, 5-α dihydrotestosterone, and 11-ketotestosterone. Although the testes are an abundant source of androgens, they can also be synthesized in other glands, including the adrenal gland and ovaries. Some androgens such as testosterone can be converted by the enzyme aromatase into estrogens. Responsible for the development and maintenance of male-typical characteristics, including development of secondary sex characteristics and behaviors.

Androgenic gland A gland near the distal portion of the sperm duct in crustaceans that secretes androgenic gland hormone.

Angiotensin II An octapeptide that plays a prominent role in the regulation of cardiovascular and body fluid homeostasis.

Angular acceptance function Photoreceptors have a limited 'field of view.' The angular extent of visual space over which a receptor receives light is described by its angular acceptance angle or function.

Anhedonia The inability to feel pleasure; a defining symptom of clinical depression.

Animal communication A behavior in which an animal produces a signal, which conveys information and influences the behavior or physiology of another animal.

Anisogamy Refers to the differences in size of the gametes of the two sexes: sperm are generally small and eggs are generally large.

Anorexia A change in eating behavior characterized by markedly reduced appetite or a total aversion to food. Anorexia is a component of sickness behavior but may refer to a behavioral change apart from febrile illnesses, such as with food allergies or psychological stress.

Anosmic animals Animals without the sense of smell.

Antagonistic pleiotropy A single gene controls for more than one trait with at least one trait being beneficial to the organism's fitness and at least one being detrimental to the organism's fitness. In analogy, a certain maternal hormone may have beneficial or detrimental effects on different offspring traits.

Anthropocentrism Regarding humans as the central element of the universe; interpreting reality exclusively in terms of human values and experience.

Anthropogenic Related to or caused by human activities (e.g., human-induced).

Anthropomorphism   Attribution of human motivation, characteristics, or behavior to animals.

Antiaphrodisiacs   Compounds transferred by the male to the female during mating that are either synthesized by the male or sequestered from the environment that render the female unattractive to rival males.

Apical gland   A gland associated with the reproductive tract of the sea hare that produces a hormone that influences egg-laying behavior.

Aposematic signal   Traits of the prey that predators can detect prior to attack and that inform the predator that the prey is defended or otherwise unattractive to attack.

Aposematism   The correlation between conspicuous signals, particularly warning coloration, and the presence of defenses in prey.

Apparent competition   When the fitness of one species is indirectly lowered by the presence of another species because of a shared parasite or predator.

Appeasement   Post-conflict interaction directed from a bystander to the aggressor to reduce the risk of being attacked.

Appetitive behavior   Behaviors that increase the probability that a particular need is satisfied. In the case of food deprivation, appetitive behavior would increase the organism's chance of locating food.

Appetitive cue   A stimulus associated with a resource (such as a food item, a host plant, or a prospective mate) that an individual would normally respond to, but which is ignored when the individual is actively migrating.

Appetitive movement   Movements that an individual makes while searching for, or in response to, appetitive cues (also called 'Trivial Movements').

Appetitive sexual behavior   A phase of reproductive behavior during which male searches, orients toward, and courts a female in preparation for copulation.

Apyrase   A calcium-dependent enzyme, found in mosquito saliva, that catalyzes the hydrolysis of ATP to ADP and inorganic phosphate.

Arginine-vasopressin   Peptide hormone involved in osmoregulation in both sexes, aggression and affiliative behavior of males.

Arginine vasotocin   Nine amino acid neuropeptide that is the homolog of arginine vasopressin found in mammals. These hormones are released at the posterior pituitary gland, but also widely in the brain where they act as neuromodulators.

Armpit effect   A system of kin recognition in which individuals learn their own phenotypic cues and use them as a template for determining the kinship status of other individuals.

Arms race   A metaphor for predator–prey coevolution wherein adaptation proceeds in an escalation/counterescalation dynamic that leads to ever exaggerated traits on both sides of the interaction.

Aromatase   The enzyme that catalyzes the transformation of androgens such as testosterone into estrogens such as estradiol.

Arthropod   Animals with an exoskeleton, a segmented body, and jointed appendages.

Artificial fruit   A device modeled on the natural problems animals deal with in opening difficult-to-process natural foods, such as fruits that needed cracking, peeling, and other forms of manipulation; typically designed to afford two or more successful opening techniques so that the fidelity of social learning about such alternatives can be objectively measured and compared.

Artificial neural network modeling   The mathematical modeling of biological nervous systems in order to simulate animal behavior and its evolution.

Asset protection   Organisms act to protect their expected future reproductive success (the asset here) from loss due to predation. Because predation would eliminate future reproduction, individuals that can expect great future success are predicted to take greater actions to protect it.

Association   (In psychology) The process of forming mental connections or bonds between sensations, ideas, or memories; two stimuli or events are associated when the experience of one leads to the effects of another, because of repeated pairing.

Associative class   A collection of objects or events signaling the same consequence or follow-up event; the members are grouped on the basis of a common association.

Assortative mating   A system in which individuals choose mates nonrandomly on the basis of a particular characteristic, selecting either mates more dissimilar to themselves than expected under random mate choice (negative assortative mating) or mates more similar to themselves than expected under random mate choice (positive assortative mating).

Asymmetric game   A subset of games in game theory, in which the differences between the contestants may affect their choice of strategies.

Attentional states   Perceptual states of the eyes that are either directed toward a viewer or directed away from them.

Attractivity   A female's ability to elicit sexual responses from a male.

Attractor   Mathematically, a set of values that a dynamical system maintains after a sufficient time, with 'sufficient' depending upon the system. An important property of an attractor is that the system returns to this set of values even after it has been slightly disturbed, that is, when it is moved a small distance away from the set of values.

Auditory template model   The model of vocal development which proposes that an animal is constrained only to copy the sounds that it hears which match a template with which it hatches or is born.

Autochthonous   Native to a place; indigenous.

Autogeny   A female ectoparasite that is able to mature a batch of eggs without an external protein meal.

Autonoesis   A special form of consciousness that allows us to be aware of being the author of the episodic memory and the episodic future imagined event.

Autotomous sting   When the sting easily tears from the body of the worker insect, sting autotomy is found in all honeybee species and some wasps.

Autotomy   The loss of a body part (generally a limb or tail) by an animal, generally as a means of escape when held by that body part.

Avoidance   The use of a habitat that has few associated natural enemies (i.e., enemy-free space).

Avpr1a gene   A gene coding the arginine-vasopressin receptor V1aR.

Awareness   Refers to the ability to perceive or feel something; awareness can refer to a wide range of sensitivity and experience, from dim perceptions to detailed conscious experience.

Bacillus   A rod-shaped bacteria cell.

Bag-cells   Neurosecretory cells of the abdominal ganglion of the sea hare that secrete egg-laying hormone.

Balanced polymorphism   The condition of having two or more alleles maintained in a population as a result of opposing evolutionary forces.

Banding (ringing)   Placing an inscribed metal or colored plastic band (ring) on the 'leg' of a bird so that the movement of the bird can be determined when recovered.

Basolateral membrane   Basal and lateral surfaces of enterocyte epithelial cells of intestinal mucosa.

Bateman gradients   Are regression lines that show the relationship between the number of mates and reproductive success for each sex in a group or population. Sometimes, they are also referred to as 'sexual selection gradients.' Generally, the steeper the slope of the regression line, the more intense will be sexual selection on that respective sex.

Batesian mimicry   Mimicry of body coloration and patterning of a toxic species by another coexisting toxic species. Sharing of the same aposematic signal between a defended species (called 'the model') and an undefended species (called 'the mimic'), such that individuals of the mimic species gain an antipredatory advantage from the shared signal.

Bathyal   Associated with benthic habitats of the continental slope between 200 and 4000 m deep.

Bayesian   Relative to animal behavior, the assumption that animals continually use new information to change their expectations of the environment (and therefore change their behavioral decisions).

Bayesian information criterion   A measure of the fit of a model to the data combining the log-likelihood of the model with a penalty term, taking into account the complexity of the model. This measure can be used to select a model among several alternatives.

Beacon   A unique marker for a location, analogous to a sign post.

Beeswax   A complex mix of hydrocarbons produced from wax glands on the abdomen of honeybees and worked into the comb structure to form the bees' nest.

Behavior sampling   Observing a group of animals and recording each occurrence of a particular behavior along with the individuals who perform it.

Behavioral deficit   A change in a behavior as a result of a contaminant or other treatment, usually having a negative effect on the animal.

Behavioral ecology   The study of the evolution and adaptive significance of behavior in the framework of recent views on the levels of action of natural selection and the importance of kin selection. There is an underlying assumption that behavior is selected for individuals to maximize the representation of their genes in the gene pools of future generations.

Behavioral fever   An increase in temperature as a response to parasites or disease. In endotherms, fevers are physiological, but in ectotherms, they may be caused by basking or the production of heat through muscular contractions.

Behavioral hierarchy   A description of interactions among behaviors, taken two at a time, that indicates which of each pair of behaviors overrides the other. Such maps are generally unidirectional (behavior A overrides B, which overrides C, which overrides D), but can have branches ($A \geq B \geq C$ or D), or feedback ($A = B \geq C \geq A$). Typically, these maps can be modified by such factors as an animal's state (e.g., hungry, sleepy, reproductive) and age.

Behavioral strategy   In game theory, a player's complete plan of action in a game, taking all other players possible actions into account.

Behavioral tradition   Nongenetic, heritable differences in behavior among groups or populations with overlapping generations, which are socially transmitted within and between generations.

Behavioristic psychology   A branch of psychology. The goal is to use the animal in an effort to understand a process of interest. Such processes include the mechanisms of learning, the prediction and control of learned behavior, and motivational processes. Some behaviorists focus on the prediction and control of behavior.

Betweenness   Centrality based on the number of shortest paths between every pair of other group members on which the focal individual lies.

Bidirectional control procedure   Manipulation of the direction of movement of an object (e.g., screen or rod) by a demonstrator to determine if an observer will manipulate the object in the same direction.

Bidirectional sex change   An individual is capable of changing sex in both directions.

Binary   Having two states; communication codes having two alternative signals.

Binding globulin   A protein molecule that binds steroids in the bloodstream and prevents both hepatic metabolism and the hormone from binding to its receptors, thus keeping the hormone in circulation. In some cases, hormones bound to binding globulins are capable of binding to receptors specific to binding-globulin-bound hormones.

Binocular stereopsis   Animals with widely separated eyes can judge the relative distance of objects because objects at different distances are imaged on slightly different parts of the retina in the two eyes. This difference is called 'retinal disparity.'

Bioassay   An appraisal of the biological activity of a substance, performed by testing its effect on an organism and comparing the result with some agreed standard.

Biodiversity   The variety of life forms at all levels of a biological system, but most often referring to the number of species.

Biogenic amine   A neurotransmitter, such as serotonin or dopamine, that can regulate behavior.

Bioindicator   A species or attribute (morphology, behavior, reproductive success) of a species or population that can be used to assess the health and well-being of an animal or plant species, a population or an ecological community.

Biological fitness   An individual's ability to survive and reproduce.

Biological model   A conceptual or mathematical description of a biological phenomenon, which generally aims to facilitate comprehension and/or to make predictions.

Biologically active (or bioactive)   Describes a substance, usually a chemical, that acts upon or influences the bodily functions of an organism.

Biologically inspired robots   Robots that are inspired by principles and mechanisms of biological systems. Bioinspired robots often share certain detailed morphological features with their biological analogs, but this is not a requirement.

Bioluminescence   Light produced by living organisms.

Biomagnification   The ability of chemicals to increase in concentration with each step in the food chain. That is, when a large fish eats a smaller fish with a given level of a contaminant, it accumulates a higher level of that contaminant in its own tissues.

Biomimesis   Mimic or imitate biological systems by artificial means (adj: biomimetic).

Biotic   Living.

Biotype   An ill-defined term generally applied to a herbivore exhibiting a specific host plant association that is noteworthy for some reason.

Bit   Contraction of *binary* digit, the uni*t* of information in the mathematical theory of communication.

Bitter pith chewing   A form of self-medication practised by chimpanzees in which an ill animal removes the outer bark and leaves of a plant, *Vernonia amygdalina*, to chew on the exposed, bitter pith. The pith has medicinal properties.

Blind to treatment   Refers to the investigator being unaware and unable to identify which animal has been treated (e.g., with a chemica) and which is a control (and has not).

Blood–brain barrier   The limited diffusion of substances from the bloodstream into the brain and cerebrospinal fluid. Small molecules and molecules that have active transport mechanisms can cross the blood brain barrier, whereas larger molecules without active transport mechanisms are prevented from crossing. The barrier makes the brain less susceptible to blood–borne substances.

Bonanza food source   A very (quantity-) rich food source that lasts for a long time.

Bouton   The enlarged terminus of a nerve cell that forms a connection, or synapse, with another nerve cell.

Bradycardia   A decrease in heart rate.

Brain size   The absolute mass or volume of the brain, based on measures of fresh tissues, brain images, or corrected endocranial volumes. The term 'relative brain size' usually refers to allometrically-corrected measures, most often residuals of log-transformed brain mass regressed against log-transformed body mass.

Branchiostegal membrane   The membrane deep to the gill operculum, connected to the small support bones of the gills (the branchiostegal bones).

Breeding dispersal   The distance between the breeding site of an individual in one year and the breeding site of the same individual in another year.

Breeding range   In migratory birds, the area in which populations reproduce.

Bridge whistle   A whistle used by animal trainers to immediately indicate to the animal that it has performed a behavior correctly. The use of the bridge whistle facilitates training if rewards cannot be given immediately after a task was performed, for example when the animal is far away from the trainer.

Bridging stimulus   A conditioned stimulus that signals the imminent delivery of reinforcement.

Broadband   A vocalization with a broad energy spectrum that often lacks sharp harmonic peaks and has an irregular pulse repetition rate.

Broodiness   Behavior of female poultry as they sit on and incubate a clutch of eggs.

Brood parasitism   Leaving eggs or young to be raised by a nonparent, usually heterospecific, host.

Brood reduction   Occurs when the number of chicks in a brood falls due to the death of one or more of them. Brood reduction is considered an adaptive process if it enhances the viability or survival of the remaining chicks or parental prospects for future reproduction.

Brush-border   Microvilli-covered apical surface of *enterocyte* epithelial cells of intestinal *mucosa*.

Budding   Mode of colony multiplication in which new colonies are founded by the departure of a relatively small force of workers accompanied by one or more queens.

Bumblefoot   Bacterial infection and inflammatory reaction of the foot.

By-product mutualism   Where two or more individuals benefit each other by investing in a cooperative behavior.

Calf   A young animal dependent on its mother.

Cameleon   A genetically engineered protein that consists of two fluorescent proteins on the N and C terminus with calmodulin and M13 domains in between. It is used to detect calcium ion concentration, using fluorescence resonance energy transfer (FRET). Cameleon remains in a linear form when no calcium ions are present. When calcium ions are present, calmodulin binds to calcium, which allows M13 to bind to calmodulin/calcium ion complex and leads to a change in confirmation (shape). The confirmation change induced by M13 and calmdulin/calcium binding pulls the two fluorescent proteins into close proximity that allows FRET to occur.

Camouflage   Concealment strategies that have evolved to reduce the chances of being detected or recognized by predators.

Candidate gene   A gene whose function suggests it may be involved in specifying variation in a quantitative trait.

Canid   Species that are dog-like, classified within the family Canidae in the order Carnivora.

Cannibalism   Feeding on conspecifics.

Capturing   A training technique that involves reinforcing a behavior that is offered spontaneously and in its final target form.

Carapace   The part of a crab's exoskeleton that covers the cephalothorax (the fused head and thorax segments).

Cardiovascular disease   Disease of the heart and/or blood vessels.

Carrier-mediated transport   Passage of glucose, amino acids, and other polar molecules through a cell membrane by 'carrier' or 'transporter' proteins in the cell membrane.

Carry-over effect   Nonfatal condition that transfers between periods of the annual cycle and influences an individual's performance through effects on the condition or timing.

Caste   Distinct social roles within a colony. Caste typically refers to reproductive caste (queen or worker), but may also refer to specialized groups within workers. These are persistent specializations in function or task.

Caste totipotency   The capacity of a social insect larva to develop into any caste within the colony.

Catadromy   The migratory pattern of fish that hatch and develop in saltwater, then migrate to fresh water for adult development, and then return to the sea and breed.

Catecholamines   The class of hormones that includes epinephrine and norepinephrine.

Categorical perception   Occurs when the continuous, variable, and confusable stimulation that reaches the sense organs is sorted by the mind into discrete, distinct categories the members of which somehow come to resemble one another more than they resemble members of other categories.

Category   (The representation of) A specifically defined, general or comprehensive division in a system of classification; often used synonymously with 'class' (see 'class').

Cation   An ion with more protons than electrons giving it a net positive charge.

Caudodorsal cells   Neurosecretory cells in a particular part of the pond snail brain that secrete a hormone that mediates egg-laying behavior.

Causal knowledge   Knowledge of causal structures or properties.

Causal properties   The properties of objects that dictate the possible ways in which they can interact with one another (e.g., solid objects cannot pass through one another).

Causal structure   The directionality of physical events (i.e., cause and effect).

cDNA library   A collection of cDNA molecules that have been inserted into host cells, typically bacteria or viruses, so that the individual cDNAs can be replicated in high numbers.

Ceilometers   A brilliant shaft of light projected on the base of the cloud layer for cloud height measurement. On misty nights, with low ceiling, birds were attracted to the light beams and collided with other birds and the ground. These devices are no longer used by the weather service.

Central pattern generator (CPG)   A neuron or neuronal circuit that produces an activity pattern that varies in time and space to produce a behavior without any need for sensory feedback. For instance, the CPG for walking in a mouse coordinates the four legs (variation in space) to produce a series of steps (variation in time), and the basic motor output pattern can be elicited in an isolated spinal cord from which all connections to sensory and motor structures have been eliminated. Although the best-studied CPGs are those that produce rhythmically patterned behaviors (e.g., walking, swimming, breathing), there are also CPGs for nonrhythmic behaviors (e.g., withdrawal, shortening, vomiting).

Centrality   A measure of an individual's structural importance in a group on the basis of its network position.

Cephalofoil   Flattened and lateral extensions of the head, typically used to describe the head of the hammerhead shark.

Cercaria (plural: cercariae)   A small free-living larval stage of the Trematoda which swims using a tail and does not feed, relying on stored glycogen for energy to find and infect the subsequent host, often a mammal.

Cerebral ganglia   Ganglia located in the head of arthropods. In insects, these are the supra and subesophageal ganglia ('brain' and SEG, respectively).

Cervical connectives   Large nerve bundles passing through an insect's neck; comparable to the spinal cord.

Cestodes   Class of parasitic flatworms, commonly called 'tapeworms,' that live in the digestive tract of vertebrates as adults and often in the bodies of various animals as juveniles.

CF–FM bats   Bats that emit a long constant frequency component terminated by a brief frequency modulated sweep for echolocation. CF–FM bats compensate for Doppler shifts in the echoes they receive.

c-fos   An immediate early gene that is expressed in cells relatively rapidly (e.g., within 30 minutes) in response to the experience of environmental stimuli or after engaging in a particular behavior. The expression of the mRNA for the c-fos gene or the protein product of this gene has been widely employed by behavioral neuroscientists to localize brain areas implicated in the expression of a given behavior.

Chagas disease   A tropical disease also known as American trypanomiasis. It is caused by the flagellate protozoan *Trypanosoea cruzi* and is transmitted by the assassin bug.

Chain migration   Where northern wintering populations breed in the northern portions of the breeding range and the southern wintering population breed in the southern parts of the range.

Channel   A communication system; the collection of alternative signals composing a communication system; the physical system conveying signals.

Channel capacity   The maximum amount of information a communication system is theoretically capable of transmitting.

Chappius effect   A phenomenon in which wavelength-specific absorption by ozone affects the spectral composition of atmospheric light; there is a relative reduction in the spectral region of 540–625 nm (yellow) and increases near 500 nm (blue-green) and 680 nm (red).

Character   A biologically transmitted attribute of a species; in behavioral phylogenetics, such attributes might include learned behaviors not encoded in the genome.

Character reconstruction   An illustration of the simplest (and putatively most likely) pattern of evolutionary changes of a trait as depicted on a phylogenetic tree.

Chase-away coevolution   A form of arms race dynamics in which reciprocal selection pushes one species to stay ahead (in terms of a trait value) of the other.

Cheating, cheater, cheat   A party in a social interaction that does not contribute its fair share. In the case of *Dictyostelium*, it would be a clone that contributed proportionally more to fertile spore cells than to sterile stalk cells during the social stage.

Chelae   Front legs of crustaceans that have been modified into claws.

Chemoreceptor   Sensillum that houses either olfactory or gustatory neurons.

Chemosensor   A sensory receptor that detects specific chemical stimuli in the environment.

Chimera   An organism that is made up of two genetically distinct lineages.

Choosy   Rejecting a particular encountered potential mate.

Chromatic aberration Optical imperfection caused by light of different wavelengths being focused in different planes by a refractive element such as a lens.

Chromophore Part or moiety of a molecule responsible for its color. In vertebrate visual pigments, the chromophore is either retinal or 3,4 dehydroretinal, aldehydes of vitamin A.

Chronesthesia The subjective awareness of the passage of time, an ability that allows us to address our own personally experienced past.

Chronic stress Either long-term exposure to a stressor or repeated exposure to an acute stressor.

Chronobiology Chronobiology, which comes from 'chrono,' meaning time, and biology, is the field of science that deals with cyclic activities in organisms and their relations to time. It is the formal study of biological rhythms.

Circadian rhythm The term 'circadian' comes from the words, 'circa,' which means about, and 'diem,' meaning day. Circadian rhythms are endogenously organized oscillations in biological processes that occur roughly with a period of about 24h and are sustained in constant conditions.

Circannual rhythm Circannual rhythms are endogenously organized oscillations in biological processes that occur each year, such as the migration patterns of some birds.

Circumboreal Occurring around the globe in the boreal, or northern regions.

Circumventricular organ A brain structure lacking a blood–brain barrier.

***cis*-Regulatory regions** DNA regions outside of the protein coding region of a gene involved in regulating transcription.

Class A collection of things sharing a common attribute, characteristic, quality, or trait.

Classic foraging theory A body of economic models concerned with prey choice and patch residence time characterized by the use of simple optimization that applies to cases without frequency-dependent payoffs and hence mostly nonsocial situations.

Classical (aka Pavlovian or Respondent) conditioning A stimulus (the unconditioned stimulus) that normally elicits a response (e.g., altered respiration) is repeatedly paired with a stimulus that does not normally elicit the response (the conditioned stimulus, e.g., light). The CS and US eventually become associated, and the organism begins to produce the behavioral response to the CS alone.

Claustral founding Colony foundation by a non-foraging queen or queens, in which energy to rear the first generation of workers comes entirely from queens stored body reserves.

Clever Hans effect The artifact that occurs when animals, including humans, may be sensitive to cues from the experimenter or the environment of which the experimenter is unaware. Double blind designs are often used to minimize Clever Hans effects.

Cloaca A single posterior opening of the gut to which the bladder and reproductive organs also join.

Cloacal protuberances Seasonally variable, occurring during the breeding season, in male birds. The protuberances are from engorgement by sperm of the storage tubules around the cloaca.

Clone A genetically identical population of cells.

Clustering coefficient (*C*) The density of the subnetwork of a focal individual's neighbors; the number of edges between neighbors is divided by the maximal possible number of edges between them.

Cnidocytes Stinging cells found in cnidarians (e.g., jellyfish) that contain cnidocysts that are fired out into potential predators, injecting venom.

Coalition formation Agonistic acts that involve at least two aggressors simultaneously joining forces to direct aggression toward the same target; such acts of coalitionary support indicate short-term cooperation between coalition partners, whereas the relationship between two individuals that repeatedly join forces over long time periods is considered to be an alliance.

Cochlear nucleus The first auditory nucleus in the brain that receives the projections from the auditory nerve. The projections from neurons in cochlear nucleus are then sent into the medulla as a series of parallel pathways that form the ascending auditory system.

Code word In certain types of codes, an ordered collection of signals making up the smallest decodable unit.

Code The way in which signals stand for their referents.

Coevolution Evolution of organisms of two or more species in which each adapts to changes in the other.

Cofoundress A female that founds a new colony in association with other females.

Cognition Psychological mechanisms that process perceptual information to enable behavioral decisions to be made, for example, learning, memory, generalization, and categorization.

Cognitive control Process in which one cognitive mechanism exerts inhibitory, excitatory, or supervisory influence on another cognitive process; executive control, executive function.

Cognitive imitation Adopting a decision rule after observing another use of that rule.

Cognitive psychology   The study of the mind's function, including perception, attention, memory, imagery, and decision-making.

Collective behavior   A phrase to describe how interactions between individuals produce group-level patterns of behavior.

Collective decision-making   The selection of one from two or more options by a group of individuals in which all members contribute to the choice, rather than following the decision of a leader.

Collective detection   Transfer of information within a group from animals that detect predation threats to others that have not detected the threats directly. Collective detection assumes that once an individual in the group has detected a threat, conspicuous signals of detection, such as alarm calls or flushing, will rapidly alert all other group members.

Collective intelligence   A group of agents that together act as a single cognitive unit to solve problems, make decisions, and carry out other complex tasks. Natural examples include social insect colonies, fish schools, and bacterial aggregations. Artificial examples include robot collectives and decentralized computer algorithms. Also known as 'swarm intelligence.'

Collective robotics   The design of groups of autonomous artificial agents that cooperate to carry out tasks. This field is strongly inspired by examples of collective behavior in animal groups.

Colonial spider   Spiders living in individual webs or nests that are interconnected by silk threads.

Colony budding   Colony founding by a group of workers and one or more queens.

Colony collapse disorder   A syndrome of unknown origin and cause afflicting beekeepers with high rates of colony mortality.

Combinatorial neurons   Neurons that respond best to signals that have two frequencies, that are harmonically or nearly harmonically related.

Command neurons   Neurons that, when stimulated individually or in small groups, can elicit a complex behavior. By strict definition, to be called a command neuron, that neuron must be active whenever the behavior occurs (correlation), stimulation of the neuron must elicit the behavior (sufficiency), and elimination of the neuron must make it impossible to trigger the behavior by its normal sensory input (necessity). Because necessity is often difficult to test, neurons that show just correlation plus sufficiency are often called command neurons.

Commensalism   An interaction between species in which at least one species is not affected, although others benefit.

Common orientation   The phenomenon whereby multiple individuals flying at high altitudes and not in visual contact of each other all take up similar flight headings, which are usually closely aligned with either the downwind direction or a seasonally preferred compass direction.

Communication   The transfer of information from a sender to a receiver by means of signals.

Communication system   An evolved network involving signal givers that produce information containing messages intended for a particular set of signal receivers; both signalers and receivers experience a net gain in reproductive success from their actions and responses.

Communicative culture   A group-specific system of signals, responses to those signals, and preferences for the class of individuals toward which signals are directed, that is socially learned and transmitted across generations.

Comparative psychology   Defined in many different ways. One approach would be the study of a great variety of behavior in a variety of species with a goal of understanding the evolutionary history, adaptive significance, development, and immediate control of behavior.

Competition   In strict biological terms, competition occurs when a necessary resource is in short supply and the use of the resource by one party denies access to that resource by another. Note that the two parties do not even have to be aware of one another, as in scramble competition, where the first party uses the resource before the second arrives and with neither necessarily aware of the other. In more general discussions of social behavior, competition may be described as a striving to outperform another where both parties are aware of the other.

Complete dimorphism   A size distribution of workers composed only of large and small individuals, with no intermediates.

Complete migration   All populations leave the breeding range of the species and move in some cases considerable distances to occupy a nonbreeding range of the species.

Components of fitness   Measures of individual fitness. Reproductive success components include the number of mates, the number of eggs laid or offspring born, the number of offspring that survive to reproductive age, the number of offspring that produce grand-offspring, and the number of grand-offspring. Survival components include age at death or lifespan.

Compound eyes   Crabs, like insects, have eyes that are composed of many repeated units called *ommatidia*, each with a *facet lens* and a transparent *crystalline cone*, which together focus light onto a narrow, elongated light-sensitive structure called the *rhabdom*. The rhabdom consists of densely packed microvilli, protruding from eight *retinula cells*, the photoreceptors. The membranes of these microvilli contain visual pigment molecules, the transmembrane protein part of which is called *opsin*.

Screening pigments in special pigment cells and in retinula cells prevent stray light from reaching the rhabdom from any other direction except through the fact lens belonging to the same ommatidium (apposition compound eyes).

Comprehension learning   Where an animal comes to extract a novel meaning from a signal as a result of experience of the usage of signals by other individuals.

Computational neuroethology   The modeling of the neural basis of animal behavior, with an emphasis on the interaction of the simulated animal with its simulated environment.

Concentrated animal-feeding operations (CAFOs)   Agricultural facilities that house a number of large animals; these operations may release waste into the environment (see http://cfpub.epa.gov/npdes/home.cfm?program_id=7 for additional information).

Concept   An abstract or general idea inferred or derived from specific instances; a mental construct or representation or idea of something formed by (mentally) combining all its characteristics or particulars (synonymously used with a general notion, a scheme or a plan).

Concerted evolution   The changes in one brain region caused as a result of changes in other associated brain regions, often thought to be due to underlying developmental mechanisms.

Condition dependence   Expression of a trait or behavior depends on the state of the organism. Many possible state variables are possible including size, age, energetic reserves, immune function, nest quality, and presence of a mate.

Conduction velocity   The speed with which action potentials travel along an axon; increases with increasing axon diameter; in vertebrates, also increased by myelin around the axon; for unmyelinated axons, typically below 20 m s^{-1}.

Conflict outcome   The result of an actual conflict in terms of the amount of conflict in the colony and the winning party, if any. For example, in honeybees, the outcome of the conflict over caste fate is that selfish individuals lose because they lack means to successful selfish behavior.

Conflict resolution   Exchange of threat and submissive signals between individuals over ownership of resources.

Conformity   A term used to define a family of biases towards high levels of fidelity in social learning, most commonly the case of copying whichever of various options is being shown by the majority of other group members.

Confound   To mingle so that the causes cannot be distinguished or separated.

Confusion effect   A reduction in capture rate by predators attacking a group resulting from their inability to single out one prey from the group.

Consciousness   In a strict medical sense, it is the state of being awake. Often refers to the state of being aware of oneself or environment.

Consensus decision   When the members of a group choose between two or more mutually exclusive actions and reach a consensus, that is, they all 'agree' on the same action.

Conservation reintroduction/benign reintroduction   An attempt to establish a species for the purpose of conservation outside its recorded distribution but within an appropriate habitat and ecogeographical area.

Consistency index   The number of character states specified in the character matrix, divided by the number of character-state transformations appearing on a phylogeny in question. The index is widely used to measure how closely the data fit a given tree.

Consolation   Postconflict affiliation from a bystander to the recipient of aggression with a stress-reducing function for the recipient of aggression. Reassuring body contact provided by a bystander to a distressed party.

Conspecific sperm precedence   Disproportional fertilization of a female by conspecific over heterospecific sperm following mating with both a con- and a heterospecific.

Conspecific   Used as either an adjective or noun to refer to another member of the same species, as contrasted with *heterospecific*, referring to a member of another species.

Constrained parents   Individuals mated to partners they do not individually prefer.

Constraints on mate preferences   Social or ecological factors that reduce the likelihood or opportunity for individuals to mate with partners they individually prefer under the assumption that mate preferences predict offspring viability.

Consummatory sexual behavior   The terminal phase of a sexual behavior sequence during which male gametes are emitted so that they can fertilize oocytes produced by the female the male is mating with.

Contagion   The unconditioned release of a predisposed behavior in one animal by the performance of the same behavior in another animal.

Contaminant   A chemical that has the potential to cause adverse effects in plants or animals.

Context   The circumstances under which a decision is made, which could be including but is not limited to, number of alternative options, nature of alternative options, temporal and spatial information.

Context-specific behavior   A behavior that occurs in one situation, but not others; the context can be defined by a social setting like aggression or mating, an environmental cue like darkness, internal condition like reproductive state or hunger, ongoing activity like flight or walking, or other factors.

Contextual fear conditioning   Pavlovian conditioning can be used to study contextual learning in which the composite properties of an experimental apparatus (e.g., its configuration, odor, illumination) frequently accompanied by an acoustical cue, act as conditioned stimuli predicting a previously experienced foot shock. Rats receiving foot shocks will typically display conditioned freezing when placed in the apparatus the following day.

Continuous (all-occurrences) sampling or recording   Observational method in which an observer records all behavioral onsets, transitions, and interactions of a single, focal animal during an observation period.

Conventional signal   A signal whose meaning could, at least theoretically, be exchanged for another within the same repertoire.

Convergent evolution   The development of similar anatomical, physiological, behavioral, or cognitive traits that may have a similar function, in two or more distantly related species; for example insects, birds, and bats all have evolved wings enabling them to fly. The traits may evolve through similar selection pressures, such as finding and processing food. Convergent evolution (analogy) is different from evolution via shared ancestry (homology), in which traits evolve because they are present in closely related species with a shared ancestor.

Cooperation   A behavior which provides a benefit to another individual (recipient), and which is selected for because of its beneficial effect on the recipient (cf. altruism which is a special case of cooperation).

Cooperative breeding   A social system in which individuals help care for young that are not their own. The parental care givers may be other reproducing adults or reproductively mature but nonreproducing adults.

Co-option or exaptation   Co-option or exaptation is the use of an ancestral adaptation (gene-trait relationship) for a new function for which the adaptation did not originally evolve. Evolutionary co-option occurs when natural selection causes traits, including behavioral traits, to assume new functions, often in new contexts. Motor patterns that are elicited in new contexts ('co-opted') can subsequently become ritualized.

Copulation   Mating; the act of inserting the male reproductive organ into the female.

Copulation solicitation display   An estrogen-dependent courtship display performed by female songbirds.

Corm   Bulblike underground part of a plant stem.

Cornicles   A pair of small upright tubes found on the hind dorsal side of aphids that are used to excrete droplets of defensive compounds.

Corpora allata   Glands near the insect brain that secrete juvenile hormone.

Corpora cardiaca   Neurohemal organs near the insect brain that store and release prothoracicotropic hormone and other neuropeptides.

Corticoids   A class of C_{21} steroid hormones secreted primarily from the adrenal cortices. There are two main types of corticoids: glucocorticoids (e.g., cortisol and corticosterone) and mineralocorticoids (e.g., aldosterone).

Corticosterone   Glucocorticoid hormones found in birds, reptiles, and mammals.

Corticotropin-releasing factor (CRF)   Forty-one amino acid polypeptides produced in the hypothalamus and extrahypothalamic sites that stimulate the release of ACTH (all vertebrates studied) and TSH (nonmammalian vertebrates) by the anterior pituitary gland. CRF-like peptides play central roles in developmental, behavioral, and physiological responses to stressors.

Cortisol   Glucocorticoid hormone most commonly found in mammals.

Corvids   Members of the crow family, which includes the rooks, ravens, magpies, jackdaws, jays, and choughs as well as crows.

Cost–benefit analysis   Cost–benefit analysis as applied to animal behavior predicts that if a behavior is adaptive, the benefits of a behavior must exceed the costs of that behavior. These costs are typically measured in terms of energy, time, and survival or reproduction.

Counterconditioning   A respondent learning technique designed to replace an undesirable response with a more desirable one. Often used to reverse fear conditioning.

Courtship   A suite of behaviors by members of one sex to attract members of the other sex for the purposes of mating.

Crepuscular   Active during periods of twilight, that is dawn and dusk.

Criterion   A rule or test on which to base a decision.

Critical flicker fusion frequency   Frequency of a flickering light at which it is perceived as steady.

Crop   A pouch-like enlargement of a bird's gullet.

Cryophilic   Having an affinity for low temperature. In behavior, cryophilic refers to animals having a tendency to move toward lower temperature.

Crypsis   Defense strategies that have specifically evolved to reduce the probability of detection.

Cryptic female choice   A type of sexual selection that can occur if a female's morphological, behavioral, or physiological traits (for instance, triggering of oviposition, ovulation, sperm transport or storage, resistance to further mating, inhibition of sperm dumping soon after copulation, etc.)

consistently biases the chances that a particular subset of conspecific mates have of siring offspring, when she copulates with more than one male. This is the postcopulatory equivalent of Darwinian female choice.

Cryptochrome   Flavoprotein ultraviolet-A receptor involved in circadian rhythm entrainment in plants, insects, and mammals.

Cue   A change in the environment made by one animal that allows another animal to acquire information, but does not benefit the animal that produced it. A source of information that can be used during orientation (e.g., a landmark).

Cue bearer   Any organism or object that carries a recognizable set of identity cues.

Cue calibration   The process of comparing compass information (i.e., directional references) derived from multiple sensory cues, such as magnetic and celestial cues, and calibrating one compass with respect to another. This can lead to a hierarchy of sensory cues, in which one particular sensory cue is being used to calibrate all the others.

Cue readers   Unintended receivers of signals (predators or parasites) using a signal to detect the location of a potential prey/host, but for which the information content of the signal is unimportant.

Culture   (a) [as commonly used by biologists]: between-group variation in behavior that owes its existence at least in part to social learning processes; (b) [as commonly used by anthropologists]: 'the complex whole which includes knowledge, belief, art, law, morals, custom, and any other capabilities and habits acquired by man as a member of society' (Tylor, 1924, p. 1).

Cupula   Gelatinous covering of the hair cells in a lateral line neuromast. The cupula forms the mechanical coupling between water movement and the displacement of the hair cell cilia.

Currency   Any quantity that can be used to evaluate the costs and benefits of different behavioral acts.

Cutaneous receptors   A cutaneous receptor is a type of sensory receptor found in the dermis or epidermis. They are a part of the somatosensory system. Cutaneous receptors include cutaneous mechanoreceptors, nociceptors (pain), and thermoreceptors (temperature).

Cysticercoids   The larval stage of many tapeworms.

Cytokine   The name literally refers to a 'moving cell,' but in this case, cytokines refer to protein and peptide molecules that act as a cell signals. Cytokines, which are secreted by immune cells that have encountered a pathogen, encompass a large and diverse family of protein and polypeptide regulators that are critical to the development and functioning of both innate and adaptive immune responses. Endogenous pyrogens, which evoke the fever reaction and sickness behavior, are a type of cytokine.

Cytoplasmic incompatibility   Differences carried within the cytoplasm of an egg or sperm prevent the formation or lead to the degradation of the zygote due to an interaction with the cytoplasm and the nuclear genetic material. The cytoplasmic effect may be due to gene products existing in the cytoplasm or cytoplasm-associated endosymbiotic organisms.

Dance language   A series of movements displayed by honeybees to recruit their nestmates to food or nest sites.

Darwinian fitness or fitness   The capability of an individual of certain genotype to reproduce, which is usually equal to the proportion of the individual's genes in all the genes of the next generation.

De novo synthesis   Produced by the organism; self-made.

Death feigning   The assumption of a false catatonic state after being captured by a predator in which the animal appears rigid and lifeless; may function to convince the predator that no further attack is necessary, allowing the prey to escape (also called: letisimulation, thanatosis, death shamming, akinesis, hypnosis, and tonic immobility).

Deception   The production of a signal that induces a receiver to behave in ways that reduce its reproductive success.

Decibel   A measurement of sound amplitude. A decibel is the ratio of two pressures on a logarithmic scale: $dB = 20 \log (p_1/p_2)$, where p_1 is the sound being measured and p_2 is a reference pressure referred to the threshold of human hearing.

Decision algorithm   A set of behavioral steps that ends with selection of one option from a choice set. The steps govern how an individual reacts to the options themselves, other aspects of the environment, and its own state. In a collective decision, they also govern interactions among group members.

Decision-making   An outcome of cognitive processes, leading to the selection of one particular course of action (or option) among several alternatives.

Declarative   In memory research, declarative memories are contrasted with nondeclarative (or implicit) memories; originally, declarative memories were those that could be explicitly talked about although today other properties may be used to characterize declarative memory; declarative memories are widely thought to depend on the temporal lobes of the brain. Nondeclarative memories control behavior without the awareness of the existence of stored information, for example, one can ride a bicycle without being able to state, in detail, how it is accomplished.

Decoding   The process of extracting information from signals.

Defeminization   A component of the sexual differentiation process during which the capacity to display female-typical behaviors is lost or reduced.

Defense call   Auditory call given by an animal standing its ground in the face of an approaching predator that may mimic the call of a species that is threatening to the predator and function to deter its further attack.

Deflective markings   Patterning on the body of a prey type that produces a fitness advantage to the bearer by manipulating the point of predatory attack on the prey's body such that successful prey capture is less likely.

Degree (k)   The number of edges a focal animal has; in an unweighted network, this is the number of other animals with which the focal individual interacts; in a weighted network, this will reflect the strength or frequency of interactions; also called *connectivity*.

Degrees of freedom   In statistical analyses it is the number of independent pieces of information upon with a statistical value is based. This along with a statistical value and the rejection criteria determine the statistical significance of a test.

Deimatic signal   A sudden change in the appearance of prey that can cause a predator to delay (or even abandon) an attack.

Delayed gratification task   Experimental situation in which rewards accumulate over time, and decision makers can choose when to stop the accumulation.

Delay-tuned neurons   Neurons in the auditory system that respond most vigorously to two brief signals that have a particular temporal separation that mimic an emitted pulse and echo that returns from a particular distance.

Demersal   Living or occurring in habitats near the bottom or seafloor.

Demographic stochasticity   The fact that some individuals fail by chance to encounter potential mates or by chance die, processes that cause fluctuations in demographic parameters.

Demography   The size and age structure of a colony.

Dendrite   Peripheral extension of a sensory neuron on which the receptor proteins are located.

Dendritic   Fingerlike, branching as a tree from a single root.

Dense cored vesicles   Small, intracellular, membrane-enclosed sacs found in neuronal terminals. Also called 'granular vesicles.'

Denticles (placoid scales)   Small outgrowths, similar in structure to teeth, which cover the skin of many cartilaginous fish including sharks. Denticles of sharks are formed of dentine with dermal papillae located in the core. The shape of a denticle varies from species to species and can be used in identification.

Dependent founding   Initiation of a new colony that requires the aid of workers. It involves colony budding or fission.

Dependent variable   A variable that is presumed to be affected or controlled by one or many independent variables.

Depth perception   The ability of animals to see the world in three dimensions.

DES   Diethylstilbestrol is a strong estrogen that was used as a preventive treatment against miscarriage.

Desensitization   The mitigation of a response to a distressing stimulus by gradual and repeated exposure to that stimulus.

Desquamation   Physical loss of skin, scales, etc.

Developmental plasticity   Environmental variation induces variation in phenotypes among individuals within populations and sometimes within individuals.

Developmental psychology   Focused on the changes in behavior as the animal matures and the interplay between genes, environment, and the organism during ontogeny.

Dewlap   A fleshy and sometimes colorful patch of skin on the throat area of some lizards. Many species have muscles that allow the dewlap to be extended as part of displays.

Dialect   The situation where acoustic communication signals form a mosaic pattern of geographic variation, with individuals within a local population producing very similar signals that are separated by relatively sharp borders from those of the neighboring groups.

Diameter (d)   The largest distance between any two vertices in the network.

Diapause   A state of arrested behavior, growth, and development that occurs at one stage in the life cycle. Quiescence accompanied by decreased metabolic rate and other physiological processes.

Diel vertical migration (DVM)   Vertical movements at sunrise and sunset, commonly used by aquatic organisms to balance feeding and predator avoidance. DVM usually involves an ascent to shallow water at sunset and descent to deeper water at sunrise, often linked to temperature and light.

DIF   Differentiation inducing factor is a chlorinated alkyl phenone produced by strong cells that induces weaker cells to become stalk, not spore.

Differential allocation hypothesis   A hypothesis about selection on parents to allocate their parental resources differently to offspring depending on the relative attractiveness of mothers versus fathers.

Differential migration   When the timing or distance of migration is different for males and females, or for young and adults, or both sex and age differences.

Diffusion chain   An experimental design for studying the serial transmission of information from model to novice, typically used to assess fidelity, corruption, and other changes, along a chain of individuals.

Dilution effect   A decrease in predation risk due to the presence of alternative targets in a group when a predator cannot capture all group members during an attack.

Dimorphism   Having two different patterns, usually referring to physical features. Males and females differ in their color patterns or sizes.

Dipsogenic   Thirst provoking.

Direct benefits   Material benefits of mate choice that accrue directly to the choosing individual as a result of the choice, such as nutrients, territory quality, or parental care provided by the mate.

Direct fitness   Fitness achieved through direct reproduction of one's own offspring. Direct reproduction is one component of inclusive fitness.

Directionality   The ability to locate the source of a stimulus in space.

Directional selection   A form of selection in which more extreme phenotypes are favored over existing phenotypes, such as larger more colorful ornaments, resulting in progressive elaboration of the phenotype over evolutionary time.

Dispersal   Movement of individuals away from an existing population or away from the parent organism.

Displacement activities   Behaviors performed in an abnormal context and in response to a seemingly unrelated motivation.

Dissociated pattern of reproduction   An annual reproductive cycle in which expression of copulatory behaviors and fertilization are not synchronized with the period of maximal activity of the gonads.

Distal   Farther from a body midline – used to describe order of segments in an appendage (e.g., a hand is distal to a shoulder).

Distractor option   A member of a choice set that is unlikely to be chosen but which may influence preferences for other options. Distractor effects exemplify the irrational decision-making often seen in humans and other animals.

Distributed cognition   Distributed cognition is an interdisciplinary branch of cognitive science that holds that cognitive processes are not confined to the brains of animals, but extend across individuals and out into the environment. An animal's 'cognitive system' consists not of its brain alone, but of its brain, body, and environment (including other animals) acting in concert. It is closely connected to the concept of embodied cognition.

Disturbance and disturbance stimulus   Disturbance is a deviation in an animal's behavior from patterns occurring without human influences. A disturbance stimulus is a human-related presence or object (e.g., birdwatcher, motorized vehicle) or sound (e.g., seismic blast) that creates a disturbance.

Diurnal   Active primarily during the daytime.

Diurnal rhythm   A biological rhythm that is synchronized to the 24 h light–dark cycle.

Diversionary display   A display performed by a parent at the approach of a predator that poses a risk to vulnerable young. If successful, the display attracts the attention of the predator causing it to move toward the parent and away from the young. These displays, most commonly described in ground-nesting birds, but also found in stickleback, incorporate elements that seem to have been co-opted and ritualized.

Division of labor   A property of a social group in which different individuals specialize in different tasks.

DNA methylation   Chemical modification of individual cytosine nucleotides in DNA that alters gene transcription.

DNQX (6,7-Dinitroquinoxaline-2,3-dione)   An AMPA and kainate antagonist. It is used in neurobiology as a tool to block AMPA and kainate type ionotropic glutamate receptors.

Domain of danger   The space closer to a focal individual than to any other group members.

Dominance   The state of having high social status in a group, often won through aggressive encounters or threats of aggressive encounters with conspecifics. Dominance is often linked to increased acquisition of resources, including food, territories, and mates.

Dominance hierarchy   A dominance hierarchy describes predictable interactions among individuals, with one giving way to another in competition for resources. A linear dominance hierarchy is transitive.

Dominance–subordinance relations   In groups of animals some individuals dominate (i.e., are higher in the 'pecking order') others that become subordinate (lower in the 'pecking order'). These relationships may be stable over many days, weeks, months, or even years, whereas in other cases they can be changing constantly (e.g., as in large groups).

Dominant frequency   The highest amplitude frequency component in a harmonic sound.

Dominant individual   High-ranking individual within a social group. This individual often has primary access to the best resources, such as food and mating partners. Dominance is often (but not always) correlated with large size and fighting ability, but also the ability to form coalitions (friendships) with other individuals.

Dopamine   A neurotransmitter occurring in both vertebrates and invertebrates. Massive loss of DA neurons in the substantia nigra in humans results in Parkinson's disease whose main characteristic is the paucity of voluntary movements, or hypokinesia. It is also associated with the pleasure system of the mammalian brain.

Doppler shifts   The increase in the frequency of a returning echo due to the difference in velocity between a bat and its target.

Dorsal root ganglion   A nodule near the spinal cord that contains cell bodies of sensory neurons in spinal nerves.

Drone   A male honeybee.

dsRNA   Double-stranded RNA.

Duration eggs   Encapsulated eggs, which can dry out or freeze and hatch once conditions are favorable again. Often found in zooplankton, especially in temporary ponds.

Dynamical system   A mathematical description of how a system behaves as a function of time. The description consists of an equation or set of equations that define the system's current state, as well as its past and predicted trajectory. The goal in considering nervous systems as dynamical systems is to characterize their oscillatory or quasi-oscillatory behavior.

Eavesdropping   The use of a signal by an animal that is not the intended receiver of the signaler.

Ecdysis   The shedding of the old, overlying exoskeleton of an arthropod; a process necessary for growth.

Ecdysteroid   A general term for a family of steroid hormones known in insects and other invertebrates. In insects, it is known as a molting hormone during larval stages, but has many other nondevelopmental effects. In most insects, the primary ecdysteroid is 20-hydroxyecdysone.

Echo ranging   The measurement of distance between a bat and its target. The acoustic cue for ranging is the time interval between the emitted pulse and the returning echo.

Echolocation   The ability to use sound waves reflected from a surface to detect objects at a distance.

Ecological determinism   Similarities among closely related species that reflect the ecological selection pressures acting on species, rather than the phylogenetic relationships between species.

Ecological time scale   A time scale of the same order of magnitude as the life span of the organisms investigated. Measured in days, months, or years, as opposed to evolutionary time scale, which is measured in thousands or millions of years.

Ecotoxicology   The study of the effect of chemicals (toxicology) on the ecology of animals or plants.

Ectoparasitoid   A parasitoid with a life-history strategy where the larva develops outside the host body by attaching or embedding in the host's tissues.

Edge   A relationship between two components of a network where the two related components are vertices in the graph model representing the network; in a social network, these can be any sort of social relationship, such as social interactions or information transfer; also called a *tie* or *link*.

Education by master apprenticeship   This is a phrase coined to describe how chimpanzees acquire new behaviors through observational learning. It is characterized by the following four aspects: (1) a long-term affectionate bond between mother and infant, (2) the mother takes on the role of the 'model' who demonstrates specific behaviors in the correct context, (3), the infant has a strong motivation to copy the model's behavior, and (4) the mother is highly tolerant toward the infant.

Effective population size   The number of breeding individuals within an idealized population, mating at random, that would have the same amount of inbreeding or of random gene frequency drift as the population under consideration.

Efficient theory   These are theories built from first principles. They are often, but not always, expressed mathematically; have few assumptions and free parameters (those that cannot be derived from a model or hypothesis); describe nature in an approximate way and is used iteratively to approach an ever-better understanding of nature. Characteristically, efficient theory has considerably fewer input variables than output variables.

Egg dumping   Occurs when a female bird lays her egg or eggs in the nest of another female and leaves that other female to care for them.

Egg pod   A capsule which encloses the egg mass of grasshoppers and which is formed through the cementing of soil particles together by secretions of the ovipositing female.

Egress   To come out or exit.

Elasmobranch   The cartilaginous fishes of the subclass Elasmobranchii including the sharks, skates, rays, and their extinct relatives.

Electric organ discharge (EOD)   The electrical signal produced by the electric organs of electric fishes. Electric organ discharges create an electrical field around the fish that can be detected by electroreceptor organs in the skin. Electric organ discharges have three functions. Extremely strong discharges (hundreds of volts) of strongly electric fish such as electric eels (*Electrophorus electricus*), electric rays (*Torpedo* spp.), and strongly electric catfish (*Malapterurus electricus*) can stun prey or potential predators. Weak electric organ discharges (typically less than a volt) of South

American knifefishes (Gymnotiformes) or African Mormyriformes are used to detect objects and prey or to communicate with conspecifics in dark, murky waters at night.

Electrocommunication   The ability of weakly electric fish to emit and receive electrical signals for the purpose of communication. Electrocommunication is limited to aquatic environments where the electrical conductivity of the medium is sufficient to transmit electric signals.

Electromyographic activity   Product of the electrical activity of muscle, which normally generates an electric current only when contracting or when its nerve is stimulated. Electrical impulses are often recorded as an electromyogram (EMG).

Electro-olfactogram   An electrical recording of the voltage across the olfactory epithelium. This type of recording allows experimenters to detect the electrical responses of olfactory sensory cells to odors.

Electroreceptor organ   Lateral-line-derived epidermal sense organs consisting of electroreceptor cells and associative structures located on the head and the trunk of certain fishes. They direct the flow of electrical current through low-resistive canals or through loosely layered patches of epithelial cells to specialized receptor cells containing membrane-bound voltage-gated ion channels, which convert outside electrical signals into sizable membrane potentials and subsequent transmitter release.

Electroretinogram (ERG)   The massed electrical response of the retina recorded by extracellular electrodes on the retinal, or more usually, corneal surface.

Embodied cognition   Embodied cognition is an interdisciplinary branch of cognitive science that argues that cognitive processes emerge from the unique manner in which an animal's morphological structure and sensorimotor capacities allow it to successfully engage with its environment. It aims to capture the way in which an animal's brain, body, and world act in concert to produce adaptive behavior, and, as such, is closely allied to the concept of distributed cognition.

Emergence   When a behavioral response to a multimodal signal is entirely different from responses elicited by any single component.

Emergency life history stage   A syndrome of physiological and behavioral traits triggered by perturbations of the environment that are designed to allow the individual to cope with the perturbation in the best condition possible until it passes.

Emergent phenomenon   Complex biological event that itself is not the target of natural or sexual selection, but which arises as the collective result of many simpler events that are under direct selection pressure.

Emergent relations   Relations between classes or class members that arise through a process of association, generalization, or inference.

Emigration   Dispersal or migration of organisms away from an area.

Emotion   A physiological and psychological state that functions to increase the survival of the organism. Basic emotions include anger, disgust, fear, happiness, sadness, and surprise.

Emotional contagion   Automatic state matching as a result of perceived emotions in others.

Empathy   The ability to recognize or understand another's state of mind or feelings (i.e., emotions).

Empirical   Evidence that can be observed.

Emulation   Recreation of the results of the efforts of another animal.

Encapsulation   A physiological immune response in host insects where a parasitoid egg, or other foreign body, is coated or engulfed by specialized cells called plasmatocytes resulting in the death of the parasitoid egg.

Encoding   The process of endowing signals with information.

Endemic   Native or restricted to a certain area.

Endocrine disruptor   A compound that interferes with the endocrine system, typically by binding to a receptor and either stimulating the effects of the receptor's hormone or blocking those effects, rendering the receptor inert. A compound produced for use as insecticides, herbicides, fungicides, or industrial applications as well as plant-produced chemicals with biological activity in living systems due to similarities in the structural and functional characteristics of native hormones, resulting in interference of endocrine systems.

Endocrine gland   A ductless gland from which hormones are released into the blood system in response to specific physiological signals. These signals can result from internal or external stimuli.

Endogenous   Phenomena arising within an organism, such as a biological rhythm.

Endogenous metabolic marker   An endogenous indicator of changes in metabolic activity within a cell.

Endogenous oscillator   An oscillator that is reset by internal stimuli. An endogenous oscillator is self-sustaining (i.e., periodic output continues after the termination of periodic input).

Endogenous pyrogens   Endogenous refers to inside the body and pyrogen refers to the generation of heat, in this case the increase in body temperature associated with a fever. Endogenous pyrogens, now commonly referred to as cytokines, evoke sickness behavior along with fever. Endogenous pyrogens are released in the body upon exposure to bacteria, bacterial cell-wall lipopolysaccharides, and viruses.

Endogenous rhythm   Internally generated, and not dependent on (but may be modified by) an external stimulus. Usually applied to seasonal processes, such as gonad growth, and migration, or diurnal processes, such as sleep.

Endoparasitoids   A parasitoid with a life-history strategy where the larva develops within the host body.

Enemy-free space   A habitat (e.g., host plant) where the herbivore is exposed to reduced rates of predation and parasitism.

Enemy release hypothesis   Hypothesizes that nonnative species become invasive because they are free from the predation and parasitism pressures of their native region.

Energy balance   Physiological adjustment of energy intake and expenditure resulting in precise maintenance of body mass; also known as 'energy homeostasis.'

Enrichment   Any aspect of enclosure design or husbandry practice that increases behavioral opportunities and promotes physical and psychological well-being in captive animals.

Enterocytes   Epithelial cells comprising the innermost layer of the gut.

Entrain (entrainment)   To adjust a rhythm so that it synchronizes with an external cycle, for example, the entraining of the internal rhythm of an organism to a light/dark cycle.

Entropy   The average amount of information encoded by signals of a system, less than or (rarely) equal to the channel capacity.

Environmental signaling   The signaling effects of environmental chemicals that directly or indirectly lead to changes in physiological functions or behaviors through interference with endocrine or exocrine mechanisms.

Environmental task specialization   Task threshold is primarily determined by the environment. Workers vary in their behavior on the basis of the environment they have experienced particularly during larval feeding.

Eph–ephrin receptors   Eph and ephrin receptors are components of cell signaling pathways involved in animal development and axon guidance. Eph receptors are classified as receptor tyrosine kinases (RTKs) and form the largest subfamily of RTKs.

Epidermis   The outermost layer of cells acting as the organism's major barrier against the environment.

Epigenetic   Originally, the term 'epigenetic' was used in a broad sense to refer to the processes of development as an interaction of genes and their products to produce the phenotype. This original definition did not imply heritability. Its definition became narrower with epigenetic being viewed as any aspect other than DNA sequence that influences the development of an organism. Modern usage of the term in molecular biology refers to the heritability of a trait over cell generations in an individual or across generations of individuals without changes in underlying DNA sequence. Epigenetic changes are preserved during the cell cycle and remain stable over the course of an individual's lifetime. For example, methylation of DNA at a Cytosine followed by a Guanine (CphosphateG or CpG site) can epigenetically switch off the adjacent gene, which may then stay 'off' in ensuing generations.

Episodic memory   The ability to remember and reexperience specific personal happenings from the past.

Epistasis   Where multiple genes interact to influence a trait.

Epistemic acts   Acts which serve to change the cognitive demands of a task so as to make it easier to solve, but which do not move an animal closer to task completion.

Epistemic engineering   The manner in which animals change their environments in order to alter the nature of the informational environments, as a means of either reducing its own cognitive load or increasing that of its enemies and rivals.

Eradication   An attempt to completely remove exotic fauna or flora from an area.

ERαKO   Knockout mice lacking a functional estrogen receptor α.

ERβKO   Knockout mice lacking a functional estrogen receptor β.

Eruption   In migration studies, a massive emigration from a particular region.

Estivation (Aestivation)   A period of dormancy over the summer that allows animals to survive an extended period of high temperatures or drought.

Estradiol   An estrogen (hormone) secreted by the ovary; it binds to estrogen receptors in many tissues including the brain.

Estradiol 17β   The most important circulating estrogen in both teleost fishes and mammals, produced in the ovaries but also other tissues including brain through the action of the enzyme aromatase.

Estrogen   A steroid hormone with 18 carbons and an aromatic ring, so named because of their estrus-generating properties in female mammals. Examples include estradiol, estriol, and estrone. Estrogens are synthesized from androgens with the help of the enzyme aromatase.

Estrus (estrous)   The period during which a female is sexually attractive, proceptive, and receptive to males and is capable of conceiving.

Ethnic marker A seemingly arbitrary cultural element that signals membership of a particular ethnic group.

Ethnopharmacology The study of the pharmacologically active compounds in plants used by traditional societies pertaining to the health care of humans and their animals.

Ethogram Inventory of behaviors of a species, with definitions.

Ethology Approach to the study of behavior developed by European zoologists that emphasizes, but is not limited to, the study of the naturally occurring behavioral patterns of free-ranging animals with particular emphasis on evolution and adaptive significance but not to the exclusion of development and immediate causation.

Ethopharmacology The study of the effects of drugs on the neurochemical mechanisms of behavior. According to some authors, ethopharmacology should include the biological variability and adaptive significance of behavior, thus explicitly relying on an evolutionary approach. Ethopharmacological studies of host–parasite interactions attempt to unravel the neuromodulatory mechanisms that underlie the behavioral alterations of a host induced by a manipulative parasite.

Euphotic zone Upper water layer of a lake or ocean to which 1% sunlight penetrates.

Euryhaline The ability to tolerate various salt concentrations, that is, describes water organisms that tolerate a wide range of salinity.

Eusocial A classification of social organization with (1) reproductive suppression, (2) overlapping generations, and (3) cooperative care of young (e.g., naked mole rat).

Eusocial (eusociality) Colonies of animals structured around in which the generations overlap and there is a division of reproductive labor with members of the older generation producing most or all of the offspring of the colony. In primitively eusocial species, the differentiation between the parental generation (queens) and their daughter workers is weak and the daughters may have the potential to reproduce. In highly eusocial species, the queen and workers are highly differentiated and workers typically lack the physical and physiological attributes required to mate and reproduce.

Eutherian mammals Eutheria are a group of mammals consisting of placental mammals plus all extinct mammals that are more closely related to living placentals (such as humans) than to living marsupials (such as kangaroos). They are distinguished from noneutherians by various features of the feet, ankles, jaws, and teeth.

Evaluator Any organism that evaluates a cue bearer and makes a decision regarding that cue-bearer's identity.

Evo-devo Evolutionary developmental biology, a field of biology that integrates studies of genetics, development, and evolution in order to understand the evolution of morphology and developmental processes.

Evolutionarily stable strategy (ESS) A strategy that, if adopted by a population of players, cannot be invaded by any alternative strategy that is initially rare.

Evolutionary algorithms Several computational techniques that use iterative progress to solve problems. Inspired by evolutionary processes, such as reproduction, mutation, recombination, and selection, the techniques are based on a population that evolves in a guided random search until the individuals who use the best solution or strategy take over.

Evolutionary game theory Evolutionary game theory is an application of the mathematical theory of games to evolutionary biology contexts, arising from the realization that frequency-dependent fitness introduces a strategic aspect. A game defines fitness of players, which reflects not only strategy of the protagonist player but also strategy of other ones. Evolutionary game theory analyzes transition of strategists' frequency in the population according to the expected fitness of each strategist, which reflects the current relative frequencies of the strategists and the game rules.

Evolutionary psychology The application of evolutionary principles to human behavior in which behavior is regarded as the product of mechanisms that evolved early in human history, possibly in the Pleistocene epoch, and may not be adaptive in the present environment. Thus, behavior need not be adaptive in the present environment. Often behavior is viewed as the product of relatively specialized modules in the brain.

Exogenous Phenomena arising outside of an organism, such as the light–dark cycle.

Exogenous metabolic marker An exogenous substance that, when introduced to an animal, can indicate changes in metabolic activity within a cell.

Exotherm An animal that depends on external sources of heat to maintain its body temperature in a viable range, as contrasted with *endotherms*, which have physiological mechanisms to generate heat and reduce heat stress.

Exotic species A species that was accidentally or deliberately transported to an area far from its native distribution range.

Expected group size The group size that is predicted on the basis of a given hypothesis for the advantage to being in groups.

Explicit In memory research, equivalent in meaning to declarative; contrasts with implicit; see declarative.

Exposure   Process or situation in which a substance in the environment, such as a chemical, gains entrance to an organism (through ingestion, inhalation, dermal, or injection).

Expression component   The production or acquisition of identity cues by a cue bearer.

External validity   How well results of a study can be generalized to other situations or conditions.

Extinction   Withholding or preventing reinforcement of a previously reinforced behavior with the goal of reducing the frequency of the behavior to baseline or eliminating it altogether.

Extracellular fluid   One of the major fluid compartments of the body comprising all fluid residing outside cells.

Extracellular recording   Monitoring the electrical activity of neurons with an electrode outside the cells; normally records the activity of many neurons simultaneously.

Extractive-foraging   Behavior aimed at accessing food embedded in a protective matrix (such as shells or spines), or that is otherwise inaccessible (such as termites in nests or insect larvae in tree holes).

Extra-pair copulations (EPCs)   Copulations with individual(s) other than a mate or social partner.

Extra-pair fertilizations (EPFs)   Fertilizations that occur when females copulate with males other than their social mate.

Extra-pair offspring (EPO)   Offspring obtained by extra-pair copulations.

Extra-pair paternity (EPP)   Occurs when a socially paired female reproduces with a male, who is not the social mate.

Extrinsic isolation   Low fitness of hybrids because of hybrid phenotypes not being adapted to the resources of either parental population.

Extrinsic marker   Tag or band affixed to an animal at the time of capture that yields data only when an individual is re-sighted or recaptured later on.

Extrinsic mortality   Mortality caused by extrinsic agents such as predators, diseases, and accidents independently of any risks taken for reproduction.

Exudate   An escape of fluid as a consequence of increased vascular permeability and inflammation.

Exuviae   The remains of a molted arthropod exoskeleton.

Facial nerve   The seventh (VII) of twelve paired cranial nerves. It emerges from the brainstem between the pons and the medulla, and controls the muscles of facial expression, and taste to the anterior regions of the tongue.

Facultative   Applies to organisms that can adopt alternative ways of living. More specifically, individual facultative migrants have the choice of whether to migrate or not.

False belief   A belief is a mental state representing knowledge about the state of the world, for example that food is hidden in a particular container. A false belief is a mental state that is contrary to reality, for example the food may have been moved without an individual witnessing the change, and therefore it will have a false belief about the location of the food. Understanding that others can have false beliefs has been suggested as the key test for theory of mind in children.

False workers   In termites, the majority of the individuals within a colony of wood-dwelling termites. They differ from the (true) workers of foraging termites as they are totipotent larvae that lack morphological differentiations. Correspondingly, they are less involved in truly altruistic working tasks, such as foraging, brood care, or building behaviors. Therefore, they may rather be regarded as large immatures that delay reproductive maturity ('hopeful reproductives').

Family group   A group of individuals that repeatedly interact, composed of one or both parents and their direct offspring; may or may not include other relatives as in 'extended family group.'

Fast mapping   A type of inference by exclusion used by psycholinguists to denote the ability of children to form quick and rough hypotheses about the meaning of a new word after only a single exposure.

Fear effects   Another term for nonconsumptive effects. This term should be avoided except in cases where it has been established that antipredator responses are driven by fear.

Fear scream   Loud, harsh auditory call emitted after being captured by a predator that may serve one or more functions, including mobbing, startling the predator, warning kin of danger, calling for help from conspecifics, and attracting other nearby predators to distract the captor (also called: distress call).

Feature learning   (In the process of categorization) The use or abstraction of common features; in contrast to feature analysis, this process is characterized by a continuous adaptation of the feature set or the feature weights in order to cope with the actual categorization task.

Fecundity   The reproductive capacity of an organism; the quantity of eggs, sperm, or offspring produced by an individual.

Fecundity selection   Selection generated by variation in the number of offspring produced among individuals of a population.

Feeling A brain construct, involving at least perceptual awareness, associated with a life-regulating system, which is recognizable by the individual when it recurs and may change behavior or act as a reinforcer when learning.

Felid Species that are docat-like, classified within the family Felidae in the order Carnivora.

Female control Refers to the idea that in species with internal insemination and fertilization that females are likely to control the fate of sperm and the likelihood of fertilization by particular sperm.

Female resistance Describes behavior, physiology, and morphology of females that decreases the likelihood that males will attempt to force them to copulate.

Fertility The number of reproductive bouts for an individual female over a season or a lifetime.

Fertilization Occurs when sperm enters an egg.

Fidelity Faithfulness, usually applied to a locality or mate.

Finder's advantage In the context of group feeding, it is the part of a clump of food that a finder gets to eat before the arrival of any other individuals at the patch.

Finder's share The fraction of the total food patch that makes up the finder's advantage.

Fisher's sex-ratio theory Sex-ratio argument predicted for diploid species that sex ratios should stabilize at 1:1 (female:male) because each offspring derives from the pairing of a female and a male, and each sex, thus, produces overall the same total number of offspring, any deviations from an even sex ratio are unstable because negative frequency-dependent selection gives the rarer sex a reproductive advantage over the more common sex, ultimately leading to equal sex ratios at the population level.

Fission Mode of colony multiplication in which new colonies are founded by one colony dividing into two relatively equal halves.

Fission–fusion society A society in which members belong to a single, permanent social group, but in which all group members are rarely observed together concurrently. Instead, individuals form temporary subgroups that change frequently in their size and composition, often in response to ecological variation.

Fitness The relative capacity of an organism to survive and transmit its genotype to reproductive offspring.

Fixed threshold model A model of task allocation in insect colonies that holds that individual workers vary in the level of stimulus required to undertake a particular task. Workers with a low threshold are likely to engage in the task. High-threshold workers will not.

Flank marking A behavior in which an animal rubs its flanks on objects to deposit contact pheromones from scent glands located on or near the flanks.

Flexible individual phenotypes Phenotypes that are induced by environmental variation; these often appear to enhance the instantaneous fitness of the individual.

Flight boundary layer The narrow layer of the atmosphere closest to the surface within which the self-powered flight speed of an individual exceeds the mean wind speed; thus within this layer, the individual can control its direction and make headway against the wind.

Flight initiation distance Distance separating a prey and an approaching predator when the prey begins to flee; synonyms: approach distance, flight distance, flush distance.

Flight zone It is the animal's personal space. The size of the flight zone is determined by how wild or tame the animal is. Animals that are trained to lead have no flight zone.

Fluctuating asymmetry Difference between the values of bilateral symmetric traits of the same individual which can be of either sign with respect to the body axis and is assumedly the product of problems during development.

Fluffing The act of shaking and loosening the feathers.

Fluorescence resonance energy transfer (FRET) also known as Forster resonance energy transfer A phenomenon in which nonradioactive transfer of energy occurs between donor and acceptor molecules when the two are in close proximity. The most common donor and acceptor pair used in molecular biology are CFP and YFP, respectively. When FRET occurs, CFP transfers its excited energy to YFP. As a result, YFP is observed instead of CFP fluorescence emission. An example of FRET application in neurobiology is using Cameleon, a genetically engineered protein, to detect temporal calcium activity inside a living cell.

Flyway A flyway is the entire range of a migratory bird species (or groups of related species or distinct populations of a single species) through which it moves on an annual basis from the breeding grounds to nonbreeding areas, including intermediate resting and feeding places as well as the area within which the birds migrate.

FM bats Bats that emit a brief pulse for echolocation where the frequencies of the emitted call sweep from high to low throughout the duration of the pulse.

Focal sampling Observational method in which an observer focuses on a single individual during a sampling period.

Follicle-stimulating hormone (FSH) Gonadotropin that supports spermatogenesis and oocyte development in the gonads; also responsible for production of the hormone, inhibin, by the gonad.

Food aversion learning A form of associative learning in which an animal associates sensory cues from a food with some deleterious consequence of eating that food and subsequently avoids the food.

Forced copulation   Contrasts with copulation that individuals seek or freely accept. Most investigators infer that copulation is forced when it is preceded by aggression or the threat of aggression, including 'violent restraint.'

Forward genetics   A phenotype-driven mutant screen.

Forward masking   Reduction of perceptual sensitivity over a given time interval following the perception of a specific stimulus.

Foundress/cofoundress   Foundresses are females that are initiating a nest, or living, on a newly established nest before the emergence of the first offspring. If more than one foundress is present in a nest, they are called cofoundresses.

Fourier analysis   A type of time series analysis that involves fitting a series of sine waves to data. The analysis identifies the amount of strength or power associated with a set of periods.

Fovea   Specifically, a depression in the center of the retina of many vertebrates, providing high-resolution vision. More generally, areas of high visual acuity in vertebrate retinas are called 'area centralis' or 'visual streak.'

Framework   A simplified conceptual structure used to solve complex problems.

Frass   The waste product from an animal's digestive tract expelled during defecation (also known as fecal material, or feces).

Free choice profiling   An experimental methodology in which observers have complete freedom to choose their own descriptive terms and apply them to the observed behavior of animal subjects.

Free-running rhythm   Free-running rhythm refers to fluctuations in physiological or behavioral responses, with a period of about 24h, that recur in the absence of environmental cues.

Freeze tolerance   The ability of an animal to survive freezing of tissues.

Freezing   Remaining motionless upon detection of a predator in hopes of avoiding detection by the predator either through cryptic morphology or habitat cover.

Frequency-dependent selection   Selection that varies depending on trait frequency in the population.

Frequency of sound   The number of cycles of vibration per second of a sound-producing object, expressed in Hz (Hertz, or cycles per second). A good set of human ears can detect frequencies of 20 Hz–20 kHz (a kHz is a kilohertz, or 1000 cycles per second). This physical property of sound is the primary determinant of our psychological experience of sound pitch.

Frequency modulation   Cyclic changes in the frequency composition of a sound over time. The process of modulation produces extra frequencies in the sound, called sidebands.

Frontal cortex   A brain region that (among other functions) plays a key role in long-term planning, executive decision-making, and impulse control.

Functional activity mapping   An analysis of the patterns of neural activity, or its correlates, during the performance of a behavior or in response to a stimulus.

Functional class   A class defined by a common (inherent) function of its members.

Fundamental frequency (f_0)   The lowest frequency component in a harmonic sound.

Future planning   The ability to imagine and preexperience specific personal scenarios that might occur in the future.

GABA–γ   Aminobutyric acid is the chief inhibitory neurotransmitter in the mammalian CNS. The binding of GABA to its receptors causes the opening of ion channels to allow the flow of either negatively charged chloride ions into the cell, or positively charged potassium ions out of the cell, to produce an inhibition of the cell. Receptors to GABA are found in both the central and peripheral nervous systems of several invertebrate phyla. Insect GABA receptors show some similarities with vertebrate GABA receptors.

Gametes   A cell that fuses with another gamete during fertilization.

Game theoretic models   These are mathematical calculations of an individual's success (fitness) in making choices when their choice depends on the choices of others.

Game theory   A mathematical technique for choosing the best strategy given the likely choice of others.

Ganglion   The CNS of insects and other invertebrates comprises a ganglion – a processing center ('brain') – for each body segment connected to the ganglia of adjacent segments by bundles of axons called 'connectives.'

Gap junctions   Specialized intercellular complexes that directly connect the cytoplasm of two cells. Gap junctions allow various molecules and ions to pass freely between cells. Between two neurons, gap junctions form electrical synapses.

Gasterosteidae   Latin name for the family of stickleback fish.

Gating neurons   A type of *command neuron* that must be active during the whole time while a behavior takes place. This term was coined in the study of leech swimming activation to distinguish these neurons from *trigger neurons*, a class of command neurons that is active only for a short time when a behavior begins.

Gene chip   A commercial microarray.

Gene flow   The transfer of alleles of genes from one population to another.

Gene regulation   Relating to the activation (expression) of genes, including both transcription and translation.

Genetically effective population size   The number of reproducing individuals in a randomly mating population; actual population size is usually larger than its genetically effective size owing to the presence of sexually immature or nonbreeding individuals.

Genetic complementarity   The potential for traits on both sides of an ecological interaction to respond evolutionarily to reciprocal selection.

Genetic diversity   The level of biodiversity within a species, in reference to its total existing number of genetic characteristics, which, importantly, provides the raw material for evolution and is critical for long-term sustainability of a population.

Genetic drift   Chance variations in gene frequencies that result from random sampling error.

Genetic monogamy   An exclusive mating relationship between a male and a female resulting in all offspring being genetically directly related to both partners.

Genetic polymorphism   A portion of the genome that is represented by numerous distinct versions in the population. The more polymorphic a given locus is, the greater the number of distinct versions that will exist in the population. Genetic polymorphisms are based on sequence variation at specific loci.

Genetic relatedness   The fraction of genes identical by descent between two individuals. Only the fraction of genes shared above background count. See piece on relatedness.

Genetic structure   The array of alleles and genotype combinations in a population.

Genetic subdivision   Reduced gene flow between populations allows them to differ in the presence and/or frequency of alleles as a result of random genetic drift or natural selection.

Genetic task specialization   Task threshold is genetically influenced. Workers of particular parentage are more likely to engage in particular tasks.

Genic selection   Selection within individual bodies between alleles at a locus.

Genomic imprinting   Form of inheritance in which the expression of a gene depends upon the parent from which the gene is inherited. Because imprinting allows genes to be silenced when inherited from one sex and not the other, it provides a potential mechanism for achieving sex-specific expression. The imprint alters the chemical structure and hence the expression of the gene, but not its nucleotide sequence. Thus, the imprint can be erased and an active gene can be passed down in the next generation.

Genomic library   A collection of fragments of genomic DNA that have been inserted into host cells, typically bacteria or viruses, so that the individual fragments can be replicated in high numbers.

Genotype   The genetic constitution of an organism or one of the loci within that organism.

Geocentric cue   A cue based on information external to the organism.

Geographic mosaic   Ecological interactions vary across space because of the specifics of biotic and abiotic local environments, leading to a spatial mosaic of coevolutionary intensity. Hotspots, where reciprocal selection is strong, and coldspots, where reciprocal selection is weak or absent, characterize the geographic mosaic.

Geolocator   A daylight-level recorder affixed to an animal at capture that can be recovered at recapture up to one year later to estimate the latitude and longitude for each day the device was attached.

Geomagnetic field   Magnetic field associated with the Earth. It is essentially dipolar (it has two poles), the northern and southern magnetic poles on the Earth's surface. Away from the surface, the field becomes distorted.

Geophagy   The ingestion of soil particles which can reduce the potency of ingested toxins.

Geotaxis   Directed movement with respect to Earth's gravitational field. Movement away from Earth is 'negative,' movement toward Earth is 'positive.'

Germinal vesicle breakdown   Dissolution of the nuclear membrane that signals continuation of meiosis.

Ghost experiment   An experiment in which the model who would normally produce some effect in the world is absent, the effect being produced instead by surreptitious ('ghostly') means, such as pulling fine fishing line, allowing a test of how much an observer will learn from this component of the display alone.

Gill operculum   The hard flaps covering the gills of a fish.

Gilliam's rule   The prediction that animals favor using patches that minimize the ratio of predation risk to either expected growth or foraging rates.

Giving-up density and time (GUD and GUT)   Giving-up density is the amount of food or prey items still remaining in the patch, when a forager leaves it. Giving-up time is the length of time a forager will go without encountering a food item before it leaves a patch. Both are important metrics for testing predictions of the marginal value theorem.

Glossopharyngeal nerve   The ninth (IX) of twelve pairs of cranial nerves. It exits the brainstem from the medulla, just rostral (closer to the nose) to the vagus nerve. The glossopharyngeal nerve is mostly sensory and is involved in tasting, swallowing, and salivary secretions.

GLU   Glutamic acid (glutamate) is the most common excitatory neurotransmitter in the mammalian brain. Receptors to GLU are found in both the central and peripheral nervous systems of several invertebrate phyla. Insect GLU receptors show some similarities with vertebrate GLU receptors.

Glucocorticoids (Glucocorticosteroids)   (1) A class of steroid hormones released from the adrenal gland, particularly in response to stress; these include cortisol and corticosterone; (2) a class of synthetic steroid hormones; these include prednisone, dexamethasone and triamcinolone.

G-matrix   A square and symmetrical matrix in which the main diagonal consists of the additive genetic variance for a series of traits, and the other elements are additive genetic covariances between pairs of traits. Additive genetic variances have values between 0 and $+1$, whereas additive genetic covariances can range between -1 and $+1$.

Gonadotropin-releasing hormone (GnRH)   One of several neuropeptides synthesized in the brain.

Gonadotropins   Peptide hormones released from the pituitary in response to gonadotropin-releasing hormone from the brain; they stimulate growth of the gonads and synthesis of gonadal steroids.

Gonochorism   A sexual pattern in which individuals mature as one sex and remain that sex.

Good genes hypotheses   Refer, collectively, to explanations of mate preferences based on information or cues about the genes in potential mates. Good genes hypotheses can refer to complementarity (dissimilarity), relative individual heterozygosity, or to traits that indicate the possession of particular genes.

Granivorous   A diet of mostly seeds.

Gravid   Ready to lay eggs, for example carrying ovulated eggs in the ovarian lumen or oviduct.

Green-beard gene   A gene that affects copies of itself via three effects: production of trait, recognition of the trait in others, and differential treatment based on that trait. Sometimes not considered as part of kin selection because benefits go not to relatives but to actual bearers of the gene.

Green leaf volatiles   A suite of chemicals released from many plants upon mechanical damage.

Gregarious   Tending to aggregate actively into groups or clusters.

Gregarization   Density-dependent behavioral phase change in locusts from mutual repulsion to attraction and aggregation.

Ground-reaction forces   The forces that are developed as an animal or robot walks by pushing against a substrate (positive) or absorb momentum (negative, braking forces).

Group foraging   The searching, handling, and consumption of food by animals in close spatial proximity, whether or not there are social interactions between them.

Group memory   Information that is stored in the properties of an entire group, rather than encoded in the nervous system of an individual animal. The distribution of honeybee waggle dancers across food sources, for example, encodes the colony's ranking of the value of these sources.

Group selection   Selection between assemblages of individuals.

Group size effect   The phenomenon that individual vigilance declines as group size increases. This is most often explained by individual adjustments to a reduced perceived predation risk.

Gustation   Sense of taste.

Gustatory receptor protein (GR)   7-transmembrane protein located on the dendrite membrane of a gustatory neuron; detects and binds specific chemicals such as sugars or minerals.

Gustatory receptor   Sensillum that houses gustatory neurons; usually a tip pore *sensillum trichodeum*.

Gymnotiform   Electric knifefish of the New World order Gymnotiformes comprising five families. All gymnotiforms are electrogenic. 'Gymnotid' refers to members of the family 'Gymnotidae' including the weakly electric genus *Gymnotus* and the strongly electric *Electrophorus* (electric eel).

Gyne   Gynes are young females who have the potential to become egg-laying foundresses.

Habituation   Often considered the most basic form of learning that is defined as a response decrease in the presence of repeated stimulation.

Hamilton's rule   Named after W.D. (Bill) Hamilton, it is an inequality ($rb-c > 0$) that predicts when a trait is favored by kin selection, where c is the fitness cost to the actor of performing the behavior, b is the benefit to the individual to which the behavior is directed, and r is a measure of the genetic relatedness between those individuals. An altruistic act by definition has positive c and positive b and so is more likely to be favored by natural selection when r is high, and requires r to be positive. A selfish act, such as cannibalizing a member of the same species, has negative c and negative b and so is more likely to be favored by natural selection when r is low, and especially when r is zero.

Handicap   A trait whose expression incurs a cost, such that the degree of trait expression reflects the quality or condition of the bearer, in the sense than only an individual of high quality or condition can afford the cost of expressing the trait. Handicaps are one type of indicator mechanism and comprise a subset of the various indirect benefit hypotheses for the evolution of sexual dimorphisms via mate choice.

Handicap principle   A hypothesis to explain honest signaling that proposes that reliable signals must be costly to the signaler in a manner that an individual with less of that trait could not afford.

Haplodiploidy   A genetic system in which females come from fertilized eggs and are diploid, while males come from unfertilized eggs and are haploid.

Haplometrosis   The founding of a eusocial insect colony by a single queen.

Haplotype   A set of alleles of closely linked loci that are usually inherited together.

Harassment   Occurs when males attempt repeatedly to copulate and in so doing impose costs on females that supposedly induce females to submit to copulation attempts.

Harderian gland   A gland found within the eye's orbit, which occurs in vertebrates that possess a nictitating membrane. In some animals, it secretes fluid that lubricates movement of the nictitating membrane.

Hardy–Weinberg law   The foundation of population genetics; the law shows that in the absence of evolutionary forces genotype and allele frequencies are stable and related to each other algebraically.

Harmonic   An integer multiple of the fundamental frequency of a sound (e.g., $2f$, $3f$, $4f$)

Harmonic sound   A complex sound consisting of multiple frequencies (sine waves), all in integer relation with each other.

Hawk–dove game   A game theory analysis of alternate strategies hawk (attack immediately) and dove (display and retreat if attacked).

Helpers/helpers-at-the-nest   Individuals, especially birds that provide care for conspecific young that are not their own offspring.

Hematophagy   The habit of feeding on blood.

Hemimetabolous   Having no pupal stage in the transition from larva to adult.

Hemoglobin   Oxygen-carrying component of red blood cells.

Hemolymph   The circulatory fluid of insects and other invertebrates, comparable to vertebrate blood.

Herbicides   Chemicals produced to kill plants/weeds; generally used in no-till agricultural operations where the previous planting and weeds are not removed prior to seeding the new crop.

Heritability   A measure of the proportion of phenotypic variation that is due to genetic variation in a population.

Hermaphroditism   A condition in which individuals have gonads of both sexes (testes and ovaries) either simultaneously or sequentially.

Heterochrony hypothesis   Proposes that an early step in the evolution of eusociality is based on simple evolutionary modification of the timing of expression of maternal care behaviors, from postreproductively towards offspring, to prereproductively, towards sibs (see reproductive groundplan hypothesis).

Heterospecific   An individual of a different species.

Heterozygosity   The proportion of genetic loci in an organism that have different alleles.

Heuristic   Is a 'rule of thumb,' educated guess or a general way to solve a problem. Often used to describe a method that rapidly leads to a solution that is good in most situations. In phylogenetics, heuristic procedures are common because exact solutions are either mathematically impossible or nearly so.

Hibernation   Dormancy during the winter. The seasonal occurrence of profound physiological changes that include strongly reduced basal rates of metabolism, heartbeat, and respiration.

Hidden Markov model   Extension of the Markov chain concept to the modeling of nonhomogeneous data. The model combines a hidden variable driven by a Markov chain and an observed variable. A different distribution of the visible variable is associated with each possible value of the hidden variable.

Hiding time   Latency between entering and emerging from refuge; synonym: emergence time.

Higher-order conditioning   This Pavlovian learning process has two phases. First-order conditioning results in the conditioned stimulus predicting the occurrence of the provocative unconditioned stimulus. Second-order conditioning involves exposing the animal to the first conditioned stimulus, which has now acquired provocative properties, in temporal association with a second, emotionally neutral conditioned stimulus. The second conditioned stimulus then becomes a predictor of both the first conditioned stimulus and the unconditioned stimulus (not employed in the second-order association) and acquires its emotionally provocative properties at even a lower level of intensity.

Highly eusocial   Eusocial society in which there are developmentally distinct specializations where some individuals are specialized for reproduction, and others have developmental differences that preclude mating and make them totally/effectively sterile under normal circumstances.

High-speed video   Allows very high time resolution for analyzing fast behaviors by using high frame rates (commonly 500–2000 frames per second); frame rate for normal video is 30 frames per second.

Hippocampus   A brain region that (among other functions), plays a critical role in learning and memory, especially spatial learning.

Historical contingency   Evolutionary changes in a characteristic are dependent on what is inherited from evolutionary ancestors and the extent to which a characteristic diverges from that historic phenotype in response to selection.

Holarctic   The northern continents of the world.

Holometabolous   Insects that undergo complete metamorphosis involving four life stages: egg, larva, pupa, and adult.

Homeostasis   The ability of or tendency for an organism or a cell to maintain ideal internal equilibrium by adjusting its physiological processes.

Homeostatic sleep regulation   A sleep regulatory mechanism that aims to keep sleep amounts unchanged over a certain period of time; for example, after an overnight sleep loss, the activation of homeostatic sleep-promoting mechanisms induces sleepiness and compensatory increases in sleep next day.

Home range   The geographic space that an individual or group utilizes over the course of a year or longer.

Homing   The ability of an animal to return to its specific territory, or home range.

Hominization   The process of human evolution. Humans (*Homo sapiens*) are a member of Hominoids, that is a group of primates, which include humans, chimpanzees, gorillas, orangutans, and gibbons.

Homolog   A gene that shares ancestry, and hence DNA sequence composition with a gene from another species.

Homology   Biological similarity due to ancestry. For example, bat wings and mammalian forelegs are homologous.

Homoplasy   Biological similarity not due to ancestry, such as convergence or parallelism, for example bat wings and insect wings.

Honest signal   A structure or behavior that conveys reliable information to a receiver.

Horizontal social influence   Social influence on behavior that occurs within a generational cohort; for example, among juveniles.

Hormone   A chemical signal produced by one gland or tissue in the body that influences the physiology of a remote tissue.

Host   An organism harboring another parasitic organism that provides nourishment and shelter for the developing parasite.

Host plant   A plant species naturally used by a herbivore for its life activities.

Host range   The suite of host plant species used by a herbivore.

Host record   Documentation from field observation that a particular herbivore naturally uses a particular plant as a host.

Host shift   An evolutionary change by a herbivore lineage from using one host plant to using another; implies the abandonment of the ancestral host.

HPG axis   Hypothalamo–pituitary–gonadal axis.

Hybridization   Nucleic acid hybridization, the annealing, or binding, of two complementary, single-stranded, nucleic acid molecules.

Hybrid vigor   The tendency of a crossbred individual to show qualities superior to those of both parents.

Hydrozoan   A class of cnidarians that includes colonial polyps such as *Hydractinia*, individual polyps such as *Hydra*, and a diverse array of jellyfish with complex life cycles that include an attached polypoid and swimming medusoid phase.

Hyperosmolarity   An abnormally high osmolarity. The osmotic concentration of a solution, normally expressed as osmoles of solute per liter of solution.

Hyperparasitoids   A type of parasitoid that uses other parasitoids as host insects (also known as secondary parasitoids).

Hyperphagia   Seasonal occurrence of excess eating to build up fat reserves.

Hyperpolarization   A change in a nerve cell's membrane potential that makes it more negative.

Hypertrophy   Growth and enlargement of tissues and organs without cell division.

Hypokinesia   Abnormally slow or diminished movement of an animal.

Hypophysectomy   Removal of the pituitary gland.

Hypothalamus   A small region in the forebrain, containing various substructures (nuclei, including the arcuate nucleus) that collectively play a role in hunger, satiety, thirst, temperature regulation, hormone release, autonomic control, and circadian rhythms.

Hypothetico-deductive method   Hypothesis testing in which a scientific hypothesis could be falsified by a test of a prediction of that hypothesis.

Hypoxia   The presence of a low oxygen environment.

Hysteresis   The dependence of a physical system's performance on its history, apparent in some emergent collective properties of animal groups. For example, the ability of a group of ants to form a pheromone recruitment trail may depend on whether it reached its current size by growth from a smaller size or reduction from a larger one.

Hysteria Uncontrollable and potentially violent episodes of extreme nervousness.

Ideal despotic distribution Expected spatial distribution of organisms that have perfect information on the relative quality of all available habitats and current residents of habitats can exclude others from entering.

Ideal free distribution (IFD) Expected spatial distribution of organisms that have perfect information on the relative quality of all available habitats and can move freely among these habitats.

Idiobiont A parasitoid life-history strategy where host development is arrested upon parasitism. Idiobiont parasitoids are typically ectoparasitoids that attack host eggs or pupae.

Imitation The reproduction of the form of a behavior produced by another animal.

Immediate early genes The first genes transcribed in a cell during a response to a stimulus; their protein products regulate the transcription of other genes.

Immigration The arrival of new individuals from elsewhere.

Immunocompetence The ability of the body to produce a normal immune response (i.e., antibody production and/or cell-mediated immunity) following exposure to an antigen, which might be an actual virus itself or an immunization shot. Immunocompetence is the opposite of immunodeficiency or immuno-incompetent or immuno-compromised.

Imposex A form of sexual abnormality in gastropods where male sex organs such as the penis and vas deferens develop in ('imposed upon') a genetic female as a result of exposure to organotin.

Impulsivity A preference for the less delayed outcome.

In situ hybridization A process in which labeled DNA or RNA probes are used to localize specific DNA or RNA sequences in sections of tissue.

In vitro Literally, 'in glass,' meaning a reaction, process, or experiment in a metaphorical test tube rather than in a living organism. As opposed to in vivo: Literally, 'in life,' meaning a reaction, process, or experiment in a living organism.

Inadvertent social information Information generated as a by-product of the behavior of other individuals.

Inbreeding Mating among close relatives.

Inclusive fitness Calculated from an individual's own reproductive success plus his/her effects on the reproductive success of his/her relatives, each one weighted by the appropriate coefficient of relatedness.

Inclusive fitness theory A synonym of kin selection theory emphasizing inclusive fitness.

Independence from irrelevant alternatives Principle of rational choice behavior. It describes the expectation that preference between a pair of options should be independent of the presence of inferior alternatives.

Independent founding Initiation of a new colony by reproductives without the help of workers.

Independent variable A variable that is presumed to affect or control the value of a dependent variable.

Indeterminate growth Growth that is not terminated in contrast to determinate growth that stops once a genetically predetermined structure has completely formed.

Index A signal whose reliability is maintained due to some physical constraint on their performance.

Indicator models A subset of indirect benefit hypotheses proposing that extravagant traits evolve via mate choice because their expression indicates the quality or condition of the bearer, which is assumed to be heritable. A handicap is an example of an indicator mechanism.

Indifference point A set of options between which agents are indifferent; that is, in preference tasks, they choose the options equally.

Indirect benefits Genetic benefits of mate choice that accrue indirectly to the choosing individual in the form of improved genetic quality of its offspring.

Indirect environmental maternal effect Indirect environmental effects occur when the mother's environment influences her own and in turn her offsprings' phenotype. With regard to hormone-mediated maternal effects, differences in the environment the mothers live in result in differences in hormonal signaling to the offspring.

Indirect fitness Indirect fitness is one component of inclusive fitness. The effects of an individual on the fitness of other individuals weighted by their genetic relatedness.

Indirect genetic maternal effect Indirect genetic maternal effects are influences on offspring phenotype due to differences in the genetic background of mothers. With regard to hormone-mediated maternal effects, genetic differences between mothers would result in, for example, the expression of certain genes that regulate hormone secretion.

Indirect reciprocity An observer C witnesses an altruistic act by A toward B, and as a result, cooperates with A in the future.

Individual comparison Direct comparison of two or more options by a single animal, allowing it to determine which option is best. Individual comparison is not necessary for a collective decision, which can emerge from interactions among individuals who have each assessed only some of the available options.

Individual- or agent-based models   Computer simulations which can be used to describe and predict the global (group or population) consequences of the local interactions of individuals.

Individual recognition   The ability to learn the phenotypes of other individuals in a population and to use that information to shape individual-specific behavioral responses during interactions.

Induced ovulation   Occurs when ovulation is tied directly to copulation or some other stimulus associated with copulation. It may have evolved as a guard against forced or coerced copulation.

Inducible defenses   Defenses that occur only when predators are present.

Induction of preference   When past experience with a plant increases the degree of preference for that plant relative to others.

Inequity aversion   An aversion to unequal distributions of resources.

Infanticide   Killing a young, relatively defenseless, member of the same species.

Infectious coryza   Acute or subacute bacterial respiratory infection in chicken, pheasant, and guinea fowl caused by *Avibacterium paragallinarum*.

Inference   (In the field of *logic*) The act of passing from one proposition, statement, or judgment considered as true to another the truth of which is believed to follow from that of the former.

Inference by exclusion   Choice of an undefined stimulus (i.e., a stimulus that does not already have a learned association with a category) over a defined one (i.e., a stimulus that is already associated) by excluding (logically rejecting) the latter, which leads to the emergence of an untrained association between the undefined stimulus and the category.

Inferential reasoning   The ability to associate a visible and an imagined event.

Inferior colliculus   The midbrain auditory nucleus where the projections from most lower centers converge and are integrated. The inferior colliculus is the nexus of the auditory system.

Infinitesimal model   A genetic model in which it is assumed that traits are determined by a large (infinite) number of loci, each with a very small (infinitesimal) effect.

Inflorescence   A group or cluster of flowers arranged on a stem.

Information   Data that, when acquired, reduces an animal's uncertainty about environmental or social conditions. A quantity in the mathematical theory of communication expressed in bits.

Information sharing   A foraging system in which all group members are instantly informed of each other's food discoveries as they search for their own food.

Information transferred   The average variety conveyed by a communicative act, less than or (commonly) equal to the entropy.

Initial phase   The first sexual phenotype seen in many protogynous species, often characterized by relatively drab colors and relatively low displays of aggression and courtship behavior.

Inka cell   Endocrine cells near the insect spiracles that secrete pre-ecdysis-triggering hormone and ecdysis-triggering hormone.

Innate behavior   A behavior that is not learnt, but inherited.

Innovation (sensu process)   A process that introduces novel behavioral variants into a population's repertoire and results in new or modified learned behavior. The introduction of a novel behavior by social learning is not considered innovation.

Innovation (sensu product)   A new or modified learned behavior not previously found in the population.

Insectivorous   A diet of mainly insects.

Insemination   Occurs when males ejaculate inside the copulatory organ of a female.

Insight   The view that problem solving occurs by sudden recognition of a solution, or 'ah-ha' experience, rather than by trial-and-error learning. It is characterized by a sudden shift in behavior with a smooth and error-free transformation, a shift before the reward is obtained, long-term retention, transfer to other, similar problems, and to appear based on a perceptual restructuring of the problem.

Instantaneous (point) sampling   Observational method in which an observer records behavior of an individual at preset intervals.

Instar   The growth stage between two successive molts.

Insulin resistance   A state in which fat cells and muscle become insensitive to insulin's signal to take up glucose from the circulation, thereby producing high blood glucose levels (hyperglycemia). This is often seen in obesity and can be a precursor to diabetes.

Integument   All components of the outer layer of an organism – includes skin, hair, feathers, scales, nails, horns, wattles, warts, etc.

Interaural time difference When sound comes from one side of the body, it reaches one ear before the other. This creates an interaural time difference (ITD) which is used to localize sound in the horizontal plane. When the ITD is zero, the source appears at the midpoint between the ears. When ITD is varied, the source shifts toward the ear at which the signal arrives earlier. ITDs depend upon head size and in some cases on an interaural canal. In general, animals with large heads have larger time differences available to them.

Interference A reversible decline in fitness with increasing competitor density.

Interleukin-1 This is one of the earliest described endogenous pyrogens or cytokines. IL-1 is also known as lymphocyte activating factor and mononuclear cell factor. IL-1 is actually composed of two distinct proteins, IL-1α and IL-1β.

Intermediate host A host which is used by a parasite during its life cycle, in which it may multiply asexually but not sexually.

Internal validity Suitability of the study design to answer the question. The extent to which an effect seen in a study can be attributed to a specific cause.

Interneuron A neuron which connects neurons to other neurons in neural circuitries and whose cell body lies in the CNS.

Interobserver reliability The extent to which two or more observers consistently score behavior in the same way.

Interommatidial angle The angle between the viewing directions of two neighboring ommatidia in compound eyes.

Intersex An individual carrying the sexual characteristics of both sexes.

Intersexual selection Selection arising from variance in mating success due to interactions between males and females, such as female preference for males with a particular trait or resource.

Interspecific competition Competition between individuals of two different species.

Intertemporal choice A choice between outcomes that yield benefits at different points in time.

Intimidation A type of male aggressive response to females' refusals to mate. It may increase the likelihood that a female will mate with a male in the future.

Intracellular fluid One of the major fluid compartments of the body comprising all fluid within cells.

Intracerebroventricular administration Injection of a substance into one of the cerebral ventricles. Drugs and hormones injected this route have a relatively direct access to the brain tissue.

Intraguild predation An interaction in which predator and prey compete for basal resources (e.g., top predators eating mesopredators as well as smaller prey eaten by mesopredators).

Intralocus sexual conflict A form of genomic conflict that occurs when males and females differ in their fitness optima for a shared trait that is coded by the same locus or set of loci. Intralocus sexual conflict arises from intrasexual genetic correlations that constrain sex-specific expression of the shared trait and it is resolved by the evolution of sex-linked inheritance or sex-limited gene expression and the subsequent evolution of sexual dimorphism.

Intraobserver reliability The extent to which an observer consistently scores behavior in the same way at successful time intervals.

Intrasexual selection An evolutionary process that favors traits which improve an individual's competitive ability against members of the same sex for access to mates. Selection arising from variance in mating success due to competitive interactions within one sex, such as male–male combat or territory defense for access to females.

Intrinsic isolation Low fitness of hybrids because of genetic incompatibilities.

Intrinsic markers Genetic material, stable isotopes, or other markers that are carried within the animal itself and require only a single capture to yield data.

Intromittant organs Male copulatory organs, which deposit sperm and other seminal fluids into the female reproductive tracts. In mammals, a very few birds (only 3% of species), lizards, and snakes, males have an intromittant organ called 'a penis.' In insects, a male's intromittant organ is called 'an eadeagus.'

Introspection Self-observation based on private mental processes; often thought to be limited to consideration of one's own conscious thoughts, feelings, and perceptions.

Invariant feature A feature (quantity or property or function) that remains unchanged under a transformation.

Invasion In migration studies, the same as irruption. More generally, the colonization of an area by a species formerly absent there.

Invasive species A nonnative species that spreads rapidly once established, with the potential to cause economic or environmental harm.

Inverse square law A mathematical formula describing the attenuation of sound as it propagates through an ideal environment. By the inverse square law, sound amplitude decreases by 6 dB per doubling of distance.

Ionospheric circulation Large-scale convection in the inner magnetosphere and the conjugated ionosphere.

Irruption   In migration studies, a massive immigration to a particular region. More generally, a form of migration in which the proportions of individuals that participate, and the distances they travel, vary greatly from year to year.

Isodar   The set of points on a plot of density of one species in different habitats at which the fitness payoffs for choosing between habitats are equal.

Isogamy   Refers to eggs and sperm that are similar, or approximately more similar, in size than is usually the case.

Isoleg   The set of points on a plot of densities of different species at which the fitness payoffs for using both (or multiple) habitats and using only one habitat are equal.

Isolume   A level of constant light intensity in the water column that is commonly represented as a line of points on a plot.

Iterated game   Contestants play a game such as the prisoner's dilemma many times, thus allowing a strategy to be contingent on past moves.

Iteroparity   The repeated or iterated cycles of reproduction, production of young, throughout the life cycle of an organism before it succumbs.

Jack   In salmon, a male that matures precociously and typically does not spend any time at sea; jacks are typically much smaller.

Juvenile hormone   A sesquiterpenoid insect hormone known to regulate many functions across insect taxa, including larval development, reproduction, and behavior.

Kairomone   Chemical signal molecule that is produced by one species and perceived by another species, resulting in altered physiology or behavior in the species perceiving the cue that benefits that species.

Kappa coefficient   An index of concordance that measures agreement between two observers in behavioral observation, taking into account the probability of agreement by chance alone.

Kendall's coefficient of concordance   A nonparametric method for measuring agreement among more than two observers in behavioral observation.

Kinematics   The characterization of a behavior in terms of the movements of the body. Most commonly, such studies involve a frame-by-frame analysis of films or videotapes of the behavior. Kinematic studies are often carried out to determine which muscles produce the movements underlying a behavior, so they are often accompanied by recording the tension or electrical activity generated by active muscles.

Kinesis   Behavior in which the organism does not move in a particular direction with reference to a stimulus but instead simply moves at an increasing or decreasing rate, or rate or turning, until it ends up farther from or closer to the object. (Contrast with taxis.)

Kinocilium   A special structure on the apex of hair cells located in the sensory epithelium of various vertebrate sensory receptors including electroreceptors.

Kin recognition   The ability to discriminate kin from nonkin, or the ability to make discriminations among kin based on degree of relatedness.

Kin selection   The process of selection as it acts through effects on relatives. Sometimes viewed as co-extensive with inclusive fitness, but sometimes viewed as excluding green-beard effects. See Hamilton's Rule.

Kleptoparasitic spiders   Spiders that live in webs of other species and steal prey from the host.

Koinobiont   A parasitoid life-history strategy where hosts continue to grow and develop after parasitism.

Labellum   Bottom part of the proboscis in flies, equipped with fine grooves to assist ingestion of liquid food.

Lag-sequential analysis   Method used for the identification of the most likely sequences of successive events appearing in a time series.

Lairage   European term for the stockyards that hold animals at a slaughter plant.

Larviposition   The act of depositing living larvae instead of eggs.

Larynx   A musculoskeletal structure that functions as a vocal organ among amphibians, reptiles, and mammals.

Laser ablation   In biology, a process of killing cells by irradiating them with a laser beam.

Latency   The amount of time until a behavior occurs. The delay between the onset of the stimulus and the beginning of the response (neural or behavioral).

Leaf swallowing   The slow and deliberate swallowing, one at a time without chewing, of whole leaves that are folded between tongue and palate, and pass through the gastrointestinal tract visibly unchanged. The behavior is known to occur in apes, some monkey species, other mammals and some birds.

Leapfrog migration   Where northern wintering populations breed in the southern portions of the breeding range and southern wintering populations breed in the northern parts of the range.

Leghorn   Breed of egg-type chickens that produce white-shelled eggs; named after the city of Leghorn, Italy, where they are considered to have originated; leghorns have provided the genetic foundation of most modern egg-type chicken strains.

Leishmaniasis   Caused by protozoan parasites in the genus *Leishmania* that are transmitted by sandflies. They can affect the skin, mucus membranes, or internal organs.

Lek Is an aggregation or cluster of male territories into arenas used for attracting, courting, and mating with females. Males that form leks provide only sperm and no other resource to the females. No lasting bonds are formed and males do not engage in any parental care.

Lek paradox The persistence of strong directional selection for exaggerated sexual ornaments or display despite the apparent lack of benefit for such choice, particularly in lek mating systems.

Lek polygyny A mating system in which individual males mate with multiple females during a breeding season and in which males aggregate at small, nonresource-containing display sites to attract females.

Leptin A type I cytokine secreted by fat cells that regulates food intake. Leptin acts on the brain to signal when the body has sufficient energy stores, thus inhibiting appetite (i.e., it is an 'adipostat'). However, leptin and its receptor are widely expressed, suggesting that leptin is much more than an 'adipostat,' and likely plays diverse roles in animal development.

Levels of organization A complex behavioral system can be broken into a hierarchy of components or networks based on their physical size and functional complexity. Causal influences operate in both a top-down and a bottom-up fashion with one-way causation characterizing the simplest interactions and two-way causation operating across multiple levels. The lowest level of organization for predator recognition is sensory input from the environment, followed by the processing of predator features in a down-stream hierarchical integration of predator features, yielding higher-order predator recognition and mediation of antipredator behavior.

Lexical syntax Structured rules for ordering semantically meaningful sound units such that their ordering carries additional meaning beyond that reflected in the units alone.

Life cycle of chemicals The passage of a compound through the environment beginning with the source of production and release; consideration of the physical/chemical properties and the migration of the chemical in various media including soil, water, and air including the production of metabolites and their activity in living systems.

Life cycle of organisms Consideration of all stages in the life of an individual with ontogeny, maturation, adult, and aging including reproductive strategy and lifespan as well as unique species characteristics.

Life-for-life relatedness Relatedness including a concept of relative sex-specific reproductive value. Life-for-life relatedness = Regression relatedness × (sex specific reproductive value of the recipient/sex-specific reproductive value of the actor).

Life history Characteristics of the growth and development of an organism, such as its length and timing of gestation, maternal dependency, sexual maturity, reproductive period, and lifespan.

Life history stage (LHS) A syndrome of morphological, physiological, and behavioral traits associated with a specific process (e.g., reproduction, nonreproduction).

Life-history traits Features of the life cycle, with particular reference to survival and reproduction (e.g., age at first reproduction, fecundity, etc.)

Lignified When something has been made hard like wood as a result of the internal deposition of *lignin*, a substance related to cellulose that provides rigidity to plant cell walls.

Linear timing The hypothesis that psychological estimates of time are linearly related to physical time.

Linkage A phenomenon whereby two genes are spatially located close to each other on a chromosome so that crossing over rarely occurs between them during meiosis. Thus, two variants in the corresponding genes are said to be in 'linkage disequilibrium' when they tend to be coinherited. If one variant is in a gene that encodes a phenotype, the linked variant acts as a marker. This is the basis for linkage studies.

Lipophilic Having an affinity for, tending to combine with, or capable of, dissolving in lipids (fats).

Lipopolysaccharides (LPS) Large molecules consisting of a lipid and a polysaccharide that are found in the outer membrane of some bacteria. The molecules, referred to as endotoxins, cause the release of endogenous pyrogens that evoke a fever, resulting in sickness behavior in the animals exposed to them. An integral component of Gram-negative bacterial cell walls that induces an acute phase response in most vertebrates.

Local enhancement Attention drawn to the location where another animal is performing a response.

Local mate competition Theory that competition for mates is stronger between related males than between related females, reducing the relative value of males; thus, sex-ratio interests of queens and workers become more closely aligned and female-biased sex ratios are considered optimal for both parties.

Local resource enhancement hypothesis The idea that related females cooperate synergistically to enhance their joint reproduction, increasing the relative value of females; thus, sex ratios should be female-biased.

Locomotor system The way an animal moves from one location to another. In primates, locomotor systems include brachiation, vertical clinging and leaping, quadrupedality, knuckle-walking and bipedality.

Logistic A logistic function or logistic curve is the most common sigmoid curve. The initial stage is approximately exponential; then, as saturation begins, the rate of increase slows and approaches an asymptote.

Log-linear model   Model for the analysis of multiway contingency tables. The principle is to first consider all possible associations between a finite set of categorical variables, and then to remove nonsignificant associations.

Longitudinal   Correlational research study that involves repeated observations of the same items over long periods of time – often many decades.

Lophotrochozoa   A major subdivision of protostome animals that includes molluscs, annelids, bryozoans, brachiopods, and other less conspicuous animal phyla. The group is named for the presence fan-like feeding structures called 'lophophores' (in the bryozoans, phoronids, and brachiopods) and trochophore larval stages found in many of the group's members. The Lophotrochozoa can be contrasted with the other group of protostomes called the 'Ecdysozoa,' which includes arthropods, nematodes, and other animal phyla.

Lordosis   A female receptive behavior exhibited by many rodents and birds, highlighted by an immobile posture with arched back and raised rump and head.

Lumen   The space within the intestinal tube.

Luminance   An indicator of brightness.

Luteinizing hormone (LH)   Gonadotropin that stimulates gonadal production of steroid hormones and supports gamete production.

Lymphatic filariasis   A tropical parasitic disease caused by thread-like filarial nematode worms that live in the lymphatic system and cause lymphedema. The worms are transmitted by mosquitoes.

Lymphocyte   This type of white blood cell makes up 25–30% of white blood cells. Lymphocytes are concentrated in central lymphoid organs and tissues, such as the spleen, tonsils, and lymph nodes. Lymphocytes determine the specificity of the immune response to infectious microorganisms. The two broad categories of lymphocytes are the large granular lymphocytes and the small lymphocytes. Large, granular lymphocytes are known as the natural killer cells and the small lymphocytes are the T cells and B cells.

Macrocyst   The sexual, diploid stage of the *Dictyostelium* life cycle.

Macroevolution   Evolutionary change that is observed as differences between species, genera, or higher taxa.

Macronutrient   Those nutrients that are needed by the body in large amounts and potentially can be used as a source of energy (proteins, carbohydrates, and fats).

Macroparasite   A parasite that does not multiply inside its definitive host.

Macrophages   Literally meaning 'big eaters,' in actuality these are white blood cells dwelling within tissues that phagocytose or engulf cellular debris and bacteria.

Macrophytes   Aquatic vegetation with roots.

Magnetic compass   A compass that provides a direction bearing, or reference, based on the polarity or inclination of the Earth's magnetic field.

Magnetic inclination angle   The angle at which field lines of Earth's magnetic field intersect the surface of the Earth.

Magnetic intensity   The strength of a magnetic field.

Magnetic map   A map based on geographic variation in the Earth's magnetic field, which could be used to determine geographic position.

Magnetite (Fe_3O_4)   One of several types of biogenically produced iron oxides. Lustrous black, magnetic mineral, Fe_3O_4. It occurs in crystals of the cubic system. A cubic mineral and member of the *spinel* structure type.

Magnetoreception   The sensory detection and use of magnetic fields, particularly the Earth's magnetic field.

Magnetoreceptor   A sensory neuron that transduces magnetic stimuli into a neural (i.e., bioelectric) signal.

Major histocompatibility complex (MHC)   A genetic region (containing > 150 genes in humans) that plays an important role in autoimmunity and immune diversity in jawed vertebrates. MHC genes products mediate self/nonself recognition in vertebrate immune systems and are involved in tissue compatibility (histocompatibility).

Male harassment of females   A type of coercion that may not be immediately associated with copulation attempts.

Mandibular gland   A salivary gland on either side of the mouth, inside the lower jaw, that discharges saliva into the oral cavity.

Mantle   Soft extensions of the body wall that in many mollusks secrete a shell. It also forms a cavity that shelters the gills.

Marginal costs   The change in costs with a change in behavior. (In economics, marginal is synonymous with the derivative from calculus.)

Marginal value theorem (MVT)   A model within optimal foraging theory that predicts whether an animal should continue to exploit a given patch based on its current (marginal) value relative to the expected gain from moving to another patch.

Marker   A trait that signals a particular genotype.

Markov chain   Stochastic process in which the value taken by a random variable X at time t is explained by the values observed for the same variable at time t- 1 (first-order model) and possibly at times t- 2, t- 3, . . . (high-order model). Transition probabilities between different values are summarized as a transition matrix.

Mark-recapture method   A method commonly used to estimate population sizes which relies on recording individually distinctive traits or making individuals, and later using these traits or marks to recognize them in future encounters. In a closed population, the proportion of animals resighted in relation to newly encountered animals allows a calculation of population size. A set of methods for estimating one or more of abundance, survival, and recruitment by recording repeated sightings or captures of animals, some of which are identifiables from marks previously placed on them. Increasingly, natural marks are used, identified from photographs or DNA fingerprinting.

Masculinization A component of the sexual differentiation process during which the capacity to display male-typical behaviors is acquired or enhanced.

Mate A social associate and need not refer to an individual with which one copulates.

Mate assessment Results from the process of evaluating potential mates; mate assessment determines an individual's preference function.

Mate choice The decision made by an individual in selecting a partner for reproduction.

Mate-choice copying A form of nonindependent mate choice whereby an individual chooses the same mate that it previously observed being chosen by another individual.

Maternal effects Nongenetic influences of the mother's phenotype (including behavior) on an individual's phenotype, especially those with evolutionary consequences.

Maternal inheritance Maternal inheritance describes the maternal inheritance of DNA and is distinct from maternal effect.

Maternal rank inheritance The process by which juveniles (e.g., cercopithecine primates, spotted hyenas) attain positions in the dominance hierarchy adjacent to those of their mothers.

Mating system The demographic pattern of breeding individuals within a group.

Matrigene In diploids, the allele inherited from the mother.

Matriline Individuals of two or more generations that are descended from the same female.

Matrotrophic A form of gestation in which the developing offspring take nutrients directly from the mother's blood through specialized embryonic structures throughout the gestation period.

Maxillary palps Sensory structures on the outer surface of the maxillae used for detecting food.

Mechanisms of heredity Ways in which information physical or otherwise – are transferred between generations. Such mechanisms include genes, culture, learning, developmental systems, and epigenetics.

Mechanisms of sexual selection Include behavioral and physiological interactions between individuals, whether male–male, female–female, or male–female that result in within-sex variance in some component of fitness.

Medulla (medulla oblongata) The lower half of the brainstem. It contains the cardiac, respiratory, and vasomotor centers and deals with autonomic functions, such as breathing, heart rate, and blood pressure.

Melanin A pigment that underlies the rusty coloration of the ventral feathers of the barn swallow and many other birds. Melanins are produced endogenously rather than acquired through diet like carotenoid pigments that add red and orange colors to the feathers of many other birds like house finches.

Melanophores A pigment cell that contains melanin.

Melatonin Melatonin is a hormone secreted by the pineal gland. Plasma levels of melatonin are low in the day and high in the night. It enables an organism to detect changes in the seasons because its expression mimics changes in day length.

Memory The retention of information from prior experience.

Memory monitoring The process of tracking or evaluating the contents of one's own memory.

Mental representation The process of internalizing a referent (external stimulus) into specific mental content. The term can also be used to refer to the content itself.

Mental state An unobservable, internal or cognitive representation of 'things' in the world (e.g., the perception of objects), the actions or plans required to interact with those objects (e.g., intentions and desires), and information about those objects' current structure, location, properties, etc. (e.g., knowledge). In theory of mind research, David Premack suggested that there are three important categories of mental states: Perceptual, Motivational, and Informational.

Mental time travel The ability to travel backwards and forwards in the mind's eye in order to reminisce about the past and imagine future scenarios.

Mentalistic psychology An approach in which the scientist attempts to understand the mental life of the animal. One posits experiences in the animal mind that are similar, at least in some respects, to those of human experience.

Mesoconsumer Intermediate consumers (i.e., herbivores and mesopredators).

Mesocosm experiments Experiments that achieve highly controlled manipulations by working at small spatial scales (e.g., experimental plots are a few square meters or smaller and invertebrate consumers often comprise the highest trophic level). They often test general principles that potentially apply to large spatial scales where experimental tests are logistically more difficult (e.g., large vertebrates in vast landscapes).

Mesopelagic   Associated with the midwater oceanic zone between 200 and 1000 m depth, a zone characterized by dim light and a steep persistent thermocline.

Mesopredator   A carnivore occupying a mid-trophic level and at risk of predation from carnivores at higher trophic levels.

Message   A decodable collection of signals transmitted as a unit; also the meaning of a communication.

Metacognition   Thinking about thinking; the ability to reflect on or think about one's own thoughts, feelings, and knowledge.

Metamemory   Knowledge of the contents and function of one's own memory; memory monitoring.

Metamorphic climax   The final and most rapid phase of morphological change when thyroid activity is at its peak.

Metamorphosis   The change in form that occurs during the postembryonic lives of insects as they transition from early feeding stages to the adult reproductive stage.

Metapopulation   A group of semi-isolated populations that are linked through exchange of individuals such that the dynamics of each subpopulation are asynchronous. A series of populations connected by dispersal; the dynamics of metapopulations involve extinction and recolonization events.

Methylation   The addition of a methyl group to a molecule; in DNA methylation, methyl groups are attached to cytosine residues and can lead to changes in gene expression including gene silencing.

Microarray   A series of microscopic spots of DNA that are attached to a solid surface; the microarray is hybridized with cDNA or RNA in order to measure differences in gene expression between two samples.

Microevolution   Evolutionary change that takes place within a population. The direct or indirect genetic response to selection or drift.

Microparasite   A parasite that reproduces inside its host.

Microsatellites   Neutral segments of DNA consisting of repeating base pairs that show a high degree of intra- and inter-specific polymorphism.

Microspectrophotometry   Measurement of the spectral composition of light that is reflected or transmitted by materials, at a microscopic scale.

Microvilli   Microscopic cellular membranous protrusions that increase the surface area of epithelial cells and are involved in a wide variety of functions, including absorption and secretion.

Migration   A seasonal, usually two-way, movement from one habitat to another to avoid unfavorable climatic conditions and/or to seek more favorable energetic conditions.

Migration syndrome   The suite of coadapted morphological, physiological, and life-history traits that enable migration and that is underlain by a genetic complex that controls the development and expression of these traits.

Migratory connectivity   Geographic linking of populations between different periods of the annual cycle, including breeding, migration, and wintering.

Miracidium (plural: miracidia)   A small free-living larval stage of the Trematoda which swims using cilia and does not feed, relying on glycogen stores to enable it to find and infect the subsequent parasite host, often a mollusc.

Mirror neurons   Brain cells that react similarly during one's own motor actions as those observed in others.

Mitochondrial DNA   An abundant single-stranded circular DNA molecule occurring in mitochondria and containing a few genes; the control region where DNA replication begins has especially high mutation rates and is valuable for population genetics studies.

Müllerian mimicry   Mimicry of body coloration, body patterning, and/or behavior of a toxic prey species by a nontoxic, coexisting species.

Müllerian ring   A group of species that are Müllerian mimics and have converged on the same aposematic signal.

Mobbing   A coordinated effort by a group (three or more) of prey in response to a predatory attack in which the prey approach, observe, harass, attack, and sometimes injure or kill the predator before it is able to attack.

Modal action pattern   An innate, relatively invariant series of behaviors, common to all members of a species, that are dependent on an external signal (sign stimulus) to trigger the sequence. Originally termed 'fixed action pattern,' George Barlow argued that because the motor pattern is not performed identically each time it is elicited 'modal action pattern' would be a more appropriate term for a recognizable motor pattern elicited by a sign stimulus.

Modules   Phenotypic units, often occurring in a repeating series that develop more or less independently of each other, which come together to form a larger whole.

Molt   The process of shedding the outer covering of the body.

Molt cycles   Replacement of skin, hair, feathers, scales, etc. usually is cyclic and occurs during restricted periods called 'molts.' There may be one to several molt cycles each year depending on the species.

Monoamines Important neural signaling molecules characterized by having an amino group connected to an aromatic ring. Important examples include serotonin, dopamine, and norepinephrine.

Monocularly With one eye only.

Monocytes This type of white blood cell changes into a macrophage. While they make up only 3–8% of all white blood cells, monocytes have two important functions related to the immune system; one is to replenish resident tissue macrophages that get used up engulfing bacteria and cell debris and the other is to move quickly to new sites of infection where they then differentiate into a new population of macrophages.

Monodomy An ant colony that occupies a single nest.

Monogamy Mating system in which males and females mate with a single partner during a particular breeding season. This typically reduces variance in mating success in both sexes, thereby limiting the opportunity for sexual selection.

Monogynous (monogyne, monogyny) Colonies having one queen.

Monomorphism Individuals of a prey species which are invariant in a specific trait, for example, color pattern.

Monophagous versus oligophagous versus polyphagous Whether a herbivore uses one versus several versus many plant taxa as hosts.

Moon watching A technique for studying nocturnal migration by observing through a 20–30× telescope birds as they pass before the disc of the moon.

Morgan's Canon States that we should not attribute behavior to higher cognitive abilities if it can be explained in terms of simpler processes.

Mormyrid African, weakly electric fishes of the family Mormyridae.

Morph A discontinuous class of morphological variation.

Morphological caste A mechanism for division of labor in which individuals vary in physical attributes, particularly size, with corresponding differences in the tasks they perform.

Morphological computation The idea that the physical body of an animal, interacting with its environment, can function in a manner that removes the need for direct neural control in the production of adaptive behavior.

Morphology The form, structure, and configuration of an organism. This includes aspects of outward appearance such as coloration as well as the form and structure of internal parts such as bones and organs.

Mosaic evolution The ability of selective pressures to produce independent changes in brain regions.

Mosquito control Many states have programs to control mosquitos, either with chemical spray or by altering habitat, such as cutting ditches in salt marshes to drain them so there is no habitat for the mosquitos to breed.

Motion parallax Motion parallax is a monocular depth cue that results from motion of the object or observer. Closer objects move farther across the visual field than distant ones.

Motoneuron (or motor neuron) A neuron located in the CNS that project its axon outside the CNS to innervated and control muscles.

Motor imitation Performing an action after seeing another perform that action.

Mucopolysaccharide Class of polysaccharide molecules, also known as 'glycosaminoglycans,' composed of amino sugars chemically linked into repeating units that give a linear unbranched polymeric compound.

Mucosa The innermost layer of the gastrointestinal tract (gut) that surrounds the lumen, or space within the intestinal tube. This layer of epithelial cells, known as *enterocytes*, comes in direct contact with food, and is the primary site of nutrient absorption.

Multifunctional neurons Neurons, particularly interneurons, that are active in – and presumably contribute to – several different behaviors.

Multiharmonic A vocalization with a nearly constant pulse repetition rate or fundamental frequency, and several prominent harmonics.

Multilevel selection theory Also known as levels of selection theory. Describes how variation in fitness can be partitioned into selection at multiple levels (e.g., between and within groups) to provide insight into how selection affects phenotypic evolution. For example, for social evolution, the balance of selection between and within social groups explains the evolution of sociality (note this can equivalently be described in terms of inclusive fitness/ kin selection theory).

Multimale groups More or less permanent social groups containing multiple, reproductively active adults of each sex.

Multimodal signal Signals produced in multiple sensory modes or channels at the same time.

Multiple messages When the individual components of multimodal signals each convey distinct information.

Multivalued Having more than two values; communication codes having three or more alternative signals.

Multivariate More than one-variable quantity; communication codes having two or more signals making up a decodable unit.

Mutant screen Organisms are exposed to a mutagenic substance and the offspring of the mutagenized organisms are then screened for mutant phenotypes.

Mutation   Any change in DNA sequence, typically caused by errors during DNA replication.

Mutual benefit/mutualism   A behavior performed by the actor that contributes to the lifetime fitness benefits of both the actor and recipient (evolutionary biology); mutualism refers to interspecies cooperation (evolutionary biology); a behavior that produces immediate benefits for both actor and recipient (social science).

Mutual gaze   Eye-to-eye contact is an important characteristic of early mother–infant relationships. Mothers look into the eyes of their infants, while the infants look back into their mothers.' This is called 'mutual gaze.' It is a truly unique feature shared by humans and chimpanzees.

Mutualism   Intra- or interspecific social interactions in which both parties benefit.

Mycophagy   Feeding on fungi.

Myelination   The development of a myelin sheath around sensory or motor neurones. Myelination improves the conduction speed of nerve impulses, enabling fast reactions and skilled movements to occur.

Narrow-sense sexual selection   Variance in fitness due entirely to variation in number of mates. It is often associated with exaggerated traits in males.

Nash equilibrium   A combination of strategies for the players of a game in which each player's strategy is the best response (i.e., one that maximizes expected payoffs) to the other players' strategies. Equilibrium point in a game at which no player can improve its payoff by changing its tactic unilaterally.

Natal   Related to ones birthplace.

Natal dispersal   The movement of an individual from birthplace to breeding place.

Natal homing   Tendency for an animal to return to reproduce in the same geographic area where it began life.

Natriorexigenic   That which provokes salt intake.

Natural selection   Nonrandom differential preservation of traits across generations, leading to changes in the distribution of traits in a population over time.

Nature–nurture controversy   Controversy over the relative importance of genetic factors (nature) and the environment (nurture) in the development of behavior. This is now regarded as supplanted by an epigenetic approach to development.

Necessity and sufficiency   Criteria required to prove causation; for instance, if a behavior disappears when a specific neuron is killed, that neuron is necessary for the behavior; if stimulation of only that single neuron elicits the behavior, it is sufficient; necessity and sufficiency can occur separately.

Necrophagy (adj. necrophagous)   Eating dead and/or decaying insects.

Necrophoresis   Movement toward dead organisms.

Nectar corridor   A series of populations of flowering plants that permit nectar-feeding bats to migrate from one area to another.

Nectarivore   An animal that eats nectar produced by flowering plants.

Negative frequency-dependent selection   A type of selection that favors rare polymorphisms in the population. Under this type of selection, the fitness of a given locus is inversely proportional to its prevalence in the population.

Negative punishment   The removal of a desirable outcome, or the opportunity for reinforcement, coincident with a behavior such that the future probability of that behavior is decreased.

Negative reinforcement   Increasing the future probability of a behavior by the removal of, or a decrease in the intensity of, an aversive stimulus.

Neighborhoods   All the groups of individuals that live within one fragment of habitat, more likely to interact with each other than with individuals from other areas; technically called a deme of a population.

Nematocytes   The stinging cells of cnidarians. These cells contain organelles called nematocysts, among the most complex intracellular structures known in animals. Nematocysts serve a variety of functions including feeding, defense, and locomotion. Nematocytes are found only in the phylum Cnidaria, although a few other noncnidarian groups possess superficially similar cells.

Nematodes (or roundworms)   Phylum of worms with an unsegmented body. Abundant in marine and freshwater habitats, in soil, and as parasites of plants and animals. Nematode species are very difficult to distinguish; over 80 000 have been described, of which over 15 000 are parasitic.

Nematomorpha   Commonly known as 'Horsehair worms' or 'Gordian worms,' parasitic animals that are morphologically and ecologically similar to nematode worms, hence the name. They range in size from 1 cm to 1 meter long, and 1–3 mm in diameter. The adult worms are free living, but the larvae are parasitic on beetles, cockroaches, Orthoptera, and crustaceans. About 326 species are known and a conservative estimate suggests that there may be about 2000 species worldwide.

Neonatal smiling   Human newborns are known to smile spontaneously with their eyes closed, a behavior known as 'neonatal smiling.'

Neophilia   A form of nonassociative learning in which novel things become more acceptable.

Neophobia   Fear of novelty. A form of nonassociative learning in which novel things become less acceptable.

Neotenic reproductives   In termites, wingless reproductives that develop within the natal colony *via* a single molt from any instar after the third larval instar. At this neotenic molt, their gonads grow and they develop some imaginal characters while maintaining an otherwise larval appearance; some characters, like wing pads, may regress. Neotenic reproductives are characterized by the absence of wings and usually by the lack of compound eyes. The cuticle is less sclerotized than in primary reproductives. They are subdivided into: (i) *replacement reproductives* if they develop after the death of the same-sex reproductive of a colony or (ii) *supplementary reproductives* if they develop in addition to other same-sex reproductive(s) already present within a colony.

Neoteny   Persistence of juvenile characteristics into adulthood.

Neotropics   An ecozone that includes Central and South America, the Mexican lowlands, the Caribbean islands, and southern Florida.

Nepotism   The preferential treatment of relatives.

Nervous system maps   A physical organization of neurons that corresponds to locations in the external world; analogous to a road map that depicts the locations of the real roads; can be sensory as in the mapping of touch sensation onto a body representation in the primate cortex or can be motor.

Nest defense   Behavior by a parent that reduces the probability that a potential predator will hurt the parent's offspring; the parent may incur some cost of defense, including increased probability of injury or death.

Neural tracer   Any substance that, when injected into brain tissue, is taken up by one part of a neuron and is transported to another part and can be used, therefore, to determine connections among brain regions.

Neuroendocrine   General interactions between the nervous and endocrine systems; specific production of endocrine signaling molecules by neurons.

Neurohemal organ   The enlarged endings of neurosecretory neurons that serve as a distinct storage and release site.

Neurohormone   A hormone that is released into the blood from a neuron rather than from endocrine tissue.

Neuromast   Functional unit of the lateral line, consisting of a cluster of hair cells with surrounding support cells, and an overlying gelatinous mass called 'a cupula.'

Neuromodulator   A chemical messenger, typically a peptide, which is released from presynaptic terminals and acts on the postsynaptic membrane to modulate the responsiveness of the postsynaptic cells to the effects of the neurotransmitter.

Neuro-muscular junction   Synapse between the motor neuron terminals and the muscle. In vertebrates, the signal passes through the neuromuscular junction via the neurotransmitter acetylcholine. In invertebrates, the transmitter is Glutamate.

Neuropeptide   Peptides found in neural tissue acting as chemical signals to communicate information (such as endorphins, or some hormones like oxytocin and vasopressin).

Neurosteroids   Some regions of the brain, especially those involved in territorial aggression, express all the enzymes needed to synthesize sex steroids such as testosterone and estradiol-17beta de novo from cholesterol. These neurosteroids are thought to act locally on neurons associated with aggressive behavior.

Neurotoxin   A toxin that acts specifically on neurons usually but not exclusively by interacting with membrane proteins such as ion channels.

Neurotransmitter   Chemicals (monoamines, ions, gases, hormones) that relay and modulate signals between a presynaptic neuron and a postsynaptic cell.

New York epigeneticists   A group of animal psychologists that developed around T. C. Schneirla and was located primarily at the American Museum of Natural History and the Institute of Animal Behavior. They generally favored nurture over nature and a 'levels' view of evolution according to which only very limited generalizations can be made across well-defined taxonomic levels.

Niche conservatism   Closely related species tending to occupy similar environments.

Niche displacement   The removal of a species from its ecological and functional space in the environment. It is most often caused by a natural catastrophe, or by interspecific interactions like predation, competition, or mating interference.

Niche (ecological niche)   The features of the environment that characterize an organism's position in the ecosystem, such as diet, preferred habitat, location within the habitat, and activity pattern. The ecological role of a species in an ecosystem encompassing abiotic, biotic, and geographical dimensions.

Nocturnal   Active at night.

Nomenclature   As subdiscipline of taxonomy, it is the naming of taxa, including species and higher level groups. In phylogenetic systematics, nomenclature must be tied to phylogeny. Formal rules governing the naming of animals are codified by the International Code of Zoological Nomenclature (ICZN).

Nonconsumptive effects   The effect of a predator's presence on the survival and reproduction of prey, not due to direct killing. In essence, nonconsumptive effects are the costs of antipredator behavior.

Nonelemental learning   Associative forms of learning in which individual events are ambiguous and only logical combinations of them can be used to solve a discrimination problem.

Nongenomic effects of steroids   Effects of steroids on behavior or physiological responses that are not mediated by their binding to their well-characterized cognate intracellular receptors that normally results in a change in gene transcription. These effects are rather thought to come about via an interaction of the steroid with the cell membrane including the binding to membrane receptors of various sorts. These nongenomic effects are observed with much shorter latencies than the traditional genomic effects and by definition do not involve the induction of their biological effects via changes in gene transcription but via changes in protein state and second messenger systems.

Nonlinear timing   The hypothesis that psychological estimates of time are nonlinearly related to physical time.

Nonrapid-eye-movement sleep   One of the two basic forms of sleep in mammals and birds; in adult humans, it constitutes about 75% of total sleep time. It is characterized by high-amplitude, low-frequency brain waves, suppressed muscle tone, and decreased metabolic rate.

Nonredundant signals   When component modes in a multimodal signal contain distinctly different kinds of information, indicated by different responses of receivers to each mode.

Norm enforcement   The infliction of harm (including gossip, shunning, and ostracism as well as physical harm) on another individual for violations of social rules and conventions (social science).

Novelty response   Sudden acceleration of the rate of EOD emitted by a pulse-type electric fish caused by the sudden appearance of a novel sensory stimulus of any modality.

Noxious   Harmful or poisonous.

Nuclear species   A species that plays an important role in the formation and maintenance of a mixed-species group, usually leading the group.

Numerical ratio effect   When comparing a set of numerical values, one's ability to discriminate the values is based on both their magnitude and the difference between them (also known as Weber's law).

Nymph   Generally, a nymph is the juvenile stage of any hemimetabolous insect; it appears similar to an adult except it is smaller and lacks wing structures and developed reproductive organs. In termites, it refers to the preadult instars that perform nonreproductive tasks in the nest. However, since they are juveniles, they can later develop into either reproductive members of the colony or into sterile adult workers. Sometimes, this term is used interchangeably with larvae; however, there is some contention to this dichotomy.

Object   Something perceptible by one or more of the senses, especially by vision or touch; also a focus of attention, feeling, thought, or action.

Object movement reenactment   Reproducing the movement of an object manipulated by another animal.

Obligate   Applies to organisms that have to behave in a particular way to survive and whose behavior is innate. More specifically, individual obligate migrants migrate every year, and do not have the option of migrating or not, their behavior being genetically fixed.

Observational conditioning   Facilitation of the acquisition of a response due to the association between an object and secondary reinforcement (the observation of the other animal making contact with the object and obtaining a reinforcer).

Occasion setting   A learning situation in which a stimulus, the occasion setter, sets the occasion for when or where a predictive relationship applies. Contextual learning is closely related to occasion setting.

Occipital nerves   One or more nerves that originate in the brain and exit the posterior end of the skull through a foramen to innervate muscles that develop from occipital somites; considered a homolog of the hypoglossal nerve of tetrapods.

Occipital somites   Embryonic segments of mesoderm in all developing vertebrates that give rise to several skeletal muscles in the head including vocal/sonic muscles associated with the larynx, syrinx, and swimbladder.

Octavolateralis system   The group of sensory systems related to the eighth, and lateral line cranial nerves. It includes the sense organs of the inner ear, the lateral line, and the electrosense.

Octopamine   A neurotransmitter found in the CNS and elsewhere in all major classes of invertebrates. Its vertebrate equivalent is considered to be noradrenaline. It has been suggested that it plays a crucial role in the flight or flight reaction in insects. In particular, OA has been suspected of having a general effect on insect arousal.

Odiferous   Producing a pungent smell, often unpleasant.

Odometry   The measurement of distance traveled.

Odorant receptor   A protein molecule situated on the membrane of the sensory neuron that recognize a particular odorant (or a class of similar odorants).

Offspring viability   A measure of the relative health of offspring and/or their survival probability.

Offspring viability selection   Occurs when variation in the number of offspring surviving to reproductive age (productivity) differs between constrained and unconstrained parents.

Oil droplets   Lipid globules located in the inner segment of the cone photoreceptor of many birds and reptiles that filter light at different wavelengths and decrease the overlap in sensitivity between cones.

Olfaction   Sense of smell.

Omnivorous   A diversified diet of plant and animal materials.

One-way migration   Movement of an organism from a location where it develops to where it breeds without returning to the natal habitat before succumbing.

One–zero sampling   A time sample that produces a proportion of periods in which the behavior occurred.

Ontogeny   The development of an organism from fertilization through maturity and adulthood, also used to refer to the development of a particular trait over the same time.

Oocyst   A zygote stage in the sporozoan life cycle that sporulates to form sporozoites.

Oogenesis   Production of eggs.

Oogenesis-flight syndrome   A kind of migration syndrome described by C. G. Johnson, and found in many insects, in which migratory activity is limited to the brief period of sexual immaturity of the adult stage that immediately follows metamorphosis to the adult form.

Ootheca   An egg case; in cockroaches, a double row of eggs enclosed by a protective outer shell.

Opaque imitation   A form of imitation in which the observer cannot see its own reproduction of the behavior that is observed (e.g., imitating a demonstrator who places his hand on his head).

Open diffusion   An experimental design for studying the social diffusion of information, in which one or more individuals proficient in a novel action pattern is introduced into, or reunited with, a group of individuals and the potential spread of the action tracked.

Operant conditioning   Associative forms of learning in which an individual learns the consequences of its own behavior. It is a form of conditioning in which the desired behavior or increasingly closer approximations to it are followed by a rewarding or reinforcing stimulus.

Operational sex ratio (OSR)   The sex ratio among individuals ready to mate (i.e., being in operation).

Opportunistic breeder   An organism that can breed at any time of year, as long as specific environmental conditions exist (can thus also be a continuous breeder under correct circumstances).

Opportunistic foragers   Animals that feed on whatever is available, and can make use of new and novel food sources.

Opportunity cost   The cost of choosing one option and foregoing the opportunity associated with another option.

Opportunity for selection   The upper bound on the rate of evolutionary change in the mean of all phenotypes in a population, which is equal to the variance in relative fitness among members of the population divided by the squared average in fitness of those individuals.

Opsin   The membrane-bound G-protein-coupled receptor protein found in photoreceptors in the retina, which when combined to the chromophore, forms a visual pigment.

Optic lobe   The portion of the insect brain that processes visual input.

Optic tectum   A portion of the vertebrate midbrain, which processes sensory information from the eyes. In mammals, the optic tectum is called 'the superior colliculus.'

Optimal foraging theory   A body of theory that predicts behavior relative to maximizing or minimizing one or a set of goals.

Optimal group size   A group size for which the net benefits of group members are at a maximum.

Optimality   The cost–benefit approach has been extended to model when the benefit-to-cost ratio is maximized so that an individual should maximize the benefit of the behavior while simultaneously minimizing any costs associated with the behavior.

Optimal outbreeding   Mating with animals that share, due to identity by descent, favorable gene combinations, while avoiding matings with first or second degree relatives (parents, sibs, offspring) that might expose deleterious lethal genetic combinations.

Optimization and trade-offs   *Optimization* is a mathematical concept in which a function is either minimized or maximized given a restricted set of alternative inputs into the function. In behavioral ecology, it applies to predicting or interpreting behavioral decisions that maximize net fitness (e.g., lifetime reproductive success) in the face of conflicting demands, such as avoiding predation, which reduces feeding rates, and foraging, which increases exposure and risk of death by predation. *Trade-offs* are the outcomes of these decisions, such as greater safety at the cost of poorer energy stores or better energy stores at the cost of higher predation risk.

Optomotor response   Innate behavior used to stabilize a moving image through movements of the eyes, head, or body.

Organizational effects   Permanent changes in morphology, physiology, and/or neural circuitry dependent on hormone exposure during development.

Oropharynx   Region including the oral cavity and pharynx.

Ortholog   A similar gene in different species, thought to be derived from a common ancestor.

Oscillator   An oscillator is a process that repeats periodically.

Osmoregulation   The homeostatic control (see homeostasis) of osmotic potential or water potential, resulting in the maintenance of a constant volume of body fluids.

Otolith   Also known as 'ear stones,' these calcium carbonate structures are attached to the sensory epithelium of subdivisions of the vertebrate inner ear that are known as the lagena, saccule, and utricle. Each subdivision may serve either a vestibular (balance) and/or an auditory (hearing) function.

Oviposition   Egg-laying.

Ovipositor   The valved egg-laying apparatus of a female insect.

Ovipositor valve   The blade-like paired structures comprising the ovipositor shaft.

Oxytocin   A peptide produced almost exclusively within the hypothalamus that is released from the posterior pituitary and from neural projections to numerous intra- and extrahypothalamic brain sites. I Involved in milk-let-down, mother–offspring, and pair-bond formation in females, and contraction of nonstriated muscles for example during parturition.

Paedomorphosis   Reproductive maturity is attained while in a larval or branchiate form.

Pain   An aversive sensation and a feeling associated with actual or potential tissue damage.

Pair bond   The temporary or permanent association formed between a female and male, potentially leading to breeding.

Palps   Lateral mouthparts of invertebrates.

Panmictic (panmixia)   When mating between individuals in a population occurs randomly.

Pan-pipes   A device designed to present a naturalistic challenge to a tool-using animal such as the chimpanzee (*Pan*). A blockage in the upper of two pipes traps a food item. In social learning experiments, the blockage is released by using the tool in either of two quite different ways, the spread of which through social learning can thus later be objectively recorded (see Whiten et al., 2005).

Paracellular solvent drag   Movement of small molecules from interior of intestine (*lumen*) to circulatory fluids by passing between *enterocyte* epithelial cells of small intestine.

Paracrine agent   A chemical messenger that is released into the extracellular fluid and diffuses to and acts on adjacent target cells without entering the systemic circulation.

Paradigm   A combination of methods used to investigate problems, or an overall model of scientific conclusions regarding a given subject (e.g., how a contaminant affects the behavior of an animal, including humans).

Paralog   A gene that duplicated from an ancestral gene.

Parasite   Something that lives in, with, or on another organism (the host) and obtains benefits from that organism. A parasite is detrimental to the host in varying degrees.

Parasite manipulation   The ability, shared by several parasite groups, to modify their hosts' behavior to their own advantage, generally through increased probability of transmission in parasitic cycles.

Parasite propagules   Life-cycle stages which enable dispersion, transmission between hosts and from which new organisms can develop.

Parasitic wasps   A number of families of wasps which lay their eggs inside or outside of the larvae, pupae, or adult-host arthropods. The eggs hatch and the wasp's larvae feed inside the host eventually killing it. The wasp's larvae then pupate inside the host and emerge as adult wasps.

Parasitoid   An organism that spends a significant portion of its life history attached to or within a single host organism that it ultimately kills.

Parasocial route   Social grouping that originates in aggregations of individuals, usually around a rich resource.

Parathyroid glands   Small endocrine glands in the neck which are involved in calcium homeostasis.

Parentage analysis   Are studies or experiments that allow the investigator to determine the parents of any given individual offspring. In modern times, this is done using molecular techniques involving DNA analyses using mostly microsatellites.

Parental distraction display   Any behavior by a parent that reduces the probability that a predator will harm the parent's offspring by means of drawing the predator's attention away from the offspring; may take the form of feigning injury, tail-flagging, explosive flight, or erratic or conspicuous running.

Parental investment (PI)   Any investment by a parent that increases offspring fitness, at the cost of investing in other offspring.

Parental manipulation   Proposes that offspring helping behavior, a fundamental characteristic of the evolution of eusociality, arises as a result of parents influencing offspring development and condition.

Parent–offspring conflict   The disparity in selective pressures arising because optima in parental investment differ between parents and offspring.

Parr   A juvenile salmon during the initial freshwater phase of life.

Parsimony   The fundamental scientific principle that assumptions (especially process assumptions) need not be inflated beyond what is necessary to explain the phenomenon. In phylogenetics, the optimal tree is one that summarizes the putative homologies in such a way that as many as possible are retained. That is, homology is maximized, and as a result, the minimum number of evolutionary changes necessary is preferred.

Parthenogenesis   Development from an unfertilized egg.

Partial migration   A situation in which some birds from a given breeding area migrate away for the nonbreeding season, while others remain in the breeding area year-round.

Passerine   A bird belonging to the order Passeriformes, also referred to as 'perching birds.' Songbirds also belong to this group.

Patch   A relatively homogeneous area that differs in some way from its surroundings.

Path integration   Estimation of the current position relative to a starting location by integrating distances traveled and changes in direction throughout the journey.

Pathogen   Any disease-causing agent, especially a microorganism.

Patience   A preference for the more delayed outcome.

Patrigene   In diploids, the allele inherited from the father.

Pavlovian conditioning   This term is used interchangeably with the term 'classical conditioning.' It is a method of learning in which animals have inescapable exposure (one or more times) to an emotionally neutral stimulus, the conditioned stimulus, in temporal association with an innately provocative stimulus, unconditioned stimulus. Because of this association, the conditioned stimulus becomes a predictor of the occurrence of the unconditioned stimulus, and it typically acquires emotionally provocative properties similar to the unconditioned stimulus but at lower intensity.

Payoff matrix   A mathematical description of the fitness benefits to one behavioral strategy when it plays other strategies.

PCR   Polymerase chain reaction, a chemical reaction that utilizes a polymerase enzyme to replicate a target DNA sequence using primers that bind to the target DNA.

Pearson coefficient   A measure of correlation between ordinal or ratio data; can be used as a measure of observer reliability.

Pecking   The act of striking with the beak.

Pectoral girdle   That part of the skeleton that connects the fins or limbs to the axial skeleton (homologous to the shoulder region in mammals).

Pelage   Soft covering of a mammal such as hair, fur, or wool.

Penetrance   A genetic term referring to the extent to which the effect of a gene is expressed.

Peptide hormones   Small proteins, typically around 100 amino acids or shorter that are released from one tissue and have their action in another. These hormones differ slightly from species to species as the result of evolutionary changes in the DNA sequence, posttranslational processing, etc.

Perception   Physical sensation interpreted in the light of experience; as a fundamental means of allowing an organism to process changes in its external environment it depends on, but is not equal, to *sensation* – the detection of a stimulus and the recognition that an event has occurred; it can be viewed as the process whereby sensory stimuli are translated into organized experience. In the human cognitive sciences, perception is the process of attaining *awareness* or understanding of sensory information.

Perception-action mechanism (PAM)   Perception of another's state or situation activates neural representations of similar states or situations that the self has experienced.

Perception component   The recognition and processing of cues and cue bearers by an evaluator.

Perceptual class   A collection of items sharing perceptual properties, that is, arrays of features or elements defined in their own absolute values; thus class membership is solely based on similarity.

Period   The amount of time taken to complete one cycle of a sinusoid is its period. Period is the reciprocal of frequency. Low-frequency sounds have long periods and high-frequency sounds have short periods. In chronobiology, period is the time it takes for a full oscillation or rhythm to occur.

Periodogram analysis   A type of time series analysis that involves combining average response rate functions assuming different underlying periodic trends. The analysis identifies the underlying periods that minimize errors of prediction.

Perspective taking   Being in a position to form a 'mental picture' of what another can see even when you cannot see it directly yourself.

Pesticides   A range of chemicals produced to kill insects; many chemical forms exist some of which are endocrine active.

Pet   A domestic or tamed animal that is individually identified, kept by a person or persons as a companion, and cared for with affection.

Phagocytosis   The cellular consumption or elimination of foreign tissues, cells, or particles.

Phagomimicry   Release of a chemical that induces feeding behavior toward the chemical (a false food stimulant) and not the animal that releases it.

Phagostimulant   Anything that triggers feeding behavior.

Pharmaceuticals   Chemicals produced for the treatment of biomedical conditions.

Pharmacology   The science of the properties of drugs and their affects on the body.

Pharynx   The part of the neck and throat situated immediately posterior to (behind) the mouth and nasal cavity.

Phase   Temporal relationship between two rhythmic processes having the same frequency.

Phase angle   Measurement of phase, expressed as the time delay between two rhythmic processes, divided by the length of the common period and multiplied by 360°.

Phase (in locusts-solitarious, gregarious, or transiens)   A combination of traits defining morphological, physiological, and behavioral state of a locust.

Phase locking   The auditory system uses phase-locked spikes to encode the timing or phase of the auditory signal. Phase-locked neurons fire spikes at, or near, particular phase angles of sinusoidal waveform. Physiological experiments measure this spike phase with respect to the stimulus period. Spike phase is plotted in a period histogram and is used to calculate the statistic vector strength (r). Each spike defines a vector of unit length with a measured phase angle. The vectors characterizing the spikes are plotted on a unit circle and the mean vector calculated. The length of the mean vector provides a measure of the degree of synchronization.

Phase shift   A phase shift is a change in the timing, or phase, of an oscillation or rhythm in response to an external cue. A widely used manipulation in the study of biological rhythms. The event that is thought to reset timing (e.g., light-dark cycle in the case of circadian rhythms) is advanced or delayed. Gradual adjustment in response to a phase shift is a characteristic feature of an endogenous oscillator.

Phenology   The repetitive sequence of events of the life cycle of plants and animals that are affected by environmental conditions.

Phenomenology   One's subjective experience or the experience from the first-person point of view.

Phenotype   Any characteristic of an organism that is the result of that individual's genotype and the interaction of the genotype with the environment during development.

Phenotype matching   The ability to learn phenotypes of group members, such as littermates, and to extend that knowledge of phenotype to make discriminations among previously unmet animals.

Phenotypic flexibility   See phenotypic plasticity.

Phenotypic interface of coevolution   The traits that mediate ecological interactions between coevolving species, such as chemical defenses of prey and resistance to those compounds by predators.

Phenotypic plasticity   The capacity of an individual organism to produce different phenotypes (morphology, physiology, behavior, etc.) in response to different environmental inputs.

Pheromone   A chemical messenger produced by an organism that influences the behavior or physiology of another organism of the same species.

Phi coefficient   A measure of correlation between nominal data; can be used as a measure of observer reliability.

Philopatric reproduction   Breeding at the natal nest.

Philopatry   The tendency of an individual to remain or return to its birthplace.

Phonological syntax   Structured rules for constructing sequences of otherwise meaningless sound units.

Phonotaxis   Locomotion towards or away from a sound source.

Photic zone   The portion of the upper water column with sufficient light for photosynthesis to occur, typically reaches between 50 and 200 m depth in oceanic waters.

Photomechanic infrared receptor   A receptor rapidly dissipating infrared energy into a micromechanical event (i.e., a brief increase in internal pressure in the core of the receptor) which is measured by a mechanoreceptor.

Photoperiod   Length of day.

Photoperiodism   Changes in reproductive physiology and behavior in response to changing day length.

Photorefractoriness   A complete shutdown of the hypothalamic-pituitary-gonad axis that terminates the breeding phase.

Photorefractoriness in birds   Physiological state in which photoperiodic birds terminate reproduction during long day lengths. Signals the end of the breeding season.

Photorefractoriness in mammals   Physiological state in which photoperiodic mammals reactivate the HPG axis after prolonged exposure to short days. Spontaneous gonadal recrudescence occurs and the short days no longer inhibit reproduction.

Phototaxis   From photos (light) and taxis (movement). It refers to movement toward or away from a light source (positive or negative phototaxis, respectively).

Phylogenetic   Relating to, or based on, evolutionary history.

Phylogenetic signal   The extent to which similarities among closely related species, such as the form of communication they use, is dependent on the phylogenetic relationships between those species.

Phylogenetic systematics (also, phylogenetics, cladistics)   The particular method of systematics proposed by Willi Hennig. Phylogenetic systematics relies on two fundamental precepts: (1) only whole character-state transformations, in the form of synapomorphies, count as evidence of relationship; and (2) taxonomic names must be applied only to natural evolutionary groups (i.e., nomenclature is united with phylogeny).

Phylogeny   A genealogy of species that reflects their evolutionary relationships.

Physiological psychology   The study of mechanisms internal to the animal that affect and are affected by behavior. Included are studies of the nervous system, endocrine function, and other internal processes.

Phytohormones   Hormone-life chemicals produced by plants that have structural characteristics that allow them to interact with steroid hormone receptors in vertebrate physiological systems; for example soy phytoestrogens.

Phytophagous   Plant eating.

Pied Piper effect   The idea (now largely discounted) that in the northern hemisphere, northward movements of insects in the spring are facilitated by favorable winds, but that the progeny of these immigrants are then trapped at high latitudes as winter approaches, leading to mass fatality.

Piloerection   This term derives from 'pilo,' meaning hair, and refers to the erection of the hair of the skin. Piloerection starts when a stimulus such as cold or a frightening stimulus causes an involuntary contraction of the small muscles that attach to the base of the hairs deep in the hair follicles. Contraction of these muscles elevates the hair follicles above the rest of the skin so the hairs seem to stand on end.

Planktivorous   Feeding primarily on organisms that drift or possess insufficient motor capabilities to overcome currents (plankton).

Plant secondary metabolite   See secondary plant compound.

Playback studies   'Playbacks' can be broadly defined as the use of broadcast signals in any sensory modality with an accompanying bioassay to address questions concerning communication and animal behavior.

Pleiotropic   (see pleiotropy).

Pleiotropy   When a gene affects multiple traits.

Pleometrosis   The founding of a eusocial insect colony by several queens.

Poikilothermic   Having a body temperature that varies with the temperature of its surroundings.

Point of balance   It is a point at the animal's shoulder that handlers can use to control animal movement. When a person stands behind the point of balance, the animal moves forward. When a person stands in front of the point of balance, the animal backs up.

Policing   Repression of selfish or competitive behavior (evolutionary biology); in social insects, inhibition of worker reproduction by aggression or destruction of eggs (evolutionary biology); impartial intervention in conflicts (social science); enforcement of societal norms and laws (social science).

Polyandry   One female has a breeding relationship with two or more males. In eusocial insects, a queen that has mated many times.

Polydomy   Of colonies having more than one nest each.

Polyembryonic   A form of reproductive in which one sexually produced embryo splits into many genetically identical offspring.

Polygamous   Having more than one partner or spouse.

Polygenic   Where several genes interact to influence a phenotype; each gene may have a varying degree of influence upon the phenotype.

Polygyny (polygynous)   Mating system in which some or all males in a population mate with more than one female per breeding season. This typically increases variance in male mating success, thereby generating sexual selection on males. In eusocial insects, a colony that has two or more queens.

Polymorphism   The existence of multiple forms within a population or species. The term can refer to morphology or alleles or physiology or behavior or any other kind of trait. In eusocial insects, size or shape variation in the worker caste.

Polymorphous class   A class in which no single feature is necessary or sufficient to determine class membership, but several features contribute to this to some degree.

Polyphagy (adj. polyphagous)   Eating many kinds of food, for example, many plant species from a range of families.

Polyphenism   Within a population, different phenotypes that arise from environmental rather than genetic causes.

Polyspermy   When more than one sperm enters the egg during fertilization.

Ponerine ants   The Ponerinae is a subfamily of ants (Hymenoptera, Formicidae). Some species within the Ponerinae are queenless, having lost the queen caste. A colony is headed by one or more mated workers that fulfill the queen's role and are sometimes known as gamergates.

Population density   The number of individuals within a specified unit of space.

Population dynamics   Study of short- and long-term changes in the size and age composition of populations, and the factors influencing those changes.

Porphyropsin   All visual pigments whose chromophore is 3,4 dehydroretinal.

Positional cloning   A technique used in molecular cloning that utilizes a set of unique genomic elements called 'genetic markers' that flank the gene of interest.

Positive punishment   The application of a stimulus immediately after a behavior which results in a reduction in the future probability of that behavior.

Positive reinforcement   The application of a stimulus coincident with or immediately after a behavior which results in an increase in the future probability of that behavior.

Postconflict bystander affiliation   Postconflict affiliative interaction between a conflict opponent and a bystander uninvolved in the conflict.

Postconflict quadratic affiliation   Postconflict affiliation between two bystanders.

Postcopulatory sexual competition   The term generally used to refer (somewhat imprecisely) to all events following the initiation of genital coupling.

Postmating-prezygotic isolation   Barriers between species or populations that result from mechanisms that prevent zygote formation after mating.

Postzygotic compensatory mechanisms   Flexible responses of constrained individuals of either sex that increase the likelihood that already produced zygotes will survive to reproductive age.

Postzygotic isolation   Barriers between species or populations that result from low fitness of hybrids.

Potential conflict   Differences in the reproductive optima of individuals or groups in a colony. For example, there is potential conflict over male production in a colony of eusocial Hymenoptera as each individual is more related to its own sons (0.5) than to the sons of the mother queen (0.25) or sister workers (full nephews 0.375, half nephews 0.125).

Potential reproductive rate (PRR)   Offspring production per unit time when unconstrained by mate availability.

Praying mantis   A predatory insect with prominent eyes and an elongated body; in the order Dictyoptera along with the cockroaches and termites.

Precedence effect   Psychophysical phenomenon in which two or more stimuli separated by a brief time interval are perceived as a single stimulus originating from the source of the first one.

Precocial   Mobile young (usually birds or mammals) that are dependent on parents for food and warmth.

Predation risk theory   The framework used to predict or interpret antipredator behavior, risk effects, and the behavioral component of trophic cascades. Fundamental to it is the assumption that prey maximize fitness (e.g., lifetime reproductive success) by making behavioral decisions that optimize trade-offs between predator avoidance and resource acquisition.

Predation sequence   The sequence of events that is necessary for a predator to kill one or more prey individuals, including search, encounter, hunting, and killing. Each of these four stages can include distinct substages.

Predator inspection   Alone or in groups, an approach toward a predator to observe and gain information about it that may function to deter attack by advertising that the predator has been detected; the behavior may also advertise ability to incur risk and escape.

Prediction   Statement of results of studies that could be performed.

Preening   The act of cleaning and trimming the feathers with the beak.

Preference function   The order in which an individual ranks potential mates.

Preference hierarchies   A ranking indicating the relative degree to which each of a set of alternative plants are preferred by a herbivore.

Premating isolation   Barriers between species or populations that result from mechanisms that prevent mating.

Premetamorphosis   Stage of amphibian larval development when the animal grows but little or no morphological change occurs; plasma thyroid hormone concentrations are low.

Preoptic area   A region of the brain just rostral to the optic chiasma where steroid action plays a key role in the activation of male sexual behavior in many vertebrate species.

Prepubescent   Prior to puberty.

Prezygotic compensatory mechanisms   Flexible or facultative responses of constrained individuals of either sex that increase the likelihood that their offspring survive to reproductive age.

Price equation   A mathematical statement of evolutionary change that partitions selection into a between- and within-group component.

Primary polygyny   Polygyny that arises through pleometrosis.

Primary predator–prey behaviors   Behaviors concerned with predators encountering prey, or prey avoiding predators, before any attack occurs.

Primary reproductive   The winged, founding members of a termite colony. A winged reproductive male and female found a new colony as the primary reproductives.

Primitively eusocial   Eusocial society in which all individuals are capable of mating and reproducing, though behaviorally specialized reproductives occur.

Prisoner's dilemma   In its simplest form, it is a two-player game in which players decide whether to cooperate (C) or defect (D). The relative sizes of the payoffs define the game, in that mutual cooperation pays more than mutual defection, but defecting while your partner cooperates provides the highest payoff, and cooperating while your partner defects provides the lowest payoff. The game captures both the temptation to defect and the low payoff for being a 'sucker' (cf. the 'tragedy of the commons,' which arises in a multiplayer version of this game).

PRKO   Knockout mice lacking a functional progesterone receptor.

Probability matching   In the study of foraging behavior, this refers to an animal's tendency to match its proportion of visits to a feeding site with the proportion of times that site produced food.

Probing motor acts (PMA)   Characteristic behaviors composed of a series of swimming movements in close proximity to an object under investigation.

Problem-solving   The use of novel means to reach a goal when direct means are unavailable.

Proboscis   Central trunk-like mouthpart of insects that feed on liquid food.

Proceptivity   Feminine behaviors that are evoked by stimuli from the male and which serve to reduce the distance between the female and the male. The extent to which a female initiates mating (i.e., a female's willingness and motivation to mate).

Producer and scrounger   Behavioral alternatives for group foragers when a resource, for example, food, is found by one individual, the producer, and then exploited by one or more animals in the group, the scroungers. The term also describes a game theory model that applies to the two alternatives.

Production learning   Where a signal is modified in form as a result of experience of the usage of signals by other individuals.

Productivity   The number of offspring that survive to reproductive age.

Progesterone   Steroid hormone produced mostly in gonads and the brain.

Progressive molt   A molt characterizing the gradual development from egg via several instars into an adult. Associated with progressive molts is an increase in body size and morphological development. This is the default developmental program in all hemimetabolous and holometabolous insects.

Progressive provisioning   Type of larval provisioning in which the larvae are fed throughout their development. In contrast to mass provisioning, in which all of the food necessary for larval development is amassed before laying an egg.

Prohormone   Precursor to the active form of a hormone.

Prometamorphosis   Stage of amphibian larval development when metamorphosis begins. Hindlimb growth and development is evident externally. The thyroid gland becomes active and secretes thyroid hormone in response to increasing plasma concentrations of pituitary thyrotropin (TSH).

Propagules   Any structure that can give rise to a new individual. This could include sexually or asexually produced zygotes, embryos, larvae, seeds, or fragments or buds.

Propolis   Plant resins collected by honeybees and used for sealing gaps and cracks in their nest.

Proprioception   The ability to sense the position and location and orientation and movement of the body and its parts.

Prosocial behavior   Tendency to help others even if this provides no immediate reward to the self.

Prosociality   The tendency to help another in a situation where there are no personal gains, and little or no personal cost.

Prostaglandins   Fatty acid hormones, such as PGF2a, that are secreted by the reproductive tract and ovary.

Protandry   A sexual pattern in which individuals mature as males and can then later change functional sex to become female.

Proteome   The set of proteins expressed by the entire genome of an organism under given environmental conditions at a given time. Proteomics is the large-scale study of the structure and function of this entire set of proteins, generally in a particular cell, tissue-type, or organ (such as the brain).

Prothoracic gland   The molting gland of the insect that secretes ecdysone, the precursor of the active form of the molting hormone, 20-hydroxyecdysone.

Protogyny   A sexual pattern in which individuals can mature as females and then later change functional sex to become secondary males. In *monandric* ('one male') protogyny, all secondary males first pass through a female stage. In *diandric* ('two males') protogyny, individuals can mature as either males or females and both can change from the initial phase (IP) to become the larger and typically colorful and aggressive terminal phase (TP) males.

Prototype   The 'best' or most typical example of a category that corresponds to the average, or central tendency, of all of the exemplars that have been experienced; it serves as the basis or standard for other members of the same category.

Protozoan   Unicellular microorganisms among eukaryotes. Comprises flagellates, ciliates, sporozoans, amoebas, foraminifers.

Proximal   Closer to a body midline (opposite of distal).

Proximate causation   Explanations of an animal's behavior based on internal and external mediators of behavior including genetic underpinnings, epigenetic forces, maternal effects on physiology, morphology, and development. Questions about proximate causes are sometimes said to be about how animal behavior is expressed or about mechanisms of animal behavior.

Proximate factors   External stimuli (such as specific daylengths) which are used as cues by an animal to trigger preparation for breeding, migration, molt, or other events, or as time keepers to set their endogenous time programs at appropriate times of the year.

Pseudergate   In termites, an alternative technical term that can be found which distinguishes workers with a flexible development and options for direct reproduction from workers with restricted developmental trajectories. Pseudergates are the 'workers' of many lower termites (including wood-dwelling and foraging species) that have broad developmental options, generally including progressive, stationary, and regressive molts. Current use of this term often lacks the precision of its original definition for individuals that develop regressively from nymphal instars to 'worker' instars without wing buds.

Pseudopregnant   Reproductive condition in which a female shows external indicators of pregnancy but is not actually pregnant.

Pseudoreciprocity   The act of increasing another individual's fitness to acquire or enhance the by-product benefits obtained from that individual.

Pseudoreplication   A statistical error in which interrelated observations or measures are treated as though they are statistically independent.

Psychoneuroimmunology   A relatively new field in medicine that explores the ability of the nervous system and psychological states to influence immune defenses, and the ability of the immune system to influence the brain and behavior.

Pterygoid teeth   Small teeth on the roof of the mouth.

Ptilochronology   The study of growth bands in feathers that indicate condition or problems during feather molt in birds.

PTT   A platform transmitter terminal (PTT) sends an ultrahigh frequency (401.650 MHz) signal to satellites.

PTTs are attached to animals in order to track their movements.

Public good   A resource that is costly to produce and provides a benefit to all the individuals in the local group. Public goods systems are often open to exploitation by cheats who benefit, but do not pay the cost.

Public information   Cues produced by animals that can potentially be used by observer animals in making behavioral decisions.

Pulse repetition rate   The rate at which individual sound pulses are produced within a single call.

Punishment   A costly behavior that is negatively reciprocal (decreases harmful behavior in the recipient) (evolutionary biology); any stimulus that reduces the frequency of a behavior (social science); behavior correction and the enforcement of social norms, typically by impartial parties; see also Third-party punishment, Policing (social science).

Pupa   A life stage in some insects that undergo complete metamorphosis that results in the transition between the larval and adult stage.

Purging selection   Mechanisms eliminating deleterious genes from the population.

Pyrophilous insects   Species strongly attracted to burning or newly burned areas, and species that have their main occurrence in burned forests 0–3 years after the fire.

Quality of life   Well-being; a multidimensional, experiential continuum that comprises an array of affective states, broadly classifiably as relating to the states of comfort–discomfort and pleasure; often equated to welfare and well-being.

Quantitative trait   A continuous trait such as body mass that is influenced by many genes and the environment.

Quantitative trait locus (QTL)   A region of DNA that is associated with a particular quantitative trait, containing a gene or genes that influence that trait. Quantitative traits typically have continuous distributions rather than discrete states, and are influenced by several or many loci, each with relatively small or large effects on the expression of the trait.

Quasi-experimental design   An experimental design where a treatment variable may be manipulated but subjects within groups are not equated or randomly assigned.

Quasiparisitism   Occurs when the female that dumps eggs in another female's nest is the resident male's extra-pair partner and her dumping is assisted by that male.

Queen   Reproductive female in a eusocial insect society. She is developmentally and/or behaviorally disposed towards performing all reproductive function for a colony.

Questing The behavior of ticks, involving an ascent on vegetation that allows for a maximum exposure of sensory receptors on the forelegs to stimuli from approaching hosts.

Quorum decision A minimum number of individuals required to perform a specific behavior (such as choosing a direction of travel) that results in all of the other members of a group adopting this behavior.

Quorum sensing A rule under which a social group member's execution of a particular act or behavioral transition is conditioned on the presence of a threshold number of fellow group members.

Radiotracking The location and tracking of a radiomarked individual from a signal emitted frequently by the radio.

Rape A legal term and includes other forms of sexual assault as well as forced copulation, including statutory rape, which may appear to be consensual copulation but with a minor; in this case women, not just men, can be rapists.

Rapid-eye-movement sleep The other basic sleep form in mammals and birds. It is often called 'paradoxical sleep' because the brain activity resembles that of the awake brain. It is characterized by the complete inhibition of muscle tone and suppressed autonomic regulation of most homeostatic functions such as thermoregulation and blood pressure.

Rate of return The ratio of the amount of food obtained to the time it took to procure the food.

Rationality A set of consistency principles that decision-makers are expected to follow if they are attempting to maximize some currency such as utility or fitness. Fitness maximization by natural selection is expected to yield rationality, but many instances of irrational choice are known in humans and other animals. Property of individual choice is used both to describe the process of making a choice and to describe the behavioral outcome of choice.

Rayleigh scatter Light scatter by particles smaller than the wavelength of light.

Reaction norm A reaction norm describes the production of a range of phenotypes by a single genotype in response to a range of an environmental parameter. Different genotypes may produce different response trajectories in response to a gradient of an environmental parameter. Reaction norms resemble dose-response curves in physiology, for example the effects of a gradient in hormone concentrations. Dose-response relationships are not necessarily monotonic but can include thresholds or show maximal (minimal) effects at low and high doses or medium doses.

Reasoning A form of logic-based thinking; the cognitive process of looking for reasons for beliefs, conclusions, actions, or feelings.

Receiver psychology Sensory capabilities of the signal receiver that affect the detectability, discriminability, and/or memorability of signals, and play a role in the evolution of signal design.

Receptivity Sexual behaviors that are necessary and sufficient for mating.

Reciprocal altruism Where individual A pays a personal cost to help individual B with the expectation that B will return the favor.

Reciprocal selection Positive feedback between selection by ecological enemies. Natural selection by predators on prey generates the evolution of increased defense, which in turn causes stronger selection by prey on predators to evolve greater exploitative abilities.

Reciprocity Delayed exchange of benefits between parties.

Recognition signals Signals that evolved to make a signaler distinctive.

Recombination In evolutionary algorithms, a process of crossover that combines elements of existing solutions in order to create at the next generation a new solution, with some of the features of each 'parent solution.' It is analogous to biological crossover.

Reconciliation Postconflict affiliative reunion between former opponents that restores their social relationship disturbed by the conflict.

Recruitment Entry of progeny into a population as reproductive adults.

Red queen Based on the quote from Lewis Carroll's Red Queen, 'It takes all the running you can do, to keep in the same place,' this metaphor describes a coevolutionary dynamic where frequencies of traits or genotypes of ecological enemies cycle through time so that as one type becomes common, it is disfavored and a rare type can spread through the population.

Redirected aggression Postconflict aggressive interaction directed from the original recipient of aggression to a bystander uninvolved in the conflict.

Redirected behavior The direction of some behavior, such as an act of aggression, away from the primary target and toward another, inappropriate target.

Redundancy reduction The reduction in the overlap of information encoded by neurons in the nervous system.

Referent The on model on which a signal is based.

Reflectance The ratio of reflected to incident light on a given area (e.g., colored patch in the plumage).

Refraction Change in direction of light caused by alteration of its velocity on obliquely entering a medium of different refractive index.

Refractive index   A measure of the speed of light in a medium.

Refractive state   The resting refractive state of an animal determines the point at which it is focused without having to expend any accommodative effort.

Regressive molt   A molt that is characterized by a decrease in body size and/or regression of morphological development, generally a reduction of wing bud size in nymphal instars. This type of development is unique to termites.

Regularity   A specific version of independence from irrelevant alternatives. It describes the expectation that the absolute preference for an option should never be increased by the addition of inferior options to the choice set.

Regurgitant   A substance produced in the gut of an insect that is excreted from the mouth as a defensive secretion.

Reinforcement   The evolution of premating isolation after secondary contact as a result of selection against hybrids or hybridization.

Reinforcement/supplementation   Addition of individuals to an existing population of conspecifics.

Reintroduction   An attempt to establish a species in an area which was once part of its historical range, but from which it has been extirpated or become extinct.

Relatedness asymmetries   A group of individuals are more closely related with a certain group of individuals than others within a colony.

Relatedness, *r*   Genetic similarity between individuals, in comparison with randomly chosen individuals in the population, that have a mean relatedness of zero by definition.

Relational class   A class defined by relations between or among its members and going beyond any perceptual similarities or functional interconnections.

Relative risk   An individual's risk of predation given the abundance of its type.

Relaxed selection   This occurs when the sources of natural selection engendering physical or behavioral traits that promote fitness diminish markedly or are no longer present in the environment. In the case of predators, prey species might be separated from their former predators by their isolation on islands. In another context, climate change tolerated by prey might diminish contact with their predators that are intolerant to climate change and eventually disappear.

Reliability   The percentage of signals of a particular type X that are accurately associated with a stimulus (X′).

REMI   Restriction enzyme mediated integration (REMI) is an ingenious method of introducing single gene knockouts in a genome in a way that allows one to identify the actual gene that is knocked out. Used in *Dictyostelium*.

Repeatability   Consistency between different measurements separated in time of a trait of a certain individual, used in population genetics as the upper limit of heritability.

Repertoire expansion   A pattern of temporal polyethism in which workers increase the types of tasks they perform as they age.

Replication   Using more than one observation per observational unit or subject per experimental treatment group.

Reproductive age   The age at which an individual becomes receptive to mating the first time.

Reproductive character displacement   The process of phenotypic evolution in a population caused by cross-species mating and which results in enhanced prezygotic reproductive isolation between sympatric species. Referred to as 'reinforcement' if postzygotic isolation is incomplete.

Reproductive compensation   Refers to any flexible response of constrained individuals that increases the likelihood that their offspring will survive to reproductive age.

Reproductive division of labor   Differentiation of individuals within a eusocial colony into those capable of reproducing, and functionally or physically sterile workers.

Reproductive effort   The proportion of available time, nutrient or energy resources that an adult invests in current reproduction, usually detracting from those available for other functions.

Reproductive groundplan hypothesis   (Originally described as ovarian groundplan hypothesis) Proposes that the evolution of eusociality is based on simple evolutionary modification of conserved reproductive and corresponding behavioral cycles so that during the course of social evolution, reproductive and nonreproductive behavioral and physiological components can be separated and used to build reproductive (queen) and nonreproductive (worker) phenotypes.

Reproductive isolation   Reduced genetic exchange between populations via reduced interbreeding and lower fitness of hybrid offspring; speciation has occurred when reproductive isolation between populations is complete.

Reproductive skew   Asymmetry in the distribution of direct reproduction among individuals within a social group.

Reproductive strategy   An organism's relative investment, behaviorally and physiologically, in offspring, including reproduction and parental care.

Reproductive success (RS)   Refers to the number of offspring an individual produces which survive and go on to reproduce in the next generation. Although 'life-time reproductive success' is the most accurate measure, logistically it is not always possible to obtain this measure.

Consequently, RS may be measured as number of eggs produced, number of young produced, number of young that fledge from the nest (e.g., birds) or survive to weaning (e.g., mammals), or number of young that survive to reproductive age.

Reproductive suppression   A mature individual does not reproduce because of physiological mechanisms that inhibit production of gametes as a direct result of communication with conspecifics.

Reproductive value   The expected reproduction of an individual from its current age onward, given that it has survived to that age. It changes with age, increasing at first and declining until death.

Residual reproductive value   The number of offspring an individual is expected to produce during its remaining lifespan.

Resource competition   A particular form of competition in which members of the same or different species compete for the same resource in an ecosystem (e.g., food, space).

Resource constraint hypothesis (Trivers–Willard effect)   Colonies should invest more in the cheaper sex (i.e., males, which are generally smaller than females in Hymenoptera) when resources are limited.

Resource holding potential   The relative fighting ability of a contestant.

Response blocking   Also called *flooding* – The process of exposing a subject to constant, high levels of a distressing stimulus, while preventing escape from the situation, in an attempt to reduce or extinguish the distress produced by the stimulus.

Retinal disparity   Difference between the images projected on the two retinas when looking at an object that serves as a binocular cue for the perception of depth.

Retinoscopy   A technique used to obtain an objective measurement of the refractive state of the eye, in which a moving light is shone into an animal's eyes and the relative motion of the reflection is observed.

Reverse genetics   A molecule-driven approach to understanding a phenotype.

Rheotaxis   Orientation or response to current flow; moving upstream is positive and downstream is negative rheotaxis.

Rhinophores   Tentacles in some gastropod mollusks that carry the olfactory organ.

Rhodopsin   All visual pigments whose chromophore is retinal, but commonly (although erroneously) used to refer only to rod visual pigments.

$R_{male-male}$ Androgen responsiveness (i.e., the change in testosterone concentrations) during aggressive interactions between territorial males.

R_{season} Seasonal androgen response, reflecting the increase from breeding baseline testosterone concentrations to maximum concentrations during specific parts of the breeding life-cycle stage, that is, during the phase of territory establishment or mate guarding.

Riparian   Interface between terrestrial and aquatic ecosystem. When intact, riparian ecosystems limit soil runoff and are characterized by high biodiversity and thus are an important buffer zone.

Risk effects   Nonconsumptive effects of predators on prey, namely the lost foraging opportunities and lower levels of growth and reproduction experienced by prey investing in antipredator behavior (also known as nonlethal effects). This term avoids the complication that prey that are not directly killed by a predator may in fact be consumed.

Risk history   The frequency, intensity, and duration of predation risk events experienced by prey in the past.

Risk threshold   The level of risk that must be exceeded for the prey to start reducing its antipredator behavior under the risk allocation hypothesis.

Ritualization   Communicative behaviors used in social interactions that evolved from other behaviors with different functions. For example, when attacked an ancestor of the wolf might have flattened the ears, crouched, and tucked the tail to avoid injury; over time these behaviors evolved to communicate submission. Evolutionary modification of a motor pattern used in communication that is thought to improve signal function, often through increased stereotypy and exaggeration.

RNA interference (RNAi)   A technique of molecular biology in which expression of a particular gene is silenced by introducing double-stranded RNA into a eukaryotic organism. RNA interference can provide conclusive proof that a particular gene influences behavior.

Roosting   The act of perching to rest or sleep.

Round-trip migration   A subcategory of migration, with seasonal to-and-fro movements between regular breeding and wintering sites, typical of many birds but rare in insects.

RT-PCR   Reverse transcription PCR, PCR that is performed on DNA that was synthesized from RNA by a reverse transcriptase enzyme.

Rule learning   The ability to infer rule information from a number of different examples connected by a logical operation 'if → then.'

Rules of thumb   Simple measures that animals can use to approximate solutions to optimal foraging problems. An example would be using the number of prey items encountered to leave patches as predicted by the marginal value theorem.

Runaway selection   A theoretical model for the evolution of extravagant traits based on female preference.

The model proposes that female preference for a male trait results in a genetic correlation between preference and trait, such that the trait evolves beyond the level favored by natural selection in a 'runaway' process fuelled by female preference. Also called the Fisher process in reference to Sir Ronald Fisher, who developed the theory.

Saccule   An otolithic subdivision of the inner ear in all vertebrates that has an auditory (hearing) function among many fishes.

Saprophagy   Feeding on dead materials.

Satellite transmitters   These tracking devices are larger than radio transmitters and emit signals that are detected by geosynchronous satellites; these devices carry substantial batteries or are solar powered and continue to transmit for relatively long periods of time (i.e., a year or more); they enable tracking to occur over substantial geographic distances.

Satiation   The feeling of fullness at the end of a meal.

Satiety   The persisting sensation of repletion that results from eating.

Scalar timing   The dominant theory of timing which assumes that the coefficient of variability (i.e., the standard deviation of time estimates divided by the mean of time estimates) is constant across a broad range of temporal estimates (i.e., a specific proposal of the linear timing hypothesis).

Scale-free power-law   A degree distribution described by $p(k) \approx k^{-\gamma}$; demonstrated by a straight line on a log–log plot.

Scan sampling   A type of instantaneous sampling in which a group of individuals is scanned at specified intervals and the behavior of each individual at that instant is recorded.

Scanning   Often synonymous to vigilance.

Scatter hoarding   Hoarding of individual food items in many different locations.

Schistosomiasis (or bilharzias)   A disease caused by a blood fluke of the genus *Schistosoma*, a type of flatworm parasite. The intermediate host is a snail, in which cercariae (larvae) develop and migrate out into water; the cercariae penetrate the skin of hosts which make contact with the water. Symptoms depend on species causing infection, but can include rash, fever, aching, cough, diarrhea, and liver and spleen enlargement.

Schnauzenorgan response   A twitching movement of the elongated chin (Schnauzenorgan) of *Gnathonemus petersii, an electric fish,* evoked by the sudden emergence of a novel object near the animal's head, which is detected through the active or passive electric sense.

Schreckstoffe   Chemical alarm signals released by aquatic injured conspecifics, which is used to warn animals about an imminent danger.

Sclerotized   The hardening of tissue.

Scolopidium   A multicellular sensory structure of arthropods used to detect stretch, vibration, or sound.

Scout   A member of a social group, such as an ant or bee colony, that searches for food sources, nest sites, or other targets of interest. It may exploit its discoveries by itself or recruit other group members to help.

Scramble competition   Organisms use up a common limiting resource but otherwise do not contest or harm each other.

Scrounging   A behavioral strategy that consists of exploiting a resource uncovered by some other individual's efforts.

Seasonal breeder   An organism that breeds only in specific seasons (i.e., not continuously).

Seasonal interaction   When events in one period of the annual cycle, such as timing or condition, of an animal to influence events in subsequent periods.

Seasonality   Changes in hormonal or behavioral status in response to change in seasons.

Secondary defenses   Traits of the prey that influence the action of the predator, subsequent to prey detection, in ways that benefit the prey. Compare with primary defenses that act prior to the predator detecting the prey.

Secondary plant compound   Molecules produced by plants, the presence of which is often characteristic of particular plant taxa and which appear not to be directly involved in primary metabolism.

Secondary polygyny   Polygyny that arises from monogyny, generally through queen adoption.

Secondary predator–prey behaviors   Behaviors concerned with predators capturing prey, or prey escaping from predators, during an attack.

Secondary reproductive   These are produced by many termite species; they are sexually capable individuals who do not have wings, and are capable of superceding sick, injured, or absent parental primary reproductives.

Secondary sexual character   A trait that differs between the sexes and is neither required for reproduction nor related to sex differences in ecology. Most such traits do not develop fully until sexual maturity, are expressed more strongly in males than in females, and are useless or costly for survival. Traits that do not differ between the sexes but share the other two qualities may also be referred to as secondary sexual characters (e.g., ornate plumage in sexually monomorphic birds).

Segregation distortion   Within-individual selection for one or another allele of a diploid body.

Selective attention   The cognitive processes of (selectively) concentrating on one aspect of the environment while ignoring others; consciously or unconsciously, the perceiving organism is focused on particular areas of the environment. This is determined by past experience and the skill being performed.

Selective differential   The difference in fitness between two or more subsets of a population subjected to different selective pressures with resulting differences in fitness.

Selective sweep   Recent and strong positive natural selection on a particular gene which leads to reduced variation in DNA sequence among individuals in a population.

Selective tidal stream transport (STST)   Vertical movements of aquatic organisms relative to tides; provides a mechanism for zooplankton and small nekton to move horizontally within and between estuaries and coastal regions.

Self-awareness (self-recognition)   Increased self-other distinction, oftentimes indicated by self-recognition in a mirror. Sensitivity to one's own thoughts and feelings; sometimes used to indicate the knowledge that one exists independent of other entities.

Self-control task   Experimental situation in which decision-makers must choose between smaller–sooner and larger–later options.

Selfish-herd effect   Bunching by foragers to decrease their relative domain of danger when facing predation threats.

Self-medicate   The use by animals of secondary plant compounds or other nonnutritional substances in preventing or treating diseases.

Self-organization   The idea that the development of complex structures and behaviors in a system can emerge from events taking place primarily within and through the system itself.

Self-propelled particle (SPP) models   Models of collective motion in which each group member is treated as a particle that responds to other group members within interaction zones. An individual moves toward or away from other individuals, or aligns itself with them, depending on which zone they occupy.

Semantic memory   The ability to acquire general factual knowledge about the world.

Semelparous (semelparity)   Reproducing once during a lifetime.

Semiclaustral founding   Colony founding procedure in which a founding queen or queens forage outside the brood cell to secure sufficient energy to rear the first generation of workers.

Semi-intact preparation   A piece of an animal, along with its nervous system, that produces a behavior or a component of a behavior. Such preparations are normally used primarily to allow access to the nervous system, but can also be used to eliminate sensory input or confounding inputs from other parts of the nervous system.

Semisociality   Social groups of same-generation adults and their offspring characterized by cooperative brood care (i.e., alloparental care occurs), and a reproductive division of labor, such that some individuals mainly reproduce while others mainly perform other tasks such as foraging and brood-care.

Senescence   The combination of biological processes of deterioration of organismic function in a living organism approaching an advanced age.

Sensillum   Hair-like structure that houses sensory neurons.

Sensitive phase   A stage of life during which the ability to learn is enhanced. Occurs most commonly early in life.

Sensory drive   The hypothesis that sensory systems and sensory conditions in the environment 'drive' evolution in particular directions.

Sensory environment   Multiple types of information – signals and cues from other animals and the physical environment – that may be perceived by an animal on the basis of its unique sensory capabilities (i.e., 'umwelt').

Sensory mode   The physical characteristics of signal production, on the basis of animal sense organs by which it is perceived (e.g., sound, patterns of light and color, vibration, etc.).

Sensory traps   In attempts to induce certain responses in other individuals, the use of stimuli whose effectiveness in inducing these responses evolved in a different context. In a sexual context, the male can produce a stimulus that elicits a particular female response; this female response exists because previous natural selection in another context favored such a response to the same (or a similar) stimulus.

Sentience   A general term for the ability to feel or perceive subjectively.

Sentinel   An individual in a group that remains vigilant and stands guard while other group members forage or carry out other activities (also called: sentry or guard).

Sentinel cells   A newly discovered cell that sweeps through a *Dictyostlium* slug mopping up toxins and bacteria, acting as a kidney, a liver, and an innate immune system.

Sequence divergence   Changes in the sequence of DNA bases in different populations or different species. Comparisons of the degree of sequence divergence are used to estimate how long ago the populations or species began to evolve independently.

Sequestering   Accumulation of a chemical in the integument or inner organs of an organism from an outside source (e.g., diet).

Serotonergic basal cells   Round cells at the base of the taste bud, which are immunoreactive to serotonin.

Serotonergic medications   Psychotropic medications that effectively increase the availability of the neurotransmitter serotonin in the brain.

Serotonin (5-HT)   A monoamine neurotransmitter that is derived from tryptophan. It is synthesized in the gut, pineal, and CNS. In the brain, 5-HT influences learning and memory as well as appetite, sleep, and muscle contraction.

Sex allocation   Sometimes used to refer to the process by, or the time at, which a parent bestows gender on offspring (see sex determination and sex allocation sequence, respectively), but more generally used to refer to how resources are apportioned to each gender (also referred to as *investment ratio*). Sex allocation can be thought of as an evolutionarily derived reproductive strategy of the parents and the sex ratio as one of its manifestations.

Sex allocation sequence   The order in which offspring of different gender are produced by a parent. Nonrandom sequences can, but do not always, imply parental control and can influence sex ratio variance.

Sex determination   The genetic basis of an individual's gender. There is an astonishing diversity of sex determination mechanisms among animals, often exerting a profound influence on reproductive behavior.

Sex-limited polymorphism   Occurrence of several discrete forms or morphs within one sex, but not the other sex.

Sex ratio   The proportion of individuals that are male, that is, males/(males + females). Sex ratios are sometimes given as the proportion females (this is not incorrect; there is no strict convention) and sometimes reported as the ratio of males to females, that is, males/females (termed sex ratio *sensu stricto*): this is not a recommended measure as it is not readily amenable to statistical analysis. The sampling unit may be indicated, for example, *population sex ratio*, *clutch sex ratio*, *parental sex ratio* (the sex ratio of offspring produced by a given parent or pair of parents). The developmental stage of offspring may also be indicated: *primary sex ratio* (the sex ratio at offspring production; this may be used to indicate the sex ratios at fertilization or at egg laying), *secondary sex ratio* (the sex ratio at some defined later stage of offspring development, for example, emergence or mating (adulthood)). Developmental mortality can mean that primary and secondary sex ratios are not equivalent.

Sex ratio variance   A measure of the diversity of sexual composition in groups of offspring (e.g., clutches, litters, etc.). Heterogametic sex determination (e.g., the XY system in mammals, the WZ system in birds) leads to the null expectation that distributions of group sex ratios conform to binomial variance. Deviations from the binomial expectation can, but do not necessarily, imply sex ratio control. Under haplodiploid sex determination, there is no particular null expectation of variance, but subbinomial variances have been observed in many haplodiploid species.

Sex-ratio conflict   Conflict between queens and workers over the investment into male versus female reproductives produced by the colony.

Sex role reversal   Occurs when males provide the majority of parental care, resulting in sexual selection on females, who can increase their reproductive success by obtaining additional mates.

Sex-role reversed species   Are those in which females compete for males and males choose among females. Typically, males take care of the young.

Sexual behavior   Behavioral interactions that facilitate the union of eggs and sperm.

Sexual coercion   Occurs when one sex, usually males, use force or the threat of force – forced copulation, harassment, intimidation, restriction of the movement of the other – to increase the probability that mating will occur.

Sexual conflict   Occurs whenever the fitness interests of individuals of different sexes conflict.

Sexual dialectics hypothesis   The idea that whenever the behavior and physiology of one sex decreases the fitness of the other, flexible individuals adaptively modify their behavior or physiology to resist the deleterious effects of interaction(s) with the other sex. Because control and resistance interactions are likely to be dynamic, changing during the lifetime of an individual, the sexual dialectics hypothesis predicts that individuals flexibly adjust resistance behavior in contemporary time.

Sexual dichromatism   A subset of sexual dimorphisms in which males and females of a species differ systematically in coloration or color pattern.

Sexual differentiation   In ontogeny, the anatomical and behavioral differentiation of males and females.

Sexual dimorphism   Refers to differences in morphology, behavior or physiology between males and females. Generally, more intense sexual selection results in greater sexual dimorphism.

Sexually antagonistic selection   A type of selection that is characterized by dynamic interactions – actions and reactions – between individuals of different sexes that can lead to a coevolutionary arms race.

Sexual reproduction   Reproduction involving gamete formation by meiosis and gamete fusion to form new individuals.

Sexual selection   Selection for traits that make individuals of one sex better able to compete for individuals of the opposite sex. As a consequence, some individuals have a mating advantage over other individuals of their own sex, such that there is nonrandom differential reproductive success among these individuals.

Sexual signals   Advertise the signaler's genetic or phenotypic quality in order to attract mates and deter rivals. Examples include conspicuous traits, such as bright colors and elaborate songs. Signals can be visual, acoustic, olfactory, tactile, or electric.

Sexual size dimorphism (SSD)   A subset of sexual dimorphisms in which males and females of a species differ systematically in body size.

Shaping   The procedure of reinforcing successive approximations of a desired behavior.

Short day breeder   An organism that enters full reproductive capability during short days of winter.

Sibling species   Anatomically similar species that are nonetheless reproductively isolated; in herbivorous insects, such species often use different host plants.

Sickness responses   The suite of adaptive behavioral and febrile reactions among vertebrate animals associated with the acute phase immune response that includes fever, iron withholding, reduced motivated behaviors such as food and water intake, and lack of sexual, parental, or other social interactions. These responses are critical to survival.

Sign   A signal; also anything that gives evidence or trace of something else; also a physical object, usually fixed in space, that is a signal when encountered by a receiver.

Sign stimulus   An external stimulus that elicits a specific motor pattern (modal or fixed action pattern).

Signal   A character or behavior that has evolved so as to provide information to other organisms.

Signal detection theory   A general model of the discrimination of signals from background noise that can be applied to data from psychophysical studies with animals and to situations where an animal must make a discrimination under conditions of uncertainty.

Signal dominance   When a multimodal signal generates a response in only one of its component modes in relation to other modes.

Signal enhancement   When receiver responses to redundant multimodal signals are increased in their intensity compared to unimodal signals.

Signal equivalence   When receiver responses to redundant multimodal signals are the same or equal to unimodal signals in their intensity (equivalence).

Signal independence   When the response to a multimodal signal includes the (different) responses to each of its unimodal components.

Signaling mode   The physical characteristics of a signal that enables it to be received by a specific type of sensory neuron in a receiver. Signaling modes include chemical, electric, sound, light, and vibration.

Signal parasite   An individual that exploits an existing communication system in a way that benefits itself at the expense of a signal giver or a signal receiver.

Signal redundancy   When individual components of a multimodal signal presented separately elicit the same response from a receiver and likely contain the same or similar kinds of information about the sender.

Significance level/criterion   In statistical analyses it is a criterion of probability below which a statistical test value is said to indicate a significant difference between populations.

Silkie   Asiatic breed of chickens characterized by fur-like plumage and dark blue flesh.

Simultaneous hermaphroditism   A sexual pattern characterized by individuals possessing both mature ovarian and spermatogenic tissue within the same functional gonad.

Single nucleotide polymorphism (SNP)   Variation in a DNA sequence that occurs when a single nucleotide – A, T, C, or G – varies between individuals of the same species.

Sinus gland   A neurohemal organ associated with the crustacean X-organ.

Siphon   Cylinder created by curling the edges of the mantle in some mollusks. It can be used to forcibly discharge the contents of the mantle cavity.

Sister groups   A pair of evolutionary lineages that share their most recent common ancestor and thus are necessarily equal in age.

Site fidelity (see philopatry).

Size constancy   The ability to determine the true size of objects despite viewing them at different distances when their images subtend various angles on the retina.

Skylight polarization   Due to scattering by particles in the earth's atmosphere, sunlight becomes polarized, with the light wave's electric field oscillating in one direction. The degree of polarization is maximal at 90°, relative to the direction of incident light.

Sloughing behavior   Specific behavior associated with sloughing off skin and associated structures such as hair, feathers, and scales. Often, this involves rhythmic movements to lift off old skin layers (e.g., in snakes), or movements allowing abrasion of skin with substrate (many birds and mammals) to break up and shed skin and its components.

Smoltification   The transformation or metamorphosis of anadromous salmonids from the parr to smolt stages, including changes in morphology, endocrinology, and behavior in preparation for saltwater entry. Some of these include increased plasma, thyroid hormone, and cortisol levels, as well as the deposition of guanine in the skin, giving the fish a silvery appearance. This impedes water loss, and with the increase of Na^+-ATPase pumps in the gills and gut, the osmoregulatory function improves as the fish enters the hyperosmotic conditions for seawater. Behaviorally, smolts leave the natal streams and migrate to open water.

Sneak spawning   Male reproductive behavior where an individual will attempt to fertilize eggs that are released during the courtship and spawning episode of another male–female pair; the individual is usually unable to defend a territory and court a female independently, and spends most of the time hiding to avoid agonistic encounters with territory-holding males.

Social cognition   Knowledge about group mates and social interactions.

Social (cooperative) spider   Spiders that share a nest, feed together, and have cooperative breeding.

Social cues   Products of the behavior of others that convey inadvertent social information.

Social dominance   The state of having high social status relative to other individuals, which react submissively during dyadic agonistic encounters. Dominant individuals have priority of access to resources over subordinate individuals.

Social eavesdropping   The extraction of social information by an individual (the eavesdropper) from a signaling interaction between other individuals (usually conspecifics) in which the eavesdropper takes no direct part.

Social facilitation or social enhancement   The effect of the mere presence of another animal on the production of a target response. The increase or initiation of a behavior already in one's behavioral repertoire when in the presence of others engaged in the same behavior.

Social foraging theory   A body of game theoretic models designed to analyze foraging decisions made under conditions of frequency-dependent payoffs.

Social hymenoptera   Meaning the eusocial Hymenoptera. Eusociality has evolved approximately nine times in the Hymenoptera, once in ants, three times in wasps, and approximately five times in bees.

Social influence   The effect of another animal on the production of a target response (e.g., contagion or social facilitation) that does not involve the acquisition of information about the to-be acquired response (e.g., imitation).

Social information   Information obtained by an individual from other animals in its social group.

Social insects (see eusocial)   Insects that live in groups in which some group members rear offspring that are not their own.

Social intelligence   An influential theory developed by Alison Jolly, Nick Humphrey, and others to explain the superior intelligence of primates, including humans. The theory is based on the idea that living in a complex social world requires cognitive abilities related to learning from others, forming social relationships in order to gain dominance, and deceiving others to gain resources normally unavailable to them. This extreme form has been named Machiavellian Intelligence.

Social interaction   A dynamic, changing sequence of social actions between individuals that modify their actions and reactions according to those of their interaction partner(s).

Sociality   Associations and interactions of individuals within a social group.

Social learning   Any process whereby the behavior of an individual is altered as a result of it either observing the behavior of another individual, interacting with it, or being exposed to its products.

Social learning strategy   An evolved psychological rule specifying under what circumstances an individual learns from others and/or from whom it learns.

Socially mediated learning   Learning that is influenced by presence and activity of conspecifics, also referred to as socially biased learning; the process by which social context contributes to learning.

Social mimicry   Imitation between species that associate among each other.

Social monogamy   A type of mating system in which one male and one female form a bonded pair for the purposes of reproduction. Typically, the pair will stay together and raise young together. However, both the male and/or the female may engage in copulations with other individuals from outside the bonded pair (EPCs).

Social network   Pattern of social connectedness, either through behavioral interactions or spatial proximity, between individuals in a population.

Social norm   A pattern of behavior that is accepted as being the normal way of behaving for a particular group of people, and to which all the group members are expected to conform.

Social organization   The size, demographic composition, and spatiotemporal coordination of individuals within a group.

Social (other-regarding) preferences   Behavior motivated out of concern for the effects it has on other individuals over and above material self-interest; these can be positive or negative (social science).

Social parasitism   The coexistence in the same colony of two species of social insects, one of which parasitizes the other.

Social selection   A type of natural selection characterized by nonrandom, differential reproductive success of individuals bearing some trait relevant to social interactions (either competitive or cooperative) for access to resources such as food, territories, allies, and mates.

Social structure   The pattern of relationships among individuals within groups, groups within demes (subpopulations), and demes within a population of a given species.

Social transmission   Transfer of information among individuals in a group or population, both within and between generations, through social learning or teaching.

Social transport   A form of recruitment used by certain ant species, in which one ant carries another to a destination, typically in a stereotyped posture. This is most commonly seen when colonies emigrate from one nest site to another.

Sociobiology   An extension of Darwinian theory and the evolutionary synthesis that developed during the 1960s and 1970s. The core principles were that natural selection works at the level of the individual or gene, not the population or species (still contested) and that the representation of one's genes in future generations could be achieved by facilitating the reproductive success of close relatives.

Sociomatrix   For a group with n members, an $n \times n$ matrix with each group member along the vertical and horizontal axes and each entry in the grid as the weight of the social relationship, if any, between the two intersecting individuals.

Soldier   Similar to workers, these are nonreproductive members of a colony (sometimes known, especially in ants, as 'major workers'). Unlike normal workers, however, they are generally larger, with specialized head structures, and primarily perform nest-defense tasks.

Solitarious   Living singly or in pairs; the term refers to behavioral, morphological, and physiological traits (especially for the solitarious phase of locusts).

Somatic   Pertaining to the body.

Somatic fusion   The process by which the nonreproductive tissues of two individuals join to form a single individual with a shared body (soma). In many taxa, this can occur either between clones or closely related individuals. Fusion between allogeneic (nonclonemates) organisms produces a genetically chimeric individual.

Somatic recombination   The process by which regions of the genome are physically edited in the nucleus, giving rise to novel genetic elements. Accounts of this process are rare and are known from a few systems where genetic diversity is of primary importance.

Somatic rejection   The process by which two individuals reject each other, often involving the formation of a physical barrier between them and the preservation of genetic individuality.

Somatosensory system   A diverse sensory system comprising the receptors and processing centers to produce the sensory modalities such as touch, temperature, proprioception (body position), and nociception (pain). The sensory receptors cover the skin and epithelia, skeletal muscles, bones and joints, internal organs, and the cardiovascular system.

Somatotropic axis   A group of hierarchically regulated hypothalamic, pituitary, and peripheral tissue hormones which are involved in the regulation of somatic growth.

Song   Loud, often complex sound usually produced by males of a species in defense of a breeding territory and/or to attract females.

Song control nuclei   Interconnected regions of the brain in songbirds that regulate the production and learning of song.

Sore footed   A type of lameness which is caused by pain in the animal's hoof.

Sparse coding   The representation of information in the nervous system by the activation of a relatively small set of neurons.

Spatial contrast sensitivity function   Plot of the contrast required to detect gratings of different spatial frequencies.

Spatial resolution (acuity)   The ability of an animal to perceive spatial detail.

Spawning   Oviposition, or the deposition of eggs, in water.

Spearman coefficient   A measure of correlation between ranked data; can be used as a measure of observer reliability.

Spectral sensitivity   The differential sensitivity of photoreceptors to different wavelengths of light.

Spectrogram   A display of the frequency components of a sound over time.

Speed/accuracy tradeoff   A fundamental decision-making constraint that captures the cost in time that must be paid to improve the accuracy with which the best available option can be chosen.

Sperm allocation   Refers to situations in which males that are running low on sperm will vary the amount of sperm in an ejaculate, so as to provide more sperm for some females and less for others. Generally, it is assumed that males will provide more sperm for females that are of higher quality or status.

Spermatheca   A small sac associated with the median oviduct of the female, in which sperm are stored following copulation.

Spermatophore   A sac produced by accessory glands of male insects and directly or indirectly transferred to the female, containing sperm and often proteinaceous material.

Sperm capacitation Changes the spermatozoa undergo to become ready to interact with the ovum and hence able to fertilize.

Sperm competition A type of sexual selection that can occur if a male or his seminal products directly reduce the changes that the sperm of other males which have mated with the same female have of fathering her offspring. This is the postcopulatory equivalent of male–male battles.

Sperm depletion Refers to the fact that males may be limited in the number of sperm that they can produce per unit time and eventually they may run out of sperm. In such cases, males need a period of time to rebuild their sperm supplies.

Sperm precedence An individual male's share of paternity when females mate with multiple partners.

Spherical aberration Optical imperfection caused by light striking a refractive surface at different points being focused in different planes.

Spite A behavior that reduces the lifetime fitness of the recipient while also reducing the fitness of the actor (evolutionary biology); harming behavior resulting from a desire for the suffering or misfortunes of another individual (social science).

Split sex ratios Population-wide bimodal sex-ratio distributions with co-occurring colonies that specialize in the production of either male or female reproductives.

Sporozoites A stage in the life cycle of apicomplexan protists that is produced by sporulation and invades host cells.

Stabilizing selection A form of selection in which deviations from a main phenotype, such as changes to a conspecific call type, are selected against maintaining the same phenotype over evolutionary time. Contrast with directional selection.

Stable group size A group size at which no individual can gain by unilaterally leaving or joining the group.

Stable isotopes Nonradioactive forms of an element having an extra neutron; stable isotopes of carbon, nitrogen, and hydrogen, among others, are very useful for ecological and behavioral studies.

Stable supine posture The ability of infants to lie on their backs on the ground or another surface is uniquely human. Chimpanzee and other non-human primate infants are unstable when they are laid on their backs – they move their limbs in an attempt to grasp and cling to something. From a developmental perspective stable supine posture enabled humans to become by far the most versatile and proficient tool users in nature. The stable supine posture provides the basis of tool use, face-to-face communication, and vocal exchange.

Stage 4 sleep The deepest stage of NREMS. It is characteristic of the first half of the night in humans; our ability to enter this stage diminishes with aging.

Startle signal See deimatic signal.

Starvation–predation risk trade-off Animals must balance the time or effort they spend feeding to prevent themselves from starving, with the time or effort they spend looking out for predators to prevent themselves from being eaten. Any animal that spends all its time looking out for, or avoiding predators will starve to death. Any animal that spends all its time feeding may not ever starve, but is more likely to be caught by a predator.

State-dependent model Models that use the techniques of stochastic dynamic optimization to predict animal behavior. Often used to model tradeoffs that animals face when having to decide between competing factors such as getting food and avoiding predators.

Stationary molt An intermittent molt that is associated with a lack of increase in body size and morphological development. This type of development occurs in several insect species and is frequently associated with periods of food shortage, when a larva or nymph is not capable of passing a critical mass threshold in an instar. In some termites, it might also be linked to the wear of mandibles.

Statocyst Inertial balance organ of aquatic invertebrates, consisting of a heavy mineral body (statolith) resting on a bed of mechanoreceptors that register the displacement of this body whenever its orientation relative to the direction of gravity changes.

Stereocilia Nonmotile tufts of secretory microvilli on the free surface of cells. Thought to be a variant of microvilli and characterized by their length (distinguishing them from microvilli) and their lack of motility (distinguishing them from cilia).

Stereotypic behavior Behavior that is repetitive, relatively invariant, and has no obvious goal or function.

Steroid hormone A class of molecules that include the sex hormones and stress hormones from the adrenal cortex that share a common biosynthetic pathway.

Steroid receptors Steroid hormones act largely through intracellular steroid receptor proteins that bind hormone and then function as 'ligand activated transcription factors' to regulate gene expression in target cells. Teleosts have multiple forms of both the estrogen and the androgen receptors.

Stimulant A substance that quickens and enlivens the physiological and metabolic activity of the body.

Stimulus (In physiology) Something that can elicit or evoke a physiological response in a (sensory) cell, a (sense) organ, or an organism; it can be internal or external; (in psychology) something that has an impact or an effect on an organism so that its behavior is modified in a detectable way.

Stimulus enhancement   The facilitation of an observer's response (e.g., through approach and manipulation) resulting from the pairing of an object with reinforcement.

Stimulus generalization and discrimination   When prey respond to the olfactory, auditory, or visual cues of a species which are similar to those of another species, prey are said to generalize their species recognition to these cues. If prey fail to respond or respond weakly to these cues because they are dissimilar, they are said to discriminate these cues from those of another species. Therefore, stimulus generalization and discrimination are reciprocal effects, with higher stimulus generalization indicating lower stimulus discrimination.

Stochastic dynamic optimization   A mathematical technique that predicts optimal behavior by having computers examine every possible set of behaviors. This produces a numerical rather than an analytical solution as found by the marginal value theorem or Gilliam's rule.

Stop-over habitats   Habitats along the migration routes of animals that allow them to feed and replenish fat stores before moving on.

Slotting   Vertical jumping in ungulates during flight away from a predator in which all four legs leave the ground at the same time, the legs being held straight while the animal is in the air; similar behaviors include pronking, spronking, bounding, and leaping; may function to deter further attack by a predator or distract the predator's attention away from vulnerable offspring.

Strategic design   Aspects of signals relating to its function, for example, brightness of plumage conveying male quality.

Strategy   A set of behavioral decisions that are highly heritable, associated with a particular genotype within the gene pool of a species.

Stratified squamous epithelium   An epithelium characterized by multiple layers of flat, scale-like cells called 'squamous cells.'

Stress   A descriptive label with varying meanings for the biological processes involved when an animal perceives a threat that challenges internal homeostasis (both motivational and physiological 'set points') and the behavioral and physiological adjustments that the organism undergoes to avoid or adapt to the stressor and return to homeostasis. An environmental effect on an individual that overtaxes its control systems and results in adverse consequences and eventually in reduced fitness.

Stress response   The physiological and behavioral responses to a sudden emergency situation.

Stressor   A challenge (whether physical or psychological) to homeostatic balance (see 'homeostasis').

Stress-response   The array of neural and endocrine adaptations that occur in the body in response to a stressful challenge.

Stretch activation   In some muscles, physical elongation by mechanical means can lead directly to contraction of the muscle, counteracting the induced stretch.

Striated muscle   Also known as 'skeletal muscles,' these muscles have alternating bands of overlap and nonoverlap between thick (myosin) and thin (actin) filaments, giving them a striated appearance.

Stridulation   The rubbing of skeletal elements against one another that is a common form of sound production in fishes and many insects.

Strong inference   A method in the cognitive structure and logic of scientific discovery in which investigators attempt to identify and test simultaneously alternative hypothetical-deductive hypotheses with crucial predictions. Crucial predictions are predictions about a phenomenon that are in opposite directions. If tested well with a crucial experiment, two hypotheses can be tested simultaneously and one hypothesis supported and another rejected.

Strong reciprocity   A propensity to reward others for cooperative, norm-abiding, behaviors coupled with a propensity to punish others for norm violations.

Stunning   A method that renders animals insensible to pain before slaughter.

Sublethal effects   Effects that are negative, but do not immediately kill the organisms, such as decreased ability to stand, walk, eat, or avoid predators.

Submission   Behavior that indicates a low probability of initiating aggressive behavior. A submissive individual, however, may respond to injurious aggression with aggression. Submissive individuals often terminate interactions by physical distancing.

Subordinance   The state of having low social status in a group, often because the individual was defeated in an aggressive encounter.

Subordinate   A low-ranking individual within the group that does not usually get access to resources. Subordinate individuals tend to be smaller and weaker and do not form close social networks.

Subsociality   Family groups consisting of parents and immature offspring, and are characterized by brood defense or brood provisioning by parents.

Subsocial route   Social grouping that originates from an extended family and restricted, or no dispersal of, young.

Suprachiasmatic nucleus (SCN)   The suprachiasmatic nucleus (SCN) of the hypothalamus is a bilateral structure that sits at the base of the mammalian brain. It serves as a central pacemaker and acts to synchronize the body with the environment. It also synchronizes endogenous circadian rhythms. The SCN can be divided into two areas, a ventral area containing cells that receive direct input from the retina, and a dorsal area which contains highly rhythmic cells that serve in output processes. Input from the ventral SCN synchronizes the rhythmic cells of the dorsal SCN.

Surprisal   In the mathematical theory of communication, the entropy or information associated with a particular signal.

Survivorship cost   Reduction in fitness in the form of decreased probability of survival.

Swimbladder   An anatomical structure comprising connective tissue, filled with a mixture of oxygen, carbon dioxide, and nitrogen (hence, also known as a 'gas bladder') that has multiple functions among fishes including control of buoyancy, sound production, and sound reception.

Symbol   Something – such as an object, picture, written word, a sound, or particular mark – that represents (or stands for) something else through association, resemblance, or convention, especially a material object used to represent something invisible.

Sympathetic nervous system   The branch of the peripheral nervous system, regulated by epinephrine and norepinephrine release, that orchestrates the immediate responses to a stressor.

Sympatric   Geographically overlapping; for example, populations on the same island with no barrier to movement between them.

Sympatric speciation   The development of isolating mechanisms while incipient species are within the same geographic area, specifically when individuals from each population are within cruising range of one another.

Synanthropic (synanthropy)   Describes a population of wild animals that lives near or within human settlements or anthropogenic habitats; usually implies some degree of dependence on humans or exploitation of human-derived resources.

Synapomorphy   A shared, derived character; the only valid character type for revealing phylogenetic affinities.

Synapse   The gap or junction between nerve cells.

Synaptic pruning   The reduction in the number and connectivity of synapses that may accompany development.

Synaptic transmission   The transmission of an electrical signal from one cell to another which occurs at the point of connection between these two nerve cells called synapse.

Syrinx   Musculoskeletal structure that functions as a vocal organ found among birds.

Systematics   The general field of researching inferring, and proposing the evolutionary relationships of organisms. One of the oldest fields of biology, its relevance today is stronger than ever, including multiple sources of data and methodological techniques.

Tachycardia   An increase in heart rate.

Tactical design   Aspects of signals relating to its effectiveness in transmission, for example, male songs with higher amplitude signals in certain frequencies get more female attention.

Tactics   A set of behavioral decisions for which the phenotype develops as a result of any combination of learned mechanisms (genetic heritability is unspecified).

Tail streamer   The elongated outer tail feathers (rectrices) of the swallow tail, giving the tail its forked appearance.

Tandem run   A form of recruitment used by certain ant species, in which one ant leads a single follower to a destination. The pair remain in contact by the exchange of pheromone signals from the leader and tactile signals from the follower.

Tangled bank theory   The idea that the world, and the challenges that it poses to organisms, is variable and complex. In such a world, the production of genetically variable offspring increases the chances of at least some of them being able to survive and reproduce.

Tapetum   Reflective layer in either the retinal pigment epithelium or choroid that reflects light not absorbed by the photoreceptors back through the retina, thus improving sensitivity in animals in low light levels.

Task   A behavior or set of behaviors that contribute to the work necessary for the function of a social group.

Task specialization   When an individual within a social group preferentially performs one task over other tasks being performed by that group.

Task threshold   The level of stimulus required to make a worker engage in a task.

Tastant   Chemical molecule that induces the sensation of taste, such as sugars or salts.

Tautologous   A circular logical argument in which the conclusion is included in the propositions.

Taxis   The movement of an organism in a particular direction with reference to a stimulus. A taxis usually involves the employment of one sense and a movement directly toward or away from the stimulus, or else the maintenance of a constant angle to it. (Contrast with kinesis.) See phototaxis and geotaxis as examples.

Taxonomy   The scientific discipline concerned with studies of taxa, including the subdisciplines of systematics and nomenclature.

Teaching Behavior modified by an experienced individual in the presence of a naïve individual, such that the naïve individual learns the behavior more quickly than it would otherwise and at some cost to the teacher.

Tegmen (pl. tegmina) A leathery, hardened forewing (usually of Orthopteroids).

Teleost Fish infraclass Teleostei within the ray-finned, bony fishes, excluding gars and bowfins. One of three infraclasses of ray-finned fishes (Actinopterygii) that includes most common fish.

Template The neurological or physical model against which cue bearers are compared and evaluated.

Temporal caste discretization A form of age-related division of labor in which workers form distinct age groups that have roles composed of sets of nonoverlapping tasks.

Temporal contrast sensitivity function Plot of the contrast required for detection of a light flickering at different frequencies.

Temporal discounting A decrease in the subjective value of a delayed benefit.

Temporal information Of, relating to, or involving an awareness of time.

Temporal information processing The sequence of computational steps that are hypothesized to occur while processing events that unfold in time.

Temporal polyethism A pattern of division of labor in eusocial insect colonies in which task performance is associated with worker age.

Temporal representation The internal format of stored information about events that unfold in time.

Temporal structure Describes the amplitude and frequency modulations of an acoustic waveform over time.

Tergal glands Glands on the dorsal surface of the abdomen; usually referring to those on males that entice females into position for copulatory engagement.

Terminal investment strategy Is a term sometimes used to refer to species in which the male usually, or always, is killed and cannibalized by the female during, or immediately after, copulation. The term implies that the male may be investing in its future offspring by providing food and nutrients (its own body) to the female.

Termites, higher Comprises only the termite species of the family Termitidae. They have bacterial gut symbionts only.

Termites, lower All termites with the exception of the Termitidae. Lower termites harbor bacteria and flagellates in their guts.

Territory Any defended space; can be for breeding, foraging, caring for young, or a combination.

Test of congruence A central component of phylogenetic systematics; it is the result of simultaneous analysis of characters. Given sufficient evidence, true synapomorphies will tend to reinforce one another guiding tree inference, and characters that do not in fact reveal phylogeny will be revealed as such. The test of congruence is therefore the primary tool in testing homology and identifying homoplasy, and it flows logically from the recognition that there is but one optimal phylogeny for a group of taxa.

Testosterone (T) Important androgenic steroid hormone in all classes of vertebrates; critically, this steroid often functions as a biosynthetic intermediate in estradiol or 11-ketotestosterone production.

Tethered flight A laboratory technique in which an insect is suspended by a wire or stick attached to its dorsal surface; with a wind blowing on the head, removal of foot (tarsal) contact triggers sustained flight.

Thanatosis An antipredator behavior in which the organism feigns death.

Thelytokous automixis A kind of parthenogenesis in which two gametes produced by meiosis fuse to produce a diploid female.

Thelytoky A form of parthenogenetic reproduction in which only female offspring are produced.

Theory of mind The ability to attribute mental (cognitive) states to others.

Thermocline A zone of rapidly changing temperature.

Thermolability See poikilothermic.

Thermoneutral ambient temperature An ambient temperature where the activities of heat-producing and the heat loss mechanisms are at a minimum level; the animal needs the least thermoregulatory effort to maintain its normal body temperature.

Thermoregulation The ability of an organism to keep its body temperature within certain boundaries, even when the temperature surrounding is very different. The regulation of body temperature.

Thiamine A water-soluble vitamin of the B complex (vitamin B_1), whose phosphate derivatives are involved in many cellular processes.

Third-party punishment Imposition of sanctions by an impartial observer on an individual for actions directed toward a third party; see also Policing punishment (social science).

Third-party relationships Relationships or interactions among conspecific group members in which the observer itself is not directly involved.

Threshold The lowest stimulus strength that reliably elicits a response; a low threshold means high sensitivity; the exact criterion for threshold differs among studies.

Thyroid hormones Iodinated tyrosine residues produced in the thyroid gland. The gland mostly secretes thyroxine (T4) that is then converted in the blood or in target organs by deiodinases to tri-iodothyronine (T3). T3 is regarded as being the biologically active form. Biological effects of thyroid hormones include regulation of metabolism (temperature regulation), development, and behavioral effects.

Thyrotropin (TSH) Glycoprotein hormone comprising two subunits produced by the anterior pituitary gland that stimulates the production of thyroid hormone by the thyroid gland.

Thyrotropin-releasing hormone (TRH) Tripeptide produced in the hypothalamus and extrahypothalamic sites that stimulates the release of TSH by the anterior pituitary gland.

Time perception The experience of time.

Time sampling Behavior is sampled periodically at a specified sample point at the end of a specific sample interval.

Time series analysis Events that unfold in time may be characterized by the periodic trends that make up the temporal structure of the events.

Timing The general ability to keep track of time.

Tonotopy The orderly mapping of frequency along the cochlea. The orderly arrangement of frequency is then preserved in each of the successively higher nuclei of the auditory system up to and including the auditory cortex.

Tool-use Directing an unattached object towards one's self or another object (animate or inanimate) in order to achieve a goal.

Tool-user A species that regularly uses tools in its natural environment.

Totipotent In eusocial insects, having the ability to express either the reproductive queen or the nonreproductive worker phenotype.

Toxic Substances that are poisonous and injurious or lethal to predators that attempt to consume it.

Toxicological effects Direct effects of chemicals that interfere with physiological processes resulting in the deterioration of function and may ultimately cause organ and system failure.

Trade-off The cost–benefit approach has been extended to model when this benefit-to-cost ratio is optimal, and states that an individual should maximize the benefit of the behavior while simultaneously minimizing any costs associated with the behavior. In other words, the benefit of any particular behavior should be considered with the costs associated with the behavior.

Trade-off theory A theory to explain the emergence of symbolic representation in humans. At a certain point in human evolution, brain capacity reached a limit and in order to accumulate new functions, old functions needed to be lost. Consequently, humans may have lost much of their ability for olfactory processing and developed instead highly sensitive visual, auditory, and crossmodal functions. A similar scenario may be applied to the trade-off between memory and symbol use, where human memory capacity may have been sacrificed in exchange for enhanced symbolic capabilities.

Tradition An enduring behavior pattern shared among members of a group that depends to a measurable degree on social contributions to learning.

Tragedy of the commons A situation in which individuals would do better if they all cooperate, compared to them all defecting, but in which cooperation is unstable because each individual gains by selfishly pursuing their own short-term interests (cf. Prisoner's dilemma).

Trained losing and winning The learning processes whereby an animal either acquires a stronger tendency to submit or yield to other individuals after losing previous agonistic encounters, or acquires a stronger tendency to attack or dominate other individuals after winning previous encounters. Modification of the tendency is in relation to other individuals generally, not limited to opponents involved in previous encounters.

Transcellular diffusion Substances travel through the cell, passing through both the *apical membrane* and the *basolateral membrane*.

Transcription factor A gene that directly affects the expression of another gene or genes.

Transcriptional Relating to transcription, the process by which DNA is converted into messenger RNA.

Transcriptome The set of all messenger RNA (mRNA) molecules produced in one cell or a population of cells, or in a given organism, under particular environmental conditions at a given time. Transcriptomics is the large-scale study of gene expression level (mRNAs) in a given cell population (such as brain cells), often using high-throughput techniques based on DNA microarray technology.

Transduction mechanism In a sensory neuron the odorant receptor-ligand complex induces a series of cellular reactions that ultimately release action potentials in the axons.

Transiens Transitional locust phase, from the solitarious to gregarious or vice versa.

Transition matrix Squared matrix in which each row is a probability distribution. This is the fundamental element of a Markov chain.

Transitive inference A form of reasoning in which given prior information a subject deduces a logical conclusion.

Specifically, the ordinal relation between two elements in a series must be inferred from information that establishes the relations of those two elements to a third.

Transitivity A fundamental principle of rational choice behavior that applies specifically to binary choices. Preferences are transitive between the three options A, B, and C if A is preferred to B, B is preferred to C, and A is preferred to C.

Translational Relating to translation, a process by which mRNA is converted into protein.

Translocation Technique used in wildlife conservation, wherein wild individuals are captured from one location and transported and introduced to another part of their range, often with the purpose of re-establishing a local population which has become extirpated.

Transmission distance Refers to the change of sound intensity with increasing distance relative to a reference point.

Transposable element A mobile piece of DNA that can insert itself into the genome.

Trematodes Groups of parasitic worms, commonly referred to as 'flukes.' Almost all trematodes infect mollusks as the first host in the life cycle, and most have a complex life cycle involving other hosts. Most trematodes are monoecious and alternately reproduce sexually and asexually. The two main exceptions to this are the Aspidogastrea, which have no asexual reproduction, and the schistosomes, which are dioecious. The Trematoda are estimated to include 18 000–24 000 species.

Triadic mother–infant–object relationships It is also called 'social referencing.' Human infants often manipulate objects within a social context. Suppose that a human infant encounters a new toy. She may look up at the mother *before* touching it. The mother may nod or smile, and only then will the infant actually start manipulating the object. While playing with the toy, the infant may often show it to the mother while smiling. The mother may smile back at her child and give social praise.

Trigeminal nerve The fifth cranial nerve in vertebrates, which is known to be both sensory and motor in function. The ophthalmic branch of the trigeminal has been shown to be sensitive to magnetic fields.

Trigger neurons A class of command neurons whose short-lasting activation (e.g., less than a second) produces a long-lasting behavioral response (e.g., for tens of seconds). This term was coined in the study of leech swimming activation to distinguish these neurons from *gating neurons*, a class of command neurons that must be active during the whole time while a behavior takes place.

Tri-trophic level interactions Interactions that take place between organisms at three different levels within a food chain, for example, a plant, herbivore, and a carnivore.

Trivial movement See Appetitive movement.

Trophic cascades The indirect effects of top predators on the population processes of plants and animal species at lower trophic levels, as mediated by the density and foraging behavior of intermediate consumers.

Trophic level An organism's feeding position in a food web, with primary producers occupying the lowest level, herbivores the second, and carnivores occupying higher trophic levels.

Trophollaxis Mouth-to-mouth transfer of food or other substances.

Tropic hormone A hormone that modulates the secretion of another hormone.

True workers In termites, workers in colonies of foraging termites. They can be considered altruistic individuals as they perform most tasks within a colony (e.g., foraging, brood care, and building behavior) except for reproduction and specialized defense. Although they sometimes, especially in lower termites, still have some reproductive options (for instance as neotenic reproductives), their morphological differentiations (especially their sclerotization) largely restrict their developmental capability. In functional terms, these true workers, often just called workers, are equivalent to the workers of the social Hymenoptera, even though the latter are imagoes, whereas the true workers here are preimaginal stages.

Trypanosomiasis The name given to several diseases of vertebrates, including man, that are endemic in parts of Africa and the American continents. They are caused by protozoan parasites of the genus *Trypanosoma*.

Tuber Enlarged area of a root (e.g., a sweet potato).

Two-action design An experimental design used in social learning studies, in which each of two different actions on the same object is modeled in either of two different experimental conditions, permitting measurement of the extent to which observers match their later behavior to the alternative they witnessed.

Two-action procedure The demonstration of a response in two distinctly different ways that results in the same effect on the environment (e.g., stepping on vs. pecking at a treadle).

Tympanum Eardrum; a thin membrane that vibrates in response to sound.

Type I error A statistical error in which the null hypothesis is rejected when it is, in fact, true.

Type II error A statistical error in which the null hypothesis is not rejected when it is, in fact, not true.

Ultimate causation Evolutionary explanations of animal behavior. Questions about ultimate causes of behavior are about why a behavior is expressed. Ultimate causes explain the adaptive significance of behavior.

Ultrasonic vocalization   A vocalization consisting only of frequencies higher than 20 kHz, that is, higher than the range of frequencies audible to human ears. Many species hear very high frequencies, well above the frequency range of human hearing.

Ultrasound   Sounds with frequencies above the limit of human hearing; normally considered to be 20 kHz and higher.

Unconstrained parents   Individuals mated to partners they do individually prefer.

Undertaking behavior   A behavioral routine found in social insects that involves collecting and removing the corpses of colony-mates from the nest.

Units   In extracellular multichannel recordings, investigators use mathematical techniques to separate differently sized and shaped action potentials from each other, calling each one a 'unit.' It is thought that these represent recordings from individual neurons. Because of the properties of extracellular recording, however, one cannot be absolutely certain that these are unique neurons. Hence, people who work in this area often use the less specific term 'unit.'

Univoltine   Having but a single generation a year.

Unpalatable   Unable to be eaten due to an unpleasant/noxious taste or toxicity.

Usage learning   Where an animal comes to use an existing signal in a new context as a result of experience of the usage of signals by other individuals.

Usurpation   Take over or adoption of nest, brood, and/or workers produced by other queens.

Vacuum activities   Behaviors, such as fly snapping, performed out of context, without an obvious stimulus.

Vagotomy   The transsection of the vagus nerve.

Vagus nerve   The tenth of the 12 pairs of cranial nerves, which originates in the brain stem and sends nerve fibers to the head, neck and viscera, including the lungs, heart, liver, and gastrointestinal tract. Most of the nerve fibers in the vagus nerve are sensory and the remainder are part of the parasympathetic nervous system. It contributes to the innervation of the viscera and conveys sensory information about the state of the body's organs to the central nervous system. The vagus is also called the *pneumogastric* nerve since it innervates both the lungs and the stomach.

Value   One of the alternative states of a variable; in communication, one of the alternative signals of a code.

Value of information   The fitness of an animal with access to information, contrasted to the fitness of an animal without access to the information.

Variable reinforcement schedule   In operant conditioning, the reinforcement of a desired behavior is given at random intervals.

Variance   A measure of statistical dispersion obtained by averaging the squared distance of its possible values from the expected value (mean). Whereas the mean is a way to describe the location of a distribution, the variance is a way to capture its scale or degree of being spread out.

Variance in fitness   A measure of deviation from mean fitness.

Variance in number of mates   Refers to a measure of the variation (deviation around the mean) in the number of mates obtained by different individuals of the same sex within a population. For example, in any given population, variance in number of mates is low when all individuals of one sex are able to obtain more or less the same number of mates. Variance in number of mates is high when some individuals mate with many members of the opposite sex, while others mate with very few or none.

Variance in reproductive success   Refers to a measure of the variation (deviation around the mean) in number of young produced by different individuals of the same sex within a population. For example, in any given population, variance in RS is low when all individuals of one sex produce more or less the same number of young. Variance in RS is high when some individuals produce most of the young, while others produce few or none.

Varroa mite   The mite species *Varroa destructor*, originally a pest of *Apis cerana* and now found on *A. mellifera*. A serious pest of honeybees and the cause of substantial colony mortality in *A. mellifera*.

Vasopressin   A peptide produced predominantly by magnocellular cells within hypothalamus, but also by centrally projecting neurons within the hypothalamus and amygdala.

Vasotocin   Peptide hormone secreted by the posterior pituitary; also released in the brain where it affects many social behaviors.

Veliger   One of the larval stages of some mollusks, including gastropods.

Venomous   Substances that are toxic and injure or kill animals, in most cases injected by biting or stinging.

Vent   External opening of the cloaca.

Ventricle   A cavity within the brain that is filled with cerebrospinal fluid. The cerebroventricular system comprises four ventricles: two lateral ventricles, the third ventricle, and the fourth ventricle. Cerebrospinal fluid flows from the lateral ventricles, to the third ventricle, then to the fourth ventricle before leaving the brain and entering the central canal of the spinal cord or into the subarachnoid space.

Vergence eye movements   Eye movements where the angle between the eyes changes.

Vertex   A component of a network with known relationships to others in the graph model representing the network; in a social network, this can be an individual animal or group; also called a *node* or *point*.

Vertical social influence   Influence by an individual on another from a different generation, such as a mother's influence on her offspring.

Vesicle   Knob-like structure on the terminal region of a nerve cell that stores and releases neurotransmitters. Also called synaptic vesicle.

Viability   Capacity for survival, more specifically used to mean a capacity for living, developing, or germinating under favorable conditions.

Viability selection   Selection generated by variation in survival among individuals of a population.

Vibrissae   Specialized hairs usually used for tactile sensation (singular: vibrissa).

Vicarious (or social) sampling   Gathering of information about the environment by observation of the behavior or products of the behavior of others.

Vigilance   Visual or auditory monitoring of the surroundings aimed at detecting threats related to predation. Vigilance can also be aimed at rivals or mates within the group.

Viral vector   A virus that is engineered to transport a specific DNA sequence into infected cells.

Viscera   The organs in the cavities of the body.

Visual acuity   Spatial visual resolution, the minimum angular separation between two objects that are perceived as different within the visual field.

Visual fields   Volume of space around an animal from which visual information can be obtained.

Vitellogenesis   Yolk deposition into the oocyte (egg).

Vitellogenin   Egg yolk precursor protein involved in regulation of behavioral maturation in social insects.

Viviparous   An animal giving birth to live young which have developed inside the body of the parent.

Vocal mimicry   Imitation by one species of sounds produced by another.

Vocal muscle   A vertebrate striated muscle used in sound production; also known as 'sonic muscle.'

Vocal production learning   Signals are modified in form as a result of experience with those of other individuals, leading to signals that are either similar or dissimilar to the model.

Vomeronasal organ (Jacobson's organ)   An accessory olfactory (odor-detecting) organ that is located in the roof of the mouth or nasal septum. The vomeronasal organ is particularly important for processing odors related to social signals.

Waders   Used in Europe and refers to shorebirds but used in North America with reference to herons and egrets.

Waiting game   A game in which both predator and prey need to decide for how long to wait when the prey entered a refuge that restricts its ability to collect information about the continued presence of the predator.

Wave refraction zone   The shallow area of ocean adjacent to a coastline where waves approaching the shore at an angle are redirected by interactions with the sea floor so that they approach directly toward shore.

Wavelength   The spatial distance between two consecutive cycles of a sine wave. Numerically, wavelength is the velocity of sound divided by its frequency. In a given medium, low-frequency sounds have long wavelengths and high-frequency sounds have short wavelengths.

Weakly electric fish   Electric fish with electric organs that produce very weak electric organ discharges that function in electrolocation and communication, but that are too weak to function in stunning predators or prey.

Weaning   The transition of young mammals from nursing to independent feeding, especially the parent's role in facilitating that transition.

Weber's law   A psychological law stating that one's ability to discriminate two quantities or intensities depends on the ratio between them.

Welfare   The health, happiness, and prosperity of an individual in its state as regards its attempts to cope with its environment; equated with 'well-being,' generally measured on a scale from very good to very poor.

Welfare illustrator grid   The assessment and two-dimensional illustration of welfare, designed to account for a temporal component and the cause of the animal's suffering.

'When' strategy   A social learning strategy specifying the circumstances under which individuals copy others.

'Who' strategy   A social learning strategy specifying from whom individuals learn.

Wild-type   The phenotypic composition of an organism as it occurs in nature.

Wing aspect ratio   The ratio of wing length to wing width; high aspect ratio wings permit fast, agile flight; low aspect ratio wings permit slow, maneuverable flight.

Wing polymorphism   Having more than one wing form within a population, for example, long-winged (migratory) and short-winged (nonmigratory) individuals may be found within the same population of many species of planthoppers (known as 'wing-dimorphic species').

Wintering area   In migratory birds, the area where populations spend the nonbreeding season, usually at lower latitudes.

Wintering dispersal   The distance between the wintering site of an individual in one year and its wintering site in another year.

Winter territory   A home range that an individual occupies and defends its boundaries against others (usually conspecifics but sometimes other species as well). This territory/home range may be held exclusively by the individual or as a pair or as a small group.

Wiring costs   The energetic costs associated with total length neural wiring (axons and dendrites).

Wisdom of crowds   The principle that the collective performance of a group of decision-makers can exceed that of a randomly chosen individual acting alone.

Within-pair offspring (WPC)   Offspring sired by the social father.

Within-sex variance in reproductive success   An operational definition of sexual selection.

Worker   Individual in a eusocial society that primarily performs all nonreproductive tasks in a colony. In primitively eusocial groups, this individual may be physically capable of reproduction; however, in highly eusocial groups, it is effectively sterile.

Xenoestrogens   Chemicals that are produced for agricultural, private, or industrial use that have estrogenic activity in living organisms.

X-organ   A group of neurosecretory neurons in the crustacean eyestalk that synthesize several peptide hormones.

Y-organ   The molting gland of crustaceans that usually secretes ecdysone, the precursor of the active form of the molting hormone, 20-hydroxyecdysone.

Zeitgeber   German word for 'time-giver'; an exogenous cue that entrains an endogenous biological rhythm.

Zoological psychology   A part of animal psychology that lies at the boundary between psychology and zoology. The approach is animal-centered in that the focus is primarily on studying the life of the animal rather than on asking arbitrary questions in a so-called animal model. The emphasis is often upon the natural behavioral repertoire of the animal rather than training the animal to engage in some arbitrary task.

Zoopharmacognosy   The study of how animals use medicinal substances. Interchangeably used by some with the term animal self-medication.

Zooplankton   Small pelagicorganisms in aquatic ecosystems that form central part of the food web. They typically eat algae (phytoplankton) and are consumed by small (planktivorous) fish.

Zugunruhe   Migratory restlessness (hopping or hovering) in caged migratory birds often oriented with respect to seasonal directions of migration (e.g., northward in spring and southward in fall).

Zygote   A newly fertilized egg.